A COMPUTER-MAPPED FLORA

A STUDY OF THE
COUNTY OF WARWICKSHIRE

A COMPUTER-MAPPED FLORA

A STUDY OF THE COUNTY OF WARWICKSHIRE

by

D. A. CADBURY, J. G. HAWKES

and

R. C. READETT

With a section on Bryophytes

by

T. LAFLIN

and contributed chapters by

F. W. SHOTTON, H. THORPE *and* G. T. WARWICK

1971

Published for the

BIRMINGHAM NATURAL HISTORY SOCIETY

by

ACADEMIC PRESS · LONDON · NEW YORK

ACADEMIC PRESS INC. (LONDON) LTD
24-28 Oval Road,
London, NW1 7DD

U.S. Edition published by
ACADEMIC PRESS INC.
111 Fifth Avenue,
New York, New York 10003

Library of Congress Catalog Card Number: 75-153532

ISBN: 0-12-333360-1

PRINTED IN GREAT BRITAIN BY
THE WHITEFRIARS PRESS LTD., LONDON AND TONBRIDGE

PREFACE

This is a County Flora with a difference. On the one hand, it attempts to follow the distinguished traditions of British Local and County Floras, giving the detailed distribution of plants in a restricted area some 50 miles long and 30 miles wide, and adding some preliminary chapters on the history of botanical study, geology, soils, climate, etc., which are designed to explain or add a gloss to the distribution data. On the other hand, it brings a completely new look to County Flora work, and indeed to Floras of any sort, attempting for the first time to make use of electronic data processing methods to sort and codify the information sent in by the recorders.

More importantly perhaps, this Flora uses computer techniques to construct distribution maps which not only rcport the presence or absence of a species in a particular area but indicate also the major habitats in which that species has been found and its approximate frequency.

We believe that this work will be of interest by reason of its methods and techniques to botanists and other biologists outside the boundaries of Warwickshire, and, indeed, of Britain. The methods we use can be applied anywhere and to any kind of distribution data, whether it be insects, birds, crop plants, human populations, blood groups, or industrial and agricultural developments, to name but a few examples.

The study is based on the unremitting efforts of our recorders who sent in some 175 000 records over a period of 15 years, from 1950 to 1965. Without these helpers, both amateur and professional, the Flora could never have been written, and we thank them here, listing their names and contributions on pp. 737–9.

We also wish to thank very particularly Mr. M. C. Clark for many years of help in recording, checking records, sorting specimens and correcting manuscripts and proofs. Our special thanks are due also to Mr. Brian Kershaw of the Birmingham University Computer Centre for his inspired and enthusiastic development of the computer programs which were used to sort and collate the data and print or draw the maps, and for his help in writing the section on computer methods.

We should also like to record our grateful thanks to Mr. T. Laflin who provided the bryophyte section, and to the following who contributed chapters and/or maps to the Flora, mostly based on original unpublished research:

Professor F. W. Shotton	Geology and Soils
Professor Harry Thorpe	Historical Geography
Dr. Gordon Warwick	Physical Background
Mr. B. D. Giles	Rainfall and Temperature Maps

We thank also the many national and international specialists (of whom a detailed list appears on pp. 740–42) for identifying difficult or critical specimens for us, and particularly Professor J. Heslop-Harrison (*Rosa*), Dr. R. Melville (*Ulmus*), and Mr. D. Meikle (*Salix*), who, in addition, led one or more special excursions for the study of the genera mentioned after their names.

It is a pleasure also to record our thanks to the following Institutes, Museums, Universities and Herbaria, who loaned us material, provided facilities for study and sent us information and references:

The Herbarium, Royal Botanic Gardens, Kew
The Department of Botany, Natural History Section, British Museum
The Herbarium, Botany School, University of Cambridge
The Herbarium, Department of Botany, University of Oxford
The Natural History Section, Birmingham Museum and Art Gallery
The Natural History Section, Coventry Museum
The County Museum, Warwick
Birmingham Central Reference Library
The Linnean Society Library, London

We should also like to thank the Botanical Society of the British Isles for helping us with some of the recording. Help was also kindly given by the Coventry, and the Warwick and Leamington Natural History Societies.

Apart from those who helped with recording and others who identified difficult material for us (whose names are listed later) we should like to record our indebtedness to the following for the special help noted after their names:

Bangerter, E. B.	(Checking through British Museum Records)
Dandy, J. E.	(Supplying marked maps and notes on the Vice-County boundary and much help on nomenclatural problems, as well as facilities for study)
Davies, K. C.	(Supplying material and information on the Coventry Museum collections)
Dawson, R. W. B.	(Help and information on the Warwick Museum specimens and records)
Good, Prof. R.	(Advice in early planning stage)
Green, P. S.	(Co-founder of the Flora, with R. C. Readett)
Griffiths, Miss S.	(Making a card index of Warwickshire specimens in the British Museum (with Miss Powell))
Haiste, J. H.	(Loan of Bloxam collection from the Library at Rugby)
McClintock, D.	(Criticisms and suggestions on the first check list)
Morris, Miss J.	(Providing facilities and help at Warwick Museum)
Perring, Dr. F.	(Provision of records from the B.S.B.I. survey; discussions on data processing)

Powell, Miss S. E.	(Making a card index of Warwickshire specimens in the British Museum (with Miss Griffiths))
Richards, Dr. A. J.	(Providing material and information from the University of Oxford Botany Department Herbarium)
Sell, P.	(Supplying material and information from the University of Cambridge Botany School Herbarium)
Spreadbury, Mrs. P. M.	(Making a card index of Warwickshire specimens at the Warwick Museum)
Warburg, Dr. E. F.	(Provision of specimens, materials and facilities at the University of Oxford Botany Department Herbarium)
Wilkins, Dr. D. A.	(Advice and comments on Chapter 7)

Our special thanks are due to Mr. J. Heyes and his staff of Benson Lehner Ltd. (now Computer Instrumentation, Ltd.) for running the tapes through the incremental plotter at the C.I. Data Centre at Aldershot and for re-running several dozen until perfection was reached, with no extra charge.

We also wish to record here our thanks to the following who worked with us at Birmingham University:

- Mrs. Peggy Kershaw (Tape punching and editing)
- Mrs. M. Cotton, Mrs. B. Hulme and Mrs. E. Pickvance (Transfer of records to species cards, etc.)
- Mrs. M. Cotton and Mrs. S. Jones (Mounting specimens and many other routine duties connected with the Flora)
- Mrs. E. Facer and Mrs. Z. Lees (Typing part of the manuscript)
- Miss Jean L. Wilson (Cartographic work)

Finally, we should like to record our indebtedness to the Birmingham Natural History Society for providing financial aid; and to the University of Birmingham for working and storage facilities, secretarial help, free computer services, and various grants, including payments for tape punching and editing.

July, 1971

D. A. C.
J. G. H.
R. C. R.

CONTENTS

1. INTRODUCTION

Bagnall's "Flora of Warwickshire" was published in 1891. Fifty years later it was still much used, but there was a growing feeling among local botanists that so many changes had taken place in habitats and plant distribution, that a new survey was needed. Nothing practical was done, however, until 4th October, 1949, when members of the Birmingham Natural History and Philosophical Society (as it was then called) and members of the staff of the Botany Department of Birmingham University met at the Birmingham Museum to discuss a proposal to begin work on a new County Flora. The President of the Natural History Society, W. Salmon, was in the chair. The Botany Department of the University was represented by P. S. Green, A. D. Skelding, and E. S. Twyman; and the Botanical Section of the Natural History Society by C. E. A. Andrews, R. C. L. Burges, and R. C. Readett. It was agreed that a new Flora of Warwickshire was needed and that the project should be undertaken jointly by the Natural History Society and the University; that the new Flora should be more ecological than that of Bagnall and should attempt to deal with frequency and distribution, especially of the commoner plants.

On 31st January, 1950, a Joint Advisory Committee was set up consisting of the representatives of the Natural History Society and of the University who attended the preliminary meeting together with D. Payler, Keeper of the Natural History Museum and his assistant, V. C. Smith. Miss D. A. Cadbury was co-opted at a meeting later in the year. A draft Recorder's Card, setting out the information to be asked for with each record and containing a list of critical genera and species in respect of which a specimen would be required with each record, and a draft Species Card for the card-indexing of records were submitted by Mr. Green and Mr. Readett, and were approved. It was agreed by the Committee that the direction of the work should be in the hands of Mr. Green and Mr. Readett. In November, 1950, when Professor Good came to Birmingham to speak to the Natural History Society on "The Story of the Dorset Flora" the plans for the Warwickshire Flora were discussed with him.

Early Work

Recording was begun as from 1st January, 1950. It was decided that, in the first place at any rate, the revision would be limited to vascular plants. The area to be surveyed was to be Vice-county 38 as delimited on a set of 1 in. O.S. maps marked for the purpose by J. E. Dandy of the British Museum (Natural History). It was intended to use a team of voluntary workers of varying degrees of competence and experience, and to require 6-figure map references with all records. As records were received they were to be entered on species cards and then plotted by means of dots on transparent outline maps. In this way, provided the county was covered evenly, there would eventually be outline maps showing the distribution of each species; and these transparent outline maps being placed over maps to the same scale showing the surface geology, river valleys, woodlands and other natural features, could be used to interpret the distributions. Records of plants in the critical groups were to be accompanied by specimens; and the specimens were to be submitted for determination to specialists of national standing so that there would be uniform and authoritative determinations.

By the end of 1952 about 15 000 records had been received from about 30 workers; and approximately 1000 specimens of plants in critical groups had been sent in. By this time considerable changes affecting the work on the Flora had taken place. First of all, Mr. Green left Birmingham to take up an appointment at the Royal Botanic Garden, Edinburgh. While retaining his interest, he could no longer take part in the day-by-day work on the Flora. Secondly, the new "Flora of the British Isles" by Clapham, Tutin and Warburg was published. (So far, nomenclature had been based mainly on Professor Clapham's Check List.) Thirdly, Professor Maskell was appointed to the Chair of Botany at Birmingham University, and Dr. J. G. Hawkes was appointed lecturer in Taxonomic Botany. Professor Maskell, Dr. Hawkes and Mr. J. F. Woolman were co-opted on to the Advisory Committee, and Dr. Hawkes and Mr. Skelding were appointed Joint Recorders with Mr. Green and Mr. Readett. Later, the committee ceased to function owing to the sad deaths of Professor Maskell, Dr. Burges and D. Pailer, as well as to the fact that Green, Skelding and Twyman left the country, and Smith left Birmingham. It was agreed by the Natural History Society and the Botany Department of the University that the work should continue under the direction of Miss D. A. Cadbury, Dr. (later Professor) Hawkes and Mr. R. C. Readett as Joint Editors, together with Mr. M. C. Clark.

A Review and Revision of Methods

By the middle of 1953 it was felt that the time had come to make a critical survey of the first three years' work on the Flora to see whether the methods adopted were producing or likely to produce the desired results. Dr. Hawkes drew up a memorandum in which a new system of recording, described by him as the "Basic Square Method", was proposed. After discussion and some modification, a pilot survey on the proposed new lines was carried out, and the new system was brought into operation on 1st January, 1954. All records previously obtained were retained so long as grid references made it possible to put them into the appropriate basic squares.

About this time T. Laflin began, independently, a survey of the Warwickshire bryophytes. He was working along similar lines to those being used for the vascular plants, and it was decided that the value of the Flora would be considerably enhanced by the addition of a section on bryophytes contributed by so experienced a worker in this field. This he agreed to provide.

The modified system of vascular plant recording referred to above required future recording to be based on 1 km squares. One square was selected at random from each "tetrad" or block of four for detailed study. This made it possible to provide an even survey over the whole county with one quarter of the effort that would have been required if every square had been surveyed. Furthermore, the random selection of a 1 km square in each tetrad ensured that no bias towards certain habitats at the expense of others would be exercised when squares were selected for recording.

Recorders were asked to make species lists on standard recording forms and to add against the name of each species information on habitats and frequencies, using the appropriate symbols provided. The lists when completed and sent in were entered on to "Map-index" Species Cards, printed in blocks of 10 × 10 km at a scale of one inch to the mile. Records were written on to these in the correct spatial relationship so that cards of any one species could be laid out in the form of a map and its distribution studied at any time during the progress of the work.

From these index cards the information for the majority of the species was conveyed to punched tape and processed by the Birmingham University Computer (English Electric KDF9). This made possible the sorting of information in a variety of ways so that statistical data could be presented in the Flora (see notes under Species and the Check List). Finally, by means of an Incremental Plotter and a Line Printer it was possible to provide computer maps of the flowering plants, vascular cryptogams and bryophytes with ease and precision.

The methods outlined here and described in detail in Chapter 6 have not been used, so far as we are aware, in any other botanical work and we believe that they are capable of very wide application to many types of map production, not necessarily confined to the biological field.

Bagnall's Flora covered not only vascular plants and bryophytes but also included a section on the Fungi. In 1965 a committee was set up to explore the possibility of bringing Bagnall's work up to date and extending it to groups of fungi which he did not attempt to cover. A fungus survey of the county is now well advanced and the intention is to publish the results in due course.

2. THE PHYSICAL BACKGROUND

G. T. WARWICK

Part I—Physiography

Introduction

A plant has to adapt itself to the environment as it finds it. From the botanical point of view it is an artificial distinction to separate natural from man-made features of the landscape, since the latter may possess properties that have just as strongly determinate an effect upon the plants that take root and thrive upon them as those of the natural environment. For this reason human modifications of the land surface such as built-up areas, spoil heaps, quarries, canals and reservoirs will be considered here, where appropriate, as variants of the natural scene.

The land surface itself is extremely complex in form, as may be seen on the relief and drainage overlay maps I, II and IX. In order to bring out the major differences the vice-county area may be divided up into a series of physiographic regions, as shown on Map 1 (see p. 21).

Although Warwickshire is one of the counties furthest from the sea and has the important Trent–Severn watershed running across it, it remains a lowland county with most of it less than 500 ft above sea level. The height varies from about 80 ft O.D. near Cleeve Mill on the Avon in the south-west to 854 ft O.D. on the summit of Ebrington Hill, some eight miles away to the south-east, both points lying on the county boundary. Essentially, the landscape is made up of a low plateau ranging between 400 and 500 ft O.D., dissected by the rivers Avon and Tame and their tributaries. Edge Hill in the south-east is exceptional, rising abruptly from the vale at its foot at about 350 ft O.D. to a crest which tops the 700 ft contour. Further to the north-east the scarp is broken by a series of low cols through which easy passage can be made into the headwaters of the Cherwell.

The alignment of the Avon, Anker and Tame valleys sets the pattern of the relief, with N–S orientation in the northern part and a NE–SW grain towards the south-east. The intervening plateaux exhibit very gentle gradients sloping towards the dissecting valleys and terminating in rounded changes of slope. Below the latter are the steeper valley-side bluffs some 50–100 ft high whose lower slopes gradually grade into flatter features cut into the solid rock or similar benches built up with gravels. In the Avon valley there are five gravel terraces and one or more rock benches, though only fragments remain as a result of erosion by the main river and its tributaries. Below the lowest step comes the floodplain, fringed by short steep slopes, against which the river occasionally impinges as it meanders across its innermost valley. In times of excess discharge after storms or the melting of snow the channel can no longer contain all the water and flooding occurs, with deposition of silt near the banks.

The variations in slope provide differing substrates for plants. In general, the steeper the slope, the thinner the soil and the greater the tendency for the soil and subsoil material to migrate downslope. It is rare to find a slope too steep to maintain plant life because of the unstable surface. This does occur along the outside bends of the river channels, and where the rivers swing against the valley-side bluffs the bank above may show shallow slips, known as "sheepwalks". The movement of fine material on the steeper slopes results in an accumulation at the base where plants may find a greater supply of soil-water. The flatter slopes also show a greater degree of leaching than the steeper ones, with consequent contrasts in base-status and availability of minerals. Besides its position in relation to the drainage, the inclination of a slope is also affected by the nature and structure of the solid rocks. However, the weathering product derived from higher up may mask the solid rocks at the base. In places these accumulations date back to conditions reigning at the end of the last glaciation, and are the result of spring thawing of the ground frozen in winter, resulting in more rapid downhill movement than occurs today, helped by a sparser vegetation cover. Such fossil hill-wash material is referred to as "head" and forms the parent material of the soils found at the foot of Edge Hill. The floodplain is perhaps the most specialised habitat, as it has a high water table and under natural conditions is poorly drained. This is because the flood deposit along the channel edge produces a gentle slope away from the stream. In places swampy vegetation can be found along the foot of the floodplain bluff e.g. by the Avon, west of Westham House, Barford and at Bradnock's Marsh in the Blythe valley. These damp conditions contrast strongly with the well-drained soils on the adjacent river terraces. On the plateau tops the soils are generally much thicker than on the bounding slopes, though their nature is affected by the parent material which may consist of deposits left by ice-sheets of the geologically recent past.

Slope has a secondary effect upon plant life since the steeper slopes are unsuitable for ploughing and may even be too steep for cattle to graze. This is probably the reason for the retention of Edge Hill under timber. In general the flatter the slope the greater is the chance of disturbance by agriculture. Also hedges and their associated plants have a greater chance of survival where they run along the edge of a steep slope. On the gravel terraces intensive horticulture has been favoured at the lower levels by the ease of working, even after heavy rain, though competition has been encountered with gravel workings which lower the surface and usually leave a heavier soil. Where these workings go below the water table, pools may result which become new aqueous habitats for plants, fish, and birds, e.g. in the Tame valley near Kingsbury. Some interference may occur, however, by the takeover of such pools for sailing. The floodplains have usually remained unploughed and been left for grazing, often in the form of permanent pasture which carries a greater variety of plants than the short leys of the higher land, especially those species able to withstand the physical effects of animal grazing.

The Physiographic Regions

The broad outlines of the physiographic regions of the West Midlands were given by Lapworth in 1913. He did not produce a map, but he noted that the 400-foot contour gave a close approximation to the plateau edge. However, the Geographical Section of the Birmingham Natural History and Philosophical Society preferred to use the 300-foot contour which was used by Martineau (1913) in editing a study of the fauna of what he called the Midland Plateau. Subsequently, the late Professor R. H. Kinvig (1928) re-named this the Birmingham Plateau. A more precise scheme based upon fieldwork and a study of 1 : 25 000 maps was introduced by the author in 1950 (Warwick, 1950) for the Birmingham region and this has now been extended to cover all the Warwickshire vice-county. The result is given on Map 1 and the key is provided by Table 1 below. In many cases these regions have local names, but some new ones have been invented, usually based upon some local settlement. For purposes of continuity features named from adjacent counties have continued to be called by "foreign" names. The regions fall into three groups consisting of valleys, plateaux and uplands or higher hills and these have been subdivided, in some cases down to three orders of magnitude. On Map 1 (p. 21), the boundaries of the second order units are shown by continuous lines and the others by hatched lines.

TABLE 1. Physiographic Regions of Warwickshire

I. The Birmingham Plateau

- (a) The South Staffordshire Plateau
  - (1) Sutton Plateau
  - (2) West Bromwich–Harborne Plateau
  - (3) Solihull Plateau
  - (4) The Ridge Way
- (b) The East Warwickshire Plateau
  - (1) Ansley Plateau
  - (2) Corley Plateau
  - (3) Wroxall Plateau
  - (4) Claverdon Plateau

II. The Trent Valley System

- (a) Tame Valley
  - (1) Lower Tame Valley
  - (2) Mid-Tame, Rea and Cole Valleys
  - (3) Blythe Valley
  - (4) Anker Valley
- (b) Mease Valley
- (c) Soar Valley

III. The Avon Valley System

- (a) Arrow–Alne Valleys
  - (1) Arrow Valley
  - (2) Alne Valley
- (b) Mid-Avon Valley
- (c) Stour Valley
- (d) Kineton–Southam Vale (Feldon)
- (e) Upper Avon Valley
- (f) Sowe Valley
- (g) Leam Valley

IV. The Fringing Hills and Plateaux

- (a) The Leicester Border Hills
- (b) Dunsmore
- (c) Gaydon Plateau
- (d) Northamptonshire Uplands
- (e) Burton Dassett Hills
- (f) Edge Hill
- (g) Whichford Hills
- (h) Ebrington Hill (Ilmington Downs)

V. The Thames Valley System

- (a) Cherwell Valley

The Birmingham Plateau. In the wider context of the West Midlands this consists of two units, the *South Staffordshire* and *East Warwickshire Plateaux* as named by Lapworth (1913). Only about a fifth of the former lies within the vice-county of Warwickshire, but the name has been retained to avoid confusion. The units consist mainly of low plateaux, largely made from Triassic rocks covered with glacial clays and gravels. In the north the *Sutton Plateau* is underlain by the conglomerates and sandstones of the Bunter Pebble Beds. Thanks to the gift of Henry VIII, Sutton Park has preserved a non-agricultural landscape over much of the region with dry heath and woodland on the upper surfaces and valley sides and wet heath in the valley bottoms, though the presence of old millpools has also introduced extensive bodies of water. Temporary disturbances during wartime and by a scout jamboree have been partially obliterated by natural growth. Only the southern part of the *West Bromwich–Harborne Plateau* lies within the vice-county; this is almost entirely covered by the streets and buildings of the city centre and northern and western suburbs of Birmingham, though it does contain the large artificial Rotton Park Reservoir and several urban parks. The *Solihull Plateau* is cut up by the Cole and its tributaries in the north, but southwards the headwaters of the Blythe only dimple the general flat surface which consists of drift-covered Keuper Marl. The southern boundary forms part of the main Severn–Trent watershed, overlooking the short steep slopes of the upper valleys of the Arrow and Alne. On the western side of the Arrow there is a narrow band of plateau surface preserved on the top of the *Ridge Way* south of Redditch.

The *East Warwickshire Plateau* might be more appropriately called the Central Warwickshire Plateau, but it has long been known by the former name in the West Midlands and it was not thought wise to confuse the issue. The coalfield of which it is largely made is also known as the East Warwickshire coalfield. It is more dissected than the South Staffordshire Plateau and it has been reduced to several narrow "necks" by minor streams.

The Bourne valley divides the northern part into the *Ansley and Corley Plateaux*. The watershed runs in a gently sinuous arc swinging from east of centre towards the west in the south. Between the valleys there is little relief except for the artificial spoil heaps of the collieries, most of which have now been closed owing to exhaustion of their reserves. The centre of the plateau is underlain by upper Coal Measures, consisting mainly of red marls with subsidiary sandstones and thin limestones, but for the most part it is free from drift. The middle Coal Measures outcrop along the northern edges of the plateau and much of the ground has been disturbed by open-cast mining. On the north-east older basement rocks appear, of hard diorite and quartzite which have been extensively quarried for roadstone. The waste heaps from the quartzite are somewhat inhospitable, but those from the diorites form interesting habitats for saxicolous and calcicolous plants. Marl holes serving old brick and tile works sometimes provide aqueous habitats, though several are now being infilled with rubbish.

The *Wroxall* and *Claverdon Plateaux* are cut out of red Keuper Marl and the latter especially has a cover of gravel and boulder clay some of which has been worked for gravel, but the pits are now being used for rubbish disposal.

Between the two halves of the Birmingham Plateau is the narrow *Kingswood Gap*, utilised by the Grand Union Canal and the Western Region railway line from Birmingham to Paddington. Here the watershed is hardly discernible on the ground, separating two minor streams that drain the floor of the gap.

The Trent Valley System. Although the Trent does not flow directly through Warwickshire, the county is drained by three of its major tributaries, though only the Tame is of any importance. The *River Tame* rises in the heart of the Black Country, then flows through the lowest parts of Birmingham running eastwards, until joined by the Cole and Blythe at Whitacre where it turns abruptly northwards along the line of the Blythe, leaving the county near Tamworth where it is joined by the Anker. The Tame has the doubtful reputation of being one of the most polluted rivers in the county, both by organic material from overloaded sewage works and by industrial effluents. This imposes a severe limit on plant and animal life, though it is hoped that new sewage works will produce a steady improvement in the future. Floodwaters also pollute the pastures alongside the river banks. The *Mid-Tame* valley is lined with canals along much of its way through Warwickshire, and they also suffer from considerable pollution in the industrial areas. Within Birmingham the valley sides of the *Tame* and *Rea* are covered with roads, houses and factories. Eastwards the river flows past Minworth sewage works where additional water supplies, originating in Wales, are discharged after treatment, to be succeeded in turn by the Hams Hall Power Station complex where some of the river water is used for cooling purposes, returning to the river slightly warmer for this reason. Over much of this stretch the channel has been artificially straightened and even lined with bricks in places. The *Cole Valley* is also heavily urbanised, though little of this lies within the vice-county.

Thanks to the protection of the provisional Green Belt, the *Lower Tame Valley* and the *Blythe Valley* remain as largely undeveloped countryside. Both are cut in Keuper Marl with a fine network of small tributaries. The Tame Valley is wider and has a much more developed floodplain with at least two gravel terraces rising above it. Flooded gravel workings opposite Kingsbury form an extensive area of water which will remain open for sailing. In the Bourne Valley joining the Tame also near Whitacre is Shustoke Reservoir which is used for drinking water storage. Downstream the Tame Valley becomes more industrialised, with mining and brickworks as well as manufacturing industry. The Blythe Valley is much more rural with a steep eastern side, the western flanks being gentler and more finely dissected. The *Anker Valley* is a fairly open one with steepish upper sides. The narrow floodplain is flanked by gravel terraces but in the lower valley some of the right bank tributaries flow over very wide spreads of alluvium, for example around Austrey. Most of the valley is cut in Keuper Marl, except for the upper part of the left flank, made up of Coal Measures, Cambrian shales and quartzite and a small area of Precambrian tuffs and ashes. However, the lowest part of the valley is cut across the Coal Measures. Here the workings have caused the valley floor to subside, forming Alvecote Pools, which are now preserved as a nature reserve. The left side of the valley is also followed by the Coventry Canal which is no longer actively used for commercial traffic, but forms a home for water plants.

Only an insignificant part of the *Mease Valley* lies in the vice-county in the extreme northern part, cut here in Keuper sandstone. Similarly, only small parts of the valleys of the headwaters of the Leicestershire *Soar* are to be found on the eastern flanks of the Leicester Border Hills (see below).

The Avon Valley System. Over half of the vice-county is drained by the river Avon and its tributaries. It is, geologically speaking, a young river, which once flowed SW–NE to join the Soar. After being dammed by ice-sheets and infilled with deposits of the ensuing lakes, the last lake in this valley overflowed westwards and the river then flowed in its present direction. This is the reason for the peculiar orientation of some of the right bank tributaries and the narrowness of the upper Avon basin. None of the tributaries is very large and the flow is supplemented by Severn water via the sewers and treatment works of Coventry. This lack of dilution from tributaries produces a high biological oxygen demand in times of drought; black sludge and long streams of *Cladophora* reinforce the evidence of pollution, though this is insufficient to suppress plant and animal life. The Avon rises near Naseby in Northamptonshire and receives no major tributary until it is joined by the Sowe at Stoneleigh. It has its confluence with the Leam between Leamington and Warwick, by which time it appears to be much wider and deeper, thanks to the presence of weirs. Other left bank tributaries are the small river Dene and the somewhat larger Stour. The only right bank tributary is the Arrow, coming in near Bidford on Avon, just on the county boundary.

The valley sides descend in a series of steps cut in the solid rock, with up to five gravel terraces below, numbered 1–5 in ascending order. Only numbers 2 and 4 are at all extensive, the former being the most widespread. The tributary valleys also possess parts of the same sequence since the terraces represent the remains of old valley floors, but these are usually smaller and often the upper flights may be missing altogether. Details of the occurrence of these terraces are to be found in the publications of Shotton (1953) and Dr. Mabel Tomlinson (1929, 1936).

The *Upper Avon Valley* is a relatively narrow strip, consisting of a small floodplain bordered by a suite of four gravel terraces. It is cut into a mass of alternating layers of soft clays, gravels and sand. This makes for very rapid alternations of dry and wet land. The gravels have been worked and several of the abandoned pits, e.g. near Wolston, have been cut down to the water table and are now flooded. Near Ryton on Dunsmore the valley floor has been affected by subsidence owing to the workings of Binley colliery, resulting in a wide "flash" (subsidence lake). The *Sowe Valley* to the north has a more complex pattern of streams and associated valleys. The lower part of the river has a valley similar in many respects to that of the Avon. Smite Brook continues the line of the lower Sowe and it has been dammed in Combe Abbey park to form an extensive pool. The upper Sowe, Sherbourne and Finham Brooks all have more easterly elements in their courses contrary to the general pattern of the Avon basin. The relief pattern is reminiscent of the upper Blythe valley with relatively little incision. The *Leam Valley* is rather wider than the upper Avon Valley with unusually wide spreads of terrace gravel stretching upstream to a little beyond the junction of the Itchen.

The Warwickshire section of the *Middle Avon Valley* is much wider than the headwater valleys and all of the sub-units are correspondingly larger. The floodplain is crossed by the meandering river in wide sweeps below considerable expanses of flat gravel lands forming No. 2 terrace, especially on the left bank. Wellesbourne airfield was constructed on this feature. The steeper slopes up to the level of No. 4 terrace mark the edge of an inner valley within which settlement is largely concentrated. Above this level is a series of steps with a marked scarp on the south-east forming the boundary of the *Kineton-Southam Vale*, which is part of the Feldon region (see Chapter 4). Much of the steepness of the scarp is due to the presence of the White Lias limestone and in places the Blue Lias limestone, which reinforce the clays occurring between them. The vale is formed by the upper parts of the rivers Dene and Itchen, and slopes are gentle, being cut in the soft Lower Lias clays. The south-eastern side of the Gaydon Plateau forms a small counter-scarp facing Edge Hill. Within the Itchen Valley are several small hills capped with sands and gravels, such as Wormleighton; these are all that remain of former widespread deposits laid down in a glacial lake (Bishop, 1958; Shotton, 1953).

The *Stour Valley* is a miniature epitome of the Middle Avon Valley, with its own reduced train of terraces, but its headwaters are marked by steeper slopes than usual, descending from the fringing uplands which are about 500 ft above the valley floor. The valley is cut for the most part in the soft Lower Lias clays though some of the minor eminences such as Honington and Idlicote Hills are capped with old gravels. The *Alne* and *Arrow Valleys* are wide in comparison with their length and cut deeply into the Keuper Marl of the Solihull Plateau. On the west is the Ridge Way, a flat-topped ridge which once marked the watershed of the old Avon–Soar system. Terraces occur only in the lower parts of the valleys, but the left side of these parts is capped by the same stiffener of White Lias noted previously which forms a distinct escarpment, frequently left as woodland, such as Withycombe Wood, south-east of Haselor.

The Fringing Hills and Plateaux. These are mainly shared with neighbouring counties, with two exceptions, Dunsmore and Gaydon Plateau. The *Leicester Border Hills* have no accepted name; they lie along the north-eastern boundary, with gently sloping tops, separating the drainage basins of the Upper Avon tributaries from those of the Soar. In strong contrast to the East Warwickshire Plateau their underlying Keuper Marl is masked by a thick cover of chalky boulder clay; even so, it contains little chalk, but large flints derived from the chalk are much commoner. Most of the upper surface is a little over 400 ft above the sea, with only one small area above 450 ft. On the southern side of the Avon lies *Dunsmore*, a small plateau between the Avon and Leam composed of relatively recent lacustrine clays and capped with porous and infertile gravels. A series of tributaries of the Avon and Leam have mildly dissected the upper surface, leaving gravel-capped ridges separated by wetter clay-floored valleys. The A45 runs along the main divide to avoid these undulations. The general level is somewhat lower than the average Warwickshire plateau, the break in slope occurring about 350 ft O.D. and rising to a maximum of only 372 ft O.D. The south-western spur of Dunsmore carries an unusually high proportion of woodland.

The *Gaydon Plateau* corresponds to the southern parts of the East Warwickshire Plateau. Its surface is just over 400 ft in altitude and Gaydon airfield is built upon it. It is largely cut out of the Blue Lias shales with limestone and a capping of boulder clay.

The south-eastern boundary of the vice-county is marked by some of the highest land, rising abruptly to over 600 ft and forming the southern watershed of the Avon basin. The higher parts are capped for the most part by some 150 ft of Middle Lias marlstones, rich in iron oxides which impart a deep rich brown colour to the soils, houses and boundary walls of the fields. Locally, even younger deposits, Upper Lias clays and Inferior Oolite are to be found, especially in the south on the highest areas. Occasional quartzite pebbles in the soils of the hilltops indicate the presence of some of the earliest glacial deposits. North-eastwards from Edge Hill the line of hills is broken by a series of gaps, some of which were used by the overflow waters of the glacial lakes. The intervening hills form the outer fringe of the *Northamptonshire Uplands*, and they are capped with the resistant marlstone. These hills, e.g. Napton and Beacon Hills, rise from a low plateau whose surface approaches 500 ft in altitude, and terminate towards the north-west with a marked escarpment. The Fenny Compton Gap is the lowest of the cols and with the slightly higher Burton Dassett Gap, isolates the elongated *Burton Dassett Hills* whose marlstone cap has been cambered over the edges as a result of flowage of the underlying clays (Edmonds *et al.*, 1965). Further south the escarpment is unbroken, with *Edge Hill* forming a very straight feature for about 2–3 miles, probably the steepest slope in Warwickshire and covered by beautiful hanging woodland. This contrasts strongly with the flat top. Further to the south-west the edge becomes more sinuous with deep valleys of the Stour headstreams gashing the scarp but not breaking the skyline as in the north-east. Since the character of the escarpment is so distinct I have differentiated the area from Edge Hill *sensu stricto*, and for want of a traditional name have called it the *Whichford Hills* after the only village on the flanks of the hills near their Warwickshire centre. Brailes Hill is isolated from the main ridge, being connected to it by a low plateau reminiscent of the northern area. The crest of this southern area is over 750 ft O.D. and is capped with Inferior Oolite.

The last of the fringing uplands is *Ebrington Hill*, the most north-easterly outlier of the Cotswolds and again capped with Inferior Oolite, forming the highest part of the vice-county. The Warwickshire part of this hill is sometimes referred to as *Ilmington Downs* after the village nestling at its foot.

The Thames Valley System. Only a very small part of Warwickshire drains towards the Thames, lying to the south-east of the Avon–Thames watershed. A few small areas of the upper basins of the *Cherwell* and its tributaries occur within the vice-county boundary at the back of Edge Hill and the Northamptonshire Uplands fringe. The valley cut into Edge Hill just avoids notching the skyline.

Part II—Climate

Introduction

Heat, light and water are fundamental to plant life and the amount and availability of these play a large part in controlling plant growth. The seasonal changes and the density of cloud cover obviously influence strongly the amount of the sun's radiant energy available for plants, in the form of both light and heat. The water comes usually in the form of rain but may also be present as dew, mist, fog, snow and hail, the last two not being available for plants until melted. Not all the precipitation is available for plants since a proportion is evaporated into the atmosphere and for the most part is only available when present in the soil. Critical temperatures, usually below freezing point, may cause severe damage to certain plants, though the duration of the extreme cold is also important. Other extremes are concerned with excessive rain or its absence for a long period, as in a drought. It is usual to take the average values for the climatic elements as a measure of the climate or average weather of a place, though such values may hide the extreme weather conditions which may be so important for the flora. Normally the observations are taken under standard conditions within a shaded Stevenson screen, over a grass surface some 4 ft from the ground. The actual value experienced by plants may differ from these conditions and within the plant cover there may be local variations, usually within somewhat narrow limits. Again, the number of recording stations is low so that the isopleths shown in the accompanying maps are generalised from a relatively small sample. These maps, drawn by Mr. B. D. Giles, have been compiled from the Monthly Weather Reports issued by the Meteorological Office (see overlay Maps IV, V and VI).

Since Warwickshire lies in the centre of England and Wales, its climate tends to be somewhat more extreme than in coastal districts, with higher summer temperatures and lower winter ones. On the other hand it lies between the wetter west and the drier east. Overall, it possesses a temperate climate with rainfall fairly well distributed throughout the year, but with a general tendency towards more in winter than in summer. The reader wishing to find out more about the climatic variations is referred to Saward (1950).

Within the vice-county the small altitudinal range does not result in any great regional variations, though the Avon Valley tends to be the warmest and driest part. The large cities such as Birmingham and Coventry produce their own local climates owing to the extra heat derived from the myriad chimneys and reflected and radiated by the buildings. Their atmospheres are also notable for a higher proportion of fine particles and sulphur dioxide. This combination also affects the receipt of energy from the sun. It is highly probable that Edge Hill and the Whichford Hills may produce local perturbations in the air flowing over them but data are not available to show this. The aspect of slopes also affects the amount of energy received, southerly and westward facing slopes being the most favoured, but again it is probably only on Edge Hill that this is at all significant.

Temperatures

Temperature exercises an important control over plant growth, and it is usually reckoned that growth only occurs when the temperature is above 6°C. Since most of the area experiences critical temperatures below this figure for 4–5 months in winter (Gregory, 1964) most plants have a dormant period during the Warwickshire winters. January is usually the coldest month, and since the lowest temperatures are critical ones a map has been prepared of the mean minimum temperatures for this month (Map V). This brings out the Birmingham heat-island, though Coventry does not appear to have much effect. Another "high" area lies around the Gaydon Plateau in contrast to the lower part of the Middle Avon Valley, probably as a result of cold air seepage from the Whichford Hills down the Stour Valley during clear cold nights. This phenomenon restricts the development of fruit growing in low-lying areas and may similarly affect the distribution of plants subject to frost damage, especially during flowering. In July the mean maximum temperatures (Map VI) again bring out the Birmingham heat island and also show up the Avon Valley as another area of high temperature. A study of the date of flowering of selected plants would probably reflect some of the slight variations between different parts of the vice-county, but phenological maps on this basis are not available.

Precipitation

There is perhaps a greater range shown by the rainfall map (Map IV), which indicates a mean annual difference of 5 in. between the Avon Valley floor and the top of Edge Hill, where the total is 30 in. on average. None of these values is very high and in summer many crops would benefit from irrigation, though there is not always sufficient water available for this purpose. The soils derived from sand and gravel are the worst hit by summer drought since they retain so little moisture that plants may wilt and the grass become scorched. Whilst this is rarely fatal for plants adapted to such conditions, it must affect those with greater water demands.

Snow is a very variable climatic element, with considerable variation both in amount and duration of lying from year to year. In general, such records as are available indicate that in most years the Avon Valley experiences the least amount of snow and that often it lies for only about half to two-thirds as long as in Edgbaston. Snow may protect plants from the effects of frost by the insulating effect of the air contained in the snow, but it also inhibits growth. Much of the meltwater sinks into the ground to replenish the soil water, though since it is close to freezing point it will also chill the soil.

Miscellaneous Climatic Elements

South-westerly winds predominate over others so that air pollution is spread towards the north-east from the sources of pollution. The strength of the winds is rarely sufficient to cause damage, though occasional strong gales may topple older trees. Probably no trees and shrubs in Warwickshire show much effect of the excessive transpiration on the windward side that causes stunted growth near the coast.

Unfortunately, there is not much information available about cloudiness and sunshine, but in general those areas with the lowest precipitation will experience better sunshine conditions, providing more heat and light for plant growth. Hence the Avon Valley is the most favoured locality within the vice-county which may affect plants growing near to their northern limits. Near the large cities pollution may stimulate fogs, but it is pleasing to record that fog incidence is being reduced as a result of clean air measures.

Hail can cause damage to crops and plants by breaking branches and stems and stripping leaves, but such occurrences are rare and do not appear to exercise any controls within the vice-county.

Further details of the distribution of climatic mean values can be obtained from the Climatological Atlas of the British Isles and the other sources listed below. The writer wishes to acknowledge the great help that he has received from his colleague B. D. Giles of the Department of Geography, University of Birmingham, for drawing the climatic maps and helping with advice on the second part of this chapter, though he should not be held responsible for any errors.

References*

Physiography

Bishop, W. W. (1958). The Pleistocene geology and geomorphology of three gaps in the Midland Jurassic Escarpment. *Phil. Trans. R. Soc.* B. **241,** 255–306.

Bloom, J. H. (1916). "Warwickshire." Cambridge, pp. 3–20.

Edmonds, E. A. *et al.* (1965). "Geology of the Country around Banbury and Edge Hill", Mem. Geol. Svy of G.B. Explanation of Sheet 201, New Series. H.M.S.O., London.

Kinvig, R. H. (1928). The North-West Midlands. *In* Ogilvie, A. G. (Ed.), "Great Britain, Essays in Regional Geography", Cambridge. Chapter XII, pp. 216–36.

Lapworth, C. (1913). The Birmingham Country. Its Geology and Physiography. *In* Auden, G. A. (Ed.), "A Handbook for Birmingham and the Neighbourhood", Birmingham. pp. 547–611.

Mackney, D., and Burnham, C. P. (1964). "The Soils of the West Midlands." Bull. No. 2. Soil Svy of G.B. H.M.S.O., London.

Martineau, P. E. (1913). "Introduction to the Fauna of the Midland Plateau", Bgham nat. Hist. phil. Soc., Birmingham. pp. 7.

McPherson, A. W. (1946). "Warwickshire." Part 62 of the Land of Britain, the Report of the Land Utilisation Survey of Great Britain, London. pp. 653–842. (Section II is especially useful for Geomorphology and Soils.)

Pickering, R. (1957). The Pleistocene geology of the South Birmingham area. *Q. Jl geol. Soc. Lond.* **113,** 223–37.

Shotton, F. W. (1929). The geology of the country around Kenilworth. *Q. Jl geol. Soc. Lond.* **85,** 165–222.

* Reference lists throughout this volume follow the system of abbreviations used in the "World List of Scientific Periodicals", 4th edition (1963), published by Butterworths, London.

Shotton, F. W. (1953). The Pleistocene deposits of the area between Coventry, Rugby and Leamington and their bearing upon the topographic development of the Midlands. *Phil. Trans. R. Soc.* B. **237,** 209–60.

Shotton, F. W. (1968). The Pleistocene succession around Brandon, Warwickshire. *Phil. Trans. R. Soc.* B. **254,** 387–400.

Simpson, C. A. (1914). The upper basin of the Warwick Avon. *Geogrl Teach.* **7,** 369–82.

Tomlinson, M. E. (1929). The drifts of the Stour–Evenlode watershed and their extension into the valleys of the Warwickshire Stour and upper Evenlode. *Proc. Bgham nat. Hist. phil. Soc.* **15,** Pt. viii, 157–96.

Tomlinson, M. E. (1936). The superficial deposits of the country north of Stratford on Avon. *Q. Jl geol. Soc. Lond.* **91,** 423–62.

Warwick, G. T. (1950). Relief and Physiographic Regions. *In* Wise, M. J. (Ed.), "Birmingham and its Regional Setting", Birmingham. pp. 3–14.

Climate

Bilham, E. G. (1938). "The Climate of the British Isles." London.

Gregory, S. (1964). Climate. *In* Wreford Watson, J. and Sissons, J. B. (Eds), "The British Isles, A Systematic Geography", London. Chapter 4, pp. 53–73.

Lamb, H. H. (1964). "The English Climate." 2nd Edn. London.

McPherson, A. W. (1946). "Warwickshire", Part 62 of the Land of Britain, the Report of the Land Utilisation Survey of Great Britain, London. pp. 674–83.

Meteorological Office. Various. *Monthly Weather Reports.* Bracknell.

(Undated.) "Rainfall over the area of the Trent River Board" 1916–50. *Hydrological Memoranda, No. 10.* Bracknell.

(Undated.) "Rainfall over the catchment area of the Severn River Board" 1916–50. *Hydrological Memoranda, No. 15.* Bracknell.

1952. "Climatological Atlas of the British Isles." H.M.S.O., London.

Saward, B. (1950). Climate. *In* Wise, M. J. (Ed.), "Birmingham and its Regional Setting", Birmingham. pp. 47–54.

3. GEOLOGY AND SOILS

F. W. SHOTTON, F.R.S.

Map XIII which is reproduced at the scale of $\frac{1}{2}$ in. to 1 mile (see pocket at end of book), and also as simplified and reduced overlays for comparison with the species distribution maps (III and X), compromises between showing geological outcrops and soils but with a stronger emphasis on geology. There is no map of the Soil Survey which covers the county and it would be quite impossible to show soils in the complexity of classification which that science uses. Soils, however, do reflect broadly the underlying geology and it is this latter large-scale pattern which appears to control the distribution of some plant species.

In preparing this map, there were available for the north of the county the "New Series" Drift edition maps of the Geological Survey, Sheets 154 (Lichfield), 155 (Atherstone), 168 (Birmingham) and 169 (Coventry). Parts of the southern end of the county come into Sheets 201 (Banbury), 217 (Moreton in Marsh) and 218 (Chipping Norton), but for the heart of the area, official government maps are all in the "Old Series", dating back to the mid-nineteenth century and showing no Quaternary deposits. They could therefore be thoroughly misleading, but fortunately there have been a number of other publications which go a long way towards dispelling our ignorance of this particular area (Bishop, 1958; Matley, 1912; Shotton, 1929, 1953, 1954, 1968; Tomlinson, 1925, 1929, 1935).

The solid geology pattern of the county is fairly simple. In the south-east, the Middle Jurassic oolitic limestones of the Cotswold Scarp just appear above Long Compton and these provide the only and very limited areas of dry calcareous upland. Close to these outcrops occurs the Middle Lias ironstone which builds the strong escarpment of Warmington, Edge and Sunrising Hills and provides outliers at Burton Dassett and Shuckburgh which might well be responsible for a few plants that favour ferruginous soils.

Apart from these restricted areas, a great proportion of the southern part of the county is floored by Lower Lias clays and cementstones, giving a very heavy but a calcareous soil. This belt of calcareous clays is supremely important in controlling the occurrence and abundance of many plants, particularly the orchids, and perhaps most important of all are the spoil heaps associated with the cement industry's quarries at Long Itchington, Stockton, Harbury and Ufton. The north-western limit of this very significant area is marked by a thin limestone (White Lias) which forms quite a prominent escarpment at Chesterton and Ufton. It gives rise to narrow areas of drained hard limestone, particularly in the spoil heaps of small old quarries and in many respects repeats the pattern of the oolitic limestones on the southern fringe of the county.

North-westward from the Lias clay belt is a large area of Keuper Marl, giving rise to the deep red clay soils which many people regard as typical of Warwickshire. Despite the name of "Marl", these soils are not calcareous. There is, however, a thin sandstone about 100 ft below the top of the Keuper Marl, known as the Arden Sandstone, which has been separately shown on the map. It often forms a conspicuous small escarpment with better drained, coarser soil, and its dolomite content could explain the occurrence of a few calcicoles which have been found upon it.

Rising from beneath the Keuper Marl and forming a much more irregular outcrop pattern are the lower members of the Trias, the Keuper Sandstone and, around Birmingham and Sutton Coldfield, the Bunter Sandstones and Conglomerates. It has been convenient to show these together, since they produce well-drained sandy and often pebbly soils.

The older rocks of the county are found principally in the Warwickshire coalfield extending from Alvecote to Kenilworth, comprising a narrow outcrop of productive Coal Measures, with plenty of old tip heaps adjacent, and a very much larger area of "Barren Coal Measures". In these, clays are dominant, but there are important outcrops of sandstone and conglomerate which typically give rise to well-drained sandy ridges (e.g. Corley Rocks; Meriden Hill; Hollyberry End; and Gibbet Hill, Coventry). There are also a number of extremely thin limestones (Spirorbis Limestones) which despite being rarely more than a foot thick, were dug in ancient times and the old spoil heaps at least might encourage unusual plants; for this reason, they have been indicated. The oldest rocks of all are the Cambrian Stockingford Shale of Dosthill and the Mancetter–Griff area, the Cambrian Hartshill Quartzite of the Hartshill ridge, and the Pre-Cambrian volcanic series of Caldecote. All these are cut by thick sills of igneous rock, camptonite, which, by their weathering, could produce soils of unusual character. Perhaps more important is the vast extent of quarrying of these outcrops, so that ruderal habitats abound.

If only "solid" formations outcropped in Warwickshire, this chapter could now close. Further complications are caused, however, by "Drift", or Quaternary deposits. These divide into two very distinct geomorphological series, the older drifts which occur as residual patches capping the higher ground, and the newer drifts which form terrace deposits in the valleys of the Avon, Leam, Tame, Arrow, Itchen and indeed all the principal rivers. These terraces stand at least a few feet above the level of the alluvial floodplains where there are water meadows and the possibility of marshes.

The older drifts include some very extensive sheets of sand or gravel, but also considerable thicknesses of glacial lake clays or boulder clays which may be non-calcareous through incorporation of much Keuper Marl, or calcareous (Chalky Boulder Clay). Where sands and gravels occur on the higher ground, they give rise to dry heathland or sandy ploughland, often bedevilled by iron-pan beneath the surface. Where these Quaternary gravels rest on the major clay formations such as the Lias or Keuper Marl, they have been separately distinguished since they produce a radical change in soil conditions, but where they lie upon equally pervious and coarse-grained strata such as the Bunter Sandstones and Conglomerates, there is no point in separating them. The importance of these high level gravelly older drifts can be judged from the numerous "heath" place-names—Dunsmore, Ryton, Brinklow, Coleshill Heaths to mention but a few, whilst in built-up Birmingham, King's Heath, Balsall Heath, Small Heath and Washwood Heath are some of the names which perpetuate a picture of an environment long since disappeared.

The clay members of the older drift have been shown only when they rest upon a contrasted geological formation. Thus any type of glacial clay has been shown when the underlying "solid" geology is sandy; but where Chalky Boulder Clay rests upon the calcareous Lower Lias clays, or where red Keuper-derived Quaternary clays rest upon Keuper Marl, it has not been thought to be useful to separate the outcrops.

Along the major river valleys there are flat terraces underlain by gravels, which give well-drained soils where otherwise very heavy waterlogged ground might occur. The detail of these terraces is often complex, too complicated to be shown in detail on a small-scale map, and in general they have been simplified to show belts of gravelly soil intruding into the clay areas. Aerial photography has shown, from crop markings, how heavily these gravel stretches in the river valleys were used for settlements by ancient people from the Neolithic onwards. It seems likely that the plant assemblages have been more modified by man's agricultural activities along the river-terrace belts than in any other part of the county.

Finally, in a county which is notably devoid of large sheets of water, it has been considered useful to show, on the geological map, the larger pools where many of the aquatic plants will have to be sought. All these are in some sense artificial—dammed fish-pools as at Combe Abbey, water supply reservoirs (Shustoke), flooded gravel pits along the Tame in particular, or even the natural flooding of land subsiding over coal workings (Alvecote Pools).

References

Maps (Drift edition) and accompanying Memoirs of the Geological Survey of Great Britain, for sheets 154, 155, 168, 169, 201, 217 and 218.

Bishop, W. W. (1958). The Pleistocene geology and geomorphology of three gaps in the midland Jurassic escarpment. *Phil. Trans. R. Soc.* B. **241,** 255–306.

Matley, C. A. (1912). The Upper Keuper (or Arden) Sandstone Group and associated rocks of Warwickshire. *Q. Jl geol. Soc. Lond.* **68,** 252.

Shotton, F. W. (1929). The geology of the country around Kenilworth, Warwickshire. *Q. Jl geol. Soc. Lond.* **85,** 167.

Shotton, F. W. (1953). The Pleistocene deposits of the area between Coventry, Rugby and Leamington and their bearing upon the topographic development of the Midlands. *Phil. Trans. R. Soc.* B. **237,** 209.

Shotton, F. W. (1954). The geology around Hams Hall, near Coleshill, Warwickshire. *Proc. Coventry Distr. nat. Hist. scient. Soc.* **2,** No. 8, 237.

Shotton, F. W. (1968). The Pleistocene succession around Brandon, Warwickshire. *Phil. Trans. R. Soc.* B. **254,** 387.
Tomlinson, M. E. (1925). River terraces of the lower valley of the Warwickshire Avon. *Q. Jl geol. Soc. Lond.* **81,** 137.
Tomlinson, M. E. (1929). The drifts of the Stour–Evenlode watershed and their extension into the valleys of the Warwickshire Stour and upper Evenlode. *Proc. Bgham nat. Hist. phil. Soc.* **15,** 157.
Tomlinson, M. E. (1935). The superficial deposits of the country north of Stratford on Avon. *Q. Jl geol. Soc. Lond.* **91,** 423.

4. HISTORICAL GEOGRAPHY
THE EVOLUTION OF SETTLEMENT AND LAND USE

H. THORPE

The previous two chapters have, as it were, set the stage across which a long and varied succession of actors has moved to fashion Warwickshire's present landscape of villages, farms and fields, of towns and industries, linked by an intricate network of communications. In this process of the slow conversion of a once natural landscape into the complex cultural landscape of the twentieth century, the greatest change has undoubtedly been in vegetation. The vast, monotonous sweep of forest that covered most of Warwickshire by the Boreal-Atlantic transition (see Godwin, 1956) around 6000–5000 B.C., when insular conditions finally prevailed, has gradually been cleared back to make way for plants and animals of man's own choosing; for homes, mines and factories; for roads, canals, railways and airfields. Yet the description, "Leafy Warwickshire", still graces the county today, though its sylvan character now stems mainly from the small tracts of remnant woodland, from the many copses and great parks, and, even more so, from a vast lattice of tree-studded hedgerows that bear the imprint of a long process of enclosure rather than from puny survivals of the mixed-oak forests of Neolithic times.

Prehistoric Warwickshire (see Map 2, p. 21)

Unfortunately, we know little in detail of the floral character of these early deciduous forests of Warwickshire, but by analogy elsewhere it would seem that oak (which features so strongly in the county today, both arboreally and in such place-names as Oakley Wood or Cryer's Oak) was prominent, being found sometimes in pure stands, sometimes mixed with other trees such as elm and lime. In this connection it is interesting that a recent excavation at Barford in the Middle Avon Valley revealed acorns in association with a Secondary Neolithic cremation (Oswald, 1966–67). Similarly, we can only hint at the sequences of secondary vegetation that sprang up as man abandoned some of his early clearings—at least until subsequent pressures of population made them ripe for the axe and plough again. The many names incorporating *Heath* (e.g. Hockley Heath, Barton-on-the-Heath, Washwood Heath, Dunsmore Heath) are an eloquent reminder of anthropogenic influences on our vegetation, including also the development of distinctive "ruderal" or wayside habitats.

Some five thousand years ago small numbers of Neolithic pastoralists and cultivators had penetrated parts of Warwickshire and, with the aid of polished stone axes and

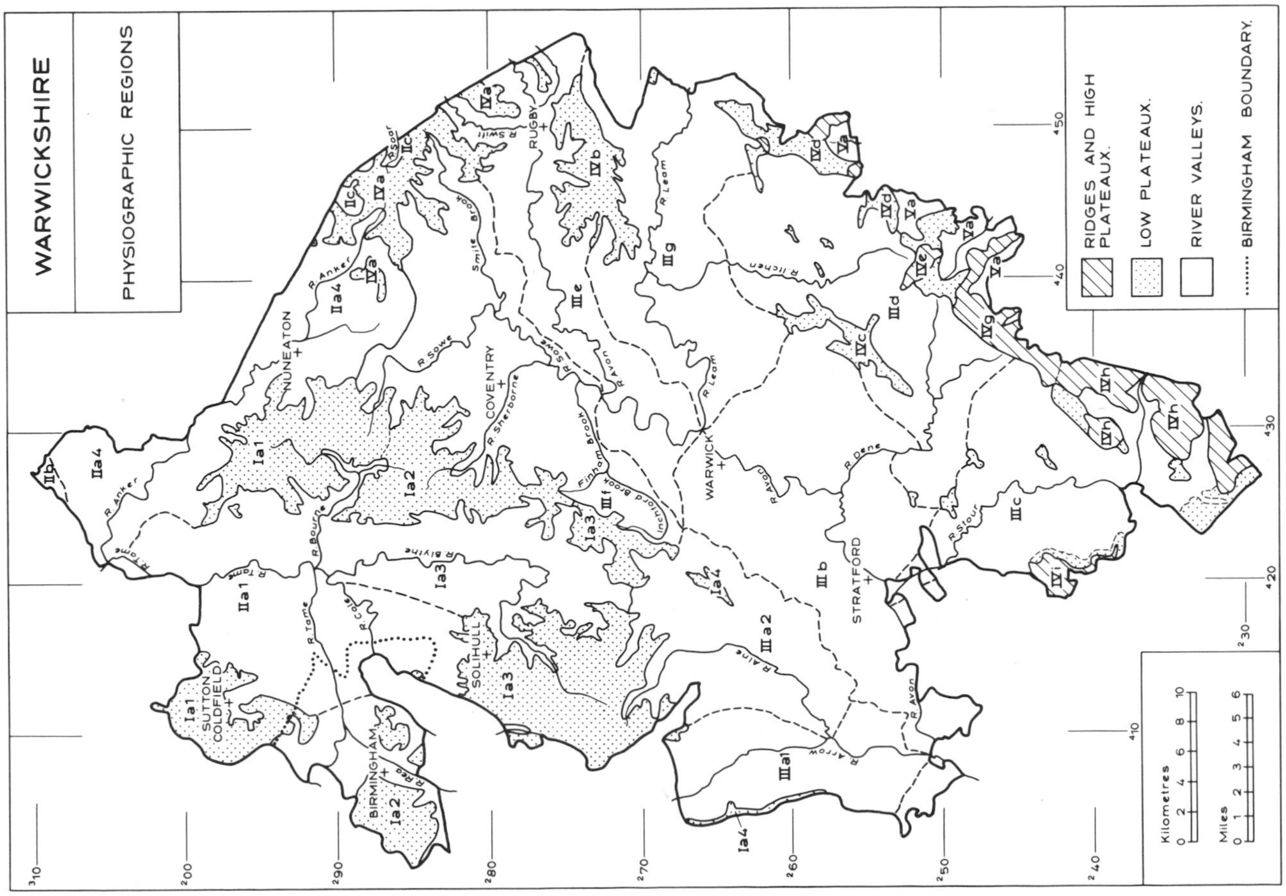

MAP. 1. Physiographic Regions

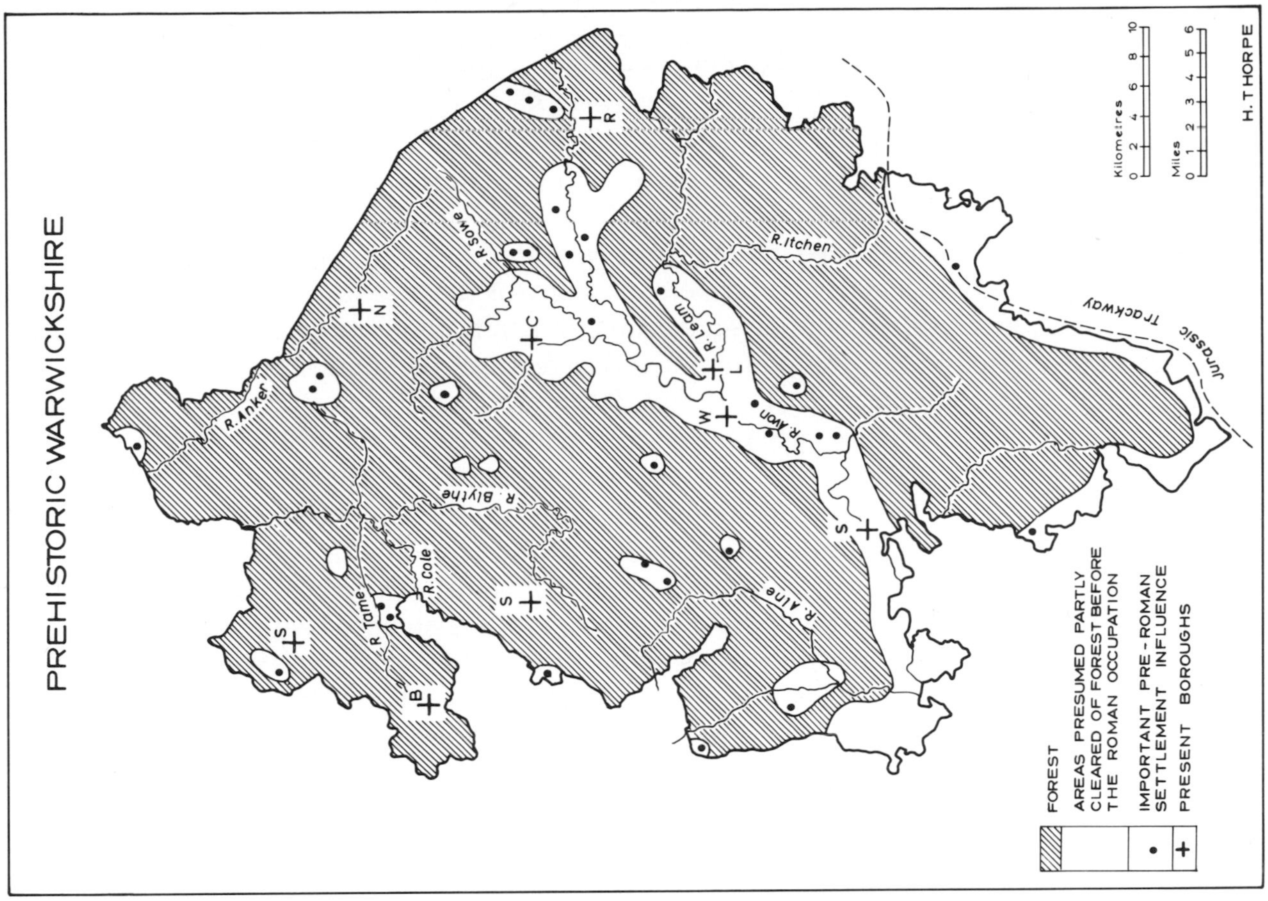

MAP. 2. Prehistoric Warwickshire

adzes, ring-barking and fire, had fashioned the first tiny, irregular clearings for grain and grass. Aerial photography in connection with the Avon–Severn Valleys Research Project has identified an intricate pattern of crop-marks on the broad terraces of sand and gravel, with their light loamy soils, that border the Avon and its major tributaries. The many sites (see Webster and Hobley, 1965), which total around one hundred in Warwickshire alone, cover collectively a wide range in date from Neolithic to Dark Ages and beyond. Continuous occupation, particularly by farmers, of many of these favourable sites right up to the present day has resulted not only in the superimposition of one cultural horizon on another, but also often in the obliteration of earlier features by later ones. The result is frequently a tangled web of piece-meal evidence. Important excavations have, however, taken place, often as "rescue digs" which have attempted to record the history and pre-history of threatened sites before gravel excavation completely destroyed all evidence, and the results have been recorded in annual Reports of the Avon–Severn Valleys Research Project and elsewhere.* Important glimpses of the economy in Neolithic times have been provided by the excavation of stockaded enclosures for animals, clay-lined pits that may have held water for stock, grain storage pits (for emmer wheat and barley) and numerous saddle querns in the area between Barford and Charlecote. Pastoral and arable produce would be supplemented by fishing and fowling along the river and by hunting. Fodder for livestock, especially cattle, sheep, pigs and goats, was available from the grass, weeds (especially plantain) and shrubs that were slowly colonising small clearings abandoned by shifting cultivators, and from forest leaves, litter and tree bark. Blackthorn, hawthorn, wild rose, rowan, bramble and holly probably spread over many such abandoned clearings.

There is also strong evidence that Neolithic folk along the Avon Terrace Belt were numerically and culturally strong enough to have an active concern with religious and ceremonial matters. Elongated rectangular enclosures of the cursus type (Anon, 1960a) have been identified, and no less than four lie between Charlecote and Barford (Webster and Hobley, 1965). Barford has also recently revealed a very fine henge monument of Dorchester type (Oswald, 1966–67), a long barrow (now destroyed) was also found at Loxley, and one cannot doubt that the influence of early man in Warwickshire from Neolithic times onwards was much stronger than was once supposed. As aerial reconnaissance and excavation continue, not only along the Avon Valley and its tributaries but also along the Tame, Blythe and Anker Valleys to the north, this is likely to become even more abundantly clear.

Much of the early archaeological evidence so far discussed must be considered as spanning Neolithic and Bronze Age times. This is especially so with many stone artefacts, a large number of which have been found in the upper Avon Valley north of Warwick; of particular interest is the concentration of such finds, ranging from Palaeolithic to Bronze Age date (and beyond), on the dry sands and gravels of the Baginton area which suggests not only continuity of occupation but also considerable movement of folk both along

*Vide *West Midland Annual Archaeological News Sheets*, produced and distributed by the Department of Extramural Studies, University of Birmingham.

and across this important river line (see Thorpe, 1950). Before 1950 very little Neolithic or Bronze Age pottery had been identified in the county, but recent excavations have revealed Neolithic pottery in association with small hovels on the Lower Keuper Sandstone meander-core at Warwick, as well as on the Avon terraces at places such as Barford and Charlecote. Bronze Age pottery was also found on these sites, while at Ryton on Dunsmore a cemetery of middle to late Bronze Age date with cremation urns was unearthed. A modest number of bronze axes, particularly from the upper Avon and its tributaries, suggests by inference that clearing was being extended, aided by fire. On such clearings circular stockades for livestock, similar to that excavated at Charlecote, might have been seen, with small arable closes nearby, while round barrows for the dead were probably more numerous than present evidence suggests. A number of these have been found between the Sowe and upper Avon, and it seems likely that where intensive arable farming has persisted down to the present day the bite of the plough has removed others from sites further down stream. For example, recent excavations at Barford have revealed two such barrows. In all, we may therefore conclude that during Neolithic and Bronze Age times very effective occupation, associated with active clearing, cultivation and stock keeping, had taken place along the major river lines, as well as on the higher ground around Hartshill on the East Warwickshire Plateau.

On the present evidence there is little sign of Neolithic or Bronze Age occupation on the forested Lower Lias Clays well south of the Avon or along the drier Cotswold Fringe itself. Similarly, north of the river there were large tracts of woodland on the Keuper Marl which had never experienced the axe or fire. Yet the presence of the Rollright Stones on the Warwickshire–Oxfordshire border suggests that there may have been more activity along the Cotswold edge than the paucity of other finds would have us suppose. For by Early Iron Age times an important branch of the Jurassic Way followed the Cotswold edge, and one may speculate whether such scarp-top routeways had indeed not come into being much earlier. The location of numerous hill forts or camps on the Oxfordshire side of the scarp, together with the presence of similar features at Meon Hill, Ilmington and Nadbury, suggests that a strong occupation of the scarp top had occurred in Early Iron Age times, if not before. The keeping of livestock was again important among these clustered communities, and the light loamy soils of the Cotswolds probably gave good arable yields for one or two years before diminishing returns set in and new land had to be cleared. The slow spread of such farming activities northward down the scarp face on to the heavy Lower Lias Clays probably marked the first major attack on the extensive forests that then extended almost to the Avon. It was perhaps at this time that the Red Horse of Tysoe was etched on the Middle Lias Marlstone outcrop, presupposing that much of the scarp face—hereabouts at least—was by then free from wood. Such a figure was intended to be a dominating man-made feature of the local landscape and was clearly best seen from some distance back in the Vale that was later to bear its name; but one can only speculate as to how far back the forest had by then been pushed.

Along the Avon at this time not only had there been continuing occupation of earlier sites, but clearing had been considerably extended both along, and away from, the river. The importance of livestock in the economy is seen in the large stockaded enclosure of Oakley Wood, while north of the river many small circular or quasi-circular enclosures had come into being as population pressed into the Arden woodland. For example, univallate enclosures, such as that at Beausale, may have served as summer cattle pounds for clustered communities strung along the Avon terraces; they were certainly too small and poorly defended to have been built hurriedly, late in Early Iron Age times, to oppose the Roman advance beyond the Avon. The excavation of crop-mark sites along the Avon has revealed many circular and rectilinear enclosures with Early Iron Age associations of pottery and the like. A good many of these may have been concerned with arable or pastoral pursuits; in this connection, one of the sites excavated at Barford revealed banks and ditches that appeared to be part of a Romano-British field system, though within Warwickshire as a whole one knows very little in detail about prehistoric field systems.

Apart from such great fortresses as Meon Hill (just outside VC 38), along the Cotswold Fringe, strongly defended camps had been established at Oldbury and Corley on the East Warwickshire Plateau, while in the valley of the Leam at Wappenbury the impressive fort guarding this river-crossing point gave evidence of Belgic penetration into our midland area. The low survival value of iron, compared with gold and bronze, probably accounts in part for the paucity of Iron Age metal finds in Warwickshire; this being so, it is remarkable both as a survival and as a token of wealth and influence that an early excavation in 1824 at Meon Hill produced almost four hundred currency bars of iron from this one site. The likelihood of continuity of occupation from Neolithic to Bronze and Early Iron Age times is suggested by evidence from numerous sites, and it is logical to suppose that the subsequent occupation of the area by the Romans and the rise of Romano-British communities was strongly conditioned by earlier clearing, the rise of trackways and river-crossing points. It is little wonder then that at places such as Alcester and Chesterton on Fosse evidence of pre-Roman occupation has been found beneath Roman horizons. Finally, an attempt has been made in Map 2 to indicate cartographically those parts of Warwickshire presumed to have been at least partly cleared of forest before the Roman occupation. In preparing this and subsequent maps, "clearing influence" in whole or in part has been assumed to have taken place within a radius of half a mile of all major occupation sites, clusters of finds, datable crop-marks and important routeways. The valley of the Avon and its tributaries stands out as the principal cleared area, followed by the Cotswold Fringe; between the two, dense forest occupied the Lower Lias and Keuper Marl until the ground fell gently to the well-settled belt of Avon Terraces. North of the river, pockets of clearing and settlement had arisen in the lower reaches of the Alne and Arrow (river names which, like the Avon, are of Celtic origin) (Mawer and Stenton, 1936); near Lapworth; around Castle Bromwich on the interfluve between the Tame and the Cole; in the vicinity of Sutton Park, and around the forts of Oldbury and Corley on the East Warwickshire Plateau. Apart from this the Forest of Arden dominated the regional personality of the land north of the Avon, extending from the Lower Lias projection north-west of Stratford across the Keuper Marl to the Coal Measures and older rocks of the East Warwickshire Plateau. Elsewhere, beyond the line of alder, reed and alluvial water-meadow bordering the main stream the ground rose gently to the first of the terraces with its small fields of grain, grassy enclosures, and sporadic settlements ranging from small farmsteads to hamlets, and even to fortified villages such as Wappenbury. Close by, on abandoned clearings along the forest fringe secondary vegetation of birch, hazel, blackthorn, hawthorn, rowan and elder could no doubt be found, with small shrubs, grass and weeds competing for space between the thickets.

Roman Warwickshire (see Map 3, p. 28)

Roman influence within Warwickshire, following the Claudian invasion of A.D. 43, was considerable and it is essential for our understanding that the area covered by our present county be viewed within a wider, national context. The main military advance from the Thames crossing at London was effected by three columns spreading fanwise. The left wing, consisting of one legion, advanced westward towards Exeter while the right wing, also of legionary strength, struck north towards Lincoln. The task of pressing north-westward across the scarplands towards the Avon and Severn Valleys, and ultimately towards the Dee, was entrusted to two legions whose great supply road, the Watling Street, now forms the north-east boundary of the county. By about A.D. 47 the right wing had reached Lincoln, the left was on the borders of Devon and Dorset, with units of the main force deployed along the Avon line and its watershed with the Trent. Following the subjugation of Celtic tribesmen occupying fortified camps along the Cotswold edge, a great diagonal road, the Fosse Way, was constructed linking the south-west with Lincoln; its straight course was cut transversely across Warwickshire south of, and roughly parallel to, the Avon to intersect Watling Street at High Cross (Venonae). Expansion beyond the Fosse saw the extension of Watling Street towards Wall (Letocetum, in south Staffordshire), lying in the col between Cannock Chase and the South Staffordshire Plateau, where it was joined by the Ryknield Street; the latter, which marked an important further stage in the advance, left the Fosse at Bourton-on-the-Water, struck northward to cross the Avon at Bidford, the Alne at Alcester, and the headwaters of the Rea and Tame near Birmingham.

In their advance across Warwickshire the Roman forces probably encountered little serious resistance for, as we have seen, the number of Celtic strongholds was small. Roman forts were probably established at such strategic points as Cave's Inn (Tripontium) and High Cross along Watling Street, and have been proved by excavation at Baginton in the densely settled upper Avon–Sowe Valleys and at Metchley on the site of the Medical School of the University of Birmingham. Along the line of the great roads and around the forts swathes of clearing must have been considerable, as also along such secondary roads as that running from Alcester to cross the Avon at Stratford, or that linking Tripontium, Baginton and Droitwich (Salinae, in what is now Worcestershire).

As peaceful conditions prevailed, important civil settlements arose on new sites such as Chesterton and Princethorpe along the Fosse, at Mancetter (Manduessedum) on Watling Street and at Tiddington on the Avon. From the very large number of coins found in the Birmingham area one would also suppose that some considerable clearing and settlement had taken place here along the Ryknield Street and in the valleys of the Tame, Rea and Bourne Brook. Romano-British farming must also have been extensive along the Avon Valley, in the lower Alne, and at points along the Fosse and Watling Streets associated with the slow spread of clearing. It is also thought that the Romans introduced some "foreign" trees, such as sweet-chestnut, walnut and sycamore, to their urban gardens and villa estates.

A comparison of Maps 2 and 3 shows that by the latter part of the fourth century there had been a considerable extension of clearing south from the Avon Terrace Belt to the line of the Fosse, with lobes following minor roads. The shrunken block of south Warwickshire woodland now lay south of the Fosse on the heavy soils of the Lower Lias Clay almost as far as the scarp foot of the Cotswolds where further extensions of deforestation for arable and pasture had taken place. North from the Avon Terrace Belt felling had also continued, with broad extensions up the valley of the Alne and along Ryknield Street and its tributary roads. The Forest of Arden was now sharply defined as a great triangular expanse of woodland bounded by the Avon Terrace Belt in the south-east, by the Ryknield Street and lobes of clearing around Alcester and Metchley (Birmingham) on the west, and by the Watling Street and its cleared flanks on the north-east. But within parts of the Forest of Arden itself considerable industrial activity could have been discerned, associated particularly with the making of pottery, tiles and glass, while around the clusters of kilns felling of timber for fuel had proceeded apace, associated also with the slow spread of subsistence farming across the clearings. Thus at Hartshill in north-east Warwickshire over thirty kilns specialising in *mortaria*, and ranging in date from the early second to the fourth century, have been found, the source of clay again being the local Keuper Marl (Anon, 1960b, 1961). At Mancetter, not very far to the north, several Romano-British pottery kilns were discovered associated with a heated drying shed for "green" pottery and/or for grain. A further find of considerable interest on the same site was a glass-making furnace (Anon, 1964), while at Chilvers Coton and Griff Hill Farm in Bedworth other tile kilns have been excavated. Elsewhere in Arden, tile kilns using Keuper Marl have been found at Chase Woods (near Kenilworth) and also at Rowington Green along the west–east road linking Cave's Inn, Baginton and Droitwich, while south of Baginton a large pottery factory has come to light at Wappenbury with other kilns at Snowford Bridge and Ryton on Dunsmore. In view of the numerous finds that have been revealed by detailed archaeological investigation in recent years, one may safely conclude that Roman influence in Warwickshire was considerable, even in parts of the Forest of Arden, long considered to be a cultural back-water. As such investigations proceed, it may well be necessary to modify further the present picture which, on the evidence of Map 3, suggests that the Romano-British contribution to the evolving regional personality of Warwickshire was profound.

Anglo-Saxon Warwickshire (see Map 4, p. 28)

We know very little about the conditions that obtained in the Warwickshire area after the withdrawal of Roman power at the turn of the fourth and fifth centuries. A resurgence of paganism among Celtic tribes on the Welsh border followed by raids on the pacified and Christianised communities within the Trent–Severn–Avon watershed must for some time have disrupted both local farming and urban organisation (see Thorpe, 1954). The reduced demand for grain, following the closure of Roman markets, probably meant that in many areas tracts of former arable were abandoned to grass, scrub and heath. These difficulties were further aggravated by repercussions of the Anglo-Saxon invasions spreading inland from the east and south coasts. Many scholars (see Gelling, 1954; Cook, 1958) now believe that the arrival of barbarian forces in our area may have been taking place by A.D. 500, which is somewhat earlier than was formerly supposed. It seems probable, too, that the first invaders, who were also essentially colonists, either took over pre-existing Celtic settlements and fields or, in areas where Celtic resistance was strong, settled on old abandoned clearings basing their livelihood initially on plunder and on pillaged cattle, supplemented by hunting and fishing. As peaceful conditions obtained, so the swing to cultivation and settled farming took place.

Three main penetration routes can be clearly recognised in our area—a well-defined Anglian route penetrating up the Trent and Tame Valleys; a second Anglian entry via the Nene and Welland to the Leam, upper Avon and Sowe; and a Saxon (Hwiccan) route following the lower and middle Avon and the Fosse Way (see Thorpe, 1950). Somewhere in the vicinity of Warwick, Anglian and Saxon elements met and came to terms concerning peaceful coexistence. Such cemeteries as those at Bidford, Tiddington, Longbridge, Emscote and Baginton show the concentration of early communities on the long-cleared and cultivated terraces bordering the Avon, or, like those at Brinklow, Marton, Offchurch, Compton Verney and Stretton on Fosse, on ribbons of clearing flanking a major Roman road. An interesting phase of expansion of population from the primary Anglo-Saxon settlements into the slowly shrinking woodland now began. South of the Avon a close stipple of Anglo-Saxon nucleated villages and hamlets arose with open fields and common pastures biting deeply into the remaining woodland, creating a landscape of extensive cleared areas, separated by shrinking forest blocks, whose personality was strongly reflected in the regional name, the *Feldon*, that came into being to describe the area between the Avon and the Cotswold edge. By contrast, north of the Avon Terrace Belt expansion into the Forest of Arden was a much more gradual and piece-meal process that continued well into post-Norman times, and was associated with a scatter of small farmsteads and tiny hamlets, with only here and there an occasional village such as Birmingham and Coleshill.

In view of the rather late character of most of the Anglo-Saxon settlement in Warwickshire, it is not surprising that authentic *-ingas* names are absent from the area, but several examples of early terminations in *-ingatun* and *-ingaham* occur in the Tame and Avon Valleys. Thus *Beormingahām*, the early form of the name for Birmingham (Gelling, 1954),

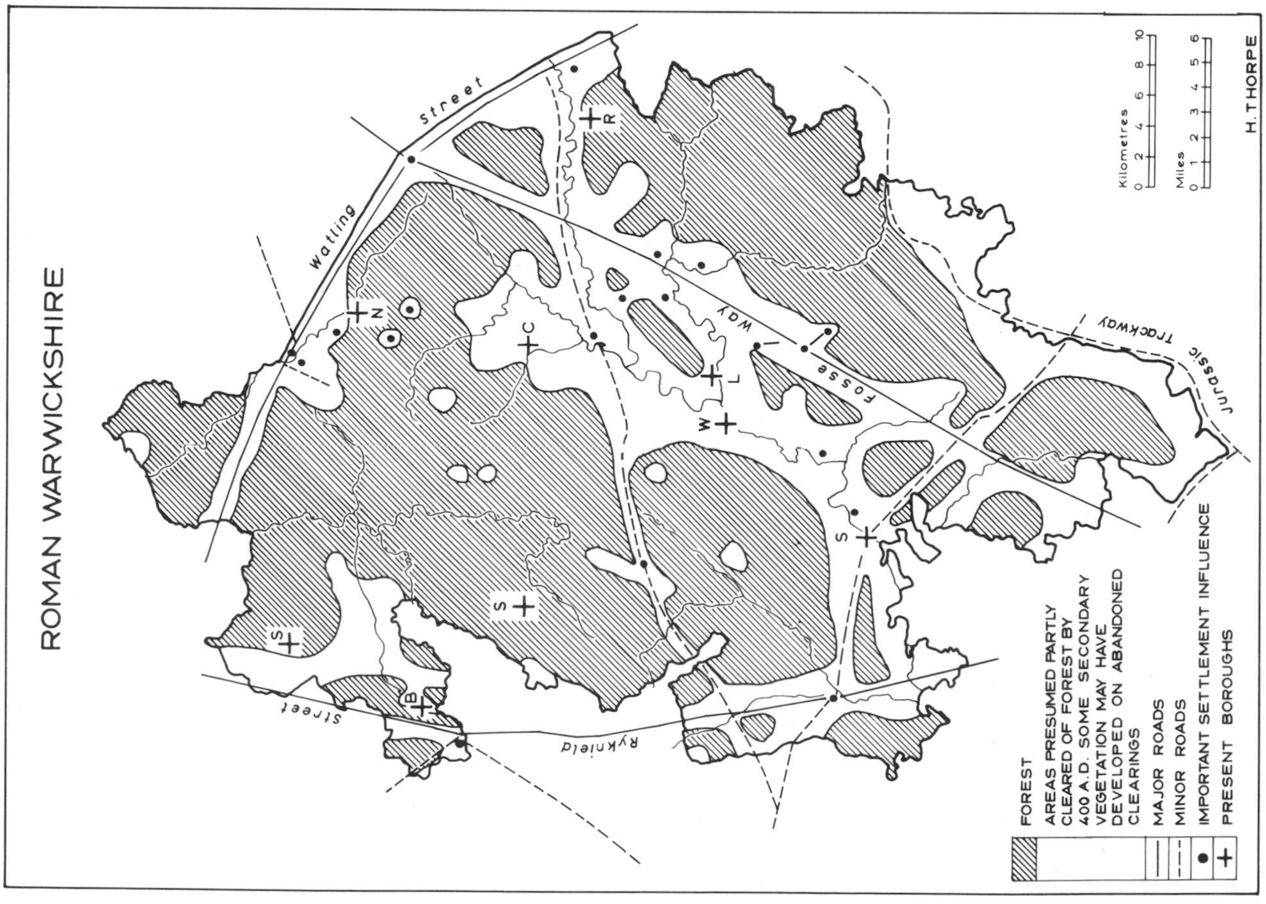

MAP 3. Roman Warwickshire

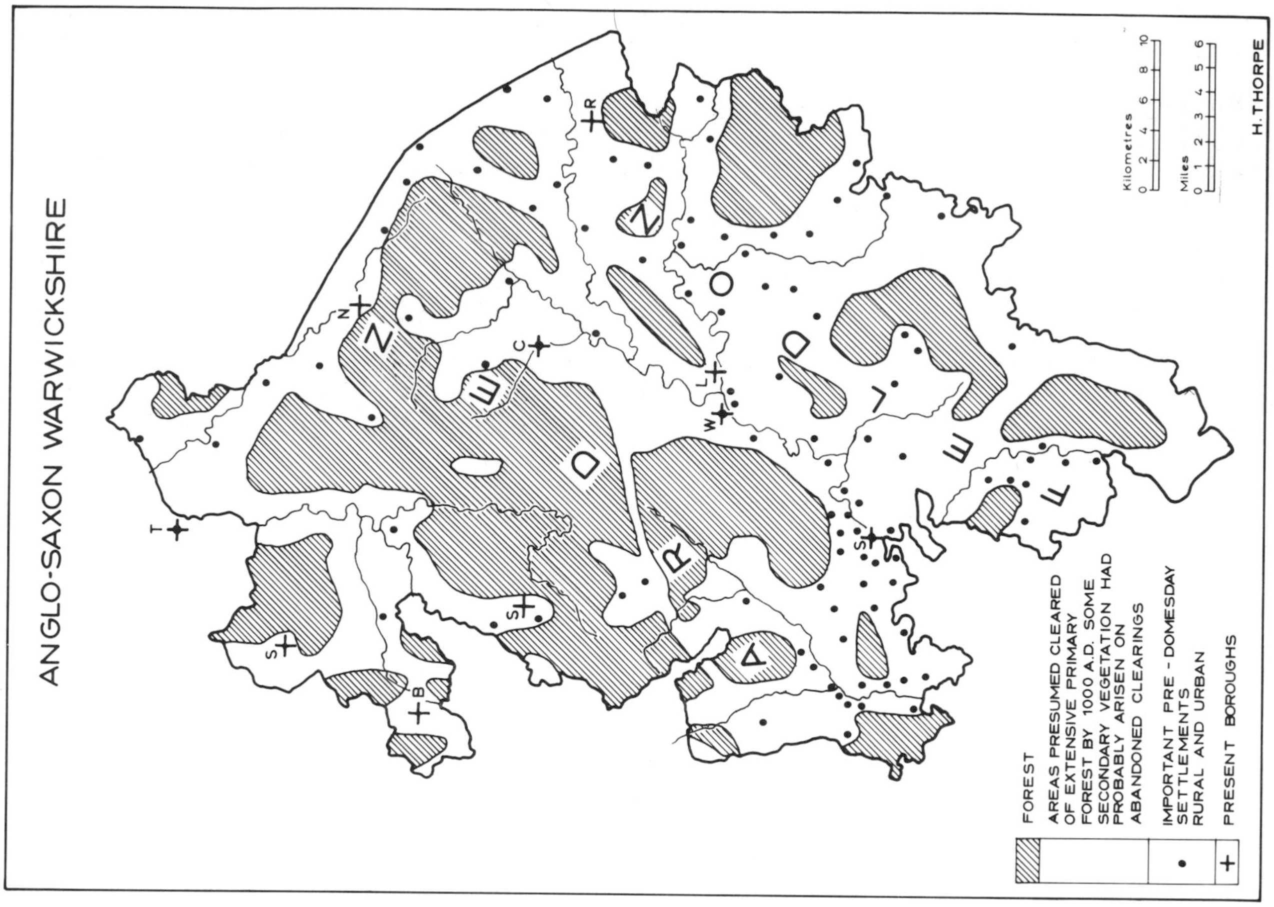

MAP 4. Anglo-Saxon Warwickshire

means "homestead of the people of *Beorma*" and is consistent with the settlement of an early Anglo-Saxon group under a leader in the valley of the Rea which, as we have already seen, had witnessed a good deal of clearing in Roman times. It is interesting, too, that whereas names in *-ton* and *-ham* are numerous along the Avon Terrace Belt and in the Feldon, late Anglo-Saxon clearing names in *-ley*, *-worth* and *-field* are concentrated almost exclusively in the late cleared tracts of Arden, scarcely any examples being found south of the Avon (Thorpe, 1950; Gelling, 1954). The occurrence of Anglo-Saxon names with pagan associations provides further evidence of early Anglian penetration of the Tame and Rea Valleys. Thus, Weoley Castle, now the name of a suburb on the south-western outskirts of Birmingham, goes back to Old English *weoh-leah* meaning "woodland clearing associated with idol worship". A little further south and just across the border into Worcestershire, lies Arrowfield Top, incorporating the element *hearg* signifying a heathen shrine. North of Birmingham the worship of *Woden* was also well established in groves within the woodland of south Staffordshire at places like Wednesbury (Woden's fortress) and Wednesfield (Woden's clearing), while the worship of *Tīw* is recorded in the name Tysmere near Alvechurch (in Worcestershire).

It is again significant that pre-Domesday place-names incorporating tree or wood elements are concentrated north of the Avon. A word of caution should be introduced here, however, since the distribution of such names in a datable context may depend on the fortuitous survival of manuscripts for particular areas. Thus Wootton Wawen ("farmstead by the wood", recorded A.D. 723–27) was the centre of an enormous forested township from which later sprang the townships of Henley in Arden and Ullenhall. In the same general area of western Arden, Nuthurst and Westgrove (Haselor parish) were both recorded in A.D. 704–9, while Aspley (A.D. 963, Tanworth parish) took its name from a clearing amid the aspens (*Populus tremula*). Mention has been made earlier of the slow spread of secondary vegetation on abandoned clearings, giving rise to tracts of heath, and this may well be reflected in the name, Broom (A.D. 710), located on sandy river terrace soils of the Arrow close to Ryknield Street.

The grid of Roman roads, with the earlier pattern of Celtic trackways, was now reinforced by a finer mesh of rough roads linking Anglo-Saxon hamlets, villages and towns. In the Feldon highly organised open field cultivation, under a two- or three-field system perhaps, was probably associated with another distinctive landscape feature, namely the holding of land in long acre strips, and in subdivisions thereof; similarly the use of yokes of oxen drawing the heavy, mould-board plough, together with the need for drainage on the heavy Lower Lias Clays, may have been accompanied by the introduction of the reversed-S curves of ridge and furrow, those symmetrical corrugations that were so typical of Feldon fields until the coming of the tractor. The reversed-S curve, or aratral curve, is attributed to the swing of the ox-team to the left when approaching the headland at each end of the strip in order to facilitate the turning of the heavy mould-board plough. The needs of livestock were met by grazing on the fallow field and on the waste, reinforced by the provision of hay from alluvial water-meadows and grassy enclosures. By contrast, within Arden irregular hedged and banked fields well protected against deer had spread, the emphasis generally being on cultivation by individuals or small groups rather than on co-aration as practised in Feldon. Hunting in the forest was probably still at this stage an important aspect of the economy.

By the end of the Anglo-Saxon period fortified towns had arisen at Warwick (the shire town) and at Tamworth (the old Mercian capital), while such places as Coventry, Stratford, Coleshill and Birmingham were probably also discharging urban functions by then. Danish settlement in Warwickshire from the latter half of the ninth century had been only slight, reflected in names such as Rugby, Willoughby and Wibtoft all lying close to the Watling Street that generally marked the western extent of the Danelaw proper. But Danish pressure on Mercia had one very important influence, in that the creation by Aethelfleda, Lady of Mercia, of the fortified *burh* at Warwick in 914 defined the territorial unit that was to be tributary to the shire town and by so doing created Warwickshire. Map 4 portrays the extent of remnant woodland in Warwickshire around A.D. 1000, and should be considered in close association with the map of Domesday woodland and clearing, Map 5. It is evident that the forest blocks of Feldon had shrunk considerably, whereas within Arden clearance had been more sporadic and much less spectacular, concentrating particularly in the valleys of the Tame, Cole, Blythe and Alne.

Domesday Warwickshire (see Map 5, p. 32)

By the end of the Anglo-Saxon and Viking periods Warwickshire had received its last bulk contribution of colonial settlement. The Norman Conquest initiated in 1066 was to contribute military strength on a national scale, overlordship, powerful organisation and a strong feudal system rather than bands of homely settlers. In this it had certain affinities with the Roman occupation, but very little in common with the Anglo-Saxon colonisation. After the Battle of Hastings Norman control was quickly established in Warwickshire, and Norman lords took over many estates and their appurtenant manors. Even the Anglo-Saxon thegn, Turchil of Arden, who had not fought against William, was not allowed to retain his position for long before a powerful Norman, Henry de Newburgh, was created Earl of Warwick with control of the shire town. That peace was quickly established in the county as a whole may be judged by the paucity of defences of the motte and bailey type within the shire, except for those in important towns such as Warwick and Tamworth, and at strategic settlements in rural areas such as Brinklow on the Fosse, Seckington and Castle Bromwich. Where lesser Saxon lords remained in control of manors, they did so only by acknowledging fealty to a Norman overlord. Once peace was firmly established, farming went on apace as new Norman lords sought to extend their revenues by expanding cultivation, increasing their flocks and herds, and by undertaking colonisation of woodland and waste. It would seem that the rabbit was also introduced at this time, initially being bred in captivity as a tasty addition to the festive board but soon escaping to colonise vast areas of woodland and waste. Its influence on smaller plants and herbs was to be considerable, while its importance as an addition to the food supply of many villeins, bordars, cottiers and serfs should by no means be overlooked.

The desire of William to assess the extent and wealth of his new domain led to the making of the Domesday Survey in 1086. This remarkable record gives a description of every manor, its population and land use, enabling us to reconstruct the major patterns of social and economic geography for the period. The pictures thus obtained are best considered as constituting a summary of the regional geography of Warwickshire at the end of the Anglo-Saxon period rather than as something distinctly Norman. Detailed analyses of the Domesday geography of the county have already been published (Kinvig, 1950; Darby and Terrett, 1954), and this is not the place to attempt another. Instead, we propose to present here the salient features of the regional differences to be observed in the county at that time, and to use these data in understanding the complex of historico-geographical factors that underlie the distinctive patterns of remnant woodland and cleared land that are shown in Map 5. Although the existence of many places is recorded, or can be proven by archaeology, for the period before the Norman Conquest, Domesday Book provides the first comprehensive list of rural and urban settlements county by county for most of England. The resulting distribution pattern for Warwickshire, which records some 280 places, is incorporated in Map 5 and clearly shows a greater density of settlement within the Feldon and the Avon Terrace Belt than existed in the Forest of Arden with its thinner, more sporadic occupation. No account is given of the actual buildings in the settlements, though from the reference to a priest in 71 settlements one would imagine that a church existed in each; one also would like to know whether yew trees graced the churchyards then (carefully walled or fenced as they must have been to prevent livestock from breaking in), providing bows for the woodmen and archers of Arden, but unfortunately Domesday is silent on such points. Though small market towns probably existed, none is mentioned apart from the single borough of Warwick which then had 248 houses. The recorded population of Warwickshire was probably around 7000 families which might mean as a very rough estimate a total population of about 35 000 or 40 000. Domesday population data are usually plotted in terms of recorded adults, but, as these were generally the male heads of families, it is better to think in terms of families. If one plots the density of population on the basis of the old Anglo-Saxon Hundred divisions within the county, most of the Feldon averages 11 families (or about 55–60 persons) per square mile and the Avon Terrace Belt about 8 families (40–45 persons) per square mile. By contrast, population densities north of the Avon Terrace Belt fell to no more than 4 families (about 20–22 persons) per square mile around Birmingham and even as low as 2–2·5 families (about 10–13 persons) per square mile in the extensive Arden woodland between Tanworth in Arden and Atherstone.

A measure of the intensity of arable farming is provided by the number of plough-teams recorded for each manor. Notionally, a plough-team would cultivate the equivalent of one hide of land, which averaged about 120 acres, although the exact dimensions varied locally according to custom, the character of soils and the size of the plough-team. Totalling these by territorial Hundreds, there is a marked gradation northward from 4 plough-teams per square mile in the Feldon, indicating intensive cultivation of the great

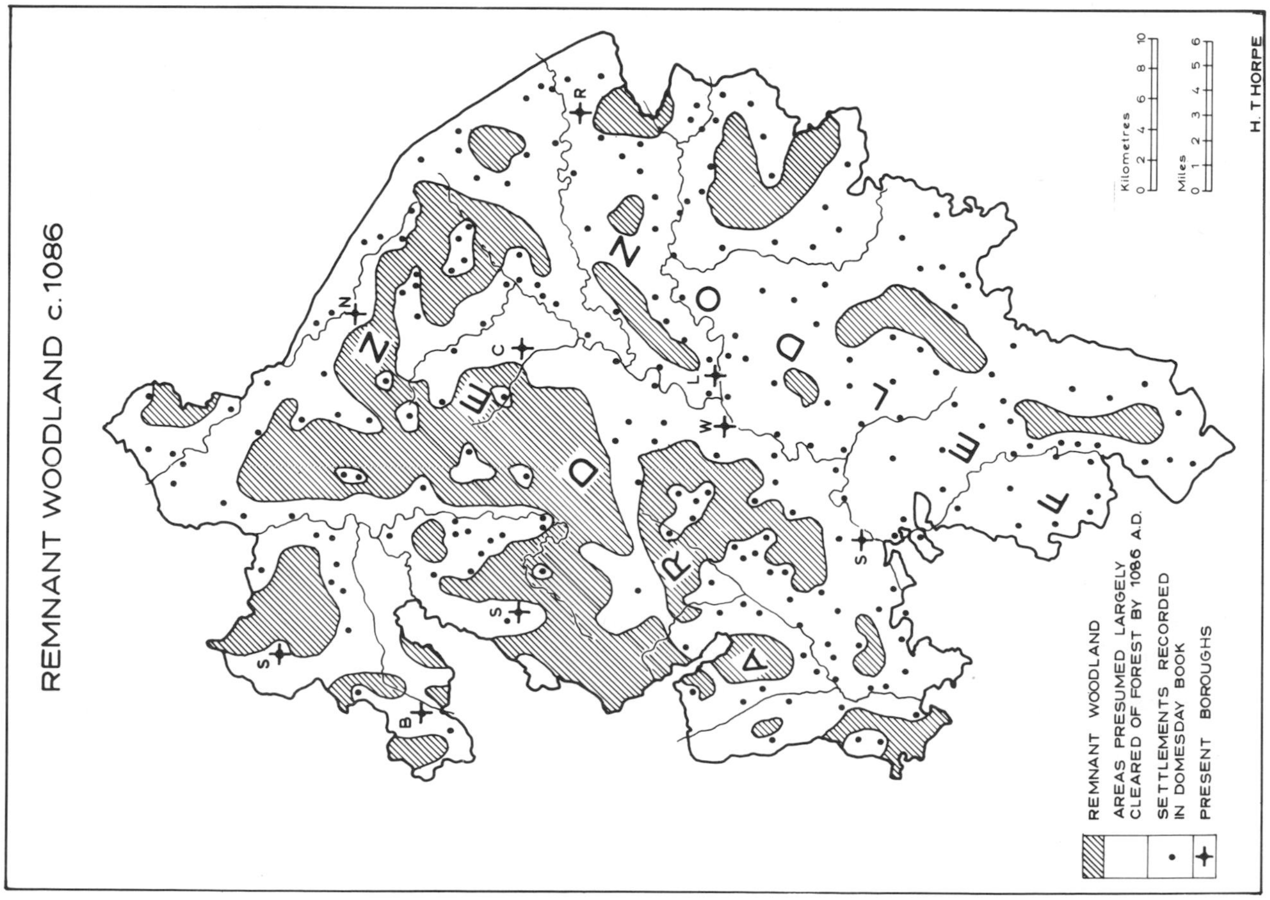

MAP 5. Remnant Woodland *c*. 1086

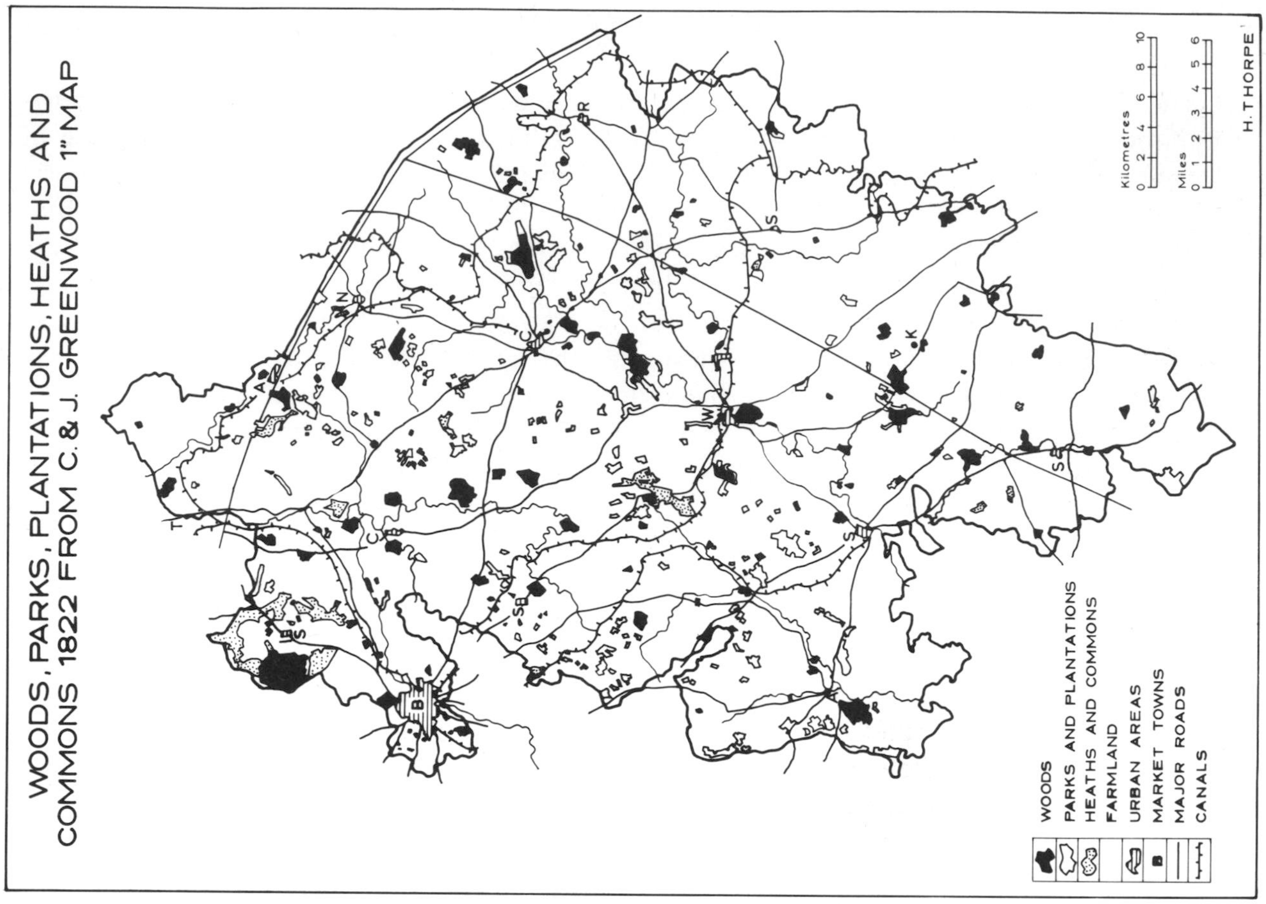

MAP 6. Woods, Parks, etc. 1822

open fields that surrounded the many nucleated villages, to 3 plough-teams per square mile among the Avon Valley settlements. Once into Arden the density dropped to 2 plough-teams, as around Birmingham, and to as low as 0·7 and 0·6 in the woodland heart north-east from Tanworth. But meadow land was of even greater value than arable land in medieval times, for it provided the important hay crop on which so many livestock depended for supplementary feed to see them through the long winter. Recorded acres of meadow were high within the Feldon and the Avon Terrace Belt, much of this being derived from alluvial water-meadow bordering streams but also often being mown from grassy enclosures. Remarkably little meadow was recorded in the area north of the Avon terraces, apart from small acreages in the Tame and Blythe Valleys, suggesting that abundant feed was available from wood and waste. By contrast, one has the impression that pressure of population in the Avon Valley and in the Feldon had increased to the point where the extent of plough land had been pushed to the township limits in many villages; where forage was scarce, the greater was the need for hay in winter to feed plough oxen, milch cows and sheep. The number of recorded mills further highlights the intensity of grain production in the Feldon and along the Avon.

It is most fortunate for this account of the spread of clearing in Warwickshire that the Domesday Survey records the amount of woodland remaining in each manor. The entries are usually given as a notional area, so many leagues (a Domesday league equalled 12 Domesday furlongs) in length and so many leagues in breadth; sometimes the dimensions are in furlongs and very occasionally in perches or even directly as acres. Supplementary information concerning the pannage capacity of the wood may also sometimes be given. Thus the large wood at Stoneleigh, which was 4 leagues long and 2 leagues broad, provided "feed for 2000 swine", whereas the smaller wood of 6 furlongs by 4 furlongs at Coughton could support only 50 swine. Warwickshire lacked the great expanses of Royal Forest to be found in many other counties, such as neighbouring Worcestershire, so that much of its woodland was ripe for clearing and exploitation, as will be seen later.

The extent of remnant woodland on the basis of Domesday data has been carefully plotted on Map 5, which provides a useful check on Map 4 for the Anglo-Saxon period—a map that had been built up by following a long succession of clearing influences beginning in distant Neolithic times. From a comparison of the two maps it is evident that clearing had continued strongly in the Feldon throughout the latter part of the Anglo-Saxon and early Norman periods, leaving only a few small blocks of woodland. North of the Avon, however, the strength of Arden woodland was only slightly diminished, clearings having been extended in the Cole and Blythe Valleys, and many new "intakes" established as small nodes of clearing in the great forest block lying north-west and north-east of Coventry. In all, one is reminded of Leland's recognition of a major regional distinction persisting within Warwickshire even as late as about 1540:

> I learnyd that the moste parte of the shire of Warwike that lyeth as Avon river descendithe on the right hand or rype of it, is in Arden . . . and the grownd in Arden is muche enclosyd, plentifull of gres . . . and woode, but no great plenty of corne. The othar part of Warwykshire that lyeth

on the lefte hond or rype of Avon river, muche to the southe, is for the muste part champion, somewhat barren of woode, but very plentifull of corne. (Toulmin Smith, 1906–10)

The term "champion" here signified open-field arable land subject to co-aration and lying unenclosed. As one would expect, the names of places recorded for the first time in Domesday Book include many Anglo-Saxon clearing names such as Haseley, Snitterfield and Kenilworth north of the Avon. Elmdon ("hill of the elms") and Whichford ("ford by the wych-elms") are also recorded, being the only major place-names in Warwickshire to include this tree as an element.

Warwickshire from the Twelfth to Seventeenth Centuries (see Map VII)

As peaceful conditions prevailed in Warwickshire after the force of conquest was spent, Norman lords, lay and ecclesiastical alike, ensured that their new estates were made as productive as possible. Not only were the areas of arable and meadow extended by encroachments on local wood and waste, but a vigorous colonisation of Arden woodland was fostered by local lords such as the Earl of Warwick who offered tracts of land for clearing on attractive monetary and service terms. The conflicting needs of cutting timber for domestic use, and of preserving large trees for the future, probably led in many places to a development of what amounted to "coppice with standards" in the underwood of which hazel featured strongly. From studies of a great many land charters, Roberts (1965) has shown how piece-meal clearing, associated with irregular fields and amorphous intakes, occurred within Arden linked with the rise of many small freemen farmers dwelling within moated homesteads that still remain strong features of the rural settlement of the area today (see also Roberts, 1962; Emery, 1962). It has been possible to date the establishment of many of the irregular banked and ditched fields at places like Umberslade and Earlswood in and around Tanworth parish, and to trace the slow creep of deforestation across the area; similarly, it is known that many of the moated homesteads arose between 1150 and 1350. Earlswood took its name from the great wood of the Earl of Warwick, the latter being overlord of the manor of Tanworth in Arden. Although well-organised open-field cultivation was practised around such villages as Tanworth and Hampton in Arden, many small scattered farmsteads now came into being, each set amid an "island" of clearing. The moat surrounding the homestead, and the large hedged banks and ditches bordering the small fields served to keep deer and other wild animals from damaging the kitchen garden and the field crops respectively. The large fishponds, such as those at Codbarow near Umberslade, which we find closely associated at this time with many moated homesteads in Arden (as also with numerous nucleated villages in Feldon) indicate the great importance of fish in the food supply then. Although most of these fishponds have since been drained, today their soggy beds favour the growth of marshy meadow dotted with meadowsweet and sedge.

As one might expect, many Arden place-names incorporating references to wood are mentioned in documents of the twelfth and thirteenth centuries, as, for example, Henwood (about 1155, Solihull), Forewood (1265, Wootton Wawen) and Waverley Wood

(1204, Stoneleigh). Sometimes the type of tree is specified, as in Hazelwood (1221, Preston Bagot) or The Hollies (1279, Allesley); occasionally, too, we find personal names such as *John le Berker* (1288, Exhall near Coventry) suggesting the use of oak for tanning bark, as the village name Tanworth in Arden (1201, "Tanner's clearing") also implies. References to *Wood* (*-wood*), and *Wood End* names continue to be plentiful throughout the fourteenth and fifteenth centuries, by the end of which some slackening in the force of clearing seems to have come about. Of particular interest, too, in this post-Norman period are the many names incorporating *heath* and *moor*, which again occur most strongly within Arden. Thus the sandy district of Moorhills in Rowington parish seems to take its name from *Robert de la More* who had his home there in 1297, while *Simon de la Heth* is recorded in Tanworth parish in 1332. It would seem that some extensive tracts of heath or moor had sprung up on abandoned clearings within Arden, particularly where the dominant Keuper Marl was covered by deposits of glacial sands and gravels. Although both the latter may have aided small-scale farming at the outset, the soils probably quickly became hungry after which they may have been rested for long periods or abandoned to secondary vegetation. On some of these man-made wastes squatters had initiated yet another phase of scattered, individual settlement, and the clustering of their hovels around a small central space or green where trackways across the heath intersected had given rise to names like Heathcote (1196, *Hethcote*) in Wasperton parish or Green End (Fillongley) where *Richard atte Grene* was living in 1332. (For further information on such squatting greens see Thorpe, 1949, 1965a.)

Grants of Arden land to ecclesiastical bodies for the founding of great religious houses, such as the Cistercian abbeys at Combe, Stoneleigh and Merevale, were also followed by widespread clearing as the economic life of the estates gained strength. Nor was emphasis placed on arable farming alone, for the demand from Birmingham and Coventry townsfolk for dairy produce, meat, wool and hides was reflected in extensive pastures and meadows. But, whereas there was still opportunity north of the Avon for the winning of more farm land from wood, south of the river many Feldon villages seem to have expanded their common fields and pastures at an early stage to the very limits of the parish. Indeed, the disappearance of woodland was even posing serious problems in the provision of timber for homes and implements, as well as of faggots for the fire. Some indication of this difficulty was given by John Spencer, admittedly towards the end of the period under discussion, when he recorded that at the time of his first visit to Wormleighton in the latter part of the fifteenth century there was "noo wood no tymber growing within xij or xiij myle" and poor folk had to "bren the strawe that theire cattell shuld lyve by" (see Thorpe, 1965b).

The probability that a period of stagnation had settled on Feldon farming by the middle of the thirteenth century, if not before, has been ably discussed by Harley (1958, 1964) using the Hundred Rolls of 1279. It would seem that a fine point of balance had now been reached on the heavy Lower Lias Clay soils of the Feldon between pressure of population and the carrying capacity of the land in terms of crops and livestock. By contrast, within Arden there was still plenty of *lebensraum* to cater for expansion of population. Although population pressure within Feldon was relieved by the Black Death of 1348–49, the succeeding century was fraught with difficulties stemming from a lack of balance and a changing emphasis within the social and economic structure (see Thorpe, 1965b for a fuller discussion). One direct outcome of the Black Death seems to have been a considerable increase in the extent of pasture, since there was initially not only a smaller labour force but also a reduced demand for grain. At the same time, too, considerable expansion of the native English woollen industry had taken place, and there was an expanding market for raw wool. Throughout the fifteenth century towns grew steadily in size, as population increased and industries developed, and there was a ready market for dairy produce and meat, as well as for hides for the leather industry. The land market had also become increasingly active as estates changed hands, and it became patently clear to certain lords that better returns were to be gained from manors given over to pasture than from those that were still worked by co-aration. Eventually, ruthless lords adopted pastoral land-use as their economic policy, putting entire parishes down to sheepwalks in charge of a very small labour force represented by a handful of shepherds, and driving the wretched peasants from their homes.

Beresford (1945–46; 1954) has already given us well-documented accounts of the great depopulations that affected England in general, and Warwickshire in particular, during the fifteenth and early sixteenth centuries and I do not propose to enlarge on them here. For our purpose, it is necessary to inquire into the distribution and intensity of such depopulation within the county, and to consider its influence on the landscape. No fewer than 130 villages and hamlets in Warwickshire suffered depopulation either wholly or in part, the majority lying within the Feldon and Avon Terrace Belt, whereas Arden, with its traditional early piece-meal enclosure, was virtually immune. As a result of almost complete depopulation, adjacent townships such as Hodnell, Upper and Lower Radbourn, Chapel Ascot and Watergall on the heavy Lower Lias Clays south of the small market town of Southam, were converted to great ranches where sheep, cattle and horses now grazed the ancient corrugations of ridge and furrow which for centuries had carried corn crops. The enclosure of former open fields as huge stock runs is well documented for the parish of Wormleighton where the final blow of depopulation came about 1499 (Thorpe, 1965b). Here and elsewhere it was common practice to enclose the great fields within double banks and ditches, set with hedges of hawthorn and blackthorn, and these can be clearly seen in Wormleighton to this day. The effect on the landscape of the creation of a great lattice of hedges was considerable, not merely in topographic geometry but also in the provision of a sheltered habitat for hedgerow plants. We know, too, that it was common practice in deforested Feldon to plant trees in the great hedgerows and John Spencer, who purchased the depopulated manor of Wormleighton in 1506, describes how he carefully planted acorns along his field boundaries. Indeed, the stumps of great trees dating from this period can still be identified along the double banks in Wormleighton today.

Within some of the great blocks of grassy fields that had now arisen were to be found smaller enclosures of use for penning ewes in the lambing season, gathering flocks for shearing, segregating stock for the market and for breaking in young horses. A second stage in enclosure began about the middle of the sixteenth century when, as the wool boom declined, some enlightened landowners began to subdivide further their great pastures and lease them at high rents as smaller holdings to venturesome farmers still on the ascendant. The good prices now being paid for corn in the growing towns meant that on some of these compact farm units land that had been under grass for a century or more was ploughed up to raise cereals as well as stock. By about 1600, too, some of the remaining open-field parishes of south Warwickshire were beginning to agitate for the enclosure of *champion* land and its subdivision into small fields for continuing arable use by individual farmers, including the lord of the manor himself. But in many other Feldon townships the grazing tradition continued strongly from the days of depopulation right through to this century. Yet, interspersed among the latter, there still remained throughout this period of flux a good many thriving villages and hamlets that had never been disrupted and whose great open fields continued to bear the imprint of co-aration until even as late as the General Inclosure Acts of 1801 and 1845.

Active buying, selling and leasing of land created a demand for large-scale manuscript maps of estates, and our knowledge of events in Wormleighton and elsewhere from the sixteenth century onward owes much to cartographic evidence (Thorpe, 1965b). Furthermore, upstart lords, some of whom had risen from prosperous butcher-graziers to the ranks of the nobility, were anxious to comprehend the extent of their estates in map form; some had used their sudden wealth to build a great house on top of the ruins of a "lost" village, and to empark the surrounding area, introducing clumps of trees and sheets of water offering new habitats for a variety of plants and animals. The distribution of deer parks (Map VII) in Warwickshire is clearly shown on the first printed map of the county produced by Christopher Saxton in 1576, and on later maps by William Smith (1603), Speed (1610) and Sir William Dugdale (1656) (see Harvey and Thorpe, 1959). In addition to deer parks associated with "lost" villages at such places as Compton Wynyates in the Feldon, a number of much older parks had arisen within Arden in Norman times; these included ancient parks of the Earls of Warwick at Wedgnock, Grove and Haseley, as well as the great Chase of Kenilworth, in all of which a large number of oaks still survived.

The steady growth of Coventry and Birmingham as industrial towns had been accompanied by extensive clearing of remnant woodland for miles around, while the provision of charcoal for the iron industry of South Staffordshire and for furnaces at such places as Aston and Perry Barr made further inroads into local forest resources (see Pelham, 1950). As such names as Newfallings Coppice (1544, Tanworth) and Hackings Wood (1544, Exhall) suggest, clearing for fuel and for farming around the shire town and around the market towns of Alcester, Stratford, Rugby, Southam, Kineton, Shipston, Henley, Kenilworth, Solihull, Coleshill, Nuneaton, Atherstone and Tamworth had also been considerable; but as Saxton, Speed, Dugdale and Map VII show, a central spine of Arden woodland including a large proportion of oak in the seventeenth century still followed the low watershed between streams draining south to the Avon and north to the

Trent. In the seventeenth century the significance of prominent trees, particularly oaks, in marking part of the boundary of certain parishes is recorded in a number of Gospel Oak names, such as that recorded in 1608 in Lapworth parish where a halt was made and a passage from the Scriptures read during the Rogation ceremony of "beating the bounds".

Warwickshire from the Eighteenth Century to the Present Day
(see Maps 6, p. 32, and VIII)

In the preceding two sections little was said about the strengthening of the road network in Warwickshire and the steady increase in traffic that was carried, requiring improvements in the width and surfacing of the routes and eventually leading to the turnpike roads that sprang up in Warwickshire during the early eighteenth century. William Smith's map of 1603, the first printed map of the county to portray roads, shows the nodal importance of Birmingham, Coventry and Warwick, while the great highway from London to Daventry, Coventry, Coleshill, Lichfield and Chester is clearly shown. Unfortunately, records which might give us a glimpse of the interesting ribbon-like "ruderal" or wayside habitats do not exist, but Ogilby's magnificent road book of 1675 at a scale of one inch to the statute mile distinguishes by pecked lines roads traversing open field areas from those, shown by continuous lines, crossing enclosed ground. Throughout the seventeenth century moves to enclose champion land had continued, and no less than 62% of the land bordering Ogilby's road strips was enclosed, much of this being associated with the "green roads" so typical of the county today. This percentage cannot be accepted as an index for the whole county, but none the less confirms that enclosed fields were dominant in Arden with a secondary scatter across the Feldon, while unenclosed land was mainly concentrated in the south. Enclosures by common agreement and by Private Act are numerous throughout the eighteenth century, and the final stages in the superimposition of a draught-board pattern of hedged and fenced fields across Warwickshire's cleared surface was completed by the General Inclosure Acts of 1801 and 1845 (see Tate, 1943–44). Regional distinctions between Arden and Feldon, so sharp during Anglo-Saxon and Norman times, had by now become less pronounced as hedges and hedgerow trees gave a leafy look to both areas. But the trained eye can still see the influence of the past on the present in the larger, more rectangular fields of Feldon, bounded by reversed-S curves that reflect former open-field patterns, while the smaller, often more ancient, fields of Arden lie irregular as fire and forester's axe fashioned them.

Supplementing the extents of heath in the county (see Maps 6 and VIII), there now arose another newer form of waste land as coal-mining spoil heaps sprang up along the thin inverted horse-shoe of the exposed coalfield in East Warwickshire (see Mitcheson, 1952). North of Coventry some extraction of coal had continued from the early thirteenth century at least, particularly around Bedworth, Chilvers Coton and Nuneaton, and this was to increase considerably during the eighteenth century as Henry Beighton's magnificent one-inch map of Warwickshire for 1722–25 shows (see Harvey and Thorpe, 1959). Some mining of ironstone on a very small scale also occurred, and blast furnaces were at work in Bedworth in 1873, but the greatest contribution to derelict land has been made here by coal mining, the extraction of clay for brick, tile and pipe manufacture, and the quarrying of Cambrian quartzite and igneous rocks. Elsewhere, the quarrying of Lower Lias Limestone, as around Lighthorne, Harbury and Bishop's Itchington, has left deep scars on the landscape, as also has to a lesser degree the extraction of Middle Lias Marlstone along the Cotswold face. But the total loss in terms of plant habitat attributable to these activities is considerably less than that associated with the rapid expansion of towns and cities within the county, particularly Birmingham and Coventry (Map VIII).

The addition of a canal network, following the opening of the important Birmingham Canal in 1769, not only assisted the flow of goods, coal and raw materials within the county but also encouraged the spread of aquatic plants. As an important secondary effect of canal development, large reservoirs were constructed at places such as Earlswood, astride the main watershed between the Avon and Trent, to supply much needed water to the Stratford on Avon canal. Along the shores of such man-made lakes one can today find communities showing all stages in hydrarch succession from submerged associations to alder-willow swamp. Similarly the coming of the railways, with the opening of the Grand Junction Railway line from Birmingham to Warrington in 1837, initiated further great changes in transport and in settlement, accompanied by the rise of rail-side plant communities on embankments and in damper cuttings where periodic burning associated with sparks from coal-fired engines was a constant hazard until the coming of diesel and electric locomotives. But with recent closures of lines following the implementation of the Beeching Plan, a new distribution of long sinuous trackway ruderal habitats is coming into being along abandoned routes (Map VIII). The coming of hard-surfaced civil and military airfields to urban and rural areas of the county this century has also made dramatic changes in local habitats, though interesting reversions have occurred where Second World War runways have been abandoned or released for farming purposes.

Returning to a consideration of the economic use of tracts of woodland remaining in the county, apart from the provision of tanbark and charcoal, to which reference has already been made, the emphasis was mainly on timber for construction purposes especially for buildings, implements, fences and carts. Probably because of its distance from the sea and the lack of Royal Forests, we find very few references to the cutting of ships' timbers within Warwickshire, though fine stands of oak were maintained by careful planting on such estates as Merevale, Stoneleigh and Ragley (Page, 1908). Indeed, so effective had been the clearing of remnant Arden woodland since the sixteenth century that C. and J. Greenwoods' one-inch map of 1822 (Map 6) showed only a thin stipple compared with the compact blocks of Domesday times and earlier. In many areas the emphasis was now on conservation, but even so the acreage of woodland recorded in 1888 was only 19 562 acres out of a total county area of 566 458 acres, which agrees remarkably closely with the figure of 19 320 acres given in the Ministry of Agriculture's official statistics for 1932–33.* Within the great parks, whose layout is so dramatically

* The County area has since been increased to 624 467 acres.

illustrated by Henry Beighton's prospects in the 1730 edition of Dugdale's *Antiquities of Warwickshire*, further changes were in progress during the nineteenth century. A deep interest in science and exploration had found one of many outlets in the study of foreign plants, and soon after the middle of the century *Rhododendron ponticum*, together with *Prunus laurocerasus*, *Sequoia sempervirens*, *Sequoiadendron giganteum*, *Araucaria araucana*, *Cedrus libani* and numerous other trees and shrubs were firmly established in many parks and grounds where they vied with *Aesculus hippocastanum* introduced from continental Europe some two centuries earlier. Within urban areas, too, the provision of public parks and recreation grounds preserved some semi-natural open space associated with formal groupings of grass, trees, gardens and ornamental lakes. For example, Adderley Park in Birmingham was established in 1856 and Calthorpe Park the following year, though Sutton Park in Sutton Coldfield goes back much further, having been originally a royal chase before being transferred to the "Warden and Society of the Royal Town" about 1528. As Map VIII shows, Sutton Park, established on infertile Bunter Pebble Beds, and on pebbly drift, is the largest tract of heath in the county today.

As we have seen, in south Warwickshire the protracted process of enclosure had produced a somewhat regimented field pattern accompanied in many cases by the rise of a scattered distribution of farmsteads as farmers moved from village centres to virgin sites amid their newly allocated fields. Allocation of land to individuals was again followed by much buying and selling of land as farmers sought to consolidate scattered holdings into compact farm units. In the drive for high yields attempts were also made within Arden to supplement the benefits of farm manure by spreading marl from holes dug at suitable spots in local fields. This practice, which is recorded during the fourteenth century and became more common during the eighteenth and nineteenth centuries, has produced a remarkably close stipple of ponds within Arden, as a cursory glance at the O.S. 1 : 25 000 sheets will show. Although the return in fertility from spreading marl was probably not very high, the flooded marl-holes made good drinking ponds for stock, and

TABLE 2. Composition of Woodland in Warwickshire in 1924

		Acres	% of Total Woodland
High Forest:	Conifers	687	3·6
	Hardwoods	7503	38·8
	Mixed	2497	12·9
Coppice		1144	5·9
Coppice with Standards		4752	24·6
Scrub		1210	6·3
Felled or devastated		962	5·0
Uneconomic		565	2·9
		19 320	100·0

supported interesting aquatic plant communities ringed around by alder and willow. The needs of shelter, hunting and shooting, and the provision of a farm timber supply also encouraged the establishment of coverts and small plantations among which, soils permitting, quick growing conifers were often strongly represented. This late use of conifers on a small scale gave more variety to Warwickshire spinneys and copses, though they have never featured strongly as landscape elements; thus out of a total woodland acreage for Warwickshire of 19 320 in 1924, only 687 acres were high forest conifers (Anon, 1928, 1952). The detailed composition of Warwickshire woodland at that date was as shown in Table 2.

Thus in 1924, woodlands occupied only about 3% of the county area, which, despite the impression conveyed by the popular description of the county as "leafy Warwickshire", was considerably less than the corresponding figure of about 5% for England as a whole. As the Land Utilisation Survey for Warwickshire (see Stamp, 1946; Rees and Skelding, 1950) undertaken in 1932–33 showed, the greater part of such Warwickshire woodland as remained lay in the old Arden area north of the Avon; this is further confirmed by Map 6 for 1822 and by Map VIII for about 1960.

Ancient pastures ribbed by ridge and furrow, some dating virtually from the period of the great depopulations of five centuries ago, remained in many "empty" Feldon parishes until the First World War when they felt again the sharp cut of the share during the plough-up campaigns. Some escaped re-conversion to arable even then, and yet again during the Second World War. The grazing tradition in such Feldon parishes as Wormleighton, Chesterton or Hodnell died hard, but in the period since 1946 the increased use of tractors and the move of new farmers into the area has brought back arable farming to many long rested pastures. Considerable acres of fields, already large from the days of early enclosure, have also recently been further consolidated by the removal of ancient hedgerows, so that in townships such as Hodnell one sees again an enormous sweep of open field arable land-use almost from one end of the parish to the other. But the old parcels of medieval strips are missing, and the relentless onset of the tractor plough has obliterated the deep corrugations of ridge and furrow. Yet in autumn and spring one can still discern these ancient patterns as alternations of light and dark stripes in the strong brown earth; one cannot help wondering, too, whether problems of drainage will not arise on some of these heavy Lower Lias Clays so recently levelled. The virtual removal of the rabbit from extensive areas following the introduction of myxomatosis has also changed the ecological balance in subtle ways, as, indeed, has the use of pesticides and weedkillers, and the grubbing up of hedgerows.

Today emphasis is placed, albeit belatedly, on conservation, but in Neolithic times when our story began the small bands of hunters, fishers, herdsmen and cultivators made little impress on landscape and neither the problem nor the underlying philosophy of conservation arose. They were hemmed in by forest, by monotonous great unknowns in which wild animals lurked and where both views and movement were restricted. In such an oppressive environment considerable effort and care was necessary to establish minority plants—grass and crops—in the man-made clearings. As settlements took shape and population grew, the forest was pushed back slowly and relentlessly at many points, until in some townships the realisation slowly dawned that it was now expedient to retain some woodland for its resources of timber, game, pannage and wild honey. So the tide of colonisation moved elsewhere, gnawing continuously at the forest fringes until eventually the coalescence of colonising movements launched from different directions carved up the forest into clearly identifiable blocks. Norman overlordship may for a short time have held up this clearing process, but very soon in Warwickshire the financial benefits to be gained from selling and leasing tracts of woodland to favoured freemen overcame an early desire to preserve it for the chase. A predominantly cleared landscape, a Feldon, had arisen early along and south of the Avon Valley, though its materialisation came later in Arden, but eventually the result in general landscape terms in both areas was to be virtually the same. The first step in the evolution of our intricate landscape, in whatever part of Warwickshire we find ourselves, has thus always been destruction of woodland, so permitting a succession of other landscapes to arise—fields, farms, villages and towns with a web of communications to serve them. So the cultural landscape has gradually become dominant, and the botanist and zoologist now strive to record survivals of plant and animal life in Warwickshire's remnant "natural" habitats which are themselves already markedly semi-natural and man-adjusted. How effective man's influence in the county has been is clearly expressed in the crude land-use statistics for 1970 (Table 3) with which it seems appropriate to conclude this account.

TABLE 3. Warwickshire Estimated Acreages, 1970
(1887 figures in parentheses)*

Land Use	Acres		%	
Forest and Woodland	16 479	(16 659)	2·6	(2·9)
Arable (including Temporary Grass)	268 760	(196 740)	43·2	(34·7)
Permanent Grass	165 346	(308 689)	26·5	(54·6)
Rough Grazing	5 505		0·9	
Orchards	1 937		0·3	
Built-up Area and Land Agriculturally Unproductive	165 148	(44 370)	26·5	(7·8)
Total Area of Warwickshire (including Birmingham, Coventry and Solihull C.B.'s)	623 175	(566 458)	100·0	(100·0)

* From Bagnall, 1891.

References

Anon (1928). Forestry Commission Report on Census of Woodlands and Census of Home Grown Timber, 1924. H.M.S.O.

Anon (1952). Census Report No. 1. Census of Woodlands 1947–49. H.M.S.O.

Anon (1960a). A Matter of Time. Royal Commission on Historical Monuments, pp. 24–7. H.M.S.O.

Anon (1960b). West Midlands Annual Archaeological News Sheet Nos 3, 4 and 5.

Anon (1961). West Midlands Annual Archaeological News Sheet Nos 4 and 5.

Anon (1964). West Midlands Annual Archaeological News Sheet Nos 7 and 8.

Bagnall, J. E. (1891). "The Flora of Warwickshire." London and Birmingham.

Beresford, M. W. (1945–46). The Deserted Villages of Warwickshire. *Trans. Bgham arch. Soc.* **66,** 49–106.

Beresford, M. W. (1954). "The Lost Villages of England", pp. 445. Lutterworth Press.

Cook, J. M. (1958). An Anglo-Saxon Cemetery at Broadway Hill, Broadway, Worcestershire. *Antiq. J.* **38**, 76–81.

Darby, H. C. and Terrett, I. B. (1954). (Eds). "The Domesday Geography of Midland England", Chapter 6, Warwickshire, by Kinvig, R. H. 270–308. Cambridge University Press.

Dugdale, W. (1656). "The Antiquities of Warwickshire", 2nd edn. by Thomas, W. (Ed.) (1730). pp. 1153. John Osborn and Thomas Longman.

Emery, F. V. (1962). Moated Settlements in England. *Geography* **47,** 4, 378–88.

Gelling, M. (1954). Some Notes on the Place Names of Birmingham and the Surrounding District. *Trans. Bgham arch. Soc.* **72,** 14–15.

Godwin, H. (1956). "The History of the British Flora", Chapter 3, 13–63. Cambridge University Press.

Harley, J. B. (1958). Population Trends and Agricultural Developments from the Warwickshire Hundred Rolls of 1279. *Econ. Hist. Rev.* **11** (No. 1), 8–18.

Harley, J. B. (1964). The Settlement Geography of Early Medieval Warwickshire. *Trans. Inst. Br. Geogr.* **34,** 115–30.

Harvey, P. D. A. and Thorpe, H. (1959). "The Printed Maps of Warwickshire 1576–1900." Warwick County Occasional Series, 1. pp. 279. Warwick.

Kinvig, R. H. (1950). The Birmingham District in Domesday Times. *In* "Birmingham and its Regional Setting", 113–34, Birmingham.

Mawer, A. and Stenton, F. M. (1936). "The Place-Names of Warwickshire." English Place-Name Society, Vol. 13, pp. 409. Cambridge University Press.

Mitcheson, J. C. (1952). The East Warwickshire Coalfield. *In* "Birmingham and its Regional Setting", 289–302. Birmingham.

Oswald, A. (1966–67). Excavations for the Avon/Severn Research Committee at Barford, Warwickshire. *Trans. Bgham arch. Soc.* **83,** 7.

Page, W. (Ed.). (1908). "The Victoria County History of Warwickshire", **2,** 287–95. Archibald Constable, London.

Pelham, R. A. (1950). The Growth of Settlement and Industry *c.* 1100–*c.* 1700. *In* "Birmingham and its Regional Setting", 135–58. Birmingham.

Rees, W. J. and Skelding, A. D. (1950). Vegetation. *In* "Birmingham and its Regional Setting", 65–76. Birmingham.

Roberts, B. K. (1962). Moated Sites. *Amat. Hist.* Winter, 1962. 34–8.

Roberts, B. K. (1965). "Settlement, Land Use and Population in the Western Portion of the Forest of Arden, Warwickshire, between 1086 and 1350." Ph.D. Thesis, University of Birmingham.

Stamp, L. D. (Ed.). (1946). "The Land of Britain: Report of the Land Utilisation Survey of Britain." Part 62, Warwickshire by McPherson, A. W. (Fig. 66 shows the distribution of woodland, heath, common and rough pasture in 1932–33). pp. 655–840. Geographical Publications, London.

Tate, W. E. (1943–44). Enclosure Acts and Awards relating to Warwickshire. *Trans. Bgham arch. Soc.* **65**, 45–104.

Thorpe, H. (1949). The Green Villages of County Durham. *Trans. Inst. Br. Geogr.* **15,** 155-80.

Thorpe, H. (1950). The Growth of Settlement before the Norman Conquest. *In* "Birmingham and its Regional Setting", 87–112. Birmingham (see especially Fig. 18 presenting pre-Roman evidence).

Thorpe, H. (1954). The City of Lichfield: A Study of its Growth and Function. *Staffordshire Historical Collections*, 1950–51. 144–5.

Thorpe, H. (1965a). The Green Village in its European Setting. *The Fourth Viking Congress, Transactions* 85–111.

Thorpe, H. (1965b). The Lord and the Landscape, illustrated through the Changing Fortunes of a Warwickshire Parish, Wormleighton. *Trans. Bgham arch. Soc.* **80,** 62.

Toulmin Smith, Lucy (Ed.). (1906–10). "The Itinerary of John Leland." **2,** 47–51; **5,** 155–6. George Bell and Sons, London.

Webster, G. and Hobley, B. (1965). Aerial Reconnaissance over the Warwickshire Avon. *Archaeol. J.* **121,** 1–22. (This includes a distribution map, a selection of large-scale plans and a summary of the character of each site.)

5. SOME WARWICKSHIRE BOTANISTS: A HISTORICAL SURVEY

"The Authors of the best local Floras of modern date have deemed it a useful and honourable task to trace the history of the discovery of the indigenous vegetation of the districts upon which they have written and to describe the successive contributions made by their predecessors to our knowledge of the present day." So William Mathews opens his "History of the County Botany of Worcester" in the *Midland Naturalist* of 1887; and, in order that the new Warwickshire Flora may qualify for a place in that select company, it may be well to follow his lead.

The subject is not without its difficulties. Botany, of course, deals with plants, not with people; and in searching for records one finds a note of a plant, a date, a location—ideally a specimen which may be examined. The person who collected it may not be mentioned at all; or, at best, may appear as a note on a herbarium sheet. Sometimes the name is of a well-known person, and then there is little difficulty in giving some account of his life and times. But often little or nothing is known of the recorder even when his name is given.

There are, in Warwickshire, cases like that of **William Cheshire,** a working printer of Stratford on Avon, who contributed notes to the *Phytologist* and specimens to Perry's Herbarium, of which 14 are first records for the county. He died about 1855–57, but the date of his birth is unrecorded and virtually nothing is known of his life. **Samuel Freeman** flourished about 1840, lived at 11, Sun Street, Birmingham, and contributed a "List of some of the Rarer Plants observed in the neighbourhood of Birmingham" to the *Phytologist* in October 1841. These include 3 first records. That seems to be all that is definitely known of him. Of the brothers **Cross,** who contributed plants to the Herbarium at Warwick Museum, we do not know certainly whether there were two or three. They, and many like them, as well as the larger figures whose lives can be sketched in some detail, all helped to gather the knowledge we have of Warwickshire plants; and we should be grateful to them. If time allowed, exhaustive research might throw more light upon some of the shadowy figures.

Nowadays, when a biographer takes up his pen to write, too often it seems that his first intention is to cut down his subject to size, and to emphasise how much better we do things today. In this short and somewhat rash attempt at biography the reverse is the case. The more one learns of the work of our predecessors in botanical investigation in the county, the fuller one must surely be with admiration for what they achieved under the conditions in which they laboured and with the equipment at their disposal.

Of course, although botanical work within a county is largely that of men and women little known beyond their home territory, it is nevertheless of great importance. It is fascinating at times to watch a distinguished lecturer drawing bold lines on a world map to delimit the range of some well-known species, and to realise that to be accurate these must be built upon detailed information given in scores of local Floras.

The people who compile local Floras may often be little known, and those who contribute to them even less distinguished: they may not usually be, to quote John Masefield,

> . . . the princes and prelates with
> periwigged charioteers
> Riding triumphantly laurelled to lap the
> fat of the years:

but, in the case of Warwickshire, we may boast of one who, but for the political accidents of his time, would undoubtedly have been a prelate, and whose intellectual endowments were princely:

John Ray was born in 1627 and died in 1705, that is, he was born in the reign of Charles I and died in the reign of Queen Anne. For a man of principle there could scarcely have been a more difficult period, and Ray was no Vicar of Bray. Neither the promise of a brilliant academic career (he was appointed lecturer in Greek at Cambridge at the age of 23, and a few years later lecturer in Mathematics and the Humanities), nor the prospect of high preferment in the Church (he was ordained in 1660, and when, in 1677, he refused the Secretaryship of the Royal Society, could say to John Aubrey, author of "Brief Lives", "Divinity is my profession") could move him from the path he felt it his duty to take. He refused to subscribe to the Act of Uniformity of 1662 and forfeited his Fellowship. Henceforward he could hold no position in the Church or the University.

There can be no doubt that Ray felt the separation very deeply, but the action was deliberate and there were no expressions of regret. Indeed, throughout all his correspondence, as Raven says, "there is literally not a word to indicate that he was living in one of the most exciting periods of English history or that he knew or cared what was happening". Now his life was to be devoted mainly to science—to ornithology, entomology, zoology but, above all, to botany; and its record to be largely a list of publications interspersed with journeys at home and on the Continent in search of material.

It was on the fourth of these itineraries, in May 1662, that Ray left Cambridge for Northampton, Coventry and Middleton, where he joined Willughby and Skippon (another pupil). (Later, from 1669 to 1675, he was to live at Middleton Hall as tutor in the family of Francis Willughby—friend, pupil, patron and fellow-naturalist—and afterwards for a short time at Coleshill and at Sutton Coldfield.) The trio proceeded from Middleton to Sutton Coldfield where they found *Botrychium lunaria*, the moonwort, which they called *Lunaria minor* "in great plenty . . . in a close in Sutton Coldfield Park". There it still is, though not in great plenty, but whether in the station where Ray saw it, it is not possible to say. According to Bagnall it was seen in his day near Longmoor Pool Mill, which is a long way from where it was recorded in 1954 and where it still flourishes.

Perhaps the most interesting other plants included in nearly 40 first records he contributed to the Flora of Warwickshire—all from Middleton and its neighbourhood—are *Cladium mariscus*, *Empetrum nigrum*, *Hypochoeris glabra*, *Schoenus nigricans*, *Scirpus cespitosus* ssp. *germanicus*, *Vaccinium oxycoccos* and *Carex pulicaris*. Of these, four are still present in Warwickshire in Sutton Park only, and three have disappeared completely from the county. These, and other records referred to later on, show convincingly how the countryside has changed in comparatively recent times, particularly through the disappearance of heathland and bog; and emphasise the urgency of keeping such areas in Sutton Park from a similar fate.

In the history of botany, and, indeed, of natural history generally, John Ray is an astonishing figure. His "Catalogue of Plants growing round Cambridge", published in 1660, was the first British local Flora. His "Synopsis of the British Flora", which appeared in 1690, remained the standard British Flora for fifty years. His *Methodus Plantarum* of 1682 marks the beginning of a natural system of classification. Though he never possessed a microscope, he separated flowering from flowerless plants and was the first to distinguish monocotyledons from dicotyledons. A quotation from Ray's correspondence might have served as a preface to this new "Flora of Warwickshire" if it had not been made use of, in part, by the editors of the "Atlas of the British Flora". It is from a letter to Willughby from Trinity College, and is dated 25th February, 1659.

> You will remember that we lately, out of Gerard, Parkinson and *Phytologia Britannica* made a collection of rare plants whose places are therein mentioned and ranked under the several counties. My intention is now to carry on and perfect that design; to which purpose I am now writing to all my friends and acquaintances who are skilful in Herbary to request them this next summer to search diligently his county for plants, and to send me a catalogue of such as they find together with the places where they grow. In divers counties I have such as are skilful and industrious: for Warwickshire . . . I must beg your assistance.

Most of Ray's Warwickshire records are from his *Catalogus Plantarum Angliae et Insularum adjacentium*, published in 1670: a few from other works and from his correspondence.

To have kept strictly to chronological order it would have been necessary to have begun this survey with mention of **Mathias de L'Obel** (1538–1616), who was born at Lille. He was a Dutchman who had practised medicine in Antwerp and had been physician to William, Prince of Orange (afterwards known as William the Silent). He came to London as a young man and was later appointed Botanographer to James I. In conjunction with **Peter Pena** he published in 1570 *Stirpium Adversaria Nova*, and recorded *Centaurea scabiosa* as *Jacea maior* from cornfields at Coventry, and *Lysimachia nemorum* as *Anagallis lutea* from woods near Coventry. The book was dedicated to Queen Elizabeth.

In the early years of the reign of William and Mary a group of undergraduates, probably at Magdalen College, Oxford, led by one **Henry Holden,** made a collection of plants which came to the Birmingham Reference Library through the family of the late A. E. Housman, distinguished Classical scholar and poet who died in 1936. The collection was brought to the knowledge of the authors of this Flora through the kindness of Dr. R. J. Hetherington. It contains a number of plants but few localities are given: only six localities for Warwickshire, three of which are for wild plants. *Astragalus glycyphyllos* is the first record for Warwickshire "in ye lane before you go up Frizwell hill from Warwick": and so is *Anthriscus caucalis* "In ye hedg. as you go to Longbridge near Warwick": the third, not a first county record, is of unusual interest because it gives the same location, in the same words, as does Ray: *Vaccinium oxycoccos* "on moorish Grounds and Quagmires v.g. in Sutton-Cofield-parke in Warwickshire". There are many queries relating to this collection which await elucidation.

And now, perhaps, a slight deviation from the strictly scientific attitude might allow us to consider **Shakespeare** as a botanical recorder. It seems a pity that our first record of *Narcissus pseudo-narcissus* should be Ray's record of 1670, when, in "The Winter's Tale", written before 1611 and published in 1623, William Shakespeare of Stratford on Avon wrote of

golden daffodils,
That come before the swallow dares, and take
The winds of March with beauty;

and, in the same place, of

pale primroses,
That die unmarried, ere they can behold
Bright Phoebus in his strength.

Who can doubt that the poet had seen and known these flowers from his own county? Our first record of *Primula vulgaris* is by Rev. W. T. Bree and published in Purton's "Midland Flora" in 1821.

Sometimes there is doubt as to the identity of some of the plants referred to; as in "Hamlet", the Queen, speaking of Ophelia, says

There with fantastic garlands did she come,
Of crow-flowers, nettles, daisies and long purples,
That liberal shepherds give a grosser name,
But our cold maids do dead men's fingers call them.

Nettles and daisies are clear enough and would give us much earlier records for *Urtica dioica* (1831) and *Bellis perennis* (1813). Long purples have been considered to be Early Purple Orchids: but what are crow-flowers?

In "King Lear" we read

Crowned with rank fumiter and furrow-weeds,
With harlocks, hemlock, nettles, cuckoo-flowers,
Darnel, and all the idle weeds that grow
In our sustaining corn.

This suggests earlier records than we have for *Fumaria officinalis* (1812) and *Conium maculatum* (1831)—but we cannot be sure of harlocks, cuckoo-flowers or even darnel.

King Henry VI, in the last of the plays Shakespeare devoted to his history, asks the question:

> Gives not the hawthorn-bush a sweeter shade
> To shepherds, looking on their silly sheep,
> Than doth a rich embroider'd canopy
> To kings that fear their subjects' treachery?

A last quotation, from "A Midsummer Night's Dream": Oberon, King of the Fairies, tells Puck:

> I know a bank whereon the wild thyme blows,
> Where ox-lips and the nodding violet grows;
> Quite over-canopied with lush woodbine,
> With sweet musk-roses, and with eglantine.

Our first record for *Crataegus monogyna* is dated 1836 and for *Thymus drucei* 1812. In this connection it is interesting to recall that Druce, in his "Flora of Northamptonshire", took a number of records from the poems of John Clare; but then Clare certainly knew his plants and wrote almost entirely of his native county.

After Ray died in 1705 there was an interval of over half a century during which little of botanical note seems to have taken place in the county. In 1762 a student of 21 went to Edinburgh to study medicine; his name was **William Withering** (1741–99). At the University, John Hope, Professor of Medical Botany and first lecturer on the Linnaean system of classification, offered each year a gold medal to his most industrious student. William wrote to his parents: "An incitement of this kind . . . will hardly have charm enough to banish the disagreeable ideas I have formed of the study of Botany." A few years later he began collecting botanical specimens for a young lady patient to sketch. She became Mrs. Withering and he became one of Britain's leading botanists, with a European reputation.

Withering was first and foremost a Doctor of Medicine. He became physician at the Birmingham General Hospital in 1779 on the recommendation of **Dr. Erasmus Darwin,** grandfather of the more famous Charles. These two men were for many years the most sought-after provincial physicians of the day, and to some extent were rivals in the field of botany; for, while Withering was preparing his "Systematic Arrangement of British Plants", Darwin was writing his "Botanic Garden", a long treatise in verse, which included the "Loves of the Plants". This was immensely popular at the time and set forth the author's views on evolution, which some knowledgeable people consider influenced the author of "The Origin of Species"; but of which Charles wrote, with all the irreverence of a grandson, as anticipating "the views and erroneous grounds of opinion of Lamarck". About this time, too, while Withering and his collaborator, **Dr. Jonathan Stokes,** were preparing the second edition of the "Arrangement" for the press, the Botanical Society at Lichfield, which, for all practical purposes, was Erasmus Darwin, was busy translating the *Species Plantarum* of Linnaeus. The Linnaean system, as is well known, arranged the flowering plants in classes according to the number of stamens, and in orders according to the number of pistils. Darwin likens the pistils to "wives" and the stamens to "husbands"; and we have the picturesque if not very scientific descriptions of flowers in which "wives" have four "husbands", two taller than the other two; of "husbands" that live with "wives" in the same house, but have different beds (monoecious), and of others where "wives" and "husbands" have different houses (dioecious). In his *Phytologia* the same author writes: "The vegetable passion of love is agreeably seen in the flower of the *Parnassia*, in which males alternately approach and recede from the female, and in the flower of *Nigella*, or devil in the bush, in which the tall females bend down to their dwarf husbands."

Withering's principal contribution to British botany was the popularisation of the Linnaean system of classification, which, although now mainly of historic interest, focused a great deal of attention on plant classification, and thus played a great part in the development of systematic botany in this country. His "Arrangement" went through three editions during his lifetime (most of our first records are from the second edition of 1787) and four more during the lifetime of his son. It was then "corrected and arranged" by W. Macgillivray, whose first edition was published in 1830 and went into numerous editions down to the end of the century. Associated with Withering in his botanical work were his son William (1775–1832) and Dr. Jonathan Stokes of Kidderminster (1755–1831) who, besides acting as editor and illustrator, himself contributed a number of records, including six first records.

Controversy seems to have sprung up almost spontaneously around Withering. First there was the quarrel with Darwin over the discovery of the use of *Digitalis* as a cure for dropsy, which led to estrangement to the end of their lives; then the often repeated allegation that the second edition of the "Arrangement" was more the work of Stokes than of Withering. Certainly Withering lent Stokes a number of valuable books to assist him in whatever he did do; and there is extant a schedule of these books and a demand for their return. Apparently they had all been badly mutilated by Stokes's cutting out illustrations.

In Birmingham, Withering lived for a time in a house in Cherry Street, which Benjamin Walker said was "possibly the first house in Birmingham to have a water-closet". At this time the cherry orchard which gave the street its name was possibly still in existence; certainly William Hutton, first Birmingham historian, who was a contemporary, wrote of the yellow corn waving at the corner of Bennetts Hill and New Street. Withering had rooms in Old Square and in 1786 went to live at Edgbaston Hall. He was a member of the famous Lunar Society, and, at one time, had even Matthew Boulton collecting plants. He is buried at Edgbaston Parish Church.

Withering contributed a number of interesting Warwickshire first records. Editions of the "Arrangement" published in his lifetime included over fifty, and about half that number were added in later editions up to 1830. He recorded *Drosera rotundifolia*, *Eriophorum angustifolium* and *Narthecium ossifragum* from Birmingham Heath; and *Eriophorum vaginatum* from "a marshy valley crossed by a foot-road to Winson Green".

In this connection it is interesting to recall a notice which appeared in the *Birmingham*

Gazette on 17th November, 1788, relating to the Crescent, behind Cambridge Street, of which few, if any, bricks now remain.

> A correspondent who has seen the design for the elegant Crescent . . . remarks that the houses will be very convenient, and the situation excellent in every respect, either for a winter or summer residence, as the houses will have both a southerly and northerly aspect. . . . And it is an additional recommendation of the plan in this growing town, that there is not the least probability of any further buildings ever excluding the inhabitants of the Crescent from a most agreeable prospect of the country.

This prospect would have included, in Withering's day, a part, at least, of Birmingham Heath.

Alongside such a drastic change it is pleasing to point out that the edition of the "Arrangement", published in 1796, records *Sambucus ebulus*, not a first record, "a few hundred yards from Knowle, by the side of the road leading to Warwick". There Bagnall saw it, and there it still flourishes.

From Edgbaston Hall to Packington Park was probably a pleasant drive in Withering's day and the doctor frequently visited his friend and fellow botanist, the charming and remarkable **Countess of Aylesford** (1760–1832), who had twelve children and whose portrait, painted by Sir Joshua Reynolds, is in the possession of the present Earl. She was the daughter of the Marquis of Bath, who, as Viscount Weymouth, had been a founder member of the Linnean Society and had given his name to the Weymouth Pine. At an early age she became interested in botany, and, in 1784, when she was 24, began a series of flower paintings which grew to the number of 2830, the last being painted in 1816. These paintings, now in the possession of the Dartmouth branch of the family (Augusta, granddaughter of the 4th Countess, painter of the pictures, married William, 5th Earl of Dartmouth), achieved a high standard of accuracy according to Druce, who added that she always gave the source of the plant. Many of these plants were left by her to her daughter Charlotte, who married Rev. Charles Palmer, rector of Lighthorne; and in turn came to her daughter Charlotte who presented them, together with her own collection, to Druce. Some of these are now in the Oxford University Herbarium.

Altogether the Countess of Aylesford provided about 30 first records to the county Flora, of which the majority were first contributed to the "Botanist's Guide through England and Wales" by Dawson Turner and Lewis Weston Dillwyn, published in 1805. The remainder are from her specimens in the Oxford University Herbarium. The most interesting are *Anagallis tenella*, *Cirsium dissectum*, *Colchicum autumnale*, *Scutellaria minor* and *Teesdalia nudicaulis* from Packington; *Limosella aquatica* and *Littorella uniflora* from Coleshill Pool; *Paris quadrifolia* from near Coleshill; *Lycopodium inundatum* from Coleshill Heath, and, of particular interest to those who live in or near Birmingham, *Myosurus minimus* from Chelmsley Wood.

From the granddaughter, **Charlotte Eden Palmer** (1830–1914), we have first records, all but one represented by specimens in the Oxford University Herbarium. They include *Blysmus compressus* collected in the seemingly unlikely neighbourhood of Lighthorne; but there is no doubt about the determination of the plant. In 1872 Charlotte Palmer and her elder sister moved to Hampshire—they contributed a number of records to the Flora of Hampshire, by Fred Townsend, brother of Elizabeth Townsend of Honington Hall, Warwickshire—and here it was they entertained Druce—a somewhat frequent visitor. Late in life the two old ladies took up stamp-collecting and tried to interest Druce. One can well imagine what a stamp collector George Claridge Druce would have been and what his annotated volume of Stanley Gibbons' Catalogue would have looked like.

In 1907 Miss Charlotte Palmer gave Druce her large herbarium, which included the collection made by her grandmother (the Countess) and specimens from the Rev. W. T. Bree, together with a collection from Miss Elizabeth Townsend, which included plants given to her by Perry and by Fred Townsend (her brother) and some from Bolton King (her nephew). It was records of these collections together with a list from Baxter which Druce sent to Bagnall too late for inclusion in his Flora, and which were the basis of the list published by Bagnall in the *Midland Naturalist* of 1892–93. It has not been possible to trace the actual manuscript.

In the year 1817 two events of importance in the history of botanical work in the county took place. There were published an abridged edition of Dugdale's "Antiquities of Warwickshire" with a "Select List of Plants" found in the county, by W. G. Perry; and "The Midland Flora" by Purton. **Thomas Purton** (1768–1833) was then 49, while **William Groves Perry** (1796–1863) was only 21; and it is striking that a large number of first records contributed to the County Flora are of plants collected and named by this young man during the year 1812 when he was 16.

Thomas Purton was a surgeon at Alcester where he began to practise in 1793. The first two volumes of his Flora, published in 1817, rejoiced in the title of "The Midland Flora: a Botanical Description of British Plants in the Midland Counties, particularly of those in the neighbourhood of Alcester, with Occasional Notes and Observations, to which is prefaced a Short Introduction to the Study of Botany and to the Knowledge of the Principal Natural Orders". It followed the Linnaean system of classification and contained over one hundred plant records new to the county. The third volume followed in 1821. All three volumes contain a number of records provided by the **Rev. W. T. Bree** (1787–1863), son of the **Rev. William Bree** (1760–1820) who also made a few contributions and drew the plates.

Purton's Flora is a comprehensive work describing over 1600 species, but, as he did not give locations for many of the commoner plants, it is not possible in many cases to trace a record to a particular county. In fact it has been possible to credit him with about 130 first records of which about 20 were sent to him by the Rev. W. T. Bree. Among the more interesting are *Cephalanthera longifolia* from Oversley Wood, where it still is; *Coeloglossum viride*, *Epipactis palustris* and *Gymnadenia conopsea* from the same area, where, however, they have not been seen recently; *Groenlandia densa*; *Leonurus cardiaca*; *Pilularia globulifera* and *Radiola linoides* both from Coleshill Pool; *Plantago coronopus* and *Vicia lathyroides*.

The Rev. W. T. Bree, who contributed to Purton's "Midland Flora", to Watson's "New Botanist's Guide" and to the *Magazine of Natural History* (1828–30), was one of a remarkable family of divines. His father, who has already been mentioned, was Rector of Allesley before him, and he was succeeded in that office by his eldest son, the Venerable William Bree, D.D., who became Archdeacon of Coventry, and died in 1917. He was a good naturalist. The family at one time held land at Beausale and at Baddesley Clinton, and still own land at Shrewley where there are still some unusual willows, which the Rev. W. T. Bree may have had a hand in planting. Bree contributed to the "New Botanist's Guide" the first Warwickshire record for *Chamaemelum nobile* from Shrewley Pool; and to Purton's "Midland Flora" first records for *Genista anglica* and *Lycopodium clavatum* from Coleshill Heath; *Lycopodium selago* from Coleshill Pool; *Pinguicula vulgaris* from Bannerley Pool; *Thelypteris palustris* and *Tulipa sylvestris* from Allesley. All these have long since disappeared except *Pinguicula vulgaris* which remains in the county—in Sutton Park. In 1850 he contributed to the *Phytologist* a paper on "A Warwickshire Habitat for *Gagea lutea*", which the Rev. J. Gorle had discovered in 1837—at Sheldon, presumably not far from the spot where it flourished until the recent housing development.

Writing to a friend from Allesley Rectory on 12th August, 1841, Bree remarks: "True it is Coleshill Heath has been greatly curtailed by enclosures and cultivation since the days I have alluded to; still, however, considerable tracts remain in *statu quo*: the drier parts purpled o'er with the three common kinds of heath, the bogs a sheet of gold in their season with the blossom of *Narthecium*, and abounding with *Oxycoccos*, *Comarum*, *Eriophorum*, *Menyanthes*, etc."

To William Groves Perry Warwickshire botany owes a great debt. Reference has been made to the important list of plants he contributed to the edition of Dugdale's "Warwickshire" published in 1817. This was followed three years later by *Plantae Varvicensis Selectae*, or Botanist's Guide through the county of Warwick. This work was the earliest exclusively devoted to Warwickshire botany, and contained 379 records, largely from the works of Ray, Withering, and Turner and Dillwyn; but about one-quarter were new to the county. It seems clear that Perry intended a complete Flora of the county at a later date, but he did not finish the project. Bagnall, who had Perry's own annotated copy of *Plantae Selectae*, says that Perry started collecting material for the larger Flora and left manuscripts; but these are lost.

Even more important than his "List" and *Plantae Selectae* is the collection of plants now part of the Warwickshire County Museum Herbarium. This collection is referred to here as the Perry Herbarium—not without good reason, and in good company, for Bagnall did the same—but might have been referred to as the Warwickshire Museum Herbarium, a title here reserved for those plants added to the collection after the death of Perry in 1863. The early history of this collection is a little confused. In 1836, when Perry was 40 years old and had already built up an extensive herbarium of plants collected by himself and a number of friends and correspondents, the Warwickshire Natural History and Archaeological Society was formed and obtained the use of rooms at the Market Hall, Warwick, now the Warwick County Museum. In 1841 or 1842 the Society commenced the collection of plants, and Perry, who was a bookseller in the High Street, and was for some time Honorary Secretary of the Society, undertook the arrangement and care of the collection. At this time, however, although the Herbarium contained a number of plants collected by Perry, it did not have the main body of Perry's own collection, which was purchased by the Society from Perry's executors about 1870. When the Society became defunct the Herbarium became part of the Museum's collection.

From the three sources—Perry's "List" of 1817, his Flora of 1820 and his herbarium—we have listed over 270 first records for Warwickshire, making him the largest contributor of first records to the Flora. From so many records it is difficult to select a few as of particular interest. From the "List" we have *Campanula glomerata* from near Pillerton, *Cynoglossum germanicum*, *Fritillaria meleagris* from Wroxall Field, *Nepeta cataria* and *Silybum marianum* from the race-course at Warwick; from *Plantae Selectae: Calamagrostis canescens*, *Hottonia palustris* and *Ophioglossum vulgatum*; from the Herbarium: *Baldellia ranunculoides*, *Chrysosplenium alternifolium*, *Dactylorhiza praetermissa*, *Jasione montana* and *Parnassia palustris*.

Perry had a number of correspondents who contributed specimens to his Herbarium. Henry Bromwich and William Ick are mentioned more fully in subsequent pages. William Cheshire has been named already. **Thomas Kirk** (1828–98) was born at Coventry, where his mother was a nursery-woman and florist, and where he worked at one time as a clerk in a sawmill. He contributed records to Bagnall's Flora, and there are specimens collected by him in the herbaria of Birmingham University and the British Museum as well as the Perry Herbarium. He contributed papers to the *Phytologist*; and in one of these in 1851 gives the first record for the county of *Elodea canadensis* as *Anacharis alsinastrum*. There is a specimen of his of the same date in the British Museum. In 1863 (when he was 35), for health reasons it is believed, he emigrated with his family to New Zealand, where, after some eventful and difficult years, he became secretary-curator of the Auckland Institute and Museum, and, eventually, Chief Conservator of Forests. Another notable contributor to the Perry herbarium was **Dr. Lloyd,** who collected around Warwick about 1835. **Dr. St. Brody,** who lived at Guy's Cliffe and was, for a time, a master at Leamington Spa Boys' College, also collected in this area about 1875; but this was after Perry's death.

Mention has been made of a number of records sent to Bagnall by Druce too late for inclusion in the "Flora of Warwickshire", and which Bagnall published, with others, in the *Midland Naturalist* of 1892–93. These included a large number of records of plants collected by **William Baxter** (1787–1871) who was born at Rugby and died at Oxford, where he was Curator of the Botanic Garden from 1813 to 1851. It was in 1831 that he revisited the place of his birth with his son to collect botanical records with a view to publishing a Flora of the district. He did not carry out this design. The records are mostly of common plants and over 60 of them are first county records.

Things have changed at Oxford since Baxter's time. Druce, in his "Flora of Oxfordshire", quoting from a writer in the *Gardeners' Chronicle*, says: "Botany at Oxford had sunk to its lowest level. . . . Dr. Williams, who was Professor of Botany, although an elegant scholar, added nothing to botanical science, and for practical instruction in botany students had recourse to the teaching of Mr. Baxter." The salary he drew as Curator of the Botanic Garden was £50 per annum when he was appointed in 1813, and had increased by 1829 to £70.

Baxter published his "British Phaenogamous Botany" in 6 volumes between 1834 and 1843. It was a very handsome production with beautiful illustrations, and was embellished, as were many such works about that time, with many verses from many poets. Perhaps it is not too irrelevant to recall that Baxter claimed connection with Shakespeare through Mrs. Hart, the poet's sister, and named his son William Hart Baxter. Druce records that during the First World War he learned that the copper plates for Baxter's Flora were to be melted down. He wrote immediately to the owner, making an offer, but the letter miscarried and the plates were lost.

In 1835 the United Committee of the Birmingham Botanical and Warwickshire Floral Societies offered a medal for the best *hortus siccus* of native plants correctly named with local habitations, collected within 10 miles of Birmingham between 1st August, 1835, and 1st August, 1836. The winner was **William Ick** (1800–44), who came to Birmingham as a child of three and became a tutor at a school near Warwick. He was for a time curator of the Birmingham Philosophical Institute and was awarded a German Ph.D. for his work in geology. His small herbarium (about 300 sheets) was lost sight of for many years, but was presented to the Birmingham Museum in 1948.

Ick's winning list of plants named 320 species and was published in the *Analyst, a Monthly Journal of Science, Literature and the Fine Arts* in July 1837. Few of his records are of particular interest but his comments on some of the habitats certainly are. He writes:

> I need scarcely observe the extensive changes which have taken place in Birmingham and its environs during the last 40 years, or since Dr. Withering published the last edition of his Systematic Arrangement of British Plants, make it the object of some interest to ascertain which of the localities of plants that gentleman's long residence near this town enabled him to point out, still exist. . . .
>
> Birmingham Heath and Washwood Heath—where in Dr. Withering's time the rambles of the botanist were rewarded by *Hypochoeris glabra*, *Vaccinium oxycoccos* or *Eriophorum vaginatum*—now exist as *heaths* only in name. Houses, canals and the murky steam engine cover the places where "once the wild flower smiled", and the busy hum of men has long succeeded "the buzzing wing of the Drowsy Dorr", even in spots as yet unconscious of the march of bricks and mortar. The labours of agriculture have swept away almost all the gleanings of the botanist: the Common Potato (*Solanum tuberosum*) now keeps, far more profitably, the place of its noxious congenors *Solanum dulcamara* and *S. nigrum;* while the slender Sea Cabbage (*Brassica oleracea*) has been doomed "neath the gardener's plastic art" to undergo more metamorphoses than ever Proteus tried or Ovid sang.

The **Rev. Andrew Bloxam** (1801–78), who contributed records to Watson's "New Botanist's Guide" and to Bagnall's Flora, has been described as the last of the all-round British Naturalists: and, if he was not exactly that, he was certainly a fine example of a type of which the nineteenth century produced quite a number. In natural history he was skilled in many branches without it being possible for anyone to say that he was master of none. He was associated for many years with the Rev. Churchill Babington with whom he contributed a list of flowering plants to Potter's "History of Charnwood Forest". With the Rev. W. H. Coleman he contributed a plant list to Potter's "History of Leicestershire". In addition to botany he was a student of conchology and ornithology. He was accepted as an expert in the study of Fungi and his collection is in the British Museum. *Agaricus bloxamii* is named after him as is *Agaricus babingtonii* after his friend. He was a critical student of both *Rosa* and *Rubus*, and here again he is remembered in *Rubus bloxamii* (and Babington in *Rubus babingtonianus*). A few years ago Dr. Mary Jones found in the Rugby Public Library a small exercise book which contained a collection of plants made by Bloxam in 1875, when he was Rector of Harborough Magna. It contained a first record of *Alchemilla xanthochlora*.

Before settling down to the church (and botany) in Warwickshire, Bloxam had led a full and colourful life. His father was a master at Rugby School and he entered his father's house in 1809, afterwards proceeding to Oxford. His mother was the sister of Sir Thomas Lawrence, the famous painter, and his uncle by marriage was Purton. At the age of 23 he was appointed naturalist on the Frigate *Blonde*, commanded by Capt. Lord Byron (cousin of the poet) which was sent by the Government to the Sandwich Islands to convey there the bodies of the King and Queen of the Islands who had died in this country. On his return he took Holy Orders, became Rector of Twycross in Leicestershire, then of Harborough Magna in Warwickshire.

There is a chalk drawing of him by his uncle, Sir Thomas Lawrence, made in 1824, and, in the National Gallery, a water-colour by Turner showing the six brothers attending the funeral of their uncle. Bloxam's wife was a descendant of Nehemiah Grew (1641–1712), the English plant anatomist who was born at Coventry and became secretary of the Royal Society. Bloxam assisted with the early reports of Rugby School Natural History Society, of which Bagnall made good use and from which we have taken a number of first records. He was a corresponding member of the Birmingham Natural History and Microscopical Society.

Hewett Cottrell Watson (1804–81) is a name important in British botany. It was he who introduced the concept of the vice-county, the area which we use today to record plant distribution instead of the geographical or administrative county, the boundaries of which have changed considerably since his time. His *Cybele Britannica*, published in four volumes between 1847 and 1859, is a treatise on the geographical distribution of plants in the counties of Britain, and was followed by his "Topographical Botany" in two volumes in 1873–74. But, as a source of first records his earlier work, published in 1835 with additions in 1837, is much more important. This is "The New Botanist's Guide to the Localities of the Rarer Plants of Britain on the Plan of Turner and Dillwyn's Botanist's Guide".

The source of a large number of these records is said to be a "Checked Catalogue of Plants found in Warwickshire, chiefly in the neighbourhood of Allesley and of Coleshill, or the intermediate places" by Rev. W. T. Bree; but there are also records from Purton's "Midland Flora", the "Botanist's Guide", the *Magazine of Natural History* and

A. Bloxam. Only about a dozen of these, all but one from Bree and Bloxam, are first records. In his introduction to the Warwickshire Section Watson refers to some records of Perry's and adds, "The latter gentleman, I am told, has published a Botanical Guide to the county; but on giving this name to my bookseller, it was marked by him as that of a country publication not met with in town".

The name of **Henry Bromwich** (1828–1907) is usually associated with that of Bagnall, though he contributed a number of specimens to the Perry Collection as well as to Bagnall's Herbarium. He also supplied Bagnall with many records for his Flora; and there are specimens of his in the Herbaria at the British Museum and Birmingham University. He seems particularly to have had an eye for aliens and casuals. To him we owe the record of *Daphne mezereum* from the canal cutting near Shrewley in 1864—a record confirmed by Dr. Burges in 1933, but not seen lately; records of *Viola canina* in 1857 from Milverton and of *Stellaria palustris* in 1844 from Leek Wootton, neither of which has turned up in these areas during the present survey: and *Scirpus tabernaemontani* and *Scirpus maritimus* in 1870 and 1872 respectively, from Southam Holt, where they still flourish. In all we owe to Bromwich nearly 80 first records. Bromwich, son of a gardener, was himself for many years at Wroxall Abbey, then owned by Miss Wren, a descendant of Sir Christopher. His home was at Milverton, and he was for some time curator of the Botanical Department of Warwick Museum. In a report to the Birmingham Natural History and Microscopical Society in 1878, Bagnall says: "Many of the districts south of Warwick I have visited in company with my friend Mr. H. Bromwich, an excellent botanist . . . who has worked with great success most of the county around Warwick."

To most of us, of course, the Flora of Warwickshire means Bagnall's, although it has been shown how considerable a part many others, notably Perry, played in making that work possible. Local botanists have been brought up on it, botanically speaking, for many years; and the more they used it the more impressed they were with the quality as well as the quantity of the work done by the indefatigable **James Eustace Bagnall** (1830–1918). And, indeed, indefatigable is the word that comes to mind again and again in considering Bagnall's work. As he himself reminds us in his Preface: "all my work, whether clerical or botanical, has been done in the scant leisure of a manufactory clerk; and . . . my knowledge of botany has been self-acquired." By contemporary standards, his Flora was a very sound piece of work, and was described by J. G. Baker, F.R.S., of the Royal Botanic Gardens, Kew, as "one of the best county floras that has been written, and will stand as a permanent memorial to his diligence and ability". He had, of course, correspondents who sent him records and whom he consulted over critical determinations; but the planning and execution of the work were his own. When one considers, in addition to his work on Flowering Plants and Ferns, that on Mosses and Liverworts and with Grove on Fungi, and his detailed recording of the microspecies of *Rosa* and *Rubus*, one feels a very humble worker in the same field. In addition, of course, one has to remember the immense amount of work he did for our society, the Birmingham Natural History and Microscopical Society, as it then was.

Bagnall was born at Aston in 1830 and lived there all his life. He was educated at

Singers Hill, where Rowland Hill, who was the originator of Britain's Penny Post, and his brother Matthew Davenport Hill, who became Birmingham's first Recorder, developed the educational innovations of their father. In 1844 he began work in his father's business, but nine years later moved to Hinks and Wells, the world-famous manufacturers of steel pen nibs, where he became chief clerk and remained until he retired from business in 1897. In 1864, at the age of 34, he had begun the study of botany. A friend lent him a compound microscope and a few slides. He collected his first botanical specimen, made a slide of its pink petals, bought a second-hand copy of Purton's "Midland Flora" and "Bentham and Hooker's Handbook" from Cornish's, and identified *Geranium robertianum.* He saw a small Herbarium at a soirée held by the Birmingham Naturalists' Union at the shop of Franklin (Taxidermist) in Suffolk Street and began his long career as a collector.

Two years later, in 1866 (just over 100 years ago), he joined the Birmingham Natural History and Microscopical Society. He was for many years Honorary Secretary of the Biological Section and held office also as Honorary Librarian and Vice-President. During his membership he was present at nearly every meeting and it was rare indeed that he did not in some way contribute to the proceedings. In the year he joined, the Society decided to undertake a survey of the flora within 10 miles of the centre of Birmingham, and Sutton Park was allotted to him. In 1870 he contributed to the *Proceedings* of the Society a classified list of the Flowering Plants and Ferns Indigenous to the Neighbourhood around Birmingham and this developed into his "Notes on Sutton Park: its Flowering Plants, Ferns and Mosses" published in 1876, and led directly to his Flora published in 1891. Meanwhile he contributed many papers, notably to the *Midland Naturalist* and to the *Journal of Botany*, on the flowering plants, mosses and fungi of Warwickshire. He was elected an Associate of the Linnean Society in 1885 and was awarded the Darwin Medal of the Midland Union of Natural History Societies in 1888.

After the publication of his "Flora of Warwickshire" he published the additional records in the *Midland Naturalist* of 1892–93 and contributed to the *Journal of Botany* the "Mosses and Hepatics of Staffordshire" (1896), an "Outline Flora of Staffordshire" (1901), and the "Mosses and Hepatics of Worcestershire" (1903), which was reprinted with additions in the "Botany of Worcestershire" by Amphlett and Rea in 1909. In 1913 he presented his collection of about 10 000 specimens (including Warwickshire Flora material), a set of British brambles and a collection of mosses and liverworts to the Birmingham Museum. He died in 1918.

There are many other names of men and women who have contributed records to this Flora. Space cannot be found for them all; and, although selection is apt to be invidious, there are a few who must be mentioned. First, **George Claridge Druce,** from whose "Comital Flora", published in 1932, but which does not give sources, we have a few first records. Then there is **Dr. R. C. L. Burges,** who died in 1959, who wished so much to see our Society produce a new Flora of Warwickshire, and from whose herbarium we have about a dozen first records. **Dr. R. Baker** of Birmingham, and later of Leamington, was one of a number of botanists, including Bagnall and Bromwich, some of whose records were first published by the Warwick Natural History Society in 1873, in what afterwards became known as the Young and Baker Catalogue. Finally, **David Cameron,** who contributed our first record for *Cystopteris fragilis,* deserves mention as the first Curator of the Birmingham Botanical Gardens opened in 1830.

Of the many still with us, who have contributed records, this is not the place to write. Their work is acknowledged elsewhere in this volume, and they can rest assured that not only is this work appreciated, but that, without it, the new County Flora would never have seen the light of day.

This, then, is the real Prologue to the new "Flora of Warwickshire": a chronicle, all too brief, of many men and women—some famous, some almost unknown—who have, each in his or her own way, contributed something to make the work possible; and for whom every user of the book will spare a kindly thought.

PRINCIPAL WORKS CONSULTED

Ray, John

Raven, C. E. (1942). "John Ray, Naturalist: His Life and Works." Cambridge.

Withering, William

Withering, William, Jun. (Ed.). (1822). "The Miscellaneous tracts of the late William Withering, M.D., F.R.S., . . . to which is prefixed a Memoir of his life, character, and writings." London.

Schofield, R. E. (1963). "The Lunar Society of Birmingham; a social history of provincial science and industry in eighteenth-century England." Oxford.

Hutton, William (1781). "An History of Birmingham to the end of the year 1780." Birmingham.

Walker, Benjamin (1932). Some 18th Century Birmingham houses and the men who lived in them. *Trans. Bgham Midl. Inst.* **56,** 1–36.

Darwin, Erasmus

Linnaeus, C. (1787). "The families of plants. Translated [by Erasmus Darwin] from the last edition of the Genera Plantarum and of the Mantissae Plantarum. . . . By a Botanical Society at Lichfield." Lichfield.

Darwin, E. (1800). "Phytologia, or the Philosophy of agriculture and gardening." London.

Aylesford, Countess of

Druce, G. C. (1926). "The Flora of Buckinghamshire." Arbroath.

Palmer, Charlotte

Druce, G. C. (1915). Obituaries. Charlotte Eden Palmer. *Rep. botl Soc. Exch. Club Br. Isl.* **4,** 48–51.

Baxter, William

Druce, G. C. (1927). "Flora of Oxfordshire", Edition 2. Oxford.

Bloxam, Rev. Andrew

Berkeley, Rev. M. J. (1878). The Rev. Andrew Bloxam: A Memoir. *Midl. Nat.* **1,** 88–90.

Perry, William Groves

[Anon.]. (1863). *Warwick and Warwickshire Advertiser and Leamington Gazette, 28th March*, pp. 2–3

Bagnall, James Eustace

Badger, E. W. (1897). "A Sketch of the Botanical Work of James E. Bagnall." Birmingham.
W. B. G[roves]. (1918). James Eustace Bagnall (1830–1918). *J. Bot., Lond.* **56,** 354–6.
[Anon.]. (1918). Obituaries. Bagnall, J. E. *Rep. botl Soc. Exch. Club Br. Isl.* **5,** 349–52.
Bagnall MSS. Birmingham Reference Library.

Miscellaneous

Bagnall, J. E. (1891). "Flora of Warwickshire." London and Birmingham.
Dent, R. K. (1879). "Old and New Birmingham." Birmingham.
"Dictionary of National Biography."
Allen, D. E. *MS Notes on Birmingham Museum botanical collections.*
Dawson, R. *MS Notes on Warwick Museum botanical collections.*
Analyst, Lond. 1834–40.
Phytologist. 1841–54.
J. Bot., Lond. Various volumes.
Rep. botl Soc. Exch. Club Br. Isl. Various volumes.
Proc. Bgham nat. Hist. phil. Soc. Various volumes.

6. METHODS OF RECORDING AND PROCESSING THE DATA

Introduction

The "basic square" method used in this Flora for the vascular plants (see Hawkes and Readett, 1954, 1963) is founded on the concept of area rather than point recording. The unit of reference is an area of 1 km^2, identical with the smallest subdivision marked on the one inch to the mile Ordnance Survey series. There are some 2450 of these basic squares in the County of Warwickshire, but only a quarter of these was selected for recording, as we shall see later.

This method owes much to the work of Good (1948) who, in his survey of the flora of the county of Dorset, recorded lists of species ("stands") at various points ("loci"), spaced as evenly as possible throughout the county. Good's selection of loci was arrived at arbitrarily, each locus being situated within a well-defined habitat (e.g. bog, pasture, wood, etc.). In about ten years he was able to complete some 7500 stands which, estimating the area of Dorset as 1000 square miles, gives an average stand or locus frequency of 7·5 to the square mile or a little less than 3 to the square kilometre.

The methods devised by Good and the information resulting from his survey are quite clearly of the greatest importance to the study of the distribution and ecology of British plant species. There are, however, one or two disadvantages, as Good himself pointed out. Firstly, the loci were situated in arbitrarily selected natural or semi-natural habitats, so that more "artificial" habitats such as waste ground, farmyards, roadsides, gardens and vegetational communities which were transitional or not clear-cut were to a large extent neglected.

Further, although the frequency of a species in the County as a whole was recorded by Good (i.e. the number of stands in which it occurred) its relative frequency within each stand was not indicated. A plant which occurred very thinly scattered over the county might by this method be considered as occurring much more frequently than one that occurred as a dominant in some parts only.

In planning the methods used in the Flora of Warwickshire we followed Good's system in which species distributions are built up from a mosaic of individual lists, and we have tried to rectify the weaknesses in his methods by substituting area recording for point or locus recording. The advantage of area recording is that the field worker is required to identify and record all habitats in the form of data connected with the species records rather than as arbitrarily selected items. In our method we also required frequency data for each species habitat record; this solved the problem referred to above with Good's methods and enabled us to give two types of frequency evaluation, which are shown in the Check List (Chapter 11).

Our area system of recording bears at first glance a great similarity to that used in the B.S.B.I. "Atlas of the British Flora" (Perring and Walters, 1962; Perring, 1968). The surveys were begun at about the same time but were thought out independently. The areas used by Perring and Walters were squares of 10 × 10 km, and no habitat or frequency data were required with the records sent in. This is understandable enough when a complete country is being surveyed. Distribution maps were printed by a tabulator on to pre-printed base maps from data on 40-column punched cards, only one kind of symbol being used. Other symbols were added to these maps by hand in certain instances.

Other County Floras published during the rather long period during which our own work has been developing have given grid records for the less common species (Cambridgeshire: Perring *et al.*, 1964; Derbyshire: Clapham, 1969; and some maps) whilst others have shown maps based on a tetrad unit (Hertfordshire: Dony, 1967; Berkshire: Bowen, 1968).

Here again, the maps do no more than indicate the presence or absence of a species within a grid square or "tetrad" of four squares, and the authors have not attempted to show habitat or frequency data except for frequency indications in the Berkshire maps. Neither have computer methods been used in any of these Floras.

Works directly derived from the methods used in the Warwickshire survey are a study of the Breckland of Norfolk, Suffolk and parts of Cambridgeshire which is being carried out by the Nature Conservancy and voluntary helpers from the counties concerned; and a new Flora of Durham by the Rev. G. G. Graham. It will be particularly valuable in due course to compare the Warwickshire results with those from these other similar surveys in different parts of Britain.

Methods

(i) *Area of Study*. The area surveyed during the present study was Vice-County 38 (Warwickshire) and not the political or administrative county, whose boundary has changed continuously over the years. The vice-county and political boundaries are compared on Map 7 (p. 63), and it can be seen that in most regions the discrepancies between the two boundaries are not very great. The largest changes occur in the vicinities of Tamworth, Stratford on Avon, and especially of Birmingham, where the western boundaries of the city have in most instances been taken formally as the boundaries of the county, though the City is itself completely distinct for administrative purposes.

The Watsonian vice-county system (see Watson, 1870a and b) has provided a stable series of areas of roughly comparable size throughout the country which were based mainly on the political boundaries of the time when it was devised. Very large counties were divided into two or more portions, and small counties or fragmented "islands" and "peninsulas" were added to the counties surrounding or adjacent to them.

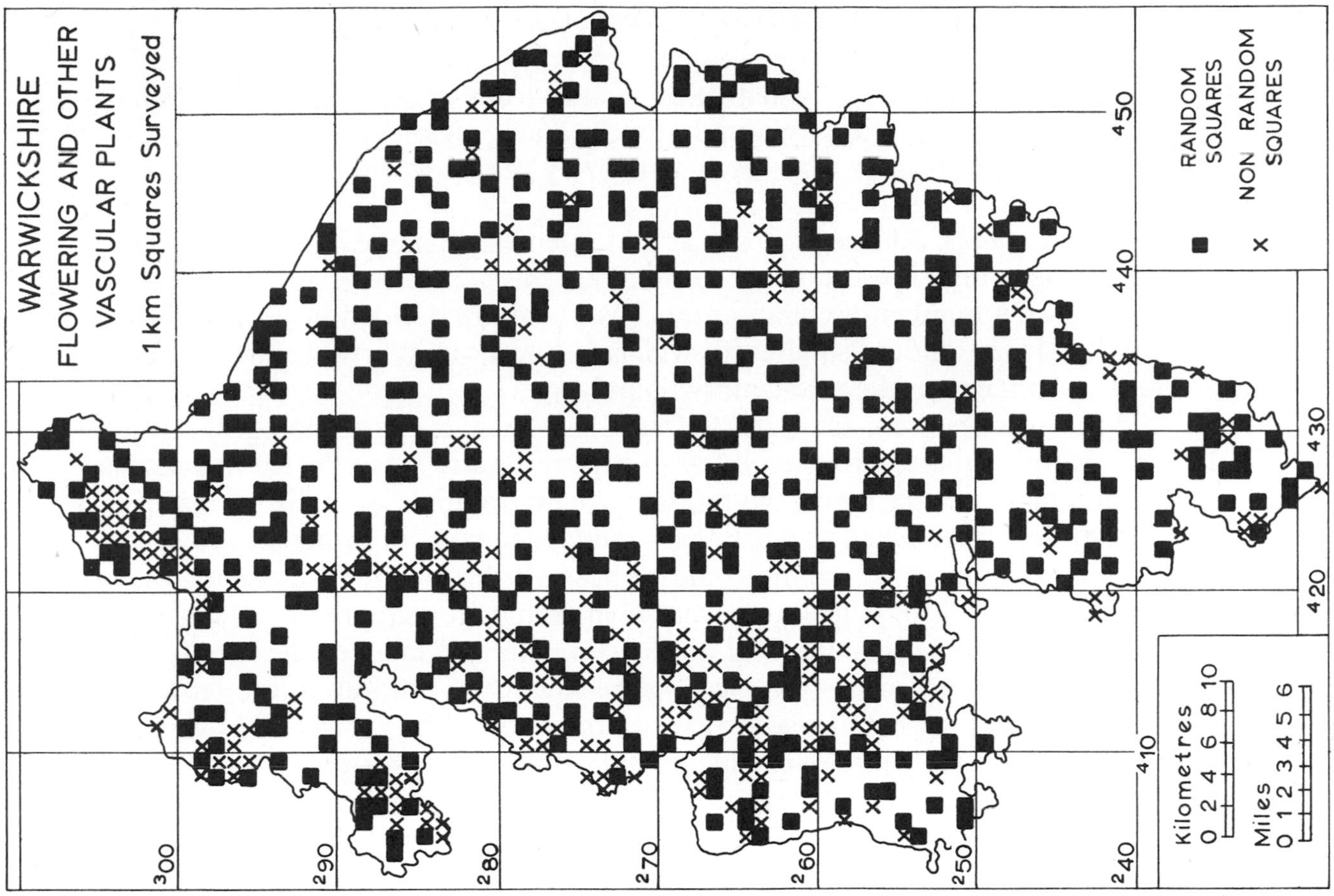

MAP 8. Vascular Plants. 1 km squares surveyed.

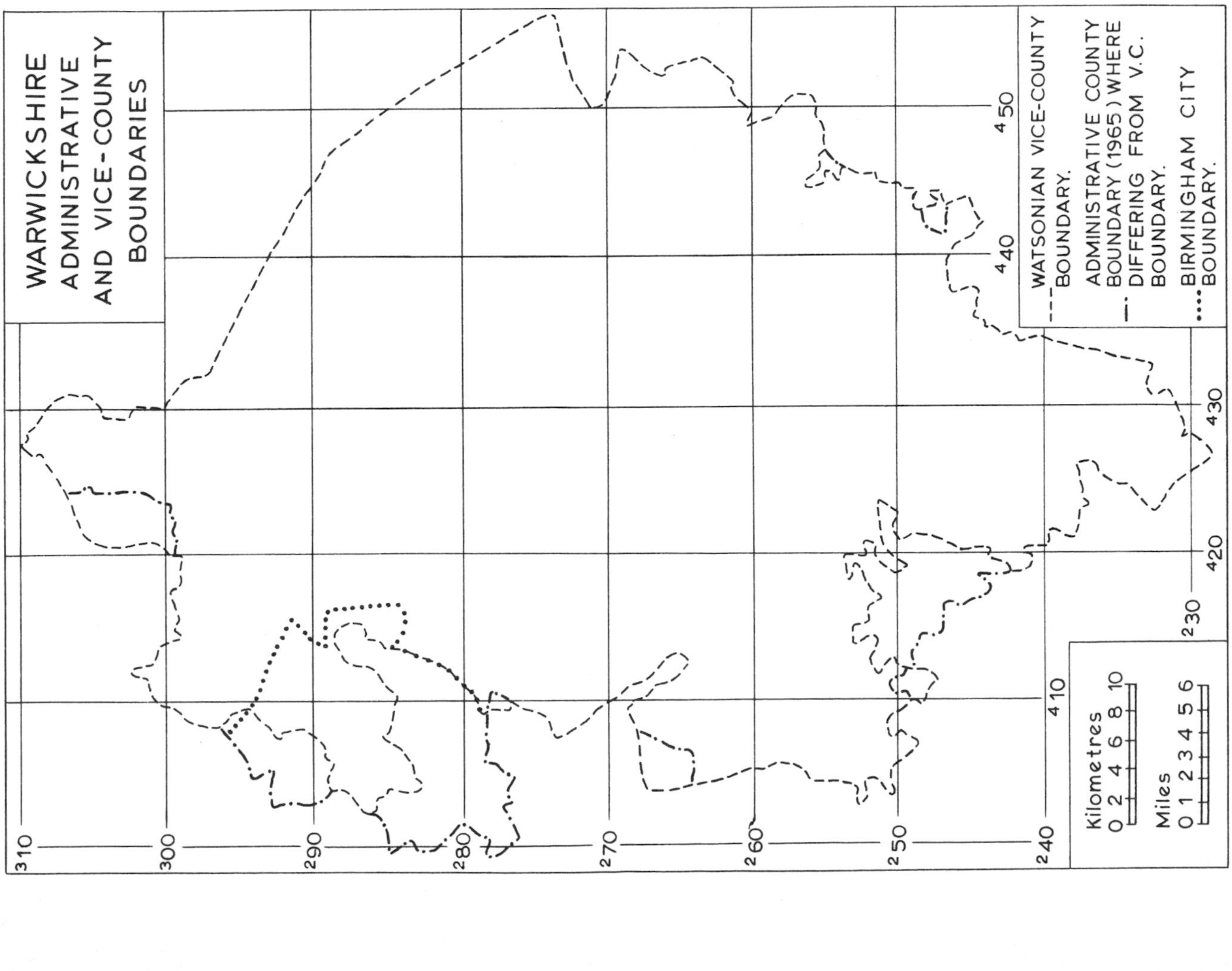

MAP 7. Administrative and VC boundaries.

The plants occurring in each could then be compared from decade to decade without introducing the confusion of varying boundaries. In many regions the vice-county and county boundaries do not differ from each other very markedly, as we have seen for most of Warwickshire.

(ii) *Selection of Squares*. Although it was at first intended to survey all the 2450 kilometre squares in the county it was quickly realised that this would be an impossible task, even with the help of a number of collaborators. We therefore decided in 1955 to record whenever possible from random squares, of which 613 occur in Warwickshire, omitting small portions of squares on the vice-county boundary.

Blocks of four squares were set out ("tetrads") so that 25 were included in each 10 × 10 km "major" square. From each tetrad a 1 km square was selected at random and marked on a master map. Later, the random square numbers were also transferred to a list. All recorders were asked to work these random squares in preference to others so that an even and unbiased recording would be possible throughout the county. This ensured that at least 25 1 km squares were worked in each major square of 100 km^2, and that a good estimate of the distribution of all the common and most of the rare species would be made. Another advantage of selecting the squares at random in the tetrads is that no bias towards certain habitats and against others would be exercised, as we shall see later when we discuss the problem of habitat analyses. Although recorders were asked to take random squares for study we did not reject lists of records sent in from non-random squares, and an extra 256 of this type were submitted, making the total squares surveyed 869. As we shall see later, these records, from both random and non-random squares, were used in the map construction, but only the random square records were utilised in the statistical work. The random and non-random squares recorded are shown on Map 8.

(iii) *Field Recording*. One or more random squares were registered with each recorder, who was asked to work over the area with the aid of a $2\frac{1}{2}$ in. or 1 in. to the mile O.S. map. Each recorder was given a card with clear instructions on the method of work and the sort of information required. He was asked to walk through his square, carefully visiting all parts of it and noting down all species of flowering plants and vascular cryptogams on a specially prepared form (see Fig. 1). A portion of a completed form is shown in Fig. 2. The square number on the form was the 4-figure grid reference for 10 km and 1 km squares, the 100 km reference being omitted, since although Warwickshire falls within two 100 km grid squares, 42 and 43, there is no overlap in the county of 10 km grid references between them.

The "Basic square locality" was merely an arbitrarily chosen place-name used as a cross-check on the grid reference in case of error in grid citation. It was also necessary to mention the name of the Flora used for identification purposes since the connotation or circumscription varied for some species in the Floras commonly used for identification. Latterly, however, the second edition of Clapham, Tutin and Warburg's "Flora of the British Isles" (1962) was used almost exclusively. The names which are used in this present Flora do not always coincide with those in Clapham, Tutin and Warburg,

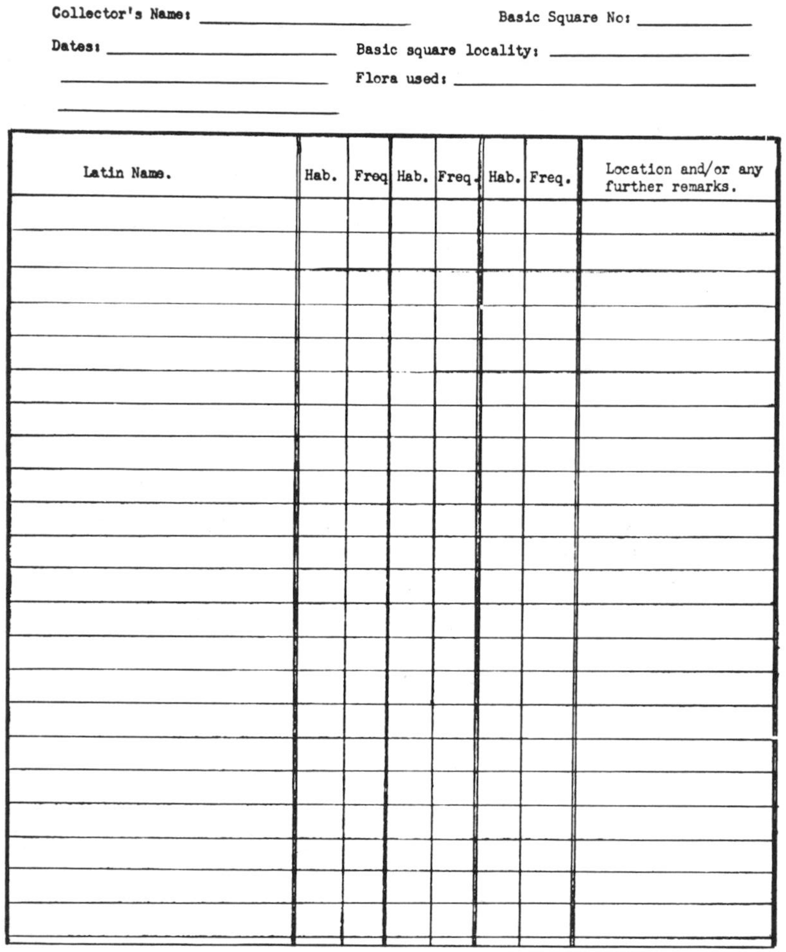

Collector's Name: ______ Basic Square No: ______

Dates: ______ Basic square locality: ______

______ Flora used: ______

Latin Name.	Hab.	Freq	Hab.	Freq.	Hab.	Freq.	Location and/or any further remarks.

FIG. 1. Recorder's form for vascular plants. See text for further explanation.

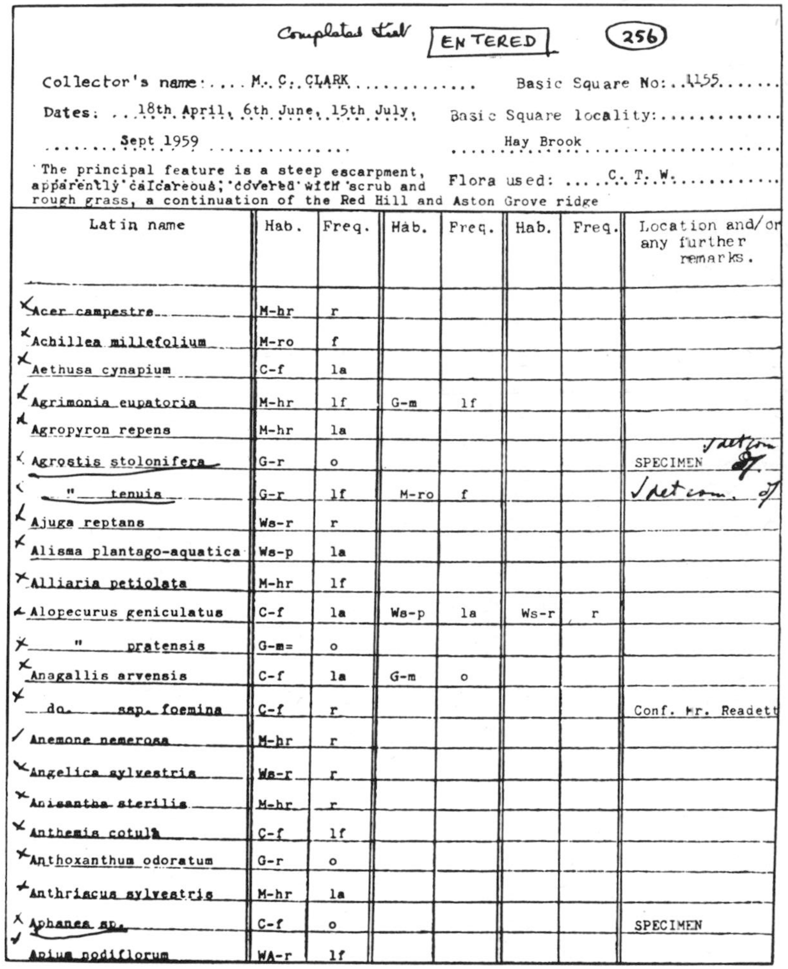

Completed list ENTERED 256

Collector's name: .... M. C. CLARK .... Basic Square No: ..1155......

Dates: ..18th April, 6th June, 15th July, Basic Square locality: ....

........Sept 1959 ....Hay Brook..................

The principal feature is a steep escarpment, apparently calcareous, covered with scrub and rough grass, a continuation of the Red Hill and Aston Grove ridge Flora used: ....C. T. W...........

Latin name	Hab.	Freq.	Hab.	Freq.	Hab.	Freq.	Location and/or any further remarks.
Acer campestre	M-hr	r					
Achillea millefolium	M-ro	f					
Aethusa cynapium	C-f	la					
Agrimonia eupatoria	M-hr	lf	G-m	lf			
Agropyron repens	M-hr	la					
Agrostis stolonifera	G-r	o					SPECIMEN
" tenuis	G-r	lf	M-ro	f			
Ajuga reptans	Ws-r	r					
Alisma plantago-aquatica	Ws-p	la					
Alliaria petiolata	M-hr	lf					
Alopecurus geniculatus	C-f	la	Ws-p	la	Ws-r	r	
" pratensis	G-m=	o					
Anagallis arvensis	C-f	la	G-m	o			
do. ssp. foemina	C-f	r					Conf. Mr. Readett
Anemone nemerosa	M-hr	r					
Angelica sylvestris	Ws-r	r					
Anisantha sterilis	M-hr	r					
Anthemis cotula	C-f	lf					
Anthoxanthum odoratum	G-r	o					
Anthriscus sylvestris	M-hr	la					
Aphanes sp.	C-f	o					SPECIMEN
Apium nodiflorum	WA-r	lf					

FIG. 2. Portion of a completed form.

however, since we have attempted to use the most recently agreed name for each species.

Four columns for habitat and frequency were provided, as well as a final column for special remarks, such as exact location, 6-figure grid reference giving the position of a plant to the nearest 100 m, further notes on habitat, degree of dominance or rarity, or whether a specimen was being sent for confirmation of the record or had already been checked by one or more members of the panel.

At least three visits were to be made at different seasons, or more than three visits if time permitted. A visit in early spring followed by one in late spring or early summer and a third in late summer or early autumn were recommended so that as complete a picture as possible of all the species occurring in the square might be obtained.

Although for each basic square each species was noted only once, a record was to be made of its presence in each distinct habitat, together with its frequency in that habitat. Most species were recorded in one or two habitats only, but occasionally records were made for certain species in three, four or even five habitats within a square. The habitats were carefully codified and given standard abbreviations. They were grouped into eight main categories ("major habitats"), each of which was subdivided into subsidiary categories ("minor habitats") (see Table 4). At a much later stage it was considered advisable to re-group the eight major categories into nine by re-allocating certain of the

TABLE 4. Habitat Categories used in Field Recording

1. WO = Woodland
   - WO-c = Conifer Wood
   - WO-m = Mixed Wood
   - WO-o = Oak Wood
   - WO-sc = Scrub or Derelict Wood

2. WS = Waterside, Marsh
   - WS-c = Canalside
   - WS-d = Ditchside
   - WS-l = Lakeside
   - WS-p = Pondside
   - WS-r = Riverside
   - WS-m = Marsh

3. WA = Immersed in Water
   - WA-c = Canal
   - WA-d = Ditch
   - WA-l = Lake
   - WA-p = Pond
   - WA-r = River

4. H = Acid Heath, Bog
   - H-b = Bog
   - H-d = Dry Heath
   - H-g = Heath Grassland

5. M = Marginal
   - M-hr = Hedgerow
   - M-ro = Roadside
   - M-ry = Railway embankment or cutting
   - M-wa = Wall

6. G = Grassland (excluding heath grassland)
   - G-m = Meadow
   - G-p = Pasture
   - G-o = Orchard
   - G-r = Rough Grassland

7. C = Cultivated Land, Arable
   - C-a = Allotment
   - C-f = Cultivated Field (Arable)
   - C-g = Garden
   - C-p = Public Park

8. RU = Ruderal, Waste Land
   - RU-b = Brick or Marl Pit
   - RU-f = Farmyard
   - RU-q = Quarry
   - RU-w = Waste Ground, Tips, Refuse Dumps

minor categories. However, this re-grouping was done entirely by the computer when the records had already been transferred to computer store. The revised arrangement appears in Table 5, and it is to these that the species text always refers, and which the maps show. It should be repeated, however, that the field recording was based on the scheme shown in Table 4. We shall discuss the reasons for this re-grouping in Chapter 7.

TABLE 5. Habitat Categories used in this Work (See also Table 4)

1. WO = Woodland
   - WO-c = Conifer Wood
   - WO-m = Mixed Wood
   - WO-o = Oak Wood
2. M = Marginal (Hedgerows, Scrub Woodland)
   - M-hr = Hedgerow
   - M-sc = Scrub or Derelict Wood (formerly WO-sc)
3. H = Acid Heath, Bog
   - H-b = Bog
   - H-d = Dry Heath
   - H-g = Heath Grassland
4. G = Grassland
   - G-m = Meadow
   - G-p = Pasture
   - G-o = Orchard
   - G-r = Rough Grassland
5. WS = Waterside, Marsh
   - WS-c = Canalside
   - WS-d = Ditchside
   - WS-l = Lakeside
   - WS-p = Pondside
   - WS-r = Riverside
   - WS-m = Marsh
6. WA = Immersed in Water
   - WA-c = Canal
   - WA-d = Ditch
   - WA-l = Lake
   - WA-p = Pond
   - WA-r = River
7. RU = Ruderal (Waysides, Waste Places)
   - RU-ro = Roadsides, pathsides (formerly M-ro)
   - RU-ry = Railway embankment or cutting (formerly M-ry)
   - RU-b = Brick or Marl Pit
   - RU-f = Farmyard
   - RU-w = Waste Ground, Tips, Refuse Dumps
8. ST = Stones, Quarries
   - ST-wa = Walls (formerly M-wa)
   - ST-qu = Quarries (formerly RU-q)
9. C = Cultivated Land, Arable
   - C-a = Allotment
   - C-f = Cultivated Field (Arable)
   - C-g = Garden
   - C-p = Public Park

The frequency of the species in any one basic square was rather difficult to estimate with accuracy. Detailed quadrat surveys in the usual ecological sense could clearly not be carried out on a county-wide scale. Since there must be an element of subjectivity in any method which falls short of detailed ecological surveys we considered it best to keep the system fairly simple, by using six conventional and well-known terms, which we give below (with code abbreviations) in descending order of frequency:

1. Abundant (a)
2. Locally abundant (1a)
3. Frequent (f)
4. Locally frequent (1f)
5. Occasional (o)
6. Rare (r)

It must be emphasised that the frequency assessments did not refer to the frequency of a species in the basic square as a whole but only to the particular habitat for which the record was made. Thus, for example, *Hedera helix* was often recorded as frequent in hedgerows but not found in any other habitat, with the exception of woodlands. Over the square as a whole it probably would not have ranked as more than occasional or, at most, locally frequent. Nevertheless the record "Hedgerows—frequent (Mhr-f)", was correct, since only the habitat hedgerow was under consideration for that record.

Difficulties with identification were to be expected, and we designated a list of "Critical Groups" (see Table 6) for which specimens were needed to support records. However, it was also possible merely to denote the genus (e.g. *Rosa*, *Rubus*, *Hieracium*, etc.), or to note the species in the wide or aggregate sense (e.g. *Rosa canina* agg. or s.l., *Rubus fruticosus* agg.). Even when no attempt at identification beyond the genus or the species aggregate was made, specimens were requested whenever possible. Further work with the specimens submitted will be dealt with in a later section.

TABLE 6. List of Critical Groups

**Ranunculus* subgenus *Batrachium*	*Symphytum*
**Fumaria*	*Calystegia*
Viola	*Rhinanthus*
**Polygala*	**Euphrasia*
**Cerastium*	*Orobanche*
†**Sagina*	*Mentha*
**Chenopodium*	*Valerianella*
**Montia*	*Arctium*
**Prunus* (except *P. spinosa*)	*Centaurea*
Rubus fruticosus agg.	*Hieracium* (except *H. pilosella*)
Alchemilla	**Taraxacum*
Aphanes	*Potamogeton*
Rosa	**Juncus*
**Sorbus* (except *S. aucuparia* and *S. torminalis*)	*Epipactis*
**Callitriche*	†*Dactylorhiza*
Epilobium (except *E. angustifolium* and *E. hirsutum*)	**Carex*
**Oenanthe*	*Glyceria*
**Polygonum aviculare* agg.	*Festuca*
**Rumex*	*Poa*
Ulmus	*Bromus*
Salix	*Agrostis*

* Specimen should be in fruit. † Fresh material desirable.

Although most basic squares were surveyed by individual collectors, working independently, many were recorded jointly. It was found to be rather difficult to allocate squares in the north-eastern and south-eastern parts of the county because of their isolation from Birmingham and other cities. To deal with this problem we organised a series of summer joint recording field meetings from 1955 to 1961. Transport was arranged and squares were allocated to pairs of recorders (or occasionally more), the less experienced worker being grouped where possible with one of more experience. Recorders met in the late afternoon at a convenient point to discuss problems of recording and identification. Spring, summer and autumn meetings were arranged to the same sets of squares, with other recorders substituting for earlier ones where necessary at the later meetings. This method, besides being of value in surveying poorly-known areas, helped to train field workers with little previous experience and to give them confidence in working over the squares which had been assigned to them individually.

We also received help in several cases from other Natural History Societies, including those of Coventry and of Warwick and Leamington, in providing lists from squares or individual records. Members of these Societies also sent us lists and records. Apart from individuals and Societies within the county, we also received help from the Botanical Society of the British Isles whose members collaborated with us at joint meetings in the Shipston, Nuneaton and Leamington areas.

We also received help from people in various parts of the county whose names were sent us by Dr. F. Perring, organiser of the B.S.B.I. maps scheme. We should note in passing that at a certain stage our records for major squares (10 × 10 km) were sent to Dr. Perring for inclusion in the Atlas of the British Flora, though we obtained many more records after that work went to press.

(iv) *Arrangement and Classification of Records*. The number of records submitted by collectors from "completed" squares varied considerably, due both to differences in the actual number of species growing and the competence of the collectors in seeing and identifying them.

In practice we were not prepared to accept as complete any list with fewer than 100 records, unless the square was situated in a built-up area where the habitat range was very restricted. Even in squares in the centre of Birmingham and Coventry we were often still able to find up to about 60 species growing along the canal towpaths, on derelict or building sites, or even in gasworks yards.

Lists of less than 100 records from squares in normal country regions were not considered to be adequate and the squares were not recorded as complete, even though the records were added to our files.

It seems probable that the actual number of species occurring in a 1 km square in Warwickshire (apart from the squares in built-up areas to which we have just referred) varies from about 150 to 350, the highest number recorded being 323. A good average number was about 200 records, and most good recorders would probably have seen 90–95% of the species in a square during their three visits in a single year. Subsequent visits apparently added only a very small extra number of records, except when another recorder visited a square that had previously been somewhat imperfectly studied.

The main reason for the rather large range of species to be found from square to square lies no doubt in the diversity of habitats. It does not necessarily follow, however, that areas of more natural vegetation will provide the greatest number of records. Parts of the semi-natural areas of Sutton Park in the north-western corner of the county are botanically very uniform, with large areas of *Nardus*, *Molinia* and *Calluna*. On the other hand, some of the very man-made environments in the centre and west of Warwickshire with their canals, railways, roads, paths, coppices and marshes possess a wide range of habitats and the species counts from these are correspondingly higher.

When the lists were received they were carefully screened and any doubtful records deleted or verified with the recorder before acceptance. Records of very rare species, new county records, or species recorded from apparently atypical habitats are examples of those which were carefully checked in this way. When specimens were collected the verification could be made by our Flora Panel. The members of this panel were the authors of this book, together with M. C. Clark, and it met nearly every Tuesday evening (apart from holiday periods) for some 15 years to identify material and put together the records of the Flora.

General identification of specimens submitted was carried out by the panel, with help from local, national and international specialists when required (see list on pp. 740–42). Towards the beginning of the survey a large voucher collection of material identified by specialists was built up. Subsequently, with the named sets available, and with an increasing knowledge, the panel was able to assume responsibility for naming the more straightforward material; it was thus necessary to send only particularly difficult specimens (hybrids, etc.) to the specialists.

The next stage in the process was to copy the basic square lists on to species cards (see Fig. 3), of which a separate set was made up for all the species in each major square (Hawkes, Readett and Skelding, 1955). These species cards were ruled with a grid of 100 one-kilometre squares at the scale of 1 in. to the mile. Thus each card represented a conventionalised map with the grid co-ordinates on the top and right-hand margins. Records of habitat and frequency were written in the correct spatial relationship, and by placing all the cards for a given species together in the correct order it was possible to construct an interim distribution map of that species for the whole of the county (see Fig. 4 for a portion of the north-west corner). Thus at various times during the survey it was possible to check the records for certain species, consider the reasons for particular distribution patterns and ask for additional records or observations from our recorders—all before the final maps were drawn up.

It was possible also, by ringing or marking in red ink, to note the records for critical species which had been confirmed by national or international specialists, or by our own panel.

(v) *Back-checking*. By the end of 1963 the collection of data for virtually all the random squares was apparently complete. However, the species cards and the lists sent in by certain collectors showed a number of gaps, and it was thought that these might be due to a lack of knowledge on the part of these collectors rather than an absence of species in some areas.

Accordingly, a back-check operation was organised during the 1964 and 1965 seasons,

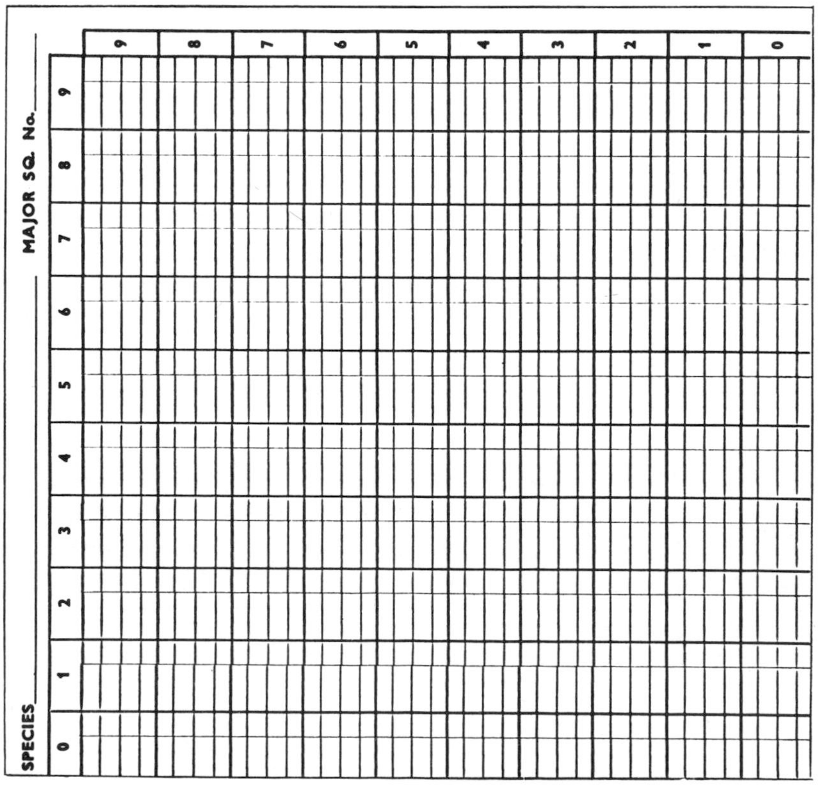

FIG. 3. Species card for vascular plants. See text for explanation.

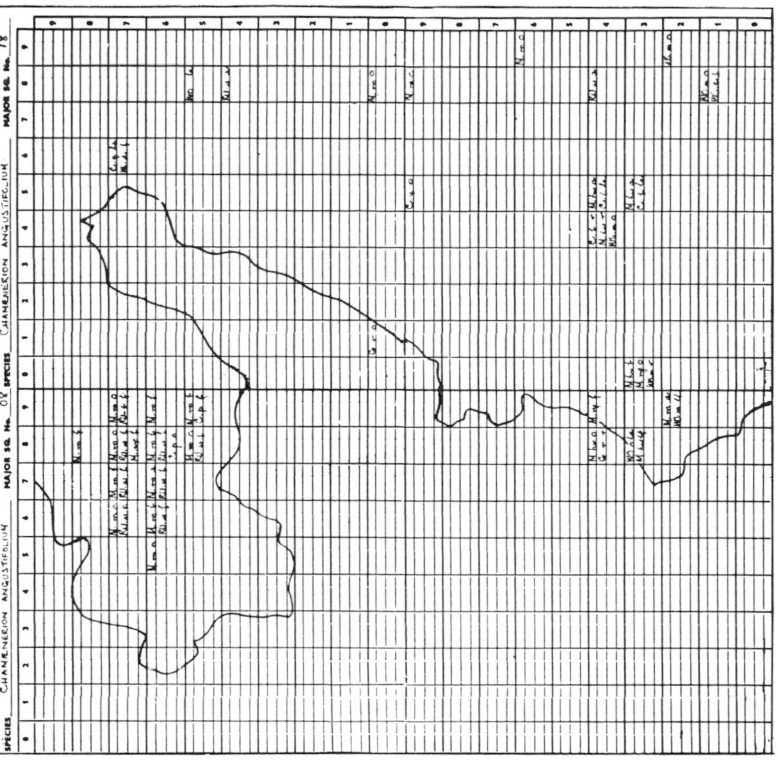

FIG. 4. A set of four species cards for *Epilobium angustifolium* from the N.W. corner of the county, with the VC boundary added. These cards could be laid out for each species to form a conventional map.

in which several recorders made themselves responsible for checking doubtful records and trying to fill obvious gaps in most of the major squares. Each recorder was provided with a detailed back-check list of the existing records for his major square and was asked to pay special attention to those 1 km squares where the number of records was low. This was quite clearly a useful operation since many thousands of extra records were added in this way.

By the end of the survey some 175 000 records had been made, representing many more individual records than had ever been gathered before in a County Flora.

(vi) *The First Check List.* As a preliminary attempt to set down the first records of plant species in the vice-county and to note their presence in surrounding vice-counties a check list of the vascular plants and bryophytes was published (Readett *et al.*, 1965). This was widely circulated both within the county and nationally, and amongst other things served the useful purpose of drawing the attention of botanists to the records available to us at that time. As a result, extra information was obtained, which we have incorporated into the present work.

(vii) *Maps and Symbols.* From the inception of the scheme it had been hoped that it would be possible to transfer the habitat data in some way on to species distribution maps by means of symbols that would give a general impression of the habitats or communities in which a species was to be found throughout the county. A number of problems had to be solved here; in the first place, none of the symbols could be of the filled-in type, since they would need to be presented in various combinations in the same square. Each must therefore be clearly visible, even when combined with several of the others. It was also necessary to devise symbols which would give easily recognisable patterns over the map as a whole. Thus, symbols which relied for their interpretation on their position within a square would clearly be valueless, since such a scheme would not enable the reader to obtain a general picture at a glance and each square would have to be examined individually. It was further decided to represent the record from each random square in the complete tetrad in which the random square occurred. This would prevent a false impression of distribution being produced as a result of "clustering" of the random squares (see Map 8). A reasonably satisfactory series of habitat symbols was devised which conformed to the needs outlined above (see Cadbury *et al.*, 1957). These have been incorporated into the present scheme with some modifications. At that stage (prior to 1964) it seemed impossible to devise a scheme for species and habitat mapping which at the same time showed frequencies within habitats, though a general idea of frequencies could of course be obtained by looking at the spread of the species over the county as a whole.

The major habitat categories (woodland, waterside, water, heath, marginal, grassland, cultivated land and ruderal), eight in number, were, as we have already mentioned, divided into minor ones; there were 34 of these minor habitat categories for Warwickshire, and it was clearly impossible to devise a system of symbols to denote all these when each symbol needed at the same time to be distinguishable in a large number of different combinations. It was therefore considered that symbols showing the eight major categories (finally nine, as we have already mentioned) would be all that could be reasonably expected for maps of this sort.

By 1964 a number of problems urgently needed attention. Since it was hoped to draw distribution maps of some 700 species of vascular plants and 200 species of bryophytes a long and expensive period of map drawing seemed to be ahead of us. Although the bryophyte data were fairly simple, the prospect of drawing by hand the habitat symbols on 700 vascular plant distribution maps was one to which we did not look forward with much pleasure. There were somewhere in the region of 1000 records for the more common species, each of which would need to be mentally interpreted into a symbol by the draughtsman and placed accurately on to the base map.

A final point might be considered here. It was planned at a fairly early stage (see Cadbury *et al.*, 1957) to include habitat analyses in the Flora; these were to show for each of the commoner species the range of habitats in which it had been recorded, expressed as the number of records for each habitat and the percentage figure, as well as the apparent "preference" of the species for a particular habitat. For the second time in a County Flora numerical data would be available for statistical analysis, the first time being Good's "Geographical Handbook of the Dorset Flora" (1948).

The information available was sorted in one or two cases by counting occurrences in the different habitats and calculating percentages, as the first stage of the process. The data were extremely interesting, but the labour involved was quite considerable and carried with it as many possibilities of error as the drawing of the maps themselves.

Electronic Data Processing

(i) *General.* In 1964 Birmingham University bought an English Electric KDF9 computer, and it was decided to process the whole of our data on this machine and at the same time to explore the possibilities of computer mapping of the distribution data. Clearly, some kind of graph plotter seemed to be the answer to the mapping problem, where X and Y co-ordinates could be used to position the symbols in the correct places on a previously printed map on continuous stationery. Unfortunately, most graph plotters incorporate a print head with very small standard symbols. These completely obscured each other if overprinted, only one symbol could generally be printed in each position, and the symbols could often only be printed in a fixed sequence. Furthermore, the symbols available were far too small for our purposes.

The methods used in the B.S.B.I. "Atlas of the British Flora" (Perring and Walters, 1962) were an excellent example of the use of data processing equipment in the production of distribution maps. Unfortunately, only a series of uniform symbols could be printed by this method, and it was therefore considered to be unsuitable for the presentation of the Warwickshire data. More sophisticated methods of distribution mapping were discussed by Soper (1964, 1966) for the vascular plants of Ontario, Canada. Soper transferred his primary data, consisting of grid co-ordinates and certain other information, on to punched tape and later to 80-column I.B.M. punched cards. He discussed the

```
2010

stellaria holostea l.    2010

05

44.womf.wosco.
54.mhro.
58.womo.mhro.
64.mror.
66.mrof.
68.gro.mhro.womo.
89.mnrf.
94.mhrla.
98.mryla.mrof.
/

06

43.womo.mhro.
44.mhro.
46.woof.
50.mhra.
51.wsdf.
53.womf.woof.
54.womf.mnrf.
56.mhrr.
60.woof.
63.gpo.mhrf.mrof.
64.mnrlf.
65.mnra.
70.mhra.
73.mhro.
76.mroo.
77.mhro.
80.mroo.
83.wsro.mnro.
84.mroa.mhro.womo.
86.mrof.
87.mnrf.
92.mrof.womf.mnrf.
93.mhrf.
98.mroa.
/

07

72.mhrf.
73.mnrf.
81.mhro.
82.mhro.mroo.
83.womo.mhrf.
84.gro.mrof.
90.mnrf.
92.mrof.womf.mhrf.
93.mrof.
94.mhra.nryf.
/

08

43.mhrr.
54.womr.
/

10

20.mrola.
/
```

FIG. 5. "Print out" of portion of data for *Stellaria holostea* L. as typed on the teleprinter, showing major square numbers heading blocks of data, minor square numbers at the start of lines, and letter strings representing individual observations (e.g. mhr-f = hedgerow, frequent). The figure 2010 at the top represents the B.S.B.I. code number for this species.

use of tabulators and digital plotters with a printing head in transferring data to a preprinted base map. These he preferred to an incremental type of plotter since he wished to provide completely blacked-in symbols which could not be drawn so satisfactorily by the latter instrument.

For the more complex system of symbols indicating habitats which we wished to portray on maps for a large number of the Warwickshire plants, the incremental plotter is clearly very advantageous. Accordingly, after several exploratory discussions with a variety of firms, we decided to use an incremental plotter with a drawing pen, marketed by Benson-Lehner Ltd. (now Computer Instrumentation, Ltd.) of Southampton. Since these instruments are relatively expensive and we only needed one for this particular project, we agreed in April, 1965 to buy time on one of Benson-Lehner's own machines. It was also decided to make some simpler maps on the teleprinter or line printer which could be used for the bryophytes and certain vascular plants, for which only a limited number of records were available. The mapping and general computer methods adopted were described by us in a preliminary paper, published in 1968 (Hawkes, Kershaw and Readett, 1968).

(ii) *Preparing the Data for the Computer.* Apart from the very necessary programming work, the first stage in preparing the data for plotting was to put it into a form which could be used by the computer. We had no punched card facilities at Birmingham when this phase of the work started and we therefore decided to use 5-hole punched tape. To this end we bought a Westrex teleprinter, which is in effect an electric typewriter with tape punching and reading facilities.

Punching began in December 1965, and we were fortunate in having the services of a skilled teleprinter operator (Mrs. M. Kershaw) for the whole period, working on a part-time basis, until Easter, 1967. This included not only the punching of the tapes, but also editing them; it must be admitted, in all fairness to the operator, that a considerable amount of the "editing" was nothing more than adding recently received records to a tape which was in fact a perfect copy of the original data.

The data were taken directly from the species cards, species by species, and in sequence of major square cards, the data being copied exactly and a "print-out" produced as a by-product of the punching operation (Fig. 5). The print-out was then used for checking against the original cards.

(ii) *Entering the Data into the Computer.* After punching and checking, the data were passed to the computer. Three programs were used, but only the first read the paper tape data. Each in turn operated on the data, taking the work a stage further and passing on partly processed intermediate results to the next stage by means of magnetic tape. The last program produced punched paper tape which was sent to Benson-Lehner and used to control the plotter to draw the final map.

The computer first checked the format of the data for four types of error:

(1) Major square number unacceptable.
(2) Minor square number unacceptable.
(3) The same 1 km square mentioned twice.
(4) Observation code unacceptable.

If an error was detected, the computer printed a message to guide the punch operator in correcting the mistake, and switched into a mode of operation in which it did not prepare intermediate results for further processing, but continued to read the data and print messages for any errors detected. In this way all errors in a particular data tape were found in one computer run and could all be corrected before the tape was re-submitted.

STELLARIA HOLOSTEA L. 2010

	A	LA	F	LF	O	R	TOTAL
MHR	17	5	49	15	59	24	169
RURO	10	5	46	12	38	7	118
WOM	1	1	16	2	23	3	46
GR		3	4	8	11	2	28
MSC	2	1	6	1	6	1	17
RURY		2	4		8	2	16
WSD			6		2	2	10
WOO			2		2	1	5
GM			2		1	1	4
RUW	1				2	1	4
STWA						2	2
GP			1			1	2
WOC	1	1					2
STQ						1	1
WSC			1				1
WSM					1		1
WSR					1		1
CF			1				1
CG			1				1
TOTAL	32	18	139	38	154	48	429

HABITAT	ABUNDANT		OCCASIONAL		TOTAL	
	OBS.	PERCENT	OBS.	PERCENT	OBS.	PERCENT
MARGINAL	96	22.4	90	21.0	186	43.4
RUDERAL	80	18.6	58	13.5	138	32.2
WOODLAND	24	5.6	29	6.8	53	12.4
GRASSLAND	18	4.2	16	3.7	34	7.9
WATERSIDE	7	1.6	6	1.4	13	3.0
STONES ETC.			3	0.7	3	0.7
CULTIVATED	2	0.5			2	0.5
TOTAL	227	52.9	202	47.1	429	100.0

FIG. 6. Data format check and simple statistics for *Stellaria holostea* L. The input tape for this species contained no errors, so the computer has printed no error messages. In the first block of data the records are listed by minor habitats in one direction and frequencies in the other; general totals are given in the right-hand column. In the second block of data the records are listed in major habitat groups on two grades of frequency. Totals and percentages are also shown.

If, however, no errors were detected, the computer assembled within its memory an "electronic image" of the required distribution map. Once this image was obtained, the rest of the computer process was directed to translating it into suitable control signals for the plotter. The first level map image was placed on magnetic tape ready to be further treated by the second program.

The first program, however, did rather more than this. Observations which were accepted and entered on the map image were counted, giving 204 totals, one for each minor habitat frequency combination. These totals were printed to form the first table in the statistics. The second table was simply the first table, recast in terms of nine major habitats and two frequencies, corresponding to the symbols drawn on the maps (see Fig. 6).

It should be noted that this first program also incorporated a "random square sieve" which allowed habitat and frequency data through for the simple statistics only if they had been recorded for random squares. Data for non-random squares were therefore suppressed, thus rendering what remained more statistically reliable. Many records had

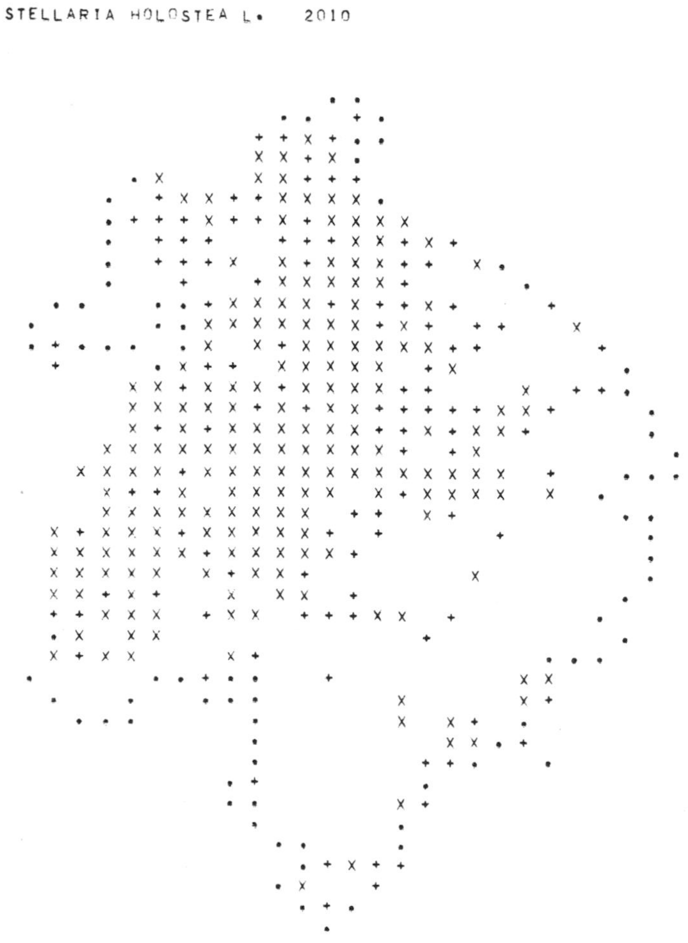

MAP 9. Sketch map of *Stellaria holostea* L., printed on the computer's fast output line printer. The map took only 3 seconds to produce

+ Occasional to rare in tetrad.
X Frequent to abundant in tetrad.
· Indicates position of VC boundary where no record exists in this tetrad.

been submitted by collectors for non-random squares as well as for random ones and all these were used for mapping. However, for showing habitat spread and preference of the species to be mapped it was considered essential not to bias the data in any way towards a preference of a collector for recording in a particular habitat, rather than a preference of a plant for growing in it. This preliminary statistical treatment of the data not only helped us to make useful and meaningful statements on habitat spread and preference in the general text of the Flora; it was also used to provide a comparative table of habitat records for each of the 718 vascular plant species mapped by the graph plotter, giving both actual and percentage records in each of the nine major habitat categories as well as other data on frequencies, numbers of records and percentage of random squares in which each species had been recorded.

A "sketch map" (Map 9) was reproduced as an interim product and used each print position to represent one tetrad. If the species had not been observed in that tetrad, the printer recorded "space" (blank); if the only observations were occasional or rare, "+" was printed; and if at least one abundant to frequent observation had been recorded, "X". There are 10 print positions to the inch across the page, but only 6 lines to the inch vertically; by inserting "space" in alternate positions we had 5 symbols to the inch horizontally and 6 to the inch vertically, which is as near as it is possible to get to equalising the horizontal and vertical scales.

The sketch maps were therefore slightly wider than they should be, but were nevertheless most useful in giving a general impression of the distribution of each species throughout the county. They do not appear in the Flora (but see bryophyte maps, etc. mentioned below) and were used merely to provide a general indication of species distribution when the textual part was being drafted.

Those tetrads through which the county boundary passed were "marked" in the computer's memory. When such a tetrad was empty, a full stop was printed for it instead of a space. These full stops built up a suggestion of the county boundary, further assisting in the interpretation of the sketch map.

(iv) *Map Images by Incremental Plotter* (Most vascular plants). The map images were taken from magnetic tape by the second program and converted into suitable control signals for the incremental plotter, which were placed on a further magnetic tape. The third program transferred from this magnetic tape to paper tape; this was done for the convenience of the computer room staff and need not be discussed further here.

The plotter consisted, effectively, of a pen moving over a sheet of paper, capable of drawing any shape required by incremental movements of 0·1 mm in any one of the four cardinal directions. Left to right movements were accomplished by the actual movement of the pen, whilst up and down movements were effected by corresponding movements of the paper. Diagonal lines were drawn by combined pen and paper movement, whilst further control signals instructed the pen to begin or to cease marking.

A reasonable solution to the problem of showing frequencies, as well as habitats, on the same map was also devised, though admittedly not all grades of frequency could be given. This was done by indicating "rare" to "occasional" by means of a thin line and "frequent" to "abundant" (including "locally frequent" and "locally abundant") by a thick one. If these ideas were to be translated into a completely automated system it would clearly be impossible to change pens every time a change of line thickness was required. We therefore decided to use a thin pen throughout and to program the plotter to re-draw the symbol after displacing the pen by two increments if a thick line was required. Since the pen thickness (0·2 mm) is the same as this displacement, the two single thin lines will ideally produce a line twice the thickness of the thin one. In theory there should not be a perfect coincidence of the two lines when diagonals and circles are drawn, but in practice they seemed to coincide quite well.

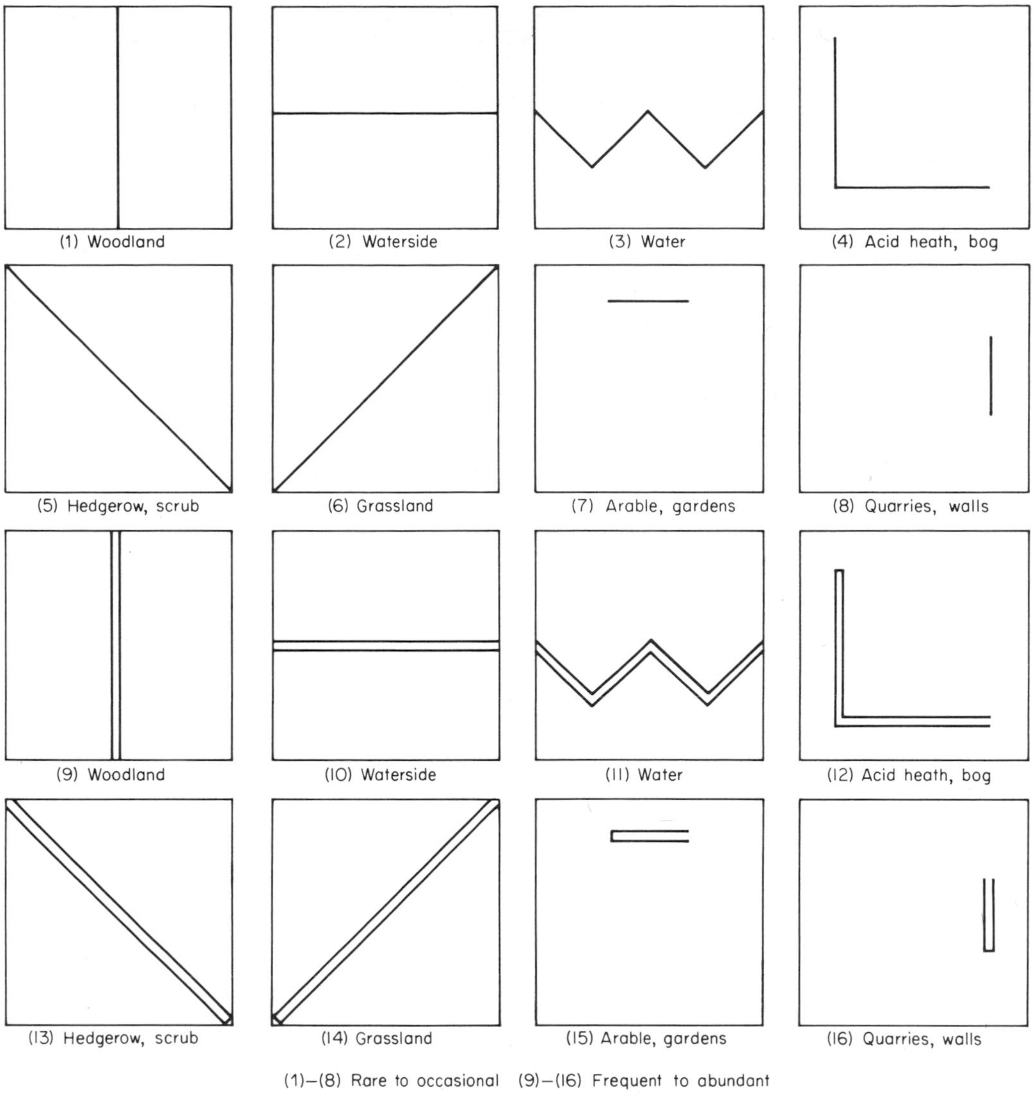

FIG. 7. Enlarged drawings of habitat symbols (other than "ruderal") showing the track of the pen centre. At the actual size at which these were drawn (in a square of 6 mm sides) the line drawn (0·2 mm) is wide enough to fill completely the space between the lines of a thick symbol.

All the symbols (except the circle, which is a special case) were made up of vertical, horizontal or diagonal straight lines (Fig. 7).

Figure 8 shows how the circles, thick and thin, were built up from short horizontal, vertical and diagonal straight lines. The exact arrangement of these lines was decided by a preliminary computer run to design the best possible circle within the limitations of the plotter.

Two problems arose in preparing the plotter control tape: moving the pen to the required map location for a symbol, and drawing the symbol once the pen was correctly located. The second problem was easier to solve, since the drawing of a symbol was identical, regardless of its position on the map: 18 standard sets of drawing instructions, one for each symbol, were held in the computer and the correct one was selected after the pen had been positioned. Positioning the pen to draw a symbol was complicated by the desire to keep pen movement, and therefore plotter time and cost, to a minimum. As each symbol was drawn it was erased from the computer's map image, and the machine then chose the nearest remaining symbol to the plotter pen's current position to be drawn next.

Besides drawing the symbols, the plotter wrote a reference number on each map so that it could be identified and given the correct title in printing; two register or "fiducial"

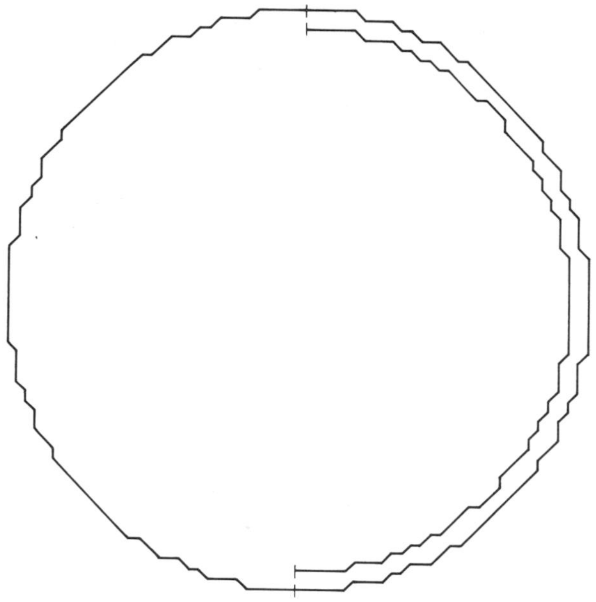

FIG. 8. The circle ("ruderal" symbol), not shown in Fig. 7. On the left is the pen track to draw a thin circle. The two tracks on the right, separated by about two steps throughout, combine to form a thick circle. The actual size of square in which the circle was drawn was of 6 mm side.

marks, one at the top left of the map and the other at the bottom right, were made so that the printers could correctly superimpose the plotted symbols on the standard outline of the county which would be overprinted on every map.

(v) *Map images by line printer* (Some flowering plants). It was mentioned above that there were insufficient records available for some of the flowering plants to make it worth producing maps by incremental plotter. Nevertheless, about 60 species were of sufficient interest to make us feel that some kind of map for each of them would be valuable.

It was therefore decided to present the data in a modified type of "sketch map" in which the marginal full stops were omitted and + signs added at top and bottom as fiducial or registration points. The type of symbol employed was the same as on the sketch map, with "X" indicating "frequent" to "abundant" (including "locally frequent" and "locally abundant"), and "+" indicating "occasional" or "rare".

The distorted dimensions of the sketch map were accepted for the same reason, that the maps were to be produced on the line printer.

These line printer maps of flowering and other vascular plants appear in a separate section after the graph-plotted maps (pp. 601–616).

(vi) *Map images by line printer* (Bryophytes). The bryophyte survey, which was planned and largely carried out by T. Laflin, was conceived on a slightly different basis from the vascular plant survey (see p. 654). In the bryophyte study, species lists without habitats but with data on "fruiting" or "sterile" were punched on the teleprinter. Thus, each record consisted of a species code reference, based on the bryophyte census catalogue number, together with a grid reference and a symbol denoting fruiting or sterile. The tapes were passed through the computer with a different program, which produced maps on the line printer in the same form as the extra flowering plant maps noted above. However, the two kinds of symbol differed in their meaning. "X" indicated "fruiting" whilst "+" indicated "sterile".

The records for these were added up and percentaged as a byproduct of the computer program. The bryophyte maps are also overprinted with a standard map in another colour, as with the flowering plant maps.

(vii) *Further use of stored data.* Once the data from a comprehensive survey of this type are in computer store the ways in which they may be used are almost limitless.

During the course of the work we have looked at the species, habitat and frequency data in various ways, some of which are shown in the Check List (pp. 617–653). Correction factors can be applied to help in reaching a truer picture of habitat spread and preference for each species, and the habitats themselves can be listed, with the species arranged in descending order of frequency in each (see Chapter 7). Furthermore, the computer can be programmed to map habitats, rather than species (see Maps 10–18, pp. 89, 97, 104, 112, 116). Other ways of looking at the data, such as comparing habitat spread and frequencies of species in different parts of the county ("Arden" and "Feldon", for example) or on different soils, have been contemplated but have had to be abandoned for the present because of lack of space available in this volume.

Finally, maps may be shown on a Visual Display Unit, operating on-line to a small computer. This allows immediate retrieval of any required species distribution map with selected habitat and frequency data (Kershaw, 1970) and would be valuable in comparison of species or habitats in rapid succession. The method has not been used in the present volume, since it did not seem essential for immediate needs and was in any case devised by B. L. Kershaw after the two methods of mapping used here had already been adopted. Nevertheless, its applications in the future may well be considerable.

(viii) *Computer Programs.* The considerable programming for this survey was done by Brian Kershaw, and we are extremely indebted to him for this work, as we have already stated in the Preface. For those interested in the detailed programs further notes can be seen in an appendix to a paper published in *Watsonia* by Hawkes, Kershaw and Readett (1968) pp. 360–4.

References

Bowen, J. J. M. (1968). "The Flora of Berkshire." Oxford.

Cadbury, D. A., Hawkes, J. G. and Readett, R. C. (1957). Flora of Warwickshire: species distribution maps and habitat analyses. *Proc. Bgham nat. Hist. phil. Soc.* **18,** 135–45.

Clapham, A. R. (1969). "Flora of Derbyshire." Derby.

Dony, J. G. (1967). "Flora of Hertfordshire." Hitchin.

Good, R. (1948). "A Geographical Handbook of the Dorset Flora." Dorchester.

Hawkes, J. G. and Readett, R. C. (1954). The Warwickshire County Flora revision: a new method of recording. *Proc. Bgham nat. Hist. phil. Soc.* **18,** 61–74.

Hawkes, J. G. and Readett, R. C. (1963). Collecting the data. A description of the methods used in the current revision of the Flora of Warwickshire. *In* Wanstall, P. J. (Ed.). "Local Floras." B.S.B.I., London.

Hawkes, J. G., Kershaw, B. L. and Readett, R. C. (1968). Computer mapping of species distribution in a County Flora. *Watsonia* **6,** 360–4.

Hawkes, J. G., Readett, R. C. and Skelding, A. D. (1955). The Warwickshire County Flora revision. Mapping distribution. *Proc. Bgham nat. Hist. phil. Soc.* **18,** 109–12.

Kershaw, B. L. (1970). "Distribution mapping by computer." M.Sc. Thesis, University of Birmingham. May, 1970.

Perring, F. H. (1968). "Critical Supplement to the Atlas of the British Flora." B.S.B.I., London.

Perring, F. H. and Walters, S. M. (1962). "Atlas of the British Flora." B.S.B.I., London.

Perring, F. H., Sell, P. D. and Walters, S. M. (1964). "A Flora of Cambridgeshire." Cambridge.

Readett, R. C., Hawkes, J. G., Cadbury, D. A. and Laflin, T. (1965). Check lists of the vascular plants and bryophytes of Warwickshire (V.C. 38) and surrounding vice-counties. *Proc. Bgham nat. Hist. phil. Soc.* (Special number) **20,** 4, pp. 64.

Soper, J. H. (1964). Mapping the distribution of plants by machine. *Can. J. Bot.* **42,** 1087–1100.

Soper, J. H. (1966). Machine-plotting of phytogeographical data. *Can. Geogr.* **10,** 1, 15–26.

Watson, H. C. (1870a). "A Compendium of the Cybele Britannica." London.

Watson, H. C. (1870b). "Topographical Botany." London.

7. HABITAT STUDIES

Introduction

The vegetation of most of the low-lying parts of Britain has been profoundly modified by man. Warwickshire is no exception to this, as Professor Thorpe has shown in considerable detail in Chapter 4. Forests have been felled and turned into arable and pasture land; roads, railways and canals have been driven through in all directions; rivers have been straightened and deepened; marshes and bogs have been drained and used for agricultural land; and reservoirs and canal feeder lakes have been constructed. In many parts, and especially in the north, urban landscapes of bricks and concrete, as well as industrial wastes, coal tips, gravel, brick, marl and lime pits, quarries and spoil heaps have spread over the land. There are countless other reminders of the hand of man in the industrial midlands of England.

Strangely enough, such changes in the natural landscape have in many ways enriched our flora. To be sure, the rarer and more sensitive plants of marsh, bog, heath and forest have diminished greatly or disappeared completely. Surprisingly, others have lingered on after they were thought to have disappeared entirely, such as the Moonwort (*Botrychium lunaria*) in Sutton Park. But the wide range of man-made habitats such as hedgerows, railways, roadsides, canal towpaths, pits and waste heaps, as well as ponds, canals and reservoirs, have provided conditions for species which might otherwise be unknown in the county. Such a richness of habitat, even though largely man-made, has provided a corresponding profusion of species that has often astonished us. Indeed, up to a point, the regions highly altered by man seem to possess the richest floras, whilst those which are more natural have provided the smaller species lists, as we have already mentioned in Chapter 6.

Warwickshire has often been compared unfavourably with other counties because of its presumed lack of rare or "interesting" plants. We can only say from our experience of the county that if one considers the sum total of species here, the recently arrived aliens as well as those presumed to be native, the common as well as the rare species, one is struck very forcibly by the richness and variety of its flora.

Furthermore, our method of recording methodically, square by square, has demonstrated time and time again the interesting species to be found in pastures and in arable fields, at any rate up to the time when the use of selective weed killers and the resowing of old pastures and meadows reduced their number in many parts of the county.

In comparing the numbers of vascular plant species present in Warwickshire at the present day with those recorded by Bagnall in 1891 and by Perry and others even earlier, we have been astonished by the vigour of plant life here. Some species, such as the Club Mosses (*Lycopodium* spp.), seem to have disappeared entirely at least 150 years ago, probably because of heath drainage. Even so, others are always appearing, and scarcely a year goes by when a new record for the county is not added to the list. Some new records are merely the discovery of plants that have been present for a long time and not previously seen or distinguished as distinct taxa, but others are new, and are often arrivals from distant continents.

Delimitation of Habitats

The co-operative system of recording in Warwickshire which we have described in Chapter 6 has made it necessary to devise a system of habitat and frequency categories that would be sufficiently complex to comprehend the variation in the environment and the plant communities. It also needed to be simple and clear enough to be used with accuracy by the amateur and professional botanist without a specialist ecological or phytogeographical background. A more complex scheme would have failed, as many have done in the past, either because it was too difficult to be easily comprehended or because its use would have been too time-consuming for the Flora ever to have been completed. We therefore make no apologies for our habitat recording system. It worked reasonably well in practice and provided easily codifiable data for computer mapping and statistical purposes.

As we have explained in Chapter 6, the 34 minor habitat categories, such as marsh, pondside, oak woodland, etc., were grouped into eight major ones for ease in handling the data, both conceptually and practically. Parts of this grouping seemed later to be ecologically indefensible, and we decided on a re-arrangement into nine major categories. This process of re-arrangement was carried out entirely by the computer.

It could perhaps be argued that the vegetation of Warwickshire, and indeed of the world as a whole, is a living continuum which can be divided into categories or units only in an arbitrary manner. However, even though there are intermediate states between many of the major habitats as we have defined them, we believe that these habitat groupings are valid and useful. Species will sometimes be found in the intergradations between two habitats, especially when they occur naturally in both (e.g. a woodland species which also occurs in hedgerows). Such species will in practice have been recorded in one or the other (or both) habitat and little harm will have been done if an error of judgement is made. In a large-scale survey of this sort such errors are of little statistical importance.

Differences between minor categories are slighter than those between major ones, and perhaps in certain instances are not very real, such as those between lakeside and pondside, or oak wood and mixed wood. To some extent we may here be considering parts of a very complex series of habitat and community differences which are impossible to quantify in a County Flora, or we may be looking at a difference of locality (lake or pond). We have felt that minor categories are of value in many instances, but for most purposes in this volume we have presented the major category data only, as representing greater and more clear cut differences than those shown by the minor ones.

Plants are clearly no respecters of man-made categories, and they may not be confined either to one or even a few clearly defined habitats. Thus, the Vascular Plant Check List and Analyses of Habitat and Frequency Data (Chapter 11, p. 617 *et seq.*) shows that most species spread widely into several habitats, with often an apparent preference for one or only a few.

Any field botanist or naturalist will have had experiences of plants that were occasionally found in the "wrong" habitats, such as lime-loving species in *Sphagnum* bogs or weeds of cultivation in woodland. There is no doubt that such recording "errors" are often not errors at all, and one must always try to avoid preconceived notions of where a plant "ought" to grow. Sometimes the apparent anomaly is due to the presence of a micro-habitat to which the species is in fact adapted; sometimes the plant is completely "out of habitat" and will not survive for long under such abnormal conditions. Our figures in Chapter 11 show these smaller percentages of occurrences for "abnormal" habitats, but we have not considered it necessary to comment further on these in Chapter 8 when we are discussing the localities and general distribution of each species. The reader himself may draw further conclusions from such figures in the light of his own experience.

There is another point that should be considered in this general section on habitats. Some of our habitats, as we have defined them, are, in fact, plant communities, whilst others are more strictly habitats. In this we follow the general example of Tansley (1939) where communities, soils, water content and other factors are used to define the categories selected. Thus, woodland, hedgerow and grassland are plant communities, and are defined by the major or common plants within them (trees, shrubs and grasses, respectively). On the other hand, water, waterside, heath, bog, cultivated land, ruderal and quarries are habitats which are conditioned by the level of the water table, the chemical and physical nature of the soil or rock, and the amount of interference by man. Even within the communities, such as woodlands, the concept of habitat must differ according to the plants under consideration. A woodland tree is part of the woodland community, but to a small shrub or herb on the woodland floor that tree forms part of its habitat. The same holds true for herbs in hedgerows and non-gramineous species in grassland. In arable land the management of the soil and the crops grown on it provide the major environmental factors for the weed species there. If we had recorded whether the soils were lime-rich or not, this would have appeared in our scheme as another habitat factor. There are, of course, well-marked differences of soil type in Warwickshire, but it was felt that these were too complex for a scheme of this sort. On the other hand, it was thought useful to record (or at least to assume) the very poor nutrient conditions of bogs and heaths. The choice was a pragmatic one and entirely influenced by what we thought could be easily recognised by the average recorder.

Habitat Preferences

A further matter which should be discussed here is the question of "habitat preference". The percentage figures of occurrence for species in the different major habitat categories shown in Chapter 11 indicate what appear to be "preferred" habitats, where the occurrences are much higher than in others. Furthermore, some species appear to be very restricted in their habitat range whilst others are tolerant of a wide spread of habitats. The whole matter of amplitude of range is a complex one for which we unfortunately have no space here.

Habitat "preference" must, however, be examined in a little more detail. If, for example, a species is recorded twice as many times in grassland as in woodland and we find that grassland itself is twice as frequent as woodland we are here seeing a false preference for grassland in a species which occurs in both habitats at equal potential frequencies. It would therefore be valuable to adjust the habitat frequency figures for each species by applying factors derived from the relative areas of each habitat in the county. We possess no actual area data of this sort from the survey; indeed, it is not necessary, since our data are based on numbers of records and not areas.

The information we actually possess is probably sufficient to supply a factor of the type required. For instance, we have the total number of observations for each major habitat and the percentages derived from these, as can be seen in Table 8. (It should be noted in passing that the total number of observations, 170 773, in the table is less than the figure of 175 000 mentioned in the Preface. This is because none of the non-mapped species was put into data store. Furthermore, records obtained before 1954 based on 100 m squares were in that year converted to 1 km square records and thus reduced in number.)

On considering further the total number of records for each habitat in an attempt to scale the actual figures, we realised that some habitats such as woodlands, hedgerows, waste places, etc., might well be richer in species, area for area, than others such as grassland. We therefore concluded that it would be better to scale our figures by a factor depending on the percentage of the total 613 km squares in which each habitat had been recorded (see Table 8). This was done by dividing 613, the total number of random squares, by the number of random squares in which that habitat had been recorded, so producing a factor for each major habitat. These factors were then used to convert the "raw" habitat occurrence figures for each species into "scaled" ones, which were then converted into percentages.

Since the Marginal, Grassland, Waterside, Ruderal and Cultivated habitats are found in nearly all squares the differences between the raw and scaled figures for these habitats are small. On the other hand, with Woodland, Heath, Water and Stones the habitat occurrences are less and the scaled figures thus show a corresponding increase.

Space prevents the inclusion of all these figures here, but Table 7 gives some examples where the habitat preferences change from the raw to the scaled figures or the preference becomes more marked (*Primula vulgaris*).

TABLE 7. Comparison of Raw and Scaled Percentage figures for major habitat frequencies of selected species

Figures in bold type indicate "preference" of a species for a particular habitat. See text for further explanation.

		WO	M	G	H	WS	WA	RU	ST	C
Agrostis canina	Raw	15·8	5·3	**36·8**	15·8	5·3	—	21·1	—	—
	Sc.	8·2	1·8	12·5	**68·5**	1·8	—	7·2	—	—
Endymion non-scriptus	Raw	37·1	**45·9**	5·7	—	1·7	—	8·9	—	0·7
	Sc.	**47·2**	38·5	4·8	—	1·4	—	7·5	—	0·6
Erigeron acer	Raw	—	—	17·6	—	—	—	**52·9**	29·4	—
	Sc.	—	—	13·9	—	—	—	41·5	**44·6**	—
Primula vulgaris	Raw	**35·9**	23·3	18·4	—	8·7	—	12·1	—	1·5
	Sc.	**45·9**	19·7	15·6	—	7·4	—	10·2	—	1·3
Ranunculus hederaceus	Raw	—	—	—	5·0	**60·0**	35·0	—	—	—
	Sc.	—	—	—	**37·7**	35·9	26·4	—	—	—
Typha latifolia	Raw	—	—	—	0·6	**50·6**	48·3	0·6	—	—
	Sc.	—	—	—	6·1	42·4	**51·0**	0·5	—	—
Viburnum opulus	Raw	32·8	**41·3**	1·1	—	21·7	—	2·6	—	0·5
	Sc.	**42·4**	35·3	0·9	—	18·7	—	2·3	—	0·5

Habitat Maps

The computer-generated habitat distribution maps (see Maps 10–18) should be mentioned here. They were produced on the fast line-printer in the format described on p. 78 for Map 9, and were developed by working over the whole of the data and recording for each tetrad every major habitat recorded in that tetrad regardless of the species observed there. A similar set of maps was printed out for minor habitats but unfortunately there were too many of these to reproduce in the present volume.

The production of habitat maps is another illustration of the very versatile use to which plant distribution data can be put, once they have been transferred to computer store.

The Major Habitats

The nature of our county-wide survey has not made it possible to give detailed analyses of the variation within each major habitat, apart from those resulting from the minor divisions. However, in the sections which follow we shall discuss local variations within communities arising from differences in soil type, present management and historical development.

The general habitat statistics show a number of interesting differences in number of species and number of records per habitat, as well as the total recording of the habitats themselves (see Table 8). From this we can see that the largest number of records (28·2%) was obtained in Ruderal, followed by Grassland (20·0%) and Marginal (18·7%), with Waterside (13·0%), Cultivated (10·5%) and Woodland (6·2%) much lower, ending with Water (1·7%), Stones (1·3%) and Heath (0·4%) right at the bottom.

TABLE 8. Total Records and Species Numbers for Major Habitats, together with Numbers of Random Squares in which each Habitat has been recorded

Major Habitat Categories	Total Records Number	%	Number of Species	Random Squares Number	%
Woodland	10 596	6·2	440	403	65·74
Marginal	31 875	18·7	550	609	99·35
Heath	640	0·4	299	48	7·83
Grassland	34 081	20·0	554	610	99·51
Waterside	22 284	13·0	503	605	98·69
Water	2907	1·7	129	480	78·30
Ruderal	48 180	28·2	610	611	99·67
Stones	2255	1·3	420	316	51·55
Cultivated	17 955	10·5	436	603	98·37
Total	170 773	100·0			

The differences between the number of *species*, rather than the *records*, in each habitat are not so extreme. Thus Ruderal (610 spp.) shows only about twice as many species as Heath (299), and the lowest number is from Water (129).

Random square habitat occurrences are also shown in Table 8. They have already been discussed, and Maps 10–18 indicate their distribution.

A word of warning should be sounded here. Although we have attempted to make the boundaries between the habitat categories as "natural" as possible, they are, even so, categories of our own construction. The ruderal category is a large one, but if we had decided, for instance, to divide it into "waste places" (ruderal) and "waysides" (viatical), the figures would obviously have been very different. Similarly, it might have been considered more "natural" to link water and waterside categories together, since the boundary between them is often difficult to define exactly. Again, the figures would not have been the same. However, if we take them for what they show, without attempting to read too much into them, they provide much interesting information, which we shall discuss further in the sections which follow.

1. Woodland (WO)

To the average field botanist woodland is amongst the most interesting of all the Warwickshire habitat categories, perhaps because it is regarded as the most "natural" and

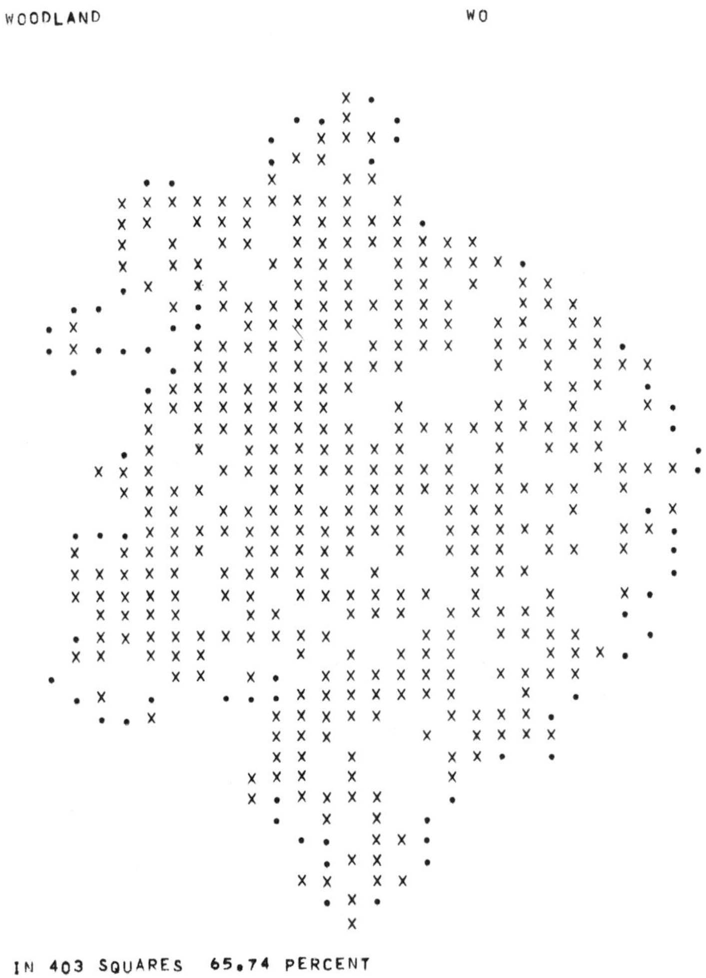

MAP. 10. Woodland habitat distribution

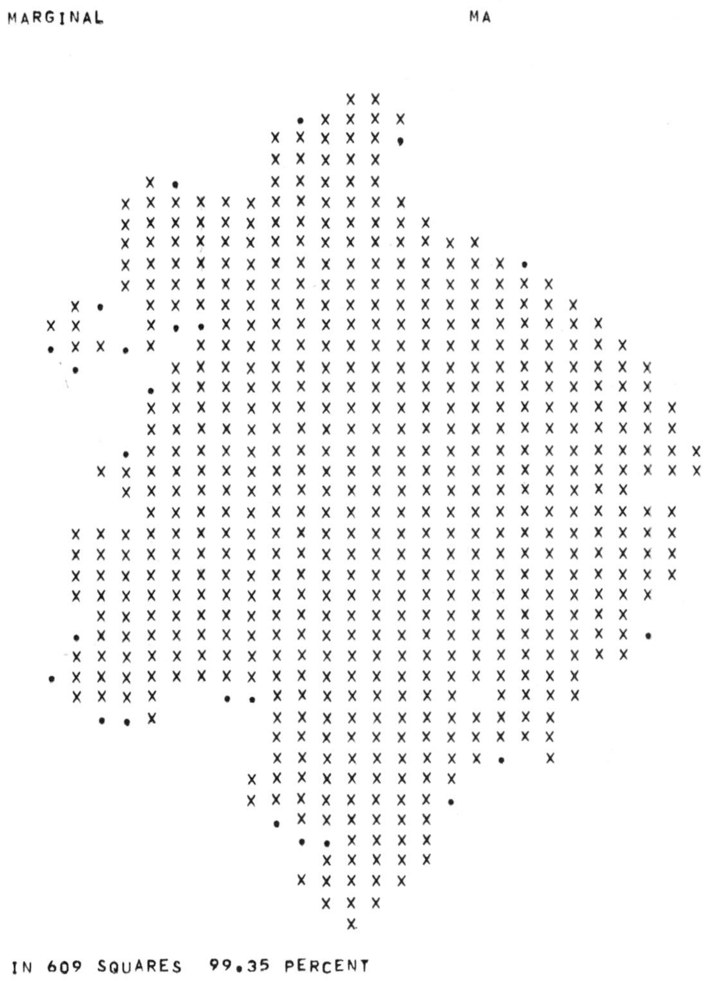

MAP. 11. Marginal habitat distribution (hedgerow and scrub)

unspoilt. Unfortunately this is very far from the case. The unbroken spread of forest mentioned by Professor Thorpe in Chapter 3 which occurred in prehistoric times began to be destroyed or modified from very early periods. The small remnants now remaining (2·6% of the total area) have been partly or completely felled, coppiced and thinned; exotic species have been planted, either completely or in part, so that no untouched or even partially preserved woods now exist. Yet the natural dominant trees, the oaks, still remain and give our woodlands their characteristic form over most of the county. The basic woodland type at present in most of Warwickshire is oak–birch woodland, but the birches represent part of a seral development from felled or coppiced woodland and they cannot be considered as part of the forest climax. Of the two oak species present, *Quercus robur* is by far the commoner, though *Q. petraea* is found to some extent in the north and west, where it frequently hybridises with *Q. robur*. Indeed, it is doubtful whether any really "pure" *Q. petraea* now exists in Warwickshire, though Rees and Skelding (1950, p. 66) speak of its being dominant in certain regions.

The birches are found principally in the northern and western parts of the county, with *Betula pendula* rather more common than *B. pubescens*. Many probable hybrids between these two species occur.

Holly (*Ilex aquifolium*) is seen mostly in the northern half of the county and forms an important component of some of the woodlands in Sutton Park. Ash (*Fraxinus excelsior*) and Wych Elm (*Ulmus glabra*) largely take the place of birch in the southern and south-eastern woods on more base-rich soils. Beech (*Fagus sylvatica*) is scattered throughout the county but was probably all originally planted.

In the north, the oak–birch woods on the poorer, more acid soils contain the following tree and shrub species:

Quercus robur
Q. petraea
Betula pendula
B. pubescens
Ilex aquifolium
(occasionally subdominant)
Sorbus aucuparia
Frangula alnus
Crataegus monogyna
Populus tremula
Fagus sylvatica
Alnus glutinosa

General oak–birch woodland on roughly neutral soils is characterised chiefly by the following species:

Quercus robur
Betula pendula
B. pubescens
Crataegus monogyna
C. laevigata
Swida sanguinea
Corylus avellana
(dominant in coppices)
Sorbus aucuparia
Fagus sylvatica
Ilex aquifolium
Acer campestre
A. pseudoplatanus
Alnus glutinosa

On fairly base-rich soils one sees fewer oaks, ash as a dominant or co-dominant and practically no birches. The following list gives some indication of the composition of this woodland type:

Fraxinus excelsior
(very common, often co-dominant)
Quercus robur
(less common, often co-dominant with ash)
Swida sanguinea
Viburnum lantana
Corylus avellana
Crataegus laevigata
Fagus sylvatica
Ulmus glabra

Scattered in neutral and alkaline woods are trees of *Tilia* × *vulgaris*, *T. cordata*, *Ulmus glabra*, *Carpinus betulus* and *Taxus baccata*. In many woodlands exotic conifers such as *Pinus sylvestris* (not native here), *P. nigra*, *Picea sitchensis*, *Larix* spp., *Tsuga* spp., *Abies* spp., *Pseudotsuga taxifolia*, *Sequoia gigantea* and *S. sempervirens* are planted.

In parts of Oversley wood near Alcester and Rough Hill Wood near Redditch, *Tilia cordata* is the dominant species. It is also frequent in other woods in south-west Warwickshire and adjoining parts of Worcestershire. Could these perhaps be a relic of the abundant forests of *Tilia platyphyllos* and *T. cordata* that were present in the Midlands some 5000 years ago? It is tempting to think so, but unfortunately the data are too scanty for us to be certain (see Kelly and Osborne, 1965).

Some more detailed work may be of interest at this point. Student studies under the direction of Dr. D. A. Wilkins of the Department of Botany, Birmingham University, were carried out during the summer of 1970. A reasonably objective analysis of tree frequency was made in a number of distinct woodland types. Since the total area occupied by a single tree varies according to species a simple count of numbers is not enough to characterise the woods. A basal area for each species was the average cross-sectional area taken from a girth measurement at 1·5 m height. From this was calculated the area × density (= basal area × number of trees per hectare) for each species. The area × density figure gives a truer idea of the importance of each species relative to the others occurring with it.

The work is still in a preliminary stage but even these results demonstrate the very large differences in woodland which are due partly to soil differences and also probably to differences of felling and management. Some simplified results are given in Table 9, from which one can see that oaks are by no means dominant numerically in every wood, even though their area × density figure often exceeds that of other species.

In making divisions of the woodland major category into minor ones, we designated three only—conifer, mixed and oak wood. In view of the very highly complex situation discovered subsequently, this proved to be a wise decision. The plantations of conifers were the easiest to identify; most of the rest was mixed woodland, with oak sometimes dominant and sometimes not. Where oak was clearly dominant the minor category WO-o was used, but there were all gradations to mixed woodland and probably nothing approaching the original oak forests of prehistoric times now exists.

TABLE 9. Percentage Occurrence and "Area × Density" for Tree Species in Certain Warwickshire Woodlands (see text for further explanation)

Locality	*Quercus*	*Betula*	*Ilex*	*Fraxinus*	*Corylus*	*Ulmus glabra*	Other Species
Sutton Park Holly Hurst	11(21)	10(4)	64(14)	—	—	—	15(12)
Sutton Park Westwood Coppice	35(15)	4(2)	—	—	—	—	61*(17)
Earlswood (Oak-birch)	38(11)	47(3)	3(0·4)	0·5(0·3)	—	—	11(0·5)
Earlswood (Ash-wood)	11(2·5)	—	—	57(28)	14(0·3)	11(1·5)	8(—)
Oversley (A)	14(3·6)	61(11)	—	—	5(—)	—	20(—)
Oversley (B)	3(—)	97(11)	—	—	—	—	—
Oversley (C)	—	34(5)	—	—	6(0·5)	—	60**(13)
Chesterton	9(10)	4(0·4)	—	66(7·5)	2(—)	9(4)	10(—)

"Area × Density" figures in parentheses.
* Mainly *Sorbus*, *Pinus* and *Castanea*.
** Mainly *Tilia cordata*.

It was not considered practicable to designate the woodlands according to the various types of management. Coppice with standards is a common enough woodland type in Britain, and much of our Warwickshire woodland is of this type. However, much woodland has been felled and allowed to regenerate in a poorly managed way, whilst some has been planted with exotic species. A detailed survey would demand very many categories such as in the one the Nature Conservancy is now carrying out. Our own scheme needed to be less complex in order to get it completed within a reasonable space of time.

(a) Oak Woodland (WO-o). Recorded in 40 squares; 6·53% of those surveyed. Total number of observations: 662. Number of species recorded: 190.

Oak woodland has been recorded chiefly in the north of the county. The following lists note some of the most frequent species in this minor category: (number of records in parenthesis, arranged in order of frequency)

Quercus robur (40)
Endymion non-scriptus (14)
Rubus fruticosus agg. (14)
Lonicera periclymenum (12)
Pteridium aquilinum (12)
Dryopteris dilatata (10)
Dryopteris filix-mas (10)
Viola riviniana (10)
Anemone nemorosa (9)
Ilex aquifolium (9)
Silene dioica (9)
Stachys sylvatica (9)
Teucrium scorodonia (9)

(*b*) *Mixed Woodland* (*WO-m*). Recorded in 396 squares; 64·6 % of those surveyed. Total number of observations: 9870. Number of species recorded: 437.

In view of the very wide range of differences in Warwickshire woodlands we can do no more here than list the averages obtained from the whole of the county. Mixed woodlands of many different types are fairly evenly scattered throughout the county but are less frequent on the calcareous clays of the south-east. The most frequent species in this minor category are as follows:

Endymion non-scriptus (204)
Quercus robur (175)
Hedera helix (156)
Dryopteris filix-mas (153)
Fraxinus excelsior (146)
Sambucus nigra (143)
Geum urbanum (136)
Brachypodium sylvaticum (135)
Corylus avellana (135)
Circaea lutetiana (132)
Acer pseudoplatanus (130)
Ranunculus ficaria (130)
Lonicera periclymenum (128)
Urtica dioica (121)
Rubus fruticosus agg. (120)

(*c*) *Coniferous Woodland* (*WO-c*). Recorded in 24 squares; 3·9 % of those surveyed. Total number of observations: 64. Number of species recorded: 41.

These are always plantations, and the ground cover is often very impoverished by reason of the dense shade. Plantations of pines, larches, spruces and silver firs have chiefly been recorded. The following list gives some of the commoner species:

Pinus sylvestris (9)
Larix decidua (6)
Rubus fruticosus agg. (3)
Deschampsia flexuosa (2)
Endymion non-scriptus (2)
Epilobium angustifolium (2)
Geum urbanum (2)
Picea abies (2)

2. *Marginal* (*M*)

The species-rich hedgerows of Warwickshire are perhaps its most dominant and characteristic feature. No doubt most of those of the centre and south-east were largely planted after the open fields of the Feldon were enclosed. In many parts of the former Arden Forest region of the centre, west and north-west (see Maps 4 and VII), however, the hedgerows are no doubt partly at least "assart", formed by the clearance of forest. Here they are rich and varied, acting very largely as refuges for woodland plants where the woodlands no longer exist. In almost all parts of the county the hedges also act as refuges for grassland plants when the old meadows and pastures have been ploughed up and re-seeded. This no doubt accounts for the high percentages of *Primula veris* in hedges, even though it is typically thought of as a pasture plant.

Hedges also act as windbreaks, as refuges for insects and other animal life, and very often as refuges for trees such as oaks and elms. However, many of the tree species occurring in hedges (*Ulmus minor* agg., *Fraxinus*, *Malus*, *Prunus domestica*, *Populus italica*, *P.* × *canadensis*, etc.) were originally planted. Where, as is often the case, the hedges follow the course of streams and ditches, they form refuges for marsh herbs. Alders and many willows are often planted here. In our grouping of minor habitats we have included hedges and scrub woodland under the category of "Marginal" on the model of Tansley (1939, 1968) who classifies as "marginal" those species of half-shade which need to grow in scrub or hedgebank. Since marginal species differ from each other in the range in which they can exist, each having a particular preference, we have considered it best to use "Marginal" as a habitat category to include not only this wide range but the shade-loving woodland and the light-loving grass and arable plants. The criterion here is merely whether the plants occur in hedgerow or scrub woodland. There seems to be very little difference in species composition between hedgerow and scrub and the distinction is perhaps more one of convenience to the botanist than of significance to the plant. In Warwickshire on the whole, scrub woodland is generally the regeneration of previously clear-felled or degraded woodland rather than a stage in seral development from grassland, even in post-myxomatosis days. However, in some areas, such as Aston Grove, the scrub woodland is developing from former limestone grassland.

We have made no attempt to try to distinguish hedgerows and scrub on acid soils in the north (with a very poor species composition, despite their apparent age) from those on neutral to slightly alkaline clays and loams in the centre and those on calcareous clays and limestones in the south and south-east. Such a distinction is impossible in a general field survey of the sort we have been engaged in and would require careful soil surveys to make it of any value.

(a) Hedgerow (M-hr). Recorded in 609 squares; 99·35% of those surveyed. Total number of observations: 28 088. Number of species recorded: 496.

Hedgerows may be considered as anthropogenic subclimax vegetation, or, in other words, thin strips of incipient woodland continually held by cutting and laying at a subclimax stage. According to Hooper (1970) most if not all the hedgerows of Britain were planted, apart from the assarts left after the felling of woodland. From a comparison of known age with shrub species numbers, he claims that they can be dated approximately by counting the number of shrubs and tree species in them and multiplying by a factor of 110. Whilst we would agree that the species richness of the Arden hedges to the north-west of the River Avon is greater than that of the Feldon ones, planted at various times after the enclosures of the Saxon open fields, this may well be due to the fact that the Arden hedges were never planted at all but represent parts of the original forest. It must not be forgotten also that much of the hedgerow plantings of the south-east will have included a variety of shrubs and trees (especially elms—see Richens, 1955, 1958, 1961) for various economic purposes. So the hedgerows, at any rate in the past, have served a dual purpose: (a) To separate livestock into easily manageable areas (b) To use for forage (elms), wild nuts and fruits (hazel, crab apples, blackberries, elderberries), cultivated fruits (plums, damsons, apples, pears, etc.), timber (oak, ash, elm, etc.) and firewood (saplings of all types). They were also used as boundary markers (Richens, 1955, 1958, 1961).

It seems clear that most of our Warwickshire hedgerows are highly artificial, planted or even cultivated, with hawthorn (*Crataegus* spp.), blackthorn (*Prunus spinosa*) and, in places, holly (*Ilex*), hazel, elm suckers and other shrubs making an impenetrable barrier if periodically cut. These are further bound together by roses, blackberries, honeysuckle, hops, clematis and ivy; and small perennial and annual herbs fill in the edges and sides to form an extremely rich community when its total area is considered.

Apart from the growth forms of the species occurring in hedgerows we can make an approximate division, as we have already mentioned, into:

1. Species also found in woodland.
2. Species also found in grassland.
3. Species also found in waste places.
4. Species also found in cultivated land.
5. Species also found in waterside and marshes.
6. Species confined to hedgerows.

From this it is clear that hedgerows form an alternative habitat for a wide range of species, as well as acting as a refuge, especially for woodland species.

A list of the most commonly occurring species is given below:

Crataegus monogyna (571)
Sambucus nigra (545)
Prunus spinosa (529)
Fraxinus excelsior (520)
Rubus fruticosus agg. (519)
Ulmus procera (496)
Acer campestre (489)
Galium aparine (478)
Hedera helix (470)
Stachys sylvatica (457)
Rosa canina agg. (453)
Tamus communis (451)
Quercus robur (439)
Malus sylvestris (430)
Anthriscus sylvestris (396)
Alliaria petiolata (392)
Solanum dulcamara (390)
Corylus avellana (388)
Urtica dioica (368)
Ligustrum vulgare (346)
Geum urbanum (326)
Ilex aquifolium (326)
Geranium robertianum (318)
Chaerophyllum temulentum (315)
Glechoma hederacea (315)
Calystegia sepium (314)

(b) Scrub or derelict woodland (M-sc). Recorded in 320 squares; 52·20% of those surveyed. Total number of observations: 3787. Number of species recorded: 455.

Scrub or derelict woodland is widely distributed in the county with some concentration in the north-west corner. Although it can be either part of the seral development of vegetation from grassland subclimax to forest, in fact very little of the Warwickshire scrub woodland is of this sort. Much more is to be found in former woods that have been clear felled or partly so, and that are now regenerating, often with the addition of exotic trees and shrubs. Scrub on grassland is also developing at Aston Grove and many of the grassy spoil banks of cement quarries, as well as on certain road verges and waste places.

The colonisers are generally elderberry, hawthorn and blackthorn, but may be gorse and birch scrub on acid soils and ash with other species on some of the limestones. With regenerating woodland the tree and shrub species of the original woodland (*Quercus*, *Corylus*, etc.) often regenerate from the cut stumps.

The species list which follows shows the predominance of herb over shrub species as compared with hedgerows, but the overall composition of these two minor habitat categories is not too dissimilar:

Hedera helix (68)
Endymion non-scriptus (62)
Sambucus nigra (59)
Brachypodium sylvaticum (58)
Arum maculatum (55)
Geum urbanum (50)
Lonicera periclymenum (49)
Rubus fruticosus agg. (49)
Crataegus monogyna (47)
Glechoma hederacea (43)
Urtica dioica (43)
Moehringia trinervia (42)
Epilobium angustifolium (41)
Ranunculus ficaria (41)
Dryopteris filix-mas (40)
Fraxinus excelsior (40)
Quercus robur (40)
Prunus spinosa (39)
Corylus avellana (35)
Silene dioica (35)
Stachys sylvatica (35)

3. Acid Heath and Bog (H)

All communities of plants growing on acid soils have been included in the major habitat category of "Heath". In fact, as we have shown earlier, heathland forms the smallest habitat in Warwickshire, both in terms of number of squares in which it was recorded and the number of records (Table 8).

From place name and historical evidence we know that heaths were once more common than they are now. Draining of bogs and ploughing, fertilising and re-seeding have turned many of the old Warwickshire heaths that were apparently so common on the gravels of the north and north-west into agricultural land; others have been overtaken by the conurbations of Birmingham and Coventry, or have been despoiled by mining and other industrial activities.

Fortunately, Sutton Park remains and its heathland communities are relatively untouched, even though drainage has turned some of the bogs into much drier grassland communities. Some parts which were ploughed up during the war have not yet returned to their natural heathland communities and have lost the podsol profile so typical of such regions. Species more typical of agricultural grassland have now become established, such as *Trifolium* spp., *Dactylis glomerata* and *Agrostis tenuis*. *Ulex* seems more frequent in these areas than in the adjacent regions which were not ploughed.

Other interesting areas of heathland are to be found at Earlswood, Coleshill, Packington and Atherstone.

We have separated the heathland communities into (a) Bogs (H-b)—areas of acid soil, chiefly obtaining their water supply from rainfall and more acid surface drainage.

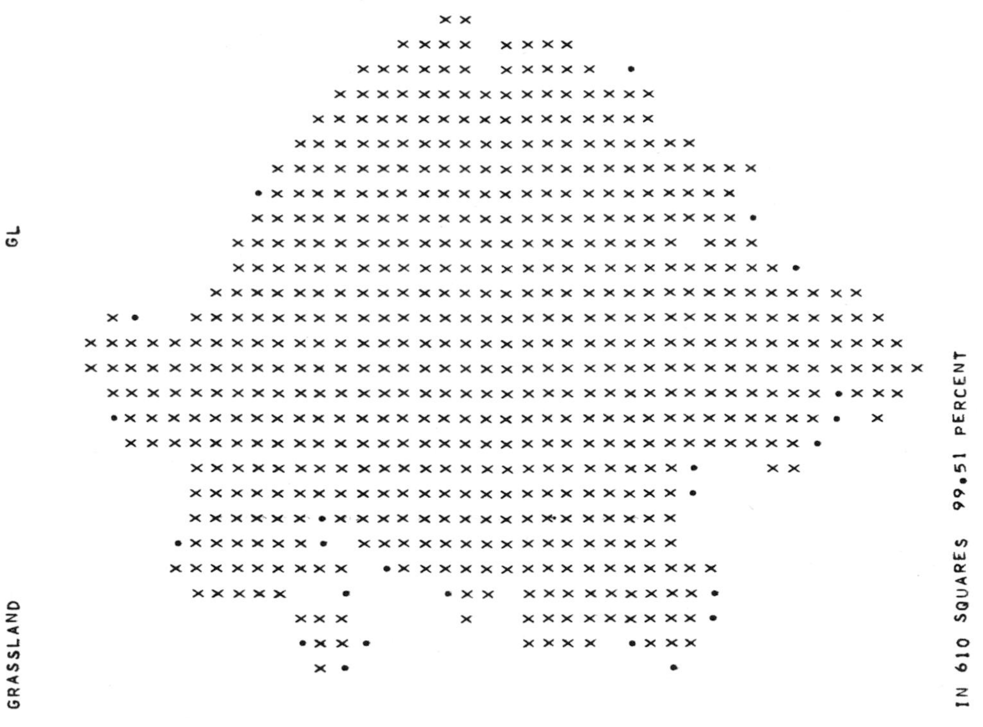

MAP 12. Grassland habitat distribution

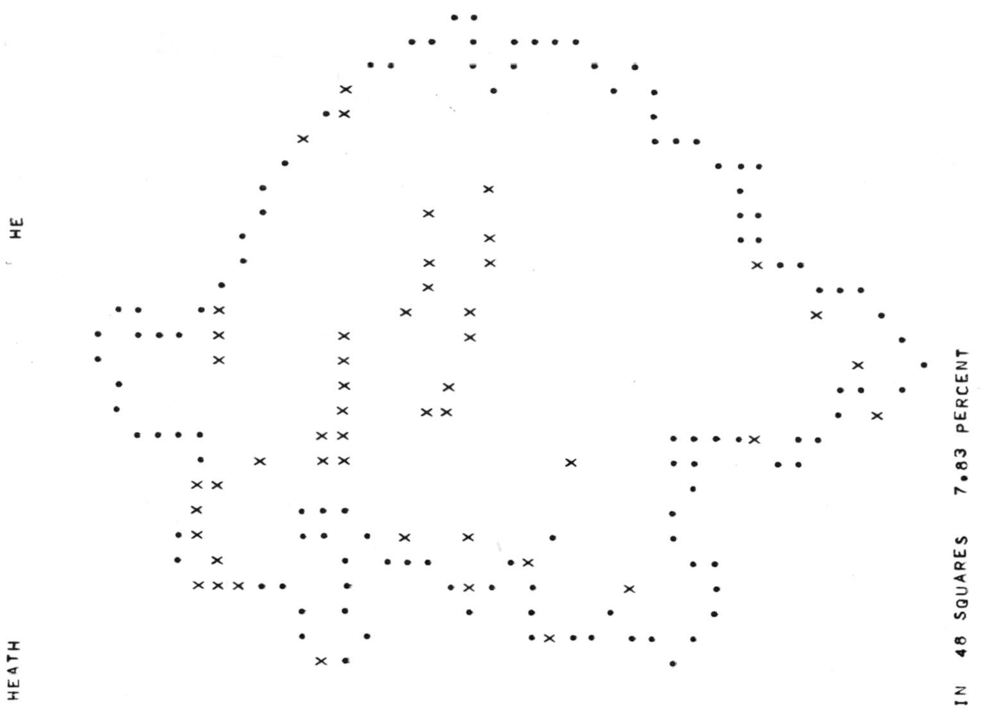

MAP 13. Heath habitat distribution (including acid bog)

(b) Dry heath (H-d) confined in general to the drier areas with heather (*Calluna*) dominant, but grading into bogs through wetter areas of *Erica tetralix* and cottongrass (*Eriophorum*). (c) Heath grassland (H-g). This separates the grassland communities of acid soils from those which occur on neutral to alkaline soils (G-m, G-p, etc.). We have not created a category for heath woodland since there is relatively very little of this in Warwickshire. Heath woodland, where it occurs, is included as "scrub woodland" (M-sc) or "mixed woodland" (WO-m). In a county with large amounts of heath woodland this should undoubtedly be distinguished as a separate category.

(a) Bog (H-b). Recorded in 18 squares; 2·94% of those surveyed. Total number of observations: 154. Number of species recorded: 101.

The true bogs, which in Warwickshire are seen chiefly in Sutton Park, are characterised by growth of bog mosses (*Sphagnum* spp., see pp. 662–3) as well as *Eriophorum angustifolium* and *E. vaginatum*. Many intergrading steps through slightly acid to neutral marsh communities are seen in Sutton Park and elsewhere, rendering the identification of this habitat difficult at times.

Outside Sutton, Coleshill Bog was a good example of former H-b communities but was largely destroyed by road development, drainage and tree growth. We give below a list of some of the more interesting bog species, most of which are still to be found in several parts of Sutton Park:

Carex dioica and *f. isogyna*
Drosera rotundifolia
Parnassia palustris
Pinguicula vulgaris
Sagina nodosa
Carex paniculata
Taraxacum spectabile agg. (*T. faeroense*)
Eleocharis quinqueflora
E. palustris
Dactylorhiza praetermissa
Vaccinium oxycoccos
Hydrocotyle vulgaris
Cirsium dissectum
Eriophorum vaginatum
E. angustifolium
Carex echinata
C. curta
Equisetum palustre
Erica tetralix
Galium uliginosum
Juncus acutiflorus
J. kochii
Potentilla palustris
Viola palustris
Vaccinium vitis-idaea

(b) Dry Heath (H-d). Recorded in 20 squares; 3·26% of those surveyed. Total number of observations: 146. Number of species recorded: 64.

Dry heath is mainly to be found in Sutton Park, grading in its wetter parts into wet heath and bog, and in other areas into heath grassland or heath woodland (see WO-m). Outside Sutton Park it is seen at Coleshill and other areas in the north and until recently at Wolford Heath in the extreme south of the county. Over most parts *Calluna vulgaris* is dominant, but in the wetter areas *Erica tetralix*, *Molinia caerulea* and various heath rushes are associated with or replace it. Scrub birch, gorse, broom and bracken are also to be found in parts of Sutton, marking the transition to heath woodland or indicating former agricultural activities (see above). A list of the commoner plants of dry heath is given below:

Calluna vulgaris (8)
Galium saxatile (8)
Rumex acetosella (6)
Ulex europaeus (6)
Ulex gallii (6)
Deschampsia flexuosa (5)
Juncus squarrosus (5)
Agrostis tenuis (4)
Festuca ovina ssp. *tenuifolia* (4)
Molinia caerulea (4)
Potentilla erecta (4)
Teucrium scorodonia (4)
Betula pubescens (3)
Centaurium erythraea (3)
Deschampsia cespitosa (3)
Erica tetralix (3)
Holcus mollis (3)
Hypericum humifusum (3)
Luzula campestris (3)
Nardus stricta (3)
Pteridium aquilinum (3)
Sieglingia decumbens (3)
Betula pendula (2)
Campanula rotundifolia (2)

(c) Heath grassland (H-g). Recorded in 37 squares; 6·04% of those surveyed. Total number of observations: 340. Number of species recorded: 157.

Heath grassland in drier areas probably in most cases represents a degradation of *Calluna* heath owing to overgrazing or burning. We have included in this the grass moors of Sutton Park and elsewhere with dominant *Nardus stricta* or *Molinia caerulea*, which grade into H-b in the wetter parts.

Apart from Sutton Park and its neighbourhood, heath grassland is to be found at Coleshill, Packington, Earlswood, Atherstone, some of the woods on gravel to the east and south of Coventry and a number of other areas in the northern half of the county.

The commoner plants are as follows:

Galium saxatile (13)
Deschampsia flexuosa (11)
Potentilla erecta (9)
Holcus lanatus (8)
Nardus stricta (8)
Agrostis tenuis (7)
Anthoxanthum odoratum (7)
Campanula rotundifolia (6)
Rumex acetosella (6)
Ulex europaeus (6)
Molinia caerulea (5)
Stellaria graminea (5)
Betula pendula (4)
Calluna vulgaris (4)
Carex pilulifera (4)
Holcus mollis (4)
Juncus squarrosus (4)
J. subuliflorus (4)
Ulex gallii (4)
Aira caryophyllea (3)
Aira praecox (3)
Betula pubescens (3)
Carex ovalis (3)
Luzula campestris (3)
Sieglingia decumbens (3)
Teucrium scorodonia (3)
Agrostis canina (2)
Festuca ovina (2)
Luzula multiflora (2)

4. Grassland (G)

We mentioned above when discussing the heath communities that heath and moor grassland were more conveniently to be grouped there, leaving the neutral and alkaline grasslands to be included under the present heading.

The Warwickshire grasslands, like the hedgerows, are largely semi-natural or man-made and can also be considered as biotic subclimaxes, kept in their present form by grazing and mowing. In pre-war days there were large expanses of old meadows and pastures, many of which had probably been in existence since the enclosures (see Chapter 3). Their richness in species was great, but improved farming methods and the ploughing up and re-seeding, together with manuring, have changed most of them beyond recognition. Since the appearance of myxomatosis much of the old close cropped grassland has disappeared, some to be colonised by hawthorn and blackthorn scrub, and some by the lush growth of the grass itself to be changed in character very greatly. However, the Burton Hills and Harrow Hill still remain as good examples of this grassland type.

After even one season's fallowing, arable land quickly turns into grassland, and some of the arable weeds noted in our lists for grassland may be from this type of succession from arable to grass, which is likely to have been returned to arable by re-ploughing shortly afterwards.

In some of the damper meadows there is a gradual change to marsh which often makes it difficult to set the boundaries between the two. Then again, much grassland can be found in open woodland, especially in the glades or rides. In these cases we instructed our field workers to record the data as "woodland", if the influence of the trees was obvious. For this reason, the woodland grasses such as *Holcus mollis*, *Deschampsia flexuosa*, *Melica uniflora*, and also those which stray quite happily into woodlands such as *Poa pratensis* and *P. trivialis* appear in the woodland lists.

It was not found practicable to ask our field workers to separate limestone grassland from neutral grassland since in Warwickshire there are so many intermediate conditions. We have therefore taken a pragmatic view, as always, and asked collectors to distinguish meadows (grasses cut for hay, not normally grazed), pastures (grasses obviously grazed), rough grassland (tussock grasses of poor run-down fields, wide road verges) and orchards (very rough grass under orchard trees, often with domestic stock, chickens, etc., and with high soil nitrogen content).

(a) Meadows (G-m). Recorded in 495 squares; 80·75% of those surveyed. Total number of observations: 7522. Number of species recorded: 376.

This habitat has not been very widely recorded for Warwickshire, probably because, on the whole, the fields are used much more for grazing and the fattening of livestock than for hay-making. Almost certainly, difficulties arose from the fact that recorders did not know enough about grassland management in general to distinguish always between pastures and meadows. Short term leys were generally not noted and the data received on these were too scanty to use. The very frequent meadow grasses *Cynosurus cristatus*, *Lolium perenne*, *Phleum pratense*, *Poa pratensis*, *P. trivialis* and *Alopecurus pratensis* are widespread.

In the list which follows only *Hordeum secalinum* could be considered as in any way a calcicole, though many calcicoles were shown on the computer print-out of species for this habitat. Unfortunately, as with all the other lists, reasons of space prevent our showing the total number of species:

Ranunculus acris (218)
Alopecurus pratensis (190)
Lolium perenne ssp. *perenne* (180)
Trifolium pratense (180)
Bellis perennis (177)
Cynosurus cristatus (173)
Anthoxanthum odoratum (158)
Dactylis glomerata (156)
Trifolium repens (156)
Phleum pratense (133)
Cerastium holosteoides (126)
Ranunculus bulbosus (123)
Cirsium arvense (121)
Holcus lanatus (119)
Rumex acetosa (119)
Taraxacum officinale agg. (114)
Chrysanthemum leucanthemum (107)
Plantago lanceolata (107)
Cardamine pratensis (103)
Festuca pratensis (101)
Ranunculus repens (93)
Luzula campestris (91)
Prunella vulgaris (91)
Trifolium dubium (90)
Cirsium vulgare (85)
Poa pratensis ssp. *pratensis* (83)
Lolium perenne ssp. *multiflorum* (80)
Achillea millefolium (78)
Agrostis tenuis (78)
Lotus corniculatus (78)
Hordeum secalinum (74)

(*b*) *Pastures* (*G-p*). Recorded in 561 squares; 91·52% of those surveyed. Total number of observations: 11 618. Number of species recorded: 387.

This category principally represents grazing land of various types and is very widely distributed. When they were unsure whether to note meadows or pastures on their lists we suspect that our recorders probably gave the benefit of the doubt to pastures.

Limestone pastures are found chiefly towards the edge of the Lower Lias where limestone strata occur quite frequently, and also on the scarp slope itself. They have also developed on the spoil heaps of the cement quarries at Rugby, Long Itchington, and Bishop's Itchington. Unfortunately, the best oolitic limestone pastures really occur outside the county in Oxfordshire and Gloucestershire, but some quite good small patches are to be found in the far south where typical *Bromus erectus—Poa angustifolia* grassland develops.

So far as the neutral grassland is concerned, *Agrostis tenuis*, *Phleum bertolonii*, *Poa pratensis*, *Festuca rubra* and *Anthoxanthum odoratum* are extremely common.

The following list is a generalised one of the most commonly occurring species and since limestone pastures are relatively uncommon in Warwickshire the species typical of them do not appear:

Bellis perennis (332)
Cynosurus cristatus (301)
Trifolium repens (268)
Lolium perenne ssp. *perenne* (261)
Ranunculus acris (261)
Trifolium pratense (235)
Prunella vulgaris (230)
Anthoxanthum odoratum (224)
Cirsium arvense (223)
Ranunculus bulbosus (219)
Luzula campestris (210)
Cirsium vulgare (204)
Alopecurus pratensis (201)
Cerastium holosteoides (198)
Lotus corniculatus (196)
Plantago lanceolata (194)
Leontodon autumnalis (190)
Dactylis glomerata (179)
Rumex acetosa (174)
Agrostis tenuis (165)
Taraxacum officinale agg. (164)
Achillea millefolium (159)
Phleum bertolonii (151)
Ranunculus repens (150)
Festuca pratensis (147)
Hypochoeris radicata (142)
Veronica serpyllifolia (141)
Phleum pratense (138)
Poa pratensis ssp. *pratensis* (129)
Festuca rubra (128)
Hordeum secalinum (127)
Holcus lanatus (125)
Trifolium dubium (123)
Plantago major (121)
Trisetum flavescens (112)
Poa annua (111)
Galium verum (108)

(*c*) *Rough Grassland* (*G-r*). Recorded in 567 squares; 92·50% of those surveyed. Total number of observations: 14 830. Number of species recorded: 529.

Included in this category is grass that is ungrazed or unmown, as well as roadside verges which may be cut from time to time. It also includes commons and neglected grassland, and is very widespread throughout the vice-county. Characteristic of this are the tussock grasses such as *Dactylis glomerata* and *Deschampsia cespitosa*, with roadside species such as *Arrhenatherum elatius* and other coarse grasses of little value to cattle such as *Holcus lanatus*. Rough grassland occurs both on neutral and alkaline soils in Warwickshire but we have not attempted to separate the species appropriate to those two soils in the list of common species that follows, most of which are in any case not typical of limestone areas:

Dactylis glomerata (210)
Centaurea nigra agg. (201)
Holcus lanatus (194)
Festuca rubra (188)
Lathyrus pratensis (185)
Agrostis tenuis (174)
Alopecurus pratensis (174)
Arrhenatherum elatius (174)
Poa pratensis ssp. *pratensis* (172)
Agrostis stolonifera (164)
Rumex acetosa (162)
Achillea millefolium (161)
Galium verum (157)
Lotus corniculatus (156)
Poa trivialis (156)
Trifolium pratense (155)
Vicia sativa (151)
Cirsium arvense (148)
Lolium perenne ssp. *perenne* (145)
Cerastium holosteoides (144)

Cirsium vulgare (143)
Taraxacum officinale agg. (141)
Trisetum flavescens (140)
Heracleum sphondylium (139)
Phleum pratense (137)
Veronica chamaedrys (134)
Plantago lanceolata (132)
Stellaria graminea (130)
Cynosurus cristatus (128)

(d) Orchards (G-o). Recorded in 33 squares; 5·38% of those surveyed. Total number of observations: 111. Number of species recorded: 75.

Orchards are probably a lot more frequent than our figures show, and perhaps are under-recorded. They are thinly scattered over the county with perhaps a little concentration in the south-west.

The obvious and typical orchard grass is *Dactylis glomerata*, though *Arrhenatherum elatius* follows it closely. One calcicolous plant occurs in the list—*Geranium pratense*.

A list of the commoner plants of orchards is given below. The planted trees are, of course, not recorded:

Dactylis glomerata (7)
Urtica dioica (5)
Anthriscus sylvestris (3)
Arrhenatherum elatius (3)
Geum urbanum (3)
Glechoma hederacea (3)
Prunella vulgaris (3)
Rumex obtusifolius (3)
Vicia tetrasperma (3)
Geranium pratense (2)

5. Waterside and Marsh (WS)

Included under this major habitat category are all areas with a high water table and with neutral to alkaline water present. Acid conditions (bogs) are placed as H-b. The category is of course a transitional one (either spatial or temporal) between aqueous habitats on the one hand and dry land on the other. The hydrosere is generally rather static, through interference by man (dredging of canals, rivers, etc., re-cutting of ditches). It would lead normally to grassland, or occasionally to hedgerows, woods or (with ponds especially) to ruderal (farmyards).

The minor habitat categories are largely defined by locality—canalside, riverside, etc., though the composition of the species lists varies enough to have made the distinctions worth while.

(a) Marsh (WS-m). Recorded in 409 squares; 66·72% of those surveyed. Total number of observations: 3983. Number of species recorded: 327.

Marshes are fairly evenly distributed but are somewhat rare in the extreme north and north-west of Warwickshire and in the Avon and Leam valleys. Where found they are never very extensive and often represent old ox-bows, silted up ponds or ends of lakes and reservoirs, very damp water meadows where the drainage has been impeded, and the like.

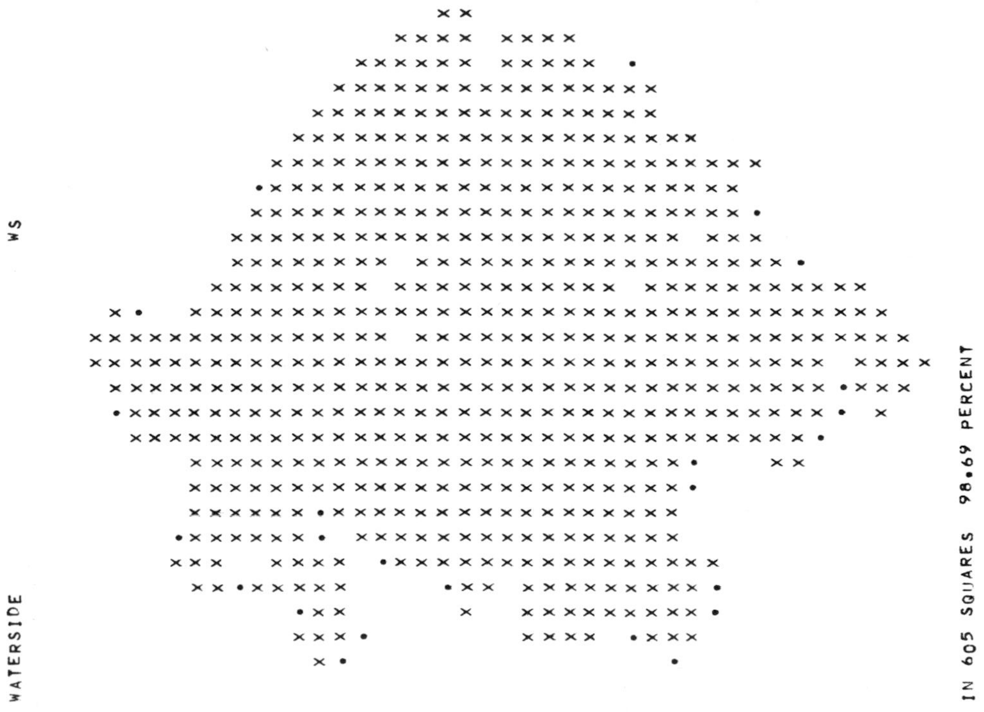

MAP 14. Waterside habitat distribution (including marsh)

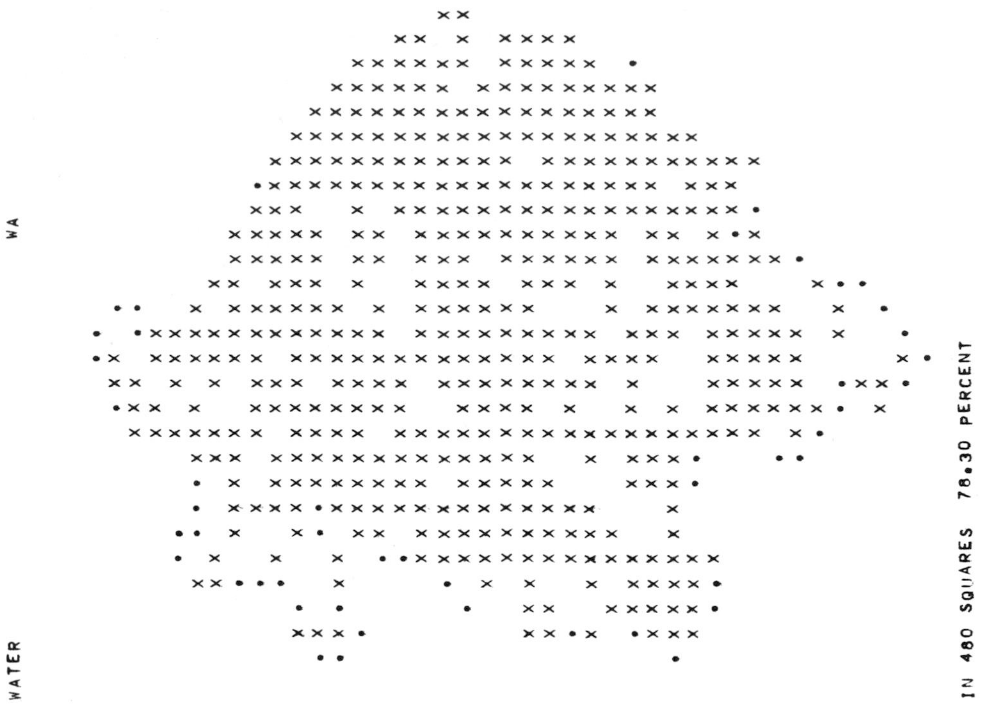

MAP 15. Water habitat distribution

In a few places in the east and south-east of Warwickshire there are a number of mildly saline springs which form small brackish marshes and ponds in their vicinities. We have not designated a special minor habitat category for these because of their rarity.

At the springs at Southam Holt (4460) and Flecknoe (5064) *Scirpus maritimus* and *S. tabernaemontani* are locally abundant; the latter species, *S. tabernaemontani*, has also been recorded near Itchington Holt (3755) and at Holt Farm (4260). Strangely enough, no other halophilous plants have been recorded in these salt marshes, in contrast to the richer salt spring vegetation of Worcestershire. We have, however, recorded *Puccinellia distans* in a very small "salt marsh" at Ansley Hall Colliery (3093) fed by water pumped out of the mine.

The floral composition of marshes is diverse and interesting and the list below gives only the more frequent species:

Alopecurus geniculatus (124)
Juncus effusus (120)
Juncus inflexus (109)
Juncus articulatus (106)
Cardamine pratensis (97)
Carex hirta (96)
Galium palustre agg. (94)
Lotus uliginosus (85)
Stellaria alsine (85)
Glyceria fluitans (82)
Deschampsia cespitosa (80)
Cirsium palustre (79)
Filipendula ulmaria (71)
Veronica beccabunga (70)
Lychnis flos-cuculi (69)
Caltha palustris (65)
Carex otrubae (61)
Ranunculus repens (60)
Juncus subuliflorus (57)
Epilobium hirsutum (55)

(*b*) *Lakeside* (*WS-l*). Recorded in 33 squares; 5·38% of those surveyed. Total number of observations: 259. Number of species recorded: 120.

To distinguish lakes from ponds we have defined them as being more than ½ hectare (1 acre) in area approximately (see note under WA-l). They are chiefly seen in the north-west and north-eastern parts of Warwickshire.

The species list for lakesides is very similar to that for canalsides:

Lycopus europaeus (8)
Alnus glutinosa (7)
Phalaris arundinacea (7)
Scutellaria galericulata (7)
Salix fragilis (6)
Angelica sylvestris (5)
Juncus articulatus (5)
Polygonum amphibium (5)
Sparganium erectum (5)
Typha latifolia (5)

(*c*) *Pondside* (*WS-p*). Recorded in 457 squares; 74·55% of those surveyed. Total number of observations: 4534. Number of species recorded: 326.

Ponds were defined as areas of water less than about ½ hectare (1 acre) in area (see note under WA-p). They are fairly evenly distributed throughout the county. Since the banks are frequently very muddy and the water polluted through the presence of cattle the species composition for pondsides differs quite considerably from that of lakesides, as the list of common species for this minor habitat shows:

Ranunculus sceleratus (184)
Glyceria fluitans (147)
Juncus effusus (146)
Solanum dulcamara (145)
Juncus inflexus (131)
Alisma plantago-aquatica (118)
Alopecurus geniculatus (117)
Salix fragilis (112)
Veronica beccabunga (103)
Epilobium hirsutum (98)
Galium palustre agg. (92)
Myosotis caespitosa (87)
Carex otrubae (79)
Rorippa nasturtium-aquaticum agg. (78)
Rumex conglomeratus (75)
Sparganium erectum (75)
Salix cinerea ssp. *oleifolia* (73)

(*d*) *Riverside (and Streamside)* (*WS-r*). Recorded in 406 squares; 66·23% of those surveyed. Total number of observations: 6246. Number of species recorded: 365.

(For distribution see notes under WA-r. See also overlay Map II.) Riverside and streamside habitats do not differ very much from marshes or lakesides, since the influence of the flowing water does not reach them. On the high banks of rivers rather distinct communities sometimes develop. However, these are influenced by the increased drainage and lower water table on the bank as distinct from the waterside itself and the more waterlogged soils in the meadow behind. Such communities have not been included under WS-r.

The list which follows shows the commoner species, which include a number of shrubby willows:

Salix fragilis (178)
Epilobium hirsutum (171)
Scrophularia auriculata (151)
Alnus glutinosa (146)
Veronica beccabunga (142)
Filipendula ulmaria (136)
Phalaris arundinacea (115)
Barbarea vulgaris (114)
Cardamine flexuosa (106)
Juncus inflexus (95)
Apium nodiflorum (94)
Juncus effusus (94)
Myosotis scorpioides (94)
Solanum dulcamara (86)
Ranunculus sceleratus (85)
Angelica sylvestris (84)
Mentha aquatica (83)
Polygonum hydropiper (80)
Myosoton aquaticum (77)
Salix alba (75)

(*e*) *Canalside* (*WS-c*). Recorded in 99 squares; 16·15% of those surveyed. Total number of observations: 1785. Number of species recorded: 250.

Canals are rather widespread in the Midlands as the overlay Map II indicates (see also notes under WA-c). On the whole, the plant list for canals is not dissimilar to that for lakes. Species growing in the stone walls of locks and obviously influenced by the water have been recorded for "canalside" and not "wall".

Lycopus europaeus (56)
Scutellaria galericulata (55)
Filipendula ulmaria (42)
Epilobium hirsutum (41)
Rumex hydrolapathum (41)
Juncus inflexus (37)
Glyceria maxima (36)
Carex otrubae (34)
Myosotis scorpioides (34)
Angelica sylvestris (31)
Galium palustre agg. (30)
Mentha aquatica (30)
Impatiens capensis (28)
Alnus glutinosa (27)
Rorippa amphibia (27)
Carex acutiformis (26)
Scrophularia auriculata (26)
Stachys palustris (26)
Potentilla anserina (25)
Bidens tripartita (24)

(*f*) ***Ditchside* (*WS-d*).** Recorded in 560 squares; 91·35% of those surveyed. Total number of observations: 5477. Number of species recorded: 345.

Ditchsides are evenly distributed over the county (see notes for WA-d). The species lists for them do not differ greatly from those of the other waterside habitats, though they are not by any means identical. The commonest species are as follows:

Epilobium hirsutum (309)
Filipendula ulmaria (201)
Juncus effusus (137)
Scrophularia auriculata (131)
Phalaris arundinacea (109)
Veronica beccabunga (105)
Angelica sylvestris (102)
Deschampsia cespitosa (99)
Solanum dulcamara (97)
Rumex sanguineus (95)
Ranunculus ficaria (94)
Carex otrubae (93)
Cardamine flexuosa (91)
Juncus inflexus (91)
Ranunculus repens (88)
Salix fragilis (80)
Equisetum arvense (76)
Barbarea vulgaris (74)
Apium nodiflorum (72)

6. Water (WA)

Warwickshire, like all other parts of lowland Britain, is threaded by thousands of small streams and ditches which drain into rivers. Some of this water also flows into canals (see Chapter 2 and Map II). There is a certain overlap between the WA and WS habitat groupings since some plants spread from one to the other, and the actual boundary between them varies as the water level rises and falls. In canals and rivers, whose levels are kept more or less constant by sluices, this need not bother us too greatly. Signs of obvious flooding and drought in lakes, ponds and ditches can also be observed and appropriate allowances made.

Certain plants are clearly to be found in one habitat only, such as *Potamogeton* species growing in open water and *Lemna* species which float on the surface (even though they can both survive if they are stranded on wet mud). Other plants possess water phenotypes and marsh phenotypes (many Batrachian *Ranunculus* species, *Polygonum amphi-*

bium) whilst others are able to grow in still or running water with different growth forms (*Sparganium erectum, Sagittaria sagittifolia*).

The division into minor categories has been, as with "Waterside", a matter largely of locality, though the species lists differ sufficiently to have made the distinction worthwhile.

(a) Lakes (WA-l). Recorded in 24 squares; 3·92% of those surveyed. Total number of observations: 78. Number of species recorded: 36.

To distinguish lakes from ponds we have included all bodies of water of more than ½ hectare (1 acre) in area approximately. They are sparsely scattered and mostly found in the north-western and north-eastern parts of the county.

No lakes occurring in the county are natural and those present are reservoirs or ornamental lakes produced by building a dam across a shallow valley. In many instances, as in Sutton Park, Edgbaston Park, etc. they represent old millpools that have been turned to other uses. Some, as for instance the Earlswood lakes, are canal feeder lakes and have now been retained chiefly for fishing and boating. Others are flooded quarries, gravel and clay pits.

The most common lake species in the county are as follows:

Polygonum amphibium (8)
Nuphar lutea (6)
Typha latifolia (6)
Elodea canadensis (5)
Myriophyllum spicatum (4)
Potamogeton natans (4)
Potamogeton pectinatus (4)
Alisma plantago-aquatica (3)
Hippuris vulgaris (3)
Potamogeton crispus (3)

(b) Ponds (WA-p). Recorded in 366 squares; 59·71% of those surveyed. Total number of observations: 1489. Number of species recorded: 90.

Ponds are widespread throughout the county and are defined as bodies of water of less than ½ hectare (1 acre) in extent. They are mostly, if not entirely, artificial, having been dug for cattle watering or for removing marl in an earlier time to spread over the land. Where cattle and ducks have had access to the ponds the vascular plants are often completely absent, but enclosed ponds are generally rich in species if not shaded by trees and bushes.

The following list shows the commoner species in ponds:

Lemna minor (259)
Potamogeton natans (135)
Alisma plantago-aquatica (107)
Lemna trisulca (94)
Callitriche stagnalis (66)
Sparganium erectum (63)
Typha latifolia (60)
Glyceria fluitans (52)
Ranunculus aquatilis (52)
Rorippa nasturtium-aquaticum agg. (47)
Polygonum amphibium (36)
Ranunculus trichophyllus (33)
Potamogeton crispus (31)

(*c*) ***Rivers (and Streams) (WA-r).*** Recorded in 213 squares; 34·75% of those surveyed. Total number of observations: 739. Number of species recorded: 80.

The main river systems have been described in Chapter 2 and are shown on overlay Map II. Because of the rather small amount of water in our rivers they are generally dammed up at various points, resulting in a series of very slow flowing "ponds". The streams also are rather slow flowing, since in any case the valley slopes are very gentle. However, the slow flow is sufficient to give a slightly different aspect to the river and stream flora as compared with that of canals.

Pollution is a very serious problem in a number of rivers, chiefly the Cole, Anker, Blythe and Rea which debouch into the Tame. In these rivers very few vascular plants are able to survive. Conditions are better in the Avon, Stour and Arrow, and their tributaries, though no doubt the range of species is not so great nowadays as 100 years ago. Certainly, some of the species recorded by Bagnall, such as *Ranunculus fluitans*, have now disappeared, though it is difficult to be certain that their absence is due solely to pollution.

The commoner species are recorded in the following list:

Nuphar lutea (62)
Apium nodiflorum (55)
Rorippa nasturtium-aquaticum agg. (53)
Sparganium erectum (50)
Lemna minor (40)
Scirpus lacustris (38)
Veronica beccabunga (34)
Callitriche stagnalis (31)
Alisma plantago-aquatica (29)
Sagittaria sagittifolia (28)
Potamogeton pectinatus (26)

(*d*) ***Canals (WA-c).*** Recorded in 60 squares; 9·79% of those surveyed. Total number of observations: 383. Number of species recorded: 57.

The canal systems are shown on Map II and are mentioned in Chapter 2. Six canals run through Warwickshire, most of them converging on Birmingham. These are the Grand Union, the Birmingham and Fazeley, the Stratford on Avon, the Coventry, the Ashby de la Zouch and the Oxford canals.

Many of the river plants are also to be found in canals, though the rather heavier rooted or rhizomatous ones such as the water lilies, *Nuphar* and *Nymphaea*, are not so common in canals because of dredging.

Pollution is also intense in the canal sections passing through Birmingham and other industrial cities, where no plants seem able to survive in the water. Even trailing tow-path herbs and grasses die off completely at water level.

All the canals we have mentioned are used for commercial and pleasure purposes. They are therefore kept reasonably well dredged and the natural hydrosere progression is kept in check.

The commonest plants of canals are found growing completely submerged (*Potamogeton* spp., etc.) though those with only their roots and basal parts of the stems in the water are by no means rare (*Butomus*, *Carex*, etc.).

The following list gives the species most frequently recorded:

Potamogeton pectinatus (44)
Potamogeton crispus (28)
Potamogeton perfoliatus (25)
Sagittaria sagittifolia (25)
Lemna minor (21)
Potamogeton natans (21)
Alisma plantago-aquatica (17)
Elodea canadensis (12)
Lemna gibba (12)
Nuphar lutea (12)
Polygonum amphibium (11)
Butomus umbellatus (10)

(*e*) *Ditches* (*WA-d*). Recorded in 115 squares; 18·76% of those surveyed. Total number of observations: 218. Number of species recorded: 52.

Ditches are fairly evenly distributed through the county. On the whole, however, ditches have not been recorded so frequently (115 squares) as ditch sides (560 squares). This is almost certainly due to the fact that during the summer when plants are flowering and the recorders are out in the field the ditch bottoms have largely dried out. Since for WA-d the plants had to be actually growing in water it is likely that they were in fact recorded for WS-d. In any case, the majority of ditch plants are really marsh plants and only the ones here noted, together with some 40 others, were actually seen growing in ditches with free water at the bottom, probably because there was a spring nearby.

The following species were the most frequently recorded:

Apium nodiflorum (36)
Rorippa nasturtium-aquaticum agg. (34)
Veronica beccabunga (26)
Lemna minor (17)
Mentha aquatica (10)
Callitriche stagnalis (8)
Sparganium erectum (8)
Epilobium hirsutum (5)
Glyceria fluitans (5)
Alisma plantago-aquatica (4)
Iris pseudacorus (4)

7. Ruderal (RU)

The flora of the areas of Britain most heavily influenced by man has been to some extent neglected, not only in County Floras but in works on British vegetation in general. Salisbury's "Weeds and Aliens" (1961) is one of the notable exceptions.

The weeds of waste places (ruderals) and of roadsides and pathsides (viaticals) are in fact of considerable interest. In a county such as Warwickshire, with its big industrial complexes of Birmingham and Coventry and the mining areas of Bedworth and Nuneaton, there are plenty of opportunities for ruderal plants to establish themselves on the highly "artificial" habitats of waste heaps, spoil banks, industrial and urban re-development areas and railway yards. Our survey spread right through such areas, and we were just as interested in these as in the more natural vegetation of woods, fields and heaths.

In this major habitat category which we have termed "Ruderal" we have also included railways, pathsides, roadsides, farmyards and brick pits. We might perhaps have separated the pathsides, roadsides and railways into a separate "Viatical" category, but the intergradations between these on the one hand and the waste areas on the other was so gradual and difficult to define that we felt it better to unite them, at least under a major habitat category.

Many of the plants found in this category are weeds of arable land, though here again many others come in from grassland and hedgerows. The arable weeds are aggressive and vigorous, and some of them have undoubtedly been with us since full and late glacial times (Godwin, 1956). Others appeared with agricultural activities in neolithic, bronze or iron age times or even later still, in Romano-British days. In fact, it would seem that there has been an almost continuous sequence of introduction of ruderal plants into Britain by man. This sequence is still in progress. Although some of our ruderal plants are almost certainly native, others have spread with farming activities from Europe, the Mediterranean and south-west Asia from quite early times. Later introductions in the eighteenth century onwards have spread to Britain either as weeds or escapes from cultivation from all parts of the world. It is therefore impossible in this continuing sequence of introduction and varying degrees of establishment to attempt to make a hard and fast distinction between truly native species and alien or introduced ones.

Ruderal habitats support secondary seres in which open areas are colonised by weeds which would in due course be replaced by a sequence of grasses, bushes and trees. Ruderal plants are vigorous and aggressive and can often withstand the poorest soil conditions or even those in which soil is completely replaced by concrete and brick rubble or stone and slag heaps. They help to form the soil and ameliorate the environment so that more exacting grass and herb species may follow them. Ruderals are able to withstand the most adverse conditions of soil or lack of it, but are not able to compete very successfully with the grasses, herbs, trees and shrubs which follow them and which therefore cause them to die out. In this respect ruderal plants conform well to the concept of "weedy plants" in the ecological sense and are probably to be considered in the same eco-physiological group as the weeds of cultivation, and, indeed, the cultivated plants themselves.

We have made five subdivisions of the ruderal category as follows: Waste places (RU-w), Roadsides and pathsides (RU-ro), Railways (RU-ry), Brick and marl pits (RU-b) and Farmyards (RU-f).

(a) Roadsides (RU-ro). Recorded in 606 squares; 98·86% of those surveyed. Total number of observations: 28 754. Number of species recorded: 553.

As might be expected in a county so well traversed by roads of all kinds as Warwickshire, this habitat category is very widespread, occurring in nearly every one of the random squares surveyed.

The roadside category (which also includes pathsides, canal towpaths, etc.) is the richest and most ubiquitous of all, with 553 species recorded, followed closely by rough grassland with 529 species. Obviously, pathsides and roadsides link up with a wide range of other habitats, and above all with rough grassland and hedgerows, as well as ditches

RUDERAL RU

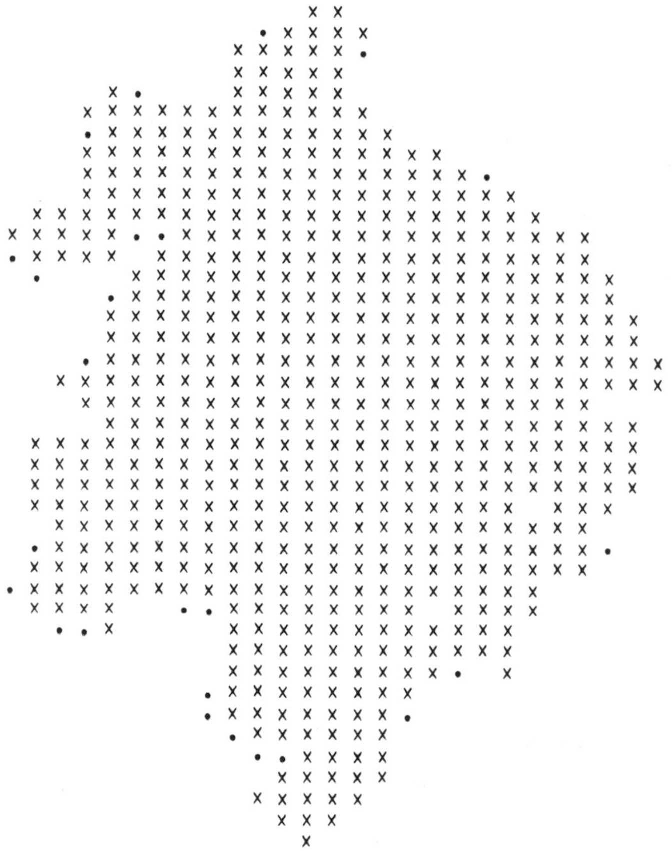

IN 611 SQUARES 99.67 PERCENT

MAP 16. Ruderal habitat distribution (including roadsides, railways and waste places)

STONE WALLS, QUARRIES ST

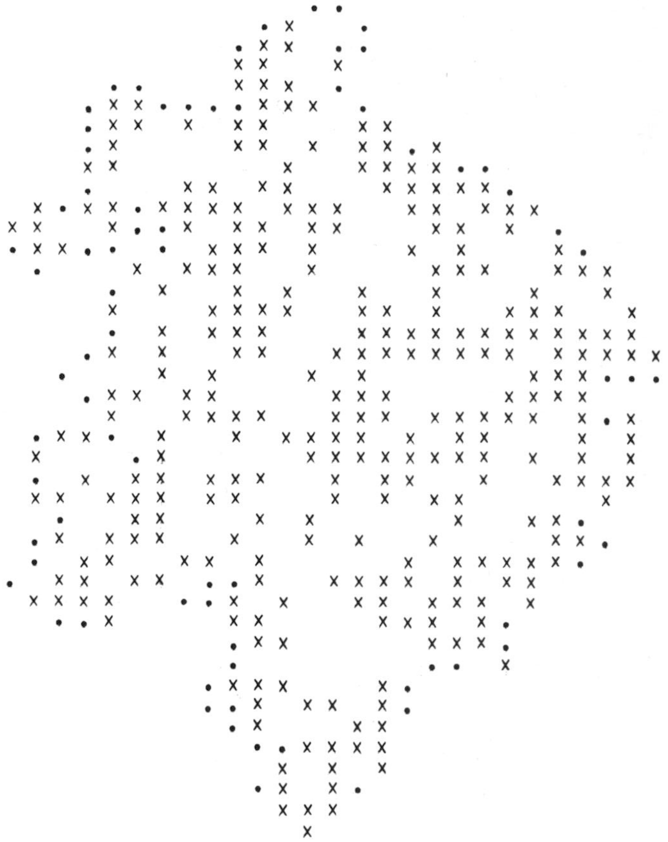

IN 316 SQUARES 51.55 PERCENT

MAP 17. Stones habitat distribution (including walls and quarries)

and even waste places. One therefore finds a range of most diverse elements on roadsides, including plants of hedges, woodlands and grassland on the one hand, with characteristic weeds of cultivated land and waste places on the other.

In our instructions to recorders the wide grassy verges of roadsides were to be recorded as rough grassland (G-r) and not roadsides; nevertheless, where the verge was narrow it was felt better to leave this in the RU-r category since the influence of the special conditions of poor soil, trampling and stones, grit, etc., would everywhere be quite important.

Of the very large number of species recorded for this habitat category the following are commonest:

Taraxacum officinale agg. (425)
Plantago major (423)
Plantago lanceolata (406)
Potentilla anserina (384)
Dactylis glomerata (372)
Achillea millefolium (368)
Trifolium repens (364)
Potentilla reptans (359)
Lolium perenne ssp. *perenne* (353)
Matricaria matricarioides (341)
Heracleum sphondylium (334)
Poa annua (324)
Arrhenatherum elatius (319)
Lamium album (298)
Alopecurus pratensis (288)
Ranunculus repens (282)
Trifolium pratense (280)
Cirsium arvense (274)
Holcus lanatus (269)
Tussilago farfara (262)

(b) Railways (RU-ry). Recorded in 165 squares; 26·92% of those surveyed. Total number of observations: 6718. Number of species recorded: 472.

This is a fairly widespread habitat category, comprising both used and derelict railways, the latter being particularly rich in species. It includes not only the railway embankments and cuttings but the sides of the tracks, station yards and sidings. In a sense it is too wide, since it would perhaps have been better to have included the grassy embankments and cuttings as rough grassland. This was done by some recorders but not consistently. The ruderal trackside plants are most characteristic, some even growing in the ballast between the sleepers in lines that are not too much used or are derelict. We find in the list a number of plants typical of rough grassland (*Achillea millefolium*), some of waysides and waste places (*Senecio squalidus*), others which are weeds of cultivation (*Convolvulus arvensis, Chaenorhinum minus*) and others, again (*Cochlearia danica*), which have advanced along the railway tracks from their original habitat of maritime shingle.

Of a very large species list the following have been recorded with the greatest frequency:

Equisetum arvense (94)
Epilobium angustifolium (89)
Tussilago farfara (85)
Centaurea nigra (83)
Achillea millefolium (77)
Hieracium pilosella (77)
Convolvulus arvensis (70)
Lathyrus pratensis (69)
Leucanthemum vulgare (66)
Senecio squalidus (65)

Vicia hirsuta (65)
Linaria vulgaris (64)
Daucus carota (63)
Chaenorhinum minus (62)
Myosotis arvensis (62)

(c) *Waste Places (RU-w).* Recorded in 562 squares; 91·68% of those surveyed. Total number of observations: 10 677. Number of species recorded: 489.

The "waste places" category is very frequent and is fairly evenly distributed in Warwickshire, though it thins out in the extreme south. It includes areas where soil, stones, gravel and other materials have been dumped, old industrial or housing sites which are partially or entirely derelict, spoil heaps from pits and quarries, coal and slag banks, industrial yards and many other types of waste land. As we have said before, this is an interesting and important habitat category which is frequent in areas of urban development and decay.

Plants growing in waste places may be divided into three main groups:

(i) The ruderal plants which seem chiefly at home in waste places and are absent or infrequent as arable weeds. Plants such as *Artemisia vulgaris*, *A. absinthium*, *Polygonum cuspidatum*, *Matricaria matricarioides* and many others fit best into this group.
(ii) Arable weeds, such as *Capsella bursa-pastoris*, *Senecio vulgaris*, *Plantago major*, *Poa annua* and *Chenopodium album*; these are annuals and seem better able to survive in areas where the soil is loose and well cultivated.
(iii) Recent escapes from cultivation such as *Aster novi-belgii.*

The more commonly recorded species of waste places are as follows:

Matricaria matricarioides (204)
Capsella bursa-pastoris (203)
Chenopodium album (189)
Polygonum aviculare (187)
Senecio vulgaris (172)
Urtica dioica (162)
Atriplex patula (153)
Epilobium angustifolium (147)
Sisymbrium officinale (145)
Tussilago farfara (145)
Plantago major (143)
Poa annua (140)
Tripleurospermum maritimum ssp. *inodorum* (134)
Stellaria media (128)
Artemisia vulgaris (120)
Sonchus oleraceus (120)
Polygonum persicaria (119)
Rumex obtusifolius (119)
Sonchus asper (112)
Lamium album (102)
Hordeum murinum (101)

(d) *Brick and Marl Pits (RU-b).* Recorded in 20 squares; 3·26% of those surveyed. Total number of observations: 241. Number of species recorded: 157.

This is a very restricted minor habitat category, since clay pits are not common and many of them have been filled with rubbish or have been turned into ponds. The plants are those of waste places, but species able to grow in clay, such as *Tussilago farfara*, are especially frequent.

The following are some of the commoner species in this habitat:

Tussilago farfara (6)
Senecio vulgaris (5)
Artemisia vulgaris (4)
Lotus corniculatus (4)
Sarothamnus scoparius (4)
Senecio squalidus (4)
Sisymbrium officinale (4)

(e) Farmyards (RU-f). Recorded in 256 squares; 41·76% of those surveyed. Total number of observations: 1790. Number of species recorded: 217.

This habitat is somewhat irregularly scattered over the county.

Since weed seeds are constantly being transferred here from the cultivated fields, the arable weed component is very high in this habitat. Two factors of special importance are the trampled ground and the high nitrogen content of the soil through the presence of manure. Plants typical of the former are *Coronopus squamatus* and *Matricaria matricarioides;* those typical of high nitrogen soils are *Urtica dioica* and *Chenopodium* spp.

The species most commonly recorded from this habitat are:

Matricaria matricarioides (80)
Coronopus squamatus (67)
Polygonum aviculare (60)
Capsella bursa-pastoris (58)
Urtica dioica (56)
Atriplex patula (41)
Plantago major (41)
Tripleurospermum maritimum ssp. *inodorum* (39)
Chenopodium album (37)
Sisymbrium officinale (36)
Senecio vulgaris (35)
Matricaria recutita (33)
Rumex obtusifolius (33)
Sinapis arvensis (33)
Poa annua (32)

8. Stones, walls and quarries (ST)

This is a reasonably well-distributed major habitat, though only recorded in some 50% of the squares surveyed, as can be seen in Table 8. The main distribution of this habitat is composed of "wall" records (see below), since quarries are much more limited. No natural cliffs occur in the county.

(a) Walls (ST-wa). Recorded in 291 squares; 47·47% of those surveyed. Total number of observations: 1144. Number of species recorded: 193.

Walls are well distributed throughout the county and tend to be composed of limestone in the southern and south-eastern calcareous areas, and of brick or sandstone in the others.

Many plants, such as various *Asplenium* species, are confined to walls; others find the

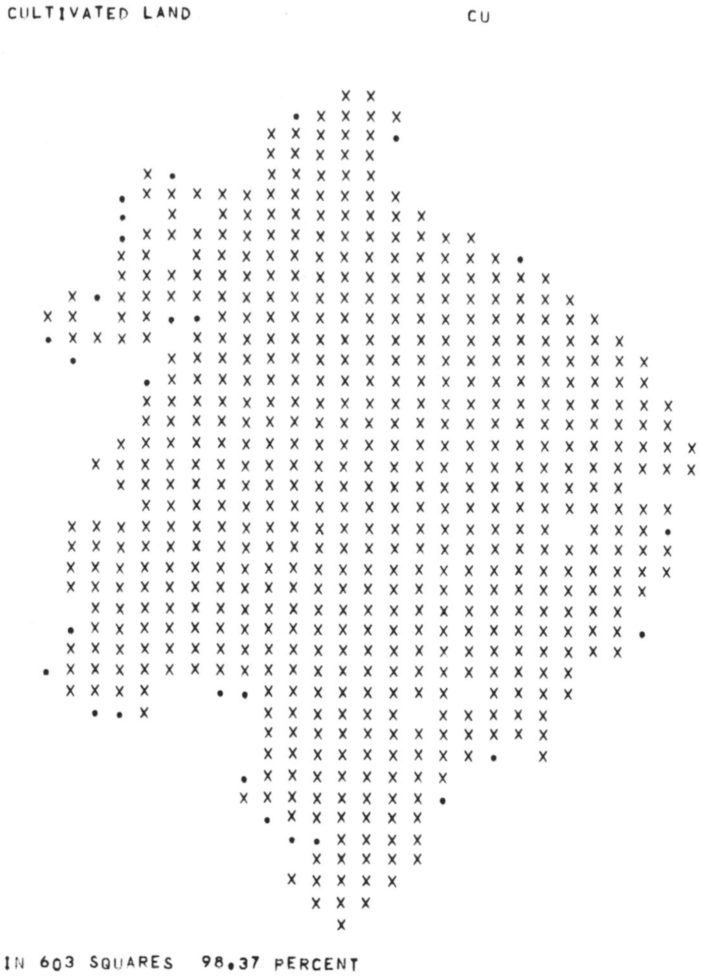

MAP 18. Cultivated land habitat distribution

dry well-drained tops or protected bases suitable habitats, whilst others are weeds or ruderals which have been recorded in and around the wall bases.

In the list that follows we have put a (C) after the species chiefly found in limestone walls. Others may of course be taking advantage of the lime mortar in walls built of other materials.

Asplenium ruta-muraria (60)
Hedera helix (58)
Cymbalaria muralis (47)
Sagina procumbens (44)
Arenaria serpyllifolia agg. (35)
Asplenium trichomanes (34)
Phyllitis scolopendrium (29)
Sedum acre (29) (C)
Dryopteris filix-mas (28)
Asplenium adiantum-nigrum (27)
Veronica arvensis (26)
Polypodium vulgare agg. (22)
Erophila verna (21)
Parietaria judaica (18)
Pteridium aquilinum (17)
Senecio squalidus (17)
Cardamine hirsuta (16)
Hordeum murinum (15)
Saxifraga tridactylites (15) (C)
Bromus sterilis (14)

Ceterach officinarum (14) (C)
Geranium robertianum (14)
Poa compressa (14)
Catapodium rigidum (8) (C)
Fragaria vesca (8) (C)
Centranthus ruber (7) (C)
Geranium lucidum (4) (C)
Myosotis ramosissima (4) (C)

(*b*) *Quarries* (*ST-qu*). Recorded in 62 squares; 10·11% of those surveyed. Total of observations: 1111. Number of species recorded: 345.

Quarries are of very different nature in different parts of the county. Thus, the Hartshill quarries of the north-east provide habitats for certain saxicolous plants and one or two that can make use of the alkaline conditions provided by weathered diorite.

Cement quarries such as those of Rugby, Stockton, Long Itchington and Bishop's Itchington provide a substrate composed of calcareous stones and clay which dries out considerably in the summer.

There are also some sandstone quarries and rock faces (hardly cliffs) in the region of Warwick, Kenilworth and Coventry where the plants are taking advantage of good drainage. Some of them which grow on the ruins of Kenilworth Castle are no doubt there for the same reason.

The Middle Lias ironstone quarries near Burton Dassett and at Edge Hill also provide well-drained conditions, whilst those of the Inferior Oolite near Little Compton are interesting areas for calcicolous species that require good drainage in addition.

We have separated the plants of quarries into three lists, so as to give some indications of the chemical as well as the physical characteristics of the rocks on which the plants grow:

General

Hieracium pilosella (15)
Vicia sativa (13)
Cirsium arvense (12)
Lotus corniculatus (12)
Senecio jacobea (11)
Tussilago farfara (11)
Crepis capillaris (10)
Plantago lanceolata (10)
Vulpia bromoides (10)
Anthyllis vulneraria (4)
Carduus nutans (4)
Centaurea scabiosa (4)
Cirsium acaule (4)
Plantago media (4)
Dactylorhiza fuchsii (4)
Blackstonia perfoliata (3)
Bromus erectus (3)
Listera ovata (3)

Limestone

Linum catharticum (7)
Pastinaca sativa (7)
Catapodium rigidum (5)
Cirsium eriophorum (5)
Erigeron acer (5)
Ophrys apifera (5)

Non-calcareous

Rumex acetosella (9)
Rubus fruticosus agg. (9)
Cerastium glomeratum (8)
Aphanes arvensis (7)
Spergularia rubra (4)
Aira praecox (4)

9. Cultivated Land (C)

Cultivated land is evenly distributed throughout Warwickshire, since gardens, allotments and parks are to be found even in the large cities.

We have included in this category the ploughed and re-dug soil of arable fields, gardens, allotments and public parks but not the planted or semi-natural grasslands, which are included in the grassland category, or the plantations and shrubberies which come within woodland. The plants recorded are therefore chiefly weeds of cultivation or garden and shrubbery plants which have become established and naturalised. Some ornamental trees, even when possibly planted, have been included in this and other categories but our general policy has been to exclude them unless they seemed to be more or less established.

The weeds of cultivated land form their own communities, of which the dominant species could be said to be the cultivated plants themselves. These cultivated plants are not recorded in our Flora unless we had reason to suppose that they were becoming naturalised. The weeds themselves, as we mentioned in the section on ruderals, have come to this county from remote periods with the neolithic, bronze age, iron age and Romano-British farmers, and in later epochs as well. There is also reason to believe that many were here even in full glacial and late glacial times (Godwin, 1956). Many of the species are found in our "ruderal" and "cultivated" habitat categories, since the open soil conditions are present in both. However, species typical of very compacted soils tend to be rare in cultivated land, and there are a number of arable weeds, such as *Myosurus minimus*, that are never found as ruderal plants, possibly because they require better soil conditions than the ruderal habitat can provide, at any rate in Warwickshire.

(a) Arable land; cultivated fields (C-f). Recorded in 565 squares; 92·17% of those surveyed. Total number of observations: 14 245. Number of species recorded: 351.

This habitat is evenly distributed in Warwickshire except in the large conurbations and their surroundings, as well as the smaller towns. It has increased greatly in the last 100 years at the expense of grassland (see Table 3, p. 42). Probably the greatest area under crops is still to be found on the calcareous soils of the south-east and the valleys of the Avon and other rivers such as the Arrow. These have been traditional farming areas since Saxon times and even before (see Chapter 4).

The composition of the weed flora of cultivated fields differs according to whether it is on the well-drained river terrace soils of the Avon and other rivers, the calcareous Lower Lias clays of the south and south-east, or the neutral or acid soils of the centre and north. Apparent calcicolous species such as *Kickxia spuria* and *K. elatine* have been recorded for the Lias clays (see distribution maps). Plants typical of more acid damp soils such as *Gnaphalium uliginosum* and *Juncus bufonius* are more confined to the centre and north, whilst those that prefer light well-drained soils (*Euphorbia exigua* for instance) occur mainly in the south and south-east of Warwickshire.

Many weed species of cultivated land were in the past brought to England as impurities in seed mixtures from more southerly countries to which they were better adapted. They grew in Britain when introduced, but did not mature a good crop of seeds except in the southern counties. Such plants as *Agrostemma githago* and *Centaurea cyanus* are of this type; they were once much more common and have now practically died out.

Many of our arable fields also had a much richer flora when the present survey was in its early years than in 1971. This is due to the use of selective weed killers which have unfortunately decimated all but the more resistant species over wide areas.

A list of the more commonly recorded species is given below:

Chenopodium album (364)
Anagallis arvensis ssp. *arvensis* (354)
Polygonum convolvulus (346)
Polygonum aviculare (339)
Sinapis arvensis (337)
Stellaria media (329)
Viola arvensis (318)
Veronica persica (311)
Aethusa cynapium (296)
Polygonum persicaria (290)
Atriplex patula (289)
Tripleurospermum maritimum ssp. *inodorum* (287)
Capsella bursa-pastoris (284)
Senecio vulgaris (281)
Spergula arvensis (239)
Raphanus raphanistrum (206)
Myosotis arvensis (201)

(b) Gardens (C-g). Recorded in 374 squares; 61·01 % of those surveyed. Total number of observations: 2933. Number of species recorded: 285.

The "gardens" category is fairly evenly distributed but is recorded as rather infrequent on the Lias clays. The weed flora is very similar in most respects to that of arable fields even though *Euphorbia peplus*, *E. helioscopia* and *Lamium purpureum* tend to be much more frequent in gardens, possibly because of the higher nitrogen content of garden soil.

Shrubs of derelict gardens, such as *Taxus baccata*, *Ligustrum vulgare* (and *L. ovalifolium*?) and *Ilex aquifolium* have also been recorded from time to time.

The list below shows the commoner species:

Euphorbia peplus (164)
Senecio vulgaris (135)
Lamium purpureum (115)
Euphorbia helioscopia (109)
Stellaria media (99)
Capsella bursa-pastoris (96)
Veronica persica (86)
Sonchus oleraceus (79)
Bellis perennis (75)
Veronica hederifolia (71)
Poa annua (68)
Aethusa cynapium (59)
Aegopodium podagraria (49)

(c) Allotments (C-a). Recorded in 59 squares; 9·62 % of those surveyed. Total number of observations: 603. Number of species recorded: 170.

Allotments have chiefly been recorded around urban areas but they are probably somewhat under-recorded. The well-cultivated ones bear a weed flora very similar to that of gardens but perhaps with some aspects of arable fields also. Those which have been neglected and are overgrown will gradually revert to grassland, scrub and woodland if left unattended, though we have not included the later stages of this succession in the allotment category. Many ornamental plants from gardens and shrubberies also become established for a time in neglected allotments, and in this respect they reflect the situation often met with in waste ground (RU-w). In fact, the two categories are sometimes hard to distinguish.

By far the commonest allotment weeds are those of arable fields and gardens, as the following list shows:

Capsella bursa-pastoris (26)
Senecio vulgaris (23)
Euphorbia helioscopia (21)
Lamium purpureum (20)
Chenopodium album (19)
Euphorbia peplus (19)
Stellaria media (19)
Poa annua (16)
Atriplex patula (15)
Veronica persica (15)
Polygonum convolvulus (11)
Sonchus oleraceus (11)
Taraxacum officinale agg. (11)
Convolvulus arvensis (10)

(d) Public Parks (C-p). Recorded in 44 squares; 7·18% of those surveyed. Total number of observations: 174. Number of species recorded: 94.

This minor category has been interpreted rather widely by our recorders, though it was intended to include only the cultivated or disturbed ground in public parks. Unfortunately it was also often interpreted as parkland (i.e. rough grass and pasture land with planted trees). It has thus been unsuccessful in its application, and we give a small list here only as an indication of the commoner plants recorded. Fortunately, perhaps, only 174 records for this minor habitat category were made.

Bellis perennis (7)
Acer pseudoplatanus (6)
Aesculus hippocastanum (6)
Fagus sylvatica (6)
Leontodon autumnalis (5)
Hypochoeris radicata (4)
Lolium perenne ssp. *perenne* (4)
Poa annua (4)
Rumex acetosa (4)
Trifolium repens (4)

References

Godwin, H. (1956). "The History of the British Flora." Cambridge University Press.
Hooper, M. D. (1970). The botanical importance of our hedgerows. *In* Perring, F. (Ed.). "The Flora of a Changing Britain." B.S.B.I., Hampton. pp. 58–62.
Kelly, M. and Osborne, P. J. (1965). Two faunas and floras from the alluvium at Shustoke, Warwickshire. *Proc. Linn. Soc. Lond.* **176,** 37–65.
Rees, W. J. and Skelding, A. D. (1950). Vegetation. *In* "Birmingham and its Regional Setting." British Association Handbook, Birmingham. pp. 65–76.
Richens, R. H. (1955). Studies on *Ulmus*. I. The range of variation in East Anglian Elms. *Watsonia* **3,** 138–53.
Richens, R. H. (1958). Studies on *Ulmus*. II. The village elms of southern Cambridgeshire. *Forestry* **31,** 132–46.

Richens, R. H. (1961). Studies on *Ulmus*. IV. The village elms of Huntingdonshire and a new method for exploring taxonomic discontinuity. *Forestry* **34,** 47–64.

Salisbury, E. J. (1961). "Wccds and Alicns." London.

Tansley, A. G. (1939). "The British Islands and their Vegetation." Cambridge University Press.

Tansley, A. G. (1968). "Britain's Green Mantle." 2nd Edition, revised by M. C. F. Proctor. (1st Edition, 1948) London.

8. FLOWERING AND OTHER VASCULAR PLANTS

General arrangement and nomenclature

The section which follows is arranged in taxonomic order, generally according to the "List of British Vascular Plants" prepared by J. E. Dandy and published in 1958. There are some exceptions. In a number of cases Dandy has indicated in correspondence changes of specific names which do not involve changes in his numerical code, and these appear here as revised. In others he has indicated changes in generic names; and here it has been necessary in using an amended name to add a letter "A" or "B" after the old generic code number for the new genus, while the original code number has been retained for the species. It is understood that these name changes will appear in the revised Plant List in preparation. In addition, there are a few plants included here which do not appear in Dandy's List, and these have been included, unnumbered, in what we consider to be the appropriate places.

Other differences arise on taxonomic grounds. In several critical groups (notably *Salix* and *Ulmus*) specialists who have determined most of the material for us have held different views from Dandy on some points of taxonomy and on the range and limits of particular taxa; and these are reflected in the names they have used, which we have felt it appropriate to follow. Finally, there are a few cases in which we ourselves have felt it more consistent with the material we have studied in VC 38 to treat as subspecies taxa which Dandy has considered it proper to treat as species; and in one or two instances the reverse is the case.

The text for each species commences with the code number from Dandy's Plant List, followed by the Latin name of the species and the common English name. This has presented some difficulties, and occasionally more than one name has been given; but the names chosen are those which we consider to be generally accepted nationally or in the county. Next appear the name of the first recorder or collector, the date of publication or collection, the name of the publication in which the record is published or the herbarium in which the specimen is located, the Latin name used by the recorder or collector when it differs from that used by us, and, where this is known, the location in the vice-county of the plant recorded. Where a distribution map appears in a later section of the book, a reference is given to that map. A short account of the distribution of the plant in VC 38 is then given, ending with a note of its presence in any of the surrounding vice-counties.

First Records

The first record given for a plant is generally the date of first publication or the earliest herbarium specimen seen, whichever is the earlier; but in all groups considered critical and whenever there did not appear to be satisfactory evidence that a published record was of a plant correctly named, the date of the earliest herbarium specimen examined has been quoted, though later in date than the published record. In certain critical groups—such as *Hieracium*, *Rubus*, *Salix*, *Taraxacum* and *Ulmus*—it has sometimes been considered advisable to include records only of material determined by national and international specialists; and this has led to some gaps, particularly in the records for surrounding VCs owing to the uncertainty of the nomenclature of older records. In a few cases a quite recent date has been given for a common plant whose existence in the vice-county has been known for some time but for which it has not been possible to trace an earlier published record or herbarium specimen.

In nearly all cases the record (publication or manuscript) quoted has been seen. The exceptions are a few manuscript notes quoted by Bagnall in his "Flora of Warwickshire" which have been accepted, and the Baxter MS of 1831 quoted by Bagnall in the *Midland Naturalist* of 1892–93 as the basis for a number of records pre-dating those given in his "Flora of Warwickshire" published in 1891. The manuscript was in the possession of Druce, who supplied Bagnall with a list of the records, but it is not with the large quantity of MSS which Druce left to Oxford University. Many of these records have been written by Bagnall into his own copy of his Flora; and, as they are generally of common plants, it has been thought well to accept them when not to have done so would have meant giving as a first record a very much later publication or herbarium specimen. Where, however, a published record or herbarium specimen has been found dated within a few years after the Baxter MS, the later record has been given as well.

Frequency and Distribution

Where a plant has been recorded in the present survey a brief statement is made as to its frequency and general distribution in the county. Where no distribution map of the species is included there follows a list of the 1 km squares from which the plant has been recorded, usually with place-names to identify the locations more easily. Where a distribution map is included the squares are not listed; but a list of the principal habitats, in descending order of "preference" by the plant, is given, each followed by a figure giving the percentage of all records (to the nearest whole number) from random-selected 1 km squares in which the plant has been recorded. Percentages have been omitted altogether when the total number of records from all habitats is small; and where percentages have been given for only some of the habitats listed, this is because the number of records from those habitats without percentages is too small to be significant. In this connection it will be recognised that a smaller percentage is significant where the number of records is large than where the total number of records is small. In certain cases, where the information is considered of particular interest, additional information is given as to earlier distribution as illustrated by herbarium specimens; and occasionally, where a plant has not been recorded in the present survey, it has been possible to give what is believed to be the date and location of the last record.

Smaller type has been used for casuals, garden escapes and obvious introductions not considered to be of particular interest. Some *Rubus* microspecies are also shown in small type; the reason for this is given in the preamble to the genus.

Keys

Keys to the identification of species are not generally appropriate in a County Flora; but those included here—*Arctium*, *Bromus*, *Hieracium*, *Rubus* and *Salix*—include, in our opinion, information not readily available elsewhere. In other groups we have occasionally added notes as to observations we have made on material from VC 38, which, as far as we know, have not been noticed elsewhere.

Surrounding Vice-counties

The vice-counties surrounding VC 38 (Warwickshire) are:

23 Oxford
32 Northampton
33 East Gloucester
37 Worcester
39 Stafford
55 Leicester with Rutland
57 Derby

The boundaries of VC 57 are nowhere actually contiguous with those of VC 38, but in one place are separated by less than a mile. It has, therefore, been considered appropriate to include it here.

CONTRIBUTORS OF FIRST RECORDS AND SOURCES OF THE RECORDS

ALLEN, D. E.

"FLORA OF THE RUGBY DISTRICT", 1957

Beers, A. G.
Dunn, S. T.
Thomas, F.

AMPHLETT, S. C. L. AND REA, B. C. L.

"BOTANY OF WORCESTERSHIRE", 1909

Lees, E.

ANALYST: 1837

Beilby, Miss M. A.
Ick, W.

BAGNALL, J. E.

"FLORA OF WARWICKSHIRE", 1891

Bagnall, J. E.
Baxter, W.
Baynes, W. W.
Bloxam, Rev. A.
Bromwich, H.
Caswell, Rev. J.
Cheshire, W.
Gorle, Rev. J.
King, Bolton
Kirk, T.
Lloyd, Dr. G.
Perry, W. G.
Power, Rev. J.
Westcott, Rev. B. T.

"FLORA OF SUTTON PARK", 1876

Bagnall, J. E.

HERBARIUM, BIRMINGHAM MUSEUM

Bagnall, J. E.
Baker, R. L.
Brody, Dr. St.
Bromwich, H.
Caswell, Rev. J.
Kirk, T.
Satchell, W.
Townsend, F.

BAXTER, W.
"BRITISH PHAENOGAMOUS BOTANY", Vol. 1, (1834), Vol. 2, (1835)
Baxter, W.
MS, 1831—See *MIDLAND NATURALIST*

BIRMINGHAM MUSEUM
HERBARIUM (additional to Bagnall Collection)
Hardaker, W. H.
Harlond, A.
Ick, W. (Ick Collection)

BIRMINGHAM NATURAL HISTORY SOCIETY PROCEEDINGS
1870: Bagnall, J. E.
1947: Andrews, C. E. A.
1947: Burges, R. C. L.

BIRMINGHAM UNIVERSITY
HERBARIUM
Anonymous
Bagnall, J. E.
Brody, Dr. St.
Bromwich, H.
Dunn, S. T.
Kirk, T.
Ley, Rev. A.
Thompson, H. S.
West, G. S.
Wilkinson, W. H.

BLOXAM, REV. A.
HERBARIUM, 1875
(RUGBY PUBLIC LIBRARY)
Bloxam, Rev. A.

BOTANICAL SOCIETY & EXCHANGE CLUB REPORTS
1922: Cumming, L.
1926: Bloomer, H. H.
1928: Bromwich, H.
1928: Lamb, J.
1945: Burges, R. C. L.
1945: Horwood, A. R.
1946–47: Burges, R. C. L.
1946–47: Wallace, E. C.

"BOTANIST'S GUIDE": See TURNER, D.

BRITISH MUSEUM
HERBARIUM
Anonymous
Bagnall, J. E.
Bailey, C.
Barker, Mrs. G. M.
Bloxam, Rev. A.
Bromwich, H.
Cadbury, Miss D. A.
Cook, C. D. K.
Edees, E. S.
Jackson, A. B.
Jacobs, V.
King, Bolton
Kirk, T.
Preston, N. C.
Riddelsdell, Rev. H. J.
Thompson, H. S.

BURGES, R. C. L.
HERBARIUM (With Birmingham Natural History Society)
Burges, R. C. L.

CADBURY, MISS D. A.
HERBARIUM
Cadbury, Miss D. A.

CAMBRIDGE UNIVERSITY
HERBARIUM
Bagnall, J. E.
Bloxam, Rev. A.
Cox, T.
Mills, W. H.

"COMPENDIUM OF CYBELE BRITANNICA": See WATSON, H. C.

COVENTRY MUSEUM:
HERBARIUM
Anonymous

DRUCE, G. C.
"COMITAL FLORA OF THE BRITISH ISLES", 1932

GIBSON, E.
EDITION OF CAMDEN'S "BRITANNIA", 1695
Ray, J.

"GLOUCESTERSHIRE, FLORA OF": 1948
Riddelsdell, Rev. H. J.

HOLDEN, H.
HERBARIUM, 1688–96
(Birmingham Reference Library)

ICK, W.
HERBARIUM:
See BIRMINGHAM MUSEUM

JOURNAL OF BOTANY
1869: Kitchener, F. E.
1896: Dunn, S. T.
1897: Townsend, F.
1901: Bromwich, H.
1904: Townsend, F.

KEW
HERBARIUM
Bromwich, H.

L'OBEL: See PENA & L'OBEL

MAGAZINE OF NATURAL HISTORY
1828: Bree, Rev. W. T.
1829: Perry, W. G.
1830: Bree, Rev. W. T.

MANCHESTER UNIVERSITY
HERBARIUM
Dashwood, I. M. Kirk, T.

"MIDLAND FLORA": See PURTON, T.

MIDLAND NATURALIST
CONTRIBUTIONS BY BAGNALL, J. E., 1892–93
Aylesford, Countess of
Baxter, W. (MS 1831)
Bromwich, H.
Palmer, Miss C. E.

NATIONAL MUSEUM OF WALES
HERBARIUM
Bagnall, J. E.
Baker, T. G.
Bloxam, Rev. A.

"NEW BOTANIST'S GUIDE":
See WATSON, H. C.

OXFORD UNIVERSITY
HERBARIUM
Aylesford, Countess of
Bromwich, H.
Palmer, Miss C. E.

PENA, P. AND L'OBEL, M. DE
"STIRPIUM ADVERSARIA NOVA", 1570
L'Obel, M. de

PERRY, W. G.
HERBARIUM, WARWICK MUSEUM
Baynes, W. W.
Bromwich, H.
Cheshire, W.
Cox, T.
Cross, F.
Cross, J.
Harris,
Ick, W.
Kirk, T.
Leighton,
Lloyd, Dr. G.
Lowe, H. E.
Murcot, J. J.
Perry, W. G.
Snape, T.

LIST: APPENDIX TO DUGDALE'S "WARWICKSHIRE", 1817
Perry, W. G.

"PLANTAE VARVICENSIS SELECTAE", 1820
Perry, W. G.

PHYTOLOGIST
EDITED BY LUXFORD, G., 1841–54
1841: Freeman, S.
1842: Freeman, S.
1842: Kirk, T.
1843: Cameron, D.
1847: Kirk, T.
1848: Bloxam, Rev. A.

PURTON, T.
"MIDLAND FLORA"
Vols. 1 & 2 (1817): Bree, Rev. W. T.
Dolben, T.
Purton, T.
Rufford, Rev. W. S.
Vols. 3 & 4 (1821): Baxter, W.
Bree, Rev. W. T.
Purton, T.

RAY, J.

"CATALOGUS PLANTARUM ANGLIAE", 1670
"CATALOGUS PLANTARUM ANGLIAE", 1677
CORRESPONDENCE (ED. BY LANKESTER), LETTER TO LISTER, 1669
"HISTORIA PLANTARUM", 1686
"HISTORIA PLANTARUM", 1688
"SYNOPSIS METHODICA STIRPIUM BRITANNICARUM", 1690
"SYNOPSIS METHODICA STIRPIUM BRITANNICARUM", 1696
"SYNOPSIS METHODICA STIRPIUM BRITANNICARUM", 1724

Ray, J. (First records in all the above works)

RUGBY SCHOOL REPORTS

1867
1868
1869
1871
1872: Bloxam, Rev. A.
1873: Kitchener, F. E.
1874
1875
1892
1923
1949

SOUTH LONDON BOTANICAL INSTITUTE

HERBARIUM

Bagnall, J. E.

"TOPOGRAPHICAL BOTANY":

See WATSON, H. C.

TURNER, DAWSON AND DILLWYN, L. W.

"BOTANIST'S GUIDE THROUGH ENGLAND & WALES", 1805

Aylesford, Countess of

TURRILL, W. B.

"BRITISH KNAPWEEDS", 1950

Turrill, W. B.

WARWICK MUSEUM

HERBARIUM (additional to Perry Collection)

Brody, Dr. St.
Bromwich, H.
Cross, F.
Townsend, F.

WATSON, H. C.

"COMPENDIUM OF CYBELE BRITANNICA", 1870 and SUPPLEMENT, 1872

"NEW BOTANIST'S GUIDE"

1835: Bree, Rev. W. T.
1837: Bloxam, Rev. A.
1837: Bree, Rev. W. T.
1837: Skipwith, Mrs. F.

"TOPOGRAPHICAL BOTANY"

1858: Kirk, T.
1873: Bree, Rev. W. T.
1873: Bromwich, H.
1873: Kirk, T.
1883: Kirk, T.

WATSONIA

1951: Ounsted, J.
1952: Allen, D. E.
1952: Falk, P.
1952: Fleetwood, T.

WITHERING, W.

"BOTANICAL ARRANGEMENT OF BRITISH PLANTS"

1787: 2nd edition, including records by Stokes, Dr. J., and Woodward, T. J.
1796: 3rd edition, including records by Stokes, Dr. J. and Miss Withering
1801: 4th edition
1812: 5th edition
1818: 6th edition
1830: 7th edition

"WORCESTERSHIRE, BOTANY OF":

See AMPHLETT, S. C. L. AND REA, B.C.L.

YOUNG, REV. J. R., AND BAKER, DR. R.

"CATALOGUE OF PLANTS COLLECTED IN WARWICKSHIRE", 1873

Baker, Dr. R.
Cox, T.

NOTE: In a few cases, listed below, specimens previously housed in the Warwick Museum have been destroyed. The labels are in existence and the records are considered quite good.

Aconitum anglicum
Lloyd: Herb. Perry 1835

Bidens tripartita
Perry: Herb. Perry 1812
Galium spurium
Dunn: Herb. Warw. Mus. 1896
Glechoma hederacea
Perry: Herb. Perry 1813
Plantago major
Lloyd: Herb. Perry 1835
Plantago media
Lloyd: Herb. Perry 1835
Rorippa nasturtium-aquaticum agg.
Perry: Herb. Perry 181–
Scabiosa columbaria
Perry: Herb. Perry 181–
Senecio vulgaris
Lloyd: Herb. Perry 1835
Sisymbrium officinale
Lloyd: Herb. Perry 1835
Succisa pratensis
Perry: Herb. Perry 1812
Tussilago farfara
Perry: Herb. Perry 1813

CONTRIBUTORS OF FIRST RECORDS DURING PRESENT SURVEY

Allen, D. E.—London
Allen, H.—Coventry
Andrews, C. E. A.—Birmingham
Arnold, G. A. & M. A.—Wilnecote
Austen, Mrs. R.—Leamington
Benoit, P. M.—Barmouth, Merioneth
Bywater, Miss J.—former student at Birmingham University
Cadbury, Miss D. A.—Birmingham
Clark, M. C.—Barnt Green, Worcestershire
Cooper, Miss G.—Warwick
Cotes, R.—Beeston, Notts.
Coultas, P. G.—Rugby
Daulman, Miss B. M.—Fillongley
Dawson, R. W. B.—Warwick Museum
Dony, Mrs. C. M. (*née* Goodman)—Luton (formerly of Birmingham)
Edees, E. S.—Newcastle, Staffs.
Evans, R. E.—Stratford on Avon
Fincher, F.—Woodcote, near Bromsgrove
Green, P. S.—Kew (formerly of Birmingham University)
Hackett, A.—Hinckley, Leics.
Hall, Mr. & Mrs. P. C.—Dartford, Kent
Hanson, F. D.—Birmingham
Hardaker, W. H.—Birmingham
Hawkes, Prof. J. G.—Birmingham University
Jacobs, V.—Australia (formerly of Birmingham)
Jeffray, D. J.—Coventry
Jones, Mrs. M. D. G.—Rugby
Kiernan, Dr. J. A.—Birmingham
Laflin, T.—Leamington
Lester, Dr. R. N.—Birmingham University
Lowe, J. H.—Warwick
Meikle, R. D.—Kew
Melville, Dr. R.—Kew
Perry, D.—former student at Birmingham University
Phipps, Dr. J. B.—London, Ontario (former student at Birmingham University)
Pickvance, Mrs. E.—Birmingham
Readett, R. C.—Solihull
Roberts, H. A.—Wellesbourne
Rushforth, Miss M.—Warwick
Shack, Dr. J.—Birmingham
Skelding, Prof. A. D.—Jamaica (formerly of Birmingham University)
Smith, Dr. T. A.—former student at Birmingham University
Thomas, C.—Birmingham
Townsend, C. C.—Kew
Tyrer, Dr. F. H.—Rochdale (formerly of Birmingham)
Ward, L. E.—Warwick
Webster, Miss M. McCallum—Forres, Moray
Woolman, J. F.—Shirley
Wright, Miss P. M.—Birmingham

PTERIDOPHYTA

LYCOPSIDA

LYCOPODIACEAE

1 Lycopodium L.

1.1 **Lycopodium selago** L. Fir Clubmoss
Bree, Rev. W. T., 1817. Mid. Flora. Bog at Coleshill Pool.
Not recorded in present survey; the last record appears to be from Coleshill in 1846; recorded from all surrounding VCs except 32.

1.2 **Lycopodium inundatum** L. Marsh Clubmoss
Countess of Aylesford, 1805. Bot. Guide. Coleshill Heath.
Not recorded in present survey; last record from Coleshill Pool in 1842; recorded from VCs 37, 39, 55 and 57.

1.4 **Lycopodium clavatum** L. Stagshorn Clubmoss
Bree, Rev. W. T., 1817. Mid. Flora. Coleshill Heath.
Not recorded in present survey; last record from Sutton Park in 1884; recorded from all surrounding VCs except 33.

SPHENOPSIDA

EQUISETACEAE

4 Equisetum L.

4.1 **Equisetum hyemale** L. Rough Horsetail
Ray, 1695. Gibson's Edition of Camden's Britannia [*E. nudum* Ger.]. "We found it in moorish ditches at Middleton, towards Drayton."
Not recorded in present survey; recorded from VCs 37, 39, 55 and 57.

4.5 **Equisetum fluviatile** L. Water Horsetail
Purton, T., 1817. Mid. Flora [*E. limosum*]. Studley. Map 19, p. 242
Frequent, but very irregularly distributed throughout the county, and almost absent from the S.W.; recorded from watersides and marshes (64%) and water (36%); recorded from all surrounding VCs.

4.6 **Equisetum palustre** L. Marsh Horsetail
Purton, T., 1817. Mid. Flora. Exhall.
Map 20, p. 242
Fairly frequent but very irregularly distributed throughout the county; recorded from marshes and watersides (90%), water (5%) and bogs (4%); recorded from all surrounding VCs.

4.7 **Equisetum sylvaticum** L. Wood Horsetail
Ray, 1670. Catalogus [*E. omnium minimum tenuifolium* Park.]. "About and in the New Park, Middleton, near Tamworth."
A rare plant of wet heath, woods and hedgerows recorded from Sutton Park (0998), Earlswood (1074) where it is abundant over a wide area, Trickley Coppice (1599) and near Wasperton (2759); formerly more widespread; recorded from all surrounding VCs.

4.9 **Equisetum arvense** L. Common Horsetail
Perry, 1813. Herb. Perry. Warwick.
Map 21, p. 243
Abundant and fairly evenly distributed throughout the county; recorded from roadsides, railway banks and waste places (40%), watersides and marshes (22%), grassland (14%), hedgerows (13%) and cultivated land (8%); recorded from all surrounding VCs.

4.10 **Equisetum telmateia** Ehrh. Great Horsetail
Perry, 1820. Plant. Varv. Sel. [*E. fluviatile*]. River Avon, Nicholas Meadow.
Map 22, p. 243
An occasional plant of a number of widely scattered areas; recorded from watersides and marshes (44%), roadsides, waste places and railway banks (24%), hedgerows (18%) and woods; recorded from all surrounding VCs.

PTEROPSIDA

POLYPODIACEAE

5 Osmunda L.

5.1 **Osmunda regalis** L. Royal Fern
Ray, 1670. Catalogus [*Filix florida sive Osmunda regalis* Ger.]. "At Middleton, Warwickshire, by the pale of the new Park."
Two records of plants naturalised by the sides of lakes at Edgbaston Park (0584) and Leek Wootton Court (2868); formerly known at Sutton Park, Coleshill Heath, Marston Green and near Rugby, but now probably extinct as a native plant; recorded from all surrounding VCs except 23 and 33.

8 Pteridium Scop.

8.1 **Pteridium aquilinum** (L.) Kuhn Bracken
Ray, 1686. Hist. Plant. [*Filix foemina* Ger.]. Without locality. Map 23, p. 244
Abundant in the N.W. half of the county, but rare in the Avon Valley and on the calcareous soils of the S. and S.E.; recorded from hedgerows and scrub (41%), woods (26%), roadsides, railway banks and waste places (19%), grassland (6%) and walls and quarries (5%); recorded from all surrounding VCs.

11 Adiantum L.

11.1 **Adiantum capillus-veneris** L.
Maidenhair Fern
Goodman, Miss C. M., 1959. Kenilworth.
Several plants with other ferns found on the railway station platform (2971) on mortar between bricks at the side of the track; recorded from VCs 32 and 37.

12 Onoclea L.

12.1 **Onoclea sensibilis** L.
Cooper, Miss G., 1964. Leek Wootton.
One record of plant naturalised at Leek Wootton (2968); not recorded from any of the surrounding VCs.

13 Blechnum L.

13.1 **Blechnum spicant** (L.) Roth Hard Fern
Withering, 1801. Arrangement. In lanes about Aston Park. Map 24, p. 244
A rare plant recorded from damp woods, hedges, ditches and heaths in the N. and W. of the county; recorded from all surrounding VCs.

14 Phyllitis Hill

14.1 **Phyllitis scolopendrium** (L.) Newm. Hartstongue Fern
Perry, 1812. Herb. Perry [*Scolopendrium vulgare* Sm.]. Without locality. Map 25, p. 245
Frequent in the centre of the county, rare in the S. and almost completely absent from the N.; recorded from walls (63%), watersides (17%), woods, scrub, railway banks and roadsides; recorded from all surrounding VCs.

15 Asplenium L.

15.1a **Asplenium adiantum-nigrum** L. ssp. **adiantum-nigrum** Black Spleenwort
Purton, T., 1817. Mid. Flora. Sambourn; Middleton; Oversley. Map 26, p. 245
An occasional plant of walls (100%), thinly distributed and very rare in the extreme N. and S. of the county; recorded from all surrounding VCs.

15.5 **Asplenium trichomanes** L. Maidenhair Spleenwort
Purton, T., 1817. Mid. Flora. On Coughton Church; Walcot. Map 27, p. 246
An occasional plant of walls (100%), thinly distributed in the county but absent from the extreme N. and S.; recorded from all surrounding VCs.

15.6 **Asplenium viride** Huds. Green Spleenwort
Palmer, Miss C., 1892–93. (Bagnall: Mid. Nat. 1892–93.) "On a brick wall at Lighthorne, not planted."
Not recorded in present survey; recorded from VCs 37, 39 and 57.

15.7 **Asplenium ruta-muraria** L. Wall-rue
Purton, T., 1817. Mid. Flora. Walcot; church porch, Gt. Alne; Wixford. Map 28, p. 246
An occasional plant of walls (100%), common in the W. of the county; recorded from all surrounding VCs.

16 Ceterach DC.

16.1 **Ceterach officinarum** DC. Rusty-back Fern
Purton, T., 1817. Mid. Flora [*Scolopendrium ceterach*]. Walcot, in Haselor parish. Map 29, p. 247
An occasional plant recorded from walls in a few scattered localities in the S. half of the county; recorded from all surrounding VCs.

17 Matteuccia Tod.

17.1 **Matteuccia struthiopteris** (L.) Tod. Ostrich Fern
Hawkes, J. G., 1961. Guy's Cliffe.
One record only of several plants naturalised in shade at base of cliff (2966); not recorded from any of the surrounding VCs.

18 Athyrium Roth

18.1 **Athyrium filix-femina** (L.) Roth Lady-fern
Bree, Rev. W. T., 1821. Mid. Flora [*Aspidium filix-foemina*]. Coleshill; Allesley. Map 30, p. 247
A moderately frequent plant almost entirely confined to the centre and N. of the county; recorded from damp woods (38%), watersides and marshes (36%) and hedgerows and scrub (20%); recorded from all surrounding VCs.

19 Cystopteris Bernh.

19.1 **Cystopteris fragilis** (L.) Bernh. Brittle Bladder-fern
Cameron, D., 1843. Phytologist. Compton Verney. Map 737, p. 602
Established on damp walls in a number of places; recorded from all surrounding VCs.

21 Dryopteris Adans.

21.1 **Dryopteris filix-mas** (L.) Schott Common Male Fern
Baxter, W., 1831. MS (Bagnall: Mid. Nat. 1892–93.) Near Rugby. Map 31, p. 248
Abundant, though a little irregularly distributed throughout most of the county, but rare on the Lower Lias clays in the S.E.; recorded from hedgerows and scrub (34%), woods (32%), watersides (14%), roadsides, railway banks and waste places (12%), walls and quarries (6%) and a few other habitats; recorded from all surrounding VCs.

21.2 **Dryopteris pseudomas** (Wollaston) Holub & Pouzar Scaly Male Fern
Baxter, W., 1831. MS (Bagnall: Mid. Nat. 1892–93.) [*Lastraea filix-mas* Presl. var. *borreri* Bab.]. Dunchurch Road, Rugby. Map 32, p. 248
A rare plant of damp woods and hedgerows in a few areas in the county; collected between 1854 and 1893 from Alveston Pastures, Bearley, Bentley Park, Binley, Cathiron, Edge Hill, Maxstoke, Trickley Coppice, near Studley, Wolford Wood and Wolvey; recorded from all surrounding VCs.

21.6 **Dryopteris carthusiana** (Vill.) H. P. Fuchs Narrow Buckler-fern
Withering, 1796. Arrangement [*Polypodium spinulosum* Muller]. Bogs on Birmingham Heath. Map 33, p. 249
An occasional plant mainly confined to the W. of the county; recorded from damp woods (54%), hedgerows and scrub (23%), watersides (12%) and several other habitats; recorded from all surrounding VCs.

21.7 **Dryopteris dilatata** (Hoffm.) A. Gray Broad Buckler-fern
Bree, Rev. W. T., 1821. Mid. Flora [*Aspidium dilatatum* Sw.]. Coleshill; Allesley; Coughton Lane; Spernall. Map 34, p. 249
Frequent but somewhat irregularly distributed in the N.W. of the county but rare on the calcareous soils in the S.E. and in the Avon Valley; recorded from damp

woods (51%), hedgerows and scrub (22%), watersides (16%), roadsides and waste places (7%) and a few other habitats; recorded from all surrounding VCs.

22 **Polystichum** Roth

22.1 **Polystichum setiferum** (Forsk.) Woynar
Soft Shield-fern
Perry, 1833. Herb. Perry [*Aspidium angulare*]. Near Warwick.
An occasional plant of woods and hedgerows recorded from Alne Wood (1060, 1061), Packwood (1772), near Stratford on Avon (1952, 1953), Knowle Hall (1976), Rowington (2069), near Windmill Hill (2141), Ufton Wood (3963) and Bretford (4277); recorded from all surrounding VCs.

22.2 **Polystichum aculeatum** (L.) Roth
Hard Shield-fern
Withering, 1801. Arrangement [*Polypodium aculeatum* L.]. In a ditch at Elmdon.
Map 35, p. 250
A rare plant of a number of widely scattered localities, with some concentration in the W. central area of the county; recorded from hedgerows, watersides, woods, scrub and railway banks; recorded from all surrounding VCs.

24 **Thelypteris** Schmidel

24.1 **Thelypteris limbosperma** (All.) H. P. Fuchs
Mountain Fern
Ray, 1696. Synopsis [*Filix maris vulgaris varietas*]. "In ericeto Dunsmore dicto agri Warwicensis prope Rugby."
A rare fern of grassland on acid soil, recorded from Sutton Park (0998), Coleshill Pool (1986), Baxterley Park (2797) and Frankton Wood (4171); previously collected from Haseley Common by Perry in 1839 and near Marston Green by Bagnall in 1859; recorded from all surrounding VCs.

24.2 **Thelypteris palustris** Schott Marsh Fern
Bree, Rev. W. T., 1817. Mid. Flora [*Aspidium thelypteris*]. Allesley.
Not recorded in present survey; collected in 1851 from near Kenilworth by Bromwich, and in 1877 from Sutton Park by Bagnall; recorded from VCs 23, 32 and 39.

24.3 **Thelypteris phegopteris** (L.) Slosson
Beech Fern
Cox, T., 1848. Herb. Perry. Berkswell.
Not recorded in present survey; recorded from VCs 37, 39 and 57.

24A **Gymnocarpium** Newm.

24.4 **Gymnocarpium dryopteris** (L.) Newm. Oak Fern
Bagnall, 1866. Flora, 1891 [*Polypodium dryopteris* L.]. Darnel Hurst, Sutton Park.
Not recorded in present survey; recorded from all surrounding VCs except 32 and 33.

24.5 **Gymnocarpium robertianum** (Hoffm.) Newm.
Limestone Fern
Bagnall, 1866. Herb. Bagnall [*Polypodium robertianum* Hoffm.]. Darnel Hurst, Sutton Park.
One record of fern on mortar and bricks of platform walls at Kenilworth railway station (2971); recorded from all surrounding VCs except 37.

25 **Polypodium** L.

25.1 **Polypodium vulgare** L. sensu lato Polypody
For first record see segregates.
Map 36, p. 250
An occasional plant of a number of widely scattered localities in the county, with a concentration in the W. central area; recorded from walls (51%), hedgerows and scrub (28%), woods (14%) and a few other habitats; recorded from all surrounding VCs.
The shapes of the fronds and of the sori do not seem to be satisfactory characters for distinguishing the segregates *P. vulgare* L. s.s. and *P. interjectum* Shivas; for certain determination, it is necessary, in our opinion, to examine the annulus of the sporangium under a magnification of at least 80 diameters.

Polypodium vulgare L. sensu stricto
Bagnall, 1874. Herb. Bagnall. Sutton Park.
Map 738, p. 602
An occasional plant of walls, hedges and tree-trunks mainly in the area of the old Forest of Arden, probably under-recorded, for the reason given under *P. interjectum*; recorded from VCs 23, 32, 33 and 57.

Polypodium interjectum Shivas
Perry, 1813. Herb. Perry [*P. vulgare*]. Warwick.
Map 739, p. 602
An occasional plant of limestone walls mainly in the S. of the county, probably under-recorded owing to material not always being in a suitable state for identification; recorded from all surrounding VCs except 39.

MARSILEACEAE

26 **Pilularia** L.

26.1 **Pilularia globulifera** L. Pillwort
Purton, T., 1817. Mid. Flora. Coleshill Pool.
Not recorded in present survey but known at Coleshill Pool till 1878 and at Sutton Park from 1875 to 1898; recorded from VCs 39 and 55.

AZOLLACEAE

27 **Azolla** Lam.

27.1 **Azolla filiculoides** Lam. Water Fern
Jacobs, V., 1949. Herb. Brit. Mus. Near Solihull.
First recorded in canal near Solihull in 1949; became locally abundant in the Stratford and Birmingham canal between Earlswood and Hockley Heath and in some local ornamental pools, but has since become extinct there; was recorded from squares 0583, 1174, 1176, 1274, 1472 and 1572; recorded from VCs 23, 33 and 37.

OPHIOGLOSSACEAE

28 Botrychium Sw.

28.1 **Botrychium lunaria** (L.) Sw. Moonwort
Ray, 1670. Catalogus [*Lunaria minor* Ger.]. "In a Close in Sutton Coldfield Park."
Recorded only from rough grassland in Sutton Park (0997, 0998); formerly recorded from Coleshill, Bickenhill, Atherstone and elsewhere, but now extinct; recorded from all surrounding VCs.

29 Ophioglossum L.

29.1a **Ophioglossum vulgatum** L. ssp. **vulgatum** Adderstongue
Perry, 1820. Plant. Varv. Sel. Plantation near Saltisford Common; field near Baly's Locks, Warwick. Map 37, p. 251
An occasional plant very irregularly distributed mainly in the S. half of the county; recorded from grassland (64%), railway banks, roadsides, waste places, woods, scrub and hedgerows; recorded from all surrounding VCs.

SPERMATOPHYTA

GYMNOSPERMAE

PINACEAE

31 Picea A. Dietr.

31.1 **Picea abies** (L.) Karst. Norway Spruce
Phipps, J. B., 1954. Forshaw Heath. Map 38, p. 251
An occasional tree in many parts of the county, always planted; recorded from woods (83%), hedgerows and scrub; recorded from VCs 23, 37 and 55.

31.2 **Picea sitchensis** (Bong.) Carrière Sitka Spruce
Phipps, J. B., 1955. Near Chadshunt.
A rare tree, always planted, recorded from squares 1097, 1656, 2272, 2332, 2333, 3652 and 4162; recorded from VC 37.

32 Larix Mill.

32.1 **Larix decidua** Mill. European Larch
Wright, Miss P. M., 1953. Broom Hall. Map 39, p. 252
A frequent tree, widespread throughout the county and probably always planted; recorded from woods (62%), scrub and hedges (27%), grassland (4%) and roadsides (4%); recorded from all surrounding VCs.

33 Pinus L.

33.1 **Pinus sylvestris** L. Scots Pine
Bagnall, 1876. Flora of Sutton Park. Map 40, p. 252
Abundant and fairly evenly distributed in the N. half of the county, but much more irregularly distributed in the S. half; probably generally planted; recorded from woods (49%), hedgerows and scrub (30%), grassland (15%) and roadsides (6%); recorded from all surrounding VCs.

Pinus nigra Arnold
Shack, J., 1952. New Park Wood.
A rare tree, always planted, recorded from Sutton Park (0997, 1097, 1098) and Keresley (3183); both ssp. *nigra* and ssp. *laricio* are in Sutton Park; *P. nigra* is recorded from VCs 23, 32, 37 and 55.

TAXACEAE

35 Taxus L.

35.1 **Taxus baccata** L. Yew
Countess of Aylesford, before 1832. Herb. Oxf. Univ. Packington. Map 41, p. 253
Frequent, mainly in the N.W. half of the county, but very rare on the Lower Lias clays of the S. and S.E.; often planted, as in churchyards and some hedges, and frequently regenerating, but possibly sometimes native in woods; recorded from woods (30%), churchyards and plantations (24%), hedges and scrub (23%), grassland (11%), roadsides and waste places (7%) and watersides (5%); recorded from all surrounding VCs.

ANGIOSPERMAE

Dicotyledones

RANUNCULACEAE

36 Caltha L.

36.1 **Caltha palustris** L. Kingcup, Marsh Marigold
Perry, 1813. Herb. Perry. Without locality. Map 42, p. 253
An abundant plant widely distributed throughout the county but somewhat rare in the extreme N. and S. particularly where the soils are likely to dry out; recorded from waterside and marsh (84%), damp grassland (8%) and a few other wet habitats; recorded from all surrounding VCs.

38 Helleborus L.

38.1 **Helleborus foetidus** L. Stinking Hellebore
Purton, T., 1817. Mid. Flora. Studley Castle; Dunnington; Arrow.
A local plant of scrub and open woodland on calcareous soil now confined to Gt. Alne (1061, 1160) and Alveston Pastures (2451); previously also known from Rowington, Morton Bagot, Holywell, Shrewley and near Allesley; recorded from all surrounding VCs.

38.2 **Helleborus viridis** L. Green Hellebore
Purton, T., 1817. Mid. Flora. In a field near Studley Castle.
A local plant of woods and hedgerows on calcareous soil recorded from Spernall Park (1062), Morton Bagot (1164), Danzey (1169), Mays Wood (1464), W. of Stratford on Avon (1654), Thistly Wood (1763), Knowles Wood (1863) and Shustoke Reservoir (2391); formerly also known from Allesley, Norton Lindsey, Edge Hill and Oldbury, near Atherstone; recorded from all surrounding VCs.

39 **Eranthis** Salisb.

39.1 **Eranthis hyemalis** (L.) Salisb. Winter Aconite
Bagnall, 1866. Herb. Bagnall. Near Curdworth.
A rare plant naturalised in grassland and woodland, recorded from Wootton Wawen (1563) since destroyed by road widening, Alveston Pastures (2356), Ettington Park (2447), near Ufton (3961), Brandon Hall (4076), Ladbroke (4159), Wolston (4175), Farnborough (4249), Frankton (4270), Dunchurch (4771) and Bilton Fields (4972); recorded from all surrounding VCs.

40 **Aconitum** L.

40.1 **Aconitum anglicum** Stapf Monkshood
Lloyd, Dr. G., 1835. Herb. Perry [*A. napellus* L.]. Without locality.
A very rare plant found only in one colony on a shaded river-bank in the S. of the county near Cherington (2836, 2837); recorded from all surrounding VCs except 32.

41 **Consolida** (DC.) S. F. Gray

41.1 **Consolida ambigua** (L.) Ball & Heywood Larkspur
Purton, T., 1817. Mid. Flora [*Delphinium consolida*]. Studley, in the Castle Field.
Not recorded in present survey; recorded from all surrounding VCs.

42 **Actaea** L.

42.1 **Actaea spicata** L. Herb Christopher
Bloxam, Rev. A., 1872. (Bagnall's Flora 1891.) Harborough Magna, near Rugby.
A casual not since recorded for Warwickshire; recorded from VCs 23 and 55.

43 **Anemone** L.

43.1 **Anemone nemorosa** L. Wood Anemone
Perry, 1817. List. Wootton Grange.
Map 43, p. 254
A frequent plant of woods (52%), hedgerows and scrub (26%), roadsides (11%) and grassland (7%) especially in the area of the old Forest of Arden in the N.W. of the county; recorded from all surrounding VCs.

45 **Clematis** L.

45.1 **Clematis vitalba** L.
Traveller's Joy, Old Man's Beard
Perry, 181–. Herb. Perry. Hedges on high ground Norbrooke to Norton Lindsey.
Map 44, p. 254
Frequent on the Lower Lias limestone in the S. part of the county with scattered occurrences elsewhere; recorded from hedges and scrub (64%), railway banks, roadsides and waste places (16%) and woods (14%); recorded from all surrounding VCs.

46 **Ranunculus** L.

46.1 **Ranunculus acris** L. Meadow Buttercup
Withering, 1796. Arrangement. Pool dam at Edgbaston.
Map 45, p. 255
Abundant and evenly distributed throughout the county; recorded from grassland (59%), roadsides, waste places and railway banks (28%), hedgerows and scrub (5%), watersides and marshes (4%) and a few other habitats; recorded from all surrounding VCs.

46.2 **Ranunculus repens** L. Creeping Buttercup
Perry, 1812. Herb. Perry. Without locality.
Map 46, p. 255
Very abundant and evenly distributed throughout the county; recorded from roadsides, waste places, railway banks and farmyards (32%), grassland (29%), watersides and marshes (21%), cultivated land (8%), hedgerows and scrub (7%) and woods (3%); recorded from all surrounding VCs.

46.3 **Ranunculus bulbosus** L. Bulbous Buttercup
Perry, 1812. Herb. Perry. Without locality.
Map 47, p. 256
Abundant and evenly distributed throughout the county; recorded from grassland (65%), roadsides, railway banks and waste places (27%), hedgerows (3%) and a few other habitats; recorded from all surrounding VCs.

46.5 **Ranunculus arvensis** L. Corn Buttercup
Perry, 1812. Herb. Perry. Without locality.
Map 48, p. 256
Frequent on the calcareous soils of the S. and E. and occasional elsewhere in the county; recorded from cultivated land (79%), roadsides and waste places (16%) and a few other habitats; recorded from all surrounding VCs.

46.7 **Ranunculus sardous** Crantz Hairy Buttercup
Bromwich, H., 1891. Bagnall's Flora. A casual in cornfields at Myton.
Not recorded in present survey; recorded from VCs 37, 39, 55 and 57.

46.9 **Ranunculus parviflorus** L.
Small-flowered Buttercup
Perry, 1819. Herb. Perry. "Oversley Hill near the turnpike on the Birmingham Road, Alcester."
One record only from a sandy bank at Hatton Rock (2257); it was collected from Hatton Rock (by W. Cheshire) and from a number of other places including the roadside between Norton Lindsey and Hampton on the Hill during the last century; recorded from all surrounding VCs.

46.10 **Ranunculus auricomus** L. Goldilocks
Perry, 1813. Herb. Perry. Without locality.
Map 49, p. 257
Abundant in W. central area of the county, frequent on the calcareous soils of the S. and S.E. and occasional elsewhere; recorded from hedgerows and scrub (32%), woods (25%), roadsides (24%), grassland (16%) and watersides (3%); recorded from all surrounding VCs.

46.11 **Ranunculus lingua** L. Greater Spearwort
Freeman, S., 1842. Phytologist. Sutton Park. Map 740, p. 602
A rare plant of shallow water at the edges of lakes, ponds and canals; recorded from all surrounding VCs.

46.12a **Ranunculus flammula** L. ssp. **flammula** Lesser Spearwort
Perry, 1812. Herb. Perry [*R. flammula* L.]. Without locality. Map 50, p. 257
Frequent, particularly in the N.W. of the county, to some extent in areas which were well-wooded in the seventeenth century; recorded from marshes and watersides (78%), damp grassland (8%), damp heaths (4%), hedgerows and scrub (4%) and woods (3%); recorded from all surrounding VCs.

46.15 **Ranunculus sceleratus** L. Celery-leaved Buttercup
Perry, 1812. Herb. Perry. Without locality. Map 51, p. 258
Abundant and evenly distributed throughout the county; recorded from watersides and marshes (91%) and water (7%); recorded from all surrounding VCs.

46.16 **Ranunculus hederaceus** L. Ivy-leaved Water Crowfoot
Perry, 1820. Plant. Varv. Sel. Rare in the neighbourhood of Warwick; roadside between Hatton and Rowington. Map 52, p. 258
An occasional plant of a number of widely scattered localities, mainly in the N. of the county; recorded from watersides and marshes, and water; recorded from all surrounding VCs.

46.17 **Ranunculus omiophyllus** Ten.
Kirk, T., 1858. Topographical Botany [*R. lenormandii*]. Without locality.
A rare plant of ditches and marshes recorded only from Sutton Park (0995, 0998, 1095, 1096, 1097) where it has been known since 1859; recorded from VCs 37, 39, 55 and 57.

46.19 **Ranunculus fluitans** Lam.
Ray, 1670. Catalogus [*Ran. aquat. alb. affine Millefolium Maratriphyllon fluitans* C.B.]. "In the River Tame about Tamworth, Middleton etc."
This has always been a very rare species in the county and var. *fluitans* has not been recorded during the present survey; records of *R. fluitans*, as such, have been made from Middleton (1898), the River Blythe near Maxstoke (2286) and the River Anker near Tamworth (2305), and as var. *bachii* (Wirtgen) Wirtgen from the River Anker near Shuttington (2405) and near Polesworth (2502); earlier specimens collected by Bagnall between 1872 and 1881 have been determined as *R. fluitans* from near Coleshill and from Hampton in Arden, as var. *fluitans* from near Coleshill, and as var. *bachii* from near Coleshill, Hampton in Arden and Witton. *R. fluitans* has been recorded from all the surrounding VCs.

Hybrids of *R. fluitans* with another species, probably *aquatilis* but perhaps *trichophyllus* or *peltatus* (according to Prof. Cook), have been recorded from a pool near Newbold on Stour (2245) and from rivers at Gt. Alne (1159), Pype Hayes Park (1292, 1392), Shelly Coppice (1576), Copt Heath (1677), near Stratford on Avon (1853), Middleton (1898), Bradnocks Marsh (2179) and Hampton in Arden (2181); similar plants were collected by Baker from Radford in 1881, and by Bagnall from Lowsonford in 1887 and from Baginton in 1888.

46.20 **Ranunculus circinatus** Sibth.
Baxter, W., 1831. MS (Bagnall: Mid. Nat. 1892–93). Near Brownsover and Rugby Mill. Map 53, p. 259
A rare plant of a number of scattered localities mainly in the N. and E. of the county; recorded from ponds and canals; recorded from all surrounding VCs.

46.21 **Ranunculus trichophyllus** Chaix
Palmer, Miss C. E., 1851. (Bagnall: Mid. Nat. 1892–93.) Lighthorne. Map 54, p. 259
An occasional plant in the S. and E. of the county, almost entirely on calcareous soils; recorded from water (82%) and watersides and marshes (18%); recorded from all surrounding VCs.

46.22a **Ranunculus aquatilis** L.
Perry, 1812. Herb. Perry [*R. heterophyllus* Weber]. Without locality. Map 55, p. 260
Frequent and widely but irregularly distributed throughout the county; recorded from water (71%) and watersides (27%); recorded from all surrounding VCs.

Ranunculus aquatilis L. × **trichophyllus** Chaix
Cadbury, Miss D. A., 1950. Near Fenny Compton.
Three records only from ponds S. of Knowle (1875), at Gospel Oak (2693) and E. of Fenny Compton (4351).

46.22b **Ranunculus peltatus** Schrank
Kirk, T. 1856. Herb. Brit. Mus. [*R. floribundus* Bab.]. Stoke, near Coventry. Map 56, p. 260
An occasional plant of a number of scattered localities almost entirely in the N. half of the county; recorded mainly from ponds and rivers; recorded from all surrounding VCs.

46.22c **Ranunculus penicillatus** (Dumort.) Bab.
(= *R. pseudofluitans* Syme)
Bagnall, 1876. Herb. Bagnall [*R. pseudofluitans* Bab.]. River Blythe near Solihull.
A rare plant of ponds and slow-moving rivers recorded from Tanners Green (0874), near Wootton Wawen (1664), near Dosthill (2099) and Alvecote Pool (2404); *R. pseudofluitans* has been recorded from all surrounding VCs.

46.24 **Ranunculus ficaria** L. Lesser Celandine
For first records see sub-species. Map 57, p. 261
Abundant and evenly distributed throughout the county; recorded from hedgerows and scrub (25%), watersides (22%), grassland (20%), roadsides and waste places (17%) and woods (15%); recorded from all surrounding VCs.
The records for the two sub-species, particulars of which follow, have been included in this map.

Ssp. **ficaria**
Perry, 1813. Herb. Perry [*R. ficaria*]. Without locality. Map 58, p. 261
Frequent but irregularly distributed throughout the county; probably under-recorded; recorded from grassland (29%), hedgerows and scrub (27%), watersides (19%), woods (13%) and roadsides (12%); recorded from all surrounding VCs.

Ssp. **bulbifer** (Marsden-Jones) Lawalrée
West, G. S., 1909. Herb. B'ham Univ. [*R. ficaria*]. Coleshill. Map 59, p. 262
Frequent but very irregularly distributed throughout the county; probably under-recorded; recorded from hedgerows and scrub (35%), watersides (20%), grassland (17%), woods (17%) and roadsides (10%), thus showing, in relation to ssp. *ficaria*, a slight preference for more shady habitats; recorded from all surrounding VCs.

48 Myosurus L.

48.1 **Myosurus minimus** L. Mousetail
Countess of Aylesford, 1805. Bot. Guide. Chelmsley Wood. Map 60, p. 262
A rare plant mainly of the lower Avon Valley; recorded chiefly from cultivated land; recorded from all surrounding VCs.

49 Aquilegia L.

49.1 **Aquilegia vulgaris** L. Columbine
Bree, Rev. W. T., 1817. Mid. Flora. Corley Woods. Map 741, p. 603
A rare plant of hedgerows, roadsides and railway banks, probably always an escape; recorded from all surrounding VCs.

50 Thalictrum L.

50.1 **Thalictrum flavum** L. Common Meadow-rue
Purton, T., 1817. Mid. Flora. Banks of the Avon at Bidford. The Arrow near Beauchamp Court. Map 61, p. 263
An occasional plant whose distribution follows the course of the Rivers Tame and Avon and their tributaries; recorded from riversides and ditchsides; recorded from all surrounding VCs.

BERBERIDACEAE

53 Berberis L.

53.1 **Berberis vulgaris** L. Barberry
Perry, 1817. List. Leek Wootton. Map 62, p. 263
A rare plant confined to the Arrow district, where it has been known since Purton's time, and a narrow belt along the N.E. and E. side of the county; considered by Purton and Bagnall to be native, at least in the Arrow district; recorded from hedgerows (100%); recorded from all surrounding VCs.

54 Mahonia Nutt.

54.1 **Mahonia aquifolium** (Pursh) Nutt. Oregon Grape
Hawkes, J. G., 1955. Ufton Wood. Map 63, p. 264
Frequently planted and sometimes escaping; naturalised in a number of scattered localities, but rare in the S. and in the extreme N. of the county; recorded from hedges and scrub (54%), woods (25%), roadsides, railway banks and waste places (11%) and one or two other habitats; recorded from all surrounding VCs.

NYMPHAEACEAE

55 Nymphaea L.

55.1 **Nymphaea alba** L. White Water-lily
Withering, 1787. Arrangement. Tamworth. Map 64, p. 264
A rare plant of a number of very widely scattered localities throughout much of the county, but almost completely absent from the calcareous soils of the S. and S.E.; recorded from ponds, rivers, canals and lakes; recorded from all surrounding VCs.

56 Nuphar Sm.

56.1 **Nuphar lutea** (L.) Sm. Yellow Water-lily, Brandy-bottle
Baxter, W., 1831. MS (Bagnall: Mid. Nat. 1892–93.) Rugby Mill and near Brownsover. Map 65, p. 265
An occasional plant of a number of scattered localities in the county with considerable concentration in the Avon Valley; recorded from rivers, canals, ponds and lakes; recorded from all surrounding VCs.

CERATOPHYLLACEAE

57 Ceratophyllum L.

57.1 **Ceratophyllum demersum** L. Spined Hornwort
Perry, 1835. Herb. Perry. Mill Pool, Chesterton. Map 66, p. 265
A rare plant recorded from water in a few scattered localities; recorded from all surrounding VCs.

PAPAVERACEAE

58 Papaver L.

58.1 **Papaver rhoeas** L. Common Poppy
Baxter, W., 1831. MS (Bagnall: Mid. Nat. 1892–93.) Clifton Road near Rugby.
Lloyd, Dr. G., 1835. Herb. Perry. Leamington. Map 67, p. 266
Abundant and fairly evenly distributed in the S.E. half of the county, but more irregularly distributed in the N.W.; recorded from roadsides, waste places, railway banks and farmyards (46%), cultivated land (42%), grassland (7%) and a few other habitats; recorded from all surrounding VCs.

58.2 **Papaver dubium** L. Long-headed Poppy
Baxter, W., 1831. MS (Bagnall: Mid. Nat. 1892–93.) Between Rugby and Sawbridge.
Ick, W., 1837. Analyst. Near Aston Church. Map 68, p. 266
Abundant and fairly evenly distributed throughout much of the county, but rather less frequent in the central N. area and in the extreme S.; recorded from roadsides, waste places, railway banks and farmyards (58%), cultivated land (33%), grassland (6%) and a few other habitats; recorded from all surrounding VCs.

58.3 **Papaver lecoqii** Lamotte Babington's Poppy
Bromwich, H., 1867. Herb. Warw. Mus. Cornfield, Ashorne. Map 69, p. 267
An occasional plant mainly on calcareous soils in the S. half of the county, but absent from the extreme S.; recorded from cultivated land (53%), waste places and roadsides (40%); recorded from all surrounding VCs.

58.5 **Papaver argemone** L. Long Prickly-headed Poppy
Bree, Rev. W. T., 1830. Mag. Nat. Hist. Claverdon. Map 70, p. 267
An occasional plant of well-drained soils, chiefly on limestone, sands and gravel, with some concentration in the Tamworth, Stratford and Rugby areas; recorded from cultivated land (50%), waste places, roadsides and railway banks (47%); recorded from all surrounding VCs.

58.6 **Papaver somniferum** L. Opium Poppy
Baxter, W., 1831. Phaen. Bot., 1834. Cornfields near the road going from Rugby to Barby, and on Jarrett's Heath near Rugby. Map 71, p. 268
An occasional plant widely distributed, generally occurring as an escape from cultivation; recorded from waste places and roadsides (73%), cultivated land (16%) and grassland (9%); recorded from all surrounding VCs.

59 **Meconopsis** Vig.

59.1 **Meconopsis cambrica** (L.) Vig. Welsh Poppy
Bagnall, 1869. Flora, 1891. Near Rowington Hall.
Not recorded in present survey; recorded from all surrounding VCs except 33.

61 **Glaucium** Mill.

61.1 **Glaucium flavum** Crantz Yellow Horned-Poppy
Lees, E., 1867. Botany of Worcestershire [*G. luteum*]. Shipston-on-Stour.
Not recorded in present survey; recorded from VC 37.

62 **Chelidonium** L.

62.1 **Chelidonium majus** L. Greater Celandine
Baxter, W., 1831. MS (Bagnall: Mid. Nat. 1892–93.) Between Hill Morton and the Engine.
Lloyd, Dr. G., 1835. Herb. Perry. Warwick. Map 72, p. 268
An abundant plant of irregular distribution throughout the county, especially near houses, but rare on the Lower Lias clays in the S.E.; recorded from roadsides and waste places (51%), hedgerows (29%), watersides (7%), walls (6%) and a few other habitats; recorded from all surrounding VCs.

FUMARIACEAE

65 **Corydalis** Medic.

65.1 **Corydalis solida** (L.) Sw.
Purton, T., 1821. Mid. Flora [*Fumaria solida*]. Near Studley Castle.
Naturalised on roadsides at Dunnington (0653) and Mappleborough Green (0866, 0867); recorded from all surrounding VCs except 57.

65.3 **Corydalis claviculata** (L.) DC. White Climbing Fumitory
Withering, 1787. Arrangement [*Fumaria claviculata*]. "In some of the least frequented roads about Birmingham."
Perry, 1817. List [*Fumaria claviculata*]. Gravelly Hill, near Birmingham.
Not recorded in present survey; recorded from all surrounding VCs except 32 and 33.

65.4 **Corydalis lutea** (L.) DC. Yellow Fumitory
Perry, 1812. Herb. Perry. Without locality. Map 73, p. 269
A garden escape naturalised on walls and by roadsides and hedges in a number of scattered localities in the county; recorded from all surrounding VCs.

66 **Fumaria** L.

66.2 **Fumaria capreolata** L. Ramping Fumitory
Baxter, W., 1831. MS (Bagnall: Mid. Nat. 1892–93.) Rugby.
Not recorded in present survey; recorded from VCs 23, 37, 39 and 57.

66.4 **Fumaria bastardii** Bor.
Bagnall, 1874. Flora, 1891 [*F. confusa* Jord.]. Without locality.
Not recorded in present survey; recorded from VCs 23 and 37.

66.6b **Fumaria muralis** Sond. ex Koch ssp. **boraei** (Jord.) Pugsl.
Anon., 1872. Herb. B'ham Univ. Honiley.
Not recorded in present survey; recorded from all surrounding VCs except 32; ssp. *muralis* is recorded from VCs 37 and 39.

66.7 **Fumaria densiflora** DC.
Roberts, H. A., 1959. Wellesbourne Hastings.
Three records from cultivated fields near Wellesbourne Hastings (2656, 2755, 2756); recorded from VCs 23 and 37.

66.8 **Fumaria officinalis** L. Common Fumitory
Perry, 1812. Herb. Perry. Without locality. Map 74, p. 269
Abundant but somewhat irregularly distributed throughout the county; recorded from cultivated land (59%), waste places, roadsides and railway sides (36%) and a few other habitats; recorded from all surrounding VCs.

CRUCIFERAE

67 **Brassica** L.

67.2 **Brassica napus** L. Rape
Purton, T., 1817. Mid. Flora. Arrow, on a hedge bank bordering a cornfield on the road leading to Cookhill. Map 75, p. 270
An occasional plant irregularly distributed throughout the county; recorded from arable land (50%), roadsides and waste places (32%) and banks of streams (11%); recorded from all surrounding VCs.

67.3 **Brassica rapa** L. Navew, Wild Turnip
Lloyd, Dr. G., 1833. Herb. Perry (Bagnall's Flora, 1891). Without locality. Map 76, p. 270
An occasional plant irregularly distributed throughout

the county; recorded from arable land (44%), waste places, farmyards, roadsides and railway banks (30%), watersides (18%) and grassland (8%); recorded from all surrounding VCs.

67.4 **Brassica nigra** (L.) Koch Black Mustard
Purton, T., 1817. Mid. Flora [*Sinapis nigra*]. In a field at Exhall near to Rosall.
Map 77, p. 271
A frequent plant on the calcareous soils in the S.W. of the county but rare elsewhere; recorded from waste places and roadsides (40%), arable land (32%), by the sides of streams and ditches (14%), grassland (8%) and hedgerows (6%); recorded from all surrounding VCs.

68 **Erucastrum** C. Presl

Erucastrum nasturtiifolium (Poir.) O. E. Schulz
Bagnall, 1877. Flora 1891 [*Brassica erucastrum* Vill.]. Sutton Park.
Not recorded in present survey; recorded from VCs 33 and 37.

69 **Rhynchosinapis** Hayek

69.3 **Rhynchosinapis cheiranthos** (Vill.) Dandy
Wallflower Cabbage
Clark, M. C. 1963. Bidford on Avon.
A casual recorded once only from the railway track near Bidford on Avon (1152); recorded from VCs 23, 33 and 37.

70 **Sinapis** L.

70.1 **Sinapis arvensis** L. Charlock, Wild Mustard
Lloyd, Dr. G., 1835. Herb. Perry. Leamington.
Map 78, p. 271
Abundant and evenly distributed throughout the county; recorded from cultivated land (52%), roadsides, waste places and farmyards (36%), grassland (8%) and a few other habitats; recorded from all surrounding VCs.

70.2 **Sinapis alba** L. White Mustard
Purton, T., 1817. Mid. Flora. Grafton. Map 79, p. 272
An occasional introduced plant of widely scattered distribution mainly in the S.E. half of the county, with some concentration on well-drained soils; recorded from cultivated land, roadsides, waste places, farmyards and rough grassland; recorded from all surrounding VCs.

72 **Diplotaxis** DC.

72.1 **Diplotaxis muralis** (L.) DC.
Annual Wall Rocket, Stinkweed
Bromwich, H., 1872. Herb. Brit. Mus. Railway embankment near Myton (New Red Sandstone); also Harbury (Lias soils). Map 80, p. 272
An introduced plant naturalised in a number of places in the middle of the county, and recorded from railway tracks, walls, waste places, gardens and roadsides; recorded from all surrounding VCs.

72.2 **Diplotaxis tenuifolia** (L.) DC.
Perennial Wall Rocket
Purton, T., 1817. Mid. Flora [*Sisymbrium tenuifolium*]. Kinwarton.
One record only from a railway bank near Wootton Wawen (1561); recorded from all surrounding VCs except 57.

73 **Eruca** Mill.

73.1 **Eruca sativa** Mill.
Bromwich, H., 1893. Herbs. Brit. Mus. and B'ham Univ. [*Brassica erucastrum* Vill.]. Railway bank, Milverton.
Not recorded in present survey; recorded from VCs 23, 32, 33, 55 and 57.

74 **Raphanus** L.

74.1 **Raphanus raphanistrum** L. Wild Radish
Baxter, W., 1831. MS (Bagnall: Mid. Nat. 1892–93.) By the side of the footpath, West Leys to Newbold on Avon, and in the Clifton Road near Rugby. Map 81, p. 273
Abundant and fairly evenly distributed throughout the county; recorded from cultivated land (52%), waste places, roadsides and farmyards (39%) and grassland (7%); yellow and white flowered variants have been recorded; recorded from all surrounding VCs.

76 **Rapistrum** Crantz

76.2 **Rapistrum rugosum** (L.) All.
Cadbury, Miss D. A., 1957. Manor Fields Farm near Stoneleigh.
A rare casual of cultivated and waste places recorded from Bearley Station Yard (1760), Manor Fields Farm (3273), Blue Lias Inn (4264) and Bilton Place (4874); recorded from all surrounding VCs.

77 **Cakile** Mill.

77.1 **Cakile maritima** Scop. Sea Rocket
Benoit, P. M., 1963. Near Warwick.
Two records, one on introduced sea-sand (2965) and one on a colliery dump (3585); not recorded from any of the surrounding VCs.

78 **Conringia** Heist. ex P.C. Fabr.

78.1 **Conringia orientalis** (L.) Dumort.
Hare's-ear Mustard
Bromwich, H., 1860. Herb. Perry [*Erysimum orientale* (L.) Cr.]. Myton.
Not recorded in present survey; recorded from all surrounding VCs except 37.

79 **Lepidium** L.

79.1 **Lepidium sativum** L. Garden Cress
Bagnall, 1877. Herb. Bagnall. Witton Road, Aston.
An occasional escape from cultivation recorded three times in present survey; recorded from VCs 23, 37 and 55.

79.2 **Lepidium campestre** (L.) R. Br.
Field Pepperwort
Perry, 181–. Herb. Perry. Whitnash Road from Warwick. Map 82, p. 273
An occasional plant almost entirely in the S. of the county with some concentration in the S.W.; recorded from cultivated fields (46%), waste places, railway banks, roadsides and farmyards (34%) and grassland (18%); recorded from all surrounding VCs.

79.3 **Lepidium heterophyllum** Benth.
Smith's Pepperwort
Bree, Rev. W. T., 1835. New Bot. Guide [*L. smithii*]. Without locality.
Perry, 1839. Herb. Perry [*L. smithii* Hook.]. Temple Balsall.

One record from rough grassland at Brandon Wood (3976) and one from farmyard at Bourton on Dunsmore (4470); recorded from all surrounding VCs.

79.4 **Lepidium ruderale** L.
Narrow-leaved Pepperwort
Bromwich, H., 1876. Bagnall: Flora, 1891. Knowle Hill, Kenilworth. Map 83, p. 274

A rare plant of a number of scattered localities mainly in the E. of the county, with a few occurrences in the W.; recorded from waste places (100%); recorded from all surrounding VCs.

Lepidium virginicum L.
Daulman, Miss B., 1961. Red Hill Farm.

One record from a farmyard at Red Hill Farm (2986); recorded from VCs 23, 33, 37 and 55.

Lepidium perfoliatum L.
Rugby School Report, 1923. Hillmorton Ballast Pits.

Not recorded in present survey; recorded from VCs 23, 33, 37 and 55.

80 Coronopus Zinn

80.1 **Coronopus squamatus** (Forsk.) Aschers.
Swine-cress, Wart-cress
Baxter, W., 1831. MS (Bagnall: Mid. Nat. 1892–93) [*Senebiera coronopus* Pers.]. Near Brownsover. Map 84, p. 274

Abundant except in the N.E. quarter of the county, where it is rare; apparently the N.W. limit of its main British distribution runs through Warwickshire; recorded from roadsides, farmyards and waste places (65%), arable land (15%), grassland (13%) and watersides (especially ponds) (6%); recorded from all surrounding VCs.

80.2 **Coronopus didymus** (L.) Sm. Lesser Swine-cress
Bromwich, H., before 1891. Bagnall's Flora [*Senebiera didyma* Pers.]. Garden weed, Myton. Map 85, p. 275

A rare plant of a number of scattered localities in the county; recorded from waste places, roadsides and farmyards (75%) and arable land; recorded from all surrounding VCs except 32.

81 Cardaria Desv.

81.1 **Cardaria draba** (L.) Desv. Hoary Cress
Bagnall, 1876. Herb. Bagnall [*Lepidium draba* L.]. Ward End. Map 86, p. 275

An occasional plant of a number of localities scattered throughout the county but almost entirely absent from the S.E.; recorded from waste places, roadsides and railway banks (59%), grassland (15%), arable land (15%) and quarries (11%); recorded from all surrounding VCs except 39.

83 Iberis L.

83.1 **Iberis amara** L. Wild Candytuft
Kirk, T., 1847. Phytologist. Exhall, near Coventry.

Not recorded in present survey; recorded from all surrounding VCs except 37.

84 Thlaspi L.

84.1 **Thlaspi arvense** L. Field Penny-cress
Ray, 1670. Catalogus [*T. Dioscoridis* Ger.]. Kingsbury. Map 87, p. 276

A plant abundant in the basins of the Avon and Arrow, with rare occurrences elsewhere in the county; recorded from cultivated land (65%), waste places, farmyards and roadsides (31%) and grassland (4%); recorded from all surrounding VCs.

85 Teesdalia R. Br.

85.1 **Teesdalia nudicaulis** (L.) R. Br.
Shepherd's Cress
Countess of Aylesford, 1805. Bot. Guide [*Iberis nudicaulis* L.]. "By the side of the road, near Packington, where it divides to Coleshill and Castle Bromwich."

Two records, one from a waste place at Wellesbourne Wood (2653) and one from the railway track at Flecknoe (5263); previously collected from Sutton Park, near Marston Green and Baddesley Common; recorded from all surrounding VCs except 33.

86 Capsella Medic.

86.1 **Capsella bursa-pastoris** (L.) Medic.
Shepherd's Purse
Baxter, W., 1831. MS (Bagnall: Mid. Nat. 1892–93.) Rugby. Map 88, p. 276

An abundant plant evenly distributed throughout the county; recorded from roadsides and waste places (50%), cultivated land (37%) and grassland (10%); recorded from all surrounding VCs.

88 Cochlearia L.

88.5 **Cochlearia danica** L. Danish Scurvy-grass
Wallace, E. C., 1946. Bot. Soc. & Exch. Club Report 1946–47. Railway track outside Rugby Station. Map 742, p. 603

A fairly recent introduction confined to railway tracks and becoming more widespread; recorded from VCs 32, 33, 39 and 55.

Neslia Desv.

Neslia paniculata (L.) Desv.
Dunn, S. T., 1897. Flora of the Rugby District; D. E. Allen, 1957. Railway cutting at Rugby.

A bird-seed alien recorded once from Rugby (4974); recorded from VCs 23, 32 and 33.

Lunaria L.

Lunaria annua L. Honesty
Hardaker, W. H., 1957. Lower Brailes.

A garden escape recorded eight times from roadsides and waste places; recorded from VCs 23, 33 and 57.

91 Alyssum L.

91.1 **Alyssum alyssoides** (L.) L. Small Alison
Bromwich, H., 1891. (Bagnall's Flora) [*A. calycinum* L.]. Myton.

A casual not found in present survey; recorded from all surrounding VCs.

92 **Lobularia** Desv.

92.1 **Lobularia maritima** (L.) Desv. Sweet Alison
Thomas, C., 1950. Gillhurst Road, Harborne, Birmingham.

An escape from cultivation recorded eight times from roadsides and waste places in the vicinities of Birmingham, Coventry, Rugby and Warwick; recorded from VCs 23, 33, 37 and 55.

93 **Berteroa** DC.

93.1 **Berteroa incana** (L.) DC. Hoary Alison
Bagnall, 1874. Flora, 1891 [*Alyssum incanum* L.]. Boldmere, near Sutton.

Not recorded during present survey; recorded from all surrounding VCs except 32.

94 **Draba** L.

94.4 **Draba muralis** L. Wall Whitlow-grass
Thomas, F. and Beers, A. G., before 1930. Flora of the Rugby District; D. E. Allen, 1957. Wall near Little Lawford Hall.

Introduced but well established on walls at Little Compton (2630); recorded from all surrounding VCs except 32 and 37.

95 **Erophila** DC.

95.1 **Erophila verna** (L.) Chevall.
Common Whitlow-grass
Countess of Aylesford, 1809. Herb. Oxf. Univ. [*Draba verna* L.]. Packington.
Map 89, p. 277

An occasional plant, mainly in the S. half of the county with particular concentration in the S.W.; recorded from railway sides and yards, roadsides and waste places (66%), walls (29%) and a few other habitats; recorded from all surrounding VCs.

95.2 **Erophila spathulata** Láng
Round-podded Whitlow-grass
Allen, H., 1952. Butlers Marston.

One record only from the railway at Butlers Marston (3250); recorded from all surrounding VCs except 37 and 55.

95.3 **Erophila praecox** (Stev.) DC.
Bagnall, 1874. Flora, 1891. Bedlam's End and Harbury Village.

Not recorded in present survey; recorded from all surrounding VCs except 33 and 39.

96 **Armoracia** Gilib.

96.1 **Armoracia rusticana** Gaertn., Mey. & Scherb.
Horse-radish
Purton, T., 1817. Mid. Flora [*Cochlearia armoracia*]. On the River Arrow near Oversley Bridge.
Map 90, p. 277

A naturalised plant abundant and evenly distributed throughout the county with the exception of the belt of Lias clay in the S.E. where it is rare; recorded from roadsides, railway banks, farmyards and waste places (72%), grassland (14%), hedgerows (7%) and cultivated land (5%); recorded from all surrounding VCs.

97 **Cardamine** L.

97.1 **Cardamine pratensis** L.
Cuckoo Flower, Lady's Smock
Baxter, W., 1831. MS (Bagnall: Mid. Nat. 1892–93.) Near Lawford.
Lloyd, Dr. G., 1835. Herb. Perry. Leamington.
Map 91, p. 278

An abundant plant evenly distributed throughout the county; recorded from watersides and marshes (49%), damp grassland (37%) and roadsides (7%); recorded from all surrounding VCs.

97.2 **Cardamine amara** L. Large Bittercress
Ray, 1670. Catalogus [*Nasturtium aquaticum amarum* Park.]. "In boggy and watery places, near Middleton."
Map 92, p. 278

An occasional plant scattered throughout the county with the exception of the calcareous soils in the S.E. where it is rare; recorded from watersides (90%) and a few other habitats; recorded from all surrounding VCs.

97.3 **Cardamine impatiens** L.
Narrow-leaved Bittercress
Bree, Rev. W. T., 1835. New Bot. Guide. Without locality.

A rare plant of diorite recorded from quarries in two localities S. of Atherstone (3095, 3096); recorded from all surrounding VCs except 23 and 55.

97.4 **Cardamine flexuosa** With. Wood Bittercress
Withering, 1796. Arrangement. "Rookery at Edgbaston, and in ditches at the tail of the Pool."
Map 93, p. 279

An abundant plant fairly evenly distributed throughout the county except on the calcareous clays in the S.E. where it is uncommon; recorded from watersides and marshes (78%), damp woods (8%), scrub and hedgerows (5%) and grassland (5%); recorded from all surrounding VCs.

97.5 **Cardamine hirsuta** L. Hairy Bittercress
Ray, 1690. Synopsis [*Cardamine impatiens altera hirsutior*]. "This is very common in Warwickshire, in gardens and moist places, flowering in the beginning of the spring."
Map 94, p. 279

A frequent plant of irregular distribution recorded from railways, roadsides and waste places (59%), gardens and cultivated fields (26%) and wall tops (10%); recorded from all surrounding VCs.

98 **Barbarea** R. Br.

98.1 **Barbarea vulgaris** R. Br. Common Wintercress
Baxter, W., 1831. MS (Bagnall: Mid. Nat. 1892–93.) Near Sawbridge.
Map 95, p. 280

An abundant plant, fairly evenly distributed throughout the county except for an area in the N.W. where it is less common; recorded from watersides (66%), roadsides and waste places (17%), hedgerows (7%) and grassland (7%); recorded from all surrounding VCs.
B. vulgaris var. *arcuata* (Opiz) Fr. (first record Bromwich, 1864. Herb. Brit. Mus. River Leam, Leamington) has also been recorded.

98.3 **Barbarea intermedia** Bor.
Medium-flowered Wintercress
Rugby School Report, 1874. Rugby district.
Map 96, p. 280
An occasional plant of arable land (57%), watersides (24%) and roadsides (14%); recorded from all surrounding VCs except 32 and 39.

98.4 **Barbarea verna** (Mill.) Aschers.
American Wintercress, Land Cress
Bloxam, Rev. A., 1837. New Bot. Guide [*B. praecox*]. Near Rugby, on rubbish probably brought from a garden. Map 97, p. 281
A rare plant of arable land, railways and waste places; recorded from all surrounding VCs.

100 Arabis L.

100.4 **Arabis hirsuta** (L.) Scop. Hairy Rockcress
Kirk, T., 1852. Herb. Perry (Bagnall: Flora 1891). Old walls, Allesley. According to Bagnall, believed to have been introduced by the Rev. W. T. Bree.
One record in limestone quarry near Little Compton (2628) and one in waste place near Wellesbourne Wood (2653); recorded from all surrounding VCs.

101 Turritis L.

101.1 **Turritis glabra** L. Tower Mustard
Ray, 1670. Catalogus [*T. vulgatior* Park.]. "Supra Dorsthill-hill prope Middleton in agro Warwicensi."
Not recorded in present survey; collected by Bagnall at Marston Green in 1868 but not seen since; recorded from all surrounding VCs except 32.

102 Rorippa Scop.

102.1 & 2 **Rorippa nasturtium-aquaticum** sensu lato
Watercress
Perry, 181–. Herb. Perry [*Nasturtium officinale*]. Without location. Map 98, p. 281
Abundant and fairly evenly distributed throughout the county; recorded from watersides and marshes (64%) and water (36%); recorded from all surrounding VCs.

102.1 **Rorippa nasturtium-aquaticum** (L.) Hayek
Edees, E. S., 1951. Herb. Brit. Mus. Marshy ground by side of pool near Blake Street station.
Probably most of the records included in the map of the aggregate are of this species but material determined as *R. nasturtium-aquaticum* sensu stricto has been collected only from the following squares: 1898, 2404, 2550, 3072, 3252, 3785, 3788, 3951, 3983 and 5165; recorded from all surrounding VCs.

102.2 **Rorippa microphylla** (Boenn.) Hyland.
One-rowed Watercress
Palmer, Miss C. E., 1852. Herb. Oxf. Univ. [*Nasturtium officinale* R. Brown]. Lighthorne.
Map 99, p. 282
An occasional plant of a number of widely scattered localities, but almost absent from the centre of the county; recorded from watersides (80%) and water (20%); possibly under-recorded; recorded from all surrounding VCs.
Records for this species are also included in the map of the aggregate species.

Rorippa microphylla × **nasturtium-aquaticum** = **R.** × **sterilis** Airy Shaw
Bromwich, H., 1874. Herb. Warw. Mus. [*Nasturtium officinale*]. Without location.
Rare and generally found in presence of parents; recorded from squares 1280, 1664, 1681, 2086, 2104, 2185, 2284, 3290, 3294, 3487, 4388, 4470, 4560, 4663, 4976 and 4977; recorded from all surrounding VCs.

102.3 **Rorippa sylvestris** (L.) Bess.
Creeping Yellow-cress
Bree, Rev. W. T., 1835. New Bot. Guide [*Nasturtium sylvestre*]. Without locality.
Map 100, p. 282
Thinly and irregularly scattered throughout the county; recorded from watersides (85%) and a few other localities; recorded from all surrounding VCs.

102.4 **Rorippa islandica** (Oeder) Borbás
Marsh Yellow-cress
Perry, 1817. List [*Sisymbrium terrestre*]. Baginton Bridge; near the race stand, Warwick. Map 101, p. 283
Frequent though irregularly distributed throughout most of the county, but almost entirely absent from the calcareous soils of the S.; recorded from watersides (86%), waste places and roadsides (11%) and a few other habitats; recorded from all surrounding VCs.

102.5 **Rorippa amphibia** (L.) Bess.
Great Yellow-cress
Withering, 1787. Arrangement [*Sisymbrium amphibium*]. Side of the river and wet ditches at Tamworth. Map 102, p. 283
Frequent though irregularly distributed throughout much of the county, but absent from the extreme S.; recorded from watersides (86%) and water (14%); recorded from all surrounding VCs.

104 Hesperis L.

104.1 **Hesperis matronalis** L. Dame's Violet
Lloyd, Dr. G., 1835. Herb. Perry. Tachbrook Mallory near Warwick. Map 103, p. 284
An occasional plant naturalised in a number of areas throughout the county; recorded from roadsides, waste places and railway banks (50%), watersides, hedgerows and a few other habitats; recorded from all surrounding VCs.

105 Erysimum L.

105.1 **Erysimum cheiranthoides** L. Treacle Mustard
Bromwich, H., 1860. Herb. Perry. Myton.
Map 104, p. 284
A rare plant of a few widely scattered localities in the county; recorded from waste places, rough grassland, arable land and quarries; Warwickshire appears to be outside the area of the main British distribution of this species; recorded from all surrounding VCs.

Erysimum repandum L.
Bromwich, H., 1900. Herb. Brit. Mus. Milverton.
Not recorded in present survey; recorded from VCs 23 and 33.

106 Cheiranthus L.

106.1 **Cheiranthus cheiri** L. Wallflower
Perry, 1820. Plant. Varv. Sel. [*C. fruticulosus*]. Walls at Warwick.
A rare plant of old walls, recorded from 0866, 1652,

2289 (Maxstoke Castle), 2772 (Kenilworth Castle), 2865 (Warwick) and 4245 (Shotteswell); recorded from all surrounding VCs.

107 **Alliaria** Heist. ex P.C. Fabr.

107.1 **Alliaria petiolata** (Bieb.) Cavara & Grande Garlic Mustard, Jack-by-the-hedge
Baxter, W., 1831. MS (Bagnall: Mid. Nat. 1892–93) [*Sisymbrium alliaria* Scop.]. Hill Morton. Map 105, p. 285
An abundant plant of hedgerows (54%), roadsides and waste places (25%), ditches and riversides (12%) and woods (4%), evenly distributed throughout the county; recorded from all surrounding VCs.

108 **Sisymbrium** L.

108.1 **Sisymbrium officinale** (L.) Scop. Hedge Mustard
Lloyd, Dr. G., 1835. Herb. Perry. Without locality. Map 106, p. 285
Abundant and fairly evenly distributed throughout the county; recorded from roadsides, waste places, farmyards and railway banks (63%), cultivated land (13%), hedgerows (12%), grassland (8%) and a few other habitats; recorded from all surrounding VCs.

108.4 **Sisymbrium orientale** L. Eastern Rocket
Burges, Dr. R. C. L., 1934. Herb. Burges. Edgbaston. Map 107, p. 286
An occasional plant of a number of widely scattered localities throughout the county but absent from the extreme S.; recorded from waste places, railway banks and roadsides (82%), cultivated land (11%) and a few other habitats; recorded from all surrounding VCs.

108.5 **Sisymbrium altissimum** L. Tall Rocket
Bromwich, H., 1894. Herb. B'ham Univ. Near Milverton Railway Station. Map 108, p. 286
An occasional plant very irregularly distributed in the N. half of the county, mainly around Birmingham and Coventry; recorded from waste places and roadsides (92%); recorded from all surrounding VCs.

109 **Arabidopsis** (DC.) Heynh.

109.1 **Arabidopsis thaliana** (L.) Heynh. Thalecress
Purton, T., 1817. Mid. Flora [*Arabis thaliana*]. Coughton; Sambourn; Oversley. Map 109, p. 287
A plant particularly associated with dry railway tracks, banks and sidings, roadsides and waste places (56%); also recorded from arable land (26%), grassland (11%) and walls and quarries (6%); frequent and well distributed but thinning out towards the N.E. and S.E. of the county; recorded from all surrounding VCs.

110 **Camelina** Crantz

110.1 **Camelina sativa** (L.) Crantz Gold of Pleasure
Bagnall, 1877. Herb. Bagnall. Sutton: railway bank near Blackroot Pool and waste land.
A rare introduction recorded from an arable field near Hatton (2265) and from a waste place at Clifford Bridge (3780); recorded from all surrounding VCs.

111 **Descurainia** Webb & Berth.

111.1 **Descurainia sophia** (L.) Webb ex Prantl Flixweed
Purton, T., 1817. Mid. Flora [*Sisymbrium sophia*]. Studley Castle and Dunnington. Map 743, p. 603
A casual which has appeared on roadsides and waste places in a few locations in the county; recorded from all surrounding VCs.

RESEDACEAE

112 **Reseda** L.

112.1 **Reseda luteola** L. Dyer's Rocket
Perry, 1817. List. Roadsides (without locality). Map 110, p. 287
Frequent but very irregularly distributed throughout the county, with some concentration in urban areas; recorded from waste places, railway banks and roadsides (77%), rough grassland (10%), quarries and walls (8%) and hedgerows (5%); recorded from all surrounding VCs.

112.2 **Reseda lutea** L. Wild Mignonette
Rugby School Report. 1868. On the Leamington railway bank near Rugby. Map 111, p. 288
An occasional plant of a number of scattered localities in the county; recorded from waste places, railway banks, roadsides and quarries; recorded from all surrounding VCs.

112.3 **Reseda alba** L. White Mignonette
Kirk, T., 1847. Phytologist [*R. suffruticulosa*]. "On the ground, from which eight or ten feet of surface soil had been removed, at the New Waterworks, Coventry."
One record only from rough grassland near Rugby (5276); recorded from all surrounding VCs except 39 and 57.

VIOLACEAE

113 **Viola** L.

113.1 **Viola odorata** L. Sweet Violet
Withering, 1796. Arrangement. "All that I have seen about Birmingham are white." Map 112, p. 288
Abundant and evenly distributed in the S. half of the county and towards the E. around Rugby, rare elsewhere; recorded from hedgerows and scrub (60%), woods (16%), roadsides and railway banks (14%), grassland (5%), watersides (4%); white and rose-purple (var. *sub-carnea*) varieties have been recorded, the white being widespread and abundant but the rose-purple very local; recorded from all surrounding VCs.

113.2a **Viola hirta** L. ssp. **hirta** Hairy Violet
Perry, 1824. Herb. Perry. On a hill between Hampton on the Hill and Norton Lindsey, near Warwick. Map 113, p. 289
A frequent plant on calcareous soils in the S. half of the county, especially on the Lower Lias limestone escarpment; recorded from roadsides and railway banks (37%), hedgerows and scrub (26%), grassland (26%), woods (7%) and a few other habitats; recorded from all surrounding VCs.

Viola hirta L. × **odorata** L.
= **V.** × **permixta** Jord.
Bagnall, 1884. Herb. Bagnall. Compton Verney.
Recorded four times from hedgerows and scrub; near Bidford on Avon (1152), near Exhall (1155), near Snitterfield (2158) and at Houndshill (2550); probably under-recorded; recorded from all surrounding VCs except 39 and 57.

113.4a **Viola riviniana** Reichb. ssp. **riviniana** Common Violet
Bagnall, 1870. Herb. Bagnall. Near Kingswood. Map 114, p. 289
Abundant but rather irregularly distributed in the N.W. half of the county, rare in the S.E. except on the Lower Lias limestone and Cotswold escarpments; recorded from hedgerows and scrub (41%), woods (29%), grassland (13%), roadsides and railway banks (13%) and watersides (4%); recorded from all surrounding VCs.

113.4b **Viola riviniana** Reichb. ssp. **minor** (Gregory) Valentine
Readett, R. C., 1951. Sutton Park.
Two authenticated records only, from Sutton Park (0998) and S. of Willey (4983); recorded from VCs 23, 33, 55 and 57.

113.5 **Viola reichenbachiana** Jord. ex Bor. Wood Violet
Kirk, T., 1861. Herb. Perry [*V. reichenbachiana* var. *sylvestris*]. Stoke Aldermoor. Map 115, p. 290
An occasional plant chiefly on the Lower Lias limestone escarpment and in the Dorridge—Shrewley—Henley in Arden area; recorded from woods, hedgerows and scrub, roadsides, rough grass and watersides; recorded from all surrounding VCs.
V. reichenbachiana × *riviniana* (first record: Lester, R., 1956. Lighthorne Hill) has been recorded three times from woods at Coldcomfort Wood (0658), near Claverdon (1864) and Lighthorne Hill (3456); it has been recorded from VCs 33, 37 and 57.

113.6a **Viola canina** L. ssp. **canina** Heath Dog-violet
Bromwich, H., 1857. Herb. Perry. Milverton.
A rare plant of heath grassland recorded from Coldcomfort Wood (0658), Coleshill Bog (2086), Packington (2384), Emscote (2965) and Hartshill (3294); collected by Bagnall from Milverton in 1872 and from Sutton Park in 1875 and 1876; recorded from all surrounding VCs.
The hybrid *V. canina* × *riviniana* has been recorded from Coleshill Bog (2086).

113.7 **Viola lactea** Sm. Pale Dog-violet
Kirk, T., 1857. Herb. Perry. Keresley Common.
Not recorded in present survey; Kirk's station is now built on; recorded from VC 23.

113.9a **Viola palustris** L. ssp. **palustris** Marsh Violet
Withering, 1787. Arrangement. Bogs, on Birmingham Heath. Map 116, p. 290
An occasional plant of a few scattered localities in the N. of the county, with some concentration in Sutton Park and the Coleshill Bog area; recorded chiefly from watersides and bogs, but also from ditches, riversides and mixed woods; recorded from all surrounding VCs except 32 and 33.

113.12a **Viola tricolor** L. ssp. **tricolor** Wild Pansy
Perry, 1845. Herb. Perry [*V. tricolor*]. Road to Briery Land from the Turnpike road. Map 117, p. 291
An occasional plant of a number of very widely scattered localities throughout the county; recorded chiefly from cultivated land, but also from waste places, railway banks, grassland and farmyards; recorded from all surrounding VCs.

113.13 **Viola arvensis** Murr. Field Pansy
Perry, 1812. Herb. Perry [*V. tricolor* var. *arvensis*]. Without locality. Map 118, p. 291
Abundant and evenly distributed throughout the county; recorded from cultivated land (74%), waste places, roadsides and railway banks (16%), grassland (6%) and a few other habitats; recorded from all surrounding VCs.

POLYGALACEAE

114 **Polygala** L.

114.1 **Polygala vulgaris** L. Common Milkwort
Perry, 1812. Herb. Perry. Without locality. Map 119, p. 292
An occasional plant mainly in the S. of the county with some concentration on the Lower Lias limestone; recorded from grassland (53%), railway banks and roadsides (26%), scrub, hedgerows, marshes and woods; recorded from all surrounding VCs.

114.2 **Polygala serpyllifolia** Hose Heath Milkwort
Countess of Aylesford, 1808. Herb. Oxf. Univ. [*P. vulgaris*]. Packington. Map 120, p. 292
A rare plant of a number of widely scattered localities in the W. of the county; recorded from railway banks, woods, roadsides, grassland, watersides, scrub, and dry and wet heath; recorded from all surrounding VCs.

HYPERICACEAE

115 **Hypericum** L.

115.1 **Hypericum androsaemum** L. Common Tutsan
Bree, Rev. W. T., 1821. Mid. Flora. Woods, Meriden.
A rare plant of woods and hedges recorded from near Maxstoke (2486), near Fillongley (2986) and near Kenilworth (3071); recorded from all surrounding VCs.

115.3 **Hypericum hircinum** L. Stinking Tutsan
Webster, Miss M. McCallum, 1958. Hillmorton Ballast Pits.
An introduction recorded once from Hillmorton (5473); recorded from VCs 32 and 33.

115.4 **Hypericum calycinum** L. Rose-of-Sharon, Aaron's Beard
Bagnall, 1891. Flora. Compton Verney.
Frequently planted in shrubberies and occasionally escaping; recorded once during present survey from Walton Hall (2852); recorded from all surrounding VCs except 39.

115.5 **Hypericum perforatum** L.
Perforate St. John's Wort
Perry, 1812. Herb. Perry. Without location.
Map 121, p. 293

Abundant and fairly evenly distributed throughout most of the county but only frequent on the calcareous soils of the S.; recorded from roadsides, railway banks and waste places (43%), hedgerows and scrub (25%), rough grassland (22%), woods (4%) and a few other habitats; recorded from all surrounding VCs.

115.6b **Hypericum maculatum** Crantz ssp. **obtusiusculum** (Tourlet) Hayek
Imperforate St. John's Wort
Withering, 1801. Arrangement [*H. dubium*]. In Mr. Digby's Plantation, Meriden.
Map 122, p. 293

Frequent in the centre of the county but rare in the Avon Valley, in the extreme N. and on the calcareous soils in the S.; recorded from roadsides, railway banks and waste places (47%), grassland (23%), watersides (13%), hedgerows and scrub (10%) and woods (6%); recorded from all surrounding VCs.

115.8 **Hypericum tetrapterum** Fr.
Square-stemmed St. John's Wort
Perry, 1812. Herb. Perry [*H. quadrangulum*]. Without locality. Map 123, p. 294

Abundant and fairly evenly distributed throughout the county; recorded from watersides and marshes (70%), damp grassland (10%), roadsides, railway banks and waste places (9%), hedgerows and scrub (7%) and woods (4%); recorded from all surrounding VCs.

115.9 **Hypericum humifusum** L.
Trailing St. John's Wort
Perry, 1817. List. Opposite Stoneleigh Lodge.
Map 124, p. 294

An occasional plant of dry heathy soils in the middle of the county, absent from the extreme N. and from the calcareous soils of the S. and S.E.; recorded from rough grassland, roadsides, waste places, dry heaths and a number of other habitats; recorded from all surrounding VCs.

115.11 **Hypericum pulchrum** L.
Slender St. John's Wort
Withering, Miss, 1796. Withering's Arrangement. On a sloping bank, near the wall, on the E. side of Edgbaston Park. Map 125, p. 295

An occasional plant of non-calcareous soils concentrated towards the W. of the county and almost entirely absent from the N. and S.; recorded from hedgerows and scrub (31%), grassland (26%), woods (23%) and a few other habitats; recorded from all surrounding VCs.

115.12 **Hypericum hirsutum** L.
Hairy St. John's Wort
Perry, 1812. Herb. Perry. Without location.
Map 126, p. 295

A frequent plant mainly of base-rich soils in the S. and S.W. of the county, but rare elsewhere; recorded from hedgerows and scrub (29%), roadsides and railway banks (26%), grassland (23%), woods (15%) and watersides (6%); recorded from all surrounding VCs.

115.14 **Hypericum elodes** L.
Marsh St. John's Wort
Withering, 1787. Arrangement. Birmingham Heath.

Two records from boggy places in the neighbourhood of Coleshill: one at Bickenhill Plantation (1884) and the other at Coleshill Pool (1986) in an area where it was at one time very abundant; recorded from VCs 37, 39, 55 and 57.

CISTACEAE

118 **Helianthemum** Mill.

118.1 **Helianthemum nummularium** (L.) Mill.
Common Rockrose
Perry, 1812. Herb. Perry [*H. vulgare* Gaertn.]. Hilly bank 2½ miles on Birmingham Road from Warwick. Map 127, p. 296

A rare plant of a few scattered localities on calcareous soils in the S. of the county; recorded from grassland and roadsides; recorded from all surrounding VCs.

ELATINACEAE

122 **Elatine** L.

122.1 **Elatine hexandra** (Lapierre) DC.
Six-stamened Waterwort
Lloyd, Dr. G., 1835. Herb. Perry. Coleshill Pool.

Recorded once only during survey for this Flora on mud in gravel pit at Coleshill (2086); previously known from Coleshill Pool, and seen there and at Olton Pool in 1893 by Bagnall and H. Stuart Thompson towards the end of a long drought; not recorded from any of the surrounding VCs.

CARYOPHYLLACEAE

123 **Silene** L.

123.1 **Silene vulgaris** (Moench) Garcke
Bladder Campion
Ick, W., 1836. Herb. Ick [*S. inflata*]. Road to Erdington 3 miles from Birmingham.
Map 128, p. 296

A frequent plant of very irregular distribution, rare in the centre of the county; recorded from roadsides and railway banks (59%), hedgerows (14%), grassland (14%) and cultivated land (9%); recorded from all surrounding VCs.

123.6 **Silene gallica** L. Small-flowered Catchfly
Kirk, T., 1860. Herb. Perry [*S. anglica* L.]. Near Brandon.

Not recorded in present survey; collected by Bagnall from Sutton Park, between Stonebridge and Coleshill, and in 1889 from Cornets End, this probably being the last record; recorded from all surrounding VCs.

123.10 **Silene nutans** L. Nottingham Catchfly
Caswell, Rev. J., 1891. Bagnall's Flora. Oscott College grounds.

Not recorded in present survey; recorded from VCs 37, 39, 55 and 57.

123.12 **Silene noctiflora** L. Night-flowering Catchfly
Baxter, W., 1831. MS (Bagnall: Flora, 1891). Rugby. Map 129, p. 297
A rare plant of a few scattered localities mainly in the S. of the county on the Cotswold escarpment; recorded from cultivated land, roadsides, waste places and quarries; recorded from all surrounding VCs.

123.13 **Silene dioica** (L.) Clairv. Red Campion
Bree, Rev. W. T., 1828. Mag. Nat. Hist. [*Lychnis dioica*]. Allesley. Map 130, p. 297
Abundant in the N.W. half of the county and in the extreme S. and S.E., but only occasional in the Avon Valley and on the Lower Lias limestone and clays; recorded from hedgerows and scrub (41%), roadsides, railway banks and waste places (22%), woods (17%), watersides and marshes (13%), grassland (4%) and a few other habitats; recorded from all surrounding VCs.

123.14 **Silene alba** (Mill.) E. H. L. Krause
White Campion
Baxter, W., 1831. MS (Bagnall: Mid. Nat. 1892–93) [*Lychnis alba* Mill.]. Near Rugby Mill. Map 131, p. 298
Abundant and fairly evenly distributed throughout the county; recorded from roadsides, waste places and railway banks (40%), hedgerows (31%), cultivated land (15%), grassland (10%) and a few other habitats; recorded from all surrounding VCs.

Silene alba × dioica
Hardaker, W. H., 1954. Bickenhill.
Map 132, p. 298
An occasional plant of a number of scattered localities mainly in the N. of the county; recorded from hedgerows, waste places, roadsides, railway banks, grassland, cultivated land and watersides; recorded from all surrounding VCs.

Silene cretica L.
Bagnall, 1877. Herb. Bagnall. Railway bank, Sutton Park.
Not recorded in present survey; not recorded from any of the surrounding VCs.

124 Lychnis L.

124.3 **Lychnis flos-cuculi** L. Ragged Robin
Bree, Rev. W. T., 1828. Mag. Nat. Hist. Coleshill. Map 133, p. 299
A frequent plant throughout most of the county but rare on the calcareous soils in the S.; recorded from marshes and watersides (63%), damp grassland (16%), woods (12%), hedgerows and scrub (6%) and a few other habitats; recorded from all surrounding VCs.

125 Agrostemma L.

125.1 **Agrostemma githago** L. Corn-cockle
Perry, 1812. Herb. Perry [*Lychnis githago* (L.) Scop.]. Without locality.
Two records, one in 1954 S. of Edge Hill (3744) and one in 1955 near Upper Brailes (2940); not seen in either locality since; formerly widespread chiefly in cornfields but now probably extinct in the county owing to the use of purer seed mixtures and chemical weed-killers; recorded from all surrounding VCs.

127 Dianthus L.

127.1 **Dianthus armeria** L. Deptford Pink
Perry, 1817. List. Hampton-on-the-Hill, between Warwick and Norton.
Not recorded in present survey; last record Bagnall, 1871 (Herb. Bagnall) from Myton; recorded from all surrounding VCs except 32 and 55.

127.7 **Dianthus gratianopolitanus** Vill. Cheddar Pink
Bromwich, H., before 1907. Herb. Warw. Mus. [*D. caesius* Sm.]. Old Wall, Barford.
Not recorded in present survey; recorded from VCs 23, 32, 33 and 37.

127.8 **Dianthus deltoides** L. Maiden Pink
Lloyd, Dr. G., 1830. Herb. Perry. Roman Camp, Chesterton.
One record as a garden escape from a roadside at Avon Dassett (4049); recorded from all surrounding VCs except 32.

128 Vaccaria Wolf

128.1 **Vaccaria pyramidata** Medic. Cowherb
Bagnall, 1877. Herb. Bagnall [*Saponaria vaccaria* L.]. Railway bank, Sutton Park.
Recorded once from a waste place at Emscote (2965) and once as a bird-seed alien at Rugby (4974); it was collected by Bromwich in 1888 from Milverton; recorded from all surrounding VCs except 39 and 57.

129 Saponaria L.

129.1 **Saponaria officinalis** L. Soapwort
Purton, T., 1817. Mid. Flora. Hedge bank, at Dunnington. Map 134, p. 299
A rare plant of a number of scattered localities in the county, generally an escape from cultivation; recorded from cultivated land, waste places, roadsides, railway banks, hedgerows and rough grassland; recorded from all surrounding VCs.

131 Cerastium L.

131.2 **Cerastium arvense** L.
Field Mouse-ear Chickweed
Kirk, T., 1860. Herb. Perry. Willenhall.
One record only from calcareous grassland near Little Compton (2628); formerly more widespread; recorded from all surrounding VCs.

131.3 **Cerastium tomentosum** L.
Snow-in-summer, Dusty-miller
Phipps, J. B., 1955. Between Hampton in Arden and Cornets End.
A garden escape occasionally becoming established and recorded from 1664, 2280, 3067, 3660, 3664, 3747, 3847 and 3873; recorded from all surrounding VCs except 39.

131.7 **Cerastium holosteoides** Fr.
Common Mouse-ear Chickweed
Perry, 1812. Herb. Perry. [*C. glomeratum*]. Without locality. Map 135, p. 300
Abundant and widely distributed throughout the county; recorded from grassland (50%), roadsides, railway banks and waste places (28%), cultivated land (15%) and a few other habitats; recorded from all surrounding VCs.

131.8 **Cerastium glomeratum** Thuill.
Sticky Mouse-ear Chickweed
Baxter, W., 1831. MS (Bagnall: Mid. Nat. 1892–93.) Hill Morton.
Lloyd, Dr. G., 1836. Herb. Perry [*C. vulgatum*]. Near Warwick. Map 136, p. 300
Abundant and fairly evenly distributed throughout the county; recorded from roadsides, railway sides and waste places (37%), dry grassland (30%), arable land (24%) and quarries and walls (5%); recorded from all surrounding VCs.

131.10 **Cerastium diffusum** Pers.
Dark Mouse-ear Chickweed
Allen, D. E., 1950. Watsonia, 1952 [*C. tetrandrum* Curt.]. Railway track near Brandon. Map 137, p. 301
An occasional plant almost entirely confined to railway ballast; recorded from all surrounding VCs except 57.

131.11 **Cerastium pumilum** Curt.
Dwarf Mouse-ear Chickweed
Jacobs, V., 1951. Terry's Green.
One record only: abundant in 1951 on ballast of disused railway track and also scattered in vicinity of Terry's Green (1073); recorded from VCs 23, 33 and 55.

131.12 **Cerastium semidecandrum** L.
Little Mouse-ear Chickweed
Purton, T., 1817. Mid. Flora. "Kinwarton, in a field near the Church." Map 744, p. 603
An occasional plant of walls, quarries, railways and dry grassland; recorded from all surrounding VCs.

132 **Myosoton** Moench

132.1 **Myosoton aquaticum** (L.) Moench
Water Chickweed
Perry, 1812. Herb. Perry [*Stellaria aquatica* (L.) Scop.]. Without locality. Map 138, p. 301
An occasional plant very irregularly distributed throughout most of the county, but with some concentration in the main river valleys; recorded from watersides and marshes (91%) and a few other habitats; recorded from all surrounding VCs.

133 **Stellaria** L.

133.2 **Stellaria media** (L.) Vill.
Common Chickweed
Lloyd, Dr. G., 1835. Herb. Perry. Leamington. Map 139, p. 302
Abundant and evenly distributed throughout the county; recorded from cultivated land (40%), roadsides, waste places, farmyards and railway banks (34%), grassland (13%), hedgerows and scrub (6%), woods (4%) and watersides (3%); recorded from all surrounding VCs.

133.3 **Stellaria pallida** (Dumort.) Piré
Lesser Chickweed
Clark, M. C., 1958. Near Broom Court.
Recorded three times, from a wall near Broom Court (0852), the edge of a cultivated field at Aston Grove (1457) and a waste place near Walton Hall (2852); recorded from all surrounding VCs except 39.

133.4 **Stellaria neglecta** Weihe Greater Chickweed
Kirk, T., 1861. Herb. Perry [*S. umbrosa* Opiz]. Willenhall. Map 140, p. 302
An occasional plant of a number of scattered localities mainly in the centre and extreme S. of the county, with a considerable concentration in the Solihull—Warwick—Coughton area; recorded from watersides and marshes (63%), hedgerows (14%), roadsides and a few other habitats; recorded from all surrounding VCs except 32.

133.5 **Stellaria holostea** L. Greater Stitchwort
Perry, 1812. Herb. Perry. Without locality. Map 141, p. 303
Abundant and evenly distributed in the N.W. of the county and occasional in the extreme S., but almost absent from the lower Avon Valley, the Lower Lias soils of the S. and S.E. and the chalky boulder clay of the N.E.; recorded from hedgerows and scrub (43%), roadsides and railway banks (32%), woods (12%), grassland (8%) and watersides (3%); recorded from all surrounding VCs.

133.6 **Stellaria palustris** Retz. Marsh Stitchwort
Bromwich, H., 1844. Herb. Perry [*S. glauca*]. Meadow near Railway Bridge, Leek Wootton.
Not recorded in present survey; a number of collections, mostly by Bromwich, were made between 1867 and 1871 from the Hill Wootton and Milverton district; recorded from all surrounding VCs.

133.7 **Stellaria graminea** L. Lesser Stitchwort
Perry, 1812. Herb. Perry. Without locality. Map 142, p. 303
Abundant and fairly evenly distributed throughout much of the county, but comparatively sparse in the lower Avon Valley and on the calcareous clays of the S. and S.E.; recorded from grassland (40%), roadsides, waste places and railway banks (23%), hedgerows and scrub (20%), watersides and marshes (12%), woods (3%) and a few other habitats; recorded from all surrounding VCs.

133.8 **Stellaria alsine** Grimm Bog Stitchwort
Perry, 1812. Herb. Perry [*S. uliginosa* Murr.]. Without locality. Map 143, p. 304
Abundant in the N.W. half of the county and frequent on the Middle Lias sandstone and ironstone in the extreme S., but almost absent from the lower Avon Valley and from the Lower Lias clays and limestone; recorded from watersides and marshes (88%), woods (8%) and a few other habitats; recorded from all surrounding VCs.

135 **Moenchia** Ehrh.

135.1 **Moenchia erecta** (L.) Gaertn., Mey. & Scherb.
Upright Chickweed
Bree, Rev. W. T., 1817. Mid. Flora [*Sagina erecta*]. Coleshill Heath.
Not recorded in present survey but records have been made from Atherstone, Berkswell, Coleshill Pool, Kenilworth Common, Sutton Park and Yarningale Common, where it was last seen in 1898; recorded from all surrounding VCs except 33.

136 Sagina L.

136.1 **Sagina apetala** Ard. Annual Pearlwort
Perry, 1812. Herb. Perry (Bagnall: Flora, 1891). Without locality. Map 144, p. 304

Thinly distributed in many parts of the county; recorded from roadsides, railways and waste places (59%), walls and quarries (20%), cultivated land (12%) and a few other habitats; recorded from all surrounding VCs.

"Flora Europaea" groups this and *S. ciliata* as one species, distinguishing as sub-species:

S. apetala ssp. *apetala* = *S. ciliata* Fr. (in present treatment)
ssp. *erecta* = *S. apetala* (auct.)
= *S. apetala* L. (in present treatment)

Warwickshire material has seemed to be distinguished readily, and we have preferred to treat them as species.

136.2 **Sagina ciliata** Fr. Ciliate Pearlwort
Bagnall, 1869. Herb. Bagnall. Coleshill Heath. Map 145, p. 305

A rare plant of a number of widely scattered localities in the county; recorded from railways, roadsides, walls, quarries and a few other habitats; recorded from all surrounding VCs.
See note under *S. apetala.*

136.4 **Sagina procumbens** L. Procumbent Pearlwort
Baxter, W., 1831. MS (Bagnall: Mid. Nat. 1892–93.) Clifton Road, near Rugby. Map 146, p. 305

Frequent though irregularly distributed throughout most of the county, but rare on the calcareous soils of the S.; recorded from roadsides, waste places and railways (48%), walls and quarries (17%), cultivated land (15%), watersides and marshes (10%), grassland (5%) and a few other habitats; recorded from all surrounding VCs.

136.10 **Sagina nodosa** (L.) Fenzl Knotted Pearlwort
Stokes, Dr. J., 1787. Withering's Arrangement. Boggy ground in Sutton Park.

A rare plant of marshes and bogs, now confined to Sutton Park (0995, 0996, 0998); previously also recorded from Bannerley Pool, Coleshill and Warwickshire Moor, Tamworth; recorded from all surrounding VCs.

137 Minuartia L.

137.4 **Minuartia hybrida** (Vill.) Schischk.
Fine-leaved Sandwort
Bloxam, Rev. A., 1837. New Bot. Guide [*Arenaria tenuifolia*]. On a heap of gravel, on the lower Hill Morton Road. Map 745, p. 604

An occasional plant of walls, quarries, railway ballast and dry waste places; recorded from all surrounding VCs.

140 Moehringia L.

140.1 **Moehringia trinervia** (L.) Clairv.
Three-nerved Sandwort
Perry, 1813. Herb. Perry. Without locality. Map 147, p. 306

Abundant though irregularly distributed throughout most of the county but only occasional on the calcareous clay in the S.; recorded from hedgerows and scrub (52%), woods (29%), roadsides, waste places and railway banks (9%), watersides (6%) and a few other habitats; recorded from all surrounding VCs.

141 Arenaria L.

Arenaria serpyllifolia sensu lato
For first record see *A. serpyllifolia* L. Map 148, p. 306

A widely but unevenly distributed plant of railways, roadsides and waste places (55%), walls and quarries (25%), grassland (11%) and arable land (8%); recorded from all surrounding VCs.
The two segregates *A. leptoclados* and *A. serpyllifolia* (q.v.) are of about equal frequency and occupy similar habitats; many recorders have not distinguished them.

141.1 **Arenaria serpyllifolia** L.
Thyme-leaved Sandwort
Perry, 1812. Herb. Perry. Without locality. Map 149, p. 307

Recorded from railways, roadsides and waste places (41%), walls and quarries (41%) and grassland (11%); recorded from all surrounding VCs.

141.2 **Arenaria leptoclados** (Reichb.) Guss.
Lesser Thyme-leaved Sandwort
Kirk, T., 1860. Herb. Perry. Coventry. Map 150, p. 307

Recorded from railways, roadsides and waste places (56%), walls and quarries (21%), grassland (13%) and arable land (10%); recorded from all surrounding VCs

141.6 **Arenaria balearica** L. Mossy Sandwort
Jones, Mrs. M. D. G., 1963. Copston Magna.

One record only of garden escape growing on a wall (4588); recorded from VCs 23 and 32.

142 Spergula L.

142.1 **Spergula arvensis** L. Corn Spurrey
Baxter, W., 1831. MS (Bagnall: Mid. Nat. 1892–93.) Near Rugby.
Lloyd, Dr. G., 1835. Herb. Perry. Leamington. Map 151, p. 308

Abundant and fairly evenly distributed throughout the county, except on the calcareous soils of the S. and S.E., where it is rare; recorded from cultivated land (74%), waste places, roadsides, farmyards and railways (20%), grassland (4%) and a few other habitats; recorded from all surrounding VCs.

var. *sativa* (Boenn.) Mert. & Koch (first record: Baxter, W., 1831. MS Bagnall: Mid. Nat. 1892–93. Blake St., near Sutton Coldfield) has been recorded once from square 2971; also recorded from all surrounding VCs.

143 Spergularia (Pers.) J. & C. Presl

143.1 **Spergularia rubra** (L.) J. & C. Presl
Common Sandspurrey
Purton, T., 1817. Mid. Flora [*Arenaria rubra*]. Turnpike Road to New Inn, Alcester Parish. Map 152, p. 308

An occasional plant of widely scattered distribution mainly in the N. of the county on light sandy and gravelly soils; recorded from waste places, roadsides and railways (72%), cultivated land and quarries; recorded from all surrounding VCs.

ILLECEBRACEAE

146 Herniaria L.

146.3 **Herniaria hirsuta** L. Hairy Rupturewort
Bagnall, 1891. Flora. Tanyards, Kenilworth.
Not recorded in present survey; recorded from VCs 37, 39 and 57.

148 Scleranthus L.

148.1 **Scleranthus annuus** L. Annual Knawel
Baxter, W., 1831. MS (Bagnall: Mid. Nat. 1892–93.) Near Richardson's Farm, Rugby.
Lloyd, Dr. G., 1835. Herb. Perry. Milverton.
Map 153, p. 309
An occasional plant widely but irregularly distributed on sandy and gravelly soils throughout much of the county; absent from the Avon Valley and the calcareous soils of the S. and S.E.; recorded from cultivated land (73%), waste places and roadsides (24%) and a few other habitats; recorded from all surrounding VCs.

PORTULACACEAE

149 Montia L.

149.1 **Montia fontana** L. Blinks
Stokes, Dr. J., 1787. Withering's Arrangement. Hockley Pool grate, near Birmingham.
Map 154, p. 309
A rare plant of very widely scattered distribution in the county, but with some concentration in Sutton Park; recorded chiefly from marshes, watersides and water; recorded from all surrounding VCs.

149.1b Ssp. **chondrosperma** (Fenzl) Walters
Bagnall, 1874. Herb. Bagnall [*M. fontana*]. Sutton Park.
Recorded from Sutton Park (0998), near Middleton (1997), Wolford Heath (2432, 2433), near Sherbourne (2562), Brandon (4076) and Frankton Wood (4171); recorded from VCs 33 and 37.

149.1c Ssp. **amporitana** Sennen
Bagnall, 1868. Herb. Bagnall [*M. fontana*]. Sutton Park.
Not recorded in present survey, but, in addition to that given above, records have been made from Marston Green in 1869 and from Kenilworth in 1875; recorded from all surrounding VCs except 23 and 32.

149.1d Ssp. **variabilis** Walters
Lowe, H., 1834. Herb. Perry [*M. fontana*]. Coleshill Pool.
Recorded only from Sutton Park (0998, 1096); recorded from VCs 32, 39, 55 and 57.

149.2 **Montia perfoliata** (Willd.) Howell
Spring-beauty
Baker, Dr. R., 1873. Young & Baker Catalogue [*Claytonia perfoliata* Don.]. Sutton Coldfield.
Map 155, p. 310
A rare introduction in a few widely scattered localities in the N. half of the county; recorded from cultivated land, watersides, roadsides, waste places and quarries; recorded from all surrounding VCs.

149.3 **Montia sibirica** (L.) Howell Pink Purslane
Thompson, H. S., c. 1900. Herb. B'ham Univ. [*Claytonia sibirica* L.]. Edgbaston Botanical Gardens (as a weed).
A plant occasionally naturalised in woods, hedgerows and roadsides and recorded from squares 1381, 2199, 2299, 2369, 2868, 2885, 2966, 2968, 3286 and 4871; recorded from VCs 37, 39, 55 and 57.

AMARANTHACEAE

153 Amaranthus L.

153.1 **Amaranthus retroflexus** L. Common Pigweed
Anon., 1872. Herb. B'ham Univ. [*A. blitum*]. Kenilworth.
A rare casual in arable fields recorded from 1254, 2756 and 3954; recorded from all surrounding VCs except 39.

153.3 **Amaranthus albus** L.
Anon., 1872. Herb. B'ham Univ. [*A. blitum*]. Kenilworth.
Not recorded in present survey; recorded from VCs 23 and 37.

153.4 **Amaranthus lividus** L.
Anon. 1874. Herb. B'ham Univ. [*A. deflexus*]. Kenilworth and Myton.
A casual not recorded in present survey; not recorded from any of the surrounding VCs.

Amaranthus deflexus L.
Bromwich, H., 1874. Herb. Brit. Mus. [*A. blitum* L.]. Waste ground, Kenilworth.
Not recorded in present survey; recorded from VC 37.

CHENOPODIACEAE

154 Chenopodium L.

154.1 **Chenopodium bonus-henricus** L.
Good King Henry
Perry, 181–. Herb. Perry. Leek Wootton Churchyard; Leamington. Map 156, p. 310
A locally frequent plant, especially near houses, found in a number of widely scattered localities; recorded from waste places, farmyards and roadsides (67%), arable land (25%) and rough grassland; recorded from all surrounding VCs.

154.2 **Chenopodium polyspermum** L.
Many-seeded Goosefoot
Perry, 1820. Herb. Perry and Plant. Varv. Sel. [*C. acutifolium*]. In a newly made garden, Saltisford, Warwick. Map 157, p. 311
A very frequent plant especially in the S. half of the county; recorded from arable land (63%), waste places, roadsides and farmyards (26%), watersides (5%) and a few other habitats; recorded from all surrounding VCs.

154.3 **Chenopodium vulvaria** L. Stinking Goosefoot
Jackson, A. B., 1897. Herb. Brit. Mus. Cultivated ground, Milverton.
One record from old tip near Sutton Coldfield (1495); collected by Bagnall in 1903 from the Corporation Tip, Solihull Lodge; recorded from all surrounding VCs.

154.4 **Chenopodium album** L. Fat Hen
Baxter, W., 1831. MS (Bagnall: Mid. Nat. 1892–93.) Near Rugby.
Cheshire, W., 1855. Herb. Perry. Great Alne.
Map 158, p. 311
Abundant and evenly distributed throughout the

county; recorded from arable land (51%), roadsides, waste places and farmyards (43%) and a number of other habitats; recorded from all surrounding VCs.

154.5 **Chenopodium suecicum** J. Murr
Kirk, T., 1854. Herb. Bagnall [*C. viride*]. Coventry.
Not recorded in present survey; recorded from all surrounding VCs.

154.7 **Chenopodium opulifolium** Schrad. ex Koch & Ziz Grey Goosefoot
Bagnall, 1891. Flora. By road sides, Sutton Park, and in waste places near Milverton.
Not recorded in present survey; recorded from all surrounding VCs except 55.

154.8 **Chenopodium hircinum** Schrad.
Bromwich, H., 1898. Bot. Soc. & Exch. Club Report 1928. Milverton.
Not recorded in present survey; recorded from all surrounding VCs except 32 and 57.

154.9 **Chenopodium ficifolium** Sm. Fig-leaved Goosefoot
St. Brody, Dr., 1876. Herb. B'ham Univ. [*C. serotinum* L.]. Kenilworth.
One record from a cultivated field at Whitehouse Farm, Idlicote (2744); recorded from all surrounding VCs.

154.11 **Chenopodium murale** L. Nettle-leaved Goosefoot
Baxter, W., 1831. MS (Bagnall: Mid. Nat. 1892–93.) Between Brownsover and Aqueduct, Rugby.
Bree, Rev. W. T., 1835. New Bot. Guide. Without locality.
One record as garden weed from Napton-on-the-Hill (4661); recorded from all surrounding VCs except 32.

154.12 **Chenopodium urbicum** L. Upright Goosefoot
Bree, Rev. W. T., 1835. New Bot. Guide. Without locality.
Not recorded in present survey; recorded from all surrounding VCs except 32.

154.13 **Chenopodium hybridum** L. Maple-leaved Goosefoot
Perry, 181–. Herb. Perry. Road between Warwick and Hampton-on-the-Hill. Map 159, p. 312
A rare plant of a few scattered localities in the S.W. of the county; recorded from waste places and arable land; recorded from all surrounding VCs except 57.

154.14 **Chenopodium rubrum** L. Red Goosefoot
Baxter, W., 1842. MS (Bagnall's Flora, 1891). Rugby. Map 160, p. 312
A very frequent plant in the S. of the county and less so in the N.W.; recorded from waste places, farmyards and roadsides (61%), watersides, especially of ponds (20%), and arable land (16%); recorded from all surrounding VCs.

Chenopodium foliosum Asch.
Perry, 1823. Herb. Perry. Stratford on Avon.
Not recorded in present survey; not recorded from any of the surrounding VCs.

156 Atriplex L.

156.2 **Atriplex patula** L. Common Orache
Baxter, W., 1831. MS (Bagnall's Flora, 1891). Near Rugby. Map 161, p. 313
An abundant plant evenly distributed throughout the county; recorded from arable land (48%), waste places, roadsides and farmyards (44%) and a few other habitats; recorded from all surrounding VCs.

156.3 **Atriplex hastata** L. Hastate Orache
Baxter, W., 1831. MS (Bagnall: Mid. Nat. 1892–93.) Near Rugby. Map 162, p. 313
A frequent plant, widely but unevenly distributed throughout the county; recorded from waste places, roadsides and farmyards (59%), arable land (25%) and watersides (12%); recorded from all surrounding VCs.

Corispermum L.

Corispermum leptopterum (Asch.) Iljin
Clark, M. C., 1962. Emscote.
Recorded growing abundantly in one location at railway sidings at Emscote (2965) on introduced sea-sand; not recorded from any of the surrounding VCs.

159 Salsola L.

159.1 **Salsola kali** L. Prickly Saltwort
Benoit, P. M., 1963. Emscote.
One record on introduced sea-sand at Emscote (2965); recorded from VCs 23, 32, 33 and 39.

TILIACEAE

162 Tilia L.

162.1 **Tilia platyphyllos** Scop. Large-leaved Lime
Watson, H. C., 1870. Comp. Cyb. Brit. [*T. grandifolia* Ehrh.]. Without locality.
A rare tree of woods and hedges, perhaps planted in Warwickshire, and recorded from Arrow (0756), Alvecote Priory (2504) and Bilton Grange (4871); recorded from all surrounding VCs.

162.2 **Tilia cordata** Mill. Small-leaved Lime
Kirk, T., before 1874. Herb. Brit. Mus. [*T. parvifolia* Ehrh.]. Whitmore Park. Map 163, p. 314
An occasional tree, widely scattered in the N.W. of the county, with a concentration in the Earlswood—Solihull—Lowsonford area; recorded from hedges, scrub and mixed woods; recorded from all surrounding VCs.

Tilia × vulgaris Hayne = **T. cordata × platyphyllos** Common Lime
Bagnall, 1868. Herb. Bagnall. Near Sutton Coldfield. Map 164, p. 314
Abundant but irregularly distributed throughout the N. half of the county, only occasional in the S. where it is almost completely absent from the calcareous clays; frequently planted; recorded from hedges and scrub (60%), mixed woods (20%), grassland (9%), roadsides and waste places (6%) and watersides (3%); recorded from all surrounding VCs.

Tilia tomentosa Moench
Jones, Mrs. M. D. G., 1960. Bilton Grange.
Two records of planted trees from Coventry (3378) and Bilton Grange (4871); recorded from VC 23.

MALVACEAE

163 **Malva** L.

163.1 **Malva moschata** L. Musk Mallow
Perry, 1812. Herb. Perry. Without locality. Map 165, p. 315
Frequent in the centre of the county but rare on calcareous clays in the S.; recorded from roadsides, railway banks and waste places (44%), grassland (28%), hedgerows (21%) and a few other habitats; recorded from all surrounding VCs.

163.2 **Malva sylvestris** L. Common Mallow
Perry, 1812. Herb. Perry. Without locality. Map 166, p. 315
Frequent but very irregularly distributed throughout most of the county; notably infrequent in the W. central area; recorded from roadsides, waste places and farmyards (72%), hedgerows (11%), grassland (10%) and cultivated land (5%); recorded from all surrounding VCs.

163.3 **Malva nicaeensis** All.
Bromwich, H., 1876. Herb. Bagnall. Milverton.
Not recorded in present survey; recorded from VCs 23, 39 and 55.

163.4 **Malva neglecta** Wallr. Dwarf Mallow
Lloyd, Dr. G., 1835. Herb. Perry [*M. rotundifolia* L.]. Leamington. Map 167, p. 316
Sparsely scattered throughout most of the county, but common in the Avon Valley; recorded from roadsides, waste places and farmyards (69%), cultivated land (20%), grassland (8%) and a few other habitats; recorded from all surrounding VCs.

163.5 **Malva pusilla** Sm. Small Mallow
Bromwich, H., 1881. Herb. Bagnall. Near Kenilworth.
Not recorded in present survey; recorded from VCs 23, 33, 37 and 55.

163.6 **Malva parviflora** L. Least Mallow
Jackson, A. B., 1902. Herb. Brit. Mus. Milverton, Warwick.
One record from a waste place at Bunkers Hill Farm (3664); recorded from all surrounding VCs except 23 and 32.

165 **Althaea** L.

165.2 **Althaea hirsuta** L. Hispid Mallow
Goodman, Miss C. M., 1957. Edgbaston.
A casual recorded once only in a waste place (0584); recorded from VCs 23, 33, 39 and 55.

LINACEAE

166 **Linum** L.

166.1 **Linum bienne** Mill. Pale Flax
Rugby School Report, 1875. [*L. angustifolium* Huds.]. Upper Hillmorton Road, near Rugby.
Not recorded in present survey; recorded from all surrounding VCs except 32.

166.2 **Linum usitatissimum** L. Common Flax
Purton, T., 1817. Mid. Flora. Broom.
An escape from cultivation recorded from arable fields (1069, 2946, 3560, 3652) and waste places (2201, 3780); recorded from all surrounding VCs.

166.4 **Linum catharticum** L. Purging Flax
Perry, 1815. Herb. Perry. Without locality. Map 168, p. 316
Frequent on calcareous soils in the S. of the county with scattered occurrences elsewhere; recorded from grassland (57%), railways, roadsides and waste places (30%), quarries (5%) and a few other habitats; recorded from all surrounding VCs.

167 **Radiola** Hill

167.1 **Radiola linoides** Roth Allseed
Purton, T., 1817. Mid. Flora [*Linum radiola*]. Coleshill Pool.
Not recorded in present survey; known from a number of stations since 1817, including Coleshill Pool and Bog, and Cornets End where the last record was made by Bagnall (Herb. Bagnall) in 1893; recorded from VCs 23, 37, 39 and 55.

GERANIACEAE

168 **Geranium** L.

168.1 **Geranium pratense** L. Meadow Cranesbill
Countess of Aylesford, 1805. Bot. Guide. Allesley. Map 169, p. 317
A frequent plant mainly confined to the S.E. half of the county; recorded from roadsides, railway banks and waste places (41%), watersides (24%), grassland (21%) and hedgerows (13%); recorded from all surrounding VCs.

168.2 **Geranium sylvaticum** L. Wood Cranesbill
Purton, T., 1817. Mid. Flora. Oversley Wood.
Not recorded in present survey; Purton's record was probably an escape from cultivation; recorded from all surrounding VCs except 23 and 32.

168.3 **Geranium endressii** Gay French Cranesbill
Druce, 1932. Com. Flora. Without locality.
A rare garden escape occasionally becoming naturalised; recorded from Edge Hill (3747) and near Princethorpe (3870); recorded from all surrounding VCs except 32 and 39.

168.4 **Geranium versicolor** L. Pencilled Cranesbill
Perry, 1829. Herb. Perry [*G. striatum* L.]. Chesford Bridge, Kenilworth.
Not recorded in present survey; recorded from VCs 23, 33, 37 and 55.

168.6 **Geranium phaeum** L. Dusky Cranesbill
Perry, 1823. MS (Bagnall's Flora, 1891). "Brought me from the rock at Woodloes by Mr. Harris."
A rare garden escape occasionally becoming naturalised in hedgerows, woodland and rough grass; recorded from Houndshill (2550), Leek Wootton (2868) and Long Compton (2931); recorded from all surrounding VCs.

168.7 **Geranium sanguineum** L. Bloody Cranesbill
Anon., 1845. Herb. Coventry Museum. Ilmington.
A rare garden escape not recorded in present survey; recorded from all surrounding VCs except 23 and 55.

168.9 **Geranium pyrenaicum** Burm. f. Hedgerow Cranesbill
Bree, Rev. W. T., 1830. Mag. Nat. Hist. Allesley. Map 170, p. 317
An occasional plant of a number of scattered localities throughout the county; recorded from roadsides and

railway banks, grassland and hedgerows; recorded from all surrounding VCs.

168.10 **Geranium columbinum** L.
Long-stalked Cranesbill
Perry, 1812. Herb. Perry. Opposite Half-way House, Stratford Road, Warwick.
Map 171, p. 318
A rare plant of several scattered localities; recorded chiefly from grassland; recorded from all surrounding VCs.

168.11 **Geranium dissectum** L.
Cut-leaved Cranesbill
Baxter, W., 1831. MS (Bagnall: Mid. Nat. 1892–93.) Near Brownsover.
Ick, W., 1836. Herb. Ick. Bank at Little Bromwich. Map 172, p. 318
Abundant and evenly distributed throughout most of the county but somewhat less abundant in the N.; recorded from roadsides, waste places and railway banks (36%), grassland (31%), cultivated land (23%) and hedgerows (8%); recorded from all surrounding VCs.

168.12 **Geranium rotundifolium** L.
Round-leaved Cranesbill
Readett, R. C. and Woolman, J. F., 1957. Newbold on Stour.
A very rare plant of rough grassland recorded from Newbold on Stour (2345) and Little Compton (2630); recorded from VCs 23, 32, 37 and 57.

168.13 **Geranium molle** L. Dovesfoot Cranesbill
Perry, 1813. Herb. Perry. Without locality.
Map 173, p. 319
Abundant and evenly distributed in the centre of the county but only frequent in the N. and extreme S.; recorded from roadsides, waste places and railway banks (41%), cultivated land (25%), grassland (22%), hedgerows (9%) and a few other habitats; recorded from all surrounding VCs.

168.14 **Geranium pusillum** L.
Small-flowered Cranesbill
Perry, 1828. Herb. Perry. Hampton Lucy at the gate leading to Bank Croft.
Map 174, p. 319
A frequent plant in the Avon Valley with a number of scattered occurrences elsewhere; recorded from roadsides and waste places (36%), cultivated land (27%), grassland (18%), hedgerows (10%) and quarries and walls (7%); recorded from all surrounding VCs.

168.15 **Geranium lucidum** L. Shining Cranesbill
Perry, 1817. List. Stank Hill; between Warwick and Longbridge. Map 175, p. 320
A rare plant of a number of scattered localities; recorded from roadsides, walls and hedgerows; recorded from all surrounding VCs.

168.16 **Geranium robertianum** L. Herb Robert
Bree, Rev. W. T., 1828. Mag. Nat. Hist. Wootton, near Warwick. Map 176, p. 320
Abundant and fairly evenly distributed throughout the county; recorded from hedgerows and scrub (46%), roadsides, railway banks and waste places (21%), watersides (14%), woods (13%) and a few other habitats; recorded from all surrounding VCs.

169 Erodium L'Hérit.

169.2 **Erodium moschatum** (L.) L'Hérit.
Musk Storksbill
Purton, T., 1817. Mid. Flora. Near Cookhill on the Ridgeway.
Not recorded in the present survey and not seen since 1884; recorded from all the surrounding VCs except 57.

169.3a **Erodium cicutarium** (L.) L'Hérit.
ssp. **cicutarium** Common Storksbill
Countess of Aylesford, 1810. Herb. Oxf. Univ. Packington. Map 177, p. 321
An occasional plant of very irregular distribution usually on well-drained sandy soils; recorded from roadsides, waste places and railway banks (53%), cultivated land (31%), quarries (9%) and grassland (7%); recorded from all surrounding VCs.

169.4 **Erodium glutinosum** Dumort. Sticky Storksbill
Benoit, P. M., 1963. Emscote.
Introduced with sea-sand at Emscote (2965); not recorded from any of the surrounding VCs.

Erodium botrys (Cav.) Bertol.
Hackett, A., 1964. Near Bedworth.
Several plants recorded from a colliery refuse tip (3686) since obliterated by further tipping; recorded from VC 37.

OXALIDACEAE

170 Oxalis L.

170.1 **Oxalis acetosella** L. Wood-sorrel
Perry, 1813. Herb. Perry. Without locality.
Map 178, p. 321
A frequent plant mainly of woods in the N. and W. of the county; almost completely absent from the Avon Valley and the Lower Lias clay in the S. and S.E.; recorded from woods (71%), hedgerows and scrub (21%), watersides (4%) and a few other habitats; recorded from all surrounding VCs.

170.2 **Oxalis corniculata** L. Procumbent Yellow Sorrel
Kirk, T., 1847. Phytologist. In a garden at Foleshill.
A plant of roadsides, gardens, walls and waste places, generally an escape, and recorded from squares 1671, 1682, 2185, 2656, 3660, 3679, 3690, 4080, 4147, 4245, 4370, 4682, 4779, 4871 and 4873; recorded from all surrounding VCs except 39.

170.4 **Oxalis europaea** Jord. Upright Yellow Sorrel
Kirk, T., 1848. Herb. Perry [*O. stricta* L.]. Arbury Hall.
Three records from canal path at Hatton (2566), waste place at Wellesbourne (2855) and gardens at Arbury Hall (3389); recorded from all surrounding VCs except 32 and 39.

BALSAMINACEAE

171 Impatiens L.

171.1 **Impatiens noli-tangere** L. Touch-me-not
Cox, T., 1852. Herb. Perry. Small wood near Berkswell Hall.
Not recorded in present survey; last record appears to be by Bagnall in 1888 (Herb. Bagnall) from a wood

near Knowle; probably always a garden escape in the Midlands; recorded from all surrounding VCs except 23.

171.2 **Impatiens capensis** Meerb. Orange Balsam
Falk, P. and Allen, D. E., 1937. Watsonia, 1952. Canal banks from Hillmorton to Cathiron, near Rugby. Map 179, p. 322
An introduced plant which has become naturalised and has spread rapidly along canal sides since it was first recorded in the Rugby area; occasionally also on river banks; recorded from all surrounding VCs.

171.3 **Impatiens parviflora** DC. Small Balsam
Bromwich, H. (Bagnall: Mid. Nat. 1892–93.) Near Milverton. Map 746, p. 604
An occasional plant of damp areas in woods, scrub and waste places; recorded from all surrounding VCs.

171.4 **Impatiens glandulifera** Royle
Indian Balsam, Policeman's Helmet
Ounsted, J., 1942. Watsonia, 1951. By River Tame and River Blythe. Map 180, p. 322
An introduced plant mainly of riversides (65%), occasionally also found in other damp places; recorded from all surrounding VCs.

ACERACEAE

173 **Acer** L.

173.1 **Acer pseudoplatanus** L. Sycamore
Purton, T., 1817. Mid. Flora. Arrow; Oversley; Alcester. Map 181, p. 323
A very common tree of hedges and scrub (57%), mixed woodland (24%) and roadsides and waste places (11%); also occasionally found in grassland; widely planted and naturalised throughout the county but less common in the S.E.; recorded from all surrounding VCs.

173.2 **Acer platanoides** L. Norway Maple
Readett, R. C., 1952. Great Alne.
Though apparently not formally recorded earlier, has been planted in woods for many years; occasionally escapes and regenerates; recorded from squares 1059, 1159, 1459, 1478, 1479, 2261, 2268, 3052, 4080, 4476, 4778, 4580 and from VCs 33, 37 and 57.

173.3 **Acer campestre** L. Common Maple
Ray, 1688. Hist. Plant. Near Middleton. Map 182, p. 323
A common tree or shrub mainly of hedges and scrub (80%), also occasionally found in mixed and oak woodland (11%), evenly distributed throughout the county; recorded from all surrounding VCs.

HIPPOCASTANACEAE

175 **Aesculus** L.

175.1 **Aesculus hippocastanum** L. Horse-chestnut
Bagnall, 1870. Herb. Bagnall. Combe Abbey Park. Map 183, p. 324
A widely distributed tree of hedges (41%), woods (21%), cultivated and waste land (20%) and grassland (16%), frequently planted and regenerating; recorded from all surrounding VCs.

Aesculus carnea Hayne Red Horse-chestnut
Phipps, J. B., 1955. Upper St. Dennis Farm.
Although not formally recorded until 1955, has been known as a planted tree in several parts of the county for many years.

AQUIFOLIACEAE

176 **Ilex** L.

176.1 **Ilex aquifolium** L. Holly
Young & Baker, 1873. Catalogue. Pinley. Map 184, p. 324
Rare on the Lias limestone in the S.E., but common in the centre and N. of the county, chiefly in hedges (67%) and woodland (23%); occasionally found on roadsides and in waste places; recorded from all surrounding VCs.

CELASTRACEAE

177 **Euonymus** L.

177.1 **Euonymus europaeus** L. Spindle-tree
Purton, T., 1817. Mid. Flora. Oversley Wood and Wetheley Wood. Map 185, p. 325
An occasional plant mainly in the S. half of the county with considerable concentration on the Jurassic limestone and clays; recorded from hedges and scrub (73%), woods (21%) and a few other habitats; Warwickshire is on the edge of the main British distribution of the species; recorded from all surrounding VCs.

BUXACEAE

178 **Buxus** L.

178.1 **Buxus sempervirens** L. Box
Withering, 1812. Arrangement. Near Sutton Coldfield. Map 186, p. 325
Scattered in the E. and centre of the county, generally planted and sometimes naturalised; recorded from hedges (61%), woods (26%) and plantations (10%); recorded from all surrounding VCs.

RHAMNACEAE

179 **Rhamnus** L.

179.1 **Rhamnus catharticus** L. Buckthorn
Countess of Aylesford, 1805. Bot. Guide. Packington. Map 187, p. 326
Abundant on the calcareous soils of the S. and S.E. of the county and occasional elsewhere; recorded from hedges (91%), mixed woods (6%) and a few other habitats; recorded from all surrounding VCs.

180 **Frangula** Mill.

180.1 **Frangula alnus** Mill. Alder Buckthorn
Purton, T., 1817. Mid. Flora [*Rhamnus frangula*]. Grafton; Great Alne and Arrow. Map 188, p. 326
A rare plant of a number of scattered areas in the W. of the county, more frequent in the N.W.; recorded from woods (67%), hedges (18%), watersides and marshes (9%); recorded from all surrounding VCs.

PAPILIONACEAE

183 **Lupinus** L.

Lupinus polyphyllus Lindl. Garden Lupin
Pickvance, Mrs. E., 1963. Edgbaston, Birmingham.
A garden escape occasionally naturalised and recorded from roadsides (1194, 2097) and waste places (0483, 0685, 1266); not recorded from any of the surrounding VCs.

184 **Laburnum** P.C. Fabr.

184.1 **Laburnum anagyroides** Medic. Golden Rain, Laburnum
Hardaker, W. H., 1953. Middle Bickenhill. Map 747, p. 604
A garden escape occasionally naturalised in woods, hedges and waste places; recorded from all surrounding VCs except 33 and 57.

185 **Genista** L.

185.1 **Genista tinctoria** L. Dyer's Greenweed
Perry, 1812. Herb. Perry. Without locality. Map 189, p. 327
An occasional plant found on the Lower Lias escarpment and in a few areas in the W. of the county; recorded chiefly from rough grassland, hedgerows and roadsides; recorded from all surrounding VCs.

185.2 **Genista anglica** L. Needle Furze, Petty Whin
Bree, Rev. W. T., 1817. Mid. Flora. Coleshill Heath.
Not recorded in present survey; formerly a rare plant of sandy heaths, recorded from Sutton Park, Coleshill Heath and similar areas mainly in the N. of the county, the last record being from Arley Wood, near Ansley (Bagnall: Herb. Bagnall, 1878); recorded from all surrounding VCs except 33.

187 **Ulex** L.

187.1 **Ulex europaeus** L. Furze, Common Gorse
Palmer, Miss C. E., 1854. Herb. Oxf. Univ. Lighthorne. Map 190, p. 327
Abundant, but unevenly distributed in the N.W. half of the county, occasional in the S.E. except on the calcareous soils, where it is very rare; recorded from hedges and scrub (36%), roadsides, waste places and railway banks (27%), grassland (25%), woods (7%) and heaths (3%); recorded from all surrounding VCs.

187.2 **Ulex gallii** Planch. Western Gorse
Perry, 1812. Herb. Perry. Without locality. Map 191, p. 328
Occasional, but very irregularly distributed in the N.W. quarter of the county, with considerable concentration in the Sutton and Tamworth area; Warwickshire is on the E. boundary of the main British distribution; recorded from heaths, hedgerows, scrub, roadsides, waste places, rough grassland and woods; recorded from all surrounding VCs except 32.

188 **Sarothamnus** Wimm.

188.1 **Sarothamnus scoparius** (L.) Wimm. ex Koch Broom
Baxter, W., 1831. MS (Bagnall: Mid. Nat. 1892–93) [*Cytisus scoparius* Link.]. Near Sawbridge.
Ick, W., 1836. Herb. Ick [*Spartium scoparium*]. A bank about a mile beyond Saltley. Map 192, p. 328
Abundant, but rather irregularly distributed throughout the N. half of the county; absent from the Avon Valley and the calcareous soils of the S. and S.E.; recorded from waste places, roadsides and railway banks (39%), hedges and scrub (31%), rough grass (15%), mixed woods (7%) and a few other habitats; recorded from all surrounding VCs.

189 **Ononis** L.

189.1 **Ononis repens** L. Common Restharrow
Baxter, W., 1831. MS (Bagnall: Mid. Nat. 1892–93.) Clifton Road, near Rugby.
Lloyd, Dr. G., 1835. Herb. Perry [*O. arvensis*]. Leamington. Map 193, p. 329
A frequent plant in the S. half of the county especially on the Lower Lias limestone; recorded from roadsides and railway banks (55%), grassland (40%) and a few other habitats; recorded from all surrounding VCs.

189.2 **Ononis spinosa** L. Spiny Restharrow
Baxter, W., 1831. MS (Bagnall: Mid. Nat. 1892–93.) Watling Street, near Rugby.
Lloyd, Dr. G., 1835. Herb. Perry. Near Tachbrook, Warwick. Map 194, p. 329
A frequent plant mainly of the Lower Lias clay in the S. and S.E. of the county, and of the boulder clay in the N.E.; recorded from grassland (66%), roadsides, railway banks and waste places (28%) and a few other habitats; recorded from all surrounding VCs.

190 **Medicago** L.

190.2 **Medicago sativa** L. Lucerne, Alfalfa
Purton, T., 1817. Mid. Flora. Grafton. Map 195, p. 330
A frequent plant in the S. of the county but rare in the N.; recorded from roadsides, railway banks and waste places (52%), cultivated land (21%), grassland (20%) and hedgerows (5%); recorded from all surrounding VCs.

190.3 **Medicago lupulina** L. Black Medick
Kirk, T., 1854. Herb. Perry. Coundon Station, Coventry. Map 196, p. 330
Abundant and fairly evenly distributed throughout most of the county but a little less common in the N.; recorded from roadsides, waste places and railway banks (49%), grassland (35%), cultivated land (9%) and a few other habitats; recorded from all surrounding VCs.

190.4 **Medicago minima** (L.) Bartal. Bur Medick
Clark, M. C., 1957. Near Bidford on Avon.
One record of several plants in railway sidings near Bidford on Avon (0952); recorded from VCs 32, 33 and 37.

190.5 **Medicago polymorpha** L. Toothed Medick
Bromwich, H., 1866. Bagnall's Flora, 1891 [*M. denticulata* Willd.]. Near Kenilworth.
One record from railway sidings near Bidford on Avon (0952); recorded from all surrounding VCs except 55.

190.6 **Medicago arabica** (L.) Huds. Spotted Medick
Perry, 1812. Herb. Perry [*M. maculata* Sibth.]. Stone Quarry, Pigwells, Warwick. Map 197, p. 331
A rare plant chiefly confined to the Lower Avon Valley; recorded from roadsides, cultivated land, grassland, railway banks and a few other habitats; recorded from all surrounding VCs except 39.

Medicago laciniata (L.) Mill.
Hackett, A., 1957. Nuneaton.
Two records from railway sidings near Bidford on Avon (0952) and a wool factory at Nuneaton (3691); recorded from VCs 37 and 39.

191 **Melilotus** Mill.

191.1 **Melilotus altissima** Thuill. Tall Melilot
Perry, 1812. Herb. Perry [*M. officinalis*]. Without locality. Map 198, p. 331
A frequent plant of the Lower Lias limestone in the S. of the county, but only occasional and widely scattered elsewhere; recorded from railway banks, roadsides and waste places (71%), grassland (12%), quarries (6%) and a few other habitats; recorded from all surrounding VCs.

191.2 **Melilotus officinalis** (L.) Pall. Ribbed Melilot
Bagnall, 1885. Herb. Bagnall [*M. officinalis* Lam.]. Bulkington. Map 199, p. 332
An occasional plant with a scattered distribution in many parts of the county; less common than *M. altissima* and absent from the Lower Lias limestone, but occasional on the calcareous clay of the S. and S.E.; recorded from waste places, roadsides and railway banks (66%), quarries (16%) and a few other habitats; recorded from all surrounding VCs.

191.3 **Melilotus alba** Medic. White Melilot
Bree, Rev. W. T., 1828. Mag. Nat. Hist. [*Trifolium officinalis*]. Coleshill. Map 200, p. 332
A rare plant of a few widely scattered areas in the county; recorded from quarries, walls, roadsides and railway banks; recorded from all surrounding VCs.

191.4 **Melilotus indica** (L.) All. Small Melilot
Bromwich, H., 1874. Herb. Warw. Mus. [*M. arvensis* Wallr.]. Kenilworth.
A rare plant naturalised in waste places at Packington Ford (2185) and Hillmorton Wharf (5473), in a quarry at Griff (3589) and in gardens near Dunchurch (4871) and Rugby (4974); recorded from all surrounding VCs.

Trigonella L.

Trigonella caerulea (L.) Ser.
Hanson, F. D., 1953. Herb. Miss C. M. Goodman. Woolman's Nurseries, Shirley.
One record only from Nursery Gardens at Shirley (1180); recorded from VCs 23, 39 and 55.

192 **Trifolium** L.

192.2 **Trifolium pratense** L. Red Clover
Lloyd, Dr. G., 1835. Herb. Perry. Wroxall Heath. Map 201, p. 333
Abundant and evenly distributed throughout the county; recorded from grassland (54%), roadsides, railway banks and waste places (34%), cultivated land (8%) and a few other habitats; recorded from all surrounding VCs.

192.4 **Trifolium medium** L. Zigzag Clover
Perry, 1813. Herb. Perry. Without locality. Map 202, p. 333
Abundant in the N.W. half of the county, very rare in the Avon Valley and only occasional on the calcareous soils of the S. and S.E.; recorded from grassland (47%), roadsides, railway banks and waste places (46%), hedgerows and scrub (5%) and a few other habitats; recorded from all surrounding VCs.

192.7 **Trifolium incarnatum** L. Crimson Clover
Bagnall, 1868. Herb. Bagnall. Castle Bromwich.
A rare plant recorded as a relic of cultivation from squares 1495, 1656, 2934 and 4170; recorded from all surrounding VCs.

192.9 **Trifolium arvense** L. Haresfoot Clover
Purton, T., 1817. Mid. Flora. Salford; Dunnington. Map 203, p. 334
An occasional plant with a scattered distribution throughout the county, mainly on sands and gravel; recorded from roadsides, waste places and railway banks (72%), grassland, walls, quarries, heaths and cultivated land; recorded from all surrounding VCs.

192.10 **Trifolium striatum** L. Soft Clover
Ray, 1670. Catalogus [*T. nodiflorum glomerulis mollioribus*]. "On Dorsthill, near Middleton." Map 204, p. 334
A rare plant, mainly confined to well-drained soils on gravels, sandstones and limestone in the S.E. half of the county; recorded from grassland and roadsides; recorded from all surrounding VCs.

192.11 **Trifolium scabrum** L. Rough Clover
Kirk, T., 1873. Top. Bot. Without locality.
Three records only from Burton Dassett: two from quarries (3951, 3952) and one from pasture (4051); recorded from all surrounding VCs except 39.

192.13 **Trifolium subterraneum** L. Subterranean Clover
Perry, 1843. Bagnall's Flora, 1891. Between Emscote and Milverton.
Two records from grassland at Salford Priors (0750) and from Sherbourne (2562) where it has been known since 1866; it was also collected in 1859 from near Stratford on Avon; recorded from all surrounding VCs except 32 and 39.

192.17 **Trifolium hybridum** L. Alsike Clover
Rugby School Report, 1867. Near Rugby. Map 205, p. 335
Abundant and fairly evenly distributed throughout the county; recorded from grassland (42%), cultivated land (32%), roadsides, waste places and railway banks (24%) and a few other habitats; recorded from all surrounding VCs.

192.18 **Trifolium repens** L. White Clover, Dutch Clover
Perry, 1822. Herb. Perry. Saltisford, Warwick. Map 206, p. 335
Abundant and evenly distributed throughout the county; recorded from grassland (47%), roadsides, waste places and railway banks (40%), cultivated land (8%) and several other habitats; recorded from all surrounding VCs.

192.19 **Trifolium fragiferum** L. Strawberry Clover
Purton, T., 1817. Mid. Flora. Kinwarton and Oversley. Map 207, p. 336
An occasional plant confined to the calcareous soils of the S. and E. of the county; recorded from roadsides (50%) and grassland (46%); recorded from all surrounding VCs except 39.

192.20 **Trifolium resupinatum** L. Reversed Clover
Caswell, Rev. J., 1891. Bagnall's Flora. Oscott College Grounds.
Not recorded in present survey; recorded from all surrounding VCs except 57.

192.21 **Trifolium campestre** Schreb. Hop Trefoil
Perry, 1812. Herb. Perry [*T. procumbens*]. Without locality. Map 208, p. 336
Abundant, though rather unevenly distributed throughout the county; recorded from roadsides, railway banks and waste places (47%), grassland (40%), cultivated land (7%), hedgerows (3%) and a few other habitats; recorded from all surrounding VCs.

192.22 **Trifolium aureum** Poll.
Rugby School Report, 1871 [*T. agrarium* L.]. Near Bilton.
One record only from a garden at Bilton Grange (4871); recorded from all surrounding VCs except 37 and 39.

192.23 **Trifolium dubium** Sibth. Lesser Trefoil
Perry, 1812. Herb. Perry. Without locality. Map 209, p. 337
Abundant and evenly distributed throughout the county; recorded from grassland (46%), roadsides, waste places and railway banks (41%), cultivated land (8%) and a few other habitats; recorded from all surrounding VCs.

192.24 **Trifolium micranthum** Viv. Slender Trefoil
Countess of Aylesford, 1810. (Bagnall: Mid. Nat. 1892–93) [*T. filiforme* L.]. Packington. Map 210, p. 337
An occasional plant mainly on light sandy soils in the valleys of the Avon and Arrow; recorded from grassland and roadsides; recorded from all surrounding VCs.

193 **Anthyllis** L.

193.1 **Anthyllis vulneraria** L. Kidney-vetch
Purton, T., 1817. Mid. Flora. Shottery; Kinwarton; Coughton Fields. Map 211, p. 338
An occasional plant of grassland on dry railway banks, roadsides, waste places and quarries, all on calcareous soil; recorded from all surrounding VCs.

195 **Lotus** L.

195.1 **Lotus corniculatus** L.
Common Birdsfoot Trefoil
Perry, 1812. Herb. Perry. Without locality. Map 212, p. 338
Abundant and evenly distributed throughout the county; recorded from grassland (53%), roadsides, railway banks and waste places (36%), hedgerows and scrub (5%) and a few other habitats; recorded from all surrounding VCs.

195.2 **Lotus tenuis** Waldst. & Kit. ex Willd.
Narrow-leaved Birdsfoot Trefoil
Lloyd, Dr. G., 1835. Herb. Perry. Wilmcote. Map 213, p. 339
An occasional plant confined to calcareous soils in the S. of the county; recorded from grassland (58%), hedgerows, roadsides, railway banks, waste places and quarries; recorded from all surrounding VCs.

195.3 **Lotus uliginosus** Schkuhr
Marsh Birdsfoot Trefoil
Baxter, W., 1831. MS (Bagnall: Mid. Nat. 1892–93) [*L. pilosus* Becke]. By Sawbrook, near Sawbridge.
Lloyd, Dr. G., 1835. Herb. Perry [*L. major*]. Leamington; Oversley Wood. Map 214, p. 339
Abundant throughout most of the county but almost absent from the Avon Valley and the Lower Lias soils; recorded from watersides and marshes (61%), damp grassland (21%), roadsides, railway banks and waste places (7%), woods (5%) and hedgerows and scrub (5%); recorded from all surrounding VCs.

197 **Galega** L.

197.1 **Galega officinalis** L. Goat's Rue
Laflin, T., 1959. Stratford on Avon.
A rare plant of roadsides and waste places naturalised S.E. of Redditch (0665), at Shirley (1178), Sharmans Cross (1289), Stratford on Avon (2055), Tile Hill (2778), near New Close Wood (4077) and at Newbold on Avon Quarry (4976); recorded from VCs 55 and 57.

198 **Robinia** L.

198.1 **Robinia pseudoacacia** L. Acacia
Arnold, G. & M., 1955. Alvecote.
Generally planted and occasionally naturalised in woodland and hedgerows, and by roadsides; recorded from squares 0650, 1194, 1196, 1197, 1956, 1986, 2196, 2504, 2849, 3163, 3690 and 4080; recorded from VCs 23, 33, 37 and 57.

200 **Astragalus** L.

200.3 **Astragalus glycyphyllos** L.
Milk-Vetch, Wild Liquorice
Holden, 1688–96. Herb. Holden [*Glaux vulgaris leguminosa s. Glycyrrhiza Sylvestris*]. "Grows in ye Lane before you go up Frizwell hill from Warwick." Map 215, p. 340
A rare plant of hedgerows (43%) and roadsides and waste places (36%), generally on calcareous soil but rare even there; recorded from all surrounding VCs.

202 **Ornithopus** L.

202.1 **Ornithopus perpusillus** L. Birdsfoot
Withering, 1787. Arrangement. Winson Green and Washwood Heath. Map 216, p. 340
A rare plant of a few scattered localities in the N. half of the county, on light sandy soils; completely absent from the calcareous soils of the S. and S.E.; recorded from waste places, grassland, quarries, railway banks, cultivated land and dry heath; recorded from all surrounding VCs.

203 **Coronilla** L.

203.1 **Coronilla varia** L. Crown Vetch
Bagnall, 1868. Herb. Bagnall. Wylde Green.
Established in a number of waste places: Dosthill (2199), near Stretton on Dunsmore (4073, 4172), Lower Shuckburgh (4862) and Rugby Cement Works (4875, 4876); recorded from all surrounding VCs.

Coronilla scorpioides Koch
Jones, Mrs. M. D. G., 1965. Rugby.
One record of bird-seed alien in garden at Rugby (4974); recorded from VCs 23, 39 and 55.

204 Hippocrepis L.

204.1 **Hippocrepis comosa** L. Horseshoe Vetch
Bromwich, H., 1864. Herb. Brit. Mus. Moreton Morrell.
A rare plant of calcareous grassland recorded from pasture at Harrow Hill (2833) and from roadsides near Weston Park (2934) and near Moreton Morrell (3154); recorded from all surrounding VCs.

205 Onobrychis Mill.

205.1 **Onobrychis viciifolia** Scop. Sainfoin
Purton, T., 1817. Mid. Flora [*Hedysarum onobrychis*]. Grafton; Billesley.
Map 217, p. 341
A rare plant, probably always a relic of cultivation, found only in a few widely scattered localities on calcareous soils in the S. and E. of the county; recorded from railway banks, roadsides, grassland and waste places; recorded from all surrounding VCs.

206 Vicia L.

206.1 **Vicia hirsuta** (L.) Gray Hairy Tare
Perry, 1812. Herb. Perry. Without locality.
Map 218, p. 341
Abundant and fairly evenly distributed throughout most of the county but rare in the extreme S.; recorded from roadsides, railway banks and waste places (42%), cultivated land (24%), grassland (20%) and hedgerows and scrub (9%); recorded from all surrounding VCs.

206.2 **Vicia tetrasperma** (L.) Schreb. Smooth Tare
Perry, 1812. Herb. Perry. Without locality.
Map 219, p. 342
Frequent, but very irregularly distributed throughout most of the county, with considerable concentration in the S.W.; recorded from cultivated land (33%), roadsides, railway banks, waste places and farmyards (27%), grassland (23%), hedgerows and scrub (13%) and a few other habitats; recorded from all surrounding VCs.

206.3 **Vicia tenuissima** (M. Bieb.) Schinz & Thell.
Slender Tare
Palmer, Miss C. E., 1854. Herb. Oxf. Univ. [prob. *V. gracilis*]. Lighthorne.
A rare plant of cultivated fields, rough grassland and quarries, recorded from near Drayton (1656), Newbold on Stour (2345), near Combrook (2950, 2952) and Lighthorne (3355); previously also collected from Honington and Moreton Morrell; recorded from VCs 23, 33 and 37.

206.4 **Vicia cracca** L. Tufted Vetch
Perry, 1812. Herb. Perry. Without locality.
Map 220, p. 342
Abundant and fairly evenly distributed throughout the county; recorded from hedgerows and scrub (38%), roadsides, railway banks and waste places (32%), grassland (20%), watersides and marshes (6%) and a few other habitats; recorded from all surrounding VCs.

206.5 **Vicia tenuifolia** Roth Fine-leaved Vetch
Dawson, R. W., 1962. Combrook.
Three records, from roadside at Binton (1453), and railway embankments near Combrook (2950) and Lower Hillmorton (5274); recorded from all surrounding VCs except 32 and 57.

206.10 **Vicia sylvatica** L. Wood Vetch
Bree, Rev. W. T., 1830. Mag. Nat. Hist. Bentley Park.
Not recorded in present survey; recorded from all surrounding VCs.

206.11 **Vicia sepium** L. Bush Vetch
Baxter, W., 1831. MS (Bagnall: Mid. Nat. 1892–93.) Near Rugby. Map 221, p. 343
Abundant but very irregularly distributed throughout most of the N.W. and the extreme S. of the county, occasional and more irregularly distributed elsewhere; recorded from roadsides, railway banks and waste places (46%), hedgerows and scrub (24%), grassland (24%), woods (3%) and a few other habitats; recorded from all surrounding VCs.

206.13 **Vicia hybrida** L. Hairy Yellow-vetch
Druce, 1932. Com. Flora. Without locality.
Not recorded in present survey; recorded from VCs 39 and 55.

206.14 **Vicia sativa** L. ssp. **sativa** Common Vetch
Ick, W., 1836. Herb. Ick [*V. sativa*]. Side of the road to Erdington 3½ miles from Birmingham.
Map 222, p. 343
Abundant and evenly distributed throughout the county, except in the extreme N. where it is comparatively rare; recorded from roadsides, railway banks and waste places (43%), grassland (31%), hedgerows and scrub (13%), cultivated land (8%) and a few other habitats; recorded from all surrounding VCs.

206.15 **Vicia sativa** L. ssp. **angustifolia** (L.) Gaud.
Narrow-leaved Vetch
Bree, Rev. W. T., 1835. New Bot. Guide [*V. angustifolia* var. *Bobartii*]. Without locality.
Map 223, p. 344
Narrow-leaved forms of *V. sativa* L. have generally been distinguished as ssp. *angustifolia*; this ssp. is frequent but very irregularly distributed throughout the county; recorded from roadsides, railway banks and waste places (45%), grassland (32%), hedgerows (12%), cultivated land (5%) and quarries (4%); recorded from all surrounding VCs.

206.16 **Vicia lathyroides** L. Spring Vetch
Purton, T., 1817. Mid. Flora. On the side of the Bridle Road from Spernall to Studley.
Not recorded in present survey; recorded from VCs 23, 37, 39 and 57.

207 Lathyrus L.

207.1 **Lathyrus aphaca** L. Yellow Vetchling
Purton, T., 1817. Mid. Flora. Alne Hills.
A rare plant of cultivated and waste places, probably not native in Warwickshire; recorded from near Wellesbourne (2653, 2756), Caludon House (3780), Long Itchington (4164), Rugby (4974) and Hillmorton (5373); recorded from all surrounding VCs except 32.

207.2 **Lathyrus nissolia** L. Grass Vetchling
Purton, T., 1817. Mid. Flora. Coughton.
Map 224, p. 344
A rare plant almost entirely confined to the S.W. of the county; recorded from grassland, roadsides, hedgerows and scrub; recorded from all surrounding VCs.

207.3 **Lathyrus hirsutus** L. Hairy Vetchling
Cadbury, Miss D. A., 1957. Near Baginton, Coventry.
One record from cultivated ground near Baginton (3473); recorded from VCs 23, 33, 39 and 57.

207.4 **Lathyrus pratensis** L. Meadow Vetchling
Perry, 1812. Herb. Perry. Without locality. Map 225, p. 345
Abundant and evenly distributed throughout the county; recorded from roadsides, railway banks and waste places (40%), grassland (37%), hedgerows and scrub (14%), watersides and marshes (5%) and a few other habitats; recorded from all surrounding VCs.

207.5 **Lathyrus tuberosus** L. Earth-nut Pea
Druce, 1932. Com. Flora. Without locality.
A very rare plant recorded from withy beds at Alveston (2456), where it was known up to 1955, and from railway bank near Studley Station (0563); recorded from all surrounding VCs except 32.

207.6 **Lathyrus sylvestris** L.
Narrow-leaved Everlasting Pea
Perry, 1812. Herb. Perry. Green's Grove, near Hatton.
A rare plant of open woods, hedges, rough grassland and quarries recorded from Spernall Park (1062), Aston Grove (1357), Gospel Oak (1858), near Coventry (3476), Edge Hill (3747) and near Flecknoe (4764, 4864); recorded from all surrounding VCs.

207.8 **Lathyrus latifolius** L.
Broad-leaved Everlasting Pea
Bagnall, 1872. Herb. Bagnall. Chesterton Wood. Map 748, p. 604
Occasionally naturalised on railway banks, in hedgerows and waste places; recorded from all surrounding VCs except 39 and 57.

207.11 **Lathyrus montanus** Bernh. Bitter Vetch
Ray, 1670. Catalogus [*Astragalus sylvaticus*]. "Warwicii frequens." Map 226, p. 345
Abundant in the W. of the county, mainly in areas which were forest in the seventeenth century; very rare elsewhere; recorded from hedgerow and scrub (43%), grassland (26%), roadsides (21%) and woods (10%); recorded from all surrounding VCs.

ROSACEAE

209 **Spiraea** L.

209.1 **Spiraea salicifolia** L. Willow-leaved Spiraea
Cadbury, Miss D. A., 1953. Edgbaston Park.
Generally planted and occasionally naturalised in woods, hedges and waste places, and recorded from Edgbaston Park (0583, 0584), Trap's Green (1069), Boldmere (1194), Temple Balsall (2075), Grove Park (2464), Warwick (2965), Kenilworth (3072) and Little Walton (4983); recorded from all surrounding VCs.

Spiraea menziesii Hook.
Arnold, G. & M., 1963. Middleton Wood Farm.
One record from a hedgerow at Middleton Wood Farm (1698); almost certainly a garden escape; not recorded from any of the surrounding VCs.

210 **Filipendula** Mill.

210.1 **Filipendula vulgaris** Moench Dropwort
Purton, T., 1817. Mid. Flora [*Spiraea filipendula*]. Spernall; Arrow. Map 227, p. 346
An occasional plant confined to the calcareous soils in the S. and S.E. of the county; recorded from grassland (79%), roadsides and railway banks (21%); recorded from all surrounding VCs.

210.2 **Filipendula ulmaria** (L.) Maxim.
Meadow-sweet
Ick, W., 1836. Herb. Ick [*Spiraea ulmaria*]. Bank near Vaughton's Hole. Map 228, p. 346
Abundant and fairly evenly distributed throughout the county; recorded from watersides and marshes (67%), hedgerows and scrub (11%), roadsides and railway banks (9%), damp grassland (8%) and damp woods (4%); recorded from all surrounding VCs.

211 **Rubus** L.

211.6 **Rubus idaeus** L. Raspberry
Ray, 1688. Hist. Plant. [*Rubus Idaeus spinosus fructu rubro* J.B.]. "Humido agro Warwicensi." Map 229, p. 347
Abundant and fairly evenly distributed in the N. of the county, but infrequent in the lower Avon Valley and on the calcareous soils of the S. and S.E.; recorded from hedges and scrub (33%), woods (29%), roadsides, waste places and railway banks (24%), watersides (6%), grassland (6%) and a few other habitats; recorded from all surrounding VCs.

211.9 **Rubus caesius** L. Dewberry
Kirk, T., 1852. Herb. Perry. Whitley Common. Map 230, p. 347
Frequent and fairly evenly distributed mainly on the calcareous soils in the S. and S.E. of the county; recorded from hedges and scrub (47%), watersides (20%), roadsides, railway banks and waste places (20%), rough grassland (8%) and woods (5%); recorded from all surrounding VCs.

211.11 **Rubus fruticosus** agg. Blackberry, Bramble
Perry, 181–. Herb. Perry [*R. fruticosus discolor*]. Without locality. Map 231, p. 348
Abundant and evenly distributed throughout the county; recorded from hedges and scrub (60%), woods (15%), waste places, railway banks and roadsides (15%), watersides (4%), grassland (4%) and a few other habitats; recorded from all surrounding VCs.

There follow some Introductory Notes and Keys to some of the Warwickshire Rubi by J. F. Woolman, which, in the opinion of the authors, form a valuable introduction to the study of this very difficult group, and which they have found useful in the identification of some of the commoner species in the county.

Introduction

The Keys which follow attempt to provide the means of identification for the commoner Warwickshire Rubi. All such Keys have their limitations, and, owing to the wide range of variation within the species, this is probably more notably the case with Rubus than with any other group. It is hoped, however, that the Keys will enable many people to identify some at least of the brambles they collect and thus stimulate their interest in the group.

Nomenclature

The names used throughout are those used by W. C. R. Watson in his "Rubi of Great Britain and Ireland", 1958, with a few exceptions to bring the nomenclature up to date.

Specimens

Great care must be taken in collecting the specimens and in assessing the characters used. Each plant consists of two types of shoot:

(a) First year's growth, or barren stems, which are the growths of the current year and do not bear inflorescences.
(b) Flowering shoots which consist of growths of previous years, from which axillary shoots have arisen bearing flowering stems with panicles.

Material required for identification consists in each case of a piece of barren stem from approximately the middle of the growth *and* an end piece or panicle from the flowering stem.

It is obviously *essential* that the two pieces are taken from the same plant though it may at times be difficult to make certain of this.

In the Keys the term *stem* (unless otherwise stated) is used to denote a suitable piece of the barren stem, and *panicle* to denote the flowering stem.

Characters used and their assessment

With the exception of the Suberecti, which are separated by their habit of growth, the Rubi are primarily separated into groups according to the nature of the armature of the barren stem, which ranges from no stalked glands, acicles or pricklets at the one extreme to a heavy growth of all three, varying in size, range and complexity, at the other (Figs 9e and f).

The stem may be round, bluntly angled or sharply angled. Its armature may include:

1. *Prickles*, which may be long or short based, weak or strong (Fig. 9a).
2. *Pricklets*, which are weaker and shorter than prickles but generally have noticeable and elongated bases (Fig. 9b).
3. *Acicles*, which are needle-like, very narrow and with little or no basal elongation (Fig. 9c).
4. *Glands*, which may be stalked or sessile; stalked glands may be of various lengths and are bristle-like, each with a round pinhead-shaped gland at the tip; sessile glands appear in the form of dark dots (Fig. 9d).

Other barren stem characters used are leaflet shape, length of leaflet stalks, leaf stipule shape and stem section. Flower characters are petal colour, shape and size, and the colours and relative lengths of stamens and styles. Petal colour may vary with age, and in assessing colour a recently opened bloom, not a semi-open or an aged one, should be used. In comparing stamen and style lengths reference should be made to the proportion apparent in a young, just fully open flower and not to the actual lengths when ascertained by dissection.

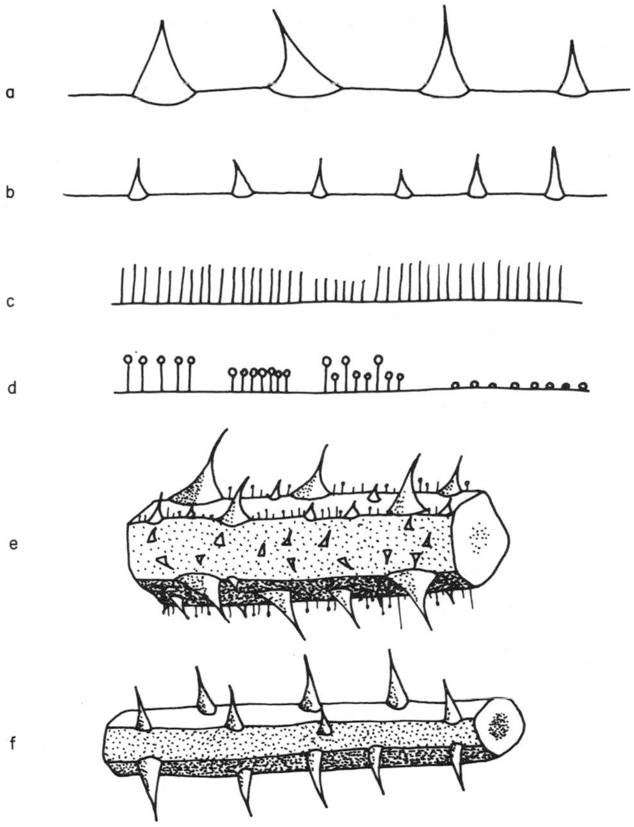

FIG. 9. Stem armature in *Rubus*.
(a) Prickles (b) Pricklets (c) Acicles (d) Glands—stalked and sessile (e) Stem with prickles, pricklets, acicles and glands (*R. fuscus*) (f) Stems with prickles and pricklets only (*R. scissus*)

In assessing some of the characters it may be useful to use a × 10 lens and, occasionally, as for examining for staminal hairs, a × 20 lens.

The Keys

These consist of:

1. A key to the main groupings (Sections and Series).
2. A further splitting of these groups into species.

At the end of some of the sections there are set out the names of a number of species coming within that group which have been collected and named, or identified as being present in the county, but which it has not proved possible for various reasons (insufficient specimens, wide range of variability etc.) to include in the Key.

At the end of the Key, there is a further list of species,

also collected and named, or identified as being present in the county, but which do not come within any of the Series included in the Key. They are also included (in smaller type) in the species list which follows the Keys, where the authorities for their inclusion are given.

Acknowledgements are gratefully made to E. S. Edees for checking specimens, supplying records and for much advice and help, and to B. A. Miles for supplying records and advice.

J. F. W.

Keys to some Warwickshire Rubi

Rubus fruticosus agg.

This is a key to the Sections and Series present in Warwickshire with the exceptions of series Apiculati and Grandifolii in Section Appendiculati, which are rather variable in the size, quantity and distribution of stalked glands and acicles on the barren stem. In this respect they lie roughly between series Radulae in Section Appendiculati and series Hystrices in Section Glandulosi. Any specimen which appears to fall between these series without readily fitting into either should bc referred to the list of additional species at the end of the Keys.

Key to Sections and Series

1. Barren stems erect or suberect, similar to those of *R. idaeus*; panicle simple, racemose or subracemose . . . . . . **Suberecti**
   Barren stems arching variously; panicle compound 2
2. Stipules of barren stem leaves linear-lanceolate to lanceolate; basal leaflets ± sessile (*R. conjungens* sometimes slightly stalked) . 3
   Stipules of barren stem leaves linear or filiform; basal leaflets stalked . . . 4
3. Barren stems pruinose-glaucous, very weak and creeping, with weak prickles; terminal flowers of the inflorescence with more styles and therefore more drupels than the lateral ones . . . . . . (See *R. caesius*)
   Barren stems not markedly pruinose-glaucous or weak and creeping; prickles usually stronger and broader based; terminal and lateral flowers not differentiated as above **Triviales**
4. Barren stems without acicles, stalked glands or pricklets 5
   Barren stems with any or all: acicles, stalked glands and pricklets 6
5. Barren stem leaves thick and leathery (except in shade forms), usually dark green and glabrous above, grey to white felted beneath; barren stems pruinose and reddish purple; prickles very strong and broad-based, pruinose; stamens in young flowers shorter than or about equal to styles. (In *R. nemoralis* [*Silvatici*] the stamens are sometimes equal to the styles, but the sepals have long leafy tips) . . . . . **Discolores**
   Barren stem leaves not thick and leathery, paler green with some hairs above, not felted beneath; prickles not noticeably very strong; stamens in young flowers clearly longer than styles . . . . **Silvatici**
6. Stamens in young flowers shorter than or equal to styles; flowers less than 2 cm diam.; petals widely separated . . . **Sprengeliani**
   Stamens in young flowers longer than styles; flower size and petal arrangement various, but not combined as above . . . 7
7. Main prickles of barren stems ± confined to the angles; acicles, stalked glands and pricklets varying in length but always shorter than and distinct from the main prickles 8
   Main prickles of the barren stems not confined to the angles and grading imperceptibly into pricklets, acicles and stalked glands **Hystrices**
8. Stalked glands and/or acicles, and/or pricklets sparse on barren stems which are pubescent, the hairs often covering the glands; terminal barren stem leaflets roundish, finely serrate **Vestiti**
   Stalked glands, acicles and pricklets plentiful on barren stems, roughly equal in length; barren stems either glabrous or hairy but hairs not covering the glands; terminal leaflets various in shape . . . **Radulae**

Key to Species

At the end of some of the Sections there are included (in small type) the names of some species recorded from Warwickshire, but not included in the Key.

Section Suberecti

Barren stems erect or suberect, similar to a raspberry thicket when seen in quantity; panicle simple, racemose or subracemose.

1. Barren stem prickles larger than 1·5 mm., flattened, long-based; stamens ± equal in length to styles, mainly reflexed and spreading; ripe fruit black . . . *R. plicatus*
   Barren stem prickles up to 1·5 mm. in length, subulate or conical; stamens shorter than styles; ripe fruit dark red . . . *R. scissus*

R arrheniiformis
R. opacus

Section Triviales

Barren stem leaves with lanceolate stipules and sessile basal leaflets (sometimes slightly stalked in *R. conjungens*); panicle compound.

1. Barren stem prickles ± equal in size and length; few or no acicles or pricklets present 2
   Barren stem prickles unequal in size and length, grading into the numerous pricklets and acicles 4
2. Barren stems round; prickles slender and short based; petals white . . . *R. sublustris*
   Barren stems not round; prickles moderate to strong, longer based; petals pink . . 3
3. Barren stem leaflets overlapping, not soft to the touch, the basal ones often slightly stalked; stems not pubescent; stalked glands, if any, on barren stem, few and short *R. conjungens*

Barren stem leaflets not noticeably overlapping, soft to the touch, the basal ones sessile; stems pubescent and often green marbled; long stalked glands clearly present *R. balfourianus*

4. Petals white . . . . *R. myriacanthus*
Petals pink or pinkish . . . . *R. scabrosus*

R. tuberculatus

Section Silvatici

Stems arching variously; no acicles, stalked glands or pricklets on barren stem; leaf stipules linear or filiform; basal leaflets of barren stem leaves stalked; leaves with upper surface medium to pale green with some pubescence (*R. lindleianus* and *R. nemoralis* often almost glabrous), greenish to grey beneath; stamens clearly longer than styles in most species.

1. Barren stem leaves with long stalked leaflets, divided into pairs of laciniate segments (A garden escape not known wild in any country) . . . . *R. laciniatus*
Not as above 2
2. Petals white or slightly pinkish . . . 3
Petals a definite pink 4
3. Barren stem prickles numerous, strong based; pedicels and panicle rachis without stalked glands . . . . . *R. carpinifolius*
Barren stem prickles moderate; leaves noticeably obovate-cuneate with wavy edges; pedicels and sometimes panicle rachis with at least some stalked glands, even if partly hidden by hairs; panicle large and showy with the branches spreading at right angles to the main axis . . . *R. lindleianus*
4. Stamen filaments noticeably longer than styles 5
Stamen filaments barely equal to or little longer than styles 6
5. Barren stem leaves glabrescent beneath . *R. gratus*
Barren stem leaves felted beneath . *R. carpinifolius*
6. Petals narrow, distinctly separated; flowers less than 2 cm diam. [*Sect. Sprengeliani*] *R. sprengelii*
Petals wider and overlapping; flowers more than 2 cm diam. . . . . *R. nemoralis*

R. albionis
R. amplificatus
R. calvatus
R. cardiophyllus
R. favonii
R. lindebergii
R. mercicus
R. mercicus var. *bracteatus*
R. polyanthemus
R. pyramidalis
R. rubritinctus

Section Discolores

No acicles, stalked glands or pricklets on barren stem, which is pruinose and reddish purple (greenish in shade forms); prickles very strong and broad based; leaf stipules filiform or linear; basal leaflets stalked; leaflets generally leathery, thick, more or less glabrous above, whitish felted beneath.

Stamens equal to or shorter than styles; leaflets dark green above (paler when young and in shade forms) . . . . *R. ulmifolius*

R. crassifolius
R. falcatus
R. winteri

Section Sprengeliani

Stalked glands, acicles and pricklets present on barren stems varying in quantity, always shorter than and distinct from the main prickles which are ± confined to the angles; flowers 2 cm diam. or less.

Barren stems roundish in section with a few pricklets and stalked glands; petals bright rose pink, narrow; stamens pink . *R. sprengelii*

Section Appendiculati

Barren stems with pricklets, acicles and stalked glands always present but varying from few (comparable in number to prickles) to many, but not grading into the prickles.

Series Vestiti

Barren stems with long stalked glands and/or pricklets, few and occasional; basal leaflets stalked.

Barren stem prickles long and subulate; leaflets hairy and soft, usually dark green above, terminal leaflet round; petals roundish, deep rose pink; stamens rose pink; styles greenish *R. vestitus*

Series Radulae

Barren stems with acicles and stalked glands plentiful and ± equal in length, but clearly shorter than the main prickles, which are ± confined to the angles of the stem; leaf stipules linear or filiform; basal leaflets of barren stem leaves stalked.

Barren stems glabrous (though acicles and glands present); panicle rachis glabrous below, not heavily felted with long hairs but some present above . . . . *R. echinatoides*
Barren stems hairy; panicle rachis ± felted with long hairs which mainly cover the stalked glands . . . . . . *R. echinatus*

R. granulatus [Series Radulae]
R. mucronulatus [Series Mucronati]

Section Glandulosi

Series Hystrices

Barren stems angled with numerous stalked glands, acicles and pricklets grading into each other and into the prickles which are distributed all round the stem, the largest being strong and broad based; inflorescence densely glandular.

Barren stem prickles noticeably declining and/or hooked, many gland-tipped; petals white; flowers 1–1·5 cm diam. . . . *R. murrayi*

Barren stem prickles not noticeably declining, not gland-tipped; petals pink; flowers 2–2·5 (–3) cm diam. . . . . . *R. dasyphyllus*

R. scabripes [Series Hystrices]
R. hylonomus [Series Euglandulosi]
R. apiculatus [Apiculati]
R. bloxamii [Apiculati]
R. diversus [Grandifolii]
R. drymophilus [Apiculati]
R. euryanthemus [Apiculati]
R. foliosus [Apiculati]
R. fuscus [Apiculati]
R. insectifolius [Apiculati]
R. leightonii [Grandifolii]
R. pallidus [Apiculati]
R. rotundifolius [Grandifolii]
R. rufescens [Apiculati]
R. scaber [Apiculati]

Rubus Species List

In compiling records of the presence of species of Rubus in Warwickshire, cognisance has been taken only of actual specimens seen and agreed by E. S. Edees, B. A. Miles or J. F. Woolman (on the basis of specimens agreed by him with E. S. Edees). A vast amount of material collected prior to the present survey exists in Herb. Bagnall, Herb. B'ham Mus. and other collections, but, owing to the complexity of the problem, the changes in the understanding of the species and the confusion of nomenclature, it has been considered advisable to refer only to specimens which have been determined as stated and which can readily be identified and referred to.

In the few cases where the name used for a species differs from that used by Watson, that used by him is added in parenthesis.

Records for the presence of a species in any of the surrounding VCs are given only for those species noted as recorded for Warwickshire.

Section Suberecti P. J. Muell.

211.11.2 **Rubus scissus** W. C. R. Wats.
Bagnall, 1873. Herb. Bagnall [*R. fissus* Lindl.]. Chelmsley Wood.
Recorded from woods and heaths in Sutton Park (0898, 1096, 1097); also collected by Bagnall between 1876 and 1888 from Bentley Park, Coughton Park, Hay Wood, Solihull, Sutton Park and Trickley Coppice; recorded from all surrounding VCs except 32 and 33.

211.11.4 **Rubus opacus** Focke ex Bertram
Bagnall, 1876. Herb. Brit. Mus. Sutton Park.
Not recorded in present survey; recorded from VCs 37, 39 and 55.

211.11.5 **Rubus plicatus** Weihe & Nees
Woolman, J. F., 1954. Sutton Park.
Recorded from dry heath and scrub in Sutton Park (0997, 1096, 1097, 1098); recorded from VCs 37, 39 and 57.

211.11.8 **Rubus arrheniiformis** W. C. R. Wats.
Bagnall, various dates 1875–91. Herbs. Brit. Mus., Camb. Univ., Nat. Mus. Wales, and S. London Bot. Inst. Sand quarry, Cornets End.
Not recorded in present survey; collected by H. Stuart Thompson in 1901 from Sutton Park; not recorded from any of the surrounding VCs.

Section Triviales P. J. Muell.

211.11.14
Rubus conjungens (Bab.) W. C. R. Wats.
Woolman, J. F., 1954. Sutton Park.
Recorded from hedgerows at Tanners Green (0874), Sutton Park (0995), Morton Bagot (1165), near Solihull (1478), Austy Wood (1662), Lowsonford (1867), near Hampton in Arden (1881, 1981), Snitterfield Bushes (1960), Temple Balsall (2076, 2176), Snitterfield (2160), Balsall Street (2276), Hornbrook Farm (2281), Alvecote (2403, 2404, 2405, 2504, 2505) and Walton (2853); not recorded from any of the surrounding VCs.

211.11.17
Rubus sublustris Lees
Kirk, T., 1852. Herb. Brit. Mus. Folly Lane, near Stoke
Recorded from hedgerows at Tanners Green (0874), Sutton Park (0995), Tidbury Green (1075), Morton Bagot (1164, 1165), Solihull (1478), near Catherine de Barnes (1881), near Barston (1978), Hampton Manor (1981), Snitterfield Bushes (2060), Lodge Farm (2160), Balsall Street (2276), near Hampton in Arden (2281), Newbold on Stour (2345, 2446), Alvecote (2403, 2504), Halford (2545) and Walton (2853); collected by Bagnall between 1873 and 1886 from lane between Bourton and Birdingbury, Curdworth Bridge, Dunchurch Road near Rugby, Meriden and near Olton; recorded from all surrounding VCs except 37.

211.11.21
Rubus balfourianus Bloxam ex Bab.
Bloxam, Rev. A., 1847. Nat. Mus. Wales. Rugby. Map 232, p. 348
The map shows a number of records from hedges and woods mainly in the Solihull–Shirley–Earlswood area, but, owing to the unevenness of the recording of *Rubus* species in the county, cannot be taken as showing the true distribution; it was collected by Bagnall near Kingwood in 1893; recorded from VC 39.

211.11.27 **Rubus tuberculatus** Bab.
Bagnall, 1885. Herb. Nat. Mus. Wales. Lane, Wetherley.
Not recorded in present survey; recorded from all surrounding VCs except 32 and 55.

211.11.30
Rubus scabrosus P. J. Muell.
Woolman, J. F., 1952. Earlswood.
Recorded mainly from hedges from Edgbaston (0486), Earlswood (0974, 1074), Morton Bagot (1165), near Salter Street (1275), Olton Reservoir (1381), near Solihull (1779), Lowsonford (1867, 1868), Catherine de Barnes (1880), Holywell (1966), Netherwood Heath (1973), Hampton Manor (1981), Balsall Street (2276), near Hampton in Arden (2281), Elder Tree Copse (2849), Lodge Clump (3559) and Willoughby (5266); recorded from VC 39.

211.11.31
Rubus myriacanthus Focke
Woolman, J. F., 1952. Cow Hayes, near Solihull.
Recorded from woods and hedgerows at Olton Reservoir (1381), Cow Hayes (1778, 1779), near Barston

(1978), Snitterfield Bushes (2060), Lodge Farm (2160) and Bentley Park (2–9–); recorded from VCs 33, 39 and 57.

Section Silvatici P. J. Muell.

211.11.34
Rubus gratus Focke
Bagnall, 1887. Herb. Nat. Mus. Wales. Birchley Hayes.
Recorded from woodland at Sutton Park (1096) and from hedgerows at Salter Street (1174), Monkspath Hill Farm (1575), Balsall Street (2276), and near Hampton in Arden (2281); collected by Bagnall in 1879 from Bentley Park, in 1880 from Kenilworth Heath and in 1881 from Ironstone Wood, Oldbury; recorded from VCs 23, 37, 39 and 57.

211.11.42 **Rubus calvatus** Lees ex Bloxam
Bagnall, 1885. Herb. Bagnall [*R. villicaulis* Koehl. var. *calvatus* Blox.]. Lane from Austrey to No Man's Heath.
One record only from Sutton Park (1–9–); recorded from VCs 23, 39, 55 and 57.

211.11.47
Rubus carpinifolius Weihe & Nees
Bagnall, 1882. Herb. Bagnall. Near Shilton, S. Warws.
Recorded from hedgerows and scrub woodland at Forshaw Heath (0873), Sutton Park (0995, 1097), Rumbush (1074), near Cheswick Green (1275), Monkspath Hill Farm (1475), Four Ashes (1575), Ravenshaw (1779), Springfield House (2076), Balsall Street (2276), Cornets End (2380), Shuttington Bridge (2405), Fernhill Farm (2570) and near Grandborough (5066); collected by Bagnall at Minworth in 1892 and at Coleshill Heath in 1893; recorded from VCs 39, 55 and 57.

211.11.52
Rubus nemoralis P. J. Muell.
Bagnall, 1881. Herb. Bagnall [*R. selmeri* Lindeb.]. Pool Hollies Wood, Sutton Park.
Recorded from rough grassland and hedgerows in Sutton Park (1097) and Alvecote (2403, 2504); also collected by Bagnall in 1885 between Alvecote Mill and Fazeley, and in 1891 from Marston Green and Water Orton; recorded from all surrounding VCs except 32 and 33.

211.11.54
Rubus laciniatus Willd.
Thomas, C., 1950. Meadow Road, Edgbaston.
A garden escape recorded from Edgbaston (0285), Shirley (1180), Small Heath (1185), Dosthill Quarries (2199) and Caludon Castle (3780); recorded from VCs 23, 33 and 57.

211.11.59
Rubus lindleianus Lees
Kirk, T., 1854. Herb. Brit. Mus. Near Stoke.
Recorded from hedgerows at Oversley Green (0956), Morton Bagot (1164), Tanworth in Arden (1170), Shirley (1180), Danzey (1268, 1269), Streetsbrook (1380), Monkspath Hill Farm (1475), Four Ashes (1575), Cow Hayes (1779), near Bearley Cross (1861), Netherwood (1973), Lodge Farm (2160), Balsall Street (2276), near Hampton in Arden (2281), Fernhill Farm (2570), Alvecote (2403, 2404, 2405, 2503, 2504, 2505, 2603, 2604) and near Willoughby (5266); between 1872 and 1892 collected by Bagnall from Arley Wood, Austy Wood, Bentley Park, near Kenilworth, Minworth, Ragley, Sutton Park and Woodloes; recorded from all surrounding VCs.

211.11.62 **Rubus mercicus** Bagnall
Bagnall, 1892. Herb. B'ham Univ. Water Orton.
Not recorded in present survey; collected by Bagnall in 1893 from Kenilworth and Minworth, and in 1905 from Water Orton; not recorded from any of the surrounding VCs.

211.11.60 **Rubus mercicus** var. **bracteatus** Bagnall
Bagnall, 1896. Herb. B'ham Univ., Hartshill Quarries.
Recorded from Snitterfield Bushes (1960, 2060) and Hartshill Quarries (3394); also collected by Bagnall from Radford Lane to Corley Moor in 1897; recorded from VCs 23, 39 and 55.

211.11.77 **Rubus amplificatus** Lees
Ley, A., 1874. Herb. B'ham Univ. [*R. affinis* W. & N.]. Hedge at Harborough Magna.
Not recorded in present survey; recorded from all surrounding VCs except 32.

211.11.80 **Rubus pyramidalis** Kalt.
Edees, E. S., 1963. Oakley Wood.
One record only from Oakley Wood (3059); recorded from VCs 32 and 39.

211.11.81 **Rubus albionis** W. C. R. Wats.
Bagnall, 1874. Herb. Bagnall [*R. macrophyllus* W. & N. var. *schlechtendalii* W. & N.]. Near Blackroot Pool, Sutton Park.
Not recorded in present survey; it was collected by Bagnall in 1882 from near Berkswell; recorded from VC 23.

211.11.107
Rubus favonii W. C. R. Wats.
Riddelsdell, H. J., 1948. Flora of Gloucestershire [*R. buttii* Bart. & Ridd.]. Copse near Four-Shire Stone.
Not recorded in present survey; recorded from VC 33.

211.11.113
Rubus polyanthemus Lindeb.
Bagnall, 1871. Herb. Bagnall [*R. pulcherrimus* Neum.]. Near Coleshill Pool.
Recorded from hedgerows at Earlswood (0974), Alvecote (2504, 2604), Bentley Park (2–9–) and Oakley Wood (3059); also collected by Bagnall in 1880 from Meriden; recorded from all surrounding VCs.

211.11.114
Rubus rubritinctus W. C. R. Wats.
Bagnall, 1897. Herb. Nat. Mus. Wales. Cubbington Woods.
Two records from hedgerows at Alvecote (2405, 2504); recorded from VCs 23, 37 and 39.

211.11.123
Rubus cardiophyllus Muell. & Lefèv.
Bagnall, 1874. Herb. Bagnall [*R. rhamnifolius* W. & N.]. Lane from Rounsel Lane, Kenilworth, to Leek Wootton.
Two records from Bentley Park Wood (2–9–) and near Oakley Wood (3059); between 1875 and 1886 collected by Bagnall in a lane from Atherstone to Bentley Park, at Burmington, Honiley and near Offchurch; recorded from all surrounding VCs.

211.11.125

Rubus lindebergii P. J. Muell.

Bagnall, 1879. Herb. Camb. Univ. [*R. carpinifolius*]. Heath lands near Kenilworth Castle.

Not recorded during present survey; recorded from VCs 39, 55 and 57.

Section Discolores P. J. Muell.

211.11.129

Rubus ulmifolius Schott

Bagnall, 1871. Herb. Bagnall [*R. rusticanus* Merc.]. Sutton Park Map 233, p. 349

A plant mainly of hedges widely distributed throughout the county though certainly under-recorded; collected by Bagnall between 1871 and 1886 from Great Alne to Spernall, Kenilworth, Knowle, Sambourne and Sutton Park; recorded from all surrounding VCs.

211.11.132

Rubus winteri P. J. Muell. ex Focke

Bloxam, prob. 184–. Herb. Camb. Univ.;
Bagnall, 1887. Herb. Brit. Mus. and Herb. Nat. Mus. Wales [*R. macroacanthus* Blox.]. Mancetter.

One record from Mancetter (3395); recorded from VCs 32 and 57.

211.11.133

Rubus crassifolius Genev. (**R. propinquus** P. J. Muell.)

Bagnall, 1879. Herb. S. Lond. Bot. Inst. Solihull.

Not recorded in present survey; recorded from VC 37.

211.11.141

Rubus falcatus Kalt.

Bagnall, 1870. Herb. Bagnall [*R. thyrsoideus* Wimm.]. Marston Green.

Not recorded in present survey; also collected by Bagnall from Atherstone on Stour in 1886 and from Bishopton in 1888; recorded from all surrounding VCs except 37.

Section Sprengeliani (Focke) W. C. R. Wats.

211.11.146

Rubus sprengelii Weihe

Bloxam, Rev. A., 1846. Herb. Camb. Univ. Near Rugby.

Recorded from woods and hedges at Earlswood (0974, 1074), Sutton Park (0896, 0897, 0898, 0997, 1097, 1098, 1196), Four Ashes (1575), Catherine de Barnes (1880), Hay Wood (2071), Balsall Street (2276), near Hampton in Arden (2281) and Alvecote Wood (2403); also collected by Bloxam in 1847 from Outwoods near Atherstone and in 1876 from near Rugby, and by Bagnall between 1873 and 1890 from Arley Wood, Berkswell, Coleshill, Earlswood, Oldbury, Rowington and Solihull; recorded from all surrounding VCs except 33.

Sections Appendiculati (Genev.) Sudre and Glandulosi P. J. Muell.

211.11.165

Rubus vestitus Weihe & Nees

Kirk, T., 1852. Herb. Brit. Mus. Coventry Park. Map 234, p. 349

The map shows a number of records from hedges and woods mainly in the W. of the county, but, owing to the unevenness of the recording of *Rubus* species in the county, cannot be taken as showing the true distribution; also collected by Bagnall between 1870 and 1887 from Bretnall Wood near Ansley, Olton, Shirley, Shuckburgh and Stivichall Common; recorded from all surrounding VCs.

211.11.182

Rubus mucronulatus Bor. (**R. mucronifer** Sudre)

Bagnall, 1876. Herb. Bagnall [*R. mucronatus* Blox.]. Bentley Park.

One record from Bentley Park (2–9–); collected by Bagnall in 1878 from Coleshill, in 1893 from Marston Green and from Atherstone (undated); recorded from VCs 39, 55 and 57.

211.11.212

Rubus echinatus Lindl. (**R. discerptus** P. J. Muell.)

Baker, T. G., 1860. Herb. Nat. Mus. Wales. Wyken.

Recorded from woods and hedges at Tanners Green (0874), Oversley Wood (0956), New End (1060), Tidbury Green (1075), Aston Grove (1458), Four Ashes (1575), Songar Grange (1861), Balsall Street (2276), Alvecote (2403, 2405, 2504), Bentley Park (2–9–) and Brinklow High Wood (4279); collected by Bagnall in 1874 from Oakley Wood and in 1877 from Oversley Wood; recorded from all surrounding VCs.

211.11.213

Rubus echinatoides (Rogers) Sudre

Woolman, J. F., 1951. Shirley.

Recorded from railway bank at Sutton Park (1097), a wood at Shirley (1280) and canal side at Catherine de Barnes (1779); recorded from VCs 23, 37, 39 and 57.

211.11.219

Rubus granulatus Muell. & Lefèv.

Bagnall, 1879. Herb. B'ham Univ. [*R. bloxamiana* Coleman]. Near Hartshill.

Two records from Bentley Park (2–9–) and Hartshill Quarries (3394); collected by Bagnall in 1885 from Austrey and Wolvey, and in 1892 and 1893 from near Mancetter; recorded from VCs 32, 39, 55 and 57.

211.11.223

Rubus foliosus Weihe & Nees

Bagnall, 1870. Herb. Bagnall. Above Blackroot Pool, Sutton Park.

One record from Bentley Park (2–9–); collected by Bagnall in 1875 in a lane between Atherstone and Bentley Park, in 1881 from Ironstone Wood, Oldbury, and in 1891 from Spernall Park; recorded from all surrounding VCs except 32.

211.11.229

Rubus bloxamii Lees

Bagnall, 1871. Herb. Bagnall. Upper Holly Hurst by Keeper's Pool, Sutton Park.

Recorded from woods and hedgerows at Forshaw Heath (0873), Sutton Park (0998, 1096, 1097), Earlswood (1074), Catherine de Barnes (1880) and Brinklow High Wood (4279); between 1871 and 1887 collected by Bagnall from Allesley, Ansley, Coleshill, Cubbington Wood, Hartshill Hayes, Maxstoke, Merevale, Moor Hall and Sutton; recorded from all surrounding VCs except 32 and 57.

211.11.234

Rubus fuscus Weihe & Nees

Edees, E. S., 1951. Sutton Park.

Two records from Sutton Park (1–9–) and Edge Hill (3747); recorded from all surrounding VCs except 32 and 57.

211.11.243
Rubus pallidus Weihe & Nees
Bagnall, 1882. Herb. B'ham Univ. [*R. humifusus*]. Crackley Wood, Kenilworth.
Recorded from scrub woodland in Sutton Park (0997, 1097); recorded from VCs 23, 39, 55 and 57.

211.11.244
Rubus drymophilus Muell. & Lefèv.
Mills, W. H., 1949. Herb. Camb. Univ. Near Four-Shire Stone.
Not recorded in present survey; not recorded from any of the surrounding VCs.

211.11.252
Rubus euryanthemus W. C. R. Wats.
Edees, E. S. 1957. Sutton Park.
Two records from Sutton Park (1096, 1097); recorded from VCs 23, 32, 39 and 55.

211.11.253
Rubus insectifolius Muell. & Lefèv.
Bagnall, 1878. Herb. Bagnall [*R. adscitus* Genev.]. Trickley Coppice.
One record only from Sutton Park (1097); collected by Bagnall in 1878 from Middleton Park and from Moor Hall, near Sutton Coldfield (undated); recorded from VCs 33, 39 and 55.

211.11.263
Rubus scaber Weihe & Nees
Bagnall, 1880. Herb. Brit. Mus. [*R. glandulosus* b. *dentatus* Blox.]. Hoare Park, Nether Whitacre.
Not recorded in present survey; also collected by Bagnall in 1887 from Atherstone Outwoods; recorded from VCs 23, 37, 39 and 55.

211.11.284
Rubus rufescens Muell. & Lefèv.
Bagnall, 1878. Herb. Bagnall [*R. rosaceus* W. & N. var. *infecundus*]. Friars Wood, Bentley Park.
Recorded from woods and hedgerows at Shirley (1176), Alvecote Wood (2403), Hollyberry End (2684) and Bentley Park Wood (2–9–); also collected by Bagnall in 1882 from Red Lane, Kenilworth; recorded from all surrounding VCs except 32.

211.11.295
Rubus apiculatus Weihe & Nees
Edees, E. S., 1963. Cornets End; Edge Hill; Hollyberry End.
Three records from Cornets End (lane to Park Farm) near Berkswell (2380), Hollyberry End (2684) and Edge Hill (3747); recorded from VCs 23, 37, 55 and 57.

211.11.310
Rubus leightonii Lees ex Leighton
Bagnall, 1885. Herb. Nat. Mus. Wales. Austrey.
Recorded from woods and hedgerows at Sutton Park (1096, 1097), Shirley (1180), Minworth (1–9–), Cornets End (2380) and Alvecote (2404, 2405, 2504); recorded from all surrounding VCs except 37.

211.11.312
Rubus diversus W. C. R. Wats.
Bagnall, 1881. Herb. Nat. Mus. Wales [*R. hirtus* var. *kaltenbachii*]. Wood near Shustoke.
Not recorded in present survey; recorded from VCs 33, 39 and 55.

211.11.322
Rubus rotundifolius (Bab.) Bloxam
Bagnall, 1890. Herb. Nat. Mus. Wales. Boultbee Wood, near Meriden.
Not recorded in present survey; recorded from VC 55.

211.11.330
Rubus murrayi Sudre
Bagnall, 1883. Herb. Bagnall [*R. rosaceus* var. *adornatus* P. J. Muell.]. Kenilworth.
Map 235, p. 350
The map shows a number of records from hedges and woods mainly in the W. of the county, but, owing to the unevenness of the recording of *Rubus* species in the county, cannot be taken as showing the true distribution; also collected by Bagnall in 1884 from outside Ironstone Wood, Oldbury and in 1886 from Bentley Park; recorded from all surrounding VCs except 32 and 57.

211.11.349
Rubus scabripes Genev.
Bagnall, 1881. Herb. Nat. Mus. Wales. Combe Park.
One record from Hollyberry End (2684); recorded from VCs 37, 39 and 55.

211.11.356
Rubus dasyphyllus (Rogers) Rogers
Bagnall, 1876. Herb. Bagnall. Lane out of Minworth to Plants Brook.
Recorded from woods and hedges at Forshaw Heath (0873), Earlswood (0973, 1074), Sutton Park (0998, 1097), Tanworth (1170), Shirley (1180, 1280), near Cheswick Green (1275), Widney Manor (1577), Songar Grange (1861), near Eastcote (1978), Hay Wood (2071), Balsall Street (2276), near Hampton in Arden (2281), Alvecote (2403, 2404, 2405, 2504, 2604) and Frankton Wood (4171); between 1872 and 1887 collected by Bagnall from near Ansley, Middleton, Minworth, Packwood and Wootton Wawen; recorded from all surrounding VCs.

211.11.387
Rubus hylonomus Muell. & Lefèv.
Riddelsdell, H. J., 1915. Herb. Brit. Mus. [*R. hostilis*]. Wolford Heath, near Four-Shire Stone.
Not recorded in present survey; recorded from VC 33.

212 Potentilla L.

212.2 **Potentilla palustris** (L.) Scop.
Marsh Cinquefoil
Withering, 1801. Arrangement [*Comarum palustre*]. S.W. side of Edgbaston Pool.
Map 236, p. 350
A rare plant of a number of scattered localities in the N. half of the county, with some concentration in the Sutton Coldfield, Coleshill and Earlswood areas; recorded from marshes, watersides and bogs; recorded from all surrounding VCs except 33.

212.3 **Potentilla sterilis** (L.) Garcke
Barren Strawberry
Bagnall, 1867. Herb. Bagnall. Small Heath.
Map 237, p. 351
Abundant mainly in areas which were well wooded in the seventeenth century; occasional on the Lower Lias limestone and the Cotswold escarpment but rare elsewhere; recorded from roadsides and railway banks (35%), hedgerows and scrub (24%), grassland (23%) and woods (17%); recorded from all surrounding VCs.

212.5 **Potentilla anserina** L. Silverweed
Lloyd, Dr. G., 1835. Herb. Perry. Near Tachbrook. Map 238, p. 351
Abundant and evenly distributed throughout the county; recorded from roadsides, waste places, farms and railway banks (60%), grassland (18%), watersides (12%), hedgerows (5%) and cultivated land (4%); recorded from all surrounding VCs.

212.6 **Potentilla argentea** L. Hoary Cinquefoil
Perry, 181–. Herb. Perry. "On a sand-rock near the Woodloes, Warwick."
A rare plant of rocks and dry banks recorded from Salford Priors (0750), the Woodloes (2766), where it has been recorded a number of times since first seen by Perry, and E. of Brandon (4176); recorded from all surrounding VCs.

212.7 **Potentilla recta** L. Sulphur Cinquefoil
Wilkinson, W. H., 1881. Herb. B'ham Univ. Roadsides, near Park Station, Sutton Coldfield.
A garden escape recorded from a hedgerow at Wellesbourne (2756) and a waste place near Rugby (4874); recorded from all surrounding VCs except 39.

212.8 **Potentilla norvegica** L. Norwegian Cinquefoil
Hardaker, W. H., 1940. Herb. B'ham Mus. Sewage Farm, Minworth.
An introduced plant recorded from waste places at Wellesbourne (2755, 2756); recorded from all surrounding VCs.

212.13 **Potentilla erecta** (L.) Räusch.
Common Tormentil
Perry, 1812. Herb. Perry [*P. tormentilla* Stokes]. Without locality. Map 239, p. 352
Abundant but somewhat irregularly distributed in the N.W. half of the county; absent from the lower Avon Valley and rare on the calcareous soils of the S. and S.E.; recorded from grassland (43%), roadsides, railway banks and waste places (27%), woods (12%), hedgerows and scrub (6%), watersides and marshes (6%) and heaths (5%); recorded from all surrounding VCs.

212.14 **Potentilla anglica** Laichard.
Trailing Tormentil
Perry, 1812. Herb. Perry [*P. tormentilla* var. *nemoralis*]. Without locality. Map 240, p. 352
An occasional plant widely scattered throughout most of the county, but absent from the Avon Valley and the calcareous soils of the S. and S.E.; recorded from grassland (57%), roadsides and railway banks (25%), hedgerows and woods; recorded from all surrounding VCs.
We have treated this species in the aggregate sense, and in view of its uncertain position and the need for further work, have preferred to map the taxa sometimes distinguished as hybrids (*P. anglica* × *erecta*, *P. anglica* × *reptans* and *P. erecta* × *reptans*) with *P. anglica* itself.

212.15 **Potentilla reptans** L. Creeping Cinquefoil
Perry, 1812. Herb. Perry. Without locality.
Map 241, p. 353
Abundant and evenly distributed throughout the county; recorded from roadsides, waste places and railway banks (56%), grassland (27%), hedgerows (8%), watersides (5%) and a few other habitats; recorded from all surrounding VCs.

215 Fragaria L.

215.1 **Fragaria vesca** L. Wild Strawberry
Perry, 1813. Herb. Perry. Without locality.
Map 242, p. 353
Abundant throughout most of the county though rare in the Birmingham–Solihull area, in the Avon Valley and on the calcareous clays in the S.E.; recorded from mixed woods (31%), roadsides and railway banks (27%), hedgerows and scrub (23%), grassland (9%), walls and quarries (5%) and a few other habitats; recorded from all surrounding VCs.

215.2 **Fragaria moschata** Duchesne
Hautbois Strawberry
Bloxam, Rev. A., 1837. New Bot. Guide [*F. elatior*] (*fide* Bagnall). Grounds round Coton House (3 miles N. of Rugby).
A rare escape from cultivation naturalised in woods at Print Wood (3864) and Newbold Revel (4580); formerly more abundant especially near Wroxall Abbey; recorded from all surrounding VCs.

215.3 **Fragaria ananassa** Duchesne Garden Strawberry
Allen, H. 1950. Fern Bank near Berkswell.
A garden escape occasionally naturalised on railway banks and in waste places; recorded from 1566, 2088, 2089, 2378, 2477, 2788, 3175, 4165, 4682, 4976 and 5374; recorded from all surrounding VCs except 37 and 39.

216 Geum L.

216.1 **Geum urbanum** L. Herb Bennet, Wood Avens
Palmer, Miss C., 1853. Herb. Oxf. Univ. Lighthorne. Map 243, p. 354
Abundant and evenly distributed throughout the county; recorded from hedgerows and scrub (51%), woods (20%), roadsides, railway banks and waste places (18%), watersides (8%) and a few other habitats; recorded from all surrounding VCs.

216.3 **Geum rivale** L. Water Avens
Ray, 1677. Catalogus. Without locality.
Map 244, p. 354
An occasional plant of a number of scattered localities throughout the county with a concentration on the Lower Lias limestone; recorded from woods, watersides and hedgerows; recorded from all surrounding VCs.

Geum rivale L. × **urbanum** L.
= **G.** × **intermedium** Ehrh.
Baynes, W. W., 1832. Bagnall: Flora, 1891. Ufton Wood.
A rare plant of woods and wood margins usually found in the presence of both parents; recorded from Gospel Oak (1858), Chesterton Wood (3457), Waverley Wood (3569), Ufton Wood (3862, 3961, 3962, 3963), Bascote Heath (4062) and High Wood, Brinklow (4280); very large hybrid swarms have appeared in Ufton Wood from time to time, and it has been found possible to arrange specimens from such a swarm in a fairly regular series with some degree of correlation in respect of petal colour, petal length, calyx colour, erect or nodding habit of the flower and stipule length; recorded from all surrounding VCs.

218 Agrimonia L.

218.1 **Agrimonia eupatoria** L. Common Agrimony
Baxter, W., 1831. MS (Bagnall: Mid. Nat. 1892–93.) Between Lawford and Newbold.
Map 245, p. 355

A plant of roadsides (51%), rough grassland (33%) and hedgerows (12%), widely distributed throughout the county; abundant in the S. but less so in the N.; recorded from all surrounding VCs.

218.2 **Agrimonia procera** Wallr. Fragrant Agrimony
Cheshire, W., 1835. Herb. Perry. [*A. odorata* (Gouan) Mill.] Snitterfield Bushes.
Map 246, p. 355

A rare plant of damp scrub, ditches and wet grassland in a few localities mainly in the W. of the county; recorded from all surrounding VCs except 33.

220 Alchemilla L.

220.3 **Alchemilla vulgaris** L.
Ray, 1669. Letter to Lister from Middleton: "plentifully in the meadows hereabouts".

Represented in Warwickshire by *A. vestita* (q.v.) and recorded from all surrounding VCs.

220.3.2 **Alchemilla vestita** (Buser) Raunk.
Lady's Mantle
Perry, 1825. Herb. Perry [*A. vulgaris*]. Claverdon. Map 247, p. 356

A plant of humus-rich grassland (58%), roadsides and railway embankments (21%) and wet grassland, marshes and watersides (14%) chiefly in the N. and W. of the county; these are areas which may have been forest in recent historical times; recorded from all surrounding VCs.

220.3.8 **Alchemilla xanthochlora** Rothm.
Bloxam, Rev. A., 1875. Herb. Bloxam [*A. vulgaris*]. Pastures near Harborough Magna.

One plant only in grassland in grounds of Dunchurch Lodge (4971) recorded in 1961 and known there since 1946; recorded from all surrounding VCs except 32.

220.3.12 **Alchemilla mollis** (Buser) Rothm.
Fincher, F., 1966. Ipsley Mill.

An escape recorded once only on waste ground at Ipsley Mill (0666); recorded from VC 33.

Alchemilla venosa Ing.
Daulman, Miss B., 1963. Hampton in Arden.

A Caucasian species originally introduced and now well established in a churchyard at Hampton in Arden (2080); not recorded from any of the surrounding VCs.

221 Aphanes L.

Aphanes arvensis *sensu lato* Parsley Piert
Baxter, W., 1831. MS (Bagnall: Mid. Nat. 1892–93) [*Alchemilla arvensis* Scop.]. Near Sawbridge. Map 248, p. 356

A frequent plant, widely distributed throughout the county, separable into two segregates with somewhat different habitat preferences; recorded from cultivated fields (67%), roadsides and waste places (19%) and poor grassland (10%); recorded from all surrounding VCs.

221.1 **Aphanes arvensis** L.
Bagnall, 1867. Herb. Bagnall [*Alchemilla arvensis* Scop.]. Wylde Green.
Map 249, p. 357

A frequent plant of dry soils, often on sandstone and gravel (like *A. microcarpa*, but more abundant), and, in addition, occasionally on the Keuper Marl and Lower Lias limestone; recorded from cultivated land (71%), roadsides and waste places (15%), poor grassland (7%) and quarries and walls (6%); recorded from all surrounding VCs.

221.2 **Aphanes microcarpa** (Boiss. & Reut.) Rothm.
Bromwich, H., 1875. Herb. Warw. Mus. [*Alchemilla arvensis*]. Milverton.
Map 250, p. 357

An occasional plant of dry soils chiefly on sandstone and gravel; recorded from roadsides and waste places (56%), poor grassland (31%) and cultivated land (13%); recorded from all surrounding VCs.

222 Sanguisorba L.

222.1 **Sanguisorba officinalis** L. Great Burnet
Perry, 1817. List. Meadows round Warwick.
Map 251, p. 358

Abundant and fairly evenly distributed in the N.W. half of the county, but less common in the Avon Valley and on the calcareous soils of the S. and S.E.; recorded from grassland (58%), railway banks, roadsides, waste places and farmyards (18%), watersides and marshes (16%), hedgerows (6%) and a few other habitats; recorded from all surrounding VCs.

223.1 **Sanguisorba minor** Scop. ssp. **minor**
Salad Burnet
Perry, 1825. Herb. Perry [*Poterium sanguisorba* L.]. Hill between Norton Lindsey and Hampton-on-the-Hill. Map 252, p. 358

Abundant in the S. of the county, mainly on calcareous and neutral soils, with a few occurrences elsewhere; recorded from grassland (55%), roadsides and railway banks (34%), hedgerows (5%) and a few other habitats; recorded from all surrounding VCs.

223.2 **Sanguisorba minor** Scop. ssp. **muricata** Aschers. & Graebn. Fodder Burnet
Kirk, T., 1849. Herb. Perry [*Poterium muricatum* Spach]. Pinley and Kenilworth.

A rare introduced plant of rough grassland recorded from Shustoke Reservoir (2391) and near Flecknoe (5064); formerly more widespread, particularly on railway banks, and collected from Binton, Knowle, Leek Wootton, Milverton, between Kenilworth and Leamington, near Stratford on Avon and Warwick; recorded from all surrounding VCs.

225 Rosa L.

225.1 **Rosa arvensis** Huds. Field Rose
Baxter, W., 1831. MS (Bagnall: Mid. Nat. 1892–93.) Near Brownsover.
Lloyd, Dr. G., 1835. Herb. Perry. Leamington.
Map 253, p. 359

Abundant and fairly evenly distributed throughout the county; recorded from hedges and scrub (75%), woods (12%), roadsides and railway banks (5%), watersides (4%) and grassland (3%); recorded from all surrounding VCs.

225.4 **Rosa pimpinellifolia** L. Burnet Rose
Perry, 1817. List [*R. spinosissima*]. Between Hatton and Warwick, and between Stratford and Warwick.
Recorded only from Aston Grove (1457), but previously widespread in the S.E. of the county; recorded from all surrounding VCs except 23.

Rosa × involuta Sm. = **R. pimpinellifolia × villosa** *sensu lato*
Purton, T., 1821. Mid. Flora [*R. sabini* and var. *doniana*]. On a high bank at Wood Bevington (Purton) and Allesley (Bree).
Not recorded in present survey; in addition to Purton's and Bree's records, it was collected by Bagnall from Woodloes, near Oakley Wood, and Hampton in Arden between 1874 and 1884, and by Bromwich from Tachbrook in 1887; recorded from all surrounding VCs except 33.

225.5 **Rosa rugosa** Thunb.
Hardaker, W. H., 1954. Packington.
A garden escape recorded three times, from a hedgerow at Short Heath, Birmingham (0993), a waste place at Packington (2183) and rough grassland at Keresley (3182); recorded from VC 37.

225.7 **Rosa stylosa** Desv.
Watson, H. C., 1872. Comp. Cyb. Brit. Supp. [*R. stylosa* var. *gallicoides*]. Without locality.
A rare plant of hedges, usually on calcareous soil, recorded from near Kinwarton (1059), near Binton (1354) and Aston Grove (1457, 1458); specimens were collected by Bagnall and Bromwich from Chesterton Wood, Drayton and near Offchurch between 1875 and 1890; recorded from all surrounding VCs except 39 and 57.

225.8, 9 & 10
Rosa canina *sensu lato*
For first record see *R. canina* L.
Map 254, p. 359
Abundant and evenly distributed throughout the county; recorded from hedges and scrub (77%), railway banks, waste places and roadsides (8%), woods (7%), watersides (4%) and a few other habitats; this map includes records for *R. canina* L. (also mapped), *R. dumalis* Bechst. (also mapped), *R. dumetorum* Thuill. (also mapped) and *R. obtusifolia* Desv., as well as records for *R. canina* in the aggregate sense; the aggregate is recorded from all surrounding VCs.

225.8 **Rosa canina** L.
Bagnall, 1871. Herb. Bagnall [*R. canina* var. *verticillacantha* Mérat]. Lane near Solihull.
Map 749, p. 605
A number of records have been made of *R. canina* L., mainly from the W. of the county, but it has almost certainly been under-recorded, many collectors having recorded this species only in the aggregate sense. *R. canina* L. has been recorded from hedgerows and scrub (84%), waste places, roadsides and railway banks (9%) and grassland (7%); recorded from all surrounding VCs.

225.9 **Rosa dumalis** Bechst.
Bromwich, H., 1871. Herb. Bagnall [*R. glauca* Vill. var. e. *subcristata* Baker]. Hedge at Hatton.
Map 750, p. 605
An occasional plant of woodland, scrub and hedges, probably under-recorded; recorded from all surrounding VCs.
This includes forms determined as *R. coriifolia* Fr.

Rosa dumetorum Thuill.
Bagnall, 1871. Herb. Bagnall [*R. dumetorum* var. q. *platyphylla* Rau]. Shustoke.
Map 751, p. 605
An occasional plant of hedges, woodland and scrub, probably under-recorded; recorded from all surrounding VCs.

225.10 **Rosa obtusifolia** Desv.
Bagnall, 1872. Herb. Bagnall [*R. borreri* Woods]. Wheyporridge Lane, Solihull.
Only two authenticated records from hedges at Grounds Farm, near Wishaw (1695) and near Bilton (4875), but almost certainly under-recorded; previously widely known throughout the county; recorded from all surrounding VCs.

225.11, 12 & 13
Rosa villosa *sensu lato*
For first record see *R. tomentosa* Sm.
Map 255, p. 360
This map includes the records, for *R. mollis* Sm., *R. sherardii* Davies and *R. tomentosa* Sm.; the aggregate is recorded from all surrounding VCs.

225.11 **Rosa tomentosa** Sm.
Bree, Rev. W. T., 1817. Mid. Flora. Allesley.
An occasional plant of hedges recorded from Earlswood (0974, 1273), near Tanworth (1071), Hockley Heath (1572), near Billesley (1656), Snitterfield (2060, 2160), near Norton Lindsey (2163), Hatton Rock (2357), near Fillongley (3086) and Stockton (4364), undoubtedly under-recorded; numerous earlier records show that it was once, at any rate, a frequent and widely distributed species; recorded from all surrounding VCs.

225.12 **Rosa sherardii** Davies
Bagnall, 1870. Herb. Bagnall [*R. andreziovii* Stev.]. Coleshill Heath, in hedges.
A rare plant of hedges and scrub recorded from Earlswood (1273), Wolford Wood (2333) and near Coventry (3683), probably under-recorded; specimens were collected by Bagnall from Alcester, Balsall Street, Claverdon, Coleshill Heath, Honiley, Over Whitacre, Shirley, Sutton and Tanworth between 1870 and 1892; recorded from all surrounding VCs except 39.

225.13 **Rosa mollis** Sm. = **R. villosa** L. *sensu* C. T. & W.
Purton, T., 1821. Mid. Flora. Pophills Lane.
A rare plant of hedges from Gilbert's Green (1071), Aston Grove (1457) and near Fillongley (2988, 3086), probably under-recorded as it seems to have been more widespread at one time; recorded from all surrounding VCs.

225.14, 15, 16 & 17
Rosa rubiginosa *sensu lato* Sweet Briar
For first record see *R. rubiginosa* L.
This includes *R. rubiginosa* L., *R. micrantha* Borrer ex Sm., *R. elliptica* Tausch and *R. agrestis* Savi (q.v.); in addition to the records set out under these species, specimens collected by Perry and Lloyd from near Coughton in 1835, and by Bagnall from Billesley in 1878 and from near Studley in 1886 have been referred to *R. rubiginosa* sensu lato, which has been recorded from all surrounding VCs.

225.14 **Rosa rubiginosa** L.
Purton, T., 1817. Mid. Flora. Alne Hills, above the village.
A rare plant of rough grassland and scrub on calcareous soil recorded from the Billesley and Wilmcote area (1356, 1357, 1457, 1557); specimens were collected by Bagnall from Solihull in 1875 and Sutton Park in 1871; recorded from all surrounding VCs.

225.15 **Rosa micrantha** Borrer ex Sm.
Purton, T., 1821. Mid. Flora [*R. rubiginosa* var. *micrantha*]. "Between Bidford and Grange; at Allesley."
Two records only, on calcareous soil from woodland at Spernall Park (1063) and scrub at Aston Grove (1457); possibly under-recorded; a number of specimens were collected by Bagnall and Bromwich from Billesley, Brinklow, Claverdon, Drayton Bushes, Grafton, Itchington Holt, Knowle, Lapworth Street, Lighthorne Rough, near Oakley Wood, Preston Bagot, Solihull, Wilmcote and Wishaw between 1867 and 1889; recorded from all surrounding VCs.

225.16 **Rosa elliptica** Tausch
Kirk, T., 1860. Herb. Perry [*R. inodora* Fries]. Binley Common Wood.
Not recorded in present survey; although Kirk's specimen is undoubtedly this species, it does not appear to have been seen since in the county; not recorded from any of the surrounding VCs.

225.17 **Rosa agrestis** Savi
Bree, Rev. W. T., 1818. Mid. Flora, 1821 [*R. rubiginosa* var. *sepium?*]. "In a small hedgerow, in a pasture field, near Bidford Grange, at the back of the brick kiln on the Stratford road."
Not recorded in present survey; a number of specimens were collected by Bagnall, Bloxam and Lloyd from Allesley, Harborough Magna, Whatcote and Wormleighton between 1835 and 1887; recorded from VCs 23, 37, 55 and 57. Druce, in his Comital Flora, described Bree's record of 1818 as the first British record.

226 Prunus L.

226.1 **Prunus spinosa** L. Blackthorn, Sloe
Lowe, H. E., 1837. Herb. Perry [*P. communis* L.]. Edgbaston. Map 256, p. 360
Abundant and evenly distributed throughout the county; recorded from hedges and scrub (80%), woods (10%), roadsides, railway banks and waste places (5%) and watersides (4%); recorded from all surrounding VCs.

226.2 **Prunus domestica** L.
For first record see ssp. *domestica.*
Map 257, p. 361
An introduced tree, frequent and widely but irregularly distributed throughout most of the county, though rare in the extreme N.; recorded from hedges and scrub (86%), woods (8%), and waste places and railway banks (3%); recorded from all surrounding VCs.
Records for the two subspecies set out below have been included on this map.

226.2a Ssp. **domestica** Plum
Baxter, W., 1831. MS. (Bagnall: Mid. Nat. 1892–93) [*P. domestica* L.]. Near Newbold on Avon and near Bilton.
Generally planted but frequently found in hedgerows where it is probably an escape; recorded from squares 0650, 0860, 1000, 1185, 1298, 1769, 2051, 2053, 2550, 4157, 4351, 4366 and 4779; recorded from all surrounding VCs except 39.

226.2b Ssp. **insititia** (L.) C. K. Schneid.
Bullace, Damson
Baxter, W., about 1837. MS (Bagnall: Flora, 1891) [*P. insititia* L.]. About Rugby and Hill Morton.
Young and Baker, 1873. Catalogue [*P. insititia* L.]. Hatton.
Frequently found in hedgerows and occasionally in woods and waste places; probably always an escape from cultivation; recorded from squares 0995, 1100, 1668, 2058, 2089, 2090, 2141, 2173, 2174, 2394, 2749, 2981, 3086, 3154, 3177, 3254, 3338, 3569, 3785, 4157, 4159, 4162, 4366, 4779 and 4871; recorded from all surrounding VCs.

226.3 **Prunus cerasifera** Ehrh. Cherry Plum
Laflin, T., 1957. Luddington.
Often planted and occasionally naturalised; recorded from hedgerows at 1066, 1368, 1652, 1888, 1956, 2058, 2255, 2759, 3371, 3672, 4471 and 4682; recorded from all surrounding VCs except 39.

226.4 **Prunus avium** (L.) L. Gean, Wild Cherry
Perry, 1821. MS (Bagnall: Flora, 1891). Lower Norton. Map 258, p. 361
An occasional plant in the N.W. half of the county, but rare in the lower Avon Valley and on the calcareous clays of the S. and S.E.; recorded from hedges and scrub (54%), woods (40%) and roadsides (4%); recorded from all surrounding VCs.

226.5 **Prunus cerasus** L. Dwarf Cherry
Baynes, W. W., 1831. Bagnall's Flora, 1891. Near Leamington. Map 752, p. 605
Often planted and occasionally naturalised, usually in hedges; recorded from all surrounding VCs.

226.6 **Prunus padus** L. Bird Cherry
Cheshire, W., 1851. Bagnall's Flora, 1891. Edge Hill.
Probably planted; recorded from squares 2055, 3569 and 4776; recorded from all surrounding VCs except 23 and 32.

226.7 **Prunus laurocerasus** L. Cherry Laurel
Bywater, Miss J., 1954. Lapworth.
Map 753, p. 606
Often planted and occasionally naturalised in woods and hedgerows; recorded from VCs 23, 32, 33 and 37.

226.8 **Prunus lusitanica** L. Portugal Laurel
Readett, R. C., 1954. Umberslade Park.
Rare and generally planted; recorded from woodland in squares 1371, 1682, 2044, 2058, 2156 and 3962; not recorded from any of the surrounding VCs.

227 **Cotoneaster** Medic.

227.4 **Cotoneaster microphyllus** Wall. ex Lindl. Small-leaved Cotoneaster
Druce, 1932. Com. Flora. Without locality.
Not recorded in present survey; recorded from VCs 32, 33, 39 and 57.

229 **Crataegus** L.

229.1 **Crataegus laevigata** (Poir.) DC. Midland Hawthorn
Baxter, W., 1831. MS (Bagnall: Flora, 1891) [*C. oxyacantha* L. a. *oxyacanthoides* Thuill.]. Between Brownsover and Coton House.
Bagnall, 1871. Herb. Bagnall [*C. oxyacantha* L. a. *oxyacantha* Jacq.]. Solihull. Map 259, p. 362
Abundant in the S. half of the county but distribution scattered in the N.; usually found singly or in small numbers whereas *C. monogyna* is usually recorded in quantity; recorded from hedges (84%), woods (9%) and a number of other habitats; apparently the N.W. border of the British distribution of this species runs through the N.W. of the county; recorded from all surrounding VCs.

229.2 **Crataegus monogyna** Jacq. Hawthorn
Ick, W., 1836. Herb. Ick [*Mespilus oxyacantha*]. Coleshill Park. Map 260, p. 362
C. monogyna and *C. laevigata* are both fairly abundant in the county though *C. laevigata* is less so than *C. monogyna* particularly in the N. half of the county. The position is complicated by the presence of a wide range of intermediates. Although many specimens have been confidently classified as one or other of the species, detailed study of a limited number of populations has shown that a surprisingly large number cannot be so distinguished by means of the characters commonly used to separate them. As a result recorders have tended to draw the line between the two species at different points, and this has to be borne in mind in assessing the figures for distribution.
C. monogyna is abundant and evenly distributed throughout the county; it is recorded from hedges and scrub (73%), woods (11%), roadsides, waste places and railway banks (7%), watersides (5%) and grassland (4%); it is recorded from all surrounding VCs.

230 **Mespilus** L.

230.1 **Mespilus germanica** L. Medlar
Bromwich, H., 1875. Herb. Warw. Mus. Without locality.
One record as an escape in a hedgerow near Rugby (4874); recorded from VCs 23, 32, 33 and 55.

232 **Sorbus** L.

232.1 **Sorbus aucuparia** L. Rowan, Mountain Ash
Bree, Rev. W. T., 1821. Mid. Flora [*Pyrus aucuparia*]. Allesley. Map 261, p. 363
Abundant in the N.W. half of the county, particularly in areas which were well wooded in the seventeenth century, though generally somewhat unevenly distributed; rare in the S. half; recorded from woods (43%), hedgerows and scrub (33%), roadsides and railway banks (8%), grassland (6%), cultivated land (6%) and a few other habitats; recorded from all surrounding VCs.

232.4 **Sorbus intermedia** (Ehrh.) Pers. *sensu stricto*
Arnold, G. & M., 1963. Kingsbury.
One record only from a hedgerow at Kingsbury (2196); recorded from all surrounding VCs except 37 and 57.

232.5 **Sorbus aria** (L.) Crantz *sensu lato* Common White Beam
Bagnall, 1873. Herb. Bagnall [*Pyrus aria* var. *rupicola*]. Billesley. Map 754, p. 606
An occasional tree, probably always planted, recorded from woods and hedges at Edgbaston (0483), Sutton Park (0997, 0998, 1097, 1098), Morton Bagot (1164), Earlswood (1273), Billesley (1456, 1557), Shrewley (2167), Little Compton (2630), Hawkes End (2982), near Fillongley (3086), Stoneleigh (3371), Bedworth (3687) and near Southam (4160); of these records the following have been determined as *S. aria* sensu stricto: Sutton Park (0997, 0998), where it has been known since 1877, Billesley (1456), which is probably Bagnall's station, where it has been known since 1873, and Little Compton (2630); *S. aria* sensu lato has been recorded from all surrounding VCs.

Sorbus aria (L.) Crantz × **aucuparia** L. = **S.** × **thuringiaca** (Ilse) Fritsch
Bagnall, 1875. Herb. Bagnall [*Pyrus semipinnata* Roth]. Planted near Leamington Station.
Not recorded in present survey; not recorded from any of the surrounding VCs.

232.7 **Sorbus torminalis** (L.) Crantz Wild Service-tree
Purton, T., 1817. Mid. Flora [*Crataegus torminalis* L.]. On the footway to Mr. Petford's, Alcester Park. Map 262, p. 363
A rare tree almost entirely confined to the W. of the county from Solihull to Alcester; recorded from hedges, woods, pondsides and scrub; recorded from all surrounding VCs.

233 **Pyrus** L.

233.1 **Pyrus communis** L. Wild Pear
Purton, T., 1817. Mid. Flora. Great Alne; Kinwarton. Map 263, p. 364
Fairly frequent in the S. half of the county but rare in the N.; doubtfully native and probably always planted in Warwickshire; recorded from hedges (82%), grassland (8%) and roadsides (5%); recorded from all surrounding VCs.

234 **Malus** Mill.

234.1 **Malus sylvestris** Mill. Crab Apple
For first record see ssp. *mitis*. Map 264, p. 364
Abundant and fairly regularly distributed throughout the county; recorded from hedges and scrub (76%), woods (13%), waste places, roadsides and railway banks (5%) and a few other habitats; recorded from all surrounding VCs.

234.1a Ssp. **sylvestris**
Bagnall, 1876. Flora of Sutton Park [*Pyrus malus* var. *acerba* DC.]. Map 265, p. 365
Recorded from hedges and scrub (82%), woods (12%) and grassland (4%), and from all surrounding VCs.

234.1b Ssp. **mitis** (Wallr.) Mansf.
Lloyd, Dr. G., about 1835. Herb. Perry (*Pyrus malus* L.]. Near Leamington.
Map 266, p. 365

The maps for this ssp. and for ssp. *sylvestris* indicate that though they are both under-recorded and ssp. *sylvestris* is a little less common than ssp. *mitis*, they have been observed in approximately the same areas. Ssp. *mitis* has been recorded from hedges (76%), woods (14%), grassland (4%) and a few other habitats; it has been recorded from all surrounding VCs.

CRASSULACEAE

235 Sedum L.

235.2 **Sedum telephium** L. Orpine, Livelong
Purton, T., 1817. Mid. Flora. Alne Hills.
Map 755, p. 606

An occasional plant of woods and hedgerows; both ssp. *purpurascens* (Koch) Syme and ssp. *fabaria* Syme have been recorded; the species has been recorded from all surrounding VCs.

235.6a **Sedum album** L. ssp. **album** White Stonecrop
Kirk, T., 1847. Phytologist [*S. album* L.]. On a wall at the back of Little Park St., Coventry.
Map 756, p. 606

An occasional garden escape found naturalised on walls, quarries, roadsides and waste places; recorded from all surrounding VCs except 39.

235.8 **Sedum acre** L. Wall-pepper, Biting Stonecrop
Purton, T., 1817. Mid. Flora. Walls at Wixford.
Map 267, p. 366

A frequent plant, widely but very irregularly distributed throughout the county; recorded from walls and quarries (62%) and waste places and roadsides (35%); recorded from all surrounding VCs.

235.9 **Sedum sexangulare** L. Tasteless Stonecrop
Cox, T., 1848. Herb. Perry. Berkswell.

Not recorded in present survey; recorded from VCs 23, 33 and 37.

235.10 **Sedum forsteranum** Sm. Rock Stonecrop
Bagnall, 1872. Herb. Bagnall [*S. rupestre*]. Oversley Mill.

Not recorded in present survey; recorded from all surrounding VCs except 23 and 55.

235.11 **Sedum reflexum** L. Reflexed Stonecrop
Purton, T., 1817. Mid. Flora. On a wall at Salford; Alcester. Map 757, p. 607

An introduced plant occasionally naturalised on walls, quarries, roadsides and waste places; recorded from all surrounding VCs.

236 Sempervivum L.

236.1 **Sempervivum tectorum** L. Houseleek
Baxter, W., before 1871. Bagnall's Flora, 1891. Roofs of houses, butchers' shambles, Rugby.

An introduced plant apparently naturalised on walls and roofs at Abbot's Salford (0650), Bidford on Avon (1051), Sheldon (1584) and Mancetter (3296); previously more widespread; recorded from all surrounding VCs.

238 Umbilicus DC.

238.1 **Umbilicus rupestris** (Salisb.) Dandy
Wall Pennywort, Navelwort
Perry, 1817. List [*Cotyledon umbilicus*]. Area, Guy's Cliffe House; Old Pound, Coten End, Warwick.

One record only from walls of Maxstoke Priory (2386); recorded from all surrounding VCs.

SAXIFRAGACEAE

239 Saxifraga L.

239.7 **Saxifraga cymbalaria** L.
Cooper, Miss G., 1961. Leek Wootton.

One record of garden escape established on tufa rock at Leek Wootton Court (2968); not recorded from any of the surrounding VCs.

239.8 **Saxifraga tridactylites** L. Rue-leaved Saxifrage
Perry, 1813. Herb. Perry. Warwick.
Map 268, p. 366

An occasional plant chiefly confined to walls on the Cotswold limestone escarpment; recorded from all surrounding VCs.

239.9 **Saxifraga granulata** L. Meadow Saxifrage
Withering, 1801. Arrangement. In the Garlick Meadows, Penn's Mill. Map 269, p. 367

An occasional plant of a number of widely scattered localities in the county, with some concentration on the central band of sandstone and gravels and on the Cotswold escarpment in the S.; recorded from grassland, roadsides, mixed woods and railway banks; recorded from all surrounding VCs.

239.15 **Saxifraga hypnoides** L. Mossy Saxifrage
Kirk, T., 1851. Herb. Perry. Coventry, railway bank.

Not recorded in present survey; probably not native in Warwickshire; recorded from VCs 23, 37, 39 and 57.

242 Chrysosplenium L.

242.1 **Chrysosplenium oppositifolium** L.
Opposite-leaved Golden Saxifrage
Perry, 1813. Herb. Perry. Crackley Wood, Kenilworth. Map 270, p. 367

A locally frequent plant mainly confined to the area of the old Forest of Arden; recorded from watersides and marshes (68%) and damp woods (30%); recorded from all surrounding VCs.

242.2 **Chrysosplenium alternifolium** L.
Alternate-leaved Golden Saxifrage
Perry, 1813. Herb. Perry. Crackley Wood.

Two records from marshy places in Whichford Wood (3034); formerly more widespread; recorded from all surrounding VCs except 32.

PARNASSIACEAE

243 Parnassia L.

243.1 **Parnassia palustris** L. Grass of Parnassus
Perry, 1812. Herb. Perry. Norbrook.

A rare plant of bogs recorded only from Sutton Park (0995, 0996, 0998), formerly also known from near Kenilworth; recorded from all surrounding VCs.

GROSSULARIACEAE

246 Ribes L.

246.1 **Ribes rubrum** L. Red Currant
Perry, 1820. Plant. Varv. Sel. Side of the Avon between Emscote and Warwick.
Map 271, p. 368
Frequent, but irregularly distributed throughout the county; often planted but possibly frequently native; recorded from woods (35%), hedgerows and scrub (29%), watersides and marshes (28%) and a number of other habitats; recorded from all surrounding VCs.

246.3 **Ribes nigrum** L. Black Currant
Ray, 1670. Catalogus [*Ribes nigrum vulgo dictum folio olente*]. "Warwicensi agro."
Map 272, p. 368
Widely but thinly distributed throughout the county; doubtfully native in Warwickshire; recorded from the sides of ponds and rivers, damp woods, waste places and railway banks, and a number of other habitats; recorded from all surrounding VCs.

246.5 **Ribes alpinum** L. Mountain Currant
Withering, 1796. Arrangement. In a wood on the S.W. side of a pool at Edgbaston.
Not recorded in present survey, but there is a record dated 1915 from Little Compton; recorded from all surrounding VCs.

246.6 **Ribes uva-crispa** L. Gooseberry
Purton, T., 1817. Mid. Flora [*R. grossularia*]. Oversley Wood, and in hedges at a distance from any house. Map 273, p. 379
Frequent, but irregularly distributed throughout the county; often introduced but possibly frequently native; recorded from hedges and scrub (51%), woods (25%), watersides (13%), roadsides and waste places (5%) and a few other habitats; recorded from all surrounding VCs.

DROSERACEAE

247 Drosera L.

247.1 **Drosera rotundifolia** L. Common Sundew
Withering, 1787. Arrangement. Birmingham Heath.
Locally abundant on *Sphagnum* bog in three areas in Sutton Park (0996, 1097); formerly more widespread but has disappeared owing to urbanisation (Birmingham Heath) and drainage (Coleshill Bog and Packington); recorded from all surrounding VCs.

LYTHRACEAE

249 Lythrum L.

249.1 **Lythrum salicaria** L. Purple Loosestrife
Perry, 1813. Herb. Perry. Without locality.
Map 274, p. 379
An occasional plant of a number of scattered localities; recorded from watersides (97%); recorded from all surrounding VCs.

Lythrum junceum Banks & Solander
Bromwich, H., 1875. Herb. Warw. Mus. Myton.
One record from waste ground at Emscote (2964); recorded from VC 23; *L. hyssopifolia* has been recorded from Warwickshire and all the surrounding VCs except 33 but the material has not been seen by us.

250 Peplis L.

250.1 **Peplis portula** L. Water Purslane
Purton, T., 1817. Mid. Flora. Coleshill Pool; at the top of Spernall Lane. Map 275, p. 370
A rare plant of scattered occurrence mainly in the W. central area of the county, but absent from the extreme N. and from the calcareous soils of the S. and S.E.; recorded from mixed woods, scrub, marshes and pondsides; recorded from all surrounding VCs.

THYMELAEACEAE

251 Daphne L.

251.1 **Daphne mezereum** L. Mezereon
Bromwich, H., 1864. Herb. Brit. Mus. Shrewley Common.
Not recorded in present survey; last seen (at Shrewley) in 1932; recorded from all surrounding VCs except 55.

251.2 **Daphne laureola** L. Spurge Laurel
Perry, 1817. List. Stank Hill Farm, near Warwick; Warwick Castle Mount; on the Stratford and Birmingham Roads near Warwick. Map 276, p. 370
An occasional plant in the S. of the county with a concentration in the S.W.; recorded from hedgerows and scrub (54%), mixed woods (40%) and roadsides; recorded from all surrounding VCs; the N.W. boundary of the main British distribution of the species passes through the middle of Warwickshire.

ONAGRACEAE

254 Epilobium L.

254.1 **Epilobium hirsutum** L.
Great Hairy Willowherb, Codlins and Cream
Baxter, W., 1831. MS (Bagnall: Mid. Nat. 1892–93.) Near Brownsover. Map 277, p. 371
Abundant and evenly distributed throughout the county; recorded from watersides (74%), roadsides, waste places, and railway banks (10%), hedgerows and scrub (9%) and a few other habitats; recorded from all surrounding VCs.

Epilobium hirsutum × montanum
= **E. × erroneum** Hausskn.
Readett, R. C., 1955. Aston Grove.
Recorded from damp woods and hedgerows at Aston Grove (1357, 1458), Austy Wood (1682) and near Priors Marston (4858); recorded from all surrounding VCs except 37 and 39.

254.2 **Epilobium parviflorum** Schreb.
Hoary Willowherb
Perry, 1812, Herb. Perry. Near Warwick.
Map 278, p. 371
Frequent but irregularly distributed throughout the county; recorded from watersides and marshes (81%), roadsides, waste places and railway banks (7%), hedgerows and scrub (4%), woods (3%) and a few other habitats; recorded from all surrounding VCs.

254.3 **Epilobium montanum** L.
Broad-leaved Willowherb
Perry, 1812. Herb. Perry. Without locality.
Map 279, p. 372
Abundant but somewhat irregularly distributed throughout the county; recorded from roadsides, railway banks and waste places (32%), watersides (21%), hedgerows and scrub (18%), woods (12%), arable land (10%) and grassland (4%); recorded from all surrounding VCs.

Epilobium montanum × **parviflorum**
= **E.** × **limosum** Schur
Green, P. S., 1950. Near Brinklow.
Recorded from damp places in woods, hedgerows and railway cuttings at Wood End (1071), Aston Grove (1457), Alveston (2356), near Compton Wynyates (3441), Stockton Locks (4365) and near Brinklow (4479); recorded from all surrounding VCs except 39.

Epilobium montanum × **roseum**
= **E.** × **mutabile** Boiss. & Reut.
Thompson, H. S., 1901. Herb. Brit. Mus. Garden weed, Edgbaston.
Not recorded in present survey; recorded from VCs 33, 37 and 55.

254.4 **Epilobium lanceolatum** Seb. & Mauri
Spear-leaved Willowherb
Clark, M. C., 1963. Stratford on Avon racecourse.
Recorded only from waste ground at Stratford on Avon racecourse (1953); recorded from VCs 23, 33 and 37.

254.5 **Epilobium roseum** Schreb. Pale Willowherb
Gorle, Rev. J., 1836. Bagnall: Flora, 1891. Sheldon. Map 280, p. 372
An occasional plant of a few widely scattered localities in the county; recorded from watersides (71%), waste places (14%) and a few other habitats; recorded from all surrounding VCs.

254.6 **Epilobium adenocaulon** Hausskn.
American Willowherb
Bagnall, 1889. Herb. Bagnall [*E. lamyi*]. Wood near Gannaway Gate. Map 281, p. 373
Abundant though unevenly distributed throughout most of the county, but almost entirely absent from the extreme S.; recorded from watersides (40%), waste places, roadsides and railway banks (26%), hedgerows and scrub (11%), grassland (8%), arable land (7%) and woods (5%); recorded from all surrounding VCs.

Epilobium adenocaulon × **tetragonum**
Benoit, P. M., 1965. Warwick.
One record only from a waste place at Warwick (2865); not recorded from any of the surrounding VCs.

Epilobium adenocaulon × **obscurum**
Cadbury, Miss D. A., 1950. Oversley Wood.
Recorded from damp places in Oversley Wood (1056), Frankton Wood (4171) and All Oaks Wood (4478) recorded from VCs 32 and 33.

254.7 **Epilobium tetragonum** L. ssp. **tetragonum**
Square-stemmed Willowherb
Cheshire, W., 1857. Herb. Perry [*E. tetragonum*]. Snitterfield Bushes. Map 282, p. 373
Frequent in the S.E. half of the county but only occasional and sparsely scattered elsewhere; recorded from watersides and marshes (40%), waste places, railway banks and roadsides (23%), hedgerows and scrub (13%), arable land (12%), grassland (6%) and woods (4%); recorded from all surrounding VCs; the N.W. limit of the main British distribution passes through Warwickshire.
There is a record of *E. tetragonum* earlier than Cheshire's given by Purton (Mid. Flora, 1817. Side of the Arrow; marshes about Bidford), but we have not seen the material.

254.8 **Epilobium tetragonum** L. ssp. **lamyi** (F. W. Schultz) Nyman
Readett, R. C., 1954. Southam Holt.
One record only from salt spring at Southam Holt (4460); recorded from all surrounding VCs except 39 and 57.

254.9 **Epilobium obscurum** Schreb.
Short-fruited Willowherb
Perry, 1845. Herb. Perry [*E. tetragonum*]. Near Warwick on the Stratford Road.
Kirk, T., 1855. Herb. Perry. Wyken.
Map 283, p. 374
Occasional and very irregularly distributed throughout the county (possibly under-recorded); recorded from watersides and marshes (65%), hedgerows (11%), waste places and roadsides (11%), grassland (7%) and a few other habitats; recorded from all surrounding VCs.

Epilobium obscurum × **parviflorum**
= **E.** × **dacicum** Borbás
Bagnall, 1874. Herb. Bagnall. Wishaw.
Not recorded in present survey; recorded from all surrounding VCs except 39.

Epilobium obscurum × **palustre**
= **E.** × **schmidtianum** Rostk.
Cadbury, Miss D. A., 1951. Sutton Park.
One record only from a streamside in Sutton Park (1096); recorded from VCs 32 and 57.

254.10 **Epilobium palustre** L. Marsh Willowherb
Purton, T., 1817. Mid. Flora. Coleshill Bog.
Map 284, p. 374
Occasional in the N.W. of the county but rare on the calcareous soils of the S.E. and almost entirely absent from the Avon Valley; recorded from marshes and watersides; recorded from all surrounding VCs.

254.13 **Epilobium nerterioides** Cunn.
New Zealand Willowherb
Evans, R., 1965. Wroxall.
Recorded only from the base of the W. wall of the church at Wroxall Abbey (2270); recorded from VCs 33, 39 and 57.

255.1 **Epilobium angustifolium** L.
Rosebay Willowherb, Fireweed
Bree, Rev. W. T., 1830. Mag. Nat. Hist. Ryton Wood. Map 285, p. 375
Very abundant and evenly distributed throughout most of the county but less common in the extreme S.; recorded from roadsides, waste places and railway banks (46%), hedgerows and scrub (24%), woods (12%), grassland (8%) and watersides (6%); recorded from all surrounding VCs.

256 Oenothera L.

256.1 **Oenothera biennis** L. Common Evening Primrose
Purton, T., 1820. Mid. Flora, 1821. On the banks of the Arrow, at a distance from any house, abundantly. Map 286, p. 375
An introduction in a number of widely scattered localities throughout most of the county, but almost entirely absent from the calcareous soils of the S. and S.E.; recorded from waste places, railway banks, cultivated land, roadsides, grassland and quarries; recorded from all surrounding VCs.

256.2 **Oenothera erythrosepala** Borbás
Large-flowered Evening Primrose
Allen, H., 1952. Whitley Common.
Map 287, p. 376
A rare introduction in a few scattered localities in the county, almost completely absent from the calcareous soils of the S. and S.E.; recorded from waste places, railway banks and cultivated land; recorded from all surrounding VCs except 39.

256.3 **Oenothera stricta** Ledeb. ex Link
Fragrant Evening Primrose
Bagnall, 1871, Herb. Bagnall [*O. odorata*]. Forge Mills Lane near Gas Works, Coleshill.
Not recorded in present survey; recorded from VCs 33, 37 and 55.

Oenothera perennis L.
Withering, 1801. Arrangement. "This elegant plant, new to the English botanist, has been found growing wild on Coleshill Common, Warwickshire, by Lord Aylesford's gardener."
Not recorded in present survey; recorded from VC 37.

258 Circaea L.

258.1 **Circaea lutetiana** L.
Common Enchanter's Nightshade
Beilby, Miss M. A., 1837. Analyst. Sheldon.
Map 288, p. 376
Abundant and widely spread except on the Lower Lias clays in the S.E. of the county; recorded from moist woodland (58%), scrub and hedgerows (25%), shaded roadsides, waste places and railway banks (9%), cultivated land (5%) and a few other habitats; recorded from all surrounding VCs.

258.2 **Circaea × intermedia** Ehrh.
Intermediate Enchanter's Nightshade
Cox, T., 1848. Herb. Perry [*C. alpina*]. Temple Balsall.
In the Mid. Flora, 1817, Rev. W. Bree jr. recorded *C. alpina* from Temple Balsall and Springfield, and it seems likely that this was the same plant as Cox's, which he named *C. alpina* but which has been determined as *C. × intermedia*, generally regarded as the hybrid between *C. alpina* and *C. lutetiana*; neither plant has been recorded in the present survey, but *C. alpina* as an aggregate species has been recorded from VCs 37 and 39, and *C. × intermedia* from 57.

HALORAGACEAE

259 Myriophyllum L.

259.1 **Myriophyllum verticillatum** L.
Whorled Water Milfoil
Countess of Aylesford, 1805. Bot. Guide. Packington.
A rare plant of still water recorded from canal near Tamworth (2204) and from pools at Alvecote (2504) and Rugby Cement Works (4876); recorded from all surrounding VCs.

259.2 **Myriophyllum spicatum** L.
Spiked Water Milfoil
Purton, T., 1817. Mid. Flora. Black Pool between Spernall and Studley Church; in a pool at Sambourn. Map 289, p. 377
An occasional plant very widely scattered throughout most of the county, but absent from the S.; recorded from ponds, canals, lakes and rivers; recorded from all surrounding VCs.

259.4 **Myriophyllum alterniflorum** DC.
Alternate-flowered Water Milfoil
Baynes, W. W., 1831. Bagnall's Flora, 1891. Blakedown Mill Pool.
One record from Bracebridge Pool, Sutton Park (0998); at one time widely distributed in the county and recorded from Arbury Park, Coleshill Pool, Allesley, Southam, Broom and other places; recorded from all surrounding VCs except 23.

HIPPURIDACEAE

261 Hippuris L.

261.1 **Hippuris vulgaris** L. Marestail
Perry, 1835. Herb. Perry. Chesterton Mill Pool.
Map 290, p. 377
A rare plant of ponds and lakes occurring in a belt in the middle of the county but absent from the N. and the extreme S.; recorded from all surrounding VCs.

CALLITRICHACEAE

262 Callitriche L.

A considerable number of records have been made of this genus without identification of species.

262.1 **Callitriche stagnalis** Scop.
Common Water Starwort
Lloyd, Dr. G., 1835. Herb. Perry [*C. platycarpa*]. Near Warwick. Map 291, p. 378
An abundant plant fairly well distributed throughout the county but less frequent in the S.; recorded from shallow water (58%), waterside (36%) and damp woodland rides (3%); recorded from VCs 23, 32, 55 and 57 and as an aggregate also (incl. *C. platycarpa*) from VCs 33, 37 and 39.

262.2 **Callitriche platycarpa** Kütz.
Long-styled Water Starwort
Kirk, T., 1850. Herb. Perry. Stoke Heath, Coventry. Map 292 p. 378
A rare plant, probably often overlooked, recorded from ponds and rivers (79%) and watersides (21%); it is recorded from VCs 23, 32, 55 and 57 and may be included in the *C. stagnalis* recorded as an aggregate from the other surrounding VCs.

262.3 **Callitriche obtusangula** Le Gall
Blunt-fruited Water Starwort
Bagnall, 1881. Herb. Bagnall. Farnborough Pool.
A rare plant of shallow water recorded from 0998

(Sutton Park), 2096 (Bodymoor Heath), 3652, 3852 and 5374; recorded from all surrounding VCs.

262.4 **Callitriche intermedia** Hoffm.
Intermediate Water Starwort
Kirk, T., 1847. Phytologist [*C. pedunculata* var. *sessilis* Bab.]. Arbury Deer Park.
The records made during the present survey have been determined as ssp. (q.v.); the species is recorded from all surrounding VCs.

Ssp. **hamulata** (Kütz. ex Koch) Clapham
Bagnall, 1882. Herb. Bagnall [*C. hamulata* Kütz.]. Banners Lane, Tile Hill.
A rare plant of shallow water, recorded from Middleton (2098), Amington (2304), Seeswood Pool (3290) and near Priors Hardwick (4855); recorded from VCs 23, 55 and 57.

Ssp. **pedunculata** (DC.) Clapham
Bagnall, 1872. Herb. Bagnall [*C. pedunculata* DC.]. Whitacre.
A very rare plant of shallow water recorded from Sutton Park (0998), May's Wood (1464), Old Milverton (2967) and near Fenny Compton (4251); recorded from VCs 23 and 55.

LORANTHACEAE

263 Viscum L.

263.1 **Viscum album** L. Mistletoe
Ray, 1688. Hist. Plant. "Viscum Middletoni, in agro Warwicensis, in Corylo."
A plant occasionally found on trees (apple, hawthorn and poplar) in hedges and orchards, and recorded from squares 1164, 1454, 2054, 2076, 2495, 2697, 2756, 2899, 3098 and 5374; recorded from all surrounding VCs.

CORNACEAE

265 Swida Opiz

265.1 **Swida sanguinea** (L.) Opiz Dogwood
Baxter, W., 1831. MS (Bagnall: Mid. Nat. 1892–93) [*Cornus sanguinea* L.]. Lane near Newbold on Avon and Rugby.
Perry, 1845. Herb. Perry [*Cornus sanguinea*]. Oversley Wood. Map 293, p. 279
Abundant on the calcareous soils of the S. and S.E. of the county and on the chalky boulder clay of the N.E.; frequent but very irregularly distributed elsewhere on non-calcareous soils; recorded from hedge and scrub (78%), woods (13%), roadsides, railway banks and waste places (5%) and watersides (3%); recorded from all surrounding VCs.

265.2 **Swida sericea** (L.) Holub
Phipps, J. B. 1955. Maxstoke.
Frequently planted and rarely naturalised; recorded once only from a wood in Maxstoke Park (2288); recorded from VC 23.

ARALIACEAE

268 Hedera L.

268.1 **Hedera helix** L. Ivy
Baxter, W., 1831. MS (Bagnall: Mid. Nat. 1892–93.) Near Dunchurch.
Lloyd, Dr. G., 1835. Herb. Perry. Leamington. Map 294, p. 379
Abundant and evenly distributed throughout the county; recorded from hedges and scrub (59%), woods (18%), roadsides, railway banks and waste places (12%), walls (7%) and a few other habitats; recorded from all surrounding VCs.

UMBELLIFERAE

269 Hydrocotyle L.

269.1 **Hydrocotyle vulgaris** L. Marsh Pennywort
Purton, T., 1817. Mid. Flora. Haselor Fields; near Hoo Mill. Map 295, p. 380
Confined to Sutton Park, the Coleshill area, Earlswood and a few other areas of acid marsh in the N. of the county; recorded from watersides and marsh (76%), bogs and a few other wet habitats; recorded from all surrounding VCs.

270 Sanicula L.

270.1 **Sanicula europaea** L. Wood Sanicle
Perry, 1813. Bagnall: Flora, 1891. Without locality. Map 296, p. 380
Frequent, mainly in the S. half of the county, on sands and gravels and on the Lower Lias and Cotswold limestones, but very rare in the Avon Valley and on the Lower Lias clays; recorded from woods (71%), scrub and hedgerows (19%) and a few other habitats; recorded from all surrounding VCs.

273 Chaerophyllum L.

273.1 **Chaerophyllum temulentum** L. Rough Chervil
Ick, W., 1836. Herb. Ick [*Myrrhis temulenta*]. Upper Saltley near Alum Rock. Map 297, p. 381
Abundant and widely distributed except in Birmingham and an area surrounding it from which it is entirely absent; recorded from hedgerows and scrub (70%), roadsides and railway banks (21%), grassland (4%) and woods (3%); recorded from all surrounding VCs.

274 Anthriscus Pers.

274.1 **Anthriscus caucalis** Bieb. Bur Chervil
Holden, H., 1688–96. Herb. Holden [*Myrrhis Nova Aequicolorum Columnae:* Small Hemlock Chervil with rough seeds]. "In ye hedg. as you go to Longbridge near Warwick. It is sometimes called Wild Cicely."
Not recorded in present survey though several records were made during the nineteenth century at Marston Green, Combe Abbey, Brandon, Hampton Lucy, Warwick and Stratford on Avon; recorded from all surrounding VCs.

274.2 **Anthriscus sylvestris** (L.) Hoffm. Cow Parsley
Lloyd, Dr. G., 1835. Herb. Perry. Warwick. Map 298, p. 381
A typical plant of hedgerows throughout the county (43%); also found in roadsides and waste places (33%), grassland (13%) and woods (6%); recorded from all surrounding VCs.

274.3 **Anthriscus cerefolium** (L.) Hoffm.
Garden Chervil
Bromwich, H., 1849. Herb. Perry. Leamington.
An escape not recorded in present survey; recorded from all surrounding VCs except 32 and 39.

275 Scandix L.

275.1 **Scandix pecten-veneris** L. Shepherd's-needle
Perry, 1812. Herb. Perry. St. Nicholas meadow, Warwick (in cornfield).
Map 299, p. 382
Frequent on the Lower Lias limestone in the S. and on the chalky boulder clay in the N.E., with scattered occurrences on the Lower Lias clays and elsewhere in the county; recorded from cultivated land (87%), waste places, railway banks and grassland; recorded from all surrounding VCs.

276 Myrrhis Mill.

276.1 **Myrrhis odorata** (L.) Scop. Sweet Cicely
Purton,T., 1817. Mid. Flora [*Scandix odorata*]. Temple Balsall; Studley Castle.
A very rare plant, almost certainly not native in Warwickshire, recorded from a railway bank at Sutton Park (1097) and hedgerow at Rugby (5173); recorded from all surrounding VCs except 32.

277 Torilis Adans.

277.1 **Torilis japonica** (Houtt.) DC.
Upright Hedge-parsley
Baxter, W., 1831. MS (Bagnall: Mid. Nat. 1892–93) [*Caucalis anthriscus* Huds.]. Near Rugby. Map 300, p. 382
Abundant and fairly evenly distributed throughout the county; recorded from hedgerows and scrub (53%), roadsides, railway banks and waste places (27%), grassland (13%), watersides (3%) and a few other habitats; recorded from all surrounding VCs.

277.2 **Torilis arvensis** (Huds.) Link
Spreading Hedge-parsley
Perry, 1815. Herb. Perry [*T. infesta*]. Without locality. Map 301, p. 383
A rare plant in the S. of the county, mainly in the lower Avon Valley; recorded from cultivated land, hedgerows and waste places; recorded from all surrounding VCs.

277.3 **Torilis nodosa** (L.) Gaertn.
Knotted Hedge-parsley
Purton, T., 1817. Mid. Flora [*Caucalis nodosa*]. On a wall at Walcote. Map 302, p. 383
A rare plant of the lower Avon Valley and some of the calcareous soils of the S. of the county; recorded from grassland, roadsides, hedgerows, quarries, cultivated land and waste places; recorded from all surrounding VCs.

278 Caucalis L.

278.1 **Caucalis platycarpos** L. Small Bur-parsley
Purton, T., 1817. Mid. Flora [*C. daucoides*]. Alne Hills (Rufford); in fields about Drayton Bushes.
Not recorded in present survey; recorded from all surrounding VCs.

279 Coriandrum L.

279.1 **Coriandrum sativum** L. Coriander
Anon., 1881. Herb. B'ham Univ. Kenilworth.
One record only from a waste place at Clifford Bridge (3780); recorded from all surrounding VCs except 57.

282 Conium L.

282.1 **Conium maculatum** L. Hemlock
Baxter, W., 1831. MS (Bagnall: Mid. Nat. 1892–93.) Rugby.
Lloyd, Dr. G., 1835. Herb. Perry (Bagnall's Flora, 1891). Without locality.
Map 303, p. 384
Frequent, mainly in the S. half of the county with scattered occurrences in the N.; recorded from watersides (especially river banks) (37%), hedgerows (34%), roadsides and waste places (22%) and grassland (5%); recorded from all surrounding VCs.

283 Bupleurum L.

283.2 **Bupleurum rotundifolium** L.
Thorow-wax, Hares-ear
Perry, 1845. Herb. Perry. Near Stratford.
Three records from arable fields near Ilmington (2242) and near Newbold on Stour (2345, 2445); formerly somewhat frequent on the Lower Lias around Aston Cantlow, Wilmcote and Moreton Morrell; recorded from all surrounding VCs except 32.

Bupleurum lancifolium Hornem.
False Thorow-wax
Bromwich, H., 1867. Herb. Warw. Mus. [*B. rotundifolium*]. Without locality.
Eight records from gardens and waste places near Stratford, Warwick, Leamington, Coventry and Rugby; recorded from VCs 23, 33, 55 and 57.

285 Apium L.

285.1 **Apium graveolens** L. Wild Celery
Baxter, W., 1831. MS (Bagnall: Mid. Nat. 1892–93.) Jarrett's Heath, near Rugby.
Baxter, W., 1835. Phaen. Botany. "In ditches by the roadside between Dunchurch and Southam; nearly opp. to the village of Leamington Hastings."
Not recorded in present survey; last seen in 1947 in the Dunchurch–Southam locality; recorded from all surrounding VCs.

285.2 **Apium nodiflorum** (L.) Lag. Fool's Watercress (incl. *A. repens* (Jacq.) Lag.)
Purton, T., 1821. Mid. Flora [*Sium repens*]. Cookhill, near Alcester. Map 304, p. 384
An abundant and fairly evenly distributed plant of watersides (69%) and water (31%), found in and by streams, ditches, ponds, marshes and canals; recorded from all surrounding VCs.

285.4 **Apium inundatum** (L.) Reichb. f.
Water Marshwort
Lowe, H. E., 1834. Herb. Perry [*Sium inundatum*]. Coleshill.
A rare plant of mud in shallow, sluggish water recorded from three locations: Earlswood (1174), near Shelly Coppice (1576) and Seeswood Pool (3290); recorded from all surrounding VCs.

286 **Petroselinum** Hill

286.1 **Petroselinum crispum** (Mill.) A. W. Hill
Wild Parsley
Perry, 1826. Herb. Perry [*P. sativum*]. On the W. gate, Warwick.
An escape from cultivation recorded from squares 2185, 2866, 3357 and 4049; recorded from all surrounding VCs.

286.2 **Petroselinum segetum** (L.) Koch
Corn Parsley
Bree, Rev. W. T., 1835. New Bot. Guide. Without locality.
A rare plant of cultivated, grassy and waste places on calcareous soil, recorded from Tattle Bank (1863), near Stratford on Avon (1953), Newbold on Stour (2345), Hampton Lucy (2557) and near Lighthorne (3357); recorded from all surrounding VCs except 57.

287 **Sison** L.

287.1 **Sison amomum** L. Stone Parsley
Baxter, W., 1831. MS (Bagnall: Mid. Nat. 1892–93.) Hill Morton Road.
Lowe, H. E., 1834. Herb. Perry. Princethorpe.
Map 305, p. 385
Abundant in the S. half of the county mainly on calcareous soils and on the chalky boulder clay in the E.; recorded from hedgerows and scrub (60%), roadsides, railway banks and waste places (21%), grassland (8%), watersides (7%) and a few other habitats; recorded from all surrounding VCs.

289 **Ammi** L.

Ammi visnaga (L.) Lamk.
Bromwich, H., 1881. Herb. Bagnall [*A. majus* L.]. Kenilworth.
A casual not recorded in present survey; recorded from VCs 33, 37 and 55.

290 **Falcaria** P. C. Fabr.

290.1 **Falcaria vulgaris** Bernh. Longleaf
Lamb, J., 1928. Bot. Soc. & Exch. Club Report [*Prionitis falcaria* (L.) Dum.]. Newbold on Stour.
Recorded from cultivated fields and field borders near Newbold on Stour (2345, 2346, 2446); recorded from VCs 23, 33 and 37.

291 **Carum** L.

291.2 **Carum carvi** L. Caraway
Kirk, T., 1860. Herb. Perry. Willenhall.
Not recorded in present survey; recorded from all surrounding VCs.

293 **Conopodium** Koch

293.1 **Conopodium majus** (Gouan) Loret
Pignut, Earthnut
Perry, 1813. Herb. Perry [*Bunium flexuosum*]. Warwick (according to Bagnall: Flora, 1891).
Map 306, p. 385
Abundant in the N. half of the county and in the extreme S.; almost entirely absent from the Lower Lias clays in the S. and S.E. and rare in the lower Avon Valley; recorded from grassland (51%), roadsides and waste places (23%), woods (12%), hedgerows (9%) and arable land (4%); recorded from all surrounding VCs.

294 **Pimpinella** L.

294.1 **Pimpinella saxifraga** L.
Lesser Burnet-saxifrage
Baxter, W., 1831. MS (Bagnall: Mid. Nat. 1892–93.) West Leys and near Clifton.
Perry, 1836. Herb. Perry. Near Stratford on Avon. Map 307, p. 386
An abundant plant in the S. half of the county with occasional occurrences elsewhere; recorded from roadsides and railway banks (49%), grassland (46%) and hedgerows (5%); recorded from all surrounding VCs.

294.2 **Pimpinella major** (L.) Huds.
Greater Burnet-saxifrage
Bree, Rev. W. T., 1830. Mag. Nat. Hist. [*P. magna*]. Allesley; Meriden. Map 308, p. 386
An abundant plant in the N.E. corner of the county, almost completely absent elsewhere; recorded from roadsides and railway banks (46%), hedgerows (19%), grassland (19%), watersides (7%) and woods (4%); the W. boundary of the main British distribution passes through Warwickshire; recorded from all surrounding VCs.

295 **Aegopodium** L.

295.1 **Aegopodium podagraria** L.
Goutweed, Ground Elder
Baxter, W., 1831. MS (Bagnall: Mid. Nat. 1892–93.) Hill Morton and Lawford Mill.
Lloyd, Dr. G., 1835. Herb. Perry. Leamington.
Map 309, p. 387
A plant mainly of roadsides and waste places (59%), hedgerows (18%), cultivated land (12%) and grassland (6%); occurs abundantly throughout the county, often as a troublesome garden weed, but less frequently on the Lias clay in the S.E.; recorded from all surrounding VCs.

297 **Berula** Koch

297.1 **Berula erecta** (Huds.) Coville
Lesser Water-parsnip
Purton, T., 1817. Mid. Flora [*Sium angustifolium*]. Washford Bridge, near Studley.
Map 310 p. 387
An occasional plant of very uneven distribution recorded from watersides (77%) and shallow water (21%); recorded from all surrounding VCs.

300 **Oenanthe** L.

300.1 **Oenanthe fistulosa** L.
Tubular Water-dropwort
Anon., 1843. Herb. B'ham Univ. Moreton Morrell. Map 311, p. 388
An occasional plant very widely scattered chiefly in the N. of the county; recorded from watersides (ponds and canals) and marshes (86%) and water; recorded from all surrounding VCs.

300.3 **Oenanthe silaifolia** Bieb.
Narrow-leaved Water-dropwort
Withering, 1818. Arrangement [*O. peucedanifolia*]. "In a gorsy field by Small Heath House near Birmingham."
Not recorded in present survey; last recorded by

Cheshire near Stratford on Avon in 1854; recorded from all surrounding VCs except 39 and 57.

300.4 **Oenanthe lachenalii** C. C. Gmel.
Parsley Water-dropwort
Cheshire, W., 1854. Herb. Perry. Near Stratford on Avon.

An occasional plant of marshes, ditches and the sides of ponds and rivers, recorded from near Temple Grafton (1255), Wilmcote Common (1557), S. of Wilmcote (1656), near Newbold on Stour (2245), near Ettington (2350), Alvecote (2404, 2504) and Houndshill (2550); recorded from all surrounding VCs except 39 and 57.

300.5 **Oenanthe crocata** L.
Hemlock Water-dropwort
Bagnall, 1867. Flora, 1891. Abundant between Witton Road and Aston Church, banks of the Tame.

An occasional plant mainly of canals and ditches; recorded from Birmingham (0586, 0588, 0686, 0787, 0888, 1090, 1290), near Shustoke (2090, 2190, 2191), Arbury Park (3389) and near Southam (4362); recorded from all surrounding VCs except 55.

300.6 **Oenanthe aquatica** (L.) Poir.
Fine-leaved Water-dropwort
Purton, T., 1817. Mid. Flora [*Phellandrium aquaticum* L.]. "In an old gravel pit full of water, at the Eden's Way (Heyden)."

Recorded once only from Chesterton Pool (3559) where it was seen by Bagnall in 1877; formerly more widespread particularly in the Avon Valley; recorded from all surrounding VCs.

300.7 **Oenanthe fluviatilis** (Bab.) Colem.
River Water-dropwort
Bromwich, H., 1870. Herb. B'ham Univ. In the River Leam at Birdingbury.

Not recorded in present survey; formerly abundant especially in the Leam Valley; recorded from all surrounding VCs except 37 and 39.

301 **Aethusa** L.

301.1 **Aethusa cynapium** L. Fool's Parsley
Baxter, W., 1831. MS (Bagnall: Mid. Nat. 1892–93.) Near Rugby.
Lloyd, Dr. G., 1835. Herb. Perry. Ufton Wood.
Map 312, p. 388

A common weed of cultivated land (60%), roadsides and waste places (30%) and grassland (6%); fairly evenly distributed throughout the county but rather less frequent in the N.; recorded from all surrounding VCs.

302 **Foeniculum** Mill.

302.1 **Foeniculum vulgare** Mill. Fennel
Cross, F., 1874. Herb. Perry. Railway bank, Emscote, Warwick. Map 313, p. 389

A rare casual found chiefly in the areas of Birmingham, Coventry, Rugby, Warwick and Leamington, and Stratford; recorded from waste places, railway banks, rough grassland and a few other habitats; recorded from all surrounding VCs.

303 **Silaum** Mill.

303.1 **Silaum silaus** (L.) Schinz & Thell.
Pepper-saxifrage
Baxter, W., 1831. MS (Bagnall: Mid. Nat. 1892–93) [*Silaus pratensis*]. By Brownsover Planks.
Bree, Rev. W. T., 1835. New Bot. Guide [*Silaus pratensis*]. Without locality.
Map 314, p. 389

Abundant in the S. half of the county especially on calcareous soils, though absent from the Avon Valley, and found in a number of widely scattered localities elsewhere; recorded from grassland (65%), roadsides and railway banks (25%), hedgerows (6%) and a few other habitats; recorded from all surrounding VCs.

307 **Angelica** L.

307.1 **Angelica sylvestris** L. Wild Angelica
Baxter, W., 1831. MS (Bagnall: Mid. Nat. 1892–93.) Near Newbold on Avon.
Map 315, p. 390

A plant of watersides and ditches (53%), roadsides and waste places (15%), damp woods (12%), hedgerows and scrub (12%) and grassland (8%); abundant and widely distributed but less frequent in the S.E. than in the rest of the county; recorded from all surrounding VCs.

310 **Pastinaca** L.

310.1 **Pastinaca sativa** L. Wild Parsnip
Baxter, W., 1831. MS (Bagnall: Mid. Nat. 1892–93) [*Peucedanum sativum*]. Newbold on Avon.
Bree, Rev. W. T., 1835. New Bot. Guide. Without locality. Map 316, p. 390

An abundant plant on calcareous soils in the S. and S.E. of the county, with scattered occurrences elsewhere; recorded from roadsides, railway banks and waste places (55%), grassland (22%), hedgerows and scrub (16%), quarries (4%) and a few other habitats; recorded from all surrounding VCs.

311 **Heracleum** L.

311.1 **Heracleum sphondylium** L.
Cow Parsnip, Hogweed
Lloyd, Dr. G., 1835. Herb. Perry. Leamington.
Map 317, p. 391

Abundant and evenly distributed throughout the county; recorded from roadsides, railway banks and waste places (43%), hedgerows and scrub (25%), grassland (21%), woods (4%) and watersides (4%); recorded from all surrounding VCs; var. *angustifolium* Huds. has also been recorded from time to time.

311.2 **Heracleum mantegazzianum** Somm. & Levier
Giant Hogweed
Cadbury, Miss D. A., 1953. Canal side, Small Heath, Birmingham.

A rare garden escape occasionally becoming naturalised by canals, rivers and roads; recorded from 0483, 0985, 1254, 2156, 2257 and 2447; recorded from all surrounding VCs.

314 Daucus L.

314.1a **Daucus carota** L. ssp. **carota** Wild Carrot
Baxter, W., 1831. MS (Bagnall: Mid. Nat. 1892–93.) Near Brownsover.
Lloyd, Dr. G., 1835. Herb. Perry (Bagnall: Flora, 1891). Without locality.
Map 318, p. 391

Abundant, mainly on the calcareous soils in the S. of the county with a scattered distribution elsewhere; recorded from railway banks, roadsides and waste places (53%), grassland (28%), hedgerows and scrub (10%), arable land (5%) and quarries (3%); recorded from all surrounding VCs; Warwickshire is near the N.W. limit of the main British distribution.

CUCURBITACEAE

315 Bryonia L.

315.1 **Bryonia dioica** Jacq. White Bryony
Perry, 1812. Herb. Perry. Warwick.
Map 319, p. 392

A fairly abundant plant throughout most of the county but rare in the E. and in the W. central area; recorded from hedges and scrub (78%), waste places, roadsides and railway banks (12%) and mixed woodland (4%); recorded from all surrounding VCs.

EUPHORBIACEAE

318 Mercurialis L.

318.1 **Mercurialis perennis** L. Dog's Mercury
Perry, 1813. Herb. Perry. Near Warwick.
Map 320, p. 392

Abundant though irregularly distributed throughout the N.W. half of the county; rare in the Avon Valley and on the calcareous clays of the S. and S.E. but frequent on the Lower Lias limestone and the Cotswold escarpment; recorded from hedgerows and scrub (49%), woods (31%), roadsides and railway banks (16%) and watersides (3%); recorded from all surrounding VCs.

318.2 **Mercurialis annua** L. Annual Mercury
Burges, Dr. R. C. L., 1946–47. Proc. B.S.B.I. Bombed site, Birmingham. Map 758, p. 607

An occasional plant of waste places, cultivated ground and roadsides; recorded from all surrounding VCs.

319 Euphorbia L.

319.2 **Euphorbia lathyris** L. Caper Spurge
Dunn, S. T., 1896. Journal of Botany. Grandborough. Map 759, p. 607

An occasional plant of gardens and waste places, probably generally an escape from cultivation; recorded from all surrounding VCs.

319.7 **Euphorbia platyphyllos** L. Broad Spurge
Bromwich, H., 1891. Bagnall's Flora. On railway banks near Myton.

Not recorded in present survey; recorded from all surrounding VCs except 39 and 57.

319.9 **Euphorbia helioscopia** L. Sun Spurge
Baxter, W., 1831. MS (Bagnall: Mid. Nat. 1892–93.) Dunchurch Road, Rugby.
Lloyd, Dr. G., 1835. Herb. Perry (Bagnall: Flora, 1891). Without locality.
Map 321, p. 393

Abundant and fairly evenly distributed throughout the county; recorded from cultivated land (69%), waste places and roadsides (28%) and a few other habitats; recorded from all surrounding VCs.

319.10 **Euphorbia peplus** L. Petty Spurge
Perry, 1812. Herb. Perry. Warwick.
Map 322, p. 393

Abundant and fairly evenly distributed throughout the county; recorded from cultivated land (68%), waste places and roadsides (28%) and a few other habitats; recorded from all surrounding VCs.

319.11 **Euphorbia exigua** L. Dwarf Spurge
Perry, 1817. List. Open field between Tachbrook and Harbury; opposite Stoneleigh Lodge. Map 323, p. 394

Abundant on calcareous soils in the S. and S.E. of the county with occasional occurrences elsewhere; recorded from cultivated land (83%), roadsides, railway sides and waste places (11%) and a few other habitats; recorded from all surrounding VCs.

319.14 & 15
Euphorbia esula L. agg. incl. **E. uralensis** Fisch. ex Link
Bromwich, H., 1871. Herb. Bagnall. Railway banks, Leek Wootton.

Not recorded in present survey; Bromwich's plant appears to be *E. esula* L.; this species was collected in the Leek Wootton–Kenilworth–Leamington area on a number of occasions between 1871 and 1897; the aggregate species has been recorded from all surrounding VCs.

319.16 **Euphorbia cyparissias** L. Cypress Spurge
Rugby School Report, 1867. Near Rugby.

Two records of plants naturalised: one in scrub at Wellesbourne Wood (2653) and one in grassland at Harborough Magna Cemetery (4779); recorded from all surrounding VCs except 57.

319.17 **Euphorbia amygdaloides** L. Wood Spurge
Lloyd, Dr. G., 1835. Herb. Perry. Oversley Wood. Map 324, p. 394

A plant chiefly confined to an area in the S.W. of the county; recorded mainly from woods; Warwickshire is on the edge of the main British distribution of the species which is recorded from all surrounding VCs.

POLYGONACEAE

320 Polygonum L.

320.1 **Polygonum aviculare** L. *sensu lato*
Common Knotgrass
Stokes, Dr. J., 1787. Withering's Arrangement. Near Coleshill. Map 325, p. 395

Abundant and evenly distributed throughout the county; recorded from roadsides, waste places, farmyards and railway banks (50%), cultivated land (37%), grassland (10%) and a few other habitats; recorded from all surrounding VCs.
The records submitted are based partly on material determined by us as *P. aviculare* sensu stricto and partly on records submitted without specimens. We are convinced that the bulk of these records are of *P. aviculare* sensu stricto.

320.1.3 **Polygonum rurivagum** Jord. ex Bor.
Coultas, P., 1957. Martin's Farm.
One record from a cultivated field at Martin's Farm, near Rugby (5072); recorded from all surrounding VCs.

Polygonum arenastrum Bor.
Small-leaved Knotgrass
Bagnall, 1873. Herb. Bagnall [*P. aviculare* var. *arenastrum* Bor.]. Road by Coles Garden, Sutton. Map 326, p. 395
An occasional plant of a number of widely scattered localities; recorded from roadsides, waste places, cultivated land, railway banks and farmyards; recorded from all surrounding VCs.

320.6 **Polygonum bistorta** L. Common Bistort
Ray, 1670. Catalogus [*Bistorta major*]. Without locality. Map 327, p. 396
An occasional plant widely distributed in the N.W. half of the county; recorded from roadsides, grassland, riversides, hedgerows and woods; recorded from all surrounding VCs.

320.7 **Polygonum amplexicaule** D. Don Red Bistort
Lowe, J. H., 1960. The Grove, Tachbrook Mallory, near Whitnash.
One record as a garden escape from Tachbrook Mallory (3162); recorded from VCs 32 and 57.

320.8 **Polygonum amphibium** L.
Amphibious Bistort
Perry, 1812. Herb. Perry. Warwick.
Map 328, p. 396
Frequent and fairly evenly distributed throughout the county; recorded from watersides and marshes (57%), water (30%), waste places and roadsides (6%), cultivated land (4%) and a few other habitats; recorded from all surrounding VCs.

320.9 **Polygonum persicaria** L. Common Persicaria
Perry, 1812. Herb. Perry. Warwick.
Map 329, p. 397
Abundant and evenly distributed throughout the county; recorded from cultivated land (41%), waste places, roadsides, farmyards and railway banks (32%), watersides (16%), grassland (8%) and a few other habitats; recorded from all surrounding VCs.

320.10 **Polygonum lapathifolium** L. (incl. **P. nodosum** Pers.) Pale Persicaria
Perry, 1812. Herb. Perry. Warwick.
Map 330, p. 397
Frequent but very irregularly distributed throughout most of the county; recorded from cultivated land (48%), waste places, roadsides and farmyards (30%), watersides (15%) and grassland (5%); recorded from all surrounding VCs.

320.12 **Polygonum hydropiper** L.
Common Water Pepper
Perry, 1812. Herb. Perry. Longbridge, near Warwick. Map 331, p. 398
Frequent but rather irregularly distributed throughout most of the county, though somewhat rarer on the calcareous soils of the S. and S.E.; recorded from watersides (71%), cultivated land (10%), grassland (6%), damp woods (5%) and waste places (4%); recorded from all surrounding VCs.

320.13 **Polygonum mite** Schrank
Tasteless Water Pepper
Rugby School Report, 1892. Bourton.
Not recorded in present survey; recorded from all surrounding VCs.

320.14 **Polygonum minus** Huds.
Small Water Pepper
Bagnall, 1880. Herb. Bagnall [*P. persicaria* var. *elatum*]. Top end of Bracebridge Pool, Sutton Park.
Recorded once only from a marsh near Little Bracebridge Pool, Sutton Park (0998), probably the same station as Bagnall's; recorded from all surrounding VCs except 32.

320.15 **Polygonum convolvulus** L. Black Bindweed
Perry, 1812. Herb. Perry. Warwick.
Map 332, p. 398
Abundant and evenly distributed throughout the county; recorded from cultivated land (67%), waste places, roadsides and railway banks (24%), hedgerows (4%) and grassland (3%); recorded from all surrounding VCs.

320.18 **Polygonum aubertii** (L. Henry) Mold. Russian Vine
Hardaker, W. H., 1959. Chadwick End.
A garden escape naturalised in hedgerows at Edgbaston (0583), Coundon (3081), Marton (4069), Long Itchington (4165) and in a wood at Chadwick End (2173); not recorded from any of the surrounding VCs.

320.19 **Polygonum cuspidatum** Sieb. & Zucc.
Japanese Knotweed
Thomas C., 1950. Gillhurst Rd., Harborne.
Map 333, p. 399
An introduced plant well established in many areas in the N. half of the county; recorded from waste places, roadsides and railway banks (60%), hedgerows (16%), cultivated land (11%), grassland (6%), quarries (4%) and watersides (3%); recorded from all surrounding VCs.

320.21 **Polygonum polystachyum** Wall. ex Meisn.
Himalayan Knotweed
Allen, H., 1960. Bombed site, Coventry.
An introduction, occasionally naturalised, recorded from a roadside near Shrewley (2267), a canal bank at Hatton (2466) and a bombed site at Coventry (3480); recorded from VCs 33 and 57.

320.22 **Polygonum campanulatum** Hook. f.
Lesser Knotweed
Burges, Dr. R. C. L., 1933. Herb. Burges [*P. bistorta* (*superba*)]. Marshy ground above Blackroot by Four Oaks Wall, Sutton Park.
One record only from a marsh in Sutton Park (1097); this is apparently the locality in which Burges saw the plant; not recorded from any of the surrounding VCs.

Polygonum arifolium L.
Bromwich, H. and Jackson, A. B., 1900. Herb. Brit. Mus. Casual in a garden at Milverton, Leamington.
A native of N. America not recorded in present survey; not recorded from any of the surrounding VCs.

Polygonum patulum Bieb.
Cadbury, Miss D. A., 1964. Marl-pit, Saltley.
One record of the plant, locally abundant, in an old brickyard (0987); habitat since destroyed; recorded from VC 39.

321 Fagopyrum Mill.

321.1 **Fagopyrum esculentum** Moench Buckwheat
Purton, T., 1817. Mid. Flora [*Polygonum fagopyrum*]. Ridgeway. Map 760, p. 607
An occasional relic of cultivation found chiefly in the Wilmcote area; recorded from all surrounding VCs.

325 Rumex L.

325.1.1 **Rumex acetosella** L. *sensu stricto* Sheep's Sorrel
Perry, 1812. Herb. Perry, Warwick. Map 334, p. 399
Abundant and fairly evenly distributed in the N. of the county, but rather rare on the calcareous soils of the S. and S.E.; recorded from roadsides, waste places and railways (43%), grassland (32%), cultivated land (13%), hedgerows and scrub (5%) and a few other habitats; recorded from all surrounding VCs.

325.1.3 **Rumex tenuifolius** (Wallr.) Löve Narrow-leaved Sheep's Sorrel
Bagnall, 1873. Herb. Bagnall [*R. acetosella* L.]. Coleshill.
A rare plant of dry sandy and stony places recorded from Yarningale Common (1865), the railway at Houndshill (2550) and a quarry near Griff (3589); earlier records are from Coleshill Bog and Sutton Park; recorded from VC 37.

325.2 **Rumex acetosa** L. Common Sorrel
Baxter, W., 1831. MS (Bagnall: Mid. Nat. 1892–93.) Near Rugby.
Lloyd, Dr. G., 1835. Herb. Perry. Leamington. Map 335, p. 400
Abundant and evenly distributed throughout the county; recorded from grassland (55%), roadsides, railway banks and waste places (33%), hedgerows and scrub (5%), cultivated land (4%) and a few other habitats; recorded from all surrounding VCs.

325.4 **Rumex hydrolapathum** Huds. Water Dock
Withering, 1787. Arrangement [*R. aquaticus* L.]. Tamworth. Map 336, p. 400
An occasional plant in many parts of the county, largely absent from the N.W. and the extreme S.; recorded from watersides (83%) and water (17%) (mainly canals and rivers); recorded from all surrounding VCs.

325.11 **Rumex crispus** L. Curled Dock
Baxter, W., 1831. MS (Bagnall: Flora, 1891). Rugby.
Lloyd, Dr. G., 1835. Herb. Perry. Leamington. Map 337, p. 401
Abundant and evenly distributed throughout the county; recorded from roadsides, waste places, farmyards and railway banks (42%), grassland (27%), cultivated land (19%), hedgerows (7%) and watersides (5%); recorded from all surrounding VCs.

Rumex crispus × obtusifolius = R. × acutus L.
Bloxam, Rev. A., 1872. Rugby School Report. Pond in Cathiron Lane, Harborough Magna.
A rare hybrid found in a number of places in the county, and recorded from squares 0385, 0454, 0583, 0664, 0852, 0885, 1484, 2238, 2806, 2986, 3254, 3449, 3493, 4479 and 5374; recorded from all surrounding VCs.

325.12 **Rumex obtusifolius** L. Broad-leaved Dock
Baxter, W., 1831. MS (Bagnall: Flora, 1891). Rugby. Map 338, p. 401
Abundant and evenly distributed throughout the county; recorded from roadsides, waste places, farmyards and railway banks (47%), grassland (24%), hedgerows (13%), cultivated land (8%), watersides and marshes (6%) and a few other habitats; recorded from all surrounding VCs.

325.13 **Rumex pulcher** L. Fiddle Dock
Cross, F., 1850. Herb. Perry (Bagnall's Flora, 1891). Harbury Village.
Two records only, from rough grassland at Salford Priors (0750) and Newbold on Stour (2446); earlier records are from Chesterton and a few other localities; recorded from all surrounding VCs.

325.14 **Rumex sanguineus** L. Wood Dock, Red-veined Dock
Perry, 181–. Herb. Perry. Leamington. Map 339, p. 402
Abundant and fairly evenly distributed throughout the county; recorded from watersides and marshes (33%), hedgerows and scrub (25%), roadsides and waste places (17%), grassland (14%), woods (9%) and a few other habitats; recorded from all surrounding VCs.
The common form in the county is var *viridis* Sibth.; var. *sanguineus* is very rare.

325.15 **Rumex conglomeratus** Murr. Clustered Dock
Baxter, W., 1831. MS (Bagnall: Mid. Nat. 1892–93.) Near Brownsover. Map 340, p. 402
Abundant and fairly evenly distributed throughout the county; recorded from watersides and marshes (64%), hedgerows (11%), grassland (10%), roadsides and waste places (9%), woods (4%); recorded from all surrounding VCs.

325.18 **Rumex maritimus** L. Golden Dock
Ray, 1686. Hist. Plant. [*Lapathum folio acuto floreo aureo* C.B.]. "Middletoni Warwicensi agro in fossis."
A rare plant of the sides of lakes, ponds and rivers recorded from Little Alne (1451, 1561), Wishaw (1794), Tamworth (2104), Alvecote (2404, 2405, 2504), Compton Verney (3052) and Moreton Morrell (3255); recorded from all surrounding VCs except 57.

URTICACEAE

326 Parietaria L.

326.1 **Parietaria judaica** L. Pellitory-of-the-wall
Purton, T., 1817. Mid. Flora [*P. officinalis*]. Haselor Churchyard. Map 341, p. 403
An occasional plant very widely scattered throughout the county; recorded from the bases of walls (72%), waste places and roadsides; recorded from all surrounding VCs.

327 Soleirolia Gaudich.

327.1 **Soleirolia soleirolii** (Req.) Dandy Mind-your-own-business
Jones, Mrs. M. D. G., 1959. Near Pailton.
An occasional garden escape recorded from squares 3660,

4671, 4682 and 4871; recorded from all surrounding VCs except 39 and 57.

328 Urtica L.

328.1 **Urtica urens** L. Small Nettle
Baxter, W., 1831. MS (Bagnall: Mid. Nat. 1892–93.) Near Newbold on Avon.
Ick, W., 1836. Herb. Ick. Washwood Heath.
Map 342, p. 403
Abundant in the Avon Valley; frequent, but very irregularly distributed elsewhere in the county; recorded from cultivated land (49%), waste places, farmyards and roadsides (38%), watersides (4%), hedgerows (4%) and a few other habitats; recorded from all surrounding VCs.

328.2 **Urtica dioica** L. Stinging Nettle
Baxter, W., 1831. MS (Bagnall: Flora, 1891). Rugby.
Lloyd, Dr. G., 1835. Herb. Perry. Leamington.
Map 343, p. 404
Very abundant, and evenly distributed throughout the county; recorded from roadsides, waste places, farmyards and railway banks (35%), hedgerows and scrub (29%), grassland (12%), watersides (10%), woods (9%) and cultivated land (5%); recorded from all surrounding VCs.

328.3 **Urtica pilulifera** L. Roman Nettle
Perry, 1823. Herb. Perry [*U. dodartii*]. Garden at Saltisford, Warwick.
Not recorded in present survey; recorded from VCs 23, 32, 39 and 55.

CANNABIACEAE

329 Humulus L.

329.1 **Humulus lupulus** L. Hop
Perry, 1817. List. Emscote Bridge.
Map 344, p. 404
Fairly frequent but widely and very irregularly distributed throughout the county; recorded from hedges (81%), waste places, roadsides and railway banks (10%) and a few other habitats; recorded from all surrounding VCs.

Cannabis L.

Cannabis sativa L. Hemp
Bagnall, 1877. Herb. Bagnall. Railway Embankment near Blackroot Pool, Sutton Park.
A rare casual recorded once only from Wilmcote Station (1658); recorded from all surrounding VCs except 33 and 39.

ULMACEAE

330 Ulmus L.

330.1 **Ulmus glabra** Huds. Wych Elm
Purton, T., 1817. Mid. Flora [*U. montana* Sm.]. Wixford Lane. Map 345, p. 405
Abundant and fairly evenly distributed throughout the county; recorded from hedges and scrub (74%), mixed woods (21%) and a number of other habitats; recorded from all surrounding VCs.
Although subspecies have not been noted generally, they have been recorded as follows:
ssp. **glabra**: (first record: Bagnall, 1873. Herb. Bagnall. Witton); recorded from squares 1076, 1275, 1467, 1561, 1695, 2641, 2742, 2744, 3286, 3654, 4483 and 4581.
ssp. **montana** (Lindq.) Tutin: (first record: Lloyd, Dr. G., 1835. Herb. Perry [*U. montana*]. Leamington); recorded from squares 1051, 1560, 2053, 2341, 2486, 2502, 2641, 2653, 2742, 3089, 3169, 3183, 3464, 3744, 4073, 4249, 4390, 4581, 4778 and 4779.

Ulmus × hollandica Mill. Dutch Elm
Bromwich, H., 1878. Herb. B'ham Univ. [*U. major*]. Myton. Map 346, p. 405
This tree, assumed to be a hybrid (*U. glabra × minor*), has been planted and recorded as occasional in the S. of the county, almost entirely in hedges though there are a few records from woods; recorded from all surrounding VCs except 37.
The two varieties have been noted:
var. **hollandica**: (first record: Bromwich, H., 1892. Herb. Bagnall [*U. campestris* Sm. var. *glabra* Mill.]. Chesterton); all records shown on the map, with the exception of two (see var. *vegeta* below) are of this variety. Recorded from all surrounding VCs except 37.
var. **vegeta** (Loud.) Rchd.: (first record: Clark, M. C., 1960. Coughton); this has been recorded from Coughton (0760) and Brandon Station (4076). Recorded from VCs 23, 32 and 33.

Ulmus minor Mill. agg. = **U. carpinifolia** agg.
(*U. angustifolia*, *U. carpinifolia* Suckow, *U. coritana*, *U. diversifolia* and *U. plotii*).
First record, *U. carpinifolia × plotii*: Bagnall, 1868. Herb. Bagnall [*U. suberosa*]. Near Marston Green. Map 347, p. 406
An occasional tree mainly confined to the chalky boulder clay in the E. of the county and the calcareous clays of the S.W. and S.E.; recorded chiefly from hedges, but also from mixed woods, roadsides and watersides; recorded from all surrounding VCs.

330.3 **Ulmus angustifolia** (Weston) Weston
Melville, R., 1963. Between Brinklow and Stretton under Fosse.
One record only from hedgerow between Brinklow and Stretton under Fosse (4480); recorded from VCs 33 and 55.

330.5 **Ulmus carpinifolia** Suckow
Green, P. S., 1951. Near Brandon.
Three records, from hedgerows at Newland Hall (3285), near Brandon (3876) and near Grandborough (4866); var. *variegata* has been recorded once as a planted tree from Bearley (1860); recorded from all surrounding VCs except 39 and 57.

Ulmus carpinifolia × plotii
Bagnall, 1868. Herb. Bagnall [*U. suberosa*]. Near Marston Green. Map 761, p. 608
An occasional tree of hedgerows mainly in the S.E. of the county; recorded from VC 23.

330.4 **Ulmus coritana** Melville
Bagnall, 1881. Herb. Bagnall [*U. nitida*]. Farnborough. Map 762, p. 608
An occasional tree, mainly of hedgerows, thinly distributed throughout the county; recorded from VCs 23 and 55.

Ulmus coritana × **plotii**
= **U.** × **diversifolia** Melville
Smith, T., 1960. Near Churchover, Rugby.
Map 763, p. 608
An occasional plant mainly of hedgerows; not recorded from any of the surrounding VCs.

330.6 **Ulmus plotii** Druce
Druce, 1932. Com. Flora. Without locality.
A rare tree of woods and hedgerows recorded from Four Oaks (1199), near Bearley (1759), Stratford on Avon racecourse (1853), Pathlow (1858), between Bacon's End and Coleshill (1888), near Ilmington (2144, 2145) and Newbold Revel (4580); recorded from all surrounding VCs except 37.

Ulmus sarniensis (Moss) Bancroft
Melville, R., 1963. Bearley Cross.
A tree frequently planted in Birmingham, Solihull and other towns, and also recorded from hedgerows at Bearley Cross (1760), Arlescote (3948) and Brinklow Station (4480, 4481); recorded from VCs 23, 32, 33 and 55.

Ulmus minor agg. × **glabra**
First record, *U. carpinifolia* × *glabra* × *plotii*: Kirk, T., 1860. Herb. Perry [*U. glabra*]. Near Bickenhill. Map 348, p. 406
Thinly and irregularly distributed throughout most of the county; recorded from hedges and scrub (74%), mixed woods (18%) and a few other habitats.

Ulmus carpinifolia × **glabra**
Clark, M. C., 1961. Saltisford Common.
An occasional tree of woods and hedgerows recorded from near Alne End (1059), near Temple Grafton (1253), Great Alne (1259), Bearley Cross (1660), near Armscote (2344), Saltisford Common (2765), Chapel Green (2785), near Monks Kirby (4483), near Napton on the Hill (4560) and Newnham Lodge Farm (4784); recorded from VCs 33, 37 and 39.

Ulmus carpinifolia × **plotii** × **glabra**
Kirk, T., 1860. Herb. Perry [*U. glabra*]. Near Bickenhill. Map 764, p. 608
An occasional tree of woods and hedgerows, mainly in the S. of the county; recorded from VC 57.

Ulmus coritana × **glabra**
Cadbury, Miss D. A., 1960. Temple Balsall.
A rare tree mainly of hedgerows recorded from Great Alne (1259), Temple Balsall (2076), near Stratford on Avon (2155), Honington (2742), Old Milverton (2967) and near Priors Marston (4757); not recorded from any of the surrounding VCs.

Ulmus coritana × **plotii** × **glabra**
Melville, R., 1963. Between Bacon's End and Coleshill.
Two records only, from woodland between Bacon's End and Coleshill (1888) and hedgerow at Pailton (4782); not recorded from any of the surrounding VCs.

Ulmus plotii × **glabra**
Rugby School Report, 1949. Brandon.
Map 765, p. 609
A frequent tree of woods and hedgerows fairly evenly distributed throughout the county; recorded from all surrounding VCs except 23 and 32.

Ulmus minor agg. × **procera**
For first record see *U. plotii* × *procera*.
For details see segregates.

Ulmus coritana × **procera**
Clark, M. C., 1963. Pathlow, near Stratford on Avon.
One record only from hedgerow at Pathlow (1858); not recorded from any of the surrounding VCs.

Ulmus plotii × **procera**
Woolman, J. F., 1958. Near Grandborough.
Three records only, from hedgerows at Drayton (1654), Stratford on Avon racecourse (1853) and Grandborough (5066); not recorded from any of the surrounding VCs.

330.2 **Ulmus procera** Salisb. English Elm
Kirk, T., 1859. Herb. Perry [*U. campestris*]. Baginton. Map 349, p. 407
Abundant and fairly evenly distributed throughout most of the county, but not as common in the N.W.; recorded from hedges (90%), scrub (5%), grassland (4%) and a few other habitats; recorded from all surrounding VCs.

JUGLANDACEAE

332 Juglans L.

332.1 **Juglans regia** L. Walnut
Phipps, J. B., 1955. Near Ilmington.
Map 350, p. 407
Though not formally recorded before 1955, has been planted for many years and occasionally has become naturalised in grassland and hedges; thinly distributed throughout the county; recorded from all surrounding VCs except 39.

PLATANACEAE

334 Platanus L.

334.1 **Platanus occidentalis** × **orientalis** = **P.** × **hybrida** Brot. London Plane
Burges, Dr. R. C. L., 1942. Herb. Burges [*P. acerifolia* Willd.]. Birmingham.
Commonly planted in streets and parks in Birmingham, Coventry, Rugby and elsewhere, and recorded several times during the present survey; probably planted in some or all of the surrounding VCs although not formally recorded.

BETULACEAE

335 Betula L.

335.1 **Betula pendula** Roth Silver Birch
Baxter, W., 1831. MS (Bagnall: Mid. Nat. 1892–93) [*Betula alba*]. Dunchurch Hill, near Rugby. Map 351, p. 408
Abundant, though a little irregular in distribution, mainly on light heathy soils in the N.W. of the county, but comparatively rare in the S.E.; recorded from woodland (40%), hedgerow and scrub (33%), grass and cultivated land (12%) and railway banks and roadsides (10%); recorded from all surrounding VCs.
Hybrids between *B. pendula* and *B. pubescens* have been recorded frequently in the presence of both parents.

335.2 **Betula pubescens** Ehrh. Downy Birch
Perry, 1846. Herb. Perry [*B. alba*]. Hay Wood.
Map 352, p. 408

Less abundant than *B. pendula*, with a similar distribution but more restricted to the somewhat damper soils of the N.W. of the county; recorded from woodland (49%), hedges and scrub (35%), railway banks and roadsides (6%) and heaths (5%); recorded from all surrounding VCs.

B. pubescens Ehrh. ssp. *odorata* (Bechst.) E. F. Warb.: four specimens have been so identified by Dr. E. F. Warburg from Windmill Naps (0972), Sutton Park (0998), Trickley Coppice (1598) and near Princethorpe (3871); probably most of the other records are for ssp. *pubescens*.

336 **Alnus** Mill.

336.1 **Alnus glutinosa** (L.) Gaertn. Alder
Baxter, W., 1831. MS (Bagnall: Mid. Nat. 1892–93.) Newbold and Lawford.
Map 353, p. 409

A tree of watersides (73%), hedges (15%) and damp woods (9%), occurring abundantly in the area of the old Forest of Arden in the N.W. of the county but thinning out towards the S.E.; recorded from all surrounding VCs.

CORYLACEAE

337 **Carpinus** L.

337.1 **Carpinus betulus** L. Hornbeam
Dolben, T., 1817. Mid. Flora. Ipsley.
Map 354, p. 409

Rather irregularly scattered over a wide area in the centre of the county; nearly always found singly or in small groups; recorded from woodland (51%), hedges and scrub (27%), roadsides and railway embankments (10%) and a few other habitats; recorded from all surrounding VCs.

338 **Corylus** L.

338.1 **Corylus avellana** L. Hazel
Baxter, W., 1831. MS (Bagnall: Mid. Nat. 1892–93.) Rugby. Map 355, p. 410

Abundant and fairly evenly distributed throughout most of the county but less common on the calcareous clays of the S. and S.E.; recorded from hedges and scrub (70%), woods (23%) and roadsides, waste places and railway banks (4%); recorded from all surrounding VCs.

FAGACEAE

339 **Fagus** L.

339.1 **Fagus sylvatica** L. Beech
Baxter, W., 1831. MS (Bagnall: Mid. Nat. 1892–93.) Near Sawbridge.
Lloyd, Dr. G., 1835. Herb. Perry. Near Leamington. Map 356, p. 410

Abundant, but very irregularly distributed throughout most of the county and almost absent from the Lower Lias clays; often planted and sometimes regenerating; recorded from woods (38%), hedges and scrub (36%), grassland (9%), roadsides (9%) and cultivated land (8%); recorded from all surrounding VCs.

340 **Castanea** Mill.

340.1 **Castanea sativa** Mill.
Sweet Chestnut, Spanish Chestnut
Purton, T., 1817. Mid. Flora [*Fagus castanea*]. Snitterfield; Ragley Woods. Map 357, p. 411

Widely but unevenly distributed throughout most of the county but almost entirely absent from the calcareous clays of the S.E.; often planted and occasionally naturalised; recorded from woodland (65%), hedges and scrub (20%), open grassland (9%) and roadsides and waste places (6%); recorded from all surrounding VCs.

341 **Quercus** L.

341.1 **Quercus cerris** L. Turkey Oak
Bagnall, 1873. Herb. Bagnall. Road to Atherstone from Coleshill by pathway through meadows to Bridge over the Cole. Map 358, p. 411

An introduced tree, occasionally naturalised, and widely scattered throughout most of the county, though rare on the calcareous soils of the S. and S.E.; recorded from hedges and scrub (44%), woods (37%), grassland (11%) and roadsides and railway banks (6%); recorded from all surrounding VCs.

341.2 **Quercus ilex** L. Evergreen Oak
Daulman, Miss B. M. and Laflin, T., 1959. Near Ilmington.

Two records of trees planted near Ilmington (2044) and near Welcombe (2156); recorded from all surrounding VCs except 39.

341.3 **Quercus robur** L. Common Oak
Withering, 1801. Arrangement. By the boathouse in Edgbaston Park. Map 359, p. 412

Abundant and fairly evenly distributed throughout the county; but on the calcareous soils of the S. and S.E. the trees are rare to occasional in most of the stations where they have been recorded, while they are frequent to abundant in stations in the N. and N.W.; recorded from hedges and scrub (61%), woods (25%), grassland (6%), roadsides, railway banks and waste places (6%) and a few other habitats; recorded from all surrounding VCs.

341.4 **Quercus petraea** (Mattuschka) Liebl.
Durmast Oak, Sessile Oak
Bree, Rev. W. T., 1817. Mid. Flora [*Q. sessiliflora*]. Allesley. Map 360, p. 412

In Warwickshire *Q. petraea* shows variation in the direction of *Q. robur*, and this may be construed as evidence of past hybridisation and introgression of *Q. robur* genes into *Q. petraea*. It may be interpreted, on the other hand, merely as evidence of wide variability in *Q. petraea*. The records mapped are for plants ranging from "good" *Q. petraea*, through those recorded as hybrids, to extreme types whose morphology seems to indicate a strong *Q. robur* influence.

The majority of *Q. petraea* records in Warwickshire are from sandstones and gravels, mainly in the N. half of the county; recorded from woods (45%), hedgerows and scrub (34%) and grassland (13%); recorded from all surrounding VCs.

The hybrid *Q. petraea* × *Q. robur* (*Q.* × *rosacea* Bechst.) was first recorded in Warwickshire by H. H. Bloomer (Proc. B.S.B.I. 1926) from Streetly Wood, Sutton Park; said to be with both parents. It has been recorded from VCs 33, 37, 55 and 57.

SALICACEAE

342 Populus L.

342.1 **Populus alba** L. White Poplar
Baxter, W., 1840. MS (Bagnall: Flora, 1891). By the river at Holbrook Grange, Rugby.
Kirk, T., 1854. Herb. Perry. Binley.
Map 361, p. 413

An occasional tree of widely scattered occurrences, probably always planted; recorded from hedgerows and scrub (57%), woods (14%), grassland (12%), waste places and roadsides (12%) and watersides (5%); recorded from all surrounding VCs.

342.2 **Populus canescens** (Ait.) Sm. Grey Poplar
Baxter, W., 1831. MS (Bagnall: Mid. Nat. 1892–93.) Near Hill Morton. Map 362, p. 413

An occasional tree, often planted, widely scattered throughout the county; recorded from hedgerows and scrub (42%), woods (40%), watersides (12%) and a few other habitats; recorded from all surrounding VCs.

342.3 **Populus tremula** L. Aspen
Baxter, W., about 1840. MS (Bagnall: Mid. Nat. 1892–93.) Without locality.
Kirk, T., 1855. Herb. Brit. Mus. Kenilworth.
Map 363, p. 414

Frequent and rather irregularly distributed throughout most of the county, but rare on the calcareous soils of the S. and S.E.; recorded from hedgerows and scrub (47%), woods (33%), watersides (11%) and roadsides and waste places (6%); recorded from all surrounding VCs.

342.4 **Populus nigra** L. Black Poplar
Bromwich, H., 1872. Herb. Bagnall. Brook side, Myton.

Three records from waterside, at Lower Binton (1553), Chapel Green (2785) and near Burton Hastings (4089), and one from roadside near Atherstone (3198); *P.* × *canadensis* has been mistaken frequently for this tree, and the records given are all the authenticated ones for *P. nigra* that we have; recorded from all the surrounding VCs except 37.

Populus nigra L. var. **italica** Duroi
Lombardy Poplar
Burges, Dr. R. C. L., 1937. Herb. Burges. Near Stratford. Map 364, p. 414

An occasional tree, always planted, widely distributed throughout the county; recorded from hedgerows (78%), woods (11%), waste places, railway banks and grassland; recorded from VCs 23, 32 and 55.

342.5 **Populus** × **canadensis** Moench
Black Italian Poplar
Druce, 1932. Com. Flora. Without locality
Map 365, p. 415

Frequent and widely distributed particularly in the N. half of the county; probably always planted; recorded from hedgerows and scrub (48%), woods (20%), watersides (11%), railway banks, waste places and roadsides (10%) and grassland (9%); recorded from all surrounding VCs.

342.6 **Populus gileadensis** Rouleau Balsam Poplar
Rugby School Report, 1949. Brandon; Dunchurch; Cosford; Crick; Straight Mile; Stockton etc.

A tree occasionally planted in hedges and woodland and recorded from squares 0486, 1471, 1597, 1865, 1998, 2993, 3252, 3476, 4062, 4371, 4476 and 4871; recorded from VCs 23, 32, 37 and 55.

Populus trichocarpa Torrey & A. Gray ex Hooker
Evans, R., 1962. Tattle Bank.

A planted tree recorded from Tattle Bank (1863) and Merevale Park (2997); not recorded from any of the surrounding VCs.

343 Salix L.

The following notes on willow collecting and the key to the identification of Warwickshire *Salix* species and hybrids have been prepared by R. D. Meikle.

For accurate determination most willows (except *Salix pentandra* and a few montane species) must be collected at two seasons, once when the catkins are in flower, and again when the leaves are fully developed. If circumstances preclude the possibility of two visits to a locality, then it is generally advisable to collect foliage rather than catkins. Immature catkins without foliage, or over-ripe catkins and immature foliage make worthless specimens and should not be collected.

Specimens should be selected from the normal twiggy branches of trees and bushes growing in open situations. Sucker and coppice shoots are uncharacteristic and misleading, as are also foliage specimens taken from plants growing in shade.

Where it is suspected that the plant is a hybrid, specimens should be collected from adjacent willows in the hope that these may throw additional light on possible parentage.

A small portion of bark should be removed from the base of each specimen to show the presence or absence of longitudinal *striae* on the surface of the underlying wood.

To ensure that catkins and foliage come from the same individual it is often necessary to mark trees and bushes in some way. No system of marking is wholly reliable, but a rough sketch map, showing the relative positions of each specimen in relation to some permanent, conspicuous landmark, generally serves best. Labels, even if weatherproof, tend to be removed or interfered with by the public, and penknife slashes soon become stained or obscured.

Key to Warwickshire Salix Species and Hybrids

1. Prostrate or decumbent shrubs with creeping branches, usually less than 1 m high; leaves normally less than 4 cm long, ovate, oblong, elliptic or lanceolate, adpressed silvery-silky on the underside or on both sides, drying black . . . . **repens**
   Trees or shrubs, usually more than 1 m high with erect, spreading or pendulous branches 2
2. Trees with slender, pendulous, yellowish branches . . . . × **chrysocoma**
   Trees or shrubs with erect or spreading branches 3
3. Catkins appearing with the leaves on short lateral shoots 4
   Catkins sessile or subsessile, appearing before the leaves; catkin-scales commonly tinged brown, red or blackish towards apex . 14
4. Leaves oblong-elliptic, ovate or obovate, 2–3 times as long as broad 5
   Leaves lanceolate or linear-lanceolate, 3–5 (or more) times as long as broad . . 6

5. Leaves resinous, shining green above, minutely and regularly serrate; twigs and buds shining as if varnished; male flowers with 5 (occasionally more) stamens; catkin scales uniformly pale greenish-yellow **pentandra**
Leaves not resinous; twigs and buds not shining as if varnished; male flowers with 2 stamens; catkin-scales tinged blackish at apex . . . . . . **nigricans**
6. Leaves adpressed silvery-silky beneath or on both sides especially when young; leaf margins very finely toothed; tall trees with ascending branches, often with narrow pyramidal crowns 7
Leaves lustrous green and glabrous above, glaucous or green beneath and glabrous or very thinly pubescent at maturity; trees or bushes with broad rounded crowns . . 9
7. Leaves glabrescent and dull green above **alba** var. **coerulea**
Leaves silvery-silky on both sides . . 8
8. Winter twigs dull olive-brown . **alba** var. **alba**
Winter twigs yellow or orange-red **alba** var. **vitellina**
9. Small trees or bushes with smooth, scaling bark; shoots distinctly angled or ridged; stipules usually conspicuous and persistent; male flowers with 3 stamens . . **triandra**
Trees with fissured bark; shoots terete or subterete; stipules often small and caducous; male flowers with 2 stamens . . 10
10. Leaves quite glabrous even when young; winter twigs ochre-coloured; catkins (male only in Britain) small and puny, 2–3 cm long . . . . **decipiens**
Leaves thinly hairy or pubescent at least when young; winter twigs olive-brown; catkins 3·5–7 cm or more long . . 11
11. Leaves 2·5–4·5 cm wide, broadly lanceolate, very coarsely toothed; male catkins often bifurcate; twigs rather lustrous. **fragilis** var. **latifolia**
Leaves 1·5–3 cm wide, narrowly lanceolate or linear-lanceolate with a long, slender acumen 12
12. Ovaries about 5 mm long at anthesis, narrowly flask-shaped, usually exceeding the subtending catkin-scale . **fragilis** var. **russelliana**
Ovaries less than 5 mm long, shortly flask-shaped, seldom exceeding the catkin-scale at anthesis 13
13. Leaves thinly hairy below at first, soon glabrous or subglabrous, margins rather coarsely and irregularly serrate. **fragilis** var. **fragilis**
Leaves persistently, though thinly hairy below even at maturity, dullish green above, margins shortly and rather regularly serrate . . . . . **alba** × **fragilis**
14. Leaves linear, lanceolate, or ovate-elliptic tapering to a slender acumen . . . 15
Leaves ovate, obovate, elliptical, oblong or oblanceolate, obtuse or with a shortly cuspidate or acute apex 22
15. Leaves narrowly linear-lanceolate, silver-white below, margins subentire, narrowly reflexed; catkins rather crowded towards the apices of the flowering twigs . **viminalis**
Leaves not as above 16
16. Leaves glabrous or subglabrous at maturity, sometimes lustrous above . . . 17
Leaves persistently hairy or pubescent on one or both surfaces 19
17. Juvenile leaves glabrous; male flowers with 1 stamen, anthers reddish; female flowers with a very small, subspherical, tomentose ovary; leaves frequently opposite . **purpurea**
Juvenile leaves pubescent or thinly tomentose on one or both surfaces, very rarely opposite 18
18. Leaves thinly tomentose at first; ovary subspherical, densely white-tomentose; catkin-scales almost orbicular, blackish; stipules linear, caducous . . . . × **forbyana**
Leaves thinly silky-pubescent at first; ovary flask-shaped, glabrous or thinly silky; catkin-scales tongue-shaped or subulate, pale greenish-yellow; stipules ovate-acuminate, often persisting . **triandra** × **viminalis**
19. Leaves somewhat wrinkled or undulate, softly pubescent below; stipules conspicuous, persistent . **aurita** × **viminalis**
Leaves not, or only very slightly, wrinkled; stipules caducous 20
20. Leaves softly pubescent below with prominent venation; wood of peeled twigs not ridged or striate . . **caprea** × **viminalis**
Leaves sparsely and shortly pubescent below; wood of peeled twigs generally with a few distinct longitudinal ridges or striae. . 21
21. Leaves commonly 3 cm wide; female catkins 3 cm long or longer at anthesis; catkin-scales densely clothed with long silky hairs; catkin-bearing twigs softly pubescent × **calodendron**
Leaves generally less than 3 cm wide; female catkins less than 3 cm long at anthesis; catkin-scales with shorter hairs; catkin-bearing twigs shortly and sparsely pubescent . . . . **cinerea** × **viminalis**
22. Leaves oblanceolate, glabrous, often opposite; catkins very small; male flowers with 1 stamen, anthers reddish . . **purpurea**
Leaves and catkins not as above . . . 23
23. Female flowers with distinct styles and bifid stigmas 24
Female flowers with short styles and undivided stigmas or with the stigmas subsessile 25
24. Ovary densely white-silky; leaves firm in texture, bright green above, conspicuously glaucous below . . . . . × **laurina**
Ovary pubescent; leaves thin in texture, pubescent below, without markedly contrasting surfaces . . **caprea** × **nigricans**
25. Leaves sparsely pubescent below and rather harsh to the touch, often clothed with scattered reddish hairs (especially noticeable in late summer and autumn); flowering twigs thinly and shortly pubescent, becoming subglabrous . . **cinerea** ssp. **oleifolia**
Leaves softly grey-pubescent or tomentose below without reddish hairs . . . 26
26. Leaves large, often 6 cm wide, broadly elliptic-oblong or almost suborbicular,

densely and softly ashy-felted below; catkins normally ovoid, commonly 1·5 cm wide, showy, with yellow or reddish bud-scales; wood of peeled twigs without longitudinal ridges or striae . . . . . **caprea**

Leaves smaller, narrower, generally less than 6 cm wide, not felted below; catkins smaller and less conspicuous; wood of peeled twigs with longitudinal ridges or striae 27

27. Leaves wrinkled, usually with an obliquely twisted apex; stipules large, persistent; catkins small, usually less than 2 cm long at anthesis, appearing late (April–May) . **aurita**

Leaves not (or slightly) wrinkled; catkins generally appearing earlier (March–April) 28

28. Leaves broadly obovate, 3 cm or more wide; flowering twigs with conspicuous (often yellowish) bud scales; female catkins commonly more than 1 cm wide **caprea × cinerea**

Leaves smaller, usually less than 3 cm wide; flowering twigs with dark reddish bud-scales; female catkins usually less than 1 cm wide 29

29. Flowering twigs closely pubescent; apex of leaf not normally twisted obliquely; peeled twigs weakly striate with relatively few ridges or striae . . **cinerea** ssp. **cinerea**

Flowering twigs subglabrous; apex of leaf often twisted obliquely; peeled twigs conspicuously striate . . **aurita × cinerea**

343.1 **Salix pentandra** L. Bay Willow
Baxter, W., 1821. Mid. Flora. On the banks of the Avon near Holbrook Grange.
Bree, Rev. W. T., 1821. Mid. Flora. Binley, near Coventry. Map 766, p. 609

A rare tree of marsh and waterside in the N.W. of the county; recorded from all surrounding VCs.

343.2 **Salix alba** L. White Willow
Perry, 1820. Plant. Varv. Sel. [*S. vitellina*]. Without location. Map 366, p. 415

Frequent, but somewhat irregularly distributed throughout most of the county; recorded from watersides (69%), hedges and scrub (23%) and a few other habitats; recorded from all surrounding VCs.

Var. *coerulea* (Sm.) Sm. has been recorded from squares 2347, 4073, 4366 and 4578; var. *vitellina* (L.) Stokes from squares 2262, 2278, 2997 and 4564. Var. *coerulea* has been recorded from all surrounding VCs except 39; var. *vitellina* has been recorded from all surrounding VCs.

Salix alba × fragilis = S. × rubens Schrank
Kirk, T., 1854. Herb. B'ham Univ. [*S. alba* var. *coerulea*]. Brandon and Pinley. Map 367, p. 416

An occasional tree widely scattered throughout the county; recorded from watersides (74%), hedges (21%) and a few other habitats; recorded from all surrounding VCs.

Salix alba var. **vitellina** (L.) Stokes × **S. babylonica** L. = **S. × chrysocoma** Dode
Daulman, Miss B. M., 1965. Fillongley.

This tree, probably always planted, was known generally, until recently, as *S. babylonica*; it has been recorded from watersides, roadsides and hedgerows in squares 1678, 2180, 2181, 2201, 2662, 2681, 2800, 2887 and 4682.

343.4 **Salix fragilis** L. Crack Willow
Kirk, T., 1849. Herb. Perry [*S. russelliana*]. Fillongley. Map 368, p. 416

Abundant and evenly distributed throughout the county; recorded from watersides and marshes (71%), hedges (22%) and several other habitats; recorded from all surrounding VCs.

var. *latifolia* Anderss. (first record: Bagnall, 1886. Herb. Bagnall. Pool Hollies, Sutton Park) has been recorded once from a pondside, near Farnborough (4451).

var. *russelliana* (Sm.) Koch (first record as above) has been recorded from watersides and marshes in squares 0483, 0862, 0954, 1073, 1674, 1867, 2053, 2071 and 2268; recorded from all surrounding VCs except 32 and 33.

Salix decipiens Hoffm.
Kirk, T., 1849. Herb. Perry. Quinton Pool, Coventry.

One record only from hedgerow at Moreton Morrell (3155); there is also a specimen collected by Bromwich from Myton, Warwick, in 1874; recorded from all surrounding VCs except 39 and 57.

343.5 **Salix triandra** L. Almond Willow
Purton, T., 1821. Mid. Flora. Osier holts near Alcester; at Broom Ford, close to the river; Wixford Bridge. Map 369, p. 417

An occasional tree mainly in the Avon Valley; recorded from watersides, hedges and a few other habitats; recorded from all surrounding VCs.

S. triandra L. × *viminalis* L. (first record: Bagnall, 1888. Herb. Bagnall [*S. lambertiana* Sm.]. Near Alvecote Mill, lane to Grendon) has been recorded once from the side of the Avon near Church Lawford (4476); recorded from VCs 23, 55 and 57.

343.6 **Salix purpurea** L. Purple Willow
Purton, T., 1817. Mid. Flora [*S. lambertiana*]. Salford; Wixford. Map 370, p. 417

An occasional tree of a number of scattered localities in the county, with some concentration in the Birmingham and Rugby areas; recorded from watersides (66%), hedges (24%) and a few other habitats; recorded from all surrounding VCs.

343.9 **Salix viminalis** L. Osier
Baxter, W., 1831. MS (Bagnall: Mid. Nat. 1892–93.) Near Brownsover and Lawford.
Bree, Rev. W. T., 1835. New Bot. Guide. Without locality. Map 371, p. 418

Frequent but irregularly distributed throughout the county; sometimes planted and in a few cases found in withy beds or as relics of such beds; recorded from watersides and marshes (61%), hedges (22%), waste places, railway banks and roadsides (11%) and a few other habitats; recorded from all surrounding VCs.

343.10 **Salix calodendron** Wimm.
Baker, R. L., 1873. Young and Baker Catalogue [*S. acuminata*]. Shrewley.
Recorded only from Shrewley Pool (2268) where it has been recorded several times since it was first seen in 1873; recorded from VC 39.

343.11a **Salix caprea** L. ssp. **caprea** Great Sallow, Goat Willow
Baxter, W., 1831. MS (Bagnall: Mid. Nat. 1892–93) [*S. caprea*]. Dunchurch Road, Rugby. Map 372, p. 418
Abundant, though somewhat irregularly distributed throughout the county; recorded from hedges and scrub (39%), watersides and marshes (28%), waste places, railway banks and roadsides (15%), woods (14%) and a few other habitats; recorded from all surrounding VCs.

Salix caprea L. ssp. **caprea** × **cinerea** L. ssp. **oleifolia** Macreight
Bromwich, H., 1873. Herb. Bagnall [*S. damascena*]. Shrewley Pool.
Recorded from scrub, waste places, hedges, mixed woods, marshes and quarries in squares 1172, 1858, 1888, 3086, 3383, 3449, 3648, 3771, 4076, 4263, 4268, 4464 and 4977.
S. caprea ssp. *caprea* × *cinerea* s.l. has been recorded from squares 1865, 2755 and 3086.
S. caprea ssp. *caprea* × *cinerea* ssp. *oleifolia* × *viminalis* has been recorded once from a pondside in square 3881.

Salix caprea L. × **nigricans** Sm.
Meikle, R. D., 1966. Shrewley Pool.
One record only from Shrewley Pool (2268), in vicinity of parents.

Salix caprea L. ssp. **caprea** × **viminalis** L.
Kirk, T., 1849. Herb. Perry [*S. ferruginea*]. Coventry Park.
Recorded from hedges, watersides and marshes in squares 0772, 1073, 1273, 1275, 1582, 1769, 1869, 1881, 2148, 2401, 2534, 2786, 2907, 2986, 3483, 3493, 3881, 4474, 4871, 4983, 5066, 5080 and 5377; recorded from VCs 32, 39, 55 and 57.

343.12 **Salix cinerea** L. Grey Willow
Purton, T., 1817. Mid. Flora [*S. aquatica*]. Low swampy places at Oversley and Ragley Woods.
This occurs generally as *S. cinerea* ssp. *oleifolia* (q.v.), but one record of ssp. *cinerea* (q.v.) has been made and a few intermediates have been noted from squares 1884, 2681, 2766 and 3971.

343.12a Ssp. **cinerea**
Baker, R. L., 1881. Herb. Bagnall [*S. cinerea* var. *aquatica*]. Kenilworth.
One record only from the canal side at Lowsonford (1867).

343.12b Ssp. **oleifolia** Macreight
Kirk, T., 1849. Herb. Perry [*S. oleifolia*]. Quinton Pool. Map 373, p. 419
Abundant and fairly evenly distributed throughout much of the county but less common in the Avon Valley and on the Lower Lias clays of the S. and S.E.; recorded from watersides and marshes (44%), hedges and scrub (35%), waste places, railway banks and roadsides (9%), woods (9%) and a few other habitats; recorded from all surrounding VCs.

Salix cinerea L. × **viminalis** L.
Kirk, T., 1854. Herb. Perry [*S. smithiana* var. *rugosa* Sm.]. Pinley.
S. cinerea L. × *viminalis* L. has been recorded from hedges and watersides in squares 1073, 1484, 1677, 1970, 2167, 2486, 2668, 2729, 2785, 4085, 4181, 4182, 4276, 4366, 4855, 4873, 4983, 5162 and 5168; it has been recorded from all surrounding VCs.
S. cinerea L. ssp. *oleifolia* Macreight × *viminalis* L. has been recorded from roadsides, railway banks and pondsides in squares 2504, 3881 and 4580.

Salix × **forbyana** Sm. = **S. cinerea** × **purpurea** × **viminalis**
Bagnall, 1885. Herb. Bagnall [*S. acuminata*]. Near Seckington.
Three records only; one from marshy woodland by Edgbaston Pool (0583), and two by the edges of pools at Shrewley (2268, 2269); previously also recorded from Alvecote Mill and Olton Reservoir; recorded from VCs 37, 39 and 57.

343.13 **Salix aurita** L. Eared Willow
Purton, T., 1821. Mid. Flora. In hedgerows near Alcester; hedges between Rugby and Dunchurch (Baxter). Map 374, p. 419
An occasional tree of a number of scattered localities mainly in the N. of the county, with a considerable concentration in the Earlswood area; recorded from watersides and marshes (42%), hedges and scrub (32%), woods (13%) and a few other habitats; recorded from all surrounding VCs.

Salix aurita L. × **cinerea** L. ssp. **oleifolia** Macreight
Burges, Dr. R. C. L., 1935. Herb. Burges [*S. cinerea*]. Windley Pool, Sutton Park.
An occasional tree of a number of widely scattered localities throughout the county; recorded from watersides, marshes, hedges and woods in squares 0558, 0873, 1073, 1076, 1152, 1381, 1561, 1881, 1953, 2071, 2181, 3449, 3654, 4147, 4190, 4283, 4285, 4478 and 5378; *S. aurita* × *cinerea* has been recorded from all surrounding VCs.

Salix aurita L. × **viminalis** L.
Evans, R., 1965. Lowsonford.
One record only from canal side Lowsonford (1867); possibly planted.

343.14 **Salix nigricans** Sm. Dark-leaved Willow
Anon., 1873. Herb. B'ham Univ. Shrewley Pool.
This rare species, probably not native in Warwickshire, has been recorded from Shrewley Pool (2268) on a number of occasions since 1873, and is still there; recorded from VC 23.

Salix × **laurina** Sm.
Bromwich, H., 1873. Herb. Brit. Mus. Shrewley Pool.
The parentage of this hybrid is doubtful: according to Meikle it may be *S. cinerea* ssp. *oleifolia* × *phylicifolia*; it has not been recorded in the present survey but was collected from Shrewley Pool in 1938; recorded from VC 39.

343.16a **Salix repens** L. ssp. **repens** Creeping Willow
Bree, Rev. W. T., 1821. Mid. Flora. Coleshill Heath.
A rare plant recorded from marshy ground at Forshaw Heath (0873) and Sutton Park (0998); also, on imported sea-sand, from Emscote (2965); since Bree's record in 1821 a number of collections have been made from Ballards Green, Coleshill Pool (including forms intermediate between ssp. *argentea* and ssp. *repens*), Honiley and Sutton Park; recorded from all surrounding VCs except 32 and 37.

ERICACEAE

345 Rhododendron L.

345.1 **Rhododendron ponticum** L. Rhododendron
Shack, J., 1952. New Park Wood.
Map 375, p. 420
Cultivated and often naturalised, spreading vegetatively and by seed; frequent in the N. of the county but almost entirely absent from the calcareous soils of the S. and S.E.; recorded from woods (70%), scrub and hedges (24%) and a few other habitats; recorded from all surrounding VCs.

356 Calluna Salisb.

356.1 **Calluna vulgaris** (L.) Hull Ling, Heather
Withering, 1787. Arrangement [*Erica vulgaris*]. Birmingham Heath. Map 376, p. 420
An occasional plant in the N. half of the county; recorded from dry heath (30%), acid woodland (28%), waste places and railway sides (18%), hedgerows (13%) and rough grassland (10%); recorded from all surrounding VCs.

357 Erica L.

357.1 **Erica tetralix** L.
Cross-leaved Heath, Bog Heather
Purton, T., 1817. Mid. Flora. Studley Common. Map 377, p. 421
A rare plant confined to acid soils in Sutton Park, the Coleshill area, Trickley Coppice, Baddesley Ensor and Forshaw Heath; recorded from bogs, dry heath, acid woods and a few other habitats; recorded from all surrounding VCs.

357.4 **Erica cinerea** L. Bell Heather
Bree, Rev. W. T., 1817. Mid. Flora. Coleshill Heath. Map 767, p. 609
A rare plant of heaths, at one time much more widespread, now confined to Sutton Park, Coleshill Bog and Wirehill; recorded from all surrounding VCs except 33.

358 Vaccinium L.

358.1 **Vaccinium vitis-idaea** L. Cowberry
Ray, 1670. Catalogus [*Vaccinia rubra* Ger.]. "Warwickshire, in Middleton Parish, by the New Park pales."
A rare plant recorded from bogs and wet heath in Sutton Park (0898, 0996, 0998, 1097) and from a clay spoil heap at Glascote (2303); recorded from VCs 37, 39, 55 and 57.

358.2 **Vaccinium myrtillus** L. Bilberry
Purton, T., 1817. Mid. Flora. Studley Woods.
Map 378, p. 421
A rare plant almost entirely confined to the Sutton Coldfield and Earlswood areas, where it is abundant; recorded chiefly from woods but also from dry heaths, hedgerows, scrub, marsh and bog; recorded from all surrounding VCs except 32 and 33.

Vaccinium myrtillus × vitis-idaea
= **V. × intermedium** Ruthe
Bagnall, 1889. Herb. Bagnall. Sutton Park.
Not recorded in present survey; recorded from VCs 39 and 57.

358.4 **Vaccinium oxycoccos** L. Common Cranberry
Ray, 1669. Correspondence [*Palustria Thymifolius*]. "*Palustria Thymifolius* . . . hereabouts we have them in great plenty." (Letter to Lister from Middleton.)
Ray, 1670. Catalogus. [*Oxycoccus seu Vaccinia palustria* J.B.]. "On moorish ground and quagmires in Sutton Park."
Recorded from boggy areas in seven squares at Sutton Park (0898, 0995, 0996, 0997, 0998, 1096, 1097); collected from Coleshill Bog by Lloyd in 1835 and by Perry in 1841; recorded from VCs 37, 39 and 57.

PYROLACEAE

359 Pyrola L.

359.1 **Pyrola minor** L. Common Wintergreen
Kirk, T., 1873. Topographical Botany. Without locality.
Not recorded in present survey; the record given above is included on the authority of H. C. Watson, who states in his Topographical Botany that he had a specimen from Kirk; recorded from all surrounding VCs except 32.

MONOTROPACEAE

362 Monotropa L.

362.1 **Monotropa hypopitys** L. *sensu lato*
Yellow Birdsnest
Satchell, W., 1848. Herb. Bagnall. Compton Verney.
Not recorded in present survey; recorded from all surrounding VCs.

EMPETRACEAE

364 Empetrum L.

364.1 **Empetrum nigrum** L. Crowberry
Ray, 1670. Catalogus [*Erica baccifera procumbens nigra* C.B.]. "On black heathy grounds, in Warwickshire."
Recorded only from heath and bog in Sutton Park (0898, 0996, 0997, 0998, 1097); recorded from VCs 39, 55 and 57.

PRIMULACEAE

367 Primula L.

367.3 **Primula veris** L. Cowslip
Lloyd, Dr. G., 1836. Herb. Perry. Lower Norton. Map 379, p. 422
Abundant on the calcareous soils of the S. and S.E. of the county, frequent on the Keuper Marl but rare on sandstones and gravels in the centre and N.; recorded

from grassland (48%), roadsides and railway banks (21%), hedgerows and scrub (20%), woods (8%) and a few other habitats; recorded from all surrounding VCs.

Primula veris × vulgaris
Perry, 1813. Herb. Perry [*P. elatior*]. Without locality. Map 380, p. 422
This rare hybrid has been recorded from woods, hedgerows and scrub, roadsides and rough grassland, generally in the presence of the parents, from a number of scattered localities the majority of which are in the S.W. of the county; recorded from all surrounding VCs except 39.

367.5 **Primula vulgaris** Huds. Primrose
Bree, Rev. W. T., 1821. Mid. Flora. Allesley. Map 381, p. 423
Frequent and widely but irregularly distributed throughout most of the county, with considerable concentration in the W. central region, to some extent in areas which were well wooded in the seventeenth century; recorded from woods (36%), hedgerows and scrub (23%), grassland (18%), roadsides and railway banks (12%) and watersides (9%); recorded from all surrounding VCs.

368 Hottonia L.

368.1 **Hottonia palustris** L. Water-violet
Perry, 1820. Plant. Varv. Sel. Tamworth.
Not recorded in present survey; last record H. C. Palmer, 1878, Allesley; recorded from all surrounding VCs.

370 Lysimachia L.

370.1 **Lysimachia nemorum** L. Yellow Pimpernel
Pena & L'Obel, 1570. Stirpium Adversaria [*Anagallis lutea*]. "In sylva Coventrive proxima." Map 382, p. 423
An occasional plant chiefly confined to areas which were well wooded in the seventeenth century; recorded from woods (78%); scrub and hedgerows (15%) and a few other habitats; recorded from all surrounding VCs.

370.2 **Lysimachia nummularia** L. Creeping Jenny
Withering, 1801. Arrangement. Near a brook which crosses the Meriden Road about 2 miles from Birmingham. Map 383, p. 424
Frequent and widely but unevenly distributed throughout the county; recorded from watersides and marshes (58%), roadsides, waste places and railway banks (13%), damp grassland (11%), woods (9%) and hedgerows and scrub (8%); recorded from all surrounding VCs.

370.3 **Lysimachia vulgaris** L. Yellow Loosestrife
Purton, T., 1817. Mid. Flora. On the side of the Avon, below Bidford Grange. Map 384, p. 424
A rare plant of a number of scattered localities in the N.W. half of the county; recorded from watersides, marshes, roadsides and waste places; recorded from all surrounding VCs.

370.4 **Lysimachia ciliata** L. Fringed Loosestrife
Cox, T., 1851. Herb. Perry. Near Berkswell.
Not recorded in present survey; not recorded from any of the surrounding VCs.

370.5 **Lysimachia punctata** L. Dotted Loosestrife
Thomas C., 1950. Edgbaston, Birmingham.
An introduced plant rarely naturalised and recorded from waste places (2766, 3686) and watersides (0386, 2466); recorded from VCs 32, 37, 55 and 57.

372 Anagallis L.

372.1 **Anagallis tenella** (L.) L. Bog Pimpernel
Countess of Aylesford, 1809. Herb. Oxf. Univ. Packington.
A rare plant of bogs in Sutton Park (0995, 0996, 0998, 1096); recorded from all surrounding VCs.

372.2 **Anagallis arvensis** L. Scarlet Pimpernel
Anon., 1829. Herb. Brit. Mus. Cornfields, Lea. Map 385, p. 425
An abundant plant, evenly distributed throughout the county, occurring mainly in arable land (71%), on roadsides and waste places (21%) and in grassland (5%); recorded from all surrounding VCs.

Anagallis arvensis L. × **foemina** Mill.
Readett, R. C., 1950. Wilmcote.
Recorded twice with parents in arable fields (1557, 2345); not recorded from any of the surrounding VCs.

372.3 **Anagallis foemina** Mill. Blue Pimpernel
Purton, T., 1817. Mid. Flora [*A. caerulea* Schreb.]. Bickmarsh; Bidford; Grafton. Map 386, p. 425
A rare plant of arable land (75%) on calcareous soil chiefly in the S.W. of the county; recorded from all surrounding VCs.

372.4 **Anagallis minima** (L.) E. H. L. Krause Chaffweed
Bagnall, 1878. Herb. Bagnall [*Centunculus minimus* L.]. Oversley Wood.
One record in woodland ride, Oversley Wood (1056); recorded from VCs 23, 37, 39 and 57.

373 Glaux L.

373.1 **Glaux maritima** L. Sea-milkwort
Ward, L. E. 1962. Stoke, Coventry.
One record from introduced Cumberland turf where it survived at least 3 years; recorded from VCs 33, 37 and 39.

374 Samolus L.

374.1 **Samolus valerandi** L. Brookweed
Purton, T., 1817. Mid. Flora. In some boggy ground near Bidford Grange; River Alne above Oversley.
A rare plant of riversides and marshes recorded from near Tredington (2443), Charlecote Park (2656), Itchington Holt (3755) and Snowford Bridge (3966); previously more widespread and collected from Harbury, near Leamington, near Stratford on Avon and Wimpstone Fields; recorded from all surrounding VCs.

BUDDLEJACEAE

375 Buddleja L.

375.1 **Buddleja davidii** Franch. Butterfly Bush
Andrews, C. E. A. & Burges, Dr. R. C. L., 1947. Proc. B'ham Nat. Hist. Soc. Bombed sites in Birmingham.
Six records from waste places, railway embankments and walls; recorded from VCs 23, 33, 37 and 55.

OLEACEAE

376 **Fraxinus** L.

376.1 **Fraxinus excelsior** L. Ash
Lloyd, Dr. G., 1835. Bagnall: Flora, 1891. Without locality. Map 387, p. 426
Abundant and evenly distributed throughout the county; recorded from hedges and scrub (67%), mixed woods (19%), roadsides, railway banks and waste places (7%) and a few other habitats; recorded from all surrounding VCs.

377 **Syringa** L.

377.1 **Syringa vulgaris** L. Lilac
Arnold, G. & M. 1955. Near Wilnecote.
Generally planted and occasionally naturalised in woods and hedges; recorded from squares 0995, 1076, 1571, 1871, 2156, 2201 and 4162; recorded from all surrounding VCs.

378 **Ligustrum** L.

378.1 **Ligustrum vulgare** L. Wild Privet
Perry, 1817. List. Hedges near Warwick. Map 388, p. 426
Abundant on calcareous soils in the S. of the county but becoming less common in the N.; recorded from hedges and scrub (77%), woods (14%), roadsides and railway banks (6%) and a few other habitats; recorded from all surrounding VCs.

378.2 **Ligustrum ovalifolium** Hassk. Garden Privet
Ick, W., 1836. Herb. Ick [*L. vulgare*]. "Road about a mile beyond Saltley towards Stitchford."
Introduced; probably always planted, and recorded fifteen times during present survey; recorded from VCs 23 and 37.

APOCYNACEAE

379 **Vinca** L.

379.1 **Vinca minor** L. Lesser Periwinkle
Withering, 1801. Arrangement. In a lane leading from the Larches to the Moseley Road, near Birmingham. Map 389, p. 427
Occasionally planted and probably generally an escape from cultivation; naturalised in a number of places with the main concentration in the W. central area of the county; recorded from hedgerows, roadsides, railway banks, woods, grassland, walls, scrub and cultivated land; recorded from all surrounding VCs.

379.2 **Vinca major** L. Greater Periwinkle
Purton, T., 1817. Mid. Flora. King's Coughton; Oversley. Map 390, p. 427
Occasionally escapes from gardens and has become naturalised in a number of places, chiefly in the S. of the county; recorded from hedgerows and scrub (58%), roadsides and railway banks (27%) and woods; recorded from all surrounding VCs.

GENTIANACEAE

382 **Centaurium** Hill

382.1 **Centaurium pulchellum** (Sw.) Druce
Lesser Centaury
Baynes, W. W., 1832. (Bagnall's Flora, 1891.) On Bascote Heath, near Southam.
Three records from damp grassy places, at Weethley (0454) and Wilmcote (1557, 1656); recorded from VCs 23, 32, 33 and 37.

382.4 **Centaurium erythraea** Rafn
Common Centaury
Purton, T., 1821. Mid. Flora [*Chironia centaurium*]. Without locality.
Map 391, p. 428
Abundant in the S.W. of the county but irregularly scattered elsewhere; recorded from grassland (44%), railway banks, roadsides and waste places (27%), scrub and hedgerows (17%) and a few other habitats; recorded from all surrounding VCs.

383 **Blackstonia** Huds.

383.1 **Blackstonia perfoliata** (L.) Huds. Yellow-wort
Perry, 1812. Herb. Perry [*Chlora perfoliata* L.]. Hill 2 or 3 miles from Stratford on Alcester Road. Map 392, p. 428
A locally frequent plant mainly of the Lower Lias limestone recorded from grassland (40%), railway banks and roadsides (36%), quarries (12%) and scrub woodland (8%); recorded from all surrounding VCs.

385 **Gentianella** Moench

385.3.1 **Gentianella amarella** (L.) Börner Felwort
Purton, T., 1817. Mid. Flora [*Gentiana amarella*]. Alne Hills. Map 393, p. 429
A rare plant of the Lower Lias limestone escarpment in the S. of the county; recorded from grassland and railway banks; recorded from all surrounding VCs.

MENYANTHACEAE

386 **Menyanthes** L.

386.1 **Menyanthes trifoliata** L. Buckbean, Bogbean
Purton, T., 1817. Mid. Flora. Coleshill Bog; in a pit on the Alne Hills; Shelfield (Rufford).
Map 394, p. 429
A rare plant mainly restricted to Sutton Park and the Earlswood area, but with one or two occurrences elsewhere in the county; recorded from watersides and marshes; recorded from all surrounding VCs.

387 **Nymphoides** Séguier

387.1 **Nymphoides peltata** (S. G. Gmel.) Kuntze
Fringed Water-lily
Freeman, S., 1841. Phytologist [*Limnanthemum peltatum* Gmel.]. Packington Park.
An introduced plant naturalised in a lake at Newbold Revel (4580) where it was first recorded by Bloxam in 1871; recorded from all surrounding VCs.

POLEMONIACEAE

388 **Polemonium** L.

388.1 **Polemonium caeruleum** L. Jacob's Ladder
Bagnall, 1871. Herb. Bagnall. Shirley Heath.
A garden escape recorded once by the canal bridge at Hatton (2466); recorded from all surrounding VCs except 32.

BORAGINACEAE

389 **Cynoglossum** L.

389.1 **Cynoglossum officinale** L. Hound's-tongue
Perry, 1821. MS (Bagnall's Flora, 1891).
Perry, 1828. Herb. Perry. On a hill above the Bank Croft, Hampton Lucy. Map 395, p. 430
Restricted to a few localities in the S.W. of the county mainly on the Lower Lias limestone; recorded from rough grassland, roadsides and waste places, hedgerows and scrub; recorded from all surrounding VCs.

389.2 **Cynoglossum germanicum** Jacq. Green Hound's-tongue
Perry, 1812. Plant. Varv. Sel. 1820 [*C. sylvaticum*]. Pigwell Lane, Warwick.
Not recorded in present survey; last recorded from Kenilworth and Stratford in 1875; recorded from VCs 23, 32, 33 and 37.

392 **Symphytum** L.

392.1 **Symphytum officinale** L. Common Comfrey
Bagnall, 1868. Herb. Bagnall. Witton. Map 396, p. 430
An occasional plant of a number of widely scattered localities; recorded from waste places, roadsides, rough grass and riversides; recorded from all surrounding VCs.

392.2 **Symphytum asperum** Lepech. Rough Comfrey
Arnold, G. & M., 1959. Alvecote.
An introduced plant occasionally naturalised, and recorded from roadside at Alvecote (2404), churchyard at Temple Balsall (2076), Hatton Railway Station (2266) and roadside near Stockton (4563); recorded from all surrounding VCs except 39 and 55.

Symphytum × uplandicum Nyman
= **S. asperum × officinale** Blue Comfrey
Perry, 1813. Herb. Perry [*S. officinale*]. Ashow Mill. Map 397, p. 431
A frequent plant of irregular distribution throughout most of the county, but rare on the calcareous clays of the S. and S.E.; recorded from roadsides and waste places (53%), watersides and marshes (19%), hedgerows and scrub (14%), grassland (9%) and cultivated land; recorded from all surrounding VCs.

392.3 **Symphytum orientale** L. White Comfrey
Bagnall, 1875. Herb. Bagnall. In grounds at Myton.
St. Brody, Dr., 1875. Herb.Warw.Mus. River banks, Guy's Cliffe.
An introduced plant occasionally naturalised on roadsides and in ditches, recorded from Alne End (1159), Aston Cantlow (1359), near Newbold on Stour (2445), The Butts, Warwick (2865), near Oakley Wood (2959), near Guy's Cliffe (2967) and near Ladbroke (4159); recorded from all surrounding VCs except 33 and 39.

392.4 **Symphytum caucasicum** Bieb.
Bailey, C., 1879. Herb. Brit. Mus. [*S. tauricum* Willd.]. Milverton.
Not recorded in present survey; recorded from VC 23.

392.5 **Symphytum tauricum** Willd.
Kirk, T., 1854. Herb. Man. Univ. Allesley.
Not recorded in present survey; not recorded from any of the surrounding VCs.

392.6 **Symphytum tuberosum** L. Tuberous Comfrey
Kirk, T., 1891. Bagnall's Flora. Allesley.
Recorded once only from a roadside and scrub woodland at Farnborough (4249) where it is locally abundant; recorded from all surrounding VCs except 23.

392.7 **Symphytum grandiflorum** DC. Creeping Comfrey
Jones, Mrs. M. D. G., 1959. Near Monks Kirby.
An introduced plant naturalised on roadside near Monks Kirby (4682); recorded from VCs 32, 33 and 37.

393 **Borago** L.

393.1 **Borago officinalis** L. Borage
Purton, T., 1817. Mid. Flora. "Among some rubbish in a field by the Arrow turnpike."
A rare casual of roadsides and hedgerows, recorded from 3240, 3338 and 3456; recorded from all surrounding VCs.

394 **Trachystemon** D. Don

394.1 **Trachystemon orientalis** (L.) G. Don
Bromwich, H., 1885. Herb. Brit. Mus. Railway bank and river bank near Warwick.
Two records: one from railway embankment near Warwick (2765) and one from a neglected garden at Hillmorton (5373); recorded from VC 33.

395 **Pentaglottis** Tausch

395.1 **Pentaglottis sempervirens** (L.) Tausch Alkanet
Withering, 1787. Arrangement [*Anchusa sempervirens*]. Near Birmingham, on the Alcester Road. Map 398, p. 431
An occasional plant very irregularly distributed throughout a good deal of the county, but almost entirely absent from the calcareous soils of the S. and S.E. and from an area in the N.E.; possibly introduced in the first place but now firmly established; recorded from roadsides and waste places (41%), hedgerows (36%), grassland (11%) and a few other habitats; recorded from all surrounding VCs.

396 **Anchusa** L.

397.1 **Anchusa arvensis** (L.) Bieb. Bugloss
Baxter, W., 1831. MS (Bagnall: Mid. Nat. 1892–93) [*Lycopsis arvensis* L.]. Clifton Road and meadow near Lawford Mill.
Ick, W., 1836. Herb. Ick [*Lycopsis arvensis* L.]. Bank at Castle Bromwich. Map 399, p. 432
A plant of light soils mainly in the Avon Valley and recorded from arable land (52%) and roadsides and waste places (37%); recorded from all surrounding VCs.

399 **Pulmonaria** L.

399.2 **Pulmonaria officinalis** L. Lungwort
Baynes, W. W., 1831. Bagnall's Flora, 1891. On the road between Leamington and Kenilworth, near Chesford Bridge.
A rare plant naturalised in woodlands near Kinwarton House (0958) and at Seeswood Pool (3290), in scrub woodland near Astley Hall Farm (3286), and in waste places at Sherbourne (2661) and at Newbold on Avon (4876); recorded from all surrounding VCs.

400 Myosotis L.

400.1 **Myosotis scorpioides** L. Water Forget-me-not
Perry, 1812. Herb. Perry [*M. palustris*]. Without locality. Map 400, p. 432
Abundant though rather unevenly distributed throughout the county; recorded from watersides and marshes (93%) and water (6%); recorded from all surrounding VCs.

400.2 **Myosotis secunda** A. Murr. Creeping Forget-me-not
Cheshire, W., 1854. Herb. Perry [*M. repens*]. Bog between Stonebridge and Coleshill.
A rare plant of bogs and wet acid soil recorded from Forshaw Heath (0873), Earlswood (1074), Sutton Park (1096), Monkspath Hill Farm (1475), Hay Wood (2071), near Packington (2084) and Coleshill Bog (2086); recorded from all surrounding VCs except 23 and 32.

400.4 **Myosotis caespitosa** K. F. Schultz Tufted Forget-me-not
Baxter, W., 1831. MS (Bagnall: Mid. Nat. 1892–93.) Near Hill Morton.
Lloyd, Dr. G., 1835. Herb. Perry. Coleshill Bog. Map 401, p. 433
Frequent and fairly evenly distributed throughout the county, except in the Avon Valley where it is rare; recorded from watersides and marshes (96%) and a few other habitats; recorded from all surrounding VCs.

Myosotis caespitosa × scorpioides
Benoit, P. M. and Jones, Mrs. M. D. G., 1963. Near Lower Shuckburgh.
One record of plant with parents from canal bank (4862); not recorded from any of the surrounding VCs.

400.7 **Myosotis sylvatica** Hoffm. Wood Forget-me-not
Perry, 1812. Herb. Perry. Without locality. Map 402, p. 433
An occasional plant of very widely scattered distribution throughout the county; recorded from scrub and hedgerows (30%), woods (27%), roadsides and railway banks (16%), watersides (14%) and grassland (13%); recorded from all surrounding VCs.

400.8 **Myosotis arvensis** (L.) Hill Field Forget-me-not
Countess of Aylesford, before 1832. Herb. Oxf. Univ. Packington. Map 403, p. 434
Abundant and evenly distributed throughout the county; recorded from roadsides, railway banks and waste places (27%), cultivated land (26%), hedgerows and scrub (20%), grassland (13%), watersides (7%) and woods (6%); recorded from all surrounding VCs.

400.9 **Myosotis discolor** Pers. Yellow and blue Forget-me-not
Perry, 1812. Herb. Perry [*M. versicolor*]. Without locality. Map 404, p. 434
An occasional plant with a very scattered distribution throughout most of the county, but very rare in the extreme N.; recorded from grassland (33%), roadsides and railway banks (25%), cultivated land (23%), woods, heaths and scrub; recorded from all surrounding VCs.

400.10 **Myosotis ramosissima** Rochel Early Forget-me-not
Lowe, H. E., 1836. Herb. Perry [*M. collina*]. Coleshill. Map 405, p. 435
A rare plant very widely scattered mainly on calcareous soils in the S. of the county; recorded from railway banks, walls, roadsides and rough grassland; recorded from all surrounding VCs.

401 Lithospermum L.

401.2 **Lithospermum officinale** L. Common Gromwell
Purton, T., 1817. Mid. Flora. Great Alne; Oversley Wood. Map 406, p. 435
An occasional plant of limestone in the S. of the county; recorded mainly from hedgerows and scrub (50%) with a few records from grassland, cultivated land, railway banks, roadsides and woods; recorded from all surrounding VCs.

401.3 **Lithospermum arvense** L. Corn Gromwell
Perry, 1825. Herb. Perry. Between Norton Lindsey and Wolverton. Map 407, p. 436
An occasional plant of calcareous soils in the S. of the county; recorded from cultivated land (79%) and a few other habitats; recorded from all surrounding VCs.

403 Echium L.

403.1 **Echium vulgare** L. Viper's Bugloss
Perry, 1812. Herb. Perry. Without locality. Map 408, p. 436
A rare plant of dry soils in a number of places in the county; recorded from waste places, railway banks, roadsides and quarries; it was collected between 1810 and 1880 from Coleshill Pool, Hatton Rock, Marston Green, Sutton Park and Wilmcote; recorded from all surrounding VCs.

Lappula Moench

Lappula myosotis Moench
Bromwich, H., 1891. Bagnall's Flora [*Echinospermum lappula* L.]. Skin yards at Kenilworth.
Not recorded in present survey; recorded from VCs 23, 32 and 55.

CONVOLVULACEAE

405 Convolvulus L.

405.1 **Convolvulus arvensis** L. Lesser Bindweed
Lloyd, Dr. G., 1835. Herb. Perry. Leamington. Map 409, p. 437
Abundant and fairly evenly distributed throughout the county and especially abundant in the S.; recorded from roadsides, waste places and railway embankments (47%), arable land (24%), grassland (14%) and hedgerows (12%); recorded from all surrounding VCs.

406 Calystegia R. Br.

406.1 **Calystegia sepium** (L.) R. Br. Hedge Bindweed
Baxter, W., 1831. MS (Bagnall: Mid. Nat. 1892–93.) Near Rugby.
Lloyd, Dr. G., 1835. Herb. Perry [*Convolvulus sepium*]. Wasperton. Map 410, p. 437
An abundant plant evenly distributed throughout the

county; more frequent than *C. silvatica* though some recorders have used the name *C. sepium* in the aggregate sense; recorded from hedges (72%), waste places, roadsides and railway banks (17%), cultivated land (4%) and a few other habitats; recorded from all surrounding VCs.
Intermediates between *C. sepium* and *C. silvatica* have been recorded.

406.2 **Calystegia pulchra** Brummitt & Heywood Hairy Bindweed
Hawkes, J. G., 1962. Long Itchington Village.
A rare plant, naturalised in hedges and waste places, recorded from Stratford (1954), Dosthill (2100), Long Itchington (4165), Wolvey (4388), Pailton (4781) and Rugby (5273); recorded from all surrounding VCs except 39 and 57.
Probable hybrids between this and *C. silvatica* have been recorded from 3286 and 3561.

406.3 **Calystegia silvatica** (Kit.) Griseb. Great Bindweed
Rugby School Report, 1949. [*C. sylvestris* (Willd.) R. & S.]. Rugby. Map 411, p. 438
A frequent plant irregularly distributed throughout the county; recorded from hedges (68%), waste places (23%) and a few other habitats; recorded from all surrounding VCs.

406.4 **Calystegia soldanella** (L.) R. Br. Sea Bindweed
Bagnall, 1895. Herb. Bagnall. Lane above Streetly Wood.
Not recorded in present survey; recorded from VC 37.

407 Cuscuta L.

407.1 **Cuscuta europaea** L. Greater Dodder
Countess of Aylesford, 1805. Bot. Guide. Flax fields about Packington.
One record of plant, probably introduced with imported seed, from Coventry (3375); also recorded by Bromwich in 1900 from a field bordering Frankton Wood; recorded from all surrounding VCs.

407.2 **Cuscuta epilinum** Weihe Flax Dodder
Lloyd, Dr. G., 1851. Herb. Perry. Budbrooke.
Not recorded in present survey; recorded from VCs 23, 32, 37 and 55.

407.3 **Cuscuta epithymum** (L.) L. Common Dodder
Perry, 1854. Herb. Perry [*C. trifolii* Bab.]. Red Hill, on the road to Grafton.
One record only on various plants in dry pasture by Slingate Coppice near Alderminster (2350); apparently more frequent up to 1875; recorded from all surrounding VCs.

Cuscuta cesatiana Bertol.
Preston, N. C., 1933. Herb. Brit. Mus. Kenilworth, on Chrysanthemums in pots.
Not recorded in present survey; not recorded from any of the surrounding VCs.

Cuscuta suaveolens Ser.
Kitchener, F. E., 1869. Journal of Botany, 1870 [*C. hassiaca* Pfeiff.]. In a field of lucerne near Rugby.
Not recorded in present survey; recorded from VCs 23 and 37.

SOLANACEAE

408 Nicandra Adans.

408.1 **Nicandra physalodes** (L.) Gaertn. Shoo-fly
Austen, Mrs. R., 1964. Warwick Sewage Farm.
An alien recorded once only (2763); recorded from VCs 23, 32, 37 and 55.

409 Lycium L.

409.1 **Lycium barbarum** L. Duke of Argyll's Tea-tree
Anon., 1874. Herb. B'ham Univ. Myton and Leamington. Map 412, p. 438
An occasional plant of a number of scattered localities in the county, with some concentration in the N.E.; generally planted and occasionally naturalised; recorded from hedges (66%), waste places and a number of other habitats; recorded from all surrounding VCs.
L. chinense Mill. has been recorded, but all specimens sent in have proved to be *L. barbarum*; further investigation may show that *L. chinense* is present in the county.

410 Atropa L.

410.1 **Atropa bella-donna** L. Deadly Nightshade
Ray, 1724. Synopsis [*Solan. manicum multis, seu Bella Donna* J.B.]. Sutton Coldfield. Map 768, p. 609
A rare plant of waste places, possibly introduced; recorded from all surrounding VCs.

411 Hyoscyamus L.

411.1 **Hyoscyamus niger** L. Henbane
Purton, T., 1817. Mid. Flora. Great Alne; Wixford. Map 769, p. 610
An occasional plant of rough grassland, farmyards and waste places; recorded from all surrounding VCs.

412 Physalis L.

412.1 **Physalis alkekengi** L. Cape Gooseberry
Kirk, T., 1842. Phytologist. Foleshill.
One record from a waste place at Binley Grange (3777); recorded from VC 23.

413 Solanum L.

413.1 **Solanum dulcamara** L. Bittersweet, Woody Nightshade
Perry, 1812. Herb. Perry. Without locality. Map 413, p. 439
Abundant and evenly distributed throughout the county; recorded from hedges and scrub (42%), watersides and marshes (37%), waste places, railway banks and roadsides (12%), woods (4%) and a number of other habitats; recorded from all surrounding VCs.

413.3 **Solanum nigrum** L. Black Nightshade
Perry, 1812. Herb. Perry. Without locality. Map 414, p. 439
Frequent in the S.E. half of the county, particularly in the Arrow and Avon basins, and occasional elsewhere; recorded from cultivated land (51%), waste places and roadsides (34%), hedgerows and scrub (8%) and a few other habitats; recorded from all surrounding VCs.

413.4 **Solanum sarrachoides** Sendtn.
Green Nightshade
Roberts, H. A., 1959. Wellesbourne.
Recorded from one colony only in cultivated and waste land at Wellesbourne (2756); recorded from VC 37.

415 **Datura** L.

415.1 **Datura stramonium** L. Thorn-apple
Purton, T., 1817. Mid. Flora. Salford; Alcester.
Map 415, p. 440
A casual plant which has appeared sporadically and has been recorded from waste places, gardens and roadsides; recorded from all surrounding VCs.

SCROPHULARIACEAE

416 **Verbascum** L.

416.1 **Verbascum thapsus** L.
Aaron's Rod, Great Mullein
Bagnall, 1869. Herb. Bagnall. Road from Stonebridge to Coleshill, near Coleshill Pool.
Map 416, p. 440
Frequent and widely but very irregularly distributed throughout most of the county; recorded from waste places, roadsides and railway banks (74%), quarries and walls (8%), cultivated land (7%), and a few other habitats; recorded from all surrounding VCs.

416.6 **Verbascum speciosum** Schrad.
Andrews, C. E. A., and Townsend, C. C., 1954. Great Alne.
One record only of plant abundantly naturalised by the disused railway station at Great Alne (1159); not recorded from any of the surrounding VCs.

416.7 **Verbascum nigrum** L. Dark Mullein
Perry, 1817. List. Between Ashow and Stoneleigh. Map 417, p. 441
A rare plant mainly of the Avon basin; recorded from grassland, waste places, hedgerows and railway banks; recorded from all surrounding VCs.

416.9 **Verbascum blattaria** L. Moth Mullein
Perry, 1860. Herb. Perry. Lapworth.
One record only from rough grassland at Halford (2545); recorded from all surrounding VCs except 57.

416.10 **Verbascum virgatum** Stokes Twiggy Mullein
Anon., 1873. Herb. B'ham Univ. Garden at Myton.
One record only of plant naturalised in waste ground near Wilnecote (2202); previously collected as a casual several times in the Warwick–Milverton–Leek Wootton area; recorded from all surrounding VCs.

417 **Misopates** Raf.

417.1 **Misopates orontium** (L.) Raf.
Lesser Snapdragon
Cross, F., 1870. Herb. Warw. Mus. [*Antirrhinum orontium* L.]. Railway Cutting, Warwick.
A rare plant of cultivated land recorded from Edgbaston (0387), near Binton (1554) and near Brandon (4176); recorded from all surrounding VCs.

418 **Antirrhinum** L.

418.1 **Antirrhinum majus** L. Snapdragon
Purton, T., 1817. Mid. Flora. Salford.
A rare escape from cultivation found in waste places (2055, 2865, 2967, 3053, 3780, 4156, 4876); occasionally naturalised on old walls (2289, 4778); recorded from all surrounding VCs.

420 **Linaria** Mill.

420.1 **Linaria pelisseriana** (L.) Mill. Jersey Toadflax
Rugby School Report, 1869. Hillmorton Road.
Not recorded in present survey; not recorded from any of the surrounding VCs.

420.2 **Linaria purpurea** (L.) Mill. Purple Toadflax
Bromwich, H., 1891. Bagnall's Flora. Walls of Warwick Castle and other old walls at Warwick.
Map 770, p. 610
An occasional escape from cultivation often naturalised on walls, railway banks, roadsides and waste places; recorded from all surrounding VCs except 39.

420.3 **Linaria repens** (L.) Mill. Pale Toadflax
Bromwich, H., 1873. Herb. Bagnall. Old walls at Claverdon. Map 771, p. 610
An occasional escape from cultivation often naturalised on walls, railway banks, roadsides and waste places; recorded from all surrounding VCs except 32.

Linaria repens × **vulgaris** = L. × **sepium** Allman
Allen, H., 1956. Near Knightcote.
One record from a railway bank near Knightcote (4154); recorded from VCs 23 and 33.

420.4 **Linaria vulgaris** Mill. Common Toadflax
Perry, 1812. Herb. Perry. Without locality.
Perry, 1817. List [*Antirrhinum linaria* L.]. Ashow and between Warwick and Longbridge.
Map 418, p. 441
A frequent plant of very irregular distribution, most common in the N. and W. of the county; recorded from railway banks, roadsides and waste places (73%), grassland (11%), hedgerows (10%) and cultivated land (4%); recorded from all surrounding VCs.

421 **Chaenorhinum** (DC.) Reichb.

421.1 **Chaenorhinum minus** (L.) Lange
Small Toadflax
Perry, 1820. Plant. Varv. Sel. [*Antirrhinum minus*]. In a field behind Union Parade, Leamington. Map 419, p. 442
A frequent plant of wide and irregular distribution, recorded from railway tracks and sides, and roadsides (76%), arable land on the Lower Lias clay (17%) and walls and quarries (5%); recorded from all surrounding VCs.

422 **Kickxia** Dumort.

422.1 **Kickxia spuria** (L.) Dumort.
Round-leaved Fluellen
Purton, T., 1817. Mid. Flora [*Antirrhinum spurium*]. Grafton. Map 420, p. 442
Frequent on calcareous soils in the S. of the county but absent elsewhere; recorded from cultivated land (96%) and railways and waste places (4%); recorded from all surrounding VCs except 57.

422.2 **Kickxia elatine** (L.) Dumort.
Sharp-leaved Fluellen
Purton, T., 1817. Mid. Flora [*Antirrhinum elatine*]. Grafton; Kinwarton.
Map 421, p. 443
Frequent on calcareous soils in the S. of the county but

very rare elsewhere; recorded from cultivated land (94%) and railways (6%); Warwickshire appears to be on the N.W. border of the main British distribution; recorded from all surrounding VCs.

423 Cymbalaria Hill

423.1 **Cymbalaria muralis** Gaertn., Mey. & Scherb. Ivy-leaved Toadflax
Perry, 1812. Herb. Perry [*Linaria cymbalaria*]. Without locality. Map 422, p. 443

An occasional plant scattered throughout the county; recorded from walls (83%), waste places (12%) and dry hedge banks (5%); recorded from all surrounding VCs.

424 Scrophularia L.

424.1 **Scrophularia nodosa** L. Common Figwort
Perry, 1812. Herb. Perry. Without locality. Map 423, p. 444

Abundant in the N.W. half of the county, particularly in areas which were well wooded in the seventeenth century, with scattered occurrences elsewhere; recorded from watersides and marshes (33%), hedgerows and scrub (21%), roadsides and railway banks (20%), woods (19%) and grassland (6%); recorded from all surrounding VCs.

424.2 **Scrophularia auriculata** L. Water Figwort
Perry, 1812. Herb. Perry [*S. aquatica* L.]. Without locality. Map 424, p. 444

Abundant and evenly distributed throughout the county; recorded from watersides and marshes (94%), woods (3%) and a few other habitats; recorded from all surrounding VCs.

424.3 **Scrophularia umbrosa** Dumort. Green Figwort
Bromwich, H., 1871. Herb. Bagnall [*S. ehrhartii* Stev.]. Brook at Tachbrook.

Not recorded in present survey; specimens were collected between 1871 and 1903 from Claverdon, Tachbrook and Whitnash; recorded from all surrounding VCs except 23 and 32.

425 Mimulus L.

425.1 **Mimulus guttatus** DC. Monkey-flower
Druce, 1932. Com. Flora. Without locality.

Recorded from the sides of streams, ponds and canals at Sutton Park (0995), Umberslade Park (1371, 1471), Alvecote Pools (2504) and near Easenhall (4578); recorded from all surrounding VCs.

426 Limosella L.

426.1 **Limosella aquatica** L. Mudwort
Countess of Aylesford, 1805. Bot. Guide. Coleshill Pool.

Recorded from mud at edge of lake at Earlswood (1174) where some years, when the level of the water is low, it is locally abundant; previously known at Arbury Hall, Coleshill Pool, Shrewley and near Coventry; recorded from all surrounding VCs.

429 Digitalis L.

429.1 **Digitalis purpurea** L. Foxglove
Perry, 1812. Herb. Perry. Guy's Cliffe, Warwick. Map 425, p. 445

Fairly abundant in the N. half of the county but almost completely absent from the calcareous soils in the S.; recorded from hedgerows and scrub (47%), woods (21%), roadsides, waste places and railway banks (19%), grassland (5%) and a few other habitats; recorded from all surrounding VCs.

430 Veronica L.

430.1 **Veronica beccabunga** L. Brooklime
Baxter, W., 1831. MS (Bagnall: Mid. Nat. 1892–93.) Near Brownsover.
Lloyd, Dr. G., 1835. Herb. Perry. Leamington. Map 426, p. 445

Abundant and evenly distributed throughout the county; recorded from watersides and marshes (82%), water (16%) and a few other wet habitats; recorded from all surrounding VCs.

430.2 **Veronica anagallis-aquatica** L. Blue Water Speedwell
Bagnall, 1875. Herb. Bagnall [*V. anagallis*]. Lane from Water Orton Station to Minworth. Map 427, p. 446

A rare plant for which only three authenticated records have been received, from riversides at Ettington Park (2447), Butlers Marston (3150) and near Mollington (4347); other records sent in as *V. anagallis-aquatica* have been assumed to be *V. catenata* (q.v.); the map includes all these records.

430.3 **Veronica catenata** Pennell Pink Water Speedwell
Perry, 1812. Herb. Perry [*V. anagallis*]. Without locality. Map 427, p. 446

A frequent plant in the S.E. half of the county especially on the Lower Lias clays and the chalky boulder clay, but rare elsewhere; recorded from watersides and marshes (84%) and water (16%); recorded from all surrounding VCs except 57.
See note under *V. anagallis-aquatica*.

430.4 **Veronica scutellata** L. Marsh Speedwell
Withering, 1796. Arrangement. Ditches about Tamworth. Map 428, p. 446

A rare plant of a number of widely scattered localities mainly in the N. half of the county; recorded from marshes and the sides of ponds; recorded from all surrounding VCs.
Var. **villosa** Schumach. (first record: Kirk, T., 1853. Herb. Perry [*V. scutellaria* var. *parmularia*]. Corley Moor) has been recorded from Bushwood Grange (1869), Maxstoke Castle Grounds (2288) and Eaves Green (2682); collected by Bagnall and others, between 1862 and 1884, from Coleshill Pool, Corley Moor and Sutton Park; recorded from VCs 37 and 55.

430.5 **Veronica officinalis** L. Heath Speedwell
Purton, T., 1817. Mid. Flora. Dunnington; Coughton. Map 429, p. 447

Frequent in the W. of the county, but rare and very irregularly distributed elsewhere; recorded from grassland (44%), roadsides and waste places (24%), woods (15%), scrub (11%) and a few other habitats; recorded from all surrounding VCs.

430.6 **Veronica montana** L. Wood Speedwell
Bree, Rev. W. T., 1830. Mag. Nat. Hist. Woods at Beausale, near Wedgnock Park.
Map 430, p. 447
Frequent but very irregularly distributed in the N.W. half of the county, and very rare in the S. and S.E.; recorded from woods (66%), hedgerows and scrub (26%), and a few other habitats; recorded from all surrounding VCs.

430.7 **Veronica chamaedrys** L. Germander Speedwell
Perry, 1812. Herb. Perry. Without locality.
Map 431, p. 448
Abundant and evenly distributed throughout the county; recorded from roadsides, railway banks and waste places (33%), grassland (30%), hedgerows and scrub (19%), cultivated land (8%), woods (7%) and a few other habitats; recorded from all surrounding VCs.

430.13a **Veronica serpyllifolia** L. ssp. **serpyllifolia**
Thyme-leaved Speedwell
Perry, 1812. Herb. Perry [*V. serpyllifolia*]. Without locality. Map 432, p. 448
Abundant and fairly evenly distributed throughout most of the county, but rare in the extreme N.; recorded from grassland (62%), cultivated land (15%), roadsides, railway banks and waste places (11%), watersides and marshes (4%) and a few other habitats; recorded from all surrounding VCs.

430.15 **Veronica arvensis** L. Wall Speedwell
Baxter, W., 1831. MS (Bagnall: Mid. Nat. 1892–93.) Near Hill Morton. Map 433, p. 449
A frequent plant fairly evenly distributed throughout most of the county but absent from the heavy clays in the N. and S.E.; recorded from cultivated land (44%), railways, roadsides and waste places (26%), grassland (16%), walls and quarries (12%) and a few other habitats; recorded from all surrounding VCs.

430.20 **Veronica hederifolia** L. Ivy-leaved Speedwell
Baxter, W., 1831. MS (Bagnall: Mid. Nat. 1892–93.) Near Sawbridge. Map 434, p. 449
A frequent plant fairly evenly distributed throughout much of the county, but rare on the heavy clays in the N. and S.E.; recorded from cultivated land (55%), roadsides, waste places and railway banks (24%), hedgerows and scrub (10%), grassland (4%) and a few other habitats; recorded from all surrounding VCs.

430.21 **Veronica persica** Poir.
Common Field Speedwell
Leighton, 1837. Herb. Perry [*V. buxbaumii*]. Without locality. Map 435, p. 450
Abundant and fairly evenly distributed throughout the county; recorded from cultivated land (68%), roadsides, waste places, railway banks and farmyards (21%), grassland (8%) and a few other habitats; recorded from all surrounding VCs.

430.22 **Veronica polita** Fr. Grey Field Speedwell
Perry, about 1840. MS (Bagnall: Flora, 1891). Saltisford. Map 436, p. 450
An occasional plant mainly in the S.E. of the county, often on calcareous soils; recorded chiefly from cultivated land, but also from waste places, roadsides and railway banks; recorded from all surrounding VCs.

430.23 **Veronica agrestis** L. Green Field Speedwell
Perry, 1813. Herb. Perry. Without locality.
Map 437, p. 451
Frequent and widely but very irregularly distributed throughout most of the county; recorded from cultivated land (74%), roadsides, waste places and railway banks (15%), grassland (8%) and a few other habitats; recorded from all surrounding VCs.

430.24 **Veronica filiformis** Sm. Slender Speedwell
Townsend, C. C., 1954. B.S.B.I. Proc. Nov. 1954. Rough Hill Wood, near Studley.
Map 438, p. 451
Introduced and naturalised in a number of places throughout much of the county, but absent from the extreme N.; recorded from grassland, roadsides, waste places, watersides, marshes and cultivated land; recorded from all surrounding VCs.

432 **Pedicularis** L.

432.1 **Pedicularis palustris** L. Marsh Lousewort
Countess of Aylesford, 1812. Herb. Oxf. Univ. Packington.
A rare plant of bogs now recorded only from Sutton Park (0995, 0996, 0998, 1097); previously known at Coleshill Pool, Sowe Waste Common, Lowsonford and near Alveston; recorded from all surrounding VCs.

432.2 **Pedicularis sylvatica** L. Lousewort
Harris, 1827. Herb. Perry. Between Whitnash and Tachbrook. Map 439, p. 452
A rare plant of widely scattered distribution mainly in the N.W. half of the county; recorded from marshes and damp grassland; recorded from all surrounding VCs.

433 **Rhinanthus** L.

433.2 **Rhinanthus minor** L. ssp. **minor** Yellow-rattle
Perry, 1812. Herb. Perry [*R. crista-galli*]. Without locality. Map 440, p. 452
Frequent and widely but irregularly distributed throughout the county mainly on clay soils; recorded from grassland (62%), roadsides and railway banks (20%), cultivated land (6%), hedgerows (4%) and a few other habitats; recorded from all surrounding VCs.

434 **Melampyrum** L.

434.3 **Melampyrum pratense** L.
Common Cow-wheat
Withering, 1796. Arrangement. Woods at Edgbaston. Map 441, p. 453
A rare plant of a few localities scattered widely throughout the county with the exception of the S. and S.E.; recorded mainly from woods (79%); recorded from all surrounding VCs.

435 **Euphrasia** L.

435.1.1 **Euphrasia micrantha** Reichb.
Bagnall, 1866. Herb. Bagnall [*E. officinalis* L. var. *gracilis* Fries]. Heathy pastures near Knowle.
One record in calcareous soil by bank of canal at Olton (1382) with some possible hybrids with *E. nemorosa*; not recorded from any of the surrounding VCs.

435.1.13 **Euphrasia nemorosa** (Pers.) Wallr.
Common Eyebright
Perry, 1812. Herb. Perry [*E. officinalis*]. Without locality. Map 442, p. 453

An occasional plant, mainly in the S. half of the county, with some concentration on the Lower Lias limestone and in the Wilmcote area; recorded from grassland (44%), roadsides and railway banks (19%), quarries (15%) and a few other habitats; recorded from all surrounding VCs.

Gatherings from S.E. of Kineton (3449) have been provisionally determined as *E. stricta* Lehm., and from Harbury (3660) as *E. pseudokerneri* Pugsl.; both need further investigation.

435.1.17 **Euphrasia borealis** Wettst. (*E. brevipila* auct.)
Townsend, F., 1896–97. Jl. of Bot. 1897 [*E. brevipila* Burnat & Gremli]. Honington.
Townsend, F., 1902. Herb. Bagnall and Brit. Mus. Honington.

Recorded from calcareous grassland at Elder Tree Copse (2849), near Kineton (3449) and Lighthorne (3455); recorded from all surrounding VCs except 37.

436 Odontites Ludw.

436.1 **Odontites verna** (Bellardi) Dumort.
Red Bartsia
Perry, 1812. Herb. Perry [*Bartsia odontites* (L.) Huds.]. Without locality. Map 443, p. 454

Abundant in the S. half of the county especially on the Lower Lias limestone; occasional and widely scattered in the N. with some concentration on the boulder clay in the N.E.; recorded from roadsides and waste places (38%), grassland (32%), cultivated land (26%) and a few other habitats; recorded from all surrounding VCs; the two subspecies (q.v.) have not generally been noted, but, where this has been done, particulars are given below, and the records are included on the map for the species.

436.1a Ssp. **verna**
Kirk, T., 1856. Herb. Brit. Mus. [*Bartsia odontites* Huds.]. Sowe Waste.

Recorded from squares 1174, 1557, 1573, 1574, 1658, 1686, 1974, 2245, 2345, 2446, 3154, 3240, 3254, 3357, 3455, 3456, 3654, 4262, 4362, 4459, 4465 and 4982; and from all surrounding VCs.

436.1b Ssp. **serotina** (Wettst.) E. F. Warb.
Bagnall, 1872. Herb. Bagnall [*Bartsia odontites* Huds. var. *serotina*]. Pathway, Steeple Hill, near Bidford; Alcester.

Recorded from squares 0554, 1073, 1457, 1458, 1561, 2344, 2550, 3156, 3449, 3648, 4164, 4375, 4464 and 4475; and from all surrounding VCs.

OROBANCHACEAE

439 Lathraea L.

439.1 **Lathraea squamaria** L. Toothwort
Power, Rev. J., 1891. Bagnall's Flora. "In a thicket at Oldbury, near Atherstone."

A very rare plant recorded on *Salix fragilis* at Rushy Flanders (2394), on hazel at Foul End (2494), on holly and elm at Kinwalsey (2585), on rotten wood in a loft between Warwick and Leamington (3064) and from a wood at Dunchurch Lodge (4871); recorded from all surrounding VCs.

440 Orobanche L.

440.3 **Orobanche rapum-genistae** Thuill.
Greater Broomrape
Countess of Aylesford, 1810. Herb. Oxf. Univ. [*O. major* L.]. Packington.

Not recorded in present survey; recorded from all surrounding VCs except 23 and 32.

440.6 **Orobanche elatior** Sutton
Knapweed Broomrape
Bree, Rev. W. T., 1830. Mag. Nat. Hist. Allesley.

One record of plant on *Centaurea scabiosa* in hedgerow at Whichford Hill (3233); recorded from all surrounding VCs except 37 and 57.

440.8 **Orobanche minor** Sm. Common Broomrape
Cheshire, W., 1852. Bagnall's Flora, 1891. Sandy field near Luddington.

An occasional plant of rough grassland, cultivated and waste places recorded from Newbold on Stour (2445, 2446), Chastleton Hill (2628, 2729), Whichford Hill (3233), near Wolston Heath (4374), Lawford Heath (4573) and S. of Rugby (5172, 5173); recorded from all surrounding VCs except 39.

LENTIBULARIACEAE

441 Pinguicula L.

441.3 **Pinguicula vulgaris** L. Common Butterwort
Bree, Rev. W. T., 1817. Mid. Flora. Bannerley Pool.

A rare plant of bogs recorded only from Sutton Park (0995, 0996, 0998); previously also known at Stivichall and Bannerley Pool; recorded from all surrounding VCs.

442 Utricularia L.

Utricularia vulgaris agg. Greater Bladderwort
Withering, 1787. Arrangement. In shallow water on Birmingham Heath.

Three records from ponds at Alvecote Pools (2503, 2504) and Tile Hill Wood (2779); it was collected from a pool on Comyns Farm, near Snitterfield, by Cheshire in 1857; recorded from all surrounding VCs.

The Alvecote Pools specimens have been identified by Dr. P. Taylor as *U. australis* R. Br. (*U. neglecta* Lehm.) who states that this species seldom flowers; in the absence of flowers it may be distinguished from *U. vulgaris* L. by the presence of small abortive bladders at the bases of some of the leaves; the presence of single or multiple bristles on the leaves is an unreliable character. Flowering specimens were collected in 1969 and the previous identification based on vegetative features was confirmed.

442.4 **Utricularia minor** L. Lesser Bladderwort
Bagnall, 1875. Herb. Bagnall. Bracebridge Pool, Sutton Park.

Not recorded in present survey; reported by Harlond as extinct at this station in 1934; recorded from VCs 23, 32, 37 and 39.

VERBENACEAE

444 Verbena L.

444.1 **Verbena officinalis** L. Vervain
Perry, 1817. List. Foot of Stank Hill, near Warwick.
Two records only; one from roadside at Morton Bagot (1164) and one from a wall at Tredington (2543); previously more widespread and collected from Chadshunt, Honington, Stratford on Avon and Wixford; recorded from all surrounding VCs.

LABIATAE

445 Mentha L.

445.2 **Mentha pulegium** L. Pennyroyal
Withering, 1787. Arrangement. Side of a pool at Erdington.
Not recorded in present survey; last known record by Bromwich in 1866 from cultivated land at Warwick; recorded from all surrounding VCs.

445.3 **Mentha arvensis** L. Corn Mint
Perry, 1824. Herb. Perry [*M. pratensis* Sole]. Near the Windmill Field, Haseley.
Map 444, p. 454
A frequent plant in the N. of the county but only occasional in the S.; recorded from cultivated land (38%), watersides and marshes (29%), grassland (13%), roadsides and waste places (9%), woods (6%) and a few other habitats; recorded from all surrounding VCs.

Mentha × gentilis L.
= **M. arvensis × spicata**
Bromwich, H., 1846. Herb. Perry. Without locality.
A rare plant recorded from streamsides at Heronfield House (1974), near Hatton Station (2265), near Leamington (3365) and from a waste place at Baginton (3474); recorded from all surrounding VCs.
Specimens from Warwickshire which have been previously identified as *M. × smithiana* should, according to Dr. R. M. Harley, be referred to *M. × gentilis.*

445.4 **Mentha aquatica** L. Water Mint
Perry, 1812. Herb. Perry. Without locality.
Map 445, p. 455
Abundant and fairly evenly distributed throughout the county; recorded from watersides and marshes (86%), water (12%) and a few other habitats; recorded from all surrounding VCs.

Mentha × verticillata L.
= **M. aquatica × arvensis**
Murcott, J. J., 1844. Herb. Perry [*M. sativa* var. *vulgaris*]. Shrewley Pool. Map 446, p. 455
An occasional plant with a scattered distribution almost entirely in the N. half of the county; recorded chiefly from watersides and marshes; recorded from all surrounding VCs.

Mentha × piperita L.
= **M. aquatica × spicata** Peppermint
Withering, 1796. Arrangement. River at Tamworth.
A rare plant of watersides and roadsides recorded from Portway (0773), near Kettlebrook (2102), near Chadwick End (2173), near Baddesley Ensor (2697), Nuthurst Heath (3090), near Southam (4160) and Calcutt Locks (4663); recorded from all surrounding VCs; var. *citrata* was recorded from Warwickshire by Bromwich in 1875.

445.5 & 6
Mentha spicata L.
(incl. *M. longifolia* auct. angl.) Spearmint
Purton, T., 1821. Mid. Flora [*M. sylvestris*]. Great Alne, on the side of the ford leading to Haselor. Map 447, p. 456
A rare plant of a few scattered localities with some concentration in the Coventry–Rugby–Southam area; recorded mainly from waste places and roadsides (69%); recorded from all surrounding VCs.

Mentha × villosa Huds. var. **alopecuroides** (Hull) Briq. = **M. spicata × suaveolens**
(**M. × niliaca** auct.)
Bromwich, H., c. 1866. Herb. Warw. Mus. [*M. alopecuroides*]. In the old moat near Chesterton Church.
A very rare hybrid recorded from roadside near Southam (4162) and a waste place near Rugby (4876); *M. × villosa* has been recorded from all the surrounding VCs except 39, but material has not been seen by us.

445.7 **Mentha suaveolens** Ehrh. Round-leaved Mint
Bagnall, 1870. Proc. B'ham N.H. Soc. [*M. rotundifolia* L.]. Abundant in a swampy field near Boldmere, Sutton.
Not recorded in present survey; a specimen collected by Bagnall in 1873 from the same locality and one collected by Bromwich at about the same date from a garden at Warwick have been determined as this species by Dr. R. M. Harley; recorded from VCs 37, 39 and 57.

446 Lycopus L.

446.1 **Lycopus europaeus** L. Gipsywort
Perry, 1812. Herb. Perry. Without locality.
Map 448, p. 456
Frequent in most of the county but rare on the Lower Lias clay in the S.; recorded from watersides and marshes (94%) and water (5%); recorded from all surrounding VCs.

447 Origanum L.

447.1 **Origanum vulgare** L. Marjoram
Bromwich, H., 1873. Topographical Botany. Without locality.
Bagnall, 1874. Herb. Bagnall. Steeple Hill, near Bidford. Map 449, p. 457
A rare plant of a number of scattered localities in the S. half of the county, often on calcareous soils; recorded from roadsides, railway banks, hedgerows and scrub; recorded from all surrounding VCs.

448 Thymus L.

448.1 **Thymus pulegioides** L. Large Thyme
Bromwich, H., 1853. Herb. Perry [*T. serpyllum*]. Hatton.
A rare plant of calcareous grassland recorded from Morton Bagot (1164) and Pailton Pastures (4982);

previously much more widespread, and collected from Billesley, Drayton Rough Moors, between Marton and Princethorpe, Preston Bagot and Yarningale Common; recorded from all surrounding VCs.

448.3 **Thymus drucei** Ronn. Wild Thyme
Perry, 1812. Herb. Perry [*T. serpyllum* L.]. Without locality. Map 450, p. 457
An occasional plant, occuring almost entirely on calcareous soils in the S. of the county; recorded from dry pastures, rough grassland and quarries; recorded from all surrounding VCs.

451 **Calamintha** Mill.

451.2 **Calamintha ascendens** Jord. Common Calamint
Withering, 1796, Arrangement [*Melissa calamintha* L.]. Near Tamworth Castle. Map 772, p. 610
An occasional plant of dry grassy banks and hedgerows, chiefly on calcareous soil; recorded from all surrounding VCs.

451.3 **Calamintha nepeta** (L.) Savi Lesser Calamint
Palmer, Miss C., 1851. Herb. Oxf. Univ. Lighthorne.
Not recorded in present survey; recorded from VCs 23, 32, 33 and 55.

452 **Acinos** Mill.

452.1 **Acinos arvensis** (Lam.) Dandy Basil-thyme
Perry, 1817. List [*Thymus acinos*]. Between Milverton and Ashow. Map 773, p. 611
A rare plant of calcareous quarries, railway cuttings and roadsides in the S.E. of the county; recorded from all surrounding VCs.

453 **Clinopodium** L.

453.1 **Clinopodium vulgare** L. Wild Basil
Perry, 1812. Herb. Perry [*Calamintha clinopodium* Benth.]. Without locality. Map 451, p. 458
An occasional plant mainly on the Lower Lias limestone in the S. of the county and especially abundant in the Wootton Wawen area, with scattered occurrences elsewhere; recorded from roadsides and railway banks (39%), hedgerows and scrub (30%), rough grassland (21%) and several other habitats; recorded from all surrounding VCs.

454 **Melissa** L.

454.1 **Melissa officinalis** L. Balm
Anon., 1873. Herb. B'ham Univ. Garden at Myton.
One record from railway bank at Wilmcote (1658); recorded from all surrounding VCs.

455 **Salvia** L.

455.2 **Salvia pratensis** L. Meadow Clary
Bolton King, 1880. Herb. Brit. Mus. Dry fields east of Kineton.
Not recorded in present survey; there are other records by Bolton King in the same area up till 1898; recorded from all surrounding VCs except 39.

455.4 **Salvia horminoides** Pourr. Wild Clary
Withering, 1796. Arrangement [*S. verbenaca* L.]. On the Castle Hill, Tamworth.
A rare plant of grassy banks near gardens, probably an escape, recorded from Bidford on Avon (1051); previously also collected from Ashorne and near Warwick; recorded from all surrounding VCs except 57.

457 **Prunella** L.

457.1 **Prunella vulgaris** L. Self-heal
Perry, 1812. Herb. Perry. Without locality. Map 452, p. 458
Abundant and evenly distributed throughout the county; recorded from grassland (56%), roadsides, railway banks and waste places (21%), cultivated land (8%), watersides and marshes (6%), hedgerows and scrub (4%) and woods (4%); recorded from all surrounding VCs.

458 **Betonica** L.

458.1 **Betonica officinalis** L. Betony
Perry, 1812. Herb. Perry [*Stachys betonica*]. Without locality. Map 453, p. 459
A moderately frequent plant in the N.W. half of the county but rare in the S.E.; recorded from grassland (46%), hedgerows and scrub (23%), roadsides (14%), watersides (8%) and woodland (6%); recorded from all surrounding VCs.

459 **Stachys** L.

459.1 **Stachys annua** (L.) L.
Bagnall, 1877. Herb. Bagnall. On railway banks above Blackroot Pool, Sutton Park.
One record as a garden weed from Hatton Railway Station (2266); recorded from VCs 23, 32, 39 and 55.

459.3 **Stachys arvensis** (L.) L. Field Woundwort
Cheshire, W., 1854. Herb. Perry. In a cornfield near Alcester. Map 454, p. 459
An occasional plant of a number of widely scattered localities in the county, mainly on light non-calcareous soils; recorded from cultivated land, waste places and hedgerows; recorded from all surrounding VCs.

459.6 **Stachys palustris** L. Marsh Woundwort
Baxter, W., 1831. MS (Bagnall: Mid. Nat. 1892–93.) Near Brownsover and Newbold on Avon.
Ick, W., 1836. Herb. Ick. Shady Lane near Nechells Green. Map 455, p. 460
Frequent and widely though very irregularly scattered through much of the county, completely absent from a belt running from Birmingham to S. of Nuneaton and rare on the calcareous clays of the S.; recorded from watersides (88%), hedgerows and scrub, cultivated land and a few other habitats; recorded from all surrounding VCs.

Stachys palustris × sylvatica
= **S. × ambigua** Sm.
Bromwich, H., 1859. Herb. Perry. Near Blacksmith's shop, Beausale.
Recorded from Catherine de Barnes (1780), S. of Bickenhill (1881), Ilmington (2143) and near Stockton Reservoir (4264); recorded from all surrounding VCs.

459.7 **Stachys sylvatica** L. Hedge Woundwort
Perry, 1812. Herb. Perry. Without locality.
Map 456, p. 460
Abundant and evenly distributed throughout the county; recorded from hedgerows and scrub (58%), roadsides, waste places and railway banks (20%), woods (12%), watersides (5%) and grassland (4%); recorded from all surrounding VCs.

460 Ballota L.

460.1 **Ballota nigra** L. Black Horehound
Perry, 1812. Herb. Perry. Without locality.
Map 457, p. 461
A frequent plant of irregular distribution, absent from a large area in the W. and smaller areas in the S.E. of the county; recorded from roadsides and waste places (49%), hedgerows (43%) and grassland (5%); recorded from all surrounding VCs.

461 Lamiastrum Heist. ex Fabr.

461.1 **Lamiastrum galeobdolon** (L.) Ehrend. & Polatsch. Yellow Archangel
Withering, 1787. Arrangement [*Galeobdolon luteum*, Huds.]. Warwickshire, frequent.
Map 458, p. 461
A frequent plant chiefly in the N. half of the county, mainly on Keuper Marl and in areas which were forests in the seventeenth century; recorded from woods (42%), hedgerows and scrub (32%), roadsides (17%), watersides (5%) and grassland (4%); recorded from all surrounding VCs.

462 Lamium L.

462.1 **Lamium amplexicaule** L. Henbit Deadnettle
Perry, 1813. Herb. Perry. Without locality.
Map 459, p. 462
An occasional plant of a number of widely scattered localities in the county; recorded from cultivated land (67%), waste places and roadsides (27%), and walls and quarries (4%); recorded from all surrounding VCs.

462.3 **Lamium hybridum** Vill. Cut-leaved Deadnettle
Baxter, W., 1840. MS (Bagnall: Flora, 1891). Oat field, near Holbrook, just above the Avon, two miles from Rugby. Map 460, p. 462
A rare plant of a number of widely scattered localities in the county with some concentration in the Rugby and Stratford on Avon areas; recorded from cultivated land, waste places and a few other habitats; recorded from all surrounding VCs.

462.4 **Lamium purpureum** L. Red Deadnettle
Perry, 1829. Mag. Nat. Hist. In a garden at Warwick. Map 461, p. 463
Abundant and fairly evenly distributed throughout the county; recorded from cultivated land (43%), roadsides, waste places, farmyards and railway banks (39%), hedgerows (10%), grassland (6%) and a few other habitats; recorded from all surrounding VCs.

462.5 **Lamium album** L. White Deadnettle
Baxter, W., 1831. MS (Bagnall: Mid. Nat. 1892–93.) Near Rugby.
Lloyd, Dr. G., 1835. Herb. Perry. Near Leamington. Map 462, p. 463
Abundant and evenly distributed throughout the county; recorded from roadsides, waste places, railway banks and farmyards (48%), hedgerows and scrub (30%), grassland (9%), cultivated land (7%) and a few other habitats; recorded from all surrounding VCs.

462.6 **Lamium maculatum** L. Spotted Deadnettle
Kirk, T., 1849. Herb. Perry [*L. rugosum*]. Naturalised at Allesley. Map 463, p. 464
Introduced and occasionally naturalised in a number of scattered localities, but rare in the extreme N. and completely absent from the extreme S. of the county; recorded from roadsides and waste places and a few other habitats; recorded from all surrounding VCs.

463 Leonurus L.

463.1 **Leonurus cardiaca** L. Motherwort
Purton, T., 1817. Mid. Flora. Kings Coughton.
One record from hedge by towpath of canal at Wilmcote (1658); recorded from all surrounding VCs except 32.

465 Galeopsis L.

465.1 **Galeopsis angustifolia** Ehrh. ex Hoffm.
Narrow-leaved Hempnettle
Perry, 1827. Herb. Perry. Near footpath from Grove Park to Pinley Abbey, near Hatton.
Map 464, p. 464
An occasional plant confined to the Lower Lias limestone and well-drained calcareous clays in the S. of the county; recorded from cultivated land and waste places; recorded from all surrounding VCs.

465.4 **Galeopsis tetrahit** L. *sensu lato*
Common Hempnettle
Baxter, W., 1831. MS (Bagnall: Mid. Nat. 1892–93.) Near Hill Morton. Map 465, p. 465
Abundant and fairly evenly distributed throughout most of the county but very rare in the Avon Valley; recorded from cultivated land (32%), roadsides and waste places (26%), hedgerows and scrub (20%), grassland (10%), watersides (7%) and woods (4%); recorded from all surrounding VCs.
G. bifida Boenn. (first record Dunn, S. T., 1896. Journal of Botany [*G. tetrahit* var. *bifida*]. Heaths by the "Straight Mile", Dunchurch) has been recorded a number of times, but, owing to the difficulty of identification, we have included only records of material determined by C. C. Townsend or seen by us. These are: near Earlswood (1273), near Nuthurst (1471), near Liveridge Hill (1569), Wootton Grange (1664), near Tile Cross (1686), Polesworth (2502), near Warton (2904), Canley (3076), Nuthurst Heath (3090), near Ryton on Dunsmore (3774), Bramcote (4089), near Withybrook (4285), near Monks Kirby (4584) and near Priors Hardwick (4855). The majority of the other records included on the map of *G. tetrahit* L. sensu lato are for *G. tetrahit* L. sensu stricto (first record Ick, W., 1836. Herb. Ick. Bank at Nechells Green), which has been recorded from all surrounding VCs; *G. bifida* from VCs 23, 32, 33 and 57.

465.5 **Galeopsis speciosa** Mill.
Large-flowered Hempnettle
Withering, 1796. Arrangement [*G. cannabina*]. "Under a moist hedge at Birches Green near Birmingham."
A rare plant of cultivated and waste places, probably introduced in this county; recorded from Birmingham

(0386, 0483, 0888), near Hampton Lucy (2458), near Austrey (2904) and near Warwick (3064); recorded from all surrounding VCs.

466 Nepeta L.

466.1 **Nepeta cataria** L. Catmint
Perry, 1817. List. Between Stratford and Warwick, opposite Welcombe Hills; between Warwick and Myton.

A rare plant of hedgerows on calcareous soil recorded from the Newbold on Stour area (2245, 2345, 2446), near Halford (2744) and on Burton Hills (4052); previously more abundant and widespread, and recorded from Binton, Stratford on Avon, Myton, Whitley Common and Great Packington; recorded from all surrounding VCs.

467 Glechoma L.

467.1 **Glechoma hederacea** L. Ground-ivy
Perry, 1813. Herb. Perry. Without locality.
Map 466, p. 465

Abundant and evenly distributed throughout the county; recorded from hedgerows and scrub (43%), roadsides, railway banks and waste places (25%), woods (14%), grassland (9%) and watersides (5%); recorded from all surrounding VCs.

468 Marrubium L.

468.1 **Marrubium vulgare** L. White Horehound
Purton, T., 1817. Mid. Flora. Near Oversley Lodge, near Alcester.

A rare plant of roadsides and waste places recorded from Earlswood (1173), Wootton Wawen (1563), Coleshill (2088) and near Norton Lindsey (2262); recorded from all surrounding VCs.

469 Scutellaria L.

469.1 **Scutellaria galericulata** L. Skullcap
Withering, 1796. Arrangement. Banks of the stews at Edgbaston. Map 467, p. 466

A frequent plant irregularly distributed throughout most of the county, very rare in the extreme S. and absent from a large area around Coventry; recorded from watersides and marshes (99%); recorded from all surrounding VCs.

S. galericulata L. × *minor* Huds. (= *S.* × *hybrida* Strail) was collected by Bagnall in 1887 (Herb. Bagnall) from the canal, Sowe Waste Common. It has not been recorded during the present survey.

469.2 **Scutellaria minor** Huds. Lesser Skullcap
Countess of Aylesford, 1805. Bot. Guide. Packington.

A rare plant of bogs and marshes recorded from Sutton Park (0996, 0998), Coleshill Bog (2086) and Ryton Wood (3872); recorded from all surrounding VCs except 32 and 33.

470 Teucrium L.

470.1 **Teucrium chamaedrys** L. Wall Germander
Cotes, R., 1962. Near Ilmington.

One record only of plant in rough grassland near Ilmington (2242); recorded from VCs 23, 33, 37 and 39.

470.4 **Teucrium scorodonia** L. Wood Sage
Perry, 1817. List. Between Stoneleigh and Wootton Field. Map 468, p. 466

Frequent and irregularly scattered throughout the N.W. of the county, but almost completely absent from the calcareous soils of the S. and S.E., the Avon Valley and the chalky boulder clay of the N.E.; recorded from hedgerows and scrub (45%), woods (23%), roadsides and railway banks (22%), grassland (5%) and heaths (3%); recorded from all surrounding VCs.

471 Ajuga L.

471.2 **Ajuga reptans** L. Bugle
Perry, 1825. Herb. Perry. Without locality.
Map 469, p. 467

A plant of heavy wet soil in grassland (28%), woods (26%), watersides (16%), hedgerows and scrub (15%) and roadsides and railway banks (14%); widely and abundantly distributed throughout the county but notably absent from the Avon Valley and urban areas; recorded from all surrounding VCs.

PLANTAGINACEAE

472 Plantago L.

472.1 **Plantago major** L. Greater Plantain
Lloyd, Dr. G., 1835. Herb. Perry. Without locality. Map 470, p. 467

Abundant and evenly distributed throughout the county; recorded from roadsides, waste places, farmyards and railway banks (57%), grassland (23%), cultivated land (15%) and a few other habitats; recorded from all surrounding VCs.

472.2 **Plantago media** L. Hoary Plantain
Lloyd, Dr. G., 1835. Herb. Perry. Without locality. Map 471, p. 468

An abundant plant in the S. of the county, particularly on calcareous soils, with occasional occurrences elsewhere; recorded from grassland (51%), roadsides and railway banks (43%) and a few other habitats; recorded from all surrounding VCs.

472.3 **Plantago lanceolata** L. Ribwort Plantain
Perry, 1813. Herb. Perry. Without locality.
Map 472, p. 468

Abundant and evenly distributed throughout the county; recorded from roadsides, waste places and railway banks (49%), grassland (39%), cultivated land (6%) and hedgerows and scrub (4%); recorded from all surrounding VCs.

472.5 **Plantago coronopus** L. Buckshorn Plantain
Purton, T., 1817. Mid. Flora. Between Crab's Cross and Headley's Cross.

A rare plant of gravelly places recorded from Sutton Park (1197), Bradnock's Marsh (2179) and near Baginton (3375); formerly more widespread and known from Sutton Park since 1859, from Bradnock's Marsh since 1882 and from Bannerley Pool and Kenilworth Heath; recorded from all surrounding VCs.

472.6 **Plantago indica** L.
Bromwich, H., 1882. Herb. Bagnall [*P. arenaria* Wildst. & Kit.]. Milverton.

A bird-seed alien recorded once from Rugby (4974); recorded from all surrounding VCs except 57.

473 **Littorella** Berg.

473.1 **Littorella uniflora** (L.) Aschers. Shoreweed
Countess of Aylesford, 1805. Bot. Guide [*L. lacustris*]. Coleshill Pool.
Recorded from edges of the lakes at Earlswood (1173, 1174) and Olton (1381); previously recorded from Coleshill Pool and Bracebridge Pool, Sutton Park, where it was seen as recently as 1934; recorded from all surrounding VCs except 33.

CAMPANULACEAE

474 **Wahlenbergia** Schrad.

474.1 **Wahlenbergia hederacea** (L.) Reichb. Ivy-leaved Bellflower
Kirk, T., 1847. Herb. Perry. "On spongy turf, with *Anagallis tenella*, near Arbury Hall; possibly planted but I think it truly indigenous to the locality."
Not recorded in present survey; the last record from Arbury Park was Bagnall's in 1872; recorded from VCs 23, 39, 55 and 57.

475 **Campanula** L.

475.1 **Campanula latifolia** L. Giant Bellflower
Countess of Aylesford, 1805. Bot. Guide. Near Packington. Map 473, p. 469
A rare plant restricted to a few damp woods (53%) and scrub woodland and hedgerows (29%); recorded from all surrounding VCs.

475.2 **Campanula trachelium** L. Nettle-leaved Bellflower
Perry, 1817. List. Stoneleigh; Pillerton; Aqueduct near Leamington. Map 474, p. 469
A rare plant of a few scattered localities with a concentration on the Lower Lias escarpment in the S.W.; recorded from hedgerow and scrub (35%), mixed woods (26%) and roadsides and waste places (26%); recorded from all surrounding VCs.

475.3 **Campanula rapunculoides** L. Creeping Bellflower
Perry, before 1863. Herb. Perry. Roadsides near Ragley. Map 774, p. 611
A garden escape sometimes becoming naturalised and persisting for some years, particularly near Wilmcote; recorded from all surrounding VCs.

475.5 **Campanula persicifolia** L. Peach-leaved Bellflower
Thomas, C., 1950. Railway cutting, Harborne.
A rare garden escape recorded from roadsides and railway banks: 0380, 1193, 3256; recorded from VCs 23, 33, 37 and 57.

475.6 **Campanula glomerata** L. Clustered Bellflower
Perry, 1817. List. Near Pillerton.
A rare plant of calcareous grassland, formerly more widespread, now recorded from Aston Grove (1357), Wilmcote (1557), Halford (2644) and Burton Hills (4051, 4052); recorded from all surrounding VCs except 39.

475:7 **Campanula rotundifolia** L. Harebell
Perry, 1820. Plant. Varv. Sel. Roadside between Leek Wootton and Wootton Grange. Map 475, p. 470
A frequent plant irregularly distributed and largely absent from the Lias clay and the Avon Valley; recorded from grassland (48%), roadsides and railway banks (29%), hedgerows (11%) and dry heaths (5%); recorded from all surrounding VCs.

475.8 **Campanula patula** L. Spreading Bellflower
Woodward, 1787. Withering's Arrangement. On the road to Coleshill.
Recorded only from three places in hedgerows near Wingate Farm (0767); recorded from all surrounding VCs except 23 and 32.

475.9 **Campanula rapunculus** L. Rampion Bellflower
Perry, 1820. Plant. Varv. Sel. "Near Guys Cliffe; by the roadside in front of the house of M. Wise, Esq. Leamington."
Not recorded in present survey; recorded from all surrounding VCs except 32 and 57.

476 **Legousia** Durande

476.1 **Legousia hybrida** (L.) Delarb. Venus's Looking-glass
Rufford, Rev. W. S., 1817. Mid. Flora [*Campanula hybrida* L.]. Alne Hills. Map 476, p. 470
A rare plant of calcareous soils in the S. of the county; recorded mainly from arable land; recorded from all surrounding VCs.

478 **Phyteuma** L.

478.2 **Phyteuma spicatum** L. Spiked Rampion
Cox, T., 1863. Herb. Camb. Univ. [*P. nigrum* Schmidt]. On an old bank between Leek Wootton and Ashow.
A garden escape (blue-flowered variety) not recorded in present survey; a specimen was collected by Bromwich (Herb. Brit. Mus.) from the railway embankment at Hill Wootton in 1863 and this may be from the same colony as Cox's specimen; recorded from VC 39.

479 **Jasione** L.

479.1 **Jasione montana** L. Sheepsbit
Perry, 181–. Herb. Perry. Without locality.
Not recorded in present survey; the last record appears to be by Bagnall in 1874 (Herb. Bagnall) from Middleton Heath; recorded from all surrounding VCs except 33.

RUBIACEAE

481 **Sherardia** L.

481.1 **Sherardia arvensis** L. Field Madder
Perry, 1812. Herb. Perry. Without locality. Map 477, p. 471
Frequent though irregularly distributed throughout much of the county, but absent from the extreme N.; recorded from cultivated land (57%), grassland (23%), roadsides and waste places (9%) and a few other habitats; recorded from all surrounding VCs.

483 Asperula L.

483.2 **Asperula cynanchica** L. Squinancywort
Bloxam, A., 1875. (Bagnall's Flora, 1891.) Cornfields near Wilmcote.
One record only in pastures near Long Compton (2833); recorded from all surrounding VCs except 39.

Asperula arvensis L.
Jones, Mrs. M. D. G., 1963. Rugby.
A casual garden weed recorded from 2662, 4974 and 5373; recorded from all surrounding VCs.

484 Cruciata Mill.

484.1 **Cruciata laevipes** Opiz Crosswort
Perry, 1812. Herb. Perry [*Galium cruciata*]. Without locality. Map 478, p. 471
Frequent in a number of fairly large areas with a wide range of soils; recorded from roadsides and railway banks (40%), grassland (31%), hedgerows and scrub (22%) and waterside (5%); recorded from all surrounding VCs.

485 Galium L.

485.1 **Galium odoratum** (L.) Scop. Woodruff
Purton, T., 1817. Mid. Flora [*Asperula odorata*]. Oversley; Spernall; Ragley Woods. Map 479, p. 472
An occasional plant of a number of scattered localities in the county; recorded chiefly from woods (73%) and hedgerows (15%); recorded from all surrounding VCs.

485.3a **Galium mollugo** L. ssp. **mollugo** Great Hedge Bedstraw
Perry, 1820. Plant. Varv. Sel. [*G. mollugo*]. Weir-break Hill and Cross-of-the-Hill near Stratford. Map 480, p. 472
Frequent on base-rich soils in the S. of the county but rare elsewhere; recorded from hedgerows (48%), roadsides and railway banks (28%), grassland (17%) and a few other habitats; recorded from all surrounding VCs.

485.3b **Galium mollugo** L. ssp. **erectum** Syme
Perry, 181–. Herb. Perry [*G. erectum*]. Near Radford Semele.
One record from calcareous grassland near Compton Verney (3053); recorded from all surrounding VCs except 39.

Galium mollugo × verum
= **G. × pomeranicum** Retz.
Bromwich, H., 1878. Herb. Bagnall [*G. verum* var. *ochroleucum* Wolf]. Near Wellesbourne.
Not recorded in present survey, but found between Wilmcote and Billesley in 1938; recorded from all surrounding VCs except 39.

485.4 **Galium verum** L. Lady's Bedstraw
Perry, 1812. Herb. Perry. Without locality. Map 481, p. 473
Abundant and evenly distributed throughout the S. of the county, but less abundant and less evenly distributed in the N.; recorded from grassland (48%), roadsides and railway banks (38%), hedgerows (10%) and a number of other habitats; recorded from all surrounding VCs.

485.5 **Galium saxatile** L. Heath Bedstraw
Perry, 1817. List. Kenilworth Heath. Map 482, p. 473
Frequent, though unevenly distributed, in the N. of the county but almost completely absent from the Avon Valley and the calcareous soils of the S. and E.; recorded from grassland (24%), woods (23%), dry heath (22%), hedgerows and scrub (15%) and roadsides and waste places (13%); recorded from all surrounding VCs.

485.6 **Galium pumilum** Murr. Slender Bedstraw
Townsend, F., 1904. Journal of Botany [*G. sylvestre* Poll.]. ¾ mile E. of Tredington.
Not recorded in present survey; recorded from all surrounding VCs except 39 and 57.

485.8 **Galium palustre** L. Marsh Bedstraw
Lloyd, Dr. G., 1835. Herb. Perry. *G. palustre* L. ssp. *palustre* [as *G. palustre*]. Wroxall Heath. Map 483, p. 474
Abundant and fairly evenly distributed throughout the county; recorded from watersides and marshes (91%), damp woods (5%) and a few other habitats; recorded from all surrounding VCs.
Ssp. *elongatum* (C. Presl) Lange (first record: Kirk, T., 1860 [*G. elongatum* Presl]. Sowe Waste, Coventry) and ssp. *palustre* (first record as above) have been recorded once or twice during the survey but their distribution is not known; both have been recorded from all surrounding VCs.

485.10 **Galium uliginosum** L. Fen Bedstraw
Perry, 1817. List. Green's Grove, Hatton. Map 484, p. 474
An occasional plant of a number of scattered localities in the county, with a concentration in the Packington–Temple Balsall–Earlswood area; recorded from watersides and marshes; recorded from all surrounding VCs.

485.11 **Galium tricornutum** Dandy Corn Goosegrass
Purton, T., 1817. Mid. Flora [*G. tricorne*]. On Alne Hills (Rufford); in a cornfield by Drayton Bushes. Map 775, p. 611
A very local plant of arable fields on calcareous soil in the S. of the county; formerly more widespread and abundant; recorded from all surrounding VCs.

485.12 **Galium aparine** L. Goosegrass, Cleavers
Lloyd, Dr. G., 1835. Herb. Perry. Leamington. Map 485, p. 475
Abundant and evenly distributed throughout the county; recorded from hedgerows and scrub (47%), roadsides, waste places and railway banks (21%), cultivated land (14%), woods (9%), grassland (5%) and watersides (4%); recorded from all surrounding VCs.

485.13 **Galium spurium** L. False Cleavers
Dunn, S. T., 1896. Herb. B'ham Univ. [*G. vaillantii* DC.]. Allotments, Long Lawford.
One record from rubbish tip near Memorial Theatre, Stratford on Avon (2054); recorded from VCs 23, 33, 39, 55 and 57.

CAPRIFOLIACEAE

487 **Sambucus** L.

487.1 **Sambucus ebulus** L. Danewort, Dwarf Elder
Withering, 1796. Arrangement. At the foot of Tamworth Castle Hill, towards the river.
A rare plant of hedgerows recorded from Temple Grafton (1254) where it has been known since 1854, Knowle (1876) where it has been known since 1830 and High Cross (4788); recorded from all surrounding VCs.

487.2 **Sambucus nigra** L. Common Elder
Withering, 1787. Arrangement. Without locality. Map 486, p. 475
Abundant and evenly distributed throughout the county; recorded from hedges and scrub (65%), mixed woods (16%), waste places, roadsides, railway banks and farmyards (13%), watersides (4%) and a few other habitats; recorded from all surrounding VCs.

488 **Viburnum** L.

488.1 **Viburnum lantana** L. Wayfaring-tree
Withering, 1830. Arrangement. Hedge between Leamington and Southam. Map 487, p. 476
Frequent in the S. of the county, mainly on calcareous soils, but very rare elsewhere; recorded from hedges and scrub (74%), woods (20%) and roadsides and railway banks (4%); recorded from all surrounding VCs except 39.

488.3 **Viburnum opulus** L. Guelder-rose
Perry, 181–. Herb. Perry. Without locality. Map 488, p. 476
Frequent and fairly evenly distributed in the N. half of the county, rare in the S. except on the Lower Lias limestone and the Cotswold escarpment; recorded from hedges and scrub (41%), woods (33%), watersides (22%) and a few other habitats; recorded from all surrounding VCs.

489 **Symphoricarpos** Duham.

489.1 **Symphoricarpos rivularis** Suksd. Snowberry
Bromwich, H., before 1907. Herb. Warw. Mus. Without locality. Map 489, p. 477
Widely planted and naturalised, this plant is frequent but irregularly distributed throughout most of the county; recorded from hedges and scrub (67%), mixed woods (12%), roadsides and waste places (10%), watersides (6%) and a few other habitats; recorded from all surrounding VCs.

491 **Lonicera** L.

491.1 **Lonicera xylosteum** L. Fly Honeysuckle
Withering, 1812. Arrangement. "In the wood S.W. side of the lake in Edgbaston Park."
Probably always planted; not recorded in present survey; recorded from all surrounding VCs.

491.3 **Lonicera periclymenum** L.
Common Honeysuckle
Baxter, W., 1831. MS (Bagnall: Mid. Nat. 1892–93.) Lawford Lane.
Lloyd, Dr. G., 1835. Herb. Perry. Ufton Wood. Map 490, p. 477
Abundant and fairly evenly distributed throughout most of the county, but less common in the Avon Valley and on the Lias clay in the S.; recorded from hedgerows and scrub (66%), woods (30%) and a few other habitats; recorded from all surrounding VCs.

491.4 **Lonicera caprifolium** L. Perfoliate Honeysuckle
Bolton King, 1891. Bagnall's Flora. Chadshunt.
Not recorded in present survey; recorded from all surrounding VCs except 57.

ADOXACEAE

493 **Adoxa** L.

493.1 **Adoxa moschatellina** L.
Moschatel, Townhall Clock
Withering, 1796. Arrangement. "In a wood on the S.W. side of the pool at Edgbaston, plentiful; in the woods N. of Aston Park." Map 491, p. 478
A locally frequent plant mainly of damp hedgerows and damp scrub (33%), watersides (24%), woodland (22%) and roadsides and waste places (15%), mainly in the N.W. of the county but also in a few localities in the extreme S.E.; recorded from all surrounding VCs.

VALERIANACEAE

494 **Valerianella** Mill.

494.1 **Valerianella locusta** (L.) Betcke
Lamb's Lettuce, Common Cornsalad
Ick, W., 1836. Herb. Ick [*Fedia olitoria*]. Lane leading from Aston Tavern. Map 492, p. 478
An occasional plant in the S. half of the county, though absent from the extreme S., and very rare in the N. half; recorded chiefly from railway banks but also from rough grassland, cultivated land, walls and waste places; recorded from all surrounding VCs.

494.3 **Valerianella rimosa** Bast.
Broad-fruited Cornsalad
Bagnall, 1873. Herb. Bagnall [*V. auricula* DC.]. Red Hill, near Alcester.
Not recorded in present survey; recorded from all surrounding VCs except 55.

494.5 **Valerianella dentata** (L.) Poll.
Narrow-fruited Cornsalad
Baynes, W. W., 1831. Herb. Perry. Whitnash.
An occasional plant, generally of calcareous soil, recorded from cultivated fields at Newbold on Stour (2345, 2445, 2446), Oldbury (3195), Ufton (3962), Bourton Heath (4371), Napton Fields (4462), Bourton on Dunsmore (4470), and from a brick pit at Lighthorne (3456); formerly more widespread and collected from near Alcester, Billesley, Birdingbury, Drayton Bushes, Hartshill Hayes, Moreton Morrell, Packington, Sowe Waste, near Stratford on Avon and Tachbrook; recorded from all surrounding VCs.

495 **Valeriana** L.

495.1 **Valeriana officinalis** L. Common Valerian
Withering, 1801. Arrangement. "In the Garlick Meadows, near Penns Mill." Map 493, p. 479
Frequent but very irregularly distributed throughout the N.W. of the county, rare in the Avon Valley and almost entirely absent from the calcareous soils of the

S. and S.E.; recorded from watersides and marshes (59%), hedgerows and scrub (15%), damp woods (8%), roadsides (8%) and grassland (5%); recorded from all surrounding VCs.

495.3 **Valeriana dioica** L. Marsh Valerian
Withering, 1801. Arrangement. Garlic Meadows, Erdington. Map 494, p. 479

An occasional plant in most of the N.W. half of the county, absent from the extreme N. and N.E. and very rare in the S.; recorded almost entirely from marshes and watersides but occasionally from grassland, woods and bogs; recorded from all surrounding VCs.

496 Centranthus DC.

496.1 **Centranthus ruber** (L.) DC. Red Valerian
Perry, 1817. List [*Valeriana rubra*]. Eastgate, Warwick. Map 495, p. 480

An occasional escape from cultivation, sometimes naturalised on walls and in waste places; recorded from all surrounding VCs.

DIPSACACEAE

497 Dipsacus L.

497.1 **Dipsacus fullonum** L. Teasel
Perry, 1817. List [*D. sylvestris*]. Roadsides, common. Map 496, p. 480

Fairly abundant in the S. of the county particularly on calcareous soils, with scattered distribution in the N.E.; rare elsewhere; recorded from watersides (41%), hedgerows and scrub (23%), roadsides, waste places and railway banks (19%) and grassland (10%); Warwickshire is near the N.W. boundary of the main British distribution.

A number of records mapped are of plants which have been identified as ssp. *fullonum*, and it is probable that ssp. *sativus* (L.) Thell. does not occur in the county.

The species is recorded from all surrounding VCs.

497.2 **Dipsacus pilosus** L. Small Teasel
Countess of Aylesford, 1805. Bot. Guide. Near Coleshill. Map 497, p. 481

A very local plant, mainly in the S.W. of the county, recorded chiefly from watersides and damp woods; recorded from all surrounding VCs.

498 Knautia L.

498.1 **Knautia arvensis** (L.) Coult. Field Scabious
Perry, 1812. Herb. Perry. Without locality. Map 498, p. 481

Abundant in the S. of the county, with concentration on the Jurassic limestones, and frequent but very irregularly distributed in the N.; recorded from roadsides and railway banks (42%), grassland (30%), hedgerows (14%), cultivated land (10%) and quarries (4%); recorded from all surrounding VCs.

499 Scabiosa L.

499.1 **Scabiosa columbaria** L. Small Scabious
Perry, 181–. Herb. Perry. Without locality.
Perry, 1835. MS (Bagnall's Flora, 1891). Corner of Whitnash Field, Fosseway, leading to Harbury. Map 776, p. 611

An occasional plant of rough grassland on calcareous soil in the S. of the county; recorded from all surrounding VCs.

500 Succisa Haller

500.1 **Succisa pratensis** Moench Devilsbit Scabious
Perry, 1812. Herb. Perry [*Scabiosa succisa*]. Without locality. Map 499, p. 482

Frequent though irregularly distributed throughout much of the N.W. half of the county and occasional in the extreme S.E., but almost absent from the lower Avon Valley, the calcareous soils of the S. and S.E. and the chalky boulder clay of the N.E.; recorded from grassland (60%), watersides and marshes (18%), roadsides and railway banks (12%), woods (5%) and a few other habitats; recorded from all surrounding VCs.

COMPOSITAE

Helianthus L.

Helianthus rigidus (Cass.) Desf.
Webster, Miss M. McCallum, 1958. Hillmorton Ballast Pits.

A garden escape occasionally becoming naturalised, recorded from 3780, 4165 and 5473; not recorded from any of the surrounding VCs.

502 Bidens L.

502.1 **Bidens cernua** L. Nodding Bur-marigold
Stokes, Dr. J., 1787. Withering's Arrangement [*B. minor* L.]. Near Birmingham. Map 500, p. 482

An occasional plant mainly in the N. of the county, recorded from watersides and marshes (88%) and shallow water (12%); recorded from all surrounding VCs.

502.2 **Bidens tripartita** L. Trifid Bur-marigold
Perry, 1812. Herb. Perry. Without locality. Map 501, p. 483

An occasional plant irregularly distributed throughout the county but absent from the extreme S.; recorded from watersides; recorded from all surrounding VCs.

502.3 **Bidens frondosa** L. Beggar-ticks
Cadbury, Miss D. A., 1952. Canal towpath between Landor St. and Saltley, Birmingham.

Four records from waste places alongside canals: 0891, 0987, 1955 and 2055; recorded from VCs 33 and 37.

503 Galinsoga Ruiz & Pav.

503.1 **Galinsoga parviflora** Cav. Gallant Soldier
Wright, Miss C. M., 1955. Arable land W. of Alcester.

A rare casual of cultivated and waste places recorded from Edgbaston, Birmingham (0588), W. of Alcester (0657) and near Tamworth (2103); recorded from all surrounding VCs.

503.2 **Galinsoga ciliata** (Raf.) Blake Shaggy Soldier
Barker, Mrs. G. M., 1953. Herb. Brit. Mus. Waste ground, Kenilworth.

A rare casual of cultivated and waste places recorded from Birmingham (0386), Monkspath (1476), Solihull (1579), Stratford on Avon (1954), Guy's Cliffe, Warwick (2866), Brandon Wood (3876) and Coton House, near Rugby (5179); recorded from VCs 23, 33, 55 and 57.

505 Xanthium L.

505.2 **Xanthium spinosum** L. Spiny Cocklebur
Cox, T., 1873. Young & Baker Catalogue. Kenilworth.

Three records from cultivated and waste places, at Abbot's Salford (0649), Bidford on Avon (0952) and near Bedworth (3686); recorded from all surrounding VCs except 57.

Guizotia Cass.

Guizotia abyssinica (L.f.) Cass.
Cadbury, Miss D. A., 1962. Stratford on Avon.

A rare casual recorded from dried sewage sludge S. of Stratford on Avon (1852) and from a garden at Stratford on Avon (2055); not recorded from any of the surrounding VCs.

506 Senecio L.

506.1 **Senecio jacobaea** L. Common Ragwort
Baxter, W., 1831. MS (Bagnall: Mid. Nat. 1892–93.) Near Rugby. Map 502, p. 483

Abundant though somewhat irregularly distributed throughout most of the county, but rare on the calcareous clays of the S. and S.E.; recorded from roadsides, railway banks and waste places (50%), grassland (31%), hedgerows and scrub (9%), quarries (3%), cultivated land (3%) and a few other habitats; recorded from all surrounding VCs.

506.2 **Senecio aquaticus** Hill Marsh Ragwort
Baxter, W., 1831. MS (Bagnall: Mid. Nat. 1892–93.) Near Brownsover. Map 503, p. 484

Abundant in the W. of the county, absent from much of the centre and from the extreme S., frequent but irregularly distributed elsewhere; recorded from watersides and marshes (93%) and damp grassland (5%); recorded from all surrounding VCs.

506.3 **Senecio erucifolius** L. Hoary Ragwort
Baxter, W., 1831. MS (Bagnall: Mid. Nat. 1892–93.) Near Sawbridge; on a bank in Lawford Lane.
Lloyd, Dr. G., 1835. Herb. Perry. Between Tachbrook and Harbury. Map 504, p. 484

Abundant in the S. half of the county especially on calcareous soils, rare in the Avon Valley and thinly but irregularly distributed elsewhere; recorded from roadsides, railway banks and waste places (49%), grassland (35%), hedgerows and scrub (12%) and a few other habitats; recorded from all surrounding VCs.

506.4 **Senecio squalidus** L. Oxford Ragwort
Bree, Rev. W. T., 1830. Mag. Nat. Hist. Allesley. Map 505, p. 485

Abundant though somewhat irregularly distributed throughout much of the county, but rare on the calcareous clays of the S. and S.E. and on the Cotswold escarpment; recorded from waste places, roadsides and railways (74%), rough grassland (10%), walls and quarries (7%), cultivated land (5%) and hedgerows and scrub (4%); recorded from all surrounding VCs.

S. squalidus L. × *vulgaris* L. (first record: Burges, Dr. R. C. L., 1945. Herb. Burges. Derelict hard tennis court, Edgbaston, Birmingham) has been recorded once from Alvecote Council tip (2404); this has been recorded from VCs 23, 33 and 37.

S. londinensis Lousley (*S. squalidus* L. × *viscosus* L.) (first record: Burges, Dr. R. C. L., 1945. B.E.C. Report. Bombed site, Birmingham) has been recorded from waste places at Birmingham University tennis courts (0583) and Alvecote Council tip (2404); this has been recorded from VC 39.

506.6 **Senecio sylvaticus** L. Heath Groundsel
Perry, 1817. List. Between Birmingham and Erdington. Map 506, p. 485

Occasional and very irregularly distributed in the N. half of the county, but completely absent from the calcareous soils of the S. and S.E.; recorded from hedgerows and scrub (39%), roadsides and waste places (29%), rough grassland (12%), cultivated land (7%) and a few other habitats; recorded from all surrounding VCs.

506.7 **Senecio viscosus** L. Sticky Groundsel
Burges, Dr. R. C. L., 1938. Herb. Burges. Railway bank, Sutton Park. Map 507, p. 486

Frequent though very irregularly distributed in the N. half of the county, but rare on the calcareous soils of the S. and S.E.; recorded from railways, waste places and roadsides (86%), walls and quarries (6%), grassland (4%) and a few other habitats; recorded from all surrounding VCs.

506.8 **Senecio vulgaris** L. Common Groundsel
Lloyd, Dr. G., 1835. Herb. Perry. Without locality. Map 508, p. 486

Abundant and evenly distributed throughout the county; recorded from roadsides, waste places, railways and farmyards (47%), cultivated land (41%), grassland (6%), hedgerows (4%) and a few other habitats; recorded from all surrounding VCs.

The rayed form (var. *hibernicus* Syme) (first record: Burges, Dr. R. C. L., 1934. Herb. Burges. Edgbaston) has been recorded from 41 squares, mainly in the N. of the county; it has been recorded from all surrounding VCs except 32 and 39.

507 Doronicum L.

507.1 **Doronicum pardalianches** L. Great Leopard's-bane
Webster, Miss M. McCallum, 1957. Hillmorton Ballast Pits.

Naturalised in three localities: woodland at Warmington (4147) and Farnborough (4249), and ballast pits at Hillmorton (5473); recorded from all surrounding VCs.

508 Tussilago L.

508.1 **Tussilago farfara** L. Coltsfoot
Perry, 1813. Herb. Perry. Without locality. Map 509, p. 487

Abundant and fairly evenly distributed throughout the county; recorded from roadsides, waste places and railway banks (61%), cultivated land (14%), grassland (10%), watersides (8%), hedgerows and scrub (4%) and a few other habitats; recorded from all surrounding VCs.

509 Petasites Mill.

509.1 **Petasites hybridus** (L.) Gaertn., Mey. & Scherb. Common Butterbur
Purton, T., 1817. Mid. Flora [*Tussilago hybrida* L.]. Hoo Mill, on a willow bed. Map 510, p. 487

An occasional plant of widely scattered occurrence

throughout most of the county; recorded from waterside and marsh (60%), roadsides and waste places (22%) and a few other habitats; in a considerable number of cases the sex of the plants has been recorded, and from these records it would appear that there is a tendency for colonies of plants of one sex to occupy areas separated from those occupied by the other sex; male plants are recorded in quantity from an area near Redditch (Major squares 06, 16) and near Shustoke (28, 29); and female plants from near Bidford on Avon (05, 15) and from a large area in the vicinity of Rugby (47, 48, 57, 58). The species has been recorded from all surrounding VCs.

509.2 **Petasites albus** (L.) Gaertn. White Butterbur
Perry, 1852. Herb. Perry. Guy's Cliffe, near Warwick.
An introduced plant established at Leek Wootton (2968), Arbury Park (3389) where it has been known since 1855, and near Brandon Castle (4075); recorded from VCs 32, 39 and 55.

509.3 **Petasites japonicus** (Sieb. & Zucc.) F. Schmidt Giant Butterbur
Rushforth, Miss M., 1957. Harbury.
An introduced plant established in damp shaded places at Edgbaston Park (0583), Woodcote Lodge, Leek Wootton (2869) and Harbury (3760); recorded from VC 37.

509.4 **Petasites fragrans** (Vill.) C. Presl Winter Heliotrope
Cheshire, W., 1855. Herb. Perry. Luddington Old Lock, near Stratford on Avon. Map 511, p. 488
An introduced plant, very rare in Bagnall's time, now firmly established in many places in a broad belt across the centre of the county; recorded from roadsides, waste places, railway banks, hedgerows and scrub; recorded from all surrounding VCs.

512 Inula L.

512.1 **Inula helenium** L. Elecampane
Purton, T., 1817. Mid. Flora. Grafton; Studley, in the Castle Field.
A rare introduced plant of hedgerows and waste places recorded from Tattle Bank (1863), Berkswell (2479) and Stoney Thorpe (4061); recorded from all surrounding VCs.

512.4 **Inula conyza** DC. Ploughman's Spikenard
Perry, 1817. List [*Conyza squarrosa*]. Near Myton; between Emscote and Leamington. Map 512, p. 488
Occasional in grassland and waste places on calcareous soil; rare on walls and in quarries; recorded from all surrounding VCs.

513 Pulicaria Gaertn.

513.1 **Pulicaria dysenterica** (L.) Bernh. Common Fleabane
Perry, 1825. Herb. Perry. Warwick. Map 513, p. 489
Frequent and widely but very irregularly distributed throughout the county; recorded from watersides and marshes (67%), grassland (17%), roadsides (12%) and hedgerows (4%); recorded from all surrounding VCs.

513.2 **Pulicaria vulgaris** Gaertn. Small Fleabane
Withering, 1796. Arrangement [*Inula cylindrica*]. About Wishaw, near Coleshill.
Not recorded in present survey; last record appears to be Ick, 1834 (Herb. Perry), Myton Road, Warwick; recorded from VCs 23, 37 and 55.

514 Filago L.

514.1 **Filago vulgaris** Lam. Common Cudweed
Perry, 1812. Herb. Perry [*F. germanica* (L.) L.]. Warwick.
A rare plant of dry banks and heathy places recorded from S. of Stratford on Avon (2053), Hay Wood (2171), Wellesbourne Hastings (2755, 2756), Waverley Wood (3571), Ryton on Dunsmore (3774), Brandon Wood (3876, 3976) and W. of Wolvey (4187); formerly much more frequent; recorded from all surrounding VCs.

514.5 **Filago minima** (Sm.) Pers. Slender Cudweed
Cross, J., 1845. Herb. Perry. Kenilworth Bush Common.
Three records from dry waste places: Bearley Railway Station Yard (1760) and Brandon Wood (3876, 3976); recorded from all surrounding VCs.

515 Gnaphalium L.

515.1 **Gnaphalium sylvaticum** L. Heath Cudweed
Stokes, J., 1787. Withering's Arrangement. "Banks of the canal in the parish of Coseley (Keresley), Warwickshire." Map 777, p. 612
An occasional plant of heathy woods, dry grassland and waste places on acid soil; recorded from all surrounding VCs.

515.4 **Gnaphalium uliginosum** L. Marsh Cudweed
Perry, 1820. Herb. Perry. Warwick. Map 514, p. 489
Abundant and fairly evenly distributed throughout the N. of the county but almost entirely absent from the calcareous soils of the S. and S.E.; recorded from cultivated land (40%), watersides and marshes (25%), roadsides and waste places (22%), grassland (8%) and a few other habitats; recorded from all surrounding VCs.

516 Anaphalis DC.

516.1 **Anaphalis margaritacea** (L.) Benth. Pearly Everlasting
Bagnall, 1879. Herb. Bagnall [*Gnaphalium margaritaceum* L.]. Aston.
A rare casual recorded from Rugby Cement Works (4875) and woodland at Coughton Park (0660); recorded from VCs 33, 37, 39 and 57.

518 Solidago L.

518.1 **Solidago virgaurea** L. Golden-rod
Perry, 1817. List. Between Wootton Fields and Stoneleigh. Map 515, p. 490
A rare plant entirely confined to the Earlswood–Solihull–Wroxall area; recorded from hedgerows, roadsides, railway banks and a wall; recorded from all surrounding VCs.

518.2 **Solidago canadensis** L. Garden Golden-rod
Burges, Dr. R. C. L., 1943. Herb. Burges. Waste ground, Cannon Hill, Birmingham. Map 516, p. 490
An occasional plant of a number of widely scattered localities mainly in the N. of the county, abundant in the Birmingham area; generally an escape from cultivation; recorded from

waste places, railway banks and roadsides (60%), grassland (28%) and hedgerows; recorded from VCs 23, 32, 37 and 55.

519 Aster L.

519.6 **Aster novi-belgii** L. Michaelmas-daisy
Thomas, C., 1950. Edgbaston. Map 517, p. 491
An occasional but increasingly common escape from cultivation found in the neighbourhood of towns; recorded from waste places and roadsides (79%) and a few other habitats; recorded from VCs 23, 33 and 37.

521 Erigeron L.

521.1 **Erigeron acer** L. Blue Fleabane
Bree, Rev. W. T., 1817. Mid. Flora [*E. acre*]. Allesley and Meriden. Map 518, p. 491
A rare plant of a few widely scattered areas in the county, with a considerable concentration in the Alcester–Wilmcote–Stratford area, and a smaller one around Rugby; recorded from railways, quarries, rough grassland and waste places; recorded from all surrounding VCs.

522 Conyza Less.

522.1 **Conyza canadensis** (L.) Cronq.
Canadian Fleabane
Andrews, C. E. A., 1946. Proc. B'ham N. H. Soc. 1947. Bombed centre of Coventry.
Map 519, p. 492
An occasional plant, mainly of waste places, railways and roadsides, recorded only from a wide band, mainly of light soils, running diagonally (N.E.–S.W.) across the middle of the county; recorded from all surrounding VCs.

524 Bellis L.

524.1 **Bellis perennis** L. Daisy
Perry, 1813. Herb. Perry. Warwick.
Map 520, p. 492
Abundant and evenly distributed throughout the county; recorded from grassland (59%), roadsides and waste places (24%) and cultivated land—gardens and parks—(11%); recorded from all surrounding VCs.

525 Eupatorium L.

525.1 **Eupatorium cannabinum** L. Hemp Agrimony
Perry, 1817. List. Lane between Pigwell's and Canal, Warwick; between Wedgenock Park and Fern Hill. Map 521, p. 493
An occasional plant of a number of widely scattered localities, with some concentration in the N. and W.; recorded from riversides and ditches (68%), hedgerows (14%) and a few other habitats; recorded from all surrounding VCs.

526 Anthemis L.

526.1 **Anthemis tinctoria** L. Yellow Chamomile
Goodman, Miss C. M., 1959. Near Kenilworth.
A casual recorded three times: from waste ground (2971), from hedgerow (2276) and from a waste place (3165); recorded from all surrounding VCs except 37.

526.2 **Anthemis cotula** L. Stinking Chamomile
Ick, W., 1836. Herb. Ick. Bank at Nechells Green. Map 522, p. 493
A frequent and widely but somewhat unevenly distributed plant of arable land (49%), roadsides, farmyards and waste places (40%) and grassland (6%); recorded from all surrounding VCs.

526.3 **Anthemis arvensis** L. Corn Chamomile
Purton, T., 1817. Mid. Flora. On the Ridgeway, on new-made earth mounds.
Map 523, p. 494
A rare plant of arable land (70%) and waste places (26%) confined to a few widely separated areas in the county; recorded from all surrounding VCs.

527 Chamaemelum Mill.

527.1 **Chamaemelum nobile** (L.) All.
Common Chamomile
Bree, Rev. W. T., 1835. New Bot. Guide [*Anthemis nobilis* L.]. Shrewley Pool (*fide* Bagnall).
Not recorded in present survey; last record believed to be by H. S. Thompson at Yarningale Common in 1901; recorded from all surrounding VCs.

528 Achillea L.

528.1 **Achillea millefolium** L. Yarrow, Milfoil
Perry, 1812. Herb. Perry. Warwick.
Map 524, p. 494
A very common species, mainly of roadsides and waste places (49%) and grassland (38%), occasionally also found in hedgerows and cultivated land; evenly distributed throughout the county; recorded from all surrounding VCs.

528.3 **Achillea ptarmica** L. Sneezewort
Perry, 1812. Herb. Perry. Warwick.
Map 525, p. 495
An occasional plant of damp humus-rich soils found in grassland (40%), marshes and watersides (38%) and roadsides and railway banks (15%), chiefly in the N.W. of the county; recorded from all surrounding VCs.

531 Tripleurospermum Schultz Bip.

531.1b **Tripleurospermum maritimum** (L.) Koch ssp. **inodorum** (L.) Hyland. ex Vaarama
Scentless Mayweed
Perry, 1820. Herb. Perry [*Pyrethrum inodorum*]. Warwick. Map 526, p. 495
Abundant and evenly distributed throughout the county; recorded from roadsides, waste places and farmyards (48%), cultivated land (38%), grassland (11%) and a few other habitats; recorded from all surrounding VCs.

532 Matricaria L.

532.1 **Matricaria recutita** L. Scented Mayweed
Perry, 1812. Herb. Perry [*M. chamomilla*]. Without locality. Map 527, p. 496
Frequent, though irregularly distributed throughout most of the county, and very rare in the extreme S.; recorded from cultivated land (47%), roadsides, waste places and farmyards (44%) and grassland (8%); recorded from all surrounding VCs.

532.2 **Matricaria matricarioides** (Less.) Porter
Pineapple Weed, Rayless Mayweed
Druce, 1932. Com. Flora [*M. suaveolens* (Pursh) Buch.]. Without locality.
Map 528, p. 496
Abundant and evenly distributed throughout the county; recorded from roadsides, waste places and farmyards (71%), cultivated land (20%), grassland (6%) and a few other habitats; recorded from all surrounding VCs.

533 Chrysanthemum L.

533.1 **Chrysanthemum segetum** L. Corn Marigold
Baxter, W., 1831. MS (Bagnall: Mid. Nat. 1892–93.) Between Rugby and Newbold on Avon. Map 529, p. 497
An occasional plant of a number of scattered localities chiefly on sandstone and gravel; recorded from arable land (56%), roadsides and waste places (26%) and grassland (16%); recorded from all surrounding VCs.

533A Leucanthemum Mill.

533.2 **Leucanthemum vulgare** Lam.
Ox-eye Daisy, Moon Daisy
Baxter, W., 1831. MS (Bagnall: Mid. Nat. 1892–93.) Near Brownsover.
Lloyd, Dr. G., 1835. Herb. Perry. Near Warwick. Map 530, p. 497
Abundant and evenly distributed throughout the county; recorded from grassland (56%), roadsides, railway banks and waste places (33%), arable land (6%) and a few other habitats; recorded from all surrounding VCs.

533.3 **Leucanthemum maximum** (Ramond) DC.
Shasta Daisy
Readett, R. C., 1954. Railway bank, Solihull.
An occasional escape from cultivation, often becoming well established on railway banks, in waste places and in quarries; recorded from 0898, 1479, 3490, 3780, 4181, 4784, 4875 and 5175; recorded from VC 55.

533B Tanacetum L.

533.4 **Tanacetum parthenium** (L.) Schultz Bip.
Feverfew
Perry, 1826. Herb. Perry [*Pyrethrum parthenium*]. Roadsides between Milverton and Emscote.
Map 531, p. 498
A frequent plant of irregular distribution often escaping from cultivation and becoming naturalised; recorded from roadsides and waste places (66%), hedgerows (8%), walls (8%), arable land (7%), grassland (5%), watersides (5%) and a few other habitats; recorded from all surrounding VCs.

533.5 **Tanacetum vulgare** L. Tansy
Perry, 1817. List. Churchyard and College walls, Warwick; Hatton Hill. Map 532, p. 498
A fairly frequent plant of irregular distribution throughout most of the county but rare on the calcareous soils in the S.E.; recorded from roadsides, waste places and railway banks (66%), rough grassland (12%), hedgerows (10%) and river-banks (7%); recorded from all surrounding VCs.

535 Artemisia L.

535.1 **Artemisia vulgaris** L. Mugwort
Baxter, W., 1831. MS (Bagnall: Mid. Nat. 1892–93.) Clifton Road near Rugby.
Map 533, p. 499
Abundant and fairly evenly distributed throughout the county except on the Lias clay in the S.E. where it is much less frequent; recorded from roadsides and waste places (79%), hedgerows (9%) and grassland (6%); recorded from all surrounding VCs.

535.6 **Artemisia absinthium** L. Common Wormwood
Bree, Rev. W. T., 1873. Topographical Botany. Without locality. Map 534, p. 499
Recorded as a very rare casual in old gardens by Bagnall, this plant is now widespread and frequent in the Birmingham and Coventry areas; recorded from waste places, roadsides and by railway tracks (89%), grassland (6%) and hedgerows (5%); recorded from all surrounding VCs.

536 Echinops L.

536.1 **Echinops sphaerocephalus** L. Globe Thistle
Arnold, M. A. & G. A., 1960. Wilnecote.
A rare escape recorded from Wilnecote (2201), Pooley Fields (2503) and railway embankment between Whitnash and Fosse Way (3361) where it is well established and spreading; recorded from VCs 23 and 39.

537 Carlina L.

537.1 **Carlina vulgaris** L. Carline Thistle
Perry, 1817. List. Between Wootton and Ashow. Map 535, p. 500
An occasional plant almost entirely confined to the Lower Lias escarpment; recorded from grassland, railway embankments and scrub; recorded from all surrounding VCs.

538 Arctium L.

These notes and the key which follows have been prepared by Dr. M. C. Lewis.

(a) Inflorescence characters refer to the arrangement of the terminal 5–7 heads on a main lateral branch of the plant.
(b) Petiole characters are most satisfactorily observed near the base of the petiole of a mature *rosette* leaf.
(c) Length of acumen of the involucral bracts is measured from the point of inflexion (at the junction of the appressed, imbricating base with the spreading limb) to the hooked apex of a bract from the median part of the involucre. This measurement is considered more useful than overall head diameter which may vary considerably between flowering and fruiting.
(d) The degree of floret protrusion may increase with fruit development and is not a reliable character for pressed material; it is preferably observed on fresh, fully flowering material.

Key

1. Heads long-stalked, widely open in fruit, in corymbose-subcorymbose clusters; acumens (8–)9–12 mm long; florets not pro-

truding beyond the involucral bracts; wider upper part of corolla distinctly shorter than the slender lower part; achenes pale; petioles solid; plant with ascending branches . . . . . **A. lappa** L.

Heads in ± racemose arrangement; wider upper part of corolla about equalling the slender lower part; achenes dark; petioles hollow. (**A. minus** Bernh.) . . . 2

2. Heads distinctly stalked, contracted at top in fruit; inflorescence racemose; acumens 4–6(–7) mm long; florets protruding beyond the involucral bracts; plant with ascending branches . . . ssp. **minus**

Heads short-stalked to sessile, open at the top in fruit; inflorescence racemose with a terminal cluster of (2–)3(–4) ± sessile heads; acumens 7–10(–12) mm long; florets not protruding beyond the involucral bracts; plants with long, arcuate lateral branches . . ssp. **nemorosum** (Lej.) Syme

Heads long-stalked (up to 15 cm), widely open in fruit, in open racemes; involucre persistently tomentose; acumens 8–10 mm long; florets not protruding beyond the involucral bracts; plants with ascending branches . . ssp. **pubens** (Bab.) J. Arènes

538.1 **Arctium lappa** L. Greater Burdock
Bagnall, 1878. Herb. Bagnall [*A. major* Bernh.]. Billesley Map 536, p. 500

A frequent plant of roadsides and waste places (34%), hedgerows and scrub (31%), watersides (15%), grassland (10%) and woods (8%), mainly in the S. of the county; recorded from all surrounding VCs.

Arctium minus Bernh. sensu lato
Lesser Burdock
Palmer, Miss C. E., 1852. Herb. Oxf. Univ. Lighthorne. Map 537, p. 501

Abundant and fairly evenly distributed throughout the county; recorded from roadsides and waste places (39%), hedgerows and scrub (29%), grassland (12%), woods (9%) and watersides (7%); recorded from all surrounding VCs.

According to Dr. M. C. Lewis, who is making a special study of the Warwickshire Arctiums, this species, which has been divided into 3 subspecies (i.e. *minus*, *nemorosum* and *pubens*) exhibits a very complex pattern of variation in Warwickshire. As suggested by Perring (Critical Suppl. Atlas British Flora. B.S.B.I., 1968), this situation is probably due to occasional hybridisation between the subspecies (and possibly also *A. lappa*) followed by numerous generations of inbreeding to produce a range of forms which are perpetuated as ± pure lines but which together exhibit a large spectrum of parental character combination. The degree of hybridisation appears to be much greater in the Midlands than elsewhere in the British Isles, and the commonest forms in this area are intermediate between "good" *minus* and "good" *nemorosum* (Perring, 1968) although *minus* characters (genes) seem to predominate. The hybridisation may have been brought about by forest clearance from Saxon times onwards, which would have broken down the ecological barriers between the woodland habitats of which *nemorosum* is characteristic and the ruderal habitats of *minus*.

Although "good" *minus* is found reasonably frequently and "good" *pubens* occasionally in Warwickshire, specimens typical of *nemorosum* have not been encountered in the present survey, although the commonest forms in the area undoubtedly incorporate *nemorosum* characters.

In addition to good ssp. *minus* (q.v.), ssp. *nemorosum* (q.v.) and ssp. *pubens* (q.v.) three intermediate forms have been distinguished by him:

Form 1: An intermediate form which is frequently encountered, especially in the more shaded habitats of woodland and hedgerows, has the typically small heads and protruding corollas of *minus* but the typically arcuate branching habit and subsessile–sessile terminally clustered heads of *nemorosum*.

Specimens have been collected from woodlands (0956, 1156, 1164, 1662, 2265, 2464), hedgerows (1266, 1456, 1471, 3472), ditchside (1465) and roadside (2160).

Form 2: A form with a typical *minus* inflorescence but with unusually large heads for this subspecies (suggesting *nemorosum* genes) has been collected at Temple Grafton (1354) in scrub woodland.

Form 3: A form with a very densely clustered inflorescence of many small heads has been recognised. This is possibly referable to *A. minus* ssp. *minus* var. *pycnanthum* J. Arènes.

Specimens have been collected from Watchbury Hill (2860) in rough grassland, and from Birdingbury Fields Farm (4366).

538.2 Ssp. **nemorosum** (Lej.) Syme
Bagnall, 1884. Herb. Bagnall [*A. intermedium*]. Near Hatton Asylum.

Not recorded in present survey; in addition to the specimen of good ssp. *nemorosum* cited above, four specimens in the Bagnall Herbarium closely approach ssp. *nemorosum*; the involucral bracts have long acumens and the plants exhibit the characteristic clustering of heads, but the peduncles, particularly the lower ones, are longer than in typical ssp. *nemorosum*; the specimens were collected between 1869 and 1889 from Alderminster, Marston Green, Maxstoke and Princethorpe; recorded from VCs 39 and 57.

538.3 Ssp. **pubens** (Bab.) J. Arènes
Kirk, T., 1856. Herb. Brit. Mus. Coventry Wood, Arbury Hall.

Two records, from roadsides near Ilmington (2145) and Warwick (2965); recorded from VC 57.

538.4 Ssp. **minus**
Bromwich, H., 1866. Herb. Brit. Mus. [*A. minus* Bernh. var. *intermedium* Lange]. Hedges at Woodloes, near Warwick.

ssp. *minus*, determined by Dr. M. C. Lewis, has been recorded from hedges and scrub (1456, 1553, 1654, 1778, 1959, 2261, 2591, 2963, 2968, 4366), roadsides and waste places (1164, 1273, 2404, 2604, 2661), rough grassland (1557, 2860), waterside (1059, 2966) and woodland (2261); recorded from all surrounding VCs.

539 Carduus L.

539.1 **Carduus tenuiflorus** Curt. Slender Thistle
Roberts, H. A., 1961. Near Wellesbourne Hastings.
Several plants recorded for the first time in the county from rough grassland near Wellesbourne Hastings (2855); recorded from all surrounding VCs except 55.

539.3 **Carduus nutans** L. Musk Thistle
Perry, 1817. List. Between Warwick and Stratford. Map 538, p. 501
An occasional plant of a few widely scattered localities; recorded from roadsides, waste places and railway banks (45%), grassland (35%) and quarries (10%); recorded from all surrounding VCs.

539.4 **Carduus acanthoides** L. Welted Thistle
Perry, 1828. MS (Bagnall's Flora, 1891). Near Hampton Lucy. Map 539, p. 502
A frequent plant in the S. and E. of the county but uncommon elsewhere; recorded from hedgerows (33%), roadsides and waste places (21%), grassland (19%) and watersides (19%); recorded from all surrounding VCs.

Carduus acanthoides × nutans
= **C. × orthocephalus** Wallr.
Bagnall, 1870. Herb. Bagnall [*C. nutans × crispus*]. Near Shustoke.
Not recorded in present survey; recorded from all surrounding VCs except 37 and 39.

540 Cirsium Mill.

540.1 **Cirsium eriophorum** (L.) Scop. Woolly Thistle
Perry, 1817. List [*Cnicus eriophorus*]. Road from Stratford to Warwick, at the turn to Snitterfield; between Hatton and Stank Hill. Map 540, p. 502
An occasional plant chiefly of the Lower Lias limestone in the S. of the county; recorded from grassland (49%), roadsides and railway banks (26%), hedgerows and scrub (12%) and quarries (10%); recorded from all surrounding VCs.

540.2 **Cirsium vulgare** (Savi) Ten. Spear Thistle
Bree, Rev. W. T., 1828. Mag. Nat. Hist. [*Cnicus lanceolatus*]. Allesley. Map 541, p. 503
Abundant and widespread throughout the county; recorded from grassland (42%), roadsides, waste places and railway banks (32%), hedgerows and scrub (14%), arable land (6%) and watersides (4%); recorded from all surrounding VCs.

540.3 **Cirsium palustre** (L.) Scop. Marsh Thistle
Bree, Rev. W. T., 1828. Mag. Nat. Hist. [*Cnicus palustris*]. Allesley. Map 542, p. 503
Abundantly and widely distributed throughout the county except in the lower Avon Valley, where it is rare; recorded from watersides and marshes (40%), damp grassland (34%), roadsides and railway banks (10%), hedgerows and scrub (8%) and mixed woodland (5%); recorded from all surrounding VCs.

540.4 **Cirsium arvense** (L.) Scop. Creeping Thistle
Bree, Rev. W. T., 1828. Mag. Nat. Hist. [*Cnicus arvensis*]. Allesley. Map 543, p. 504
Abundant and evenly distributed throughout the county; recorded from grassland (39%), roadsides, waste places and railway banks (33%), arable land (17%), and hedgerows (7%); recorded from all surrounding VCs.

540.6 **Cirsium acaule** Scop. Stemless Thistle
Perry, 1817. List [*Cnicus acaulis*]. Long Compton Hill. Map 544, p. 504
An occasional plant almost entirely confined to the calcareous soils in the S. of the county; recorded from grassland (83%), railway banks (8%) and quarries (5%); recorded from all surrounding VCs.

540.8 **Cirsium dissectum** (L.) Hill
Meadow Thistle, Marsh Plume Thistle
Countess of Aylesford, 1805. Bot. Guide [*Carduus pratensis*]. Packington. Map 545, p. 505
A rare plant recorded from a few heathy and marshy areas in the N.W. of the county; recorded from all surrounding VCs.
The hybrid *C. dissectum* × *C. palustre* [*C.* × *forsteri* (Sm.) Loud.] was collected in 1937 by Dr. R. C. L. Burges from a moist pasture above Little Bracebridge Pool, Sutton Park.

541 Silybum Adans.

541.1 **Silybum marianum** (L.) Gaertn. Milk Thistle
Perry, 1817. List [*Carduus marianus*]. Racecourse, Warwick.
One record only from waste ground by farm near Kineton (3449); collected by Lloyd in 1835 and by Cheshire in 1854 from the Scar Bank, Hampton Lucy; recorded from all surrounding VCs.

542 Onopordum L.

542.1 **Onopordum acanthium** L.
Scotch Thistle, Cotton Thistle
Purton, T., 1817. Mid. Flora. Bidford. Map 546, p. 505
A rare plant of a few scattered localities mainly in the Avon Valley; recorded from waste places, hedgerows, roadsides, grassland, cultivated land and watersides; recorded from all surrounding VCs.

544 Centaurea L.

544.1 **Centaurea scabiosa** L. Greater Knapweed
Pena & L'Obel, 1570. Stirpium Adversaria [*Jacea maior*]. "In Angliae segetibus Coventriae conterminus abunde provenit." Map 547, p. 506
An occasional plant in the S. of the county mainly on calcareous soils; recorded from roadsides and railway banks (59%), grassland (17%), hedgerows (14%) and arable land (7%); recorded from all surrounding VCs.

544.3 **Centaurea cyanus** L. Cornflower, Bluebottle
Bagnall, 1868. Herb. Bagnall. Witton. Map 548, p. 506
A rare weed of cornfields, grassland and waste places; recorded from all surrounding VCs.

544.5 **Centaurea jacea** L. Brown-rayed Knapweed
Bromwich, H., 1898. Herb. Brit. Mus. Milverton, amongst lucerne.
Not recorded in present survey; recorded from VCs 23, 33 and 39.

544.6 & 7
Centaurea nigra L.
Lesser Knapweed, Hardheads
Baxter, W., 1831. MS (Bagnall: Mid. Nat. 1892–93.) Near Brownsover Planks, Rugby.
Map 549, p. 507
C. nigra in Warwickshire is a highly variable species which shows no correlation between the characters customarily used to distinguish ssp. *nigra* from ssp. *nemoralis*. Although certain specimens can be classified as one or other of the subspecies the majority cannot be so placed, and the range of variation for all characters extends up to and beyond the type specimens of each subspecies. No clustering of specimens is evident around the Linnean and Jordanean types, and no obvious soil or pH preferences can be correlated with morphological characters.
The species is widely and evenly distributed throughout the county and has been recorded from roadsides, railway banks and waste places (46%), grassland (41%) and hedgerows (9%); it is recorded from all surrounding VCs.
Ssp. *nemoralis* (Jord.) Gugl. (first record Bromwich, H., 1847. Herb. Perry [*C. nigra* flor. alba]. Myton) and ssp. *nigra* (first record Bagnall, 1866. Herb. Bagnall [*C. nigra* L. a. *vulgaris*]. Acocks Green) have both been recorded frequently; both are also recorded from all surrounding VCs.

Centaurea nigra ssp. **nemoralis** × **jacea**
Turrill, W. B., 1950. British Knapweeds. New End.
One record in grassland (0560); not recorded from any of the surrounding VCs.

Centaurea nigra ssp. **nigra** × **jacea**
Cadbury, Miss D. A., 1951. Edgbaston.
One record from rough grassland at side of canal (0685); not recorded from any of the surrounding VCs.

544.8 **Centaurea aspera** L. Rough Star Thistle
Druce, 1932. Com. Flora. Without locality.
A rare casual recorded once from disturbed ground at Warwick (2964); not recorded from any of the surrounding VCs.

544.10 **Centaurea solstitialis** L. St. Barnaby's Thistle
Beilby, Miss M. A., 1837. Analyst. Edgbaston.
One record only (four plants) in 1959 along path in cultivated field at Dosthill (2100); recorded from all surrounding VCs except 39 and 57.

Centaurea diluta Ait.
Evans, R., 1961. Wellesbourne Wood.
A casual recorded once from scrub woodland (2653); recorded from VC 37.

545 Serratula L.

545.1 **Serratula tinctoria** L. Saw-wort
Perry, 181–. Herb. Perry. Hatton Wood.
Map 550, p. 507
Occasional and very irregularly distributed throughout the county, with some concentration in a belt from Earlswood to Nuneaton; recorded from damp grassland (34%), hedgerows (34%), watersides and marshes (22%) and a few other habitats; recorded from all surrounding VCs.

546 Cichorium L.

546.1 **Cichorium intybus** L. Chicory, Wild Succory
Withering, 1796. Arrangement Tamworth Castle. Map 551, p. 508
An occasional plant thinly distributed on the calcareous soils of the S. and S.E. but found only in a few scattered localities in the N. half of the county; recorded from cultivated land (43%), grassland (32%), roadsides (17%) and a few other habitats; recorded from all surrounding VCs.

547 Lapsana L.

547.1 **Lapsana communis** L. Nipplewort
Baxter, W., 1831. MS (Bagnall: Mid. Nat. 1892–93.) Near Dunchurch.
Lloyd, Dr. G., 1835. Herb. Perry. Near Leamington. Map 552, p. 508
Abundant and evenly distributed throughout the county; recorded from roadsides, waste places and railway banks (36%), hedgerows (33%), cultivated land (14%), grassland (7%), watersides (5%) and woods (4%); recorded from all surrounding VCs.

549 Hypochoeris L.

549.1 **Hypochoeris radicata** L. Common Catsear
Perry, 1845. Herb. Perry. Canal bank, Saltisford, Warwick. Map 553, p. 509
Abundant and evenly distributed throughout the county; recorded from grassland (48%), roadsides, waste places and railway banks (37%), hedgerows and scrub (7%), cultivated land (4%) and a few other habitats; recorded from all surrounding VCs.

549.2 **Hypochoeris glabra** L. Smooth Catsear
Ray, 1670. Catalogus [*Hieracium parvum in arenosis nascens feminum pappis densius radiatis*]. "This was found on the gravelly heath-grounds near Middleton." According to Druce (Com. Flora, 1932) this is the first British record.
Not recorded in present survey; recorded from VCs 23, 32, 37 and 55.

550 Leontodon L.

550.1 **Leontodon autumnalis** L. Autumnal Hawkbit
Baxter, W., 1831. MS (Bagnall: Mid. Nat. 1892–93.) Near Brownsover.
Lloyd, Dr. G., 1835. Herb. Perry [*Apargia autumnalis*]. Near Leamington.
Map 554, p. 509
Abundant and fairly evenly distributed throughout the county; recorded from grassland (55%), roadsides, waste places and railway banks (36%), cultivated land (3%), hedgerows and scrub (3%) and a few other habitats; recorded from all surrounding VCs.

550.2 **Leontodon hispidus** L. Rough Hawkbit
Perry, 1817. List [*Apargia hispida* Willd.]. Swan Meadow, Warwick. Map 555, p. 510
Abundant on calcareous soils in the S. of the county but becoming less frequent towards the N.; recorded from grassland (58%), roadsides and railway banks (39%) and a few other habitats; recorded from all surrounding VCs.

550.3 **Leontodon taraxacoides** (Vill.) Mérat
Lesser Hawkbit
Baxter, W., 1831. MS (Bagnall: Mid. Nat. 1892–93) [*L. hirtus* L.]. Between Sawbridge and Hill Morton. Map 556, p. 510
A frequent plant in the S. of the county but only occasional in the N.; recorded from grassland (76%), roadsides, waste places and railway banks (19%) and a few other habitats; recorded from all surrounding VCs.

551 Picris L.

551.1 **Picris echioides** L. Bristly Oxtongue
Baxter, W., 1831. MS (Bagnall: Mid. Nat. 1892–93.) Near Lawford Mill.
Lloyd, Dr. G., 1835. Herb. Perry [*Helminthia echioides* (L.) Gaertn.]. Bidford.
Map 557, p. 511
An occasional plant in the S. of the county, with a marked concentration on the Lower Lias limestone and clay in the S.W.; recorded from roadsides, waste places and railway banks (31%), grassland (29%), cultivated land (25%) and hedgerow and scrub (14%); recorded from all surrounding VCs except 39.

551.2 **Picris hieracioides** L. Hawkweed Oxtongue
Lloyd, Dr. G., 1835. Herb. Perry. Near Stratford on Avon. Map 558, p. 511
An occasional plant in the S. of the county, with some concentration on the calcareous soils of the S.W.; recorded from railway banks, roadsides and waste places (58%), grassland (26%) and hedgerows; recorded from all surrounding VCs.

552 Tragopogon L.

552.1 **Tragopogon pratensis** L. Goatsbeard
For first record see ssp. *minor*.
Map 559, p. 512
Abundant and fairly evenly distributed throughout the county; recorded from roadsides, railway banks and waste places (63%), grassland (33%) and a few other habitats; recorded from all surrounding VCs.

552.1a Ssp. **pratensis**
Bromwich, H., 1885. Herb. Bagnall [*T. pratense*]. Leek Wootton.
This subspecies has been recorded eight times in the present survey, but the only material we have seen is from Sutton Park (0997) and Henley in Arden (1465, 1466); recorded from VCs 32, 33, 37 and 57.

552.1b Ssp. **minor** (Mill.) Wahlenb.
Perry, 1824. Herb. Perry [*T. pratensis*]. Hill, Lammas Fields, Warwick.
This ssp. has been recorded many times and the records are included in the map of the species; all the material collected as *T. pratensis* which we have seen (except the three specimens detailed under ssp. *pratensis*) has proved to be ssp. *minor*, and we believe that the bulk of the records mapped as *T. pratensis* are of this ssp. and that the map may be taken as representing the distribution of ssp. *minor*. *T. pratensis* ssp. *minor* has been recorded from all surrounding VCs except 32.

Ssp. **orientalis** (L.) Vollmann
Bromwich, H., 1892. Herb. B'ham Univ. Railway bank at Burton Green, near Berkswell.
Not recorded in present survey but collected by Bromwich from Burton Green in 1893 and 1894; recorded from VC 23.

552.2 **Tragopogon porrifolius** L. Salsify
Purton, T., 1817. Mid. Flora. Gorcot Hall.
One record only from a garden at Fitzjohns, near Rugby (5173); recorded from all surrounding VCs.

553 Scorzonera L.

553.1 **Scorzonera humilis** L. Viper's-grass
Phipps, J. B., 1954. Near Portway.
A plant of pasture land, possibly introduced with seed, recorded from only one station near Portway (0871) [see Proc. B.S.B.I. Nov. 1954], locally frequent when first recorded in 1954 but gradually becoming scarcer, and not seen since 1965; not recorded from any of the surrounding VCs.

554 Lactuca L.

554.1 **Lactuca serriola** L. Prickly Lettuce
Goodman, Miss C. M., 1959. Knowle Hill, Kenilworth. Map 560, p. 512
A rare plant of a number of localities in the centre of the county; recorded mainly from waste places; recorded from all surrounding VCs except 39 and 57.

554.2 **Lactuca virosa** L. Great Lettuce
Bree, Rev. W. T., 1817. Mid. Flora. Roadsides, Stonebridge.
A very rare plant of rough grassland and waste places recorded from Stratford on Avon (1955), Maxstoke Priory Walls (2386) and Caludon House (3780); recorded from all surrounding VCs.

555 Mycelis Cass.

555.1 **Mycelis muralis** (L.) Dumort. Wall Lettuce
Perry, 1817. List [*Prenanthes muralis* L.]. Mellos' Lane and Castle Wall in Vineyard Lane, Warwick. Map 561, p. 513
A rare plant of a number of widely scattered localities mainly in the N.W. half of the county; recorded from hedgerows and scrub (46%), walls (22%), roadsides and railway banks (19%) and woods (13%); recorded from all surrounding VCs.

556 Sonchus L.

556.2 **Sonchus arvensis** L. Corn Sow-thistle
Perry, 1817. List. Whitnash Road to Warwick.
Map 562, p. 513
Abundant and evenly distributed throughout the county; recorded from roadsides, waste places, railway banks and farmyards (38%), cultivated land (36%), hedgerows and scrub (11%), grassland (10%) and watersides (5%); recorded from all surrounding VCs.

556.3 **Sonchus oleraceus** L. Smooth Sow-thistle
Baxter, W., 1831. MS (Bagnall: Mid. Nat. 1892–93.) On the Barby Road, near Rugby.
Lloyd, Dr. G., 1835. Herb. Perry. Near Stratford on Avon. Map 563, p. 514
Abundant and evenly distributed throughout the county; recorded from roadsides, waste places, railway banks and farmyards (49%), cultivated land (33%),

hedgerows (8%), grassland (6%) and a few other habitats; recorded from all surrounding VCs.

556.4 **Sonchus asper** (L.) Hill Prickly Sow-thistle
Rugby School Report, 1868. Near Rugby.
Map 564, p. 514

Abundant and evenly distributed throughout the county; recorded from roadsides, waste places, railway banks and farmyards (44%), cultivated land (31%), grassland (11%), hedgerows and scrub (7%) and watersides (6%); recorded from all surrounding VCs.

557 **Cicerbita** Wallr.

557.3 **Cicerbita macrophylla** (Willd.) Wallr.
Arnold, G. A. & M. A., 1958. Arbury Park.
Map 778, p. 612

A recent introduction becoming more widespread in waste places and rough grassland near Birmingham, Warwick, Leamington, Kenilworth and elsewhere; recorded from VCs 33 and 57.

558 **Hieracium** L.

These notes on Warwickshire Hawkweeds and the Key which follows have been prepared by C. E. A. Andrews.

In using the Key it must be borne in mind that this genus contains a number of exceedingly variable micro-species. The collector should use his best judgement in taking only what appear to be specimens of normal height and growth from a population.

Care should be taken to see that there has been no damage by grazing animals, by insect attack or by cutting, and this applies particularly to the tall leafy plants of the **Aphyllopoda** Section; these flower late in the season and may have had their early growth trimmed off in the process of mowing roadside verges or by the application of weedkillers. Secondary growth following such mutilation can be so atypical that exact identification becomes impossible.

The time of gathering is also important, and in general Hawkweeds should be collected as soon as they commence to flower. The rosette-bearing species (**Phyllopoda**) may no longer be typical after the end of June, depending on the season, or the **Aphyllopoda** after the end of August.

In naming plants from Warwickshire, special features to examine are the peduncles and phyllaries (involucral bracts) which may be glabrous, or clothed with three kinds of hair in varying proportion and not all necessarily present on the same plant. These are: (a) Simple hairs, of the usual type, which may be spreading or appressed; (b) Glandular hairs, which are like simple hairs but are tipped with black or greenish glands and are sometimes very short (microglands); (c) Stellate hairs, which are very short and branched and if sufficiently numerous may produce a felt-like tomentum or floccum.

Other important features for identification are:

(1) The colour of the styles (yellow, discoloured or black); this should be noted in fresh material as it may change in drying.
(2) The presence or absence of a basal rosette which may be of only 2 or 3 leaves; plants with rosettes have normally 0–8 stem leaves and flower from early May and are grouped as **Phyllopoda**; plants with basal leaves absent or withered at the time of flowering normally have 8–30 or more stem leaves, are usually tall plants flowering from July to September and are grouped as **Aphyllopoda.**
(3) Leaf colour (pale green, mid-green, deep green or glaucous; this also should be noted before drying).
(4) The presence or absence of brown or black spots or blotches on the leaves (these usually occur in *H. maculatum* and sometimes in *H. diaphanum* and must be distinguished from the less clearly defined reddish spots which are sometimes produced by insect attack or when plants are growing in very dry situations such as walls).
(5) The shape of the leaves: this is sometimes helpful, especially whether or not they are tapered towards the base; leaf shape can vary considerably in the same species, as can marginal toothing.

N.B. In gathering, plants should be *cut* off at ground level just below the basal leaves, and the rootstock left to regenerate. *Hieracia* have often been uprooted and as a result are much less abundant than formerly in many places, and one suspects they have suffered extinction in some cases by over-collection.

Key to Subgenera

Plant not stoloniferous; ligules always yellow . . . Subgenus **Hieracium**
Plant stoloniferous; ligules reddish-brown or yellow, often with reddish streak on the back . . . Subgenus **Pilosella**

Subgenus Hieracium

1. Rosette present with few to many leaves; stem leaves 0–8; flowering from May to September . . . 2
   No basal rosette, or rosette leaves withered at time of flowering; normally tall plants (up to 90 cm) usually with more than 8 stem leaves; not flowering before end of July . . . 6
2. Stem leaves 0–1(–2), petiolate; phyllaries (involucral bracts) densely glandular with no simple hairs . . . . **exotericum** agg.
   Stem leaves 2–8; phyllaries with glandular hairs, with or without simple hairs . . . 3
3. Phyllaries with simple hairs predominant and some shorter, often obscure, fine glandular hairs; leaves lanceolate . . . **vulgatum**
   Phyllaries with glandular hairs predominant and few or no simple hairs; leaves normally lanceolate to ovate 4
4. Leaves, at least the radical, spotted or blotched with brown; phyllaries with short, more or less equal glandular hairs . . . **maculatum**
   Leaves usually unspotted; phyllaries with long and short, often fine glandular hairs . . . 5
5. Phyllaries acute or obtuse with fine unequal glandular hairs and no simple hairs, efflocose or slightly floccose towards the base; leaves sometimes spotted with brown and sharply toothed, especially the upper ones . . . . **diaphanum**
   Phyllaries, at least the inner, acute with strong glandular hairs and normally a few simple

hairs, floccose throughout; leaves unspotted, denticulate or toothed . . . **strumosum**

6. Stem leaves 8–40 or more, more or less linear, nearly entire or with 2 or 3 teeth on either side; phyllaries more or less glabrous with recurved tips; inflorescence apically sub-umbellate . . . . . **umbellatum**
   Leaves numerous, lanceolate to ovate, toothed or the lower denticulate; inflorescence many headed, paniculate with numerous branches from the leaf axils 7
7. Plant with long white spreading simple hairs in all parts; leaves 20–40(–70), lower long lanceolate, upper ovate, sessile; phyllaries blackish-green, usually with many white spreading simple hairs and numerous fine short, more or less appressed glandular hairs **perpropinquum**
   Plant hairy below, more or less floccose only above; leaves 10–30, lanceolate, the upper broader to rhomboid and all narrowing towards the base, the lower gradually, the upper often abruptly, with a few unequal narrow teeth on both sides; phyllaries olive green with more or less ascending simple hairs and shorter glandular hairs, more or less floccose . . . . . . **eboracense**
   Plant usually sub-glabrous, at least above; phyllaries and peduncles glabrous or nearly so but often slightly floccose . . . 8
8. Leaves glaucous green with long coarse ascending teeth; phyllaries usually with a few obscure glands on the midrib . **salticola**
   Leaves dark green, denticulate or with spreading teeth; phyllaries glabrous or with slightly floccose margins . . . . **vagum**

Subgenus Pilosella

1. Ligules brownish-red; inflorescence sub-umbellate . . . . **brunneocroceum**
   Ligules yellow, normally with red streak on back; inflorescence never sub-umbellate . 2
2. Flower heads several on a stem with one or two more or less bract-like leaves . **flagellare**
   Flower heads solitary on a leafless scape . 3
3. Phyllaries with glandular and simple hairs **pilosella** var. **pilosella**
   Phyllaries with glandular hairs only **pilosella** var. **concinnatum**

558.1.98–103

Hieracium exotericum sensu lato
Bagnall, 1898. Herb. Bagnall [*H. murorum* L.]. Chelmsley Wood.

Recorded on wall and railway bank at Kenilworth (2971); recorded from all surrounding VCs except 32 and 55.

558.1.149

Hieracium vulgatum Fr. Common Hawkweed
Andrews, C. E. A., 1954. Earlswood Station. Map 779, p. 612

A rare plant of rough grassland and railway banks; recorded from all surrounding VCs except 23 and 32.

558.1.154

Hieracium maculatum Sm. Spotted Hawkweed
Bagnall, 1883. Herb. Bagnall. Hampton in Arden.

Two records from railway banks at City Road, Birmingham (0387) and Leek Wootton (2968); recorded from all surrounding VCs except 23.

558.1.158

Hieracium diaphanum Fr.
Ick, W., 1836. Herb. B'ham Mus. [*H. sabaudum*]. Green Lanes, Bordesley Green. Map 565, p. 515

An occasional plant, very irregularly distributed, mainly in the W. of the county; the S.E. border of the main British distribution appears to run through Warwickshire; recorded from railway banks and waste places (58%), rough grassland, walls and quarries and a few other habitats; recorded from all surrounding VCs.

558.1.163

Hieracium strumosum (W. R. Linton) A. Ley
Bagnall, 1867. Herb. Bagnall [*H. silvaticum*]. Earlswood. Map 566, p. 515

An occasional plant chiefly confined to the N.W. half of the county; recorded from railway banks, waste places and roadsides (50%), rough grassland (18%), hedgerows (13%), quarries and walls (9%) and woods (7%); recorded from all surrounding VCs.

558.1.203

Hieracium eboracense Pugsl.
Bagnall, 1886. Herb. Bagnall [*H. tridentatum* Fr.]. White House Warren, near Brailes.

Not recorded in present survey; recorded from VCs 37 and 57.

558.1.217

Hieracium umbellatum L.
Perry, 1817. List. St. Mary's Churchyard Wall and Castle Wall in Vineyard Lane, Warwick. Map 780, p. 612

A rare plant of woods, railways and waste places; recorded from all surrounding VCs.

558.1.219

Hieracium perpropinquum (Zahn) Druce
Bagnall, 1869. Herb. Bagnall [*H. sabaudum* var. *boreale* Fr.]. Westwood Coppice, Sutton Park. Map 567. p. 516

Frequent in the N. half of the county, mainly towards the W., but almost completely absent from the calcareous soils in the S.; recorded from hedgerows and scrub (39%), grassland (19%), roadsides, railway banks and waste places (19%) and woods (15%); recorded from all surrounding VCs.

558.1.222

Hieracium salticola (Sudre) Sell & West
Andrews, C. E. A., 1952. Herb. Andrews. Edgbaston Park. Map 568, p. 517

An occasional plant in the N. half of the county, particularly in the Birmingham–Sutton–Tamworth area; recorded from railway banks, waste places, rough grassland and roadsides; recorded from all surrounding VCs except 23, 33 and 57.

558.1.223 **Hieracium vagum** Jord.
Bagnall, 1873. Herb. Bagnall [*H. sabaudum* var. *boreale* Fr.]. Stone quarry between Atherstone and Ansley. Map 569, p. 517
A rare plant of a number of widely scattered localities in the N.W. half of the county, absent from the calcareous soils of the S.E.; recorded from railway banks, rough grassland, hedgerows, roadsides and a few other habitats; recorded from all surrounding VCs.

558.2.1 **Hieracium pilosella** L. Mouse-ear Hawkweed
Baxter, W., 1831. MS (Bagnall: Mid. Nat. 1892–93.) In a lane near Rugby. Map 570, p. 517
Abundant and fairly evenly distributed throughout the county; recorded from railway banks, roadsides and waste places (43%), dry grassland (39%), quarries and walls (7%), hedgerows (5%) and a few other habitats; recorded from all surrounding VCs.
Var. *concinnatum* F. J. Hanb. was recorded in 1967 by C. E. A. Andrews from Sutton Park (0998) and Edge Hill (3747).

558.2.5 **Hieracium flagellare** Willd.
Jones, Mrs. M. D. G., 1963. Ryton.
One record only from a waste place at Ryton (3774); recorded from VCs 23, 32 and 55.

558.2.8 **Hieracium brunneocroceum** Pugsl.
Andrews, C. E. A., 1955. Herb. Andrews. Compton Wynyates. Map 571, p. 518
An occasional plant naturalised in a number of areas in the N. half of the county; recorded from roadsides and railway banks, rough grassland and a few other habitats; recorded from all surrounding VCs except 32.

559 Crepis L.

559.2 **Crepis vesicaria** L. ssp. **taraxacifolia** (Thuill.) Thell. Beaked Hawksbeard
Bagnall, 1884. Herb. Bagnall [*C. taraxacifolia*]. Edge Hill. Map 572, p. 518
Frequent and irregularly distributed in the S.E. half of the county but only occasional and widely scattered in the N.W.; Warwickshire appears to be on the N.W. limit of its main distribution in Britain; recorded from roadsides, waste places and railway banks (64%), grassland (14%), arable land (13%), hedgerows and scrub (5%) and quarries (3%); recorded from all surrounding VCs.

559.3 **Crepis setosa** Haller f. Bristly Hawksbeard
Bromwich, H., 1866. Herb. Brit. Mus. Cornfield, Woodloes, Warwick.
One record, introduced in lawn at Rugby (4974); recorded from all surrounding VCs except 39 and 57.

559.5 **Crepis biennis** L. Rough Hawksbeard
Bromwich, H., 1872. Herb. Bagnall. Hill Wootton. Map 781, p. 613
A rare species confined to rough calcareous grassland in the S.E. of the county; recorded from all surrounding VCs except 39.

559.6 **Crepis capillaris** (L.) Wallr. Smooth Hawksbeard
Perry, 1829. Mag. Nat. Hist. [*C. tectorum*]. Near Warwick. Map 573, p. 519
Abundant and fairly evenly distributed throughout the county; recorded from roadsides, waste places and railway banks (49%), grassland (37%), arable land (5%) and hedgerows (5%); recorded from all surrounding VCs.

559.7 **Crepis nicaeensis** Balb. French Hawksbeard
Bromwich, H. (Bagnall: Mid. Nat. 1892–93). "Abundant in a marly cutting, Burton Green, near Berkswell."
Not recorded in present survey; recorded from VCs 23, 37 and 57.

559.8 **Crepis paludosa** (L.) Moench Marsh Hawksbeard
Bagnall, 1866. Flora, 1891. Sutton Park.
Recorded only from a wet place near the railway at Stratford on Avon (2053); recorded from VCs 37, 39, 55 and 57.

560 Taraxacum Weber

560.1 **Taraxacum officinale** Weber Common Dandelion
Ick, W., 1836. Herb. Ick [*Leontodon taraxacum*]. Aston. Map 574, p. 519
Abundant and evenly distributed throughout the county; recorded from roadsides, waste places and railway banks (48%), grassland (35%), cultivated land (9%), hedgerows (5%) and a few other habitats; recorded from all surrounding VCs.

Taraxacum alatum Lindb. f. (**T. officinale** agg.)
Bagnall, not later than 1913. Herb. Bagnall [*T. palustris*]. Sutton Park.
Not recorded in present survey.

Taraxacum duplidentifrons Dahlst. (**T. officinale** agg.)
Cadbury, Miss D. A., 1952. Sutton Park.
One record only from a marsh at Sutton Park (1096).

Taraxacum hamatum Raunk. (**T. officinale** agg.)
Bagnall, 1873. Herb. Bagnall [*T. officinale* var. *erythrospermum*]. Lane from Hampton to Meriden.
Recorded from hedgerow at Studley (0963), and from marshes at Monkspath Hill Farm (1475) and Bodymoor Heath (2096); also collected by Bagnall in 1875 from Marston Green.

Taraxacum polyodon Dahlst. (**T. officinale** agg.)
Evans, R., 1963. Fosse Way.
One record only from a roadside—Fosse Way (3562).

Taraxacum subhamatum Dahlst. (**T. officinale** agg.)
Evans, R., 1963. Fosse Way.
One record only from a roadside—Fosse Way (3562).

560.3 **Taraxacum spectabile** Dahlst. Red-veined Marsh Dandelion
Bromwich, H., 1866. Herb. Warw. Mus. [*T. officinale* Wigg. var. *palustre* Sm.]. Boggy meadow, Wroxall. Map 575, p. 520
An occasional plant of a number of scattered localities mainly in the N. of the county, with a considerable concentration in Sutton Park and the Earlswood–

Dorridge area; recorded from marshes, watersides, damp grassland, roadsides and bogs; recorded from all surrounding VCs except 55.

Taraxacum britannicum Dahlst.
(T. spectabile agg.)
Clark, M. C., 1963. Guy's Cliffe.
One record only from the riverside at Guy's Cliffe (2966).

Taraxacum euryphyllum Dahlst.
(T. spectabile agg.)
Hawkes, J. G., 1963. Merevale Park.
A rare plant of marsh and damp grassland recorded from squares 2997, 3076, 3572, 3771, 4171 and 4679.

Taraxacum faeroense Dahlst.
(T. spectabile agg.)
Bagnall, 1874. Herb. Bagnall [*T. palustre*]. Sutton Park. Map 576, p. 520
An occasional plant of damp grassland, marshes and bogs.

Taraxacum firmum Dahlst.
(T. spectabile agg.)
Jones, Mrs. M. D. G., 1963. N.W. of Pailton.
A rare plant of marshes and damp grassland recorded from squares 0995, 1196, 3777, 4682 and 5374.

Taraxacum nordstedtii Dahlst.
(T. spectabile agg.)
Bagnall, 1874. Herb. Bagnall [*T. palustre*]. Sutton Park.
An occasional plant of bogs, marshes and rough grassland recorded from squares 0998, 1272, 1664, 3771, 4171, 4276, 4470 and 4679; also collected by Bromwich in 1893 from pasture at Chesterton.

Taraxacum unguilobum Dahlst.
(T. spectabile agg.)
Green, P. S., 1951. Sutton Park.
Recorded from heath at Sutton Park (0998) and from grassland at Merevale Park (2997).

560.4 **Taraxacum laevigatum** (Willd.) DC.
Lesser Dandelion
For first record see *T. oxoniense* Dahlst.
Map 577, p. 521
An occasional plant of very scattered distribution mainly in the S. of the county, with a considerable concentration in the Stratford on Avon–Morton Bagot–Warwick area; recorded from dry pastures and rough grassland (46%), roadsides, railway banks and waste places (39%) and walls and quarries (12%); recorded from all surrounding VCs.

Taraxacum brachyglossum Dahlst.
(T. laevigatum agg.)
Bagnall, 1875, Herb. Bagnall [*T. erythrospermum*]. Sutton Park.
An occasional plant of grassland, dry banks and paths, recorded from squares 0650, 0895, 1058, 1456, 1664, 2199, 2366, 3449, 3951 and 4174; recorded from all surrounding VCs except 33 and 57.

Taraxacum excellens Dahlst.
(T. laevigatum agg.)
Bagnall, 1879. Herb. Bagnall [*T. udum*]. Sutton Park, near entrance.
Not recorded in present survey; not recorded from any of the surrounding VCs.

Taraxacum fulviforme Dahlst.
(T. laevigatum agg.)
Bagnall, 1879. Herb. Bagnall [*T. laevigatum*]. Sutton Park, near main entrance.
One record only from roadside by Wellesbourne Wood (2653); recorded from VC 23.

Taraxacum glauciniforme Dahlst.
(T. laevigatum agg.)
Bagnall, 1875. Herb. Bagnall [*T. erythrospermum*]. Sutton Park.
One record from an old quarry between Wilmcote and Aston Cantlow (1559); recorded from VC 23.

Taraxacum lacistophyllum Dahlst.
(T. laevigatum agg.)
Bagnall, 1875. Herb. Bagnall [*T. officinale* var. *erythrospermum*]. Marston Green and Sutton Park.
An occasional plant of pasture, rough grassland and roadsides, recorded from squares 1554, 1652, 1659, 1941, 1950, 2054, 2336, 2729, 2830, 3474, 3951, 3976, 4076, 4182 and 5374; recorded from VCs 23, 32, 37 and 57.

Taraxacum oxoniense Dahlst.
(T. laevigatum agg.)
Bagnall, 1872. Herb. Bagnall [*T. officinale* Wigg. b. *erythrospermum* Andrz.]. Hedgebank between Stonebridge and Hampton Station.
An occasional plant of rough grass, roadsides and waste places, recorded from squares 0558, 1164, 1553, 1763, 2042, 2653, 2830, 3052, 3096, 4677 and 4683; recorded from VCs 23, 37 and 57.

Taraxacum proximum Dahlst.
(T. laevigatum agg.)
Bromwich, H., 1893. Herb. Oxf. Univ. [*T. officinale* Weber d. *udum* Jord.]. Pasture at Chesterton.
Not recorded in present survey; not recorded from any of the surrounding VCs.

Monocotyledones

ALISMATACEAE

561 Baldellia Parl.

561.1 **Baldellia ranunculoides** (L.) Parl. Lesser Water-plantain
Perry, 1831. Herb. Perry (Bagnall's Flora, 1891) [*Alisma ranunculoides* L.]. Tamworth.
A very rare plant of pond margins at Earlswood (1174), Newbold on Stour (2245, 2345), and near Monks Kirby (4483, 4583); recorded from all surrounding VCs.

563 Alisma L.

563.1 **Alisma plantago-aquatica** L. Common Water-plantain
Baxter, W., 1831. MS (Bagnall: Mid. Nat. 1892–93.) Near Hill Morton. Map 578, p. 521
A frequent plant of damp ground (55%) and shallow water at the edges of canals, ponds and slow-flowing rivers (45%), evenly distributed throughout the county; recorded from all surrounding VCs.

563.2 **Alisma lanceolatum** With. Narrow-leaved Water-plantain
Young & Baker, 1873. Catalogue [*A. plantago* L. var. *lanceolata* With.]. Birdingbury; Myton. Map 782, p. 613
A rare plant of canals, lakes and rivers, which needs further investigation; recorded from all surrounding VCs.

565 Sagittaria L.

565.1 **Sagittaria sagittifolia** L. Arrowhead
Perry, 1812. Herb. Perry. Warwick. Map 579, p. 522
An occasional plant throughout most of the county, but very rare in the extreme N. and the extreme S.; recorded from rivers and canals; recorded from all surrounding VCs.

BUTOMACEAE

566 Butomus L.

566.1 **Butomus umbellatus** L. Flowering Rush
Stokes, Dr. J., 1787. Withering's Arrangement. Tamworth. Map 580, p. 522
An occasional plant mainly in the E. and S.W. of the county; recorded from canals, slow-flowing rivers and lakes (56%) and the sides of canals, rivers and lakes (41%); recorded from all surrounding VCs.

HYDROCHARITACEAE

570 Elodea Michx.

570.1 **Elodea canadensis** Michx. Canadian Pondweed
Kirk, T., 1851. Herb. Brit. Mus. [*Anacharis alsinastrum* Bab.]. Canal, Sowe Waste, near Coventry. Map 581, p. 523
An occasional plant, scattered irregularly throughout the county, recorded from ponds, canals, slow-flowing rivers and lakes; recorded from all surrounding VCs.

JUNCAGINACEAE

574 Triglochin L.

574.1 **Triglochin palustris** L. Marsh Arrow-grass
Withering, 1787. Arrangement. Tamworth. Map 582, p. 523
An occasional plant widely but very irregularly scattered throughout much of the county, absent from the extreme S.; recorded from watersides and marshes; recorded from all surrounding VCs.

POTAMOGETONACEAE

577 Potamogeton L.

577.1 **Potamogeton natans** L. Broad-leaved Pondweed
Baxter, W., 1831. MS (Bagnall: Mid. Nat. 1892–93.) Pond near Sawbridge.
Ick, W., 1836. Herb. Perry. Mill Pool near the River Rea, Fazeley Street, Birmingham. Map 583, p. 524
Abundant and fairly evenly distributed throughout most of the county, but rare in the Avon Valley and in the extreme S.; recorded chiefly from ponds and canals; recorded from all surrounding VCs.

577.2 **Potamogeton polygonifolius** Pourr. Bog Pondweed Map 783, p. 613
Withering, 1787. Arrangement [*P. natans* var. 2 *paludosum*]. Boggy ground on Birmingham Heath (Withering) and Sutton Park in places where stagnant water has been dried up or drained off (Stokes). Druce, in his Comital Flora, describes this as the first British record.
An occasional plant of bogs, acid pools and ditches; recorded from all surrounding VCs.

577.4 **Potamogeton nodosus** Poir. Loddon Pondweed
Cheshire, W., 1856. Herb. Perry [*P. lucens*]. River Stour at Alderminster.
Not recorded in present survey; no evidence of its having been seen before or after the Cheshire record; recorded from VC 23.

577.5 **Potamogeton lucens** L. Shining Pondweed
Stokes, Dr. J., 1787. Withering's Arrangement. River at Tamworth.
Purton, T., 1817. Mid. Flora. River Avon and ponds about Bidford. Map 584, p. 524
A rare plant of a number of scattered localities in the county; recorded chiefly fram canals and lakes; recorded from all surrounding VCs.

Potamogeton lucens × perfoliatus
= **P. × salicifolius** Wolfg.
Bromwich, H., c. 1870. Bagnall's Flora, 1891 [*P. decipiens* Nolte]. Canal, Warwick.
One record from canal at Coventry (3580); recorded from VCs 23, 32 and 55.

Potamogeton gramineus × lucens
= **P. × zizii** Koch ex Roth
Bagnall, 1893. Herb. Bagnall [*P. heterophyllus* Schreb.]. Earlswood Reservoir.
Still recorded from Earlswood Reservoir (1174), the

only Warwickshire station; recorded from VCs 39 and 55.

577.7 **Potamogeton alpinus** Balb.
Reddish Pondweed
Kirk, T., 1842. Phytologist [*P. rufescens* Schrad.]. Arbury Deer Park.
One record from a pool in old sand pits at Newton near Rugby (5378); formerly somewhat widespread; recorded from all surrounding VCs.

577.8 **Potamogeton praelongus** Wulf.
Long-stalked Pondweed
West, G. S., 1909. Herb. B'ham Univ. Sutton Park.
One record from canal at Brownsover Wharf, Rugby (5076); it was last recorded from Sutton Park in 1947; recorded from VCs 23, 32, 39 and 57.

577.9 **Potamogeton perfoliatus** L.
Perfoliate Pondweed
Kirk, T., 1850. Herb. Perry. Canal, Sowe Waste, Coventry. Map 585, p. 525
An occasional plant of a number of widely scattered localities mainly in the E. half of the county, and recorded chiefly from canals and rivers; recorded from all surrounding VCs.

577.11 **Potamogeton friesii** Rupr.
Flat-stalked Pondweed
Kirk, T., 1848. Herb. Perry [*P. compressus*]. Oxford Canal, Stoke Heath, Coventry.
Map 586, p. 525
A rare plant of a few widely scattered localities in the county, almost entirely confined to canals; recorded from all surrounding VCs.

577.13 **Potamogeton pusillus** L. Lesser Pondweed
Withering, 1787. Arrangement. About Tamworth. Map 587, p. 526
A rare plant of a small number of widely scattered localities, with some concentration in the Earlswood and Middleton–Kingsbury areas; recorded from canals and ponds; recorded from all surrounding VCs.

577.14 **Potamogeton obtusifolius** Mert. & Koch
Blunt-leaved Pondweed
Dashwood, I. M., 1844. Herb. Man. Univ. Sutton Park.
A rare plant of lakes and ponds in the N.W. of the county, recorded from Edgbaston Pool (0583), Windley Pool, Sutton Park (1195), S. of Coleshill (2086) and Packington Park (2284); recorded from all surrounding VCs except 23 and 33.

577.15 **Potamogeton berchtoldii** Fieb.
Small Pondweed
St. Brody, Dr., 1875. Herb. Warw. Mus. Canal, Warwick. Map 784, p. 613
An occasional plant of lakes and ponds in the N. half of the county; recorded from all surrounding VCs.

577.16 **Potamogeton trichoides** Cham. & Schlecht.
Hair-like Pondweed
St. Brody, Dr., 1875. Herb. Warw. Mus. Canal, Warwick.
Three records, one from Turner's Pool. Glascote (2102), one from a pond at Lighthorne (3356) and one from a canal at Rugby (5076); recorded from VCs 32, 33 and 37.

577.17 **Potamogeton compressus** L.
Grass-wrack Pondweed
Bloxam, Rev. A., 1837. New Bot. Guide [*P. zosteraefolius*]. "Abundant in the old Oxford Canal opposite Mr. Walker's, Newbold-upon-Avon."
An occasional plant of canals recorded from near Shirley (1078), Stratford on Avon (1955), Lowsonford (1967), near Tamworth (2102), near Alvecote (2404, 2504), near Farnborough (4451), Wormleighton (4454) and near Napton on the Hill (4560); recorded from all surrounding VCs.

577.18 **Potamogeton acutifolius** Link
Sharp-leaved Pondweed
Bloxam, Rev. A., before 1859. Herb. Brit. Mus. Canal near Rugby.
Not recorded in present survey; recorded from VC 32.

577.19 **Potamogeton crispus** L. Curled Pondweed
Perry, 181–. Herb. Perry. St. Nicholas Meadow, Warwick. Map 588, p. 526
A frequent plant very irregularly distributed throughout most of the county, but rare in the N.W. and in the extreme S.; recorded mainly from ponds, canals and rivers; recorded from all surrounding VCs.

Potamogeton crispus × lucens
= **P. × cadburyae** Dandy & Taylor
Cadbury, Miss D. A., 1948. Herb. Brit. Mus. Seeswood Pool, Nuneaton.
Not recorded in present survey; not recorded from any of the surrounding VCs.
This is the first record of this hybrid. J. E. Dandy writes (B.S.B.I. Proc. July, 1958). "A most surprising cross, but there is no doubt about the parentage, as examination of the material will quickly show. The plant looks rather like a narrow-leaved form of *P. lucens* but the influence of *P. crispus* can be seen in the obtuse leaf-apex and in the leaf-base, which though narrowed is truly sessile; other characters show intermediacy. Subsequent visits to the Warwickshire locality have failed to reveal more of the plants."

Potamogeton crispus × friesii
= **P. × lintonii** Fryer
Kiernan, J., 1961. Herb. Brit. Mus. Edgbaston.
Two records from canals at Edgbaston (0484) and Hockley Heath (1572); recorded from VCs 37, 39, 55 and 57.

577.21 **Potamogeton pectinatus** L.
Fennel-leaved Pondweed
Perry, 181–. Herb. Perry. St. Nicholas Meadow, Warwick. Map 589, p. 527
A frequent plant, widely but very irregularly distributed throughout the county, and recorded chiefly from canals, rivers and ponds; recorded from all surrounding VCs.

578 **Groenlandia** Gay

578.1 **Groenlandia densa** (L.) Fourr.
Opposite-leaved Pondweed
Purton, T., 1817. Mid. Flora [*Potamogeton densus* L.]. "In ponds and ditches on each side of the road between Stratford on Avon and Red Hill."
A plant recorded once from a pond S. of Brandon Wood (3976); formerly a more frequent plant of ponds, canals and streams; recorded from all surrounding VCs except 39.

ZANNICHELLIACEAE

580 **Zannichellia** L.

580.1 **Zannichellia palustris** L. Horned Pondweed
Purton, T., 1817. Mid. Flora. Kinwarton; Oversley. Map 590, p. 527
An occasional plant of a number of widely scattered localities throughout the county; recorded from ponds, slowly flowing rivers and canals; recorded from all surrounding VCs.

LILIACEAE

584 **Narthecium** Huds.

584.1 **Narthecium ossifragum** (L.) Huds.
Bog Asphodel
Withering, 1787. Arrangement. Birmingham Heath.
A very rare plant recorded once only from Coleshill Bog (2086); during the latter half of the nineteenth century it was abundant in a marsh between Bickenhill and Stonebridge and was also recorded from Sutton Park, probably becoming extinct in all but the Coleshill Bog area due to drainage; recorded from VCs 37, 39 and 57.

588 **Convallaria** L.

588.1 **Convallaria majalis** L. Lily-of-the-Valley
Bree, Rev. W. T., 1817. Mid. Flora. Haywood. Map 591, p. 528
Abundant in a few scattered woodland habitats in the N.W. of the county; previously more widespread and collected between 1835 and 1879 from Allesley, Hampton in Arden, Haseley, Hatton and Oversley Wood; recorded from all surrounding VCs.

589 **Polygonatum** Mill.

589.2 **Polygonatum odoratum** (Mill.) Druce
Angular Solomon's-seal
Westcott, Rev., B. F., 1836. Bagnall's Flora, 1891 [*P. officinale* All.]. Hedgerow, Erdington.
Not recorded in present survey; recorded from VCs 23, 33, 39 and 57.

589.3 **Polygonatum multiflorum** (L.) All.
Common Solomon's-seal
Cheshire, W., 1857. Herb. Perry. Wayfield Lane, Snitterfield.
Still to be found at Wayfield Lane, Snitterfield (2058) where it is probably native; also recorded from Edgbaston Park (0583, 0584) and Olton Reservoir (1381); recorded from all surrounding VCs.

Polygonatum multiflorum × **odoratum** = **P.** × **hybridum** Brügger
Hawkes, J. G., 1961. Guy's Cliffe, near Warwick.
A garden escape recorded from Sherbourne (2661) and Guy's Cliffe (2966); not recorded from any of the surrounding VCs.

590 **Maianthemum** Weber

590.1 **Maianthemum bifolium** (L.) Schmidt May Lily
Cooper, Miss G., 1961. Leek Wootton.
One record only of plant naturalised in woodland (2868); recorded from VC 23.

591 **Asparagus** L.

591.1a **Asparagus officinalis** L. ssp. **officinalis**
Asparagus
Perry, 1852. Herb. Perry [*A. officinalis*]. Coten End, Warwick. Map 592, p. 528
Naturalised and chiefly confined to a small area between Alcester and Stratford on Avon; recorded from roadsides and railway banks (37%), hedgerows (26%) and grassland (21%); recorded from all surrounding VCs except 39.

592 **Ruscus** L.

592.1 **Ruscus aculeatus** L. Butcher's Broom
Rugby School Report, 1867. Near Rugby.
A rare escape from cultivation naturalised in woods near Newbold on Stour (2347, 2447), Charlecote Park (2556), Fillongley Hall (2786) and Little Fosse (2951); recorded from all surrounding VCs except 39.

593 **Lilium** L.

593.1 **Lilium martagon** L. Martagon Lily
Rugby School Report, 1867. Newbold Grange Spinney.
A garden escape rarely naturalised in woods and recorded from Bearley (1861), Caldecote Hall (3595) and Newbold Grange (4976); recorded from all surrounding VCs except 39.

594 **Fritillaria** L.

594.1 **Fritillaria meleagris** L.
Snake's Head, Fritillary
Perry, 1817. List. Wroxall Field.
Not recorded in present survey; formerly known from near Wroxall Abbey, Godfrey's Lammas at Warwick and the Fritillary Fields, near Tamworth, where it seems to have been seen last in 1879. (It is still present a few yards over the county boundary near Tamworth in Staffordshire.) Recorded from all surrounding VCs except 57.

595 **Tulipa** L.

595.1 **Tulipa sylvestris** L. Wild Tulip
Bree, Rev. W. T., 1817. Mid. Flora. Allesley.
Not recorded in present survey; recorded from all surrounding VCs.

597 **Gagea** Salisb.

597.1 **Gagea lutea** (L.) Ker-Gawl.
Yellow Star-of-Bethlehem
Gorle, Rev. J., 1837. Bagnall's Flora, 1891 [*G. fascicularis* Salisb.]. Sheldon.
A plant of damp pastures and woods especially near streams, recorded from Edge Hill (3747) and in the Marston Green–Chelmsley Wood area (1786, 1887); in this latter area it was formerly more widespread, apparently extending along the banks of the Kingshurst

Brook and the River Cole from near Coleshill to near the Cock Inn at Elmdon. It is probably now extinct in this area due to urban development, but an effort has been made to establish it in Packington Park and at Whichford and Knavenhill woods. It is recorded from all surrounding VCs.

598 **Ornithogalum** L.

598.1 **Ornithogalum umbellatum** L.
Common Star-of-Bethlehem
Perry, 1817. List. Godfrey's Lammas, Warwick.
Map 593, p. 529
An occasional plant naturalised in a number of scattered localities throughout most of the county; recorded from roadsides, waste places, grassland, woods and hedgerows; recorded from all surrounding VCs.

598.2 **Ornithogalum nutans** L.
Drooping Star-of-Bethlehem
Young and Baker, 1873. Catalogue. Offchurch.
Naturalised in some quantity in Aston Old Churchyard (0889); recorded from all surrounding VCs except 39.

600 **Endymion** Dumort.

600.1 **Endymion non-scriptus** (L.) Garcke
Bluebell, Wild Hyacinth
Perry, 181–. Herb. Perry [*Agraphis nutans*]. Warwick. Map 594, p. 529
Abundant and fairly evenly distributed in the N.W. half of the county but only occasional in the S.E. half; recorded from hedgerows and scrub (46%), woods (37%), roadsides and waste places (9%) and grassland (6%); recorded from all surrounding VCs.

602 **Colchicum** L.

602.1 **Colchicum autumnale** L.
Meadow Saffron, Autumn Crocus
Countess of Aylesford, 1805. Bot. Guide. Near Packington. Map 595, p. 530
A rare plant mainly in the Stratford on Avon–Alcester area; previously known farther afield and collected between 1812 and 1912 from Barford, Castle Bromwich to Minworth, Darley Green, Norton Lindsey–Wolverton, Water Orton and Wroxall; recorded from woodland, grassland, waterside and hedgerow; recorded from all surrounding VCs.

603 **Paris** L.

603.1 **Paris quadrifolia** L. Herb Paris
Countess of Aylesford, 1805. Bot. Guide. Locke's (Loache's) Rough, near Coleshill.
Map 596, p. 530
A rare plant of scattered occurrence in damp woods generally on calcareous soil; recorded from all surrounding VCs.

JUNCACEAE

605 **Juncus** L.

605.1 **Juncus squarrosus** L. Heath Rush
Purton, T., 1817. Mid. Flora. Coleshill Heath. Map 597, p. 531
A locally frequent plant of Sutton Park and a few other localities, but completely absent elsewhere; recorded from heaths, acid woods and a few other acid habitats; recorded from all surrounding VCs.

605.2 **Juncus tenuis** Willd. Slender Rush
Cadbury, Miss D.A., 1944. Herb. Cadbury. Edgbaston Pool. Map 785, p. 614
A rare plant of marshes and damp waste places; recorded from all surrounding VCs except 32.

605.4 **Juncus compressus** Jacq. Round-fruited Rush
Baxter, W., 1831. MS (Bagnall: Mid. Nat. 1892–93.) Road, Rugby to Coton House.
Map 598, p. 531
An occasional plant of a number of scattered localities mainly on calcareous soils in the S. and E. of the county; recorded from marshes, damp roadsides, grassland and waste places; recorded from all surrounding VCs.

605.7 **Juncus bufonius** L. Toad Rush
Perry, 1815. Herb. Perry. Green's Grove, Hatton. Map 599, p. 532
Abundant in most parts of the county except in the Avon Valley and on the Lower Lias where it is infrequent; found chiefly by watersides (52%), roadsides and waste places (20%), grassland (10%) and cultivated ground (10%); recorded from all surrounding VCs.

605.8 **Juncus inflexus** L. Hard Rush
Perry, 1826. Herb. Perry [*J. glaucus* Sibth.]. Warwick. Map 600, p. 532
Abundant and fairly evenly distributed throughout the county; recorded from watersides and marshes (71%), damp grassland (16%), roadsides and waste places (7%) and a number of other habitats; recorded from all surrounding VCs.

605.9 **Juncus effusus** L. Soft Rush
Perry, 1826. Herb. Perry. Warwick.
Map 601, p. 533
Abundant and evenly distributed throughout most of the county though a little less common in the S.; seventeen of the records included are for var. *compactus* Hoppe, which is probably under-recorded as a number of collectors did not distinguish it; recorded from watersides and marshes (77%), grassland (9%), damp woods (5%) and a number of other habitats; recorded from all surrounding VCs.

Juncus effusus × inflexus
= **J. × diffusus** Hoppe
Bloxam, Rev. A., 1848. Phytologist [*J. diffusus*]. Baxterley Common.
Map 786, p. 614
An occasional plant usually found with parents; recorded from all surrounding VCs.

605.10 **Juncus subuliflorus** Drejer Compact Rush
Baxter, W., about 1840. Bagnall: Flora, 1891 [*J. conglomeratus* L.]. Near Rugby.
Map 602, p. 533
Abundant in the N.W. half of the county but rare in the Avon Valley and on the calcareous soils of the S.; recorded from watersides and marshes (61%), grassland (15%), damp woods (8%), waste places and roadsides (6%), scrub and hedgerows (6%) and heaths (3%); recorded from all surrounding VCs.

605.17 **Juncus subnodulosus** Schrank
Blunt-flowered Rush
Purton, T., 1817. Mid. Flora [*J. obtusiflorus*]. "In some boggy ground near Bidford Grange and in a stream at Broom." Map 787, p. 614
A rare plant of watersides and marshes; recorded from all surrounding VCs.

605.18 **Juncus acutiflorus** Ehrh. ex Hoffm.
Sharp-flowered Rush
Perry, 1815. Herb. Perry. Green's Grove, Hatton. Map 603, p. 534
A plant mainly of watersides (marshes and bogs 76%) and damp places in woods, grassland and other habitats in a number of areas in the county but more particularly in the N.W.; more confined to acid soil than *J. articulatus*.
The hybrid *J. acutiflorus* × *articulatus* (first record Rugby School Report, 1949, Newbold Quarry) has been recorded from 1076, 1096, 1273, 1372, 4364 and 4479, and from VC 37.
J. acutiflorus has been recorded from all surrounding VCs.

605.19 **Juncus articulatus** L. Jointed Rush
Perry, 181–. Herb. Perry [*J. lampocarpus* Ehrh.]. Near Whitnash. Map 604, p. 534
Abundant in most parts of the county but rare in the Avon Valley; chiefly found by watersides (81%) but also in grassland (10%) and rarely in other habitats; recorded from all surrounding VCs.

605.22 **Juncus bulbosus** L. sensu lato
Purton, T., 1817. Mid. Flora [*J. uliginosus*]. Coleshill Pool. Map 605, p. 535
This is Dandy's *J. bulbosus* L. including *J. kochii* F. W. Schultz; an occasional plant in the N.W. part of the county, recorded from bogs, marshes and wet heaths; probably all the recent records of the species are for *J. kochii* (q.v.); recorded from all surrounding VCs.

Juncus bulbosus L. sensu stricto
Bulbous Rush
Perry, 1835. Herb. Perry [*J. uliginosus*]. Coleshill Pool.
Although this has been recorded from time to time the only material seen by us and determined as *J. bulbosus* sensu stricto, is that collected by Perry in 1835; all the collections made by Bagnall and others, and seen by us, we have determined as *J. kochii* (q.v.).

Juncus kochii F. W. Schultz
Kirk, T., 1847. Herb. Perry [*J. supinus*]. Stoke Heath.
An occasional plant of marshes and bogs; this is the segregate of *J. bulbosus* sensu lato usually occurring in Warwickshire, the distribution being essentially that shown on the map for the aggregate; recorded from VCs 23, 55 and 57.

606 **Luzula** DC.

606.1 **Luzula pilosa** (L.) Willd. Hairy Woodrush
Perry, 1813. Herb. Perry. Near Warwick.
Map 606, p. 535
An occasional plant in the N.W. of the county, but almost entirely absent from the calcareous soils of the S.; recorded from woods (74%), scrub, railway banks, hedgerows and grassland; recorded from all surrounding VCs.

606.3 **Luzula sylvatica** (Huds.) Gaudin
Great Woodrush
Bagnall, 1869. Flora, 1891 [*L. maxima*]. Without locality. Map 607, p. 536
A plant of a few scattered localities mainly in a belt running diagonally N.E. to S.W. across the county; recorded from woods and scrub; recorded from all surrounding VCs.

606.4 **Luzula luzuloides** (Lam.) Dandy & Wilmott
White Woodrush
Druce, 1932. Com. Flora. Without locality.
Not recorded in present survey; recorded from VCs 33 and 37.

606.8 **Luzula campestris** (L.) DC. Field Woodrush
Perry, 1813. Herb. Perry. Without locality.
Map 608, p. 537
Abundant and evenly distributed throughout the county; recorded from grassland (78%), roadsides and railway banks (15%) and a few other habitats; recorded from all surrounding VCs.

606.9 **Luzula multiflora** (Retz.) Lejeune
Heath Woodrush
Purton, T., 1817. Mid. Flora [*Juncus liniger*]. On the road from Coughton to Sambourn.
Map 609, p. 537
Frequent, though irregularly distributed in the N.W. of the county and almost entirely absent from the calcareous soils of the S.; recorded from woods (35%), grassland (26%), scrub (13%) and a few other habitats; recorded from all surrounding VCs.
Var. **congesta** (DC.) Lej. (first record Ick, W., 1836. Herb. Ick [*Luciola congesta*]. Coleshill Bog) has been recorded from squares 0974, 1097, 2086, 2598. This also has been recorded from all surrounding VCs.

AMARYLLIDACEAE

607 **Allium** L.

607.5 **Allium vineale** L. Crow Garlic
Bromwich, H., 1867. Herb. Warw. Mus. Moreton Morrell. Map 610, p. 537
A locally frequent plant of grassy roadsides and railway banks (54%), other types of rough grassland (30%), cultivated land (8%) and hedgerows (7%), all on calcareous soil in the S. of the county especially around Stratford on Avon; recorded from all surrounding VCs.

607.6 **Allium oleraceum** L. Field Garlic
Perry, 1817. List. Near Leamington.
A rare plant of rough grass on railway banks and roadsides on calcareous soil; recorded from near Bearley (1759), S. of Alveston Pasture (2350) and Elder Tree Coppice (2849); recorded from all surrounding VCs except 23.

607.12 **Allium ursinum** L. Ramsons
Withering, 1801. Arrangement. "Several pastures, near Penn's Mill at Erdington, abound so much with the plant as to be called 'Garlick Meadows'." Map 611, p. 538
A plant of shaded riversides and ditches (46%), hedge-

rows and scrub (26%) and woods (25%), mainly confined to the area of the old Forest of Arden in the N.W. of the county; it frequently forms dense communities; recorded from all surrounding VCs.

611 **Leucojum** L.

611.2 **Leucojum aestivum** L.
Loddon Lily, Summer Snowflake
Purton, T., 1817. Mid. Flora. By the side of the Avon, near Stratford on Avon.
Not recorded in present survey; recorded from VCs 23, 33, 37 and 55.

612 **Galanthus** L.

612.1 **Galanthus nivalis** L. Snowdrop
Countess of Aylesford, 1805. Bot. Guide. Packington. Map 612, p. 538
Frequently planted and naturalised in woods, coppices and hedgerows; recorded mainly from the Avon Valley and a few scattered localities elsewhere; recorded from all surrounding VCs.

614 **Narcissus** L.

614.1 **Narcissus pseudonarcissus** L. Wild Daffodil
Ray, 1670. Catalogus [*N. sylvestris pallidus calcye luteo*]. Near Sutton Coldfield, towards Middleton. Map 613, p. 539
Abundant in the Packington area and Hay Wood; frequent in the Earlswood area, where it is diminishing; has also been recorded from one or two other isolated areas; recorded from woods, grassland and roadsides; recorded from all surrounding VCs.

614.4 **Narcissus × incomparabilis** Mill.
Bagnall, 1876. Herb. Bagnall. Guy's Cliffe.
Not recorded in present survey; recorded from VC 37.

614.6 **Narcissus majalis** Curt. Pheasant's Eye
Bree, Rev. W. T., 1837. New Bot. Guide [*N. poeticus* L.]. "A field in the parish of Fillongley is full of it. Some of the flowers are single, others semi-double; probably not truly native."
A garden plant occasionally becoming naturalised, and recorded in woods at Brandon (4076) and All Oaks Wood (4478); it has been seen in the Fillongley area on a number of occasions down to 1933; recorded from all surrounding VCs.

614.7 **Narcissus × medioluteus** Mill. = **N. poeticus × tazetta** Primrose Peerless
Perry, 1820. Plant. Varv. Sel. In the Lammas Fields, Warwick.
Not recorded in the present survey; recorded from all surrounding VCs except 55.

IRIDACEAE

616 **Iris** L.

616.3 **Iris foetidissima** L. Stinking Iris
Rufford, Rev. W. S., 1817. Mid. Flora. Alne Hills. Map 614, p. 539
An occasional plant of woods and hedgerows on calcareous soil in the S. of the county; abundant in the Billesley area; recorded from all surrounding VCs except 39.

616.4 **Iris pseudacorus** L. Yellow Iris
Baxter, W., 1831. MS (Bagnall: Mid. Nat. 1892–93.) Near Lawford and Newbold on Avon.
Lloyd, Dr. G., 1835. Herb. Perry. Near Princethorpe. Map 615, p. 540
A plant of watersides (72%) and water (24%), fairly evenly distributed throughout the county; recorded from all surrounding VCs.

618 **Crocus** L.

618.1 **Crocus nudiflorus** Sm. Autumnal Crocus
Perry, 1817. List. Pigwell Fields and Lammas Fields, Warwick.
Recorded from two areas of pasture near Warwick: Lammas Field (2764) and Priory Park (2865), where it has been known for many years and was at one time much more abundant; recorded from VCs 37, 39 and 57.

618.2 **Crocus purpureus** Weston Spring Crocus
Gorle, Rev. J., Bagnall's Flora, 1891 [*C. vernus* All.]. Sheldon and Marston Green.
Not recorded in present survey; recorded from all surrounding VCs except 32 and 55.

620 **Crocosmia** Planch.

620.1 **Crocosmia × crocosmiflora** (Lemoine) N. E. Br. Montbretia
Hall, Mr. & Mrs. P. C., 1960. Hartshill.
A garden escape naturalised on roadsides and waste places, and recorded from 0685, 1172, 3293 and 3487; recorded from VCs 37 and 57.

DIOSCOREACEAE

622 **Tamus** L.

622.1 **Tamus communis** L. Black Bryony
Perry, 1812. Herb. Perry. Warwick. Map 616, p. 540
Abundant and fairly evenly distributed throughout the county; recorded from hedges and scrub (86%), woods (10%), roadsides and railway banks (3%) and a few other habitats; recorded from all surrounding VCs.

ORCHIDACEAE

624 **Cephalanthera** Rich.

624.1 **Cephalanthera damasonium** (Mill.) Druce White Helleborine
Bagnall, 1887. Flora, 1891 [*C. pallens* Rich.]. In a coppice, near Farnborough.
Not recorded in present survey; recorded from all surrounding VCs except 39 and 55.

624.2 **Cephalanthera longifolia** (L.) Fritsch Long-leaved Helleborine
Purton, T., 1817. Mid. Flora [*Serapias ensifolia*]. Oversley Wood; Ragley Woods.
Still known from two km. squares in Oversley Wood; recorded from all surrounding VCs except 32 and 55.

625 **Epipactis** Zinn

625.1 **Epipactis palustris** (L.) Crantz Marsh Helleborine
Purton, T., 1817. Mid. Flora [*Serapias longifolia*]. Oversley Wood.
Not recorded in present survey; recorded from all surrounding VCs.

625.2 **Epipactis helleborine** (L.) Crantz
Broad Helleborine
Withering, 1796. Arrangement [*Serapias latifolia*]. In the Red Rock Plantation, Edgbaston.
Map 617, p. 541
An occasional plant of a few widely scattered localities; recorded from mixed woods, hedgerows and scrub; recorded from all surrounding VCs.

625.3 **Epipactis purpurata** Sm. Violet Helleborine
Kirk, T., 1883. Top. Bot. [*E. violacea* Bor.]. Without locality.
Bagnall, 1886. Herb. Bagnall [*E. violacea*]. Butlers Wood, Maxstoke.
A rare plant of damp woods recorded from Coughton Park (0560, 0660), Rough Hill Wood (0563) and Edgbaston Park (0583); recorded from all surrounding VCs except 39 and 57.

625.6 **Epipactis phyllanthes** G. E. Sm.
Green-flowered Helleborine
Fleetwood, T., 1857. Watsonia, 1952. Charlecote.
Not recorded in present survey; recorded from VCs 23, 33, 55 and 57.

627 **Spiranthes** Rich.

627.1 **Spiranthes spiralis** (L.) Chevall.
Autumn Lady's-tresses
Perry, 1817. List [*Neottia spiralis* Sw.]. "In a field crossed by the footroad from Warwick to Hampton-on-the-Hill."
Not recorded in present survey; collected at Badger's Wood, near Stratford on Avon, at Old Park, Warwick, and at Kenilworth, this last being in 1873; there is no record of its being seen since; recorded from all surrounding VCs.

628 **Listera** R. Br.

628.1 **Listera ovata** (L.) R. Br. Common Twayblade
Perry, 1813. Herb. Perry. Warwick.
Map 618, p. 541
A frequent plant on limestone in the S. of the county but rare elsewhere; recorded from woods (45%), roadsides and railway banks (22%), grassland (14%) and scrub (12%); recorded from all surrounding VCs.

629 **Neottia** Guett.

629.1 **Neottia nidus-avis** (L.) Rich. Birdsnest Orchid
Purton, T., 1817. Mid. Flora [*Ophrys nidus-avis*]. Middleton Woods; Ragley and Oversley Wood (Bree). Map 619, p. 542
A rare plant of a few scattered areas mainly on the Lower Lias limestone in the S. of the county; recorded from woods (100%); recorded from all surrounding VCs.

635 **Coeloglossum** Hartm.

635.1 **Coeloglossum viride** (L.) Hartm. Frog Orchid
Purton, T., 1817. Mid. Flora [*Satyrium viride*]. Meadows about Cold Comfort; Oversley Hill.
Not recorded in present survey but a number of records were made between 1843 and 1878 at Ipsley, and near Coventry, Kenilworth and Stratford on Avon; recorded from all surrounding VCs.

636 **Gymnadenia** R. Br.

636.1 **Gymnadenia conopsea** (L.) R. Br.
Fragrant Orchid
Purton, T., 1817. Mid. Flora [*Orchis conopsea* L.]. Chelmsley Wood; Bannerley Pool (Bree); Cold Comfort; Oversley.
A very rare plant of pastures and marshes on calcareous soil reported once only from Rugby Cement Works (4876), but as no specimen has been seen further investigation is necessary; formerly considerably more frequent; recorded from all surrounding VCs.

638 **Platanthera** Rich.

638.1 **Platanthera chlorantha** (Custer) Reichb.
Greater Butterfly Orchid
Perry, 1817. List [*Orchis bifolia* L.]. Plantation near Saltisford Common, Warwick.
Map 620, p. 542
A locally frequent plant mainly found on the Lower Lias limestone in the S. and S.E. of the county; recorded from woods (71%), roadsides, scrub, railway banks and quarries; recorded from all surrounding VCs.

638.2 **Platanthera bifolia** (L.) Rich.
Lesser Butterfly Orchid
Cheshire, W., 1857. Herb. Perry [*Habenaria bifolia* R. Br.]. Comyns Farm, near Stratford on Avon.
One record only of a plant from Ufton Wood (3962) which V. A. Summerhayes described as "a large-flowered specimen of the calcareous woodland form"; recorded from all surrounding VCs.

640 **Ophrys** L.

640.1 **Ophrys apifera** Huds. Bee Orchid
Skipworth, Mrs. F., 1837. New Bot. Guide. Near Anstey. Map 621, p. 543
A rare plant of calcareous soils in the S. and S.E. of the county, with some concentration on the Lower Lias limestone; recorded from quarries, claypits, roadsides, waste places and rough grassland; recorded from all surrounding VCs.

640.4 **Ophrys insectifera** L. Fly Orchid
Caswell, Rev. J., 1880. Herb. Bagnall [*O. muscifera* Huds.]. Oscott College Grounds.
Not recorded in present survey; recorded from all surrounding VCs except 55.

642 **Orchis** L.

642.5 **Orchis morio** L. Green-winged Orchid
Bree, Rev. W. T., 1828. Mag. Nat. Hist. Coleshill. Map 622, p. 543
A rare plant of scattered distribution in the centre of the county; absent from the extreme N. and very rare in the extreme S.; recorded from meadows and pastures; recorded from all surrounding VCs.

642.7 **Orchis mascula** (L.) L. Early Purple Orchid
Perry, 1825. Herb. Perry. Between Norton Lindsey and Wolverton. Map 623, p. 544
An occasional plant in the S. half of the county with some concentration on the Lower Lias limestone; recorded from woods (71%), hedgerows, scrub, grassland and waste places; recorded from all surrounding VCs.

643 Dactylorhiza Nevski

643.1a **Dactylorhiza fuchsii** (Druce) Soó ssp. **fuchsii**
Common Spotted Orchid
Perry, 181–. Herb. Perry [*Orchis latifolia*]. Warwick. Map 624, p. 544
Abundant on the Lower Lias limestone and scattered sparsely and very irregularly elsewhere; recorded from woods (33%), grassland (20%), railway banks and roadsides (17%), marshes (14%), scrub and hedgerows (12%) and quarries (4%); recorded from all surrounding VCs.

Dactylorhiza fuchsii ssp. **fuchsii** × **D. praetermissa**
Allen, H., 1953. Bishop's Itchington.
Recorded from two adjoining squares on the spoil heaps of the Lias quarries at Bishop's Itchington (3858, 3958); recorded from VCs 23, 32, 33, 55 and 57.

643.2b **Dactylorhiza maculata** (L.) Soó ssp. **ericetorum** (E. F. Linton) Hunt & Summerh.
Heath Spotted Orchid
Perry, 1813. Herb. Perry [*Orchis latifolia*]. Warwick. Map 625, p. 545
A rare plant with a sparsely scattered distribution in the W. part of the county; recorded from marshes, bogs, damp grassland and woods; recorded from all surrounding VCs.

643.3b **Dactylorhiza incarnata** (L.) Soó ssp. **pulchella** (Druce) Soó
Early Marsh Orchid
Bagnall, 1884. Flora, 1891 [*Orchis incarnata* L.]. Without locality.
Found in a marshy meadow near Halford (2645) with *D. praetermissa* in what is probably Bagnall's station; recorded from VC 55; *D. incarnata* (L.) Soó is recorded from all surrounding VCs except 37.

643.4 **Dactylorhiza praetermissa** (Druce) Soó
Southern Marsh Orchid
Perry, 1813. Herb. Perry [*Orchis latifolia*]. Warwick. Map 788, p. 614
Recorded from a few widely-separated boggy and marshy areas in the county; recorded from all surrounding VCs.

644 Aceras R. Br.

644.1 **Aceras anthropophorum** (L.) Ait. f.
Man Orchid
Jeffray, D. J., 1968. Ufton Fields.
One record only from rough grassland on limestone at Ufton Fields (3861); recorded from VCs 23, 32, 55 and 57.

645 Anacamptis Rich.

645.1 **Anacamptis pyramidalis** (L.) Rich.
Pyramidal Orchid
Perry, 1817. List [*Orchis pyramidalis* L.]. Near Pillerton. Map 789, p. 615
A rare plant of quarries and rough grassland on calcareous soil in the S.E. of the county; recorded from all surrounding VCs.

ARACEAE

646 Acorus L.

646.1 **Acorus calamus** L. Sweet Flag
Withering, 1787. Arrangement. "Tamworth, at the bottom of Mr. Oldershaw's garden." Map 626, p. 545
A rare plant introduced and now naturalised in shallow water at the margins of lakes, canals and rivers; recorded from all surrounding VCs.

649 Arum L.

649.1 **Arum maculatum** L.
Lords-and-Ladies, Cuckoo-pint
Withering, 1801. Arrangement. "Plentiful in the dingle at Edgbaston." Map 627, p. 546
Abundant and evenly distributed throughout the county except on the Lias clay in the E. where it is infrequent; recorded from hedgerows and scrub (57%), woodland (19%) and roadsides (17%); recorded from all surrounding VCs.

LEMNACEAE

650 Lemna L.

650.1 **Lemna polyrhiza** L. Greater Duckweed
Ick, W., 1836. Herb. Ick. "A pit near Saltley Hall."
A rare plant of ponds and canals recorded from Maxstoke (2288), River Anker backwater at Polesworth (2502, 2602), Old Milverton (2967), near Stretton on Dunsmore (4373), Newbold on Avon (4977) and Rugby (5076); recorded from all surrounding VCs.

650.2 **Lemna trisulca** L. Ivy-leaved Duckweed
Bree, Rev. W. T., 1835. New Bot. Guide. Without locality.
Baynes, W. W., 1835. MS (Bagnall: Flora, 1891). In a pool between Offchurch and Radford Semele. Map 628, p. 546
A frequent plant with a wide but very irregular distribution in the county; recorded from water (chiefly ponds) (95%) and watersides (5%); recorded from all surrounding VCs.

650.3 **Lemna minor** L. Common Duckweed
Baxter, W., 1831. MS (Bagnall: Mid. Nat. 1892–93.) Near Lawford.
Lowe, H. E., 1836. Herb. Perry. Pool near Erdington. Map 629, p. 547
Abundant and fairly evenly distributed throughout most of the county, but a little less common in the S.; recorded from water (chiefly ponds) (90%) and watersides (10%); recorded from all surrounding VCs.

650.4 **Lemna gibba** L. Gibbous Duckweed
Perry, 1820. Plant. Varv. Sel. Mill pond near St. Nicholas' Church, and in a brook in Baly's Lammas, Warwick. Map 630, p. 547
An occasional plant mainly in the centre and N. of the county, almost completely absent from the calcareous soils of the S.; recorded from water (chiefly ponds and canals) (92%) and watersides (8%); recorded from all surrounding VCs.

SPARGANIACEAE

652 Sparganium L.

652.1 **Sparganium erectum L.** Branched Bur-reed
Perry, 1813. Herb. Perry [*S. ramosum*]. Warwick. Map 631, p. 548
Abundant and fairly evenly distributed throughout the county; recorded from watersides (57%) and water (43%); recorded from all surrounding VCs.
Ssp. **erectum** (first record: Druce, 1932. Comital Flora). It is probable that most of the records mapped as *S. erectum* L. are for this ssp., although in most cases collectors have only recorded *S. erectum* L. and we have not seen the material: ssp. *erectum* has been recorded from squares 0584, 0867, 1173, 1174, 1275, 1378, 1392, 1479, 1597, 1678, 1690, 1779, 1977, 2071, 2183, 2184, 2245, 2393, 4459 and 5066; recorded from VCs 23, 55 and 57.
Ssp. **microcarpum** (Neuman) Hylander (first and only record: Clark, M. C., 1965. Near Lapworth, square 1569); recorded from VCs 23, 39, 55 and 57.
Ssp. **neglectum** (Beeby) Schinz & Thell. (first record: Bagnall, 1885. Herb. Bagnall [*S. neglectum*]. Anker Bridge near Wolvey); recorded from Edstone (1861), between Kineton and Tysoe (3447) and Dunsmore Heath (4373); recorded from VCs 23, 37, 55 and 57.
Ssp. **oocarpum** (Čelak) C. Cook (first and only record: Cook, C. 1947. Herb. Brit. Mus. Welford on Avon, square 1–5–); recorded from VCs 23, 55 and 57.

652.2 **Sparganium emersum** Rehm. Unbranched Bur-reed
Purton, T., 1817. Mid. Flora [*S. simplex*]. Near the Lodge Farm, Snitterfield. Map 632, p. 548
An occasional plant of a number of scattered localities throughout most of the county, absent from the extreme S.; recorded from water (71%) and watersides (29%); recorded from all surrounding VCs.

652.4 **Sparganium minimum** Wallr. Least Bur-reed
Countess of Aylesford, 1805. Bot. Guide [*S. natans*]. Packington.
Not recorded in present survey; previously collected by Kirk from Arbury Hall in 1849 and 1854, and by Cheshire near the road from Coton House to Cave's Inn in 1854; recorded from VCs 32, 37, 39 and 55.

TYPHACEAE

653 Typha L.

653.1 **Typha latifolia** L. Bulrush
Countess of Aylesford, 1818. Herb. Oxf. Univ. Packington. Map 633, p. 549
Frequent and fairly evenly distributed except in the S. of the county, where it is rare; recorded from watersides and marshes (51%) and water (48%); recorded from all surrounding VCs.

653.2 **Typha angustifolia** L. Lesser Bulrush
Ray, 1670. Catalogus [*T. palustris media* J.B.]. "Comitis Warwicensis." Map 634, p. 549
A rare plant of a few scattered localities in the extreme N.W. and the extreme E. of the county; recorded from the edges of ponds and lakes, and from marshes; recorded from all surrounding VCs.

CYPERACEAE

654 Eriophorum L.

654.1 **Eriophorum angustifolium** Honck. Common Cotton-grass
Withering, 1787. Arrangement [*E. polystachion* L.]. Birmingham Heath. Map 635, p. 550
A rare plant with a considerable concentration in Sutton Park and a few occurrences elsewhere; recorded from marshes and bogs; recorded from all surrounding VCs.

654.4 **Eriophorum vaginatum** L. Harestail Cotton-grass
Withering, 1787. Arrangement. "In a marshy valley, crossed by a foot-road to Winson Green."
A plant of bogs now confined to Sutton Park (0896, 0897, 0898, 0995, 0996, 0997, 0998, 1095, 1096, 1097), where it is often locally frequent; formerly known also at Coleshill and Packington; recorded from VCs 37, 39, 55 and 57.

655 Scirpus L.

655.2b **Scirpus cespitosus** L. ssp. **germanicus** (Palla) Broddesson Deergrass
Ray, 1670. Catalogus [*Juncus parvus montanus cum parvis capitulis luteis* J.B.]. "Circa Middleton & alibi in agro Warwicensi."
A rare plant of bogs now found only in two stations in Sutton Park (0996, 0998); collected by Lloyd at Coleshill Pool in 1835 and by Kirk at Coleshill Bog in 1848; recorded from all surrounding VCs except 23 and 32.

655.3 **Scirpus maritimus** L. Sea Clubrush
Bromwich, H., 1872. Herb. Bagnall [*S. lacustris* var. *glaucus*]. Itchington Holt.
Two records only: from a salt marsh at Southam Holt (4460) where it has been known since 1873, and from a cattle-pool near Flecknoe (5064) where it has been known since 1883; recorded from VCs 33, 37 and 39.

655.4 **Scirpus sylvaticus** L. Wood Clubrush
Ray, 1670. Catalogus [*Cyperus gramineus* J.B.]. "In several places in Warwickshire, as in ditches about Solihull; by the River Tame-side under Dorsthill-hill near Tamworth etc." Map 636, p. 550
An occasional plant of a number of widely scattered localities in the county, with some concentration in the Earlswood–Henley–Shrewley area; recorded from watersides and marshes; recorded from all surrounding VCs.

655.8 **Scirpus lacustris** L. Common Clubrush
Perry, 1825. Herb. Perry. On the side of the Avon, Warwick. Map 637, p. 551
A locally frequent plant in the centre of the county, mainly in the Avon Valley; less common elsewhere; recorded from water (69%) and watersides (31%); recorded from all surrounding VCs.

655.9 **Scirpus tabernaemontani** C. C. Gmel. Grey Clubrush
Bromwich, H., 1870. Herb. Bagnall. Southam Holt.
A very rare plant of salt marshes recorded from Itchington Holt (3755) where it has been known since

1870, Southam Holt (4260, 4460) where it has been known since 1873, and near Flecknoe (5064) where it has been known since 1874; recorded from all surrounding VCs except 32.

655.10 **Scirpus setaceus** L. Bristle Clubrush
Power, Rev. J., about 1816. MS (Bagnall: Flora, 1891). Polesworth Common.
Perry, 1843. Herb. Perry. Bishops Itchington.
Map 638, p. 551
An occasional plant of a number of localities in the W. half of the county; recorded from watersides and marshes, and a few other places on acid soil; recorded from all surrounding VCs.

655.12 **Scirpus fluitans** L. Floating Clubrush
Countess of Aylesford, 1819. Herb. Oxf. Univ. Packington.
One record only from a small bog near Packington (2084); previously somewhat widespread and collected from Beausale Common, Hill Bickenhill, Coleshill Heath and Bog, Lower Eatington, Haseley Common, Sutton Park and Wolford Heath; recorded from all surrounding VCs except 33 and 57.

656 Eleocharis R. Br.

656.2 **Eleocharis acicularis** (L.) Roem. & Schult.
Needle Spike-rush
Freeman, S., 1841. Phytologist [*E. acicularis* Sm.]. Sutton.
Recorded from the mud and shallow water at the margins of the canal at Stockton Locks (4264, 4364) and lakes at Earlswood (1174), Seeswood Pool (3290) and Newbold on Avon Quarry (4977); formerly known at a number of other canal-side localities including Stratford on Avon and Catherine de Barnes as well as at Coleshill Pool and Olton Reservoir; recorded from all surrounding VCs.

656.3 **Eleocharis quinqueflora** (F. X. Hartmann) Schwarz Few-flowered Spike-rush
Kirk, T., 1861. Herb. Perry [*Scirpus pauciflorus* Lightf.]. Stivichall, Coventry.
Recorded from two boggy areas in Sutton Park (0996, 0998); recorded from all surrounding VCs.

656.4 **Eleocharis multicaulis** (Sm.) Sm.
Many-stalked Spike-rush
Lloyd, Dr. G., 1835. Herb. Perry. Coleshill Bog.
Not recorded in present survey; previously known from Coleshill Bog, Bickenhill and Packington; last known record from near Bickenhill in 1945; recorded from all surrounding VCs except 39.

656.5 **Eleocharis palustris** (L.) Roem. & Schult.
Common Spike-rush
Baxter, W., 1831. MS (Bagnall: Flora, 1891). Pond near Sawbridge. Map 639, p. 552
A frequent plant, fairly evenly distributed throughout the county, which has been divided into ssp. *palustris* and ssp. *vulgaris* (q.v.); the species in the wider sense has been recorded from watersides and marshes (72%), water (21%) and damp grassland (4%); it has been recorded from all surrounding VCs.

Ssp. **palustris**
Lloyd, Dr. G., 1835. Herb. Perry [*E. palustris*]. Coleshill Pool.
Rare in Warwickshire; recorded from Sutton Park (1096), Idlecote (2845), Lighthorne (3356) and Long Itchington (4165); recorded from VCs 23, 33 and 37.

Ssp. **vulgaris** S. M. Walters
Lloyd, Dr. G., 1835. Herb. Perry [*E. palustris*]. Coleshill Bog.
This is evidently the commoner of the two ssps. in the county and a considerable number of the records mapped (of the species in the wider sense) have been so identified; probably the majority of the others are also of this sub-species; the ssp. is recorded from all surrounding VCs.

656.6 **Eleocharis uniglumis** (Link) Schult.
Slender Spike-rush
Readett, R. C., 1955. Southam Holt.
A rare plant of marshes recorded from Kinwarton (1058), near Henley in Arden (1465), Elder Tree Copse (2849) and Southam Holt salt spring (4460); recorded from VCs 23, 32 and 55.

657 Blysmus Panz.

657.1 **Blysmus compressus** (L.) Panz. ex Link
Broad Blysmus
Palmer, Miss C., 1854. Herb. Oxf. Univ. Near Lighthorne.
Two records from marshy places by rivers: Binton (1453) and Tamworth (2104); recorded from all surrounding VCs.

659 Schoenus L.

659.1 **Schoenus nigricans** L. Black Bogrush
Ray, 1670. Catalogus [*Juncus laevis panicula glomerata nigricante* C.C.]. "In meadows by the River Tame-side under Dorst-Hill near Middleton."
Not recorded in present survey; the last record appears to be by Purton from Coleshill Bog (Mid. Flora, 1817); recorded from all surrounding VCs.

660 Rhynchospora Vahl

660.1 **Rhynchospora alba** (L.) Vahl White Beaksedge
Withering, 1787. Arrangement [*Schoenus alba* L.]. Birmingham Heath.
Not recorded in present survey; recorded several times, mainly in the Packington and Coleshill areas, until 1885, but not seen since; recorded from VCs 23, 37, 39 and 55.

661 Cladium Browne

661.1 **Cladium mariscus** (L.) Pohl Great Fen-sedge
Ray, 1670. Catalogus [*Cyperus longus inodorus sylvestris* Ger.]. "In the boggy closes under Dorst-Hill, near Tamworth."
Not recorded in present survey; recorded from VCs 32, 37 and 39.

663 Carex L.

663.1 **Carex laevigata** Sm. Smooth-stalked Sedge
Beilby, Miss M. A., 1837. Analyst. "Moist field at Highgate, not far from the Rea."
A rare plant of damp woods and marshes in the Earlswood area (0873, 0973, 0974, 1073, 1074) and Yarningale Common (1865); recorded from all surrounding VCs except 23 and 33.

663.2 **Carex distans** L. Distant Sedge
Bromwich, H., 1873. Herb. Bagnall. Southam Holt. Map 640, p. 552
Recorded from a few scattered localities, mainly on the Lower Lias clay, in marshes and damp pasture; recorded from all surrounding VCs except 57.

663.4 **Carex hostiana** DC. Tawny Sedge
Kirk, T., 1852. Herb. Perry (Bagnall's Flora, 1891) [*C. fulva* Good.]. Stivichall.
A very rare plant of marshes and bogs in Sutton Park (0996, 0998), Brown's Coppice (1380) and Hatton (2266); recorded from all surrounding VCs.

663.5 **Carex binervis** Sm. Green-ribbed Sedge
Purton, T., 1817. Mid. Flora [*C. distans*]. Oversley; Coughton. Map 790, p. 615
A rare plant of damp heathy places and woods on acid soil; recorded from all surrounding VCs.

663.7 **Carex lepidocarpa** Tausch
Long-stalked Yellow Sedge
Bagnall, 1870. Herb. Bagnall [*C. flava* L.]. Bannerley Pool, near Coleshill.
A very rare plant, formerly more frequent, now recorded only from a rough pasture near Close Wood, Meriden (2081); recorded from all surrounding VCs.

663.8 **Carex demissa** Hornem.
Common Yellow Sedge
Lloyd, Dr. G., 1835. Herb. Perry [*C. flava*]. Coleshill Pool. Map 641, p. 553
A rare plant recorded from a few scattered marshes and bogs; recorded from all surrounding VCs.

Carex demissa × hostiana
Readett, R. C., 1951. Sutton Park.
One record, with parents, from a bog in Sutton Park (0998); recorded from VCs 33 and 57.

663.12 **Carex sylvatica** Huds. Wood Sedge
Bromwich, H., 1850. Herb. Perry. Hay Wood. Map 642, p. 553
A frequent plant mainly of the heavier soils in the S. of the county but rare on the Lower Lias clays; recorded from woods (73%), scrub and hedgerow (19%) and watersides (4%); recorded from all surrounding VCs.

663.15 **Carex pseudocyperus** L. Cyperus Sedge
Purton, 1817. Mid. Flora. On the edge of a pool at Kinwarton. Map 643, p. 554
An occasional plant somewhat unevenly distributed throughout the N. half of the county but absent from the Avon Valley and the Lower Lias clays in the S.; recorded from watersides (85%) and water (11%); recorded from all surrounding VCs.

663.16 **Carex rostrata** Stokes Bottle Sedge
Ray, 1670. Catalogus [*Gramen cyperoides polystachyon majus, spicis teretibus erectis*]. In several pools about Middleton. Map 644, p. 554
A rare plant mainly of Sutton Park and the Coleshill–Meriden area; according to Druce (Com. Flora, 1932) Ray's is the first British record; recorded from ponds and marshes (83%) and bogs (17%); recorded from all surrounding VCs.

663.17 **Carex vesicaria** L. Bladder Sedge
Withering, 1801. Arrangement. Edgbaston Pool. Map 645, p. 555
A rare plant of a few scattered localities in the W. of the county; recorded from marshes and riversides (100%); recorded from all surrounding VCs.

663.20 **Carex riparia** Curt. Greater Pond Sedge
Baxter, W., 1831. MS (Bagnall: Mid. Nat. 1892–93.) Banks of Avon near Rugby Mill. Map 646, p. 555
A frequent but unevenly distributed plant except in the N. of the county, where it is only occasional, and in the S., where it is rare; recorded from watersides and marshes (78%) and shallow water (22%); recorded from all surrounding VCs.

Carex riparia × vesicaria
= **C. × csomadensis** Simonk.
Cumming, L., 1922. Bot. Exch. Club Report. Canal by Brinklow Station.
Not recorded in present survey; recorded from VC 23. The Warwickshire record is on the authority of Druce but Salmon considered it *C. riparia.*

663.21 **Carex acutiformis** Ehrh. Lesser Pond Sedge
Baxter, W., 1831. MS (Bagnall: Mid. Nat. 1892–93) [*C. paludosa*]. Near Newbold on Avon. Map 647, p. 556
A frequent plant of widespread distribution but less common in the S. of the county; recorded from watersides (89%) and water (10%); recorded from all surrounding VCs.

663.22 **Carex pendula** Huds. Pendulous Sedge
Countess of Aylesford, 1805. Bot. Guide. Packington. Map 648, p. 556
A frequent plant mainly in the area of the old Forest of Arden in the N.W. of the county, but with a heavier concentration in the W. and a few scattered occurrences on the Lower Lias limestone escarpment; recorded from watersides (54%), damp woods (35%) and hedgerows and scrub (9%); recorded from all surrounding VCs.

663.23 **Carex strigosa** Huds.
Thin-spiked Wood Sedge
Kitchener, F. E., 1873. Rugby School Report. Princethorpe Wood.
Not recorded in present survey; recorded from all surrounding VCs.

663.24 **Carex pallescens** L. Pale Sedge
Ray, 1670. Catalogus [*Gramen cyperoides polystachion flavicans*]. "In pratis circa Middleton agri Warwicensis." Map 649, p. 557
An occasional plant mainly in the W. of the county;

rare in the E. and almost entirely absent from the extreme N. and S.; according to Druce (Com. Flora, 1932) Ray's is the first British record; recorded from watersides (46%), woodland and scrub (31%) and rough grassland and roadsides; recorded from all surrounding VCs.

663.26 **Carex panicea** L. Carnation Sedge
Perry, 1824. Herb. Perry. Norbrooke Meadows. Map 650, p. 557
Frequent in the W. half of the county but thinning towards the N. and occasional on the Lower Lias clays in the S. and S.E.; recorded from marshes and watersides (49%) and damp grassland (45%); recorded from all surrounding VCs.

663.31 **Carex flacca** Schreb. Glaucous Sedge
Snape, T., 1825. Herb. Perry [*C. recurva*]. Hampton-on-the-Hill. Map 651, p. 558
An abundant plant widely distributed throughout the county with a heavy concentration on the calcareous soils of the Lower Lias in the S.; recorded from grassland (45%), watersides and marshes (26%), roadsides and railway banks (20%) and scrub and hedgerows (5%); recorded from all surrounding VCs.

663.32 **Carex hirta** L. Hairy Sedge
Withering, 1801. Arrangement. Stew at Edgbaston. Map 652, p. 558
Abundant and evenly distributed throughout the county; recorded from watersides and marshes (45%), grassland (33%) and roadsides, railway banks and waste places (18%); recorded from all surrounding VCs.

663.34 **Carex pilulifera** L. Pill-headed Sedge
Lowe, H. E., 1836. Herb. Perry. Coleshill Heath. Map 653, p. 559
A rare plant of a few dry heathy places in the N. and W. of the county; recorded from all surrounding VCs.

663.36 **Carex caryophyllea** Latourr. Spring Sedge
Perry, 1813. Herb. Perry [*C. praecox*]. Near Warwick. Map 654, p. 559
An occasional plant, generally of dry grassland, sparsely and unevenly distributed throughout the county; recorded from grassland (80%) and heath (10%); recorded from all surrounding VCs.

663.47 **Carex acuta** L. Slender Tufted Sedge
Purton, T., 1817. Mid. Flora. Blacklands near Oversley Bridge. Map 655, p. 560
A rare plant of a few scattered waterside localities; recorded from all surrounding VCs.

663.50 **Carex nigra** (L.) Reichard Common Sedge
Baxter, W., 1831. MS (Bagnall: Mid. Nat. 1892–93) [*C. goodenovii*]. Between Hill Morton and Sawbridge. Map 656, p. 560
Frequent mainly in the N. half of the county; recorded from marshes and watersides (62%), damp grassland (26%) and bog and heath grassland (10%); recorded from all surrounding VCs.

663.54 **Carex paniculata** L. Greater Tussock Sedge
Beilby, Miss M. A., 1837. Analyst. Pond at Edgbaston near Strawberry Vale. Map 657, p. 561
An occasional plant of a few scattered localities mainly in the N. of the county; recorded from water and watersides (84%) and wet woodland (8%); recorded from all surrounding VCs.

663.56 **Carex diandra** Schrank Lesser Tussock Sedge
Bagnall, 1869. Flora, 1891 [*C. teretiuscula* Good. var. *ehrhartiana*]. Sutton Park.
A very rare plant recorded only from a bog in Sutton Park (0996) and a marsh at Packington Park (2284); recorded from VCs 37, 39, 55 and 57.

663.57 **Carex otrubae** Podp. False Fox Sedge
Perry, 181–. Herb. Perry [*C. vulpina*]. Without locality. Map 658, p. 561
Abundant and fairly evenly distributed throughout the county; recorded from watersides and marshes (80%), damp grassland (7%), roadsides (5%), hedgerows (4%) and a few other habitats; recorded from all surrounding VCs.

Carex otrubae × remota
= **C. × pseudoaxillaris** K. Richt.
Bromwich, H., 1865. Herb. Brit. Mus. [*C. axillaris* Good. non L.]. Roundshill Lane, Kenilworth.
A rare hybrid recorded with parents from ditches at three locations: Earlswood (1174), Snitterfield (2160) and Nunhold Grange (2265); recorded from all surrounding VCs except 39 and 57.

663.60 **Carex disticha** Huds. Brown Sedge
Purton, T., 1817. Mid. Flora [*C. intermedia*]. "In a thicket in the road from Dunnington to Abbot's Morton." Map 659, p. 562
An occasional plant chiefly in the N.E. and W. of the county and almost entirely absent from the calcareous soils in the S.E.; recorded from marshes and watersides (78%) and damp grassland (18%); recorded from all surrounding VCs.

663.61 **Carex arenaria** L. Sand Sedge
Phipps, J. B., 1955. Hatton Station.
A rare introduction recorded from waste places at Galley Common (3192) and E. of Kineton (3650); and from introduced sea-sand on railways at Hatton Station (2266) and Emscote, Warwick (2965); not recorded from any of the surrounding VCs.

663.65 **Carex divulsa** Stokes Grey Sedge
Purton, T., 1817. Mid. Flora. On a hedgebank between Wixford and Pophills. Map 660, p. 562
A rare plant mainly confined to the W. of the county; recorded from hedgerows, scrub, grassland, roadsides and woodland; recorded from all surrounding VCs.

663.67 **Carex spicata** Huds. Spiked Sedge
Baxter, W., 1831. MS (Bagnall: Mid. Nat. 1892–93.) Near Hill Morton.
Lloyd, Dr. G., 1835. Herb. Perry [*C. muricata*]. Warwick Castle Mount. Map 661, p. 563
Rare in the N. and N.W. of the county; abundant elsewhere especially on the Lower Lias clays in the S.; recorded from grassland (41%), roadsides and railway banks (33%), marshes and watersides (16%) and hedges and scrub (8%); recorded from all surrounding VCs.

663.68 **Carex muricata** L. ssp. **muricata** Prickly Sedge
Green, P. S., 1950. Brandon Wood.
Map 791, p. 615
A rare plant of dry grassy roadsides and railway banks; recorded from all surrounding VCs.

663.66 **Carex muricata** L. ssp. **leersii** Aschers. & Graebn. (**C. polyphylla** auct., non Kar. & Kir.)
Horwood, A. R., 1945. Proc. B.S.B.I. Claverdon.
Not recorded in present survey; recorded from all surrounding VCs.

663.69 **Carex elongata** L. Elongated Sedge
Bagnall, 1870. Herb. Bagnall. Hampton in Arden.
A very rare plant growing beside still and slowly running water in the Earlswood area (0974, 1076, 1173, 1174, 1177); recorded from VC 37.

663.70 **Carex echinata** Murr. Star Sedge
Ray, 1670. Catalogus [*Gramen sylvaticum parvum tenuifolium cum spica aculeata* J.B.]. "In moist meadows and pastures about Middleton." Map 792, p. 615
A rare plant of marshes and bogs in Sutton Park, Coleshill and other areas in the N.W. of the county; recorded from all surrounding VCs.

663.71 **Carex remota** L. Remote Sedge
Perry, 1829. Herb. Perry. "Road from Warwick to Hatton—opposite the 2nd milestone."
Map 662, p. 563
A frequent plant in the N. and W. of the county but almost entirely absent from the Avon Valley and the Lower Lias clays in the S. and S.E.; recorded from watersides (68%), damp woods (26%) and hedgerows and scrub (6%); recorded from all surrounding VCs.

663.72 **Carex curta** Gooden. White Sedge
Ray, 1670. Catalogus [*Gramen Cyperoides elegans spica composita* R.]. "In a pool not far from Middleton, towards Coleshill, in Warwickshire." Druce, in his Comital Flora, describes this as the First British Record.
Map 793, p. 616
A rare plant of marshes and bogs recorded from Sutton Park and a few other places in the N.W. of the county; recorded from VCs 37, 39 and 57.

663.74 **Carex ovalis** Gooden. Oval Sedge
Baxter, W., 1831. MS (Bagnall: Mid. Nat. 1892–93.) Near Dunchurch. Map 663, p. 564
A frequent plant fairly evenly distributed in the N. and W. of the county but very rare in the Avon Valley and on the calcareous clays in the S. and S.E.; recorded from marshes and watersides (46%), damp grassland (34%), scrub and woodland (10%) and roadsides and heaths; recorded from all surrounding VCs.

663.80 **Carex pulicaris** L. Flea Sedge
Ray, 1670. Catalogus [*Gramen cyperoides pulicare* R.]. "Flea Grass. This was so denominated by Mr. Goodyer, because the seeds, which turn downward on the stalk, do in shape and colour somewhat resemble fleas; about Middleton in Warwickshire."
A very rare sedge, formerly more frequent, now recorded only from two bogs in Sutton Park (0996, 0998); recorded from all surrounding VCs.

663.81 **Carex dioica** L. Dioecious Sedge
Baynes, W. W., 1833. MS (Bagnall's Flora, 1891). Honiley Heath.
A very rare sedge of bogs in Sutton Park (0995, 0996, 0998); recorded from all surrounding VCs; *C. dioica* f. *isogyna* (first record: Harlond, A., 1935. Herb. B'ham Mus.) is also recorded from Sutton Park (0996, 0998) but not from any of the surrounding VCs.

GRAMINEAE

665 **Phragmites** Adans.

665.1 **Phragmites australis** (Cav.) Trin. ex Steud.
Common Reed
Lloyd, Dr. G., 1835. Herb. Perry. Wasperton, near Warwick. Map 664, p. 564
A frequent plant widely but very irregularly distributed throughout the county; recorded from watersides and marshes (72%) and water (25%); recorded from all surrounding VCs.

667 **Molinia** Schrank

667.1 **Molinia caerulea** (L.) Moench
Purple Moor-grass
Purton, T., 1817. Mid. Flora [*Melica caerulea* L.]. Coleshill Bog. Map 665, p. 565
A rare plant mainly of Sutton Park and the Coleshill area, with a few scattered occurrences elsewhere in the N.W. of the county; recorded from heaths, bogs and acid grassland (54%) as well as marshes and woods; recorded from all surrounding VCs.

668 **Sieglingia** Bernh.

668.1 **Sieglingia decumbens** (L.) Bernh. Heath Grass
Purton, T., 1817. Mid. Flora [*Poa decumbens* With.]. On Tippins Hill, near to the footpath leading to Wetherley (from Alcester).
Map 666, p. 565
An occasional plant of a number of scattered localities mainly in the N.W. half of the county, with some concentration in the Sutton Coldfield and Solihull areas; recorded from heaths, grassland, marshes, hedgerows and railway banks; recorded from all surrounding VCs.

669 **Glyceria** R. Br.

669.1 **Glyceria fluitans** (L.) R. Br.
Floating Sweet-grass
Baxter, W., 1831. MS (Bagnall: Mid. Nat. 1892–93.) Near Rugby. Map 667, p. 566
Abundant and fairly evenly distributed throughout the county; recorded from watersides (84%) and water (16%); recorded from all surrounding VCs.

Glyceria × **pedicellata** Townsend
= **G. fluitans** × **plicata**
Anon. 1862. Herb. B'ham Univ. Warwick Canal. Map 668, p. 566
An occasional plant of a number of widely scattered localities throughout the county; recorded from watersides (90%) and water (10%); recorded from all surrounding VCs.

669.2 **Glyceria plicata** Fr. Plicate Sweet-grass
Perry, 1820. Herb. Perry. Without locality.
Map 669, p. 567
A frequent plant widely but irregularly distributed throughout the county; recorded from watersides (86%) and water (13%); recorded from all surrounding VCs.

669.3 **Glyceria declinata** Bréb. Small Sweet-grass
Bagnall, 1883. Herb. Bagnall [*G. fluitans* var. *declinata*]. Near Rounsel Lane, Kenilworth.
Map 670, p. 567
A fairly frequent plant with a very irregular distribution throughout the county; recorded from watersides (85%) and water (15%); recorded from all surrounding VCs.

669.4 **Glyceria maxima** (Hartm.) Holmberg
Reed Sweet-grass
Perry, 1829. Herb. Perry [*G. aquatica*]. Banks of River Leam near Leam House.
Map 671, p. 568
A fairly frequent plant with a very irregular distribution throughout the county; recorded from watersides (83%) and water (16%); recorded from all surrounding VCs.

670 Festuca L.

670.1 **Festuca pratensis** Huds. Meadow Fescue
Perry, 181–. Herb. Perry. Without locality.
Map 672, p. 568
Abundant and evenly distributed throughout the county; recorded from grassland (65%), roadsides, waste places and railway banks (27%) and a number of other habitats; recorded from all surrounding VCs.

670.2 **Festuca arundinacea** Schreb. Tall Fescue
Lloyd, Dr. G., 1835. Herb. Perry [*F. pratensis*]. Wasperton. Map 673, p. 569
Abundant in the S. half of the county particularly on the calcareous soils, but only occasional and widely scattered in the N.; recorded from roadsides and railway banks (37%), grassland (34%), watersides and marshes (17%), hedgerows and scrub (10%) and a few other habitats; recorded from all surrounding VCs.

670.3 **Festuca gigantea** (L.) Vill. Giant Fescue
Purton, T., 1817. Mid. Flora [*Bromus giganteus* L.]. Wixford Lane. Map 674, p. 569
Abundant and fairly evenly distributed throughout the county with a greater concentration in the W.; recorded from hedgerows and scrub (36%), woods (27%), watersides (24%), roadsides (9%) and grassland (4%); recorded from all surrounding VCs.

670.5 **Festuca heterophylla** Lam. Various-leaved Fescue
Hardaker, W. H., 1957. Hagley Road Railway Station.
One record only from railway track near Hagley Road Station (0385); recorded from VC 23.

670.6 **Festuca rubra** L. Creeping Fescue
Lloyd, Dr. G., 1835. Herb. Perry [*F. duriuscula*]. Wroxhall Heath. Map 675, p. 570
Abundant and evenly distributed throughout the county; recorded from grassland (54%), roadsides, railway banks and waste places (38%) and a number of other habitats; recorded from all surrounding VCs.

Ssp. **commutata** Gaudin (first record Bagnall, 1883. Herb. Bagnall [*F. rubra* var. *fallax* Th.]. Earlswood Reservoir) has been recorded from squares 0588, 0649, 0652, 0773, 0954, 1073, 1371, 1577, 1677, 1682, 1763, 1863, 2053, 2482, 2532, 2550, 2628, 3156, 3191 and 3277; recorded from VCs 23, 33, 55 and 57.

Ssp. **rubra** (first record Bagnall, 1868. Herb. Bagnall [*F. duriuscula* L.]. Marston Green) has been recorded from squares 0464, 1071, 1371, 1677, 2160, 2532, 3158, 3293, 3383, 3490, 3652, 3688 and 4369 (this last including *f. barbata* and *f. grandiflora*); recorded from VCs 23, 33, 55 and 57.

The records for the two subspecies are included in the map, but the majority of the records made have been as *F. rubra* L.

Festuca ovina L. Sheep's Fescue
For first record see ssp. *tenuifolia*
Map 676, p. 570
Scattered throughout the county and frequent in a number of areas; recorded from rough grassland and pasture (48%), roadsides and railway banks (33%), walls and quarries (8%), dry heaths (7%) and woods (3%); recorded from all surrounding VCs.

670.8 Ssp. **ovina**
St. Brody, Dr., 1875. Herb. Warw. Mus. [*F. ovina*]. Guy's Cliffe, Warwick.
Although most of the records (see *F. ovina* L. and map) have been made as *F. ovina* L., it is probable that the majority refer to ssp. *ovina*, since material so identified is widespread throughout the county; the subspecies has been recorded from all surrounding VCs except 37.

670.9 Ssp. **tenuifolia** (Sibth.) Peterm.
Fine-leaved Sheep's Fescue
Ick, W., 1836. Herb. Ick [*F. ovina*]. Sutton Coldfield.
A rare plant of dry heathy grassland confined to Sutton Park, where it is abundant, and a few other areas in the N.W. of the county; recorded from Sutton Park (0895, 0896, 0897, 0898, 0995, 0996, 0997, 0998, 1095, 1096, 1097, 1195, 1197), Solihull (1378), Thistly Wood (1763), Temple Balsall (2076), Bannerley (2185), Hawkeswell Farm (2187) and Wootton Court (2868); recorded from all surrounding VCs except 37.

670.11 **Festuca longifolia** Thuill. Hard Fescue
Bromwich, H., 1879. Herb. Kew [*F. ovina* var. *major*]. Railway banks, Hill Wootton.
Not recorded in present survey; recorded from VCs 23, 32, 55 and 57.

× Festulolium Aschers & Graebn.

× **Festulolium loliaceum** (Huds.) P. Fourn.
Cheshire, W., before 1857. Herb. Perry [*F. pratensis* Huds.]. Hatton Rock.
Map 677, p. 571
An occasional plant usually found with parents and recorded from grassland, roadsides and waste places; recorded from all surrounding VCs.

671 **Lolium** L.

671.1 **Lolium perenne** L. ssp. **perenne** Perennial Rye-grass
Lloyd, Dr. G., 1835. Herb. Perry [*L. perenne*]. Near Warwick. Map 678, p. 571
Abundant and evenly distributed throughout the county; recorded from grassland (52%), roadsides, waste places, railway banks and farmyards (38%), cultivated land (5%) and hedgerows (4%); recorded from all surrounding VCs.

671.2 **Lolium perenne** L. ssp. **multiflorum** (Lam.) Husnot Italian Rye-grass
Kirk, T., 1848. Herb. Perry [*L. perenne* var. *aristatum*]. Without locality. Map 679, p. 572
Abundant and fairly evenly distributed throughout the county; recorded from grassland (37%), roadsides, waste places and farmyards (33%), cultivated land (27%) and a few other habitats; recorded from all surrounding VCs.

671.3 **Lolium temulentum** L. Darnel
Bromwich, H., 1870. Bagnall's Flora, 1891. Near Kenilworth.
A casual recorded from waste places (1253, 2054, 4974) and roadside (3379); recorded from all surrounding VCs.

672 **Vulpia** C. C. Gmel.

672.2 **Vulpia bromoides** (L.) Gray Barren Fescue
Withering, 1796. Arrangement [*Festuca bromoides* L.]. Dry pastures near the Mass House, Edgbaston. Map 680, p. 572
Frequent but very irregularly distributed throughout most of the county; recorded from railway tracks and sidings, waste places and roadsides (64%), quarries and walls (17%), dry grassland (12%), cultivated land (3%) and a few other habitats; recorded from all surrounding VCs.

672.3 **Vulpia myuros** (L.) C. C. Gmel. Rat's-tail Fescue
Lloyd, Dr. G., 1835. Herb. Perry [*Festuca myuros*]. Warwick. Map 794, p. 616
An occasional plant of railway tracks and yards, waste places and walls, especially in the S. and E. of the county; formerly also collected in Birmingham, Sutton Park and Coleshill; recorded from all the surrounding VCs.

673 **Puccinellia** Parl.

673.2 **Puccinellia distans** (L.) Parl. Reflexed Saltmarsh-grass
Bagnall, 1890. Herb. Bagnall [*Sclerochloa distans* Bab.]. Waste ground, Milverton.
One record of plant from a ditch with effluent from a coal mine at Ansley Hall Colliery (3093); recorded from all surrounding VCs.

674 **Catapodium** Link

674.1 **Catapodium rigidum** (L.) C. E. Hubbard Fern-grass
Purton, T., 1817. Mid. Flora [*Poa rigida* L.]. Wall at Oversley Green Bridge. Map 681, p. 573
A frequent plant in the S. of the county particularly on calcareous soils; recorded from railway sides and roadsides (58%), walls and quarries (27%), dry rough grassland and a few other habitats; recorded from all surrounding VCs.

676 **Poa** L.

676.1 **Poa annua** L. Annual Poa
Lloyd, Dr. G., 1835. Herb. Perry. Leamington. Map 682, p. 573
Abundant and evenly distributed throughout the county; recorded from roadsides and waste places (48%), cultivated land (24%), grassland (22%) and a few other habitats; recorded from all surrounding VCs.

676.6 **Poa nemoralis** L. Wood Meadow-grass
Perry, 1819. Herb. Perry. Oversley Lane. Map 683, p. 574
An occasional plant, mostly in areas which were well wooded in the seventeenth century; recorded from hedgerows and scrub (49%), woods (43%) and a few other habitats; recorded from all surrounding VCs.

676.9 **Poa compressa** L. Flattened Meadow-grass
Perry, 1826. Herb. Perry. Walls of St. Mary's Churchyard, Warwick. Map 684, p. 574
A rare plant of a number of widely scattered localities, generally, but not exclusively, on dry soils; recorded from walls and quarries (68%), railway banks and roadsides (24%) and a few other habitats; recorded from all surrounding VCs.

Poa pratensis L. Meadow-grass
For first record see sub-species.
Abundant and evenly distributed throughout the county; details of distribution are given under the sub-species; recorded from all surrounding VCs.

676.10 Ssp. **pratensis** Smooth Meadow-grass
Lloyd, Dr. G., 1835. Herb. Perry [*P. pratensis*]. Leamington. Map 685, p. 575
Abundant and evenly distributed throughout the county; recorded from grassland (47%), roadsides, railway banks and waste places (38%), hedgerows (7%) and a number of other habitats; recorded from all surrounding VCs.

676.11 Ssp. **angustifolia** (L.) Gaudin Narrow-leaved Meadow-grass
Kirk, T., 1854. Herb. Perry [*P. pratensis* var. *angustifolia*]. Binley. Map 686, p. 575
An occasional plant, probably under-recorded, of a number of widely distributed localities chiefly on calcareous soils; recorded from grassland (48%), roadsides, railway banks and waste places (37%), quarries, walls and hedgerows; recorded from all surrounding VCs.

676.12 Ssp. **irrigata** (Lindm.) Lindb. f. Spreading Meadow-grass
Lloyd, Dr. G., 1835. Herb. Perry [*P. pratensis* var. *subcaerulea*]. Coleshill Hill.
A rare plant, possibly under-recorded, found generally in damp habitats; recorded from near Ipsley (0665), Birmingham (0788), Warwick (2965) and Wappenbury Wood (3771); recorded from all surrounding VCs.

676.13 **Poa trivialis** L. Rough Meadow-grass
Perry, 1820. Herb. Perry. Without locality. Map 687, p. 576
Abundant and fairly evenly distributed throughout the

county; recorded from grassland (32%), roadsides, waste places and railway banks (31%), watersides and marshes (11%), hedgerows and scrub (11%), cultivated land (8%) and woods (6%); recorded from all surrounding VCs.

676.14 **Poa palustris** L. Swamp Meadow-grass
Druce, 1932. Com. Flora. Without locality.
Not recorded in present survey; recorded from all surrounding VCs except 32.

676.15 **Poa chaixii** Vill. Broad-leaved Meadow-grass
Bromwich, H., 1875. Herb. Bagnall [*P. sudetica* Haenke]. Wood bordering Roundshill Lane, Leek Wootton.
Recorded once from a coppice near Leek Wootton (2769) where it is abundant and has persisted presumably since Bromwich recorded it in 1875; recorded from VCs 23 and 39.

677 **Catabrosa** Beauv.

677.1 **Catabrosa aquatica** (L.) Beauv.
Water Whorl-grass
Withering, 1801. Arrangement [*Aira aquatica* L.]. Edgbaston Pool. Map 688, p. 576
A rare plant of a few scattered localities recorded from watersides and marshes (67%) and water (33%); recorded from all surrounding VCs.

678 **Dactylis** L.

678.1 **Dactylis glomerata** L. Cock's-foot
Baxter, W., 1831. MS (Bagnall: Mid. Nat. 1892–93.) Near Rugby.
Lloyd, Dr. G., 1835. Herb. Perry. Leamington. Map 689, p. 577
Abundant and evenly distributed throughout the county; recorded from grassland (43%), roadsides, waste places and railway banks (39%), hedgerows and scrub (10%) and a few other habitats; recorded from all surrounding VCs.

679 **Cynosurus** L.

679.1 **Cynosurus cristatus** L. Crested Dogstail
Perry, 1818. Herb. Perry. Side of Warwick canal near Leamington. Map 690, p. 577
Abundant and evenly distributed throughout the county; recorded from grassland (77%), roadsides (17%) and a few other habitats; recorded from all surrounding VCs.

679.2 **Cynosurus echinatus** L. Rough Dogstail
Druce, 1932. Com. Flora. Without locality.
Not recorded in present survey; last recorded by C. E. A. Andrews and Dr. R. C. L. Burges in 1946 from a bombed site in Summer Lane, Birmingham; recorded from all surrounding VCs.

680 **Briza** L.

680.1 **Briza media** L. Common Quaking-grass
Ick, W., 1836. Herb. Perry. Meadow near Small Heath Turnpike. Map 691, p. 578
An abundant plant fairly evenly distributed throughout much of the county but rare in the Avon Valley and the N.W.; recorded from grassland (71%), roadsides and railway banks (18%) and marshes and watersides (5%); recorded from all surrounding VCs.

681 **Melica** L.

681.1 **Melica uniflora** Retz. Wood Melick
Purton, T., 1817. Mid. Flora. Oversley Lane. Map 692, p. 578
A frequent plant in the N.W. half of the county in areas which were well wooded in the seventeenth century; very rare elsewhere; recorded from woods (49%), hedgerows and scrub (35%), roadsides (10%) and watersides (6%); recorded from all surrounding VCs.

683 **Bromus** L.

Section **Pnigma** Dumort. (**Zerna**)

683.1 **Bromus erectus** Huds. Upright Brome
Lloyd, Dr. G., 1835. Herb. Perry. Ufton Wood. Map 693, p. 579
A frequent plant of grassland in the S. of the county, mainly preferring calcareous soil; recorded from roadsides and railway banks (56%), grassland (35%) and hedgerows (5%); recorded from all surrounding VCs.

683.2 **Bromus ramosus** Huds. Hairy Brome
Baxter, W., 1831. MS (Bagnall: Mid. Nat. 1892–93) [*B. asper* Murr.]. Near Bilton.
Lloyd, Dr. G., 1835. Herb. Perry [*B. asper*]. Ufton Wood. Map 694, p. 579
An abundant plant fairly evenly distributed throughout the county; recorded from hedgerows and scrub woodland (58%), mixed woods (19%), roadsides and railway banks (11%) and watersides (10%); recorded from all surrounding VCs.

683.4 **Bromus inermis** Leyss. Hungarian Brome
Tyrer, Dr. F. H., 1961. Near Flecknoe. Map 795, p. 616
A recent introduction occasionally naturalised on roadsides, railways and waste places; recorded from VCs 23 and 55.

Section **Genea** Dumort. (**Anisantha**)

683.5 **Bromus sterilis** L. Barren Brome
Lloyd, Dr. G., 1835. Herb. Perry. Wall, in Warwick. Map 695, p. 580
An abundant plant evenly distributed throughout the county; recorded from roadsides and waste places (46%), hedgerows (31%), grassland (12%) and a few other habitats; recorded from all surrounding VCs.

683.8 **Bromus rigidus** Roth Ripgut Brome
Bromwich, H., 1886. Herb. Bagnall [*B. maximus* Desf.]. Kenilworth.
Not recorded in present survey; recorded from VC 33.

Section **Bromus** (**Serrafalcus**)

The following account of the section, the key for identifying the species and the notes on the taxonomy of the individual species have been prepared by Dr. Philip Smith.

Bromus (sect. *Bromus*) is well represented in Warwickshire, only one native British species, *B. interruptus*, not so far having been recorded. The whole group is critical, largely because of the variability of some species, the incidence of hybridisation, and the somewhat subtle, though reliable, characters used to distinguish certain species. The most reliable characters are those of the spikelet and, to a lesser extent, the panicle. Spikelet and floret size are important, and they correlate with differences in the shape and texture of the lemmas, the relative length of paleas, the length of anthers and the robustness of awns. Caryopses display taxonomically useful variation in size and inrolling of the margins. Panicles vary in size and laxness.

Smith (*Watsonia*, **6**, 327–44, 1968) separated as *B.* × *pseudothominii* inland material previously referred to as "*B. thominii*" (sensu Tutin, in C.T.W.). The coastal populations to which the latter epithet correctly applies do not occur in Warwickshire. *B. hordeaceus* subsp. *thominii* (Hard.) Hylander, as it should now be called, occurs in a few exposed inland stations, but is typically a plant of maritime sand dunes. It may be distinguished by its short culms (to 12 cm) which are normally prostrate or procumbent, its erect, dense, often simple panicles, the grain being shorter than the palea, and the awns usually at least weakly divaricate.

Key

1. Awn divaricate, even when young; lemmas 7 mm broad . . . . . . **squarrosus**
   Awn straight, erect; lemmas narrower . . 2
2. Lemma margins inrolled in fr.; florets divaricate in fr., slow to break free 3
   Lemma with overlapping margins in fl. and in fr.; florets imbricate even in fr. soon breaking free 4
3. Lemma 7–9 mm, hairy or glabrous; panicle lax, subsecund; lower sheaths usually glabrous . . . . . . **secalinus**
   Lemma 5–6 mm, glabrous; panicle narrow, erect or nodding; lower sheaths hairy **pseudosecalinus**
4. Panicle erect, often dense, with short branches; most pedicels shorter than spikelets; lemma and glumes papery, with raised nerves . 5
   Panicle lax, with long, spreading or erect, nodding branches; most pedicels longer than their spikelets; lemma and glumes horny, without raised nerves . . . 7
5. Lemma 4·5–5·5 mm long, usually glabrous, with broadly hyaline, sharply angled margins; awns weak, setaceous; grain exceeding palea **lepidus**
   Lemma 6·5–11 mm long, hairy or glabrous, with narrowly hyaline, bluntly angled margins; awns stout; grain shorter than or equalling palea 6
6. Lemma 6·5–8 mm long, usually glabrous; grain equalling palea . . . × **pseudothominii**
   Lemma 8–11 mm long, very rarely glabrous; grain shorter than palea . . . **hordeaceus**
7. Palea equalling or exceeding lemma; lemma 7–9 mm long, obscurely angled; anthers 3–4·5 mm long . . . . . **arvensis**
   Palea shorter than lemma; lemma 6·5–11 mm long; anthers not more than 3 mm long . 8
8. Panicle narrow, erect, nodding; lemma 6·5–8 mm long, without a marginal angle; rhachilla internode 1 mm long; anthers 1·75–3 mm long . . . . **racemosus**
   Panicle spreading or drooping; lemma 8–11 mm long, its margin with a prominent blunt angle 2/3 up; rhachilla internode 1·5 mm long; anthers 1–1·5 mm long . . **commutatus**

Bromus hordeaceus L. sensu lato (**B. hordeaceus** L. ssp. **hordeaceus** and **B.** × **pseudothominii** P. Smith)
For first records see segregates.
Map 696, p. 580

In Warwickshire there are two taxa of this aggregate of closely related plants represented: *B. hordeaceus* L. ssp. *hordeaceus* and *B.* × *pseudothominii* P. Smith. Large mixed populations of these occur throughout the county, and are often associated with sown grassland. Frequently the boundary between these two taxa is hard to define, and further difficulty may result when *B. lepidus* is present in addition.

Abundant and fairly evenly distributed throughout most of the county, but a little less frequent in the N.W.; recorded from roadsides, waste places and railway banks (53%), grassland (37%), arable land (4%) and hedgerows (4%); recorded from all surrounding VCs.

This map includes records of plants which have been identified either as *B. hordeaceus* L. ssp. *hordeaceus* or *B.* × *pseudothominii* P. Smith, in addition to those identified as *B. hordeaceus* L. sensu lato.

683.10 Ssp. **hordeaceus** (**B. mollis** L.) Soft Brome
Lloyd, Dr. G., 1835. Herb. Perry [*B. mollis*]. Near Warwick. Map 697, p. 581

A fairly abundant plant well distributed throughout the county except in the N.W., where it is rare; recorded from roadsides, waste places and railway banks (48%), grassland (39%), arable land (7%) and hedgerows (4%); on roadsides it may be found in populations including *B. lepidus* and *B.* × *pseudothominii* and may be difficult to distinguish, but pure populations are to be found in established grassland as well as sometimes on roadsides and railway embankments; it seldom persists long as a weed of arable land; recorded from all surrounding VCs.

B. hordeaceus L. var. *leiostachys* Hartm. (first record Jones, Mrs. M. D. G., 1962, Hatton Rock Farm), a variety with glabrous spikelets, is rare, and is generally found in the mixed populations described above; it may result from hybridisation; it has been recorded from squares 1354, 2149, 2167, 2265, 2357, 2405, 4584, 4683 and 4875; it has been recorded from VCs 23, 33 and 37.

683.12 **Bromus** × **pseudothominii** P. Smith (**B. thominii** Hard. sensu Tutin)
Lesser Soft Brome
Bagnall, 1881. Herb. Bagnall [*B. mollis* var. *glabrescens* Coss.]. Embankment, Bracebridge Pool, Sutton Park. Map 698, p. 581

A frequent plant of the centre and S.E. of the county; recorded from roadsides (43%), grassland (41%) and arable land (10%); recorded from all surrounding VCs.

B. × *pseudothominii* comprises variants produced by the hybridisation of *B. hordeaceus* and *B. lepidus*. Some of these are similar to *B. hordeaceus*, others closely resemble *B. lepidus*, but the majority are plainly intermediate in morphological characters. *B.* × *pseudothominii* occurs usually in the presence of one or both parents, in highly variable populations on ground where grass seeds have been artificially sown. Some hybridisation may occur in these populations, but most of the variability seems to come from brome seeds contaminating the grass seed which was sown. Both *B.* × *pseudothominii* and *B. lepidus* have seeds of similar size to those of the main herbage grasses (*Festuca* spp.; *Lolium* spp.) and so are difficult to remove from herbage grass crops by normal mechanical cleaning processes. Even if the grass seed is slightly contaminated, a mixed population of bromes may arise on the sown site, e.g. roadside verge or pasture field.

B. × *pseudothominii* and *B. lepidus* occur more typically in pasture land (often sown) than *B. hordeaceus*. *B. hordeaceus* and *B.* × *pseudothominii* occur more commonly in rough, established grassland than *B. lepidus*. The hybrid forms thus seem to have a wider ecological range than either parent.

Forms of *B.* × *pseudothominii* with hairy lemmas have been named var. *hirsutus* Holmb., and occur throughout the area of distribution of the species; they have been recorded from squares 0583, 0958, 1467, 1656, 1954, 1973, 2053, 2194, 2199, 2242, 2257, 2345, 2433, 2443, 2454, 2465, 2756, 2766, 2863, 3064, 3074, 3078, 3273, 3366, 3368, 3449, 3552, 3660, 3664, 3672, 3693, 3777, 3862, 3876, 3877, 3947, 3951, 3961, 3972, 4076, 4090, 4147, 4170, 4181, 4182, 4352, 4462, 4464, 4470, 4672, 4964, 5162 and 5264.

683.13 **Bromus lepidus** Holmberg — Slender Brome
Andrews, C. E. A. and Townsend, C. C., 1953. Newbold on Stour. Map 699, p. 582

An occasional plant widely scattered throughout the county; recorded from grassland (43%), roadsides and waste places (43%) and arable land (9%); in grassland, like *B.* × *pseudothominii*, it is usually found in areas where grass seed with contaminating bromes has been sown; recorded from all surrounding VCs.

B. lepidus var. *micromollis* (Krösche) C. E. Hubbard (first record Perry, D., 1954. Fulford Heath) with hairy spikelets, may be a genuine infraspecific variant or result from the introgression of genes from *B. hordeaceus*; it has been recorded from squares 0974, 1259, 2167, 3245, 3672, 4388, 4486, 4764 and 4974; it is also recorded from VC 33.

683.14 **Bromus racemosus** L. — Smooth Brome
Perry, 1829. Herb. Perry. Hatton Farm, Hampton Lucy. Map 700, p. 582

An occasional plant largely confined to grassland on calcareous soil in the S. and S.E. of the county (73%); also recorded from grassy roadsides (14%) and watersides and marshes (9%); recorded from all surrounding VCs.

The species seldom occurs in abundance, and does not seem able to compete very successfully as a ruderal plant, unlike its close relative *B. commutatus*. The rather narrow, erect panicle of *B. racemosus* is easily overlooked in roadside populations of *B. hordeaceus*, and so the species is probably under-recorded in this habitat. Bagnall recorded an essentially similar distribution in 1891, though the species seems to have extended farther north at the time.

683.15 **Bromus commutatus** Schrad. — Meadow Brome
Lloyd, Dr. G., 1835. Herb. Perry. Field on road to Harbury from Tachbrook. Map 701, p. 583

An occasional plant on calcareous soils in the S. and S.E. of the county; recorded from grassland (45%), roadsides and waste places (34%) and arable land (17%); it seems to persist in low-lying pastures, old water meadows and hayfields where it is particularly common; it is recorded from all surrounding VCs.

B. commutatus is rarely found in great abundance but occasionally may cover large areas of wet meadow, often mixed with *Festuca* spp., and with *B. ramosus*, with which it may sometimes hybridise.

B. commutatus Schrad. var. *pubens* Wats. (first record Bagnall, 1875. Herb. Bagnall [*B. commutatus* Schrad. var. *pubescens*]. Waysides near Radway; Edge Hills and Binton), a variety with hairy spikelets, is less common than the typical variety with glabrous spikelets; it is recorded from squares 0656, 1858, 2341, 2345, 2651, 2653, 2852, 2946, 3154, 3243, 3245, 3449, 4165, 4464 and 4855; it is also recorded from VCs 23 and 33.

The hybrid *B. commutatus* Schrad. × *B. racemosus* L. has been recorded once (Evans, R., 1963) from Russell Farm, near Southam (4362).

683.17 **Bromus arvensis** L. — Field Brome
Hawkes, J. G., 1964. Elder Tree Copse.

One record only from arable ground (2849); recorded from all surrounding VCs except 57.

This is a diploid species widespread in Eurasia, which occurs in Britain only as a casual, often in ballast dumps at ports, and in waste places, particularly in the S. of England. It rarely becomes established.

683.18 **Bromus secalinus** L. — Rye Brome
Ray, 1670. Catalogus [*Festuca graminea glumis hirsutis* C. B.]. "Upon Dorsthill, not far from Tamworth."

One record only from a waste place near Stratford on Avon (2154); the name given by Ray would suggest that his plant is one of the varieties *hirsutus* or *hirtus* but as no specimen is available we have preferred to treat this as the first record for the species; recorded from all surrounding VCs.

This tetraploid species has spread through Europe as a contaminant of wheat and rye seeds because of similar seed size and shape; it is therefore hard to remove by mechanical screening. The species seems to have been much commoner at the time Bagnall wrote.

Var. **hirtus** (F. Schultz) Aschers. & Graebn.
Skelding, A. D., 1955. Near Southam.

Two records from arable fields near Haselor (1356) and roadside near Southam (4160); recorded from VCs 32, 33, 37 and 55.

Bromus pseudosecalinus P. Smith
Daulman, Miss B., 1960. Edge Hill.

This plant is rather more slender than *B. secalinus*, and has smaller floral parts. Unlike *B. secalinus* it is a diploid. It seems to occur as a contaminant in sown grass seed, perhaps especially *Lolium multiflorum*, which it resembles in seed size and shape. It is almost certainly introduced into Britain and its distribution has not yet been worked out.

In the present survey there have been five records from rough grassland, meadow and arable land: near Studley (0863), near Grove Park (2365), near Upper Tysoe (3243) and near Edge Hill (3447, 3646); recorded from VC 37.

Bromus squarrosus L. var. **villosus** Koch
Evans, R., 1961. Waverley Wood.
One record only from a rubbish dump at the edge of Waverley Wood (3470); recorded from VC 23.

This is a diploid species characteristic of S. Europe and the Mediterranean area; it is sometimes found in Britain as a casual in waste places.

Section **Ceratochloa** (Beauv.) Griseb.

683.19 **Bromus carinatus** Hook. & Arn. Californian Brome
Goodman, Miss C. M., 1961. Near Cannon Hill Park, Birmingham.
One record only; found by D. E. Allen in 1955 in Cannon Hill Park (VC 37), just outside the boundary of VC 38; in 1961 Miss Goodman noticed that it had spread across the boundary (0683); recorded from VCs 23 and 37.

683.20 **Bromus willdenowii** Kunth.
Bromwich, H., 1877. Herb. Bagnall [*Bromus unioloides* Willd.]. Warwick Old Park.
One record only, introduced with bird-seed at Rugby (4974); recorded from all surrounding VCs except 39 and 57.

684 **Brachypodium** Beauv.

684.1 **Brachypodium sylvaticum** (Huds.) Beauv. Wood False-brome
Purton, T., 1817. Mid. Flora [*Festuca sylvatica*]. Kinwarton; Grafton. Map 702, p. 583
An abundant plant fairly evenly distributed throughout the county; recorded from hedgerows and scrub woodland (56%), mixed woodland (24%), roadsides and railway banks (11%) and watersides (5%); recorded from all surrounding VCs.

684.2 **Brachypodium pinnatum** (L.) Beauv. Chalk False-brome
Purton, T., 1817. Mid. Flora [*Festuca pinnata*]. Grafton; Great Alne. Map 703, p. 584
A frequent plant mainly of grassland on calcareous soil in the S. of the county; recorded from rough grassland and pasture (38%), roadsides and railway banks (33%) and hedgerows and scrub woodland (25%); recorded from all surrounding VCs.

685 **Agropyron** Gaertn.

685.1 **Agropyron caninum** (L.) Beauv. Bearded Couch-grass
Perry, 181–. Herb. Perry [*Triticum caninum* Huds.]. Warwick. Map 704, p. 584
A plant of hedgerows (39%), roadsides (24%), watersides (17%), woods (8%) and rough grassland (8%), thinly but evenly distributed throughout the county; recorded from all surrounding VCs.

685.3 **Agropyron repens** (L.) Beauv. Couch-grass, Squitch
Perry, 181–. Herb. Perry [*Triticum repens*]. Without locality. Map 705, p. 585
A plant of roadsides and waste places (30%), hedgerows (28%), grassland (21%) and cultivated land (16%); abundant and evenly spread throughout the county; recorded from all surrounding VCs.

685.5 **Agropyron junceiforme** (A. & D. Löve) A. & D. Löve Sand Couch-grass
Clark, M. C., 1962. Emscote, Warwick.
Two records, one on introduced sea-sand (2965), the other on a colliery dump (3585); not recorded for any of the surrounding VCs.

687 **Hordeum** L.

687.1 **Hordeum secalinum** Schreb. Meadow Barley
Baxter, W., 1831. MS (Bagnall: Mid. Nat. 1892–93) [*H. pratense* Huds.]. Between West Leys and Newbold Road, Rugby.
Bree, Rev. W. T., 1835. New Bot. Guide [*H. pratense*]. Without locality. Map 706, p. 585
Abundant and fairly evenly distributed in the S.E. of the county, particularly on calcareous soils, and rare elsewhere; Warwickshire is near the W. border of the main British distribution; recorded from grassland (73%), roadsides and waste places (21%) and hedgerows (4%); recorded from all surrounding VCs.

687.2 **Hordeum murinum** L. Wall Barley
Baxter, W., 1842. MS (Bagnall: Flora, 1891). Near Rugby. Map 707, p. 586
Abundant and fairly evenly distributed throughout most of the county, but rare in the Solihull–Kenilworth–Henley in Arden area and only occasional in the extreme S.; recorded from roadsides, waste places and farmyards (75%), dry grassland (12%), hedgerows (5%), cultivated land (4%) and walls and quarries (4%); recorded from all surrounding VCs.

688 **Hordelymus** (Jessen) Harz

688.1 **Hordelymus europaeus** (L.) Harz Wood Barley
Hawkes, J. G., 1955. Aston Grove.
One record from wood on calcareous soil at Aston Grove (1458); recorded from all surrounding VCs except 39 and 55.

689 **Koeleria** Pers.

689.1 **Koeleria cristata** (L.) Pers. Crested Grass
Bromwich, H., 1853. Herb. Perry. Whitnash. Map 708, p. 586
An occasional plant confined to calcareous soils in the S. of the county; recorded from grassland (88%) and roadsides, railway banks and quarries; recorded from all surrounding VCs.

691 **Trisetum** Pers.

691.1 **Trisetum flavescens** (L.) Beauv. Yellow Oat-grass
Perry, 181–. Herb. Perry. Priory Fields, Warwick. Map 709, p. 587
Abundant and evenly distributed throughout the N. half of the county, but less common in the S. half; recorded from grassland (57%), roadsides and railway banks (36%), hedgerows (4%) and a number of other habitats; recorded from all surrounding VCs.

692 **Avena** L.

692.1 **Avena fatua** L. Common Wild Oat
Bree, Rev. W. T., 1835. New Bot. Guide. Without locality. Map 710, p. 587
A frequent and somewhat unevenly distributed plant very rare in the N.W. of the county; recorded from arable land (53%), roadsides, waste places and farmyards (33%), grassland (7%) and hedgerows (6%); recorded from all surrounding VCs.

692.2 **Avena ludoviciana** Durieu Winter Wild Oat
Anon., 1874. Herb. B'ham Univ. [*A. fatua*]. Myton.
Map 711, p. 588
A rare plant mainly confined to the Lias clay in the S.E. of the county; recorded chiefly from arable land (78%); recorded from VCs 23, 32, 33 and 37.

692.3 **Avena strigosa** Schreb. Bristle Oat
Bromwich, H., 1878. Herb. Bagnall. Moreton Morrell.
Not recorded in present survey; recorded from all surrounding VCs.

Avena sativa L. Cultivated Oat
St. Brody, Dr., 1875. Herb. Warw. Mus. Kenilworth.
A frequent relic of cultivation, or result of accidental introduction, scattered throughout the county; recorded in 34 1 km. squares; recorded from all surrounding VCs except 37 and 39.

693 **Helictotrichon** Bess.

693.1 **Helictotrichon pratense** (L.) Pilg.
Meadow Oat-grass
Lloyd, Dr. G., 1835. Herb. Perry [*Avena pratensis* L.]. Blakelow Hill. Map 712, p. 588
An occasional plant, confined to calcareous soils in the S. of the county; recorded from grassland, roadsides and railway banks; recorded from all surrounding VCs.

693.2 **Helictotrichon pubescens** (Huds.) Pilg.
Downy Oat-grass
Baynes, W. W., 1832. MS (Bagnall: Flora, 1891) [*Avena pubescens* L.]. Between Offchurch and Bascote.
Lloyd, Dr. G., 1835. Herb. Perry [*Avena pubescens* Huds.]. Coleshill Heath.
Map 713, p. 589
A frequent plant in the S. and E. of the county where it prefers base-rich soils; recorded from grassland (70%) and roadsides and railway banks (28%); recorded from all surrounding VCs.

694 **Arrhenatherum** Beauv.

694.1 **Arrhenatherum elatius** (L.) Beauv. ex J. & C. Presl False Oat-grass
Perry, 1820. Herb. Perry [*A. avenaceum*]. Without locality. Map 714, p. 589
Abundant and evenly distributed throughout the county; recorded from roadsides, railway banks and waste places (42%), grassland (26%) and hedgerows (26%); recorded from all surrounding VCs.

695 **Holcus** L.

695.1 **Holcus lanatus** L. Yorkshire Fog
Perry, 1820. Herb. Perry. Without location.
Map 715, p. 590
Abundant and evenly distributed throughout the county; recorded from grassland (44%), roadsides, waste places and railway banks (36%), hedgerows and scrub (8%), woods (5%), cultivated land (4%) and watersides (3%); recorded from all surrounding VCs.

695.2 **Holcus mollis** L. Creeping Soft-grass
Baxter, W., about 1844. Bagnall: Flora, 1891. Without locality. Map 716, p. 590
Abundant and evenly distributed throughout most of the county but distinctly less so on the calcareous soils in the S. and S.E.; recorded from roadsides, waste places and railway banks (27%), hedgerows and scrub (25%), grassland (24%), woods (17%) and a few other habitats; recorded from all surrounding VCs.

696 **Deschampsia** Beauv.

696.1 **Deschampsia cespitosa** (L.) Beauv.
Tufted Hair-grass
Perry, 1824. Herb. Perry [*Aira cespitosa*]. Without locality. Map 717, p. 591
Abundant and fairly evenly distributed throughout the county; recorded from watersides and marshes (31%), damp grassland (31%), roadsides, railway banks and waste places (16%), woods (11%) and hedgerows and scrub (9%); recorded from all surrounding VCs.

696.3 **Deschampsia flexuosa** (L.) Trin.
Wavy Hair-grass
Withering, 1796. Arrangement. Dry woods in Sutton Park. Map 718, p. 591
Frequent in light acid soils in the N.W. of the county but rare or absent elsewhere; recorded from woods (30%), roadsides, waste places and railway banks (19%), scrub and hedgerows (16%), grassland (14%), dry heath (13%) and watersides (6%); recorded from all surrounding VCs.

697 **Aira** L.

697.1 **Aira praecox** L. Early Hair-grass
Perry, 1827. Herb. Perry. Kenilworth Heath.
Map 719, p. 592
A local plant mainly restricted to dry sandy and gravelly places on railways and in grassland, waste places and quarries in a few areas in the N.W. of the county; recorded from all surrounding VCs.

697.2 **Aira caryophyllea** L. Silvery Hair-grass
Purton, T., 1817. Mid. Flora. Oversley Wood, on a sandy bank. Map 720, p. 592
A rare plant recorded from the sides of railway tracks (75%) and dry sandy and gravelly places in grassland; widely but unevenly distributed throughout the county; recorded from all surrounding VCs.

699 **Ammophila** Host

699.1 **Ammophila arenaria** (L.) Link Marram Grass
Phipps, J. B., 1956. Hatton.
A rare casual recorded twice in introduced sea-sand (2266, 2965) and once on a colliery tip (3585); not recorded from any of the surrounding VCs.

700 **Calamagrostis** Adans.

700.1 **Calamagrostis epigejos** (L.) Roth
Wood Smallreed
Purton, T., 1817. Mid. Flora [*Arundo epigejos*]. Salford; Dunnington; Wetheley.
Map 721, p. 593
An occasional plant recorded from ditches and watersides (29%), scrub woodland and hedgerows (24%), rough grassland (15%), woods (12%) and a few other habitats; recorded from all surrounding VCs.

700.2 **Calamagrostis canescens** (Weber) Roth
Purple Smallreed
Perry, 1820. Plant. Varv. Sel. [*Arundo calamagrostis*]. Without locality.

One record only from Binley Bogs, Brandon (3876); formerly much more frequent and recorded from Sutton, Olton Pool, Ufton Wood and near Griff; recorded from all surrounding VCs except 23 and 33.

701 Agrostis L.

701.2 **Agrostis canina** L. Brown Bent
Perry, 1829. Herb. Perry. Hill Wootton.
Map 722, p. 593

An occasional plant, mainly of acid grassland but also occurring in woods and on roadsides and railway banks in the N.W. of the county.

The two sub-species: ssp. *canina* (first record Perry, 1829, Herb. Perry, as above) and ssp. *montana* (Hartm.) Hartm. (first record Kirk, 1859. Herb. Perry, Kenilworth) have been recorded several times.

The species has been recorded from all surrounding VCs.

701.3 **Agrostis tenuis** Sibth. Common Bent
Perry, 1820. Herb. Perry [*A. vulgaris*]. Without locality. Map 723, p. 594

A plant mainly of grassland (57%), often being the main constituent, and roadsides and waste places (30%); abundant and widely distributed throughout the county though less common in the S.E.; recorded from all surrounding VCs.

701.4 **Agrostis gigantea** Roth Black Bent
Bagnall, 1881. Herb. Bagnall [*A. nigra* With.]. Allesley. Map 724, p. 594

A plant of cultivated land (38%), roadsides and waste places (31%) and grassland (27%), thinly and unevenly distributed throughout the county but more frequent in the W.; recorded from all surrounding VCs.

The hybrid *A. gigantea* × *stolonifera* has been recorded (Laflin, T., 1962) in pasture near Darlingscott (2242).

701.5 **Agrostis stolonifera** L. Creeping Bent
Baxter, W., 1831. MS (Bagnall: Mid. Nat. 1892–93) [*A. alba* L.]. Near Newbold on Avon.
Map 725, p. 595

Abundant and widely distributed throughout the county; recorded from grassland (38%), roadsides and waste places (34%), cultivated land (13%), waterside (8%) and a few other habitats; recorded from all surrounding VCs.

701.6 **Agrostis avenacea** J. F. Gmel.
Hackett, A., 1964. Bedworth.

A small colony of many plants on a colliery dump at Bedworth (3585); recorded from VC 37.

701.8 **Agrostis semiverticillata** (Forsk.) C. Chr.
Water Bent
Druce, 1932. Com. Flora. Without locality.

Not recorded in the county during the present survey; recorded from VCs 33, 37 and 39.

702 Apera Adans.

702.1 **Apera spica-venti** (L.) Beauv. Dense Silky-bent
Perry, 1820. Herb. Perry and Plant. Varv. Sel. [*Agrostis spica-venti* L.]. "In a newly-made garden, near the Brick Yard, Saltisford, Warwick."

A casual not recorded in the present survey; recorded from all surrounding VCs except 32.

703 Polypogon Desf.

703.1 **Polypogon monspeliensis** (L.) Desf.
Annual Beardgrass
St. Brody, Dr., 1875. Herb. Bagnall. Kenilworth Heath.

An introduction recorded from a coal-dump near Bedworth (3585) where it is abundant and apparently firmly established; recorded from all surrounding VCs except 55 and 57.

705 Gastridium Beauv.

705.1 **Gastridium ventricosum** (Gouan) Schinz & Thell. Nitgrass
Bloxam, Rev. A., 1837. New Bot. Guide [*G. lendigerum*]. Cornfield near Alcester.

A very rare plant of sandy fields and calcareous pastures recorded only once from an arable field W. of Alcester (0554); formerly more widespread in similar areas in the lower Avon basin; recorded from VCs 33 and 37.

707 Phleum L.

707.1 **Phleum bertolonii** DC. Small Catstail
Perry, 1823. Herb. Perry [*P. pratense* var. *nodosum*]. Footroad leading from Leamington to Emscote. Map 726, p. 595

Abundant and fairly evenly distributed throughout the county; recorded from grassland (60%), roadsides, waste places and railway banks (30%), cultivated land (4%) and hedgerows (4%); recorded from all surrounding VCs.

707.2 **Phleum pratense** L. Timothy Grass
Perry, 1845. Herb. Perry [*P. pratense*]. Wedgenock Deer Park. Map 727, p. 596

Abundant and evenly distributed throughout the county; recorded from grassland (51%), roadsides and waste places (27%), cultivated land (14%) and hedgerows (6%); recorded from all surrounding VCs.

707.5 **Phleum arenarium** L. Sand Catstail
Clark, M. C., 1962. Emscote.

One record on introduced sea-sand at Emscote (2965); recorded from VC 33.

708 Alopecurus L.

708.1 **Alopecurus myosuroides** Huds. Slender Foxtail
Perry, 1820. Herb. Perry [*A. agrestis* L.]. Without locality. Map 728, p. 596

A fairly frequent plant of arable land (54%), roadsides and waste places (23%) and grassland (15%), mainly in the S. part of the county; recorded from all surrounding VCs.

708.2 **Alopecurus pratensis** L. Meadow Foxtail
Bagnall, 1867. Herb. Bagnall. Marston Green.
Map 729, p. 597

A very abundant plant throughout the county, recorded from grassland (55%), roadsides and waste places (33%), hedgerows (7%) and most other habitats; recorded from all surrounding VCs.

708.3 **Alopecurus geniculatus** L. Marsh Foxtail
Perry 181–. Herb. Perry. Without locality.
Map 730, p. 597

A frequent and widely distributed plant of watersides

(65%), damp grassland (17%), damp arable fields (10%) and roadsides and waste places (6%); recorded from all surrounding VCs.

Alopecurus geniculatus × pratensis
= **A. × hybridus** Wimm.
Bromwich, H., 1899. Journal of Botany, 1901. By the River Avon at Chesford Bridge, Kenilworth.
Not recorded in present survey; recorded from VCs 23 and 55.

708.4 **Alopecurus aequalis** Sobol. Orange Foxtail
Withering, 1796. Arrangement [*A. geniculatus* L. var. 4]. "In a marshy place by the Stews in Edgbaston Park." Map 731, p. 598
A rare plant of a few scattered localities in the county chiefly at the edges of ponds; recorded from all surrounding VCs. According to Druce (Com. Flora, 1932) the Withering record is the first British record.

709 **Milium** L.

709.1 **Milium effusum** L. Wood Millet
Purton, T., 1817. Mid. Flora. Ragley and Oversley Wood. Map 732, p. 598
Frequent in the N.W. of the county mainly in areas which were well wooded in the seventeenth century; recorded from woods (78%), scrub and hedgerows (20%); recorded from all surrounding VCs.

712 **Anthoxanthum** L.

712.1 **Anthoxanthum odoratum** L.
Sweet Vernal-grass
Perry, 1813. Herb. Perry. Without locality.
Map 733, p. 599
Abundant and widely distributed throughout all types of grassland: meadow, pasture and rough grass (70%), roadsides, railway banks and waste places (20%); occasionally found in other habitats; recorded from all surrounding VCs.

712.2 **Anthoxanthum puelii** Lecoq & Lamotte
Annual Vernal-grass
Bromwich, H., 1891. (Bagnall's Flora.) Allotments near Leamington.
A casual not recorded in present survey; recorded from VCs 23, 32, 37 and 39.

713 **Phalaris** L.

713.1 **Phalaris arundinacea** L. Reed-grass
Perry, 1820. Plant. Varv. Sel. Near Leamington. Map 734, p. 599
Abundant and fairly evenly distributed throughout most of the county; recorded from watersides and marshes (90%), water (5%) and a few other habitats; recorded from all surrounding VCs.

713.2 **Phalaris canariensis** L. Canary Grass
Baxter, W., 1834. Phaen. Botany. Near Rugby, on the road to Bilton. Map 735, p. 600
An introduced plant occurring occasionally in a number of places widely scattered throughout the county; recorded from waste places and roadsides (88%) and a number of other habitats; recorded from all surrounding VCs.

714 **Parapholis** C. E. Hubbard

714.2 **Parapholis incurva** (L.) C. E. Hubbard
Curved Hard-grass
Jackson, A. B., 1902. Herb. Brit. Mus. Waste ground near skinyards at Kenilworth.
Not recorded in present survey; not recorded from any of the surrounding VCs.

715 **Nardus** L.

715.1 **Nardus stricta** L. Mat-grass
Stokes, Dr. J., 1787. Withering's Arrangement. Birmingham Heath. Map 736, p. 600
Abundant in Sutton Park and frequent in the Packington area, with occasional occurrences elsewhere, mainly in the N.W. half of the county; recorded from acid heaths (50%), railway banks, waste places, woods and grassland; recorded from all surrounding VCs.

718 **Echinochloa** Beauv.

718.1 **Echinochloa crus-galli** (L.) Beauv. Cockspur
Anon., 1874. Herb. B'ham Univ. Kenilworth.
A rare introduction which appeared in one cultivated field at Luddington (1552); recorded from all surrounding VCs except 57.

719 **Digitaria** Heist. ex P. C. Fabr.

719.2 **Digitaria sanguinalis** (L.) Scop.
Hairy Finger-grass
Bromwich, H., 1874. Herb. B'ham Univ. [*Panicum sanguinale* Scop.]. Near Kenilworth Railway Station.
Not recorded in present survey; last record from Milverton by Bromwich in 1899; recorded from VCs 23, 37 and 39.

720 **Setaria** Beauv.

720.1 **Setaria viridis** (L.) Beauv. Green Bristle-grass
Cheshire, W., 1856. Herb. Perry. Island below the Mill at Stratford on Avon.
Four records: from cultivated ground at Birmingham (0588) and Wellesbourne (2756, 2854), and from a waste place at Newbold on Avon (4876); previously also collected from Kenilworth, Leamington and Milverton; recorded from all surrounding VCs.

720.2 **Setaria verticillata** (L.) Beauv.
Rough Bristle-grass
Jackson, A. B., 1897. Herb. Brit. Mus. Milverton.
One record from a garden at Edgbaston (0583); recorded from VCs 23, 33 and 37.

720.3 **Setaria lutescens** (Weigel) Hubbard
Yellow Bristle-grass
Bromwich, H., 1884. Herb. Bagnall [*S. glauca*]. Milverton Station.
One record from a cultivated field at Edge Hill (3847); recorded from all surrounding VCs except 57.

Setaria italica (L.) Beauv. Italian Millet
Bromwich, H., 1901. Herb. Brit. Mus. Milverton.
A rare plant of a cultivated field near Luddington (1652) and waste places at Clifford Bridge (3780), W. of Newbold Grange (4876) and Hillmorton Ballast Pits (5473); recorded from VCs 23, 33 and 55.

Panicum L.

Panicum miliaceum L. Millet
Bromwich, H., 1884. Herb. Bagnall. Waste ground, Milverton.
A casual recorded from waste places at Temple Grafton (1253), near Coventry (3780), W. of Newbold Grange (4876), Rugby (4974) and Hillmorton Ballast Pits (5473); recorded from all surrounding VCs.

9 and 10. DISTRIBUTION MAPS OF FLOWERING AND OTHER VASCULAR PLANTS

The distribution maps are presented in two series, the first being drawn by an incremental graph plotter (Maps 19–736) and the second by a line printer (Maps 737–95). Both series are produced by means of computer-generated punched tape as explained in Chapter 6. In both series of maps each symbol or combination of symbols is plotted on a "tetrad" (2 × 2 km), irrespective of which 1 km square or squares within the tetrad the record comes from. All records from the present survey, whether from random or non-random squares, are included on these series of maps. The species in each are arranged in the same taxonomic sequence as the text, though many species have of course not been mapped at all, where insufficient records were available. In general, graph-plotter maps have been made of all species for which about 15 records or more were available and line-printer maps for all species with about 7–15 records. However, there are some line-printer maps with more than 15 records where we have considered that the habitat range of the species was so limited that a graph-plotted map with habitat symbols would hardly be appropriate.

We have excluded certain species from the map sequences where we were in doubt about the maps showing anything approaching a true record of the distribution (certain *Rubus* microspecies) or where the plants occurred completely sporadically as weeds of cultivation or as casuals.

In the upper right-hand panel of each map are given the Dandy Check List number, the Latin name with the authority, and the English name.

It must be remembered that the absence of a record for a species in a 2 × 2 km square does not necessarily mean that the species does not occur there. It may have been very rare and hence overlooked; the recorder for that square may not have known it; or it may have occurred in one or more of the non-random squares of the tetrad that were not surveyed.

Apart from the fact that the presence of some species may have been missed from occasional squares within the general area of their distribution, we are confident that the maps give a reasonably accurate picture of the distribution of the species recorded during the present survey.

9. GRAPH PLOTTED MAPS (Nos. 19-736)

The following series of maps (2 per page) are drawn by an incremental graph plotter and give 3 main classes of information:

(1) The presence or absence of a particular species in each tetrad. This indicates general distribution and gives some idea of frequency for the species in different parts of the county.

(2) The particular major habitat (or habitats) in which the species has been recorded for each tetrad. It will be seen that the commoner species have often been recorded for 2–4 or even 5 habitats. A distinctive symbol has been used for each habitat, and this is shown in the key to the symbols in the bottom right-hand corner of each map.

(3) A very approximate idea of frequency for each habitat within the tetrad. This has been shown by a thick line for all habitat records in which the species has been recorded as "frequent", "locally frequent", "abundant" or "locally abundant"; and a thin line for all cases where the species has been recorded as "occasional" or "rare". Where a species has been given 2 or more frequency ratings for different minor habitats or within 2 or more different 1 km squares in the tetrad the computer has been programmed to select the "most frequent" reading in the sequence "r, o, lf, f, la, a".

Transparent overlays are available in a pocket at the back of the book to be used in conjunction with these maps, as follows:

Map I	Relief
Map II	Rivers, canals and lakes
Map III	Surface geology
Map IV	Rainfall (inches). Mean annual (1951–60)
Map V	January mean minimum temperature
Map VI	July mean maximum temperature
Map VII	Well-wooded areas (1650)
Map VIII	Woods, parks, plantations, urbanisation, etc., 1960.

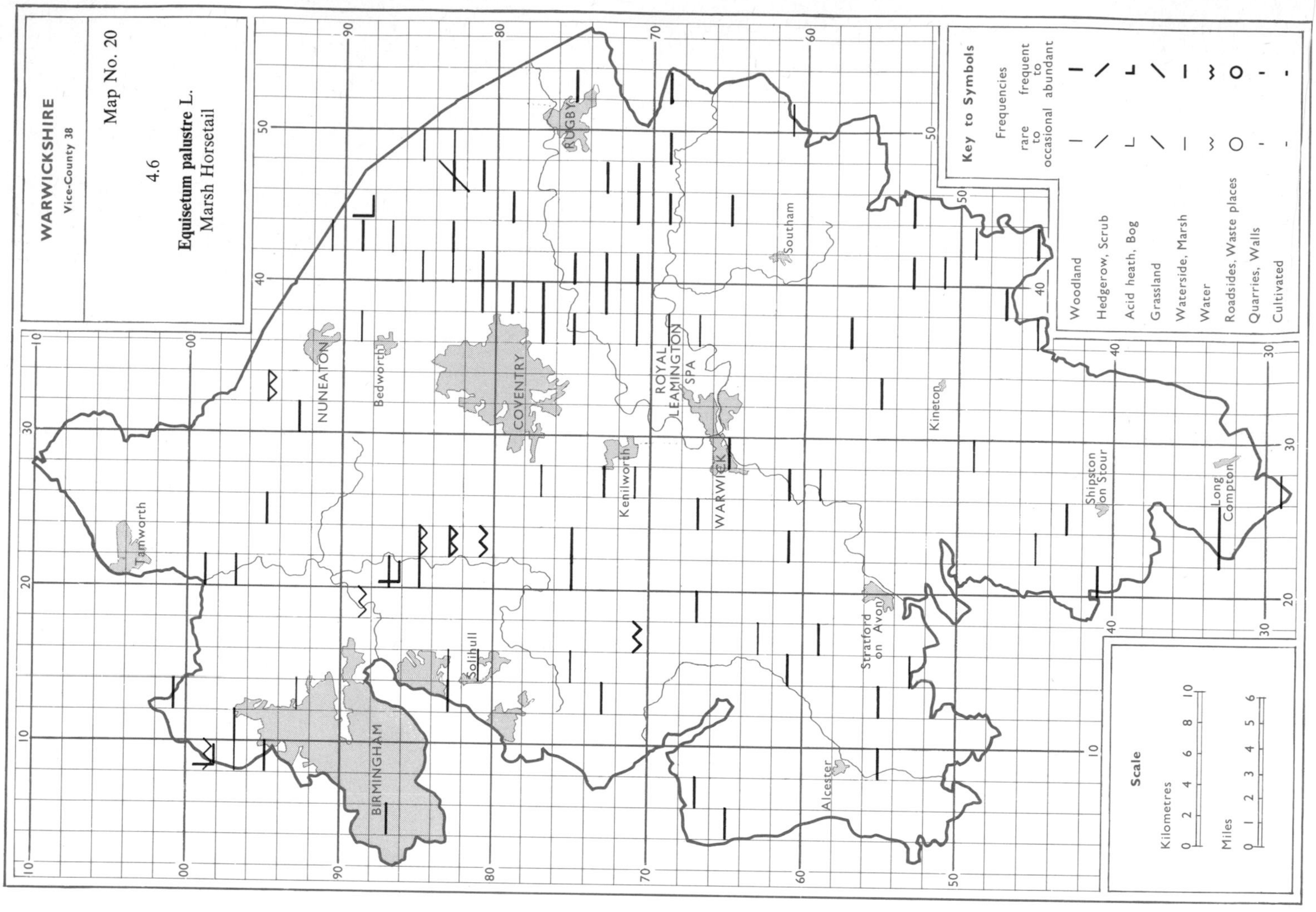
WARWICKSHIRE
Vice-County 38
Map No. 20
4.6
Equisetum palustre L.
Marsh Horsetail
Key to Symbols
Frequencies
rare to occasional
frequent to abundant
Woodland
Hedgerow, Scrub
Acid heath, Bog
Grassland
Waterside, Marsh
Water
Roadsides, Waste places
Quarries, Walls
Cultivated
Scale
Kilometres
0 2 4 6 8 10
Miles
0 1 2 3 4 5 6
Tamworth
NUNEATON
Bedworth
COVENTRY
RUGBY
BIRMINGHAM
Solihull
Kenilworth
ROYAL LEAMINGTON SPA
WARWICK
Southam
Kineton
Stratford on Avon
Alcester
Shipston on Stour
Long Compton

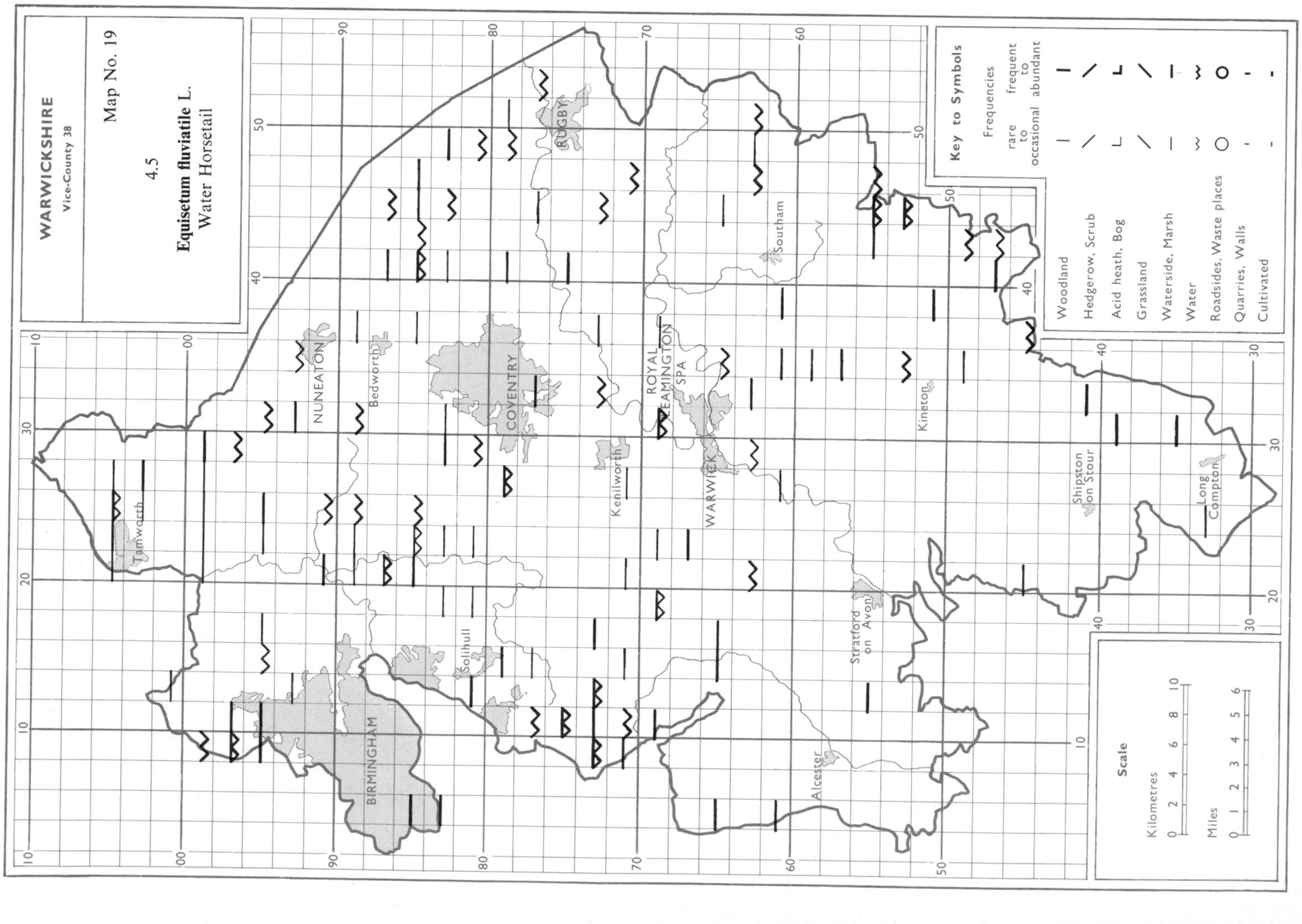
WARWICKSHIRE
Vice-County 38
Map No. 19
4.5
Equisetum fluviatile L.
Water Horsetail
Key to Symbols
Frequencies
rare to occasional
frequent to abundant
Woodland
Hedgerow, Scrub
Acid heath, Bog
Grassland
Waterside, Marsh
Water
Roadsides, Waste places
Quarries, Walls
Cultivated
Scale
Kilometres
0 2 4 6 8 10
Miles
0 1 2 3 4 5 6
Tamworth
NUNEATON
Bedworth
COVENTRY
RUGBY
BIRMINGHAM
Solihull
Kenilworth
ROYAL LEAMINGTON SPA
WARWICK
Southam
Kineton
Stratford on Avon
Alcester
Shipston on Stour
Long Compton

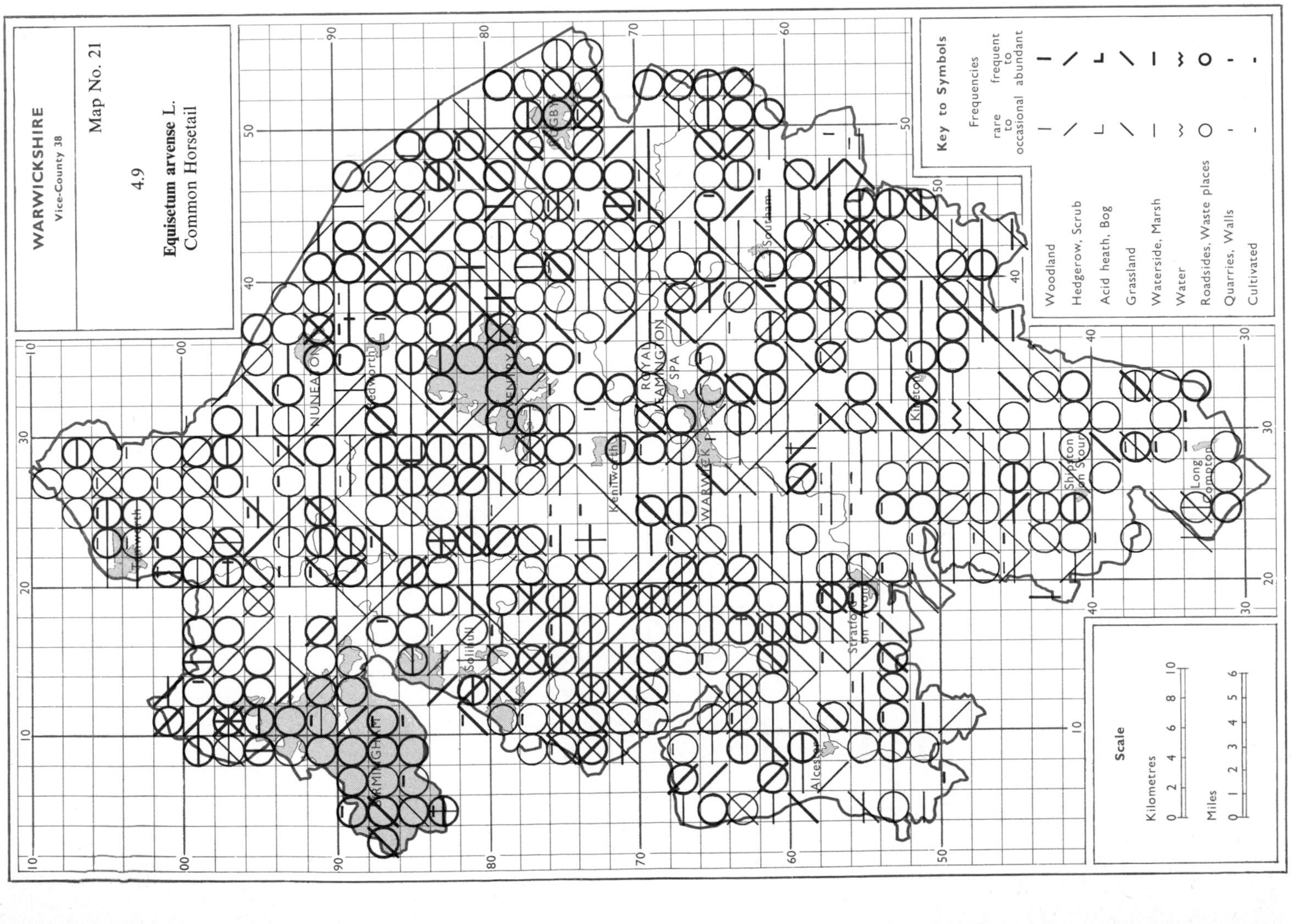
WARWICKSHIRE
Vice-County 38
Map No. 21
4.9
Equisetum arvense L.
Common Horsetail
Key to Symbols
Frequencies
rare to occasional
frequent to abundant
Woodland
Hedgerow, Scrub
Acid heath, Bog
Grassland
Waterside, Marsh
Water
Roadsides, Waste places
Quarries, Walls
Cultivated
Scale
Kilometres
Miles
NUNEATON
Bedworth
COVENTRY
RUGBY
Southam
ROYAL LEAMINGTON SPA
WARWICK
Kenilworth
Kineton
Shipston on Stour
Long Compton
Tamworth
Solihull
BIRMINGHAM
Stratford on Avon
Alcester

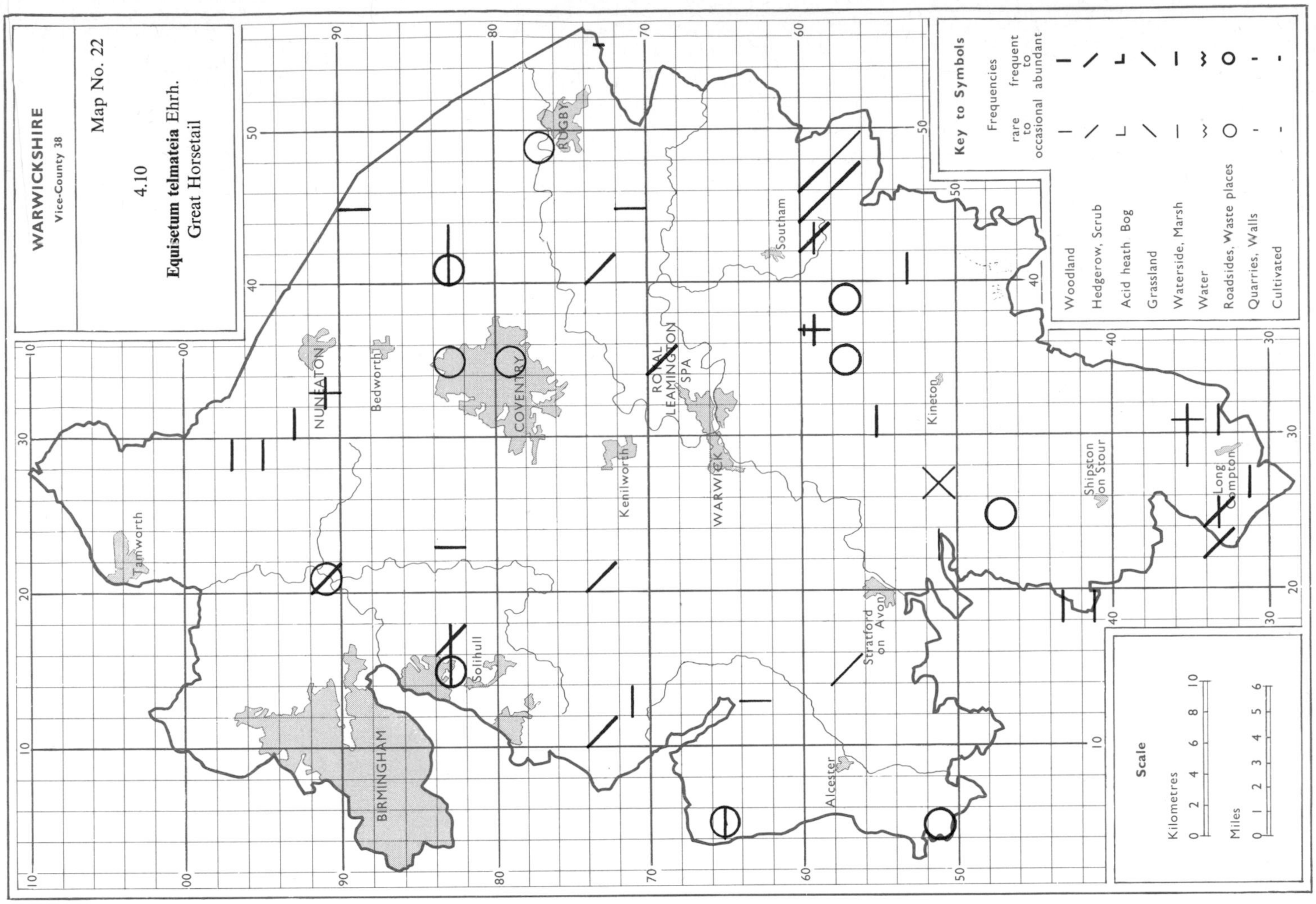
WARWICKSHIRE
Vice-County 38
Map No. 22
4.10
Equisetum telmateia Ehrh.
Great Horsetail
Key to Symbols
Frequencies
rare to occasional
frequent to abundant
Woodland
Hedgerow, Scrub
Acid heath Bog
Grassland
Waterside, Marsh
Water
Roadsides, Waste places
Quarries, Walls
Cultivated
Scale
Kilometres
Miles
NUNEATON
Bedworth
COVENTRY
RUGBY
Southam
ROYAL LEAMINGTON SPA
WARWICK
Kenilworth
Kineton
Shipston on Stour
Long Compton
Tamworth
Solihull
BIRMINGHAM
Stratford on Avon
Alcester

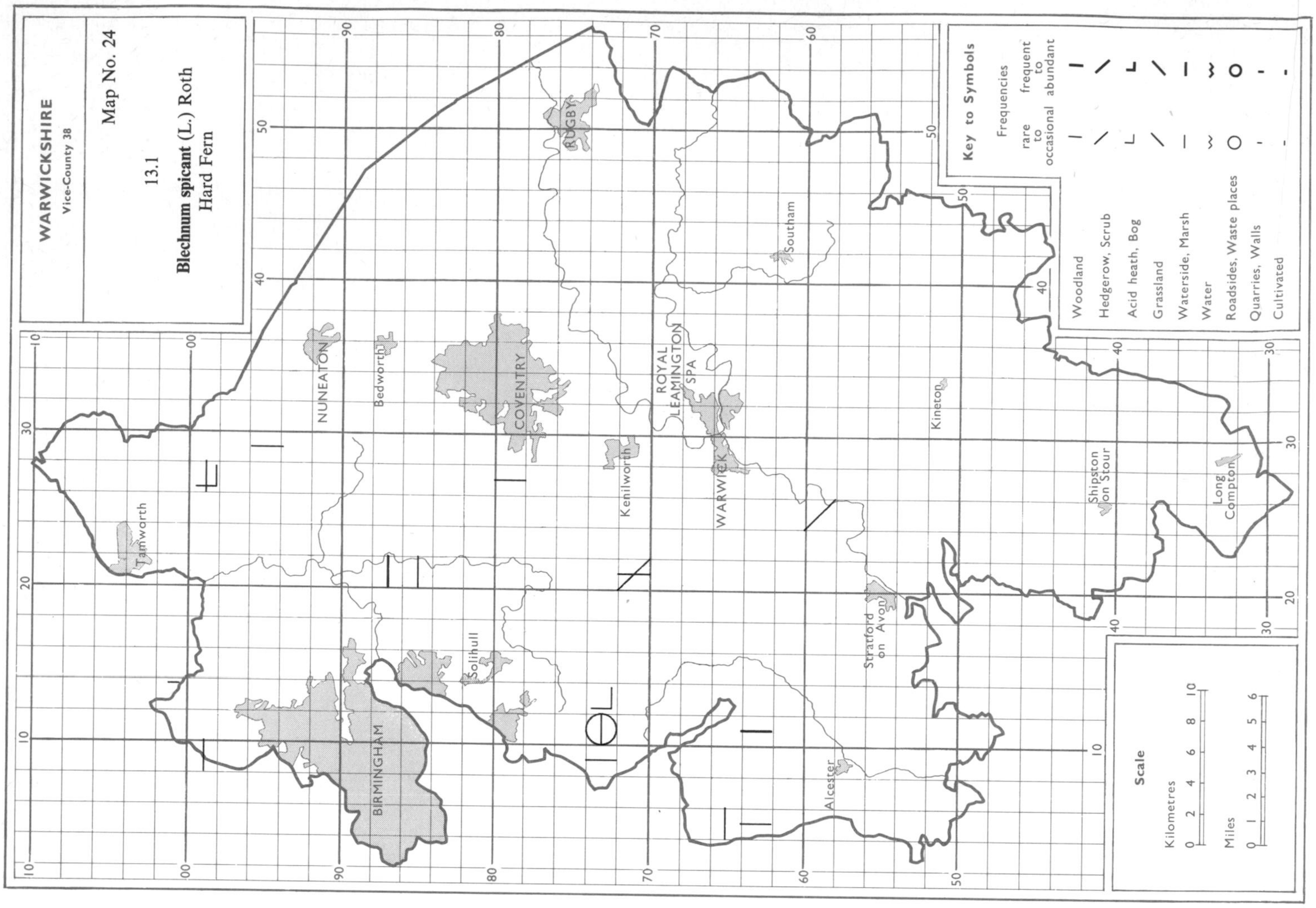
WARWICKSHIRE
Vice-County 38
Map No. 24
13.1
Blechnum spicant (L.) Roth
Hard Fern
Key to Symbols
Frequencies
rare to occasional
frequent to abundant
Woodland
Hedgerow, Scrub
Acid heath, Bog
Grassland
Waterside, Marsh
Water
Roadsides, Waste places
Quarries, Walls
Cultivated
Scale
Kilometres
0 2 4 6 8 10
Miles
0 1 2 3 4 5 6
NUNEATON
Bedworth
COVENTRY
RUGBY
Southam
ROYAL LEAMINGTON SPA
Kenilworth
WARWICK
Kineton
Shipston on Stour
Long Compton
Tamworth
Solihull
BIRMINGHAM
Stratford on Avon
Alcester

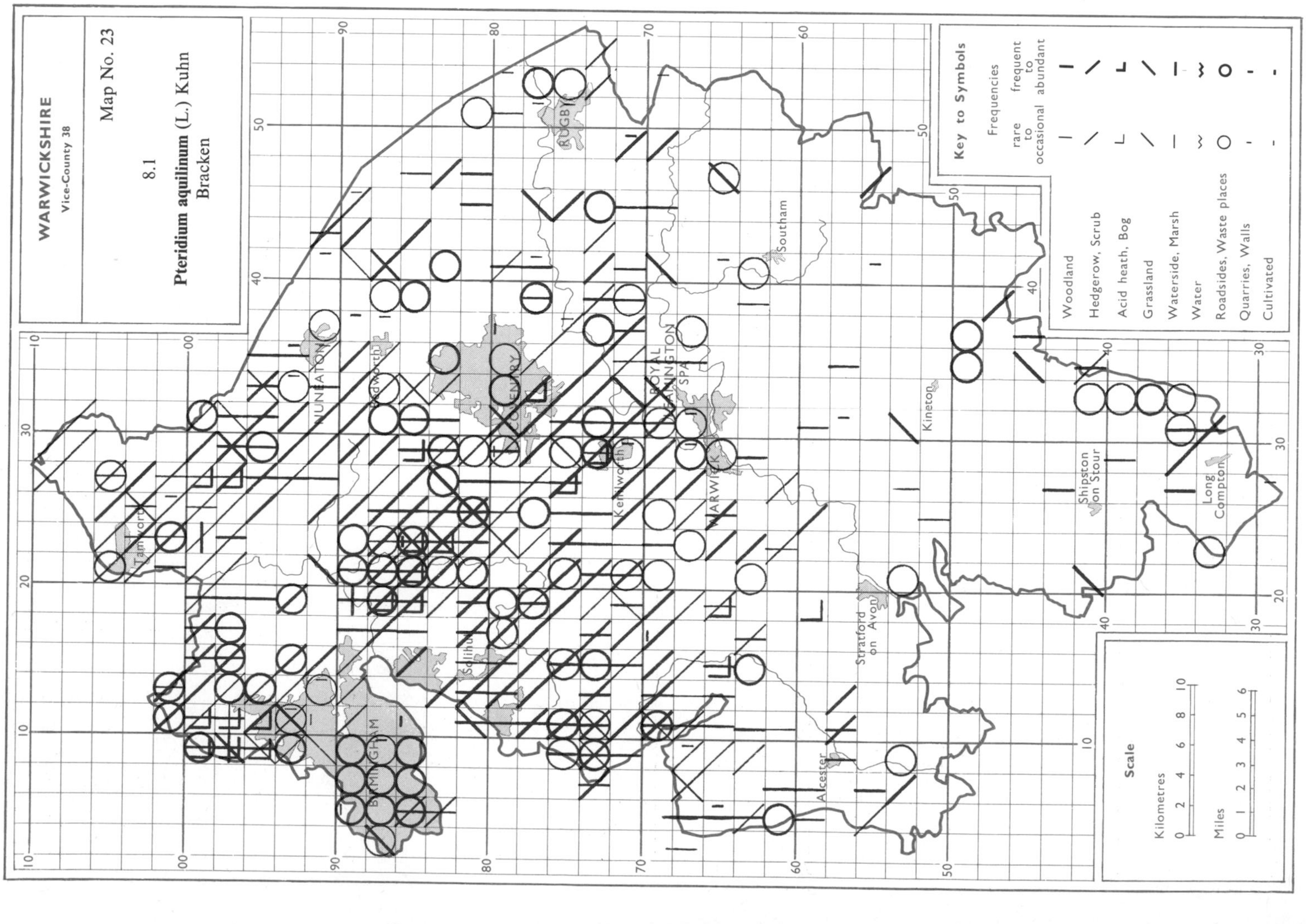
WARWICKSHIRE
Vice-County 38
Map No. 23
8.1
Pteridium aquilinum (L.) Kuhn
Bracken
Key to Symbols
Frequencies
rare to occasional
frequent to abundant
Woodland
Hedgerow, Scrub
Acid heath, Bog
Grassland
Waterside, Marsh
Water
Roadsides, Waste places
Quarries, Walls
Cultivated
Scale
Kilometres
0 2 4 6 8 10
Miles
0 1 2 3 4 5 6
NUNEATON
Bedworth
COVENTRY
RUGBY
Southam
ROYAL LEAMINGTON SPA
Kenilworth
WARWICK
Kineton
Shipston on Stour
Long Compton
Tamworth
Solihull
BIRMINGHAM
Stratford on Avon
Alcester

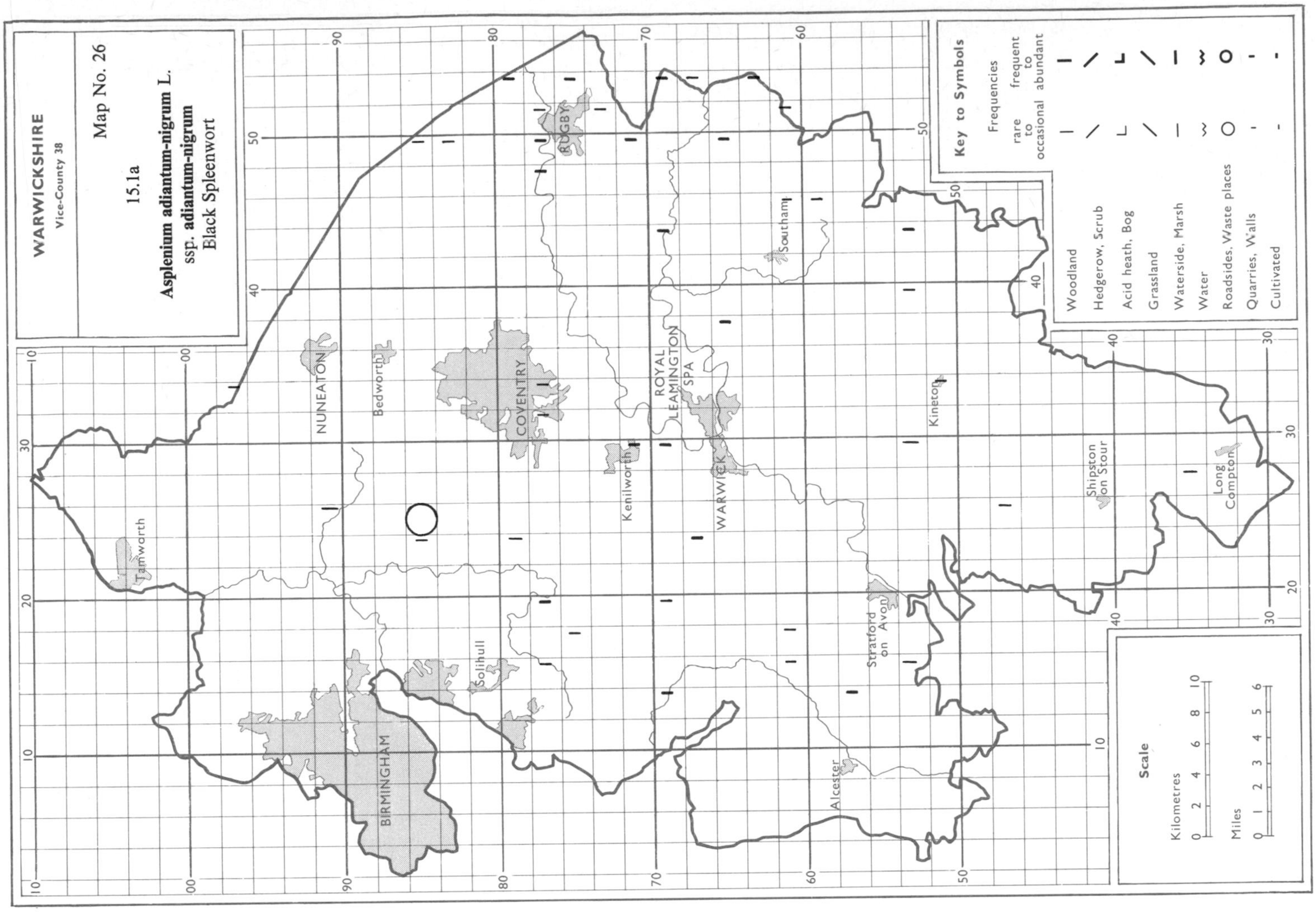
WARWICKSHIRE
Vice-County 38
Map No. 26
15.1a
Asplenium adiantum-nigrum L.
ssp. adiantum-nigrum
Black Spleenwort
Key to Symbols
Frequencies
rare to occasional
frequent to abundant
Woodland
Hedgerow, Scrub
Acid heath, Bog
Grassland
Waterside, Marsh
Water
Roadsides, Waste places
Quarries, Walls
Cultivated
Scale
Kilometres
Miles
NUNEATON
Bedworth
COVENTRY
RUGBY
ROYAL LEAMINGTON SPA
WARWICK
Kenilworth
Southam
Kineton
Shipston on Stour
Long Compton
Tamworth
Solihull
BIRMINGHAM
Stratford on Avon
Alcester

WARWICKSHIRE
Vice-County 38
Map No. 25
14.1
Phyllitis scolopendrium (L.) Newm.
Hartstongue Fern
Key to Symbols
Frequencies
rare to occasional
frequent to abundant
Woodland
Hedgerow, Scrub
Acid heath, Bog
Grassland
Waterside, Marsh
Water
Roadsides, Waste places
Quarries, Walls
Cultivated
Scale
Kilometres
Miles
NUNEATON
Bedworth
COVENTRY
RUGBY
ROYAL LEAMINGTON SPA
WARWICK
Kenilworth
Southam
Kineton
Shipston on Stour
Long Compton
Tamworth
Solihull
BIRMINGHAM
Stratford on Avon
Alcester

WARWICKSHIRE
Vice-County 38
Map No. 27
15.5
Asplenium trichomanes L.
Maidenhair Spleenwort
Tamworth
NUNEATON
Bedworth
BIRMINGHAM
Solihull
COVENTRY
RUGBY
Kenilworth
ROYAL LEAMINGTON SPA
WARWICK
Southam
Alcester
Stratford on Avon
Kineton
Shipston on Stour
Long Compton
Scale
Kilometres
Miles
Key to Symbols
Frequencies
rare to occasional
frequent to abundant
Woodland
Hedgerow, Scrub
Acid heath, Bog
Grassland
Waterside, Marsh
Water
Roadsides, Waste places
Quarries, Walls
Cultivated

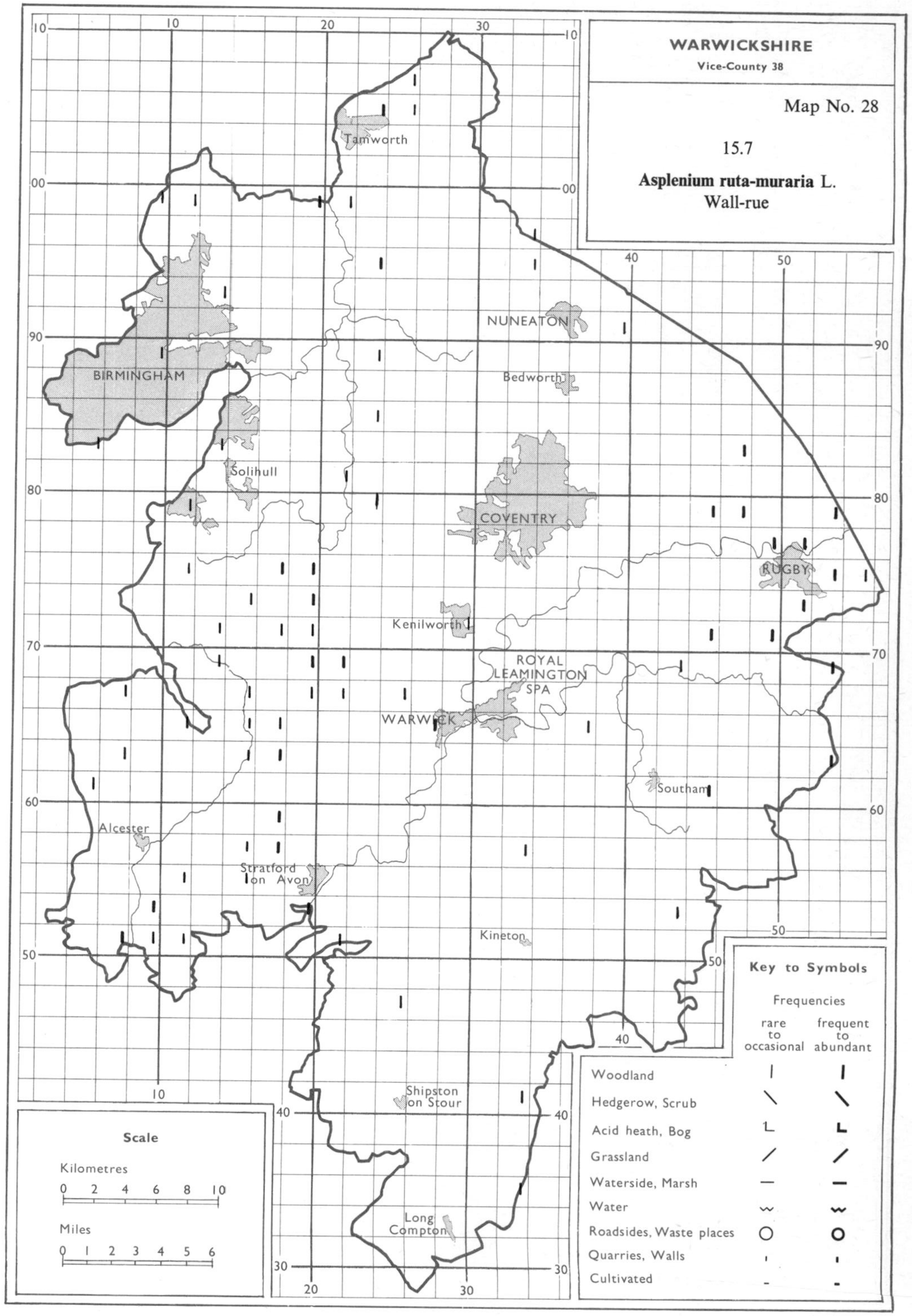
WARWICKSHIRE
Vice-County 38
Map No. 28
15.7
Asplenium ruta-muraria L.
Wall-rue
Tamworth
NUNEATON
Bedworth
BIRMINGHAM
Solihull
COVENTRY
RUGBY
Kenilworth
ROYAL LEAMINGTON SPA
WARWICK
Southam
Alcester
Stratford on Avon
Kineton
Shipston on Stour
Long Compton
Scale
Kilometres
Miles
Key to Symbols
Frequencies
rare to occasional
frequent to abundant
Woodland
Hedgerow, Scrub
Acid heath, Bog
Grassland
Waterside, Marsh
Water
Roadsides, Waste places
Quarries, Walls
Cultivated

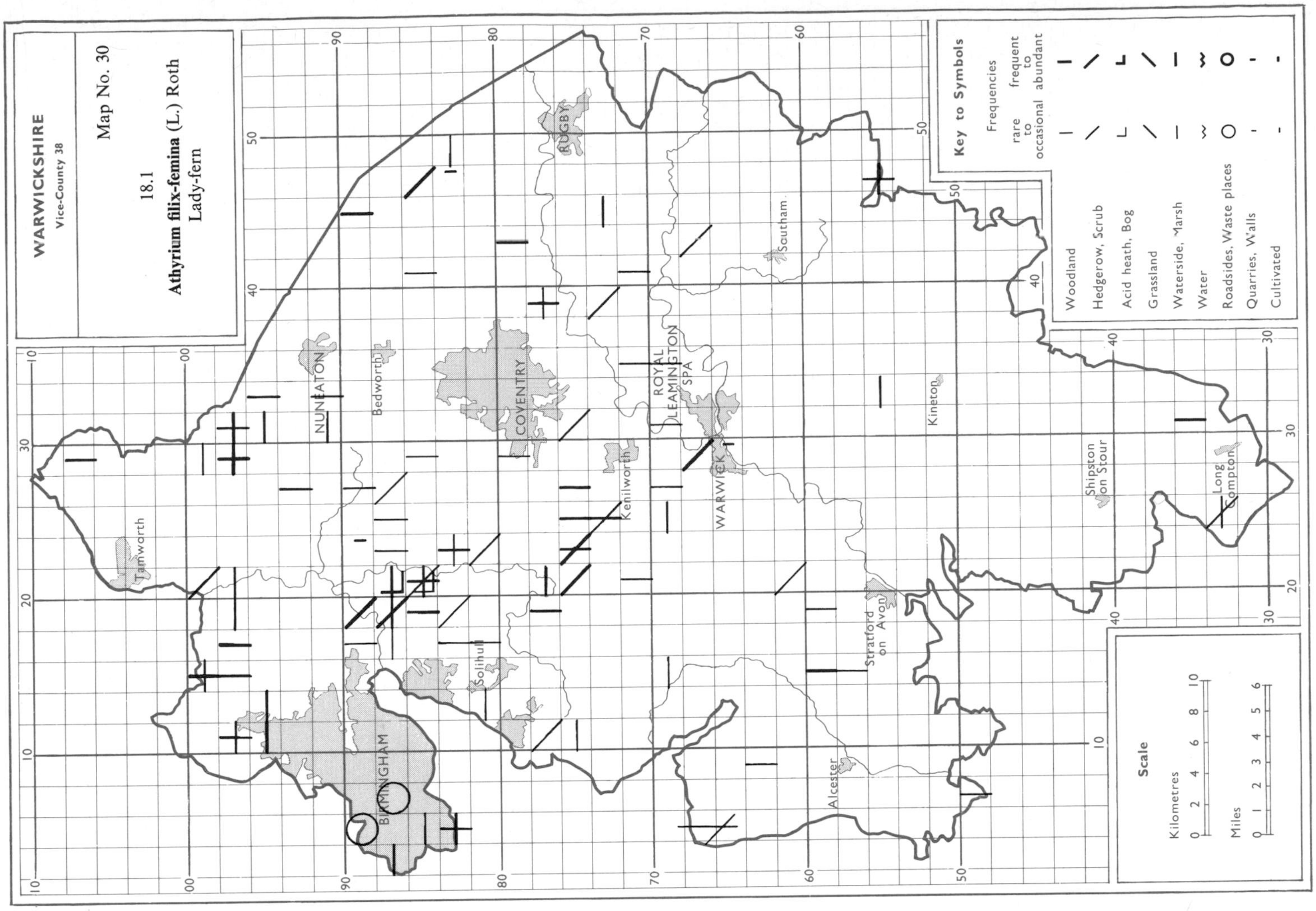
WARWICKSHIRE
Vice-County 38
Map No. 30
18.1
Athyrium filix-femina (L.) Roth
Lady-fern
Key to Symbols
Frequencies
rare to occasional
frequent to abundant
Woodland
Hedgerow, Scrub
Acid heath, Bog
Grassland
Waterside, Marsh
Water
Roadsides, Waste places
Quarries, Walls
Cultivated
Scale
Kilometres
Miles
NUNEATON
Bedworth
COVENTRY
RUGBY
ROYAL LEAMINGTON SPA
Southam
Kineton
Kenilworth
WARWICK
Shipston on Stour
Long Compton
Tamworth
Solihull
BIRMINGHAM
Stratford on Avon
Alcester

WARWICKSHIRE
Vice-County 38
Map No. 29
16.1
Ceterach officinarum DC.
Rusty-back Fern
Key to Symbols
Frequencies
rare to occasional
frequent to abundant
Woodland
Hedgerow, Scrub
Acid heath, Bog
Grassland
Waterside, Marsh
Water
Roadsides, Waste places
Quarries, Walls
Cultivated
Scale
Kilometres
Miles
NUNEATON
Bedworth
COVENTRY
RUGBY
ROYAL LEAMINGTON SPA
Southam
Kineton
Kenilworth
WARWICK
Shipston on Stour
Long Compton
Tamworth
Solihull
BIRMINGHAM
Stratford on Avon
Alcester

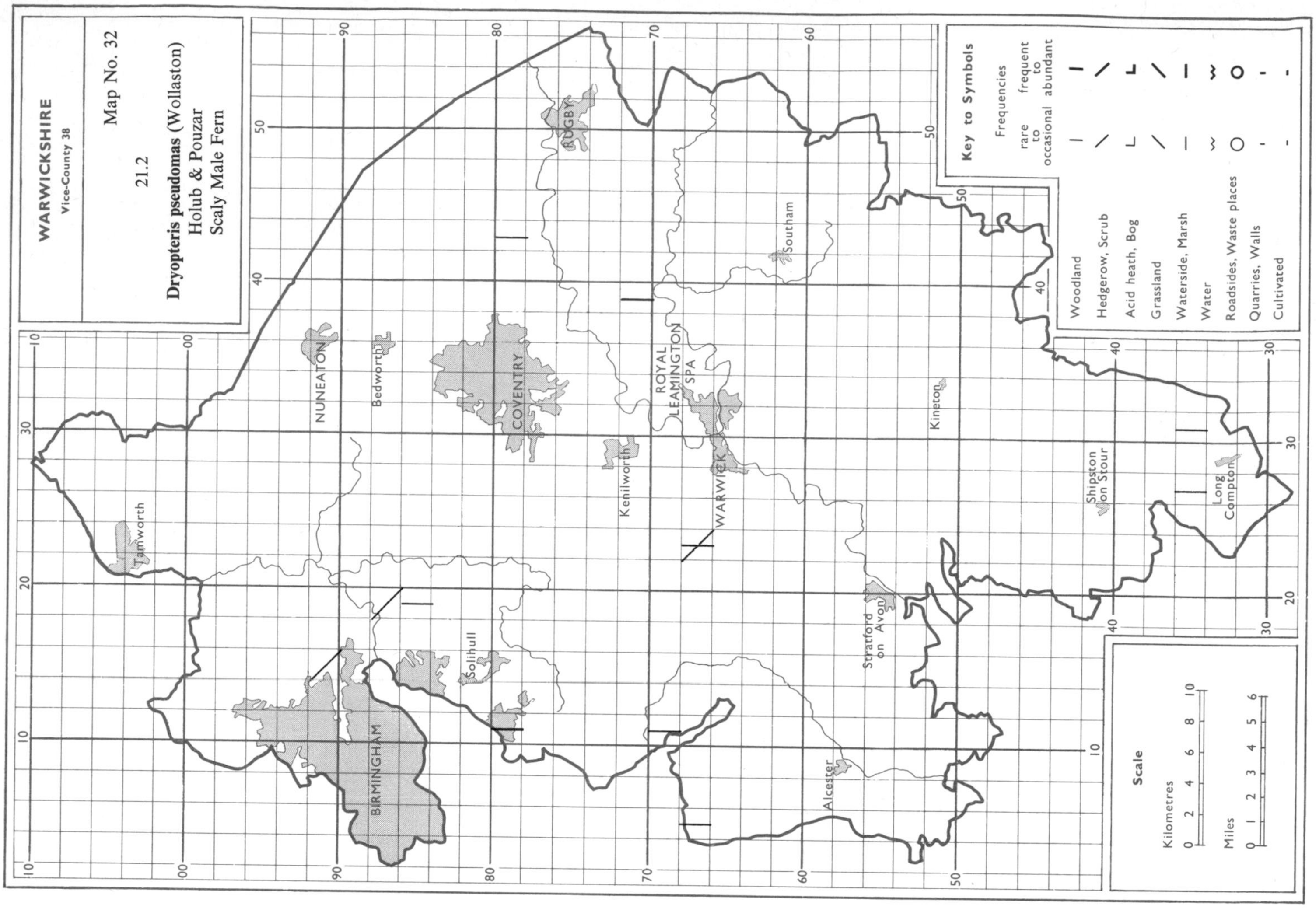

WARWICKSHIRE
Vice-County 38
Map No. 32
21.2
Dryopteris pseudomas (Wollaston) Holub & Pouzar
Scaly Male Fern
Key to Symbols
Frequencies
rare to occasional
frequent to abundant
Woodland
Hedgerow, Scrub
Acid heath, Bog
Grassland
Waterside, Marsh
Water
Roadsides, Waste places
Quarries, Walls
Cultivated
Scale
Kilometres
Miles
NUNEATON
Bedworth
COVENTRY
RUGBY
Southam
ROYAL LEAMINGTON SPA
Kenilworth
WARWICK
Kineton
Shipston on Stour
Long Compton
Tamworth
Solihull
BIRMINGHAM
Stratford on Avon
Alcester

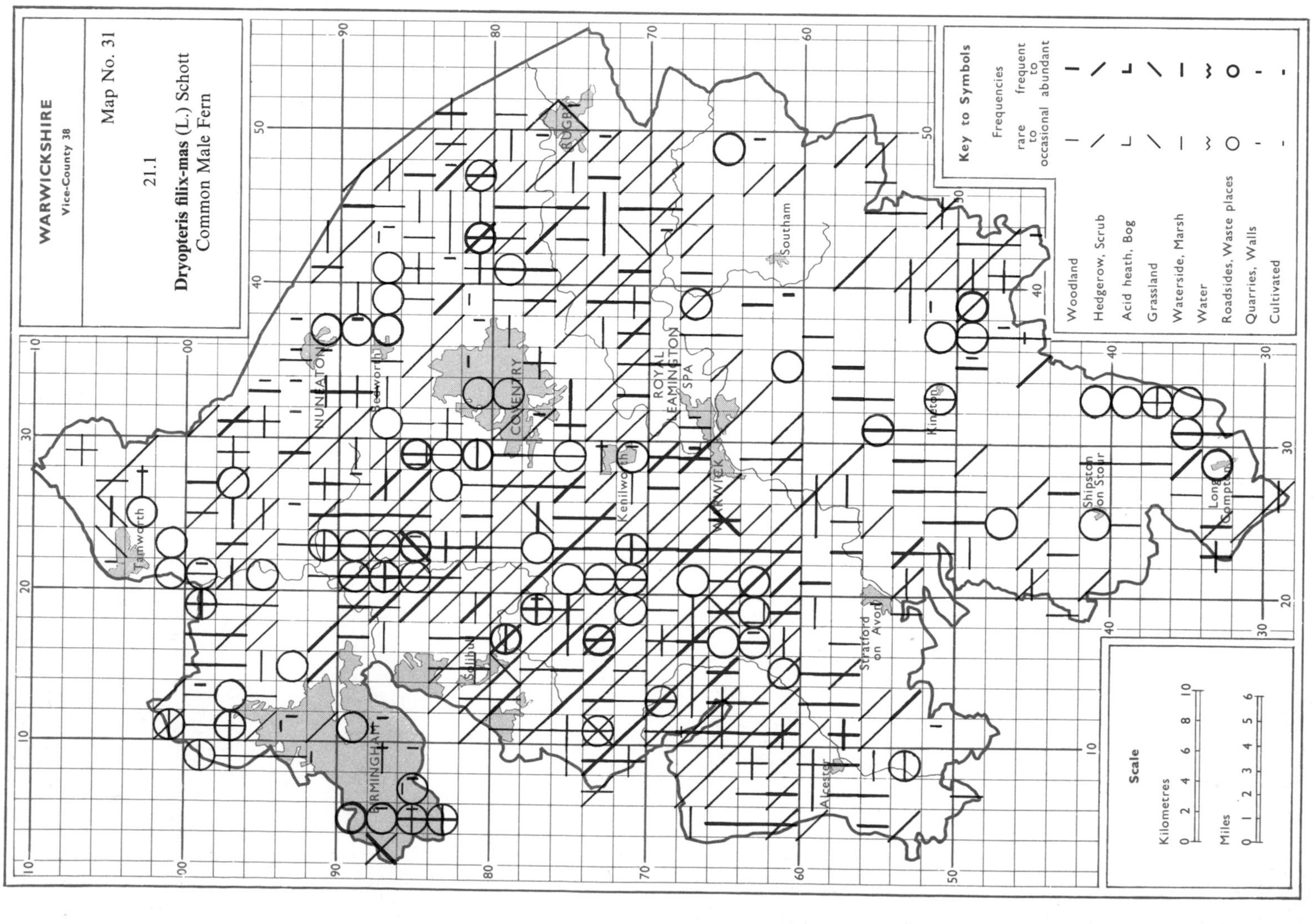

WARWICKSHIRE
Vice-County 38
Map No. 31
21.1
Dryopteris filix-mas (L.) Schott
Common Male Fern
Key to Symbols
Frequencies
rare to occasional
frequent to abundant
Woodland
Hedgerow, Scrub
Acid heath, Bog
Grassland
Waterside, Marsh
Water
Roadsides, Waste places
Quarries, Walls
Cultivated
Scale
Kilometres
Miles
NUNEATON
Bedworth
COVENTRY
RUGBY
Southam
ROYAL LEAMINGTON SPA
Kenilworth
WARWICK
Kineton
Shipston on Stour
Long Compton
Tamworth
Solihull
BIRMINGHAM
Stratford on Avon
Alcester

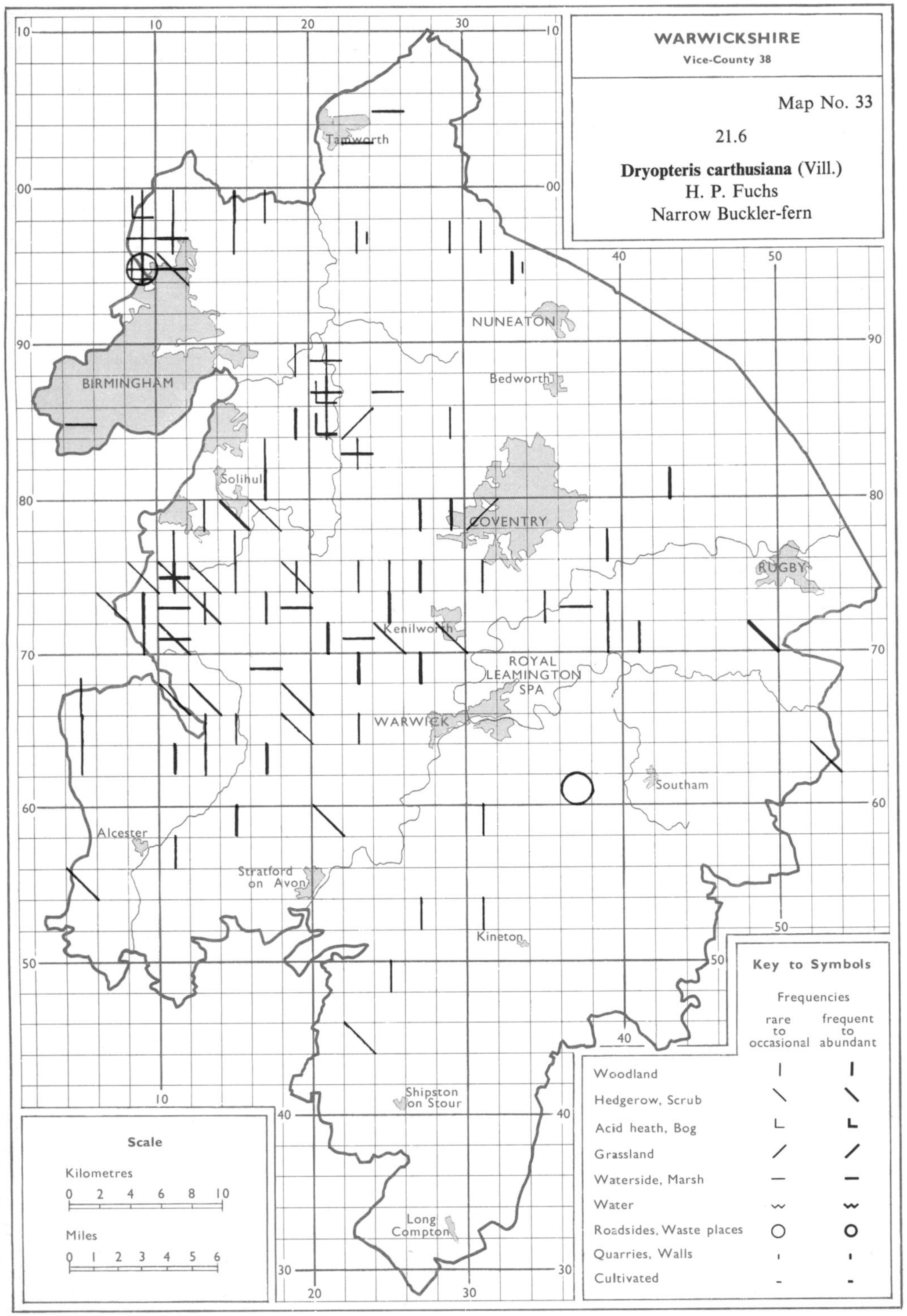
WARWICKSHIRE
Vice-County 38
Map No. 33
21.6
Dryopteris carthusiana (Vill.) H. P. Fuchs
Narrow Buckler-fern
Tamworth
NUNEATON
Bedworth
BIRMINGHAM
Solihull
COVENTRY
RUGBY
Kenilworth
ROYAL LEAMINGTON SPA
WARWICK
Southam
Alcester
Stratford on Avon
Kineton
Shipston on Stour
Long Compton
Key to Symbols
Frequencies
rare to occasional
frequent to abundant
Woodland
Hedgerow, Scrub
Acid heath, Bog
Grassland
Waterside, Marsh
Water
Roadsides, Waste places
Quarries, Walls
Cultivated
Scale
Kilometres
0 2 4 6 8 10
Miles
0 1 2 3 4 5 6

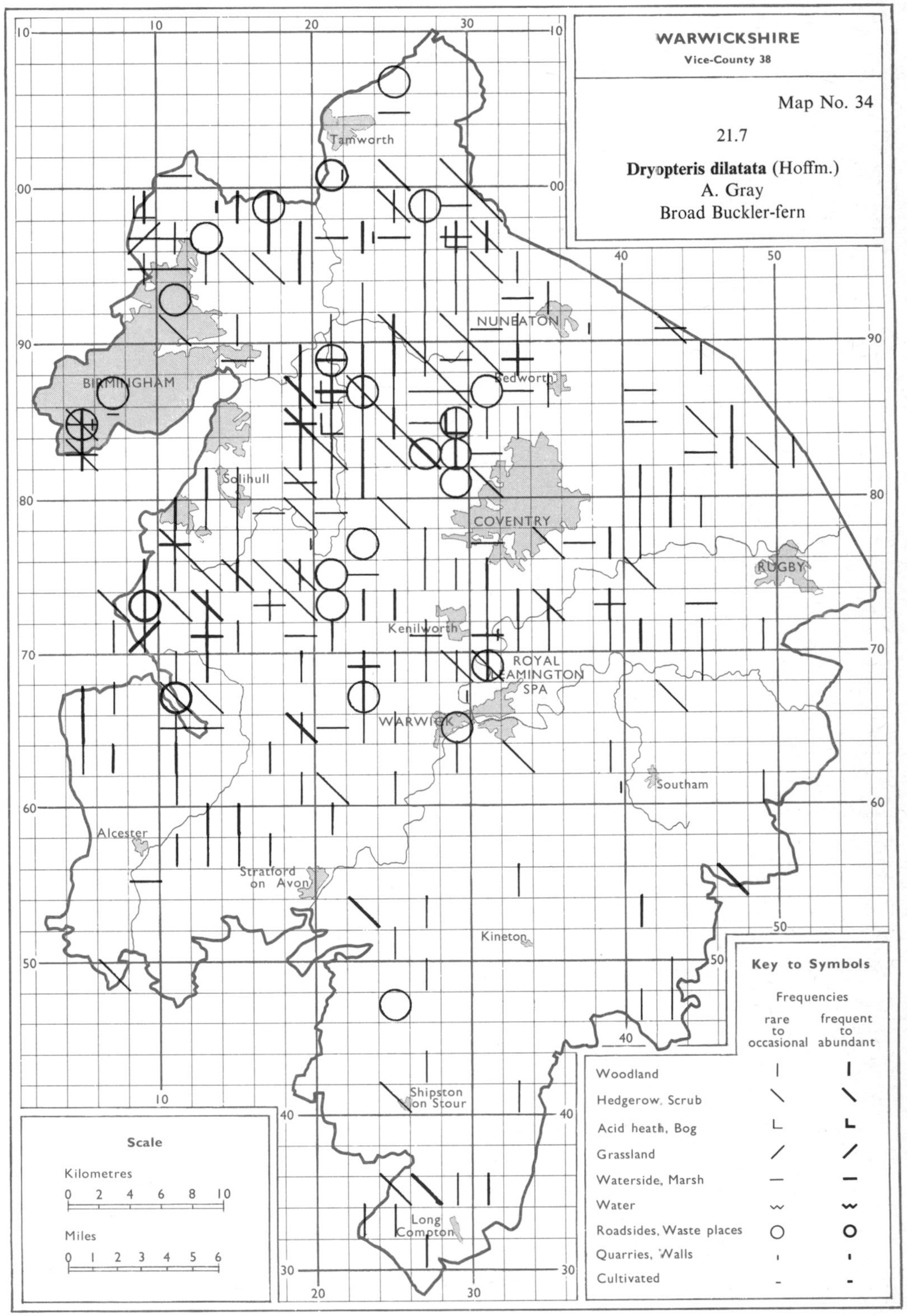
WARWICKSHIRE
Vice-County 38
Map No. 34
21.7
Dryopteris dilatata (Hoffm.) A. Gray
Broad Buckler-fern
Tamworth
NUNEATON
Bedworth
BIRMINGHAM
Solihull
COVENTRY
RUGBY
Kenilworth
ROYAL LEAMINGTON SPA
WARWICK
Southam
Alcester
Stratford on Avon
Kineton
Shipston on Stour
Long Compton
Key to Symbols
Frequencies
rare to occasional
frequent to abundant
Woodland
Hedgerow, Scrub
Acid heath, Bog
Grassland
Waterside, Marsh
Water
Roadsides, Waste places
Quarries, Walls
Cultivated
Scale
Kilometres
0 2 4 6 8 10
Miles
0 1 2 3 4 5 6

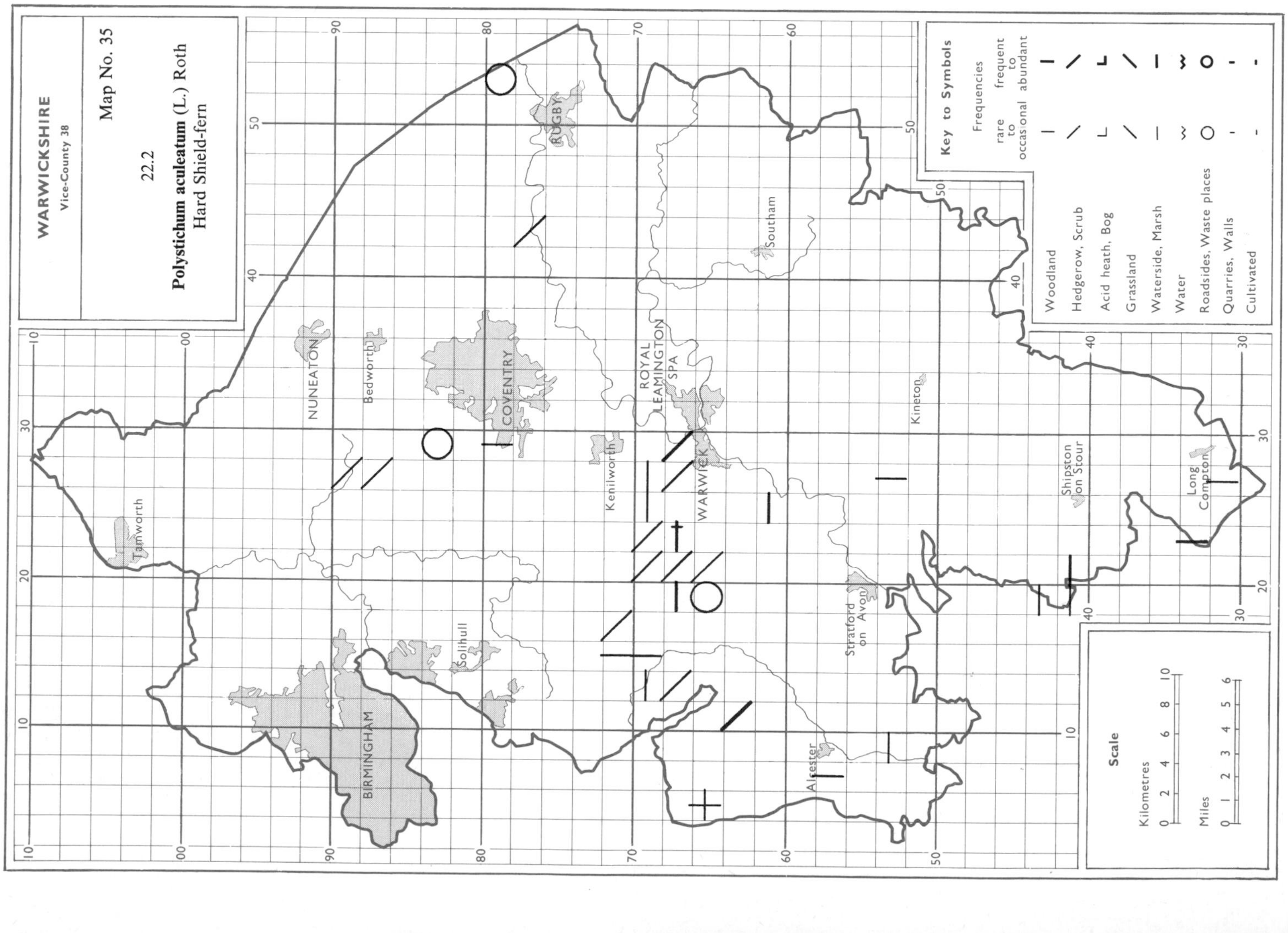
WARWICKSHIRE
Vice-County 38
Map No. 35
22.2
Polystichum aculeatum (L.) Roth
Hard Shield-fern
Key to Symbols
Frequencies
rare to occasional
frequent to abundant
Woodland
Hedgerow, Scrub
Acid heath, Bog
Grassland
Waterside, Marsh
Water
Roadsides, Waste places
Quarries, Walls
Cultivated
Scale
Kilometres
0 2 4 6 8 10
Miles
0 1 2 3 4 5 6
NUNEATON
Bedworth
COVENTRY
RUGBY
Tamworth
Kenilworth
ROYAL LEAMINGTON SPA
WARWICK
Southam
Kineton
Shipston on Stour
Long Compton
Stratford on Avon
Alcester
Solihull
BIRMINGHAM

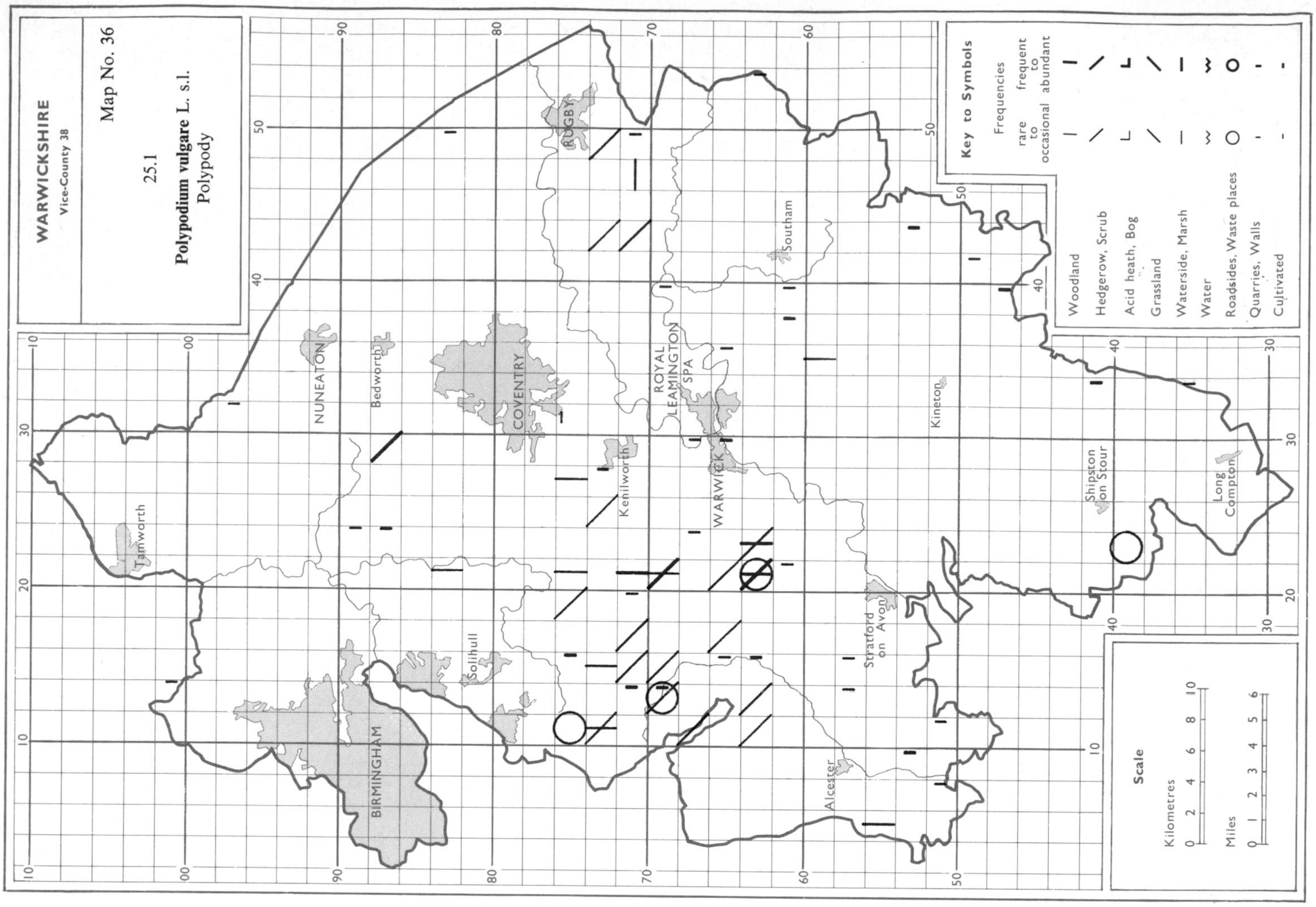
WARWICKSHIRE
Vice-County 38
Map No. 36
25.1
Polypodium vulgare L. s.l.
Polypody
Key to Symbols
Frequencies
rare to occasional
frequent to abundant
Woodland
Hedgerow, Scrub
Acid heath, Bog
Grassland
Waterside, Marsh
Water
Roadsides, Waste places
Quarries, Walls
Cultivated
Scale
Kilometres
0 2 4 6 8 10
Miles
0 1 2 3 4 5 6
NUNEATON
Bedworth
COVENTRY
RUGBY
Tamworth
Kenilworth
ROYAL LEAMINGTON SPA
WARWICK
Southam
Kineton
Shipston on Stour
Long Compton
Stratford on Avon
Alcester
Solihull
BIRMINGHAM

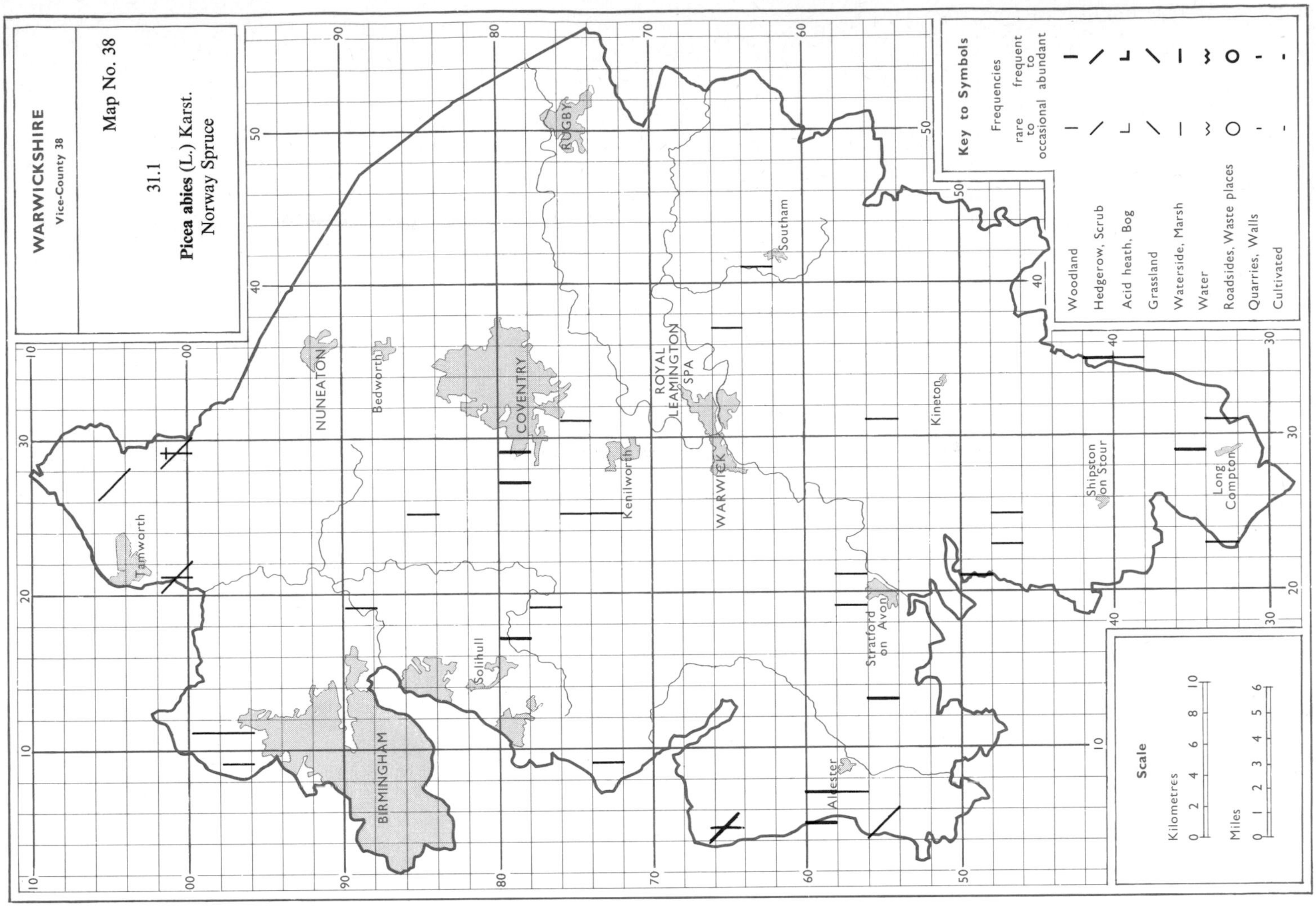

WARWICKSHIRE
Vice-County 38
Map No. 38
31.1
Picea abies (L.) Karst.
Norway Spruce
Key to Symbols
Frequencies
rare to occasional
frequent to abundant
Woodland
Hedgerow, Scrub
Acid heath, Bog
Grassland
Waterside, Marsh
Water
Roadsides, Waste places
Quarries, Walls
Cultivated
Scale
Kilometres
0 2 4 6 8 10
Miles
0 1 2 3 4 5 6
NUNEATON
Bedworth
COVENTRY
RUGBY
Southam
ROYAL LEAMINGTON SPA
WARWICK
Kenilworth
Kineton
Shipston on Stour
Long Compton
Tamworth
Solihull
BIRMINGHAM
Stratford on Avon
Alcester

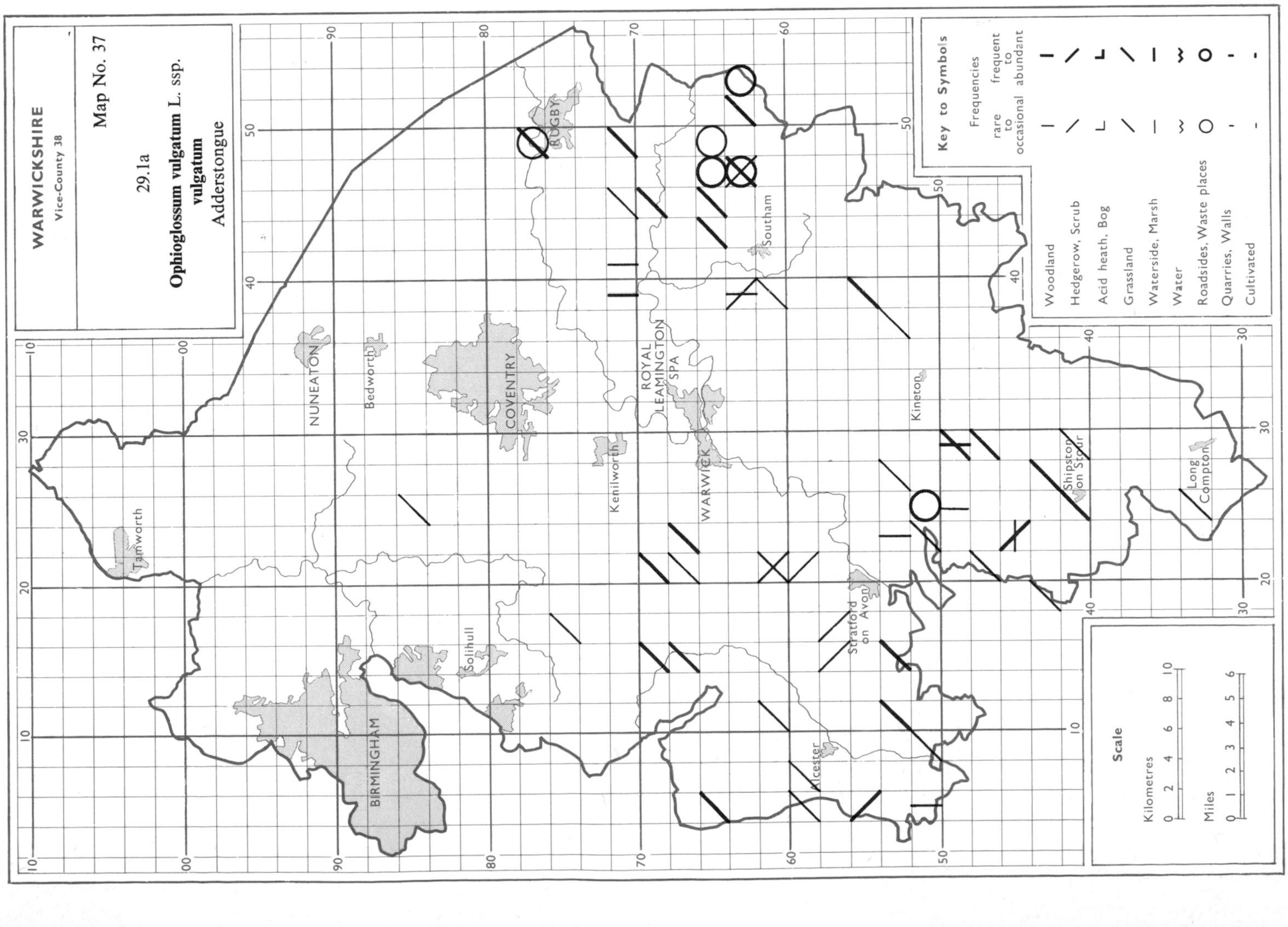

WARWICKSHIRE
Vice-County 38
Map No. 37
29.1a
Ophioglossum vulgatum L. ssp. vulgatum
Adderstongue
Key to Symbols
Frequencies
rare to occasional
frequent to abundant
Woodland
Hedgerow, Scrub
Acid heath, Bog
Grassland
Waterside, Marsh
Water
Roadsides, Waste places
Quarries, Walls
Cultivated
Scale
Kilometres
0 2 4 6 8 10
Miles
0 1 2 3 4 5 6
NUNEATON
Bedworth
COVENTRY
RUGBY
Southam
ROYAL LEAMINGTON SPA
WARWICK
Kenilworth
Kineton
Shipston on Stour
Long Compton
Tamworth
Solihull
BIRMINGHAM
Stratford on Avon
Alcester

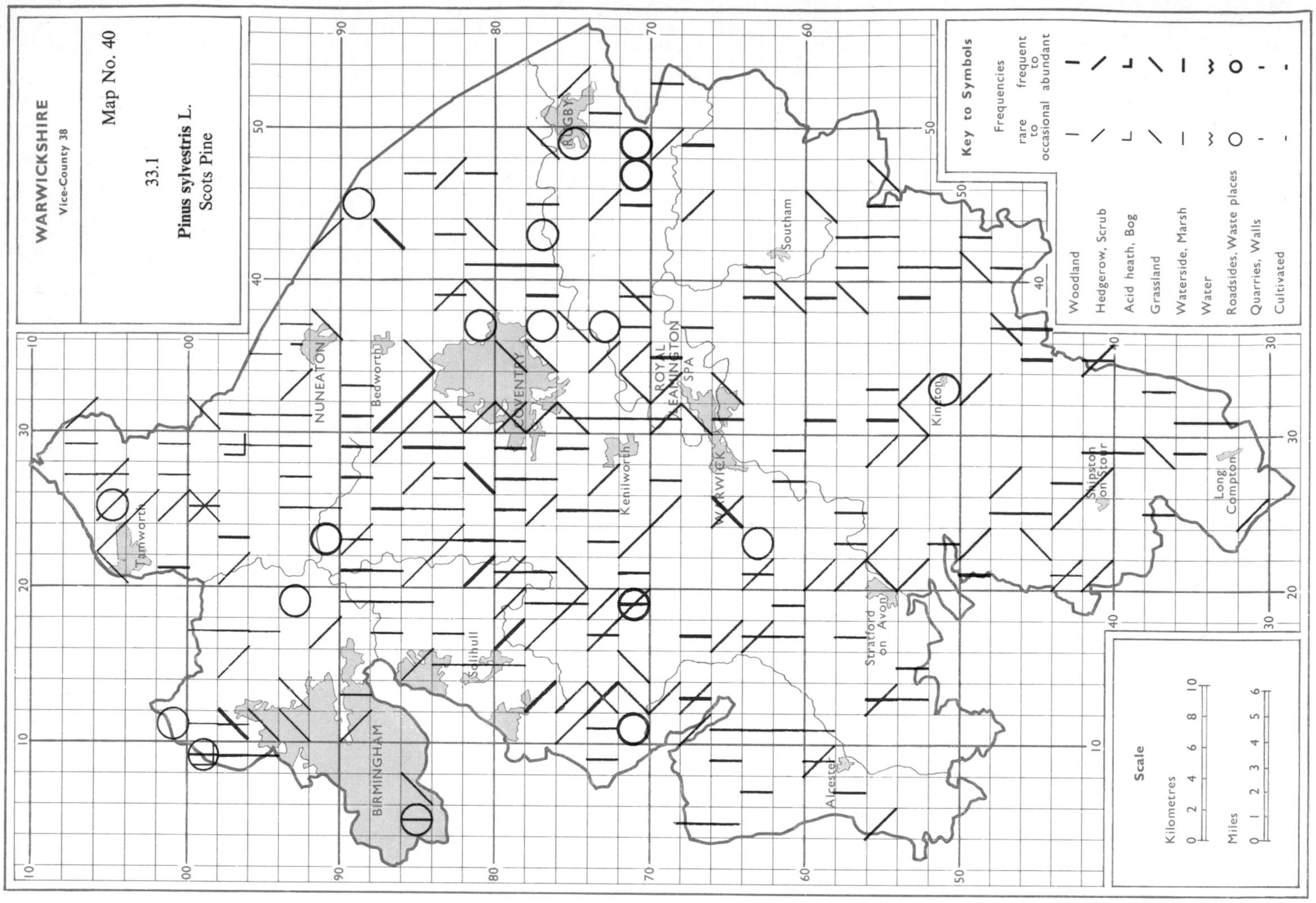

WARWICKSHIRE
Vice-County 38
Map No. 40
33.1
Pinus sylvestris L.
Scots Pine
Key to Symbols
Frequencies
rare to occasional
frequent to abundant
Woodland
Hedgerow, Scrub
Acid heath, Bog
Grassland
Waterside, Marsh
Water
Roadsides, Waste places
Quarries, Walls
Cultivated
Scale
Kilometres
0 2 4 6 8 10
Miles
0 1 2 3 4 5 6
NUNEATON
Bedworth
COVENTRY
RUGBY
ROYAL LEAMINGTON SPA
WARWICK
Kenilworth
Southam
Kineton
Shipston on Stour
Long Compton
Tamworth
Solihull
BIRMINGHAM
Stratford on Avon
Alcester

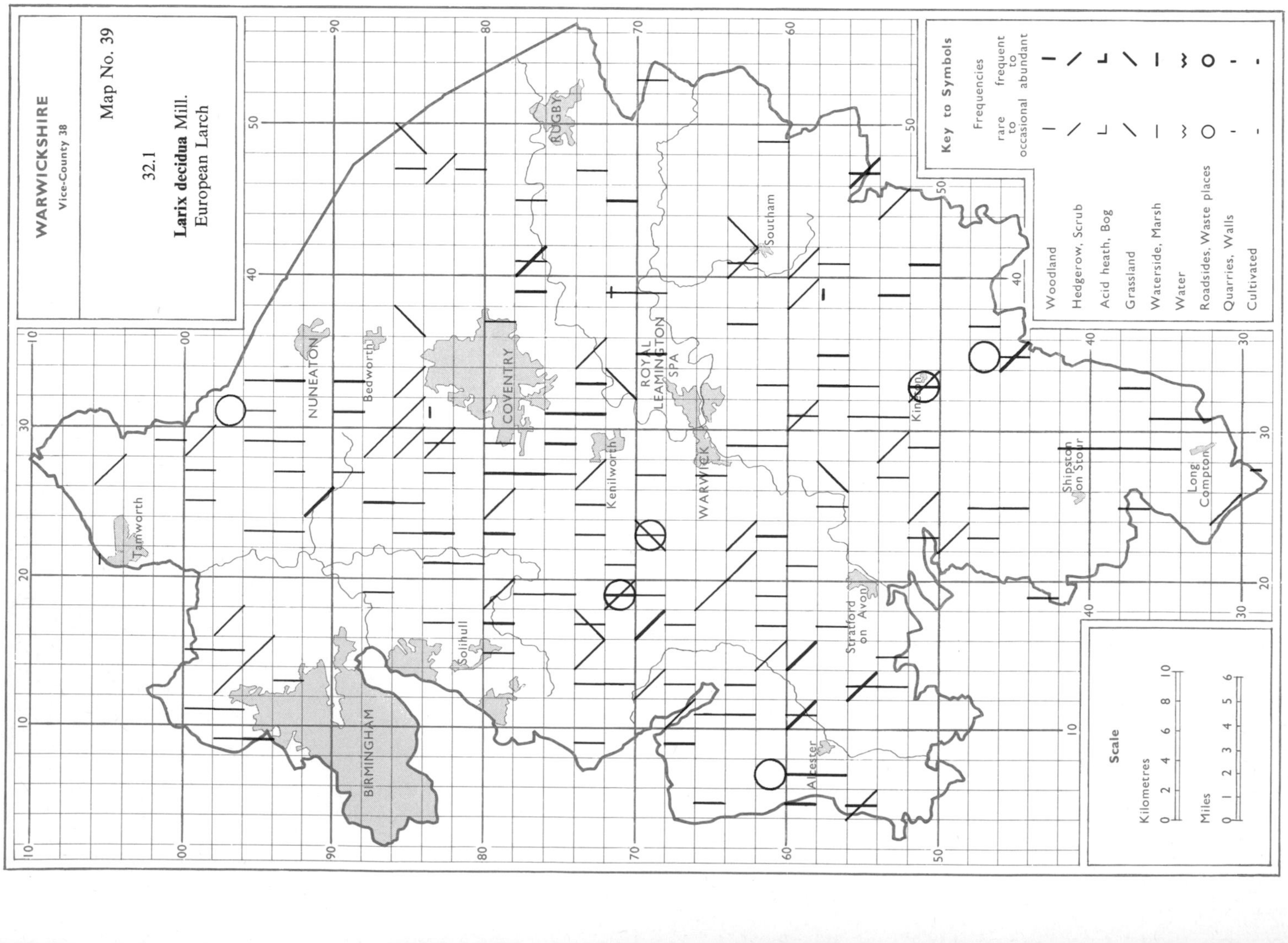

WARWICKSHIRE
Vice-County 38
Map No. 39
32.1
Larix decidua Mill.
European Larch
Key to Symbols
Frequencies
rare to occasional
frequent to abundant
Woodland
Hedgerow, Scrub
Acid heath, Bog
Grassland
Waterside, Marsh
Water
Roadsides, Waste places
Quarries, Walls
Cultivated
Scale
Kilometres
0 2 4 6 8 10
Miles
0 1 2 3 4 5 6
NUNEATON
Bedworth
COVENTRY
RUGBY
ROYAL LEAMINGTON SPA
WARWICK
Kenilworth
Southam
Kineton
Shipston on Stour
Long Compton
Tamworth
Solihull
BIRMINGHAM
Stratford on Avon
Alcester

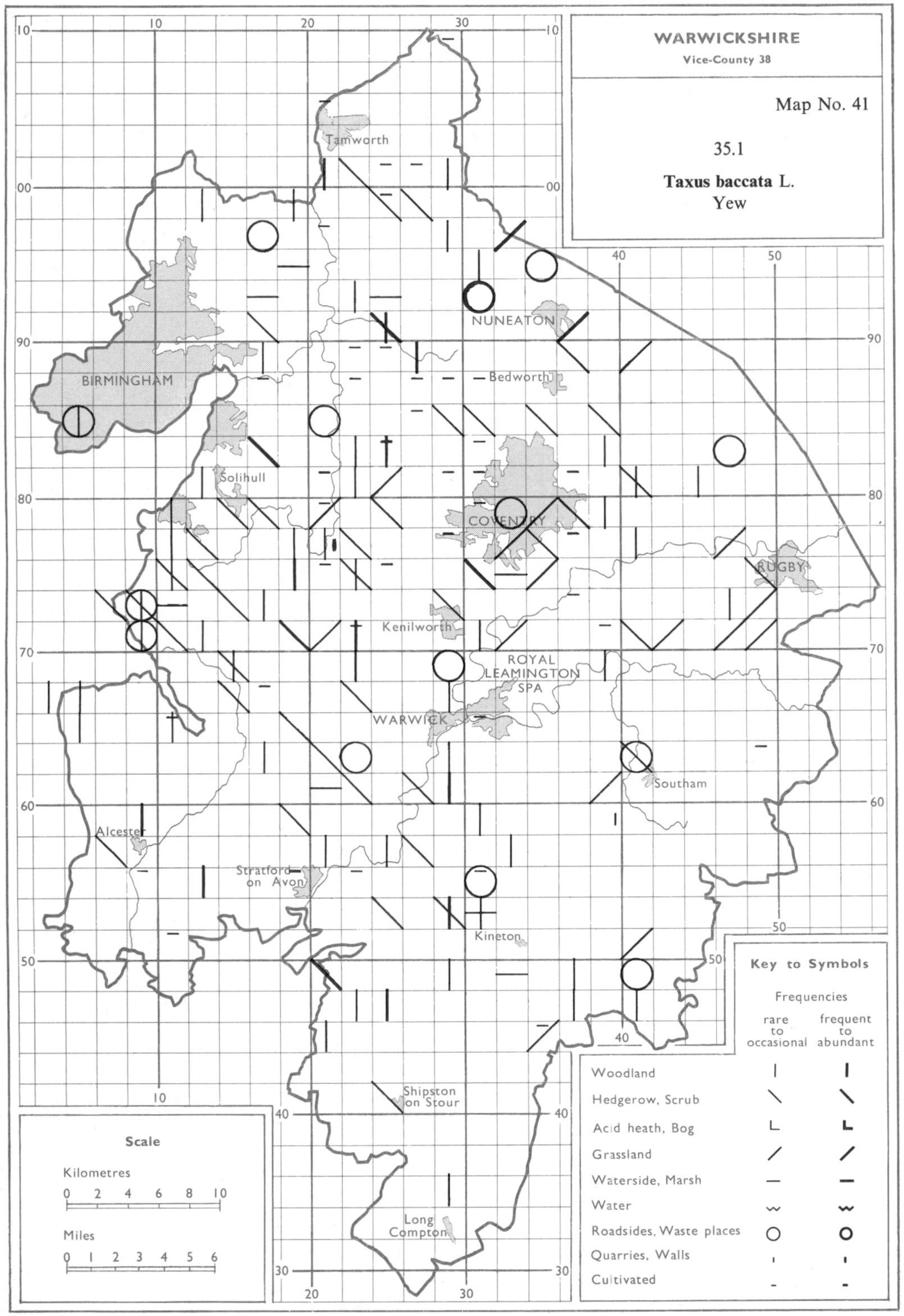
WARWICKSHIRE
Vice-County 38
Map No. 41
35.1
Taxus baccata L.
Yew
Tamworth
NUNEATON
Bedworth
BIRMINGHAM
Solihull
COVENTRY
RUGBY
Kenilworth
ROYAL LEAMINGTON SPA
WARWICK
Southam
Alcester
Stratford on Avon
Kineton
Shipston on Stour
Long Compton
Scale
Kilometres
0 2 4 6 8 10
Miles
0 1 2 3 4 5 6
Key to Symbols
Frequencies
rare to occasional
frequent to abundant
Woodland
Hedgerow, Scrub
Acid heath, Bog
Grassland
Waterside, Marsh
Water
Roadsides, Waste places
Quarries, Walls
Cultivated

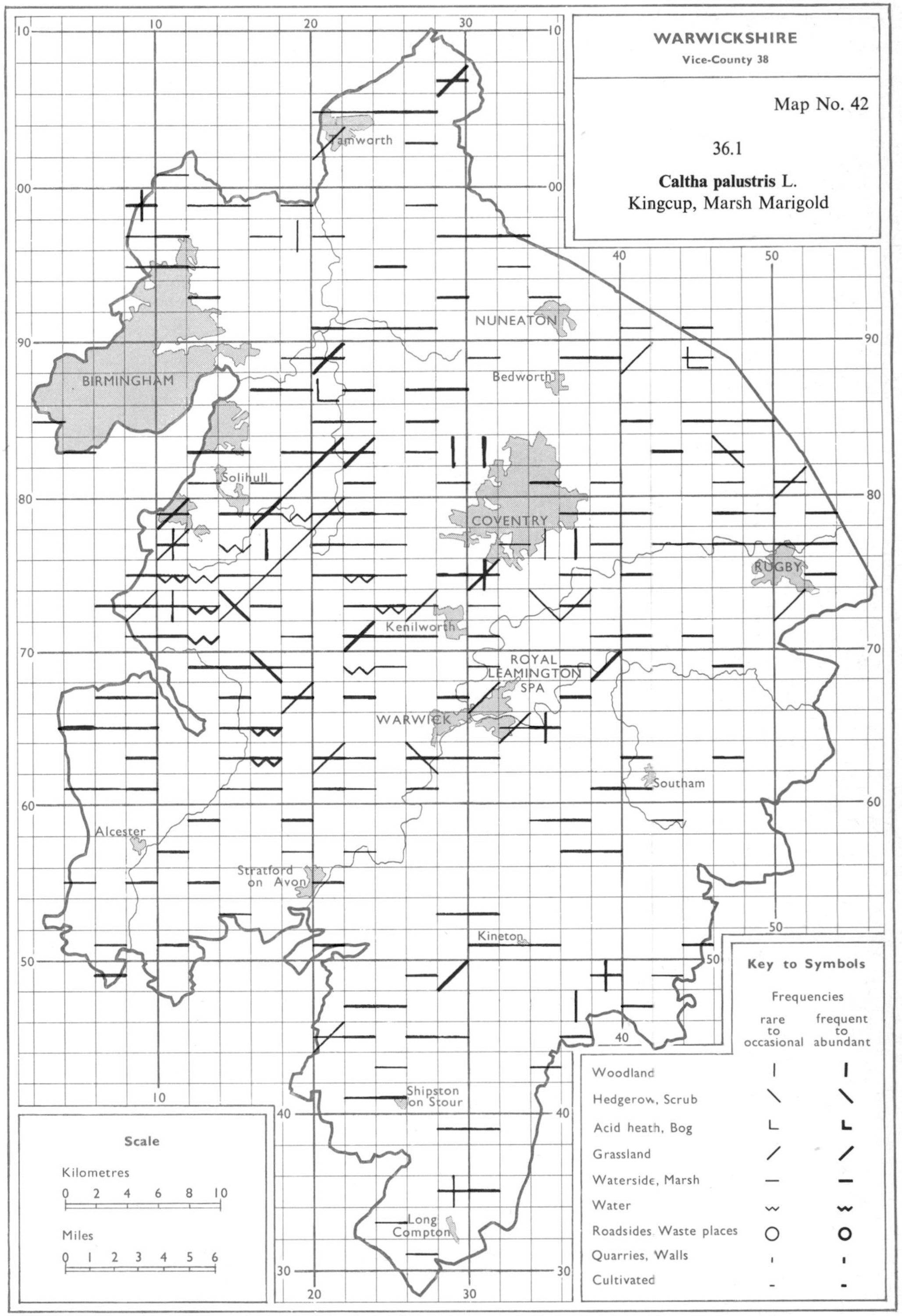
WARWICKSHIRE
Vice-County 38
Map No. 42
36.1
Caltha palustris L.
Kingcup, Marsh Marigold
Tamworth
NUNEATON
Bedworth
BIRMINGHAM
Solihull
COVENTRY
RUGBY
Kenilworth
ROYAL LEAMINGTON SPA
WARWICK
Southam
Alcester
Stratford on Avon
Kineton
Shipston on Stour
Long Compton
Scale
Kilometres
0 2 4 6 8 10
Miles
0 1 2 3 4 5 6
Key to Symbols
Frequencies
rare to occasional
frequent to abundant
Woodland
Hedgerow, Scrub
Acid heath, Bog
Grassland
Waterside, Marsh
Water
Roadsides, Waste places
Quarries, Walls
Cultivated

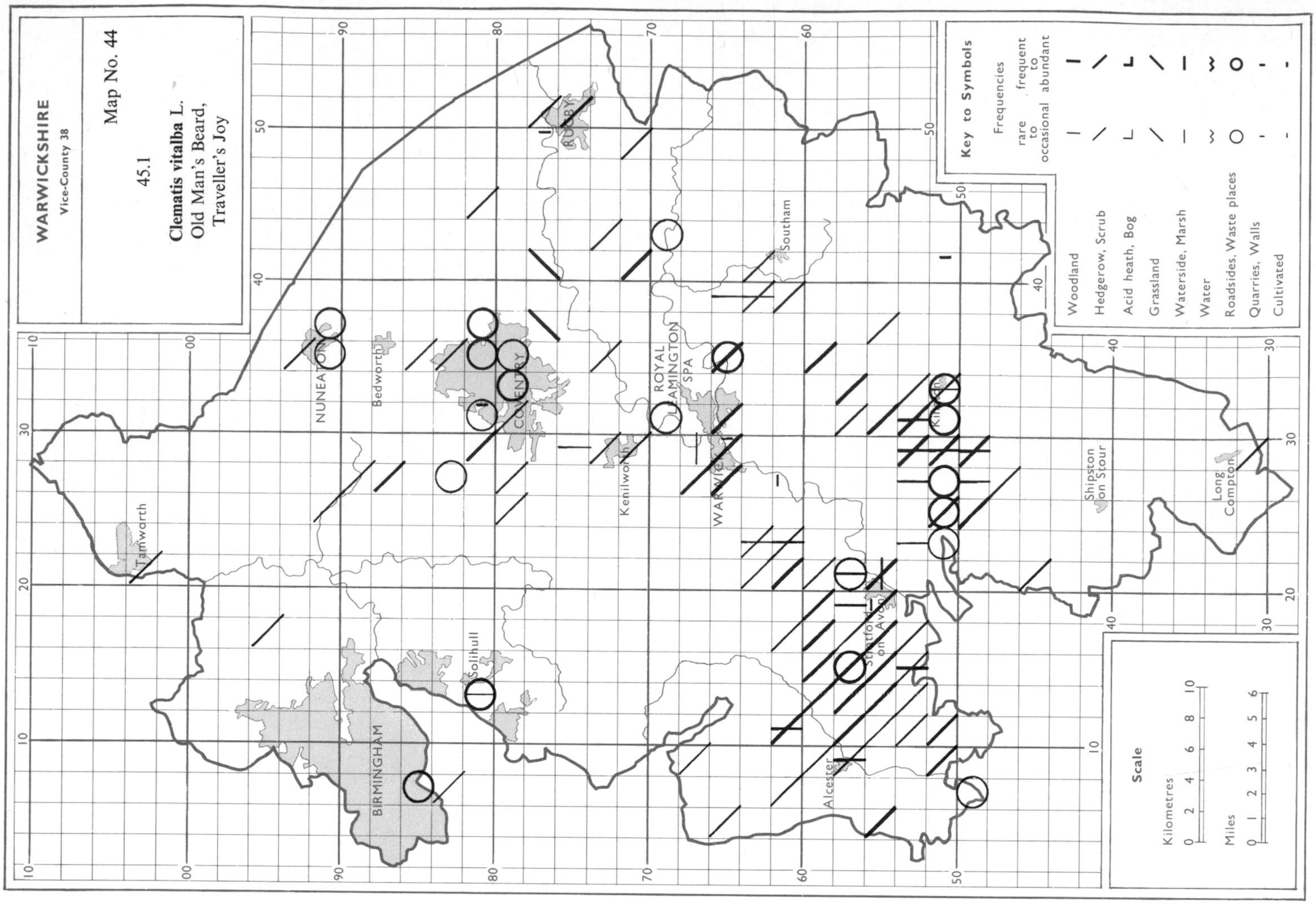
WARWICKSHIRE
Vice-County 38
Map No. 44
45.1
Clematis vitalba L.
Old Man's Beard,
Traveller's Joy
Key to Symbols
Frequencies
rare to occasional
frequent to abundant
Woodland
Hedgerow, Scrub
Acid heath, Bog
Grassland
Waterside, Marsh
Water
Roadsides, Waste places
Quarries, Walls
Cultivated
Scale
Kilometres
Miles
Tamworth
NUNEATON
Bedworth
COVENTRY
RUGBY
Southam
ROYAL LEAMINGTON SPA
Kenilworth
WARWICK
Kineton
Shipston on Stour
Long Compton
Solihull
BIRMINGHAM
Stratford on Avon
Alcester

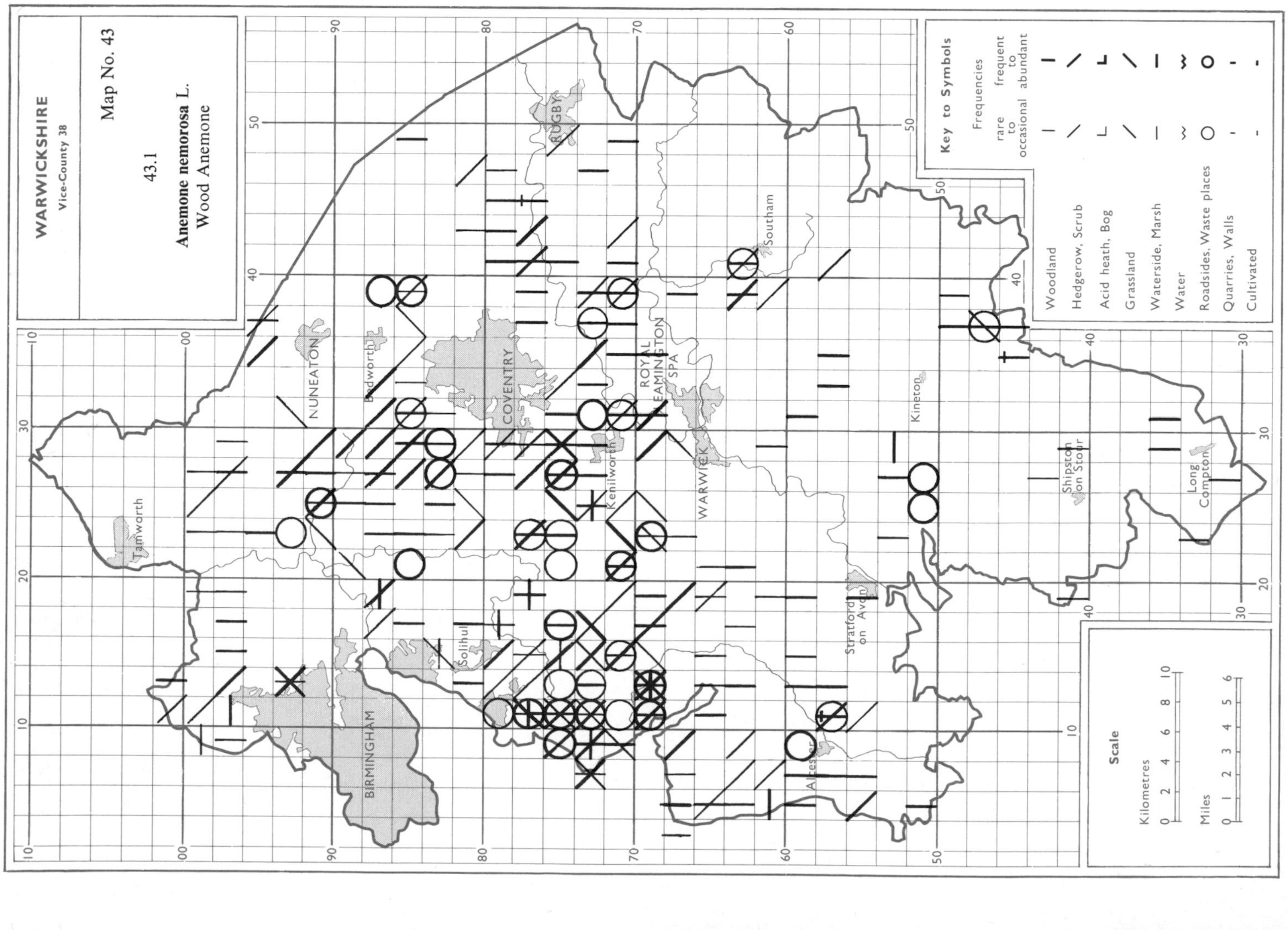
WARWICKSHIRE
Vice-County 38
Map No. 43
43.1
Anemone nemorosa L.
Wood Anemone
Key to Symbols
Frequencies
rare to occasional
frequent to abundant
Woodland
Hedgerow, Scrub
Acid heath, Bog
Grassland
Waterside, Marsh
Water
Roadsides, Waste places
Quarries, Walls
Cultivated
Scale
Kilometres
Miles
Tamworth
NUNEATON
Bedworth
COVENTRY
RUGBY
Southam
ROYAL LEAMINGTON SPA
Kenilworth
WARWICK
Kineton
Shipston on Stour
Long Compton
Solihull
BIRMINGHAM
Stratford on Avon
Alcester

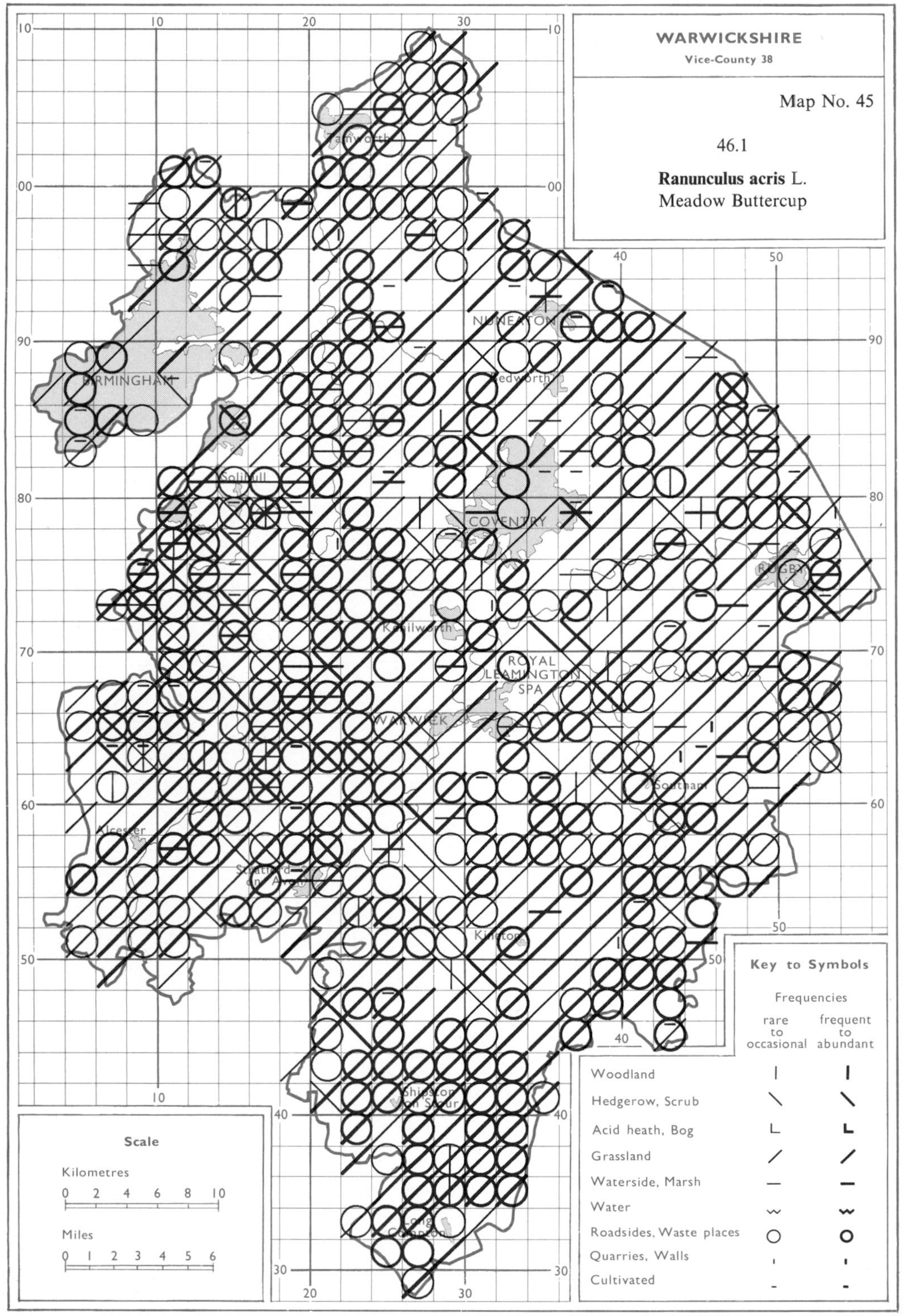
WARWICKSHIRE
Vice-County 38
Map No. 45
46.1
Ranunculus acris L.
Meadow Buttercup
BIRMINGHAM
Tamworth
Solihull
NUNEATON
Bedworth
COVENTRY
RUGBY
Kenilworth
ROYAL LEAMINGTON SPA
WARWICK
Southam
Alcester
Stratford on Avon
Kineton
Shipston on Stour
Long Compton
Key to Symbols
Frequencies
rare to occasional
frequent to abundant
Woodland
Hedgerow, Scrub
Acid heath, Bog
Grassland
Waterside, Marsh
Water
Roadsides, Waste places
Quarries, Walls
Cultivated
Scale
Kilometres
0 2 4 6 8 10
Miles
0 1 2 3 4 5 6

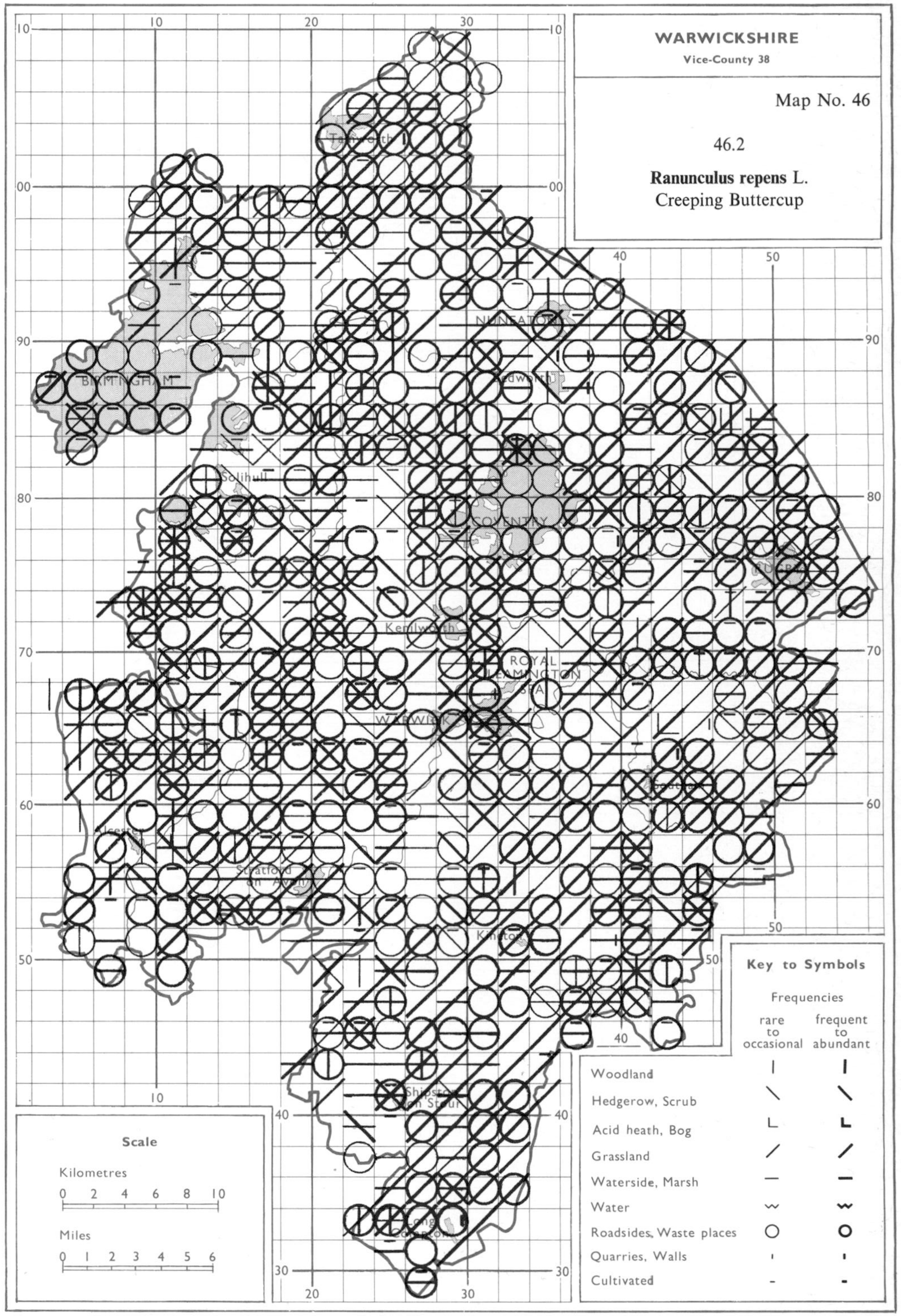
WARWICKSHIRE
Vice-County 38
Map No. 46
46.2
Ranunculus repens L.
Creeping Buttercup
BIRMINGHAM
Tamworth
Solihull
NUNEATON
Bedworth
COVENTRY
Kenilworth
ROYAL LEAMINGTON SPA
WARWICK
Alcester
Stratford on Avon
Kineton
Shipston on Stour
Long Compton
Key to Symbols
Frequencies
rare to occasional
frequent to abundant
Woodland
Hedgerow, Scrub
Acid heath, Bog
Grassland
Waterside, Marsh
Water
Roadsides, Waste places
Quarries, Walls
Cultivated
Scale
Kilometres
0 2 4 6 8 10
Miles
0 1 2 3 4 5 6

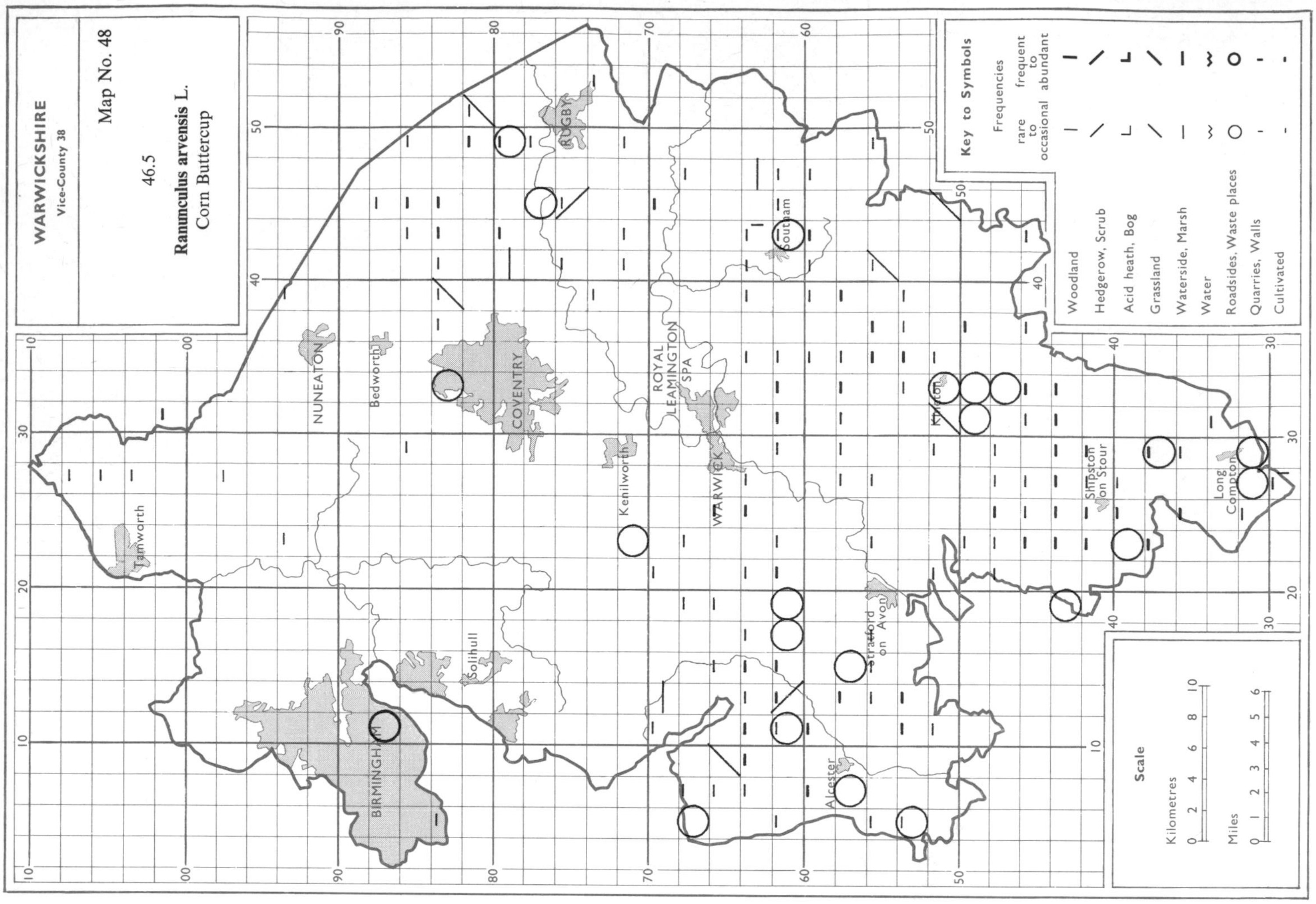
WARWICKSHIRE
Vice-County 38
Map No. 48
46.5
Ranunculus arvensis L.
Corn Buttercup
Key to Symbols
Frequencies
rare to occasional
frequent to abundant
Woodland
Hedgerow, Scrub
Acid heath, Bog
Grassland
Waterside, Marsh
Water
Roadsides, Waste places
Quarries, Walls
Cultivated
Scale
Kilometres
Miles
NUNEATON
Bedworth
COVENTRY
RUGBY
ROYAL LEAMINGTON SPA
Kenilworth
WARWICK
Southam
Kineton
Shipston on Stour
Long Compton
Tamworth
Solihull
BIRMINGHAM
Stratford on Avon
Alcester

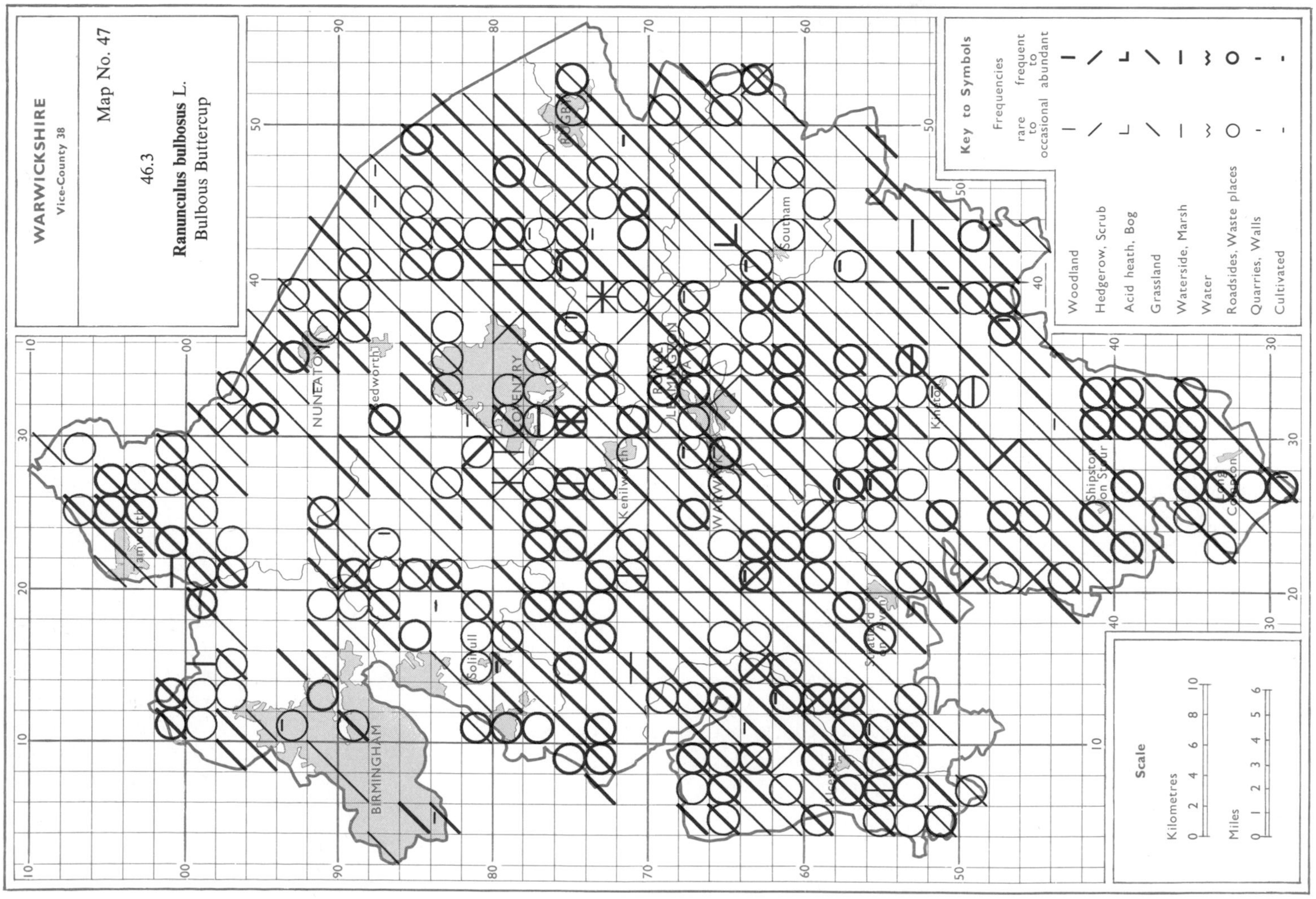
WARWICKSHIRE
Vice-County 38
Map No. 47
46.3
Ranunculus bulbosus L.
Bulbous Buttercup
Key to Symbols
Frequencies
rare to occasional
frequent to abundant
Woodland
Hedgerow, Scrub
Acid heath, Bog
Grassland
Waterside, Marsh
Water
Roadsides, Waste places
Quarries, Walls
Cultivated
Scale
Kilometres
Miles

WARWICKSHIRE
Vice-County 38
Map No. 50
46.12a
Ranunculus flammula L. ssp. flammula
Lesser Spearwort
Key to Symbols
Frequencies
rare to occasional
frequent to abundant
Woodland
Hedgerow, Scrub
Acid heath, Bog
Grassland
Waterside, Marsh
Water
Roadsides, Waste places
Quarries, Walls
Cultivated
Scale
Kilometres
Miles
Tamworth
NUNEATON
Bedworth
COVENTRY
RUGBY
Kenilworth
ROYAL LEAMINGTON SPA
WARWICK
Southam
Kineton
Shipston on Stour
Long Compton
Stratford on Avon
Alcester
Solihull
BIRMINGHAM

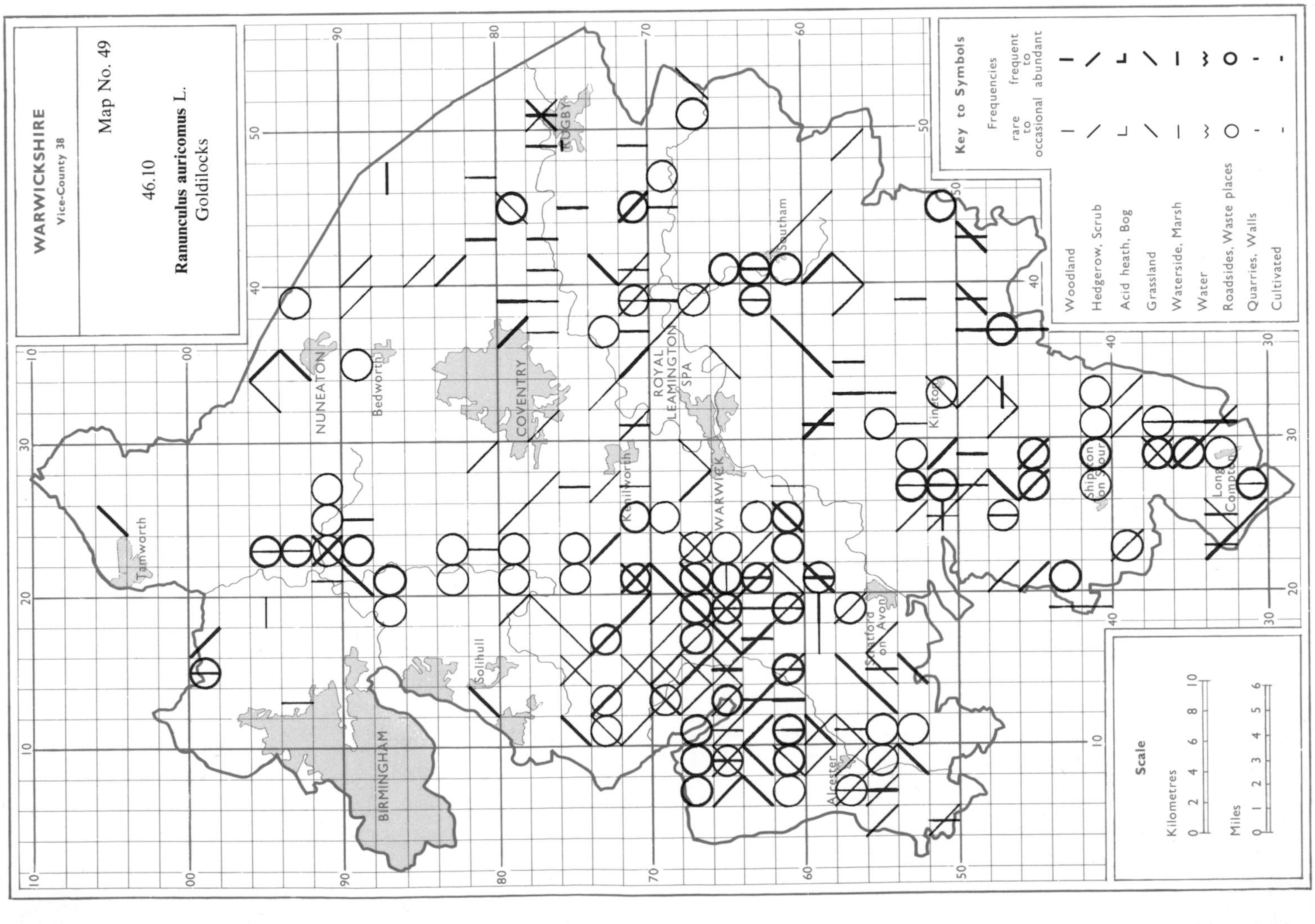
WARWICKSHIRE
Vice-County 38
Map No. 49
46.10
Ranunculus auricomus L.
Goldilocks
Key to Symbols
Frequencies
rare to occasional
frequent to abundant
Woodland
Hedgerow, Scrub
Acid heath, Bog
Grassland
Waterside, Marsh
Water
Roadsides, Waste places
Quarries, Walls
Cultivated
Scale
Kilometres
Miles
Tamworth
NUNEATON
Bedworth
COVENTRY
RUGBY
Kenilworth
ROYAL LEAMINGTON SPA
WARWICK
Southam
Kineton
Shipston on Stour
Long Compton
Stratford on Avon
Alcester
Solihull
BIRMINGHAM

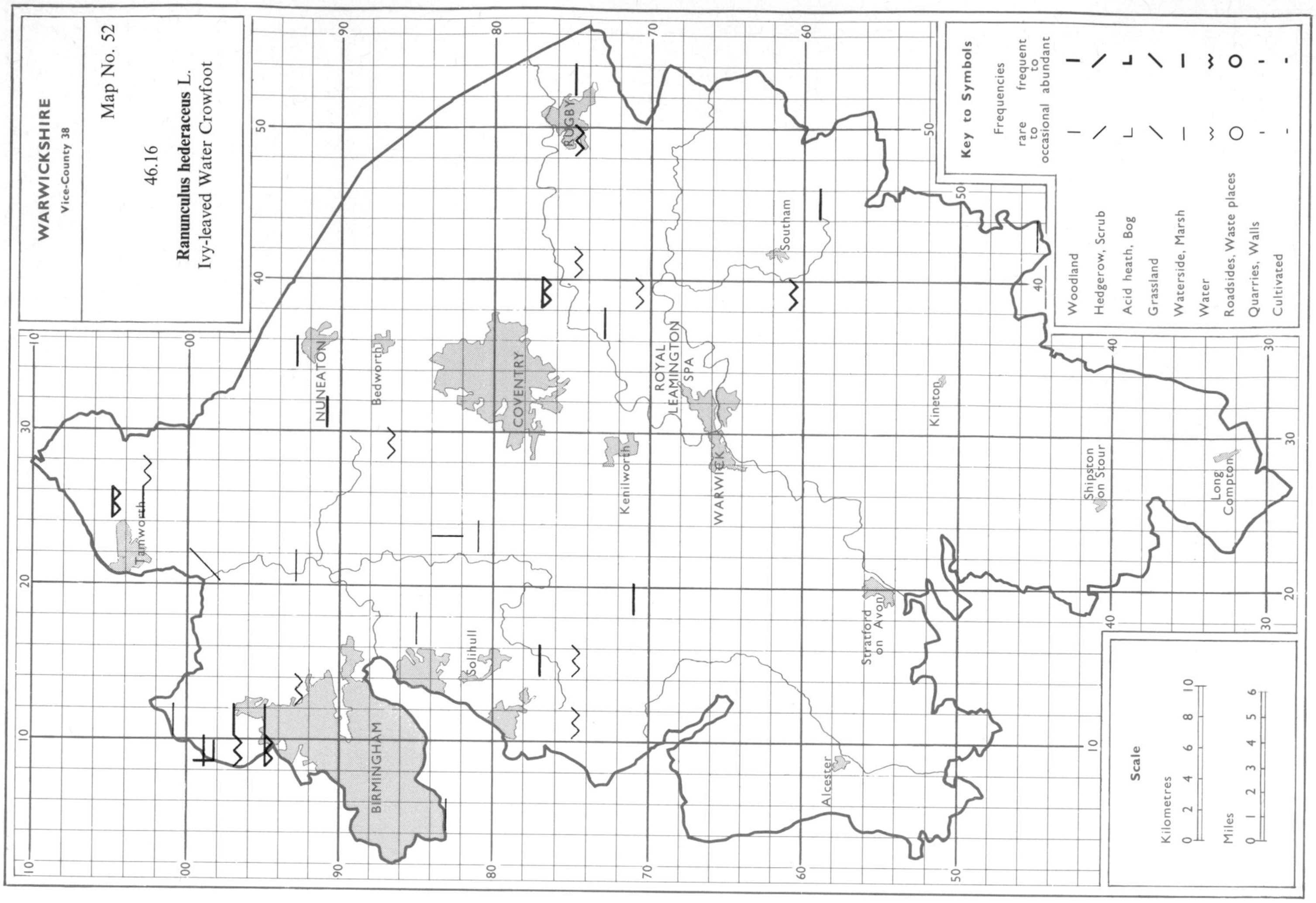
WARWICKSHIRE
Vice-County 38
Map No. 52
46.16
Ranunculus hederaceus L.
Ivy-leaved Water Crowfoot
Key to Symbols
Frequencies
rare to occasional
frequent to abundant
Woodland
Hedgerow, Scrub
Acid heath, Bog
Grassland
Waterside, Marsh
Water
Roadsides, Waste places
Quarries, Walls
Cultivated
Scale
Kilometres 0 2 4 6 8 10
Miles 0 1 2 3 4 5 6
Tamworth
NUNEATON
Bedworth
COVENTRY
RUGBY
Southam
ROYAL LEAMINGTON SPA
Kenilworth
WARWICK
Kineton
Shipston on Stour
Long Compton
Solihull
BIRMINGHAM
Stratford on Avon
Alcester

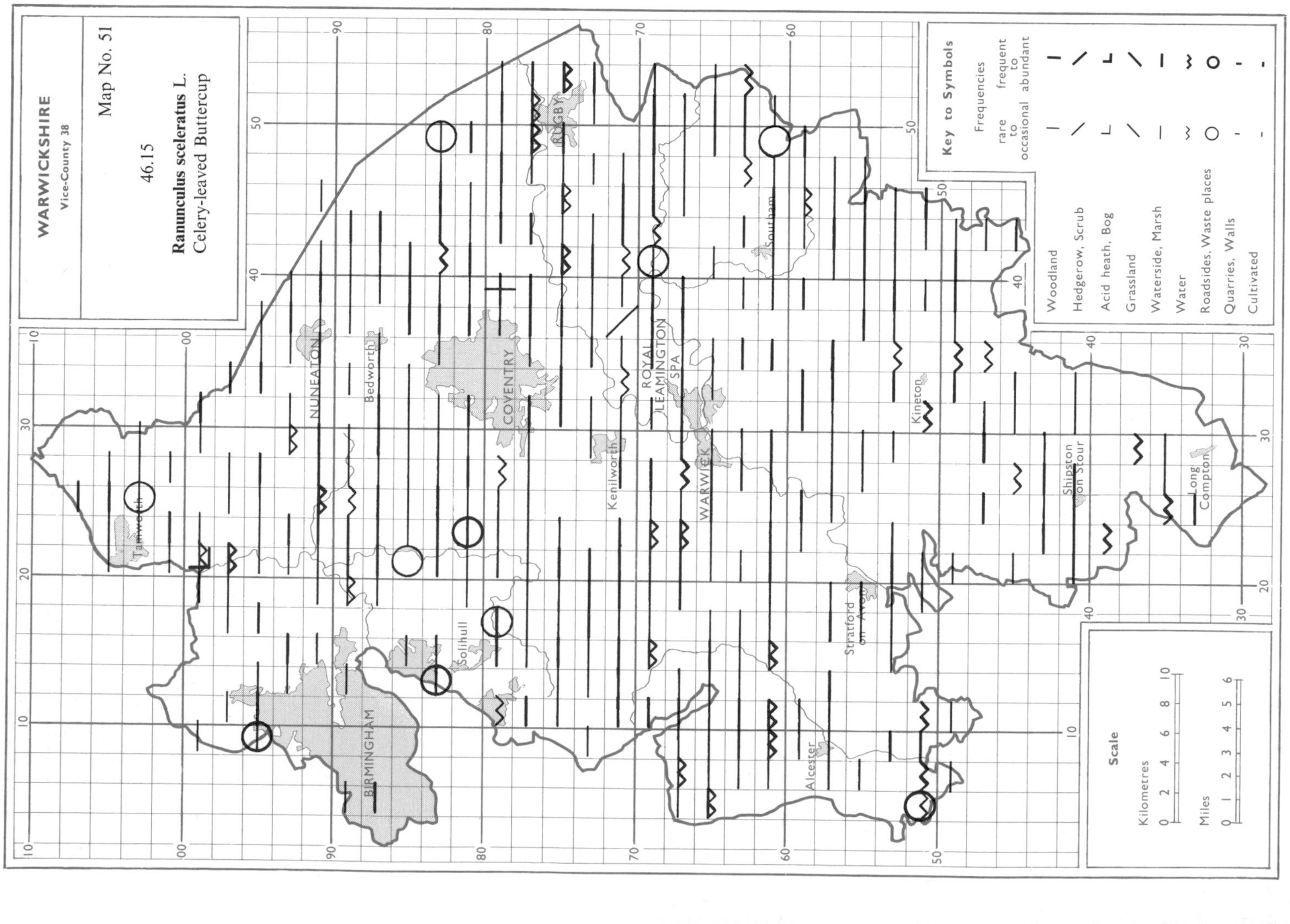
WARWICKSHIRE
Vice-County 38
Map No. 51
46.15
Ranunculus sceleratus L.
Celery-leaved Buttercup
Key to Symbols
Frequencies
rare to occasional
frequent to abundant
Woodland
Hedgerow, Scrub
Acid heath, Bog
Grassland
Waterside, Marsh
Water
Roadsides, Waste places
Quarries, Walls
Cultivated
Scale
Kilometres 0 2 4 6 8 10
Miles 0 1 2 3 4 5 6
Tamworth
NUNEATON
Bedworth
COVENTRY
RUGBY
Southam
ROYAL LEAMINGTON SPA
Kenilworth
WARWICK
Kineton
Shipston on Stour
Long Compton
Solihull
BIRMINGHAM
Stratford on Avon
Alcester

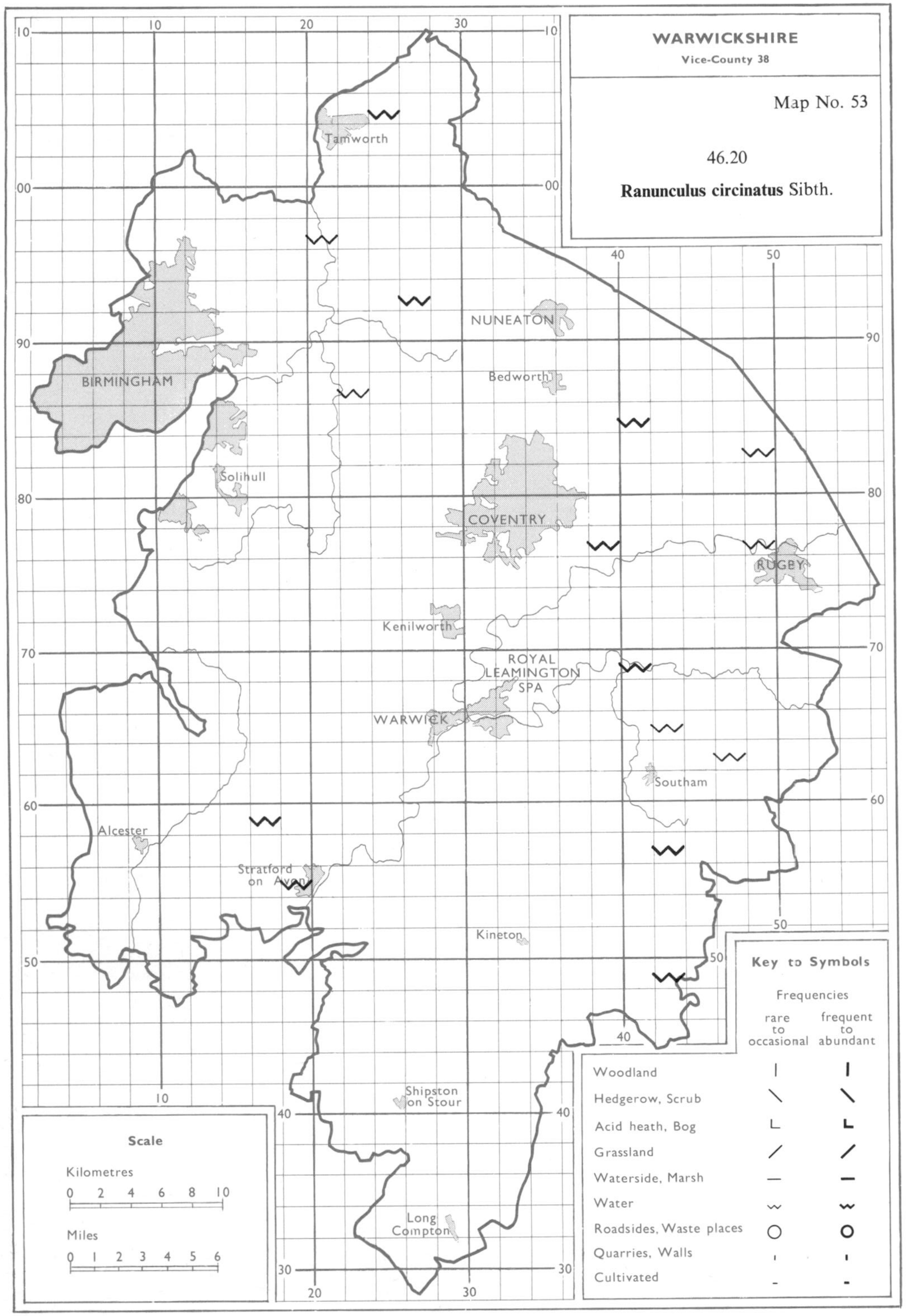
WARWICKSHIRE
Vice-County 38
Map No. 53
46.20
Ranunculus circinatus Sibth.
Tamworth
NUNEATON
Bedworth
BIRMINGHAM
Solihull
COVENTRY
RUGBY
Kenilworth
ROYAL LEAMINGTON SPA
WARWICK
Southam
Alcester
Stratford on Avon
Kineton
Shipston on Stour
Long Compton
Key to Symbols
Frequencies
rare to occasional
frequent to abundant
Woodland
Hedgerow, Scrub
Acid heath, Bog
Grassland
Waterside, Marsh
Water
Roadsides, Waste places
Quarries, Walls
Cultivated
Scale
Kilometres
0 2 4 6 8 10
Miles
0 1 2 3 4 5 6

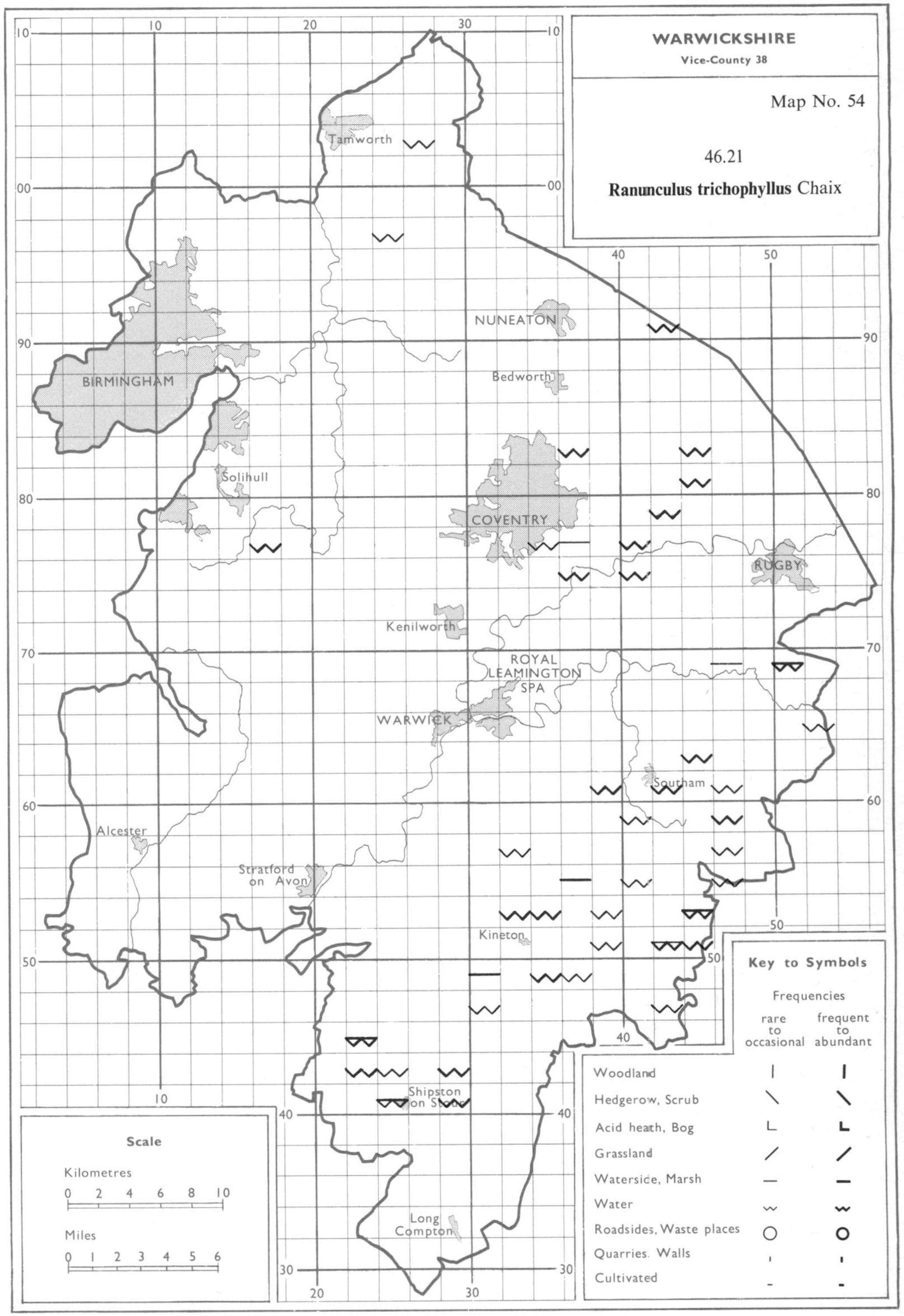
WARWICKSHIRE
Vice-County 38
Map No. 54
46.21
Ranunculus trichophyllus Chaix
Tamworth
NUNEATON
Bedworth
BIRMINGHAM
Solihull
COVENTRY
RUGBY
Kenilworth
ROYAL LEAMINGTON SPA
WARWICK
Southam
Alcester
Stratford on Avon
Kineton
Shipston on Stour
Long Compton
Key to Symbols
Frequencies
rare to occasional
frequent to abundant
Woodland
Hedgerow, Scrub
Acid heath, Bog
Grassland
Waterside, Marsh
Water
Roadsides, Waste places
Quarries, Walls
Cultivated
Scale
Kilometres
0 2 4 6 8 10
Miles
0 1 2 3 4 5 6

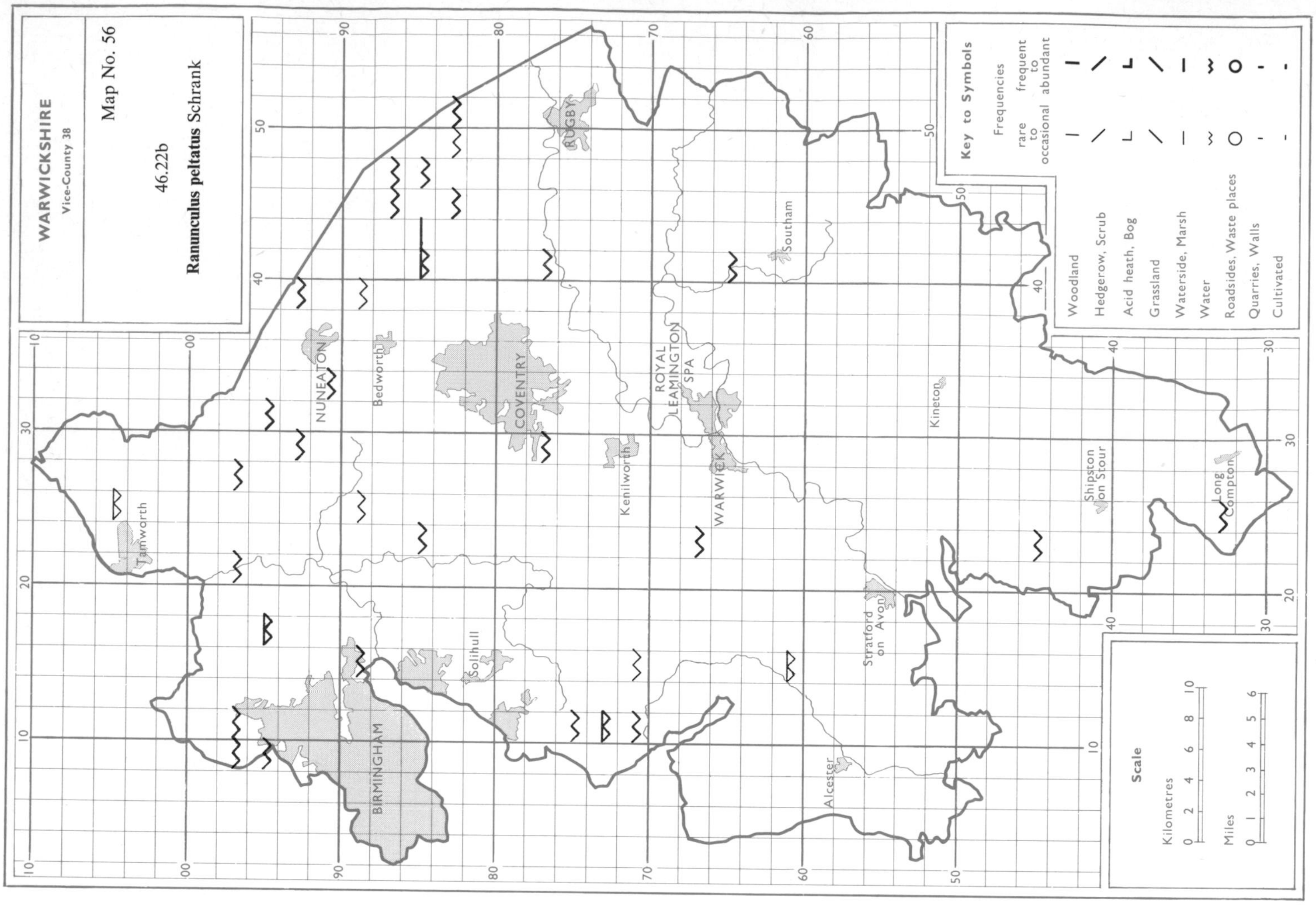
WARWICKSHIRE
Vice-County 38
Map No. 56
46.22b
Ranunculus peltatus Schrank
Key to Symbols
Frequencies
rare to occasional
frequent to abundant
Woodland
Hedgerow, Scrub
Acid heath, Bog
Grassland
Waterside, Marsh
Water
Roadsides, Waste places
Quarries, Walls
Cultivated
Scale
Kilometres
0 2 4 6 8 10
Miles
0 1 2 3 4 5 6
Tamworth
NUNEATON
Bedworth
COVENTRY
RUGBY
BIRMINGHAM
Solihull
Kenilworth
ROYAL LEAMINGTON SPA
WARWICK
Southam
Kineton
Stratford on Avon
Alcester
Shipston on Stour
Long Compton

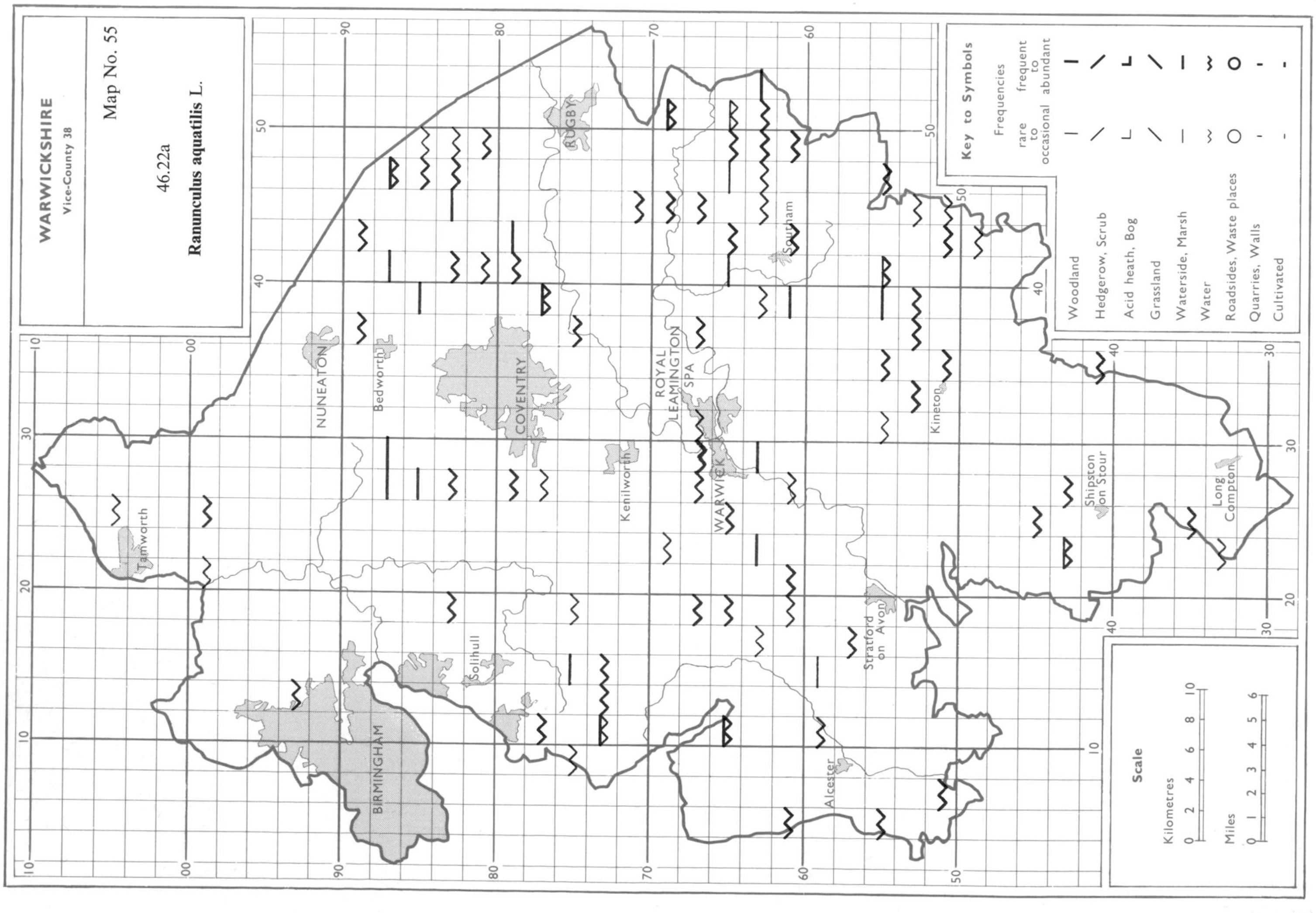
WARWICKSHIRE
Vice-County 38
Map No. 55
46.22a
Ranunculus aquatilis L.
Key to Symbols
Frequencies
rare to occasional
frequent to abundant
Woodland
Hedgerow, Scrub
Acid heath, Bog
Grassland
Waterside, Marsh
Water
Roadsides, Waste places
Quarries, Walls
Cultivated
Scale
Kilometres
0 2 4 6 8 10
Miles
0 1 2 3 4 5 6
Tamworth
NUNEATON
Bedworth
COVENTRY
RUGBY
BIRMINGHAM
Solihull
Kenilworth
ROYAL LEAMINGTON SPA
WARWICK
Southam
Kineton
Stratford on Avon
Alcester
Shipston on Stour
Long Compton

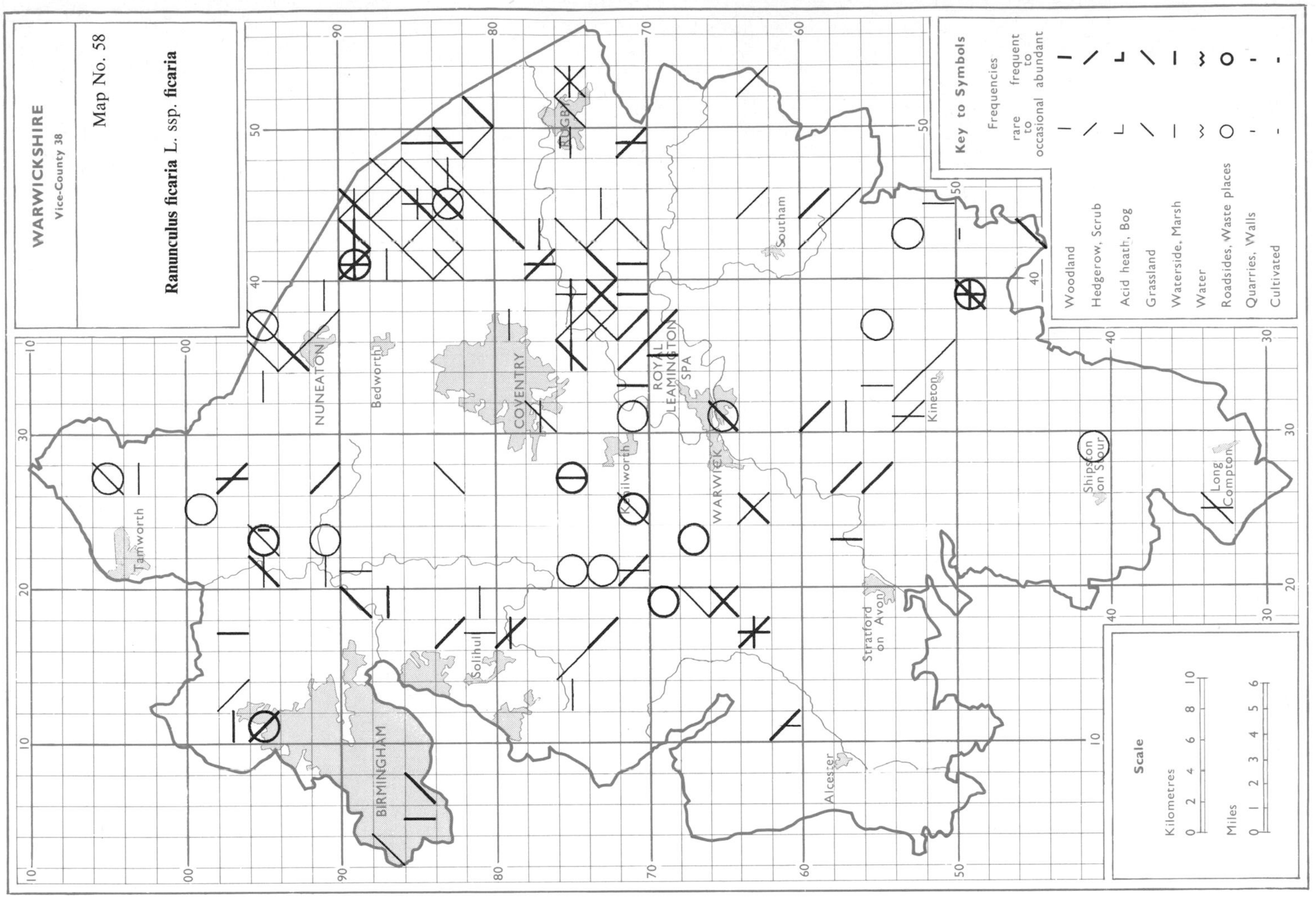
WARWICKSHIRE
Vice-County 38
Map No. 58
Ranunculus ficaria L. ssp. ficaria
Key to Symbols
Frequencies
rare to occasional
frequent to abundant
Woodland
Hedgerow, Scrub
Acid heath, Bog
Grassland
Waterside, Marsh
Water
Roadsides, Waste places
Quarries, Walls
Cultivated
Scale
Kilometres
0 2 4 6 8 10
Miles
0 1 2 3 4 5 6
NUNEATON
Bedworth
COVENTRY
RUGBY
ROYAL LEAMINGTON SPA
Southam
Kineton
WARWICK
Kenilworth
Tamworth
Solihull
BIRMINGHAM
Stratford on Avon
Alcester
Shipston on Stour
Long Compton

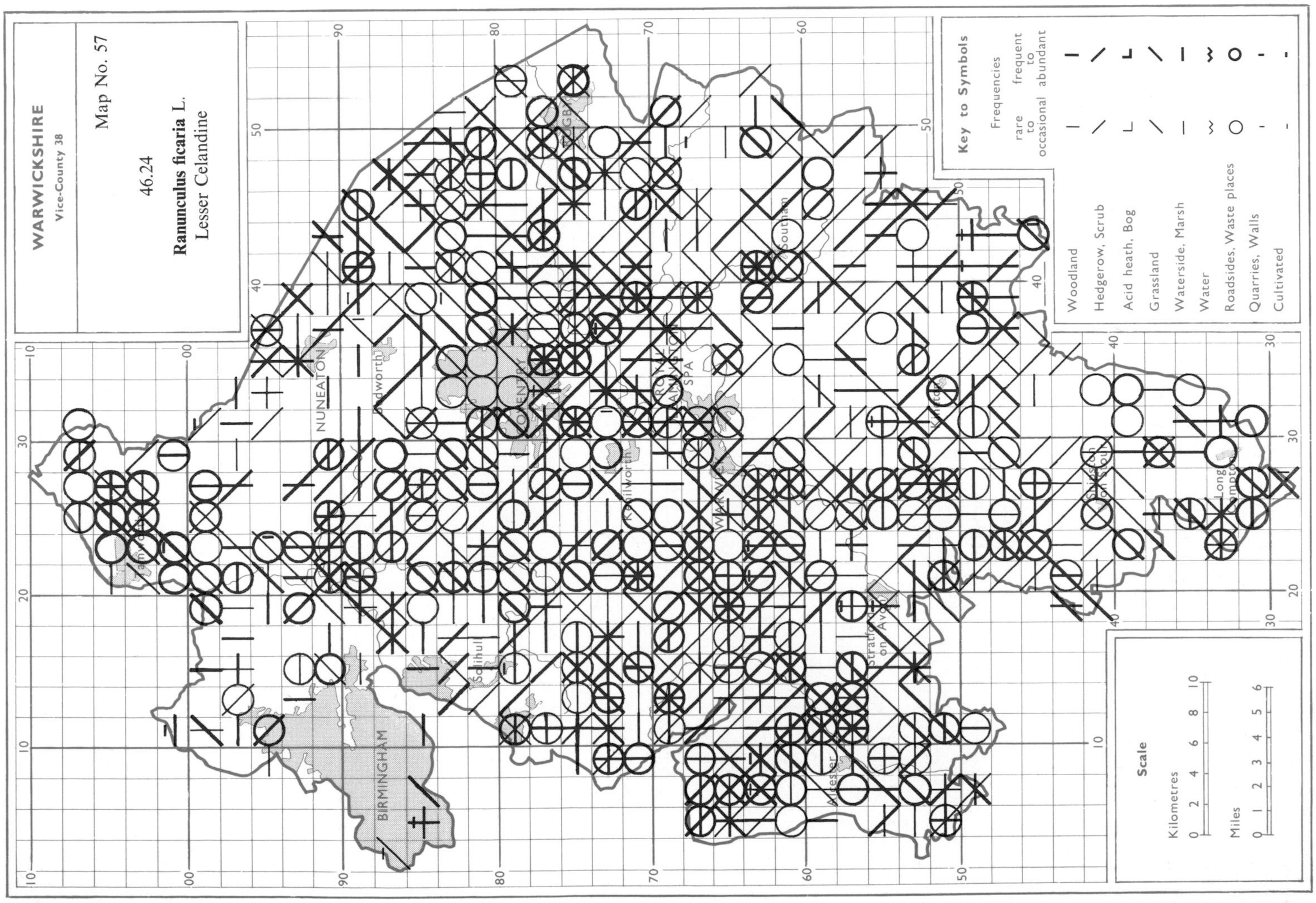
WARWICKSHIRE
Vice-County 38
Map No. 57
46.24
Ranunculus ficaria L.
Lesser Celandine
Key to Symbols
Frequencies
rare to occasional
frequent to abundant
Woodland
Hedgerow, Scrub
Acid heath, Bog
Grassland
Waterside, Marsh
Water
Roadsides, Waste places
Quarries, Walls
Cultivated
Scale
Kilometres
0 2 4 6 8 10
Miles
0 1 2 3 4 5 6
NUNEATON
Southam
Solihull
BIRMINGHAM
Kenilworth

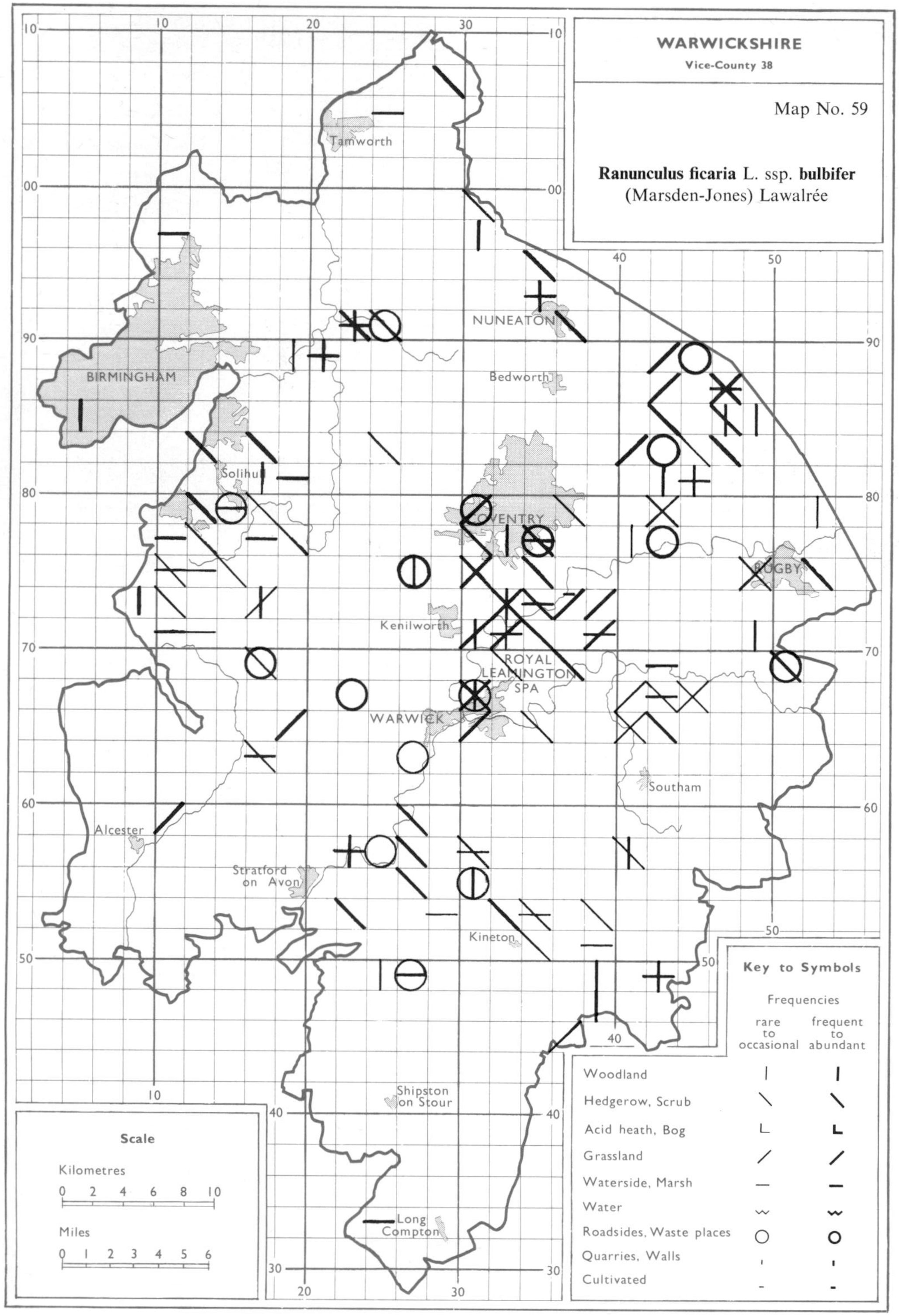
WARWICKSHIRE
Vice-County 38
Map No. 59
Ranunculus ficaria L. ssp. bulbifer (Marsden-Jones) Lawalrée
Tamworth
BIRMINGHAM
NUNEATON
Bedworth
Solihull
COVENTRY
RUGBY
Kenilworth
ROYAL LEAMINGTON SPA
WARWICK
Southam
Alcester
Stratford on Avon
Kineton
Shipston on Stour
Long Compton
Key to Symbols
Frequencies
rare to occasional
frequent to abundant
Woodland
Hedgerow, Scrub
Acid heath, Bog
Grassland
Waterside, Marsh
Water
Roadsides, Waste places
Quarries, Walls
Cultivated
Scale
Kilometres
0 2 4 6 8 10
Miles
0 1 2 3 4 5 6

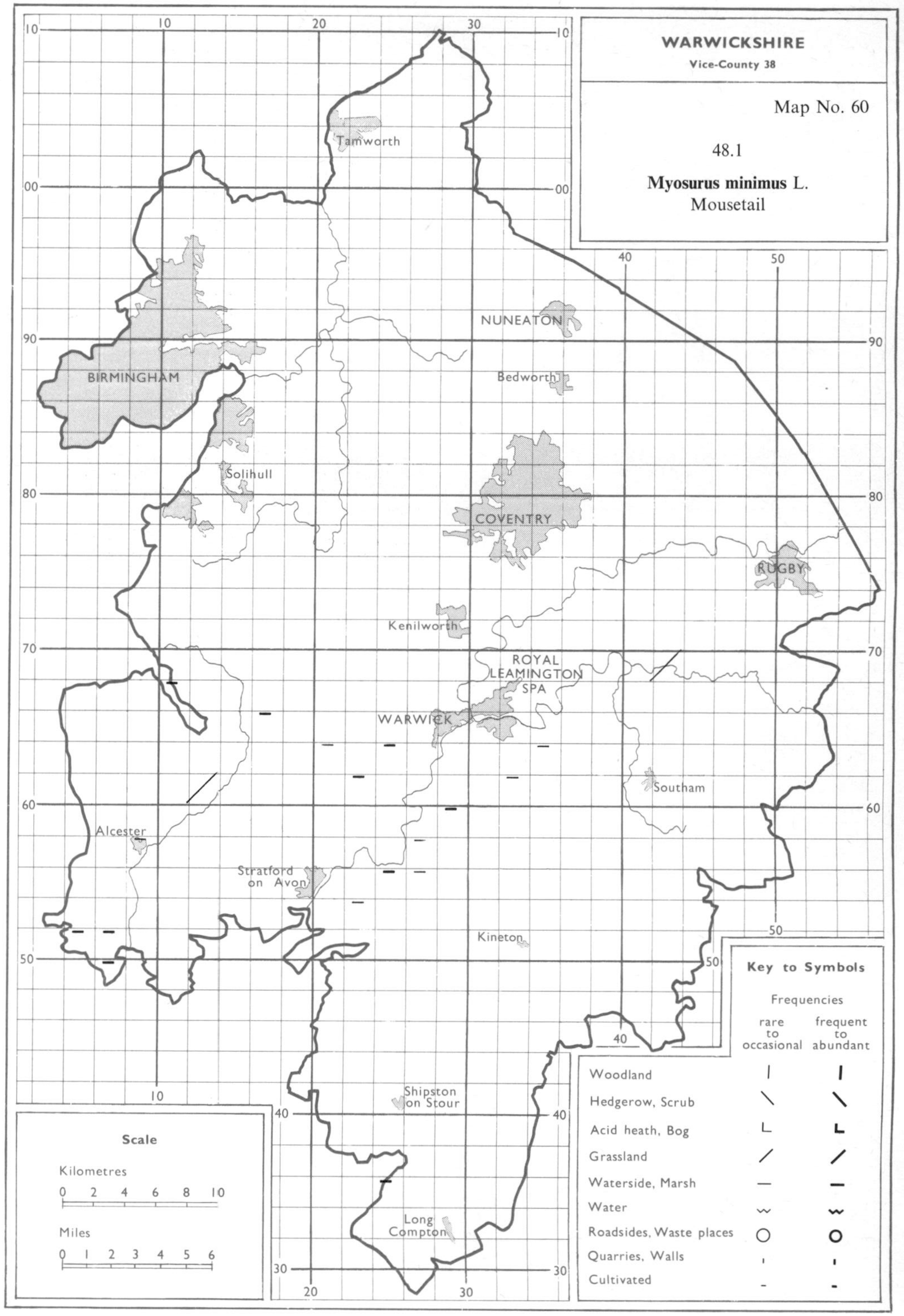
WARWICKSHIRE
Vice-County 38
Map No. 60
48.1
Myosurus minimus L.
Mousetail
Tamworth
BIRMINGHAM
NUNEATON
Bedworth
Solihull
COVENTRY
RUGBY
Kenilworth
ROYAL LEAMINGTON SPA
WARWICK
Southam
Alcester
Stratford on Avon
Kineton
Shipston on Stour
Long Compton
Key to Symbols
Frequencies
rare to occasional
frequent to abundant
Woodland
Hedgerow, Scrub
Acid heath, Bog
Grassland
Waterside, Marsh
Water
Roadsides, Waste places
Quarries, Walls
Cultivated
Scale
Kilometres
0 2 4 6 8 10
Miles
0 1 2 3 4 5 6

WARWICKSHIRE
Vice-County 38
Map No. 61
50.1
Thalictrum flavum L.
Common Meadow-rue
Key to Symbols
Frequencies
rare to occasional
frequent to abundant
Woodland
Hedgerow, Scrub
Acid heath, Bog
Grassland
Waterside, Marsh
Water
Roadsides, Waste places
Quarries, Walls
Cultivated
Scale
Kilometres
0 2 4 6 8 10
Miles
0 1 2 3 4 5 6
NUNEATON
Bedworth
COVENTRY
RUGBY
Tamworth
BIRMINGHAM
Solihull
Kenilworth
ROYAL LEAMINGTON SPA
WARWICK
Southam
Kineton
Shipston on Stour
Long Compton
Stratford on Avon
Alcester

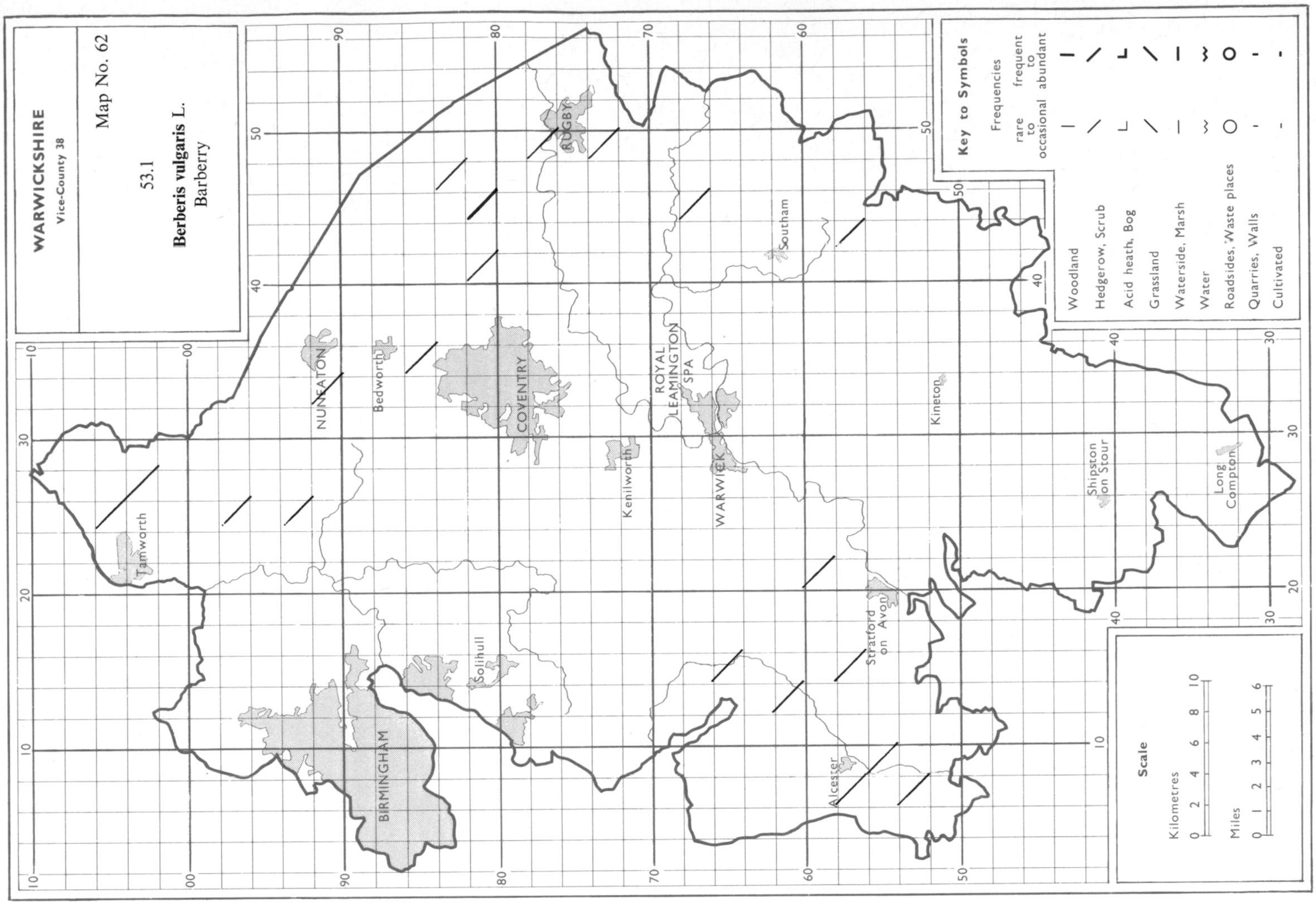
WARWICKSHIRE
Vice-County 38
Map No. 62
53.1
Berberis vulgaris L.
Barberry
Key to Symbols
Frequencies
rare to occasional
frequent to abundant
Woodland
Hedgerow, Scrub
Acid heath, Bog
Grassland
Waterside, Marsh
Water
Roadsides, Waste places
Quarries, Walls
Cultivated
Scale
Kilometres
0 2 4 6 8 10
Miles
0 1 2 3 4 5 6
NUNEATON
Bedworth
COVENTRY
RUGBY
Tamworth
BIRMINGHAM
Solihull
Kenilworth
ROYAL LEAMINGTON SPA
WARWICK
Southam
Kineton
Shipston on Stour
Long Compton
Stratford on Avon
Alcester

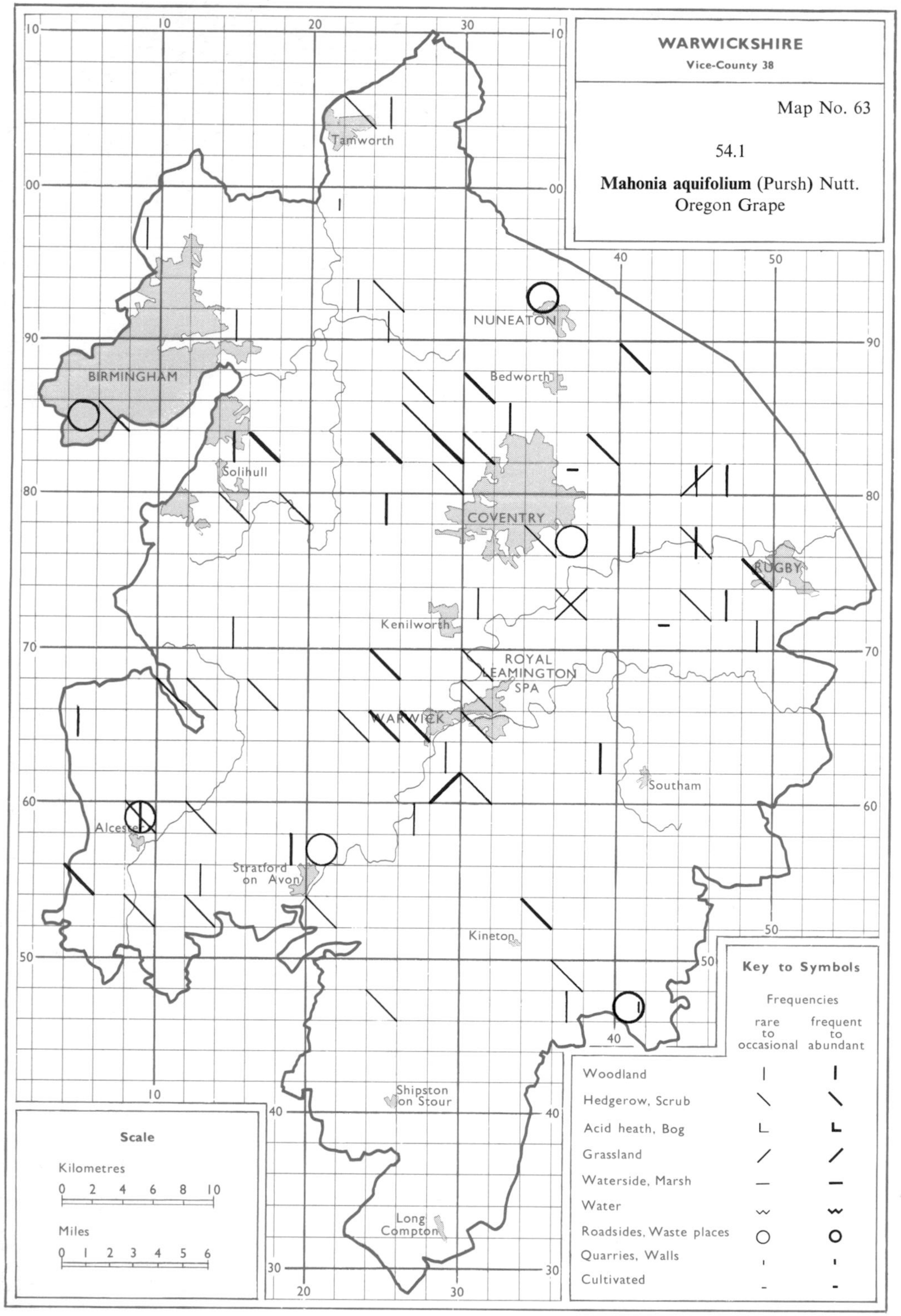
WARWICKSHIRE
Vice-County 38
Map No. 63
54.1
Mahonia aquifolium (Pursh) Nutt.
Oregon Grape
Tamworth
NUNEATON
Bedworth
BIRMINGHAM
Solihull
COVENTRY
RUGBY
Kenilworth
ROYAL LEAMINGTON SPA
WARWICK
Southam
Alcester
Stratford on Avon
Kineton
Shipston on Stour
Long Compton
Scale
Kilometres
0 2 4 6 8 10
Miles
0 1 2 3 4 5 6
Key to Symbols
Frequencies
rare to occasional
frequent to abundant
Woodland
Hedgerow, Scrub
Acid heath, Bog
Grassland
Waterside, Marsh
Water
Roadsides, Waste places
Quarries, Walls
Cultivated

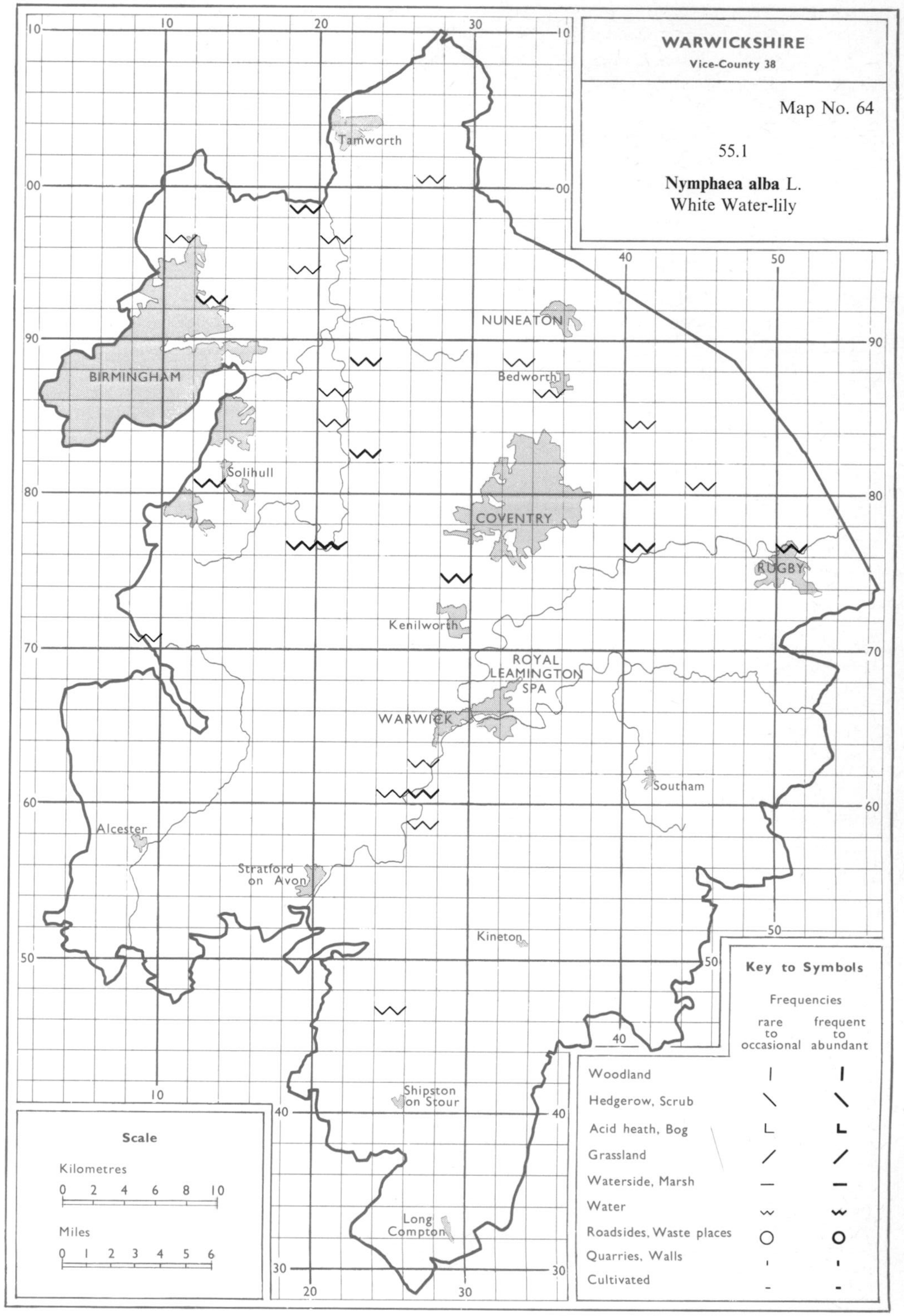
WARWICKSHIRE
Vice-County 38
Map No. 64
55.1
Nymphaea alba L.
White Water-lily
Tamworth
NUNEATON
Bedworth
BIRMINGHAM
Solihull
COVENTRY
RUGBY
Kenilworth
ROYAL LEAMINGTON SPA
WARWICK
Southam
Alcester
Stratford on Avon
Kineton
Shipston on Stour
Long Compton
Scale
Kilometres
0 2 4 6 8 10
Miles
0 1 2 3 4 5 6
Key to Symbols
Frequencies
rare to occasional
frequent to abundant
Woodland
Hedgerow, Scrub
Acid heath, Bog
Grassland
Waterside, Marsh
Water
Roadsides, Waste places
Quarries, Walls
Cultivated

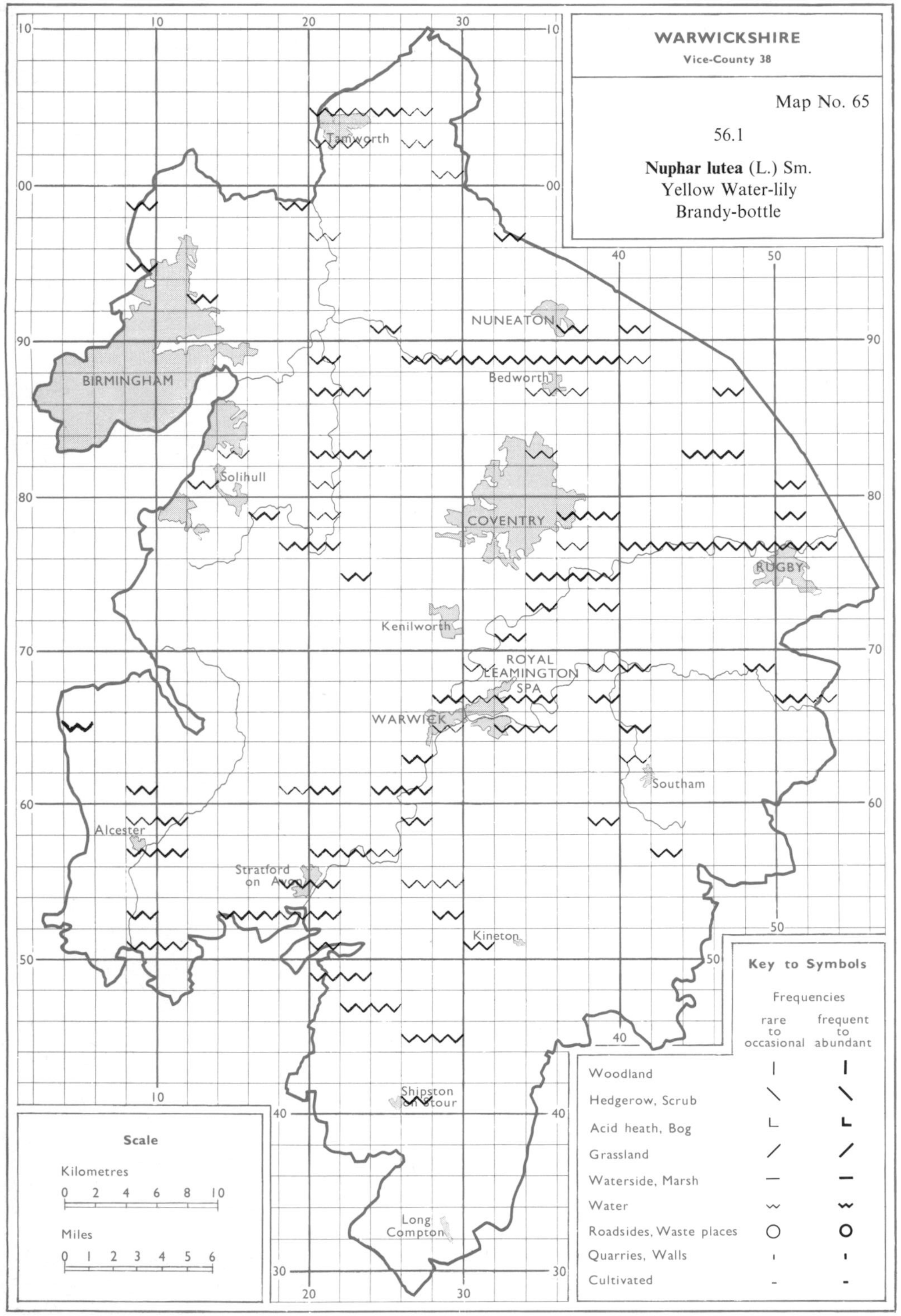
WARWICKSHIRE
Vice-County 38
Map No. 65
56.1
Nuphar lutea (L.) Sm.
Yellow Water-lily
Brandy-bottle
Tamworth
BIRMINGHAM
Solihull
NUNEATON
Bedworth
COVENTRY
RUGBY
Kenilworth
ROYAL LEAMINGTON SPA
WARWICK
Southam
Alcester
Stratford on Avon
Kineton
Shipston on Stour
Long Compton
Scale
Kilometres
0 2 4 6 8 10
Miles
0 1 2 3 4 5 6
Key to Symbols
Frequencies
rare to occasional
frequent to abundant
Woodland
Hedgerow, Scrub
Acid heath, Bog
Grassland
Waterside, Marsh
Water
Roadsides, Waste places
Quarries, Walls
Cultivated

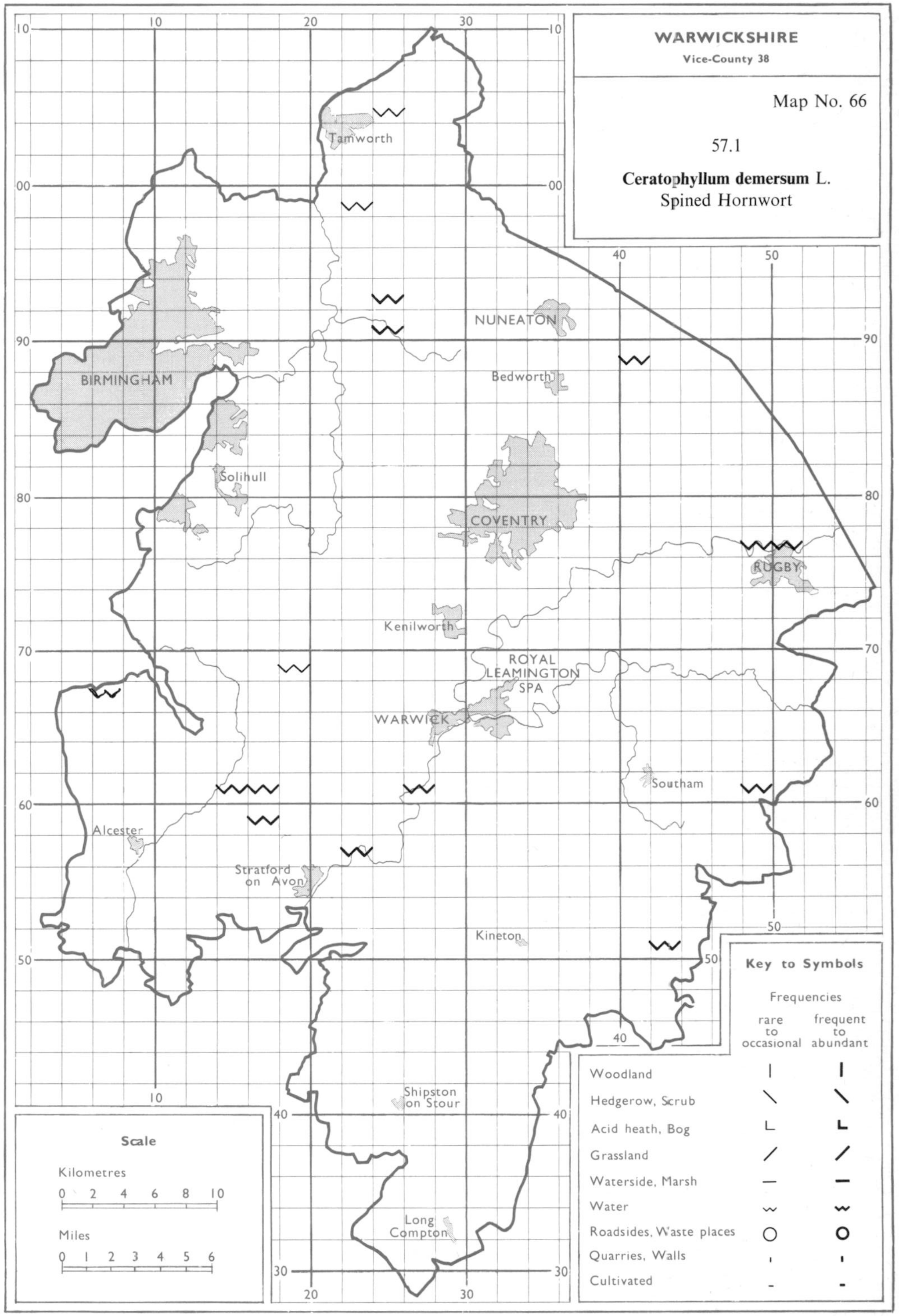
WARWICKSHIRE
Vice-County 38
Map No. 66
57.1
Ceratophyllum demersum L.
Spined Hornwort
Tamworth
BIRMINGHAM
Solihull
NUNEATON
Bedworth
COVENTRY
RUGBY
Kenilworth
ROYAL LEAMINGTON SPA
WARWICK
Southam
Alcester
Stratford on Avon
Kineton
Shipston on Stour
Long Compton
Scale
Kilometres
0 2 4 6 8 10
Miles
0 1 2 3 4 5 6
Key to Symbols
Frequencies
rare to occasional
frequent to abundant
Woodland
Hedgerow, Scrub
Acid heath, Bog
Grassland
Waterside, Marsh
Water
Roadsides, Waste places
Quarries, Walls
Cultivated

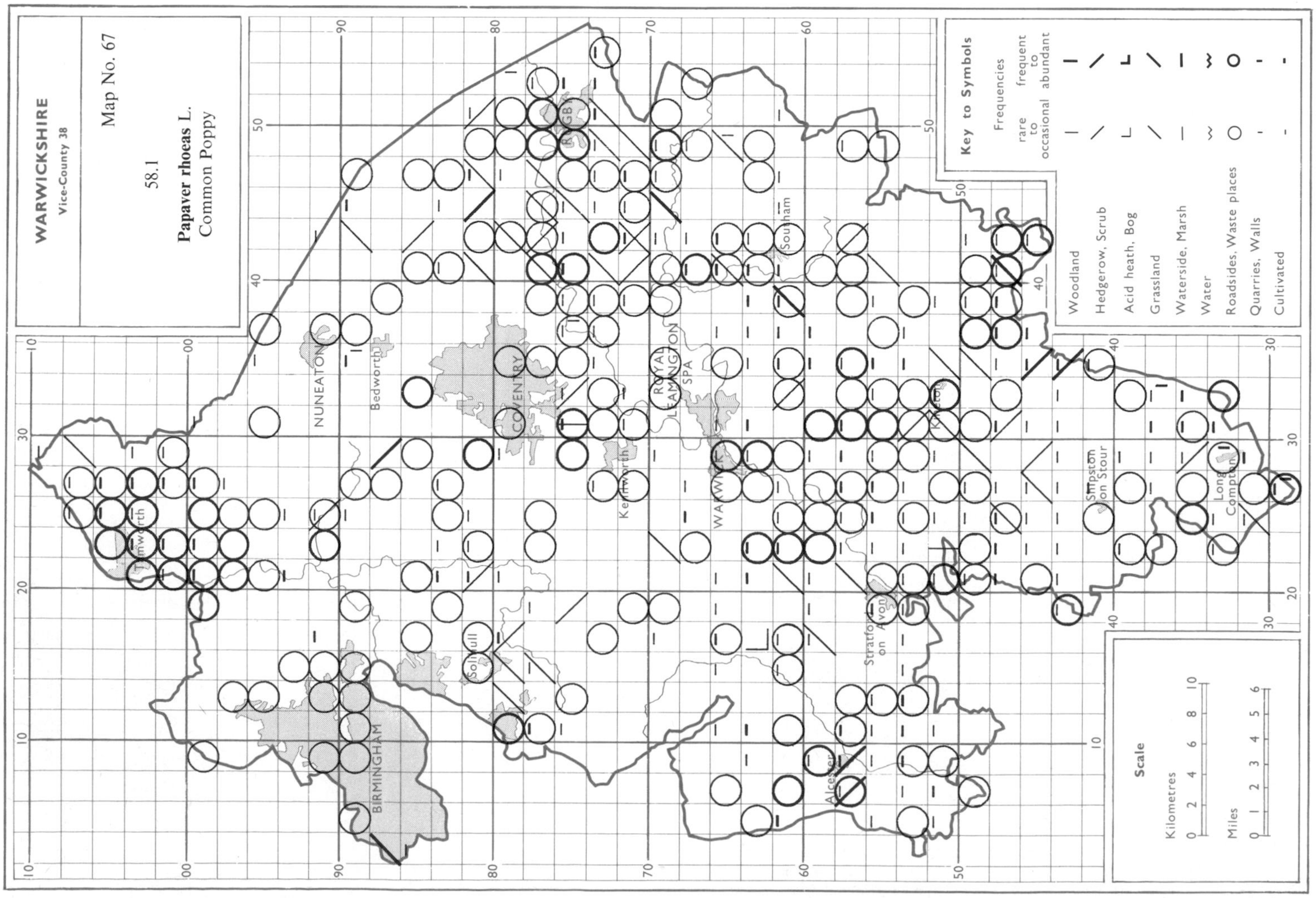
WARWICKSHIRE
Vice-County 38
Map No. 67
58.1
Papaver rhoeas L.
Common Poppy
Key to Symbols
Frequencies
rare to occasional
frequent to abundant
Woodland
Hedgerow, Scrub
Acid heath, Bog
Grassland
Waterside, Marsh
Water
Roadsides, Waste places
Quarries, Walls
Cultivated
Scale
Kilometres
0 2 4 6 8 10
Miles
0 1 2 3 4 5 6
NUNEATON
Bedworth
COVENTRY
RUGBY
Tamworth
BIRMINGHAM
Solihull
Kenilworth
ROYAL LEAMINGTON SPA
WARWICK
Southam
Stratford on Avon
Alcester
Kineton
Shipston on Stour
Long Compton

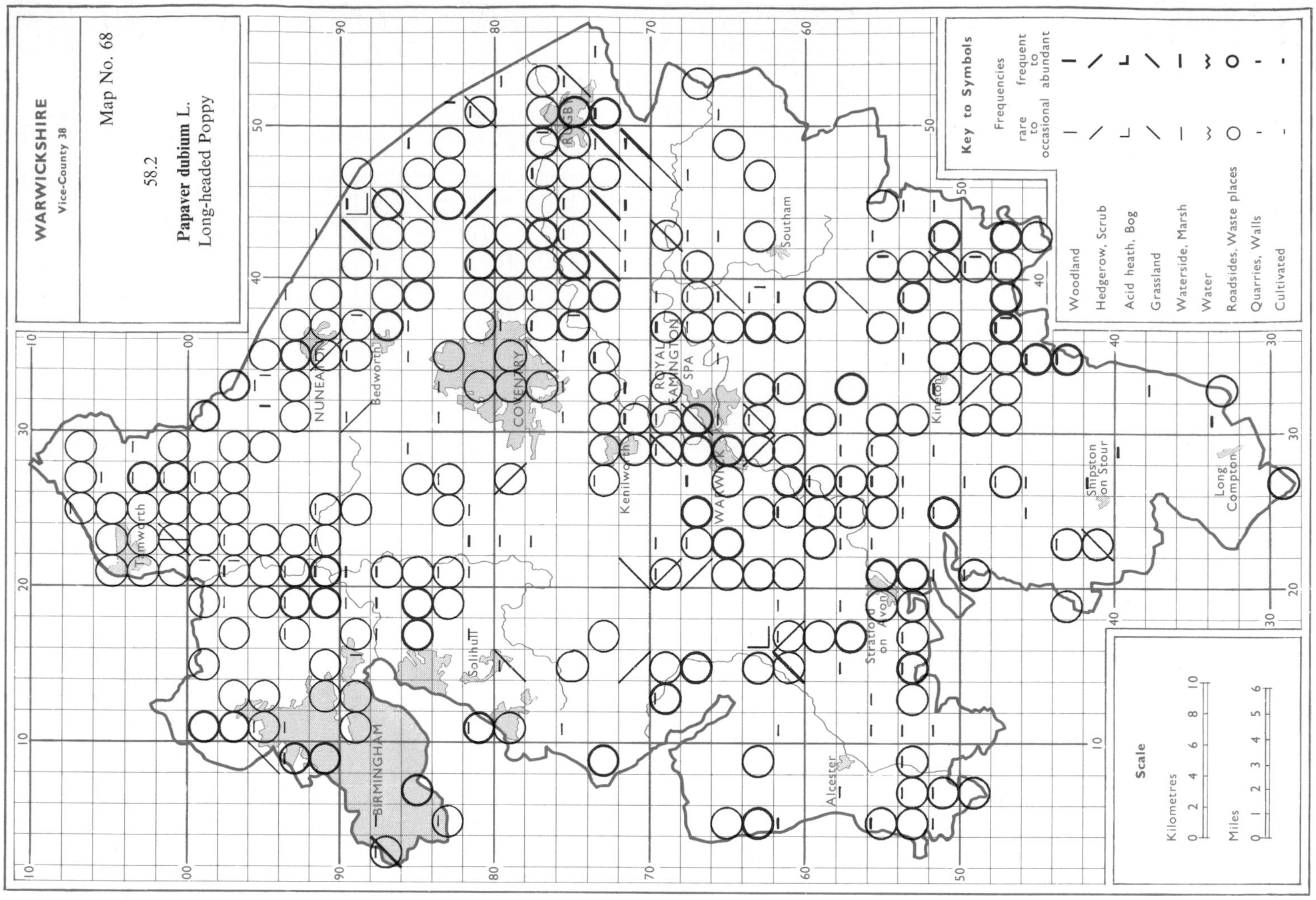
WARWICKSHIRE
Vice-County 38
Map No. 68
58.2
Papaver dubium L.
Long-headed Poppy
Key to Symbols
Frequencies
rare to occasional
frequent to abundant
Woodland
Hedgerow, Scrub
Acid heath, Bog
Grassland
Waterside, Marsh
Water
Roadsides, Waste places
Quarries, Walls
Cultivated
Scale
Kilometres
0 2 4 6 8 10
Miles
0 1 2 3 4 5 6
NUNEATON
Bedworth
COVENTRY
RUGBY
Tamworth
BIRMINGHAM
Solihull
Kenilworth
ROYAL LEAMINGTON SPA
WARWICK
Southam
Stratford on Avon
Alcester
Kineton
Shipston on Stour
Long Compton

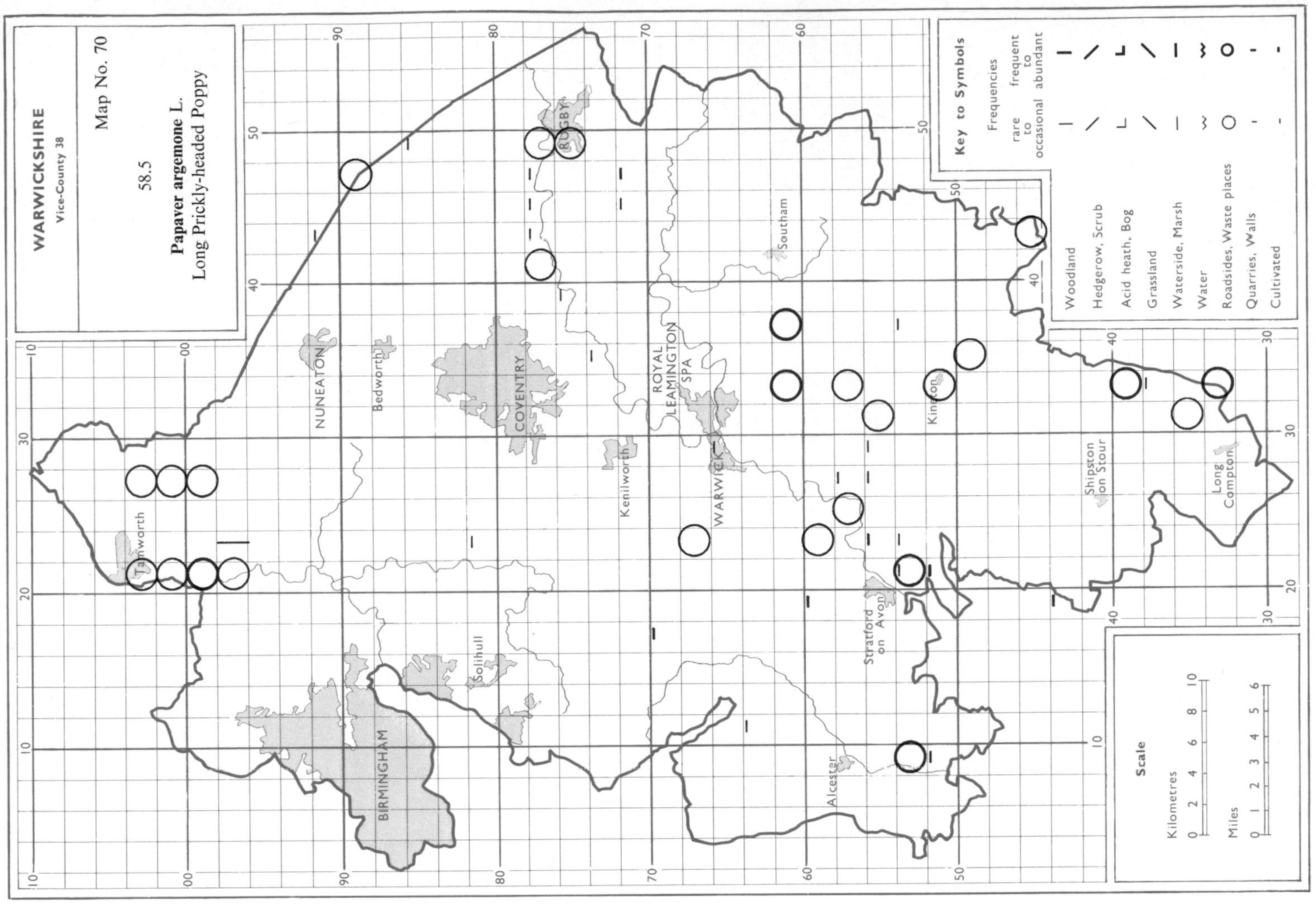
WARWICKSHIRE
Vice-County 38
Map No. 70
58.5
Papaver argemone L.
Long Prickly-headed Poppy
Key to Symbols
Frequencies
rare to occasional
frequent to abundant
Woodland
Hedgerow, Scrub
Acid heath, Bog
Grassland
Waterside, Marsh
Water
Roadsides, Waste places
Quarries, Walls
Cultivated
Scale
Kilometres 0 2 4 6 8 10
Miles 0 1 2 3 4 5 6
Tamworth
NUNEATON
Bedworth
COVENTRY
RUGBY
Southam
ROYAL LEAMINGTON SPA
Kenilworth
WARWICK
Solihull
BIRMINGHAM
Stratford on Avon
Alcester
Kineton
Shipston on Stour
Long Compton

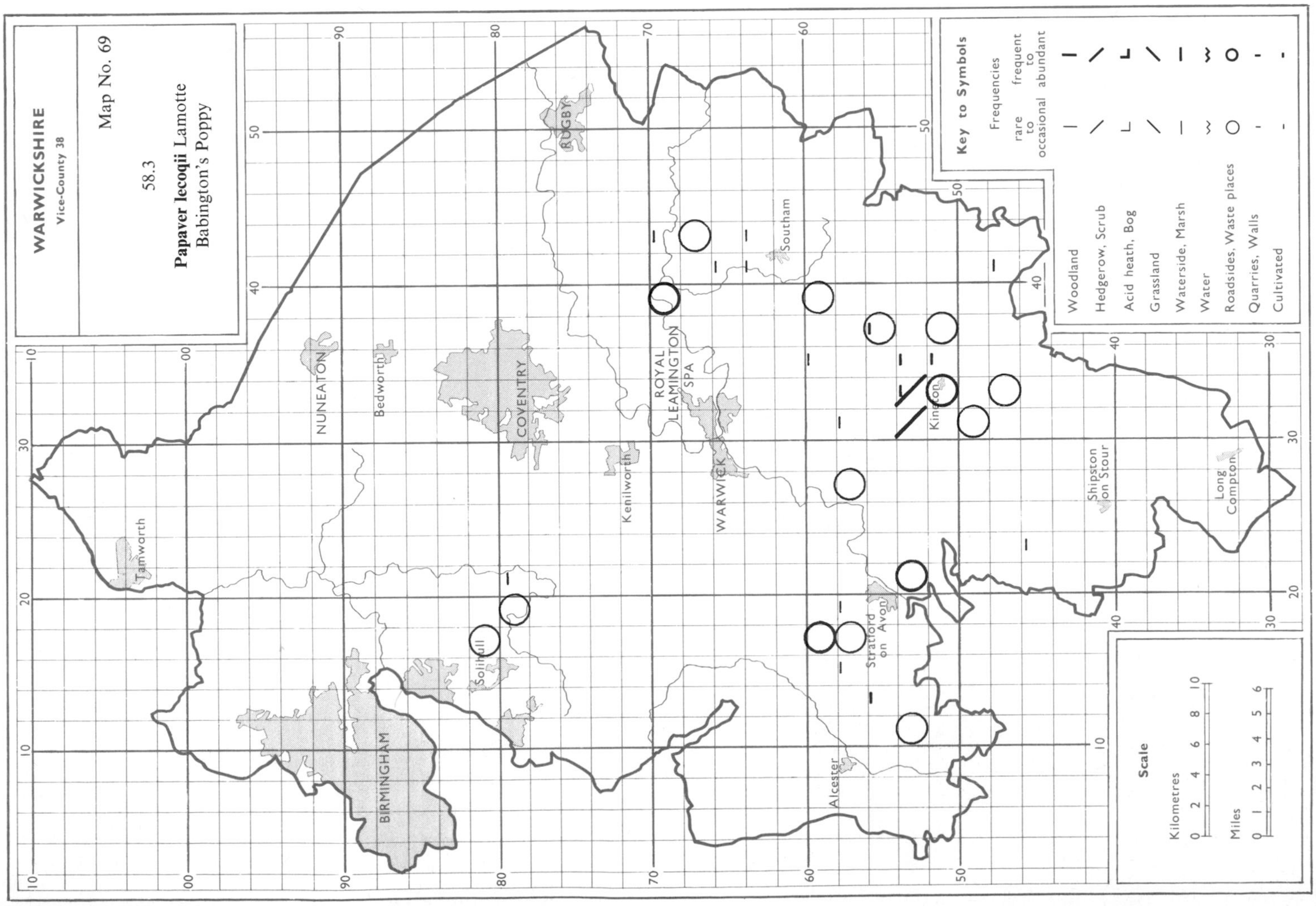
WARWICKSHIRE
Vice-County 38
Map No. 69
58.3
Papaver lecoqii Lamotte
Babington's Poppy
Key to Symbols
Frequencies
rare to occasional
frequent to abundant
Woodland
Hedgerow, Scrub
Acid heath, Bog
Grassland
Waterside, Marsh
Water
Roadsides, Waste places
Quarries, Walls
Cultivated
Scale
Kilometres 0 2 4 6 8 10
Miles 0 1 2 3 4 5 6
Tamworth
NUNEATON
Bedworth
COVENTRY
RUGBY
Southam
ROYAL LEAMINGTON SPA
Kenilworth
WARWICK
Solihull
BIRMINGHAM
Stratford on Avon
Alcester
Kineton
Shipston on Stour
Long Compton

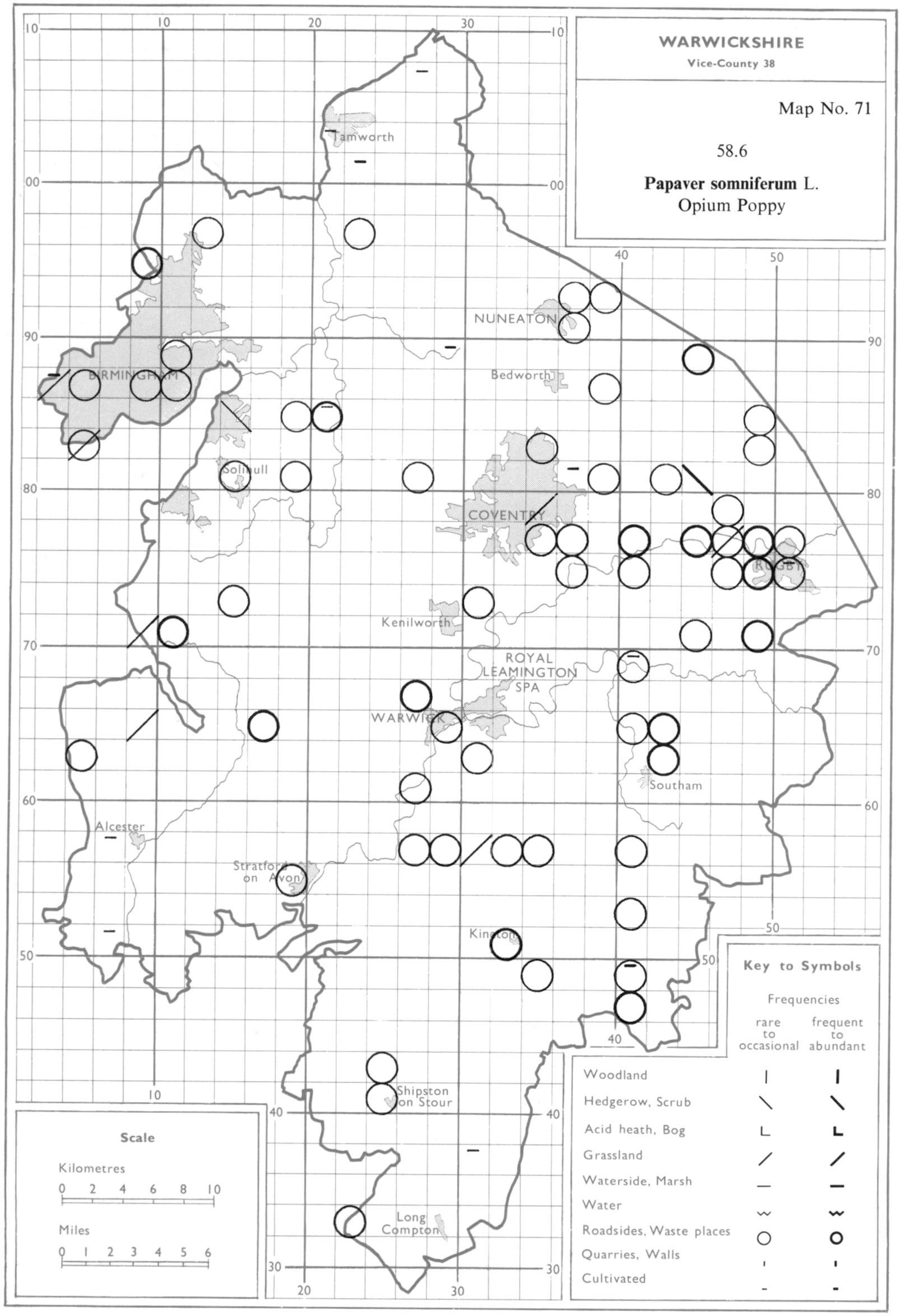
WARWICKSHIRE
Vice-County 38
Map No. 71
58.6
Papaver somniferum L.
Opium Poppy
Tamworth
NUNEATON
Bedworth
BIRMINGHAM
Solihull
COVENTRY
RUGBY
Kenilworth
ROYAL LEAMINGTON SPA
WARWICK
Southam
Alcester
Stratford on Avon
Kineton
Shipston on Stour
Long Compton
Key to Symbols
Frequencies
rare to occasional
frequent to abundant
Woodland
Hedgerow, Scrub
Acid heath, Bog
Grassland
Waterside, Marsh
Water
Roadsides, Waste places
Quarries, Walls
Cultivated
Scale
Kilometres
0 2 4 6 8 10
Miles
0 1 2 3 4 5 6

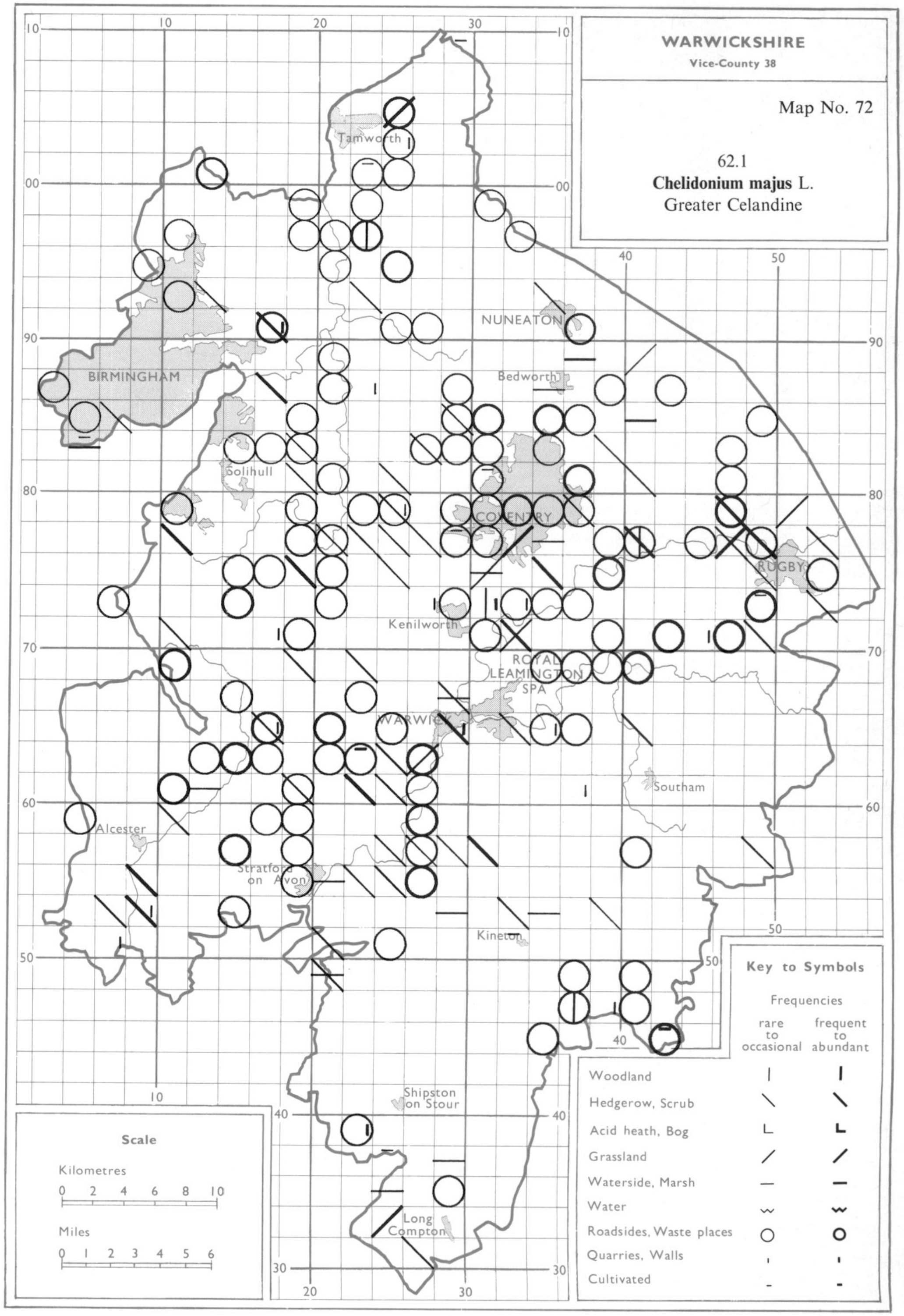
WARWICKSHIRE
Vice-County 38
Map No. 72
62.1
Chelidonium majus L.
Greater Celandine
Tamworth
NUNEATON
Bedworth
BIRMINGHAM
Solihull
COVENTRY
RUGBY
Kenilworth
ROYAL LEAMINGTON SPA
WARWICK
Southam
Alcester
Stratford on Avon
Kineton
Shipston on Stour
Long Compton
Key to Symbols
Frequencies
rare to occasional
frequent to abundant
Woodland
Hedgerow, Scrub
Acid heath, Bog
Grassland
Waterside, Marsh
Water
Roadsides, Waste places
Quarries, Walls
Cultivated
Scale
Kilometres
0 2 4 6 8 10
Miles
0 1 2 3 4 5 6

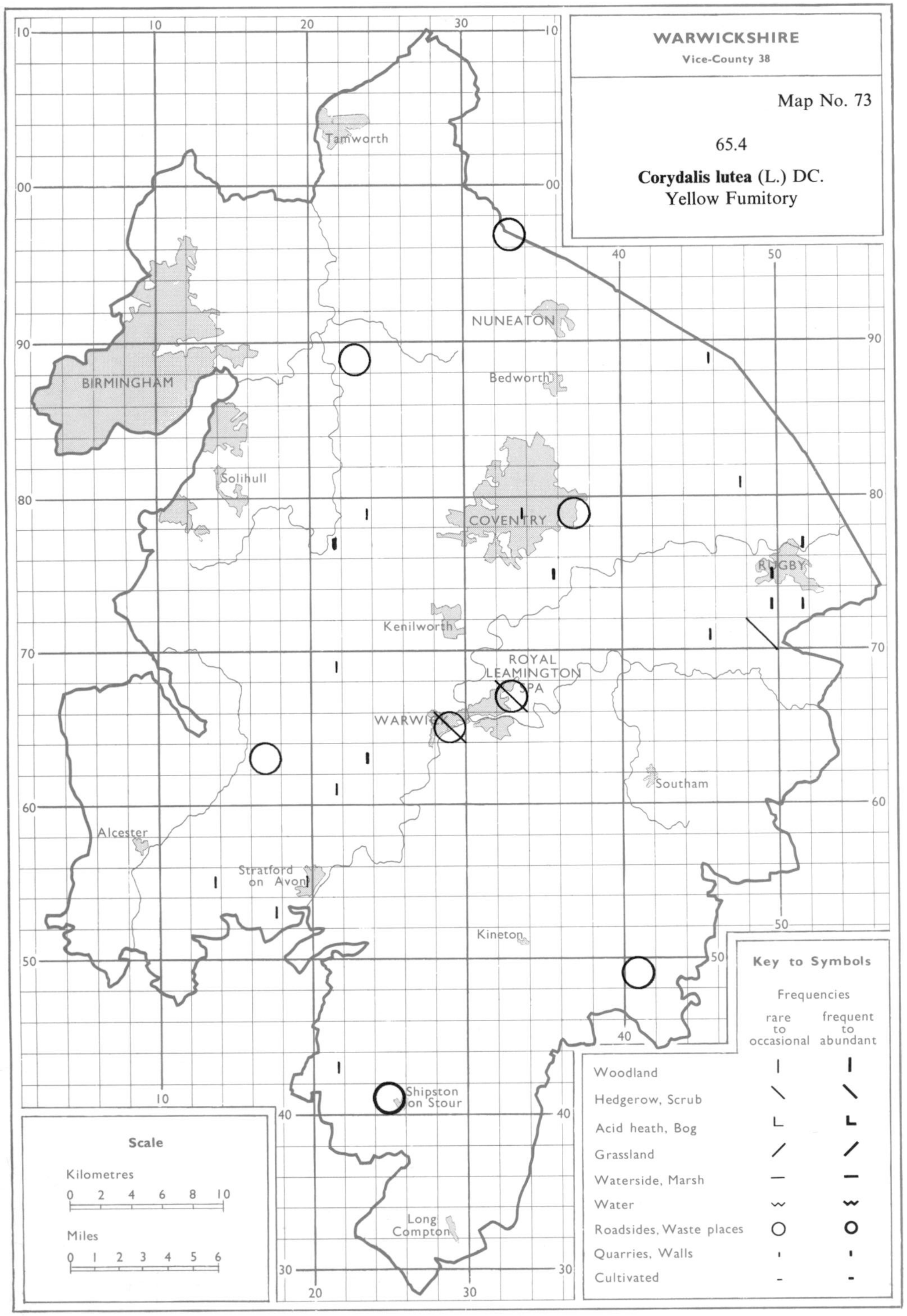
WARWICKSHIRE
Vice-County 38
Map No. 73
65.4
Corydalis lutea (L.) DC.
Yellow Fumitory
Tamworth
BIRMINGHAM
Solihull
NUNEATON
Bedworth
COVENTRY
RUGBY
Kenilworth
ROYAL LEAMINGTON SPA
WARWICK
Southam
Alcester
Stratford on Avon
Kineton
Shipston on Stour
Long Compton
Scale
Kilometres
0 2 4 6 8 10
Miles
0 1 2 3 4 5 6
Key to Symbols
Frequencies
rare to occasional
frequent to abundant
Woodland
Hedgerow, Scrub
Acid heath, Bog
Grassland
Waterside, Marsh
Water
Roadsides, Waste places
Quarries, Walls
Cultivated

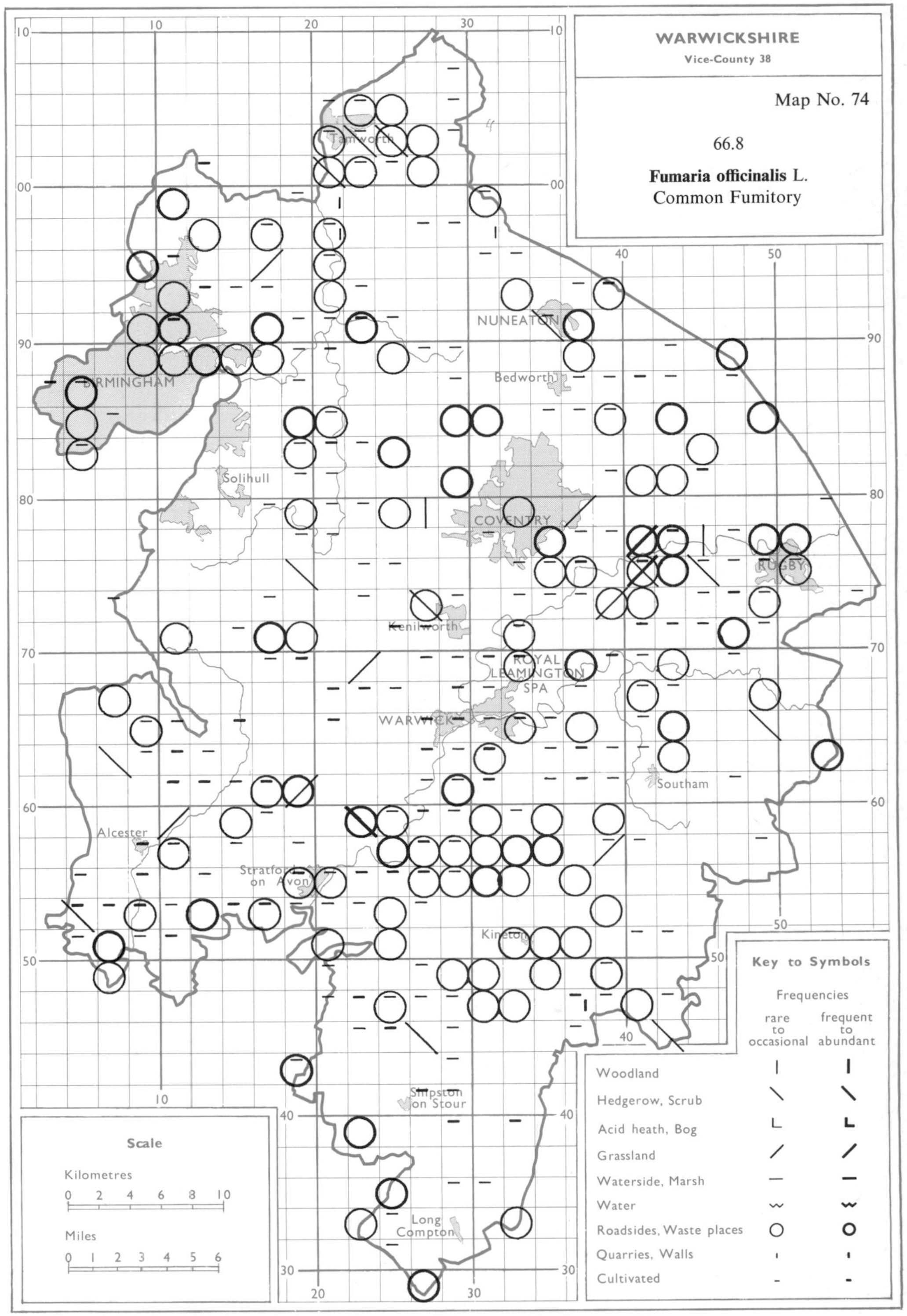
WARWICKSHIRE
Vice-County 38
Map No. 74
66.8
Fumaria officinalis L.
Common Fumitory
Tamworth
BIRMINGHAM
Solihull
NUNEATON
Bedworth
COVENTRY
RUGBY
Kenilworth
ROYAL LEAMINGTON SPA
WARWICK
Southam
Alcester
Stratford on Avon
Kineton
Shipston on Stour
Long Compton
Scale
Kilometres
0 2 4 6 8 10
Miles
0 1 2 3 4 5 6
Key to Symbols
Frequencies
rare to occasional
frequent to abundant
Woodland
Hedgerow, Scrub
Acid heath, Bog
Grassland
Waterside, Marsh
Water
Roadsides, Waste places
Quarries, Walls
Cultivated

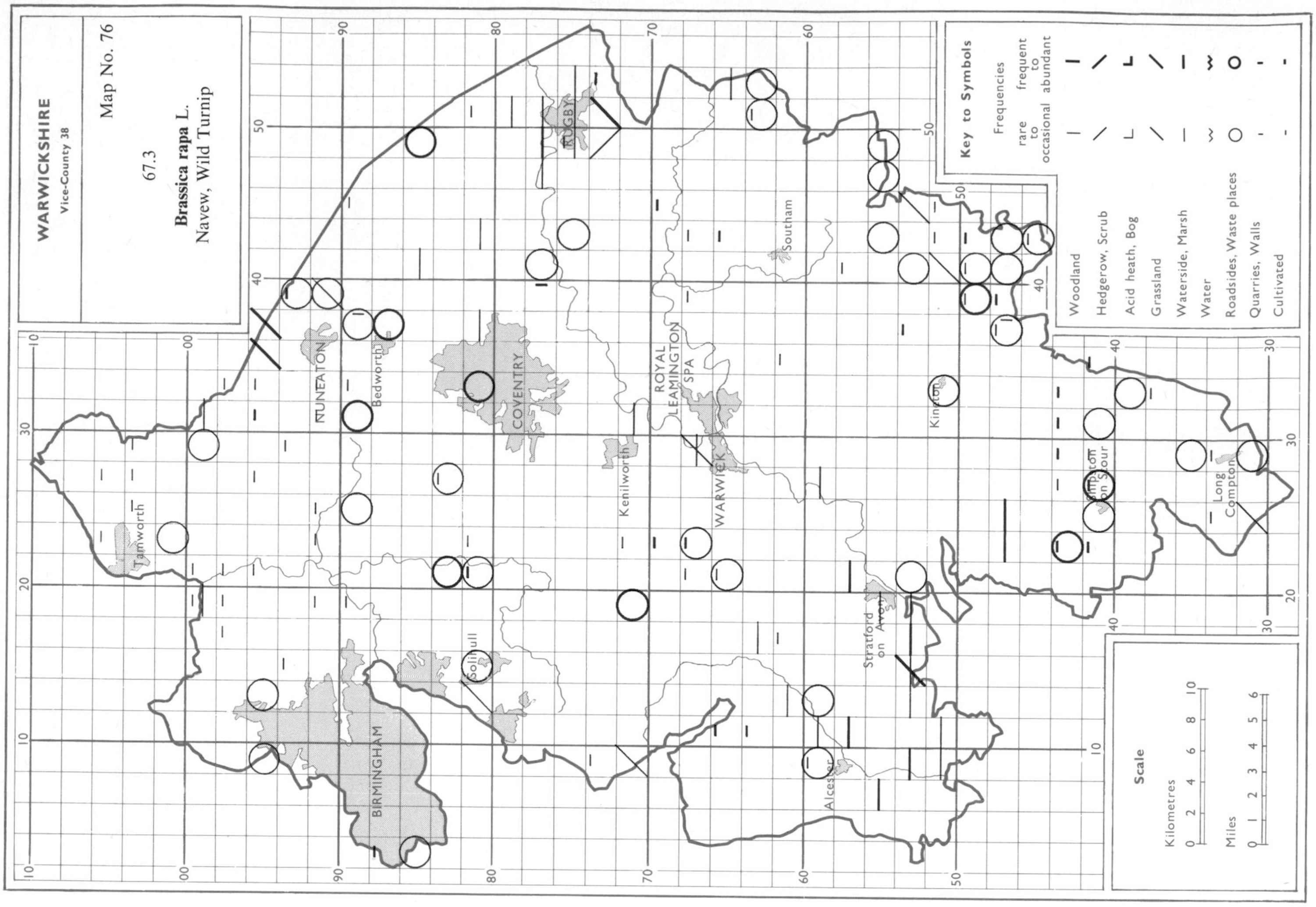
WARWICKSHIRE
Vice-County 38
Map No. 76
67.3
Brassica rapa L.
Navew, Wild Turnip
Key to Symbols
Frequencies
rare to occasional
frequent to abundant
Woodland
Hedgerow, Scrub
Acid heath, Bog
Grassland
Waterside, Marsh
Water
Roadsides, Waste places
Quarries, Walls
Cultivated
Scale
Kilometres
0 2 4 6 8 10
Miles
0 1 2 3 4 5 6
TAMWORTH
NUNEATON
Bedworth
RUGBY
COVENTRY
BIRMINGHAM
Solihull
Kenilworth
ROYAL LEAMINGTON SPA
WARWICK
Southam
Kineton
Stratford on Avon
Alcester
Shipston on Stour
Long Compton

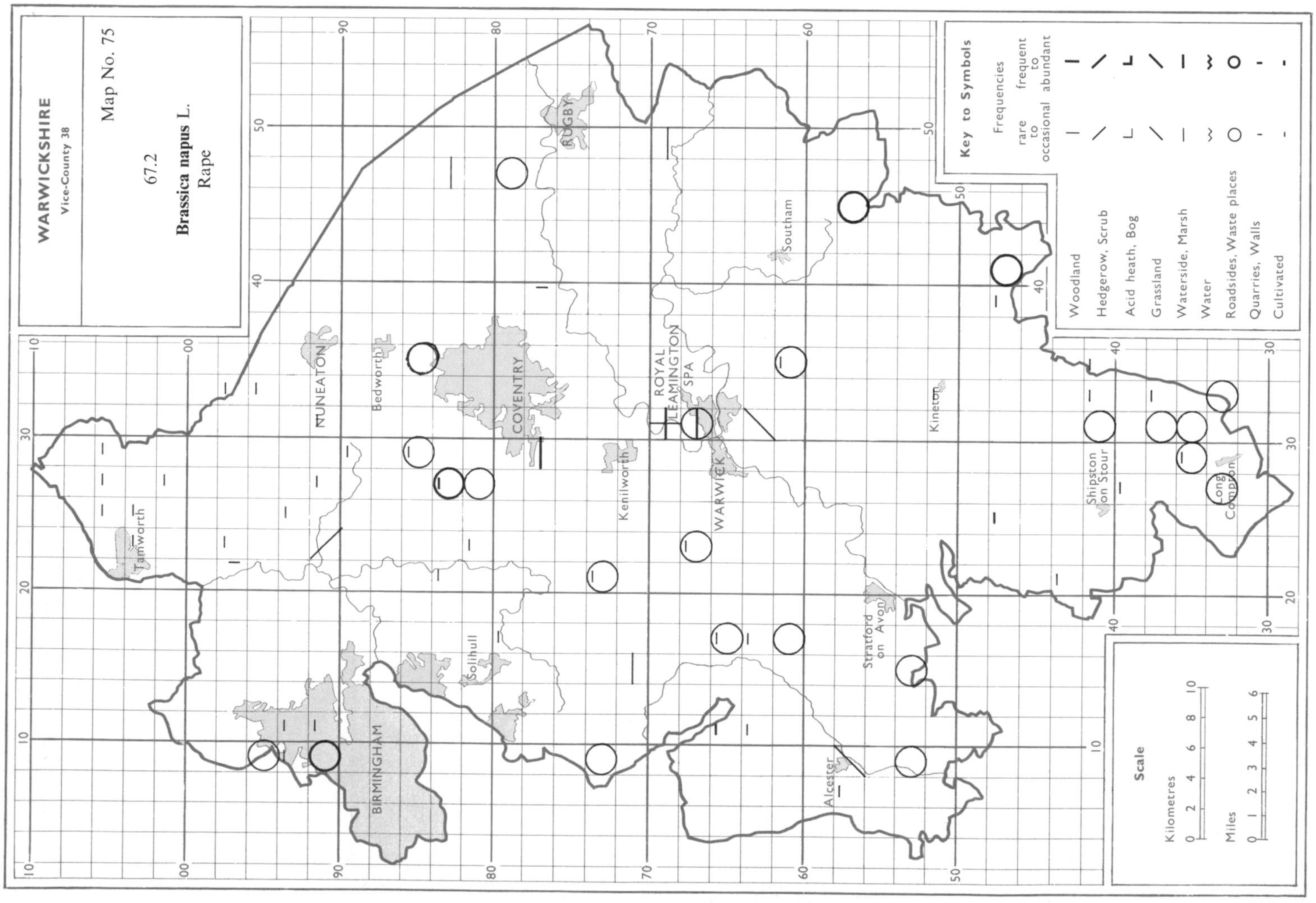
WARWICKSHIRE
Vice-County 38
Map No. 75
67.2
Brassica napus L.
Rape
Key to Symbols
Frequencies
rare to occasional
frequent to abundant
Woodland
Hedgerow, Scrub
Acid heath, Bog
Grassland
Waterside, Marsh
Water
Roadsides, Waste places
Quarries, Walls
Cultivated
Scale
Kilometres
0 2 4 6 8 10
Miles
0 1 2 3 4 5 6
Tamworth
NUNEATON
Bedworth
RUGBY
COVENTRY
BIRMINGHAM
Solihull
Kenilworth
ROYAL LEAMINGTON SPA
WARWICK
Southam
Kineton
Stratford on Avon
Alcester
Shipston on Stour
Long Compton

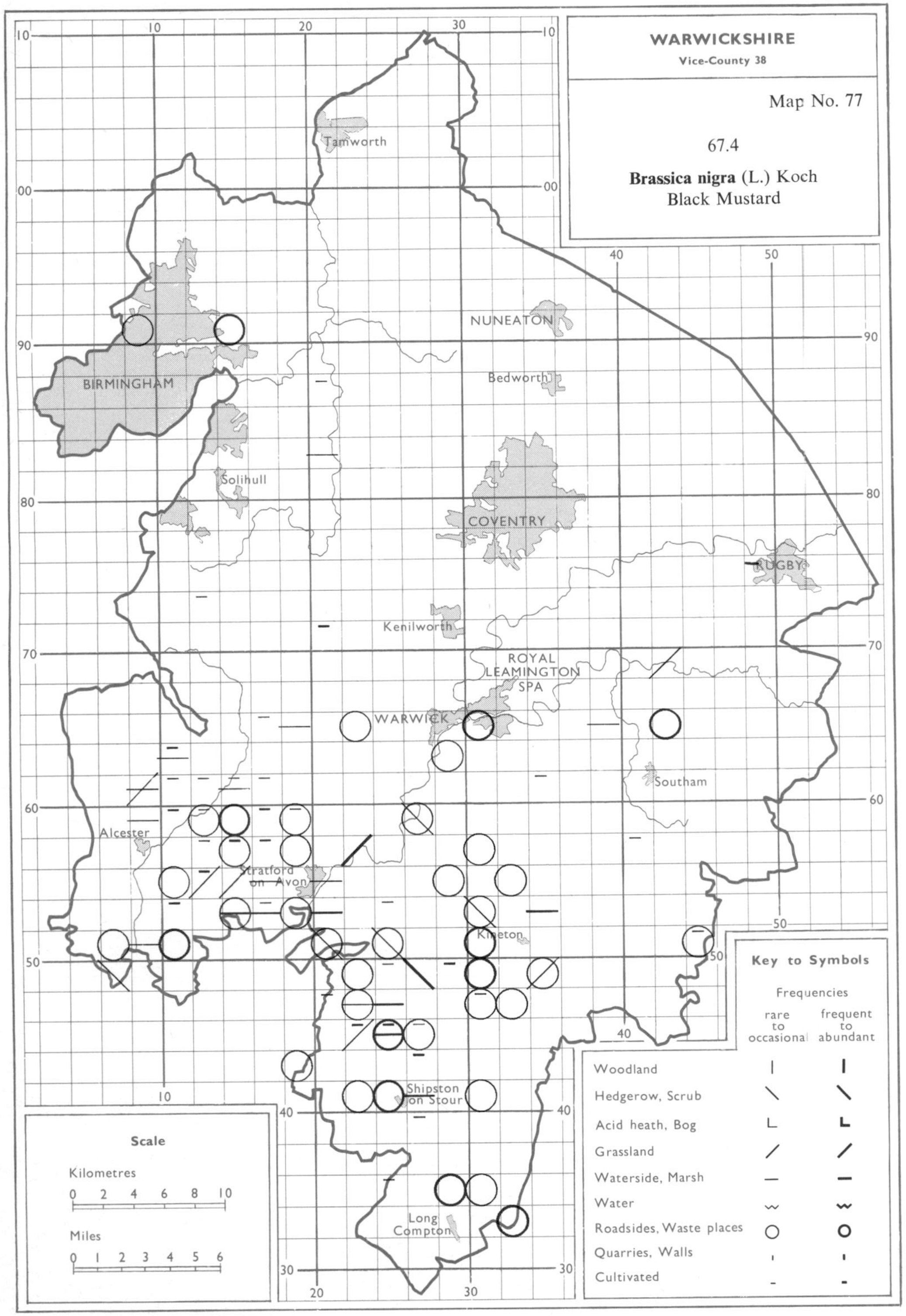
WARWICKSHIRE
Vice-County 38
Map No. 77
67.4
Brassica nigra (L.) Koch
Black Mustard
Tamworth
BIRMINGHAM
Solihull
NUNEATON
Bedworth
COVENTRY
RUGBY
Kenilworth
ROYAL LEAMINGTON SPA
WARWICK
Southam
Alcester
Stratford on Avon
Kineton
Shipston on Stour
Long Compton
Key to Symbols
Frequencies
rare to occasional
frequent to abundant
Woodland
Hedgerow, Scrub
Acid heath, Bog
Grassland
Waterside, Marsh
Water
Roadsides, Waste places
Quarries, Walls
Cultivated
Scale
Kilometres
0 2 4 6 8 10
Miles
0 1 2 3 4 5 6

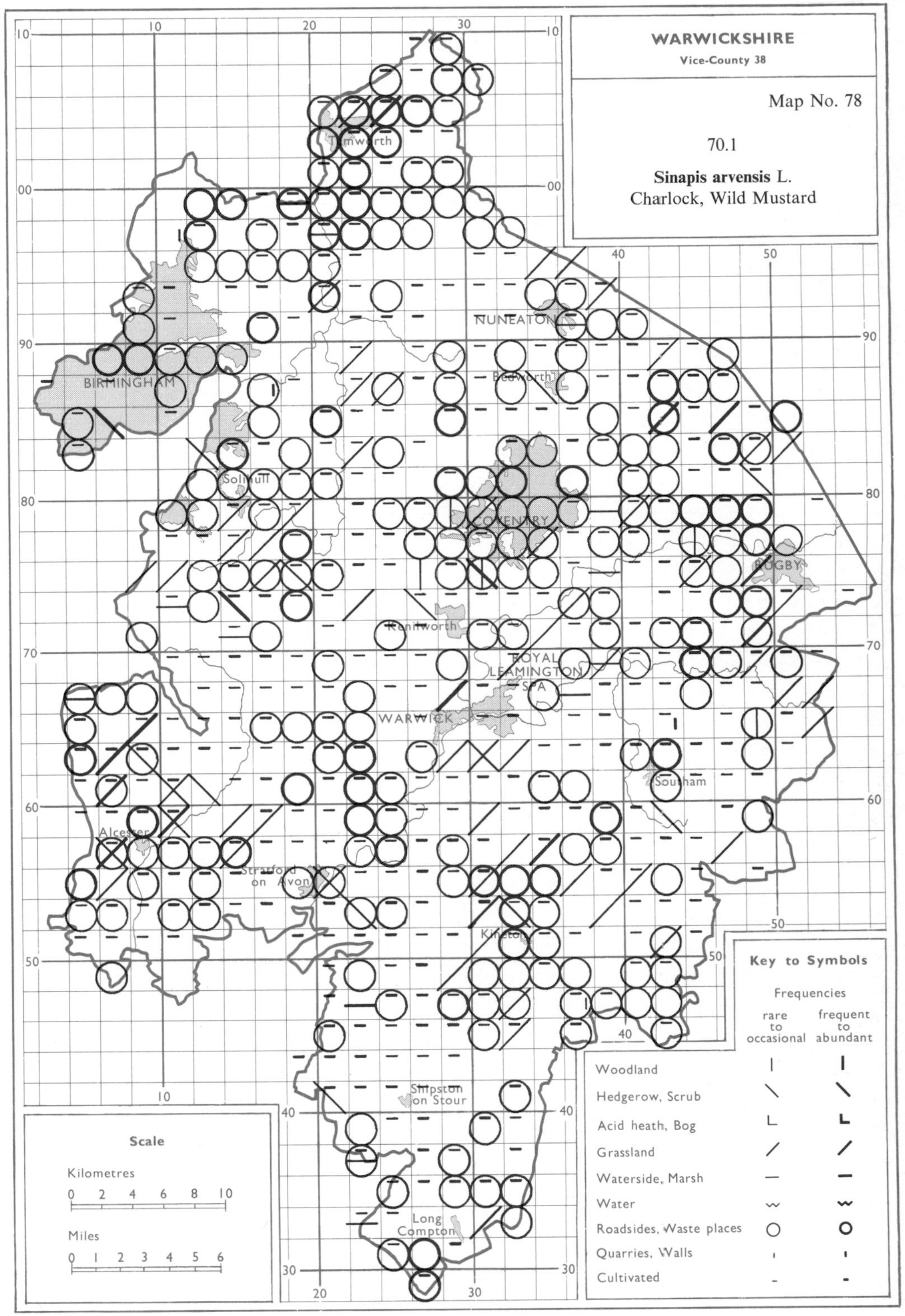
WARWICKSHIRE
Vice-County 38
Map No. 78
70.1
Sinapis arvensis L.
Charlock, Wild Mustard
Tamworth
BIRMINGHAM
Solihull
NUNEATON
Bedworth
COVENTRY
RUGBY
Kenilworth
ROYAL LEAMINGTON SPA
WARWICK
Southam
Alcester
Stratford on Avon
Kineton
Shipston on Stour
Long Compton
Key to Symbols
Frequencies
rare to occasional
frequent to abundant
Woodland
Hedgerow, Scrub
Acid heath, Bog
Grassland
Waterside, Marsh
Water
Roadsides, Waste places
Quarries, Walls
Cultivated
Scale
Kilometres
0 2 4 6 8 10
Miles
0 1 2 3 4 5 6

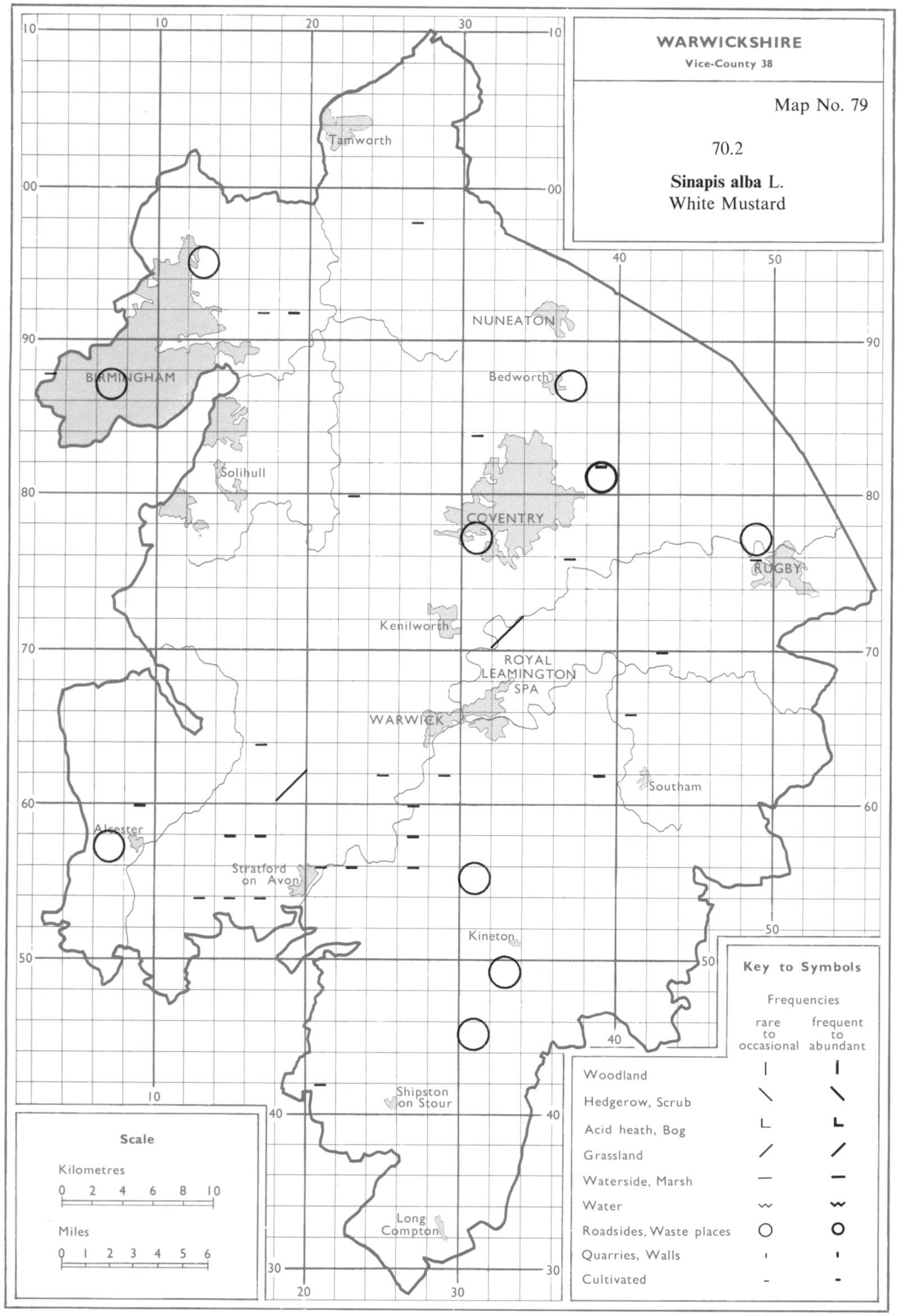

WARWICKSHIRE
Vice-County 38

Map No. 79

70.2

Sinapis alba L.
White Mustard

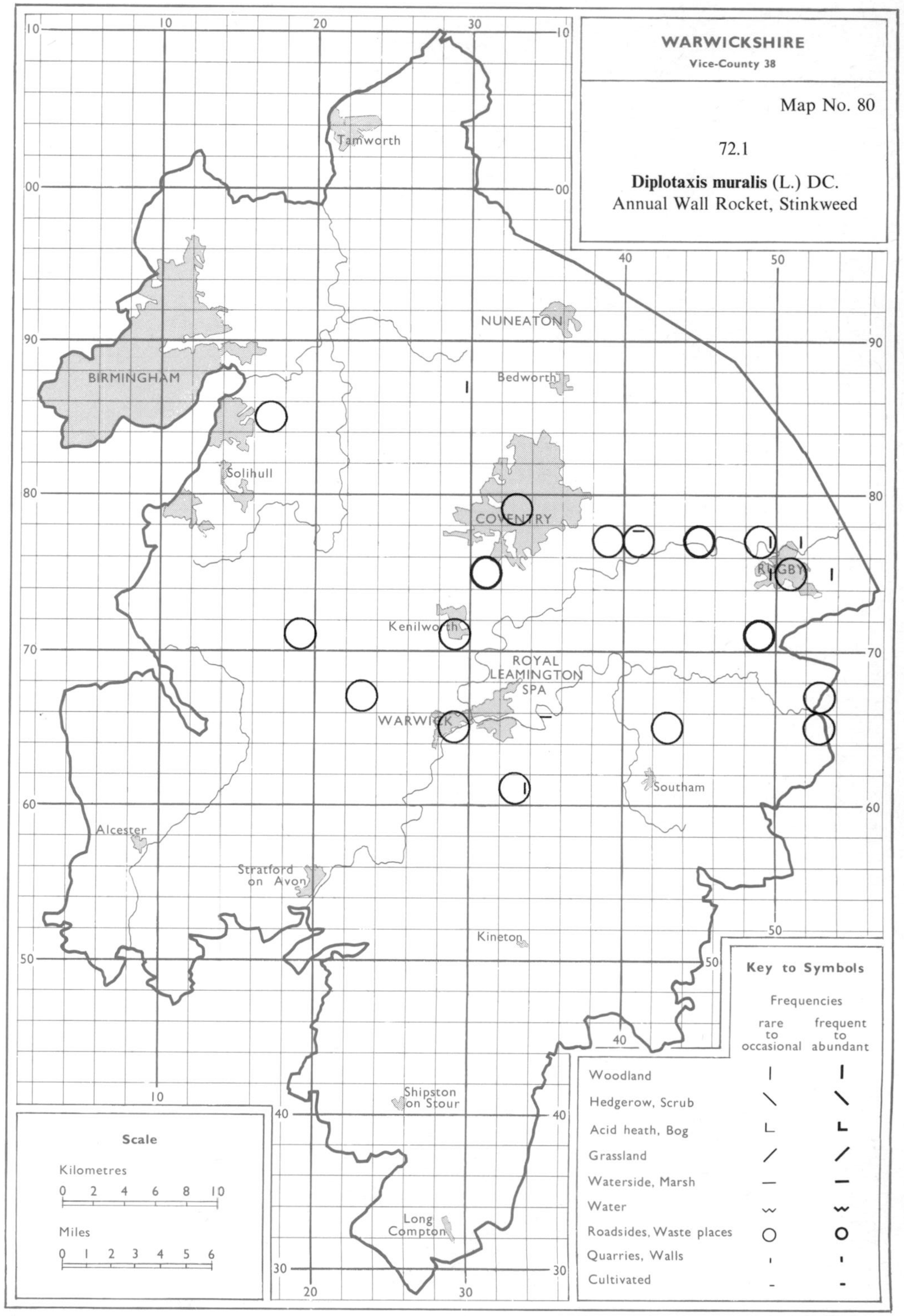

WARWICKSHIRE
Vice-County 38

Map No. 80

72.1

Diplotaxis muralis (L.) DC.
Annual Wall Rocket, Stinkweed

WARWICKSHIRE
Vice-County 38
Map No. 82
79.2
Lepidium campestre (L.) R. Br.
Field Pepperwort
Key to Symbols
Frequencies
rare to occasional
frequent to abundant
Woodland
Hedgerow, Scrub
Acid heath, Bog
Grassland
Waterside, Marsh
Water
Roadsides, Waste places
Quarries, Walls
Cultivated
Scale
Kilometres
0 2 4 6 8 10
Miles
0 1 2 3 4 5 6
NUNEATON
Bedworth
COVENTRY
RUGBY
Tamworth
Solihull
BIRMINGHAM
Kenilworth
ROYAL LEAMINGTON SPA
WARWICK
Southam
Kineton
Shipston on Stour
Long Compton
Stratford on Avon
Alcester

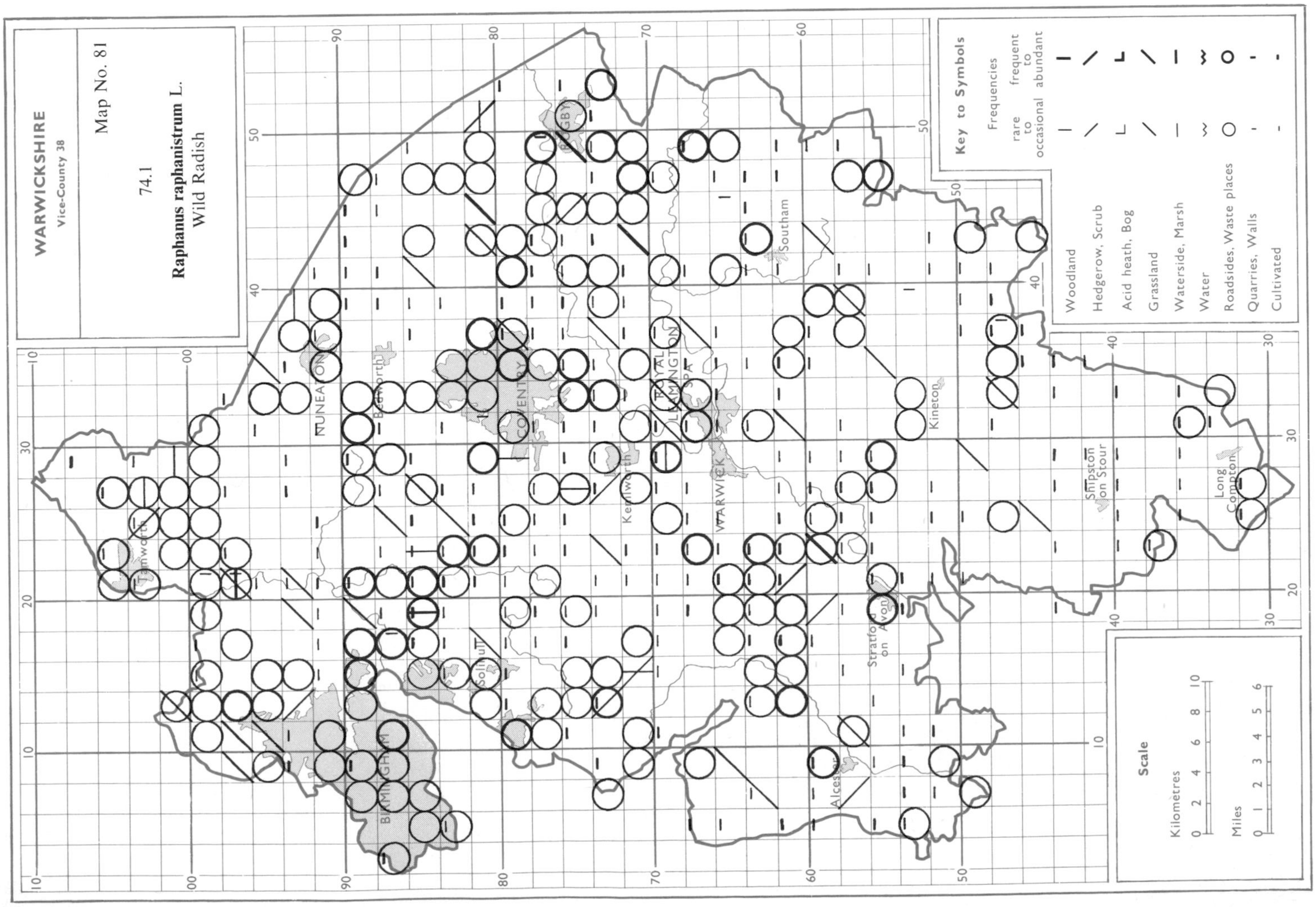

WARWICKSHIRE
Vice-County 38
Map No. 81
74.1
Raphanus raphanistrum L.
Wild Radish
Key to Symbols
Frequencies
rare to occasional
frequent to abundant
Woodland
Hedgerow, Scrub
Acid heath, Bog
Grassland
Waterside, Marsh
Water
Roadsides, Waste places
Quarries, Walls
Cultivated
Scale
Kilometres
0 2 4 6 8 10
Miles
0 1 2 3 4 5 6
NUNEATON
Bedworth
COVENTRY
RUGBY
Tamworth
Solihull
BIRMINGHAM
Kenilworth
ROYAL LEAMINGTON SPA
WARWICK
Southam
Kineton
Shipston on Stour
Long Compton
Stratford on Avon
Alcester

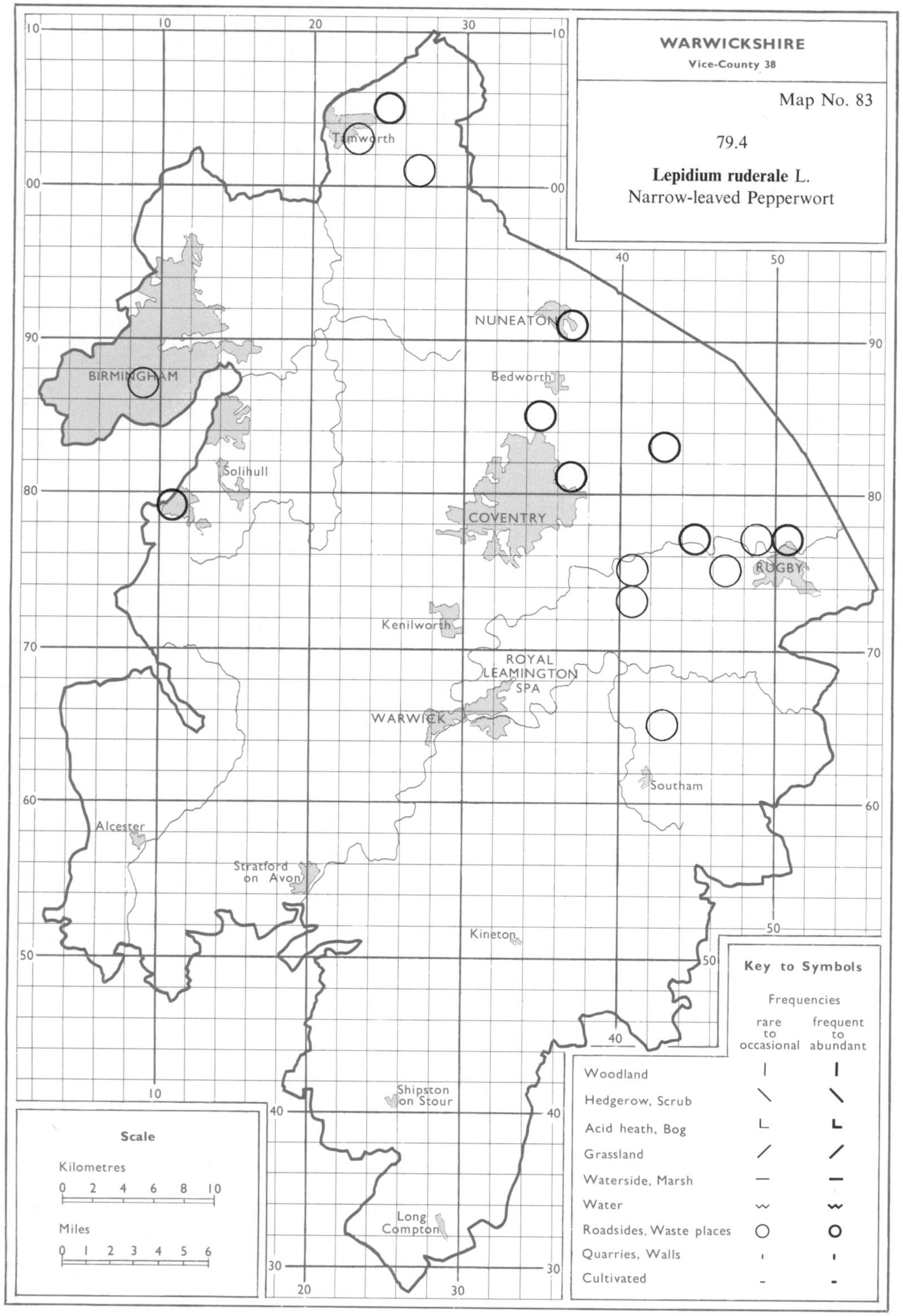
WARWICKSHIRE
Vice-County 38
Map No. 83
79.4
Lepidium ruderale L.
Narrow-leaved Pepperwort
Key to Symbols
Frequencies
rare to occasional
frequent to abundant
Woodland
Hedgerow, Scrub
Acid heath, Bog
Grassland
Waterside, Marsh
Water
Roadsides, Waste places
Quarries, Walls
Cultivated
Scale
Kilometres
Miles
Tamworth
BIRMINGHAM
Solihull
NUNEATON
Bedworth
COVENTRY
RUGBY
Kenilworth
ROYAL LEAMINGTON SPA
WARWICK
Southam
Alcester
Stratford on Avon
Kineton
Shipston on Stour
Long Compton

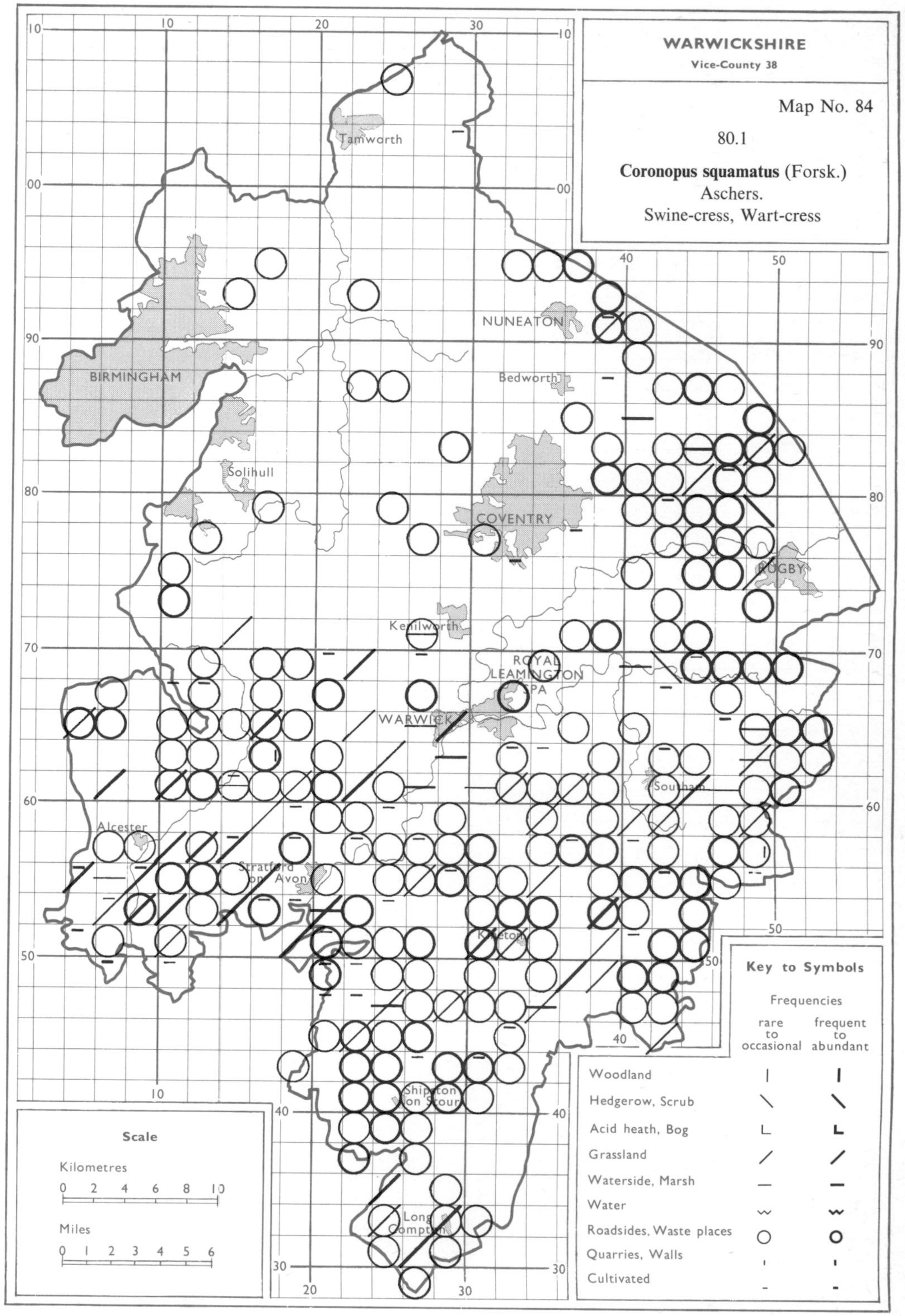
WARWICKSHIRE
Vice-County 38
Map No. 84
80.1
Coronopus squamatus (Forsk.) Aschers.
Swine-cress, Wart-cress
Key to Symbols
Frequencies
rare to occasional
frequent to abundant
Woodland
Hedgerow, Scrub
Acid heath, Bog
Grassland
Waterside, Marsh
Water
Roadsides, Waste places
Quarries, Walls
Cultivated
Scale
Kilometres
Miles
Tamworth
BIRMINGHAM
Solihull
NUNEATON
Bedworth
COVENTRY
RUGBY
Kenilworth
ROYAL LEAMINGTON SPA
WARWICK
Southam
Alcester
Stratford on Avon
Kineton
Shipston on Stour
Long Compton

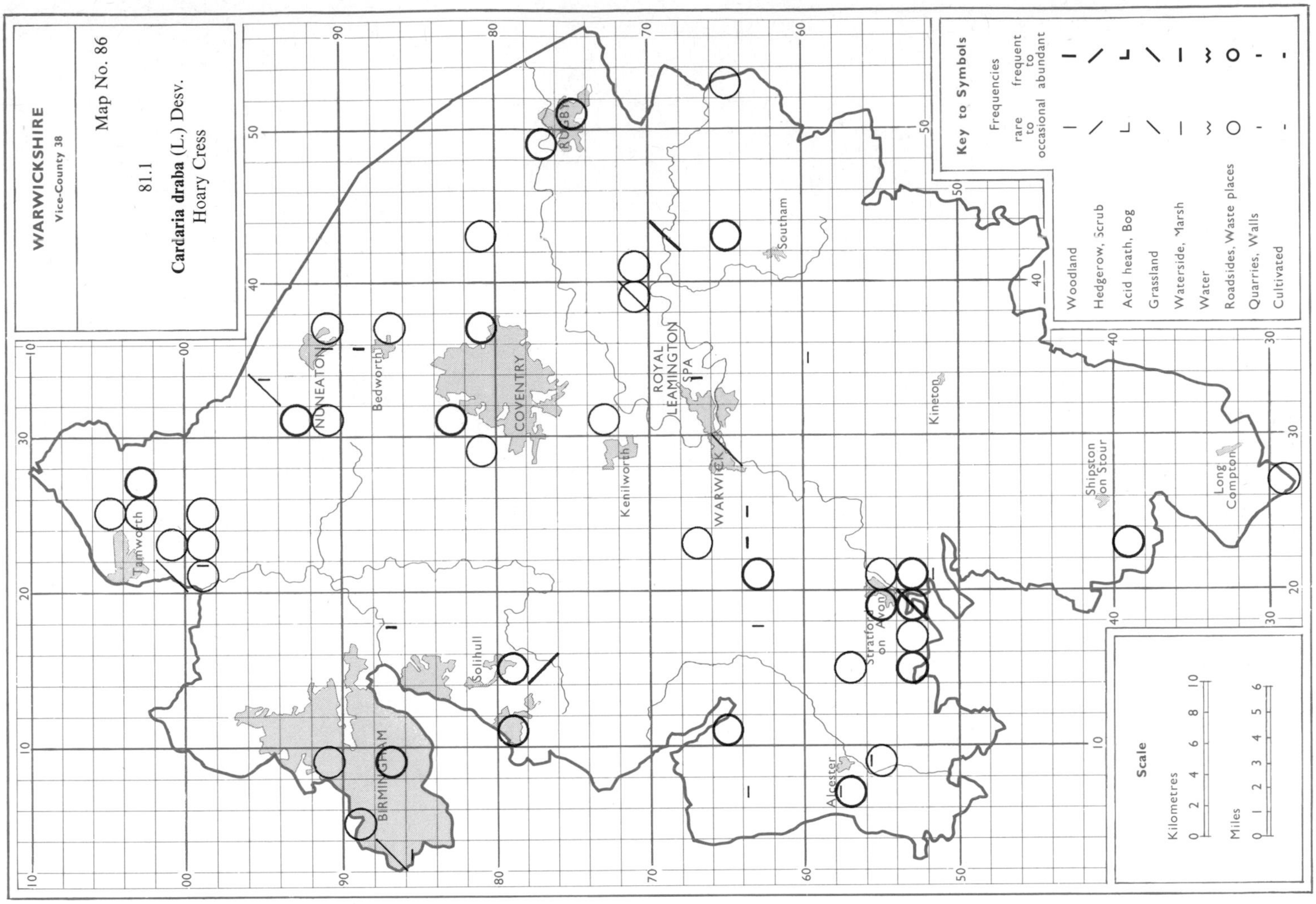
WARWICKSHIRE
Vice-County 38
Map No. 86
81.1
Cardaria draba (L.) Desv.
Hoary Cress
Key to Symbols
Frequencies
rare to occasional
frequent to abundant
Woodland
Hedgerow, Scrub
Acid heath, Bog
Grassland
Waterside, Marsh
Water
Roadsides, Waste places
Quarries, Walls
Cultivated
Scale
Kilometres
0 2 4 6 8 10
Miles
0 1 2 3 4 5 6
Tamworth
NUNEATON
Bedworth
COVENTRY
RUGBY
BIRMINGHAM
Solihull
Kenilworth
ROYAL LEAMINGTON SPA
WARWICK
Southam
Kineton
Stratford on Avon
Alcester
Shipston on Stour
Long Compton

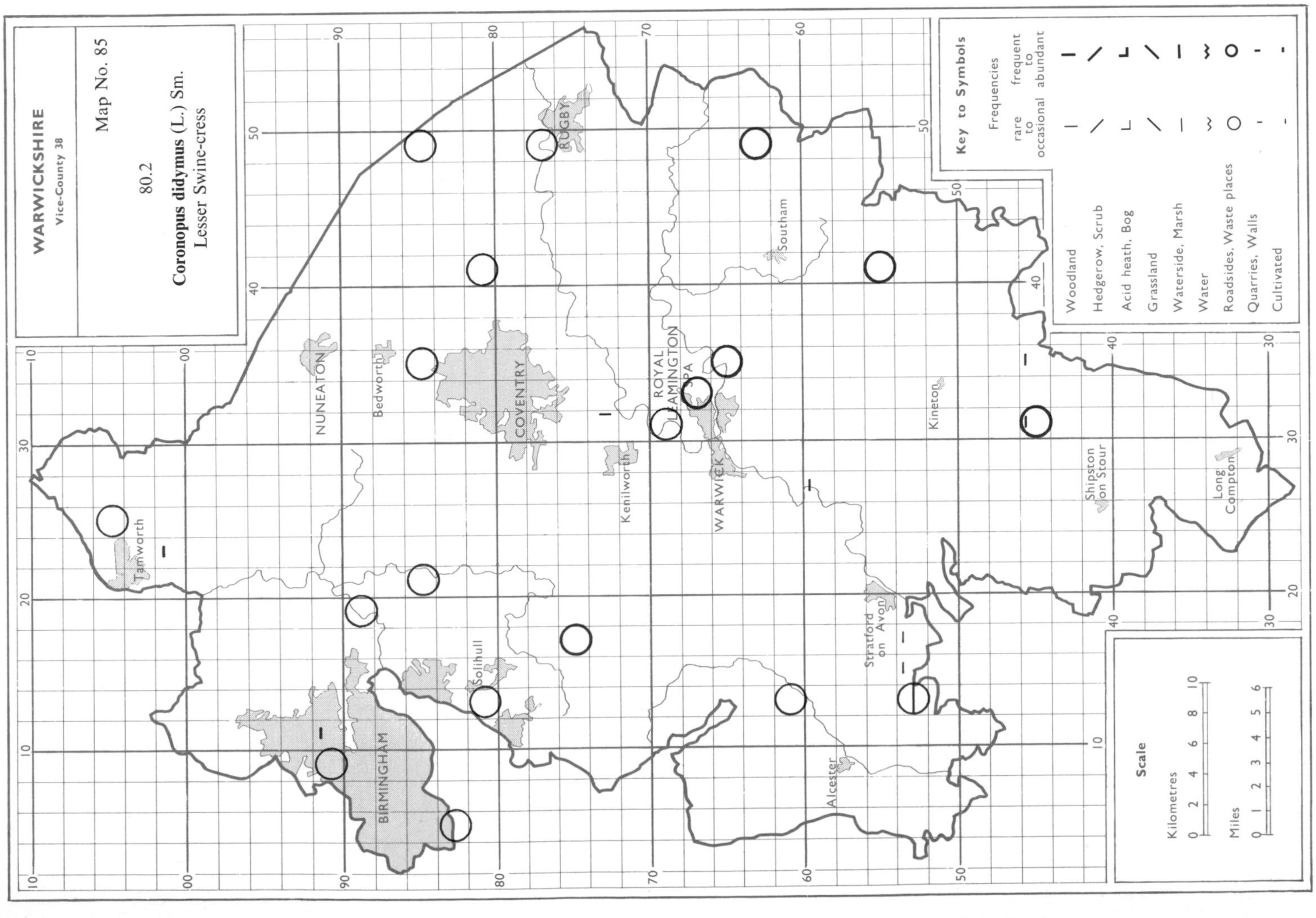
WARWICKSHIRE
Vice-County 38
Map No. 85
80.2
Coronopus didymus (L.) Sm.
Lesser Swine-cress
Key to Symbols
Frequencies
rare to occasional
frequent to abundant
Woodland
Hedgerow, Scrub
Acid heath, Bog
Grassland
Waterside, Marsh
Water
Roadsides, Waste places
Quarries, Walls
Cultivated
Scale
Kilometres
0 2 4 6 8 10
Miles
0 1 2 3 4 5 6
Tamworth
NUNEATON
Bedworth
COVENTRY
RUGBY
BIRMINGHAM
Solihull
Kenilworth
ROYAL LEAMINGTON SPA
WARWICK
Southam
Kineton
Stratford on Avon
Alcester
Shipston on Stour
Long Compton

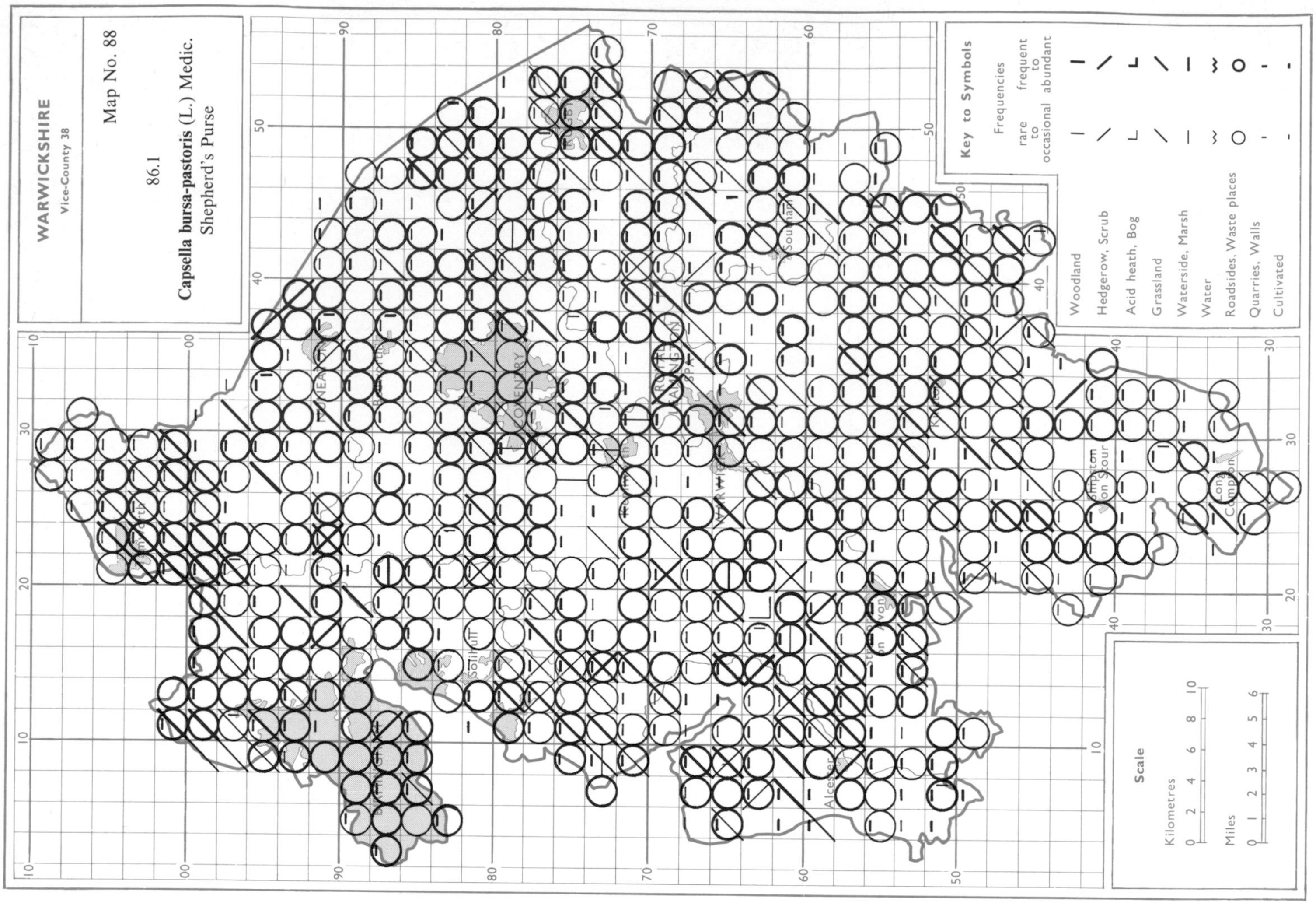
WARWICKSHIRE
Vice-County 38
Map No. 88
86.1
Capsella bursa-pastoris (L.) Medic.
Shepherd's Purse
Key to Symbols
Frequencies
rare to occasional
frequent to abundant
Woodland
Hedgerow, Scrub
Acid heath, Bog
Grassland
Waterside, Marsh
Water
Roadsides, Waste places
Quarries, Walls
Cultivated
Scale
Kilometres
Miles
COVENTRY
Solihull
Southam
Alcester
Long Compton

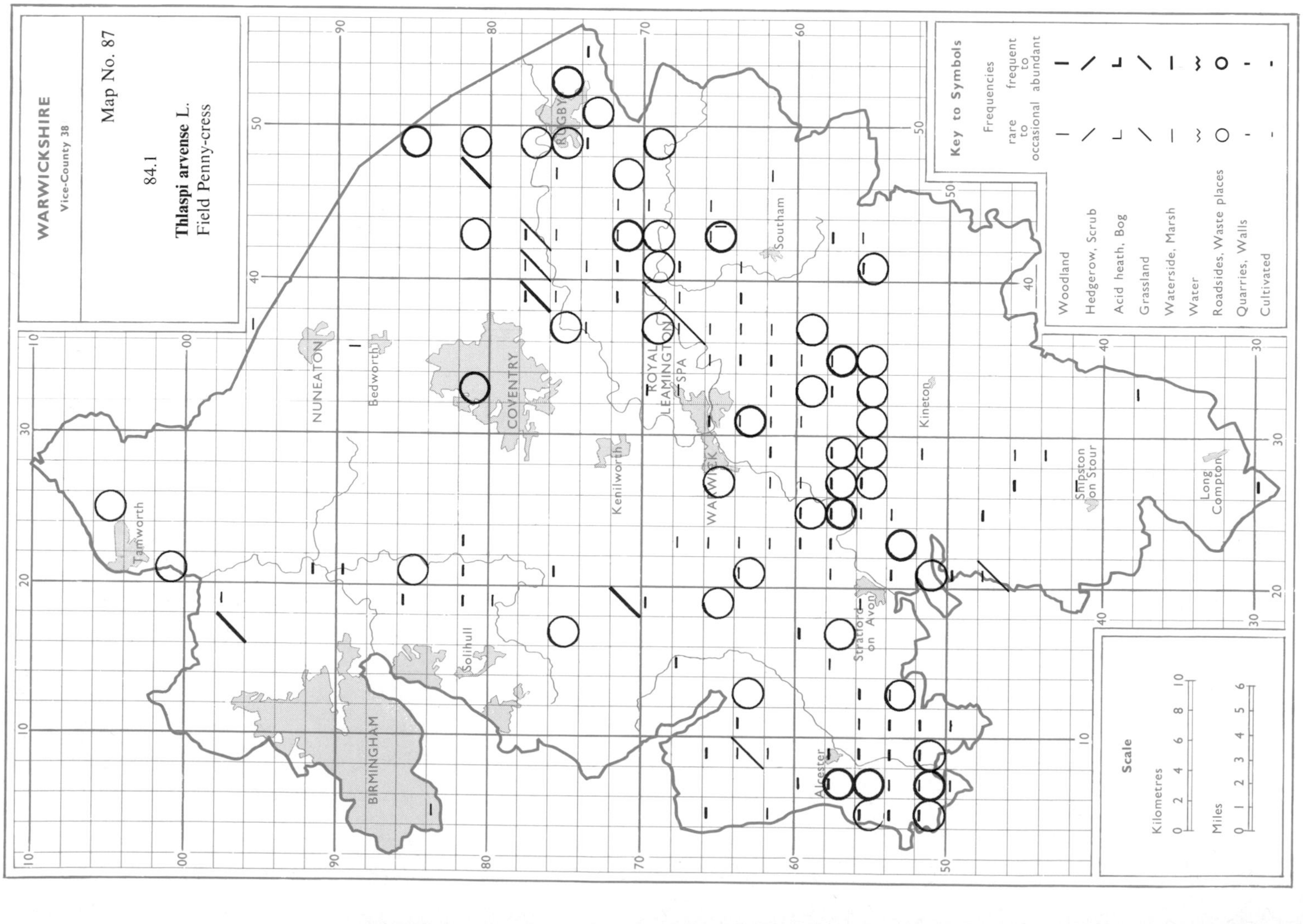
WARWICKSHIRE
Vice-County 38
Map No. 87
84.1
Thlaspi arvense L.
Field Penny-cress
Key to Symbols
Frequencies
rare to occasional
frequent to abundant
Woodland
Hedgerow, Scrub
Acid heath, Bog
Grassland
Waterside, Marsh
Water
Roadsides, Waste places
Quarries, Walls
Cultivated
Scale
Kilometres
Miles
NUNEATON
Bedworth
COVENTRY
RUGBY
ROYAL LEAMINGTON SPA
Southam
Kenilworth
WARWICK
Kineton
Shipston on Stour
Long Compton
Tamworth
Solihull
BIRMINGHAM
Stratford on Avon
Alcester

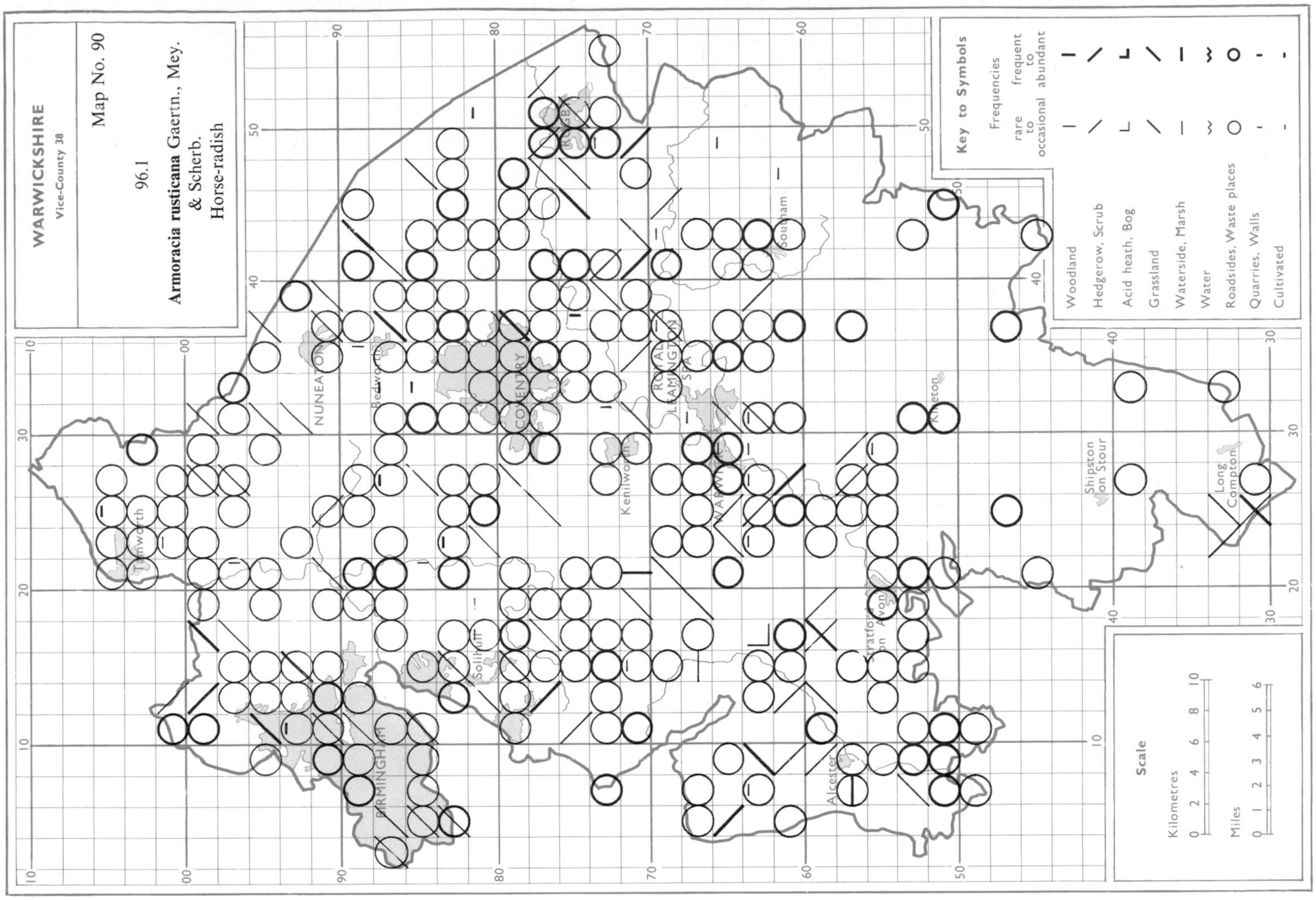
WARWICKSHIRE
Vice-County 38
Map No. 90
96.1
Armoracia rusticana Gaertn., Mey. & Scherb.
Horse-radish
Key to Symbols
Frequencies
rare to occasional
frequent to abundant
Woodland
Hedgerow, Scrub
Acid heath, Bog
Grassland
Waterside, Marsh
Water
Roadsides, Waste places
Quarries, Walls
Cultivated
Scale
Kilometres
Miles
NUNEATON
Bedworth
COVENTRY
RUGBY
Southam
ROYAL LEAMINGTON SPA
Kineton
Kenilworth
WARWICK
Tamworth
Solihull
BIRMINGHAM
Stratford on Avon
Alcester
Shipston on Stour
Long Compton

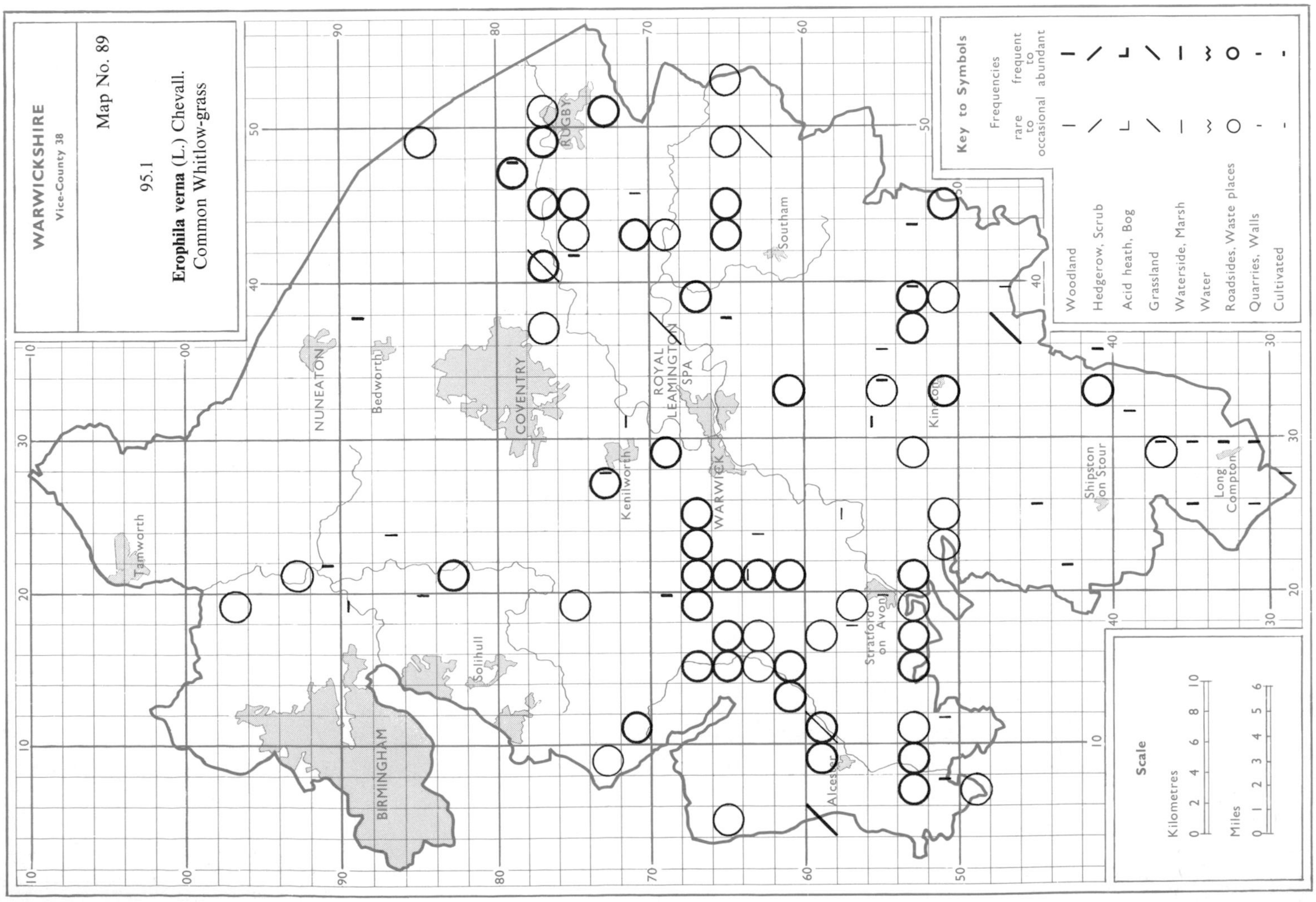
WARWICKSHIRE
Vice-County 38
Map No. 89
95.1
Erophila verna (L.) Chevall.
Common Whitlow-grass
Key to Symbols
Frequencies
rare to occasional
frequent to abundant
Woodland
Hedgerow, Scrub
Acid heath, Bog
Grassland
Waterside, Marsh
Water
Roadsides, Waste places
Quarries, Walls
Cultivated
Scale
Kilometres
Miles
NUNEATON
Bedworth
COVENTRY
RUGBY
Southam
ROYAL LEAMINGTON SPA
Kineton
Kenilworth
WARWICK
Tamworth
Solihull
BIRMINGHAM
Stratford on Avon
Alcester
Shipston on Stour
Long Compton

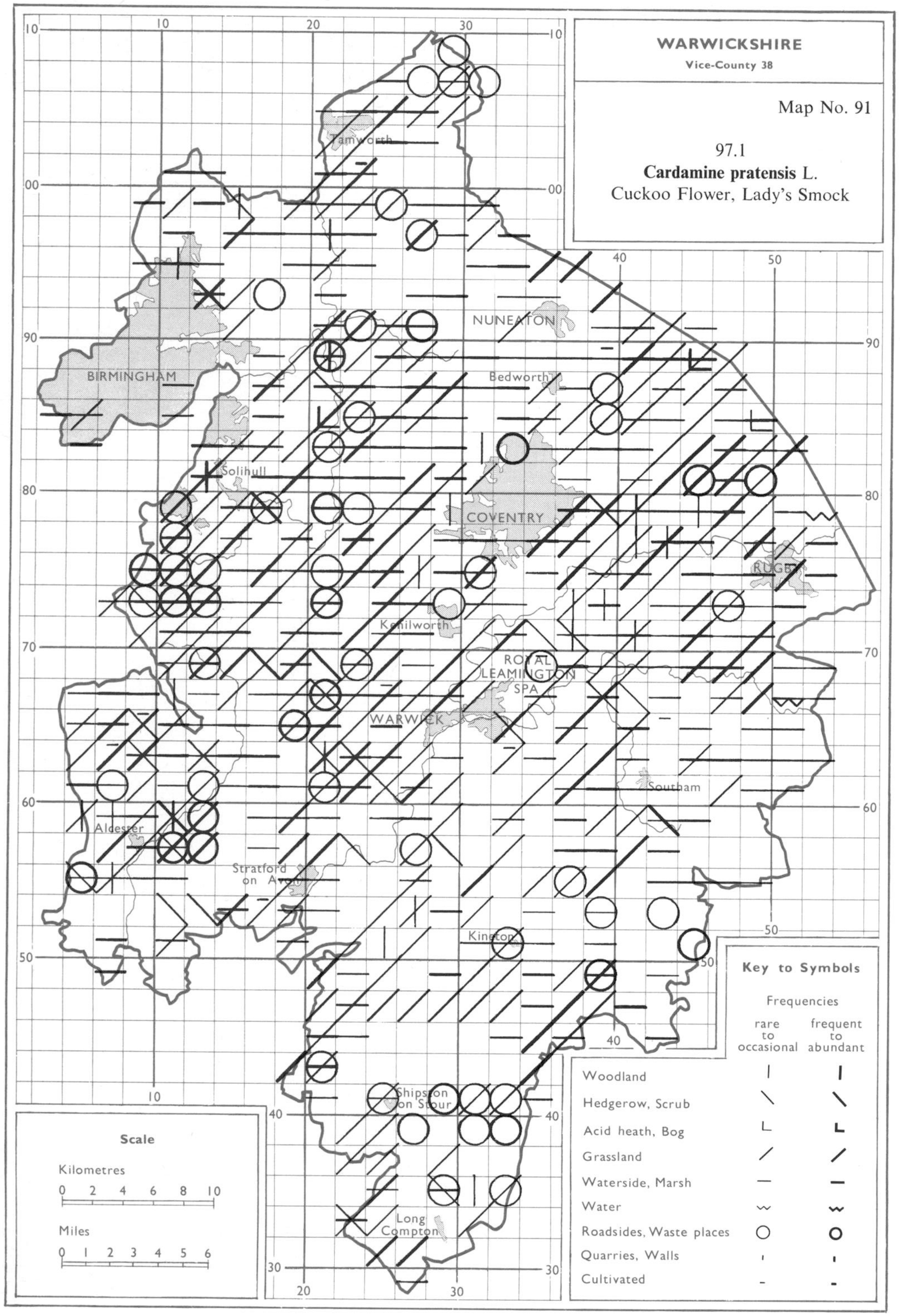
WARWICKSHIRE
Vice-County 38
Map No. 91
97.1
Cardamine pratensis L.
Cuckoo Flower, Lady's Smock
Tamworth
BIRMINGHAM
NUNEATON
Bedworth
Solihull
COVENTRY
RUGBY
Kenilworth
ROYAL LEAMINGTON SPA
WARWICK
Southam
Alcester
Stratford on Avon
Kineton
Shipston on Stour
Long Compton
Key to Symbols
Frequencies
rare to occasional
frequent to abundant
Woodland
Hedgerow, Scrub
Acid heath, Bog
Grassland
Waterside, Marsh
Water
Roadsides, Waste places
Quarries, Walls
Cultivated
Scale
Kilometres
0 2 4 6 8 10
Miles
0 1 2 3 4 5 6

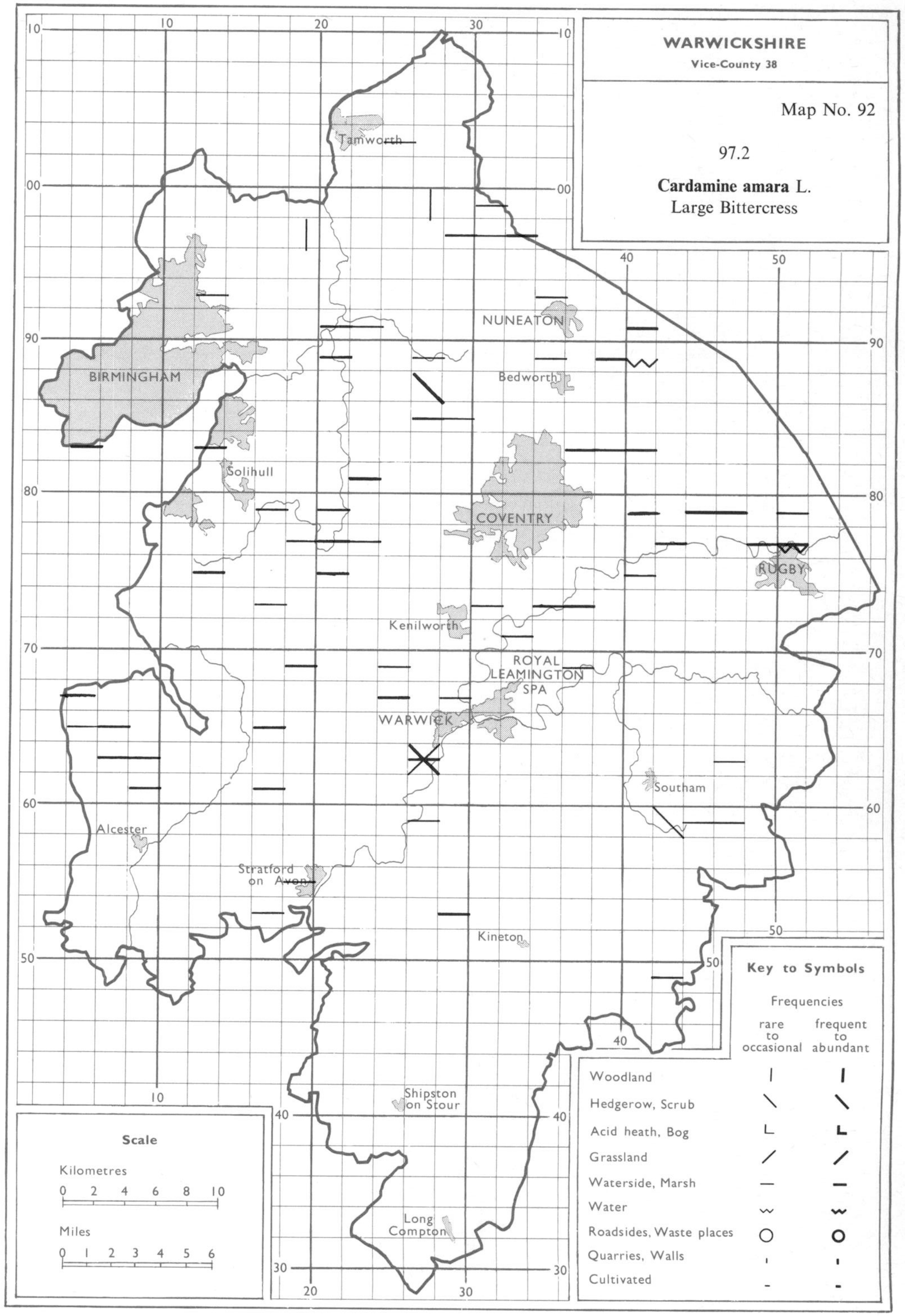
WARWICKSHIRE
Vice-County 38
Map No. 92
97.2
Cardamine amara L.
Large Bittercress
Tamworth
BIRMINGHAM
NUNEATON
Bedworth
Solihull
COVENTRY
RUGBY
Kenilworth
ROYAL LEAMINGTON SPA
WARWICK
Southam
Alcester
Stratford on Avon
Kineton
Shipston on Stour
Long Compton
Key to Symbols
Frequencies
rare to occasional
frequent to abundant
Woodland
Hedgerow, Scrub
Acid heath, Bog
Grassland
Waterside, Marsh
Water
Roadsides, Waste places
Quarries, Walls
Cultivated
Scale
Kilometres
0 2 4 6 8 10
Miles
0 1 2 3 4 5 6

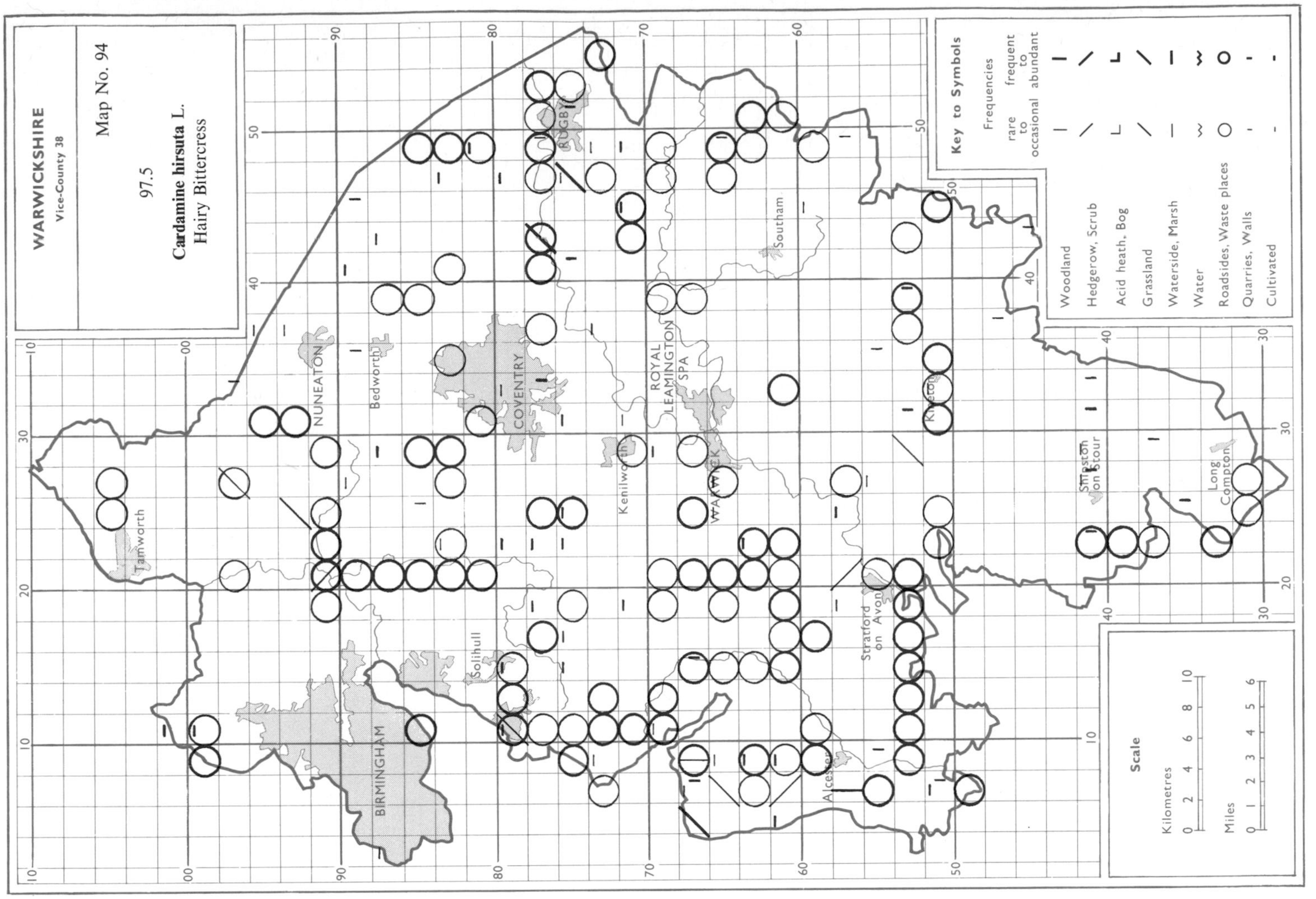
WARWICKSHIRE
Vice-County 38
Map No. 94
97.5
Cardamine hirsuta L.
Hairy Bittercress
Key to Symbols
Frequencies
rare to occasional
frequent to abundant
Woodland
Hedgerow, Scrub
Acid heath, Bog
Grassland
Waterside, Marsh
Water
Roadsides, Waste places
Quarries, Walls
Cultivated
Scale
Kilometres
Miles
Tamworth
NUNEATON
Bedworth
COVENTRY
RUGBY
Southam
ROYAL LEAMINGTON SPA
WARWICK
Kenilworth
Kineton
Shipston on Stour
Long Compton
Solihull
BIRMINGHAM
Stratford on Avon
Alcester

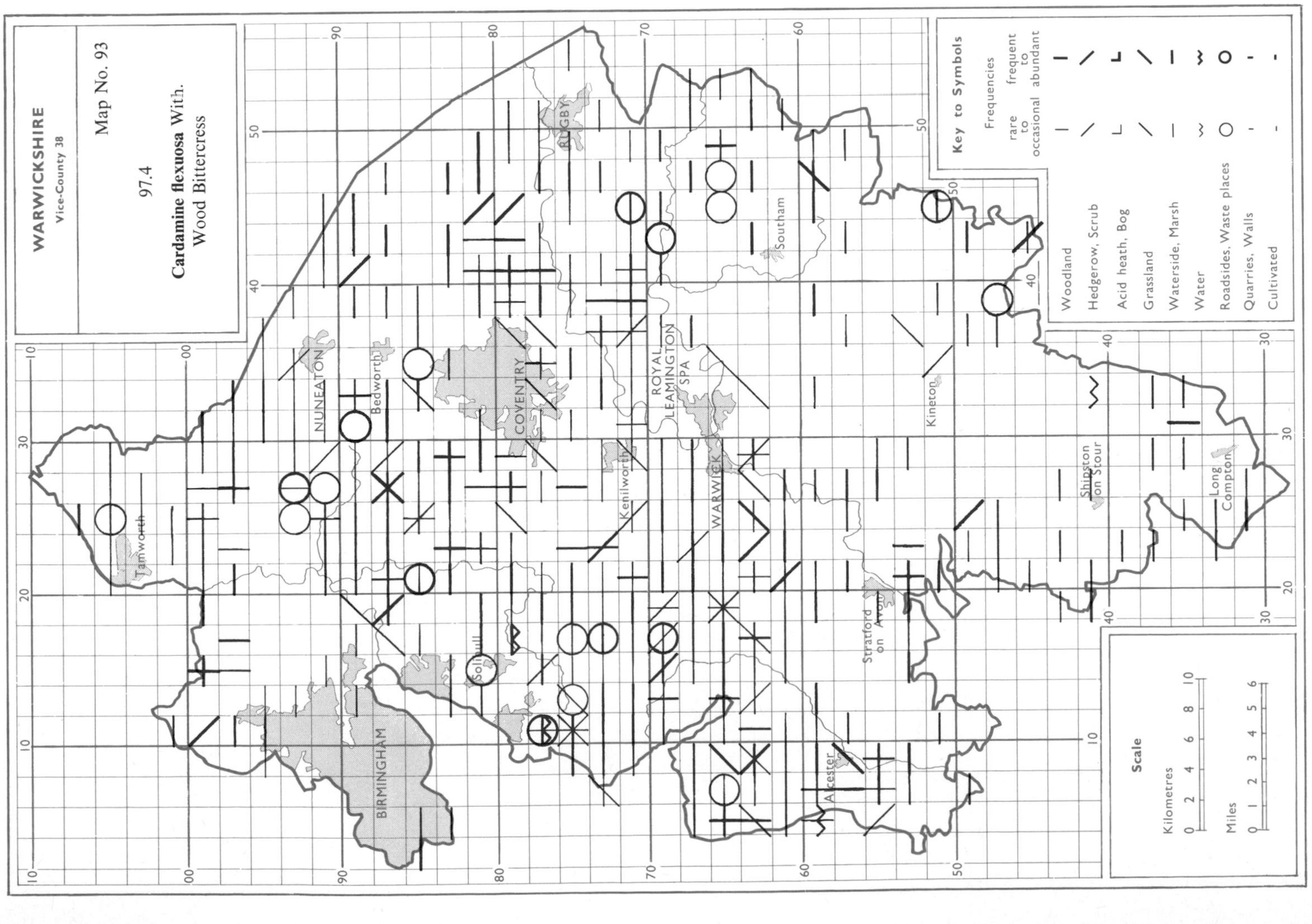
WARWICKSHIRE
Vice-County 38
Map No. 93
97.4
Cardamine flexuosa With.
Wood Bittercress
Key to Symbols
Frequencies
rare to occasional
frequent to abundant
Woodland
Hedgerow, Scrub
Acid heath, Bog
Grassland
Waterside, Marsh
Water
Roadsides, Waste places
Quarries, Walls
Cultivated
Scale
Kilometres
Miles
Tamworth
NUNEATON
Bedworth
COVENTRY
RUGBY
Southam
ROYAL LEAMINGTON SPA
WARWICK
Kenilworth
Kineton
Shipston on Stour
Long Compton
Solihull
BIRMINGHAM
Stratford on Avon
Alcester

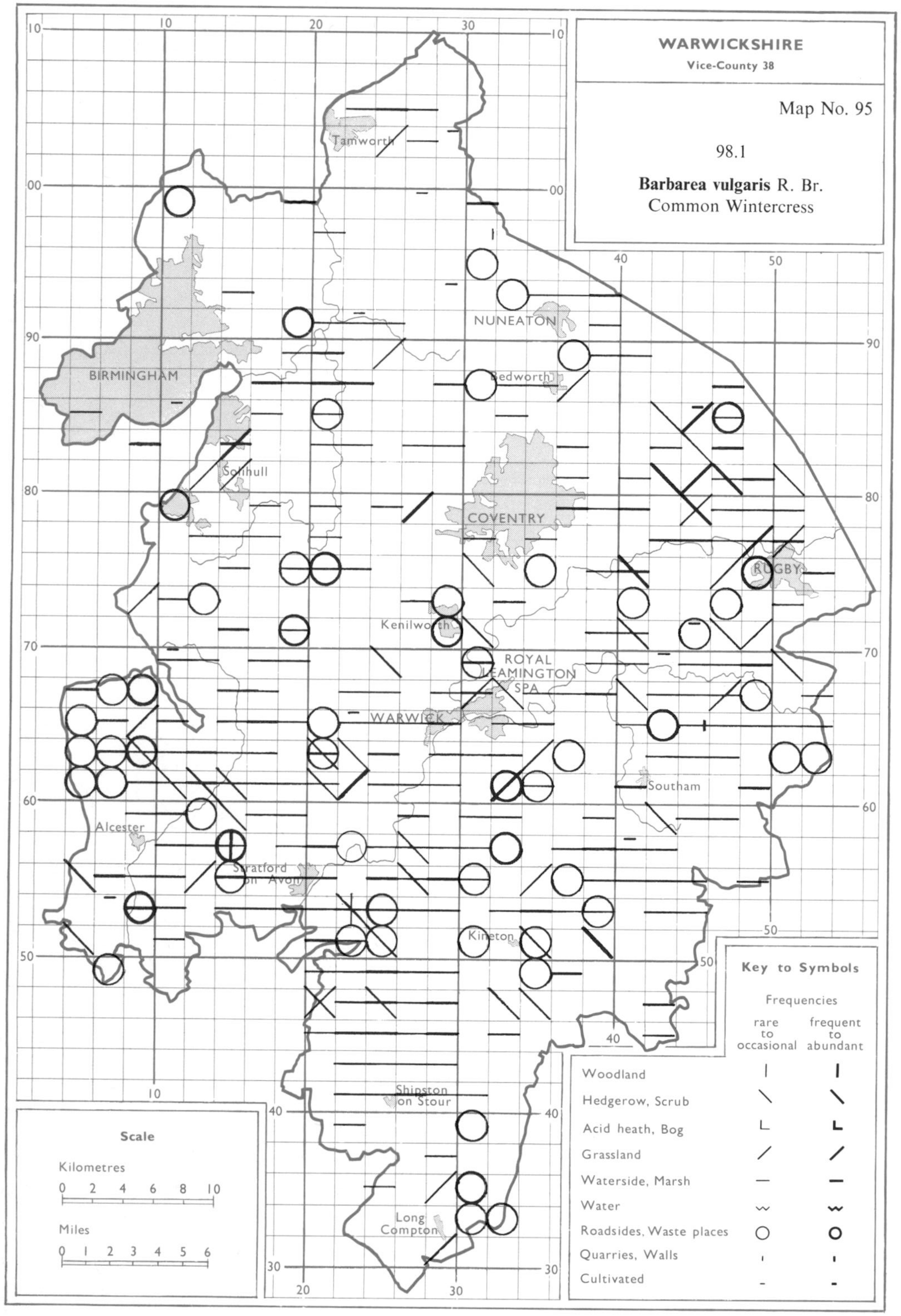
WARWICKSHIRE
Vice-County 38
Map No. 95
98.1
Barbarea vulgaris R. Br.
Common Wintercress
Tamworth
NUNEATON
Bedworth
BIRMINGHAM
Solihull
COVENTRY
RUGBY
Kenilworth
ROYAL LEAMINGTON SPA
WARWICK
Southam
Alcester
Stratford on Avon
Kineton
Shipston on Stour
Long Compton
Key to Symbols
Frequencies
rare to occasional
frequent to abundant
Woodland
Hedgerow, Scrub
Acid heath, Bog
Grassland
Waterside, Marsh
Water
Roadsides, Waste places
Quarries, Walls
Cultivated
Scale
Kilometres
0 2 4 6 8 10
Miles
0 1 2 3 4 5 6

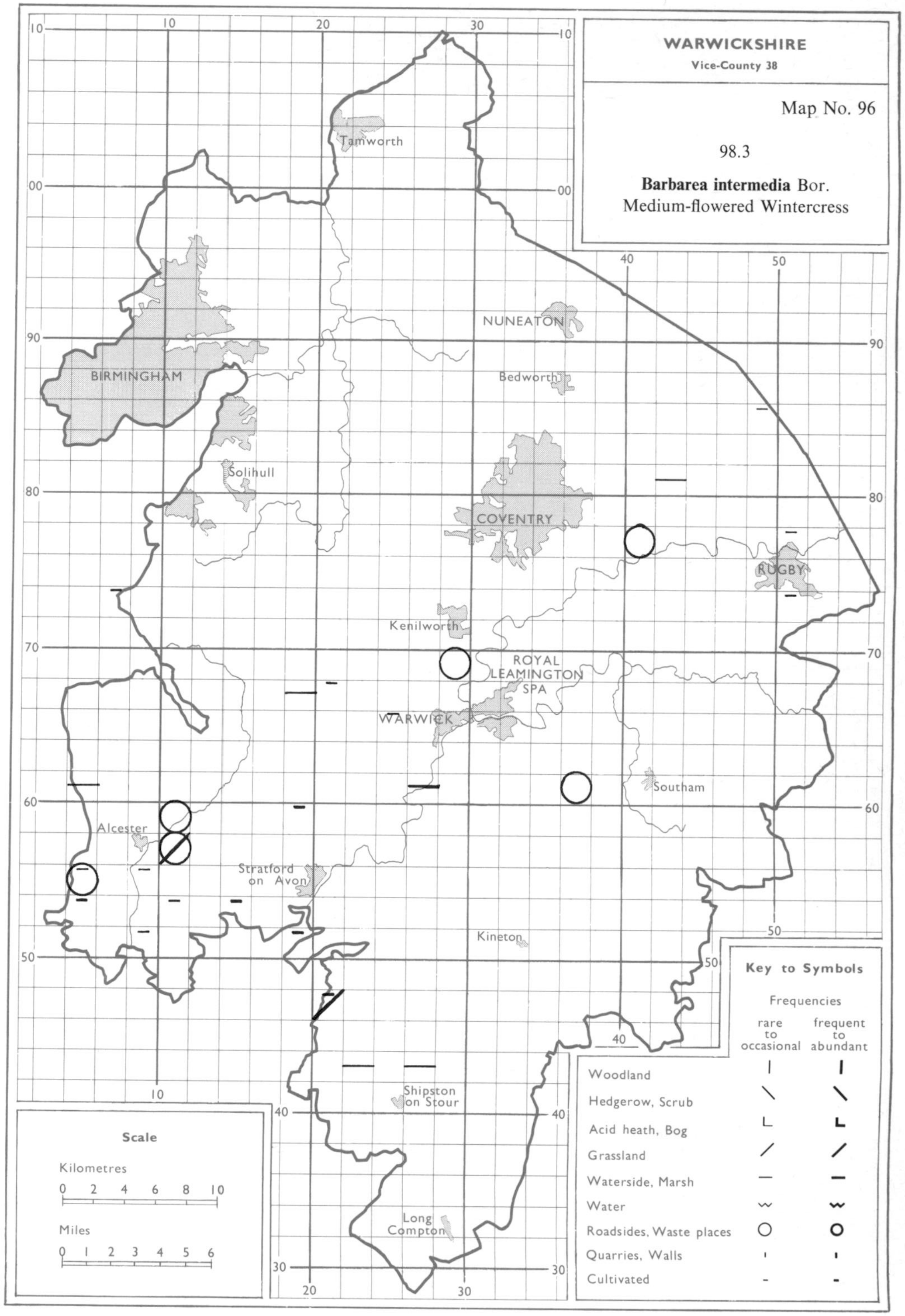
WARWICKSHIRE
Vice-County 38
Map No. 96
98.3
Barbarea intermedia Bor.
Medium-flowered Wintercress
Tamworth
NUNEATON
Bedworth
BIRMINGHAM
Solihull
COVENTRY
RUGBY
Kenilworth
ROYAL LEAMINGTON SPA
WARWICK
Southam
Alcester
Stratford on Avon
Kineton
Shipston on Stour
Long Compton
Key to Symbols
Frequencies
rare to occasional
frequent to abundant
Woodland
Hedgerow, Scrub
Acid heath, Bog
Grassland
Waterside, Marsh
Water
Roadsides, Waste places
Quarries, Walls
Cultivated
Scale
Kilometres
0 2 4 6 8 10
Miles
0 1 2 3 4 5 6

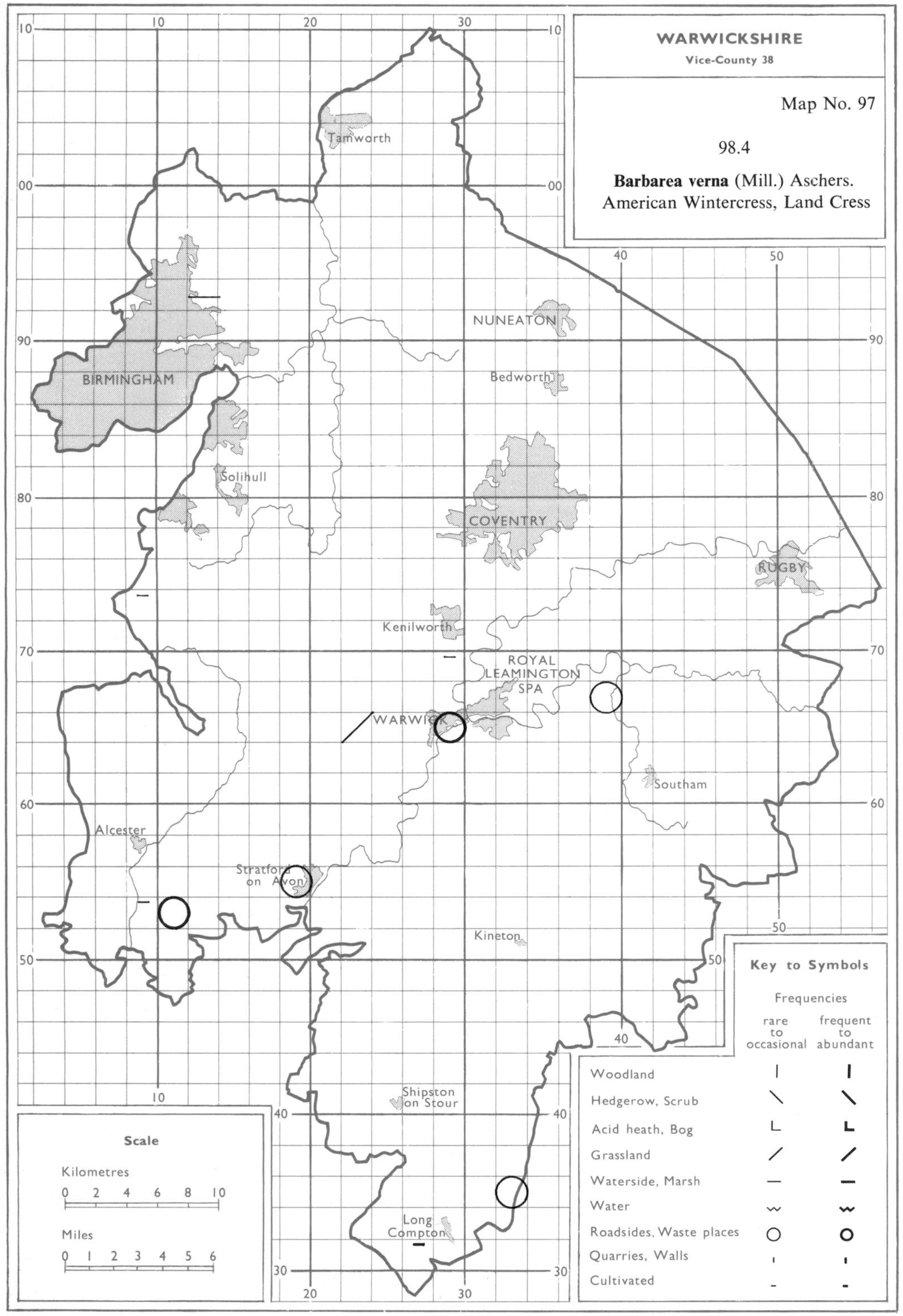
WARWICKSHIRE
Vice-County 38
Map No. 97
98.4
Barbarea verna (Mill.) Aschers.
American Wintercress, Land Cress
Tamworth
NUNEATON
Bedworth
BIRMINGHAM
Solihull
COVENTRY
RUGBY
Kenilworth
ROYAL LEAMINGTON SPA
WARWICK
Southam
Alcester
Stratford on Avon
Kineton
Shipston on Stour
Long Compton
Key to Symbols
Frequencies
rare to occasional
frequent to abundant
Woodland
Hedgerow, Scrub
Acid heath, Bog
Grassland
Waterside, Marsh
Water
Roadsides, Waste places
Quarries, Walls
Cultivated
Scale
Kilometres
0 2 4 6 8 10
Miles
0 1 2 3 4 5 6

WARWICKSHIRE
Vice-County 38
Map No. 98
102.1 & 2
Rorippa nasturtium-aquaticum s.l.
Watercress
Tamworth
NUNEATON
Bedworth
BIRMINGHAM
Solihull
COVENTRY
RUGBY
Kenilworth
ROYAL LEAMINGTON SPA
WARWICK
Southam
Alcester
Stratford on Avon
Kineton
Shipston on Stour
Long Compton
Key to Symbols
Frequencies
rare to occasional
frequent to abundant
Woodland
Hedgerow, Scrub
Acid heath, Bog
Grassland
Waterside, Marsh
Water
Roadsides, Waste places
Quarries, Walls
Cultivated
Scale
Kilometres
0 2 4 6 8 10
Miles
0 1 2 3 4 5 6

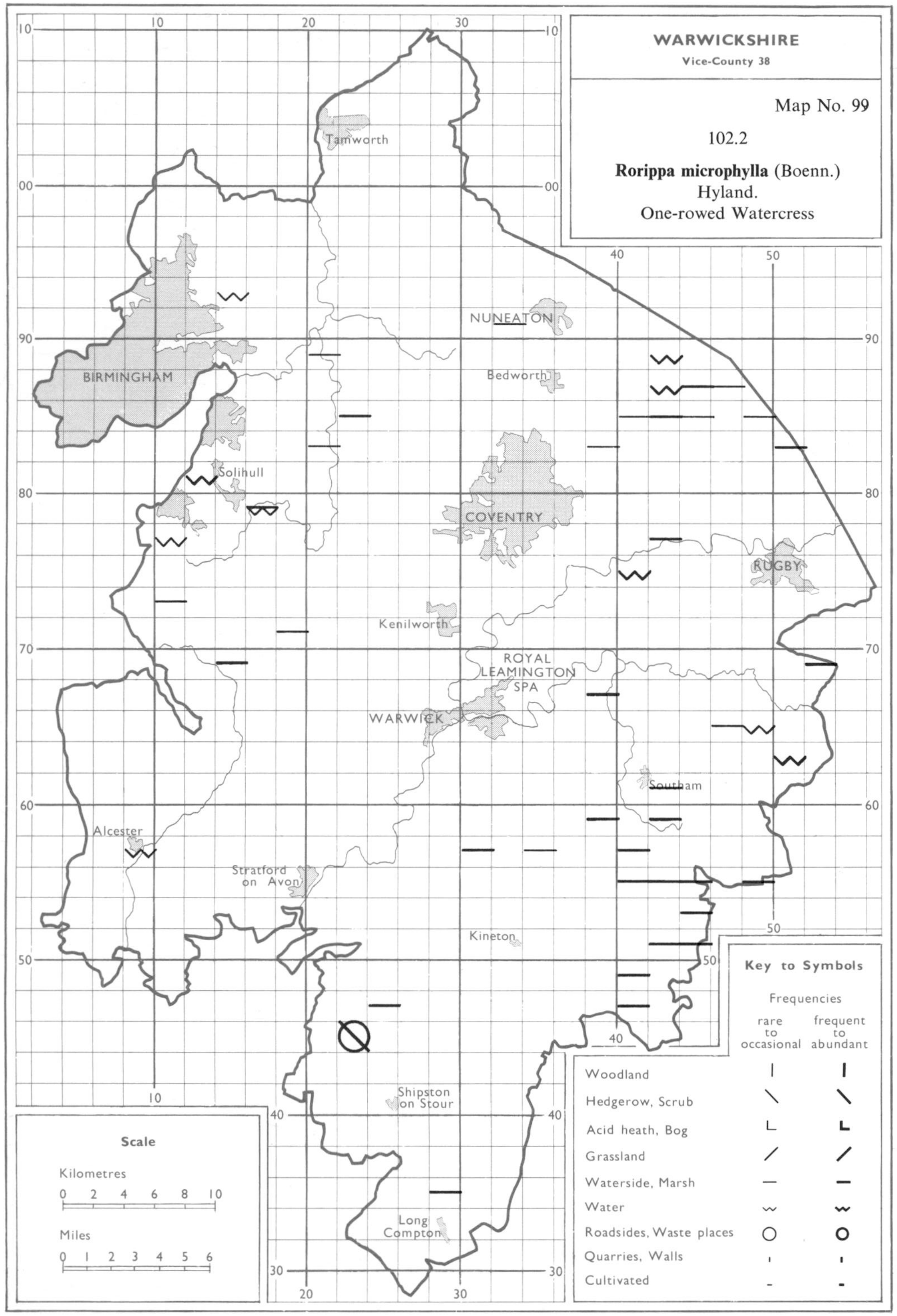
WARWICKSHIRE
Vice-County 38
Map No. 99
102.2
Rorippa microphylla (Boenn.) Hyland.
One-rowed Watercress
Tamworth
NUNEATON
Bedworth
BIRMINGHAM
Solihull
COVENTRY
RUGBY
Kenilworth
ROYAL LEAMINGTON SPA
WARWICK
Southam
Alcester
Stratford on Avon
Kineton
Shipston on Stour
Long Compton
Scale
Kilometres
0 2 4 6 8 10
Miles
0 1 2 3 4 5 6
Key to Symbols
Frequencies
rare to occasional
frequent to abundant
Woodland
Hedgerow, Scrub
Acid heath, Bog
Grassland
Waterside, Marsh
Water
Roadsides, Waste places
Quarries, Walls
Cultivated

WARWICKSHIRE
Vice-County 38
Map No. 100
102.3
Rorippa sylvestris (L.) Bess.
Creeping Yellow-cress
Tamworth
NUNEATON
Bedworth
BIRMINGHAM
Solihull
COVENTRY
RUGBY
Kenilworth
ROYAL LEAMINGTON SPA
WARWICK
Southam
Alcester
Stratford on Avon
Kineton
Shipston on Stour
Long Compton
Scale
Kilometres
0 2 4 6 8 10
Miles
0 1 2 3 4 5 6
Key to Symbols
Frequencies
rare to occasional
frequent to abundant
Woodland
Hedgerow, Scrub
Acid heath, Bog
Grassland
Waterside, Marsh
Water
Roadsides, Waste places
Quarries, Walls
Cultivated

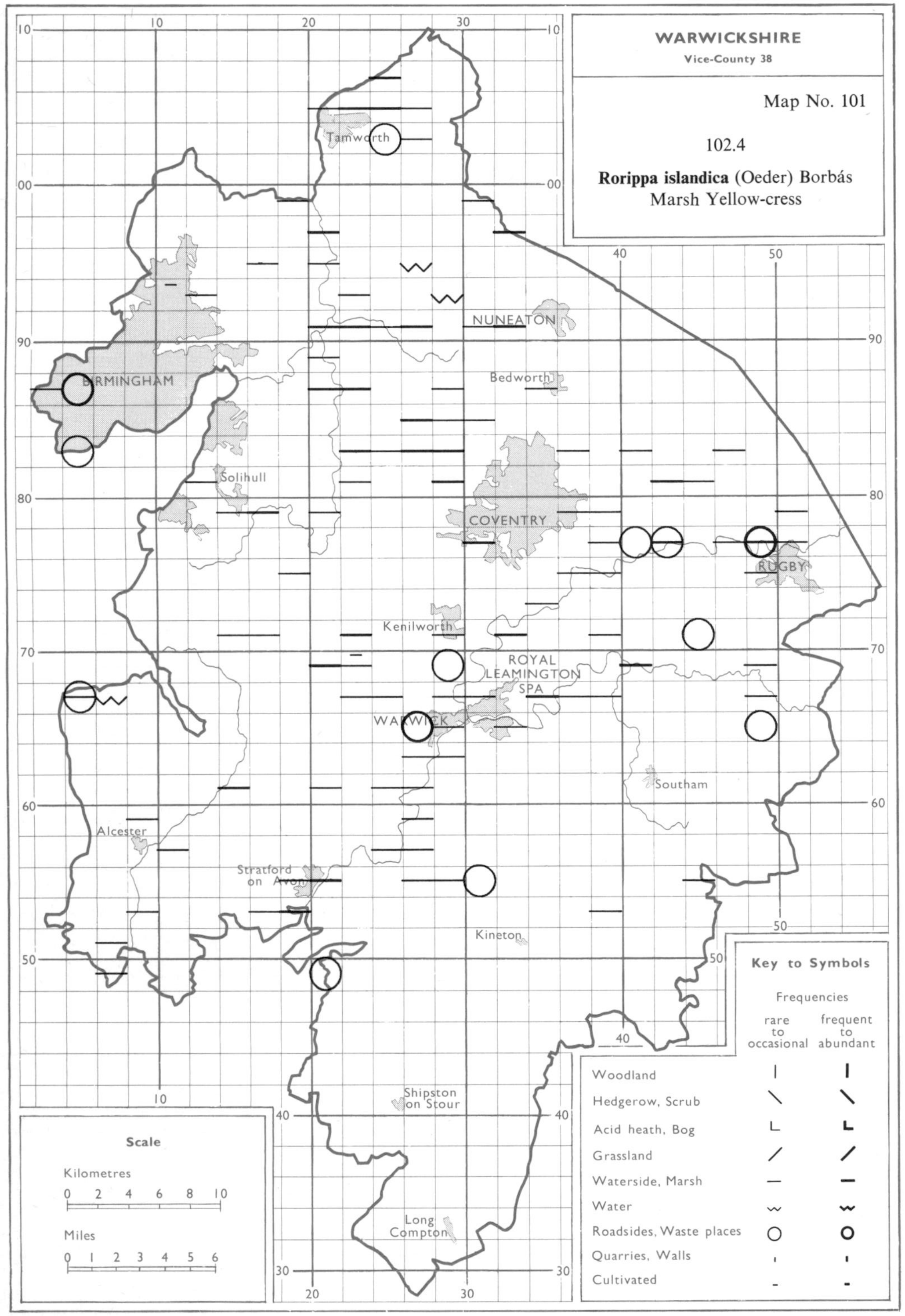
WARWICKSHIRE
Vice-County 38
Map No. 101
102.4
Rorippa islandica (Oeder) Borbás
Marsh Yellow-cress
Tamworth
NUNEATON
Bedworth
BIRMINGHAM
Solihull
COVENTRY
RUGBY
Kenilworth
ROYAL LEAMINGTON SPA
WARWICK
Southam
Alcester
Stratford on Avon
Kineton
Shipston on Stour
Long Compton
10
20
30
40
50
00
90
80
70
60
Key to Symbols
Frequencies
rare to occasional
frequent to abundant
Woodland
Hedgerow, Scrub
Acid heath, Bog
Grassland
Waterside, Marsh
Water
Roadsides, Waste places
Quarries, Walls
Cultivated
Scale
Kilometres
0 2 4 6 8 10
Miles
0 1 2 3 4 5 6

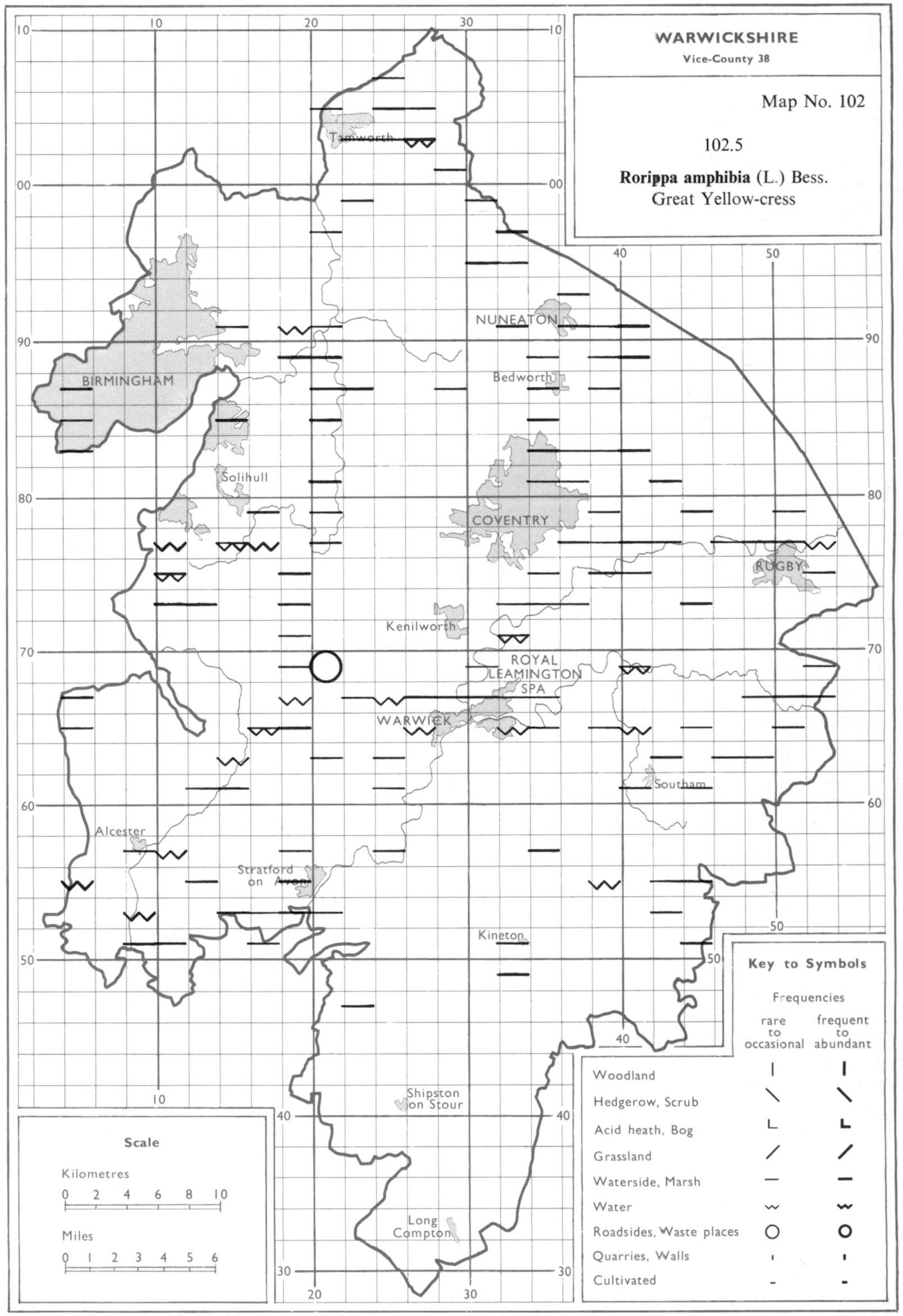
WARWICKSHIRE
Vice-County 38
Map No. 102
102.5
Rorippa amphibia (L.) Bess.
Great Yellow-cress
Tamworth
NUNEATON
Bedworth
BIRMINGHAM
Solihull
COVENTRY
RUGBY
Kenilworth
ROYAL LEAMINGTON SPA
WARWICK
Southam
Alcester
Stratford on Avon
Kineton
Shipston on Stour
Long Compton
10
20
30
40
50
00
90
80
70
60
Key to Symbols
Frequencies
rare to occasional
frequent to abundant
Woodland
Hedgerow, Scrub
Acid heath, Bog
Grassland
Waterside, Marsh
Water
Roadsides, Waste places
Quarries, Walls
Cultivated
Scale
Kilometres
0 2 4 6 8 10
Miles
0 1 2 3 4 5 6

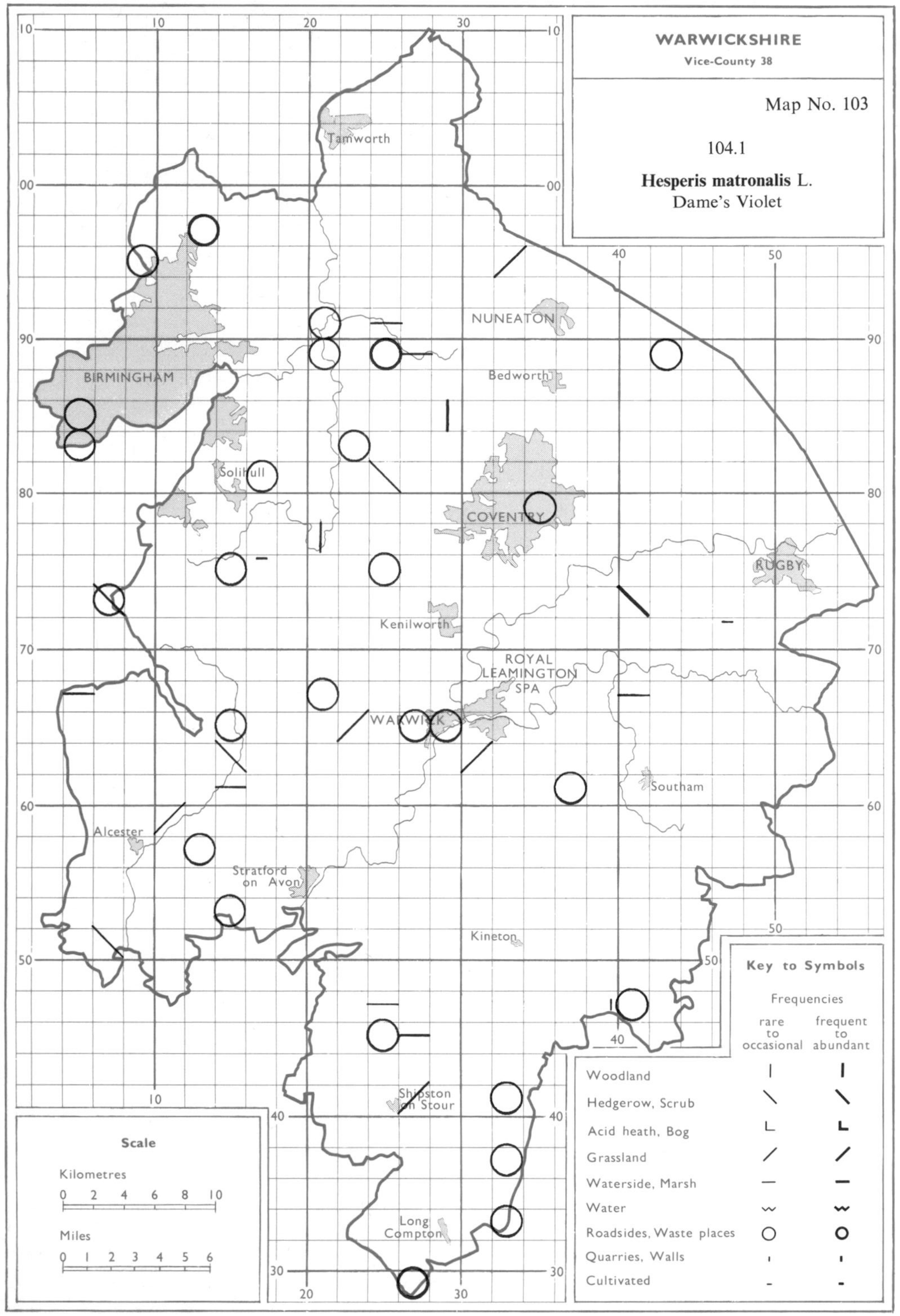
WARWICKSHIRE
Vice-County 38
Map No. 103
104.1
Hesperis matronalis L.
Dame's Violet
Tamworth
NUNEATON
Bedworth
BIRMINGHAM
Solihull
COVENTRY
RUGBY
Kenilworth
ROYAL LEAMINGTON SPA
WARWICK
Southam
Alcester
Stratford on Avon
Kineton
Shipston on Stour
Long Compton
Key to Symbols
Frequencies
rare to occasional
frequent to abundant
Woodland
Hedgerow, Scrub
Acid heath, Bog
Grassland
Waterside, Marsh
Water
Roadsides, Waste places
Quarries, Walls
Cultivated
Scale
Kilometres
0 2 4 6 8 10
Miles
0 1 2 3 4 5 6

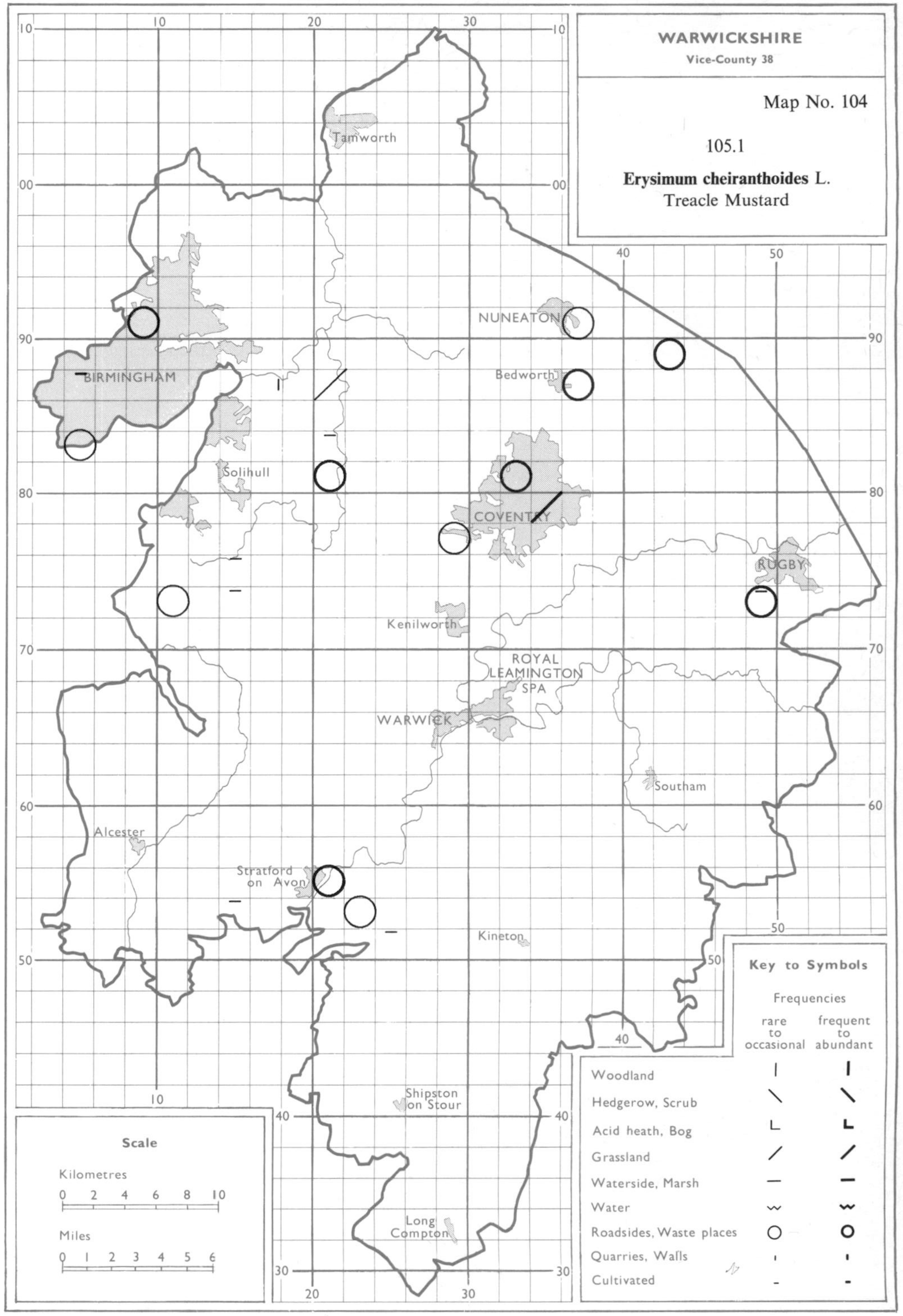
WARWICKSHIRE
Vice-County 38
Map No. 104
105.1
Erysimum cheiranthoides L.
Treacle Mustard
Tamworth
NUNEATON
Bedworth
BIRMINGHAM
Solihull
COVENTRY
RUGBY
Kenilworth
ROYAL LEAMINGTON SPA
WARWICK
Southam
Alcester
Stratford on Avon
Kineton
Shipston on Stour
Long Compton
Key to Symbols
Frequencies
rare to occasional
frequent to abundant
Woodland
Hedgerow, Scrub
Acid heath, Bog
Grassland
Waterside, Marsh
Water
Roadsides, Waste places
Quarries, Walls
Cultivated
Scale
Kilometres
0 2 4 6 8 10
Miles
0 1 2 3 4 5 6

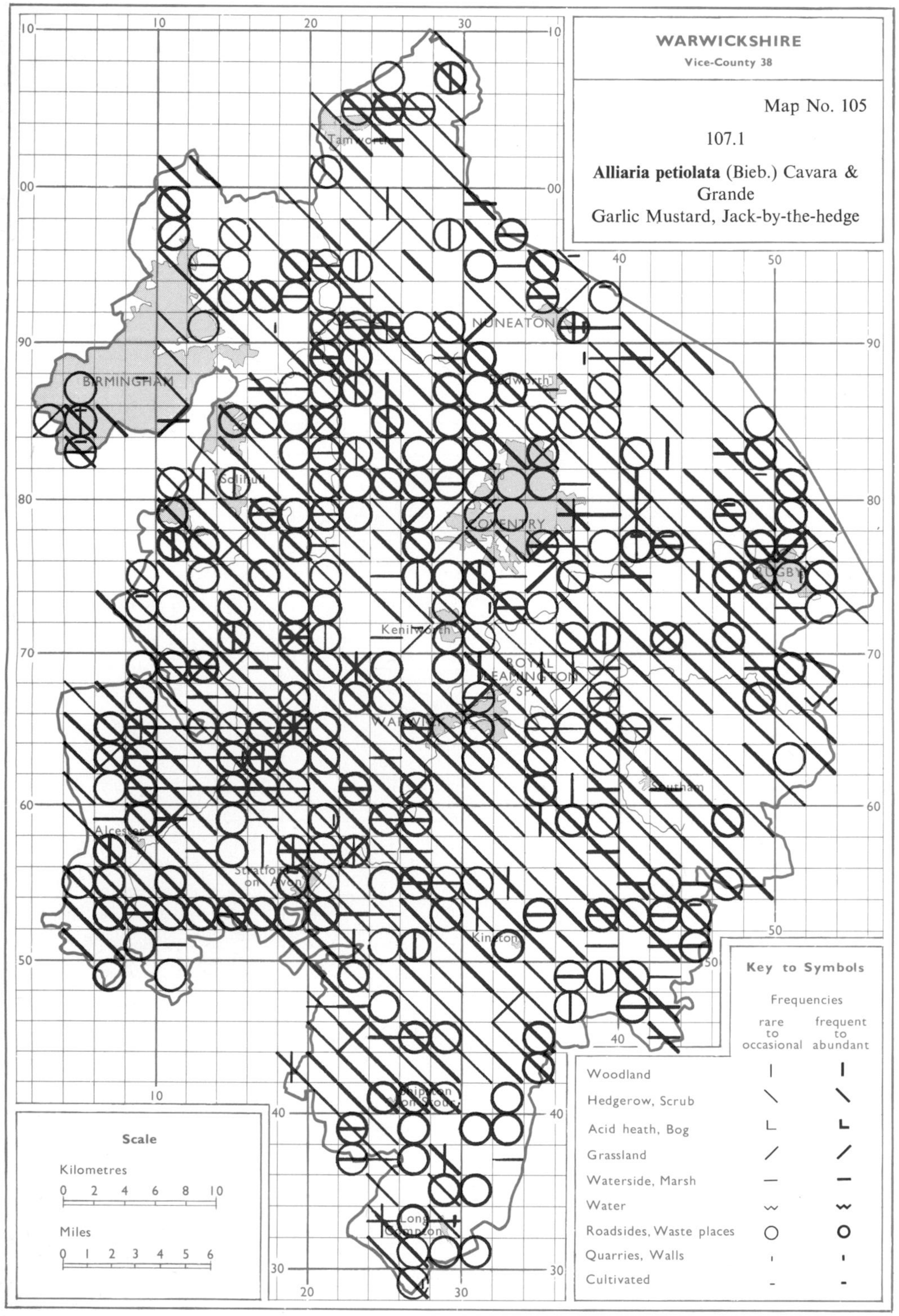
WARWICKSHIRE
Vice-County 38
Map No. 105
107.1
Alliaria petiolata (Bieb.) Cavara & Grande
Garlic Mustard, Jack-by-the-hedge
Tamworth
NUNEATON
Bedworth
BIRMINGHAM
Solihull
COVENTRY
RUGBY
Kenilworth
ROYAL LEAMINGTON SPA
WARWICK
Southam
Alcester
Stratford on Avon
Kineton
Shipston on Stour
Long Compton
Key to Symbols
Frequencies
rare to occasional
frequent to abundant
Woodland
Hedgerow, Scrub
Acid heath, Bog
Grassland
Waterside, Marsh
Water
Roadsides, Waste places
Quarries, Walls
Cultivated
Scale
Kilometres
0 2 4 6 8 10
Miles
0 1 2 3 4 5 6

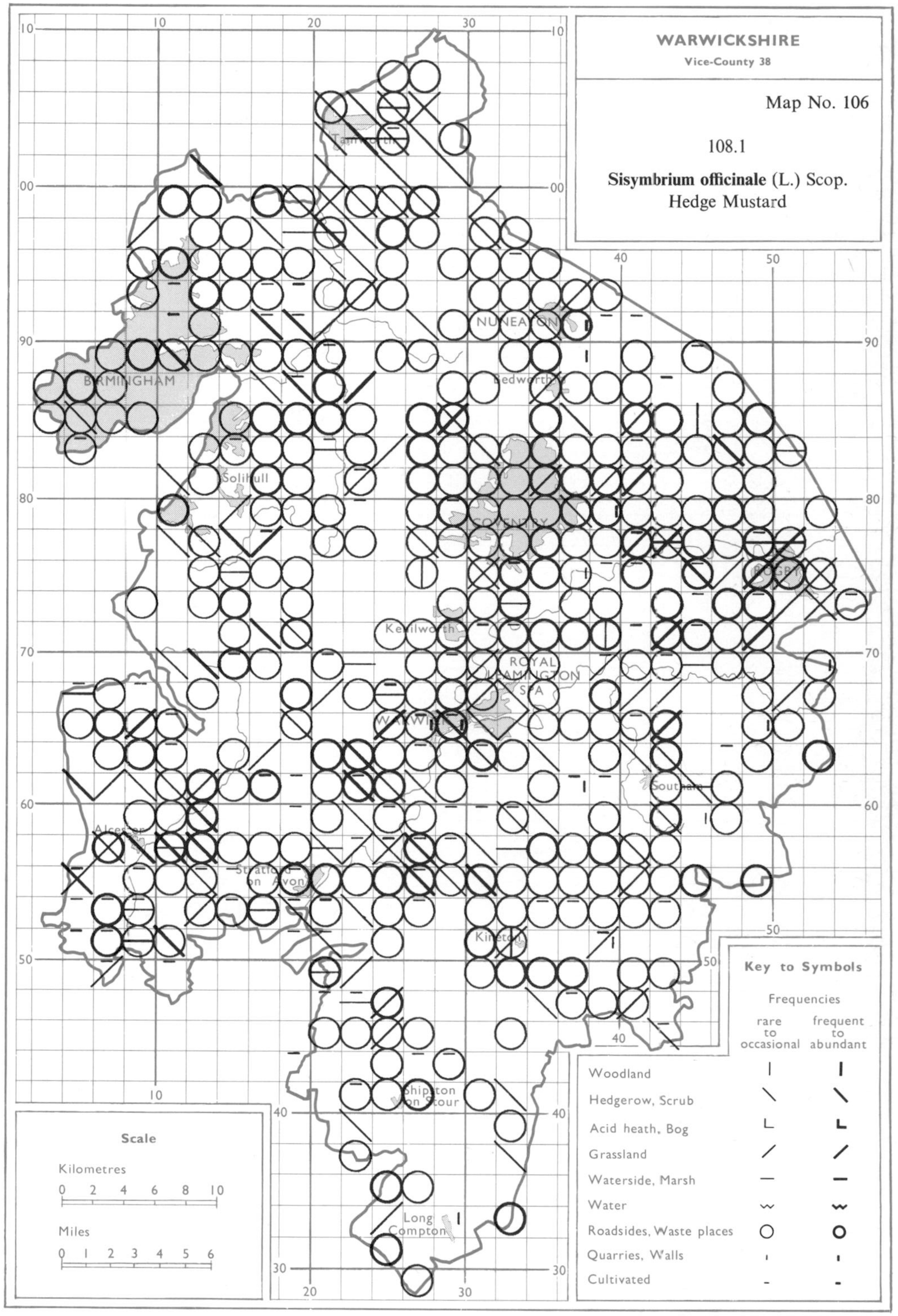
WARWICKSHIRE
Vice-County 38
Map No. 106
108.1
Sisymbrium officinale (L.) Scop.
Hedge Mustard
Tamworth
NUNEATON
Bedworth
BIRMINGHAM
Solihull
COVENTRY
RUGBY
Kenilworth
ROYAL LEAMINGTON SPA
WARWICK
Southam
Alcester
Stratford on Avon
Kineton
Shipston on Stour
Long Compton
Key to Symbols
Frequencies
rare to occasional
frequent to abundant
Woodland
Hedgerow, Scrub
Acid heath, Bog
Grassland
Waterside, Marsh
Water
Roadsides, Waste places
Quarries, Walls
Cultivated
Scale
Kilometres
0 2 4 6 8 10
Miles
0 1 2 3 4 5 6

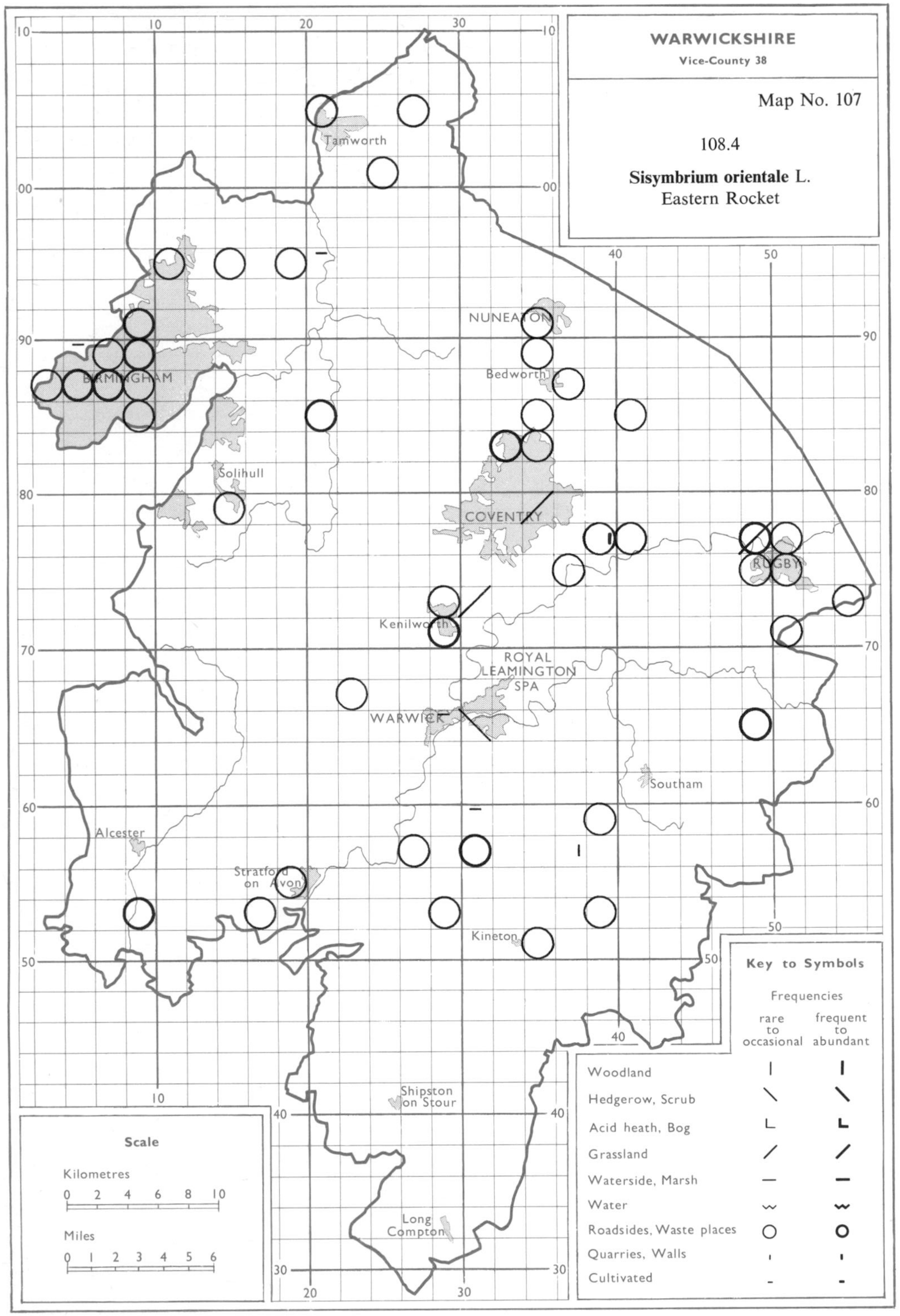

WARWICKSHIRE
Vice-County 38
Map No. 107
108.4
Sisymbrium orientale L.
Eastern Rocket
Tamworth
NUNEATON
Bedworth
BIRMINGHAM
Solihull
COVENTRY
RUGBY
Kenilworth
ROYAL LEAMINGTON SPA
WARWICK
Southam
Alcester
Stratford on Avon
Kineton
Shipston on Stour
Long Compton
Scale
Kilometres
0 2 4 6 8 10
Miles
0 1 2 3 4 5 6
Key to Symbols
Frequencies
rare to occasional
frequent to abundant
Woodland
Hedgerow, Scrub
Acid heath, Bog
Grassland
Waterside, Marsh
Water
Roadsides, Waste places
Quarries, Walls
Cultivated

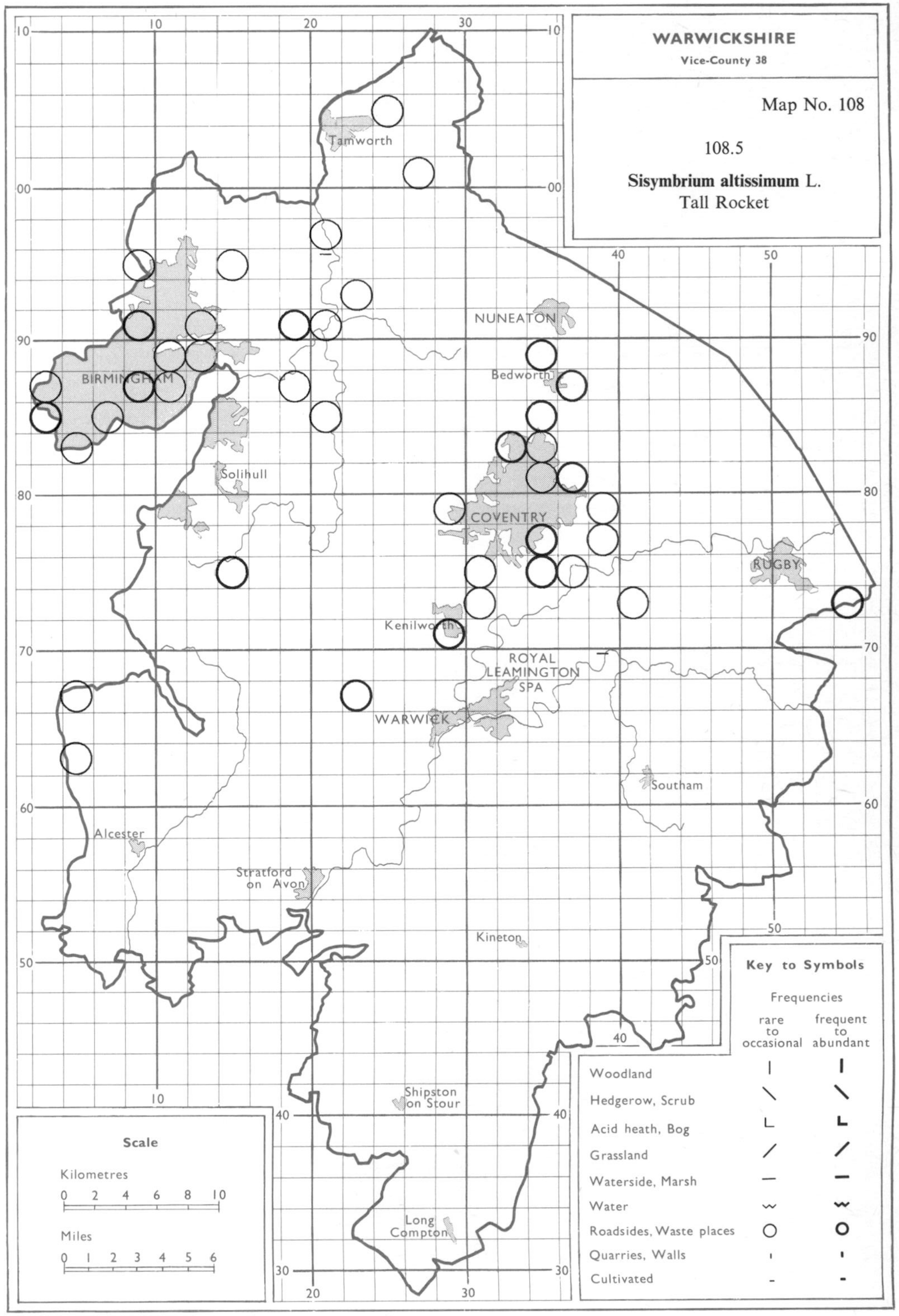

WARWICKSHIRE
Vice-County 38
Map No. 108
108.5
Sisymbrium altissimum L.
Tall Rocket
Tamworth
NUNEATON
Bedworth
BIRMINGHAM
Solihull
COVENTRY
RUGBY
Kenilworth
ROYAL LEAMINGTON SPA
WARWICK
Southam
Alcester
Stratford on Avon
Kineton
Shipston on Stour
Long Compton
Scale
Kilometres
0 2 4 6 8 10
Miles
0 1 2 3 4 5 6
Key to Symbols
Frequencies
rare to occasional
frequent to abundant
Woodland
Hedgerow, Scrub
Acid heath, Bog
Grassland
Waterside, Marsh
Water
Roadsides, Waste places
Quarries, Walls
Cultivated

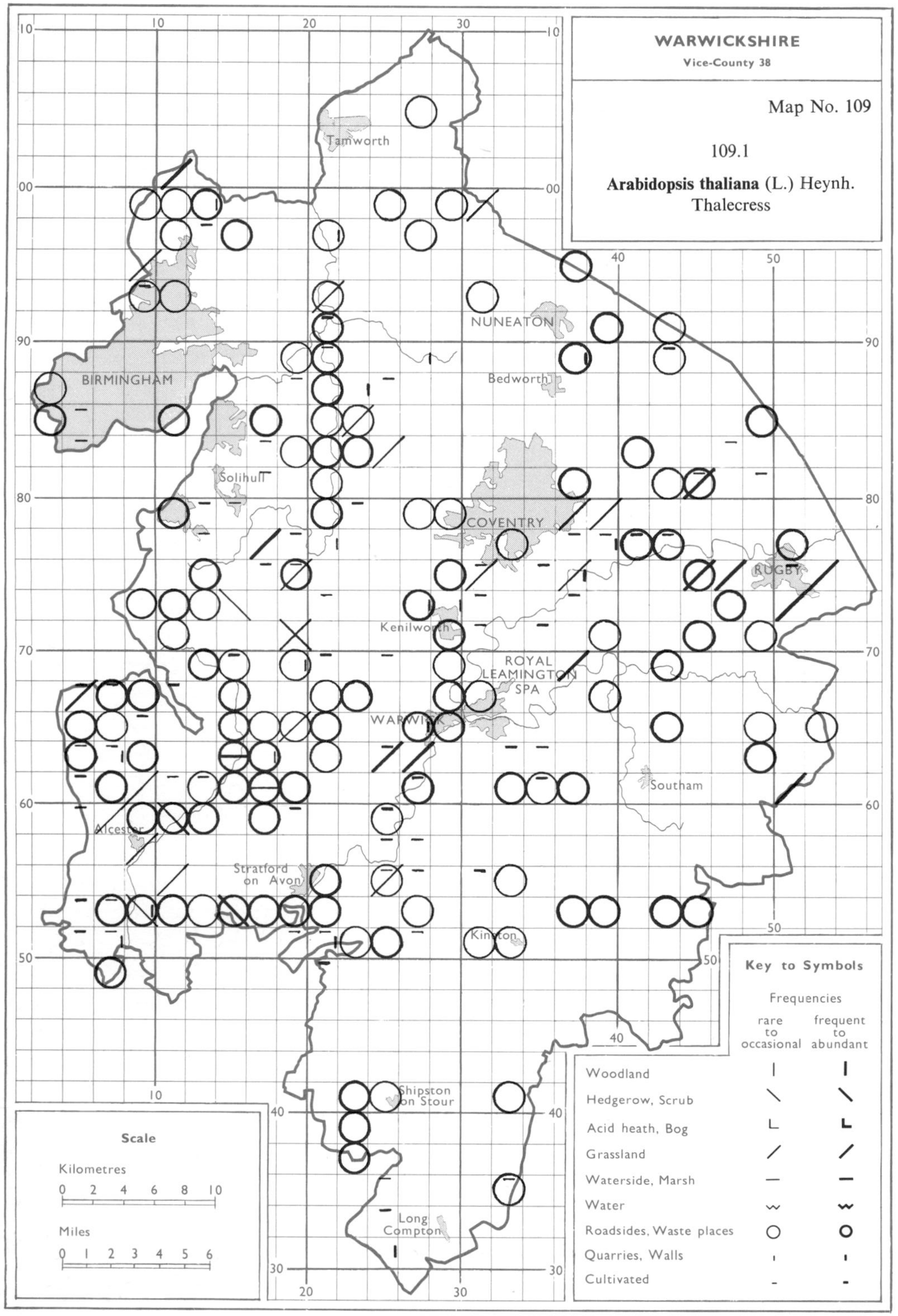
WARWICKSHIRE
Vice-County 38
Map No. 109
109.1
Arabidopsis thaliana (L.) Heynh.
Thalecress
Tamworth
NUNEATON
Bedworth
BIRMINGHAM
Solihull
COVENTRY
RUGBY
Kenilworth
ROYAL LEAMINGTON SPA
WARWICK
Southam
Alcester
Stratford on Avon
Kineton
Shipston on Stour
Long Compton
Key to Symbols
Frequencies
rare to occasional
frequent to abundant
Woodland
Hedgerow, Scrub
Acid heath, Bog
Grassland
Waterside, Marsh
Water
Roadsides, Waste places
Quarries, Walls
Cultivated
Scale
Kilometres
0 2 4 6 8 10
Miles
0 1 2 3 4 5 6

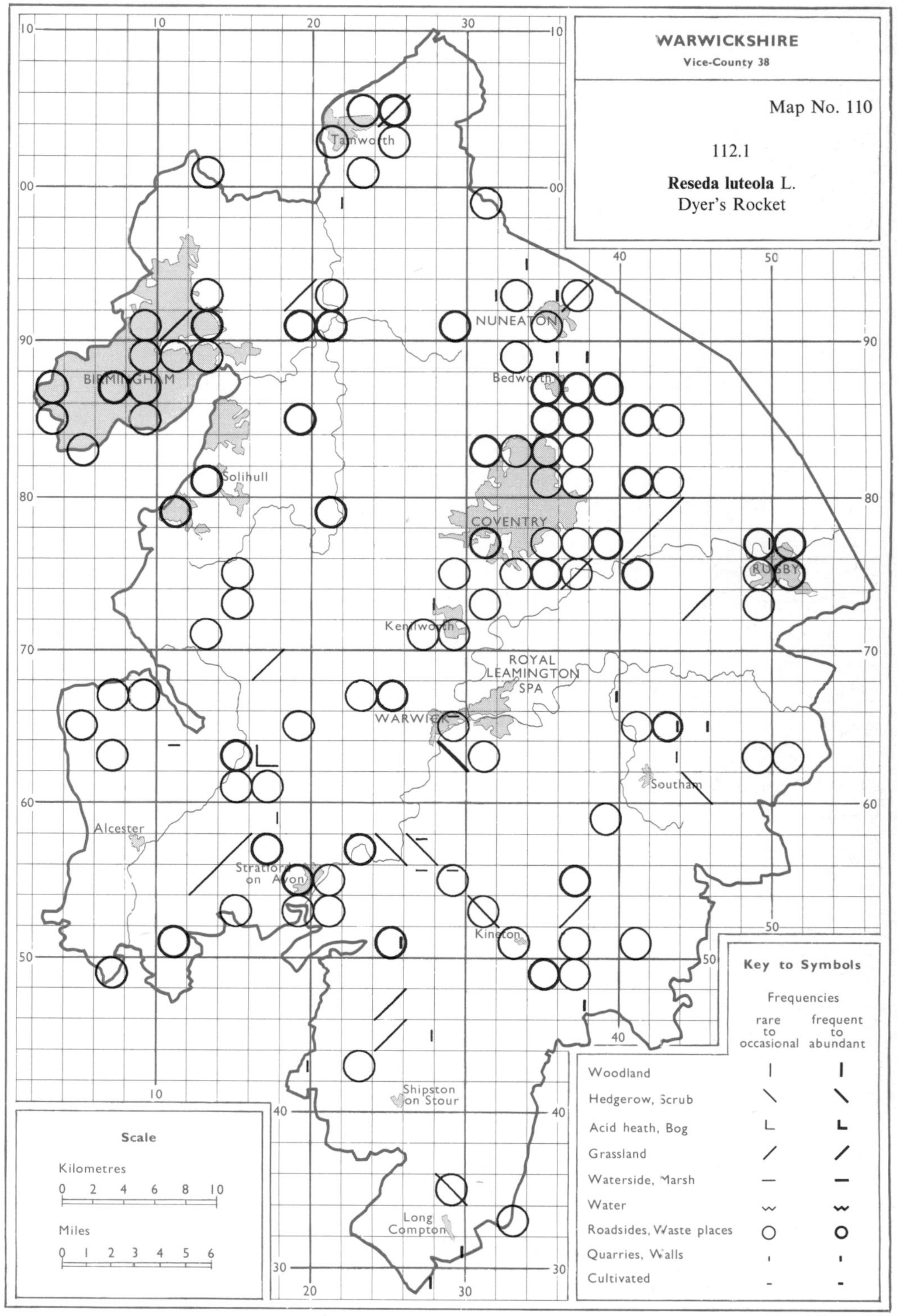
WARWICKSHIRE
Vice-County 38
Map No. 110
112.1
Reseda luteola L.
Dyer's Rocket
Tamworth
NUNEATON
Bedworth
BIRMINGHAM
Solihull
COVENTRY
RUGBY
Kenilworth
ROYAL LEAMINGTON SPA
WARWICK
Southam
Alcester
Stratford on Avon
Kineton
Shipston on Stour
Long Compton
Key to Symbols
Frequencies
rare to occasional
frequent to abundant
Woodland
Hedgerow, Scrub
Acid heath, Bog
Grassland
Waterside, Marsh
Water
Roadsides, Waste places
Quarries, Walls
Cultivated
Scale
Kilometres
0 2 4 6 8 10
Miles
0 1 2 3 4 5 6

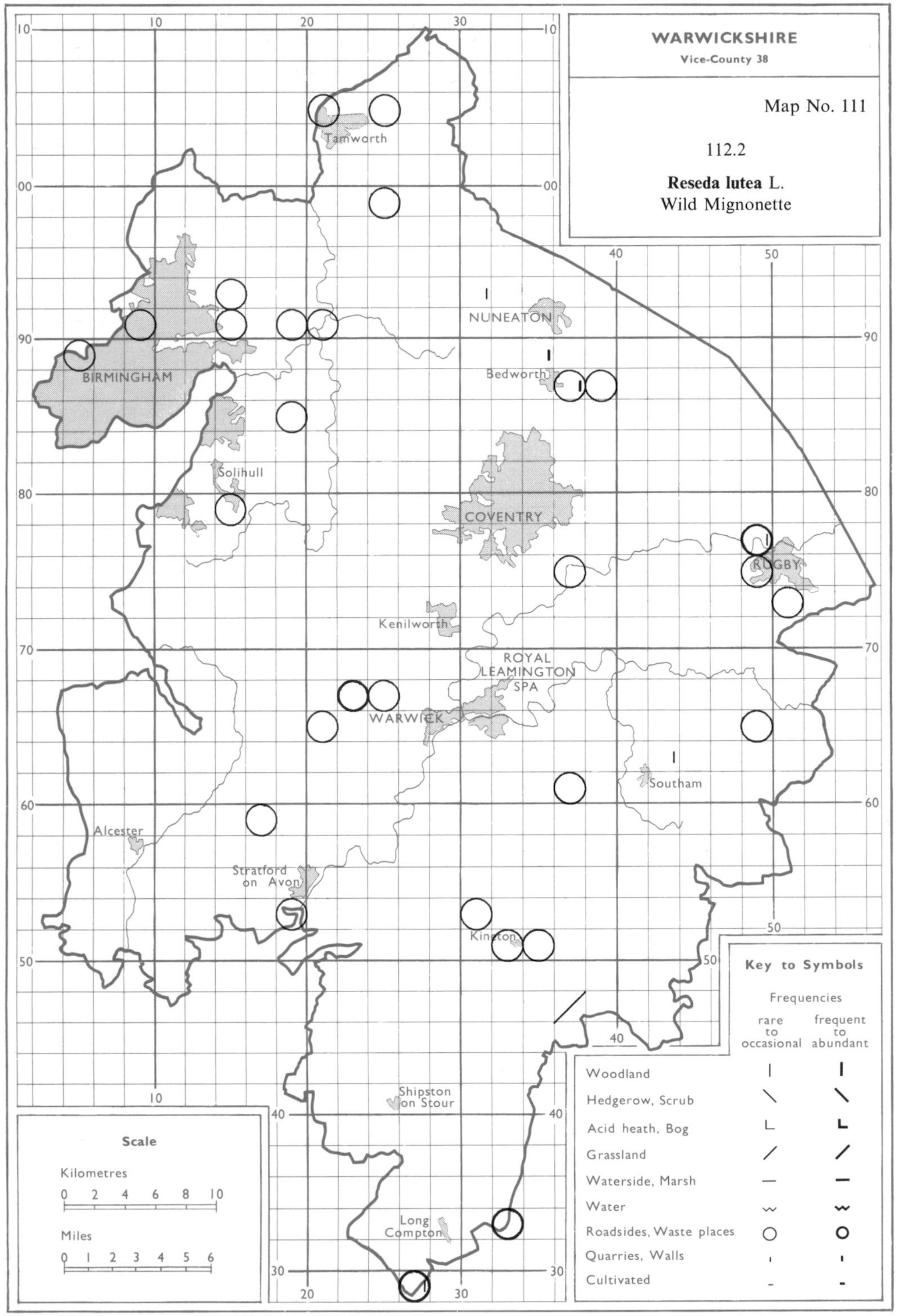
WARWICKSHIRE
Vice-County 38
Map No. 111
112.2
Reseda lutea L.
Wild Mignonette
Tamworth
NUNEATON
Bedworth
BIRMINGHAM
Solihull
COVENTRY
RUGBY
Kenilworth
ROYAL LEAMINGTON SPA
WARWICK
Southam
Alcester
Stratford on Avon
Kineton
Shipston on Stour
Long Compton
Key to Symbols
Frequencies
rare to occasional
frequent to abundant
Woodland
Hedgerow, Scrub
Acid heath, Bog
Grassland
Waterside, Marsh
Water
Roadsides, Waste places
Quarries, Walls
Cultivated
Scale
Kilometres
0 2 4 6 8 10
Miles
0 1 2 3 4 5 6

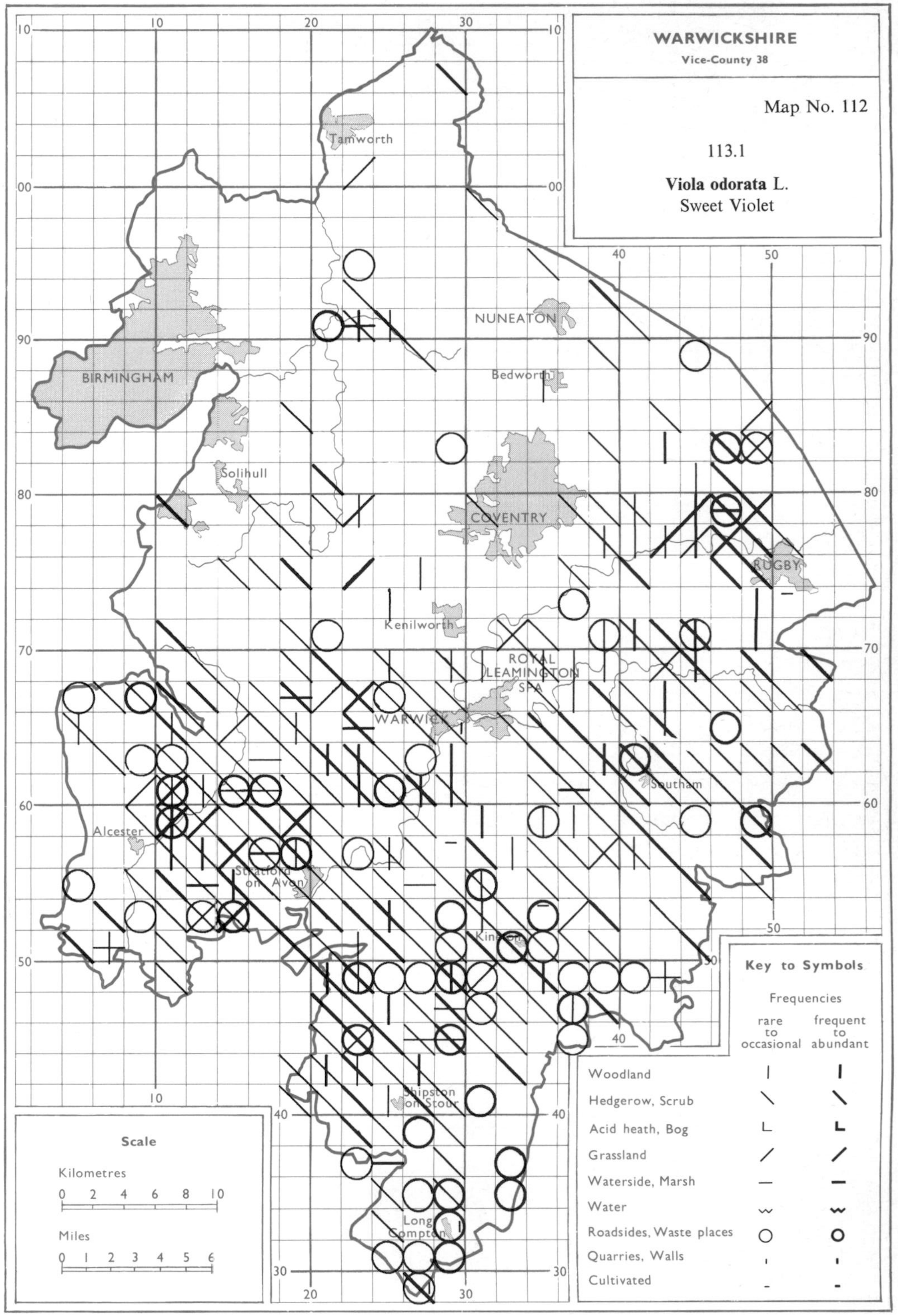
WARWICKSHIRE
Vice-County 38
Map No. 112
113.1
Viola odorata L.
Sweet Violet
Tamworth
NUNEATON
Bedworth
BIRMINGHAM
Solihull
COVENTRY
RUGBY
Kenilworth
ROYAL LEAMINGTON SPA
WARWICK
Southam
Alcester
Stratford on Avon
Kineton
Shipston on Stour
Long Compton
Key to Symbols
Frequencies
rare to occasional
frequent to abundant
Woodland
Hedgerow, Scrub
Acid heath, Bog
Grassland
Waterside, Marsh
Water
Roadsides, Waste places
Quarries, Walls
Cultivated
Scale
Kilometres
0 2 4 6 8 10
Miles
0 1 2 3 4 5 6

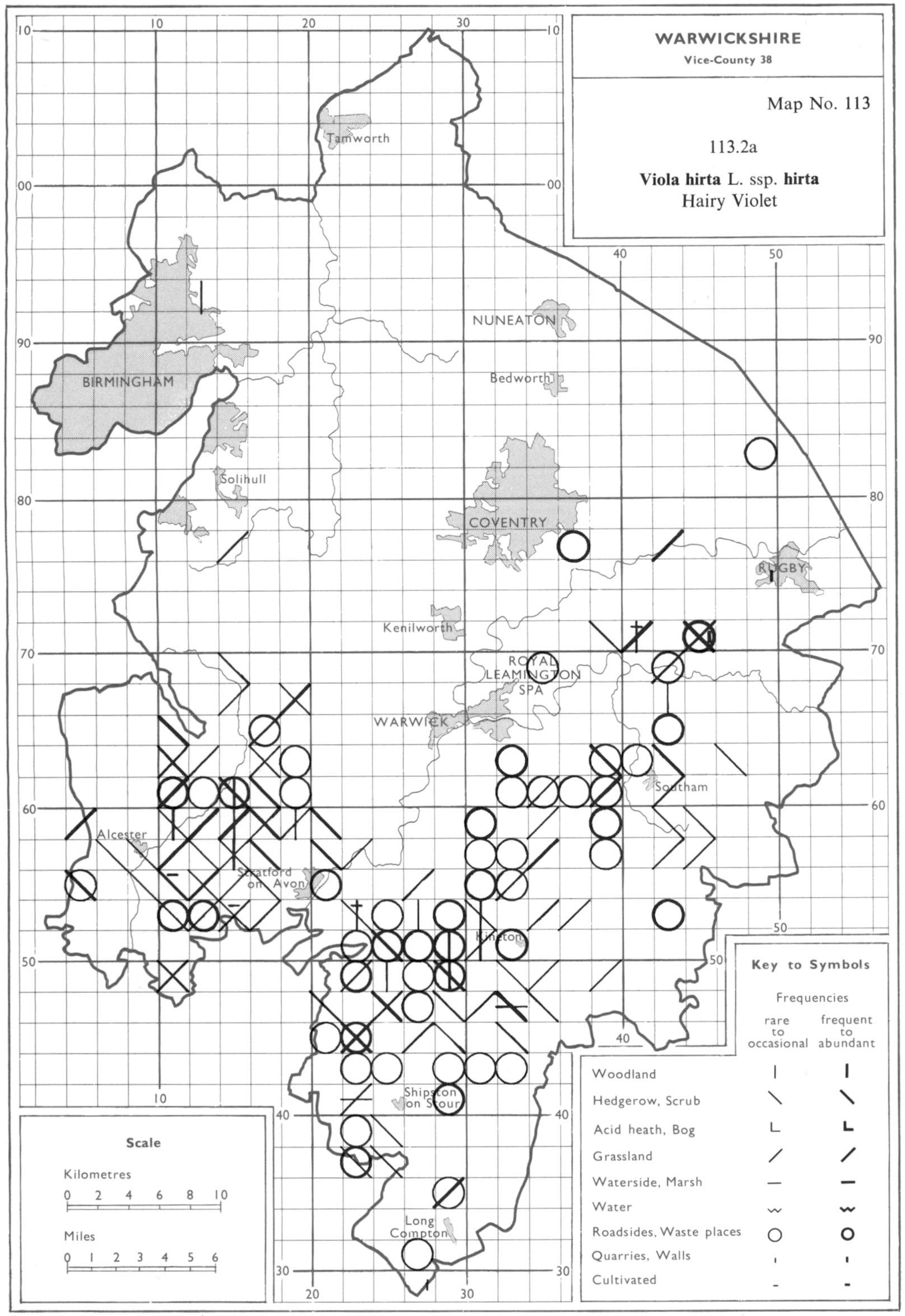
WARWICKSHIRE
Vice-County 38
Map No. 113
113.2a
Viola hirta L. ssp. hirta
Hairy Violet
Tamworth
BIRMINGHAM
NUNEATON
Bedworth
Solihull
COVENTRY
RUGBY
Kenilworth
ROYAL LEAMINGTON SPA
WARWICK
Southam
Alcester
Stratford on Avon
Kineton
Shipston on Stour
Long Compton
Key to Symbols
Frequencies
rare to occasional
frequent to abundant
Woodland
Hedgerow, Scrub
Acid heath, Bog
Grassland
Waterside, Marsh
Water
Roadsides, Waste places
Quarries, Walls
Cultivated
Scale
Kilometres
Miles

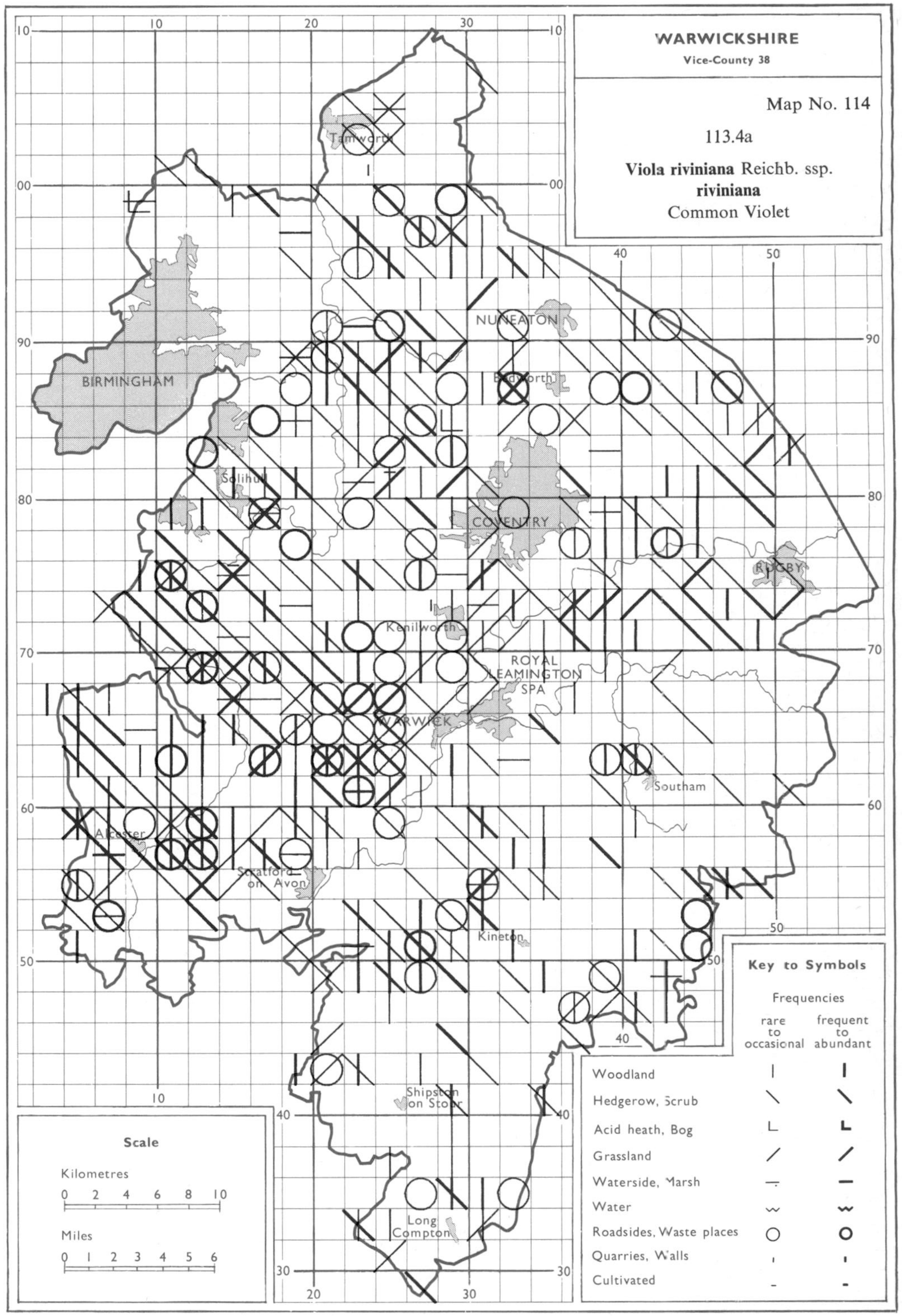
WARWICKSHIRE
Vice-County 38
Map No. 114
113.4a
Viola riviniana Reichb. ssp. riviniana
Common Violet
Tamworth
BIRMINGHAM
NUNEATON
Bedworth
Solihull
COVENTRY
RUGBY
Kenilworth
ROYAL LEAMINGTON SPA
WARWICK
Southam
Alcester
Stratford on Avon
Kineton
Shipston on Stour
Long Compton
Key to Symbols
Frequencies
rare to occasional
frequent to abundant
Woodland
Hedgerow, Scrub
Acid heath, Bog
Grassland
Waterside, Marsh
Water
Roadsides, Waste places
Quarries, Walls
Cultivated
Scale
Kilometres
Miles

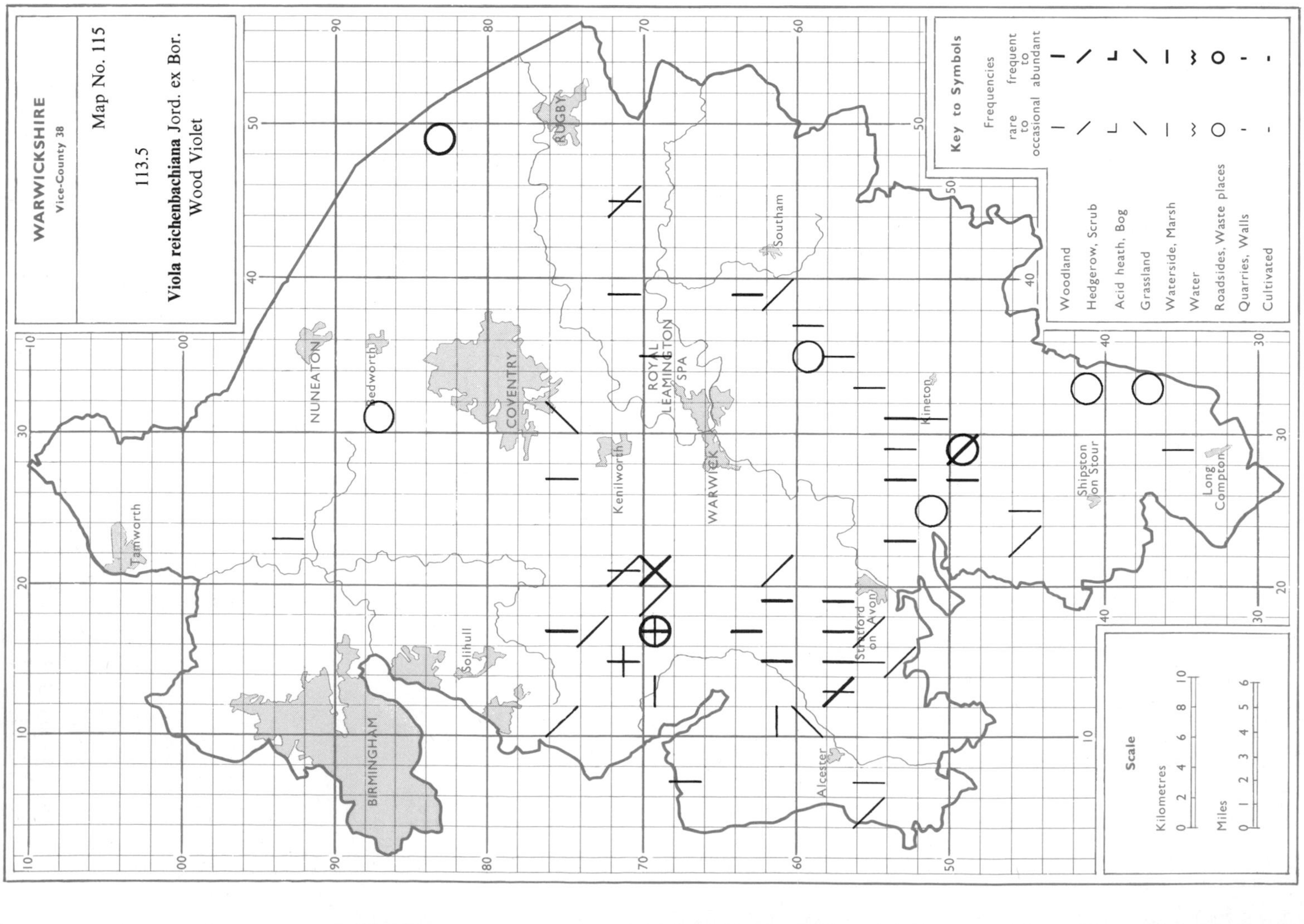
WARWICKSHIRE
Vice-County 38
Map No. 115
113.5
Viola reichenbachiana Jord. ex Bor.
Wood Violet
Key to Symbols
Frequencies
rare to occasional
frequent to abundant
Woodland
Hedgerow, Scrub
Acid heath, Bog
Grassland
Waterside, Marsh
Water
Roadsides, Waste places
Quarries, Walls
Cultivated
Scale
Kilometres
Miles
NUNEATON
Bedworth
COVENTRY
RUGBY
Southam
ROYAL LEAMINGTON SPA
WARWICK
Kenilworth
Kineton
Tamworth
BIRMINGHAM
Solihull
Stratford on Avon
Alcester
Shipston on Stour
Long Compton

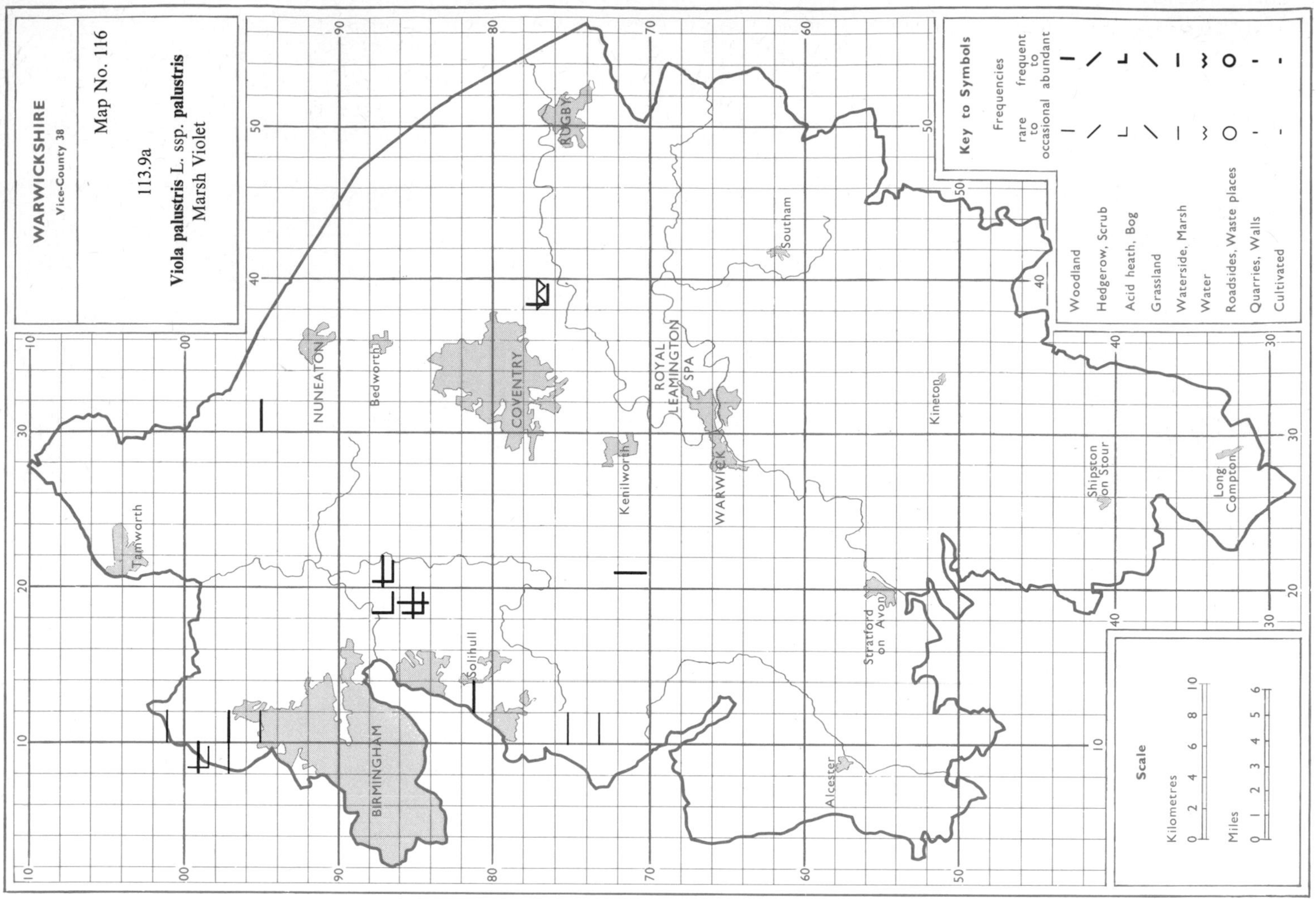
WARWICKSHIRE
Vice-County 38
Map No. 116
113.9a
Viola palustris L. ssp. palustris
Marsh Violet
Key to Symbols
Frequencies
rare to occasional
frequent to abundant
Woodland
Hedgerow, Scrub
Acid heath, Bog
Grassland
Waterside, Marsh
Water
Roadsides, Waste places
Quarries, Walls
Cultivated
Scale
Kilometres
Miles
NUNEATON
Bedworth
COVENTRY
RUGBY
Southam
ROYAL LEAMINGTON SPA
WARWICK
Kenilworth
Kineton
Tamworth
BIRMINGHAM
Solihull
Stratford on Avon
Alcester
Shipston on Stour
Long Compton

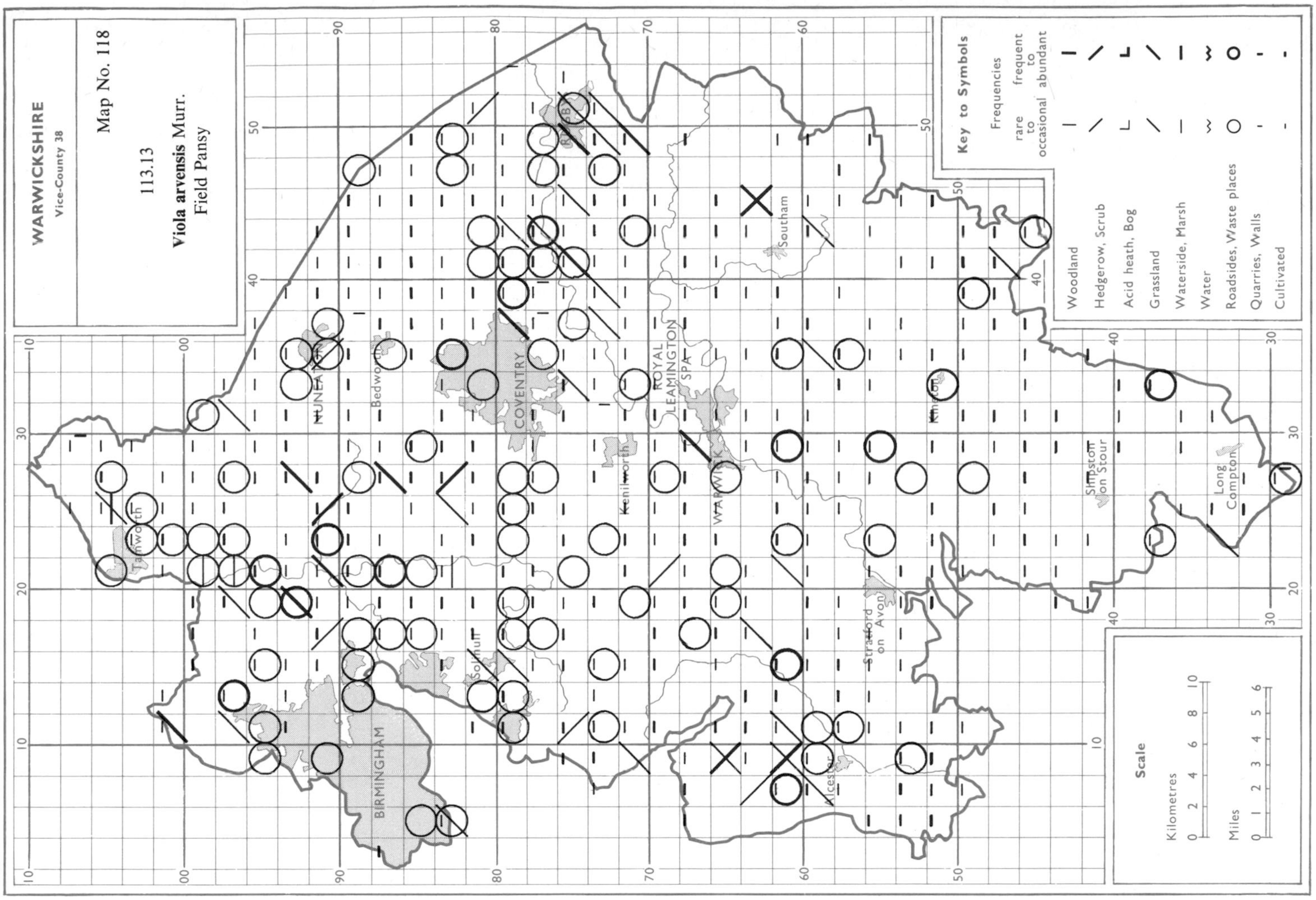
WARWICKSHIRE
Vice-County 38
Map No. 118
113.13
Viola arvensis Murr.
Field Pansy
Key to Symbols
Frequencies
rare to occasional
frequent to abundant
Woodland
Hedgerow, Scrub
Acid heath, Bog
Grassland
Waterside, Marsh
Water
Roadsides, Waste places
Quarries, Walls
Cultivated
Scale
Kilometres 0 2 4 6 8 10
Miles 0 1 2 3 4 5 6
Tamworth
NUNEATON
Bedworth
COVENTRY
RUGBY
BIRMINGHAM
Solihull
Kenilworth
ROYAL LEAMINGTON SPA
WARWICK
Southam
Kineton
Stratford on Avon
Alcester
Shipston on Stour
Long Compton

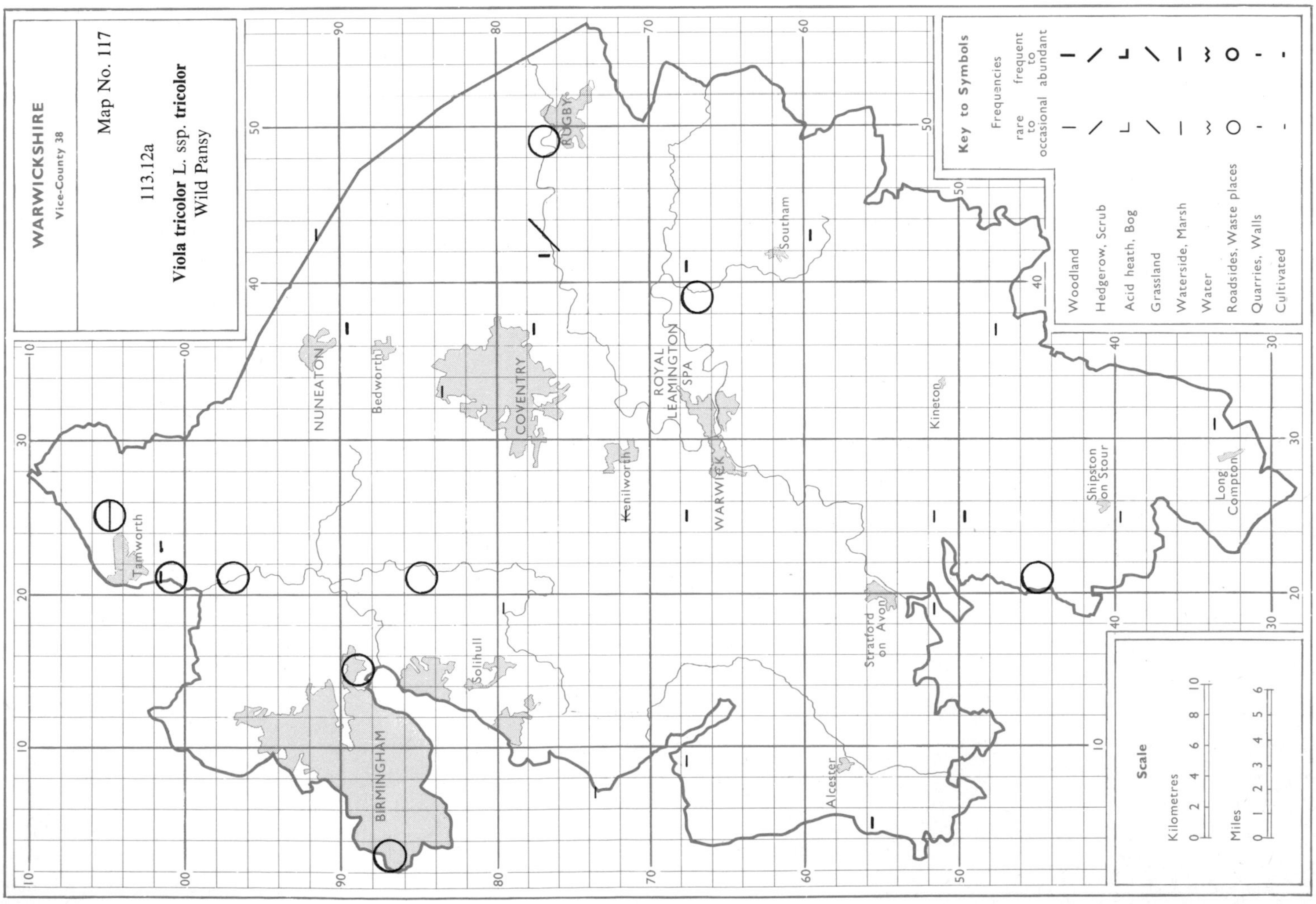
WARWICKSHIRE
Vice-County 38
Map No. 117
113.12a
Viola tricolor L. ssp. tricolor
Wild Pansy
Key to Symbols
Frequencies
rare to occasional
frequent to abundant
Woodland
Hedgerow, Scrub
Acid heath, Bog
Grassland
Waterside, Marsh
Water
Roadsides, Waste places
Quarries, Walls
Cultivated
Scale
Kilometres 0 2 4 6 8 10
Miles 0 1 2 3 4 5 6
Tamworth
NUNEATON
Bedworth
COVENTRY
RUGBY
BIRMINGHAM
Solihull
Kenilworth
ROYAL LEAMINGTON SPA
WARWICK
Southam
Kineton
Stratford on Avon
Alcester
Shipston on Stour
Long Compton

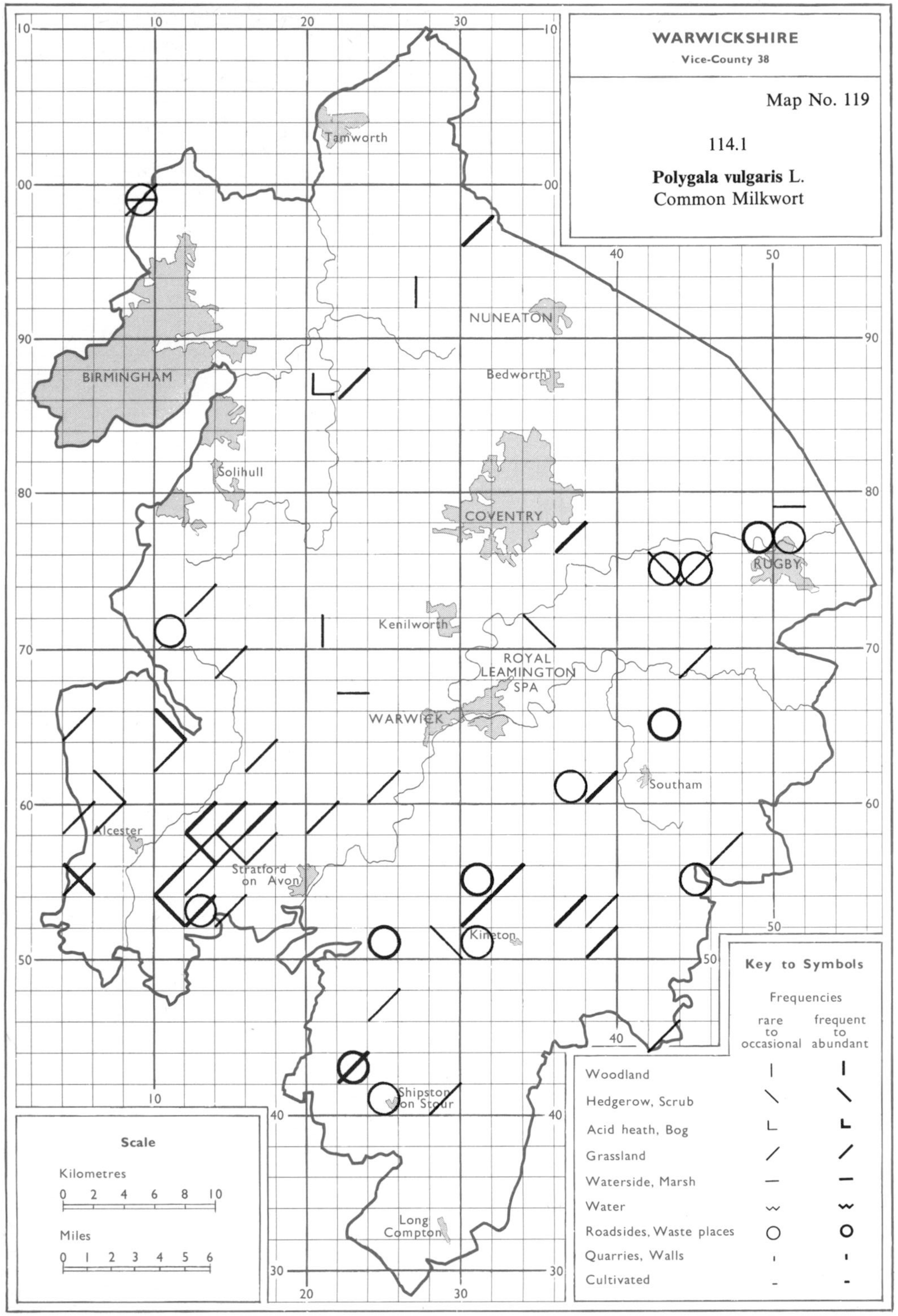
WARWICKSHIRE
Vice-County 38
Map No. 119
114.1
Polygala vulgaris L.
Common Milkwort
Tamworth
NUNEATON
Bedworth
BIRMINGHAM
Solihull
COVENTRY
RUGBY
Kenilworth
ROYAL LEAMINGTON SPA
WARWICK
Southam
Alcester
Stratford on Avon
Kineton
Shipston on Stour
Long Compton
Key to Symbols
Frequencies
rare to occasional
frequent to abundant
Woodland
Hedgerow, Scrub
Acid heath, Bog
Grassland
Waterside, Marsh
Water
Roadsides, Waste places
Quarries, Walls
Cultivated
Scale
Kilometres
0 2 4 6 8 10
Miles
0 1 2 3 4 5 6

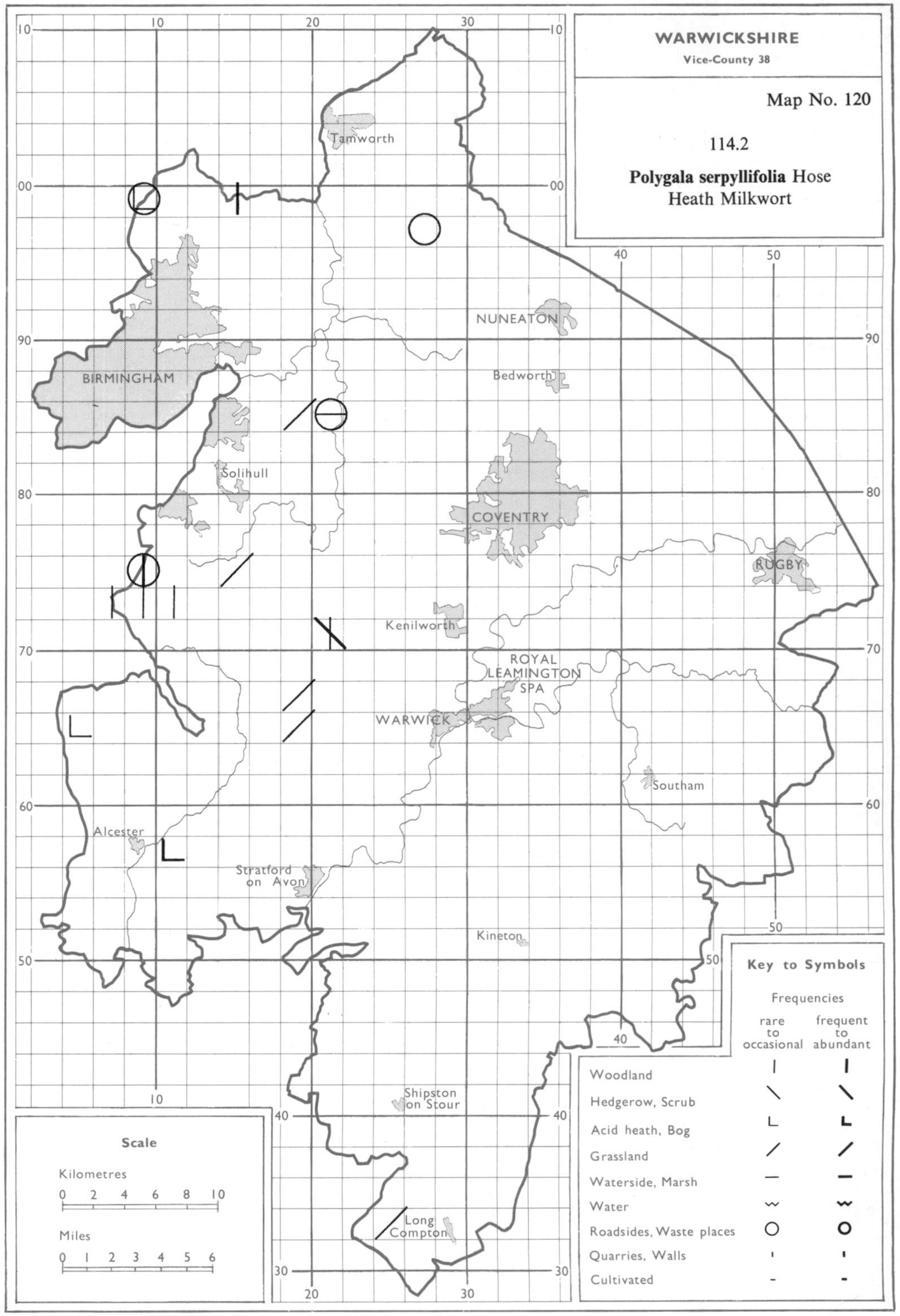
WARWICKSHIRE
Vice-County 38
Map No. 120
114.2
Polygala serpyllifolia Hose
Heath Milkwort
Tamworth
NUNEATON
Bedworth
BIRMINGHAM
Solihull
COVENTRY
RUGBY
Kenilworth
ROYAL LEAMINGTON SPA
WARWICK
Southam
Alcester
Stratford on Avon
Kineton
Shipston on Stour
Long Compton
Key to Symbols
Frequencies
rare to occasional
frequent to abundant
Woodland
Hedgerow, Scrub
Acid heath, Bog
Grassland
Waterside, Marsh
Water
Roadsides, Waste places
Quarries, Walls
Cultivated
Scale
Kilometres
0 2 4 6 8 10
Miles
0 1 2 3 4 5 6

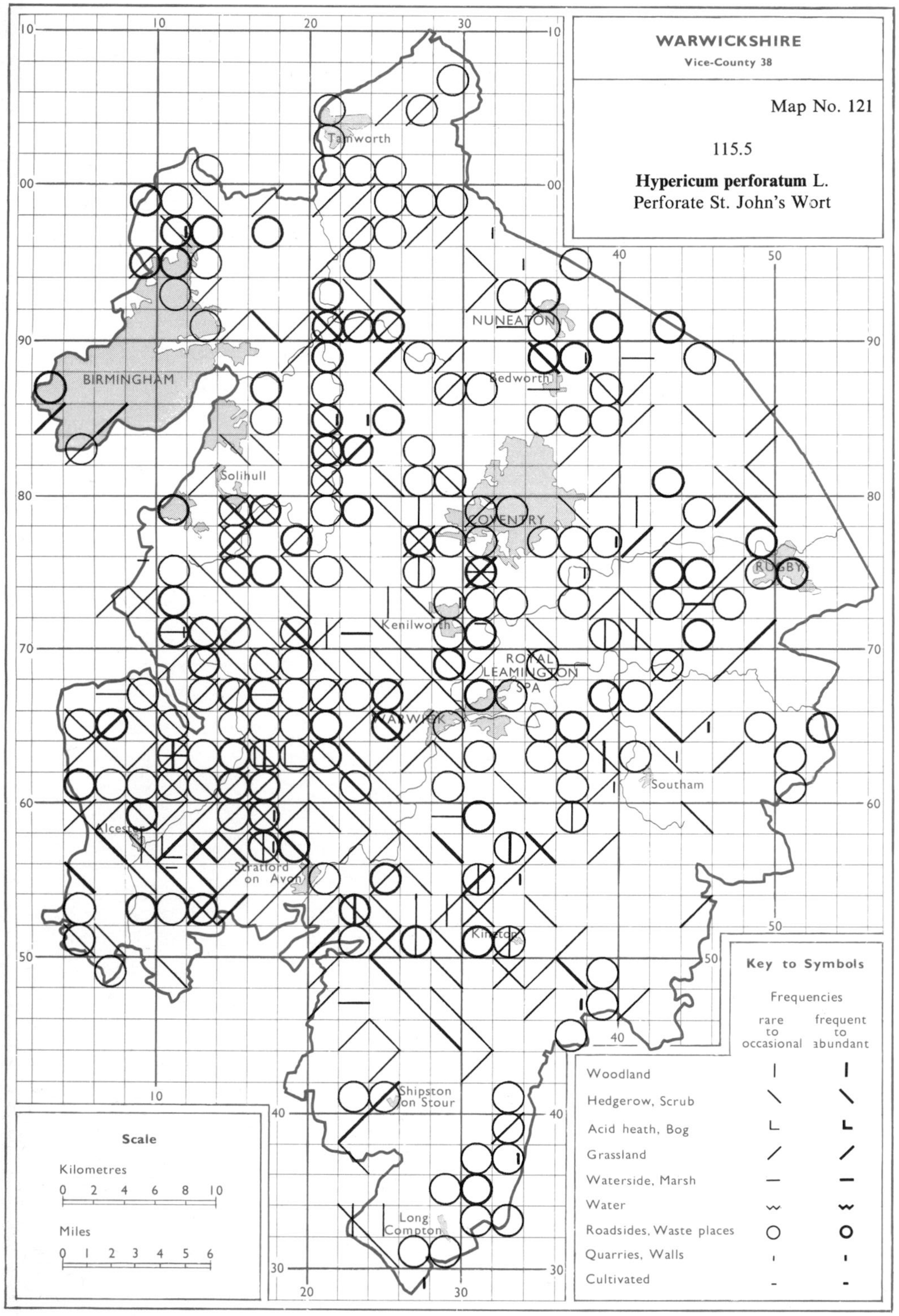
WARWICKSHIRE
Vice-County 38
Map No. 121
115.5
Hypericum perforatum L.
Perforate St. John's Wort
Tamworth
NUNEATON
Bedworth
BIRMINGHAM
Solihull
COVENTRY
RUGBY
Kenilworth
ROYAL LEAMINGTON SPA
WARWICK
Southam
Alcester
Stratford on Avon
Kineton
Shipston on Stour
Long Compton
Scale
Kilometres 0 2 4 6 8 10
Miles 0 1 2 3 4 5 6
Key to Symbols
Frequencies
rare to occasional
frequent to abundant
Woodland
Hedgerow, Scrub
Acid heath, Bog
Grassland
Waterside, Marsh
Water
Roadsides, Waste places
Quarries, Walls
Cultivated

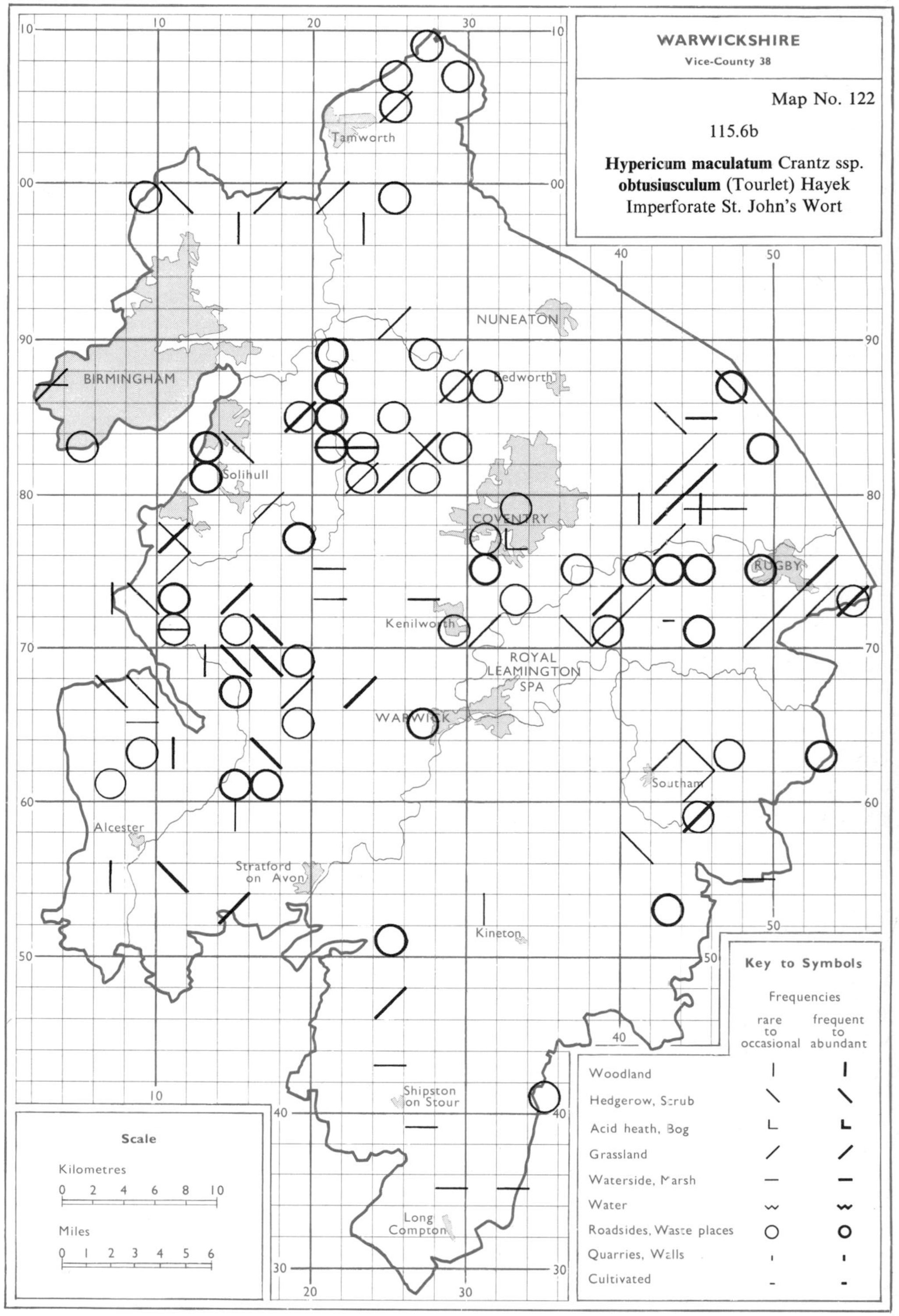
WARWICKSHIRE
Vice-County 38
Map No. 122
115.6b
Hypericum maculatum Crantz ssp. obtusiusculum (Tourlet) Hayek
Imperforate St. John's Wort
Tamworth
NUNEATON
Bedworth
BIRMINGHAM
Solihull
COVENTRY
RUGBY
Kenilworth
ROYAL LEAMINGTON SPA
WARWICK
Southam
Alcester
Stratford on Avon
Kineton
Shipston on Stour
Long Compton
Scale
Kilometres 0 2 4 6 8 10
Miles 0 1 2 3 4 5 6
Key to Symbols
Frequencies
rare to occasional
frequent to abundant
Woodland
Hedgerow, Scrub
Acid heath, Bog
Grassland
Waterside, Marsh
Water
Roadsides, Waste places
Quarries, Walls
Cultivated

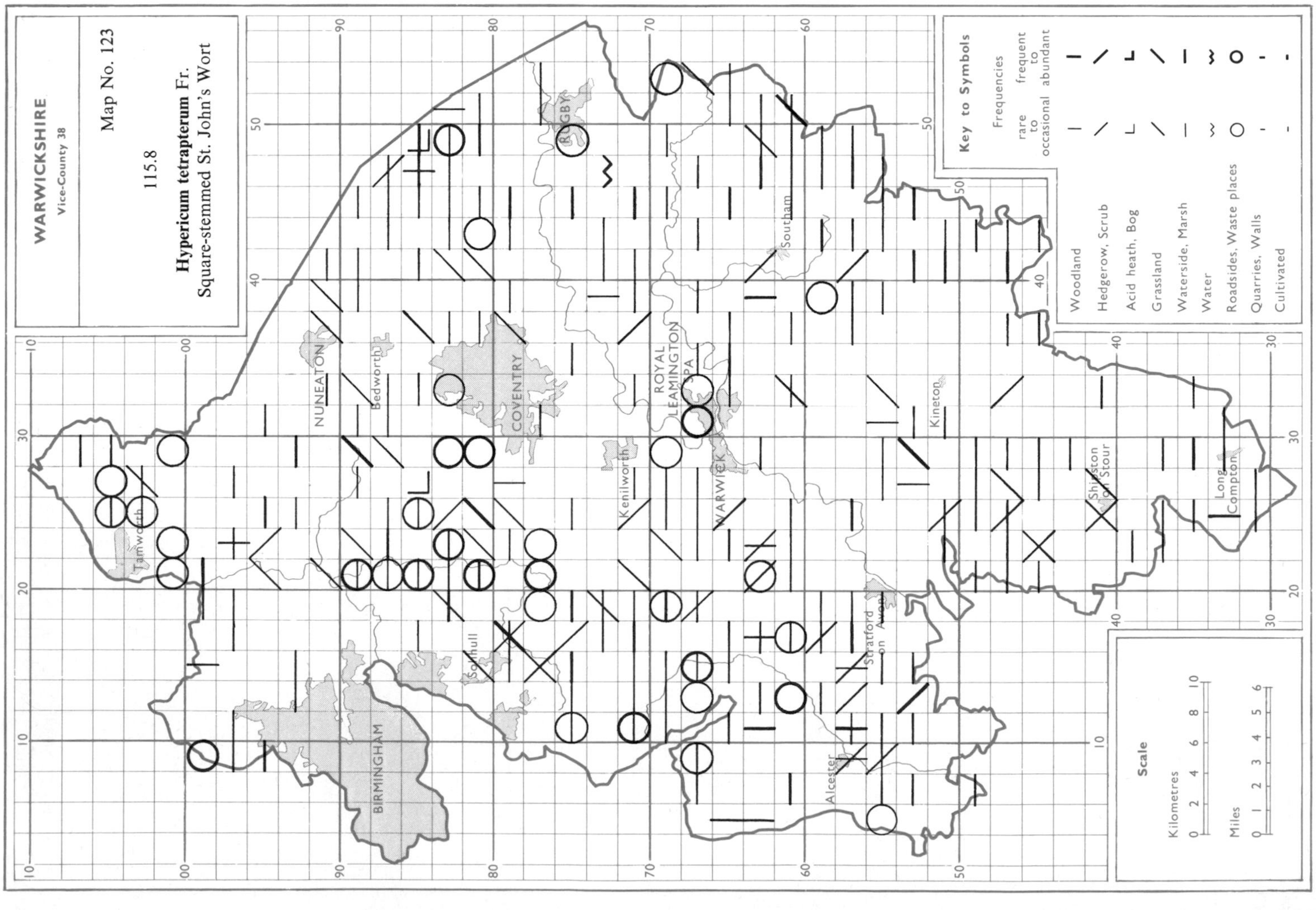
WARWICKSHIRE
Vice-County 38
Map No. 123
115.8
Hypericum tetrapterum Fr.
Square-stemmed St. John's Wort
Key to Symbols
Frequencies
rare to occasional
frequent to abundant
Woodland
Hedgerow, Scrub
Acid heath, Bog
Grassland
Waterside, Marsh
Water
Roadsides, Waste places
Quarries, Walls
Cultivated
Scale
Kilometres
0 2 4 6 8 10
Miles
0 1 2 3 4 5 6
NUNEATON
Bedworth
COVENTRY
RUGBY
Southam
ROYAL LEAMINGTON SPA
WARWICK
Kenilworth
Kineton
Shipston on Stour
Long Compton
Tamworth
Solihull
BIRMINGHAM
Stratford on Avon
Alcester

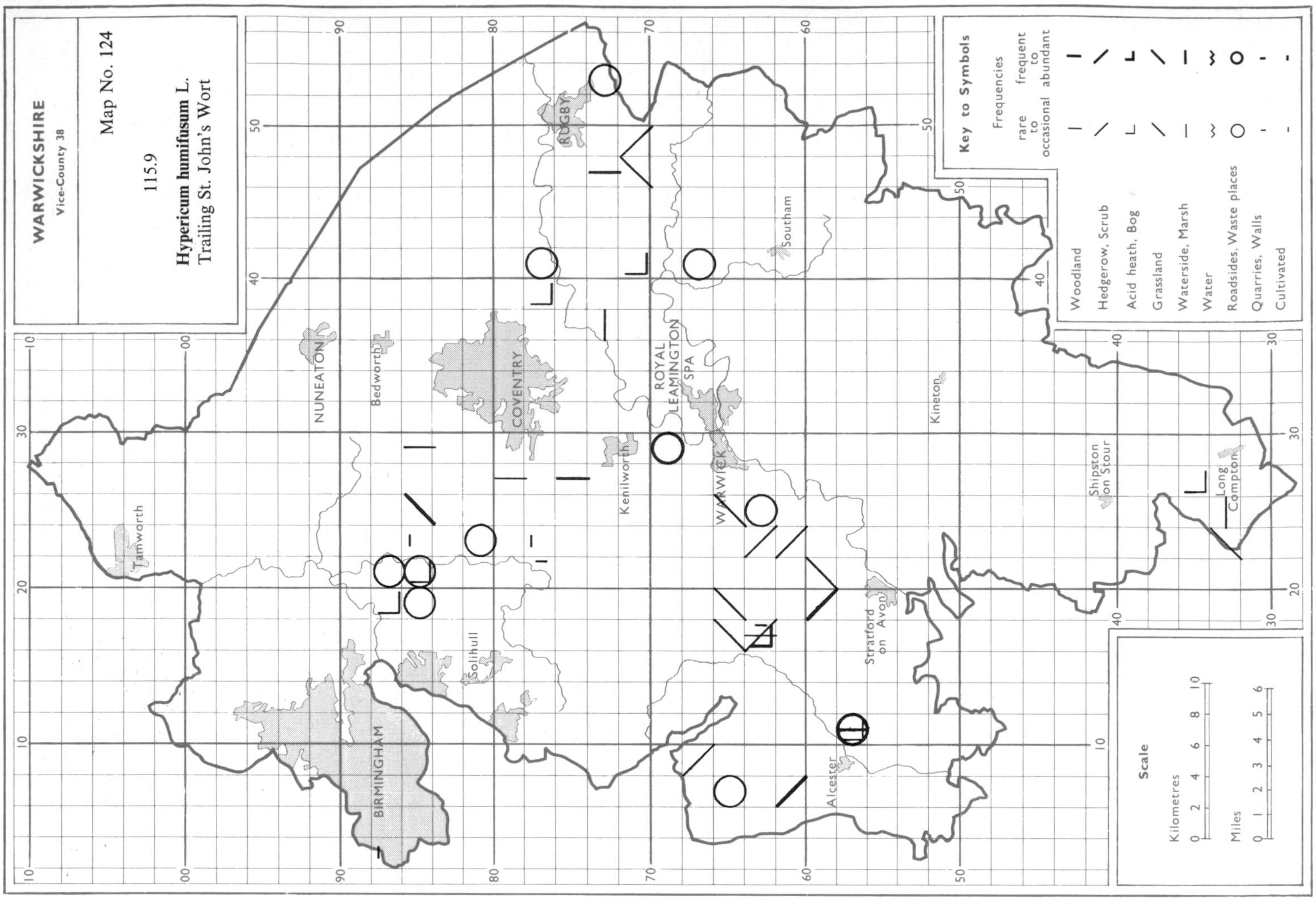
WARWICKSHIRE
Vice-County 38
Map No. 124
115.9
Hypericum humifusum L.
Trailing St. John's Wort
Key to Symbols
Frequencies
rare to occasional
frequent to abundant
Woodland
Hedgerow, Scrub
Acid heath, Bog
Grassland
Waterside, Marsh
Water
Roadsides, Waste places
Quarries, Walls
Cultivated
Scale
Kilometres
0 2 4 6 8 10
Miles
0 1 2 3 4 5 6
NUNEATON
Bedworth
COVENTRY
RUGBY
Southam
ROYAL LEAMINGTON SPA
WARWICK
Kenilworth
Kineton
Shipston on Stour
Long Compton
Tamworth
Solihull
BIRMINGHAM
Stratford on Avon
Alcester

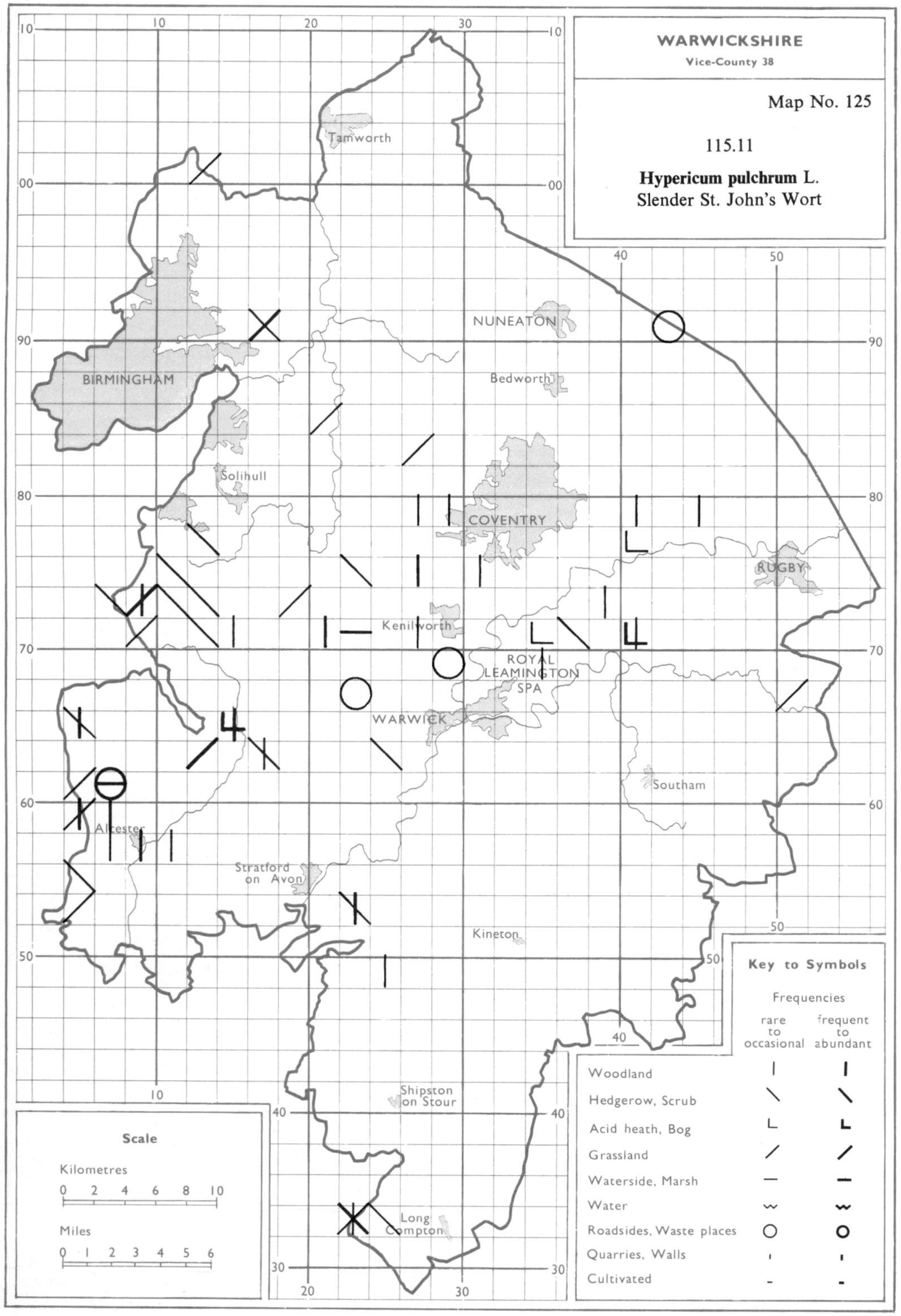
WARWICKSHIRE
Vice-County 38
Map No. 125
115.11
Hypericum pulchrum L.
Slender St. John's Wort
Tamworth
NUNEATON
Bedworth
BIRMINGHAM
Solihull
COVENTRY
RUGBY
Kenilworth
ROYAL LEAMINGTON SPA
WARWICK
Southam
Alcester
Stratford on Avon
Kineton
Shipston on Stour
Long Compton
Scale
Kilometres
0 2 4 6 8 10
Miles
0 1 2 3 4 5 6
Key to Symbols
Frequencies
rare to occasional
frequent to abundant
Woodland
Hedgerow, Scrub
Acid heath, Bog
Grassland
Waterside, Marsh
Water
Roadsides, Waste places
Quarries, Walls
Cultivated

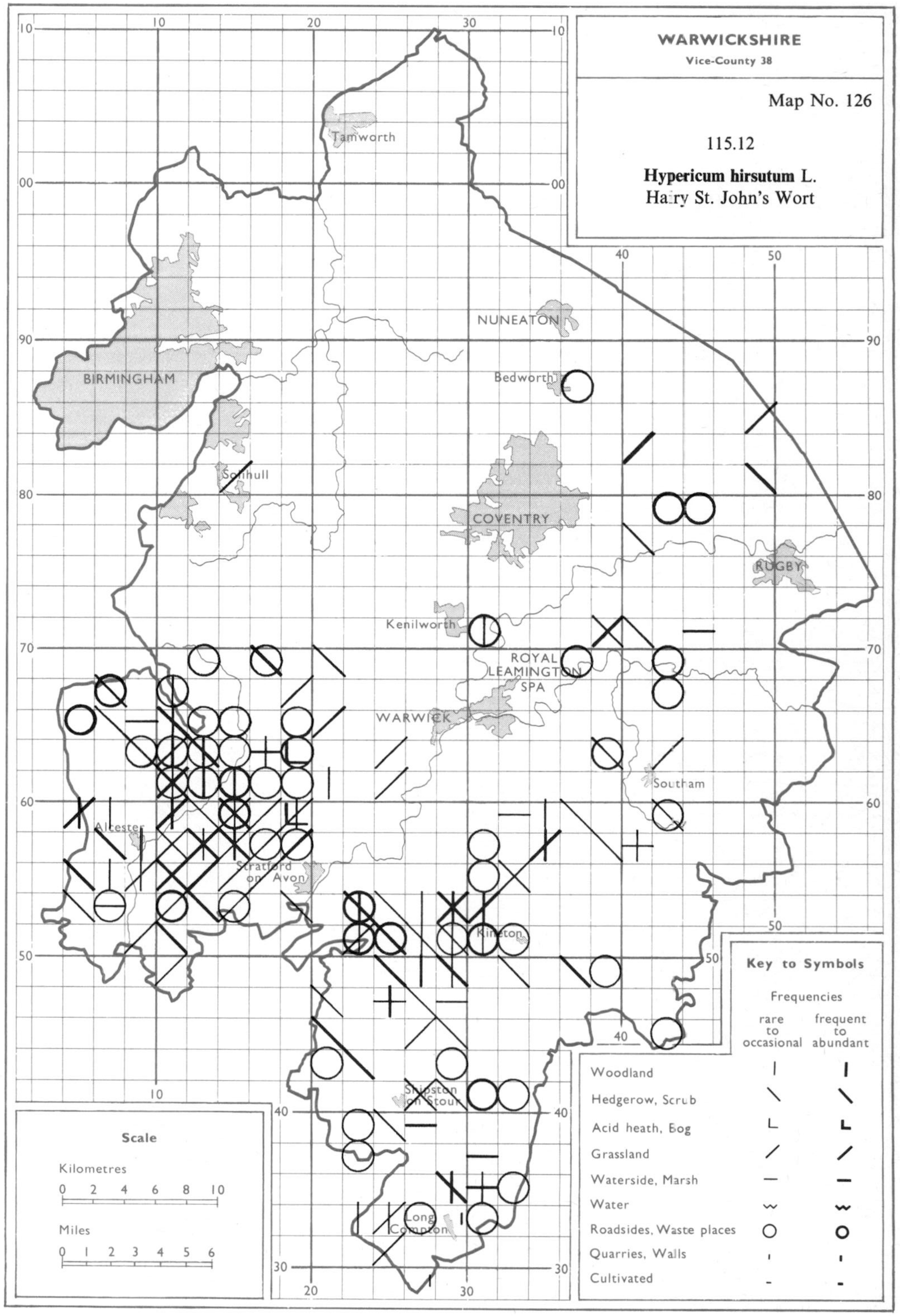
WARWICKSHIRE
Vice-County 38
Map No. 126
115.12
Hypericum hirsutum L.
Hairy St. John's Wort
Tamworth
NUNEATON
Bedworth
BIRMINGHAM
Solihull
COVENTRY
RUGBY
Kenilworth
ROYAL LEAMINGTON SPA
WARWICK
Southam
Alcester
Stratford on Avon
Kineton
Shipston on Stour
Long Compton
Scale
Kilometres
0 2 4 6 8 10
Miles
0 1 2 3 4 5 6
Key to Symbols
Frequencies
rare to occasional
frequent to abundant
Woodland
Hedgerow, Scrub
Acid heath, Bog
Grassland
Waterside, Marsh
Water
Roadsides, Waste places
Quarries, Walls
Cultivated

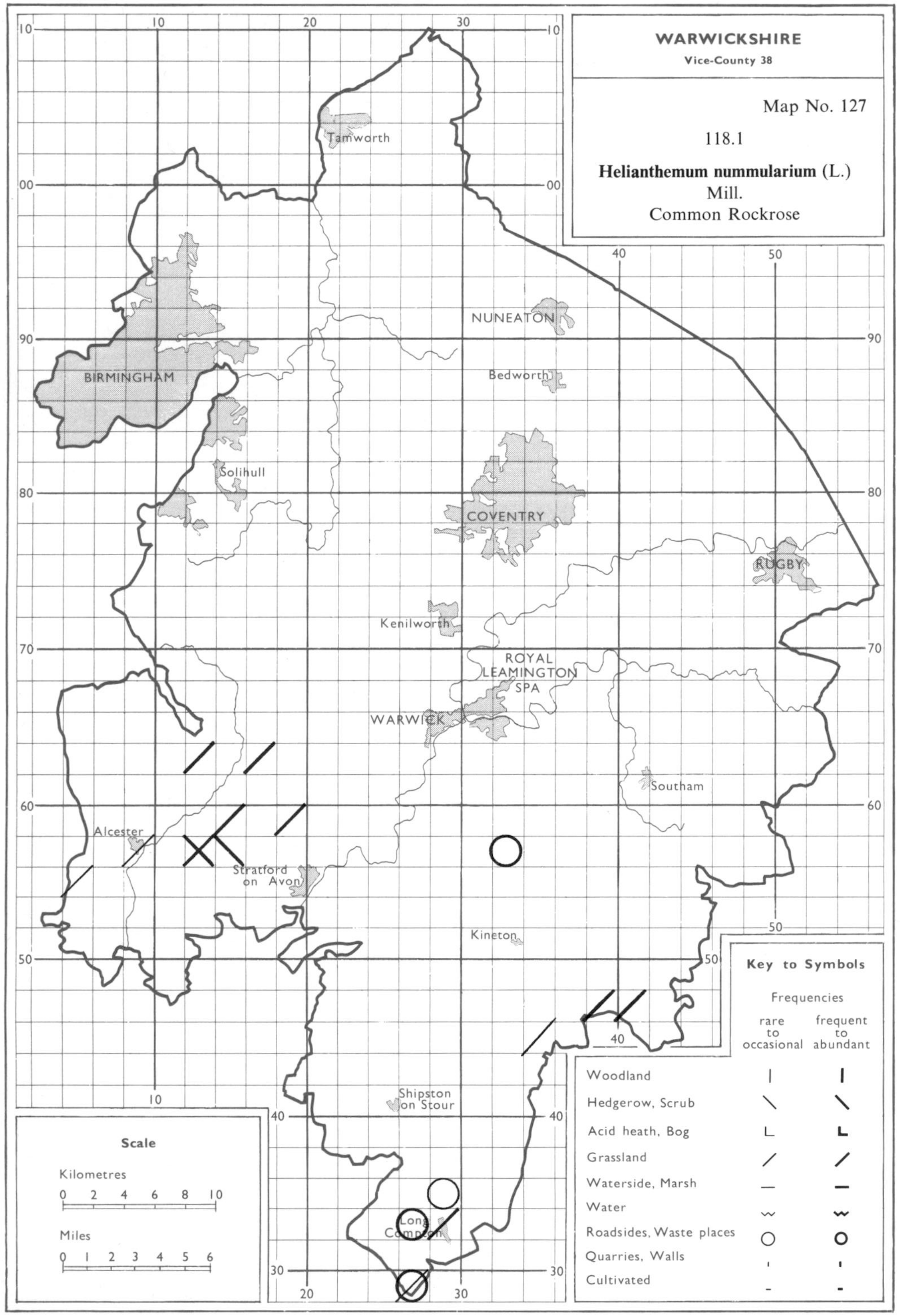
WARWICKSHIRE
Vice-County 38
Map No. 127
118.1
Helianthemum nummularium (L.) Mill.
Common Rockrose
Tamworth
NUNEATON
Bedworth
BIRMINGHAM
Solihull
COVENTRY
RUGBY
Kenilworth
ROYAL LEAMINGTON SPA
WARWICK
Southam
Alcester
Stratford on Avon
Kineton
Shipston on Stour
Long Compton
Key to Symbols
Frequencies
rare to occasional
frequent to abundant
Woodland
Hedgerow, Scrub
Acid heath, Bog
Grassland
Waterside, Marsh
Water
Roadsides, Waste places
Quarries, Walls
Cultivated
Scale
Kilometres
0 2 4 6 8 10
Miles
0 1 2 3 4 5 6

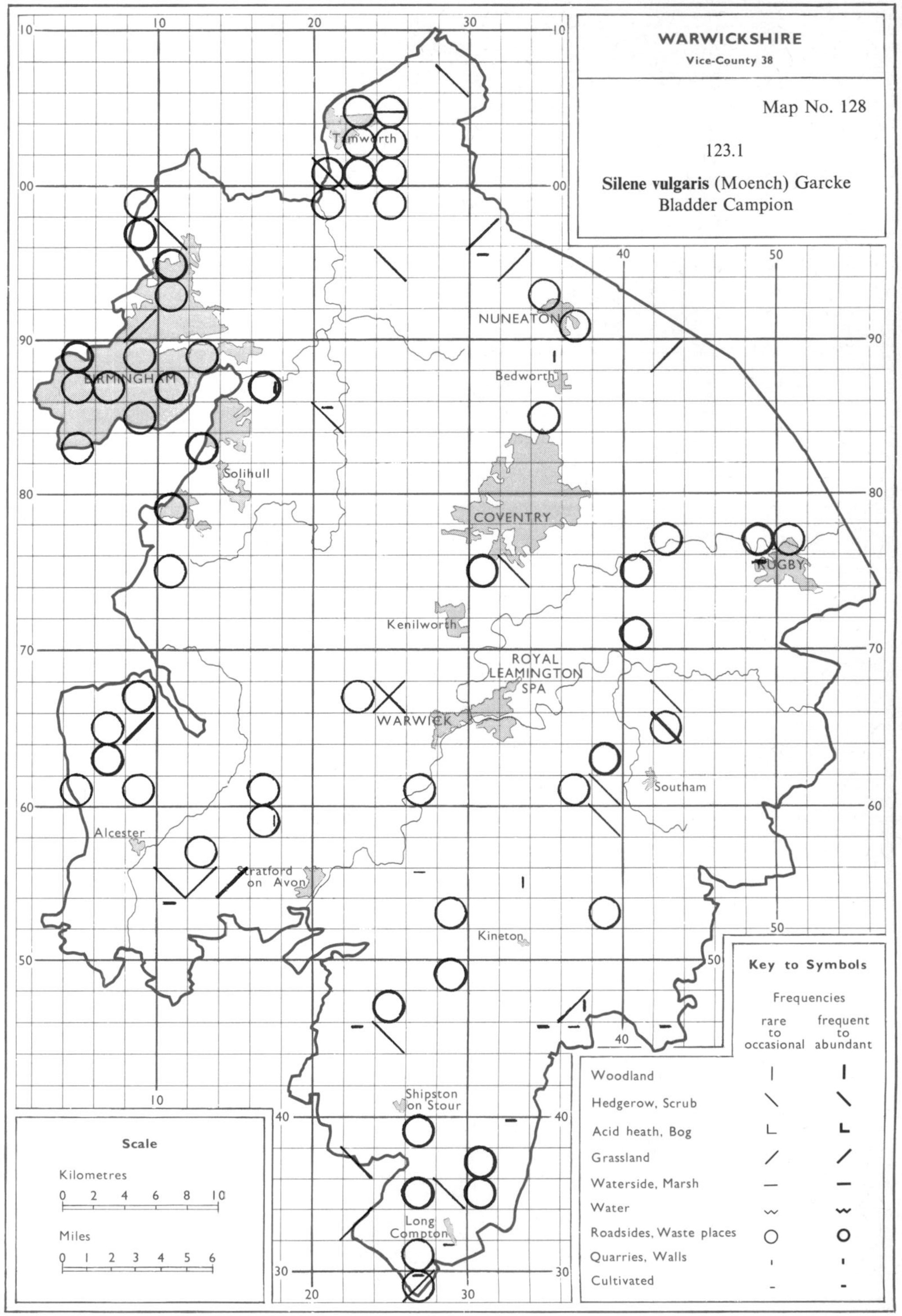
WARWICKSHIRE
Vice-County 38
Map No. 128
123.1
Silene vulgaris (Moench) Garcke
Bladder Campion
Tamworth
NUNEATON
Bedworth
BIRMINGHAM
Solihull
COVENTRY
RUGBY
Kenilworth
ROYAL LEAMINGTON SPA
WARWICK
Southam
Alcester
Stratford on Avon
Kineton
Shipston on Stour
Long Compton
Key to Symbols
Frequencies
rare to occasional
frequent to abundant
Woodland
Hedgerow, Scrub
Acid heath, Bog
Grassland
Waterside, Marsh
Water
Roadsides, Waste places
Quarries, Walls
Cultivated
Scale
Kilometres
0 2 4 6 8 10
Miles
0 1 2 3 4 5 6

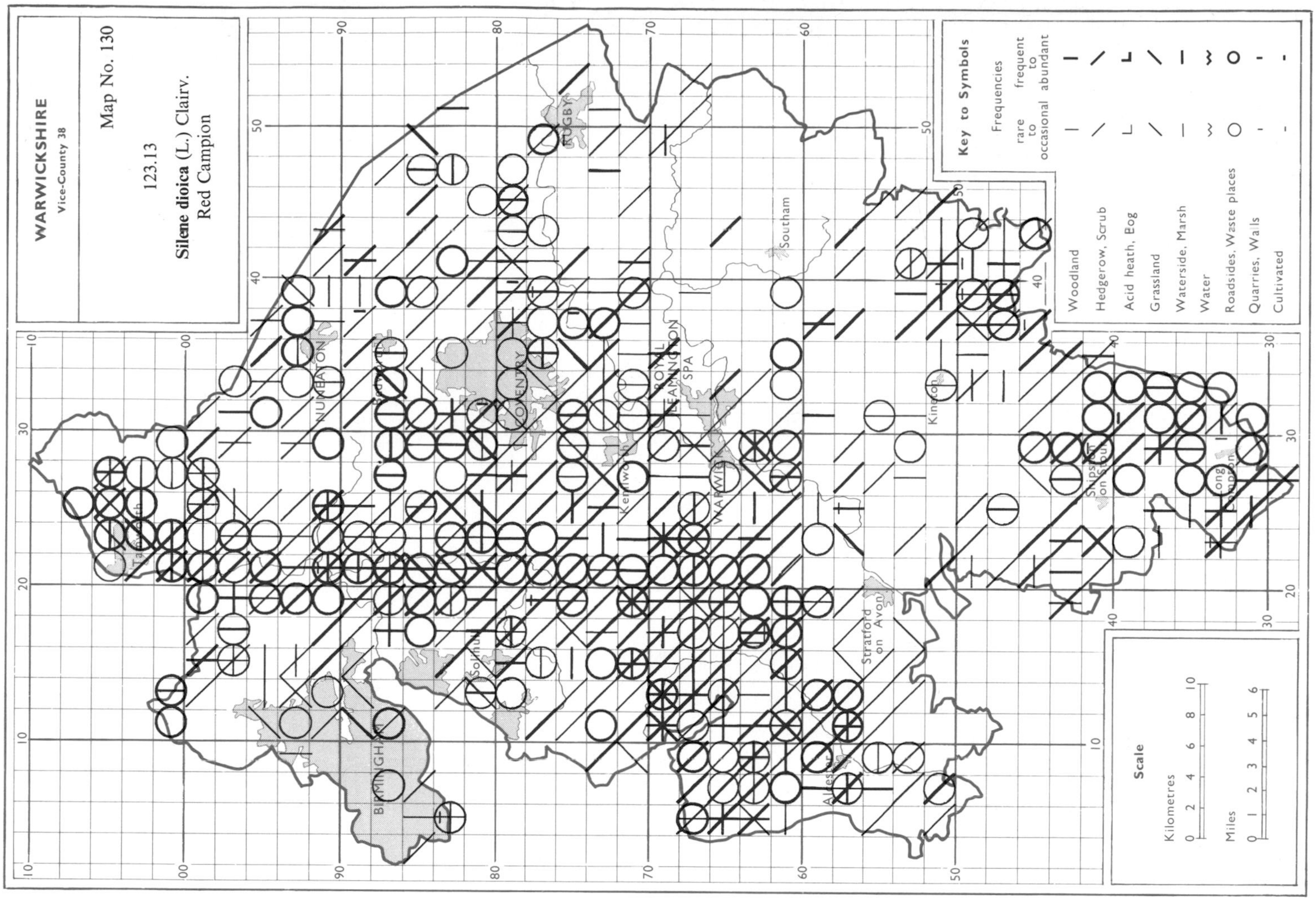
WARWICKSHIRE
Vice-County 38
Map No. 130
123.13
Silene dioica (L.) Clairv.
Red Campion
Key to Symbols
Frequencies
rare to occasional
frequent to abundant
Woodland
Hedgerow, Scrub
Acid heath, Bog
Grassland
Waterside, Marsh
Water
Roadsides, Waste places
Quarries, Walls
Cultivated
Scale
Kilometres
Miles
NUNEATON
COVENTRY
RUGBY
ROYAL LEAMINGTON SPA
Kenilworth
WARWICK
Southam
Kineton
Shipston on Stour
Long Compton
Stratford on Avon
Alcester
Solihull
BIRMINGHAM
Tamworth

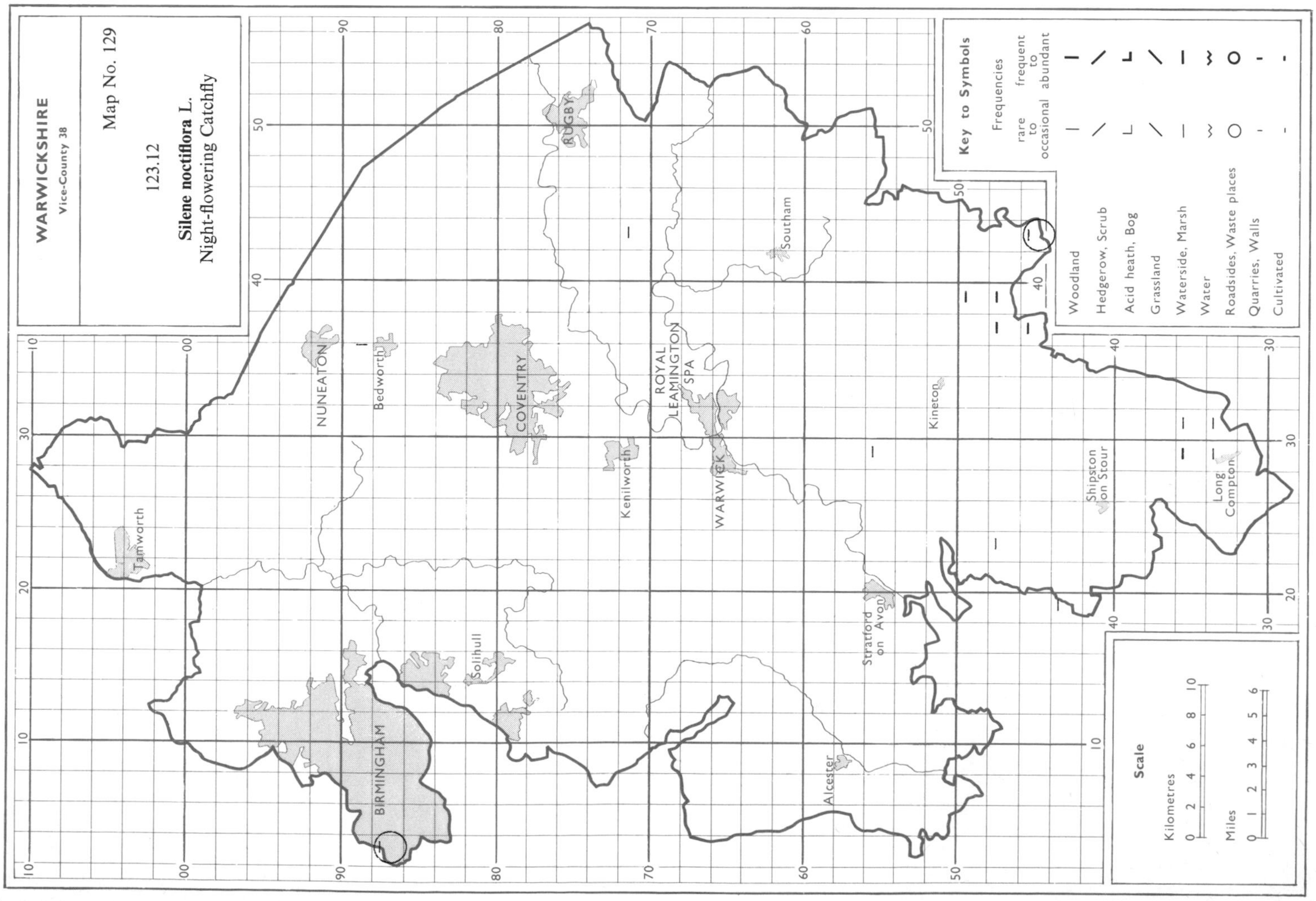
WARWICKSHIRE
Vice-County 38
Map No. 129
123.12
Silene noctiflora L.
Night-flowering Catchfly
Key to Symbols
Frequencies
rare to occasional
frequent to abundant
Woodland
Hedgerow, Scrub
Acid heath, Bog
Grassland
Waterside, Marsh
Water
Roadsides, Waste places
Quarries, Walls
Cultivated
Scale
Kilometres
Miles
NUNEATON
Bedworth
COVENTRY
RUGBY
ROYAL LEAMINGTON SPA
Kenilworth
WARWICK
Southam
Kineton
Shipston on Stour
Long Compton
Stratford on Avon
Alcester
Solihull
BIRMINGHAM
Tamworth

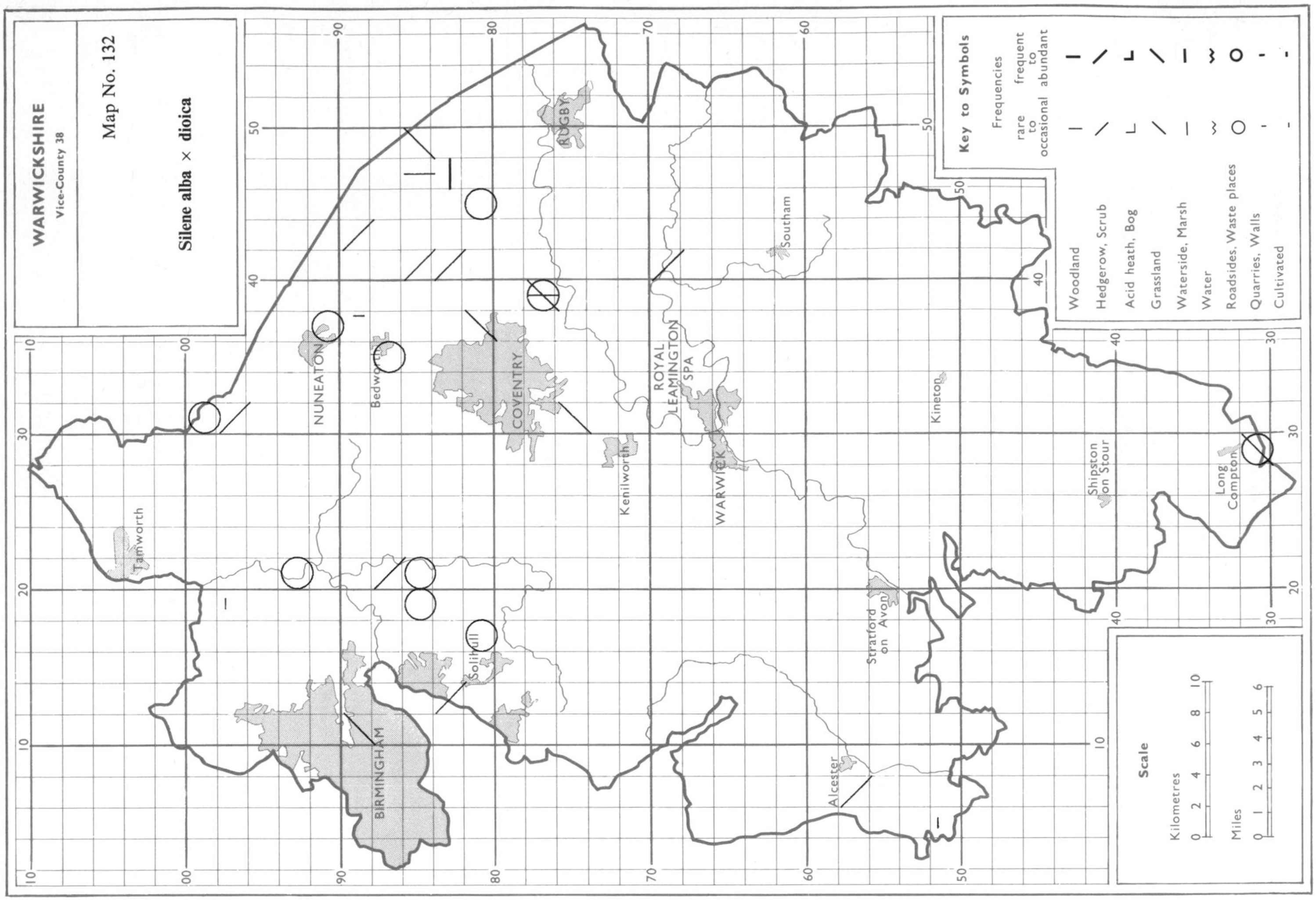
WARWICKSHIRE
Vice-County 38
Map No. 132
Silene alba × dioica
Key to Symbols
Frequencies
rare to occasional
frequent to abundant
Woodland
Hedgerow, Scrub
Acid heath, Bog
Grassland
Waterside, Marsh
Water
Roadsides, Waste places
Quarries, Walls
Cultivated
Scale
Kilometres 0 2 4 6 8 10
Miles 0 1 2 3 4 5 6
NUNEATON
Bedworth
COVENTRY
RUGBY
ROYAL LEAMINGTON SPA
WARWICK
Kenilworth
Southam
Kineton
Shipston on Stour
Long Compton
Stratford on Avon
Alcester
Solihull
BIRMINGHAM
Tamworth

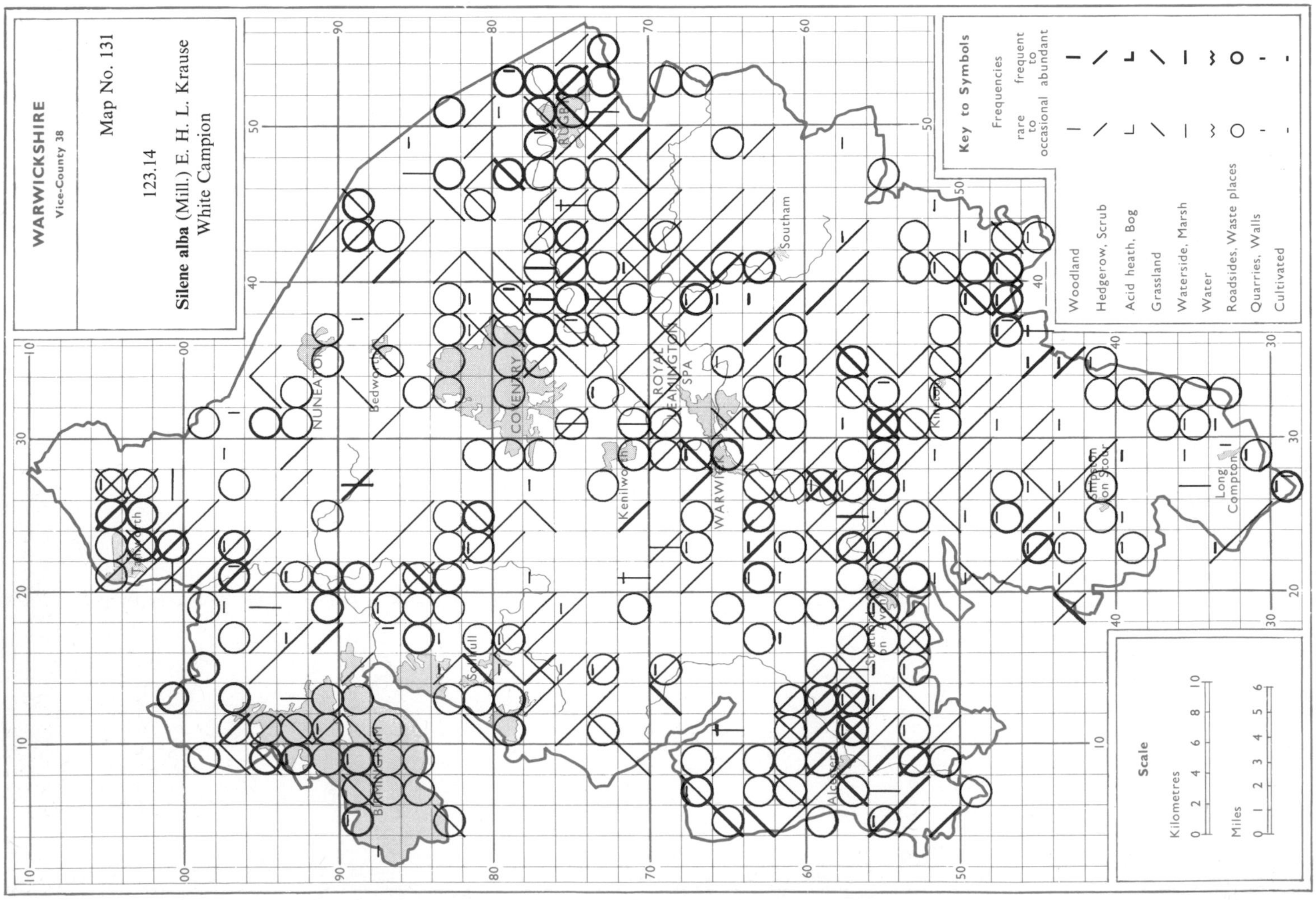
WARWICKSHIRE
Vice-County 38
Map No. 131
123.14
Silene alba (Mill.) E. H. L. Krause
White Campion
Key to Symbols
Frequencies
rare to occasional
frequent to abundant
Woodland
Hedgerow, Scrub
Acid heath, Bog
Grassland
Waterside, Marsh
Water
Roadsides, Waste places
Quarries, Walls
Cultivated
Scale
Kilometres 0 2 4 6 8 10
Miles 0 1 2 3 4 5 6
NUNEATON
Bedworth
COVENTRY
RUGBY
ROYAL LEAMINGTON SPA
WARWICK
Kenilworth
Southam
Shipston on Stour
Long Compton
Solihull
BIRMINGHAM

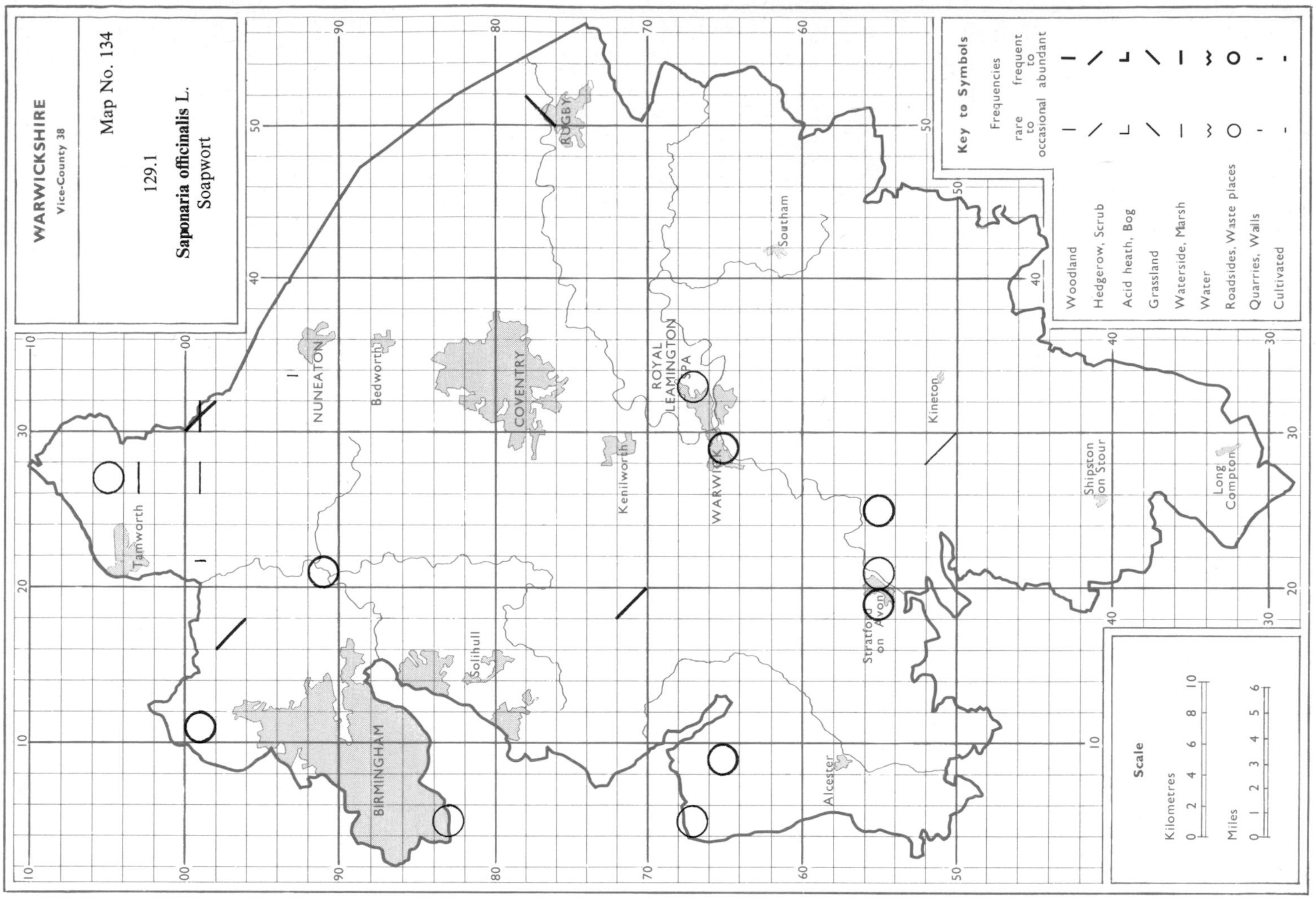
WARWICKSHIRE
Vice-County 38
Map No. 134
129.1
Saponaria officinalis L.
Soapwort
Key to Symbols
Frequencies
rare to occasional
frequent to abundant
Woodland
Hedgerow, Scrub
Acid heath, Bog
Grassland
Waterside, Marsh
Water
Roadsides, Waste places
Quarries, Walls
Cultivated
Scale
Kilometres
0 2 4 6 8 10
Miles
0 1 2 3 4 5 6
TAMWORTH
NUNEATON
Bedworth
COVENTRY
RUGBY
BIRMINGHAM
Solihull
Kenilworth
ROYAL LEAMINGTON SPA
WARWICK
Southam
Kineton
Stratford on Avon
Alcester
Shipston on Stour
Long Compton

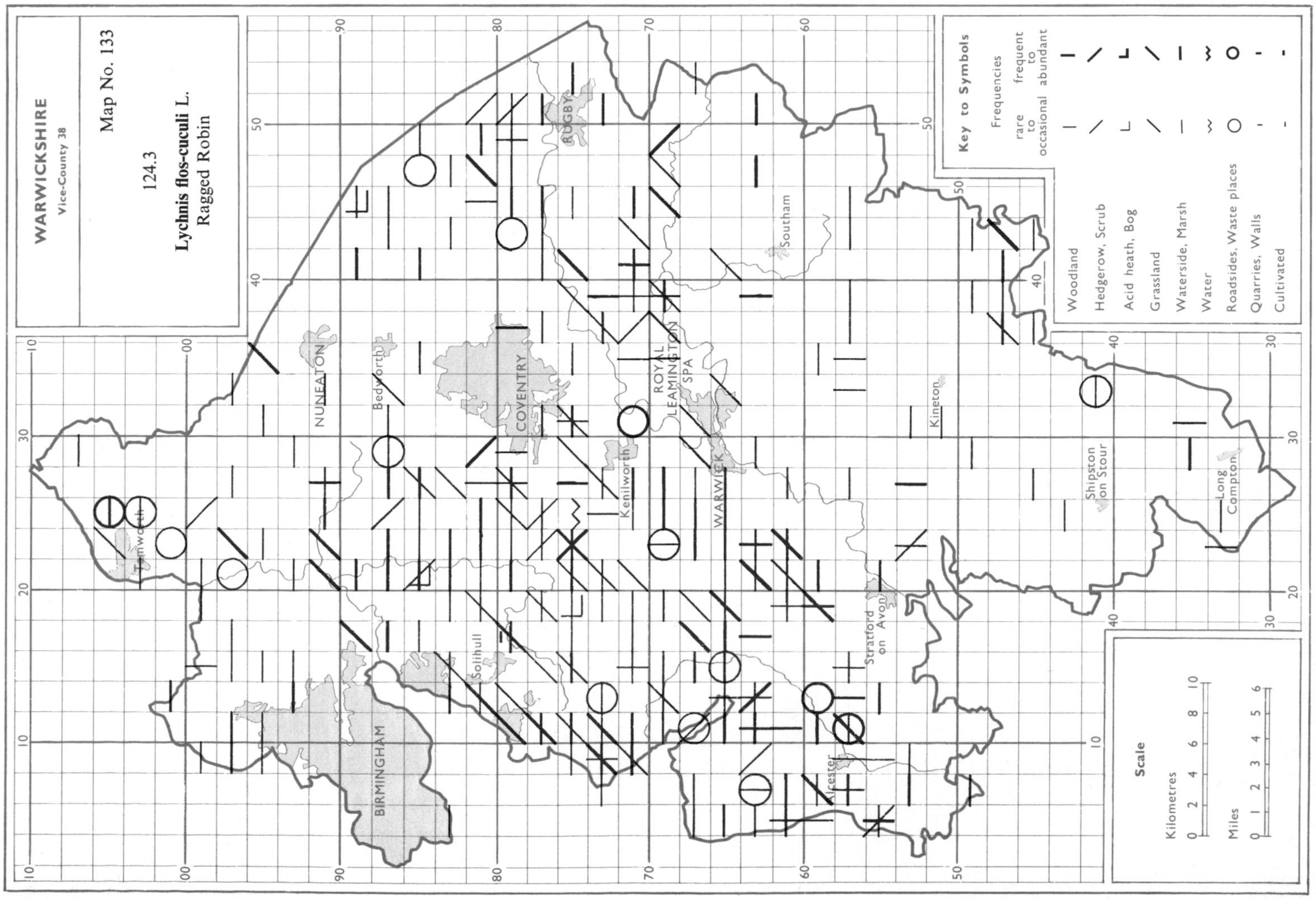
WARWICKSHIRE
Vice-County 38
Map No. 133
124.3
Lychnis flos-cuculi L.
Ragged Robin
Key to Symbols
Frequencies
rare to occasional
frequent to abundant
Woodland
Hedgerow, Scrub
Acid heath, Bog
Grassland
Waterside, Marsh
Water
Roadsides, Waste places
Quarries, Walls
Cultivated
Scale
Kilometres
0 2 4 6 8 10
Miles
0 1 2 3 4 5 6
Tamworth
NUNEATON
Bedworth
COVENTRY
RUGBY
BIRMINGHAM
Solihull
Kenilworth
ROYAL LEAMINGTON SPA
WARWICK
Southam
Kineton
Stratford on Avon
Alcester
Shipston on Stour
Long Compton

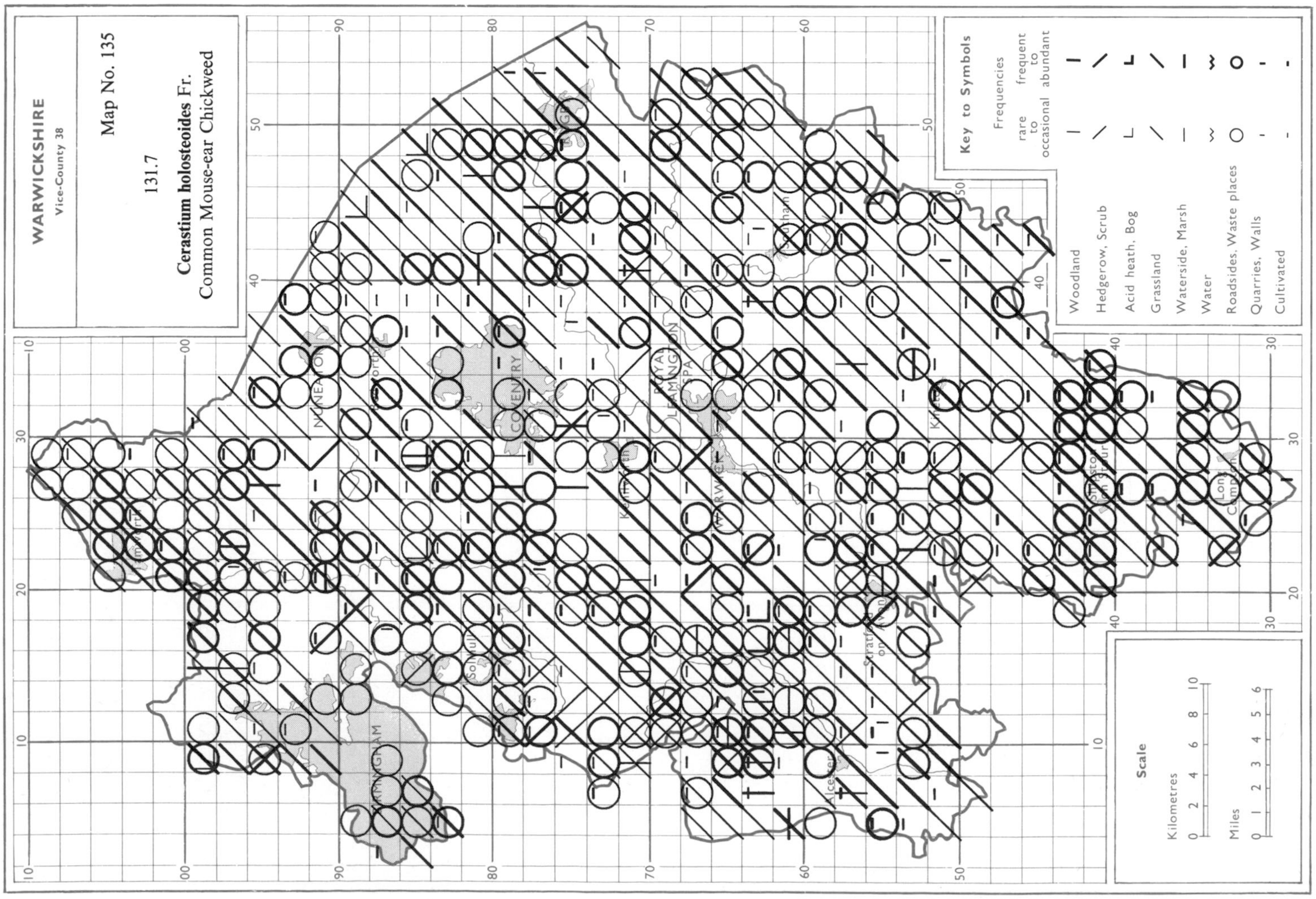
WARWICKSHIRE
Vice-County 38
Map No. 135
131.7
Cerastium holosteoides Fr.
Common Mouse-ear Chickweed
Key to Symbols
Frequencies
rare to occasional
frequent to abundant
Woodland
Hedgerow, Scrub
Acid heath, Bog
Grassland
Waterside, Marsh
Water
Roadsides, Waste places
Quarries, Walls
Cultivated
Scale
Kilometres
Miles
BIRMINGHAM
NUNEATON
COVENTRY
ROYAL LEAMINGTON SPA
Solihull
Southam
Alcester
Long Compton

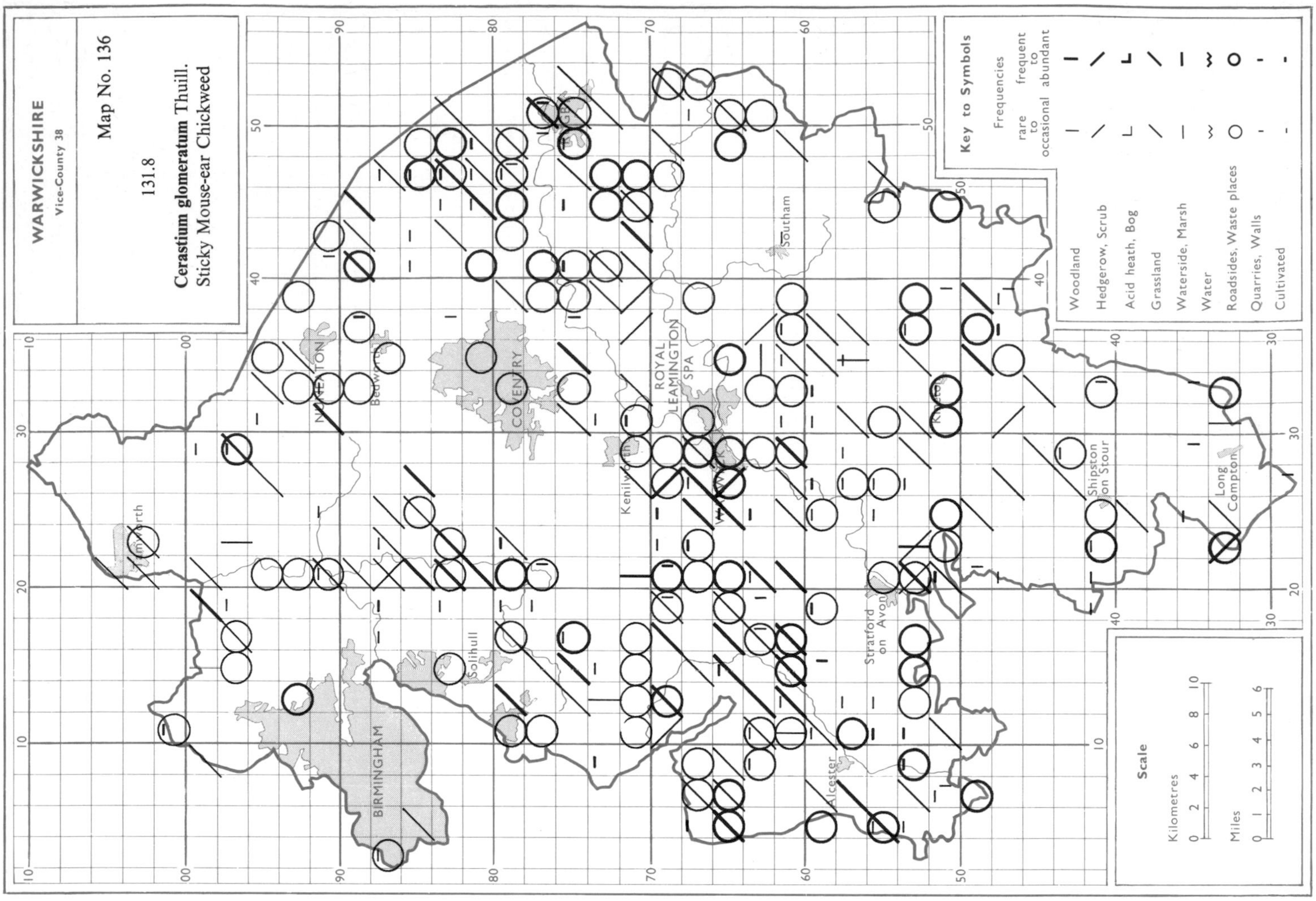
WARWICKSHIRE
Vice-County 38
Map No. 136
131.8
Cerastium glomeratum Thuill.
Sticky Mouse-ear Chickweed
Key to Symbols
Frequencies
rare to occasional
frequent to abundant
Woodland
Hedgerow, Scrub
Acid heath, Bog
Grassland
Waterside, Marsh
Water
Roadsides, Waste places
Quarries, Walls
Cultivated
Scale
Kilometres
Miles
BIRMINGHAM
NUNEATON
Bedworth
COVENTRY
ROYAL LEAMINGTON SPA
Kenilworth
Solihull
Tamworth
Southam
Stratford on Avon
Alcester
Shipston on Stour
Long Compton

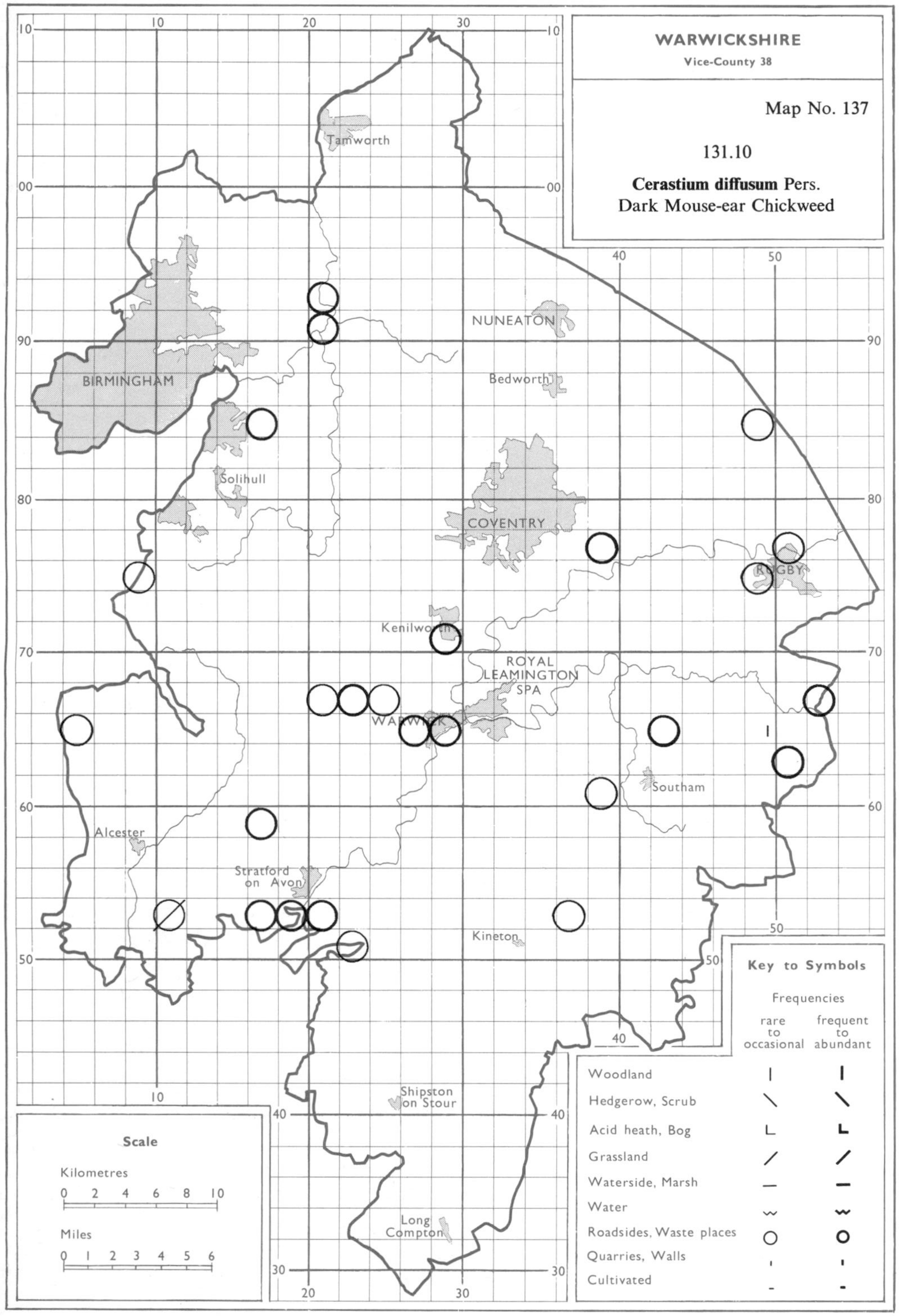
WARWICKSHIRE
Vice-County 38
Map No. 137
131.10
Cerastium diffusum Pers.
Dark Mouse-ear Chickweed
Tamworth
NUNEATON
Bedworth
BIRMINGHAM
Solihull
COVENTRY
RUGBY
Kenilworth
ROYAL LEAMINGTON SPA
WARWICK
Southam
Alcester
Stratford on Avon
Kineton
Shipston on Stour
Long Compton
Scale
Kilometres
Miles
Key to Symbols
Frequencies
rare to occasional
frequent to abundant
Woodland
Hedgerow, Scrub
Acid heath, Bog
Grassland
Waterside, Marsh
Water
Roadsides, Waste places
Quarries, Walls
Cultivated

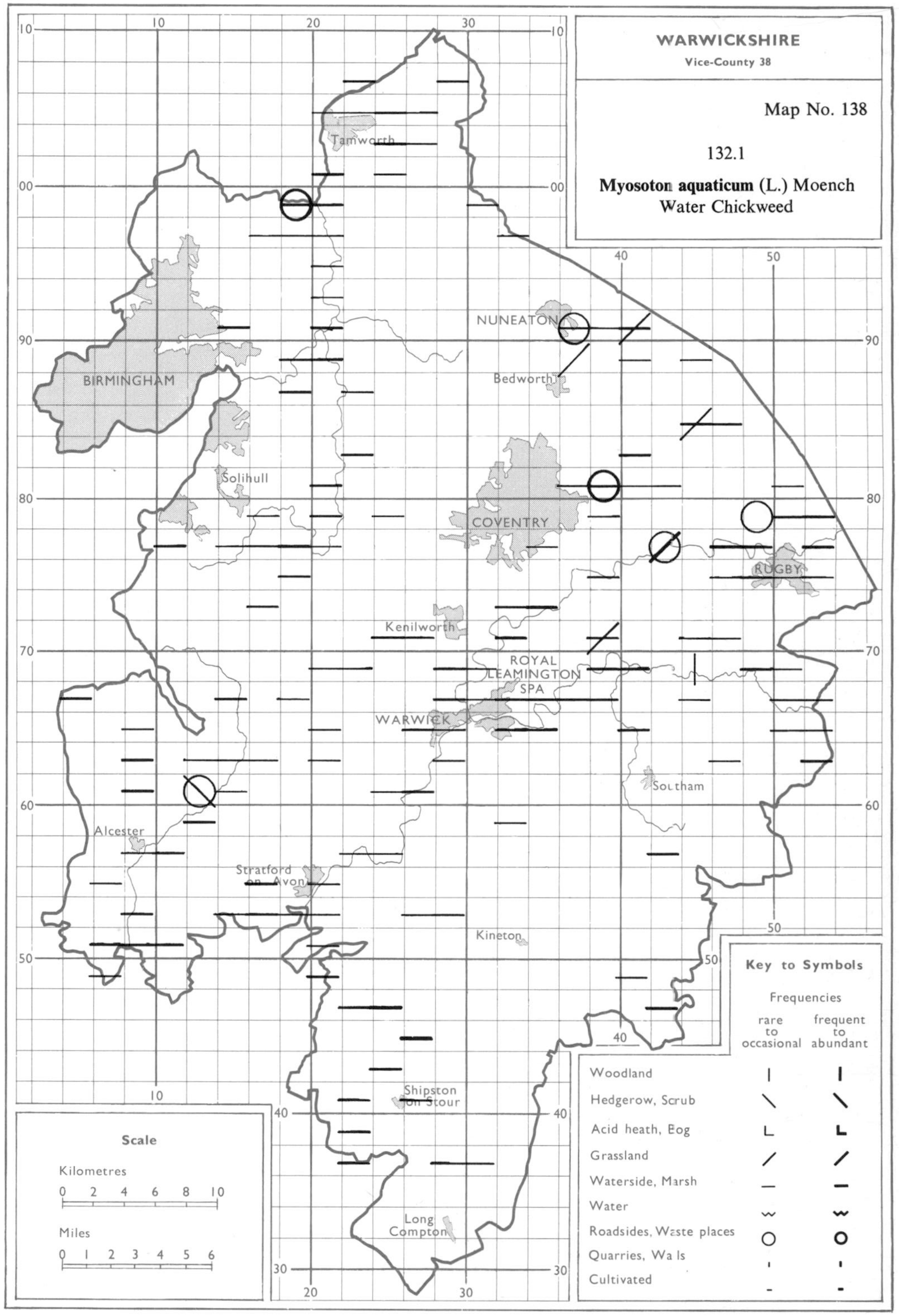
WARWICKSHIRE
Vice-County 38
Map No. 138
132.1
Myosoton aquaticum (L.) Moench
Water Chickweed
Tamworth
NUNEATON
Bedworth
BIRMINGHAM
Solihull
COVENTRY
RUGBY
Kenilworth
ROYAL LEAMINGTON SPA
WARWICK
Southam
Alcester
Stratford on Avon
Kineton
Shipston on Stour
Long Compton
Scale
Kilometres
Miles
Key to Symbols
Frequencies
rare to occasional
frequent to abundant
Woodland
Hedgerow, Scrub
Acid heath, Bog
Grassland
Waterside, Marsh
Water
Roadsides, Waste places
Quarries, Walls
Cultivated

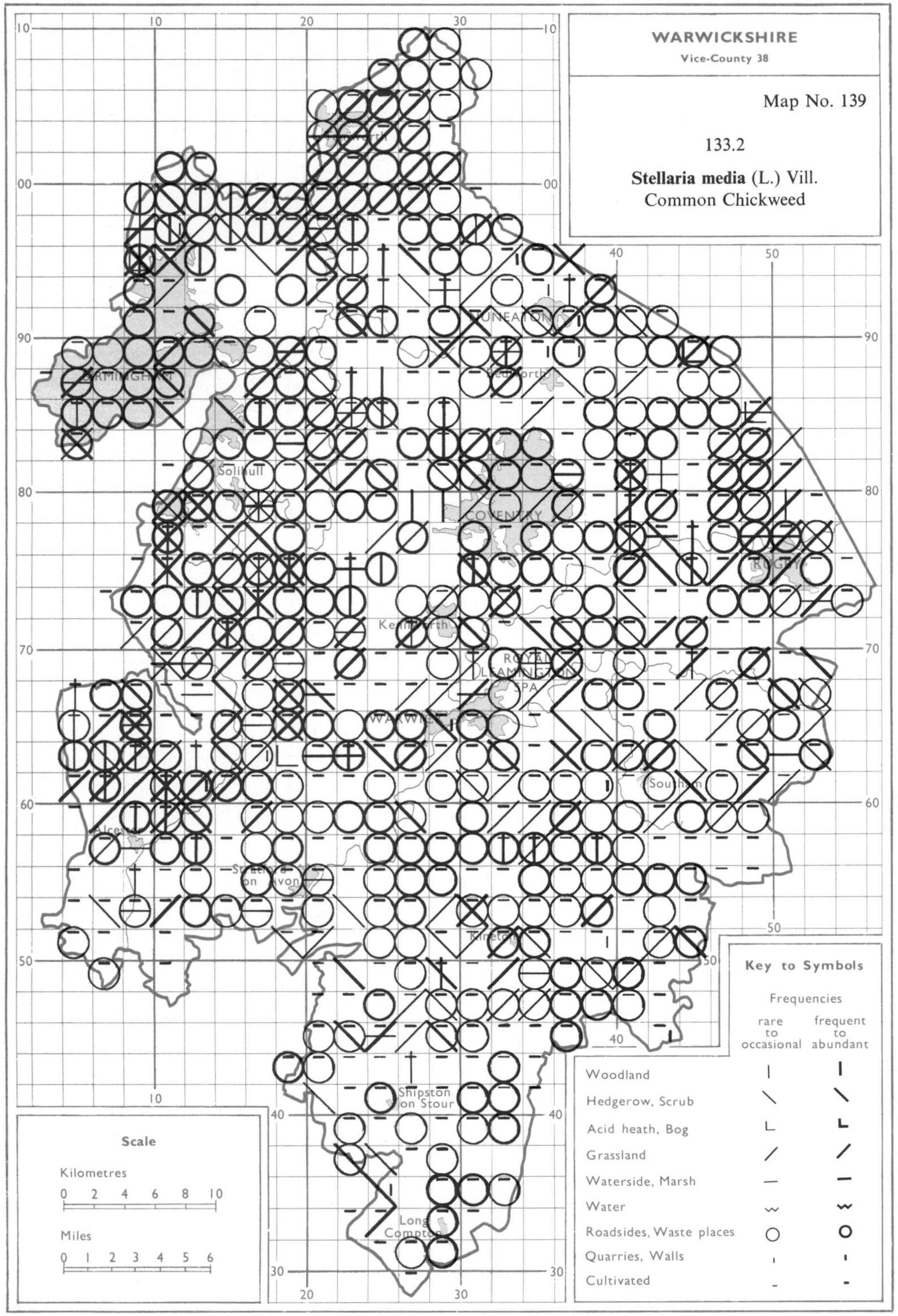
WARWICKSHIRE
Vice-County 38
Map No. 139
133.2
Stellaria media (L.) Vill.
Common Chickweed
Solihull
COVENTRY
NUNEATON
Bedworth
Kenilworth
ROYAL LEAMINGTON SPA
WARWICK
Southam
Stratford on Avon
Kineton
Shipston on Stour
Long Compton
Key to Symbols
Frequencies
rare to occasional
frequent to abundant
Woodland
Hedgerow, Scrub
Acid heath, Bog
Grassland
Waterside, Marsh
Water
Roadsides, Waste places
Quarries, Walls
Cultivated
Scale
Kilometres
0 2 4 6 8 10
Miles
0 1 2 3 4 5 6

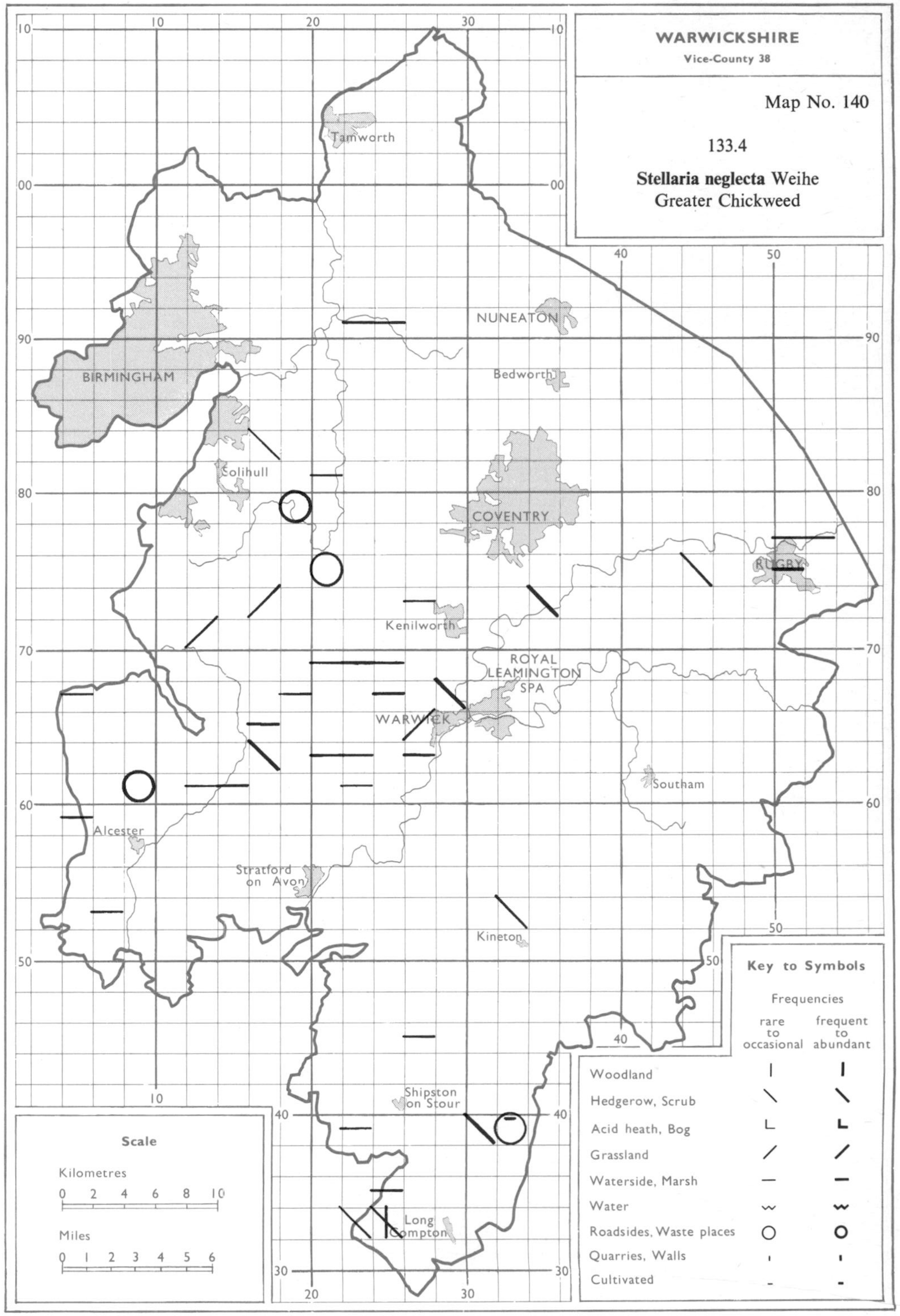
WARWICKSHIRE
Vice-County 38
Map No. 140
133.4
Stellaria neglecta Weihe
Greater Chickweed
Tamworth
BIRMINGHAM
Solihull
NUNEATON
Bedworth
COVENTRY
RUGBY
Kenilworth
ROYAL LEAMINGTON SPA
WARWICK
Southam
Alcester
Stratford on Avon
Kineton
Shipston on Stour
Long Compton
Key to Symbols
Frequencies
rare to occasional
frequent to abundant
Woodland
Hedgerow, Scrub
Acid heath, Bog
Grassland
Waterside, Marsh
Water
Roadsides, Waste places
Quarries, Walls
Cultivated
Scale
Kilometres
0 2 4 6 8 10
Miles
0 1 2 3 4 5 6

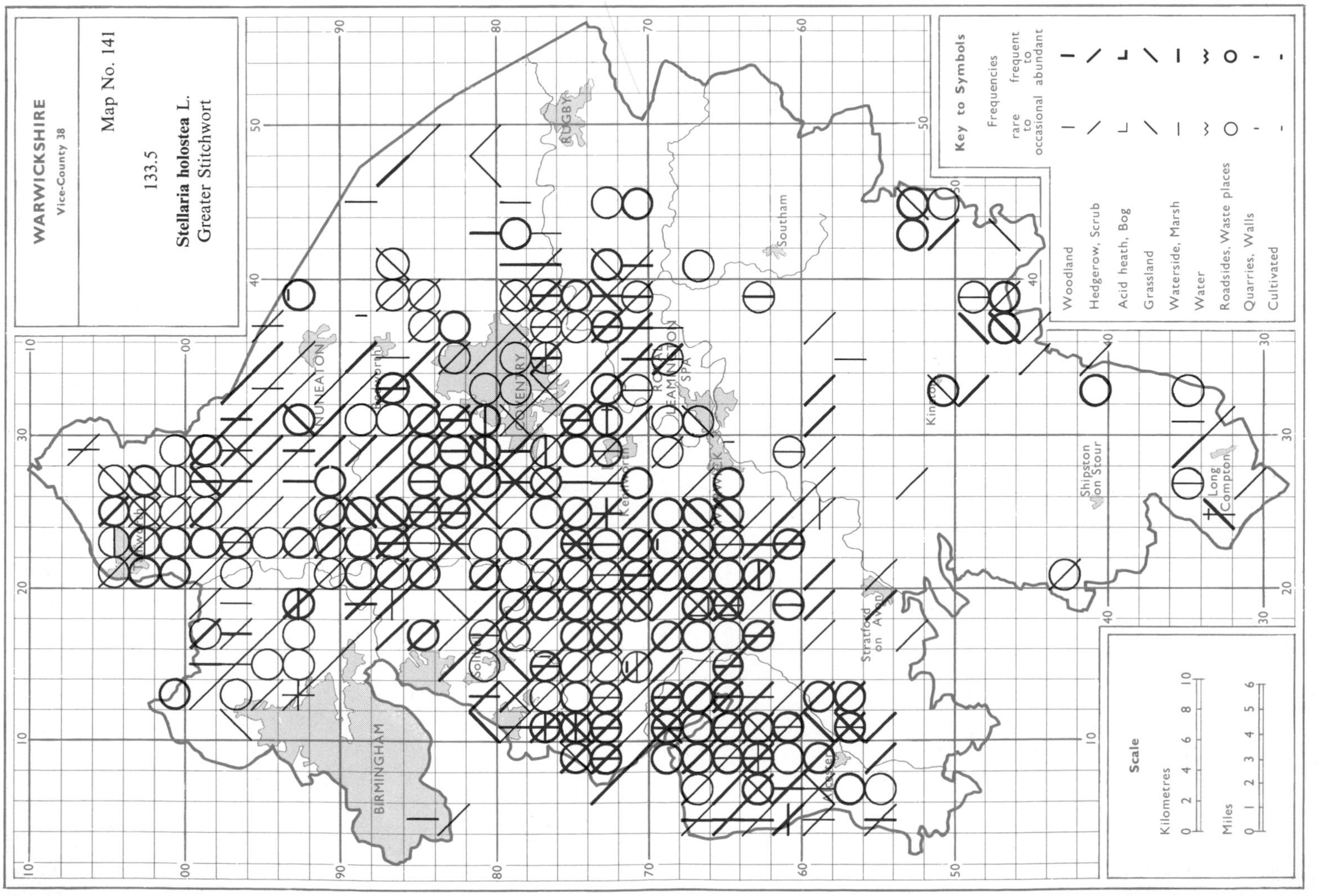
WARWICKSHIRE
Vice-County 38
Map No. 141
133.5
Stellaria holostea L.
Greater Stitchwort
Key to Symbols
Frequencies
rare to occasional
frequent to abundant
Woodland
Hedgerow, Scrub
Acid heath, Bog
Grassland
Waterside, Marsh
Water
Roadsides, Waste places
Quarries, Walls
Cultivated
Scale
Kilometres
0 2 4 6 8 10
Miles
0 1 2 3 4 5 6
NUNEATON
COVENTRY
RUGBY
BIRMINGHAM
Southam
ROYAL LEAMINGTON SPA
WARWICK
Kenilworth
Kineton
Shipston on Stour
Long Compton
Stratford on Avon
Alcester

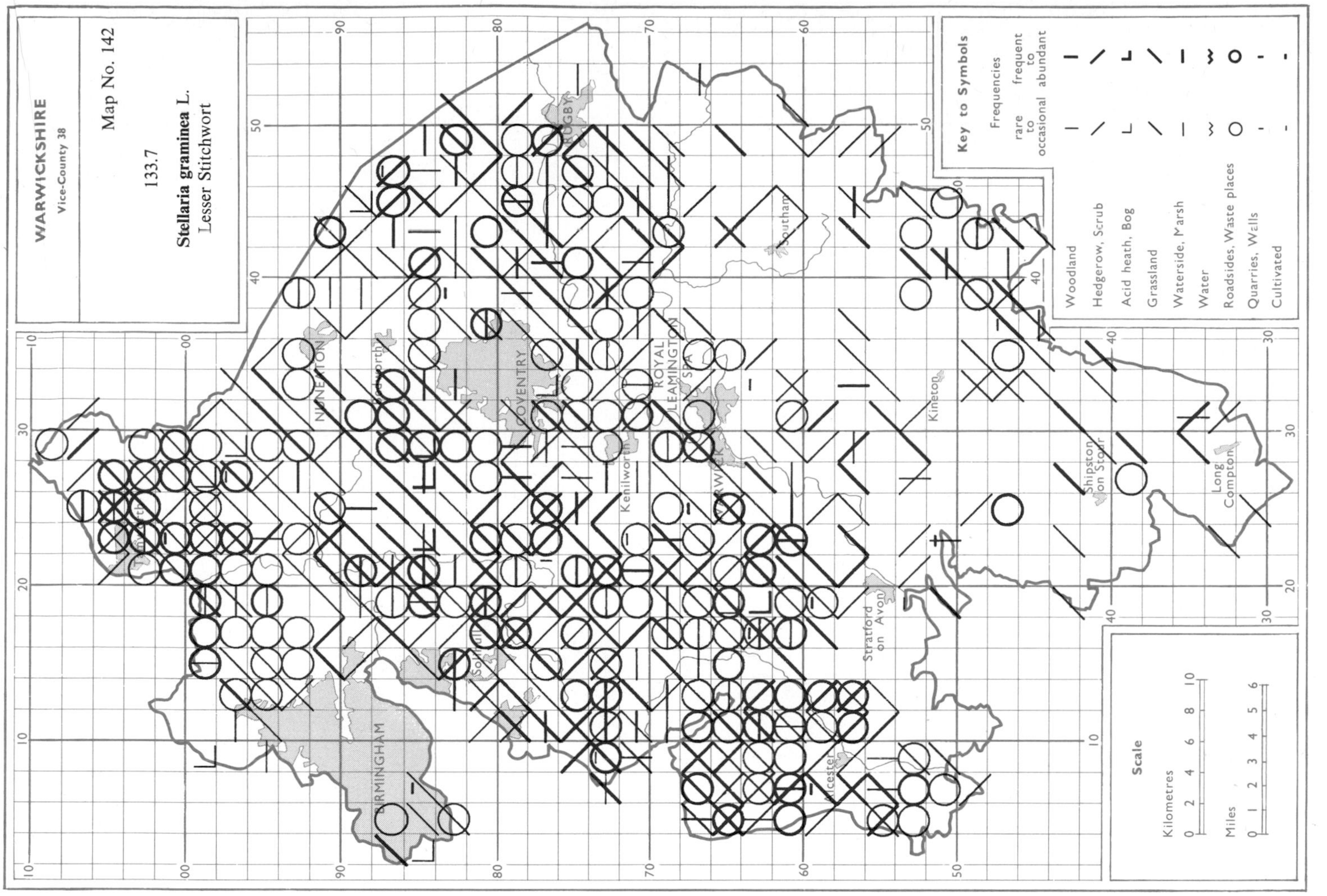
WARWICKSHIRE
Vice-County 38
Map No. 142
133.7
Stellaria graminea L.
Lesser Stitchwort
Key to Symbols
Frequencies
rare to occasional
frequent to abundant
Woodland
Hedgerow, Scrub
Acid heath, Bog
Grassland
Waterside, Marsh
Water
Roadsides, Waste places
Quarries, Walls
Cultivated
Scale
Kilometres
0 2 4 6 8 10
Miles
0 1 2 3 4 5 6
NUNEATON
COVENTRY
RUGBY
BIRMINGHAM
Southam
ROYAL LEAMINGTON SPA
WARWICK
Kenilworth
Kineton
Shipston on Stour
Long Compton
Stratford on Avon
Alcester

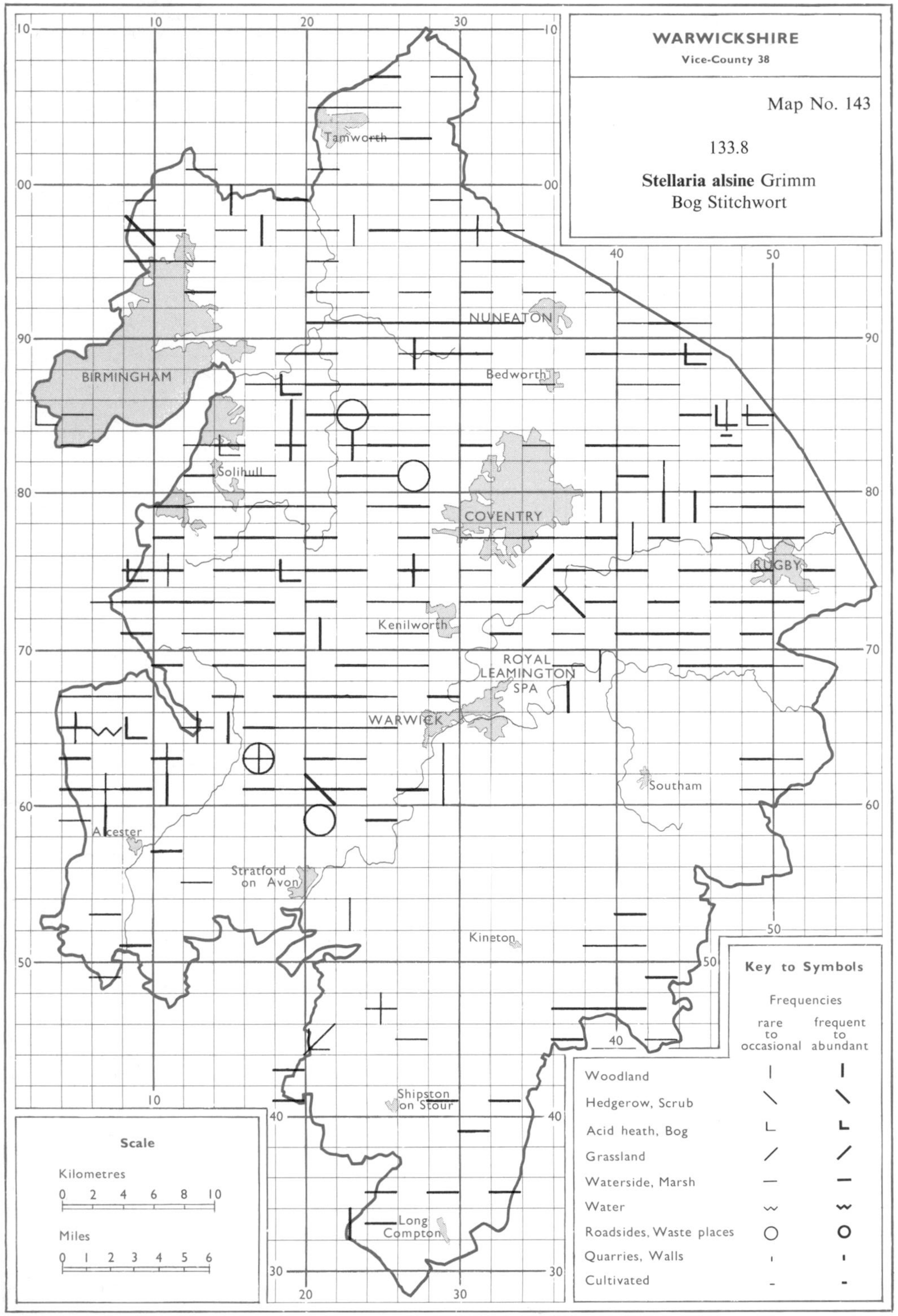
WARWICKSHIRE
Vice-County 38
Map No. 143
133.8
Stellaria alsine Grimm
Bog Stitchwort
Tamworth
NUNEATON
Bedworth
BIRMINGHAM
Solihull
COVENTRY
RUGBY
Kenilworth
ROYAL LEAMINGTON SPA
WARWICK
Southam
Alcester
Stratford on Avon
Kineton
Shipston on Stour
Long Compton
Key to Symbols
Frequencies
rare to occasional
frequent to abundant
Woodland
Hedgerow, Scrub
Acid heath, Bog
Grassland
Waterside, Marsh
Water
Roadsides, Waste places
Quarries, Walls
Cultivated
Scale
Kilometres
0 2 4 6 8 10
Miles
0 1 2 3 4 5 6

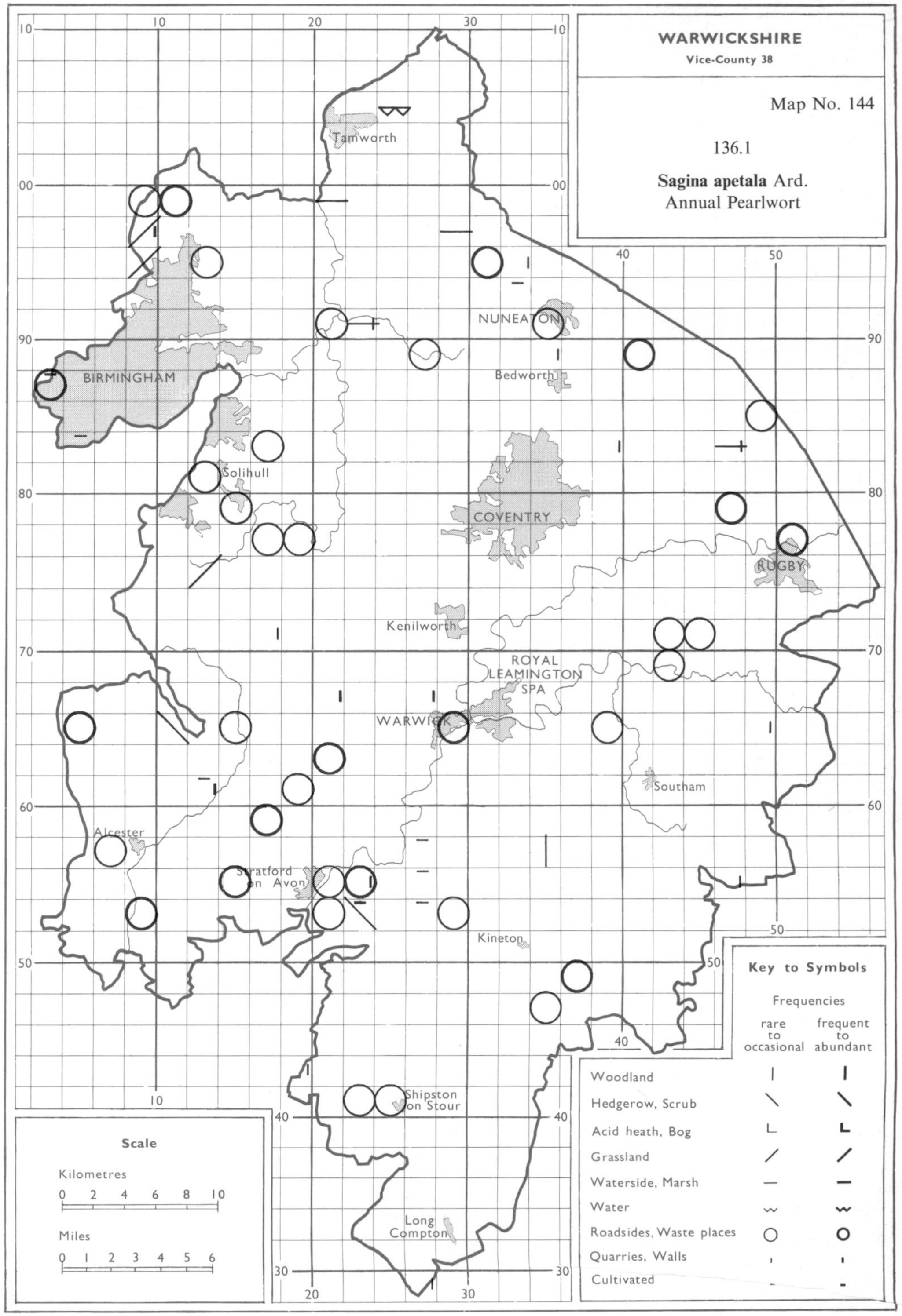
WARWICKSHIRE
Vice-County 38
Map No. 144
136.1
Sagina apetala Ard.
Annual Pearlwort
Tamworth
NUNEATON
Bedworth
BIRMINGHAM
Solihull
COVENTRY
RUGBY
Kenilworth
ROYAL LEAMINGTON SPA
WARWICK
Southam
Alcester
Stratford on Avon
Kineton
Shipston on Stour
Long Compton
Key to Symbols
Frequencies
rare to occasional
frequent to abundant
Woodland
Hedgerow, Scrub
Acid heath, Bog
Grassland
Waterside, Marsh
Water
Roadsides, Waste places
Quarries, Walls
Cultivated
Scale
Kilometres
0 2 4 6 8 10
Miles
0 1 2 3 4 5 6

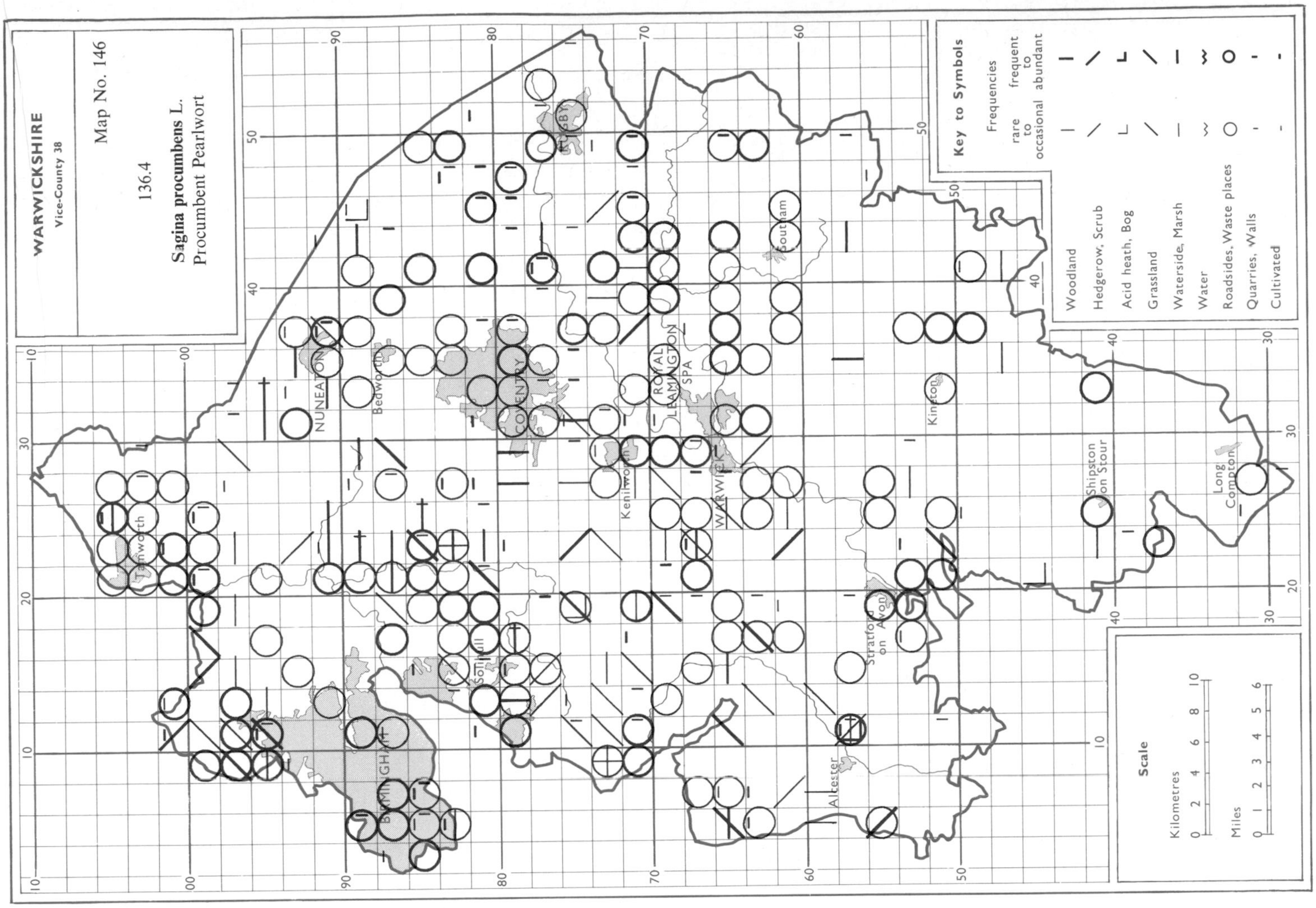
WARWICKSHIRE
Vice-County 38
Map No. 146
136.4
Sagina procumbens L.
Procumbent Pearlwort
Key to Symbols
Frequencies
rare to occasional
frequent to abundant
Woodland
Hedgerow, Scrub
Acid heath, Bog
Grassland
Waterside, Marsh
Water
Roadsides, Waste places
Quarries, Walls
Cultivated
Scale
Kilometres
Miles
NUNEATON
Bedworth
COVENTRY
RUGBY
Tamworth
BIRMINGHAM
Solihull
Kenilworth
ROYAL LEAMINGTON SPA
WARWICK
Southam
Kineton
Shipston on Stour
Long Compton
Stratford on Avon
Alcester

WARWICKSHIRE
Vice-County 38
Map No. 145
136.2
Sagina ciliata Fr.
Ciliate Pearlwort
Key to Symbols
Frequencies
rare to occasional
frequent to abundant
Woodland
Hedgerow, Scrub
Acid heath, Bog
Grassland
Waterside, Marsh
Water
Roadsides, Waste places
Quarries, Walls
Cultivated
Scale
Kilometres
Miles
NUNEATON
Bedworth
COVENTRY
RUGBY
Tamworth
BIRMINGHAM
Solihull
Kenilworth
ROYAL LEAMINGTON SPA
WARWICK
Southam
Kineton
Shipston on Stour
Long Compton
Stratford on Avon
Alcester

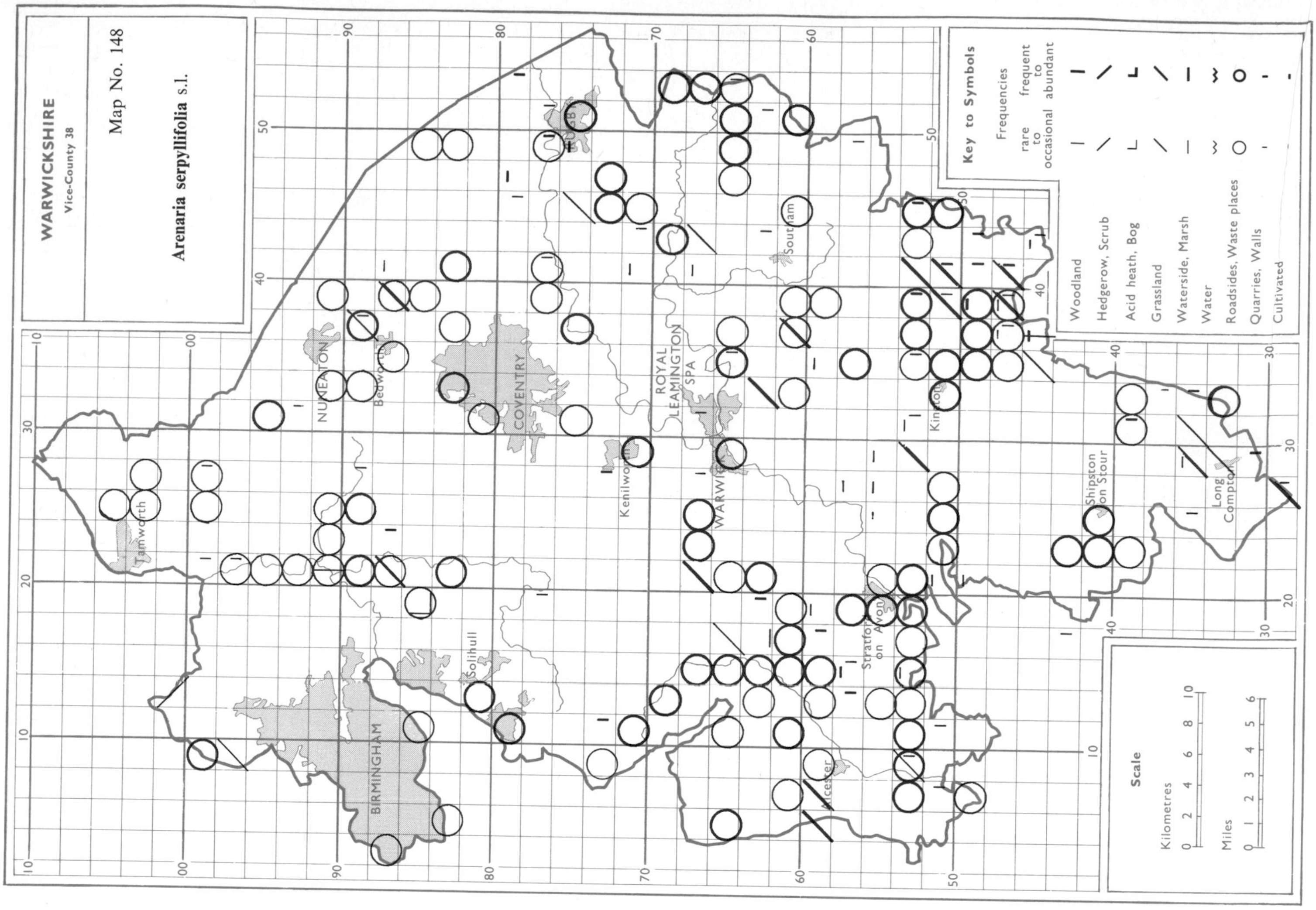
WARWICKSHIRE
Vice-County 38
Map No. 148
Arenaria serpyllifolia s.l.
Key to Symbols
Frequencies
rare to occasional
frequent to abundant
Woodland
Hedgerow, Scrub
Acid heath, Bog
Grassland
Waterside, Marsh
Water
Roadsides, Waste places
Quarries, Walls
Cultivated
Scale
Kilometres
Miles
NUNEATON
Bedworth
COVENTRY
RUGBY
ROYAL LEAMINGTON SPA
Kenilworth
WARWICK
Southam
Kineton
Shipston on Stour
Long Compton
Tamworth
Solihull
BIRMINGHAM
Stratford on Avon
Alcester

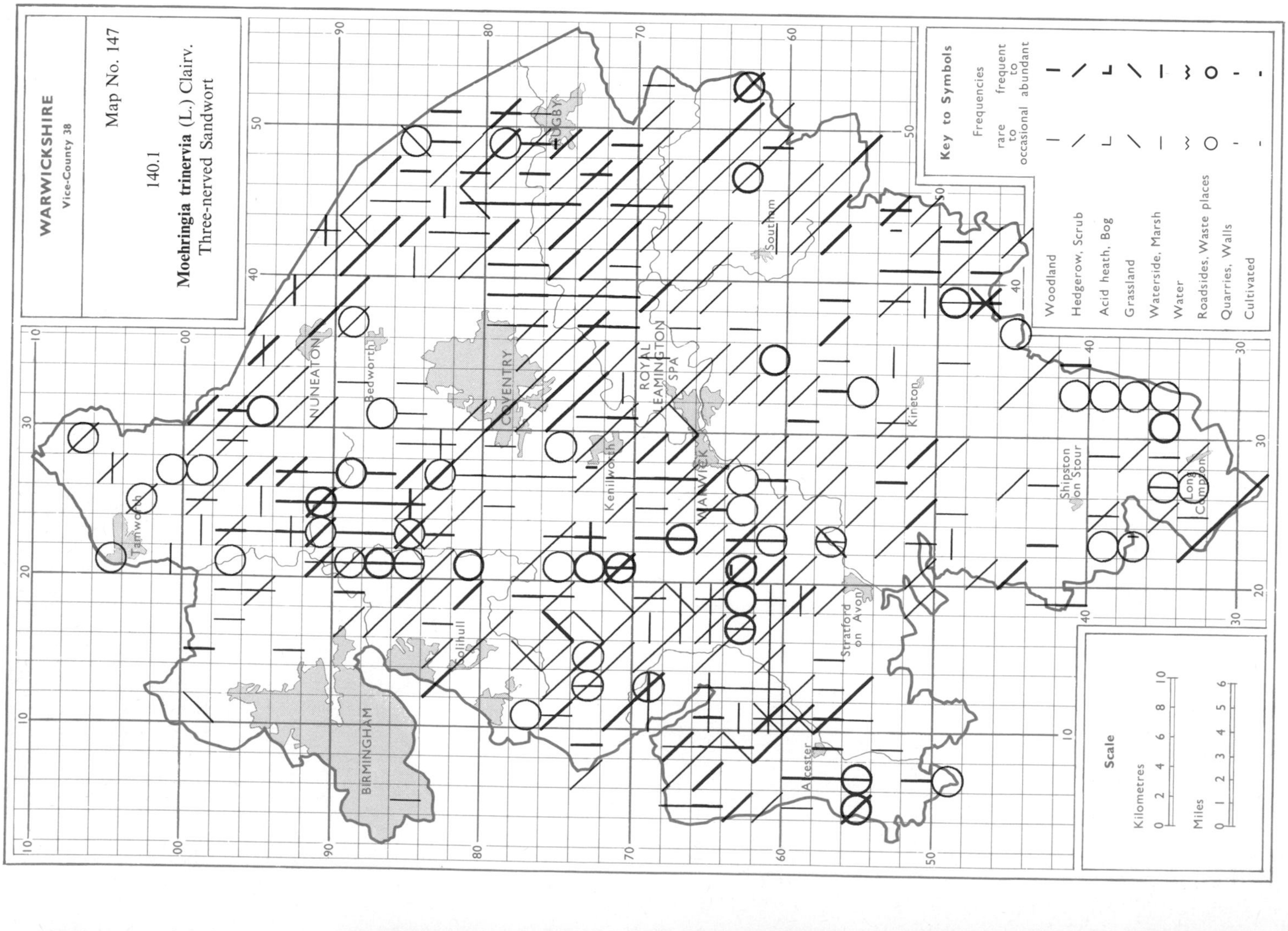
WARWICKSHIRE
Vice-County 38
Map No. 147
140.1
Moehringia trinervia (L.) Clairv.
Three-nerved Sandwort
Key to Symbols
Frequencies
rare to occasional
frequent to abundant
Woodland
Hedgerow, Scrub
Acid heath, Bog
Grassland
Waterside, Marsh
Water
Roadsides, Waste places
Quarries, Walls
Cultivated
Scale
Kilometres
Miles
NUNEATON
Bedworth
COVENTRY
RUGBY
ROYAL LEAMINGTON SPA
Kenilworth
WARWICK
Southam
Kineton
Shipston on Stour
Long Compton
Tamworth
Solihull
BIRMINGHAM
Stratford on Avon
Alcester

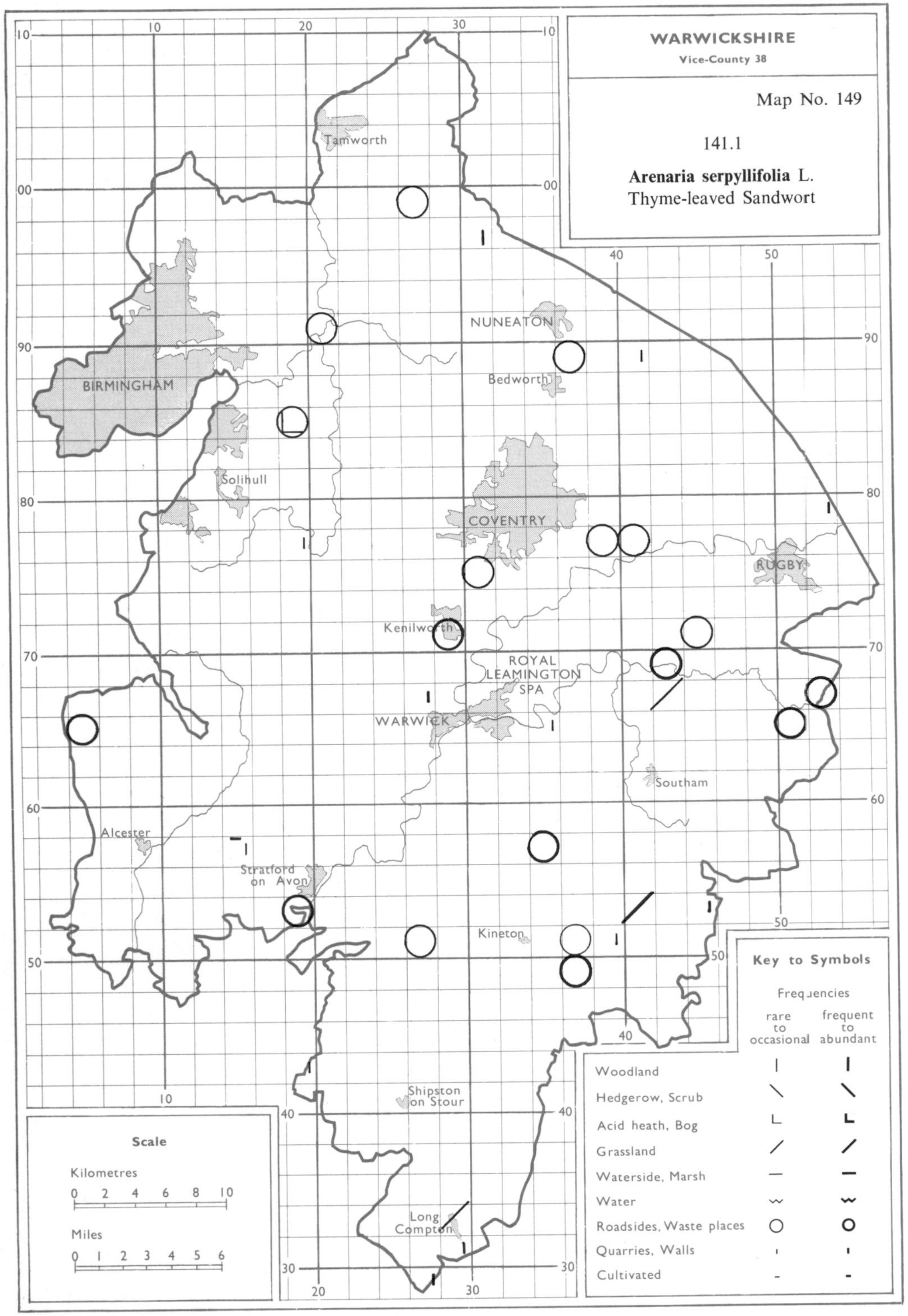
WARWICKSHIRE
Vice-County 38
Map No. 149
141.1
Arenaria serpyllifolia L.
Thyme-leaved Sandwort
Tamworth
NUNEATON
Bedworth
BIRMINGHAM
Solihull
COVENTRY
RUGBY
Kenilworth
ROYAL LEAMINGTON SPA
WARWICK
Southam
Alcester
Stratford on Avon
Kineton
Shipston on Stour
Long Compton
Key to Symbols
Frequencies
rare to occasional
frequent to abundant
Woodland
Hedgerow, Scrub
Acid heath, Bog
Grassland
Waterside, Marsh
Water
Roadsides, Waste places
Quarries, Walls
Cultivated
Scale
Kilometres
0 2 4 6 8 10
Miles
0 1 2 3 4 5 6

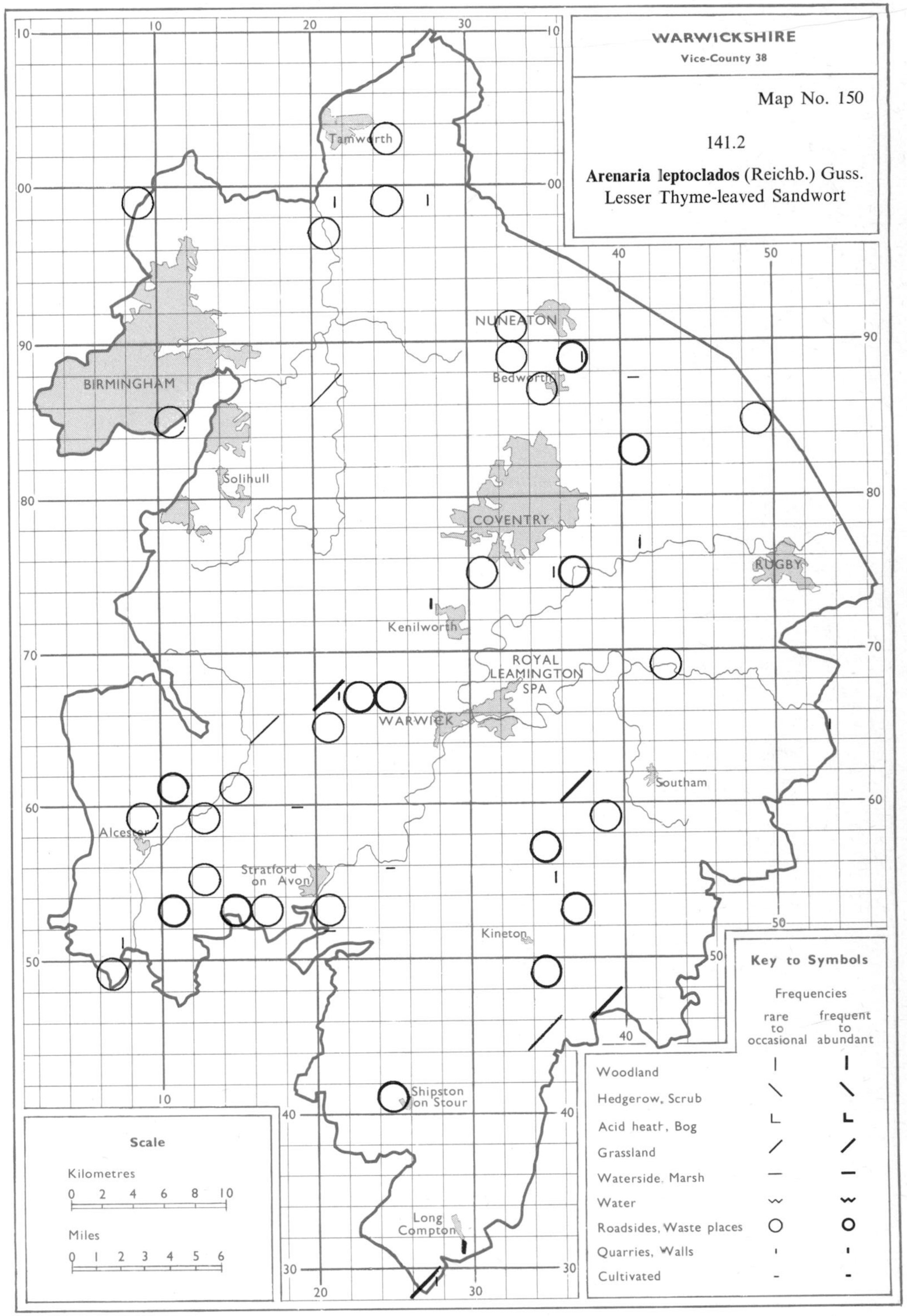
WARWICKSHIRE
Vice-County 38
Map No. 150
141.2
Arenaria leptoclados (Reichb.) Guss.
Lesser Thyme-leaved Sandwort
Tamworth
NUNEATON
Bedworth
BIRMINGHAM
Solihull
COVENTRY
RUGBY
Kenilworth
ROYAL LEAMINGTON SPA
WARWICK
Southam
Alcester
Stratford on Avon
Kineton
Shipston on Stour
Long Compton
Key to Symbols
Frequencies
rare to occasional
frequent to abundant
Woodland
Hedgerow, Scrub
Acid heath, Bog
Grassland
Waterside, Marsh
Water
Roadsides, Waste places
Quarries, Walls
Cultivated
Scale
Kilometres
0 2 4 6 8 10
Miles
0 1 2 3 4 5 6

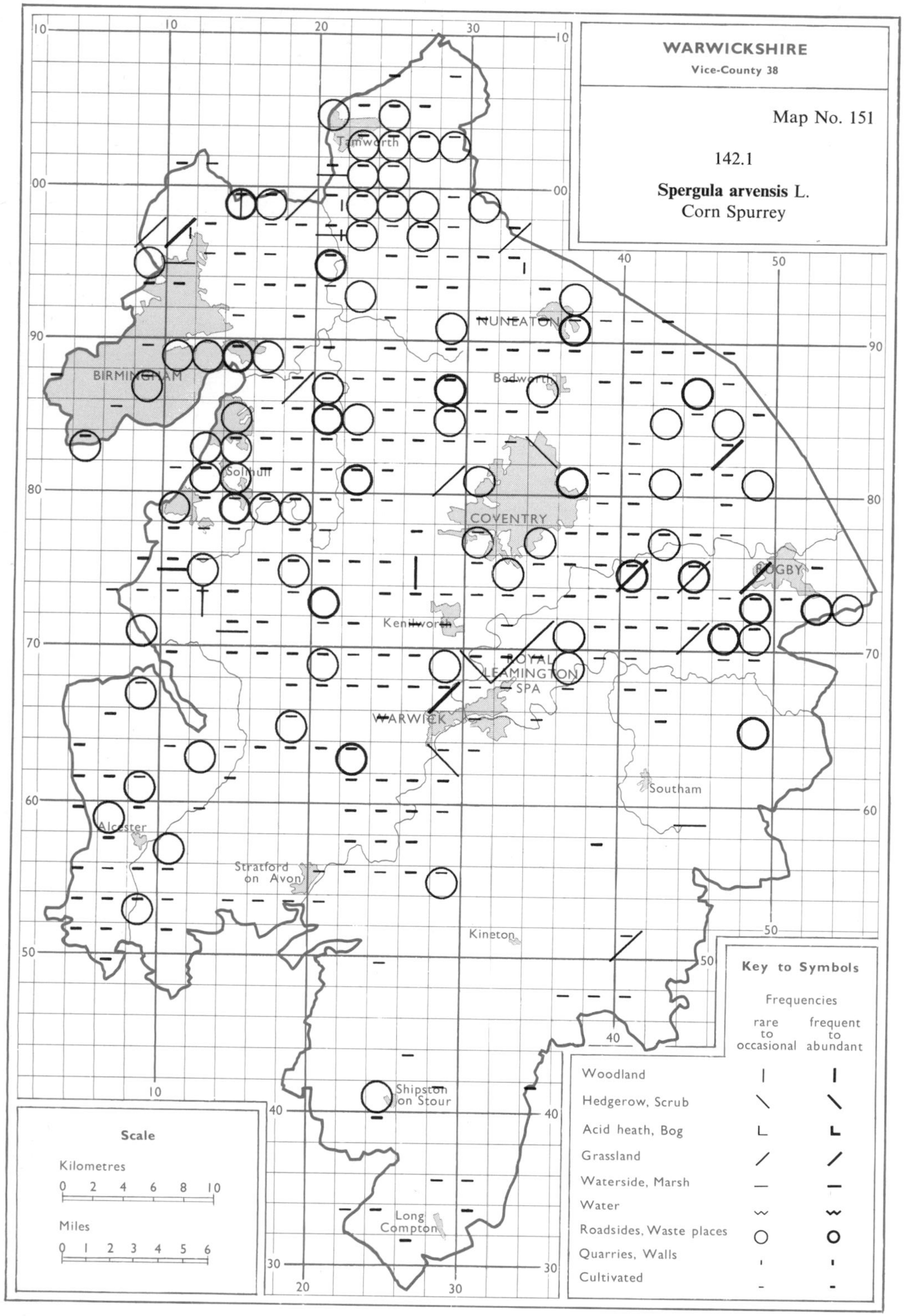
WARWICKSHIRE
Vice-County 38
Map No. 151
142.1
Spergula arvensis L.
Corn Spurrey
Tamworth
BIRMINGHAM
Solihull
NUNEATON
Bedworth
COVENTRY
Kenilworth
ROYAL LEAMINGTON SPA
WARWICK
RUGBY
Southam
Alcester
Stratford on Avon
Kineton
Shipston on Stour
Long Compton
Scale
Kilometres
0 2 4 6 8 10
Miles
0 1 2 3 4 5 6
Key to Symbols
Frequencies
rare to occasional
frequent to abundant
Woodland
Hedgerow, Scrub
Acid heath, Bog
Grassland
Waterside, Marsh
Water
Roadsides, Waste places
Quarries, Walls
Cultivated

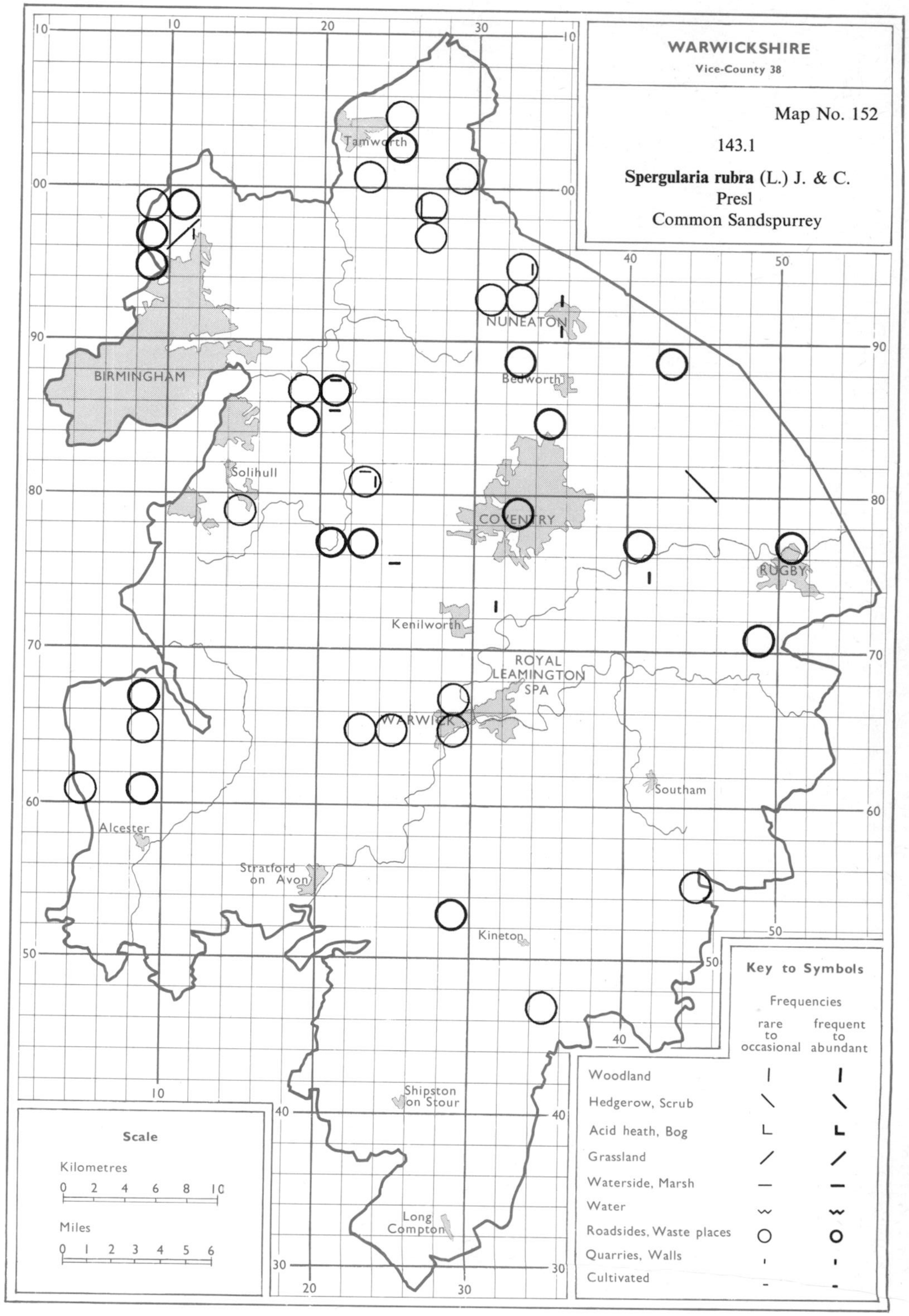
WARWICKSHIRE
Vice-County 38
Map No. 152
143.1
Spergularia rubra (L.) J. & C. Presl
Common Sandspurrey
Tamworth
BIRMINGHAM
Solihull
NUNEATON
Bedworth
COVENTRY
Kenilworth
ROYAL LEAMINGTON SPA
WARWICK
RUGBY
Southam
Alcester
Stratford on Avon
Kineton
Shipston on Stour
Long Compton
Scale
Kilometres
0 2 4 6 8 10
Miles
0 1 2 3 4 5 6
Key to Symbols
Frequencies
rare to occasional
frequent to abundant
Woodland
Hedgerow, Scrub
Acid heath, Bog
Grassland
Waterside, Marsh
Water
Roadsides, Waste places
Quarries, Walls
Cultivated

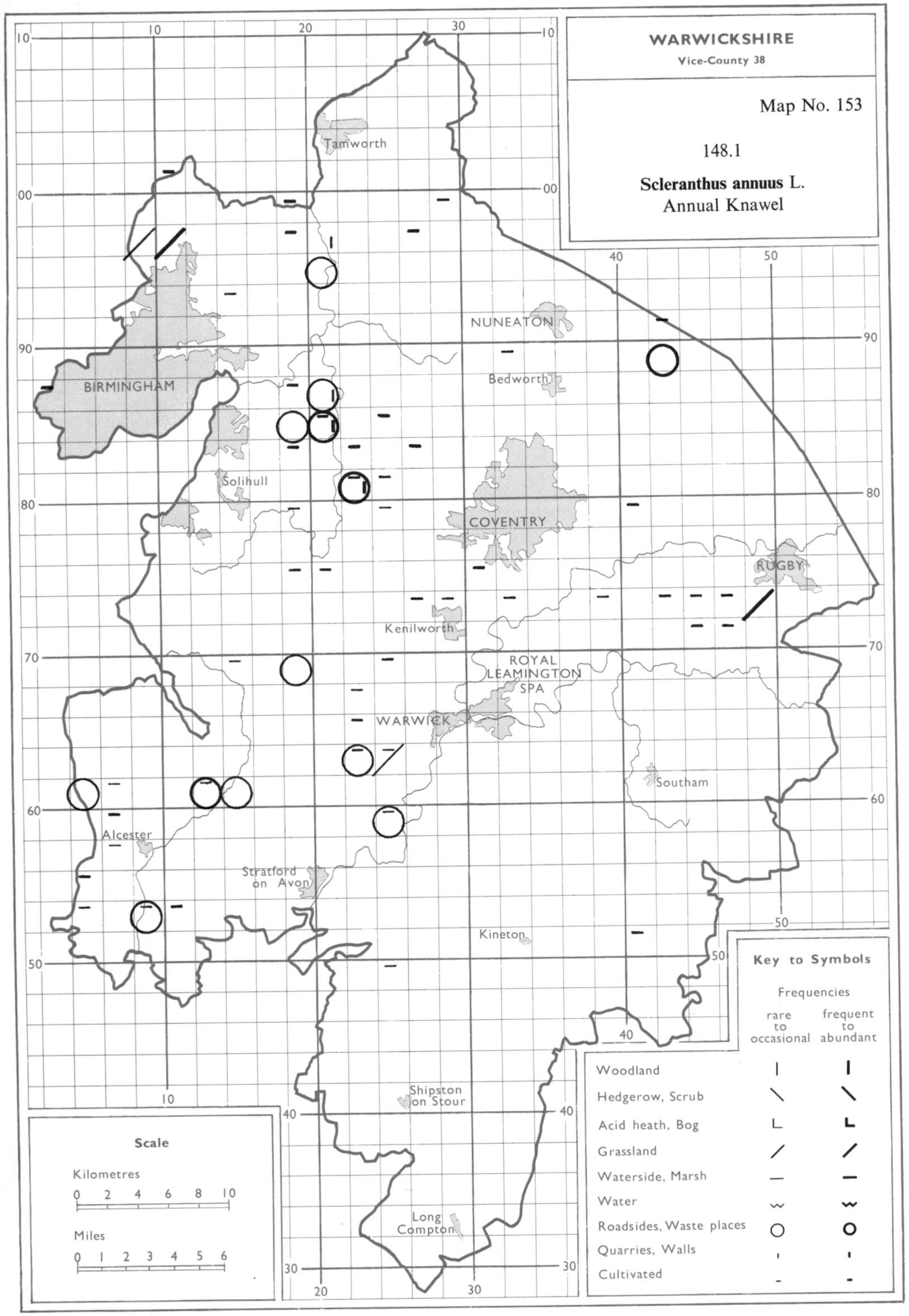
WARWICKSHIRE
Vice-County 38
Map No. 153
148.1
Scleranthus annuus L.
Annual Knawel
Tamworth
NUNEATON
Bedworth
BIRMINGHAM
Solihull
COVENTRY
RUGBY
Kenilworth
ROYAL LEAMINGTON SPA
WARWICK
Southam
Alcester
Stratford on Avon
Kineton
Shipston on Stour
Long Compton
Scale
Kilometres
0 2 4 6 8 10
Miles
0 1 2 3 4 5 6
Key to Symbols
Frequencies
rare to occasional
frequent to abundant
Woodland
Hedgerow, Scrub
Acid heath, Bog
Grassland
Waterside, Marsh
Water
Roadsides, Waste places
Quarries, Walls
Cultivated

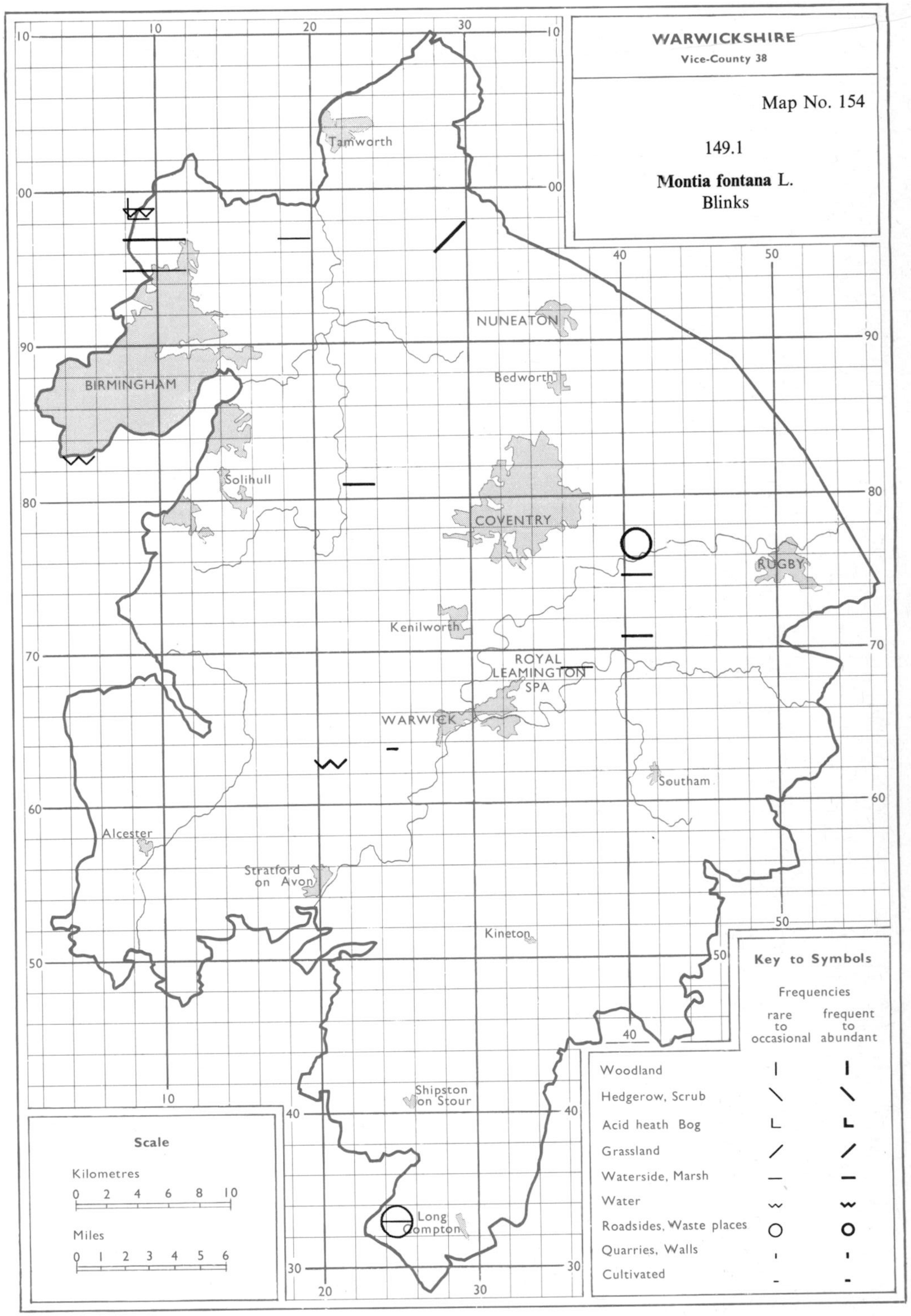
WARWICKSHIRE
Vice-County 38
Map No. 154
149.1
Montia fontana L.
Blinks
Tamworth
NUNEATON
Bedworth
BIRMINGHAM
Solihull
COVENTRY
RUGBY
Kenilworth
ROYAL LEAMINGTON SPA
WARWICK
Southam
Alcester
Stratford on Avon
Kineton
Shipston on Stour
Long Compton
Scale
Kilometres
0 2 4 6 8 10
Miles
0 1 2 3 4 5 6
Key to Symbols
Frequencies
rare to occasional
frequent to abundant
Woodland
Hedgerow, Scrub
Acid heath Bog
Grassland
Waterside, Marsh
Water
Roadsides, Waste places
Quarries, Walls
Cultivated

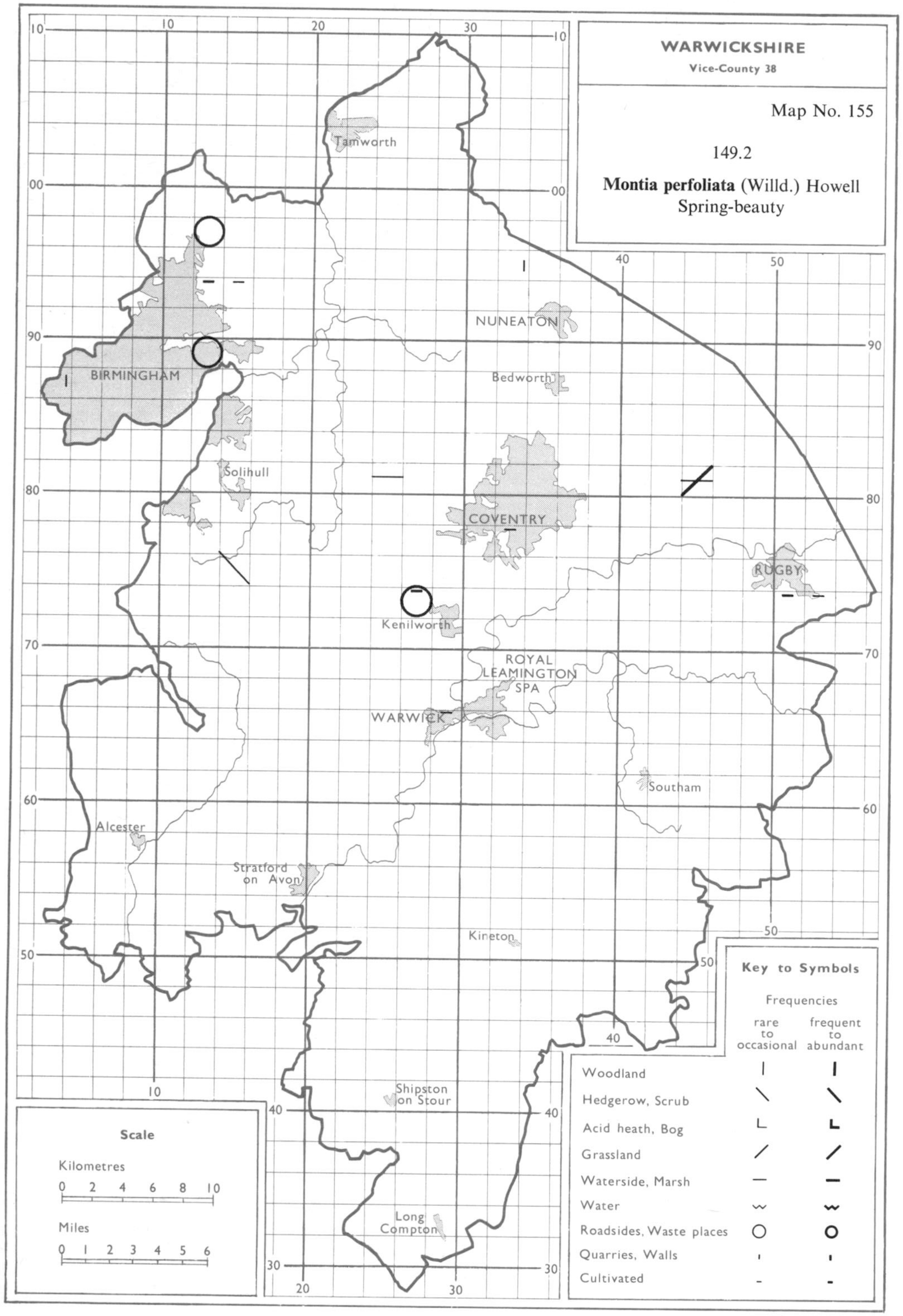

WARWICKSHIRE
Vice-County 38
Map No. 155
149.2
Montia perfoliata (Willd.) Howell
Spring-beauty
Key to Symbols
Frequencies
rare to occasional
frequent to abundant
Woodland
Hedgerow, Scrub
Acid heath, Bog
Grassland
Waterside, Marsh
Water
Roadsides, Waste places
Quarries, Walls
Cultivated
Scale
Kilometres 0 2 4 6 8 10
Miles 0 1 2 3 4 5 6
NUNEATON
Bedworth
COVENTRY
RUGBY
Southam
ROYAL LEAMINGTON SPA
WARWICK
Kenilworth
Kineton
Tamworth
Solihull
BIRMINGHAM
Stratford on Avon
Alcester
Shipston on Stour
Long Compton

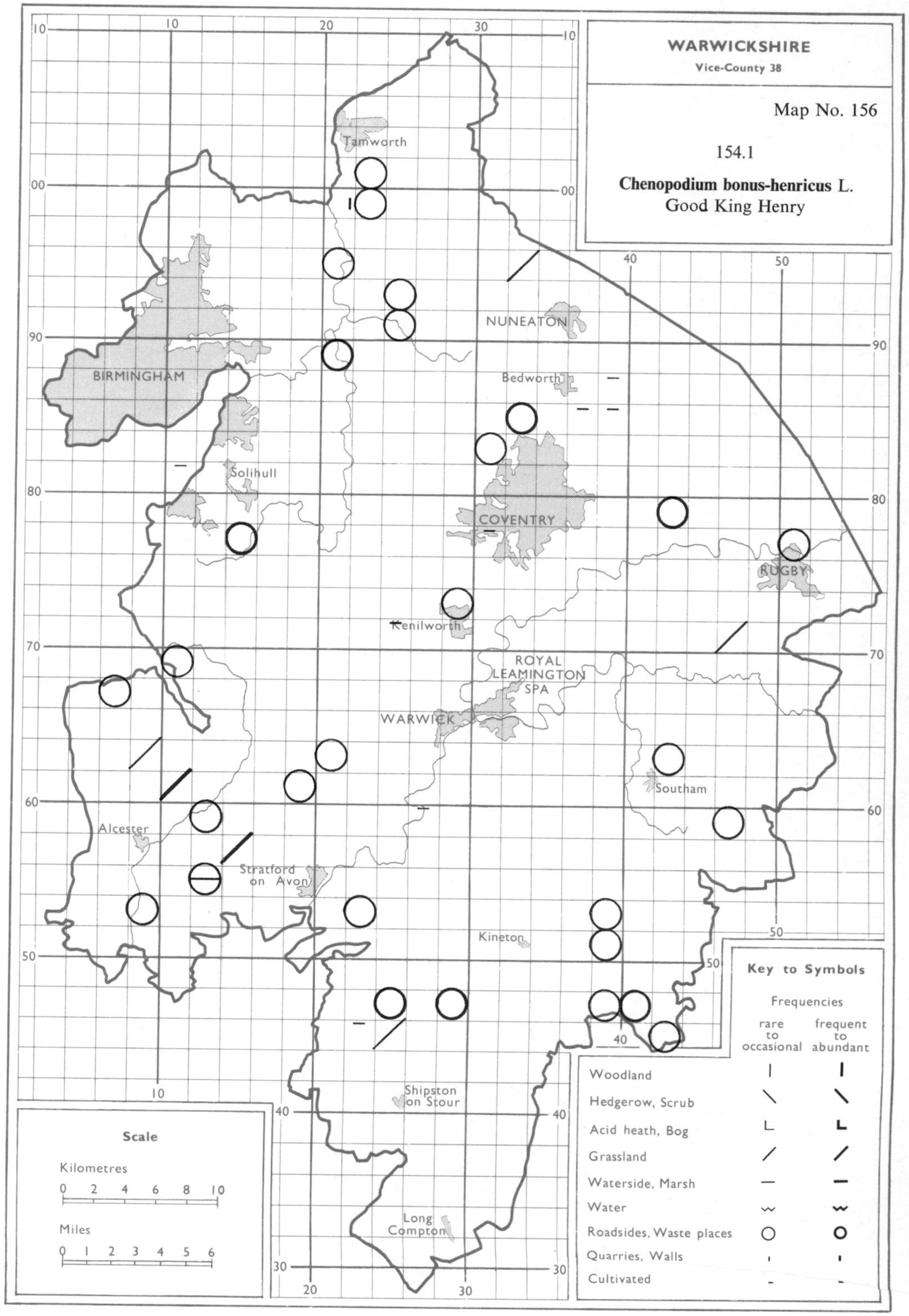

WARWICKSHIRE
Vice-County 38
Map No. 156
154.1
Chenopodium bonus-henricus L.
Good King Henry
Key to Symbols
Frequencies
rare to occasional
frequent to abundant
Woodland
Hedgerow, Scrub
Acid heath, Bog
Grassland
Waterside, Marsh
Water
Roadsides, Waste places
Quarries, Walls
Cultivated
Scale
Kilometres 0 2 4 6 8 10
Miles 0 1 2 3 4 5 6
NUNEATON
Bedworth
COVENTRY
RUGBY
Southam
ROYAL LEAMINGTON SPA
WARWICK
Kenilworth
Kineton
Tamworth
Solihull
BIRMINGHAM
Stratford on Avon
Alcester
Shipston on Stour
Long Compton

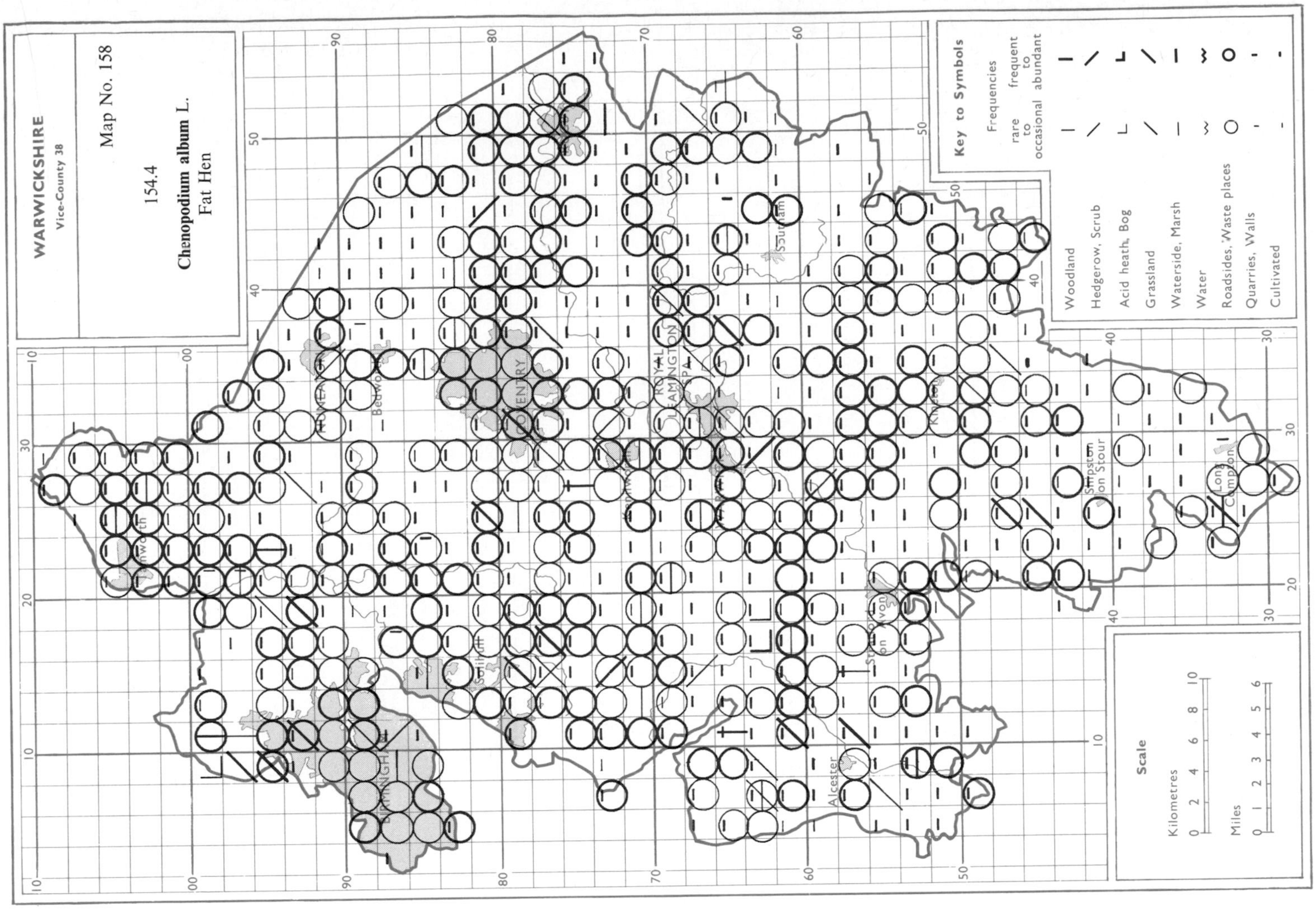
WARWICKSHIRE
Vice-County 38
Map No. 158
154.4
Chenopodium album L.
Fat Hen
Key to Symbols
Frequencies
rare to occasional
frequent to abundant
Woodland
Hedgerow, Scrub
Acid heath, Bog
Grassland
Waterside, Marsh
Water
Roadsides, Waste places
Quarries, Walls
Cultivated
Scale
Kilometres
Miles
ROYAL LEAMINGTON SPA
COVENTRY
Bedworth
Southam
Kineton
Shipston on Stour
Long Compton
Stratford on Avon
Alcester
Solihull
Tamworth

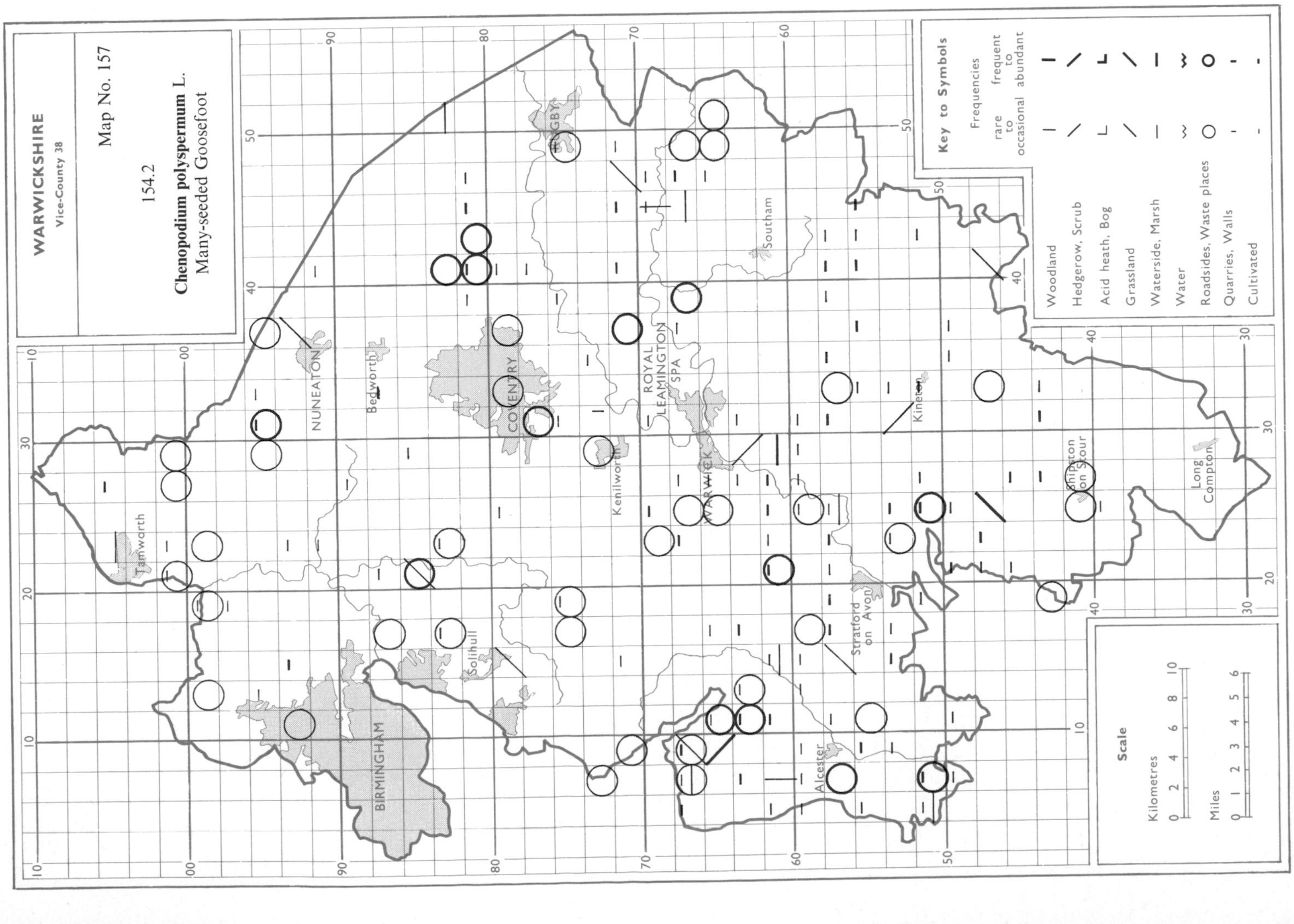
WARWICKSHIRE
Vice-County 38
Map No. 157
154.2
Chenopodium polyspermum L.
Many-seeded Goosefoot
Key to Symbols
Frequencies
rare to occasional
frequent to abundant
Woodland
Hedgerow, Scrub
Acid heath, Bog
Grassland
Waterside, Marsh
Water
Roadsides, Waste places
Quarries, Walls
Cultivated
Scale
Kilometres
Miles
NUNEATON
Bedworth
COVENTRY
ROYAL LEAMINGTON SPA
WARWICK
Kenilworth
RUGBY
Southam
Kineton
Shipston on Stour
Long Compton
Stratford on Avon
Alcester
Solihull
BIRMINGHAM
Tamworth

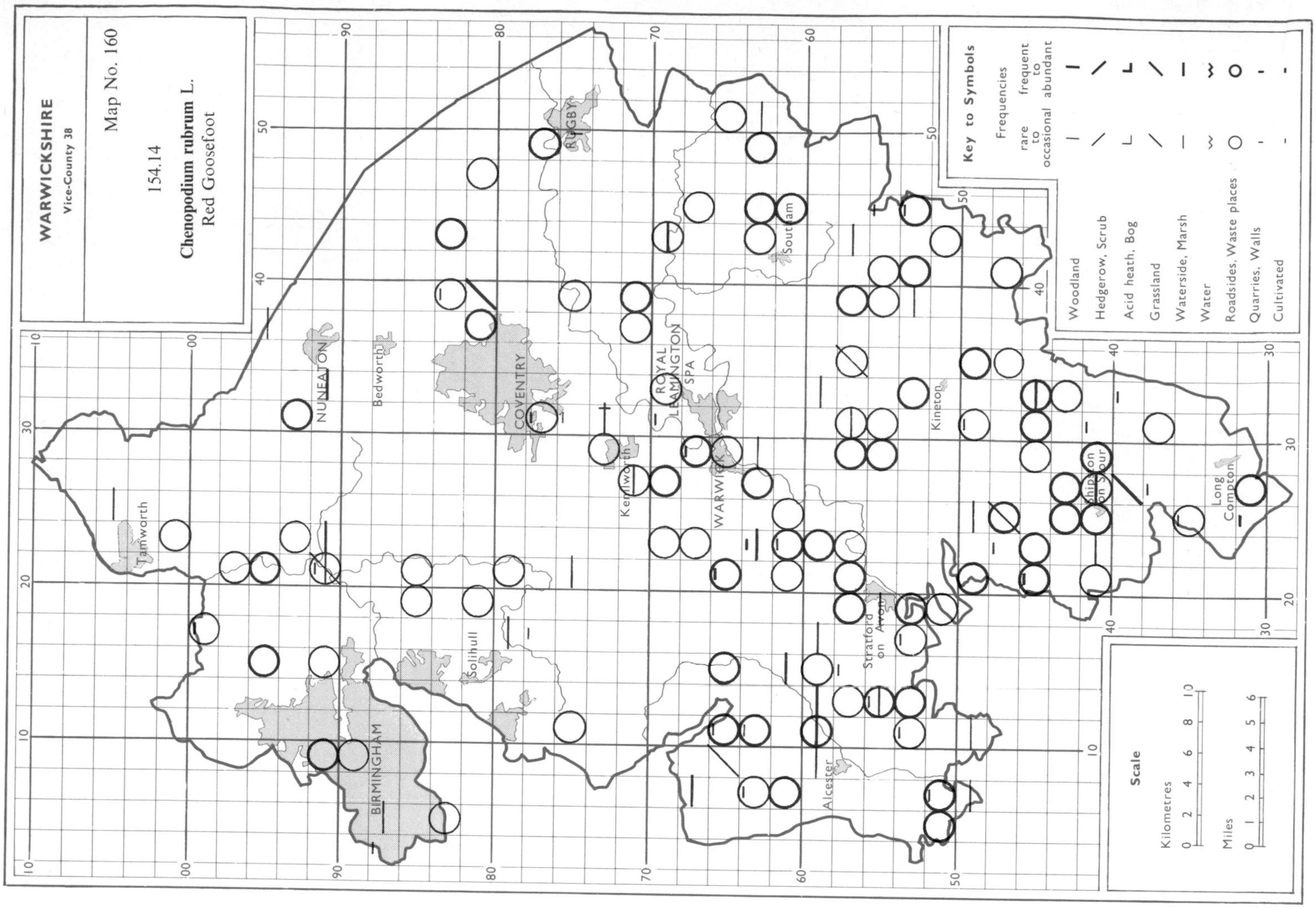
WARWICKSHIRE
Vice-County 38
Map No. 160
154.14
Chenopodium rubrum L.
Red Goosefoot
Key to Symbols
Frequencies
rare to occasional
frequent to abundant
Woodland
Hedgerow, Scrub
Acid heath, Bog
Grassland
Waterside, Marsh
Water
Roadsides, Waste places
Quarries, Walls
Cultivated
Scale
Kilometres
0 2 4 6 8 10
Miles
0 1 2 3 4 5 6
NUNEATON
Bedworth
COVENTRY
RUGBY
Southam
ROYAL LEAMINGTON SPA
Kenilworth
WARWICK
Kineton
Tamworth
Solihull
BIRMINGHAM
Stratford on Avon
Alcester
Shipston on Stour
Long Compton

WARWICKSHIRE
Vice-County 38
Map No. 159
154.13
Chenopodium hybridum L.
Maple-leaved Goosefoot
Key to Symbols
Frequencies
rare to occasional
frequent to abundant
Woodland
Hedgerow, Scrub
Acid heath, Bog
Grassland
Waterside, Marsh
Water
Roadsides, Waste places
Quarries, Walls
Cultivated
Scale
Kilometres
0 2 4 6 8 10
Miles
0 1 2 3 4 5 6
NUNEATON
Bedworth
COVENTRY
RUGBY
Southam
ROYAL LEAMINGTON SPA
Kenilworth
WARWICK
Kineton
Tamworth
Solihull
BIRMINGHAM
Stratford on Avon
Alcester
Shipston on Stour
Long Compton

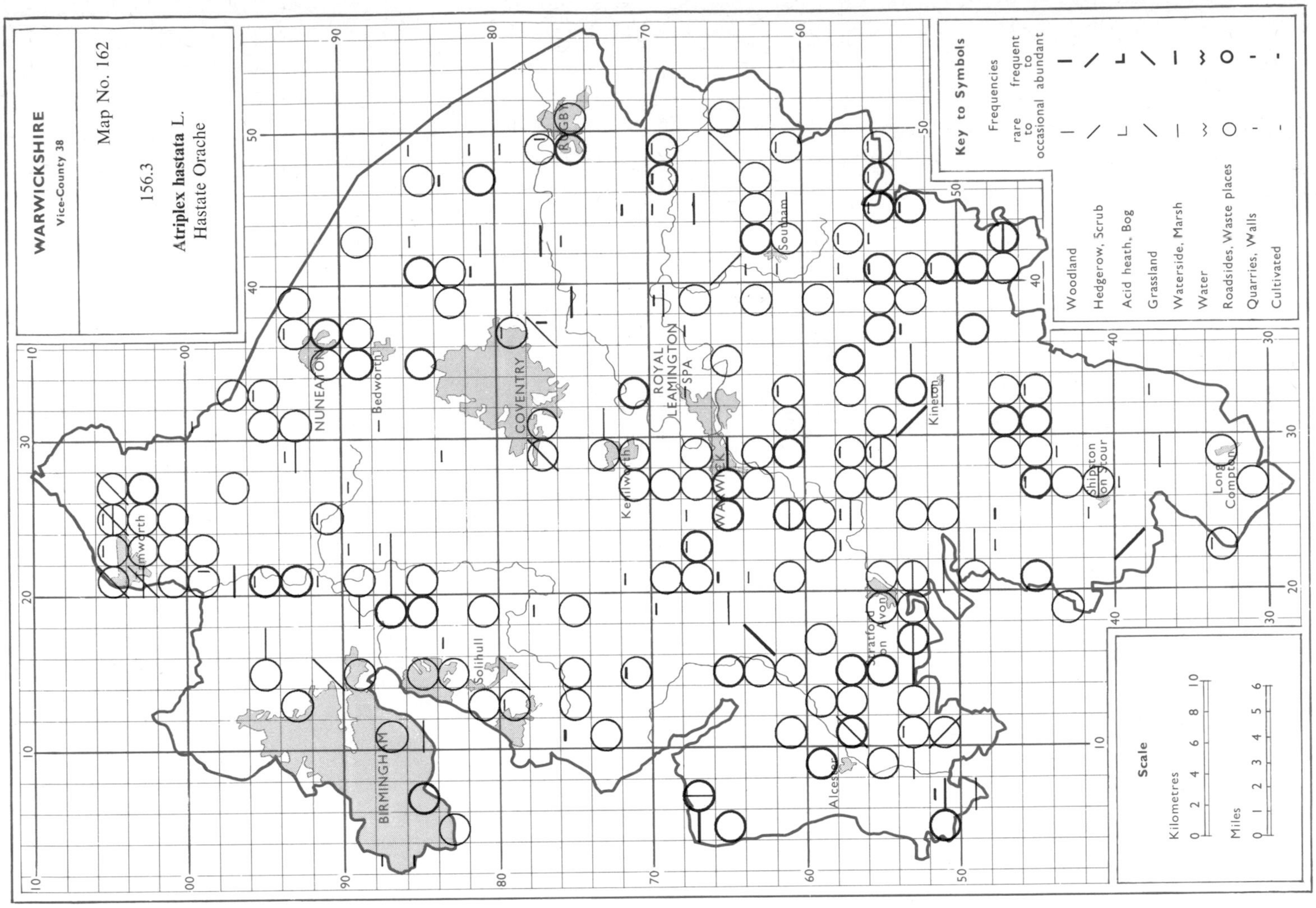
WARWICKSHIRE
Vice-County 38
Map No. 162
156.3
Atriplex hastata L.
Hastate Orache
Key to Symbols
Frequencies
rare to occasional
frequent to abundant
Woodland
Hedgerow, Scrub
Acid heath, Bog
Grassland
Waterside, Marsh
Water
Roadsides, Waste places
Quarries, Walls
Cultivated
Scale
Kilometres
0 2 4 6 8 10
Miles
0 1 2 3 4 5 6
NUNEATON
Bedworth
COVENTRY
RUGBY
ROYAL LEAMINGTON SPA
Kenilworth
WARWICK
Southam
Kineton
Shipston on Stour
Long Compton
Stratford on Avon
Alcester
Solihull
BIRMINGHAM
Tamworth

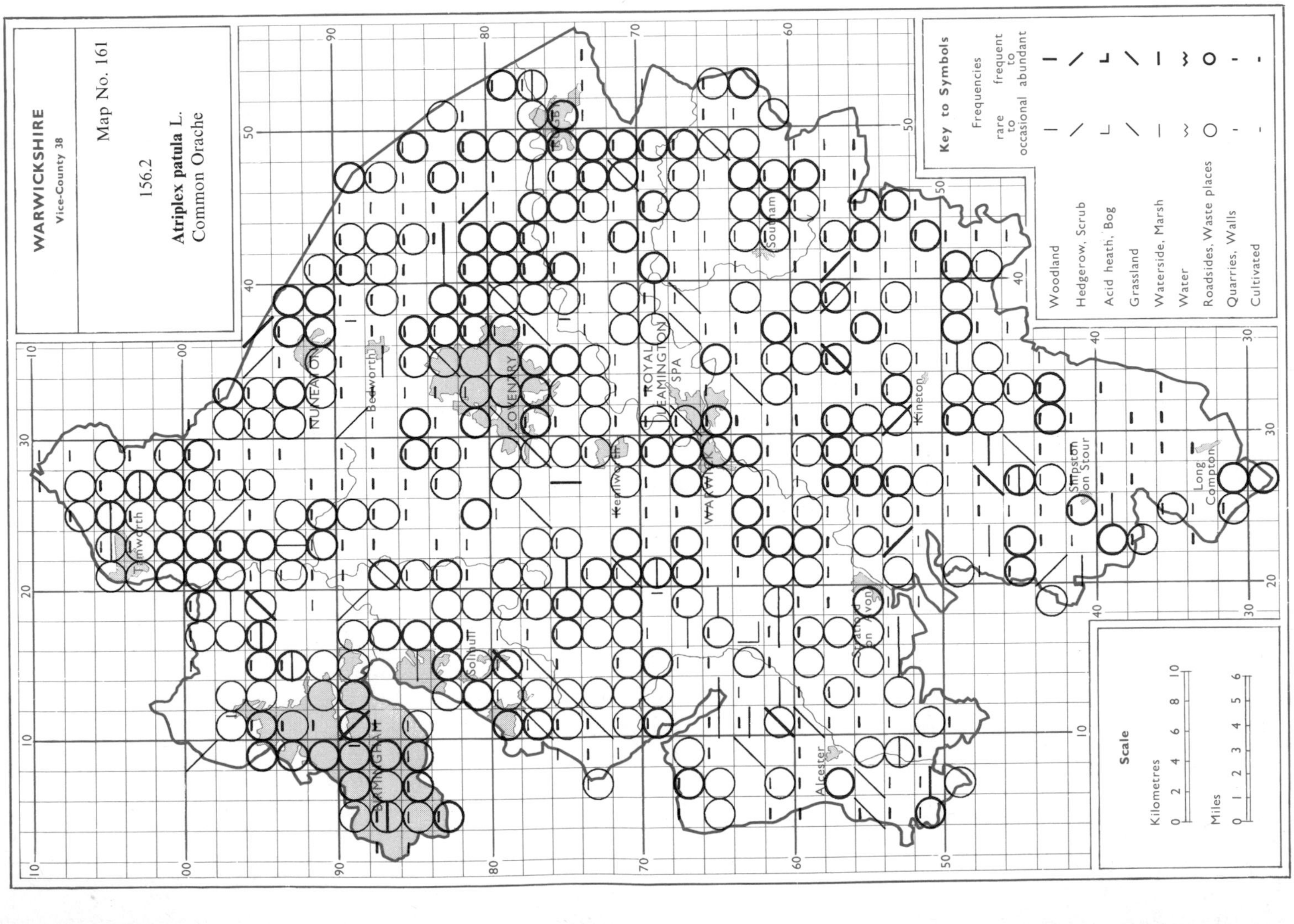
WARWICKSHIRE
Vice-County 38
Map No. 161
156.2
Atriplex patula L.
Common Orache
Key to Symbols
Frequencies
rare to occasional
frequent to abundant
Woodland
Hedgerow, Scrub
Acid heath, Bog
Grassland
Waterside, Marsh
Water
Roadsides, Waste places
Quarries, Walls
Cultivated
Scale
Kilometres
0 2 4 6 8 10
Miles
0 1 2 3 4 5 6
NUNEATON
Bedworth
COVENTRY
RUGBY
ROYAL LEAMINGTON SPA
Kenilworth
WARWICK
Southam
Kineton
Shipston on Stour
Long Compton
Stratford on Avon
Alcester
Solihull
BIRMINGHAM
Tamworth

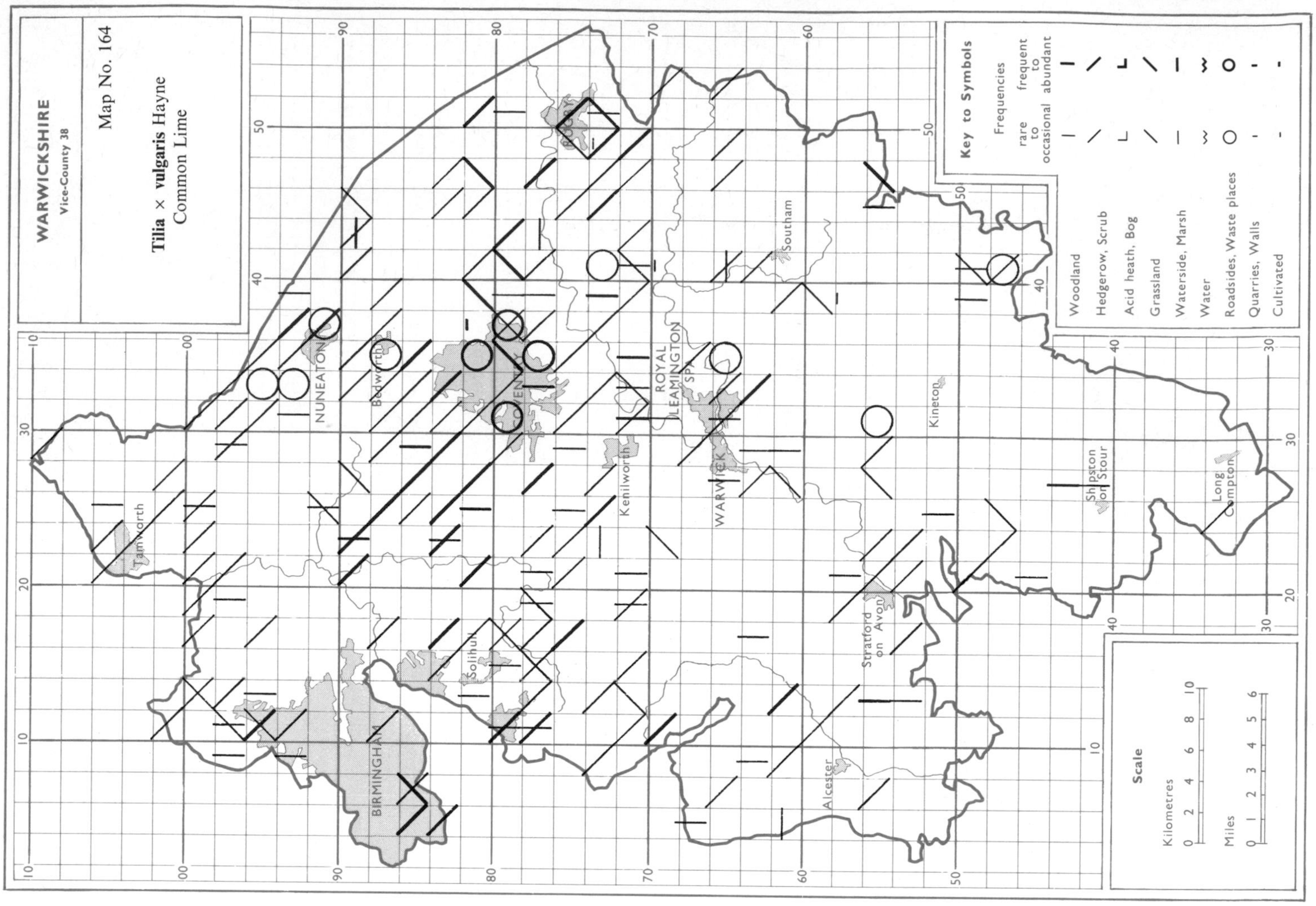
WARWICKSHIRE
Vice-County 38
Map No. 164
Tilia × vulgaris Hayne
Common Lime
Key to Symbols
Frequencies
rare to occasional
frequent to abundant
Woodland
Hedgerow, Scrub
Acid heath, Bog
Grassland
Waterside, Marsh
Water
Roadsides, Waste places
Quarries, Walls
Cultivated
Scale
Kilometres
Miles
NUNEATON
Bedworth
COVENTRY
RUGBY
ROYAL LEAMINGTON SPA
WARWICK
Kenilworth
Southam
Kineton
Shipston on Stour
Long Compton
Stratford on Avon
Alcester
Solihull
BIRMINGHAM
Tamworth

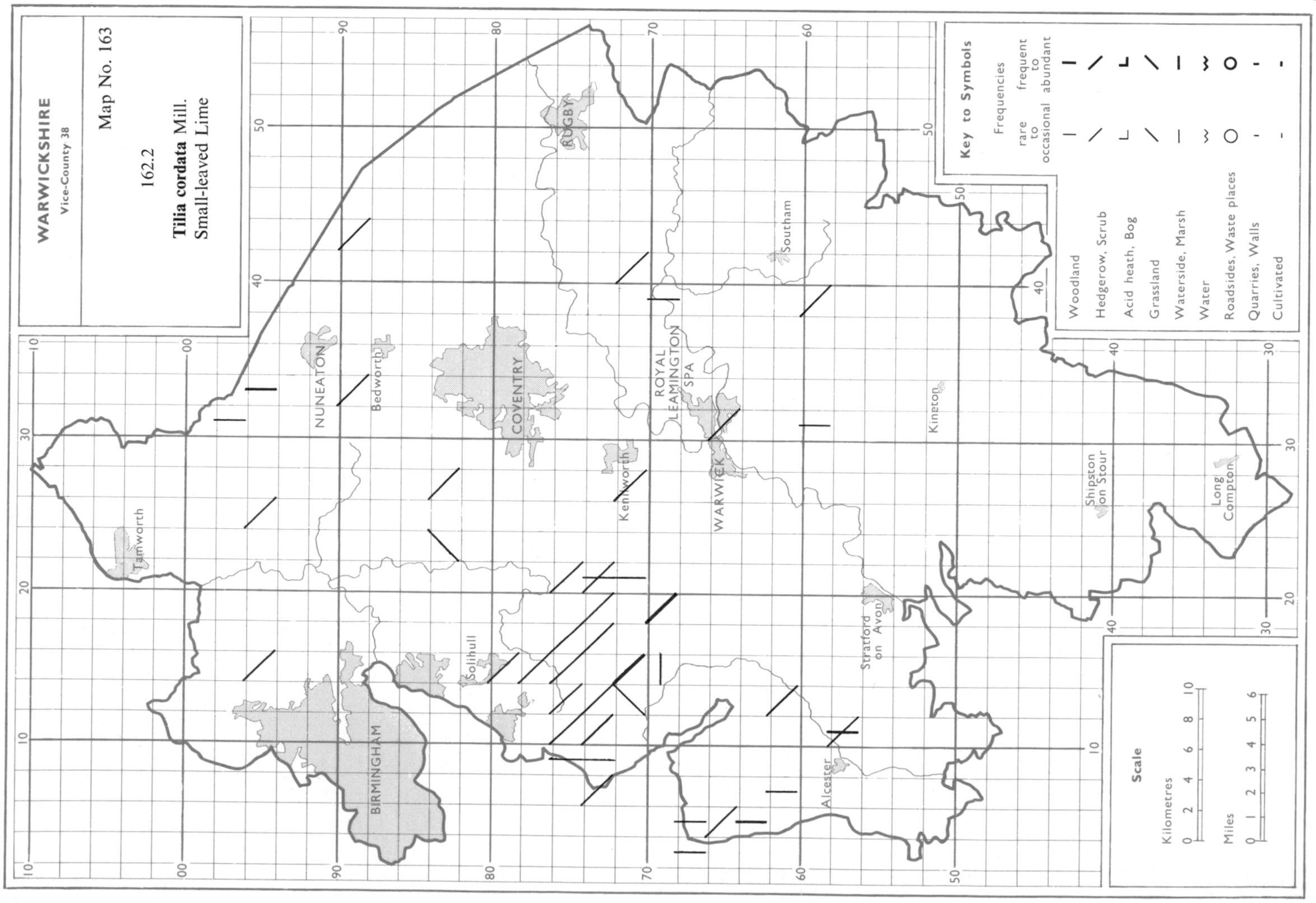
WARWICKSHIRE
Vice-County 38
Map No. 163
162.2
Tilia cordata Mill.
Small-leaved Lime
Key to Symbols
Frequencies
rare to occasional
frequent to abundant
Woodland
Hedgerow, Scrub
Acid heath, Bog
Grassland
Waterside, Marsh
Water
Roadsides, Waste places
Quarries, Walls
Cultivated
Scale
Kilometres
Miles
NUNEATON
Bedworth
COVENTRY
RUGBY
ROYAL LEAMINGTON SPA
WARWICK
Kenilworth
Southam
Kineton
Shipston on Stour
Long Compton
Stratford on Avon
Alcester
Solihull
BIRMINGHAM
Tamworth

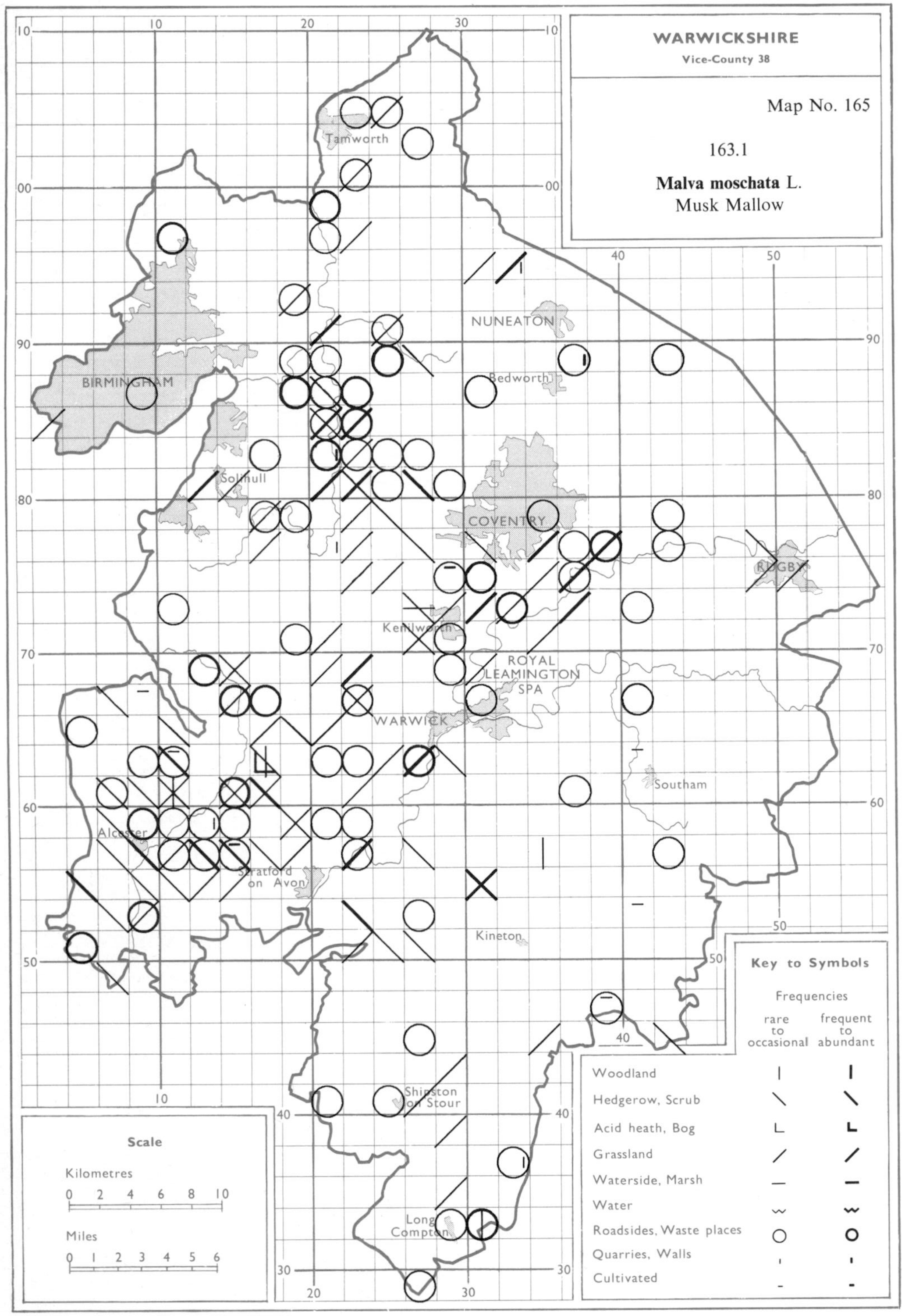
WARWICKSHIRE
Vice-County 38
Map No. 165
163.1
Malva moschata L.
Musk Mallow
Tamworth
NUNEATON
Bedworth
BIRMINGHAM
Solihull
COVENTRY
RUGBY
Kenilworth
ROYAL LEAMINGTON SPA
WARWICK
Southam
Alcester
Stratford on Avon
Kineton
Shipston on Stour
Long Compton
Key to Symbols
Frequencies
rare to occasional
frequent to abundant
Woodland
Hedgerow, Scrub
Acid heath, Bog
Grassland
Waterside, Marsh
Water
Roadsides, Waste places
Quarries, Walls
Cultivated
Scale
Kilometres
0 2 4 6 8 10
Miles
0 1 2 3 4 5 6

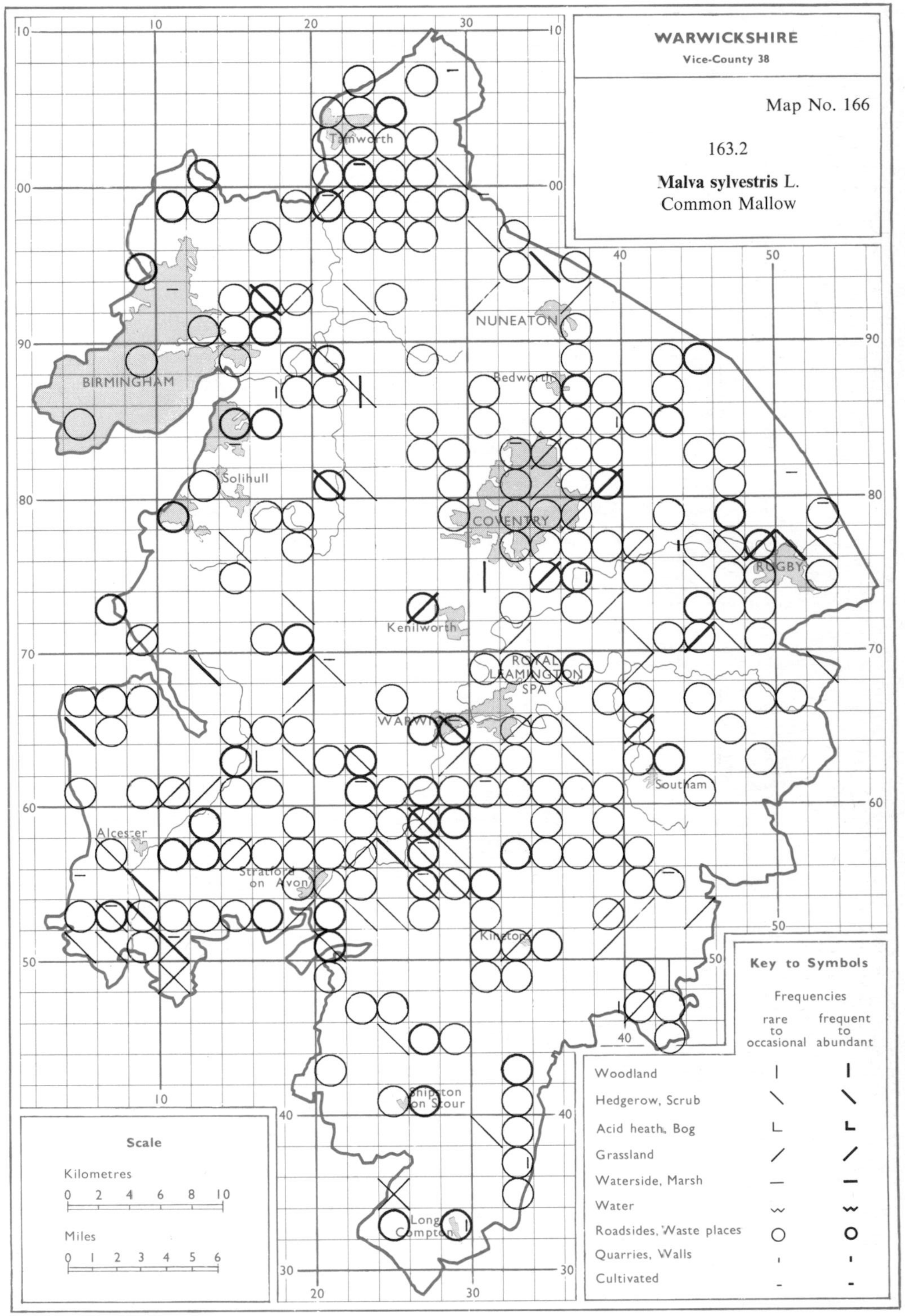
WARWICKSHIRE
Vice-County 38
Map No. 166
163.2
Malva sylvestris L.
Common Mallow
Tamworth
NUNEATON
Bedworth
BIRMINGHAM
Solihull
COVENTRY
RUGBY
Kenilworth
ROYAL LEAMINGTON SPA
WARWICK
Southam
Alcester
Stratford on Avon
Kineton
Shipston on Stour
Long Compton
Key to Symbols
Frequencies
rare to occasional
frequent to abundant
Woodland
Hedgerow, Scrub
Acid heath, Bog
Grassland
Waterside, Marsh
Water
Roadsides, Waste places
Quarries, Walls
Cultivated
Scale
Kilometres
0 2 4 6 8 10
Miles
0 1 2 3 4 5 6

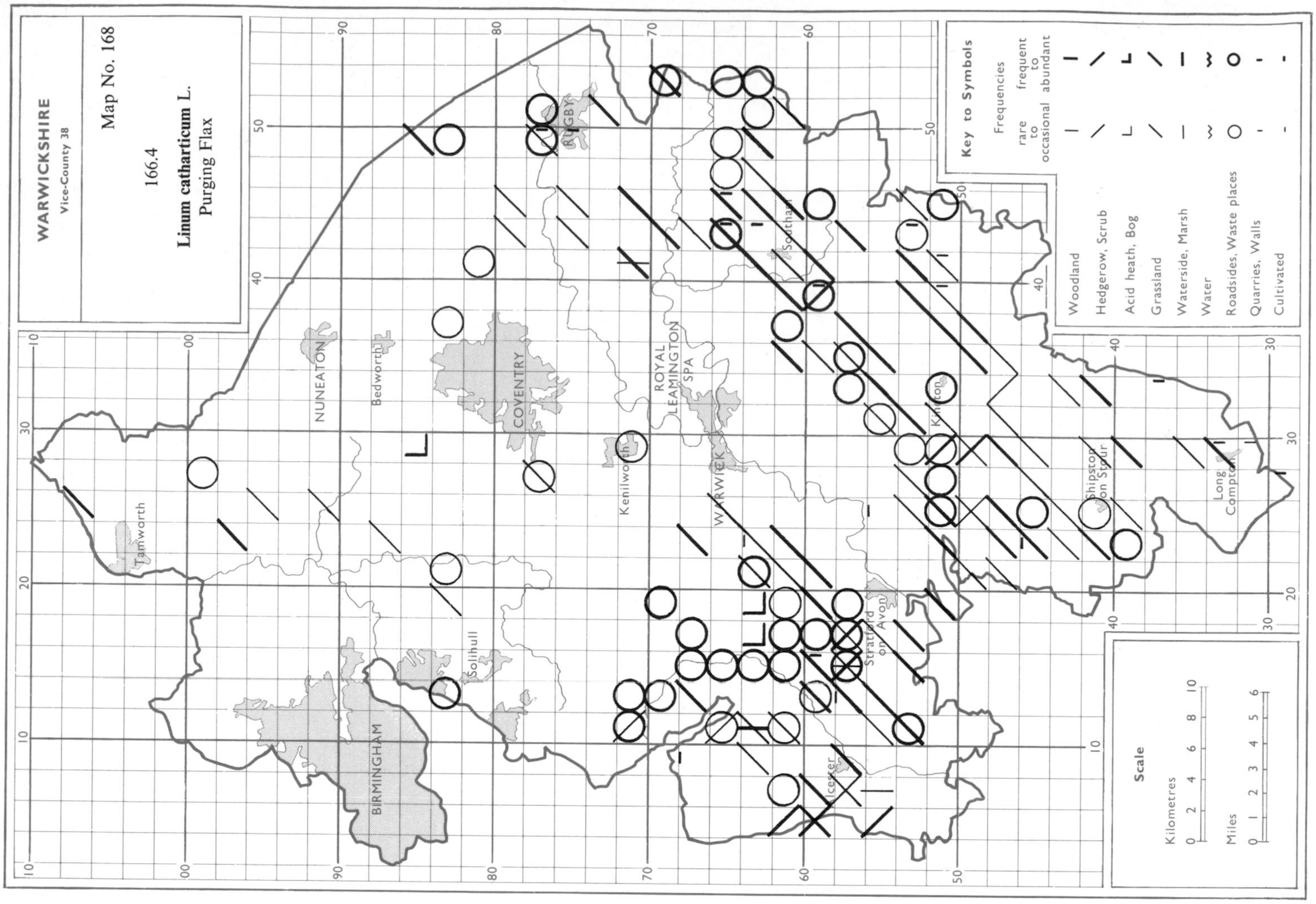
WARWICKSHIRE
Vice-County 38
Map No. 168
166.4
Linum catharticum L.
Purging Flax
Key to Symbols
Frequencies
rare to occasional
frequent to abundant
Woodland
Hedgerow, Scrub
Acid heath, Bog
Grassland
Waterside, Marsh
Water
Roadsides, Waste places
Quarries, Walls
Cultivated
Scale
Kilometres
Miles
NUNEATON
Bedworth
COVENTRY
RUGBY
ROYAL LEAMINGTON SPA
Kenilworth
WARWICK
Southam
Kineton
Shipston on Stour
Long Compton
Stratford on Avon
Alcester
Solihull
BIRMINGHAM
Tamworth

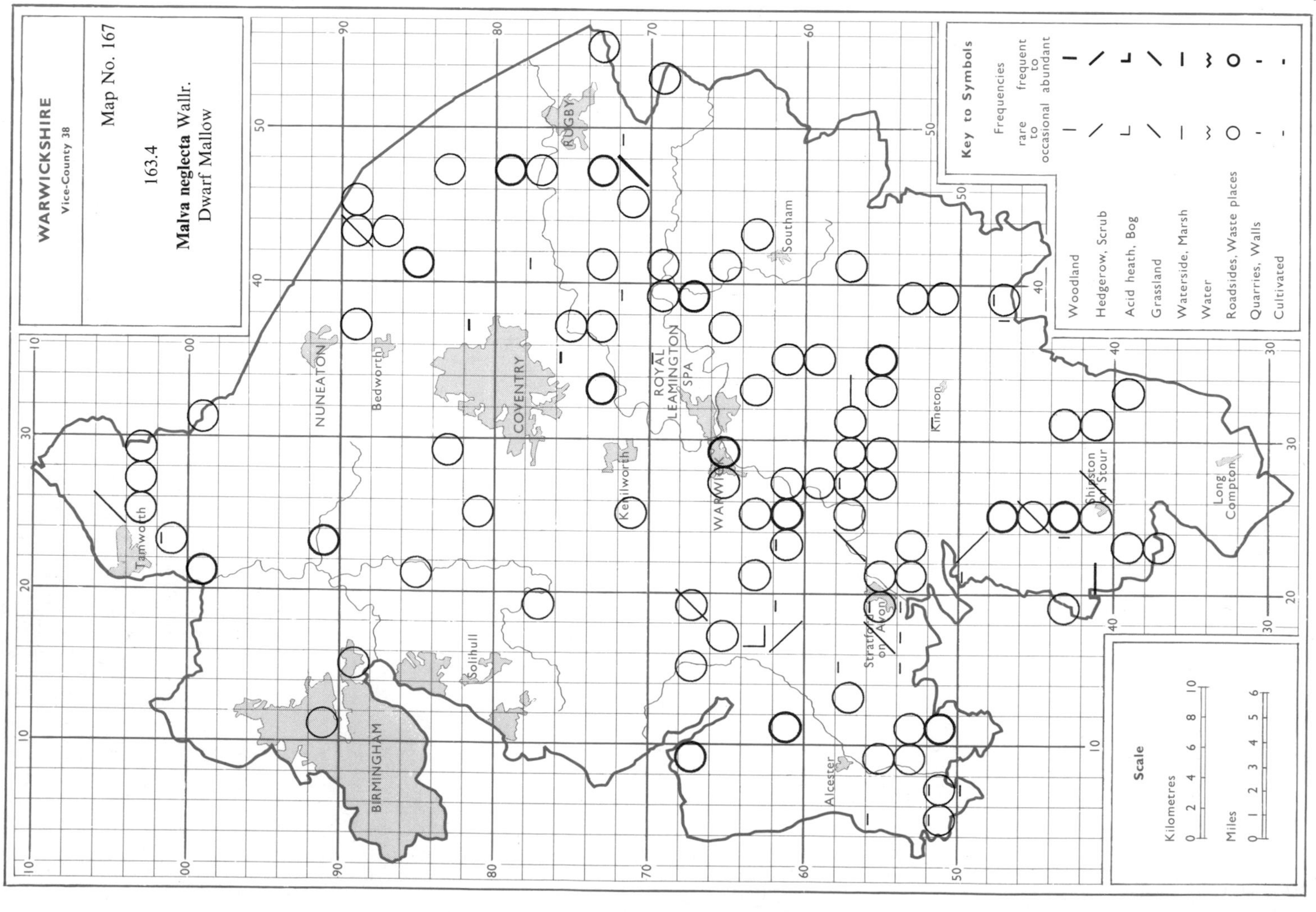
WARWICKSHIRE
Vice-County 38
Map No. 167
163.4
Malva neglecta Wallr.
Dwarf Mallow
Key to Symbols
Frequencies
rare to occasional
frequent to abundant
Woodland
Hedgerow, Scrub
Acid heath, Bog
Grassland
Waterside, Marsh
Water
Roadsides, Waste places
Quarries, Walls
Cultivated
Scale
Kilometres
Miles
NUNEATON
Bedworth
COVENTRY
RUGBY
ROYAL LEAMINGTON SPA
Kenilworth
WARWICK
Southam
Kineton
Shipston on Stour
Long Compton
Stratford on Avon
Alcester
Solihull
BIRMINGHAM
Tamworth

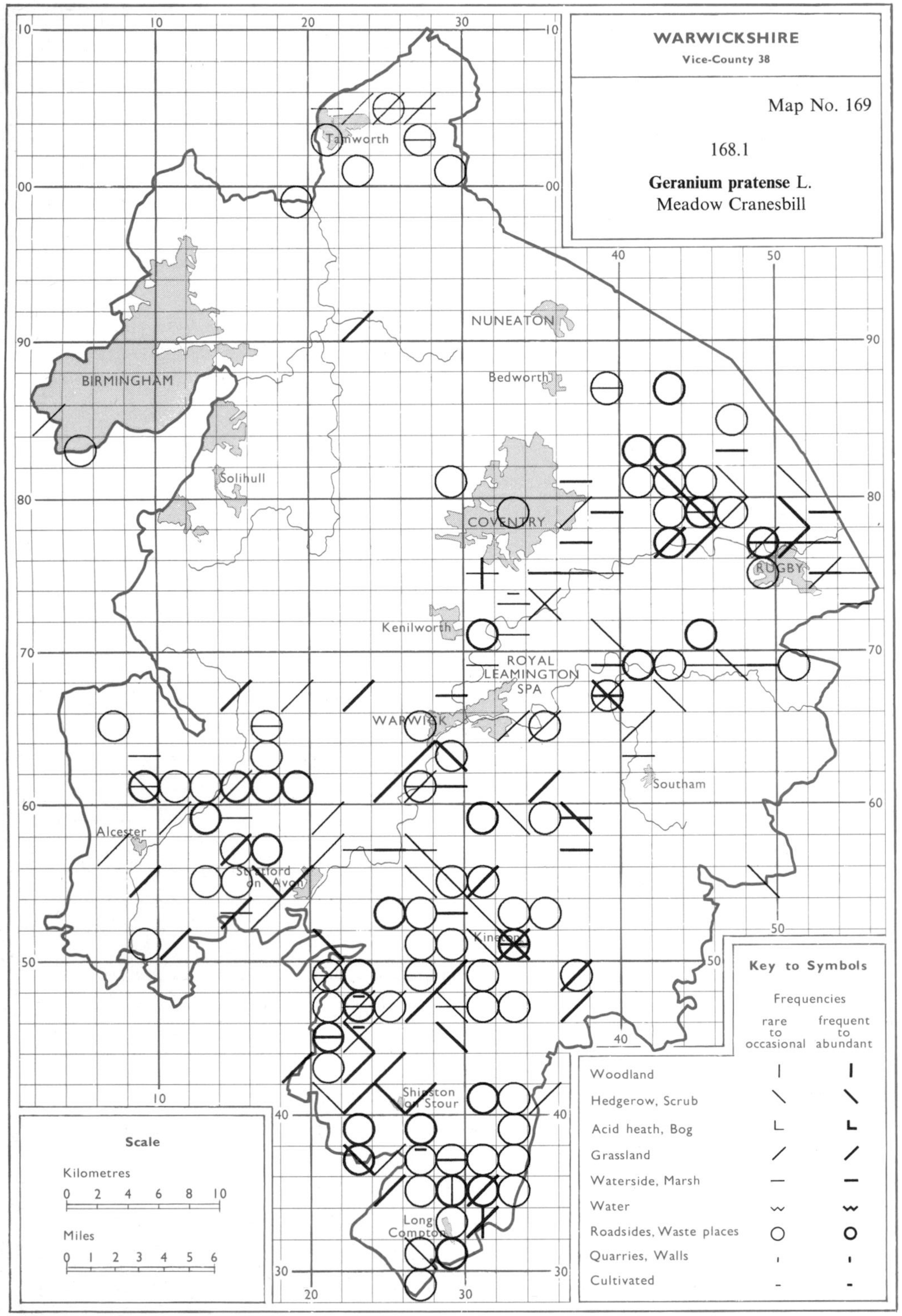
WARWICKSHIRE
Vice-County 38
Map No. 169
168.1
Geranium pratense L.
Meadow Cranesbill
Tamworth
NUNEATON
Bedworth
BIRMINGHAM
Solihull
COVENTRY
RUGBY
Kenilworth
ROYAL LEAMINGTON SPA
WARWICK
Southam
Alcester
Stratford on Avon
Kineton
Shipston on Stour
Long Compton
Scale
Kilometres
0 2 4 6 8 10
Miles
0 1 2 3 4 5 6
Key to Symbols
Frequencies
rare to occasional
frequent to abundant
Woodland
Hedgerow, Scrub
Acid heath, Bog
Grassland
Waterside, Marsh
Water
Roadsides, Waste places
Quarries, Walls
Cultivated

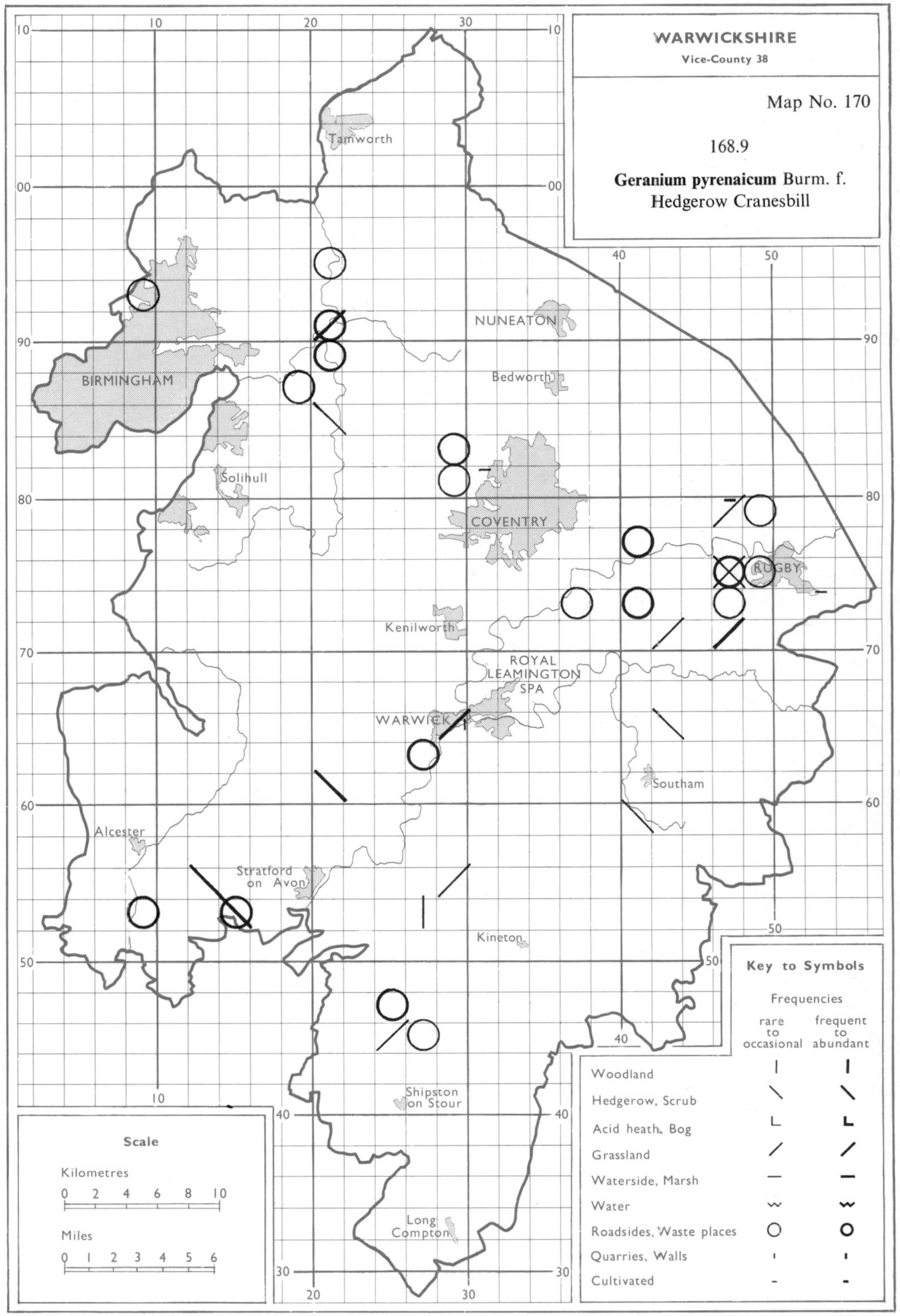
WARWICKSHIRE
Vice-County 38
Map No. 170
168.9
Geranium pyrenaicum Burm. f.
Hedgerow Cranesbill
Tamworth
NUNEATON
Bedworth
BIRMINGHAM
Solihull
COVENTRY
RUGBY
Kenilworth
ROYAL LEAMINGTON SPA
WARWICK
Southam
Alcester
Stratford on Avon
Kineton
Shipston on Stour
Long Compton
Scale
Kilometres
0 2 4 6 8 10
Miles
0 1 2 3 4 5 6
Key to Symbols
Frequencies
rare to occasional
frequent to abundant
Woodland
Hedgerow, Scrub
Acid heath, Bog
Grassland
Waterside, Marsh
Water
Roadsides, Waste places
Quarries, Walls
Cultivated

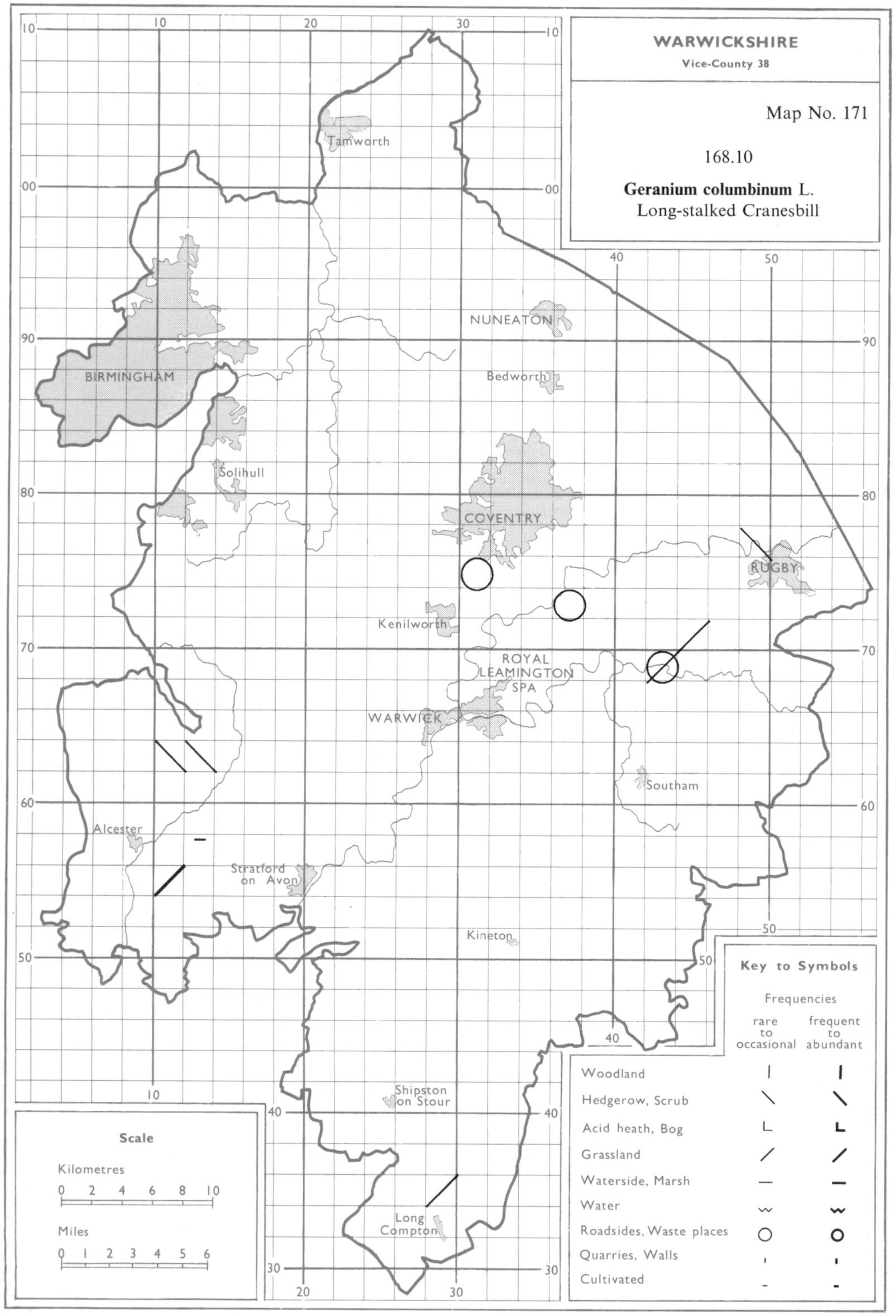
WARWICKSHIRE
Vice-County 38
Map No. 171
168.10
Geranium columbinum L.
Long-stalked Cranesbill
Key to Symbols
Frequencies
rare to occasional
frequent to abundant
Woodland
Hedgerow, Scrub
Acid heath, Bog
Grassland
Waterside, Marsh
Water
Roadsides, Waste places
Quarries, Walls
Cultivated
Scale
Kilometres
Miles
Tamworth
NUNEATON
Bedworth
BIRMINGHAM
Solihull
COVENTRY
RUGBY
Kenilworth
ROYAL LEAMINGTON SPA
WARWICK
Southam
Alcester
Stratford on Avon
Kineton
Shipston on Stour
Long Compton

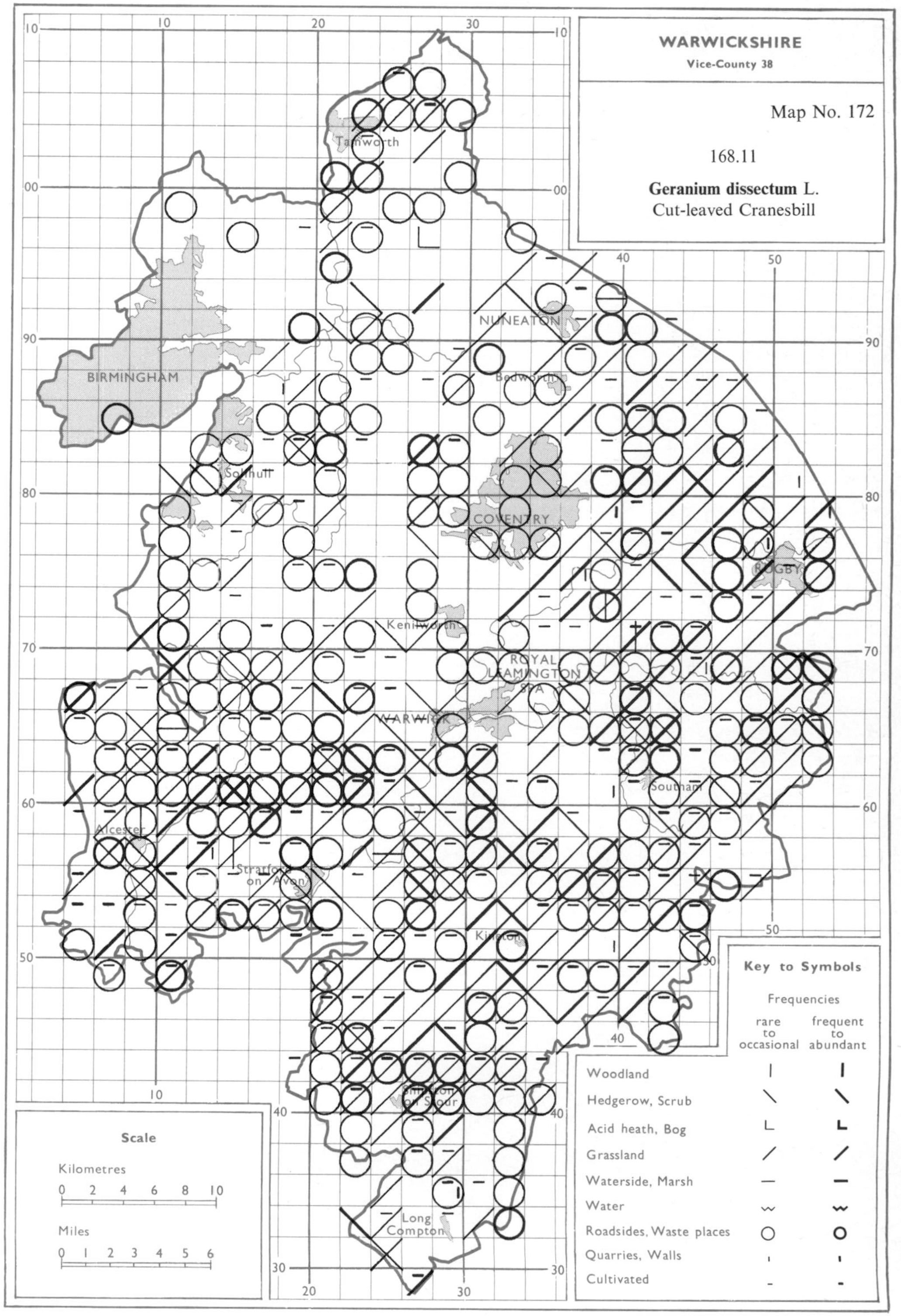
WARWICKSHIRE
Vice-County 38
Map No. 172
168.11
Geranium dissectum L.
Cut-leaved Cranesbill
Key to Symbols
Frequencies
rare to occasional
frequent to abundant
Woodland
Hedgerow, Scrub
Acid heath, Bog
Grassland
Waterside, Marsh
Water
Roadsides, Waste places
Quarries, Walls
Cultivated
Scale
Kilometres
Miles

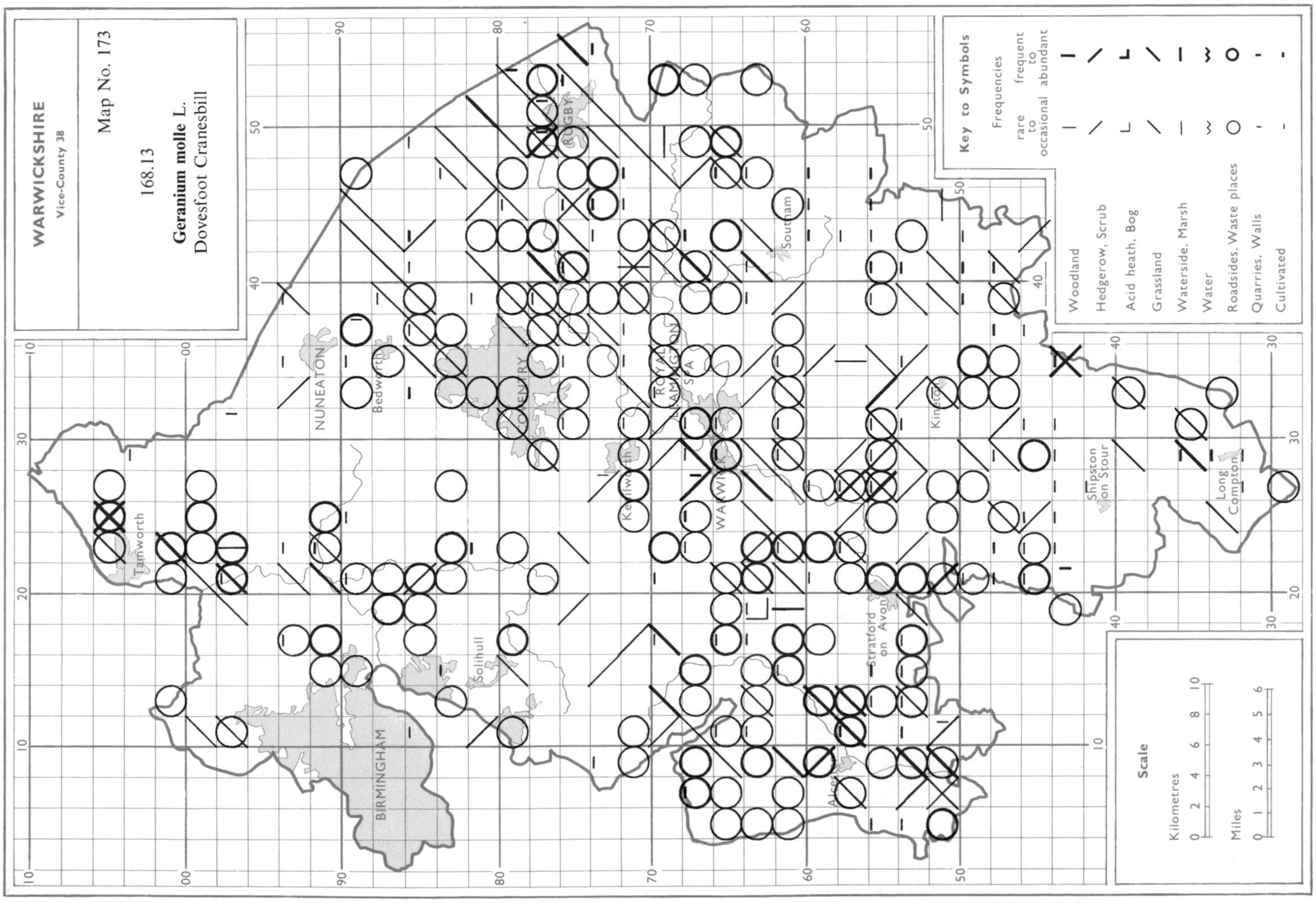
WARWICKSHIRE
Vice-County 38
Map No. 173
168.13
Geranium molle L.
Dovesfoot Cranesbill
Key to Symbols
Frequencies
rare to occasional
frequent to abundant
Woodland
Hedgerow, Scrub
Acid heath, Bog
Grassland
Waterside, Marsh
Water
Roadsides, Waste places
Quarries, Walls
Cultivated
Scale
Kilometres 0 2 4 6 8 10
Miles 0 1 2 3 4 5 6
Tamworth
NUNEATON
Bedworth
COVENTRY
RUGBY
BIRMINGHAM
Solihull
Kenilworth
ROYAL LEAMINGTON SPA
WARWICK
Southam
Kineton
Stratford on Avon
Alcester
Shipston on Stour
Long Compton

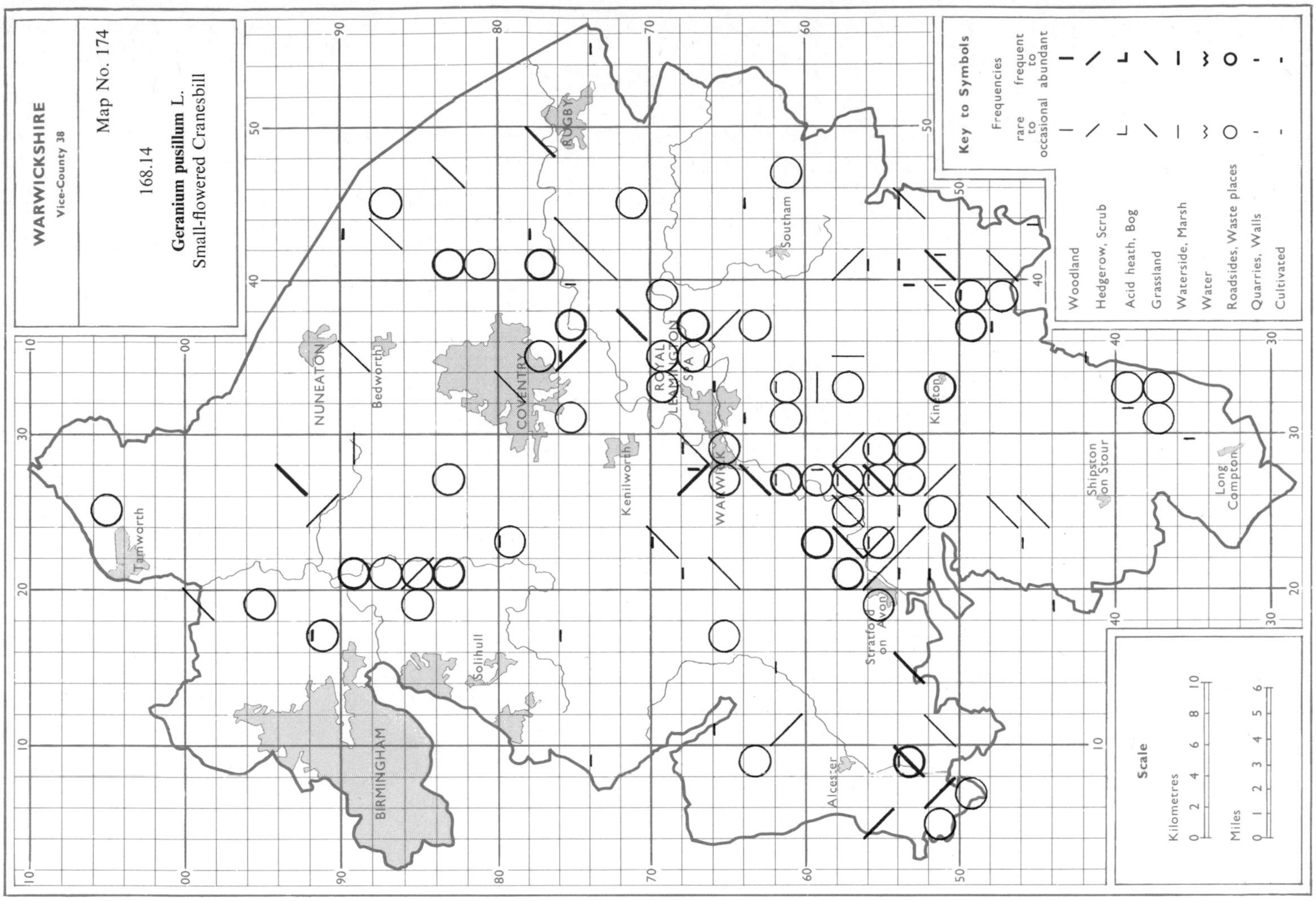
WARWICKSHIRE
Vice-County 38
Map No. 174
168.14
Geranium pusillum L.
Small-flowered Cranesbill
Key to Symbols
Frequencies
rare to occasional
frequent to abundant
Woodland
Hedgerow, Scrub
Acid heath, Bog
Grassland
Waterside, Marsh
Water
Roadsides, Waste places
Quarries, Walls
Cultivated
Scale
Kilometres 0 2 4 6 8 10
Miles 0 1 2 3 4 5 6
Tamworth
NUNEATON
Bedworth
COVENTRY
RUGBY
BIRMINGHAM
Solihull
Kenilworth
ROYAL LEAMINGTON SPA
WARWICK
Southam
Kineton
Stratford on Avon
Alcester
Shipston on Stour
Long Compton

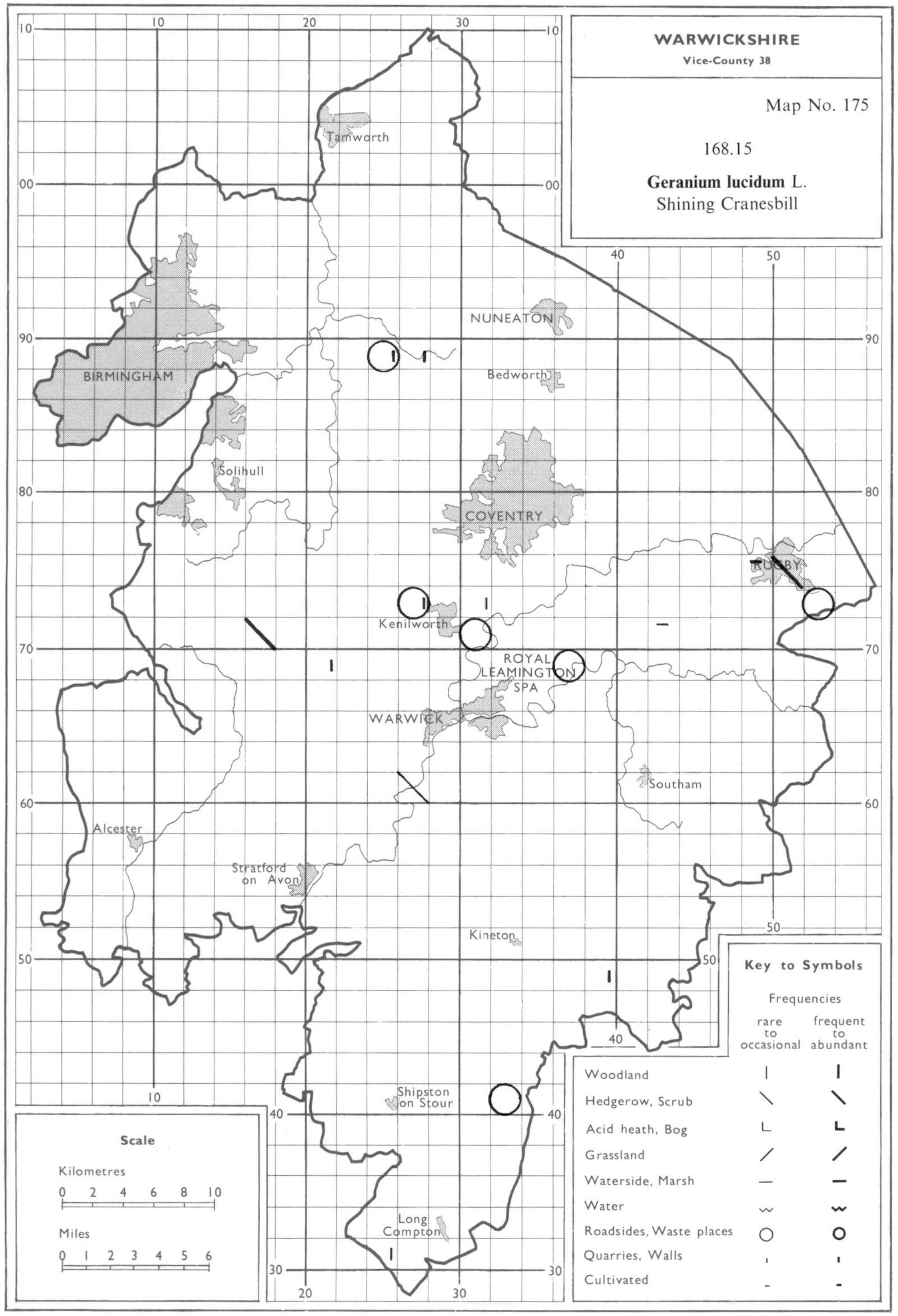
WARWICKSHIRE
Vice-County 38
Map No. 175
168.15
Geranium lucidum L.
Shining Cranesbill
Tamworth
NUNEATON
Bedworth
BIRMINGHAM
Solihull
COVENTRY
RUGBY
Kenilworth
ROYAL LEAMINGTON SPA
WARWICK
Southam
Alcester
Stratford on Avon
Kineton
Shipston on Stour
Long Compton
Key to Symbols
Frequencies
rare to occasional
frequent to abundant
Woodland
Hedgerow, Scrub
Acid heath, Bog
Grassland
Waterside, Marsh
Water
Roadsides, Waste places
Quarries, Walls
Cultivated
Scale
Kilometres
0 2 4 6 8 10
Miles
0 1 2 3 4 5 6

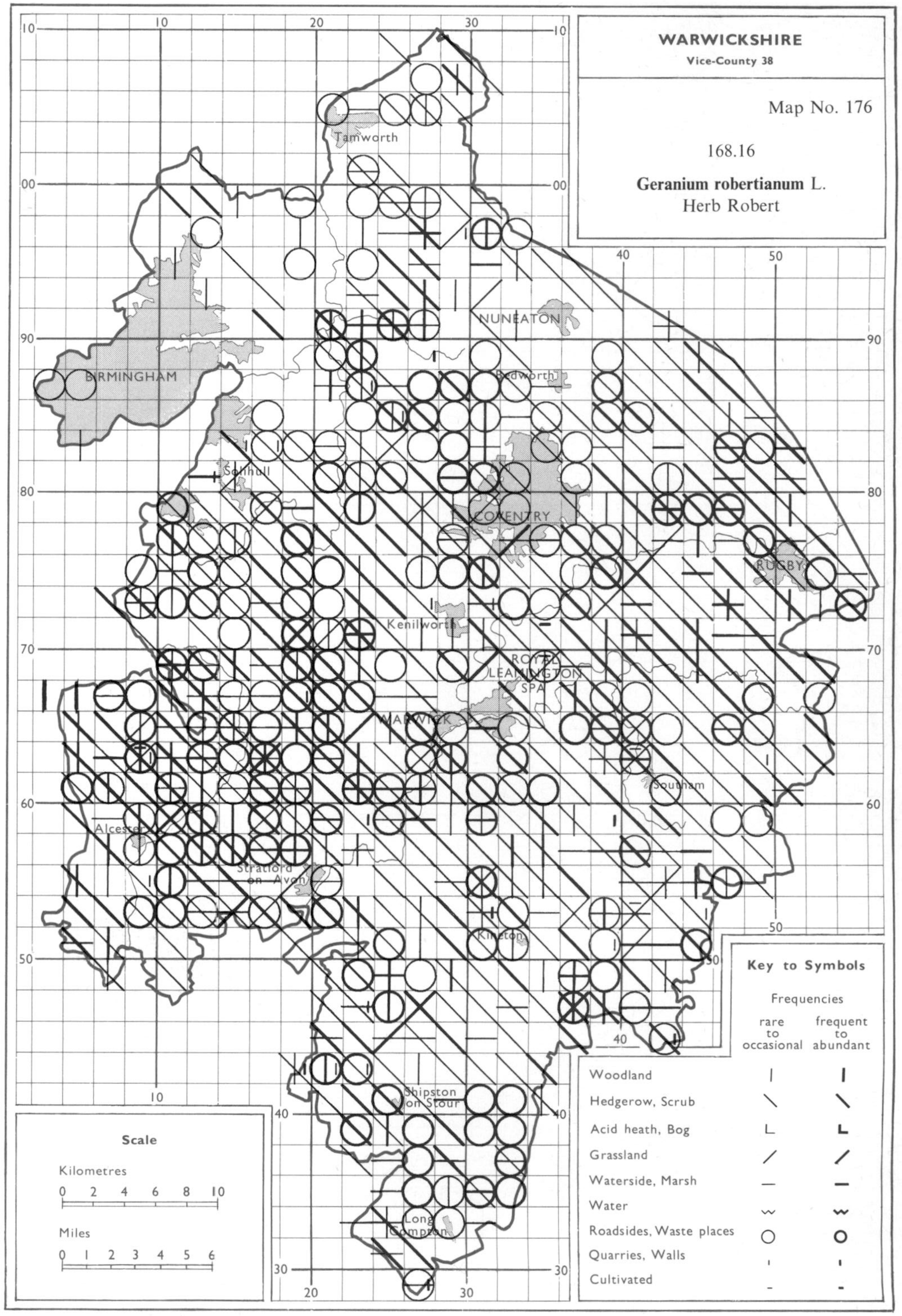
WARWICKSHIRE
Vice-County 38
Map No. 176
168.16
Geranium robertianum L.
Herb Robert
Tamworth
NUNEATON
Bedworth
BIRMINGHAM
Solihull
COVENTRY
RUGBY
Kenilworth
ROYAL LEAMINGTON SPA
WARWICK
Southam
Alcester
Stratford on Avon
Kineton
Shipston on Stour
Long Compton
Key to Symbols
Frequencies
rare to occasional
frequent to abundant
Woodland
Hedgerow, Scrub
Acid heath, Bog
Grassland
Waterside, Marsh
Water
Roadsides, Waste places
Quarries, Walls
Cultivated
Scale
Kilometres
0 2 4 6 8 10
Miles
0 1 2 3 4 5 6

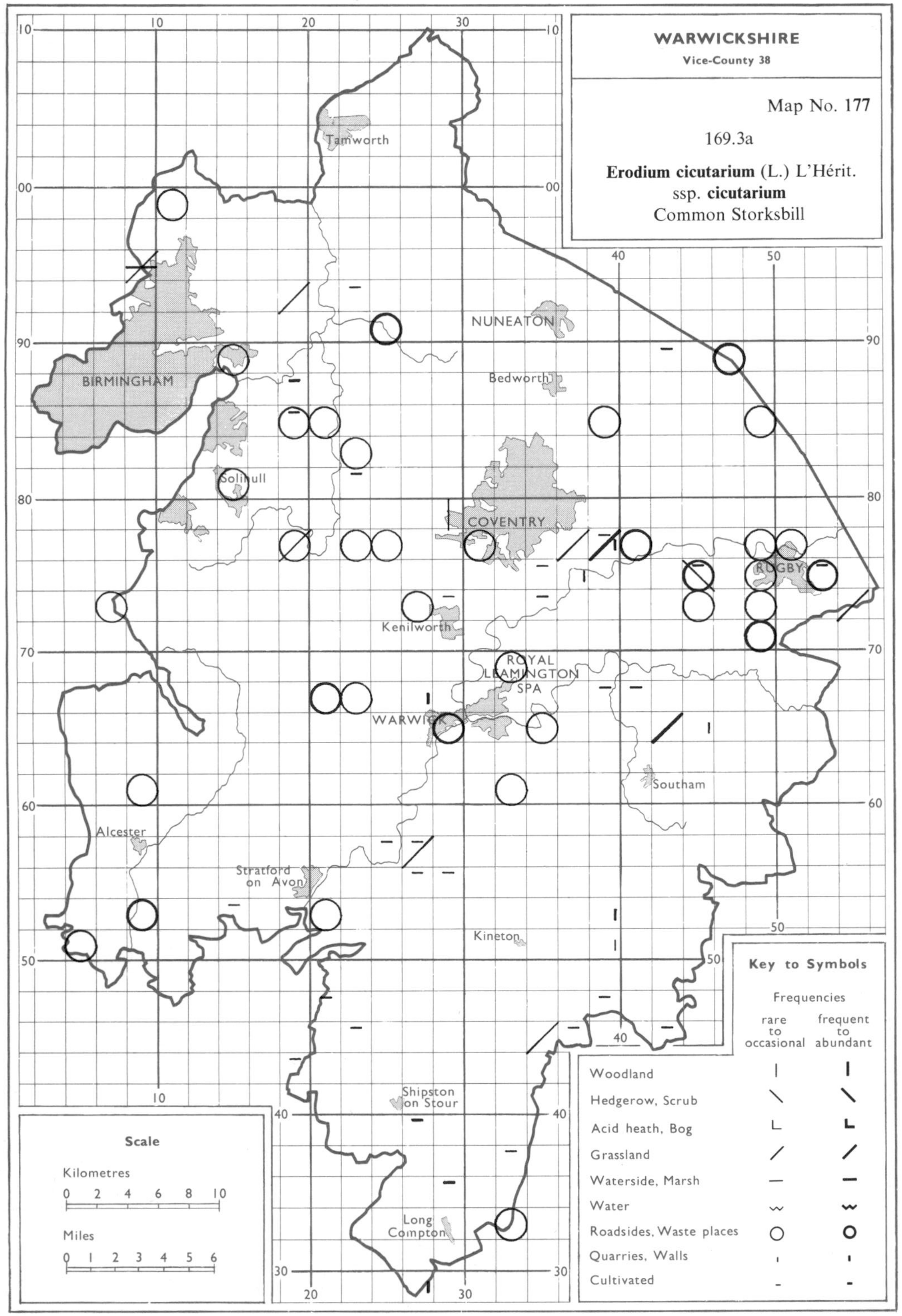
WARWICKSHIRE
Vice-County 38
Map No. 177
169.3a
Erodium cicutarium (L.) L'Hérit.
ssp. **cicutarium**
Common Storksbill
Tamworth
BIRMINGHAM
Solihull
NUNEATON
Bedworth
COVENTRY
RUGBY
Kenilworth
ROYAL LEAMINGTON SPA
WARWICK
Southam
Alcester
Stratford on Avon
Kineton
Shipston on Stour
Long Compton
Key to Symbols
Frequencies
rare to occasional
frequent to abundant
Woodland
Hedgerow, Scrub
Acid heath, Bog
Grassland
Waterside, Marsh
Water
Roadsides, Waste places
Quarries, Walls
Cultivated
Scale
Kilometres
0 2 4 6 8 10
Miles
0 1 2 3 4 5 6

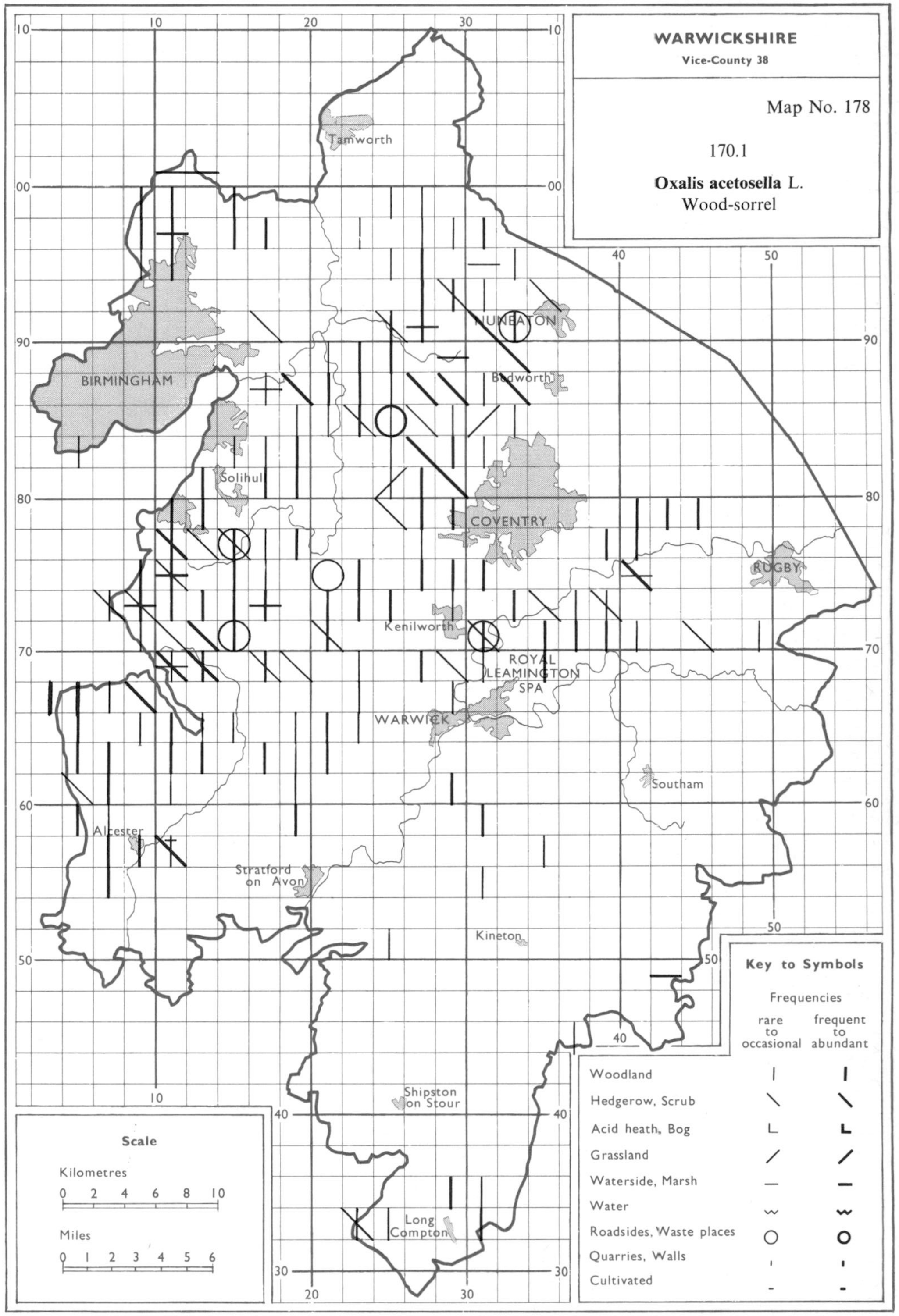
WARWICKSHIRE
Vice-County 38
Map No. 178
170.1
Oxalis acetosella L.
Wood-sorrel
Tamworth
BIRMINGHAM
Solihull
NUNEATON
Bedworth
COVENTRY
RUGBY
Kenilworth
ROYAL LEAMINGTON SPA
WARWICK
Southam
Alcester
Stratford on Avon
Kineton
Shipston on Stour
Long Compton
Key to Symbols
Frequencies
rare to occasional
frequent to abundant
Woodland
Hedgerow, Scrub
Acid heath, Bog
Grassland
Waterside, Marsh
Water
Roadsides, Waste places
Quarries, Walls
Cultivated
Scale
Kilometres
0 2 4 6 8 10
Miles
0 1 2 3 4 5 6

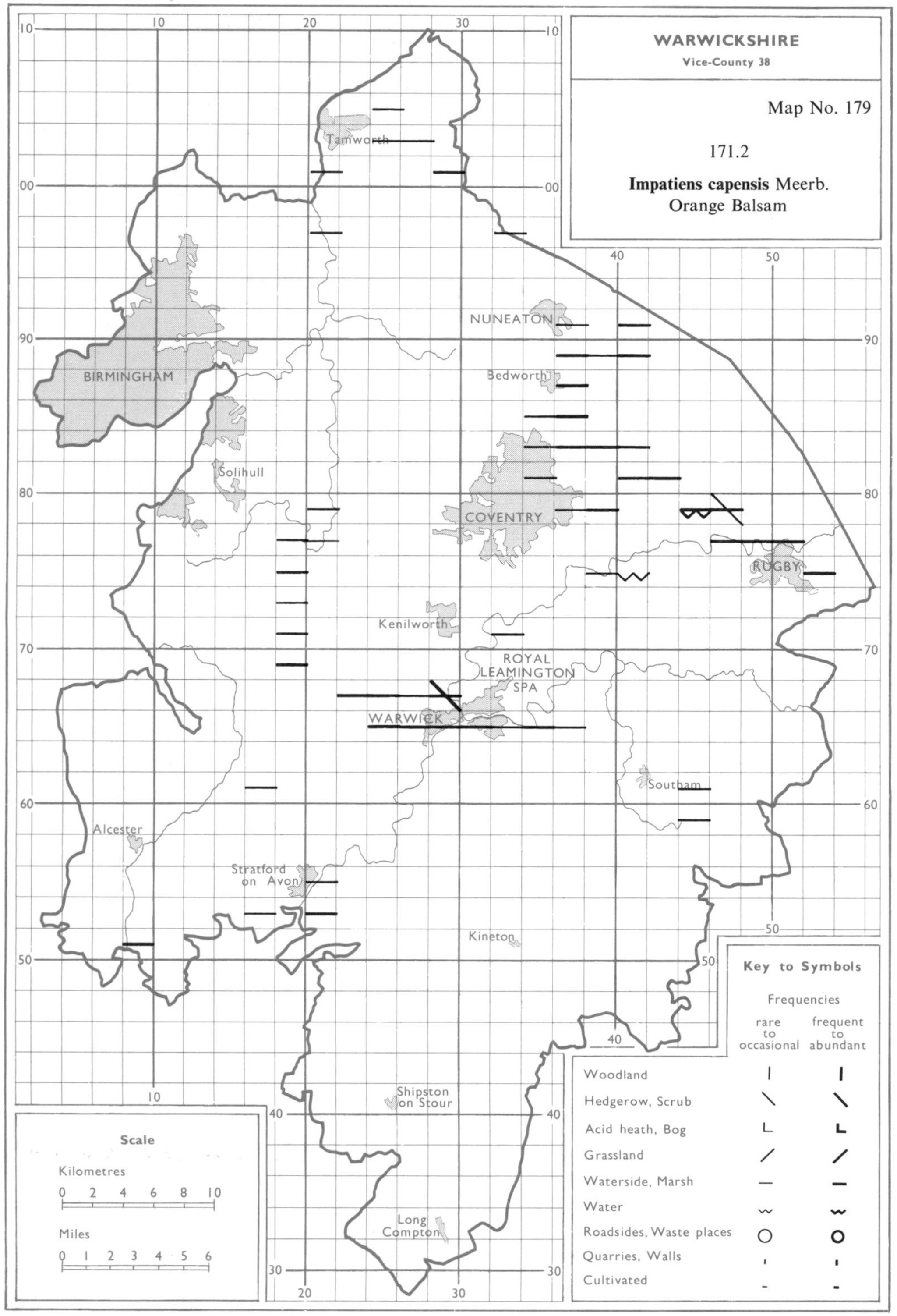
WARWICKSHIRE
Vice-County 38
Map No. 179
171.2
Impatiens capensis Meerb.
Orange Balsam
Tamworth
BIRMINGHAM
Solihull
NUNEATON
Bedworth
COVENTRY
RUGBY
Kenilworth
ROYAL LEAMINGTON SPA
WARWICK
Southam
Alcester
Stratford on Avon
Kineton
Shipston on Stour
Long Compton
Key to Symbols
Frequencies
rare to occasional
frequent to abundant
Woodland
Hedgerow, Scrub
Acid heath, Bog
Grassland
Waterside, Marsh
Water
Roadsides, Waste places
Quarries, Walls
Cultivated
Scale
Kilometres
0 2 4 6 8 10
Miles
0 1 2 3 4 5 6

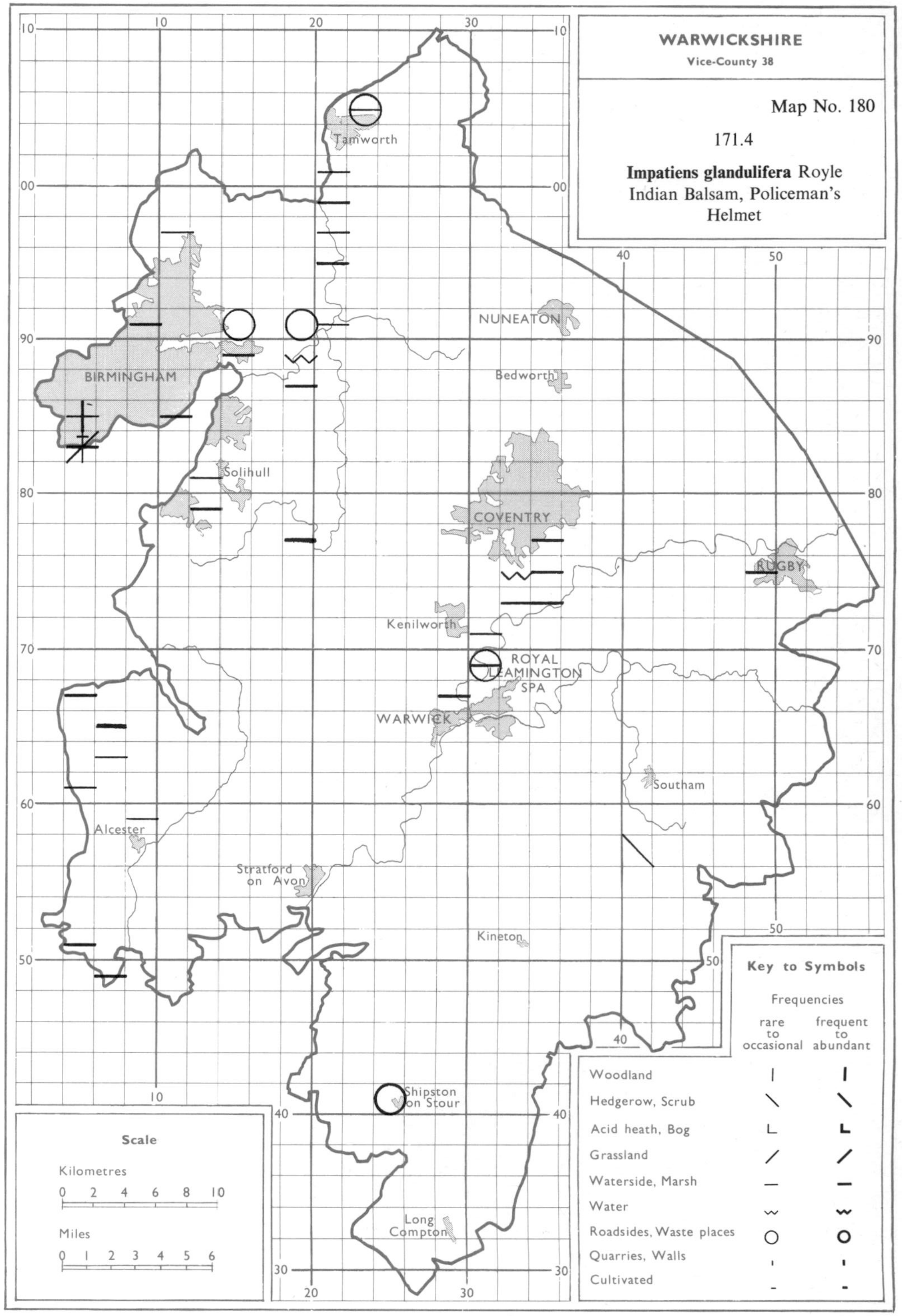
WARWICKSHIRE
Vice-County 38
Map No. 180
171.4
Impatiens glandulifera Royle
Indian Balsam, Policeman's Helmet
Tamworth
BIRMINGHAM
Solihull
NUNEATON
Bedworth
COVENTRY
RUGBY
Kenilworth
ROYAL LEAMINGTON SPA
WARWICK
Southam
Alcester
Stratford on Avon
Kineton
Shipston on Stour
Long Compton
Key to Symbols
Frequencies
rare to occasional
frequent to abundant
Woodland
Hedgerow, Scrub
Acid heath, Bog
Grassland
Waterside, Marsh
Water
Roadsides, Waste places
Quarries, Walls
Cultivated
Scale
Kilometres
0 2 4 6 8 10
Miles
0 1 2 3 4 5 6

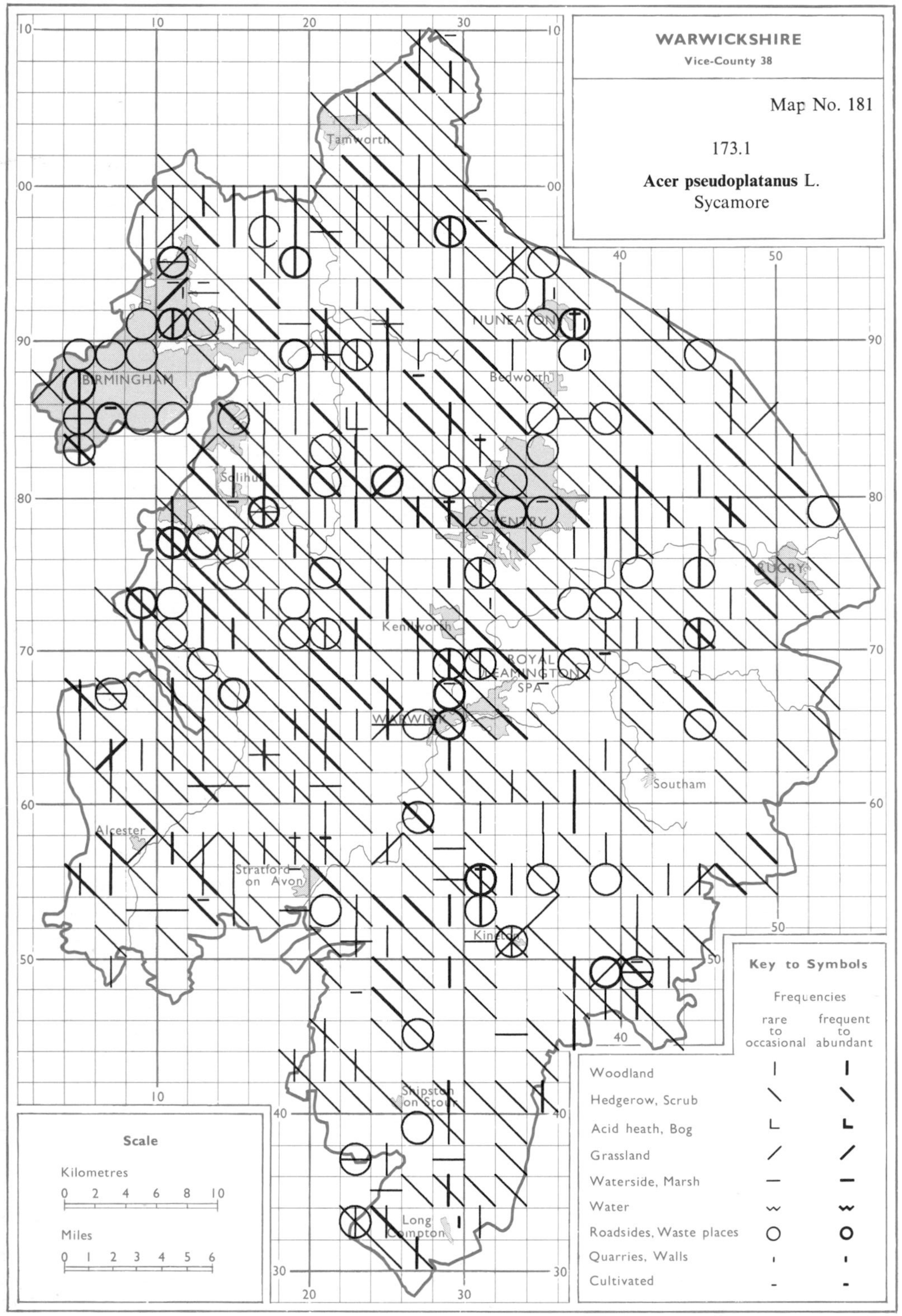
WARWICKSHIRE
Vice-County 38
Map No. 181
173.1
Acer pseudoplatanus L.
Sycamore
Tamworth
BIRMINGHAM
NUNEATON
Bedworth
Solihull
COVENTRY
RUGBY
Kenilworth
ROYAL LEAMINGTON SPA
WARWICK
Southam
Alcester
Stratford on Avon
Kineton
Shipston on Stour
Long Compton
Key to Symbols
Frequencies
rare to occasional
frequent to abundant
Woodland
Hedgerow, Scrub
Acid heath, Bog
Grassland
Waterside, Marsh
Water
Roadsides, Waste places
Quarries, Walls
Cultivated
Scale
Kilometres
0 2 4 6 8 10
Miles
0 1 2 3 4 5 6

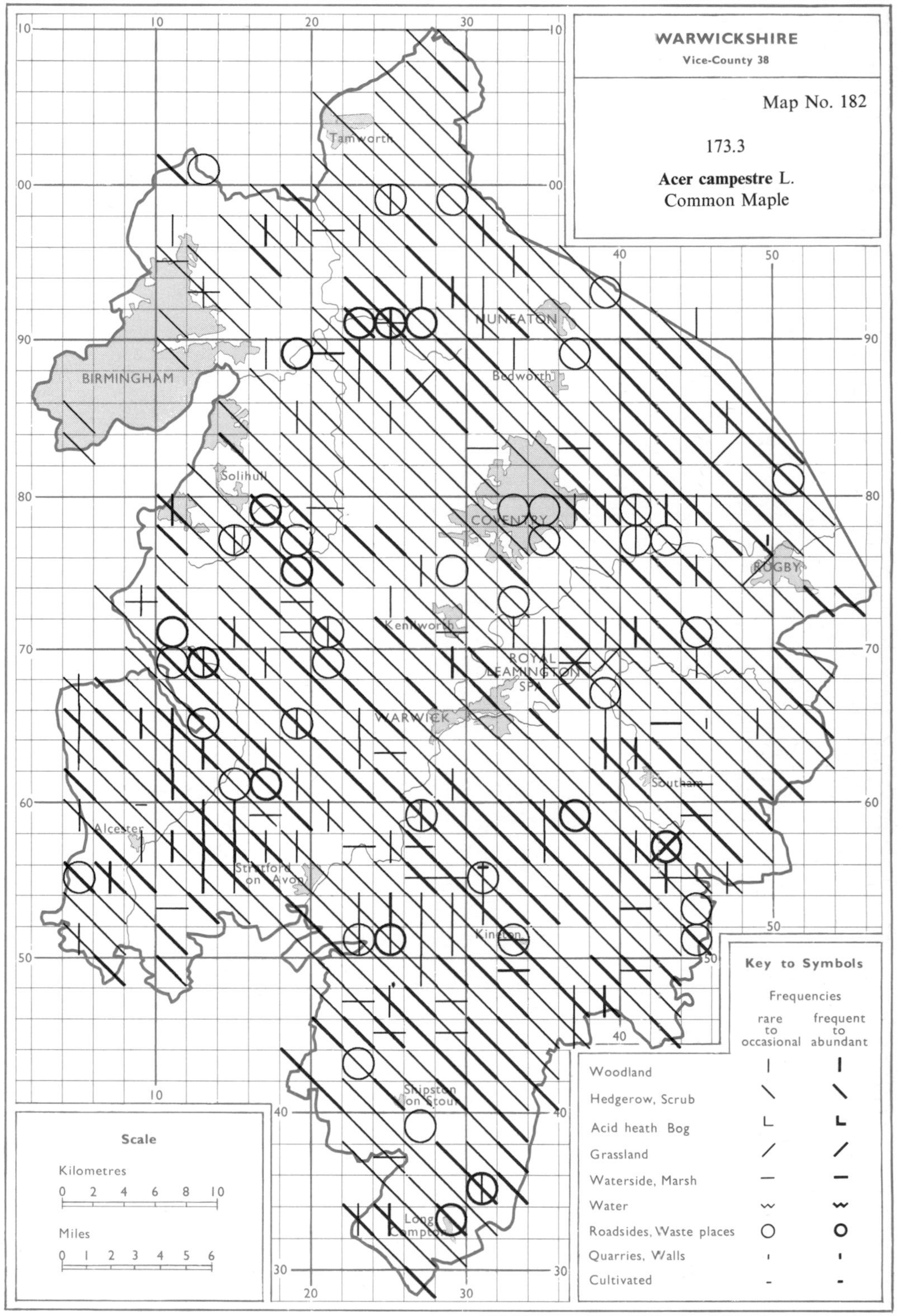
WARWICKSHIRE
Vice-County 38
Map No. 182
173.3
Acer campestre L.
Common Maple
Tamworth
BIRMINGHAM
NUNEATON
Bedworth
Solihull
COVENTRY
RUGBY
Kenilworth
ROYAL LEAMINGTON SPA
WARWICK
Southam
Alcester
Stratford on Avon
Kineton
Shipston on Stour
Long Compton
Key to Symbols
Frequencies
rare to occasional
frequent to abundant
Woodland
Hedgerow, Scrub
Acid heath Bog
Grassland
Waterside, Marsh
Water
Roadsides, Waste places
Quarries, Walls
Cultivated
Scale
Kilometres
0 2 4 6 8 10
Miles
0 1 2 3 4 5 6

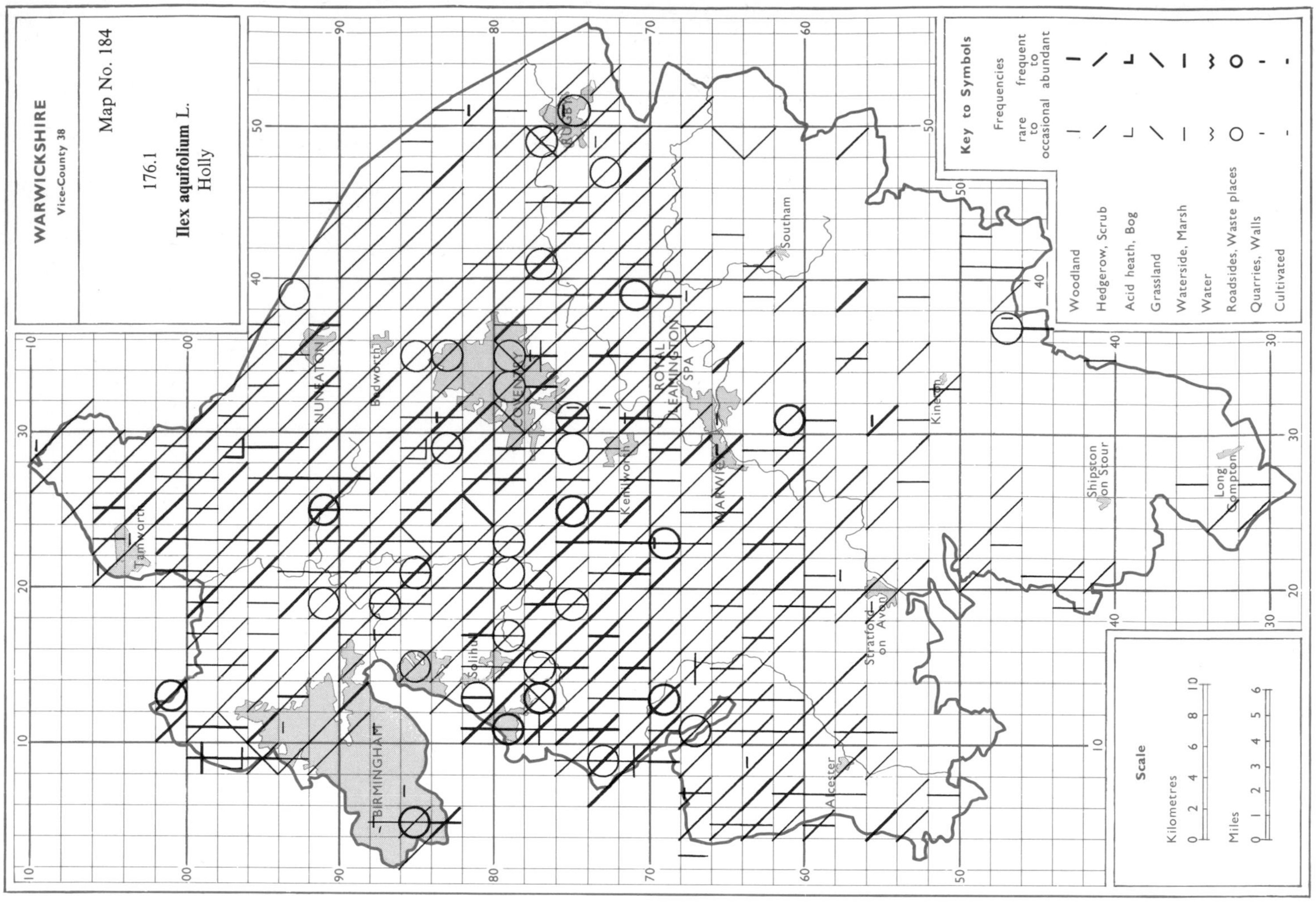
WARWICKSHIRE
Vice-County 38
Map No. 184
176.1
Ilex aquifolium L.
Holly
Key to Symbols
Frequencies
rare to occasional
frequent to abundant
Woodland
Hedgerow, Scrub
Acid heath, Bog
Grassland
Waterside, Marsh
Water
Roadsides, Waste places
Quarries, Walls
Cultivated
Scale
Kilometres 0 2 4 6 8 10
Miles 0 1 2 3 4 5 6
Tamworth
NUNEATON
Bedworth
COVENTRY
RUGBY
Solihull
BIRMINGHAM
Kenilworth
ROYAL LEAMINGTON SPA
WARWICK
Southam
Kineton
Stratford on Avon
Alcester
Shipston on Stour
Long Compton

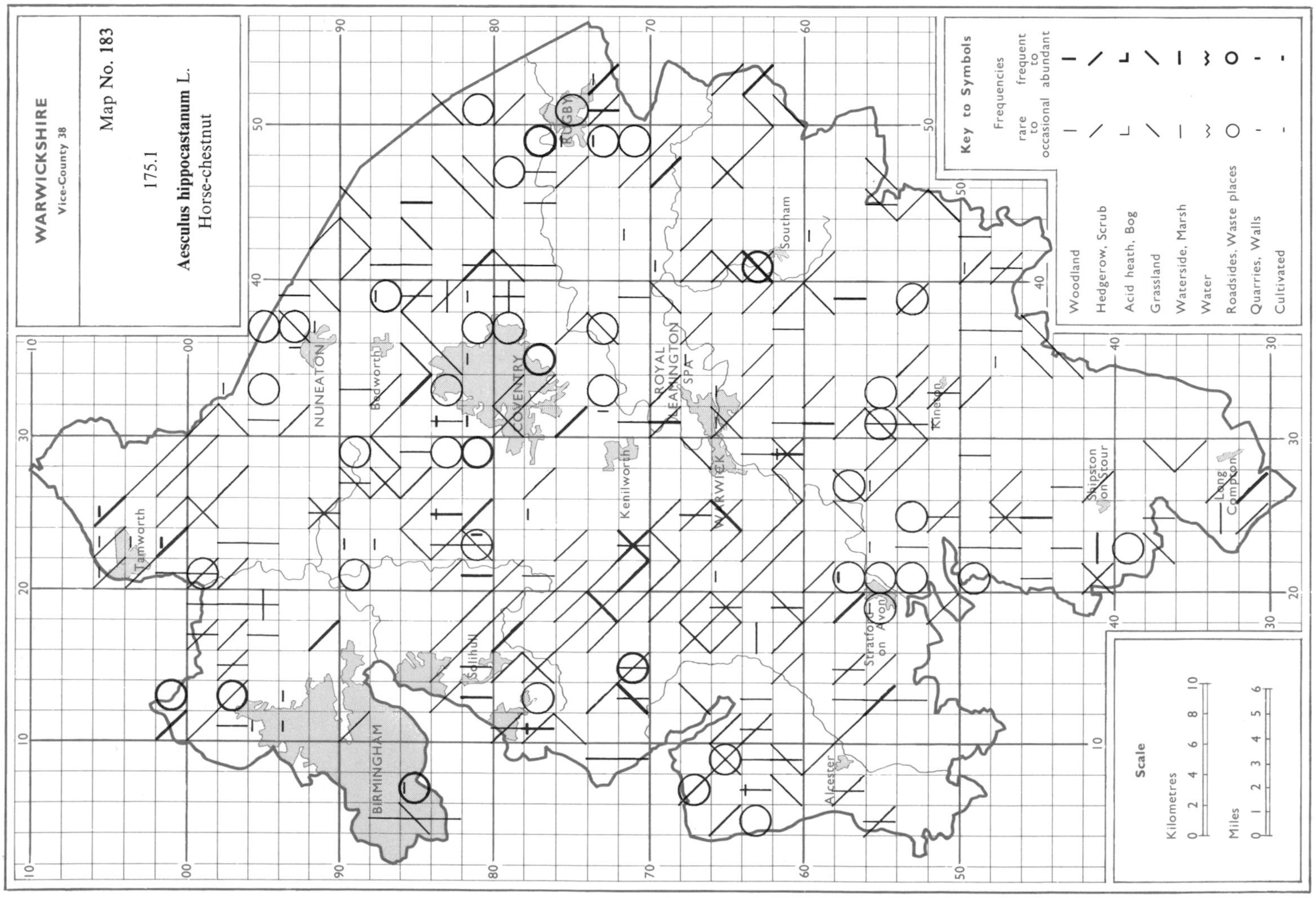
WARWICKSHIRE
Vice-County 38
Map No. 183
175.1
Aesculus hippocastanum L.
Horse-chestnut
Key to Symbols
Frequencies
rare to occasional
frequent to abundant
Woodland
Hedgerow, Scrub
Acid heath, Bog
Grassland
Waterside, Marsh
Water
Roadsides, Waste places
Quarries, Walls
Cultivated
Scale
Kilometres 0 2 4 6 8 10
Miles 0 1 2 3 4 5 6
Tamworth
NUNEATON
Bedworth
COVENTRY
RUGBY
Solihull
BIRMINGHAM
Kenilworth
ROYAL LEAMINGTON SPA
WARWICK
Southam
Kineton
Stratford on Avon
Alcester
Shipston on Stour
Long Compton

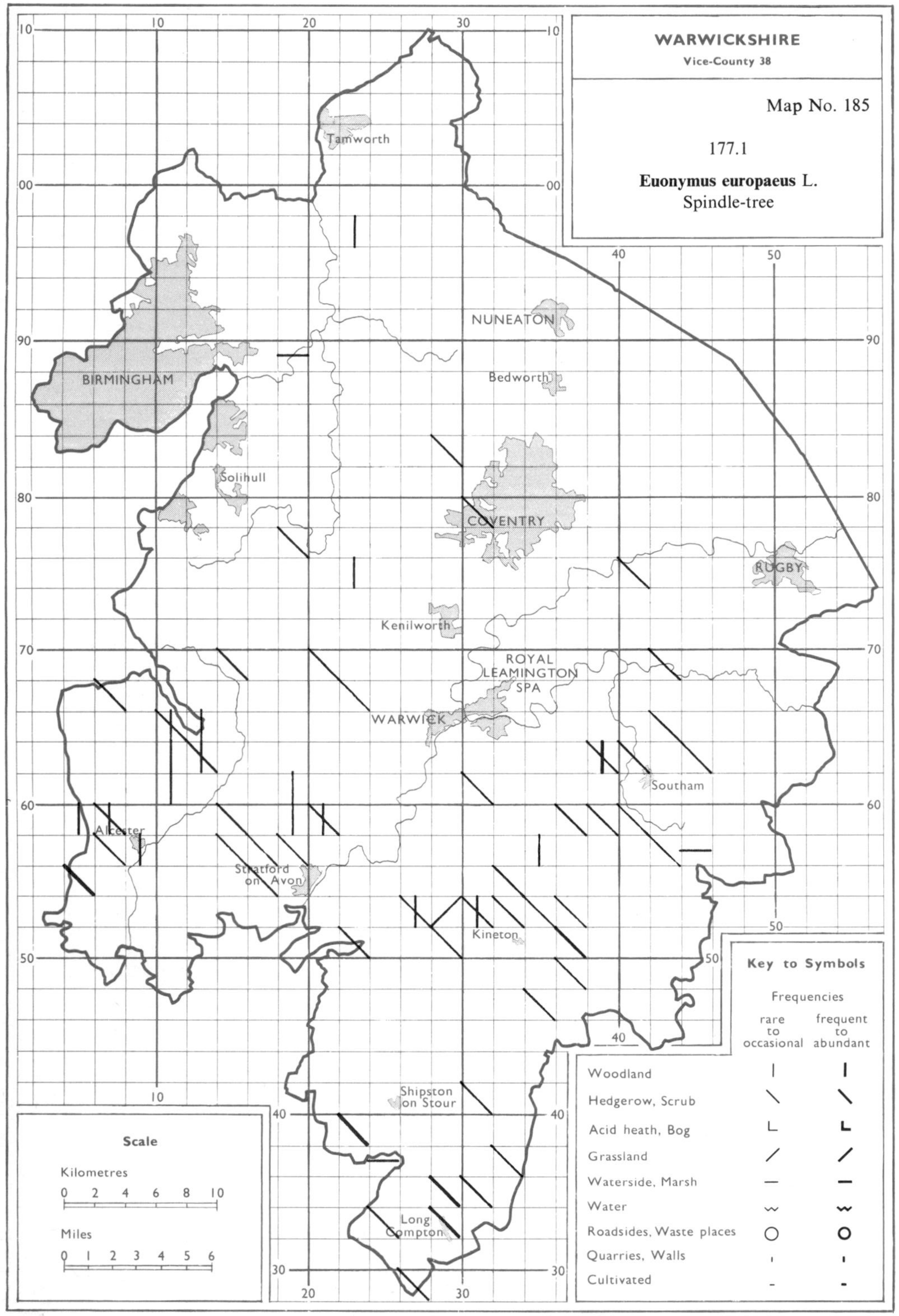
WARWICKSHIRE
Vice-County 38
Map No. 185
177.1
Euonymus europaeus L.
Spindle-tree
Tamworth
NUNEATON
Bedworth
BIRMINGHAM
Solihull
COVENTRY
RUGBY
Kenilworth
ROYAL LEAMINGTON SPA
WARWICK
Southam
Alcester
Stratford on Avon
Kineton
Shipston on Stour
Long Compton
Key to Symbols
Frequencies
rare to occasional
frequent to abundant
Woodland
Hedgerow, Scrub
Acid heath, Bog
Grassland
Waterside, Marsh
Water
Roadsides, Waste places
Quarries, Walls
Cultivated
Scale
Kilometres
0 2 4 6 8 10
Miles
0 1 2 3 4 5 6

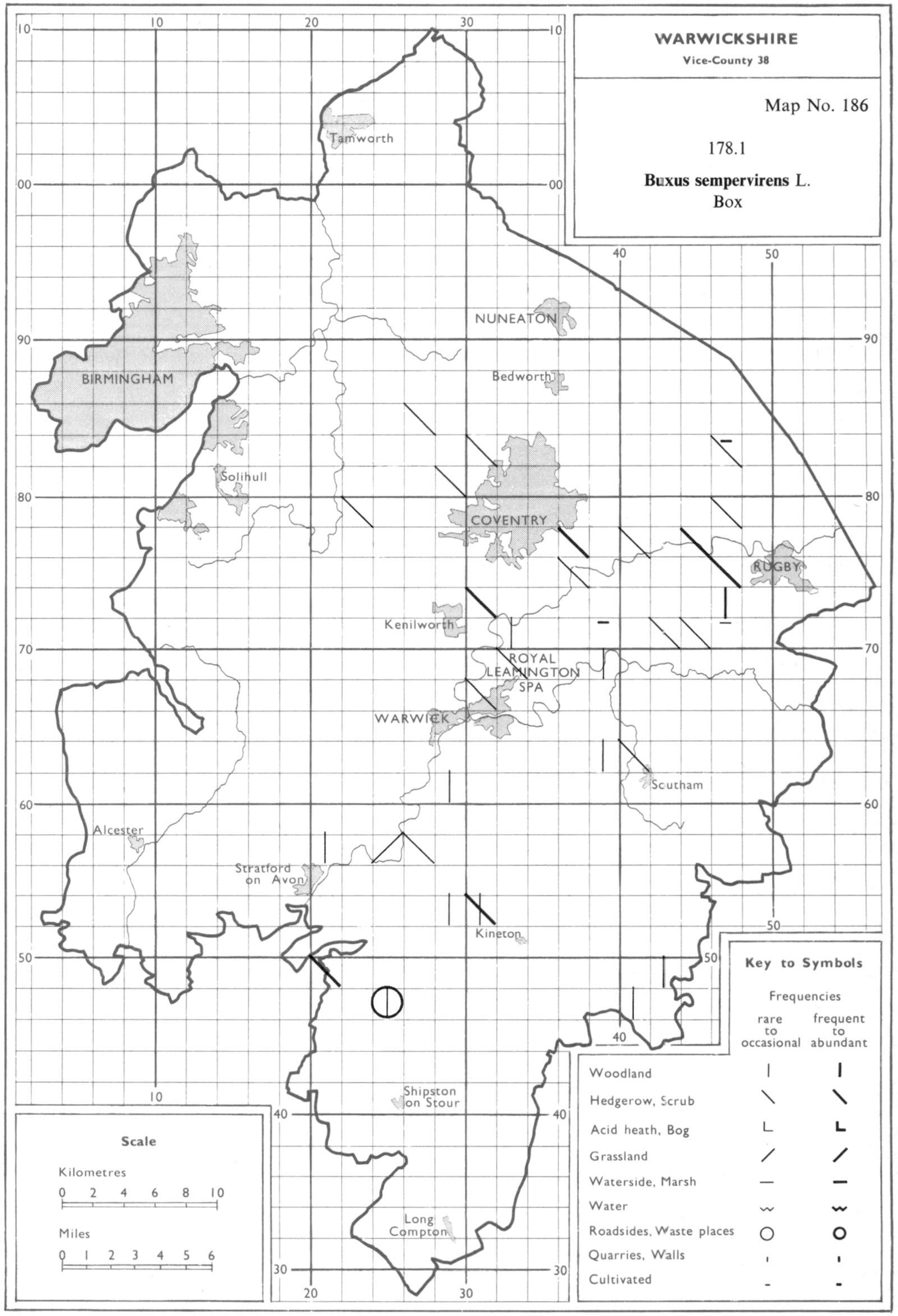
WARWICKSHIRE
Vice-County 38
Map No. 186
178.1
Buxus sempervirens L.
Box
Tamworth
NUNEATON
Bedworth
BIRMINGHAM
Solihull
COVENTRY
RUGBY
Kenilworth
ROYAL LEAMINGTON SPA
WARWICK
Scutham
Alcester
Stratford on Avon
Kineton
Shipston on Stour
Long Compton
Key to Symbols
Frequencies
rare to occasional
frequent to abundant
Woodland
Hedgerow, Scrub
Acid heath, Bog
Grassland
Waterside, Marsh
Water
Roadsides, Waste places
Quarries, Walls
Cultivated
Scale
Kilometres
0 2 4 6 8 10
Miles
0 1 2 3 4 5 6

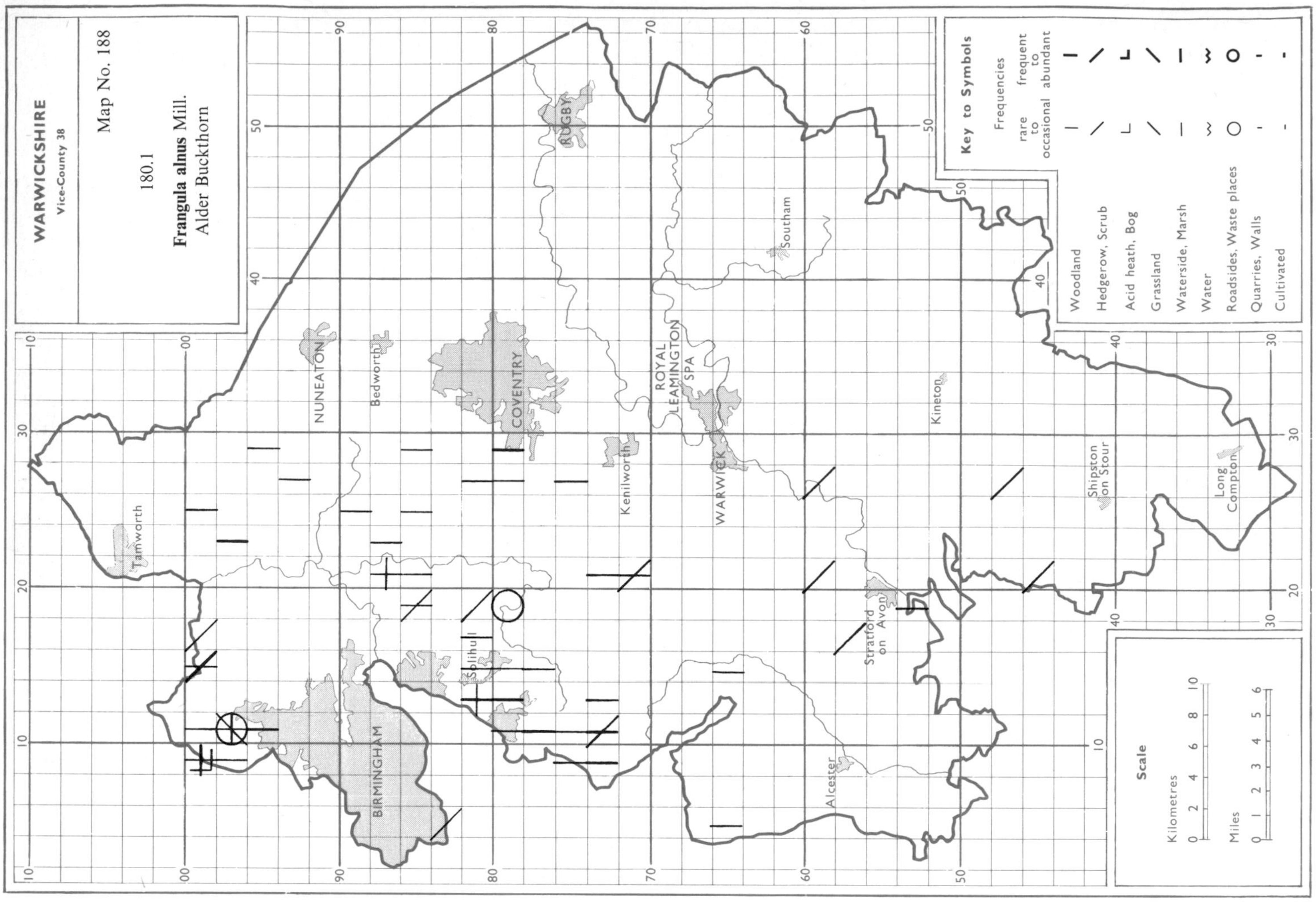
WARWICKSHIRE
Vice-County 38
Map No. 188
180.1
Frangula alnus Mill.
Alder Buckthorn
Key to Symbols
Frequencies
rare to occasional
frequent to abundant
Woodland
Hedgerow, Scrub
Acid heath, Bog
Grassland
Waterside, Marsh
Water
Roadsides, Waste places
Quarries, Walls
Cultivated
Scale
Kilometres
0 2 4 6 8 10
Miles
0 1 2 3 4 5 6
NUNEATON
Bedworth
COVENTRY
RUGBY
Southam
ROYAL LEAMINGTON SPA
Kenilworth
WARWICK
Kineton
Tamworth
Solihull
BIRMINGHAM
Stratford on Avon
Alcester
Shipston on Stour
Long Compton

WARWICKSHIRE
Vice-County 38
Map No. 187
179.1
Rhamnus catharticus L.
Buckthorn
Key to Symbols
Frequencies
rare to occasional
frequent to abundant
Woodland
Hedgerow, Scrub
Acid heath, Bog
Grassland
Waterside, Marsh
Water
Roadsides, Waste places
Quarries, Walls
Cultivated
Scale
Kilometres
0 2 4 6 8 10
Miles
0 1 2 3 4 5 6
NUNEATON
Bedworth
COVENTRY
RUGBY
Southam
ROYAL LEAMINGTON SPA
Kenilworth
WARWICK
Kineton
Tamworth
Solihull
BIRMINGHAM
Stratford on Avon
Alcester
Shipston on Stour
Long Compton

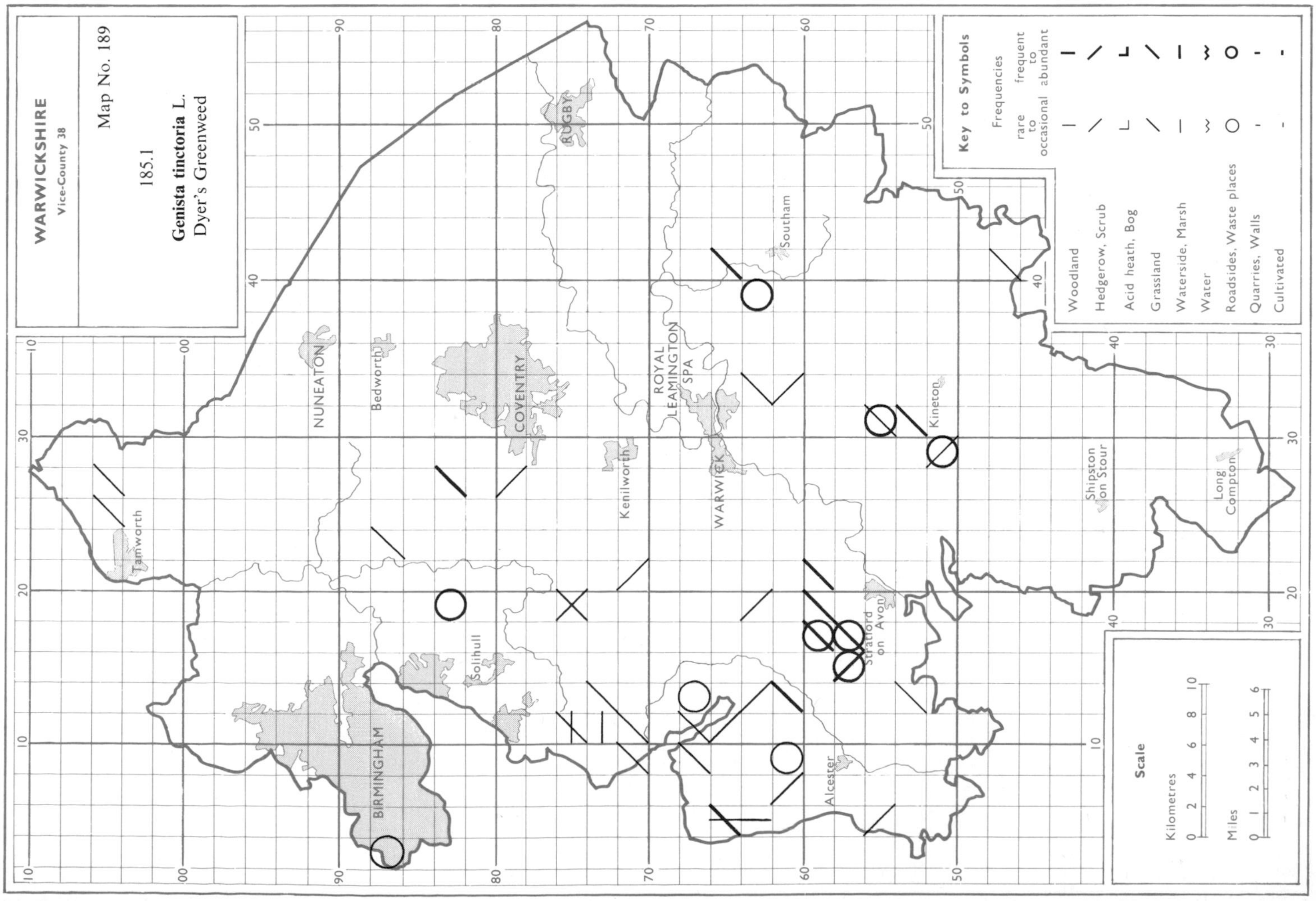
WARWICKSHIRE
Vice-County 38
Map No. 189
185.1
Genista tinctoria L.
Dyer's Greenweed
Key to Symbols
Frequencies
rare to occasional
frequent to abundant
Woodland
Hedgerow, Scrub
Acid heath, Bog
Grassland
Waterside, Marsh
Water
Roadsides, Waste places
Quarries, Walls
Cultivated
Scale
Kilometres
Miles
NUNEATON
Bedworth
COVENTRY
RUGBY
ROYAL LEAMINGTON SPA
Southam
Kenilworth
WARWICK
Kineton
Tamworth
Solihull
BIRMINGHAM
Stratford on Avon
Alcester
Shipston on Stour
Long Compton

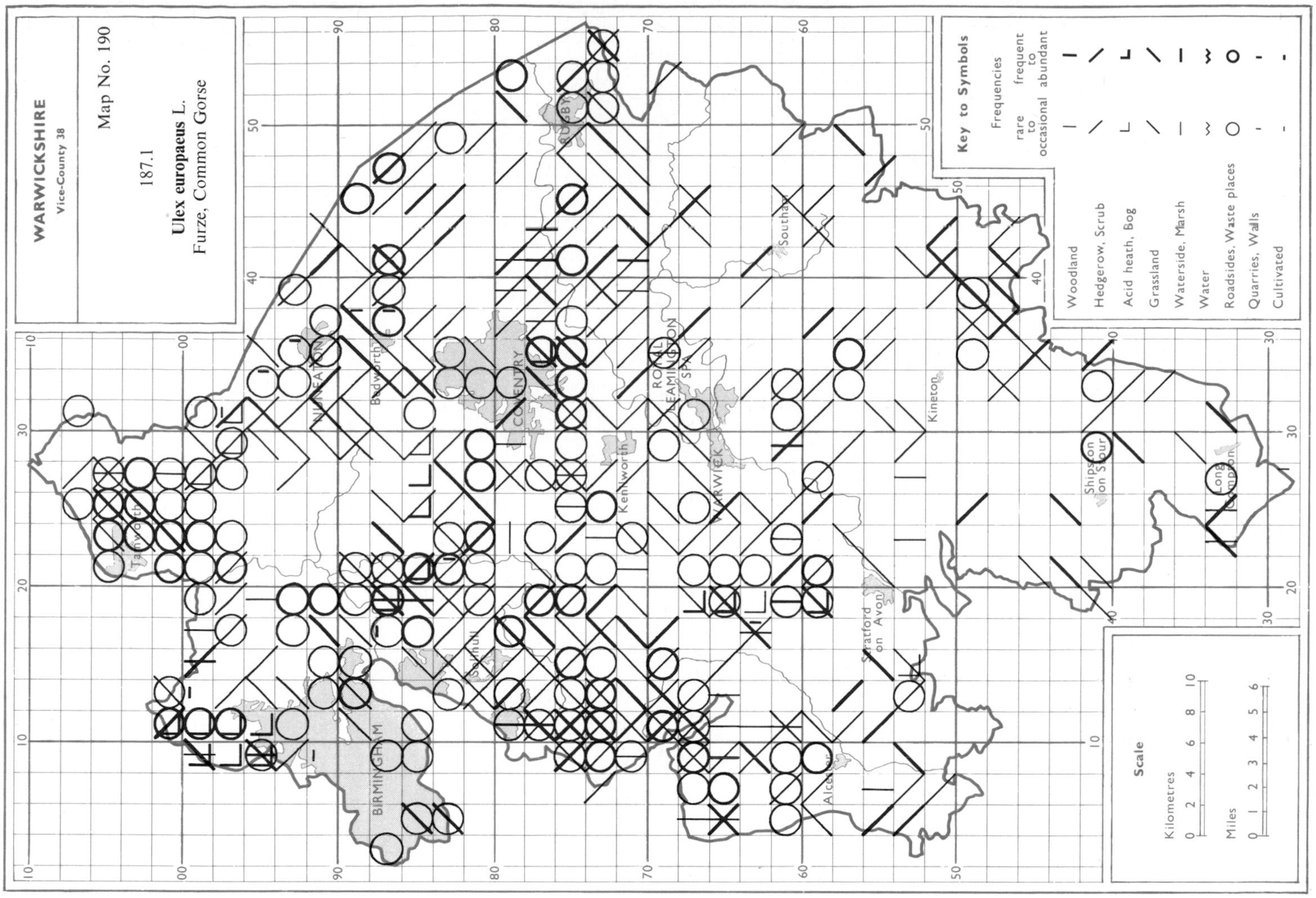
WARWICKSHIRE
Vice-County 38
Map No. 190
187.1
Ulex europaeus L.
Furze, Common Gorse
Key to Symbols
Frequencies
rare to occasional
frequent to abundant
Woodland
Hedgerow, Scrub
Acid heath, Bog
Grassland
Waterside, Marsh
Water
Roadsides, Waste places
Quarries, Walls
Cultivated
Scale
Kilometres
Miles
NUNEATON
Bedworth
COVENTRY
RUGBY
ROYAL LEAMINGTON SPA
Southam
Kenilworth
WARWICK
Kineton
Tamworth
Solihull
BIRMINGHAM
Stratford on Avon
Alcester
Shipston on Stour
Long Compton

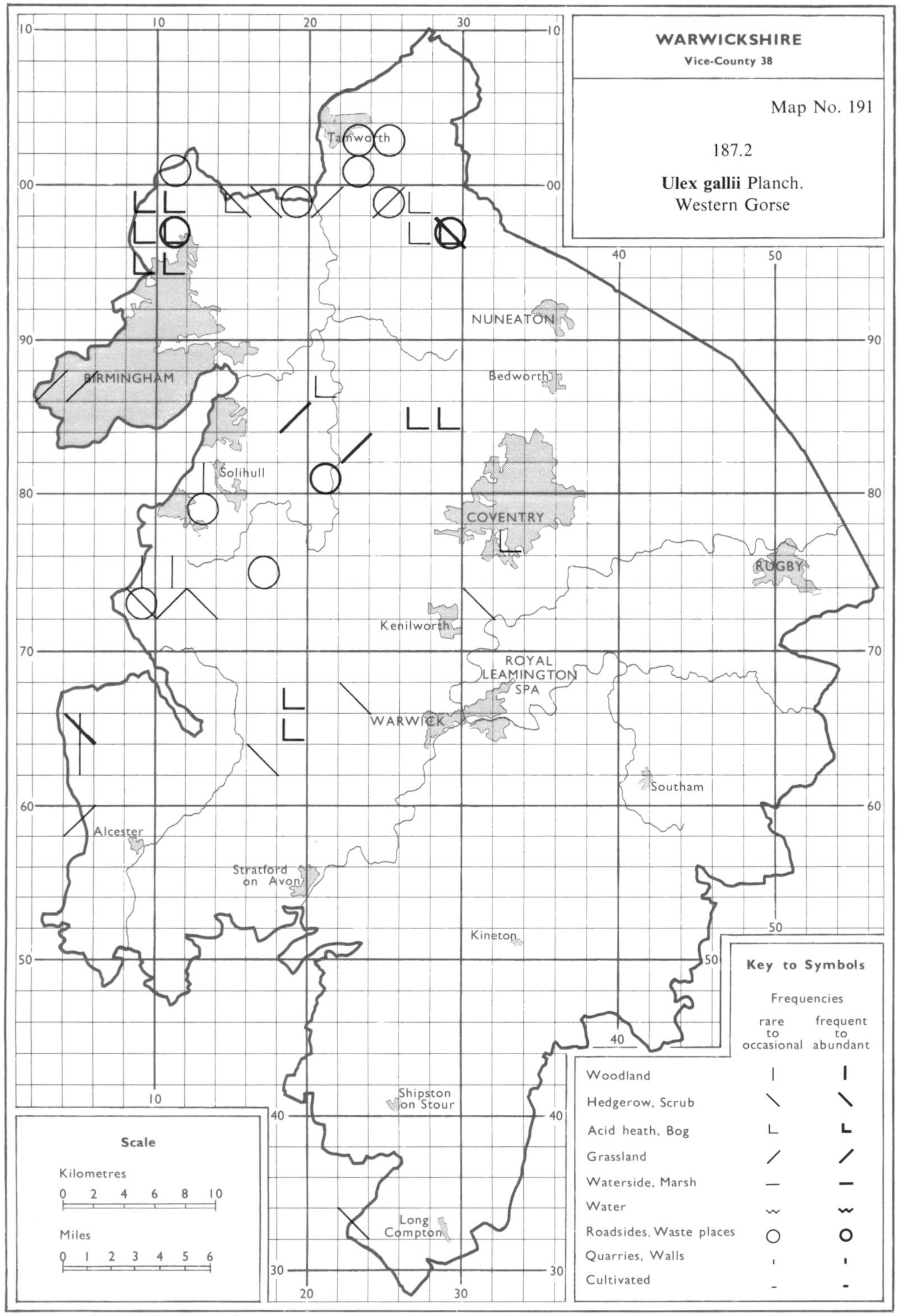
WARWICKSHIRE
Vice-County 38
Map No. 191
187.2
Ulex gallii Planch.
Western Gorse
Tamworth
BIRMINGHAM
Solihull
NUNEATON
Bedworth
COVENTRY
RUGBY
Kenilworth
ROYAL LEAMINGTON SPA
WARWICK
Southam
Alcester
Stratford on Avon
Kineton
Shipston on Stour
Long Compton
Key to Symbols
Frequencies
rare to occasional
frequent to abundant
Woodland
Hedgerow, Scrub
Acid heath, Bog
Grassland
Waterside, Marsh
Water
Roadsides, Waste places
Quarries, Walls
Cultivated
Scale
Kilometres
0 2 4 6 8 10
Miles
0 1 2 3 4 5 6

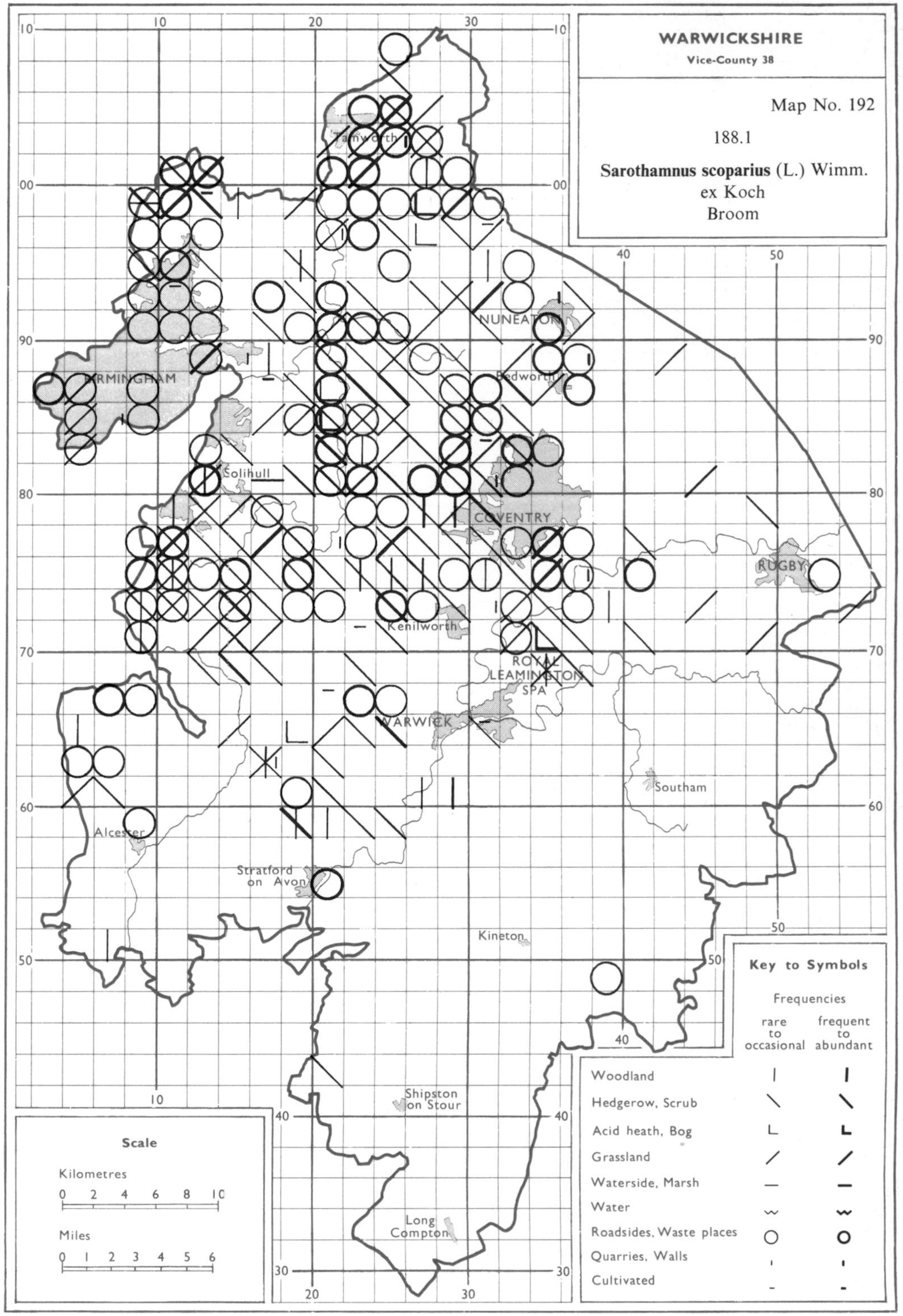
WARWICKSHIRE
Vice-County 38
Map No. 192
188.1
Sarothamnus scoparius (L.) Wimm. ex Koch
Broom
Tamworth
BIRMINGHAM
Solihull
NUNEATON
Bedworth
COVENTRY
RUGBY
Kenilworth
ROYAL LEAMINGTON SPA
WARWICK
Southam
Alcester
Stratford on Avon
Kineton
Shipston on Stour
Long Compton
Key to Symbols
Frequencies
rare to occasional
frequent to abundant
Woodland
Hedgerow, Scrub
Acid heath, Bog
Grassland
Waterside, Marsh
Water
Roadsides, Waste places
Quarries, Walls
Cultivated
Scale
Kilometres
0 2 4 6 8 10
Miles
0 1 2 3 4 5 6

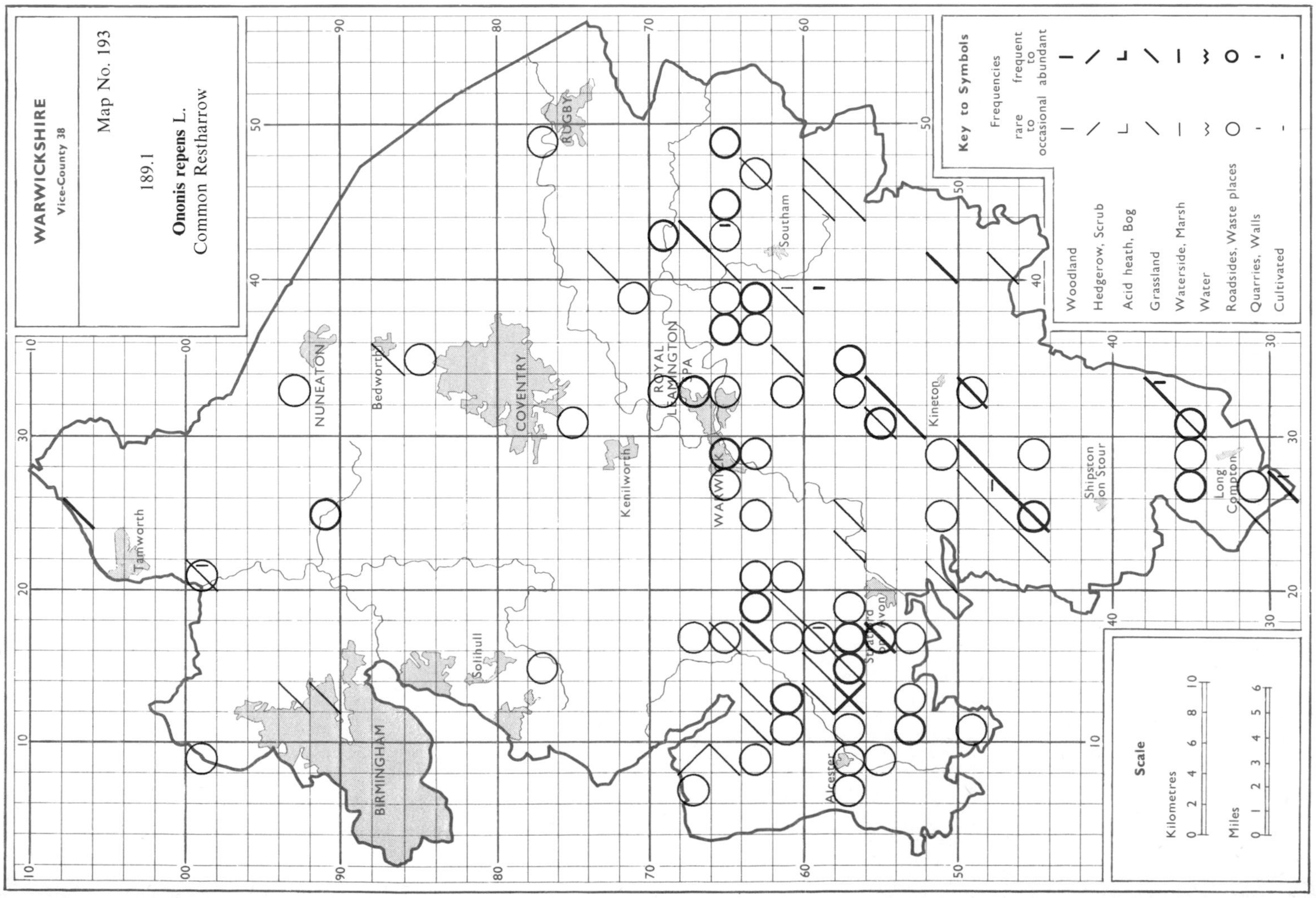
WARWICKSHIRE
Vice-County 38
Map No. 193
189.1
Ononis repens L.
Common Restharrow
Key to Symbols
Frequencies
rare to occasional
frequent to abundant
Woodland
Hedgerow, Scrub
Acid heath, Bog
Grassland
Waterside, Marsh
Water
Roadsides, Waste places
Quarries, Walls
Cultivated
Scale
Kilometres
0 2 4 6 8 10
Miles
0 1 2 3 4 5 6
Tamworth
NUNEATON
Bedworth
COVENTRY
RUGBY
Southam
ROYAL LEAMINGTON SPA
Kenilworth
WARWICK
Kineton
Shipston on Stour
Long Compton
Solihull
BIRMINGHAM
Stratford on Avon
Alcester

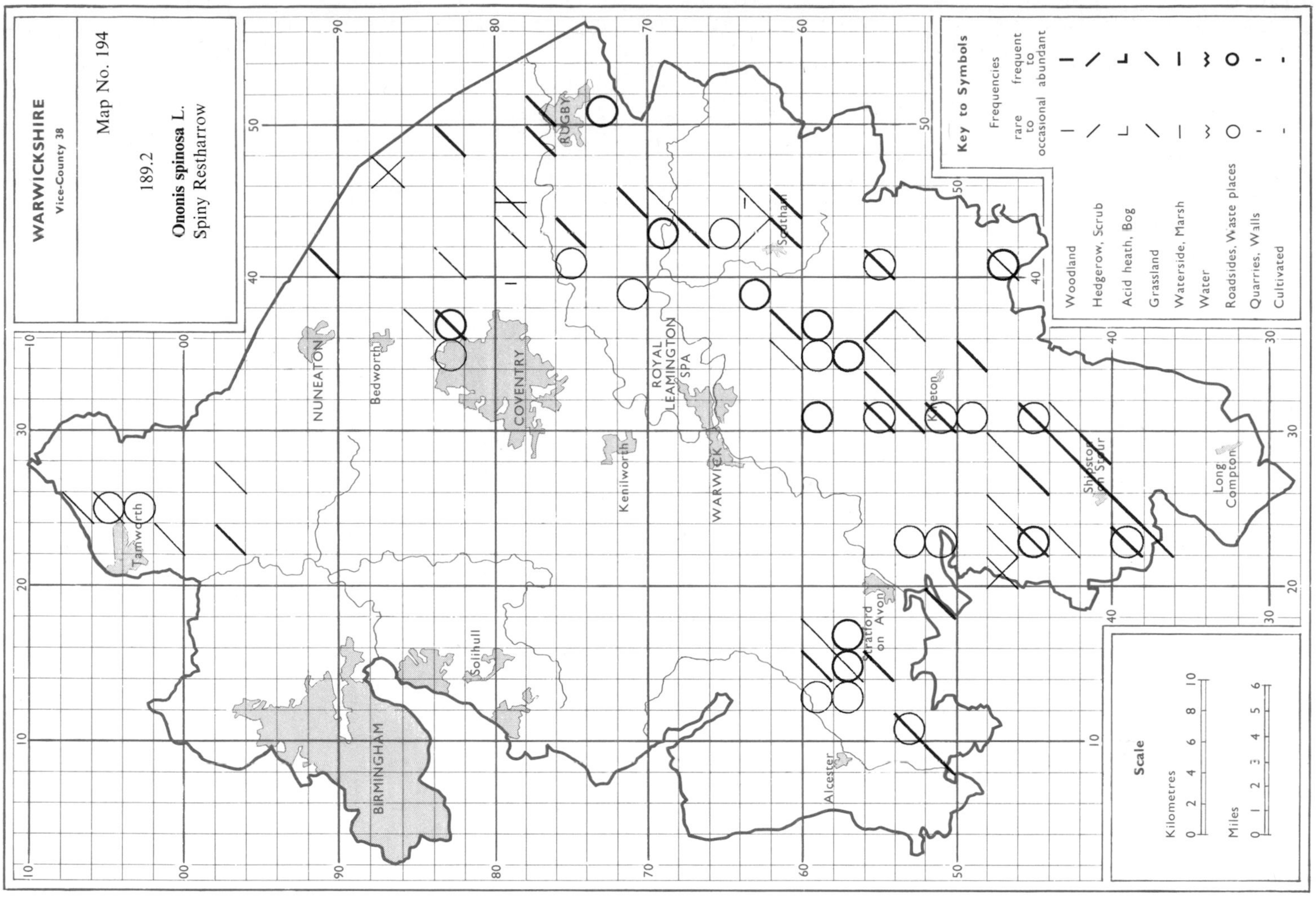
WARWICKSHIRE
Vice-County 38
Map No. 194
189.2
Ononis spinosa L.
Spiny Restharrow
Key to Symbols
Frequencies
rare to occasional
frequent to abundant
Woodland
Hedgerow, Scrub
Acid heath, Bog
Grassland
Waterside, Marsh
Water
Roadsides, Waste places
Quarries, Walls
Cultivated
Scale
Kilometres
0 2 4 6 8 10
Miles
0 1 2 3 4 5 6
Tamworth
NUNEATON
Bedworth
COVENTRY
RUGBY
Southam
ROYAL LEAMINGTON SPA
Kenilworth
WARWICK
Kineton
Shipston on Stour
Long Compton
Solihull
BIRMINGHAM
Stratford on Avon
Alcester

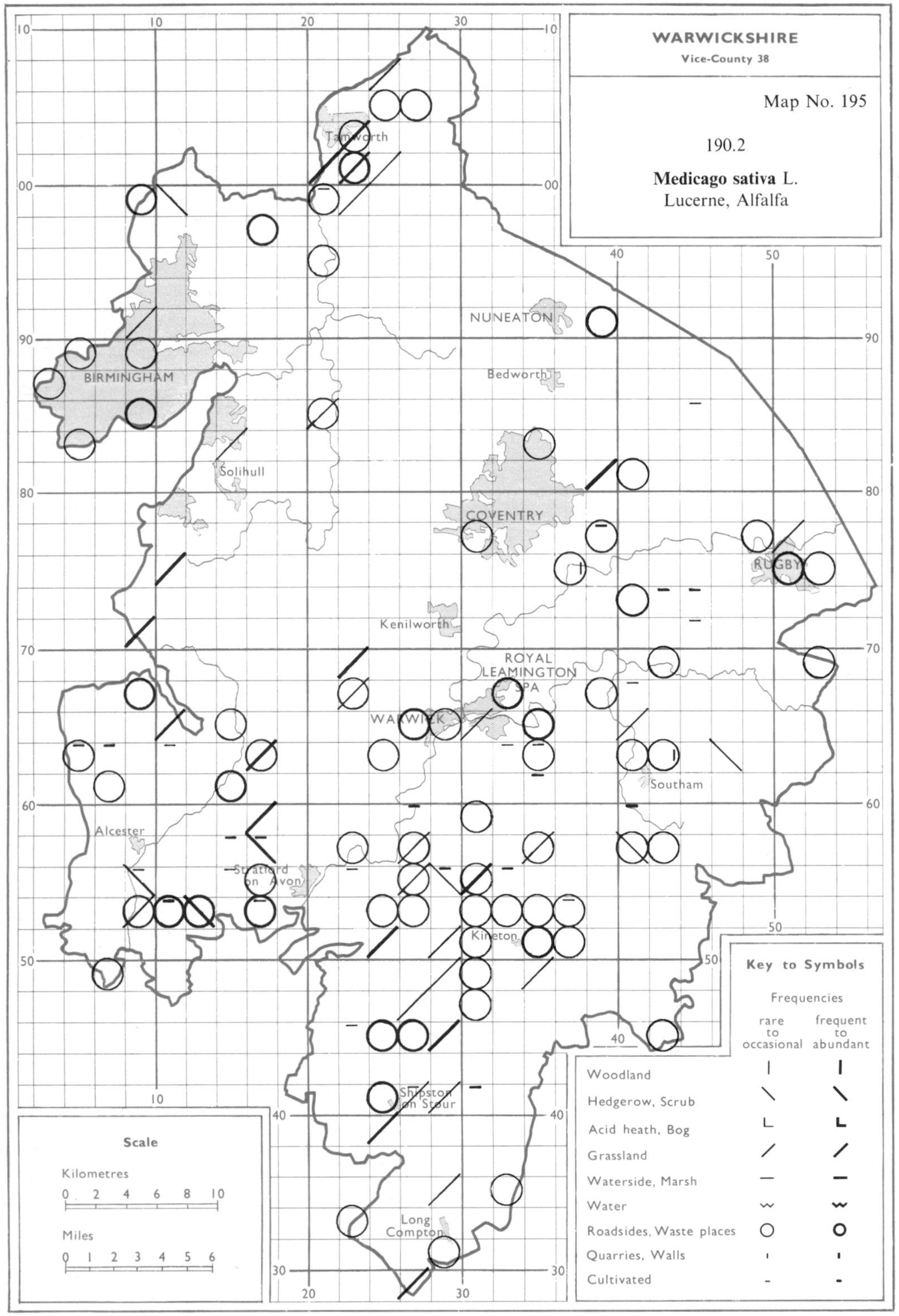

WARWICKSHIRE
Vice-County 38
Map No. 195
190.2
Medicago sativa L.
Lucerne, Alfalfa
Key to Symbols
Frequencies
rare to occasional
frequent to abundant
Woodland
Hedgerow, Scrub
Acid heath, Bog
Grassland
Waterside, Marsh
Water
Roadsides, Waste places
Quarries, Walls
Cultivated
Scale
Kilometres
Miles

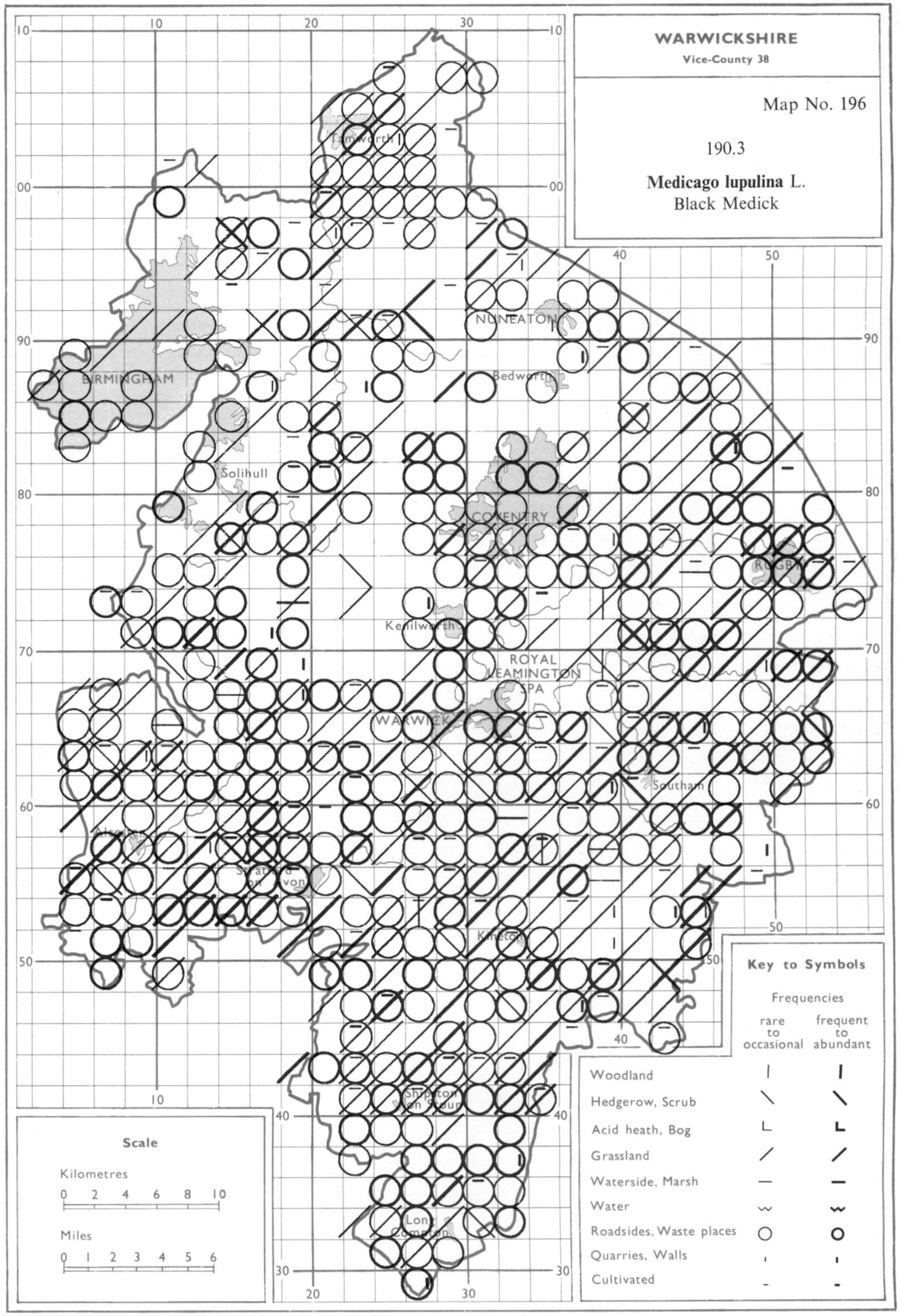

WARWICKSHIRE
Vice-County 38
Map No. 196
190.3
Medicago lupulina L.
Black Medick
Key to Symbols
Frequencies
rare to occasional
frequent to abundant
Woodland
Hedgerow, Scrub
Acid heath, Bog
Grassland
Waterside, Marsh
Water
Roadsides, Waste places
Quarries, Walls
Cultivated
Scale
Kilometres
Miles

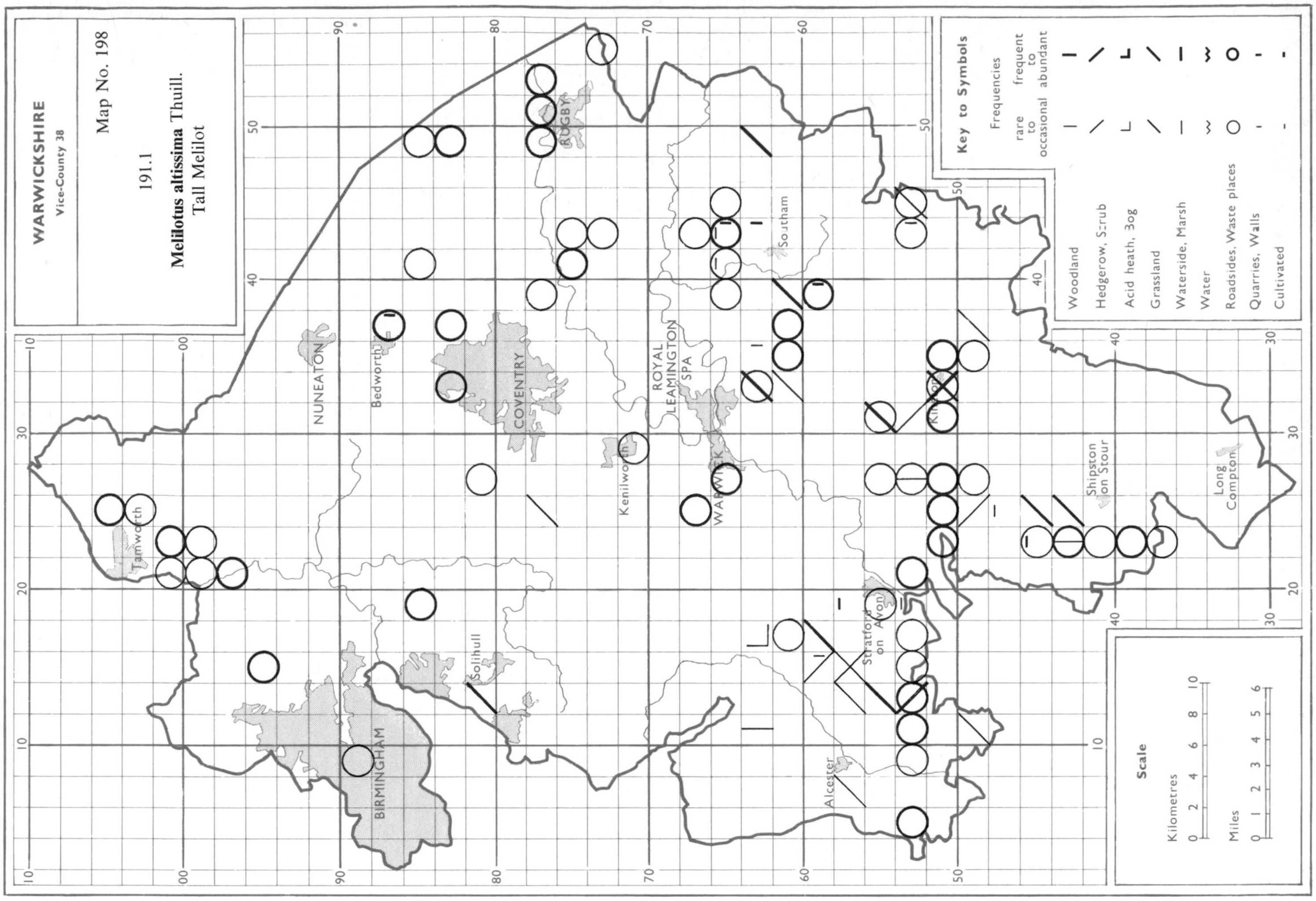
WARWICKSHIRE
Vice-County 38
Map No. 198
191.1
Melilotus altissima Thuill.
Tall Melilot
Key to Symbols
Frequencies
rare to occasional
frequent to abundant
Woodland
Hedgerow, Scrub
Acid heath, Bog
Grassland
Waterside, Marsh
Water
Roadsides, Waste places
Quarries, Walls
Cultivated
Scale
Kilometres
Miles
NUNEATON
Bedworth
COVENTRY
RUGBY
ROYAL LEAMINGTON SPA
Southam
Kenilworth
WARWICK
Kineton
Shipston on Stour
Long Compton
Tamworth
Solihull
BIRMINGHAM
Stratford on Avon
Alcester

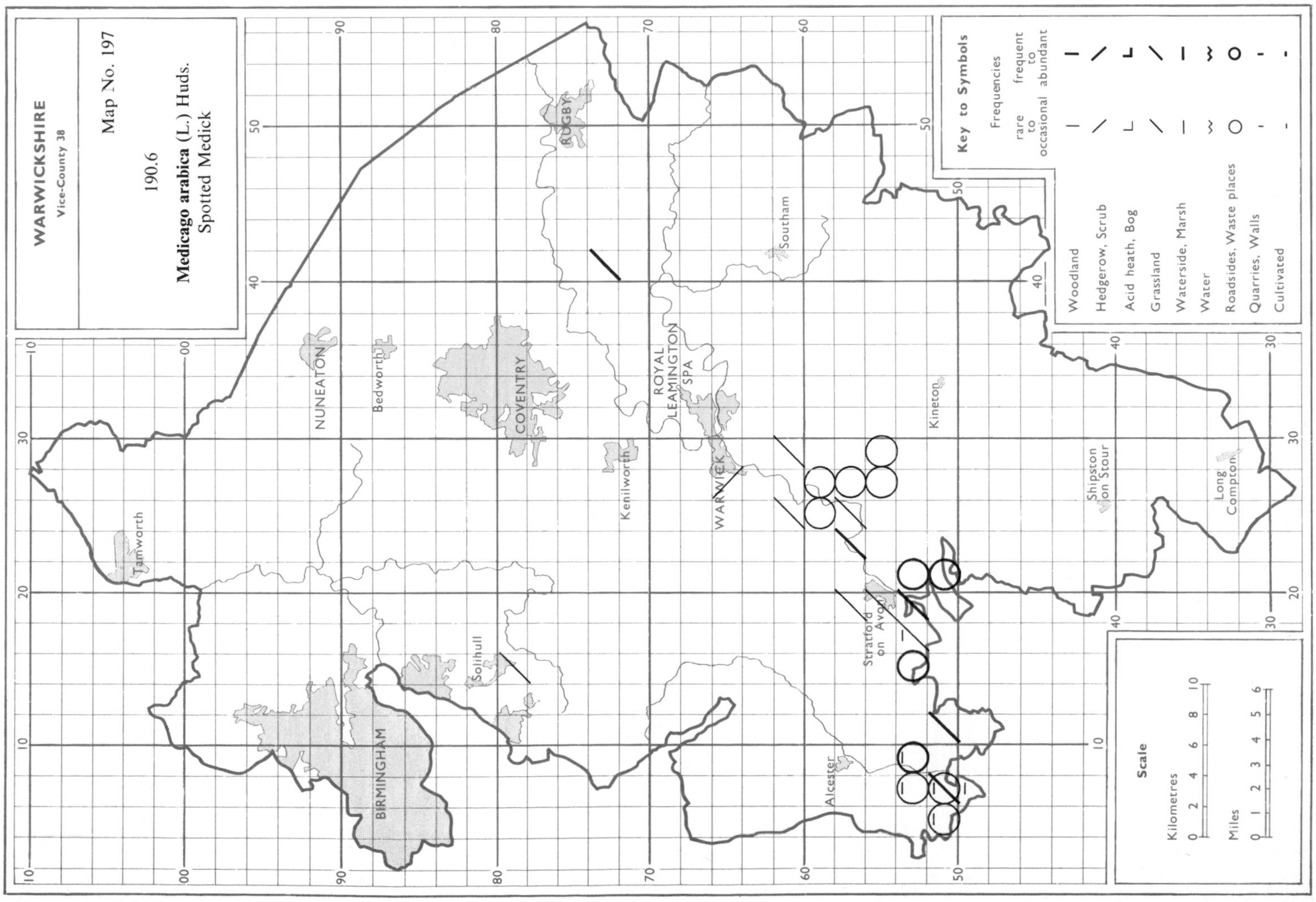
WARWICKSHIRE
Vice-County 38
Map No. 197
190.6
Medicago arabica (L.) Huds.
Spotted Medick
Key to Symbols
Frequencies
rare to occasional
frequent to abundant
Woodland
Hedgerow, Scrub
Acid heath, Bog
Grassland
Waterside, Marsh
Water
Roadsides, Waste places
Quarries, Walls
Cultivated
Scale
Kilometres
Miles
NUNEATON
Bedworth
COVENTRY
RUGBY
ROYAL LEAMINGTON SPA
Southam
Kenilworth
WARWICK
Kineton
Shipston on Stour
Long Compton
Tamworth
Solihull
BIRMINGHAM
Stratford on Avon
Alcester

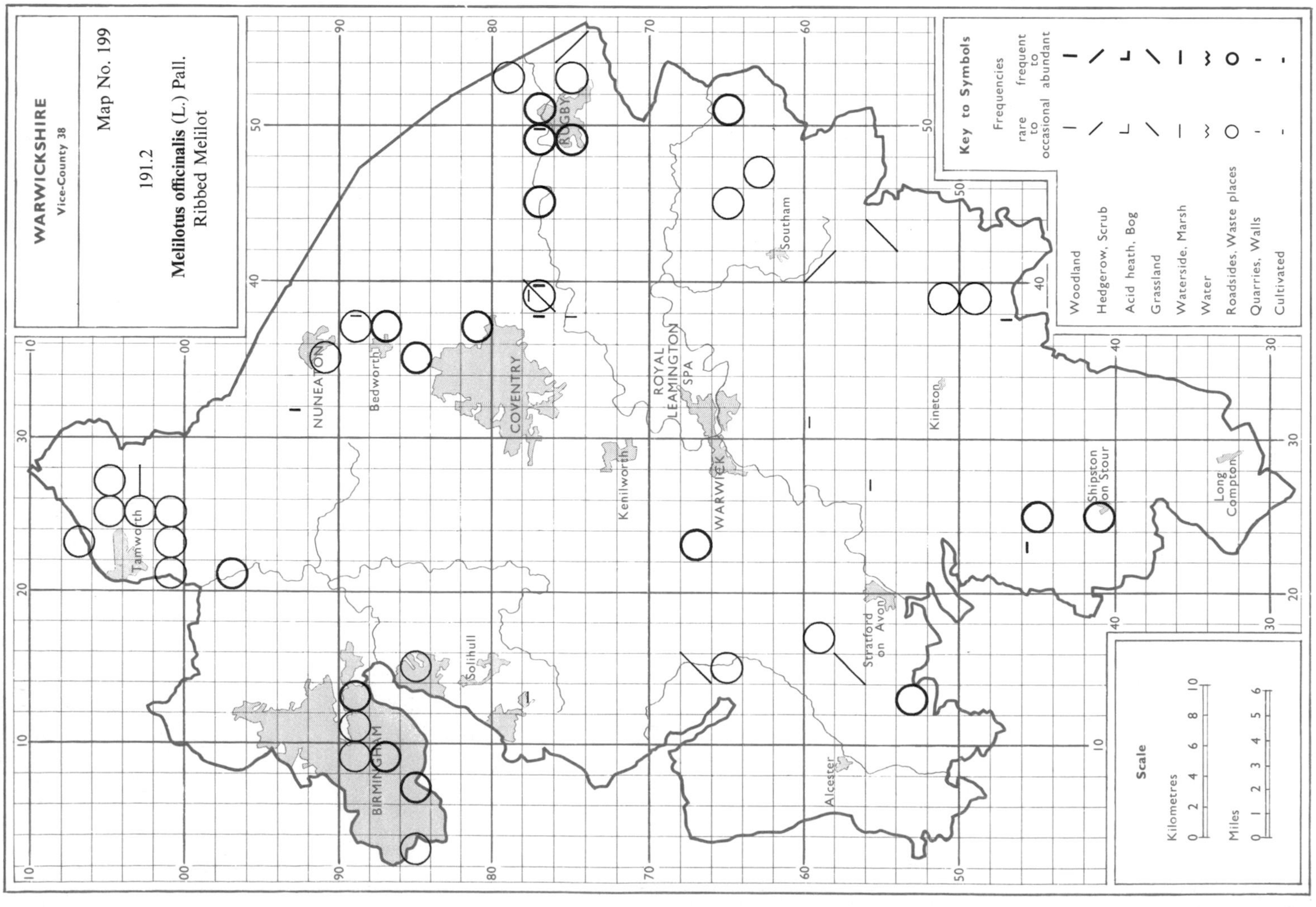
WARWICKSHIRE
Vice-County 38
Map No. 199
191.2
Melilotus officinalis (L.) Pall.
Ribbed Melilot
Key to Symbols
Frequencies
rare to occasional
frequent to abundant
Woodland
Hedgerow, Scrub
Acid heath, Bog
Grassland
Waterside, Marsh
Water
Roadsides, Waste places
Quarries, Walls
Cultivated
Scale
Kilometres
0 2 4 6 8 10
Miles
0 1 2 3 4 5 6
NUNEATON
Bedworth
COVENTRY
RUGBY
Tamworth
BIRMINGHAM
Solihull
Kenilworth
ROYAL LEAMINGTON SPA
WARWICK
Southam
Kineton
Stratford on Avon
Alcester
Shipston on Stour
Long Compton

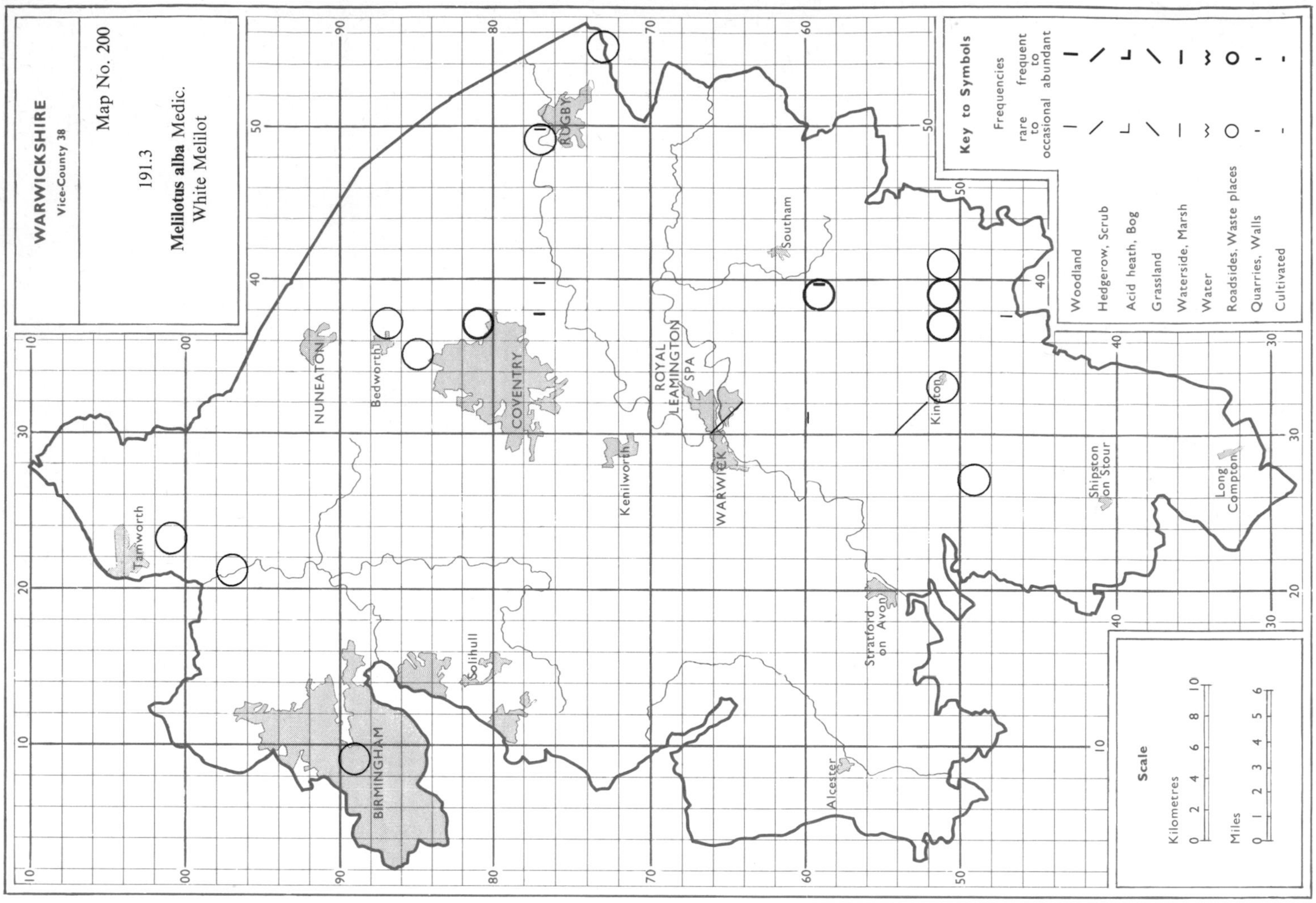
WARWICKSHIRE
Vice-County 38
Map No. 200
191.3
Melilotus alba Medic.
White Melilot
Key to Symbols
Frequencies
rare to occasional
frequent to abundant
Woodland
Hedgerow, Scrub
Acid heath, Bog
Grassland
Waterside, Marsh
Water
Roadsides, Waste places
Quarries, Walls
Cultivated
Scale
Kilometres
0 2 4 6 8 10
Miles
0 1 2 3 4 5 6
NUNEATON
Bedworth
COVENTRY
RUGBY
Tamworth
BIRMINGHAM
Solihull
Kenilworth
ROYAL LEAMINGTON SPA
WARWICK
Southam
Kineton
Stratford on Avon
Alcester
Shipston on Stour
Long Compton

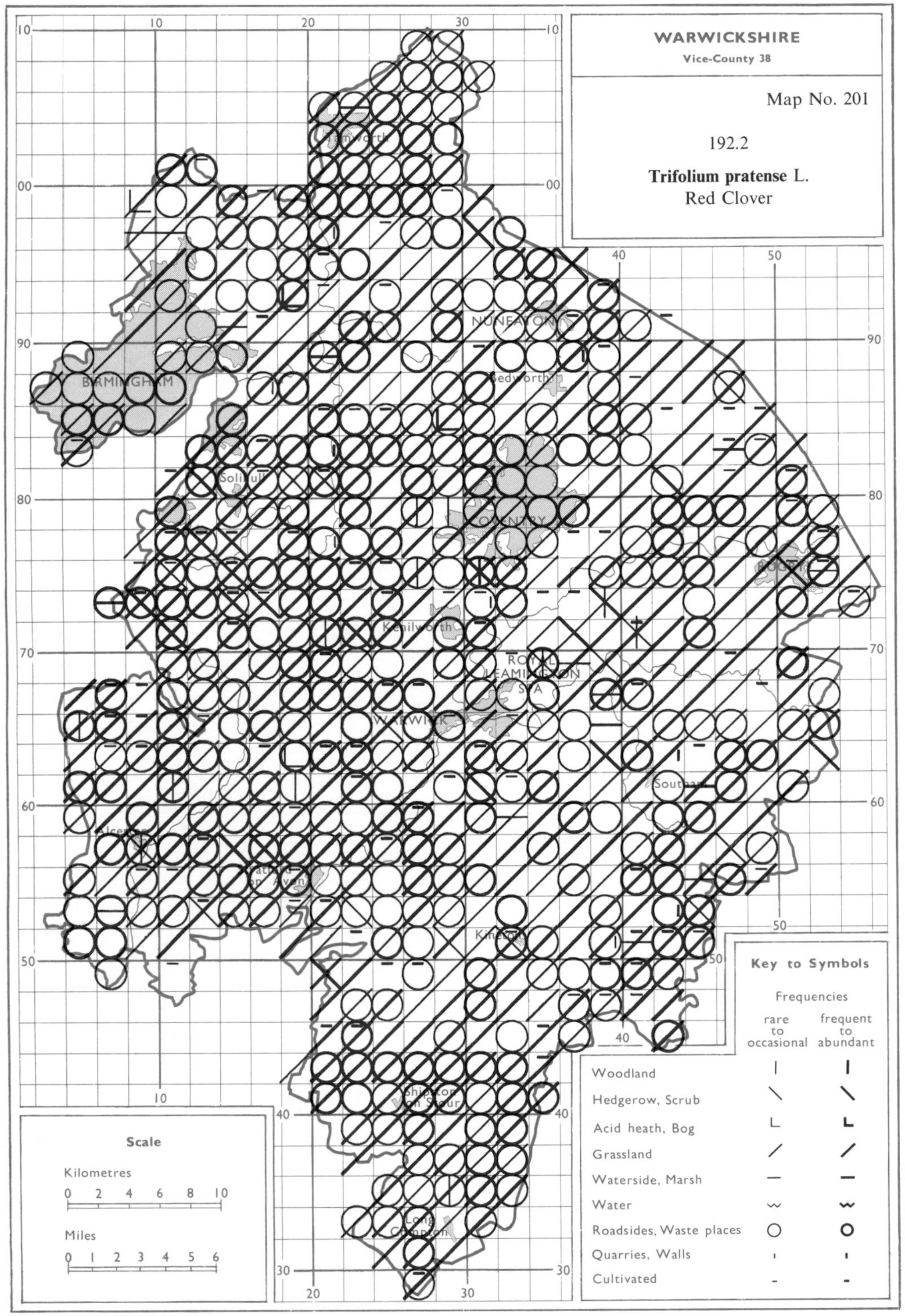
WARWICKSHIRE
Vice-County 38
Map No. 201
192.2
Trifolium pratense L.
Red Clover
Key to Symbols
Frequencies
rare to occasional
frequent to abundant
Woodland
Hedgerow, Scrub
Acid heath, Bog
Grassland
Waterside, Marsh
Water
Roadsides, Waste places
Quarries, Walls
Cultivated
Scale
Kilometres
0 2 4 6 8 10
Miles
0 1 2 3 4 5 6
BIRMINGHAM
NUNEATON
Bedworth
Solihull
COVENTRY
Kenilworth
WARWICK
Southam
Alcester
Kineton
Long Compton

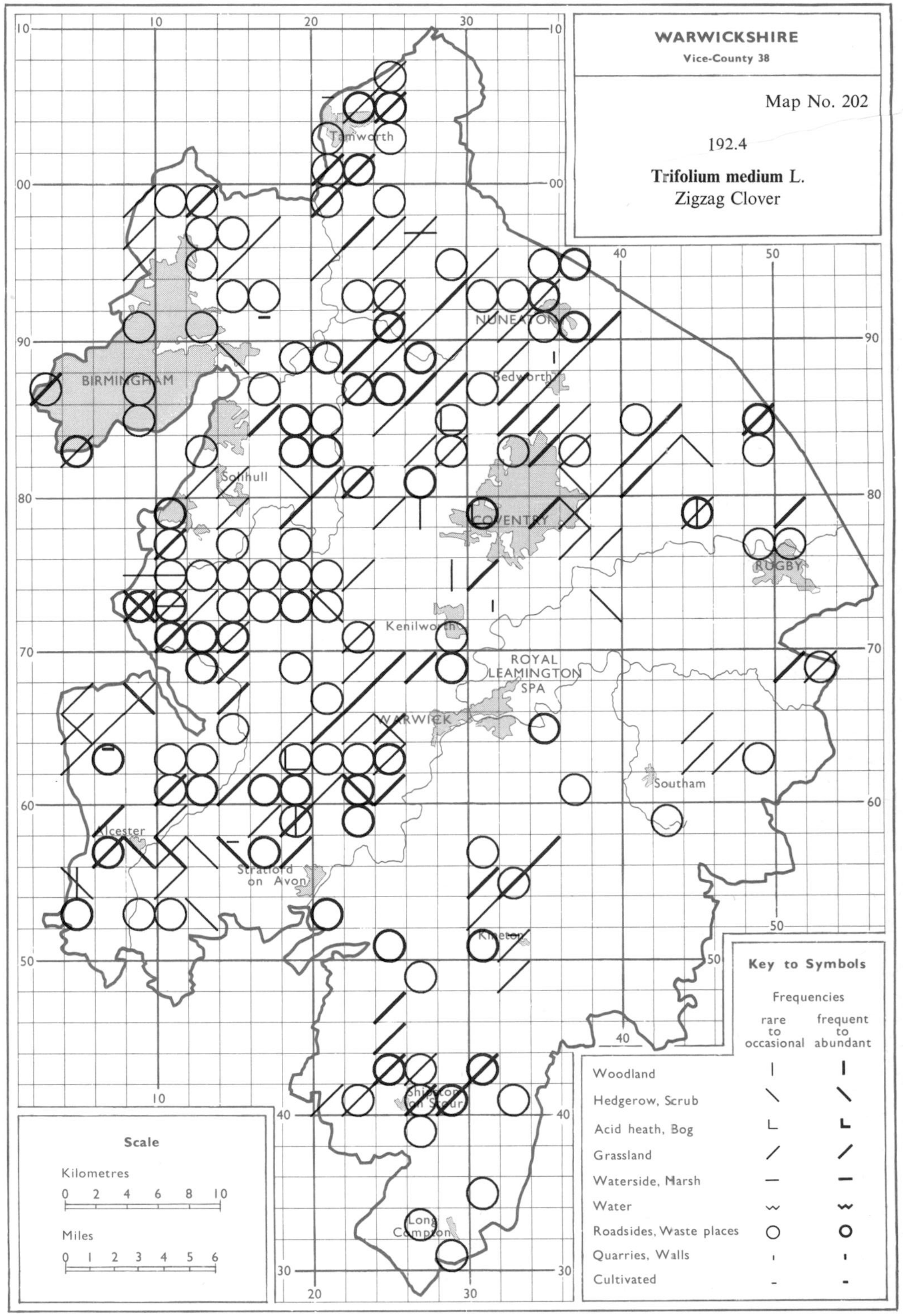
WARWICKSHIRE
Vice-County 38
Map No. 202
192.4
Trifolium medium L.
Zigzag Clover
Key to Symbols
Frequencies
rare to occasional
frequent to abundant
Woodland
Hedgerow, Scrub
Acid heath, Bog
Grassland
Waterside, Marsh
Water
Roadsides, Waste places
Quarries, Walls
Cultivated
Scale
Kilometres
0 2 4 6 8 10
Miles
0 1 2 3 4 5 6
Tamworth
BIRMINGHAM
NUNEATON
Bedworth
Solihull
COVENTRY
RUGBY
Kenilworth
ROYAL LEAMINGTON SPA
WARWICK
Southam
Alcester
Stratford on Avon
Kineton
Shipston on Stour
Long Compton

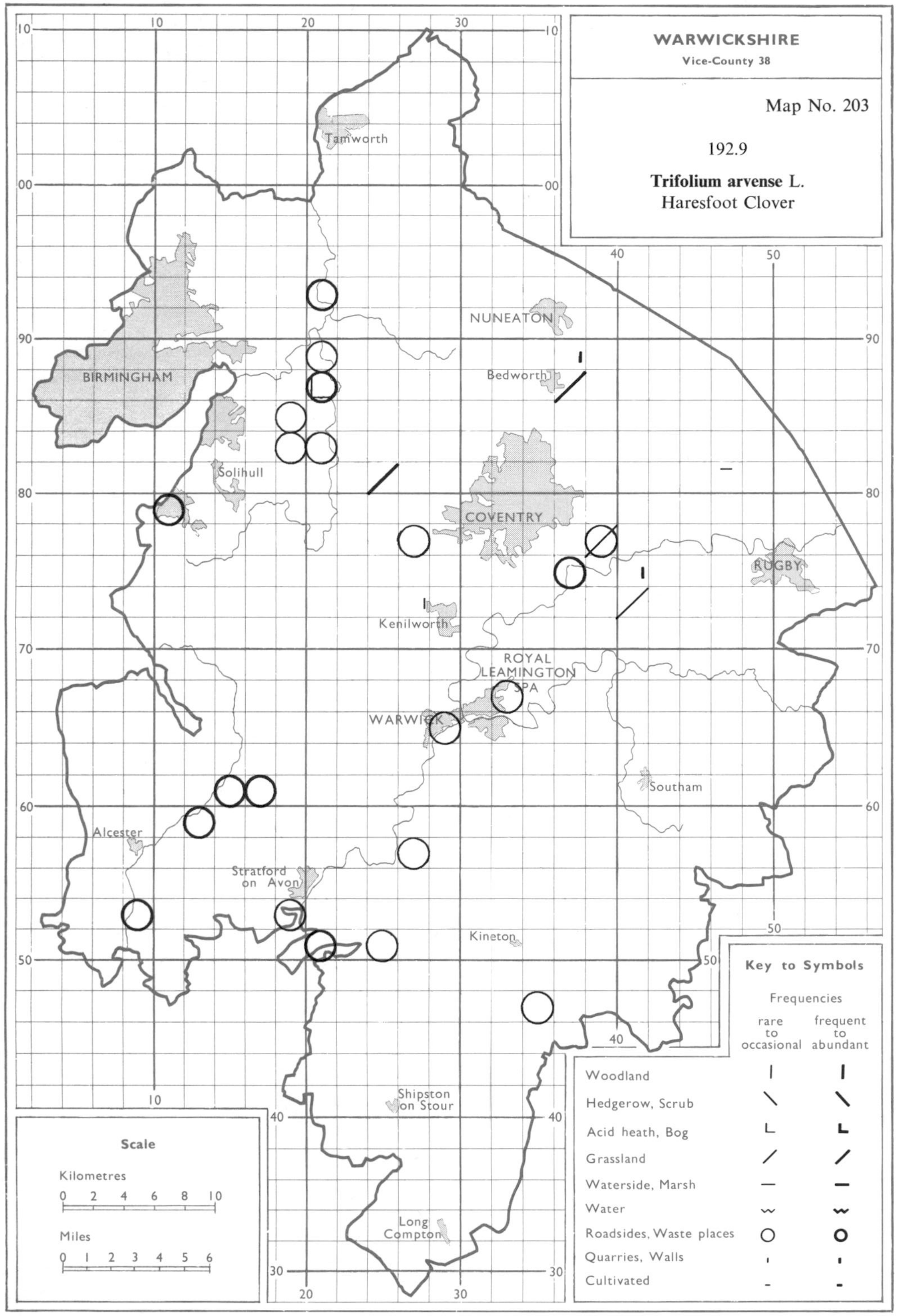
WARWICKSHIRE
Vice-County 38
Map No. 203
192.9
Trifolium arvense L.
Haresfoot Clover
Tamworth
NUNEATON
Bedworth
BIRMINGHAM
Solihull
COVENTRY
RUGBY
Kenilworth
ROYAL LEAMINGTON SPA
WARWICK
Southam
Alcester
Stratford on Avon
Kineton
Shipston on Stour
Long Compton
Key to Symbols
Frequencies
rare to occasional
frequent to abundant
Woodland
Hedgerow, Scrub
Acid heath, Bog
Grassland
Waterside, Marsh
Water
Roadsides, Waste places
Quarries, Walls
Cultivated
Scale
Kilometres
0 2 4 6 8 10
Miles
0 1 2 3 4 5 6

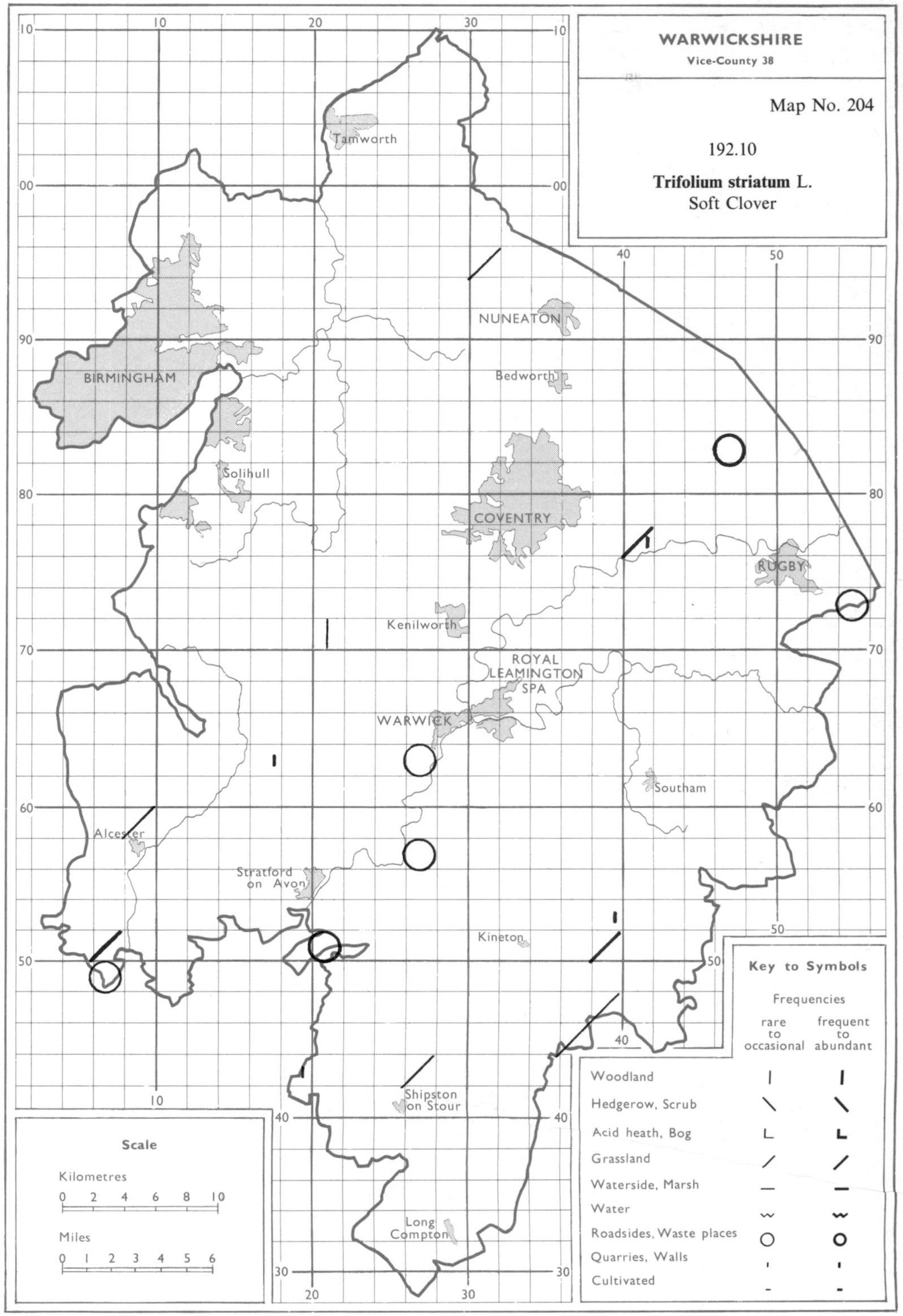
WARWICKSHIRE
Vice-County 38
Map No. 204
192.10
Trifolium striatum L.
Soft Clover
Tamworth
NUNEATON
Bedworth
BIRMINGHAM
Solihull
COVENTRY
RUGBY
Kenilworth
ROYAL LEAMINGTON SPA
WARWICK
Southam
Alcester
Stratford on Avon
Kineton
Shipston on Stour
Long Compton
Key to Symbols
Frequencies
rare to occasional
frequent to abundant
Woodland
Hedgerow, Scrub
Acid heath, Bog
Grassland
Waterside, Marsh
Water
Roadsides, Waste places
Quarries, Walls
Cultivated
Scale
Kilometres
0 2 4 6 8 10
Miles
0 1 2 3 4 5 6

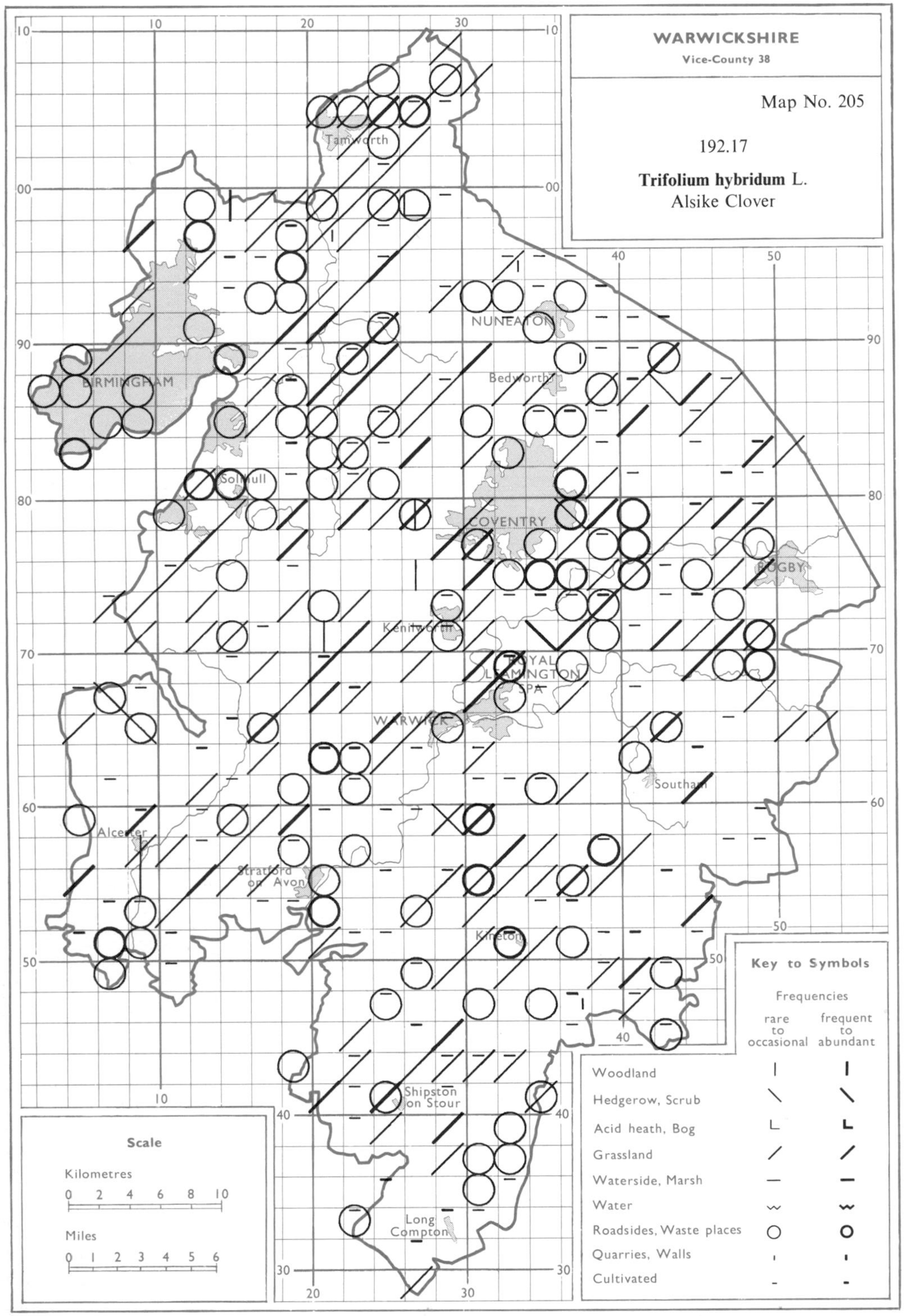
WARWICKSHIRE
Vice-County 38
Map No. 205
192.17
Trifolium hybridum L.
Alsike Clover
Tamworth
BIRMINGHAM
Solihull
NUNEATON
Bedworth
COVENTRY
RUGBY
Kenilworth
ROYAL LEAMINGTON SPA
WARWICK
Southam
Alcester
Stratford on Avon
Kineton
Shipston on Stour
Long Compton
Key to Symbols
Frequencies
rare to occasional
frequent to abundant
Woodland
Hedgerow, Scrub
Acid heath, Bog
Grassland
Waterside, Marsh
Water
Roadsides, Waste places
Quarries, Walls
Cultivated
Scale
Kilometres
0 2 4 6 8 10
Miles
0 1 2 3 4 5 6

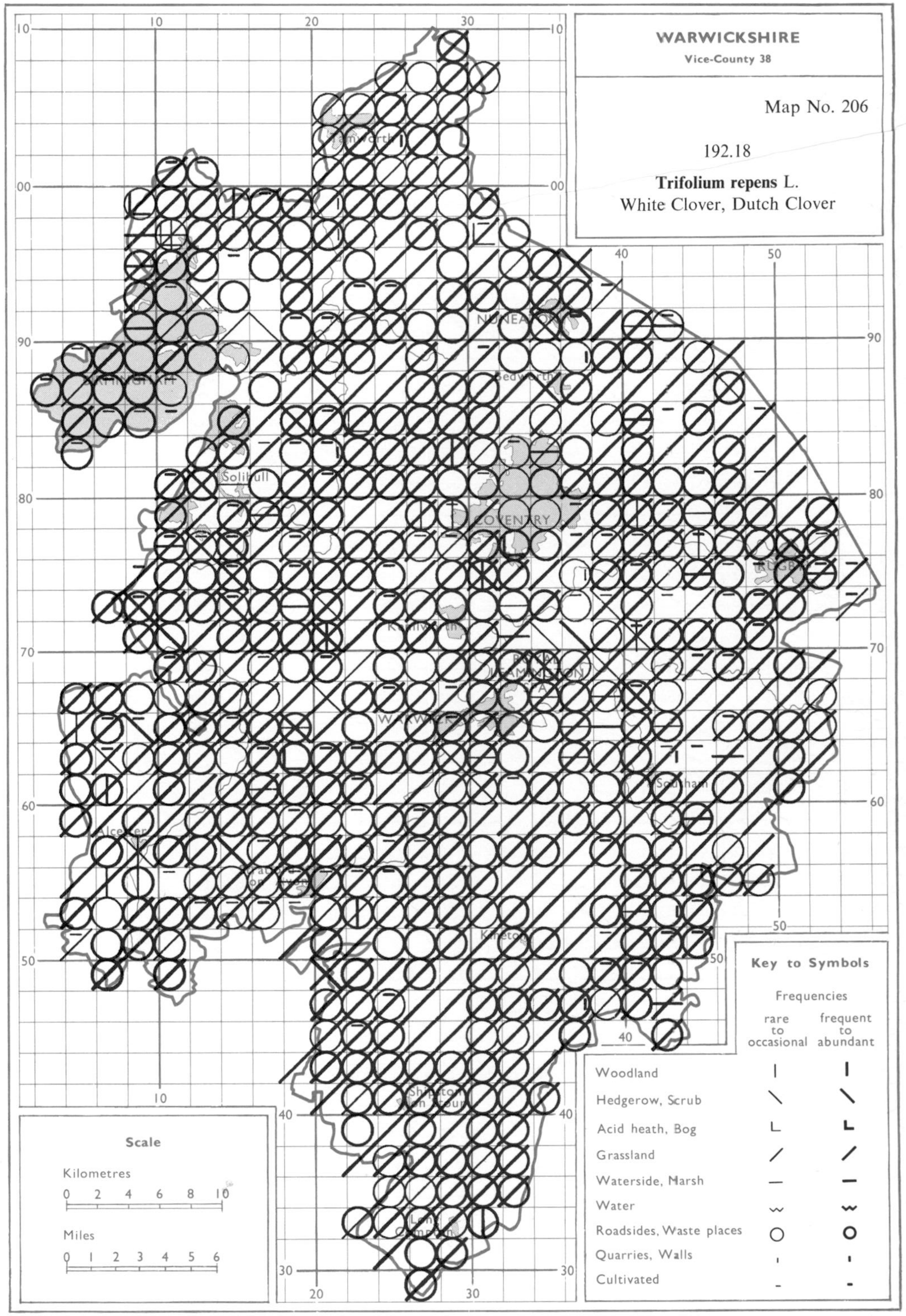
WARWICKSHIRE
Vice-County 38
Map No. 206
192.18
Trifolium repens L.
White Clover, Dutch Clover
Key to Symbols
Frequencies
rare to occasional
frequent to abundant
Woodland
Hedgerow, Scrub
Acid heath, Bog
Grassland
Waterside, Marsh
Water
Roadsides, Waste places
Quarries, Walls
Cultivated
Scale
Kilometres
0 2 4 6 8 10
Miles
0 1 2 3 4 5 6

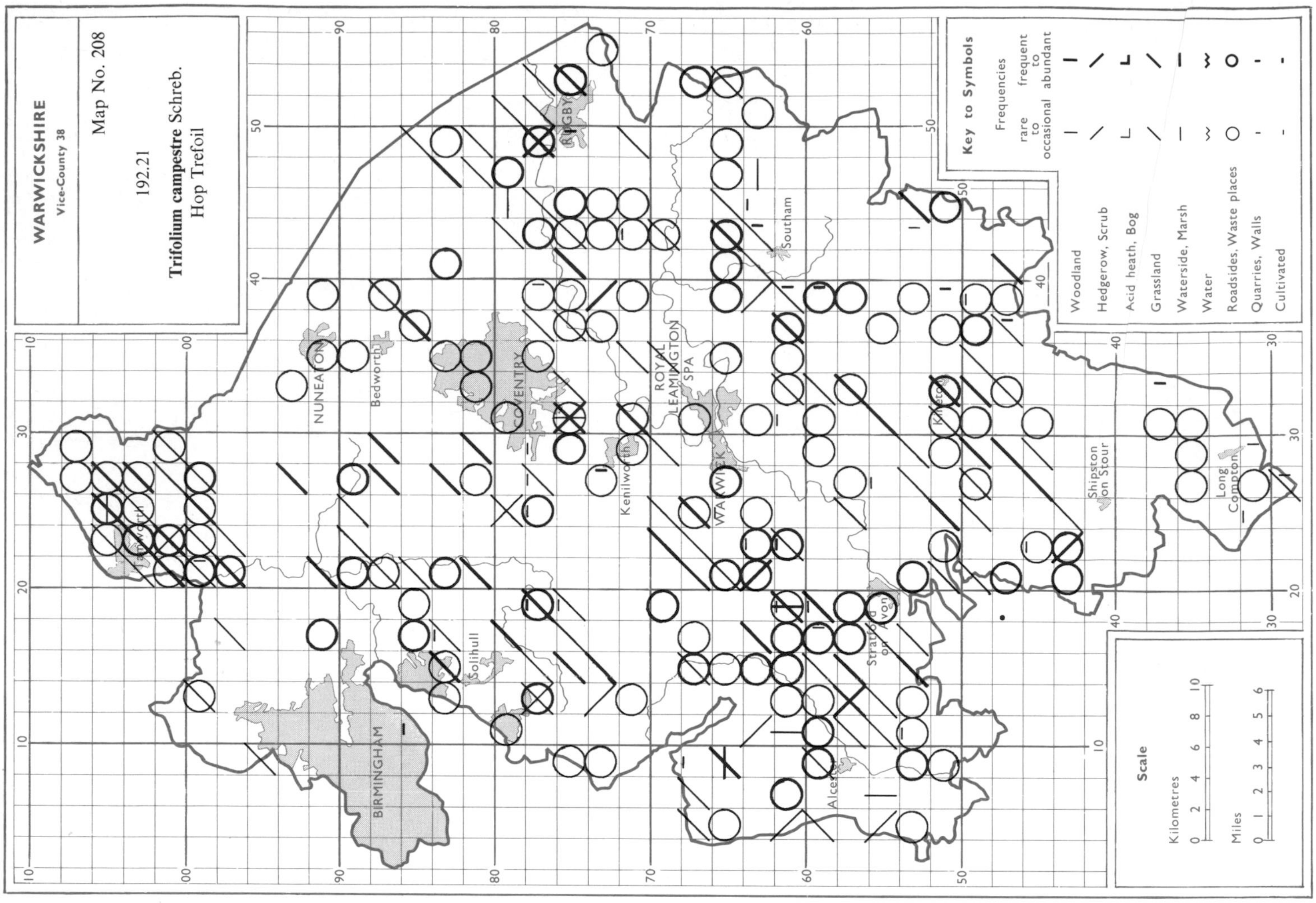
WARWICKSHIRE
Vice-County 38
Map No. 208
192.21
Trifolium campestre Schreb.
Hop Trefoil
Key to Symbols
Frequencies
rare to occasional
frequent to abundant
Woodland
Hedgerow, Scrub
Acid heath, Bog
Grassland
Waterside, Marsh
Water
Roadsides, Waste places
Quarries, Walls
Cultivated
Scale
Kilometres
Miles
NUNEATON
Bedworth
COVENTRY
RUGBY
ROYAL LEAMINGTON SPA
Kenilworth
WARWICK
Southam
Kineton
Shipston on Stour
Long Compton
Stratford on Avon
Alcester
Solihull
BIRMINGHAM
Tamworth

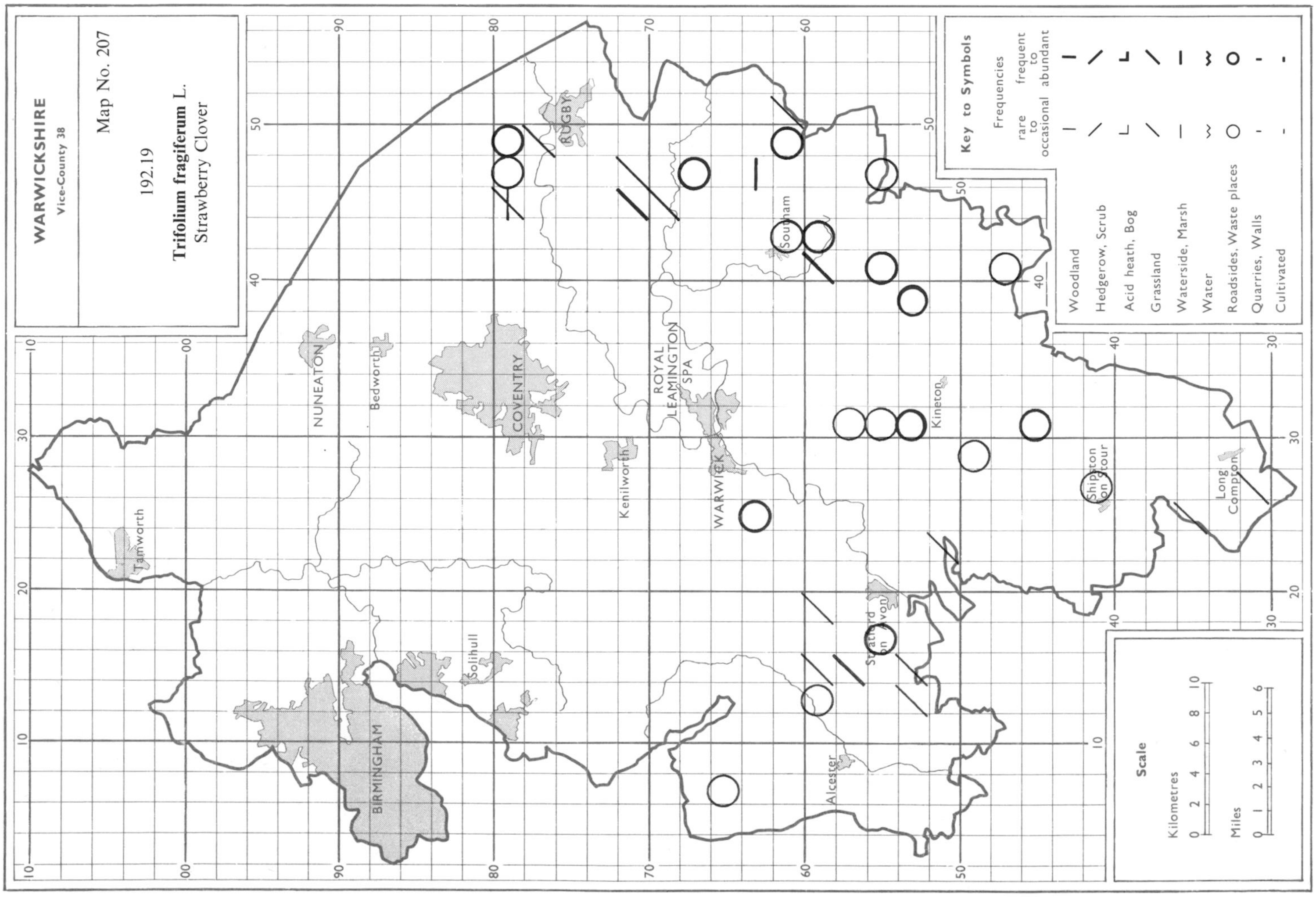
WARWICKSHIRE
Vice-County 38
Map No. 207
192.19
Trifolium fragiferum L.
Strawberry Clover
Key to Symbols
Frequencies
rare to occasional
frequent to abundant
Woodland
Hedgerow, Scrub
Acid heath, Bog
Grassland
Waterside, Marsh
Water
Roadsides, Waste places
Quarries, Walls
Cultivated
Scale
Kilometres
Miles
NUNEATON
Bedworth
COVENTRY
RUGBY
ROYAL LEAMINGTON SPA
Kenilworth
WARWICK
Southam
Kineton
Shipston on Stour
Long Compton
Stratford on Avon
Alcester
Solihull
BIRMINGHAM
Tamworth

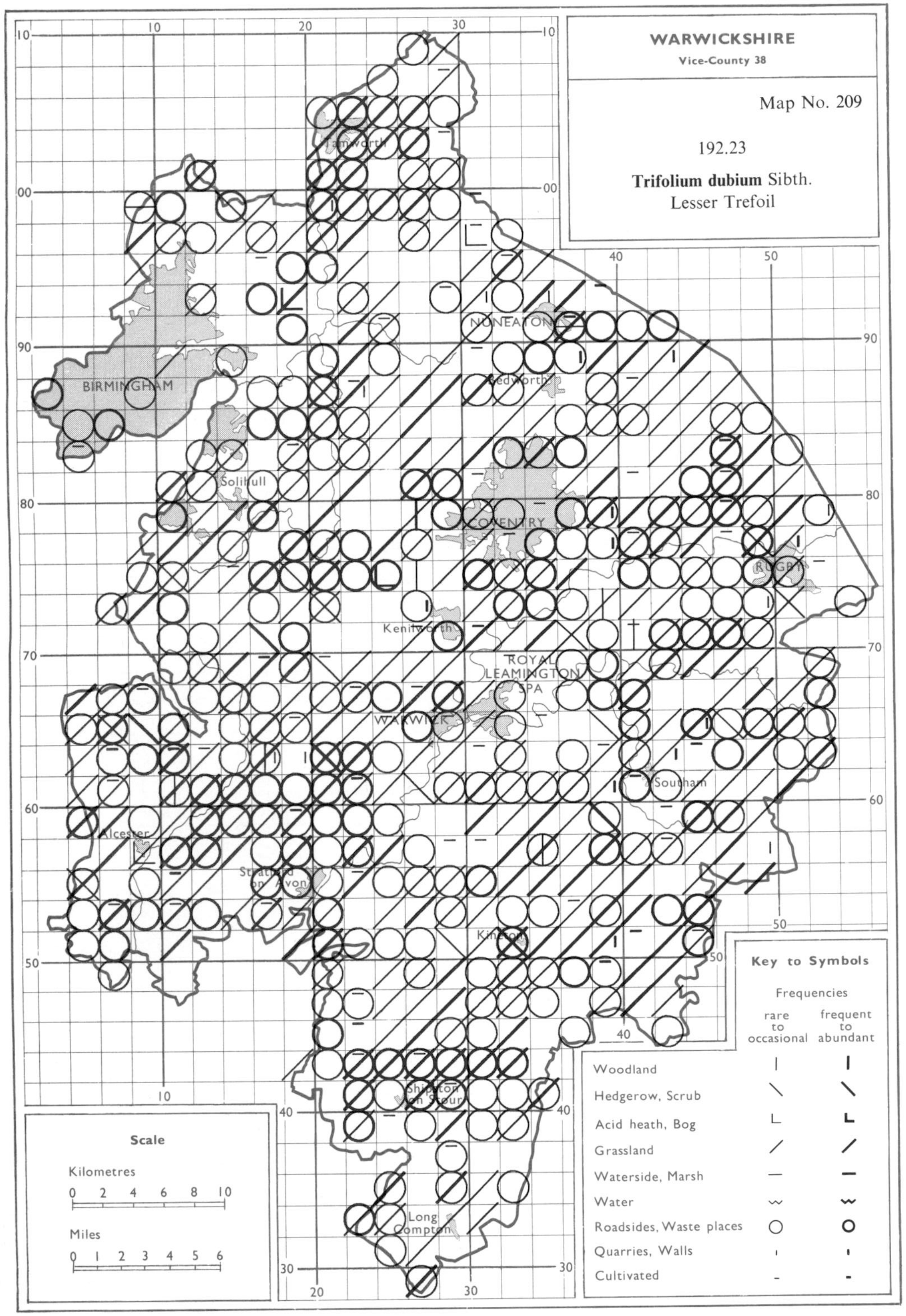
WARWICKSHIRE
Vice-County 38
Map No. 209
192.23
Trifolium dubium Sibth.
Lesser Trefoil
Tamworth
NUNEATON
Bedworth
BIRMINGHAM
Solihull
COVENTRY
RUGBY
Kenilworth
ROYAL LEAMINGTON SPA
WARWICK
Southam
Alcester
Stratford on Avon
Kineton
Shipston on Stour
Long Compton
Key to Symbols
Frequencies
rare to occasional
frequent to abundant
Woodland
Hedgerow, Scrub
Acid heath, Bog
Grassland
Waterside, Marsh
Water
Roadsides, Waste places
Quarries, Walls
Cultivated
Scale
Kilometres
0 2 4 6 8 10
Miles
0 1 2 3 4 5 6

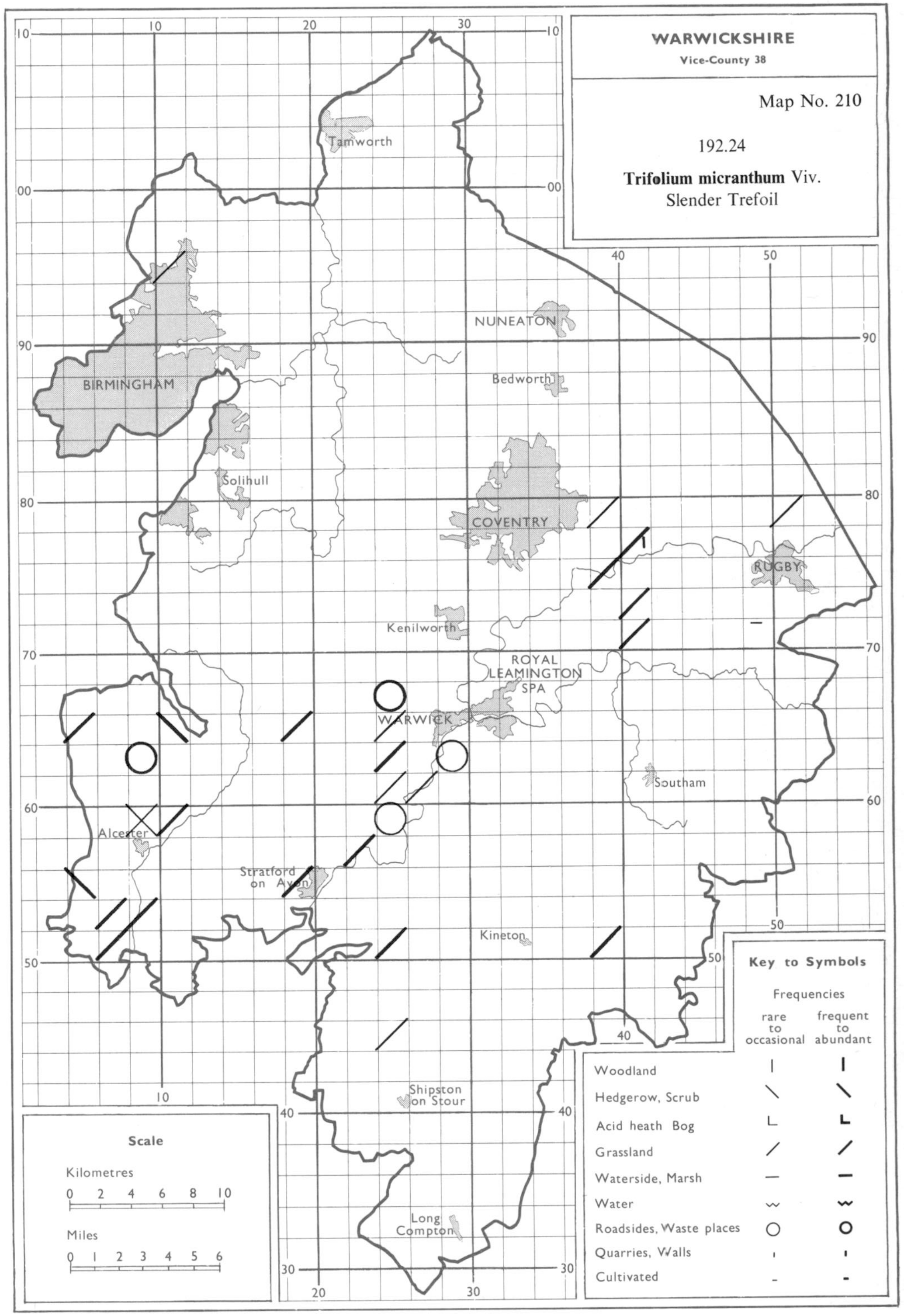
WARWICKSHIRE
Vice-County 38
Map No. 210
192.24
Trifolium micranthum Viv.
Slender Trefoil
Tamworth
NUNEATON
Bedworth
BIRMINGHAM
Solihull
COVENTRY
RUGBY
Kenilworth
ROYAL LEAMINGTON SPA
WARWICK
Southam
Alcester
Stratford on Avon
Kineton
Shipston on Stour
Long Compton
Key to Symbols
Frequencies
rare to occasional
frequent to abundant
Woodland
Hedgerow, Scrub
Acid heath Bog
Grassland
Waterside, Marsh
Water
Roadsides, Waste places
Quarries, Walls
Cultivated
Scale
Kilometres
0 2 4 6 8 10
Miles
0 1 2 3 4 5 6

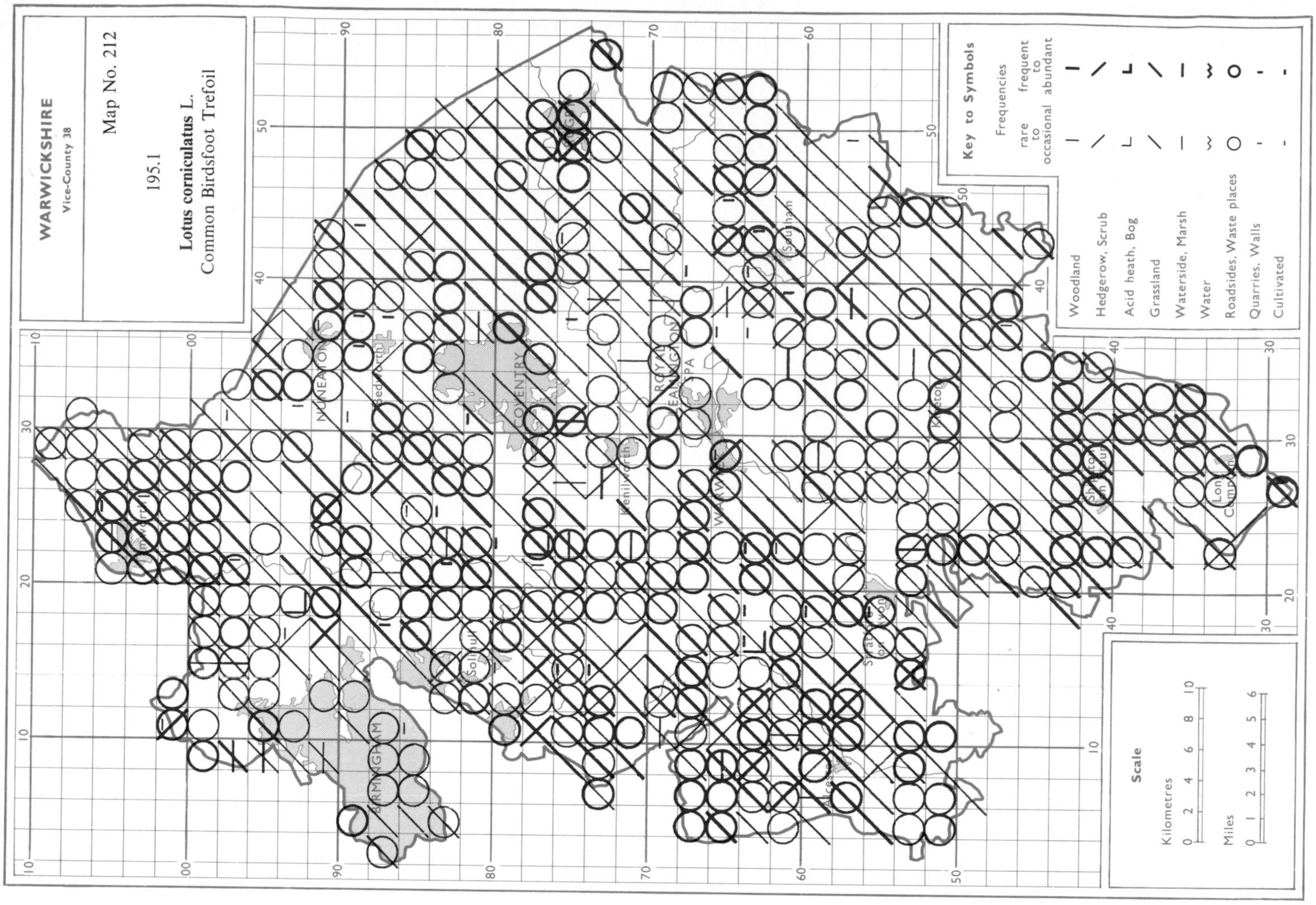
WARWICKSHIRE
Vice-County 38
Map No. 212
195.1
Lotus corniculatus L.
Common Birdsfoot Trefoil
Key to Symbols
Frequencies
rare to occasional
frequent to abundant
Woodland
Hedgerow, Scrub
Acid heath, Bog
Grassland
Waterside, Marsh
Water
Roadsides, Waste places
Quarries, Walls
Cultivated
Scale
Kilometres
0 2 4 6 8 10
Miles
0 1 2 3 4 5 6
NUNEATON
Bedworth
COVENTRY
Kenilworth
ROYAL LEAMINGTON SPA
WARWICK
Southam
Kineton
Shipston on Stour
Long Compton
Tamworth
BIRMINGHAM
Solihull
Stratford on Avon
Alcester
RUGBY

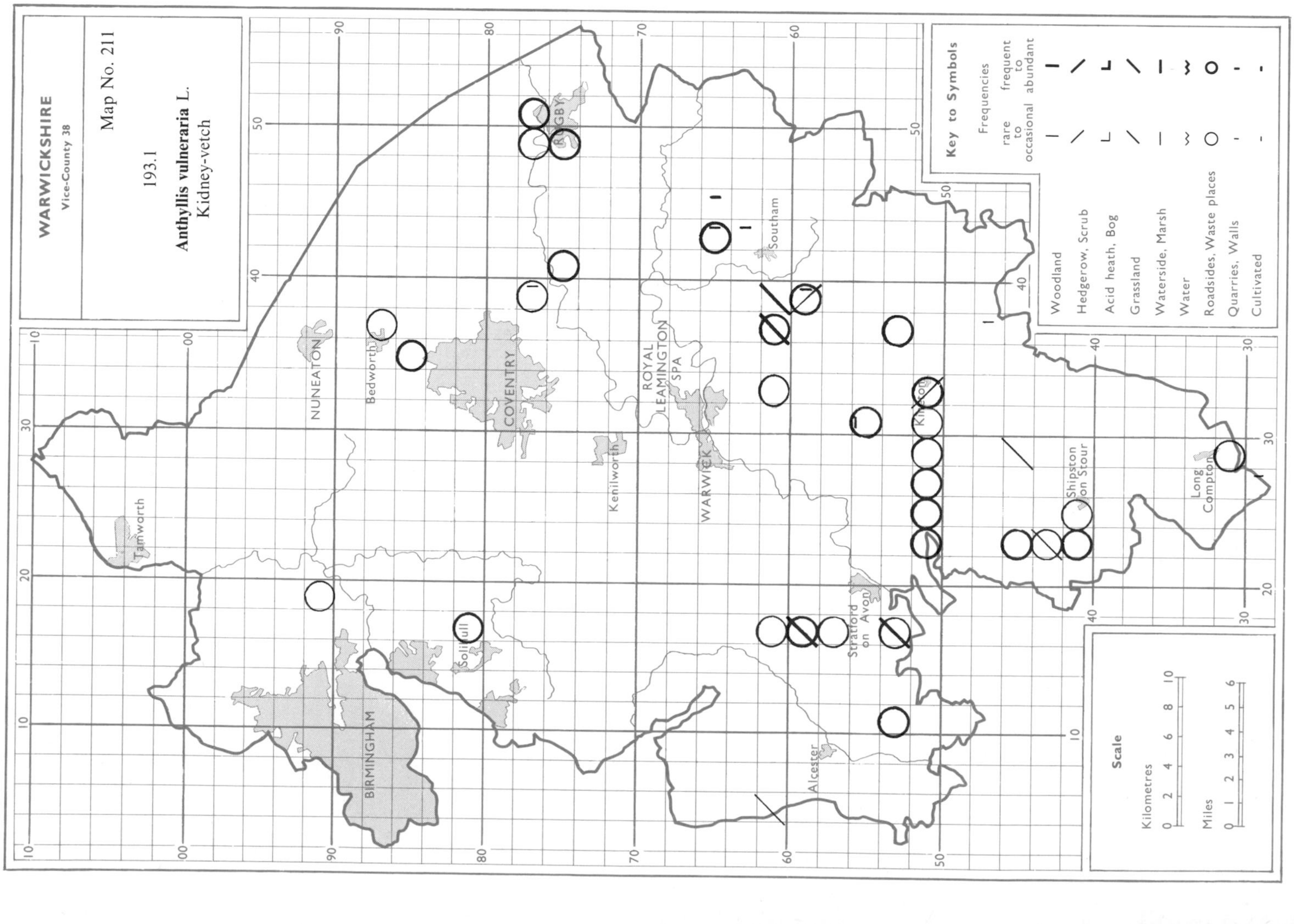
WARWICKSHIRE
Vice-County 38
Map No. 211
193.1
Anthyllis vulneraria L.
Kidney-vetch
Key to Symbols
Frequencies
rare to occasional
frequent to abundant
Woodland
Hedgerow, Scrub
Acid heath, Bog
Grassland
Waterside, Marsh
Water
Roadsides, Waste places
Quarries, Walls
Cultivated
Scale
Kilometres
0 2 4 6 8 10
Miles
0 1 2 3 4 5 6
NUNEATON
Bedworth
COVENTRY
Kenilworth
ROYAL LEAMINGTON SPA
WARWICK
Southam
Kineton
Shipston on Stour
Long Compton
Tamworth
BIRMINGHAM
Solihull
Stratford on Avon
Alcester
RUGBY

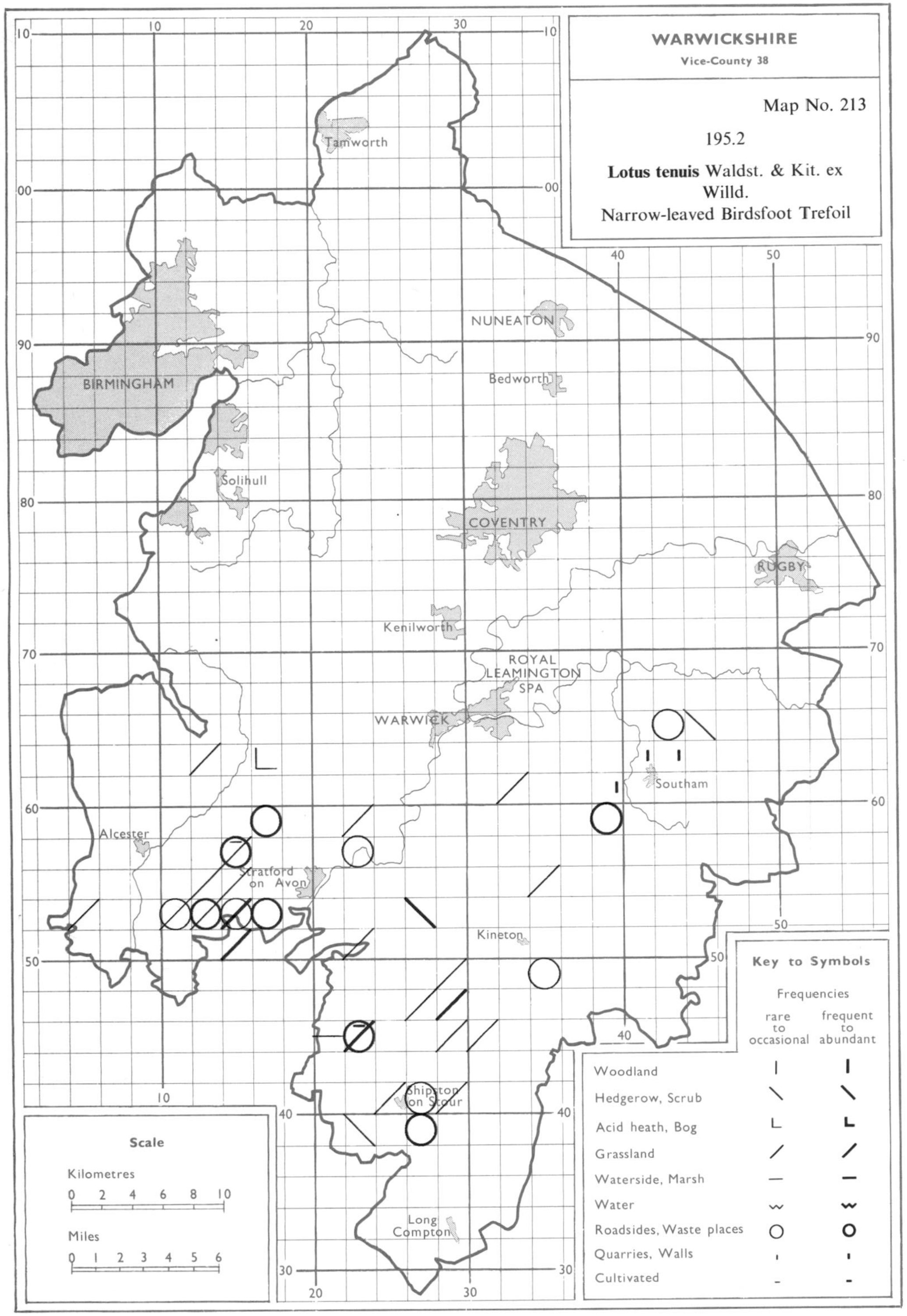
WARWICKSHIRE
Vice-County 38
Map No. 213
195.2
Lotus tenuis Waldst. & Kit. ex Willd.
Narrow-leaved Birdsfoot Trefoil
Tamworth
BIRMINGHAM
Solihull
NUNEATON
Bedworth
COVENTRY
RUGBY
Kenilworth
ROYAL LEAMINGTON SPA
WARWICK
Southam
Alcester
Stratford on Avon
Kineton
Shipston on Stour
Long Compton
Key to Symbols
Frequencies
rare to occasional
frequent to abundant
Woodland
Hedgerow, Scrub
Acid heath, Bog
Grassland
Waterside, Marsh
Water
Roadsides, Waste places
Quarries, Walls
Cultivated
Scale
Kilometres
0 2 4 6 8 10
Miles
0 1 2 3 4 5 6

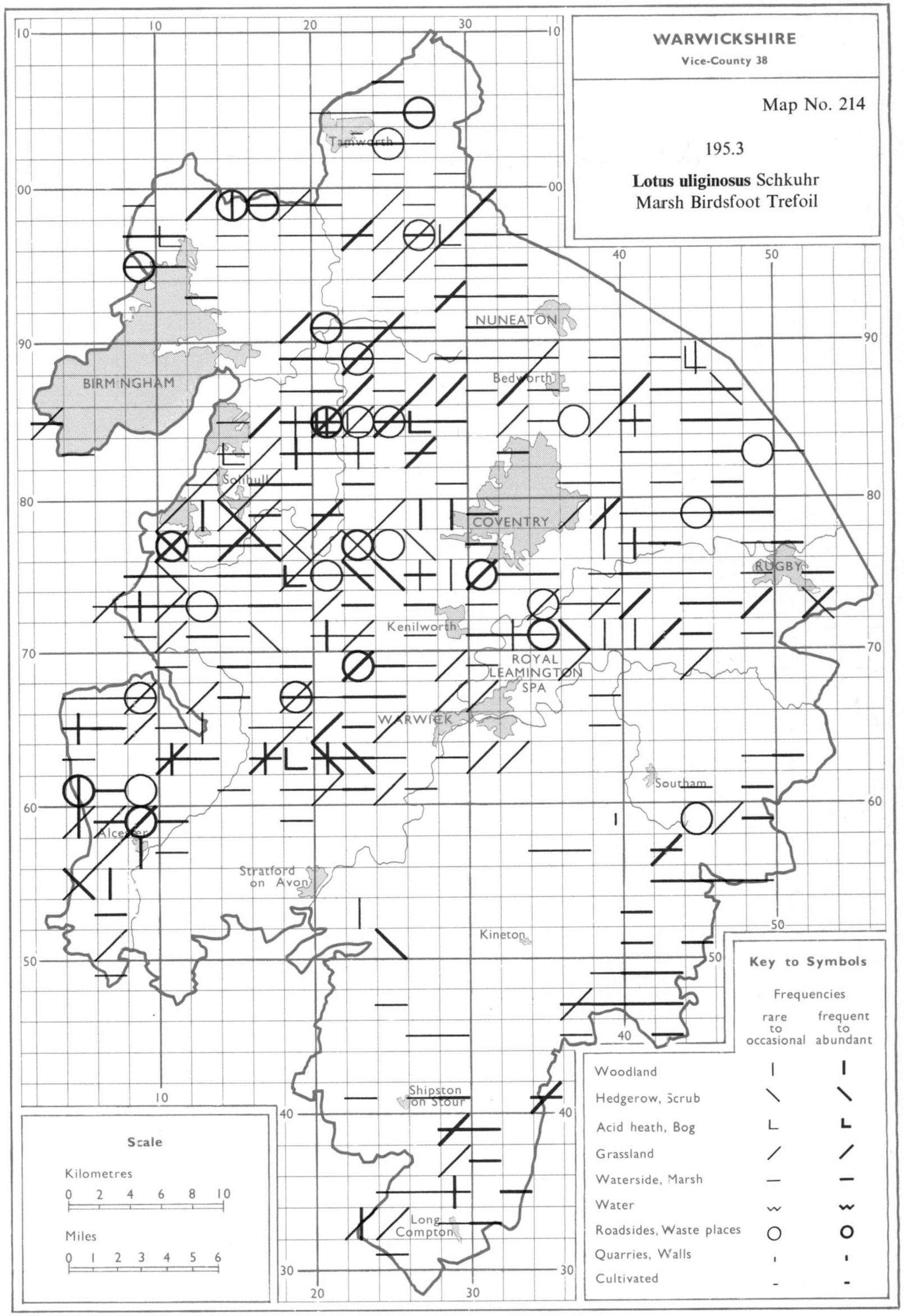
WARWICKSHIRE
Vice-County 38
Map No. 214
195.3
Lotus uliginosus Schkuhr
Marsh Birdsfoot Trefoil
Tamworth
BIRMINGHAM
Solihull
NUNEATON
Bedworth
COVENTRY
RUGBY
Kenilworth
ROYAL LEAMINGTON SPA
WARWICK
Southam
Alcester
Stratford on Avon
Kineton
Shipston on Stour
Long Compton
Key to Symbols
Frequencies
rare to occasional
frequent to abundant
Woodland
Hedgerow, Scrub
Acid heath, Bog
Grassland
Waterside, Marsh
Water
Roadsides, Waste places
Quarries, Walls
Cultivated
Scale
Kilometres
0 2 4 6 8 10
Miles
0 1 2 3 4 5 6

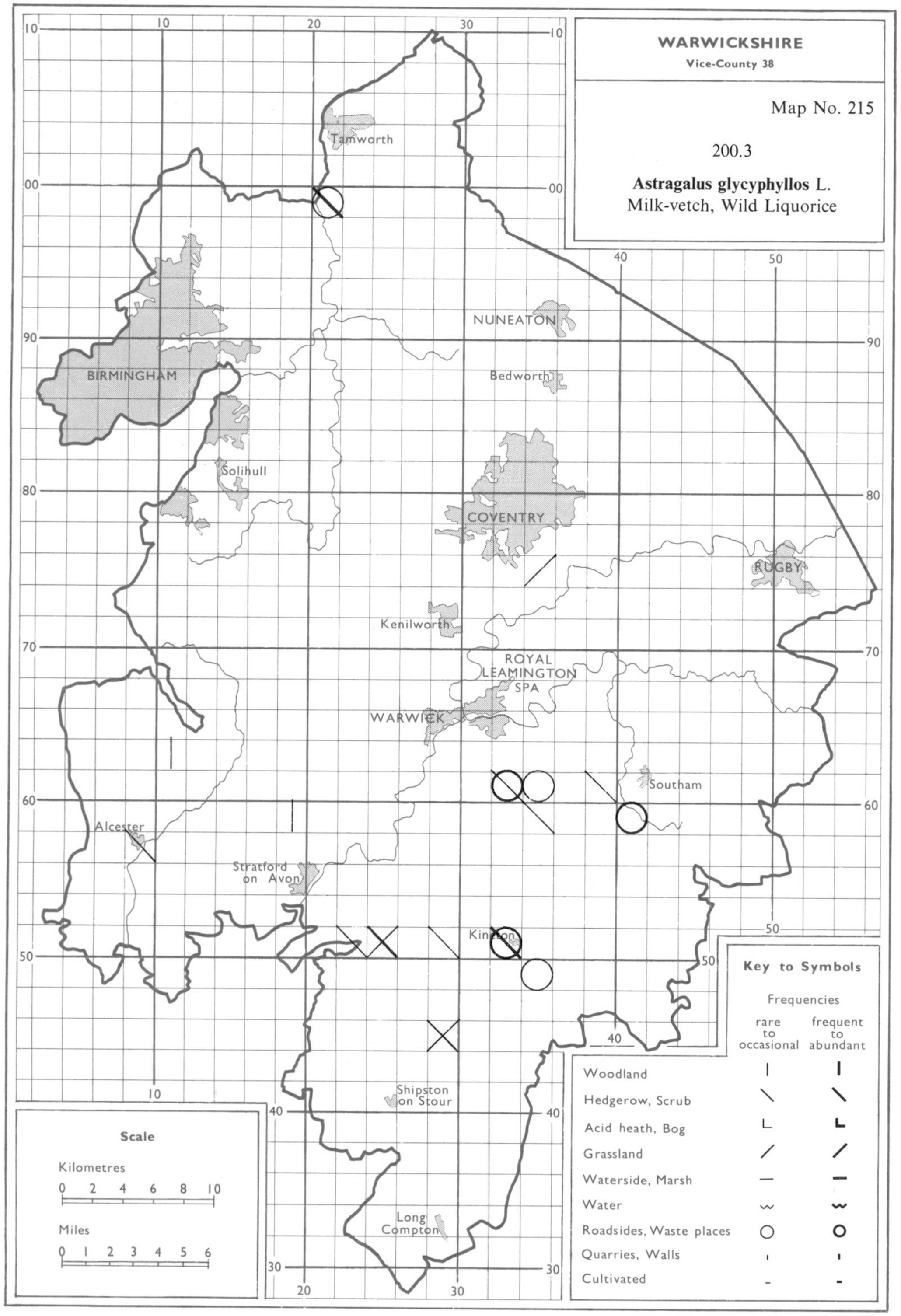

WARWICKSHIRE
Vice-County 38

Map No. 215

200.3

Astragalus glycyphyllos L.
Milk-vetch, Wild Liquorice

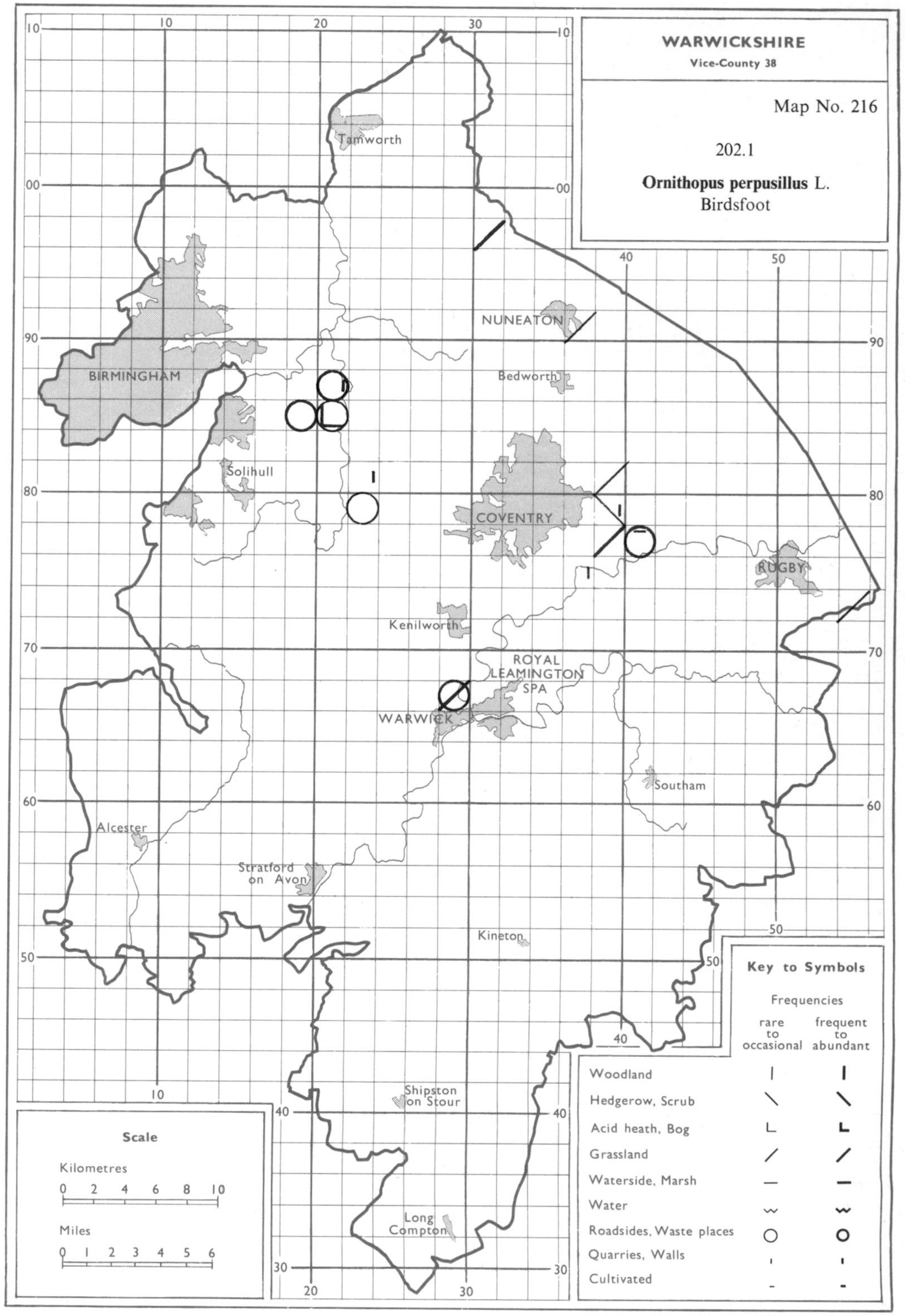

WARWICKSHIRE
Vice-County 38

Map No. 216

202.1

Ornithopus perpusillus L.
Birdsfoot

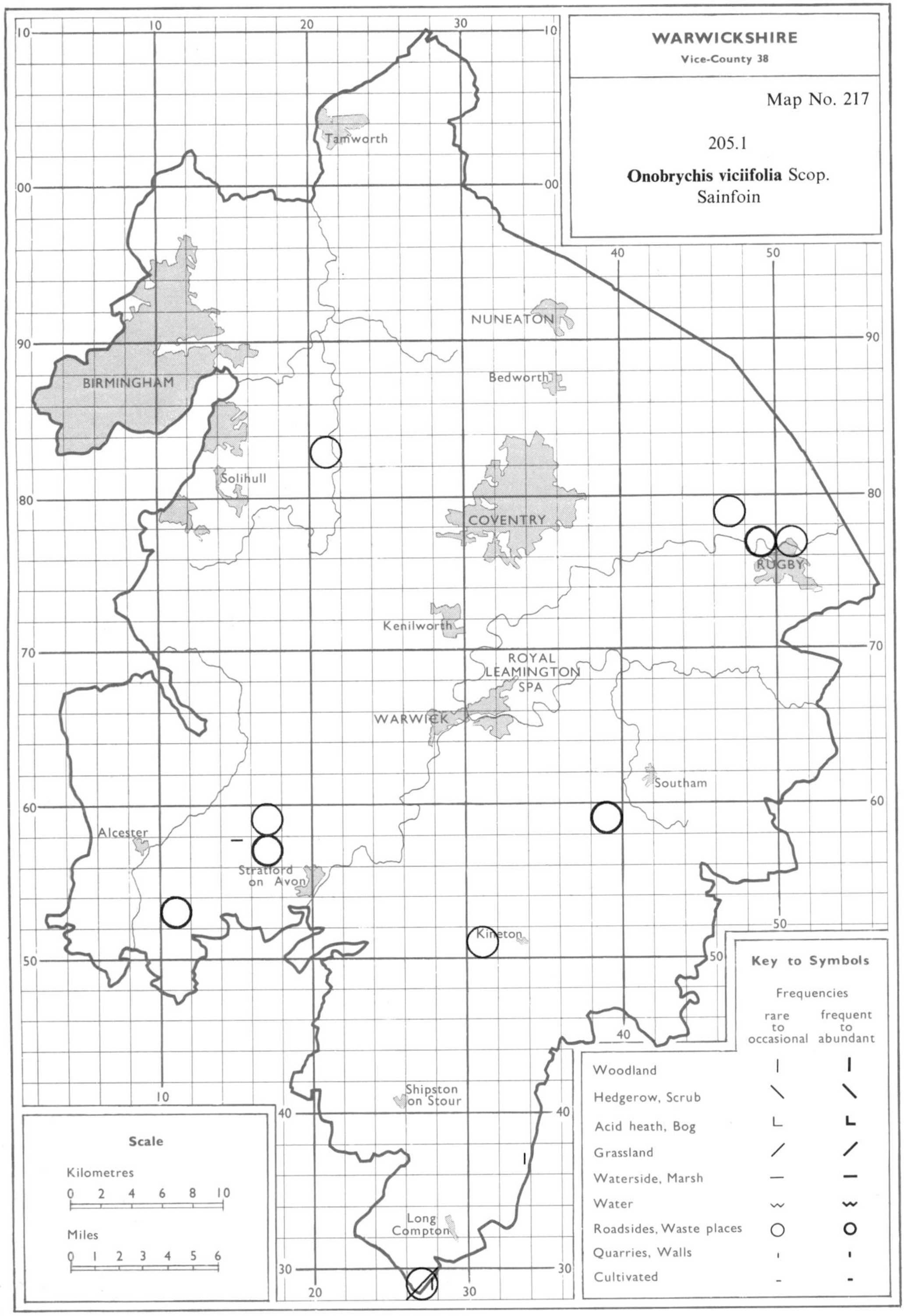
WARWICKSHIRE
Vice-County 38
Map No. 217
205.1
Onobrychis viciifolia Scop.
Sainfoin
Key to Symbols
Frequencies
rare to occasional
frequent to abundant
Woodland
Hedgerow, Scrub
Acid heath, Bog
Grassland
Waterside, Marsh
Water
Roadsides, Waste places
Quarries, Walls
Cultivated
Scale
Kilometres 0 2 4 6 8 10
Miles 0 1 2 3 4 5 6
Tamworth
BIRMINGHAM
Solihull
NUNEATON
Bedworth
COVENTRY
RUGBY
Kenilworth
ROYAL LEAMINGTON SPA
WARWICK
Southam
Alcester
Stratford on Avon
Kineton
Shipston on Stour
Long Compton

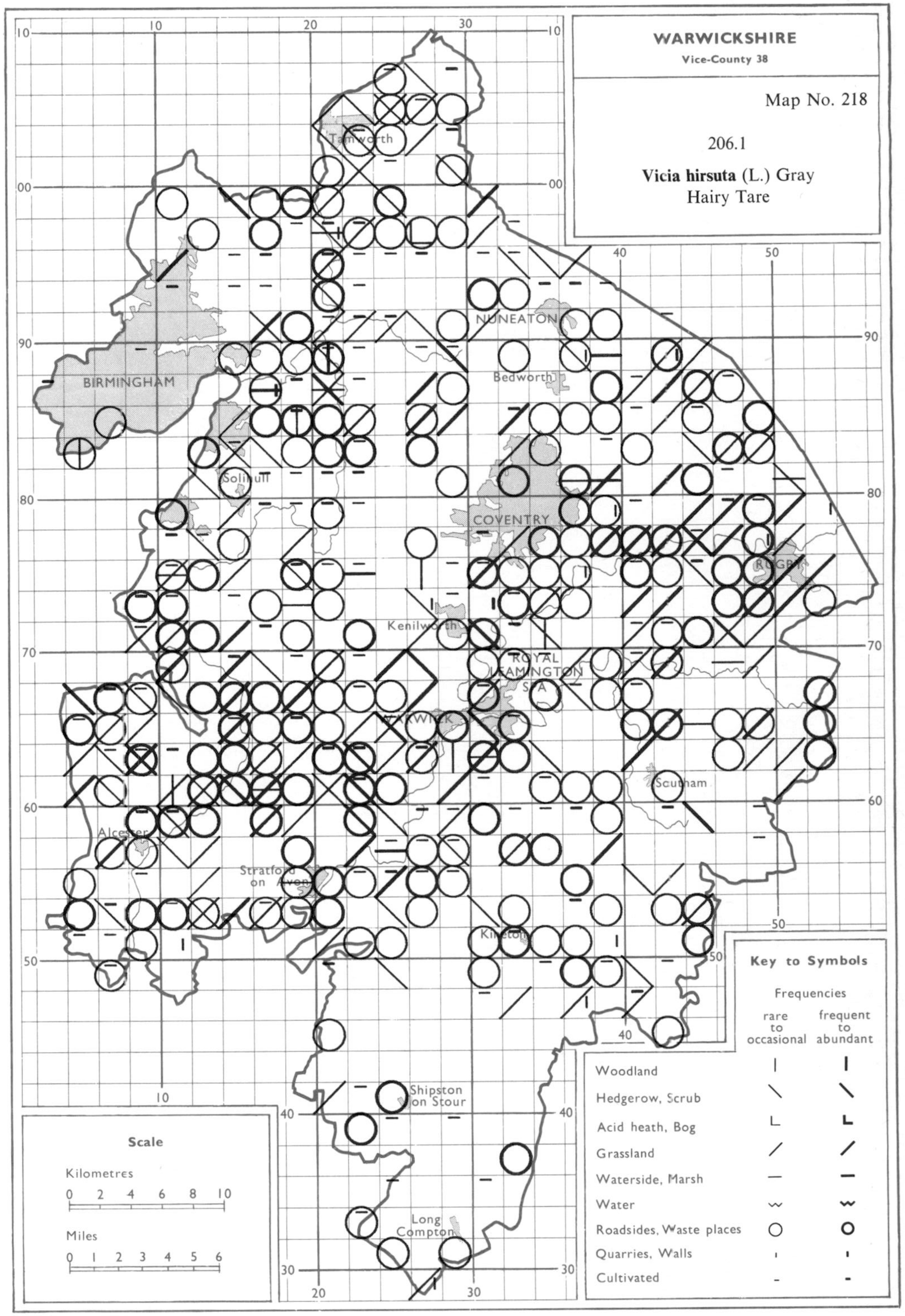
WARWICKSHIRE
Vice-County 38
Map No. 218
206.1
Vicia hirsuta (L.) Gray
Hairy Tare
Key to Symbols
Frequencies
rare to occasional
frequent to abundant
Woodland
Hedgerow, Scrub
Acid heath, Bog
Grassland
Waterside, Marsh
Water
Roadsides, Waste places
Quarries, Walls
Cultivated
Scale
Kilometres 0 2 4 6 8 10
Miles 0 1 2 3 4 5 6
Tamworth
BIRMINGHAM
Solihull
NUNEATON
Bedworth
COVENTRY
RUGBY
Kenilworth
ROYAL LEAMINGTON SPA
WARWICK
Southam
Alcester
Stratford on Avon
Kineton
Shipston on Stour
Long Compton

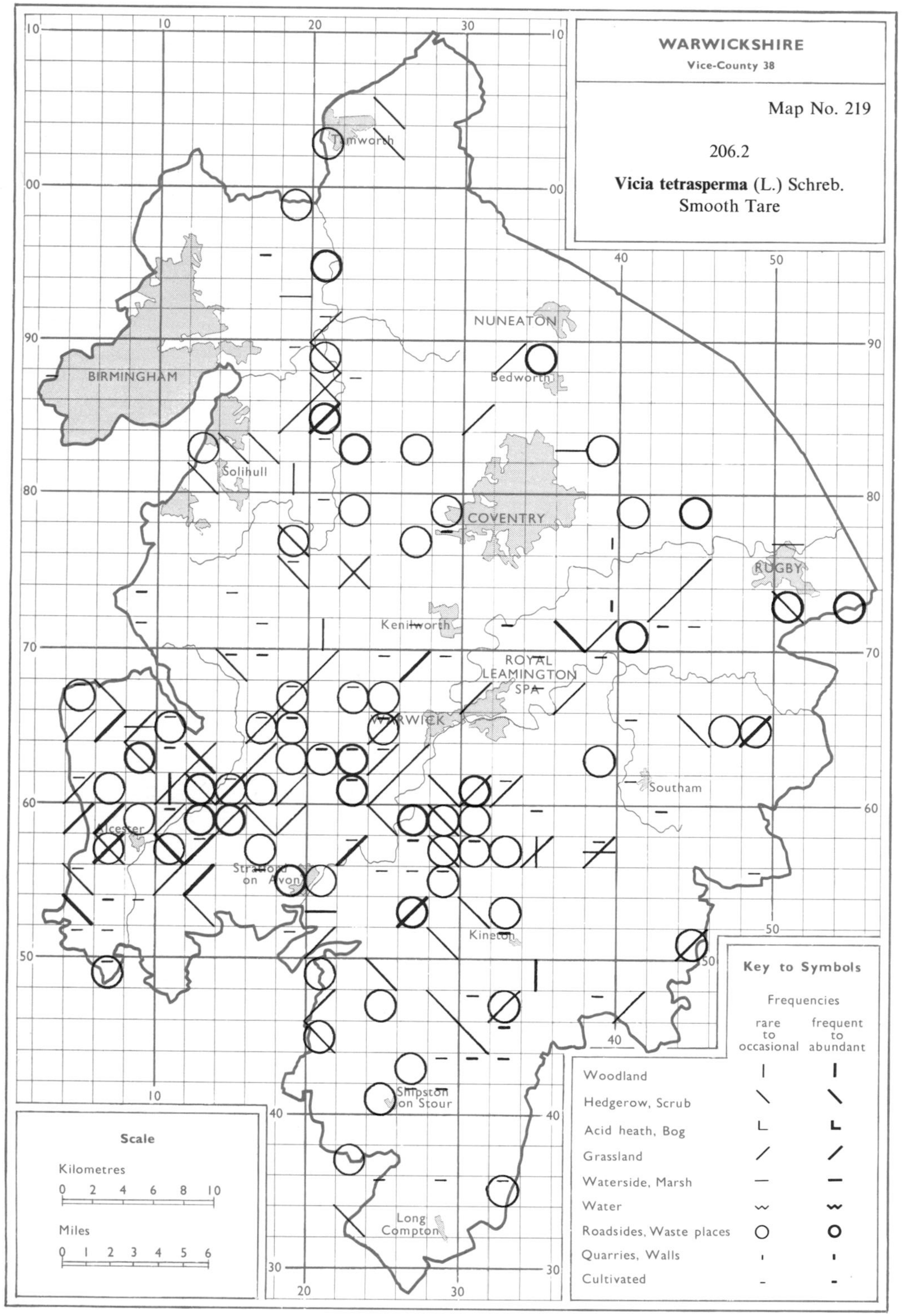
WARWICKSHIRE
Vice-County 38
Map No. 219
206.2
Vicia tetrasperma (L.) Schreb.
Smooth Tare
BIRMINGHAM
Tamworth
Solihull
NUNEATON
Bedworth
COVENTRY
RUGBY
Kenilworth
ROYAL LEAMINGTON SPA
WARWICK
Southam
Alcester
Stratford on Avon
Kineton
Shipston on Stour
Long Compton
Key to Symbols
Frequencies
rare to occasional
frequent to abundant
Woodland
Hedgerow, Scrub
Acid heath, Bog
Grassland
Waterside, Marsh
Water
Roadsides, Waste places
Quarries, Walls
Cultivated
Scale
Kilometres
0 2 4 6 8 10
Miles
0 1 2 3 4 5 6

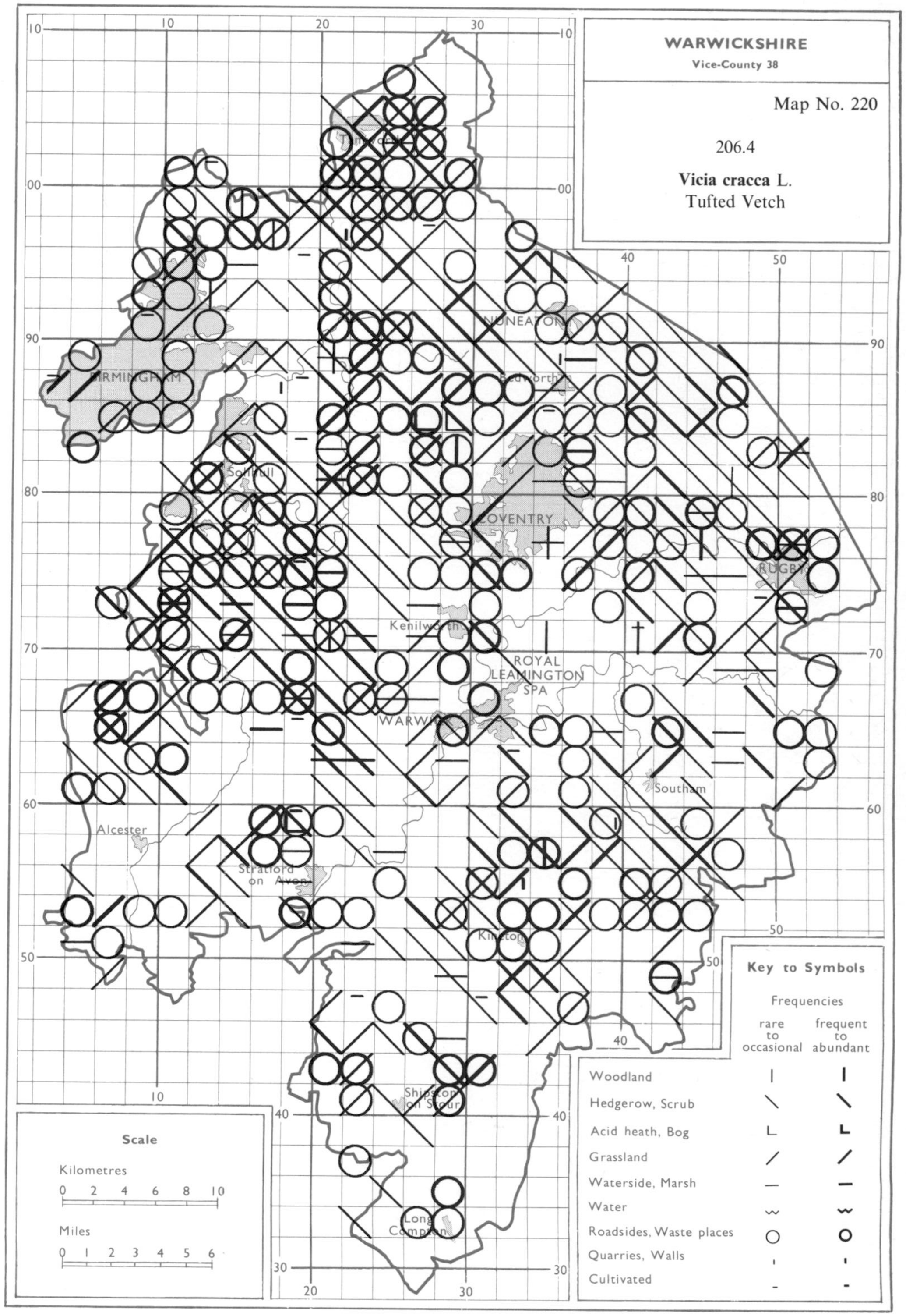
WARWICKSHIRE
Vice-County 38
Map No. 220
206.4
Vicia cracca L.
Tufted Vetch
BIRMINGHAM
Solihull
NUNEATON
Bedworth
COVENTRY
RUGBY
Kenilworth
ROYAL LEAMINGTON SPA
Southam
Alcester
Stratford on Avon
Shipston on Stour
Key to Symbols
Frequencies
rare to occasional
frequent to abundant
Woodland
Hedgerow, Scrub
Acid heath, Bog
Grassland
Waterside, Marsh
Water
Roadsides, Waste places
Quarries, Walls
Cultivated
Scale
Kilometres
0 2 4 6 8 10
Miles
0 1 2 3 4 5 6

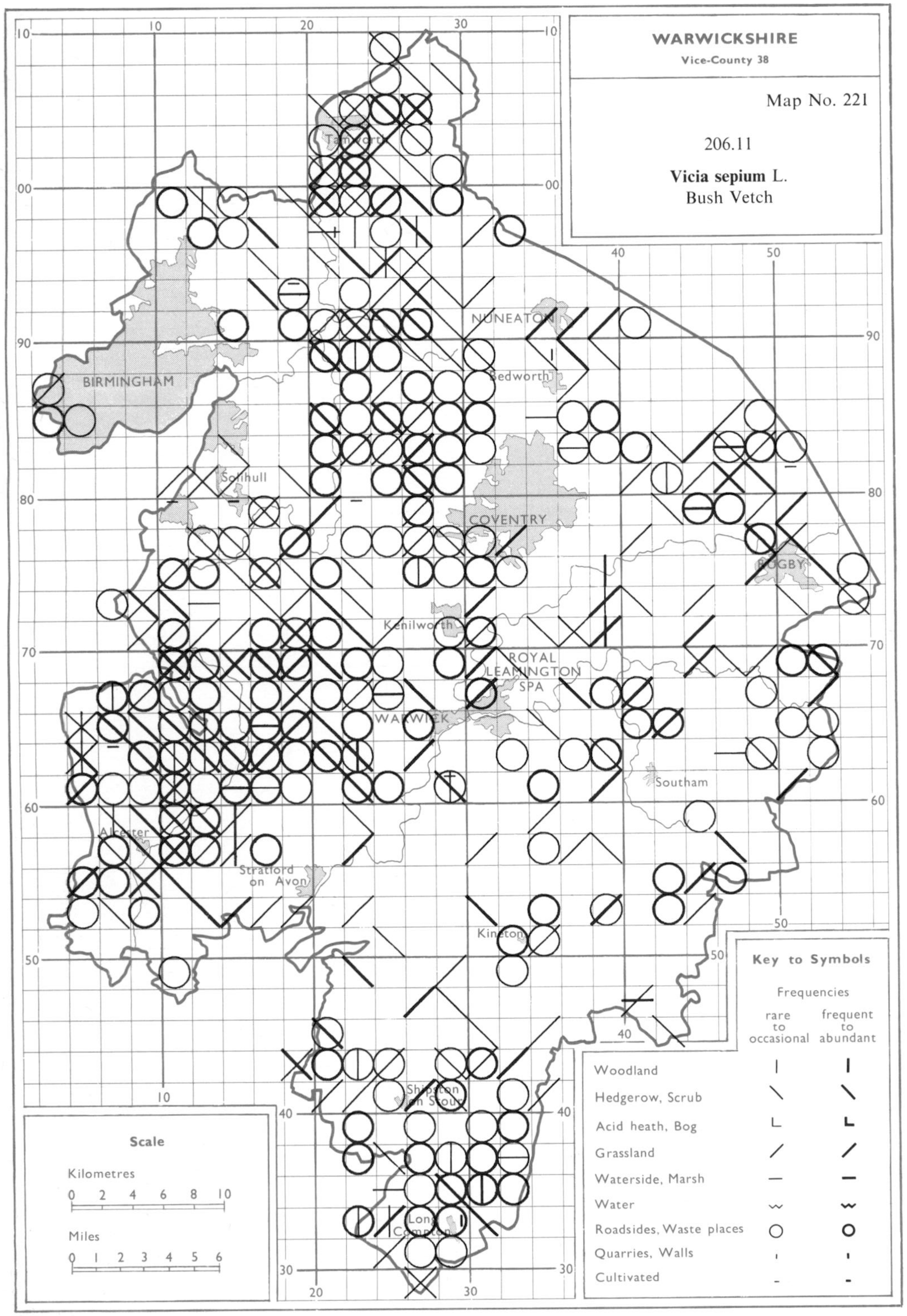
WARWICKSHIRE
Vice-County 38
Map No. 221
206.11
Vicia sepium L.
Bush Vetch
Tamworth
BIRMINGHAM
NUNEATON
Bedworth
Solihull
COVENTRY
RUGBY
Kenilworth
ROYAL LEAMINGTON SPA
WARWICK
Southam
Alcester
Stratford on Avon
Kineton
Shipston on Stour
Long Compton
Key to Symbols
Frequencies
rare to occasional
frequent to abundant
Woodland
Hedgerow, Scrub
Acid heath, Bog
Grassland
Waterside, Marsh
Water
Roadsides, Waste places
Quarries, Walls
Cultivated
Scale
Kilometres
0 2 4 6 8 10
Miles
0 1 2 3 4 5 6

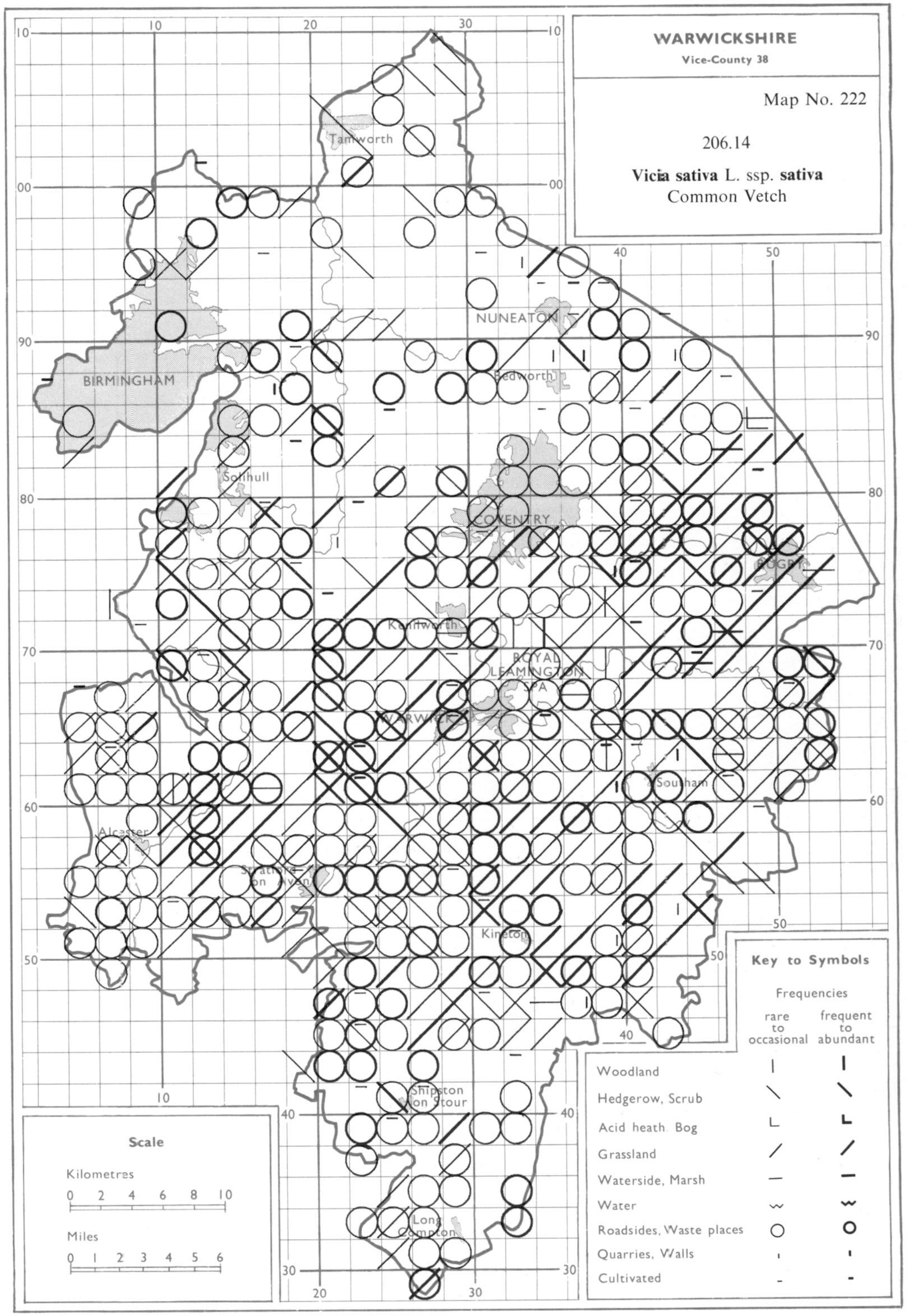
WARWICKSHIRE
Vice-County 38
Map No. 222
206.14
Vicia sativa L. ssp. sativa
Common Vetch
Tamworth
BIRMINGHAM
NUNEATON
Bedworth
Solihull
COVENTRY
RUGBY
Kenilworth
ROYAL LEAMINGTON SPA
WARWICK
Southam
Alcester
Stratford on Avon
Kineton
Shipston on Stour
Long Compton
Key to Symbols
Frequencies
rare to occasional
frequent to abundant
Woodland
Hedgerow, Scrub
Acid heath Bog
Grassland
Waterside, Marsh
Water
Roadsides, Waste places
Quarries, Walls
Cultivated
Scale
Kilometres
0 2 4 6 8 10
Miles
0 1 2 3 4 5 6

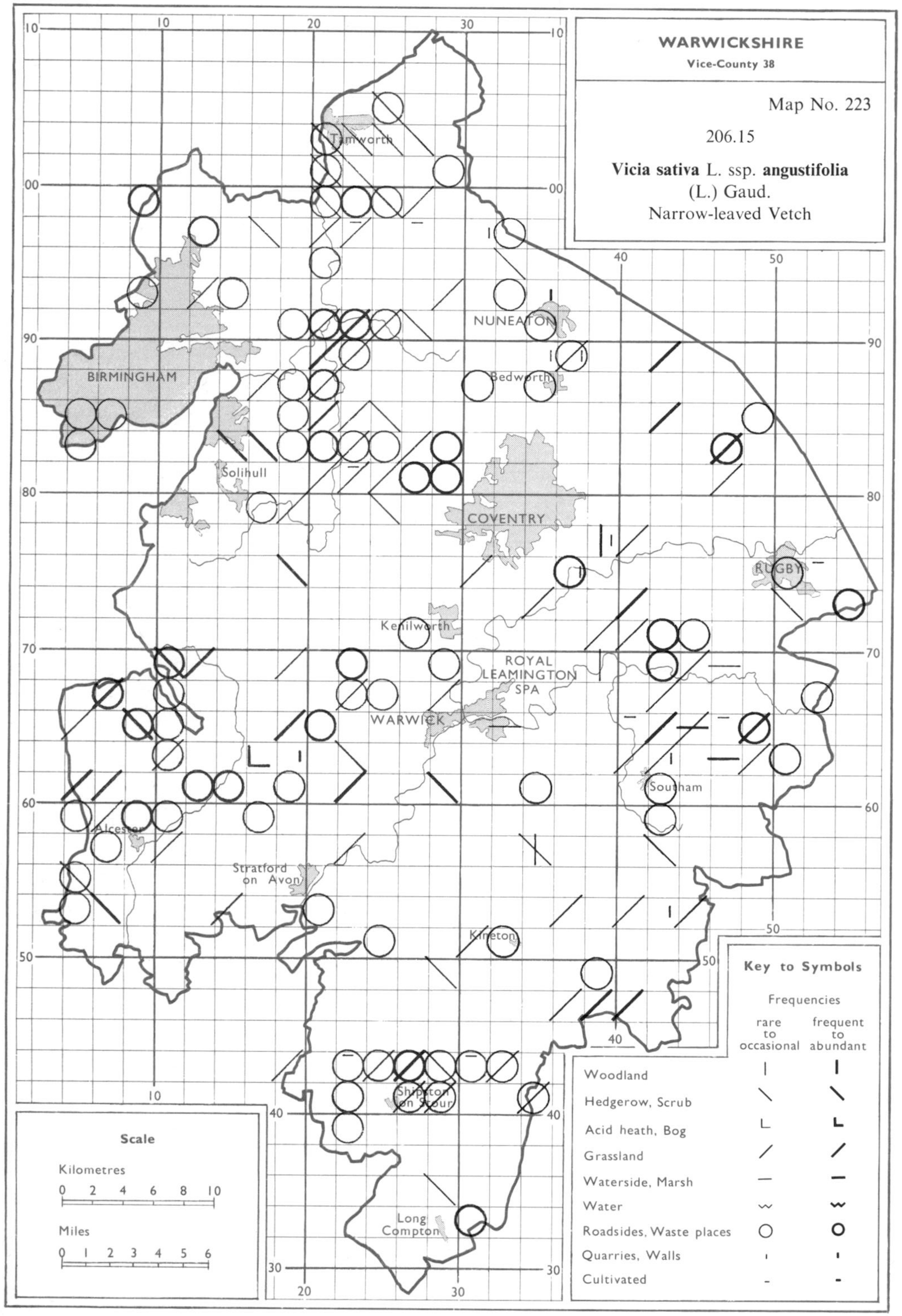
WARWICKSHIRE
Vice-County 38
Map No. 223
206.15
Vicia sativa L. ssp. angustifolia (L.) Gaud.
Narrow-leaved Vetch
Tamworth
BIRMINGHAM
Solihull
NUNEATON
Bedworth
COVENTRY
RUGBY
Kenilworth
ROYAL LEAMINGTON SPA
WARWICK
Southam
Alcester
Stratford on Avon
Kineton
Shipston on Stour
Long Compton
Key to Symbols
Frequencies
rare to occasional
frequent to abundant
Woodland
Hedgerow, Scrub
Acid heath, Bog
Grassland
Waterside, Marsh
Water
Roadsides, Waste places
Quarries, Walls
Cultivated
Scale
Kilometres
0 2 4 6 8 10
Miles
0 1 2 3 4 5 6

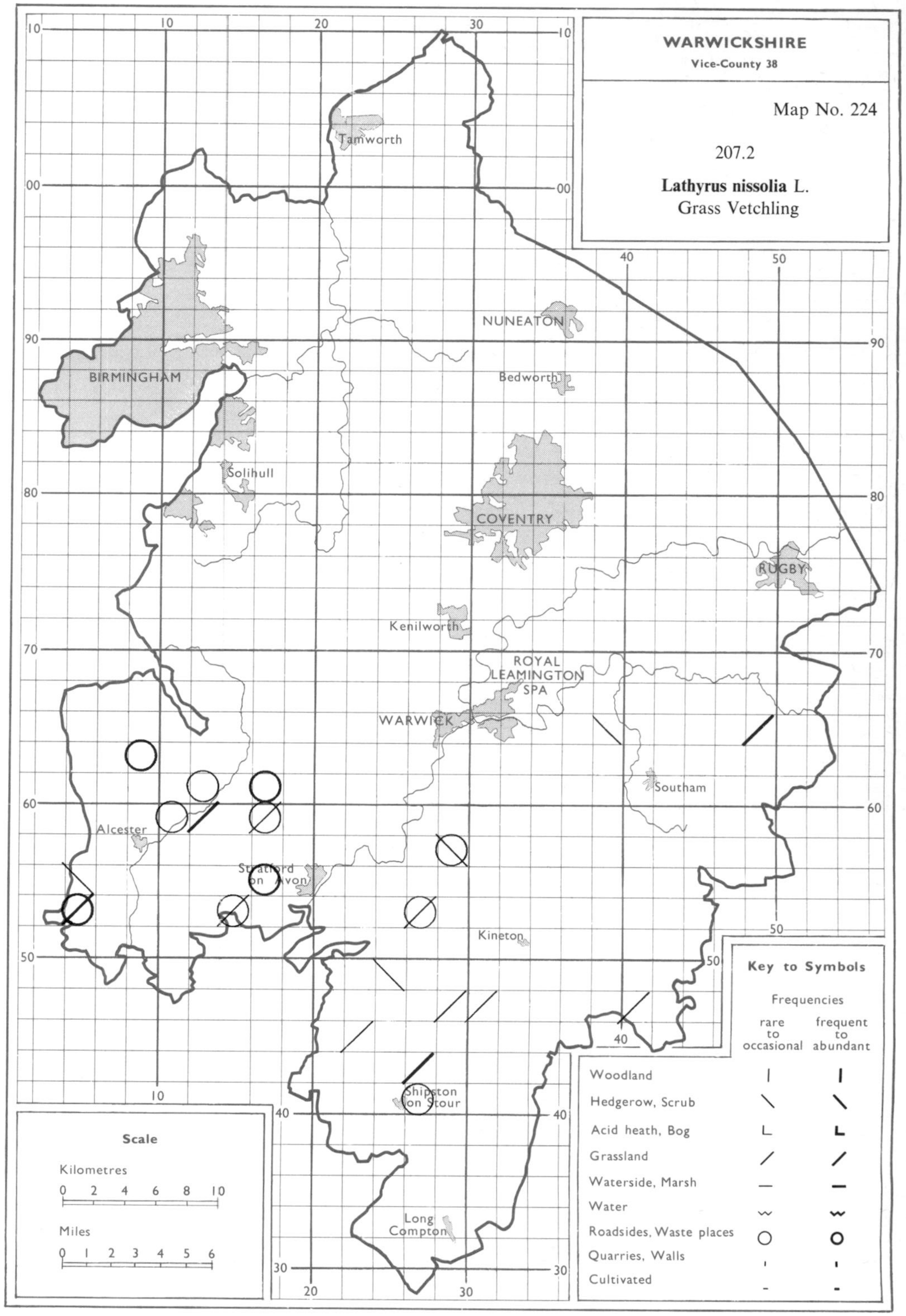
WARWICKSHIRE
Vice-County 38
Map No. 224
207.2
Lathyrus nissolia L.
Grass Vetchling
Tamworth
BIRMINGHAM
Solihull
NUNEATON
Bedworth
COVENTRY
RUGBY
Kenilworth
ROYAL LEAMINGTON SPA
WARWICK
Southam
Alcester
Stratford on Avon
Kineton
Shipston on Stour
Long Compton
Key to Symbols
Frequencies
rare to occasional
frequent to abundant
Woodland
Hedgerow, Scrub
Acid heath, Bog
Grassland
Waterside, Marsh
Water
Roadsides, Waste places
Quarries, Walls
Cultivated
Scale
Kilometres
0 2 4 6 8 10
Miles
0 1 2 3 4 5 6

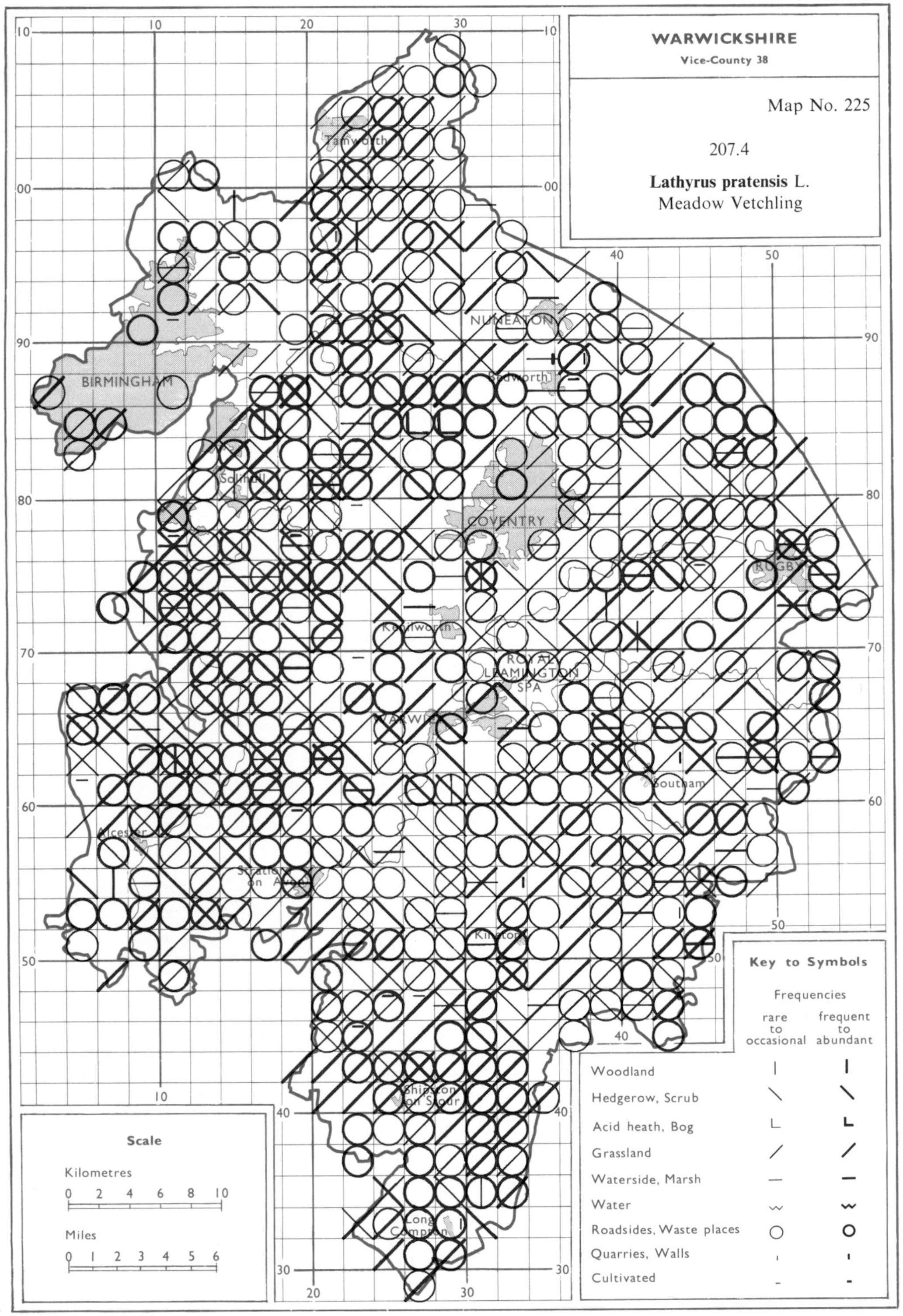
WARWICKSHIRE
Vice-County 38
Map No. 225
207.4
Lathyrus pratensis L.
Meadow Vetchling
Tamworth
NUNEATON
Bedworth
BIRMINGHAM
Solihull
COVENTRY
RUGBY
Kenilworth
ROYAL LEAMINGTON SPA
WARWICK
Southam
Alcester
Stratford on Avon
Kineton
Shipston on Stour
Long Compton
Key to Symbols
Frequencies
rare to occasional
frequent to abundant
Woodland
Hedgerow, Scrub
Acid heath, Bog
Grassland
Waterside, Marsh
Water
Roadsides, Waste places
Quarries, Walls
Cultivated
Scale
Kilometres
0 2 4 6 8 10
Miles
0 1 2 3 4 5 6

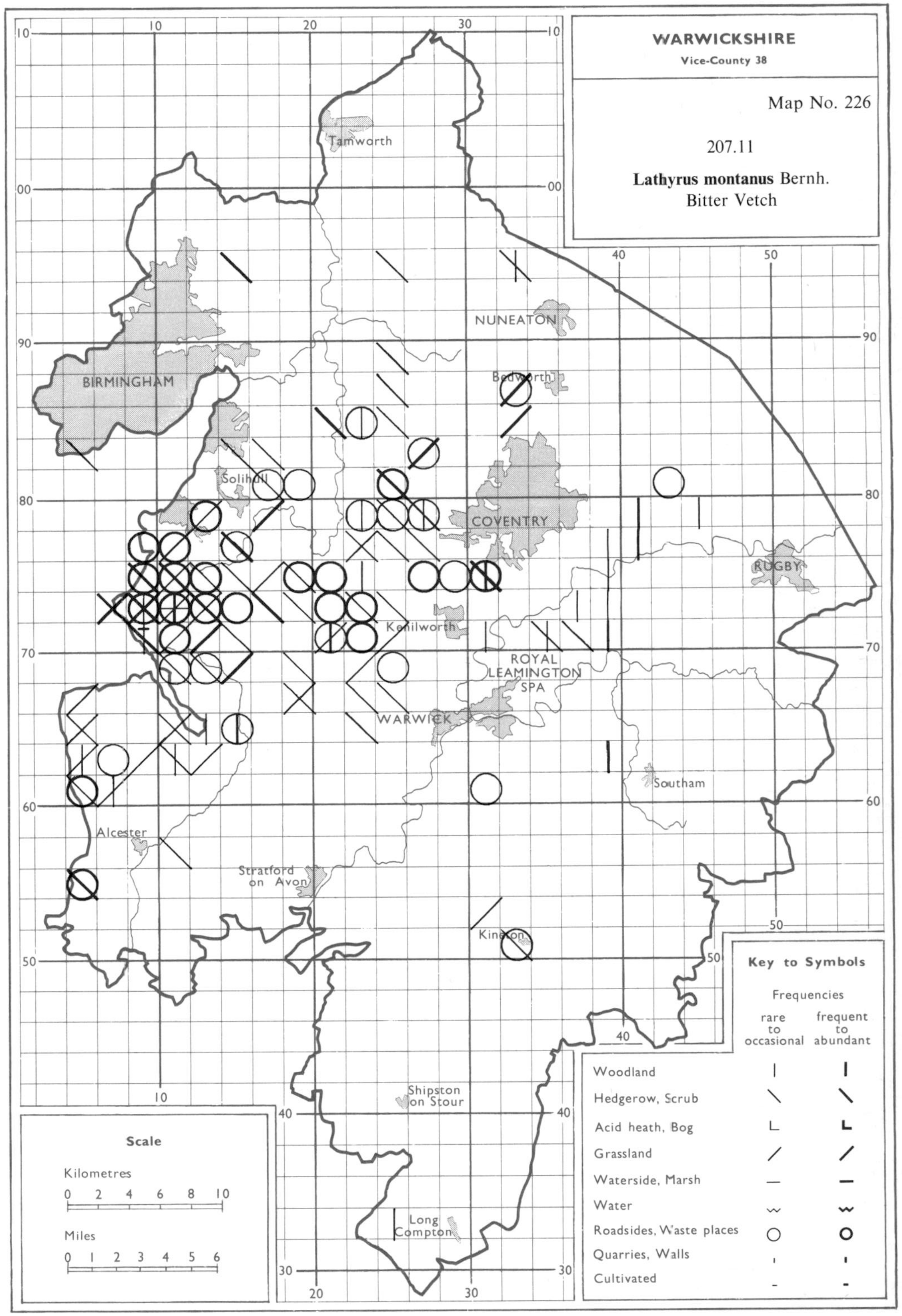
WARWICKSHIRE
Vice-County 38
Map No. 226
207.11
Lathyrus montanus Bernh.
Bitter Vetch
Tamworth
NUNEATON
Bedworth
BIRMINGHAM
Solihull
COVENTRY
RUGBY
Kenilworth
ROYAL LEAMINGTON SPA
WARWICK
Southam
Alcester
Stratford on Avon
Kineton
Shipston on Stour
Long Compton
Key to Symbols
Frequencies
rare to occasional
frequent to abundant
Woodland
Hedgerow, Scrub
Acid heath, Bog
Grassland
Waterside, Marsh
Water
Roadsides, Waste places
Quarries, Walls
Cultivated
Scale
Kilometres
0 2 4 6 8 10
Miles
0 1 2 3 4 5 6

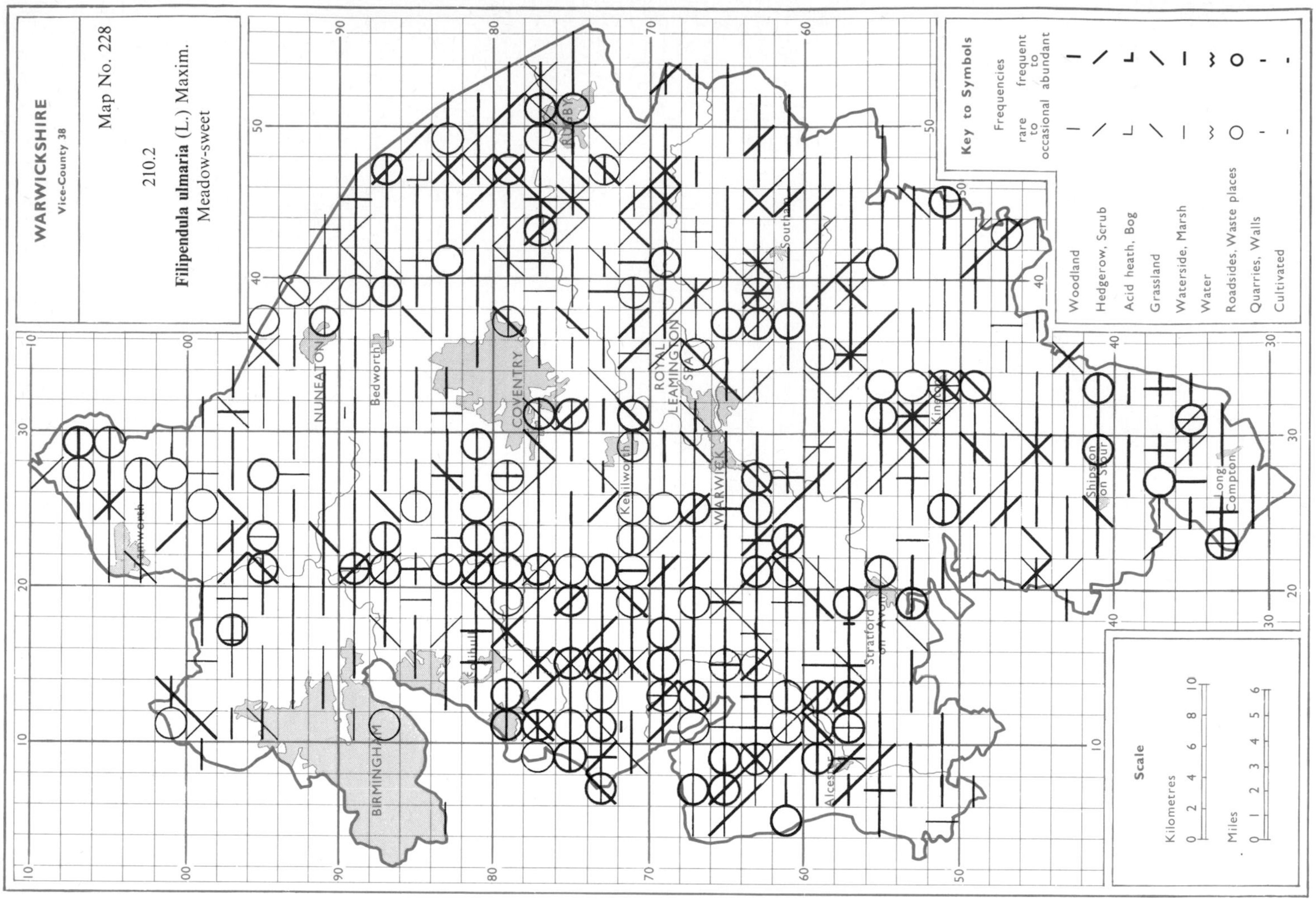
WARWICKSHIRE
Vice-County 38
Map No. 228
210.2
Filipendula ulmaria (L.) Maxim.
Meadow-sweet
Key to Symbols
Frequencies
rare to occasional
frequent to abundant
Woodland
Hedgerow, Scrub
Acid heath, Bog
Grassland
Waterside, Marsh
Water
Roadsides, Waste places
Quarries, Walls
Cultivated
Scale
Kilometres 0 2 4 6 8 10
Miles 0 1 2 3 4 5 6
NUNEATON
Bedworth
COVENTRY
RUGBY
Southam
ROYAL LEAMINGTON SPA
WARWICK
Kenilworth
Tamworth
Solihull
BIRMINGHAM
Alcester
Stratford on Avon
Kineton
Shipston on Stour
Long Compton

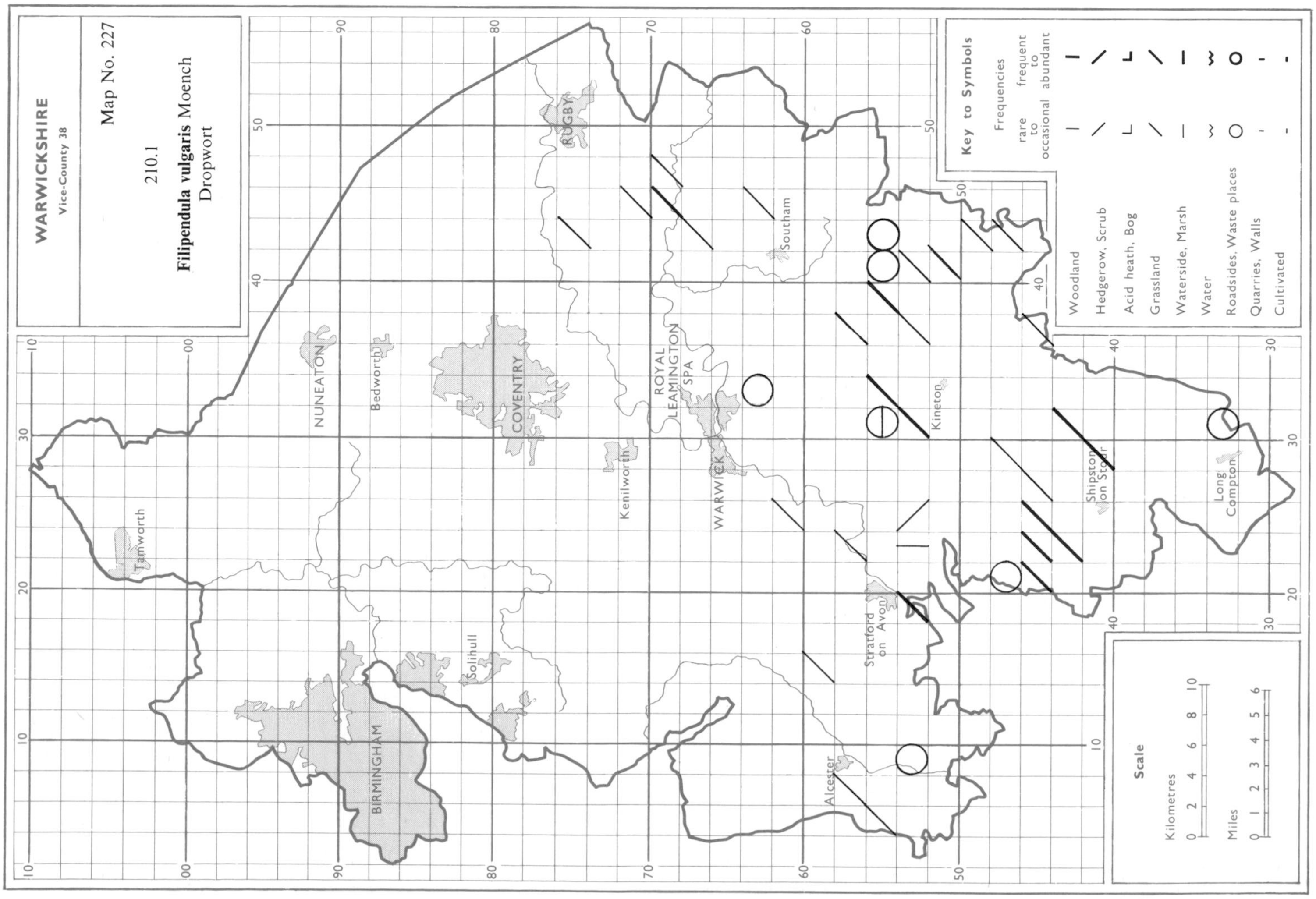
WARWICKSHIRE
Vice-County 38
Map No. 227
210.1
Filipendula vulgaris Moench
Dropwort
Key to Symbols
Frequencies
rare to occasional
frequent to abundant
Woodland
Hedgerow, Scrub
Acid heath, Bog
Grassland
Waterside, Marsh
Water
Roadsides, Waste places
Quarries, Walls
Cultivated
Scale
Kilometres 0 2 4 6 8 10
Miles 0 1 2 3 4 5 6
NUNEATON
Bedworth
COVENTRY
RUGBY
Southam
ROYAL LEAMINGTON SPA
WARWICK
Kenilworth
Tamworth
Solihull
BIRMINGHAM
Alcester
Stratford on Avon
Kineton
Shipston on Stour
Long Compton

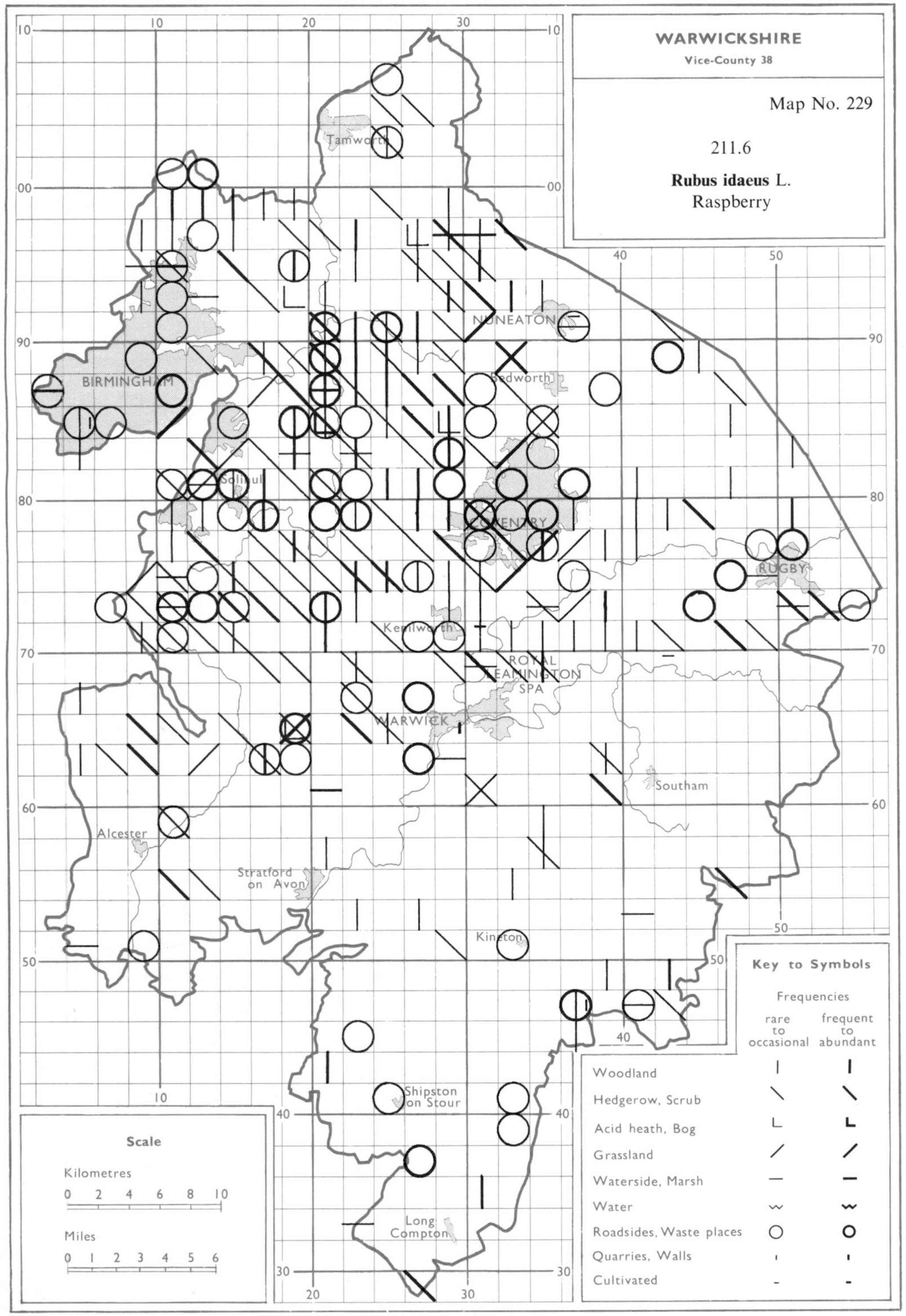
WARWICKSHIRE
Vice-County 38
Map No. 229
211.6
Rubus idaeus L.
Raspberry
Tamworth
NUNEATON
Bedworth
BIRMINGHAM
Solihull
COVENTRY
RUGBY
Kenilworth
ROYAL LEAMINGTON SPA
WARWICK
Southam
Alcester
Stratford on Avon
Kineton
Shipston on Stour
Long Compton
Scale
Kilometres
0 2 4 6 8 10
Miles
0 1 2 3 4 5 6
Key to Symbols
Frequencies
rare to occasional
frequent to abundant
Woodland
Hedgerow, Scrub
Acid heath, Bog
Grassland
Waterside, Marsh
Water
Roadsides, Waste places
Quarries, Walls
Cultivated

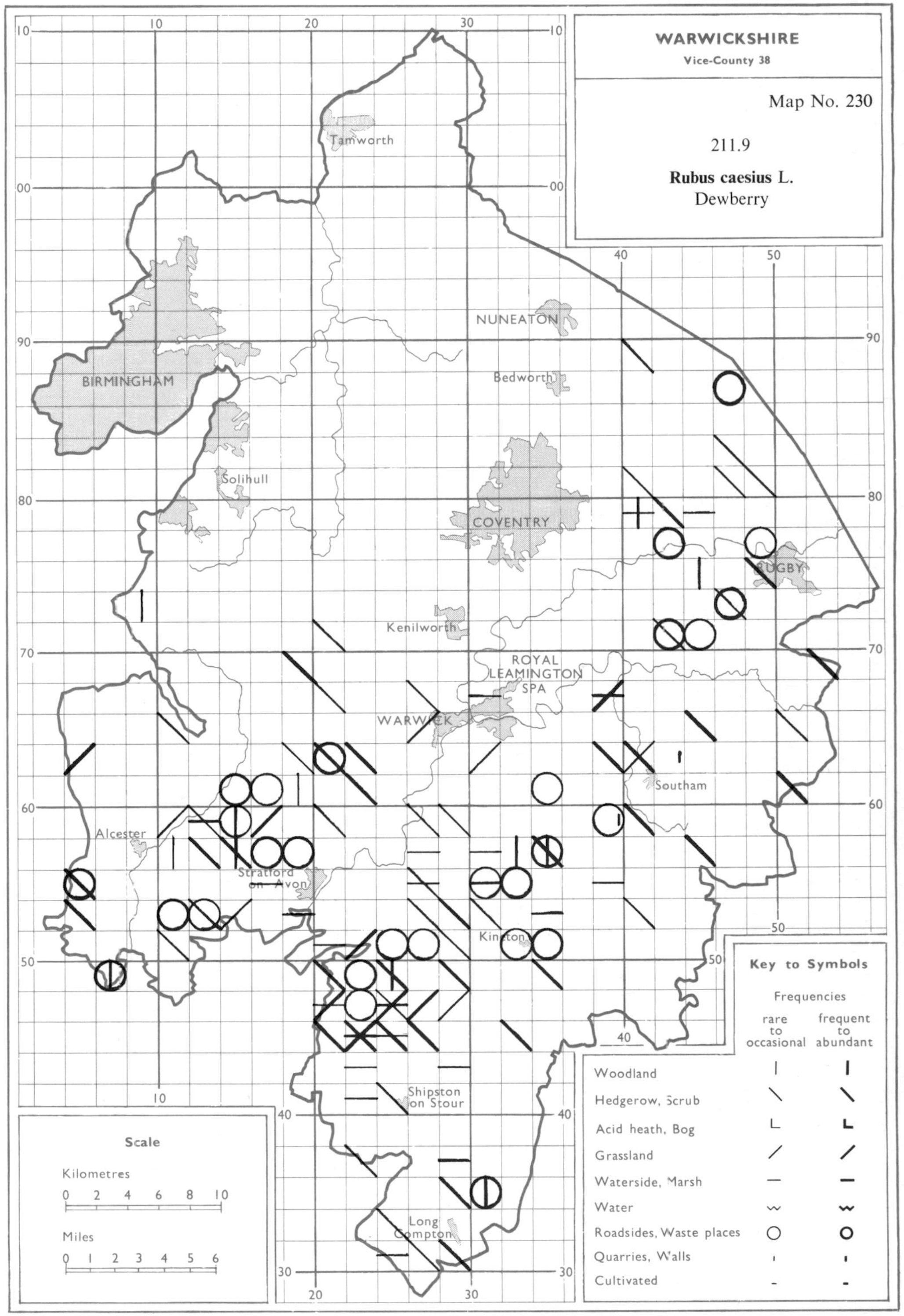
WARWICKSHIRE
Vice-County 38
Map No. 230
211.9
Rubus caesius L.
Dewberry
Tamworth
NUNEATON
Bedworth
BIRMINGHAM
Solihull
COVENTRY
RUGBY
Kenilworth
ROYAL LEAMINGTON SPA
WARWICK
Southam
Alcester
Stratford on Avon
Kineton
Shipston on Stour
Long Compton
Scale
Kilometres
0 2 4 6 8 10
Miles
0 1 2 3 4 5 6
Key to Symbols
Frequencies
rare to occasional
frequent to abundant
Woodland
Hedgerow, Scrub
Acid heath, Bog
Grassland
Waterside, Marsh
Water
Roadsides, Waste places
Quarries, Walls
Cultivated

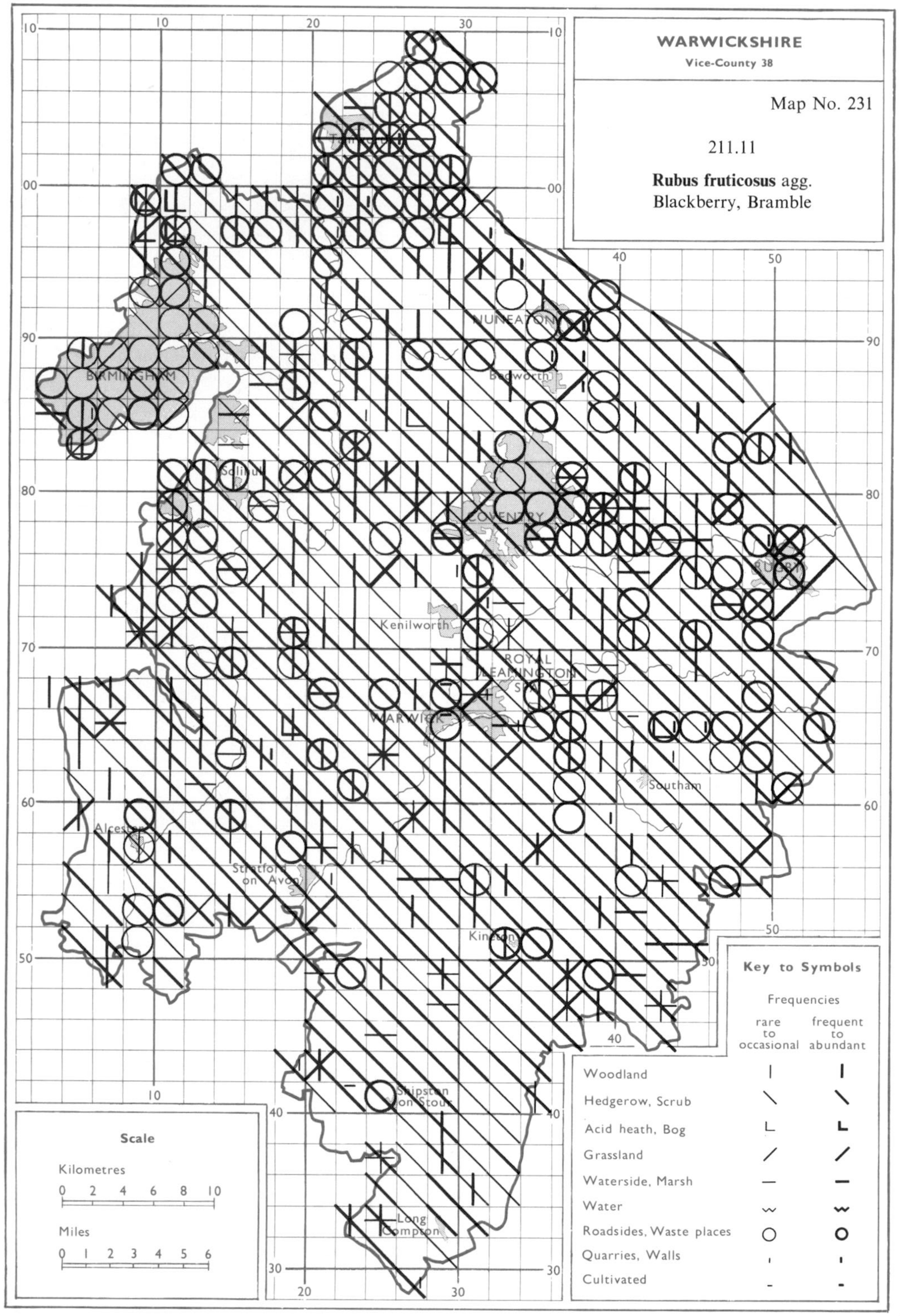
WARWICKSHIRE
Vice-County 38
Map No. 231
211.11
Rubus fruticosus agg.
Blackberry, Bramble
NUNEATON
Bedworth
BIRMINGHAM
Solihull
COVENTRY
RUGBY
Kenilworth
ROYAL LEAMINGTON SPA
WARWICK
Southam
Alcester
Stratford on Avon
Kineton
Shipston on Stour
Long Compton
Key to Symbols
Frequencies
rare to occasional
frequent to abundant
Woodland
Hedgerow, Scrub
Acid heath, Bog
Grassland
Waterside, Marsh
Water
Roadsides, Waste places
Quarries, Walls
Cultivated
Scale
Kilometres
0 2 4 6 8 10
Miles
0 1 2 3 4 5 6

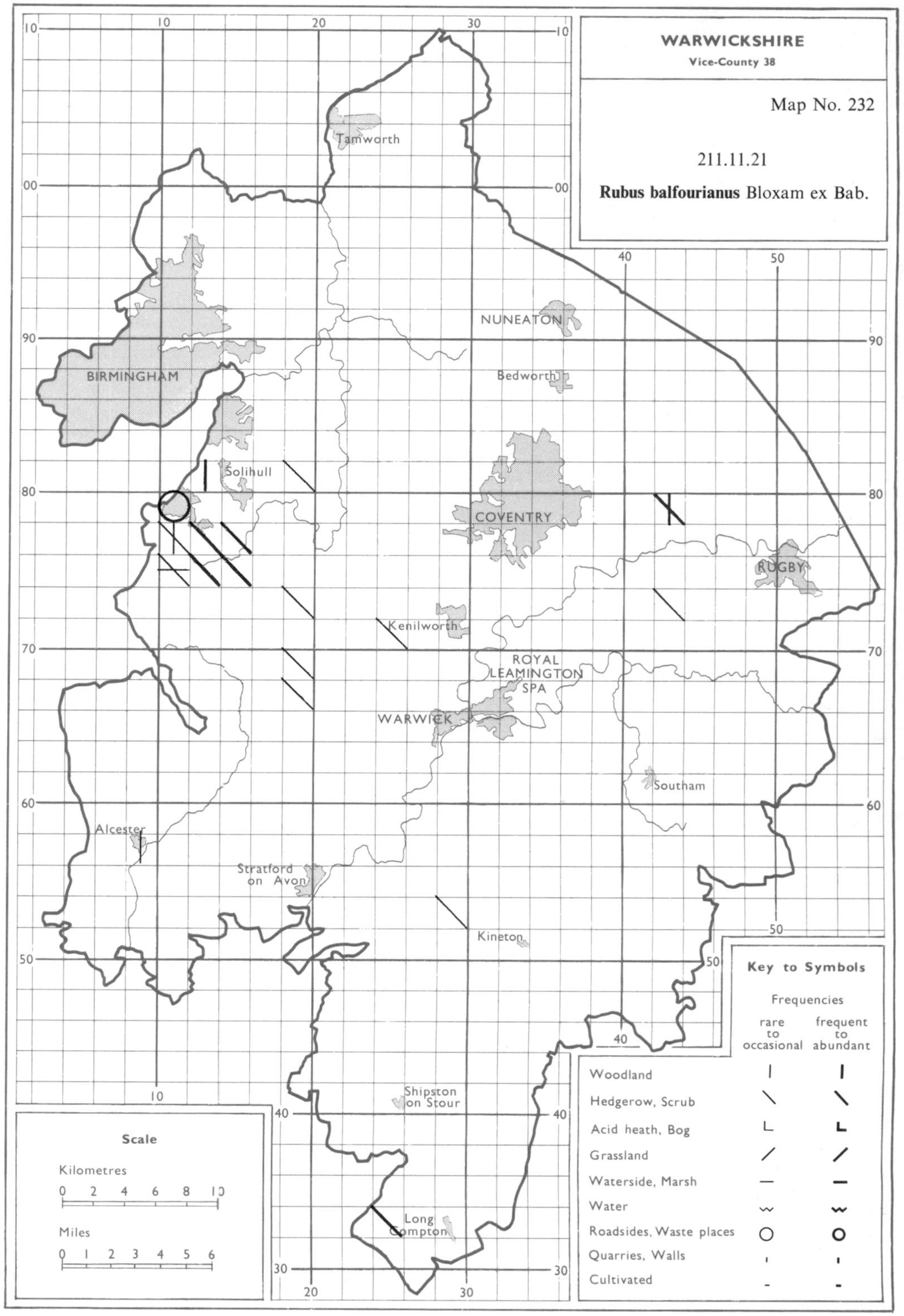
WARWICKSHIRE
Vice-County 38
Map No. 232
211.11.21
Rubus balfourianus Bloxam ex Bab.
Tamworth
NUNEATON
Bedworth
BIRMINGHAM
Solihull
COVENTRY
RUGBY
Kenilworth
ROYAL LEAMINGTON SPA
WARWICK
Southam
Alcester
Stratford on Avon
Kineton
Shipston on Stour
Long Compton
Key to Symbols
Frequencies
rare to occasional
frequent to abundant
Woodland
Hedgerow, Scrub
Acid heath, Bog
Grassland
Waterside, Marsh
Water
Roadsides, Waste places
Quarries, Walls
Cultivated
Scale
Kilometres
0 2 4 6 8 10
Miles
0 1 2 3 4 5 6

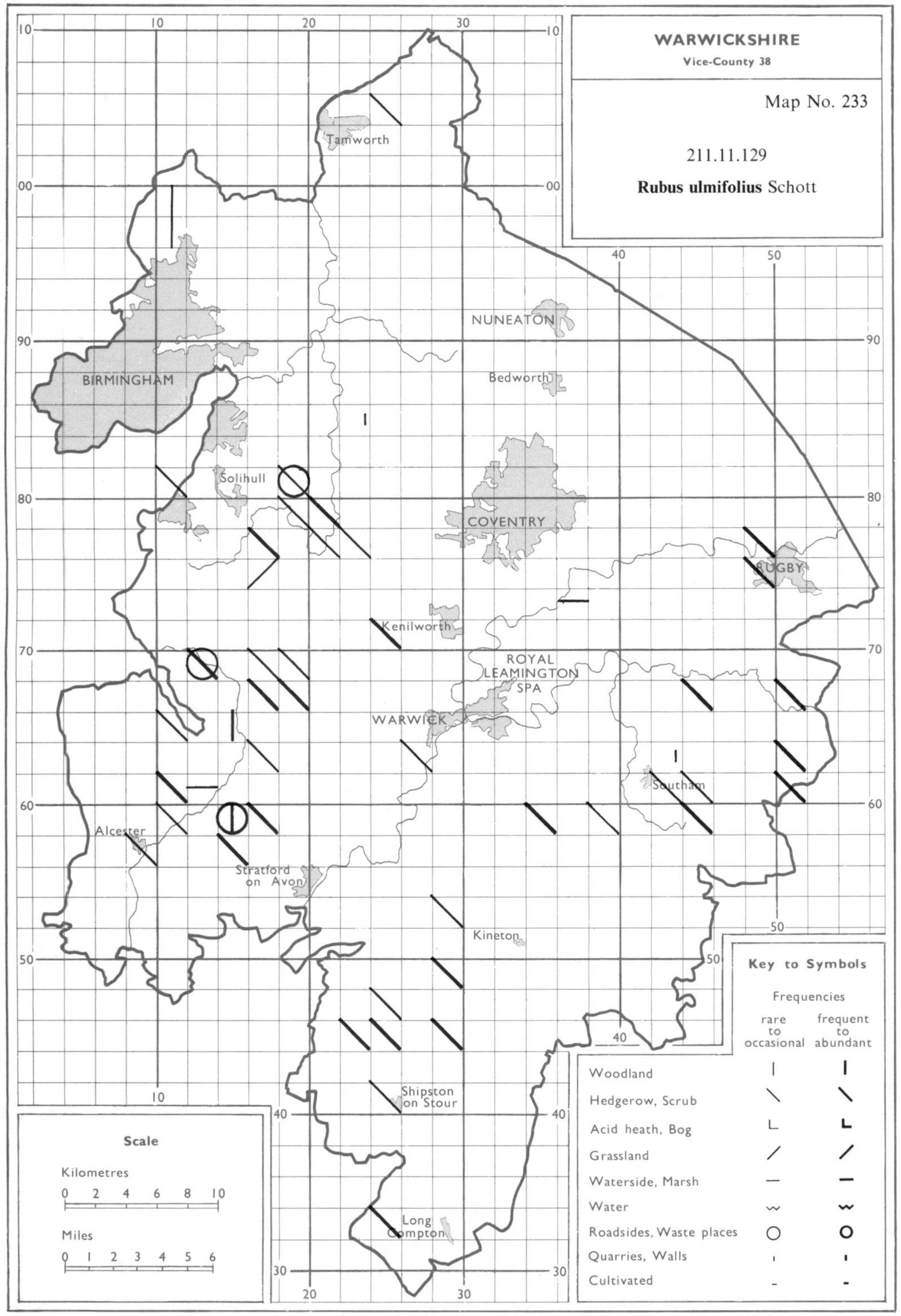
WARWICKSHIRE
Vice-County 38
Map No. 233
211.11.129
Rubus ulmifolius Schott
Tamworth
NUNEATON
Bedworth
BIRMINGHAM
Solihull
COVENTRY
RUGBY
Kenilworth
ROYAL LEAMINGTON SPA
WARWICK
Southam
Alcester
Stratford on Avon
Kineton
Shipston on Stour
Long Compton
Key to Symbols
Frequencies
rare to occasional
frequent to abundant
Woodland
Hedgerow, Scrub
Acid heath, Bog
Grassland
Waterside, Marsh
Water
Roadsides, Waste places
Quarries, Walls
Cultivated
Scale
Kilometres
0 2 4 6 8 10
Miles
0 1 2 3 4 5 6

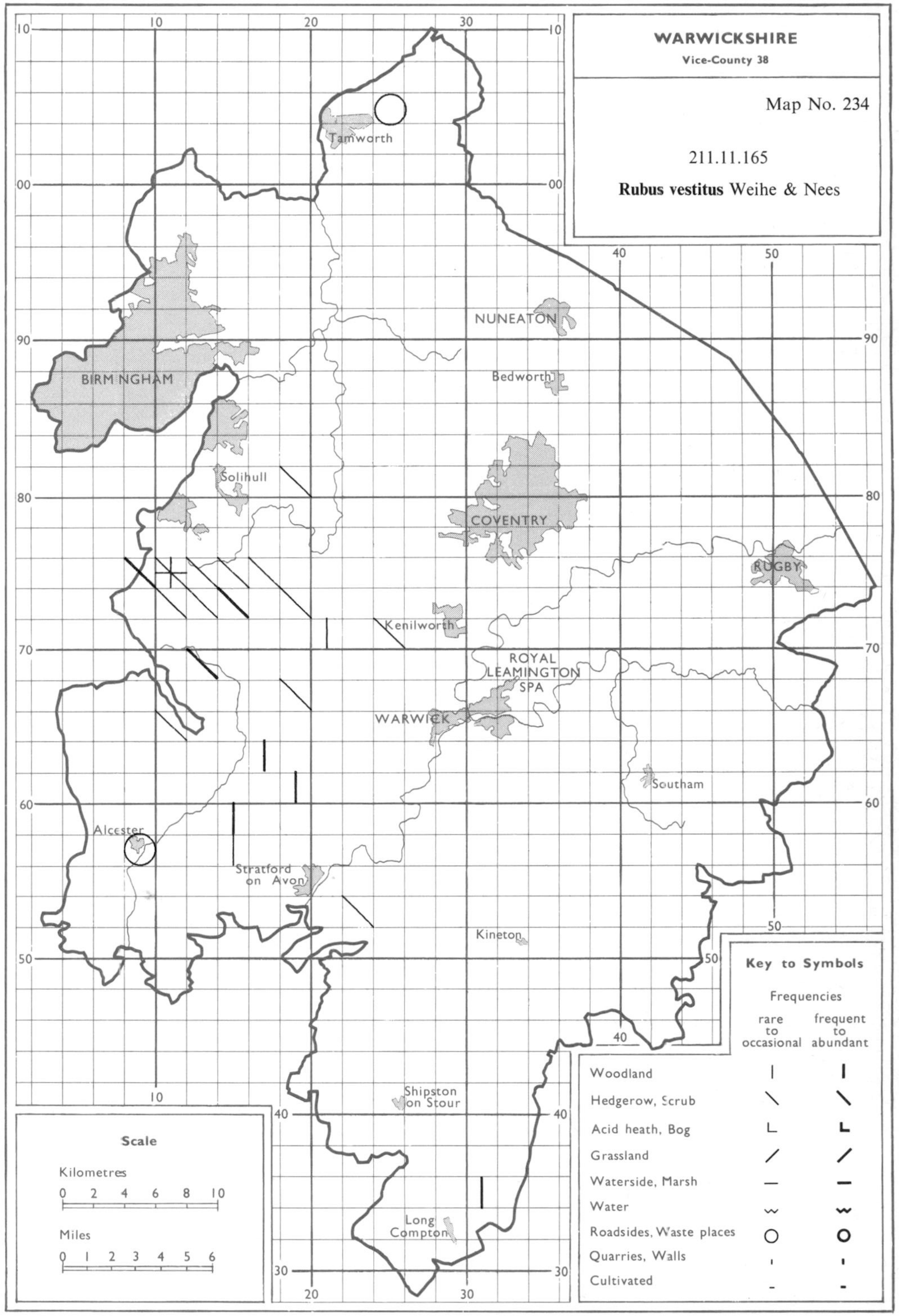
WARWICKSHIRE
Vice-County 38
Map No. 234
211.11.165
Rubus vestitus Weihe & Nees
Tamworth
NUNEATON
Bedworth
BIRMNGHAM
Solihull
COVENTRY
RUGBY
Kenilworth
ROYAL LEAMINGTON SPA
WARWICK
Southam
Alcester
Stratford on Avon
Kineton
Shipston on Stour
Long Compton
Key to Symbols
Frequencies
rare to occasional
frequent to abundant
Woodland
Hedgerow, Scrub
Acid heath, Bog
Grassland
Waterside, Marsh
Water
Roadsides, Waste places
Quarries, Walls
Cultivated
Scale
Kilometres
0 2 4 6 8 10
Miles
0 1 2 3 4 5 6

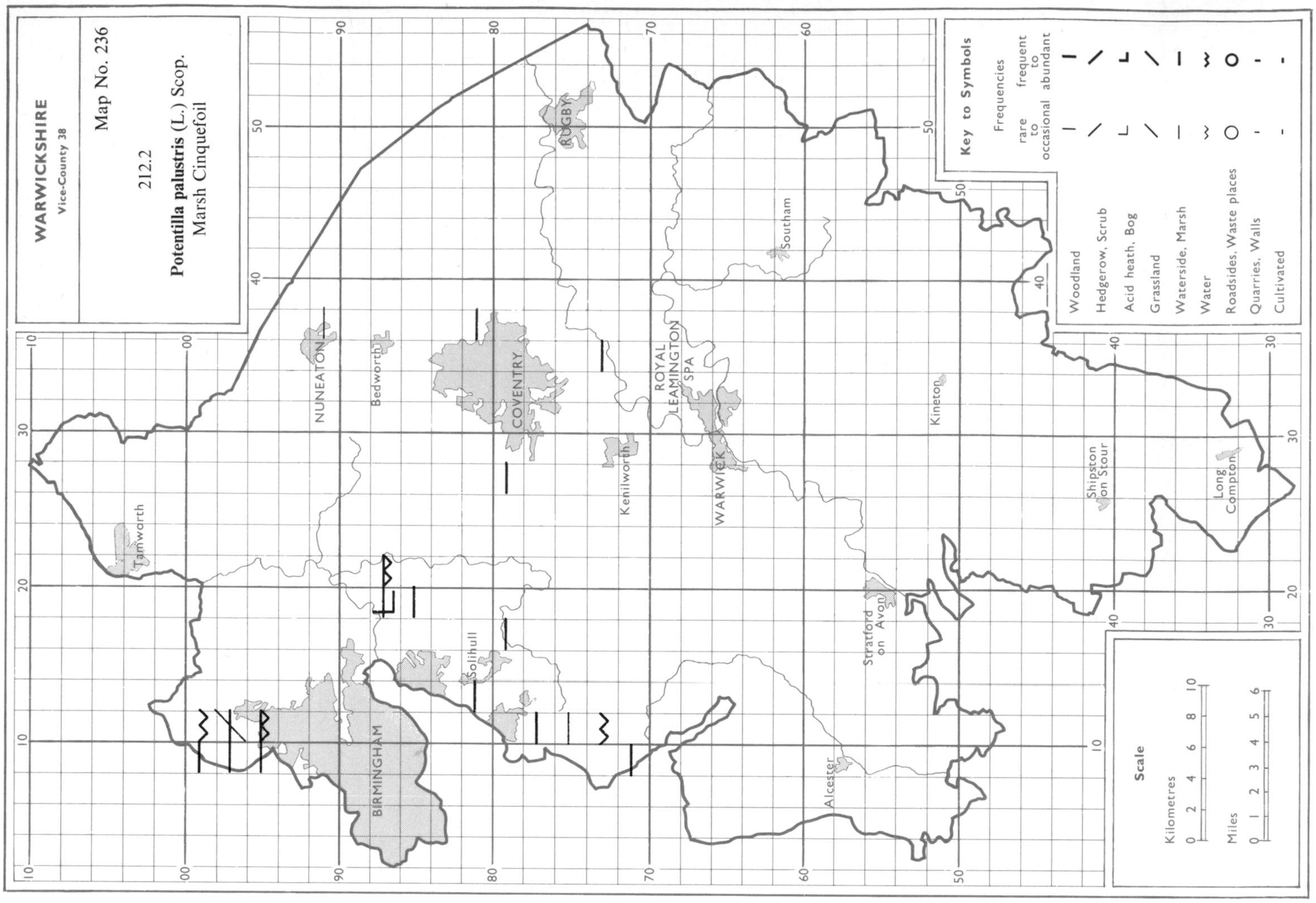
WARWICKSHIRE
Vice-County 38
Map No. 236
212.2
Potentilla palustris (L.) Scop.
Marsh Cinquefoil
Key to Symbols
Frequencies
rare to occasional
frequent to abundant
Woodland
Hedgerow, Scrub
Acid heath, Bog
Grassland
Waterside, Marsh
Water
Roadsides, Waste places
Quarries, Walls
Cultivated
Scale
Kilometres
Miles
Tamworth
NUNEATON
Bedworth
COVENTRY
RUGBY
BIRMINGHAM
Solihull
Kenilworth
ROYAL LEAMINGTON SPA
WARWICK
Southam
Kineton
Stratford on Avon
Alcester
Shipston on Stour
Long Compton

WARWICKSHIRE
Vice-County 38
Map No. 235
211.11.330
Rubus murrayi Sudre
Key to Symbols
Frequencies
rare to occasional
frequent to abundant
Woodland
Hedgerow, Scrub
Acid heath, Bog
Grassland
Waterside, Marsh
Water
Roadsides, Waste places
Quarries, Walls
Cultivated
Scale
Kilometres
Miles
Tamworth
NUNEATON
Bedworth
COVENTRY
RUGBY
BIRMINGHAM
Solihull
Kenilworth
ROYAL LEAMINGTON SPA
WARWICK
Southam
Kineton
Stratford on Avon
Alcester
Shipston on Stour
Long Compton

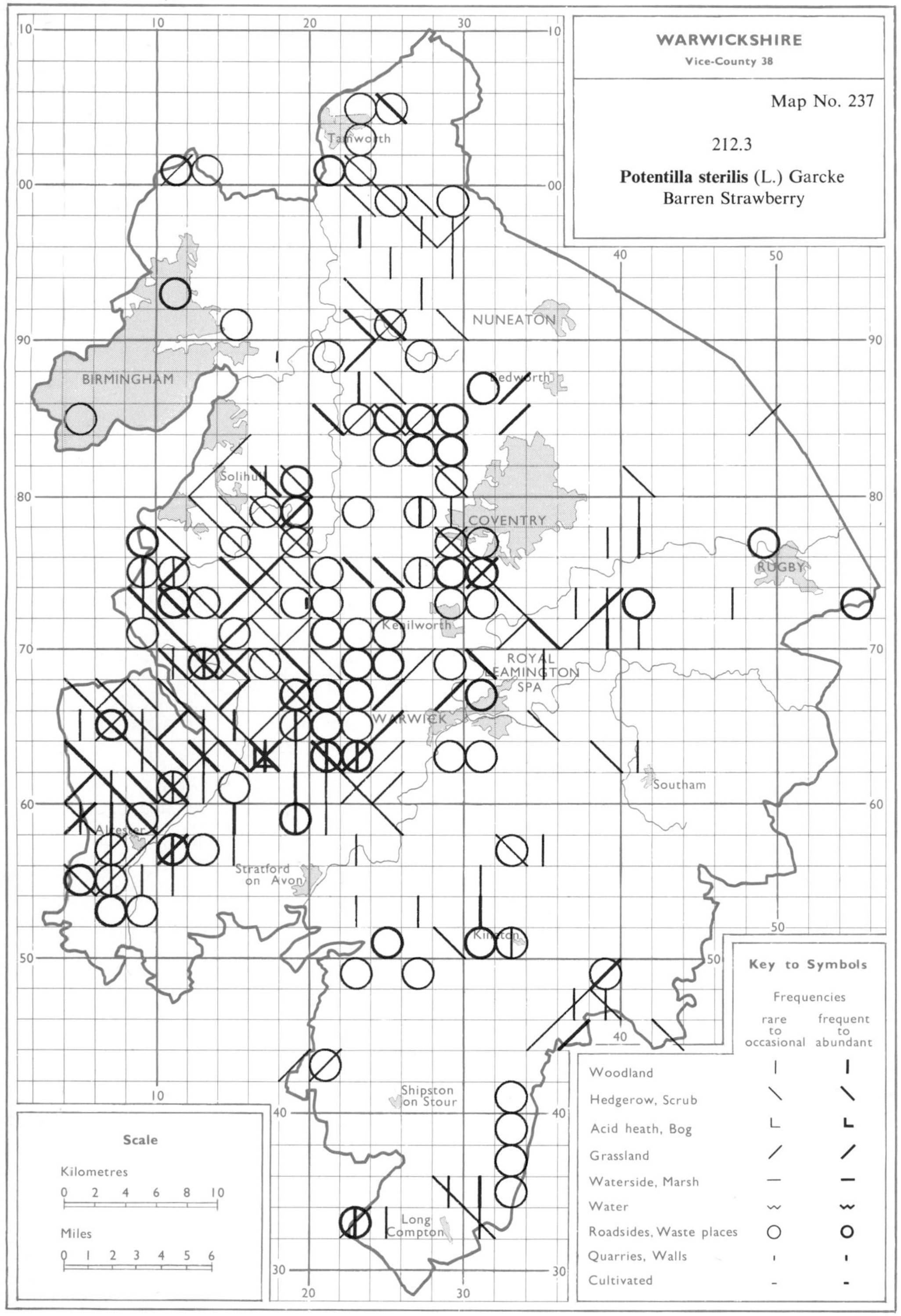
WARWICKSHIRE
Vice-County 38
Map No. 237
212.3
Potentilla sterilis (L.) Garcke
Barren Strawberry
Tamworth
BIRMINGHAM
NUNEATON
Bedworth
Solihull
COVENTRY
RUGBY
Kenilworth
ROYAL LEAMINGTON SPA
WARWICK
Southam
Alcester
Stratford on Avon
Kineton
Shipston on Stour
Long Compton
Key to Symbols
Frequencies
rare to occasional
frequent to abundant
Woodland
Hedgerow, Scrub
Acid heath, Bog
Grassland
Waterside, Marsh
Water
Roadsides, Waste places
Quarries, Walls
Cultivated
Scale
Kilometres
0 2 4 6 8 10
Miles
0 1 2 3 4 5 6

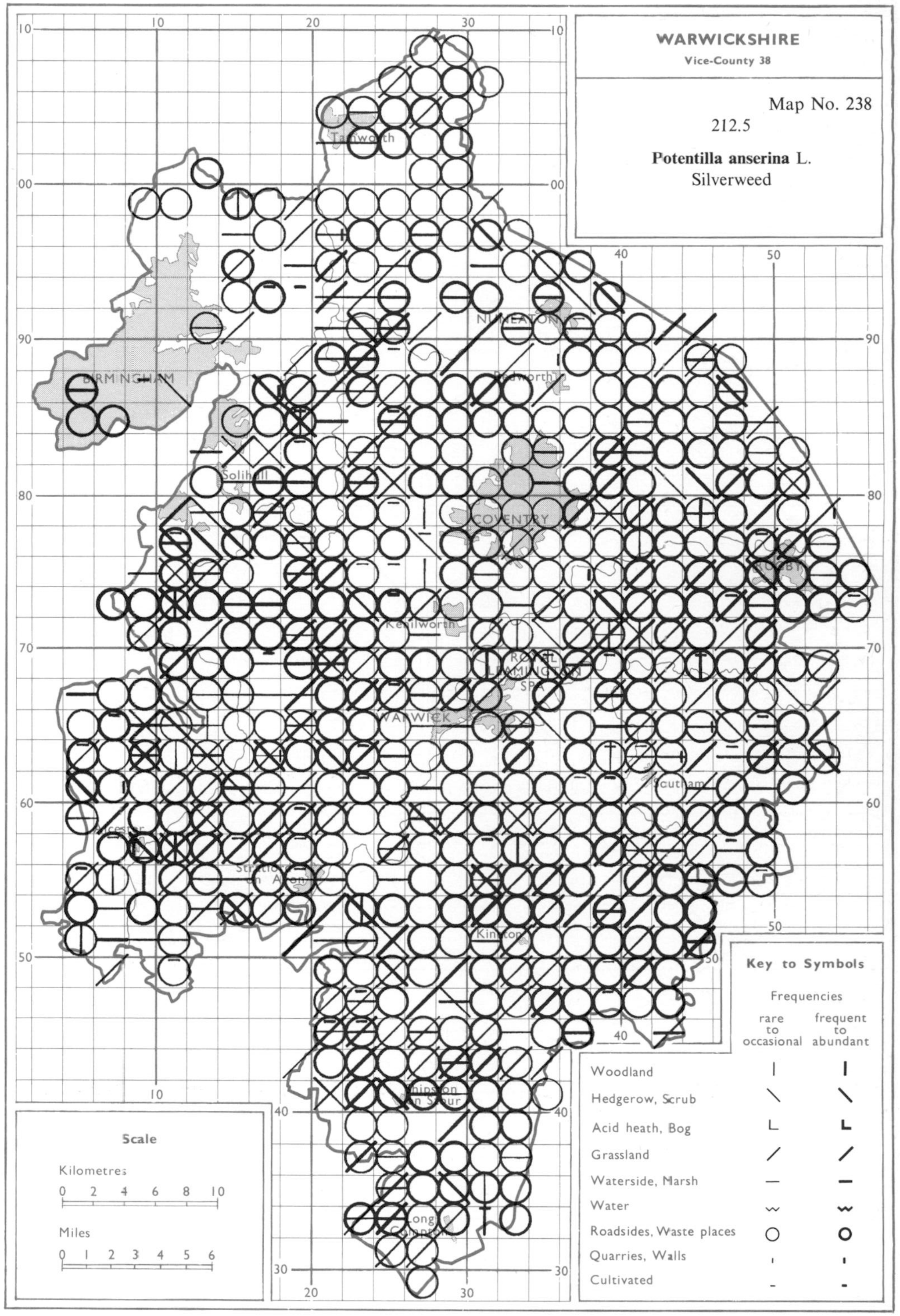
WARWICKSHIRE
Vice-County 38
Map No. 238
212.5
Potentilla anserina L.
Silverweed
Tamworth
BIRMINGHAM
NUNEATON
Bedworth
Solihull
COVENTRY
RUGBY
Kenilworth
ROYAL LEAMINGTON SPA
WARWICK
Southam
Alcester
Stratford on Avon
Kineton
Shipston on Stour
Long Compton
Key to Symbols
Frequencies
rare to occasional
frequent to abundant
Woodland
Hedgerow, Scrub
Acid heath, Bog
Grassland
Waterside, Marsh
Water
Roadsides, Waste places
Quarries, Walls
Cultivated
Scale
Kilometres
0 2 4 6 8 10
Miles
0 1 2 3 4 5 6

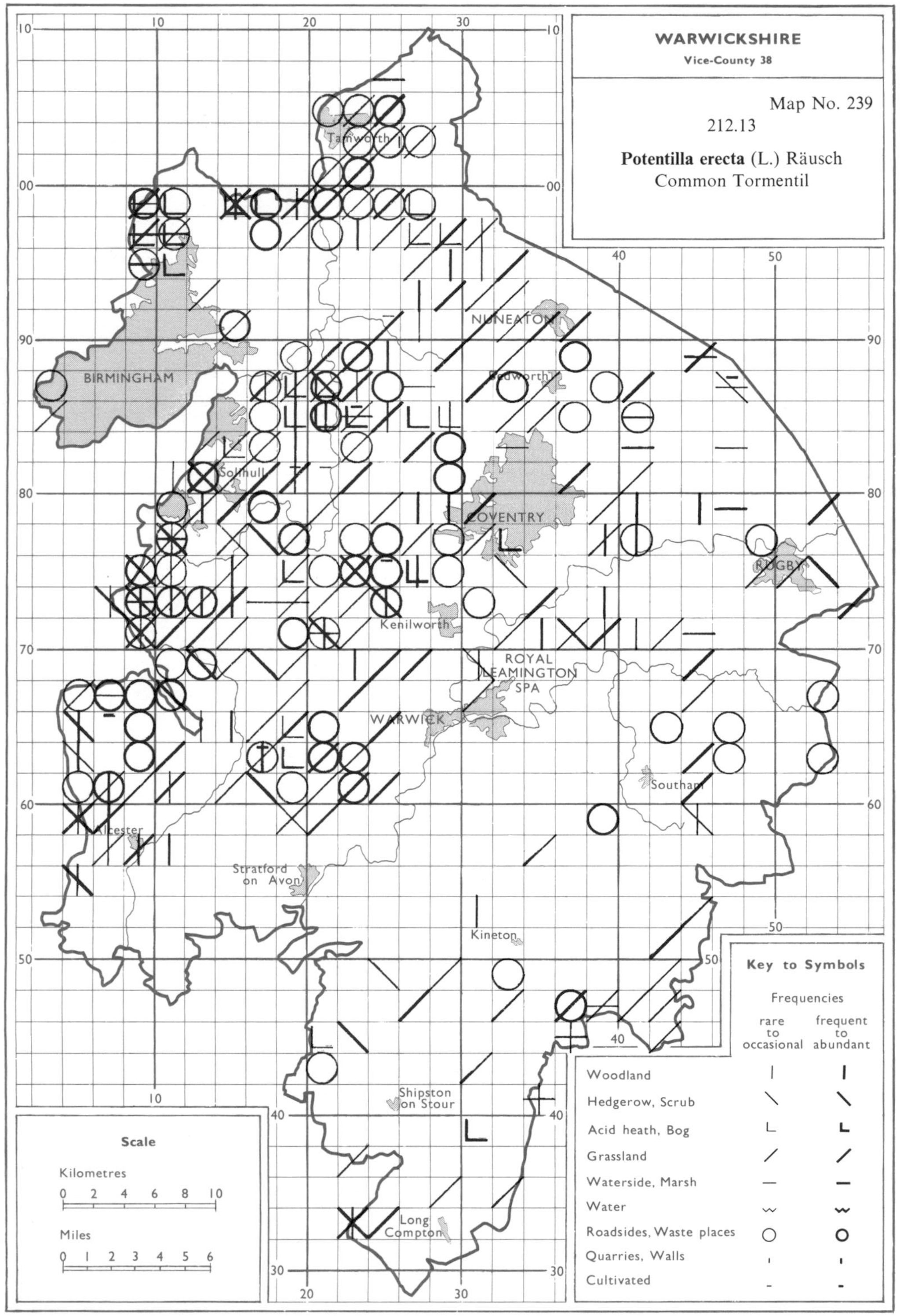
WARWICKSHIRE
Vice-County 38
Map No. 239
212.13
Potentilla erecta (L.) Räusch
Common Tormentil
Key to Symbols
Frequencies
rare to occasional
frequent to abundant
Woodland
Hedgerow, Scrub
Acid heath, Bog
Grassland
Waterside, Marsh
Water
Roadsides, Waste places
Quarries, Walls
Cultivated
Scale
Kilometres
0 2 4 6 8 10
Miles
0 1 2 3 4 5 6
BIRMINGHAM
Tamworth
NUNEATON
Bedworth
Solihull
COVENTRY
RUGBY
Kenilworth
ROYAL LEAMINGTON SPA
WARWICK
Southam
Alcester
Stratford on Avon
Kineton
Shipston on Stour
Long Compton

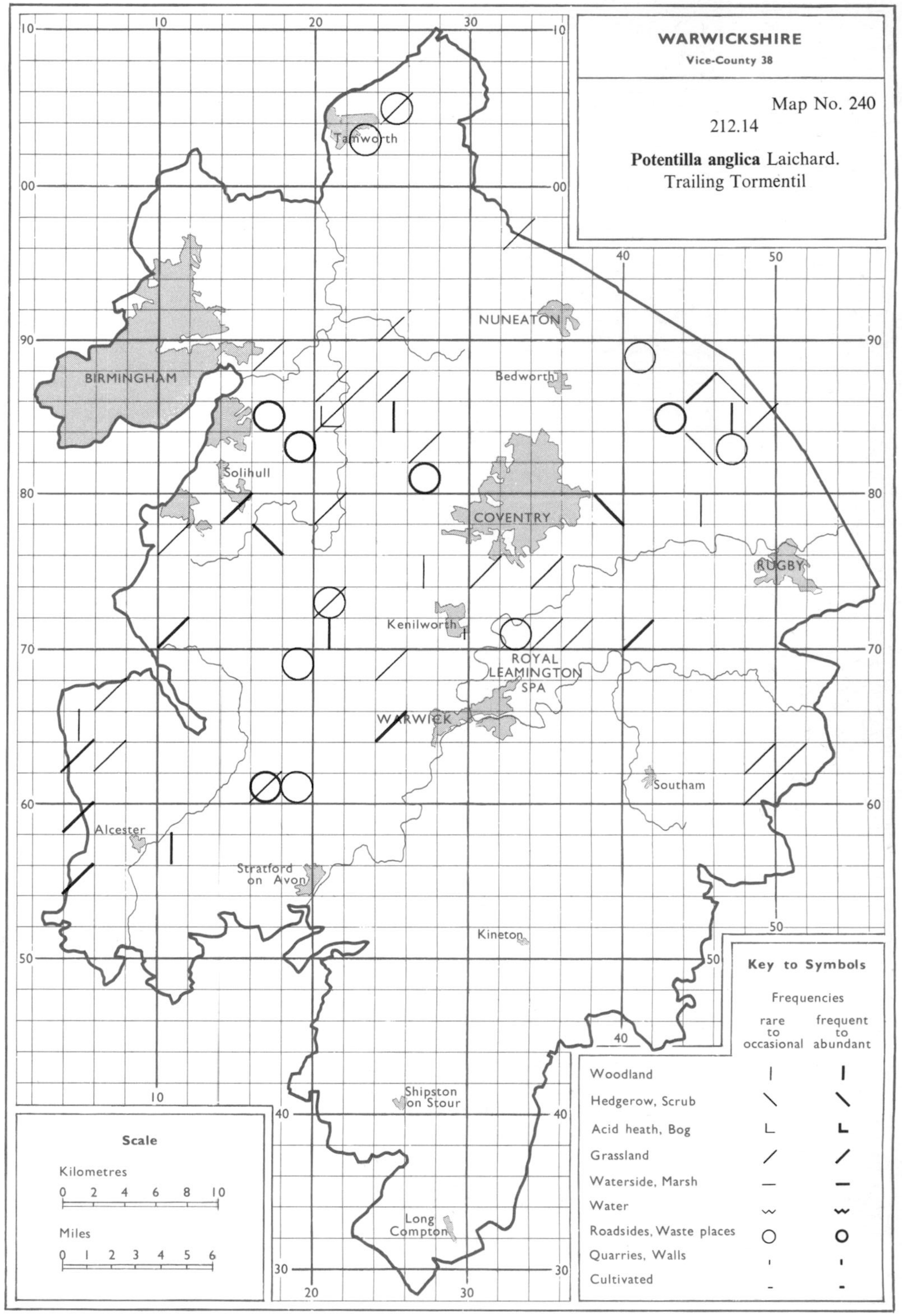
WARWICKSHIRE
Vice-County 38
Map No. 240
212.14
Potentilla anglica Laichard.
Trailing Tormentil
Key to Symbols
Frequencies
rare to occasional
frequent to abundant
Woodland
Hedgerow, Scrub
Acid heath, Bog
Grassland
Waterside, Marsh
Water
Roadsides, Waste places
Quarries, Walls
Cultivated
Scale
Kilometres
0 2 4 6 8 10
Miles
0 1 2 3 4 5 6
BIRMINGHAM
Tamworth
NUNEATON
Bedworth
Solihull
COVENTRY
RUGBY
Kenilworth
ROYAL LEAMINGTON SPA
WARWICK
Southam
Alcester
Stratford on Avon
Kineton
Shipston on Stour
Long Compton

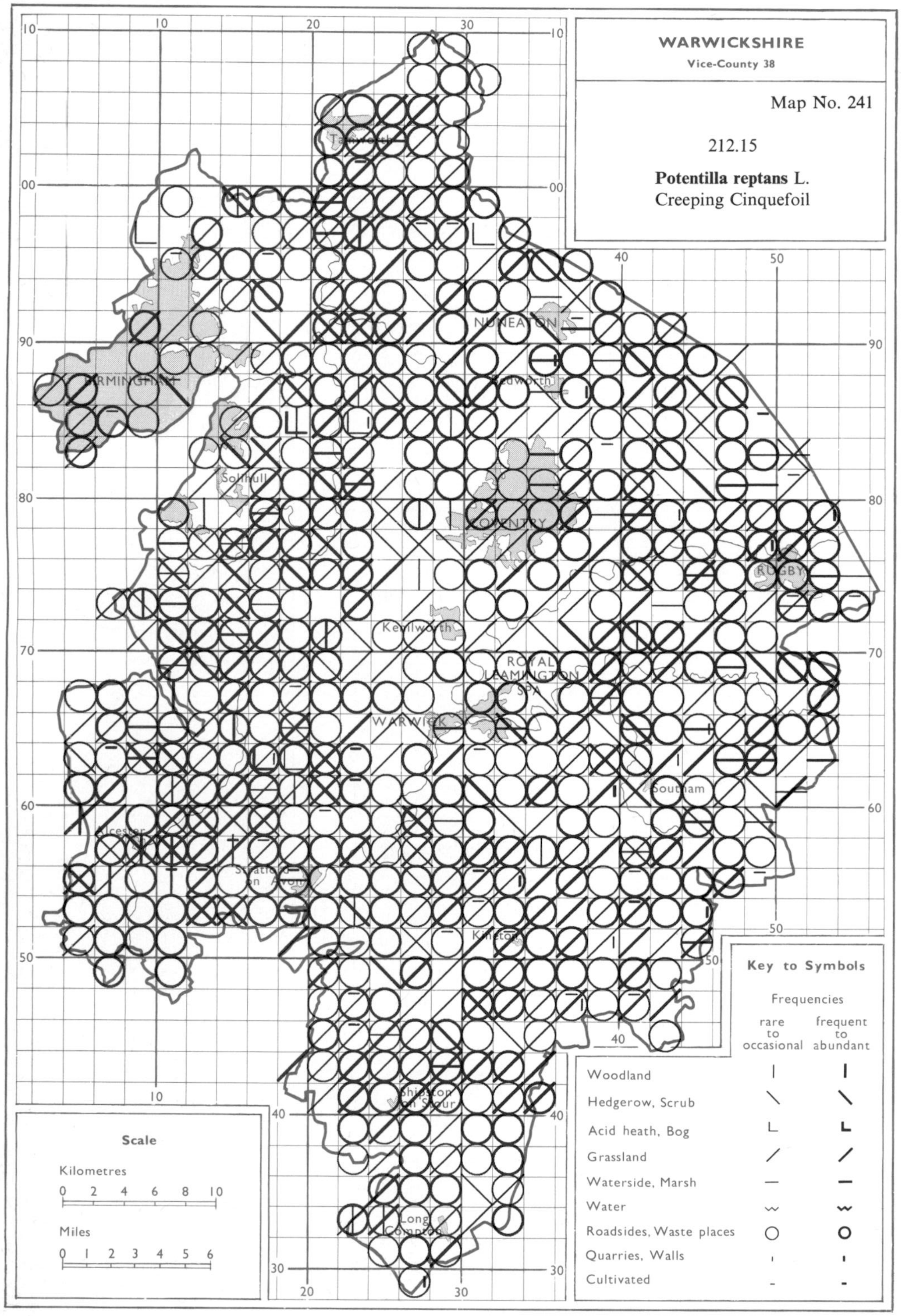
WARWICKSHIRE
Vice-County 38
Map No. 241
212.15
Potentilla reptans L.
Creeping Cinquefoil
Tamworth
BIRMINGHAM
NUNEATON
Bedworth
Solihull
COVENTRY
RUGBY
Kenilworth
ROYAL LEAMINGTON SPA
WARWICK
Southam
Alcester
Stratford on Avon
Kineton
Shipston on Stour
Long Compton
Key to Symbols
Frequencies
rare to occasional
frequent to abundant
Woodland
Hedgerow, Scrub
Acid heath, Bog
Grassland
Waterside, Marsh
Water
Roadsides, Waste places
Quarries, Walls
Cultivated
Scale
Kilometres
0 2 4 6 8 10
Miles
0 1 2 3 4 5 6

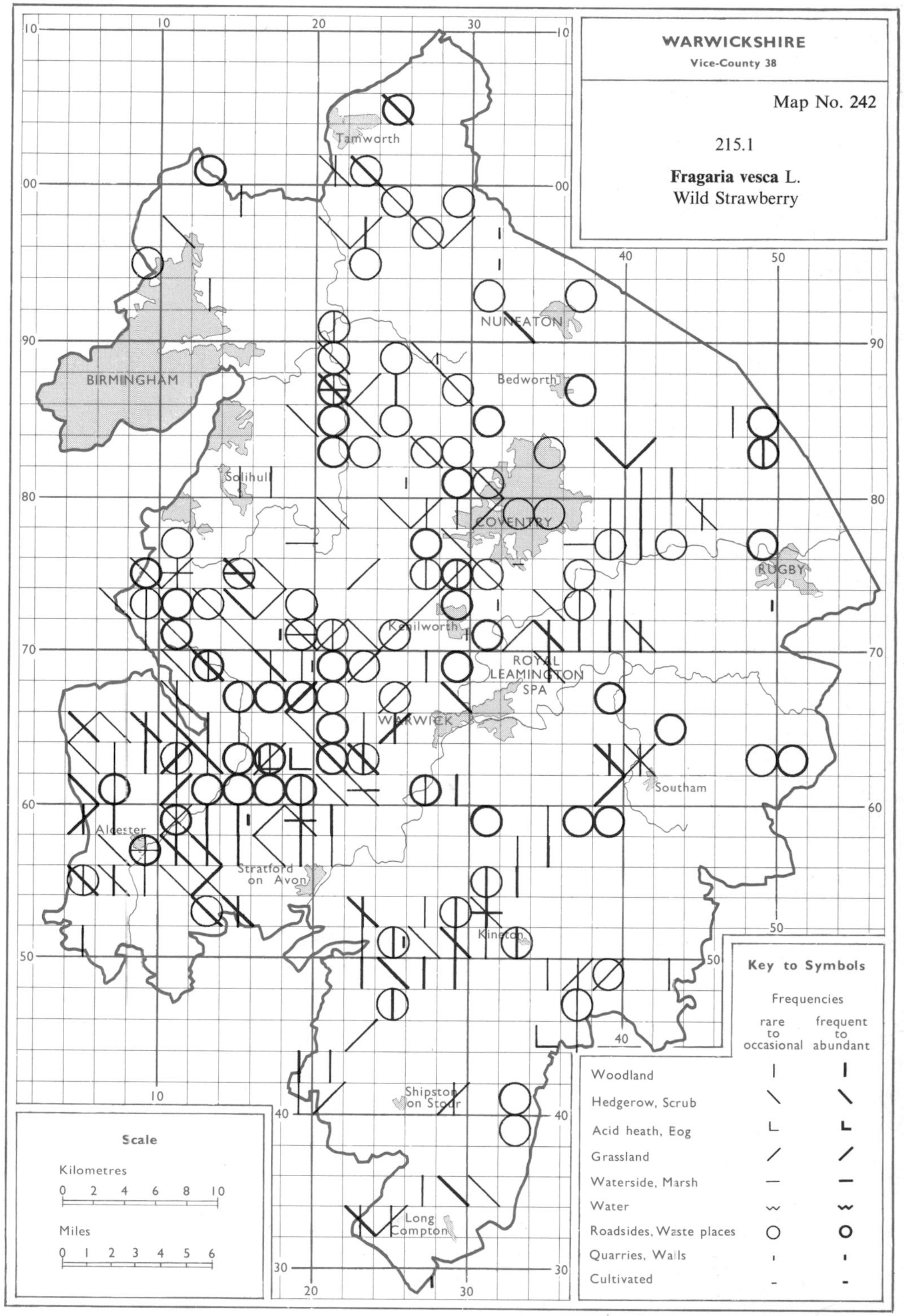
WARWICKSHIRE
Vice-County 38
Map No. 242
215.1
Fragaria vesca L.
Wild Strawberry
Tamworth
BIRMINGHAM
NUNEATON
Bedworth
Solihull
COVENTRY
RUGBY
Kenilworth
ROYAL LEAMINGTON SPA
WARWICK
Southam
Alcester
Stratford on Avon
Kineton
Shipston on Stour
Long Compton
Key to Symbols
Frequencies
rare to occasional
frequent to abundant
Woodland
Hedgerow, Scrub
Acid heath, Bog
Grassland
Waterside, Marsh
Water
Roadsides, Waste places
Quarries, Walls
Cultivated
Scale
Kilometres
0 2 4 6 8 10
Miles
0 1 2 3 4 5 6

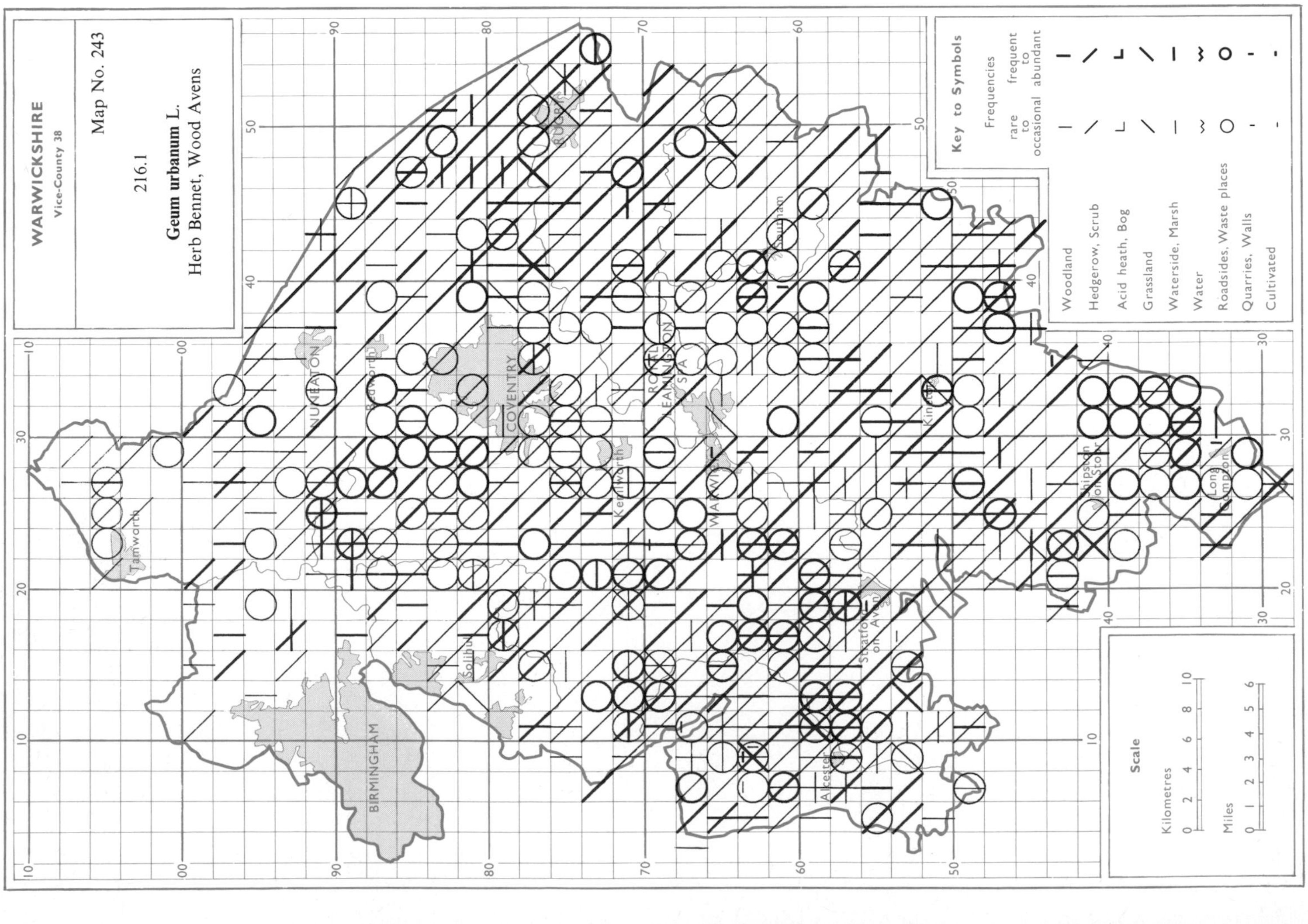
WARWICKSHIRE
Vice-County 38
Map No. 243
216.1
Geum urbanum L.
Herb Bennet, Wood Avens
Key to Symbols
Frequencies
rare to occasional
frequent to abundant
Woodland
Hedgerow, Scrub
Acid heath, Bog
Grassland
Waterside, Marsh
Water
Roadsides, Waste places
Quarries, Walls
Cultivated
Scale
Kilometres 0 2 4 6 8 10
Miles 0 1 2 3 4 5 6
NUNEATON
Bedworth
COVENTRY
RUGBY
Southam
ROYAL LEAMINGTON SPA
WARWICK
Kenilworth
Kineton
Shipston on Stour
Long Compton
Tamworth
Solihull
BIRMINGHAM
Stratford on Avon
Alcester

WARWICKSHIRE
Vice-County 38
Map No. 244
216.3
Geum rivale L.
Water Avens
Key to Symbols
Frequencies
rare to occasional
frequent to abundant
Woodland
Hedgerow, Scrub
Acid heath, Bog
Grassland
Waterside, Marsh
Water
Roadsides, Waste places
Quarries, Walls
Cultivated
Scale
Kilometres 0 2 4 6 8 10
Miles 0 1 2 3 4 5 6
NUNEATON
Bedworth
COVENTRY
RUGBY
Southam
ROYAL LEAMINGTON SPA
WARWICK
Kenilworth
Kineton
Shipston on Stour
Long Compton
Tamworth
Solihull
BIRMINGHAM
Stratford on Avon
Alcester

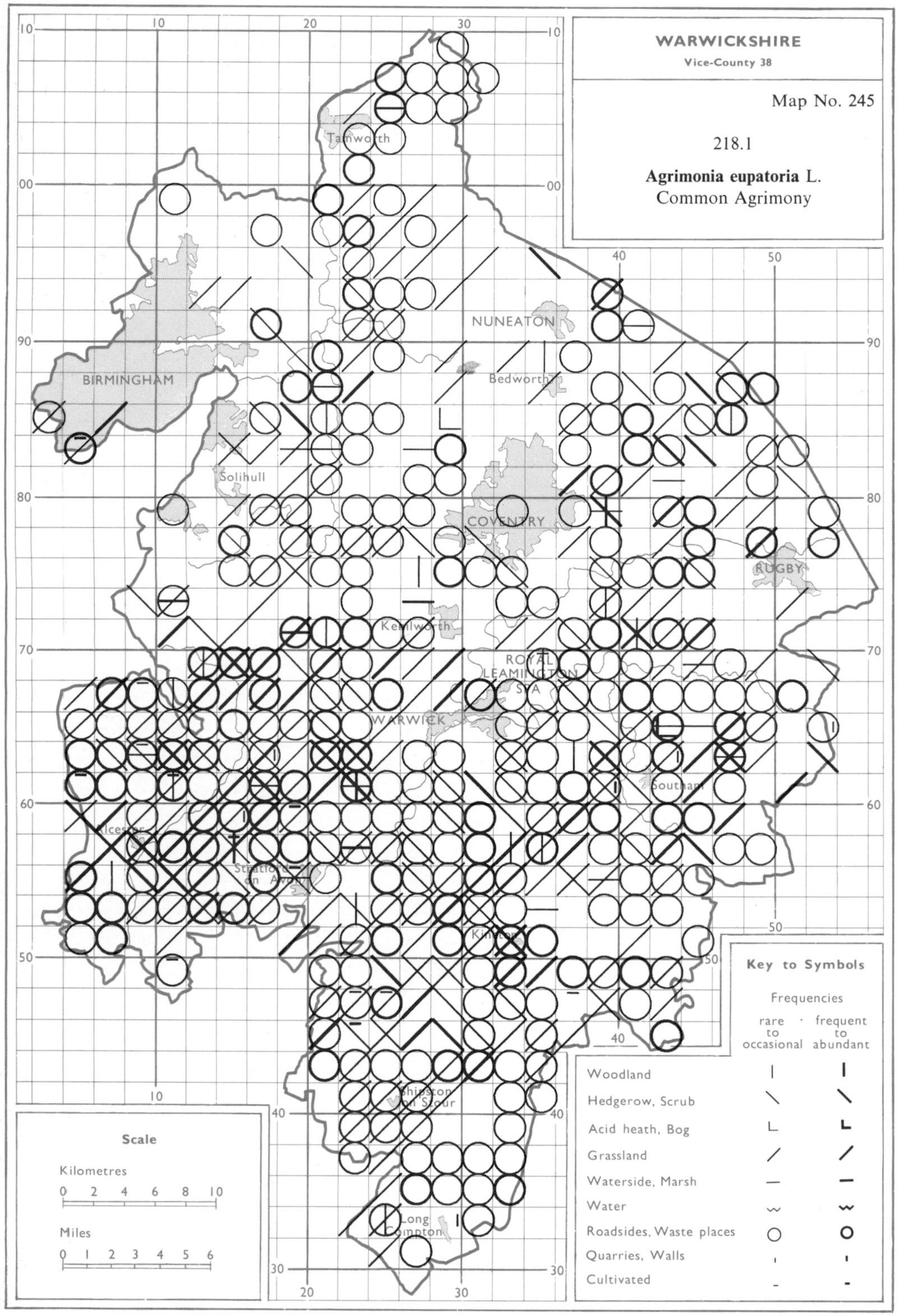
WARWICKSHIRE
Vice-County 38
Map No. 245
218.1
Agrimonia eupatoria L.
Common Agrimony
Tamworth
BIRMINGHAM
NUNEATON
Bedworth
Solihull
COVENTRY
RUGBY
Kenilworth
ROYAL LEAMINGTON SPA
WARWICK
Southam
Alcester
Stratford on Avon
Kineton
Shipston on Stour
Long Compton
Key to Symbols
Frequencies
rare to occasional
frequent to abundant
Woodland
Hedgerow, Scrub
Acid heath, Bog
Grassland
Waterside, Marsh
Water
Roadsides, Waste places
Quarries, Walls
Cultivated
Scale
Kilometres
0 2 4 6 8 10
Miles
0 1 2 3 4 5 6

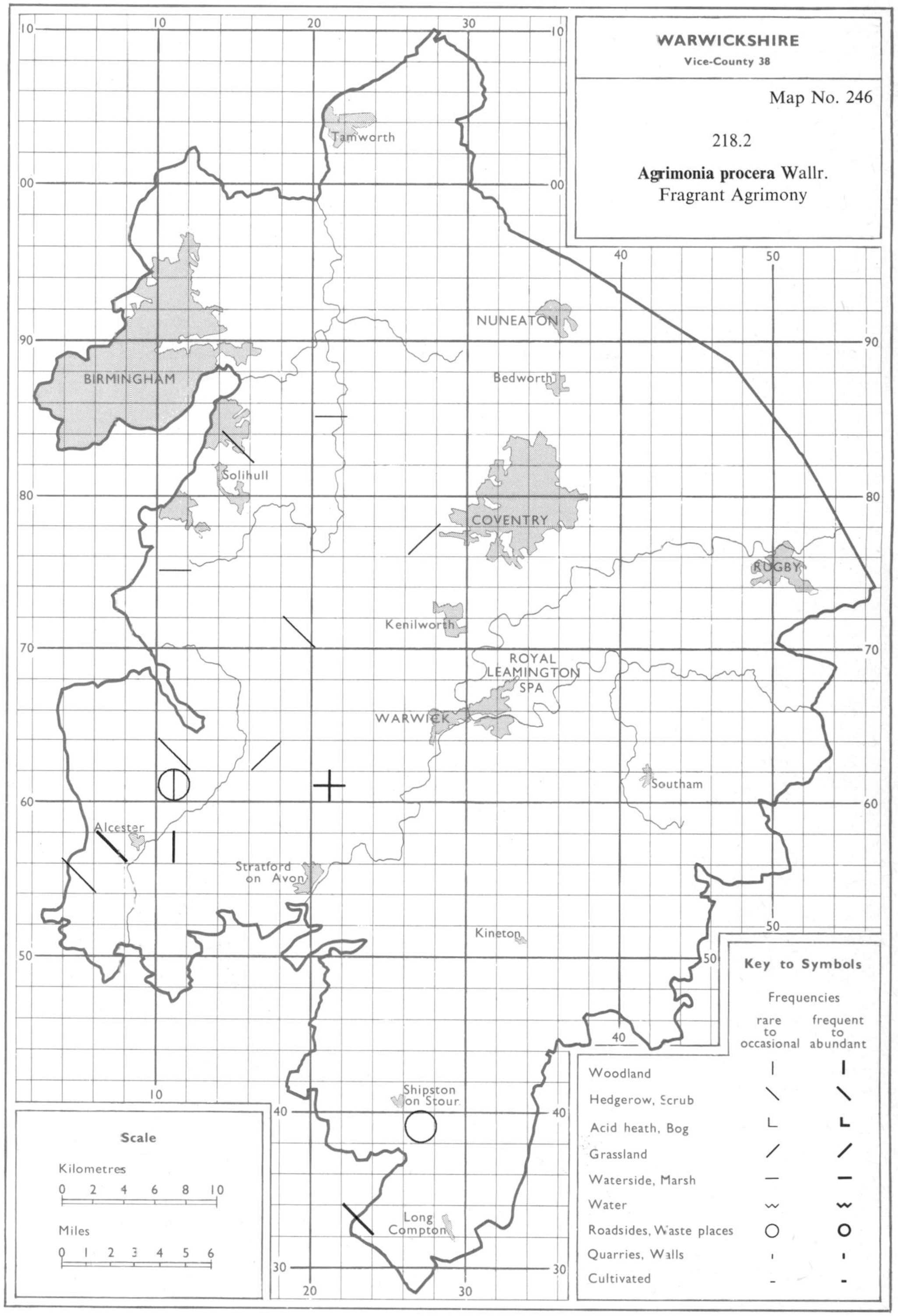
WARWICKSHIRE
Vice-County 38
Map No. 246
218.2
Agrimonia procera Wallr.
Fragrant Agrimony
Tamworth
BIRMINGHAM
NUNEATON
Bedworth
Solihull
COVENTRY
RUGBY
Kenilworth
ROYAL LEAMINGTON SPA
WARWICK
Southam
Alcester
Stratford on Avon
Kineton
Shipston on Stour
Long Compton
Key to Symbols
Frequencies
rare to occasional
frequent to abundant
Woodland
Hedgerow, Scrub
Acid heath, Bog
Grassland
Waterside, Marsh
Water
Roadsides, Waste places
Quarries, Walls
Cultivated
Scale
Kilometres
0 2 4 6 8 10
Miles
0 1 2 3 4 5 6

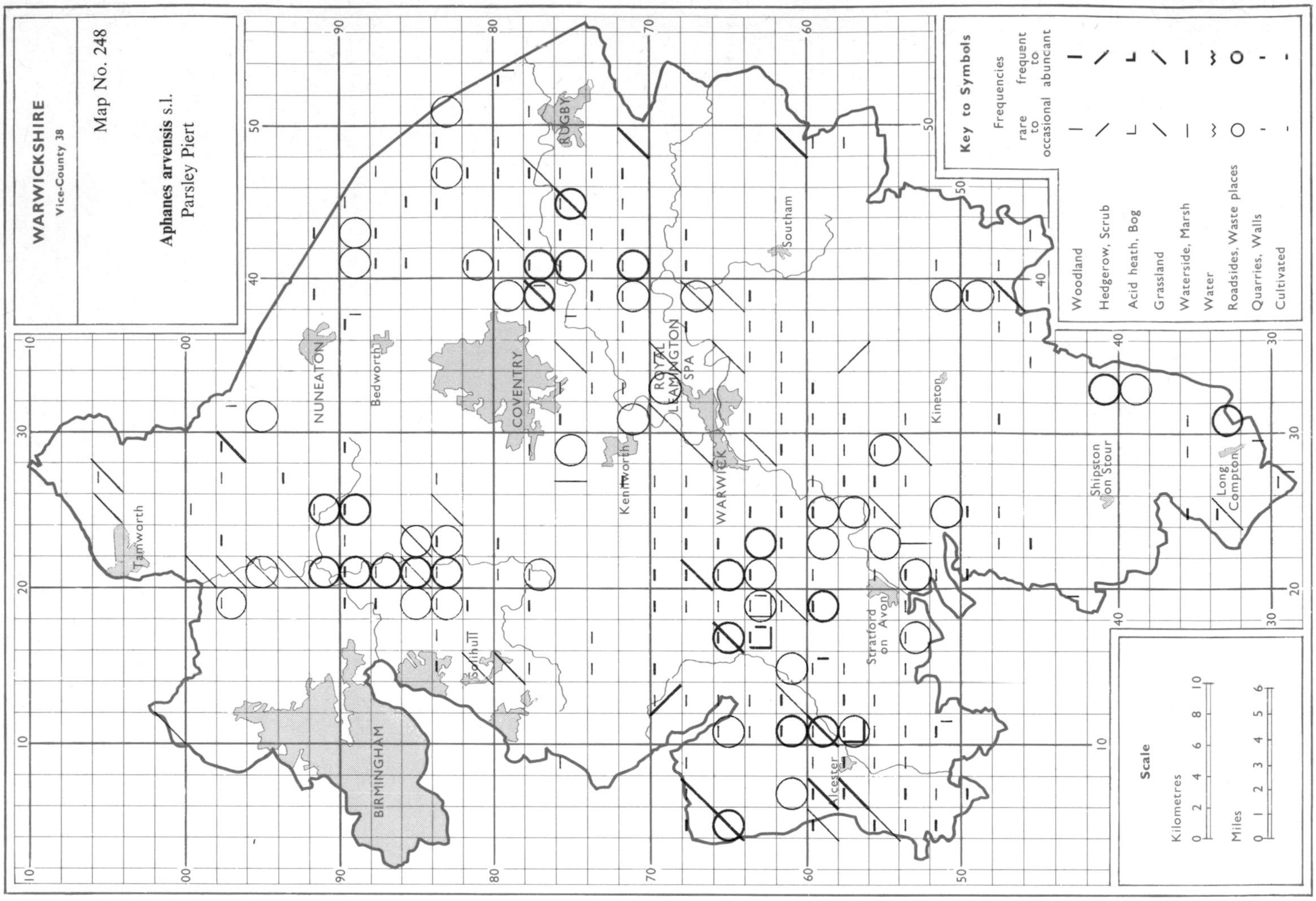
WARWICKSHIRE
Vice-County 38
Map No. 248
Aphanes arvensis s.l.
Parsley Piert
Key to Symbols
Frequencies
rare to occasional
frequent to abundant
Woodland
Hedgerow, Scrub
Acid heath, Bog
Grassland
Waterside, Marsh
Water
Roadsides, Waste places
Quarries, Walls
Cultivated
Scale
Kilometres
0 2 4 6 8 10
Miles
0 1 2 3 4 5 6
NUNEATON
Bedworth
COVENTRY
RUGBY
Southam
ROYAL LEAMINGTON SPA
Kenilworth
WARWICK
Kineton
Shipston on Stour
Long Compton
Tamworth
Solihull
BIRMINGHAM
Stratford on Avon
Alcester

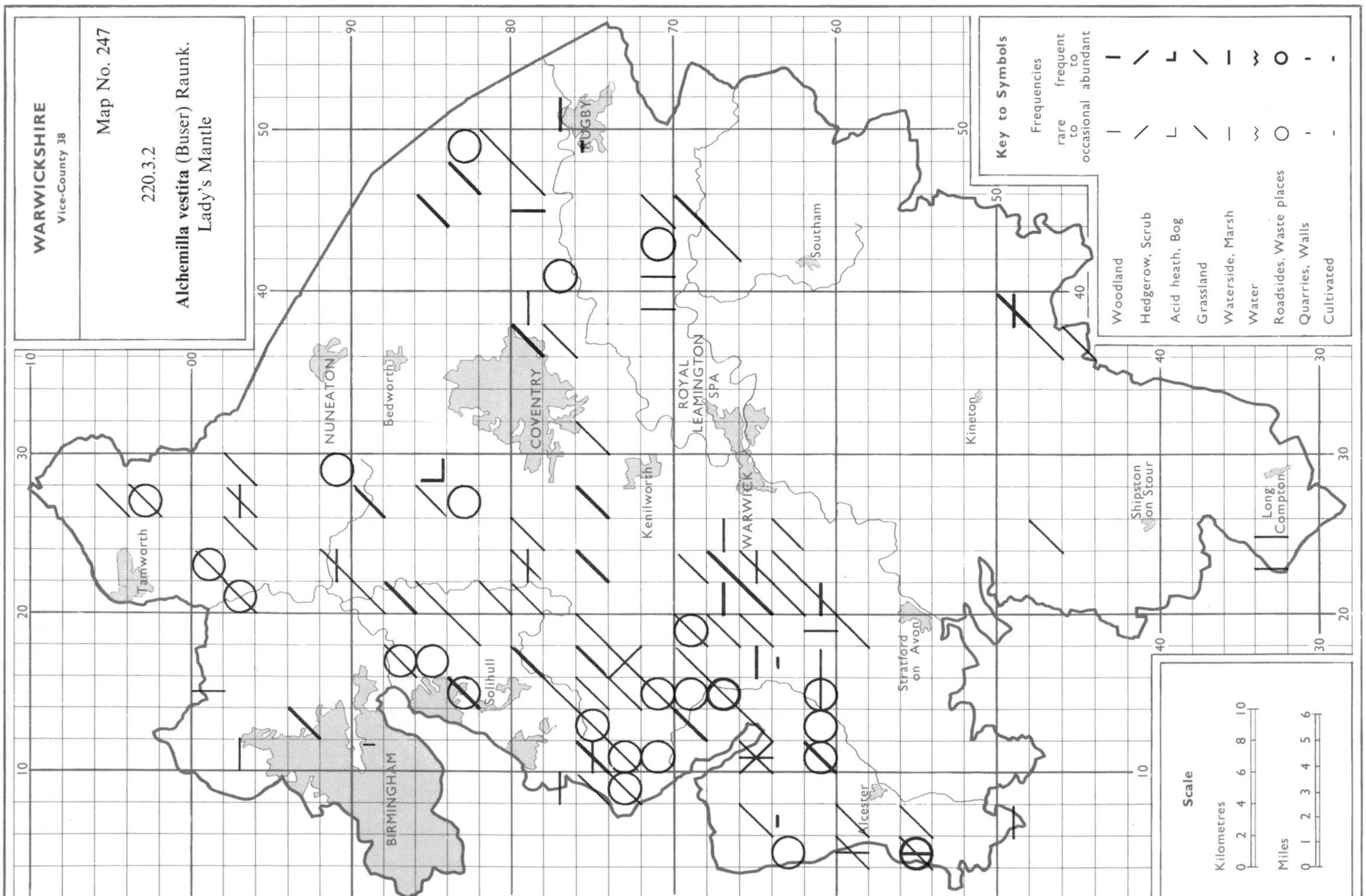
WARWICKSHIRE
Vice-County 38
Map No. 247
220.3.2
Alchemilla vestita (Buser) Raunk.
Lady's Mantle
Key to Symbols
Frequencies
rare to occasional
frequent to abundant
Woodland
Hedgerow, Scrub
Acid heath, Bog
Grassland
Waterside, Marsh
Water
Roadsides, Waste places
Quarries, Walls
Cultivated
Scale
Kilometres
0 2 4 6 8 10
Miles
0 1 2 3 4 5 6
NUNEATON
Bedworth
COVENTRY
RUGBY
Southam
ROYAL LEAMINGTON SPA
Kenilworth
WARWICK
Kineton
Shipston on Stour
Long Compton
Tamworth
Solihull
BIRMINGHAM
Stratford on Avon
Alcester

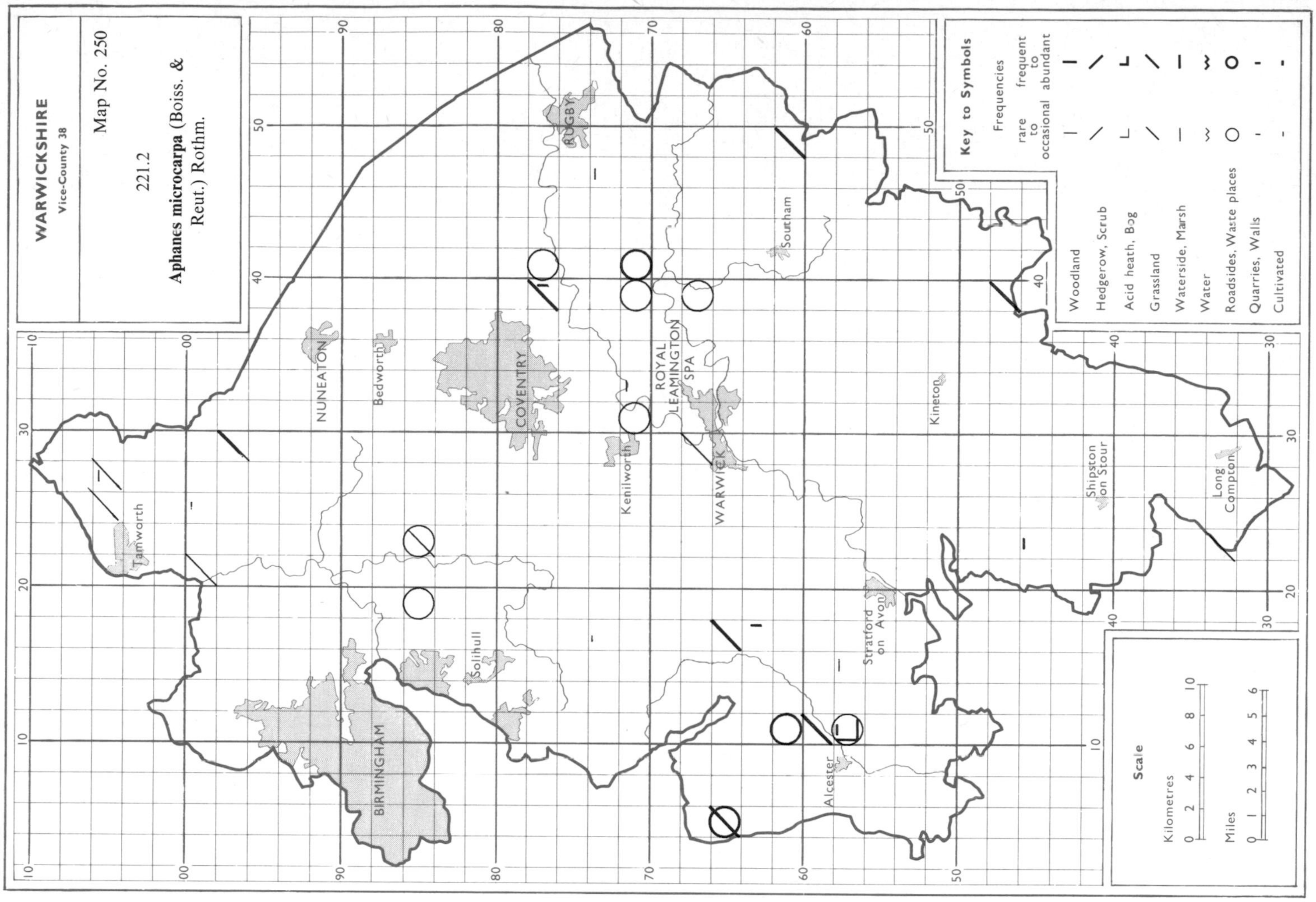

WARWICKSHIRE
Vice-County 38
Map No. 250
221.2
Aphanes microcarpa (Boiss. & Reut.) Rothm.
Key to Symbols
Frequencies
rare to occasional
frequent to abundant
Woodland
Hedgerow, Scrub
Acid heath, Bog
Grassland
Waterside, Marsh
Water
Roadsides, Waste places
Quarries, Walls
Cultivated
Scale
Kilometres
Miles
NUNEATON
Bedworth
COVENTRY
RUGBY
Southam
ROYAL LEAMINGTON SPA
Kenilworth
WARWICK
Kineton
Shipston on Stour
Long Compton
Tamworth
Solihull
BIRMINGHAM
Stratford on Avon
Alcester

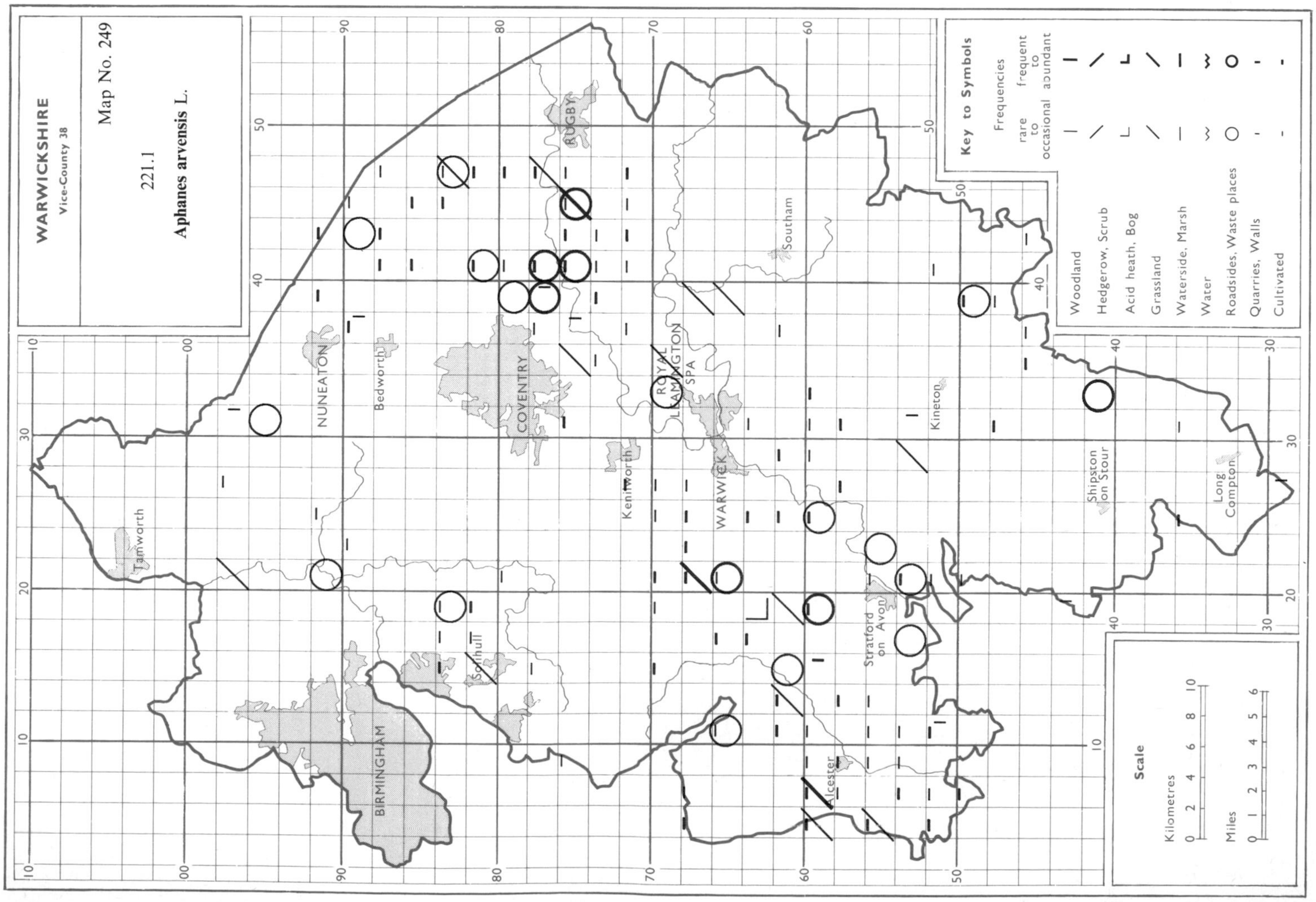

WARWICKSHIRE
Vice-County 38
Map No. 249
221.1
Aphanes arvensis L.
Key to Symbols
Frequencies
rare to occasional
frequent to abundant
Woodland
Hedgerow, Scrub
Acid heath, Bog
Grassland
Waterside, Marsh
Water
Roadsides, Waste places
Quarries, Walls
Cultivated
Scale
Kilometres
Miles
NUNEATON
Bedworth
COVENTRY
RUGBY
Southam
ROYAL LEAMINGTON SPA
Kenilworth
WARWICK
Kineton
Shipston on Stour
Long Compton
Tamworth
Solihull
BIRMINGHAM
Stratford on Avon
Alcester

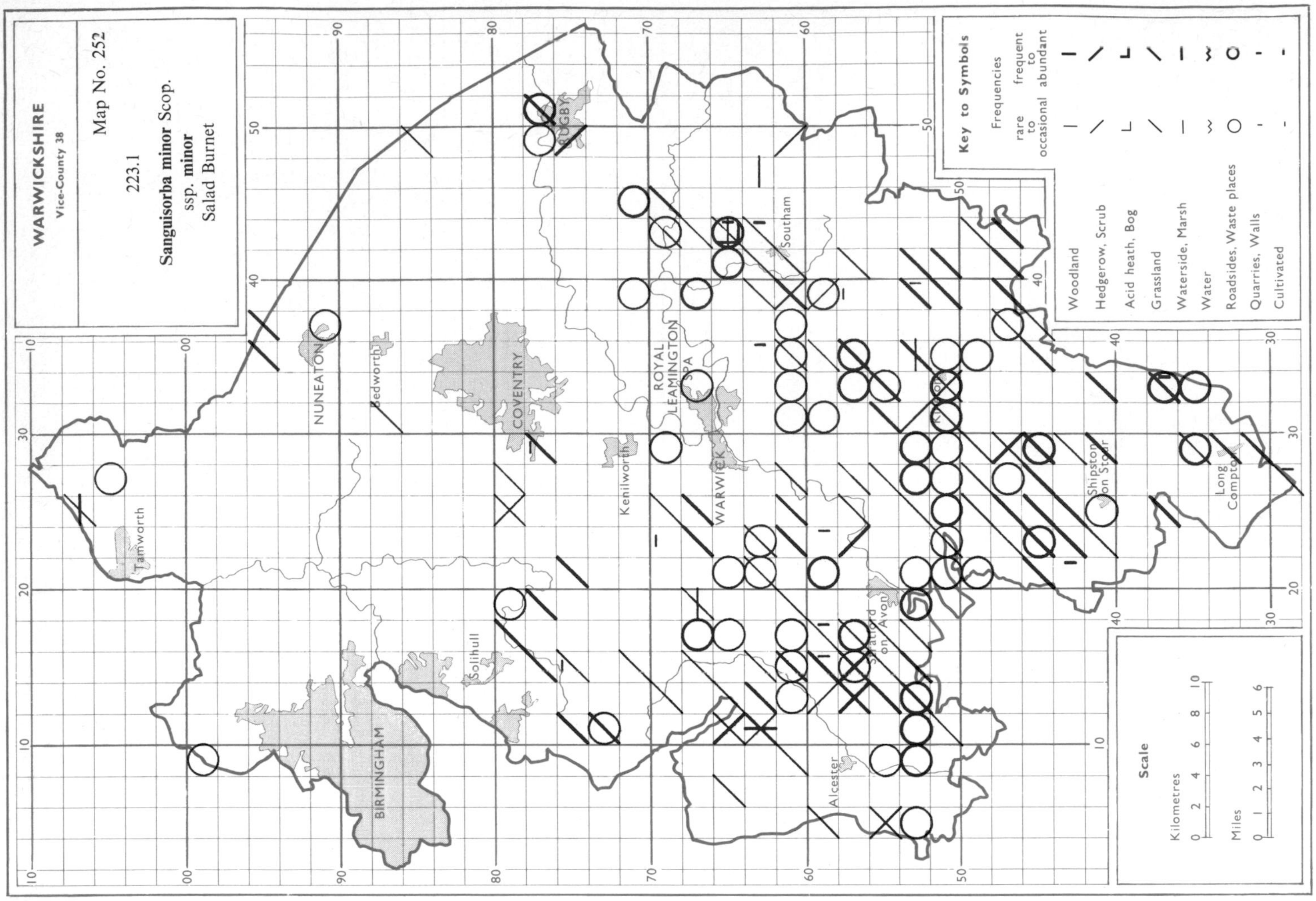
WARWICKSHIRE
Vice-County 38
Map No. 252
223.1
Sanguisorba minor Scop.
ssp. minor
Salad Burnet
Key to Symbols
Frequencies
rare to occasional
frequent to abundant
Woodland
Hedgerow, Scrub
Acid heath, Bog
Grassland
Waterside, Marsh
Water
Roadsides, Waste places
Quarries, Walls
Cultivated
Scale
Kilometres
0 2 4 6 8 10
Miles
0 1 2 3 4 5 6
Tamworth
NUNEATON
Bedworth
COVENTRY
RUGBY
BIRMINGHAM
Solihull
Kenilworth
ROYAL LEAMINGTON SPA
WARWICK
Southam
Stratford on Avon
Alcester
Kineton
Shipston on Stour
Long Compton

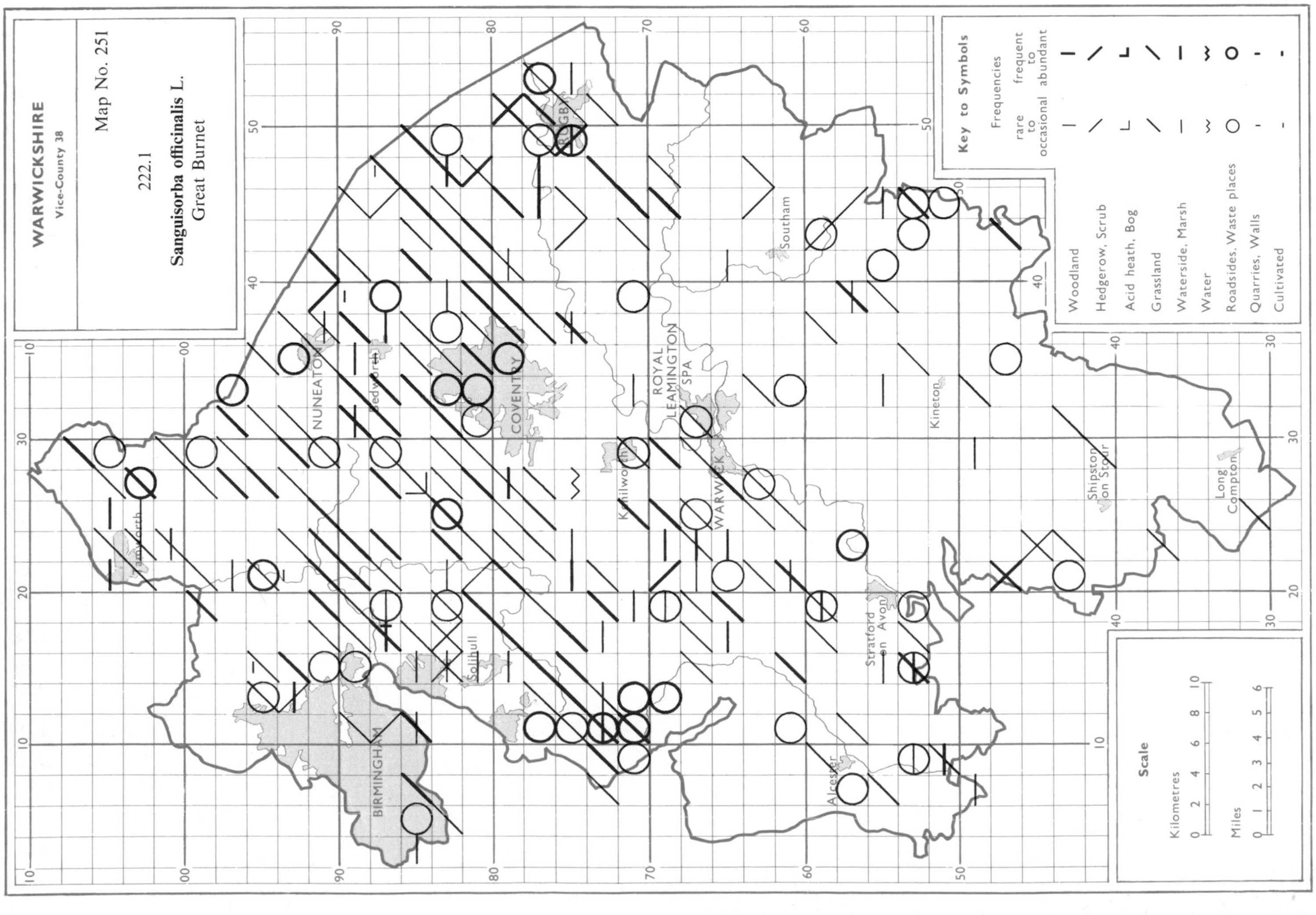
WARWICKSHIRE
Vice-County 38
Map No. 251
222.1
Sanguisorba officinalis L.
Great Burnet
Key to Symbols
Frequencies
rare to occasional
frequent to abundant
Woodland
Hedgerow, Scrub
Acid heath, Bog
Grassland
Waterside, Marsh
Water
Roadsides, Waste places
Quarries, Walls
Cultivated
Scale
Kilometres
0 2 4 6 8 10
Miles
0 1 2 3 4 5 6
Tamworth
NUNEATON
Bedworth
COVENTRY
RUGBY
BIRMINGHAM
Solihull
Kenilworth
ROYAL LEAMINGTON SPA
WARWICK
Southam
Stratford on Avon
Alcester
Kineton
Shipston on Stour
Long Compton

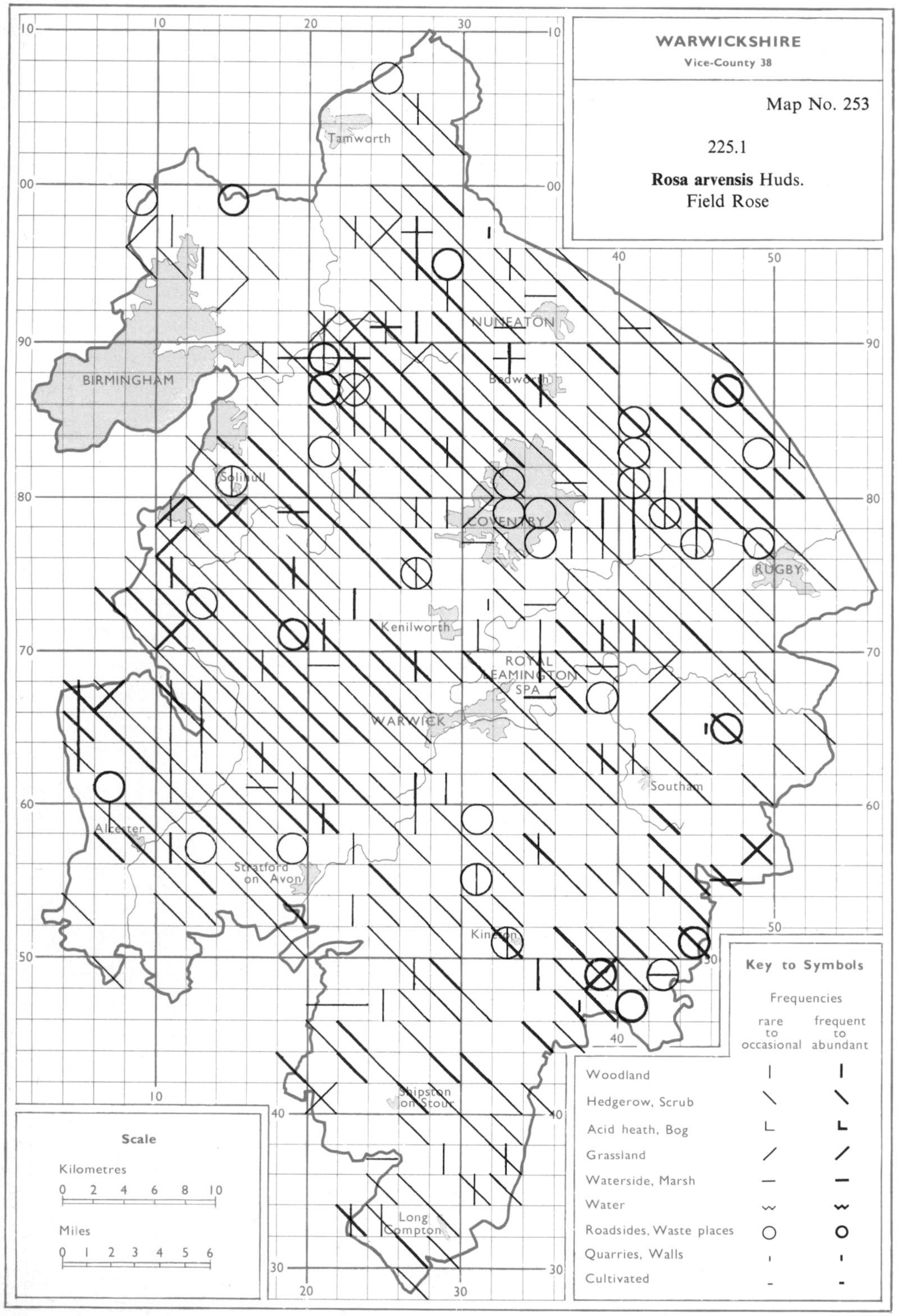
WARWICKSHIRE
Vice-County 38
Map No. 253
225.1
Rosa arvensis Huds.
Field Rose
Tamworth
BIRMINGHAM
NUNEATON
Bedworth
Solihull
COVENTRY
RUGBY
Kenilworth
ROYAL LEAMINGTON SPA
WARWICK
Southam
Alcester
Stratford on Avon
Kineton
Shipston on Stour
Long Compton
Key to Symbols
Frequencies
rare to occasional
frequent to abundant
Woodland
Hedgerow, Scrub
Acid heath, Bog
Grassland
Waterside, Marsh
Water
Roadsides, Waste places
Quarries, Walls
Cultivated
Scale
Kilometres
0 2 4 6 8 10
Miles
0 1 2 3 4 5 6

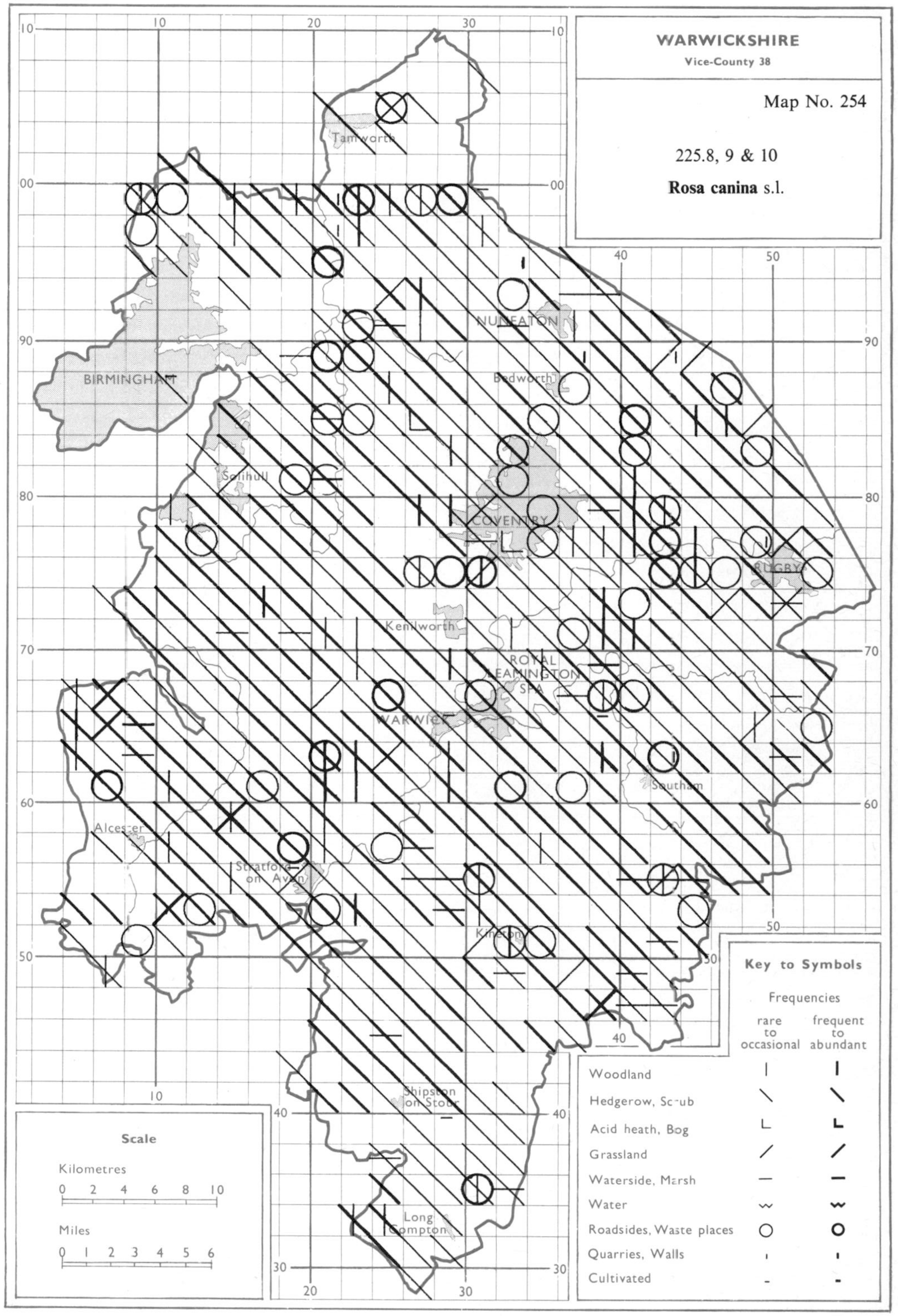
WARWICKSHIRE
Vice-County 38
Map No. 254
225.8, 9 & 10
Rosa canina s.l.
Tamworth
BIRMINGHAM
NUNEATON
Bedworth
Solihull
COVENTRY
RUGBY
Kenilworth
ROYAL LEAMINGTON SPA
WARWICK
Southam
Alcester
Stratford on Avon
Kineton
Shipston on Stour
Long Compton
Key to Symbols
Frequencies
rare to occasional
frequent to abundant
Woodland
Hedgerow, Scrub
Acid heath, Bog
Grassland
Waterside, Marsh
Water
Roadsides, Waste places
Quarries, Walls
Cultivated
Scale
Kilometres
0 2 4 6 8 10
Miles
0 1 2 3 4 5 6

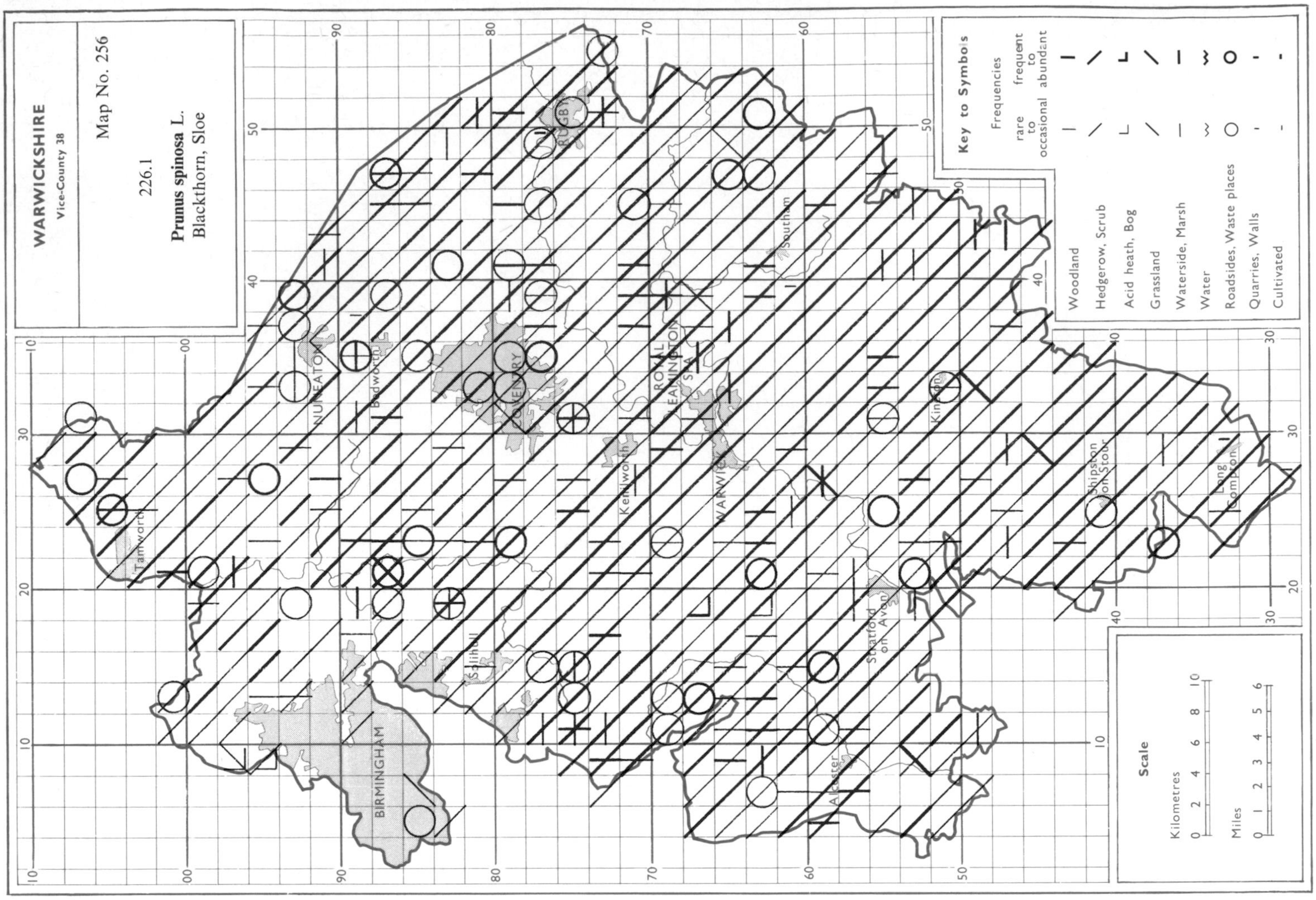
WARWICKSHIRE
Vice-County 38
Map No. 256
226.1
Prunus spinosa L.
Blackthorn, Sloe
Key to Symbols
Frequencies
rare to occasional
frequent to abundant
Woodland
Hedgerow, Scrub
Acid heath, Bog
Grassland
Waterside, Marsh
Water
Roadsides, Waste places
Quarries, Walls
Cultivated
Scale
Kilometres
0 2 4 6 8 10
Miles
0 1 2 3 4 5 6
NUNEATON
Bedworth
COVENTRY
RUGBY
ROYAL LEAMINGTON SPA
Kenilworth
WARWICK
Southam
Kineton
Shipston on Stour
Long Compton
Stratford on Avon
Alcester
Solihull
BIRMINGHAM
Tamworth

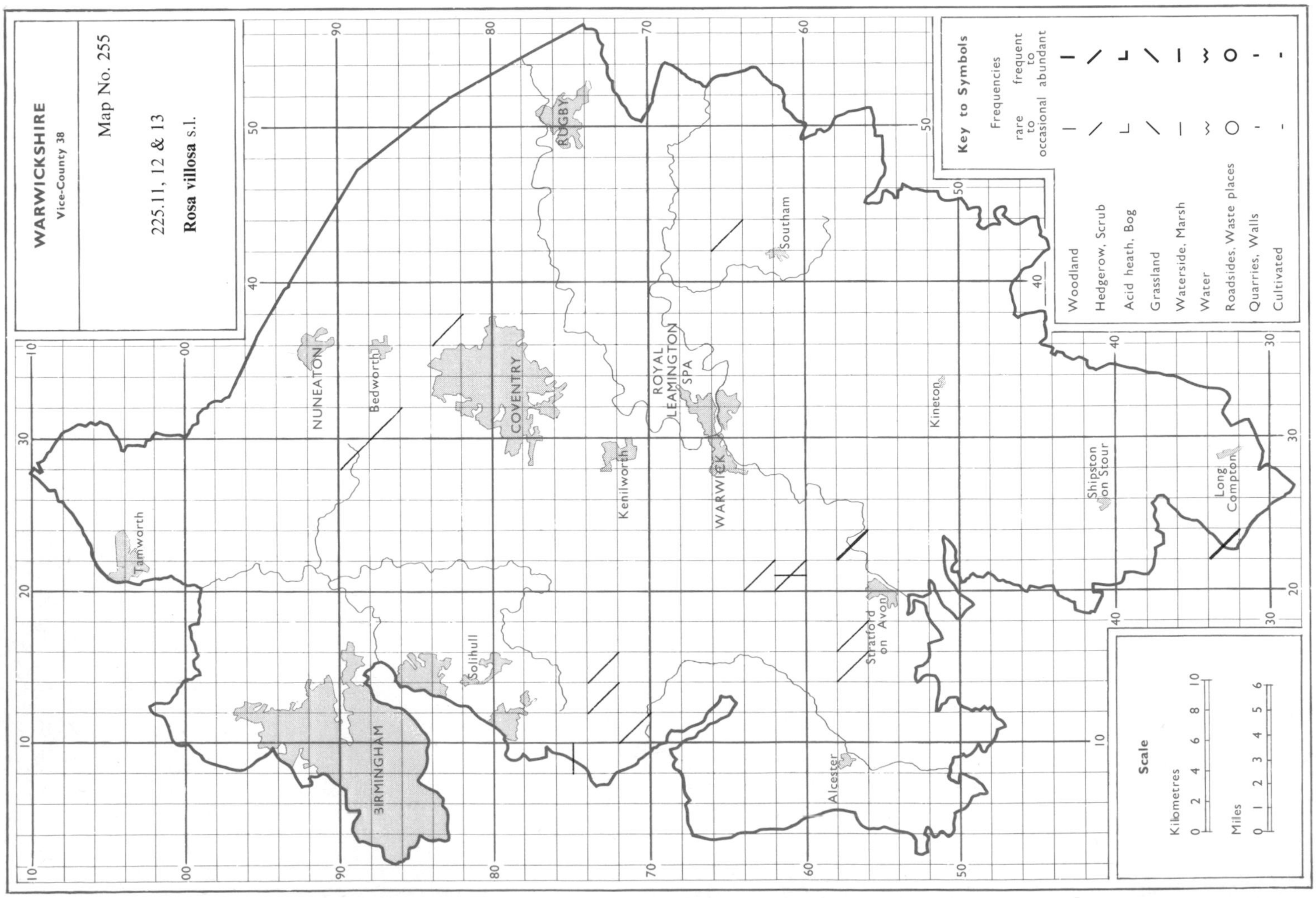
WARWICKSHIRE
Vice-County 38
Map No. 255
225.11, 12 & 13
Rosa villosa s.l.
Key to Symbols
Frequencies
rare to occasional
frequent to abundant
Woodland
Hedgerow, Scrub
Acid heath, Bog
Grassland
Waterside, Marsh
Water
Roadsides, Waste places
Quarries, Walls
Cultivated
Scale
Kilometres
0 2 4 6 8 10
Miles
0 1 2 3 4 5 6
NUNEATON
Bedworth
COVENTRY
RUGBY
ROYAL LEAMINGTON SPA
Kenilworth
WARWICK
Southam
Kineton
Shipston on Stour
Long Compton
Stratford on Avon
Alcester
Solihull
BIRMINGHAM
Tamworth

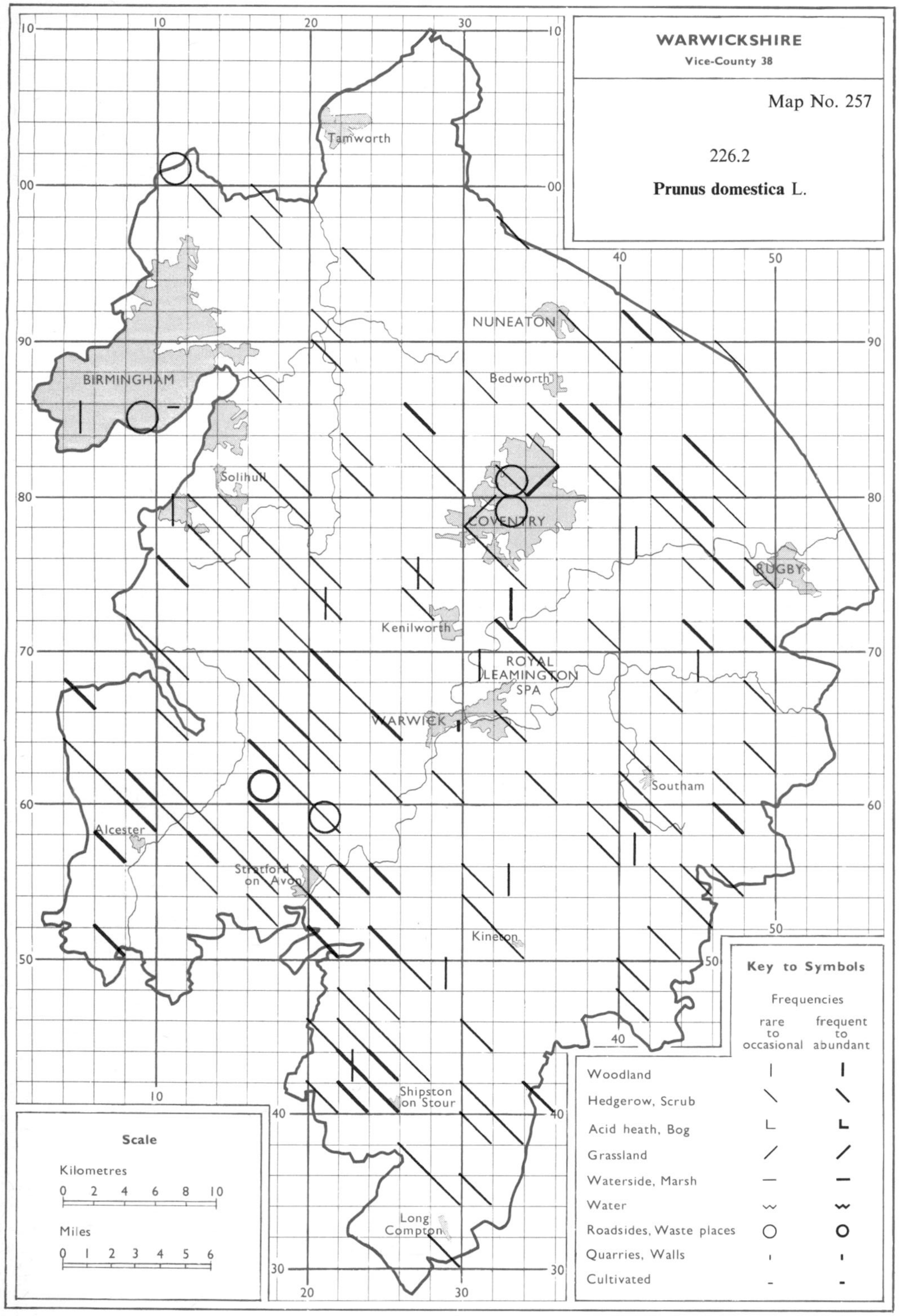
WARWICKSHIRE
Vice-County 38
Map No. 257
226.2
Prunus domestica L.
Tamworth
NUNEATON
Bedworth
BIRMINGHAM
Solihull
COVENTRY
RUGBY
Kenilworth
ROYAL LEAMINGTON SPA
WARWICK
Southam
Alcester
Stratford on Avon
Kineton
Shipston on Stour
Long Compton
Scale
Kilometres
0 2 4 6 8 10
Miles
0 1 2 3 4 5 6
Key to Symbols
Frequencies
rare to occasional
frequent to abundant
Woodland
Hedgerow, Scrub
Acid heath, Bog
Grassland
Waterside, Marsh
Water
Roadsides, Waste places
Quarries, Walls
Cultivated

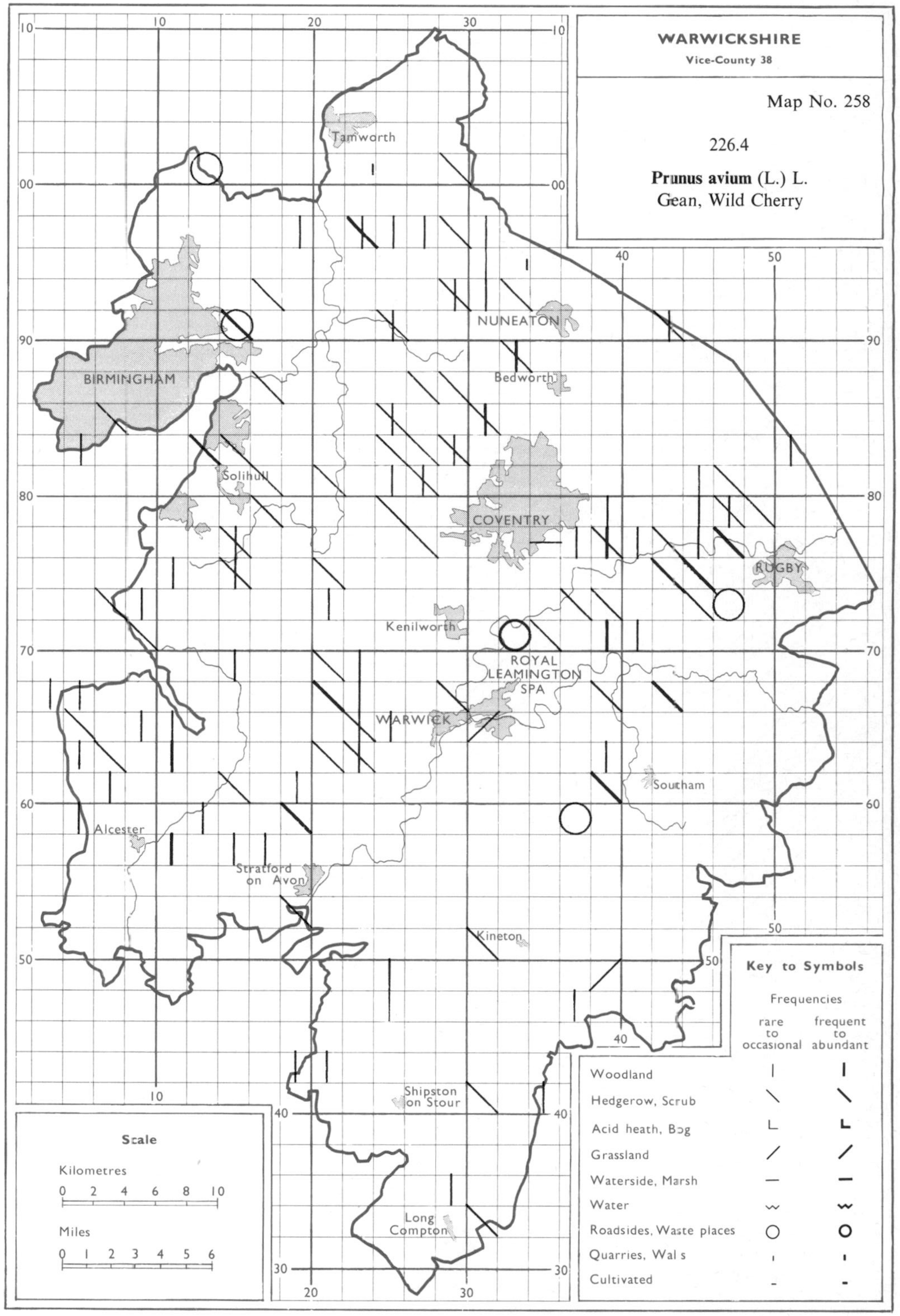
WARWICKSHIRE
Vice-County 38
Map No. 258
226.4
Prunus avium (L.) L.
Gean, Wild Cherry
Tamworth
NUNEATON
Bedworth
BIRMINGHAM
Solihull
COVENTRY
RUGBY
Kenilworth
ROYAL LEAMINGTON SPA
WARWICK
Southam
Alcester
Stratford on Avon
Kineton
Shipston on Stour
Long Compton
Scale
Kilometres
0 2 4 6 8 10
Miles
0 1 2 3 4 5 6
Key to Symbols
Frequencies
rare to occasional
frequent to abundant
Woodland
Hedgerow, Scrub
Acid heath, Bog
Grassland
Waterside, Marsh
Water
Roadsides, Waste places
Quarries, Wal s
Cultivated

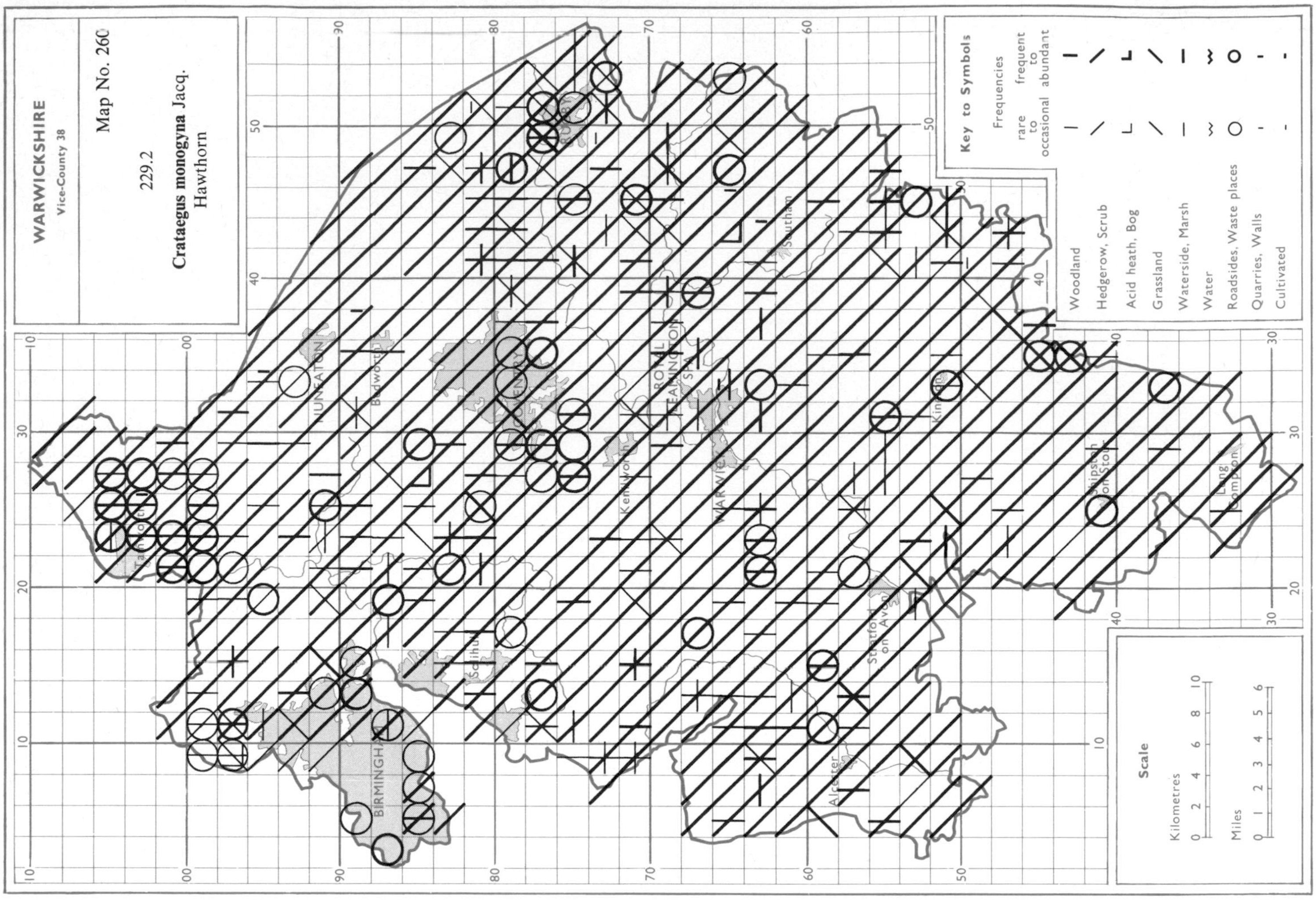
WARWICKSHIRE
Vice-County 38
Map No. 260
229.2
Crataegus monogyna Jacq.
Hawthorn
Key to Symbols
Frequencies
rare to occasional
frequent to abundant
Woodland
Hedgerow, Scrub
Acid heath, Bog
Grassland
Waterside, Marsh
Water
Roadsides, Waste places
Quarries, Walls
Cultivated
Scale
Kilometres
Miles
TAMWORTH
NUNEATON
Bedworth
COVENTRY
RUGBY
Southam
ROYAL LEAMINGTON SPA
Kenilworth
WARWICK
Kineton
Shipston on Stour
Long Compton
Stratford on Avon
Alcester
Solihull
BIRMINGHAM

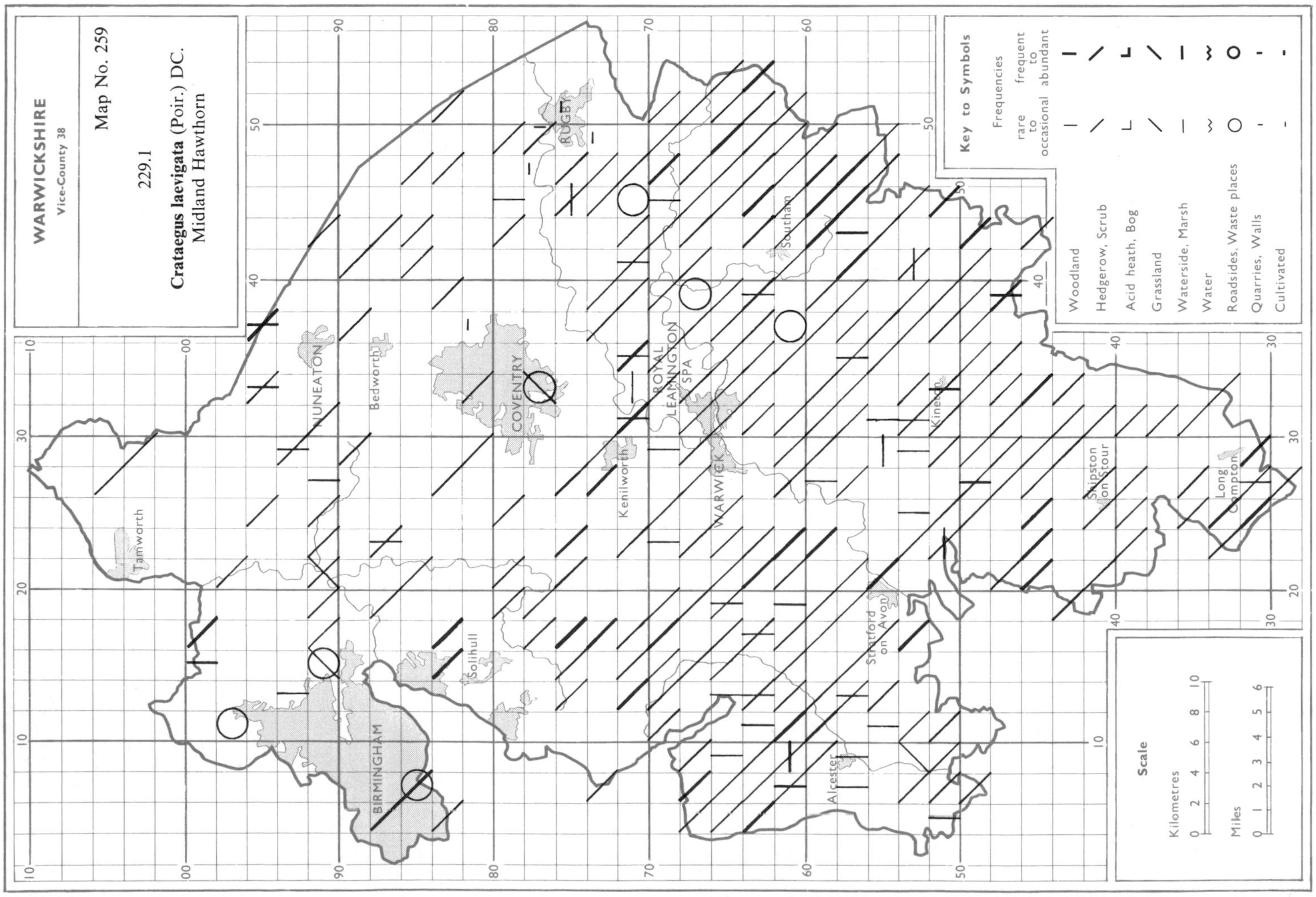
WARWICKSHIRE
Vice-County 38
Map No. 259
229.1
Crataegus laevigata (Poir.) DC.
Midland Hawthorn
Key to Symbols
Frequencies
rare to occasional
frequent to abundant
Woodland
Hedgerow, Scrub
Acid heath, Bog
Grassland
Waterside, Marsh
Water
Roadsides, Waste places
Quarries, Walls
Cultivated
Scale
Kilometres
Miles
Tamworth
NUNEATON
Bedworth
COVENTRY
RUGBY
Southam
ROYAL LEAMINGTON SPA
Kenilworth
WARWICK
Kineton
Shipston on Stour
Long Compton
Stratford on Avon
Alcester
Solihull
BIRMINGHAM

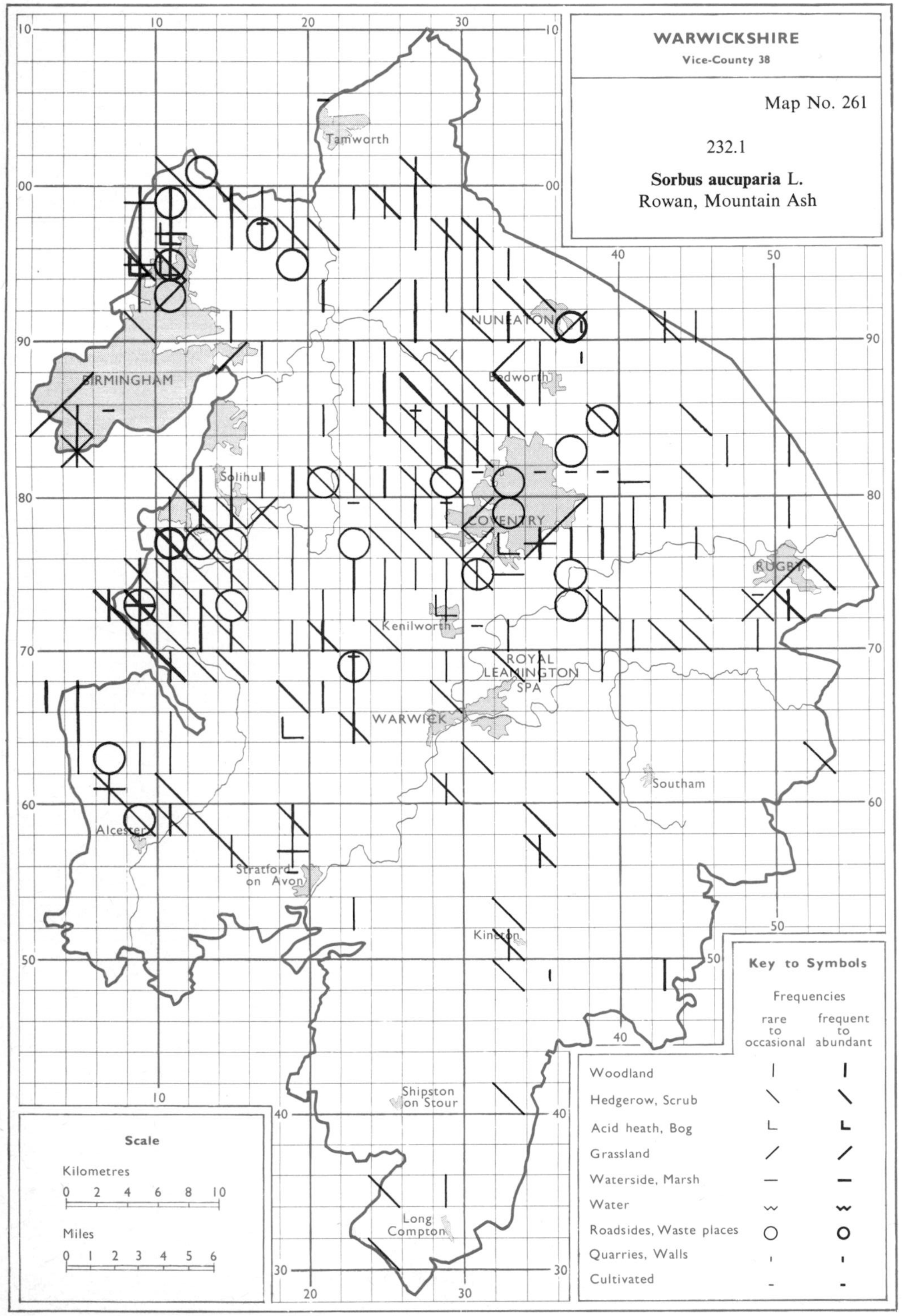
WARWICKSHIRE
Vice-County 38
Map No. 261
232.1
Sorbus aucuparia L.
Rowan, Mountain Ash
Tamworth
NUNEATON
Bedworth
BIRMINGHAM
Solihull
COVENTRY
RUGBY
Kenilworth
ROYAL LEAMINGTON SPA
WARWICK
Southam
Alcester
Stratford on Avon
Kineton
Shipston on Stour
Long Compton
Key to Symbols
Frequencies
rare to occasional
frequent to abundant
Woodland
Hedgerow, Scrub
Acid heath, Bog
Grassland
Waterside, Marsh
Water
Roadsides, Waste places
Quarries, Walls
Cultivated
Scale
Kilometres
0 2 4 6 8 10
Miles
0 1 2 3 4 5 6

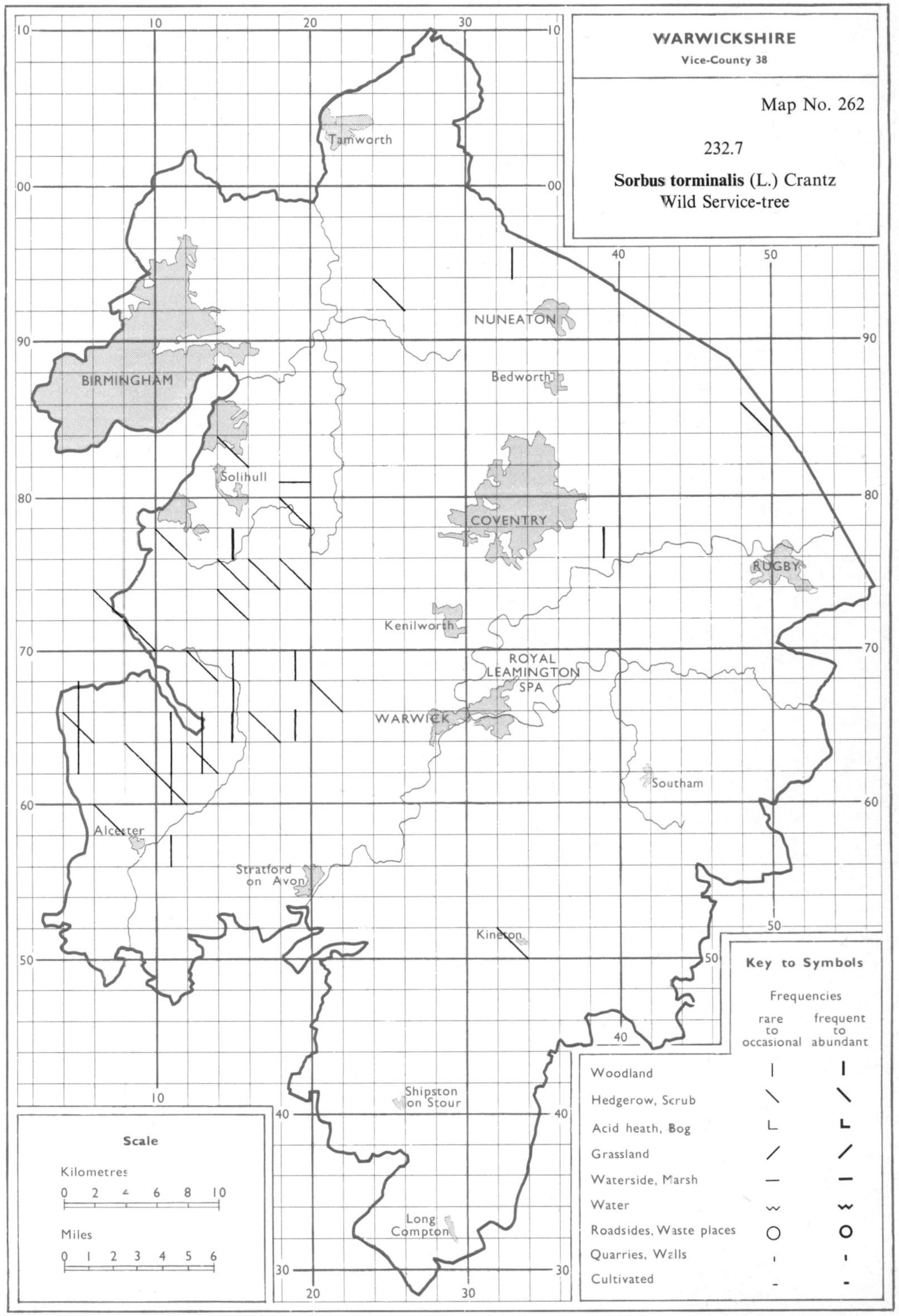
WARWICKSHIRE
Vice-County 38
Map No. 262
232.7
Sorbus torminalis (L.) Crantz
Wild Service-tree
Tamworth
NUNEATON
Bedworth
BIRMINGHAM
Solihull
COVENTRY
RUGBY
Kenilworth
ROYAL LEAMINGTON SPA
WARWICK
Southam
Alcester
Stratford on Avon
Kineton
Shipston on Stour
Long Compton
Key to Symbols
Frequencies
rare to occasional
frequent to abundant
Woodland
Hedgerow, Scrub
Acid heath, Bog
Grassland
Waterside, Marsh
Water
Roadsides, Waste places
Quarries, Walls
Cultivated
Scale
Kilometres
0 2 4 6 8 10
Miles
0 1 2 3 4 5 6

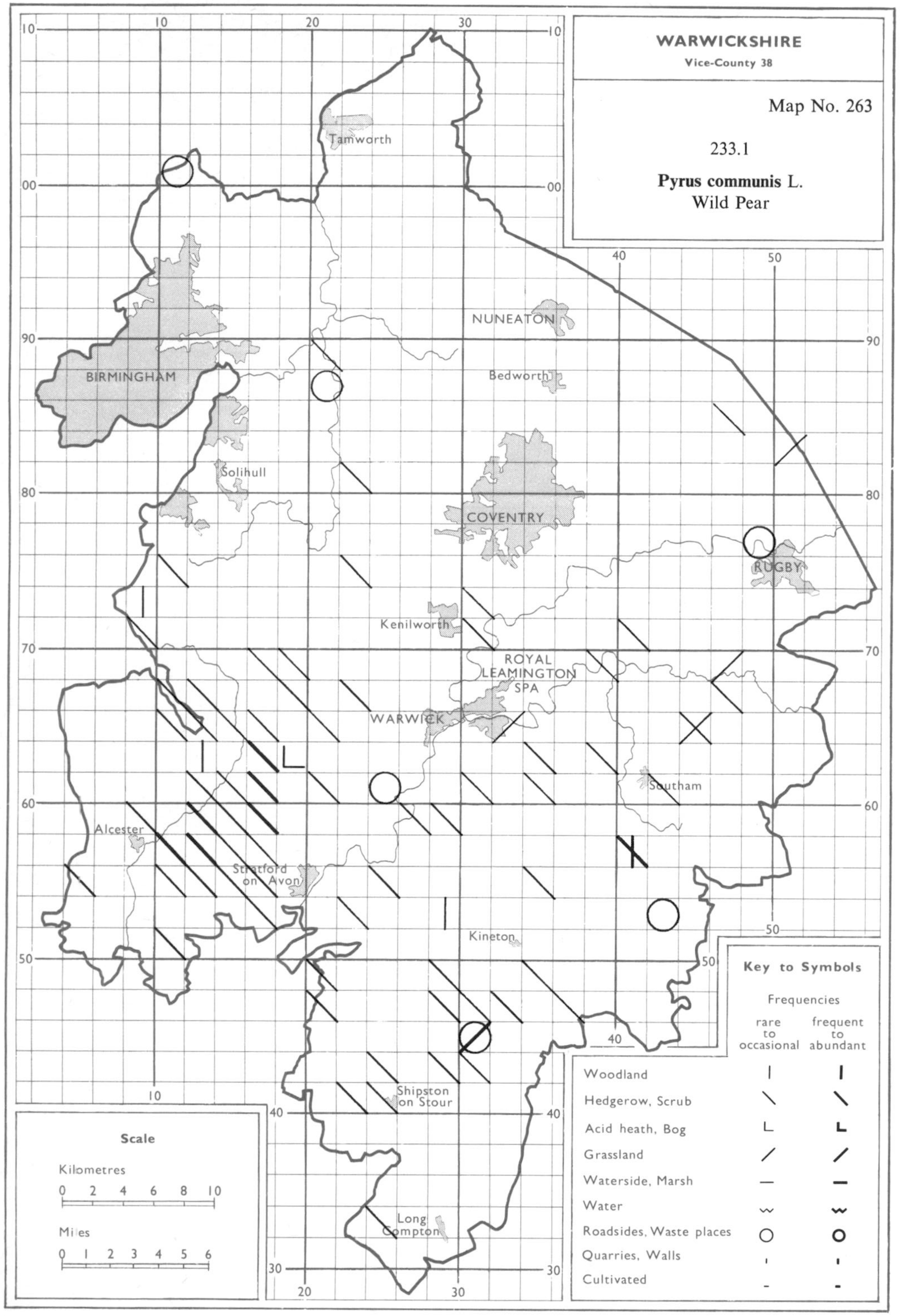
WARWICKSHIRE
Vice-County 38
Map No. 263
233.1
Pyrus communis L.
Wild Pear
Tamworth
NUNEATON
Bedworth
BIRMINGHAM
Solihull
COVENTRY
RUGBY
Kenilworth
ROYAL LEAMINGTON SPA
WARWICK
Southam
Alcester
Stratford on Avon
Kineton
Shipston on Stour
Long Compton
Key to Symbols
Frequencies
rare to occasional
frequent to abundant
Woodland
Hedgerow, Scrub
Acid heath, Bog
Grassland
Waterside, Marsh
Water
Roadsides, Waste places
Quarries, Walls
Cultivated
Scale
Kilometres
0 2 4 6 8 10
Miles
0 1 2 3 4 5 6

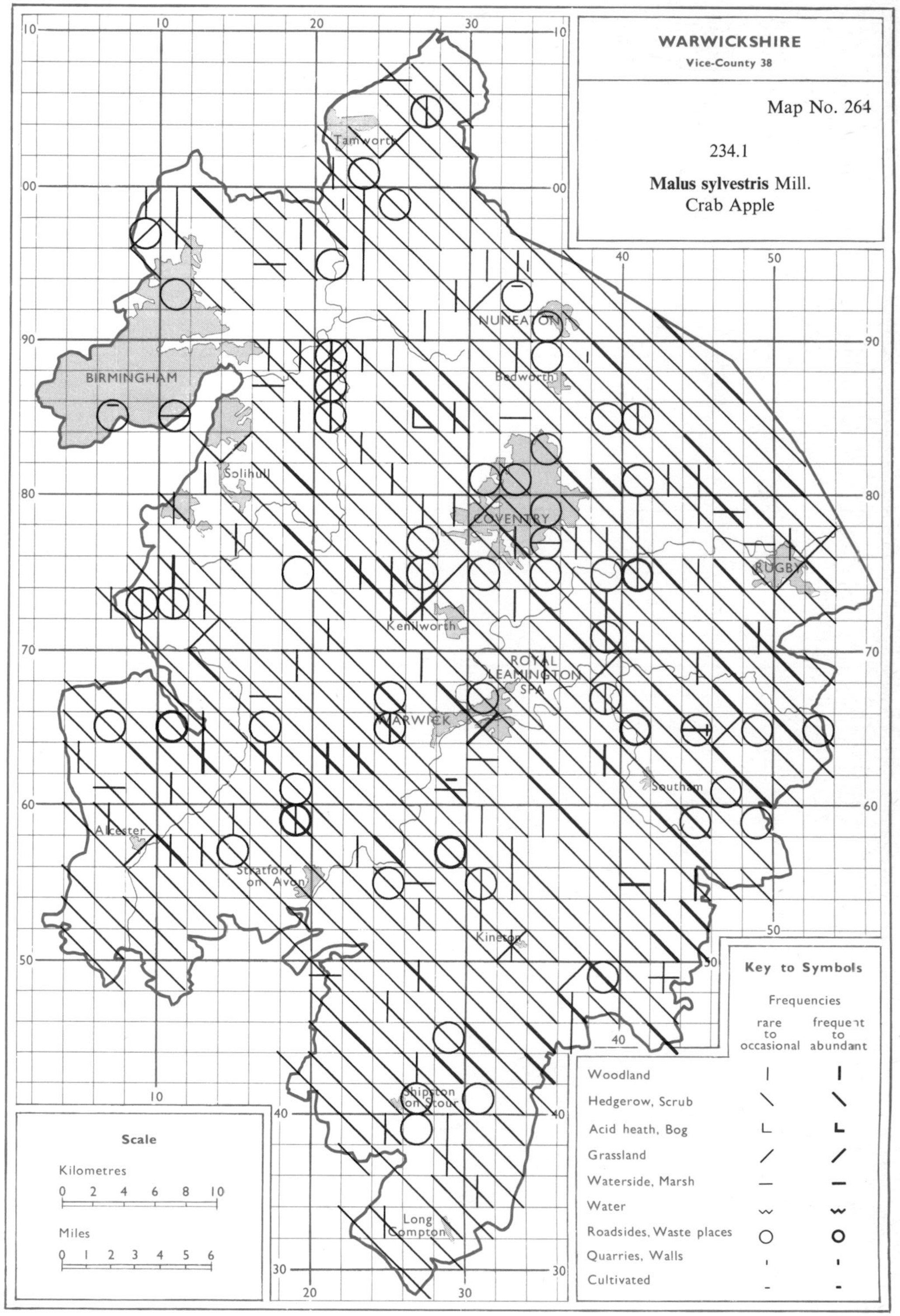
WARWICKSHIRE
Vice-County 38
Map No. 264
234.1
Malus sylvestris Mill.
Crab Apple
Tamworth
NUNEATON
Bedworth
BIRMINGHAM
Solihull
COVENTRY
RUGBY
Kenilworth
ROYAL LEAMINGTON SPA
WARWICK
Southam
Alcester
Stratford on Avon
Kineton
Shipston on Stour
Long Compton
Key to Symbols
Frequencies
rare to occasional
frequent to abundant
Woodland
Hedgerow, Scrub
Acid heath, Bog
Grassland
Waterside, Marsh
Water
Roadsides, Waste places
Quarries, Walls
Cultivated
Scale
Kilometres
0 2 4 6 8 10
Miles
0 1 2 3 4 5 6

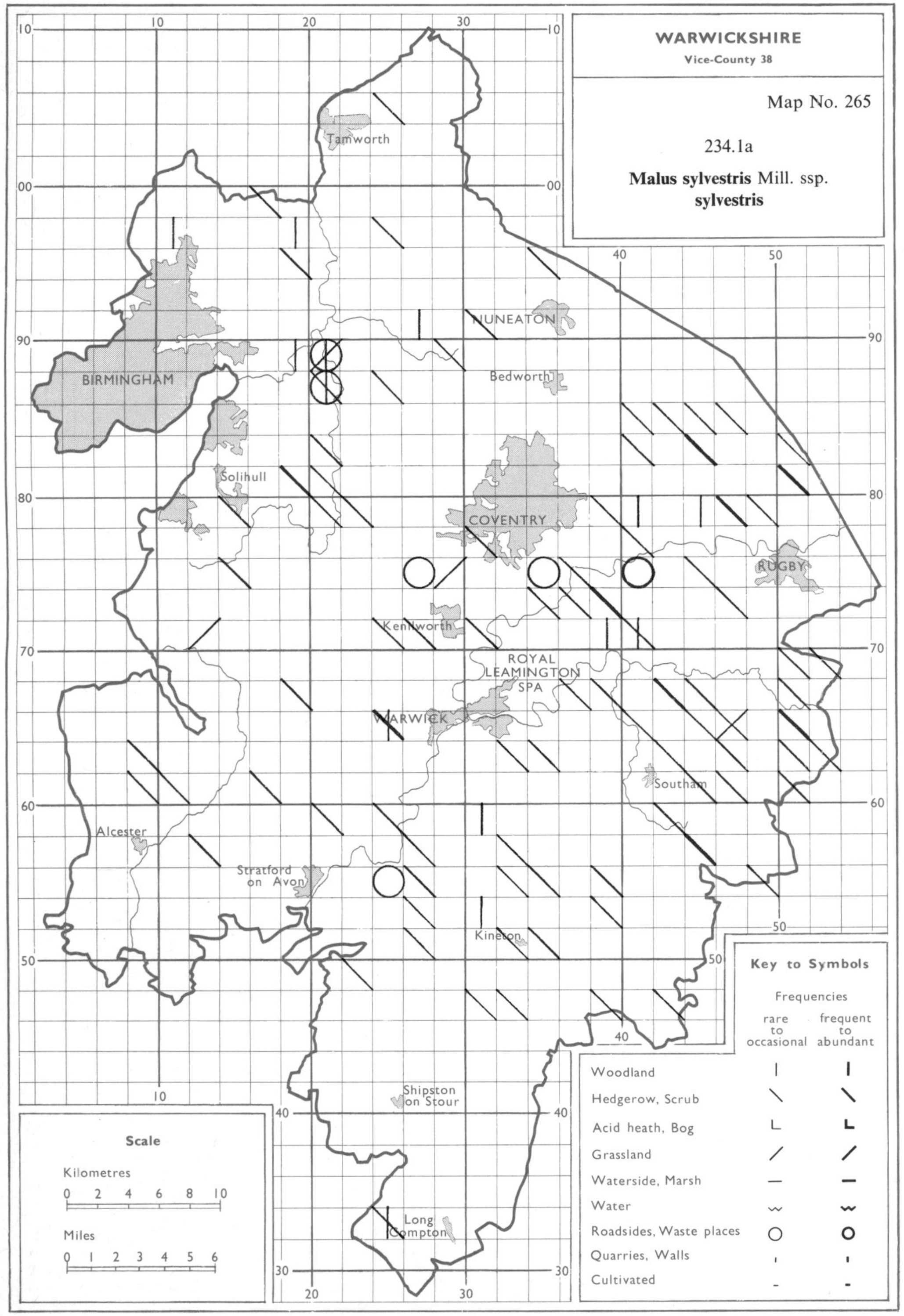
WARWICKSHIRE
Vice-County 38
Map No. 265
234.1a
Malus sylvestris Mill. ssp. **sylvestris**
Tamworth
NUNEATON
Bedworth
BIRMINGHAM
Solihull
COVENTRY
RUGBY
Kenilworth
ROYAL LEAMINGTON SPA
WARWICK
Southam
Alcester
Stratford on Avon
Kineton
Shipston on Stour
Long Compton
Key to Symbols
Frequencies
rare to occasional
frequent to abundant
Woodland
Hedgerow, Scrub
Acid heath, Bog
Grassland
Waterside, Marsh
Water
Roadsides, Waste places
Quarries, Walls
Cultivated
Scale
Kilometres
0 2 4 6 8 10
Miles
0 1 2 3 4 5 6

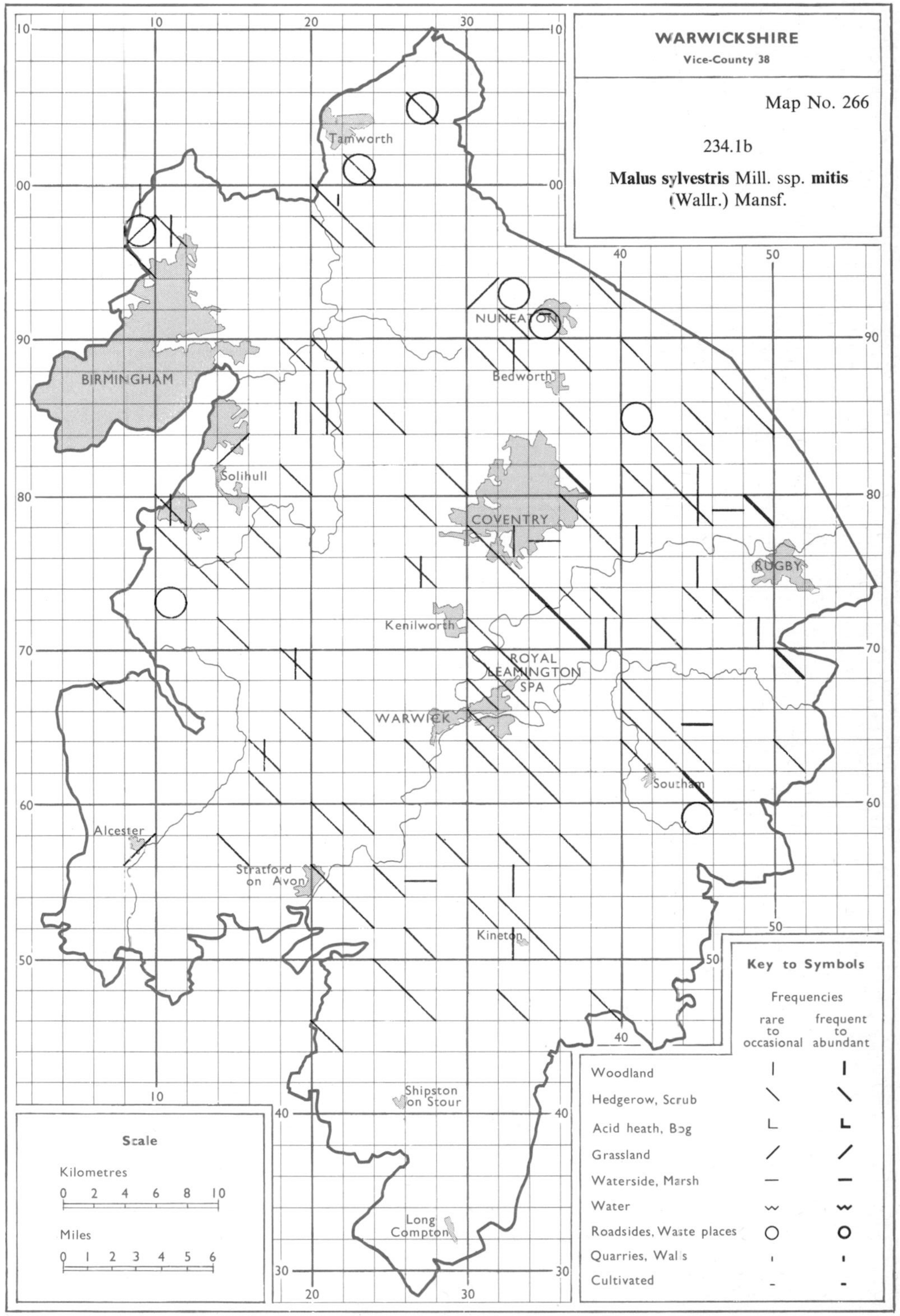
WARWICKSHIRE
Vice-County 38
Map No. 266
234.1b
Malus sylvestris Mill. ssp. **mitis** (Wallr.) Mansf.
Tamworth
NUNEATON
Bedworth
BIRMINGHAM
Solihull
COVENTRY
RUGBY
Kenilworth
ROYAL LEAMINGTON SPA
WARWICK
Southam
Alcester
Stratford on Avon
Kineton
Shipston on Stour
Long Compton
Key to Symbols
Frequencies
rare to occasional
frequent to abundant
Woodland
Hedgerow, Scrub
Acid heath, Bog
Grassland
Waterside, Marsh
Water
Roadsides, Waste places
Quarries, Walls
Cultivated
Scale
Kilometres
0 2 4 6 8 10
Miles
0 1 2 3 4 5 6

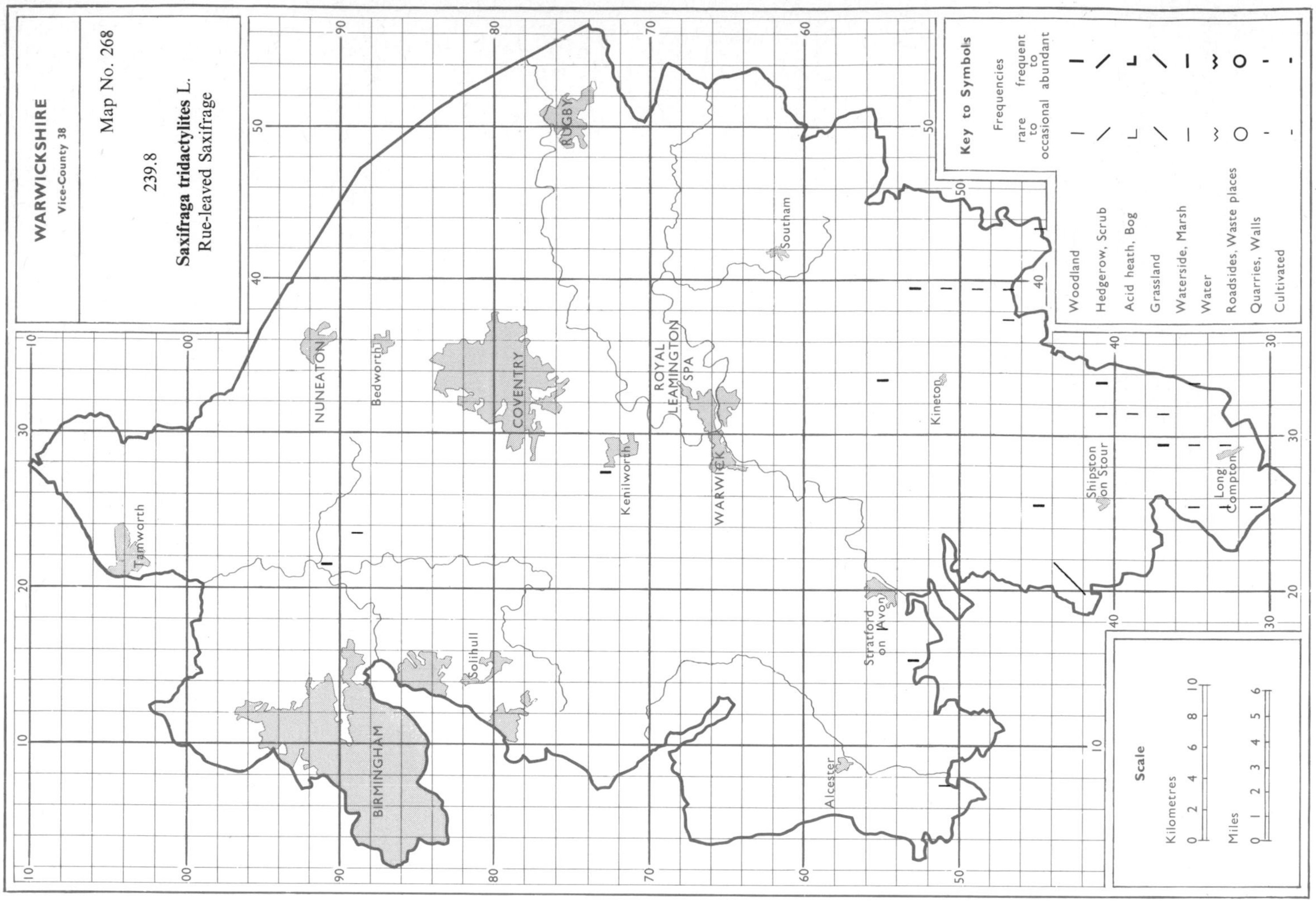
WARWICKSHIRE
Vice-County 38
Map No. 268
239.8
Saxifraga tridactylites L.
Rue-leaved Saxifrage
Key to Symbols
Frequencies
rare to occasional
frequent to abundant
Woodland
Hedgerow, Scrub
Acid heath, Bog
Grassland
Waterside, Marsh
Water
Roadsides, Waste places
Quarries, Walls
Cultivated
Scale
Kilometres
Miles
NUNEATON
Bedworth
COVENTRY
RUGBY
ROYAL LEAMINGTON SPA
Kenilworth
WARWICK
Southam
Kineton
Shipston on Stour
Long Compton
Tamworth
Solihull
BIRMINGHAM
Stratford on Avon
Alcester

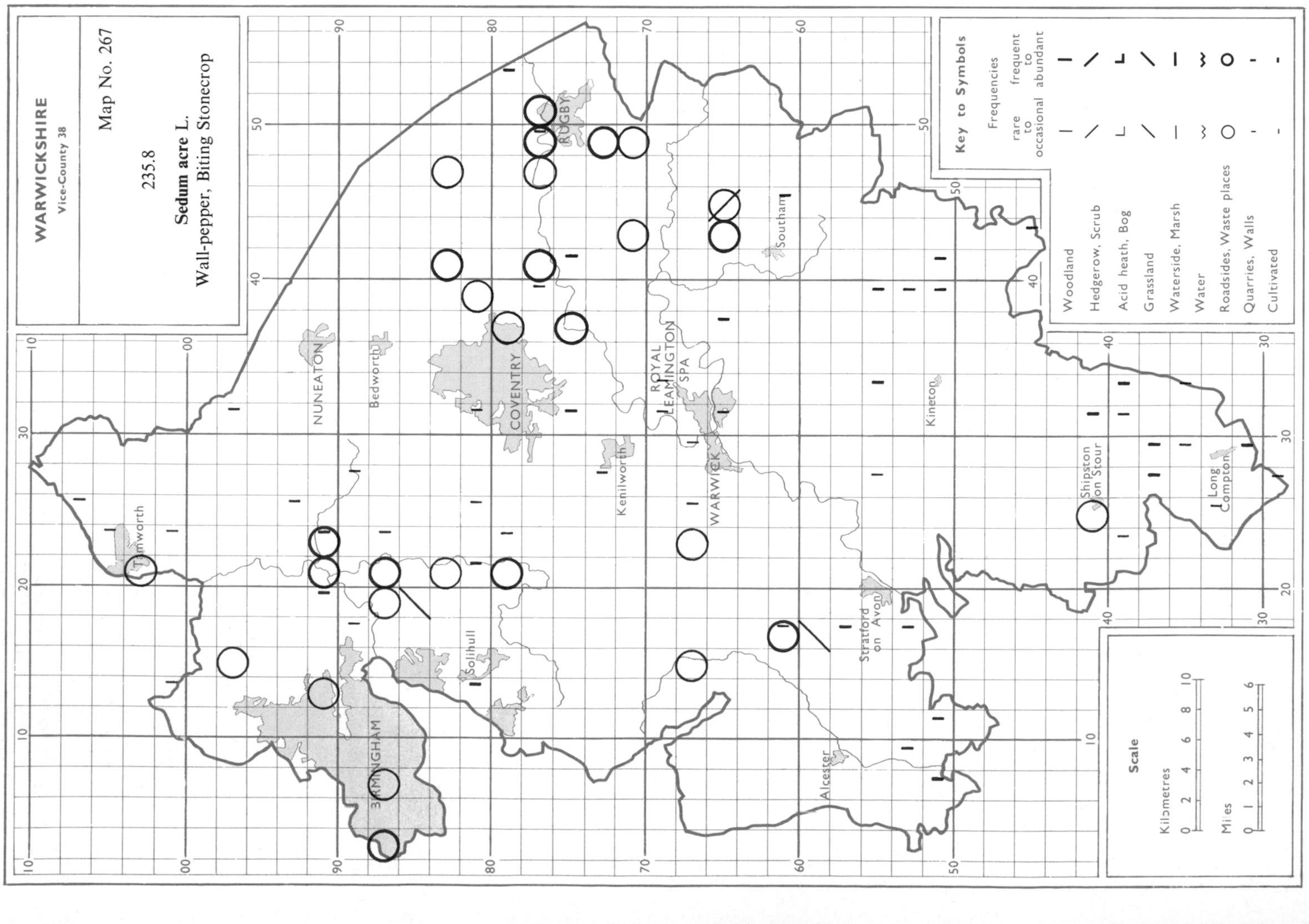
WARWICKSHIRE
Vice-County 38
Map No. 267
235.8
Sedum acre L.
Wall-pepper, Biting Stonecrop
Key to Symbols
Frequencies
rare to occasional
frequent to abundant
Woodland
Hedgerow, Scrub
Acid heath, Bog
Grassland
Waterside, Marsh
Water
Roadsides, Waste places
Quarries, Walls
Cultivated
Scale
Kilometres
Miles
NUNEATON
Bedworth
COVENTRY
RUGBY
ROYAL LEAMINGTON SPA
Kenilworth
WARWICK
Southam
Kineton
Shipston on Stour
Long Compton
Tamworth
Solihull
BIRMINGHAM
Stratford on Avon
Alcester

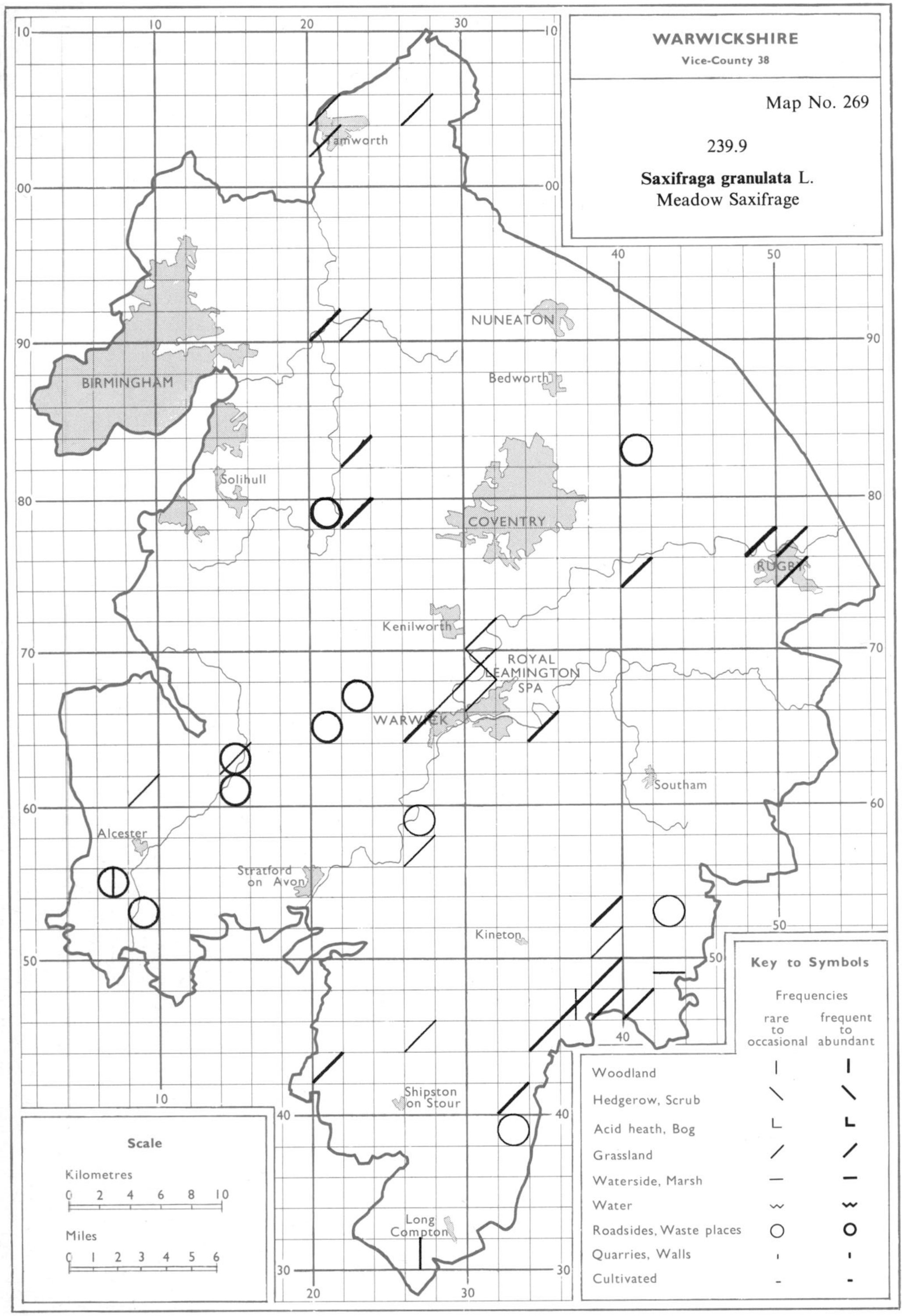
WARWICKSHIRE
Vice-County 38
Map No. 269
239.9
Saxifraga granulata L.
Meadow Saxifrage
Tamworth
NUNEATON
Bedworth
BIRMINGHAM
Solihull
COVENTRY
RUGBY
Kenilworth
ROYAL LEAMINGTON SPA
WARWICK
Southam
Alcester
Stratford on Avon
Kineton
Shipston on Stour
Long Compton
Scale
Kilometres
Miles
Key to Symbols
Frequencies
rare to occasional
frequent to abundant
Woodland
Hedgerow, Scrub
Acid heath, Bog
Grassland
Waterside, Marsh
Water
Roadsides, Waste places
Quarries, Walls
Cultivated

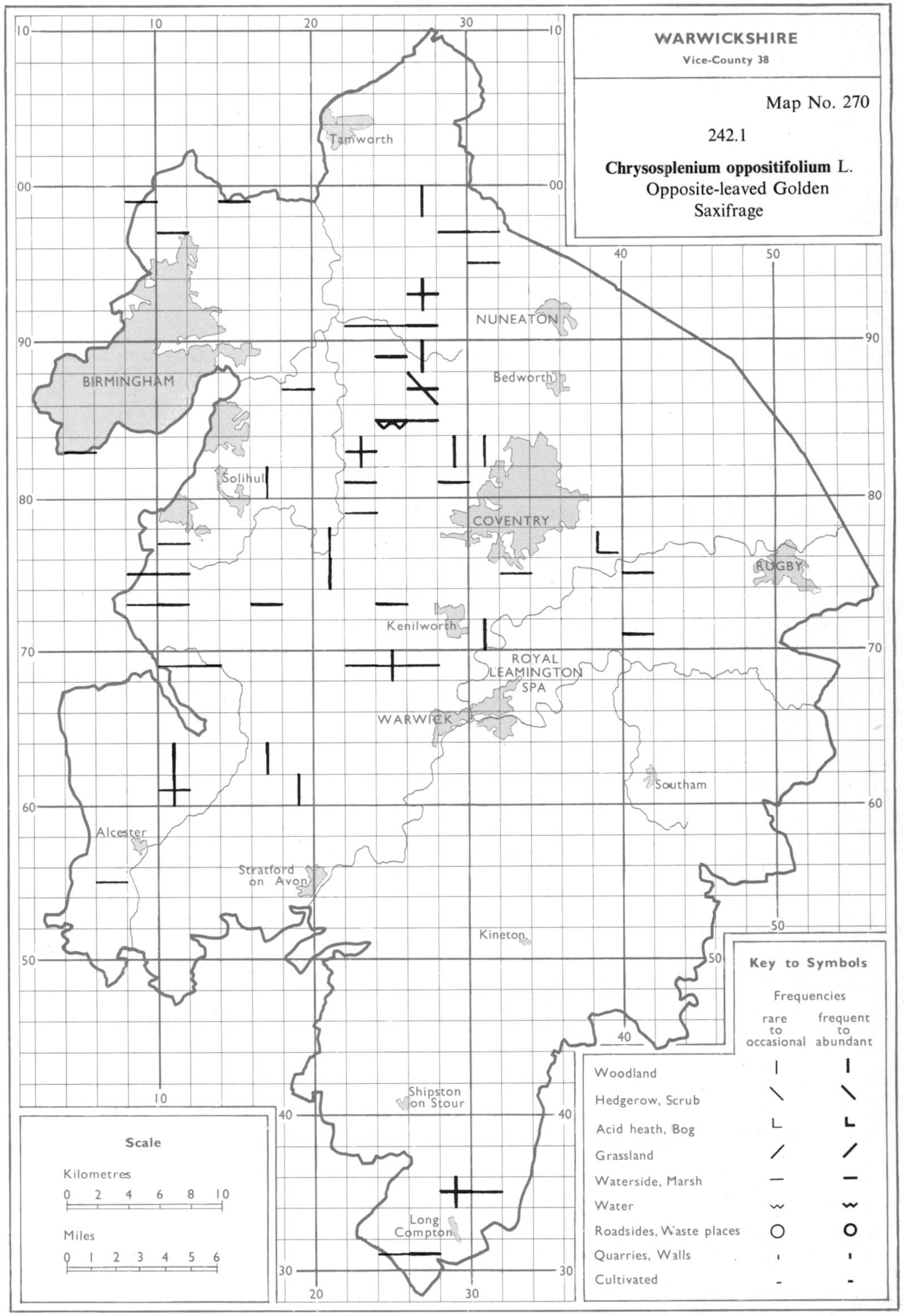
WARWICKSHIRE
Vice-County 38
Map No. 270
242.1
Chrysosplenium oppositifolium L.
Opposite-leaved Golden Saxifrage
Tamworth
NUNEATON
Bedworth
BIRMINGHAM
Solihull
COVENTRY
RUGBY
Kenilworth
ROYAL LEAMINGTON SPA
WARWICK
Southam
Alcester
Stratford on Avon
Kineton
Shipston on Stour
Long Compton
Scale
Kilometres
Miles
Key to Symbols
Frequencies
rare to occasional
frequent to abundant
Woodland
Hedgerow, Scrub
Acid heath, Bog
Grassland
Waterside, Marsh
Water
Roadsides, Waste places
Quarries, Walls
Cultivated

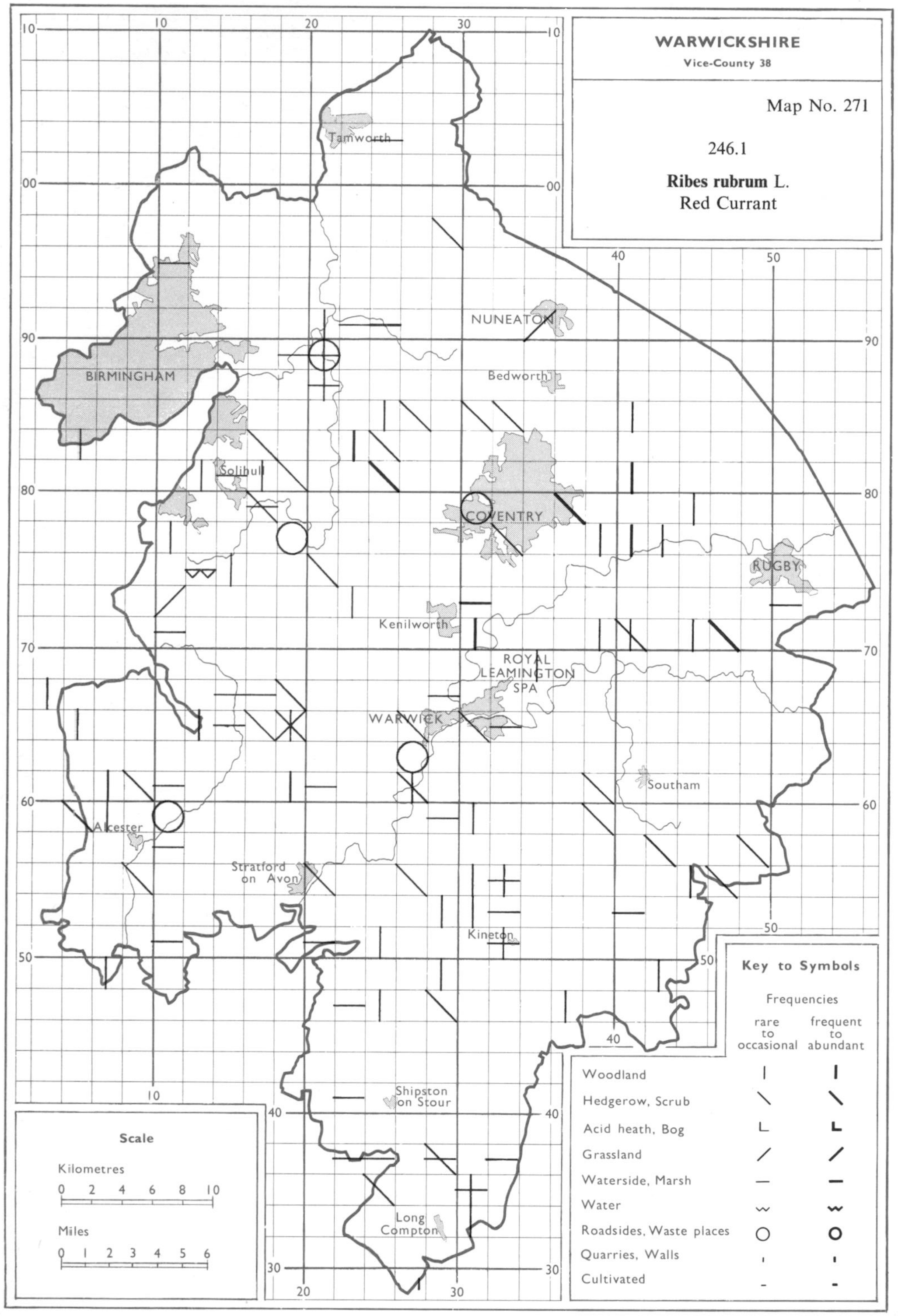
WARWICKSHIRE
Vice-County 38
Map No. 271
246.1
Ribes rubrum L.
Red Currant
Tamworth
NUNEATON
Bedworth
BIRMINGHAM
Solihull
COVENTRY
RUGBY
Kenilworth
ROYAL LEAMINGTON SPA
WARWICK
Southam
Alcester
Stratford on Avon
Kineton
Shipston on Stour
Long Compton
Key to Symbols
Frequencies
rare to occasional
frequent to abundant
Woodland
Hedgerow, Scrub
Acid heath, Bog
Grassland
Waterside, Marsh
Water
Roadsides, Waste places
Quarries, Walls
Cultivated
Scale
Kilometres
0 2 4 6 8 10
Miles
0 1 2 3 4 5 6

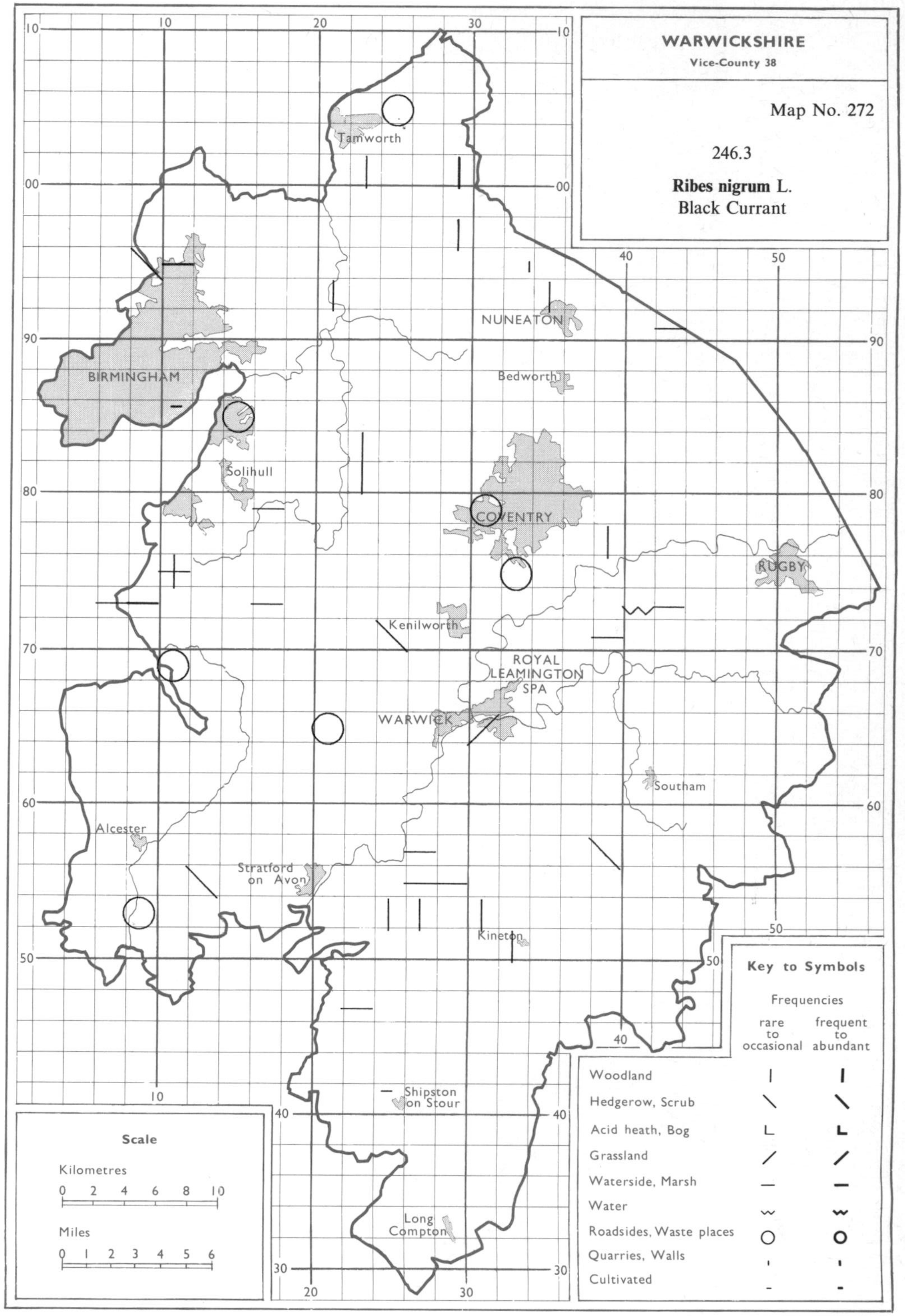
WARWICKSHIRE
Vice-County 38
Map No. 272
246.3
Ribes nigrum L.
Black Currant
Tamworth
NUNEATON
Bedworth
BIRMINGHAM
Solihull
COVENTRY
RUGBY
Kenilworth
ROYAL LEAMINGTON SPA
WARWICK
Southam
Alcester
Stratford on Avon
Kineton
Shipston on Stour
Long Compton
Key to Symbols
Frequencies
rare to occasional
frequent to abundant
Woodland
Hedgerow, Scrub
Acid heath, Bog
Grassland
Waterside, Marsh
Water
Roadsides, Waste places
Quarries, Walls
Cultivated
Scale
Kilometres
0 2 4 6 8 10
Miles
0 1 2 3 4 5 6

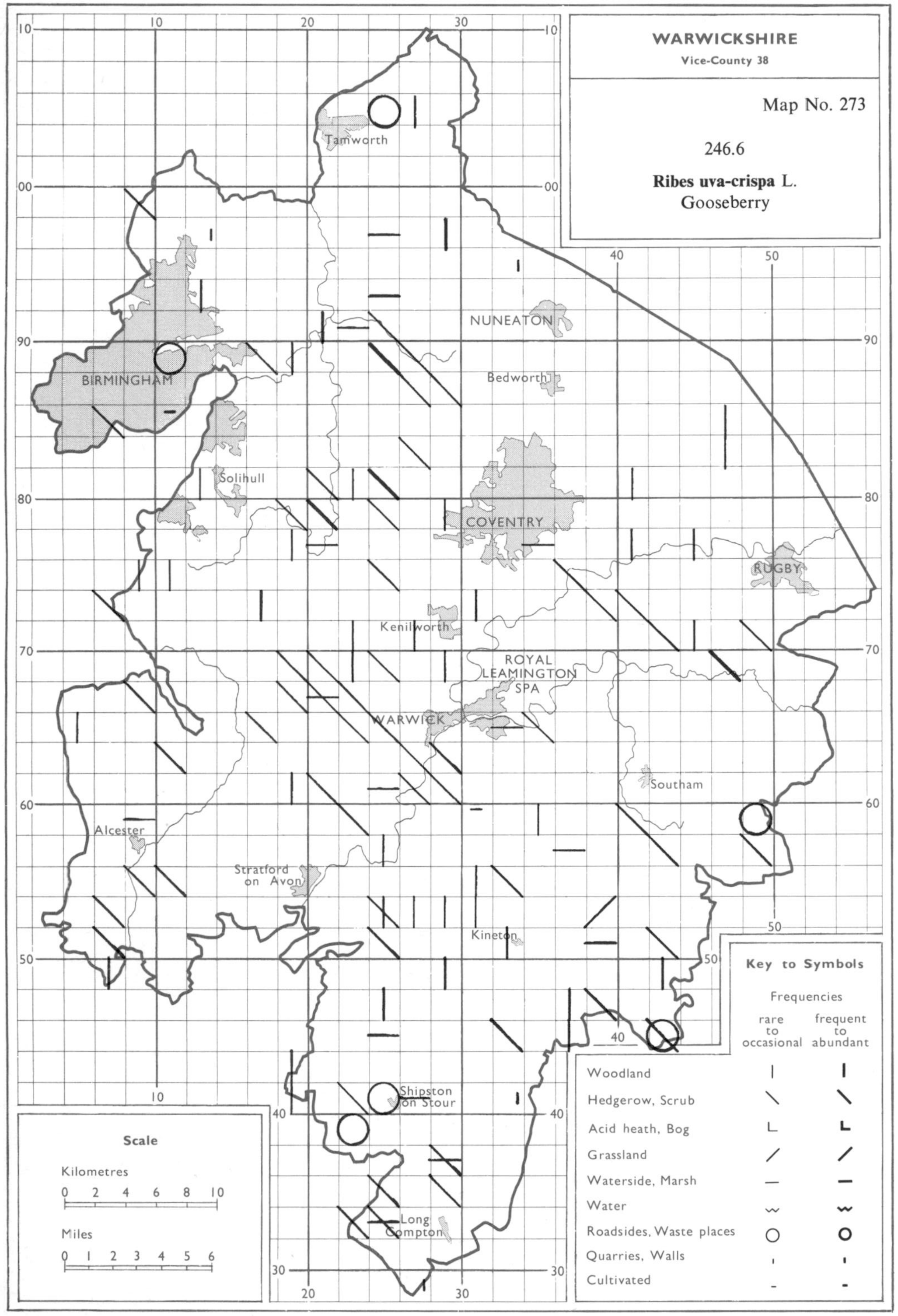
WARWICKSHIRE
Vice-County 38
Map No. 273
246.6
Ribes uva-crispa L.
Gooseberry
Tamworth
NUNEATON
Bedworth
BIRMINGHAM
Solihull
COVENTRY
RUGBY
Kenilworth
ROYAL LEAMINGTON SPA
WARWICK
Southam
Alcester
Stratford on Avon
Kineton
Shipston on Stour
Long Compton
Scale
Kilometres
0 2 4 6 8 10
Miles
0 1 2 3 4 5 6
Key to Symbols
Frequencies
rare to occasional
frequent to abundant
Woodland
Hedgerow, Scrub
Acid heath, Bog
Grassland
Waterside, Marsh
Water
Roadsides, Waste places
Quarries, Walls
Cultivated

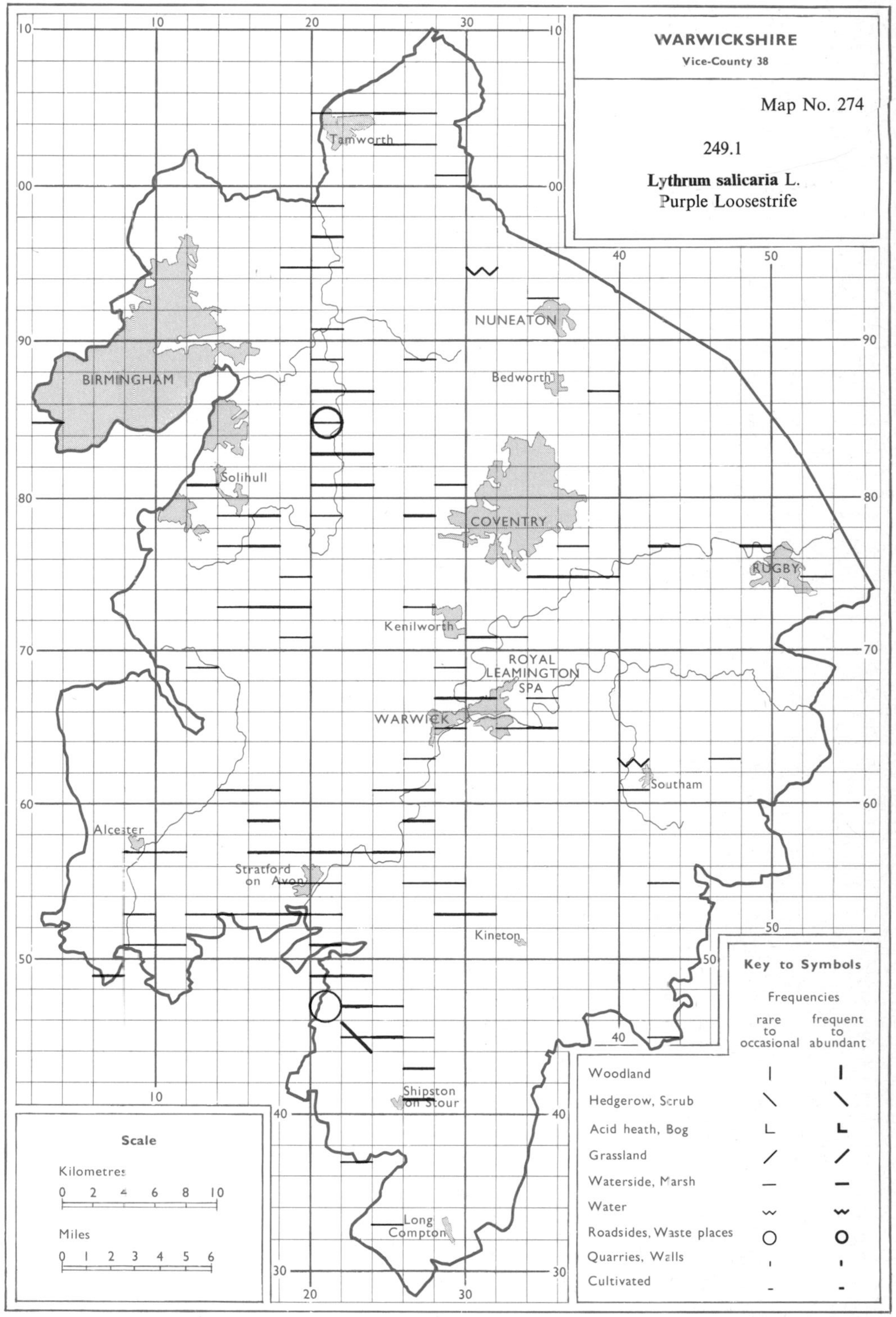
WARWICKSHIRE
Vice-County 38
Map No. 274
249.1
Lythrum salicaria L.
Purple Loosestrife
Tamworth
NUNEATON
Bedworth
BIRMINGHAM
Solihull
COVENTRY
RUGBY
Kenilworth
ROYAL LEAMINGTON SPA
WARWICK
Southam
Alcester
Stratford on Avon
Kineton
Shipston on Stour
Long Compton
Scale
Kilometres
0 2 4 6 8 10
Miles
0 1 2 3 4 5 6
Key to Symbols
Frequencies
rare to occasional
frequent to abundant
Woodland
Hedgerow, Scrub
Acid heath, Bog
Grassland
Waterside, Marsh
Water
Roadsides, Waste places
Quarries, Walls
Cultivated

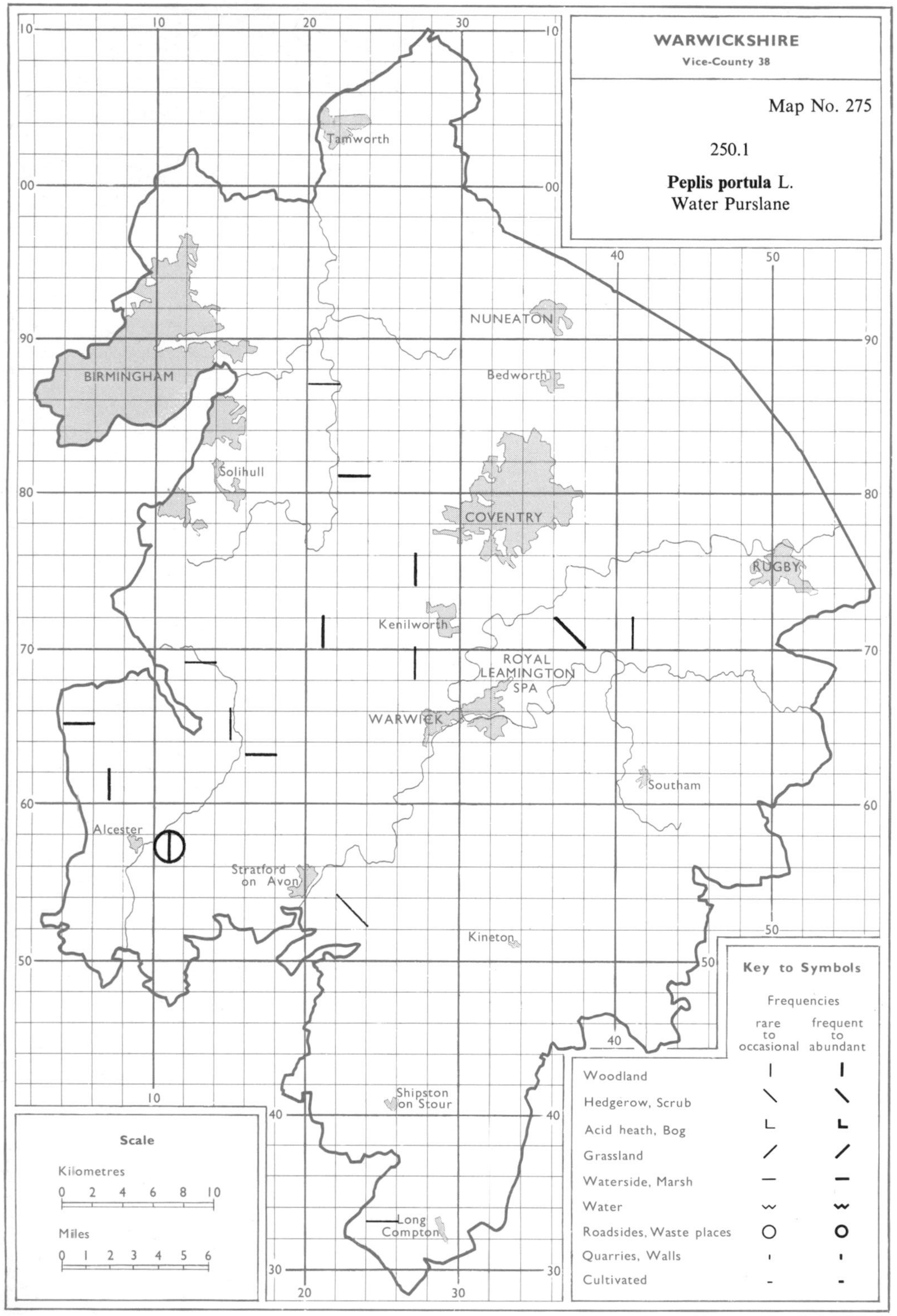
WARWICKSHIRE
Vice-County 38
Map No. 275
250.1
Peplis portula L.
Water Purslane
Tamworth
BIRMINGHAM
NUNEATON
Bedworth
Solihull
COVENTRY
RUGBY
Kenilworth
ROYAL LEAMINGTON SPA
WARWICK
Southam
Alcester
Stratford on Avon
Kineton
Shipston on Stour
Long Compton
Scale
Kilometres
0 2 4 6 8 10
Miles
0 1 2 3 4 5 6
Key to Symbols
Frequencies
rare to occasional
frequent to abundant
Woodland
Hedgerow, Scrub
Acid heath, Bog
Grassland
Waterside, Marsh
Water
Roadsides, Waste places
Quarries, Walls
Cultivated

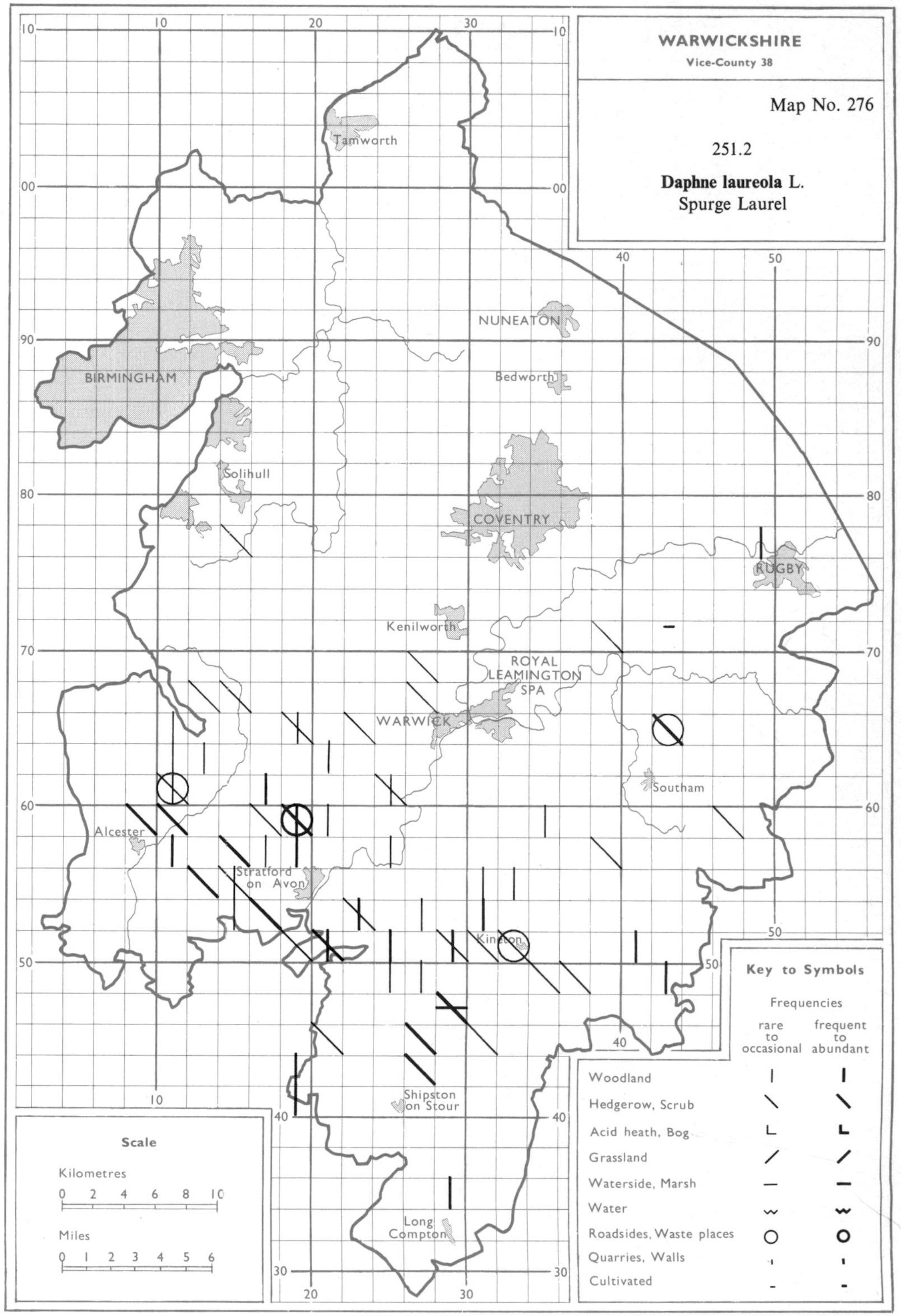
WARWICKSHIRE
Vice-County 38
Map No. 276
251.2
Daphne laureola L.
Spurge Laurel
Tamworth
BIRMINGHAM
NUNEATON
Bedworth
Solihull
COVENTRY
RUGBY
Kenilworth
ROYAL LEAMINGTON SPA
WARWICK
Southam
Alcester
Stratford on Avon
Kineton
Shipston on Stour
Long Compton
Scale
Kilometres
0 2 4 6 8 10
Miles
0 1 2 3 4 5 6
Key to Symbols
Frequencies
rare to occasional
frequent to abundant
Woodland
Hedgerow, Scrub
Acid heath, Bog
Grassland
Waterside, Marsh
Water
Roadsides, Waste places
Quarries, Walls
Cultivated

WARWICKSHIRE
Vice-County 38
Map No. 278
254.2
Epilobium parviflorum Schreb.
Hoary Willowherb
Key to Symbols
Frequencies
rare to occasional
frequent to abundant
Woodland
Hedgerow, Scrub
Acid heath, Bog
Grassland
Waterside, Marsh
Water
Roadsides, Waste places
Quarries, Walls
Cultivated
Scale
Kilometres
0 2 4 6 8 10
Miles
0 1 2 3 4 5 6
NUNEATON
Bedworth
COVENTRY
RUGBY
Southam
ROYAL LEAMINGTON SPA
WARWICK
Kenilworth
Kineton
Shipston on Stour
Long Compton
Tamworth
Solihull
BIRMINGHAM
Stratford on Avon
Alcester

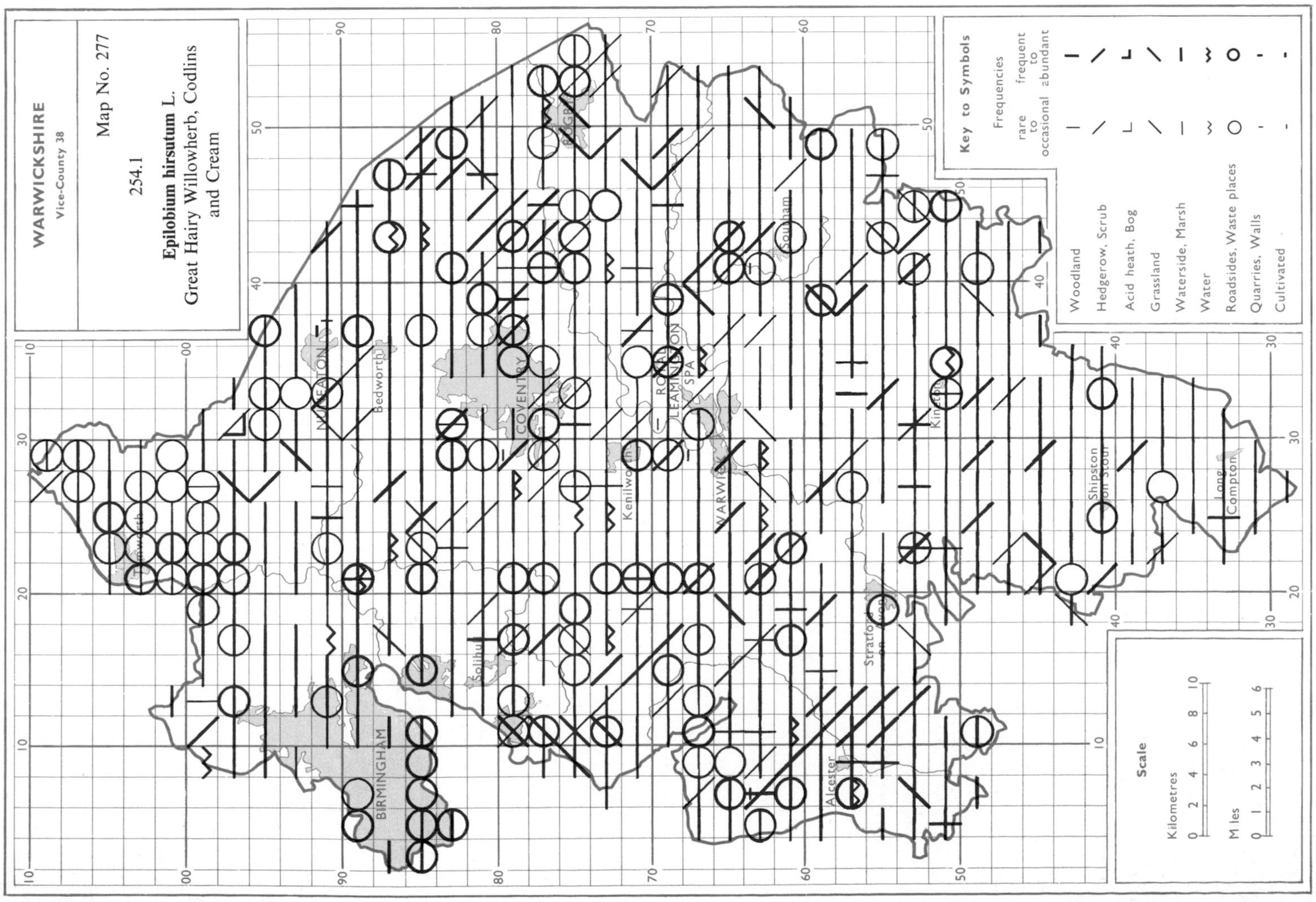
WARWICKSHIRE
Vice-County 38
Map No. 277
254.1
Epilobium hirsutum L.
Great Hairy Willowherb, Codlins and Cream
Key to Symbols
Frequencies
rare to occasional
frequent to abundant
Woodland
Hedgerow, Scrub
Acid heath, Bog
Grassland
Waterside, Marsh
Water
Roadsides, Waste places
Quarries, Walls
Cultivated
Scale
Kilometres
0 2 4 6 8 10
Miles
0 1 2 3 4 5 6
NUNEATON
Bedworth
COVENTRY
RUGBY
Southam
ROYAL LEAMINGTON SPA
WARWICK
Kenilworth
Kineton
Shipston on Stour
Long Compton
Tamworth
Solihull
BIRMINGHAM
Stratford on Avon
Alcester

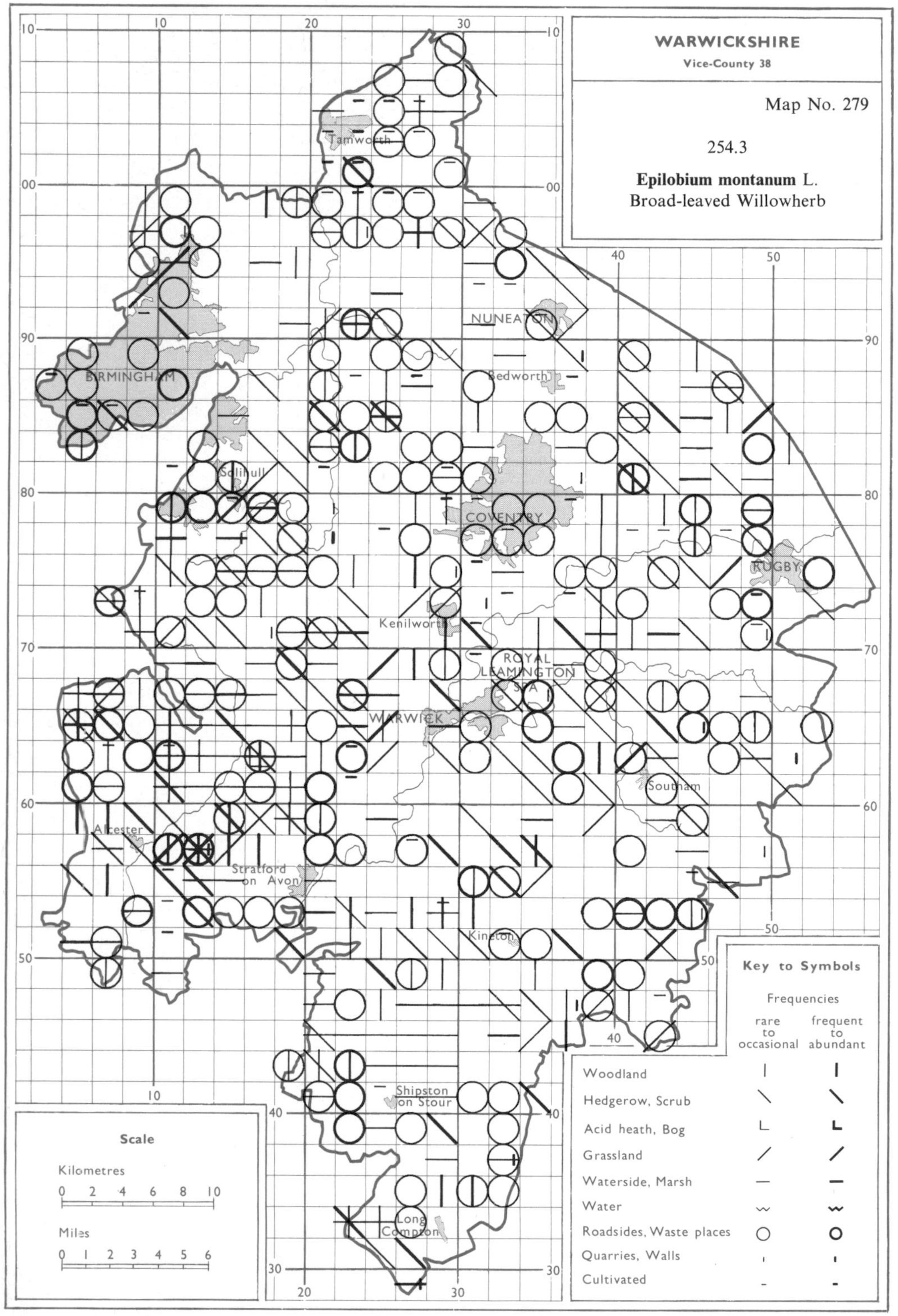
WARWICKSHIRE
Vice-County 38
Map No. 279
254.3
Epilobium montanum L.
Broad-leaved Willowherb
Key to Symbols
Frequencies
rare to occasional
frequent to abundant
Woodland
Hedgerow, Scrub
Acid heath, Bog
Grassland
Waterside, Marsh
Water
Roadsides, Waste places
Quarries, Walls
Cultivated
Scale
Kilometres
0 2 4 6 8 10
Miles
0 1 2 3 4 5 6
Tamworth
Nuneaton
Bedworth
Birmingham
Solihull
Coventry
Rugby
Kenilworth
Royal Leamington Spa
Warwick
Southam
Alcester
Stratford on Avon
Kineton
Shipston on Stour
Long Compton

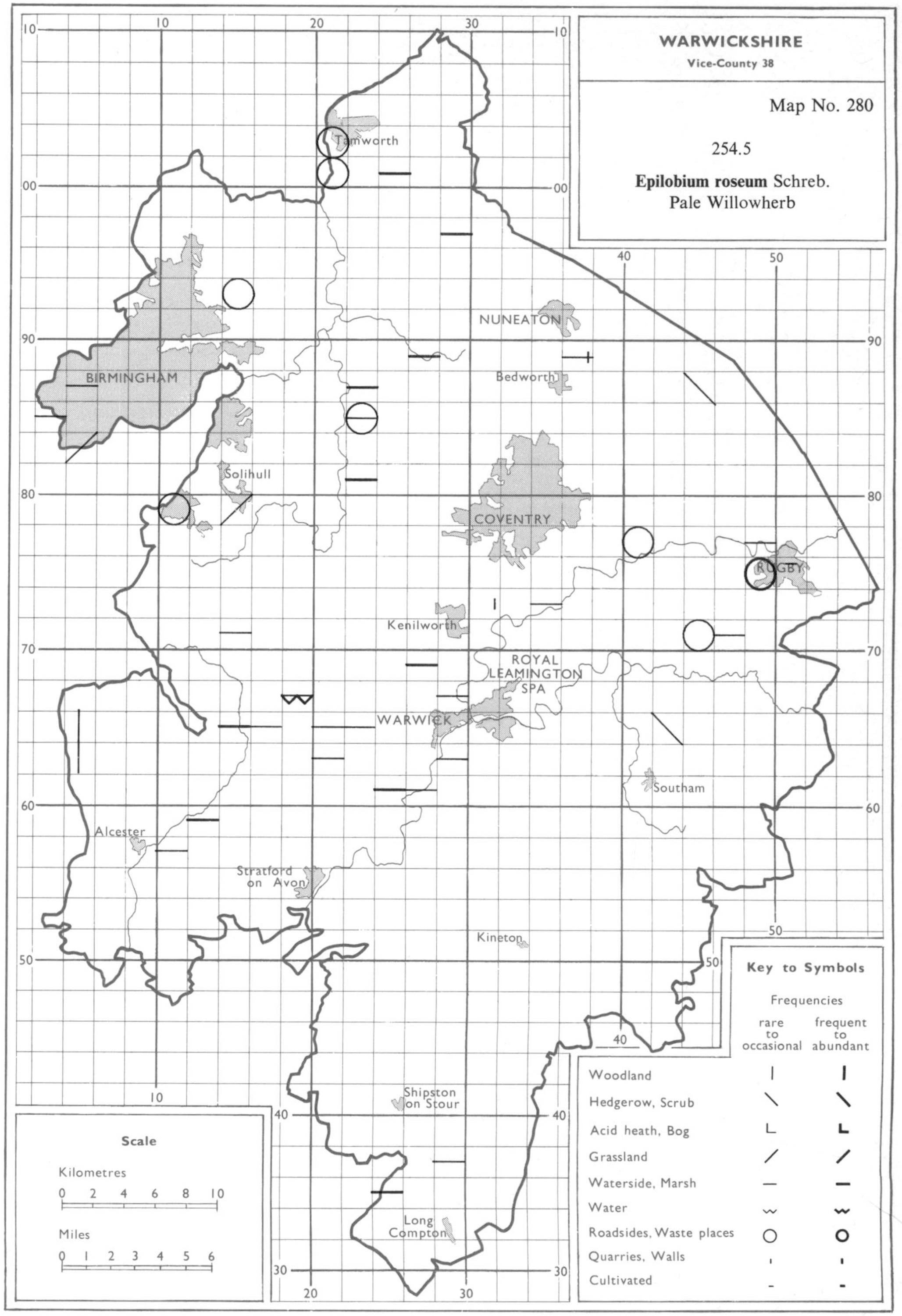
WARWICKSHIRE
Vice-County 38
Map No. 280
254.5
Epilobium roseum Schreb.
Pale Willowherb
Key to Symbols
Frequencies
rare to occasional
frequent to abundant
Woodland
Hedgerow, Scrub
Acid heath, Bog
Grassland
Waterside, Marsh
Water
Roadsides, Waste places
Quarries, Walls
Cultivated
Scale
Kilometres
0 2 4 6 8 10
Miles
0 1 2 3 4 5 6
Tamworth
Nuneaton
Bedworth
Birmingham
Solihull
Coventry
Rugby
Kenilworth
Royal Leamington Spa
Warwick
Southam
Alcester
Stratford on Avon
Kineton
Shipston on Stour
Long Compton

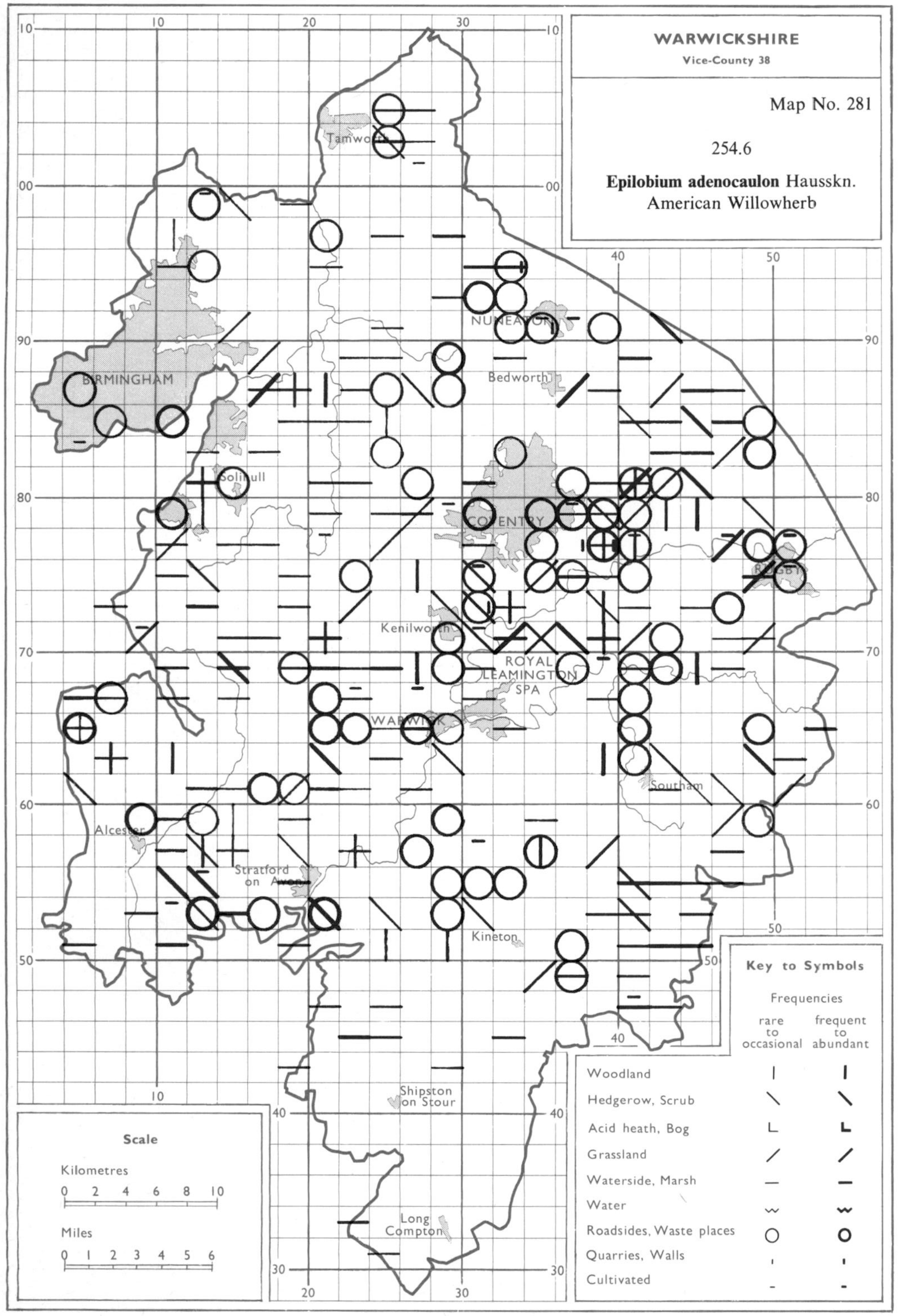
WARWICKSHIRE
Vice-County 38
Map No. 281
254.6
Epilobium adenocaulon Hausskn.
American Willowherb
Tamworth
NUNEATON
Bedworth
BIRMINGHAM
Solihull
COVENTRY
RUGBY
Kenilworth
ROYAL LEAMINGTON SPA
WARWICK
Southam
Alcester
Stratford on Avon
Kineton
Shipston on Stour
Long Compton
Key to Symbols
Frequencies
rare to occasional
frequent to abundant
Woodland
Hedgerow, Scrub
Acid heath, Bog
Grassland
Waterside, Marsh
Water
Roadsides, Waste places
Quarries, Walls
Cultivated
Scale
Kilometres
0 2 4 6 8 10
Miles
0 1 2 3 4 5 6

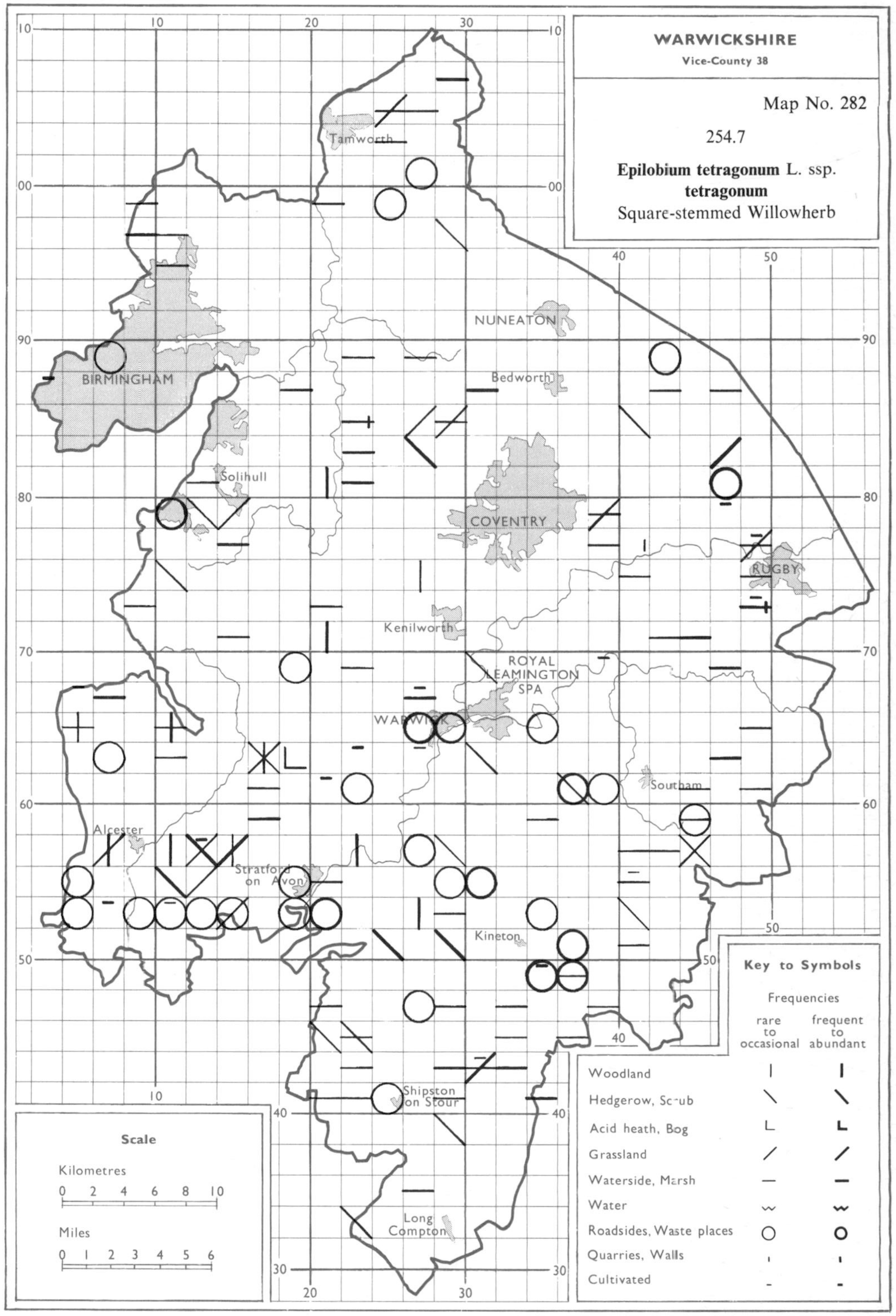
WARWICKSHIRE
Vice-County 38
Map No. 282
254.7
Epilobium tetragonum L. ssp. tetragonum
Square-stemmed Willowherb
Tamworth
NUNEATON
Bedworth
BIRMINGHAM
Solihull
COVENTRY
RUGBY
Kenilworth
ROYAL LEAMINGTON SPA
WARWICK
Southam
Alcester
Stratford on Avon
Kineton
Shipston on Stour
Long Compton
Key to Symbols
Frequencies
rare to occasional
frequent to abundant
Woodland
Hedgerow, Scrub
Acid heath, Bog
Grassland
Waterside, Marsh
Water
Roadsides, Waste places
Quarries, Walls
Cultivated
Scale
Kilometres
0 2 4 6 8 10
Miles
0 1 2 3 4 5 6

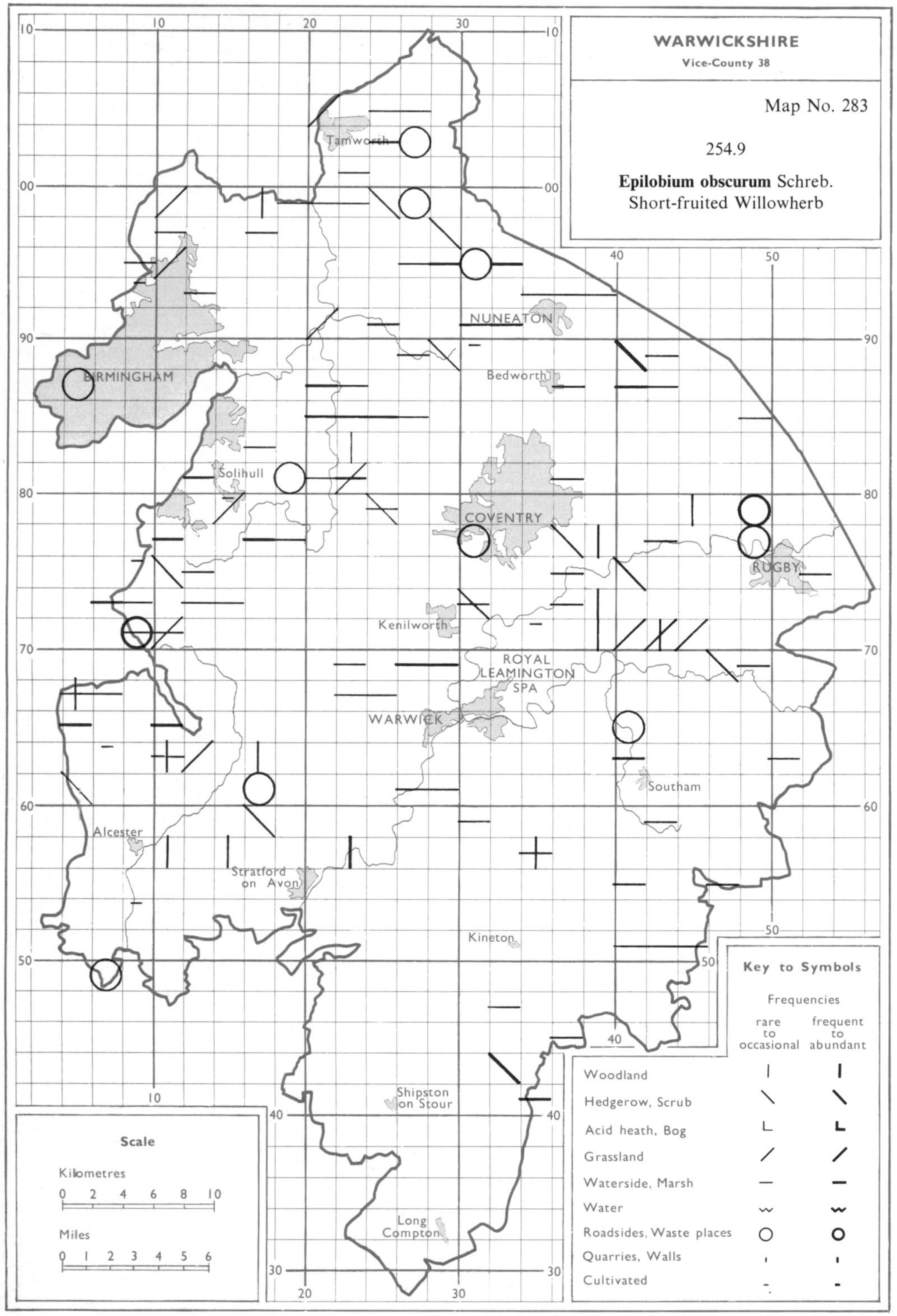
WARWICKSHIRE
Vice-County 38
Map No. 283
254.9
Epilobium obscurum Schreb.
Short-fruited Willowherb
Tamworth
NUNEATON
Bedworth
BIRMINGHAM
Solihull
COVENTRY
RUGBY
Kenilworth
ROYAL LEAMINGTON SPA
WARWICK
Southam
Alcester
Stratford on Avon
Kineton
Shipston on Stour
Long Compton
Key to Symbols
Frequencies
rare to occasional
frequent to abundant
Woodland
Hedgerow, Scrub
Acid heath, Bog
Grassland
Waterside, Marsh
Water
Roadsides, Waste places
Quarries, Walls
Cultivated
Scale
Kilometres
0 2 4 6 8 10
Miles
0 1 2 3 4 5 6

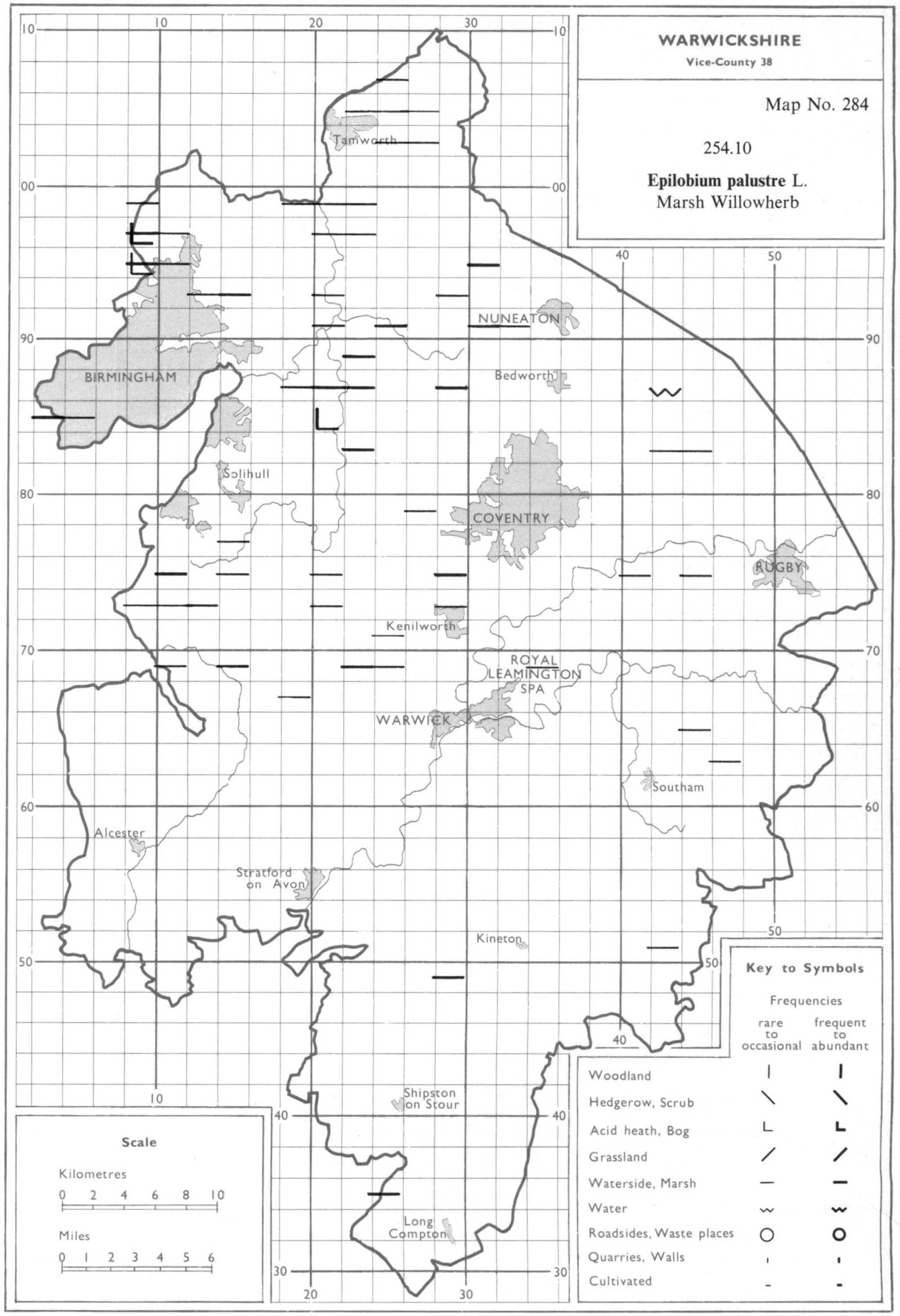
WARWICKSHIRE
Vice-County 38
Map No. 284
254.10
Epilobium palustre L.
Marsh Willowherb
Tamworth
NUNEATON
Bedworth
BIRMINGHAM
Solihull
COVENTRY
RUGBY
Kenilworth
ROYAL LEAMINGTON SPA
WARWICK
Southam
Alcester
Stratford on Avon
Kineton
Shipston on Stour
Long Compton
Key to Symbols
Frequencies
rare to occasional
frequent to abundant
Woodland
Hedgerow, Scrub
Acid heath, Bog
Grassland
Waterside, Marsh
Water
Roadsides, Waste places
Quarries, Walls
Cultivated
Scale
Kilometres
0 2 4 6 8 10
Miles
0 1 2 3 4 5 6

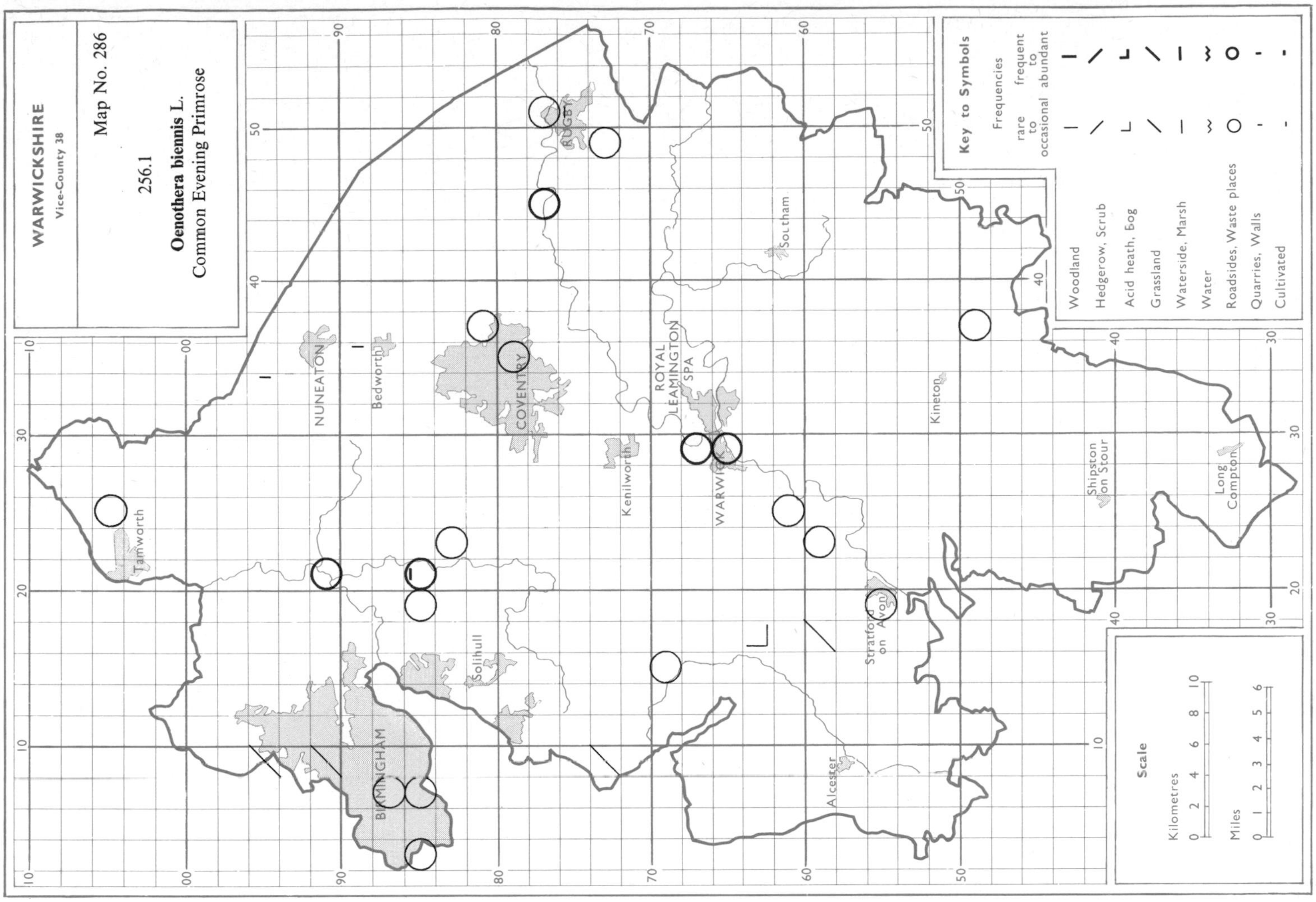
WARWICKSHIRE
Vice-County 38
Map No. 286
256.1
Oenothera biennis L.
Common Evening Primrose
Key to Symbols
Frequencies
rare to occasional
frequent to abundant
Woodland
Hedgerow, Scrub
Acid heath, Bog
Grassland
Waterside, Marsh
Water
Roadsides, Waste places
Quarries, Walls
Cultivated
Scale
Kilometres
Miles
Tamworth
NUNEATON
Bedworth
COVENTRY
RUGBY
Southam
ROYAL LEAMINGTON SPA
Kenilworth
WARWICK
Kineton
Shipston on Stour
Long Compton
Stratford on Avon
Solihull
BIRMINGHAM
Alcester

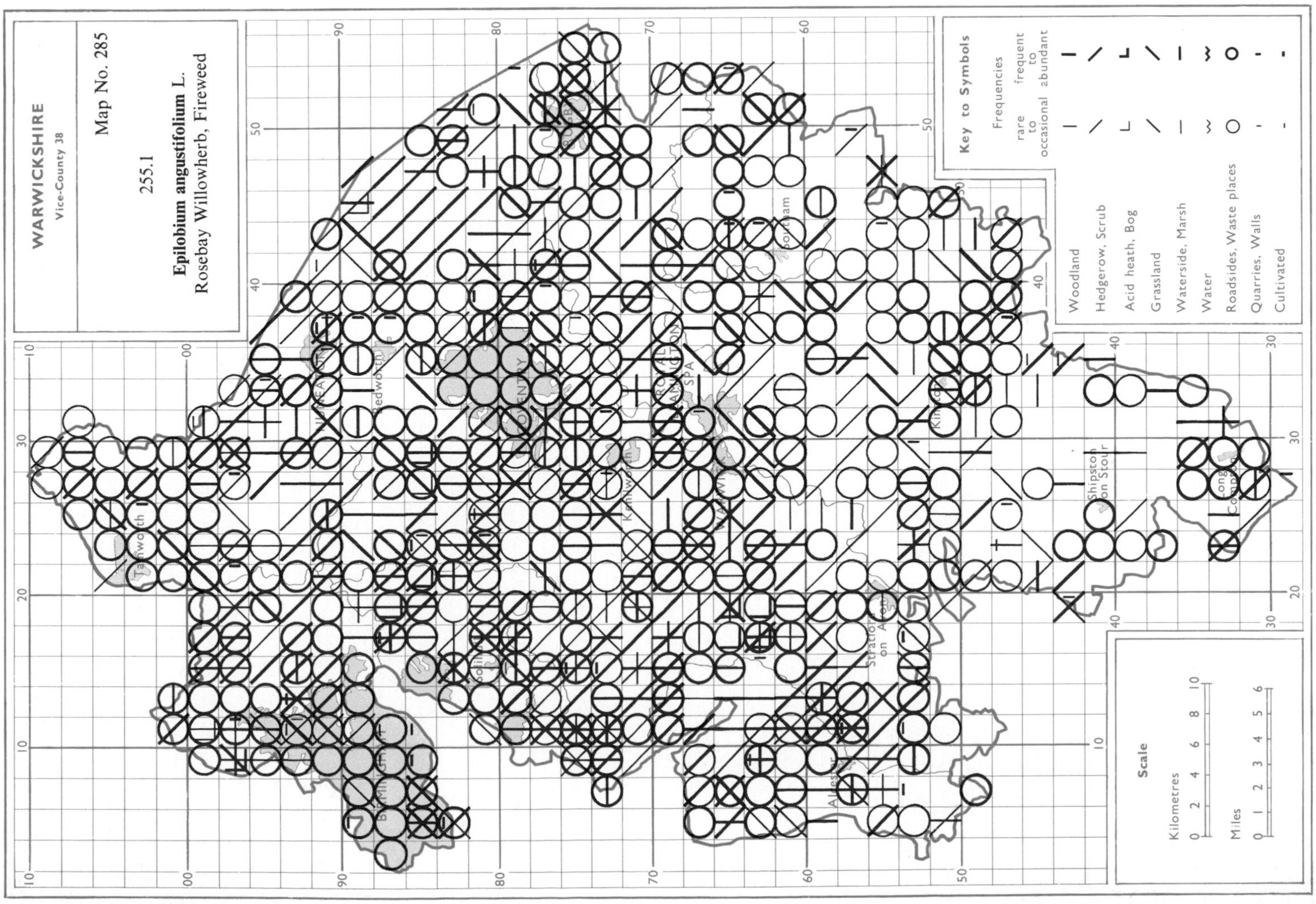
WARWICKSHIRE
Vice-County 38
Map No. 285
255.1
Epilobium angustifolium L.
Rosebay Willowherb, Fireweed
Key to Symbols
Frequencies
rare to occasional
frequent to abundant
Woodland
Hedgerow, Scrub
Acid heath, Bog
Grassland
Waterside, Marsh
Water
Roadsides, Waste places
Quarries, Walls
Cultivated
Scale
Kilometres
Miles

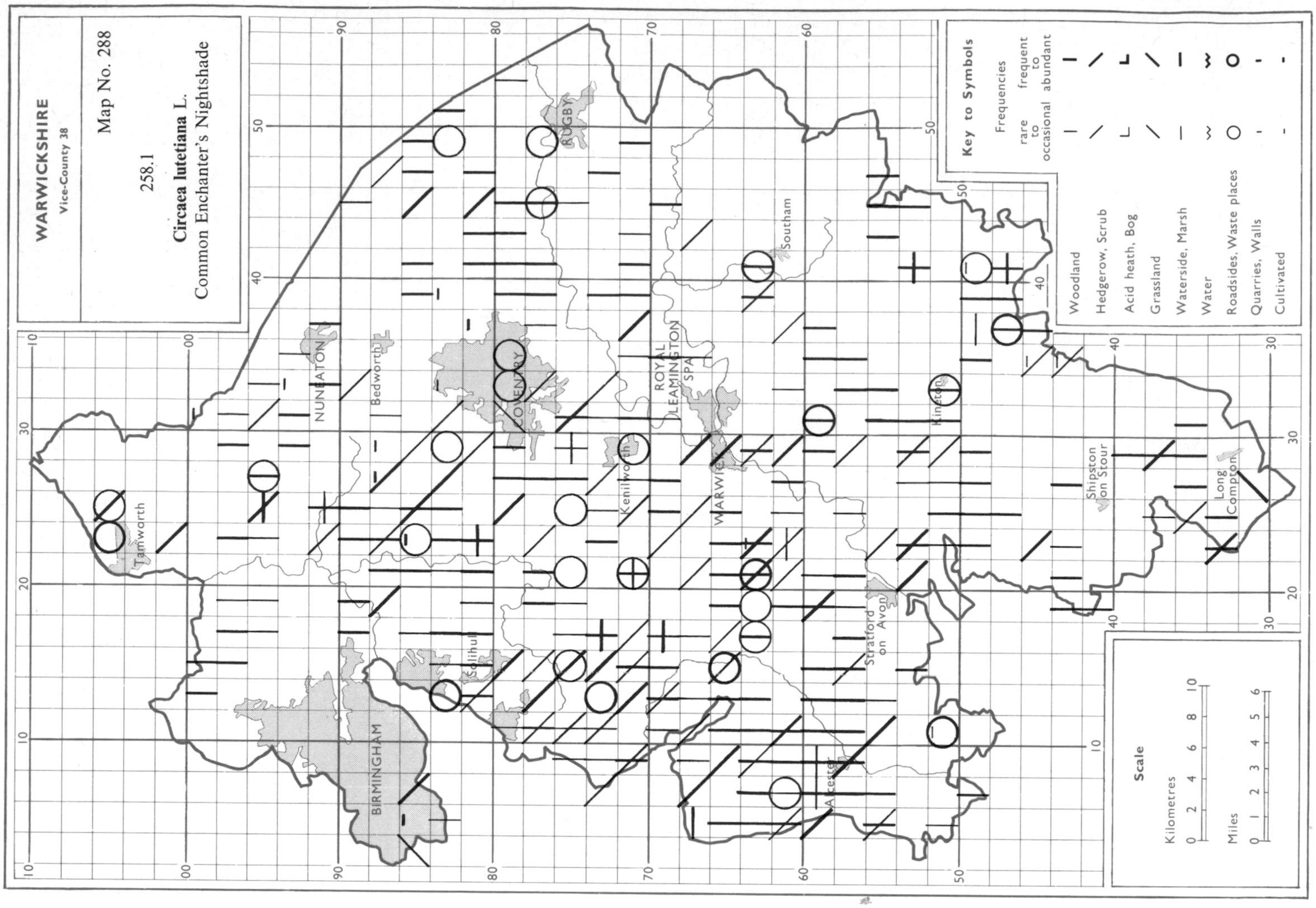
WARWICKSHIRE
Vice-County 38
Map No. 288
258.1
Circaea lutetiana L.
Common Enchanter's Nightshade
Key to Symbols
Frequencies
rare to occasional
frequent to abundant
Woodland
Hedgerow, Scrub
Acid heath, Bog
Grassland
Waterside, Marsh
Water
Roadsides, Waste places
Quarries, Walls
Cultivated
Scale
Kilometres
Miles
TAMWORTH
NUNEATON
Bedworth
COVENTRY
RUGBY
Southam
ROYAL LEAMINGTON SPA
Kenilworth
WARWICK
Kineton
Shipston on Stour
Long Compton
Solihull
BIRMINGHAM
Stratford on Avon
Alcester

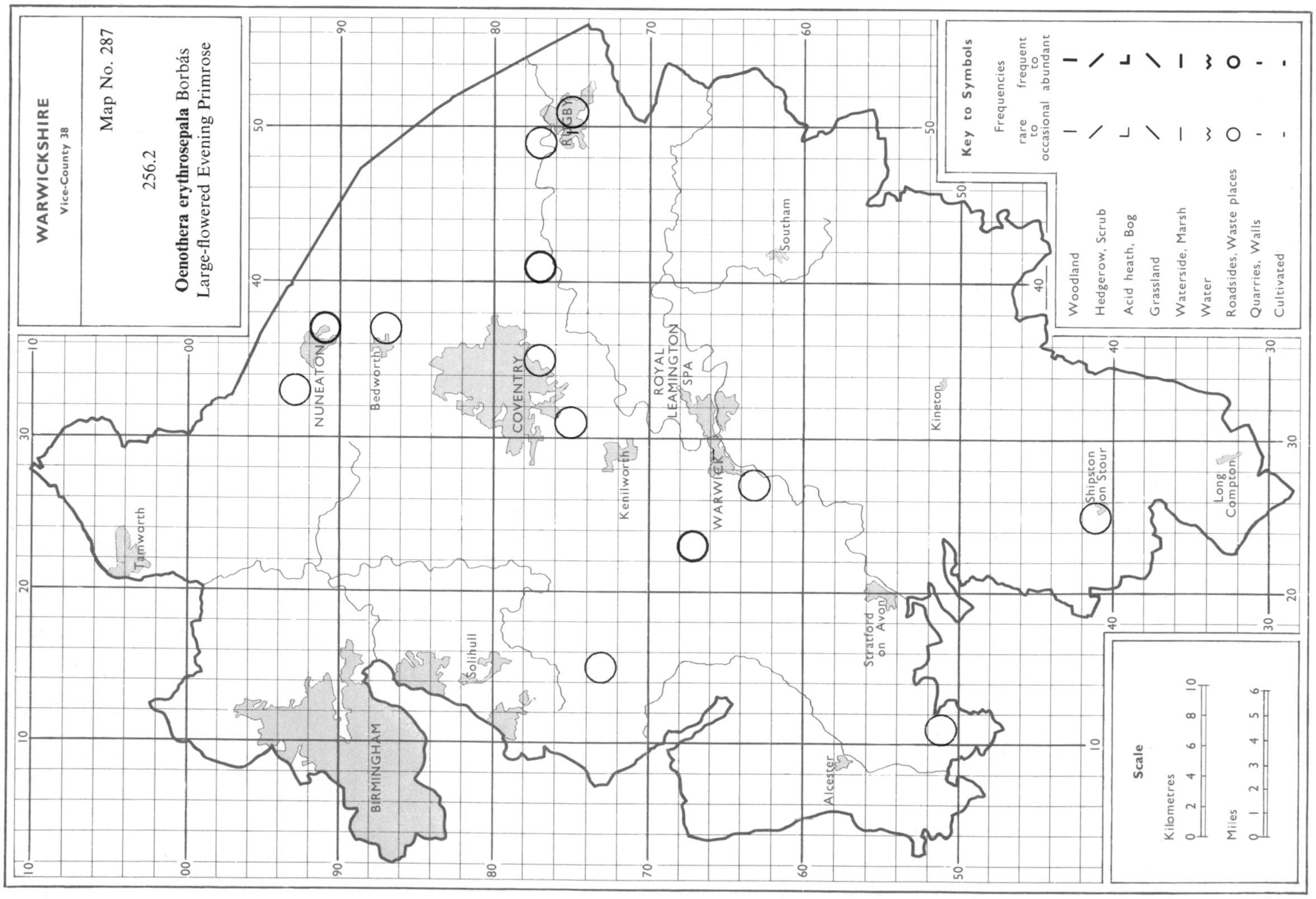
WARWICKSHIRE
Vice-County 38
Map No. 287
256.2
Oenothera erythrosepala Borbás
Large-flowered Evening Primrose
Key to Symbols
Frequencies
rare to occasional
frequent to abundant
Woodland
Hedgerow, Scrub
Acid heath, Bog
Grassland
Waterside, Marsh
Water
Roadsides, Waste places
Quarries, Walls
Cultivated
Scale
Kilometres
Miles
Tamworth
NUNEATON
Bedworth
COVENTRY
RUGBY
Southam
ROYAL LEAMINGTON SPA
Kenilworth
WARWICK
Kineton
Shipston on Stour
Long Compton
Solihull
BIRMINGHAM
Stratford on Avon
Alcester

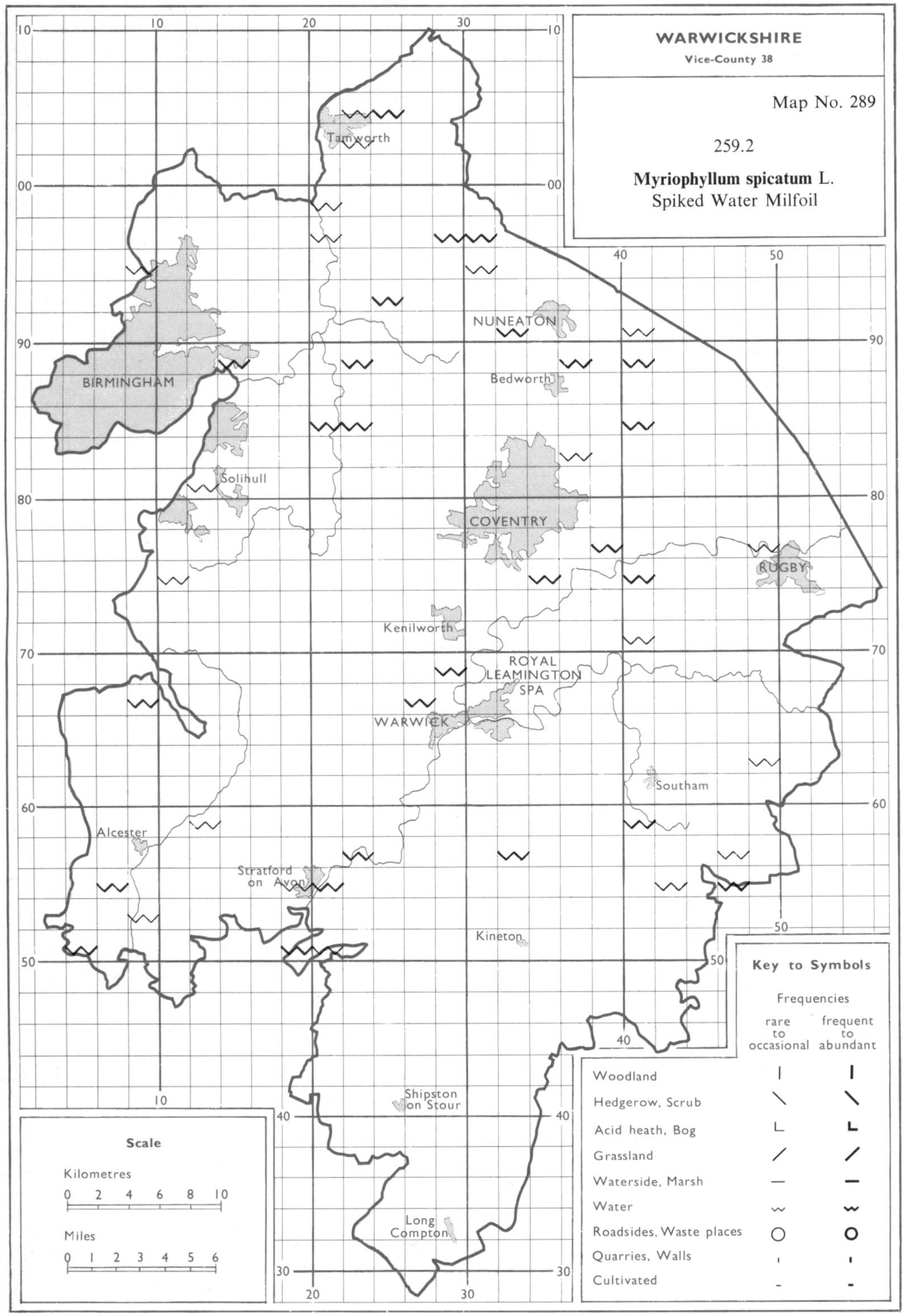
WARWICKSHIRE
Vice-County 38
Map No. 289
259.2
Myriophyllum spicatum L.
Spiked Water Milfoil
Tamworth
NUNEATON
Bedworth
BIRMINGHAM
Solihull
COVENTRY
RUGBY
Kenilworth
ROYAL LEAMINGTON SPA
WARWICK
Southam
Alcester
Stratford on Avon
Kineton
Shipston on Stour
Long Compton
Key to Symbols
Frequencies
rare to occasional
frequent to abundant
Woodland
Hedgerow, Scrub
Acid heath, Bog
Grassland
Waterside, Marsh
Water
Roadsides, Waste places
Quarries, Walls
Cultivated
Scale
Kilometres
0 2 4 6 8 10
Miles
0 1 2 3 4 5 6

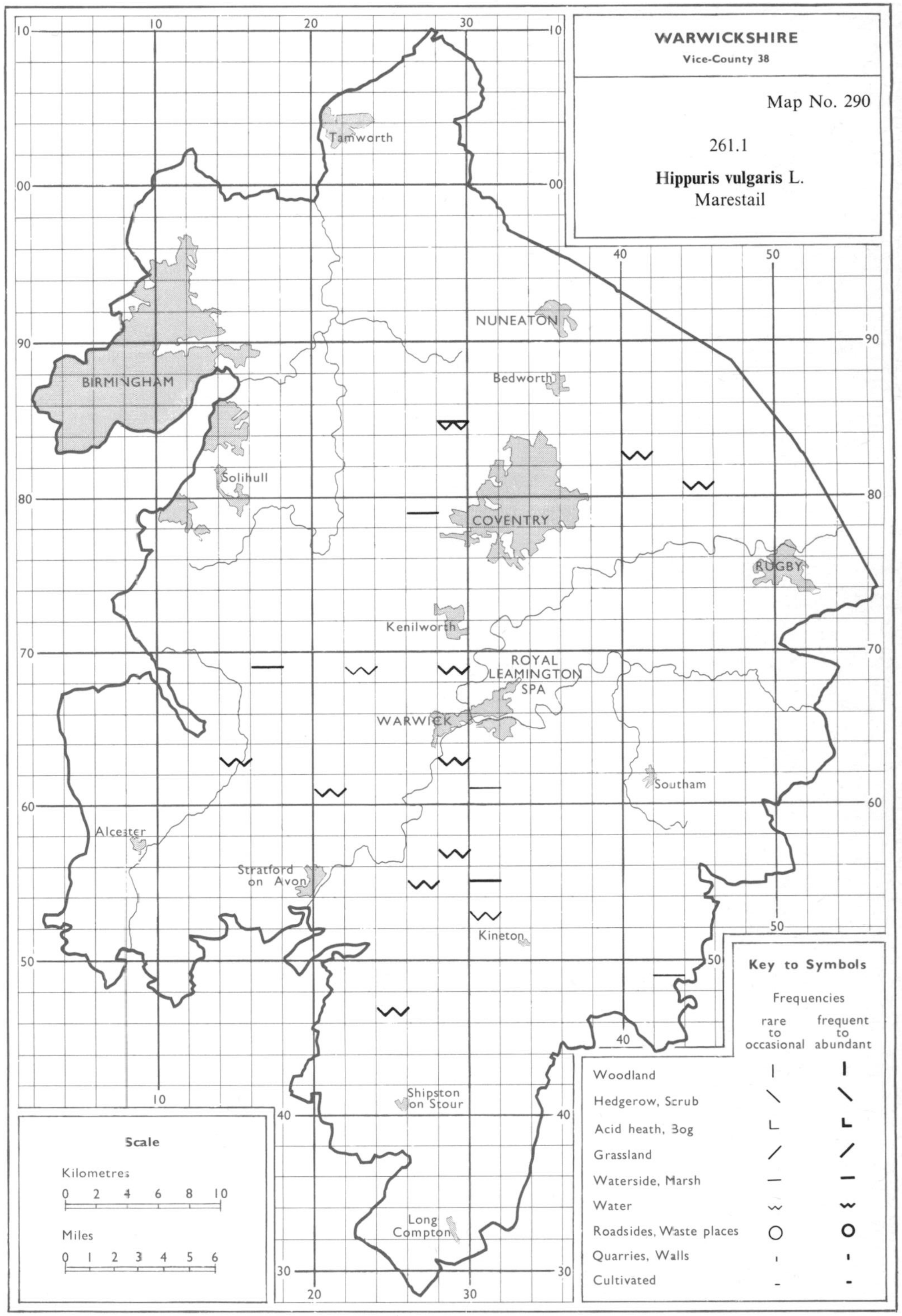
WARWICKSHIRE
Vice-County 38
Map No. 290
261.1
Hippuris vulgaris L.
Marestail
Tamworth
NUNEATON
Bedworth
BIRMINGHAM
Solihull
COVENTRY
RUGBY
Kenilworth
ROYAL LEAMINGTON SPA
WARWICK
Southam
Alcester
Stratford on Avon
Kineton
Shipston on Stour
Long Compton
Key to Symbols
Frequencies
rare to occasional
frequent to abundant
Woodland
Hedgerow, Scrub
Acid heath, Bog
Grassland
Waterside, Marsh
Water
Roadsides, Waste places
Quarries, Walls
Cultivated
Scale
Kilometres
0 2 4 6 8 10
Miles
0 1 2 3 4 5 6

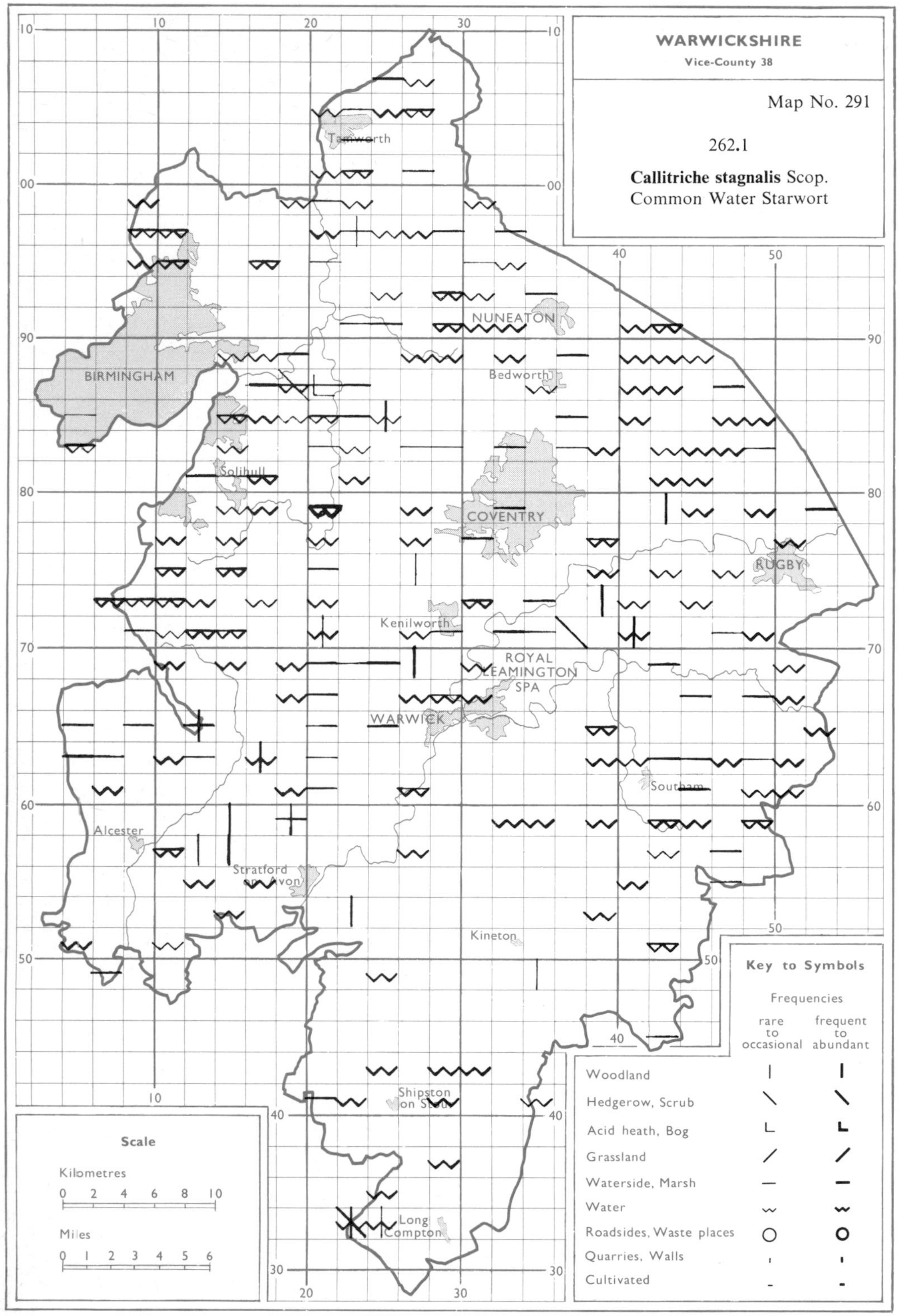

WARWICKSHIRE
Vice-County 38
Map No. 291
262.1
Callitriche stagnalis Scop.
Common Water Starwort
Key to Symbols
Frequencies
rare to occasional
frequent to abundant
Woodland
Hedgerow, Scrub
Acid heath, Bog
Grassland
Waterside, Marsh
Water
Roadsides, Waste places
Quarries, Walls
Cultivated
Scale
Kilometres 0 2 4 6 8 10
Miles 0 1 2 3 4 5 6
Tamworth
NUNEATON
Bedworth
BIRMINGHAM
Solihull
COVENTRY
RUGBY
Kenilworth
ROYAL LEAMINGTON SPA
WARWICK
Southam
Alcester
Stratford on Avon
Kineton
Shipston on Stour
Long Compton

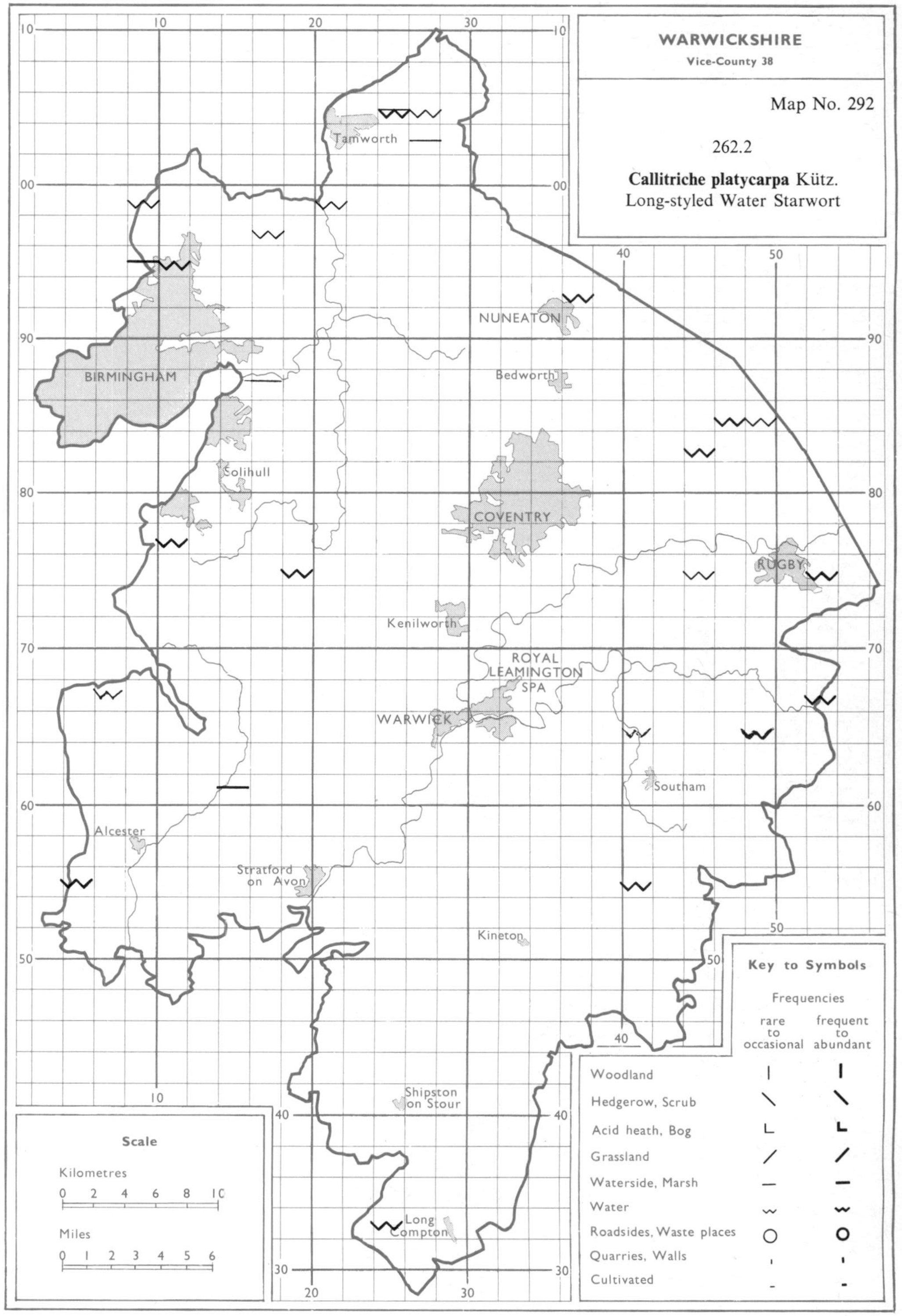

WARWICKSHIRE
Vice-County 38
Map No. 292
262.2
Callitriche platycarpa Kütz.
Long-styled Water Starwort
Key to Symbols
Frequencies
rare to occasional
frequent to abundant
Woodland
Hedgerow, Scrub
Acid heath, Bog
Grassland
Waterside, Marsh
Water
Roadsides, Waste places
Quarries, Walls
Cultivated
Scale
Kilometres 0 2 4 6 8 10
Miles 0 1 2 3 4 5 6
Tamworth
NUNEATON
Bedworth
BIRMINGHAM
Solihull
COVENTRY
RUGBY
Kenilworth
ROYAL LEAMINGTON SPA
WARWICK
Southam
Alcester
Stratford on Avon
Kineton
Shipston on Stour
Long Compton

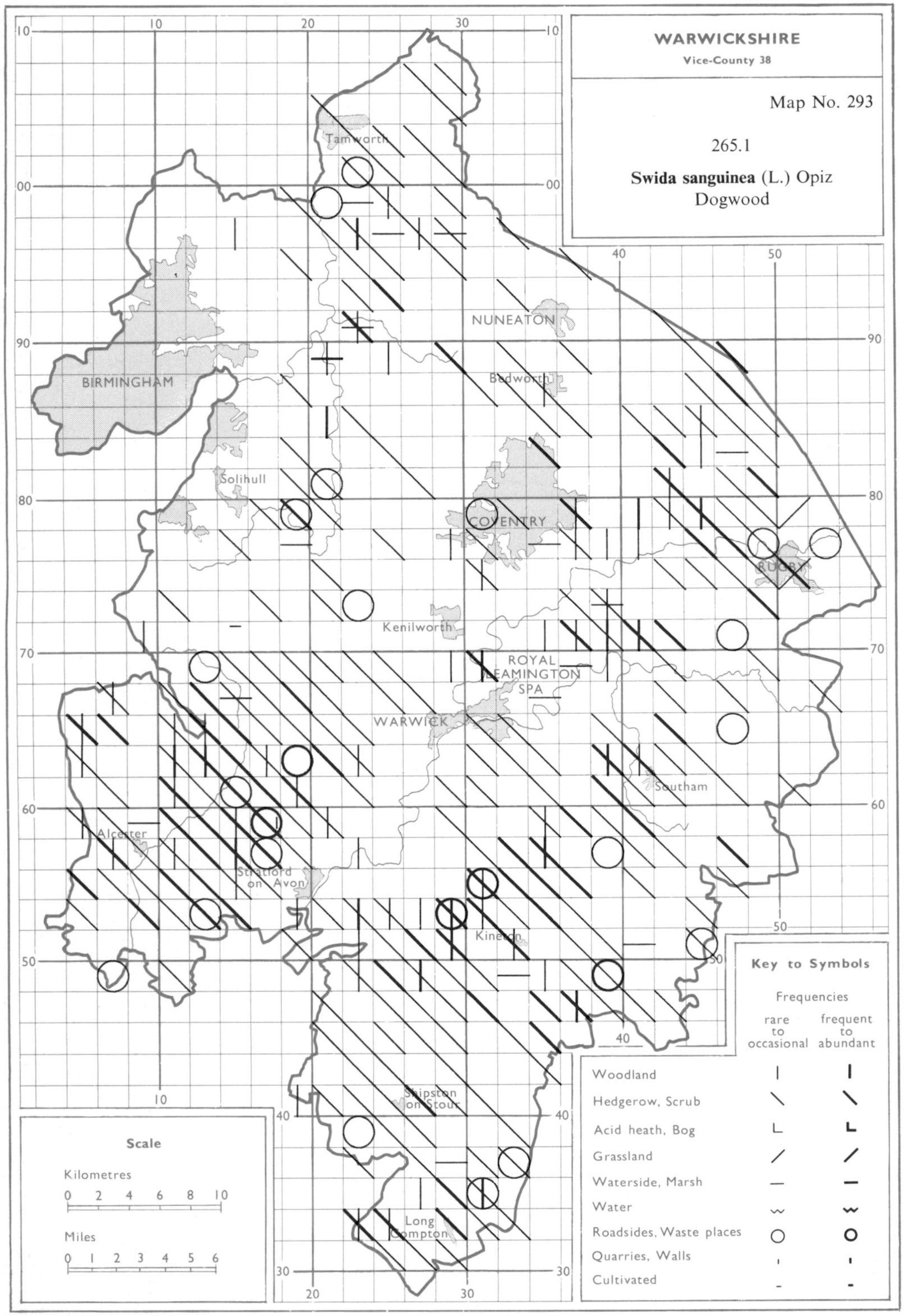

Map No. 293

265.1

Swida sanguinea (L.) Opiz

Dogwood

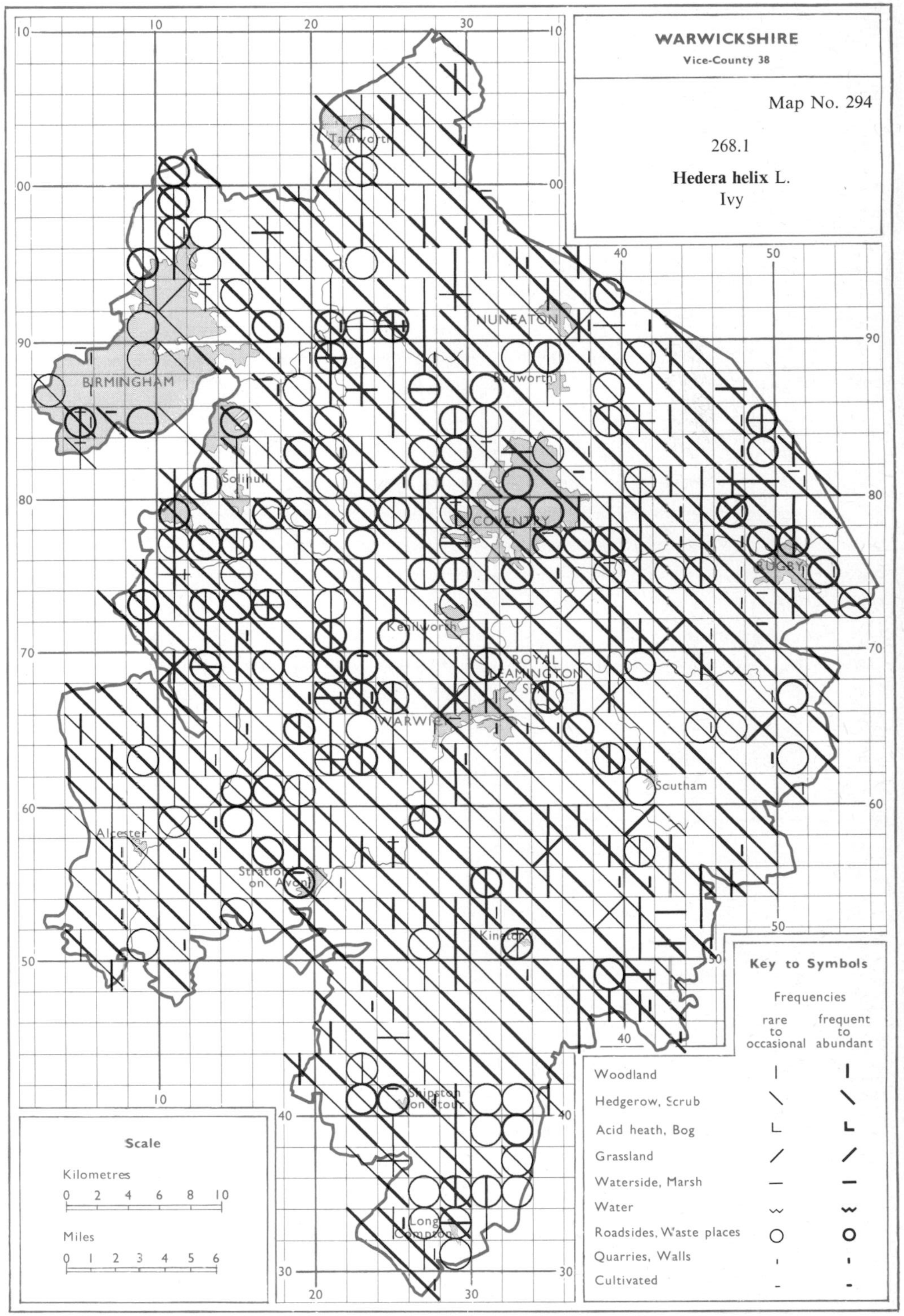

Map No. 294

268.1

Hedera helix L.

Ivy

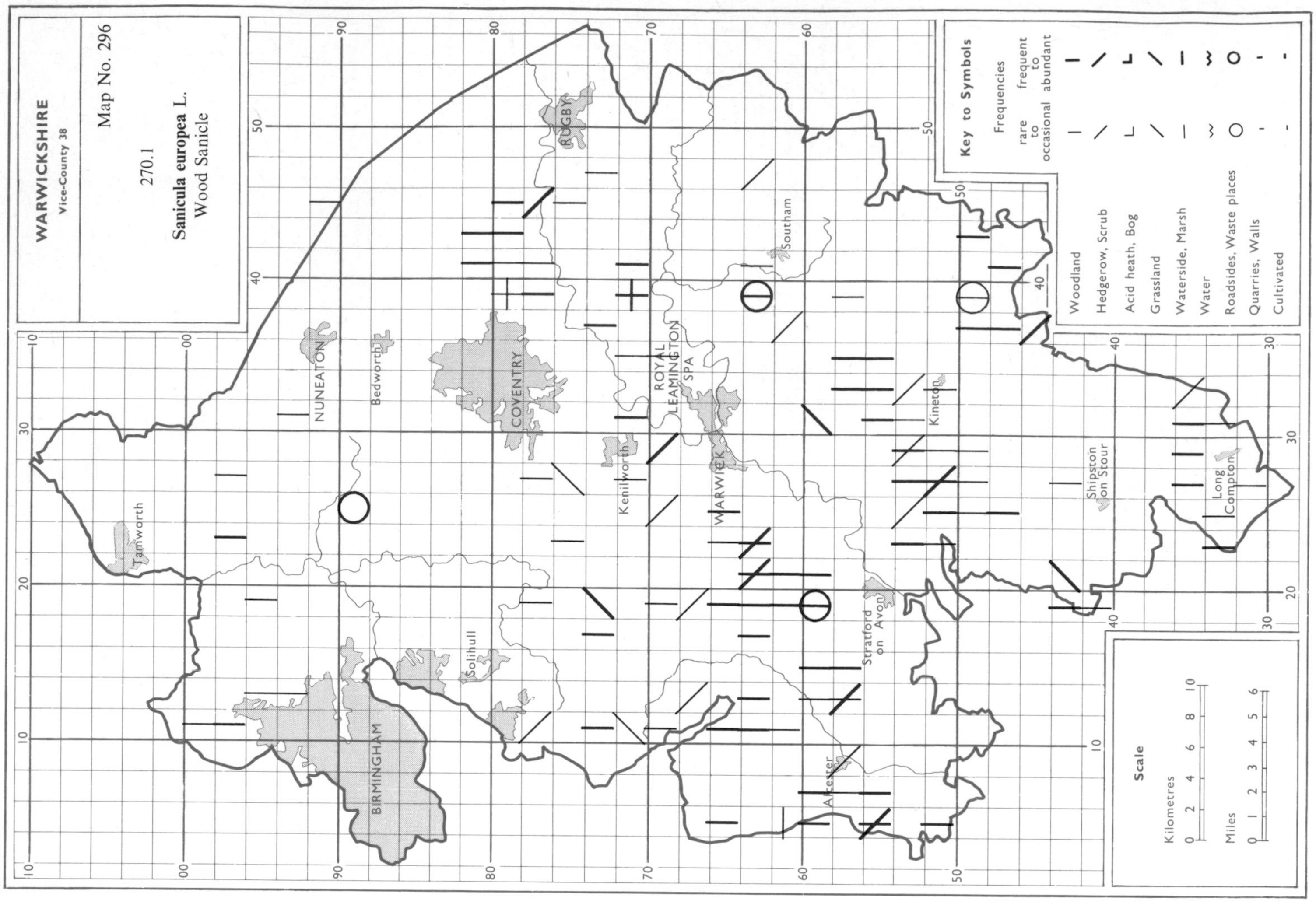
WARWICKSHIRE
Vice-County 38
Map No. 296
270.1
Sanicula europea L.
Wood Sanicle
Key to Symbols
Frequencies
rare to occasional
frequent to abundant
Woodland
Hedgerow, Scrub
Acid heath, Bog
Grassland
Waterside, Marsh
Water
Roadsides, Waste places
Quarries, Walls
Cultivated
Scale
Kilometres
0 2 4 6 8 10
Miles
0 1 2 3 4 5 6
NUNEATON
Bedworth
COVENTRY
RUGBY
Tamworth
Kenilworth
ROYAL LEAMINGTON SPA
WARWICK
Southam
Kineton
Shipston on Stour
Long Compton
Stratford on Avon
Solihull
BIRMINGHAM
Alcester

WARWICKSHIRE
Vice-County 38
Map No. 295
269.1
Hydrocotyle vulgaris L.
Marsh Pennywort
Key to Symbols
Frequencies
rare to occasional
frequent to abundant
Woodland
Hedgerow, Scrub
Acid heath, Bog
Grassland
Waterside, Marsh
Water
Roadsides, Waste places
Quarries, Walls
Cultivated
Scale
Kilometres
0 2 4 6 8 10
Miles
0 1 2 3 4 5 6
NUNEATON
Bedworth
COVENTRY
RUGBY
Tamworth
Kenilworth
ROYAL LEAMINGTON SPA
WARWICK
Southam
Kineton
Shipston on Stour
Long Compton
Stratford on Avon
Solihull
BIRMINGHAM
Alcester

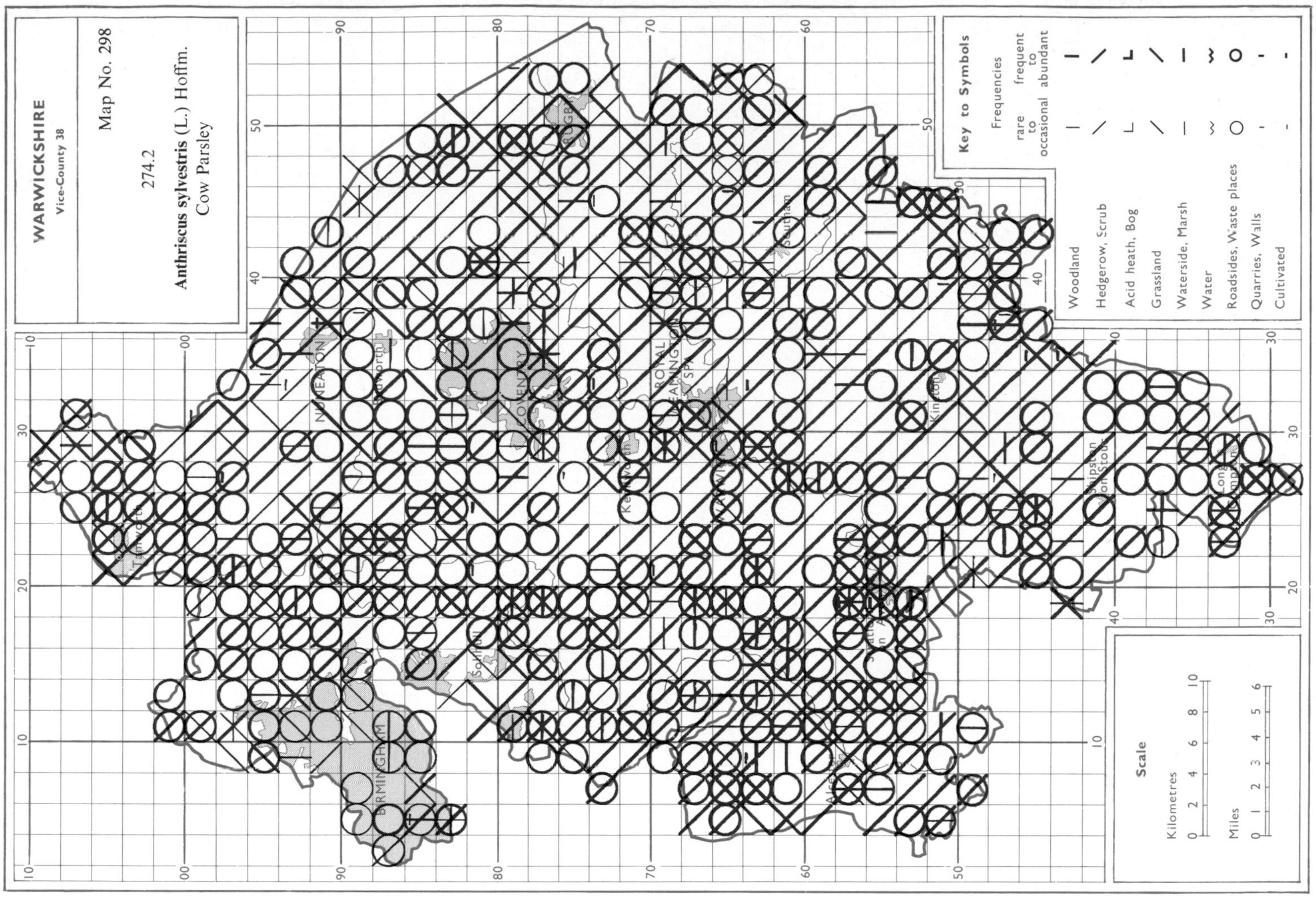
WARWICKSHIRE
Vice-County 38
Map No. 298
274.2
Anthriscus sylvestris (L.) Hoffm.
Cow Parsley
Key to Symbols
Frequencies
rare to occasional
frequent to abundant
Woodland
Hedgerow, Scrub
Acid heath, Bog
Grassland
Waterside, Marsh
Water
Roadsides, Waste places
Quarries, Walls
Cultivated
Scale
Kilometres 0 2 4 6 8 10
Miles 0 1 2 3 4 5 6
TAMWORTH
NUNEATON
Bedworth
COVENTRY
RUGBY
BIRMINGHAM
Solihull
Kenilworth
ROYAL LEAMINGTON SPA
WARWICK
Southam
Kineton
Shipston on Stour
Long Compton
Alcester
Stratford upon Avon

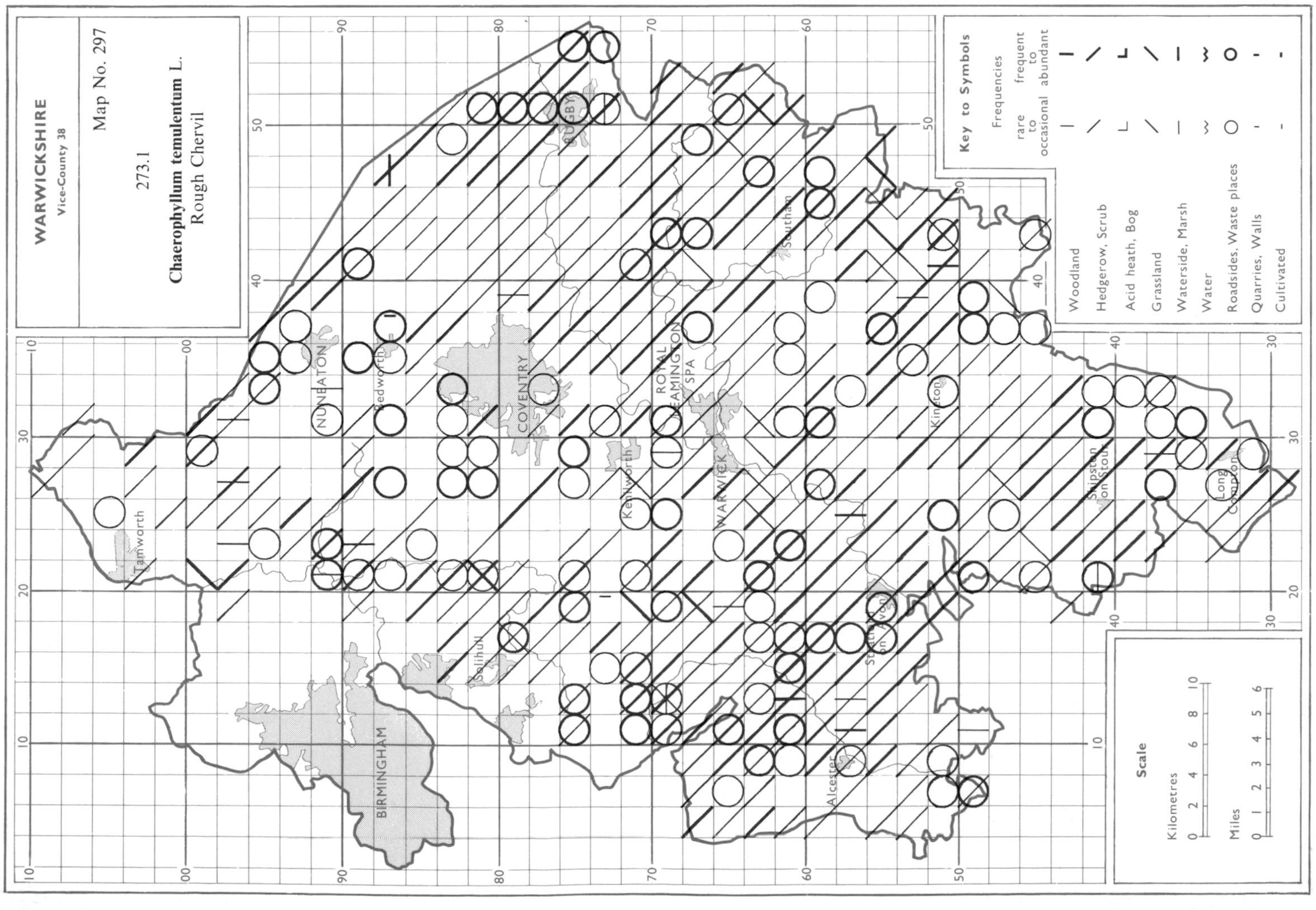
WARWICKSHIRE
Vice-County 38
Map No. 297
273.1
Chaerophyllum temulentum L.
Rough Chervil
Key to Symbols
Frequencies
rare to occasional
frequent to abundant
Woodland
Hedgerow, Scrub
Acid heath, Bog
Grassland
Waterside, Marsh
Water
Roadsides, Waste places
Quarries, Walls
Cultivated
Scale
Kilometres 0 2 4 6 8 10
Miles 0 1 2 3 4 5 6
Tamworth
NUNEATON
Bedworth
COVENTRY
RUGBY
BIRMINGHAM
Solihull
Kenilworth
ROYAL LEAMINGTON SPA
WARWICK
Southam
Kineton
Shipston on Stour
Long Compton
Alcester
Stratford upon Avon

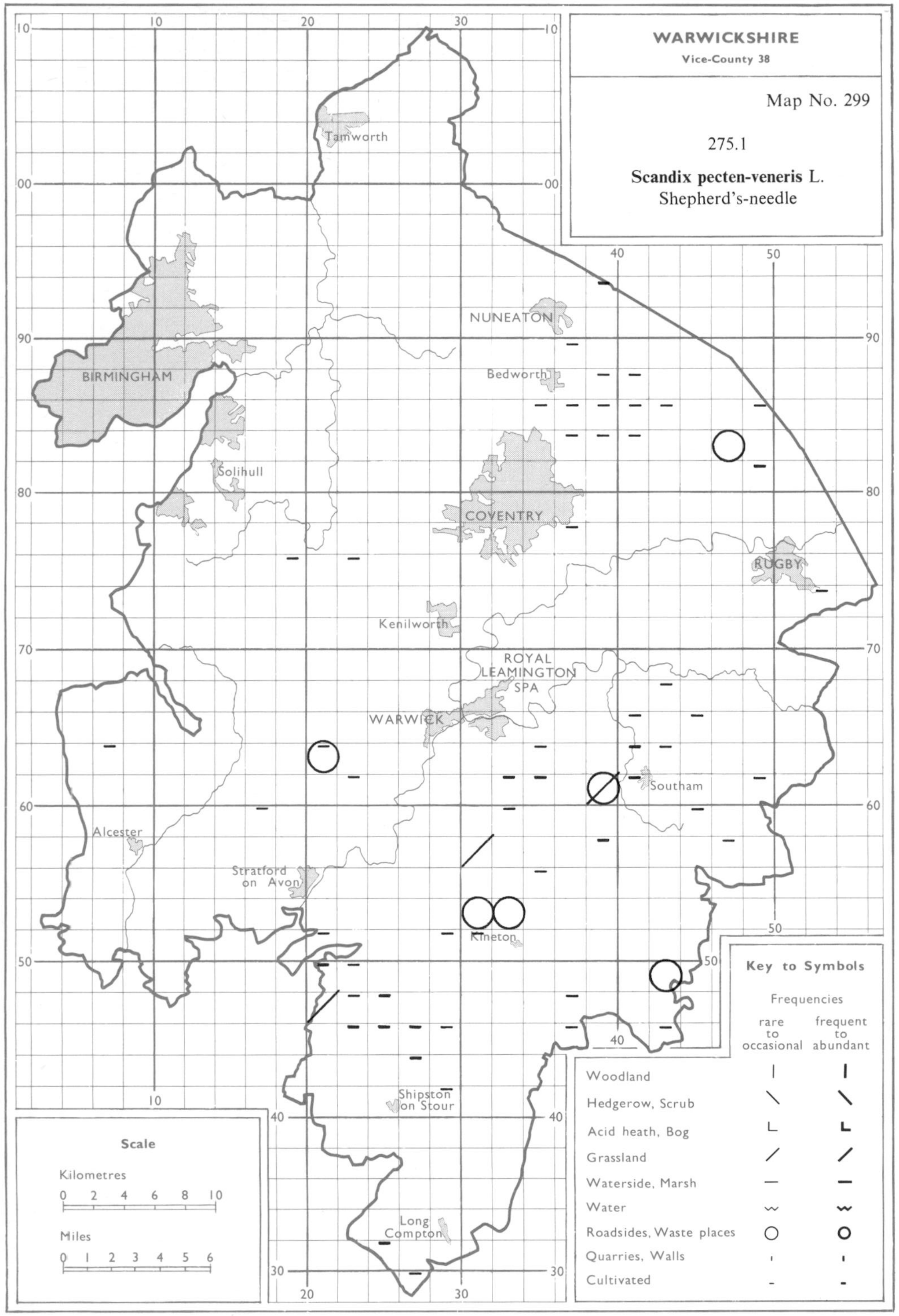
WARWICKSHIRE
Vice-County 38
Map No. 299
275.1
Scandix pecten-veneris L.
Shepherd's-needle
Tamworth
NUNEATON
Bedworth
BIRMINGHAM
Solihull
COVENTRY
RUGBY
Kenilworth
ROYAL LEAMINGTON SPA
WARWICK
Southam
Alcester
Stratford on Avon
Kineton
Shipston on Stour
Long Compton
Key to Symbols
Frequencies
rare to occasional
frequent to abundant
Woodland
Hedgerow, Scrub
Acid heath, Bog
Grassland
Waterside, Marsh
Water
Roadsides, Waste places
Quarries, Walls
Cultivated
Scale
Kilometres
0 2 4 6 8 10
Miles
0 1 2 3 4 5 6

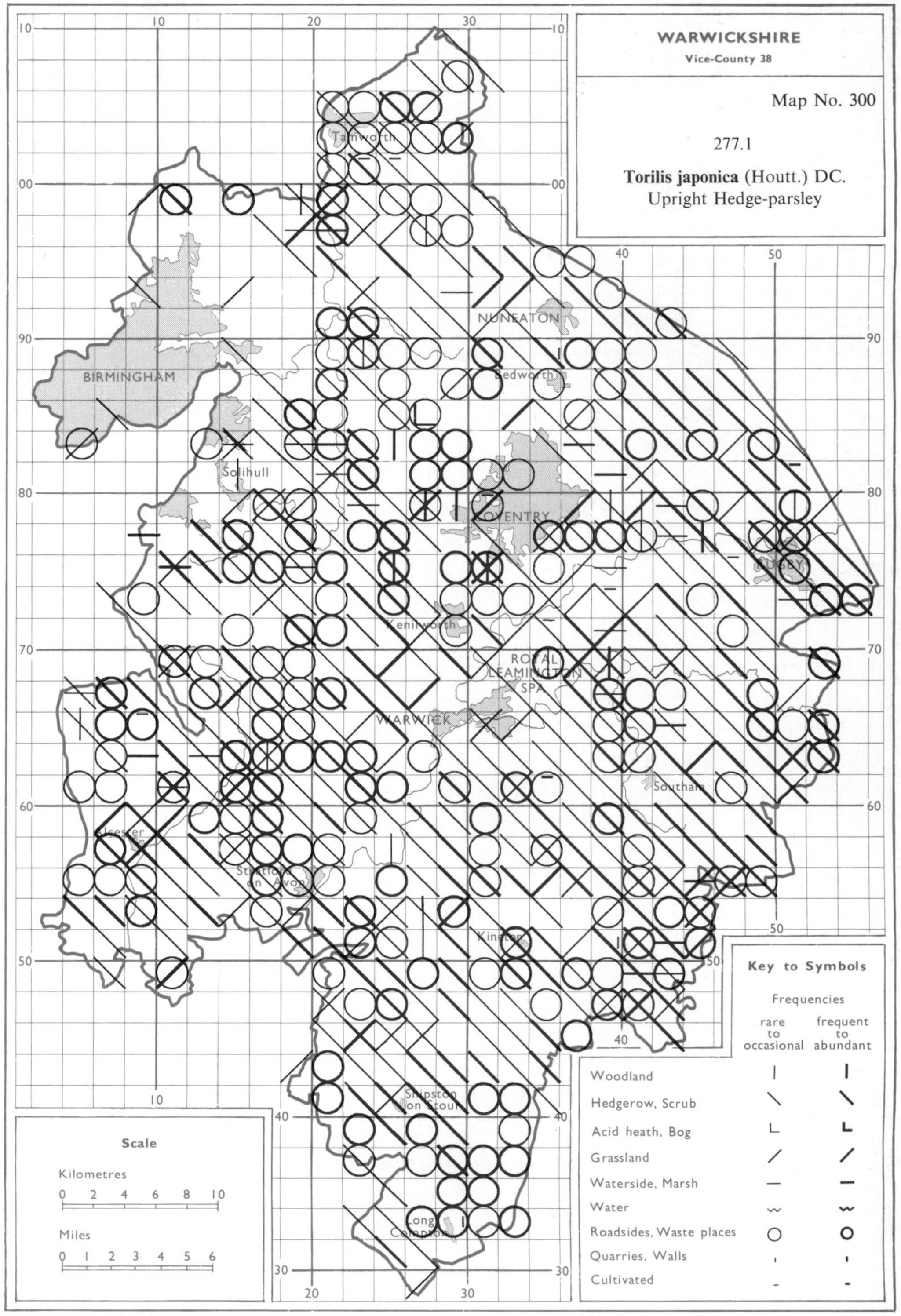
WARWICKSHIRE
Vice-County 38
Map No. 300
277.1
Torilis japonica (Houtt.) DC.
Upright Hedge-parsley
Tamworth
NUNEATON
Bedworth
BIRMINGHAM
Solihull
COVENTRY
RUGBY
Kenilworth
ROYAL LEAMINGTON SPA
WARWICK
Southam
Alcester
Stratford on Avon
Kineton
Shipston on Stour
Long Compton
Key to Symbols
Frequencies
rare to occasional
frequent to abundant
Woodland
Hedgerow, Scrub
Acid heath, Bog
Grassland
Waterside, Marsh
Water
Roadsides, Waste places
Quarries, Walls
Cultivated
Scale
Kilometres
0 2 4 6 8 10
Miles
0 1 2 3 4 5 6

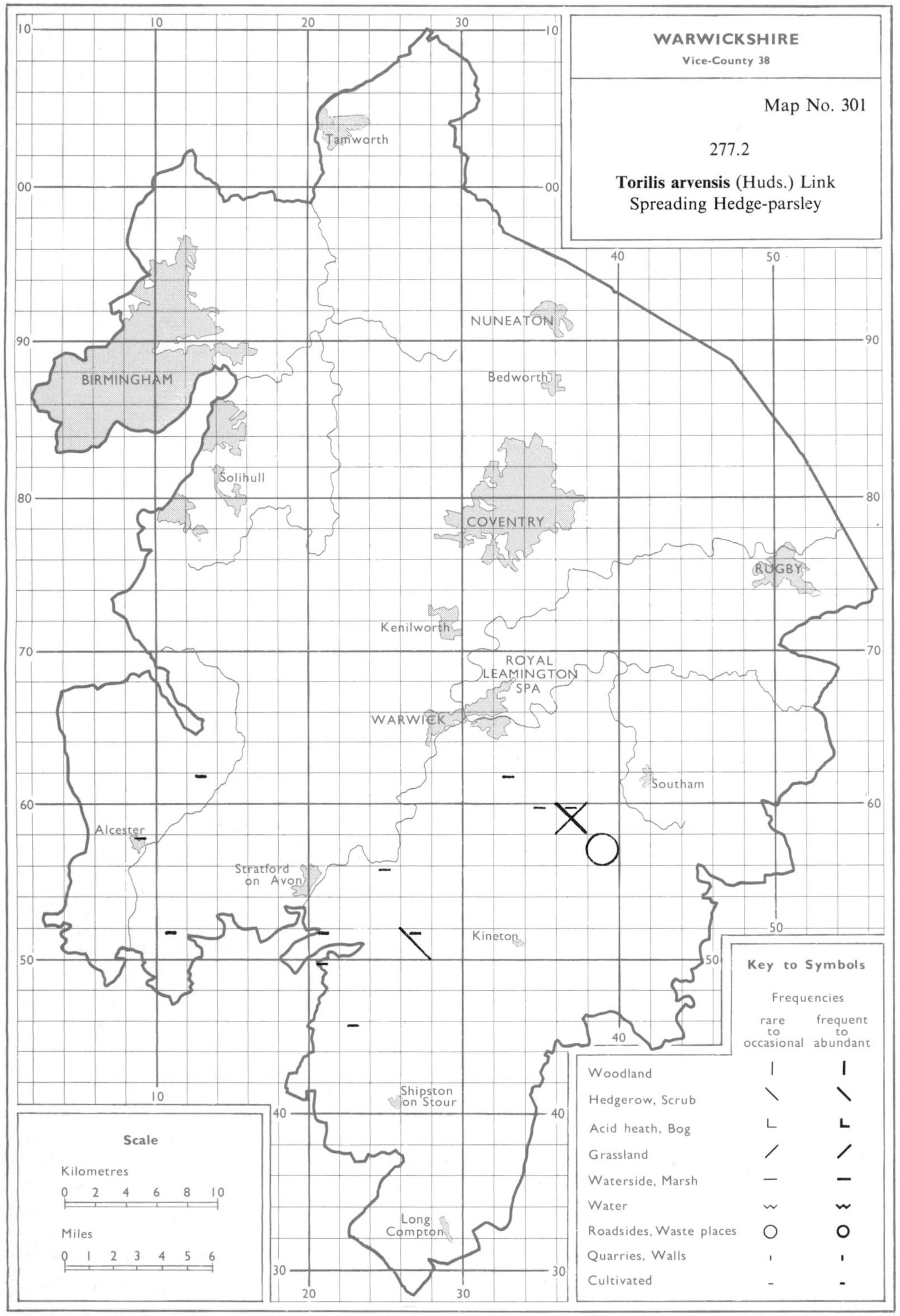

WARWICKSHIRE
Vice-County 38
Map No. 301
277.2
Torilis arvensis (Huds.) Link
Spreading Hedge-parsley
Tamworth
NUNEATON
Bedworth
BIRMINGHAM
Solihull
COVENTRY
RUGBY
Kenilworth
ROYAL LEAMINGTON SPA
WARWICK
Southam
Alcester
Stratford on Avon
Kineton
Shipston on Stour
Long Compton
Scale
Kilometres
0 2 4 6 8 10
Miles
0 1 2 3 4 5 6
Key to Symbols
Frequencies
rare to occasional
frequent to abundant
Woodland
Hedgerow, Scrub
Acid heath, Bog
Grassland
Waterside, Marsh
Water
Roadsides, Waste places
Quarries, Walls
Cultivated

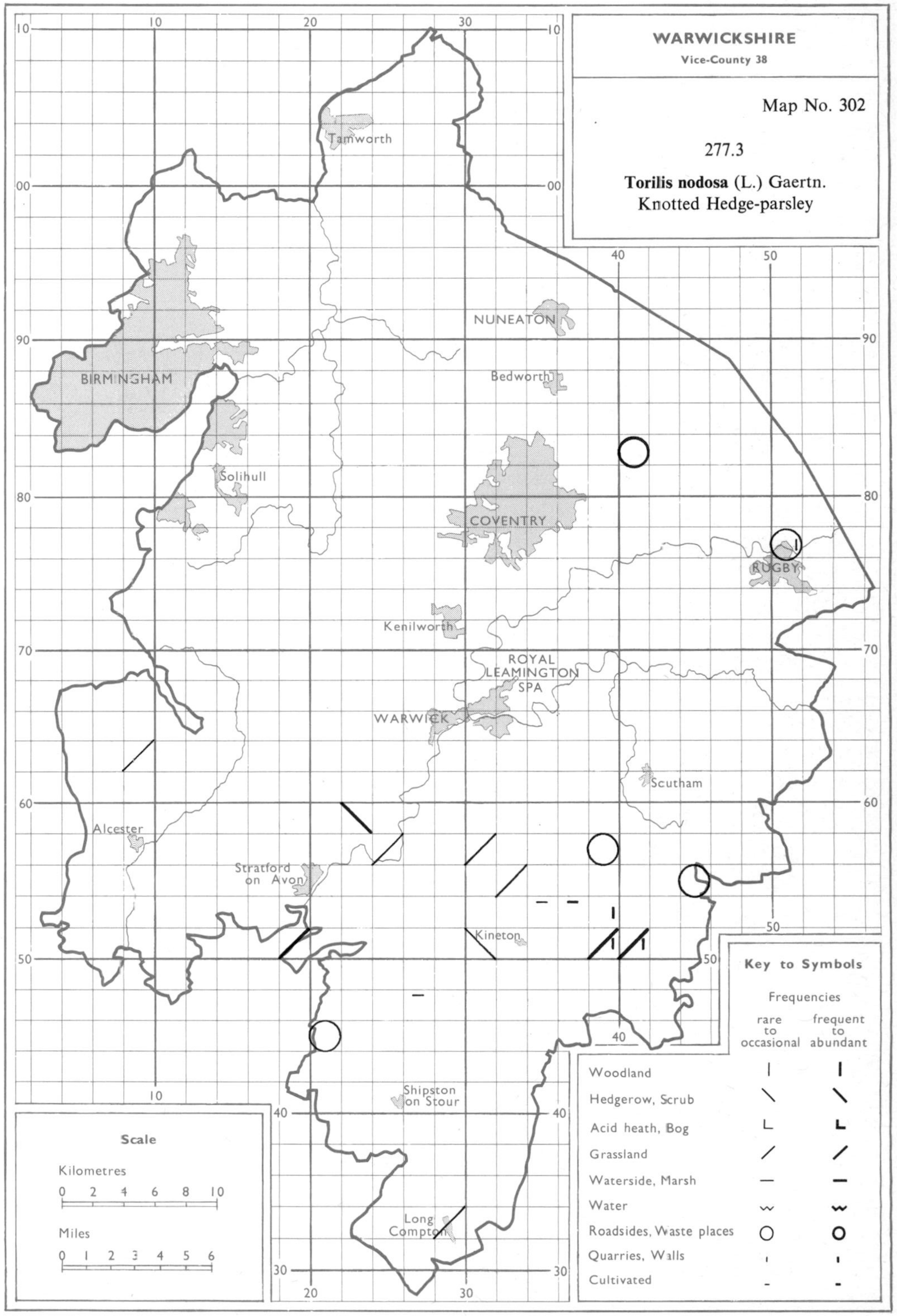

WARWICKSHIRE
Vice-County 38
Map No. 302
277.3
Torilis nodosa (L.) Gaertn.
Knotted Hedge-parsley
Tamworth
NUNEATON
Bedworth
BIRMINGHAM
Solihull
COVENTRY
RUGBY
Kenilworth
ROYAL LEAMINGTON SPA
WARWICK
Southam
Alcester
Stratford on Avon
Kineton
Shipston on Stour
Long Compton
Scale
Kilometres
0 2 4 6 8 10
Miles
0 1 2 3 4 5 6
Key to Symbols
Frequencies
rare to occasional
frequent to abundant
Woodland
Hedgerow, Scrub
Acid heath, Bog
Grassland
Waterside, Marsh
Water
Roadsides, Waste places
Quarries, Walls
Cultivated

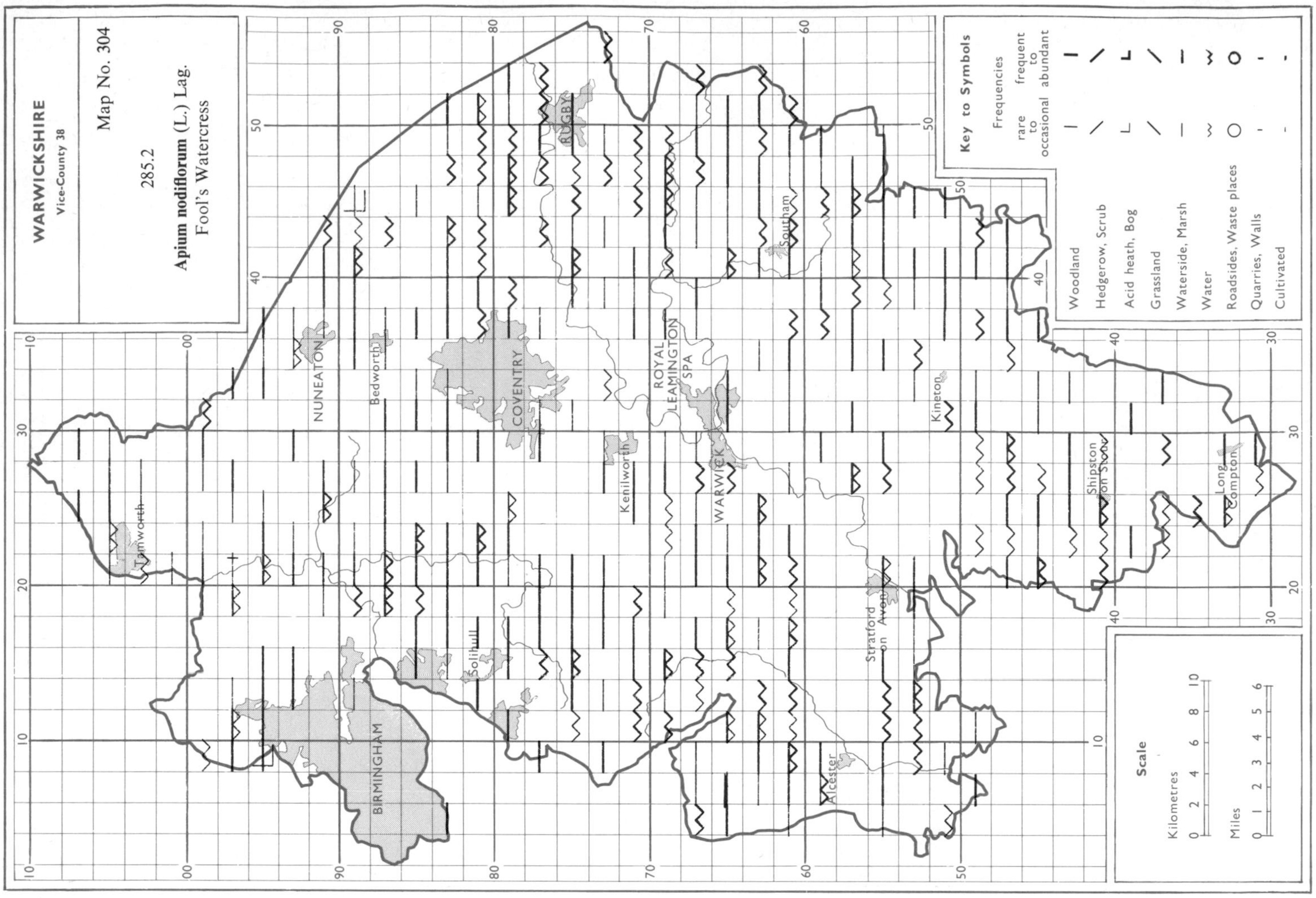
WARWICKSHIRE
Vice-County 38
Map No. 304
285.2
Apium nodiflorum (L.) Lag.
Fool's Watercress
Key to Symbols
Frequencies
rare to occasional
frequent to abundant
Woodland
Hedgerow, Scrub
Acid heath, Bog
Grassland
Waterside, Marsh
Water
Roadsides, Waste places
Quarries, Walls
Cultivated
Scale
Kilometres 0 2 4 6 8 10
Miles 0 1 2 3 4 5 6
NUNEATON
Bedworth
COVENTRY
RUGBY
ROYAL LEAMINGTON SPA
Kenilworth
WARWICK
Southam
Kineton
Shipston on Stour
Long Compton
Tamworth
Solihull
BIRMINGHAM
Stratford on Avon
Alcester

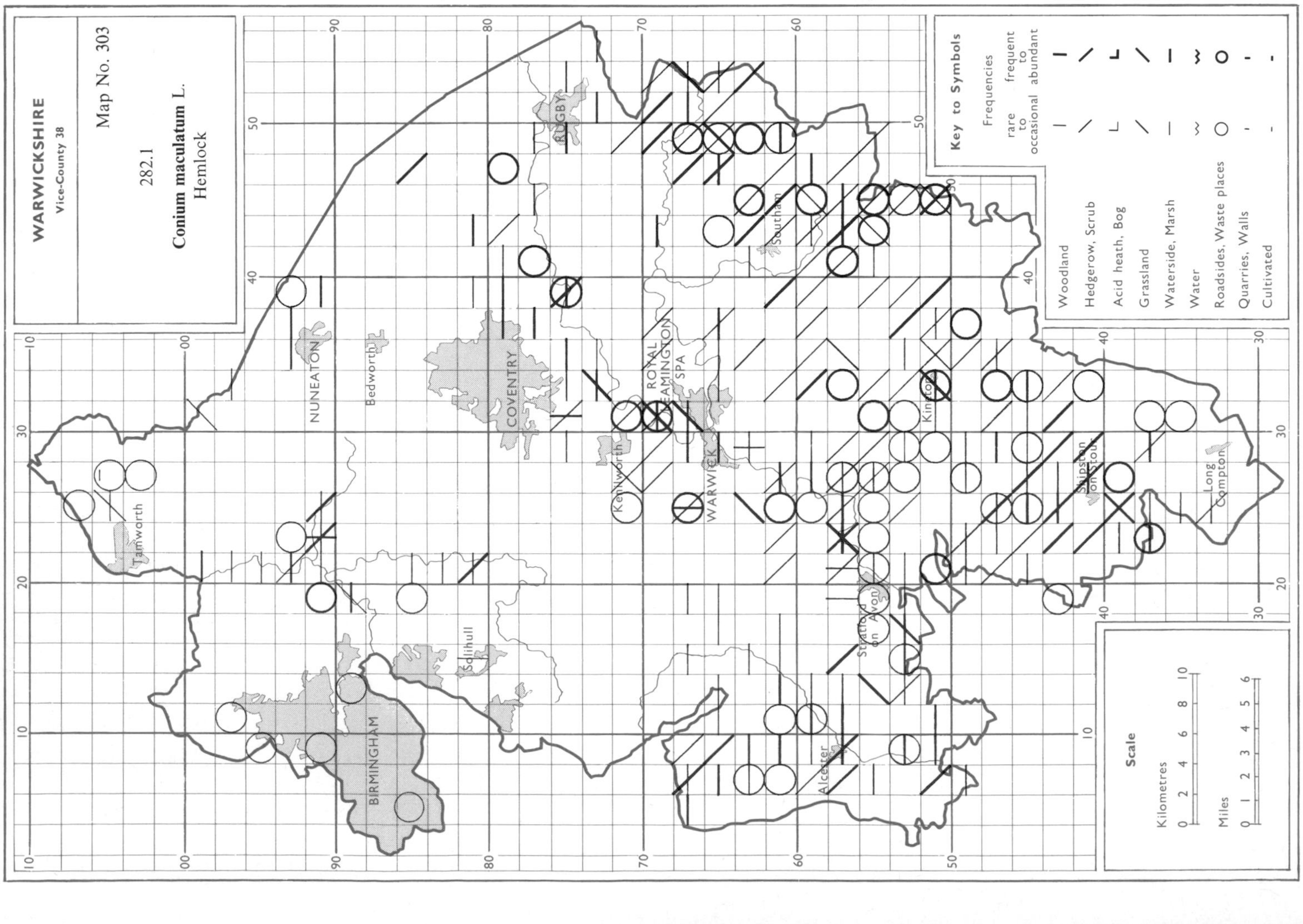
WARWICKSHIRE
Vice-County 38
Map No. 303
282.1
Conium maculatum L.
Hemlock
Key to Symbols
Frequencies
rare to occasional
frequent to abundant
Woodland
Hedgerow, Scrub
Acid heath, Bog
Grassland
Waterside, Marsh
Water
Roadsides, Waste places
Quarries, Walls
Cultivated
Scale
Kilometres 0 2 4 6 8 10
Miles 0 1 2 3 4 5 6
NUNEATON
Bedworth
COVENTRY
RUGBY
ROYAL LEAMINGTON SPA
Kenilworth
WARWICK
Southam
Kineton
Shipston on Stour
Long Compton
Tamworth
Solihull
BIRMINGHAM
Stratford on Avon
Alcester

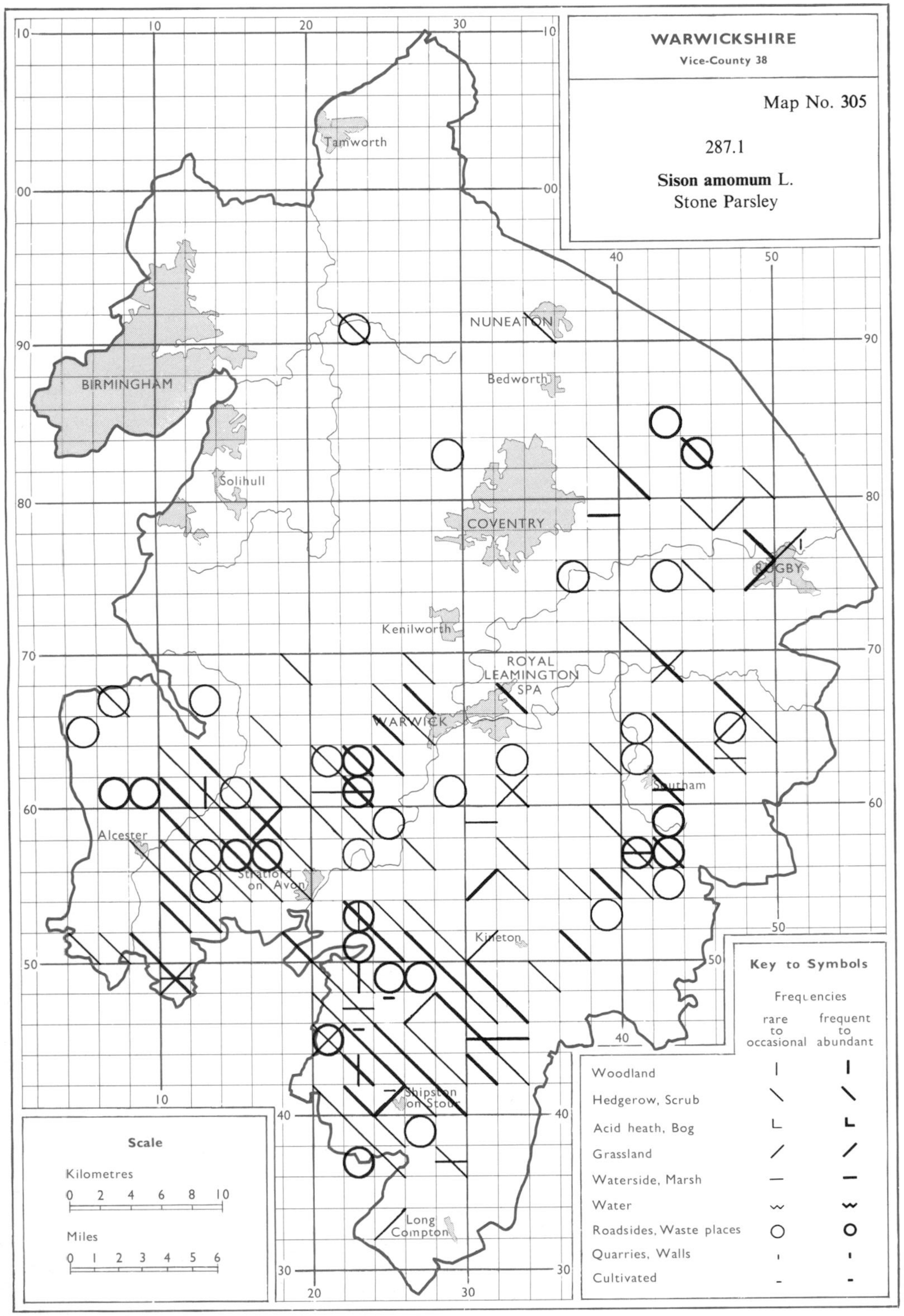
WARWICKSHIRE
Vice-County 38
Map No. 305
287.1
Sison amomum L.
Stone Parsley
Tamworth
NUNEATON
Bedworth
BIRMINGHAM
Solihull
COVENTRY
RUGBY
Kenilworth
ROYAL LEAMINGTON SPA
WARWICK
Southam
Alcester
Stratford on Avon
Kineton
Shipston on Stour
Long Compton
Key to Symbols
Frequencies
rare to occasional
frequent to abundant
Woodland
Hedgerow, Scrub
Acid heath, Bog
Grassland
Waterside, Marsh
Water
Roadsides, Waste places
Quarries, Walls
Cultivated
Scale
Kilometres
0 2 4 6 8 10
Miles
0 1 2 3 4 5 6

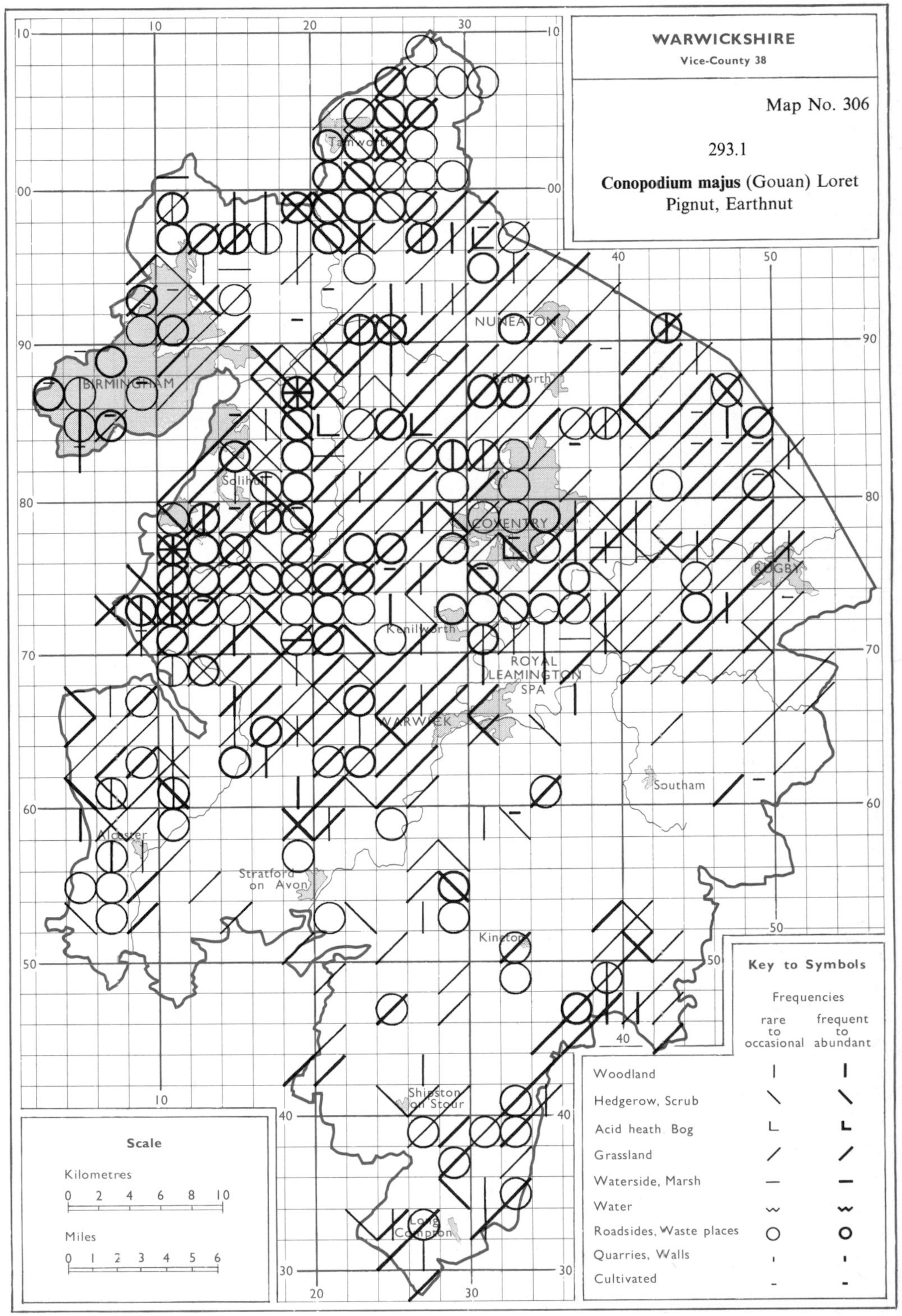
WARWICKSHIRE
Vice-County 38
Map No. 306
293.1
Conopodium majus (Gouan) Loret
Pignut, Earthnut
Tamworth
NUNEATON
Bedworth
BIRMINGHAM
Solihull
COVENTRY
RUGBY
Kenilworth
ROYAL LEAMINGTON SPA
WARWICK
Southam
Alcester
Stratford on Avon
Kineton
Shipston on Stour
Long Compton
Key to Symbols
Frequencies
rare to occasional
frequent to abundant
Woodland
Hedgerow, Scrub
Acid heath Bog
Grassland
Waterside, Marsh
Water
Roadsides, Waste places
Quarries, Walls
Cultivated
Scale
Kilometres
0 2 4 6 8 10
Miles
0 1 2 3 4 5 6

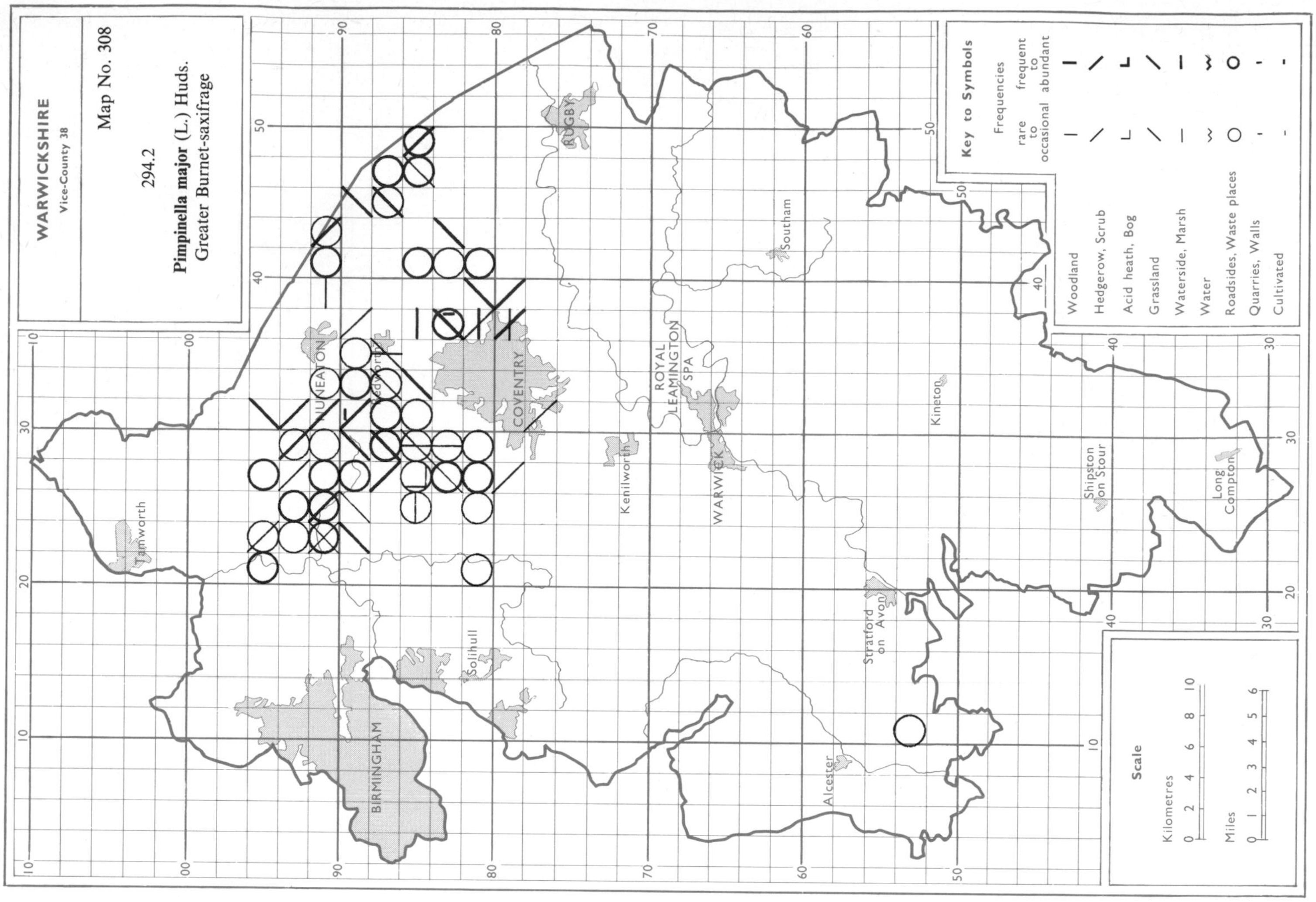
WARWICKSHIRE
Vice-County 38
Map No. 308
294.2
Pimpinella major (L.) Huds.
Greater Burnet-saxifrage
Key to Symbols
Frequencies
rare to occasional
frequent to abundant
Woodland
Hedgerow, Scrub
Acid heath, Bog
Grassland
Waterside, Marsh
Water
Roadsides, Waste places
Quarries, Walls
Cultivated
Scale
Kilometres
Miles
NUNEATON
Bedworth
COVENTRY
RUGBY
Southam
ROYAL LEAMINGTON SPA
Kineton
Kenilworth
WARWICK
Shipston on Stour
Long Compton
Tamworth
Solihull
BIRMINGHAM
Stratford on Avon
Alcester

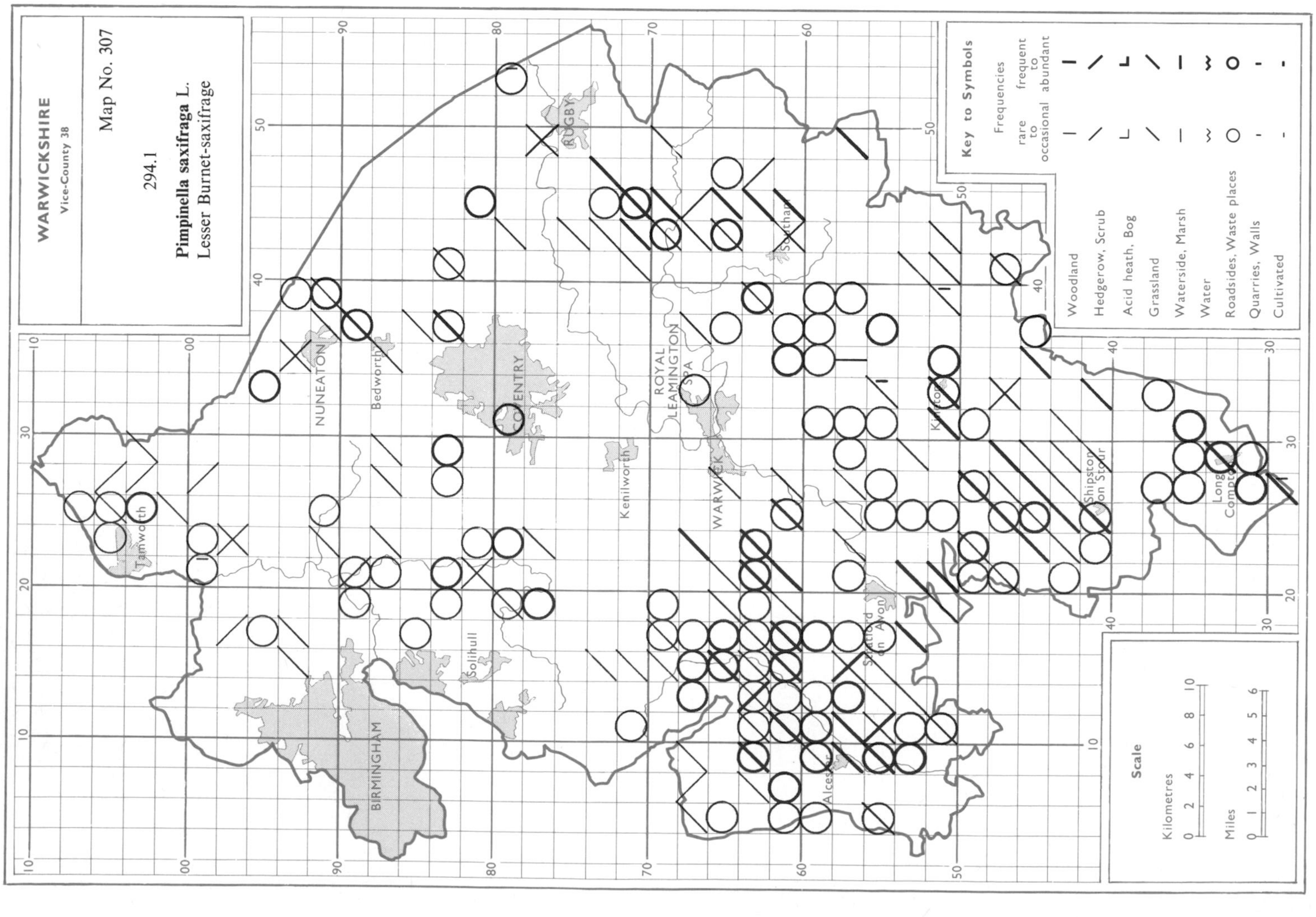
WARWICKSHIRE
Vice-County 38
Map No. 307
294.1
Pimpinella saxifraga L.
Lesser Burnet-saxifrage
Key to Symbols
Frequencies
rare to occasional
frequent to abundant
Woodland
Hedgerow, Scrub
Acid heath, Bog
Grassland
Waterside, Marsh
Water
Roadsides, Waste places
Quarries, Walls
Cultivated
Scale
Kilometres
Miles
NUNEATON
Bedworth
COVENTRY
RUGBY
Southam
ROYAL LEAMINGTON SPA
Kineton
Kenilworth
WARWICK
Shipston on Stour
Long Compton
Tamworth
Solihull
BIRMINGHAM
Stratford on Avon
Alcester

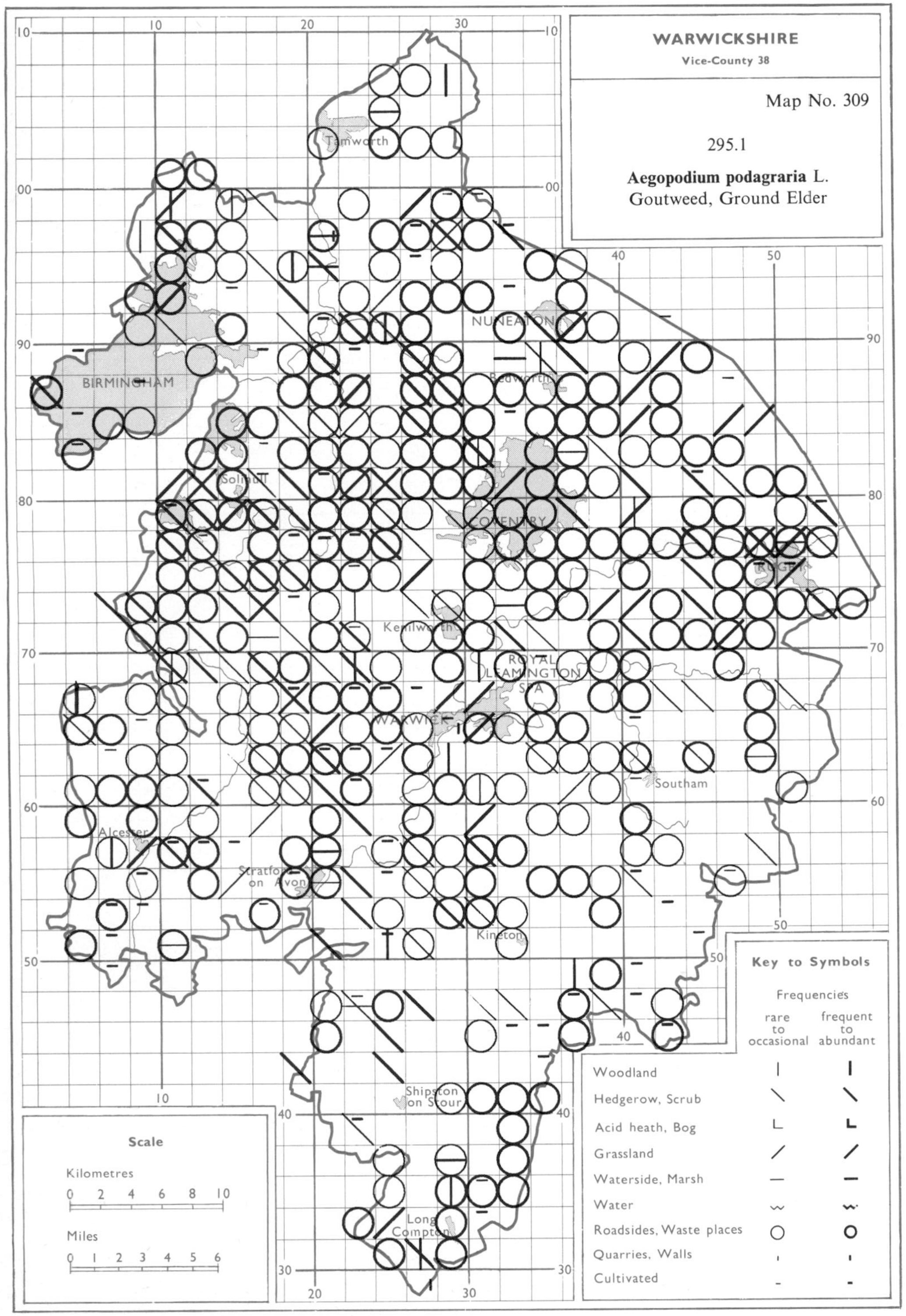

WARWICKSHIRE
Vice-County 38
Map No. 309
295.1
Aegopodium podagraria L.
Goutweed, Ground Elder
Tamworth
NUNEATON
Bedworth
BIRMINGHAM
Solihull
COVENTRY
RUGBY
Kenilworth
ROYAL LEAMINGTON SPA
WARWICK
Southam
Alcester
Stratford on Avon
Kineton
Shipston on Stour
Long Compton
Key to Symbols
Frequencies
rare to occasional
frequent to abundant
Woodland
Hedgerow, Scrub
Acid heath, Bog
Grassland
Waterside, Marsh
Water
Roadsides, Waste places
Quarries, Walls
Cultivated
Scale
Kilometres
0 2 4 6 8 10
Miles
0 1 2 3 4 5 6

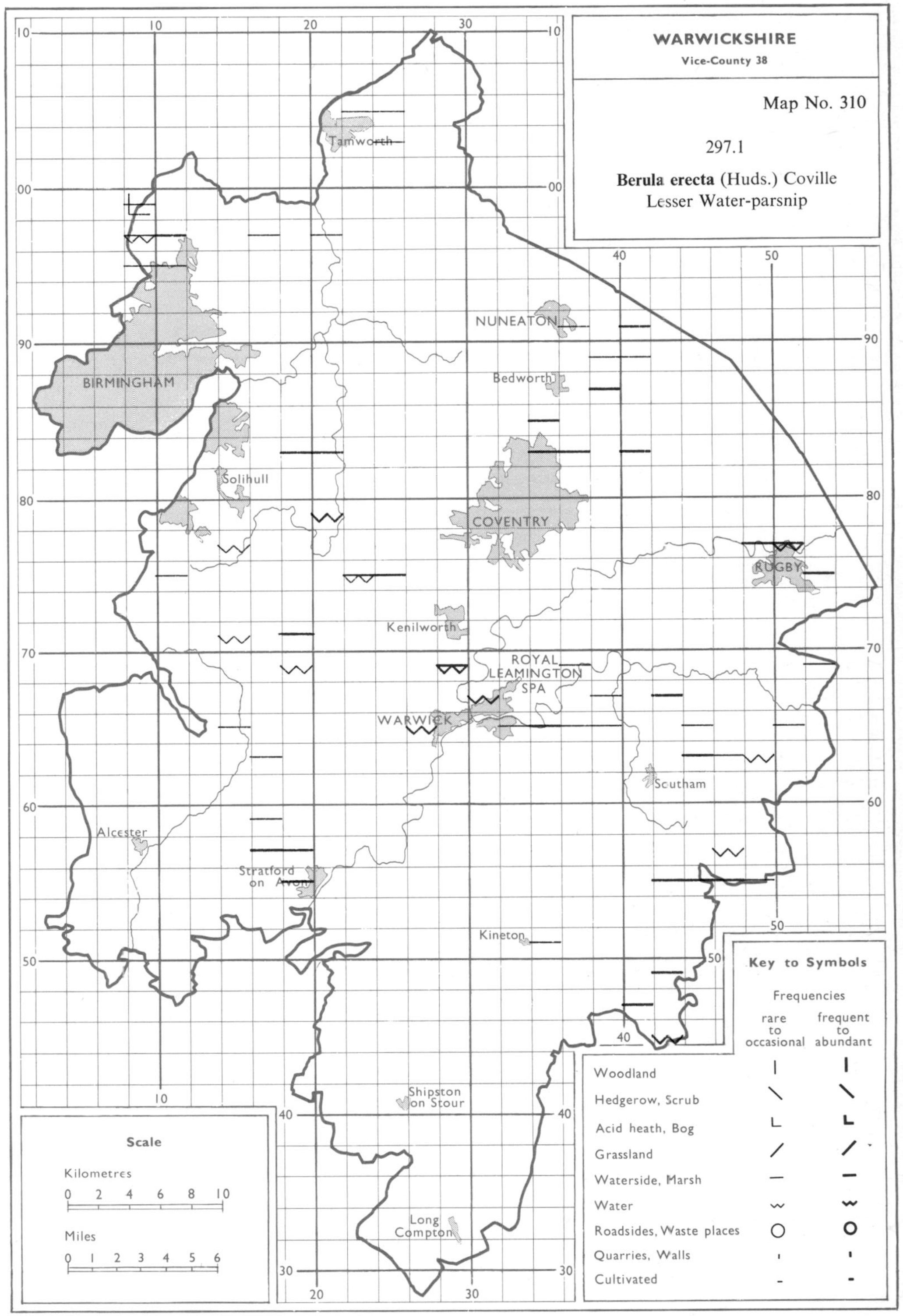

WARWICKSHIRE
Vice-County 38
Map No. 310
297.1
Berula erecta (Huds.) Coville
Lesser Water-parsnip
Tamworth
NUNEATON
Bedworth
BIRMINGHAM
Solihull
COVENTRY
RUGBY
Kenilworth
ROYAL LEAMINGTON SPA
WARWICK
Southam
Alcester
Stratford on Avon
Kineton
Shipston on Stour
Long Compton
Key to Symbols
Frequencies
rare to occasional
frequent to abundant
Woodland
Hedgerow, Scrub
Acid heath, Bog
Grassland
Waterside, Marsh
Water
Roadsides, Waste places
Quarries, Walls
Cultivated
Scale
Kilometres
0 2 4 6 8 10
Miles
0 1 2 3 4 5 6

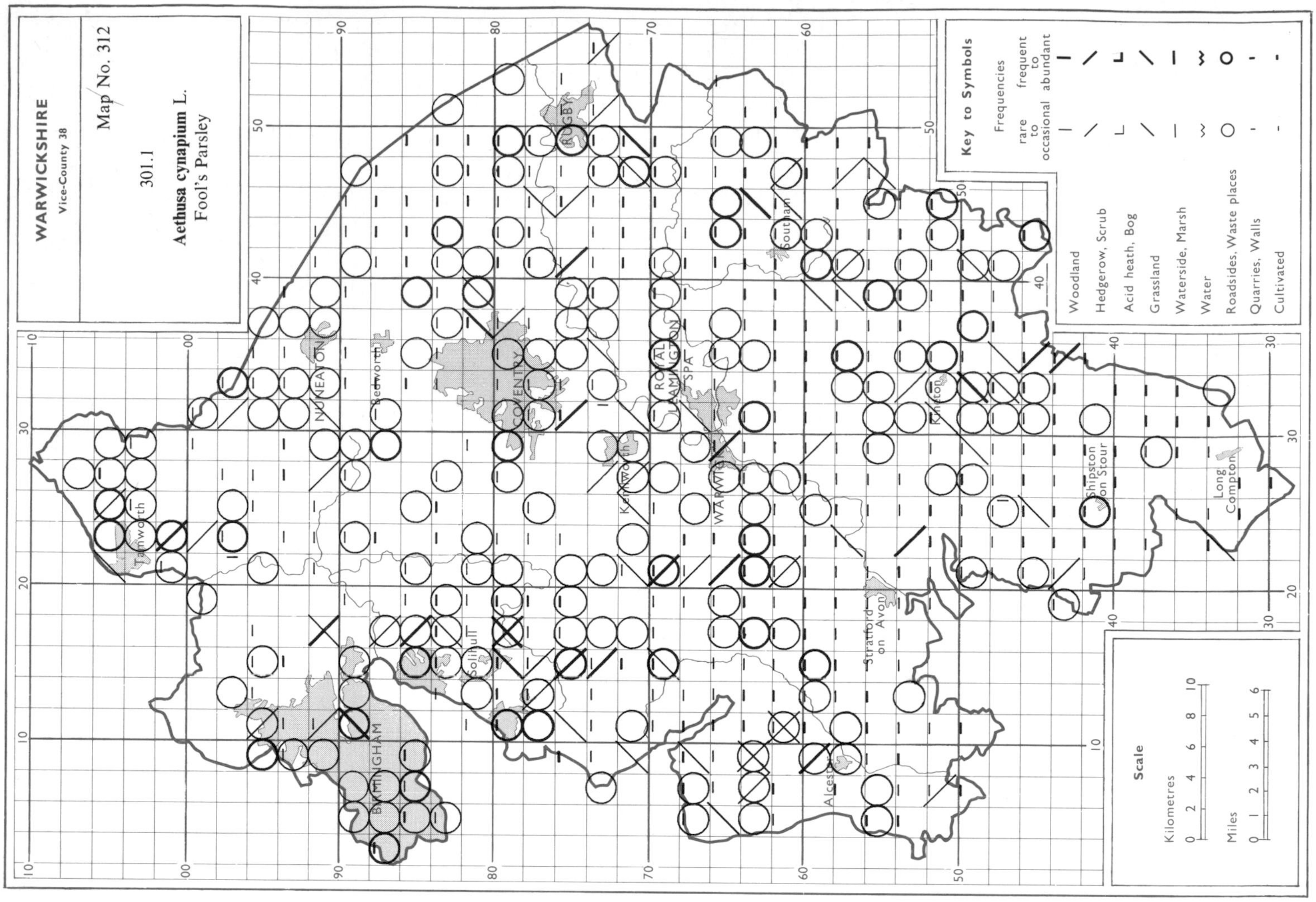
WARWICKSHIRE
Vice-County 38
Map No. 312
301.1
Aethusa cynapium L.
Fool's Parsley
Key to Symbols
Frequencies
rare to occasional
frequent to abundant
Woodland
Hedgerow, Scrub
Acid heath, Bog
Grassland
Waterside, Marsh
Water
Roadsides, Waste places
Quarries, Walls
Cultivated
Scale
Kilometres
0 2 4 6 8 10
Miles
0 1 2 3 4 5 6
NUNEATON
Bedworth
COVENTRY
RUGBY
ROYAL LEAMINGTON SPA
Kenilworth
WARWICK
Southam
Kineton
Tamworth
BIRMINGHAM
Solihull
Stratford on Avon
Alcester
Shipston on Stour
Long Compton

WARWICKSHIRE
Vice-County 38
Map No. 311
300.1
Oenanthe fistulosa L.
Tubular Water-dropwort
Key to Symbols
Frequencies
rare to occasional
frequent to abundant
Woodland
Hedgerow, Scrub
Acid heath, Bog
Grassland
Waterside, Marsh
Water
Roadsides, Waste places
Quarries, Walls
Cultivated
Scale
Kilometres
0 2 4 6 8 10
Miles
0 1 2 3 4 5 6
NUNEATON
Bedworth
COVENTRY
RUGBY
ROYAL LEAMINGTON SPA
Kenilworth
WARWICK
Southam
Kineton
Tamworth
BIRMINGHAM
Solihull
Stratford on Avon
Alcester
Shipston on Stour
Long Compton

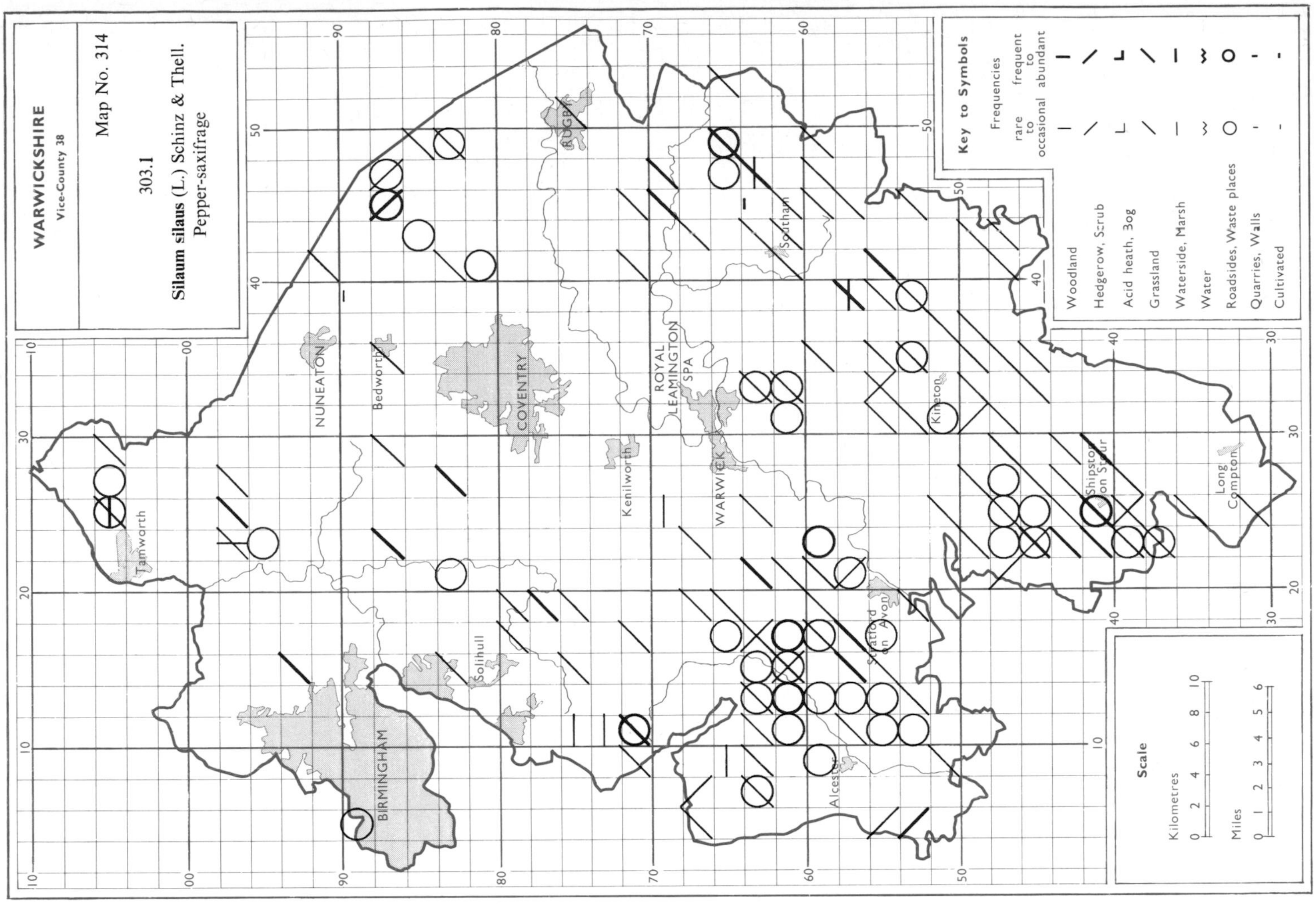
WARWICKSHIRE
Vice-County 38
Map No. 314
303.1
Silaum silaus (L.) Schinz & Thell.
Pepper-saxifrage
Key to Symbols
Frequencies
rare to occasional
frequent to abundant
Woodland
Hedgerow, Scrub
Acid heath, Bog
Grassland
Waterside, Marsh
Water
Roadsides, Waste places
Quarries, Walls
Cultivated
Scale
Kilometres
Miles
NUNEATON
Bedworth
COVENTRY
RUGBY
ROYAL LEAMINGTON SPA
Southam
Kineton
Shipston on Stour
Long Compton
Tamworth
Kenilworth
WARWICK
Stratford on Avon
Solihull
BIRMINGHAM
Alcester

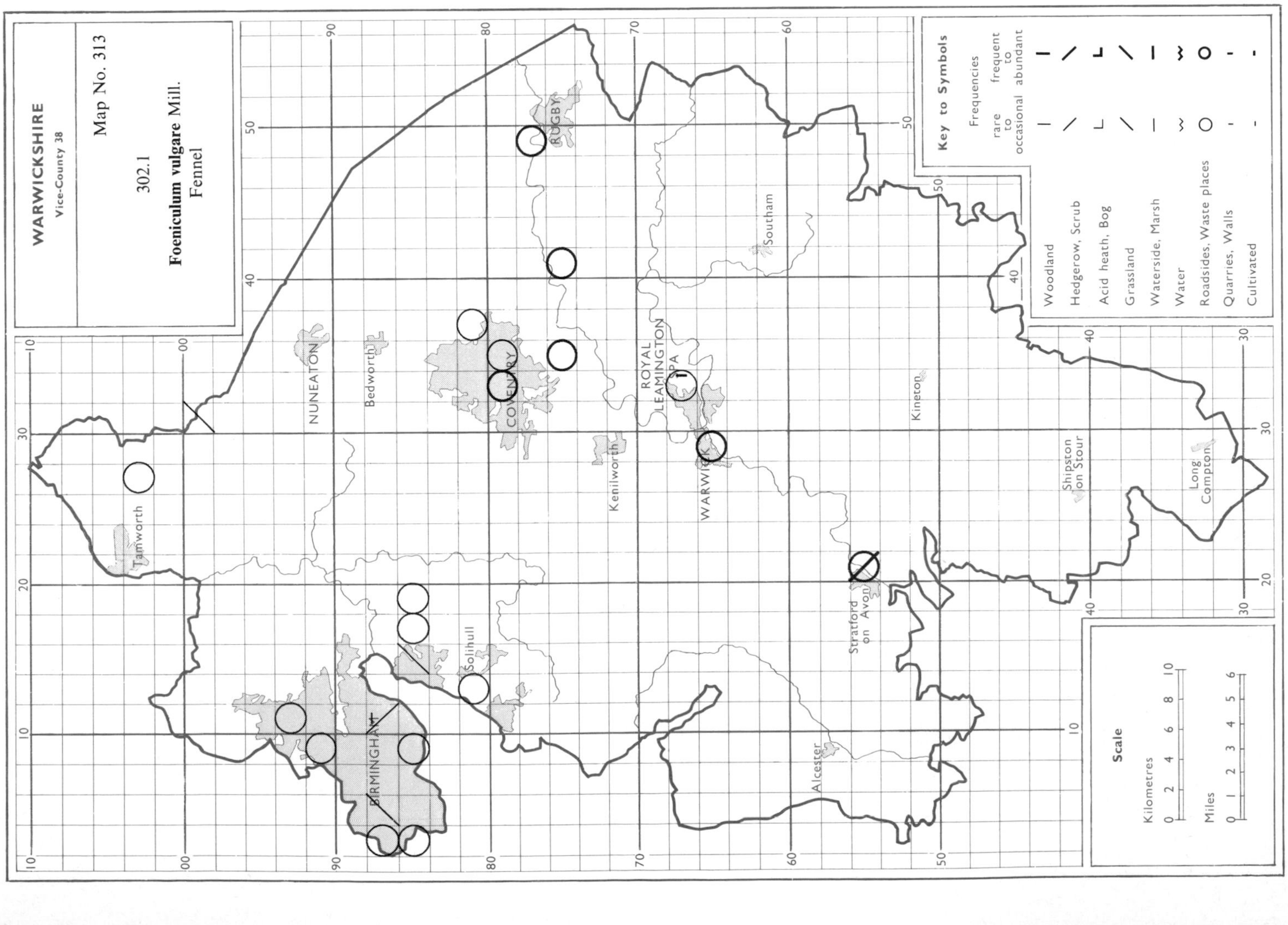
WARWICKSHIRE
Vice-County 38
Map No. 313
302.1
Foeniculum vulgare Mill.
Fennel
Key to Symbols
Frequencies
rare to occasional
frequent to abundant
Woodland
Hedgerow, Scrub
Acid heath, Bog
Grassland
Waterside, Marsh
Water
Roadsides, Waste places
Quarries, Walls
Cultivated
Scale
Kilometres
Miles
NUNEATON
Bedworth
COVENTRY
RUGBY
ROYAL LEAMINGTON SPA
Southam
Kineton
Shipston on Stour
Long Compton
Tamworth
Kenilworth
WARWICK
Stratford on Avon
Solihull
BIRMINGHAM
Alcester

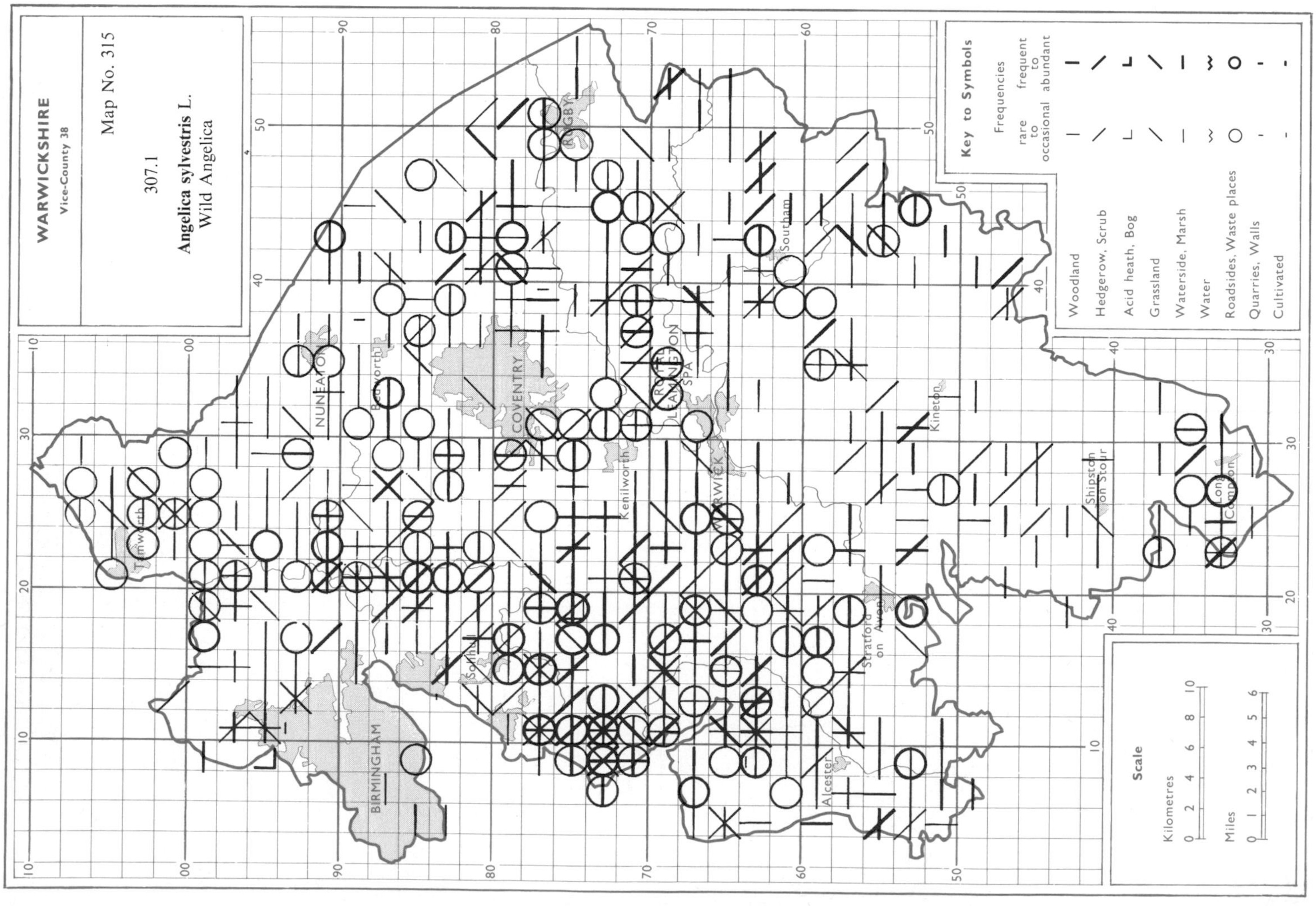
WARWICKSHIRE
Vice-County 38
Map No. 315
307.1
Angelica sylvestris L.
Wild Angelica
Key to Symbols
Frequencies
rare to occasional
frequent to abundant
Woodland
Hedgerow, Scrub
Acid heath, Bog
Grassland
Waterside, Marsh
Water
Roadsides, Waste places
Quarries, Walls
Cultivated
Scale
Kilometres
0 2 4 6 8 10
Miles
0 1 2 3 4 5 6
NUNEATON
Bedworth
COVENTRY
RUGBY
ROYAL LEAMINGTON SPA
Kenilworth
WARWICK
Southam
Kineton
Shipston on Stour
Long Compton
Tamworth
BIRMINGHAM
Solihull
Stratford on Avon
Alcester

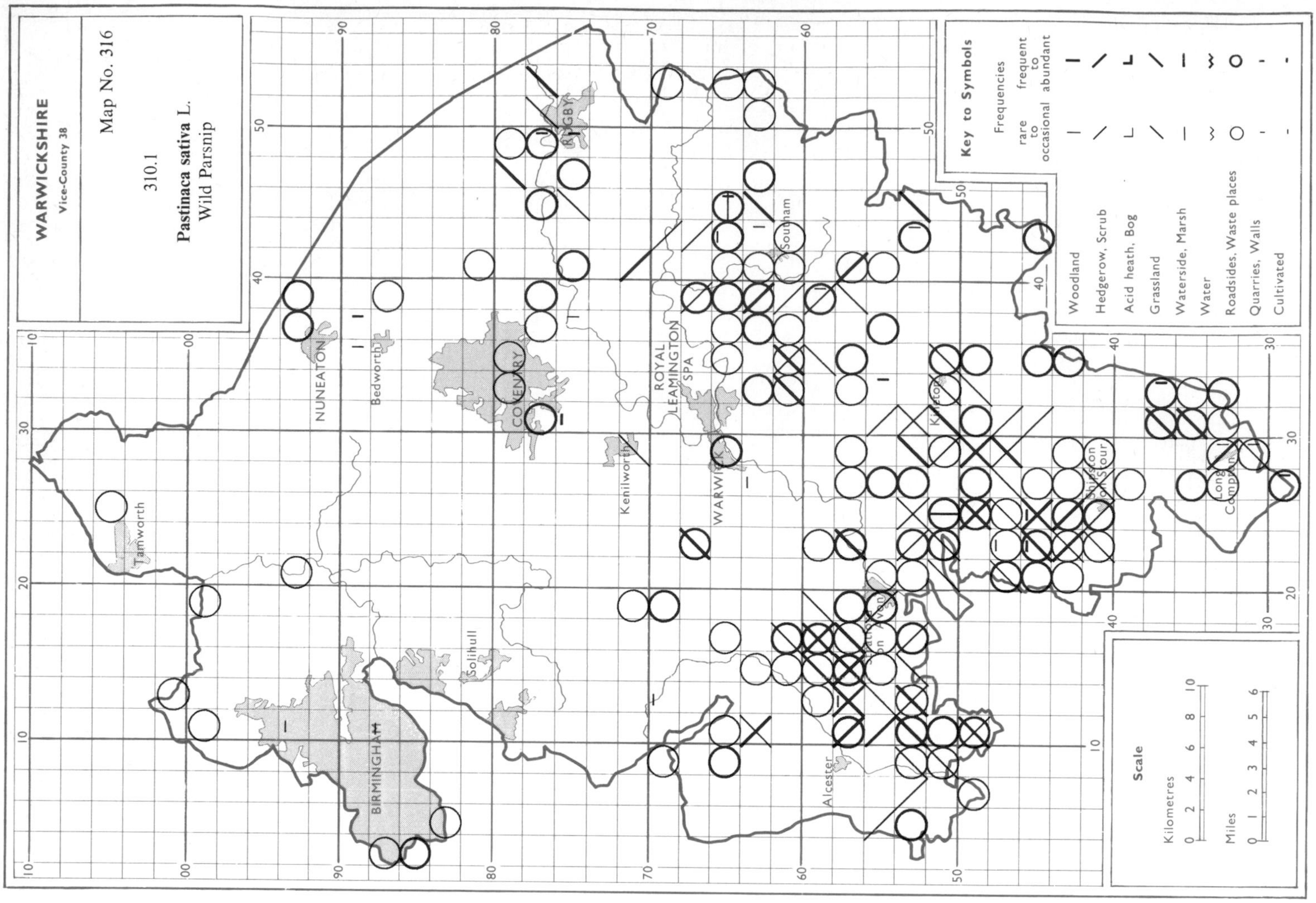
WARWICKSHIRE
Vice-County 38
Map No. 316
310.1
Pastinaca sativa L.
Wild Parsnip
Key to Symbols
Frequencies
rare to occasional
frequent to abundant
Woodland
Hedgerow, Scrub
Acid heath, Bog
Grassland
Waterside, Marsh
Water
Roadsides, Waste places
Quarries, Walls
Cultivated
Scale
Kilometres
0 2 4 6 8 10
Miles
0 1 2 3 4 5 6
NUNEATON
Bedworth
COVENTRY
RUGBY
ROYAL LEAMINGTON SPA
Kenilworth
WARWICK
Southam
Kineton
Shipston on Stour
Long Compton
Tamworth
BIRMINGHAM
Solihull
Stratford on Avon
Alcester

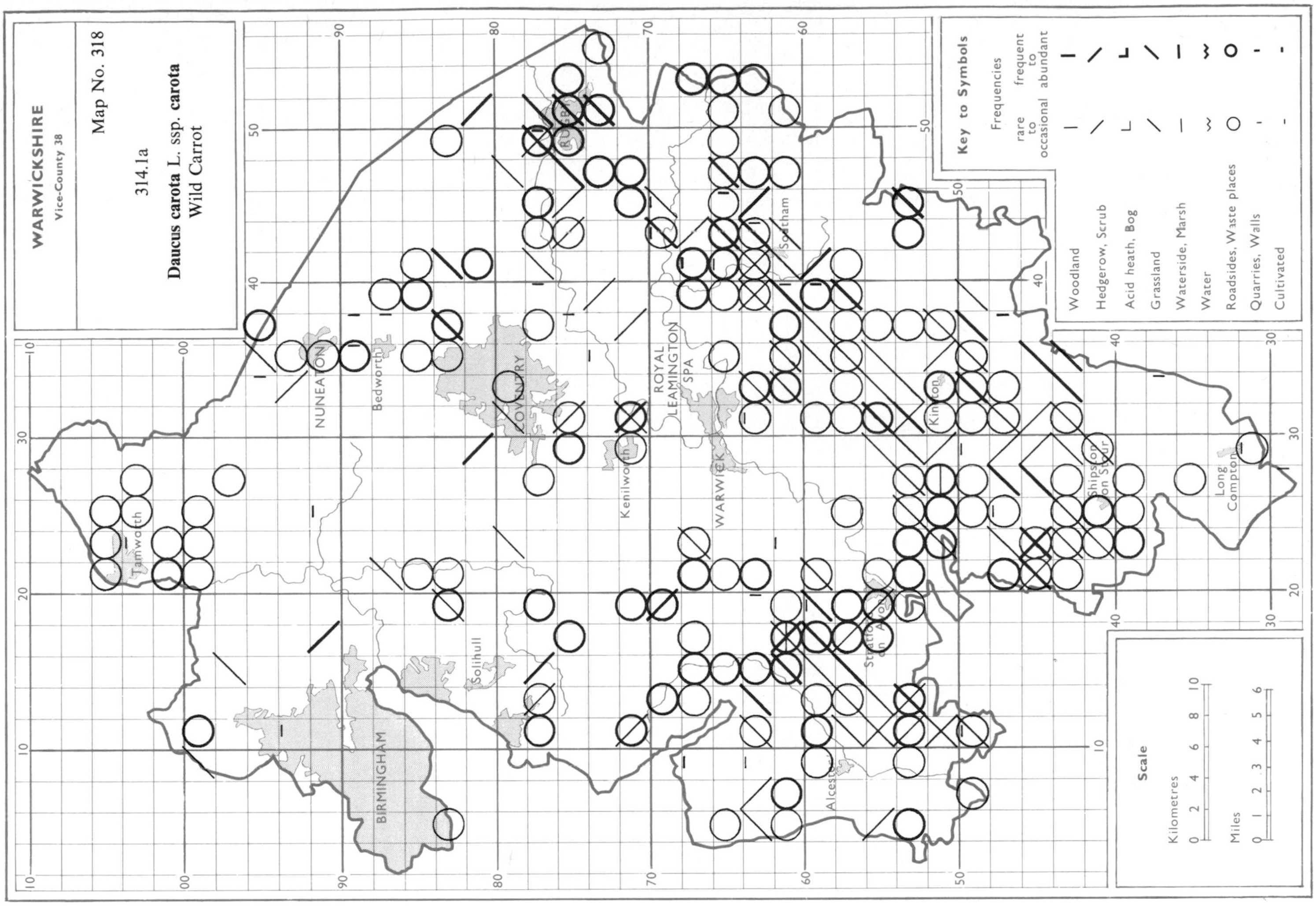
WARWICKSHIRE
Vice-County 38
Map No. 318
314.1a
Daucus carota L. ssp. carota
Wild Carrot
Key to Symbols
Frequencies
rare to occasional
frequent to abundant
Woodland
Hedgerow, Scrub
Acid heath, Bog
Grassland
Waterside, Marsh
Water
Roadsides, Waste places
Quarries, Walls
Cultivated
Scale
Kilometres
Miles
NUNEATON
Bedworth
COVENTRY
Rugby
Tamworth
Solihull
BIRMINGHAM
Kenilworth
ROYAL LEAMINGTON SPA
WARWICK
Southam
Kineton
Shipston on Stour
Long Compton
Stratford on Avon
Alcester

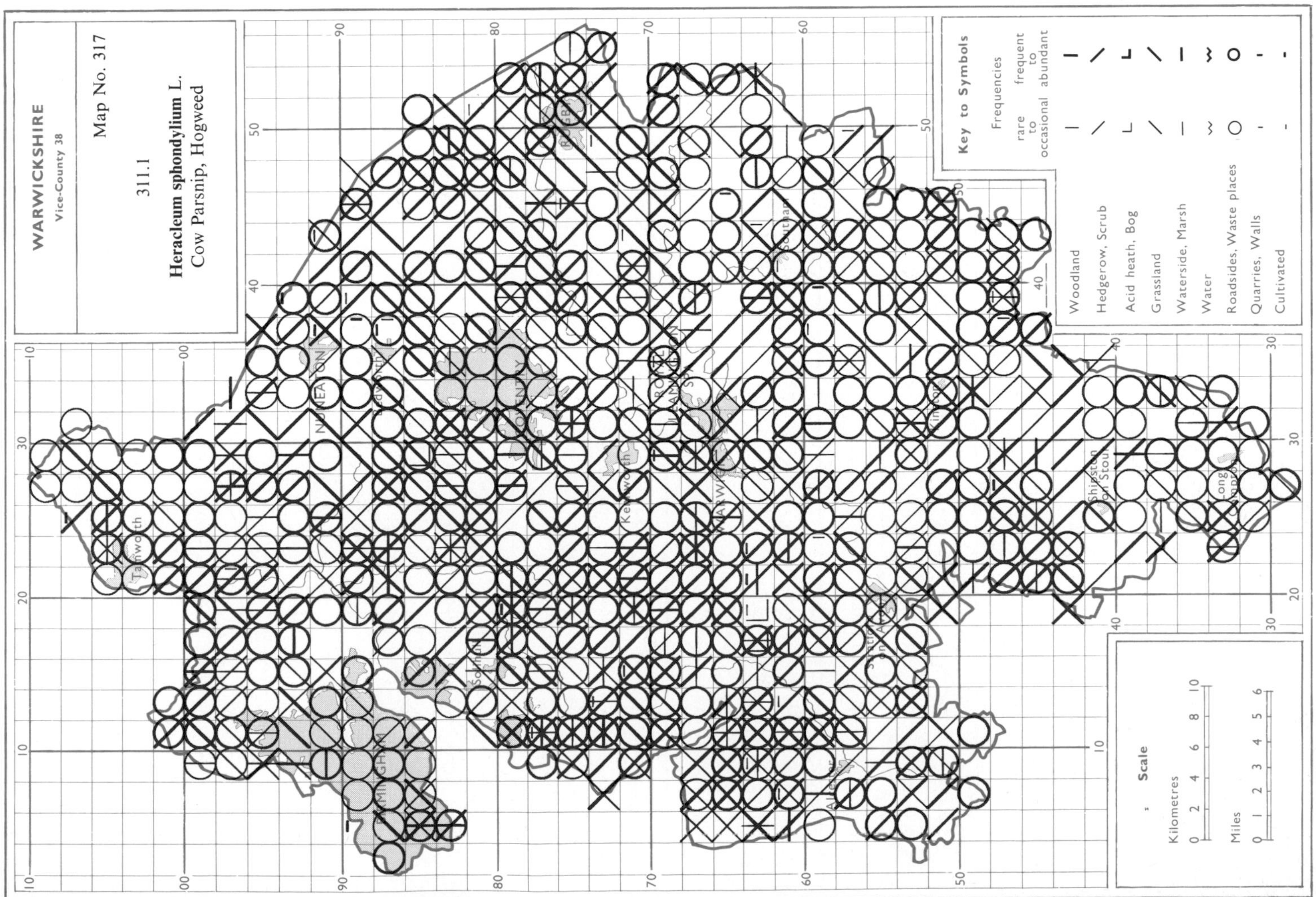
WARWICKSHIRE
Vice-County 38
Map No. 317
311.1
Heracleum sphondylium L.
Cow Parsnip, Hogweed
Key to Symbols
Frequencies
rare to occasional
frequent to abundant
Woodland
Hedgerow, Scrub
Acid heath, Bog
Grassland
Waterside, Marsh
Water
Roadsides, Waste places
Quarries, Walls
Cultivated
Scale
Kilometres
Miles

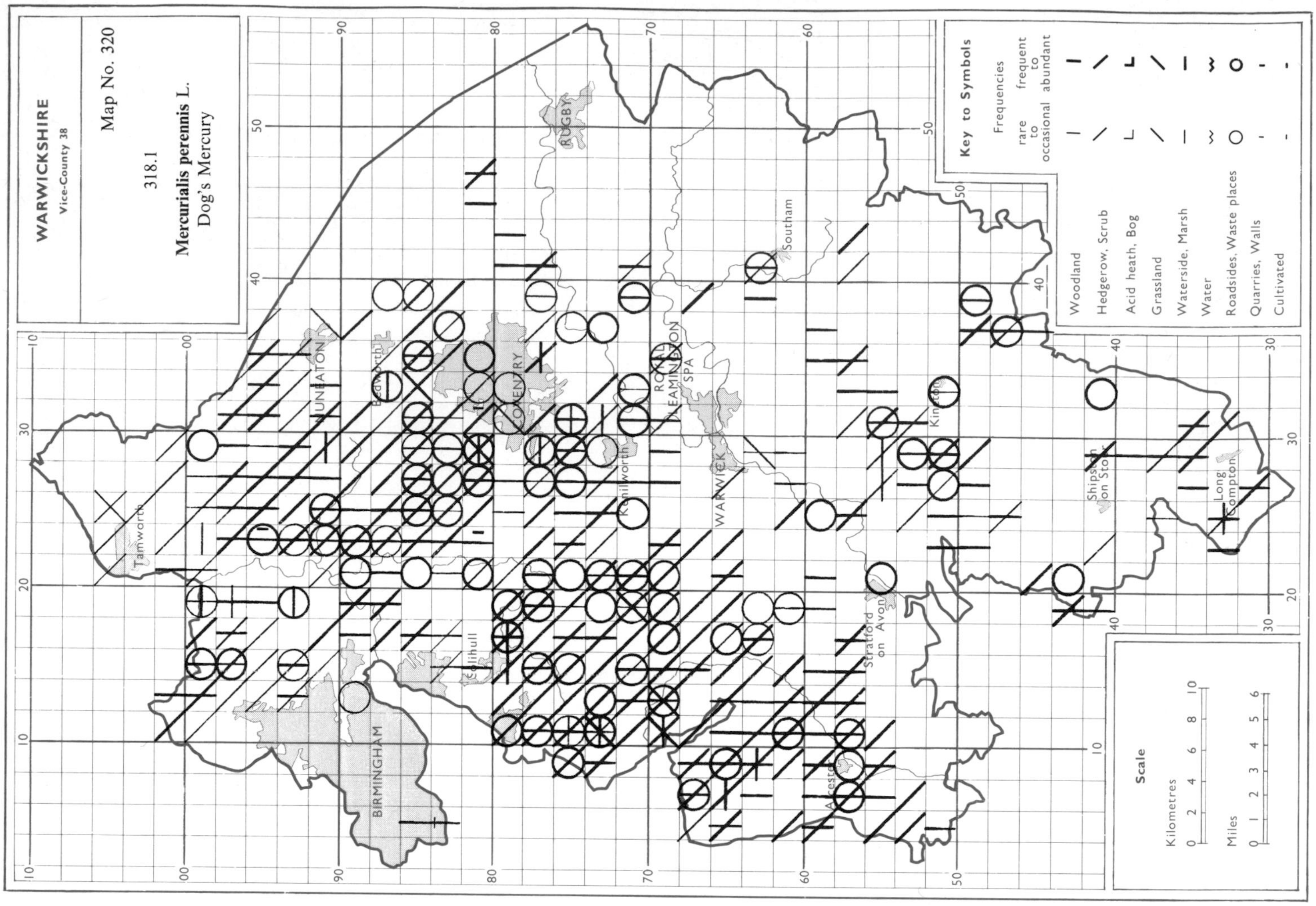
WARWICKSHIRE
Vice-County 38
Map No. 320
318.1
Mercurialis perennis L.
Dog's Mercury
Key to Symbols
Frequencies
rare to occasional
frequent to abundant
Woodland
Hedgerow, Scrub
Acid heath, Bog
Grassland
Waterside, Marsh
Water
Roadsides, Waste places
Quarries, Walls
Cultivated
Scale
Kilometres
0 2 4 6 8 10
Miles
0 1 2 3 4 5 6
NUNEATON
Bedworth
COVENTRY
RUGBY
Southam
ROYAL LEAMINGTON SPA
Kenilworth
WARWICK
Kineton
Shipston on Stour
Long Compton
Tamworth
Solihull
BIRMINGHAM
Stratford on Avon
Alcester

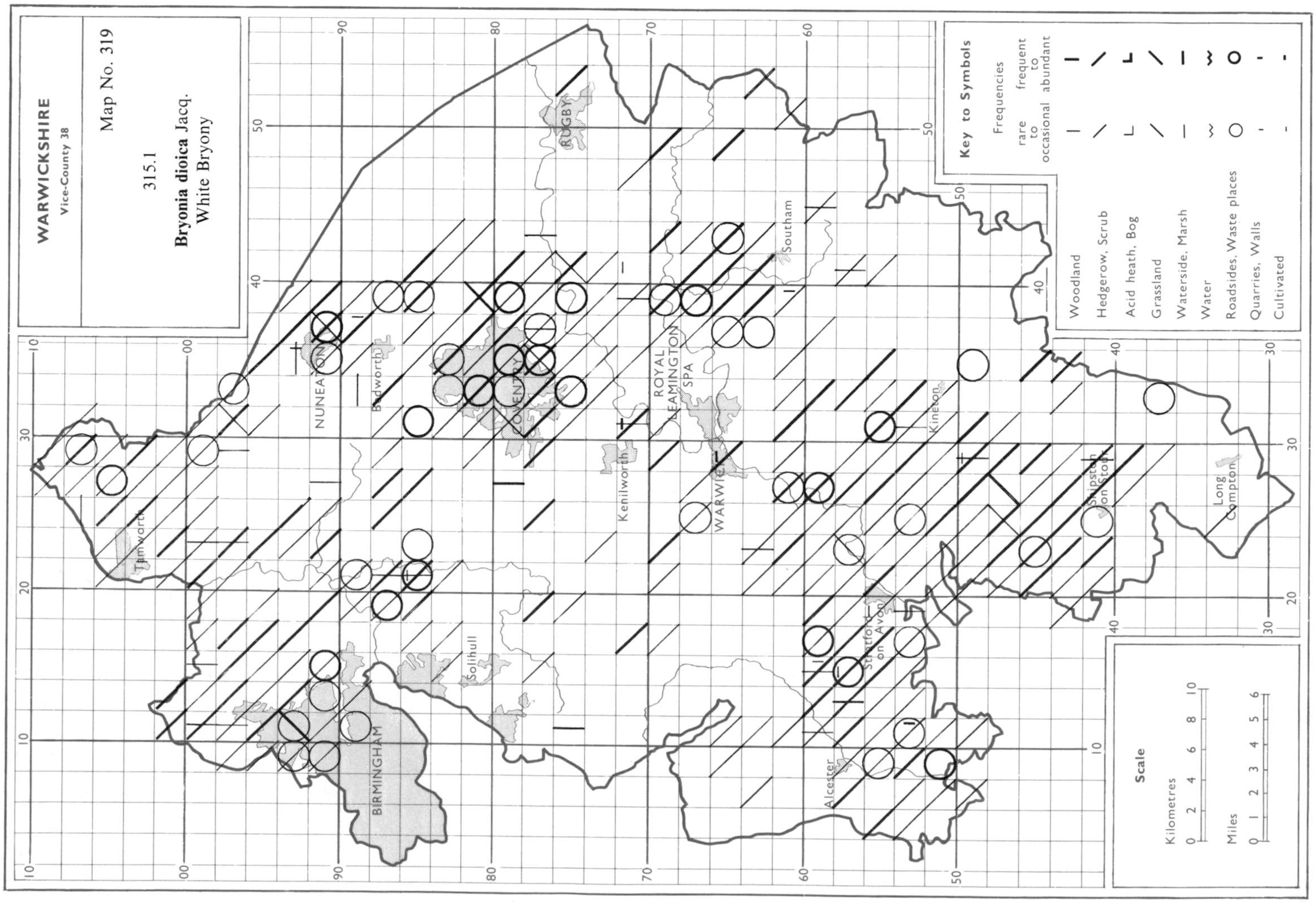
WARWICKSHIRE
Vice-County 38
Map No. 319
315.1
Bryonia dioica Jacq.
White Bryony
Key to Symbols
Frequencies
rare to occasional
frequent to abundant
Woodland
Hedgerow, Scrub
Acid heath, Bog
Grassland
Waterside, Marsh
Water
Roadsides, Waste places
Quarries, Walls
Cultivated
Scale
Kilometres
0 2 4 6 8 10
Miles
0 1 2 3 4 5 6
NUNEATON
Bedworth
COVENTRY
RUGBY
Southam
ROYAL LEAMINGTON SPA
Kenilworth
WARWICK
Kineton
Shipston on Stour
Long Compton
Tamworth
Solihull
BIRMINGHAM
Stratford on Avon
Alcester

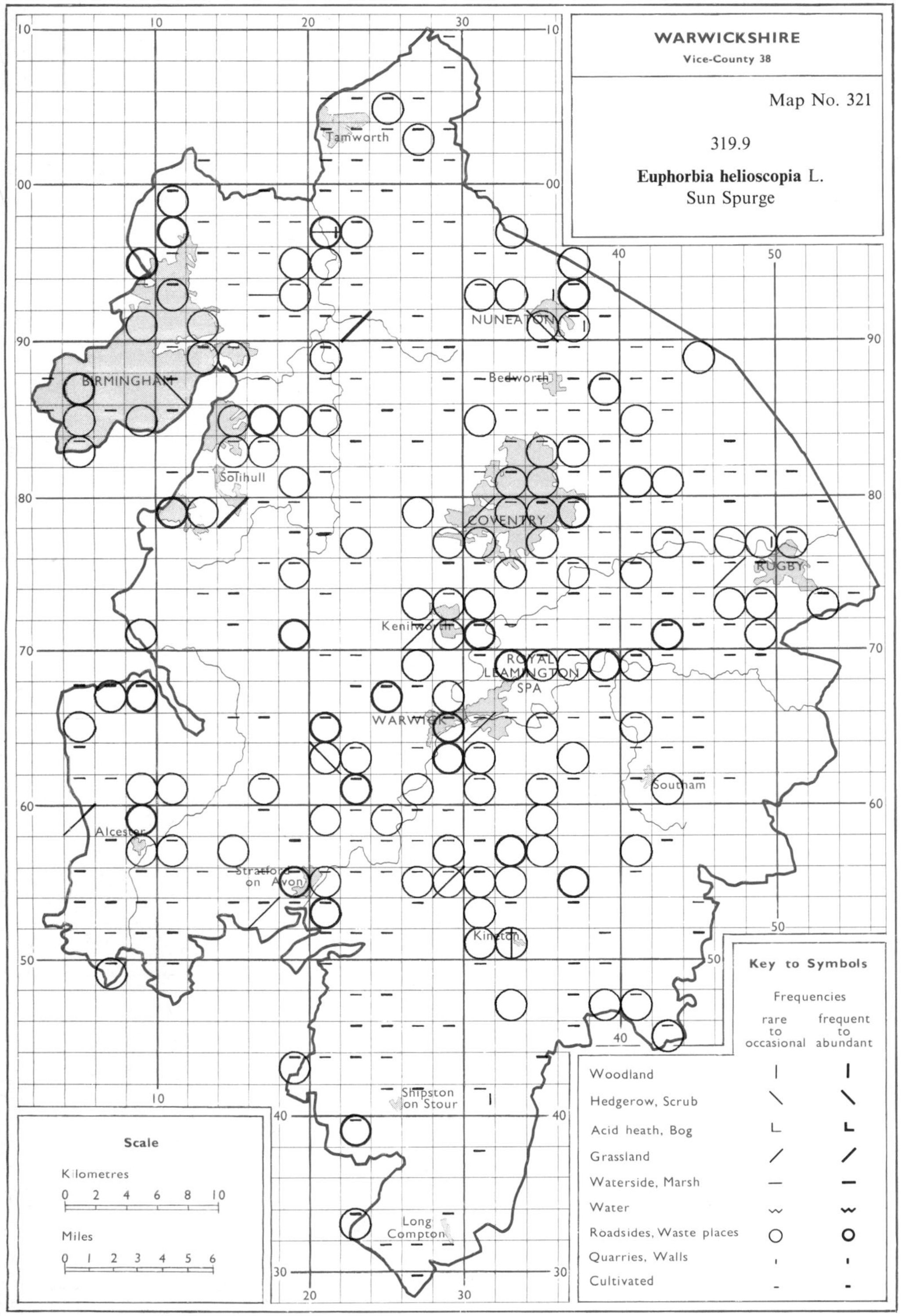
WARWICKSHIRE
Vice-County 38
Map No. 321
319.9
Euphorbia helioscopia L.
Sun Spurge
Key to Symbols
Frequencies
rare to occasional
frequent to abundant
Woodland
Hedgerow, Scrub
Acid heath, Bog
Grassland
Waterside, Marsh
Water
Roadsides, Waste places
Quarries, Walls
Cultivated
Scale
Kilometres
Miles
Tamworth
Nuneaton
Bedworth
Coventry
Rugby
Birmingham
Solihull
Kenilworth
Royal Leamington Spa
Warwick
Southam
Kineton
Shipston on Stour
Long Compton
Stratford on Avon
Alcester

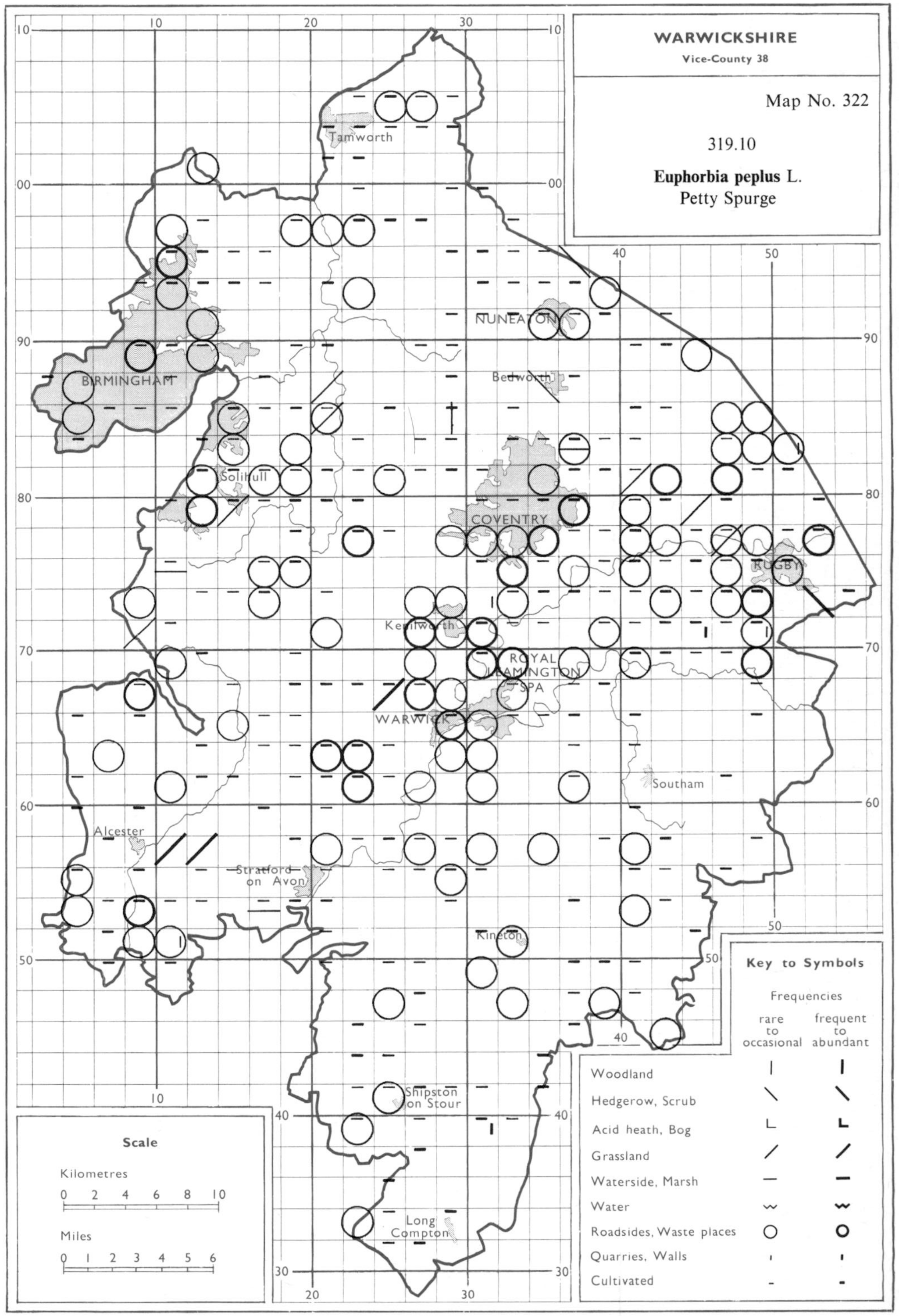
WARWICKSHIRE
Vice-County 38
Map No. 322
319.10
Euphorbia peplus L.
Petty Spurge
Key to Symbols
Frequencies
rare to occasional
frequent to abundant
Woodland
Hedgerow, Scrub
Acid heath, Bog
Grassland
Waterside, Marsh
Water
Roadsides, Waste places
Quarries, Walls
Cultivated
Scale
Kilometres
Miles
Tamworth
Nuneaton
Bedworth
Coventry
Rugby
Birmingham
Solihull
Kenilworth
Royal Leamington Spa
Warwick
Southam
Kineton
Shipston on Stour
Long Compton
Stratford on Avon
Alcester

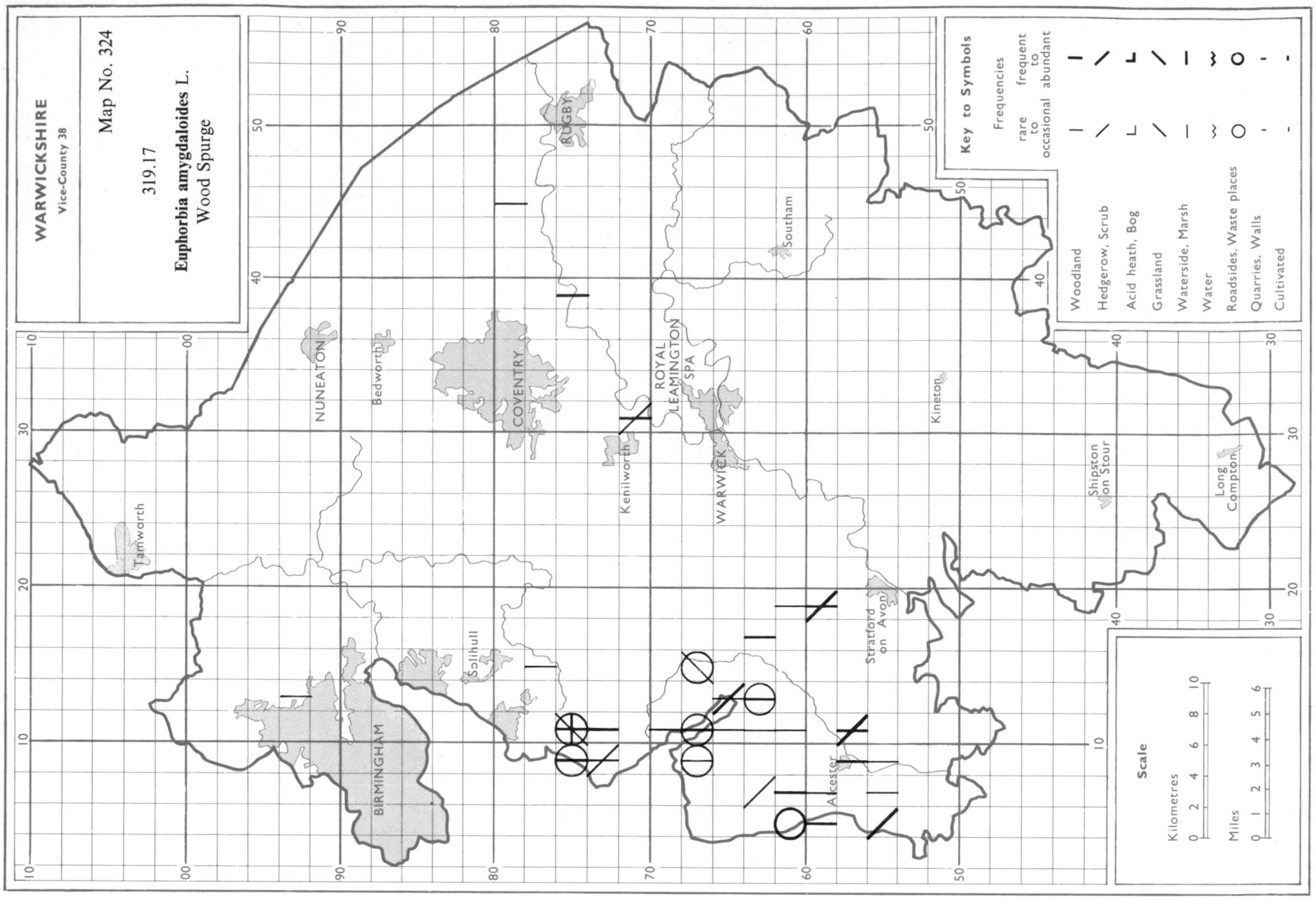
WARWICKSHIRE
Vice-County 38
Map No. 324
319.17
Euphorbia amygdaloides L.
Wood Spurge
Key to Symbols
Frequencies
rare to occasional
frequent to abundant
Woodland
Hedgerow, Scrub
Acid heath, Bog
Grassland
Waterside, Marsh
Water
Roadsides, Waste places
Quarries, Walls
Cultivated
Scale
Kilometres
Miles
NUNEATON
Bedworth
COVENTRY
RUGBY
Southam
ROYAL LEAMINGTON SPA
Kenilworth
WARWICK
Kineton
Shipston on Stour
Long Compton
Tamworth
Solihull
BIRMINGHAM
Stratford on Avon
Alcester

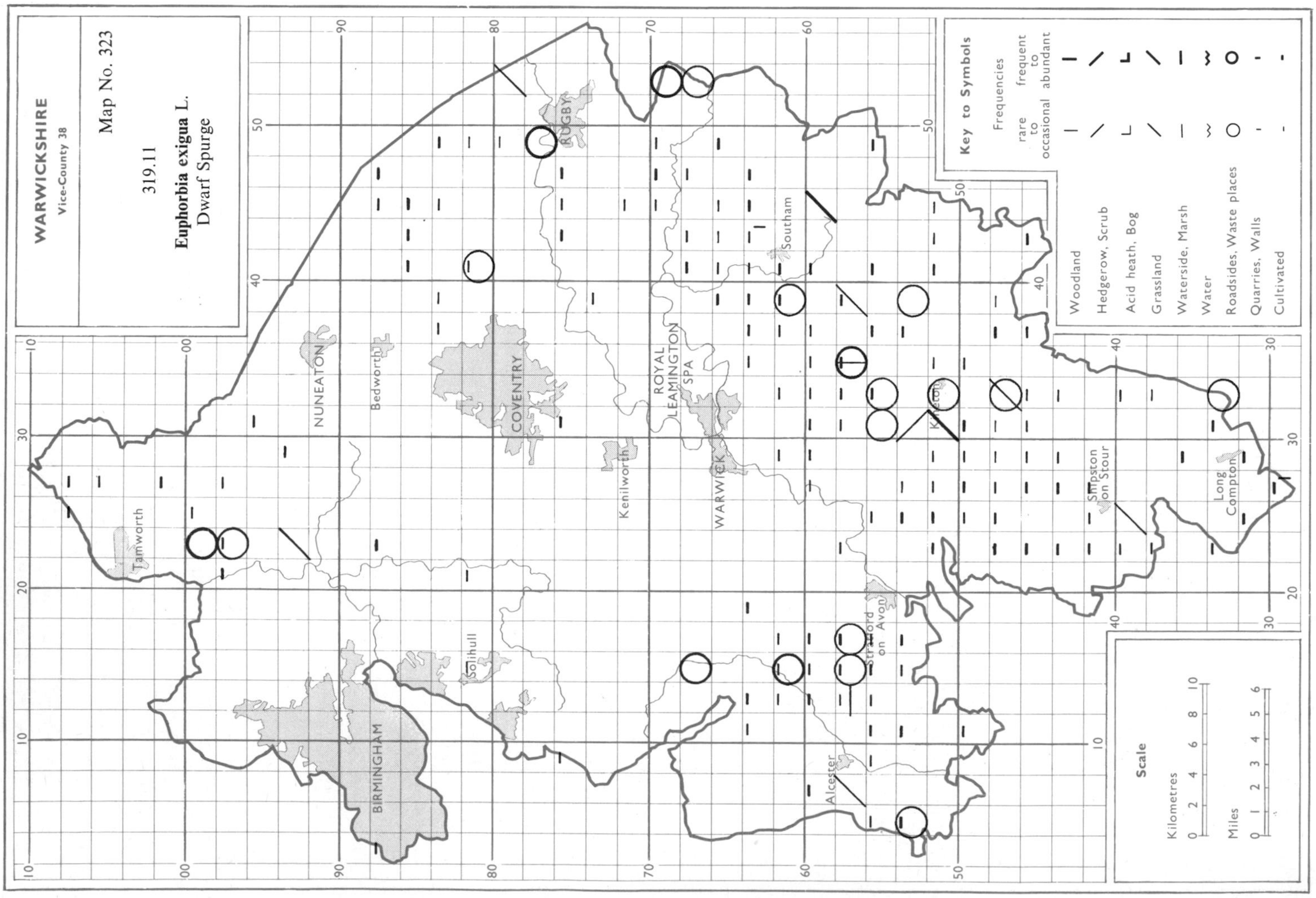
WARWICKSHIRE
Vice-County 38
Map No. 323
319.11
Euphorbia exigua L.
Dwarf Spurge
Key to Symbols
Frequencies
rare to occasional
frequent to abundant
Woodland
Hedgerow, Scrub
Acid heath, Bog
Grassland
Waterside, Marsh
Water
Roadsides, Waste places
Quarries, Walls
Cultivated
Scale
Kilometres
Miles
NUNEATON
Bedworth
COVENTRY
RUGBY
Southam
ROYAL LEAMINGTON SPA
Kenilworth
WARWICK
Kineton
Shipston on Stour
Long Compton
Tamworth
Solihull
BIRMINGHAM
Stratford on Avon
Alcester

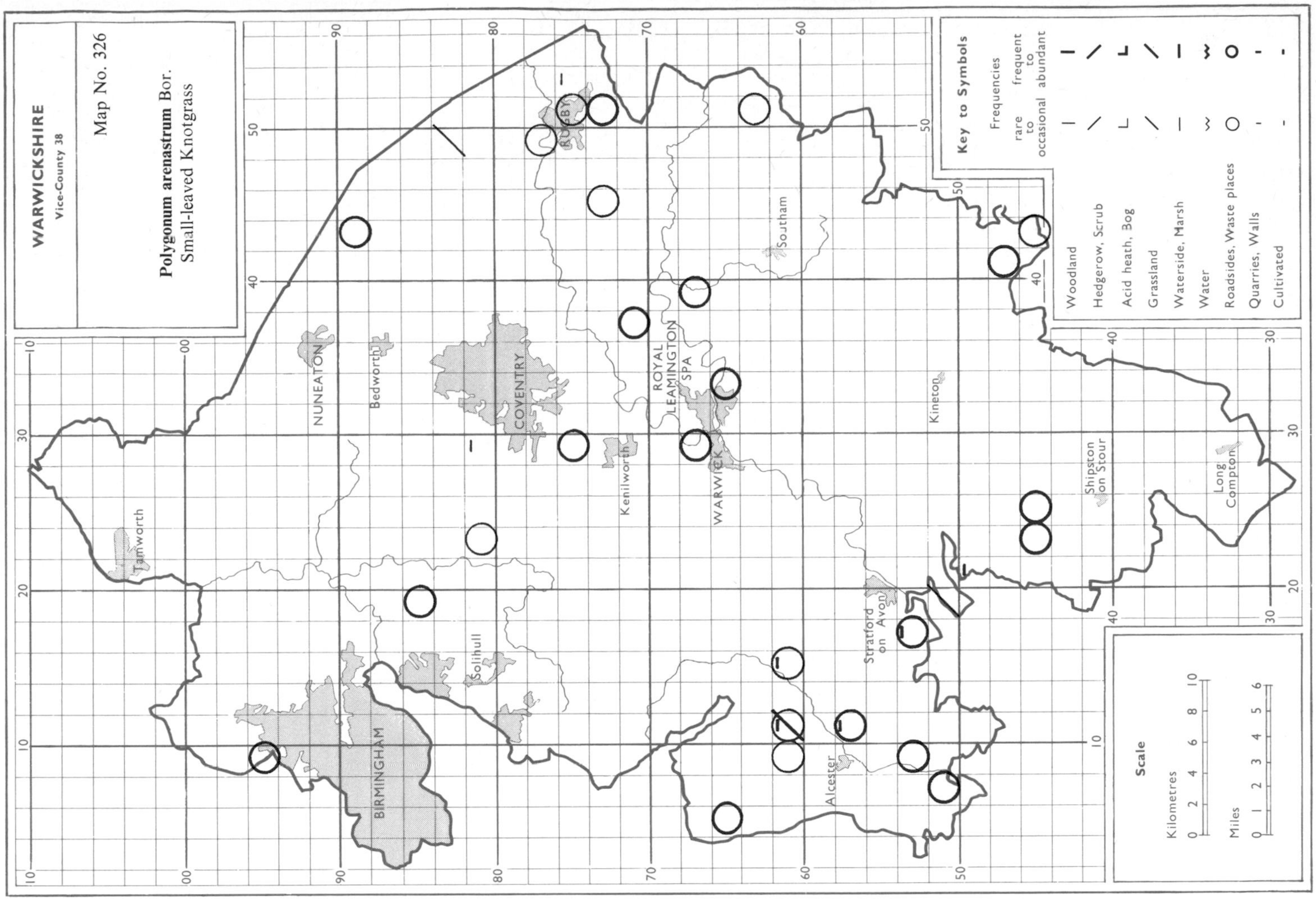
WARWICKSHIRE
Vice-County 38
Map No. 326
Polygonum arenastrum Bor.
Small-leaved Knotgrass
Key to Symbols
Frequencies
rare to occasional
frequent to abundant
Woodland
Hedgerow, Scrub
Acid heath, Bog
Grassland
Waterside, Marsh
Water
Roadsides, Waste places
Quarries, Walls
Cultivated
Scale
Kilometres 0 2 4 6 8 10
Miles 0 1 2 3 4 5 6
NUNEATON
Bedworth
COVENTRY
RUGBY
ROYAL LEAMINGTON SPA
WARWICK
Kenilworth
Southam
Kineton
Tamworth
Solihull
BIRMINGHAM
Stratford on Avon
Alcester
Shipston on Stour
Long Compton

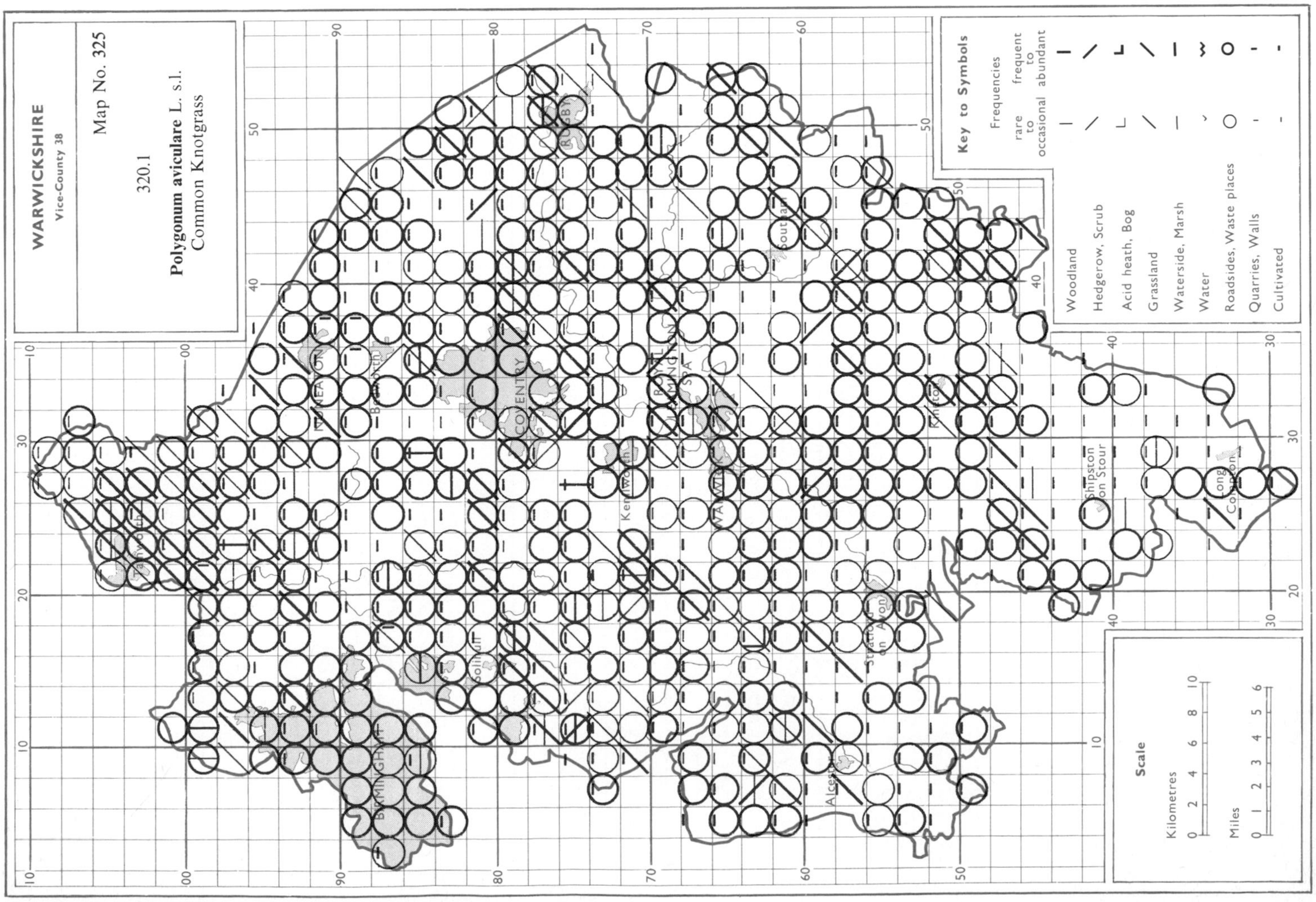
WARWICKSHIRE
Vice-County 38
Map No. 325
320.1
Polygonum aviculare L. s.l.
Common Knotgrass
Key to Symbols
Frequencies
rare to occasional
frequent to abundant
Woodland
Hedgerow, Scrub
Acid heath, Bog
Grassland
Waterside, Marsh
Water
Roadsides, Waste places
Quarries, Walls
Cultivated
Scale
Kilometres 0 2 4 6 8 10
Miles 0 1 2 3 4 5 6
Shipston on Stour
Long Compton
Stratford on Avon
Alcester
Kineton
Southam

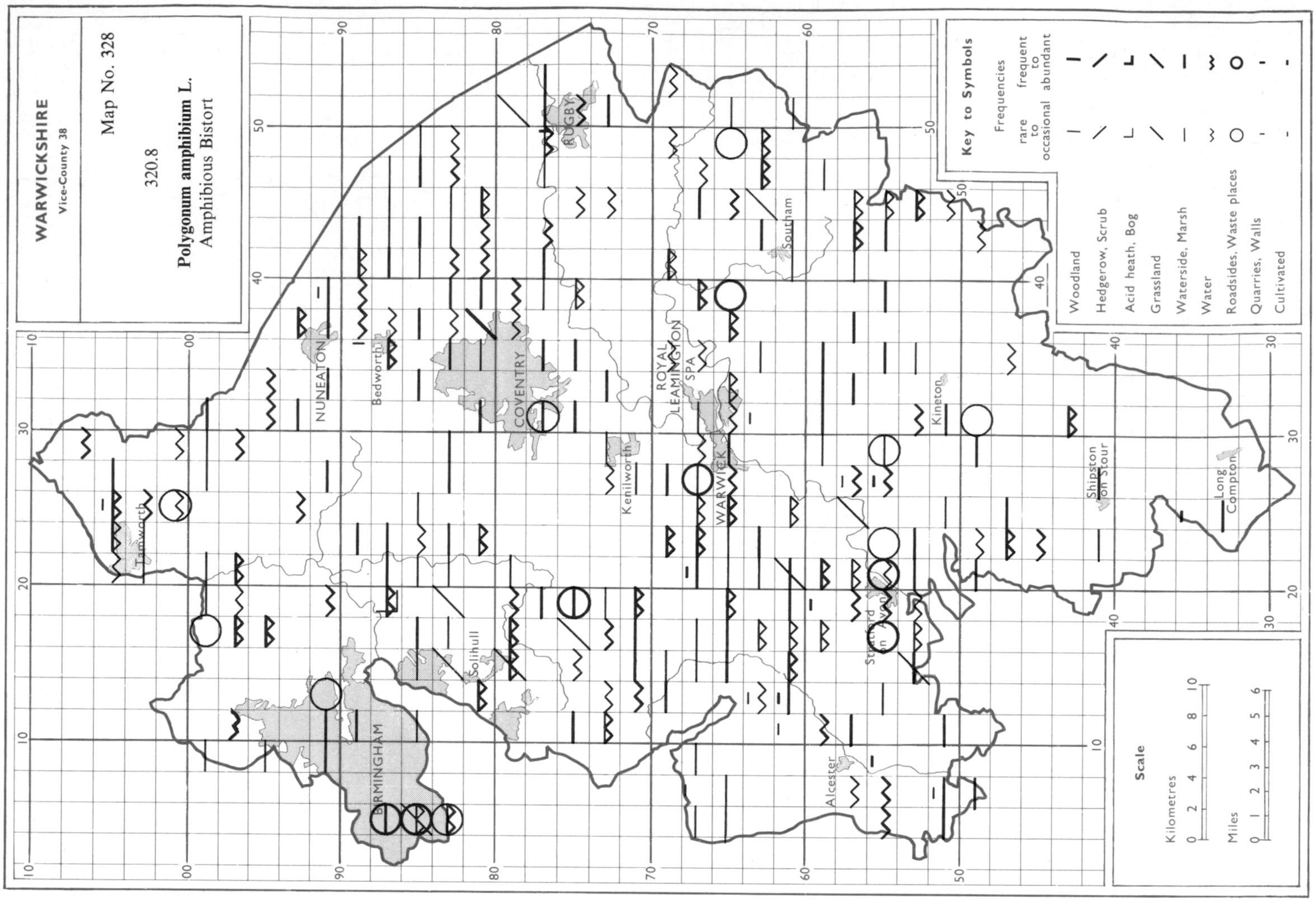
WARWICKSHIRE
Vice-County 38
Map No. 328
320.8
Polygonum amphibium L.
Amphibious Bistort
Key to Symbols
Frequencies
rare to occasional
frequent to abundant
Woodland
Hedgerow, Scrub
Acid heath, Bog
Grassland
Waterside, Marsh
Water
Roadsides, Waste places
Quarries, Walls
Cultivated
Scale
Kilometres 0 2 4 6 8 10
Miles 0 1 2 3 4 5 6
NUNEATON
Bedworth
COVENTRY
RUGBY
Tamworth
ROYAL LEAMINGTON SPA
WARWICK
Kenilworth
Southam
Kineton
Shipston on Stour
Long Compton
Stratford on Avon
Alcester
Solihull
BIRMINGHAM

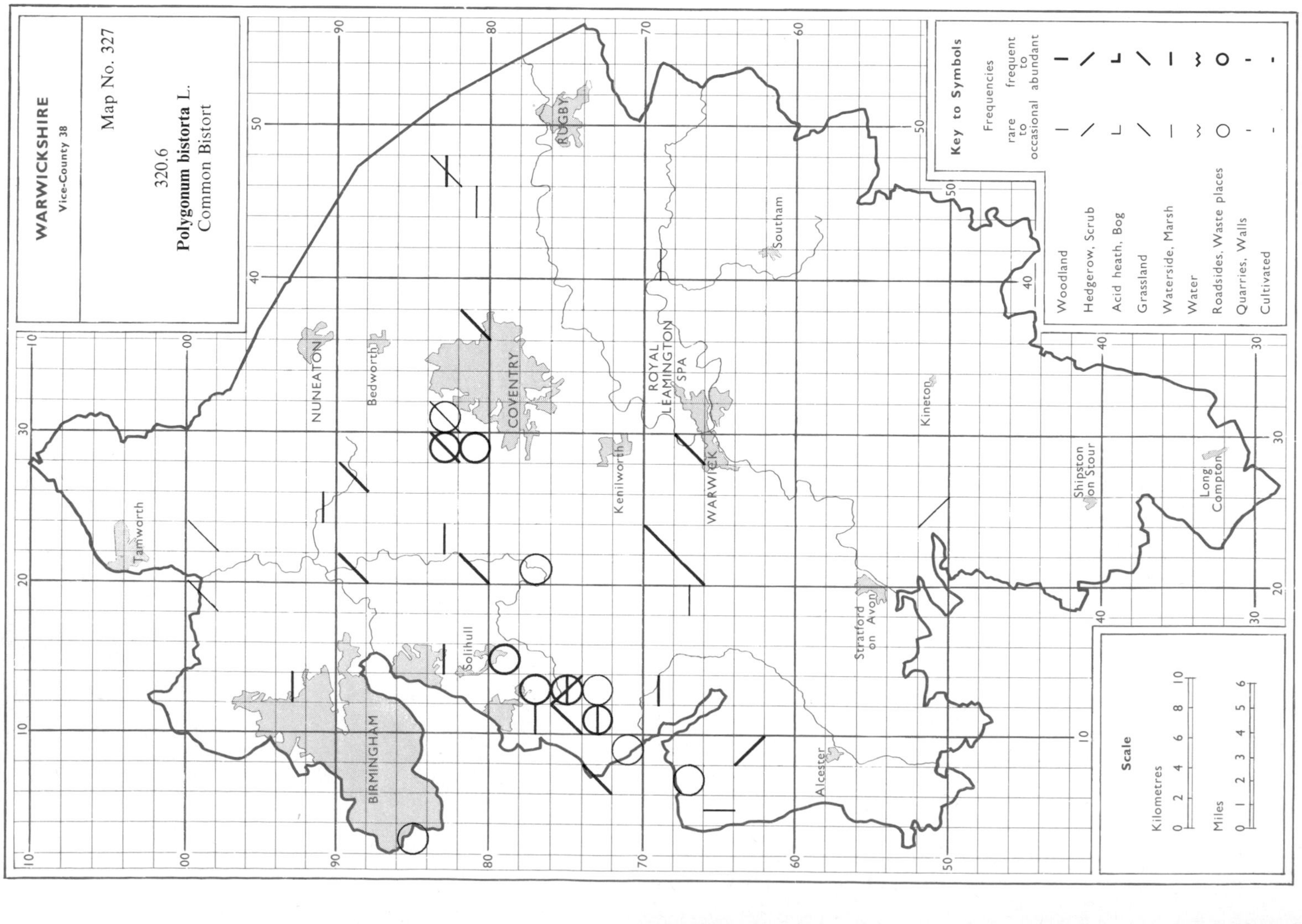
WARWICKSHIRE
Vice-County 38
Map No. 327
320.6
Polygonum bistorta L.
Common Bistort
Key to Symbols
Frequencies
rare to occasional
frequent to abundant
Woodland
Hedgerow, Scrub
Acid heath, Bog
Grassland
Waterside, Marsh
Water
Roadsides, Waste places
Quarries, Walls
Cultivated
Scale
Kilometres 0 2 4 6 8 10
Miles 0 1 2 3 4 5 6
NUNEATON
Bedworth
COVENTRY
RUGBY
Tamworth
ROYAL LEAMINGTON SPA
WARWICK
Kenilworth
Southam
Kineton
Shipston on Stour
Long Compton
Stratford on Avon
Alcester
Solihull
BIRMINGHAM

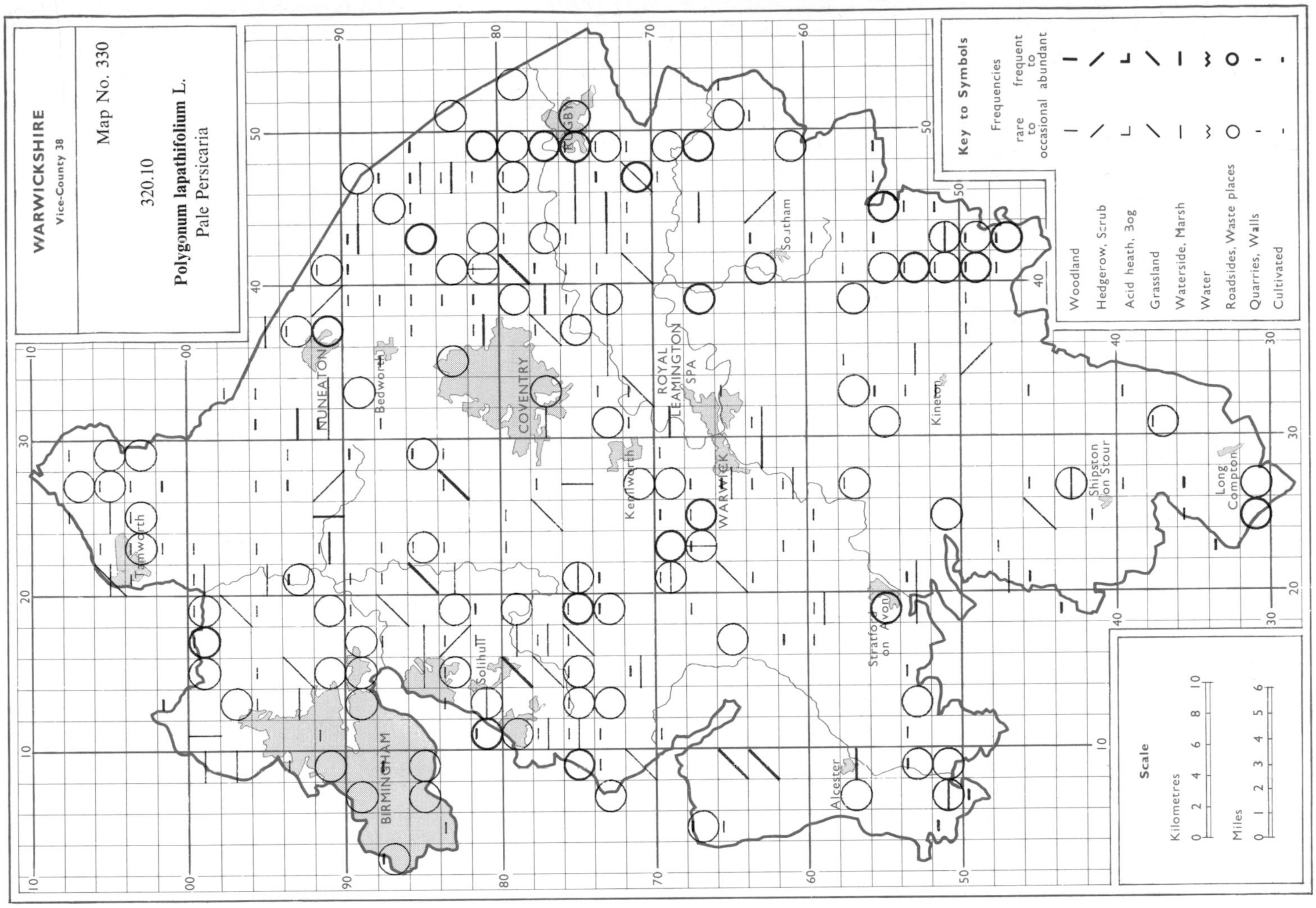
WARWICKSHIRE
Vice-County 38
Map No. 330
320.10
Polygonum lapathifolium L.
Pale Persicaria
Key to Symbols
Frequencies
rare to occasional
frequent to abundant
Woodland
Hedgerow, Scrub
Acid heath, Bog
Grassland
Waterside, Marsh
Water
Roadsides, Waste places
Quarries, Walls
Cultivated
Scale
Kilometres
Miles
TAMWORTH
NUNEATON
Bedworth
COVENTRY
RUGBY
BIRMINGHAM
Solihull
Kenilworth
ROYAL LEAMINGTON SPA
WARWICK
Southam
Kineton
Stratford on Avon
Alcester
Shipston on Stour
Long Compton

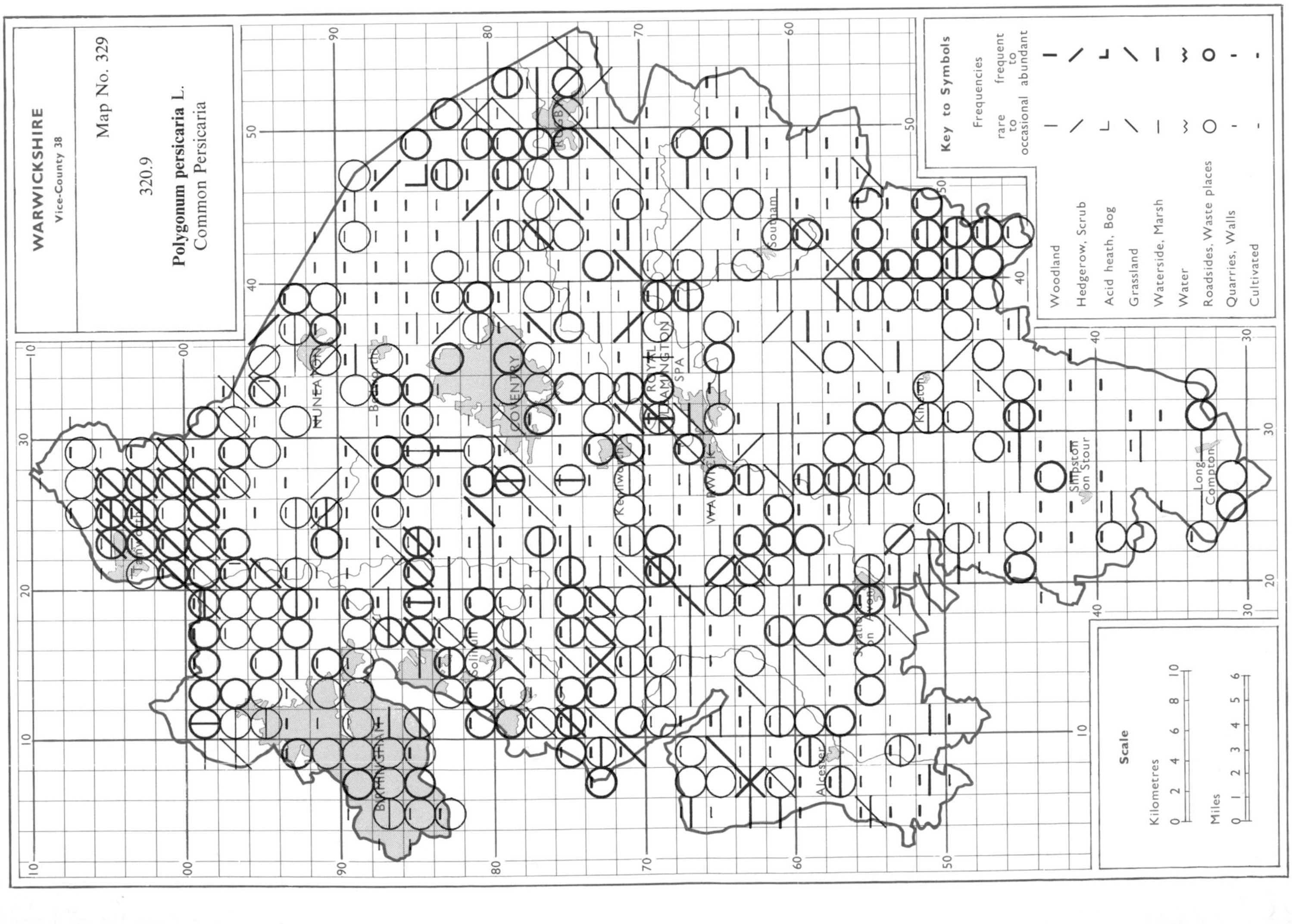
WARWICKSHIRE
Vice-County 38
Map No. 329
320.9
Polygonum persicaria L.
Common Persicaria
Key to Symbols
Frequencies
rare to occasional
frequent to abundant
Woodland
Hedgerow, Scrub
Acid heath, Bog
Grassland
Waterside, Marsh
Water
Roadsides, Waste places
Quarries, Walls
Cultivated
Scale
Kilometres
Miles
NUNEATON
COVENTRY
ROYAL LEAMINGTON SPA
Southam
Kineton
Kenilworth
Solihull
Shipston on Stour
Long Compton
Alcester

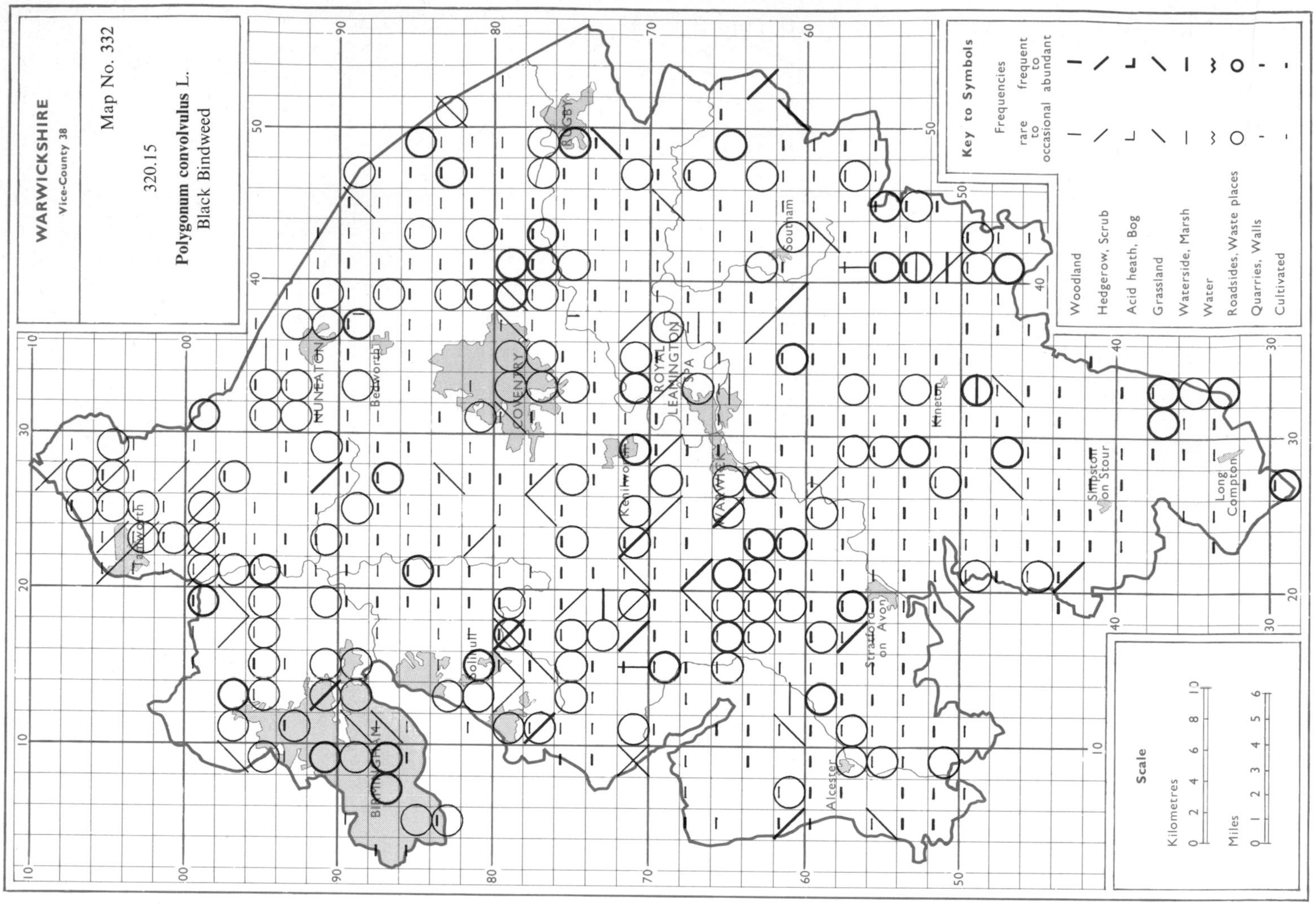
WARWICKSHIRE
Vice-County 38
Map No. 332
320.15
Polygonum convolvulus L.
Black Bindweed
Key to Symbols
Frequencies
rare to occasional
frequent to abundant
Woodland
Hedgerow, Scrub
Acid heath, Bog
Grassland
Waterside, Marsh
Water
Roadsides, Waste places
Quarries, Walls
Cultivated
Scale
Kilometres
0 2 4 6 8 10
Miles
0 1 2 3 4 5 6
NUNEATON
Bedworth
COVENTRY
RUGBY
ROYAL LEAMINGTON SPA
Southam
Kineton
Tamworth
Kenilworth
WARWICK
Solihull
BIRMINGHAM
Stratford on Avon
Alcester
Shipston on Stour
Long Compton

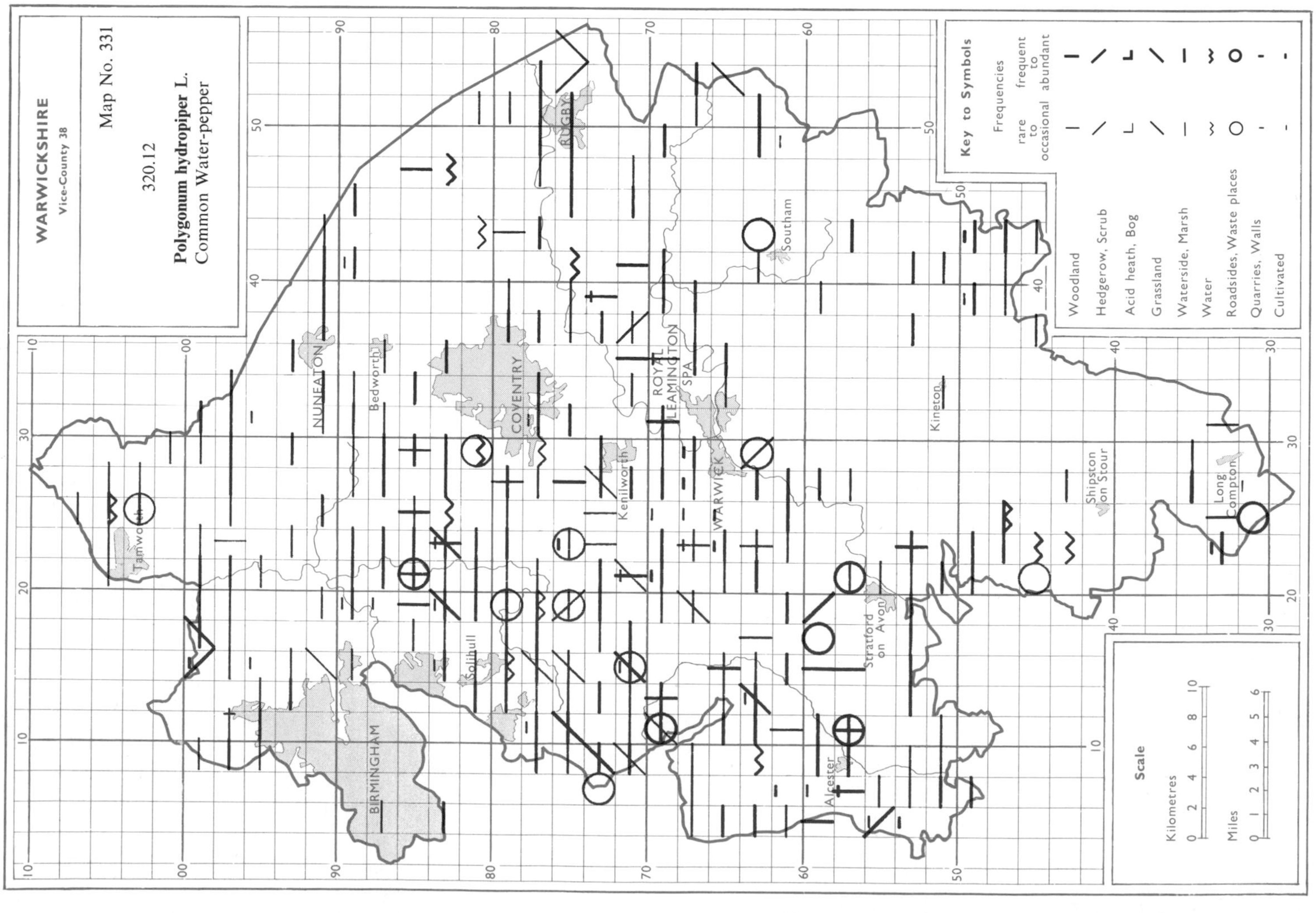
WARWICKSHIRE
Vice-County 38
Map No. 331
320.12
Polygonum hydropiper L.
Common Water-pepper
Key to Symbols
Frequencies
rare to occasional
frequent to abundant
Woodland
Hedgerow, Scrub
Acid heath, Bog
Grassland
Waterside, Marsh
Water
Roadsides, Waste places
Quarries, Walls
Cultivated
Scale
Kilometres
0 2 4 6 8 10
Miles
0 1 2 3 4 5 6
NUNEATON
Bedworth
COVENTRY
RUGBY
ROYAL LEAMINGTON SPA
Southam
Kineton
Tamworth
Kenilworth
WARWICK
Solihull
BIRMINGHAM
Stratford on Avon
Alcester
Shipston on Stour
Long Compton

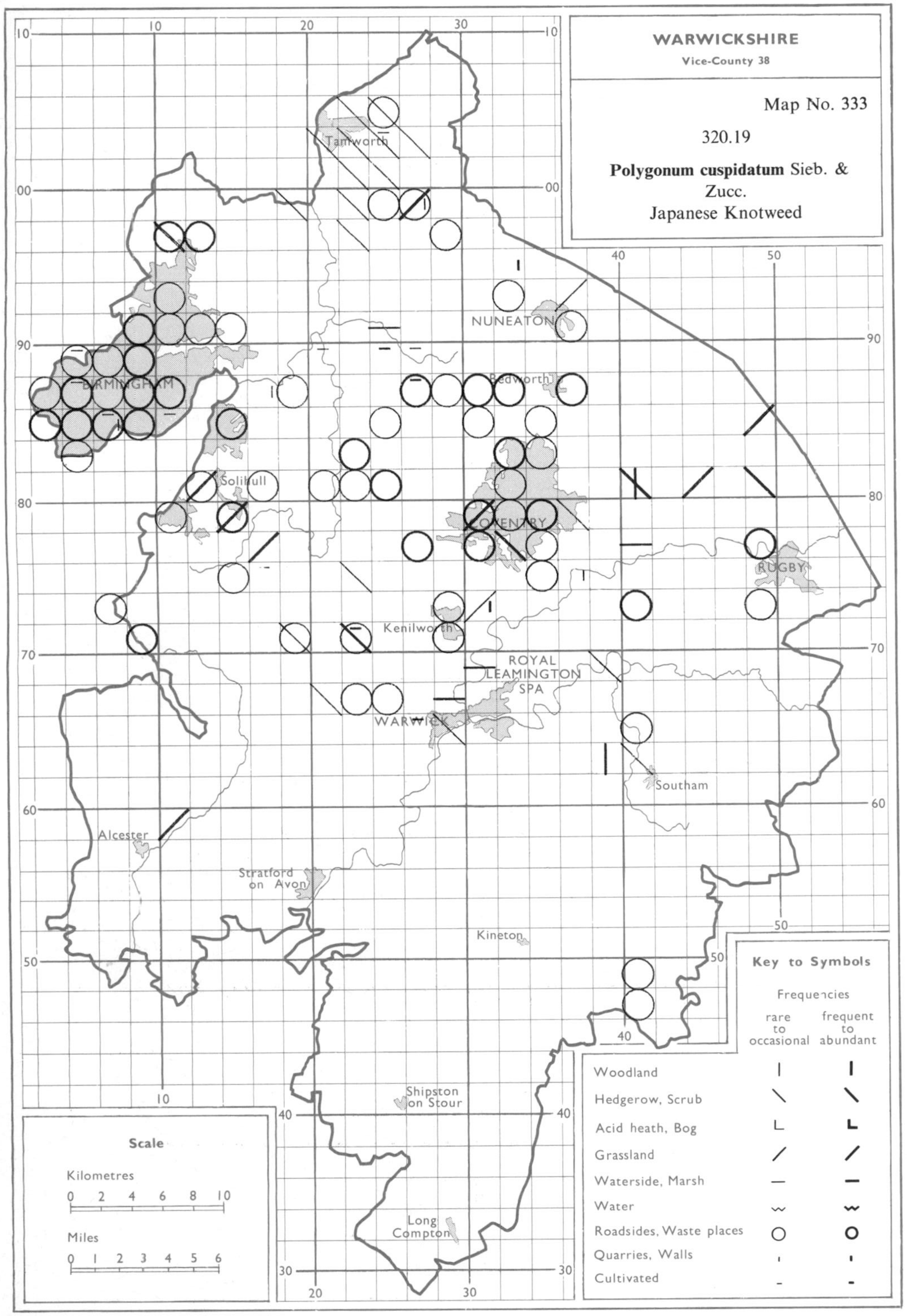
WARWICKSHIRE
Vice-County 38
Map No. 333
320.19
Polygonum cuspidatum Sieb. & Zucc.
Japanese Knotweed
Tamworth
Birmingham
Solihull
Nuneaton
Bedworth
Coventry
Rugby
Kenilworth
Royal Leamington Spa
Warwick
Southam
Alcester
Stratford on Avon
Kineton
Shipston on Stour
Long Compton
Key to Symbols
Frequencies
rare to occasional
frequent to abundant
Woodland
Hedgerow, Scrub
Acid heath, Bog
Grassland
Waterside, Marsh
Water
Roadsides, Waste places
Quarries, Walls
Cultivated
Scale
Kilometres
0 2 4 6 8 10
Miles
0 1 2 3 4 5 6

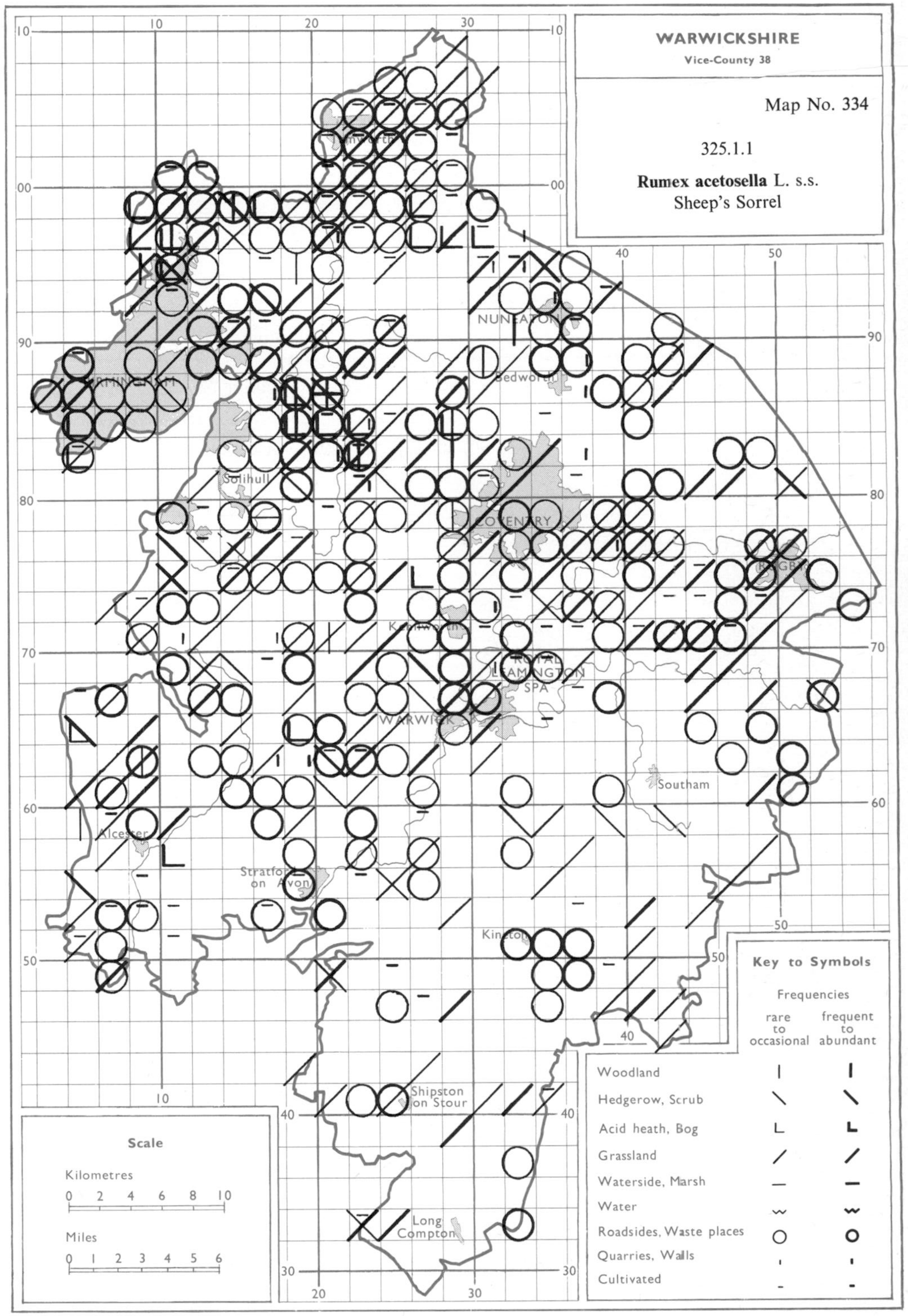
WARWICKSHIRE
Vice-County 38
Map No. 334
325.1.1
Rumex acetosella L. s.s.
Sheep's Sorrel
Tamworth
Birmingham
Solihull
Nuneaton
Bedworth
Coventry
Rugby
Kenilworth
Royal Leamington Spa
Warwick
Southam
Alcester
Stratford on Avon
Kineton
Shipston on Stour
Long Compton
Key to Symbols
Frequencies
rare to occasional
frequent to abundant
Woodland
Hedgerow, Scrub
Acid heath, Bog
Grassland
Waterside, Marsh
Water
Roadsides, Waste places
Quarries, Walls
Cultivated
Scale
Kilometres
0 2 4 6 8 10
Miles
0 1 2 3 4 5 6

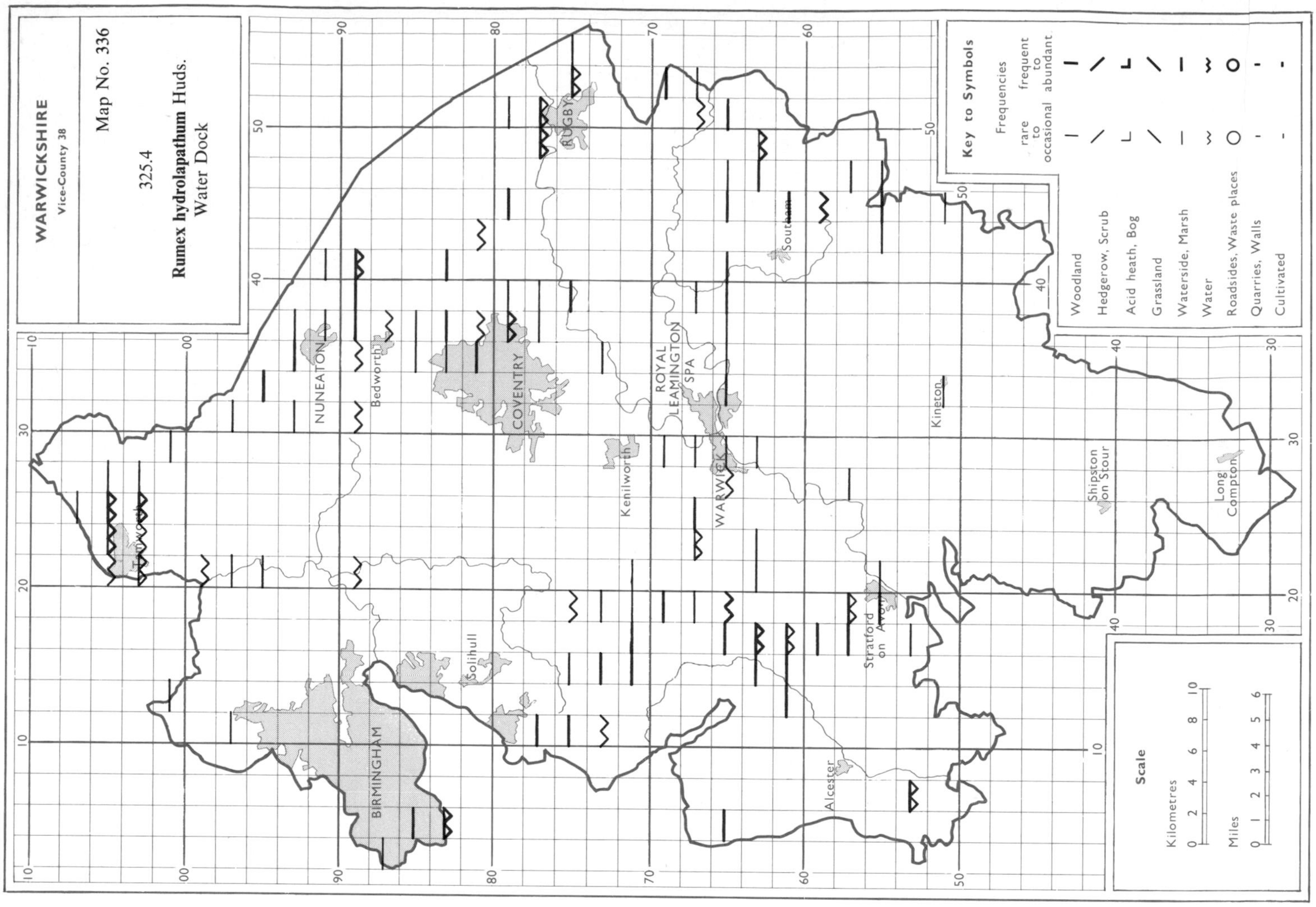
WARWICKSHIRE
Vice-County 38
Map No. 336
325.4
Rumex hydrolapathum Huds.
Water Dock
Key to Symbols
Frequencies
rare to occasional
frequent to abundant
Woodland
Hedgerow, Scrub
Acid heath, Bog
Grassland
Waterside, Marsh
Water
Roadsides, Waste places
Quarries, Walls
Cultivated
Scale
Kilometres
Miles
NUNEATON
Bedworth
COVENTRY
RUGBY
ROYAL LEAMINGTON SPA
Kenilworth
WARWICK
Southam
Kineton
Shipston on Stour
Long Compton
Tamworth
Solihull
BIRMINGHAM
Stratford on Avon
Alcester

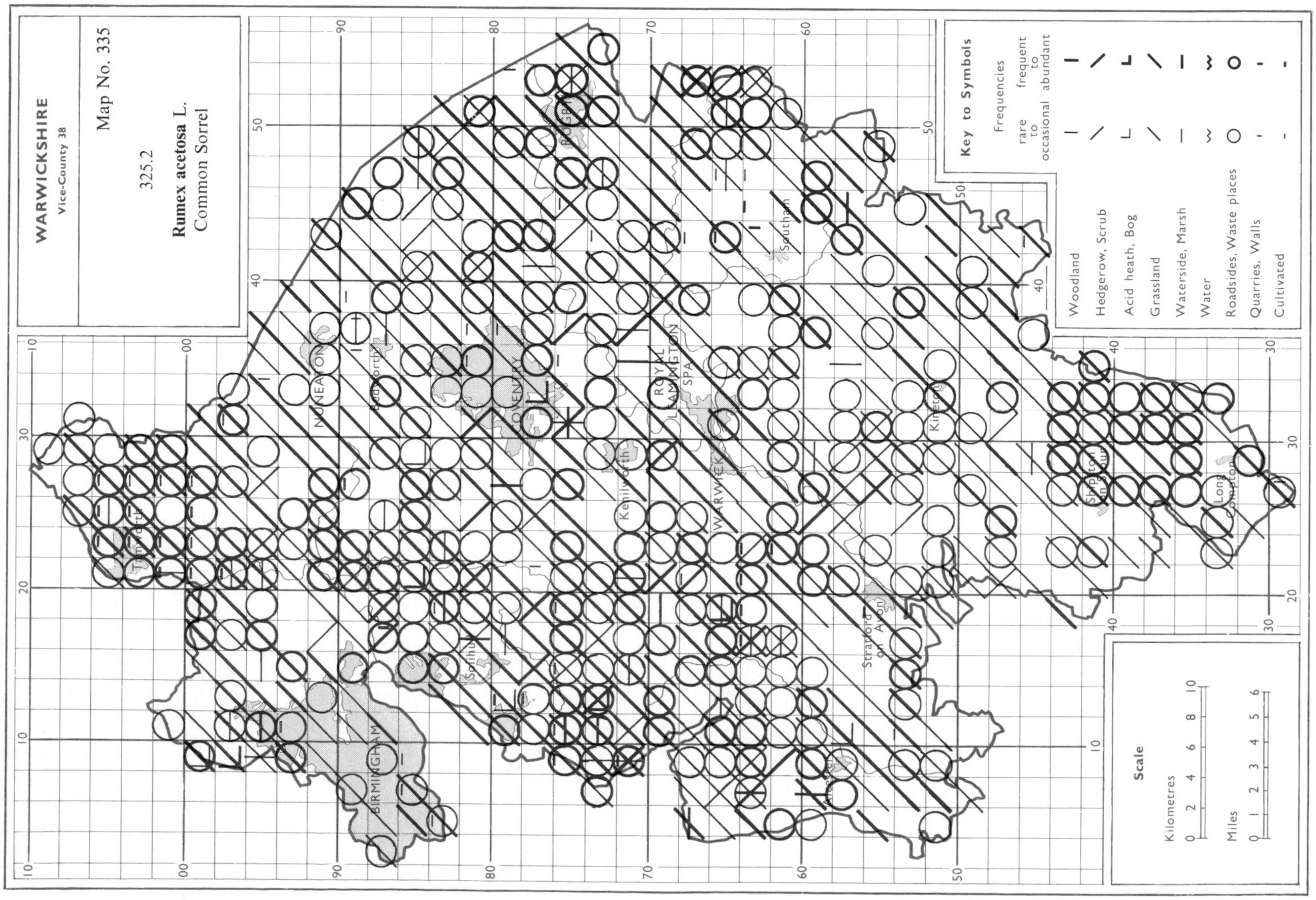
WARWICKSHIRE
Vice-County 38
Map No. 335
325.2
Rumex acetosa L.
Common Sorrel
Key to Symbols
Frequencies
rare to occasional
frequent to abundant
Woodland
Hedgerow, Scrub
Acid heath, Bog
Grassland
Waterside, Marsh
Water
Roadsides, Waste places
Quarries, Walls
Cultivated
Scale
Kilometres
Miles
NUNEATON
Bedworth
COVENTRY
RUGBY
ROYAL LEAMINGTON SPA
Kenilworth
WARWICK
Southam
Kineton
Shipston on Stour
Long Compton
Tamworth
Solihull
BIRMINGHAM
Stratford on Avon

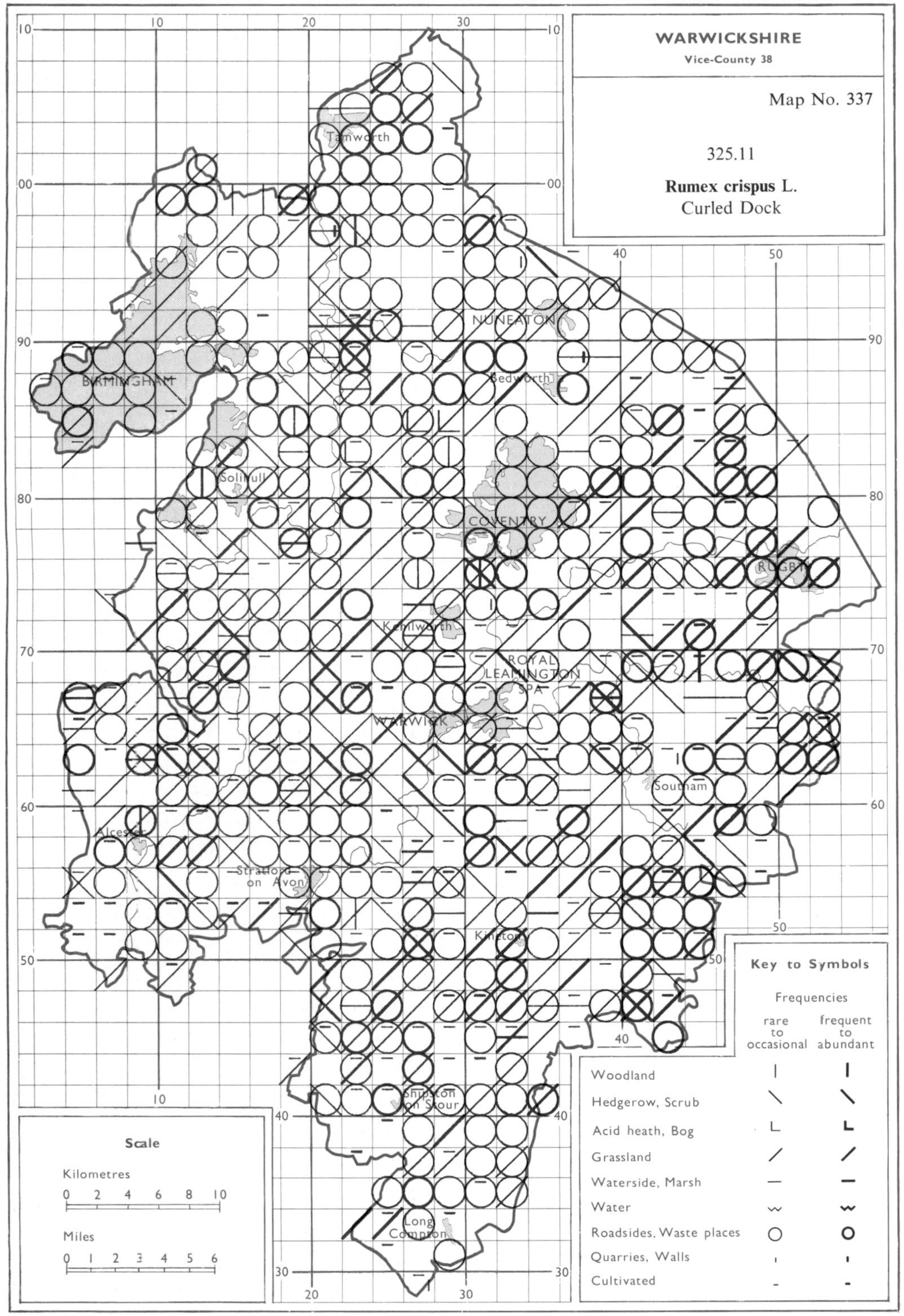
WARWICKSHIRE
Vice-County 38
Map No. 337
325.11
Rumex crispus L.
Curled Dock
Key to Symbols
Frequencies
rare to occasional
frequent to abundant
Woodland
Hedgerow, Scrub
Acid heath, Bog
Grassland
Waterside, Marsh
Water
Roadsides, Waste places
Quarries, Walls
Cultivated
Scale
Kilometres
Miles
Tamworth
NUNEATON
Bedworth
BIRMINGHAM
Solihull
COVENTRY
RUGBY
Kenilworth
ROYAL LEAMINGTON SPA
WARWICK
Southam
Alcester
Stratford on Avon
Kineton
Shipston on Stour
Long Compton

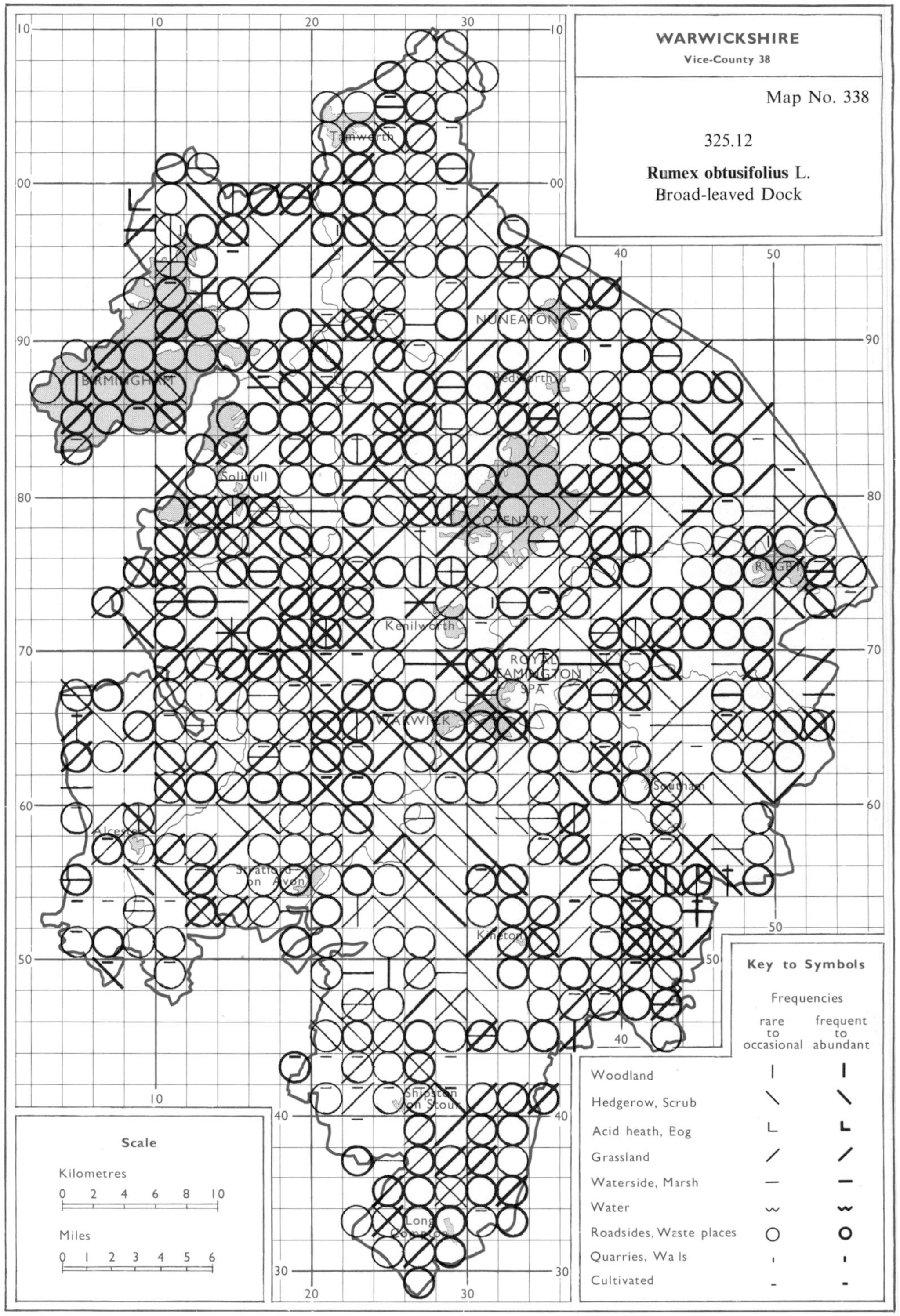
WARWICKSHIRE
Vice-County 38
Map No. 338
325.12
Rumex obtusifolius L.
Broad-leaved Dock
Key to Symbols
Frequencies
rare to occasional
frequent to abundant
Woodland
Hedgerow, Scrub
Acid heath, Eog
Grassland
Waterside, Marsh
Water
Roadsides, Waste places
Quarries, Wa ls
Cultivated
Scale
Kilometres
Miles
Tamworth
NUNEATON
Bedworth
BIRMINGHAM
Solihull
COVENTRY
RUGBY
Kenilworth
ROYAL LEAMINGTON SPA
WARWICK
Southam
Alcester
Stratford on Avon
Kineton
Shipston on Stour
Long Compton

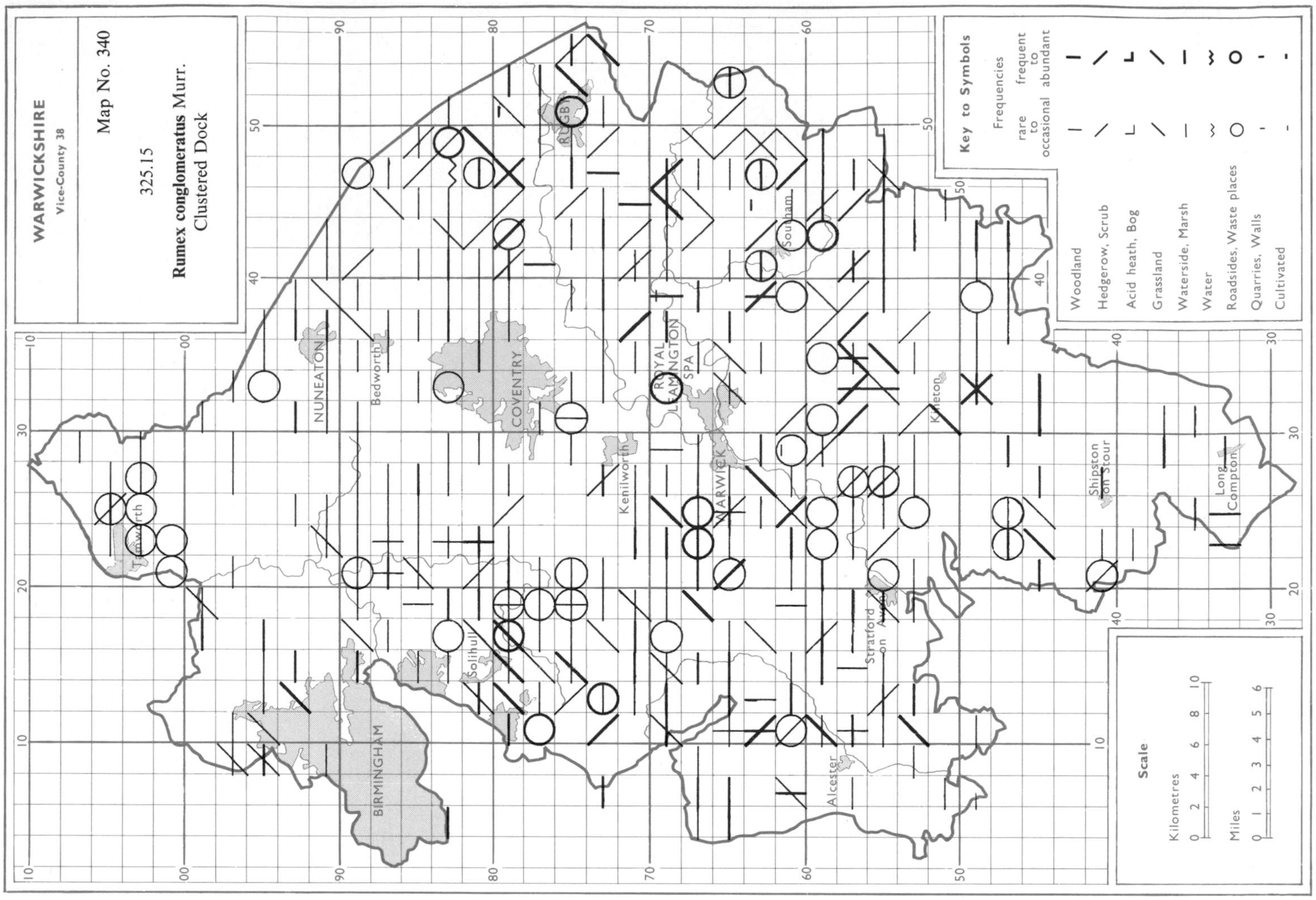
WARWICKSHIRE
Vice-County 38
Map No. 340
325.15
Rumex conglomeratus Murr.
Clustered Dock
Key to Symbols
Frequencies
rare to occasional
frequent to abundant
Woodland
Hedgerow, Scrub
Acid heath, Bog
Grassland
Waterside, Marsh
Water
Roadsides, Waste places
Quarries, Walls
Cultivated
Scale
Kilometres
0 2 4 6 8 10
Miles
0 1 2 3 4 5 6
NUNEATON
Bedworth
COVENTRY
RUGBY
Southam
ROYAL LEAMINGTON SPA
Kenilworth
WARWICK
Kineton
Shipston on Stour
Long Compton
Tamworth
Solihull
BIRMINGHAM
Stratford on Avon
Alcester

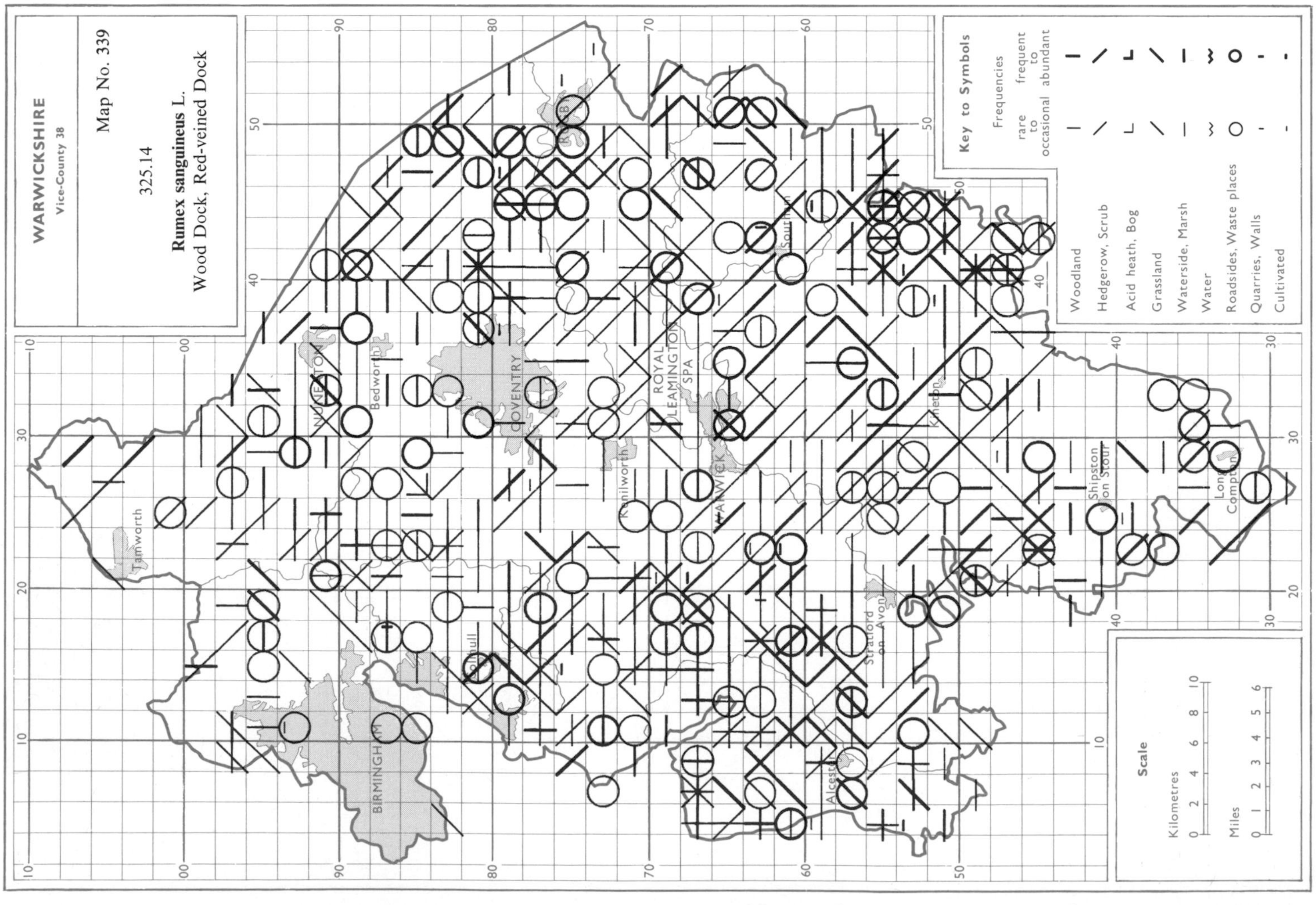
WARWICKSHIRE
Vice-County 38
Map No. 339
325.14
Rumex sanguineus L.
Wood Dock, Red-veined Dock
Key to Symbols
Frequencies
rare to occasional
frequent to abundant
Woodland
Hedgerow, Scrub
Acid heath, Bog
Grassland
Waterside, Marsh
Water
Roadsides, Waste places
Quarries, Walls
Cultivated
Scale
Kilometres
0 2 4 6 8 10
Miles
0 1 2 3 4 5 6
NUNEATON
Bedworth
COVENTRY
RUGBY
Southam
ROYAL LEAMINGTON SPA
Kenilworth
WARWICK
Kineton
Shipston on Stour
Long Compton
Tamworth
Solihull
BIRMINGHAM
Stratford on Avon
Alcester

WARWICKSHIRE
Vice-County 38
Map No. 341
326.1
Parietaria judaica L.
Pellitory-of-the-wall
Tamworth
NUNEATON
Bedworth
BIRMINGHAM
Solihull
COVENTRY
RUGBY
Kenilworth
ROYAL LEAMINGTON SPA
WARWICK
Southam
Alcester
Stratford on Avon
Kineton
Shipston on Stour
Long Compton
Scale
Kilometres
Miles
Key to Symbols
Frequencies
rare to occasional
frequent to abundant
Woodland
Hedgerow, Scrub
Acid heath, Bog
Grassland
Waterside, Marsh
Water
Roadsides, Waste places
Quarries, Walls
Cultivated

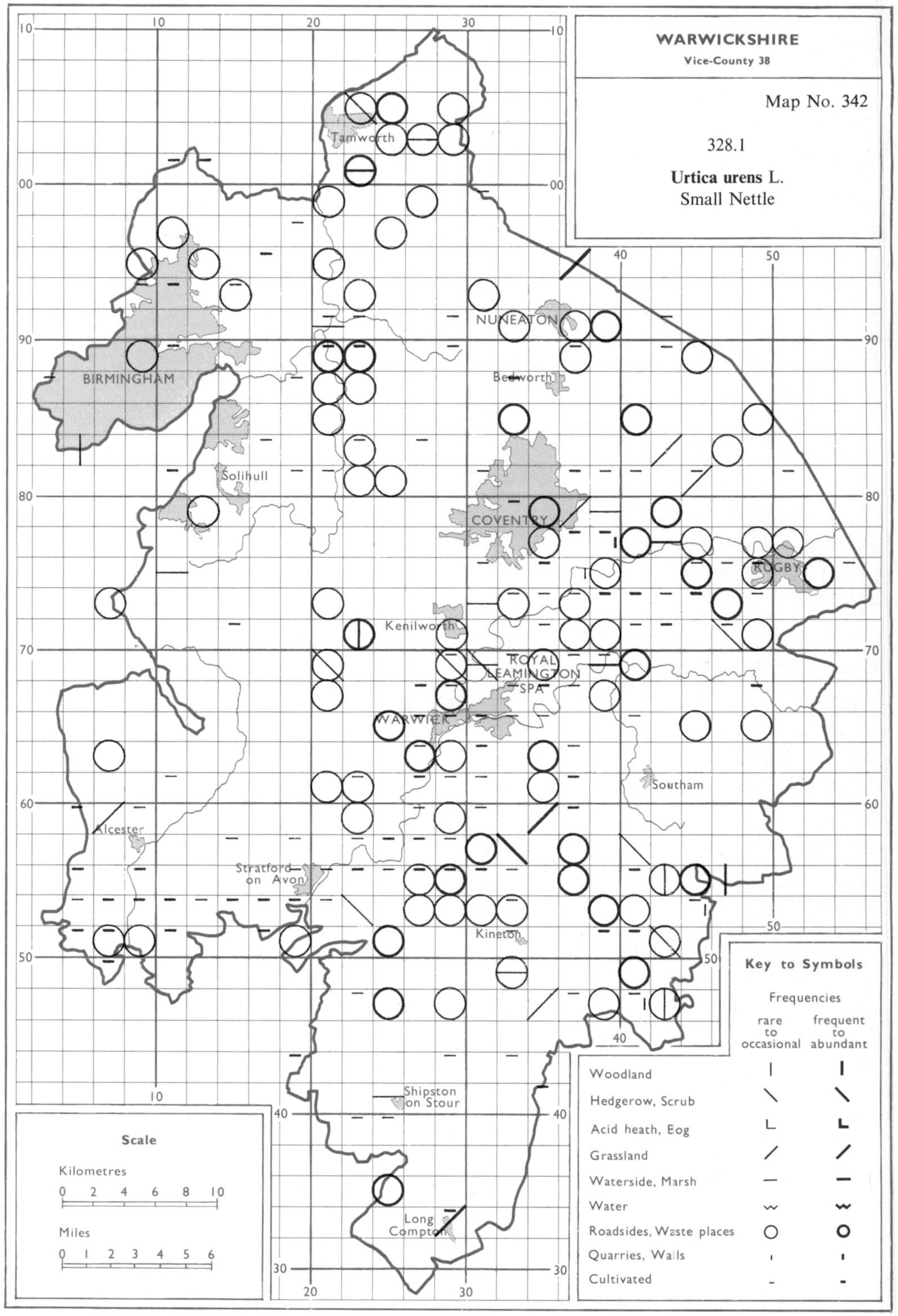
WARWICKSHIRE
Vice-County 38
Map No. 342
328.1
Urtica urens L.
Small Nettle
Tamworth
NUNEATON
Bedworth
BIRMINGHAM
Solihull
COVENTRY
RUGBY
Kenilworth
ROYAL LEAMINGTON SPA
WARWICK
Southam
Alcester
Stratford on Avon
Kineton
Shipston on Stour
Long Compton
Scale
Kilometres
Miles
Key to Symbols
Frequencies
rare to occasional
frequent to abundant
Woodland
Hedgerow, Scrub
Acid heath, Bog
Grassland
Waterside, Marsh
Water
Roadsides, Waste places
Quarries, Walls
Cultivated

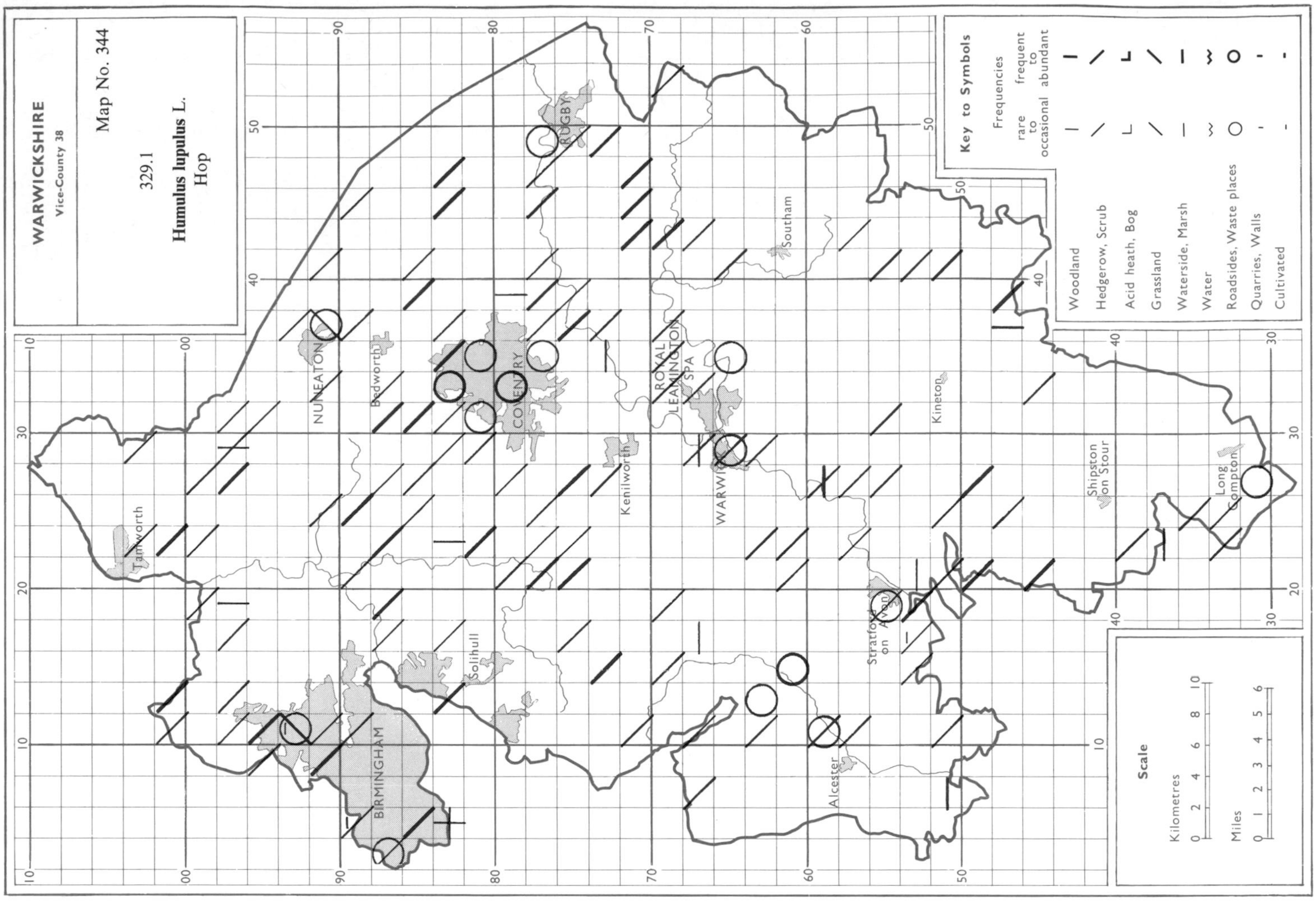
WARWICKSHIRE
Vice-County 38
Map No. 344
329.1
Humulus lupulus L.
Hop
Key to Symbols
Frequencies
rare to occasional
frequent to abundant
Woodland
Hedgerow, Scrub
Acid heath, Bog
Grassland
Waterside, Marsh
Water
Roadsides, Waste places
Quarries, Walls
Cultivated
Scale
Kilometres
Miles
RUGBY
Southam
NUNEATON
Bedworth
COVENTRY
ROYAL LEAMINGTON SPA
Kineton
Kenilworth
WARWICK
Shipston on Stour
Long Compton
Tamworth
Solihull
Stratford on Avon
BIRMINGHAM
Alcester

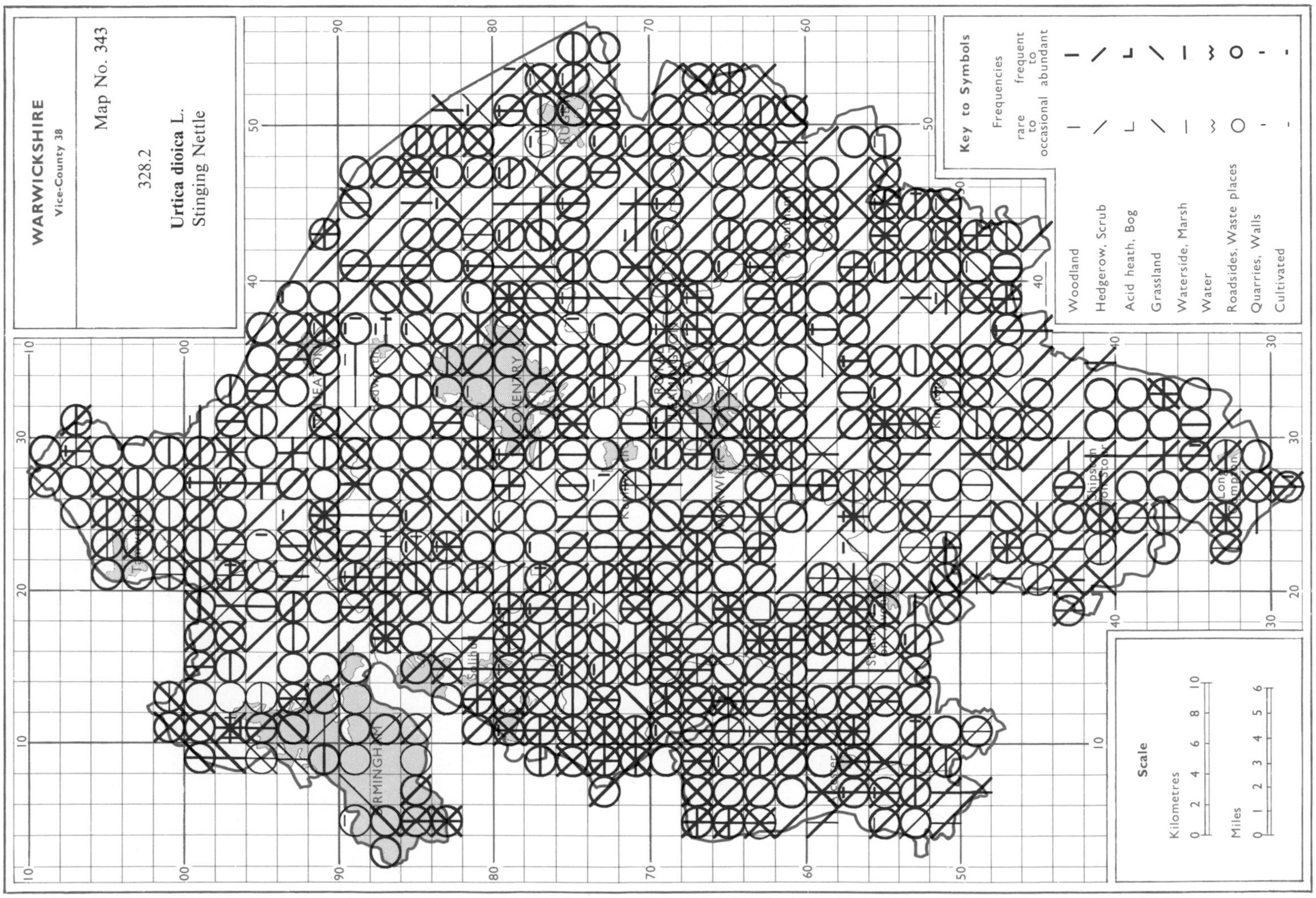
WARWICKSHIRE
Vice-County 38
Map No. 343
328.2
Urtica dioica L.
Stinging Nettle
Key to Symbols
Frequencies
rare to occasional
frequent to abundant
Woodland
Hedgerow, Scrub
Acid heath, Bog
Grassland
Waterside, Marsh
Water
Roadsides, Waste places
Quarries, Walls
Cultivated
Scale
Kilometres
Miles

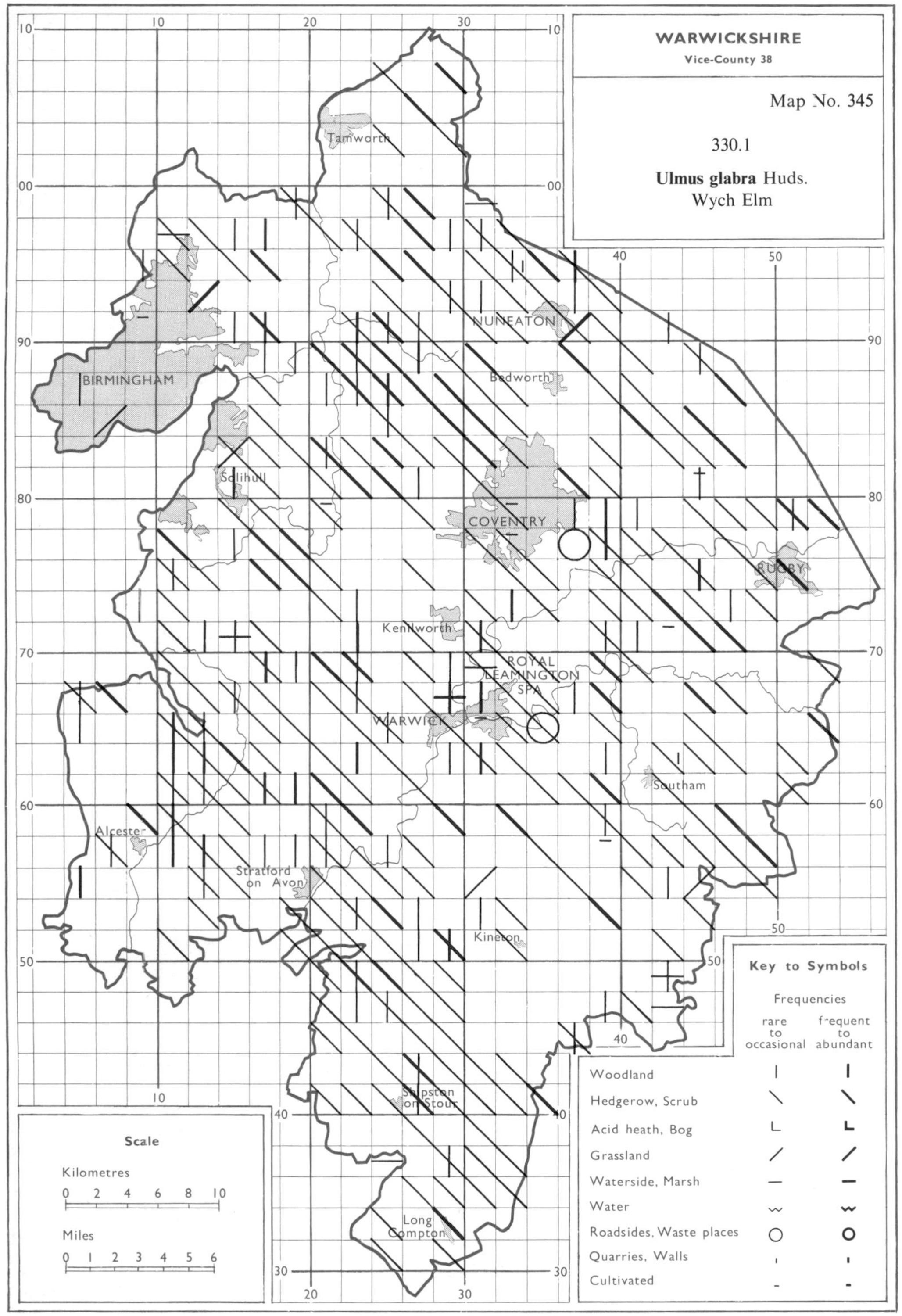
WARWICKSHIRE
Vice-County 38
Map No. 345
330.1
Ulmus glabra Huds.
Wych Elm
Tamworth
NUNEATON
Bedworth
BIRMINGHAM
Solihull
COVENTRY
RUGBY
Kenilworth
ROYAL LEAMINGTON SPA
WARWICK
Southam
Alcester
Stratford on Avon
Kineton
Shipston on Stour
Long Compton
Key to Symbols
Frequencies
rare to occasional
frequent to abundant
Woodland
Hedgerow, Scrub
Acid heath, Bog
Grassland
Waterside, Marsh
Water
Roadsides, Waste places
Quarries, Walls
Cultivated
Scale
Kilometres
0 2 4 6 8 10
Miles
0 1 2 3 4 5 6

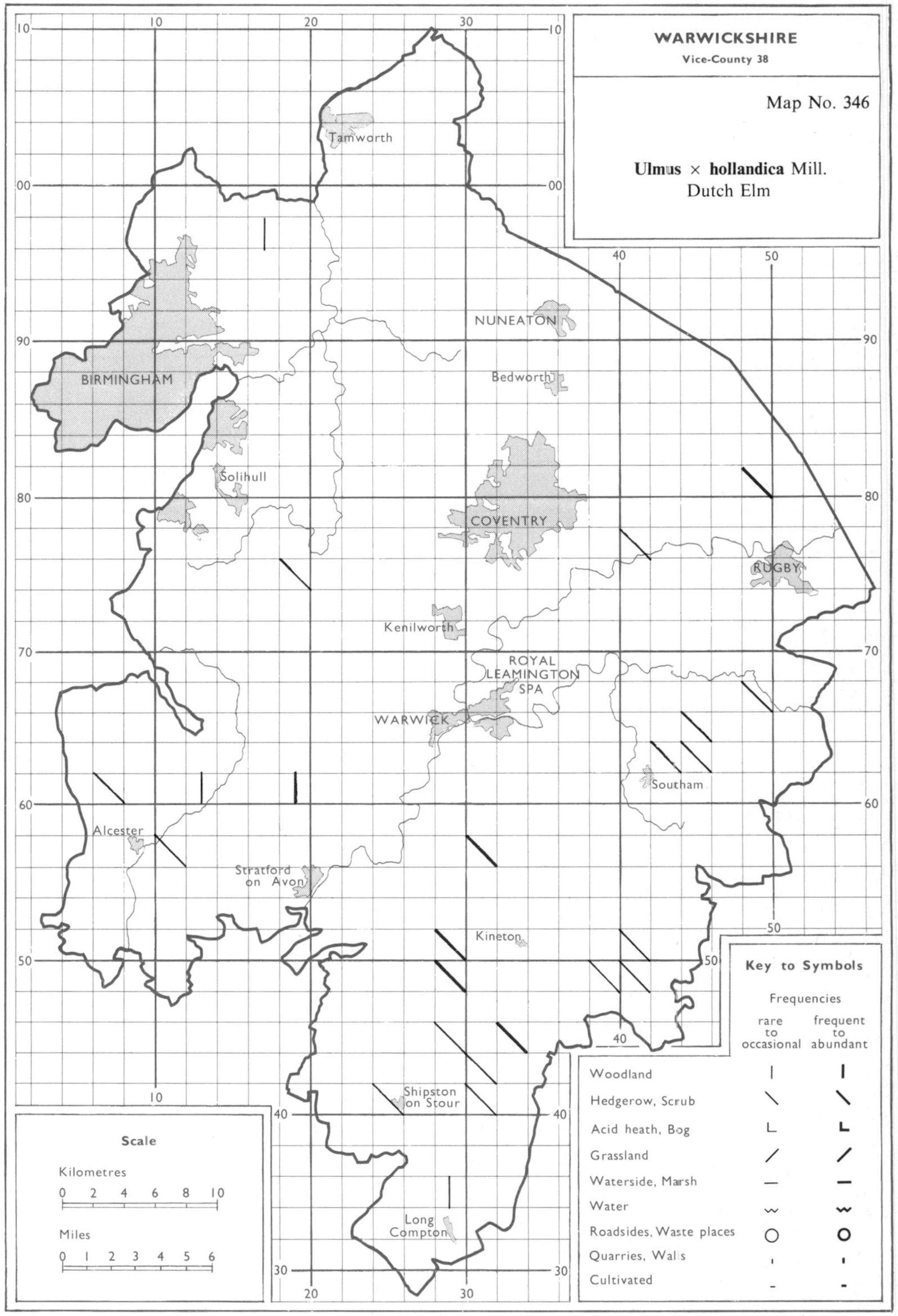
WARWICKSHIRE
Vice-County 38
Map No. 346
Ulmus × hollandica Mill.
Dutch Elm
Tamworth
NUNEATON
Bedworth
BIRMINGHAM
Solihull
COVENTRY
RUGBY
Kenilworth
ROYAL LEAMINGTON SPA
WARWICK
Southam
Alcester
Stratford on Avon
Kineton
Shipston on Stour
Long Compton
Key to Symbols
Frequencies
rare to occasional
frequent to abundant
Woodland
Hedgerow, Scrub
Acid heath, Bog
Grassland
Waterside, Marsh
Water
Roadsides, Waste places
Quarries, Walls
Cultivated
Scale
Kilometres
0 2 4 6 8 10
Miles
0 1 2 3 4 5 6

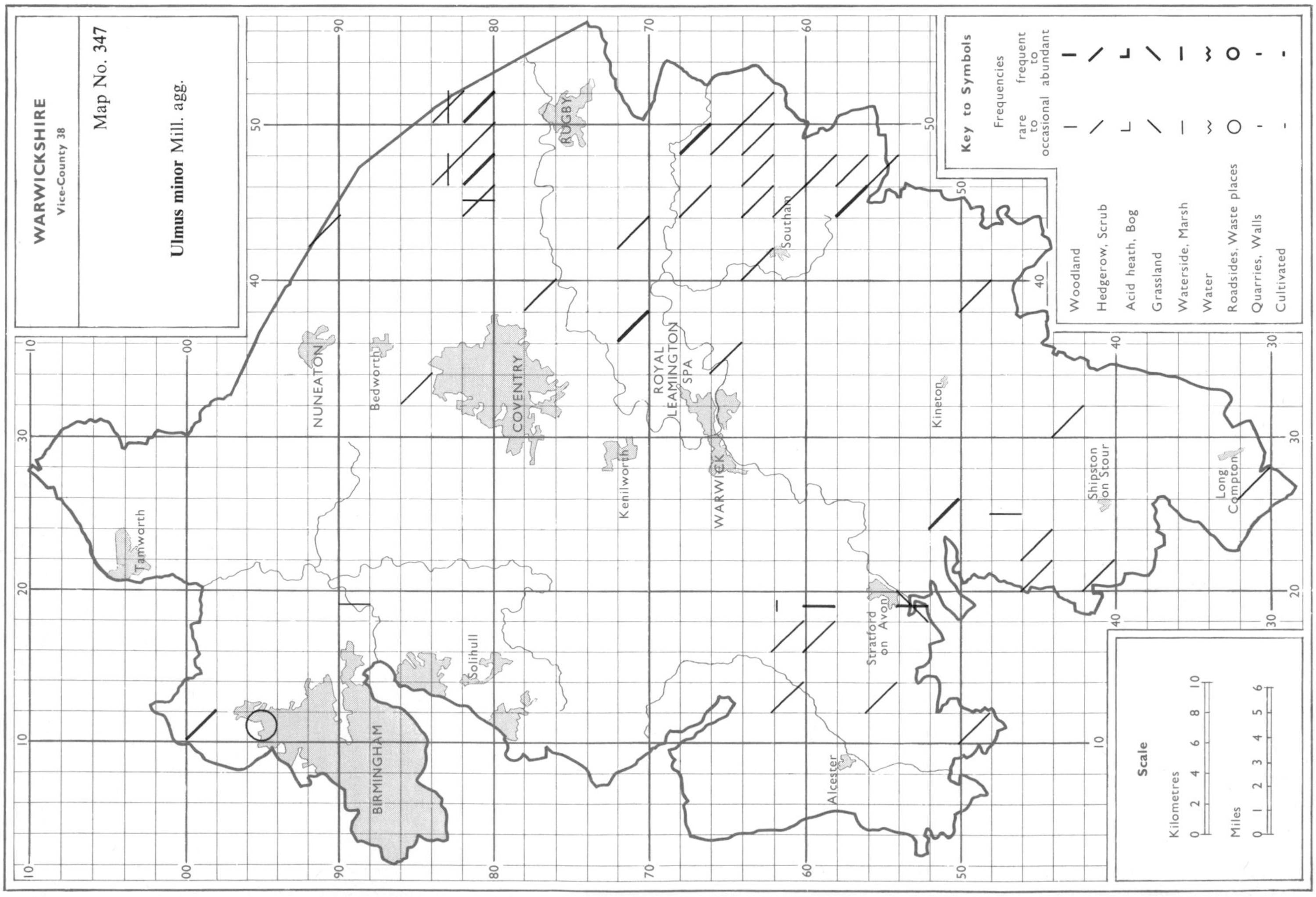
WARWICKSHIRE
Vice-County 38
Map No. 347
Ulmus minor Mill. agg.
Key to Symbols
Frequencies
rare to occasional
frequent to abundant
Woodland
Hedgerow, Scrub
Acid heath, Bog
Grassland
Waterside, Marsh
Water
Roadsides, Waste places
Quarries, Walls
Cultivated
Scale
Kilometres 0 2 4 6 8 10
Miles 0 1 2 3 4 5 6
NUNEATON
Bedworth
COVENTRY
RUGBY
Tamworth
Kenilworth
ROYAL LEAMINGTON SPA
WARWICK
Southam
Kineton
Shipston on Stour
Long Compton
Solihull
BIRMINGHAM
Stratford on Avon
Alcester

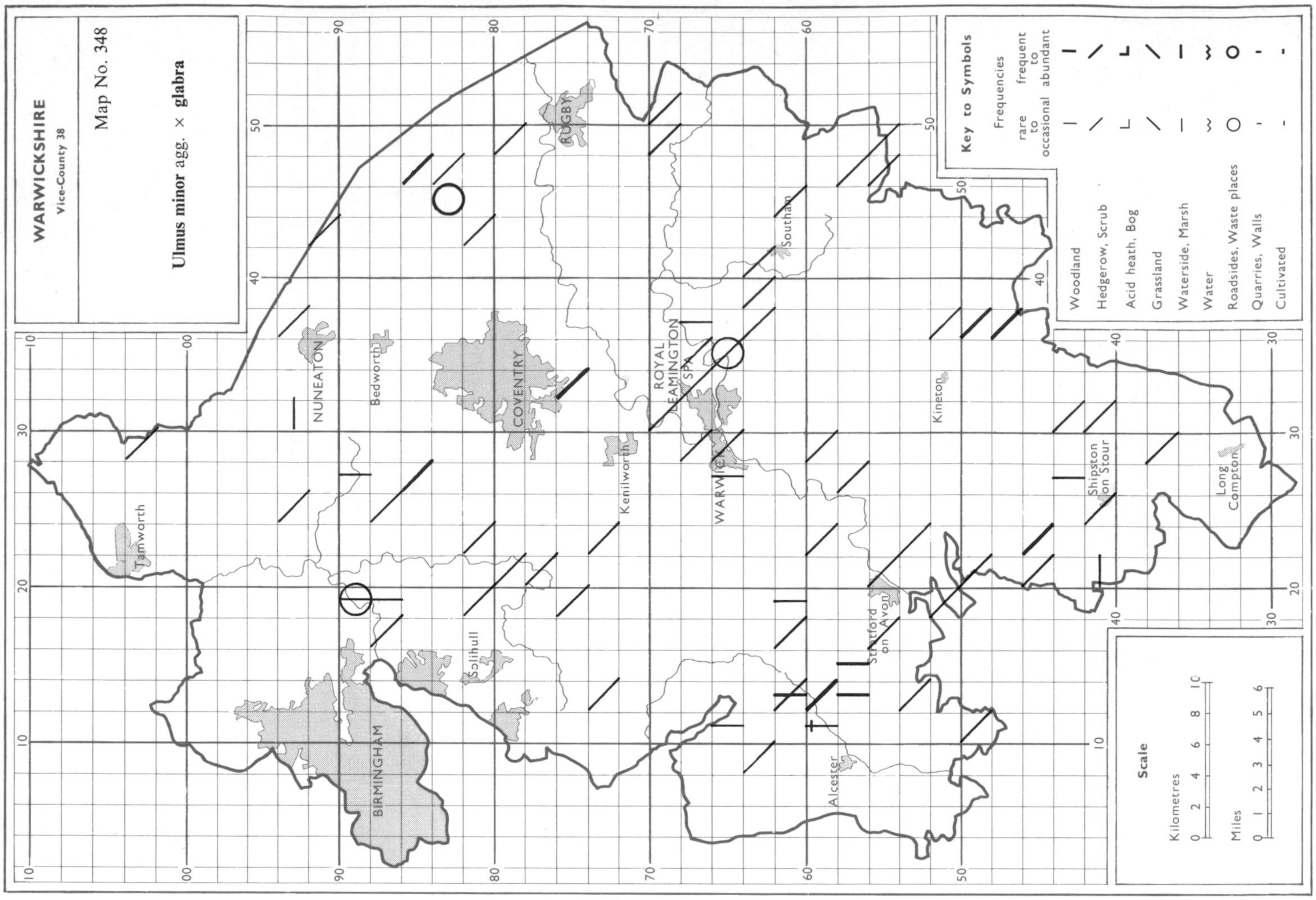
WARWICKSHIRE
Vice-County 38
Map No. 348
Ulmus minor agg. × glabra
Key to Symbols
Frequencies
rare to occasional
frequent to abundant
Woodland
Hedgerow, Scrub
Acid heath, Bog
Grassland
Waterside, Marsh
Water
Roadsides, Waste places
Quarries, Walls
Cultivated
Scale
Kilometres 0 2 4 6 8 10
Miles 0 1 2 3 4 5 6
NUNEATON
Bedworth
COVENTRY
RUGBY
Tamworth
Kenilworth
ROYAL LEAMINGTON SPA
WARWICK
Southam
Kineton
Shipston on Stour
Long Compton
Solihull
BIRMINGHAM
Stratford on Avon
Alcester

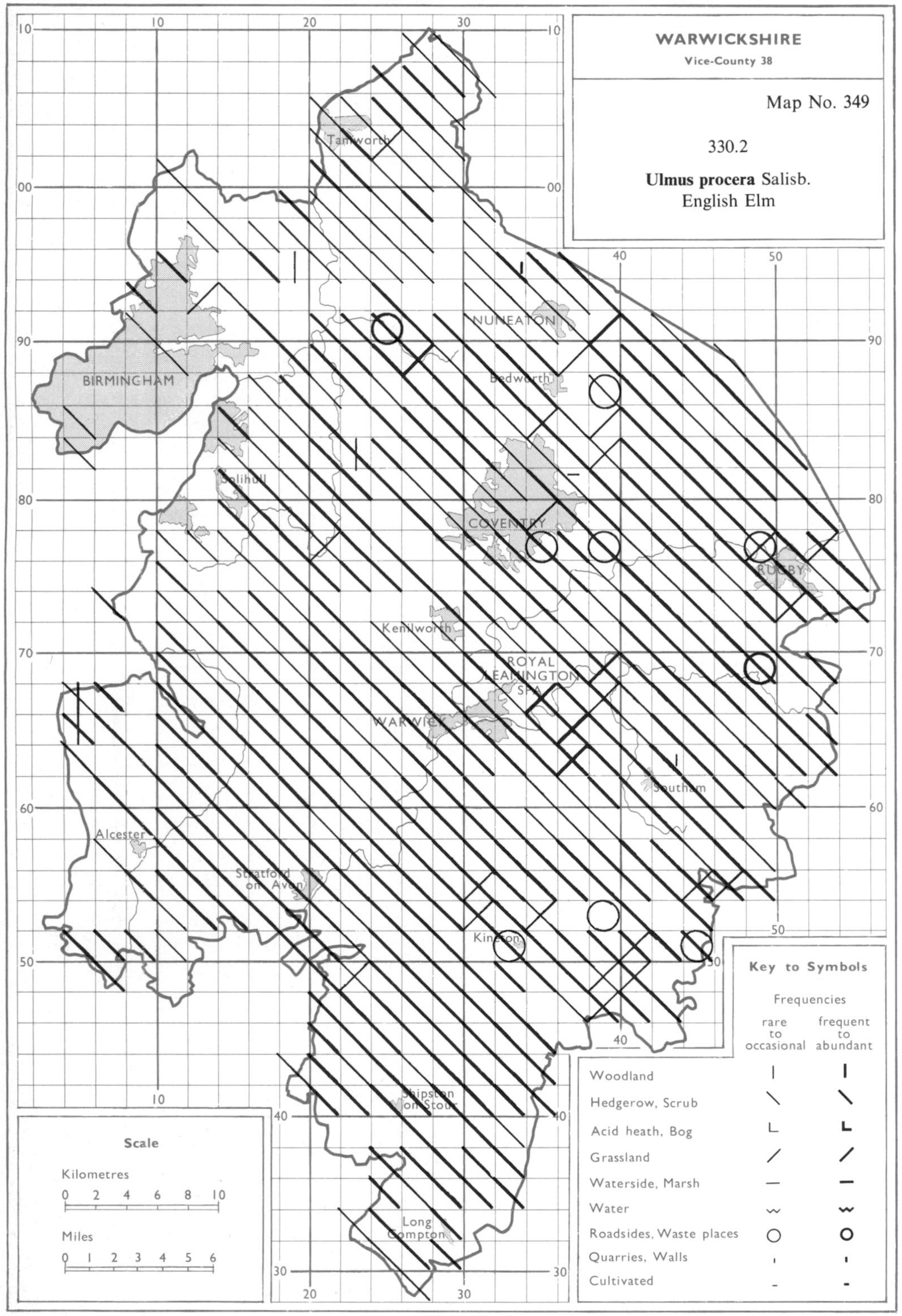
WARWICKSHIRE
Vice-County 38
Map No. 349
330.2
Ulmus procera Salisb.
English Elm
Tamworth
NUNEATON
Bedworth
BIRMINGHAM
Solihull
COVENTRY
RUGBY
Kenilworth
ROYAL LEAMINGTON SPA
WARWICK
Southam
Alcester
Stratford on Avon
Kineton
Shipston on Stour
Long Compton
Scale
Kilometres
0 2 4 6 8 10
Miles
0 1 2 3 4 5 6
Key to Symbols
Frequencies
rare to occasional
frequent to abundant
Woodland
Hedgerow, Scrub
Acid heath, Bog
Grassland
Waterside, Marsh
Water
Roadsides, Waste places
Quarries, Walls
Cultivated

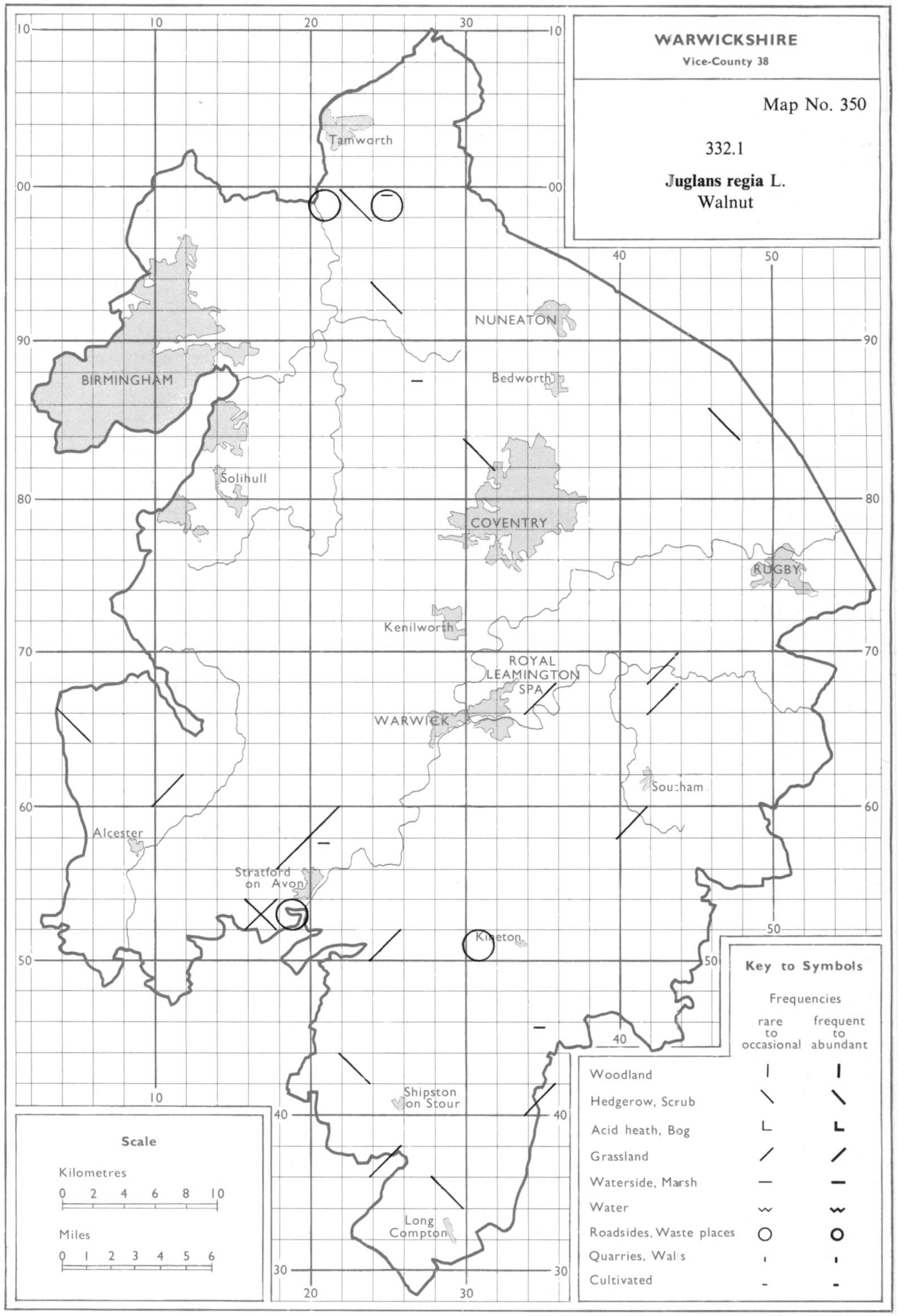
WARWICKSHIRE
Vice-County 38
Map No. 350
332.1
Juglans regia L.
Walnut
Tamworth
NUNEATON
Bedworth
BIRMINGHAM
Solihull
COVENTRY
RUGBY
Kenilworth
ROYAL LEAMINGTON SPA
WARWICK
Southam
Alcester
Stratford on Avon
Kineton
Shipston on Stour
Long Compton
Scale
Kilometres
0 2 4 6 8 10
Miles
0 1 2 3 4 5 6
Key to Symbols
Frequencies
rare to occasional
frequent to abundant
Woodland
Hedgerow, Scrub
Acid heath, Bog
Grassland
Waterside, Marsh
Water
Roadsides, Waste places
Quarries, Walls
Cultivated

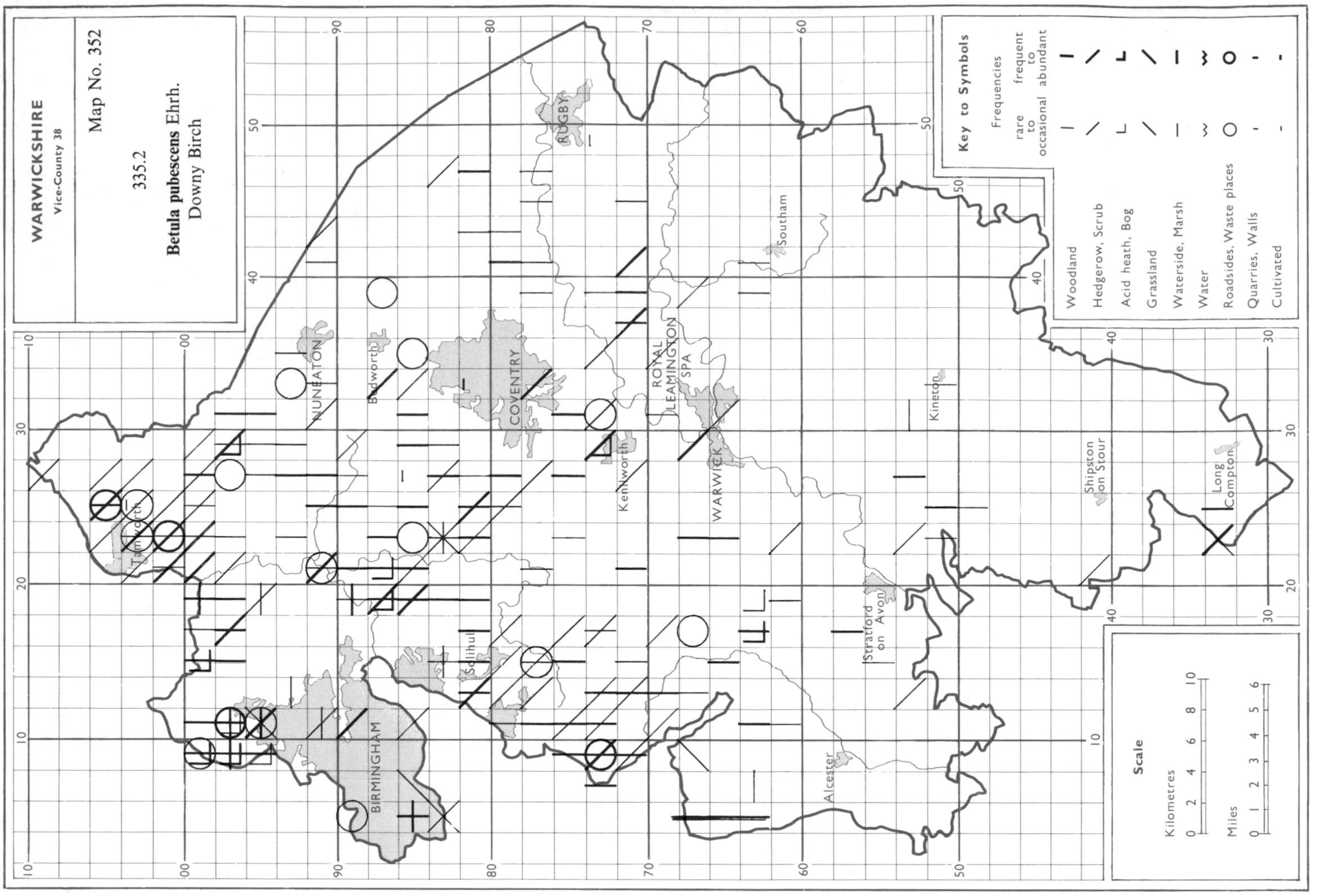
WARWICKSHIRE
Vice-County 38
Map No. 352
335.2
Betula pubescens Ehrh.
Downy Birch
Key to Symbols
Frequencies
rare to occasional
frequent to abundant
Woodland
Hedgerow, Scrub
Acid heath, Bog
Grassland
Waterside, Marsh
Water
Roadsides, Waste places
Quarries, Walls
Cultivated
Scale
Kilometres
0 2 4 6 8 10
Miles
0 1 2 3 4 5 6
NUNEATON
Bedworth
COVENTRY
RUGBY
Southam
ROYAL LEAMINGTON SPA
WARWICK
Kenilworth
Kineton
Shipston on Stour
Long Compton
Stratford on Avon
Alcester
Solihull
BIRMINGHAM
Tamworth

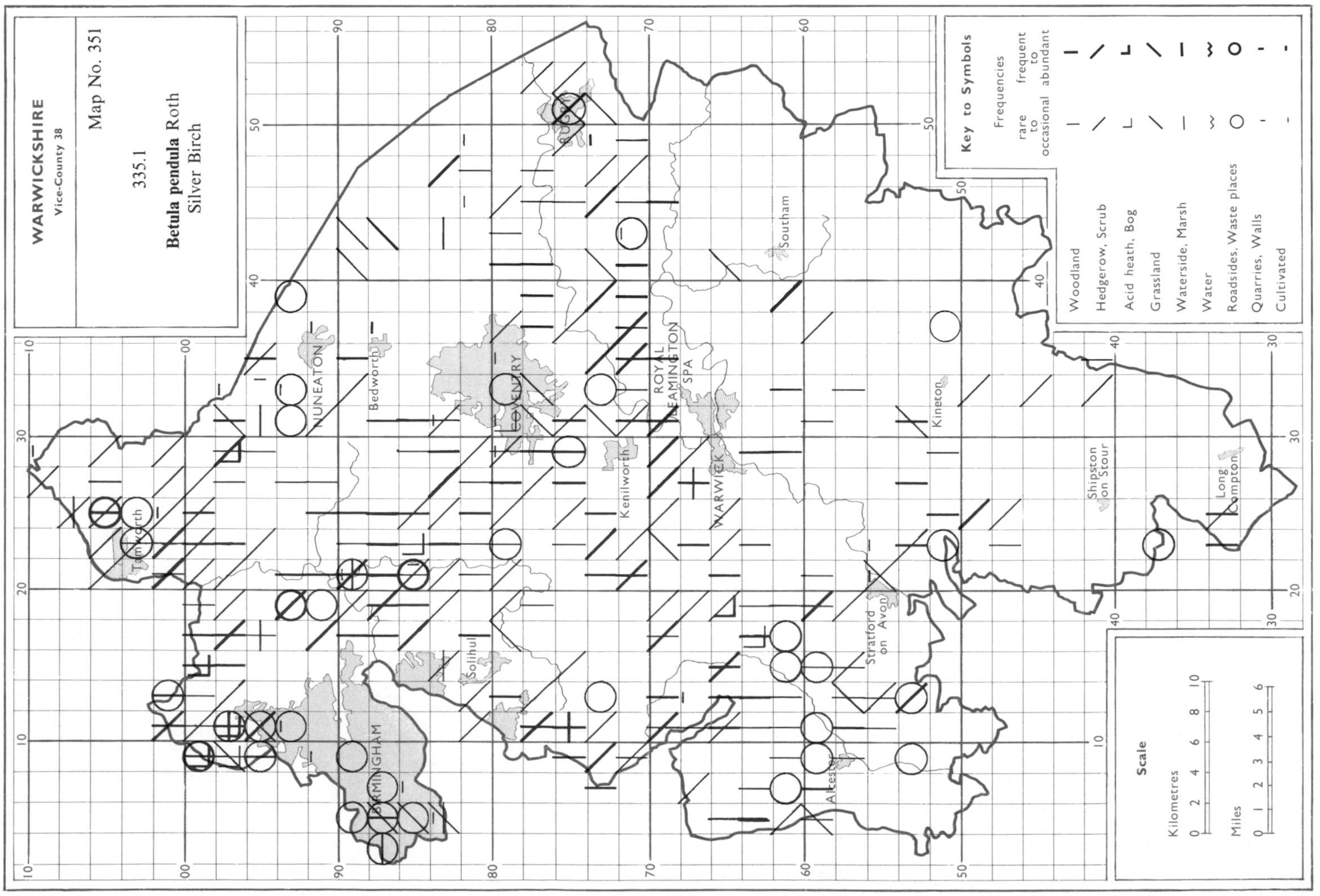
WARWICKSHIRE
Vice-County 38
Map No. 351
335.1
Betula pendula Roth
Silver Birch
Key to Symbols
Frequencies
rare to occasional
frequent to abundant
Woodland
Hedgerow, Scrub
Acid heath, Bog
Grassland
Waterside, Marsh
Water
Roadsides, Waste places
Quarries, Walls
Cultivated
Scale
Kilometres
0 2 4 6 8 10
Miles
0 1 2 3 4 5 6
NUNEATON
Bedworth
COVENTRY
RUGBY
Southam
ROYAL LEAMINGTON SPA
WARWICK
Kenilworth
Kineton
Shipston on Stour
Long Compton
Stratford on Avon
Alcester
Solihull
BIRMINGHAM
Tamworth

WARWICKSHIRE
Vice-County 38
Map No. 354
337.1
Carpinus betulus L.
Hornbeam
Key to Symbols
Frequencies
rare to occasional
frequent to abundant
Woodland
Hedgerow, Scrub
Acid heath, Bog
Grassland
Waterside, Marsh
Water
Roadsides, Waste places
Quarries, Walls
Cultivated
Scale
Kilometres
0 2 4 6 8 10
Miles
0 1 2 3 4 5 6
NUNEATON
Bedworth
COVENTRY
RUGBY
ROYAL LEAMINGTON SPA
WARWICK
Kenilworth
Southam
Kineton
Shipston on Stour
Long Compton
Tamworth
Solihull
BIRMINGHAM
Stratford on Avon
Alcester

WARWICKSHIRE
Vice-County 38
Map No. 353
336.1
Alnus glutinosa (L.) Gaertn.
Alder
Key to Symbols
Frequencies
rare to occasional
frequent to abundant
Woodland
Hedgerow, Scrub
Acid heath, Bog
Grassland
Waterside, Marsh
Water
Roadsides, Waste places
Quarries, Walls
Cultivated
Scale
Kilometres
0 2 4 6 8 10
Miles
0 1 2 3 4 5 6
NUNEATON
Bedworth
COVENTRY
RUGBY
ROYAL LEAMINGTON SPA
WARWICK
Kenilworth
Southam
Kineton
Shipston on Stour
Long Compton
Tamworth
BIRMINGHAM
Stratford on Avon
Alcester

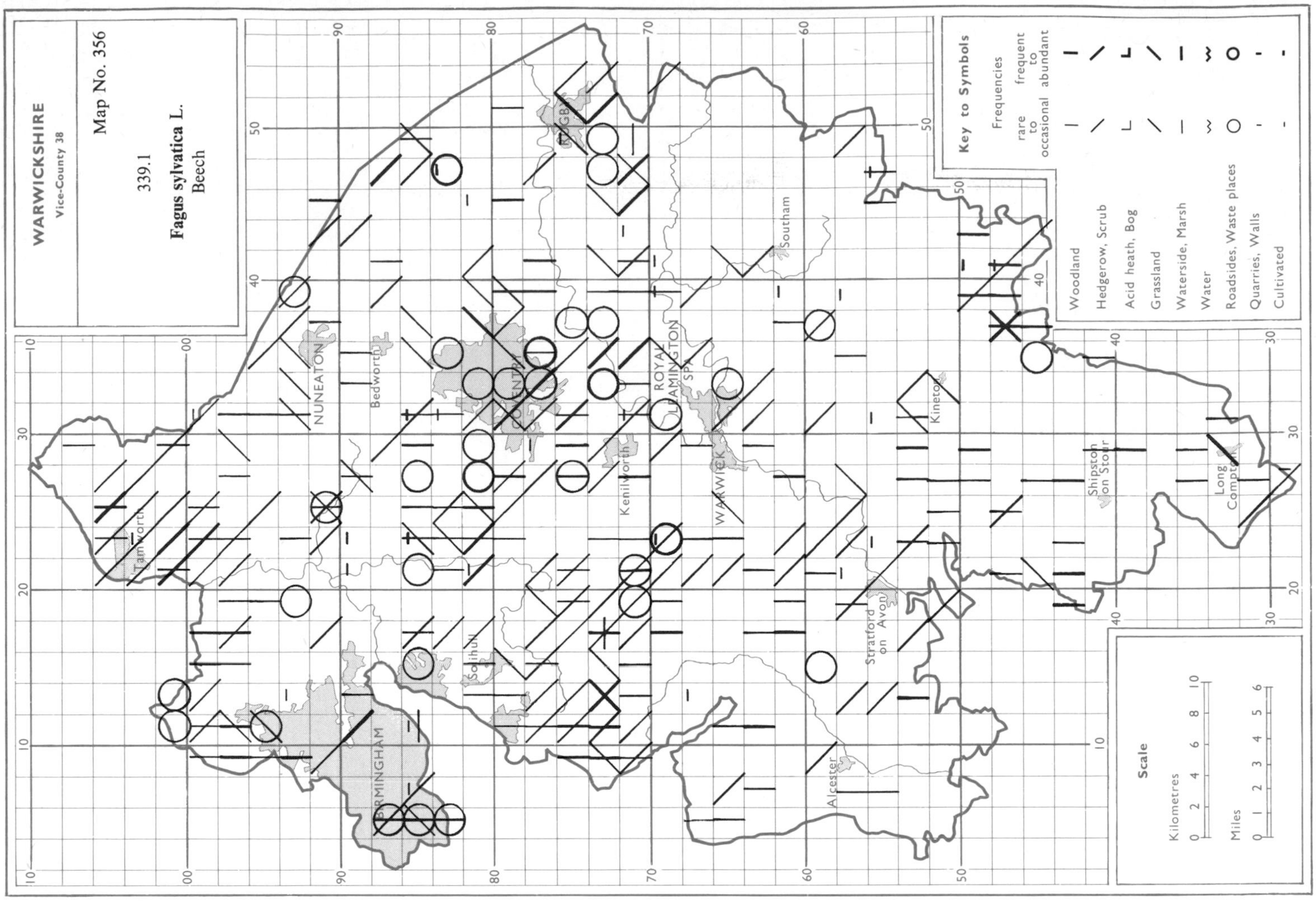
WARWICKSHIRE
Vice-County 38
Map No. 356
339.1
Fagus sylvatica L.
Beech
Key to Symbols
Frequencies
rare to occasional
frequent to abundant
Woodland
Hedgerow, Scrub
Acid heath, Bog
Grassland
Waterside, Marsh
Water
Roadsides, Waste places
Quarries, Walls
Cultivated
Scale
Kilometres
0 2 4 6 8 10
Miles
0 1 2 3 4 5 6
NUNEATON
Bedworth
COVENTRY
RUGBY
ROYAL LEAMINGTON SPA
Southam
Kineton
Shipston on Stour
Long Compton
WARWICK
Kenilworth
Tamworth
Solihull
BIRMINGHAM
Stratford on Avon
Alcester

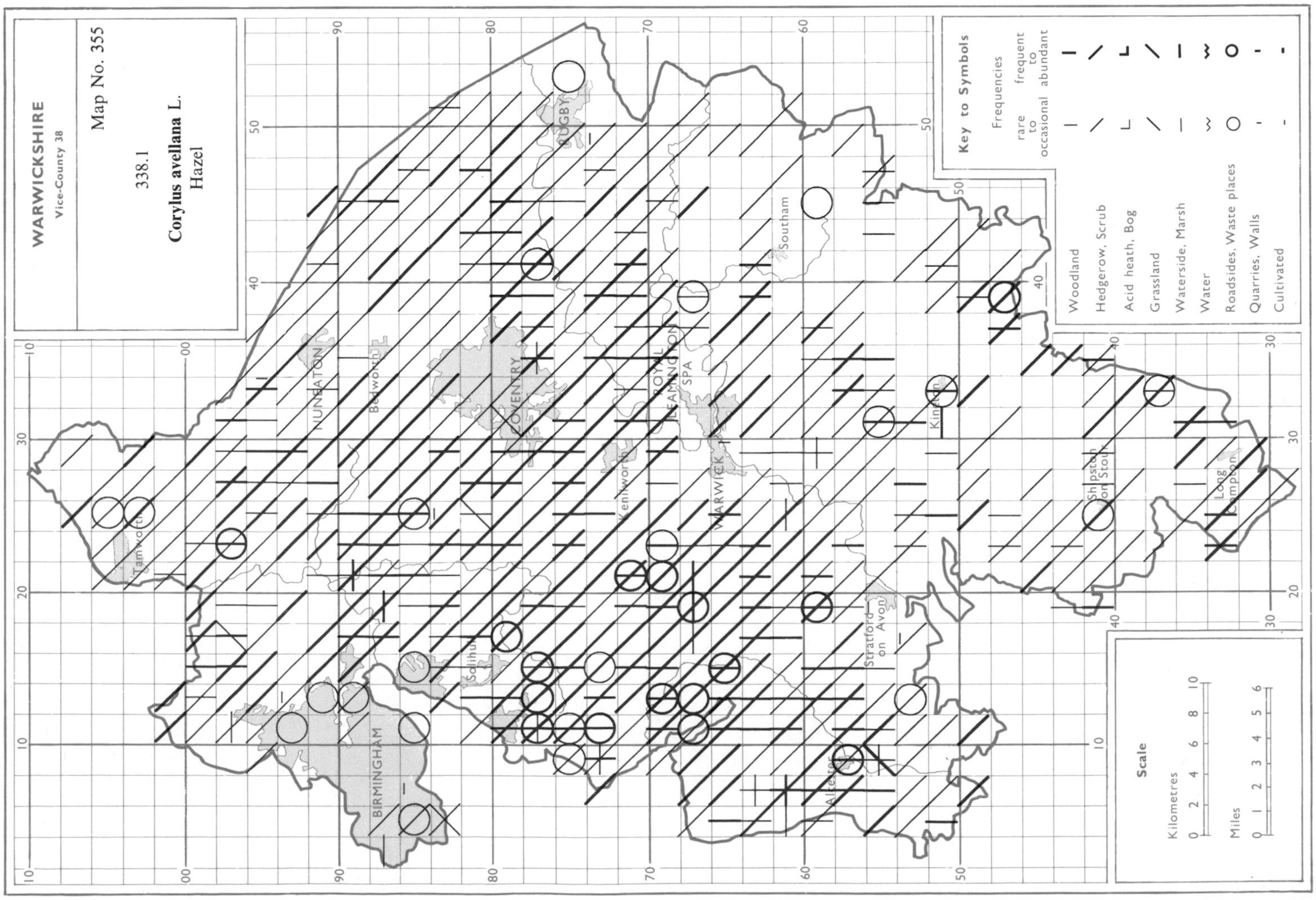
WARWICKSHIRE
Vice-County 38
Map No. 355
338.1
Corylus avellana L.
Hazel
Key to Symbols
Frequencies
rare to occasional
frequent to abundant
Woodland
Hedgerow, Scrub
Acid heath, Bog
Grassland
Waterside, Marsh
Water
Roadsides, Waste places
Quarries, Walls
Cultivated
Scale
Kilometres
0 2 4 6 8 10
Miles
0 1 2 3 4 5 6
NUNEATON
Bedworth
COVENTRY
RUGBY
ROYAL LEAMINGTON SPA
Southam
Kineton
Shipston on Stour
Long Compton
WARWICK
Kenilworth
Tamworth
Solihull
BIRMINGHAM
Stratford on Avon
Alcester

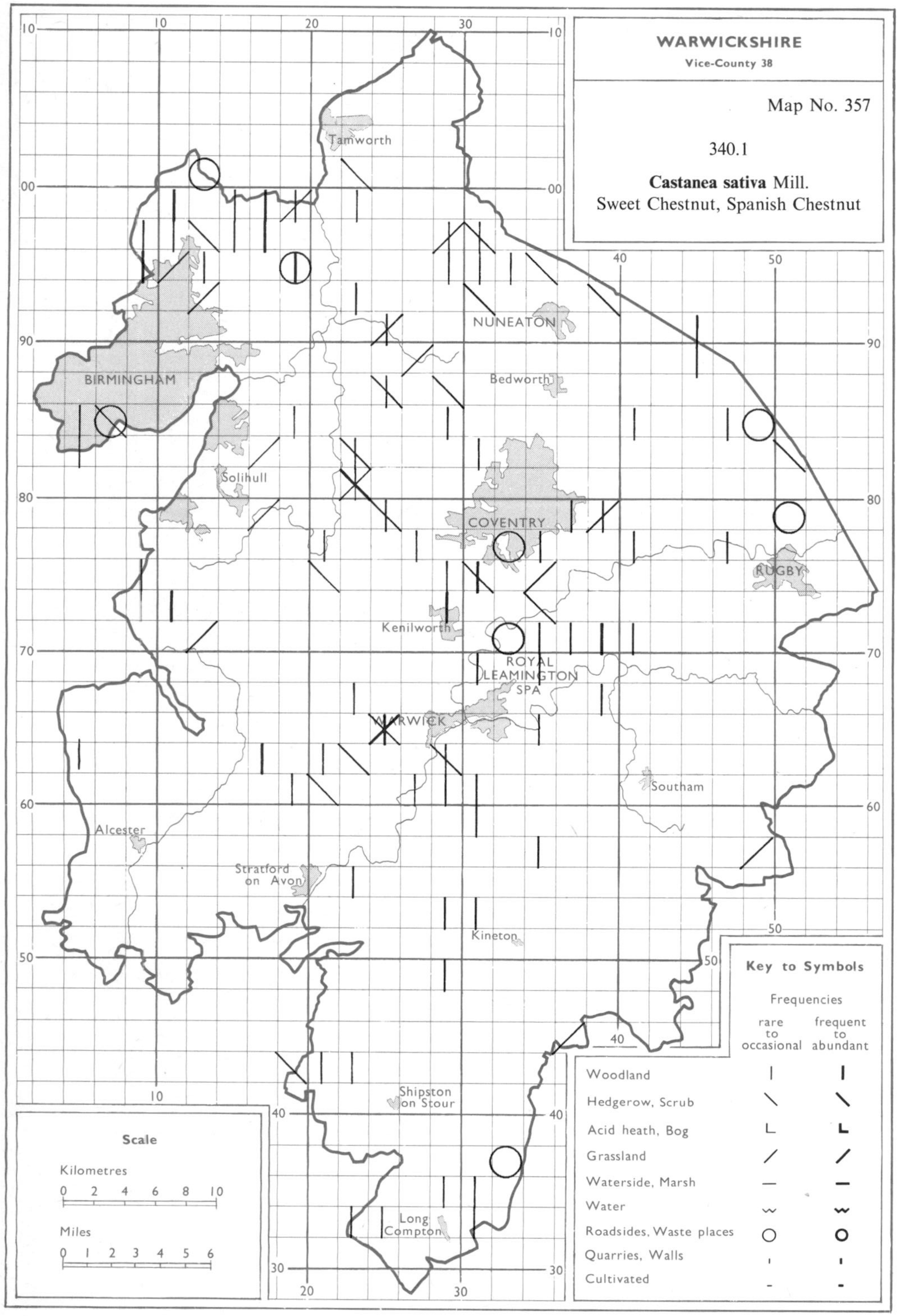
WARWICKSHIRE
Vice-County 38
Map No. 357
340.1
Castanea sativa Mill.
Sweet Chestnut, Spanish Chestnut
Tamworth
NUNEATON
Bedworth
BIRMINGHAM
Solihull
COVENTRY
RUGBY
Kenilworth
ROYAL LEAMINGTON SPA
WARWICK
Southam
Alcester
Stratford on Avon
Kineton
Shipston on Stour
Long Compton
Key to Symbols
Frequencies
rare to occasional
frequent to abundant
Woodland
Hedgerow, Scrub
Acid heath, Bog
Grassland
Waterside, Marsh
Water
Roadsides, Waste places
Quarries, Walls
Cultivated
Scale
Kilometres
0 2 4 6 8 10
Miles
0 1 2 3 4 5 6

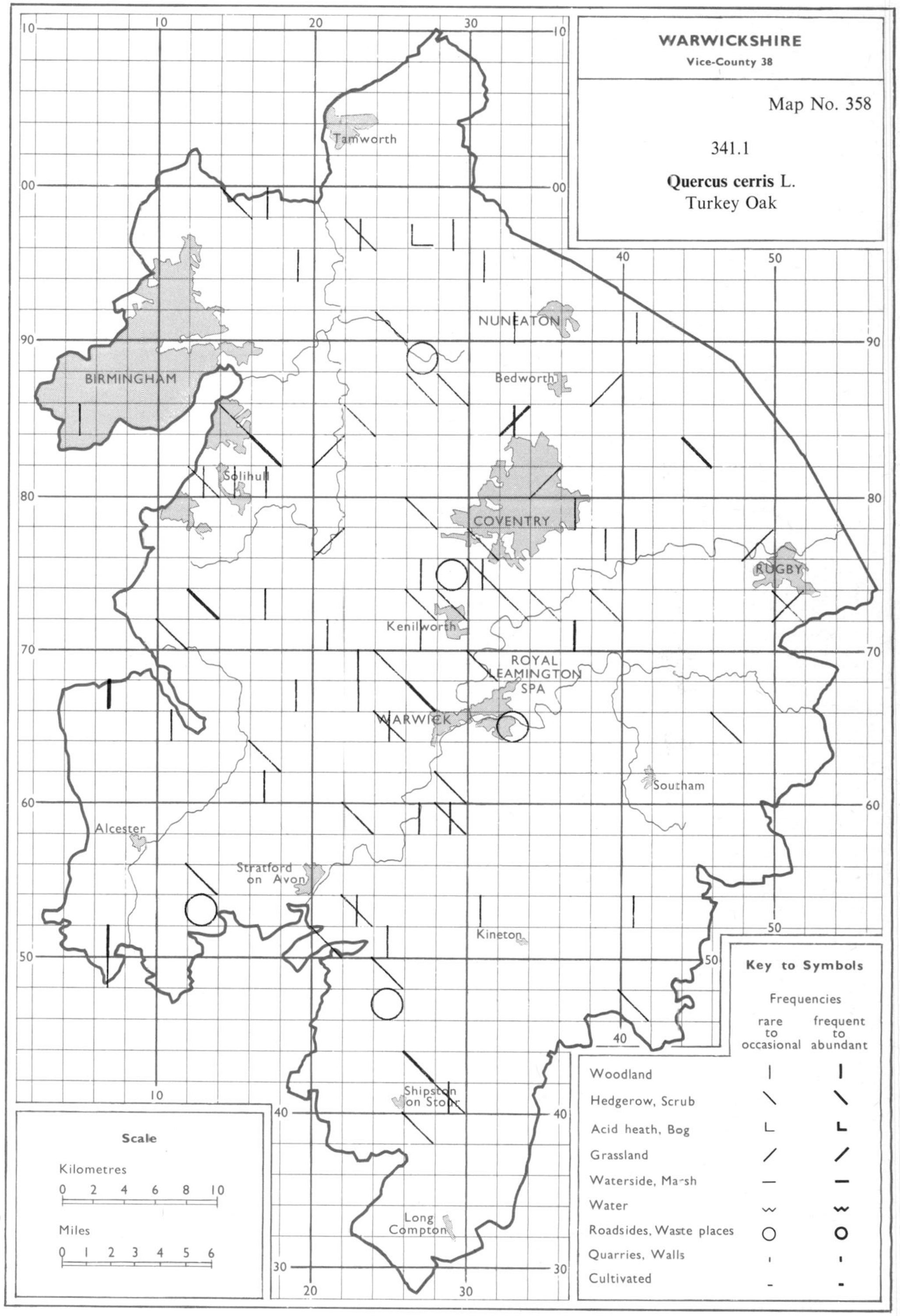
WARWICKSHIRE
Vice-County 38
Map No. 358
341.1
Quercus cerris L.
Turkey Oak
Tamworth
NUNEATON
Bedworth
BIRMINGHAM
Solihull
COVENTRY
RUGBY
Kenilworth
ROYAL LEAMINGTON SPA
WARWICK
Southam
Alcester
Stratford on Avon
Kineton
Shipston on Stour
Long Compton
Key to Symbols
Frequencies
rare to occasional
frequent to abundant
Woodland
Hedgerow, Scrub
Acid heath, Bog
Grassland
Waterside, Marsh
Water
Roadsides, Waste places
Quarries, Walls
Cultivated
Scale
Kilometres
0 2 4 6 8 10
Miles
0 1 2 3 4 5 6

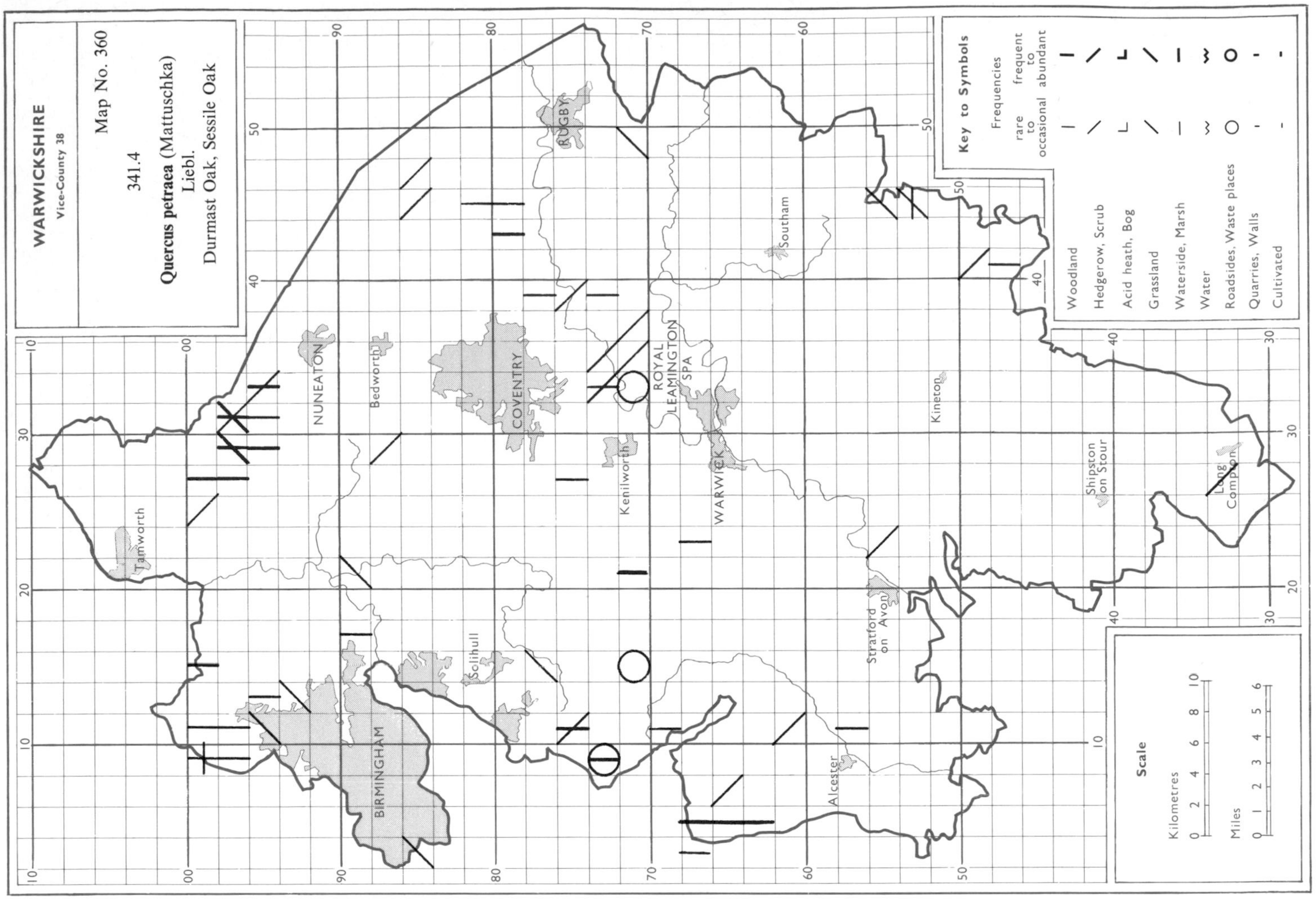
WARWICKSHIRE
Vice-County 38
Map No. 360
341.4
Quercus petraea (Mattuschka) Liebl.
Durmast Oak, Sessile Oak
Key to Symbols
Frequencies
rare to occasional
frequent to abundant
Woodland
Hedgerow, Scrub
Acid heath, Bog
Grassland
Waterside, Marsh
Water
Roadsides, Waste places
Quarries, Walls
Cultivated
Scale
Kilometres
Miles
NUNEATON
Bedworth
COVENTRY
RUGBY
Southam
ROYAL LEAMINGTON SPA
Kenilworth
WARWICK
Kineton
Tamworth
Solihull
BIRMINGHAM
Stratford on Avon
Alcester
Shipston on Stour
Long Compton

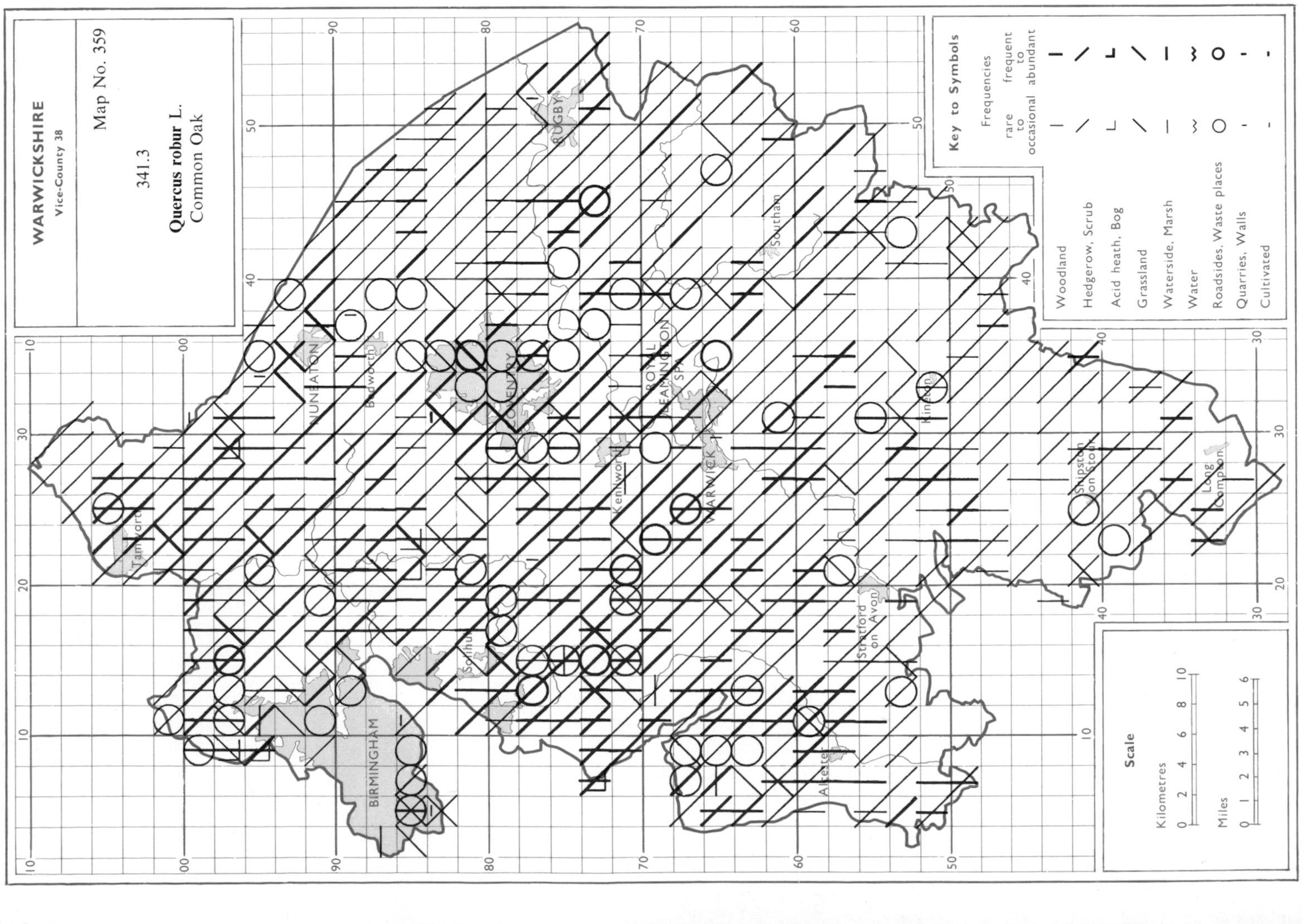
WARWICKSHIRE
Vice-County 38
Map No. 359
341.3
Quercus robur L.
Common Oak
Key to Symbols
Frequencies
rare to occasional
frequent to abundant
Woodland
Hedgerow, Scrub
Acid heath, Bog
Grassland
Waterside, Marsh
Water
Roadsides, Waste places
Quarries, Walls
Cultivated
Scale
Kilometres
Miles

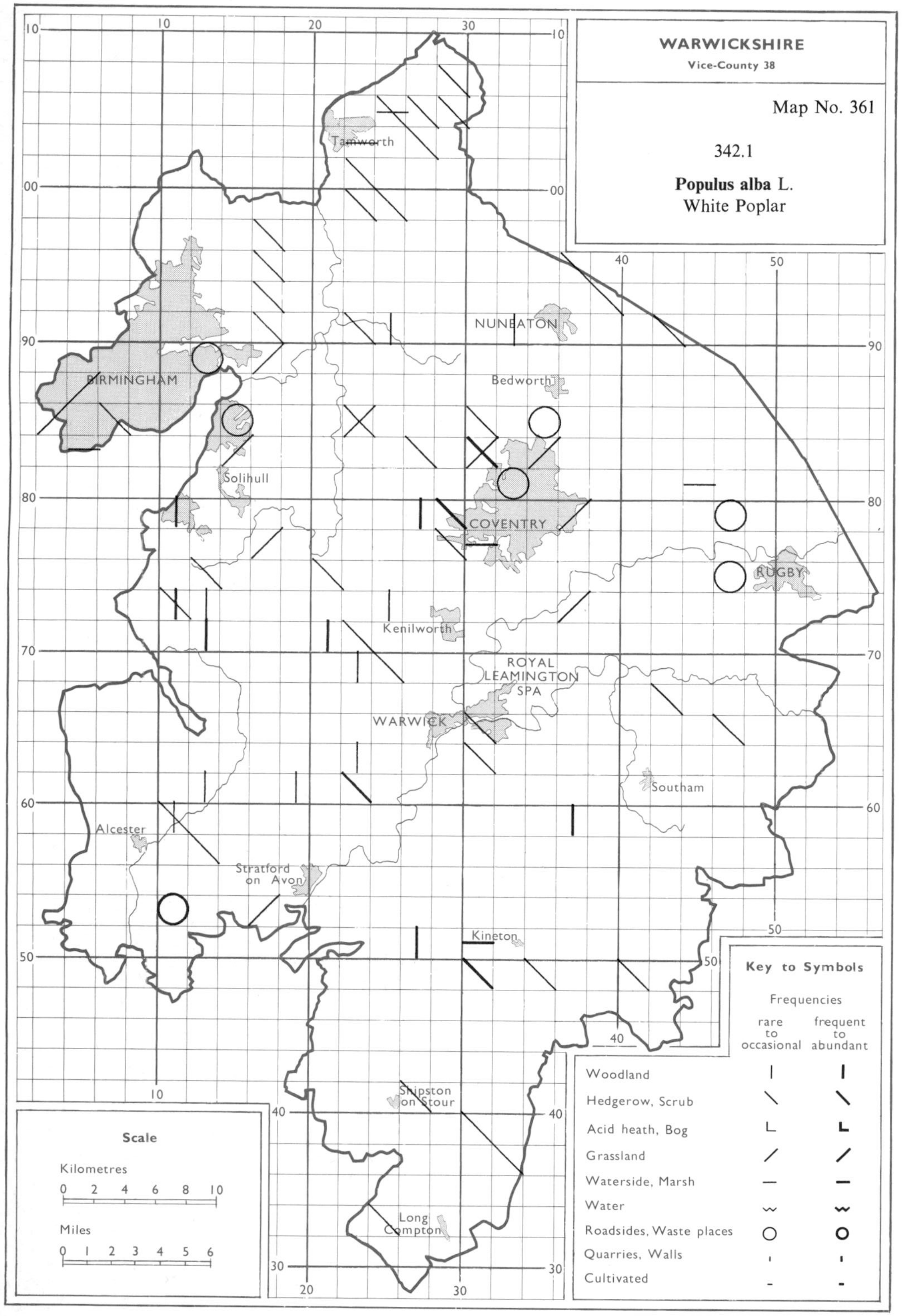
WARWICKSHIRE
Vice-County 38
Map No. 361
342.1
Populus alba L.
White Poplar
Tamworth
NUNEATON
Bedworth
BIRMINGHAM
Solihull
COVENTRY
RUGBY
Kenilworth
ROYAL LEAMINGTON SPA
WARWICK
Southam
Alcester
Stratford on Avon
Kineton
Shipston on Stour
Long Compton
Scale
Kilometres
0 2 4 6 8 10
Miles
0 1 2 3 4 5 6
Key to Symbols
Frequencies
rare to occasional
frequent to abundant
Woodland
Hedgerow, Scrub
Acid heath, Bog
Grassland
Waterside, Marsh
Water
Roadsides, Waste places
Quarries, Walls
Cultivated

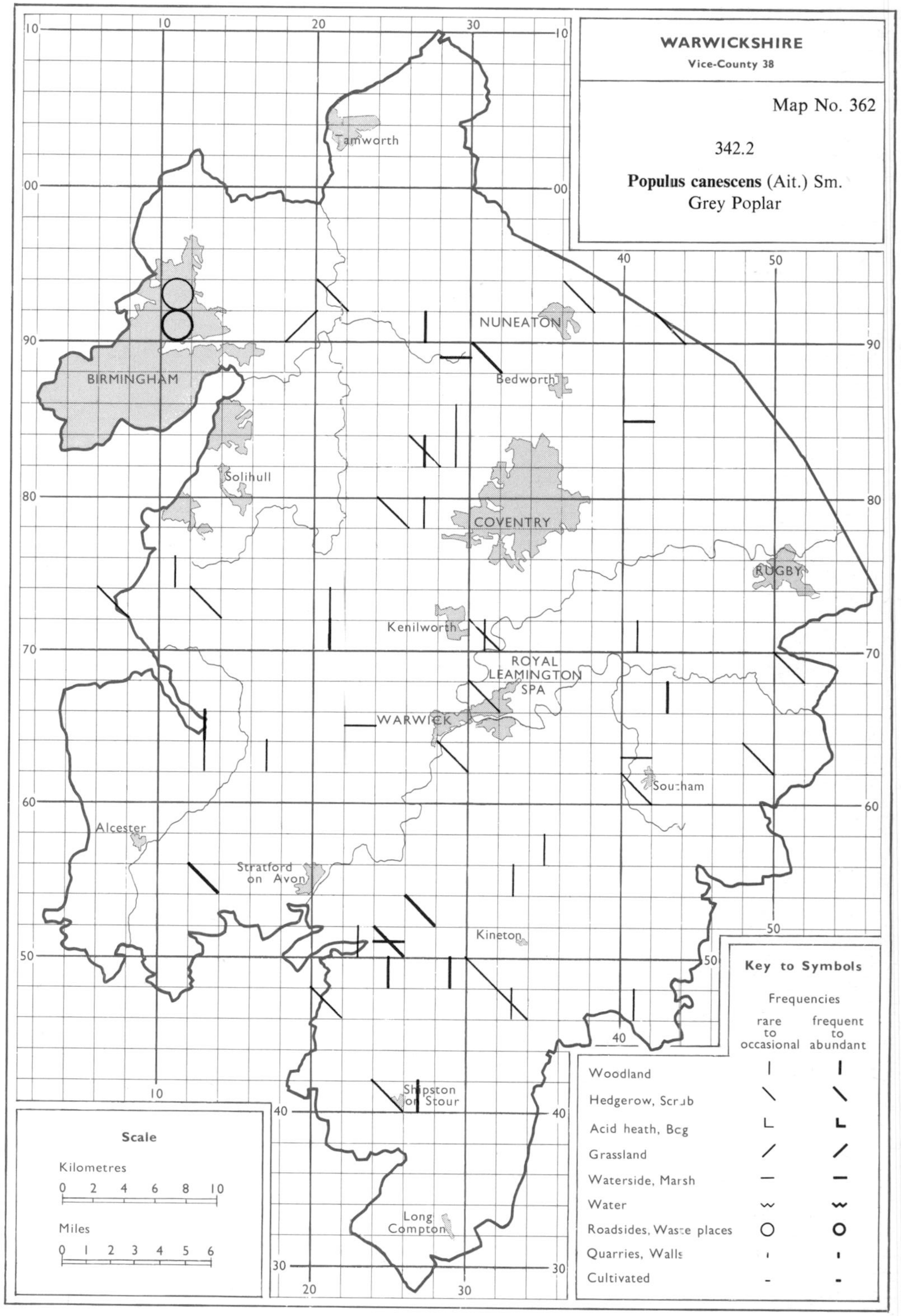
WARWICKSHIRE
Vice-County 38
Map No. 362
342.2
Populus canescens (Ait.) Sm.
Grey Poplar
Tamworth
NUNEATON
Bedworth
BIRMINGHAM
Solihull
COVENTRY
RUGBY
Kenilworth
ROYAL LEAMINGTON SPA
WARWICK
Southam
Alcester
Stratford on Avon
Kineton
Shipston on Stour
Long Compton
Scale
Kilometres
0 2 4 6 8 10
Miles
0 1 2 3 4 5 6
Key to Symbols
Frequencies
rare to occasional
frequent to abundant
Woodland
Hedgerow, Scrub
Acid heath, Bog
Grassland
Waterside, Marsh
Water
Roadsides, Waste places
Quarries, Walls
Cultivated

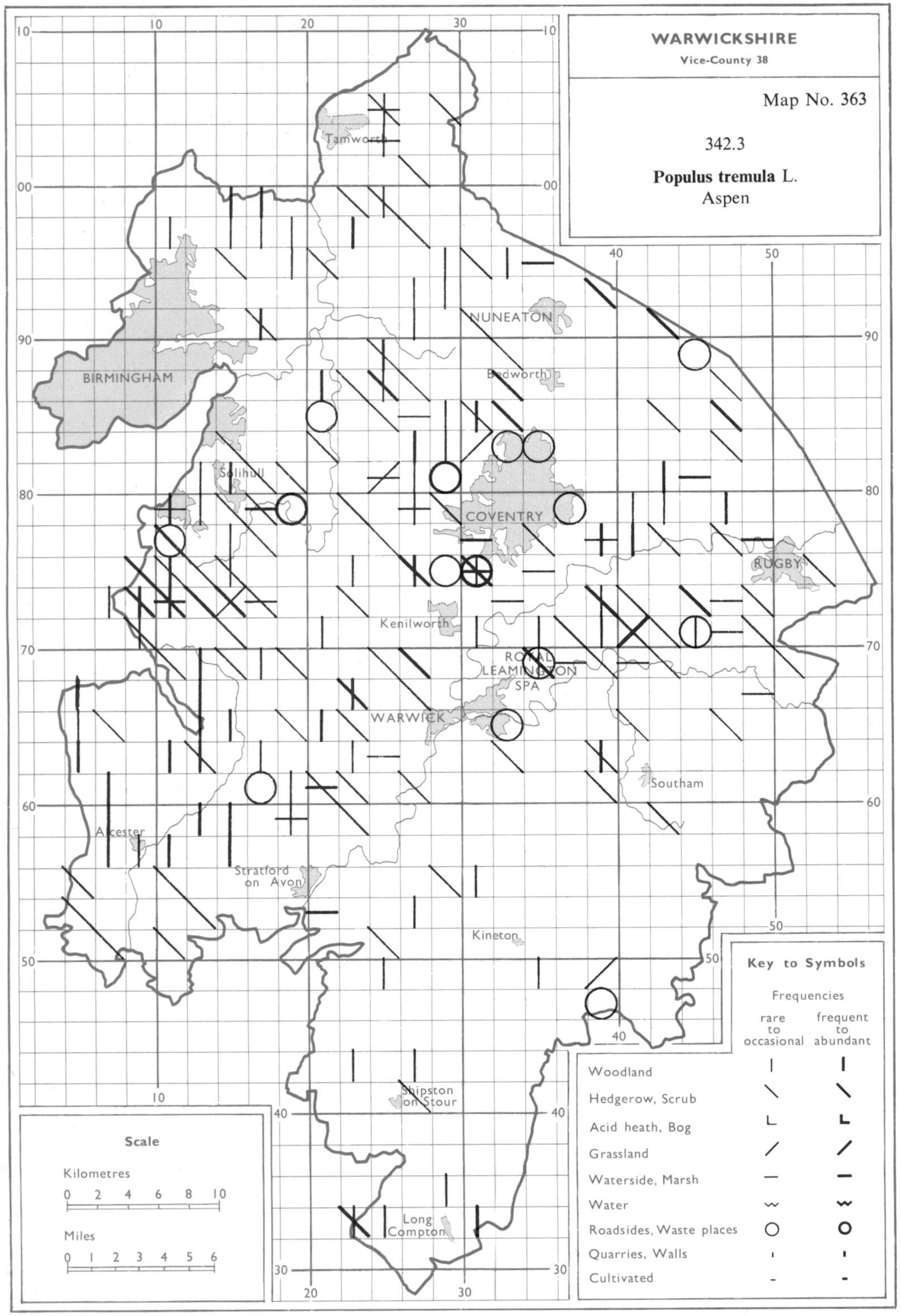
WARWICKSHIRE
Vice-County 38
Map No. 363
342.3
Populus tremula L.
Aspen
Tamworth
BIRMINGHAM
NUNEATON
Bedworth
Solihull
COVENTRY
RUGBY
Kenilworth
ROYAL LEAMINGTON SPA
WARWICK
Southam
Alcester
Stratford on Avon
Kineton
Shipston on Stour
Long Compton
Key to Symbols
Frequencies
rare to occasional
frequent to abundant
Woodland
Hedgerow, Scrub
Acid heath, Bog
Grassland
Waterside, Marsh
Water
Roadsides, Waste places
Quarries, Walls
Cultivated
Scale
Kilometres
0 2 4 6 8 10
Miles
0 1 2 3 4 5 6

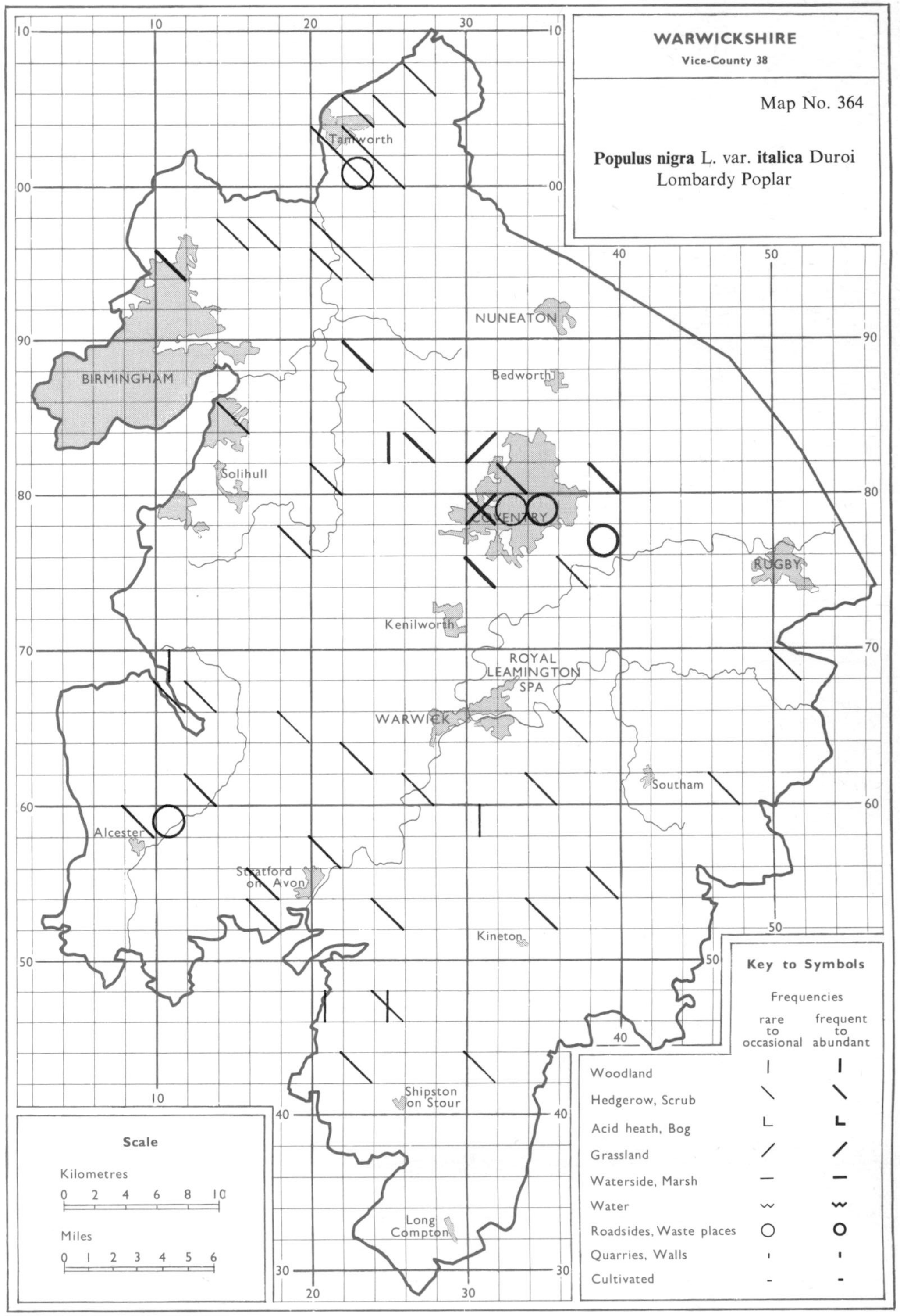
WARWICKSHIRE
Vice-County 38
Map No. 364
Populus nigra L. var. italica Duroi
Lombardy Poplar
Tamworth
BIRMINGHAM
NUNEATON
Bedworth
Solihull
COVENTRY
RUGBY
Kenilworth
ROYAL LEAMINGTON SPA
WARWICK
Southam
Alcester
Stratford on Avon
Kineton
Shipston on Stour
Long Compton
Key to Symbols
Frequencies
rare to occasional
frequent to abundant
Woodland
Hedgerow, Scrub
Acid heath, Bog
Grassland
Waterside, Marsh
Water
Roadsides, Waste places
Quarries, Walls
Cultivated
Scale
Kilometres
0 2 4 6 8 10
Miles
0 1 2 3 4 5 6

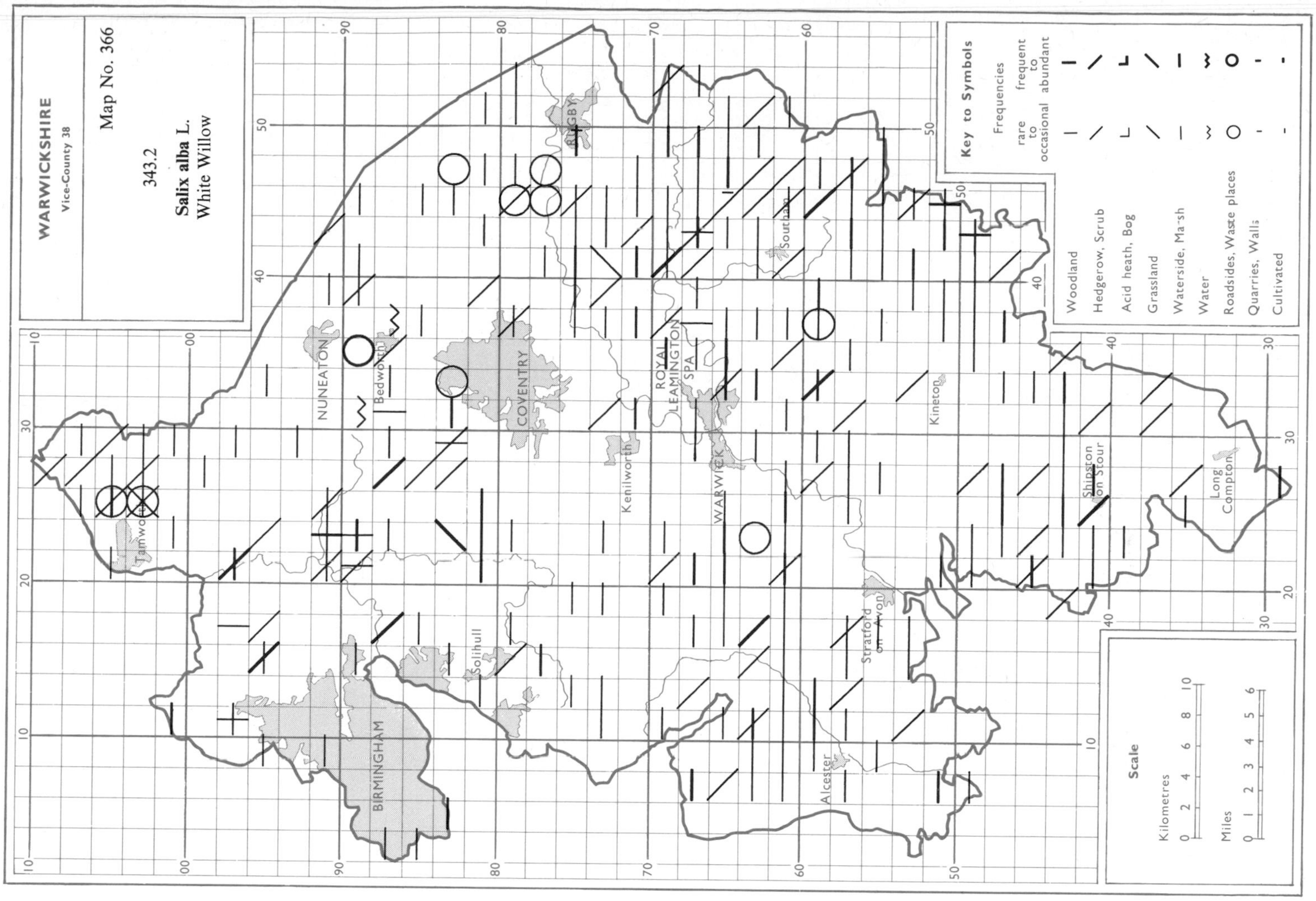
WARWICKSHIRE
Vice-County 38
Map No. 366
343.2
Salix alba L.
White Willow
Key to Symbols
Frequencies
rare to occasional
frequent to abundant
Woodland
Hedgerow, Scrub
Acid heath, Bog
Grassland
Waterside, Marsh
Water
Roadsides, Waste places
Quarries, Walls
Cultivated
Scale
Kilometres
Miles
NUNEATON
Bedworth
COVENTRY
RUGBY
ROYAL LEAMINGTON SPA
WARWICK
Kenilworth
Southam
Kineton
Shipston on Stour
Long Compton
Tamworth
Solihull
BIRMINGHAM
Stratford on Avon
Alcester

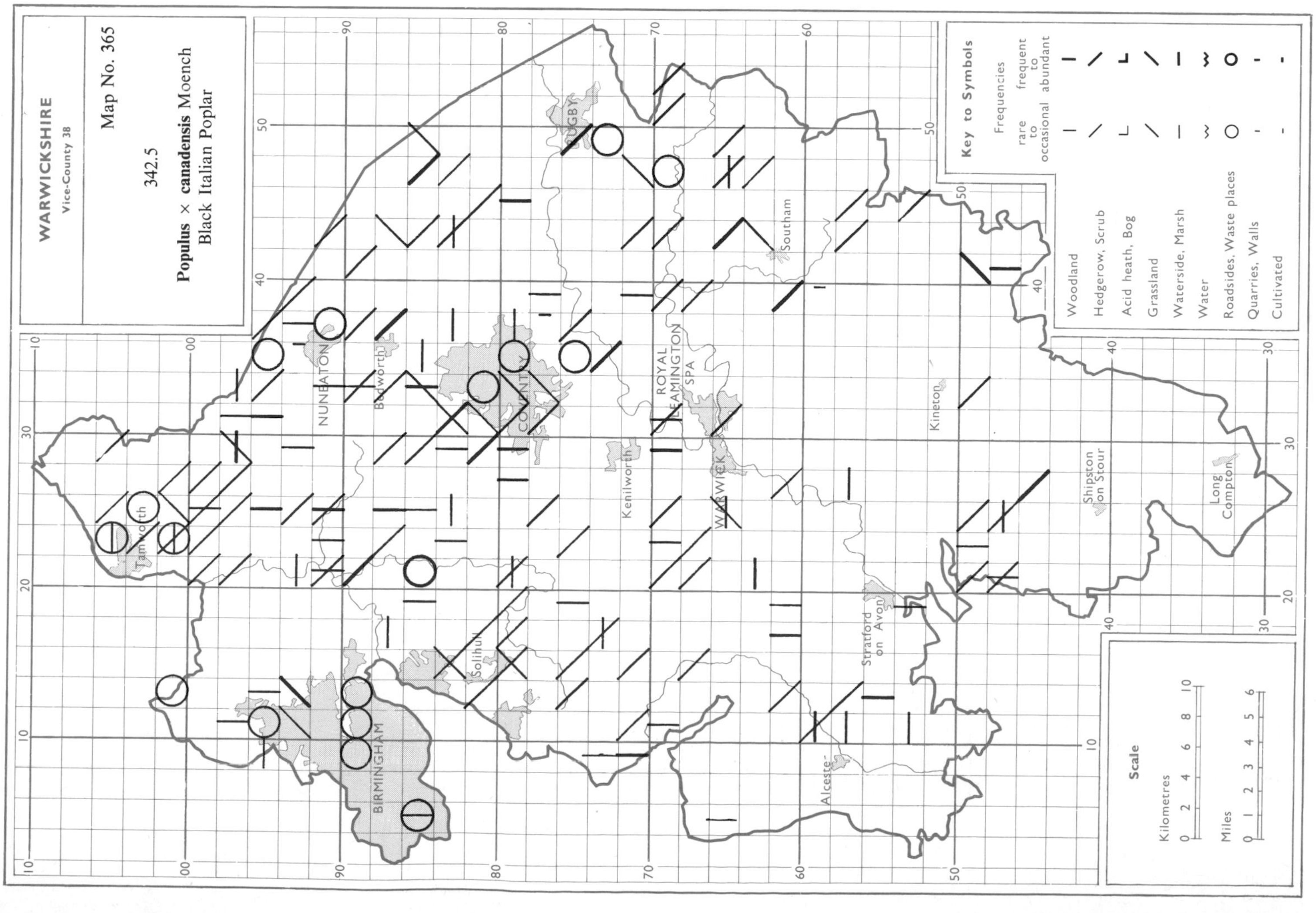
WARWICKSHIRE
Vice-County 38
Map No. 365
342.5
Populus × canadensis Moench
Black Italian Poplar
Key to Symbols
Frequencies
rare to occasional
frequent to abundant
Woodland
Hedgerow, Scrub
Acid heath, Bog
Grassland
Waterside, Marsh
Water
Roadsides, Waste places
Quarries, Walls
Cultivated
Scale
Kilometres
Miles
NUNEATON
Bedworth
COVENTRY
RUGBY
ROYAL LEAMINGTON SPA
WARWICK
Kenilworth
Southam
Kineton
Shipston on Stour
Long Compton
Tamworth
Solihull
BIRMINGHAM
Stratford on Avon
Alcester

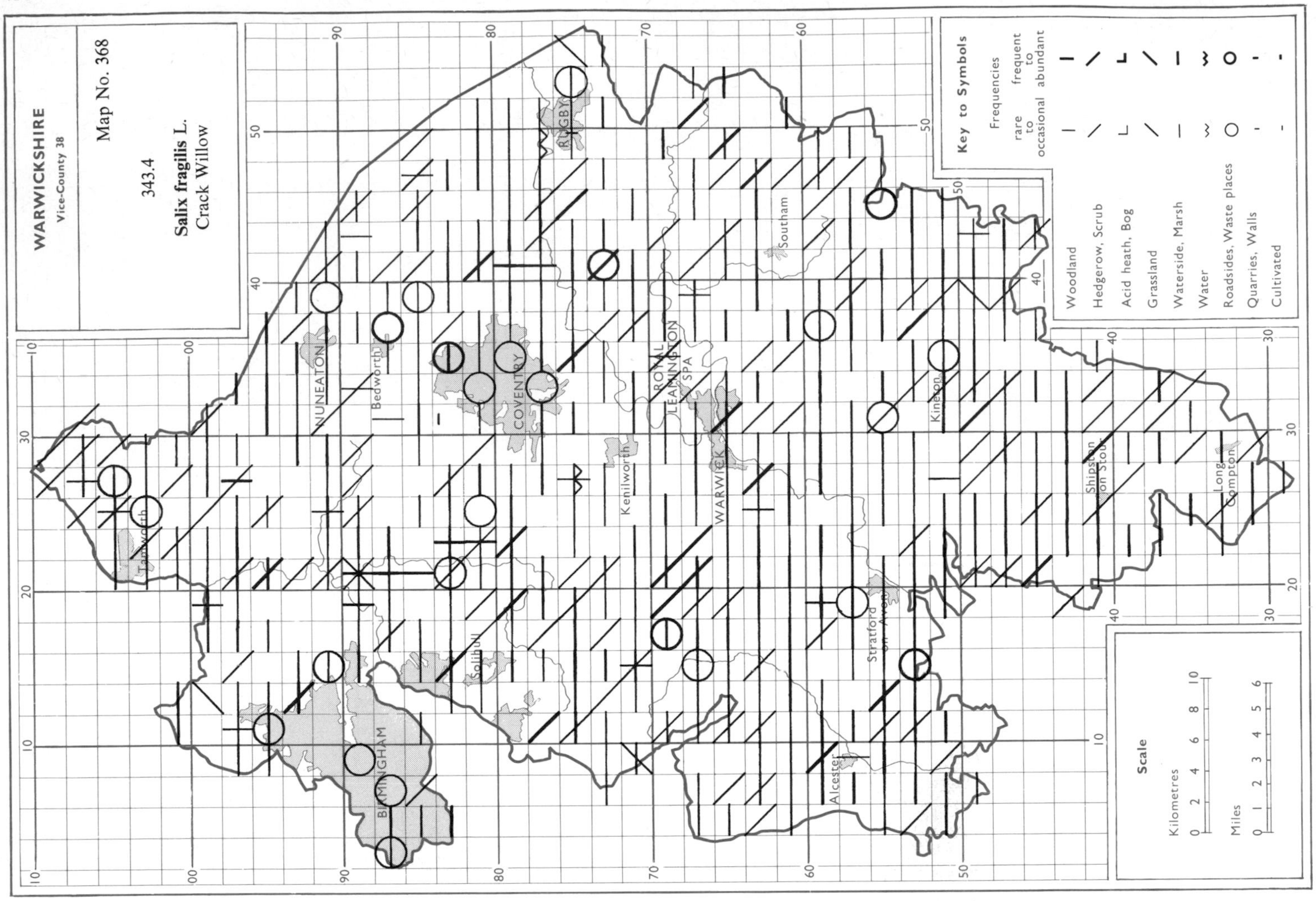
WARWICKSHIRE
Vice-County 38
Map No. 368
343.4
Salix fragilis L.
Crack Willow
Key to Symbols
Frequencies
rare to occasional
frequent to abundant
Woodland
Hedgerow, Scrub
Acid heath, Bog
Grassland
Waterside, Marsh
Water
Roadsides, Waste places
Quarries, Walls
Cultivated
Scale
Kilometres
0 2 4 6 8 10
Miles
0 1 2 3 4 5 6
NUNEATON
Bedworth
COVENTRY
RUGBY
Southam
ROYAL LEAMINGTON SPA
Kineton
Tamworth
Kenilworth
WARWICK
Shipston on Stour
Long Compton
Solihull
Stratford on Avon
BIRMINGHAM
Alcester

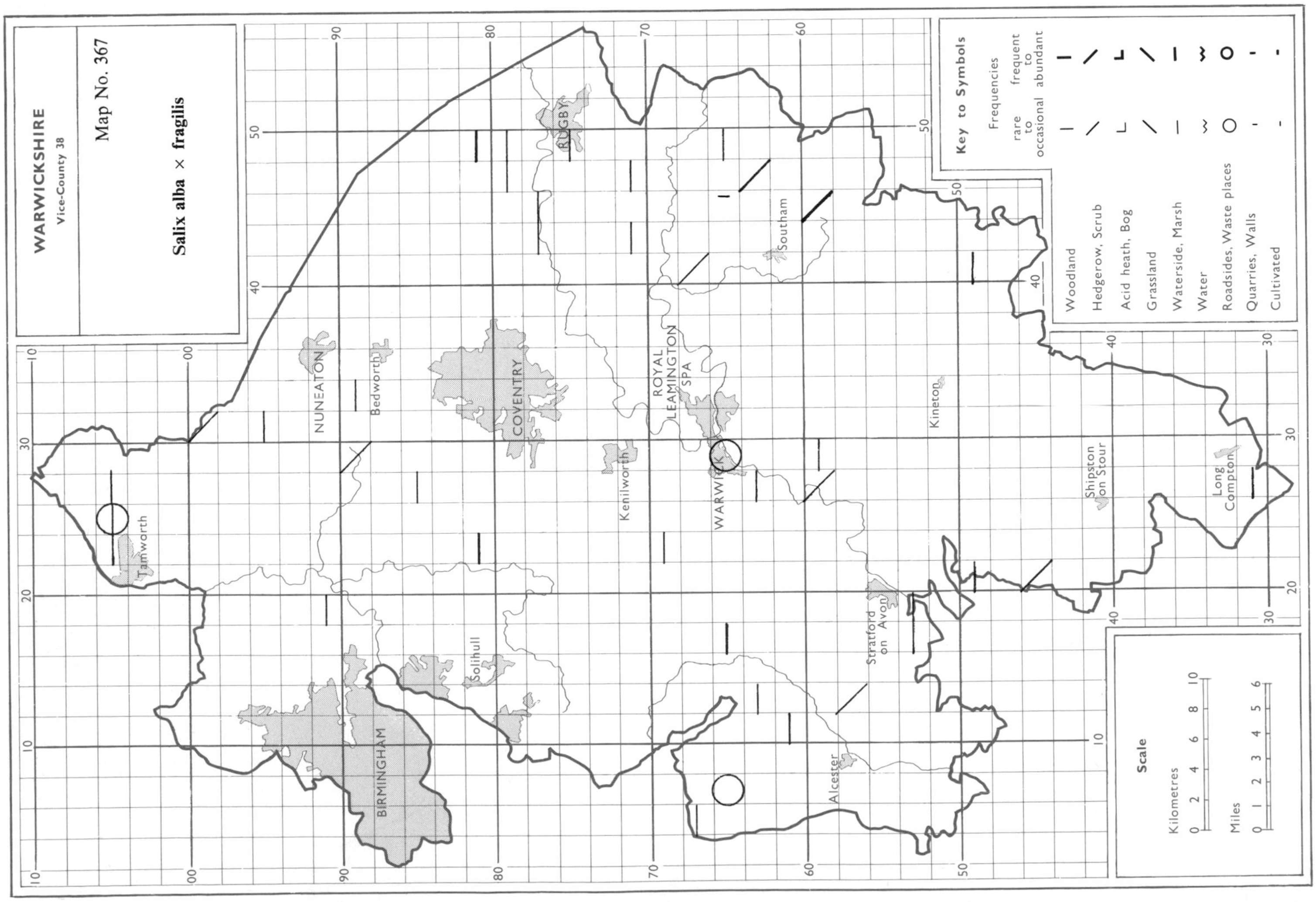
WARWICKSHIRE
Vice-County 38
Map No. 367
Salix alba × fragilis
Key to Symbols
Frequencies
rare to occasional
frequent to abundant
Woodland
Hedgerow, Scrub
Acid heath, Bog
Grassland
Waterside, Marsh
Water
Roadsides, Waste places
Quarries, Walls
Cultivated
Scale
Kilometres
0 2 4 6 8 10
Miles
0 1 2 3 4 5 6
NUNEATON
Bedworth
COVENTRY
RUGBY
Southam
ROYAL LEAMINGTON SPA
Kineton
Tamworth
Kenilworth
WARWICK
Shipston on Stour
Long Compton
Solihull
Stratford on Avon
BIRMINGHAM
Alcester

WARWICKSHIRE
Vice-County 38
Map No. 370
343.6
Salix purpurea L.
Purple Willow
Key to Symbols
Frequencies
rare to occasional
frequent to abundant
Woodland
Hedgerow, Scrub
Acid heath, Bog
Grassland
Waterside, Marsh
Water
Roadsides, Waste places
Quarries, Walls
Cultivated
Scale
Kilometres
Miles
Tamworth
NUNEATON
Bedworth
COVENTRY
RUGBY
Kenilworth
ROYAL LEAMINGTON SPA
WARWICK
Southam
Kineton
Shipston on Stour
Long Compton
Stratford on Avon
Alcester
Solihull
BIRMINGHAM

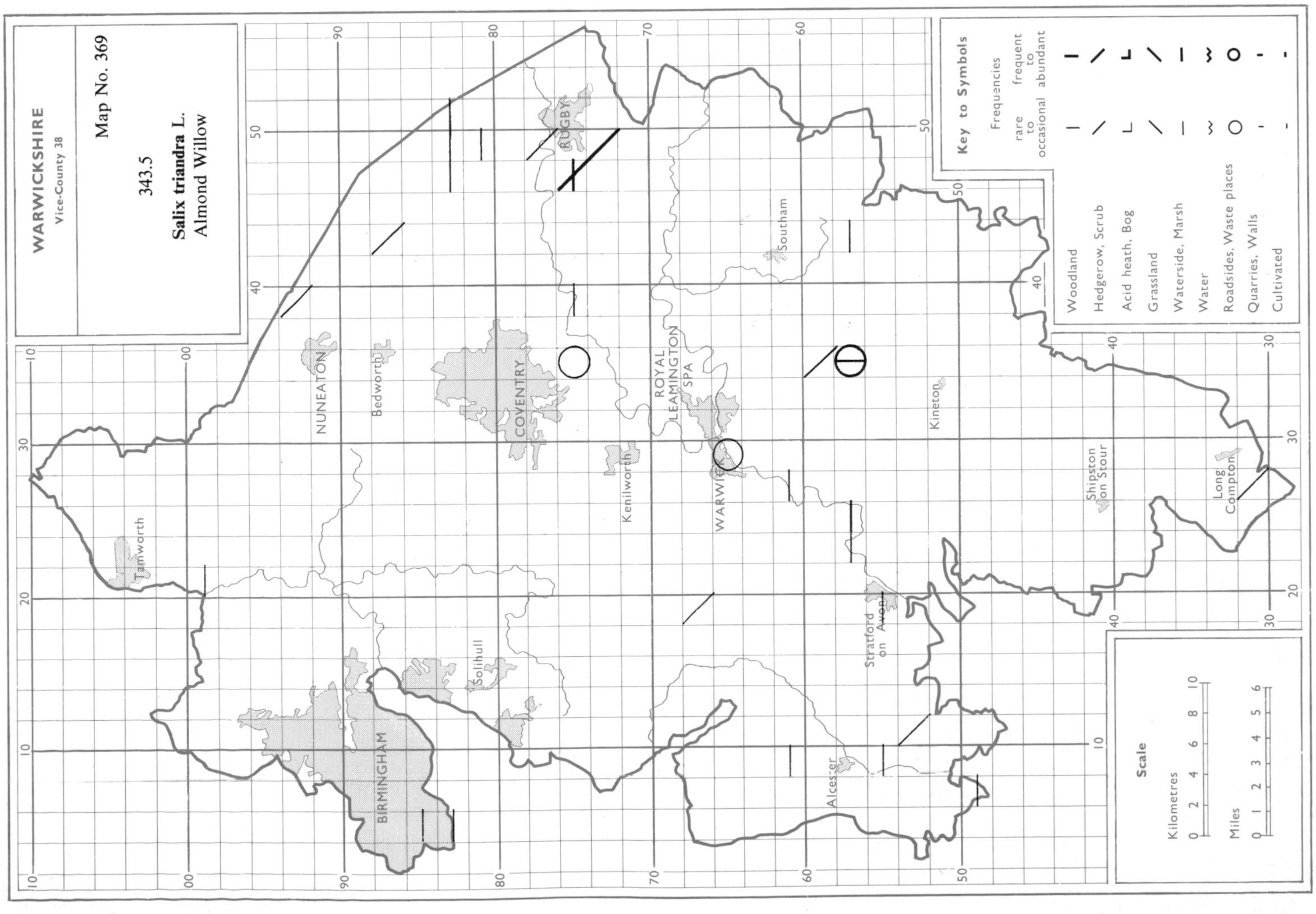
WARWICKSHIRE
Vice-County 38
Map No. 369
343.5
Salix triandra L.
Almond Willow
Key to Symbols
Frequencies
rare to occasional
frequent to abundant
Woodland
Hedgerow, Scrub
Acid heath, Bog
Grassland
Waterside, Marsh
Water
Roadsides, Waste places
Quarries, Walls
Cultivated
Scale
Kilometres
Miles
Tamworth
NUNEATON
Bedworth
COVENTRY
RUGBY
Kenilworth
ROYAL LEAMINGTON SPA
WARWICK
Southam
Kineton
Shipston on Stour
Long Compton
Stratford on Avon
Alcester
Solihull
BIRMINGHAM

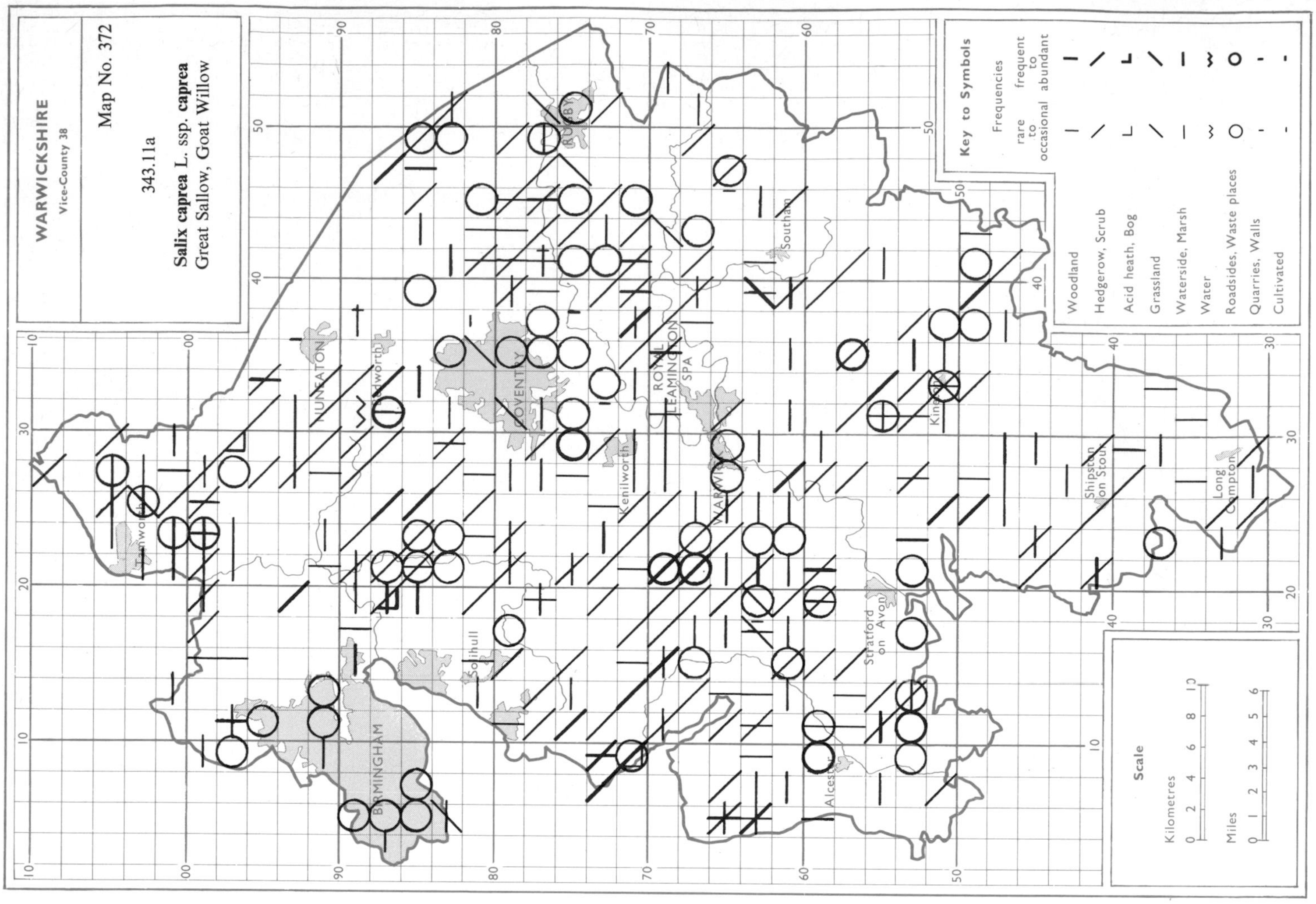
WARWICKSHIRE
Vice-County 38
Map No. 372
343.11a
Salix caprea L. ssp. caprea
Great Sallow, Goat Willow
Key to Symbols
Frequencies
rare to occasional
frequent to abundant
Woodland
Hedgerow, Scrub
Acid heath, Bog
Grassland
Waterside, Marsh
Water
Roadsides, Waste places
Quarries, Walls
Cultivated
Scale
Kilometres 0 2 4 6 8 10
Miles 0 1 2 3 4 5 6
NUNEATON
Bedworth
COVENTRY
RUGBY
Southam
ROYAL LEAMINGTON SPA
Kenilworth
WARWICK
Tamworth
Solihull
BIRMINGHAM
Kineton
Shipston on Stour
Long Compton
Stratford on Avon
Alcester

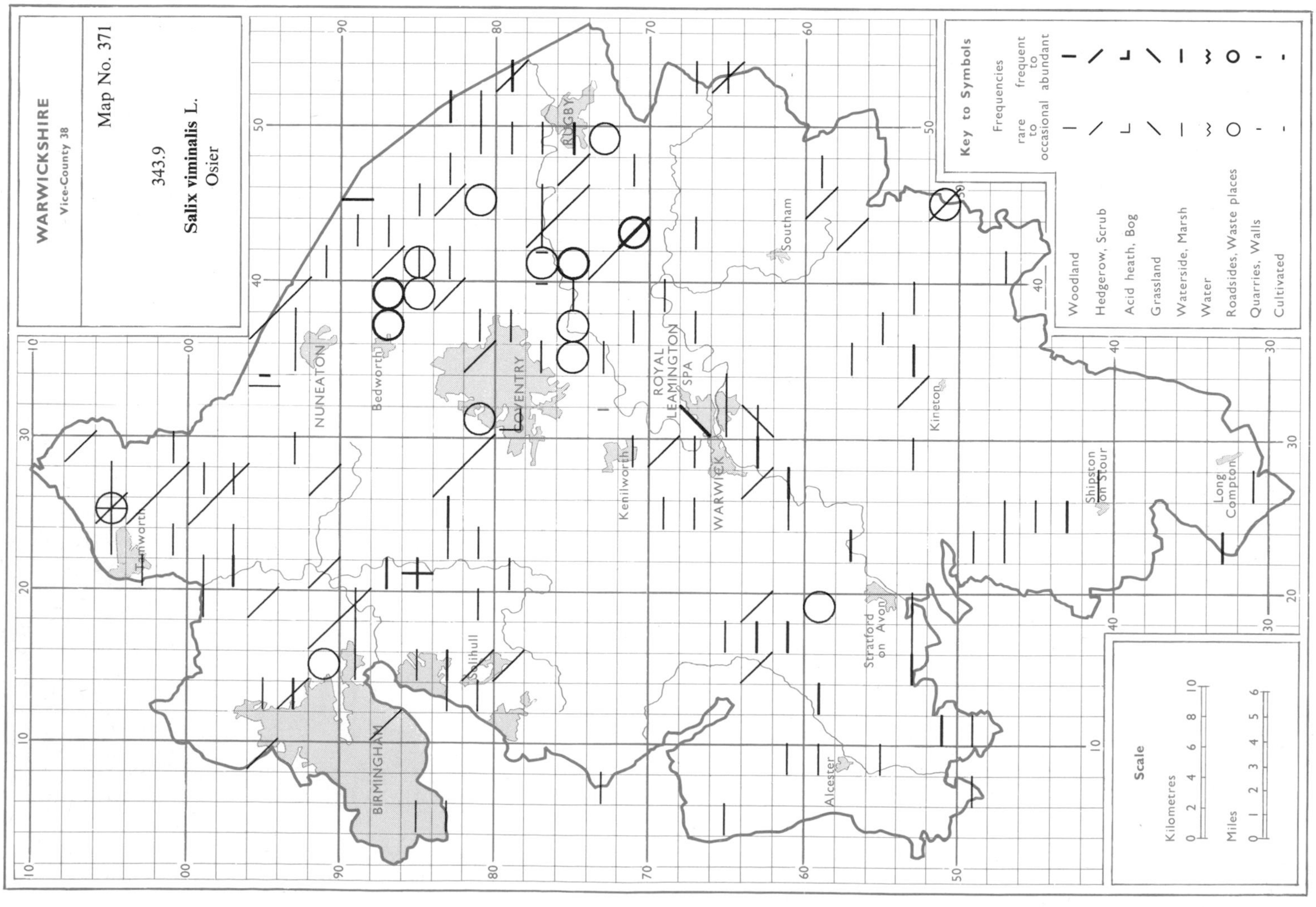
WARWICKSHIRE
Vice-County 38
Map No. 371
343.9
Salix viminalis L.
Osier
Key to Symbols
Frequencies
rare to occasional
frequent to abundant
Woodland
Hedgerow, Scrub
Acid heath, Bog
Grassland
Waterside, Marsh
Water
Roadsides, Waste places
Quarries, Walls
Cultivated
Scale
Kilometres 0 2 4 6 8 10
Miles 0 1 2 3 4 5 6
NUNEATON
Bedworth
COVENTRY
RUGBY
Southam
ROYAL LEAMINGTON SPA
Kenilworth
WARWICK
Tamworth
Solihull
BIRMINGHAM
Kineton
Shipston on Stour
Long Compton
Stratford on Avon
Alcester

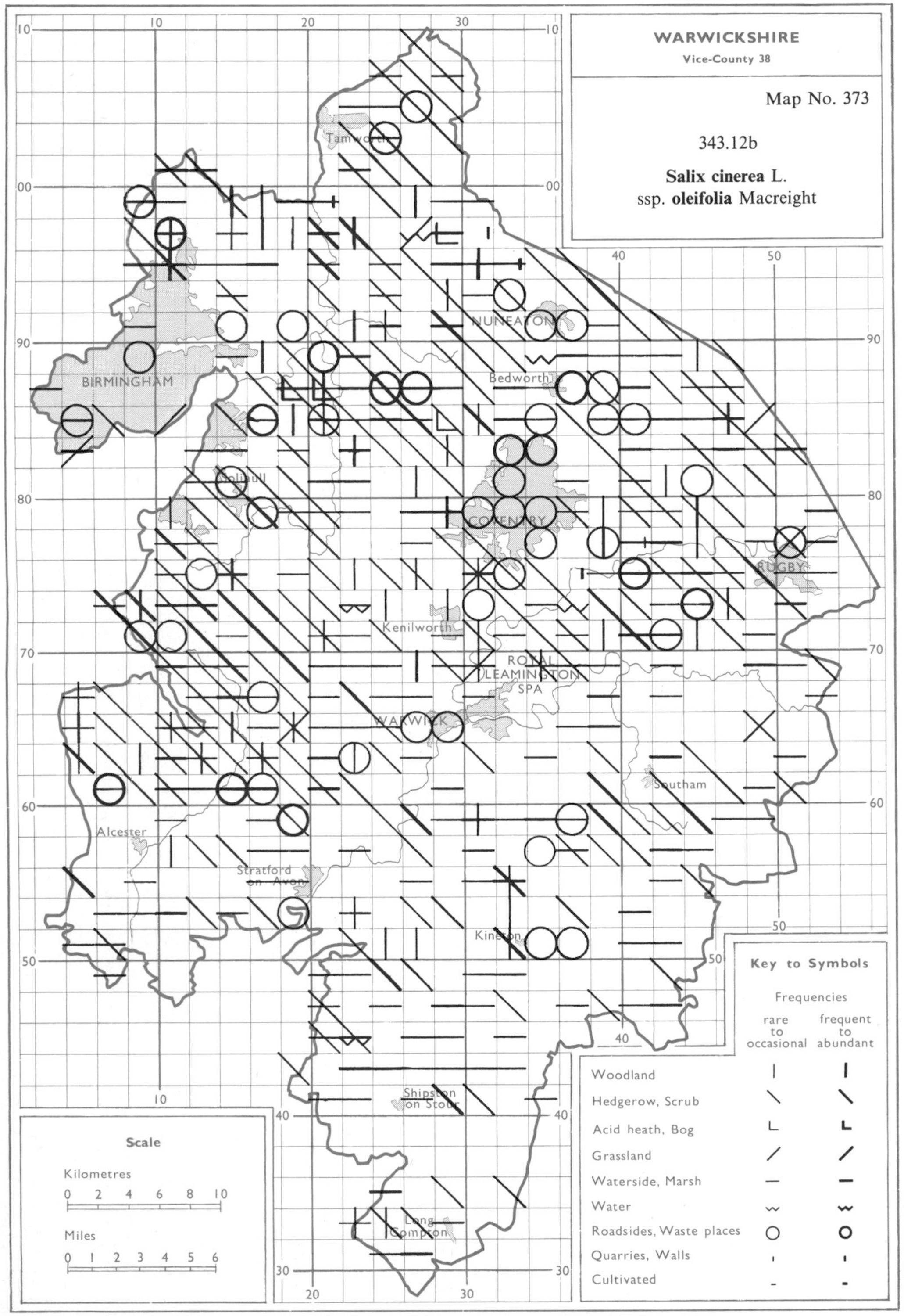
WARWICKSHIRE
Vice-County 38
Map No. 373
343.12b
Salix cinerea L.
ssp. **oleifolia** Macreight
Tamworth
NUNEATON
Bedworth
BIRMINGHAM
Solihull
COVENTRY
RUGBY
Kenilworth
ROYAL LEAMINGTON SPA
WARWICK
Southam
Alcester
Stratford on Avon
Kineton
Shipston on Stour
Long Compton
Scale
Kilometres
0 2 4 6 8 10
Miles
0 1 2 3 4 5 6
Key to Symbols
Frequencies
rare to occasional
frequent to abundant
Woodland
Hedgerow, Scrub
Acid heath, Bog
Grassland
Waterside, Marsh
Water
Roadsides, Waste places
Quarries, Walls
Cultivated

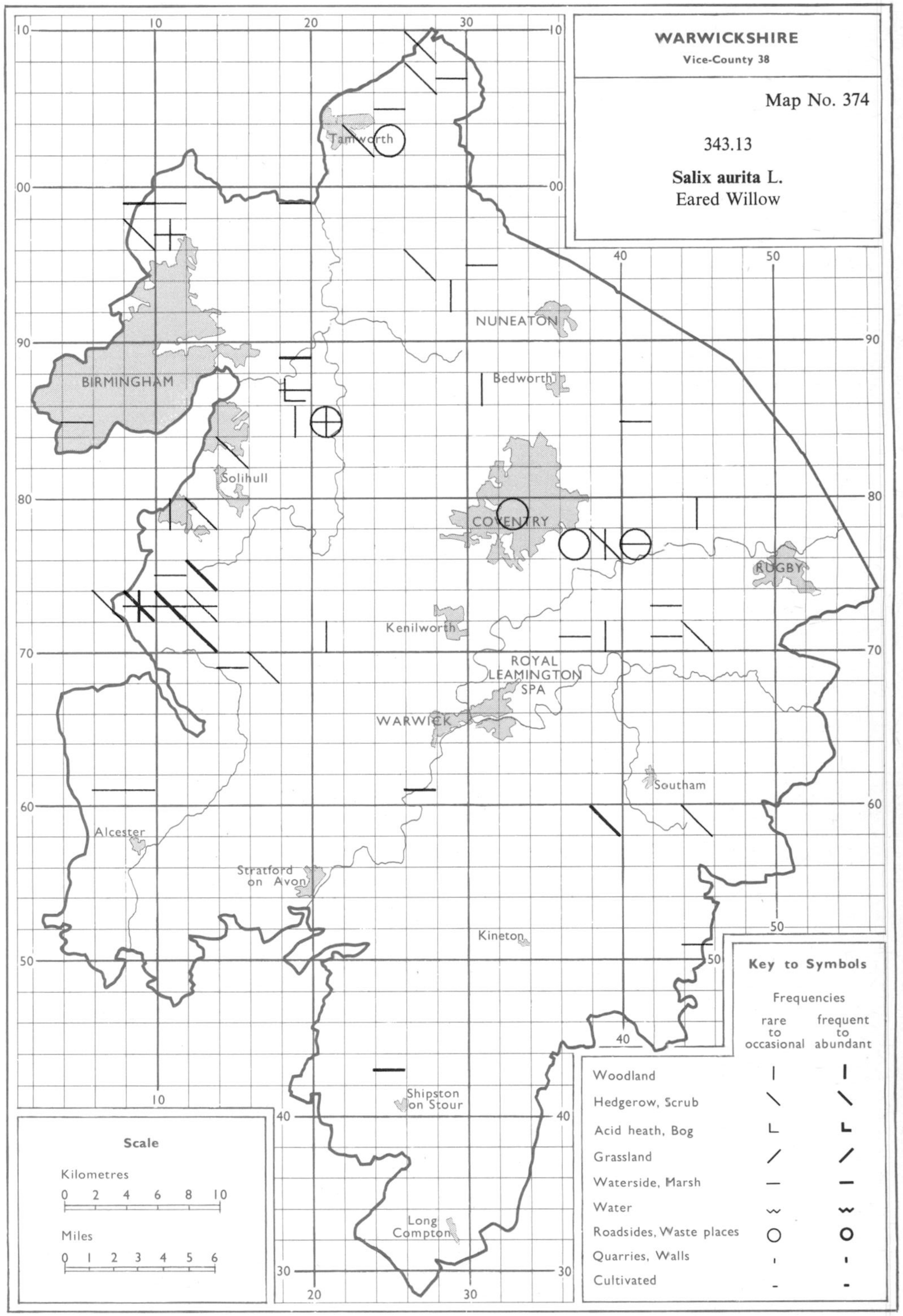
WARWICKSHIRE
Vice-County 38
Map No. 374
343.13
Salix aurita L.
Eared Willow
Tamworth
NUNEATON
Bedworth
BIRMINGHAM
Solihull
COVENTRY
RUGBY
Kenilworth
ROYAL LEAMINGTON SPA
WARWICK
Southam
Alcester
Stratford on Avon
Kineton
Shipston on Stour
Long Compton
Scale
Kilometres
0 2 4 6 8 10
Miles
0 1 2 3 4 5 6
Key to Symbols
Frequencies
rare to occasional
frequent to abundant
Woodland
Hedgerow, Scrub
Acid heath, Bog
Grassland
Waterside, Marsh
Water
Roadsides, Waste places
Quarries, Walls
Cultivated

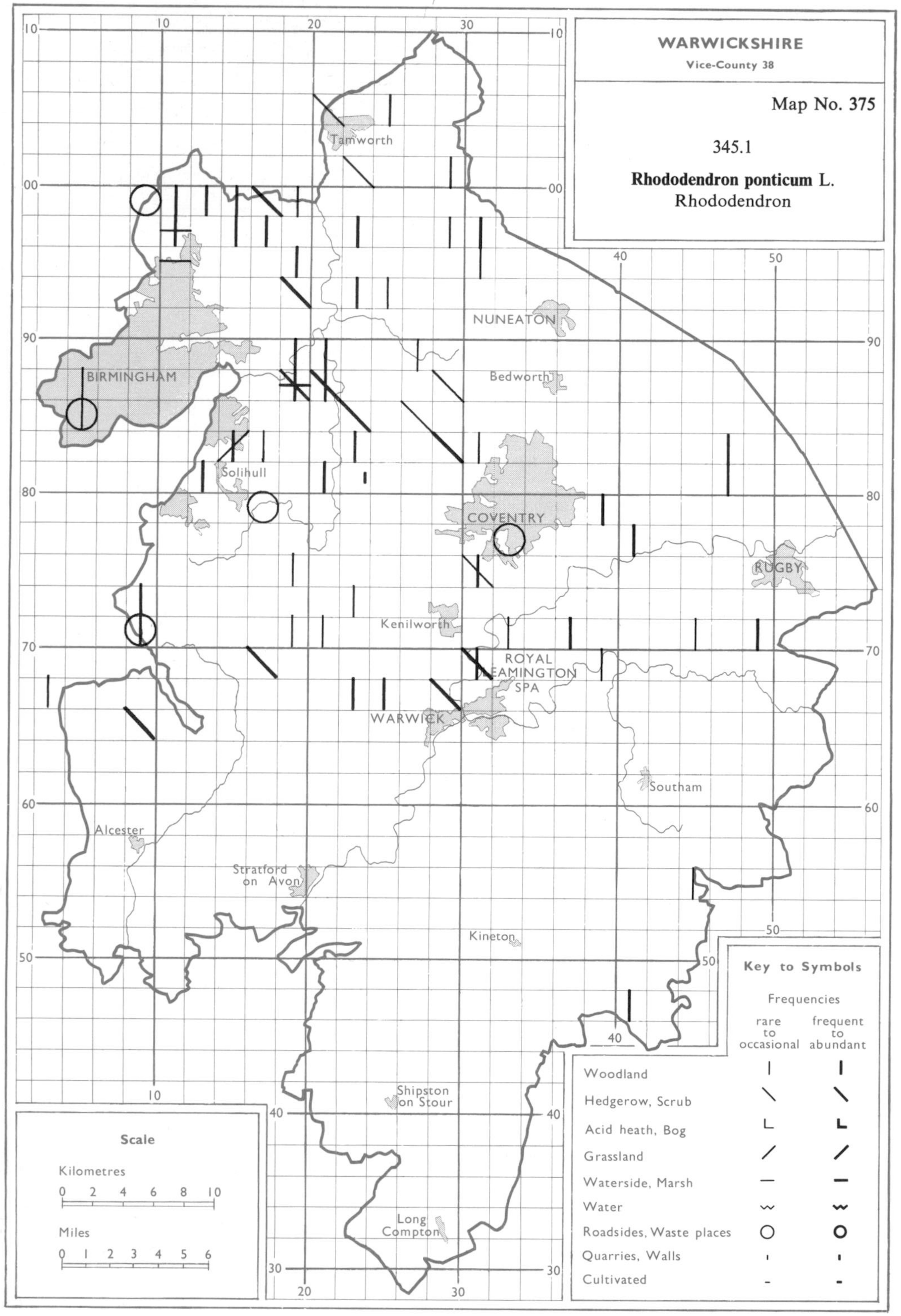
WARWICKSHIRE
Vice-County 38
Map No. 375
345.1
Rhododendron ponticum L.
Rhododendron
Tamworth
NUNEATON
Bedworth
BIRMINGHAM
Solihull
COVENTRY
RUGBY
Kenilworth
ROYAL LEAMINGTON SPA
WARWICK
Southam
Alcester
Stratford on Avon
Kineton
Shipston on Stour
Long Compton
Key to Symbols
Frequencies
rare to occasional
frequent to abundant
Woodland
Hedgerow, Scrub
Acid heath, Bog
Grassland
Waterside, Marsh
Water
Roadsides, Waste places
Quarries, Walls
Cultivated
Scale
Kilometres
0 2 4 6 8 10
Miles
0 1 2 3 4 5 6

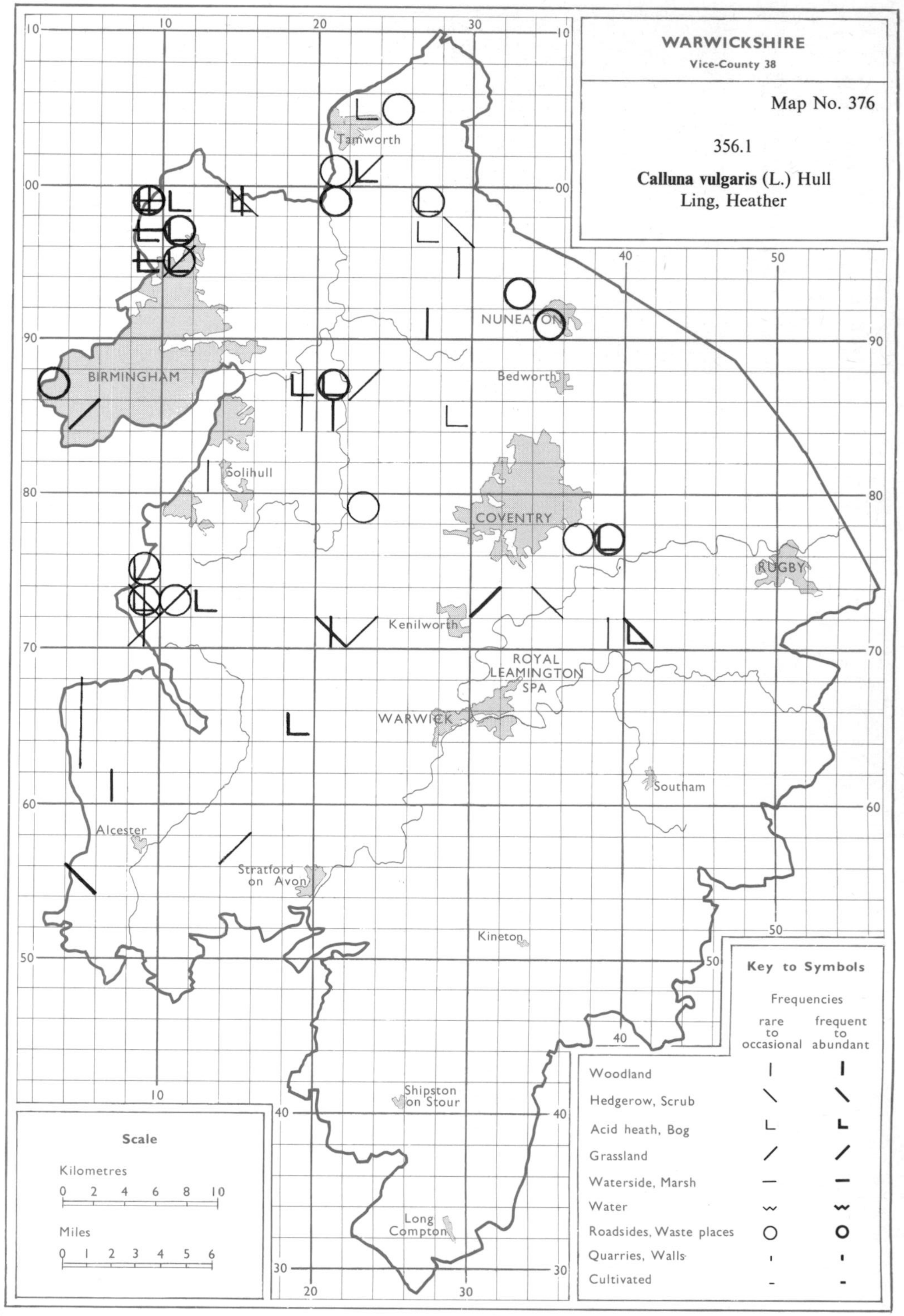
WARWICKSHIRE
Vice-County 38
Map No. 376
356.1
Calluna vulgaris (L.) Hull
Ling, Heather
Tamworth
NUNEATON
Bedworth
BIRMINGHAM
Solihull
COVENTRY
RUGBY
Kenilworth
ROYAL LEAMINGTON SPA
WARWICK
Southam
Alcester
Stratford on Avon
Kineton
Shipston on Stour
Long Compton
Key to Symbols
Frequencies
rare to occasional
frequent to abundant
Woodland
Hedgerow, Scrub
Acid heath, Bog
Grassland
Waterside, Marsh
Water
Roadsides, Waste places
Quarries, Walls
Cultivated
Scale
Kilometres
0 2 4 6 8 10
Miles
0 1 2 3 4 5 6

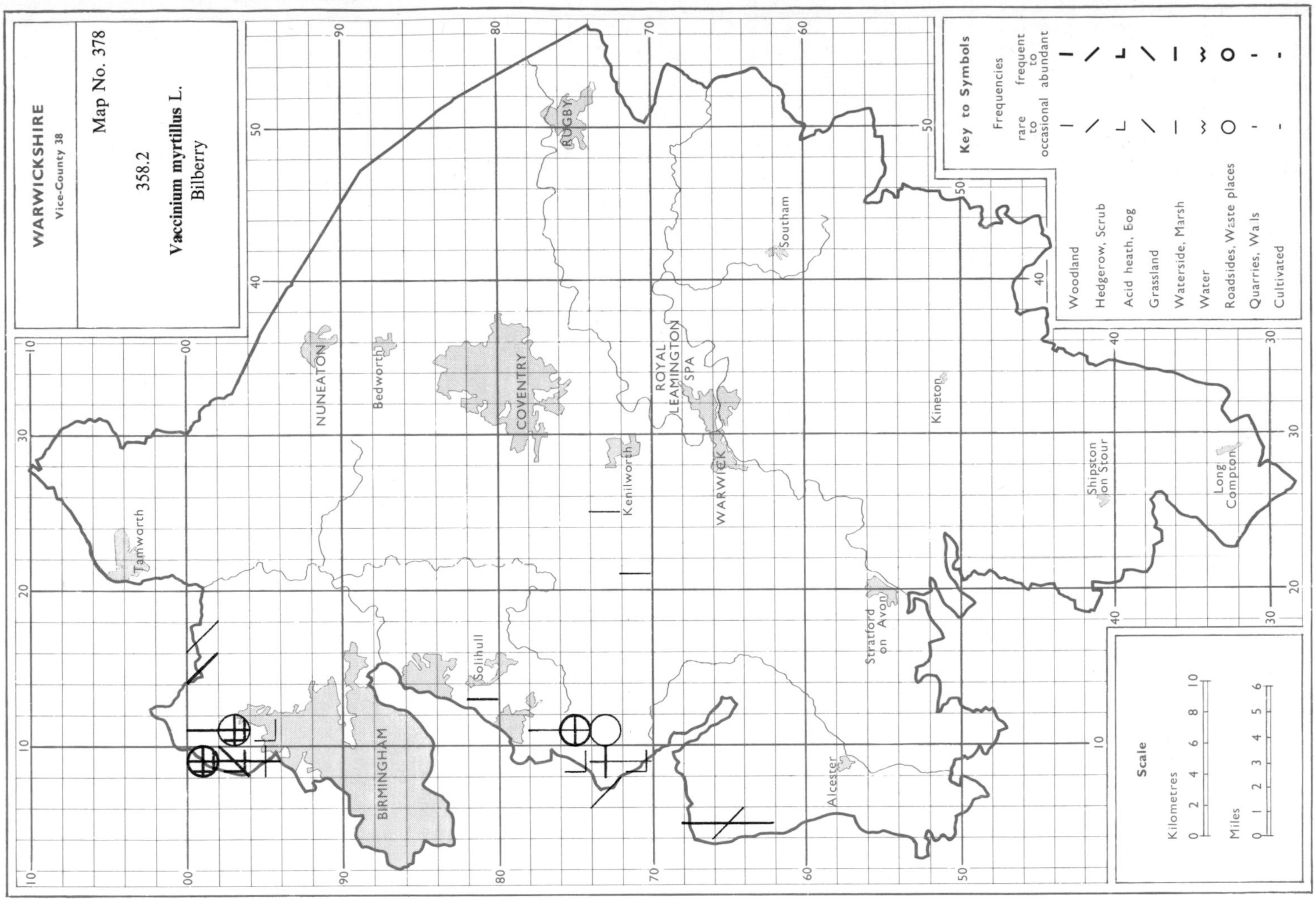
WARWICKSHIRE
Vice-County 38
Map No. 378
358.2
Vaccinium myrtillus L.
Bilberry
Key to Symbols
Frequencies
rare to occasional
frequent to abundant
Woodland
Hedgerow, Scrub
Acid heath, Bog
Grassland
Waterside, Marsh
Water
Roadsides, Waste places
Quarries, Walls
Cultivated
Scale
Kilometres
0 2 4 6 8 10
Miles
0 1 2 3 4 5 6
Tamworth
NUNEATON
Bedworth
COVENTRY
RUGBY
BIRMINGHAM
Solihull
Kenilworth
ROYAL LEAMINGTON SPA
WARWICK
Southam
Kineton
Stratford on Avon
Alcester
Shipston on Stour
Long Compton

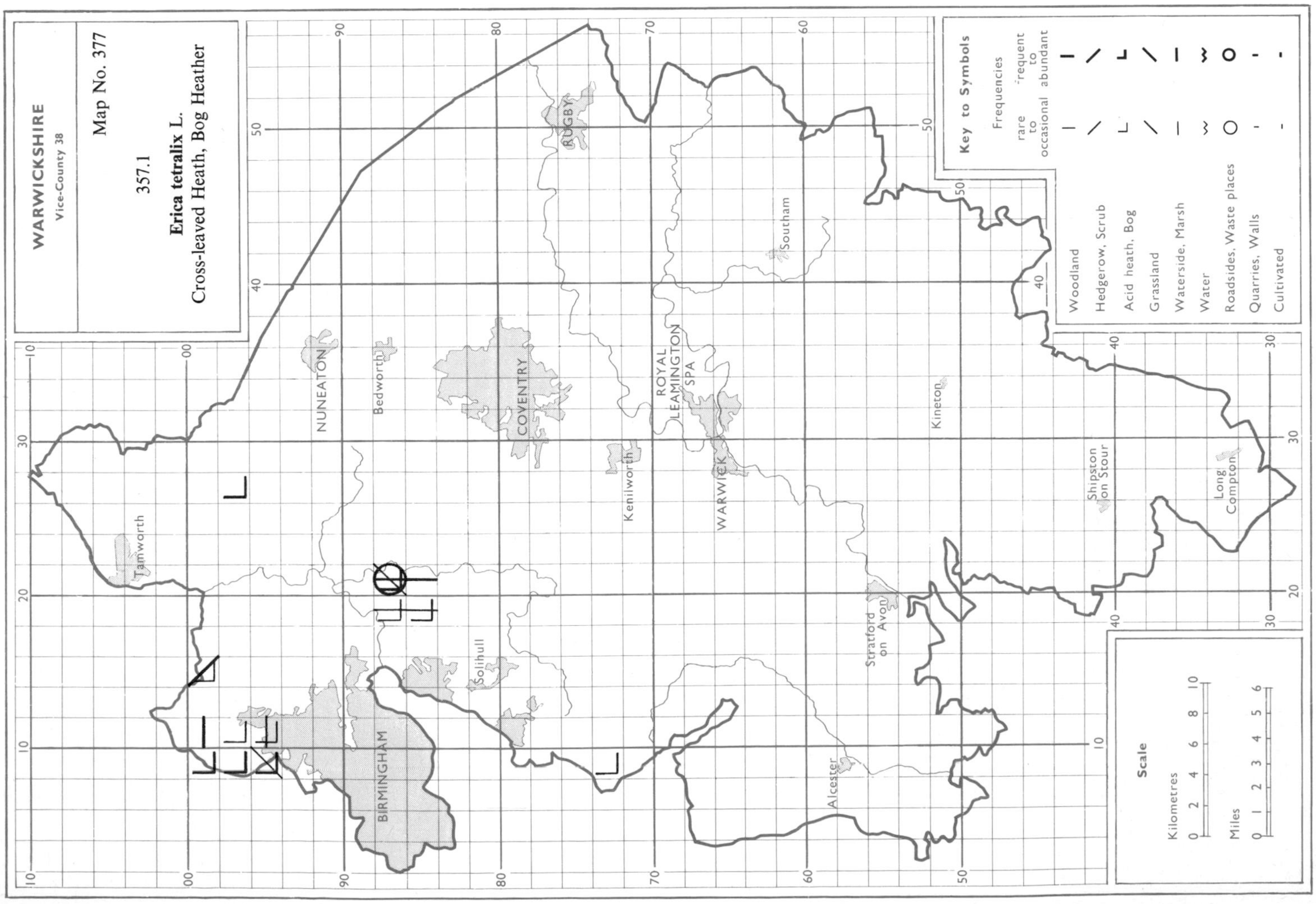
WARWICKSHIRE
Vice-County 38
Map No. 377
357.1
Erica tetralix L.
Cross-leaved Heath, Bog Heather
Key to Symbols
Frequencies
rare to occasional
frequent to abundant
Woodland
Hedgerow, Scrub
Acid heath, Bog
Grassland
Waterside, Marsh
Water
Roadsides, Waste places
Quarries, Walls
Cultivated
Scale
Kilometres
0 2 4 6 8 10
Miles
0 1 2 3 4 5 6
Tamworth
NUNEATON
Bedworth
COVENTRY
RUGBY
BIRMINGHAM
Solihull
Kenilworth
ROYAL LEAMINGTON SPA
WARWICK
Southam
Kineton
Stratford on Avon
Alcester
Shipston on Stour
Long Compton

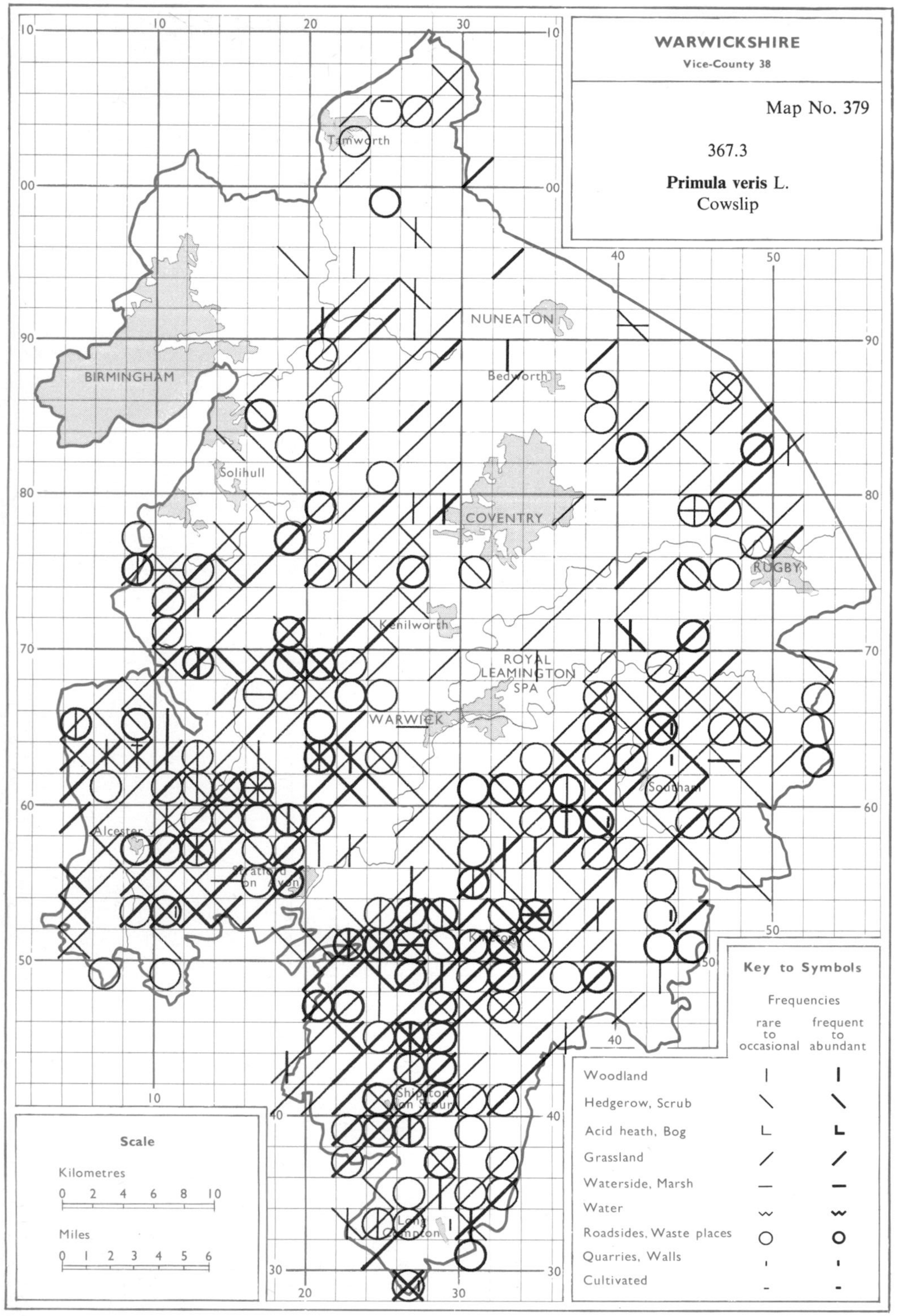
WARWICKSHIRE
Vice-County 38
Map No. 379
367.3
Primula veris L.
Cowslip
Tamworth
NUNEATON
Bedworth
BIRMINGHAM
Solihull
COVENTRY
RUGBY
Kenilworth
ROYAL LEAMINGTON SPA
WARWICK
Southam
Alcester
Stratford on Avon
Kineton
Shipston on Stour
Long Compton
Scale
Kilometres
0 2 4 6 8 10
Miles
0 1 2 3 4 5 6
Key to Symbols
Frequencies
rare to occasional
frequent to abundant
Woodland
Hedgerow, Scrub
Acid heath, Bog
Grassland
Waterside, Marsh
Water
Roadsides, Waste places
Quarries, Walls
Cultivated

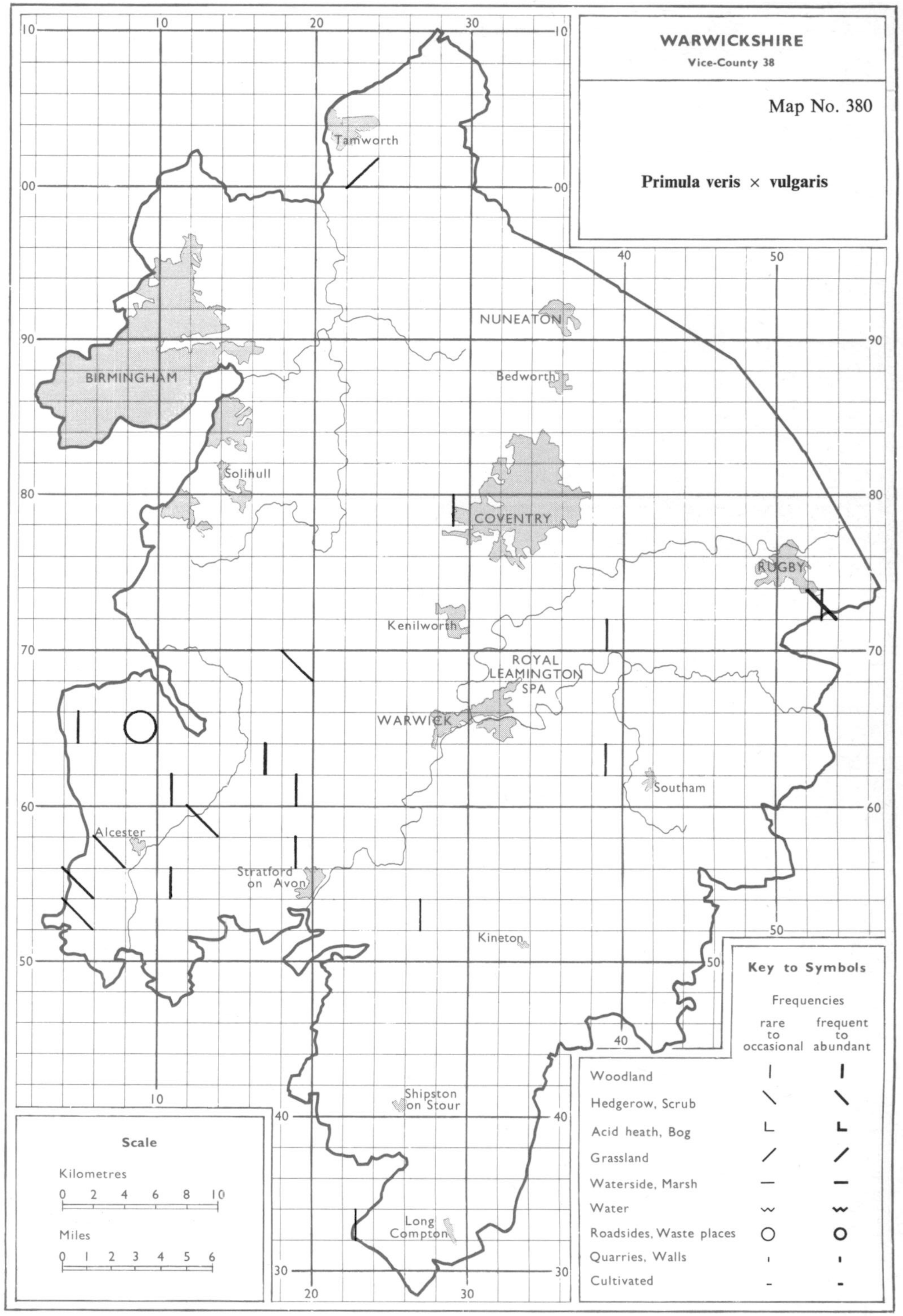
WARWICKSHIRE
Vice-County 38
Map No. 380
Primula veris × vulgaris
Tamworth
NUNEATON
Bedworth
BIRMINGHAM
Solihull
COVENTRY
RUGBY
Kenilworth
ROYAL LEAMINGTON SPA
WARWICK
Southam
Alcester
Stratford on Avon
Kineton
Shipston on Stour
Long Compton
Scale
Kilometres
0 2 4 6 8 10
Miles
0 1 2 3 4 5 6
Key to Symbols
Frequencies
rare to occasional
frequent to abundant
Woodland
Hedgerow, Scrub
Acid heath, Bog
Grassland
Waterside, Marsh
Water
Roadsides, Waste places
Quarries, Walls
Cultivated

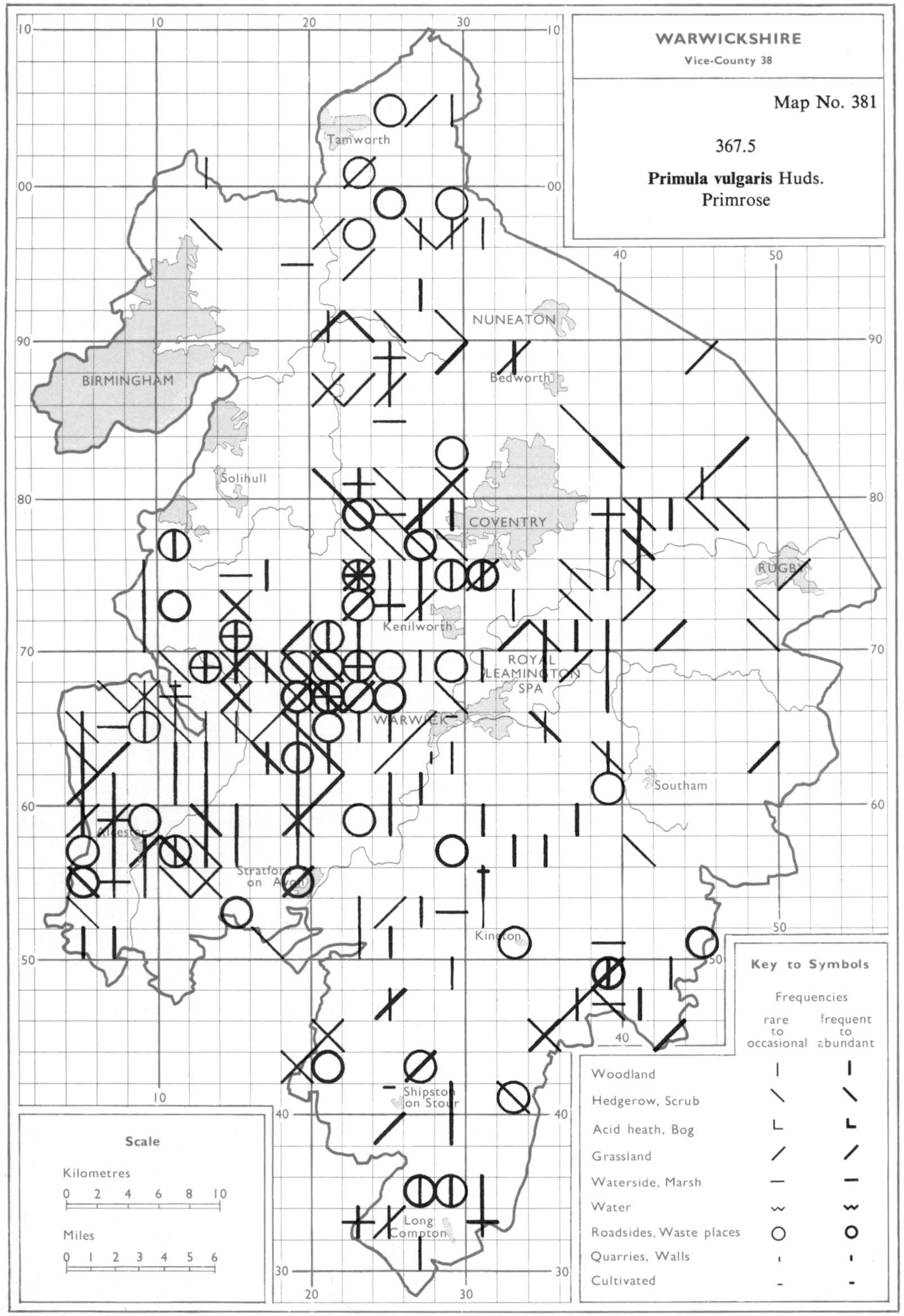

WARWICKSHIRE
Vice-County 38
Map No. 381
367.5
Primula vulgaris Huds.
Primrose
Tamworth
NUNEATON
Bedworth
BIRMINGHAM
Solihull
COVENTRY
RUGBY
Kenilworth
ROYAL LEAMINGTON SPA
WARWICK
Southam
Alcester
Stratford on Avon
Kineton
Shipston on Stour
Long Compton
Scale
Kilometres
0 2 4 6 8 10
Miles
0 1 2 3 4 5 6
Key to Symbols
Frequencies
rare to occasional
frequent to abundant
Woodland
Hedgerow, Scrub
Acid heath, Bog
Grassland
Waterside, Marsh
Water
Roadsides, Waste places
Quarries, Walls
Cultivated

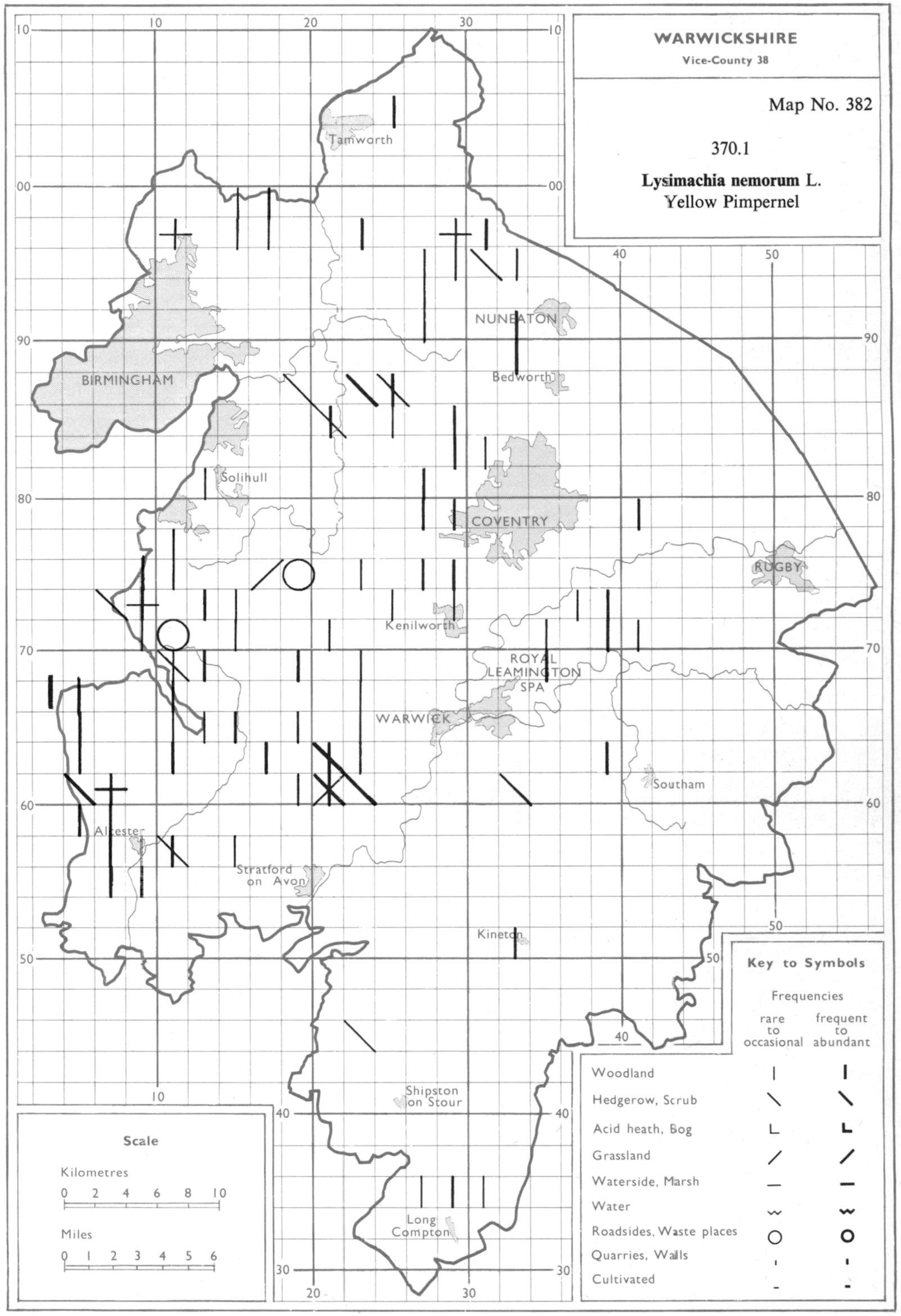

WARWICKSHIRE
Vice-County 38
Map No. 382
370.1
Lysimachia nemorum L.
Yellow Pimpernel
Tamworth
NUNEATON
Bedworth
BIRMINGHAM
Solihull
COVENTRY
RUGBY
Kenilworth
ROYAL LEAMINGTON SPA
WARWICK
Southam
Alcester
Stratford on Avon
Kineton
Shipston on Stour
Long Compton
Scale
Kilometres
0 2 4 6 8 10
Miles
0 1 2 3 4 5 6
Key to Symbols
Frequencies
rare to occasional
frequent to abundant
Woodland
Hedgerow, Scrub
Acid heath, Bog
Grassland
Waterside, Marsh
Water
Roadsides, Waste places
Quarries, Walls
Cultivated

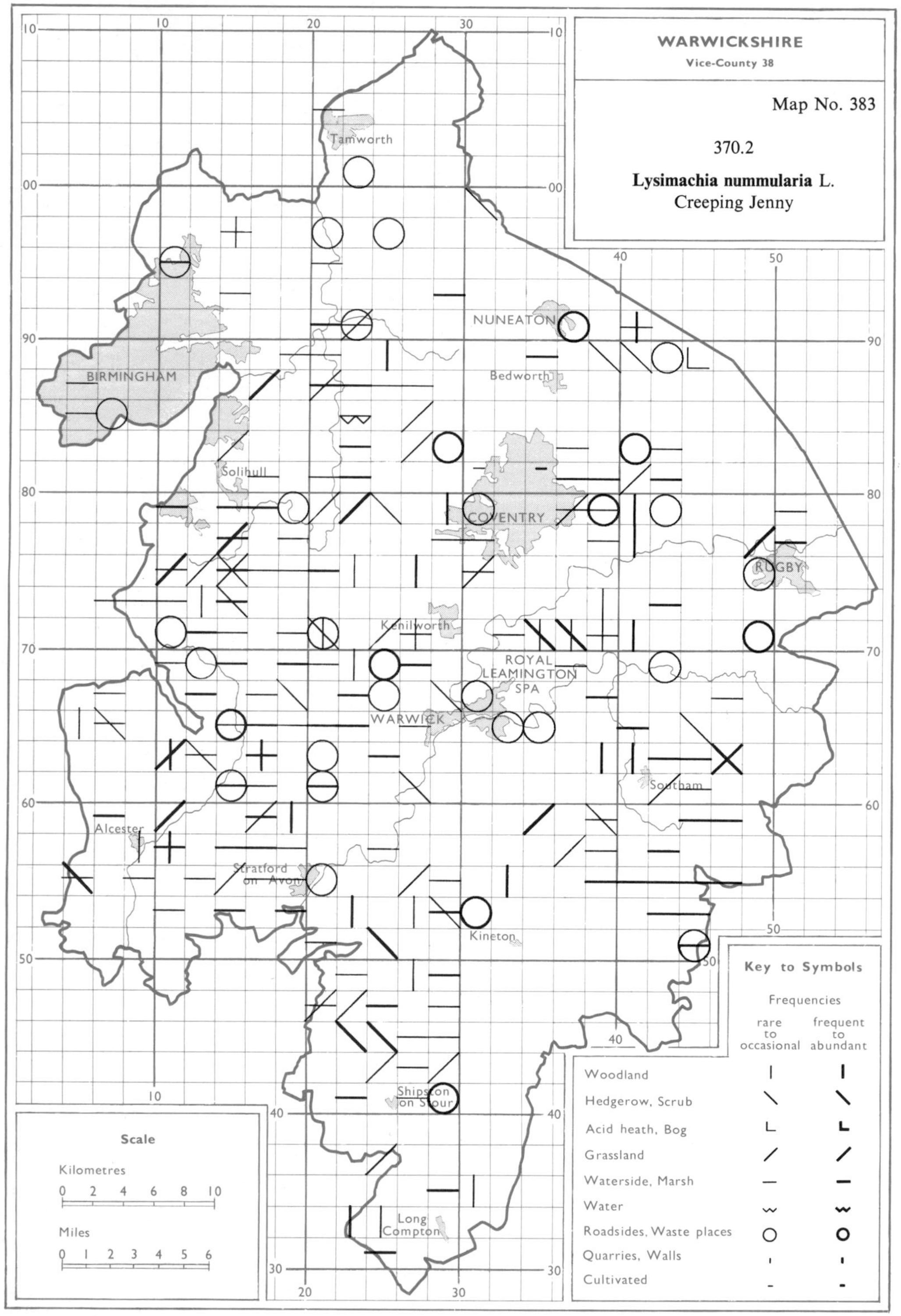
WARWICKSHIRE
Vice-County 38
Map No. 383
370.2
Lysimachia nummularia L.
Creeping Jenny
Tamworth
NUNEATON
Bedworth
BIRMINGHAM
Solihull
COVENTRY
RUGBY
Kenilworth
ROYAL LEAMINGTON SPA
WARWICK
Southam
Alcester
Stratford on Avon
Kineton
Shipston on Stour
Long Compton
Key to Symbols
Frequencies
rare to occasional
frequent to abundant
Woodland
Hedgerow, Scrub
Acid heath, Bog
Grassland
Waterside, Marsh
Water
Roadsides, Waste places
Quarries, Walls
Cultivated
Scale
Kilometres
0 2 4 6 8 10
Miles
0 1 2 3 4 5 6

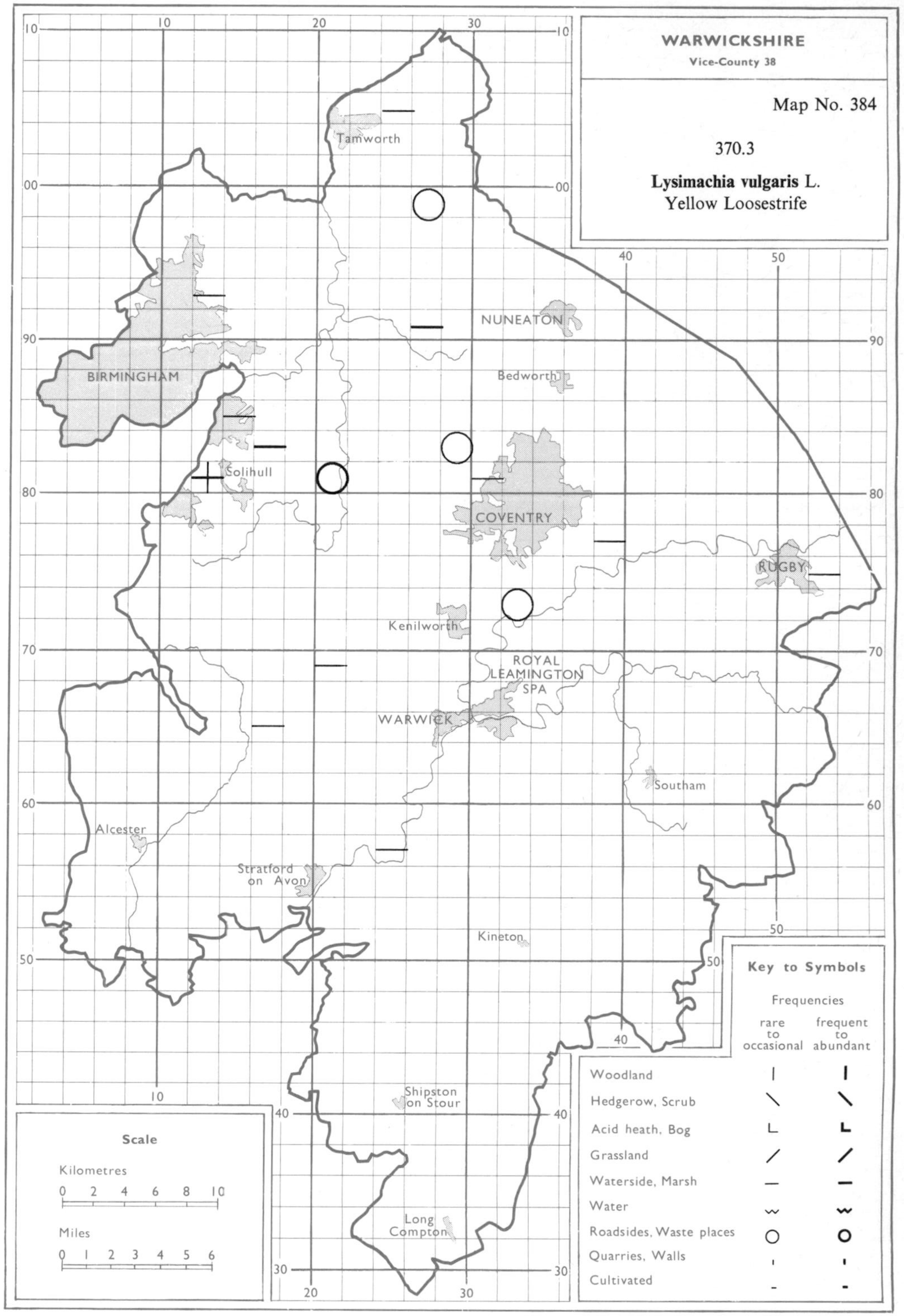
WARWICKSHIRE
Vice-County 38
Map No. 384
370.3
Lysimachia vulgaris L.
Yellow Loosestrife
Tamworth
NUNEATON
Bedworth
BIRMINGHAM
Solihull
COVENTRY
RUGBY
Kenilworth
ROYAL LEAMINGTON SPA
WARWICK
Southam
Alcester
Stratford on Avon
Kineton
Shipston on Stour
Long Compton
Key to Symbols
Frequencies
rare to occasional
frequent to abundant
Woodland
Hedgerow, Scrub
Acid heath, Bog
Grassland
Waterside, Marsh
Water
Roadsides, Waste places
Quarries, Walls
Cultivated
Scale
Kilometres
0 2 4 6 8 10
Miles
0 1 2 3 4 5 6

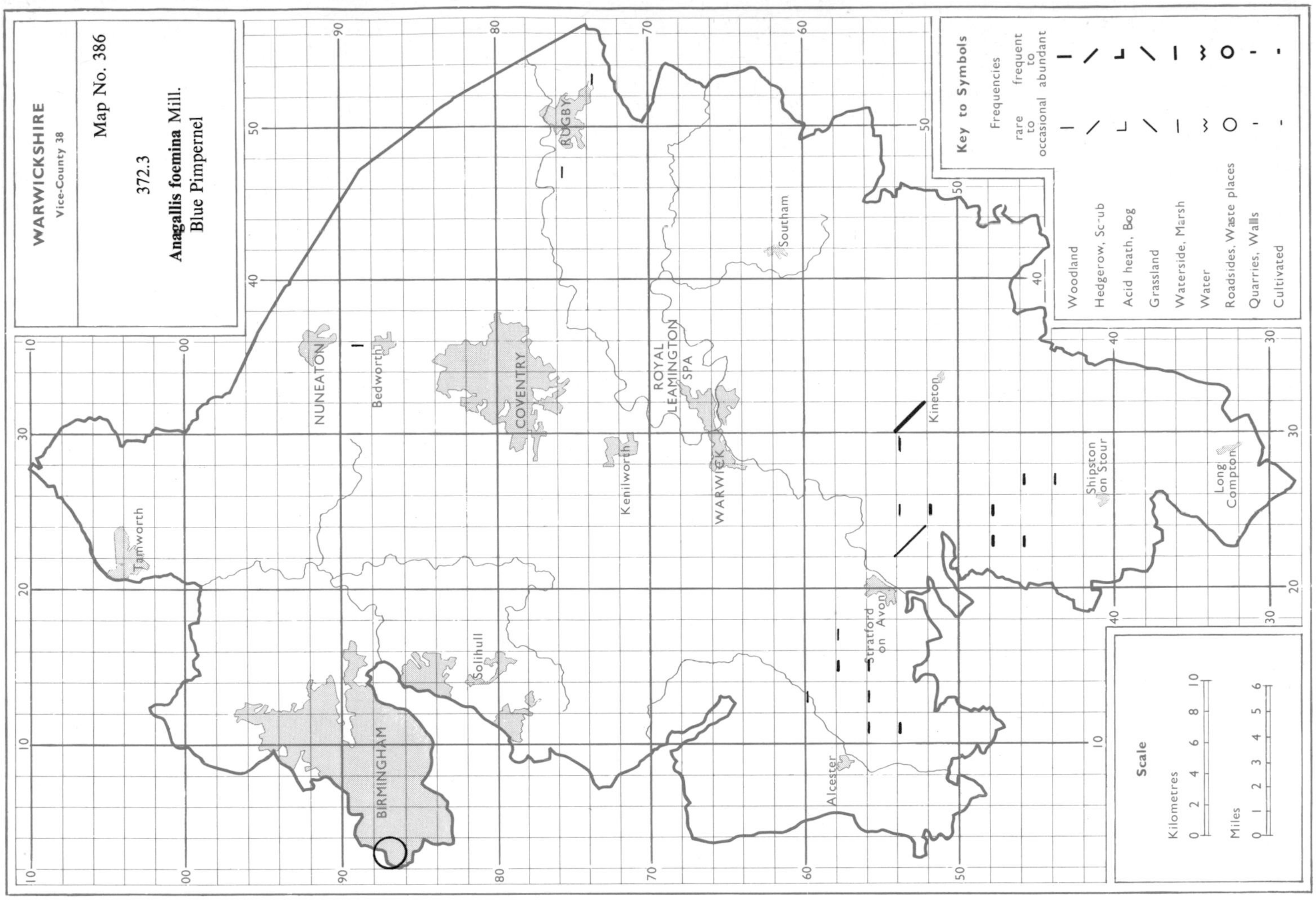
WARWICKSHIRE
Vice-County 38
Map No. 386
372.3
Anagallis foemina Mill.
Blue Pimpernel
Key to Symbols
Frequencies
rare to occasional
frequent to abundant
Woodland
Hedgerow, Scrub
Acid heath, Bog
Grassland
Waterside, Marsh
Water
Roadsides, Waste places
Quarries, Walls
Cultivated
Scale
Kilometres
Miles
NUNEATON
Bedworth
COVENTRY
RUGBY
Southam
ROYAL LEAMINGTON SPA
WARWICK
Kenilworth
Kineton
Shipston on Stour
Long Compton
Stratford on Avon
Alcester
Solihull
BIRMINGHAM
Tamworth

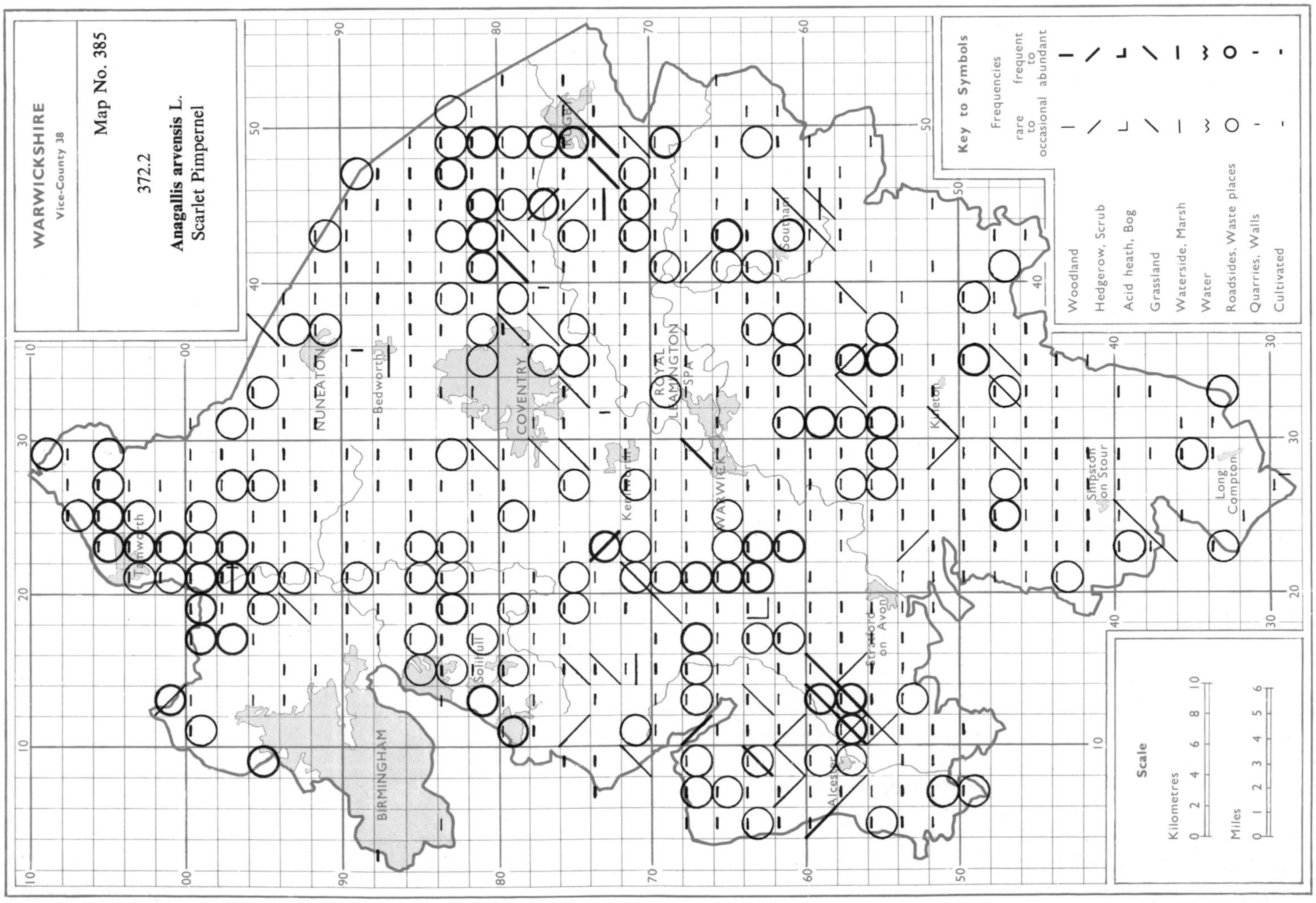
WARWICKSHIRE
Vice-County 38
Map No. 385
372.2
Anagallis arvensis L.
Scarlet Pimpernel
Key to Symbols
Frequencies
rare to occasional
frequent to abundant
Woodland
Hedgerow, Scrub
Acid heath, Bog
Grassland
Waterside, Marsh
Water
Roadsides, Waste places
Quarries, Walls
Cultivated
Scale
Kilometres
Miles
NUNEATON
Bedworth
COVENTRY
Southam
ROYAL LEAMINGTON SPA
WARWICK
Kenilworth
Kineton
Shipston on Stour
Long Compton
Stratford on Avon
Alcester
Solihull
BIRMINGHAM
Tamworth

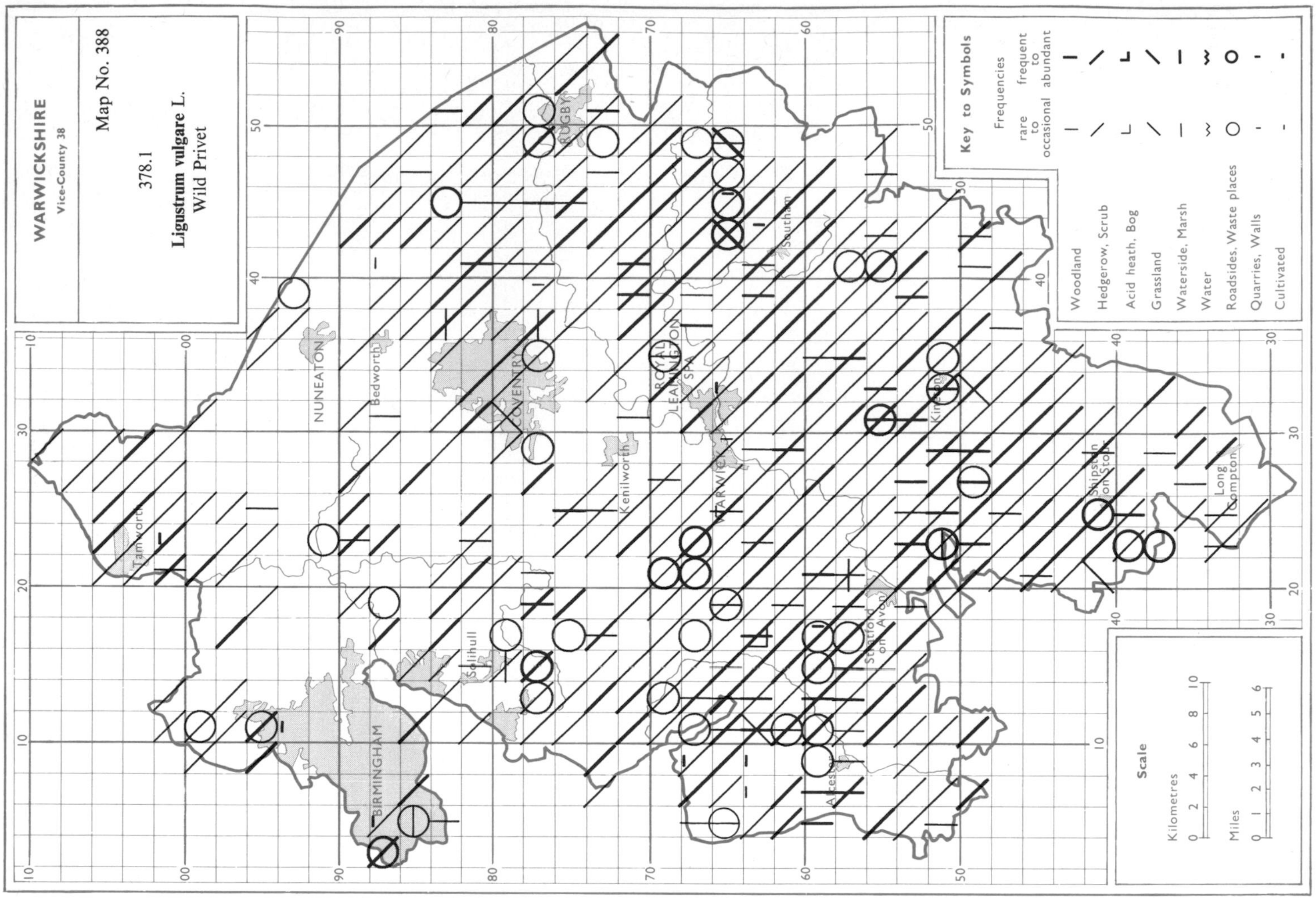
WARWICKSHIRE
Vice-County 38
Map No. 388
378.1
Ligustrum vulgare L.
Wild Privet
Key to Symbols
Frequencies
rare to occasional
frequent to abundant
Woodland
Hedgerow, Scrub
Acid heath, Bog
Grassland
Waterside, Marsh
Water
Roadsides, Waste places
Quarries, Walls
Cultivated
Scale
Kilometres
0 2 4 6 8 10
Miles
0 1 2 3 4 5 6
NUNEATON
Bedworth
COVENTRY
RUGBY
ROYAL LEAMINGTON SPA
Southam
Kineton
Shipston on Stour
Long Compton
WARWICK
Kenilworth
Tamworth
Solihull
BIRMINGHAM
Stratford on Avon
Alcester

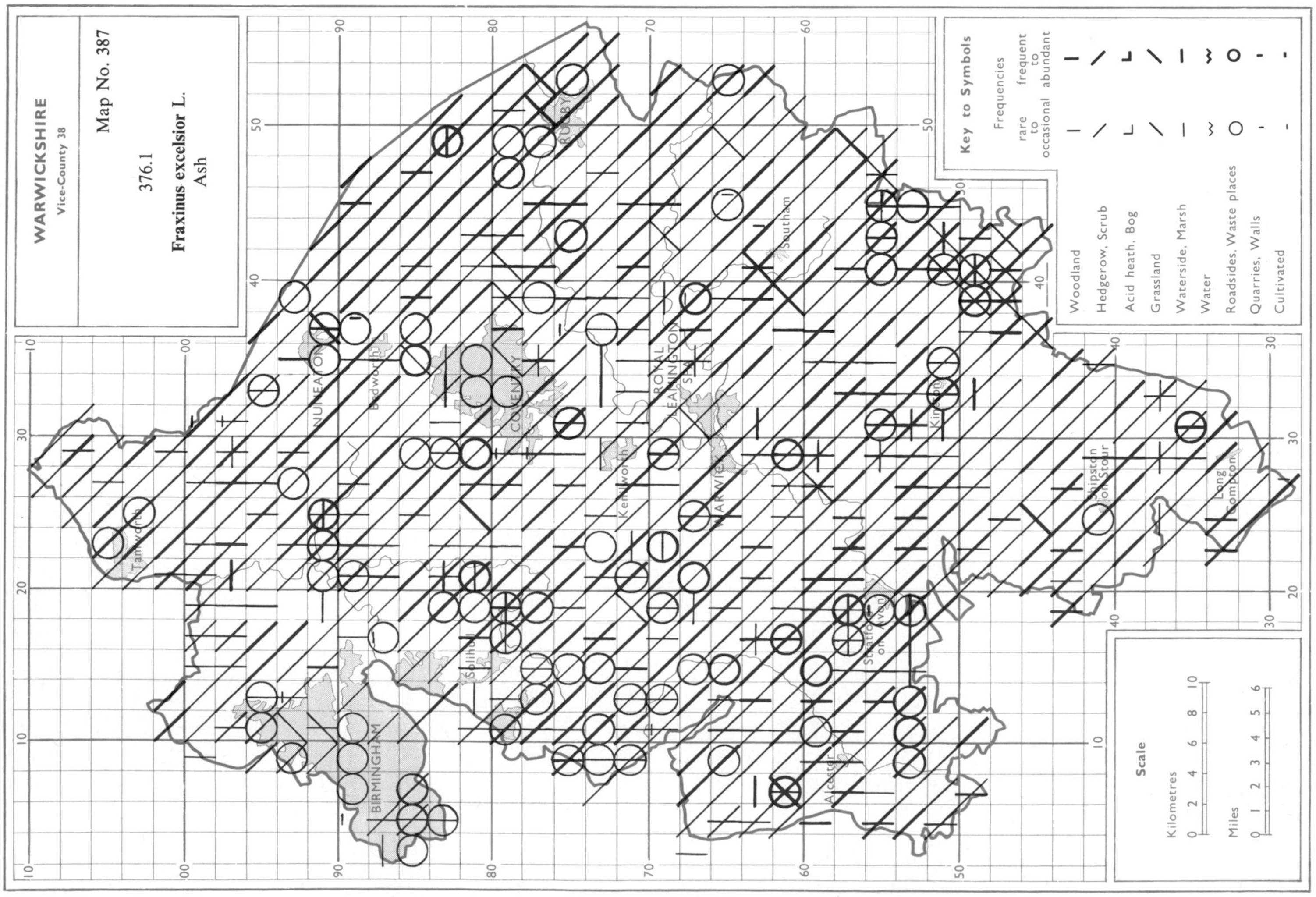
WARWICKSHIRE
Vice-County 38
Map No. 387
376.1
Fraxinus excelsior L.
Ash
Key to Symbols
Frequencies
rare to occasional
frequent to abundant
Woodland
Hedgerow, Scrub
Acid heath, Bog
Grassland
Waterside, Marsh
Water
Roadsides, Waste places
Quarries, Walls
Cultivated
Scale
Kilometres
0 2 4 6 8 10
Miles
0 1 2 3 4 5 6
NUNEATON
Bedworth
COVENTRY
RUGBY
ROYAL LEAMINGTON SPA
Southam
Kineton
Shipston on Stour
Long Compton
WARWICK
Kenilworth
Tamworth
Solihull
BIRMINGHAM
Stratford on Avon
Alcester

WARWICKSHIRE
Vice-County 38
Map No. 390
379.2
Vinca major L.
Greater Periwinkle
Key to Symbols
Frequencies
rare to occasional
frequent to abundant
Woodland
Hedgerow, Scrub
Acid heath, Bog
Grassland
Waterside, Marsh
Water
Roadsides, Waste places
Quarries, Walls
Cultivated
Scale
Kilometres
0 2 4 6 8 10
Miles
0 1 2 3 4 5 6
NUNEATON
Bedworth
COVENTRY
RUGBY
Tamworth
Solihull
BIRMINGHAM
Kenilworth
ROYAL LEAMINGTON SPA
WARWICK
Southam
Kineton
Stratford on Avon
Alcester
Shipston on Stour
Long Compton

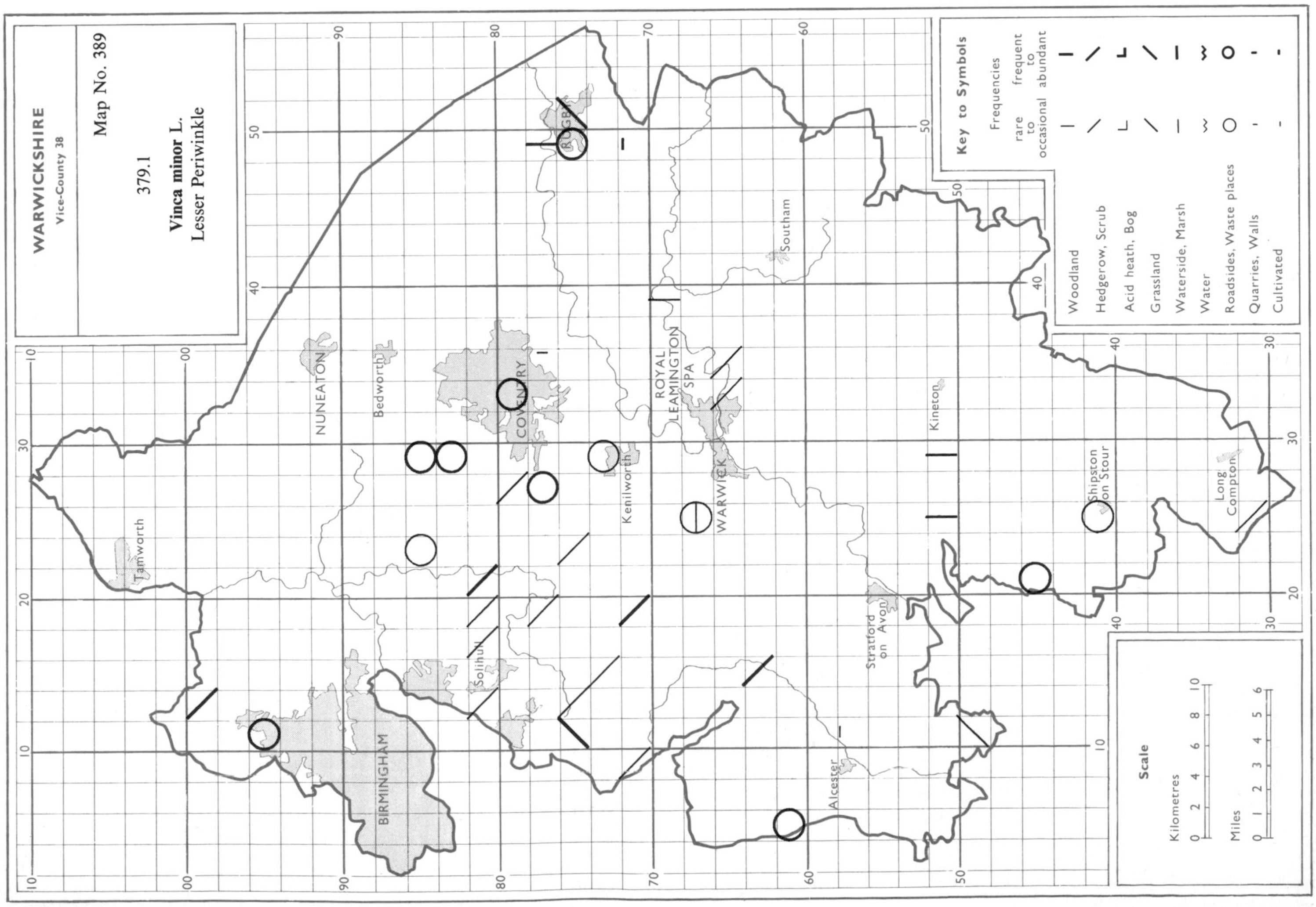
WARWICKSHIRE
Vice-County 38
Map No. 389
379.1
Vinca minor L.
Lesser Periwinkle
Key to Symbols
Frequencies
rare to occasional
frequent to abundant
Woodland
Hedgerow, Scrub
Acid heath, Bog
Grassland
Waterside, Marsh
Water
Roadsides, Waste places
Quarries, Walls
Cultivated
Scale
Kilometres
0 2 4 6 8 10
Miles
0 1 2 3 4 5 6
NUNEATON
Bedworth
COVENTRY
RUGBY
Tamworth
Solihull
BIRMINGHAM
Kenilworth
ROYAL LEAMINGTON SPA
WARWICK
Southam
Kineton
Stratford on Avon
Alcester
Shipston on Stour
Long Compton

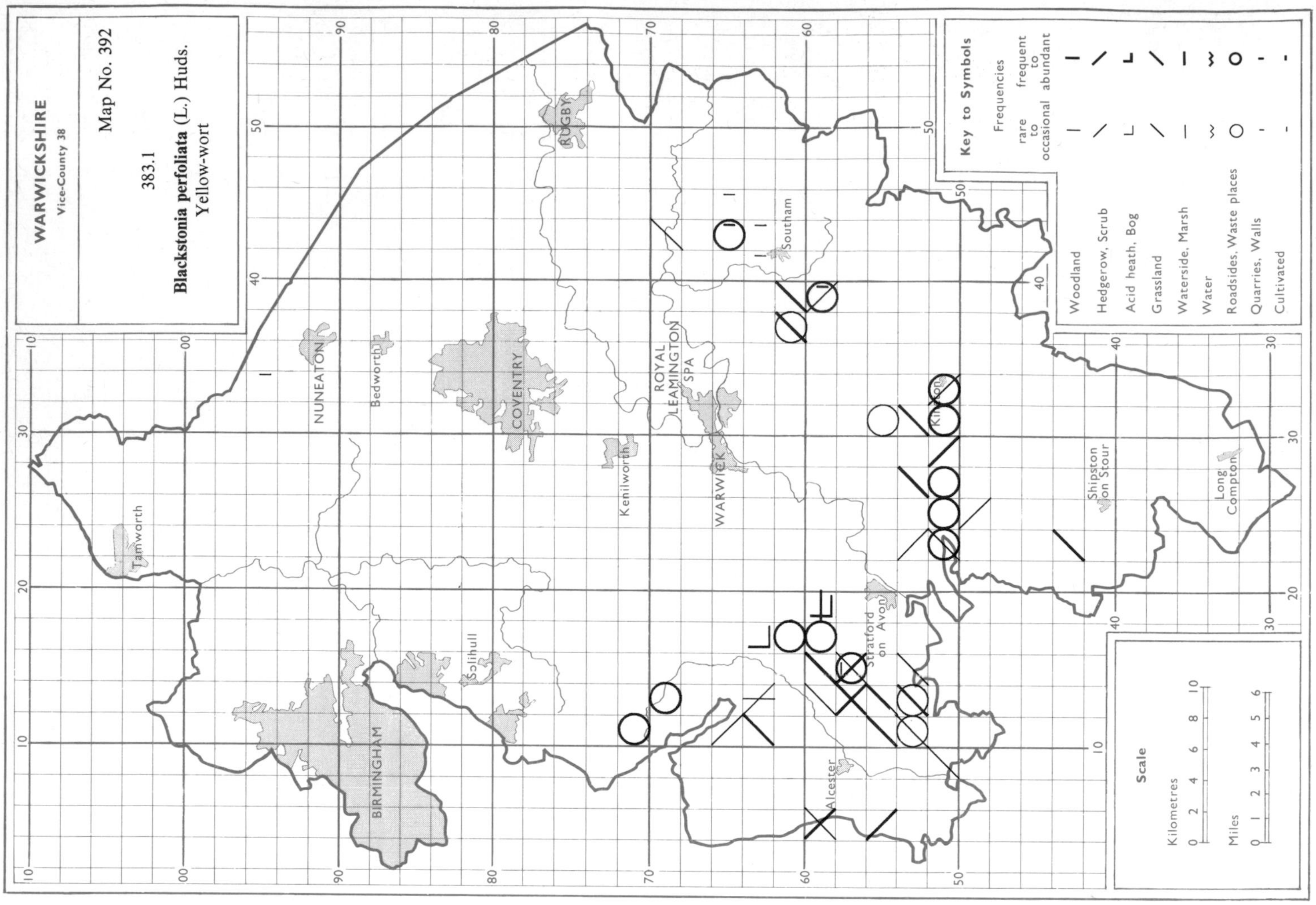
WARWICKSHIRE
Vice-County 38
Map No. 392
383.1
Blackstonia perfoliata (L.) Huds.
Yellow-wort
Key to Symbols
Frequencies
rare to occasional
frequent to abundant
Woodland
Hedgerow, Scrub
Acid heath, Bog
Grassland
Waterside, Marsh
Water
Roadsides, Waste places
Quarries, Walls
Cultivated
Scale
Kilometres 0 2 4 6 8 10
Miles 0 1 2 3 4 5 6
NUNEATON
Bedworth
COVENTRY
RUGBY
ROYAL LEAMINGTON SPA
Kenilworth
WARWICK
Southam
Kineton
Tamworth
Solihull
BIRMINGHAM
Stratford on Avon
Alcester
Shipston on Stour
Long Compton

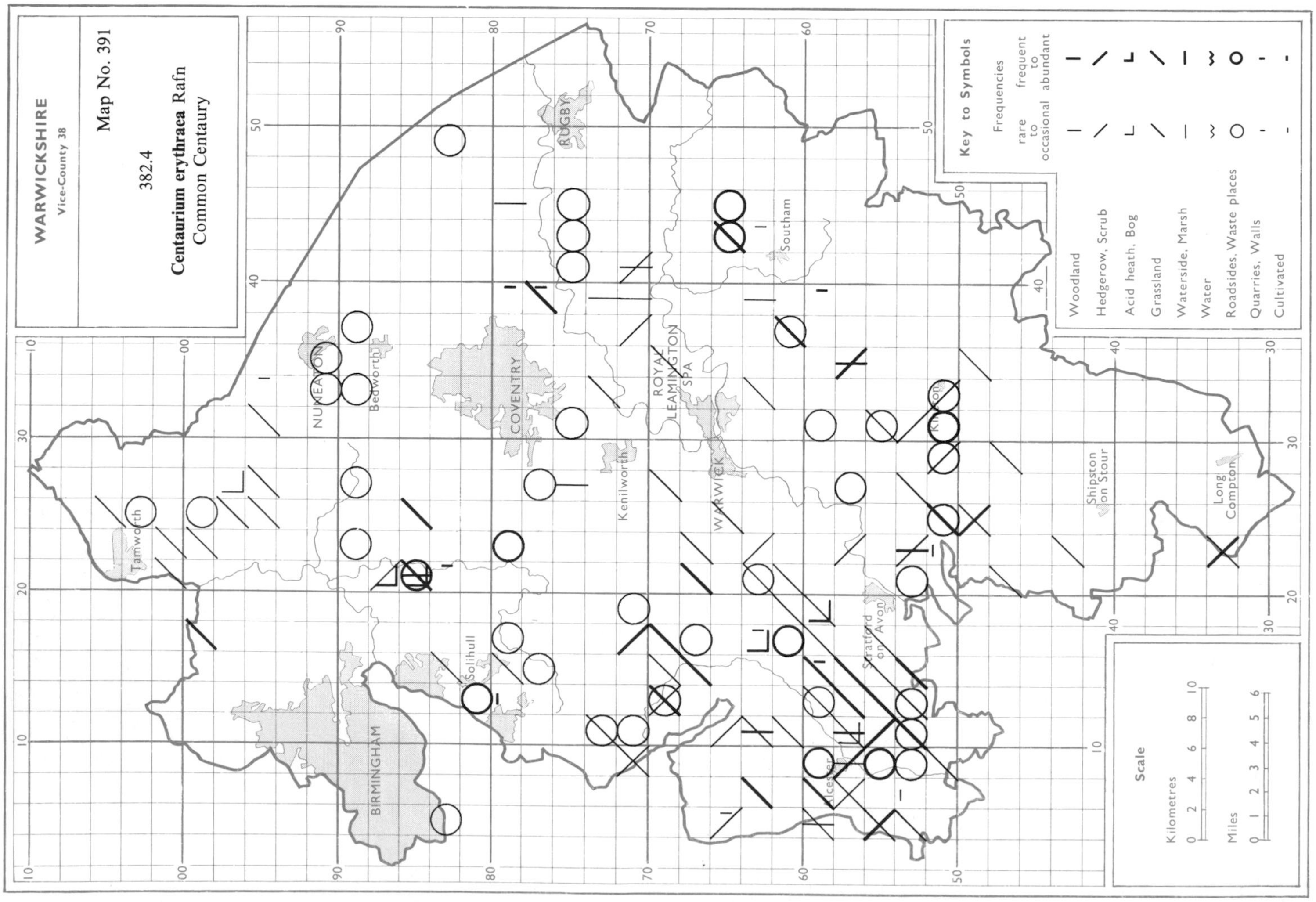
WARWICKSHIRE
Vice-County 38
Map No. 391
382.4
Centaurium erythraea Rafn
Common Centaury
Key to Symbols
Frequencies
rare to occasional
frequent to abundant
Woodland
Hedgerow, Scrub
Acid heath, Bog
Grassland
Waterside, Marsh
Water
Roadsides, Waste places
Quarries, Walls
Cultivated
Scale
Kilometres 0 2 4 6 8 10
Miles 0 1 2 3 4 5 6
NUNEATON
Bedworth
COVENTRY
RUGBY
ROYAL LEAMINGTON SPA
Kenilworth
WARWICK
Southam
Kineton
Tamworth
Solihull
BIRMINGHAM
Stratford on Avon
Alcester
Shipston on Stour
Long Compton

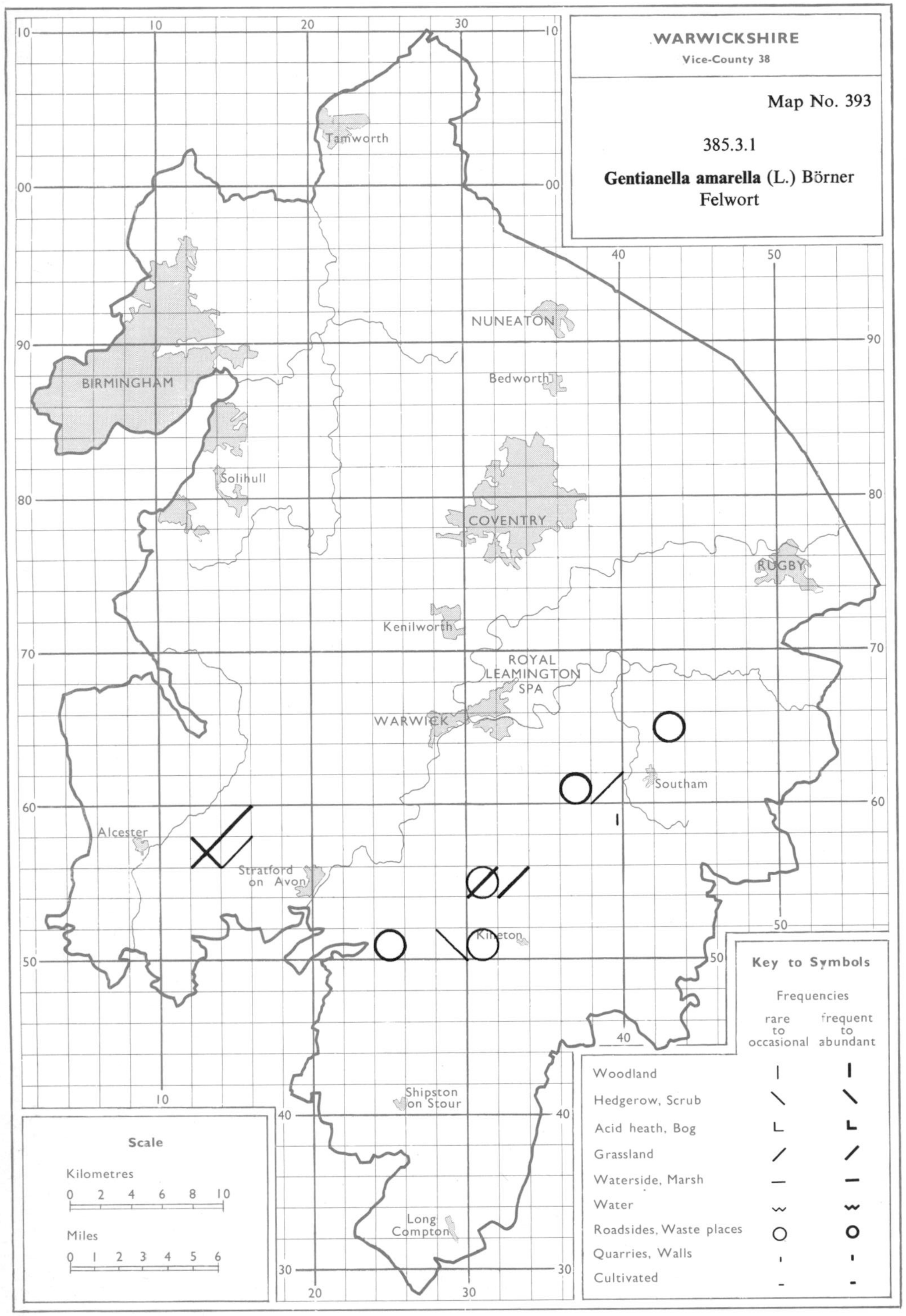
WARWICKSHIRE
Vice-County 38
Map No. 393
385.3.1
Gentianella amarella (L.) Börner
Felwort
Tamworth
NUNEATON
Bedworth
BIRMINGHAM
Solihull
COVENTRY
RUGBY
Kenilworth
ROYAL LEAMINGTON SPA
WARWICK
Southam
Alcester
Stratford on Avon
Kineton
Shipston on Stour
Long Compton
10 20 30 40 50
00 90 80 70 60 50 40 30
Scale
Kilometres
0 2 4 6 8 10
Miles
0 1 2 3 4 5 6
Key to Symbols
Frequencies
rare to occasional
frequent to abundant
Woodland
Hedgerow, Scrub
Acid heath, Bog
Grassland
Waterside, Marsh
Water
Roadsides, Waste places
Quarries, Walls
Cultivated

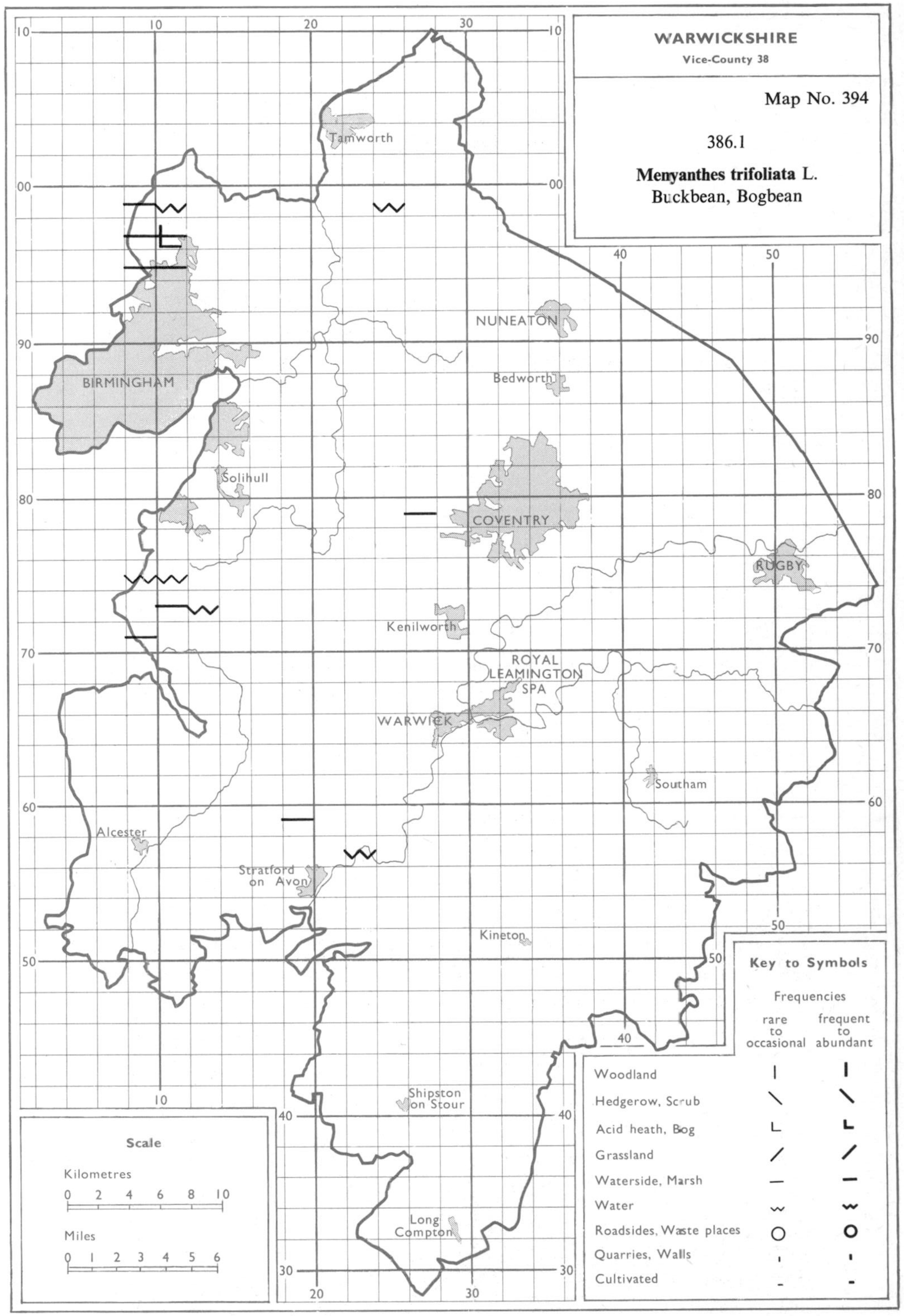
WARWICKSHIRE
Vice-County 38
Map No. 394
386.1
Menyanthes trifoliata L.
Buckbean, Bogbean
Tamworth
NUNEATON
Bedworth
BIRMINGHAM
Solihull
COVENTRY
RUGBY
Kenilworth
ROYAL LEAMINGTON SPA
WARWICK
Southam
Alcester
Stratford on Avon
Kineton
Shipston on Stour
Long Compton
10 20 30 40 50
00 90 80 70 60 50 40 30
Scale
Kilometres
0 2 4 6 8 10
Miles
0 1 2 3 4 5 6
Key to Symbols
Frequencies
rare to occasional
frequent to abundant
Woodland
Hedgerow, Scrub
Acid heath, Bog
Grassland
Waterside, Marsh
Water
Roadsides, Waste places
Quarries, Walls
Cultivated

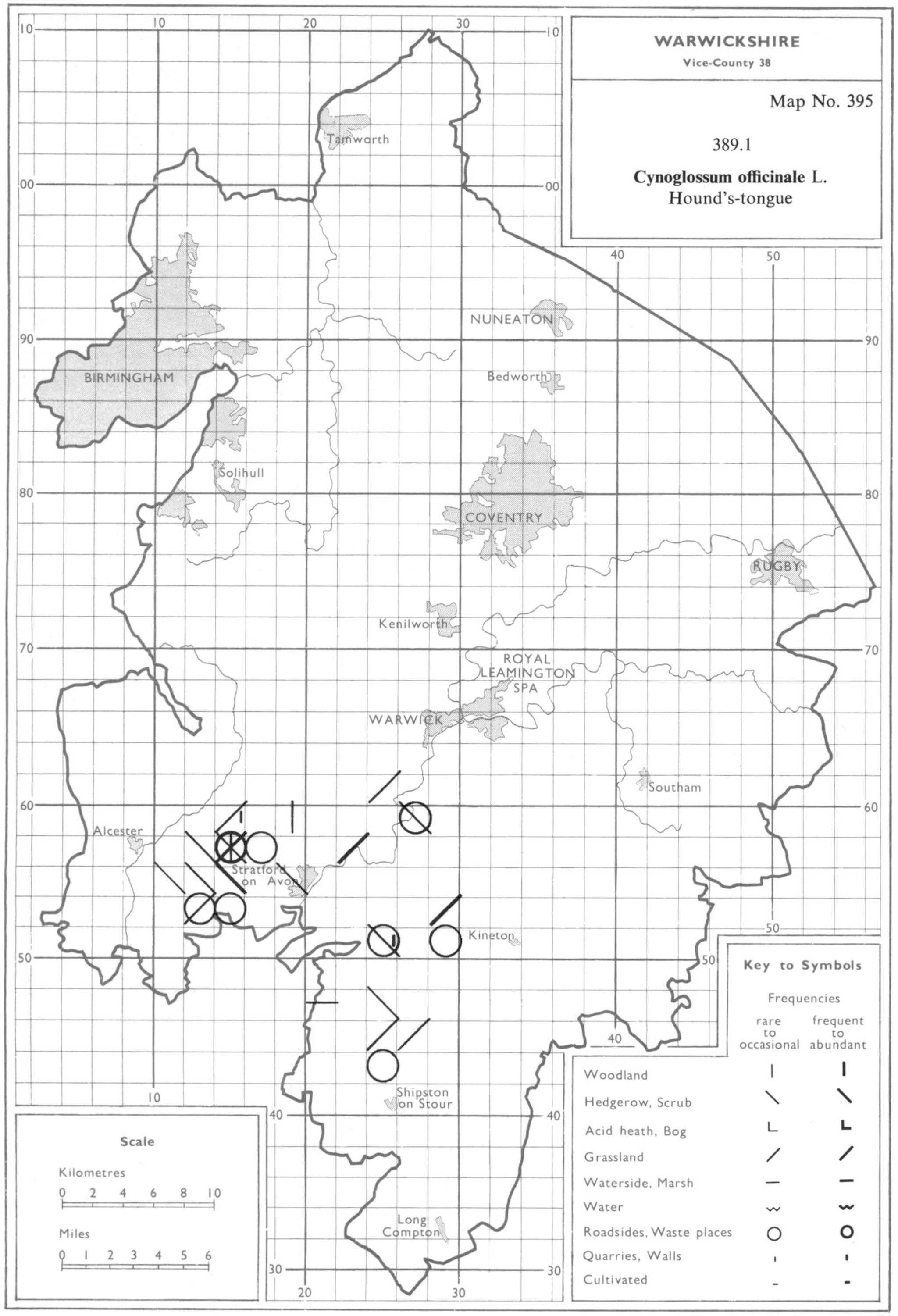
WARWICKSHIRE
Vice-County 38
Map No. 395
389.1
Cynoglossum officinale L.
Hound's-tongue
Tamworth
BIRMINGHAM
Solihull
NUNEATON
Bedworth
COVENTRY
RUGBY
Kenilworth
ROYAL LEAMINGTON SPA
WARWICK
Southam
Alcester
Stratford on Avon
Kineton
Shipston on Stour
Long Compton
Key to Symbols
Frequencies
rare to occasional
frequent to abundant
Woodland
Hedgerow, Scrub
Acid heath, Bog
Grassland
Waterside, Marsh
Water
Roadsides, Waste places
Quarries, Walls
Cultivated
Scale
Kilometres
0 2 4 6 8 10
Miles
0 1 2 3 4 5 6

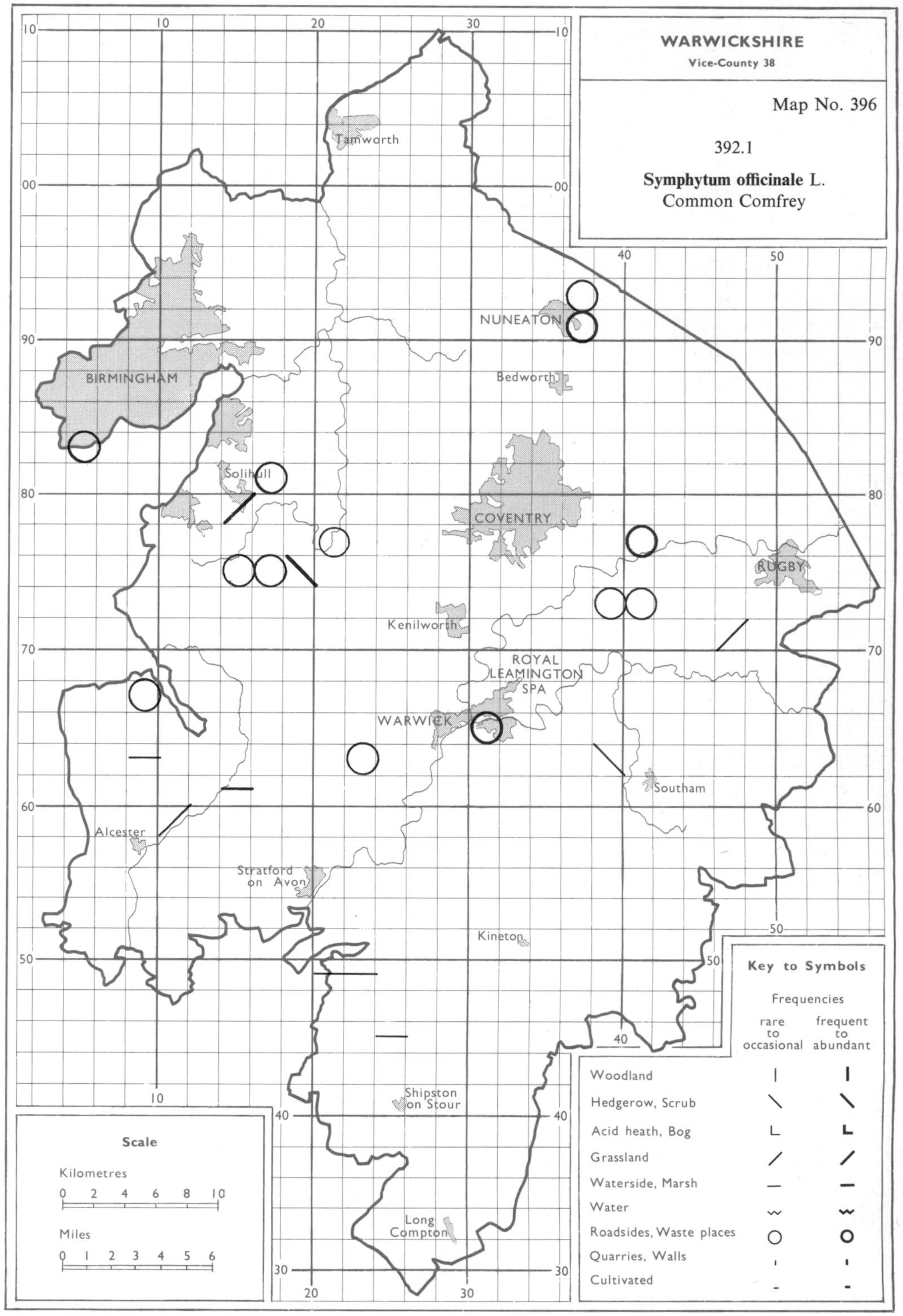
WARWICKSHIRE
Vice-County 38
Map No. 396
392.1
Symphytum officinale L.
Common Comfrey
Tamworth
BIRMINGHAM
Solihull
NUNEATON
Bedworth
COVENTRY
RUGBY
Kenilworth
ROYAL LEAMINGTON SPA
WARWICK
Southam
Alcester
Stratford on Avon
Kineton
Shipston on Stour
Long Compton
Key to Symbols
Frequencies
rare to occasional
frequent to abundant
Woodland
Hedgerow, Scrub
Acid heath, Bog
Grassland
Waterside, Marsh
Water
Roadsides, Waste places
Quarries, Walls
Cultivated
Scale
Kilometres
0 2 4 6 8 10
Miles
0 1 2 3 4 5 6

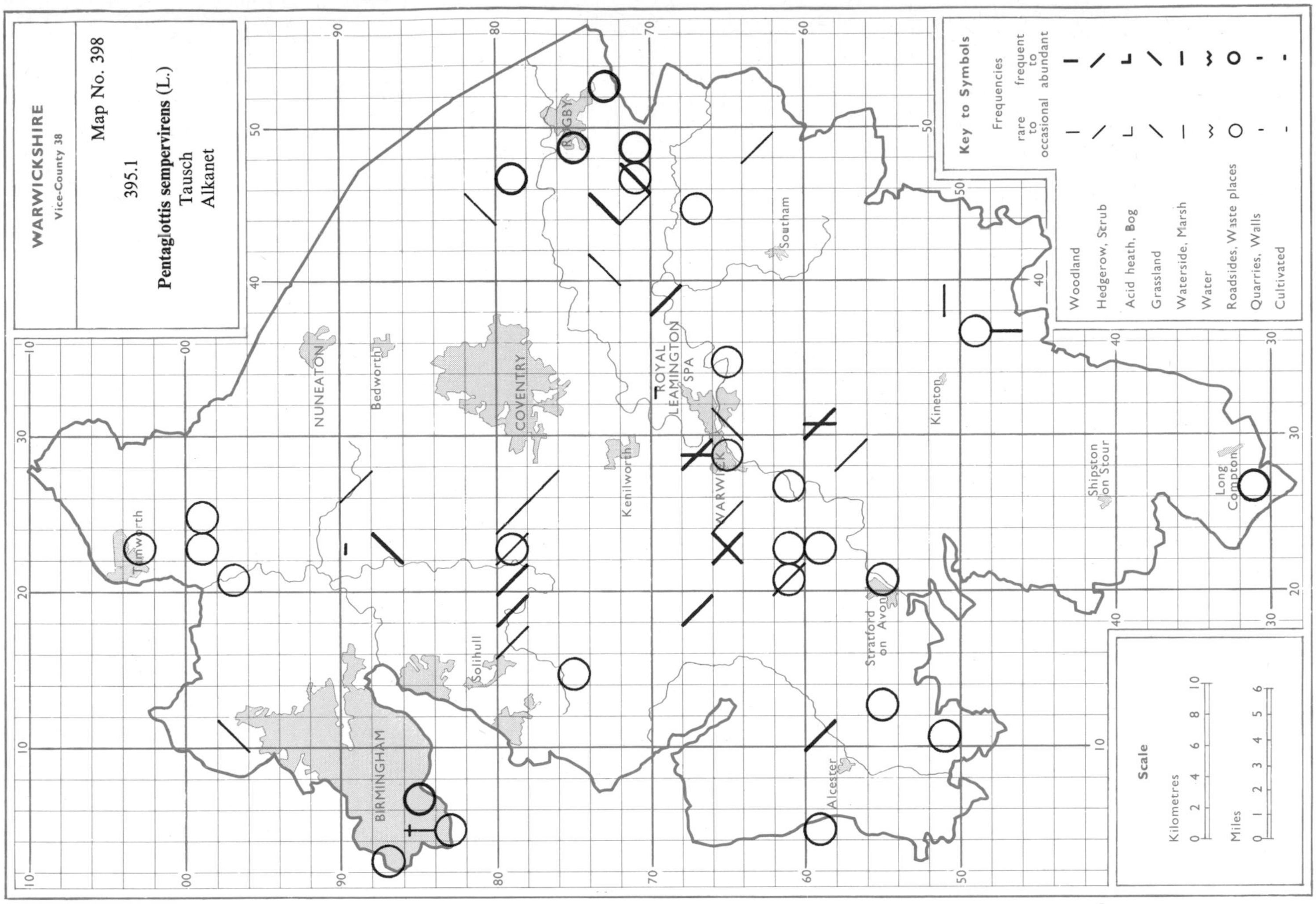
WARWICKSHIRE
Vice-County 38
Map No. 398
395.1
Pentaglottis sempervirens (L.) Tausch
Alkanet
Key to Symbols
Frequencies
rare to occasional
frequent to abundant
Woodland
Hedgerow, Scrub
Acid heath, Bog
Grassland
Waterside, Marsh
Water
Roadsides, Waste places
Quarries, Walls
Cultivated
Scale
Kilometres
Miles
NUNEATON
Bedworth
COVENTRY
RUGBY
Southam
ROYAL LEAMINGTON SPA
Kenilworth
WARWICK
Kineton
Shipston on Stour
Long Compton
Tamworth
Solihull
BIRMINGHAM
Stratford on Avon
Alcester

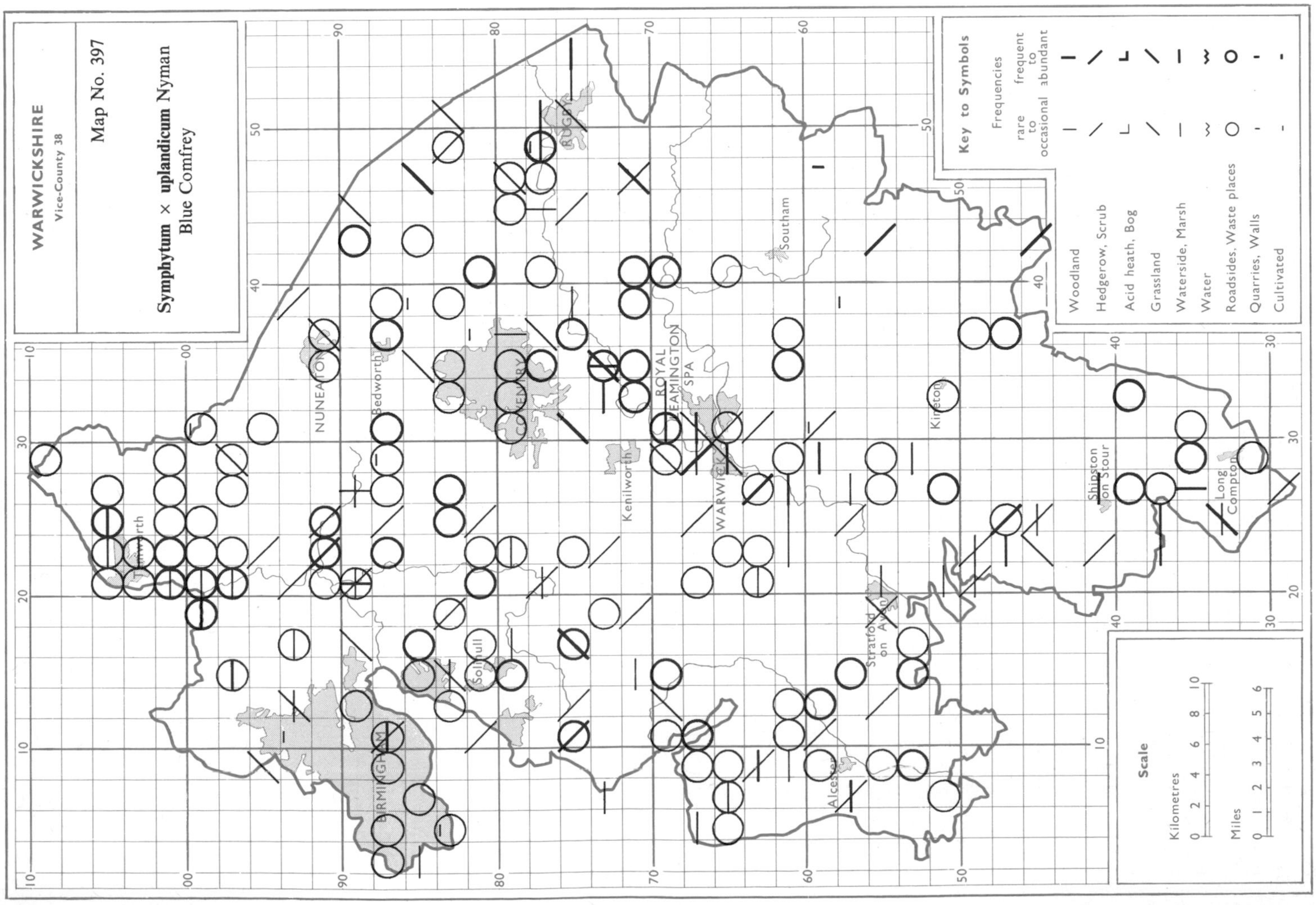
WARWICKSHIRE
Vice-County 38
Map No. 397
Symphytum × uplandicum Nyman
Blue Comfrey
Key to Symbols
Frequencies
rare to occasional
frequent to abundant
Woodland
Hedgerow, Scrub
Acid heath, Bog
Grassland
Waterside, Marsh
Water
Roadsides, Waste places
Quarries, Walls
Cultivated
Scale
Kilometres
Miles
NUNEATON
Bedworth
COVENTRY
RUGBY
Southam
ROYAL LEAMINGTON SPA
Kenilworth
WARWICK
Kineton
Shipston on Stour
Long Compton
Tamworth
Solihull
BIRMINGHAM
Stratford on Avon
Alcester

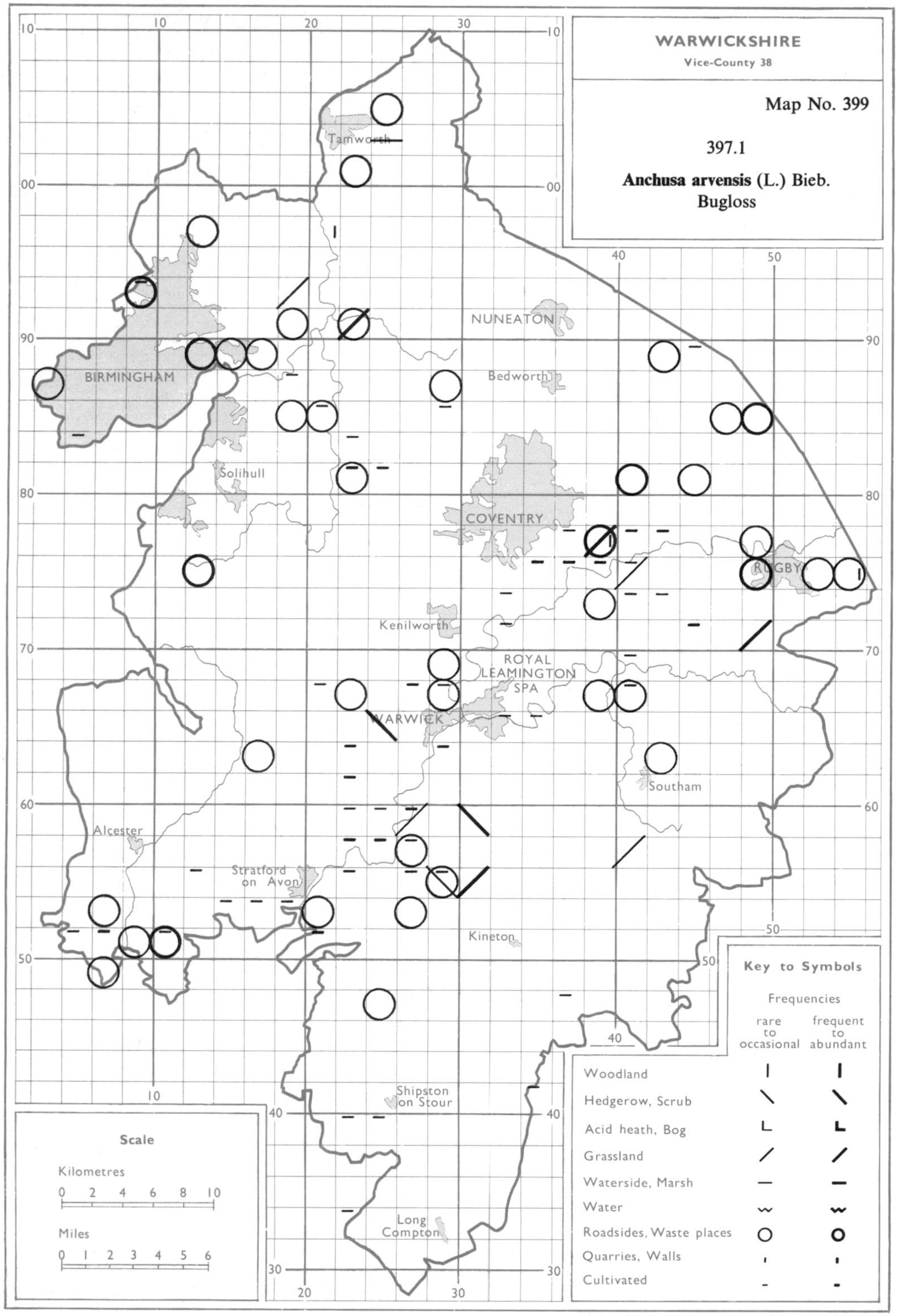

WARWICKSHIRE
Vice-County 38
Map No. 399
397.1
Anchusa arvensis (L.) Bieb.
Bugloss

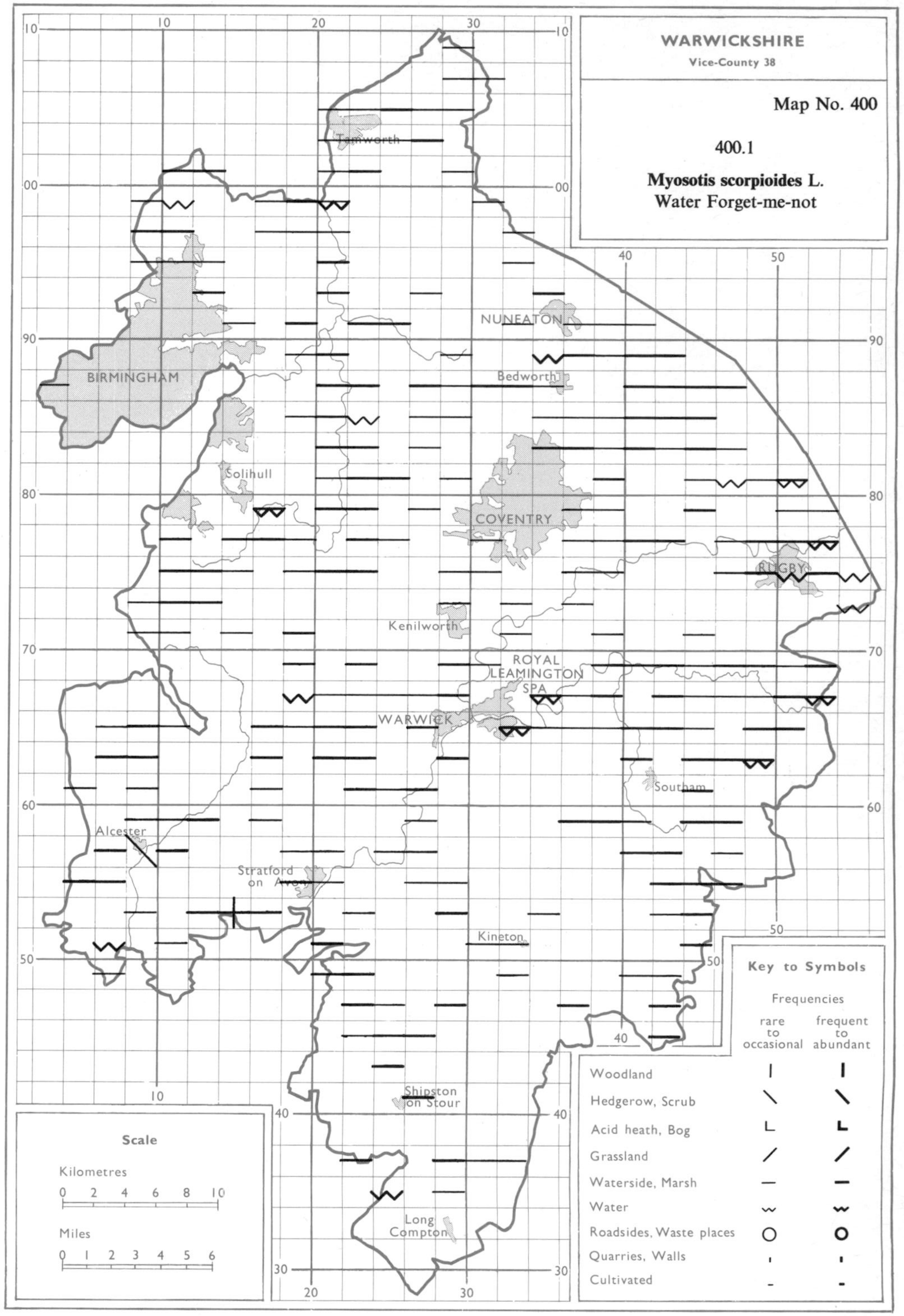

WARWICKSHIRE
Vice-County 38
Map No. 400
400.1
Myosotis scorpioides L.
Water Forget-me-not

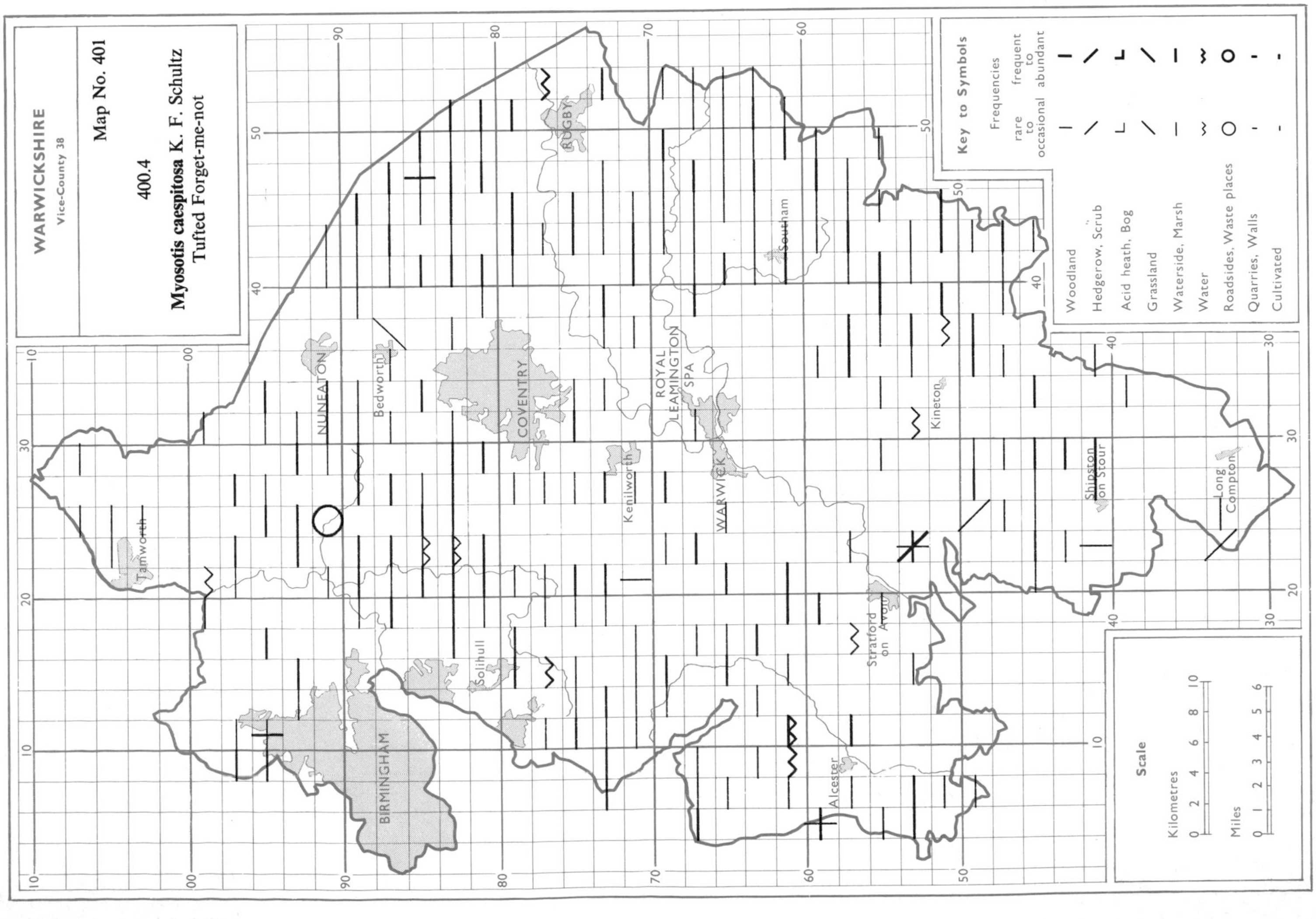
WARWICKSHIRE
Vice-County 38
Map No. 401
400.4
Myosotis caespitosa K. F. Schultz
Tufted Forget-me-not
Key to Symbols
Frequencies
rare to occasional
frequent to abundant
Woodland
Hedgerow, Scrub
Acid heath, Bog
Grassland
Waterside, Marsh
Water
Roadsides, Waste places
Quarries, Walls
Cultivated
Scale
Kilometres
Miles
NUNEATON
Bedworth
COVENTRY
RUGBY
ROYAL LEAMINGTON SPA
Southam
Kenilworth
WARWICK
Kineton
Shipston on Stour
Long Compton
Tamworth
Solihull
BIRMINGHAM
Stratford on Avon
Alcester

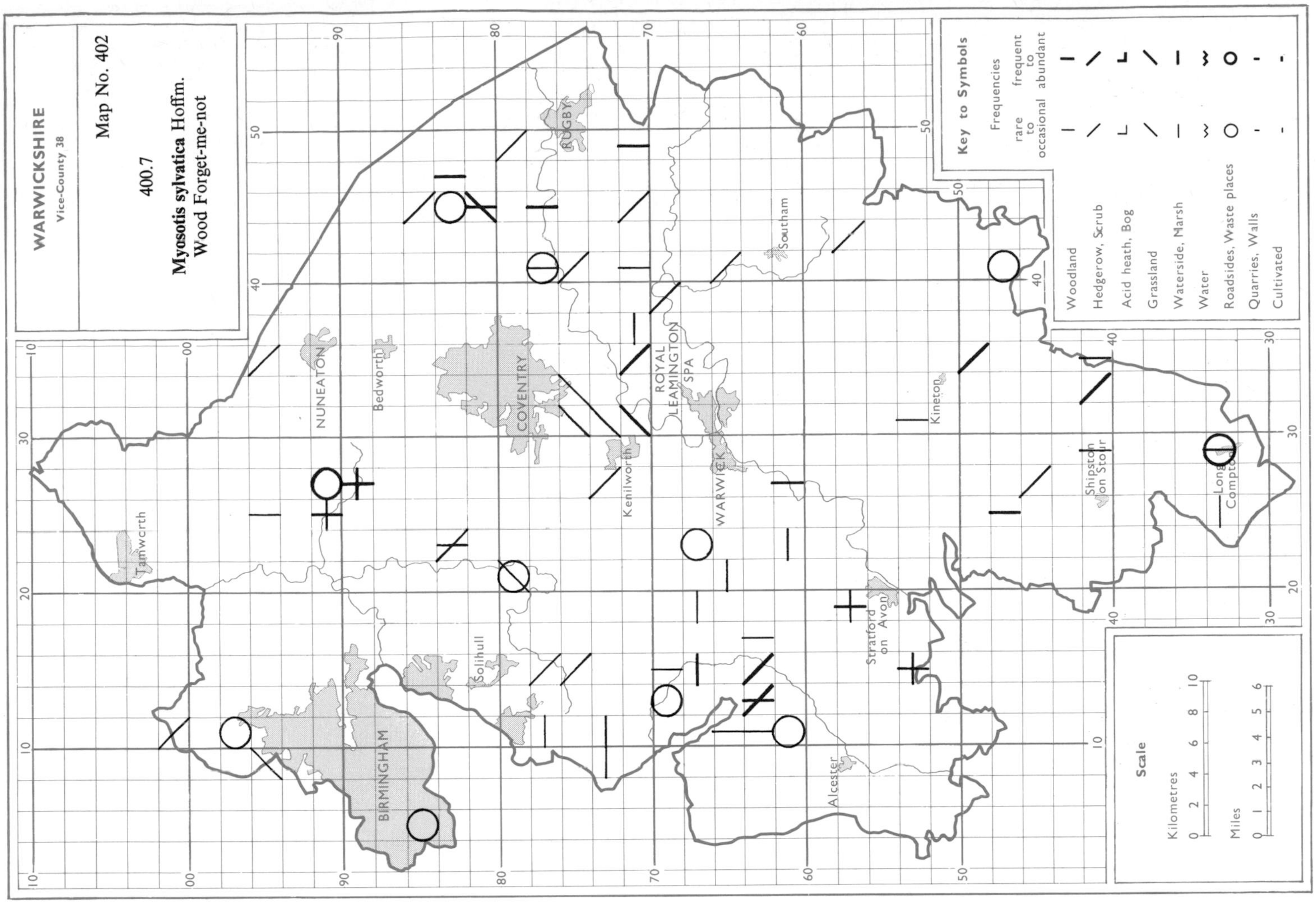
WARWICKSHIRE
Vice-County 38
Map No. 402
400.7
Myosotis sylvatica Hoffm.
Wood Forget-me-not
Key to Symbols
Frequencies
rare to occasional
frequent to abundant
Woodland
Hedgerow, Scrub
Acid heath, Bog
Grassland
Waterside, Marsh
Water
Roadsides, Waste places
Quarries, Walls
Cultivated
Scale
Kilometres
Miles
NUNEATON
Bedworth
COVENTRY
RUGBY
ROYAL LEAMINGTON SPA
Southam
Kenilworth
WARWICK
Kineton
Shipston on Stour
Long Compton
Tamworth
Solihull
BIRMINGHAM
Stratford on Avon
Alcester

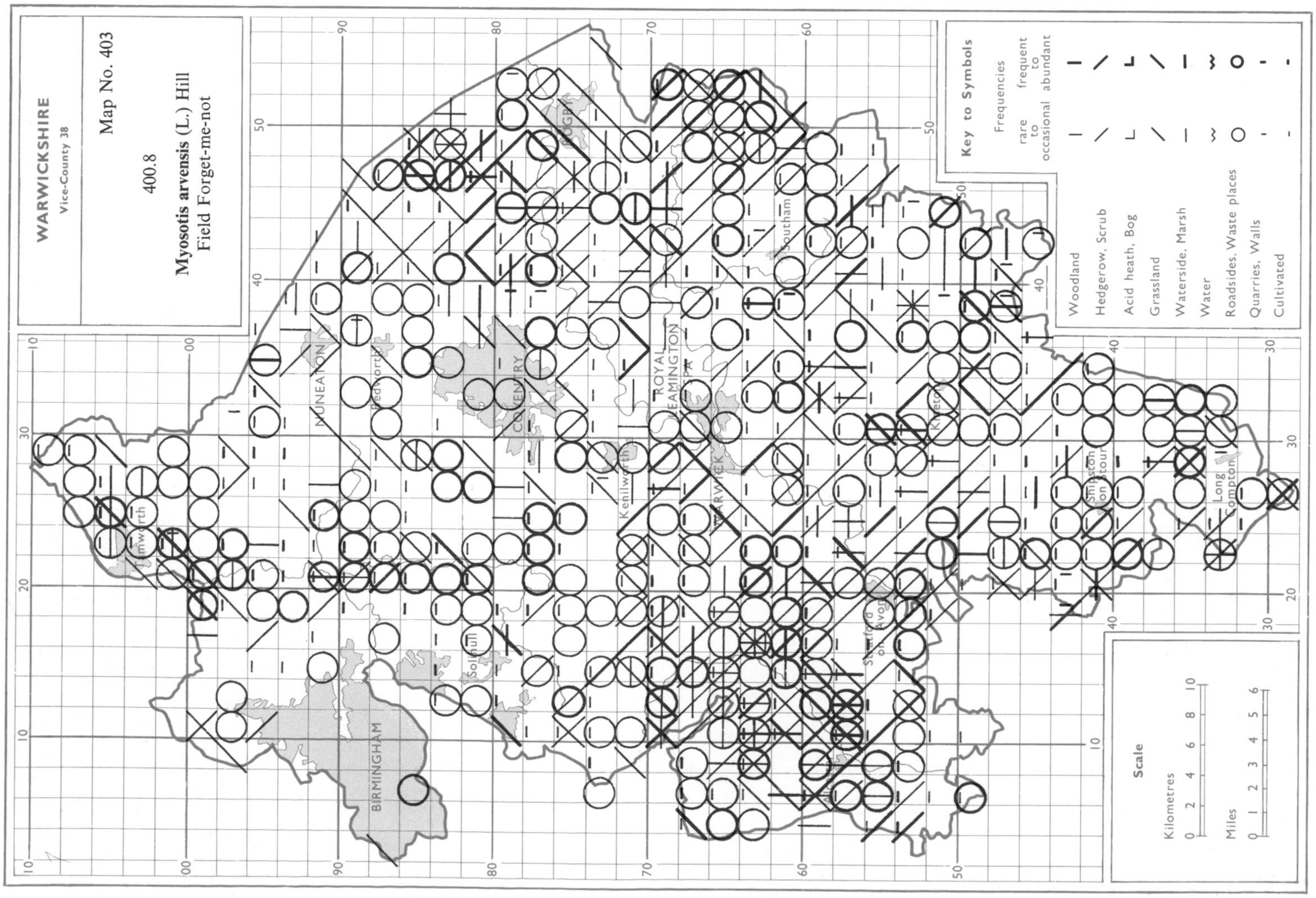
WARWICKSHIRE
Vice-County 38
Map No. 403
400.8
Myosotis arvensis (L.) Hill
Field Forget-me-not
Key to Symbols
Frequencies
rare to occasional
frequent to abundant
Woodland
Hedgerow, Scrub
Acid heath, Bog
Grassland
Waterside, Marsh
Water
Roadsides, Waste places
Quarries, Walls
Cultivated
Scale
Kilometres
0 2 4 6 8 10
Miles
0 1 2 3 4 5 6
NUNEATON
Bedworth
COVENTRY
RUGBY
Southam
ROYAL LEAMINGTON SPA
Kenilworth
WARWICK
Kineton
Shipston on Stour
Long Compton
Tamworth
Solihull
BIRMINGHAM
Stratford on Avon
Alcester

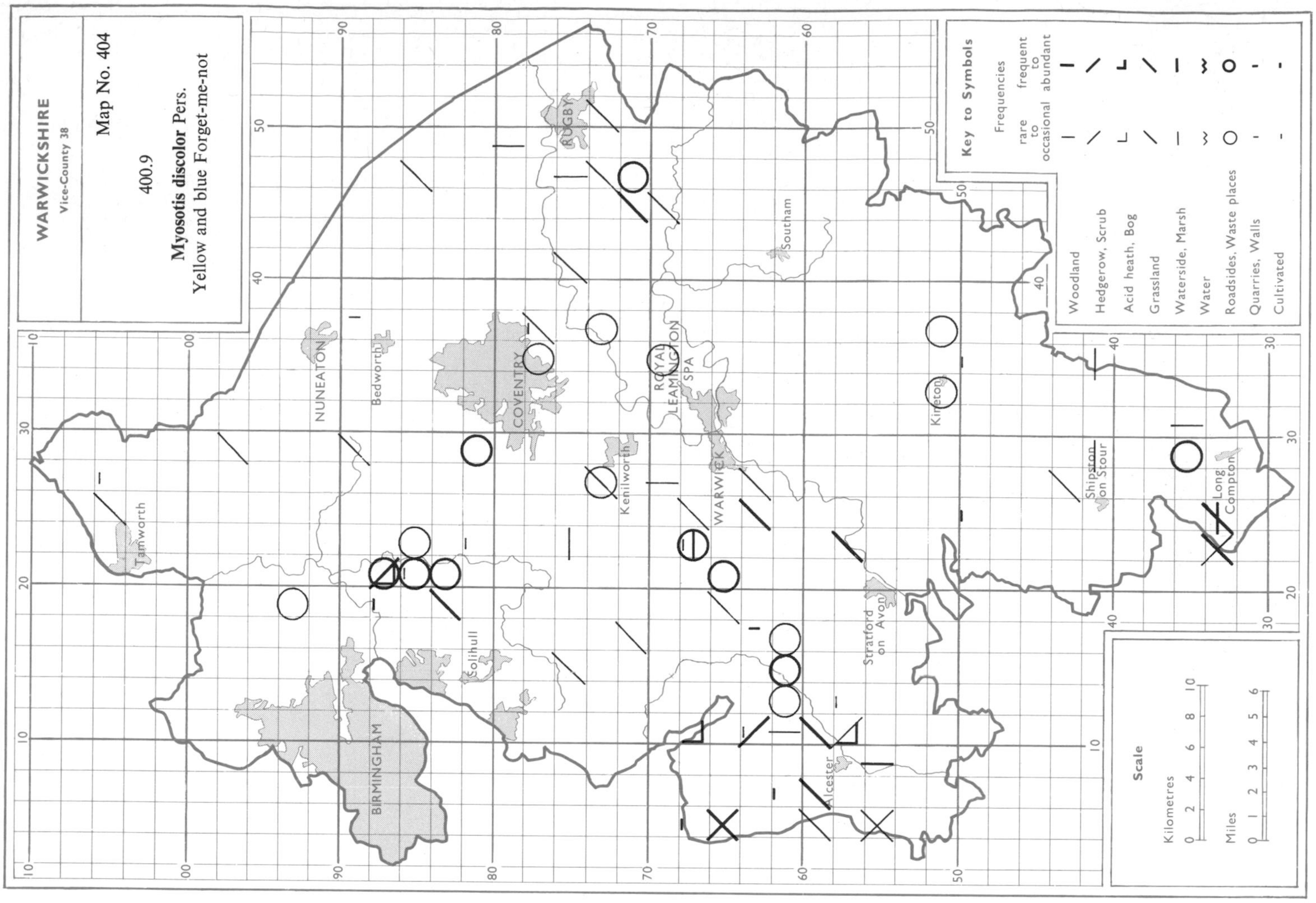
WARWICKSHIRE
Vice-County 38
Map No. 404
400.9
Myosotis discolor Pers.
Yellow and blue Forget-me-not
Key to Symbols
Frequencies
rare to occasional
frequent to abundant
Woodland
Hedgerow, Scrub
Acid heath, Bog
Grassland
Waterside, Marsh
Water
Roadsides, Waste places
Quarries, Walls
Cultivated
Scale
Kilometres
0 2 4 6 8 10
Miles
0 1 2 3 4 5 6
NUNEATON
Bedworth
COVENTRY
RUGBY
Southam
ROYAL LEAMINGTON SPA
Kenilworth
WARWICK
Kineton
Shipston on Stour
Long Compton
Tamworth
Solihull
BIRMINGHAM
Stratford on Avon
Alcester

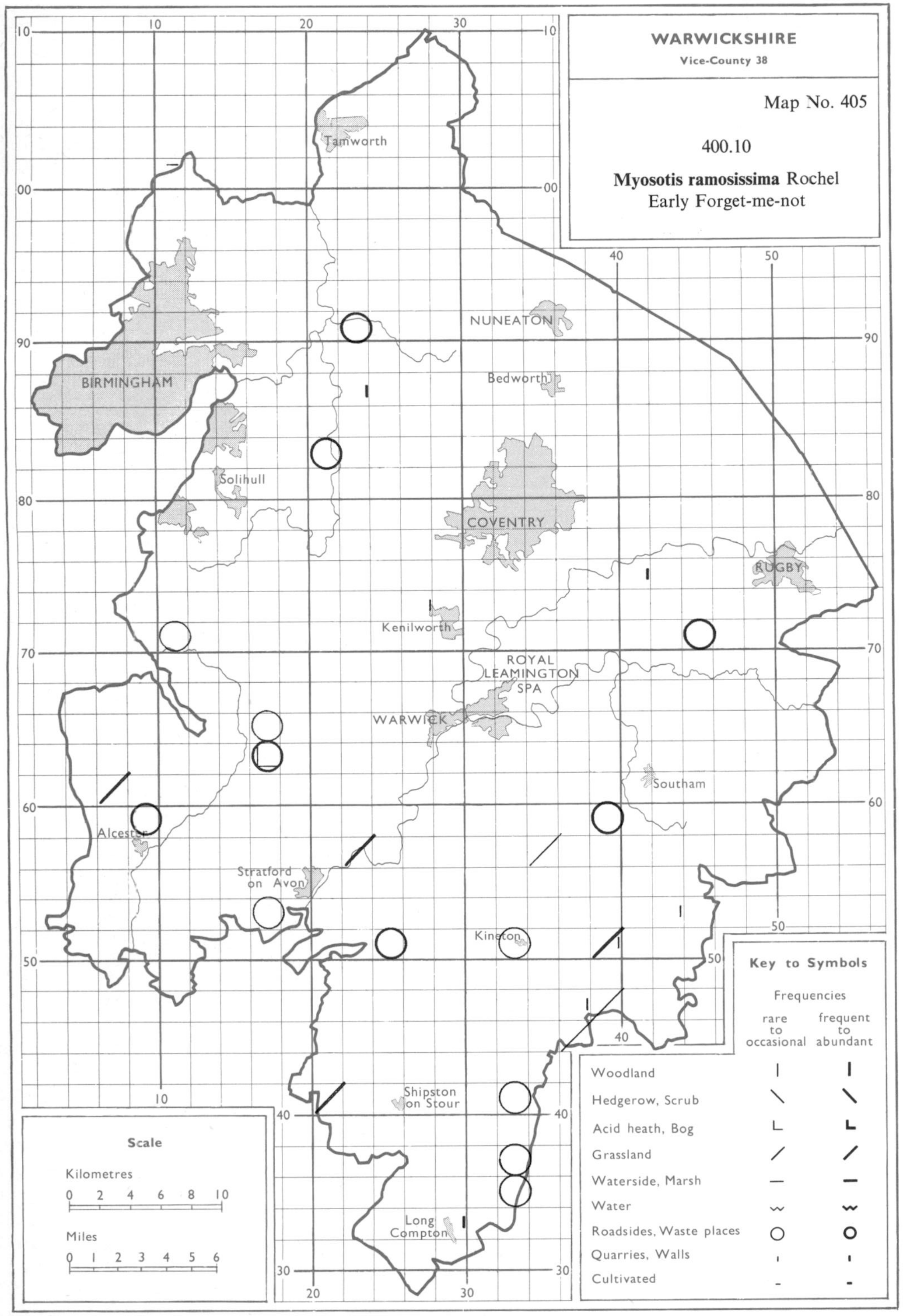
WARWICKSHIRE
Vice-County 38
Map No. 405
400.10
Myosotis ramosissima Rochel
Early Forget-me-not
Tamworth
NUNEATON
Bedworth
BIRMINGHAM
Solihull
COVENTRY
RUGBY
Kenilworth
ROYAL LEAMINGTON SPA
WARWICK
Southam
Alcester
Stratford on Avon
Kineton
Shipston on Stour
Long Compton
Key to Symbols
Frequencies
rare to occasional
frequent to abundant
Woodland
Hedgerow, Scrub
Acid heath, Bog
Grassland
Waterside, Marsh
Water
Roadsides, Waste places
Quarries, Walls
Cultivated
Scale
Kilometres
0 2 4 6 8 10
Miles
0 1 2 3 4 5 6

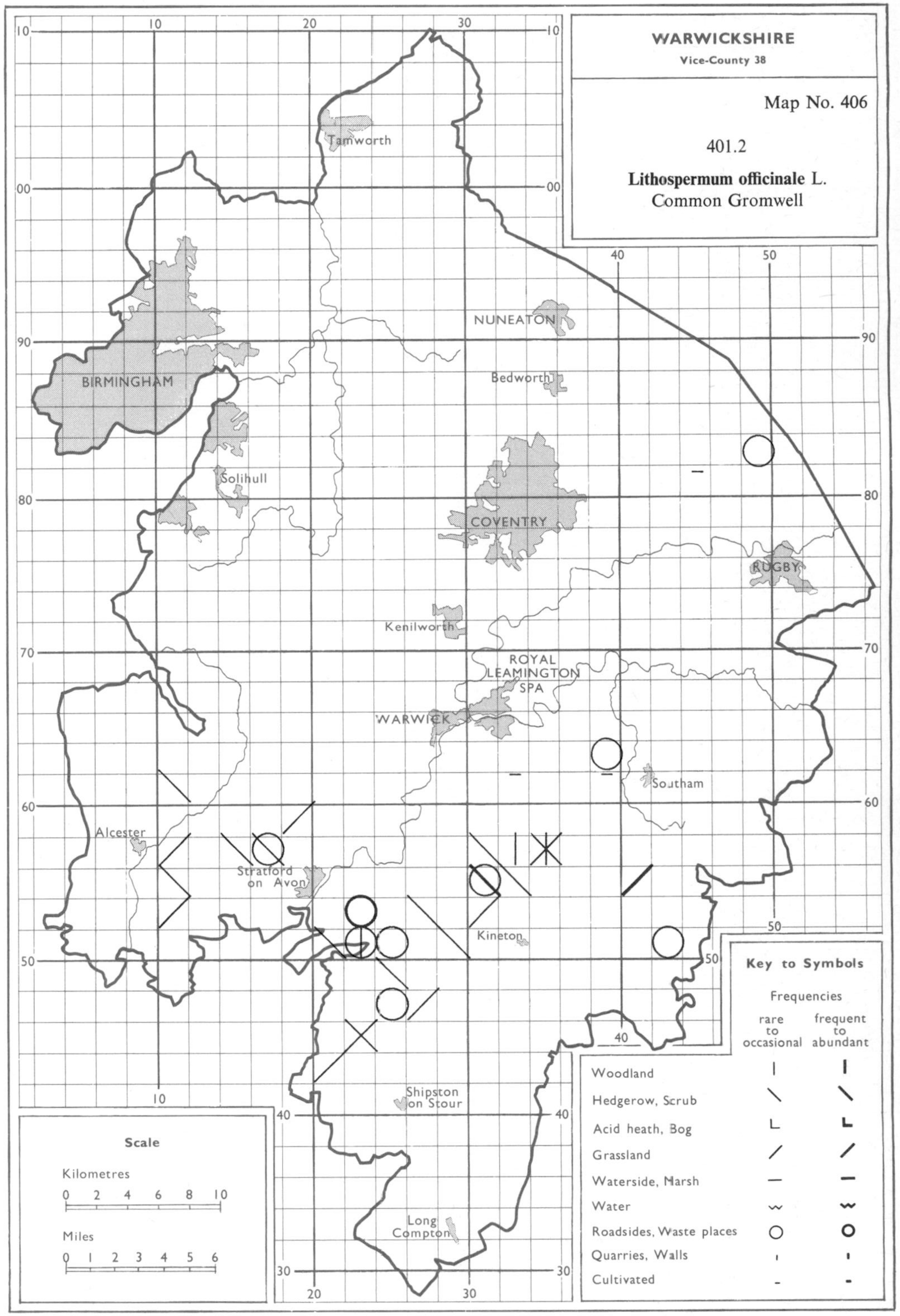
WARWICKSHIRE
Vice-County 38
Map No. 406
401.2
Lithospermum officinale L.
Common Gromwell
Tamworth
NUNEATON
Bedworth
BIRMINGHAM
Solihull
COVENTRY
RUGBY
Kenilworth
ROYAL LEAMINGTON SPA
WARWICK
Southam
Alcester
Stratford on Avon
Kineton
Shipston on Stour
Long Compton
Key to Symbols
Frequencies
rare to occasional
frequent to abundant
Woodland
Hedgerow, Scrub
Acid heath, Bog
Grassland
Waterside, Marsh
Water
Roadsides, Waste places
Quarries, Walls
Cultivated
Scale
Kilometres
0 2 4 6 8 10
Miles
0 1 2 3 4 5 6

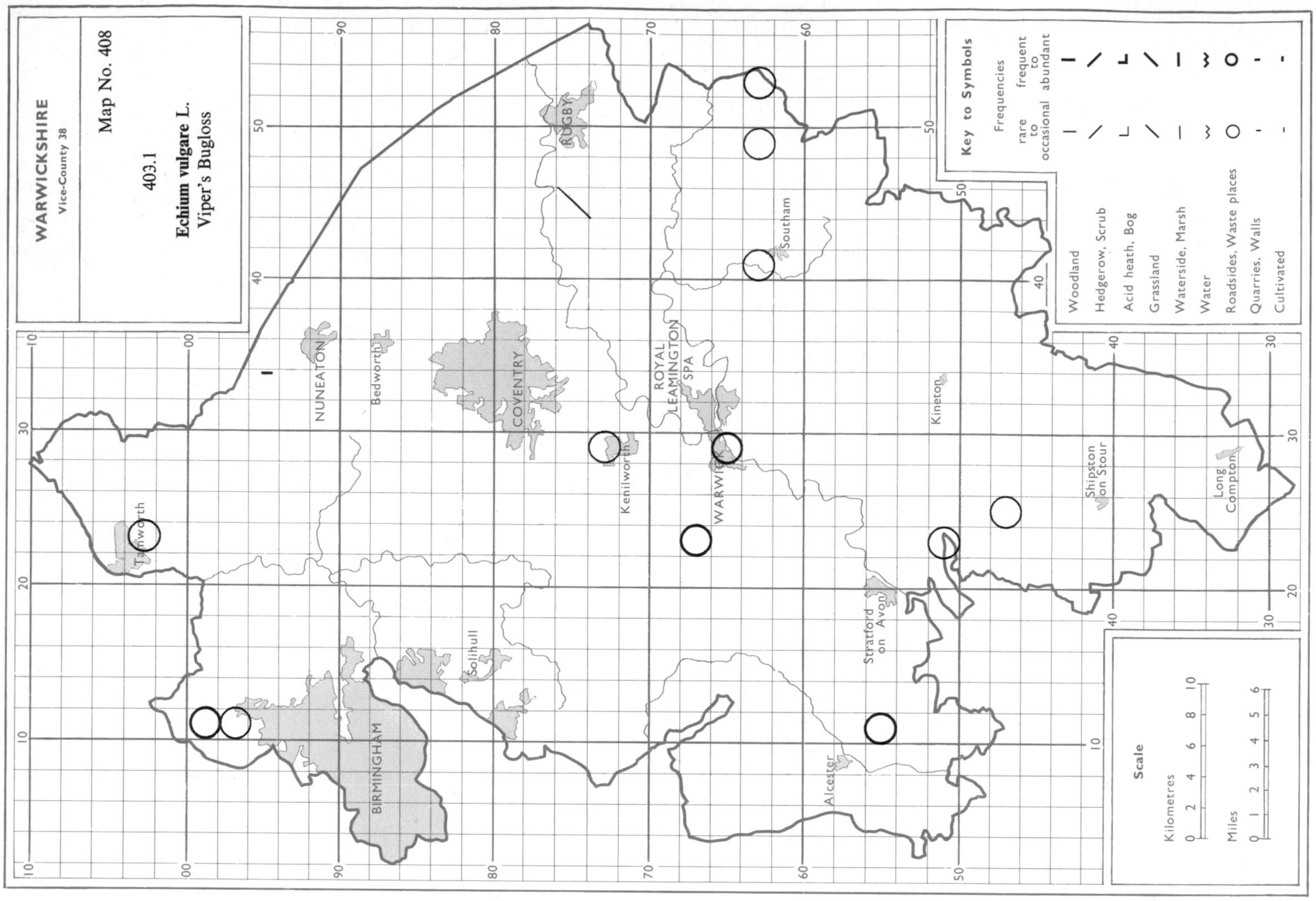
WARWICKSHIRE
Vice-County 38
Map No. 408
403.1
Echium vulgare L.
Viper's Bugloss
Key to Symbols
Frequencies
rare to occasional
frequent to abundant
Woodland
Hedgerow, Scrub
Acid heath, Bog
Grassland
Waterside, Marsh
Water
Roadsides, Waste places
Quarries, Walls
Cultivated
Scale
Kilometres
Miles
Tamworth
NUNEATON
Bedworth
COVENTRY
RUGBY
BIRMINGHAM
Solihull
Kenilworth
ROYAL LEAMINGTON SPA
WARWICK
Southam
Kineton
Stratford on Avon
Alcester
Shipston on Stour
Long Compton

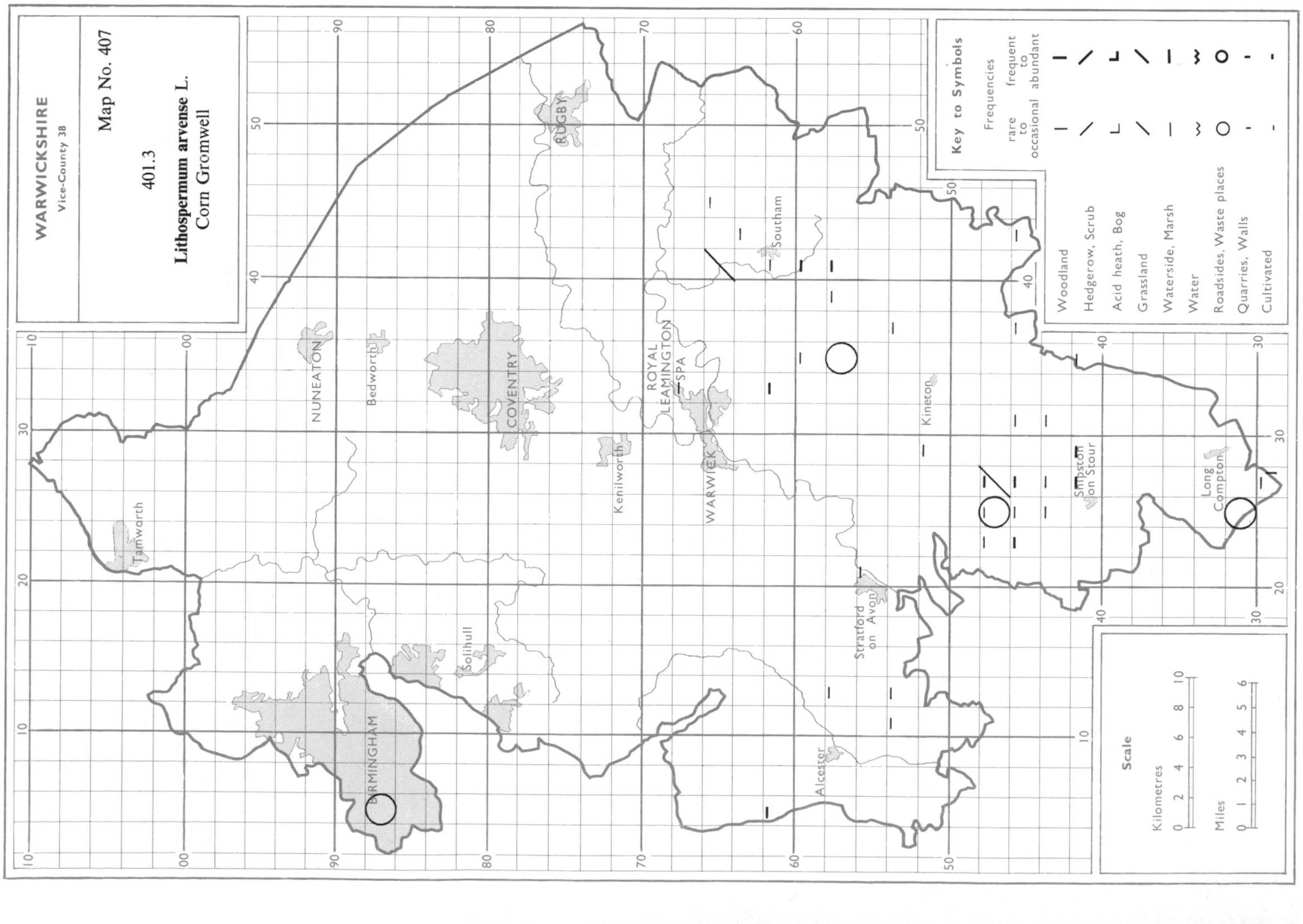
WARWICKSHIRE
Vice-County 38
Map No. 407
401.3
Lithospermum arvense L.
Corn Gromwell
Key to Symbols
Frequencies
rare to occasional
frequent to abundant
Woodland
Hedgerow, Scrub
Acid heath, Bog
Grassland
Waterside, Marsh
Water
Roadsides, Waste places
Quarries, Walls
Cultivated
Scale
Kilometres
Miles
Tamworth
NUNEATON
Bedworth
COVENTRY
RUGBY
BIRMINGHAM
Solihull
Kenilworth
ROYAL LEAMINGTON SPA
WARWICK
Southam
Kineton
Stratford on Avon
Alcester
Shipston on Stour
Long Compton

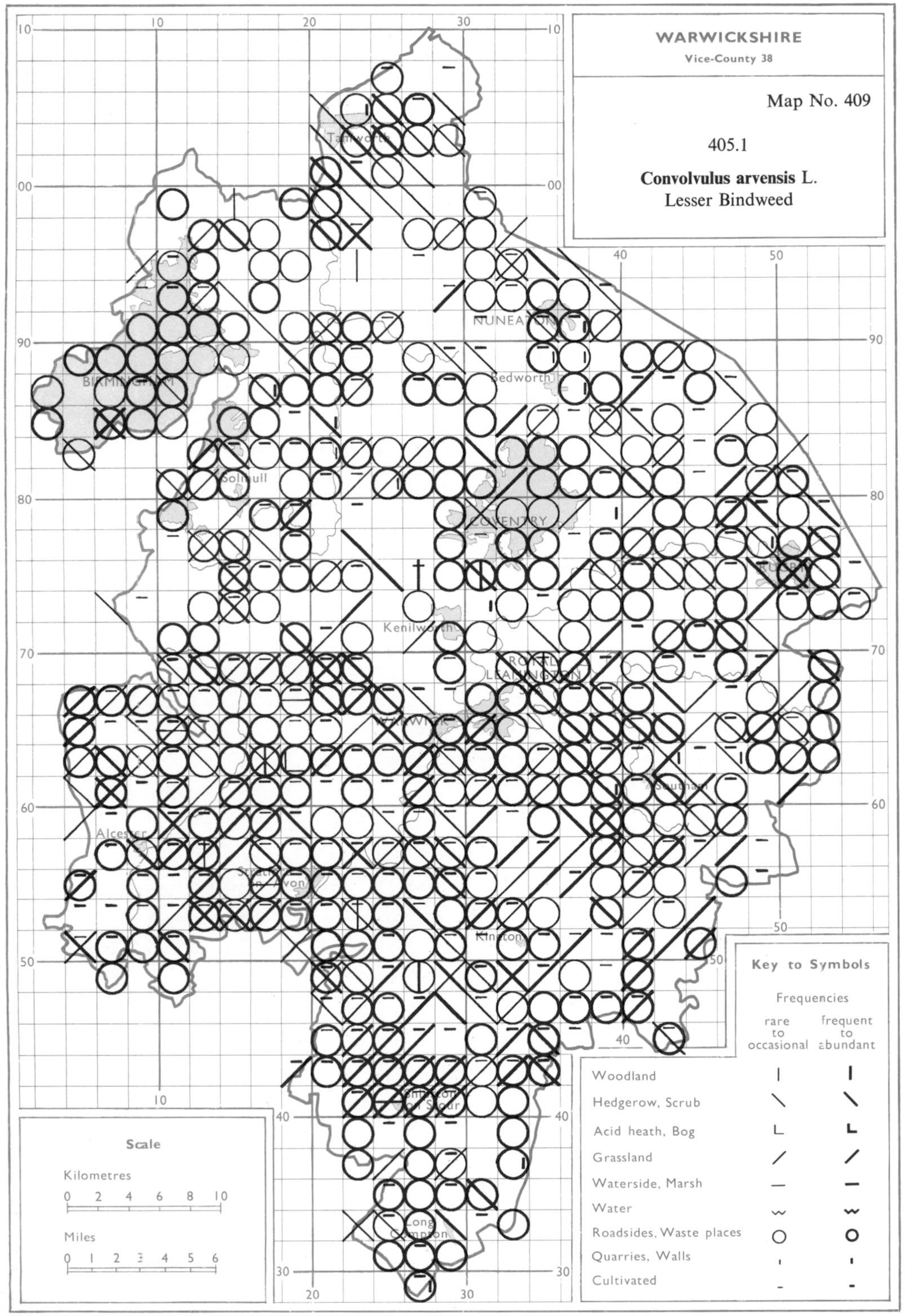
WARWICKSHIRE
Vice-County 38
Map No. 409
405.1
Convolvulus arvensis L.
Lesser Bindweed
Key to Symbols
Frequencies
rare to occasional
frequent to abundant
Woodland
Hedgerow, Scrub
Acid heath, Bog
Grassland
Waterside, Marsh
Water
Roadsides, Waste places
Quarries, Walls
Cultivated
Scale
Kilometres
Miles
Tamworth
Nuneaton
Bedworth
Birmingham
Solihull
Coventry
Rugby
Kenilworth
Royal Leamington
Warwick
Southam
Alcester
Stratford on Avon
Kineton
Shipston on Stour
Long Compton

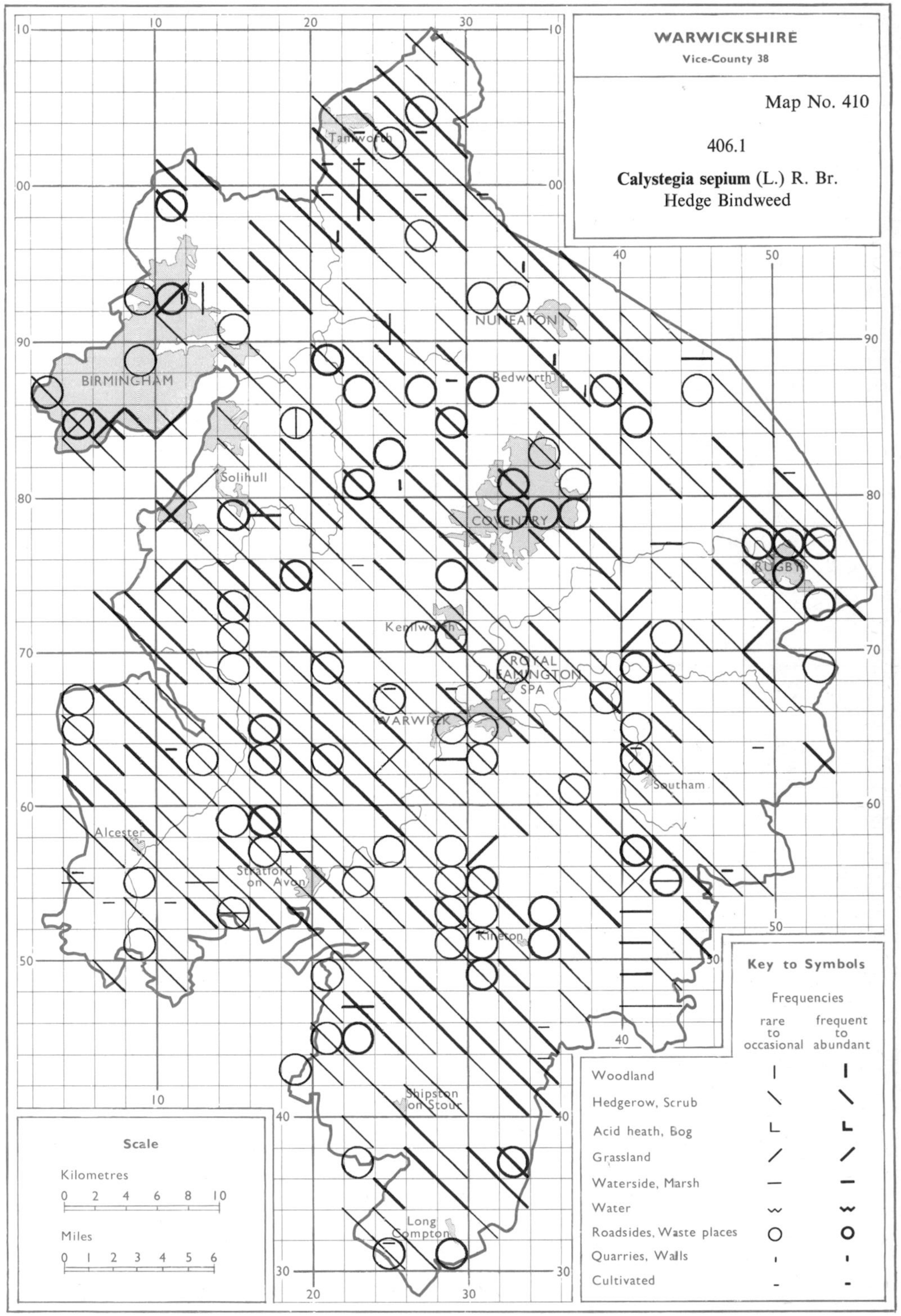
WARWICKSHIRE
Vice-County 38
Map No. 410
406.1
Calystegia sepium (L.) R. Br.
Hedge Bindweed
Key to Symbols
Frequencies
rare to occasional
frequent to abundant
Woodland
Hedgerow, Scrub
Acid heath, Bog
Grassland
Waterside, Marsh
Water
Roadsides, Waste places
Quarries, Walls
Cultivated
Scale
Kilometres
Miles
Tamworth
Nuneaton
Bedworth
Birmingham
Solihull
Coventry
Rugby
Kenilworth
Royal Leamington Spa
Warwick
Southam
Alcester
Stratford on Avon
Kineton
Shipston on Stour
Long Compton

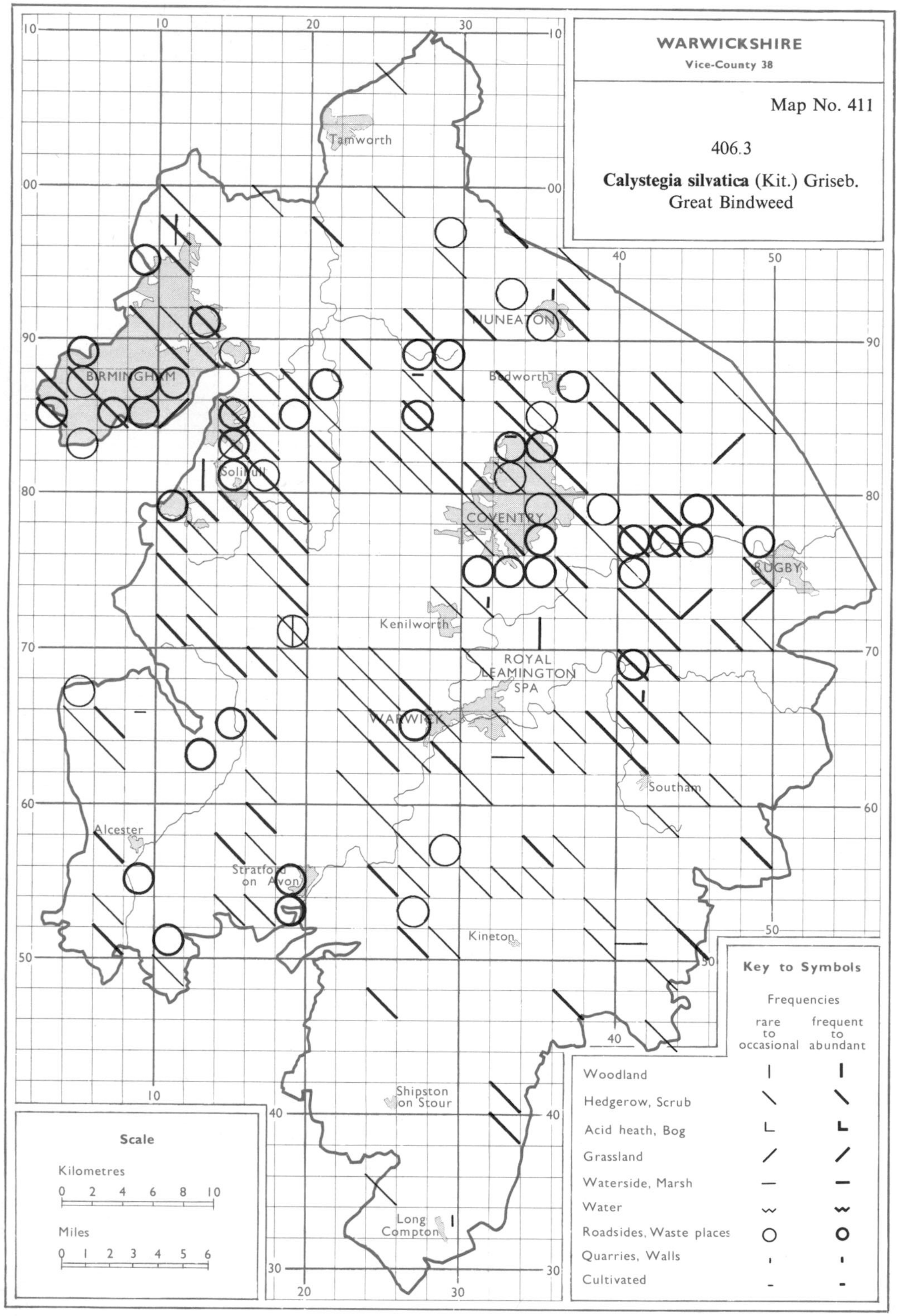
WARWICKSHIRE
Vice-County 38
Map No. 411
406.3
Calystegia silvatica (Kit.) Griseb.
Great Bindweed
Tamworth
NUNEATON
Bedworth
BIRMINGHAM
Solihull
COVENTRY
RUGBY
Kenilworth
ROYAL LEAMINGTON SPA
WARWICK
Southam
Alcester
Stratford on Avon
Kineton
Shipston on Stour
Long Compton
Key to Symbols
Frequencies
rare to occasional
frequent to abundant
Woodland
Hedgerow, Scrub
Acid heath, Bog
Grassland
Waterside, Marsh
Water
Roadsides, Waste places
Quarries, Walls
Cultivated
Scale
Kilometres
0 2 4 6 8 10
Miles
0 1 2 3 4 5 6

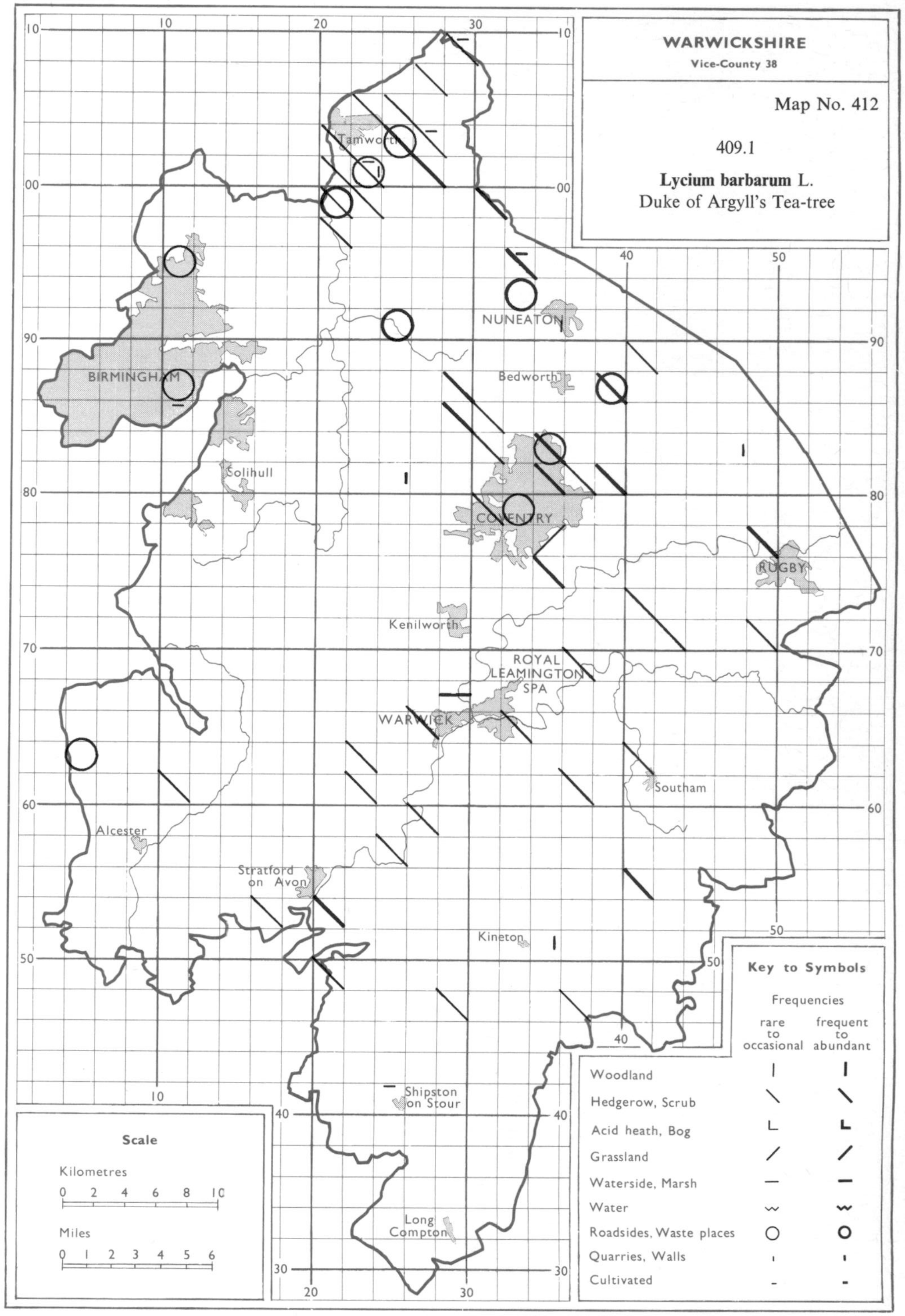
WARWICKSHIRE
Vice-County 38
Map No. 412
409.1
Lycium barbarum L.
Duke of Argyll's Tea-tree
Tamworth
NUNEATON
Bedworth
BIRMINGHAM
Solihull
COVENTRY
RUGBY
Kenilworth
ROYAL LEAMINGTON SPA
WARWICK
Southam
Alcester
Stratford on Avon
Kineton
Shipston on Stour
Long Compton
Key to Symbols
Frequencies
rare to occasional
frequent to abundant
Woodland
Hedgerow, Scrub
Acid heath, Bog
Grassland
Waterside, Marsh
Water
Roadsides, Waste places
Quarries, Walls
Cultivated
Scale
Kilometres
0 2 4 6 8 10
Miles
0 1 2 3 4 5 6

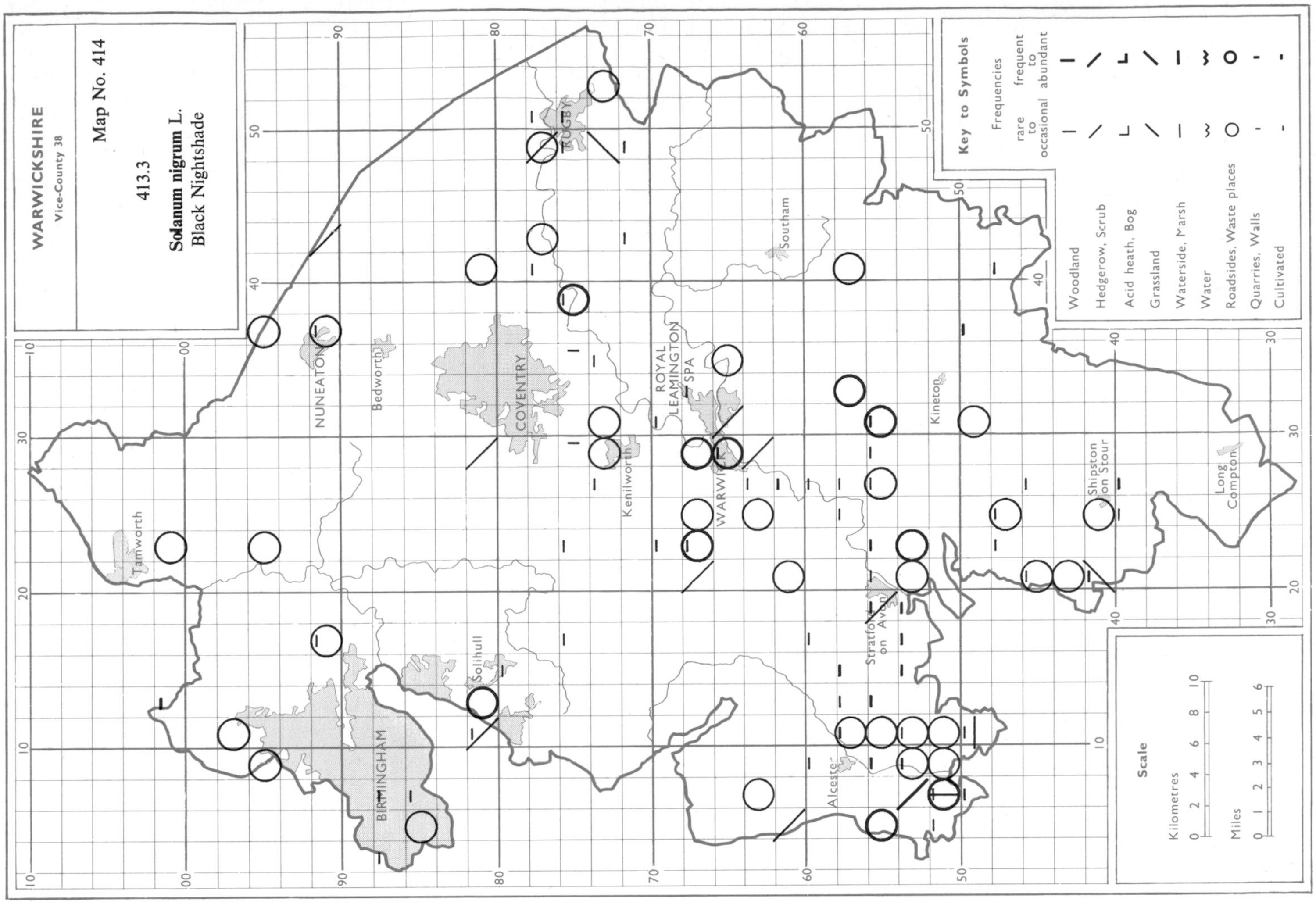
WARWICKSHIRE
Vice-County 38
Map No. 414
413.3
Solanum nigrum L.
Black Nightshade
Key to Symbols
Frequencies
rare to occasional
frequent to abundant
Woodland
Hedgerow, Scrub
Acid heath, Bog
Grassland
Waterside, Marsh
Water
Roadsides, Waste places
Quarries, Walls
Cultivated
Scale
Kilometres
0 2 4 6 8 10
Miles
0 1 2 3 4 5 6
Tamworth
NUNEATON
Bedworth
COVENTRY
RUGBY
Southam
ROYAL LEAMINGTON SPA
Kenilworth
WARWICK
Kineton
Shipston on Stour
Long Compton
Solihull
BIRMINGHAM
Stratford on Avon
Alcester

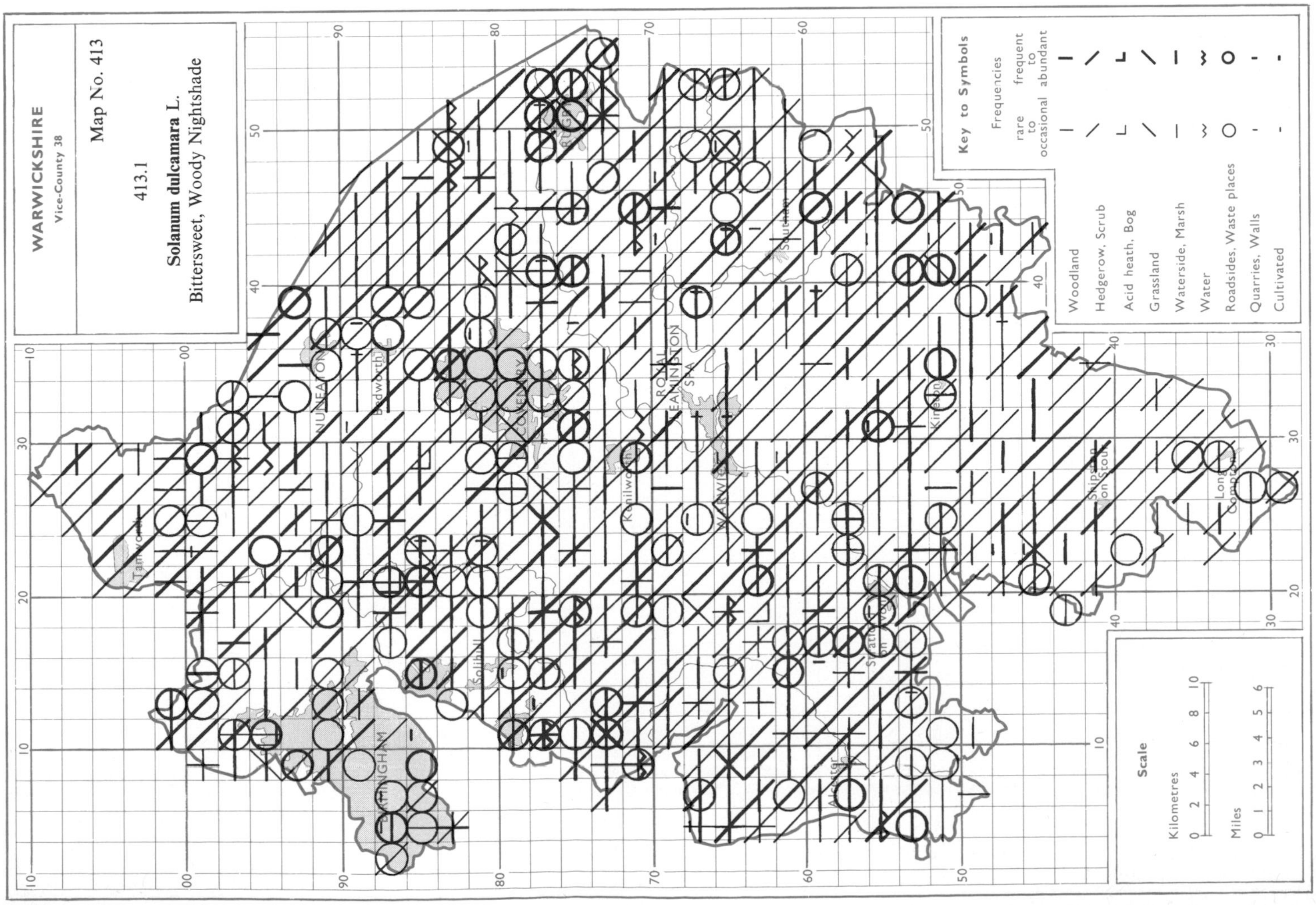
WARWICKSHIRE
Vice-County 38
Map No. 413
413.1
Solanum dulcamara L.
Bittersweet, Woody Nightshade
Key to Symbols
Frequencies
rare to occasional
frequent to abundant
Woodland
Hedgerow, Scrub
Acid heath, Bog
Grassland
Waterside, Marsh
Water
Roadsides, Waste places
Quarries, Walls
Cultivated
Scale
Kilometres
0 2 4 6 8 10
Miles
0 1 2 3 4 5 6

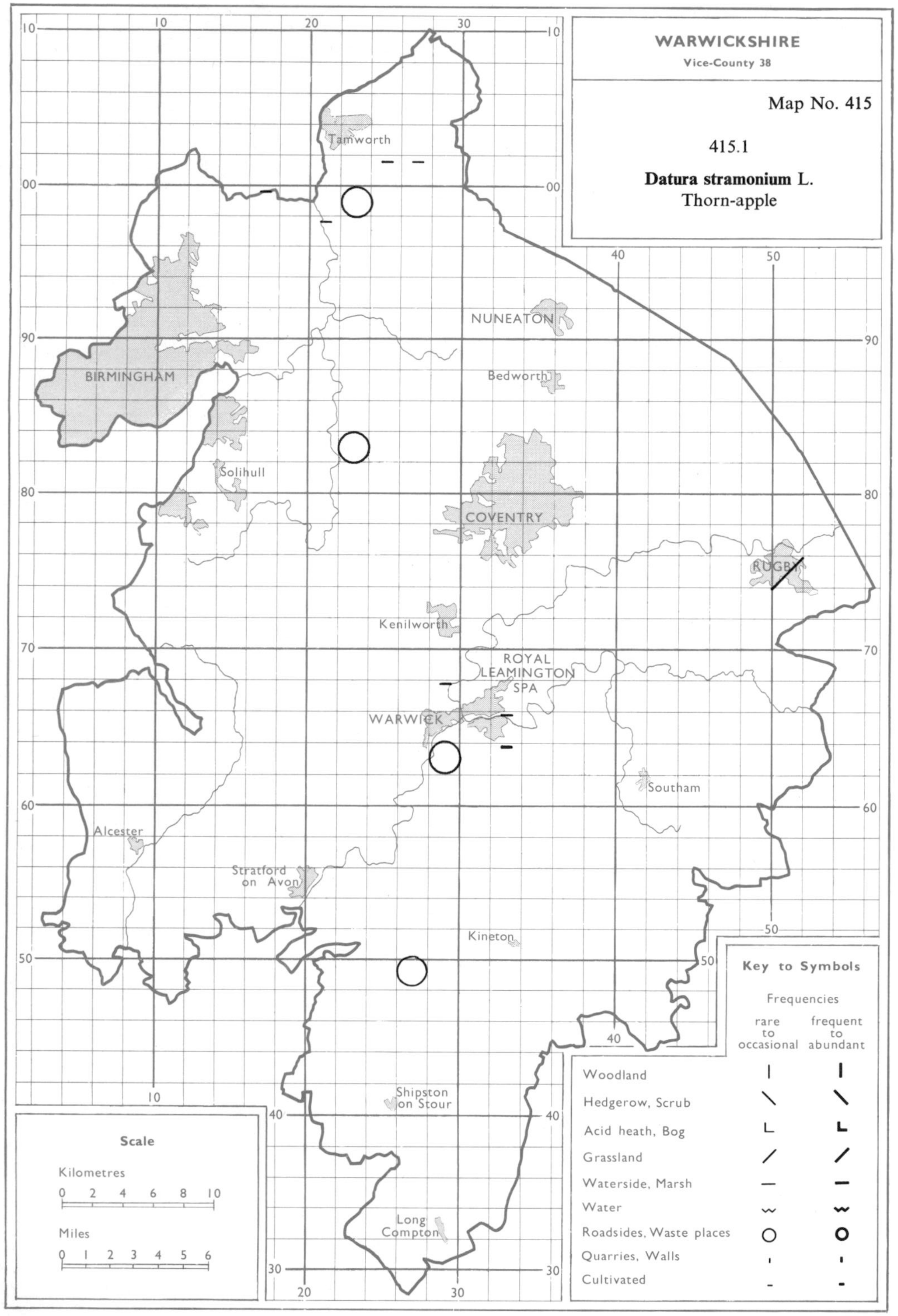
WARWICKSHIRE
Vice-County 38
Map No. 415
415.1
Datura stramonium L.
Thorn-apple
Tamworth
NUNEATON
Bedworth
BIRMINGHAM
Solihull
COVENTRY
RUGBY
Kenilworth
ROYAL LEAMINGTON SPA
WARWICK
Southam
Alcester
Stratford on Avon
Kineton
Shipston on Stour
Long Compton
Key to Symbols
Frequencies
rare to occasional
frequent to abundant
Woodland
Hedgerow, Scrub
Acid heath, Bog
Grassland
Waterside, Marsh
Water
Roadsides, Waste places
Quarries, Walls
Cultivated
Scale
Kilometres
0 2 4 6 8 10
Miles
0 1 2 3 4 5 6

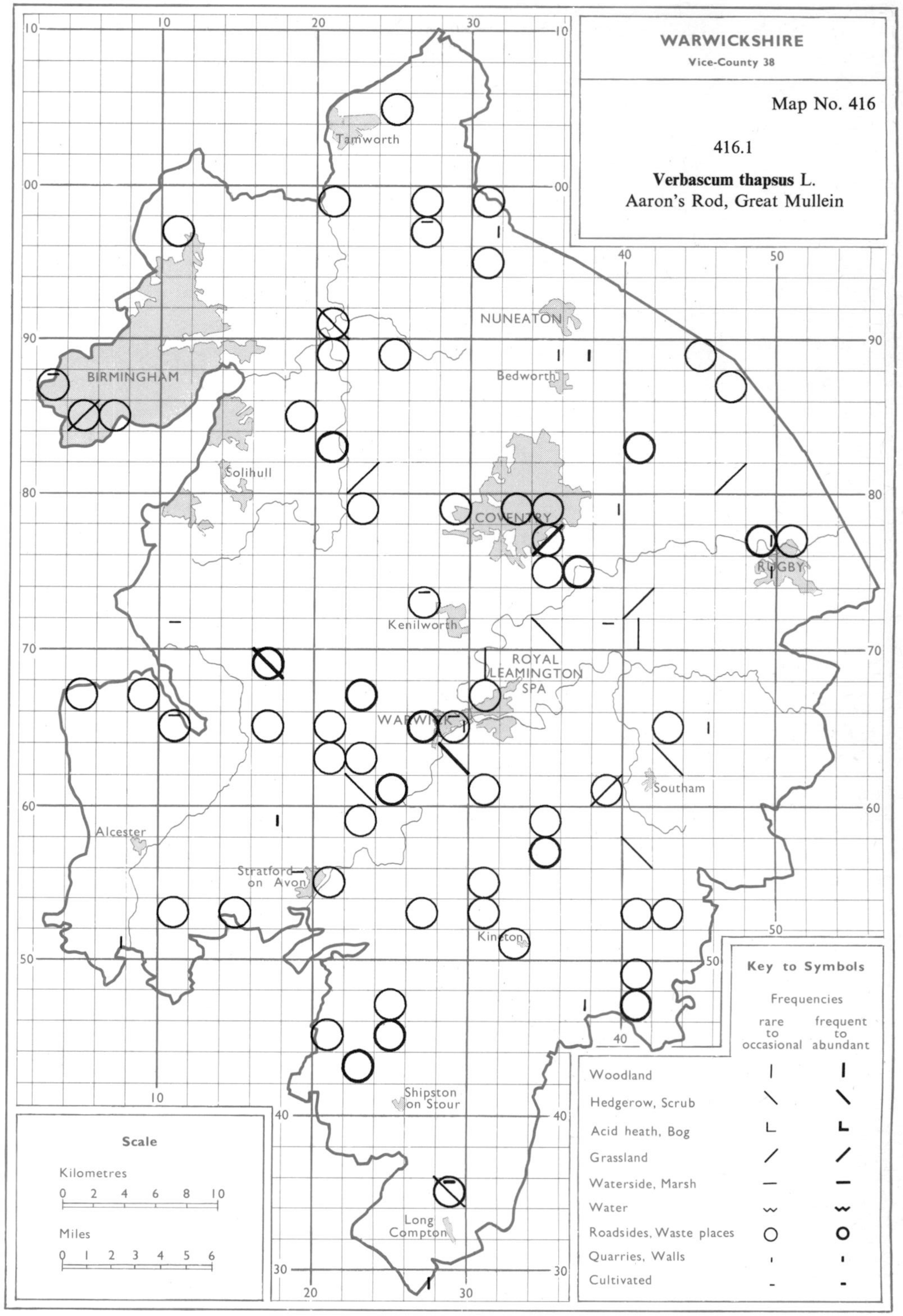
WARWICKSHIRE
Vice-County 38
Map No. 416
416.1
Verbascum thapsus L.
Aaron's Rod, Great Mullein
Tamworth
NUNEATON
Bedworth
BIRMINGHAM
Solihull
COVENTRY
RUGBY
Kenilworth
ROYAL LEAMINGTON SPA
WARWICK
Southam
Alcester
Stratford on Avon
Kineton
Shipston on Stour
Long Compton
Key to Symbols
Frequencies
rare to occasional
frequent to abundant
Woodland
Hedgerow, Scrub
Acid heath, Bog
Grassland
Waterside, Marsh
Water
Roadsides, Waste places
Quarries, Walls
Cultivated
Scale
Kilometres
0 2 4 6 8 10
Miles
0 1 2 3 4 5 6

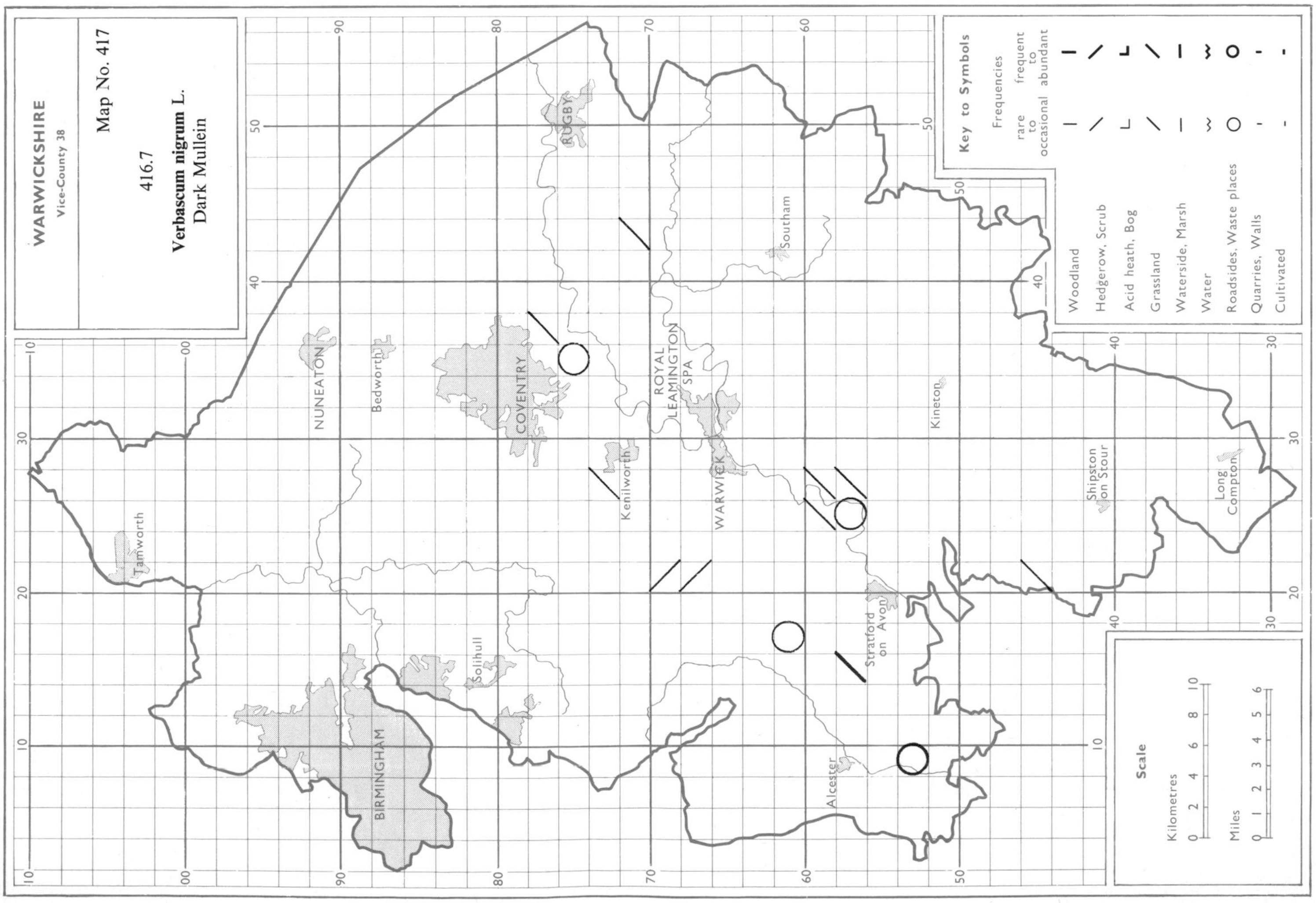
WARWICKSHIRE
Vice-County 38
Map No. 417
416.7
Verbascum nigrum L.
Dark Mullein
Key to Symbols
Frequencies
rare to occasional
frequent to abundant
Woodland
Hedgerow, Scrub
Acid heath, Bog
Grassland
Waterside, Marsh
Water
Roadsides, Waste places
Quarries, Walls
Cultivated
Scale
Kilometres
0 2 4 6 8 10
Miles
0 1 2 3 4 5 6
NUNEATON
Bedworth
COVENTRY
RUGBY
ROYAL LEAMINGTON SPA
WARWICK
Kenilworth
Southam
Kineton
Shipston on Stour
Long Compton
Tamworth
Solihull
BIRMINGHAM
Alcester
Stratford on Avon

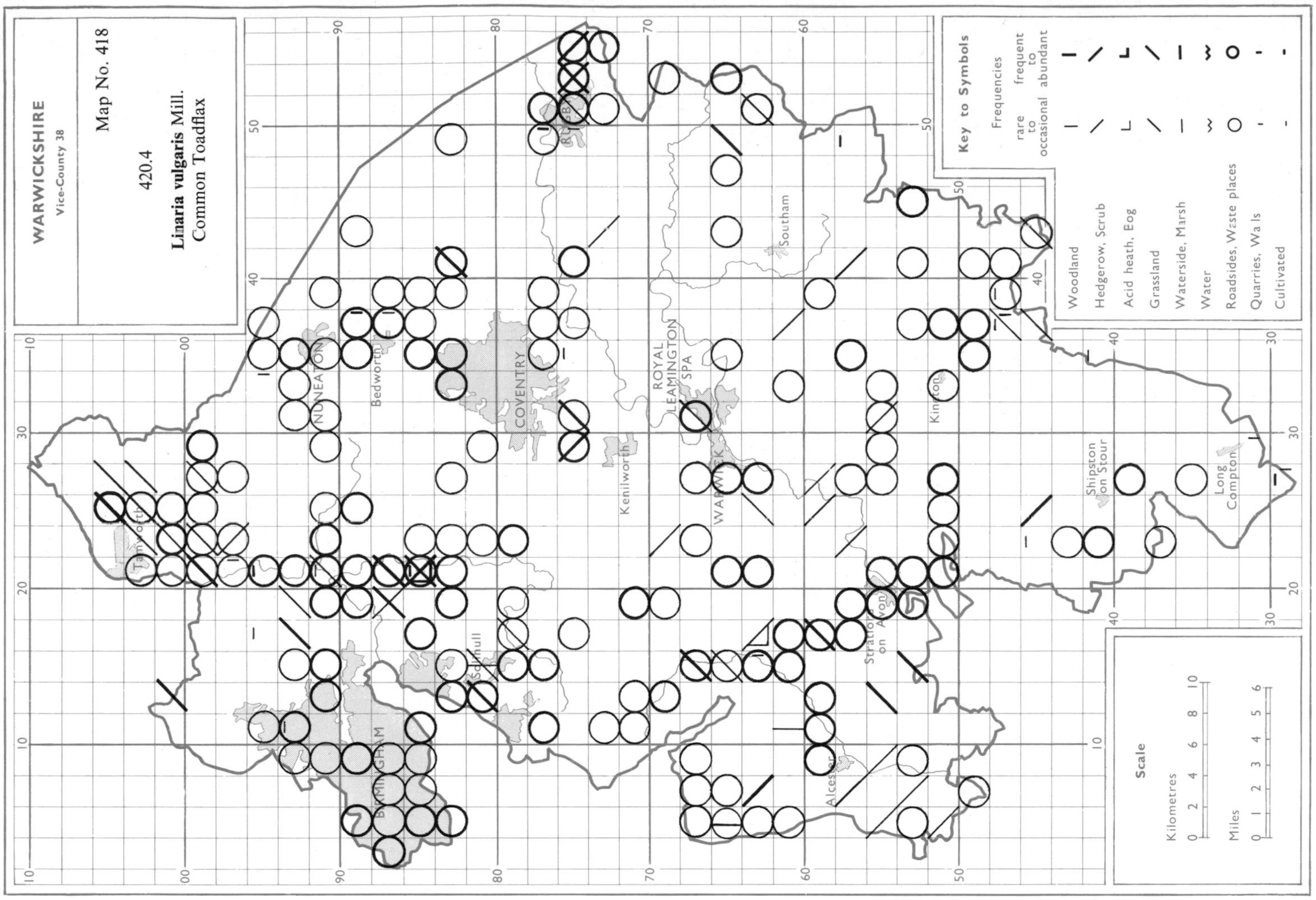
WARWICKSHIRE
Vice-County 38
Map No. 418
420.4
Linaria vulgaris Mill.
Common Toadflax
Key to Symbols
Frequencies
rare to occasional
frequent to abundant
Woodland
Hedgerow, Scrub
Acid heath, Bog
Grassland
Waterside, Marsh
Water
Roadsides, Waste places
Quarries, Walls
Cultivated
Scale
Kilometres
0 2 4 6 8 10
Miles
0 1 2 3 4 5 6
NUNEATON
Bedworth
COVENTRY
RUGBY
ROYAL LEAMINGTON SPA
WARWICK
Kenilworth
Southam
Kineton
Shipston on Stour
Long Compton
Tamworth
Solihull
BIRMINGHAM
Alcester
Stratford on Avon

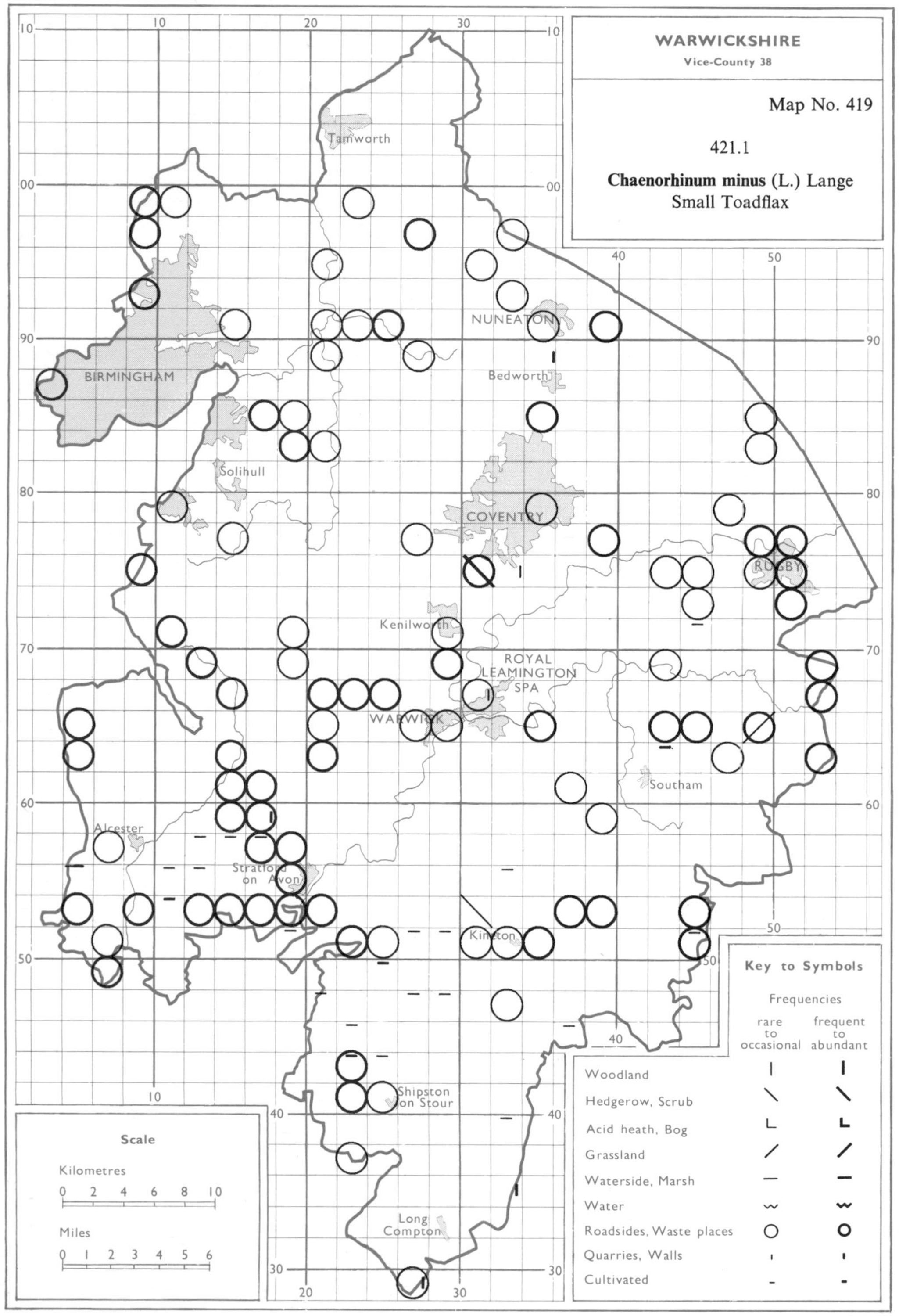
WARWICKSHIRE
Vice-County 38
Map No. 419
421.1
Chaenorhinum minus (L.) Lange
Small Toadflax
Tamworth
NUNEATON
Bedworth
BIRMINGHAM
Solihull
COVENTRY
RUGBY
Kenilworth
ROYAL LEAMINGTON SPA
WARWICK
Southam
Alcester
Stratford on Avon
Kineton
Shipston on Stour
Long Compton
Key to Symbols
Frequencies
rare to occasional
frequent to abundant
Woodland
Hedgerow, Scrub
Acid heath, Bog
Grassland
Waterside, Marsh
Water
Roadsides, Waste places
Quarries, Walls
Cultivated
Scale
Kilometres
0 2 4 6 8 10
Miles
0 1 2 3 4 5 6

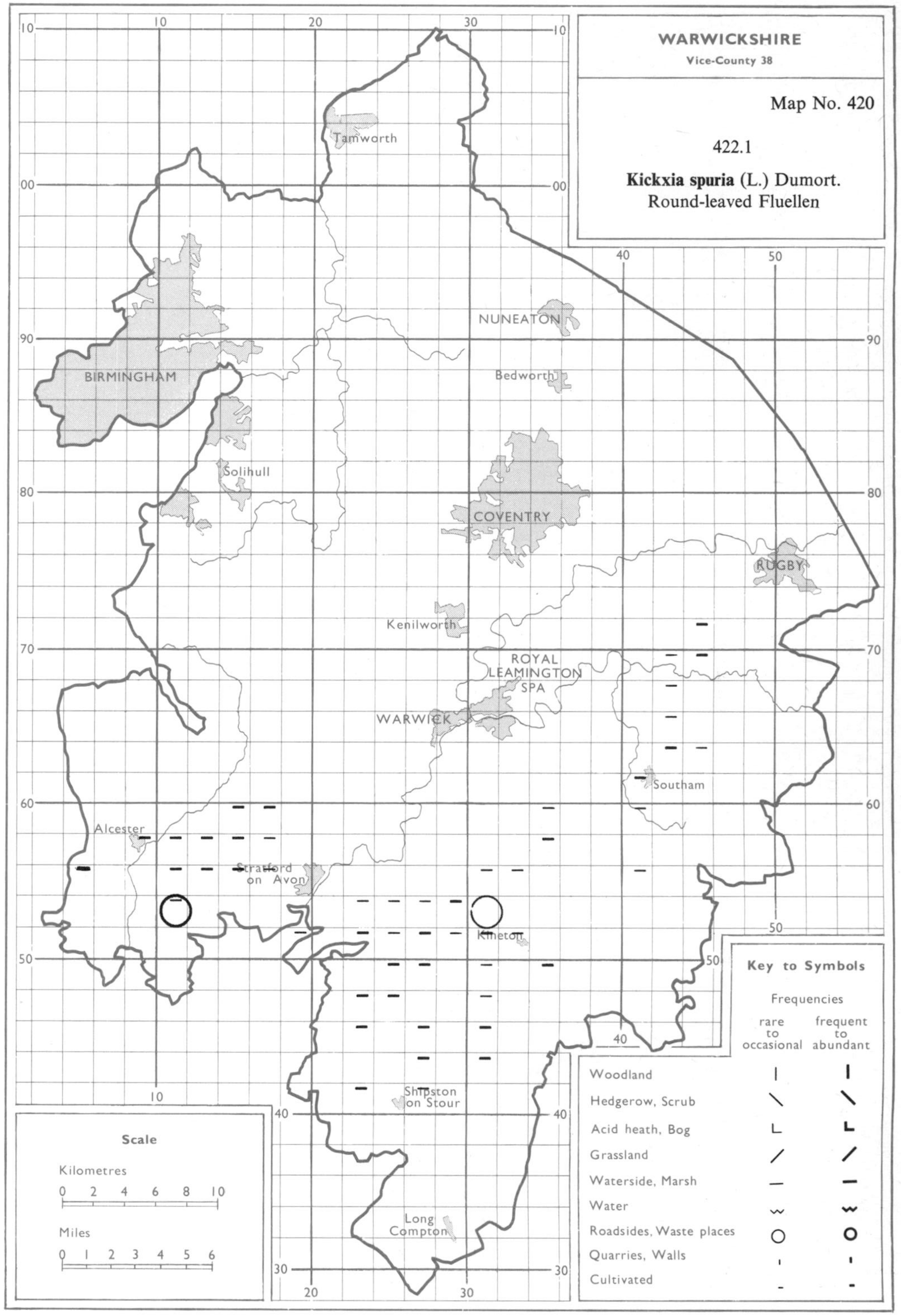
WARWICKSHIRE
Vice-County 38
Map No. 420
422.1
Kickxia spuria (L.) Dumort.
Round-leaved Fluellen
Tamworth
NUNEATON
Bedworth
BIRMINGHAM
Solihull
COVENTRY
RUGBY
Kenilworth
ROYAL LEAMINGTON SPA
WARWICK
Southam
Alcester
Stratford on Avon
Kineton
Shipston on Stour
Long Compton
Key to Symbols
Frequencies
rare to occasional
frequent to abundant
Woodland
Hedgerow, Scrub
Acid heath, Bog
Grassland
Waterside, Marsh
Water
Roadsides, Waste places
Quarries, Walls
Cultivated
Scale
Kilometres
0 2 4 6 8 10
Miles
0 1 2 3 4 5 6

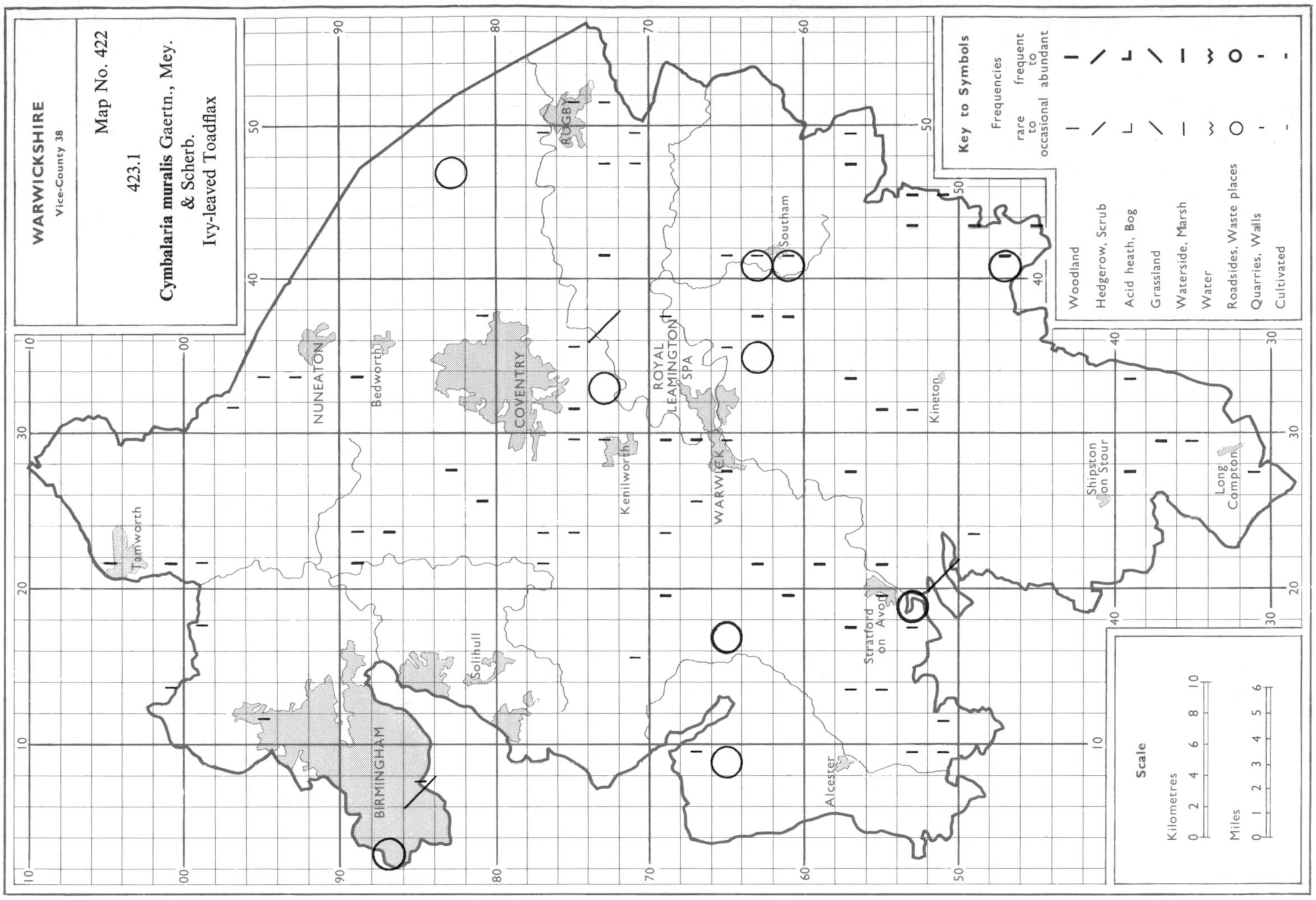
WARWICKSHIRE
Vice-County 38
Map No. 422
423.1
Cymbalaria muralis Gaertn., Mey. & Scherb.
Ivy-leaved Toadflax
Key to Symbols
Frequencies
rare to occasional
frequent to abundant
Woodland
Hedgerow, Scrub
Acid heath, Bog
Grassland
Waterside, Marsh
Water
Roadsides, Waste places
Quarries, Walls
Cultivated
Scale
Kilometres
Miles
Tamworth
NUNEATON
Bedworth
COVENTRY
RUGBY
BIRMINGHAM
Solihull
Kenilworth
ROYAL LEAMINGTON SPA
WARWICK
Southam
Kineton
Stratford on Avon
Alcester
Shipston on Stour
Long Compton

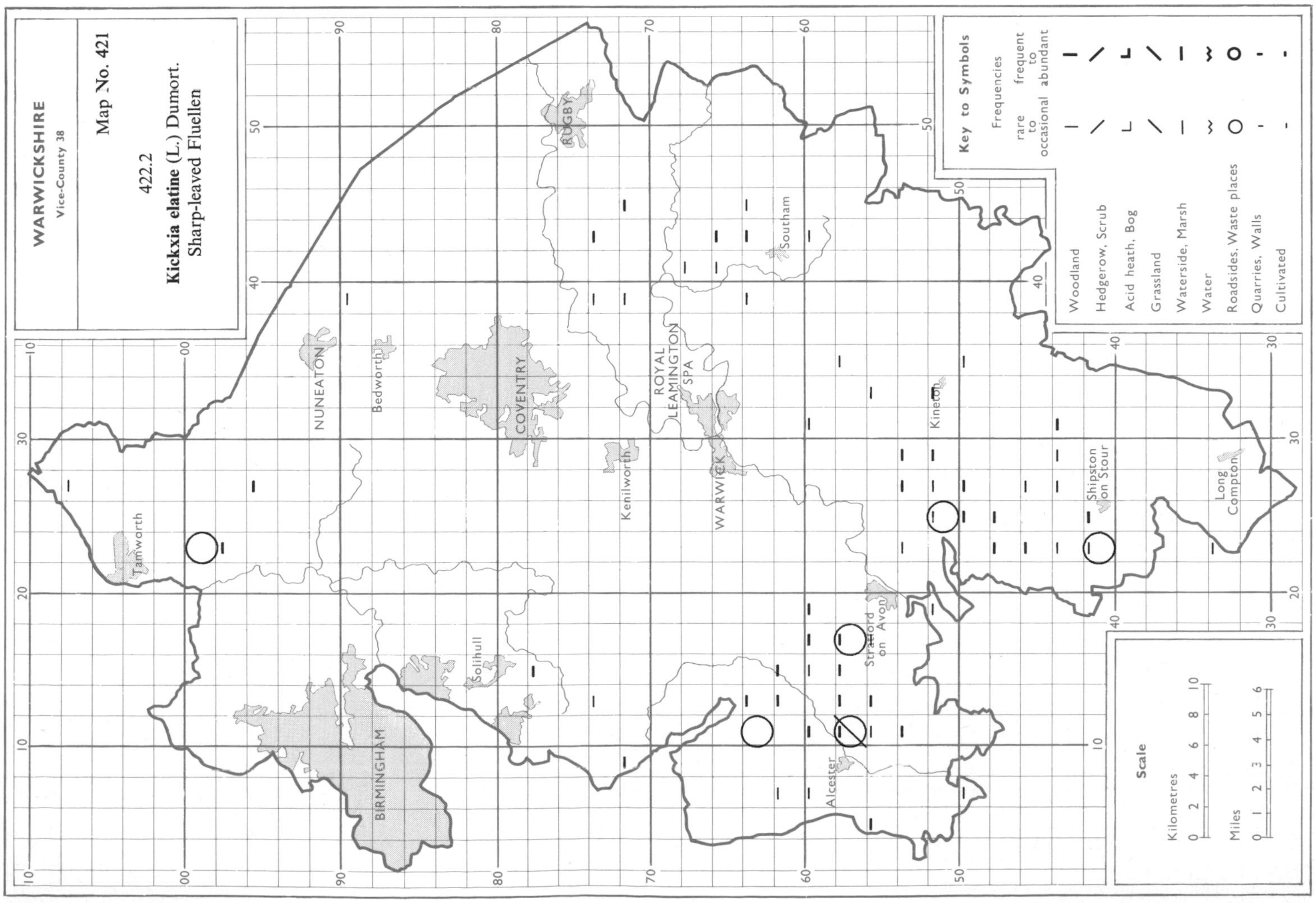
WARWICKSHIRE
Vice-County 38
Map No. 421
422.2
Kickxia elatine (L.) Dumort.
Sharp-leaved Fluellen
Key to Symbols
Frequencies
rare to occasional
frequent to abundant
Woodland
Hedgerow, Scrub
Acid heath, Bog
Grassland
Waterside, Marsh
Water
Roadsides, Waste places
Quarries, Walls
Cultivated
Scale
Kilometres
Miles
Tamworth
NUNEATON
Bedworth
COVENTRY
RUGBY
BIRMINGHAM
Solihull
Kenilworth
ROYAL LEAMINGTON SPA
WARWICK
Southam
Kineton
Stratford on Avon
Alcester
Shipston on Stour
Long Compton

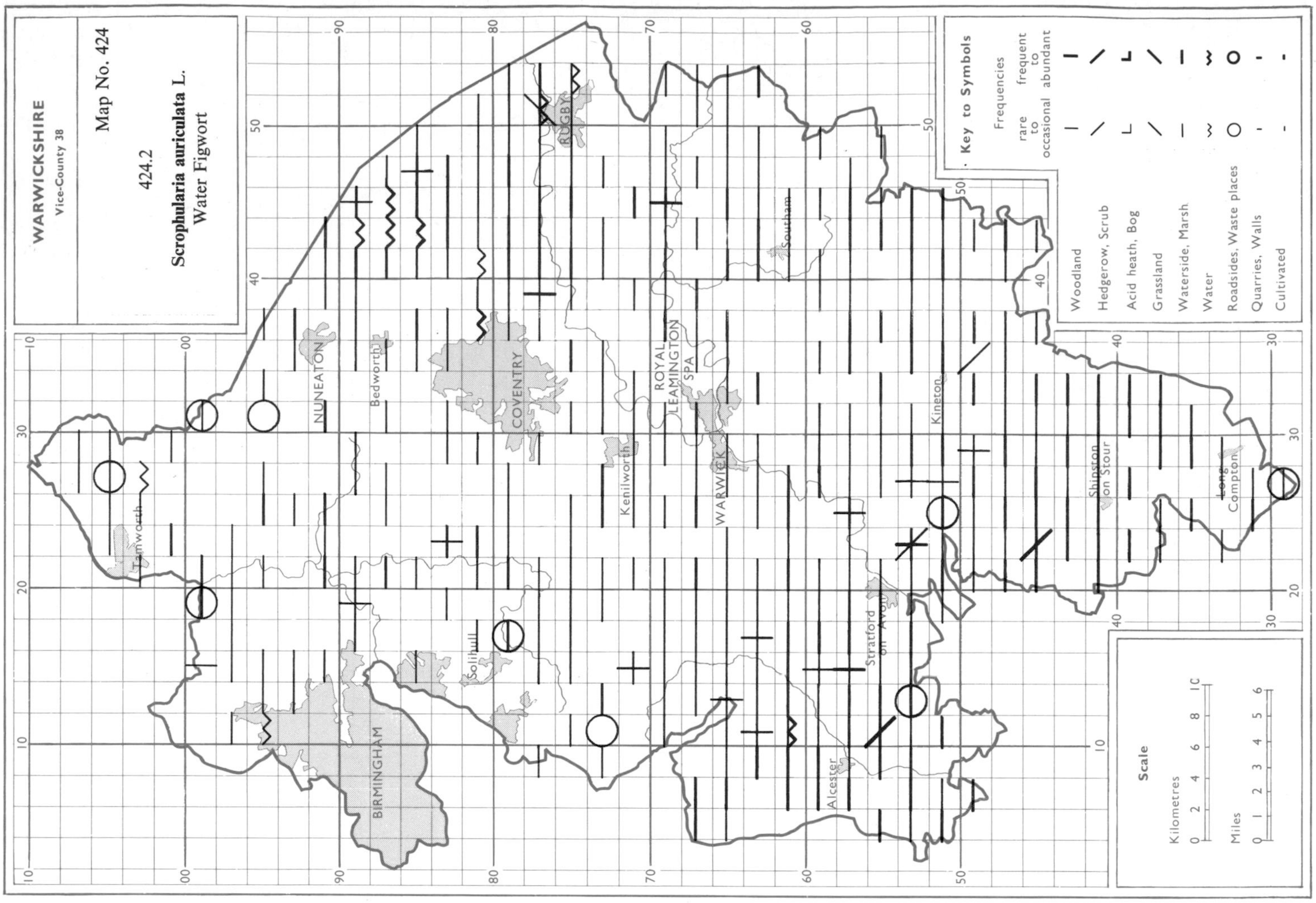
WARWICKSHIRE
Vice-County 38
Map No. 424
424.2
Scrophularia auriculata L.
Water Figwort
Key to Symbols
Frequencies
rare to occasional
frequent to abundant
Woodland
Hedgerow, Scrub
Acid heath, Bog
Grassland
Waterside, Marsh
Water
Roadsides, Waste places
Quarries, Walls
Cultivated
Scale
Kilometres
0 2 4 6 8 10
Miles
0 1 2 3 4 5 6
NUNEATON
Bedworth
COVENTRY
RUGBY
Southam
ROYAL LEAMINGTON SPA
Kenilworth
WARWICK
Kineton
Shipston on Stour
Long Compton
Tamworth
Solihull
BIRMINGHAM
Stratford on Avon
Alcester

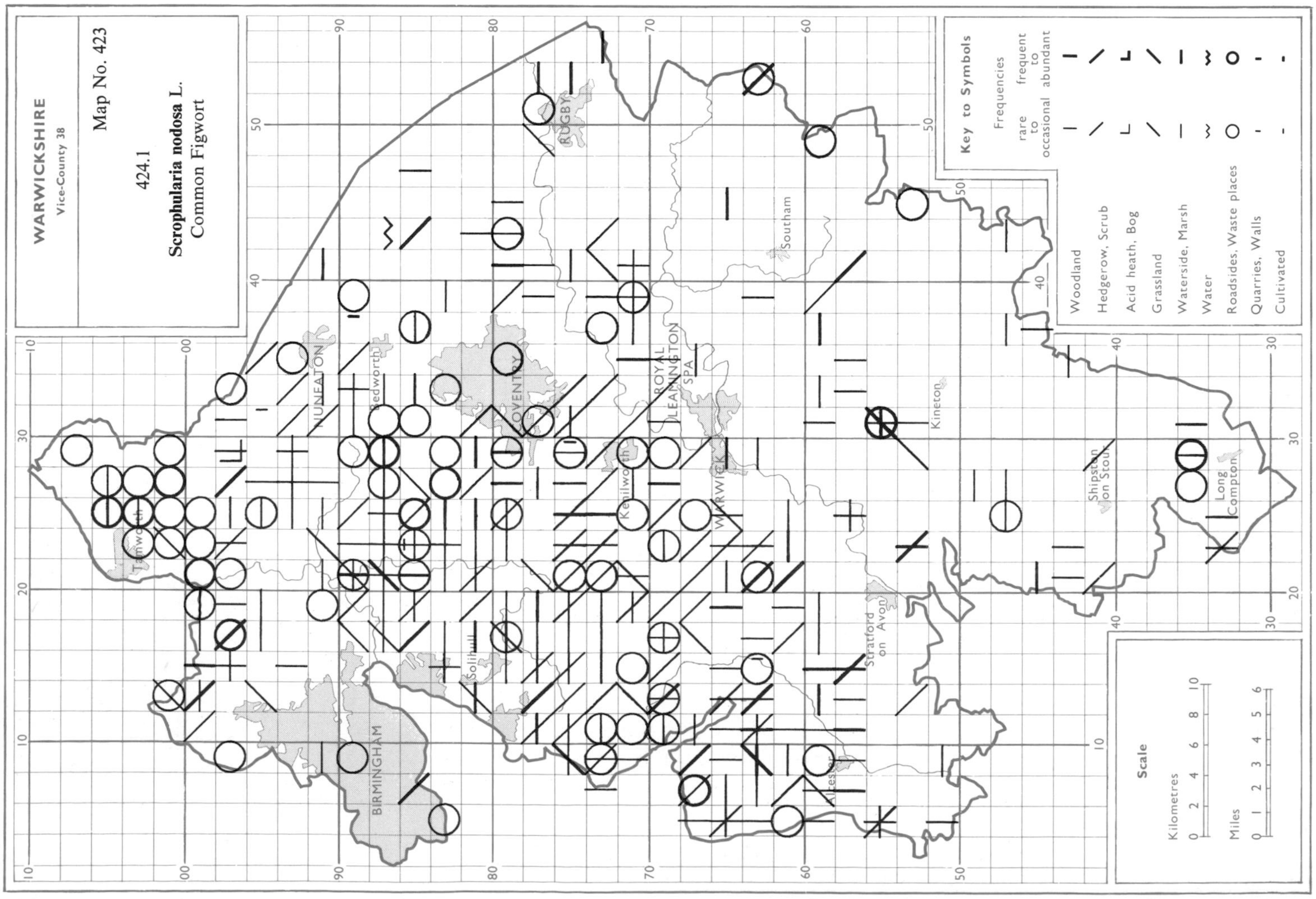
WARWICKSHIRE
Vice-County 38
Map No. 423
424.1
Scrophularia nodosa L.
Common Figwort
Key to Symbols
Frequencies
rare to occasional
frequent to abundant
Woodland
Hedgerow, Scrub
Acid heath, Bog
Grassland
Waterside, Marsh
Water
Roadsides, Waste places
Quarries, Walls
Cultivated
Scale
Kilometres
0 2 4 6 8 10
Miles
0 1 2 3 4 5 6
NUNEATON
Bedworth
COVENTRY
RUGBY
Southam
ROYAL LEAMINGTON SPA
Kenilworth
WARWICK
Kineton
Shipston on Stour
Long Compton
Tamworth
Solihull
BIRMINGHAM
Stratford on Avon
Alcester

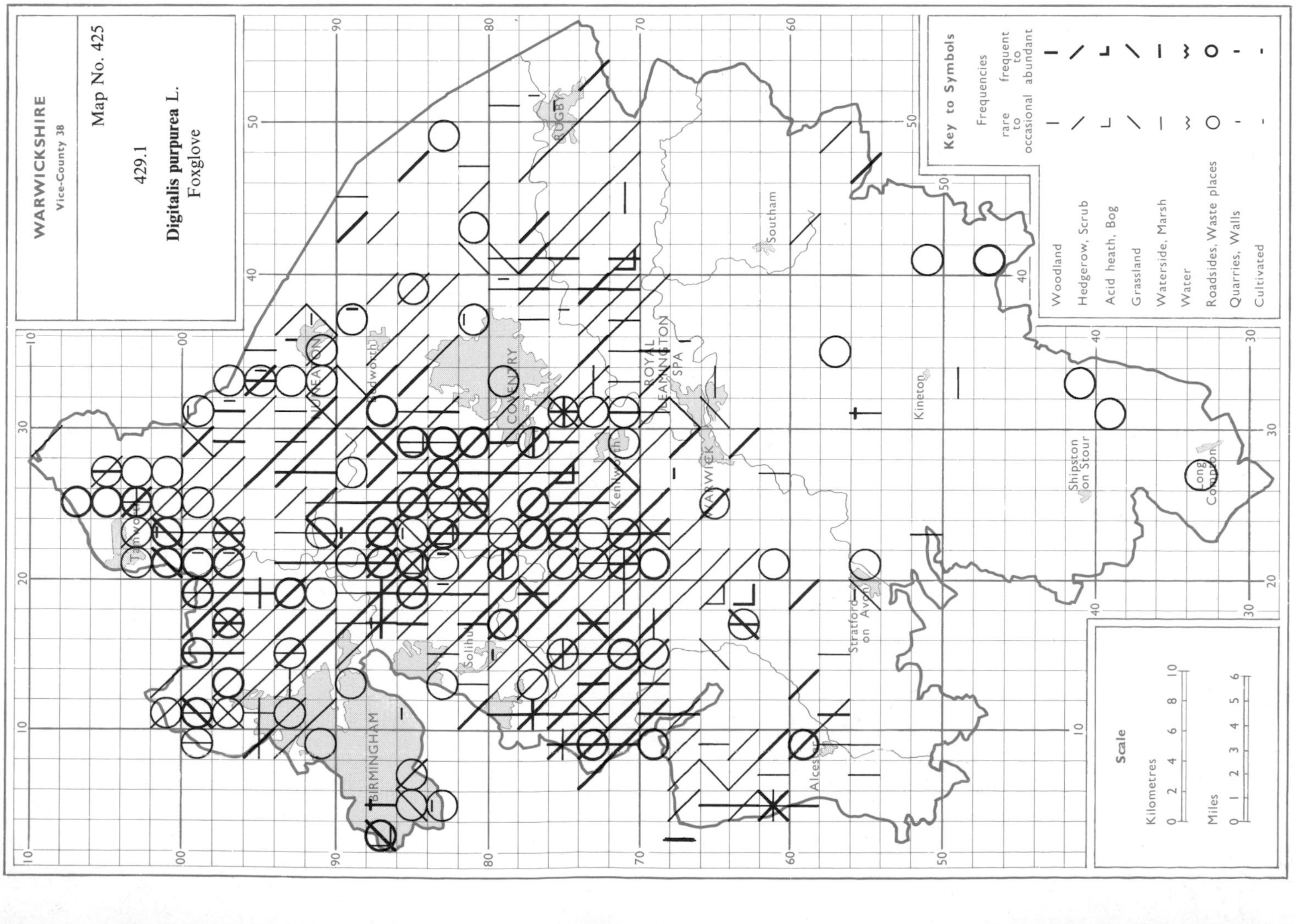
WARWICKSHIRE
Vice-County 38
Map No. 425
429.1
Digitalis purpurea L.
Foxglove
Key to Symbols
Frequencies
rare to occasional
frequent to abundant
Woodland
Hedgerow, Scrub
Acid heath, Bog
Grassland
Waterside, Marsh
Water
Roadsides, Waste places
Quarries, Walls
Cultivated
Scale
Kilometres
Miles
Tamworth
NUNEATON
Bedworth
COVENTRY
RUGBY
Southam
ROYAL LEAMINGTON SPA
WARWICK
Kenilworth
Kineton
Shipston on Stour
Long Compton
Solihull
BIRMINGHAM
Stratford on Avon
Alcester

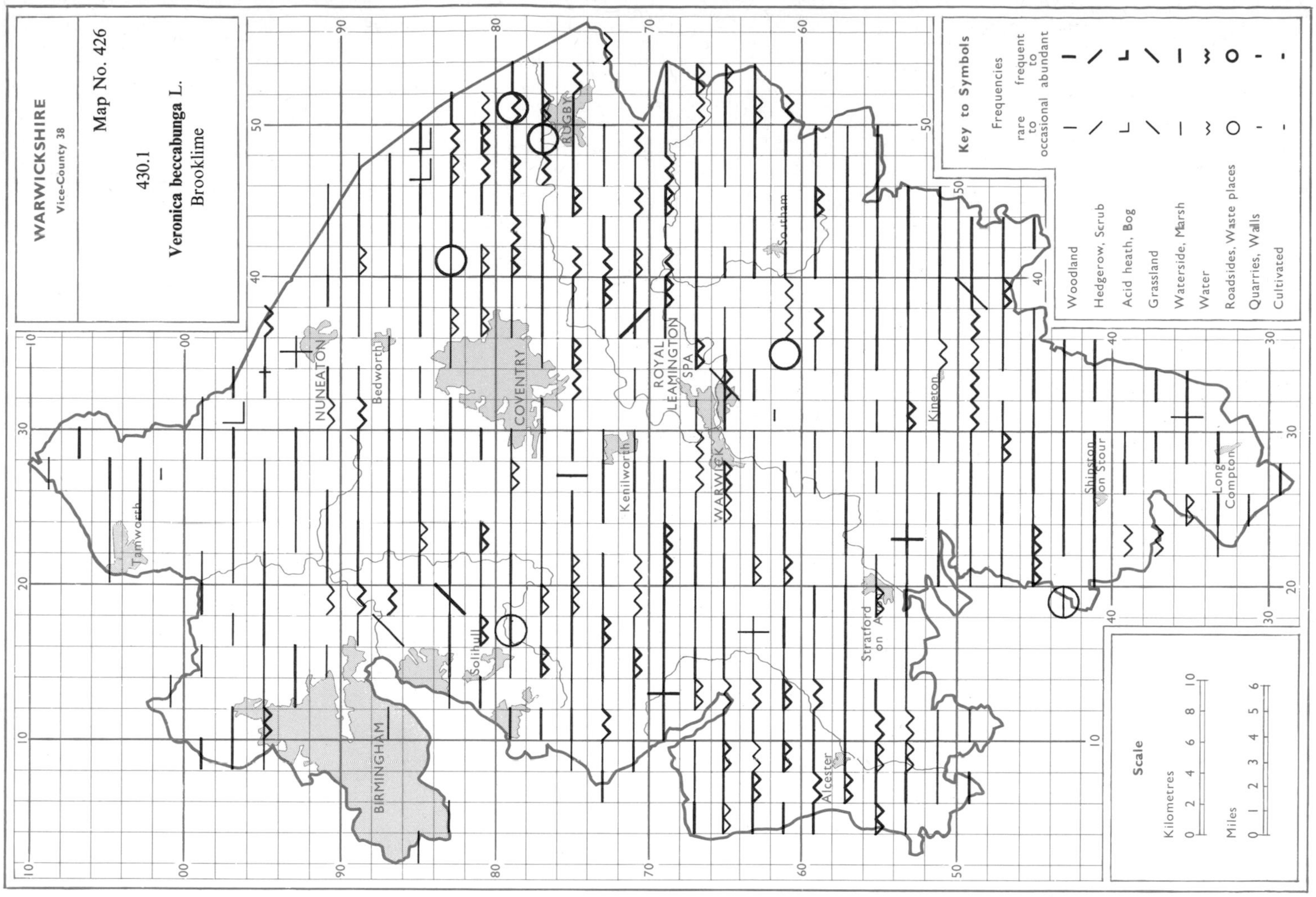
WARWICKSHIRE
Vice-County 38
Map No. 426
430.1
Veronica beccabunga L.
Brooklime
Key to Symbols
Frequencies
rare to occasional
frequent to abundant
Woodland
Hedgerow, Scrub
Acid heath, Bog
Grassland
Waterside, Marsh
Water
Roadsides, Waste places
Quarries, Walls
Cultivated
Scale
Kilometres
Miles
Tamworth
NUNEATON
Bedworth
COVENTRY
RUGBY
Southam
ROYAL LEAMINGTON SPA
WARWICK
Kenilworth
Kineton
Shipston on Stour
Long Compton
Solihull
BIRMINGHAM
Stratford on Avon
Alcester

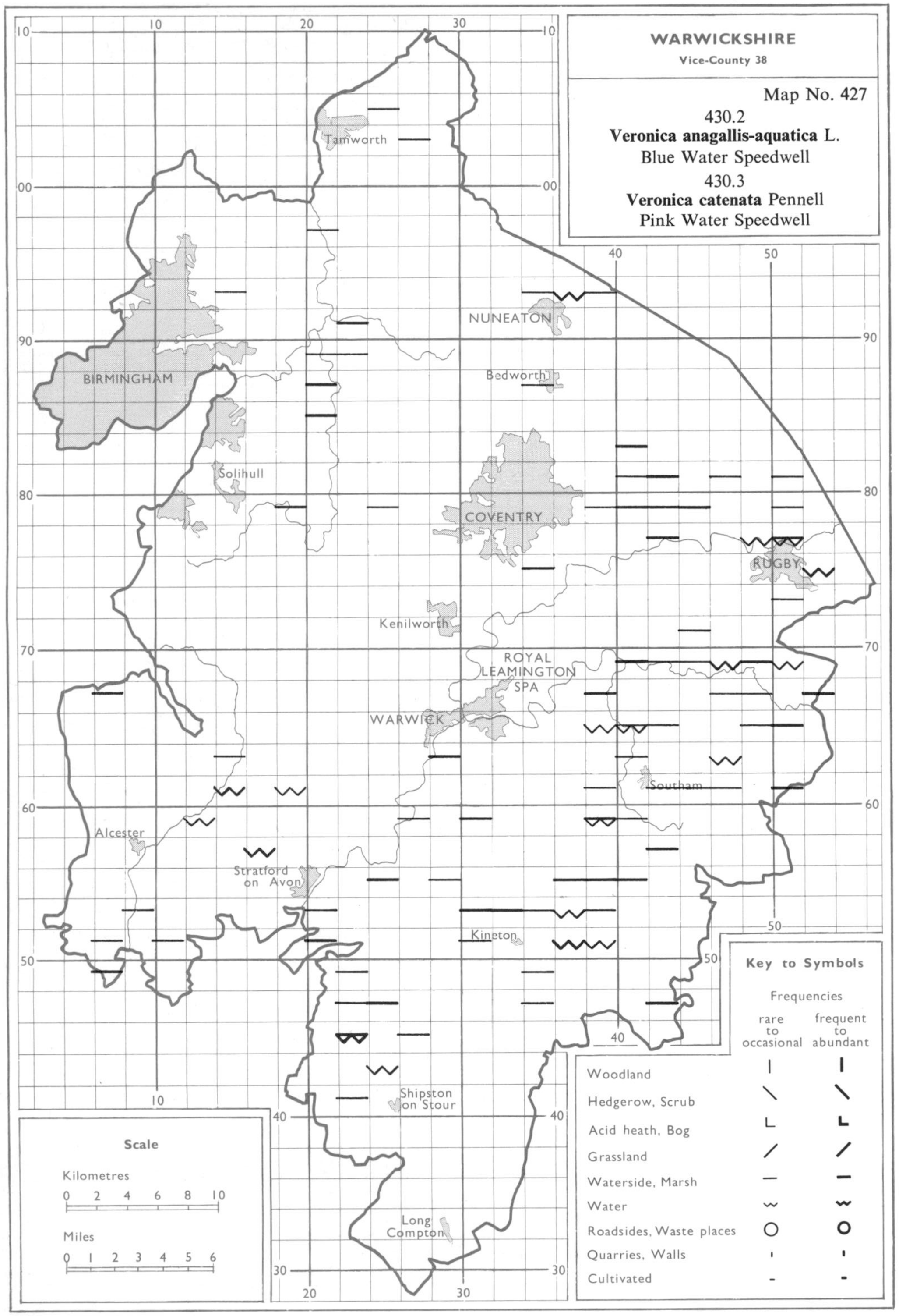
WARWICKSHIRE
Vice-County 38
Map No. 427
430.2
Veronica anagallis-aquatica L.
Blue Water Speedwell
430.3
Veronica catenata Pennell
Pink Water Speedwell
Tamworth
NUNEATON
Bedworth
BIRMINGHAM
Solihull
COVENTRY
RUGBY
Kenilworth
ROYAL LEAMINGTON SPA
WARWICK
Southam
Alcester
Stratford on Avon
Kineton
Shipston on Stour
Long Compton
Key to Symbols
Frequencies
rare to occasional
frequent to abundant
Woodland
Hedgerow, Scrub
Acid heath, Bog
Grassland
Waterside, Marsh
Water
Roadsides, Waste places
Quarries, Walls
Cultivated
Scale
Kilometres
0 2 4 6 8 10
Miles
0 1 2 3 4 5 6

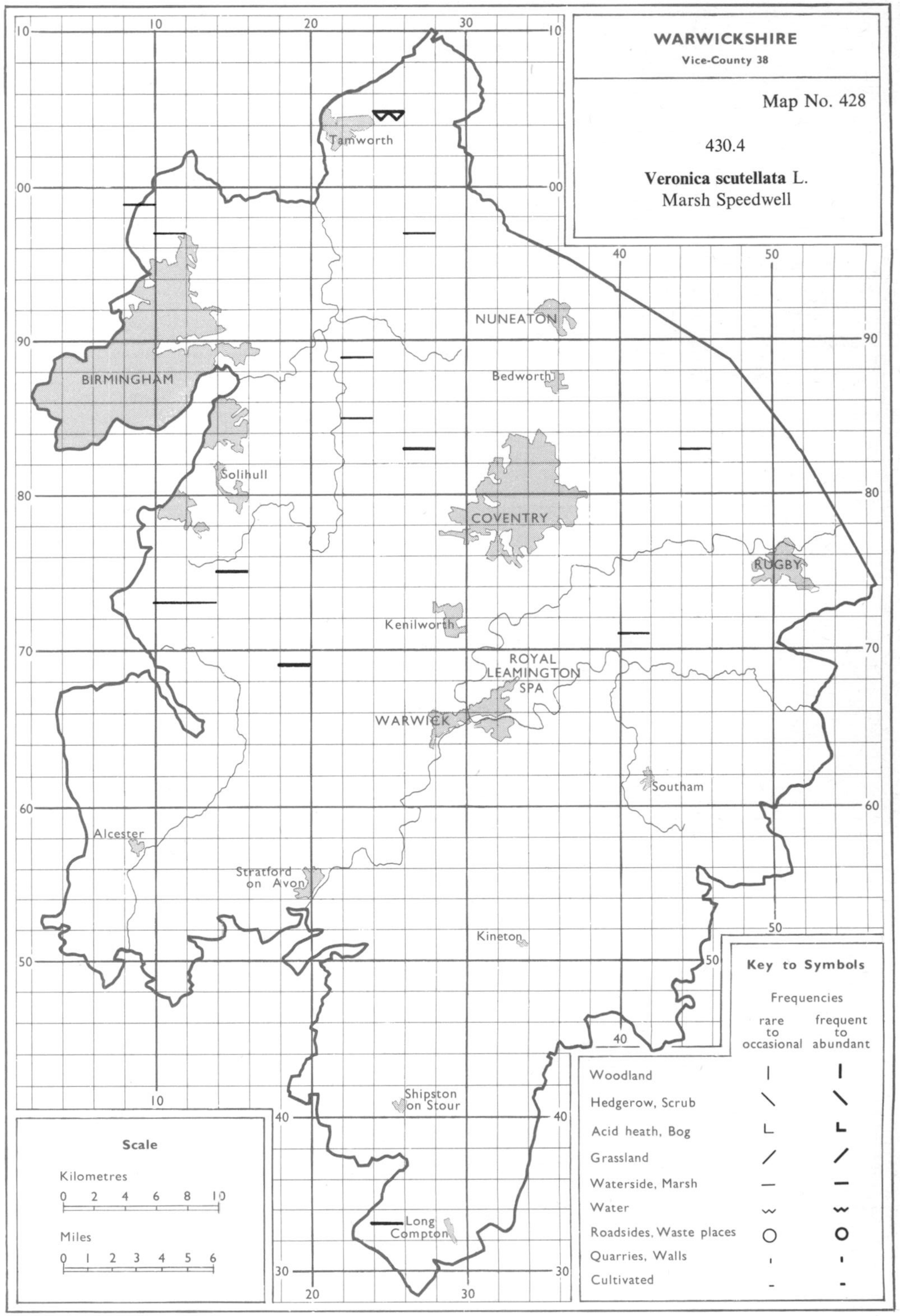
WARWICKSHIRE
Vice-County 38
Map No. 428
430.4
Veronica scutellata L.
Marsh Speedwell
Tamworth
NUNEATON
Bedworth
BIRMINGHAM
Solihull
COVENTRY
RUGBY
Kenilworth
ROYAL LEAMINGTON SPA
WARWICK
Southam
Alcester
Stratford on Avon
Kineton
Shipston on Stour
Long Compton
Key to Symbols
Frequencies
rare to occasional
frequent to abundant
Woodland
Hedgerow, Scrub
Acid heath, Bog
Grassland
Waterside, Marsh
Water
Roadsides, Waste places
Quarries, Walls
Cultivated
Scale
Kilometres
0 2 4 6 8 10
Miles
0 1 2 3 4 5 6

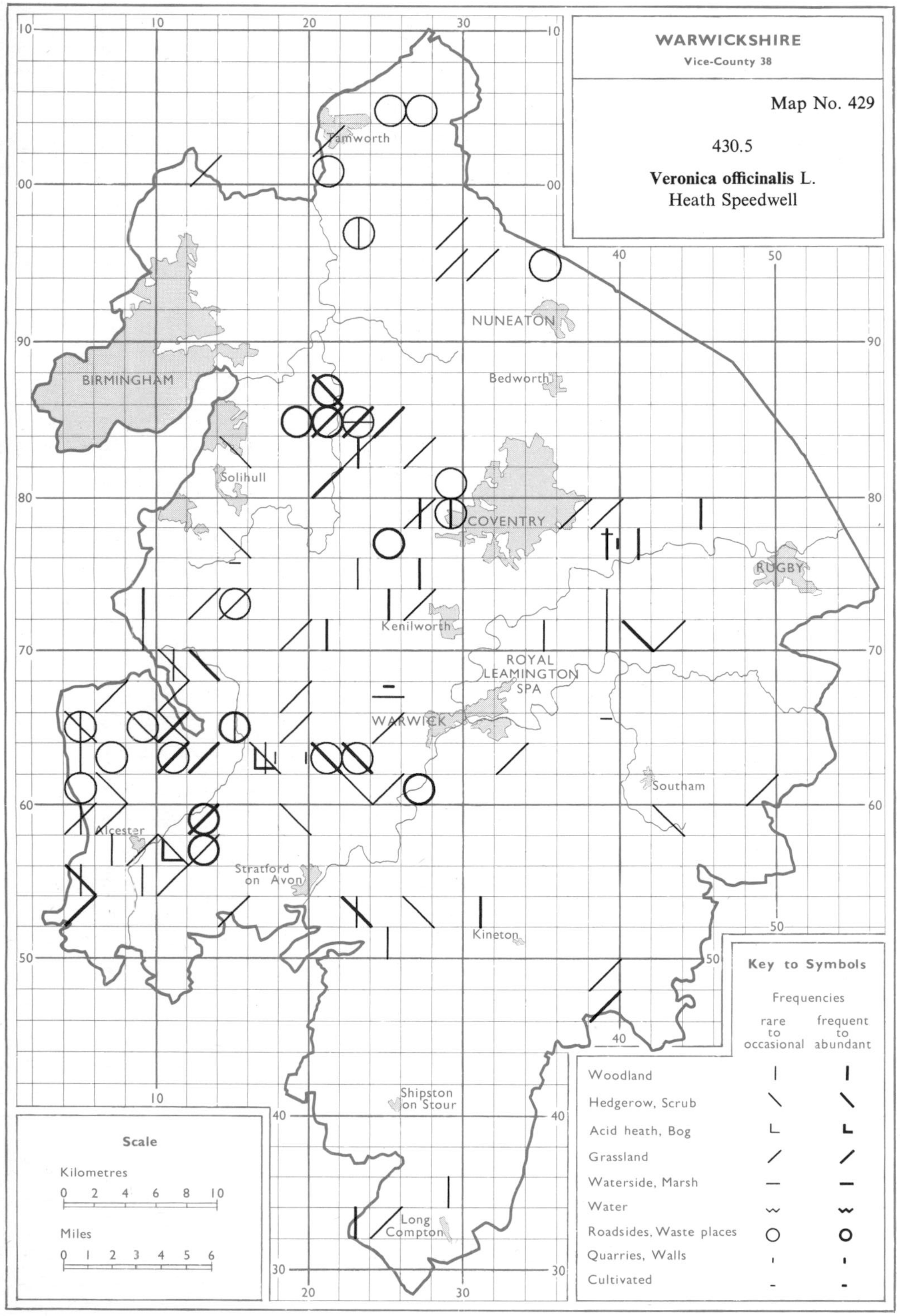
WARWICKSHIRE
Vice-County 38
Map No. 429
430.5
Veronica officinalis L.
Heath Speedwell
Tamworth
NUNEATON
Bedworth
BIRMINGHAM
Solihull
COVENTRY
RUGBY
Kenilworth
ROYAL LEAMINGTON SPA
WARWICK
Southam
Alcester
Stratford on Avon
Kineton
Shipston on Stour
Long Compton
Key to Symbols
Frequencies
rare to occasional
frequent to abundant
Woodland
Hedgerow, Scrub
Acid heath, Bog
Grassland
Waterside, Marsh
Water
Roadsides, Waste places
Quarries, Walls
Cultivated
Scale
Kilometres
0 2 4 6 8 10
Miles
0 1 2 3 4 5 6

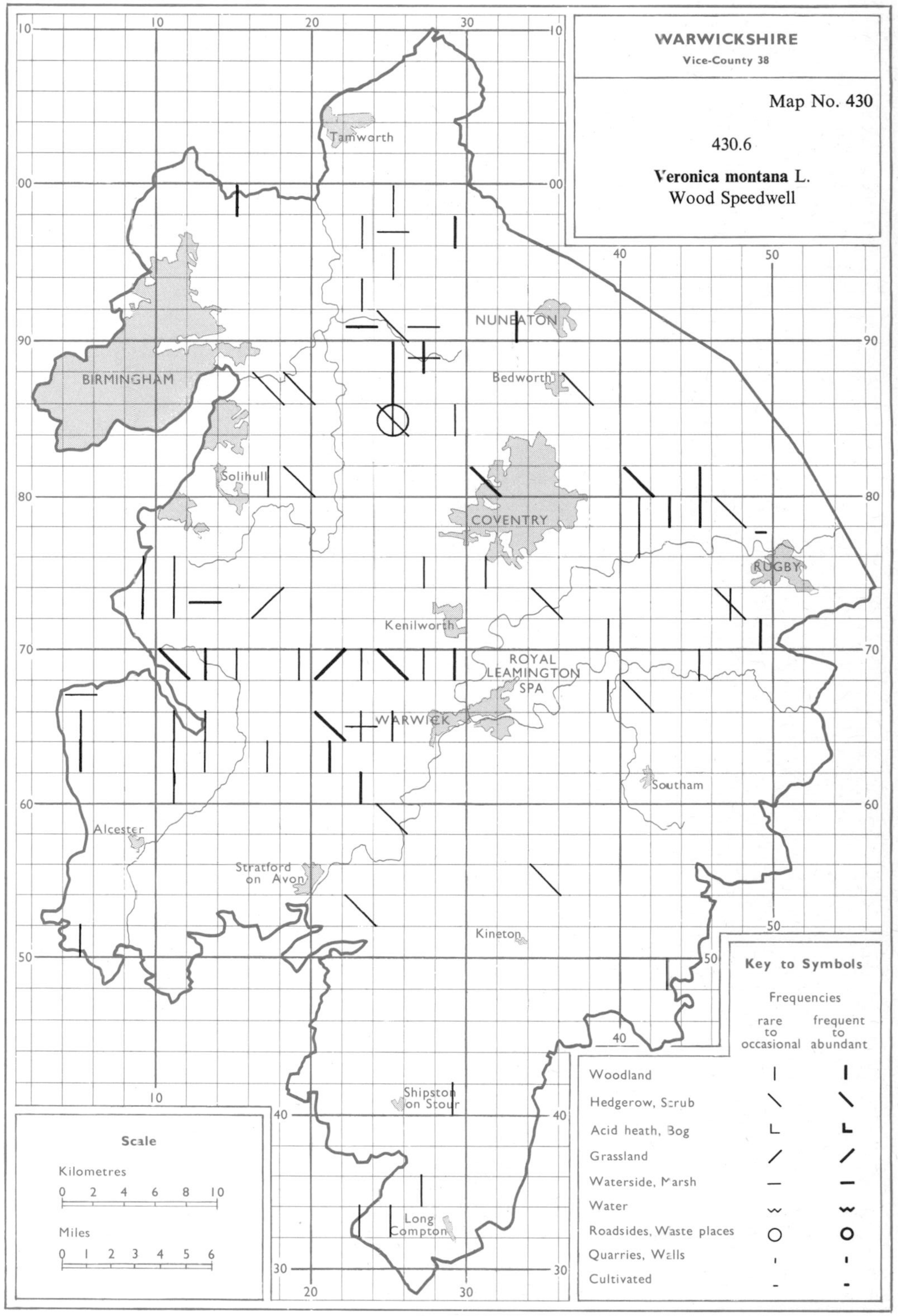
WARWICKSHIRE
Vice-County 38
Map No. 430
430.6
Veronica montana L.
Wood Speedwell
Tamworth
NUNEATON
Bedworth
BIRMINGHAM
Solihull
COVENTRY
RUGBY
Kenilworth
ROYAL LEAMINGTON SPA
WARWICK
Southam
Alcester
Stratford on Avon
Kineton
Shipston on Stour
Long Compton
Key to Symbols
Frequencies
rare to occasional
frequent to abundant
Woodland
Hedgerow, Scrub
Acid heath, Bog
Grassland
Waterside, Marsh
Water
Roadsides, Waste places
Quarries, Walls
Cultivated
Scale
Kilometres
0 2 4 6 8 10
Miles
0 1 2 3 4 5 6

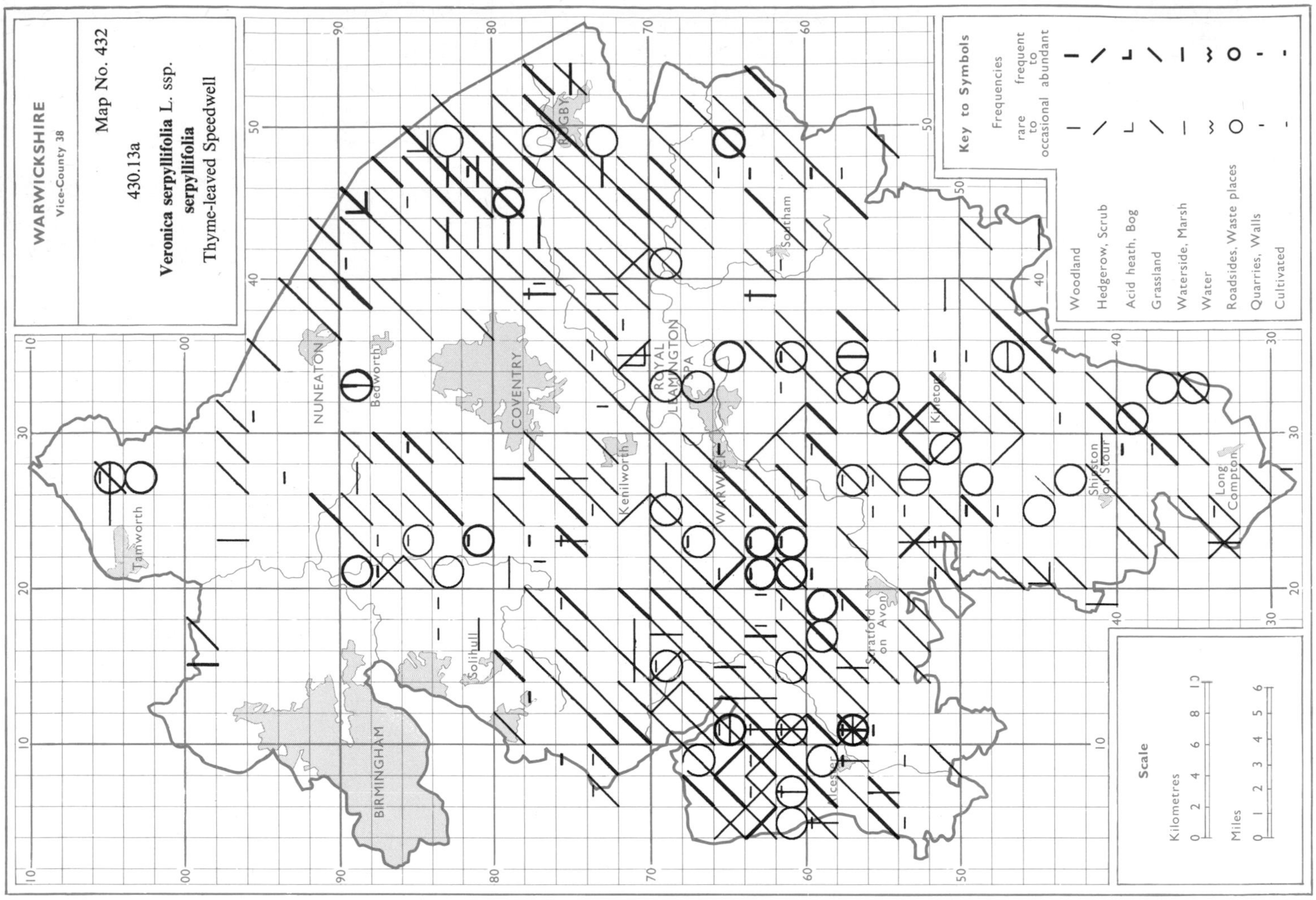
WARWICKSHIRE
Vice-County 38
Map No. 432
430.13a
Veronica serpyllifolia L. ssp. serpyllifolia
Thyme-leaved Speedwell
Key to Symbols
Frequencies
rare to occasional
frequent to abundant
Woodland
Hedgerow, Scrub
Acid heath, Bog
Grassland
Waterside, Marsh
Water
Roadsides, Waste places
Quarries, Walls
Cultivated
Scale
Kilometres
Miles
Tamworth
NUNEATON
Bedworth
COVENTRY
RUGBY
Kenilworth
ROYAL LEAMINGTON SPA
WARWICK
Southam
Solihull
BIRMINGHAM
Stratford on Avon
Alcester
Kineton
Shipston on Stour
Long Compton

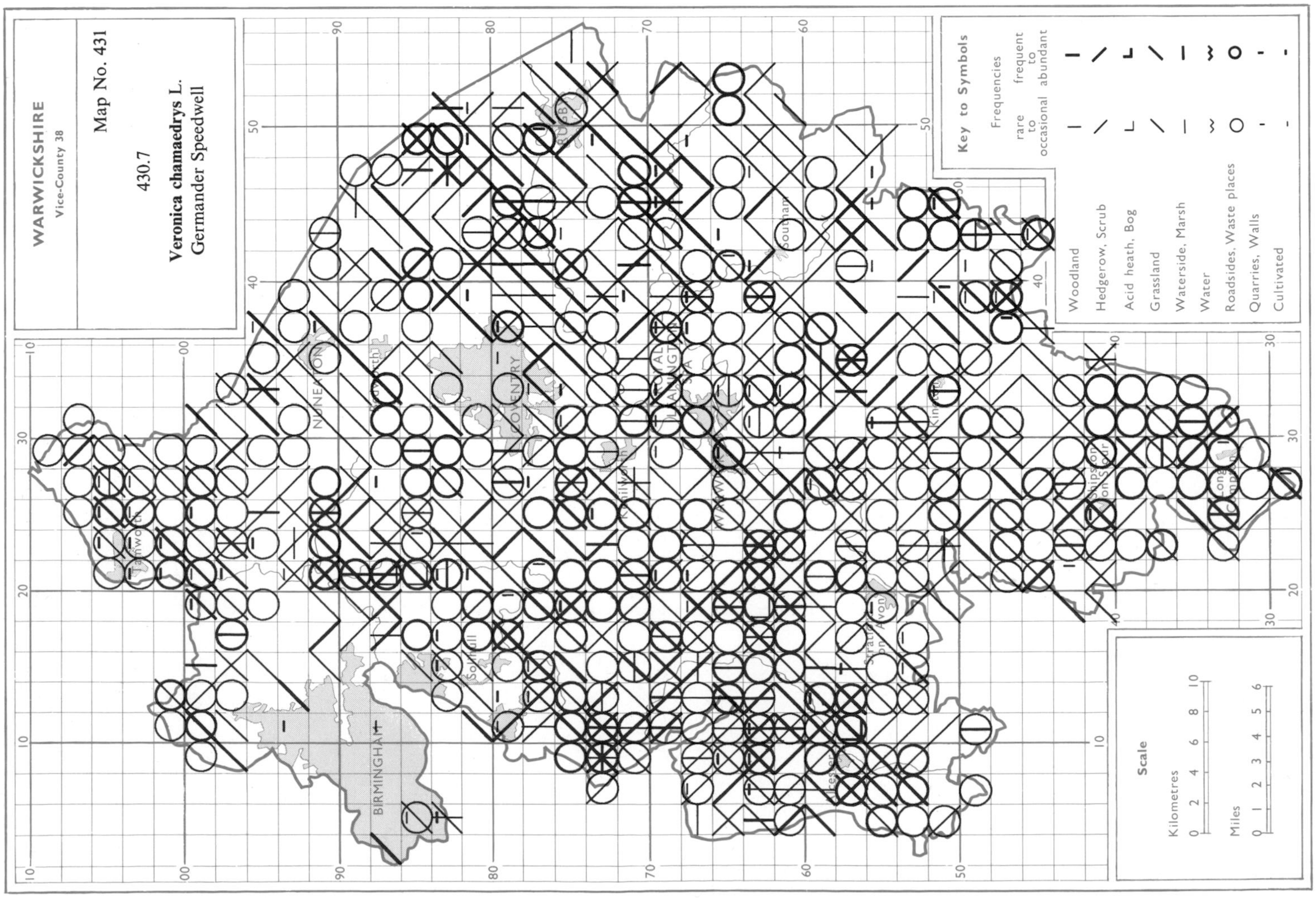
WARWICKSHIRE
Vice-County 38
Map No. 431
430.7
Veronica chamaedrys L.
Germander Speedwell
Key to Symbols
Frequencies
rare to occasional
frequent to abundant
Woodland
Hedgerow, Scrub
Acid heath, Bog
Grassland
Waterside, Marsh
Water
Roadsides, Waste places
Quarries, Walls
Cultivated
Scale
Kilometres
Miles
BIRMINGHAM
Southam

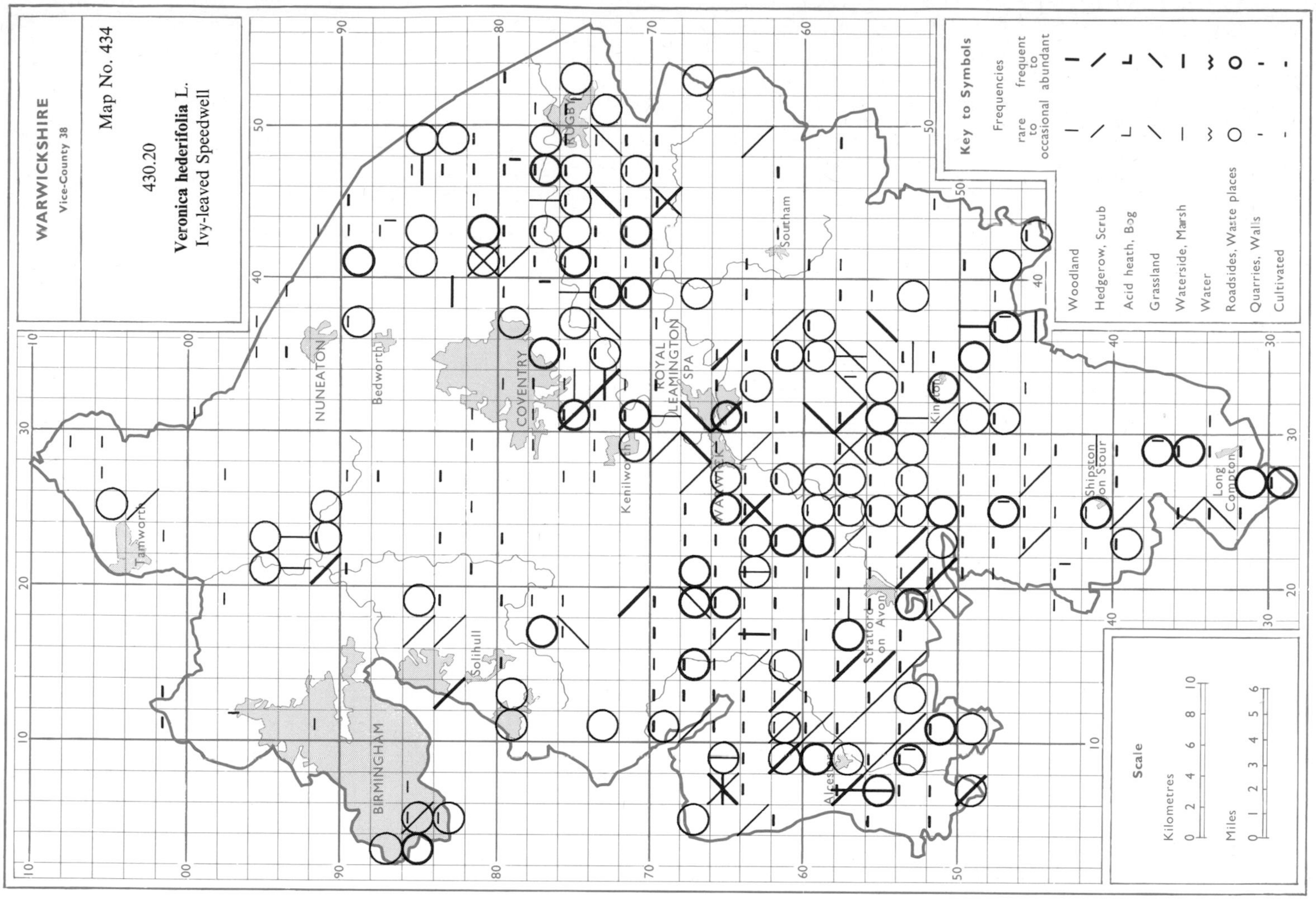
WARWICKSHIRE
Vice-County 38
Map No. 434
430.20
Veronica hederifolia L.
Ivy-leaved Speedwell
Key to Symbols
Frequencies
rare to occasional
frequent to abundant
Woodland
Hedgerow, Scrub
Acid heath, Bog
Grassland
Waterside, Marsh
Water
Roadsides, Waste places
Quarries, Walls
Cultivated
Scale
Kilometres
0 2 4 6 8 10
Miles
0 1 2 3 4 5 6
NUNEATON
Bedworth
COVENTRY
RUGBY
ROYAL LEAMINGTON SPA
Kenilworth
WARWICK
Southam
Kineton
Shipston on Stour
Long Compton
Stratford on Avon
Alcester
Solihull
BIRMINGHAM
Tamworth

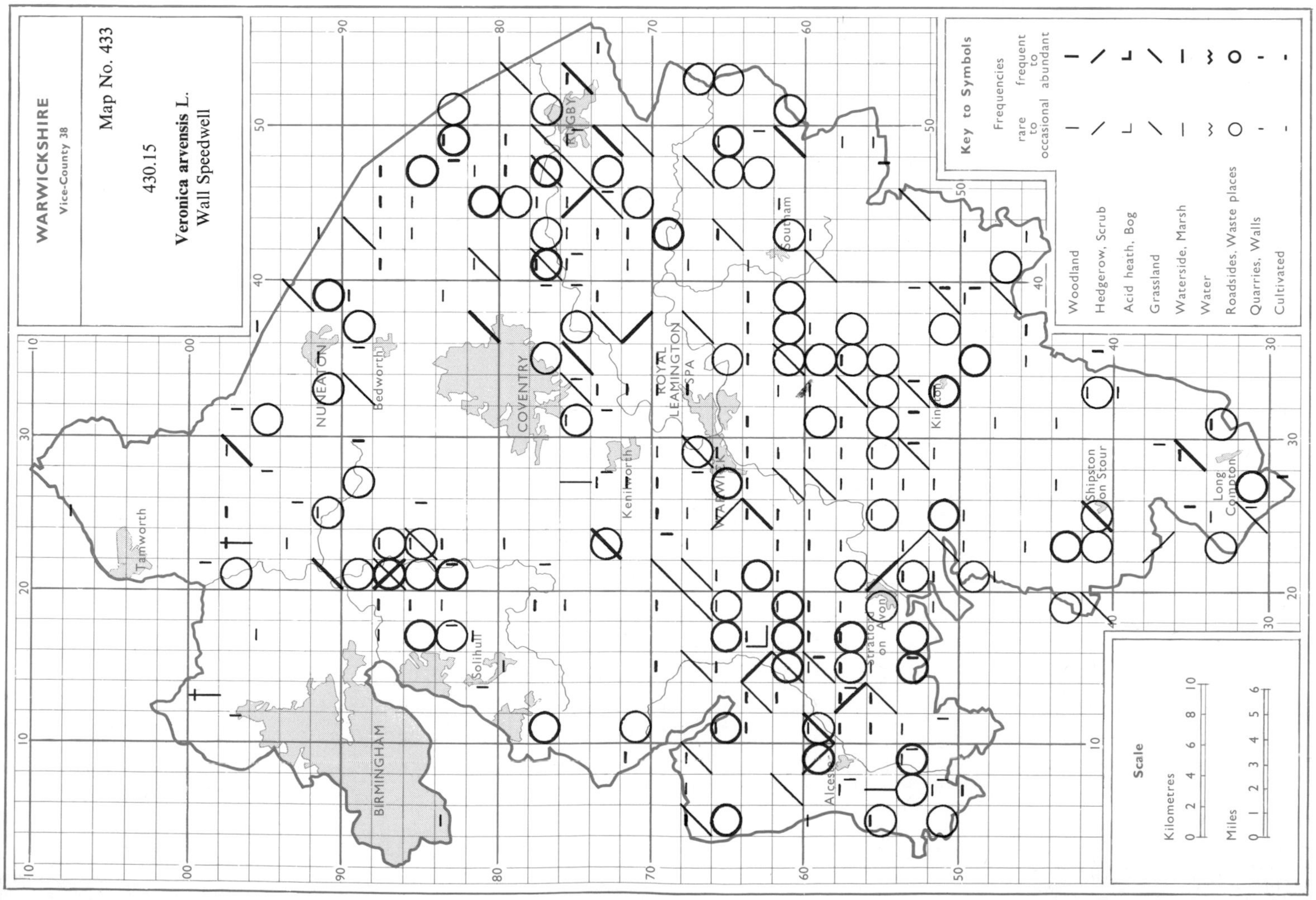
WARWICKSHIRE
Vice-County 38
Map No. 433
430.15
Veronica arvensis L.
Wall Speedwell
Key to Symbols
Frequencies
rare to occasional
frequent to abundant
Woodland
Hedgerow, Scrub
Acid heath, Bog
Grassland
Waterside, Marsh
Water
Roadsides, Waste places
Quarries, Walls
Cultivated
Scale
Kilometres
0 2 4 6 8 10
Miles
0 1 2 3 4 5 6
NUNEATON
Bedworth
COVENTRY
RUGBY
ROYAL LEAMINGTON SPA
Kenilworth
WARWICK
Southam
Kineton
Shipston on Stour
Long Compton
Stratford on Avon
Alcester
Solihull
BIRMINGHAM
Tamworth

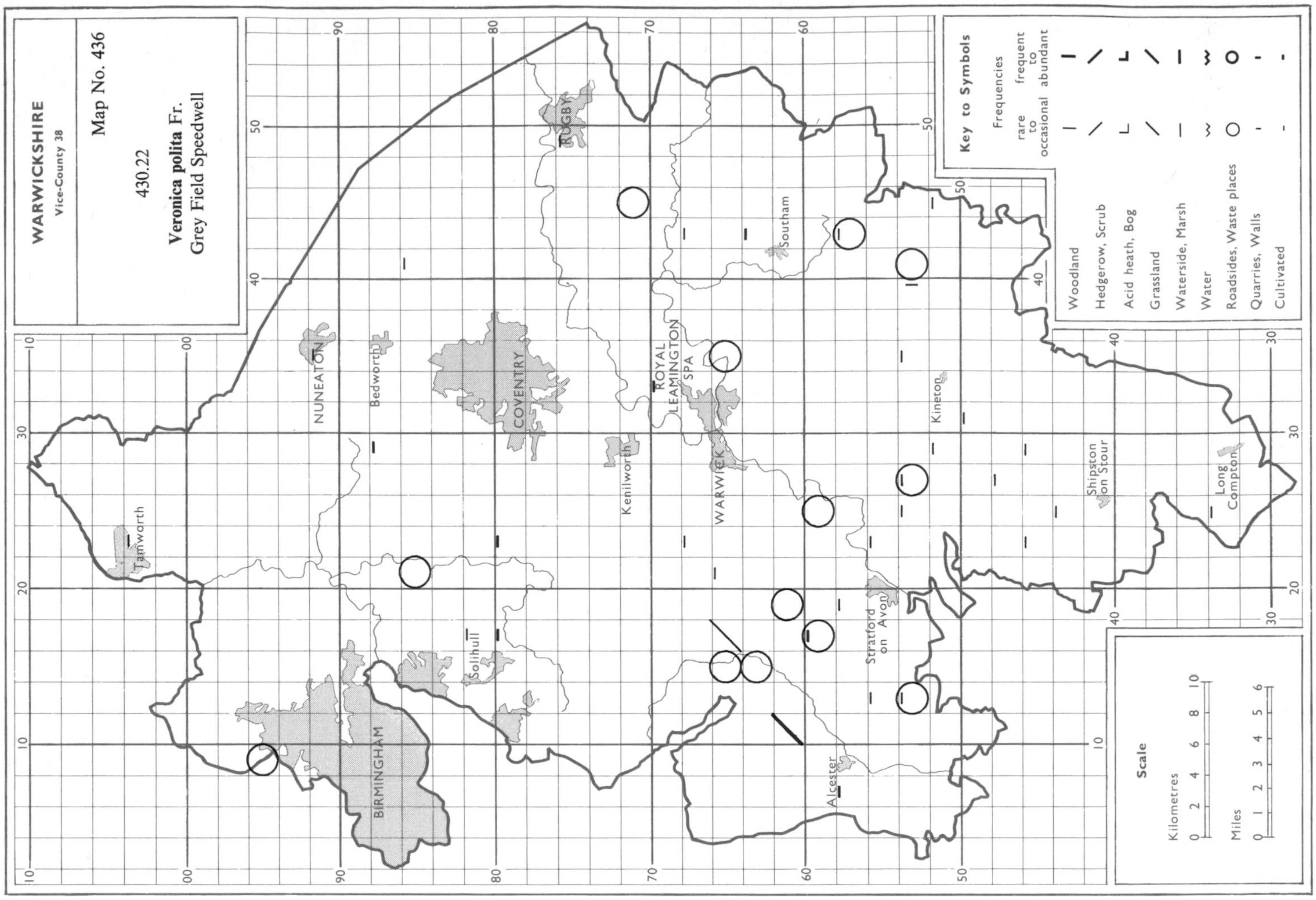
WARWICKSHIRE
Vice-County 38
Map No. 436
430.22
Veronica polita Fr.
Grey Field Speedwell
Key to Symbols
Frequencies
rare to occasional
frequent to abundant
Woodland
Hedgerow, Scrub
Acid heath, Bog
Grassland
Waterside, Marsh
Water
Roadsides, Waste places
Quarries, Walls
Cultivated
Scale
Kilometres
0 2 4 6 8 10
Miles
0 1 2 3 4 5 6
NUNEATON
Bedworth
COVENTRY
RUGBY
Southam
ROYAL LEAMINGTON SPA
Kenilworth
WARWICK
Kineton
Tamworth
Solihull
BIRMINGHAM
Stratford on Avon
Alcester
Shipston on Stour
Long Compton

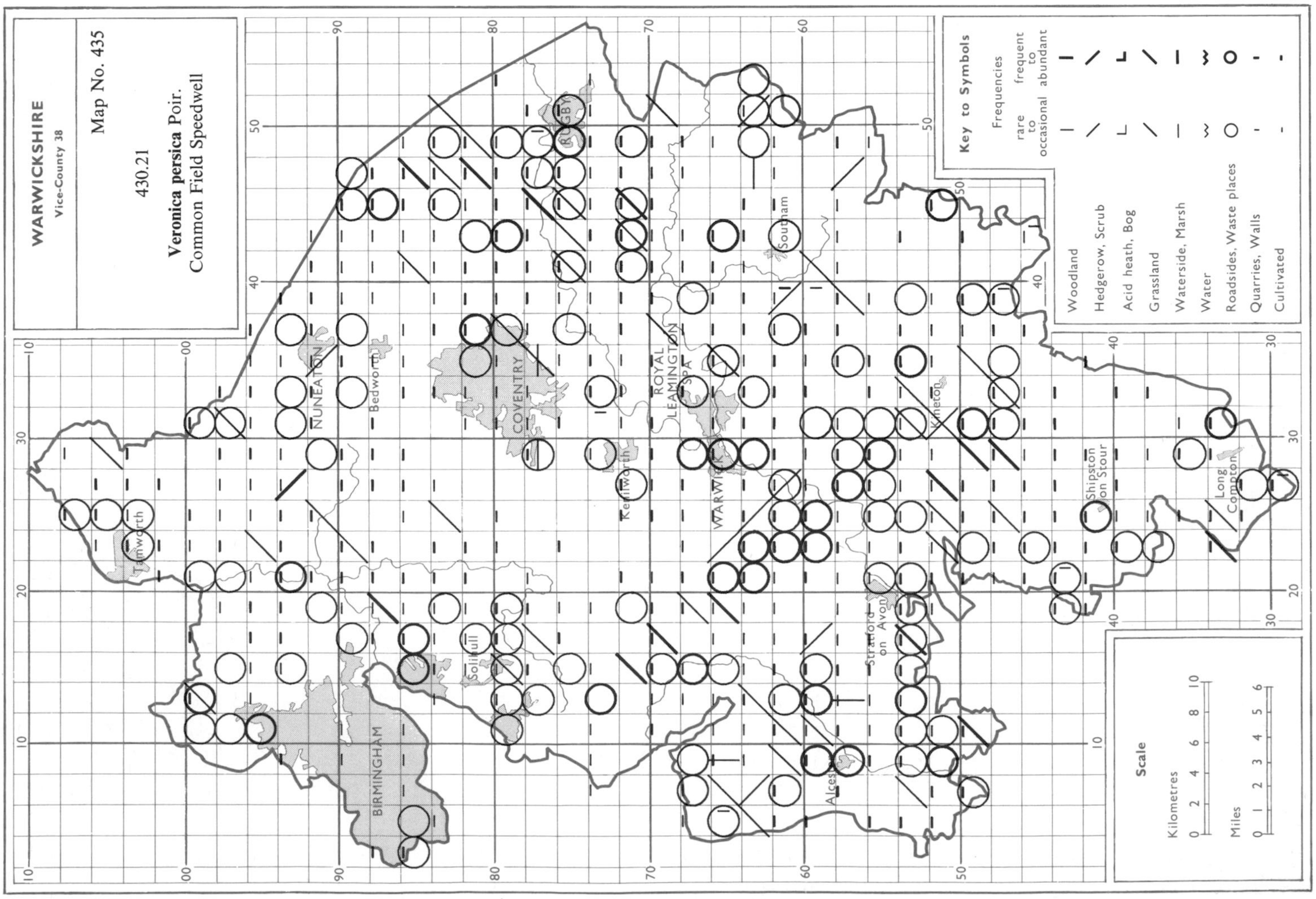
WARWICKSHIRE
Vice-County 38
Map No. 435
430.21
Veronica persica Poir.
Common Field Speedwell
Key to Symbols
Frequencies
rare to occasional
frequent to abundant
Woodland
Hedgerow, Scrub
Acid heath, Bog
Grassland
Waterside, Marsh
Water
Roadsides, Waste places
Quarries, Walls
Cultivated
Scale
Kilometres
0 2 4 6 8 10
Miles
0 1 2 3 4 5 6
NUNEATON
Bedworth
COVENTRY
RUGBY
Southam
ROYAL LEAMINGTON SPA
Kenilworth
WARWICK
Kineton
Tamworth
Solihull
BIRMINGHAM
Stratford on Avon
Alcester
Shipston on Stour
Long Compton

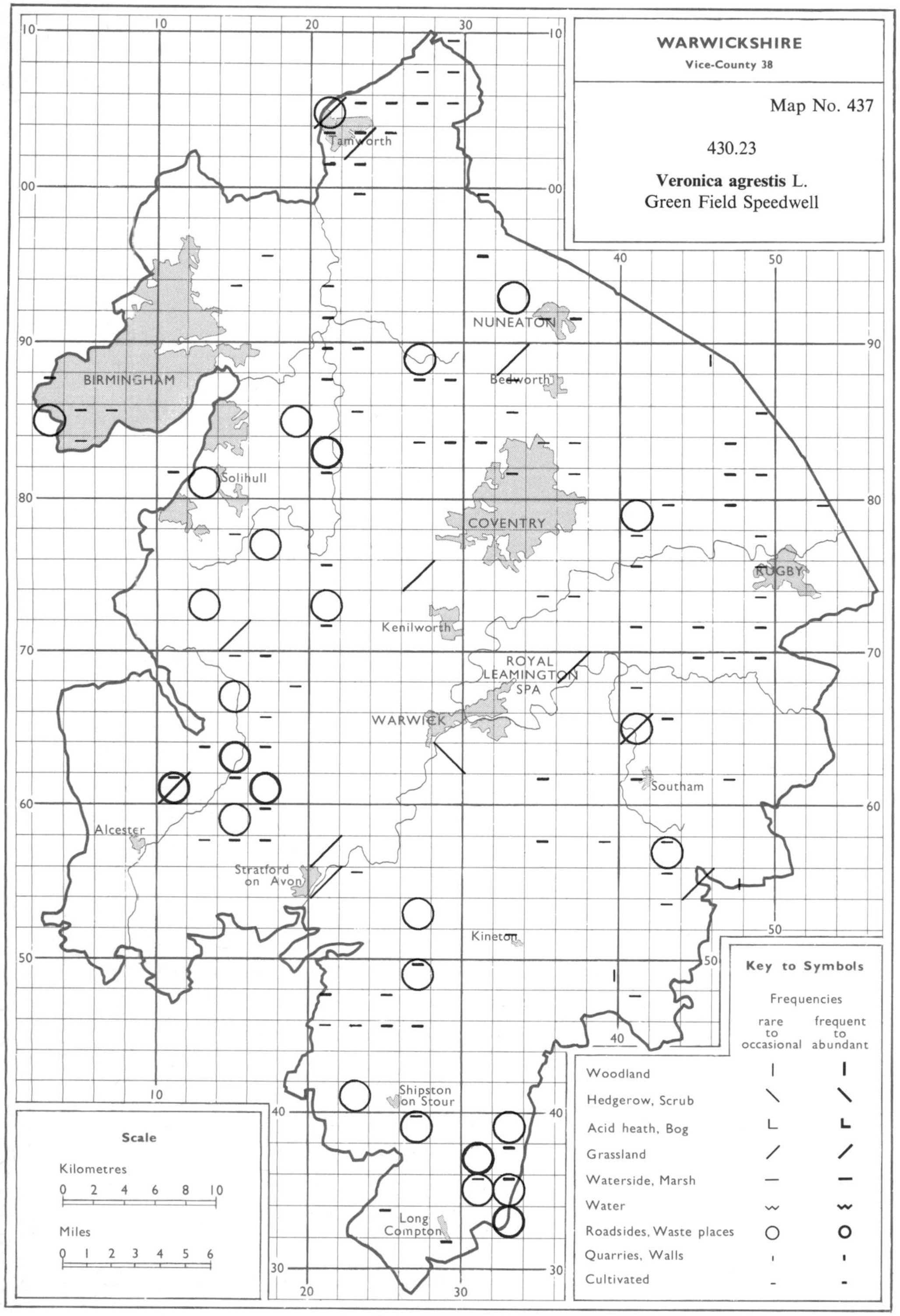

WARWICKSHIRE
Vice-County 38
Map No. 437
430.23
Veronica agrestis L.
Green Field Speedwell
Tamworth
NUNEATON
Bedworth
BIRMINGHAM
Solihull
COVENTRY
RUGBY
Kenilworth
ROYAL LEAMINGTON SPA
WARWICK
Southam
Alcester
Stratford on Avon
Kineton
Shipston on Stour
Long Compton
Key to Symbols
Frequencies
rare to occasional
frequent to abundant
Woodland
Hedgerow, Scrub
Acid heath, Bog
Grassland
Waterside, Marsh
Water
Roadsides, Waste places
Quarries, Walls
Cultivated
Scale
Kilometres
0 2 4 6 8 10
Miles
0 1 2 3 4 5 6

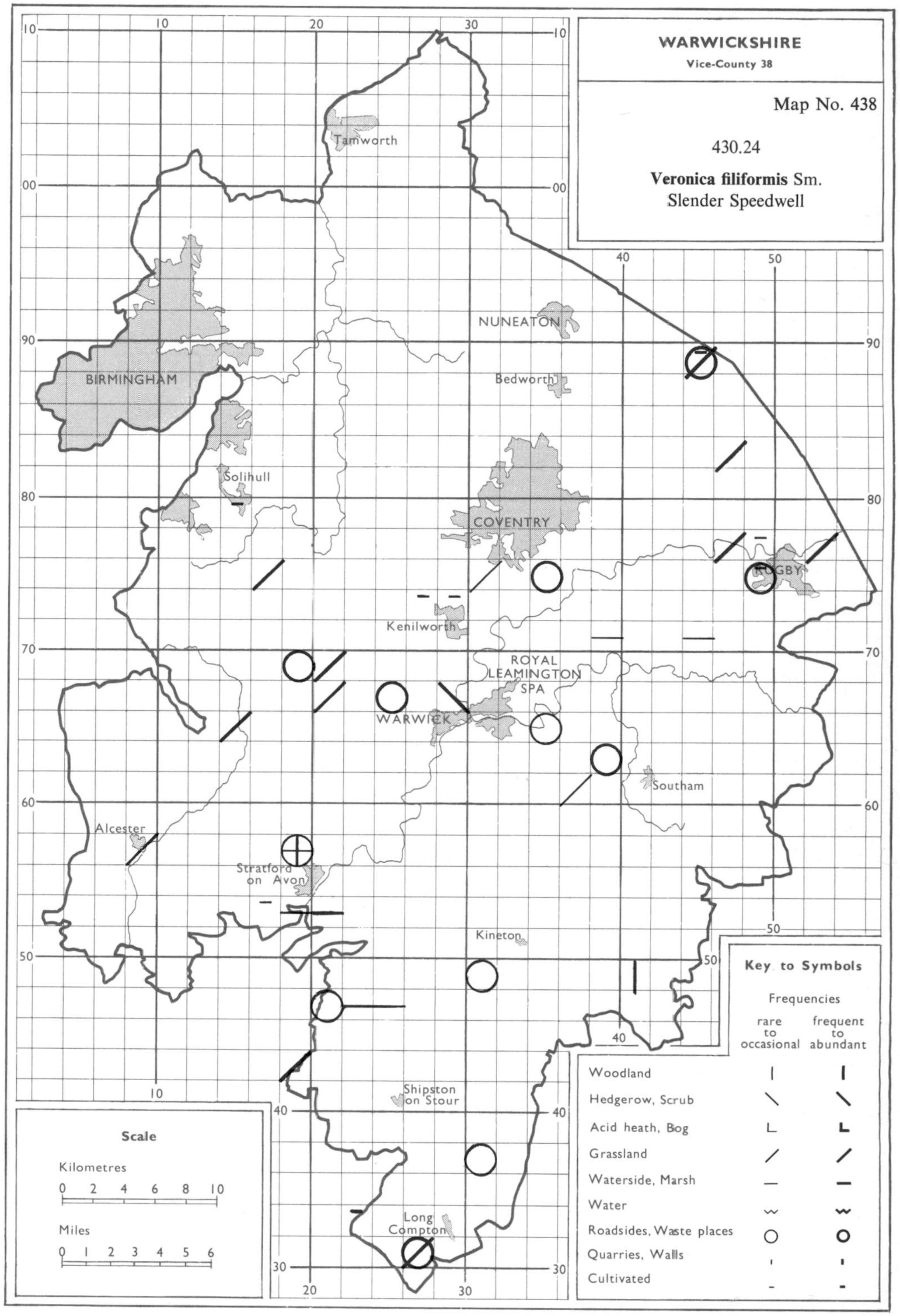

WARWICKSHIRE
Vice-County 38
Map No. 438
430.24
Veronica filiformis Sm.
Slender Speedwell
Tamworth
NUNEATON
Bedworth
BIRMINGHAM
Solihull
COVENTRY
RUGBY
Kenilworth
ROYAL LEAMINGTON SPA
WARWICK
Southam
Alcester
Stratford on Avon
Kineton
Shipston on Stour
Long Compton
Key to Symbols
Frequencies
rare to occasional
frequent to abundant
Woodland
Hedgerow, Scrub
Acid heath, Bog
Grassland
Waterside, Marsh
Water
Roadsides, Waste places
Quarries, Walls
Cultivated
Scale
Kilometres
0 2 4 6 8 10
Miles
0 1 2 3 4 5 6

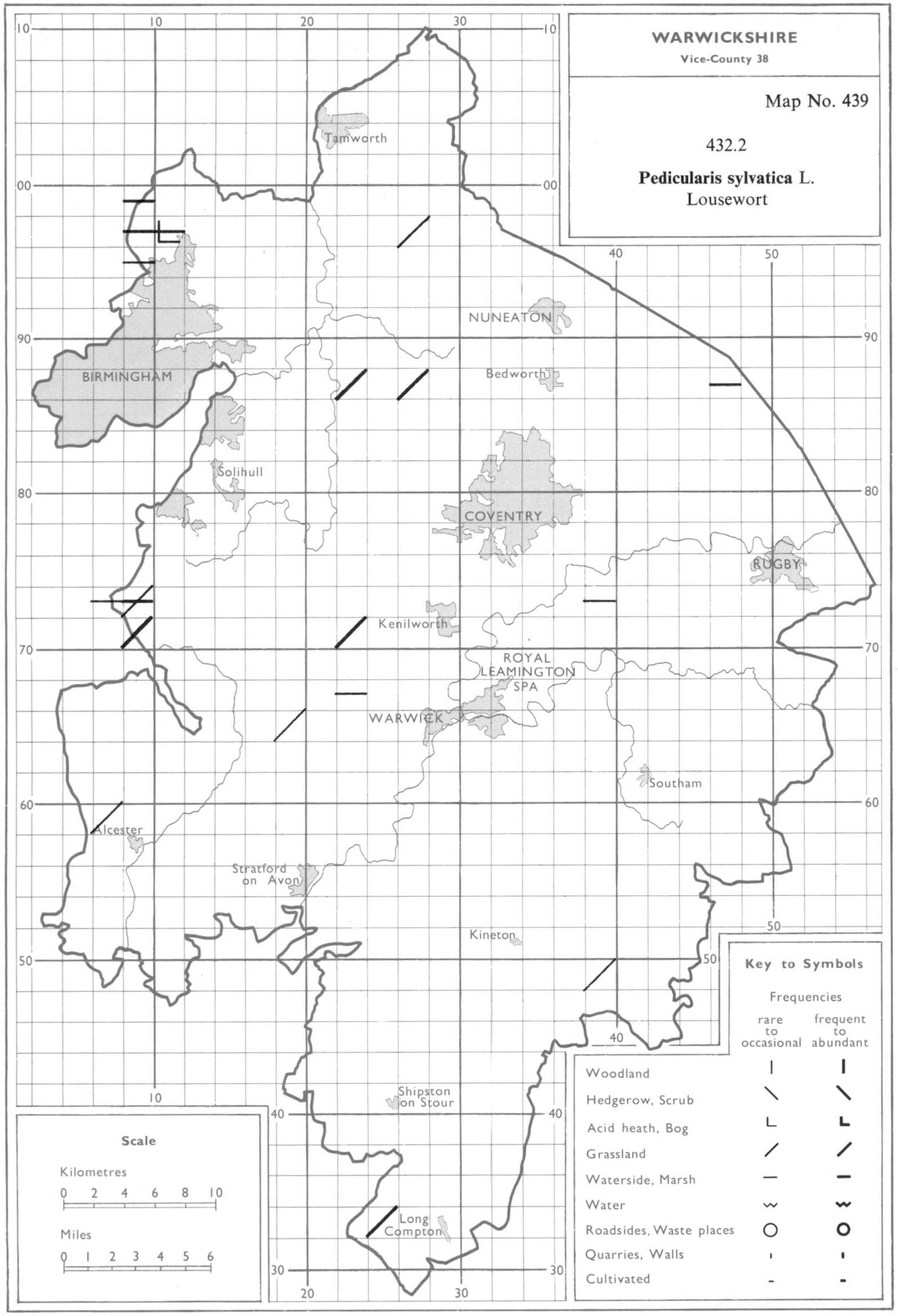
WARWICKSHIRE
Vice-County 38
Map No. 439
432.2
Pedicularis sylvatica L.
Lousewort
Tamworth
NUNEATON
Bedworth
BIRMINGHAM
Solihull
COVENTRY
RUGBY
Kenilworth
ROYAL LEAMINGTON SPA
WARWICK
Southam
Alcester
Stratford on Avon
Kineton
Shipston on Stour
Long Compton
Scale
Kilometres
0 2 4 6 8 10
Miles
0 1 2 3 4 5 6
Key to Symbols
Frequencies
rare to occasional
frequent to abundant
Woodland
Hedgerow, Scrub
Acid heath, Bog
Grassland
Waterside, Marsh
Water
Roadsides, Waste places
Quarries, Walls
Cultivated

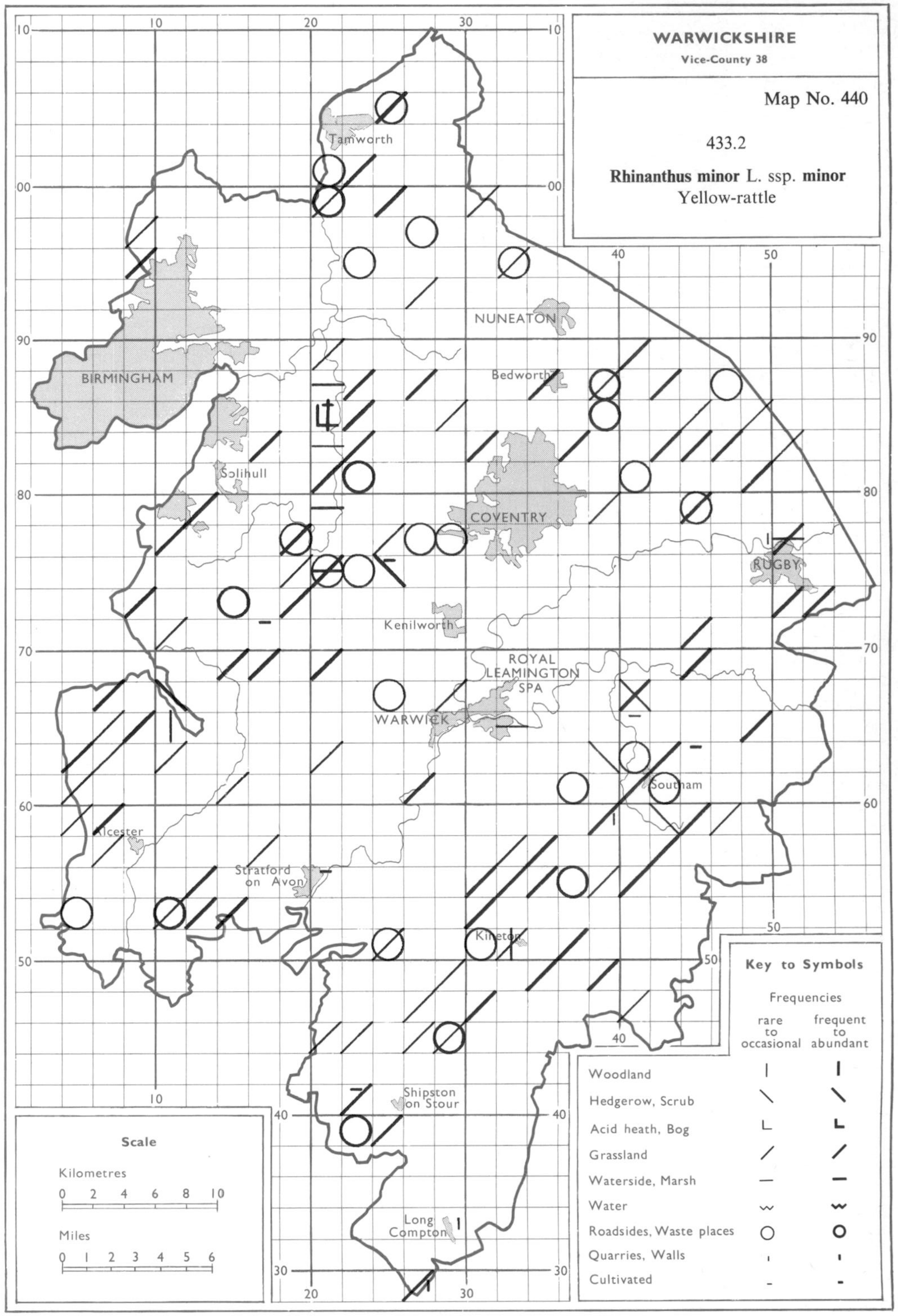
WARWICKSHIRE
Vice-County 38
Map No. 440
433.2
Rhinanthus minor L. ssp. minor
Yellow-rattle
Tamworth
NUNEATON
Bedworth
BIRMINGHAM
Solihull
COVENTRY
RUGBY
Kenilworth
ROYAL LEAMINGTON SPA
WARWICK
Southam
Alcester
Stratford on Avon
Kineton
Shipston on Stour
Long Compton
Scale
Kilometres
0 2 4 6 8 10
Miles
0 1 2 3 4 5 6
Key to Symbols
Frequencies
rare to occasional
frequent to abundant
Woodland
Hedgerow, Scrub
Acid heath, Bog
Grassland
Waterside, Marsh
Water
Roadsides, Waste places
Quarries, Walls
Cultivated

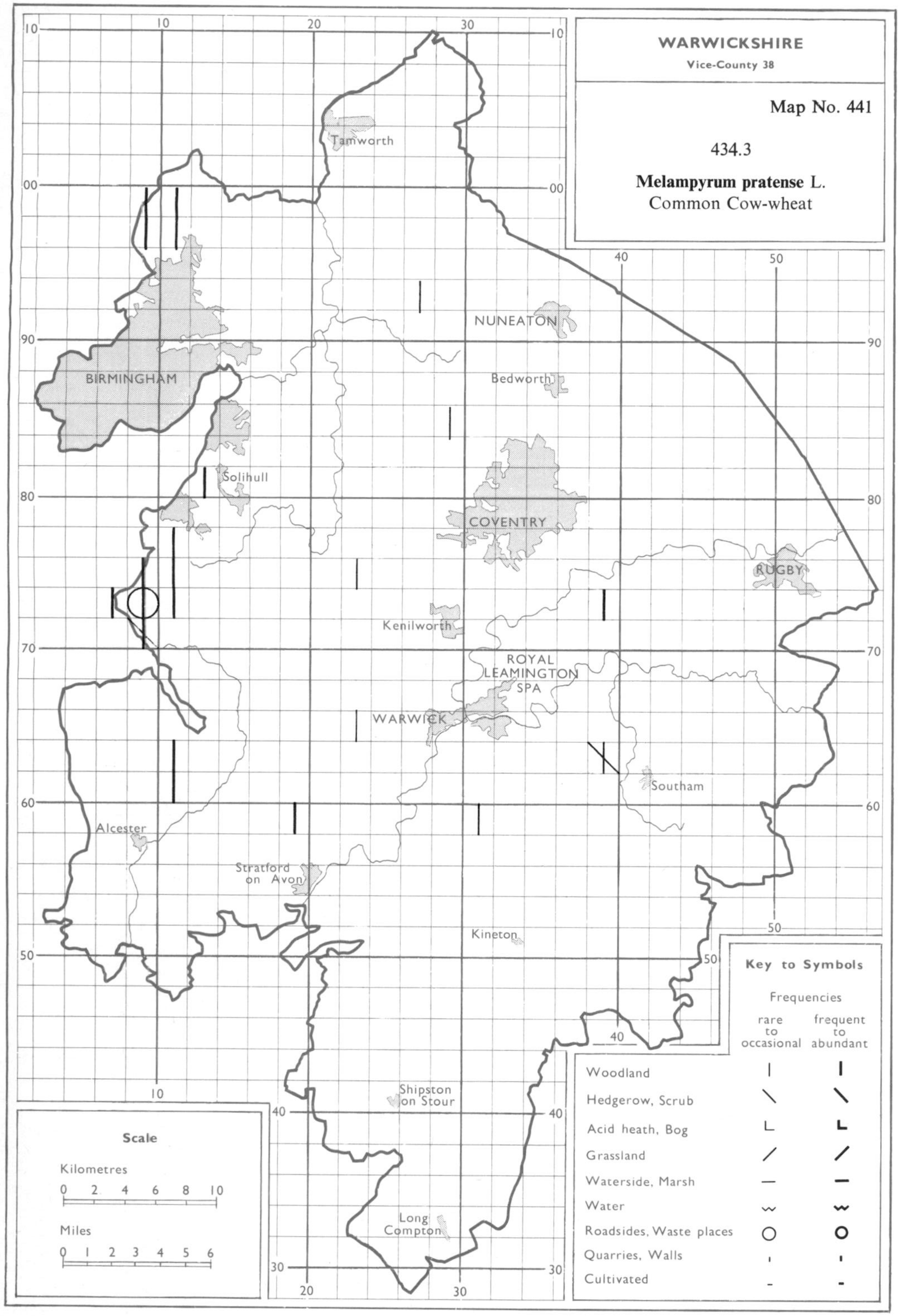
WARWICKSHIRE
Vice-County 38
Map No. 441
434.3
Melampyrum pratense L.
Common Cow-wheat
Tamworth
NUNEATON
Bedworth
BIRMINGHAM
Solihull
COVENTRY
RUGBY
Kenilworth
ROYAL LEAMINGTON SPA
WARWICK
Southam
Alcester
Stratford on Avon
Kineton
Shipston on Stour
Long Compton
Key to Symbols
Frequencies
rare to occasional
frequent to abundant
Woodland
Hedgerow, Scrub
Acid heath, Bog
Grassland
Waterside, Marsh
Water
Roadsides, Waste places
Quarries, Walls
Cultivated
Scale
Kilometres
0 2 4 6 8 10
Miles
0 1 2 3 4 5 6

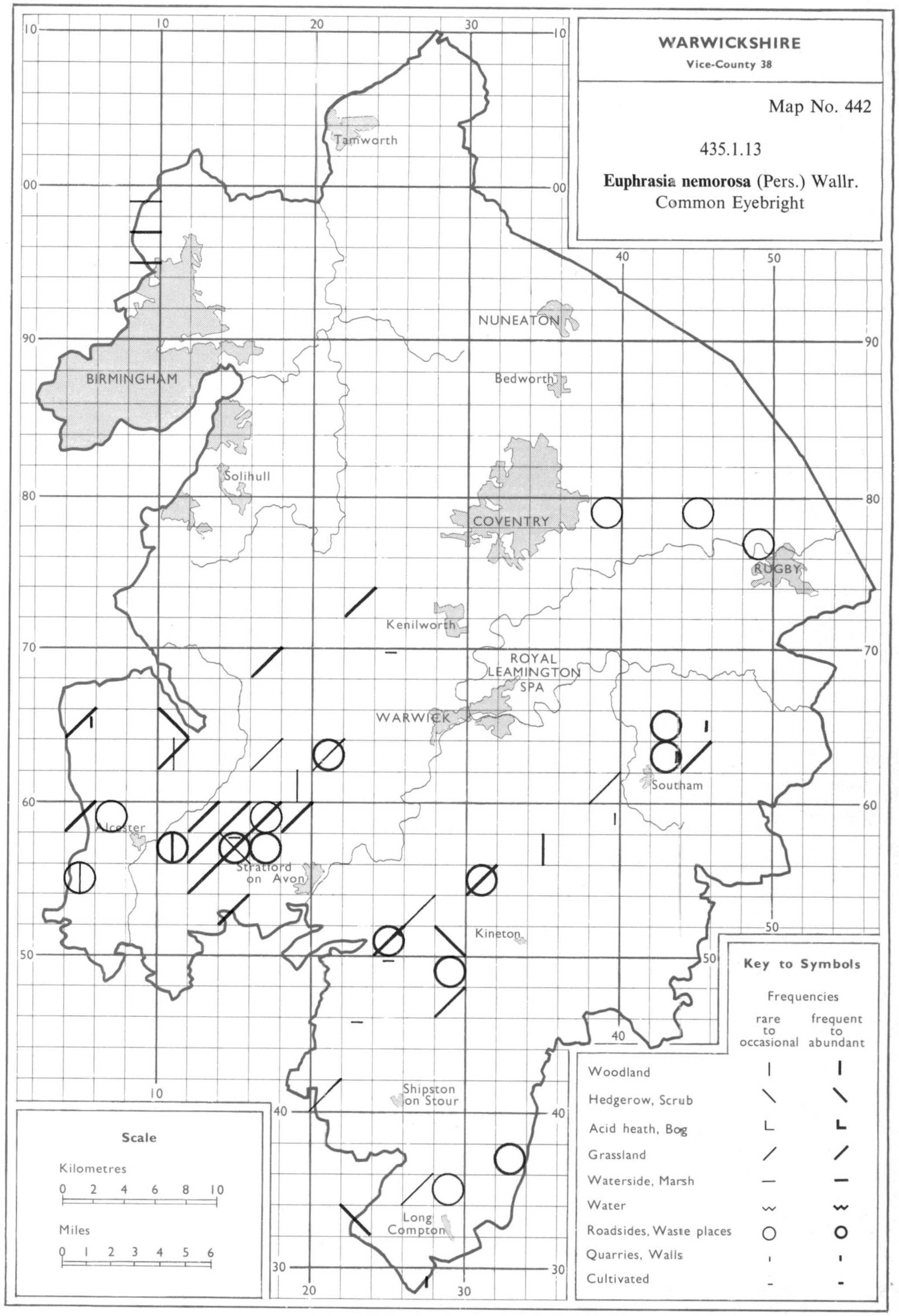
WARWICKSHIRE
Vice-County 38
Map No. 442
435.1.13
Euphrasia nemorosa (Pers.) Wallr.
Common Eyebright
Tamworth
NUNEATON
Bedworth
BIRMINGHAM
Solihull
COVENTRY
RUGBY
Kenilworth
ROYAL LEAMINGTON SPA
WARWICK
Southam
Alcester
Stratford on Avon
Kineton
Shipston on Stour
Long Compton
Key to Symbols
Frequencies
rare to occasional
frequent to abundant
Woodland
Hedgerow, Scrub
Acid heath, Bog
Grassland
Waterside, Marsh
Water
Roadsides, Waste places
Quarries, Walls
Cultivated
Scale
Kilometres
0 2 4 6 8 10
Miles
0 1 2 3 4 5 6

WARWICKSHIRE
Vice-County 38
Map No. 444
445.3
Mentha arvensis L.
Corn Mint
Key to Symbols
Frequencies
rare to occasional
frequent to abundant
Woodland
Hedgerow, Scrub
Acid heath, Bog
Grassland
Waterside, Marsh
Water
Roadsides, Waste places
Quarries, Walls
Cultivated
Scale
Kilometres
0 2 4 6 8 10
Miles
0 1 2 3 4 5 6
NUNEATON
Bedworth
COVENTRY
RUGBY
Tamworth
BIRMINGHAM
Solihull
Kenilworth
ROYAL LEAMINGTON SPA
WARWICK
Southam
Kineton
Stratford on Avon
Alcester
Shipston on Stour
Long Compton

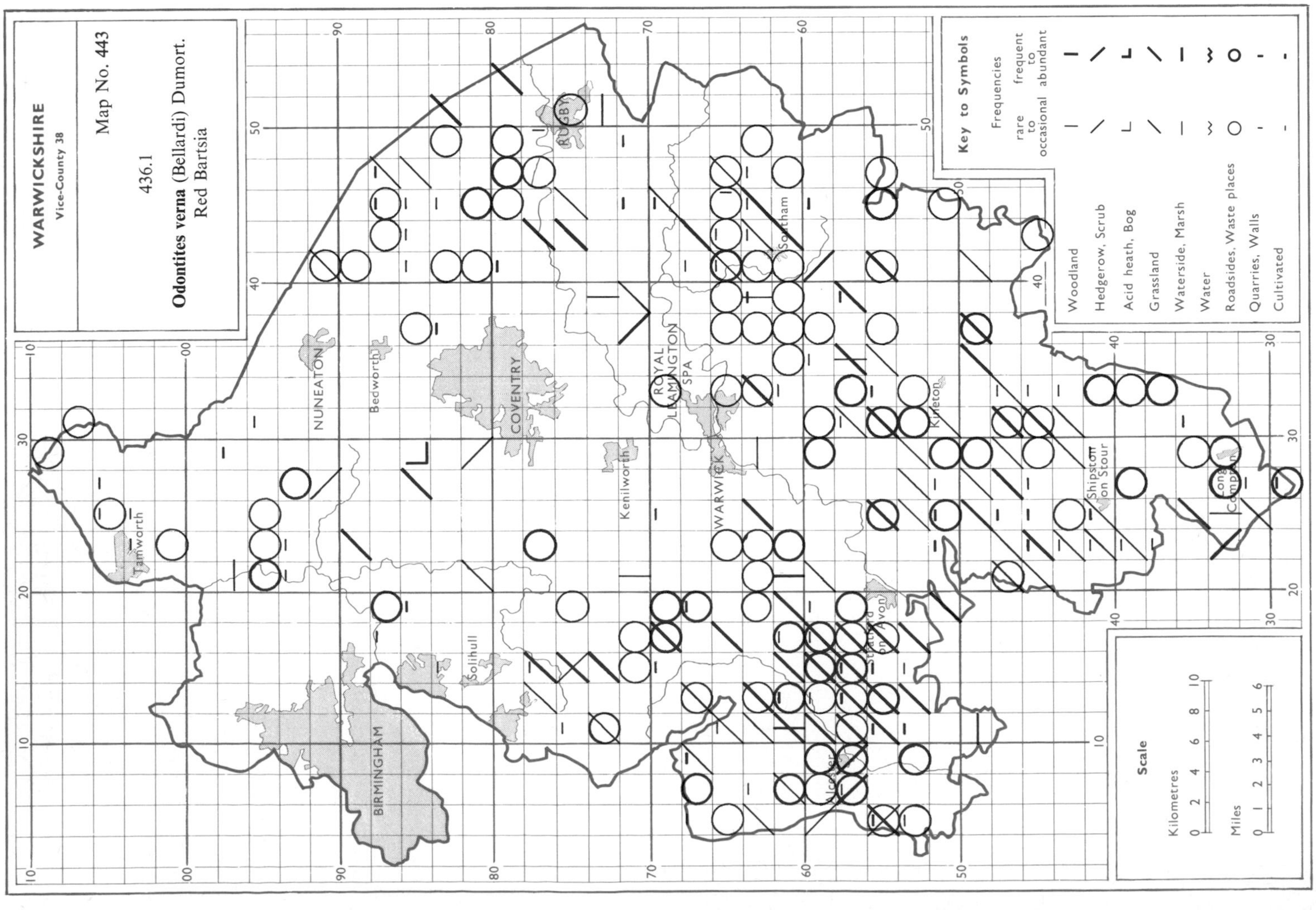
WARWICKSHIRE
Vice-County 38
Map No. 443
436.1
Odontites verna (Bellardi) Dumort.
Red Bartsia
Key to Symbols
Frequencies
rare to occasional
frequent to abundant
Woodland
Hedgerow, Scrub
Acid heath, Bog
Grassland
Waterside, Marsh
Water
Roadsides, Waste places
Quarries, Walls
Cultivated
Scale
Kilometres
0 2 4 6 8 10
Miles
0 1 2 3 4 5 6
NUNEATON
Bedworth
COVENTRY
RUGBY
Tamworth
BIRMINGHAM
Solihull
Kenilworth
ROYAL LEAMINGTON SPA
WARWICK
Southam
Kineton
Stratford on Avon
Alcester
Shipston on Stour
Long Compton

WARWICKSHIRE
Vice-County 38
Map No. 445
445.4
Mentha aquatica L.
Water Mint
Tamworth
NUNEATON
Bedworth
BIRMINGHAM
Solihull
COVENTRY
RUGBY
Kenilworth
ROYAL LEAMINGTON SPA
WARWICK
Southam
Alcester
Stratford on Avon
Kineton
Shipston on Stour
Long Compton
Key to Symbols
Frequencies
rare to occasional
frequent to abundant
Woodland
Hedgerow, Scrub
Acid heath, Bog
Grassland
Waterside, Marsh
Water
Roadsides, Waste places
Quarries, Walls
Cultivated
Scale
Kilometres
0 2 4 6 8 10
Miles
0 1 2 3 4 5 6

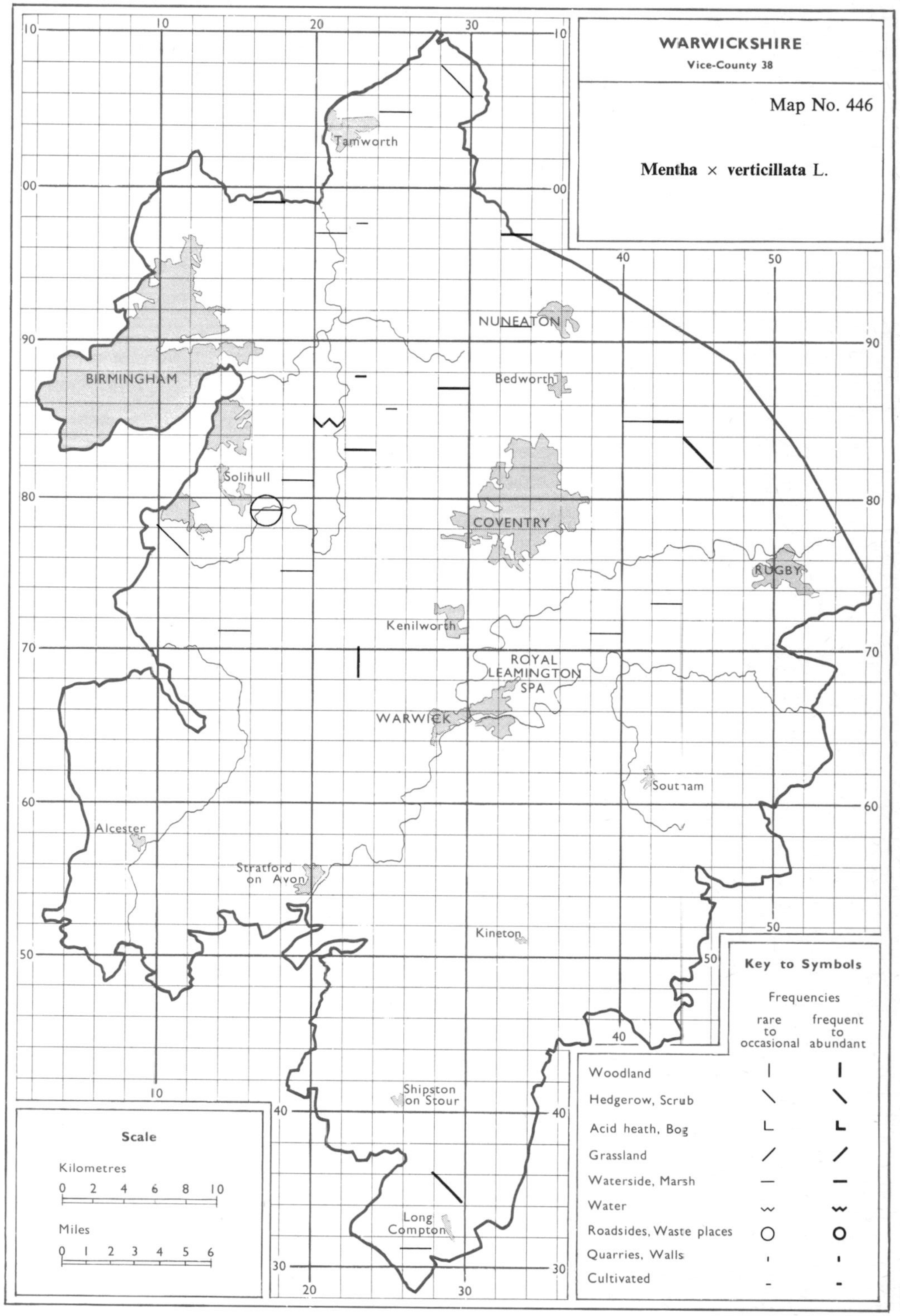
WARWICKSHIRE
Vice-County 38
Map No. 446
Mentha × verticillata L.
Tamworth
NUNEATON
Bedworth
BIRMINGHAM
Solihull
COVENTRY
RUGBY
Kenilworth
ROYAL LEAMINGTON SPA
WARWICK
Southam
Alcester
Stratford on Avon
Kineton
Shipston on Stour
Long Compton
Key to Symbols
Frequencies
rare to occasional
frequent to abundant
Woodland
Hedgerow, Scrub
Acid heath, Bog
Grassland
Waterside, Marsh
Water
Roadsides, Waste places
Quarries, Walls
Cultivated
Scale
Kilometres
0 2 4 6 8 10
Miles
0 1 2 3 4 5 6

WARWICKSHIRE
Vice-County 38
Map No. 447
445.5 & 6
Mentha spicata L.
Spearmint
Key to Symbols
Frequencies
rare to occasional
frequent to abundant
Woodland
Hedgerow, Scrub
Acid heath, Bog
Grassland
Waterside, Marsh
Water
Roadsides, Waste places
Quarries, Walls
Cultivated
Scale
Kilometres
Miles
NUNEATON
Bedworth
COVENTRY
RUGBY
Southam
ROYAL LEAMINGTON SPA
WARWICK
Kenilworth
Kineton
Shipston on Stour
Long Compton
Tamworth
Solihull
BIRMINGHAM
Stratford on Avon
Alcester

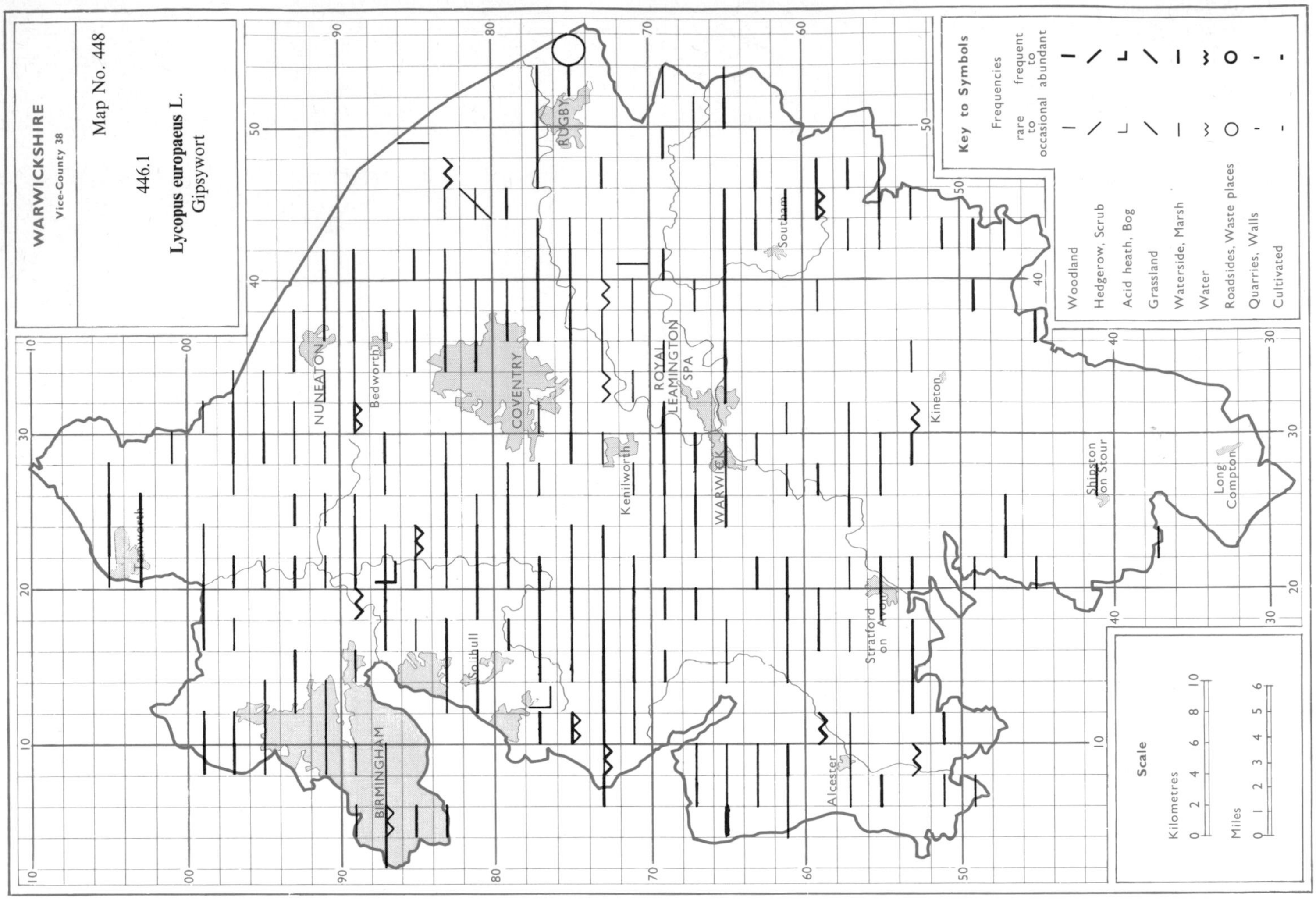
WARWICKSHIRE
Vice-County 38
Map No. 448
446.1
Lycopus europaeus L.
Gipsywort
Key to Symbols
Frequencies
rare to occasional
frequent to abundant
Woodland
Hedgerow, Scrub
Acid heath, Bog
Grassland
Waterside, Marsh
Water
Roadsides, Waste places
Quarries, Walls
Cultivated
Scale
Kilometres
Miles
NUNEATON
Bedworth
COVENTRY
RUGBY
Southam
ROYAL LEAMINGTON SPA
WARWICK
Kenilworth
Kineton
Shipston on Stour
Long Compton
Tamworth
Solihull
BIRMINGHAM
Stratford on Avon
Alcester

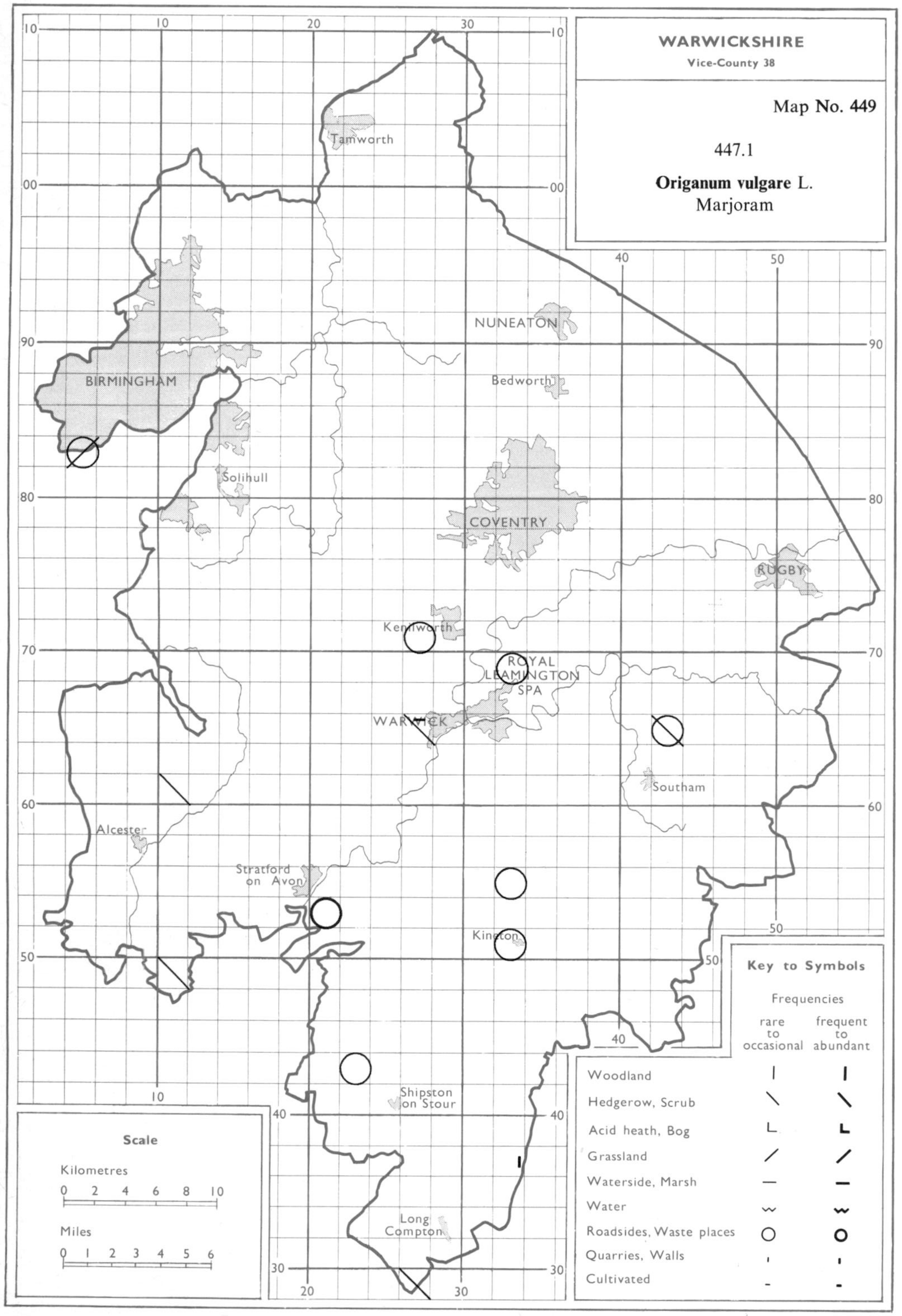
WARWICKSHIRE
Vice-County 38
Map No. 449
447.1
Origanum vulgare L.
Marjoram
Tamworth
NUNEATON
Bedworth
BIRMINGHAM
Solihull
COVENTRY
RUGBY
Kenilworth
ROYAL LEAMINGTON SPA
WARWICK
Southam
Alcester
Stratford on Avon
Kineton
Shipston on Stour
Long Compton
Key to Symbols
Frequencies
rare to occasional
frequent to abundant
Woodland
Hedgerow, Scrub
Acid heath, Bog
Grassland
Waterside, Marsh
Water
Roadsides, Waste places
Quarries, Walls
Cultivated
Scale
Kilometres
0 2 4 6 8 10
Miles
0 1 2 3 4 5 6

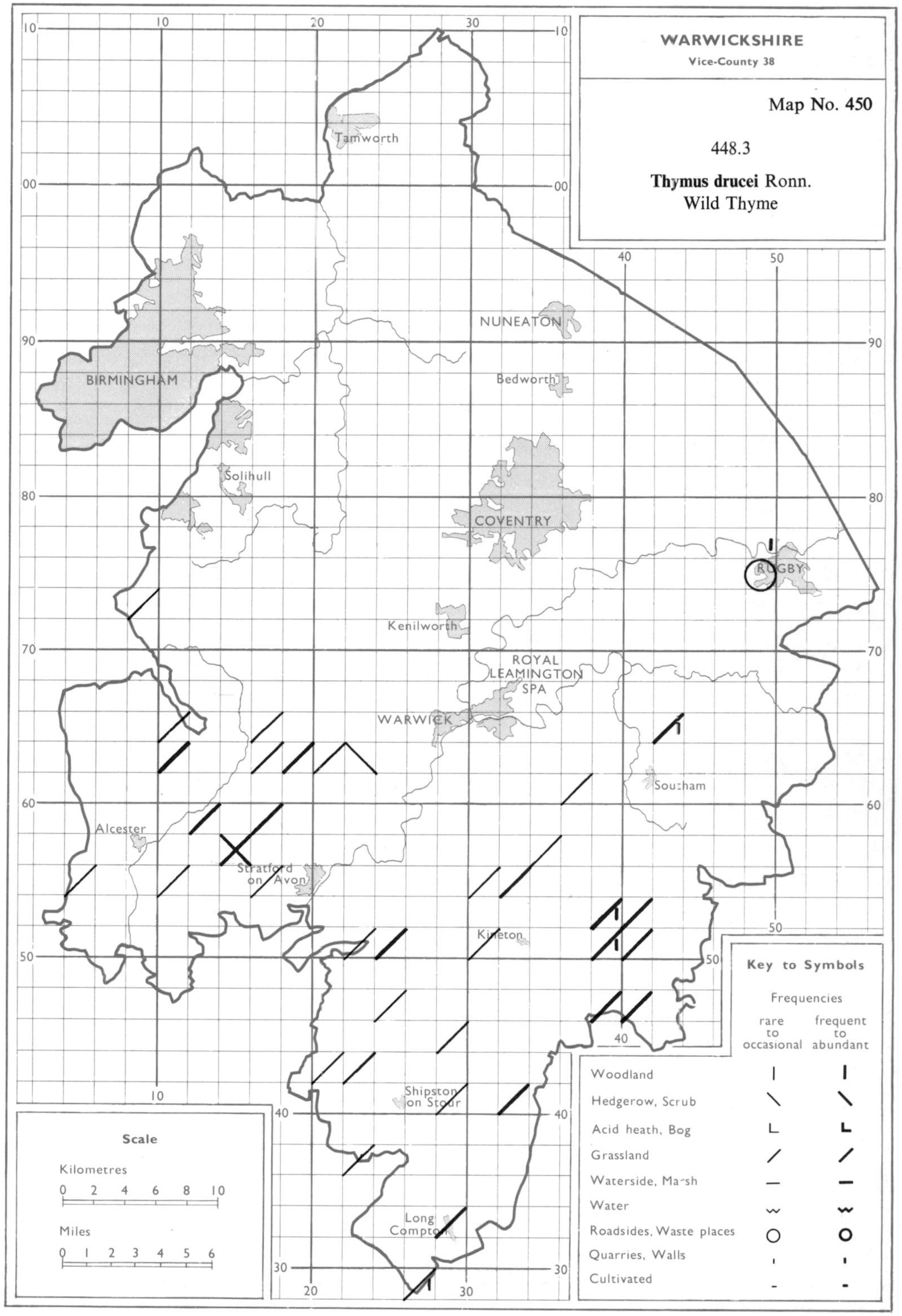
WARWICKSHIRE
Vice-County 38
Map No. 450
448.3
Thymus drucei Ronn.
Wild Thyme
Tamworth
NUNEATON
Bedworth
BIRMINGHAM
Solihull
COVENTRY
RUGBY
Kenilworth
ROYAL LEAMINGTON SPA
WARWICK
Southam
Alcester
Stratford on Avon
Kineton
Shipston on Stour
Long Compton
Key to Symbols
Frequencies
rare to occasional
frequent to abundant
Woodland
Hedgerow, Scrub
Acid heath, Bog
Grassland
Waterside, Ma-sh
Water
Roadsides, Waste places
Quarries, Walls
Cultivated
Scale
Kilometres
0 2 4 6 8 10
Miles
0 1 2 3 4 5 6

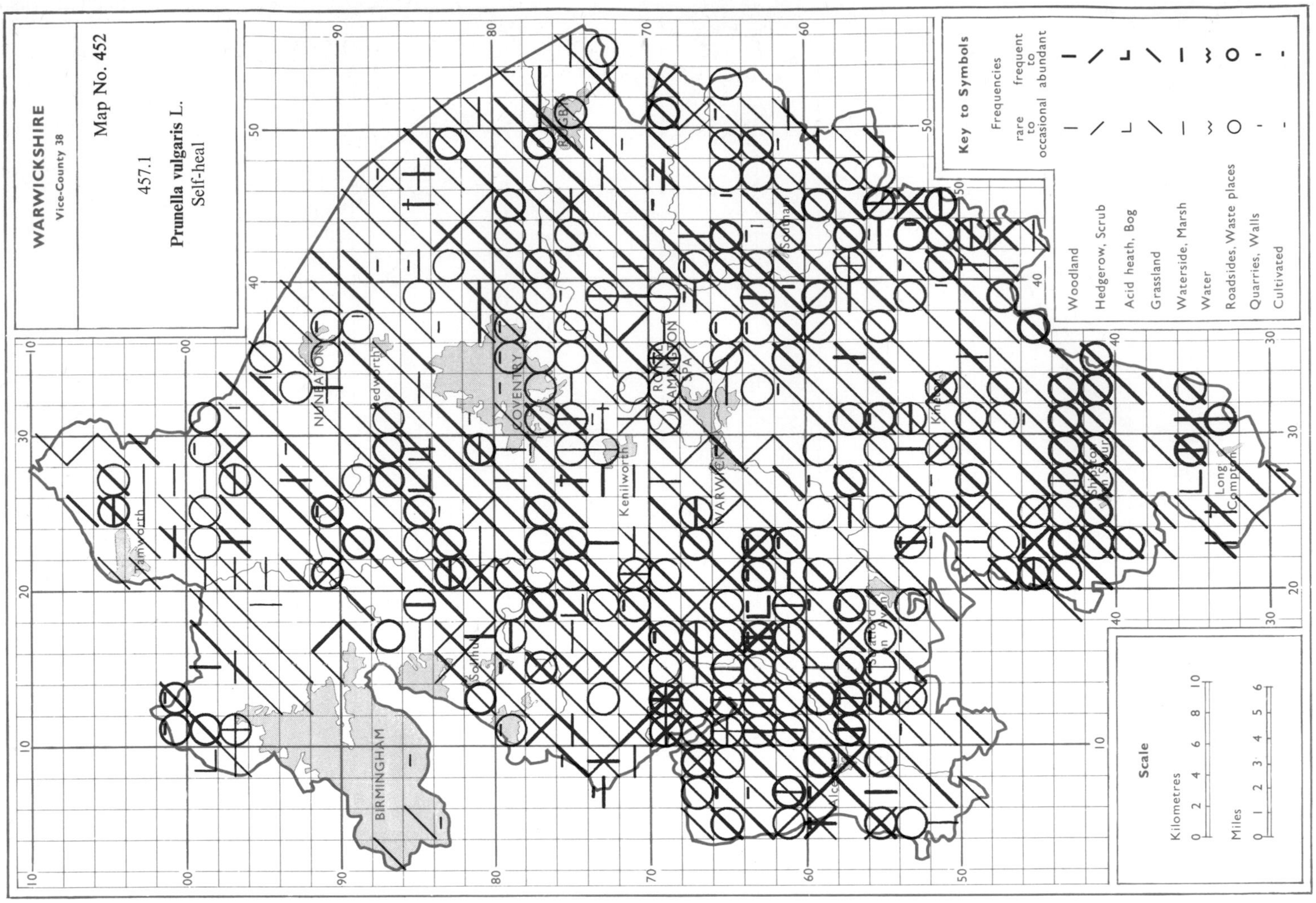
WARWICKSHIRE
Vice-County 38
Map No. 452
457.1
Prunella vulgaris L.
Self-heal
Key to Symbols
Frequencies
rare to occasional
frequent to abundant
Woodland
Hedgerow, Scrub
Acid heath, Bog
Grassland
Waterside, Marsh
Water
Roadsides, Waste places
Quarries, Walls
Cultivated
Scale
Kilometres
Miles
NUNEATON
Bedworth
COVENTRY
RUGBY
Southam
ROYAL LEAMINGTON SPA
Kenilworth
WARWICK
Kineton
Shipston on Stour
Long Compton
Tamworth
Solihull
BIRMINGHAM
Stratford on Avon
Alcester

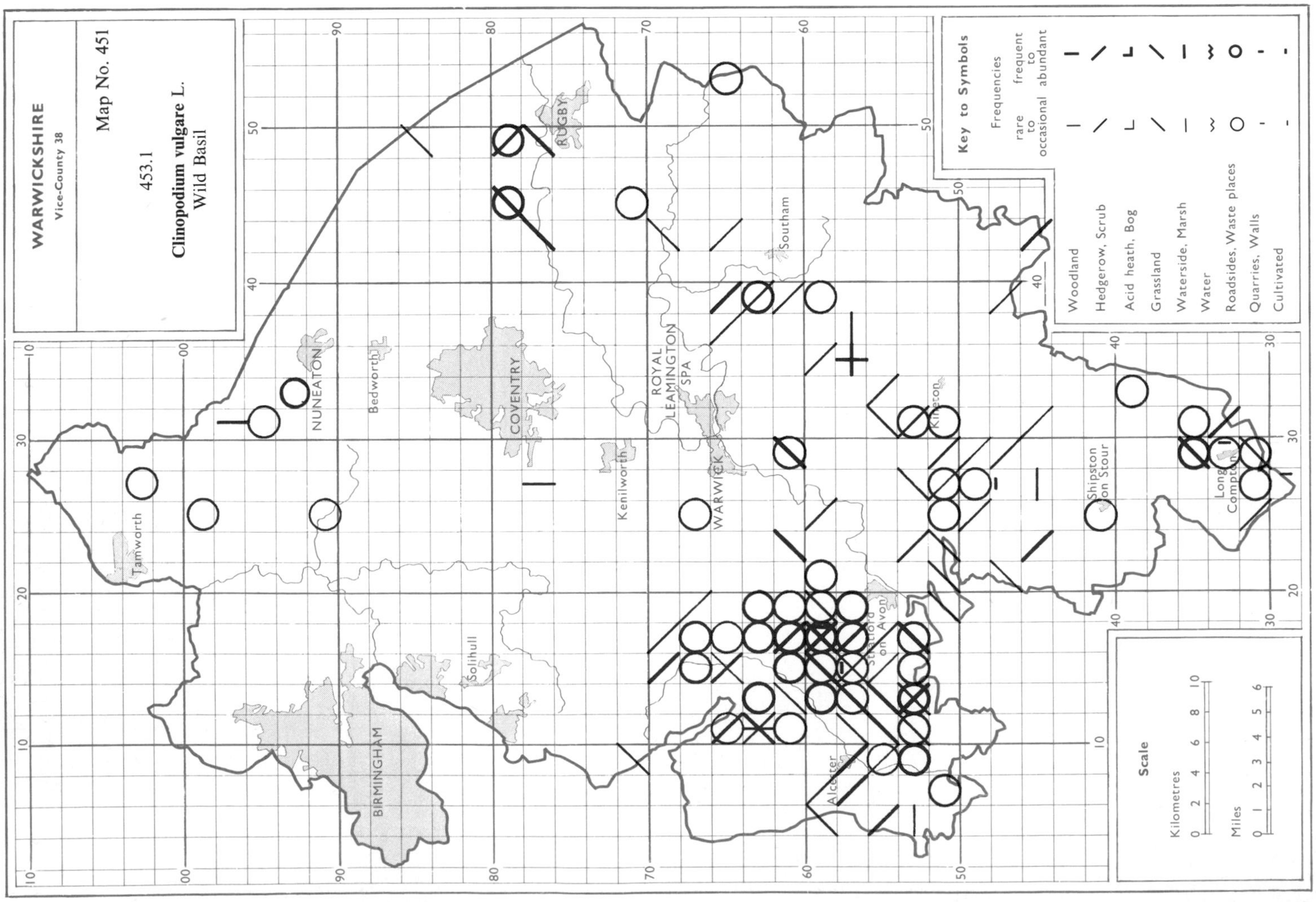
WARWICKSHIRE
Vice-County 38
Map No. 451
453.1
Clinopodium vulgare L.
Wild Basil
Key to Symbols
Frequencies
rare to occasional
frequent to abundant
Woodland
Hedgerow, Scrub
Acid heath, Bog
Grassland
Waterside, Marsh
Water
Roadsides, Waste places
Quarries, Walls
Cultivated
Scale
Kilometres
Miles
NUNEATON
Bedworth
COVENTRY
RUGBY
Southam
ROYAL LEAMINGTON SPA
Kenilworth
WARWICK
Kineton
Shipston on Stour
Long Compton
Tamworth
Solihull
BIRMINGHAM
Stratford on Avon
Alcester

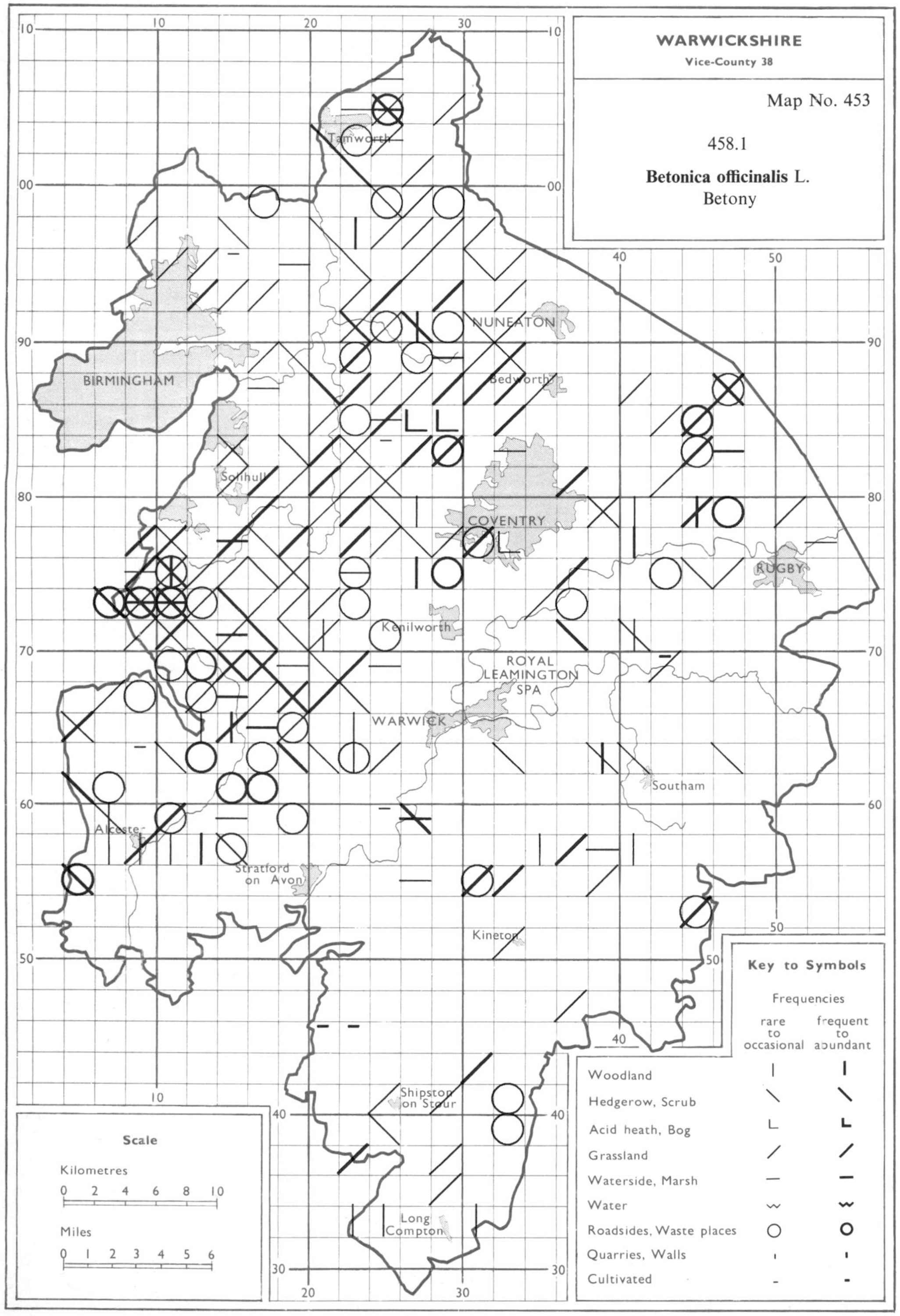
WARWICKSHIRE
Vice-County 38
Map No. 453
458.1
Betonica officinalis L.
Betony
Tamworth
BIRMINGHAM
Solihull
NUNEATON
Bedworth
COVENTRY
RUGBY
Kenilworth
ROYAL LEAMINGTON SPA
WARWICK
Southam
Alcester
Stratford on Avon
Kineton
Shipston on Stour
Long Compton
Key to Symbols
Frequencies
rare to occasional
frequent to abundant
Woodland
Hedgerow, Scrub
Acid heath, Bog
Grassland
Waterside, Marsh
Water
Roadsides, Waste places
Quarries, Walls
Cultivated
Scale
Kilometres
0 2 4 6 8 10
Miles
0 1 2 3 4 5 6

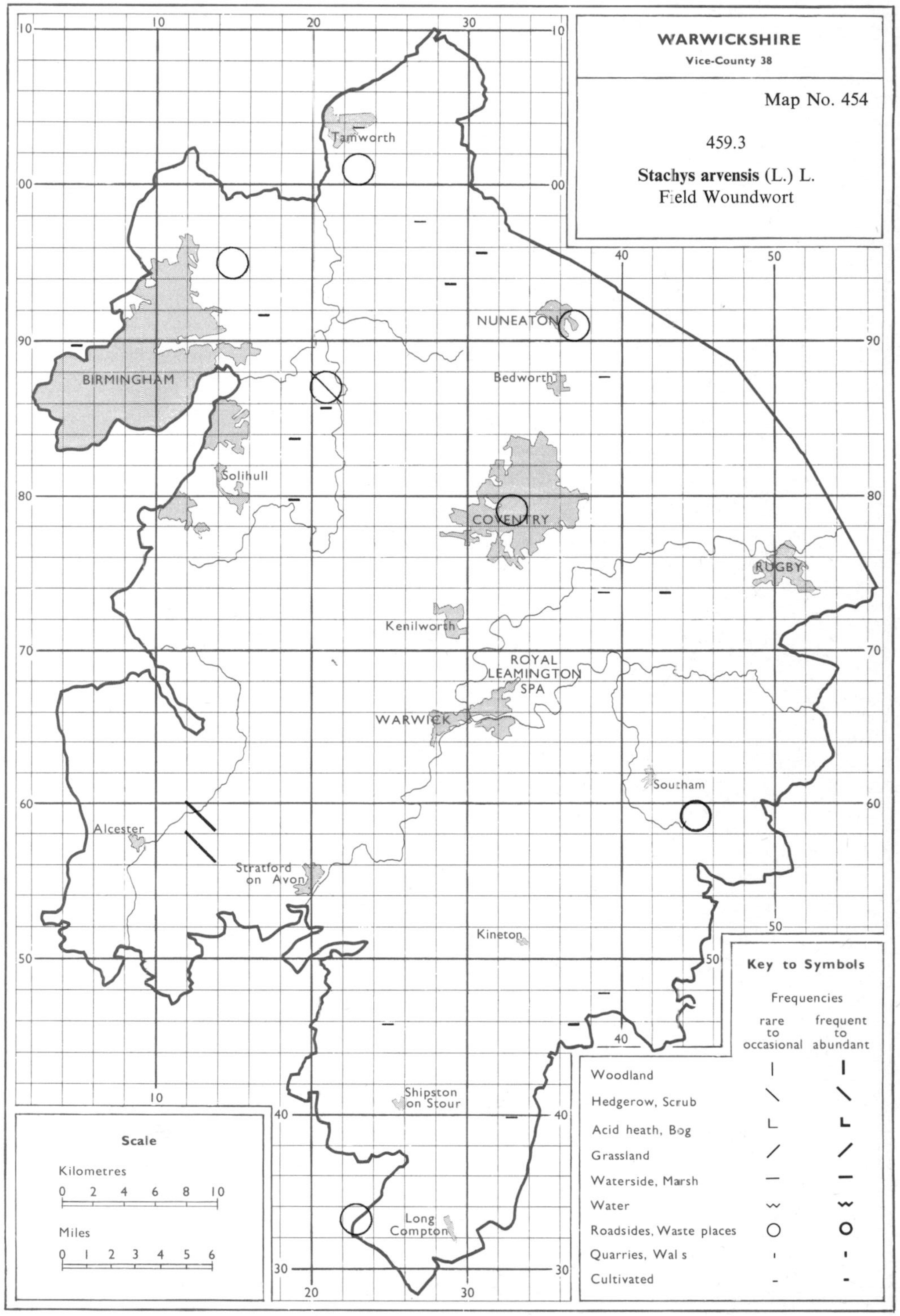
WARWICKSHIRE
Vice-County 38
Map No. 454
459.3
Stachys arvensis (L.) L.
Field Woundwort
Tamworth
BIRMINGHAM
Solihull
NUNEATON
Bedworth
COVENTRY
RUGBY
Kenilworth
ROYAL LEAMINGTON SPA
WARWICK
Southam
Alcester
Stratford on Avon
Kineton
Shipston on Stour
Long Compton
Key to Symbols
Frequencies
rare to occasional
frequent to abundant
Woodland
Hedgerow, Scrub
Acid heath, Bog
Grassland
Waterside, Marsh
Water
Roadsides, Waste places
Quarries, Walls
Cultivated
Scale
Kilometres
0 2 4 6 8 10
Miles
0 1 2 3 4 5 6

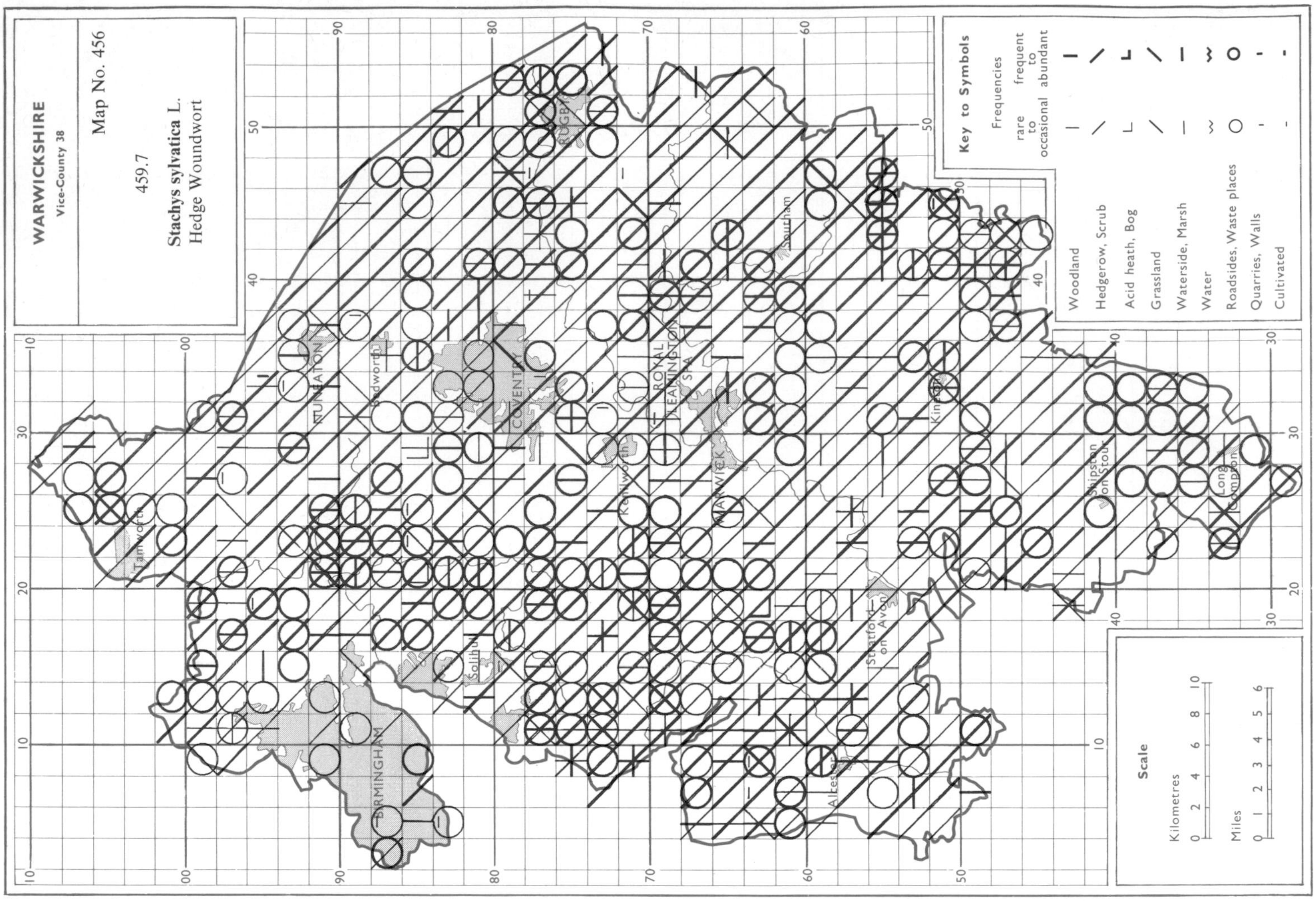
WARWICKSHIRE
Vice-County 38
Map No. 456
459.7
Stachys sylvatica L.
Hedge Woundwort
Key to Symbols
Frequencies
rare to occasional
frequent to abundant
Woodland
Hedgerow, Scrub
Acid heath, Bog
Grassland
Waterside, Marsh
Water
Roadsides, Waste places
Quarries, Walls
Cultivated
Scale
Kilometres
Miles
Tamworth
NUNEATON
Bedworth
COVENTRY
RUGBY
Southam
ROYAL LEAMINGTON SPA
Kenilworth
WARWICK
Kineton
Shipston on Stour
Long Compton
Solihull
BIRMINGHAM
Stratford on Avon
Alcester

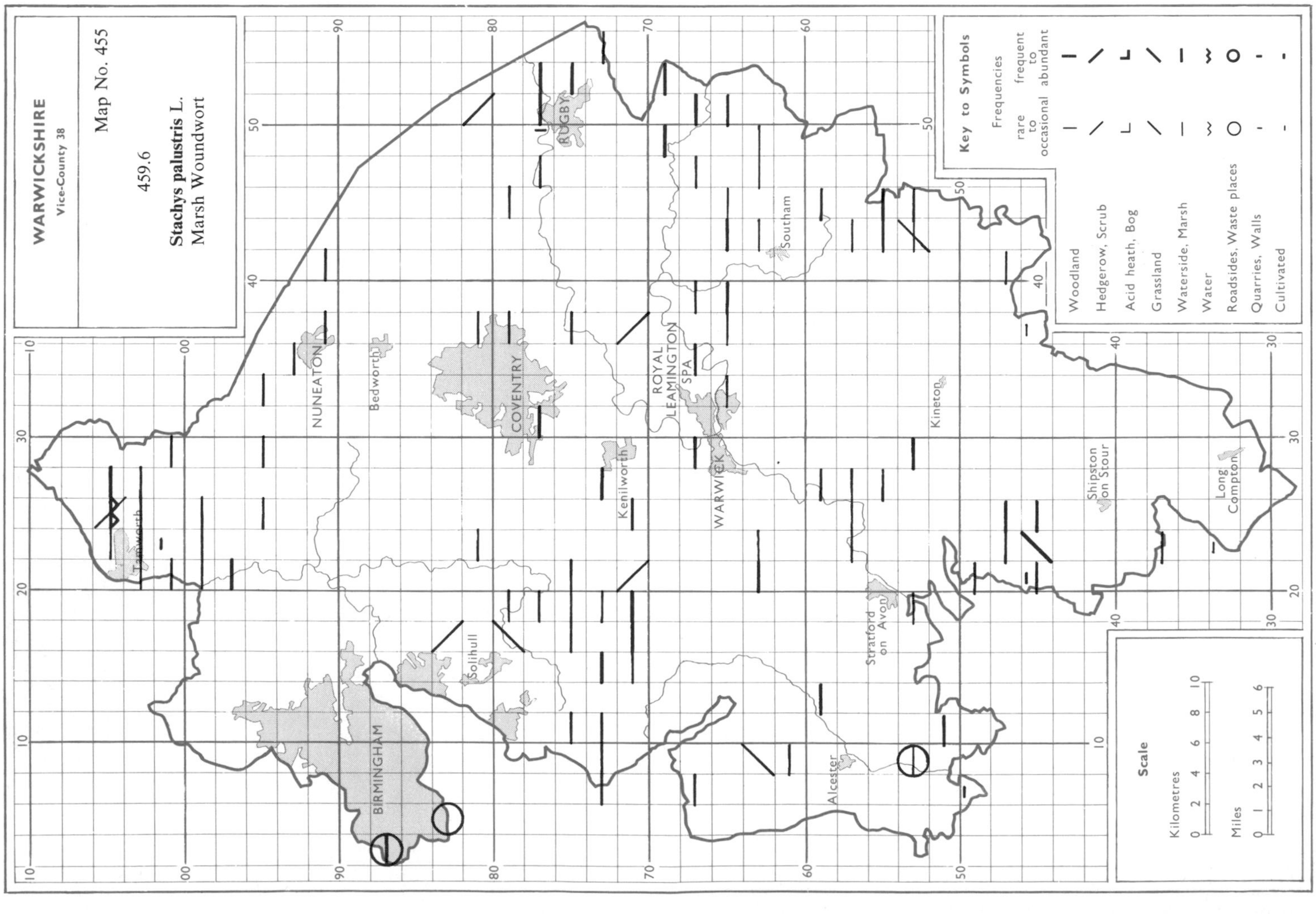
WARWICKSHIRE
Vice-County 38
Map No. 455
459.6
Stachys palustris L.
Marsh Woundwort
Key to Symbols
Frequencies
rare to occasional
frequent to abundant
Woodland
Hedgerow, Scrub
Acid heath, Bog
Grassland
Waterside, Marsh
Water
Roadsides, Waste places
Quarries, Walls
Cultivated
Scale
Kilometres
Miles
Tamworth
NUNEATON
Bedworth
COVENTRY
RUGBY
Southam
ROYAL LEAMINGTON SPA
Kenilworth
WARWICK
Kineton
Shipston on Stour
Long Compton
Solihull
BIRMINGHAM
Stratford on Avon
Alcester

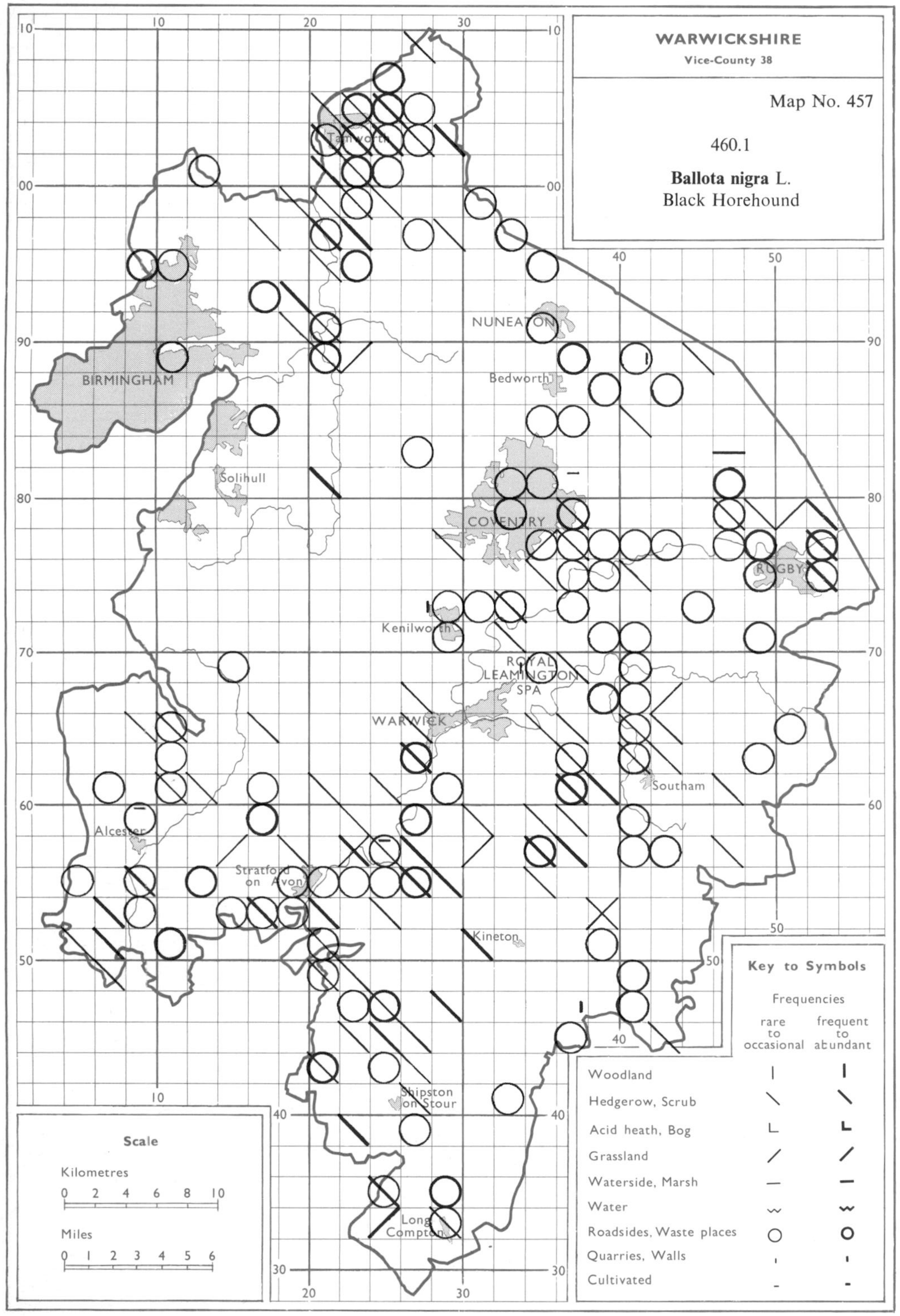

WARWICKSHIRE
Vice-County 38

Map No. 457

460.1

Ballota nigra L.
Black Horehound

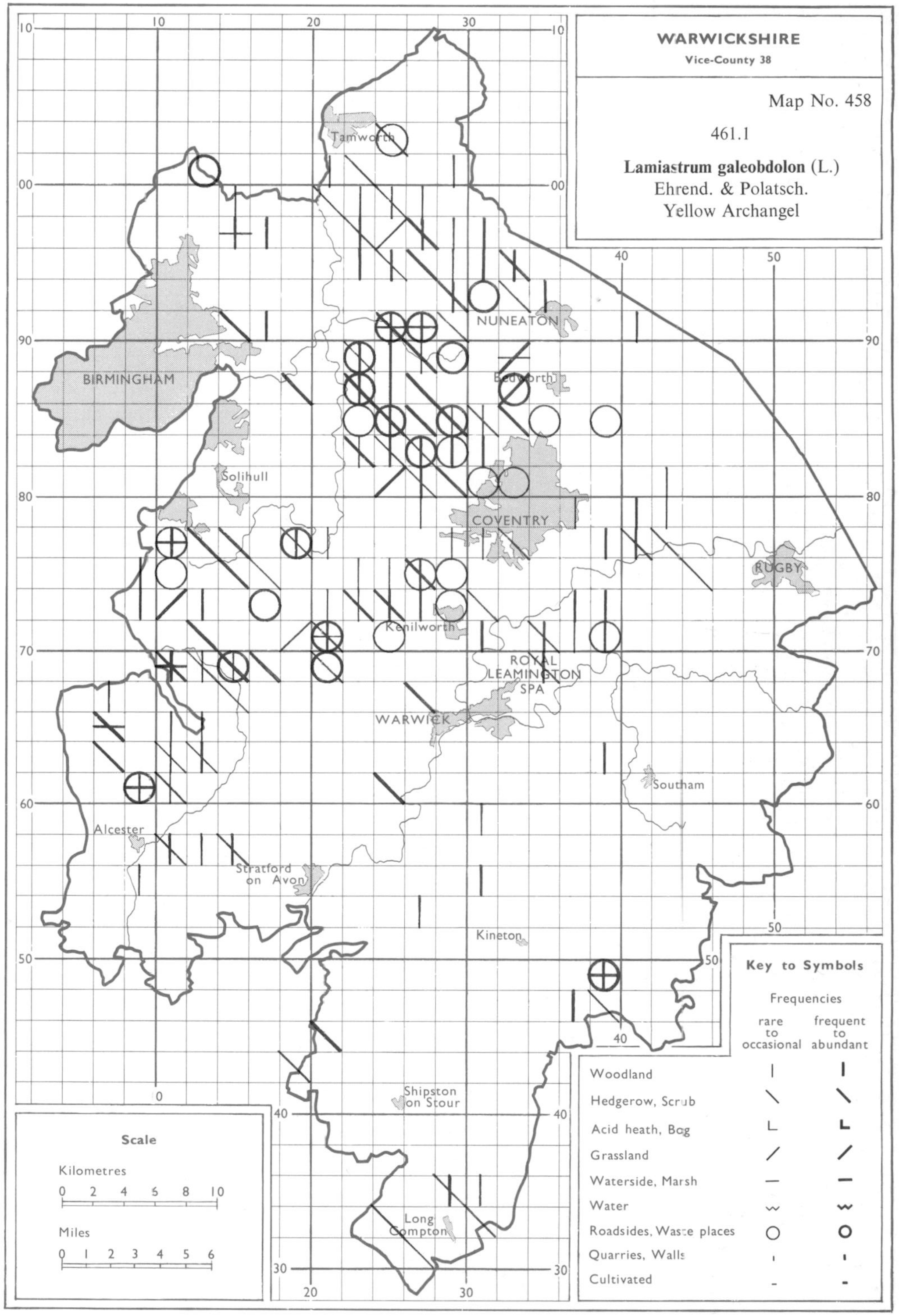

WARWICKSHIRE
Vice-County 38

Map No. 458

461.1

Lamiastrum galeobdolon (L.)
Ehrend. & Polatsch.
Yellow Archangel

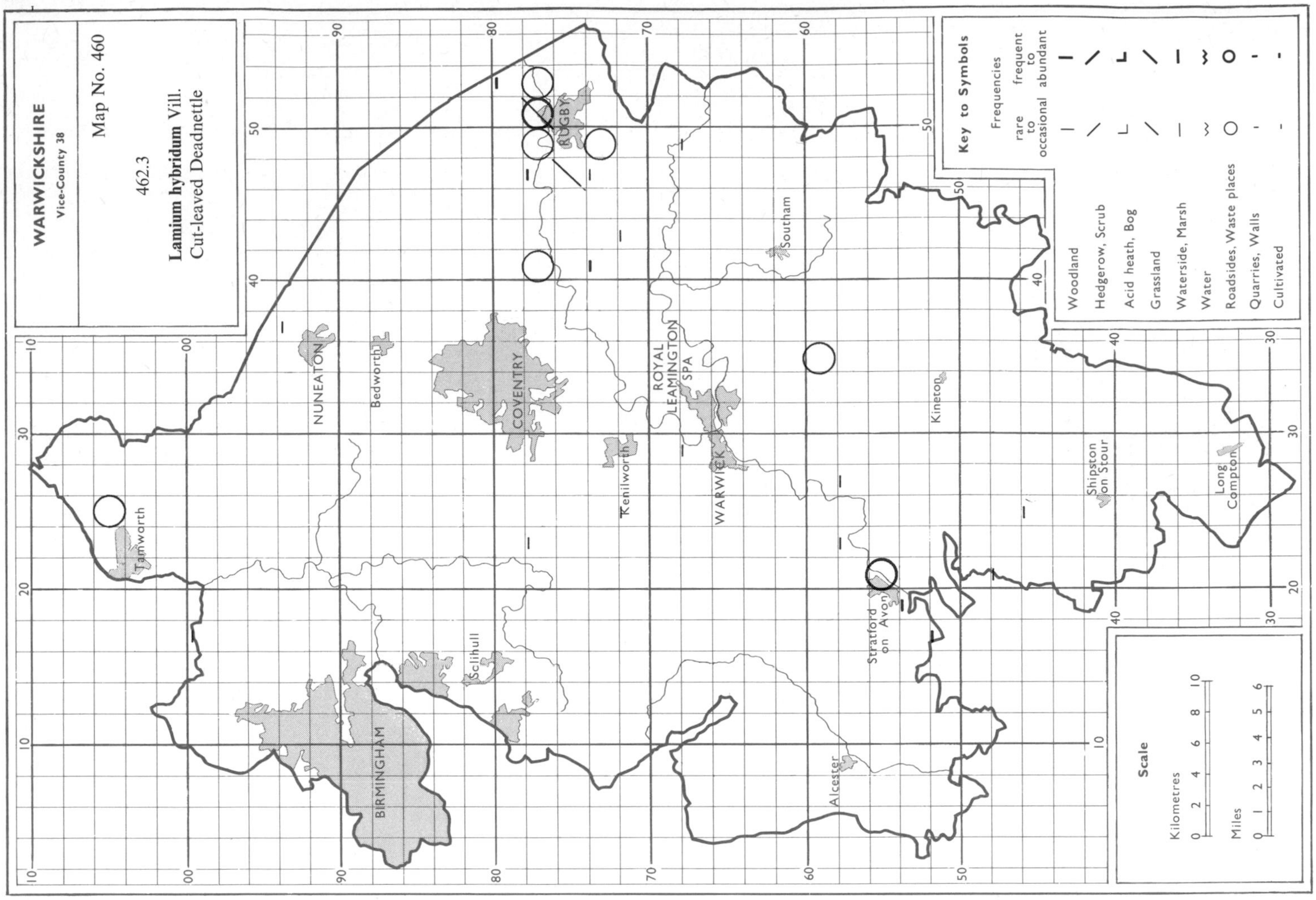

WARWICKSHIRE
Vice-County 38
Map No. 460
462.3
Lamium hybridum Vill.
Cut-leaved Deadnettle
Key to Symbols
Frequencies
rare to occasional
frequent to abundant
Woodland
Hedgerow, Scrub
Acid heath, Bog
Grassland
Waterside, Marsh
Water
Roadsides, Waste places
Quarries, Walls
Cultivated
Scale
Kilometres
0 2 4 6 8 10
Miles
0 1 2 3 4 5 6
Tamworth
NUNEATON
Bedworth
COVENTRY
RUGBY
Southam
ROYAL LEAMINGTON SPA
Kenilworth
WARWICK
Kineton
Shipston on Stour
Long Compton
Stratford on Avon
Solihull
BIRMINGHAM
Alcester

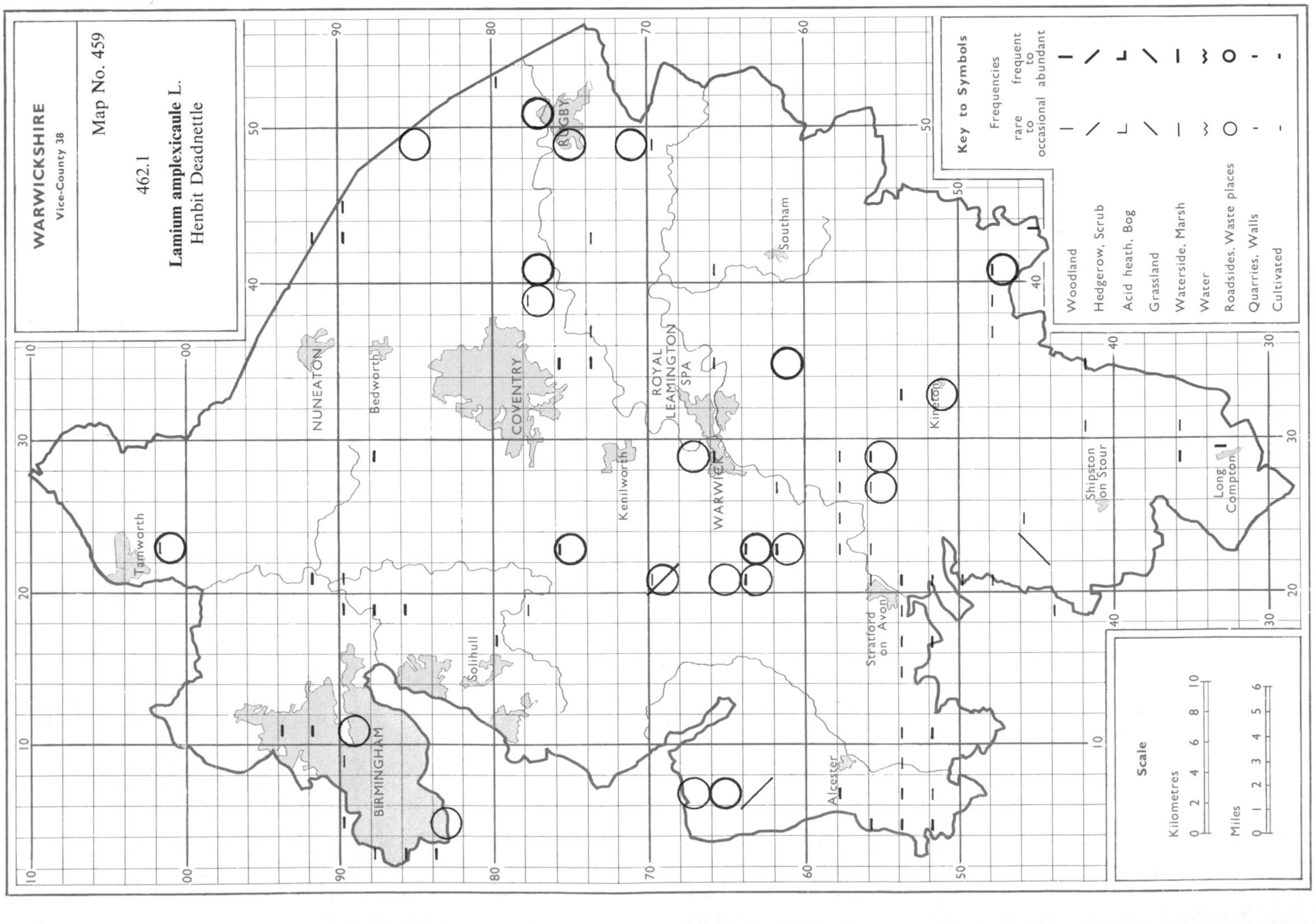

WARWICKSHIRE
Vice-County 38
Map No. 459
462.1
Lamium amplexicaule L.
Henbit Deadnettle
Key to Symbols
Frequencies
rare to occasional
frequent to abundant
Woodland
Hedgerow, Scrub
Acid heath, Bog
Grassland
Waterside, Marsh
Water
Roadsides, Waste places
Quarries, Walls
Cultivated
Scale
Kilometres
0 2 4 6 8 10
Miles
0 1 2 3 4 5 6
Tamworth
NUNEATON
Bedworth
COVENTRY
RUGBY
Southam
ROYAL LEAMINGTON SPA
Kenilworth
WARWICK
Kineton
Shipston on Stour
Long Compton
Stratford on Avon
Solihull
BIRMINGHAM
Alcester

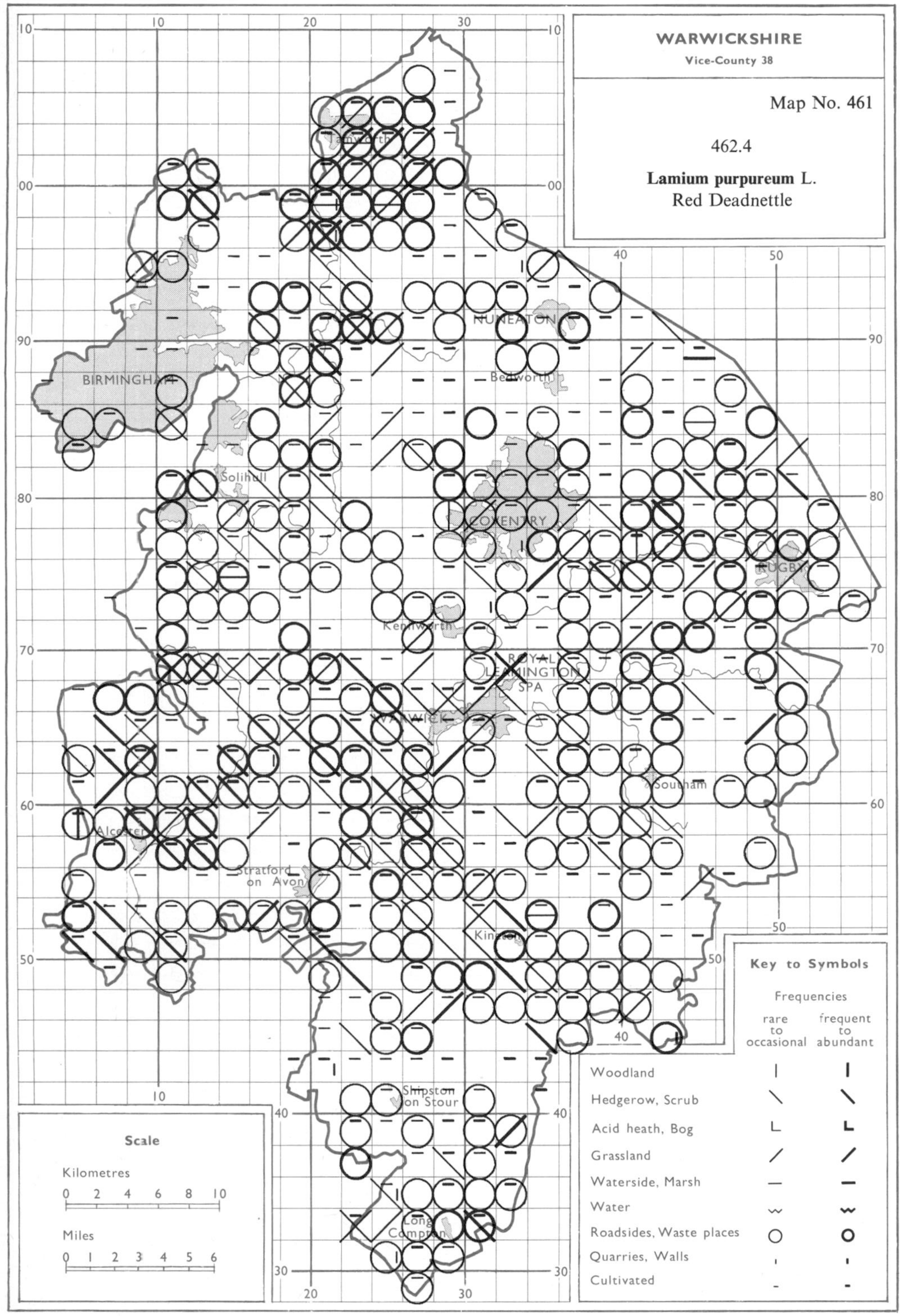
WARWICKSHIRE
Vice-County 38
Map No. 461
462.4
Lamium purpureum L.
Red Deadnettle
BIRMINGHAM
Solihull
NUNEATON
Bedworth
COVENTRY
RUGBY
Kenilworth
ROYAL LEAMINGTON SPA
WARWICK
Southam
Alcester
Stratford on Avon
Kineton
Shipston on Stour
Long Compton
Key to Symbols
Frequencies
rare to occasional
frequent to abundant
Woodland
Hedgerow, Scrub
Acid heath, Bog
Grassland
Waterside, Marsh
Water
Roadsides, Waste places
Quarries, Walls
Cultivated
Scale
Kilometres
0 2 4 6 8 10
Miles
0 1 2 3 4 5 6

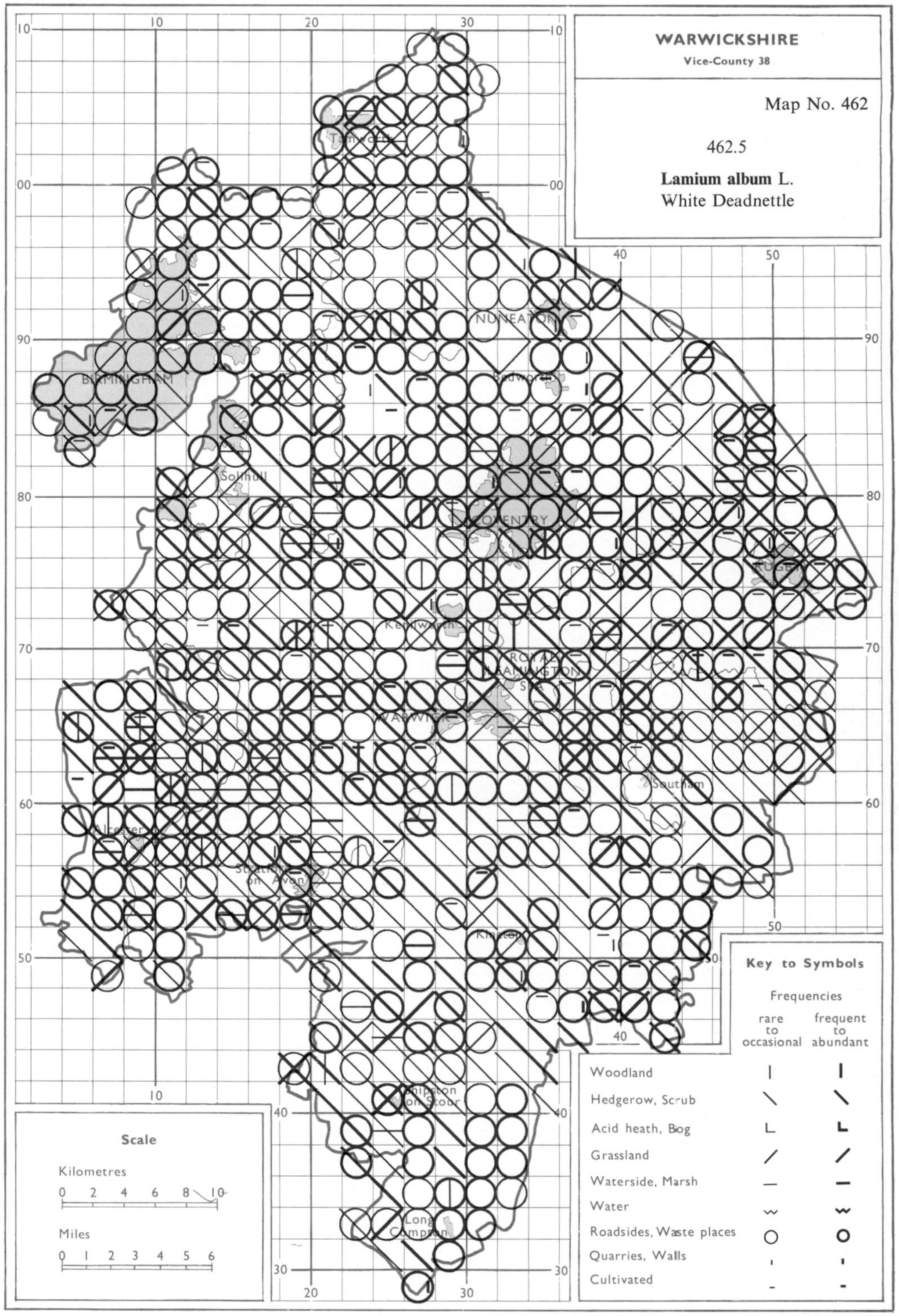
WARWICKSHIRE
Vice-County 38
Map No. 462
462.5
Lamium album L.
White Deadnettle
BIRMINGHAM
Solihull
NUNEATON
Bedworth
COVENTRY
RUGBY
Kenilworth
ROYAL LEAMINGTON SPA
WARWICK
Southam
Alcester
Stratford on Avon
Kineton
Shipston on Stour
Long Compton
Key to Symbols
Frequencies
rare to occasional
frequent to abundant
Woodland
Hedgerow, Scrub
Acid heath, Bog
Grassland
Waterside, Marsh
Water
Roadsides, Waste places
Quarries, Walls
Cultivated
Scale
Kilometres
0 2 4 6 8 10
Miles
0 1 2 3 4 5 6

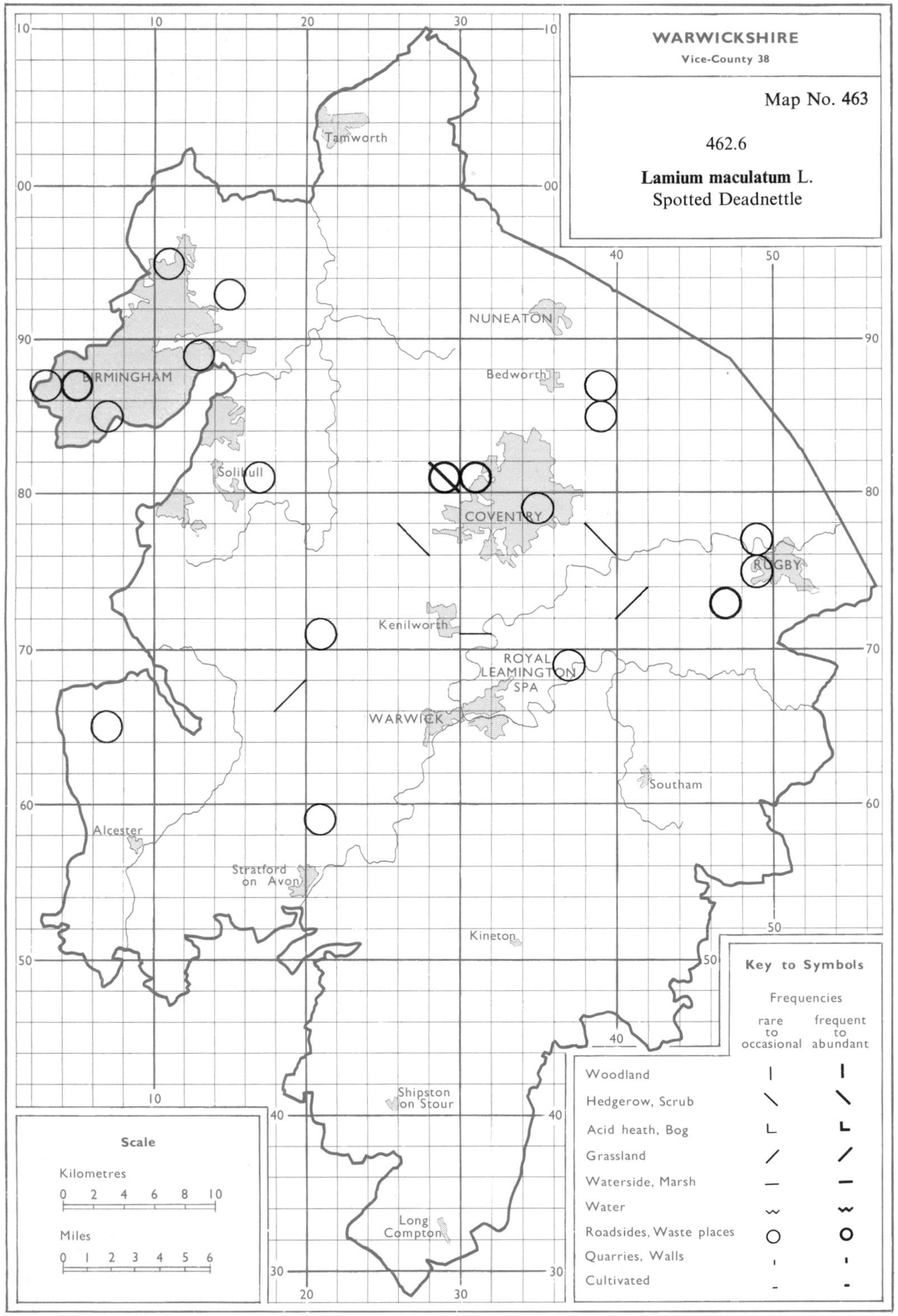
WARWICKSHIRE
Vice-County 38
Map No. 463
462.6
Lamium maculatum L.
Spotted Deadnettle
Tamworth
NUNEATON
Bedworth
BIRMINGHAM
Solihull
COVENTRY
RUGBY
Kenilworth
ROYAL LEAMINGTON SPA
WARWICK
Southam
Alcester
Stratford on Avon
Kineton
Shipston on Stour
Long Compton
Scale
Kilometres
0 2 4 6 8 10
Miles
0 1 2 3 4 5 6
Key to Symbols
Frequencies
rare to occasional
frequent to abundant
Woodland
Hedgerow, Scrub
Acid heath, Bog
Grassland
Waterside, Marsh
Water
Roadsides, Waste places
Quarries, Walls
Cultivated

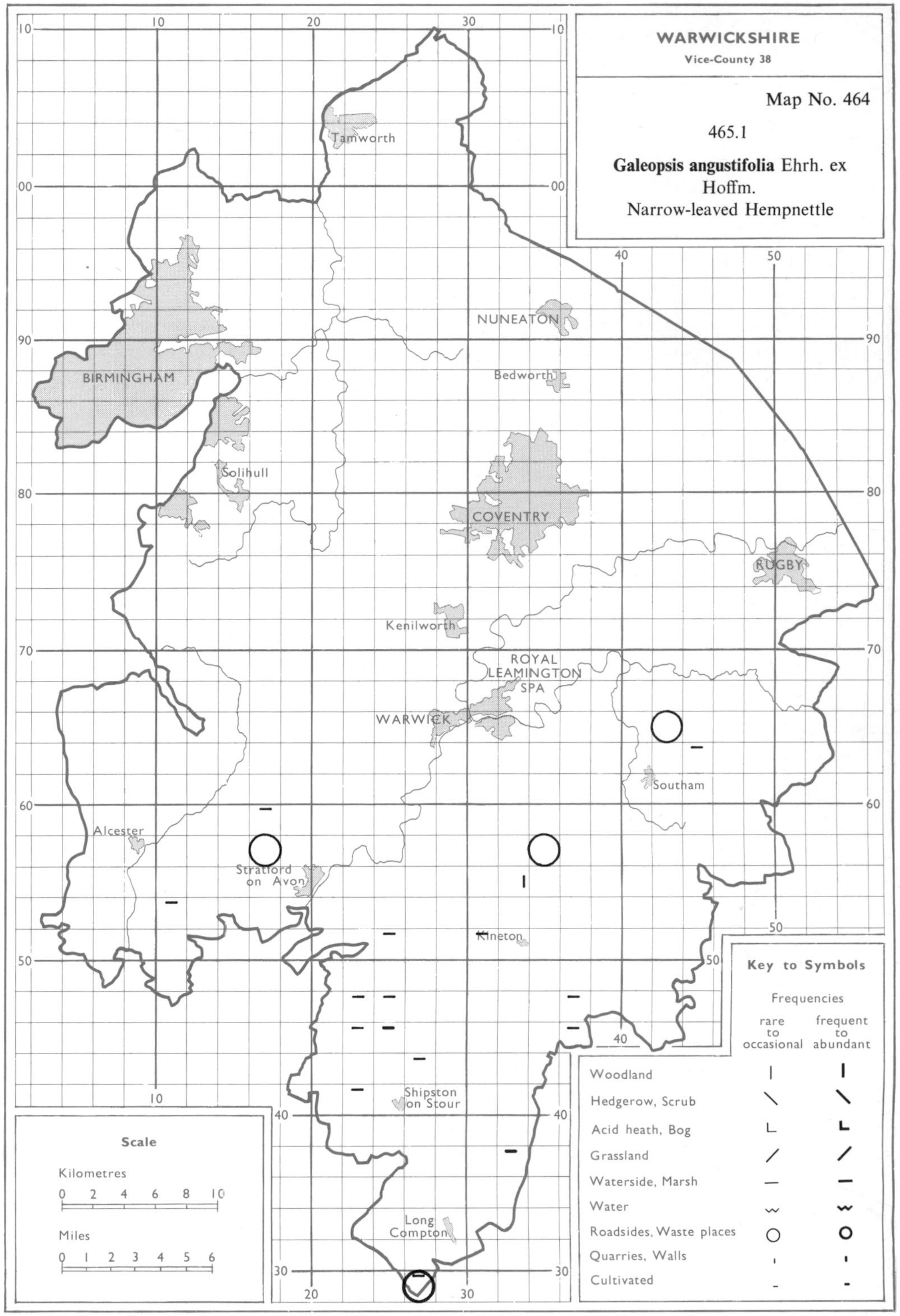
WARWICKSHIRE
Vice-County 38
Map No. 464
465.1
Galeopsis angustifolia Ehrh. ex Hoffm.
Narrow-leaved Hempnettle
Tamworth
NUNEATON
Bedworth
BIRMINGHAM
Solihull
COVENTRY
RUGBY
Kenilworth
ROYAL LEAMINGTON SPA
WARWICK
Southam
Alcester
Stratford on Avon
Kineton
Shipston on Stour
Long Compton
Scale
Kilometres
0 2 4 6 8 10
Miles
0 1 2 3 4 5 6
Key to Symbols
Frequencies
rare to occasional
frequent to abundant
Woodland
Hedgerow, Scrub
Acid heath, Bog
Grassland
Waterside, Marsh
Water
Roadsides, Waste places
Quarries, Walls
Cultivated

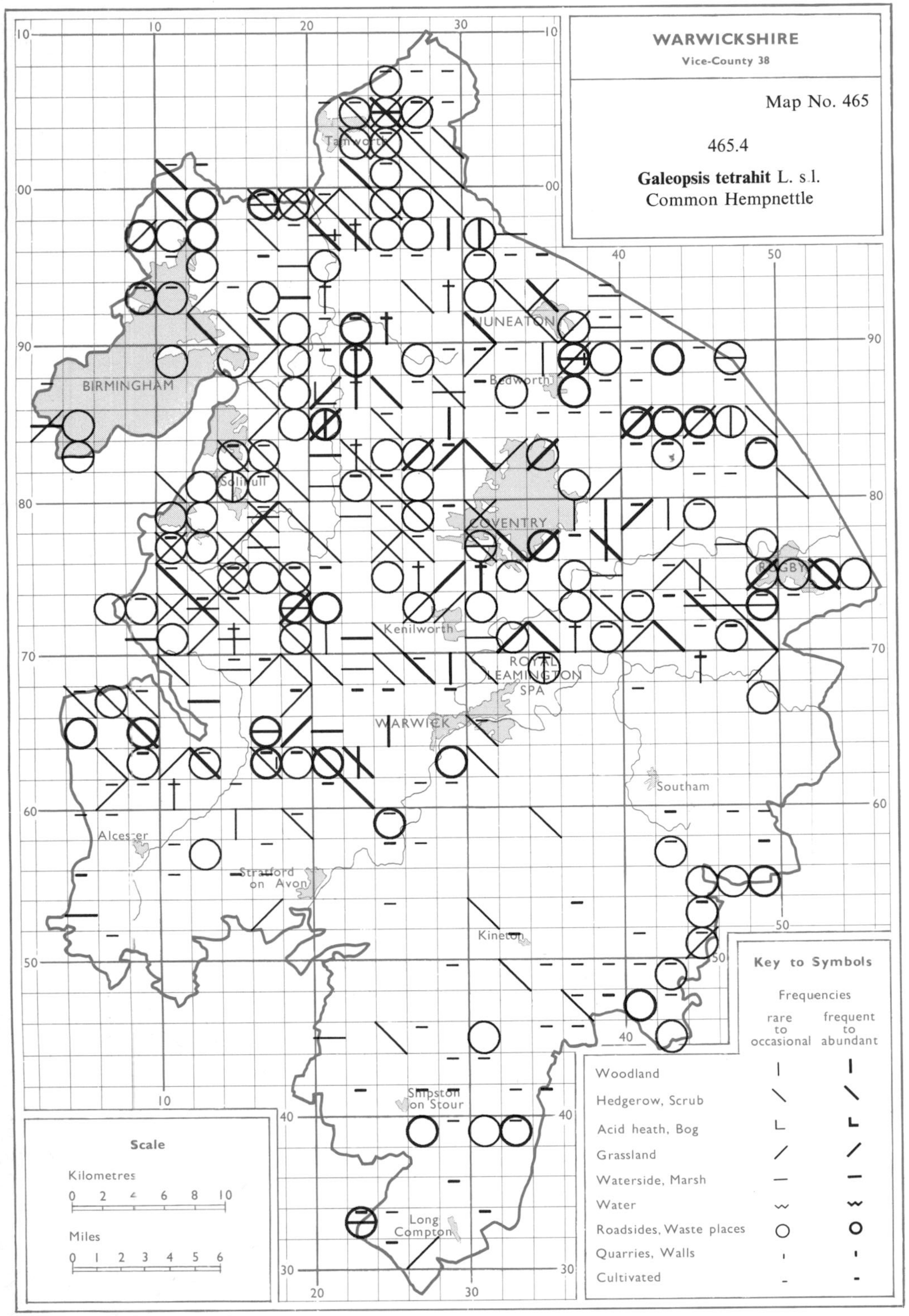
WARWICKSHIRE
Vice-County 38
Map No. 465
465.4
Galeopsis tetrahit L. s.l.
Common Hempnettle
Key to Symbols
Frequencies
rare to occasional
frequent to abundant
Woodland
Hedgerow, Scrub
Acid heath, Bog
Grassland
Waterside, Marsh
Water
Roadsides, Waste places
Quarries, Walls
Cultivated
Scale
Kilometres
Miles
Tamworth
BIRMINGHAM
NUNEATON
Bedworth
Solihull
COVENTRY
RUGBY
Kenilworth
ROYAL LEAMINGTON SPA
WARWICK
Southam
Alcester
Stratford on Avon
Kineton
Shipston on Stour
Long Compton

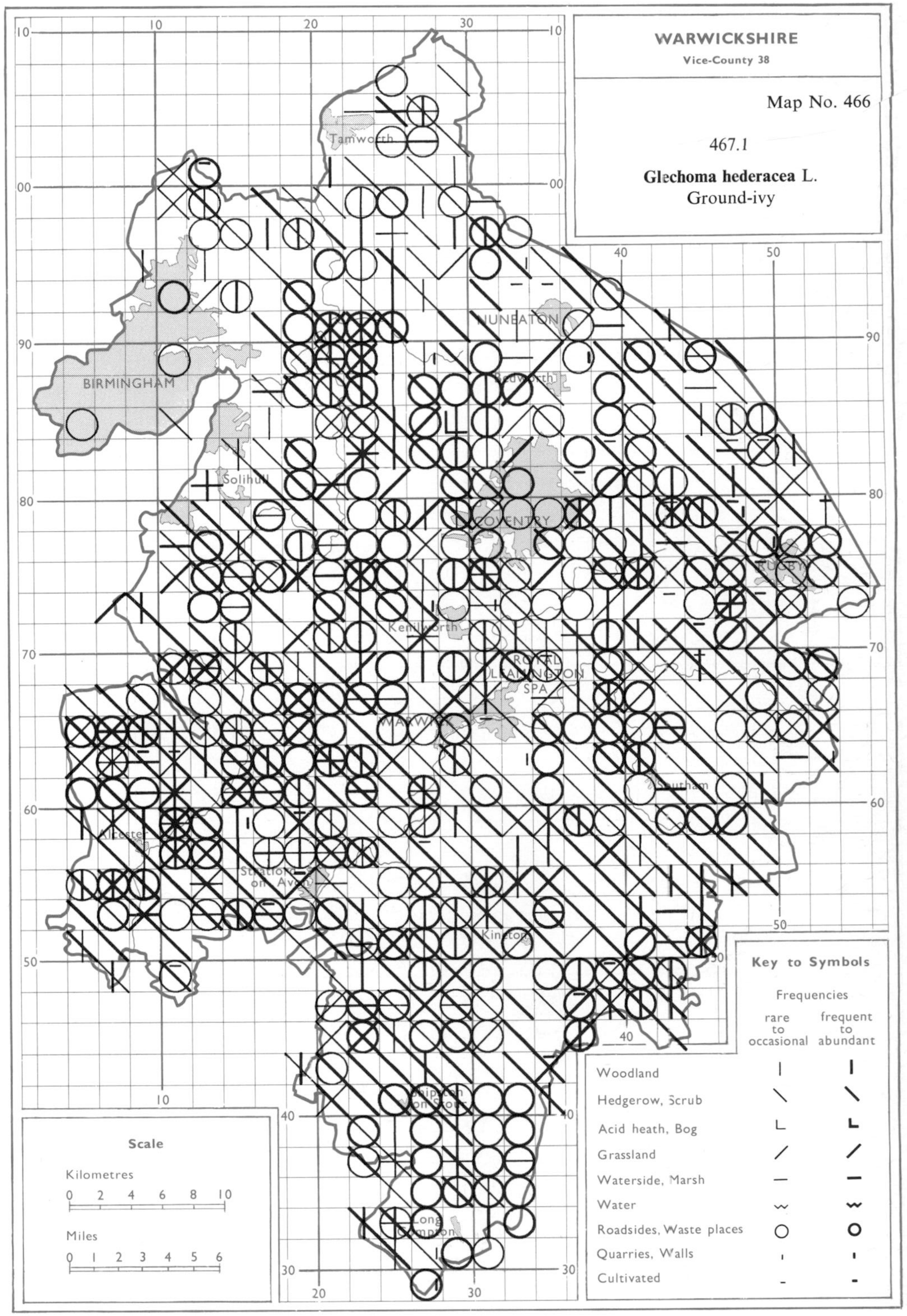
WARWICKSHIRE
Vice-County 38
Map No. 466
467.1
Glechoma hederacea L.
Ground-ivy
Key to Symbols
Frequencies
rare to occasional
frequent to abundant
Woodland
Hedgerow, Scrub
Acid heath, Bog
Grassland
Waterside, Marsh
Water
Roadsides, Waste places
Quarries, Walls
Cultivated
Scale
Kilometres
Miles
Tamworth
BIRMINGHAM
NUNEATON
Bedworth
Solihull
COVENTRY
RUGBY
Kenilworth
ROYAL LEAMINGTON SPA
WARWICK
Southam
Alcester
Stratford on Avon
Kineton
Shipston on Stour
Long Compton

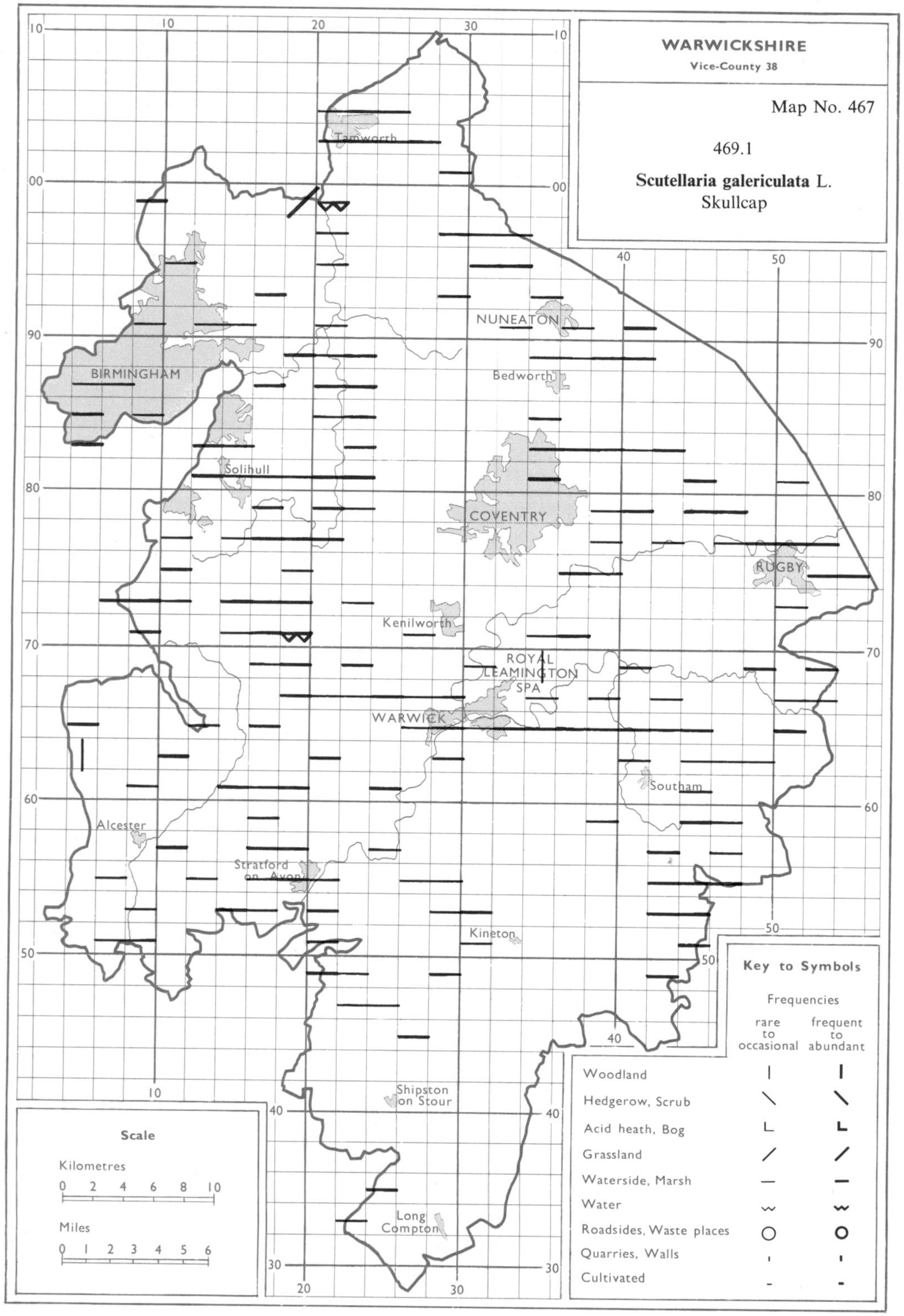
WARWICKSHIRE
Vice-County 38
Map No. 467
469.1
Scutellaria galericulata L.
Skullcap
Tamworth
NUNEATON
Bedworth
BIRMINGHAM
Solihull
COVENTRY
RUGBY
Kenilworth
ROYAL LEAMINGTON SPA
WARWICK
Southam
Alcester
Stratford on Avon
Kineton
Shipston on Stour
Long Compton
Key to Symbols
Frequencies
rare to occasional
frequent to abundant
Woodland
Hedgerow, Scrub
Acid heath, Bog
Grassland
Waterside, Marsh
Water
Roadsides, Waste places
Quarries, Walls
Cultivated
Scale
Kilometres
0 2 4 6 8 10
Miles
0 1 2 3 4 5 6

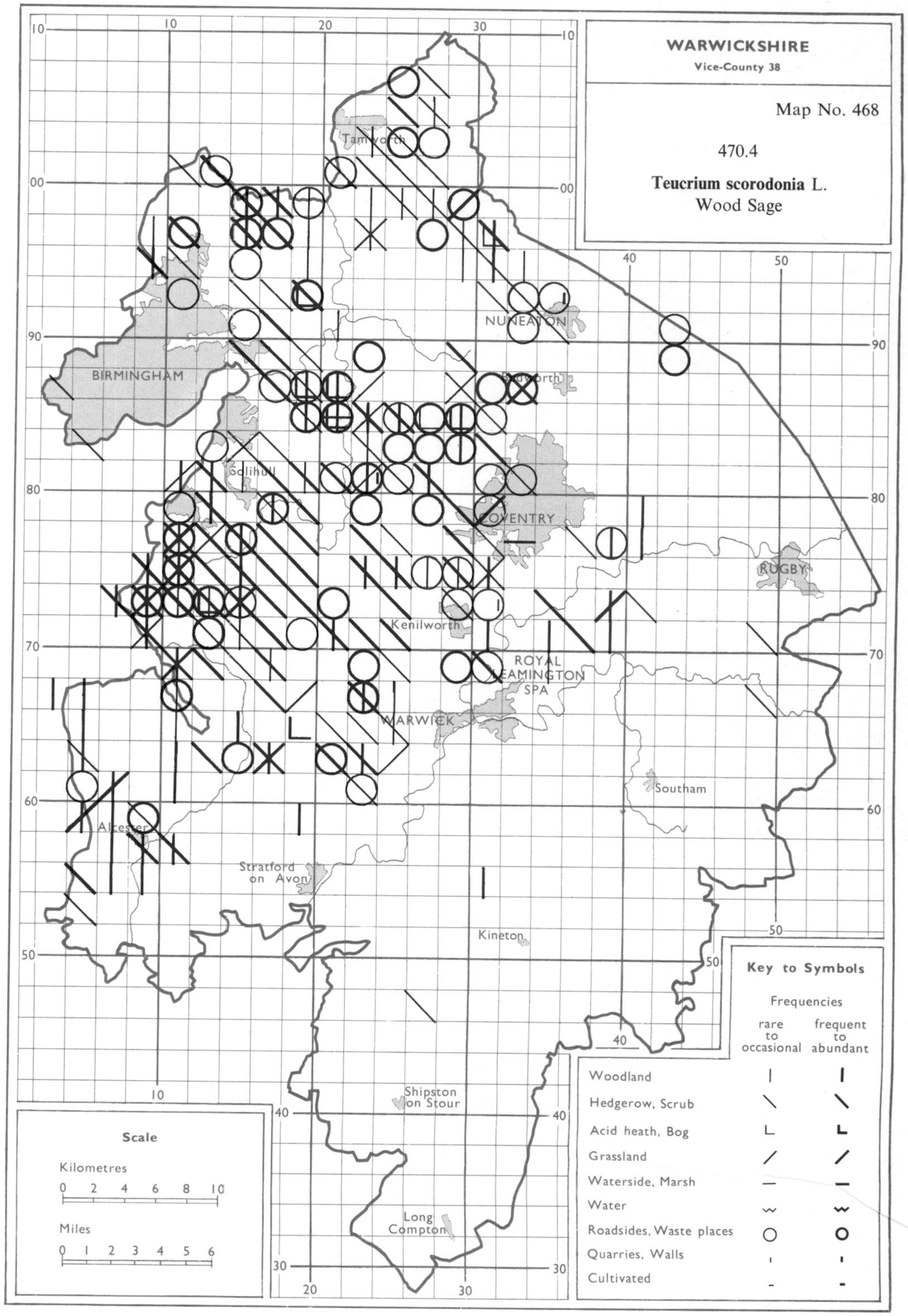
WARWICKSHIRE
Vice-County 38
Map No. 468
470.4
Teucrium scorodonia L.
Wood Sage
Tamworth
NUNEATON
Bedworth
BIRMINGHAM
Solihull
COVENTRY
RUGBY
Kenilworth
ROYAL LEAMINGTON SPA
WARWICK
Southam
Alcester
Stratford on Avon
Kineton
Shipston on Stour
Long Compton
Key to Symbols
Frequencies
rare to occasional
frequent to abundant
Woodland
Hedgerow, Scrub
Acid heath, Bog
Grassland
Waterside, Marsh
Water
Roadsides, Waste places
Quarries, Walls
Cultivated
Scale
Kilometres
0 2 4 6 8 10
Miles
0 1 2 3 4 5 6

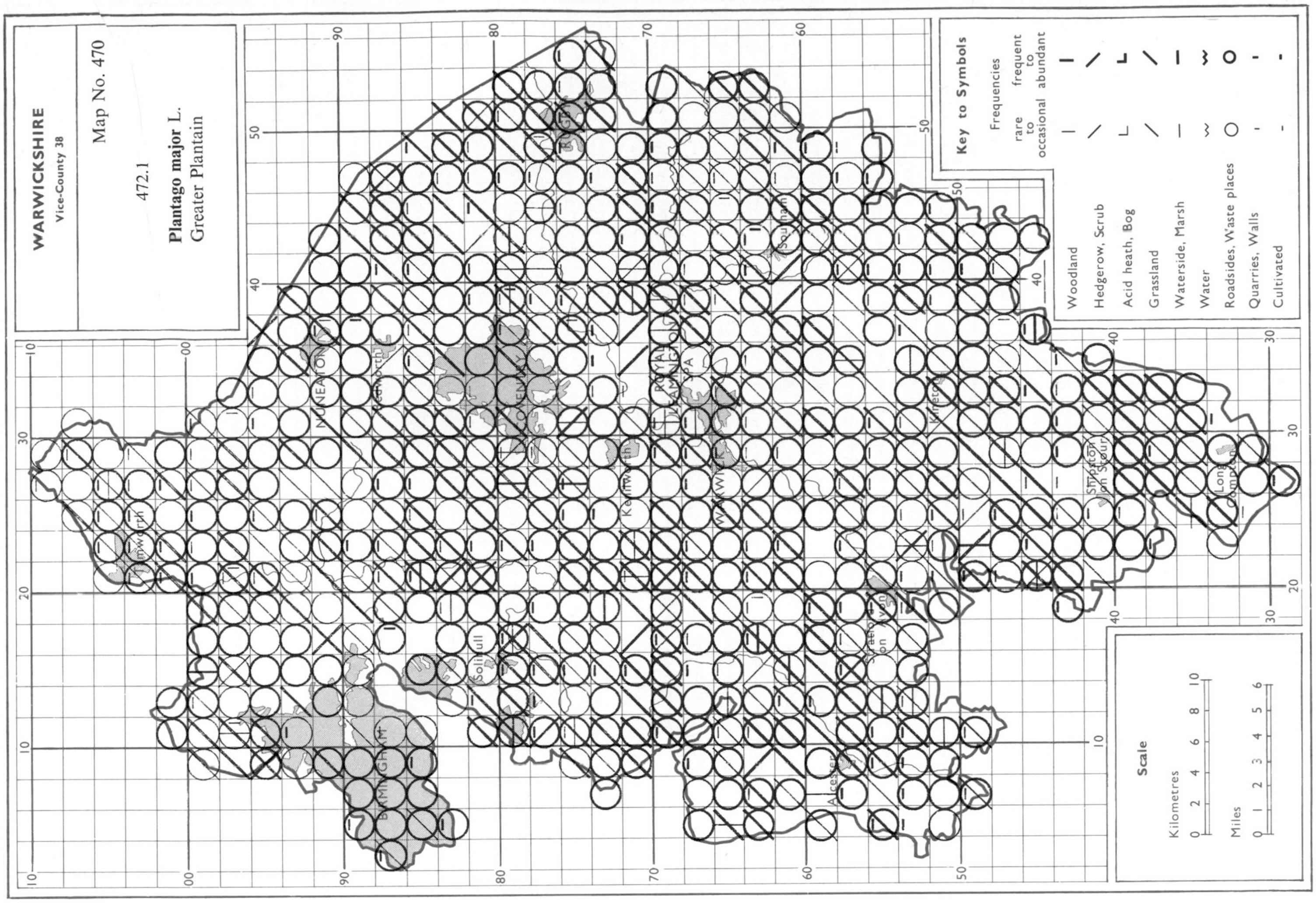
WARWICKSHIRE
Vice-County 38
Map No. 470
472.1
Plantago major L.
Greater Plantain
Key to Symbols
Frequencies
rare to occasional
frequent to abundant
Woodland
Hedgerow, Scrub
Acid heath, Bog
Grassland
Waterside, Marsh
Water
Roadsides, Waste places
Quarries, Walls
Cultivated
Scale
Kilometres
Miles
Tamworth
NUNEATON
Bedworth
COVENTRY
RUGBY
Kenilworth
ROYAL LEAMINGTON SPA
WARWICK
Southam
Kineton
Shipston on Stour
Long Compton
Solihull
BIRMINGHAM
Stratford on Avon
Alcester

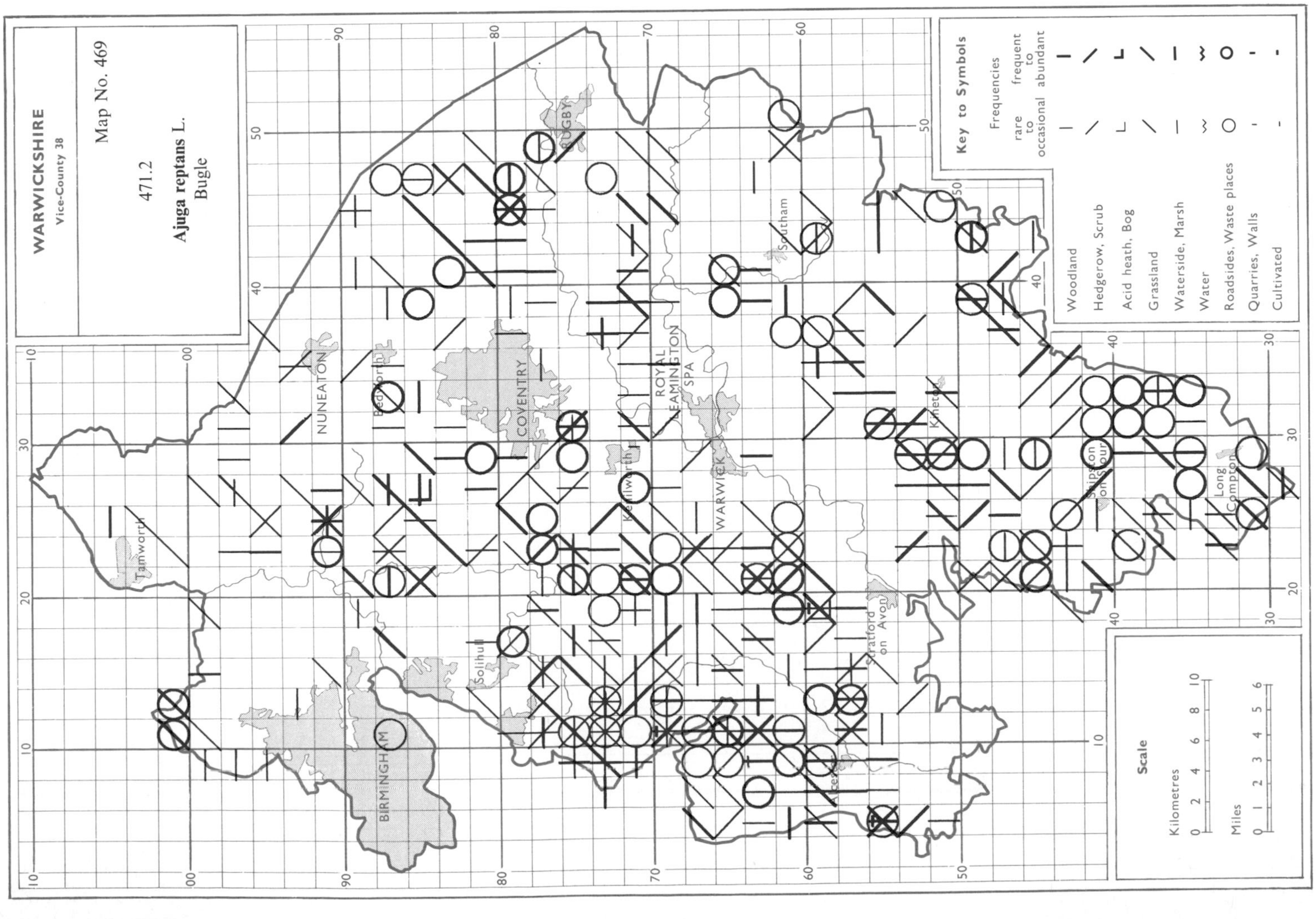
WARWICKSHIRE
Vice-County 38
Map No. 469
471.2
Ajuga reptans L.
Bugle
Key to Symbols
Frequencies
rare to occasional
frequent to abundant
Woodland
Hedgerow, Scrub
Acid heath, Bog
Grassland
Waterside, Marsh
Water
Roadsides, Waste places
Quarries, Walls
Cultivated
Scale
Kilometres
Miles
Tamworth
NUNEATON
Bedworth
COVENTRY
RUGBY
Kenilworth
ROYAL LEAMINGTON SPA
WARWICK
Southam
Kineton
Shipston on Stour
Long Compton
Solihull
BIRMINGHAM
Stratford on Avon

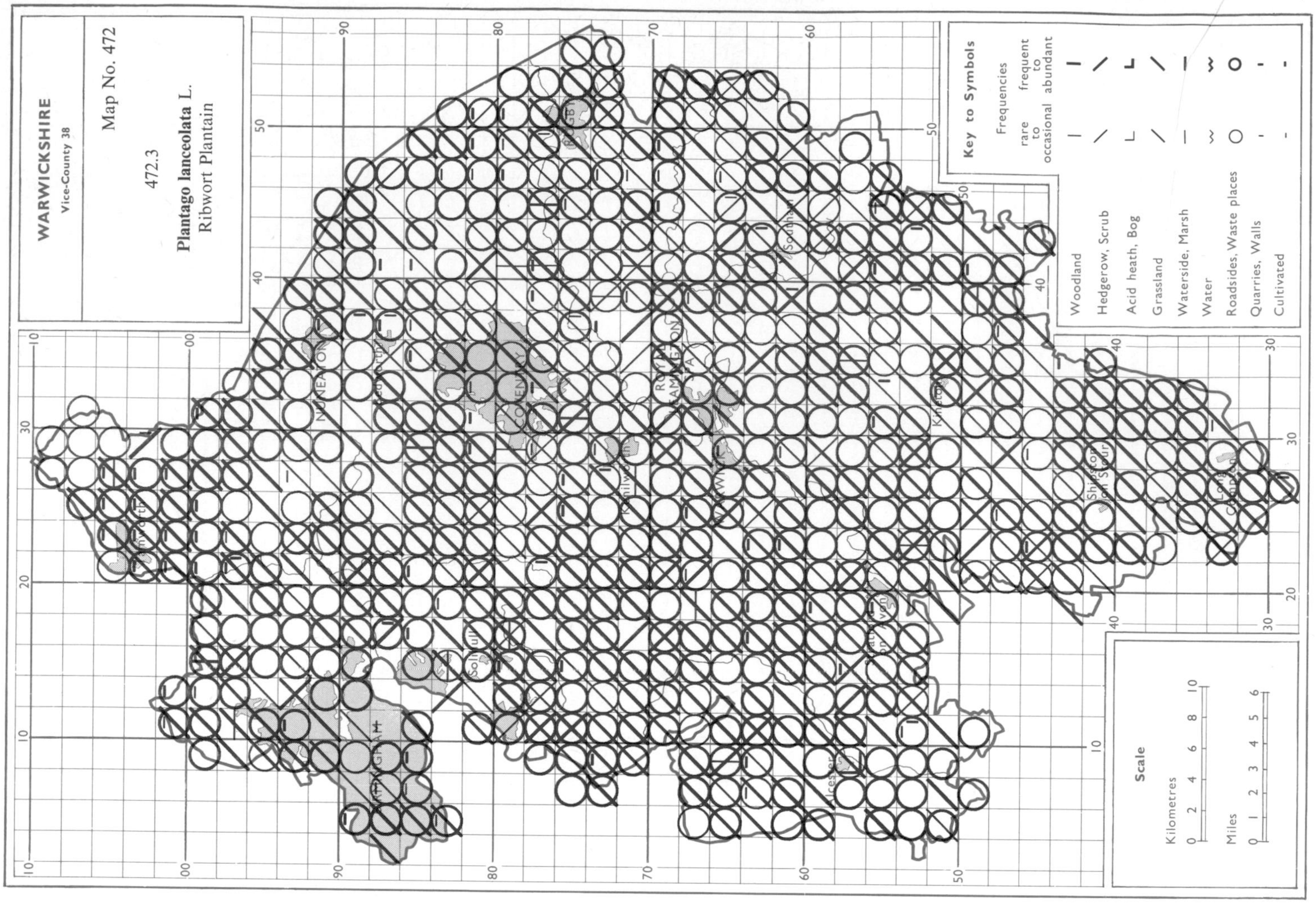
WARWICKSHIRE
Vice-County 38
Map No. 472
472.3
Plantago lanceolata L.
Ribwort Plantain
Key to Symbols
Frequencies
rare to occasional
frequent to abundant
Woodland
Hedgerow, Scrub
Acid heath, Bog
Grassland
Waterside, Marsh
Water
Roadsides, Waste places
Quarries, Walls
Cultivated
Scale
Kilometres 0 2 4 6 8 10
Miles 0 1 2 3 4 5 6

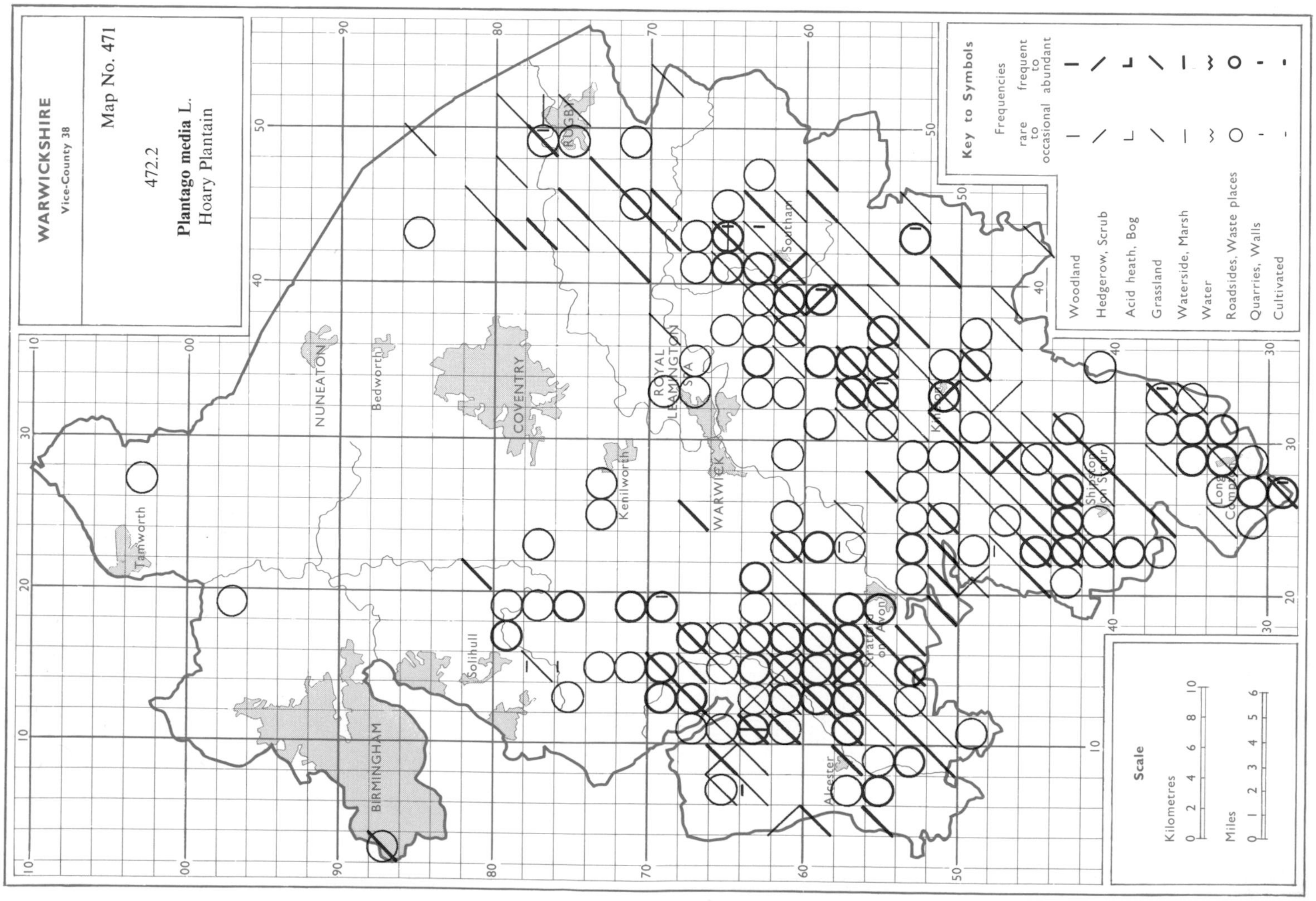
WARWICKSHIRE
Vice-County 38
Map No. 471
472.2
Plantago media L.
Hoary Plantain
Key to Symbols
Frequencies
rare to occasional
frequent to abundant
Woodland
Hedgerow, Scrub
Acid heath, Bog
Grassland
Waterside, Marsh
Water
Roadsides, Waste places
Quarries, Walls
Cultivated
Scale
Kilometres 0 2 4 6 8 10
Miles 0 1 2 3 4 5 6

WARWICKSHIRE
Vice-County 38
Map No. 474
475.2
Campanula trachelium L.
Nettle-leaved Bellflower
Key to Symbols
Frequencies
rare to occasional
frequent to abundant
Woodland
Hedgerow, Scrub
Acid heath, Bog
Grassland
Waterside, Marsh
Water
Roadsides, Waste places
Quarries, Walls
Cultivated
Scale
Kilometres
Miles
Tamworth
NUNEATON
Bedworth
COVENTRY
RUGBY
Southam
ROYAL LEAMINGTON SPA
Kenilworth
WARWICK
Kineton
Shipston on Stour
Long Compton
Stratford on Avon
Alcester
Solihull
BIRMINGHAM

WARWICKSHIRE
Vice-County 38
Map No. 473
475.1
Campanula latifolia L.
Giant Bellflower
Key to Symbols
Frequencies
rare to occasional
frequent to abundant
Woodland
Hedgerow, Scrub
Acid heath, Bog
Grassland
Waterside, Marsh
Water
Roadsides, Waste places
Quarries, Walls
Cultivated
Scale
Kilometres
Miles
Tamworth
NUNEATON
Bedworth
COVENTRY
RUGBY
Southam
ROYAL LEAMINGTON SPA
Kenilworth
WARWICK
Kineton
Shipston on Stour
Long Compton
Stratford on Avon
Alcester
Solihull
BIRMINGHAM

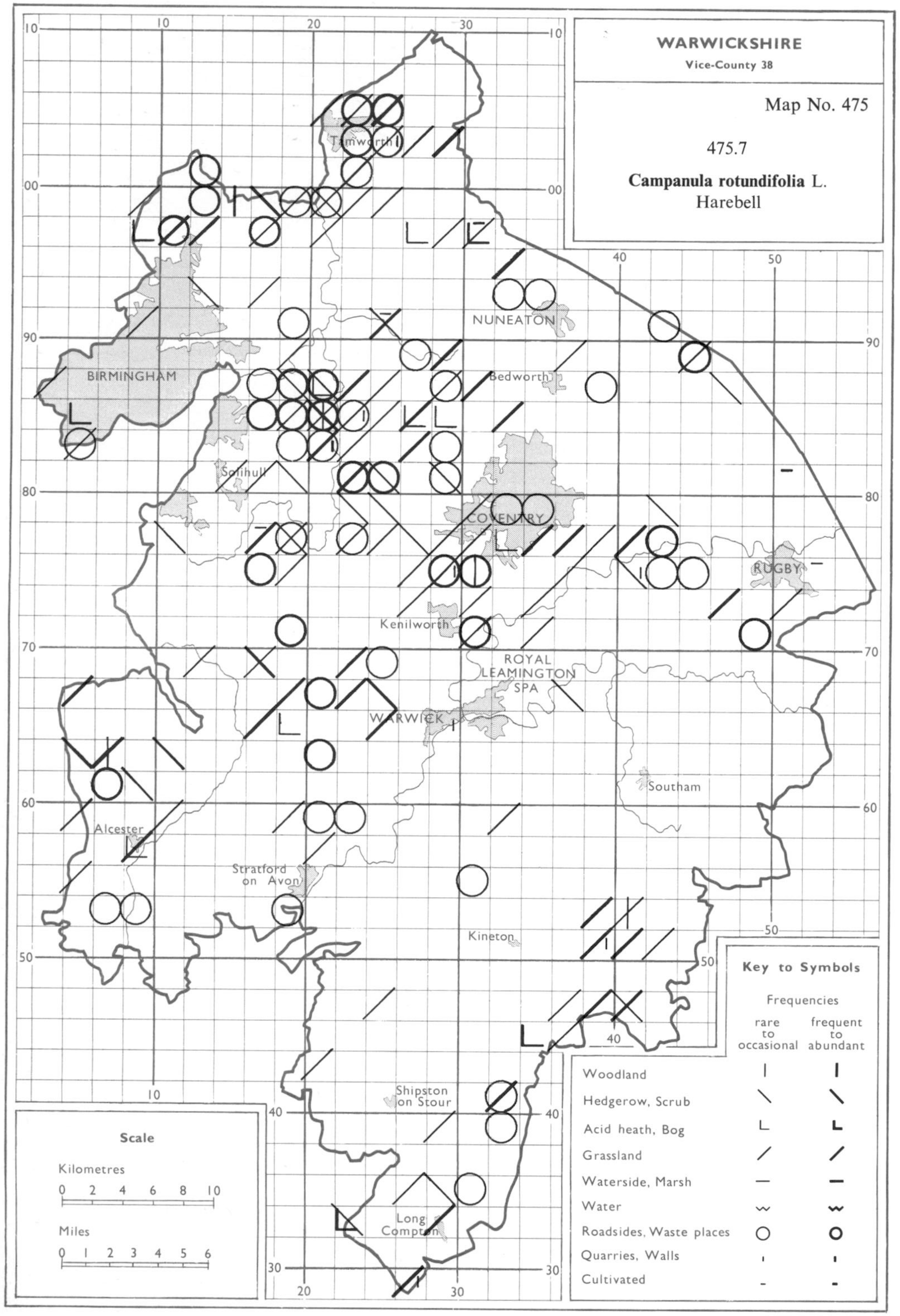
WARWICKSHIRE
Vice-County 38
Map No. 475
475.7
Campanula rotundifolia L.
Harebell
Key to Symbols
Frequencies
rare to occasional
frequent to abundant
Woodland
Hedgerow, Scrub
Acid heath, Bog
Grassland
Waterside, Marsh
Water
Roadsides, Waste places
Quarries, Walls
Cultivated
Scale
Kilometres
0 2 4 6 8 10
Miles
0 1 2 3 4 5 6
Tamworth
NUNEATON
Bedworth
BIRMINGHAM
Solihull
COVENTRY
RUGBY
Kenilworth
ROYAL LEAMINGTON SPA
WARWICK
Southam
Alcester
Stratford on Avon
Kineton
Shipston on Stour
Long Compton

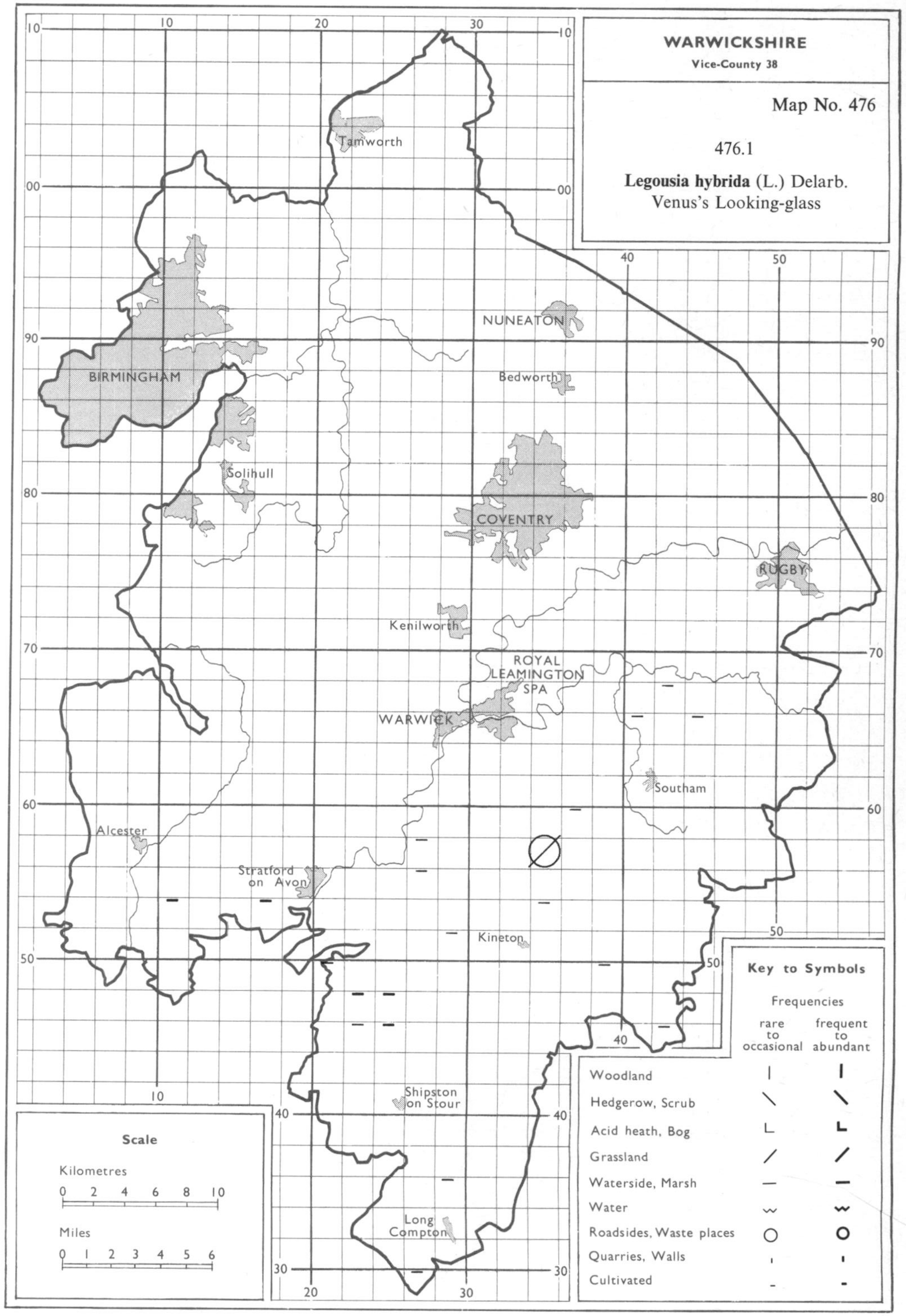
WARWICKSHIRE
Vice-County 38
Map No. 476
476.1
Legousia hybrida (L.) Delarb.
Venus's Looking-glass
Key to Symbols
Frequencies
rare to occasional
frequent to abundant
Woodland
Hedgerow, Scrub
Acid heath, Bog
Grassland
Waterside, Marsh
Water
Roadsides, Waste places
Quarries, Walls
Cultivated
Scale
Kilometres
0 2 4 6 8 10
Miles
0 1 2 3 4 5 6
Tamworth
NUNEATON
Bedworth
BIRMINGHAM
Solihull
COVENTRY
RUGBY
Kenilworth
ROYAL LEAMINGTON SPA
WARWICK
Southam
Alcester
Stratford on Avon
Kineton
Shipston on Stour
Long Compton

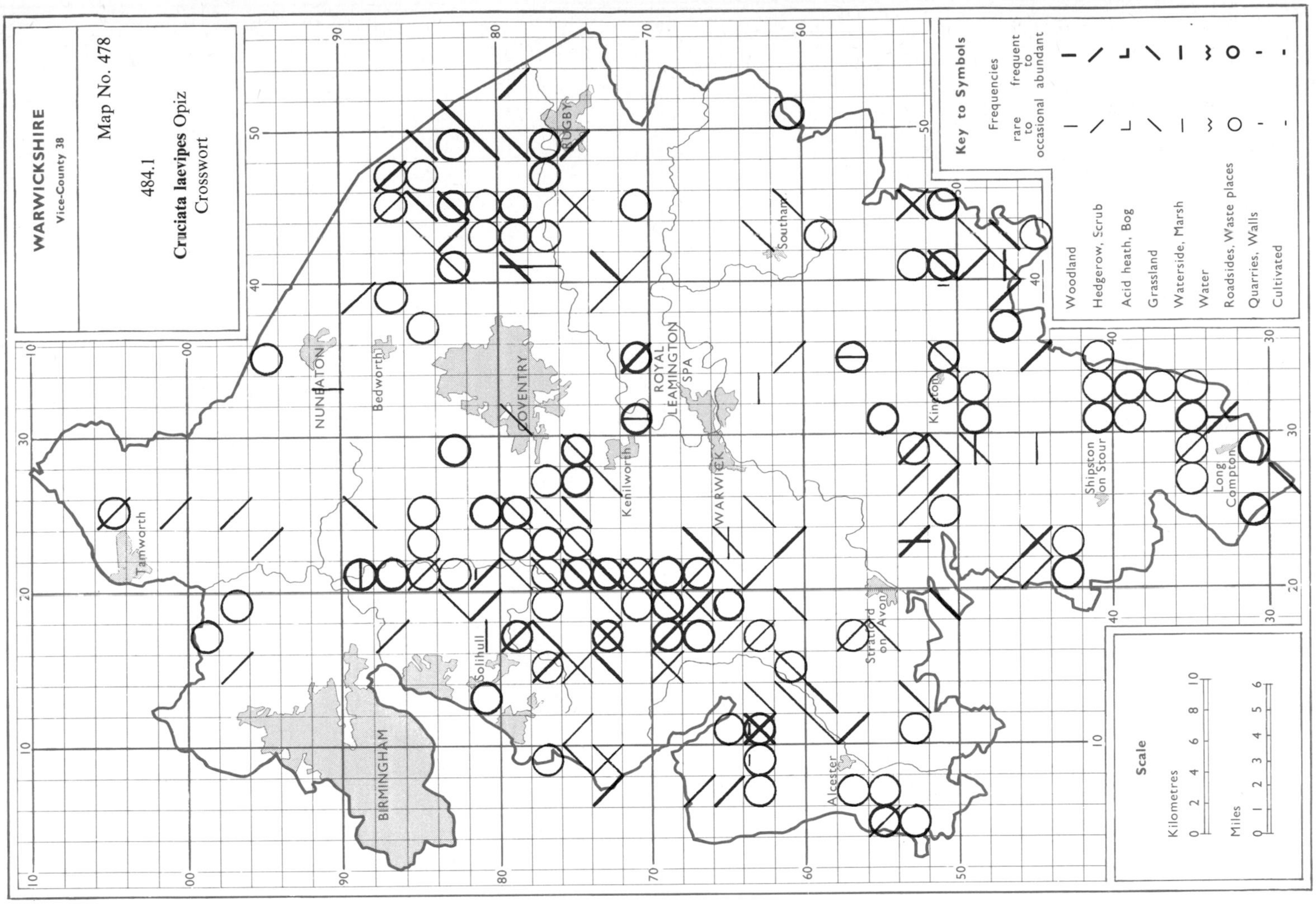
WARWICKSHIRE
Vice-County 38
Map No. 478
484.1
Cruciata laevipes Opiz
Crosswort
Key to Symbols
Frequencies
rare to occasional
frequent to abundant
Woodland
Hedgerow, Scrub
Acid heath, Bog
Grassland
Waterside, Marsh
Water
Roadsides, Waste places
Quarries, Walls
Cultivated
Scale
Kilometres
Miles
Tamworth
NUNEATON
Bedworth
COVENTRY
RUGBY
BIRMINGHAM
Solihull
Kenilworth
ROYAL LEAMINGTON SPA
WARWICK
Southam
Stratford on Avon
Alcester
Kineton
Shipston on Stour
Long Compton

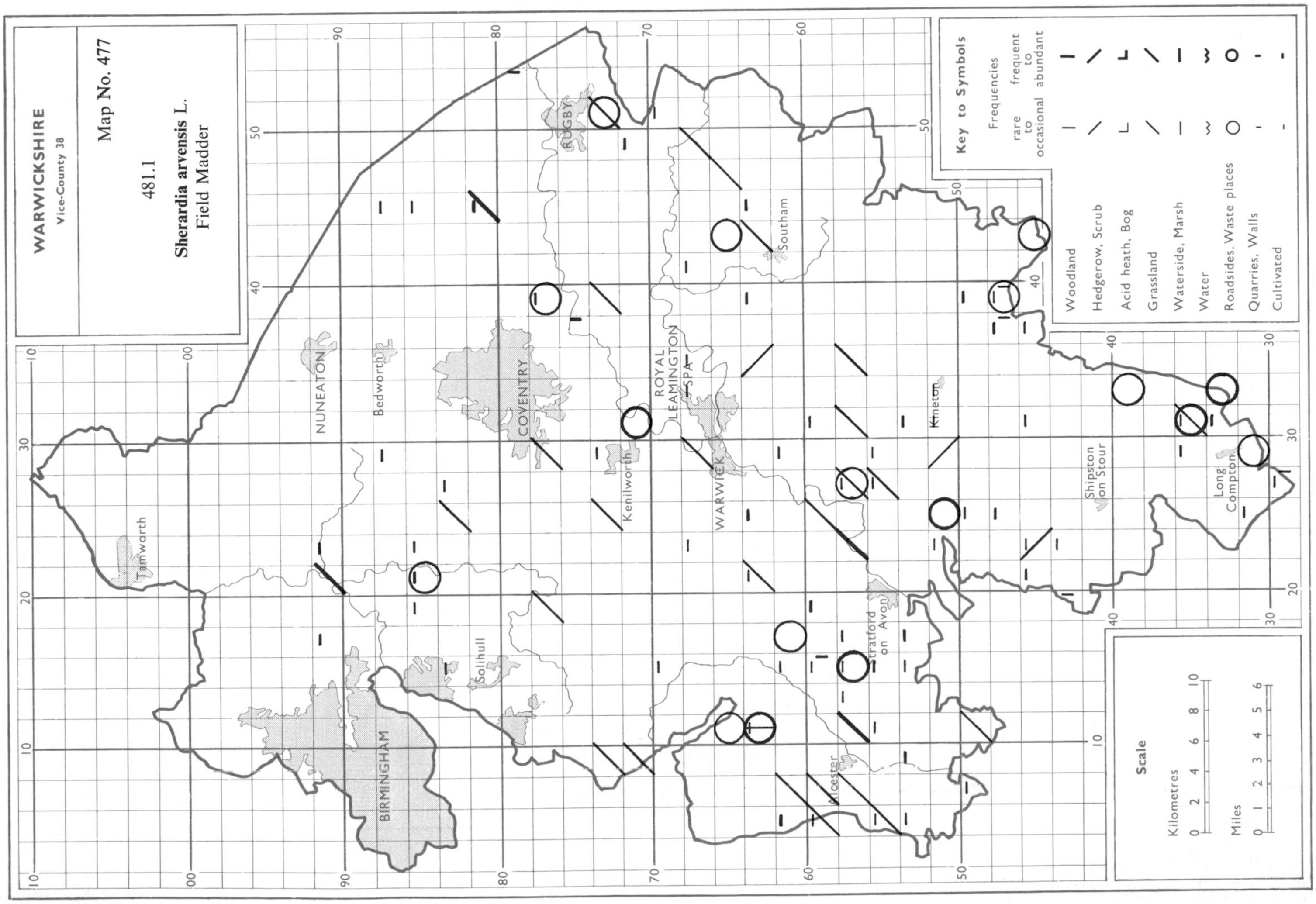
WARWICKSHIRE
Vice-County 38
Map No. 477
481.1
Sherardia arvensis L.
Field Madder
Key to Symbols
Frequencies
rare to occasional
frequent to abundant
Woodland
Hedgerow, Scrub
Acid heath, Bog
Grassland
Waterside, Marsh
Water
Roadsides, Waste places
Quarries, Walls
Cultivated
Scale
Kilometres
Miles
Tamworth
NUNEATON
Bedworth
COVENTRY
RUGBY
BIRMINGHAM
Solihull
Kenilworth
ROYAL LEAMINGTON SPA
WARWICK
Southam
Stratford on Avon
Alcester
Kineton
Shipston on Stour
Long Compton

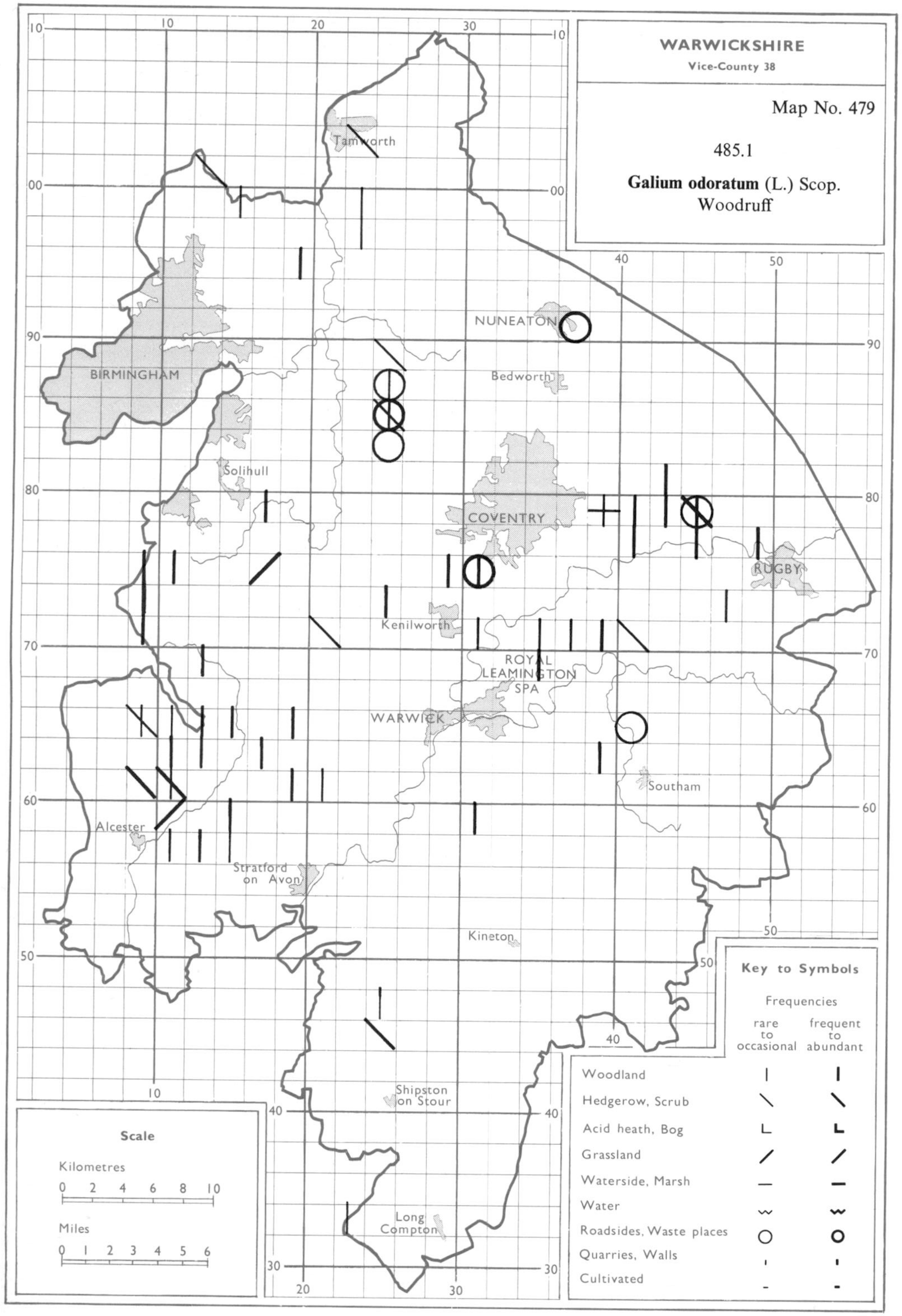
WARWICKSHIRE
Vice-County 38
Map No. 479
485.1
Galium odoratum (L.) Scop.
Woodruff
Tamworth
BIRMINGHAM
Solihull
NUNEATON
Bedworth
COVENTRY
RUGBY
Kenilworth
ROYAL LEAMINGTON SPA
WARWICK
Southam
Alcester
Stratford on Avon
Kineton
Shipston on Stour
Long Compton
Key to Symbols
Frequencies
rare to occasional
frequent to abundant
Woodland
Hedgerow, Scrub
Acid heath, Bog
Grassland
Waterside, Marsh
Water
Roadsides, Waste places
Quarries, Walls
Cultivated
Scale
Kilometres
0 2 4 6 8 10
Miles
0 1 2 3 4 5 6

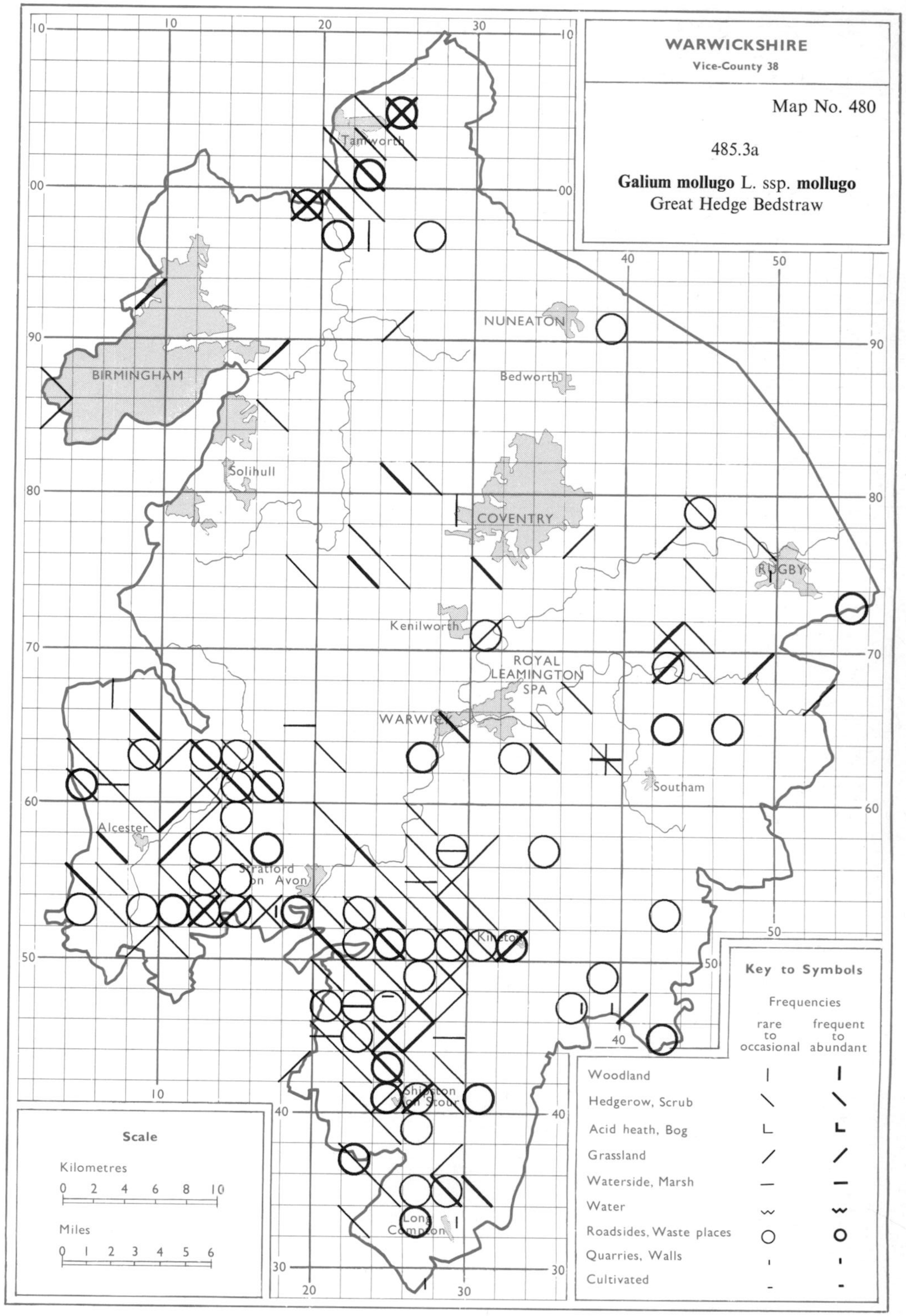
WARWICKSHIRE
Vice-County 38
Map No. 480
485.3a
Galium mollugo L. ssp. mollugo
Great Hedge Bedstraw
Tamworth
BIRMINGHAM
Solihull
NUNEATON
Bedworth
COVENTRY
RUGBY
Kenilworth
ROYAL LEAMINGTON SPA
WARWICK
Southam
Alcester
Stratford on Avon
Kineton
Shipston on Stour
Long Compton
Key to Symbols
Frequencies
rare to occasional
frequent to abundant
Woodland
Hedgerow, Scrub
Acid heath, Bog
Grassland
Waterside, Marsh
Water
Roadsides, Waste places
Quarries, Walls
Cultivated
Scale
Kilometres
0 2 4 6 8 10
Miles
0 1 2 3 4 5 6

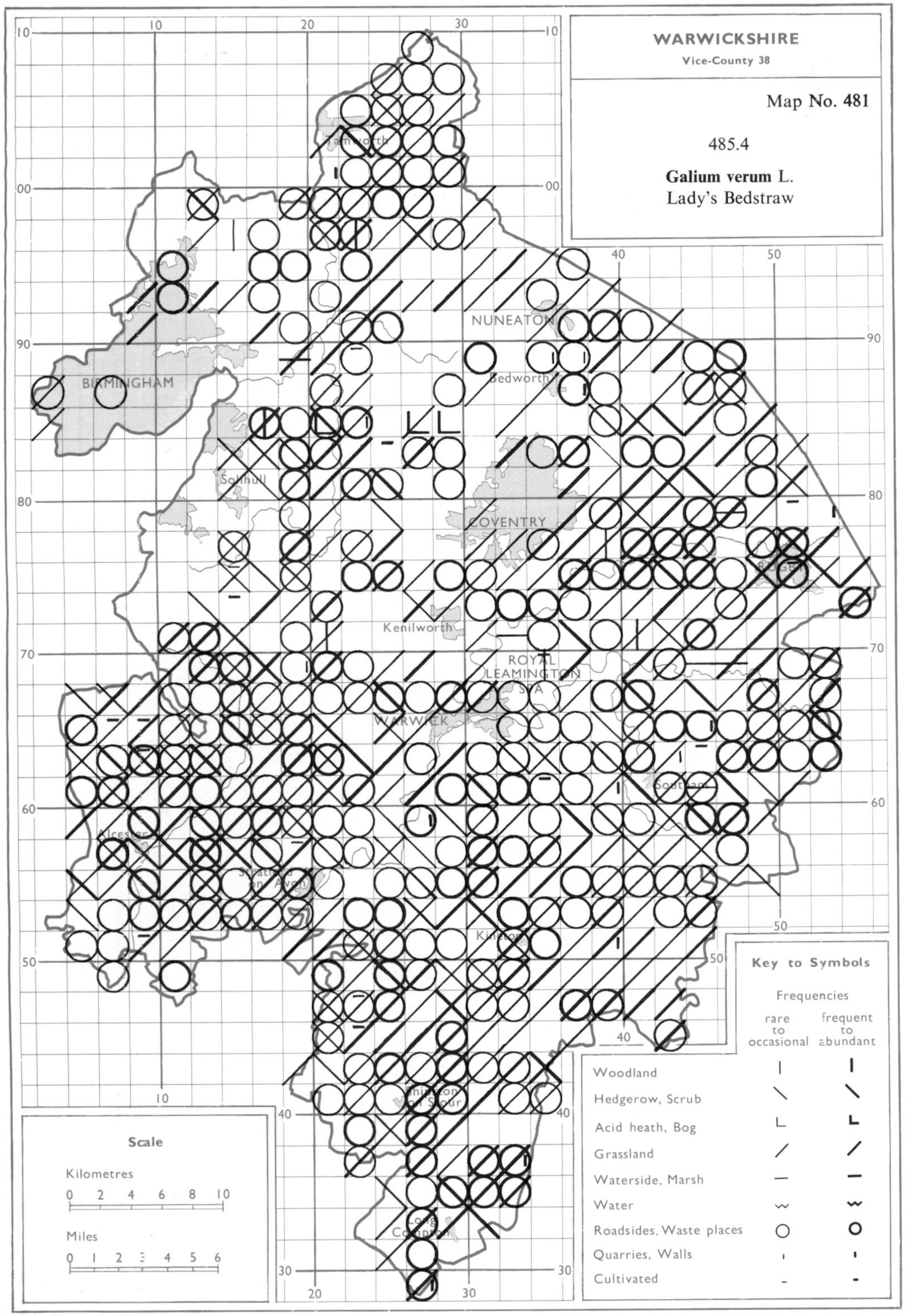
WARWICKSHIRE
Vice-County 38
Map No. 481
485.4
Galium verum L.
Lady's Bedstraw
NUNEATON
Bedworth
BIRMINGHAM
Solihull
COVENTRY
Kenilworth
ROYAL LEAMINGTON SPA
WARWICK
Southam
Alcester
Stratford on Avon
Kineton
Shipston on Stour
Long Compton
Tamworth
Rugby
Key to Symbols
Frequencies
rare to occasional
frequent to abundant
Woodland
Hedgerow, Scrub
Acid heath, Bog
Grassland
Waterside, Marsh
Water
Roadsides, Waste places
Quarries, Walls
Cultivated
Scale
Kilometres
0 2 4 6 8 10
Miles
0 1 2 3 4 5 6

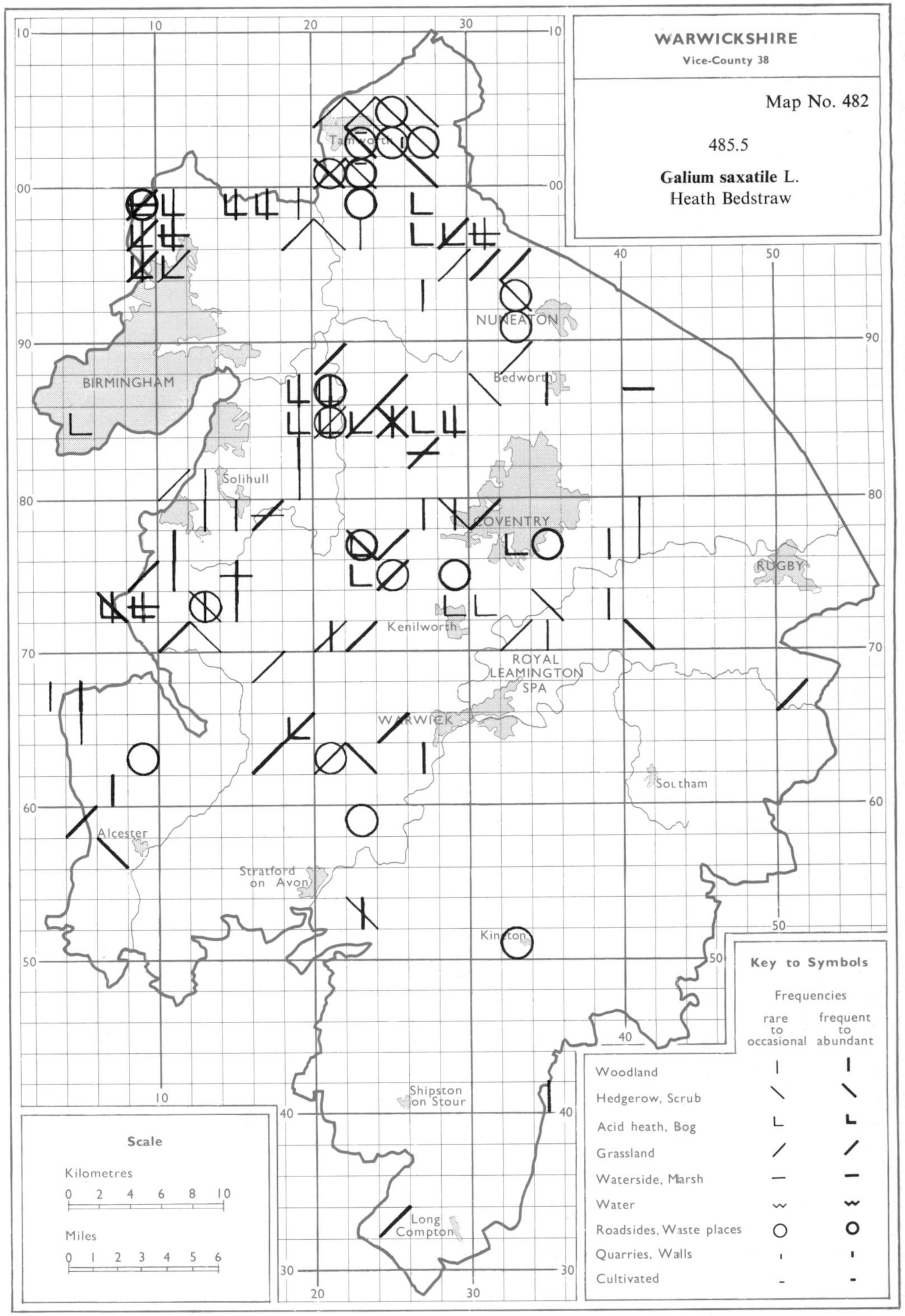
WARWICKSHIRE
Vice-County 38
Map No. 482
485.5
Galium saxatile L.
Heath Bedstraw
Tamworth
NUNEATON
Bedworth
BIRMINGHAM
Solihull
COVENTRY
RUGBY
Kenilworth
ROYAL LEAMINGTON SPA
WARWICK
Southam
Alcester
Stratford on Avon
Kineton
Shipston on Stour
Long Compton
Key to Symbols
Frequencies
rare to occasional
frequent to abundant
Woodland
Hedgerow, Scrub
Acid heath, Bog
Grassland
Waterside, Marsh
Water
Roadsides, Waste places
Quarries, Walls
Cultivated
Scale
Kilometres
0 2 4 6 8 10
Miles
0 1 2 3 4 5 6

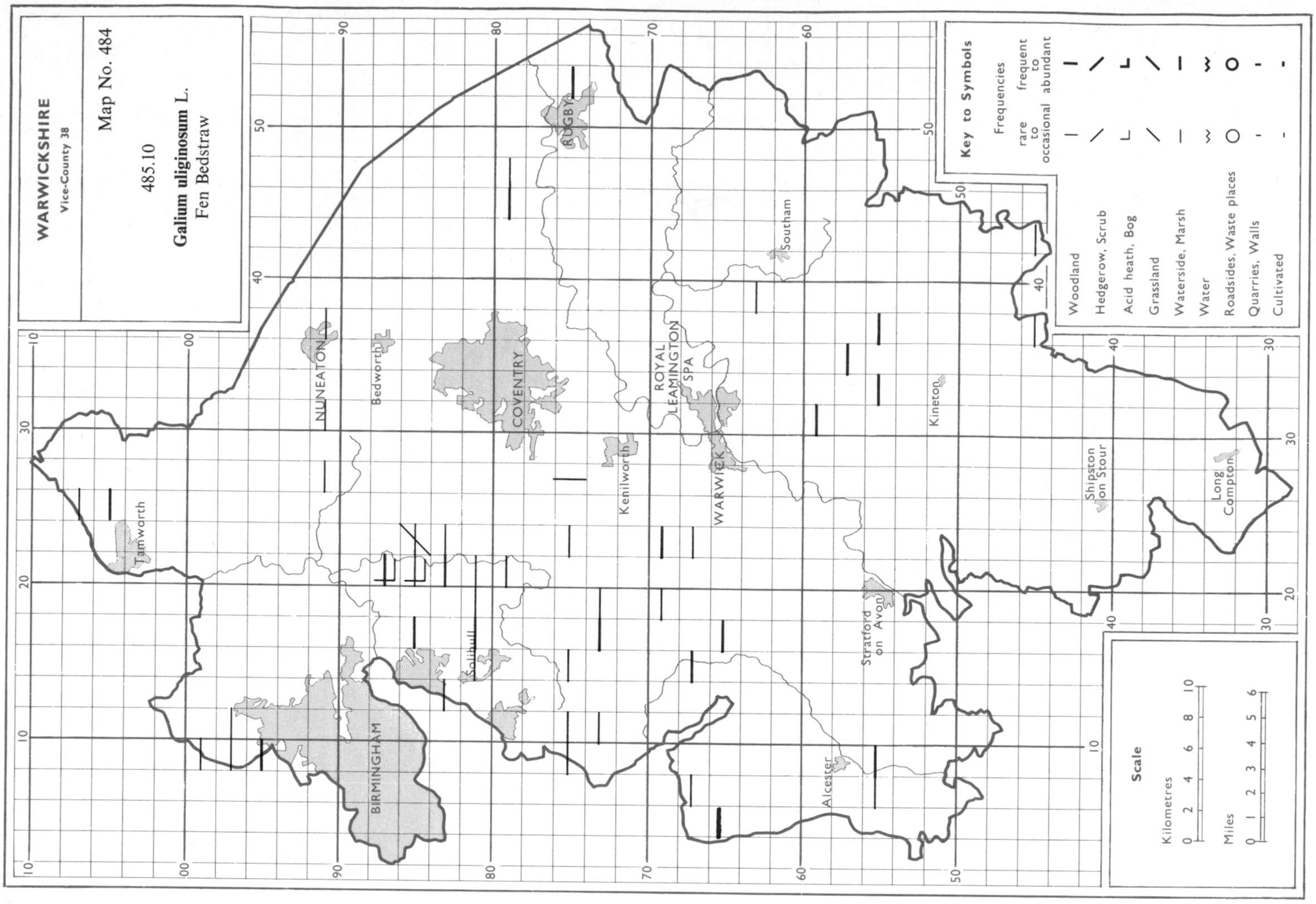
WARWICKSHIRE
Vice-County 38
Map No. 484
485.10
Galium uliginosum L.
Fen Bedstraw
Key to Symbols
Frequencies
rare to occasional
frequent to abundant
Woodland
Hedgerow, Scrub
Acid heath, Bog
Grassland
Waterside, Marsh
Water
Roadsides, Waste places
Quarries, Walls
Cultivated
Scale
Kilometres
Miles
RUGBY
Southam
NUNEATON
Bedworth
COVENTRY
ROYAL LEAMINGTON SPA
Kineton
Kenilworth
WARWICK
Tamworth
Shipston on Stour
Long Compton
Solihull
BIRMINGHAM
Stratford on Avon
Alcester

WARWICKSHIRE
Vice-County 38
Map No. 483
485.8
Galium palustre L.
Marsh Bedstraw
Key to Symbols
Frequencies
rare to occasional
frequent to abundant
Woodland
Hedgerow, Scrub
Acid heath, Bog
Grassland
Waterside, Marsh
Water
Roadsides, Waste places
Quarries, Walls
Cultivated
Scale
Kilometres
Miles
RUGBY
Southam
NUNEATON
Bedworth
COVENTRY
ROYAL LEAMINGTON SPA
Kineton
Kenilworth
WARWICK
Tamworth
Shipston on Stour
Long Compton
Solihull
BIRMINGHAM
Stratford on Avon
Alcester

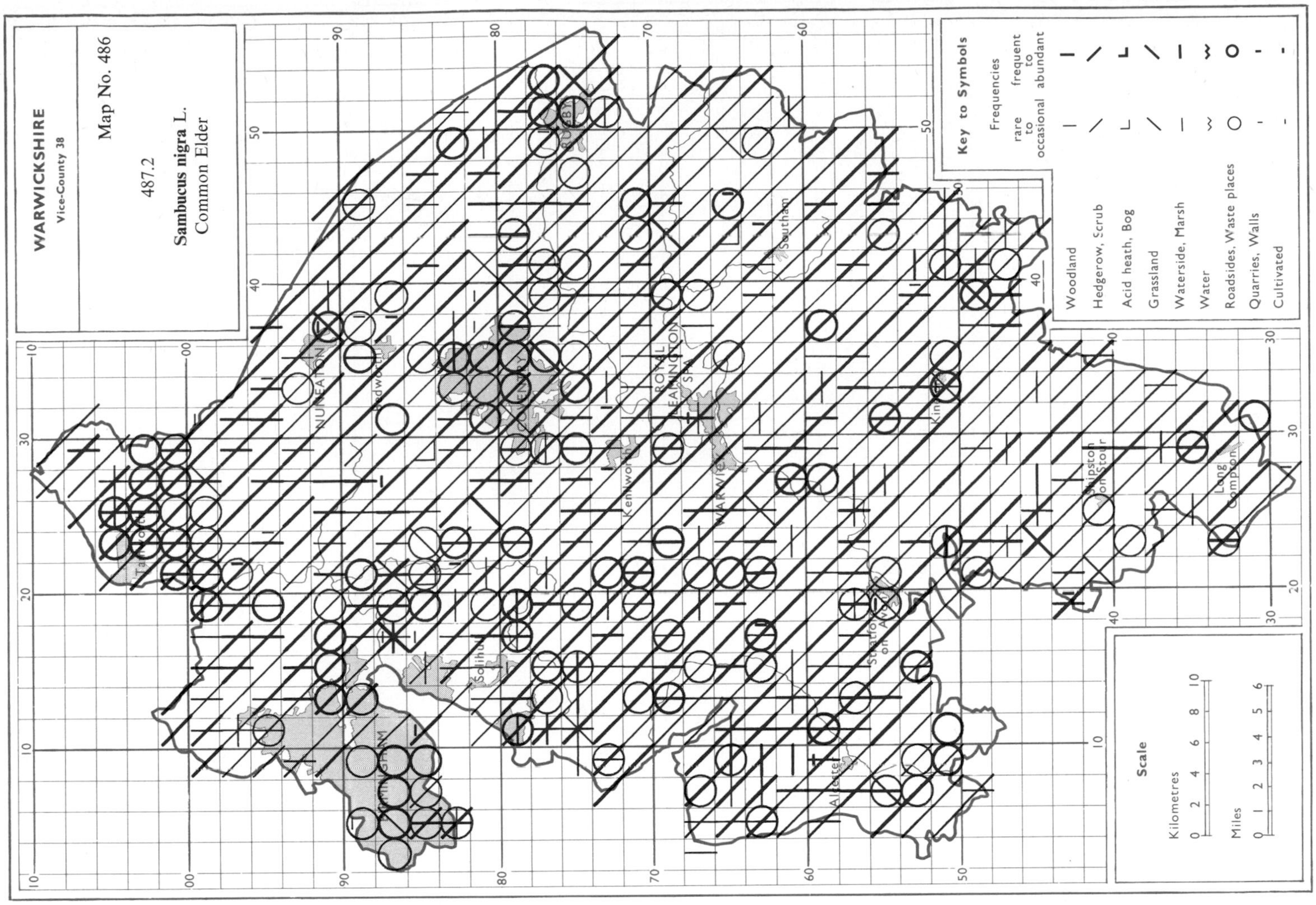
WARWICKSHIRE
Vice-County 38
Map No. 486
487.2
Sambucus nigra L.
Common Elder
Key to Symbols
Frequencies
rare to occasional
frequent to abundant
Woodland
Hedgerow, Scrub
Acid heath, Bog
Grassland
Waterside, Marsh
Water
Roadsides, Waste places
Quarries, Walls
Cultivated
Scale
Kilometres
Miles
NUNEATON
COVENTRY
RUGBY
Southam
ROYAL LEAMINGTON SPA
Kenilworth
WARWICK
Solihull
Tamworth
Alcester
Stratford on Avon
Shipston on Stour
Long Compton
BIRMINGHAM

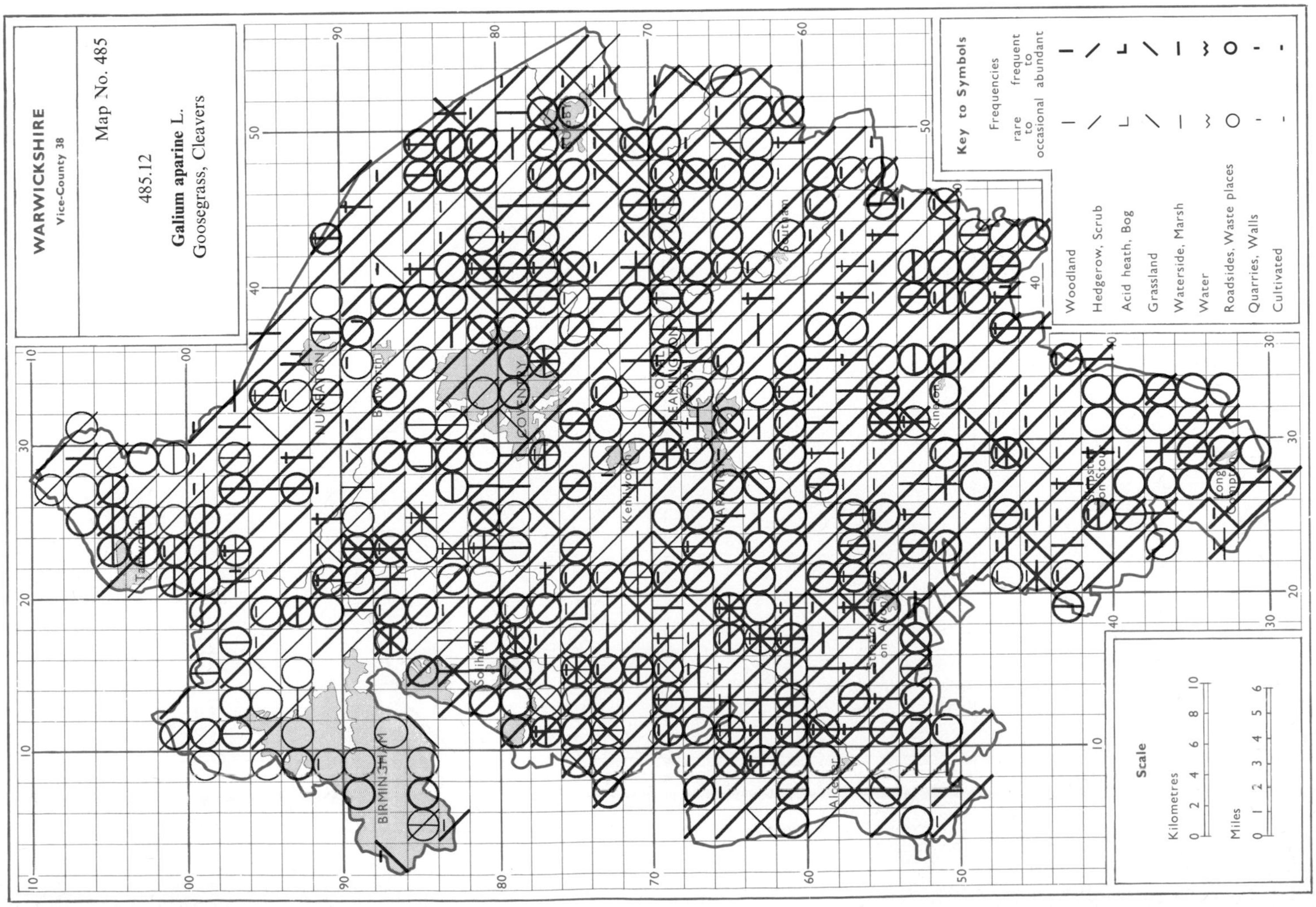
WARWICKSHIRE
Vice-County 38
Map No. 485
485.12
Galium aparine L.
Goosegrass, Cleavers
Key to Symbols
Frequencies
rare to occasional
frequent to abundant
Woodland
Hedgerow, Scrub
Acid heath, Bog
Grassland
Waterside, Marsh
Water
Roadsides, Waste places
Quarries, Walls
Cultivated
Scale
Kilometres
Miles
NUNEATON
Bedworth
COVENTRY
Southam
ROYAL LEAMINGTON SPA
Kenilworth
WARWICK
Solihull
BIRMINGHAM
Alcester
Stratford on Avon
Shipston on Stour
Long Compton

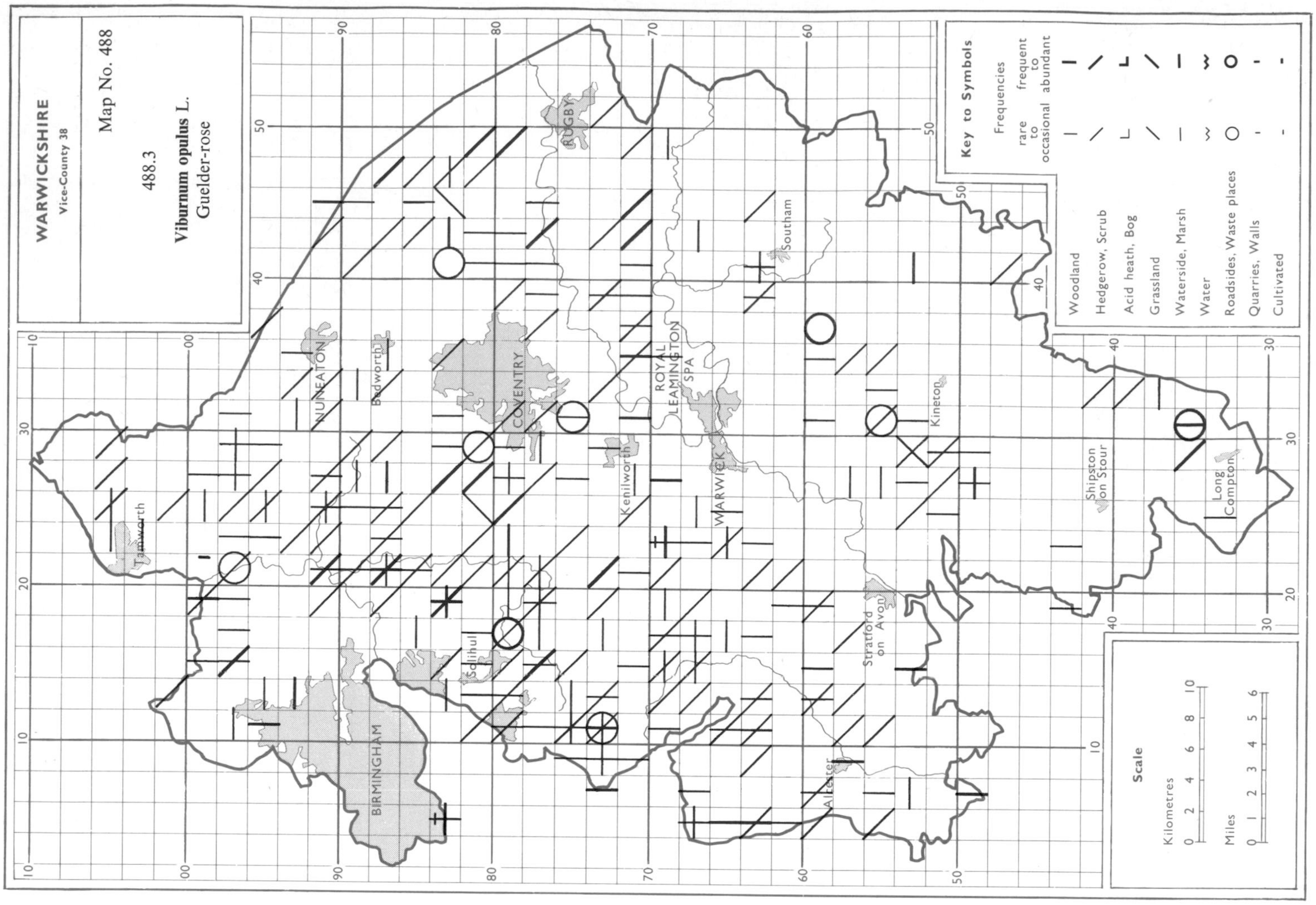
WARWICKSHIRE
Vice-County 38
Map No. 488
488.3
Viburnum opulus L.
Guelder-rose
Key to Symbols
Frequencies
rare to occasional
frequent to abundant
Woodland
Hedgerow, Scrub
Acid heath, Bog
Grassland
Waterside, Marsh
Water
Roadsides, Waste places
Quarries, Walls
Cultivated
Scale
Kilometres 0 2 4 6 8 10
Miles 0 1 2 3 4 5 6
TAMWORTH
NUNEATON
Bedworth
RUGBY
COVENTRY
Southam
ROYAL LEAMINGTON SPA
Kenilworth
WARWICK
Kineton
Solihull
BIRMINGHAM
Stratford on Avon
Shipston on Stour
Long Compton
Alcester

WARWICKSHIRE
Vice-County 38
Map No. 487
488.1
Viburnum lantana L.
Wayfaring-tree
Key to Symbols
Frequencies
rare to occasional
frequent to abundant
Woodland
Hedgerow, Scrub
Acid heath, Bog
Grassland
Waterside, Marsh
Water
Roadsides, Waste places
Quarries, Walls
Cultivated
Scale
Kilometres 0 2 4 6 8 10
Miles 0 1 2 3 4 5 6
Tamworth
NUNEATON
Bedworth
RUGBY
COVENTRY
Southam
ROYAL LEAMINGTON SPA
Kenilworth
WARWICK
Solihull
BIRMINGHAM
Stratford on Avon
Shipston on Stour
Long Compton
Alcester

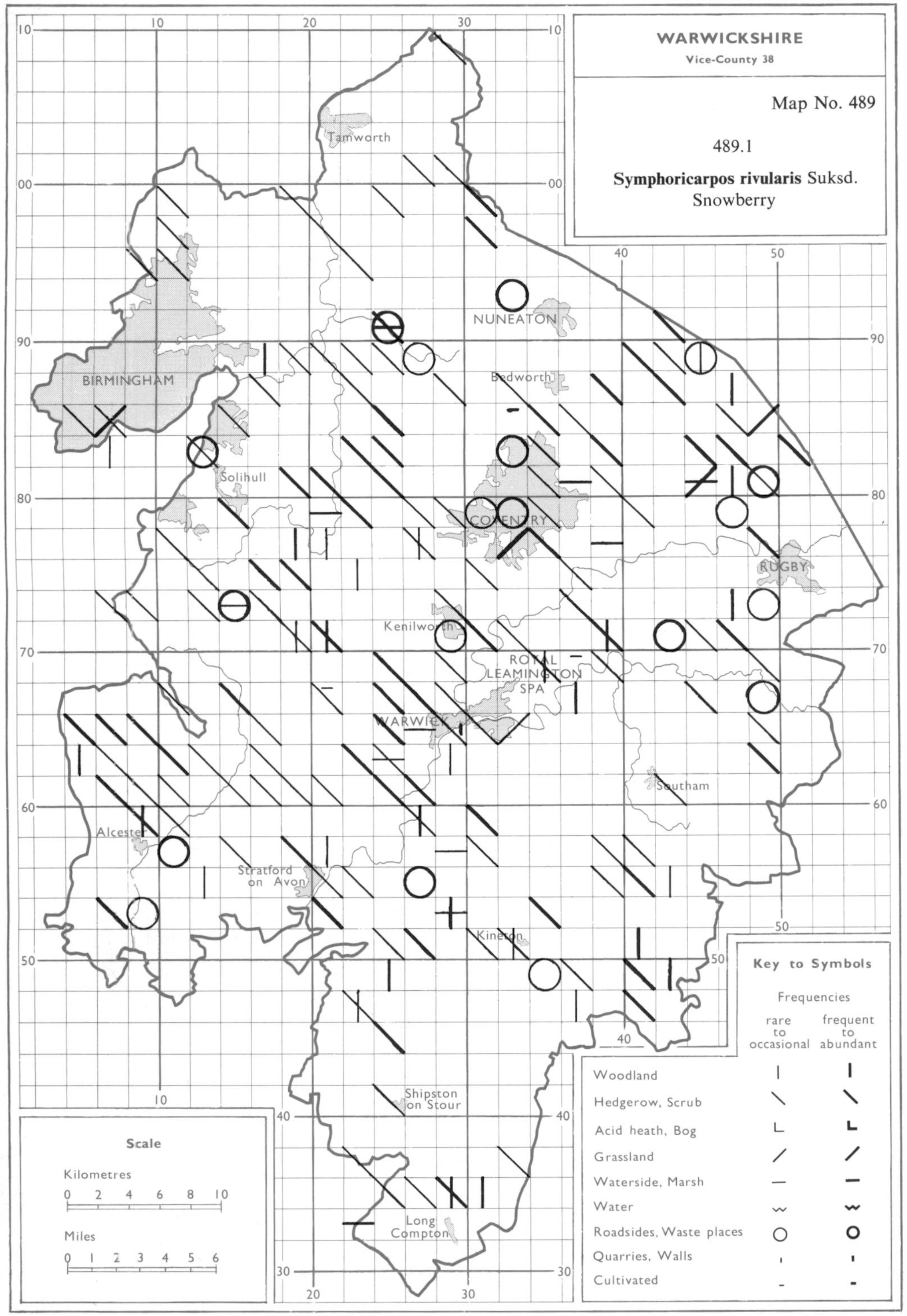

WARWICKSHIRE
Vice-County 38
Map No. 489
489.1
Symphoricarpos rivularis Suksd.
Snowberry
Tamworth
NUNEATON
Bedworth
BIRMINGHAM
Solihull
COVENTRY
RUGBY
Kenilworth
ROYAL LEAMINGTON SPA
WARWICK
Southam
Alcester
Stratford on Avon
Kineton
Shipston on Stour
Long Compton
Key to Symbols
Frequencies
rare to occasional
frequent to abundant
Woodland
Hedgerow, Scrub
Acid heath, Bog
Grassland
Waterside, Marsh
Water
Roadsides, Waste places
Quarries, Walls
Cultivated
Scale
Kilometres
Miles

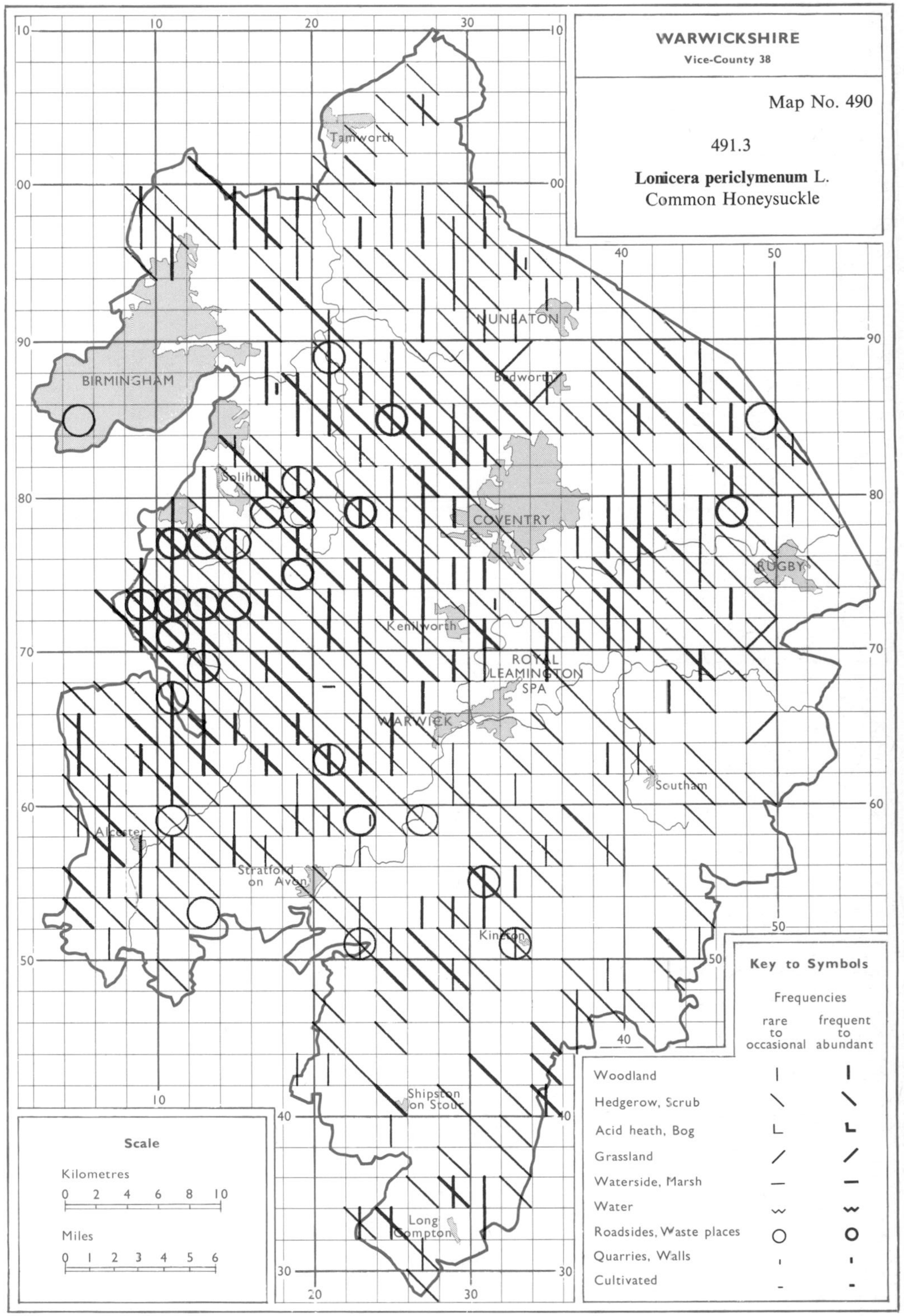

WARWICKSHIRE
Vice-County 38
Map No. 490
491.3
Lonicera periclymenum L.
Common Honeysuckle
Tamworth
NUNEATON
Bedworth
BIRMINGHAM
Solihull
COVENTRY
RUGBY
Kenilworth
ROYAL LEAMINGTON SPA
WARWICK
Southam
Alcester
Stratford on Avon
Kineton
Shipston on Stour
Long Compton
Key to Symbols
Frequencies
rare to occasional
frequent to abundant
Woodland
Hedgerow, Scrub
Acid heath, Bog
Grassland
Waterside, Marsh
Water
Roadsides, Waste places
Quarries, Walls
Cultivated
Scale
Kilometres
Miles

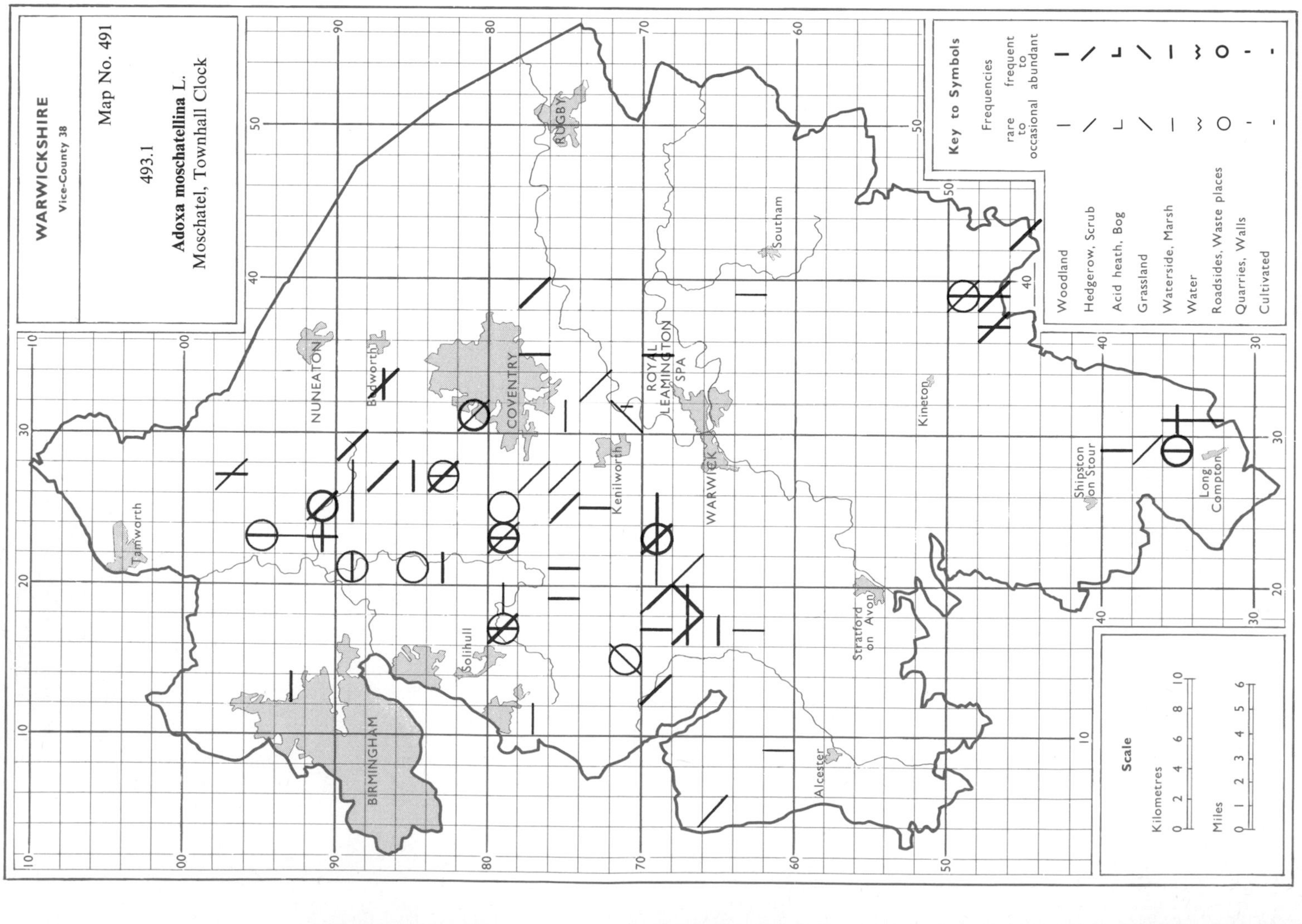
WARWICKSHIRE
Vice-County 38
Map No. 491
493.1
Adoxa moschatellina L.
Moschatel, Townhall Clock
Key to Symbols
Frequencies
rare to occasional
frequent to abundant
Woodland
Hedgerow, Scrub
Acid heath, Bog
Grassland
Waterside, Marsh
Water
Roadsides, Waste places
Quarries, Walls
Cultivated
Scale
Kilometres
Miles
NUNEATON
Bedworth
COVENTRY
RUGBY
Tamworth
Kenilworth
ROYAL LEAMINGTON SPA
WARWICK
Southam
Kineton
Shipston on Stour
Long Compton
Solihull
BIRMINGHAM
Stratford on Avon
Alcester

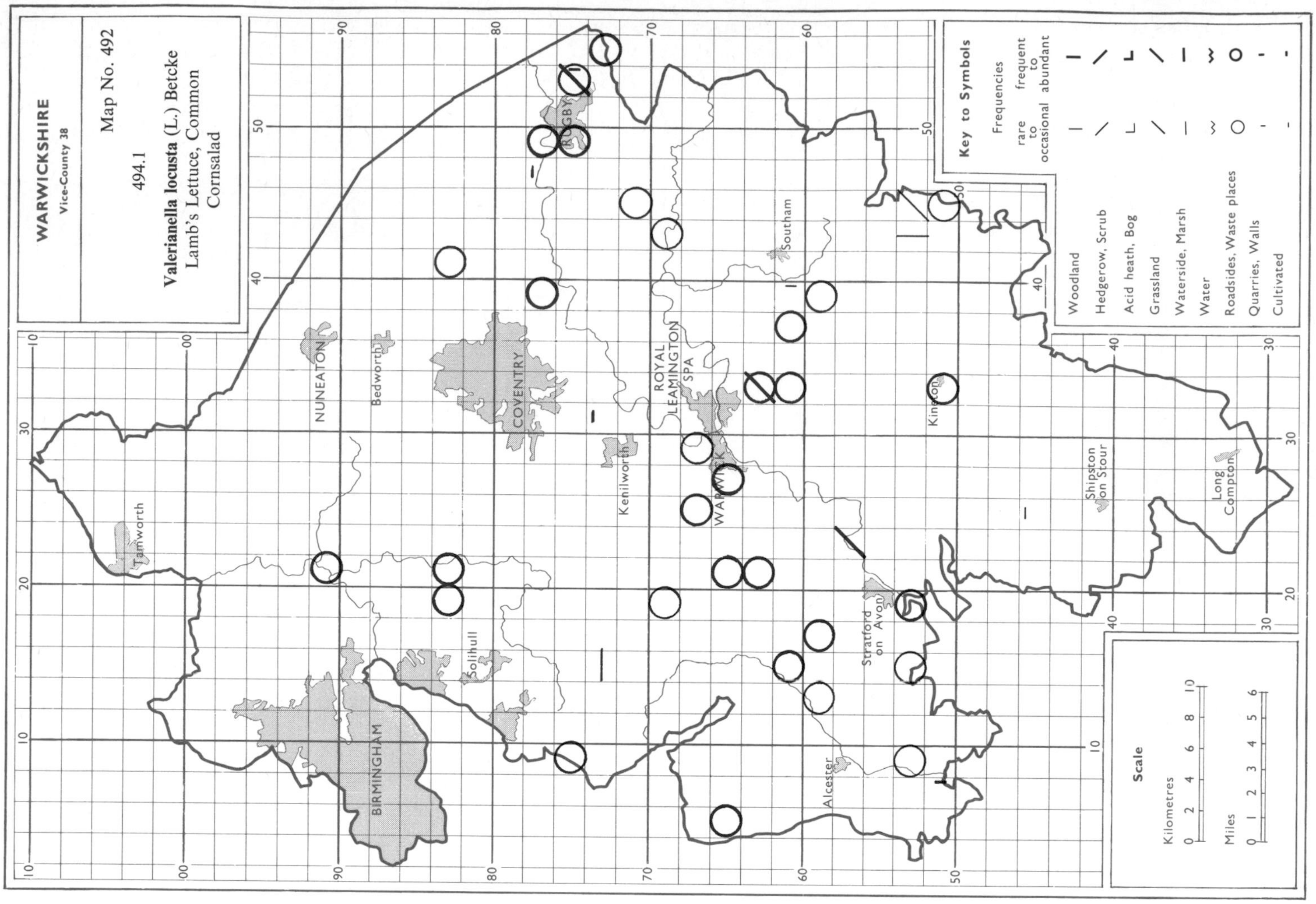
WARWICKSHIRE
Vice-County 38
Map No. 492
494.1
Valerianella locusta (L.) Betcke
Lamb's Lettuce, Common Cornsalad
Key to Symbols
Frequencies
rare to occasional
frequent to abundant
Woodland
Hedgerow, Scrub
Acid heath, Bog
Grassland
Waterside, Marsh
Water
Roadsides, Waste places
Quarries, Walls
Cultivated
Scale
Kilometres
Miles
NUNEATON
Bedworth
COVENTRY
RUGBY
Tamworth
Kenilworth
ROYAL LEAMINGTON SPA
WARWICK
Southam
Kineton
Shipston on Stour
Long Compton
Solihull
BIRMINGHAM
Stratford on Avon
Alcester

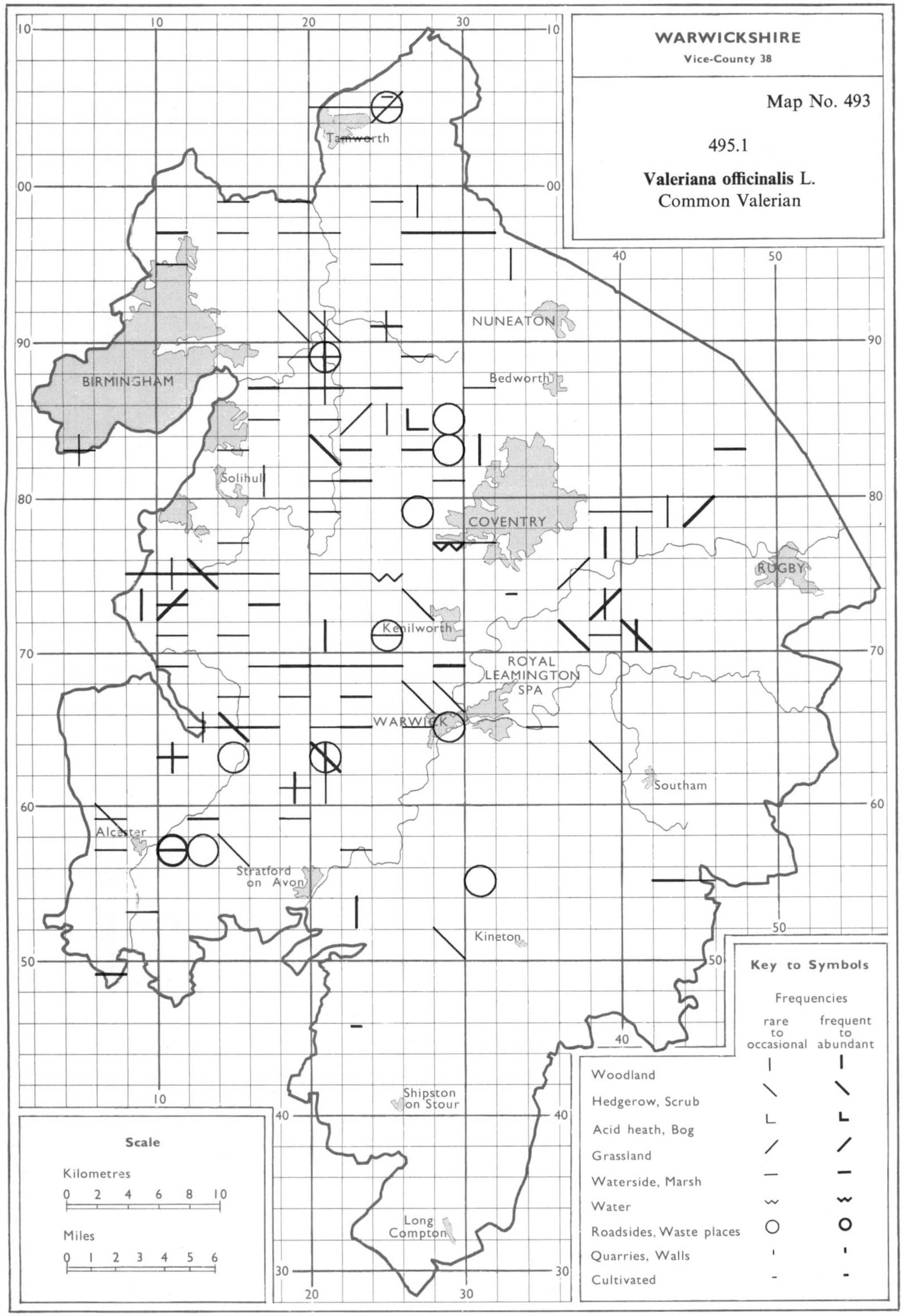
WARWICKSHIRE
Vice-County 38
Map No. 493
495.1
Valeriana officinalis L.
Common Valerian
Tamworth
NUNEATON
Bedworth
BIRMINGHAM
Solihull
COVENTRY
RUGBY
Kenilworth
ROYAL LEAMINGTON SPA
WARWICK
Southam
Alcester
Stratford on Avon
Kineton
Shipston on Stour
Long Compton
Key to Symbols
Frequencies
rare to occasional
frequent to abundant
Woodland
Hedgerow, Scrub
Acid heath, Bog
Grassland
Waterside, Marsh
Water
Roadsides, Waste places
Quarries, Walls
Cultivated
Scale
Kilometres
0 2 4 6 8 10
Miles
0 1 2 3 4 5 6

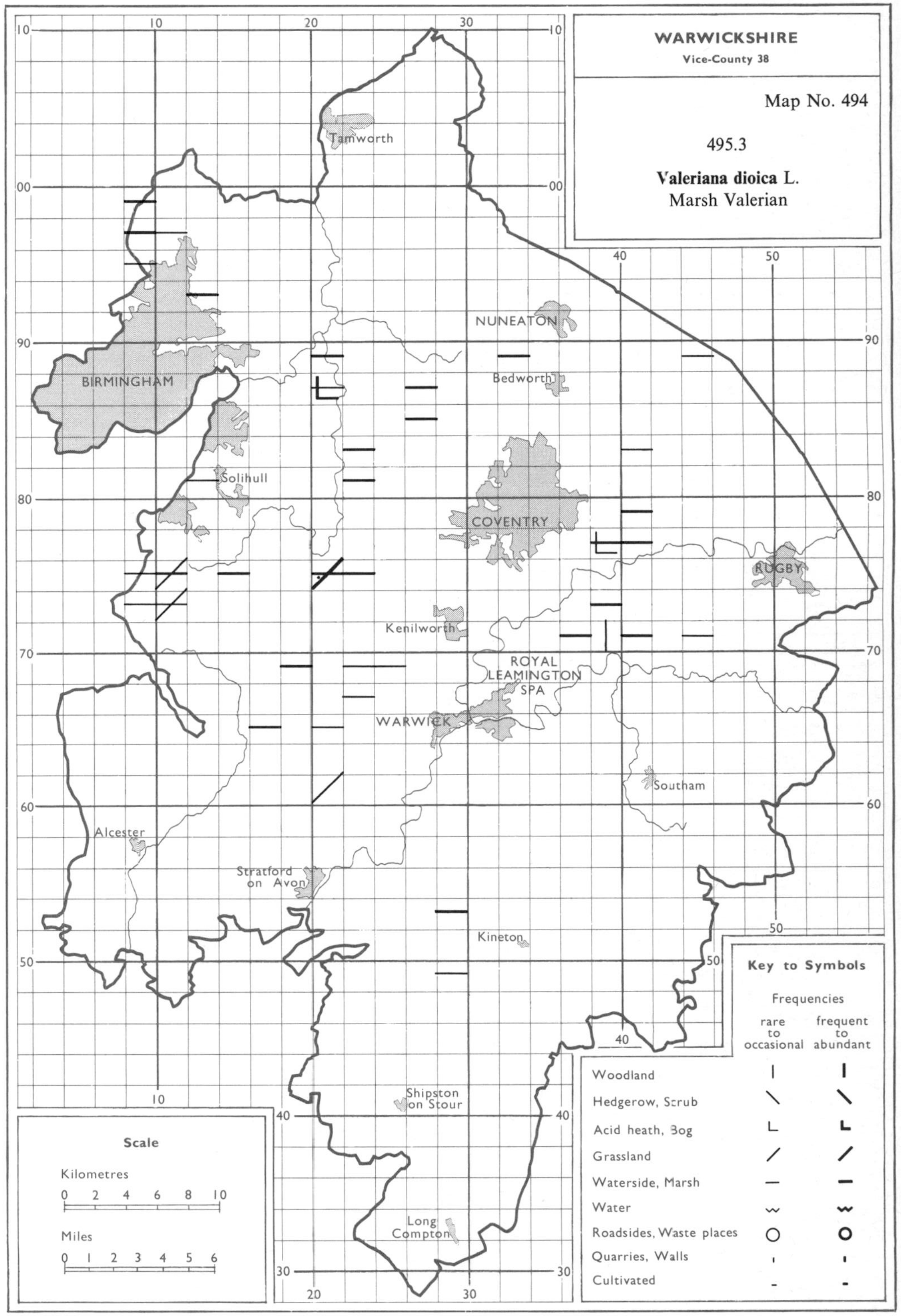
WARWICKSHIRE
Vice-County 38
Map No. 494
495.3
Valeriana dioica L.
Marsh Valerian
Tamworth
NUNEATON
Bedworth
BIRMINGHAM
Solihull
COVENTRY
RUGBY
Kenilworth
ROYAL LEAMINGTON SPA
WARWICK
Southam
Alcester
Stratford on Avon
Kineton
Shipston on Stour
Long Compton
Key to Symbols
Frequencies
rare to occasional
frequent to abundant
Woodland
Hedgerow, Scrub
Acid heath, Bog
Grassland
Waterside, Marsh
Water
Roadsides, Waste places
Quarries, Walls
Cultivated
Scale
Kilometres
0 2 4 6 8 10
Miles
0 1 2 3 4 5 6

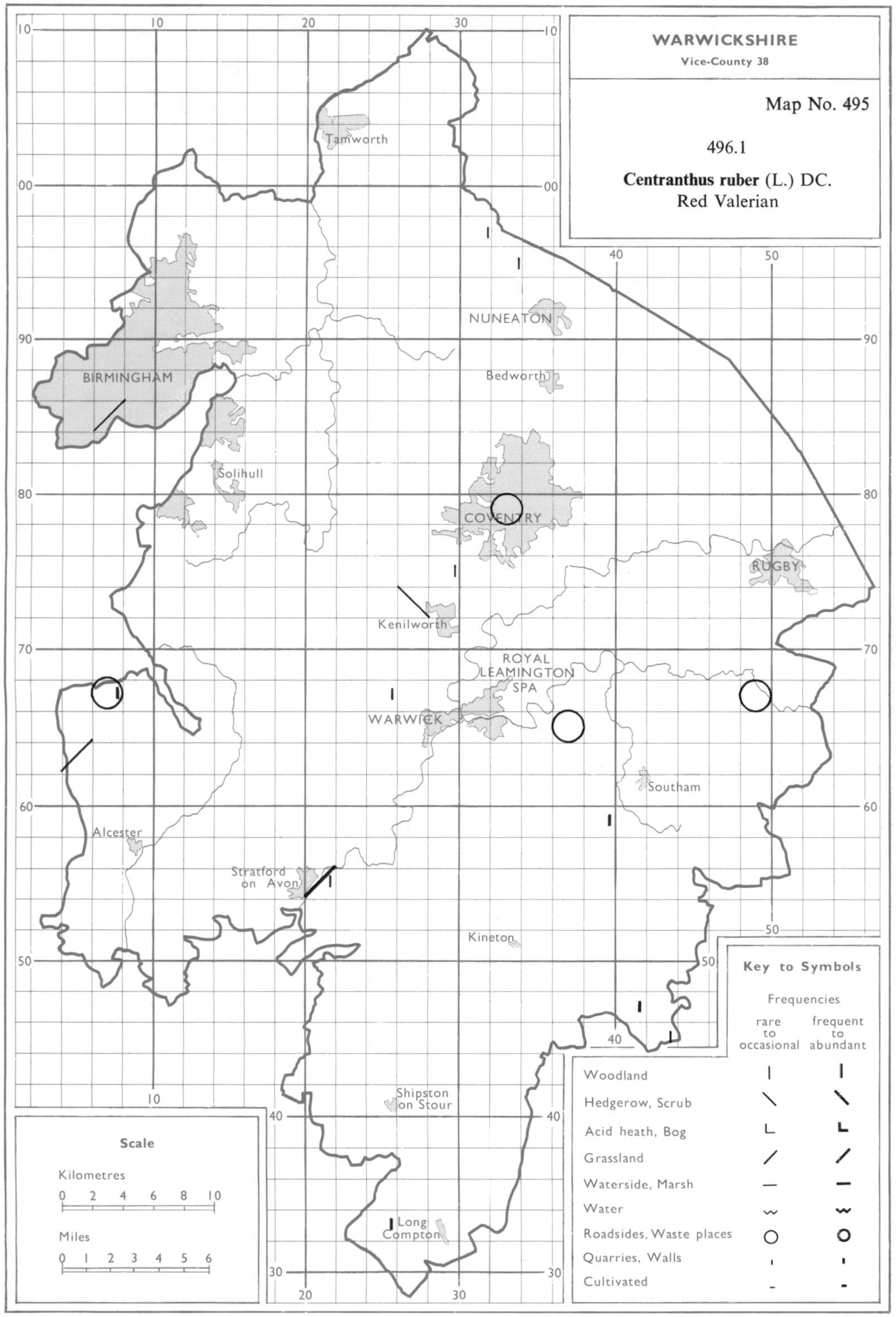
WARWICKSHIRE
Vice-County 38
Map No. 495
496.1
Centranthus ruber (L.) DC.
Red Valerian
Tamworth
NUNEATON
Bedworth
BIRMINGHAM
Solihull
COVENTRY
RUGBY
Kenilworth
ROYAL LEAMINGTON SPA
WARWICK
Southam
Alcester
Stratford on Avon
Kineton
Shipston on Stour
Long Compton
Key to Symbols
Frequencies
rare to occasional
frequent to abundant
Woodland
Hedgerow, Scrub
Acid heath, Bog
Grassland
Waterside, Marsh
Water
Roadsides, Waste places
Quarries, Walls
Cultivated
Scale
Kilometres
0 2 4 6 8 10
Miles
0 1 2 3 4 5 6

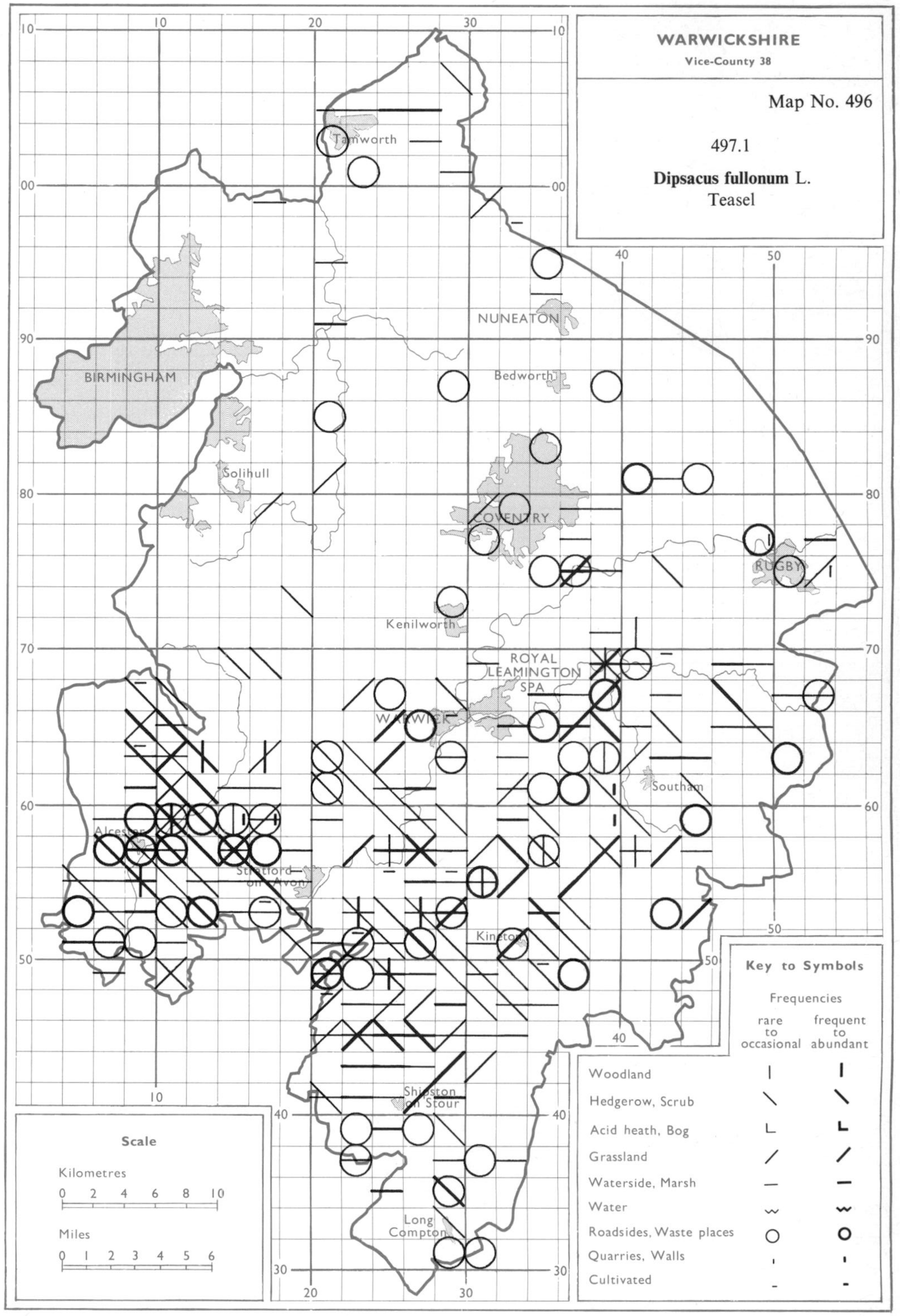
WARWICKSHIRE
Vice-County 38
Map No. 496
497.1
Dipsacus fullonum L.
Teasel
Tamworth
NUNEATON
Bedworth
BIRMINGHAM
Solihull
COVENTRY
RUGBY
Kenilworth
ROYAL LEAMINGTON SPA
WARWICK
Southam
Alcester
Stratford on Avon
Kineton
Shipston on Stour
Long Compton
Key to Symbols
Frequencies
rare to occasional
frequent to abundant
Woodland
Hedgerow, Scrub
Acid heath, Bog
Grassland
Waterside, Marsh
Water
Roadsides, Waste places
Quarries, Walls
Cultivated
Scale
Kilometres
0 2 4 6 8 10
Miles
0 1 2 3 4 5 6

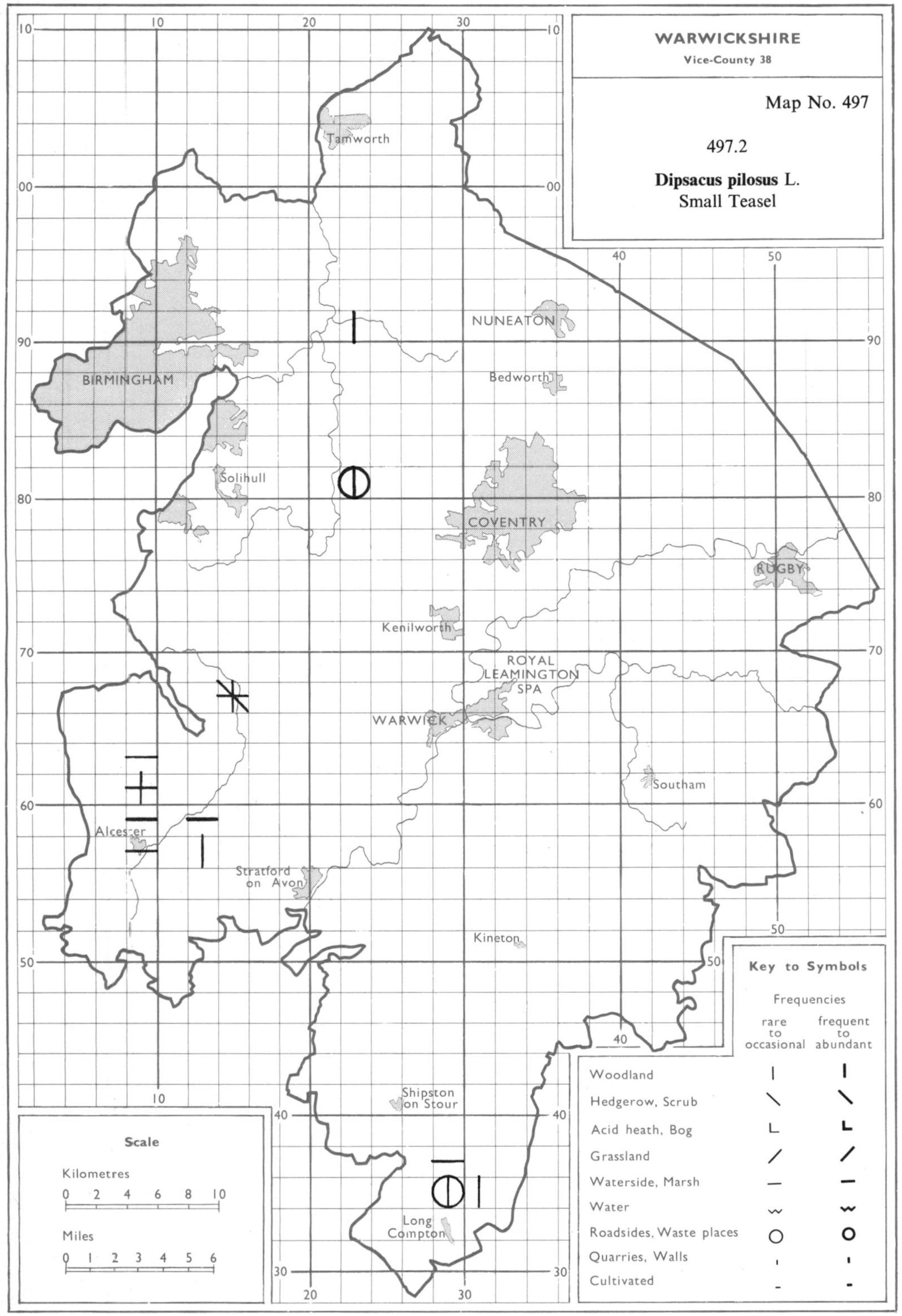
WARWICKSHIRE
Vice-County 38
Map No. 497
497.2
Dipsacus pilosus L.
Small Teasel
Tamworth
NUNEATON
Bedworth
BIRMINGHAM
Solihull
COVENTRY
RUGBY
Kenilworth
ROYAL LEAMINGTON SPA
WARWICK
Southam
Alcester
Stratford on Avon
Kineton
Shipston on Stour
Long Compton
Key to Symbols
Frequencies
rare to occasional
frequent to abundant
Woodland
Hedgerow, Scrub
Acid heath, Bog
Grassland
Waterside, Marsh
Water
Roadsides, Waste places
Quarries, Walls
Cultivated
Scale
Kilometres
0 2 4 6 8 10
Miles
0 1 2 3 4 5 6

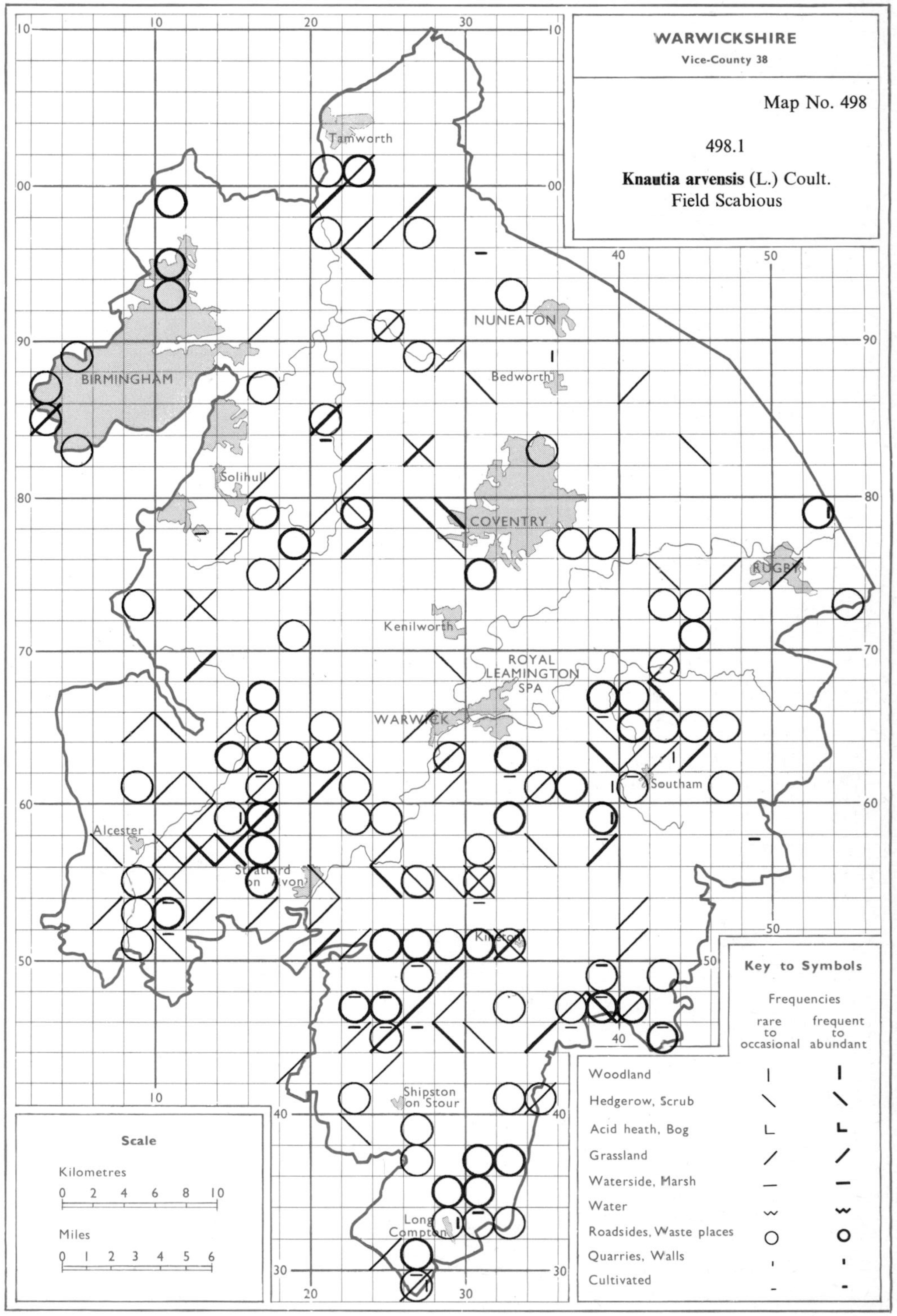
WARWICKSHIRE
Vice-County 38
Map No. 498
498.1
Knautia arvensis (L.) Coult.
Field Scabious
Tamworth
NUNEATON
Bedworth
BIRMINGHAM
Solihull
COVENTRY
RUGBY
Kenilworth
ROYAL LEAMINGTON SPA
WARWICK
Southam
Alcester
Stratford on Avon
Kineton
Shipston on Stour
Long Compton
Key to Symbols
Frequencies
rare to occasional
frequent to abundant
Woodland
Hedgerow, Scrub
Acid heath, Bog
Grassland
Waterside, Marsh
Water
Roadsides, Waste places
Quarries, Walls
Cultivated
Scale
Kilometres
0 2 4 6 8 10
Miles
0 1 2 3 4 5 6

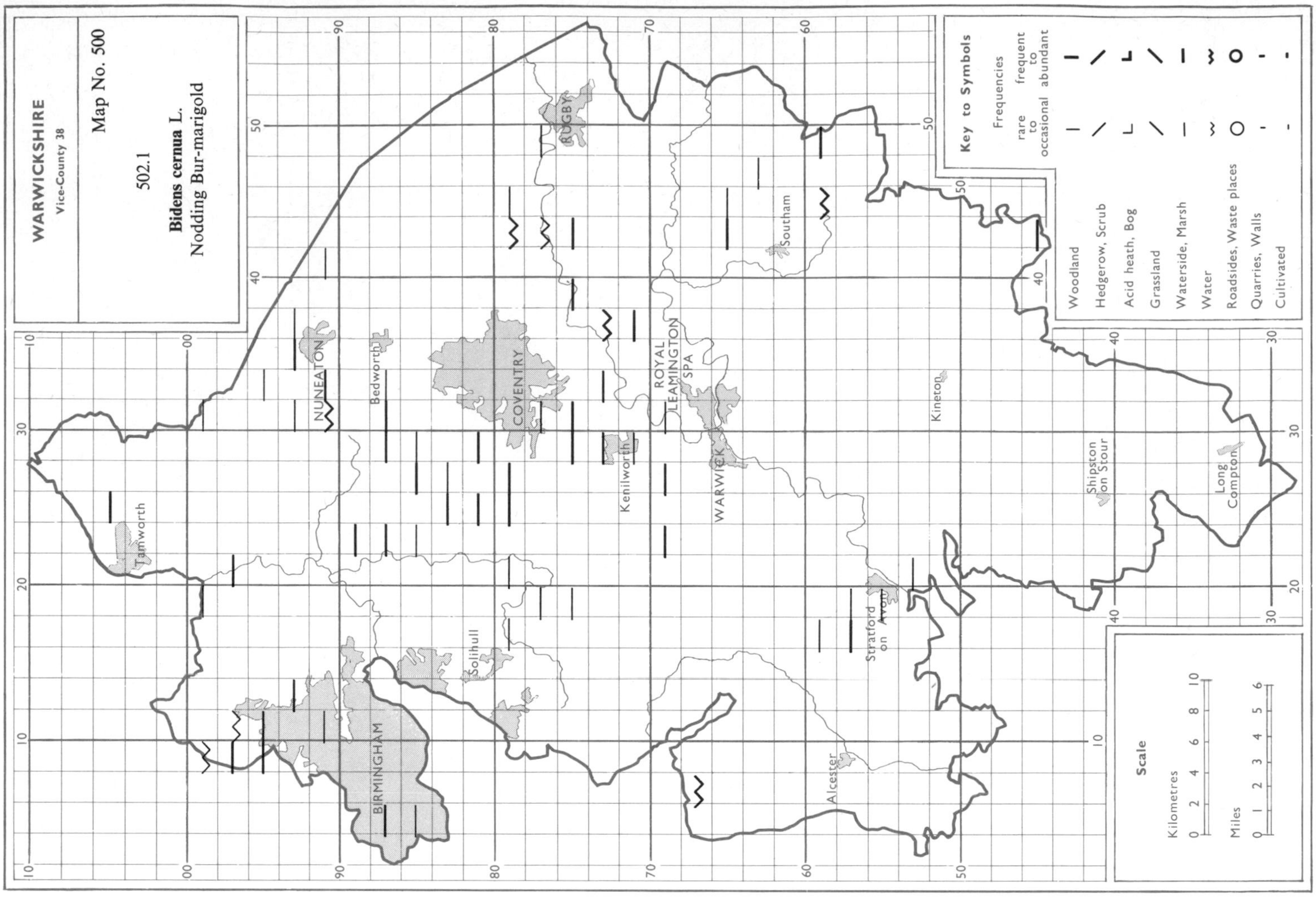
WARWICKSHIRE
Vice-County 38
Map No. 500
502.1
Bidens cernua L.
Nodding Bur-marigold
Key to Symbols
Frequencies
rare to occasional
frequent to abundant
Woodland
Hedgerow, Scrub
Acid heath, Bog
Grassland
Waterside, Marsh
Water
Roadsides, Waste places
Quarries, Walls
Cultivated
Scale
Kilometres 0 2 4 6 8 10
Miles 0 1 2 3 4 5 6
Tamworth
NUNEATON
Bedworth
COVENTRY
RUGBY
Southam
ROYAL LEAMINGTON SPA
Kenilworth
WARWICK
Kineton
Shipston on Stour
Long Compton
Solihull
BIRMINGHAM
Stratford on Avon
Alcester

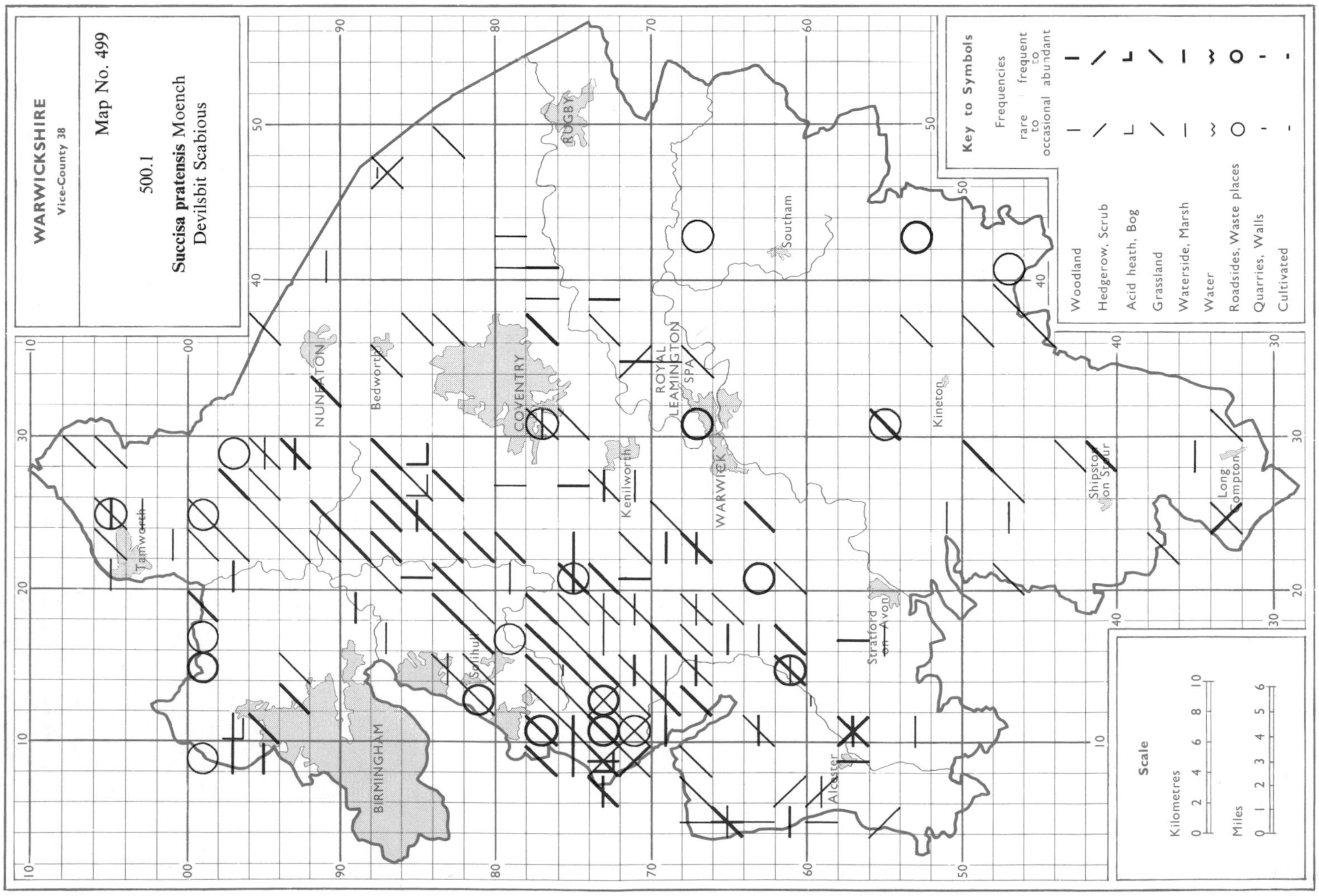
WARWICKSHIRE
Vice-County 38
Map No. 499
500.1
Succisa pratensis Moench
Devilsbit Scabious
Key to Symbols
Frequencies
rare to occasional
frequent to abundant
Woodland
Hedgerow, Scrub
Acid heath, Bog
Grassland
Waterside, Marsh
Water
Roadsides, Waste places
Quarries, Walls
Cultivated
Scale
Kilometres 0 2 4 6 8 10
Miles 0 1 2 3 4 5 6
Tamworth
NUNEATON
Bedworth
COVENTRY
RUGBY
Southam
ROYAL LEAMINGTON SPA
Kenilworth
WARWICK
Kineton
Shipston on Stour
Long Compton
Solihull
BIRMINGHAM
Stratford on Avon
Alcester

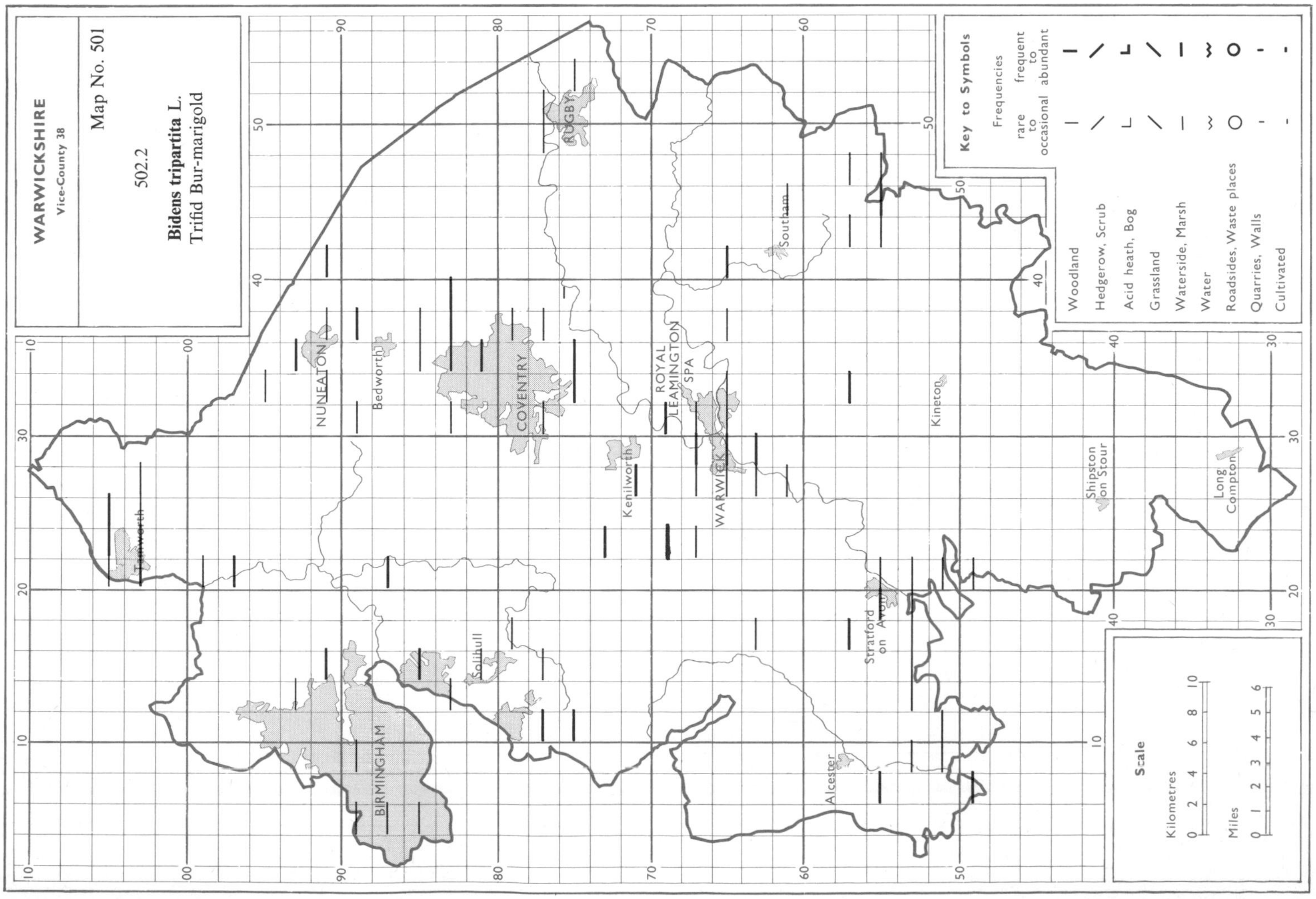
WARWICKSHIRE
Vice-County 38
Map No. 501
502.2
Bidens tripartita L.
Trifid Bur-marigold
Key to Symbols
Frequencies
rare to occasional
frequent to abundant
Woodland
Hedgerow, Scrub
Acid heath, Bog
Grassland
Waterside, Marsh
Water
Roadsides, Waste places
Quarries, Walls
Cultivated
Scale
Kilometres 0 2 4 6 8 10
Miles 0 1 2 3 4 5 6
NUNEATON
Bedworth
COVENTRY
RUGBY
Southam
ROYAL LEAMINGTON SPA
WARWICK
Kenilworth
Kineton
Shipston on Stour
Long Compton
Tamworth
BIRMINGHAM
Solihull
Stratford on Avon
Alcester

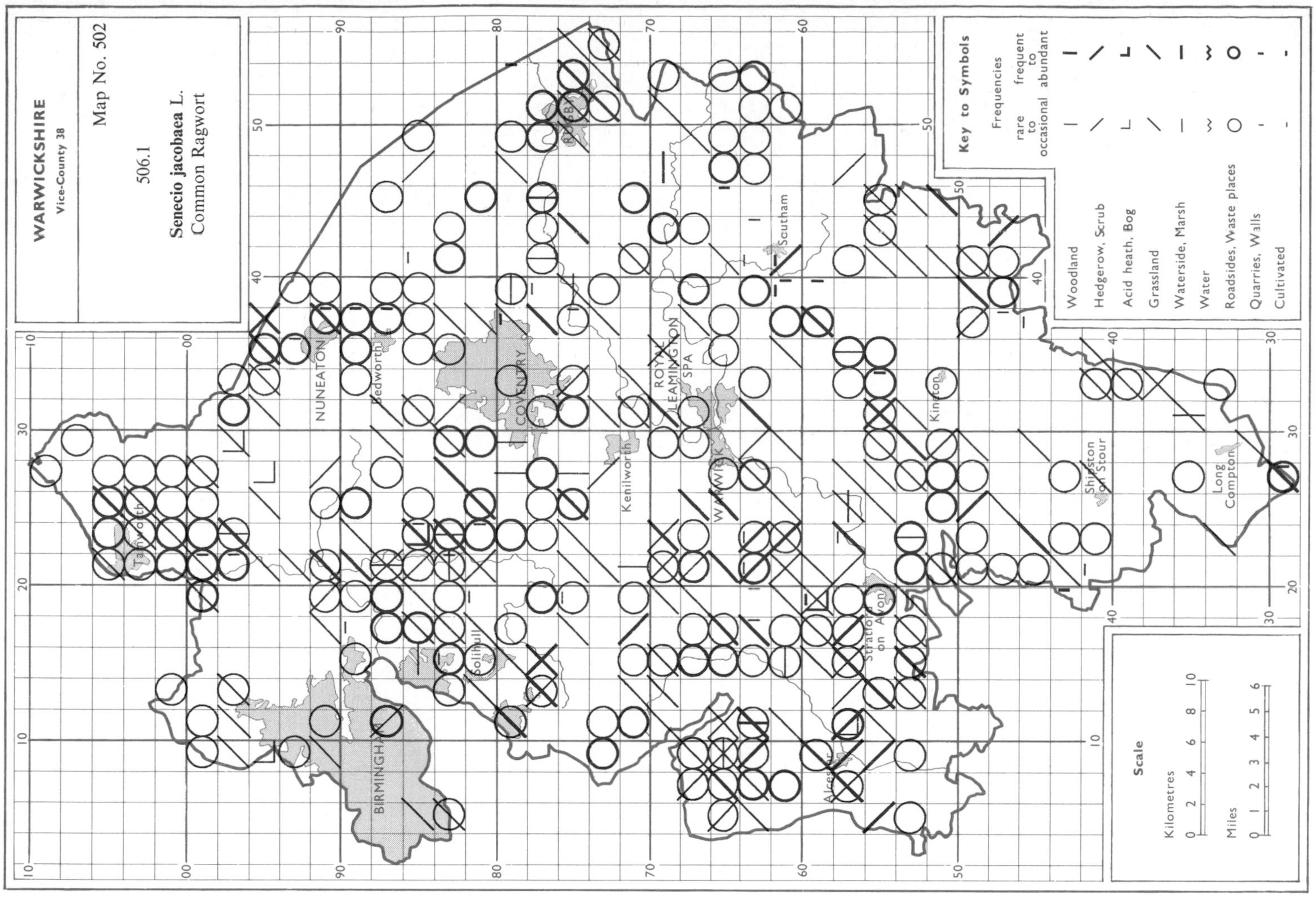
WARWICKSHIRE
Vice-County 38
Map No. 502
506.1
Senecio jacobaea L.
Common Ragwort
Key to Symbols
Frequencies
rare to occasional
frequent to abundant
Woodland
Hedgerow, Scrub
Acid heath, Bog
Grassland
Waterside, Marsh
Water
Roadsides, Waste places
Quarries, Walls
Cultivated
Scale
Kilometres 0 2 4 6 8 10
Miles 0 1 2 3 4 5 6
NUNEATON
Bedworth
COVENTRY
RUGBY
Southam
ROYAL LEAMINGTON SPA
WARWICK
Kenilworth
Kineton
Shipston on Stour
Long Compton
Tamworth
BIRMINGHAM
Solihull
Stratford on Avon
Alcester

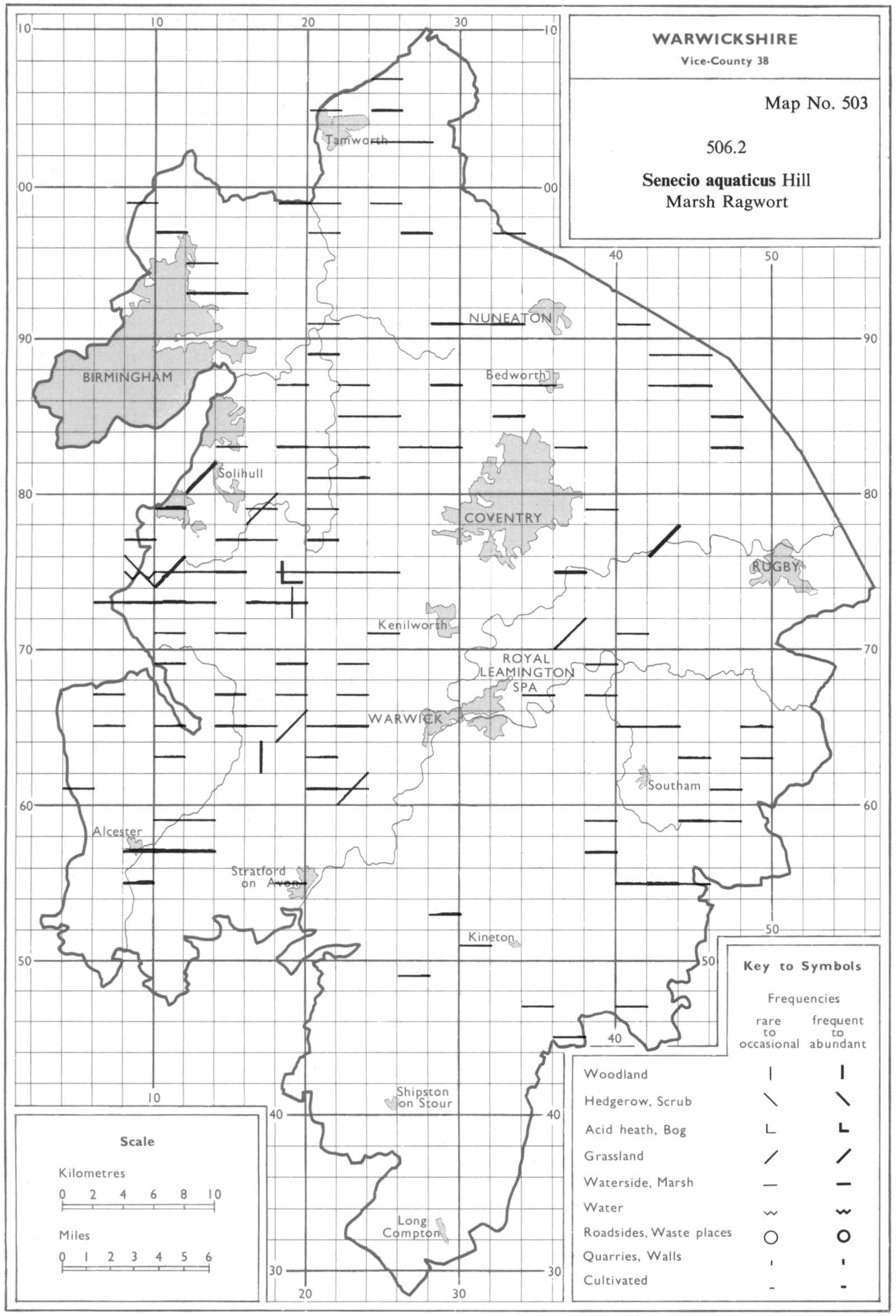
WARWICKSHIRE
Vice-County 38
Map No. 503
506.2
Senecio aquaticus Hill
Marsh Ragwort
Tamworth
NUNEATON
Bedworth
BIRMINGHAM
Solihull
COVENTRY
RUGBY
Kenilworth
ROYAL LEAMINGTON SPA
WARWICK
Southam
Alcester
Stratford on Avon
Kineton
Shipston on Stour
Long Compton
Key to Symbols
Frequencies
rare to occasional
frequent to abundant
Woodland
Hedgerow, Scrub
Acid heath, Bog
Grassland
Waterside, Marsh
Water
Roadsides, Waste places
Quarries, Walls
Cultivated
Scale
Kilometres
0 2 4 6 8 10
Miles
0 1 2 3 4 5 6

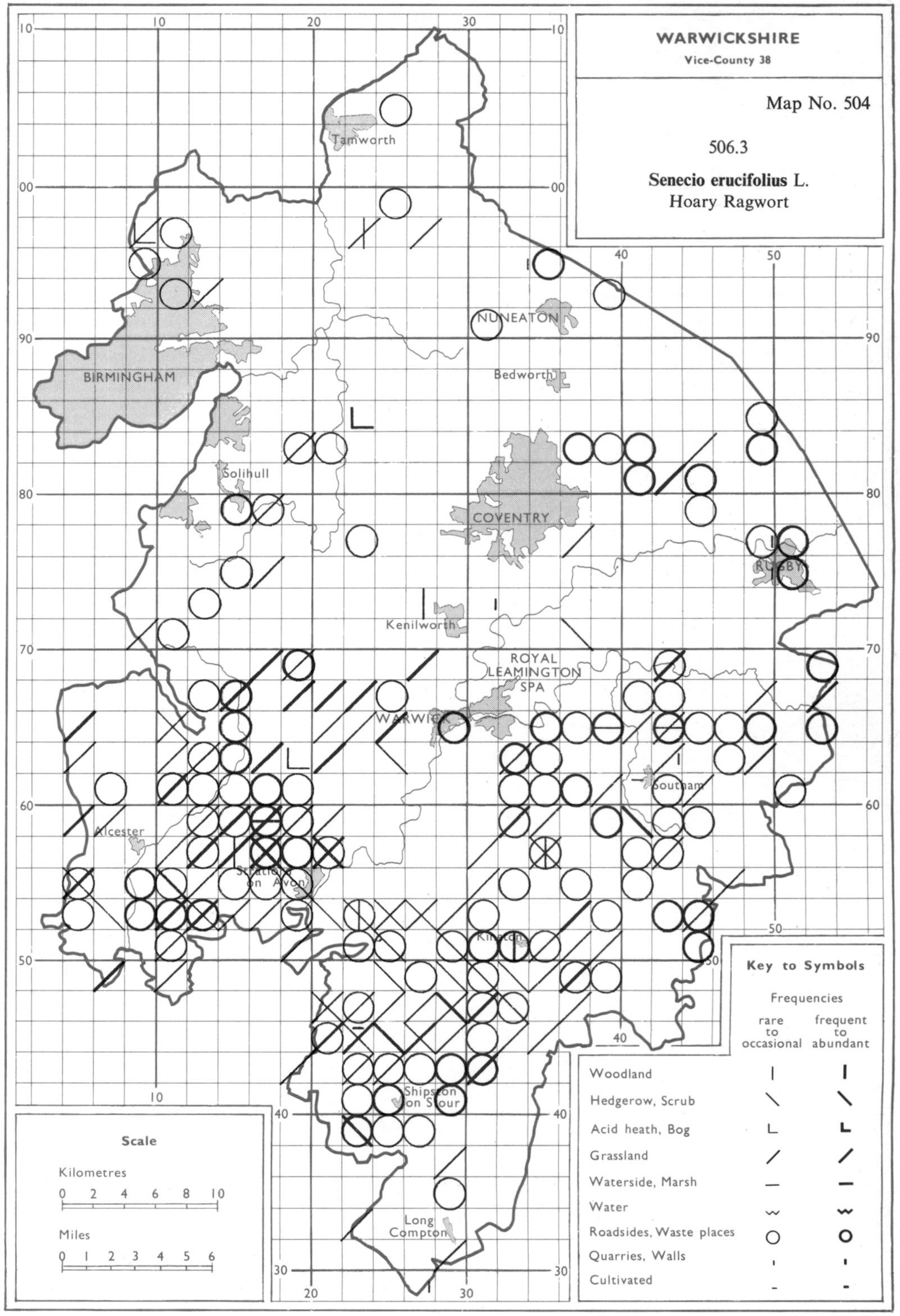
WARWICKSHIRE
Vice-County 38
Map No. 504
506.3
Senecio erucifolius L.
Hoary Ragwort
Tamworth
NUNEATON
Bedworth
BIRMINGHAM
Solihull
COVENTRY
RUGBY
Kenilworth
ROYAL LEAMINGTON SPA
WARWICK
Southam
Alcester
Stratford on Avon
Kineton
Shipston on Stour
Long Compton
Key to Symbols
Frequencies
rare to occasional
frequent to abundant
Woodland
Hedgerow, Scrub
Acid heath, Bog
Grassland
Waterside, Marsh
Water
Roadsides, Waste places
Quarries, Walls
Cultivated
Scale
Kilometres
0 2 4 6 8 10
Miles
0 1 2 3 4 5 6

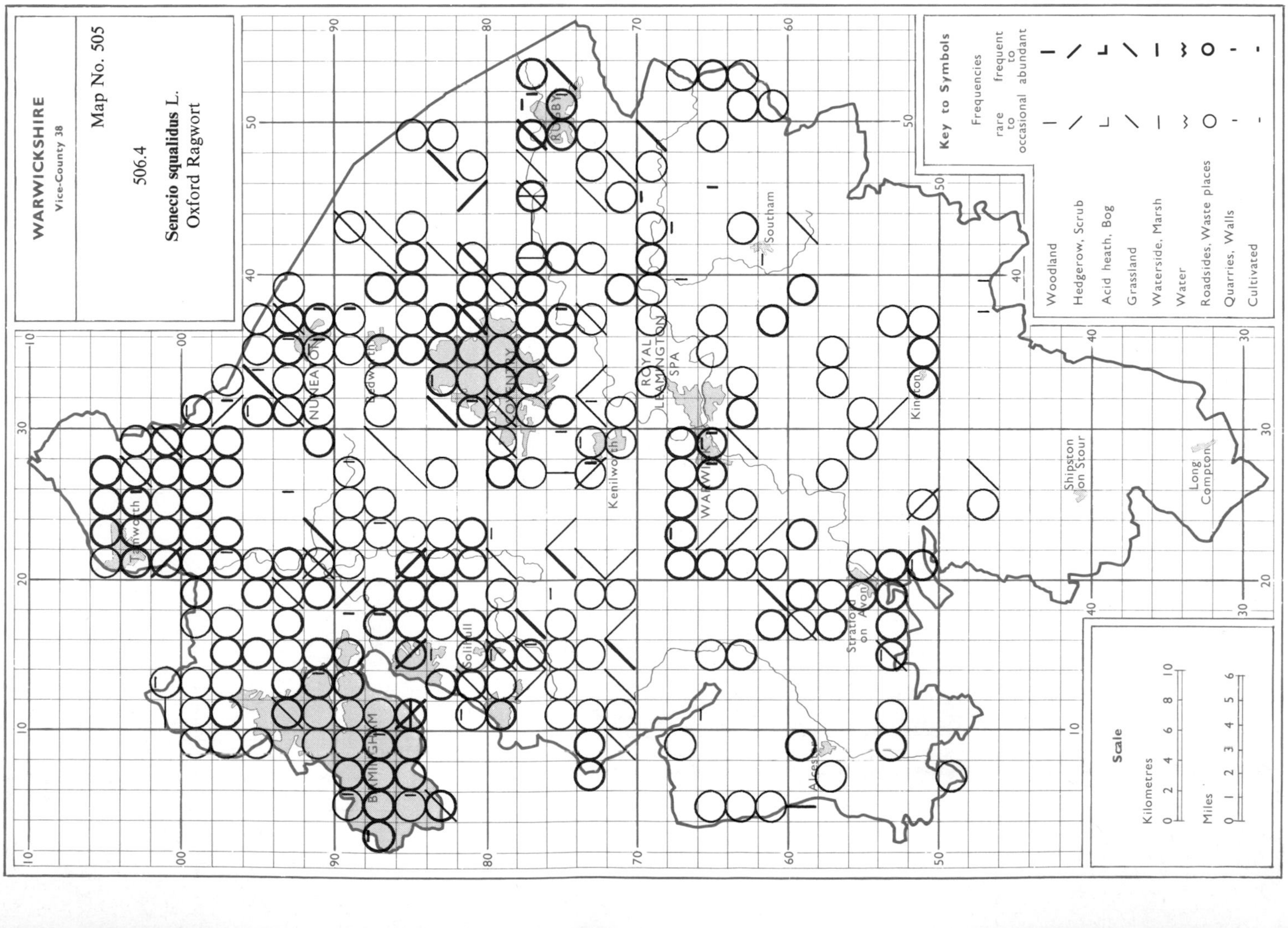
WARWICKSHIRE
Vice-County 38
Map No. 505
506.4
Senecio squalidus L.
Oxford Ragwort
Key to Symbols
Frequencies
rare to occasional
frequent to abundant
Woodland
Hedgerow, Scrub
Acid heath, Bog
Grassland
Waterside, Marsh
Water
Roadsides, Waste places
Quarries, Walls
Cultivated
Scale
Kilometres
Miles
Tamworth
Nuneaton
Bedworth
Coventry
Rugby
Southam
Royal Leamington Spa
Warwick
Kenilworth
Kineton
Shipston on Stour
Long Compton
Stratford on Avon
Solihull
Birmingham
Alcester

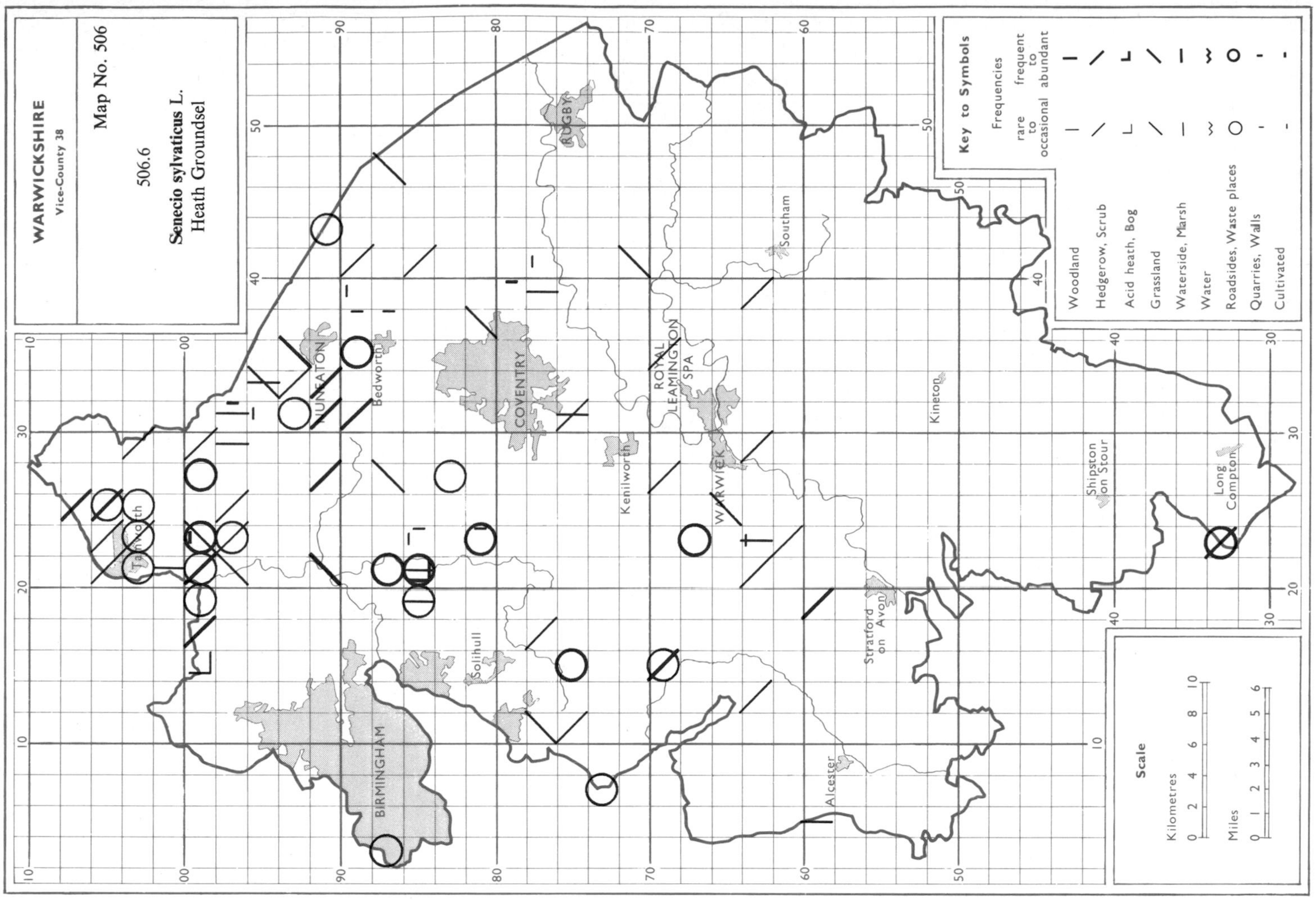
WARWICKSHIRE
Vice-County 38
Map No. 506
506.6
Senecio sylvaticus L.
Heath Groundsel
Key to Symbols
Frequencies
rare to occasional
frequent to abundant
Woodland
Hedgerow, Scrub
Acid heath, Bog
Grassland
Waterside, Marsh
Water
Roadsides, Waste places
Quarries, Walls
Cultivated
Scale
Kilometres
Miles
Tamworth
Nuneaton
Bedworth
Coventry
Rugby
Southam
Royal Leamington Spa
Warwick
Kenilworth
Kineton
Shipston on Stour
Long Compton
Stratford on Avon
Solihull
Birmingham
Alcester

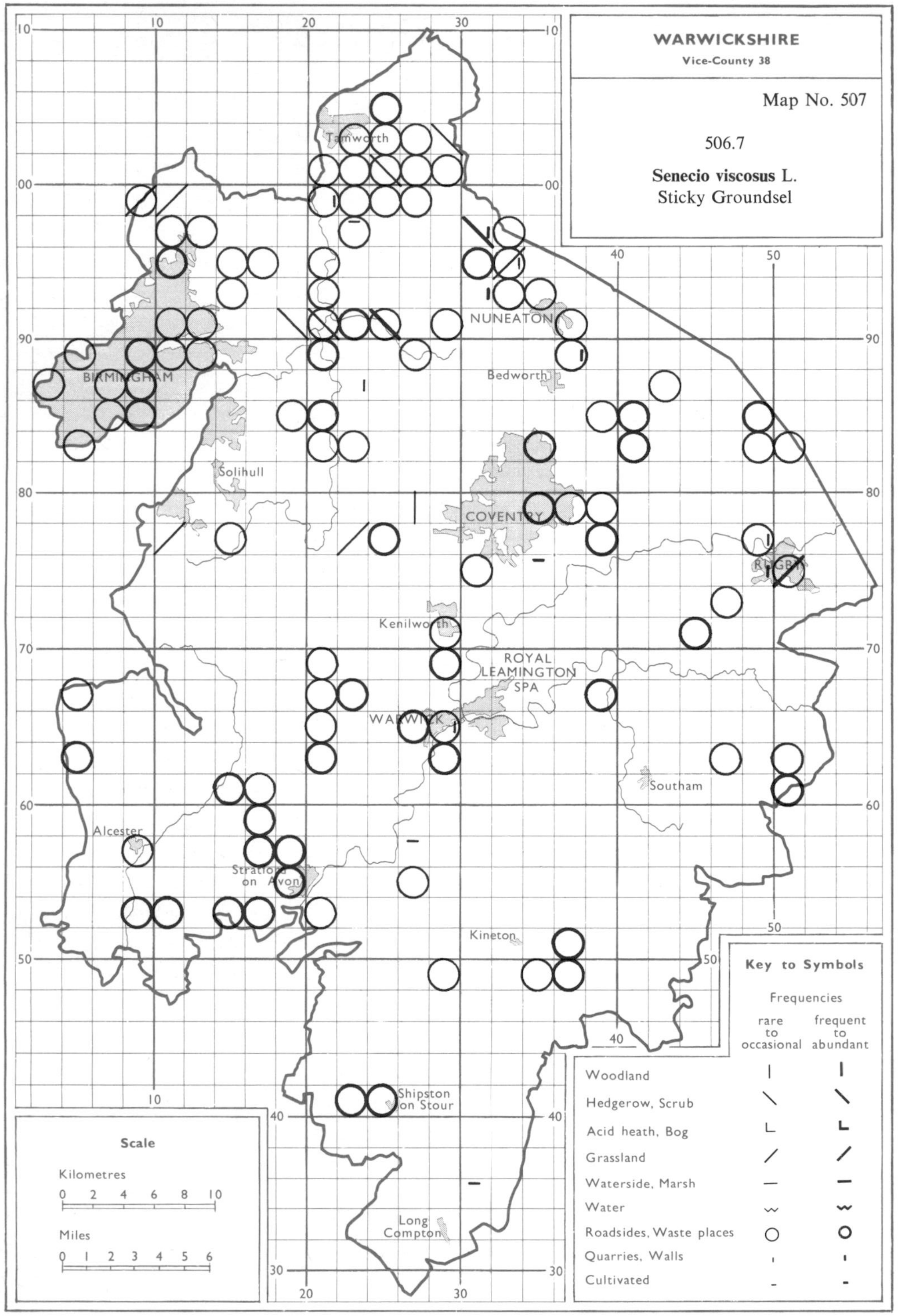
WARWICKSHIRE
Vice-County 38
Map No. 507
506.7
Senecio viscosus L.
Sticky Groundsel
Tamworth
NUNEATON
Bedworth
BIRMINGHAM
Solihull
COVENTRY
RUGBY
Kenilworth
ROYAL LEAMINGTON SPA
WARWICK
Southam
Alcester
Stratford on Avon
Kineton
Shipston on Stour
Long Compton
Key to Symbols
Frequencies
rare to occasional
frequent to abundant
Woodland
Hedgerow, Scrub
Acid heath, Bog
Grassland
Waterside, Marsh
Water
Roadsides, Waste places
Quarries, Walls
Cultivated
Scale
Kilometres
0 2 4 6 8 10
Miles
0 1 2 3 4 5 6

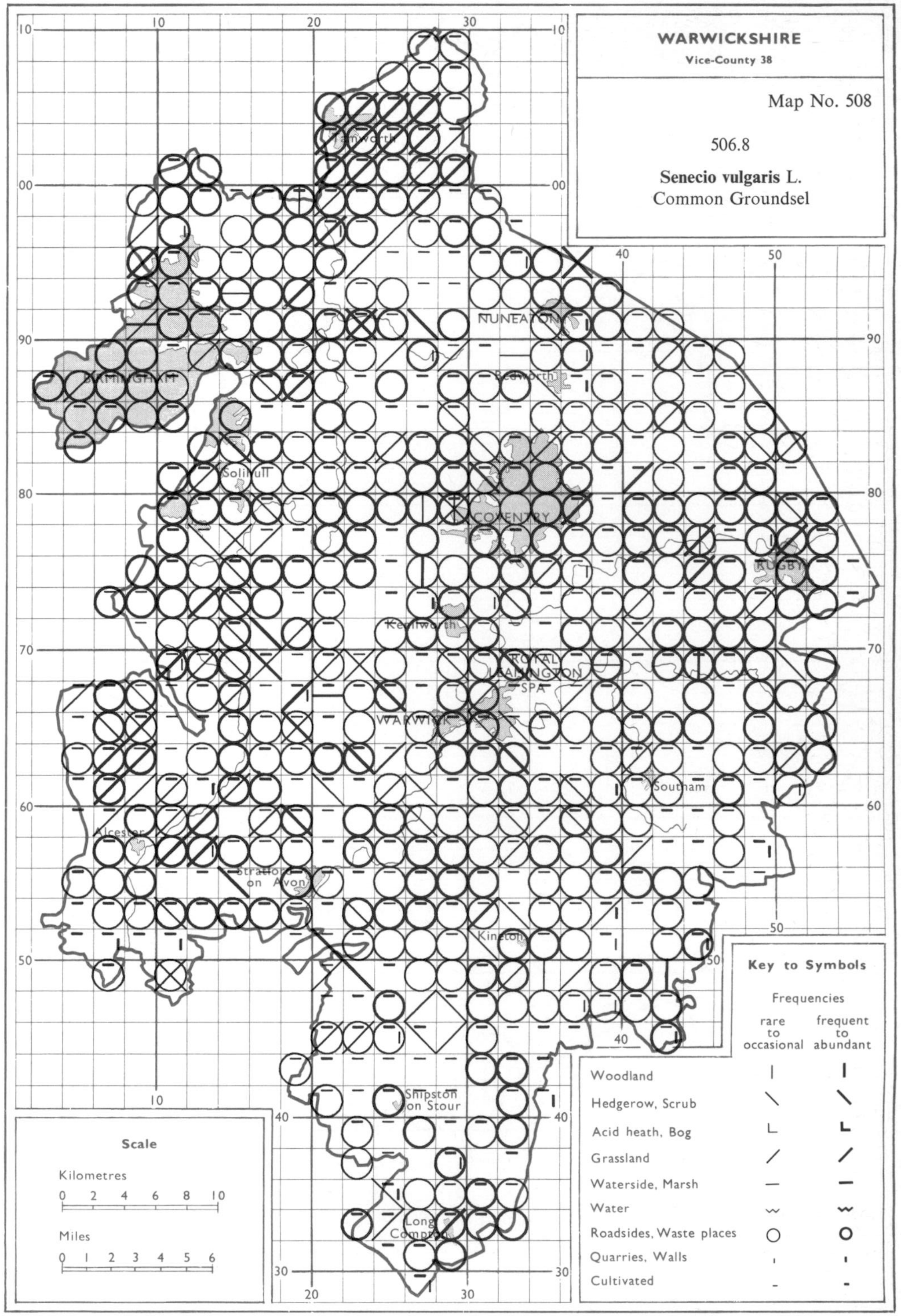
WARWICKSHIRE
Vice-County 38
Map No. 508
506.8
Senecio vulgaris L.
Common Groundsel
Tamworth
NUNEATON
Bedworth
BIRMINGHAM
Solihull
COVENTRY
RUGBY
Kenilworth
ROYAL LEAMINGTON SPA
WARWICK
Southam
Alcester
Stratford on Avon
Kineton
Shipston on Stour
Long Compton
Key to Symbols
Frequencies
rare to occasional
frequent to abundant
Woodland
Hedgerow, Scrub
Acid heath, Bog
Grassland
Waterside, Marsh
Water
Roadsides, Waste places
Quarries, Walls
Cultivated
Scale
Kilometres
0 2 4 6 8 10
Miles
0 1 2 3 4 5 6

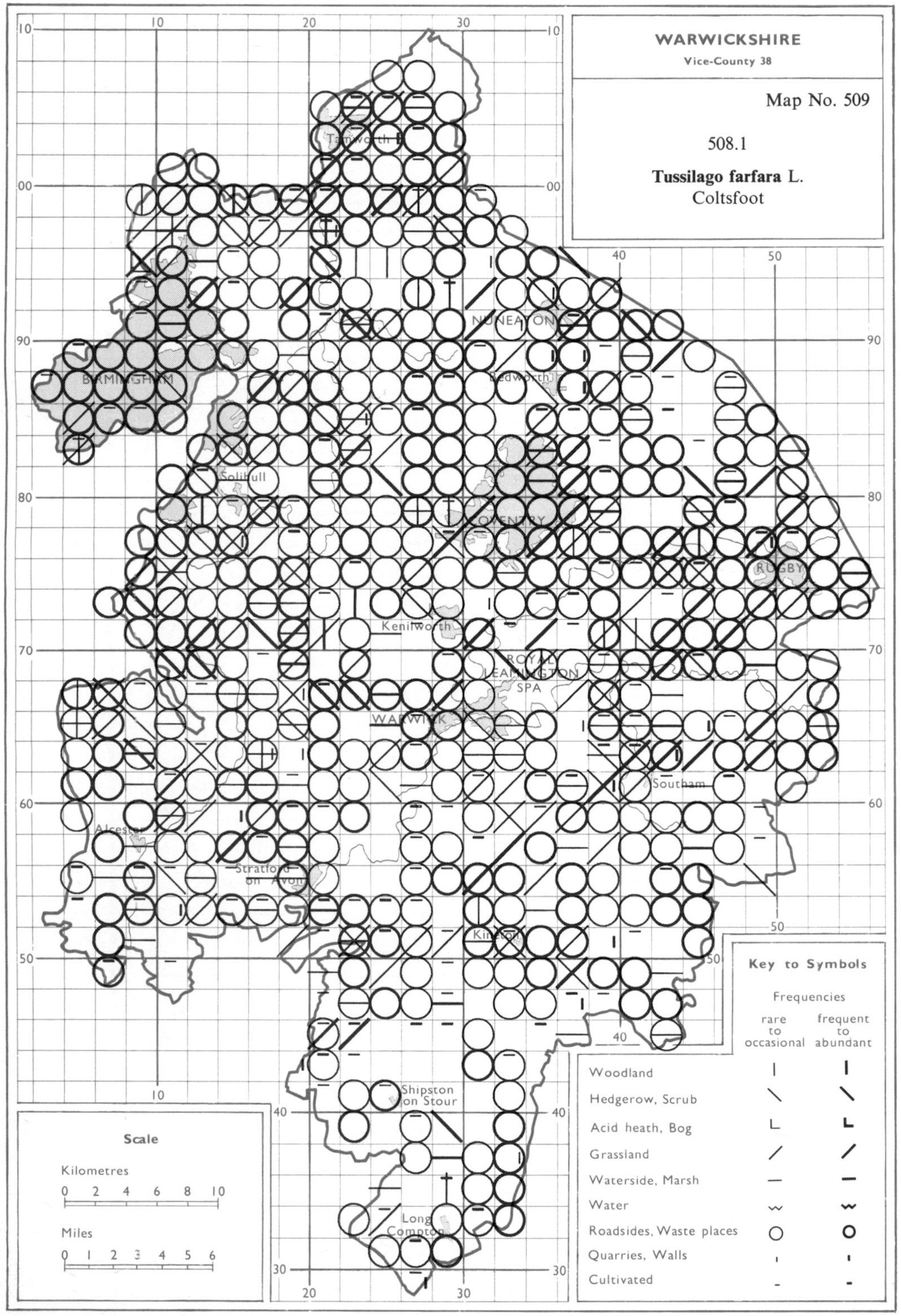
WARWICKSHIRE
Vice-County 38
Map No. 509
508.1
Tussilago farfara L.
Coltsfoot
Tamworth
NUNEATON
Bedworth
BIRMINGHAM
Solihull
COVENTRY
RUGBY
Kenilworth
ROYAL LEAMINGTON SPA
WARWICK
Southam
Alcester
Stratford on Avon
Kineton
Shipston on Stour
Long Compton
Key to Symbols
Frequencies
rare to occasional
frequent to abundant
Woodland
Hedgerow, Scrub
Acid heath, Bog
Grassland
Waterside, Marsh
Water
Roadsides, Waste places
Quarries, Walls
Cultivated
Scale
Kilometres
0 2 4 6 8 10
Miles
0 1 2 3 4 5 6

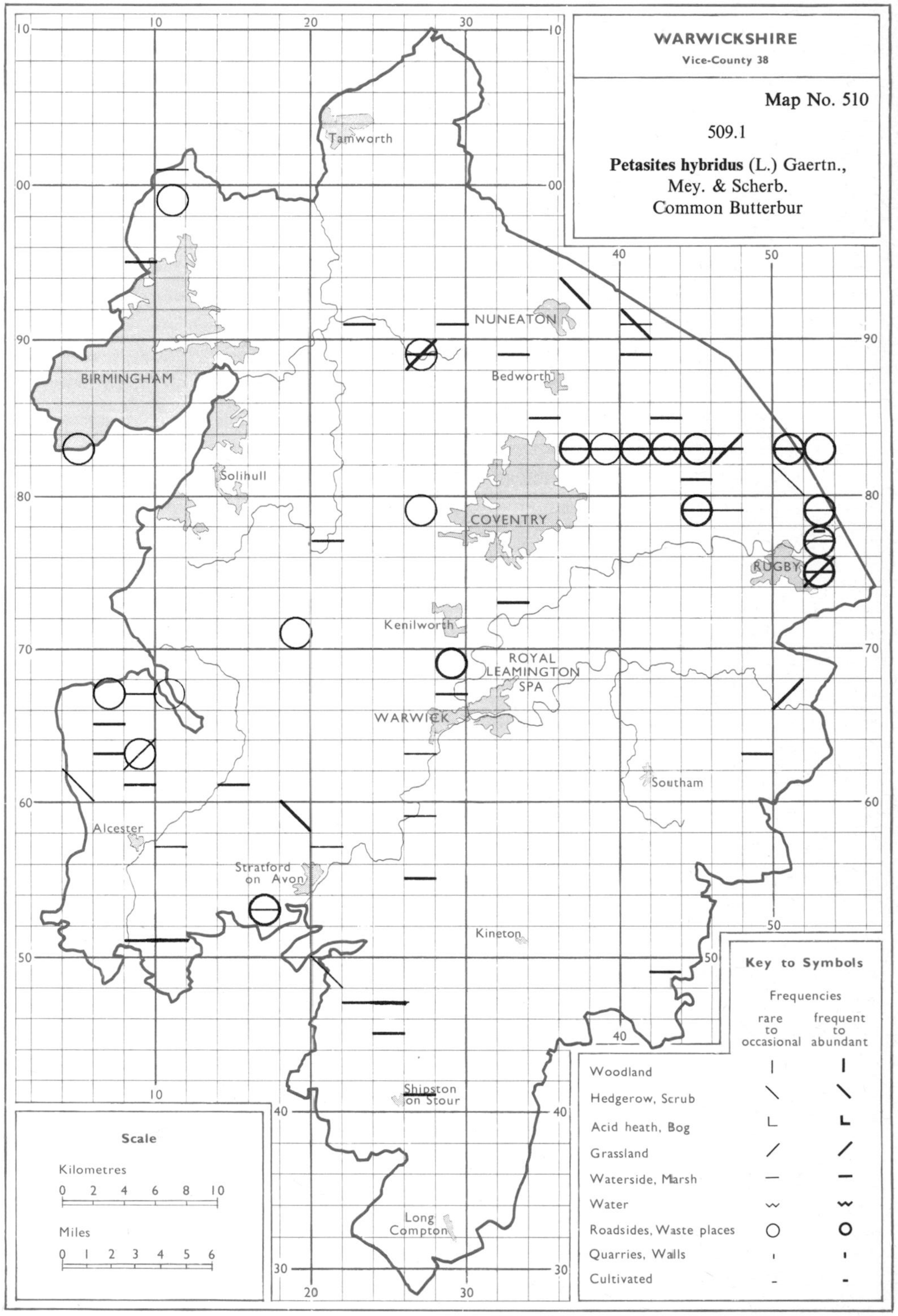
WARWICKSHIRE
Vice-County 38
Map No. 510
509.1
Petasites hybridus (L.) Gaertn., Mey. & Scherb.
Common Butterbur
Tamworth
NUNEATON
Bedworth
BIRMINGHAM
Solihull
COVENTRY
RUGBY
Kenilworth
ROYAL LEAMINGTON SPA
WARWICK
Southam
Alcester
Stratford on Avon
Kineton
Shipston on Stour
Long Compton
Key to Symbols
Frequencies
rare to occasional
frequent to abundant
Woodland
Hedgerow, Scrub
Acid heath, Bog
Grassland
Waterside, Marsh
Water
Roadsides, Waste places
Quarries, Walls
Cultivated
Scale
Kilometres
0 2 4 6 8 10
Miles
0 1 2 3 4 5 6

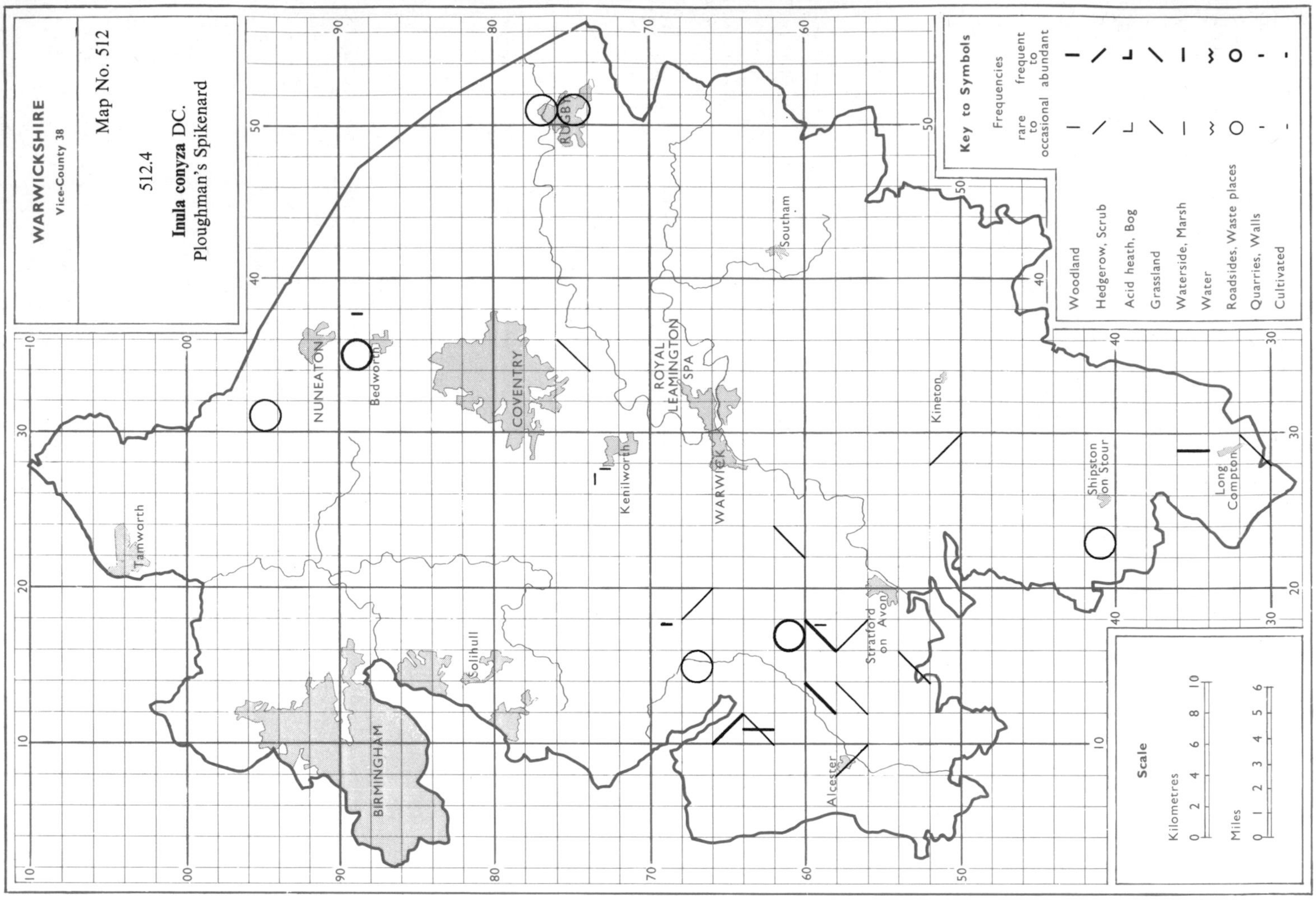
WARWICKSHIRE
Vice-County 38
Map No. 512
512.4
Inula conyza DC.
Ploughman's Spikenard
Key to Symbols
Frequencies
rare to occasional
frequent to abundant
Woodland
Hedgerow, Scrub
Acid heath, Bog
Grassland
Waterside, Marsh
Water
Roadsides, Waste places
Quarries, Walls
Cultivated
Scale
Kilometres
Miles
NUNEATON
Bedworth
COVENTRY
ROYAL LEAMINGTON SPA
WARWICK
Kenilworth
Tamworth
Solihull
BIRMINGHAM
RUGBY
Southam
Kineton
Shipston on Stour
Long Compton
Stratford on Avon
Alcester

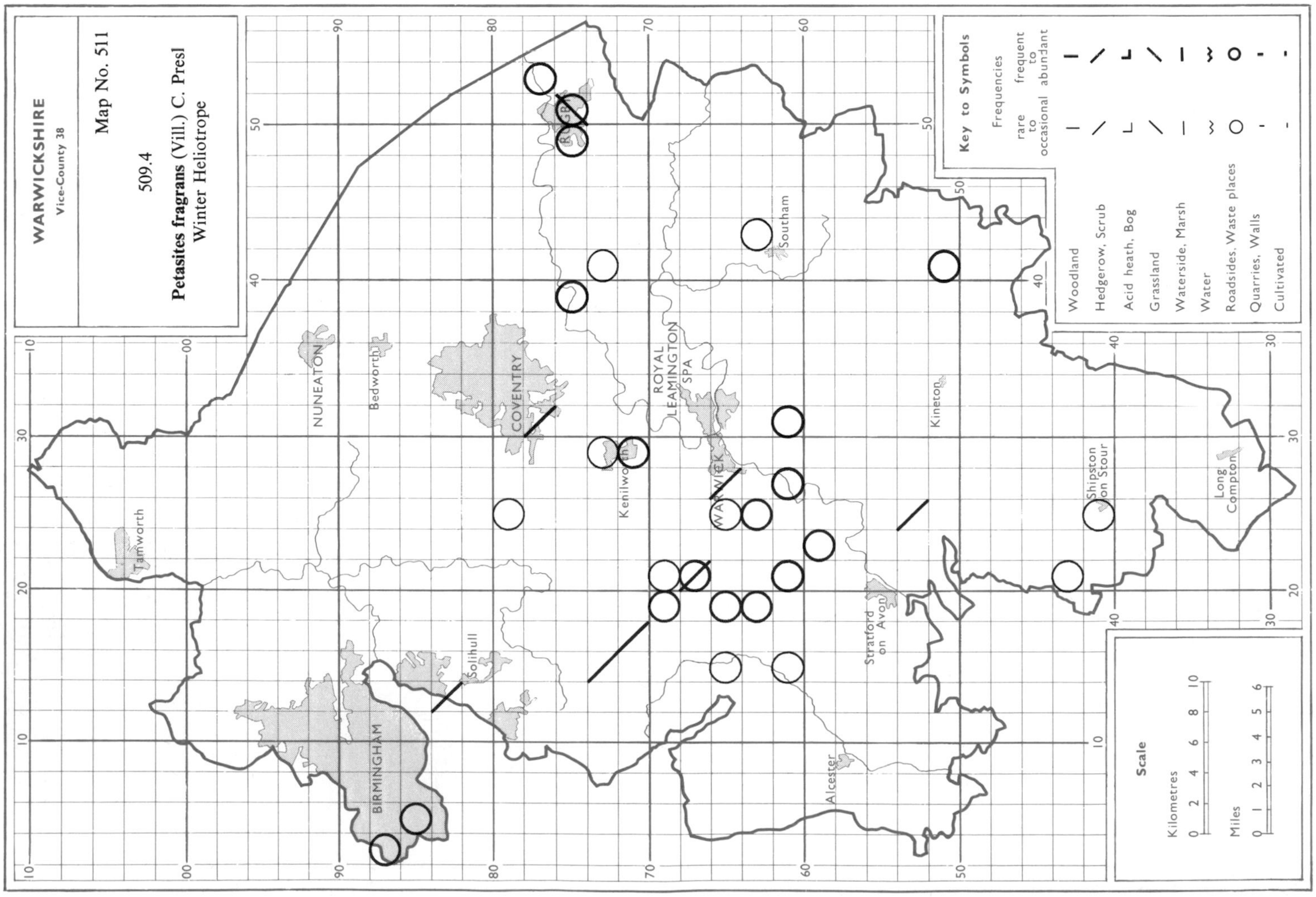
WARWICKSHIRE
Vice-County 38
Map No. 511
509.4
Petasites fragrans (Vill.) C. Presl
Winter Heliotrope
Key to Symbols
Frequencies
rare to occasional
frequent to abundant
Woodland
Hedgerow, Scrub
Acid heath, Bog
Grassland
Waterside, Marsh
Water
Roadsides, Waste places
Quarries, Walls
Cultivated
Scale
Kilometres
Miles
NUNEATON
Bedworth
COVENTRY
ROYAL LEAMINGTON SPA
WARWICK
Kenilworth
Tamworth
Solihull
BIRMINGHAM
RUGBY
Southam
Kineton
Shipston on Stour
Long Compton
Stratford on Avon
Alcester

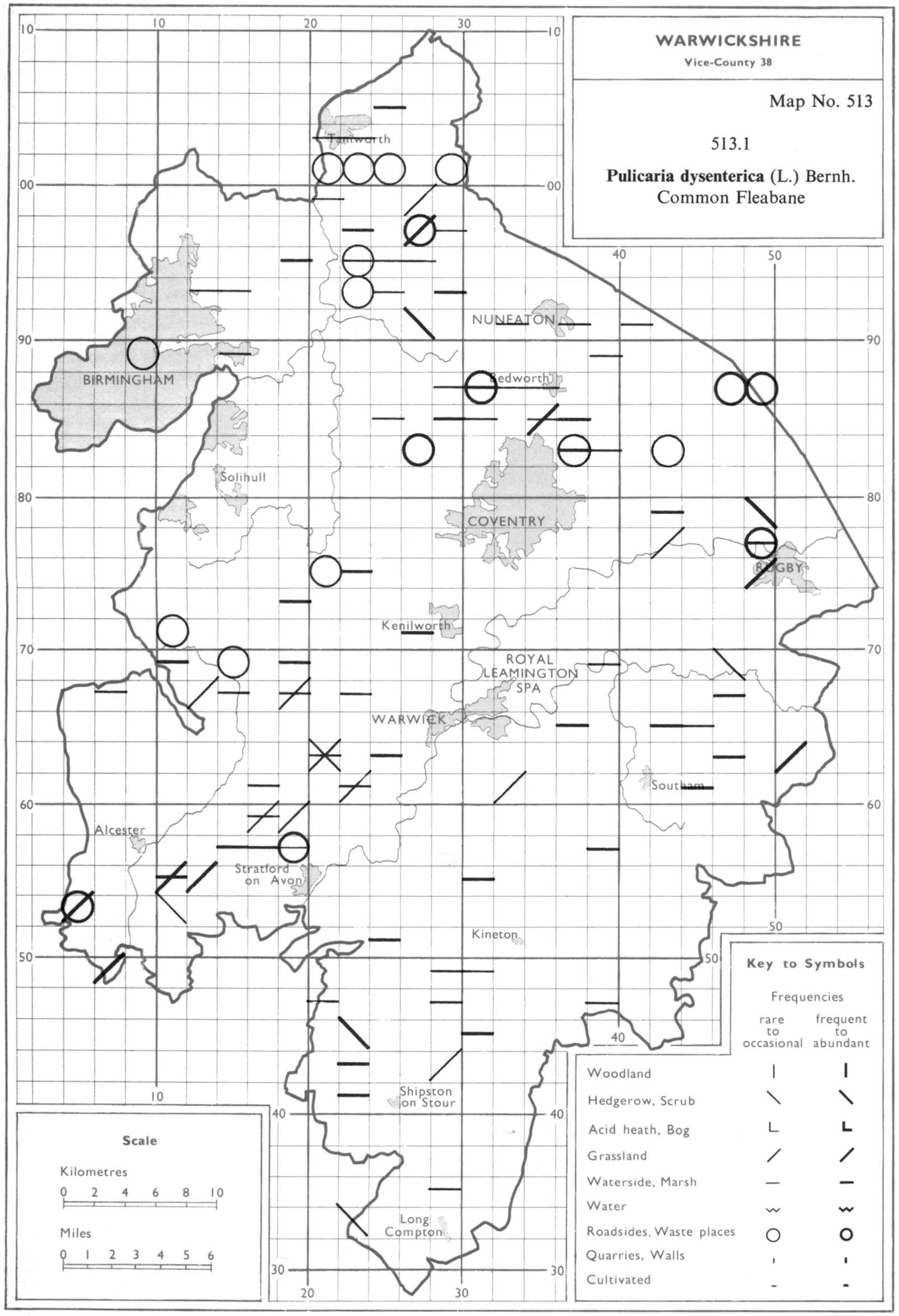
WARWICKSHIRE
Vice-County 38
Map No. 513
513.1
Pulicaria dysenterica (L.) Bernh.
Common Fleabane
Tamworth
NUNEATON
Bedworth
BIRMINGHAM
Solihull
COVENTRY
RUGBY
Kenilworth
ROYAL LEAMINGTON SPA
WARWICK
Southam
Alcester
Stratford on Avon
Kineton
Shipston on Stour
Long Compton
Key to Symbols
Frequencies
rare to occasional
frequent to abundant
Woodland
Hedgerow, Scrub
Acid heath, Bog
Grassland
Waterside, Marsh
Water
Roadsides, Waste places
Quarries, Walls
Cultivated
Scale
Kilometres
0 2 4 6 8 10
Miles
0 1 2 3 4 5 6

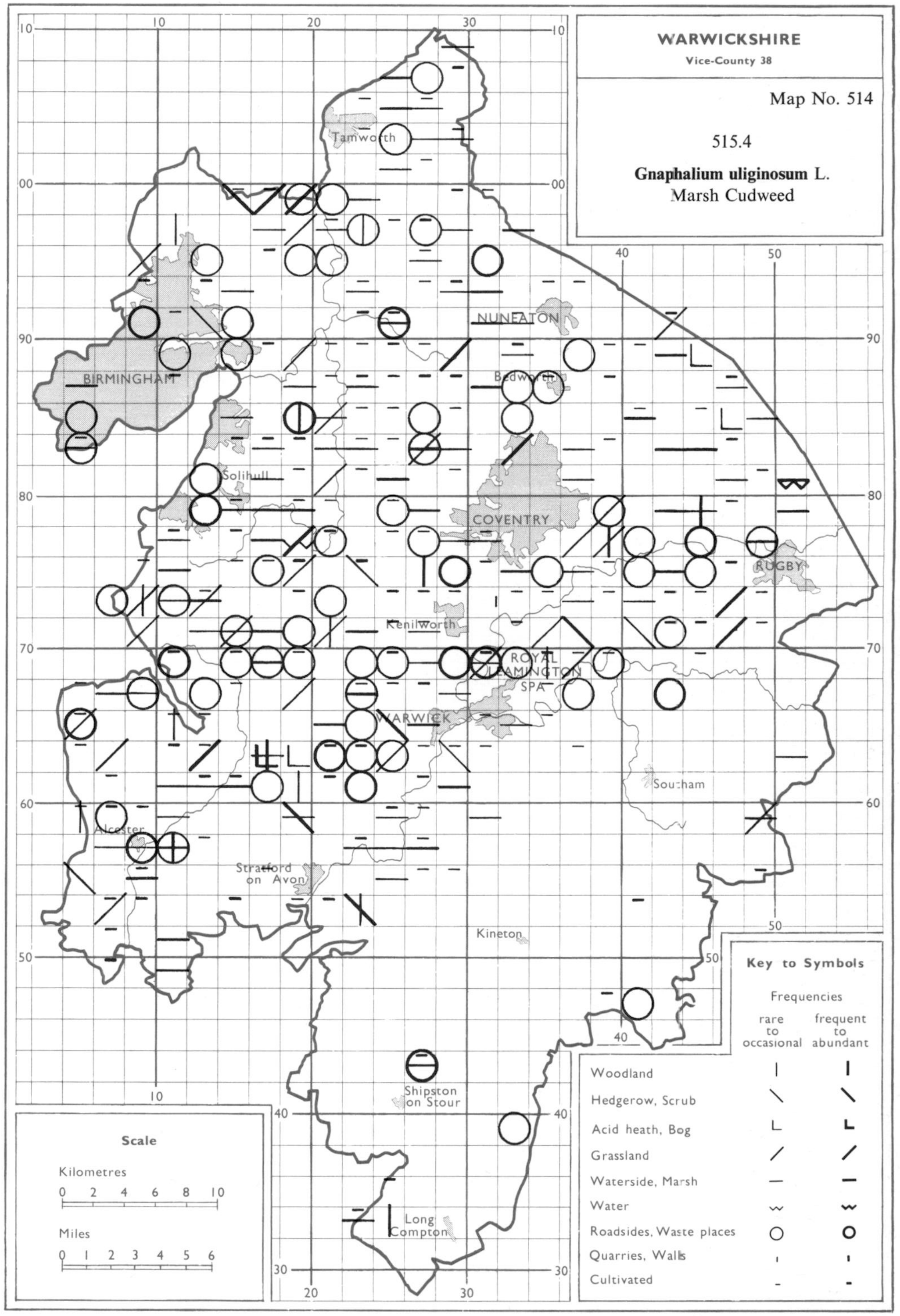
WARWICKSHIRE
Vice-County 38
Map No. 514
515.4
Gnaphalium uliginosum L.
Marsh Cudweed
Tamworth
NUNEATON
Bedworth
BIRMINGHAM
Solihull
COVENTRY
RUGBY
Kenilworth
ROYAL LEAMINGTON SPA
WARWICK
Southam
Alcester
Stratford on Avon
Kineton
Shipston on Stour
Long Compton
Key to Symbols
Frequencies
rare to occasional
frequent to abundant
Woodland
Hedgerow, Scrub
Acid heath, Bog
Grassland
Waterside, Marsh
Water
Roadsides, Waste places
Quarries, Walls
Cultivated
Scale
Kilometres
0 2 4 6 8 10
Miles
0 1 2 3 4 5 6

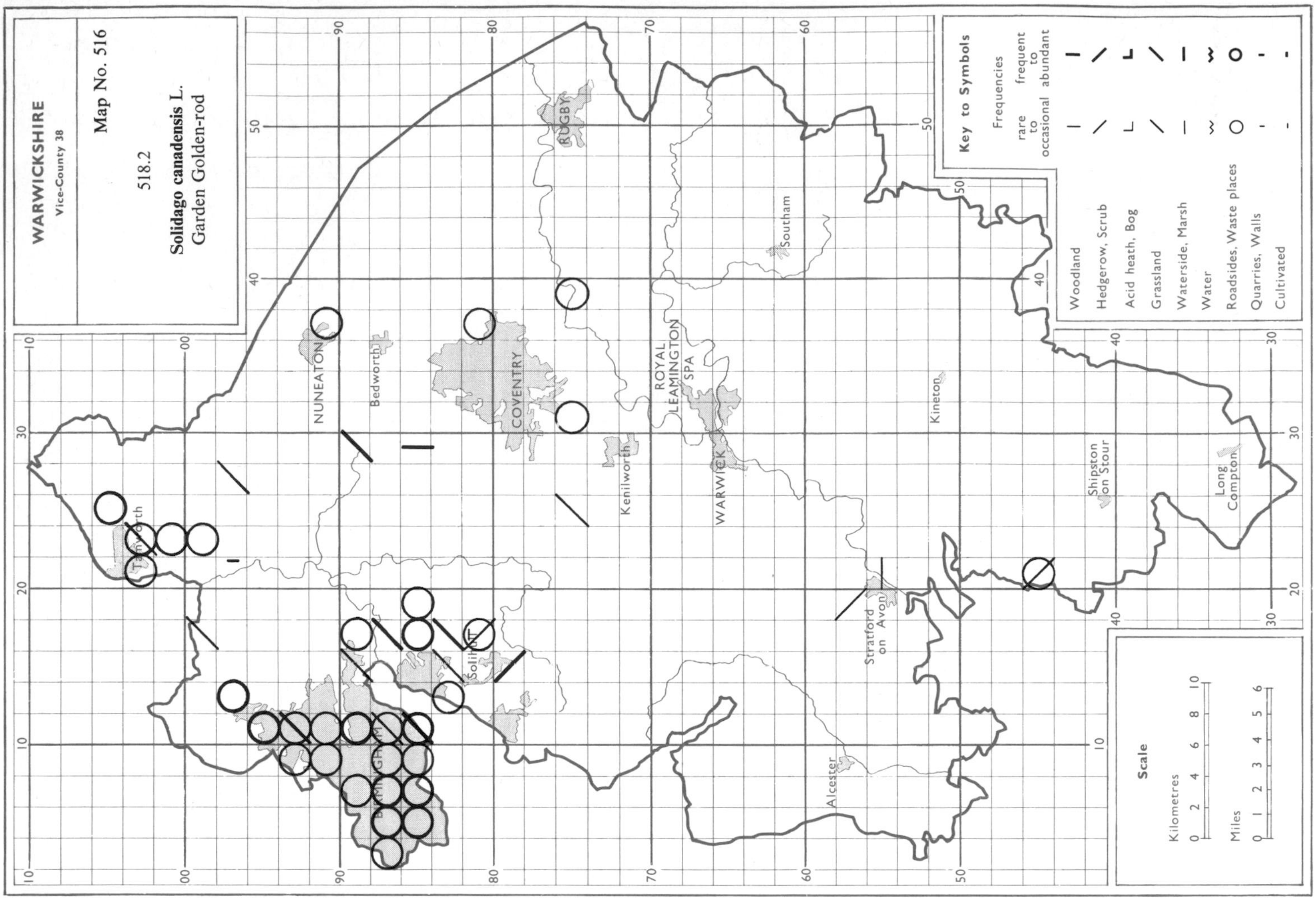
WARWICKSHIRE
Vice-County 38
Map No. 516
518.2
Solidago canadensis L.
Garden Golden-rod
Key to Symbols
Frequencies
rare to occasional
frequent to abundant
Woodland
Hedgerow, Scrub
Acid heath, Bog
Grassland
Waterside, Marsh
Water
Roadsides, Waste places
Quarries, Walls
Cultivated
Scale
Kilometres
0 2 4 6 8 10
Miles
0 1 2 3 4 5 6
Tamworth
NUNEATON
Bedworth
COVENTRY
RUGBY
Southam
ROYAL LEAMINGTON SPA
Kenilworth
WARWICK
Kineton
Shipston on Stour
Long Compton
Stratford on Avon
Alcester
Solihull
BIRMINGHAM

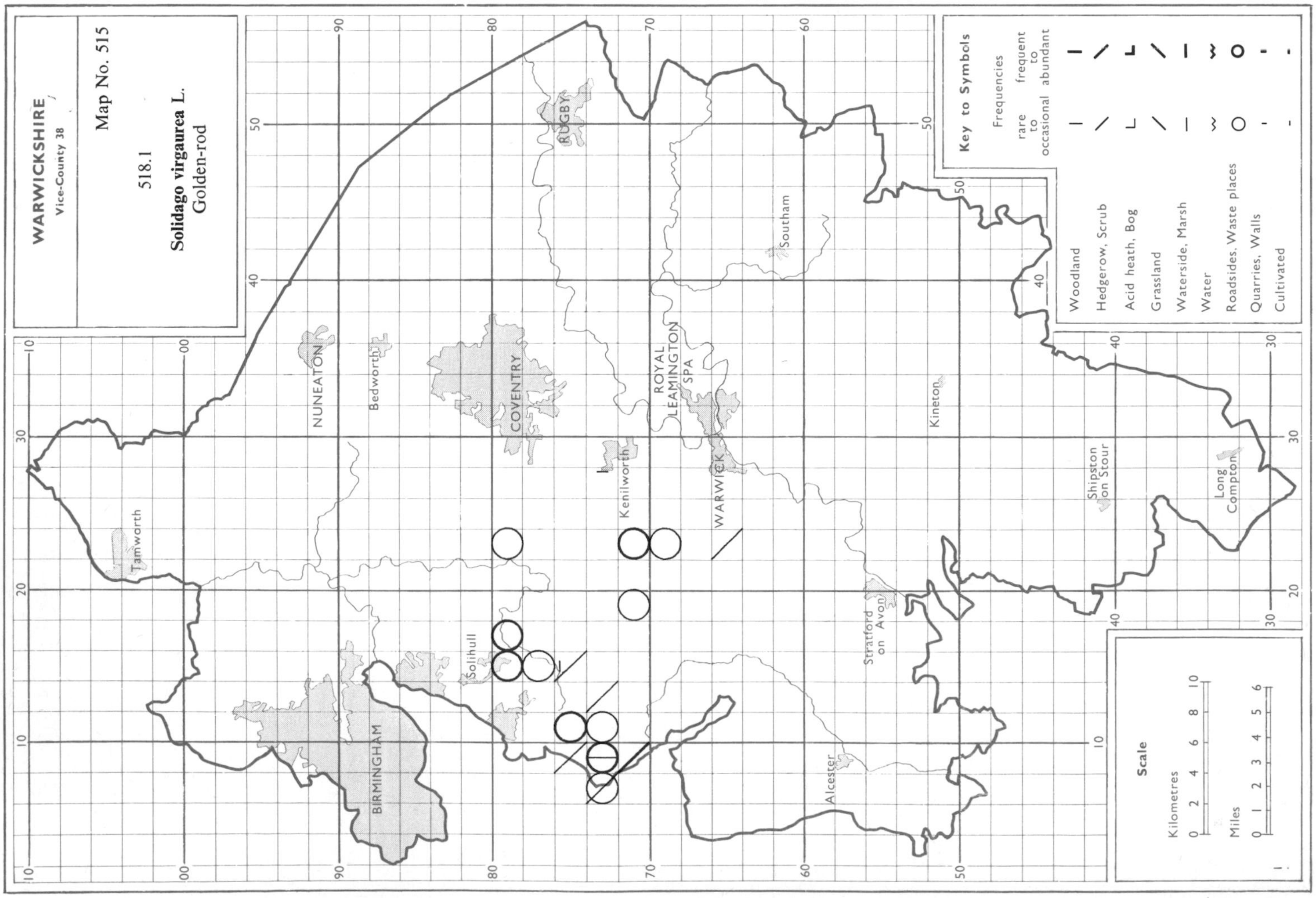
WARWICKSHIRE
Vice-County 38
Map No. 515
518.1
Solidago virgaurea L.
Golden-rod
Key to Symbols
Frequencies
rare to occasional
frequent to abundant
Woodland
Hedgerow, Scrub
Acid heath, Bog
Grassland
Waterside, Marsh
Water
Roadsides, Waste places
Quarries, Walls
Cultivated
Scale
Kilometres
0 2 4 6 8 10
Miles
0 1 2 3 4 5 6
Tamworth
NUNEATON
Bedworth
COVENTRY
RUGBY
Southam
ROYAL LEAMINGTON SPA
Kenilworth
WARWICK
Kineton
Shipston on Stour
Long Compton
Stratford on Avon
Alcester
Solihull
BIRMINGHAM

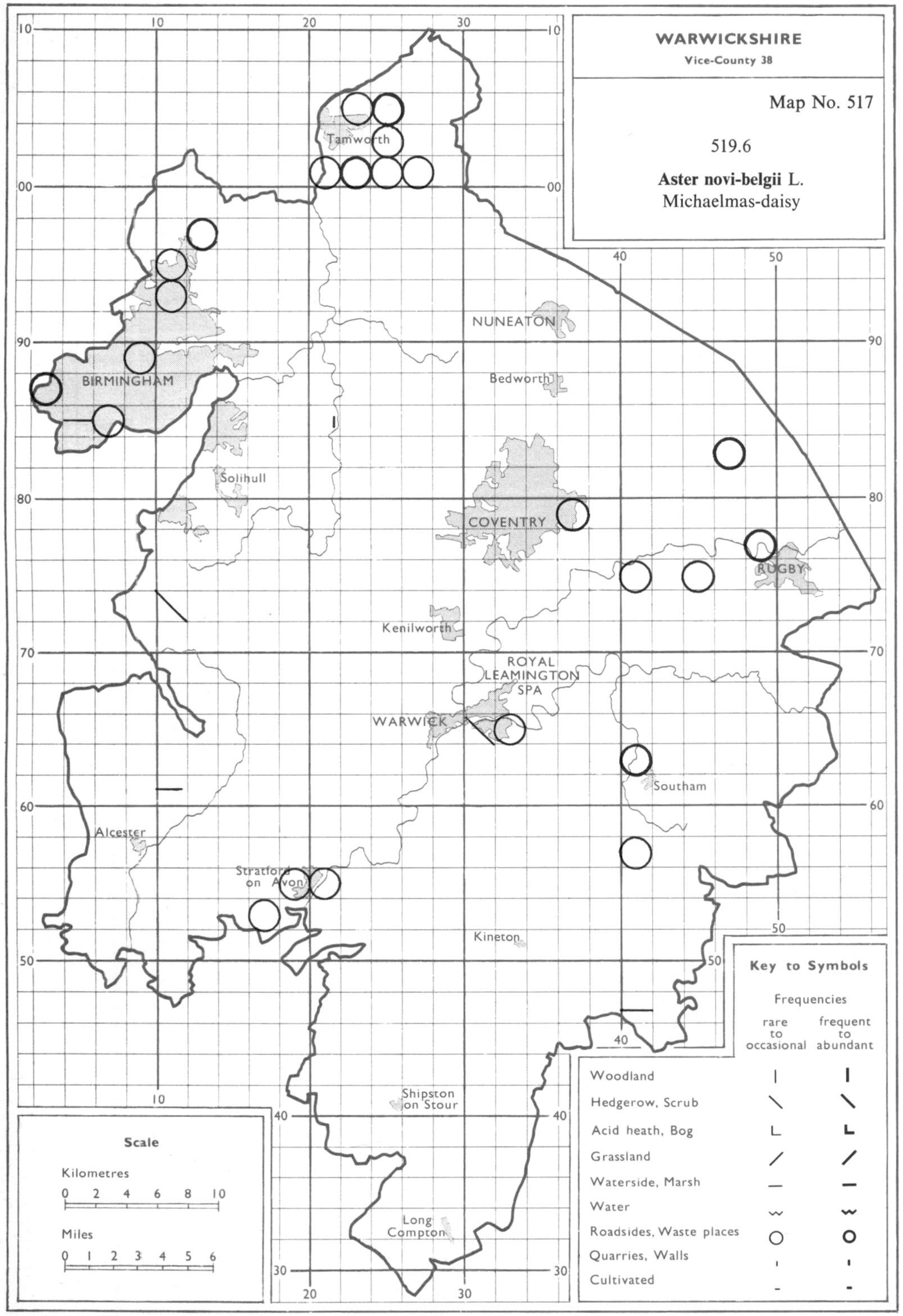
WARWICKSHIRE
Vice-County 38
Map No. 517
519.6
Aster novi-belgii L.
Michaelmas-daisy
Tamworth
BIRMINGHAM
Solihull
NUNEATON
Bedworth
COVENTRY
RUGBY
Kenilworth
ROYAL LEAMINGTON SPA
WARWICK
Southam
Alcester
Stratford on Avon
Kineton
Shipston on Stour
Long Compton
Scale
Kilometres
0 2 4 6 8 10
Miles
0 1 2 3 4 5 6
Key to Symbols
Frequencies
rare to occasional
frequent to abundant
Woodland
Hedgerow, Scrub
Acid heath, Bog
Grassland
Waterside, Marsh
Water
Roadsides, Waste places
Quarries, Walls
Cultivated

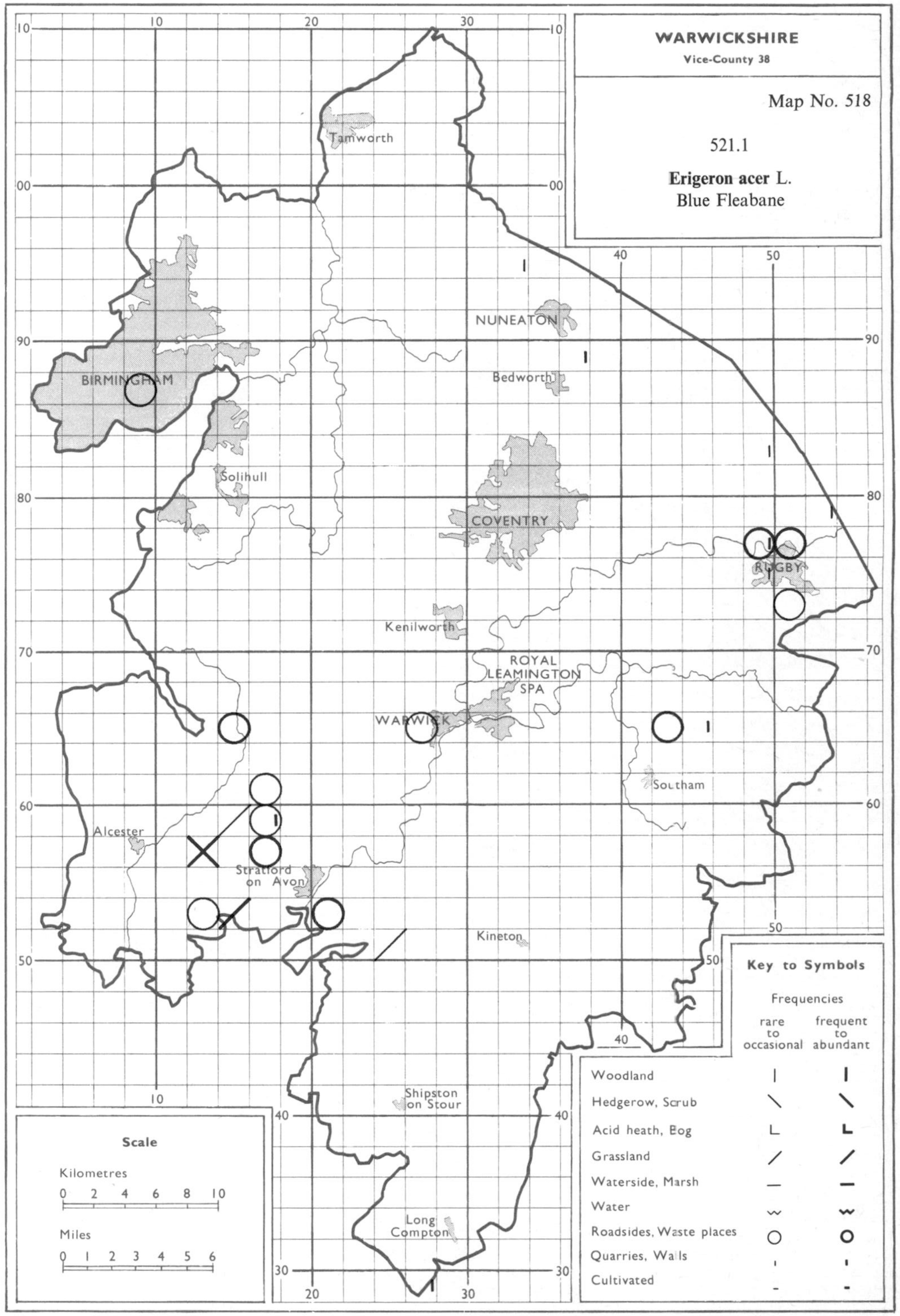
WARWICKSHIRE
Vice-County 38
Map No. 518
521.1
Erigeron acer L.
Blue Fleabane
Tamworth
BIRMINGHAM
Solihull
NUNEATON
Bedworth
COVENTRY
RUGBY
Kenilworth
ROYAL LEAMINGTON SPA
WARWICK
Southam
Alcester
Stratford on Avon
Kineton
Shipston on Stour
Long Compton
Scale
Kilometres
0 2 4 6 8 10
Miles
0 1 2 3 4 5 6
Key to Symbols
Frequencies
rare to occasional
frequent to abundant
Woodland
Hedgerow, Scrub
Acid heath, Bog
Grassland
Waterside, Marsh
Water
Roadsides, Waste places
Quarries, Walls
Cultivated

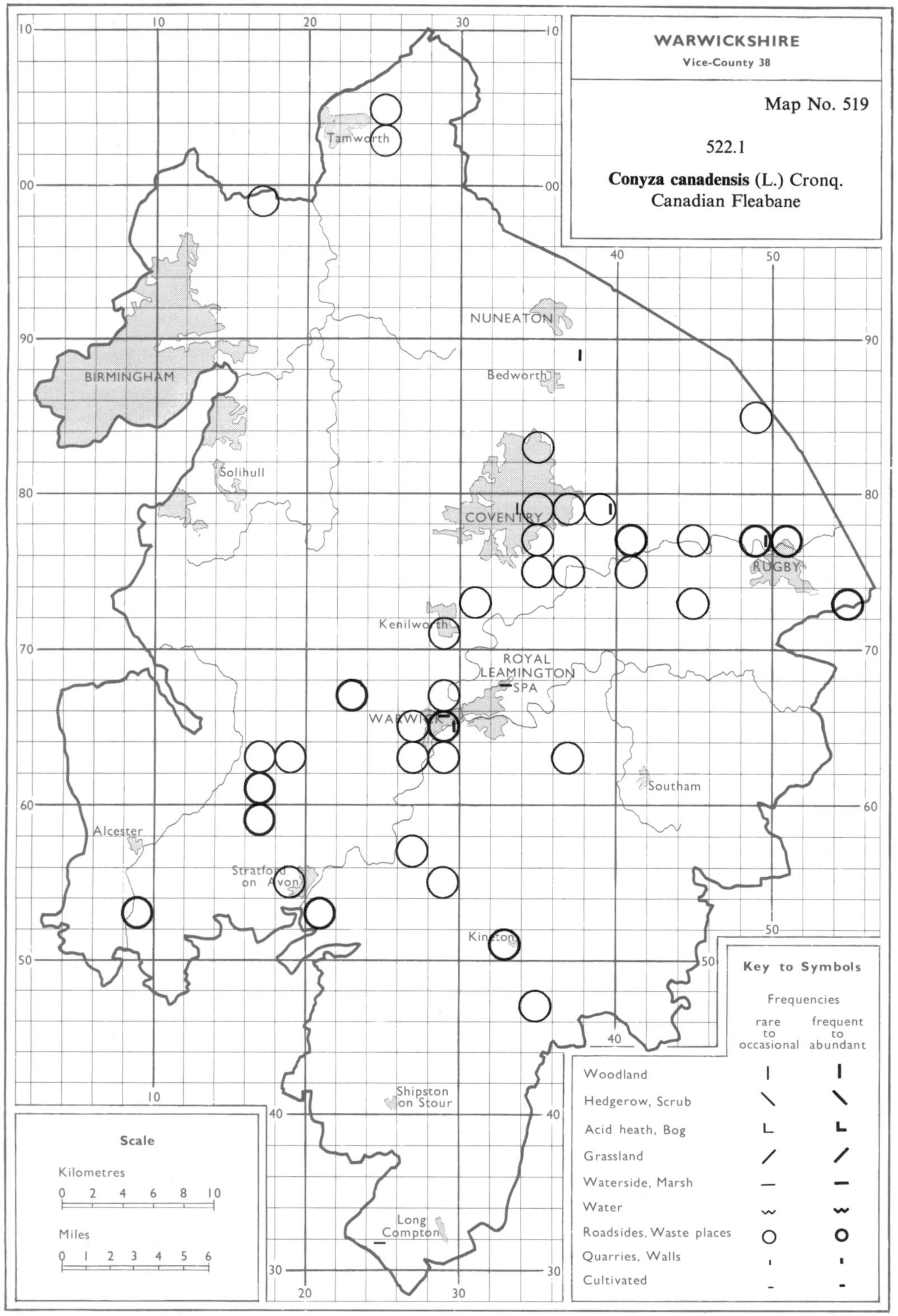
WARWICKSHIRE
Vice-County 38
Map No. 519
522.1
Conyza canadensis (L.) Cronq.
Canadian Fleabane
Key to Symbols
Frequencies
rare to occasional
frequent to abundant
Woodland
Hedgerow, Scrub
Acid heath, Bog
Grassland
Waterside, Marsh
Water
Roadsides, Waste places
Quarries, Walls
Cultivated
Scale
Kilometres
Miles

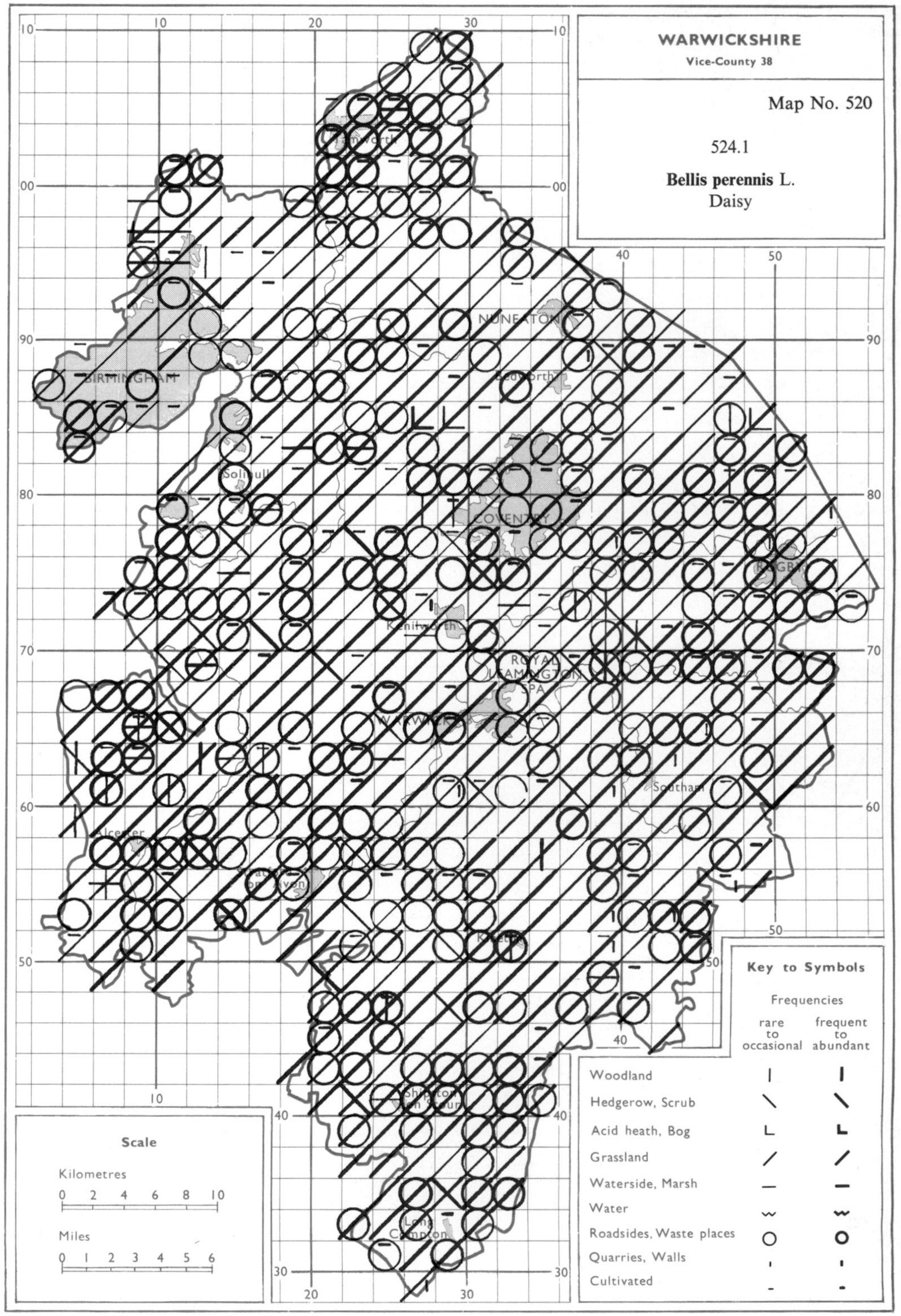
WARWICKSHIRE
Vice-County 38
Map No. 520
524.1
Bellis perennis L.
Daisy
Key to Symbols
Frequencies
rare to occasional
frequent to abundant
Woodland
Hedgerow, Scrub
Acid heath, Bog
Grassland
Waterside, Marsh
Water
Roadsides, Waste places
Quarries, Walls
Cultivated
Scale
Kilometres
Miles

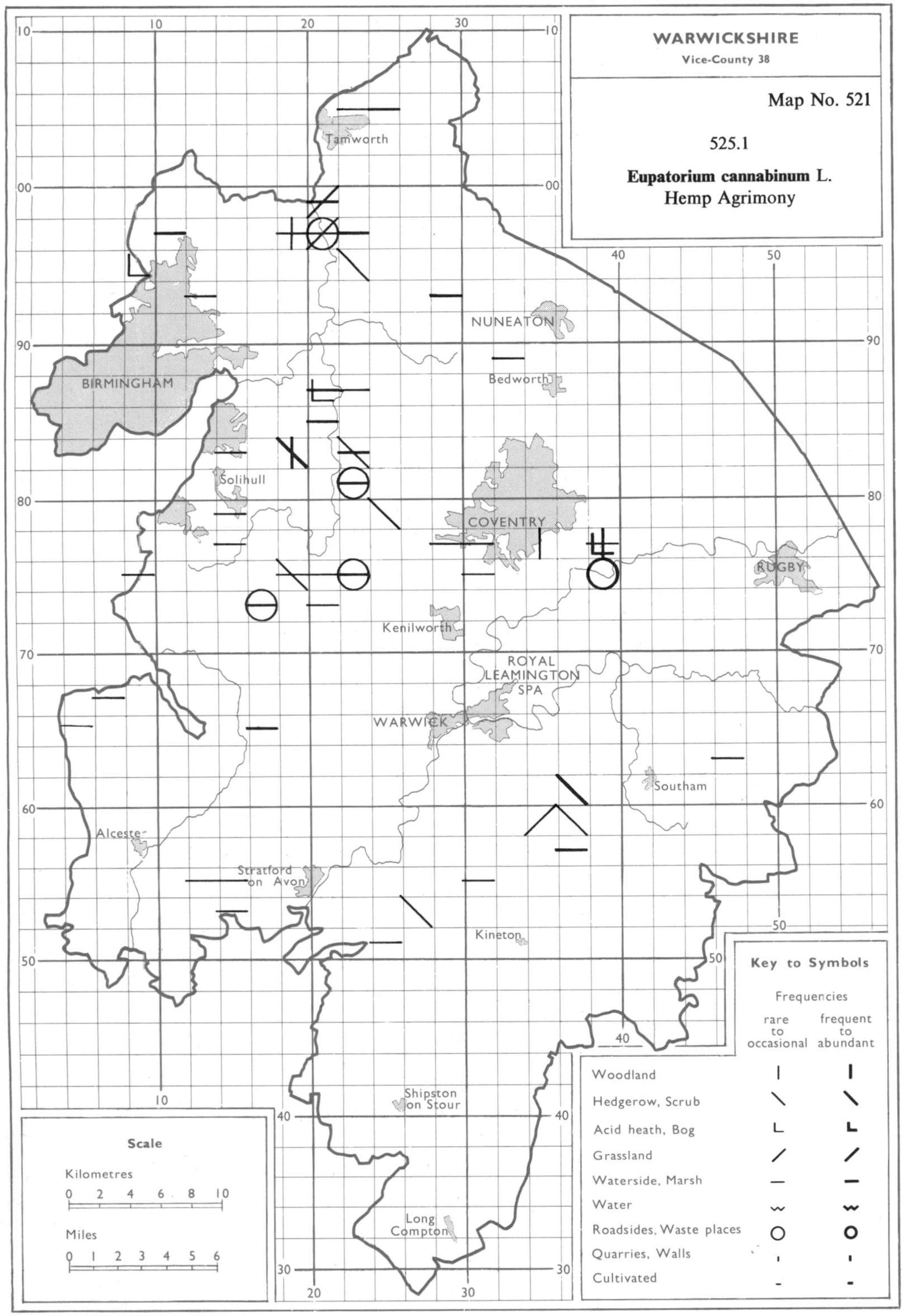
WARWICKSHIRE
Vice-County 38
Map No. 521
525.1
Eupatorium cannabinum L.
Hemp Agrimony
BIRMINGHAM
Tamworth
NUNEATON
Bedworth
Solihull
COVENTRY
RUGBY
Kenilworth
ROYAL LEAMINGTON SPA
WARWICK
Southam
Alcester
Stratford on Avon
Kineton
Shipston on Stour
Long Compton
Key to Symbols
Frequencies
rare to occasional
frequent to abundant
Woodland
Hedgerow, Scrub
Acid heath, Bog
Grassland
Waterside, Marsh
Water
Roadsides, Waste places
Quarries, Walls
Cultivated
Scale
Kilometres
0 2 4 6 8 10
Miles
0 1 2 3 4 5 6

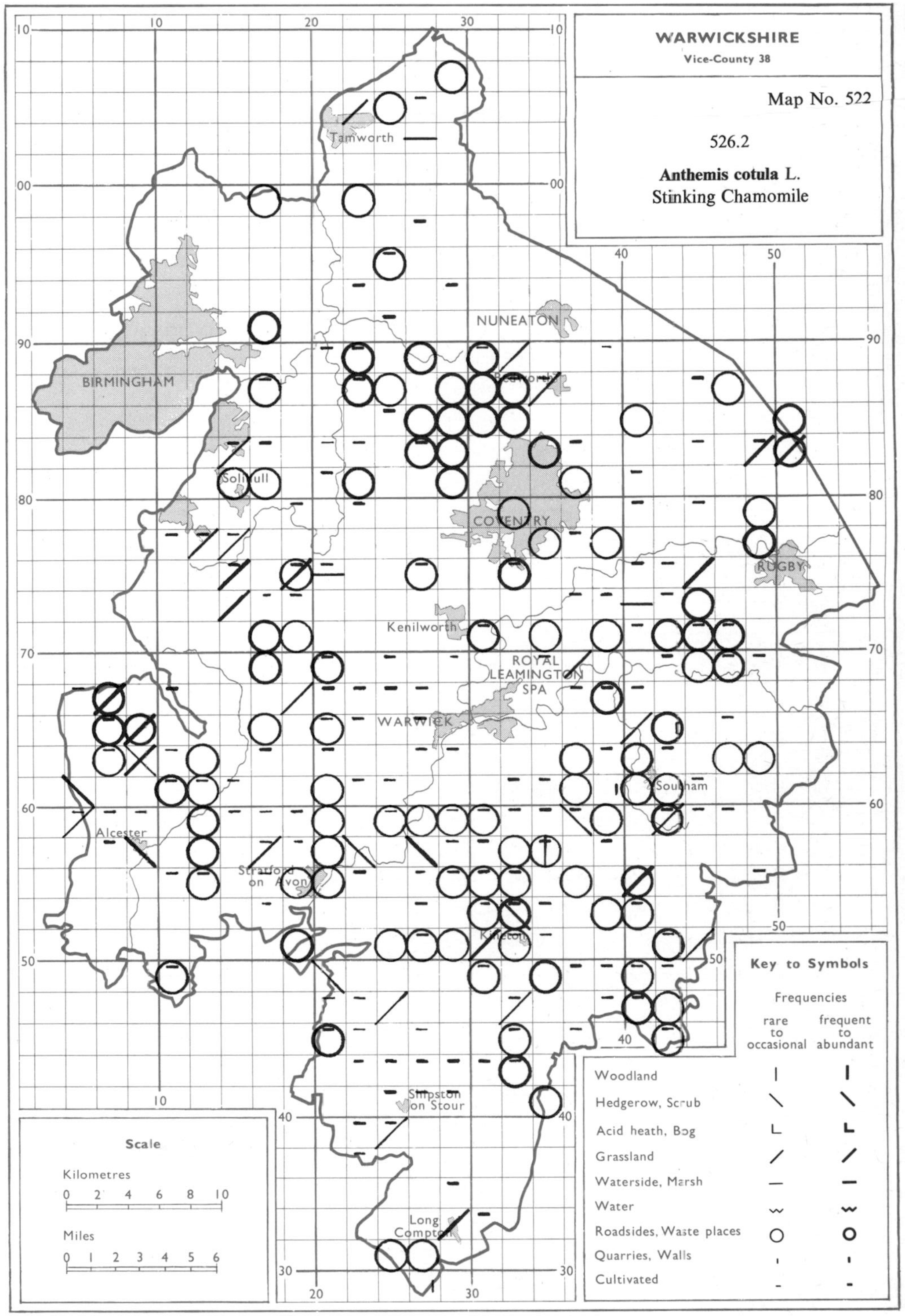
WARWICKSHIRE
Vice-County 38
Map No. 522
526.2
Anthemis cotula L.
Stinking Chamomile
BIRMINGHAM
Tamworth
NUNEATON
Bedworth
Solihull
COVENTRY
RUGBY
Kenilworth
ROYAL LEAMINGTON SPA
WARWICK
Southam
Alcester
Stratford on Avon
Kineton
Shipston on Stour
Long Compton
Key to Symbols
Frequencies
rare to occasional
frequent to abundant
Woodland
Hedgerow, Scrub
Acid heath, Bog
Grassland
Waterside, Marsh
Water
Roadsides, Waste places
Quarries, Walls
Cultivated
Scale
Kilometres
0 2 4 6 8 10
Miles
0 1 2 3 4 5 6

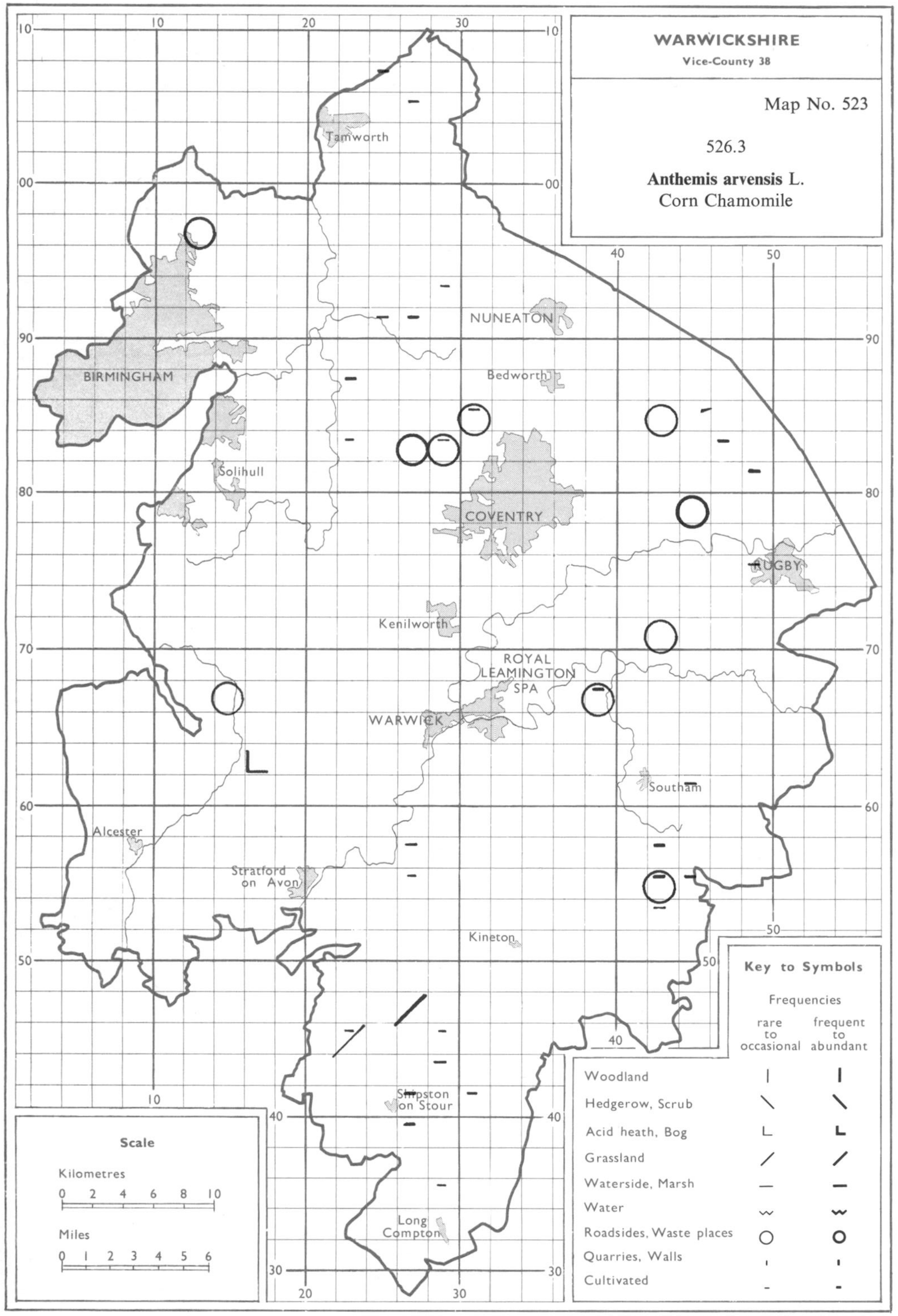
WARWICKSHIRE
Vice-County 38
Map No. 523
526.3
Anthemis arvensis L.
Corn Chamomile
Tamworth
NUNEATON
Bedworth
BIRMINGHAM
Solihull
COVENTRY
RUGBY
Kenilworth
ROYAL LEAMINGTON SPA
WARWICK
Southam
Alcester
Stratford on Avon
Kineton
Shipston on Stour
Long Compton
Key to Symbols
Frequencies
rare to occasional
frequent to abundant
Woodland
Hedgerow, Scrub
Acid heath, Bog
Grassland
Waterside, Marsh
Water
Roadsides, Waste places
Quarries, Walls
Cultivated
Scale
Kilometres
0 2 4 6 8 10
Miles
0 1 2 3 4 5 6

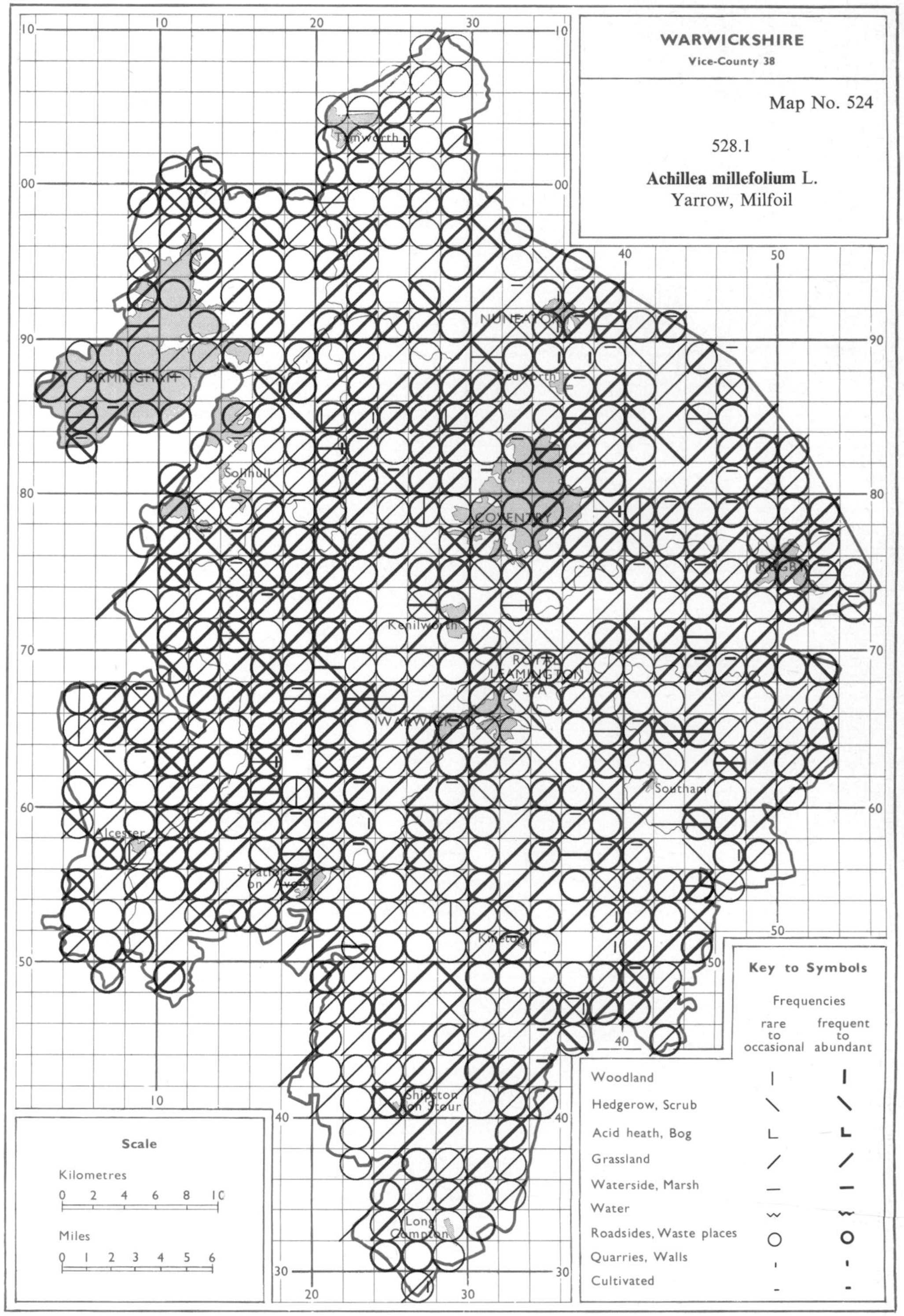
WARWICKSHIRE
Vice-County 38
Map No. 524
528.1
Achillea millefolium L.
Yarrow, Milfoil
Key to Symbols
Frequencies
rare to occasional
frequent to abundant
Woodland
Hedgerow, Scrub
Acid heath, Bog
Grassland
Waterside, Marsh
Water
Roadsides, Waste places
Quarries, Walls
Cultivated
Scale
Kilometres
0 2 4 6 8 10
Miles
0 1 2 3 4 5 6

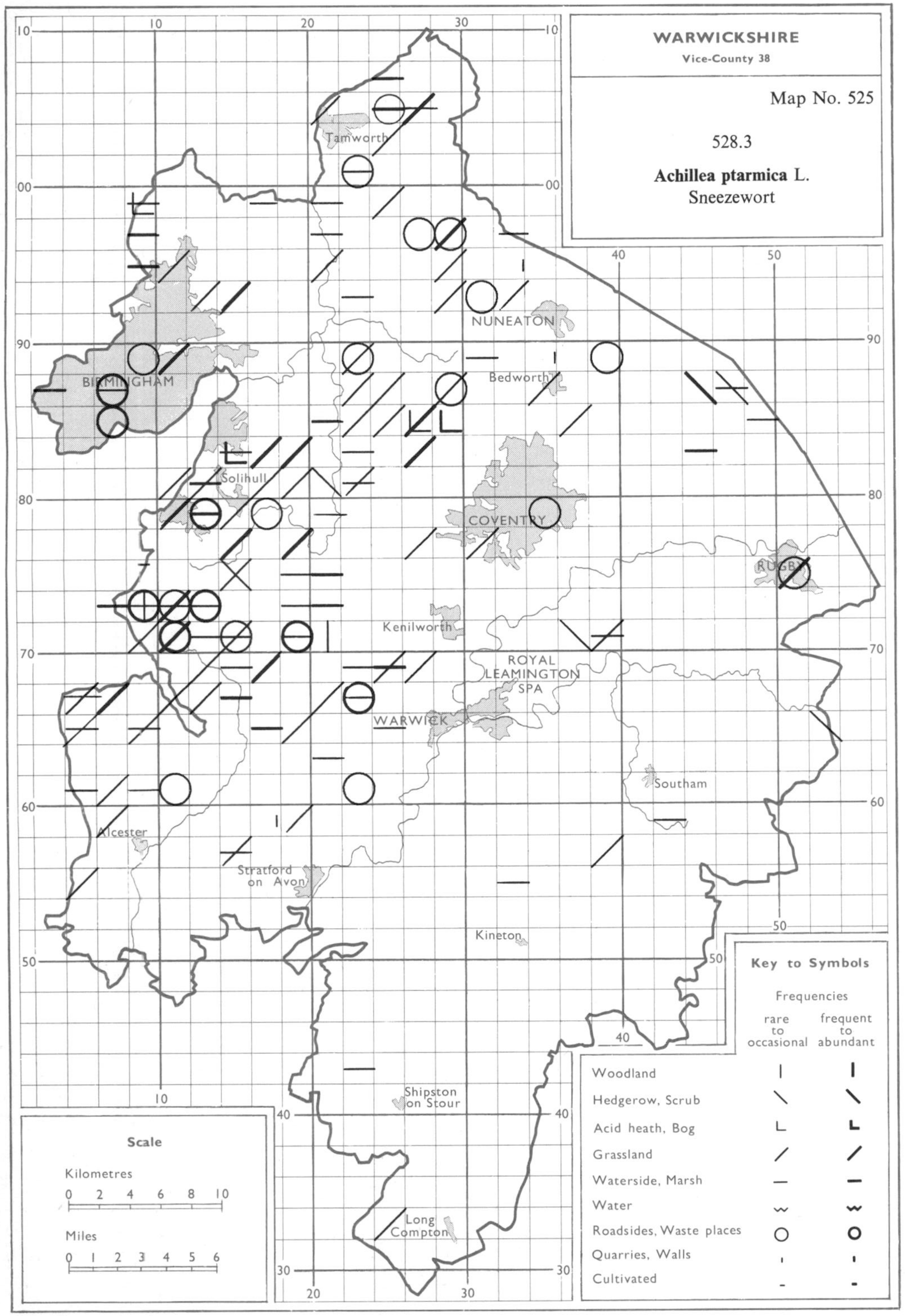
WARWICKSHIRE
Vice-County 38
Map No. 525
528.3
Achillea ptarmica L.
Sneezewort
Tamworth
BIRMINGHAM
NUNEATON
Bedworth
Solihull
COVENTRY
RUGBY
Kenilworth
ROYAL LEAMINGTON SPA
WARWICK
Southam
Alcester
Stratford on Avon
Kineton
Shipston on Stour
Long Compton
Key to Symbols
Frequencies
rare to occasional
frequent to abundant
Woodland
Hedgerow, Scrub
Acid heath, Bog
Grassland
Waterside, Marsh
Water
Roadsides, Waste places
Quarries, Walls
Cultivated
Scale
Kilometres
0 2 4 6 8 10
Miles
0 1 2 3 4 5 6

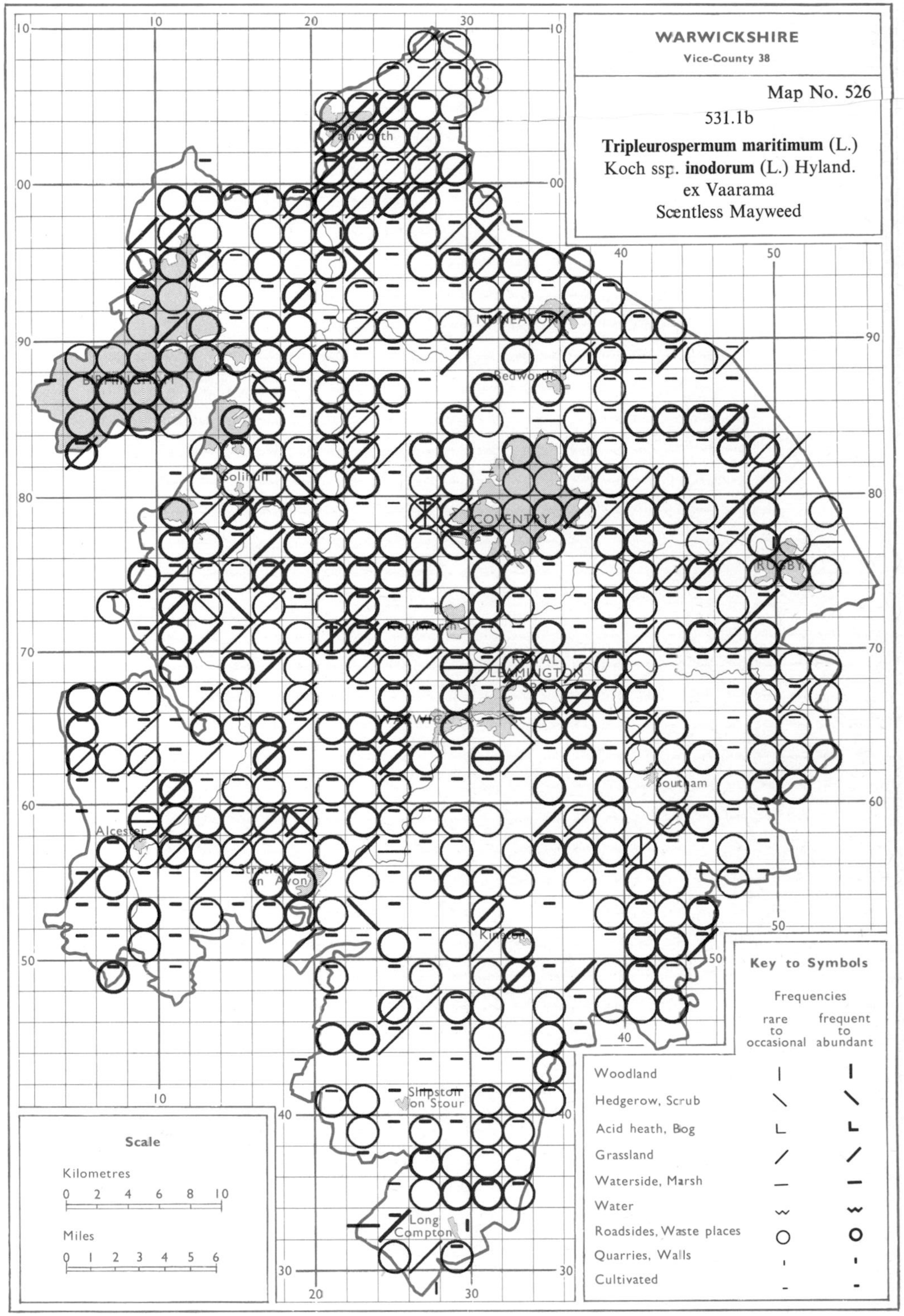
WARWICKSHIRE
Vice-County 38
Map No. 526
531.1b
Tripleurospermum maritimum (L.) Koch ssp. inodorum (L.) Hyland. ex Vaarama
Scentless Mayweed
Key to Symbols
Frequencies
rare to occasional
frequent to abundant
Woodland
Hedgerow, Scrub
Acid heath, Bog
Grassland
Waterside, Marsh
Water
Roadsides, Waste places
Quarries, Walls
Cultivated
Scale
Kilometres
0 2 4 6 8 10
Miles
0 1 2 3 4 5 6

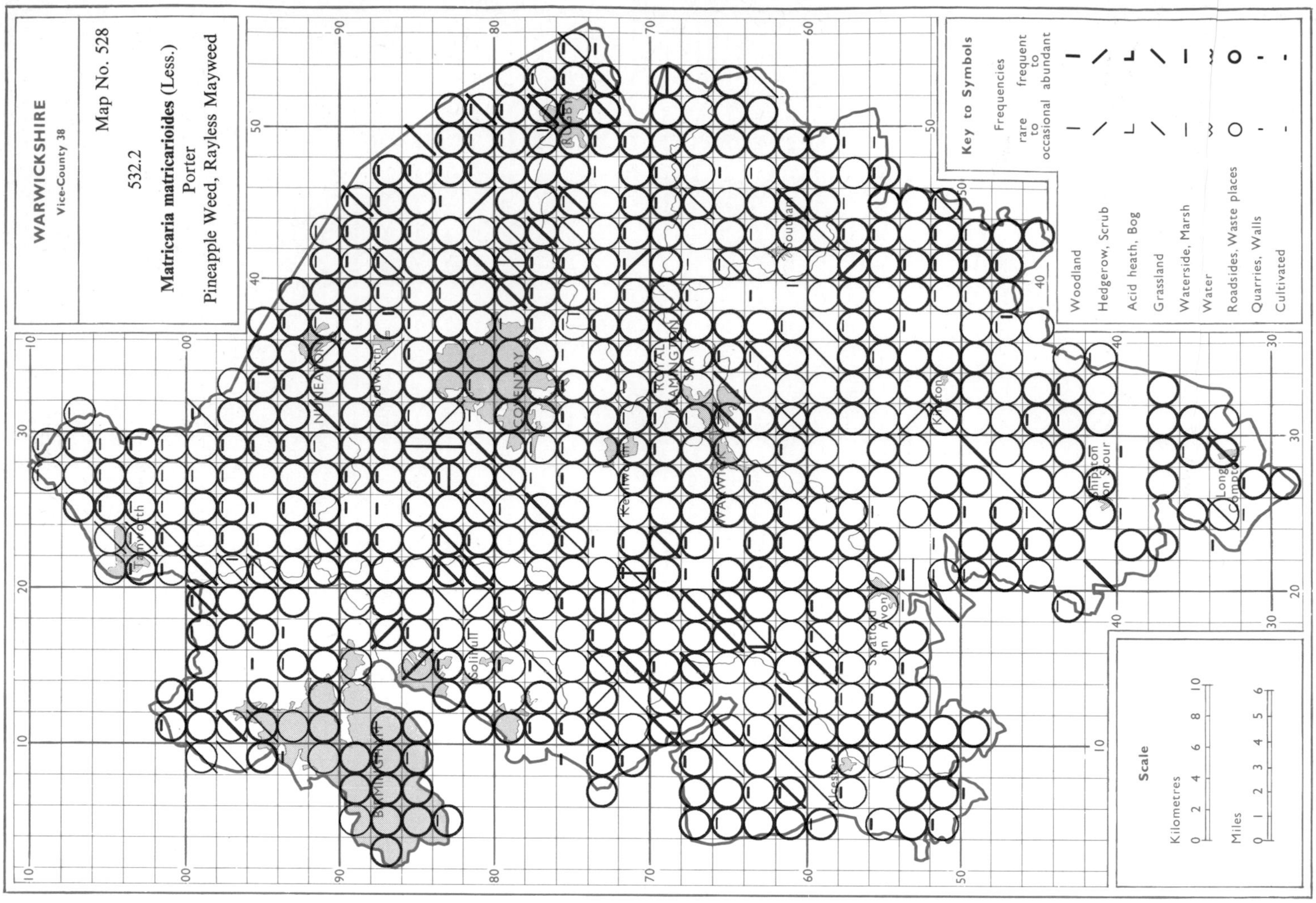
WARWICKSHIRE
Vice-County 38
Map No. 528
532.2
Matricaria matricarioides (Less.) Porter
Pineapple Weed, Rayless Mayweed
Key to Symbols
Frequencies
rare to occasional
frequent to abundant
Woodland
Hedgerow, Scrub
Acid heath, Bog
Grassland
Waterside, Marsh
Water
Roadsides, Waste places
Quarries, Walls
Cultivated
Scale
Kilometres
0 2 4 6 8 10
Miles
0 1 2 3 4 5 6
Tamworth
NUNEATON
Bedworth
COVENTRY
RUGBY
ROYAL LEAMINGTON SPA
Kenilworth
WARWICK
Southam
Kineton
Shipston on Stour
Long Compton
Solihull
BIRMINGHAM
Alcester
Stratford on Avon

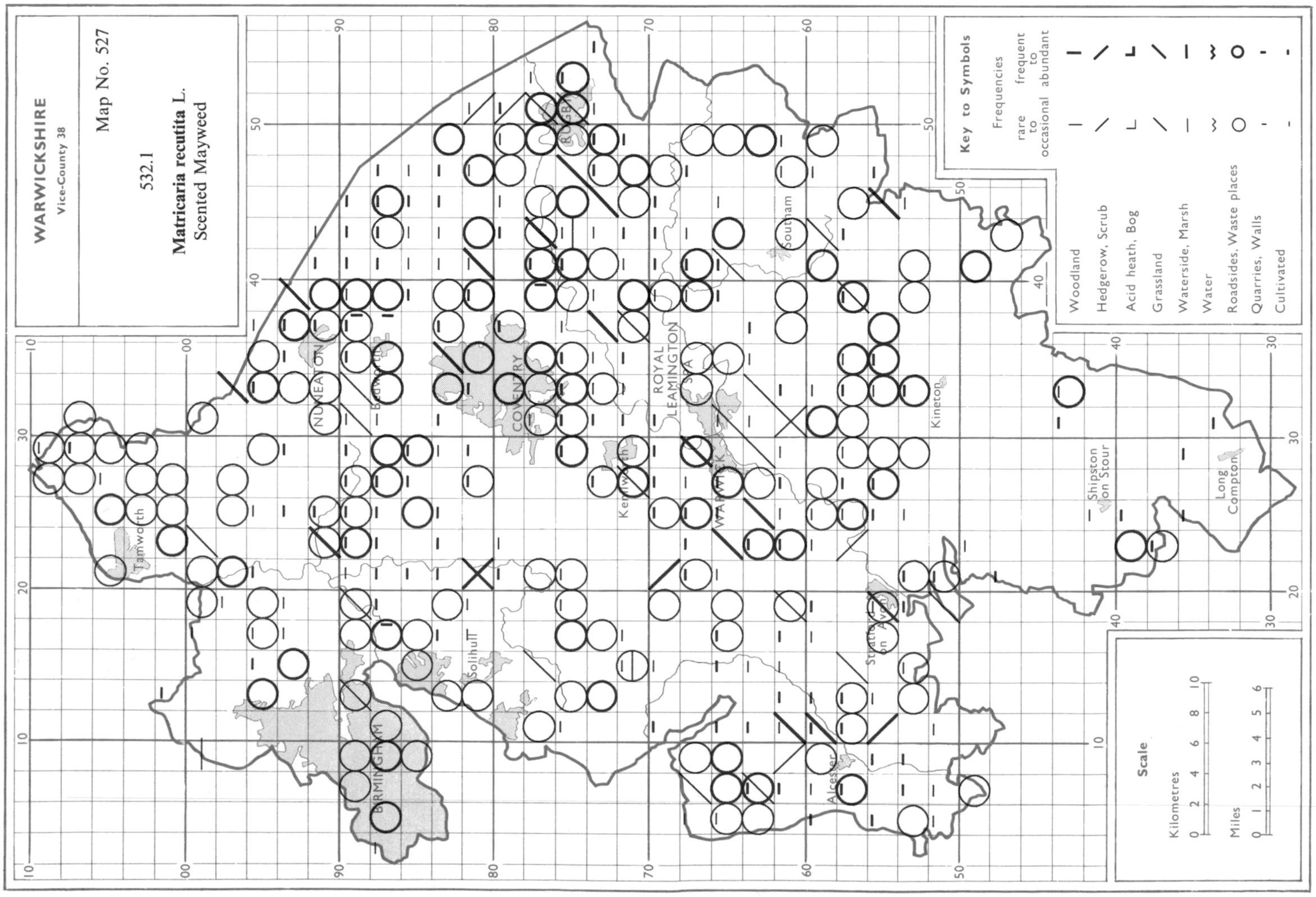
WARWICKSHIRE
Vice-County 38
Map No. 527
532.1
Matricaria recutita L.
Scented Mayweed
Key to Symbols
Frequencies
rare to occasional
frequent to abundant
Woodland
Hedgerow, Scrub
Acid heath, Bog
Grassland
Waterside, Marsh
Water
Roadsides, Waste places
Quarries, Walls
Cultivated
Scale
Kilometres
0 2 4 6 8 10
Miles
0 1 2 3 4 5 6
Tamworth
NUNEATON
Bedworth
COVENTRY
RUGBY
ROYAL LEAMINGTON SPA
Kenilworth
WARWICK
Southam
Kineton
Shipston on Stour
Long Compton
Solihull
BIRMINGHAM
Alcester
Stratford on Avon

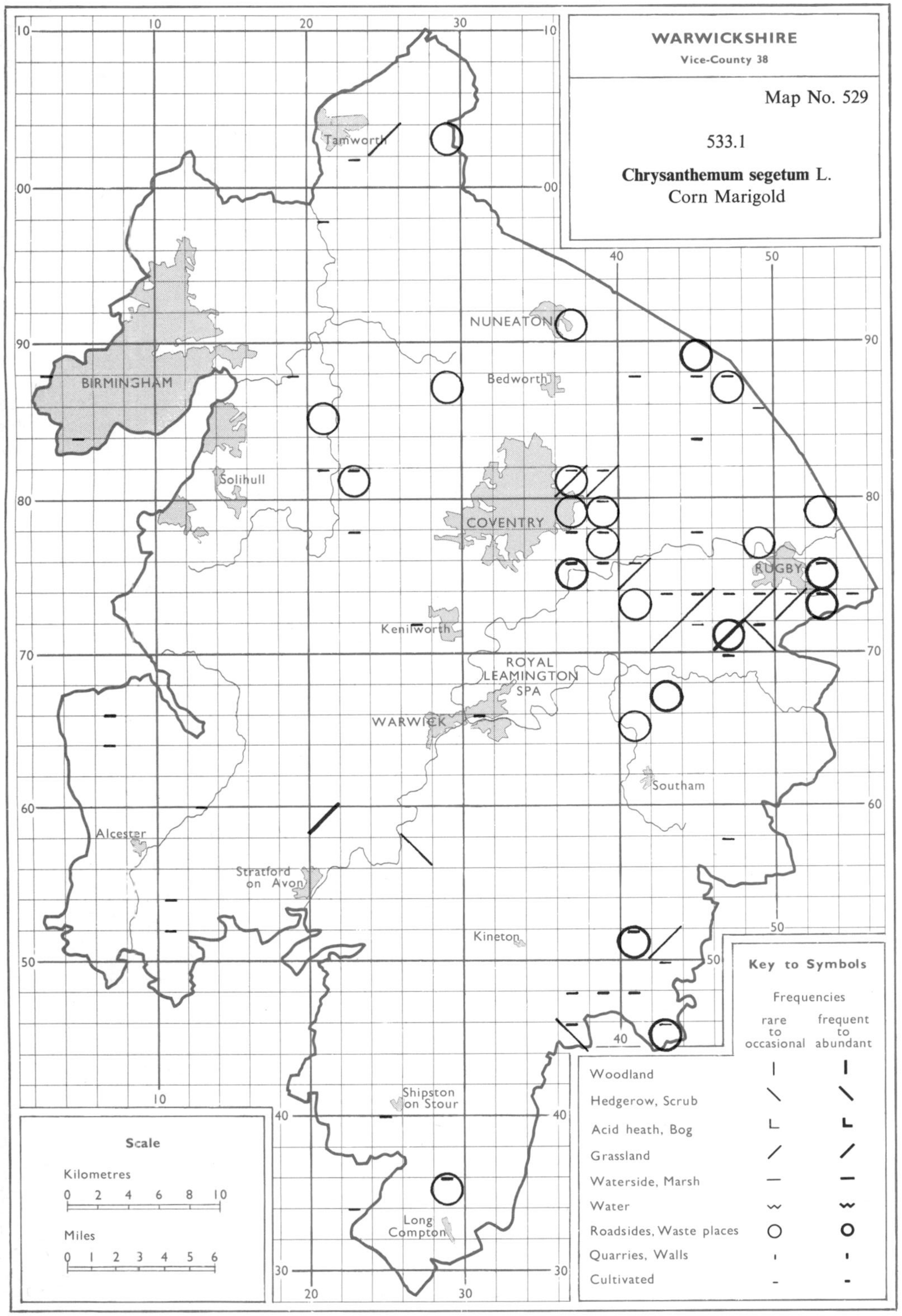
WARWICKSHIRE
Vice-County 38
Map No. 529
533.1
Chrysanthemum segetum L.
Corn Marigold
Tamworth
NUNEATON
Bedworth
BIRMINGHAM
Solihull
COVENTRY
RUGBY
Kenilworth
ROYAL LEAMINGTON SPA
WARWICK
Southam
Alcester
Stratford on Avon
Kineton
Shipston on Stour
Long Compton
Key to Symbols
Frequencies
rare to occasional
frequent to abundant
Woodland
Hedgerow, Scrub
Acid heath, Bog
Grassland
Waterside, Marsh
Water
Roadsides, Waste places
Quarries, Walls
Cultivated
Scale
Kilometres
0 2 4 6 8 10
Miles
0 1 2 3 4 5 6

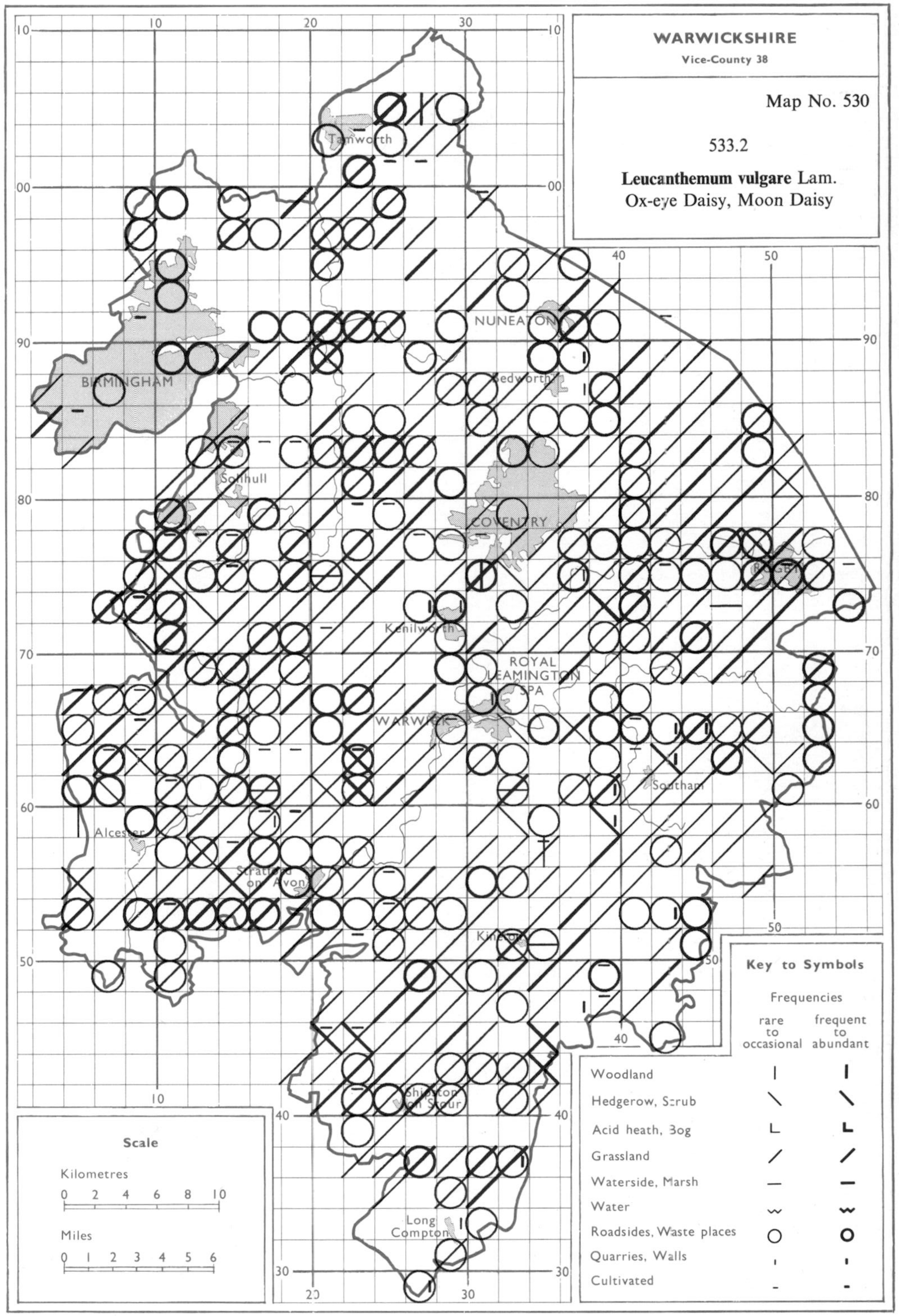
WARWICKSHIRE
Vice-County 38
Map No. 530
533.2
Leucanthemum vulgare Lam.
Ox-eye Daisy, Moon Daisy
Tamworth
NUNEATON
Bedworth
BIRMINGHAM
Solihull
COVENTRY
RUGBY
Kenilworth
ROYAL LEAMINGTON SPA
WARWICK
Southam
Alcester
Stratford on Avon
Kineton
Shipston on Stour
Long Compton
Key to Symbols
Frequencies
rare to occasional
frequent to abundant
Woodland
Hedgerow, Scrub
Acid heath, Bog
Grassland
Waterside, Marsh
Water
Roadsides, Waste places
Quarries, Walls
Cultivated
Scale
Kilometres
0 2 4 6 8 10
Miles
0 1 2 3 4 5 6

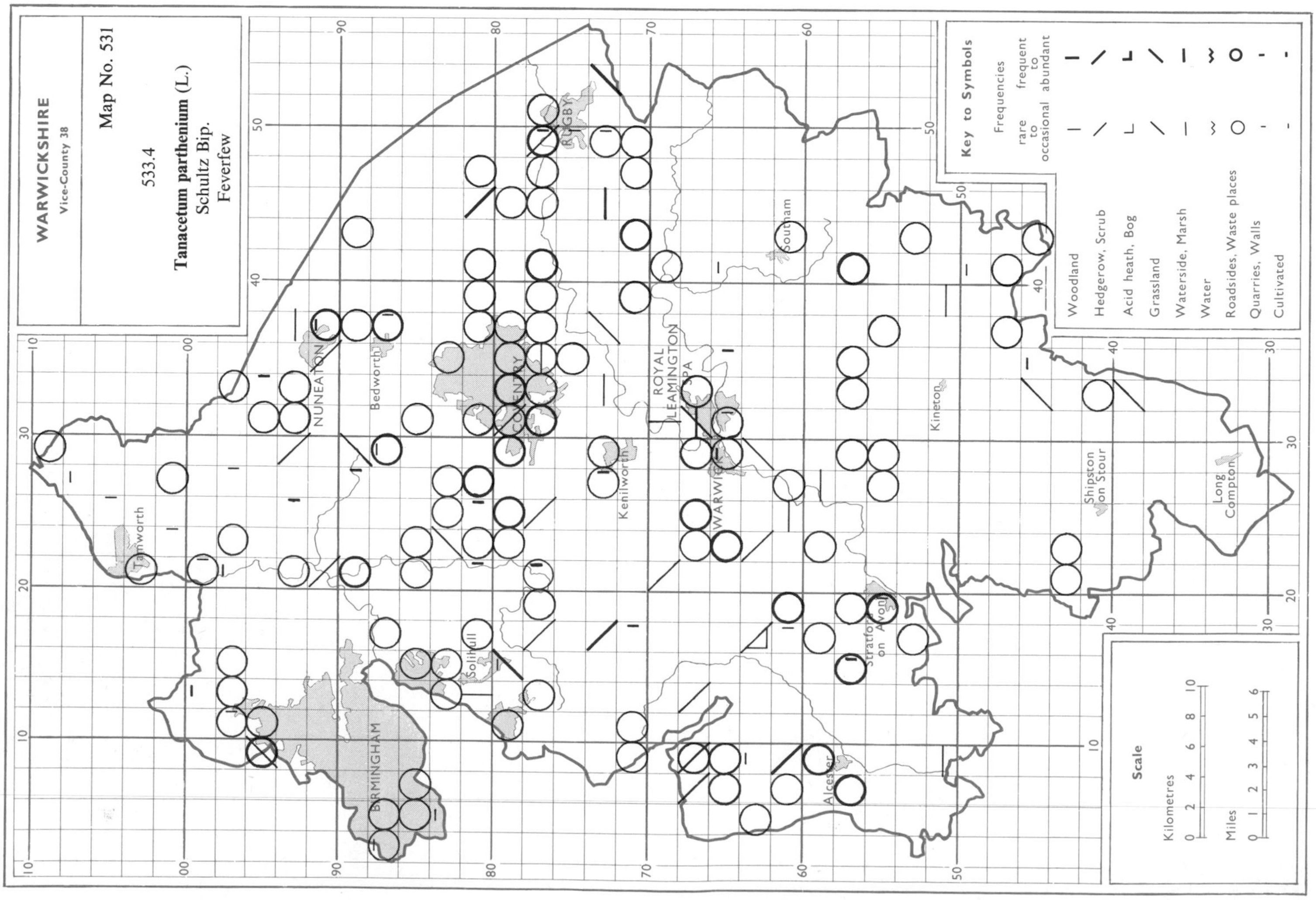
WARWICKSHIRE
Vice-County 38
Map No. 531
533.4
Tanacetum parthenium (L.) Schultz Bip.
Feverfew
Key to Symbols
Frequencies
rare to occasional
frequent to abundant
Woodland
Hedgerow, Scrub
Acid heath, Bog
Grassland
Waterside, Marsh
Water
Roadsides, Waste places
Quarries, Walls
Cultivated
Scale
Kilometres
0 2 4 6 8 10
Miles
0 1 2 3 4 5 6
BIRMINGHAM
Solihull
Tamworth
NUNEATON
Bedworth
COVENTRY
RUGBY
Kenilworth
ROYAL LEAMINGTON SPA
WARWICK
Southam
Kineton
Stratford on Avon
Alcester
Shipston on Stour
Long Compton

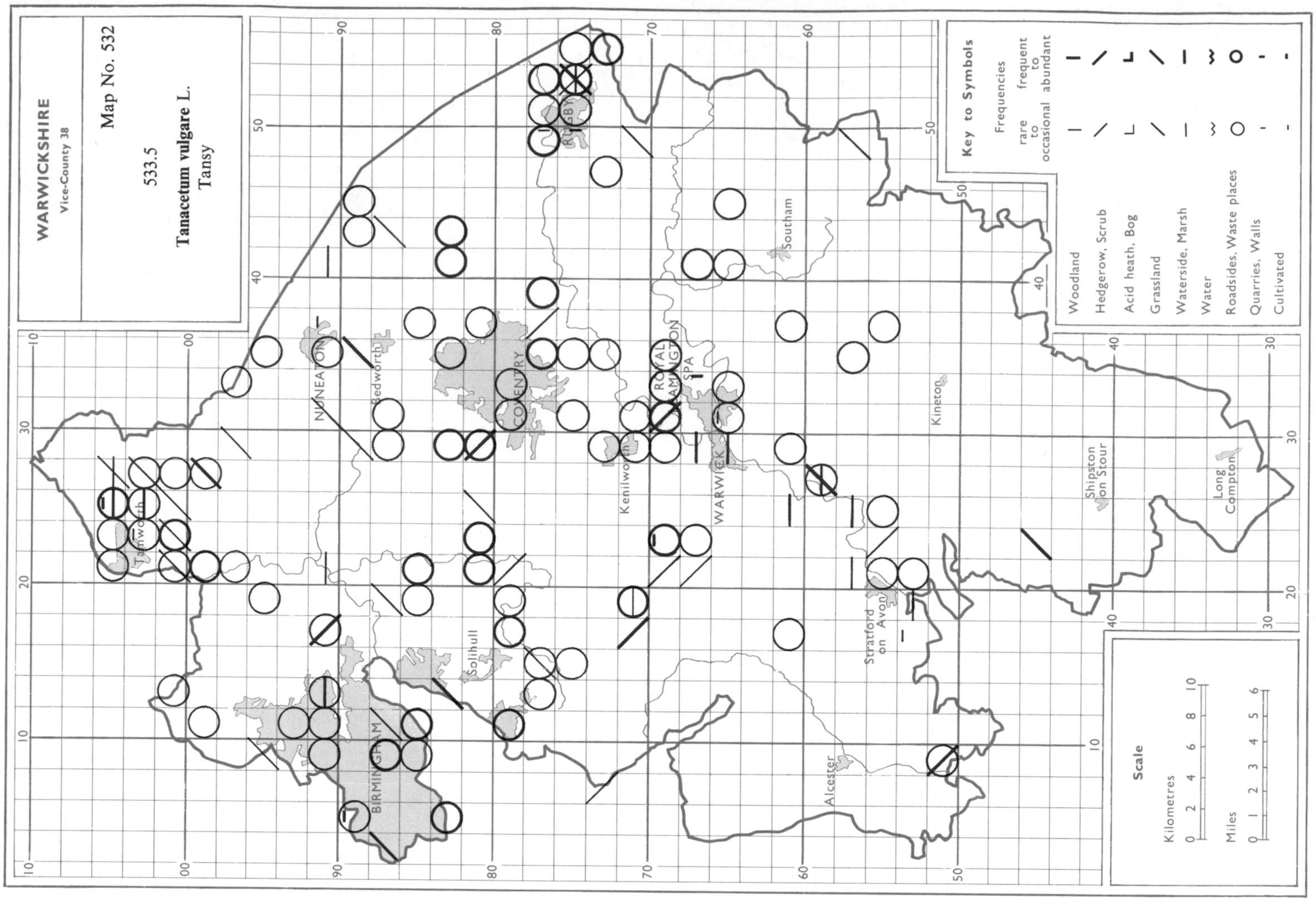
WARWICKSHIRE
Vice-County 38
Map No. 532
533.5
Tanacetum vulgare L.
Tansy
Key to Symbols
Frequencies
rare to occasional
frequent to abundant
Woodland
Hedgerow, Scrub
Acid heath, Bog
Grassland
Waterside, Marsh
Water
Roadsides, Waste places
Quarries, Walls
Cultivated
Scale
Kilometres
0 2 4 6 8 10
Miles
0 1 2 3 4 5 6
BIRMINGHAM
Solihull
Tamworth
NUNEATON
Bedworth
COVENTRY
RUGBY
Kenilworth
ROYAL LEAMINGTON SPA
WARWICK
Southam
Kineton
Stratford on Avon
Alcester
Shipston on Stour
Long Compton

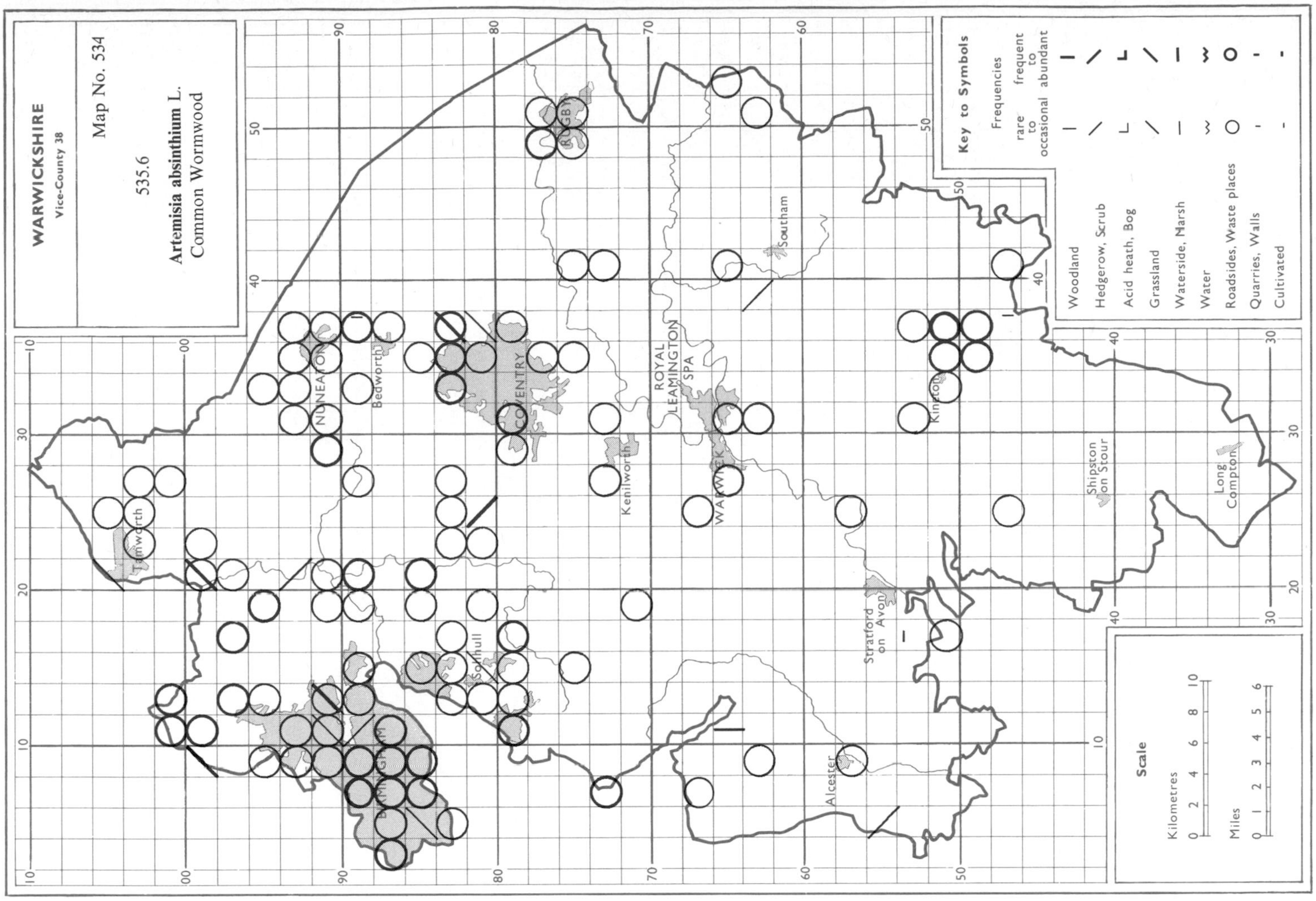
WARWICKSHIRE
Vice-County 38
Map No. 534
535.6
Artemisia absinthium L.
Common Wormwood
Key to Symbols
Frequencies
rare to occasional
frequent to abundant
Woodland
Hedgerow, Scrub
Acid heath, Bog
Grassland
Waterside, Marsh
Water
Roadsides, Waste places
Quarries, Walls
Cultivated
Scale
Kilometres
0 2 4 6 8 10
Miles
0 1 2 3 4 5 6
Tamworth
NUNEATON
Bedworth
COVENTRY
RUGBY
Southam
ROYAL LEAMINGTON SPA
WARWICK
Kenilworth
Kineton
Shipston on Stour
Long Compton
Stratford on Avon
Solihull
BIRMINGHAM
Alcester

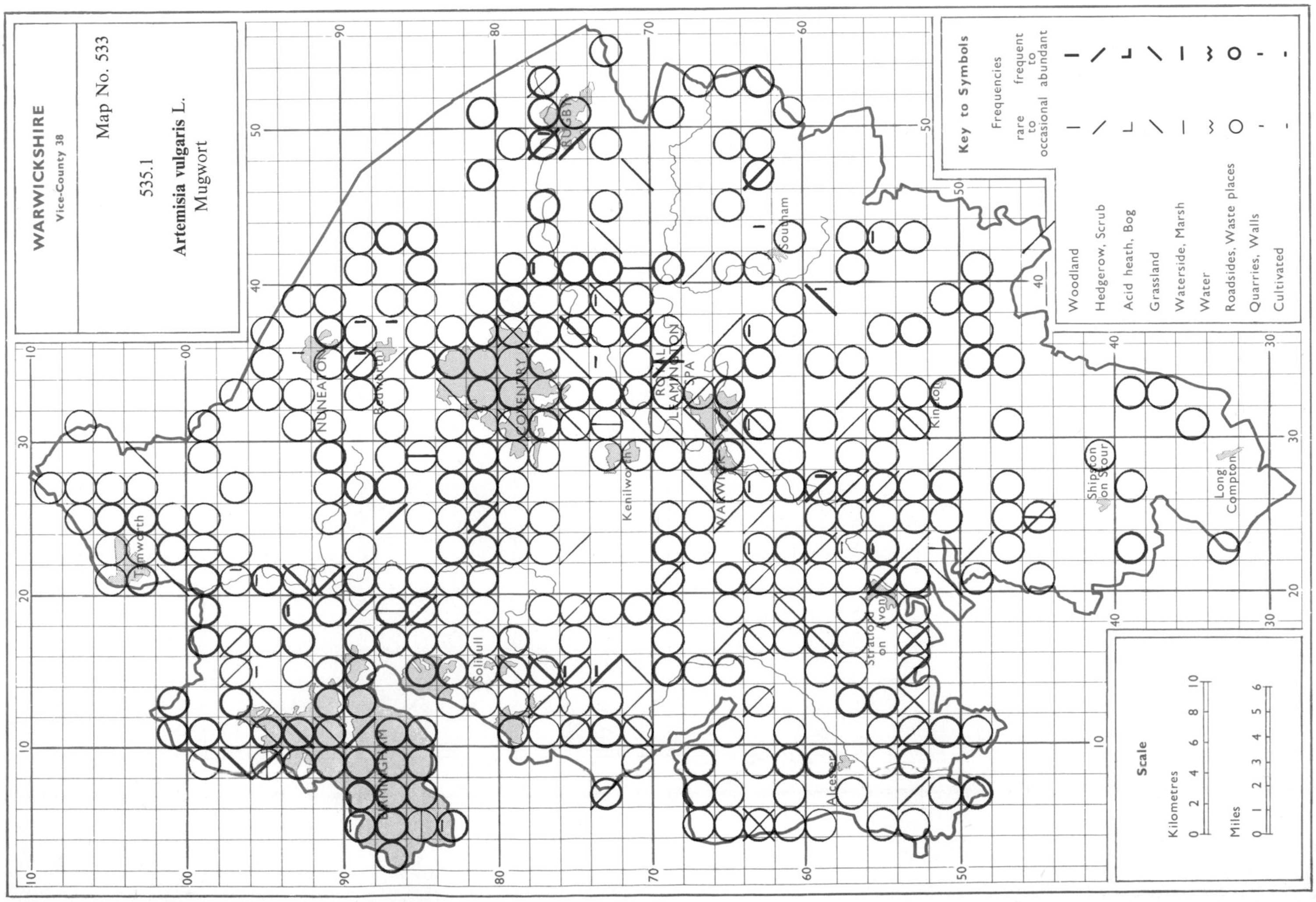
WARWICKSHIRE
Vice-County 38
Map No. 533
535.1
Artemisia vulgaris L.
Mugwort
Key to Symbols
Frequencies
rare to occasional
frequent to abundant
Woodland
Hedgerow, Scrub
Acid heath, Bog
Grassland
Waterside, Marsh
Water
Roadsides, Waste places
Quarries, Walls
Cultivated
Scale
Kilometres
0 2 4 6 8 10
Miles
0 1 2 3 4 5 6
Tamworth
NUNEATON
Bedworth
COVENTRY
RUGBY
Southam
ROYAL LEAMINGTON SPA
WARWICK
Kenilworth
Kineton
Shipston on Stour
Long Compton
Stratford on Avon
Solihull
BIRMINGHAM
Alcester

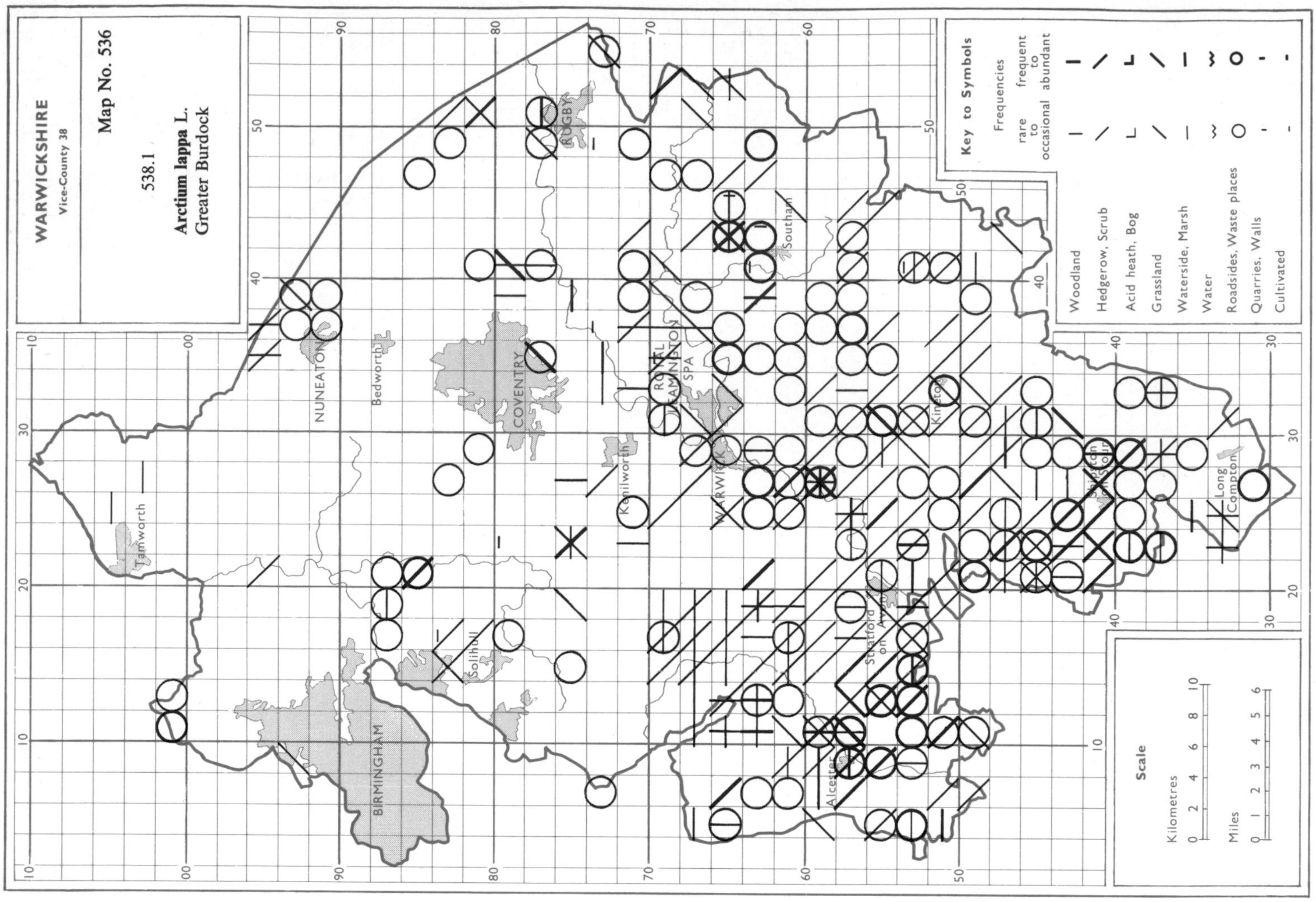
WARWICKSHIRE
Vice-County 38
Map No. 536
538.1
Arctium lappa L.
Greater Burdock
Key to Symbols
Frequencies
rare to occasional
frequent to abundant
Woodland
Hedgerow, Scrub
Acid heath, Bog
Grassland
Waterside, Marsh
Water
Roadsides, Waste places
Quarries, Walls
Cultivated
Scale
Kilometres 0 2 4 6 8 10
Miles 0 1 2 3 4 5 6
NUNEATON
Bedworth
COVENTRY
RUGBY
Southam
ROYAL LEAMINGTON SPA
Kenilworth
WARWICK
Kineton
Tamworth
Solihull
BIRMINGHAM
Stratford on Avon
Shipston on Stour
Long Compton
Alcester

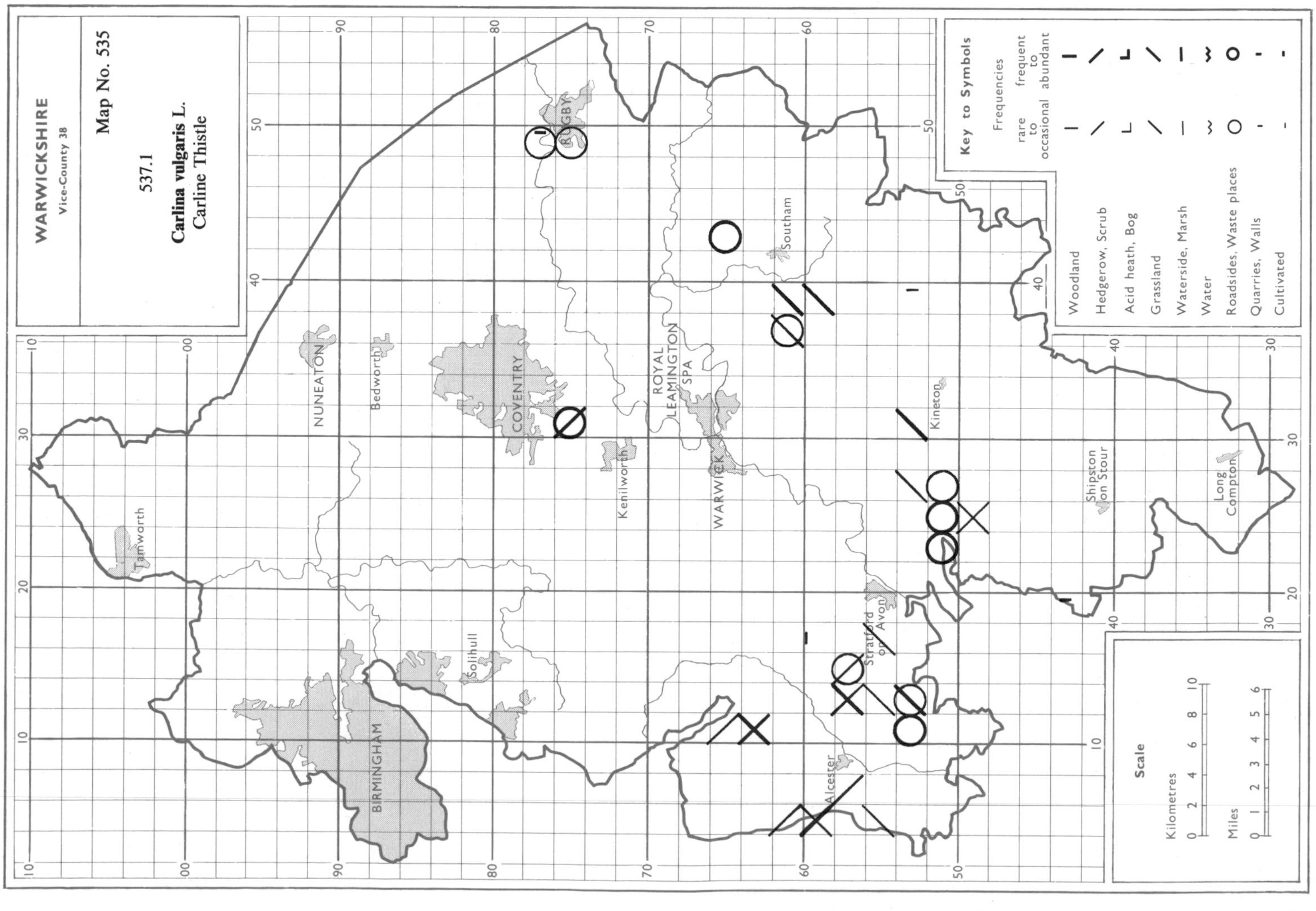
WARWICKSHIRE
Vice-County 38
Map No. 535
537.1
Carlina vulgaris L.
Carline Thistle
Key to Symbols
Frequencies
rare to occasional
frequent to abundant
Woodland
Hedgerow, Scrub
Acid heath, Bog
Grassland
Waterside, Marsh
Water
Roadsides, Waste places
Quarries, Walls
Cultivated
Scale
Kilometres 0 2 4 6 8 10
Miles 0 1 2 3 4 5 6
NUNEATON
Bedworth
COVENTRY
RUGBY
Southam
ROYAL LEAMINGTON SPA
Kenilworth
WARWICK
Kineton
Tamworth
Solihull
BIRMINGHAM
Stratford on Avon
Shipston on Stour
Long Compton
Alcester

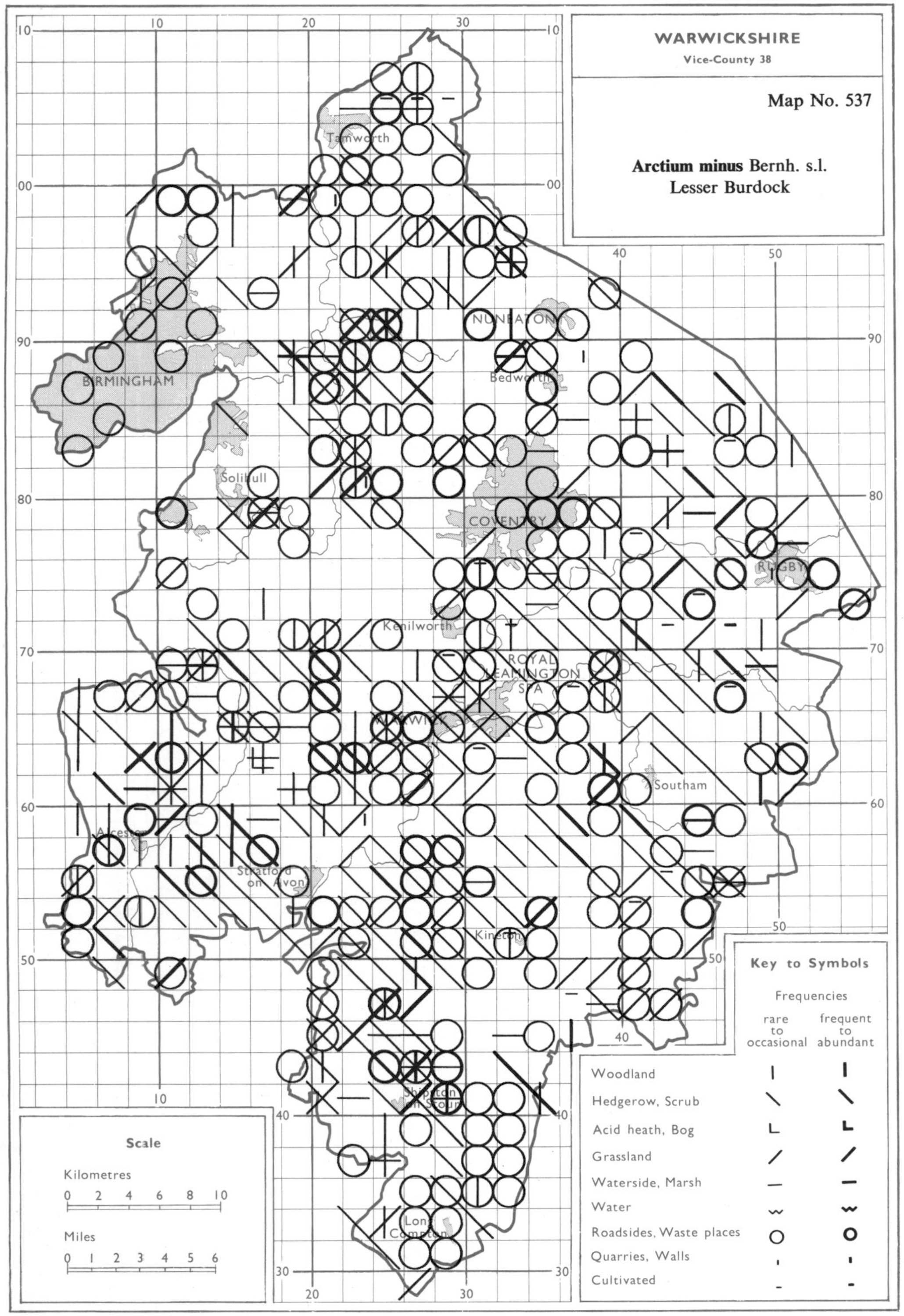
WARWICKSHIRE
Vice-County 38
Map No. 537
Arctium minus Bernh. s.l.
Lesser Burdock
Tamworth
NUNEATON
Bedworth
BIRMINGHAM
Solihull
COVENTRY
RUGBY
Kenilworth
ROYAL LEAMINGTON SPA
WARWICK
Southam
Alcester
Stratford on Avon
Kineton
Shipston on Stour
Long Compton
Key to Symbols
Frequencies
rare to occasional
frequent to abundant
Woodland
Hedgerow, Scrub
Acid heath, Bog
Grassland
Waterside, Marsh
Water
Roadsides, Waste places
Quarries, Walls
Cultivated
Scale
Kilometres
0 2 4 6 8 10
Miles
0 1 2 3 4 5 6

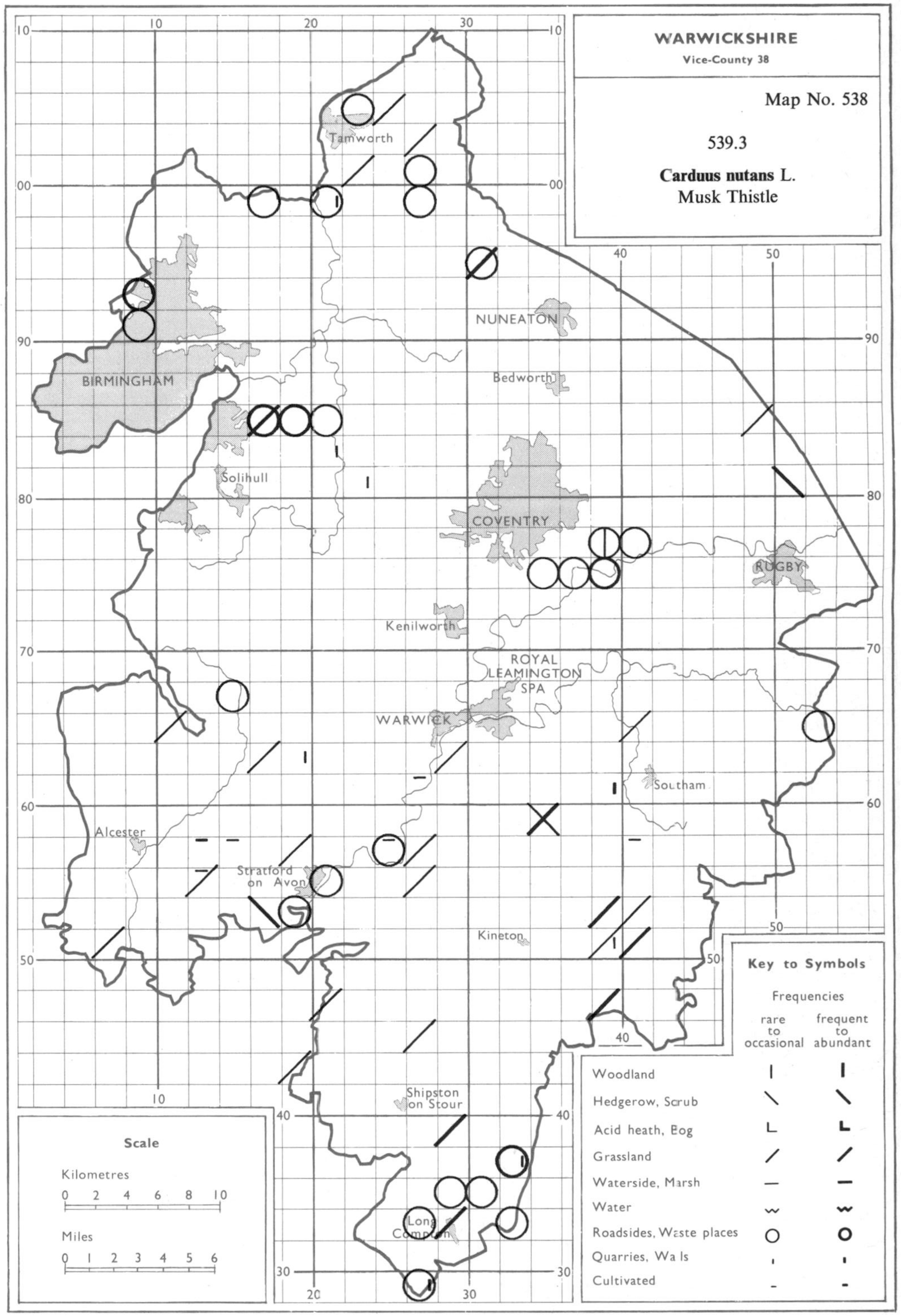
WARWICKSHIRE
Vice-County 38
Map No. 538
539.3
Carduus nutans L.
Musk Thistle
Tamworth
NUNEATON
Bedworth
BIRMINGHAM
Solihull
COVENTRY
RUGBY
Kenilworth
ROYAL LEAMINGTON SPA
WARWICK
Southam
Alcester
Stratford on Avon
Kineton
Shipston on Stour
Long Compton
Key to Symbols
Frequencies
rare to occasional
frequent to abundant
Woodland
Hedgerow, Scrub
Acid heath, Bog
Grassland
Waterside, Marsh
Water
Roadsides, Waste places
Quarries, Walls
Cultivated
Scale
Kilometres
0 2 4 6 8 10
Miles
0 1 2 3 4 5 6

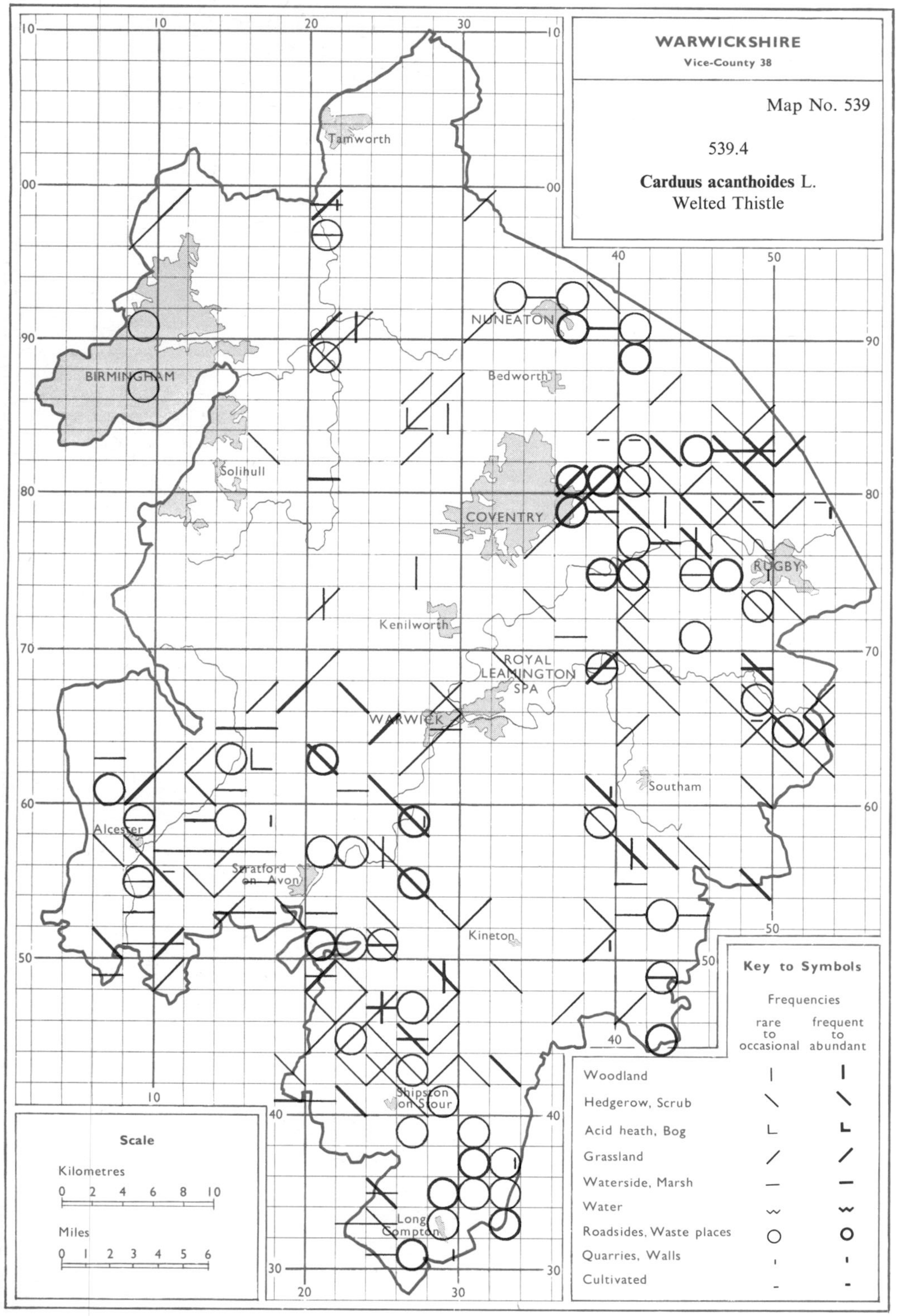
WARWICKSHIRE
Vice-County 38
Map No. 539
539.4
Carduus acanthoides L.
Welted Thistle
Tamworth
NUNEATON
Bedworth
BIRMINGHAM
Solihull
COVENTRY
RUGBY
Kenilworth
ROYAL LEAMINGTON SPA
WARWICK
Southam
Alcester
Stratford on Avon
Kineton
Shipston on Stour
Long Compton
Key to Symbols
Frequencies
rare to occasional
frequent to abundant
Woodland
Hedgerow, Scrub
Acid heath, Bog
Grassland
Waterside, Marsh
Water
Roadsides, Waste places
Quarries, Walls
Cultivated
Scale
Kilometres
0 2 4 6 8 10
Miles
0 1 2 3 4 5 6

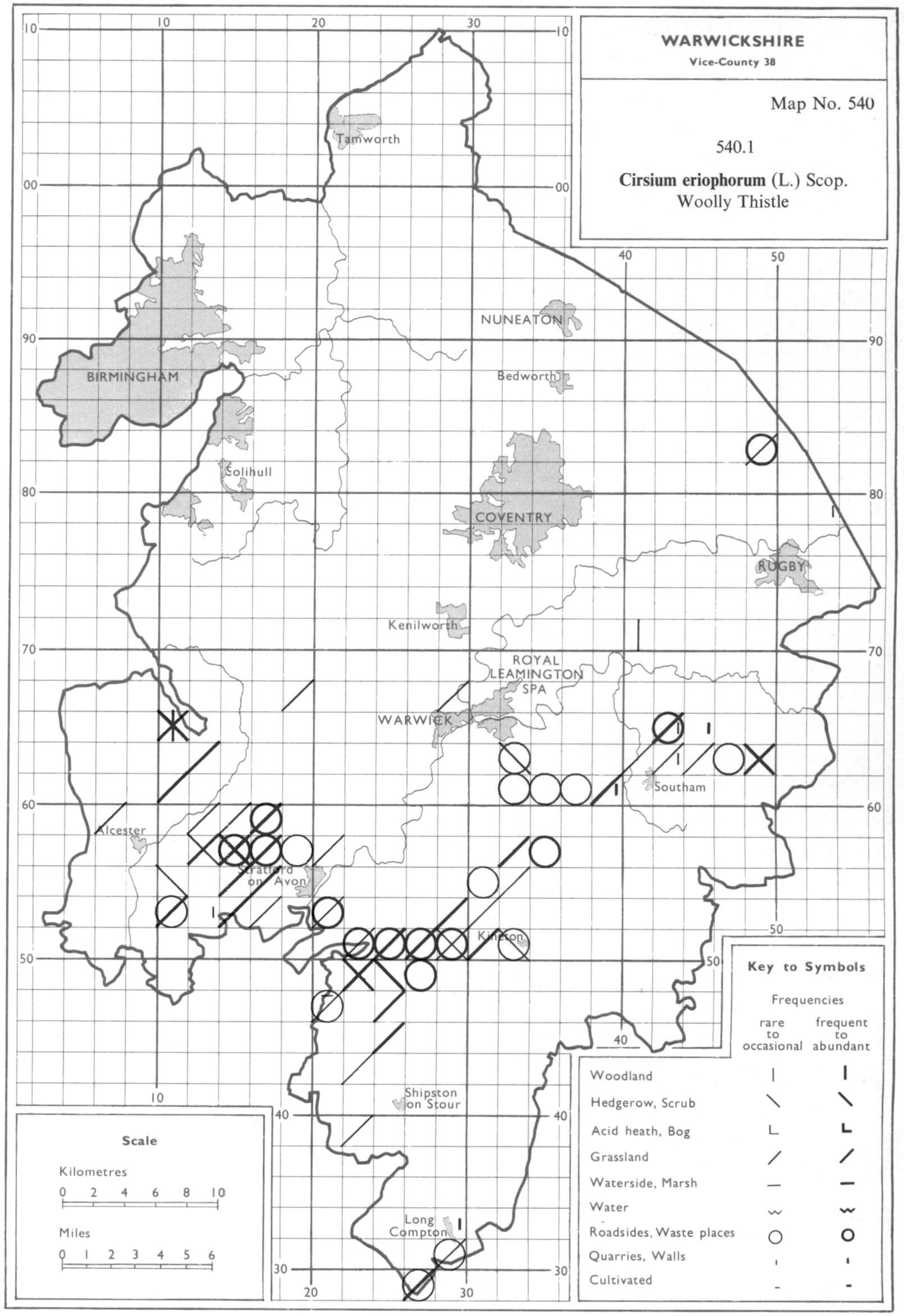
WARWICKSHIRE
Vice-County 38
Map No. 540
540.1
Cirsium eriophorum (L.) Scop.
Woolly Thistle
Tamworth
NUNEATON
Bedworth
BIRMINGHAM
Solihull
COVENTRY
RUGBY
Kenilworth
ROYAL LEAMINGTON SPA
WARWICK
Southam
Alcester
Stratford on Avon
Kineton
Shipston on Stour
Long Compton
Key to Symbols
Frequencies
rare to occasional
frequent to abundant
Woodland
Hedgerow, Scrub
Acid heath, Bog
Grassland
Waterside, Marsh
Water
Roadsides, Waste places
Quarries, Walls
Cultivated
Scale
Kilometres
0 2 4 6 8 10
Miles
0 1 2 3 4 5 6

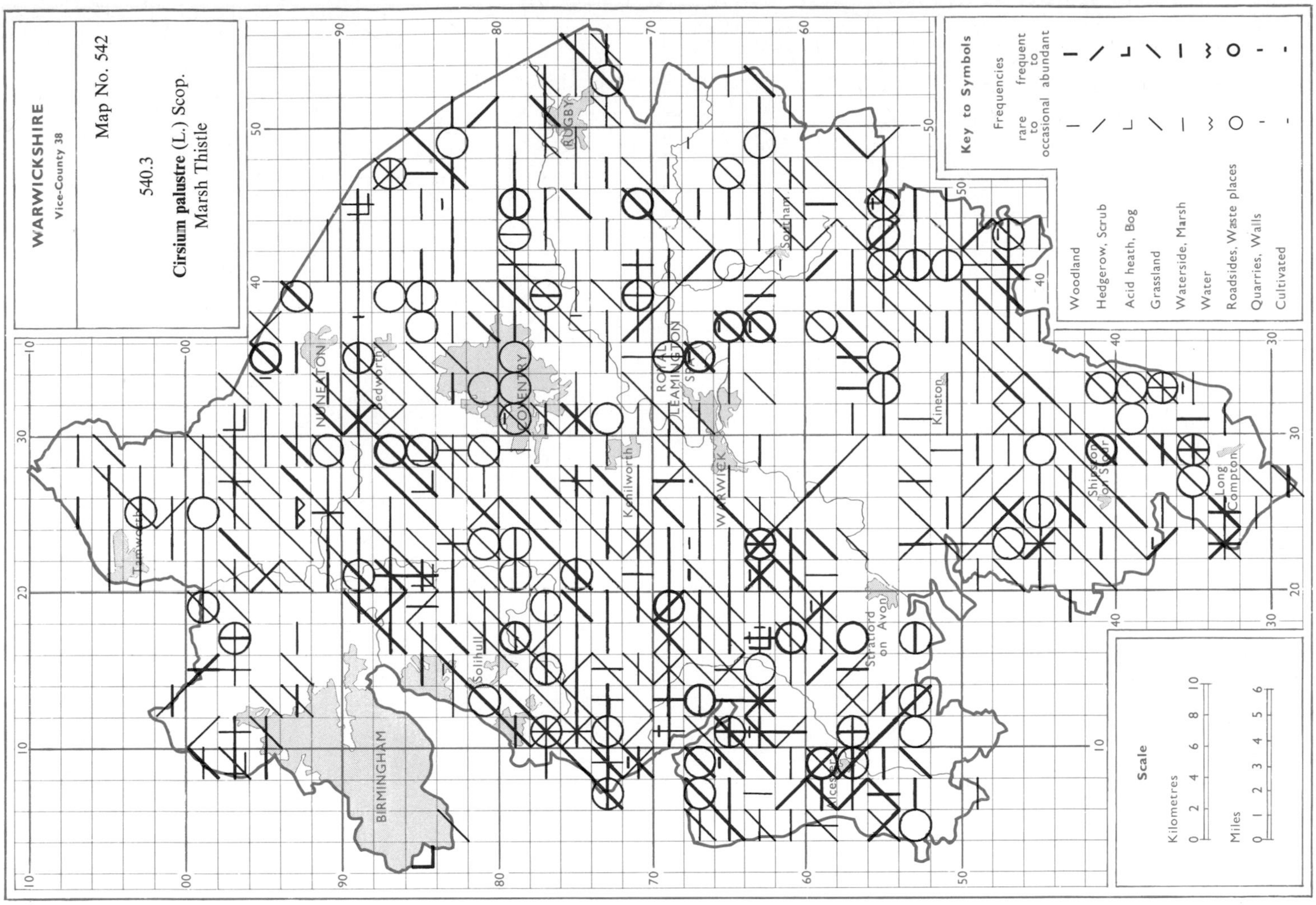
WARWICKSHIRE
Vice-County 38
Map No. 542
540.3
Cirsium palustre (L.) Scop.
Marsh Thistle
Key to Symbols
Frequencies
rare to occasional
frequent to abundant
Woodland
Hedgerow, Scrub
Acid heath, Bog
Grassland
Waterside, Marsh
Water
Roadsides, Waste places
Quarries, Walls
Cultivated
Scale
Kilometres
0 2 4 6 8 10
Miles
0 1 2 3 4 5 6
NUNEATON
Bedworth
COVENTRY
RUGBY
Tamworth
BIRMINGHAM
Solihull
Kenilworth
WARWICK
ROYAL LEAMINGTON SPA
Southam
Kineton
Stratford on Avon
Alcester
Shipston on Stour
Long Compton

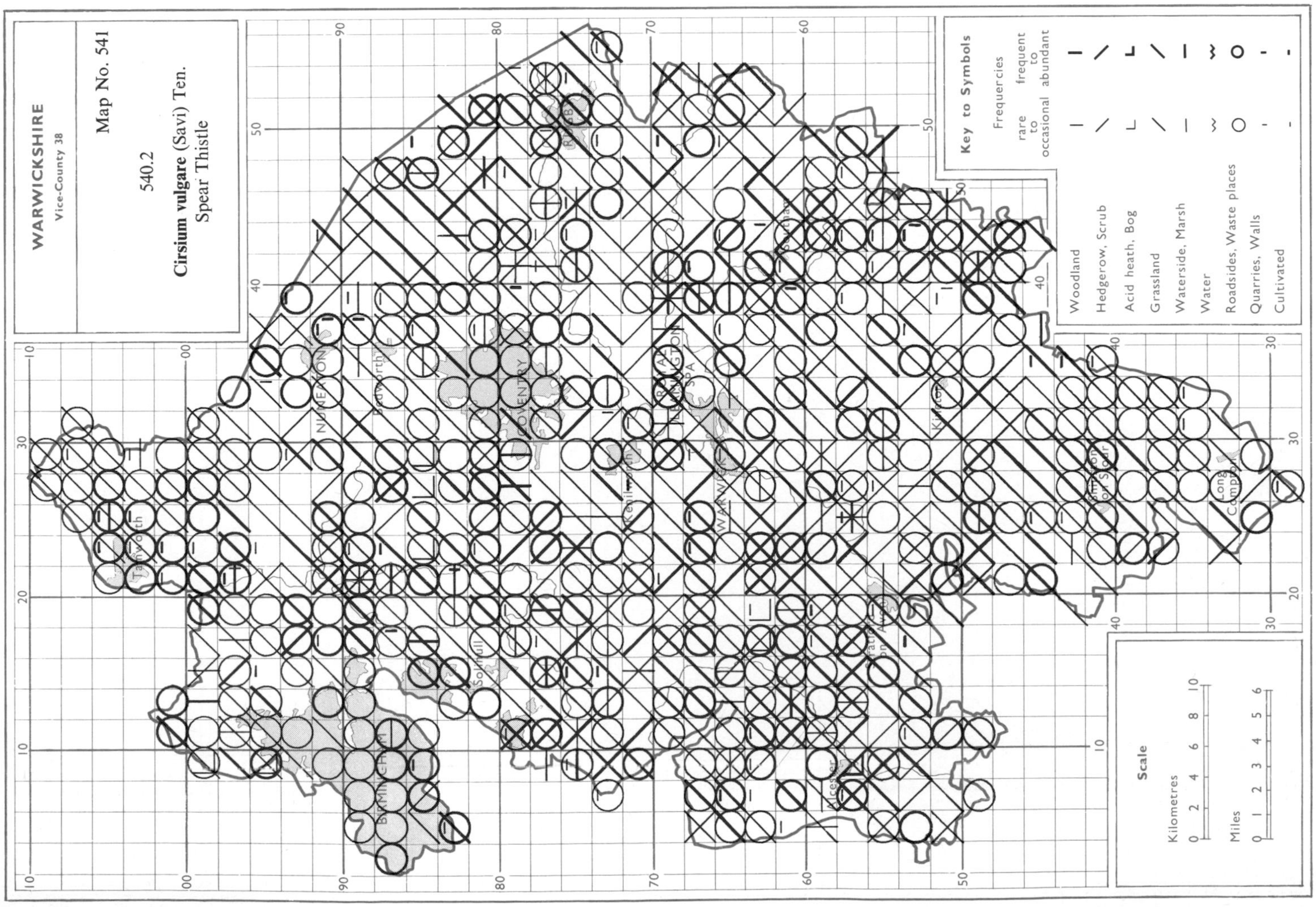
WARWICKSHIRE
Vice-County 38
Map No. 541
540.2
Cirsium vulgare (Savi) Ten.
Spear Thistle
Key to Symbols
Frequencies
rare to occasional
frequent to abundant
Woodland
Hedgerow, Scrub
Acid heath, Bog
Grassland
Waterside, Marsh
Water
Roadsides, Waste places
Quarries, Walls
Cultivated
Scale
Kilometres
0 2 4 6 8 10
Miles
0 1 2 3 4 5 6
NUNEATON
Bedworth
COVENTRY
RUGBY
Tamworth
BIRMINGHAM
Solihull
Kenilworth
WARWICK
ROYAL LEAMINGTON SPA
Kineton
Stratford on Avon
Alcester
Shipston on Stour
Long Compton

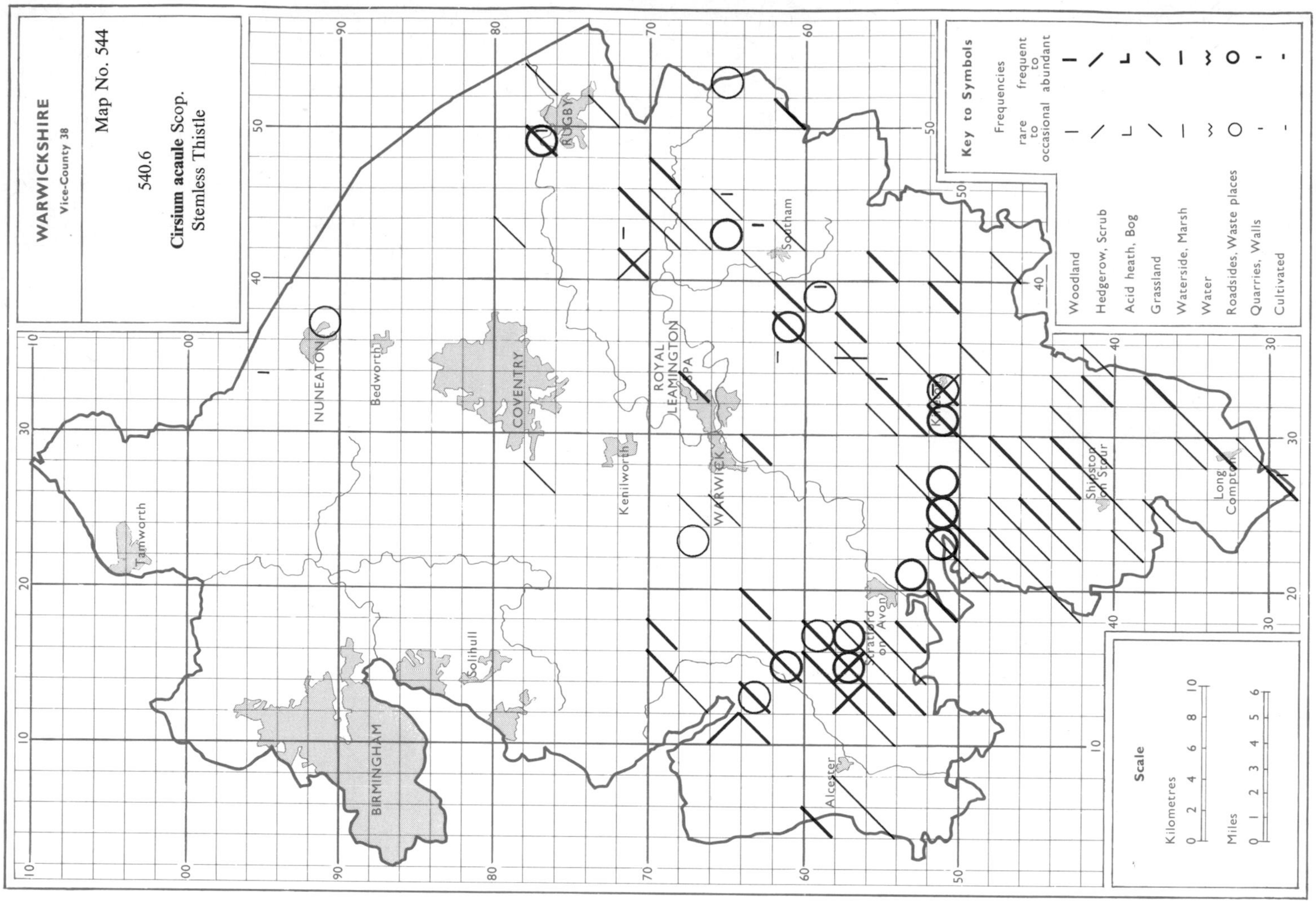
WARWICKSHIRE
Vice-County 38
Map No. 544
540.6
Cirsium acaule Scop.
Stemless Thistle
Key to Symbols
Frequencies
rare to occasional
frequent to abundant
Woodland
Hedgerow, Scrub
Acid heath, Bog
Grassland
Waterside, Marsh
Water
Roadsides, Waste places
Quarries, Walls
Cultivated
Scale
Kilometres
Miles
RUGBY
NUNEATON
Bedworth
COVENTRY
ROYAL LEAMINGTON SPA
WARWICK
Kenilworth
Southam
Kineton
Shipston on Stour
Long Compton
Stratford on Avon
Alcester
Solihull
BIRMINGHAM
Tamworth

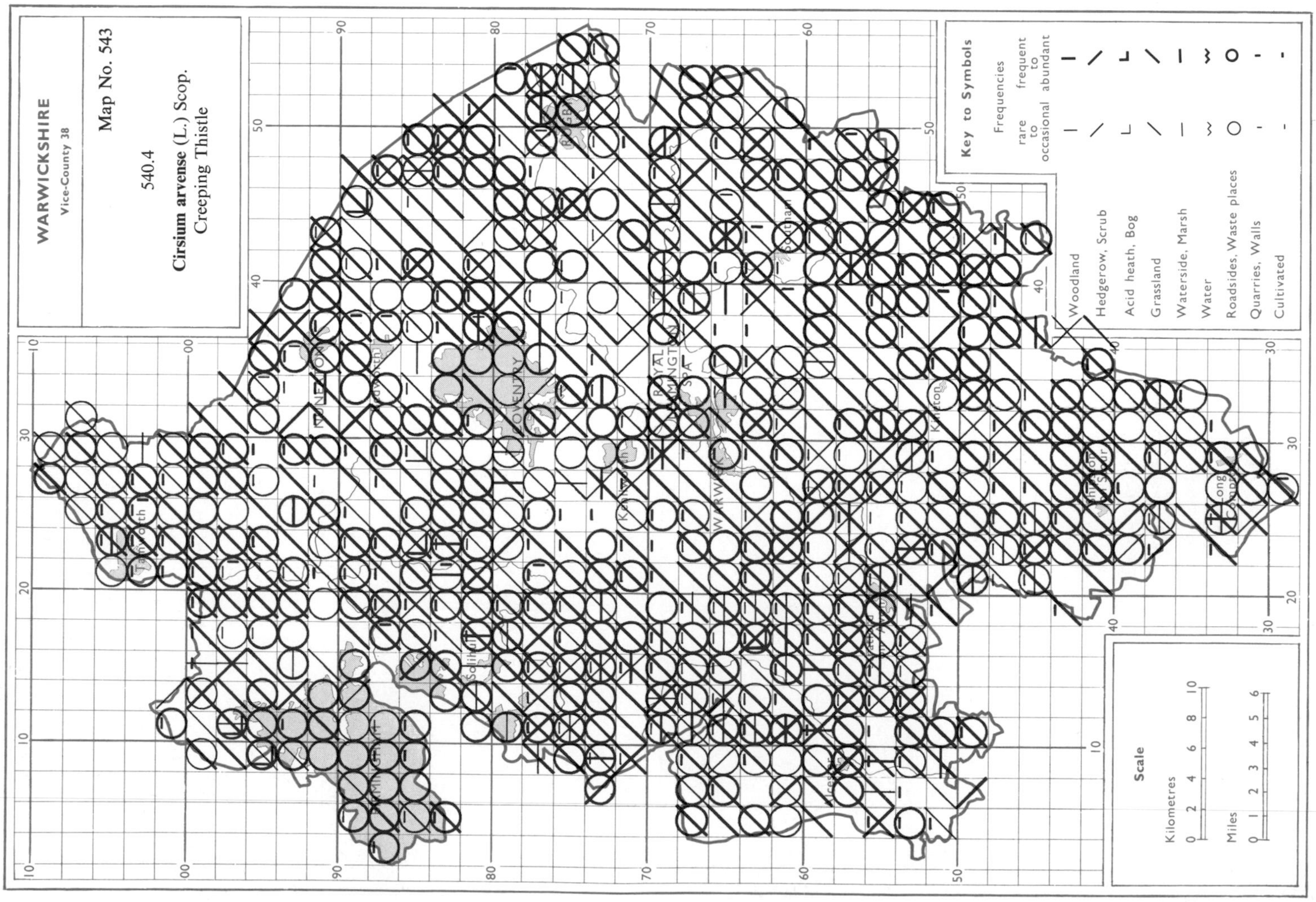
WARWICKSHIRE
Vice-County 38
Map No. 543
540.4
Cirsium arvense (L.) Scop.
Creeping Thistle
Key to Symbols
Frequencies
rare to occasional
frequent to abundant
Woodland
Hedgerow, Scrub
Acid heath, Bog
Grassland
Waterside, Marsh
Water
Roadsides, Waste places
Quarries, Walls
Cultivated
Scale
Kilometres
Miles

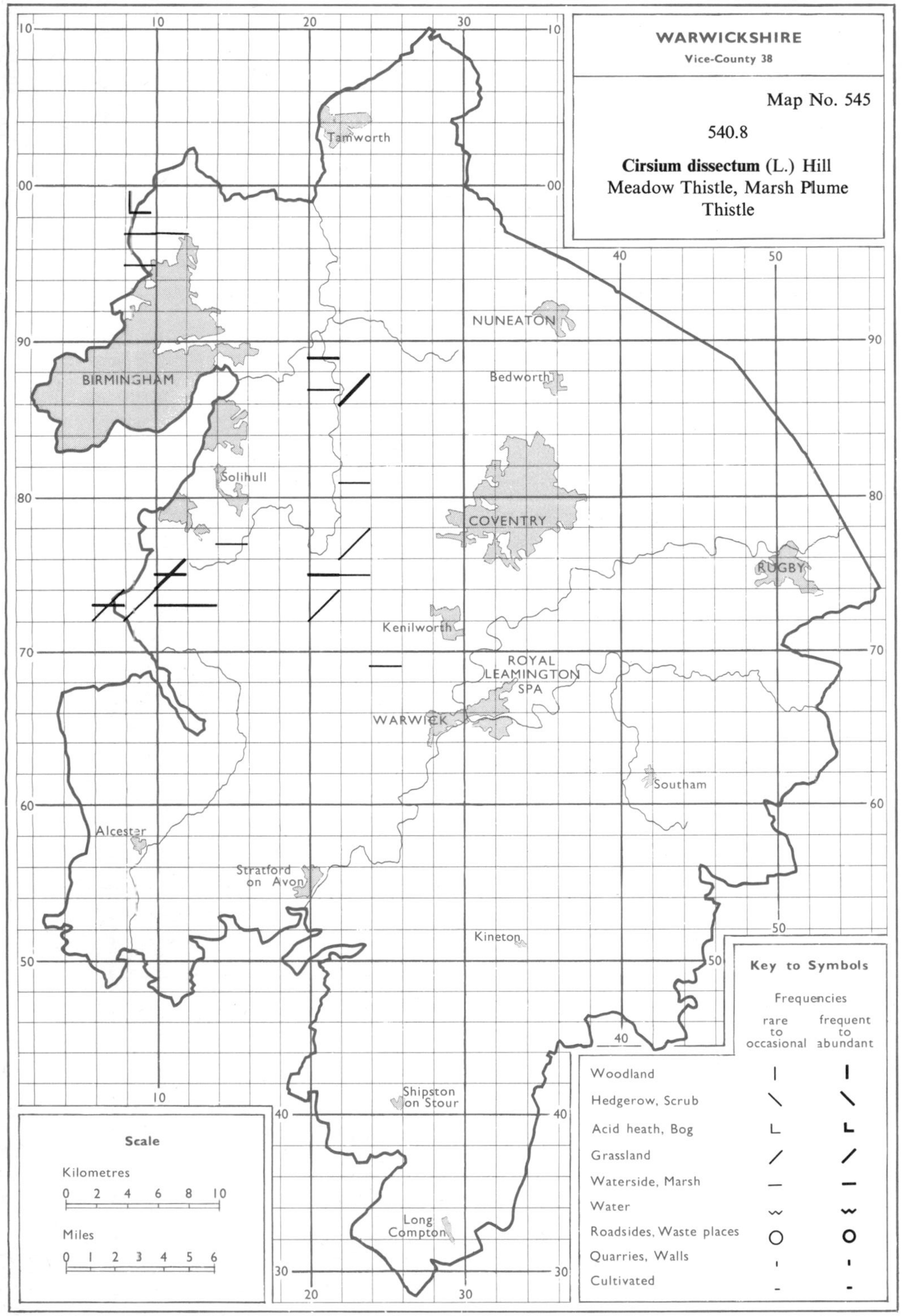
WARWICKSHIRE
Vice-County 38
Map No. 545
540.8
Cirsium dissectum (L.) Hill
Meadow Thistle, Marsh Plume Thistle
Tamworth
NUNEATON
Bedworth
BIRMINGHAM
Solihull
COVENTRY
RUGBY
Kenilworth
ROYAL LEAMINGTON SPA
WARWICK
Southam
Alcester
Stratford on Avon
Kineton
Shipston on Stour
Long Compton
Key to Symbols
Frequencies
rare to occasional
frequent to abundant
Woodland
Hedgerow, Scrub
Acid heath, Bog
Grassland
Waterside, Marsh
Water
Roadsides, Waste places
Quarries, Walls
Cultivated
Scale
Kilometres
0 2 4 6 8 10
Miles
0 1 2 3 4 5 6

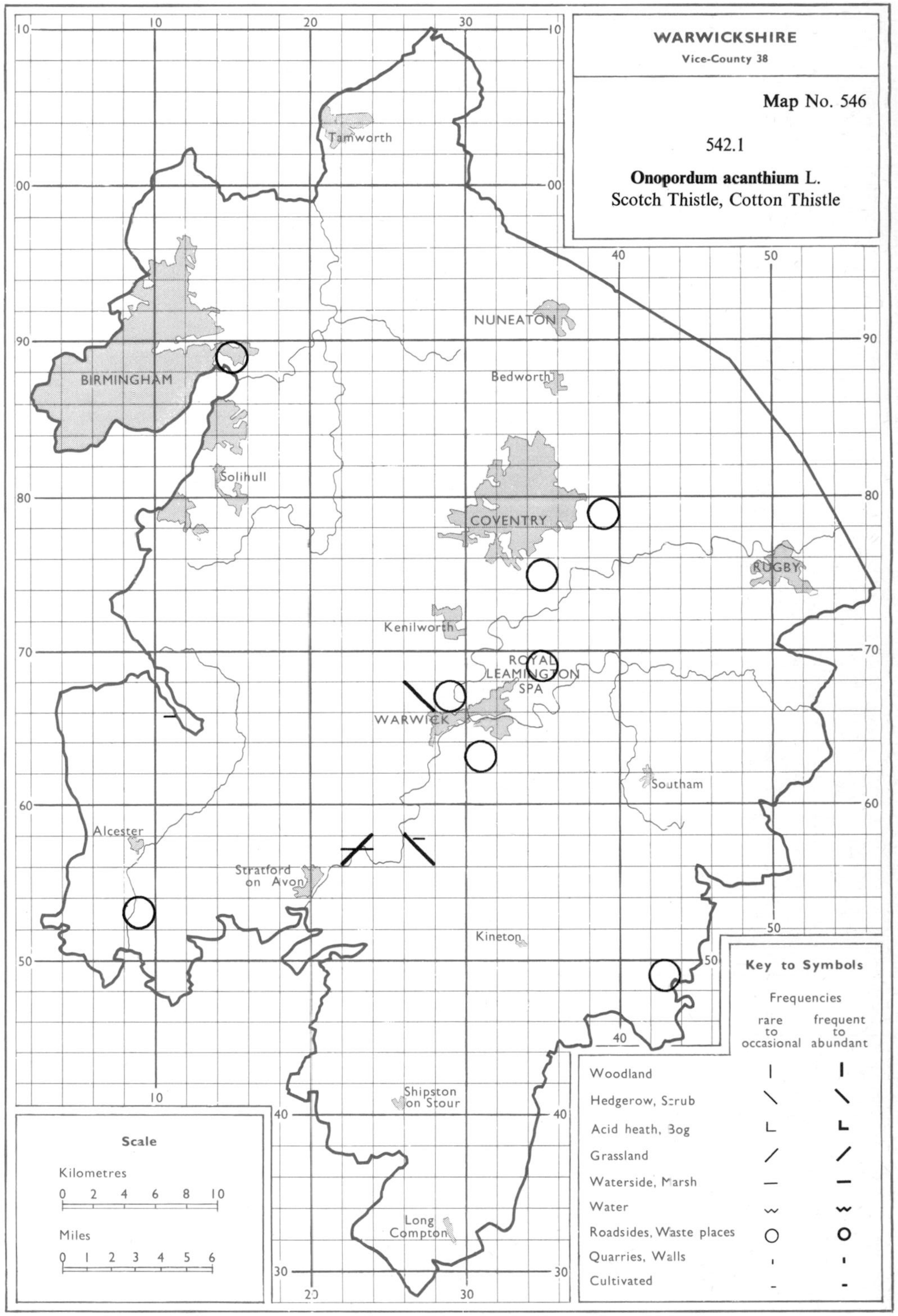
WARWICKSHIRE
Vice-County 38
Map No. 546
542.1
Onopordum acanthium L.
Scotch Thistle, Cotton Thistle
Tamworth
NUNEATON
Bedworth
BIRMINGHAM
Solihull
COVENTRY
RUGBY
Kenilworth
ROYAL LEAMINGTON SPA
WARWICK
Southam
Alcester
Stratford on Avon
Kineton
Shipston on Stour
Long Compton
Key to Symbols
Frequencies
rare to occasional
frequent to abundant
Woodland
Hedgerow, Scrub
Acid heath, Bog
Grassland
Waterside, Marsh
Water
Roadsides, Waste places
Quarries, Walls
Cultivated
Scale
Kilometres
0 2 4 6 8 10
Miles
0 1 2 3 4 5 6

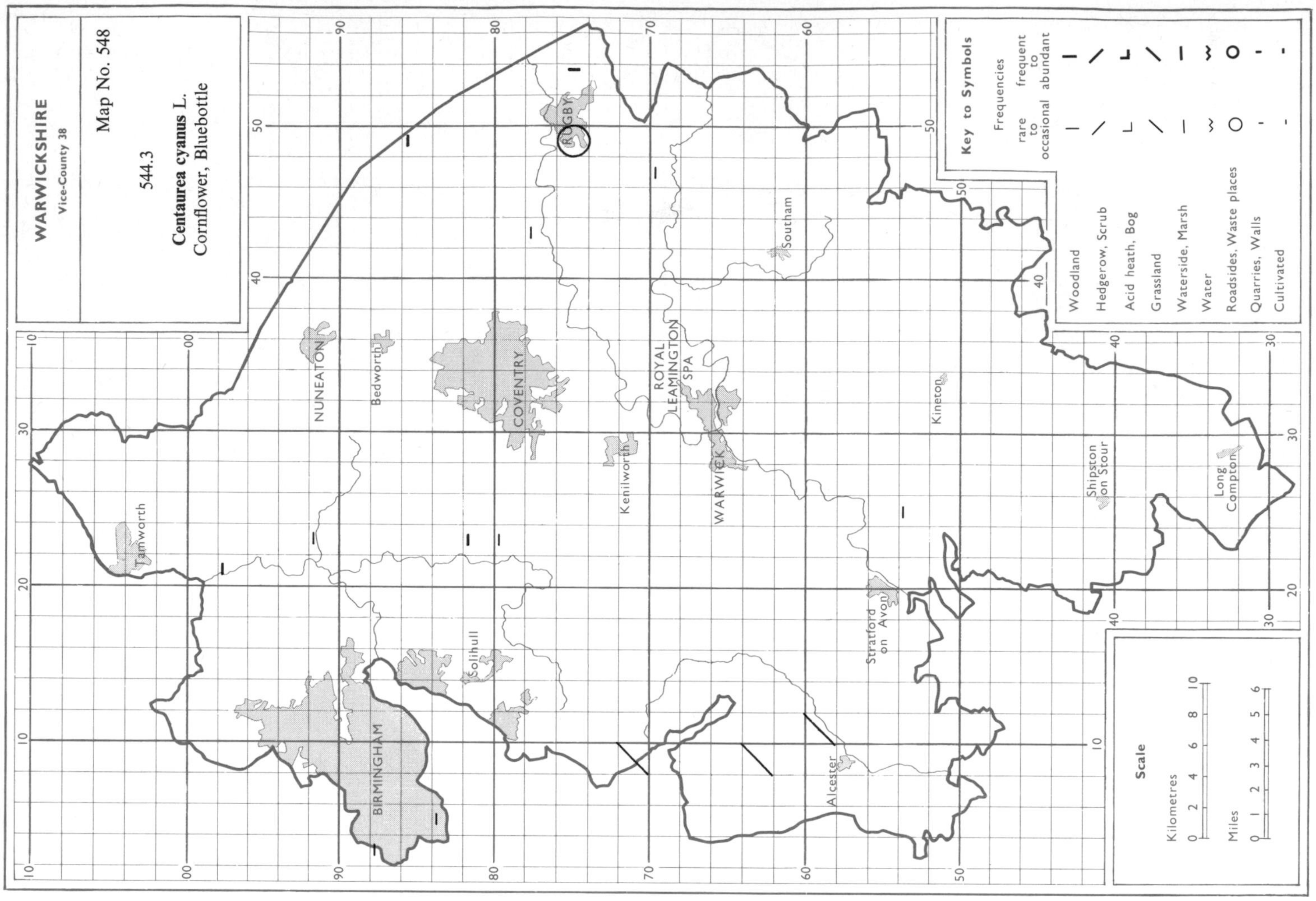
WARWICKSHIRE
Vice-County 38
Map No. 548
544.3
Centaurea cyanus L.
Cornflower, Bluebottle
Key to Symbols
Frequencies
rare to occasional
frequent to abundant
Woodland
Hedgerow, Scrub
Acid heath, Bog
Grassland
Waterside, Marsh
Water
Roadsides, Waste places
Quarries, Walls
Cultivated
Scale
Kilometres
Miles
NUNEATON
Bedworth
COVENTRY
RUGBY
Southam
ROYAL LEAMINGTON SPA
Kenilworth
WARWICK
Kineton
Tamworth
Shipston on Stour
Long Compton
Solihull
BIRMINGHAM
Stratford on Avon
Alcester

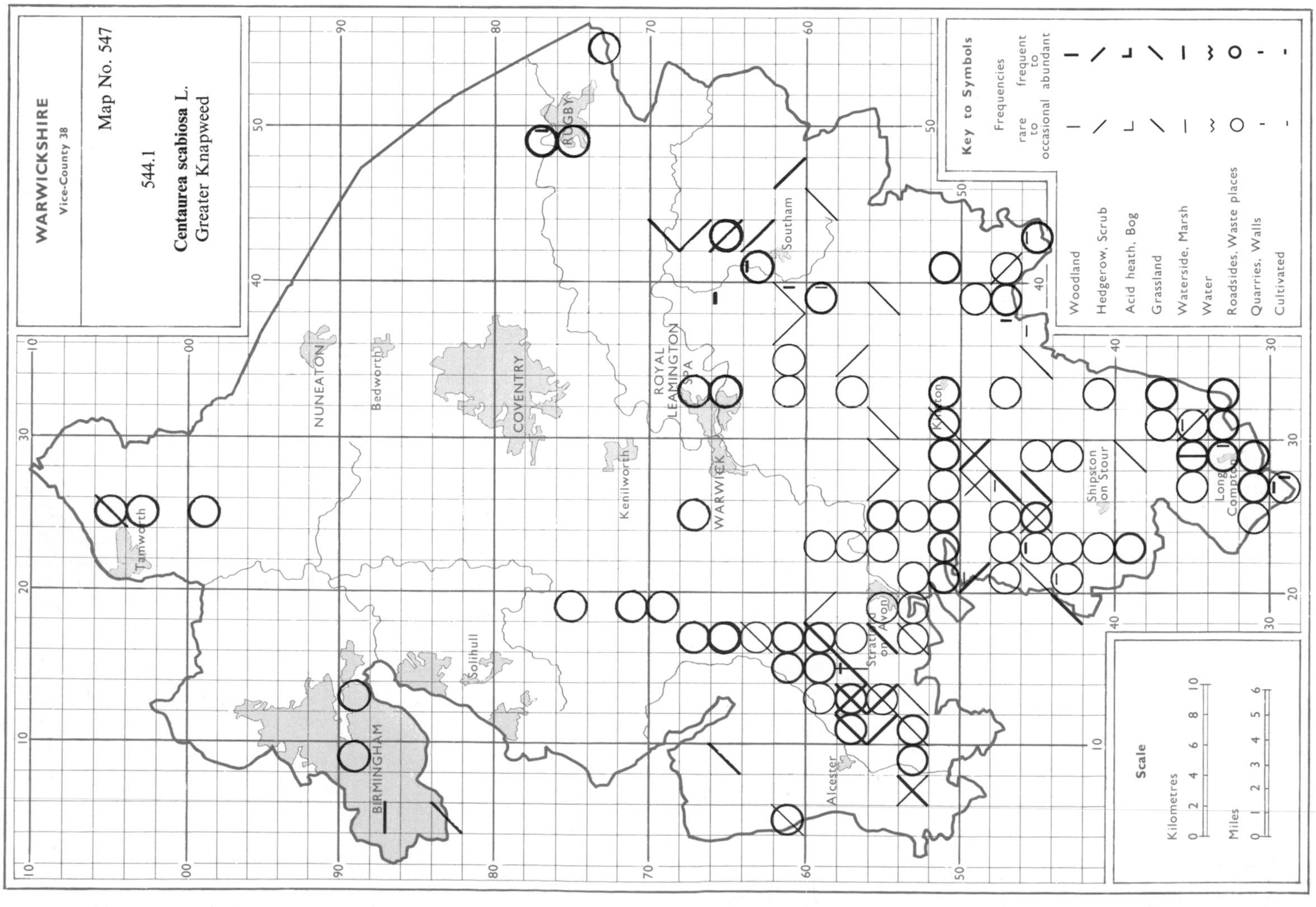
WARWICKSHIRE
Vice-County 38
Map No. 547
544.1
Centaurea scabiosa L.
Greater Knapweed
Key to Symbols
Frequencies
rare to occasional
frequent to abundant
Woodland
Hedgerow, Scrub
Acid heath, Bog
Grassland
Waterside, Marsh
Water
Roadsides, Waste places
Quarries, Walls
Cultivated
Scale
Kilometres
Miles
NUNEATON
Bedworth
COVENTRY
RUGBY
Southam
ROYAL LEAMINGTON SPA
Kenilworth
WARWICK
Kineton
Tamworth
Shipston on Stour
Long Compton
Solihull
BIRMINGHAM
Stratford on Avon
Alcester

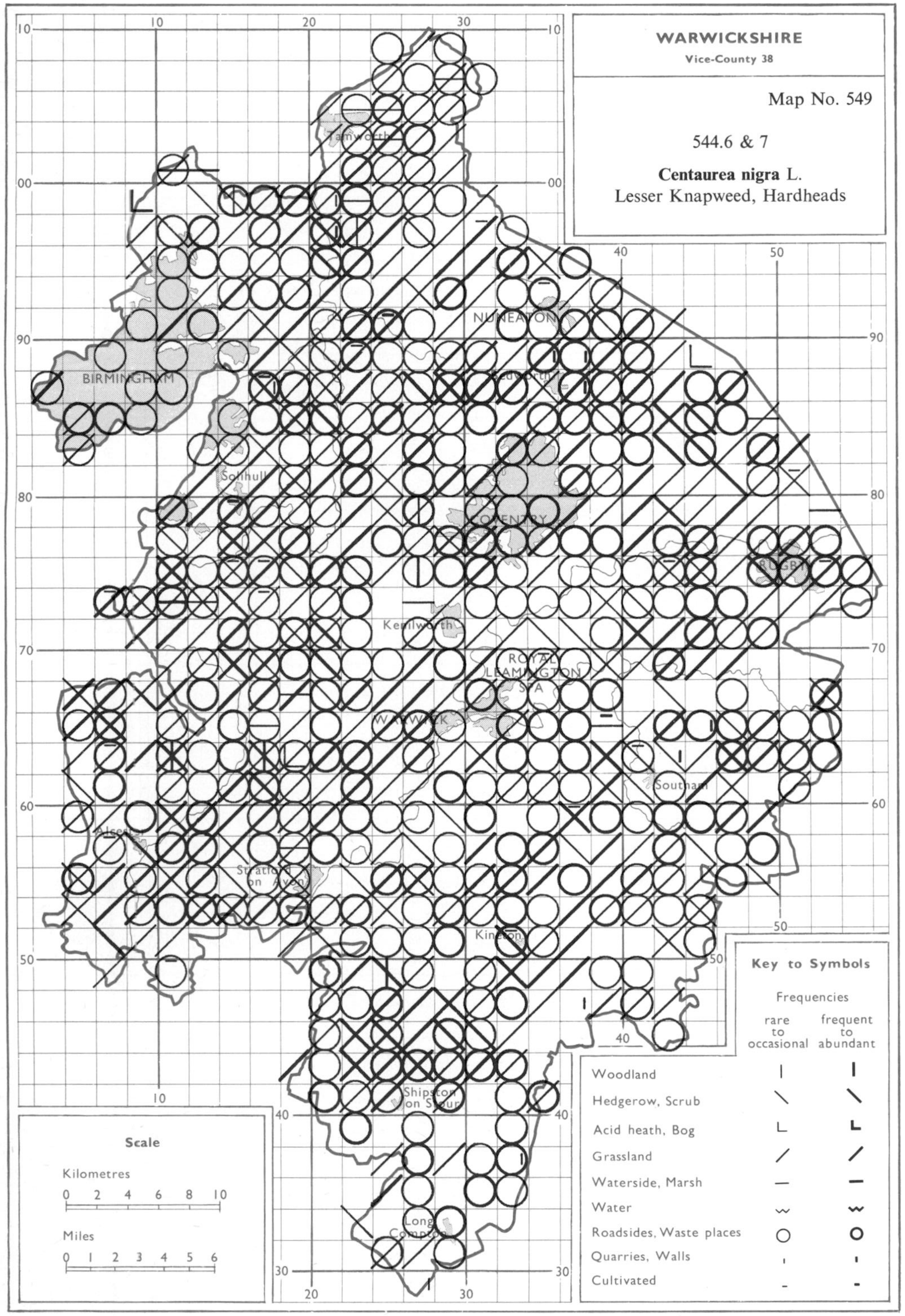
WARWICKSHIRE
Vice-County 38
Map No. 549
544.6 & 7
Centaurea nigra L.
Lesser Knapweed, Hardheads
Tamworth
BIRMINGHAM
NUNEATON
Bedworth
Solihull
COVENTRY
RUGBY
Kenilworth
ROYAL LEAMINGTON SPA
WARWICK
Southam
Alcester
Stratford on Avon
Kineton
Shipston on Stour
Long Compton
Key to Symbols
Frequencies
rare to occasional
frequent to abundant
Woodland
Hedgerow, Scrub
Acid heath, Bog
Grassland
Waterside, Marsh
Water
Roadsides, Waste places
Quarries, Walls
Cultivated
Scale
Kilometres
0 2 4 6 8 10
Miles
0 1 2 3 4 5 6

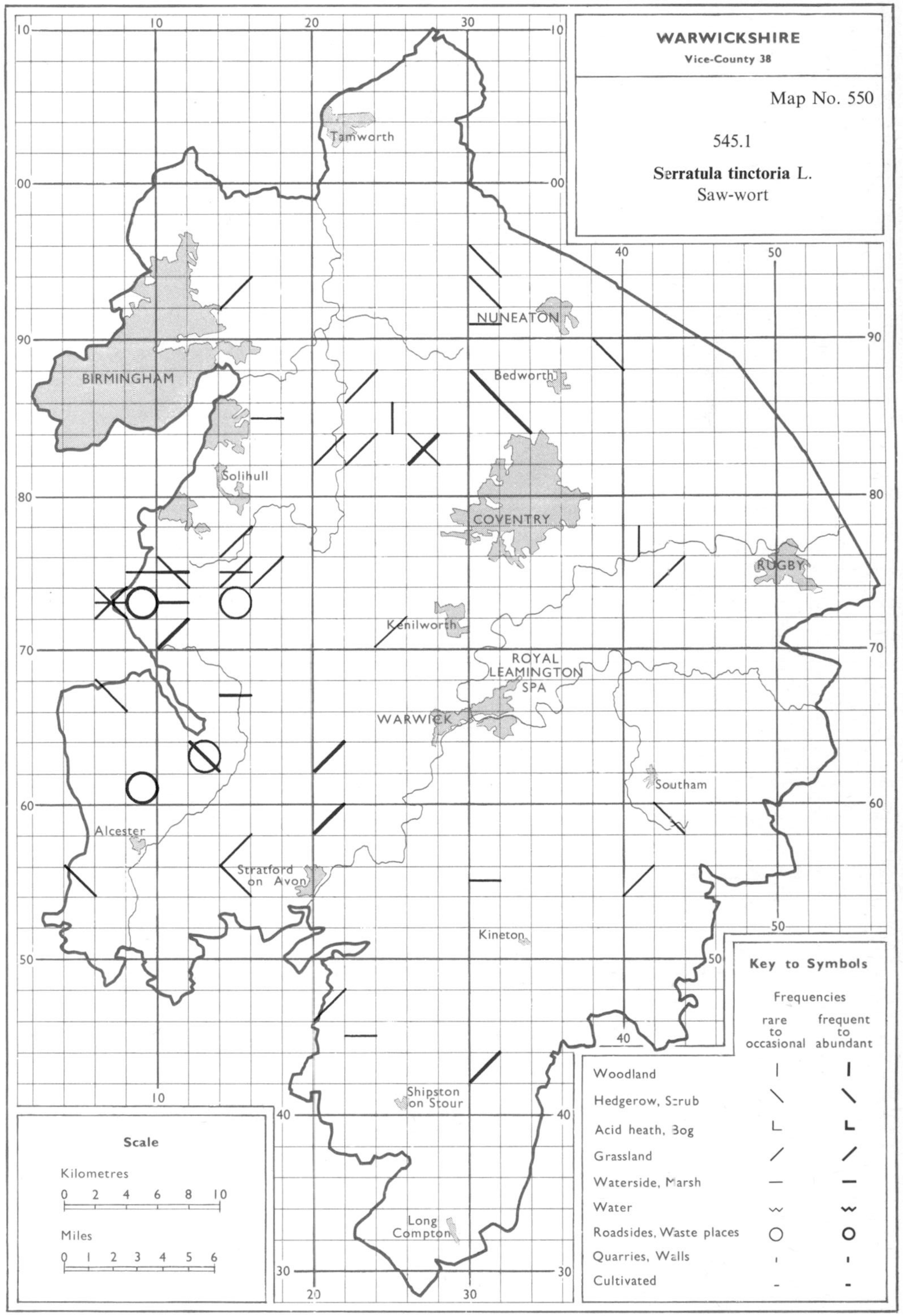
WARWICKSHIRE
Vice-County 38
Map No. 550
545.1
Serratula tinctoria L.
Saw-wort
Tamworth
BIRMINGHAM
NUNEATON
Bedworth
Solihull
COVENTRY
RUGBY
Kenilworth
ROYAL LEAMINGTON SPA
WARWICK
Southam
Alcester
Stratford on Avon
Kineton
Shipston on Stour
Long Compton
Key to Symbols
Frequencies
rare to occasional
frequent to abundant
Woodland
Hedgerow, Scrub
Acid heath, Bog
Grassland
Waterside, Marsh
Water
Roadsides, Waste places
Quarries, Walls
Cultivated
Scale
Kilometres
0 2 4 6 8 10
Miles
0 1 2 3 4 5 6

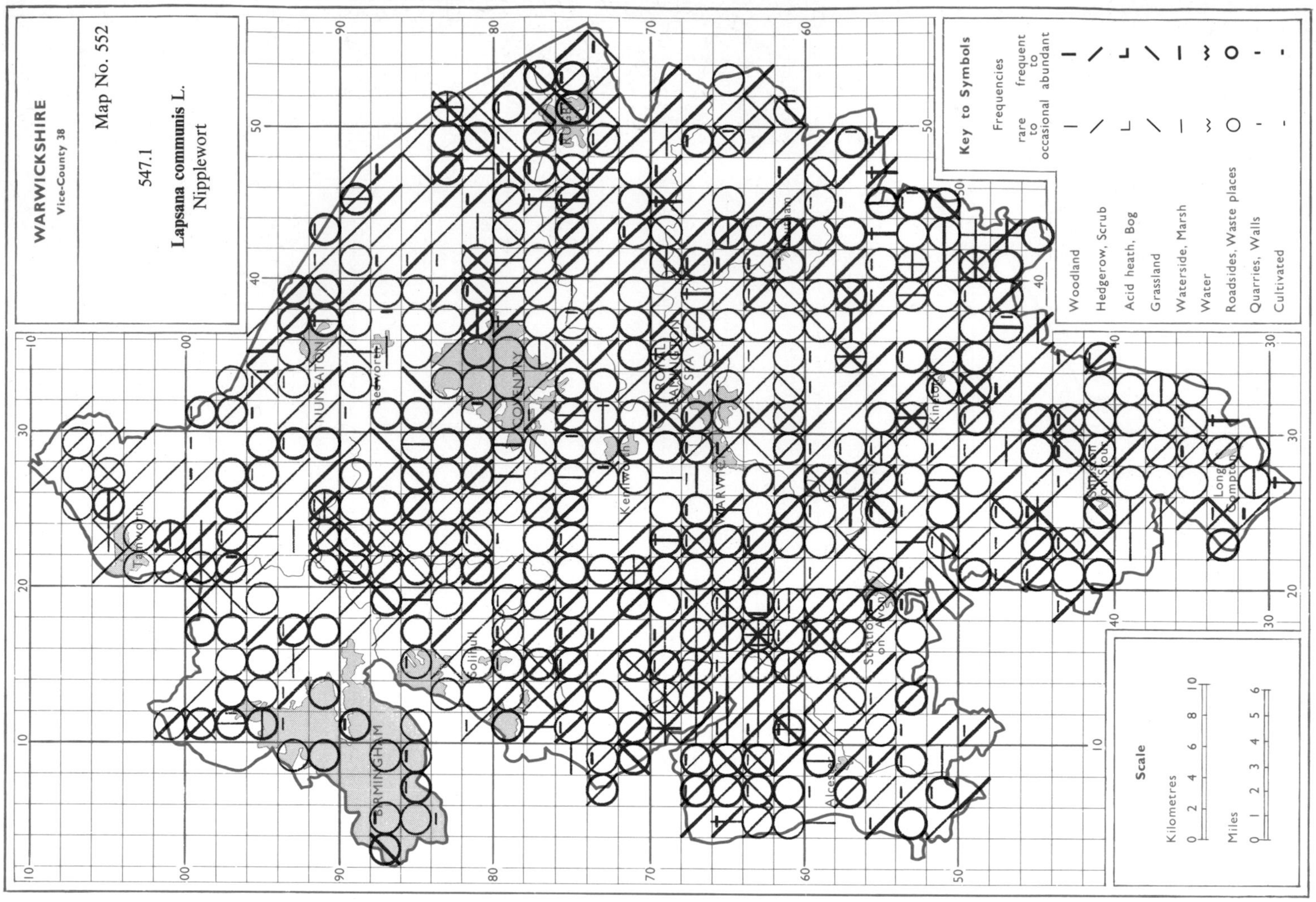
WARWICKSHIRE
Vice-County 38
Map No. 552
547.1
Lapsana communis L.
Nipplewort
Key to Symbols
Frequencies
rare to occasional
frequent to abundant
Woodland
Hedgerow, Scrub
Acid heath, Bog
Grassland
Waterside, Marsh
Water
Roadsides, Waste places
Quarries, Walls
Cultivated
Scale
Kilometres
Miles
NUNEATON
Bedworth
COVENTRY
RUGBY
ROYAL LEAMINGTON SPA
Southam
Kineton
Shipston on Stour
Long Compton
Kenilworth
WARWICK
Tamworth
Solihull
BIRMINGHAM
Stratford on Avon
Alcester

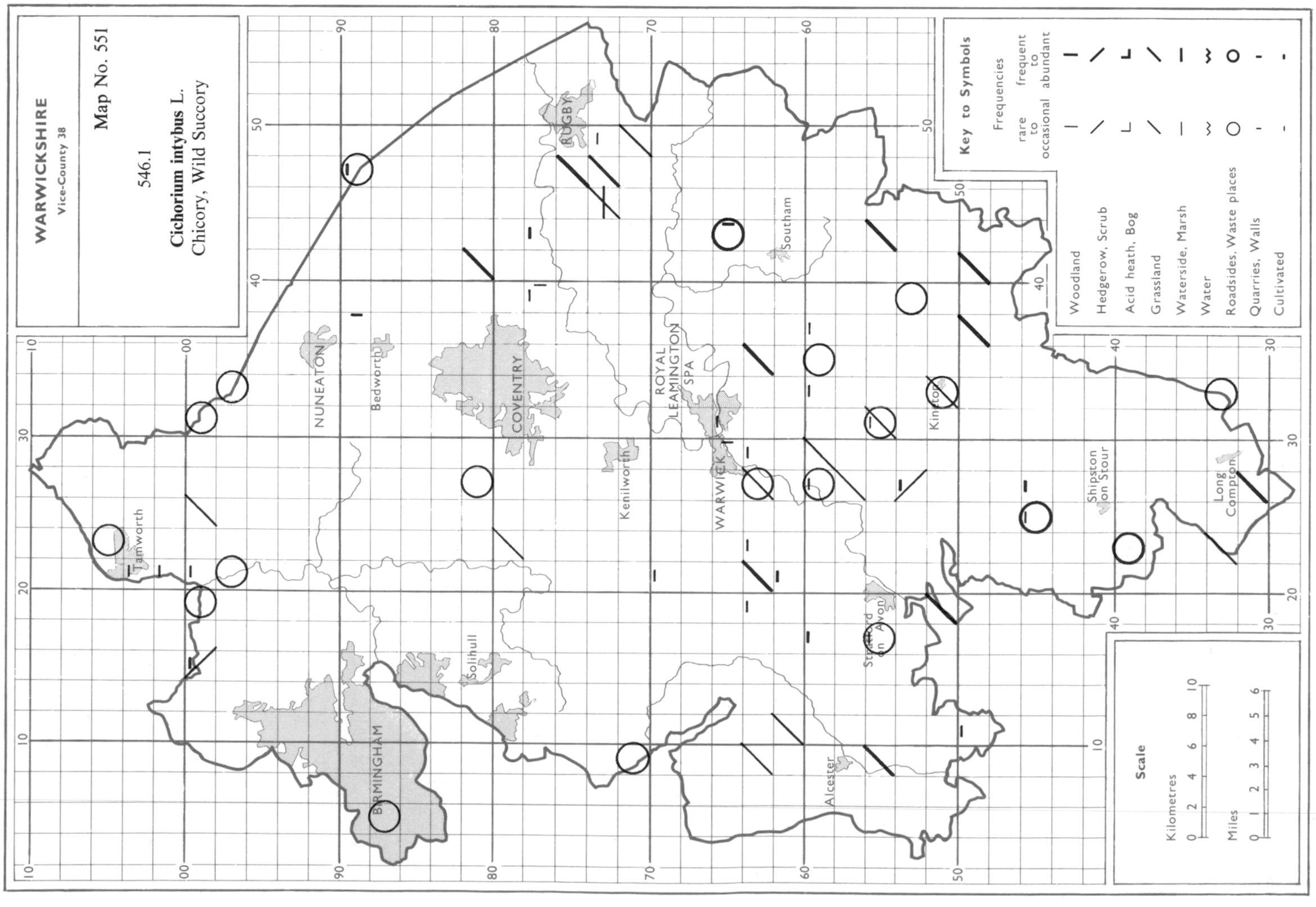
WARWICKSHIRE
Vice-County 38
Map No. 551
546.1
Cichorium intybus L.
Chicory, Wild Succory
Key to Symbols
Frequencies
rare to occasional
frequent to abundant
Woodland
Hedgerow, Scrub
Acid heath, Bog
Grassland
Waterside, Marsh
Water
Roadsides, Waste places
Quarries, Walls
Cultivated
Scale
Kilometres
Miles
NUNEATON
Bedworth
COVENTRY
RUGBY
ROYAL LEAMINGTON SPA
Southam
Kineton
Shipston on Stour
Long Compton
Kenilworth
WARWICK
Tamworth
Solihull
BIRMINGHAM
Stratford on Avon
Alcester

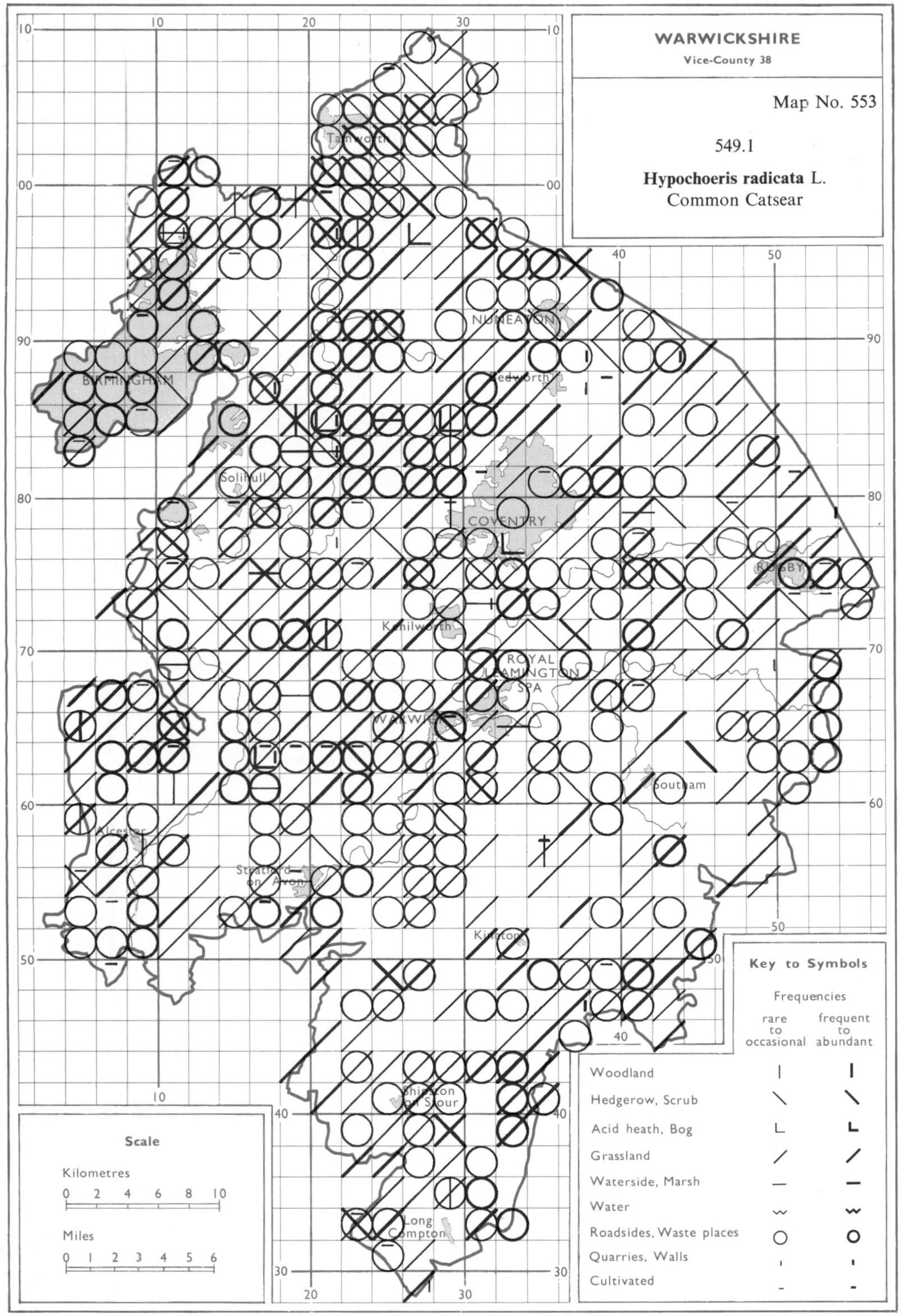
WARWICKSHIRE
Vice-County 38
Map No. 553
549.1
Hypochoeris radicata L.
Common Catsear
Tamworth
Birmingham
Solihull
Nuneaton
Bedworth
Coventry
Rugby
Kenilworth
Royal Leamington Spa
Warwick
Southam
Alcester
Stratford on Avon
Kineton
Shipston on Stour
Long Compton
Scale
Kilometres
Miles
Key to Symbols
Frequencies
rare to occasional
frequent to abundant
Woodland
Hedgerow, Scrub
Acid heath, Bog
Grassland
Waterside, Marsh
Water
Roadsides, Waste places
Quarries, Walls
Cultivated

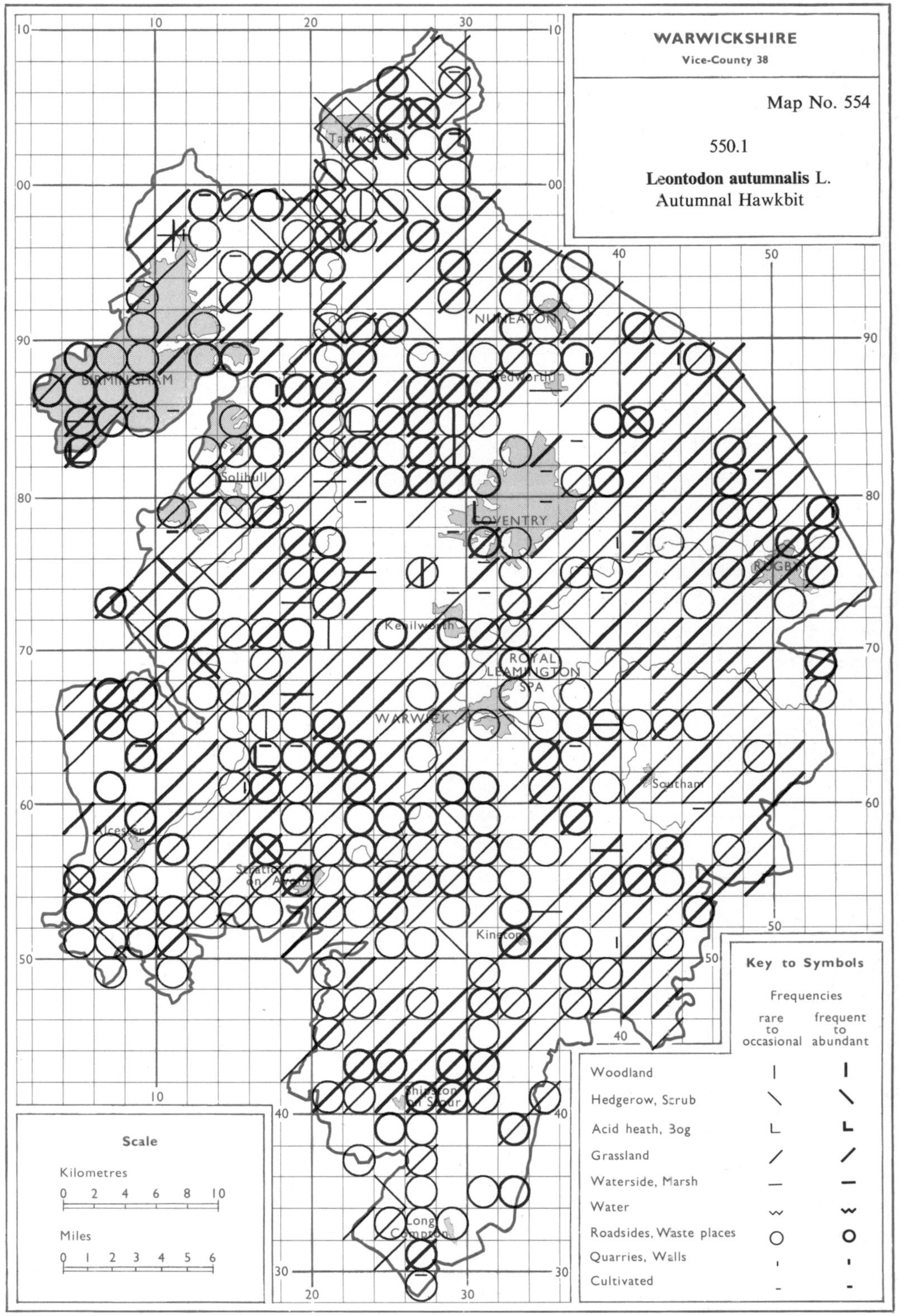
WARWICKSHIRE
Vice-County 38
Map No. 554
550.1
Leontodon autumnalis L.
Autumnal Hawkbit
Tamworth
Birmingham
Solihull
Nuneaton
Bedworth
Coventry
Rugby
Kenilworth
Royal Leamington Spa
Warwick
Southam
Alcester
Stratford on Avon
Kineton
Shipston on Stour
Long Compton
Scale
Kilometres
Miles
Key to Symbols
Frequencies
rare to occasional
frequent to abundant
Woodland
Hedgerow, Scrub
Acid heath, Bog
Grassland
Waterside, Marsh
Water
Roadsides, Waste places
Quarries, Walls
Cultivated

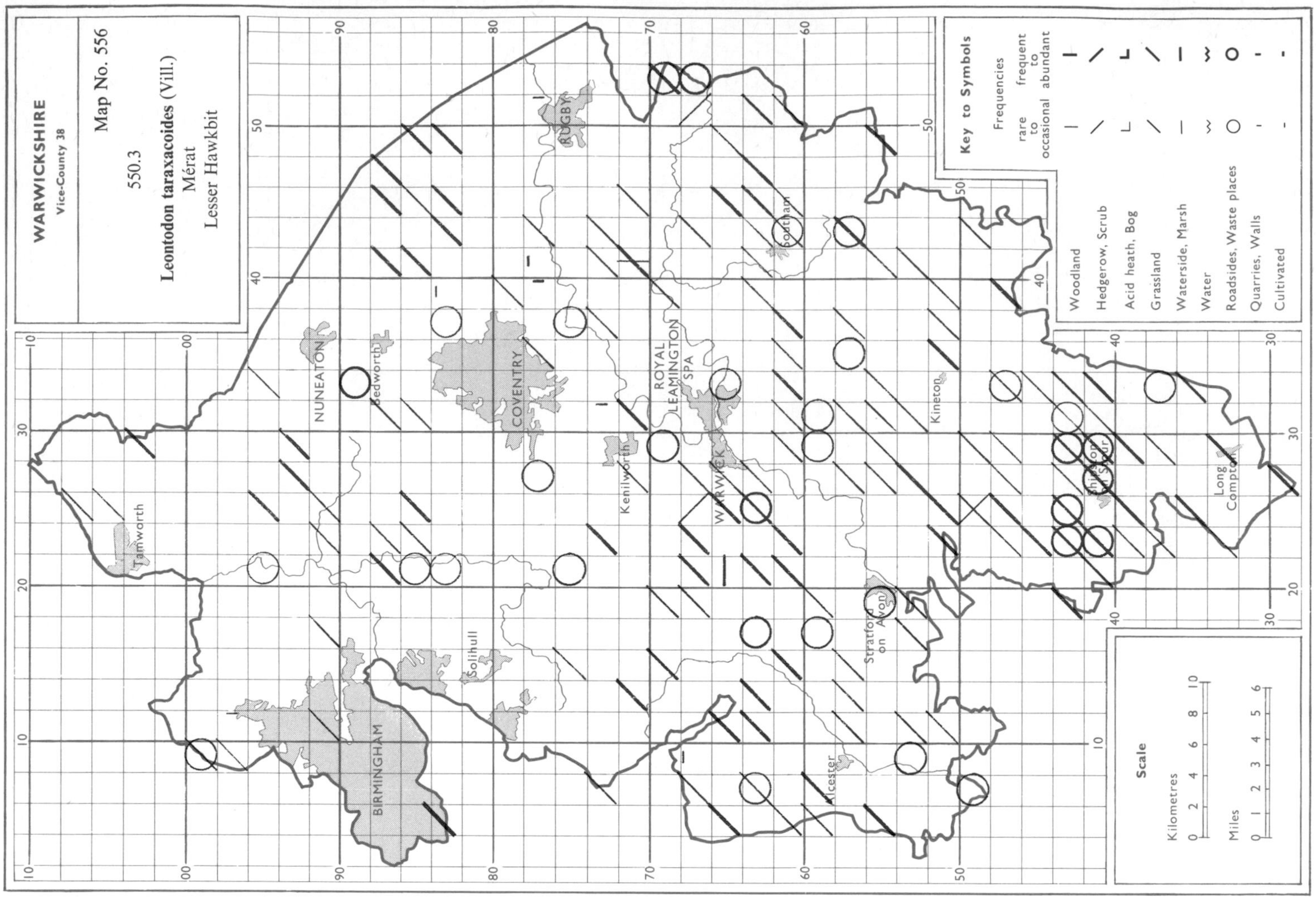
WARWICKSHIRE
Vice-County 38
Map No. 556
550.3
Leontodon taraxacoides (Vill.) Mérat
Lesser Hawkbit
Key to Symbols
Frequencies
rare to occasional
frequent to abundant
Woodland
Hedgerow, Scrub
Acid heath, Bog
Grassland
Waterside, Marsh
Water
Roadsides, Waste places
Quarries, Walls
Cultivated
Scale
Kilometres
0 2 4 6 8 10
Miles
0 1 2 3 4 5 6
NUNEATON
Bedworth
COVENTRY
RUGBY
Tamworth
Solihull
BIRMINGHAM
Kenilworth
ROYAL LEAMINGTON SPA
WARWICK
Southam
Kineton
Stratford on Avon
Alcester
Shipston on Stour
Long Compton

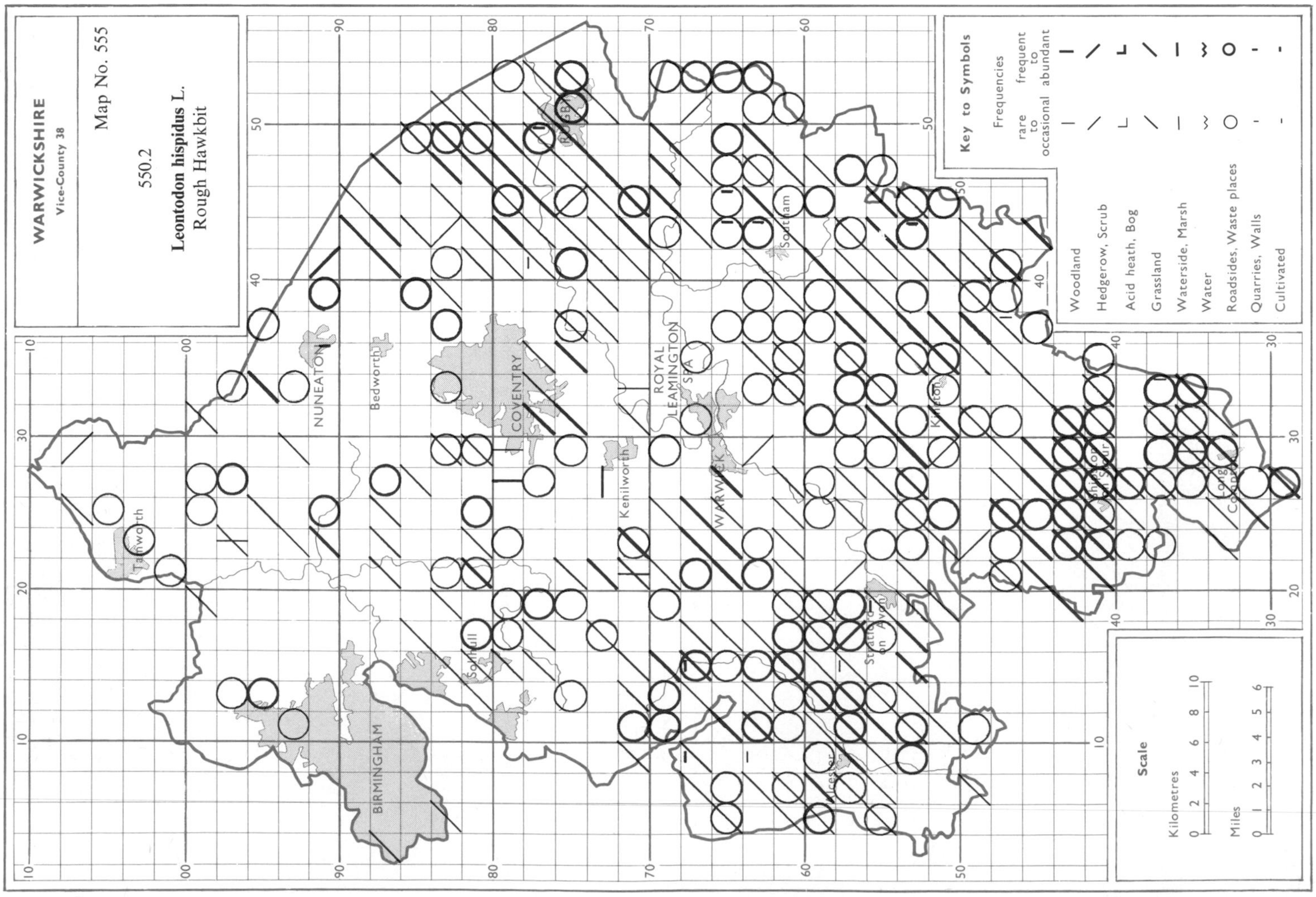
WARWICKSHIRE
Vice-County 38
Map No. 555
550.2
Leontodon hispidus L.
Rough Hawkbit
Key to Symbols
Frequencies
rare to occasional
frequent to abundant
Woodland
Hedgerow, Scrub
Acid heath, Bog
Grassland
Waterside, Marsh
Water
Roadsides, Waste places
Quarries, Walls
Cultivated
Scale
Kilometres
0 2 4 6 8 10
Miles
0 1 2 3 4 5 6
NUNEATON
Bedworth
COVENTRY
RUGBY
Tamworth
Solihull
BIRMINGHAM
Kenilworth
ROYAL LEAMINGTON SPA
WARWICK
Southam
Kineton
Stratford on Avon
Alcester
Shipston on Stour
Long Compton

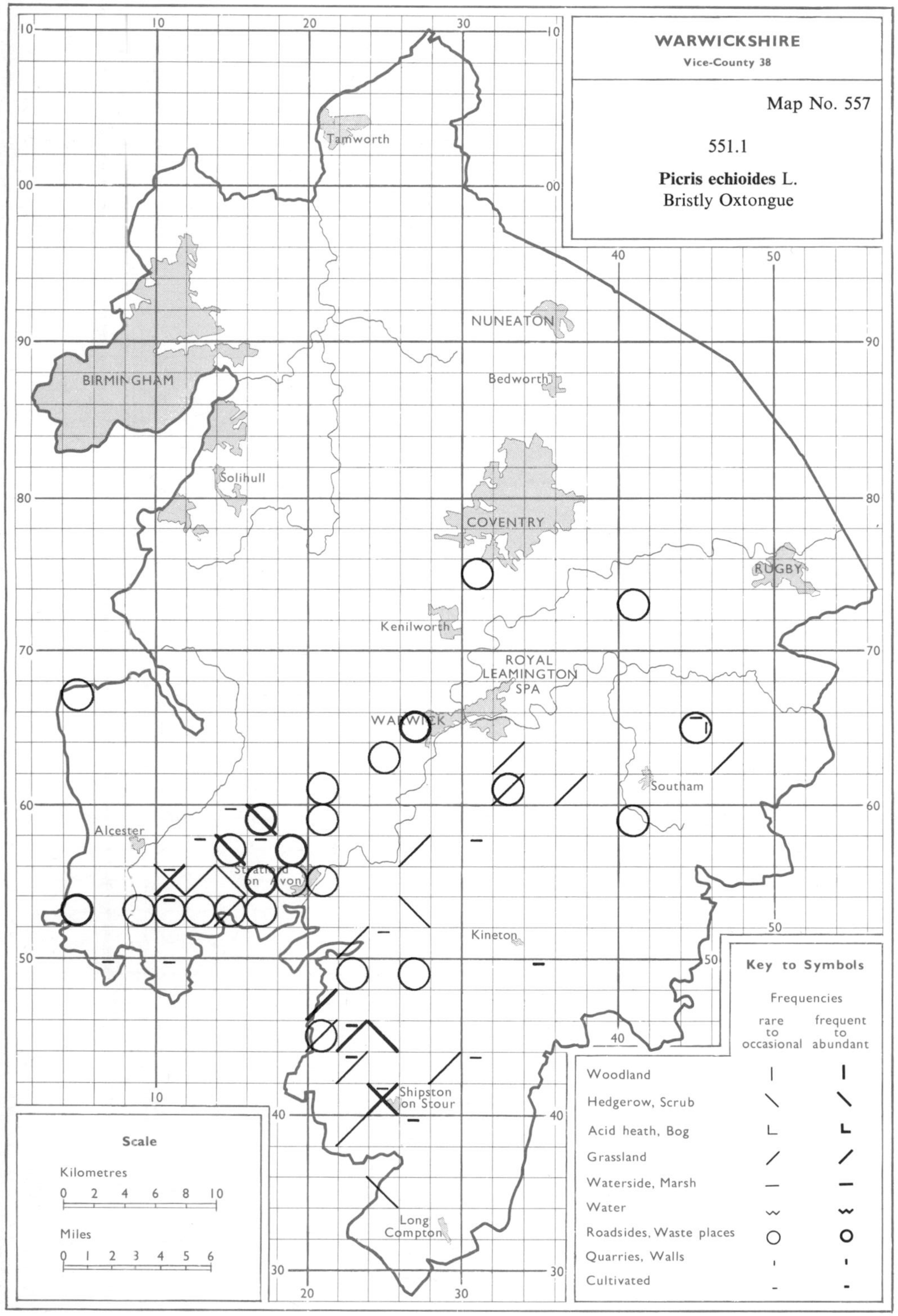
WARWICKSHIRE
Vice-County 38
Map No. 557
551.1
Picris echioides L.
Bristly Oxtongue
Tamworth
NUNEATON
Bedworth
BIRMINGHAM
Solihull
COVENTRY
RUGBY
Kenilworth
ROYAL LEAMINGTON SPA
WARWICK
Southam
Alcester
Stratford on Avon
Kineton
Shipston on Stour
Long Compton
Key to Symbols
Frequencies
rare to occasional
frequent to abundant
Woodland
Hedgerow, Scrub
Acid heath, Bog
Grassland
Waterside, Marsh
Water
Roadsides, Waste places
Quarries, Walls
Cultivated
Scale
Kilometres
0 2 4 6 8 10
Miles
0 1 2 3 4 5 6

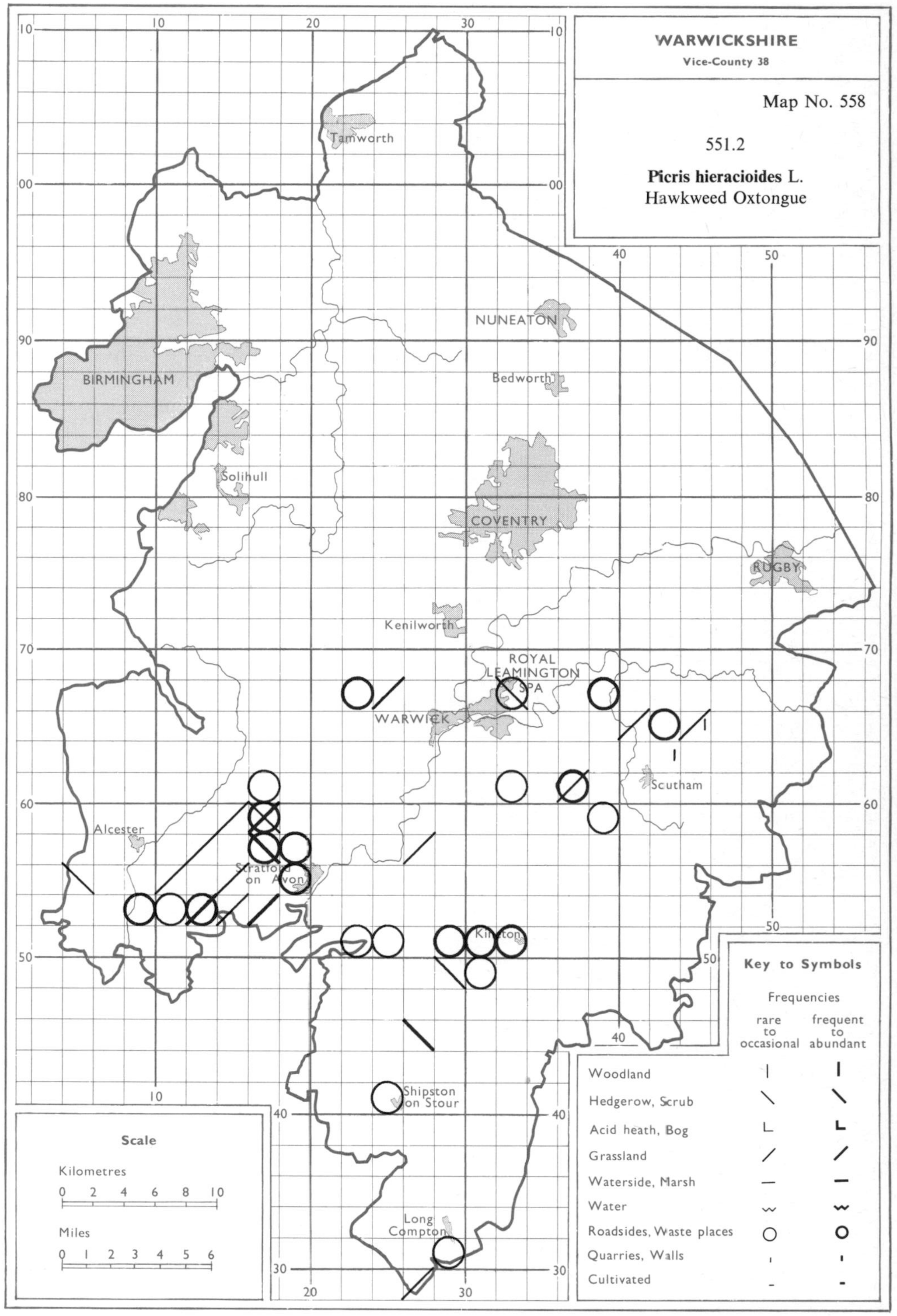
WARWICKSHIRE
Vice-County 38
Map No. 558
551.2
Picris hieracioides L.
Hawkweed Oxtongue
Tamworth
NUNEATON
Bedworth
BIRMINGHAM
Solihull
COVENTRY
RUGBY
Kenilworth
ROYAL LEAMINGTON SPA
WARWICK
Scutham
Alcester
Stratford on Avon
Kineton
Shipston on Stour
Long Compton
Key to Symbols
Frequencies
rare to occasional
frequent to abundant
Woodland
Hedgerow, Scrub
Acid heath, Bog
Grassland
Waterside, Marsh
Water
Roadsides, Waste places
Quarries, Walls
Cultivated
Scale
Kilometres
0 2 4 6 8 10
Miles
0 1 2 3 4 5 6

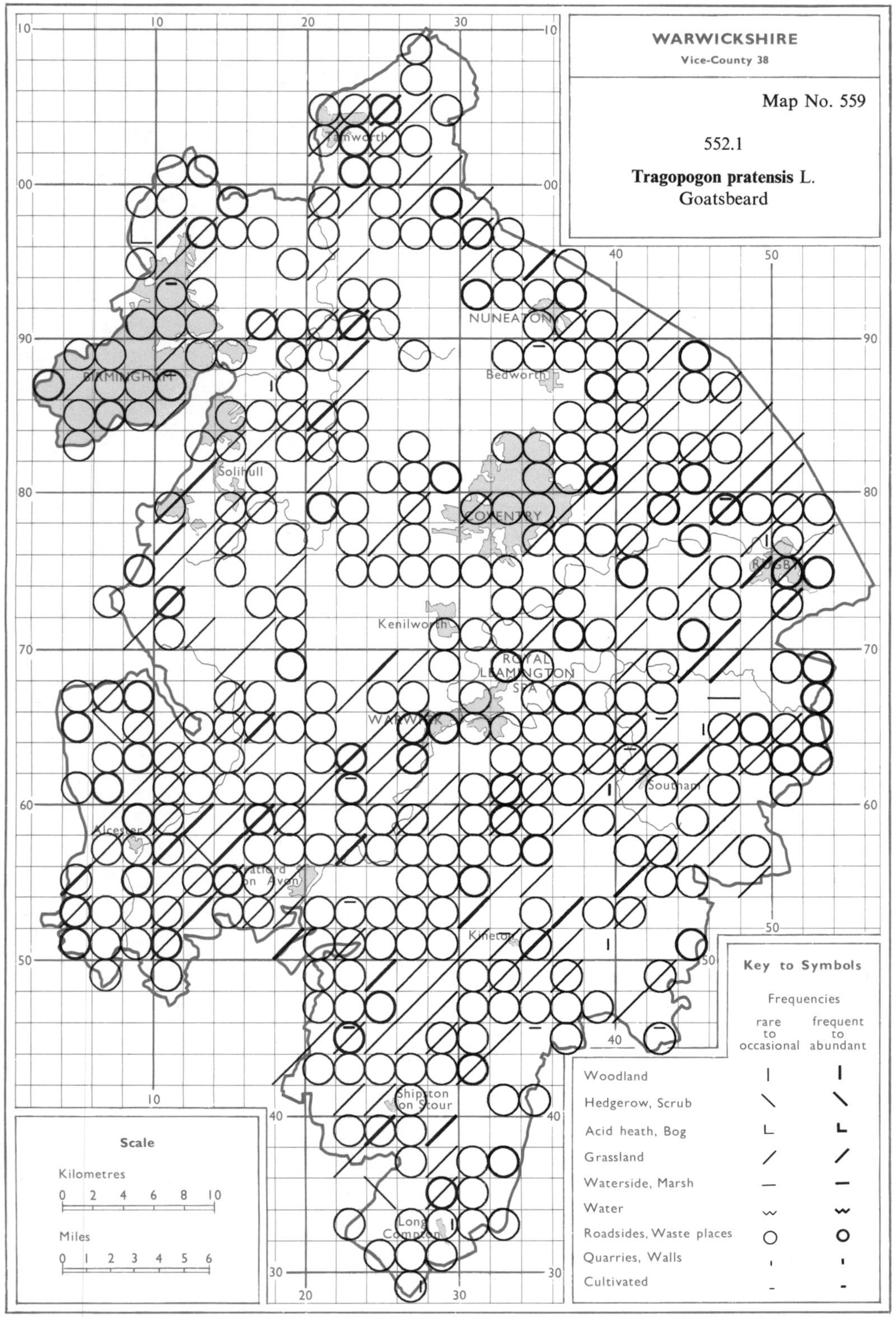
WARWICKSHIRE
Vice-County 38
Map No. 559
552.1
Tragopogon pratensis L.
Goatsbeard
Tamworth
NUNEATON
Bedworth
BIRMINGHAM
Solihull
COVENTRY
RUGBY
Kenilworth
ROYAL LEAMINGTON SPA
WARWICK
Southam
Alcester
Stratford on Avon
Kineton
Shipston on Stour
Long Compton
Key to Symbols
Frequencies
rare to occasional
frequent to abundant
Woodland
Hedgerow, Scrub
Acid heath, Bog
Grassland
Waterside, Marsh
Water
Roadsides, Waste places
Quarries, Walls
Cultivated
Scale
Kilometres
0 2 4 6 8 10
Miles
0 1 2 3 4 5 6

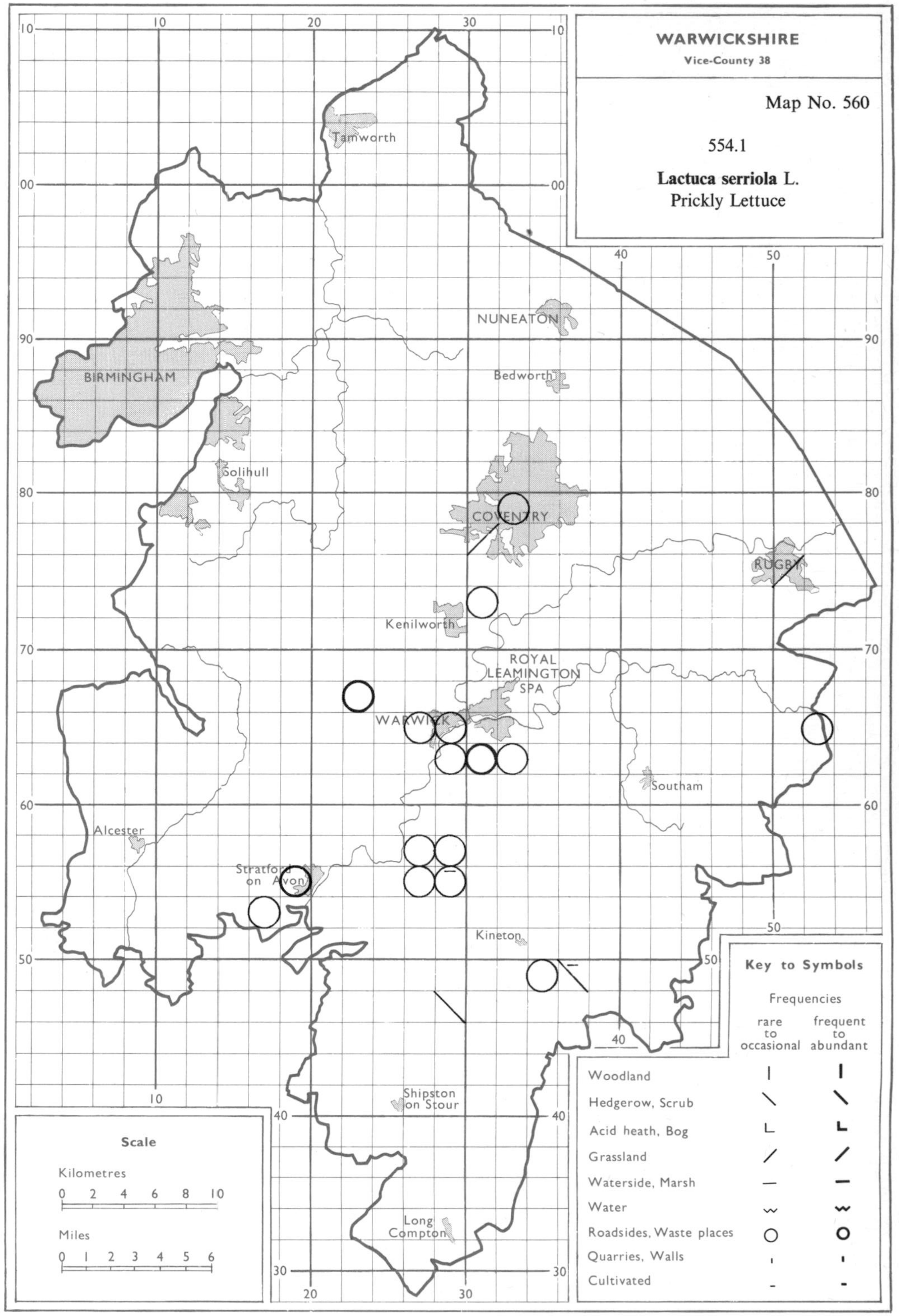
WARWICKSHIRE
Vice-County 38
Map No. 560
554.1
Lactuca serriola L.
Prickly Lettuce
Tamworth
NUNEATON
Bedworth
BIRMINGHAM
Solihull
COVENTRY
RUGBY
Kenilworth
ROYAL LEAMINGTON SPA
WARWICK
Southam
Alcester
Stratford on Avon
Kineton
Shipston on Stour
Long Compton
Key to Symbols
Frequencies
rare to occasional
frequent to abundant
Woodland
Hedgerow, Scrub
Acid heath, Bog
Grassland
Waterside, Marsh
Water
Roadsides, Waste places
Quarries, Walls
Cultivated
Scale
Kilometres
0 2 4 6 8 10
Miles
0 1 2 3 4 5 6

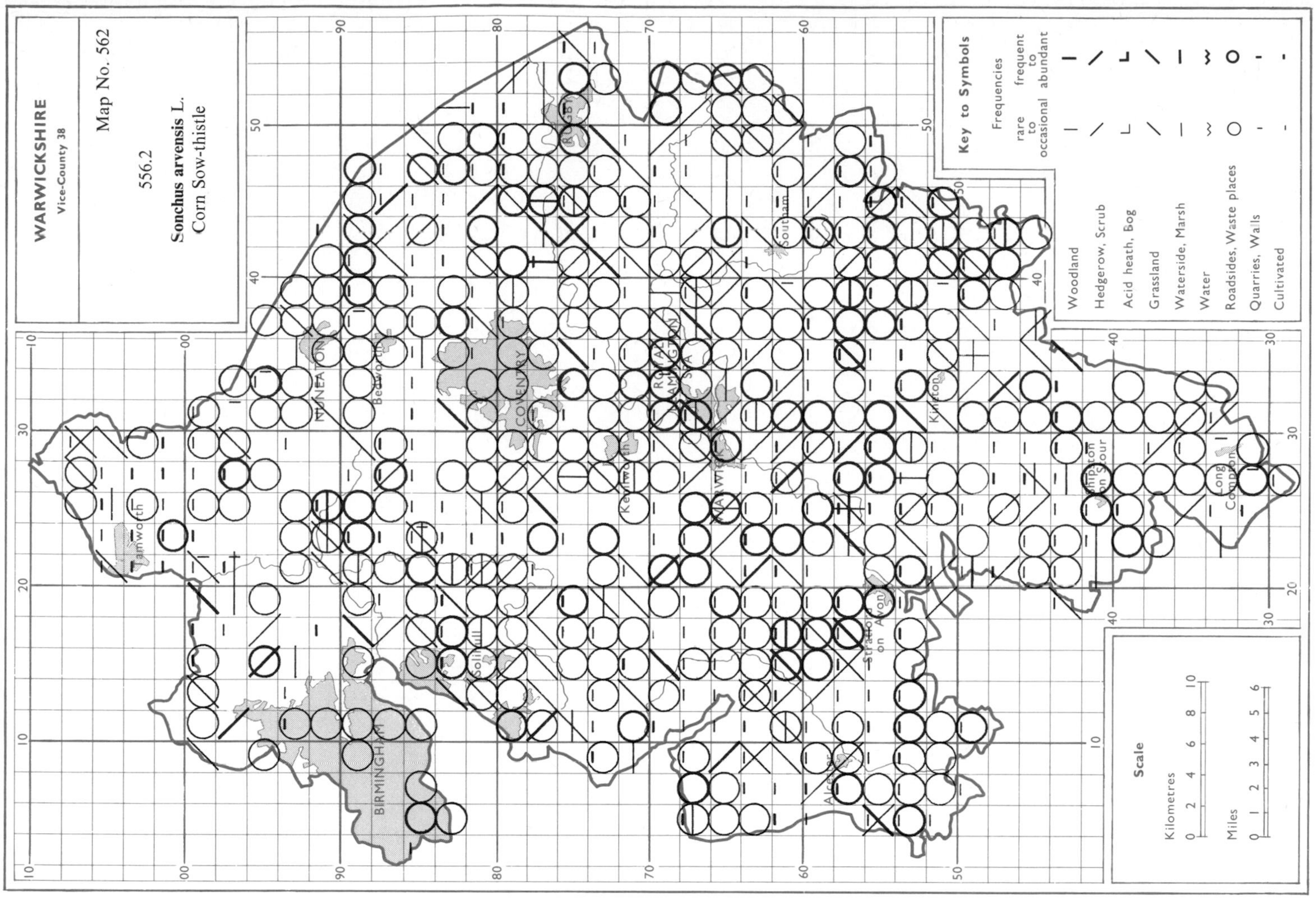
WARWICKSHIRE
Vice-County 38
Map No. 562
556.2
Sonchus arvensis L.
Corn Sow-thistle
Key to Symbols
Frequencies
rare to occasional
frequent to abundant
Woodland
Hedgerow, Scrub
Acid heath, Bog
Grassland
Waterside, Marsh
Water
Roadsides, Waste places
Quarries, Walls
Cultivated
Scale
Kilometres
Miles
TAMWORTH
NUNEATON
Bedworth
COVENTRY
RUGBY
Southam
ROYAL LEAMINGTON SPA
WARWICK
Kenilworth
Solihull
BIRMINGHAM
Kineton
Shipston on Stour
Long Compton
Stratford on Avon
Alcester

WARWICKSHIRE
Vice-County 38
Map No. 561
555.1
Mycelis muralis (L.) Dumort.
Wall Lettuce
Key to Symbols
Frequencies
rare to occasional
frequent to abundant
Woodland
Hedgerow, Scrub
Acid heath, Bog
Grassland
Waterside, Marsh
Water
Roadsides, Waste places
Quarries, Walls
Cultivated
Scale
Kilometres
Miles
Tamworth
NUNEATON
Bedworth
COVENTRY
RUGBY
Southam
ROYAL LEAMINGTON SPA
WARWICK
Kenilworth
Solihull
BIRMINGHAM
Kineton
Shipston on Stour
Long Compton
Stratford on Avon
Alcester

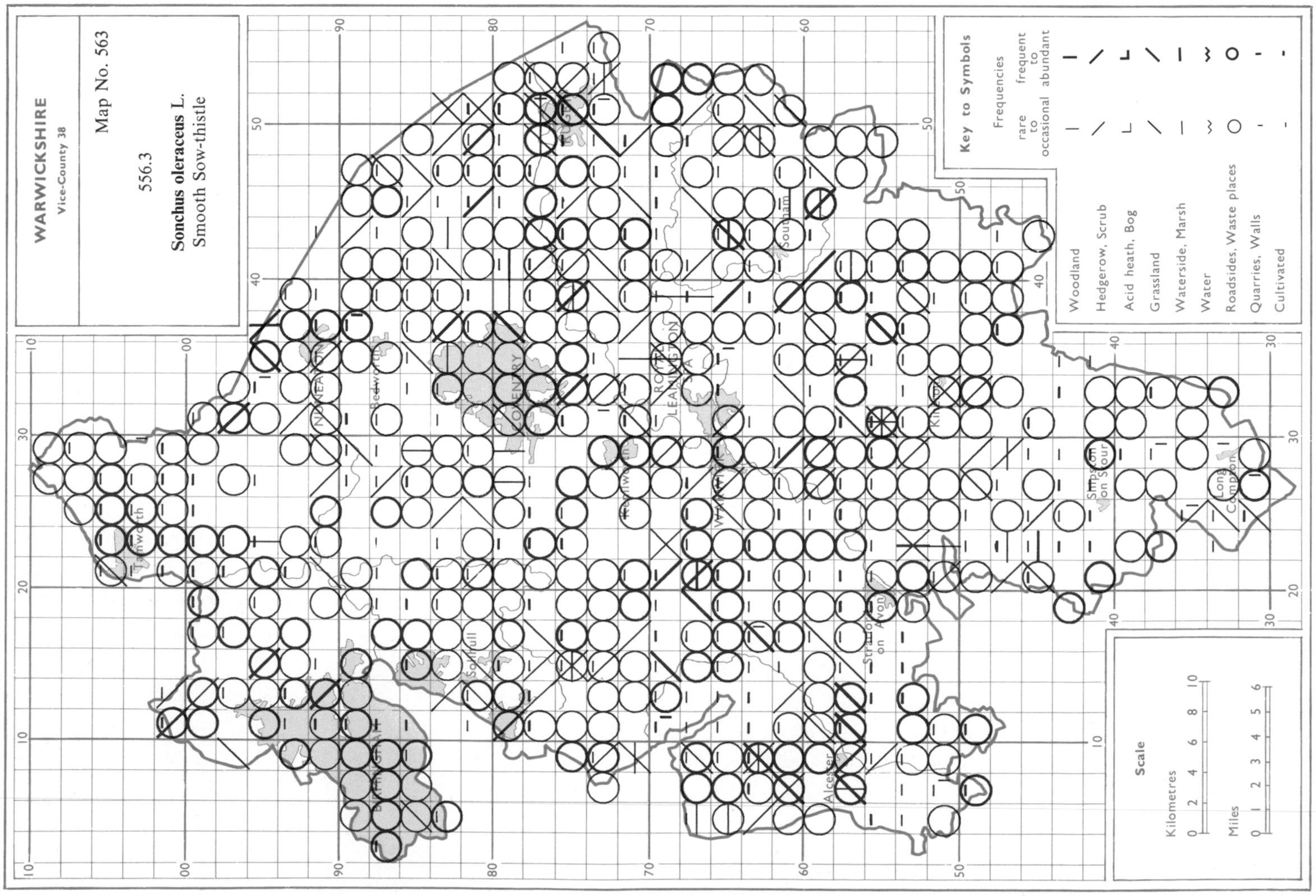
WARWICKSHIRE
Vice-County 38
Map No. 563
556.3
Sonchus oleraceus L.
Smooth Sow-thistle
Key to Symbols
Frequencies
rare to occasional
frequent to abundant
Woodland
Hedgerow, Scrub
Acid heath, Bog
Grassland
Waterside, Marsh
Water
Roadsides, Waste places
Quarries, Walls
Cultivated
Scale
Kilometres 0 2 4 6 8 10
Miles 0 1 2 3 4 5 6
Tamworth
Nuneaton
Bedworth
Coventry
Rugby
Birmingham
Solihull
Kenilworth
Royal Leamington Spa
Warwick
Southam
Kineton
Stratford on Avon
Alcester
Shipston on Stour
Long Compton

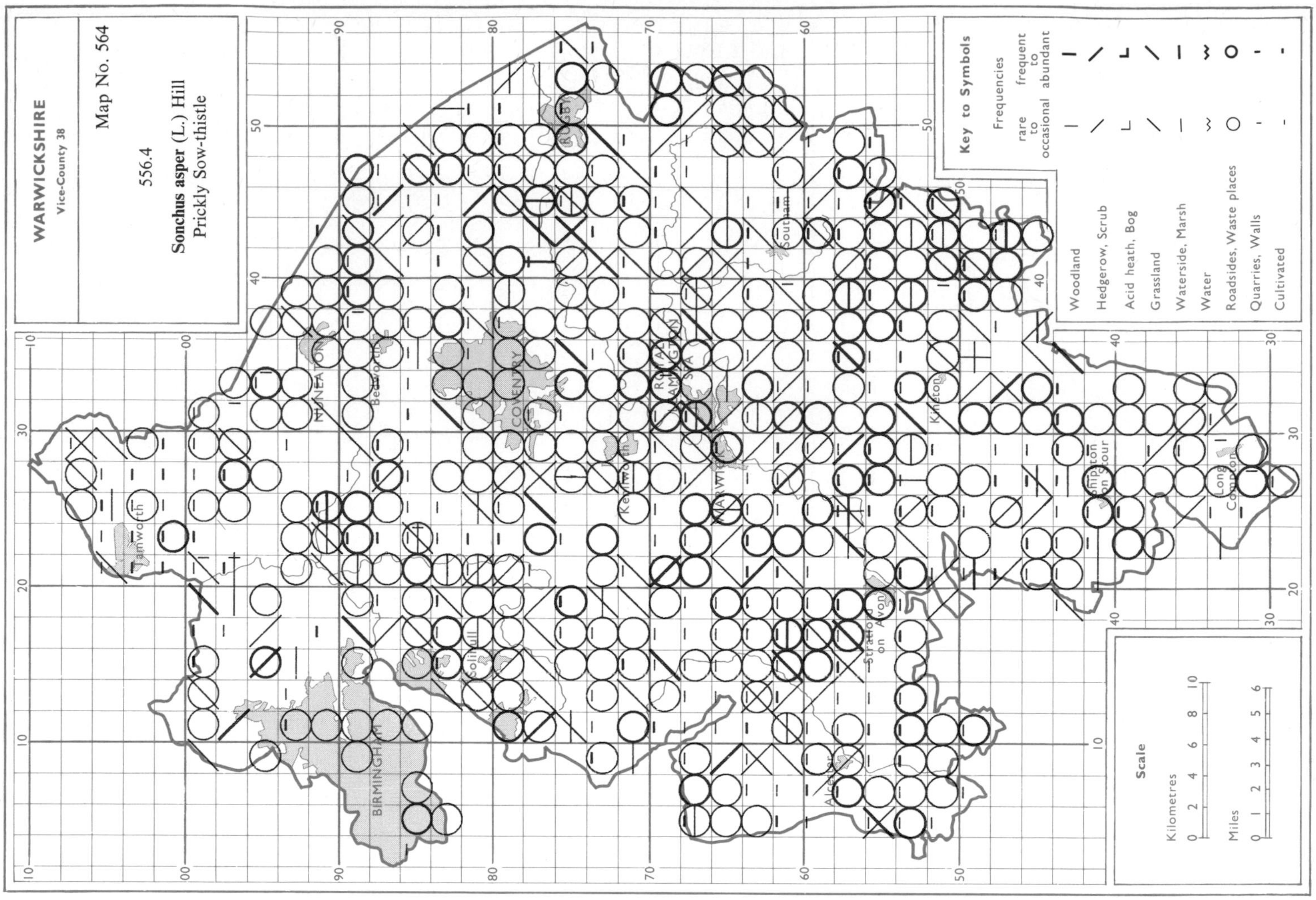
WARWICKSHIRE
Vice-County 38
Map No. 564
556.4
Sonchus asper (L.) Hill
Prickly Sow-thistle
Key to Symbols
Frequencies
rare to occasional
frequent to abundant
Woodland
Hedgerow, Scrub
Acid heath, Bog
Grassland
Waterside, Marsh
Water
Roadsides, Waste places
Quarries, Walls
Cultivated
Scale
Kilometres 0 2 4 6 8 10
Miles 0 1 2 3 4 5 6
Tamworth
Nuneaton
Bedworth
Coventry
Rugby
Birmingham
Solihull
Kenilworth
Royal Leamington Spa
Warwick
Southam
Kineton
Stratford on Avon
Alcester
Shipston on Stour
Long Compton

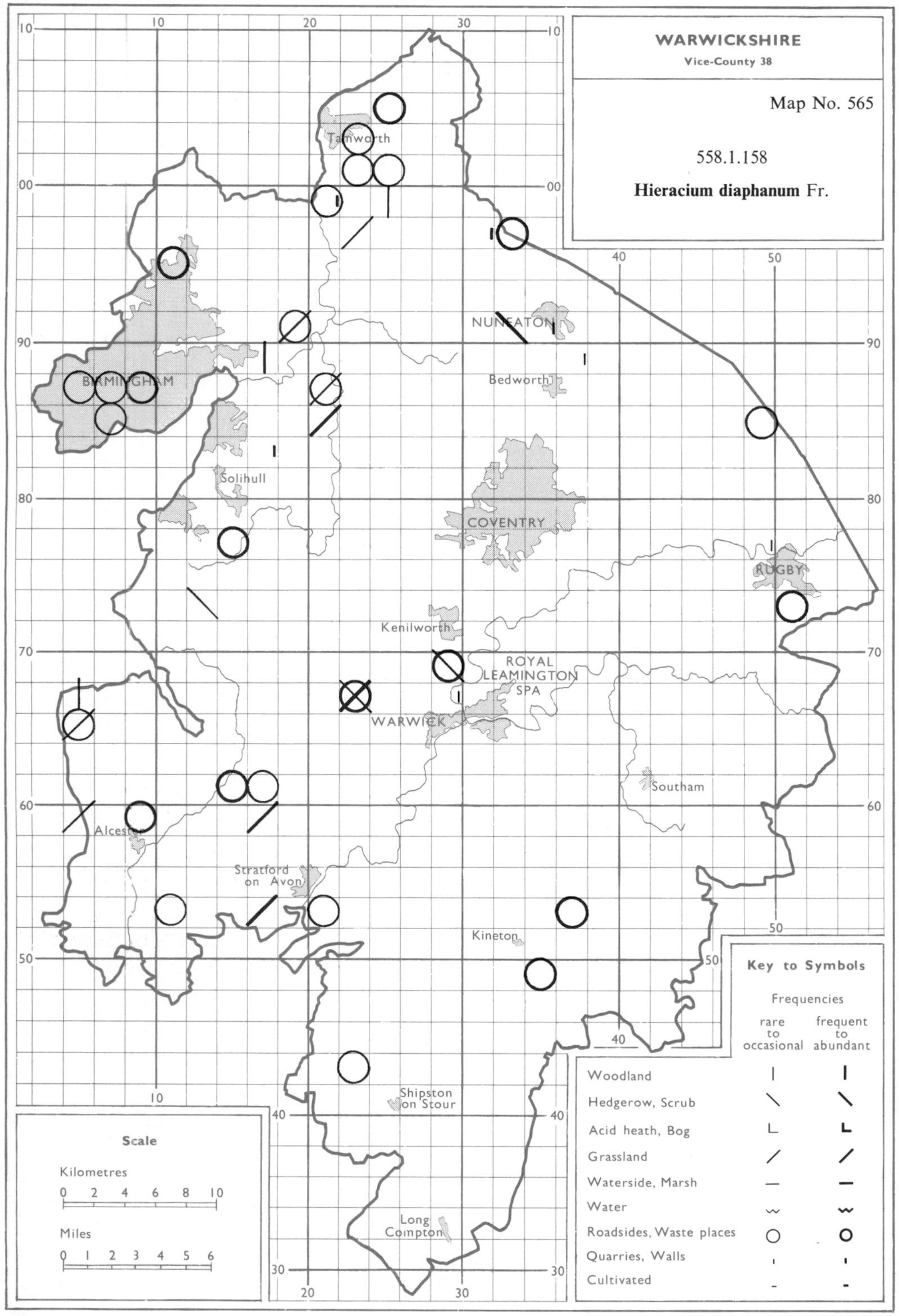
WARWICKSHIRE
Vice-County 38
Map No. 565
558.1.158
Hieracium diaphanum Fr.
Tamworth
Birmingham
Solihull
Nuneaton
Bedworth
Coventry
Rugby
Kenilworth
Royal Leamington Spa
Warwick
Southam
Alcester
Stratford on Avon
Kineton
Shipston on Stour
Long Compton
Key to Symbols
Frequencies
rare to occasional
frequent to abundant
Woodland
Hedgerow, Scrub
Acid heath, Bog
Grassland
Waterside, Marsh
Water
Roadsides, Waste places
Quarries, Walls
Cultivated
Scale
Kilometres
0 2 4 6 8 10
Miles
0 1 2 3 4 5 6

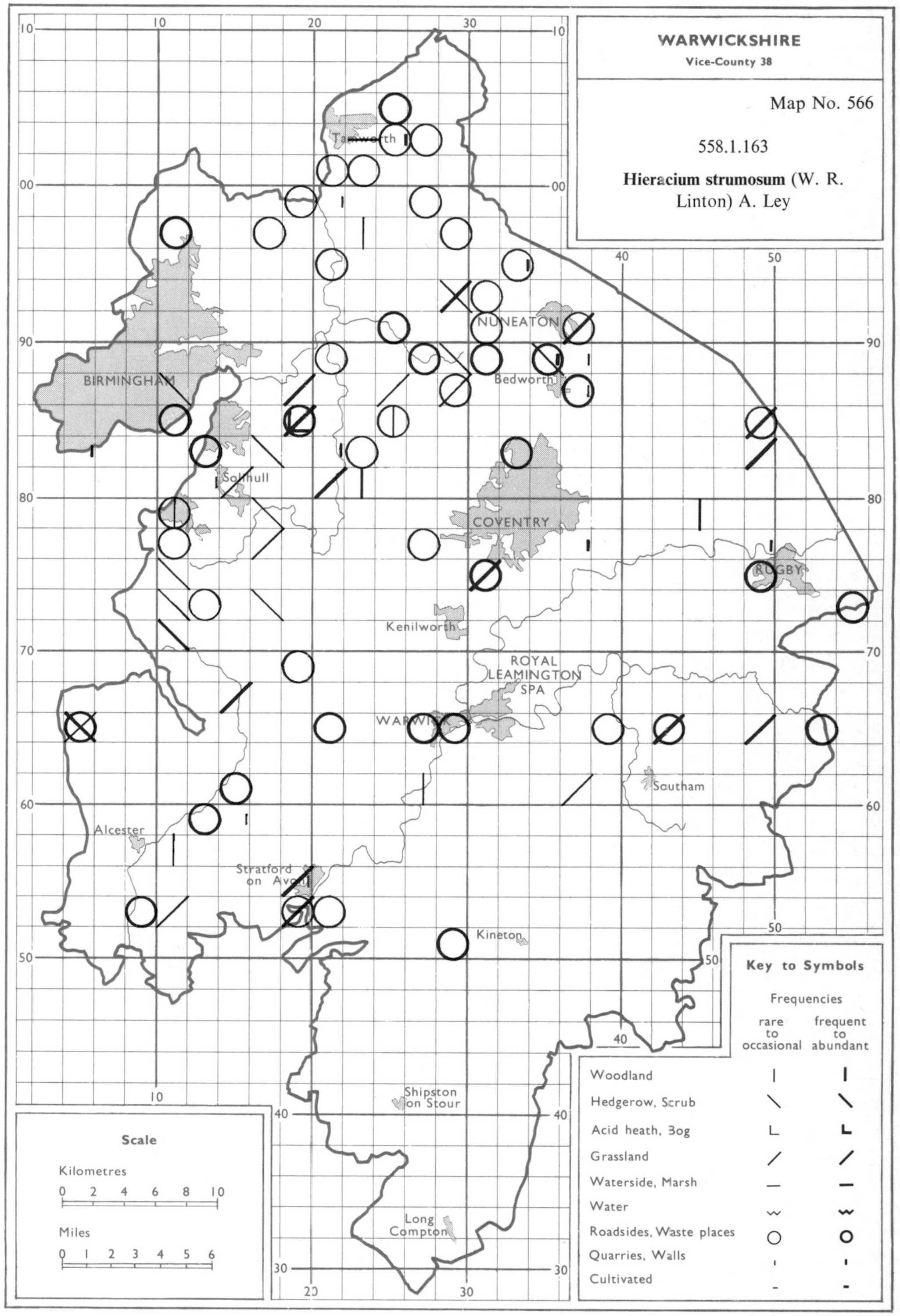
WARWICKSHIRE
Vice-County 38
Map No. 566
558.1.163
Hieracium strumosum (W. R. Linton) A. Ley
Tamworth
Birmingham
Solihull
Nuneaton
Bedworth
Coventry
Rugby
Kenilworth
Royal Leamington Spa
Warwick
Southam
Alcester
Stratford on Avon
Kineton
Shipston on Stour
Long Compton
Key to Symbols
Frequencies
rare to occasional
frequent to abundant
Woodland
Hedgerow, Scrub
Acid heath, Bog
Grassland
Waterside, Marsh
Water
Roadsides, Waste places
Quarries, Walls
Cultivated
Scale
Kilometres
0 2 4 6 8 10
Miles
0 1 2 3 4 5 6

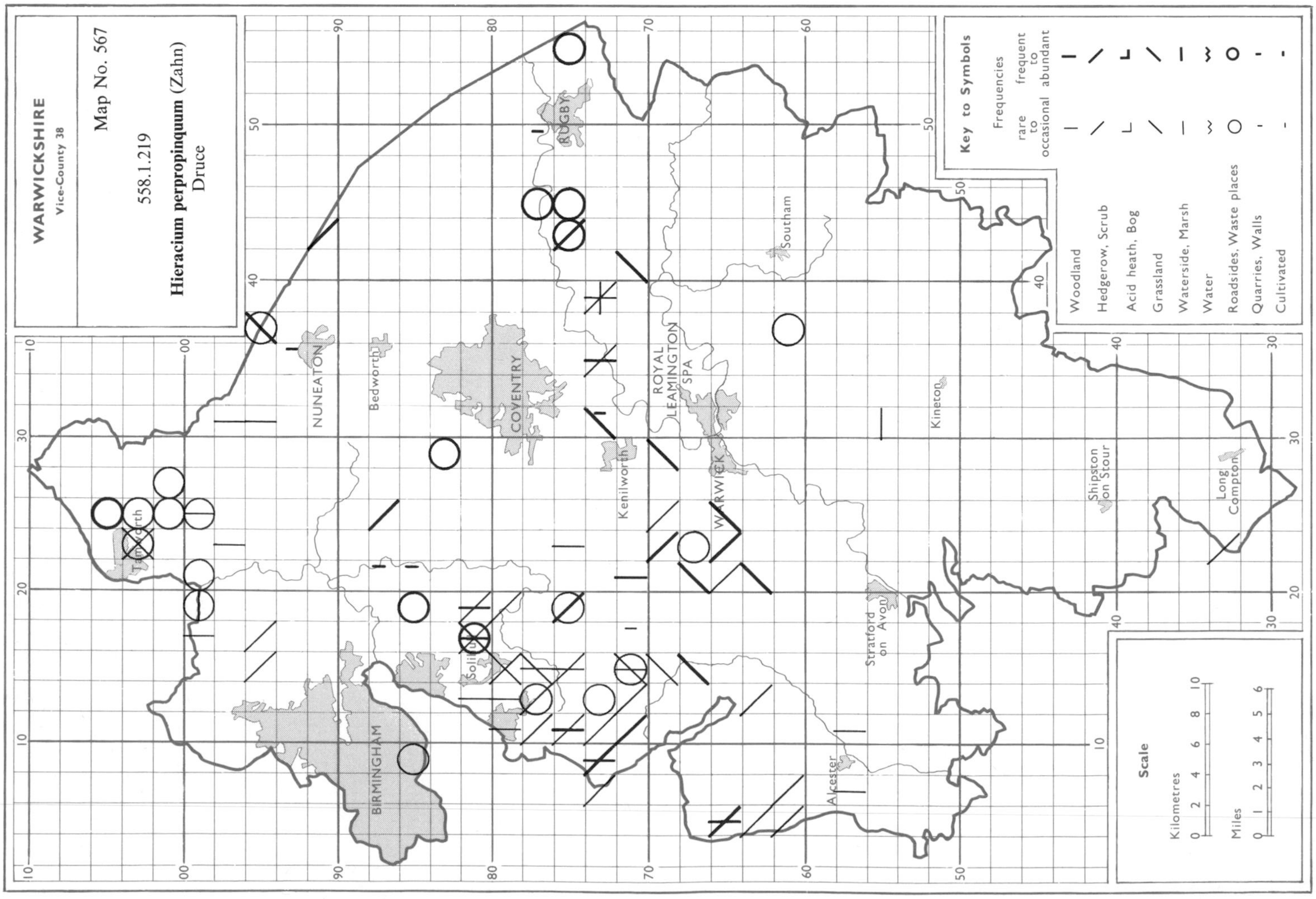
WARWICKSHIRE
Vice-County 38
Map No. 567
558.1.219
Hieracium perpropinquum (Zahn) Druce
Key to Symbols
Frequencies
rare to occasional
frequent to abundant
Woodland
Hedgerow, Scrub
Acid heath, Bog
Grassland
Waterside, Marsh
Water
Roadsides, Waste places
Quarries, Walls
Cultivated
Scale
Kilometres
Miles
NUNEATON
Bedworth
COVENTRY
RUGBY
Southam
ROYAL LEAMINGTON SPA
WARWICK
Kenilworth
Kineton
Tamworth
BIRMINGHAM
Solihull
Stratford on Avon
Alcester
Shipston on Stour
Long Compton

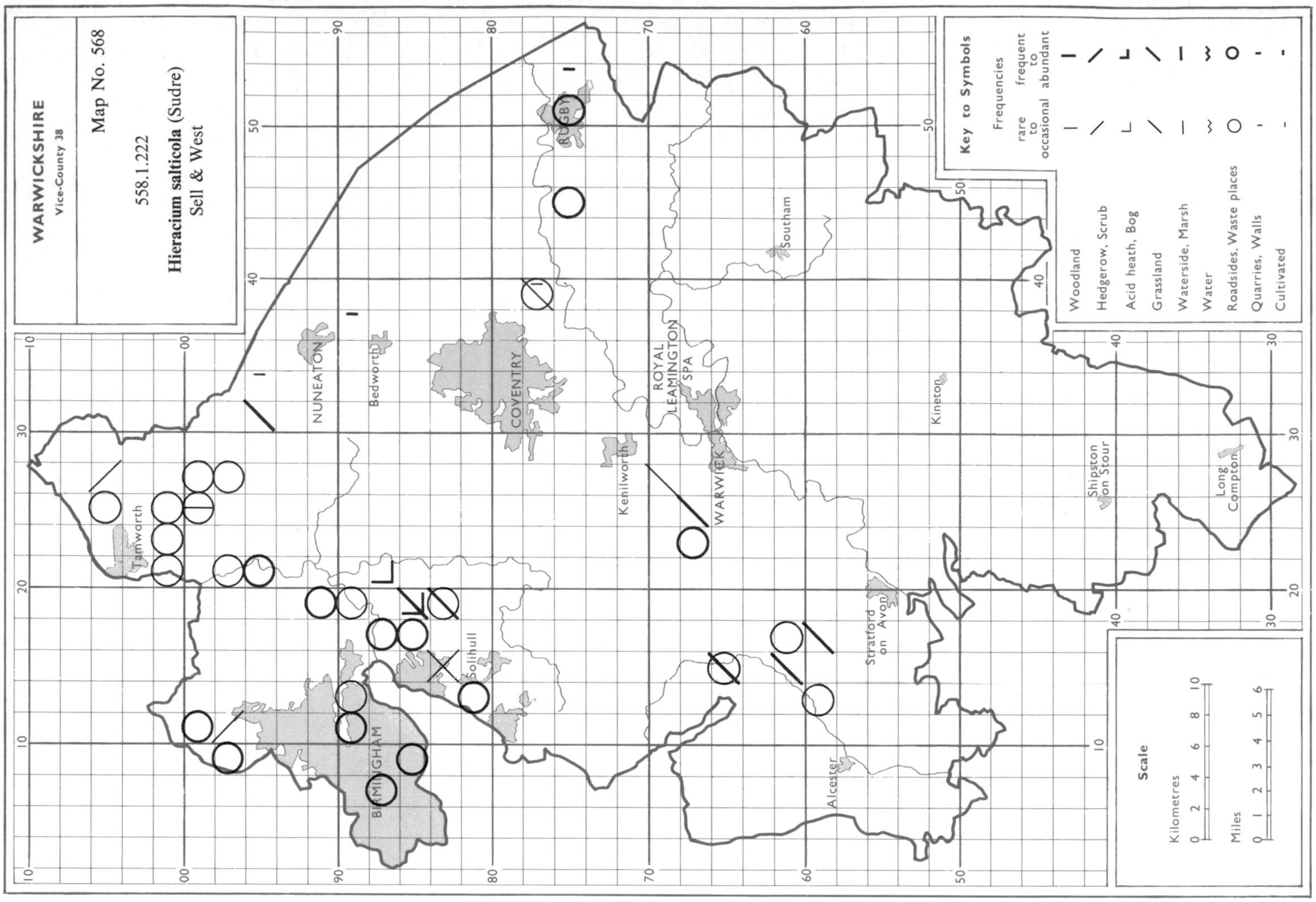
WARWICKSHIRE
Vice-County 38
Map No. 568
558.1.222
Hieracium salticola (Sudre) Sell & West
Key to Symbols
Frequencies
rare to occasional
frequent to abundant
Woodland
Hedgerow, Scrub
Acid heath, Bog
Grassland
Waterside, Marsh
Water
Roadsides, Waste places
Quarries, Walls
Cultivated
Scale
Kilometres
Miles
NUNEATON
Bedworth
COVENTRY
RUGBY
Southam
ROYAL LEAMINGTON SPA
WARWICK
Kenilworth
Kineton
Tamworth
BIRMINGHAM
Solihull
Stratford on Avon
Alcester
Shipston on Stour
Long Compton

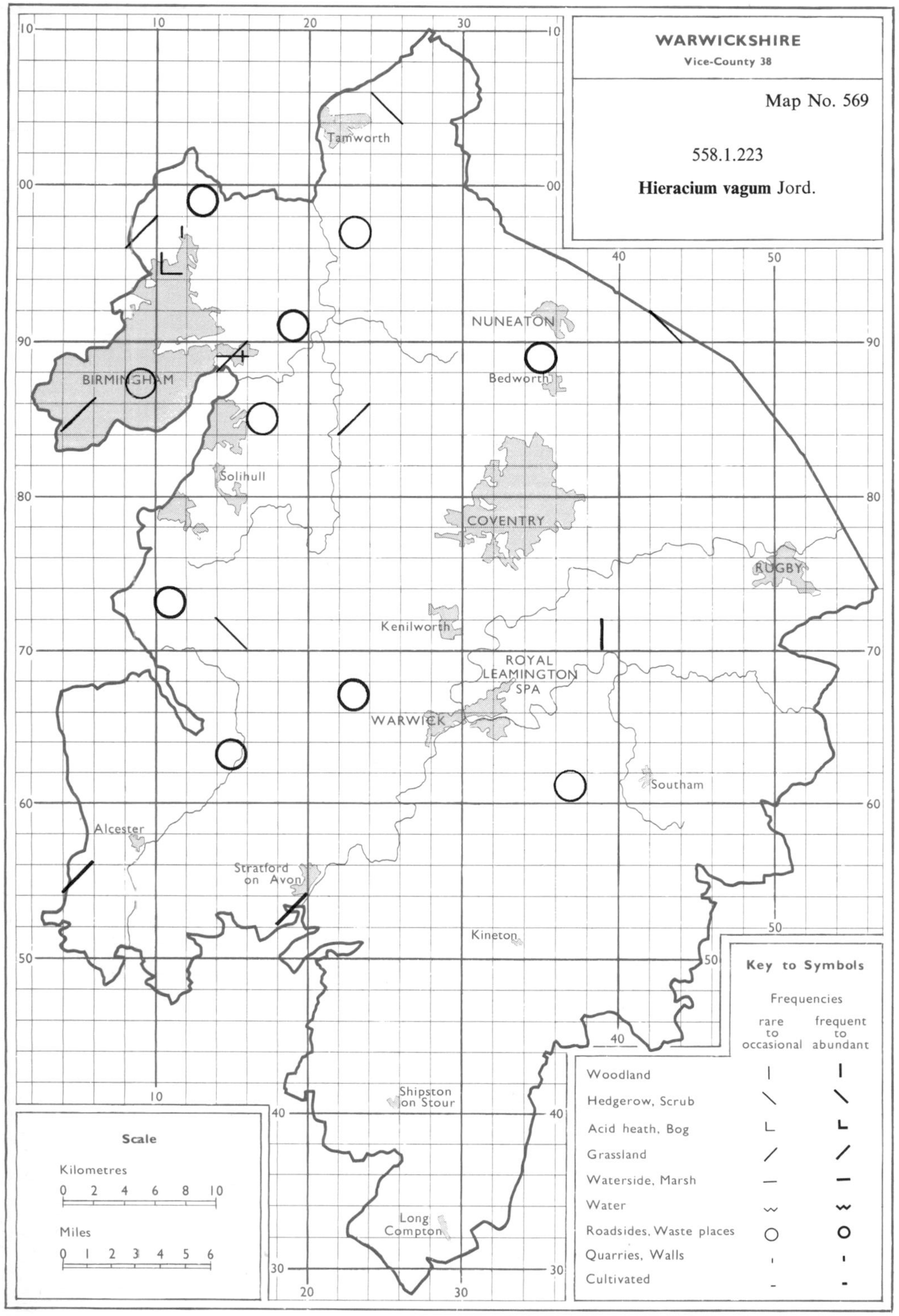

WARWICKSHIRE
Vice-County 38

Map No. 569

558.1.223

Hieracium vagum Jord.

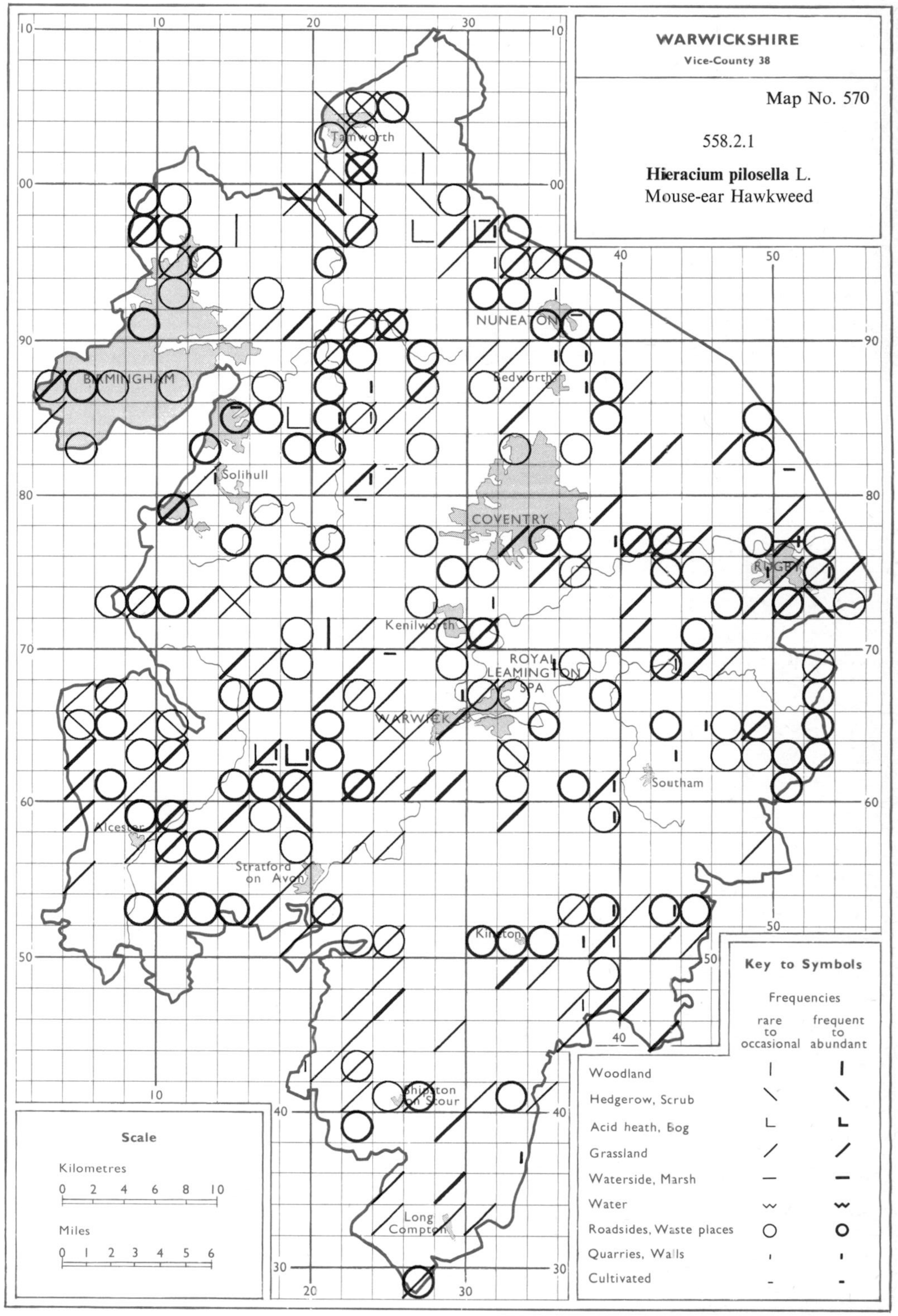

WARWICKSHIRE
Vice-County 38

Map No. 570

558.2.1

Hieracium pilosella L.
Mouse-ear Hawkweed

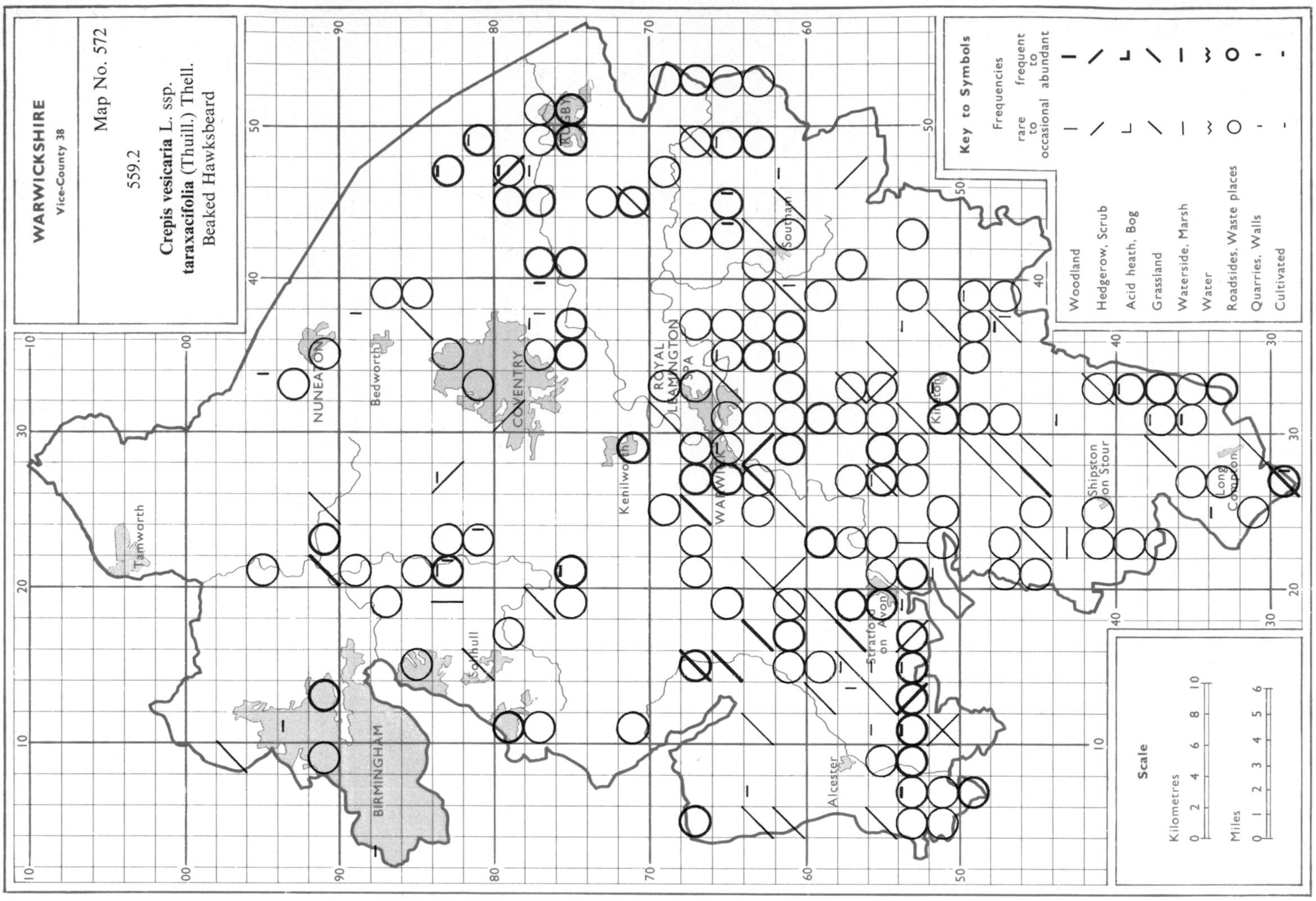

WARWICKSHIRE
Vice-County 38
Map No. 572
559.2
Crepis vesicaria L. ssp. taraxacifolia (Thuill.) Thell.
Beaked Hawksbeard
Key to Symbols
Frequencies
rare to occasional
frequent to abundant
Woodland
Hedgerow, Scrub
Acid heath, Bog
Grassland
Waterside, Marsh
Water
Roadsides, Waste places
Quarries, Walls
Cultivated
Scale
Kilometres
Miles
TAMWORTH
NUNEATON
Bedworth
COVENTRY
RUGBY
BIRMINGHAM
Solihull
Kenilworth
ROYAL LEAMINGTON SPA
WARWICK
Southam
Stratford on Avon
Alcester
Kineton
Shipston on Stour
Long Compton

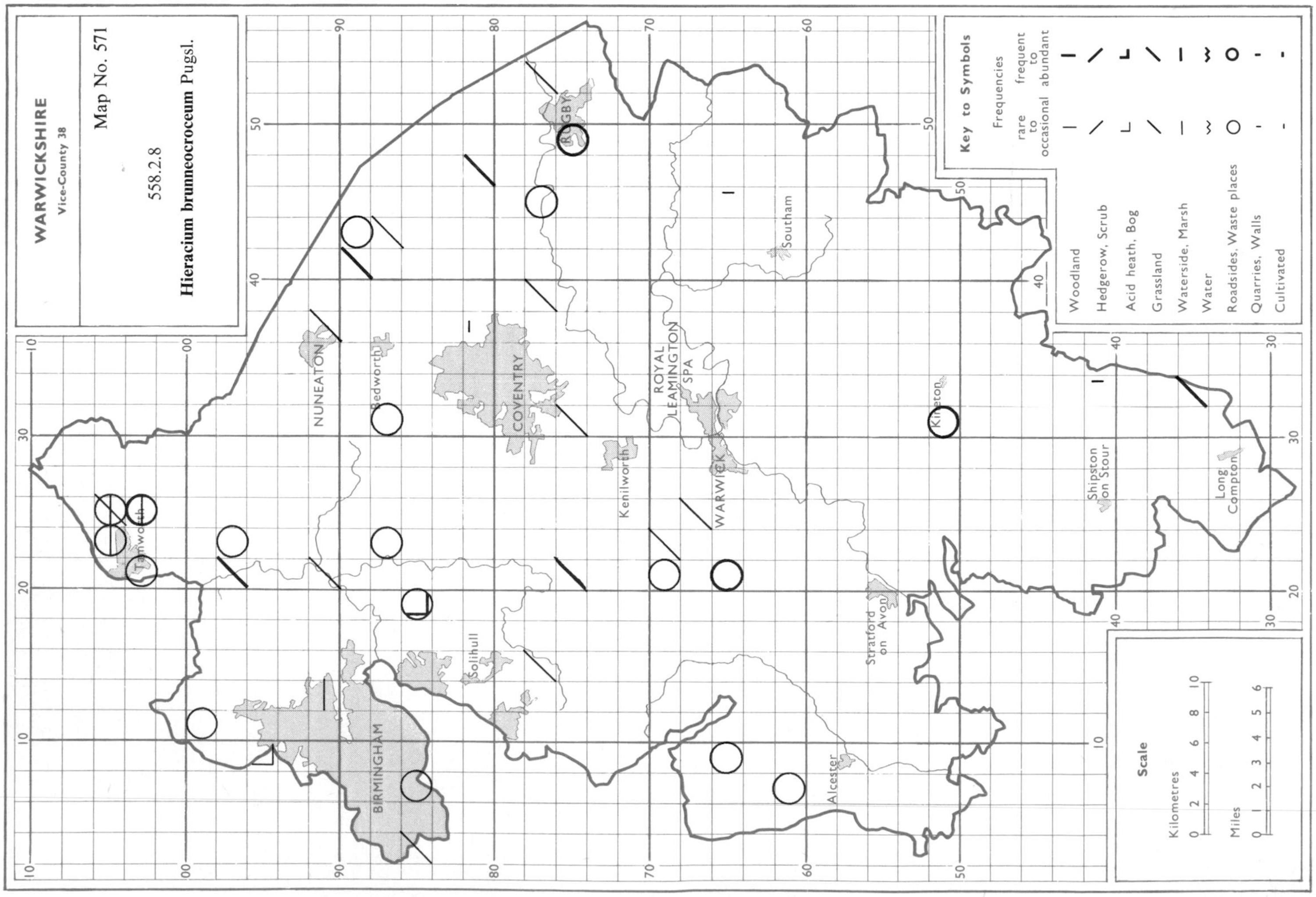

WARWICKSHIRE
Vice-County 38
Map No. 571
558.2.8
Hieracium brunneocroceum Pugsl.
Key to Symbols
Frequencies
rare to occasional
frequent to abundant
Woodland
Hedgerow, Scrub
Acid heath, Bog
Grassland
Waterside, Marsh
Water
Roadsides, Waste places
Quarries, Walls
Cultivated
Scale
Kilometres
Miles
Tamworth
NUNEATON
Bedworth
COVENTRY
RUGBY
BIRMINGHAM
Solihull
Kenilworth
ROYAL LEAMINGTON SPA
WARWICK
Southam
Stratford on Avon
Alcester
Kineton
Shipston on Stour
Long Compton

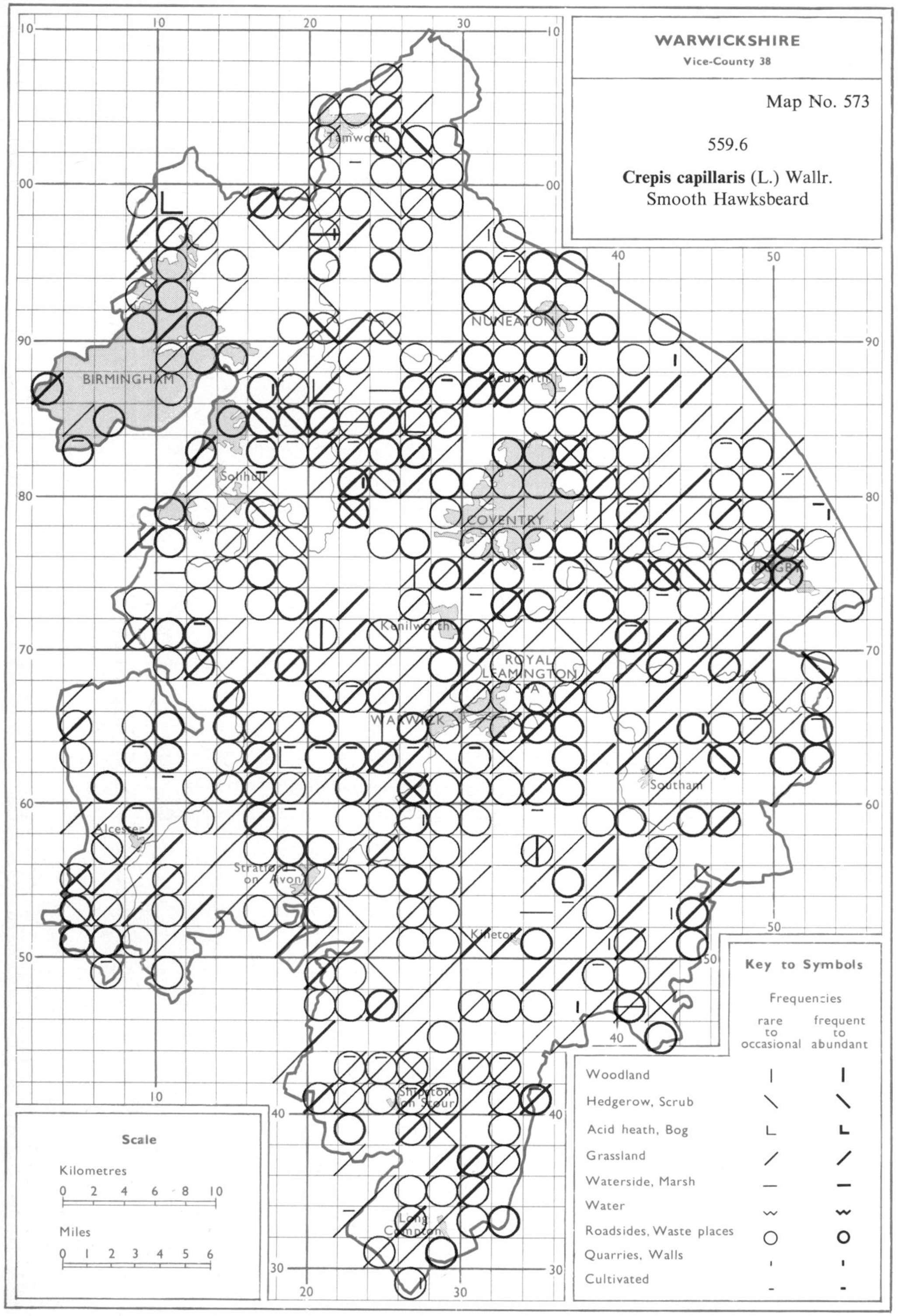
WARWICKSHIRE
Vice-County 38
Map No. 573
559.6
Crepis capillaris (L.) Wallr.
Smooth Hawksbeard
Tamworth
NUNEATON
BIRMINGHAM
Solihull
COVENTRY
Rugby
Kenilworth
ROYAL LEAMINGTON SPA
WARWICK
Southam
Alcester
Stratford on Avon
Kineton
Shipston on Stour
Long Compton
Key to Symbols
Frequencies
rare to occasional
frequent to abundant
Woodland
Hedgerow, Scrub
Acid heath, Bog
Grassland
Waterside, Marsh
Water
Roadsides, Waste places
Quarries, Walls
Cultivated
Scale
Kilometres
0 2 4 6 8 10
Miles
0 1 2 3 4 5 6

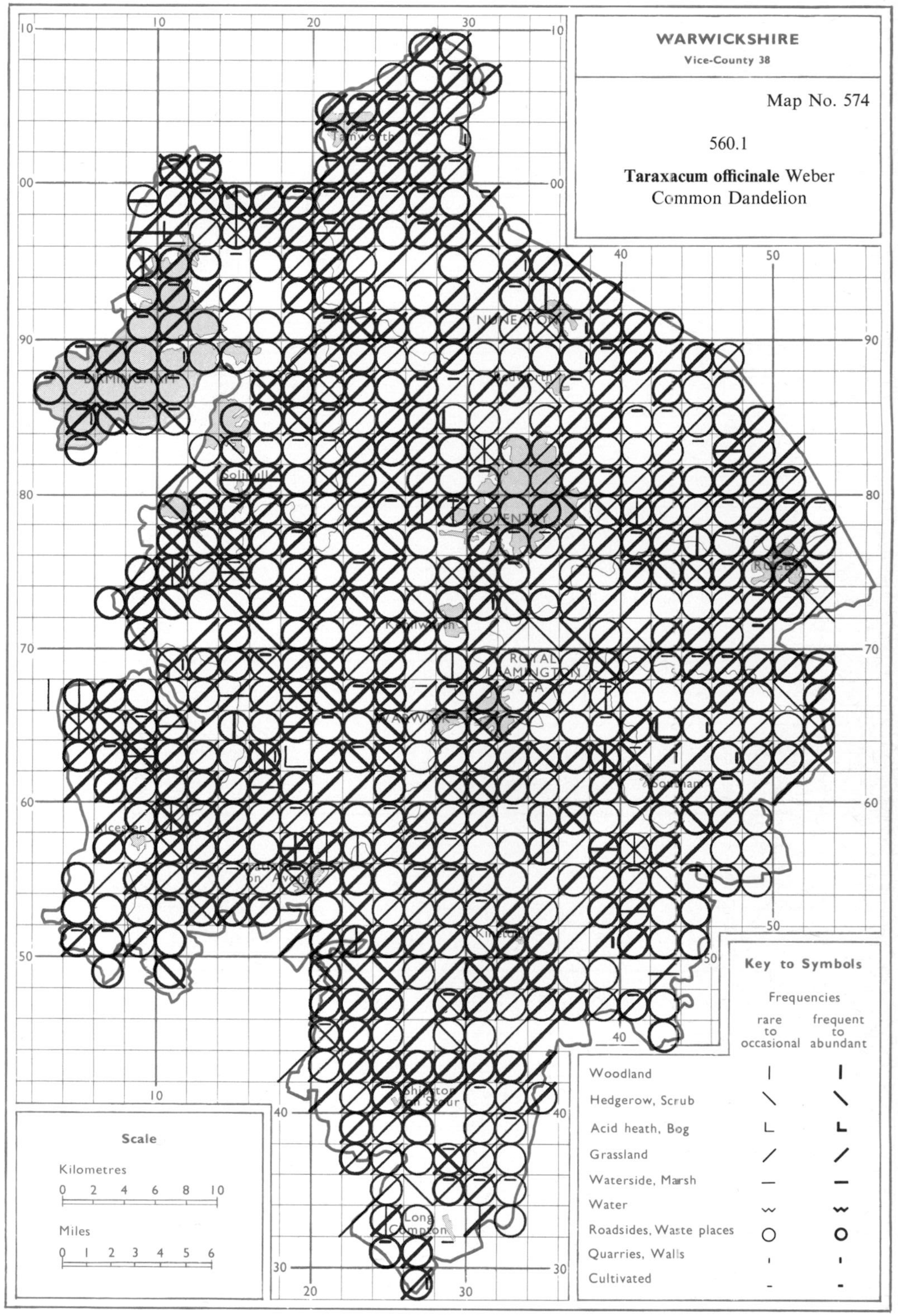
WARWICKSHIRE
Vice-County 38
Map No. 574
560.1
Taraxacum officinale Weber
Common Dandelion
Key to Symbols
Frequencies
rare to occasional
frequent to abundant
Woodland
Hedgerow, Scrub
Acid heath, Bog
Grassland
Waterside, Marsh
Water
Roadsides, Waste places
Quarries, Walls
Cultivated
Scale
Kilometres
0 2 4 6 8 10
Miles
0 1 2 3 4 5 6

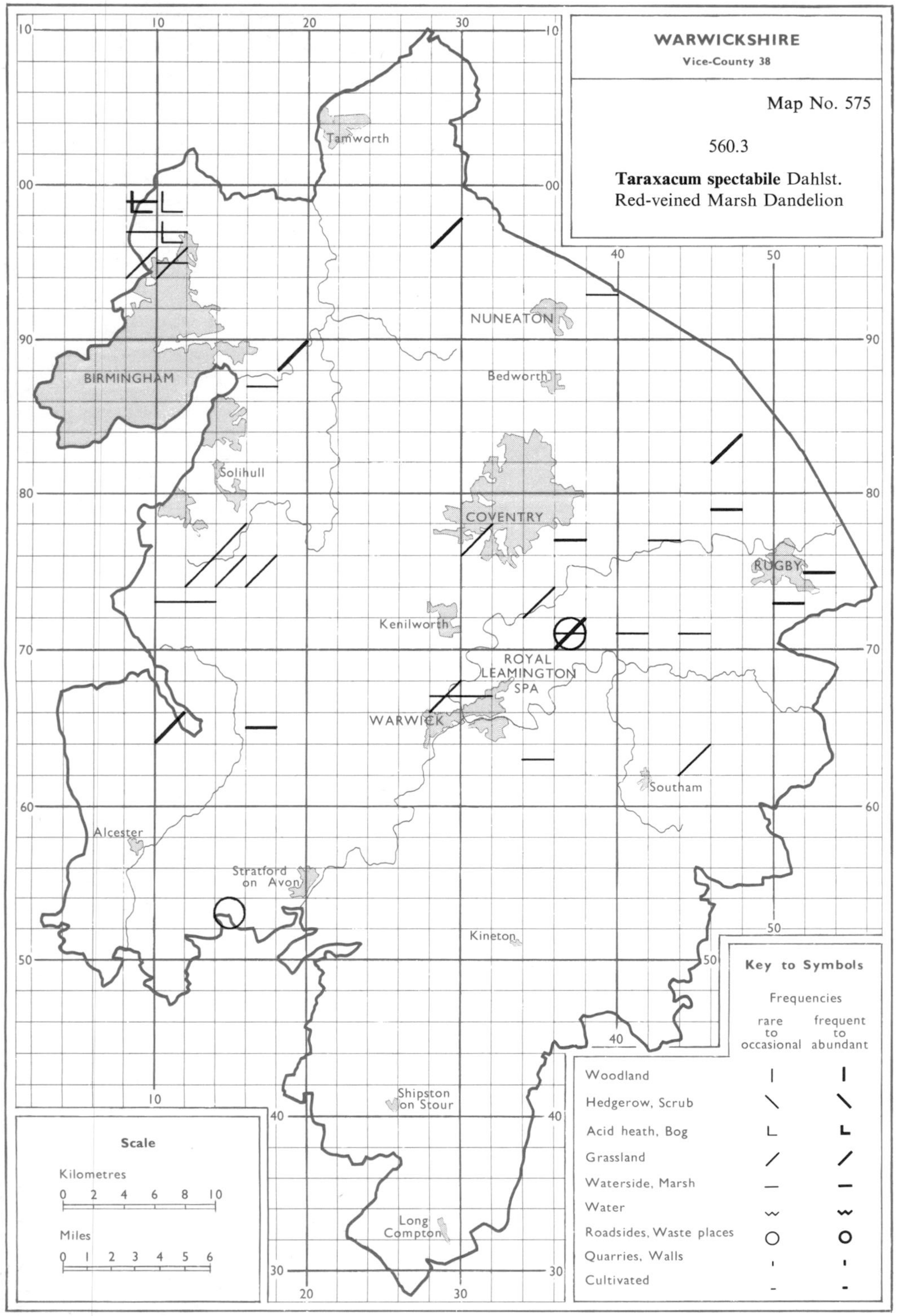
WARWICKSHIRE
Vice-County 38
Map No. 575
560.3
Taraxacum spectabile Dahlst.
Red-veined Marsh Dandelion
Tamworth
NUNEATON
Bedworth
BIRMINGHAM
Solihull
COVENTRY
RUGBY
Kenilworth
ROYAL LEAMINGTON SPA
WARWICK
Southam
Alcester
Stratford on Avon
Kineton
Shipston on Stour
Long Compton
Key to Symbols
Frequencies
rare to occasional
frequent to abundant
Woodland
Hedgerow, Scrub
Acid heath, Bog
Grassland
Waterside, Marsh
Water
Roadsides, Waste places
Quarries, Walls
Cultivated
Scale
Kilometres
0 2 4 6 8 10
Miles
0 1 2 3 4 5 6

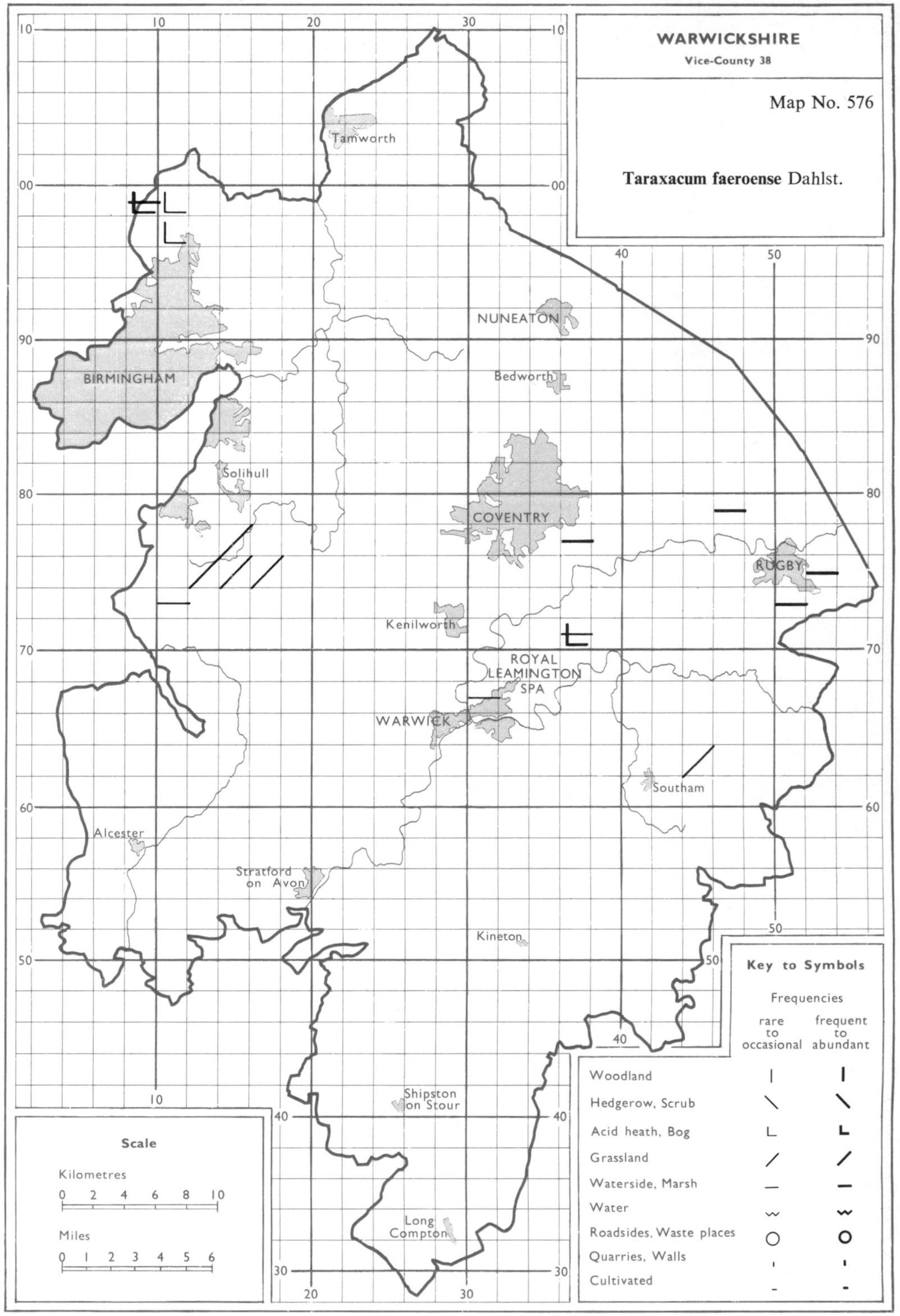
WARWICKSHIRE
Vice-County 38
Map No. 576
Taraxacum faeroense Dahlst.
Tamworth
NUNEATON
Bedworth
BIRMINGHAM
Solihull
COVENTRY
RUGBY
Kenilworth
ROYAL LEAMINGTON SPA
WARWICK
Southam
Alcester
Stratford on Avon
Kineton
Shipston on Stour
Long Compton
Key to Symbols
Frequencies
rare to occasional
frequent to abundant
Woodland
Hedgerow, Scrub
Acid heath, Bog
Grassland
Waterside, Marsh
Water
Roadsides, Waste places
Quarries, Walls
Cultivated
Scale
Kilometres
0 2 4 6 8 10
Miles
0 1 2 3 4 5 6

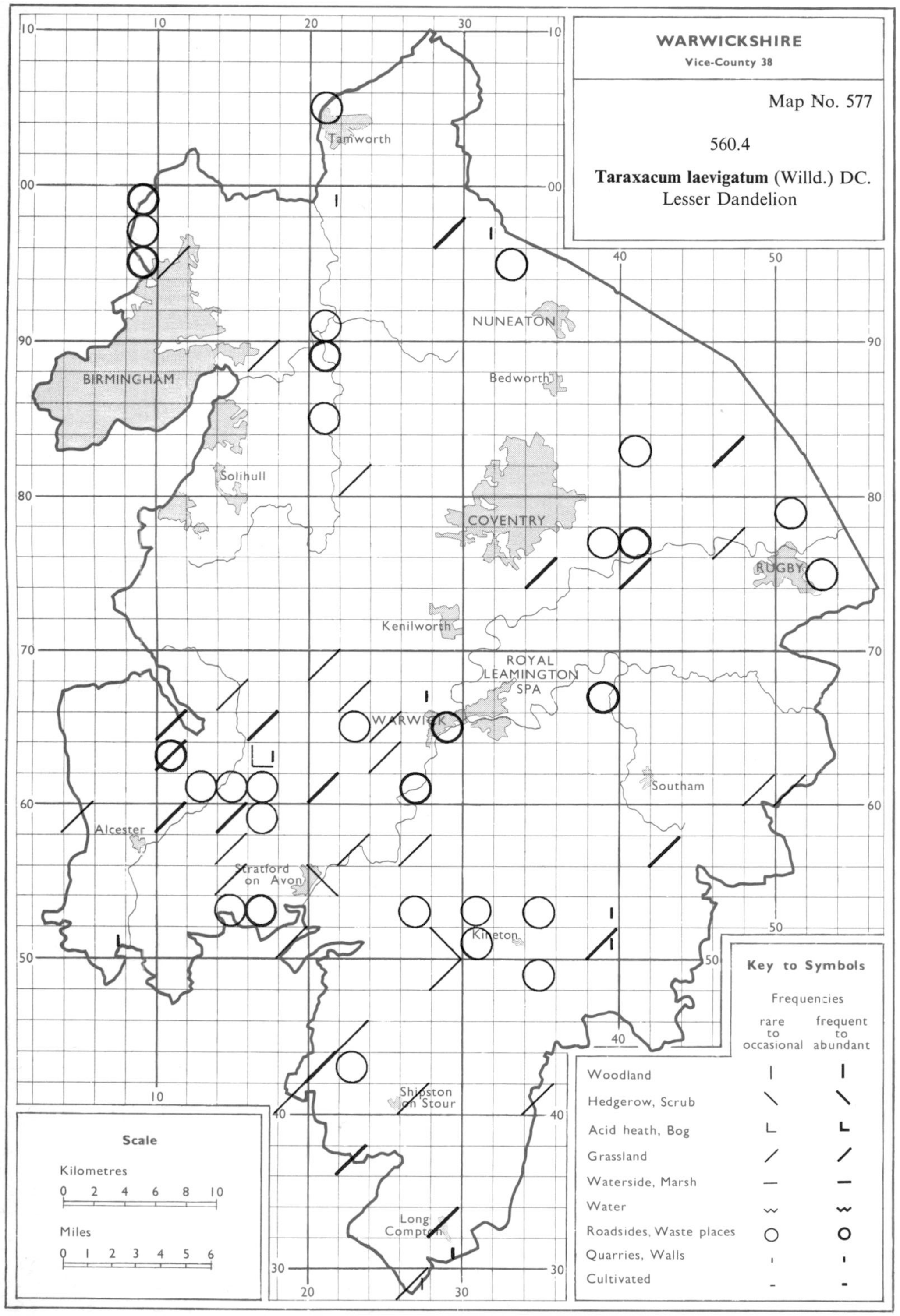

WARWICKSHIRE
Vice-County 38
Map No. 577
560.4
Taraxacum laevigatum (Willd.) DC.
Lesser Dandelion
Tamworth
NUNEATON
Bedworth
BIRMINGHAM
Solihull
COVENTRY
RUGBY
Kenilworth
ROYAL LEAMINGTON SPA
WARWICK
Southam
Alcester
Stratford on Avon
Kineton
Shipston on Stour
Long Compton
Key to Symbols
Frequencies
rare to occasional
frequent to abundant
Woodland
Hedgerow, Scrub
Acid heath, Bog
Grassland
Waterside, Marsh
Water
Roadsides, Waste places
Quarries, Walls
Cultivated
Scale
Kilometres
0 2 4 6 8 10
Miles
0 1 2 3 4 5 6

WARWICKSHIRE
Vice-County 38
Map No. 578
563.1
Alisma plantago-aquatica L.
Common Water-plantain
Tamworth
NUNEATON
Bedworth
BIRMINGHAM
Solihull
COVENTRY
RUGBY
Kenilworth
ROYAL LEAMINGTON SPA
WARWICK
Southam
Alcester
Stratford on Avon
Kineton
Shipston on Stour
Long Compton
Key to Symbols
Frequencies
rare to occasional
frequent to abundant
Woodland
Hedgerow, Scrub
Acid heath, Bog
Grassland
Waterside, Marsh
Water
Roadsides, Waste places
Quarries, Walls
Cultivated
Scale
Kilometres
0 2 4 6 8 10
Miles
0 1 2 3 4 5 6

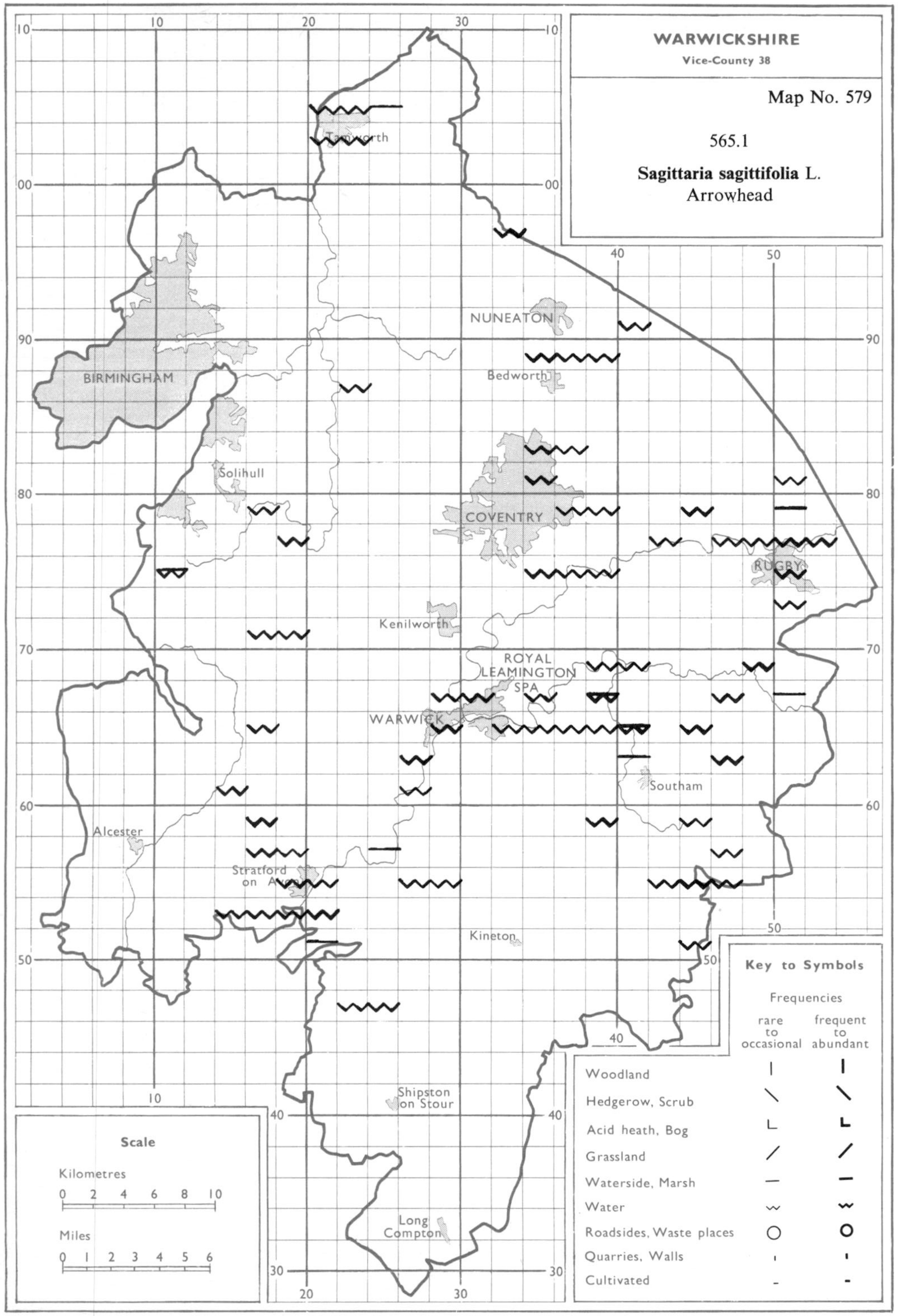
WARWICKSHIRE
Vice-County 38
Map No. 579
565.1
Sagittaria sagittifolia L.
Arrowhead
Tamworth
NUNEATON
Bedworth
BIRMINGHAM
Solihull
COVENTRY
RUGBY
Kenilworth
ROYAL LEAMINGTON SPA
WARWICK
Southam
Alcester
Stratford on Avon
Kineton
Shipston on Stour
Long Compton
Key to Symbols
Frequencies
rare to occasional
frequent to abundant
Woodland
Hedgerow, Scrub
Acid heath, Bog
Grassland
Waterside, Marsh
Water
Roadsides, Waste places
Quarries, Walls
Cultivated
Scale
Kilometres
0 2 4 6 8 10
Miles
0 1 2 3 4 5 6

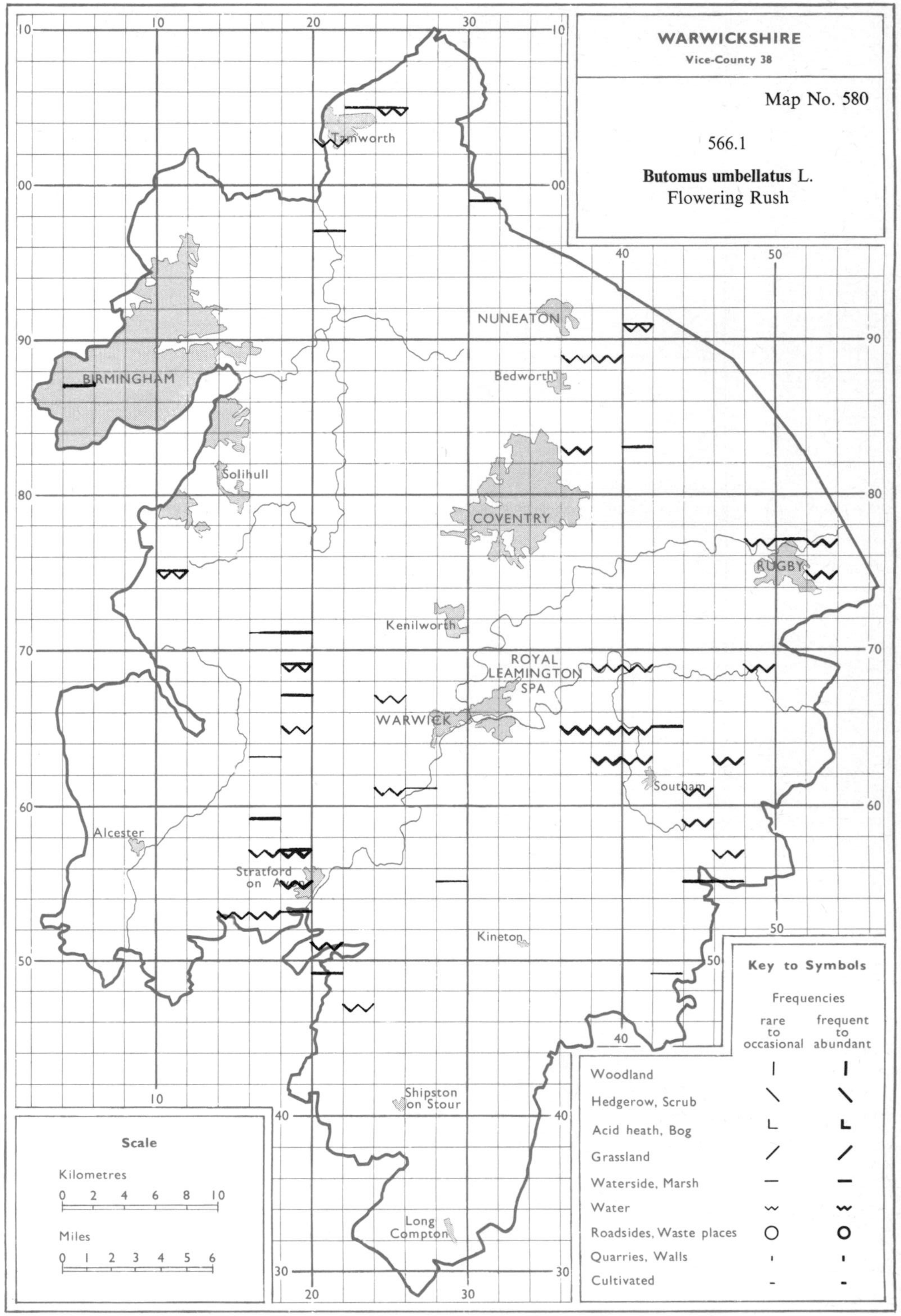
WARWICKSHIRE
Vice-County 38
Map No. 580
566.1
Butomus umbellatus L.
Flowering Rush
Tamworth
NUNEATON
Bedworth
BIRMINGHAM
Solihull
COVENTRY
RUGBY
Kenilworth
ROYAL LEAMINGTON SPA
WARWICK
Southam
Alcester
Stratford on Avon
Kineton
Shipston on Stour
Long Compton
Key to Symbols
Frequencies
rare to occasional
frequent to abundant
Woodland
Hedgerow, Scrub
Acid heath, Bog
Grassland
Waterside, Marsh
Water
Roadsides, Waste places
Quarries, Walls
Cultivated
Scale
Kilometres
0 2 4 6 8 10
Miles
0 1 2 3 4 5 6

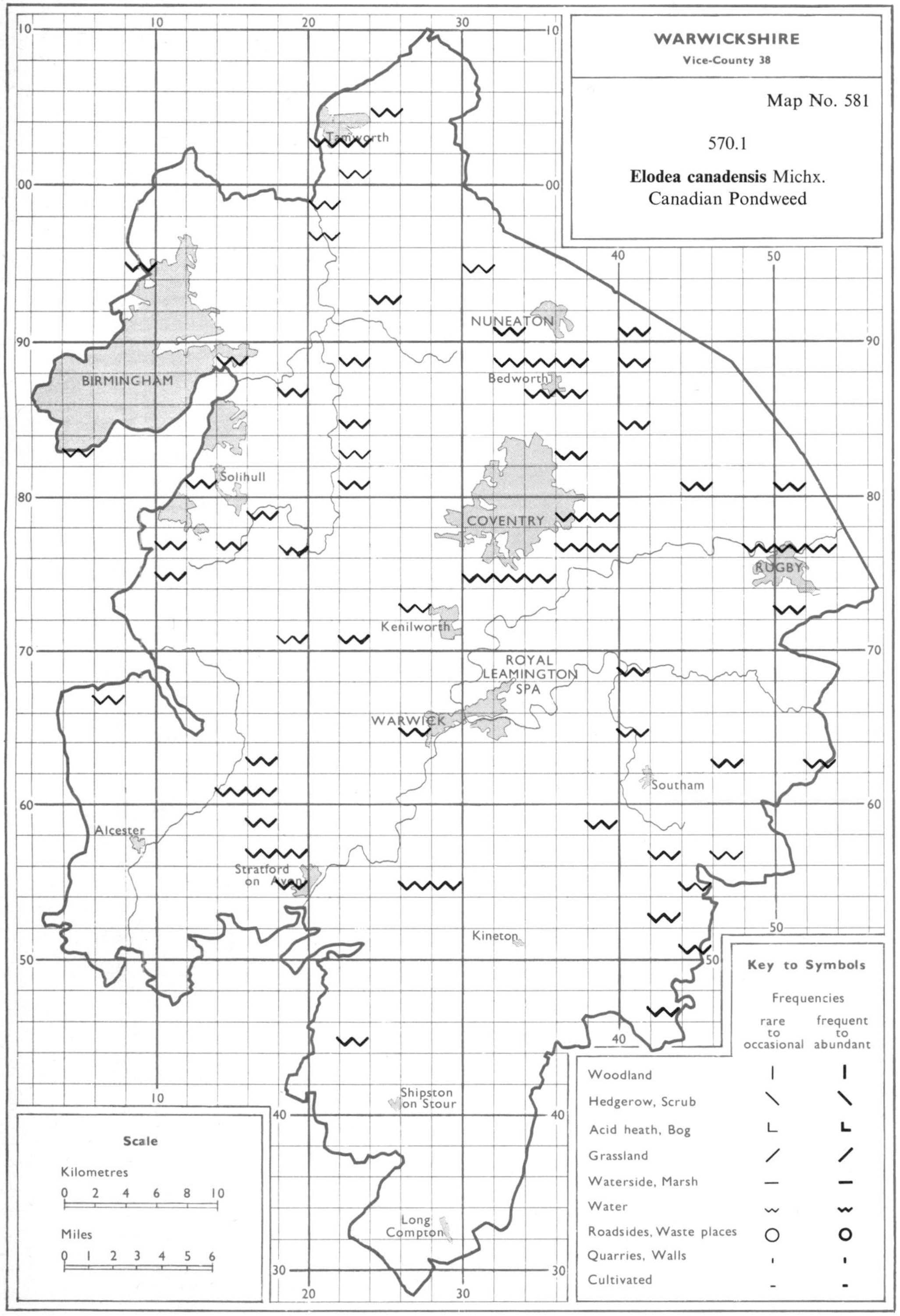
WARWICKSHIRE
Vice-County 38
Map No. 581
570.1
Elodea canadensis Michx.
Canadian Pondweed
Tamworth
NUNEATON
Bedworth
BIRMINGHAM
Solihull
COVENTRY
RUGBY
Kenilworth
ROYAL LEAMINGTON SPA
WARWICK
Southam
Alcester
Stratford on Avon
Kineton
Shipston on Stour
Long Compton
Key to Symbols
Frequencies
rare to occasional
frequent to abundant
Woodland
Hedgerow, Scrub
Acid heath, Bog
Grassland
Waterside, Marsh
Water
Roadsides, Waste places
Quarries, Walls
Cultivated
Scale
Kilometres
0 2 4 6 8 10
Miles
0 1 2 3 4 5 6

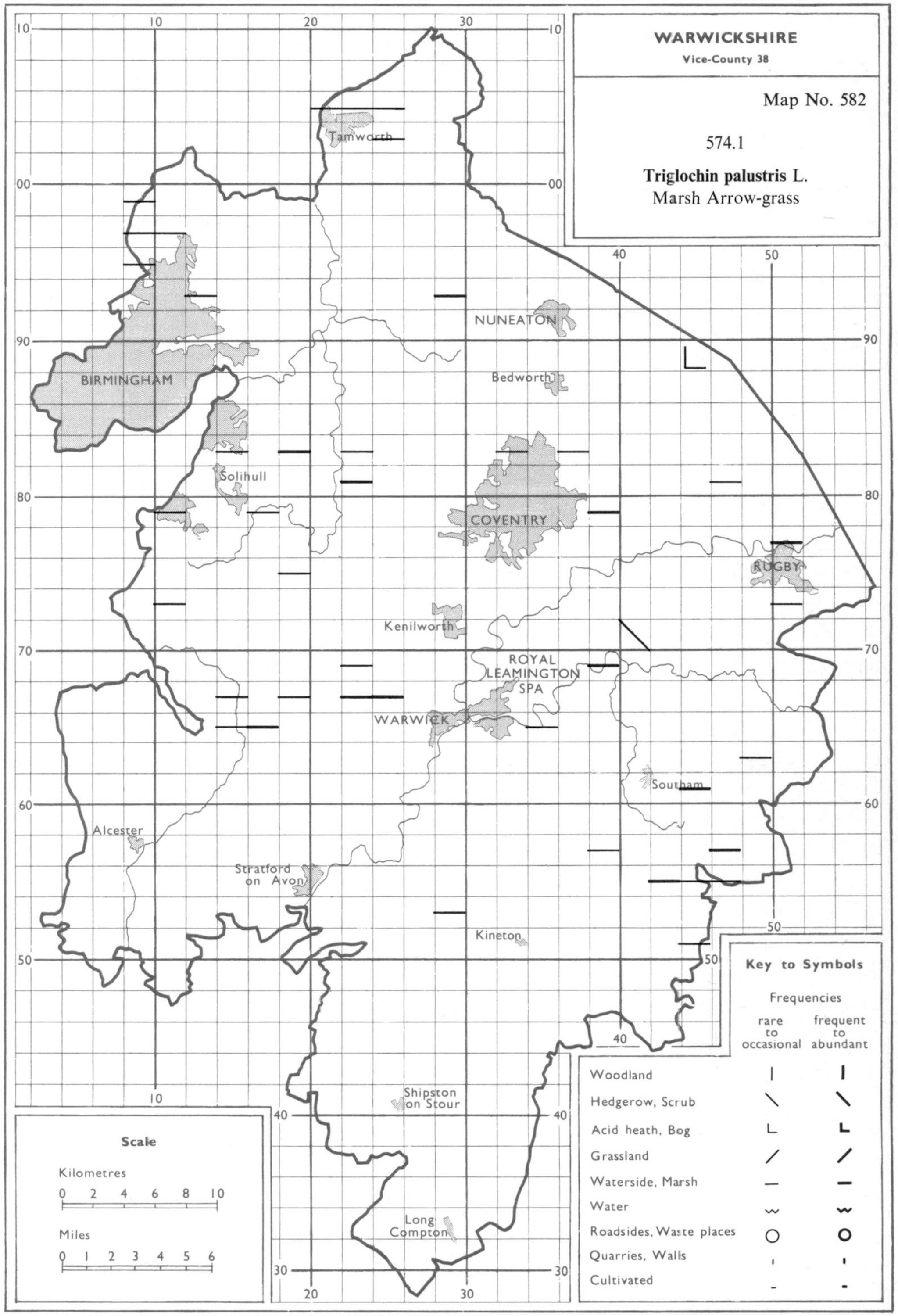
WARWICKSHIRE
Vice-County 38
Map No. 582
574.1
Triglochin palustris L.
Marsh Arrow-grass
Tamworth
NUNEATON
Bedworth
BIRMINGHAM
Solihull
COVENTRY
RUGBY
Kenilworth
ROYAL LEAMINGTON SPA
WARWICK
Southam
Alcester
Stratford on Avon
Kineton
Shipston on Stour
Long Compton
Key to Symbols
Frequencies
rare to occasional
frequent to abundant
Woodland
Hedgerow, Scrub
Acid heath, Bog
Grassland
Waterside, Marsh
Water
Roadsides, Waste places
Quarries, Walls
Cultivated
Scale
Kilometres
0 2 4 6 8 10
Miles
0 1 2 3 4 5 6

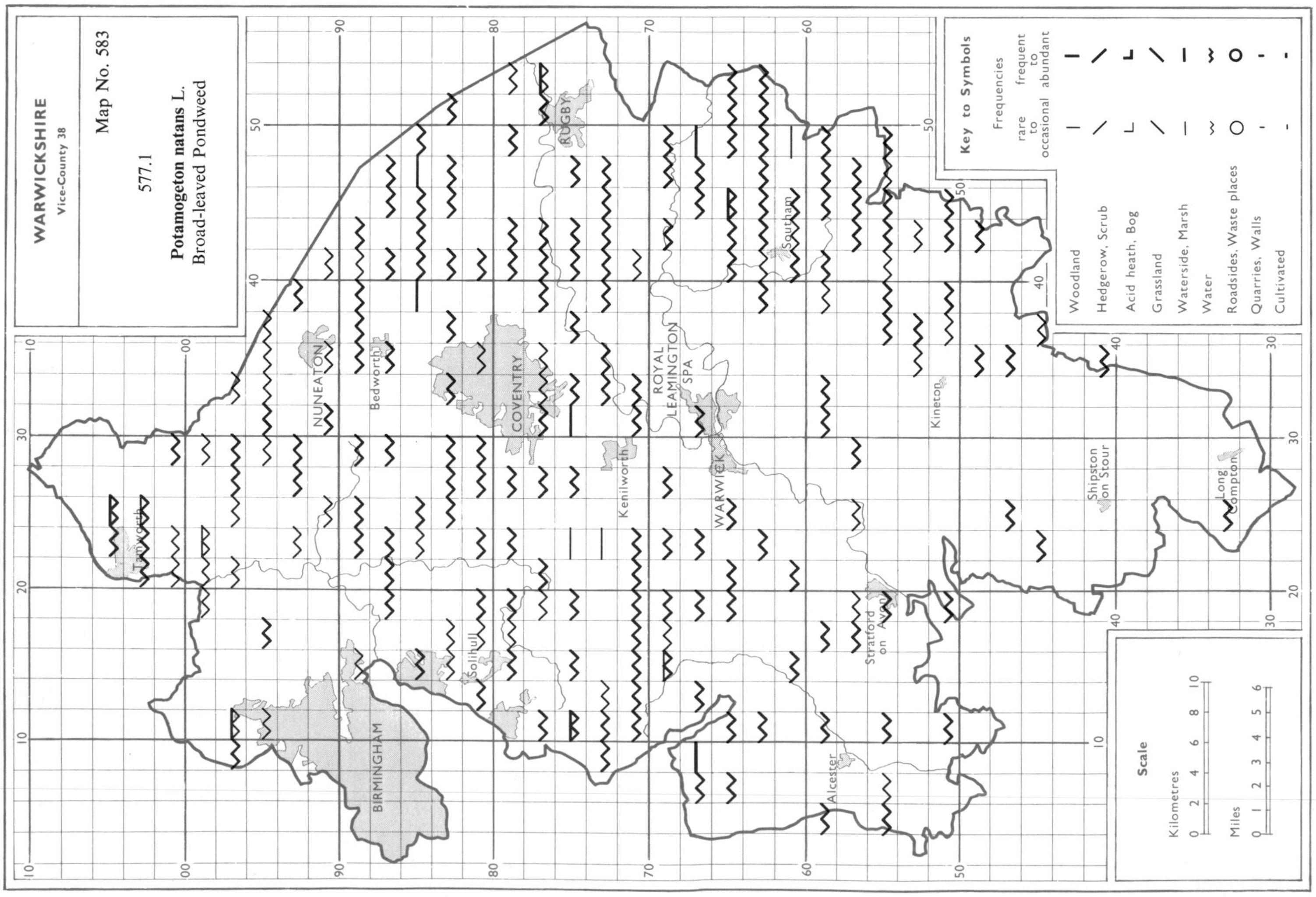
WARWICKSHIRE
Vice-County 38
Map No. 583
577.1
Potamogeton natans L.
Broad-leaved Pondweed
Key to Symbols
Frequencies
rare to occasional
frequent to abundant
Woodland
Hedgerow, Scrub
Acid heath, Bog
Grassland
Waterside, Marsh
Water
Roadsides, Waste places
Quarries, Walls
Cultivated
Scale
Kilometres
0 2 4 6 8 10
Miles
0 1 2 3 4 5 6
NUNEATON
Bedworth
COVENTRY
RUGBY
Southam
ROYAL LEAMINGTON SPA
WARWICK
Kenilworth
Kineton
Shipston on Stour
Long Compton
Tamworth
Solihull
BIRMINGHAM
Stratford on Avon
Alcester

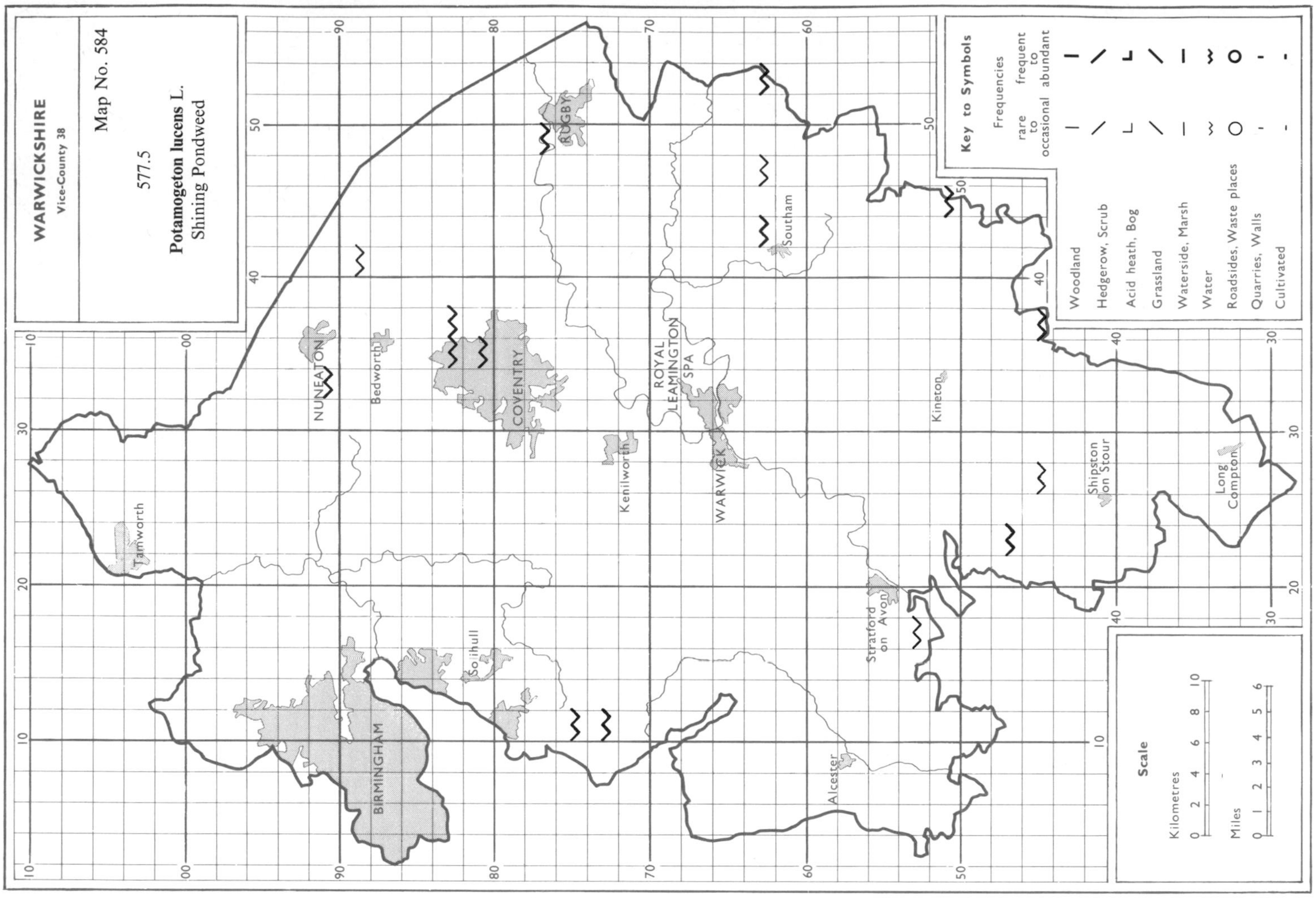
WARWICKSHIRE
Vice-County 38
Map No. 584
577.5
Potamogeton lucens L.
Shining Pondweed
Key to Symbols
Frequencies
rare to occasional
frequent to abundant
Woodland
Hedgerow, Scrub
Acid heath, Bog
Grassland
Waterside, Marsh
Water
Roadsides, Waste places
Quarries, Walls
Cultivated
Scale
Kilometres
0 2 4 6 8 10
Miles
0 1 2 3 4 5 6
NUNEATON
Bedworth
COVENTRY
RUGBY
Southam
ROYAL LEAMINGTON SPA
WARWICK
Kenilworth
Kineton
Shipston on Stour
Long Compton
Tamworth
Solihull
BIRMINGHAM
Stratford on Avon
Alcester

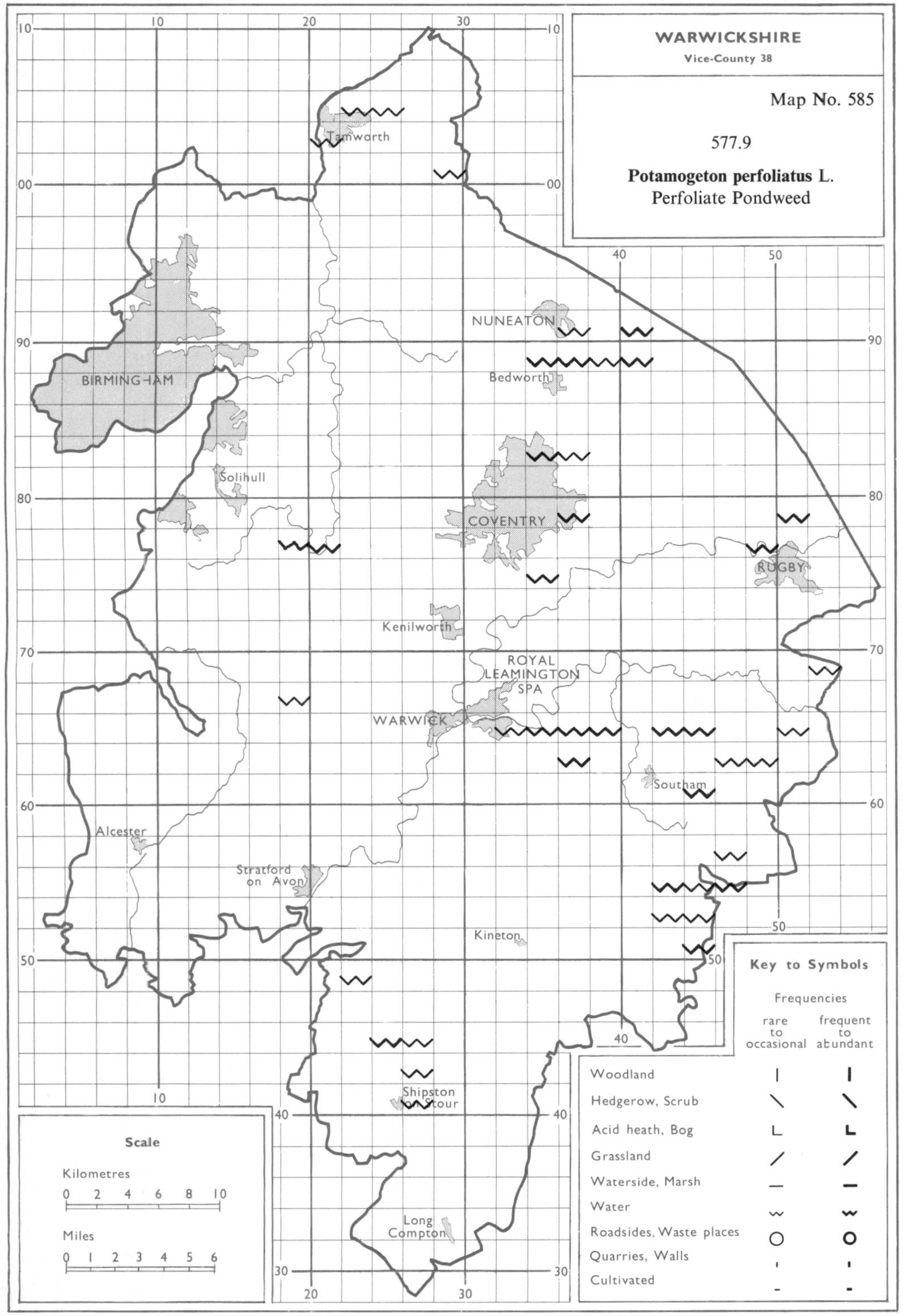
WARWICKSHIRE
Vice-County 38
Map No. 585
577.9
Potamogeton perfoliatus L.
Perfoliate Pondweed
Tamworth
NUNEATON
Bedworth
BIRMINGHAM
Solihull
COVENTRY
RUGBY
Kenilworth
ROYAL LEAMINGTON SPA
WARWICK
Southam
Alcester
Stratford on Avon
Kineton
Shipston on Stour
Long Compton
Key to Symbols
Frequencies
rare to occasional
frequent to abundant
Woodland
Hedgerow, Scrub
Acid heath, Bog
Grassland
Waterside, Marsh
Water
Roadsides, Waste places
Quarries, Walls
Cultivated
Scale
Kilometres
0 2 4 6 8 10
Miles
0 1 2 3 4 5 6

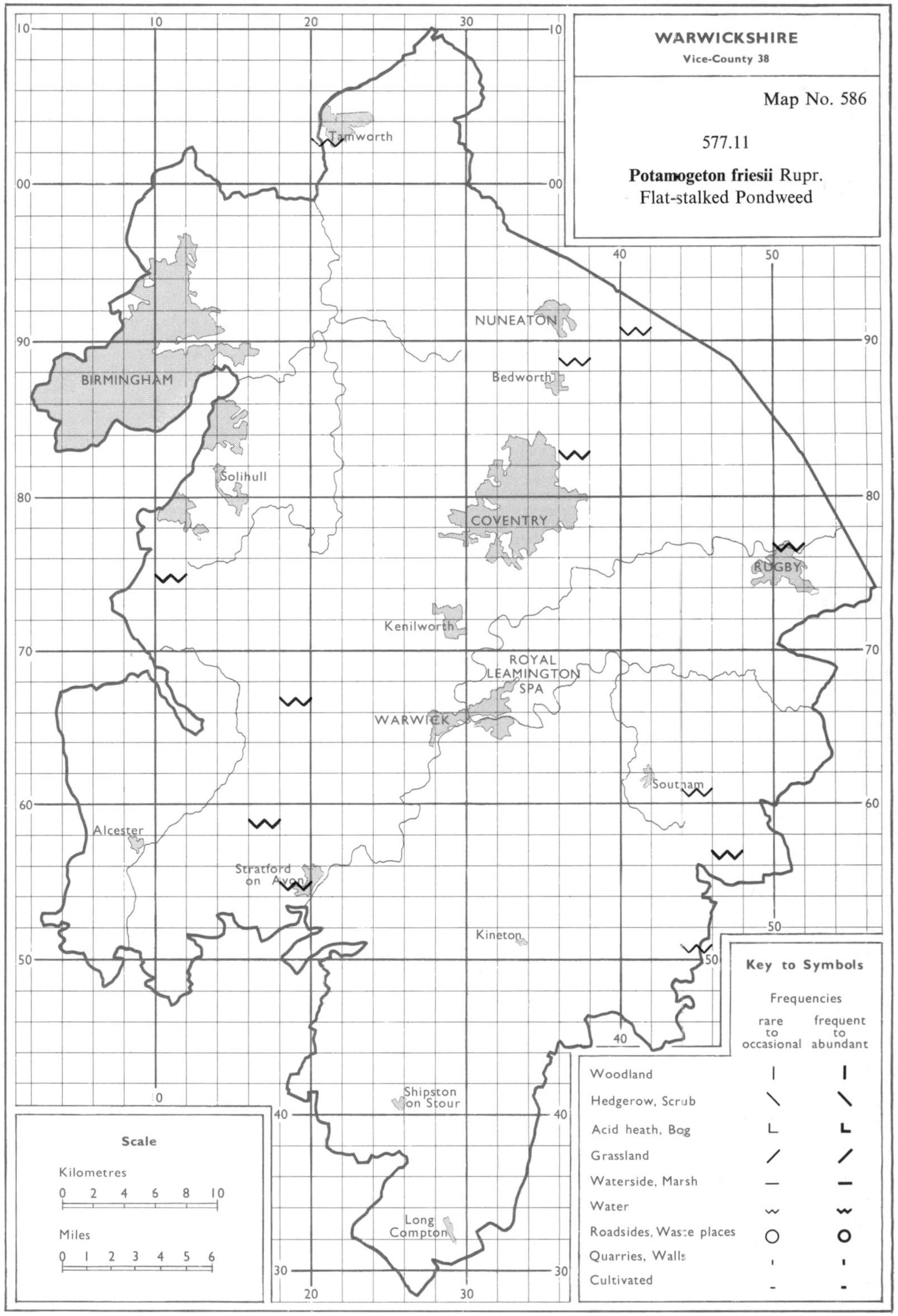
WARWICKSHIRE
Vice-County 38
Map No. 586
577.11
Potamogeton friesii Rupr.
Flat-stalked Pondweed
Tamworth
NUNEATON
Bedworth
BIRMINGHAM
Solihull
COVENTRY
RUGBY
Kenilworth
ROYAL LEAMINGTON SPA
WARWICK
Southam
Alcester
Stratford on Avon
Kineton
Shipston on Stour
Long Compton
Key to Symbols
Frequencies
rare to occasional
frequent to abundant
Woodland
Hedgerow, Scrub
Acid heath, Bog
Grassland
Waterside, Marsh
Water
Roadsides, Waste places
Quarries, Walls
Cultivated
Scale
Kilometres
0 2 4 6 8 10
Miles
0 1 2 3 4 5 6

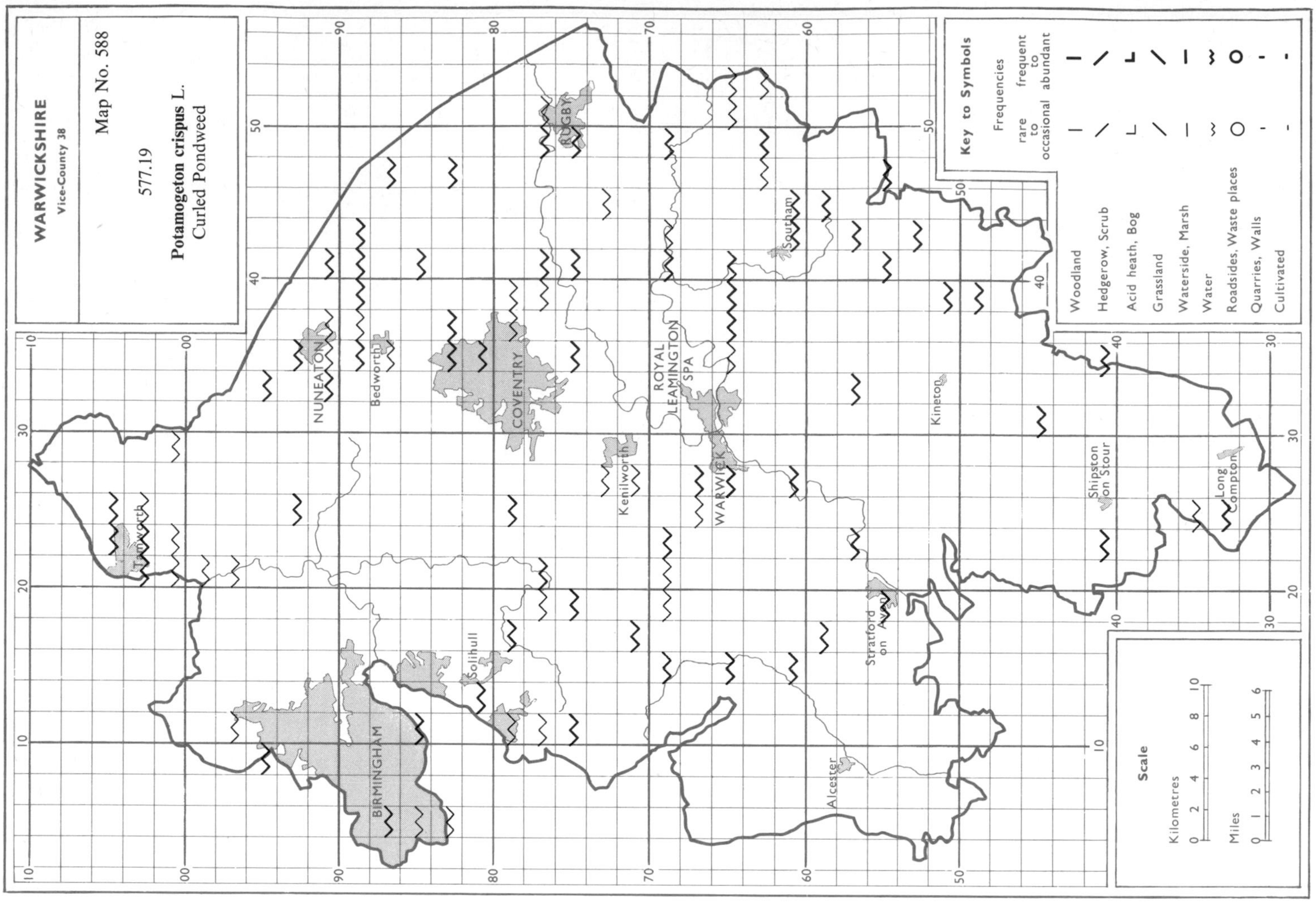
WARWICKSHIRE
Vice-County 38
Map No. 588
577.19
Potamogeton crispus L.
Curled Pondweed
Key to Symbols
Frequencies
rare to occasional
frequent to abundant
Woodland
Hedgerow, Scrub
Acid heath, Bog
Grassland
Waterside, Marsh
Water
Roadsides, Waste places
Quarries, Walls
Cultivated
Scale
Kilometres
0 2 4 6 8 10
Miles
0 1 2 3 4 5 6
NUNEATON
Bedworth
COVENTRY
RUGBY
Southam
ROYAL LEAMINGTON SPA
Kineton
Kenilworth
WARWICK
Shipston on Stour
Long Compton
Tamworth
Solihull
BIRMINGHAM
Stratford on Avon
Alcester

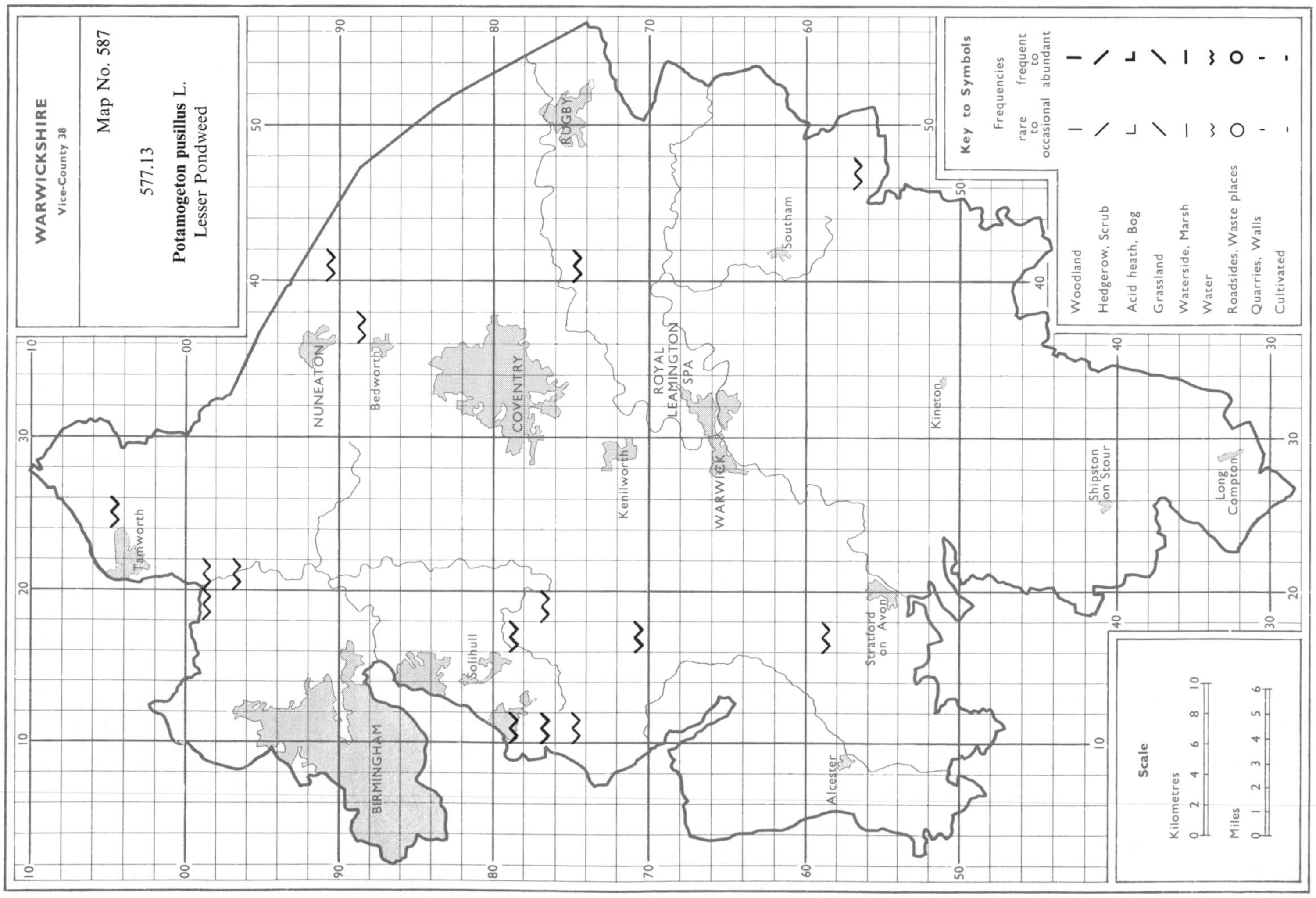
WARWICKSHIRE
Vice-County 38
Map No. 587
577.13
Potamogeton pusillus L.
Lesser Pondweed
Key to Symbols
Frequencies
rare to occasional
frequent to abundant
Woodland
Hedgerow, Scrub
Acid heath, Bog
Grassland
Waterside, Marsh
Water
Roadsides, Waste places
Quarries, Walls
Cultivated
Scale
Kilometres
0 2 4 6 8 10
Miles
0 1 2 3 4 5 6
NUNEATON
Bedworth
COVENTRY
RUGBY
Southam
ROYAL LEAMINGTON SPA
Kineton
Kenilworth
WARWICK
Shipston on Stour
Long Compton
Tamworth
Solihull
BIRMINGHAM
Stratford on Avon
Alcester

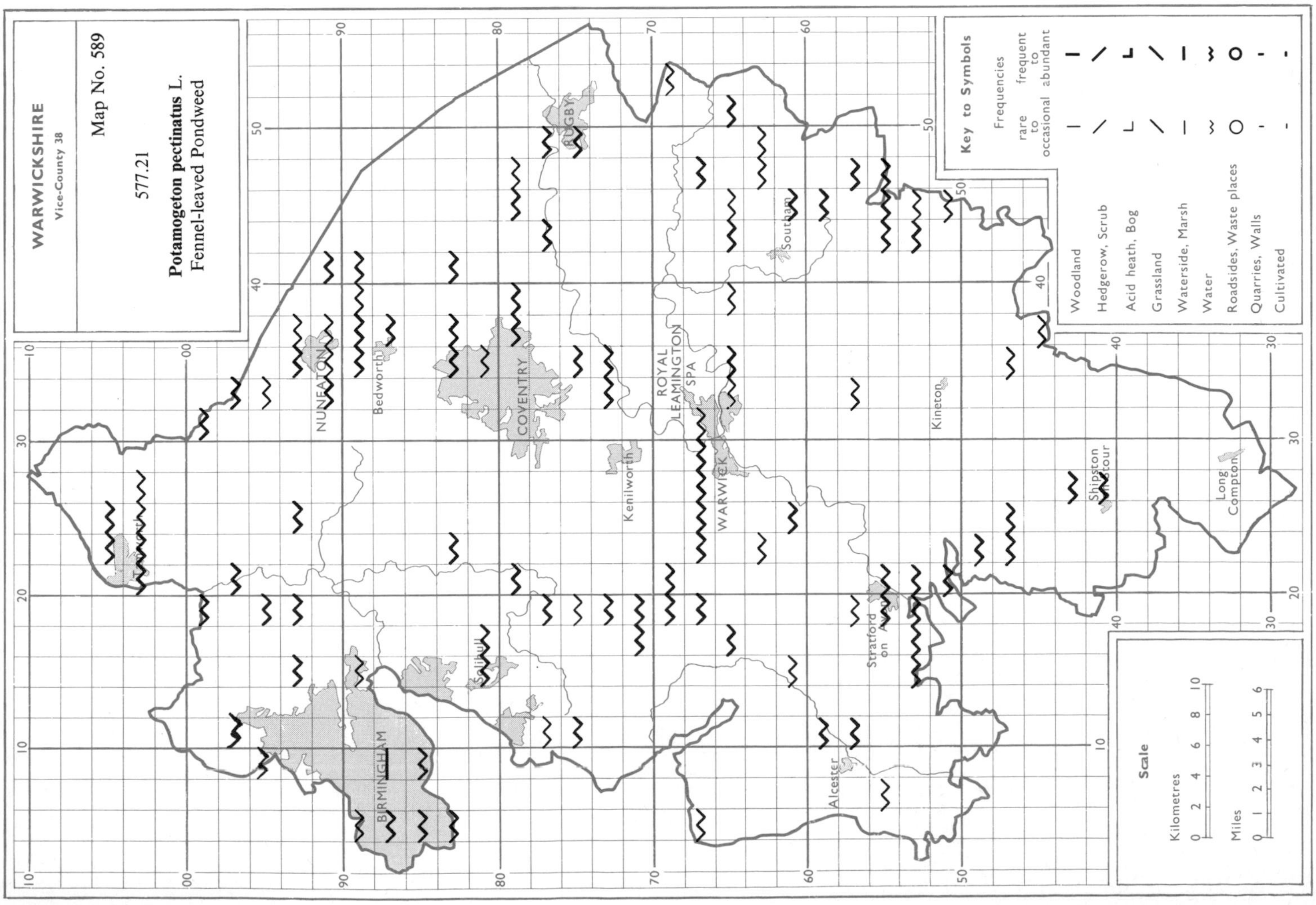
WARWICKSHIRE
Vice-County 38
Map No. 589
577.21
Potamogeton pectinatus L.
Fennel-leaved Pondweed
Key to Symbols
Frequencies
rare to occasional
frequent to abundant
Woodland
Hedgerow, Scrub
Acid heath, Bog
Grassland
Waterside, Marsh
Water
Roadsides, Waste places
Quarries, Walls
Cultivated
Scale
Kilometres
0 2 4 6 8 10
Miles
0 1 2 3 4 5 6
NUNEATON
Bedworth
COVENTRY
RUGBY
ROYAL LEAMINGTON SPA
WARWICK
Kenilworth
Southam
Kineton
Shipston on Stour
Long Compton
Stratford on Avon
Alcester
Solihull
BIRMINGHAM
Tamworth

WARWICKSHIRE
Vice-County 38
Map No. 590
580.1
Zannichellia palustris L.
Horned Pondweed
Key to Symbols
Frequencies
rare to occasional
frequent to abundant
Woodland
Hedgerow, Scrub
Acid heath, Bog
Grassland
Waterside, Marsh
Water
Roadsides, Waste places
Quarries, Walls
Cultivated
Scale
Kilometres
0 2 4 6 8 10
Miles
0 1 2 3 4 5 6
NUNEATON
Bedworth
COVENTRY
RUGBY
ROYAL LEAMINGTON SPA
WARWICK
Kenilworth
Southam
Kineton
Shipston on Stour
Long Compton
Stratford on Avon
Alcester
Solihull
BIRMINGHAM
Tamworth

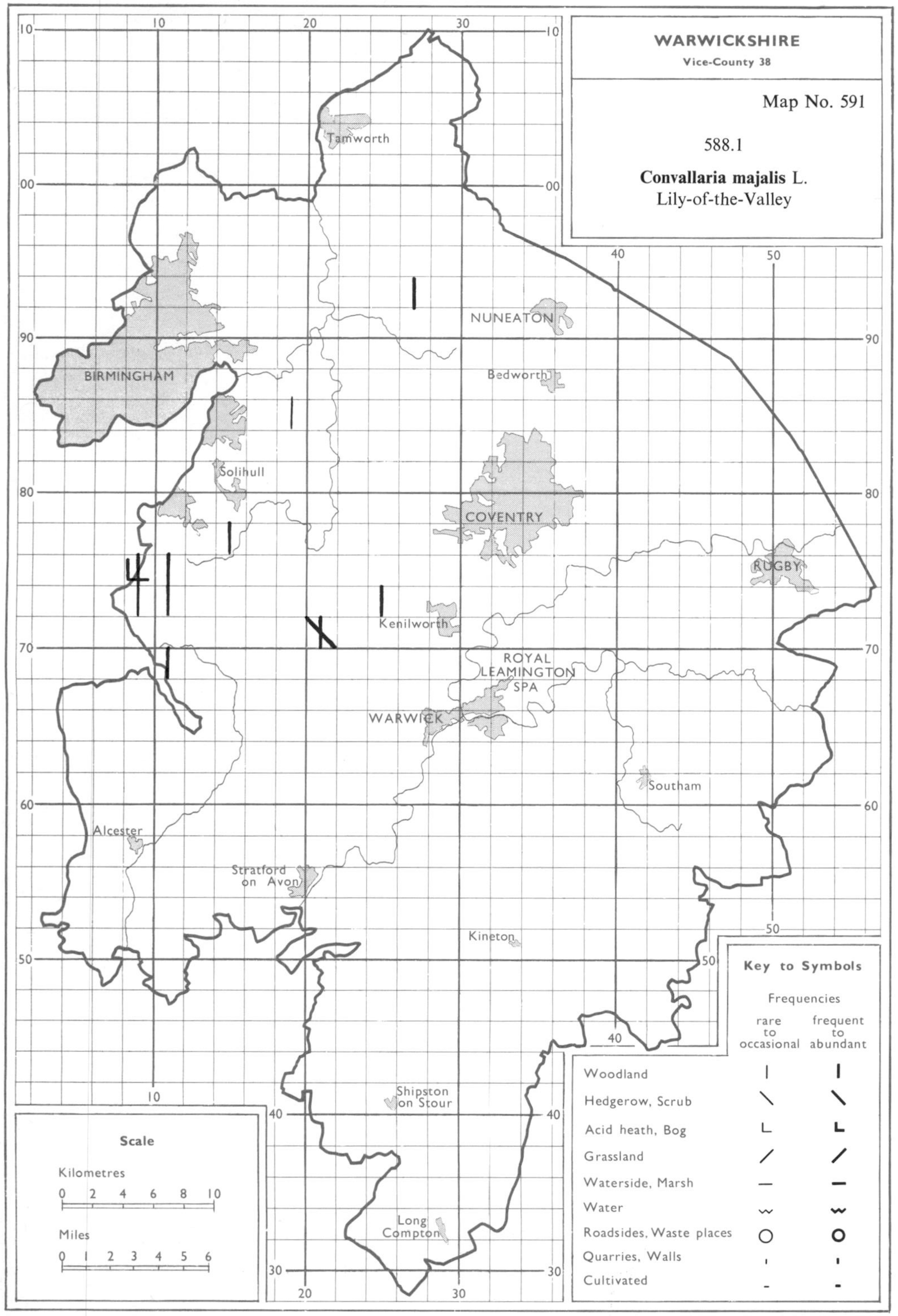
WARWICKSHIRE
Vice-County 38
Map No. 591
588.1
Convallaria majalis L.
Lily-of-the-Valley
Tamworth
BIRMINGHAM
NUNEATON
Bedworth
Solihull
COVENTRY
RUGBY
Kenilworth
ROYAL LEAMINGTON SPA
WARWICK
Southam
Alcester
Stratford on Avon
Kineton
Shipston on Stour
Long Compton
Key to Symbols
Frequencies
rare to occasional
frequent to abundant
Woodland
Hedgerow, Scrub
Acid heath, Bog
Grassland
Waterside, Marsh
Water
Roadsides, Waste places
Quarries, Walls
Cultivated
Scale
Kilometres
0 2 4 6 8 10
Miles
0 1 2 3 4 5 6

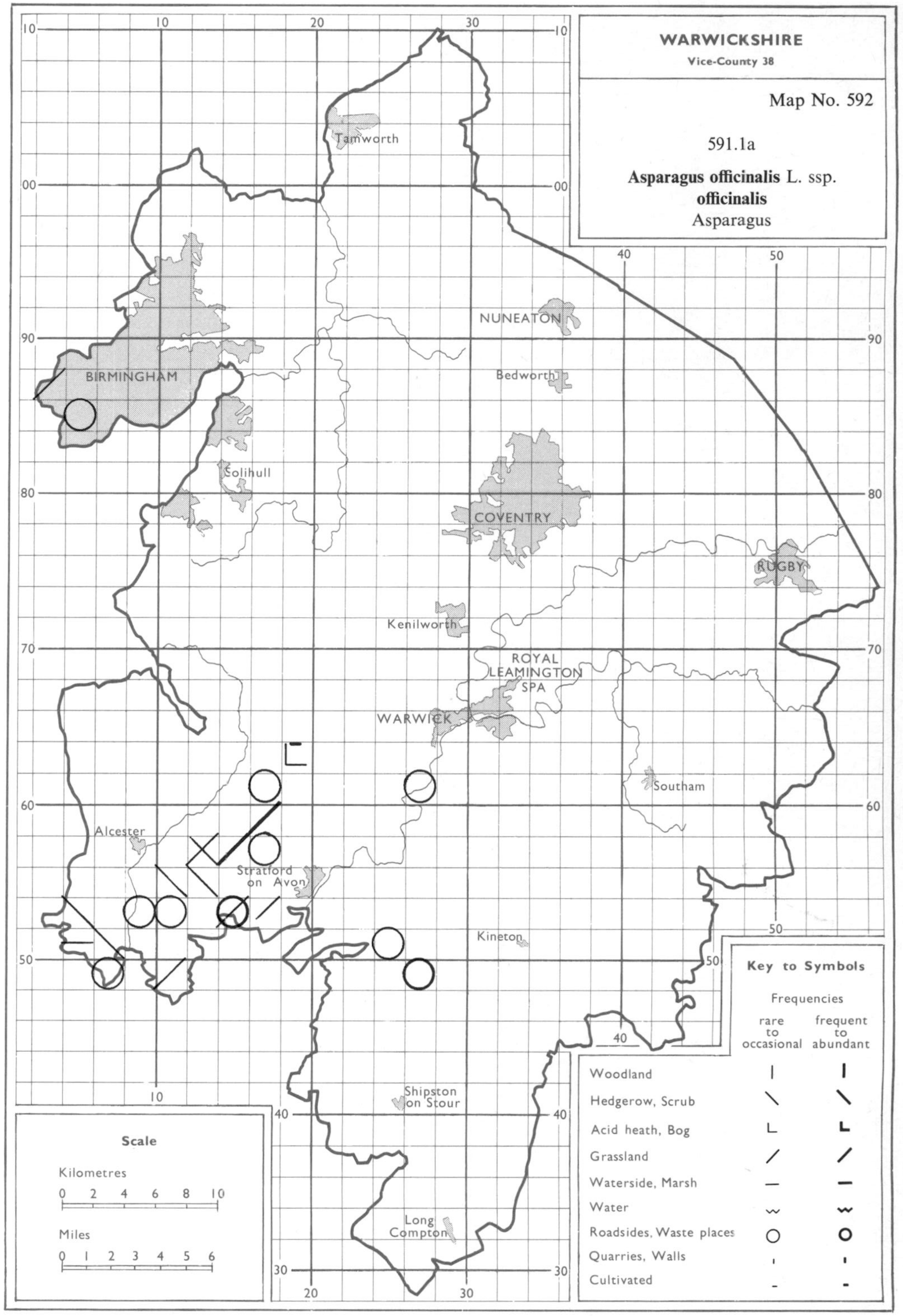
WARWICKSHIRE
Vice-County 38
Map No. 592
591.1a
Asparagus officinalis L. ssp. officinalis
Asparagus
Tamworth
BIRMINGHAM
NUNEATON
Bedworth
Solihull
COVENTRY
RUGBY
Kenilworth
ROYAL LEAMINGTON SPA
WARWICK
Southam
Alcester
Stratford on Avon
Kineton
Shipston on Stour
Long Compton
Key to Symbols
Frequencies
rare to occasional
frequent to abundant
Woodland
Hedgerow, Scrub
Acid heath, Bog
Grassland
Waterside, Marsh
Water
Roadsides, Waste places
Quarries, Walls
Cultivated
Scale
Kilometres
0 2 4 6 8 10
Miles
0 1 2 3 4 5 6

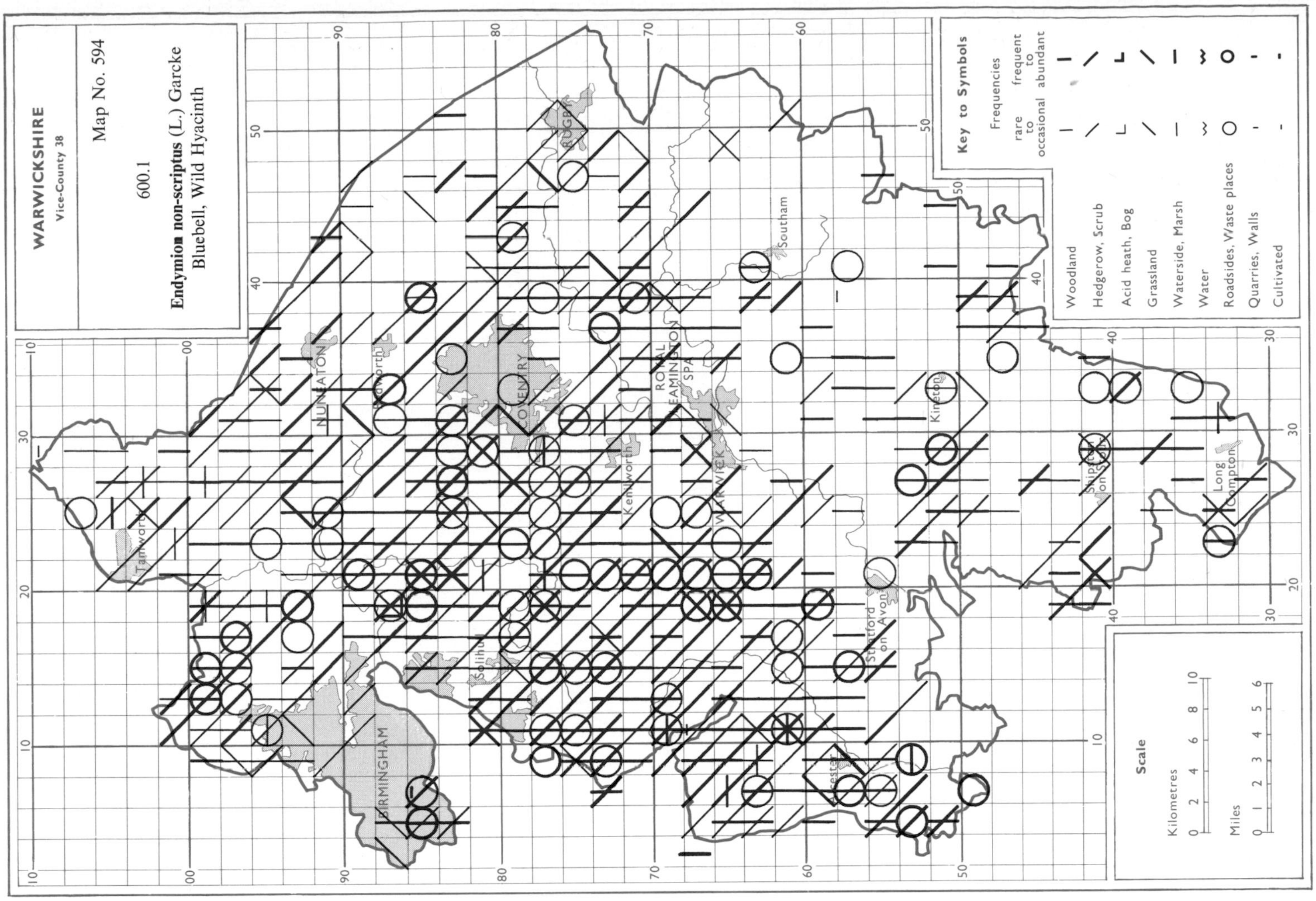
WARWICKSHIRE
Vice-County 38
Map No. 594
600.1
Endymion non-scriptus (L.) Garcke
Bluebell, Wild Hyacinth
Key to Symbols
Frequencies
rare to occasional
frequent to abundant
Woodland
Hedgerow, Scrub
Acid heath, Bog
Grassland
Waterside, Marsh
Water
Roadsides, Waste places
Quarries, Walls
Cultivated
Scale
Kilometres
Miles
Tamworth
NUNEATON
Bedworth
COVENTRY
RUGBY
BIRMINGHAM
Solihull
Kenilworth
ROYAL LEAMINGTON SPA
WARWICK
Southam
Kineton
Stratford on Avon
Shipston on Stour
Long Compton
Alcester

WARWICKSHIRE
Vice-County 38
Map No. 593
598.1
Ornithogalum umbellatum L.
Common Star-of-Bethlehem
Key to Symbols
Frequencies
rare to occasional
frequent to abundant
Woodland
Hedgerow, Scrub
Acid heath, Bog
Grassland
Waterside, Marsh
Water
Roadsides, Waste places
Quarries, Walls
Cultivated
Scale
Kilometres
Miles
Tamworth
NUNEATON
Bedworth
COVENTRY
RUGBY
BIRMINGHAM
Solihull
Kenilworth
ROYAL LEAMINGTON SPA
WARWICK
Southam
Kineton
Stratford on Avon
Shipston on Stour
Long Compton
Alcester

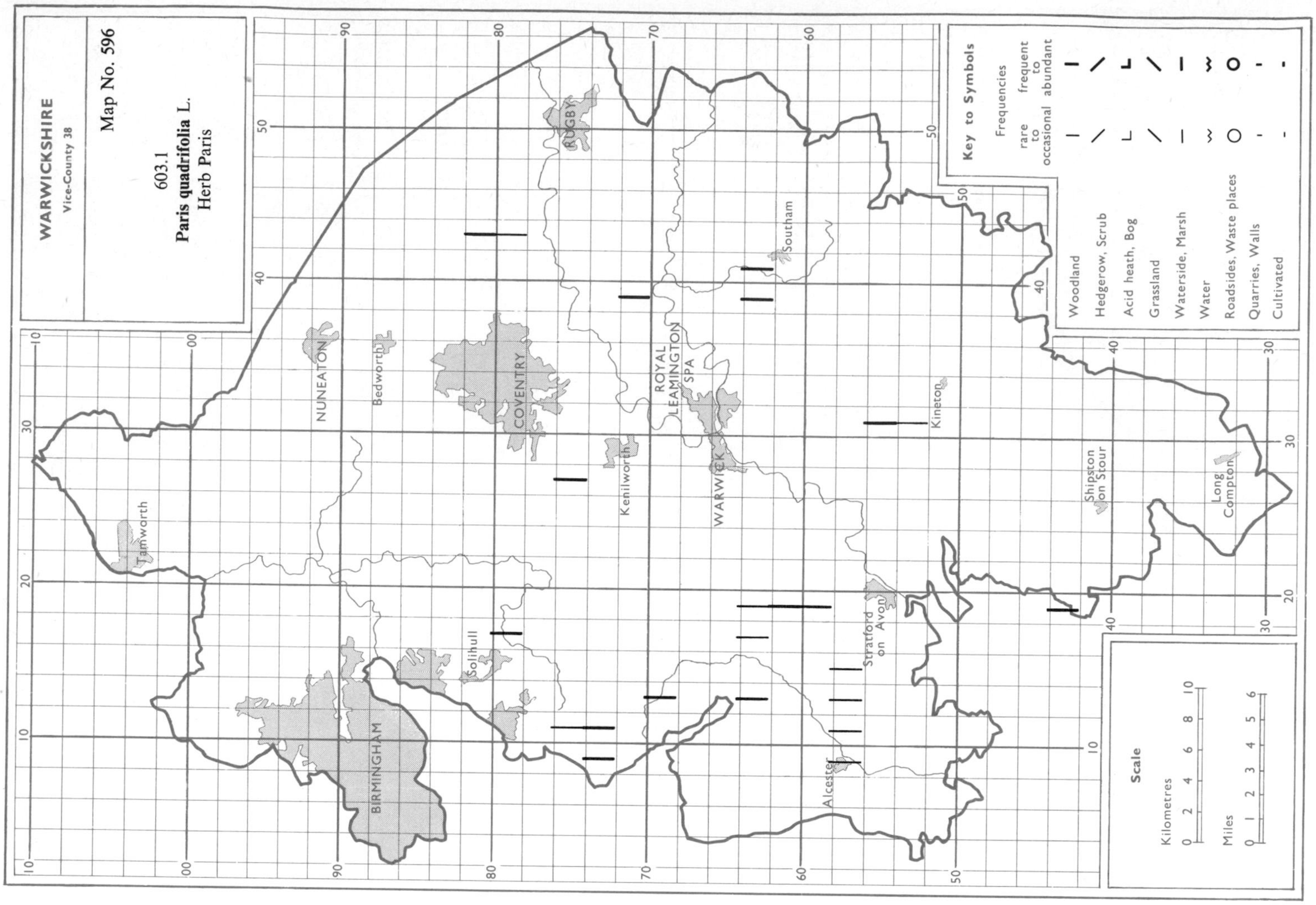
WARWICKSHIRE
Vice-County 38
Map No. 596
603.1
Paris quadrifolia L.
Herb Paris
Key to Symbols
Frequencies
rare to occasional
frequent to abundant
Woodland
Hedgerow, Scrub
Acid heath, Bog
Grassland
Waterside, Marsh
Water
Roadsides, Waste places
Quarries, Walls
Cultivated
Scale
Kilometres
0 2 4 6 8 10
Miles
0 1 2 3 4 5 6
NUNEATON
Bedworth
COVENTRY
RUGBY
Southam
ROYAL LEAMINGTON SPA
WARWICK
Kenilworth
Kineton
Shipston on Stour
Long Compton
Tamworth
Solihull
BIRMINGHAM
Stratford on Avon
Alcester

WARWICKSHIRE
Vice-County 38
Map No. 595
602.1
Colchicum autumnale L.
Meadow Saffron, Autumn Crocus
Key to Symbols
Frequencies
rare to occasional
frequent to abundant
Woodland
Hedgerow, Scrub
Acid heath, Bog
Grassland
Waterside, Marsh
Water
Roadsides, Waste places
Quarries, Walls
Cultivated
Scale
Kilometres
0 2 4 6 8 10
Miles
0 1 2 3 4 5 6
NUNEATON
Bedworth
COVENTRY
RUGBY
Southam
ROYAL LEAMINGTON SPA
WARWICK
Kenilworth
Kineton
Shipston on Stour
Long Compton
Tamworth
Solihull
BIRMINGHAM
Stratford on Avon
Alcester

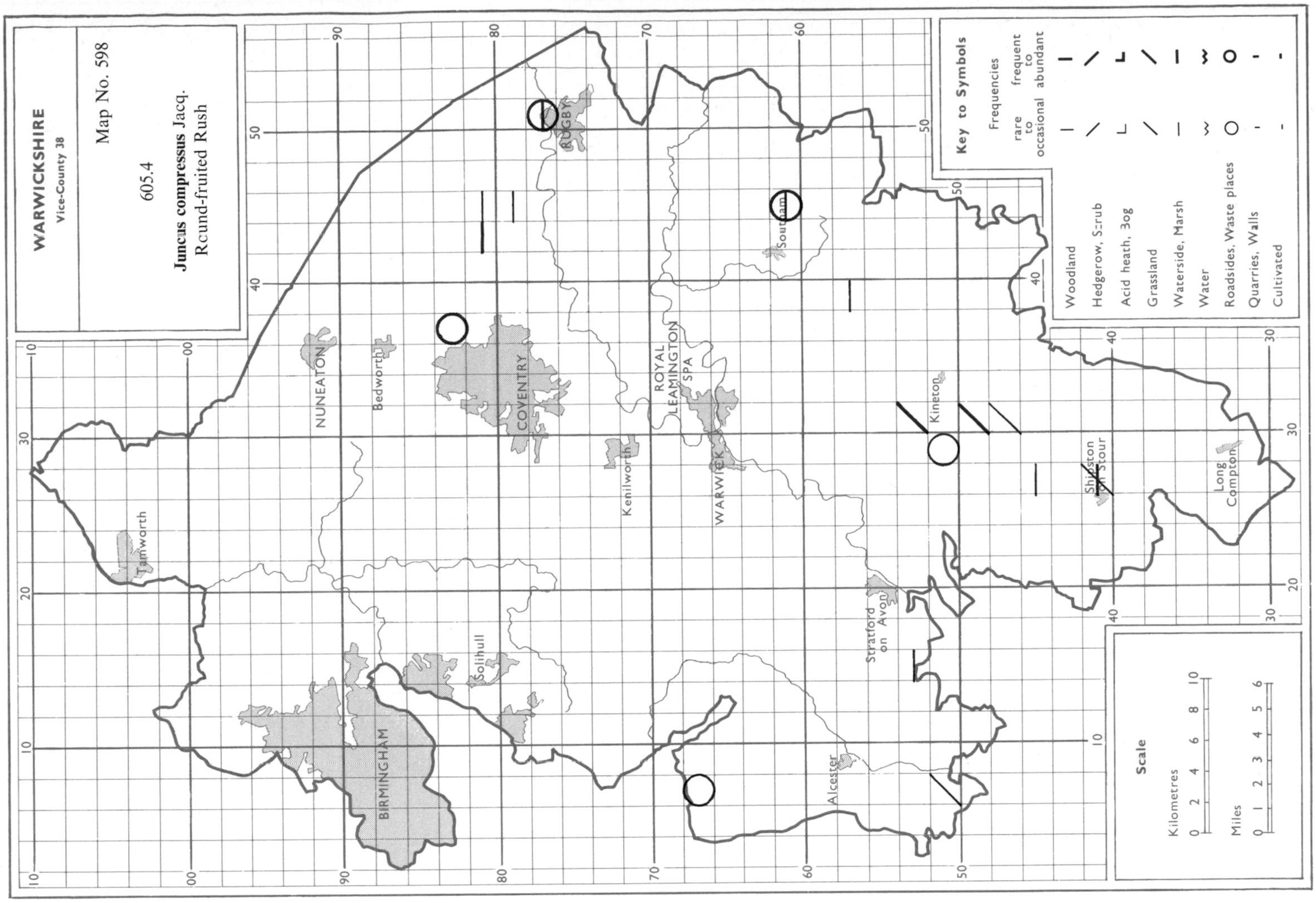

WARWICKSHIRE
Vice-County 38
Map No. 598
605.4
Juncus compressus Jacq.
Round-fruited Rush
Key to Symbols
Frequencies
rare to occasional
frequent to abundant
Woodland
Hedgerow, Scrub
Acid heath, Bog
Grassland
Waterside, Marsh
Water
Roadsides, Waste places
Quarries, Walls
Cultivated
Scale
Kilometres
Miles
NUNEATON
Bedworth
COVENTRY
RUGBY
Southam
ROYAL LEAMINGTON SPA
Kenilworth
WARWICK
Kineton
Shipston on Stour
Long Compton
Tamworth
Solihull
BIRMINGHAM
Stratford on Avon
Alcester

WARWICKSHIRE
Vice-County 38
Map No. 597
605.1
Juncus squarrosus L.
Heath Rush
Key to Symbols
Frequencies
rare to occasional
frequent to abundant
Woodland
Hedgerow, Scrub
Acid heath, Bog
Grassland
Waterside, Marsh
Water
Roadsides, Waste places
Quarries, Walls
Cultivated
Scale
Kilometres
Miles
NUNEATON
Bedworth
COVENTRY
RUGBY
Southam
ROYAL LEAMINGTON SPA
Kenilworth
WARWICK
Kineton
Shipston on Stour
Long Compton
Tamworth
Solihull
BIRMINGHAM
Stratford on Avon
Alcester

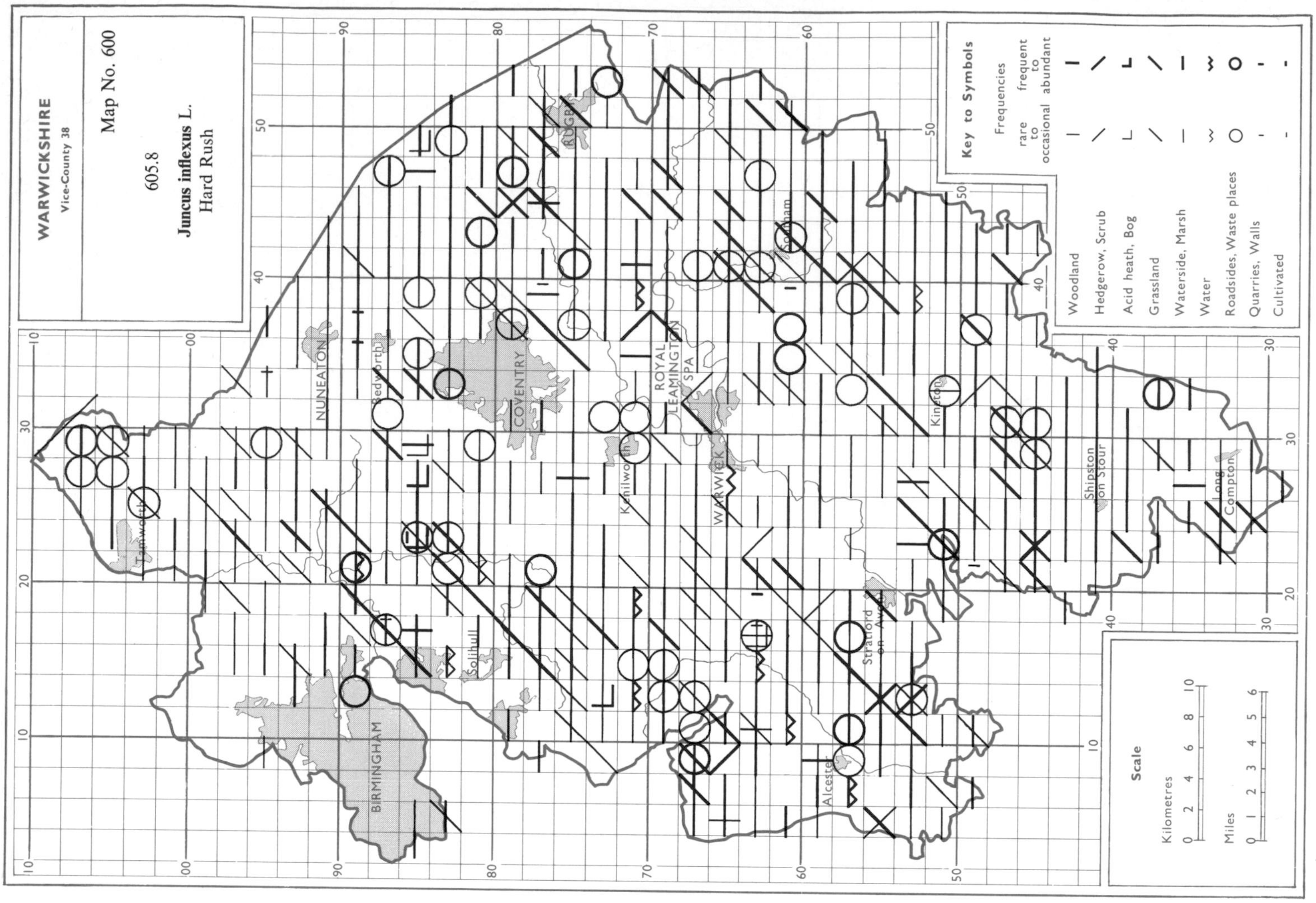
WARWICKSHIRE
Vice-County 38
Map No. 600
605.8
Juncus inflexus L.
Hard Rush
Key to Symbols
Frequencies
rare to occasional
frequent to abundant
Woodland
Hedgerow, Scrub
Acid heath, Bog
Grassland
Waterside, Marsh
Water
Roadsides, Waste places
Quarries, Walls
Cultivated
Scale
Kilometres 0 2 4 6 8 10
Miles 0 1 2 3 4 5 6
Tamworth
NUNEATON
Bedworth
COVENTRY
RUGBY
BIRMINGHAM
Solihull
Kenilworth
ROYAL LEAMINGTON SPA
WARWICK
Southam
Kineton
Shipston on Stour
Long Compton
Stratford on Avon
Alcester

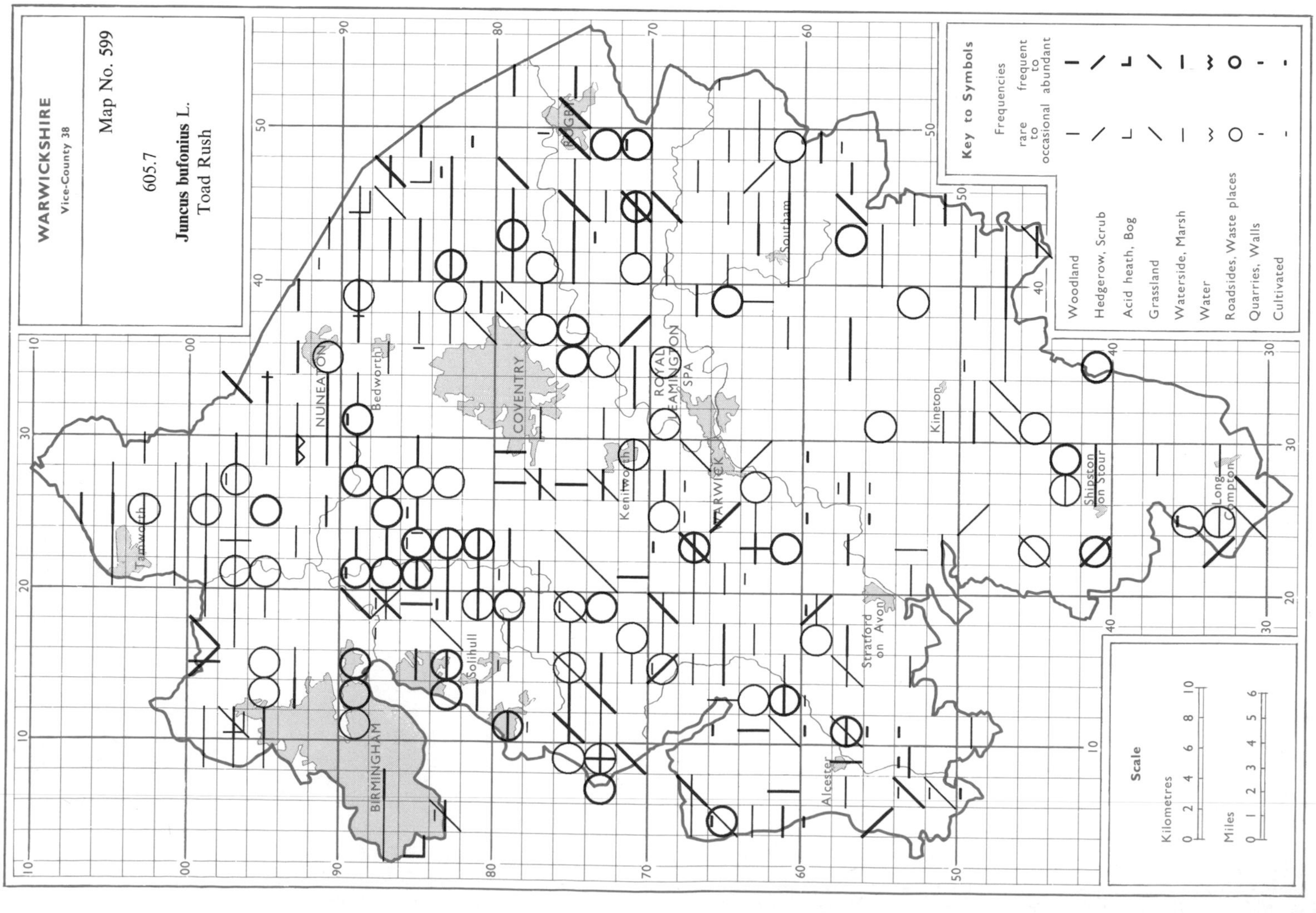
WARWICKSHIRE
Vice-County 38
Map No. 599
605.7
Juncus bufonius L.
Toad Rush
Key to Symbols
Frequencies
rare to occasional
frequent to abundant
Woodland
Hedgerow, Scrub
Acid heath, Bog
Grassland
Waterside, Marsh
Water
Roadsides, Waste places
Quarries, Walls
Cultivated
Scale
Kilometres 0 2 4 6 8 10
Miles 0 1 2 3 4 5 6
Tamworth
NUNEATON
Bedworth
COVENTRY
RUGBY
BIRMINGHAM
Solihull
Kenilworth
ROYAL LEAMINGTON SPA
WARWICK
Southam
Kineton
Shipston on Stour
Long Compton
Stratford on Avon
Alcester

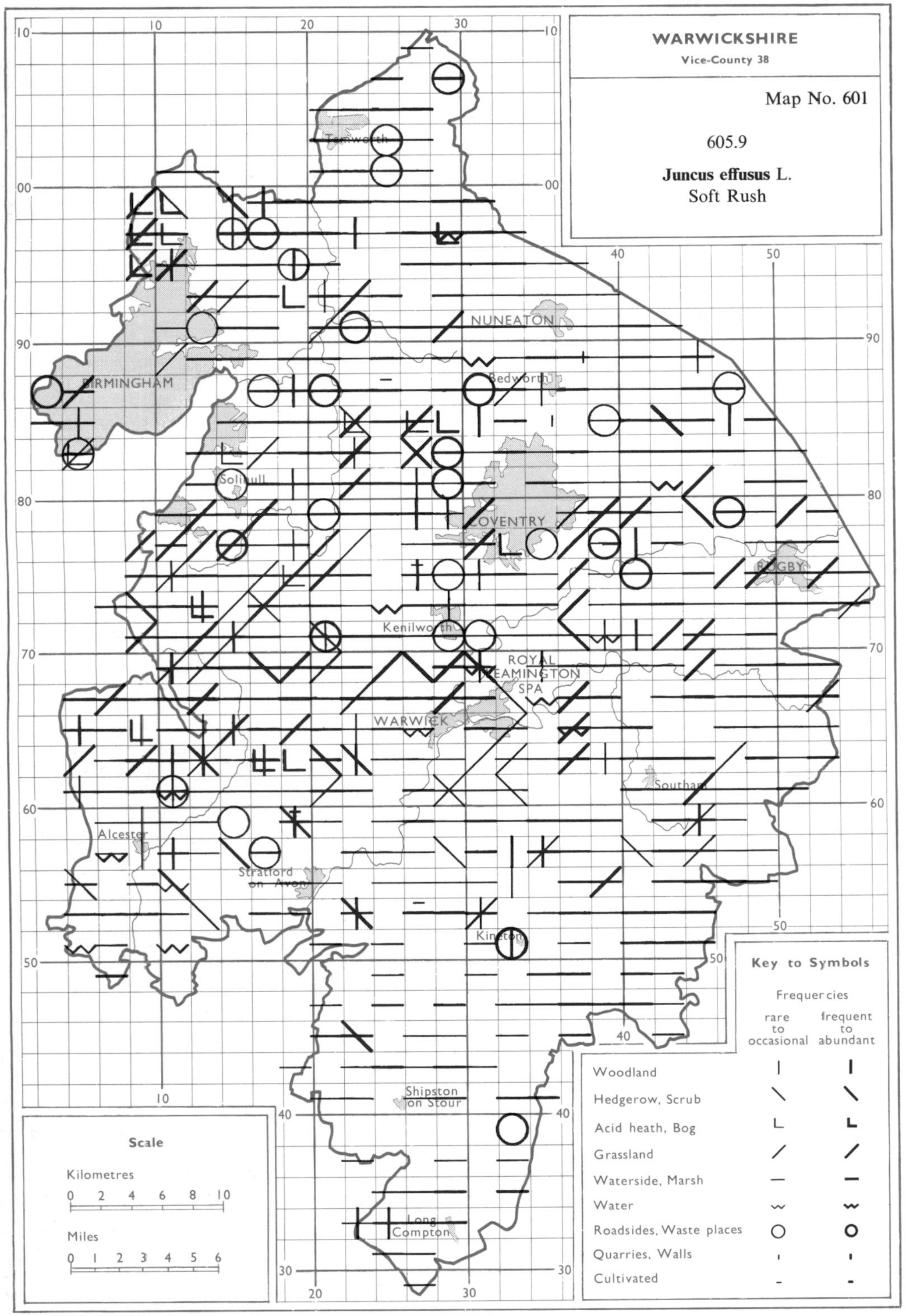
WARWICKSHIRE
Vice-County 38
Map No. 601
605.9
Juncus effusus L.
Soft Rush
NUNEATON
Bedworth
Tamworth
BIRMINGHAM
Solihull
COVENTRY
RUGBY
Kenilworth
ROYAL LEAMINGTON SPA
WARWICK
Southam
Alcester
Stratford on Avon
Kineton
Shipston on Stour
Long Compton
Key to Symbols
Frequencies
rare to occasional
frequent to abundant
Woodland
Hedgerow, Scrub
Acid heath, Bog
Grassland
Waterside, Marsh
Water
Roadsides, Waste places
Quarries, Walls
Cultivated
Scale
Kilometres
0 2 4 6 8 10
Miles
0 1 2 3 4 5 6

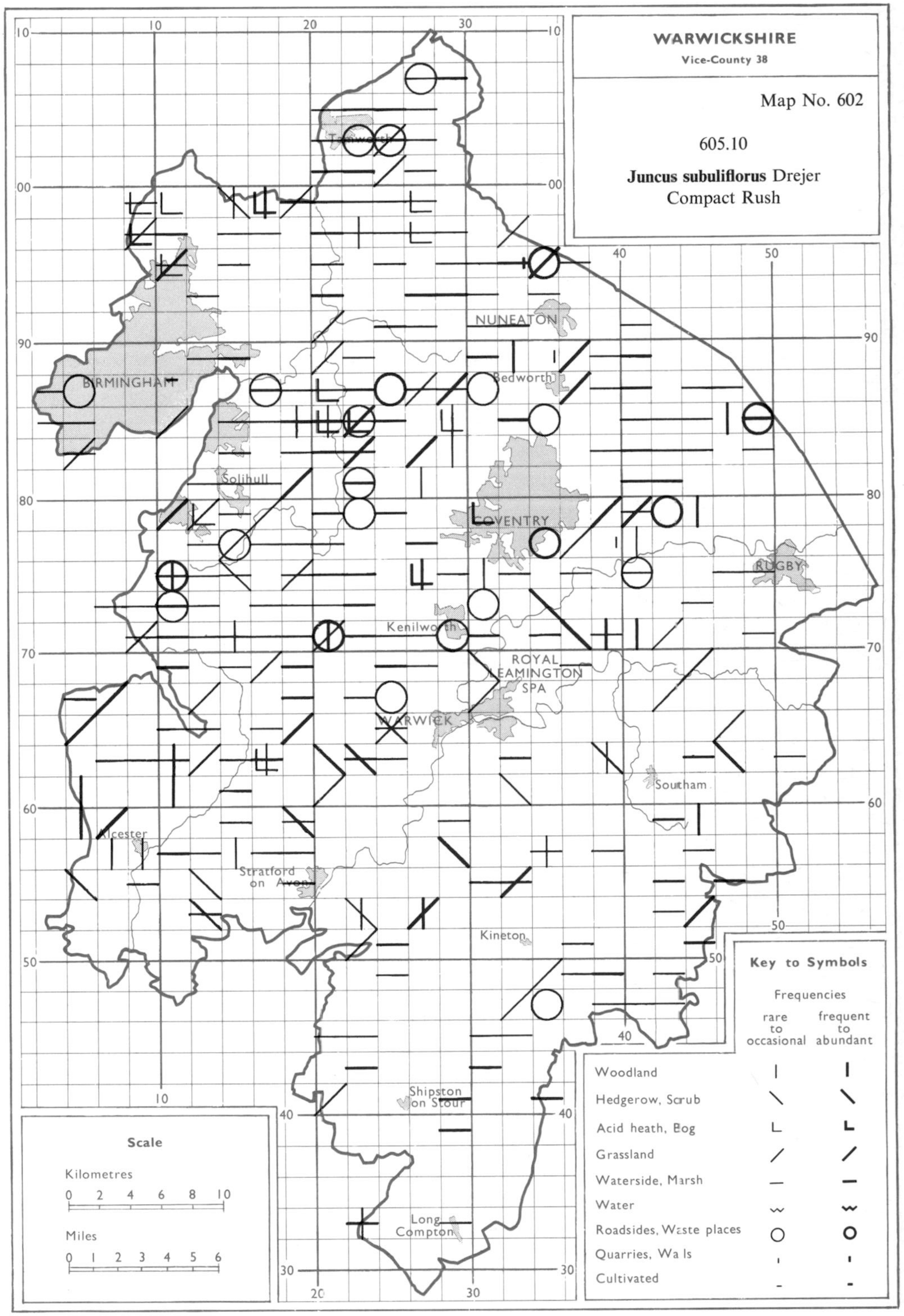
WARWICKSHIRE
Vice-County 38
Map No. 602
605.10
Juncus subuliflorus Drejer
Compact Rush
NUNEATON
Bedworth
Tamworth
BIRMINGHAM
Solihull
COVENTRY
RUGBY
Kenilworth
ROYAL LEAMINGTON SPA
WARWICK
Southam
Alcester
Stratford on Avon
Kineton
Shipston on Stour
Long Compton
Key to Symbols
Frequencies
rare to occasional
frequent to abundant
Woodland
Hedgerow, Scrub
Acid heath, Bog
Grassland
Waterside, Marsh
Water
Roadsides, Waste places
Quarries, Wa ls
Cultivated
Scale
Kilometres
0 2 4 6 8 10
Miles
0 1 2 3 4 5 6

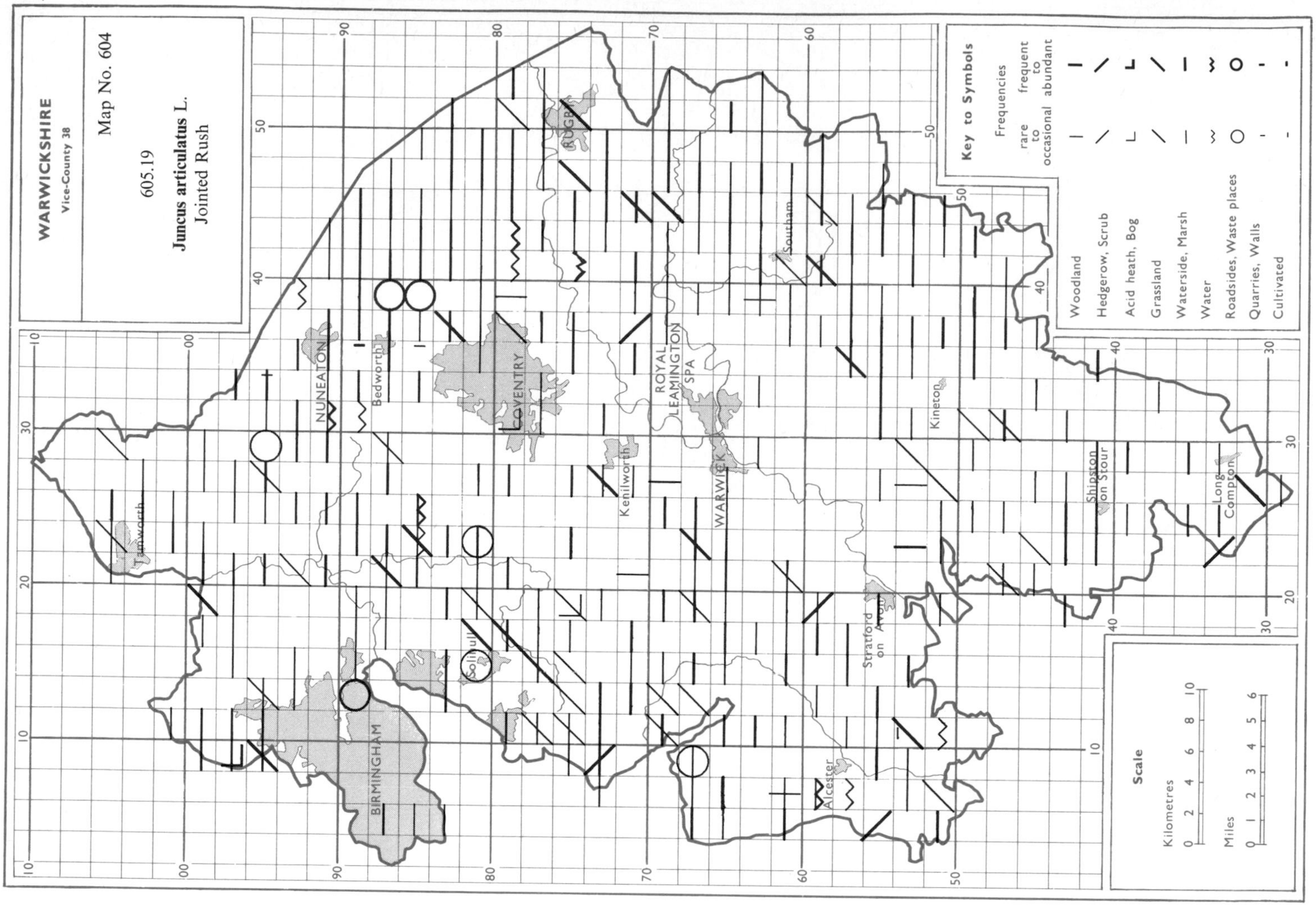
WARWICKSHIRE
Vice-County 38
Map No. 604
605.19
Juncus articulatus L.
Jointed Rush
Key to Symbols
Frequencies
rare to occasional
frequent to abundant
Woodland
Hedgerow, Scrub
Acid heath, Bog
Grassland
Waterside, Marsh
Water
Roadsides, Waste places
Quarries, Walls
Cultivated
Scale
Kilometres
Miles
Tamworth
NUNEATON
Bedworth
COVENTRY
RUGBY
Southam
ROYAL LEAMINGTON SPA
Kenilworth
WARWICK
Kineton
Shipston on Stour
Long Compton
Solihull
BIRMINGHAM
Stratford on Avon
Alcester

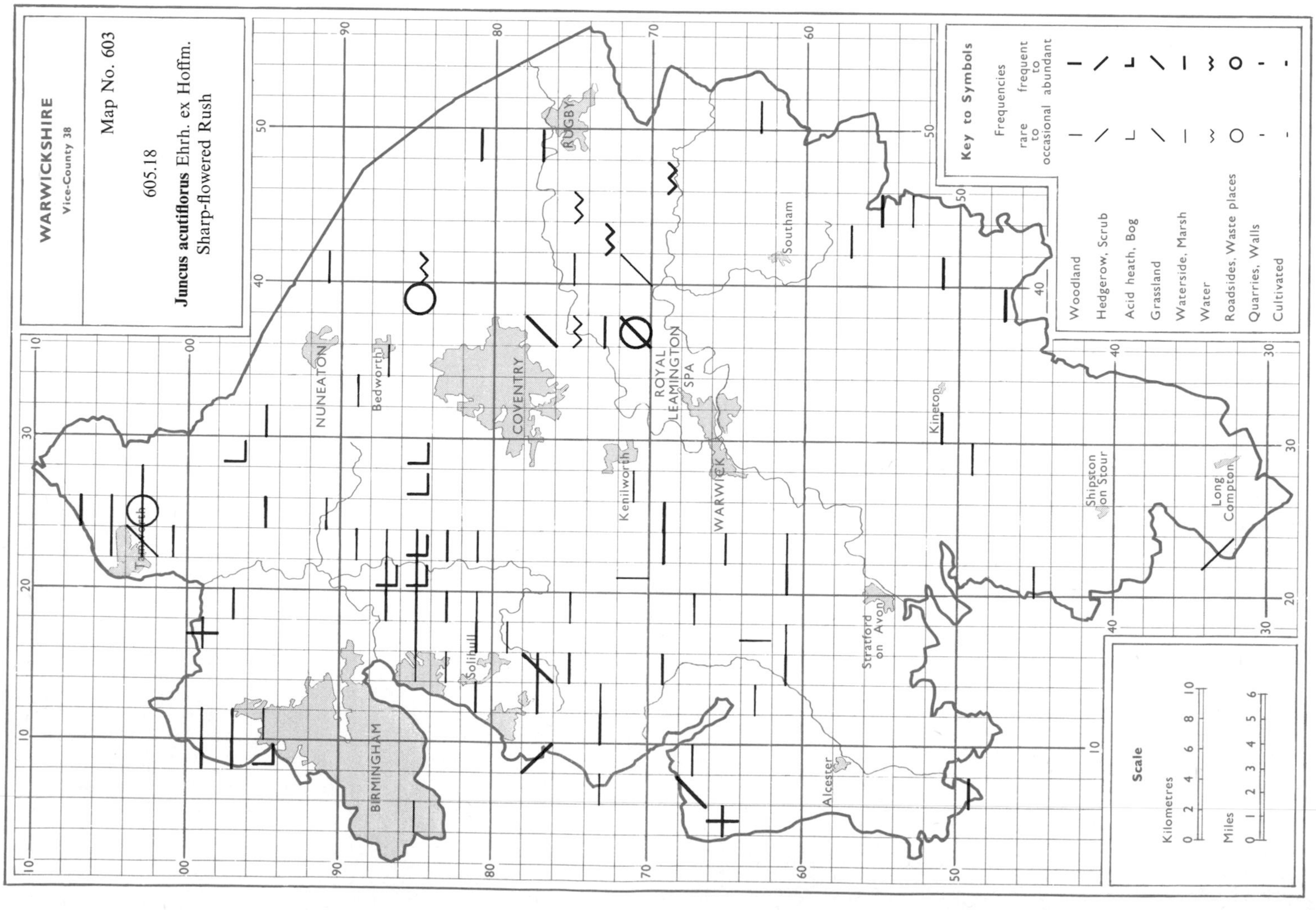
WARWICKSHIRE
Vice-County 38
Map No. 603
605.18
Juncus acutiflorus Ehrh. ex Hoffm.
Sharp-flowered Rush
Key to Symbols
Frequencies
rare to occasional
frequent to abundant
Woodland
Hedgerow, Scrub
Acid heath, Bog
Grassland
Waterside, Marsh
Water
Roadsides, Waste places
Quarries, Walls
Cultivated
Scale
Kilometres
Miles
Tamworth
NUNEATON
Bedworth
COVENTRY
RUGBY
Southam
ROYAL LEAMINGTON SPA
Kenilworth
WARWICK
Kineton
Shipston on Stour
Long Compton
Solihull
BIRMINGHAM
Stratford on Avon
Alcester

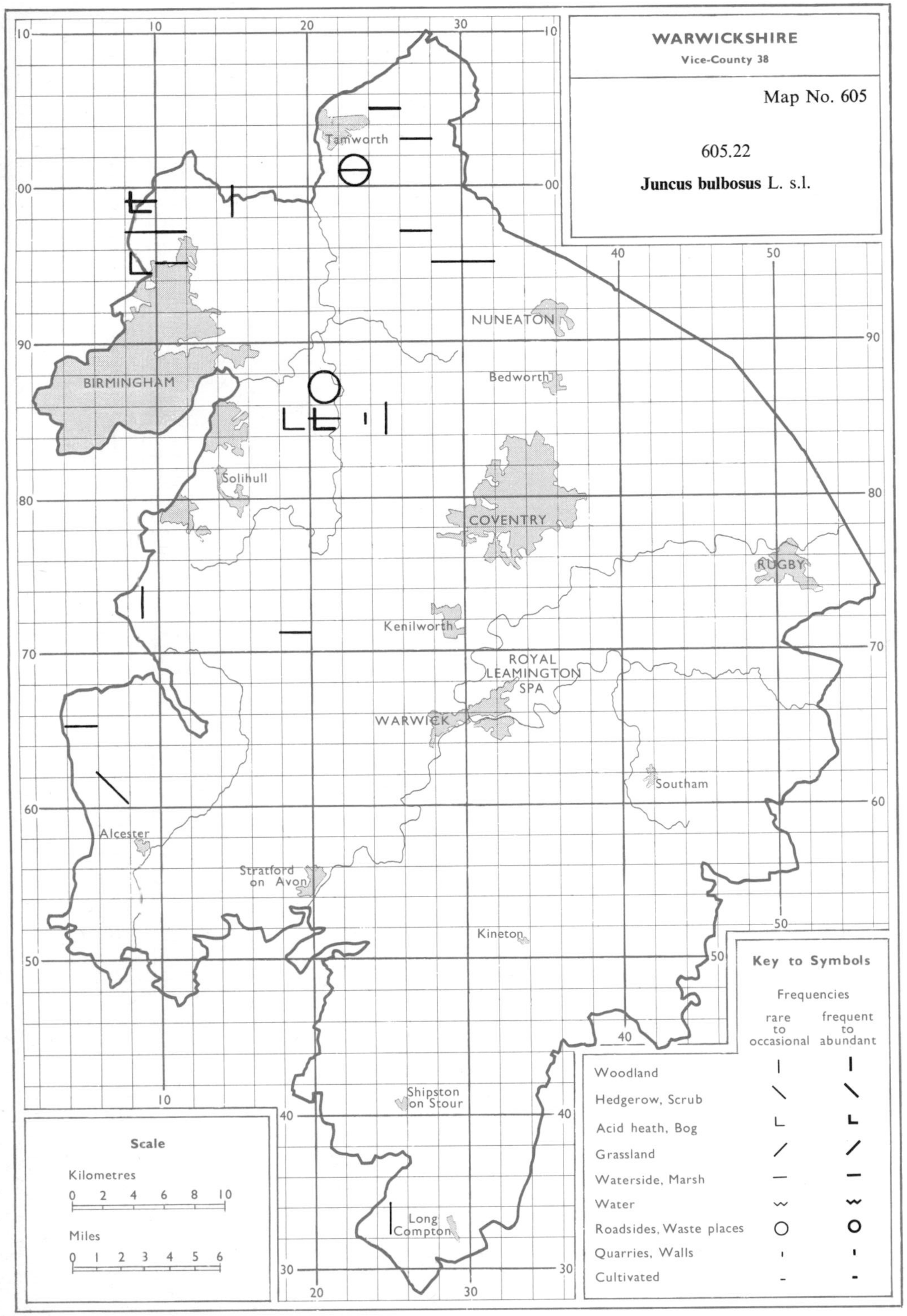
WARWICKSHIRE
Vice-County 38
Map No. 605
605.22
Juncus bulbosus L. s.l.
Tamworth
NUNEATON
Bedworth
BIRMINGHAM
Solihull
COVENTRY
RUGBY
Kenilworth
ROYAL LEAMINGTON SPA
WARWICK
Southam
Alcester
Stratford on Avon
Kineton
Shipston on Stour
Long Compton
Key to Symbols
Frequencies
rare to occasional
frequent to abundant
Woodland
Hedgerow, Scrub
Acid heath, Bog
Grassland
Waterside, Marsh
Water
Roadsides, Waste places
Quarries, Walls
Cultivated
Scale
Kilometres
0 2 4 6 8 10
Miles
0 1 2 3 4 5 6

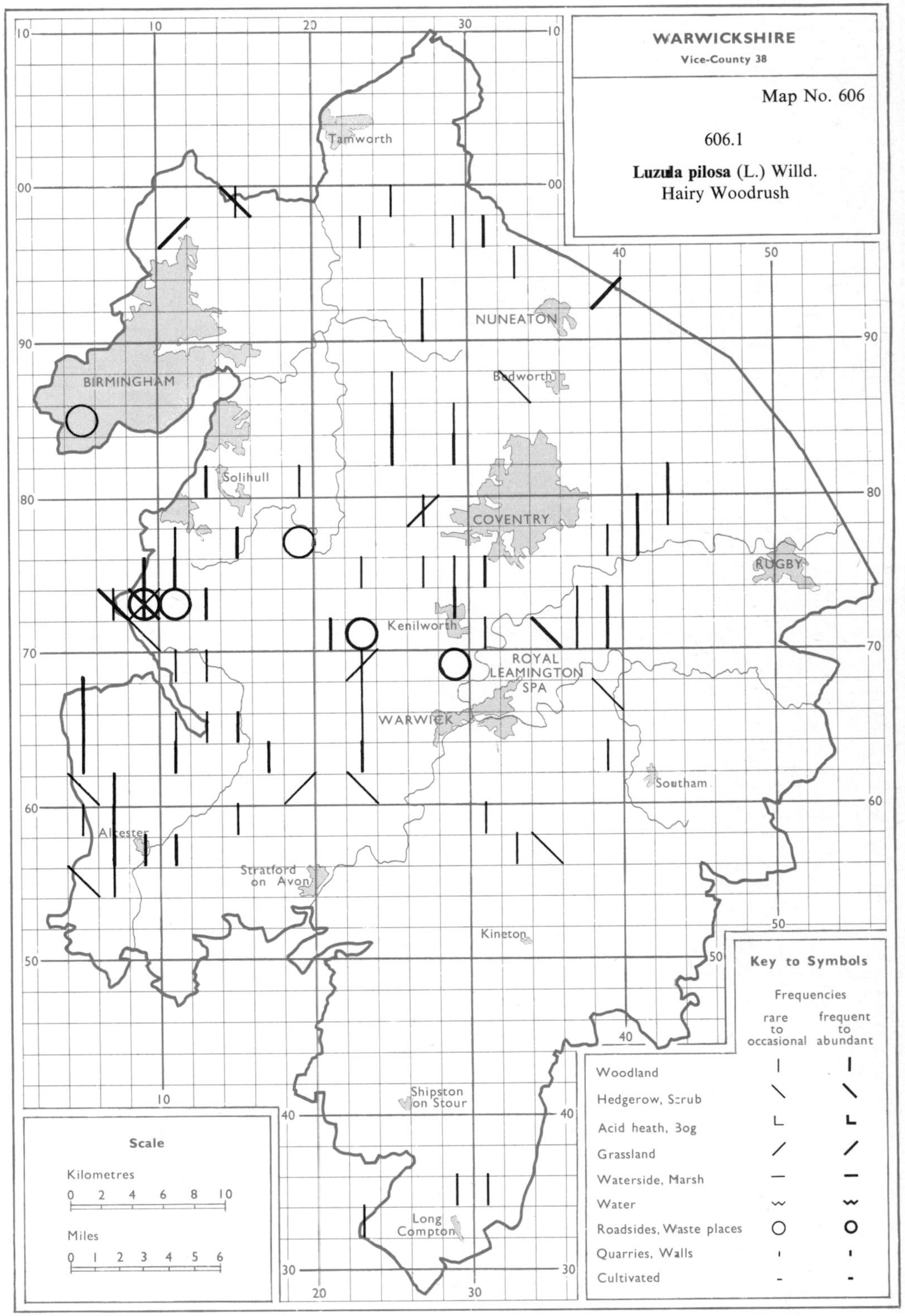
WARWICKSHIRE
Vice-County 38
Map No. 606
606.1
Luzula pilosa (L.) Willd.
Hairy Woodrush
Tamworth
NUNEATON
Bedworth
BIRMINGHAM
Solihull
COVENTRY
RUGBY
Kenilworth
ROYAL LEAMINGTON SPA
WARWICK
Southam
Alcester
Stratford on Avon
Kineton
Shipston on Stour
Long Compton
Key to Symbols
Frequencies
rare to occasional
frequent to abundant
Woodland
Hedgerow, Scrub
Acid heath, Bog
Grassland
Waterside, Marsh
Water
Roadsides, Waste places
Quarries, Walls
Cultivated
Scale
Kilometres
0 2 4 6 8 10
Miles
0 1 2 3 4 5 6

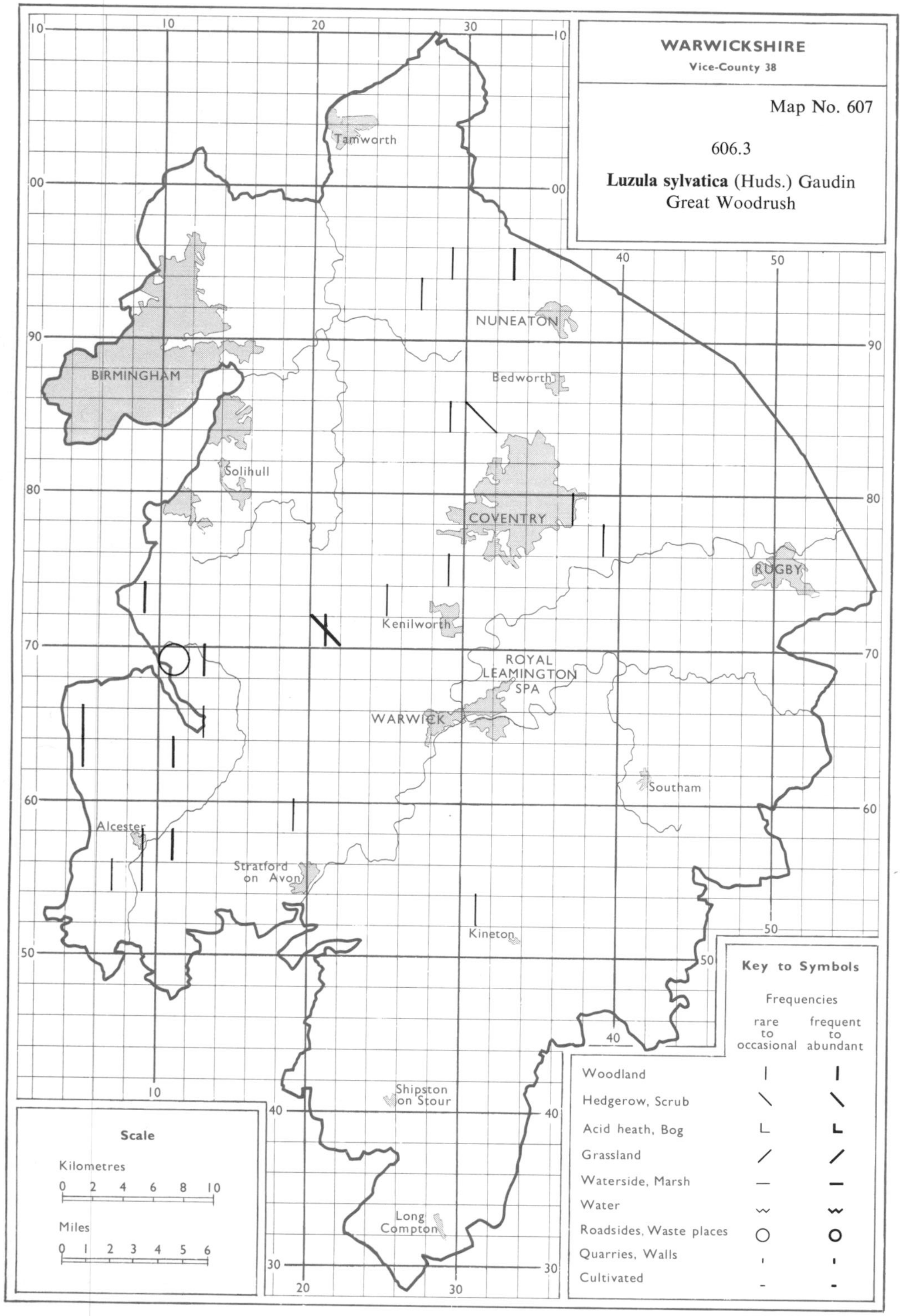
WARWICKSHIRE
Vice-County 38
Map No. 607
606.3
Luzula sylvatica (Huds.) Gaudin
Great Woodrush
Tamworth
Birmingham
Solihull
Nuneaton
Bedworth
Coventry
Rugby
Kenilworth
Royal Leamington Spa
Warwick
Southam
Alcester
Stratford on Avon
Kineton
Shipston on Stour
Long Compton
Key to Symbols
Frequencies
rare to occasional
frequent to abundant
Woodland
Hedgerow, Scrub
Acid heath, Bog
Grassland
Waterside, Marsh
Water
Roadsides, Waste places
Quarries, Walls
Cultivated
Scale
Kilometres
Miles

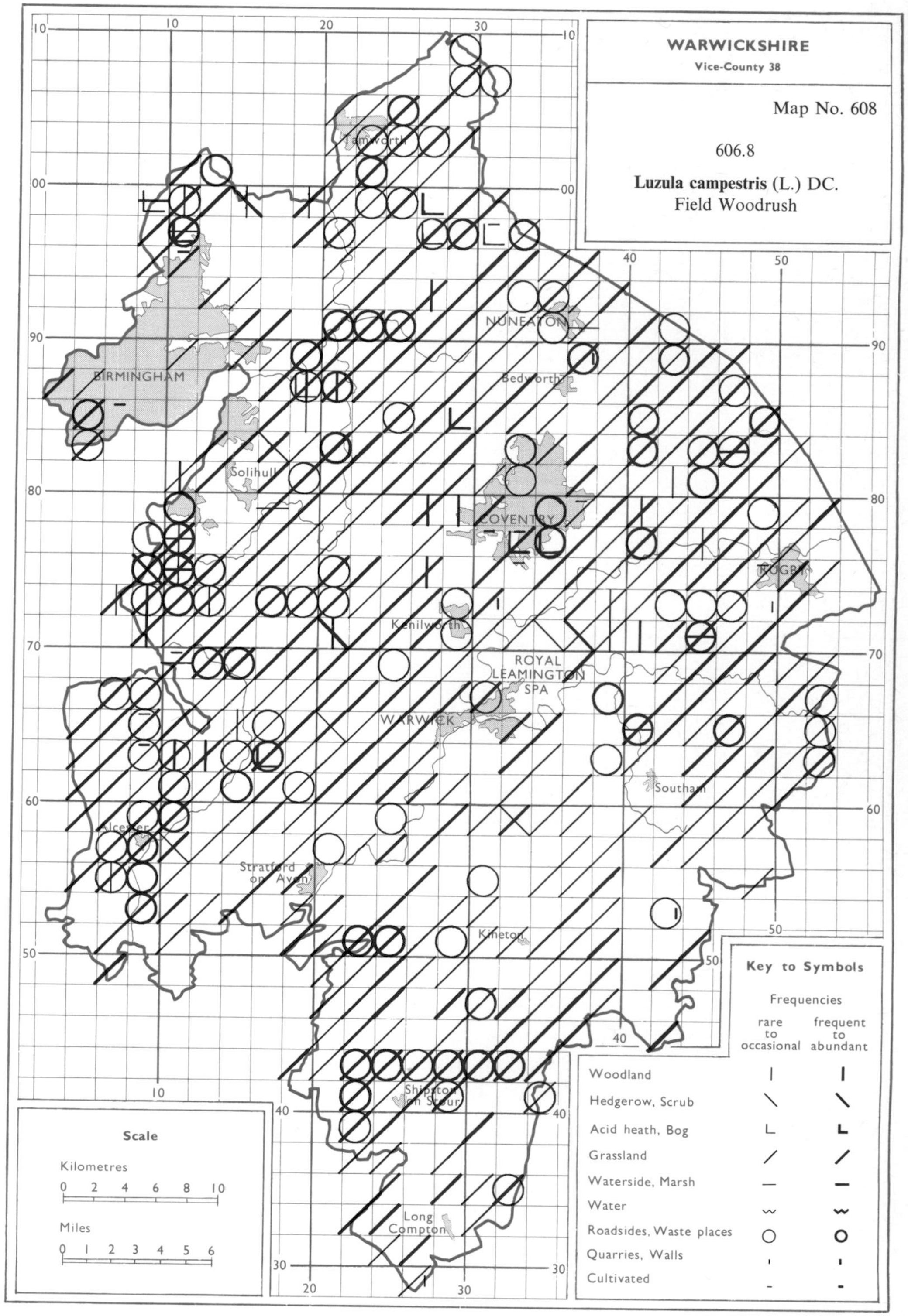
WARWICKSHIRE
Vice-County 38
Map No. 608
606.8
Luzula campestris (L.) DC.
Field Woodrush
Tamworth
Birmingham
Solihull
Nuneaton
Bedworth
Coventry
Rugby
Kenilworth
Royal Leamington Spa
Warwick
Southam
Alcester
Stratford on Avon
Kineton
Shipston on Stour
Long Compton
Key to Symbols
Frequencies
rare to occasional
frequent to abundant
Woodland
Hedgerow, Scrub
Acid heath, Bog
Grassland
Waterside, Marsh
Water
Roadsides, Waste places
Quarries, Walls
Cultivated
Scale
Kilometres
Miles

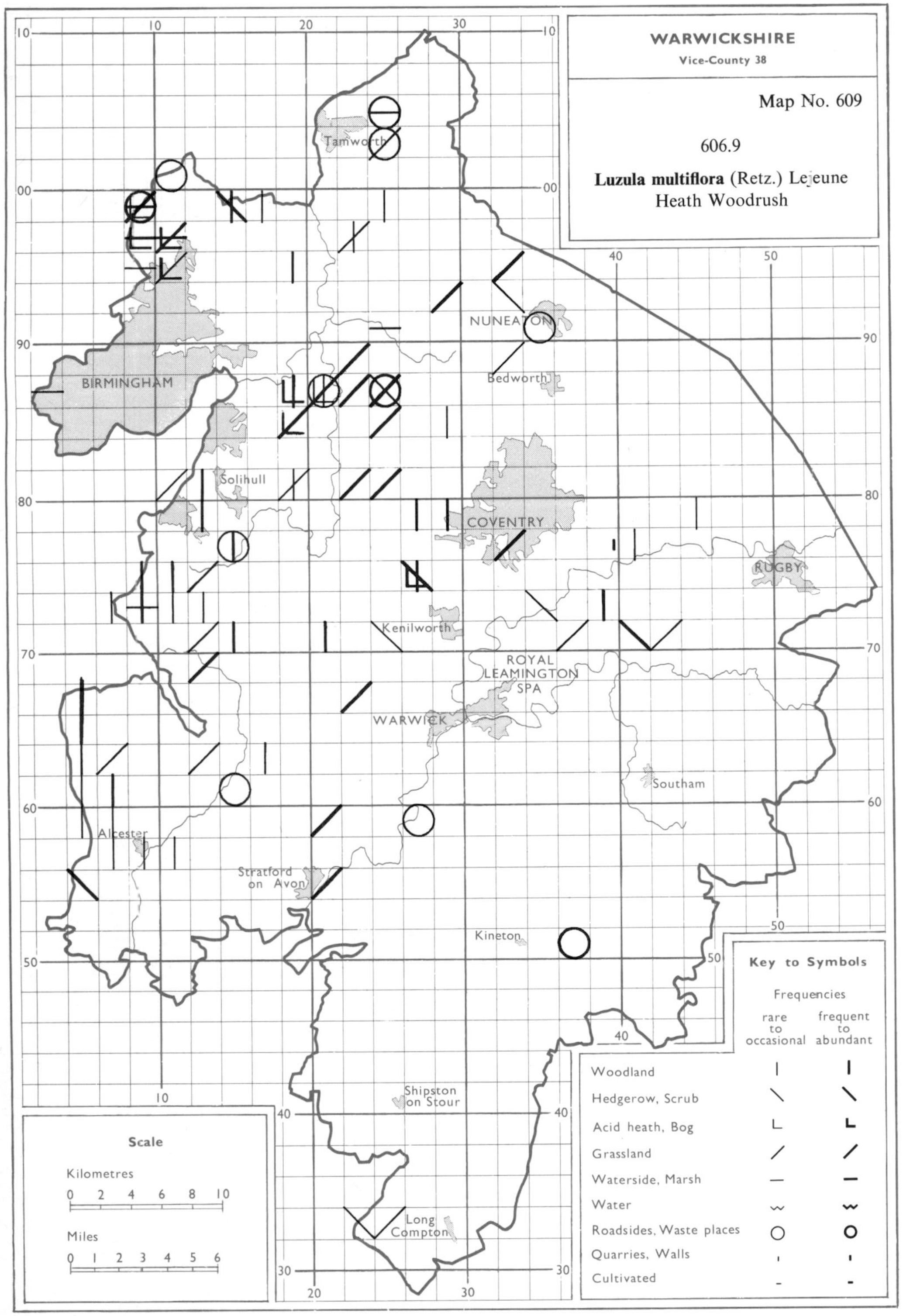
WARWICKSHIRE
Vice-County 38
Map No. 609
606.9
Luzula multiflora (Retz.) Lejeune
Heath Woodrush
Tamworth
BIRMINGHAM
NUNEATON
Bedworth
Solihull
COVENTRY
RUGBY
Kenilworth
ROYAL LEAMINGTON SPA
WARWICK
Southam
Alcester
Stratford on Avon
Kineton
Shipston on Stour
Long Compton
Key to Symbols
Frequencies
rare to occasional
frequent to abundant
Woodland
Hedgerow, Scrub
Acid heath, Bog
Grassland
Waterside, Marsh
Water
Roadsides, Waste places
Quarries, Walls
Cultivated
Scale
Kilometres
0 2 4 6 8 10
Miles
0 1 2 3 4 5 6

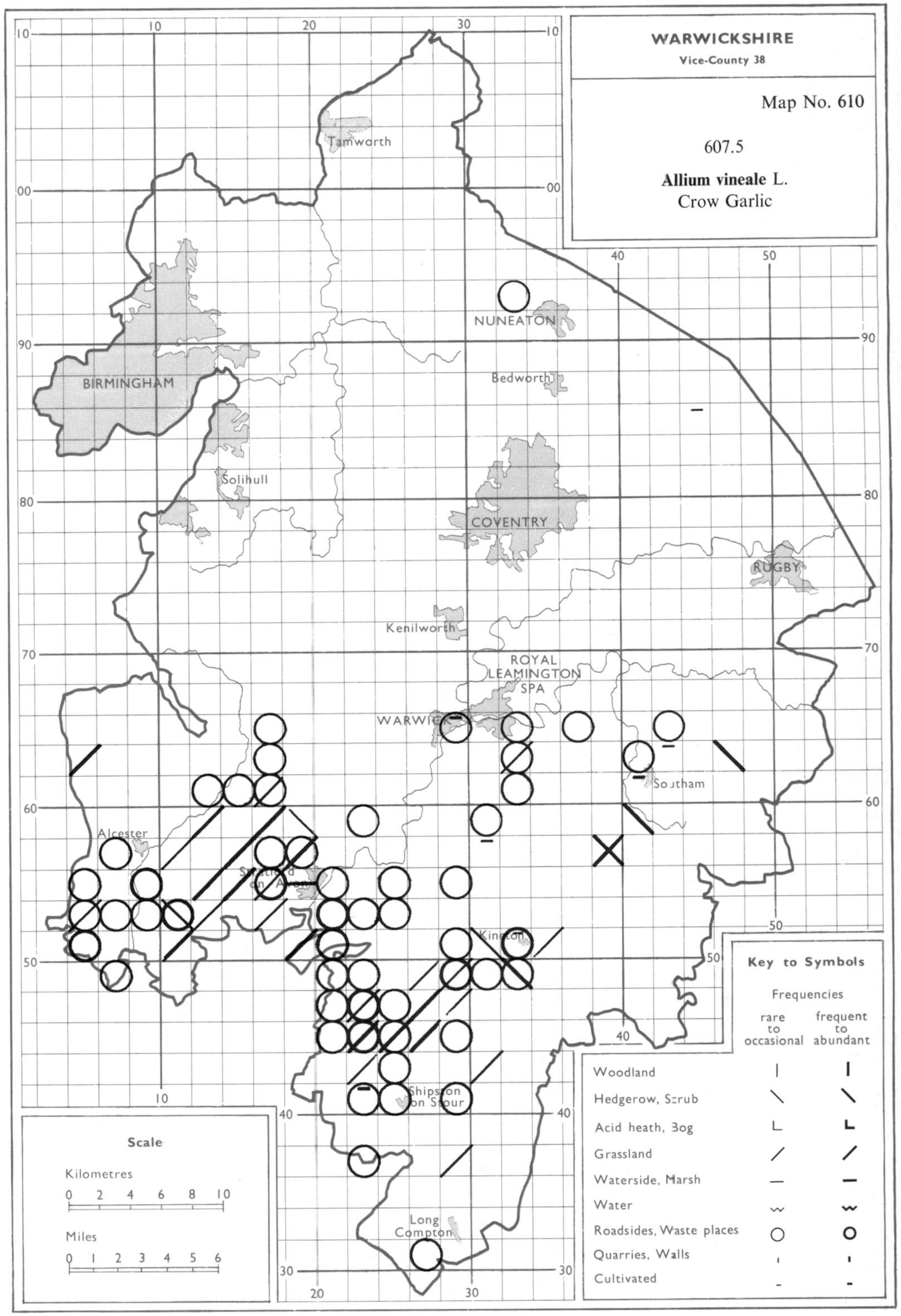
WARWICKSHIRE
Vice-County 38
Map No. 610
607.5
Allium vineale L.
Crow Garlic
Tamworth
BIRMINGHAM
NUNEATON
Bedworth
Solihull
COVENTRY
RUGBY
Kenilworth
ROYAL LEAMINGTON SPA
WARWICK
Southam
Alcester
Stratford on Avon
Kineton
Shipston on Stour
Long Compton
Key to Symbols
Frequencies
rare to occasional
frequent to abundant
Woodland
Hedgerow, Scrub
Acid heath, Bog
Grassland
Waterside, Marsh
Water
Roadsides, Waste places
Quarries, Walls
Cultivated
Scale
Kilometres
0 2 4 6 8 10
Miles
0 1 2 3 4 5 6

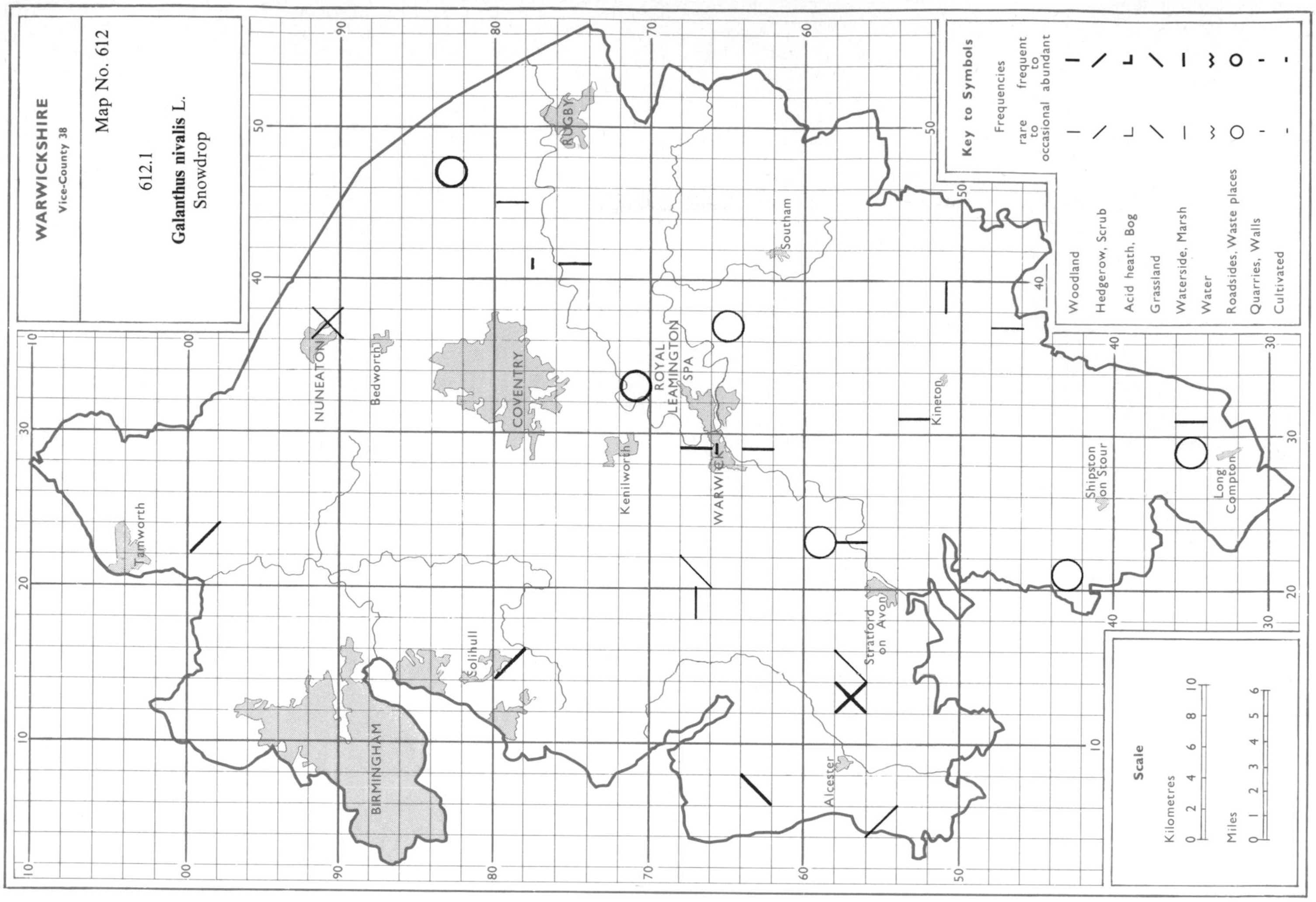
WARWICKSHIRE
Vice-County 38
Map No. 612
612.1
Galanthus nivalis L.
Snowdrop
Key to Symbols
Frequencies
rare to occasional
frequent to abundant
Woodland
Hedgerow, Scrub
Acid heath, Bog
Grassland
Waterside, Marsh
Water
Roadsides, Waste places
Quarries, Walls
Cultivated
Scale
Kilometres
0 2 4 6 8 10
Miles
0 1 2 3 4 5 6
NUNEATON
Bedworth
COVENTRY
RUGBY
Southam
ROYAL LEAMINGTON SPA
Kenilworth
WARWICK
Kineton
Shipston on Stour
Long Compton
Tamworth
Solihull
BIRMINGHAM
Stratford on Avon
Alcester

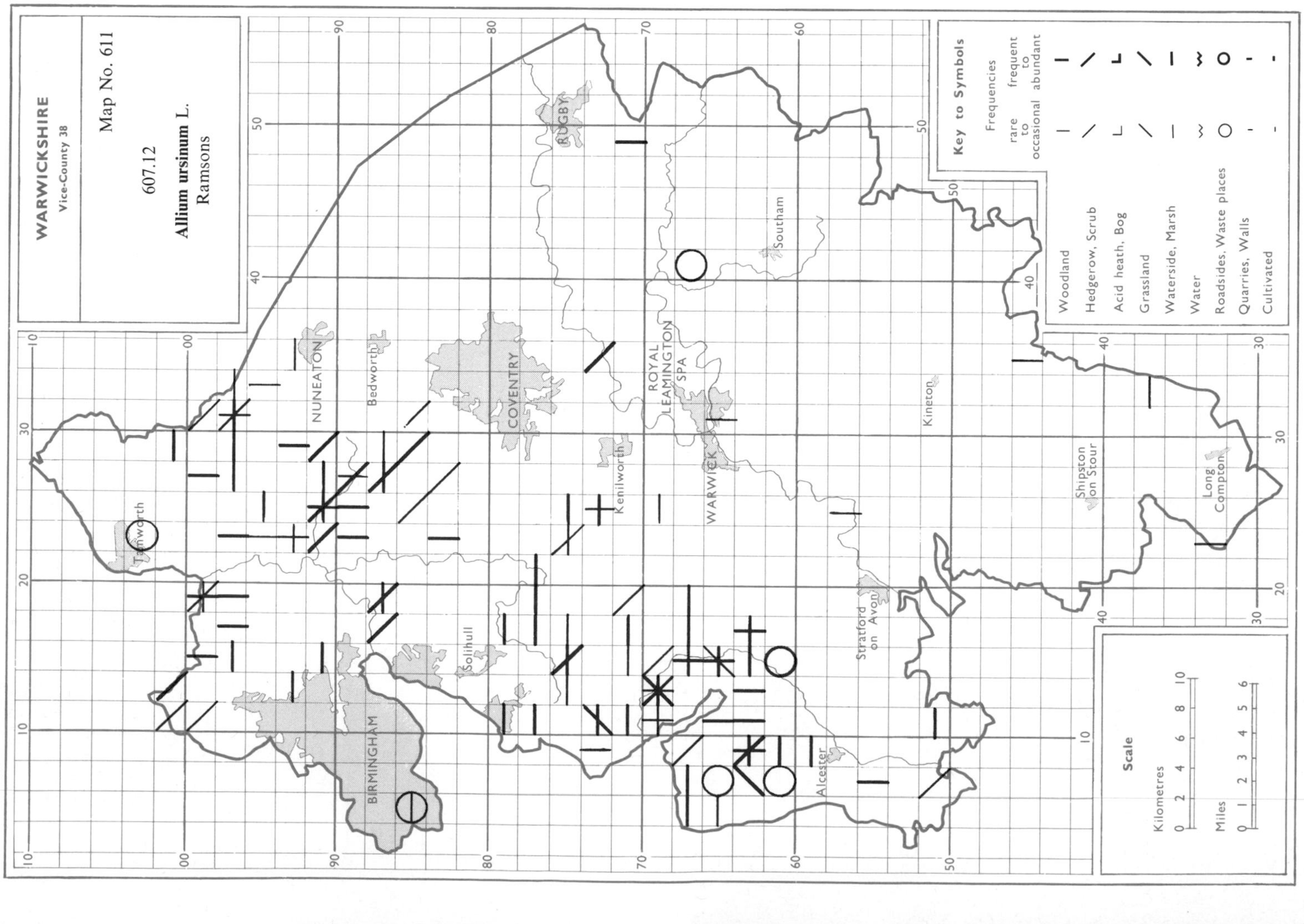
WARWICKSHIRE
Vice-County 38
Map No. 611
607.12
Allium ursinum L.
Ramsons
Key to Symbols
Frequencies
rare to occasional
frequent to abundant
Woodland
Hedgerow, Scrub
Acid heath, Bog
Grassland
Waterside, Marsh
Water
Roadsides, Waste places
Quarries, Walls
Cultivated
Scale
Kilometres
0 2 4 6 8 10
Miles
0 1 2 3 4 5 6
NUNEATON
Bedworth
COVENTRY
RUGBY
Southam
ROYAL LEAMINGTON SPA
Kenilworth
WARWICK
Kineton
Shipston on Stour
Long Compton
Tamworth
Solihull
BIRMINGHAM
Stratford on Avon
Alcester

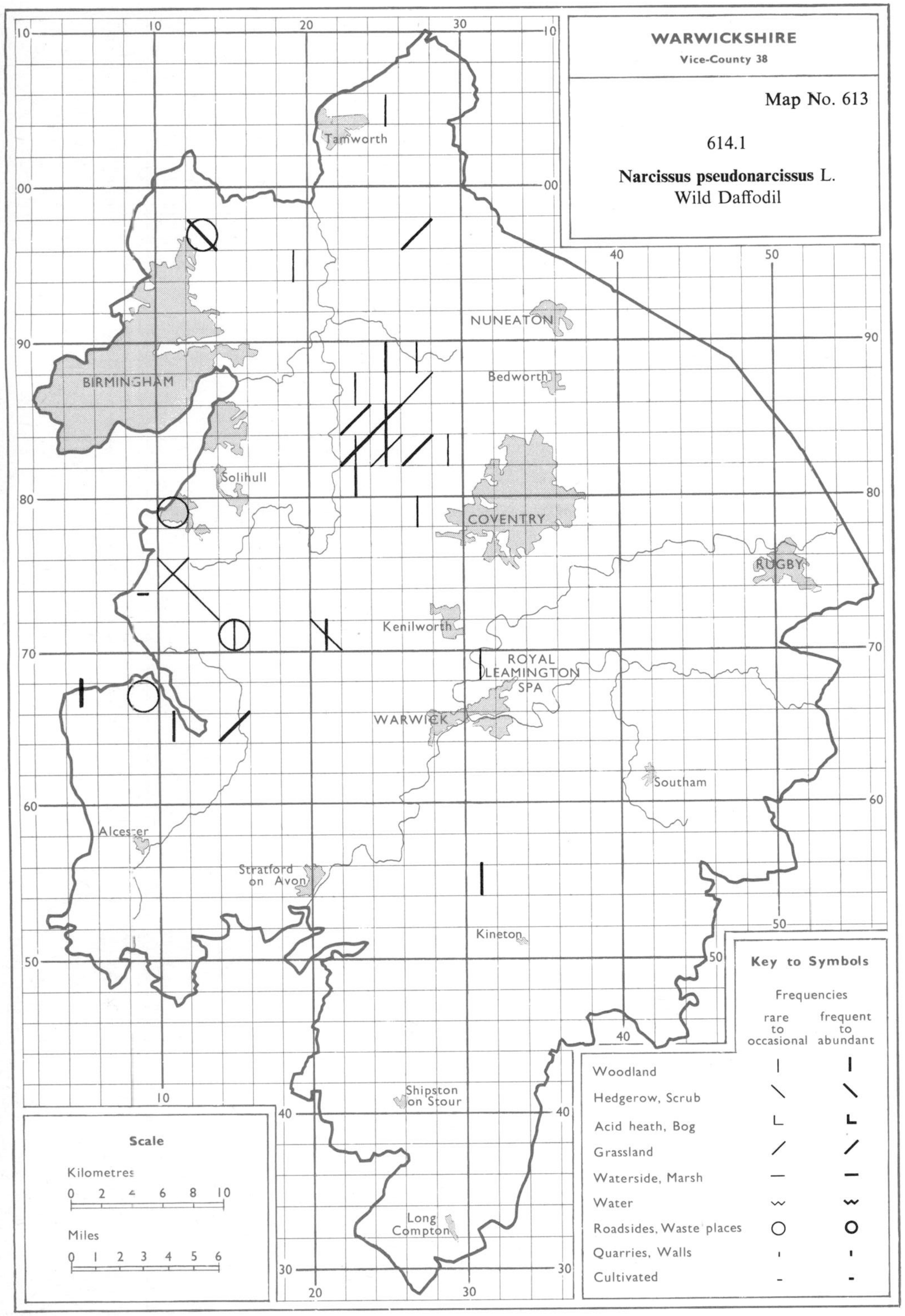
WARWICKSHIRE
Vice-County 38
Map No. 613
614.1
Narcissus pseudonarcissus L.
Wild Daffodil
Tamworth
Birmingham
Solihull
Nuneaton
Bedworth
Coventry
Rugby
Kenilworth
Royal Leamington Spa
Warwick
Southam
Alcester
Stratford on Avon
Kineton
Shipston on Stour
Long Compton
Key to Symbols
Frequencies
rare to occasional
frequent to abundant
Woodland
Hedgerow, Scrub
Acid heath, Bog
Grassland
Waterside, Marsh
Water
Roadsides, Waste places
Quarries, Walls
Cultivated
Scale
Kilometres
0 2 4 6 8 10
Miles
0 1 2 3 4 5 6

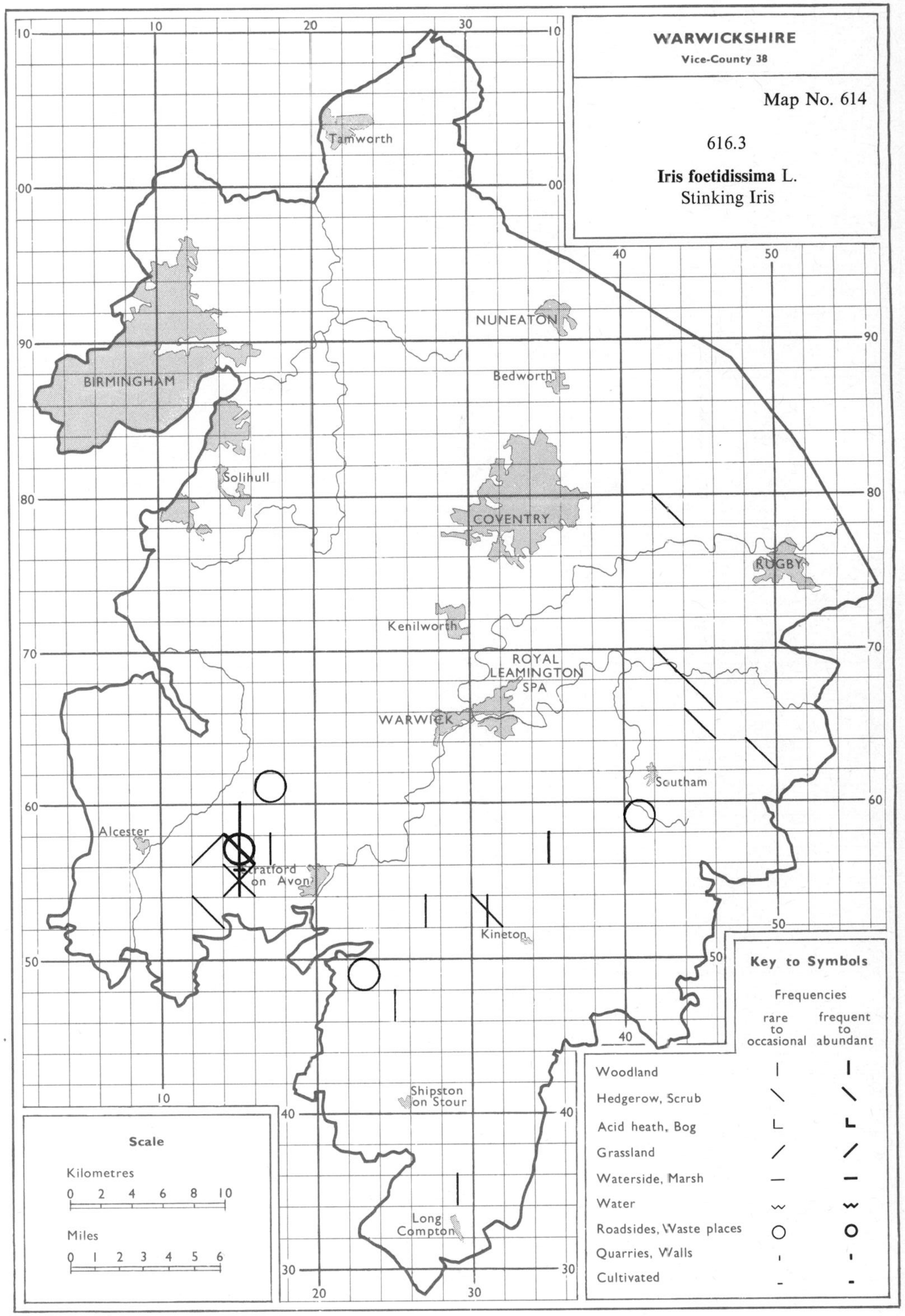
WARWICKSHIRE
Vice-County 38
Map No. 614
616.3
Iris foetidissima L.
Stinking Iris
Tamworth
Birmingham
Solihull
Nuneaton
Bedworth
Coventry
Rugby
Kenilworth
Royal Leamington Spa
Warwick
Southam
Alcester
Stratford on Avon
Kineton
Shipston on Stour
Long Compton
Key to Symbols
Frequencies
rare to occasional
frequent to abundant
Woodland
Hedgerow, Scrub
Acid heath, Bog
Grassland
Waterside, Marsh
Water
Roadsides, Waste places
Quarries, Walls
Cultivated
Scale
Kilometres
0 2 4 6 8 10
Miles
0 1 2 3 4 5 6

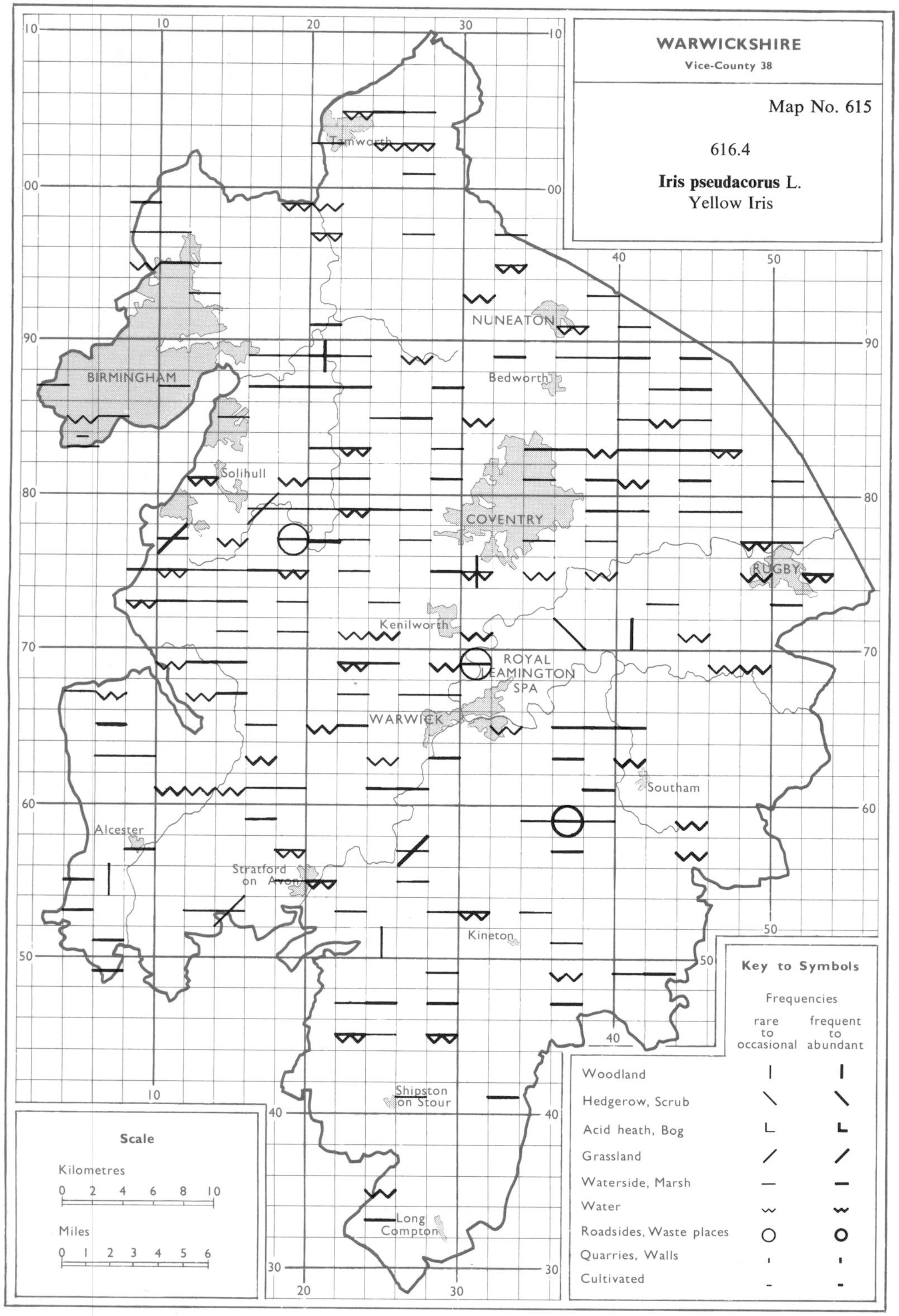
WARWICKSHIRE
Vice-County 38
Map No. 615
616.4
Iris pseudacorus L.
Yellow Iris
Tamworth
NUNEATON
Bedworth
BIRMINGHAM
Solihull
COVENTRY
RUGBY
Kenilworth
ROYAL LEAMINGTON SPA
WARWICK
Southam
Alcester
Stratford on Avon
Kineton
Shipston on Stour
Long Compton
Scale
Kilometres
0 2 4 6 8 10
Miles
0 1 2 3 4 5 6
Key to Symbols
Frequencies
rare to occasional
frequent to abundant
Woodland
Hedgerow, Scrub
Acid heath, Bog
Grassland
Waterside, Marsh
Water
Roadsides, Waste places
Quarries, Walls
Cultivated

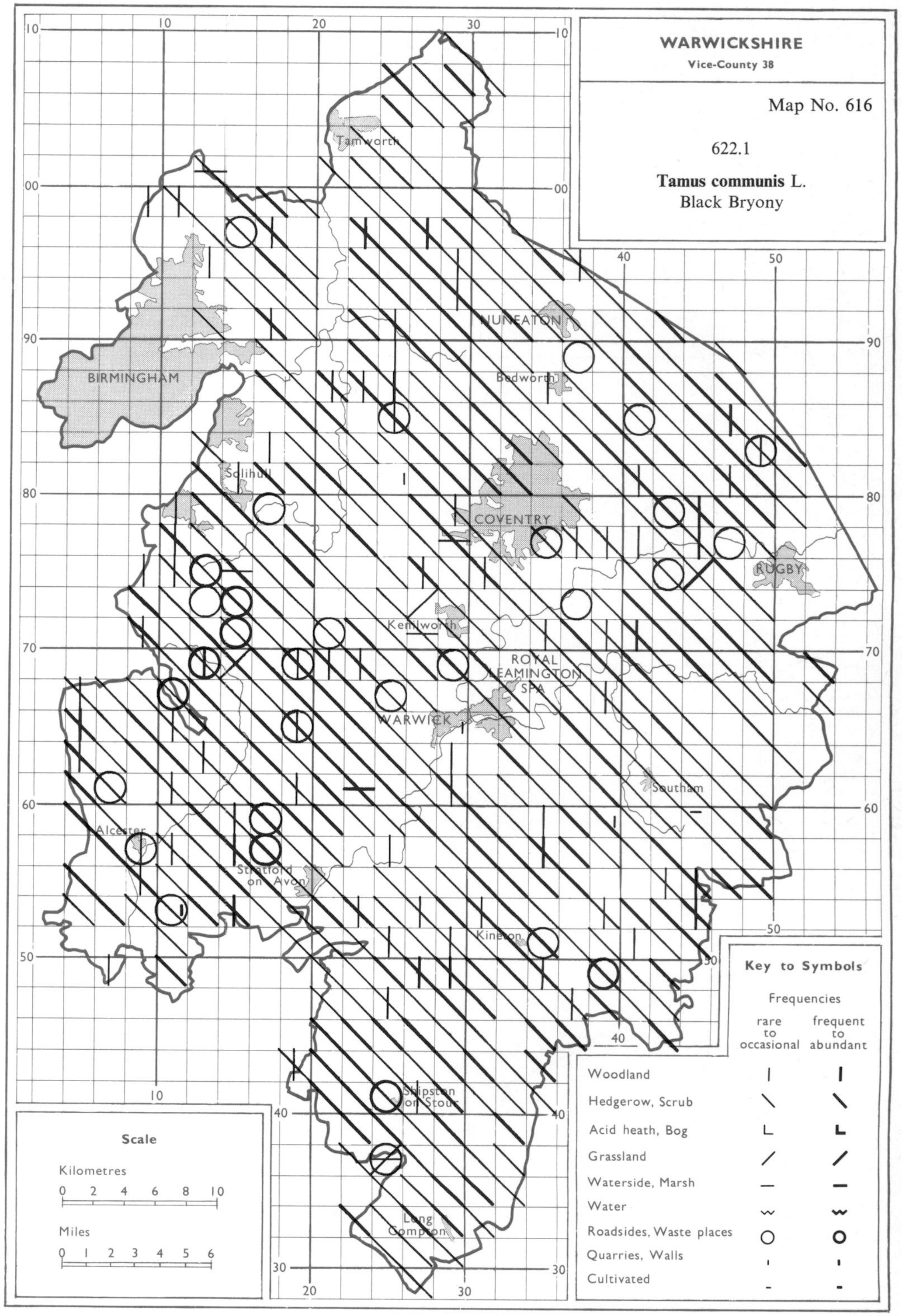
WARWICKSHIRE
Vice-County 38
Map No. 616
622.1
Tamus communis L.
Black Bryony
Tamworth
NUNEATON
Bedworth
BIRMINGHAM
Solihull
COVENTRY
RUGBY
Kenilworth
ROYAL LEAMINGTON SPA
WARWICK
Southam
Alcester
Stratford on Avon
Kineton
Shipston on Stour
Long Compton
Scale
Kilometres
0 2 4 6 8 10
Miles
0 1 2 3 4 5 6
Key to Symbols
Frequencies
rare to occasional
frequent to abundant
Woodland
Hedgerow, Scrub
Acid heath, Bog
Grassland
Waterside, Marsh
Water
Roadsides, Waste places
Quarries, Walls
Cultivated

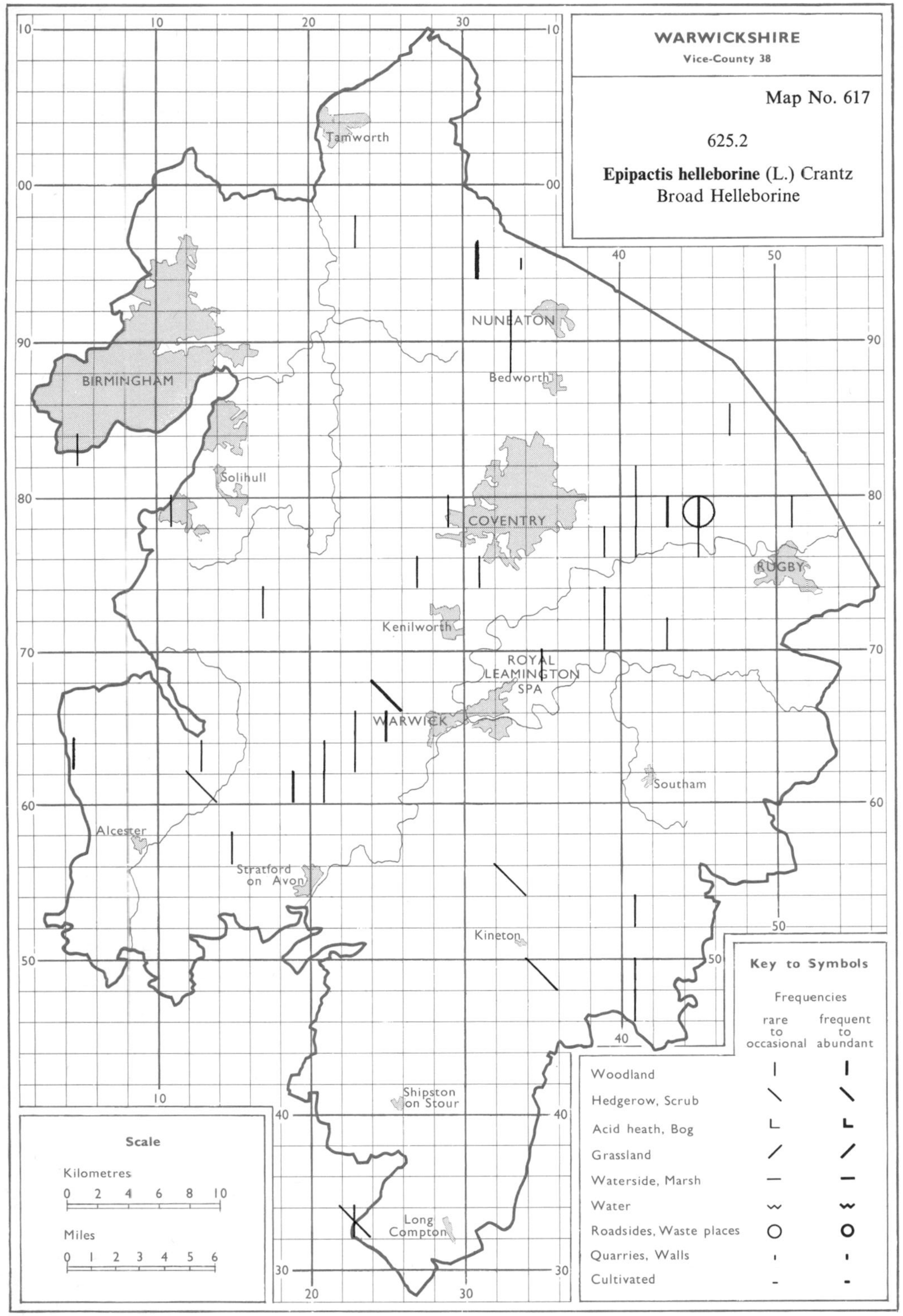
WARWICKSHIRE
Vice-County 38
Map No. 617
625.2
Epipactis helleborine (L.) Crantz
Broad Helleborine
Tamworth
NUNEATON
Bedworth
BIRMINGHAM
Solihull
COVENTRY
RUGBY
Kenilworth
ROYAL LEAMINGTON SPA
WARWICK
Southam
Alcester
Stratford on Avon
Kineton
Shipston on Stour
Long Compton
Key to Symbols
Frequencies
rare to occasional
frequent to abundant
Woodland
Hedgerow, Scrub
Acid heath, Bog
Grassland
Waterside, Marsh
Water
Roadsides, Waste places
Quarries, Walls
Cultivated
Scale
Kilometres
0 2 4 6 8 10
Miles
0 1 2 3 4 5 6

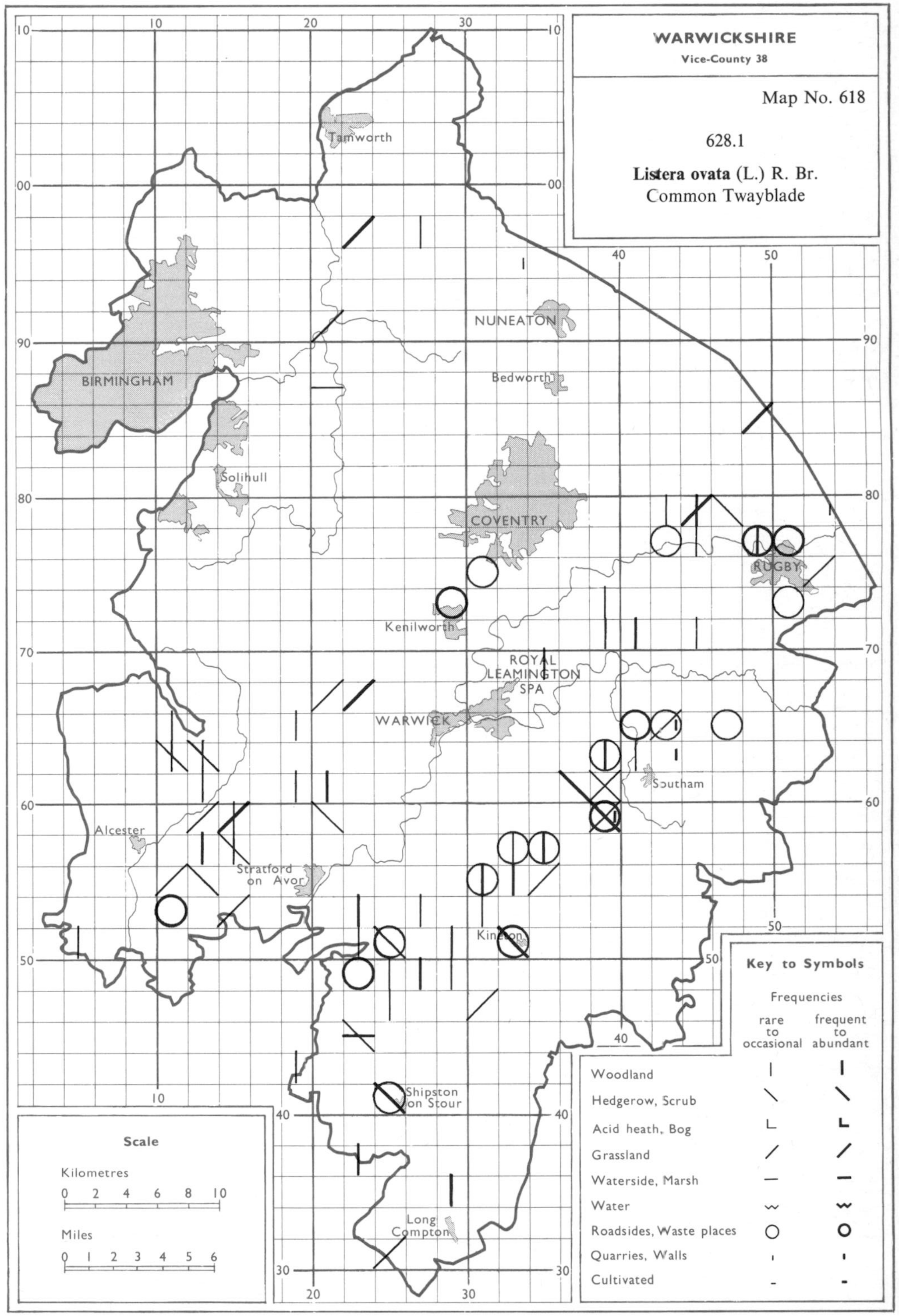
WARWICKSHIRE
Vice-County 38
Map No. 618
628.1
Listera ovata (L.) R. Br.
Common Twayblade
Tamworth
NUNEATON
Bedworth
BIRMINGHAM
Solihull
COVENTRY
RUGBY
Kenilworth
ROYAL LEAMINGTON SPA
WARWICK
Southam
Alcester
Stratford on Avon
Kineton
Shipston on Stour
Long Compton
Key to Symbols
Frequencies
rare to occasional
frequent to abundant
Woodland
Hedgerow, Scrub
Acid heath, Bog
Grassland
Waterside, Marsh
Water
Roadsides, Waste places
Quarries, Walls
Cultivated
Scale
Kilometres
0 2 4 6 8 10
Miles
0 1 2 3 4 5 6

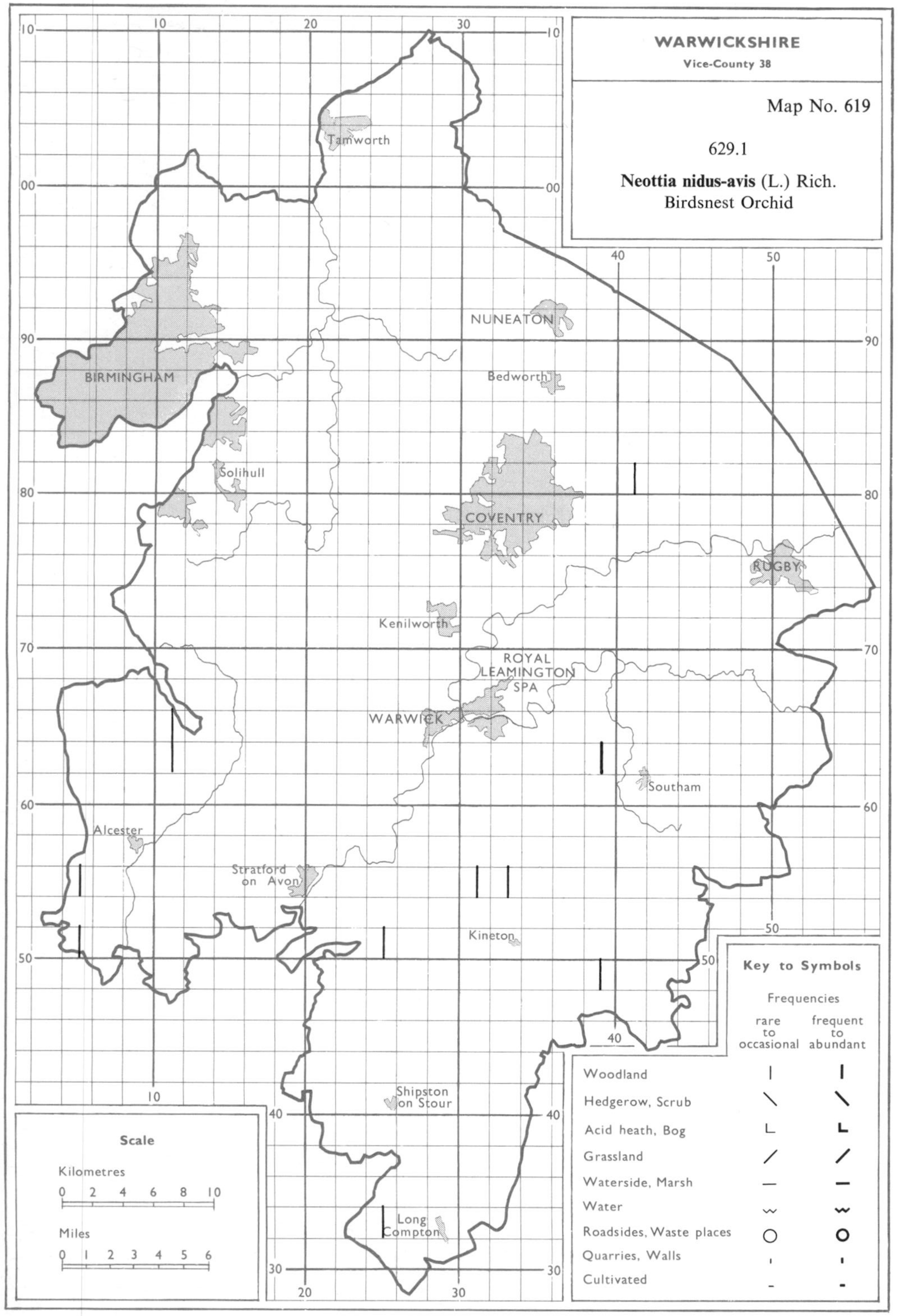
WARWICKSHIRE
Vice-County 38
Map No. 619
629.1
Neottia nidus-avis (L.) Rich.
Birdsnest Orchid
Tamworth
NUNEATON
Bedworth
BIRMINGHAM
Solihull
COVENTRY
RUGBY
Kenilworth
ROYAL LEAMINGTON SPA
WARWICK
Southam
Alcester
Stratford on Avon
Kineton
Shipston on Stour
Long Compton
Key to Symbols
Frequencies
rare to occasional
frequent to abundant
Woodland
Hedgerow, Scrub
Acid heath, Bog
Grassland
Waterside, Marsh
Water
Roadsides, Waste places
Quarries, Walls
Cultivated
Scale
Kilometres
0 2 4 6 8 10
Miles
0 1 2 3 4 5 6

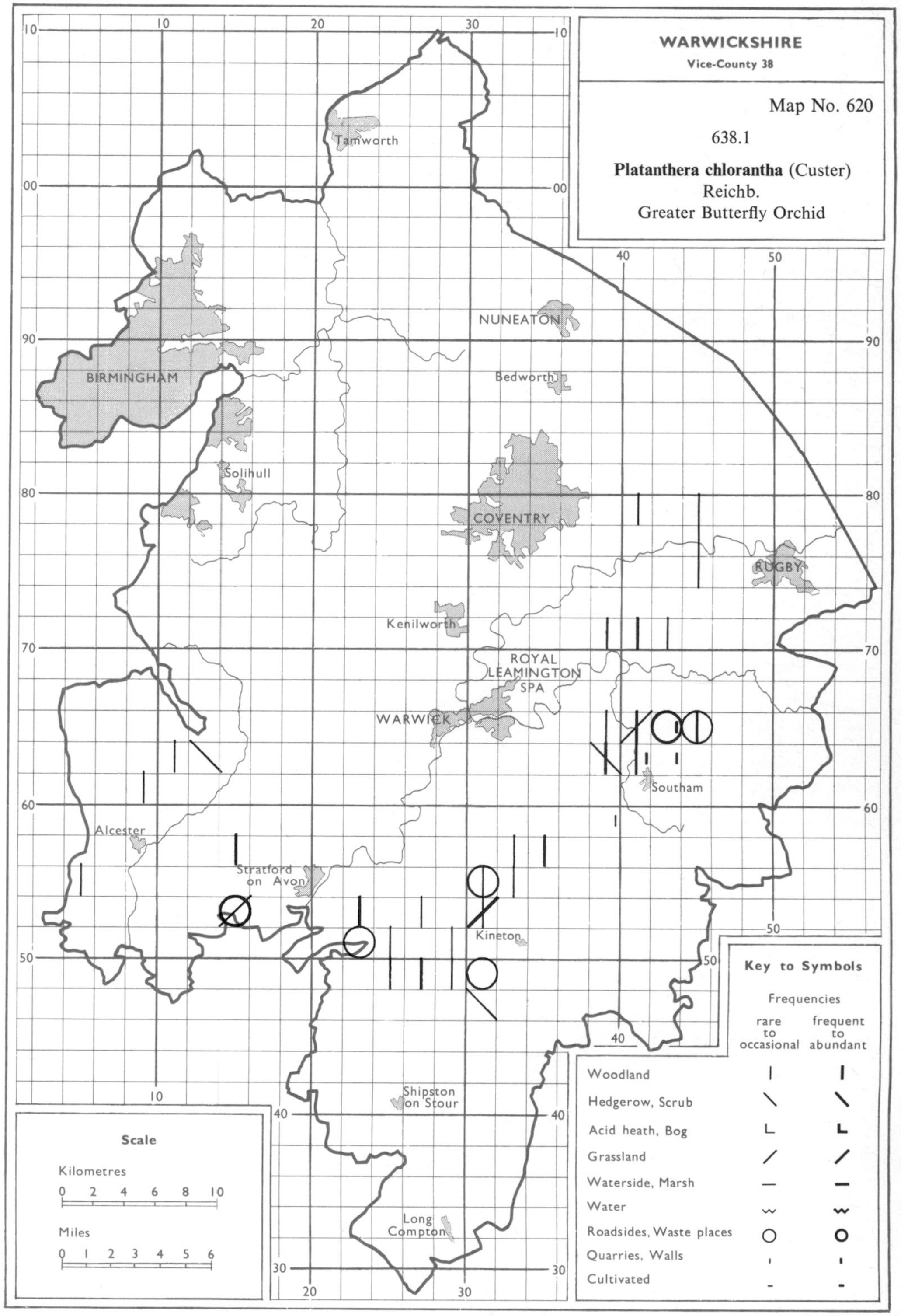
WARWICKSHIRE
Vice-County 38
Map No. 620
638.1
Platanthera chlorantha (Custer) Reichb.
Greater Butterfly Orchid
Tamworth
NUNEATON
Bedworth
BIRMINGHAM
Solihull
COVENTRY
RUGBY
Kenilworth
ROYAL LEAMINGTON SPA
WARWICK
Southam
Alcester
Stratford on Avon
Kineton
Shipston on Stour
Long Compton
Key to Symbols
Frequencies
rare to occasional
frequent to abundant
Woodland
Hedgerow, Scrub
Acid heath, Bog
Grassland
Waterside, Marsh
Water
Roadsides, Waste places
Quarries, Walls
Cultivated
Scale
Kilometres
0 2 4 6 8 10
Miles
0 1 2 3 4 5 6

WARWICKSHIRE
Vice-County 38
Map No. 622
642.5
Orchis morio L.
Green-winged Orchid
Key to Symbols
Frequencies
rare to occasional
frequent to abundant
Woodland
Hedgerow, Scrub
Acid heath, Bog
Grassland
Waterside, Marsh
Water
Roadsides, Waste places
Quarries, Walls
Cultivated
Scale
Kilometres
Miles
NUNEATON
Bedworth
COVENTRY
RUGBY
ROYAL LEAMINGTON SPA
Southam
Kineton
Tamworth
Kenilworth
WARWICK
Shipston on Stour
Long Compton
Solihull
BIRMINGHAM
Stratford on Avon
Alcester

WARWICKSHIRE
Vice-County 38
Map No. 621
640.1
Ophrys apifera Huds.
Bee Orchid
Key to Symbols
Frequencies
rare to occasional
frequent to abundant
Woodland
Hedgerow, Scrub
Acid heath, Bog
Grassland
Waterside, Marsh
Water
Roadsides, Waste places
Quarries, Walls
Cultivated
Scale
Kilometres
Miles
NUNEATON
Bedworth
COVENTRY
RUGBY
ROYAL LEAMINGTON SPA
Southam
Kineton
Tamworth
Kenilworth
WARWICK
Shipston on Stour
Long Compton
Solihull
BIRMINGHAM
Stratford on Avon
Alcester

WARWICKSHIRE
Vice-County 38
Map No. 623
642.7
Orchis mascula (L.) L.
Early Purple Orchid
Tamworth
NUNEATON
Bedworth
BIRMINGHAM
Solihull
COVENTRY
RUGBY
Kenilworth
ROYAL LEAMINGTON SPA
WARWICK
Southam
Alcester
Stratford on Avon
Kineton
Shipston on Stour
Long Compton
Key to Symbols
Frequencies
rare to occasional
frequent to abundant
Woodland
Hedgerow, Scrub
Acid heath, Bog
Grassland
Waterside, Marsh
Water
Roadsides, Waste places
Quarries, Walls
Cultivated
Scale
Kilometres
0 2 4 6 8 10
Miles
0 1 2 3 4 5 6

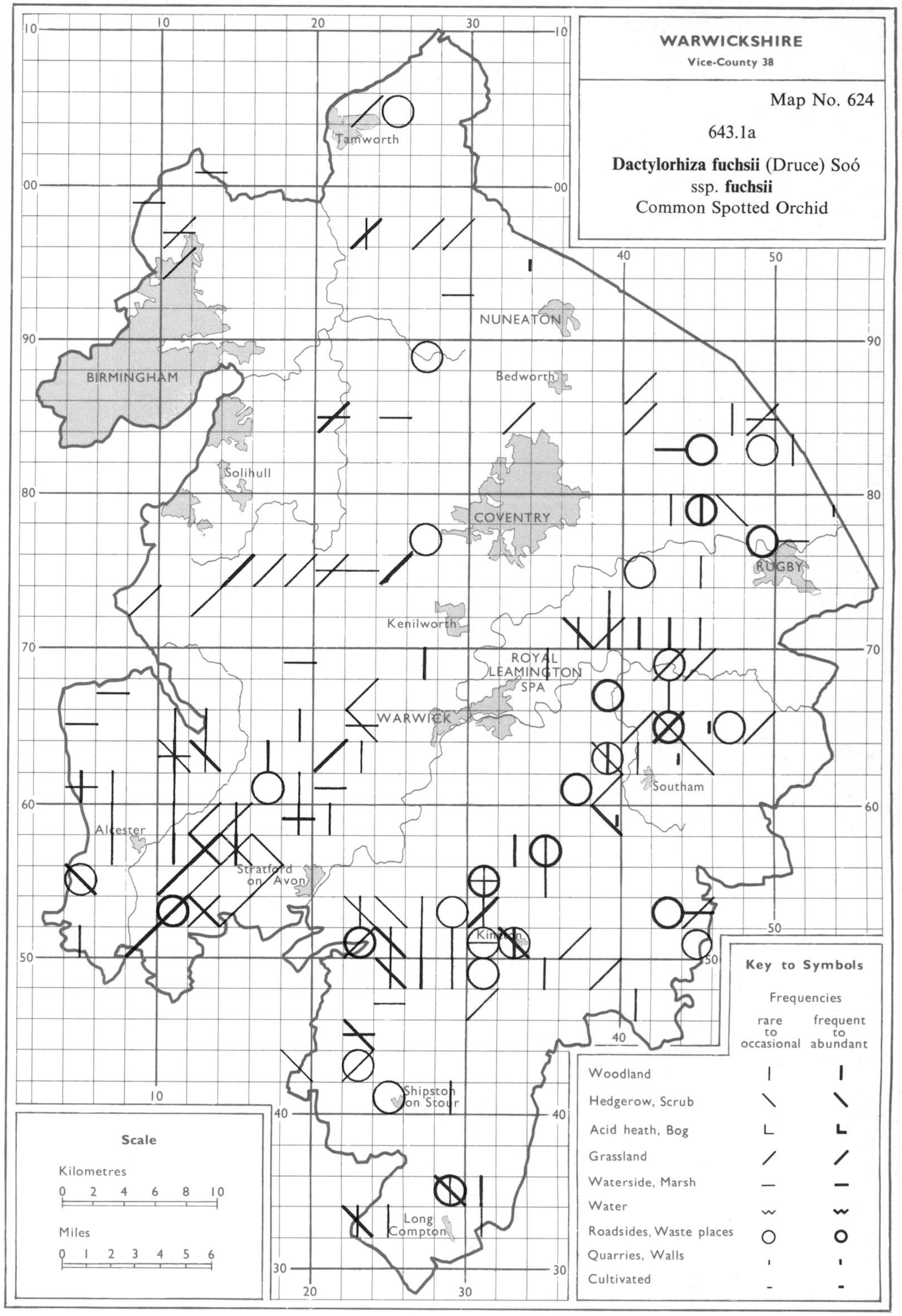
WARWICKSHIRE
Vice-County 38
Map No. 624
643.1a
Dactylorhiza fuchsii (Druce) Soó
ssp. **fuchsii**
Common Spotted Orchid
Tamworth
NUNEATON
Bedworth
BIRMINGHAM
Solihull
COVENTRY
RUGBY
Kenilworth
ROYAL LEAMINGTON SPA
WARWICK
Southam
Alcester
Stratford on Avon
Kineton
Shipston on Stour
Long Compton
Key to Symbols
Frequencies
rare to occasional
frequent to abundant
Woodland
Hedgerow, Scrub
Acid heath, Bog
Grassland
Waterside, Marsh
Water
Roadsides, Waste places
Quarries, Walls
Cultivated
Scale
Kilometres
0 2 4 6 8 10
Miles
0 1 2 3 4 5 6

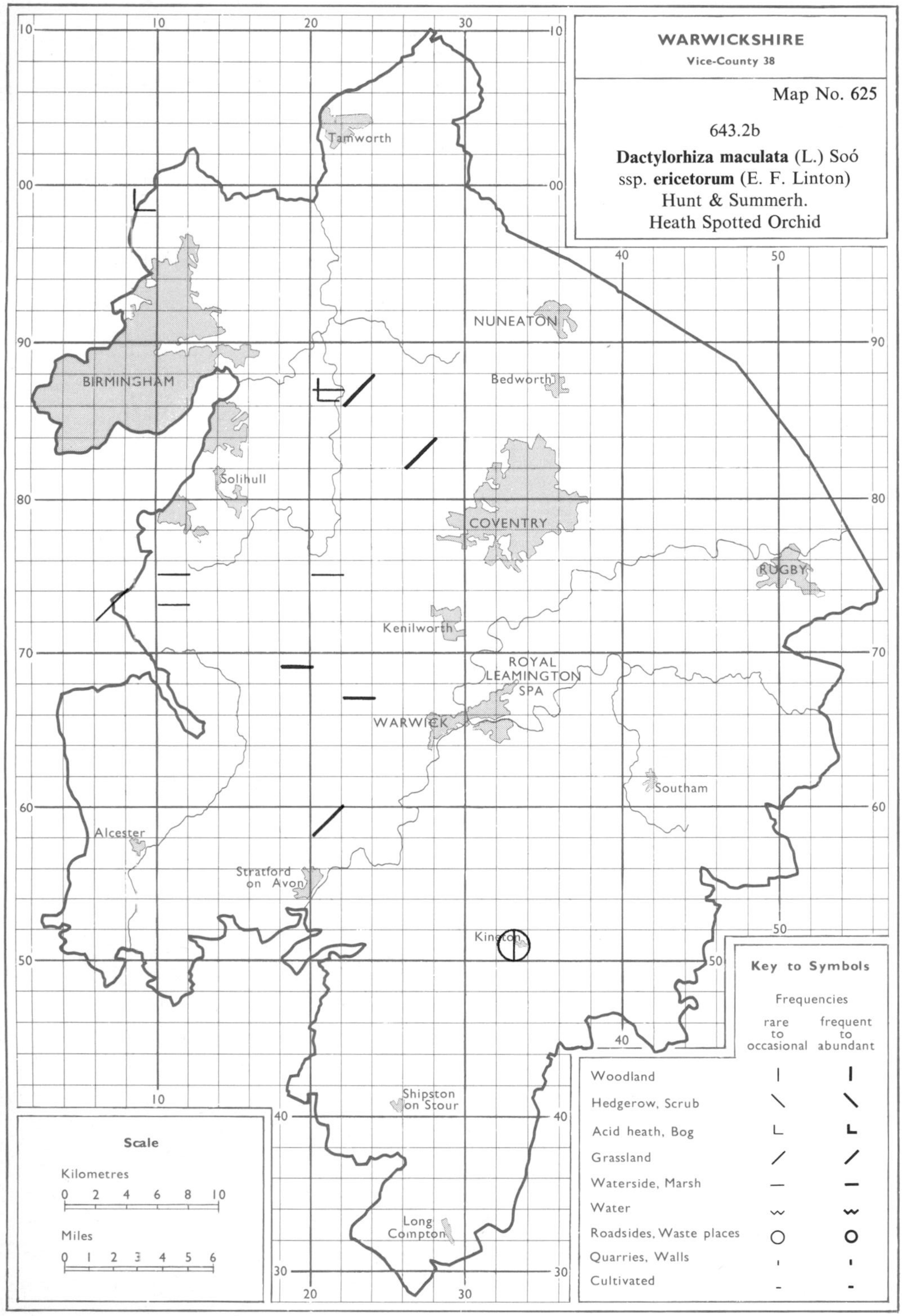
WARWICKSHIRE
Vice-County 38
Map No. 625
643.2b
Dactylorhiza maculata (L.) Soó
ssp. **ericetorum** (E. F. Linton)
Hunt & Summerh.
Heath Spotted Orchid
Tamworth
NUNEATON
Bedworth
BIRMINGHAM
Solihull
COVENTRY
RUGBY
Kenilworth
ROYAL
LEAMINGTON
SPA
WARWICK
Southam
Alcester
Stratford
on Avon
Kineton
Shipston
on Stour
Long
Compton
Key to Symbols
Frequencies
rare to occasional
frequent to abundant
Woodland
Hedgerow, Scrub
Acid heath, Bog
Grassland
Waterside, Marsh
Water
Roadsides, Waste places
Quarries, Walls
Cultivated
Scale
Kilometres
0 2 4 6 8 10
Miles
0 1 2 3 4 5 6

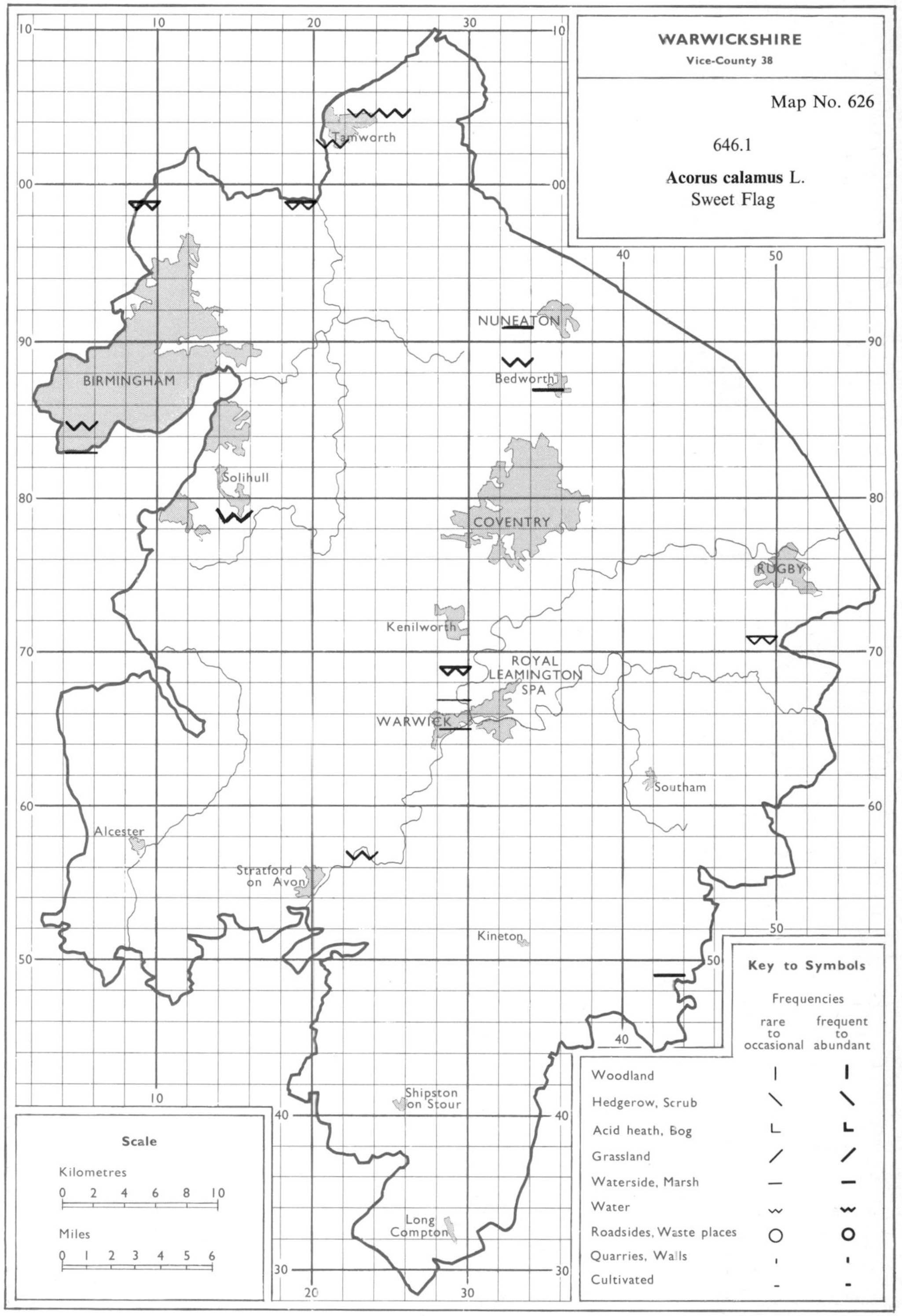
WARWICKSHIRE
Vice-County 38
Map No. 626
646.1
Acorus calamus L.
Sweet Flag
Tamworth
NUNEATON
Bedworth
BIRMINGHAM
Solihull
COVENTRY
RUGBY
Kenilworth
ROYAL
LEAMINGTON
SPA
WARWICK
Southam
Alcester
Stratford
on Avon
Kineton
Shipston
on Stour
Long
Compton
Key to Symbols
Frequencies
rare to occasional
frequent to abundant
Woodland
Hedgerow, Scrub
Acid heath, Bog
Grassland
Waterside, Marsh
Water
Roadsides, Waste places
Quarries, Walls
Cultivated
Scale
Kilometres
0 2 4 6 8 10
Miles
0 1 2 3 4 5 6

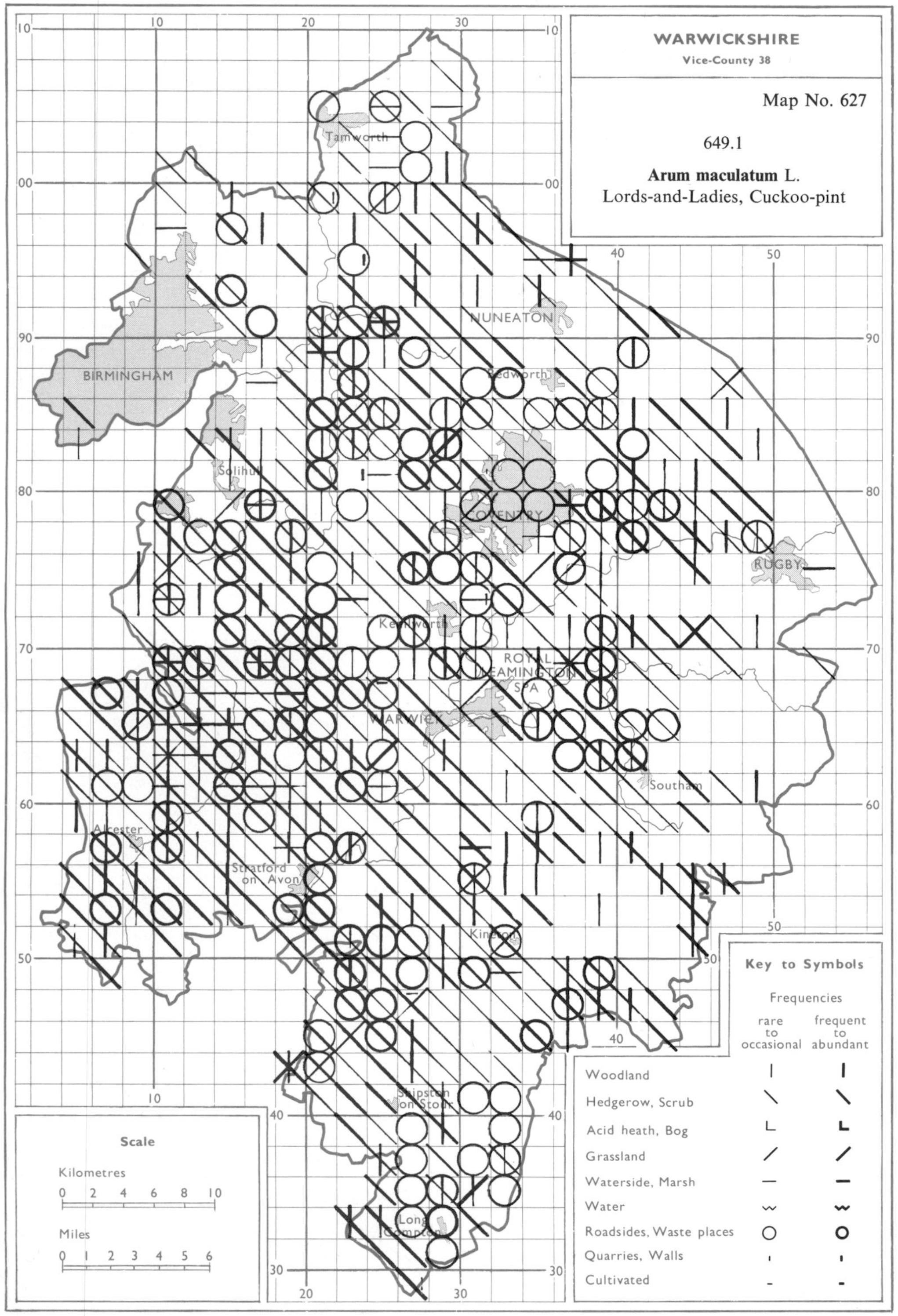
WARWICKSHIRE
Vice-County 38
Map No. 627
649.1
Arum maculatum L.
Lords-and-Ladies, Cuckoo-pint
Tamworth
NUNEATON
Bedworth
BIRMINGHAM
Solihull
COVENTRY
RUGBY
Kenilworth
ROYAL LEAMINGTON SPA
WARWICK
Southam
Alcester
Stratford on Avon
Kineton
Shipston on Stour
Long Compton
Key to Symbols
Frequencies
rare to occasional
frequent to abundant
Woodland
Hedgerow, Scrub
Acid heath, Bog
Grassland
Waterside, Marsh
Water
Roadsides, Waste places
Quarries, Walls
Cultivated
Scale
Kilometres
0 2 4 6 8 10
Miles
0 1 2 3 4 5 6

WARWICKSHIRE
Vice-County 38
Map No. 628
650.2
Lemna trisulca L.
Ivy-leaved Duckweed
Tamworth
NUNEATON
Bedworth
BIRMINGHAM
Solihull
COVENTRY
RUGBY
Kenilworth
ROYAL LEAMINGTON SPA
WARWICK
Southam
Alcester
Stratford on Avon
Kineton
Shipston on Stour
Long Compton
Key to Symbols
Frequencies
rare to occasional
frequent to abundant
Woodland
Hedgerow, Scrub
Acid heath, Bog
Grassland
Waterside, Marsh
Water
Roadsides, Waste places
Quarries, Walls
Cultivated
Scale
Kilometres
0 2 4 6 8 10
Miles
0 1 2 3 4 5 6

WARWICKSHIRE
Vice-County 38
Map No. 629
650.3
Lemna minor L.
Common Duckweed
Tamworth
NUNEATON
Bedworth
BIRMINGHAM
Solihull
COVENTRY
RUGBY
Kenilworth
ROYAL LEAMINGTON SPA
WARWICK
Southam
Alcester
Stratford on Avon
Kineton
Shipston on Stour
Long Compton
Scale
Kilometres
0 2 4 6 8 10
Miles
0 1 2 3 4 5 6
Key to Symbols
Frequencies
rare to occasional
frequent to abundant
Woodland
Hedgerow, Scrub
Acid heath, Bog
Grassland
Waterside, Marsh
Water
Roadsides, Waste places
Quarries, Walls
Cultivated

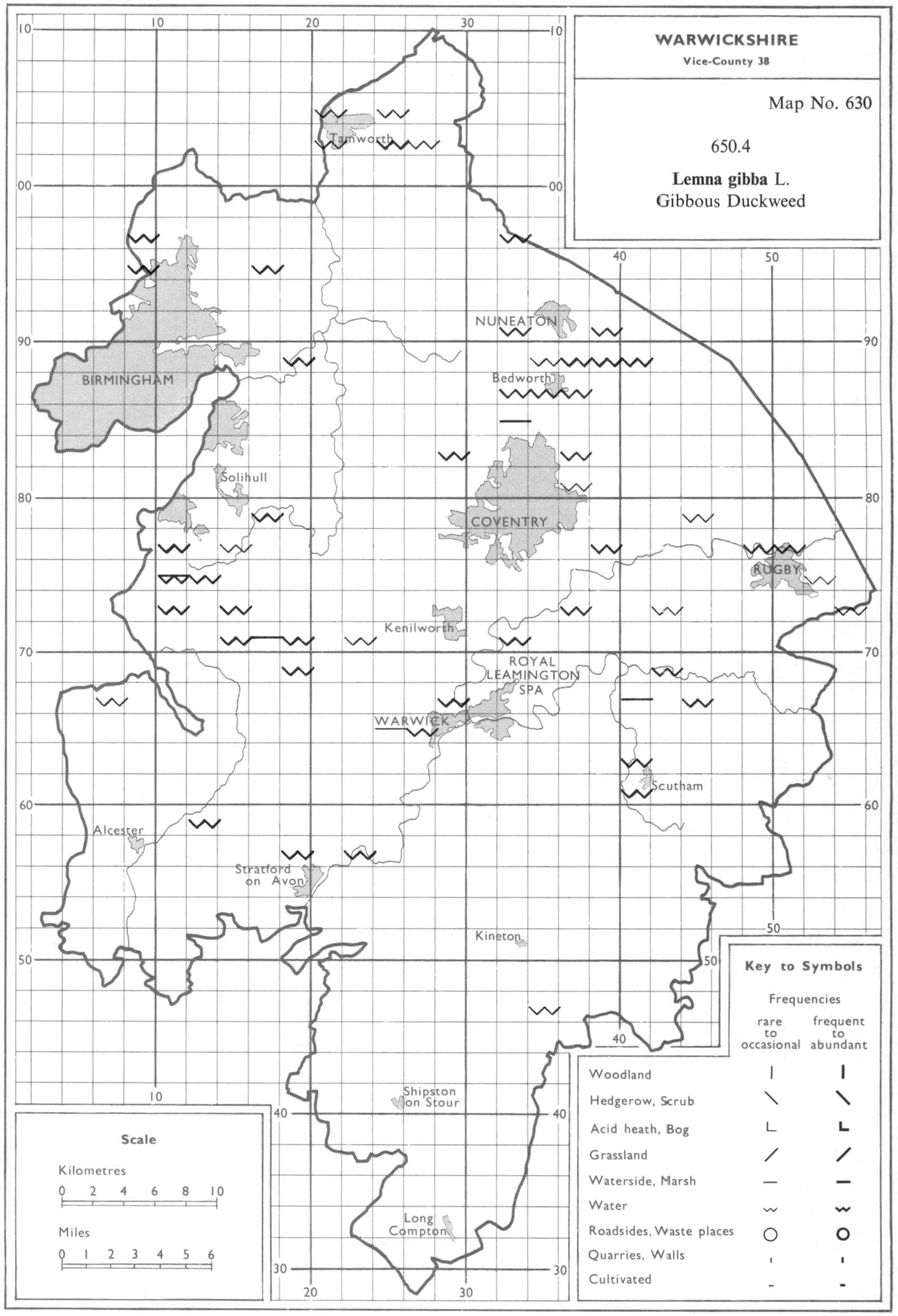
WARWICKSHIRE
Vice-County 38
Map No. 630
650.4
Lemna gibba L.
Gibbous Duckweed
Tamworth
NUNEATON
Bedworth
BIRMINGHAM
Solihull
COVENTRY
RUGBY
Kenilworth
ROYAL LEAMINGTON SPA
WARWICK
Scutham
Alcester
Stratford on Avon
Kineton
Shipston on Stour
Long Compton
Scale
Kilometres
0 2 4 6 8 10
Miles
0 1 2 3 4 5 6
Key to Symbols
Frequencies
rare to occasional
frequent to abundant
Woodland
Hedgerow, Scrub
Acid heath, Bog
Grassland
Waterside, Marsh
Water
Roadsides, Waste places
Quarries, Walls
Cultivated

WARWICKSHIRE
Vice-County 38
Map No. 631
652.1
Sparganium erectum L.
Branched Bur-reed
Tamworth
NUNEATON
Bedworth
BIRMINGHAM
Solihull
COVENTRY
RUGBY
Kenilworth
ROYAL LEAMINGTON SPA
WARWICK
Southam
Alcester
Stratford on Avon
Kineton
Shipston on Stour
Long Compton
Scale
Kilometres
0 2 4 6 8 10
Miles
0 1 2 3 4 5 6
Key to Symbols
Frequencies
rare to occasional
frequent to abundant
Woodland
Hedgerow, Scrub
Acid heath, Bog
Grassland
Waterside, Marsh
Water
Roadsides, Waste places
Quarries, Walls
Cultivated

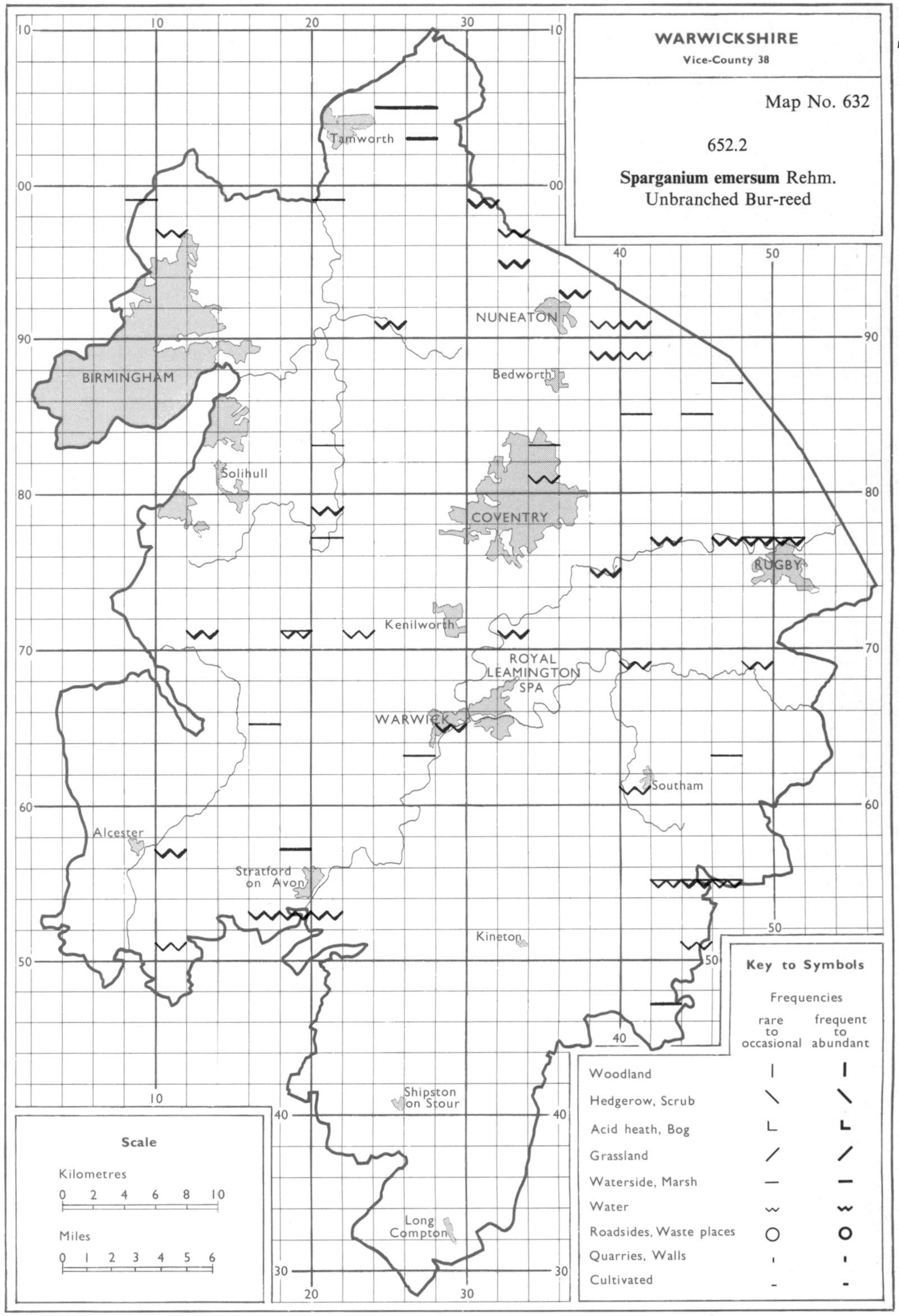
WARWICKSHIRE
Vice-County 38
Map No. 632
652.2
Sparganium emersum Rehm.
Unbranched Bur-reed
Tamworth
NUNEATON
Bedworth
BIRMINGHAM
Solihull
COVENTRY
RUGBY
Kenilworth
ROYAL LEAMINGTON SPA
WARWICK
Southam
Alcester
Stratford on Avon
Kineton
Shipston on Stour
Long Compton
Scale
Kilometres
0 2 4 6 8 10
Miles
0 1 2 3 4 5 6
Key to Symbols
Frequencies
rare to occasional
frequent to abundant
Woodland
Hedgerow, Scrub
Acid heath, Bog
Grassland
Waterside, Marsh
Water
Roadsides, Waste places
Quarries, Walls
Cultivated

WARWICKSHIRE
Vice-County 38
Map No. 634
653.2
Typha angustifolia L.
Lesser Bulrush
Key to Symbols
Frequencies
rare to occasional
frequent to abundant
Woodland
Hedgerow, Scrub
Acid heath, Bog
Grassland
Waterside, Marsh
Water
Roadsides, Waste places
Quarries, Walls
Cultivated
Scale
Kilometres 0 2 4 6 8 10
Miles 0 1 2 3 4 5 6
Tamworth
NUNEATON
Bedworth
COVENTRY
RUGBY
Southam
ROYAL LEAMINGTON SPA
Kenilworth
WARWICK
Kineton
Shipston on Stour
Long Compton
Stratford on Avon
Solihull
BIRMINGHAM
Alcester

WARWICKSHIRE
Vice-County 38
Map No. 633
653.1
Typha latifolia L.
Bulrush
Key to Symbols
Frequencies
rare to occasional
frequent to abundant
Woodland
Hedgerow, Scrub
Acid heath, Bog
Grassland
Waterside, Marsh
Water
Roadsides, Waste places
Quarries, Walls
Cultivated
Scale
Kilometres 0 2 4 6 8 10
Miles 0 1 2 3 4 5 6
Tamworth
NUNEATON
Bedworth
COVENTRY
RUGBY
Southam
ROYAL LEAMINGTON SPA
Kenilworth
WARWICK
Kineton
Shipston on Stour
Long Compton
Stratford on Avon
Solihull
BIRMINGHAM
Alcester

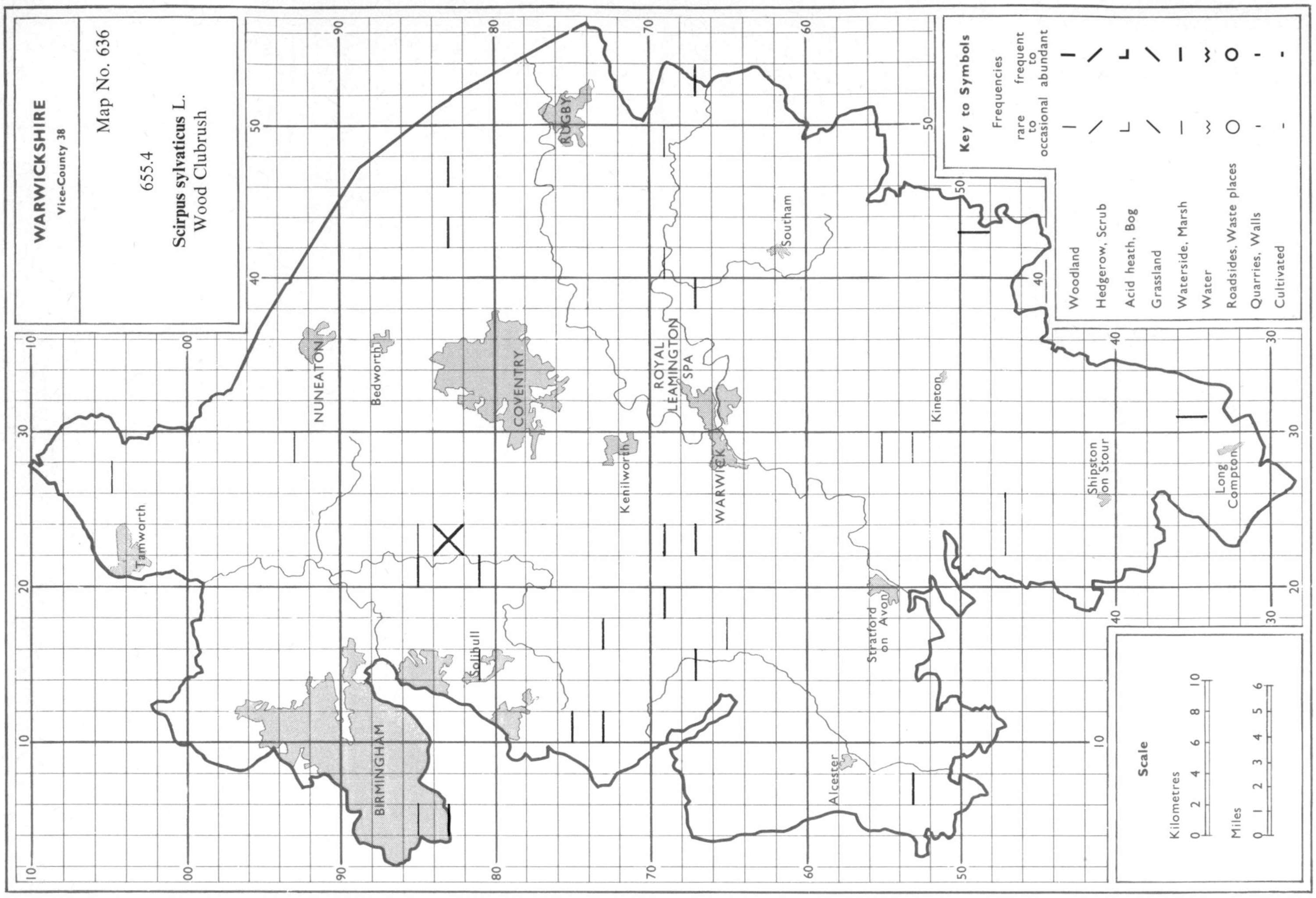
WARWICKSHIRE
Vice-County 38
Map No. 636
655.4
Scirpus sylvaticus L.
Wood Clubrush
Key to Symbols
Frequencies
rare to occasional
frequent to abundant
Woodland
Hedgerow, Scrub
Acid heath, Bog
Grassland
Waterside, Marsh
Water
Roadsides, Waste places
Quarries, Walls
Cultivated
Scale
Kilometres
Miles
NUNEATON
Bedworth
COVENTRY
RUGBY
Tamworth
Kenilworth
ROYAL LEAMINGTON SPA
WARWICK
Southam
Kineton
Shipston on Stour
Long Compton
Stratford on Avon
Solihull
BIRMINGHAM
Alcester

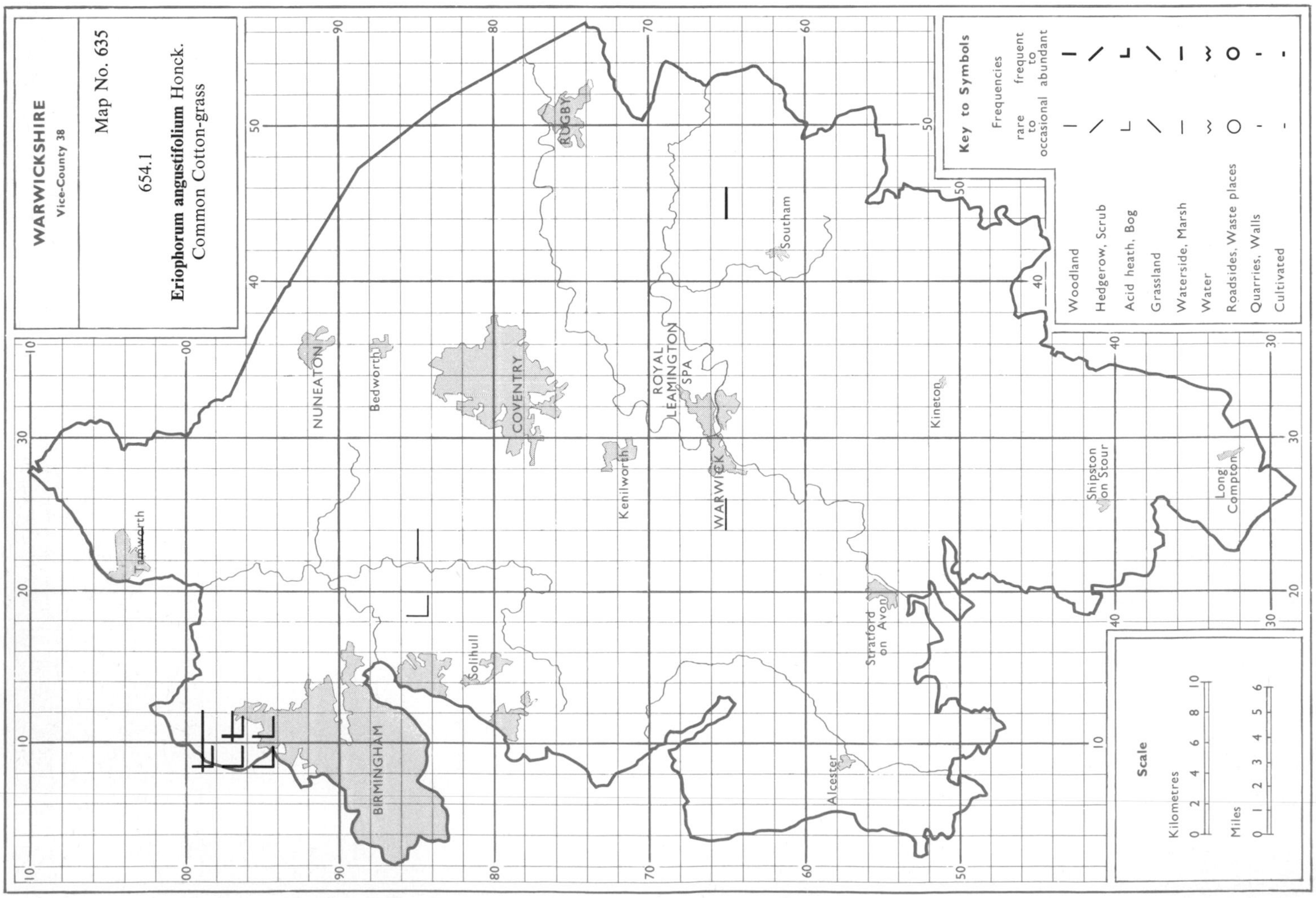
WARWICKSHIRE
Vice-County 38
Map No. 635
654.1
Eriophorum angustifolium Honck.
Common Cotton-grass
Key to Symbols
Frequencies
rare to occasional
frequent to abundant
Woodland
Hedgerow, Scrub
Acid heath, Bog
Grassland
Waterside, Marsh
Water
Roadsides, Waste places
Quarries, Walls
Cultivated
Scale
Kilometres
Miles
NUNEATON
Bedworth
COVENTRY
RUGBY
Tamworth
Kenilworth
ROYAL LEAMINGTON SPA
WARWICK
Southam
Kineton
Shipston on Stour
Long Compton
Stratford on Avon
Solihull
BIRMINGHAM
Alcester

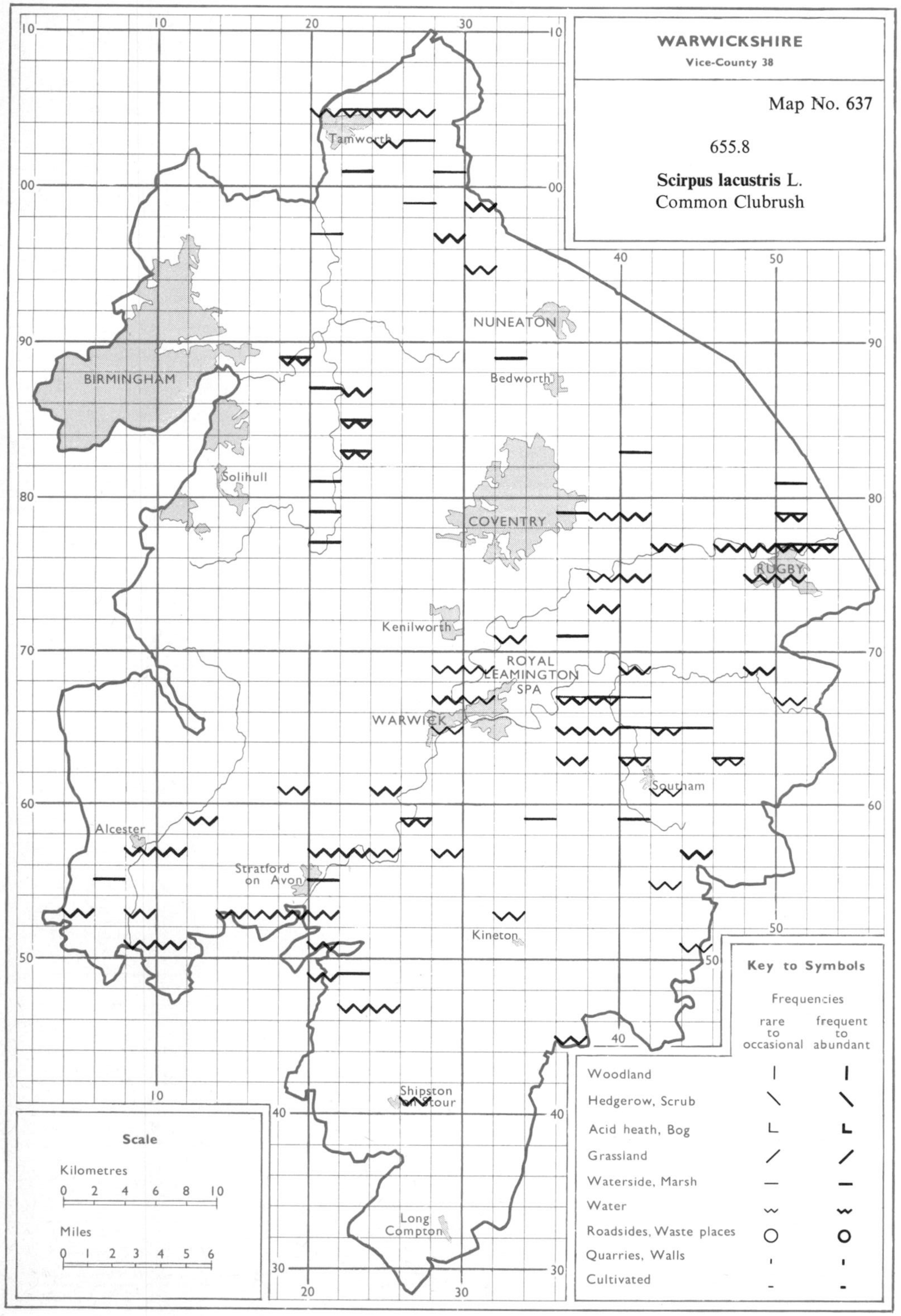
WARWICKSHIRE
Vice-County 38
Map No. 637
655.8
Scirpus lacustris L.
Common Clubrush
Tamworth
NUNEATON
Bedworth
BIRMINGHAM
Solihull
COVENTRY
RUGBY
Kenilworth
ROYAL LEAMINGTON SPA
WARWICK
Southam
Alcester
Stratford on Avon
Kineton
Shipston on Stour
Long Compton
Key to Symbols
Frequencies
rare to occasional
frequent to abundant
Woodland
Hedgerow, Scrub
Acid heath, Bog
Grassland
Waterside, Marsh
Water
Roadsides, Waste places
Quarries, Walls
Cultivated
Scale
Kilometres 0 2 4 6 8 10
Miles 0 1 2 3 4 5 6

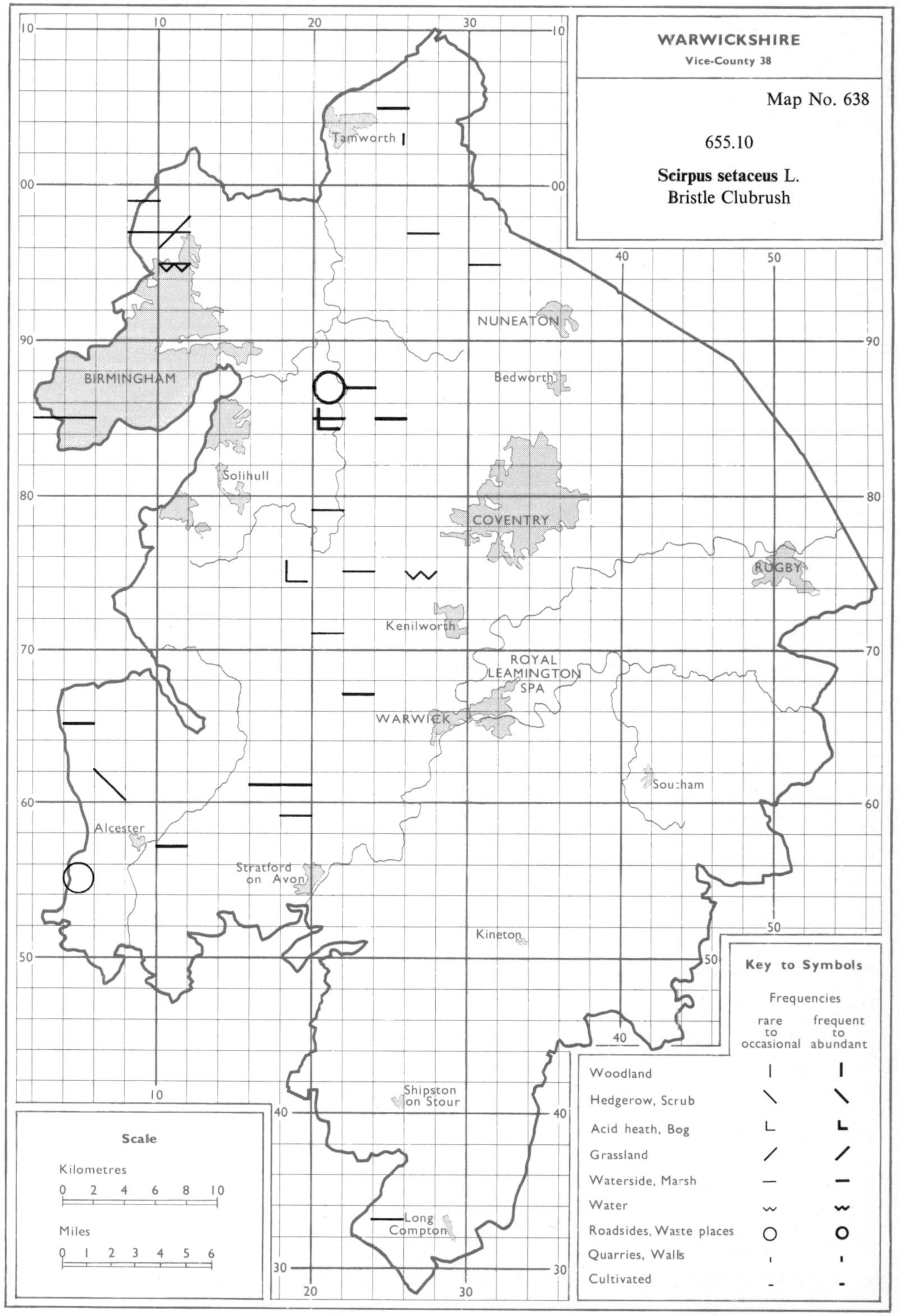
WARWICKSHIRE
Vice-County 38
Map No. 638
655.10
Scirpus setaceus L.
Bristle Clubrush
Tamworth
NUNEATON
Bedworth
BIRMINGHAM
Solihull
COVENTRY
RUGBY
Kenilworth
ROYAL LEAMINGTON SPA
WARWICK
Southam
Alcester
Stratford on Avon
Kineton
Shipston on Stour
Long Compton
Key to Symbols
Frequencies
rare to occasional
frequent to abundant
Woodland
Hedgerow, Scrub
Acid heath, Bog
Grassland
Waterside, Marsh
Water
Roadsides, Waste places
Quarries, Walls
Cultivated
Scale
Kilometres 0 2 4 6 8 10
Miles 0 1 2 3 4 5 6

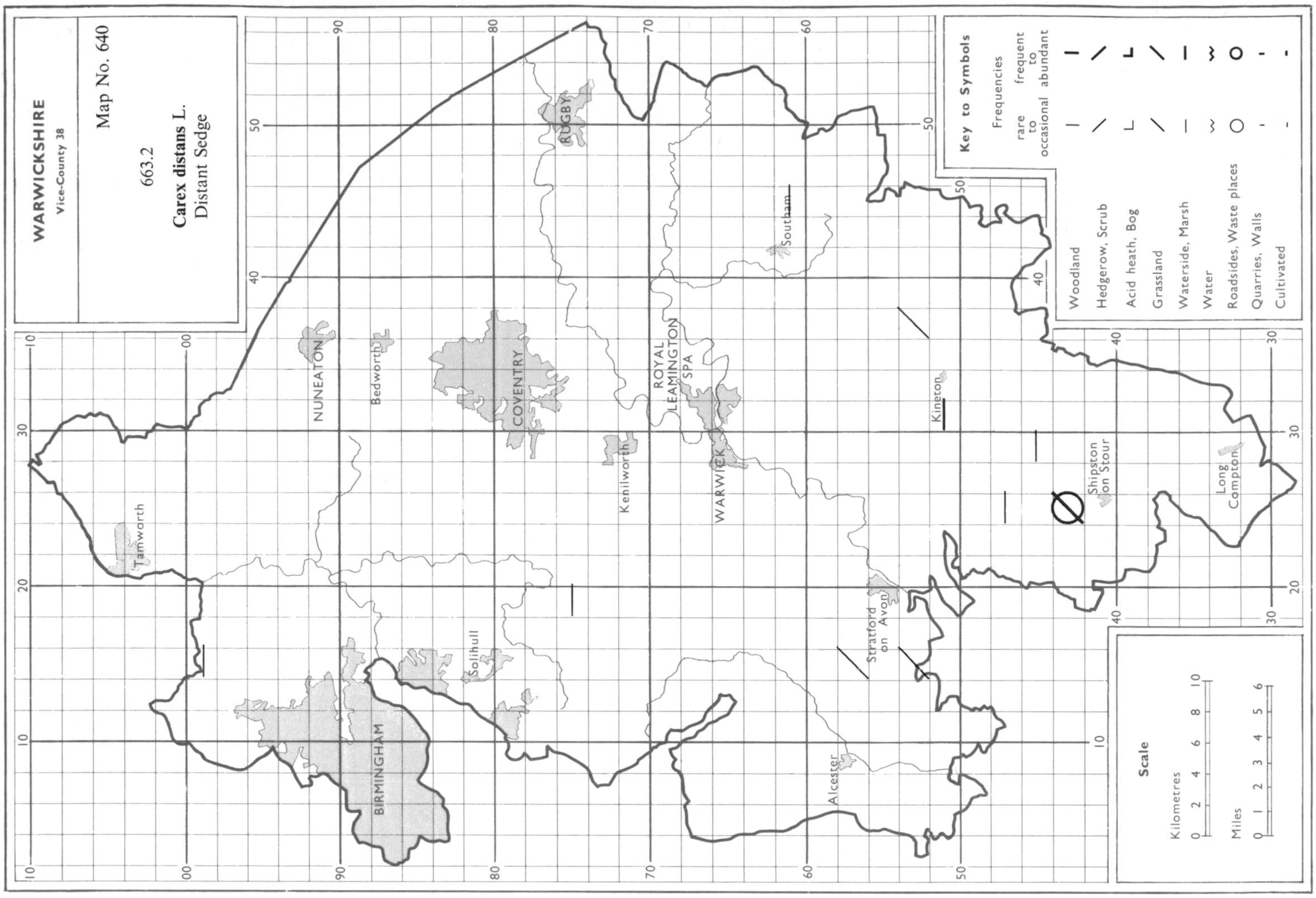
WARWICKSHIRE
Vice-County 38
Map No. 640
663.2
Carex distans L.
Distant Sedge
Key to Symbols
Frequencies
rare to occasional
frequent to abundant
Woodland
Hedgerow, Scrub
Acid heath, Bog
Grassland
Waterside, Marsh
Water
Roadsides, Waste places
Quarries, Walls
Cultivated
Scale
Kilometres
Miles
NUNEATON
Bedworth
COVENTRY
RUGBY
Southam
ROYAL LEAMINGTON SPA
WARWICK
Kenilworth
Kineton
Shipston on Stour
Long Compton
Tamworth
Solihull
BIRMINGHAM
Stratford on Avon
Alcester

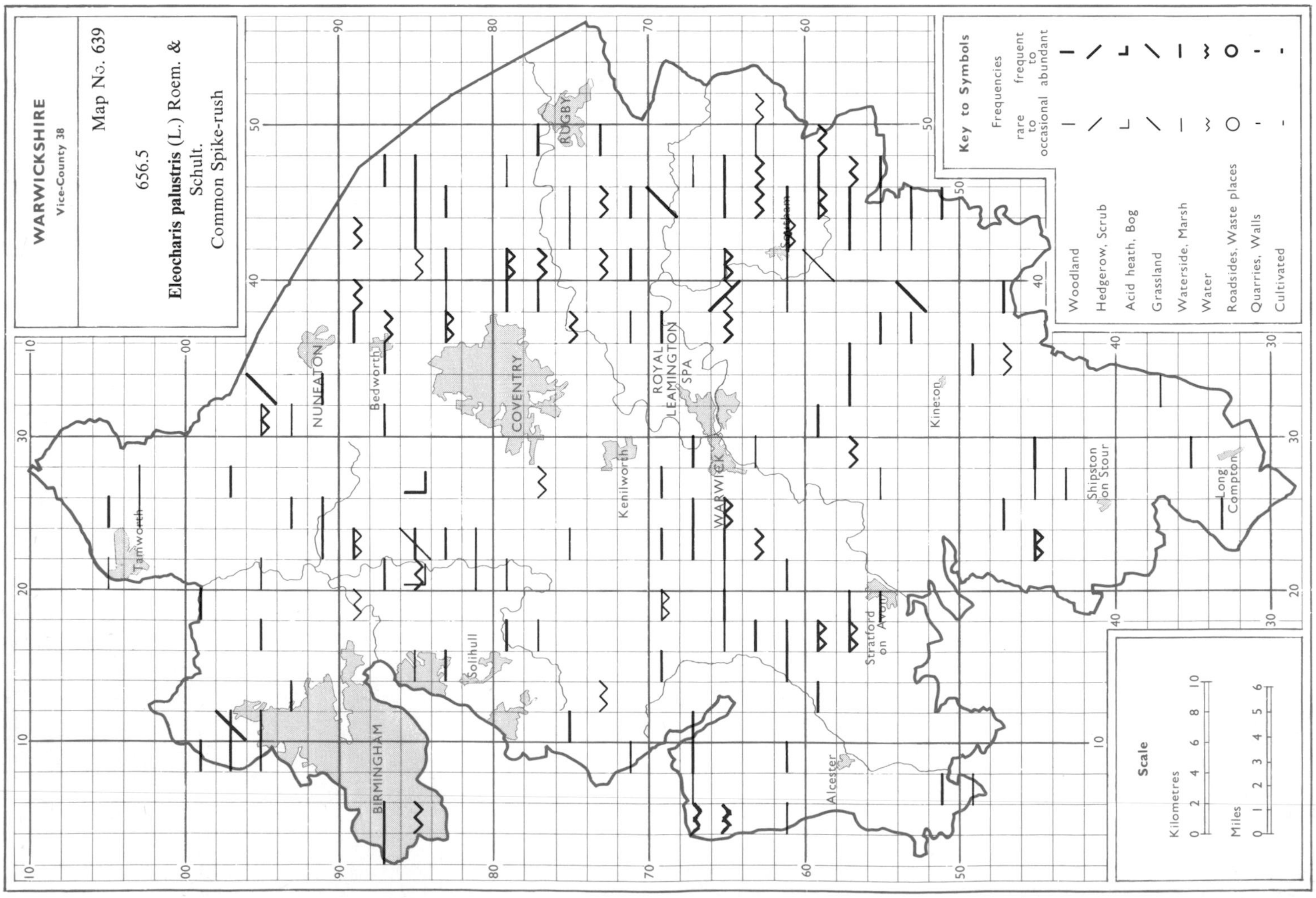
WARWICKSHIRE
Vice-County 38
Map No. 639
656.5
Eleocharis palustris (L.) Roem. & Schult.
Common Spike-rush
Key to Symbols
Frequencies
rare to occasional
frequent to abundant
Woodland
Hedgerow, Scrub
Acid heath, Bog
Grassland
Waterside, Marsh
Water
Roadsides, Waste places
Quarries, Walls
Cultivated
Scale
Kilometres
Miles
NUNEATON
Bedworth
COVENTRY
RUGBY
Southam
ROYAL LEAMINGTON SPA
WARWICK
Kenilworth
Kineton
Shipston on Stour
Long Compton
Tamworth
Solihull
BIRMINGHAM
Stratford on Avon
Alcester

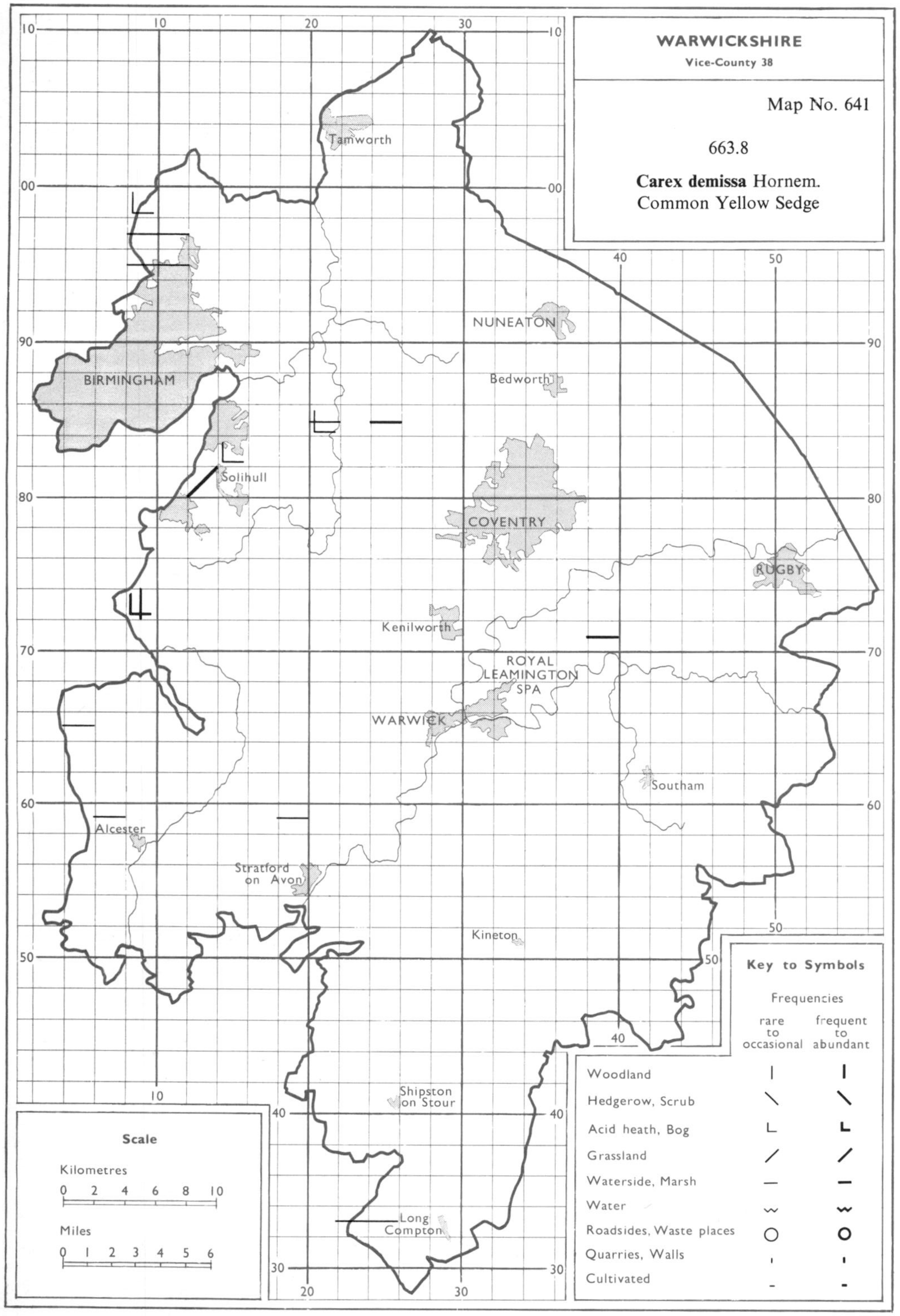
WARWICKSHIRE
Vice-County 38
Map No. 641
663.8
Carex demissa Hornem.
Common Yellow Sedge
Tamworth
NUNEATON
Bedworth
BIRMINGHAM
Solihull
COVENTRY
RUGBY
Kenilworth
ROYAL LEAMINGTON SPA
WARWICK
Southam
Alcester
Stratford on Avon
Kineton
Shipston on Stour
Long Compton
Key to Symbols
Frequencies
rare to occasional
frequent to abundant
Woodland
Hedgerow, Scrub
Acid heath, Bog
Grassland
Waterside, Marsh
Water
Roadsides, Waste places
Quarries, Walls
Cultivated
Scale
Kilometres
0 2 4 6 8 10
Miles
0 1 2 3 4 5 6

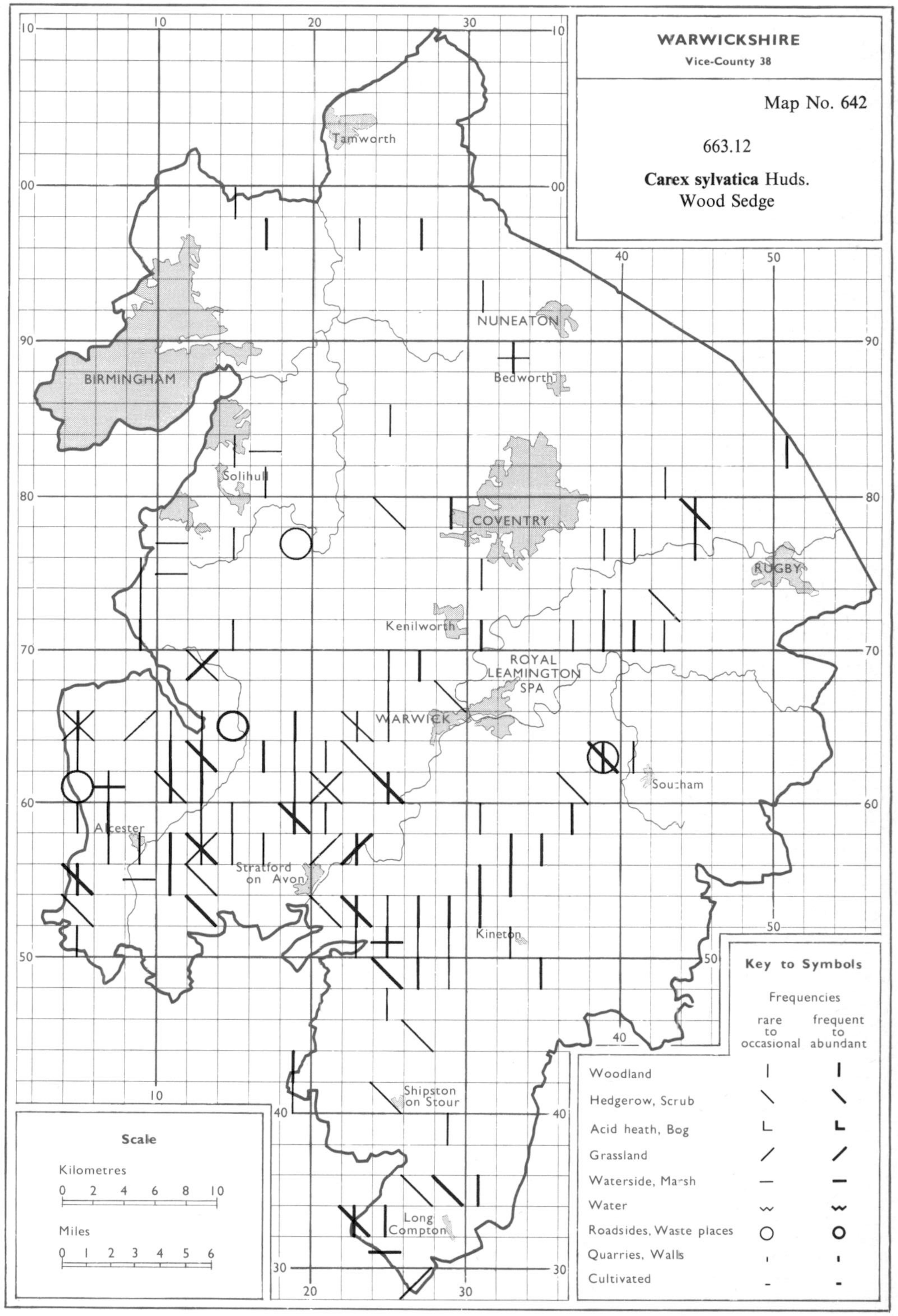
WARWICKSHIRE
Vice-County 38
Map No. 642
663.12
Carex sylvatica Huds.
Wood Sedge
Tamworth
NUNEATON
Bedworth
BIRMINGHAM
Solihull
COVENTRY
RUGBY
Kenilworth
ROYAL LEAMINGTON SPA
WARWICK
Southam
Alcester
Stratford on Avon
Kineton
Shipston on Stour
Long Compton
Key to Symbols
Frequencies
rare to occasional
frequent to abundant
Woodland
Hedgerow, Scrub
Acid heath, Bog
Grassland
Waterside, Marsh
Water
Roadsides, Waste places
Quarries, Walls
Cultivated
Scale
Kilometres
0 2 4 6 8 10
Miles
0 1 2 3 4 5 6

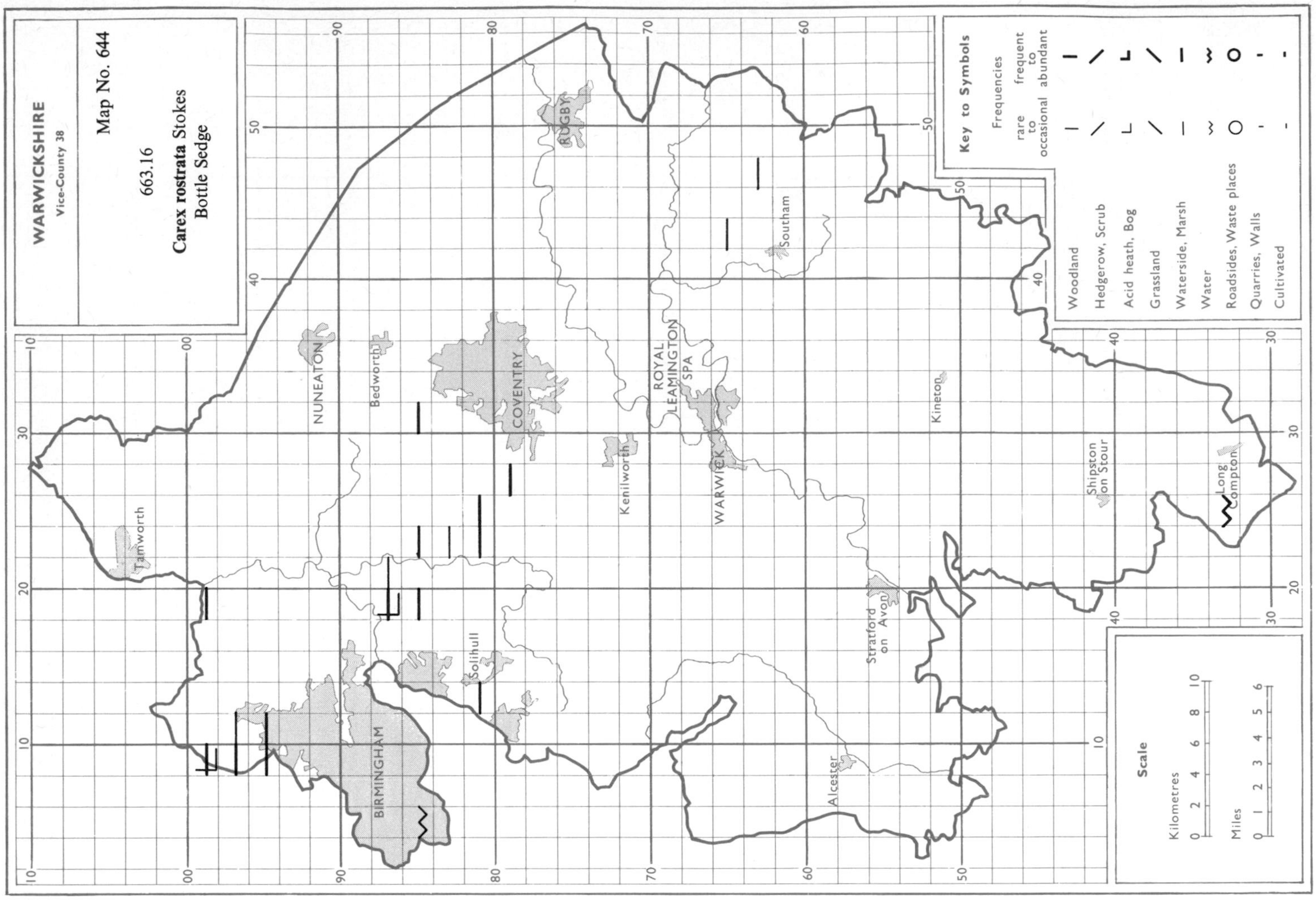
WARWICKSHIRE
Vice-County 38
Map No. 644
663.16
Carex rostrata Stokes
Bottle Sedge
Key to Symbols
Frequencies
rare to occasional
frequent to abundant
Woodland
Hedgerow, Scrub
Acid heath, Bog
Grassland
Waterside, Marsh
Water
Roadsides, Waste places
Quarries, Walls
Cultivated
Scale
Kilometres
0 2 4 6 8 10
Miles
0 1 2 3 4 5 6
TAMWORTH
NUNEATON
Bedworth
COVENTRY
RUGBY
Southam
ROYAL LEAMINGTON SPA
Kenilworth
WARWICK
Kineton
Shipston on Stour
Long Compton
BIRMINGHAM
Solihull
Stratford on Avon
Alcester

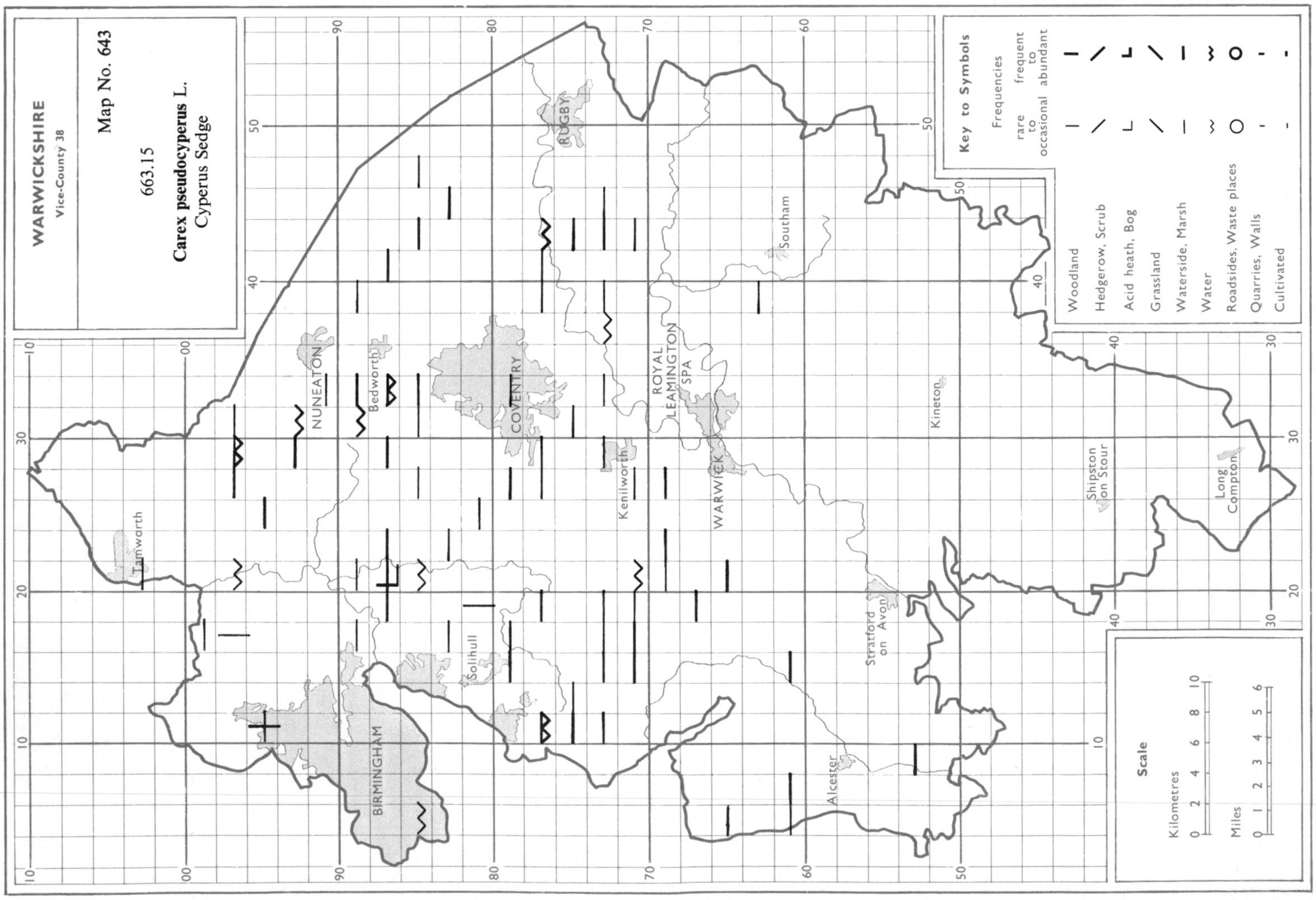
WARWICKSHIRE
Vice-County 38
Map No. 643
663.15
Carex pseudocyperus L.
Cyperus Sedge
Key to Symbols
Frequencies
rare to occasional
frequent to abundant
Woodland
Hedgerow, Scrub
Acid heath, Bog
Grassland
Waterside, Marsh
Water
Roadsides, Waste places
Quarries, Walls
Cultivated
Scale
Kilometres
0 2 4 6 8 10
Miles
0 1 2 3 4 5 6
Tamworth
NUNEATON
Bedworth
COVENTRY
RUGBY
Southam
ROYAL LEAMINGTON SPA
Kenilworth
WARWICK
Kineton
Shipston on Stour
Long Compton
BIRMINGHAM
Solihull
Stratford on Avon
Alcester

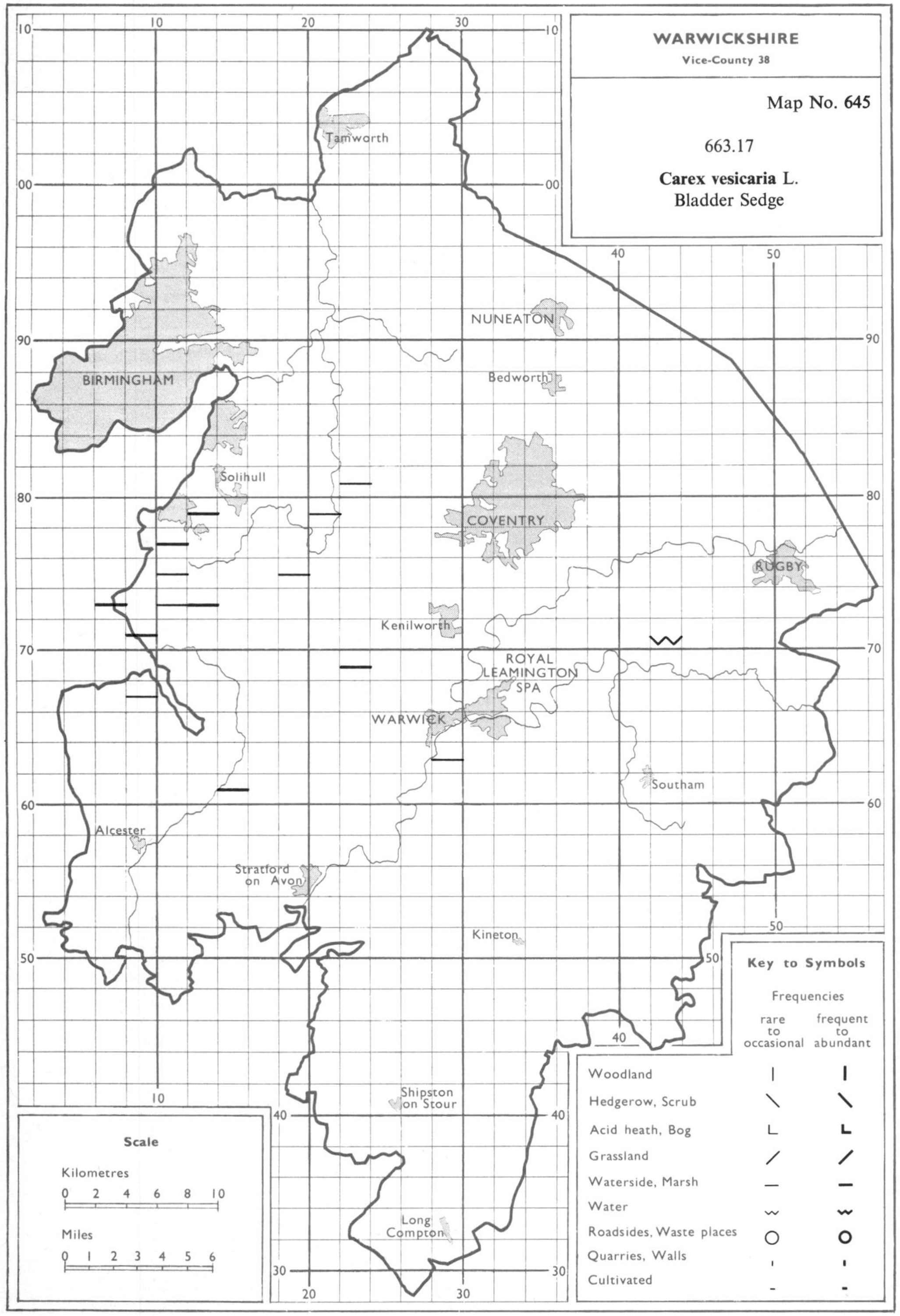

WARWICKSHIRE
Vice-County 38
Map No. 645
663.17
Carex vesicaria L.
Bladder Sedge
Tamworth
NUNEATON
Bedworth
BIRMINGHAM
Solihull
COVENTRY
RUGBY
Kenilworth
ROYAL LEAMINGTON SPA
WARWICK
Southam
Alcester
Stratford on Avon
Kineton
Shipston on Stour
Long Compton
Key to Symbols
Frequencies
rare to occasional
frequent to abundant
Woodland
Hedgerow, Scrub
Acid heath, Bog
Grassland
Waterside, Marsh
Water
Roadsides, Waste places
Quarries, Walls
Cultivated
Scale
Kilometres
0 2 4 6 8 10
Miles
0 1 2 3 4 5 6

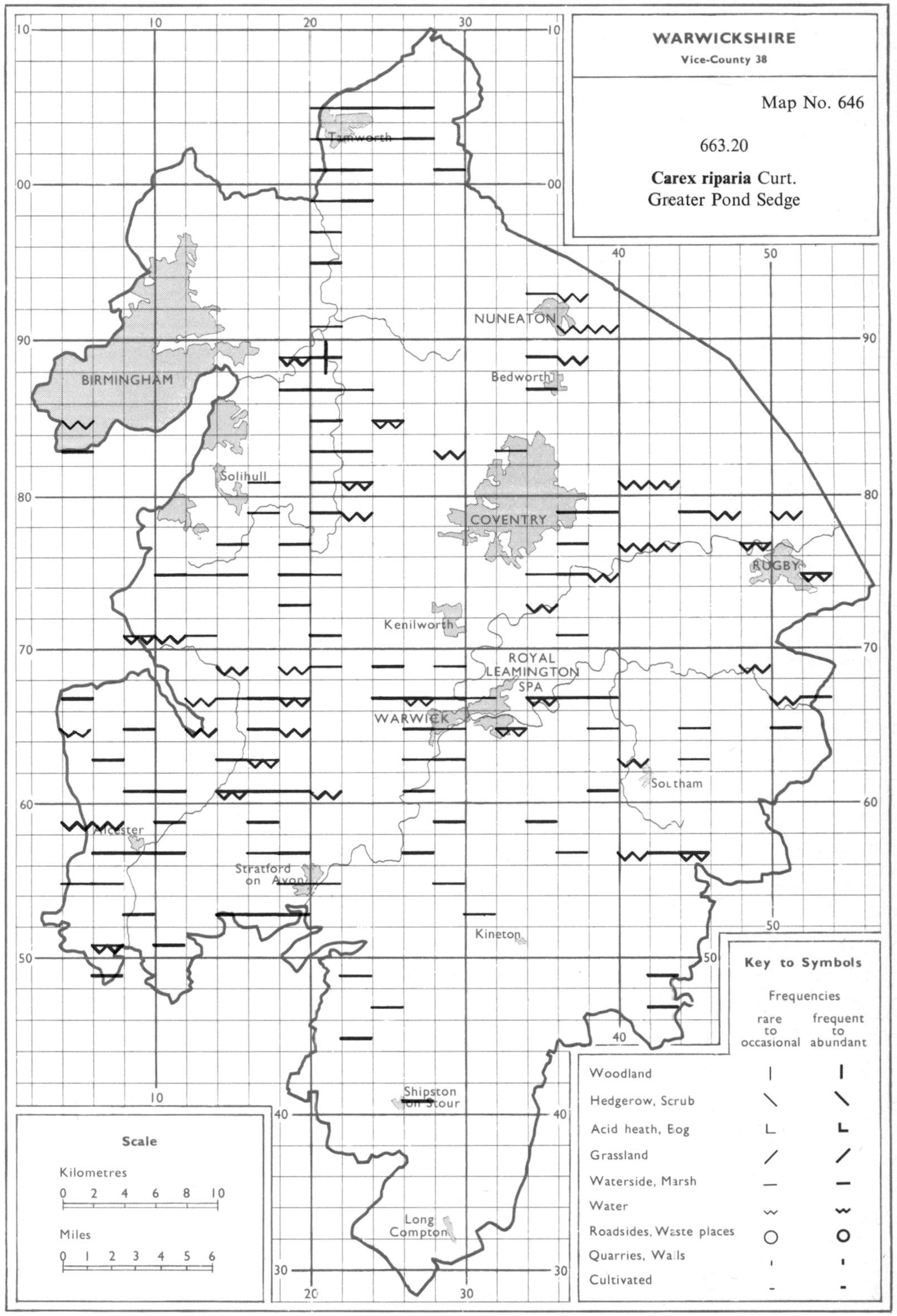

WARWICKSHIRE
Vice-County 38
Map No. 646
663.20
Carex riparia Curt.
Greater Pond Sedge
Tamworth
NUNEATON
Bedworth
BIRMINGHAM
Solihull
COVENTRY
RUGBY
Kenilworth
ROYAL LEAMINGTON SPA
WARWICK
Southam
Alcester
Stratford on Avon
Kineton
Shipston on Stour
Long Compton
Key to Symbols
Frequencies
rare to occasional
frequent to abundant
Woodland
Hedgerow, Scrub
Acid heath, Bog
Grassland
Waterside, Marsh
Water
Roadsides, Waste places
Quarries, Walls
Cultivated
Scale
Kilometres
0 2 4 6 8 10
Miles
0 1 2 3 4 5 6

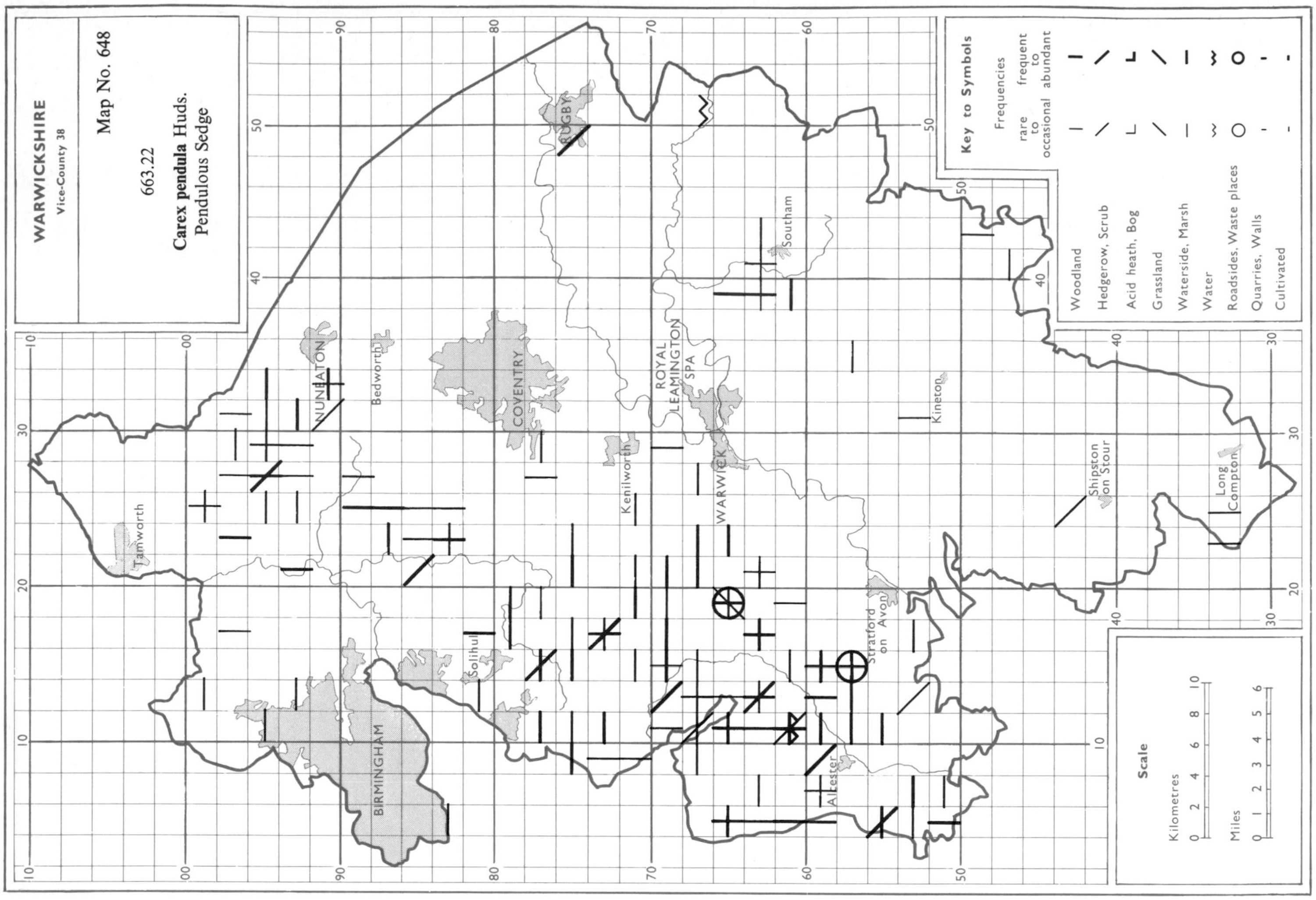
WARWICKSHIRE
Vice-County 38
Map No. 648
663.22
Carex pendula Huds.
Pendulous Sedge
Key to Symbols
Frequencies
rare to occasional
frequent to abundant
Woodland
Hedgerow, Scrub
Acid heath, Bog
Grassland
Waterside, Marsh
Water
Roadsides, Waste places
Quarries, Walls
Cultivated
Scale
Kilometres
0 2 4 6 8 10
Miles
0 1 2 3 4 5 6
Tamworth
NUNEATON
Bedworth
COVENTRY
RUGBY
Southam
ROYAL LEAMINGTON SPA
Kenilworth
WARWICK
Kineton
Shipston on Stour
Long Compton
Solihull
BIRMINGHAM
Stratford on Avon
Alcester

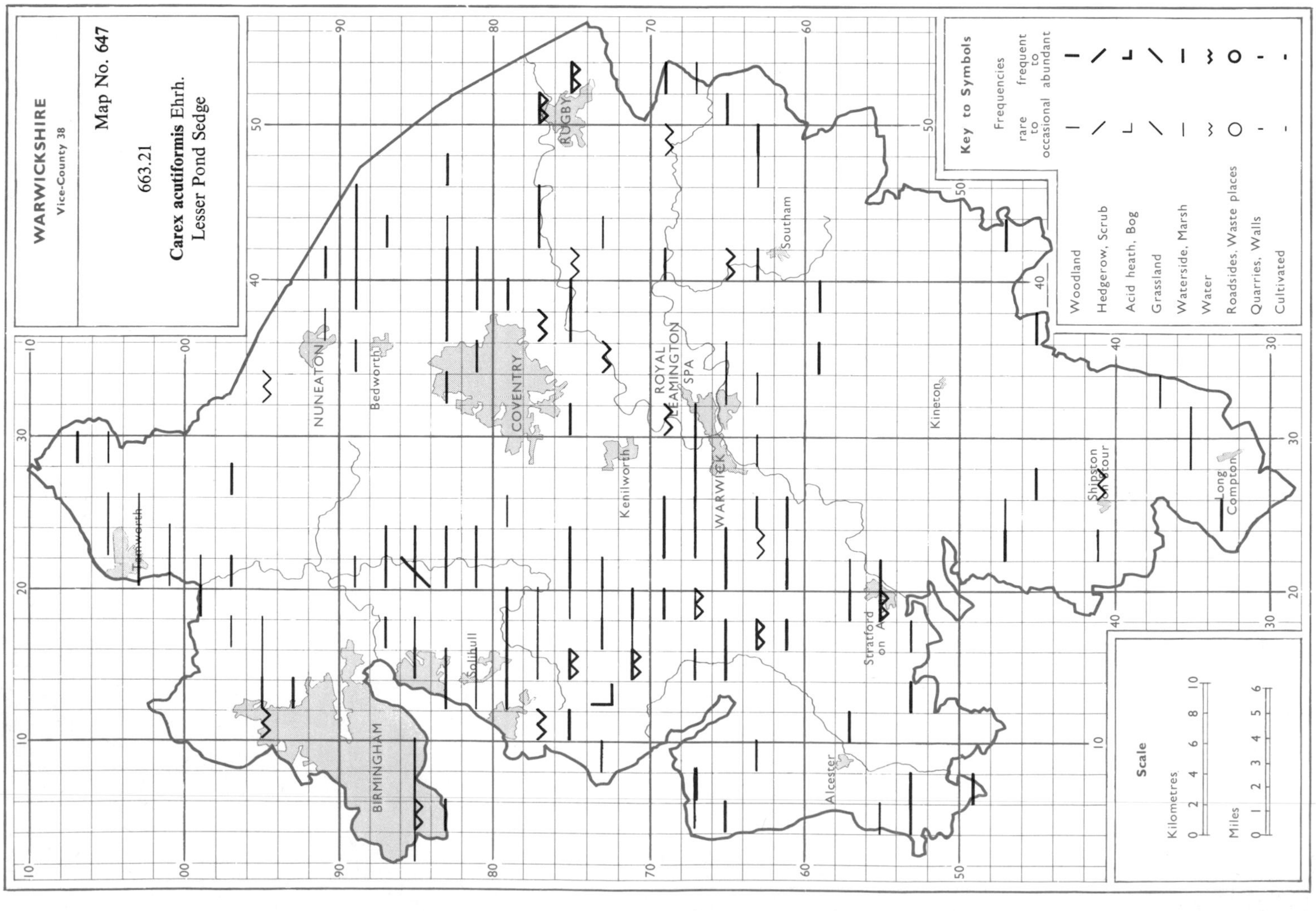
WARWICKSHIRE
Vice-County 38
Map No. 647
663.21
Carex acutiformis Ehrh.
Lesser Pond Sedge
Key to Symbols
Frequencies
rare to occasional
frequent to abundant
Woodland
Hedgerow, Scrub
Acid heath, Bog
Grassland
Waterside, Marsh
Water
Roadsides, Waste places
Quarries, Walls
Cultivated
Scale
Kilometres
0 2 4 6 8 10
Miles
0 1 2 3 4 5 6
Tamworth
NUNEATON
Bedworth
COVENTRY
RUGBY
Southam
ROYAL LEAMINGTON SPA
Kenilworth
WARWICK
Kineton
Shipston on Stour
Long Compton
Solihull
BIRMINGHAM
Stratford on Avon
Alcester

WARWICKSHIRE
Vice-County 38
Map No. 649
663.24
Carex pallescens L.
Pale Sedge
Tamworth
BIRMINGHAM
Solihull
NUNEATON
Bedworth
COVENTRY
RUGBY
Kenilworth
ROYAL LEAMINGTON SPA
WARWICK
Southam
Alcester
Stratford on Avon
Kineton
Shipston on Stour
Long Compton
Scale
Kilometres
0 2 4 6 8 10
Miles
0 1 2 3 4 5 6
Key to Symbols
Frequencies
rare to occasional
frequent to abundant
Woodland
Hedgerow, Scrub
Acid heath, Bog
Grassland
Waterside, Marsh
Water
Roadsides, Waste places
Quarries, Walls
Cultivated

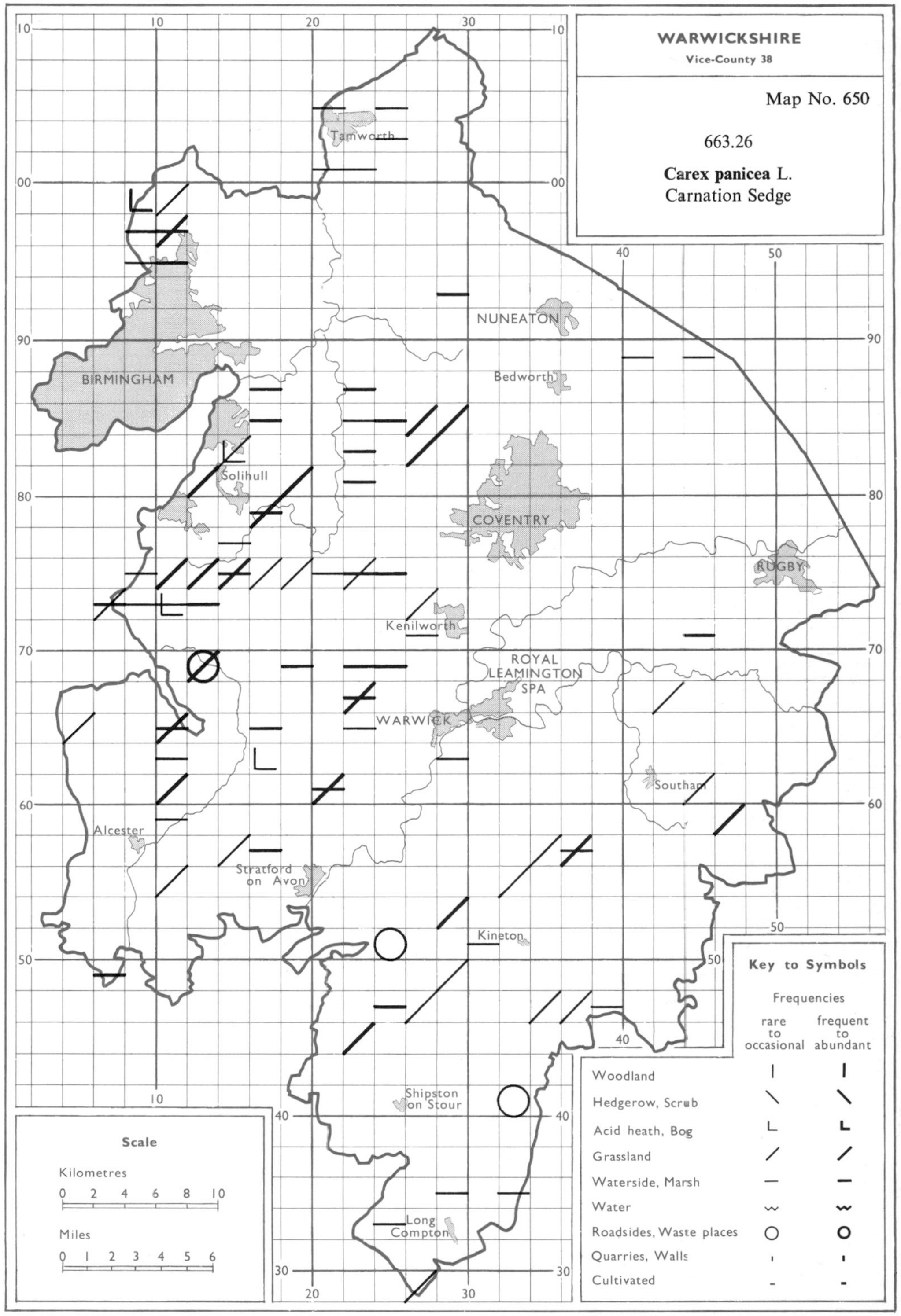
WARWICKSHIRE
Vice-County 38
Map No. 650
663.26
Carex panicea L.
Carnation Sedge
Tamworth
BIRMINGHAM
Solihull
NUNEATON
Bedworth
COVENTRY
RUGBY
Kenilworth
ROYAL LEAMINGTON SPA
WARWICK
Southam
Alcester
Stratford on Avon
Kineton
Shipston on Stour
Long Compton
Scale
Kilometres
0 2 4 6 8 10
Miles
0 1 2 3 4 5 6
Key to Symbols
Frequencies
rare to occasional
frequent to abundant
Woodland
Hedgerow, Scrub
Acid heath, Bog
Grassland
Waterside, Marsh
Water
Roadsides, Waste places
Quarries, Walls
Cultivated

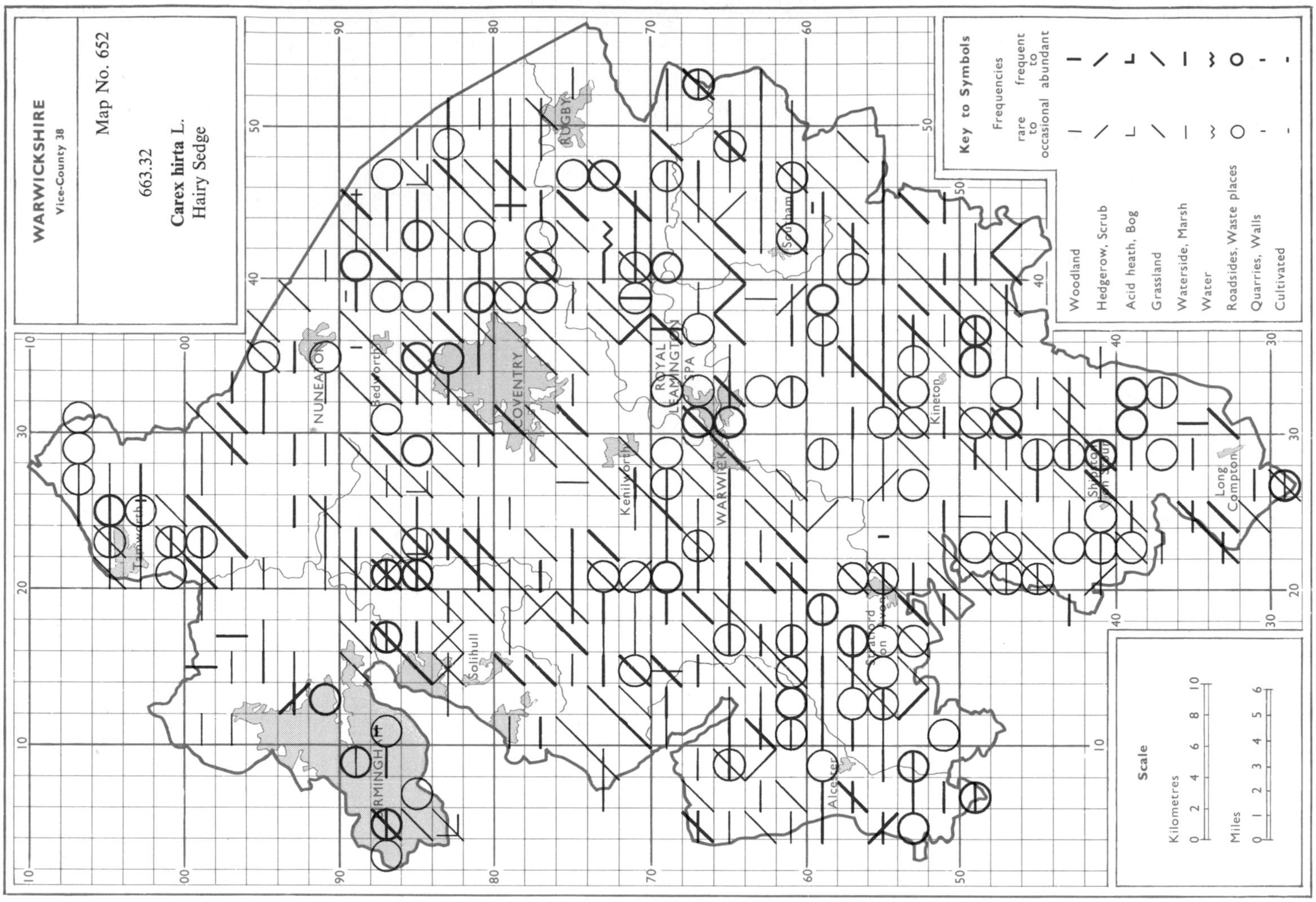
WARWICKSHIRE
Vice-County 38
Map No. 652
663.32
Carex hirta L.
Hairy Sedge
Key to Symbols
Frequencies
rare to occasional
frequent to abundant
Woodland
Hedgerow, Scrub
Acid heath, Bog
Grassland
Waterside, Marsh
Water
Roadsides, Waste places
Quarries, Walls
Cultivated
Scale
Kilometres
Miles
NUNEATON
Bedworth
COVENTRY
RUGBY
Tamworth
BIRMINGHAM
Solihull
Kenilworth
ROYAL LEAMINGTON SPA
WARWICK
Southam
Kineton
Stratford on Avon
Alcester
Shipston on Stour
Long Compton

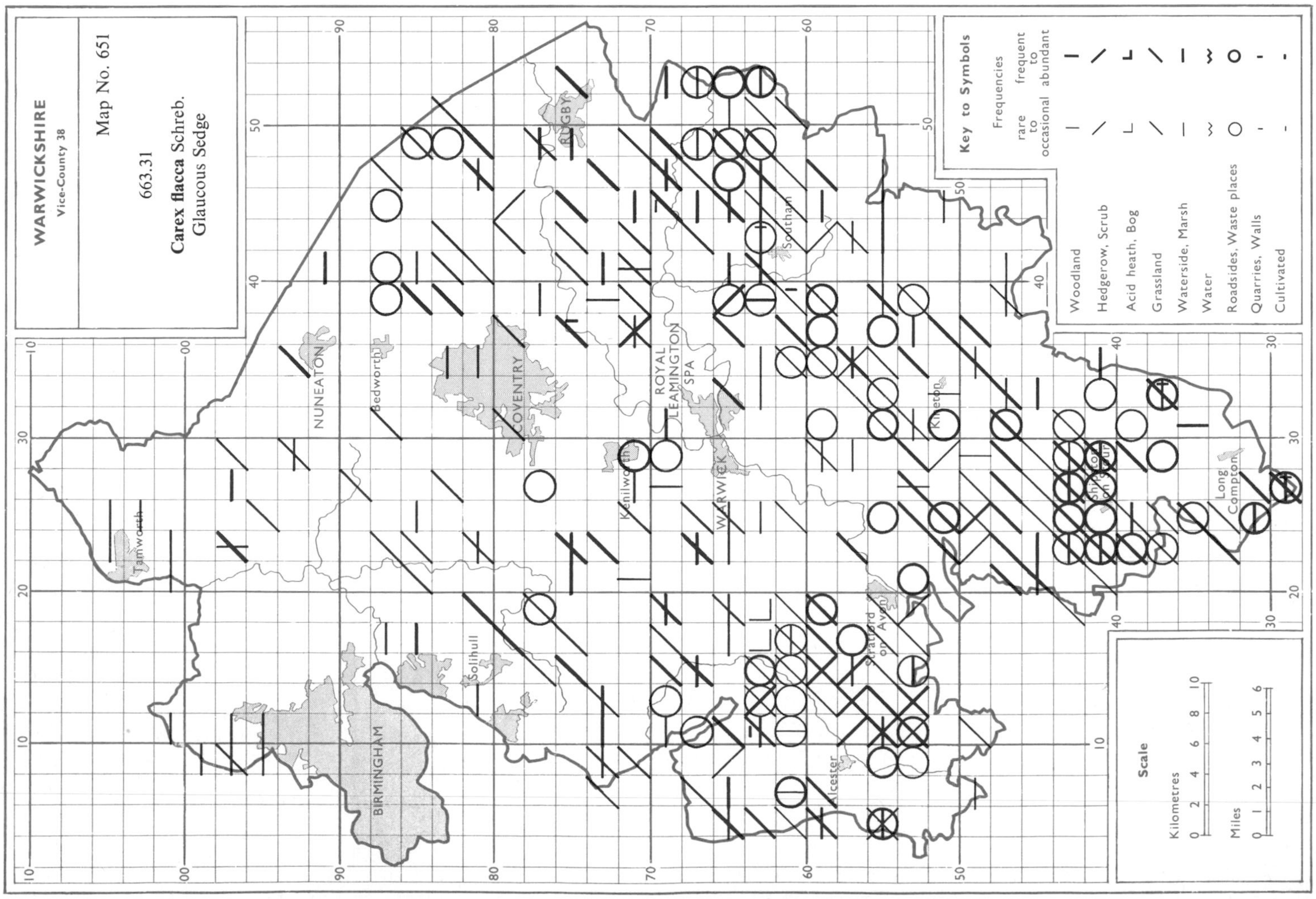
WARWICKSHIRE
Vice-County 38
Map No. 651
663.31
Carex flacca Schreb.
Glaucous Sedge
Key to Symbols
Frequencies
rare to occasional
frequent to abundant
Woodland
Hedgerow, Scrub
Acid heath, Bog
Grassland
Waterside, Marsh
Water
Roadsides, Waste places
Quarries, Walls
Cultivated
Scale
Kilometres
Miles
NUNEATON
Bedworth
COVENTRY
RUGBY
Tamworth
BIRMINGHAM
Solihull
Kenilworth
ROYAL LEAMINGTON SPA
WARWICK
Southam
Kineton
Stratford on Avon
Alcester
Shipston on Stour
Long Compton

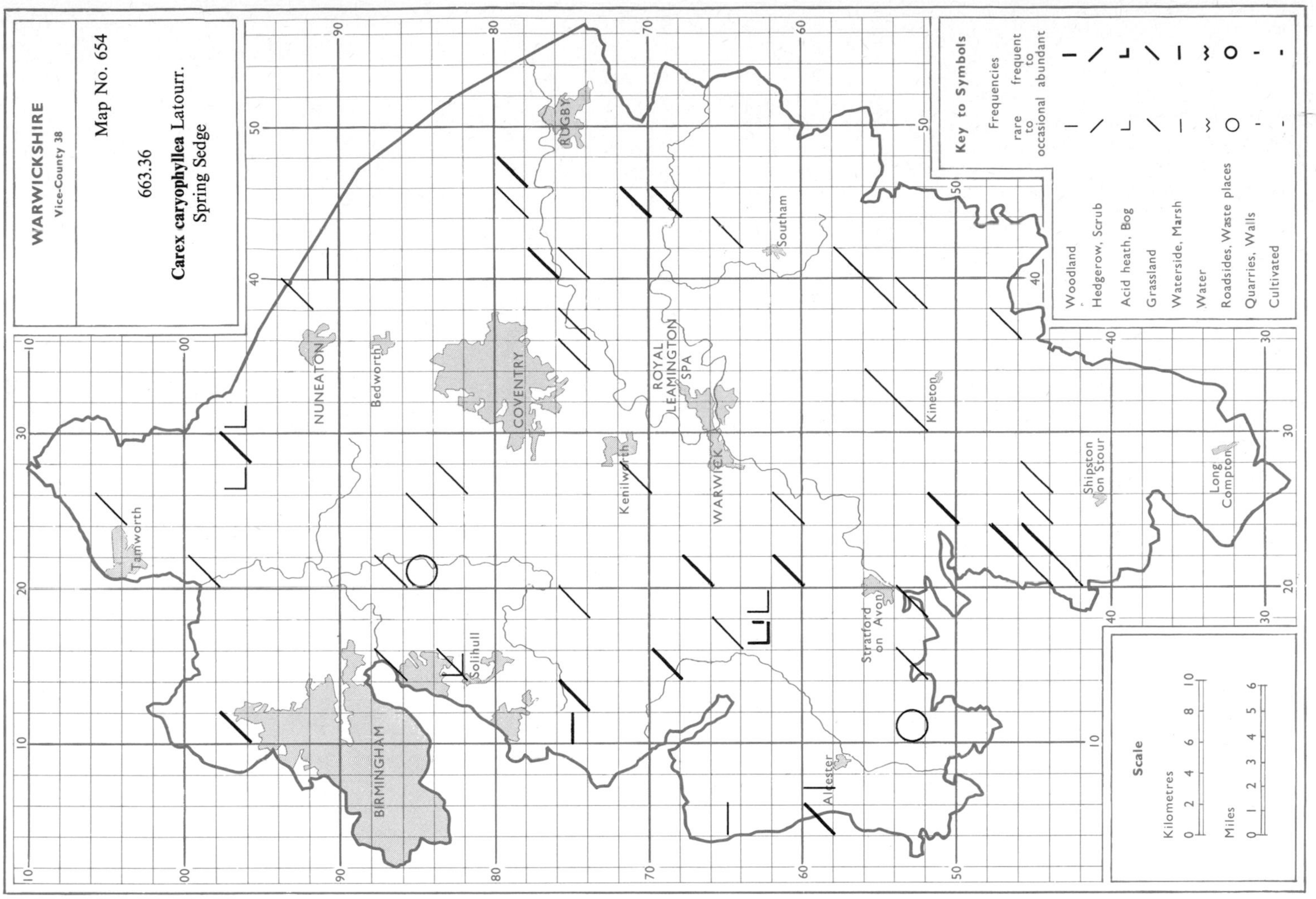
WARWICKSHIRE
Vice-County 38
Map No. 654
663.36
Carex caryophyllea Latourr.
Spring Sedge
NUNEATON
Bedworth
COVENTRY
RUGBY
Southam
ROYAL LEAMINGTON SPA
Kineton
Tamworth
Kenilworth
WARWICK
Shipston on Stour
Long Compton
Solihull
BIRMINGHAM
Stratford on Avon
Alcester
Key to Symbols
Frequencies
rare to occasional
frequent to abundant
Woodland
Hedgerow, Scrub
Acid heath, Bog
Grassland
Waterside, Marsh
Water
Roadsides, Waste places
Quarries, Walls
Cultivated
Scale
Kilometres
0 2 4 6 8 10
Miles
0 1 2 3 4 5 6

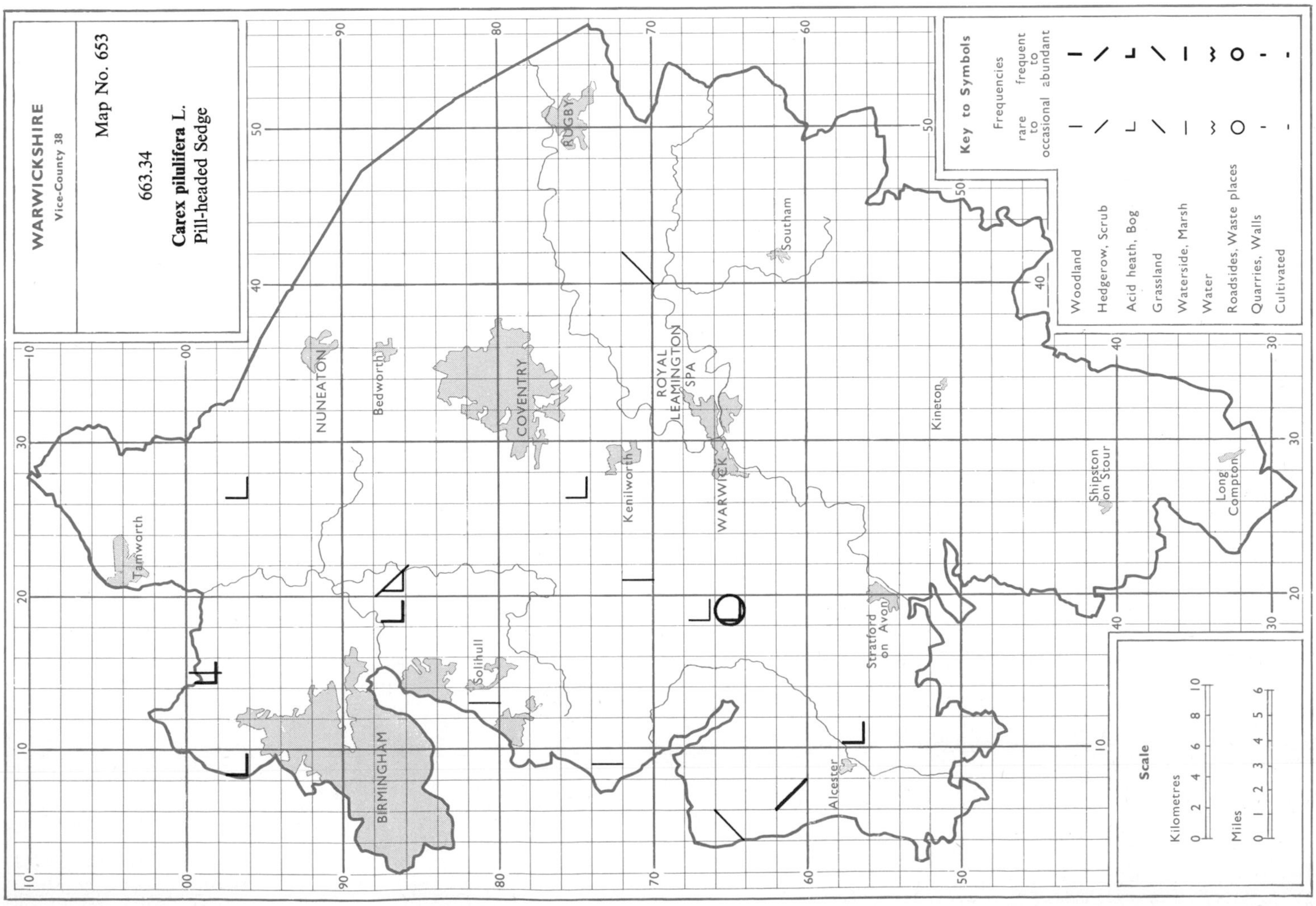
WARWICKSHIRE
Vice-County 38
Map No. 653
663.34
Carex pilulifera L.
Pill-headed Sedge
NUNEATON
Bedworth
COVENTRY
RUGBY
Southam
ROYAL LEAMINGTON SPA
Kineton
Tamworth
Kenilworth
WARWICK
Shipston on Stour
Long Compton
Solihull
BIRMINGHAM
Stratford on Avon
Alcester
Key to Symbols
Frequencies
rare to occasional
frequent to abundant
Woodland
Hedgerow, Scrub
Acid heath, Bog
Grassland
Waterside, Marsh
Water
Roadsides, Waste places
Quarries, Walls
Cultivated
Scale
Kilometres
0 2 4 6 8 10
Miles
0 1 2 3 4 5 6

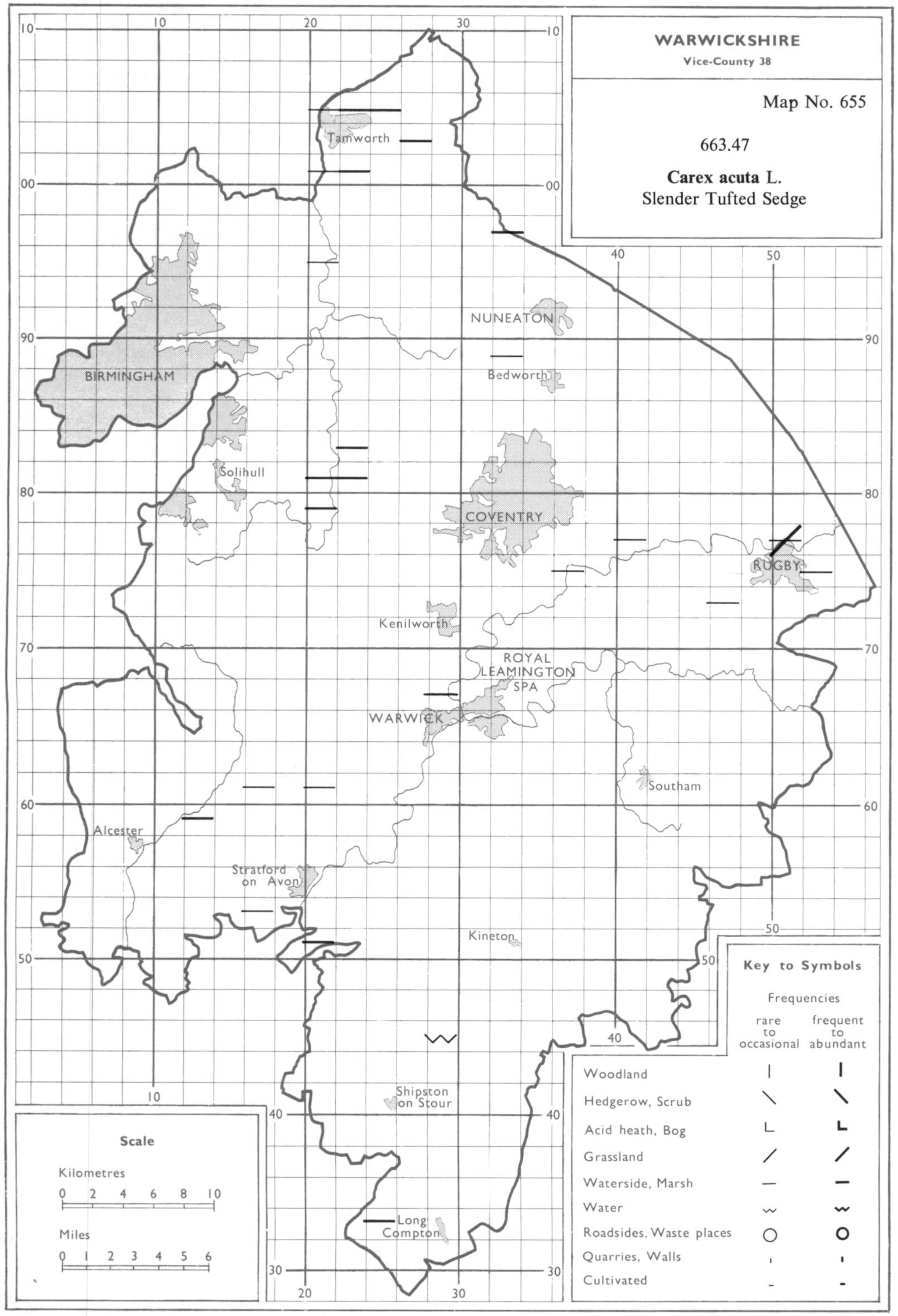
WARWICKSHIRE
Vice-County 38
Map No. 655
663.47
Carex acuta L.
Slender Tufted Sedge
Tamworth
NUNEATON
Bedworth
BIRMINGHAM
Solihull
COVENTRY
RUGBY
Kenilworth
ROYAL LEAMINGTON SPA
WARWICK
Southam
Alcester
Stratford on Avon
Kineton
Shipston on Stour
Long Compton
Key to Symbols
Frequencies
rare to occasional
frequent to abundant
Woodland
Hedgerow, Scrub
Acid heath, Bog
Grassland
Waterside, Marsh
Water
Roadsides, Waste places
Quarries, Walls
Cultivated
Scale
Kilometres
0 2 4 6 8 10
Miles
0 1 2 3 4 5 6

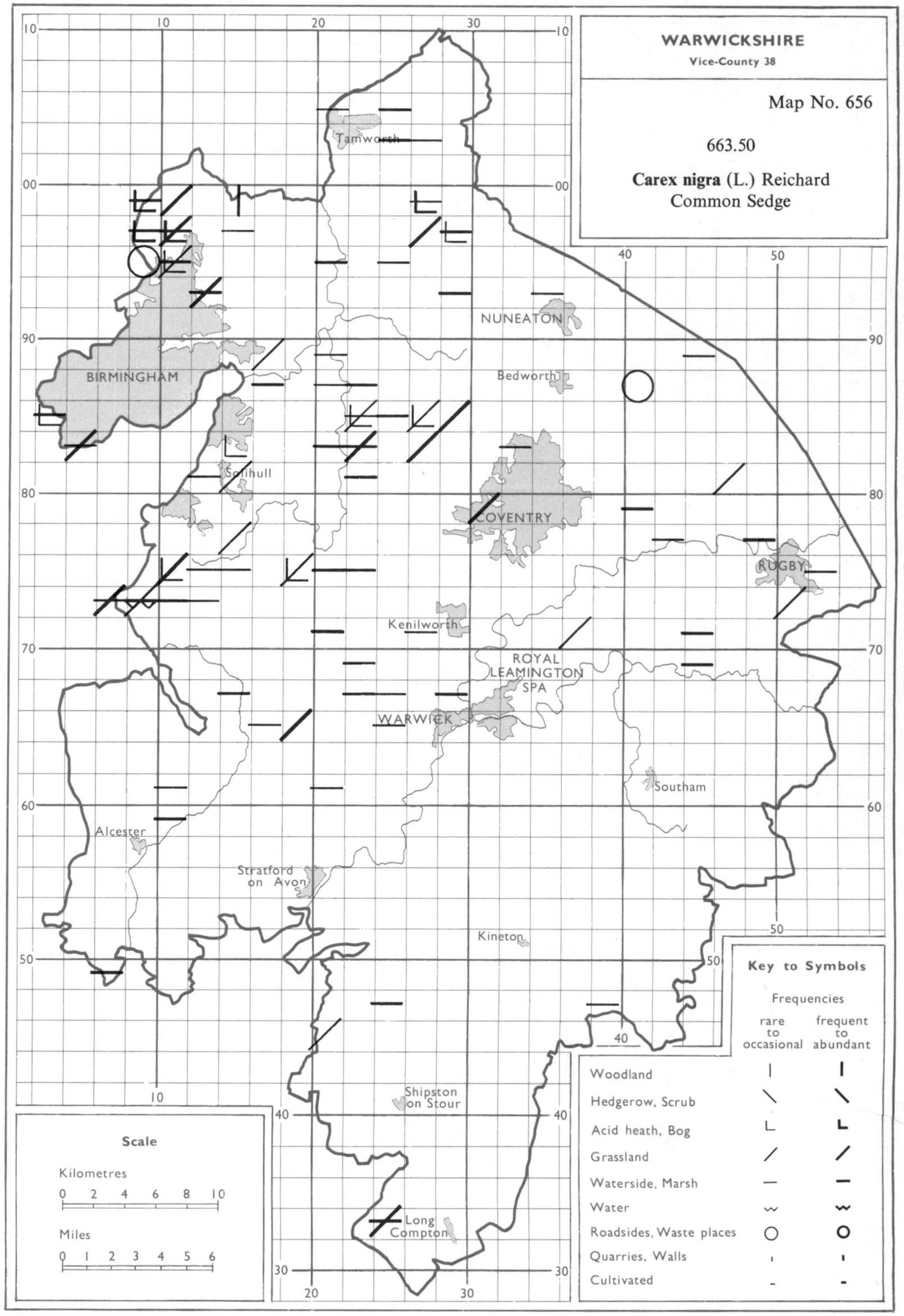
WARWICKSHIRE
Vice-County 38
Map No. 656
663.50
Carex nigra (L.) Reichard
Common Sedge
Tamworth
NUNEATON
Bedworth
BIRMINGHAM
Solihull
COVENTRY
RUGBY
Kenilworth
ROYAL LEAMINGTON SPA
WARWICK
Southam
Alcester
Stratford on Avon
Kineton
Shipston on Stour
Long Compton
Key to Symbols
Frequencies
rare to occasional
frequent to abundant
Woodland
Hedgerow, Scrub
Acid heath, Bog
Grassland
Waterside, Marsh
Water
Roadsides, Waste places
Quarries, Walls
Cultivated
Scale
Kilometres
0 2 4 6 8 10
Miles
0 1 2 3 4 5 6

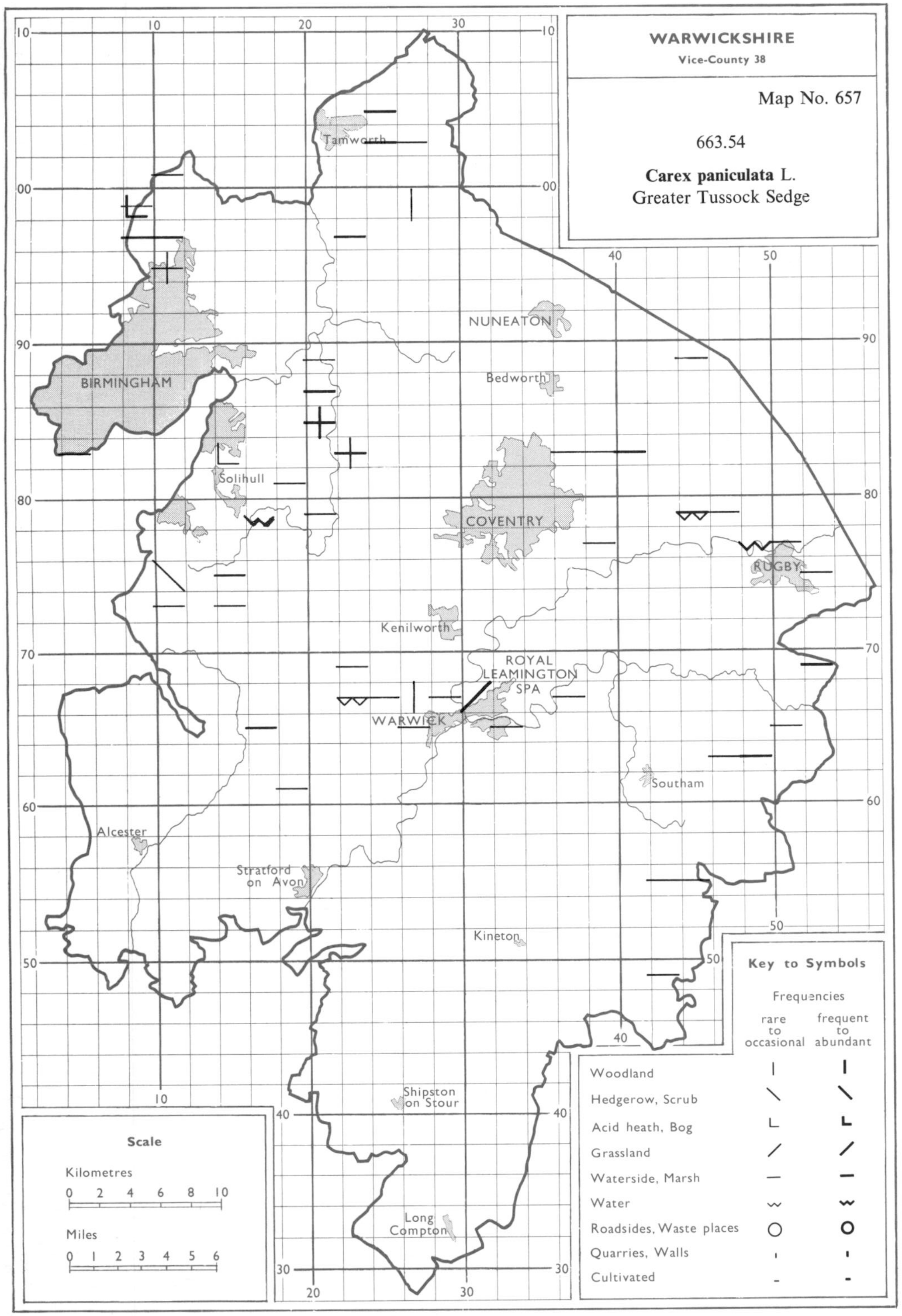
WARWICKSHIRE
Vice-County 38
Map No. 657
663.54
Carex paniculata L.
Greater Tussock Sedge
Tamworth
NUNEATON
Bedworth
BIRMINGHAM
Solihull
COVENTRY
RUGBY
Kenilworth
ROYAL LEAMINGTON SPA
WARWICK
Southam
Alcester
Stratford on Avon
Kineton
Shipston on Stour
Long Compton
Key to Symbols
Frequencies
rare to occasional
frequent to abundant
Woodland
Hedgerow, Scrub
Acid heath, Bog
Grassland
Waterside, Marsh
Water
Roadsides, Waste places
Quarries, Walls
Cultivated
Scale
Kilometres
0 2 4 6 8 10
Miles
0 1 2 3 4 5 6

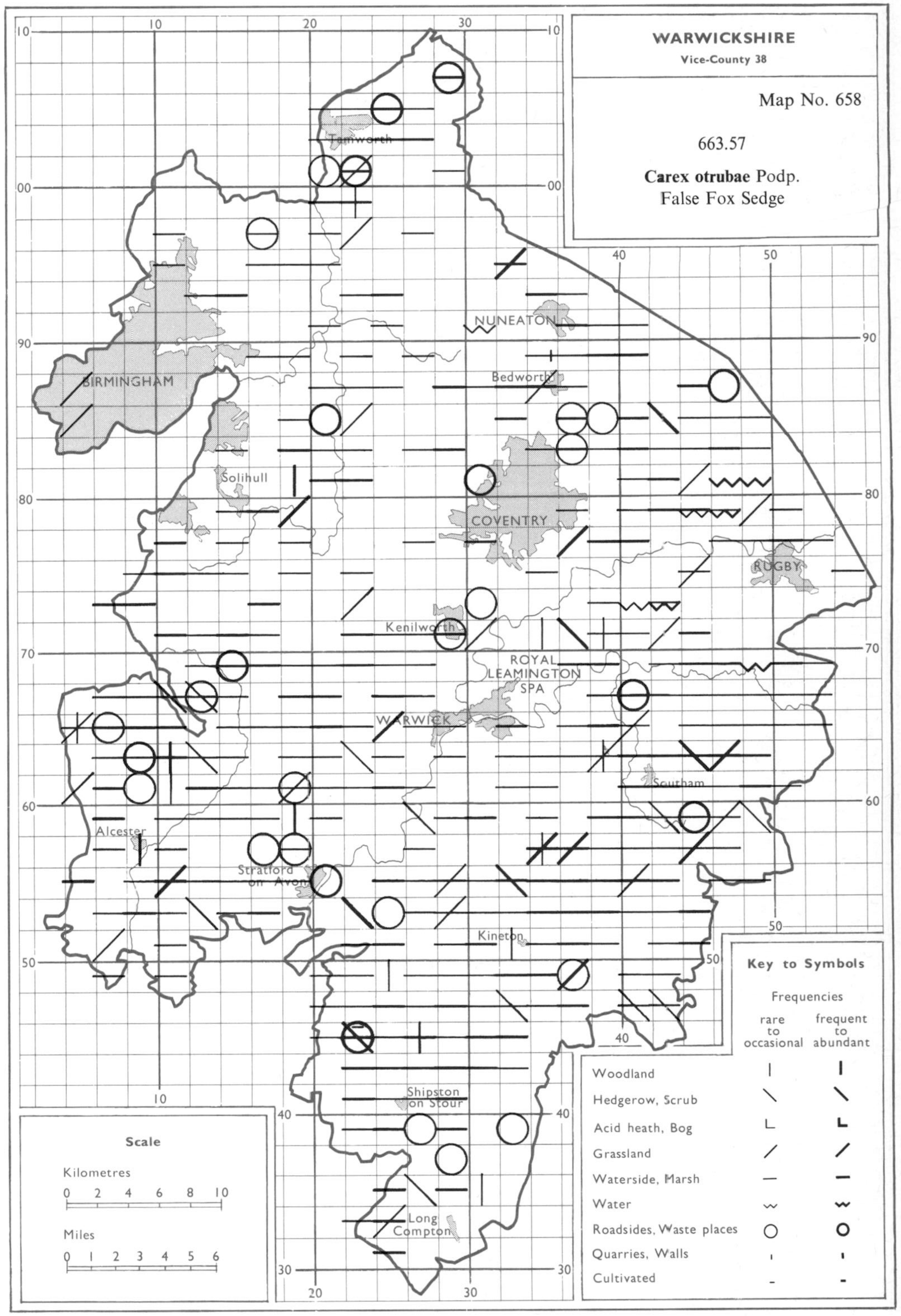
WARWICKSHIRE
Vice-County 38
Map No. 658
663.57
Carex otrubae Podp.
False Fox Sedge
Tamworth
NUNEATON
Bedworth
BIRMINGHAM
Solihull
COVENTRY
RUGBY
Kenilworth
ROYAL LEAMINGTON SPA
WARWICK
Southam
Alcester
Stratford on Avon
Kineton
Shipston on Stour
Long Compton
Key to Symbols
Frequencies
rare to occasional
frequent to abundant
Woodland
Hedgerow, Scrub
Acid heath, Bog
Grassland
Waterside, Marsh
Water
Roadsides, Waste places
Quarries, Walls
Cultivated
Scale
Kilometres
0 2 4 6 8 10
Miles
0 1 2 3 4 5 6

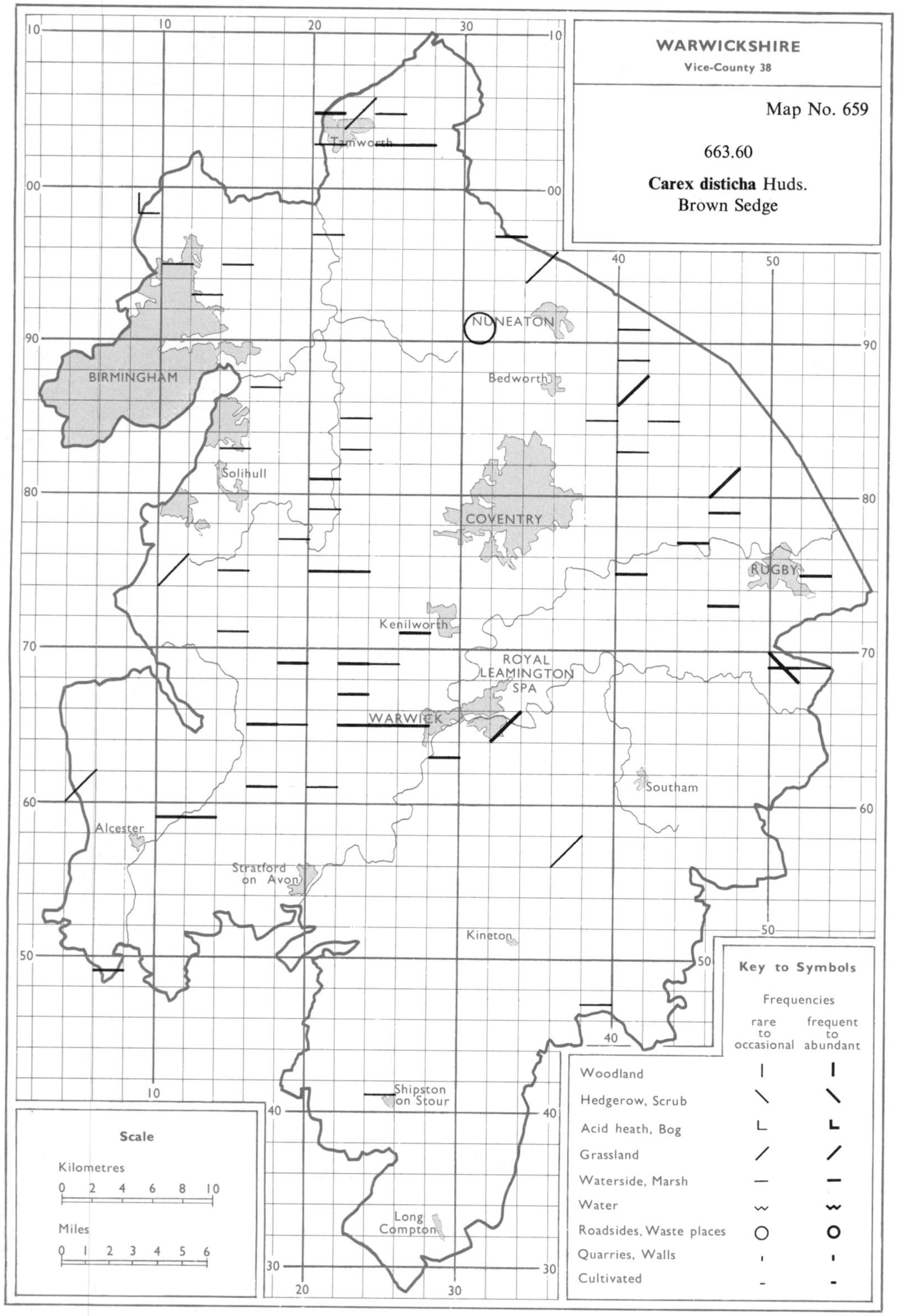

WARWICKSHIRE
Vice-County 38

Map No. 659

663.60

Carex disticha Huds.
Brown Sedge

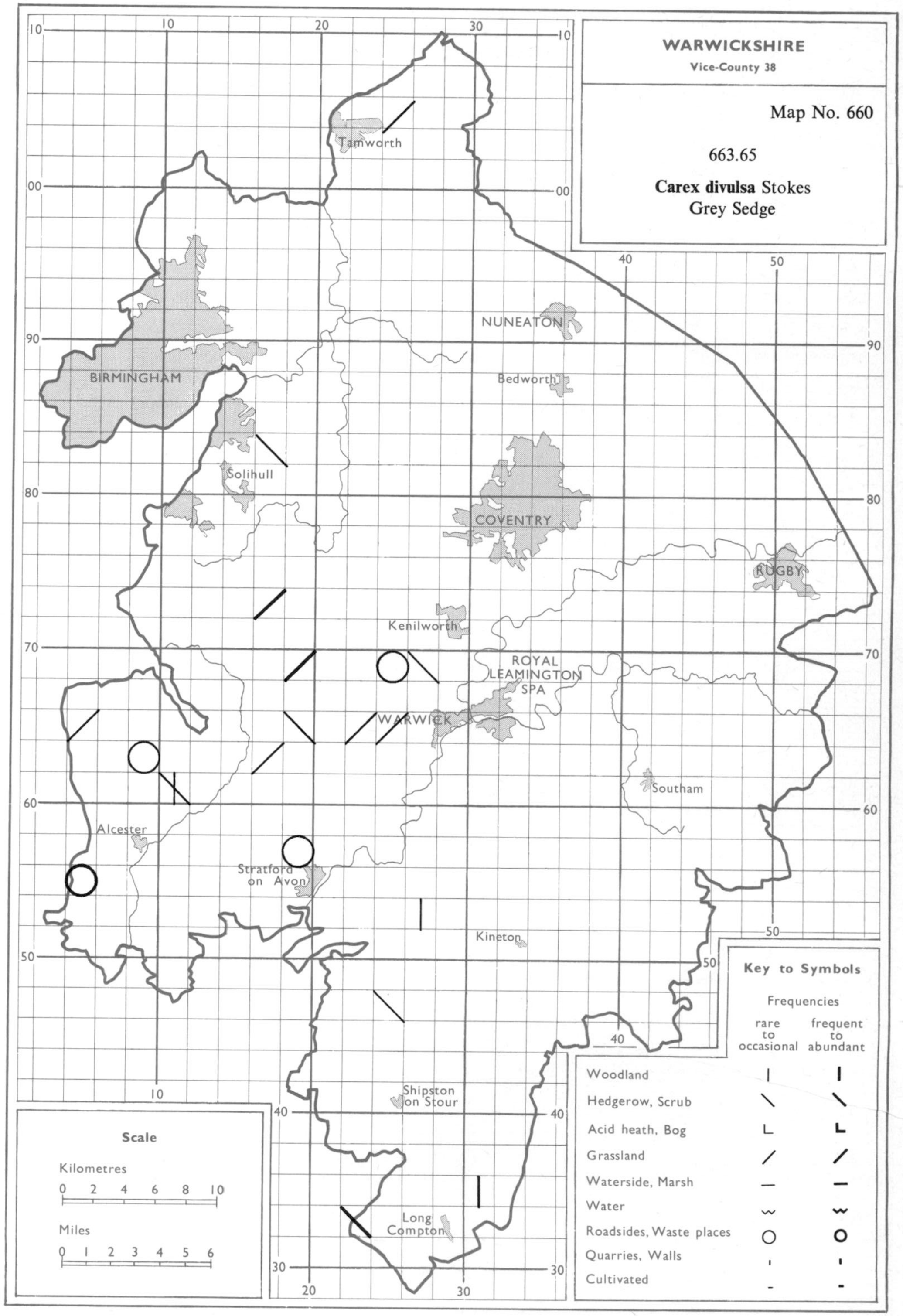

WARWICKSHIRE
Vice-County 38

Map No. 660

663.65

Carex divulsa Stokes
Grey Sedge

WARWICKSHIRE
Vice-County 38
Map No. 662
663.71
Carex remota L.
Remote Sedge
Key to Symbols
Frequencies
rare to occasional
frequent to abundant
Woodland
Hedgerow, Scrub
Acid heath, Bog
Grassland
Waterside, Marsh
Water
Roadsides, Waste places
Quarries, Walls
Cultivated
Scale
Kilometres 0 2 4 6 8 10
Miles 0 1 2 3 4 5 6
NUNEATON
Bedworth
COVENTRY
RUGBY
Southam
ROYAL LEAMINGTON SPA
Kineton
Tamworth
Kenilworth
WARWICK
Shipston on Stour
Long Compton
Stratford on Avon
Solihull
BIRMINGHAM
Alcester

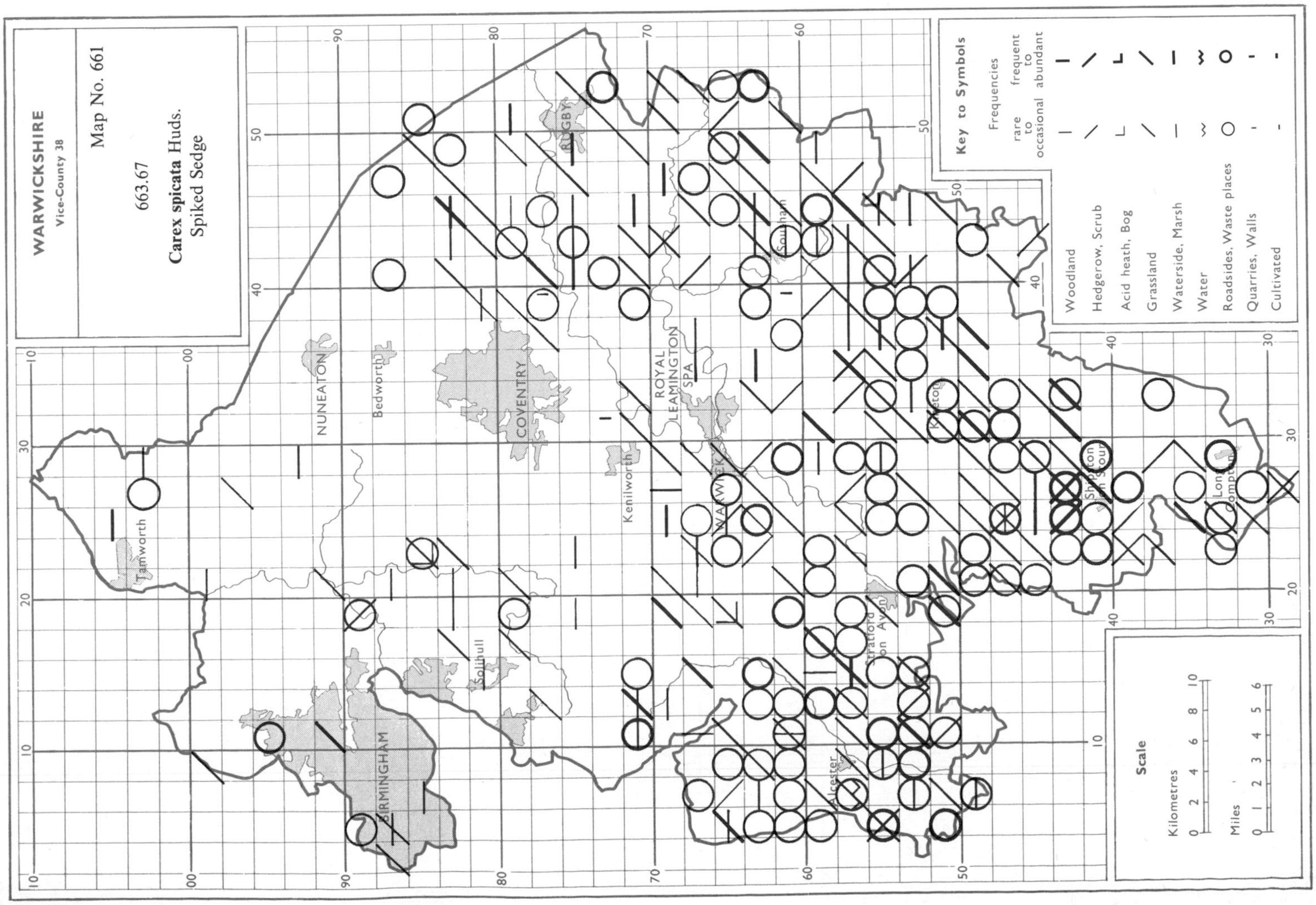
WARWICKSHIRE
Vice-County 38
Map No. 661
663.67
Carex spicata Huds.
Spiked Sedge
Key to Symbols
Frequencies
rare to occasional
frequent to abundant
Woodland
Hedgerow, Scrub
Acid heath, Bog
Grassland
Waterside, Marsh
Water
Roadsides, Waste places
Quarries, Walls
Cultivated
Scale
Kilometres 0 2 4 6 8 10
Miles 0 1 2 3 4 5 6
NUNEATON
Bedworth
COVENTRY
RUGBY
Southam
ROYAL LEAMINGTON SPA
Kineton
Tamworth
Kenilworth
WARWICK
Shipston on Stour
Long Compton
Stratford on Avon
Solihull
BIRMINGHAM
Alcester

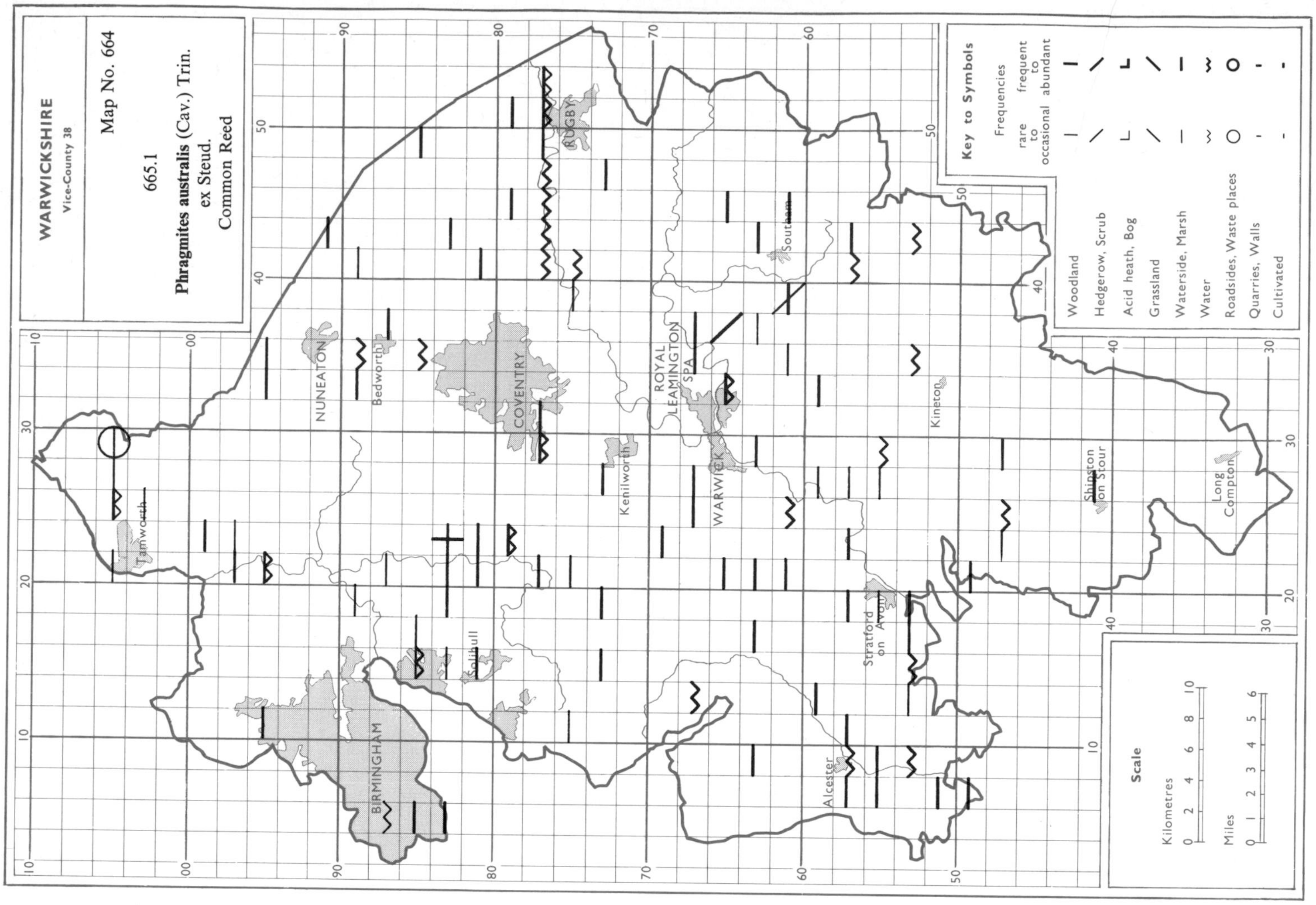
WARWICKSHIRE
Vice-County 38
Map No. 664
665.1
Phragmites australis (Cav.) Trin. ex Steud.
Common Reed
Key to Symbols
Frequencies
rare to occasional
frequent to abundant
Woodland
Hedgerow, Scrub
Acid heath, Bog
Grassland
Waterside, Marsh
Water
Roadsides, Waste places
Quarries, Walls
Cultivated
Scale
Kilometres 0 2 4 6 8 10
Miles 0 1 2 3 4 5 6
Tamworth
NUNEATON
Bedworth
RUGBY
COVENTRY
Kenilworth
ROYAL LEAMINGTON SPA
WARWICK
Southam
Kineton
Shipston on Stour
Long Compton
Stratford on Avon
Alcester
Solihull
BIRMINGHAM

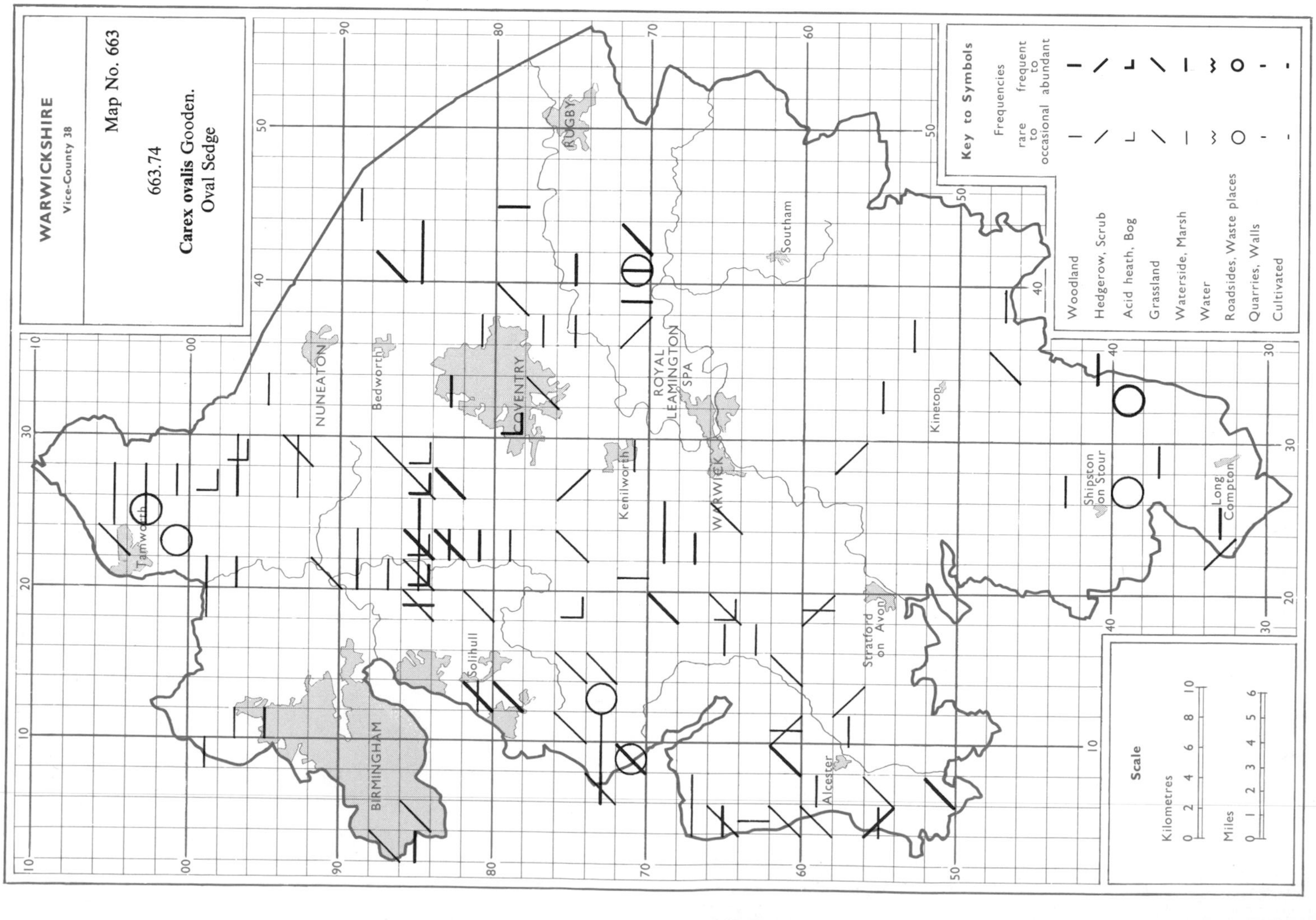
WARWICKSHIRE
Vice-County 38
Map No. 663
663.74
Carex ovalis Gooden.
Oval Sedge
Key to Symbols
Frequencies
rare to occasional
frequent to abundant
Woodland
Hedgerow, Scrub
Acid heath, Bog
Grassland
Waterside, Marsh
Water
Roadsides, Waste places
Quarries, Walls
Cultivated
Scale
Kilometres 0 2 4 6 8 10
Miles 0 1 2 3 4 5 6
Tamworth
NUNEATON
Bedworth
RUGBY
COVENTRY
Kenilworth
ROYAL LEAMINGTON SPA
WARWICK
Southam
Kineton
Shipston on Stour
Long Compton
Stratford on Avon
Alcester
Solihull
BIRMINGHAM

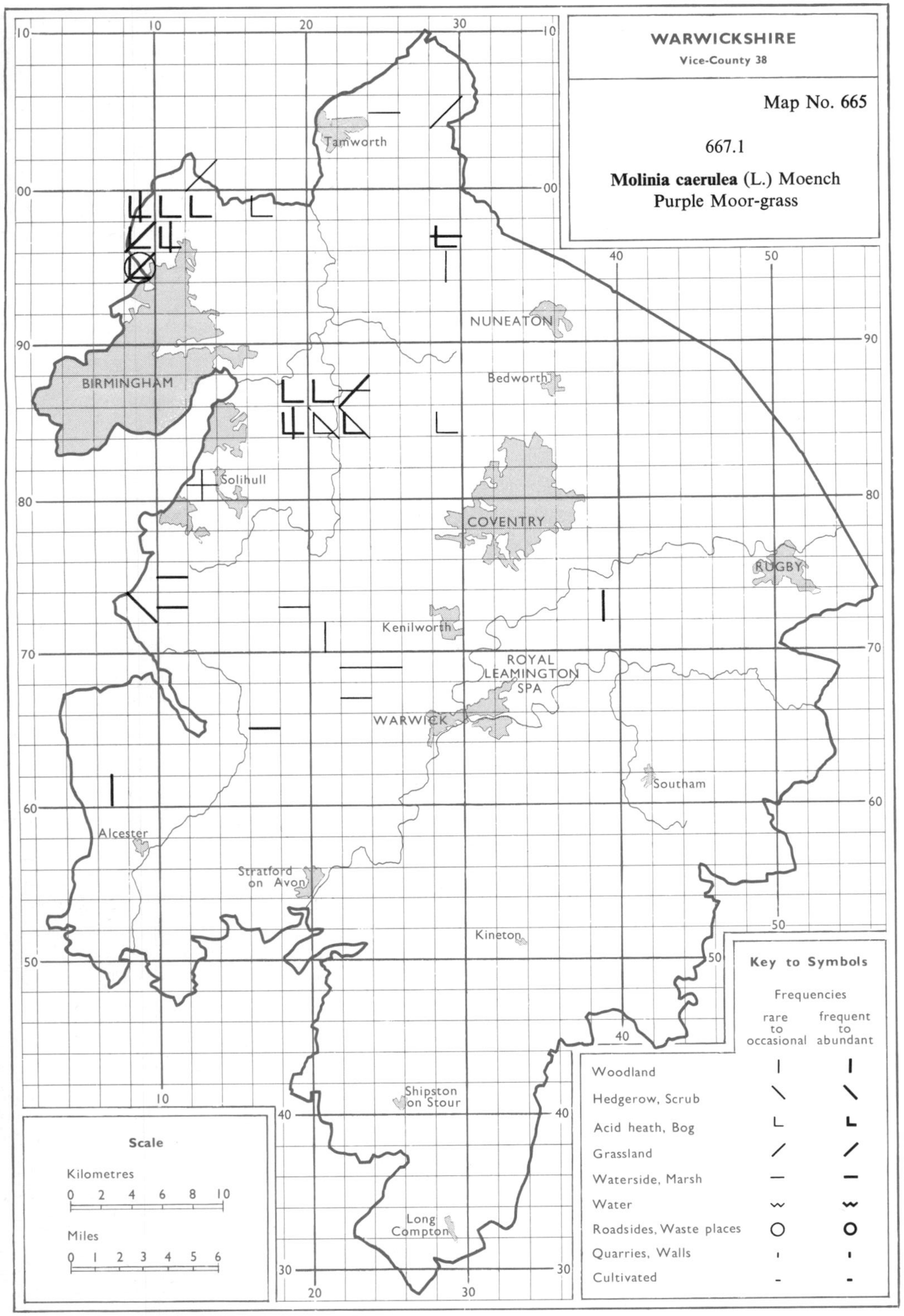
WARWICKSHIRE
Vice-County 38
Map No. 665
667.1
Molinia caerulea (L.) Moench
Purple Moor-grass
Tamworth
NUNEATON
Bedworth
BIRMINGHAM
Solihull
COVENTRY
RUGBY
Kenilworth
ROYAL LEAMINGTON SPA
WARWICK
Southam
Alcester
Stratford on Avon
Kineton
Shipston on Stour
Long Compton
Key to Symbols
Frequencies
rare to occasional
frequent to abundant
Woodland
Hedgerow, Scrub
Acid heath, Bog
Grassland
Waterside, Marsh
Water
Roadsides, Waste places
Quarries, Walls
Cultivated
Scale
Kilometres
0 2 4 6 8 10
Miles
0 1 2 3 4 5 6

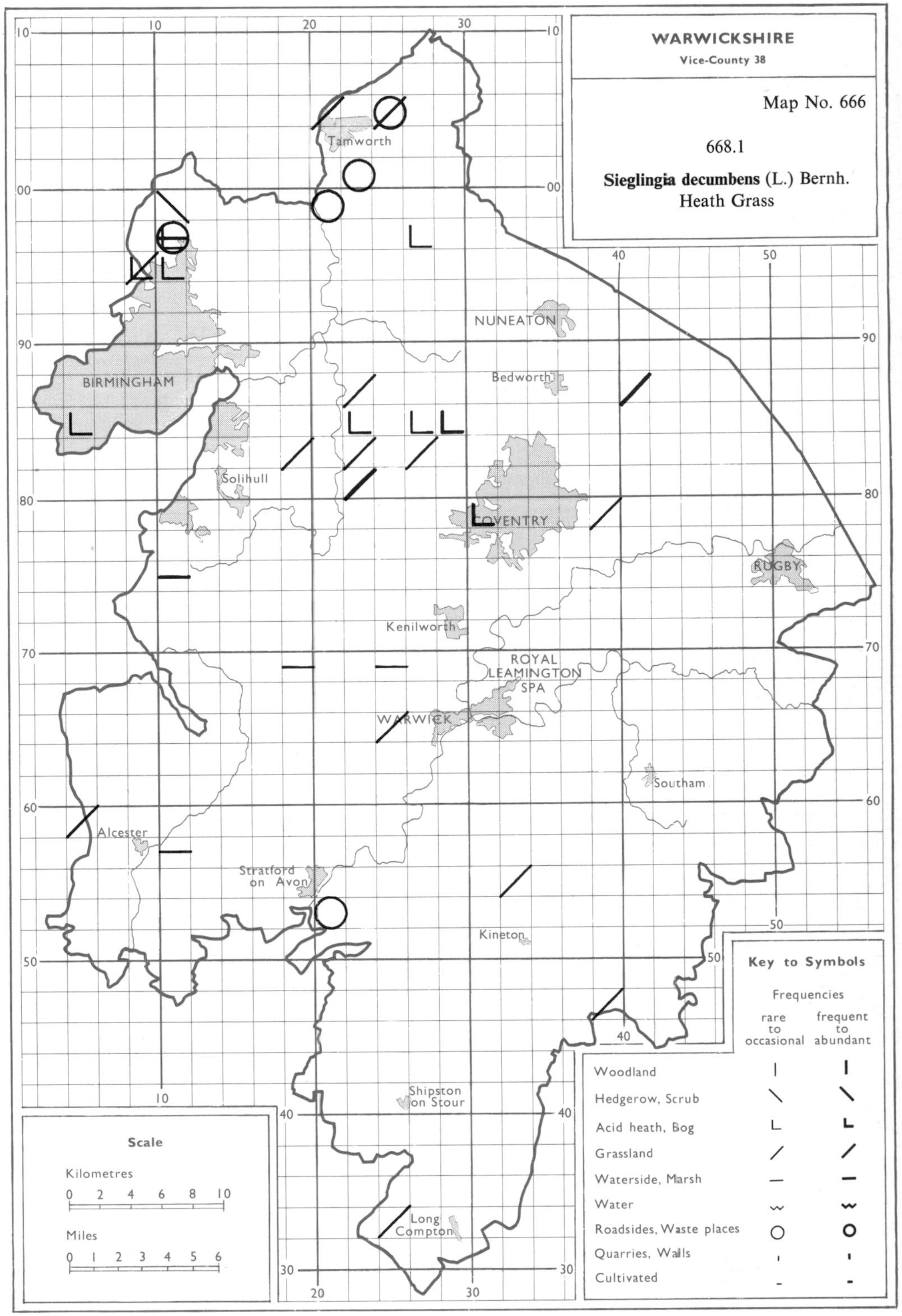
WARWICKSHIRE
Vice-County 38
Map No. 666
668.1
Sieglingia decumbens (L.) Bernh.
Heath Grass
Tamworth
NUNEATON
Bedworth
BIRMINGHAM
Solihull
COVENTRY
RUGBY
Kenilworth
ROYAL LEAMINGTON SPA
WARWICK
Southam
Alcester
Stratford on Avon
Kineton
Shipston on Stour
Long Compton
Key to Symbols
Frequencies
rare to occasional
frequent to abundant
Woodland
Hedgerow, Scrub
Acid heath, Bog
Grassland
Waterside, Marsh
Water
Roadsides, Waste places
Quarries, Walls
Cultivated
Scale
Kilometres
0 2 4 6 8 10
Miles
0 1 2 3 4 5 6

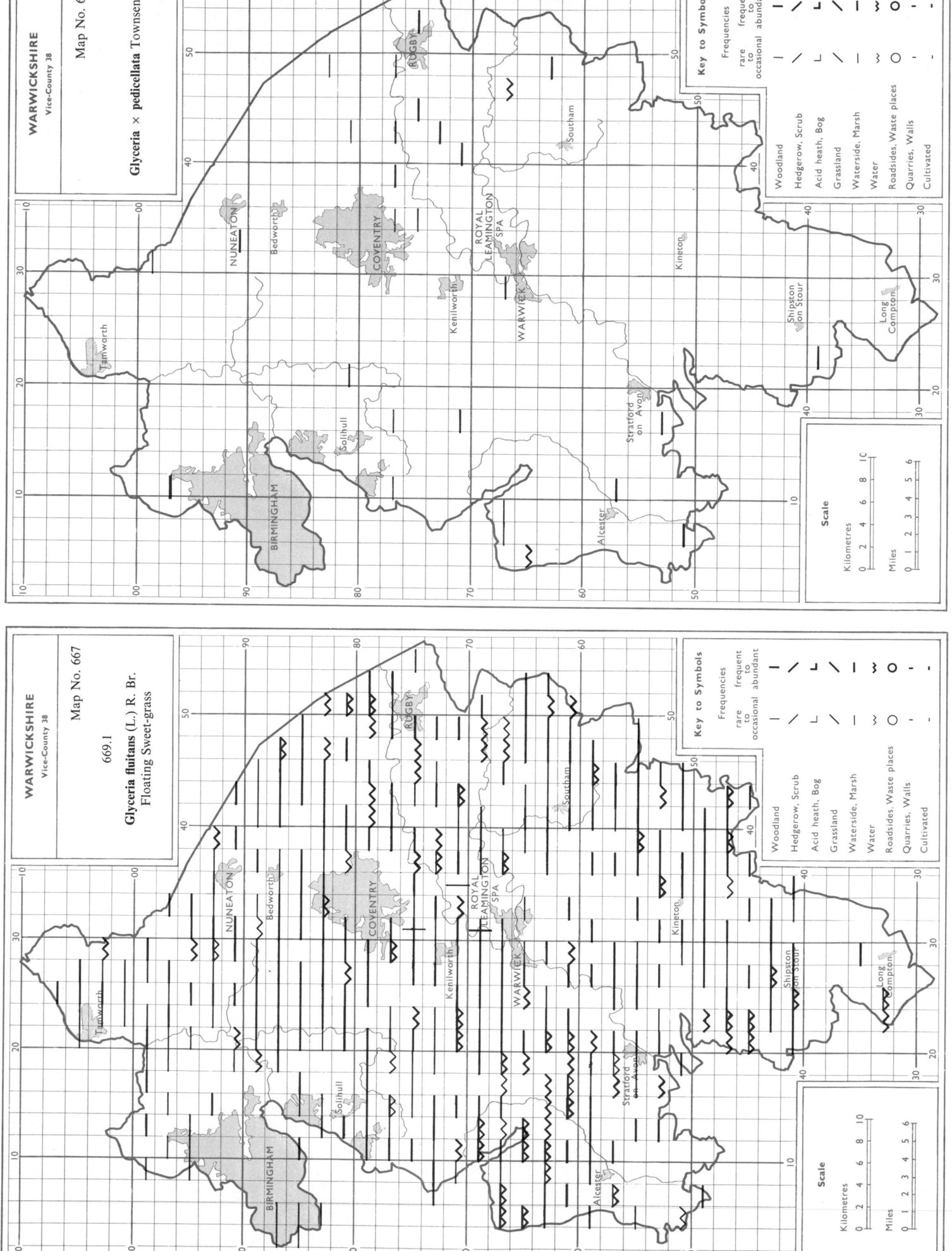

WARWICKSHIRE
Vice-County 38

Map No. 667

669.1

Glyceria fluitans (L.) R. Br.
Floating Sweet-grass

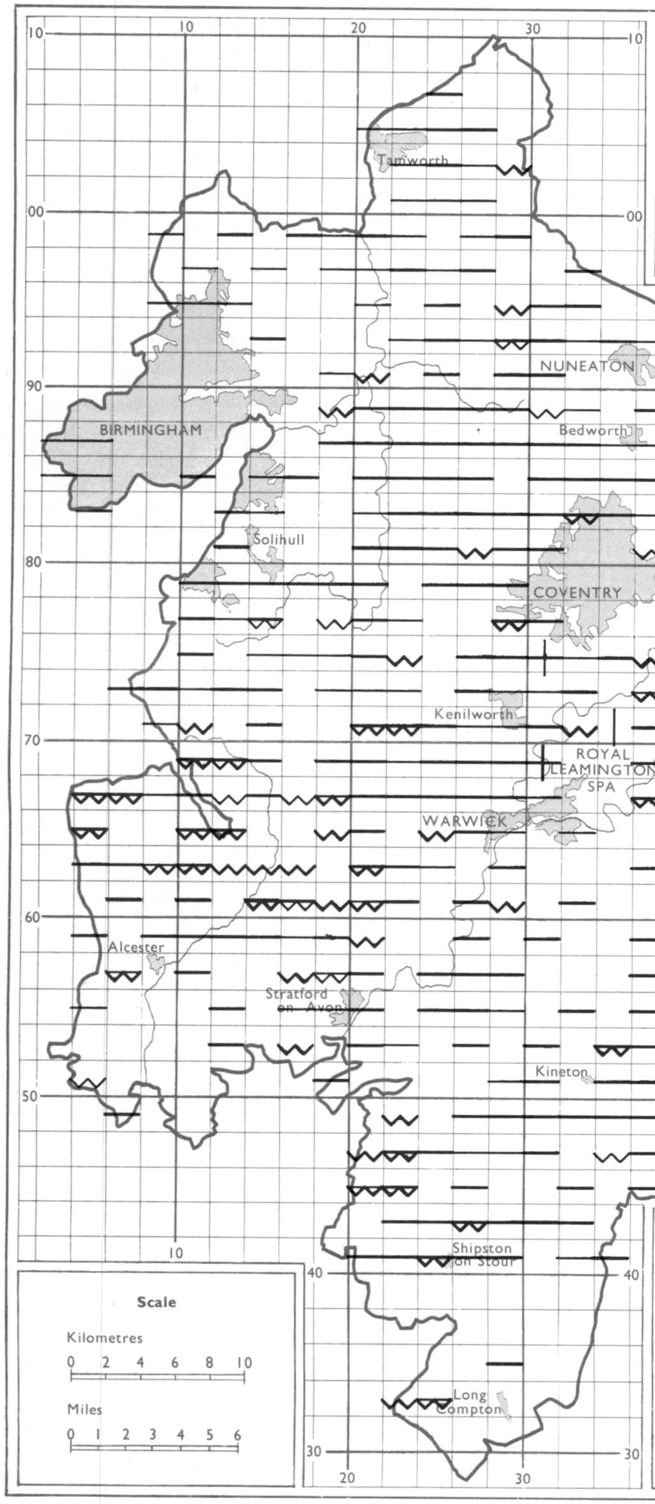

WARWICKSHIRE
Vice-County 38

Map No. 668

Glyceria × pedicellata Townsend

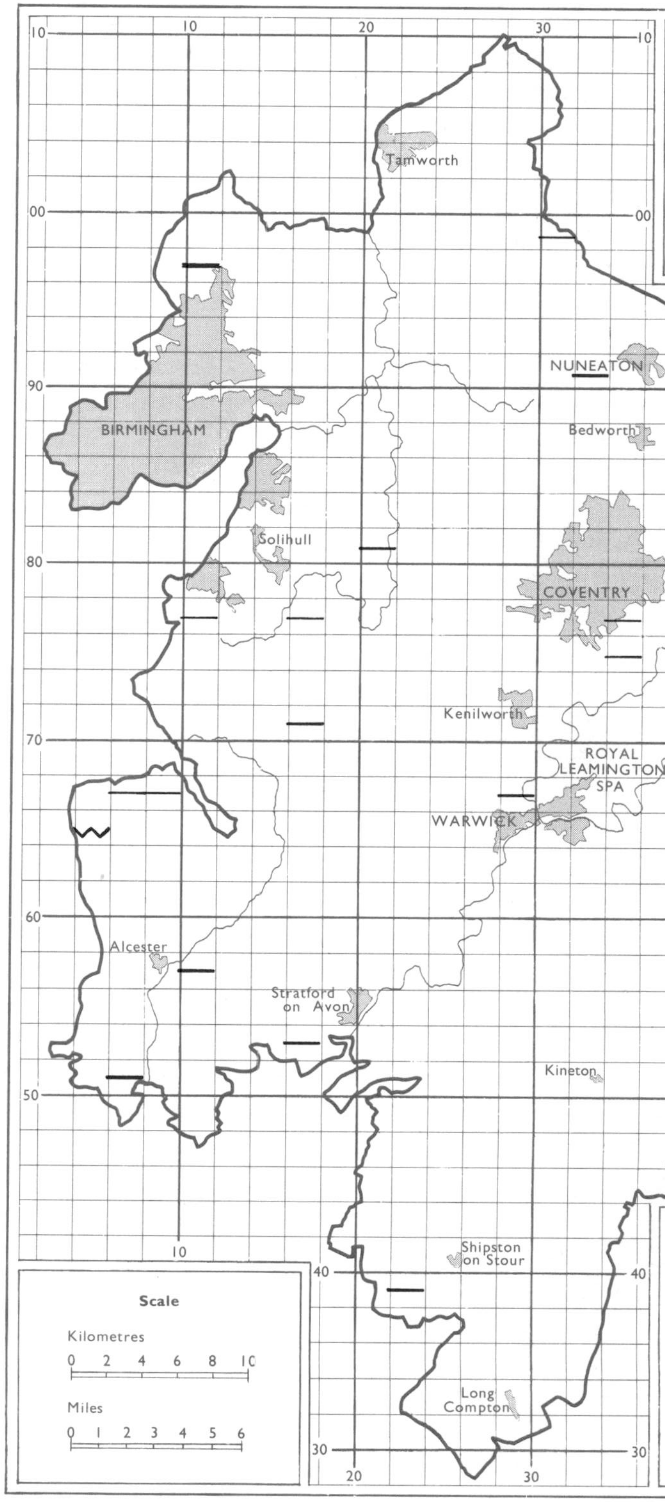

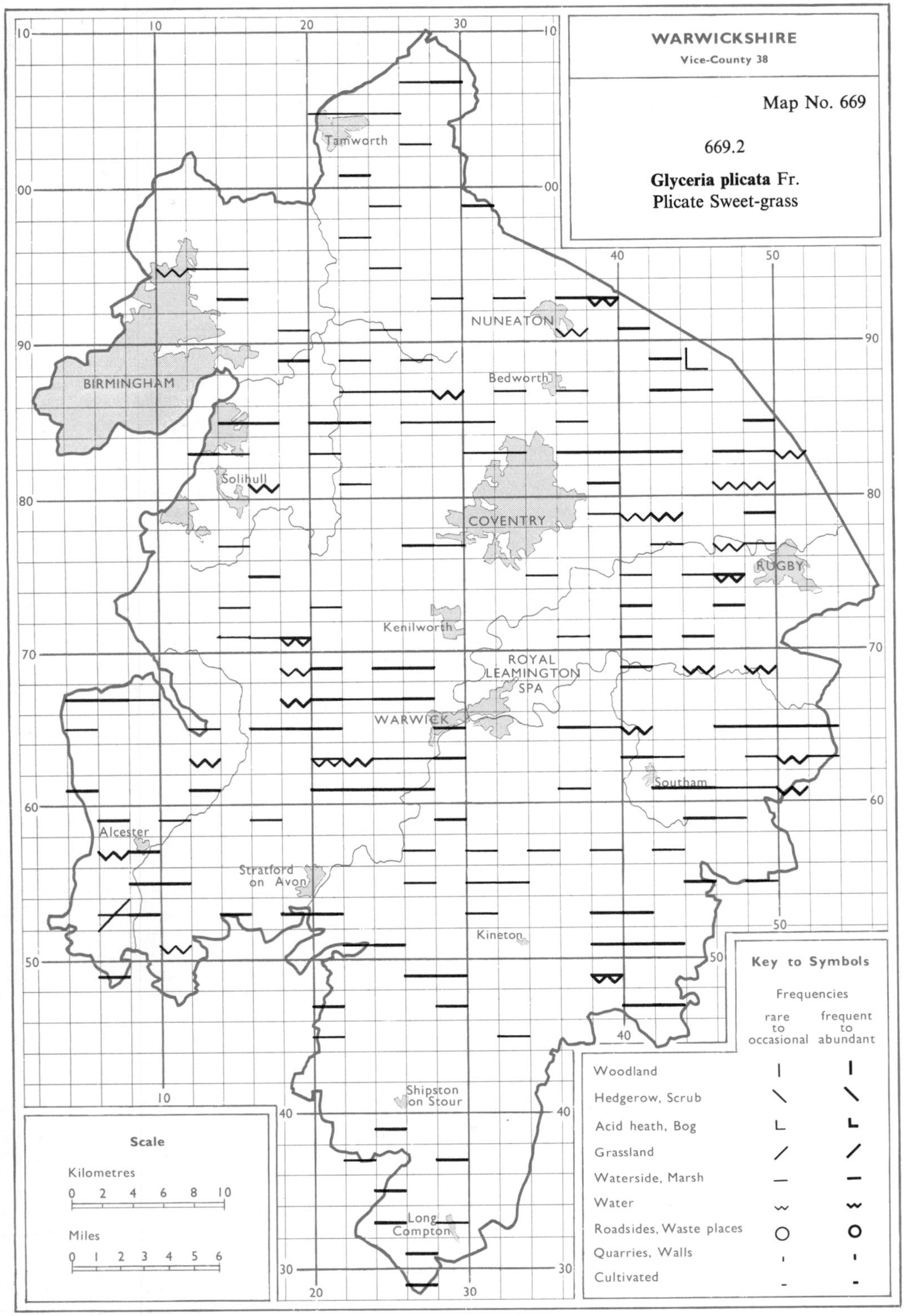
WARWICKSHIRE
Vice-County 38
Map No. 669
669.2
Glyceria plicata Fr.
Plicate Sweet-grass
Tamworth
NUNEATON
Bedworth
BIRMINGHAM
Solihull
COVENTRY
RUGBY
Kenilworth
ROYAL LEAMINGTON SPA
WARWICK
Southam
Alcester
Stratford on Avon
Kineton
Shipston on Stour
Long Compton
Key to Symbols
Frequencies
rare to occasional
frequent to abundant
Woodland
Hedgerow, Scrub
Acid heath, Bog
Grassland
Waterside, Marsh
Water
Roadsides, Waste places
Quarries, Walls
Cultivated
Scale
Kilometres
0 2 4 6 8 10
Miles
0 1 2 3 4 5 6

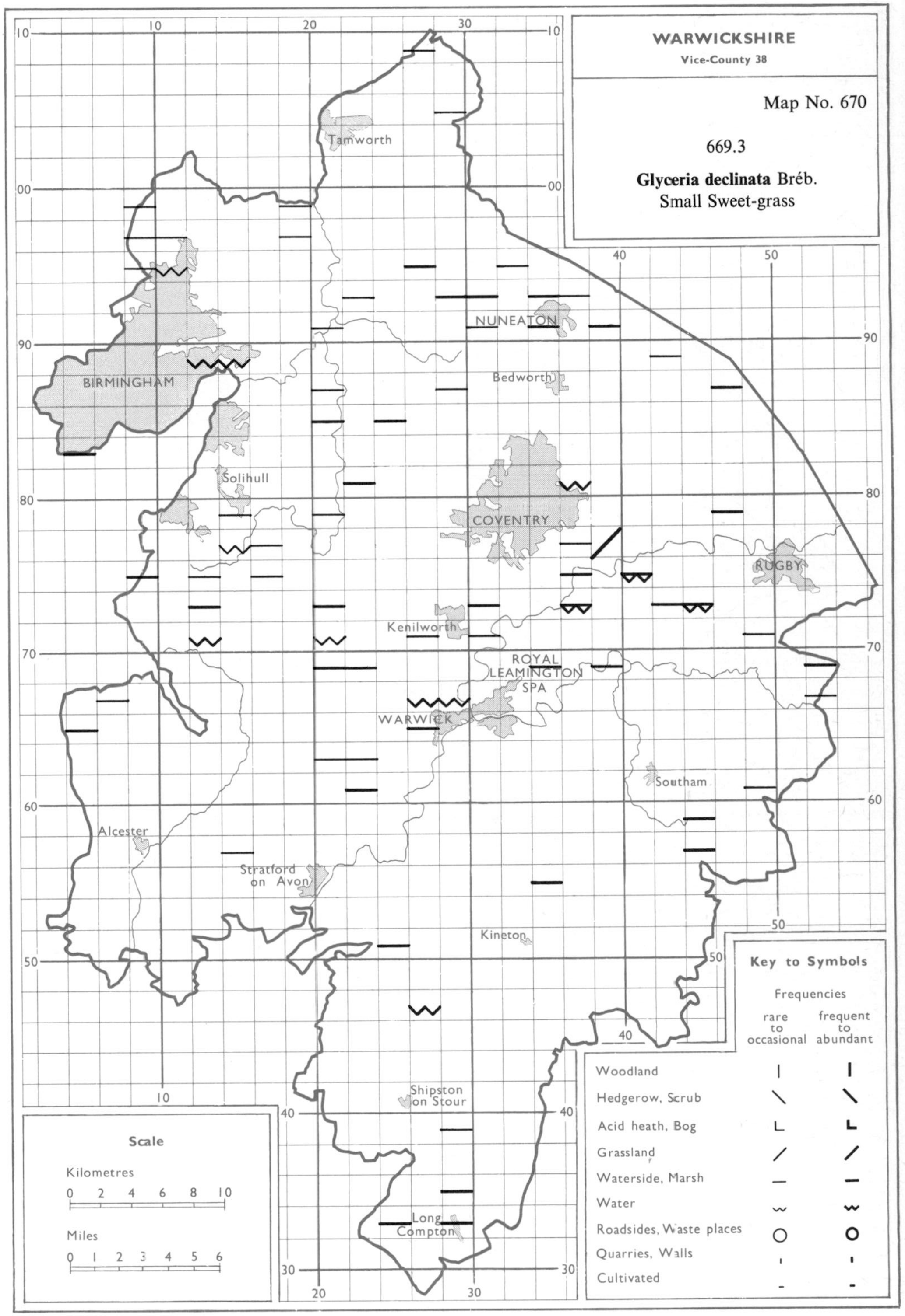
WARWICKSHIRE
Vice-County 38
Map No. 670
669.3
Glyceria declinata Bréb.
Small Sweet-grass
Tamworth
NUNEATON
Bedworth
BIRMINGHAM
Solihull
COVENTRY
RUGBY
Kenilworth
ROYAL LEAMINGTON SPA
WARWICK
Southam
Alcester
Stratford on Avon
Kineton
Shipston on Stour
Long Compton
Key to Symbols
Frequencies
rare to occasional
frequent to abundant
Woodland
Hedgerow, Scrub
Acid heath, Bog
Grassland
Waterside, Marsh
Water
Roadsides, Waste places
Quarries, Walls
Cultivated
Scale
Kilometres
0 2 4 6 8 10
Miles
0 1 2 3 4 5 6

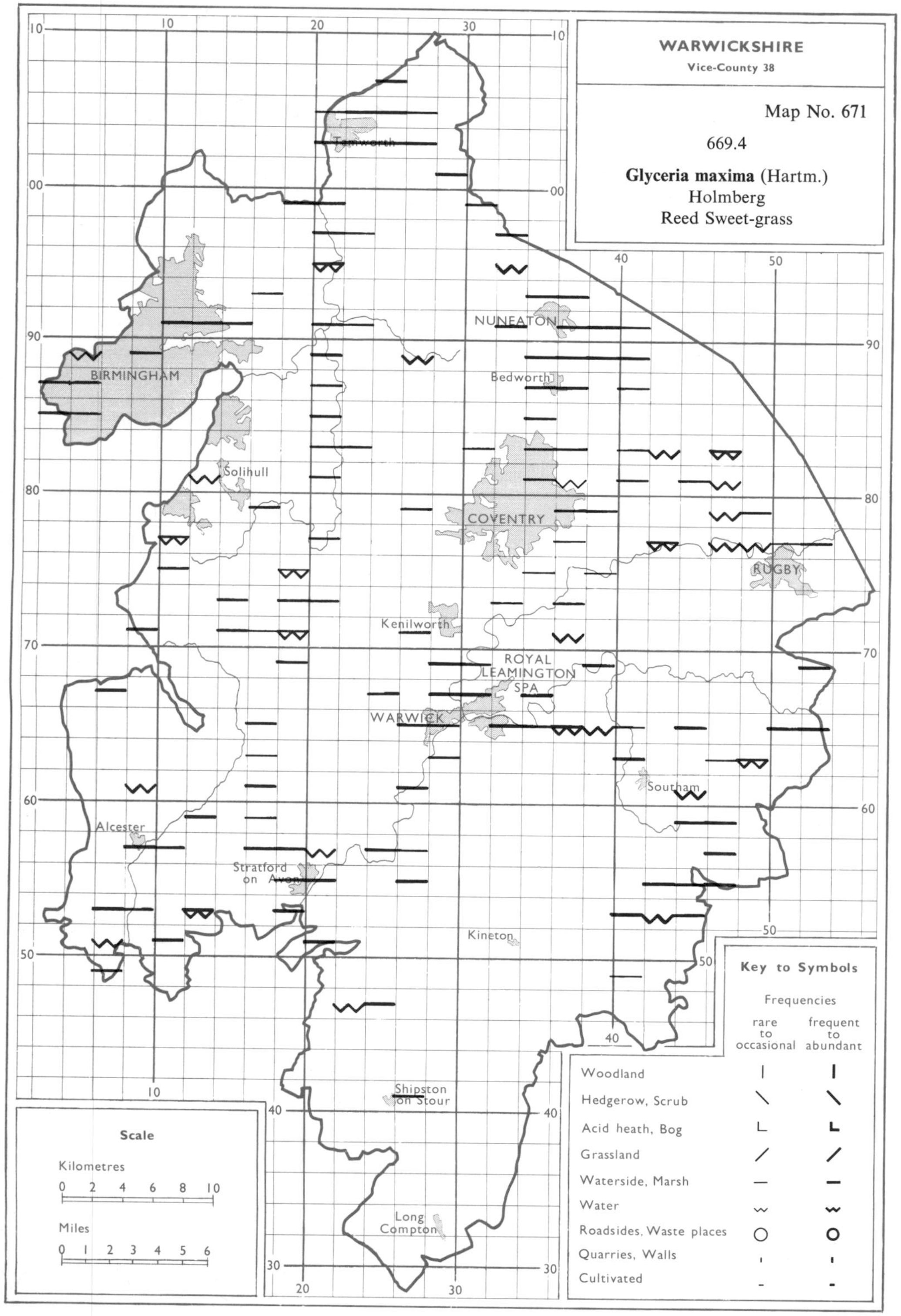
WARWICKSHIRE
Vice-County 38
Map No. 671
669.4
Glyceria maxima (Hartm.) Holmberg
Reed Sweet-grass
Tamworth
NUNEATON
Bedworth
BIRMINGHAM
Solihull
COVENTRY
RUGBY
Kenilworth
ROYAL LEAMINGTON SPA
WARWICK
Southam
Alcester
Stratford on Avon
Kineton
Shipston on Stour
Long Compton
Key to Symbols
Frequencies
rare to occasional
frequent to abundant
Woodland
Hedgerow, Scrub
Acid heath, Bog
Grassland
Waterside, Marsh
Water
Roadsides, Waste places
Quarries, Walls
Cultivated
Scale
Kilometres
0 2 4 6 8 10
Miles
0 1 2 3 4 5 6

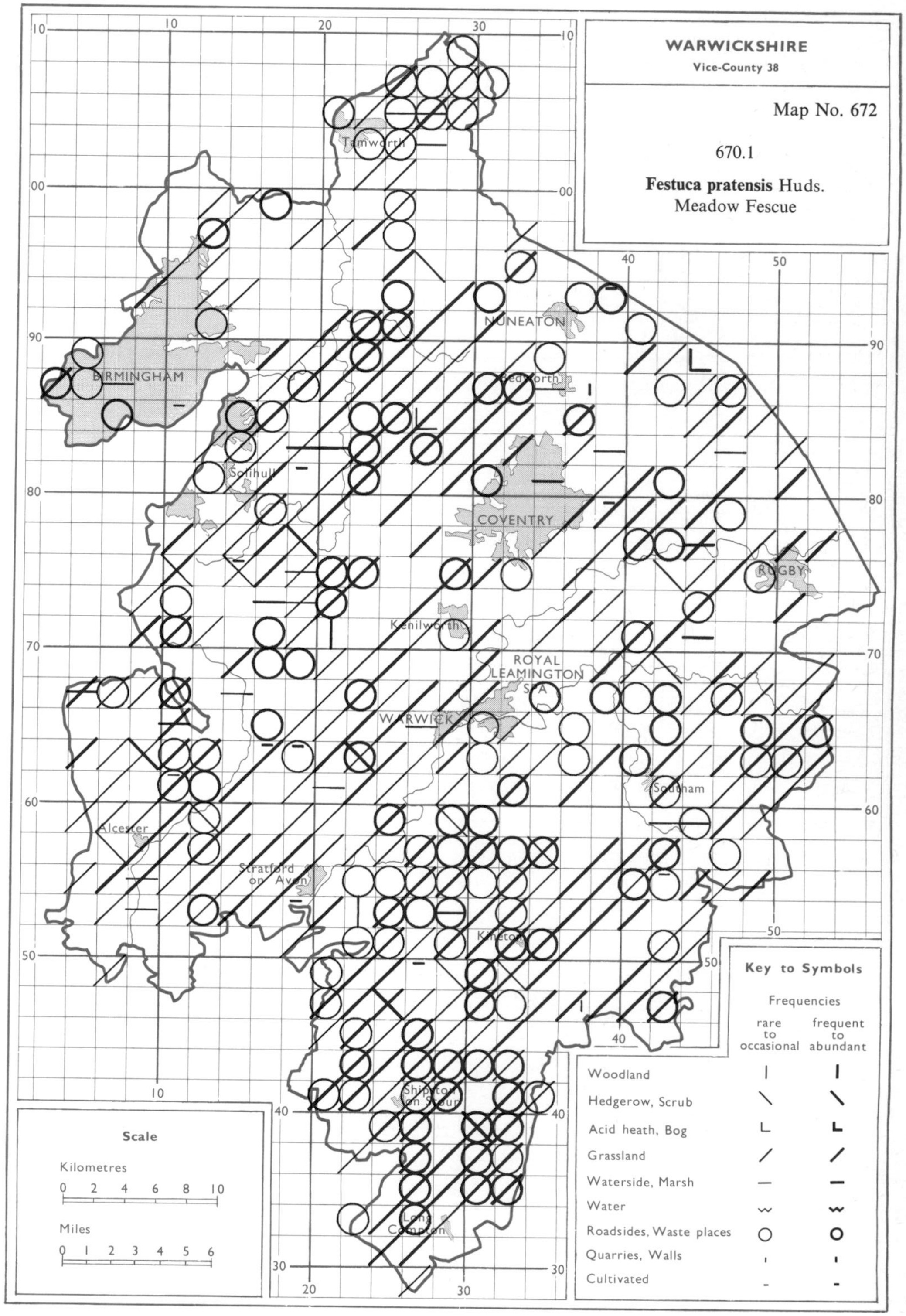
WARWICKSHIRE
Vice-County 38
Map No. 672
670.1
Festuca pratensis Huds.
Meadow Fescue
Tamworth
NUNEATON
Bedworth
BIRMINGHAM
Solihull
COVENTRY
RUGBY
Kenilworth
ROYAL LEAMINGTON SPA
WARWICK
Southam
Alcester
Stratford on Avon
Kineton
Shipston on Stour
Long Compton
Key to Symbols
Frequencies
rare to occasional
frequent to abundant
Woodland
Hedgerow, Scrub
Acid heath, Bog
Grassland
Waterside, Marsh
Water
Roadsides, Waste places
Quarries, Walls
Cultivated
Scale
Kilometres
0 2 4 6 8 10
Miles
0 1 2 3 4 5 6

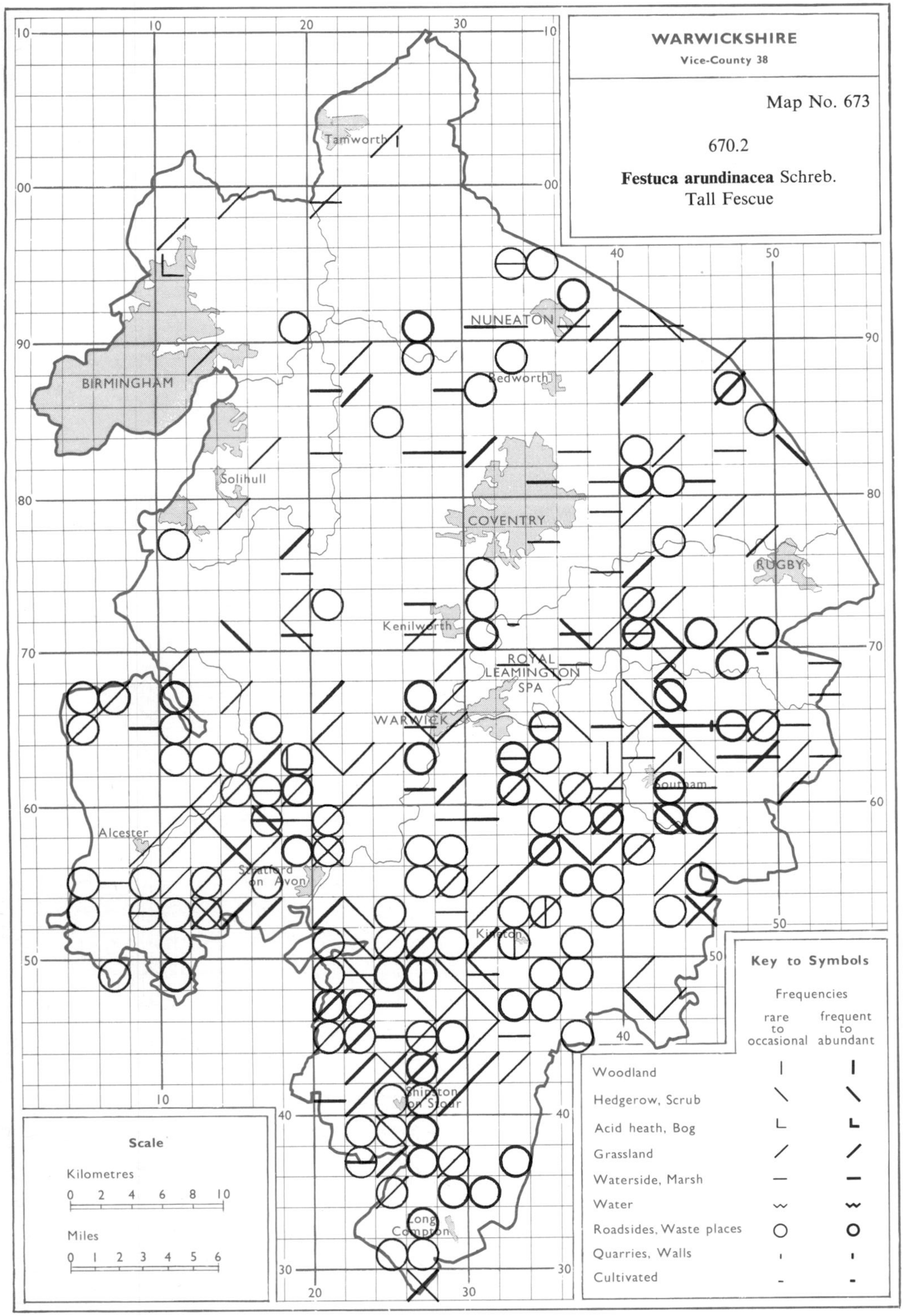

WARWICKSHIRE
Vice-County 38

Map No. 673

670.2

Festuca arundinacea Schreb.
Tall Fescue

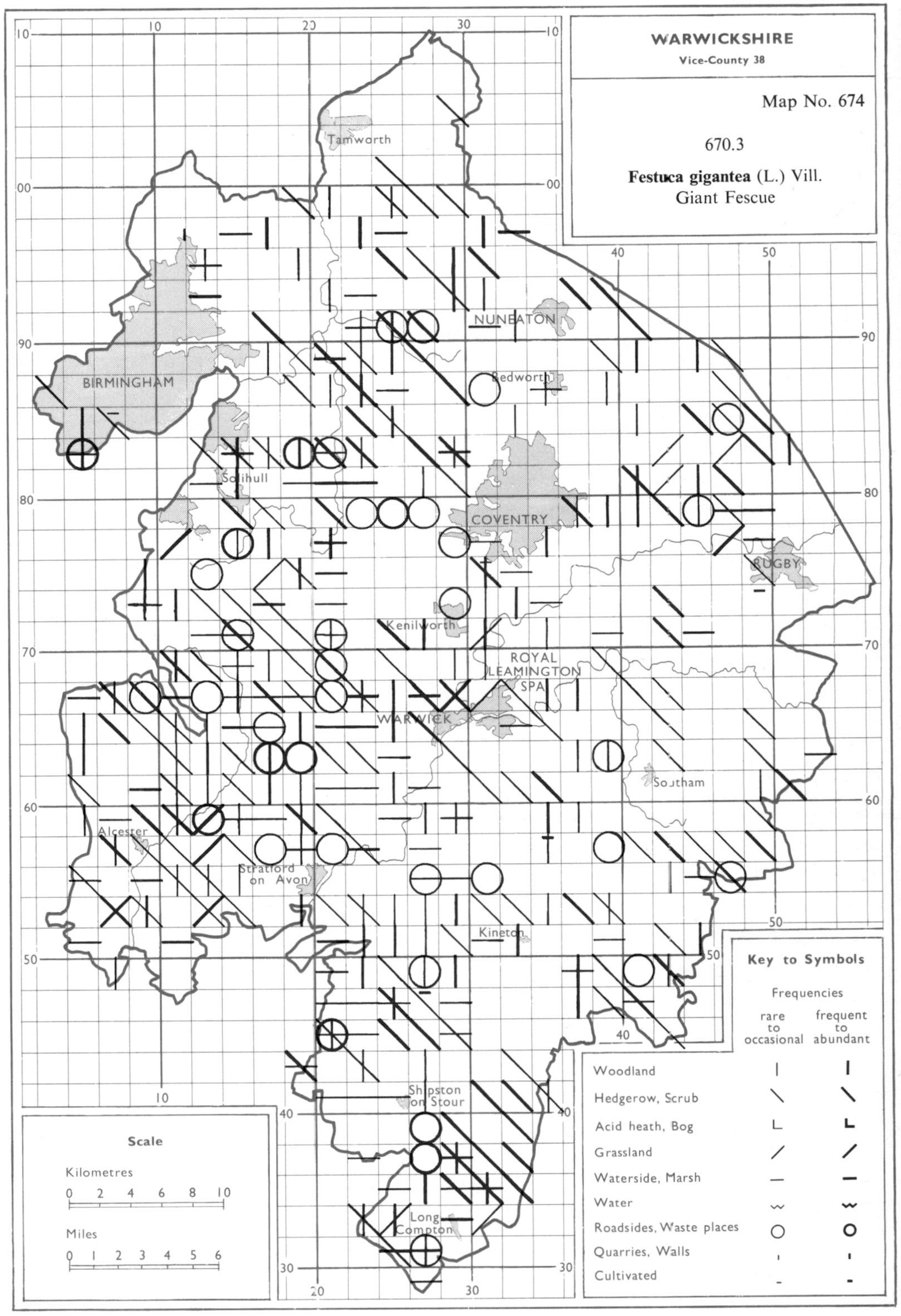

WARWICKSHIRE
Vice-County 38

Map No. 674

670.3

Festuca gigantea (L.) Vill.
Giant Fescue

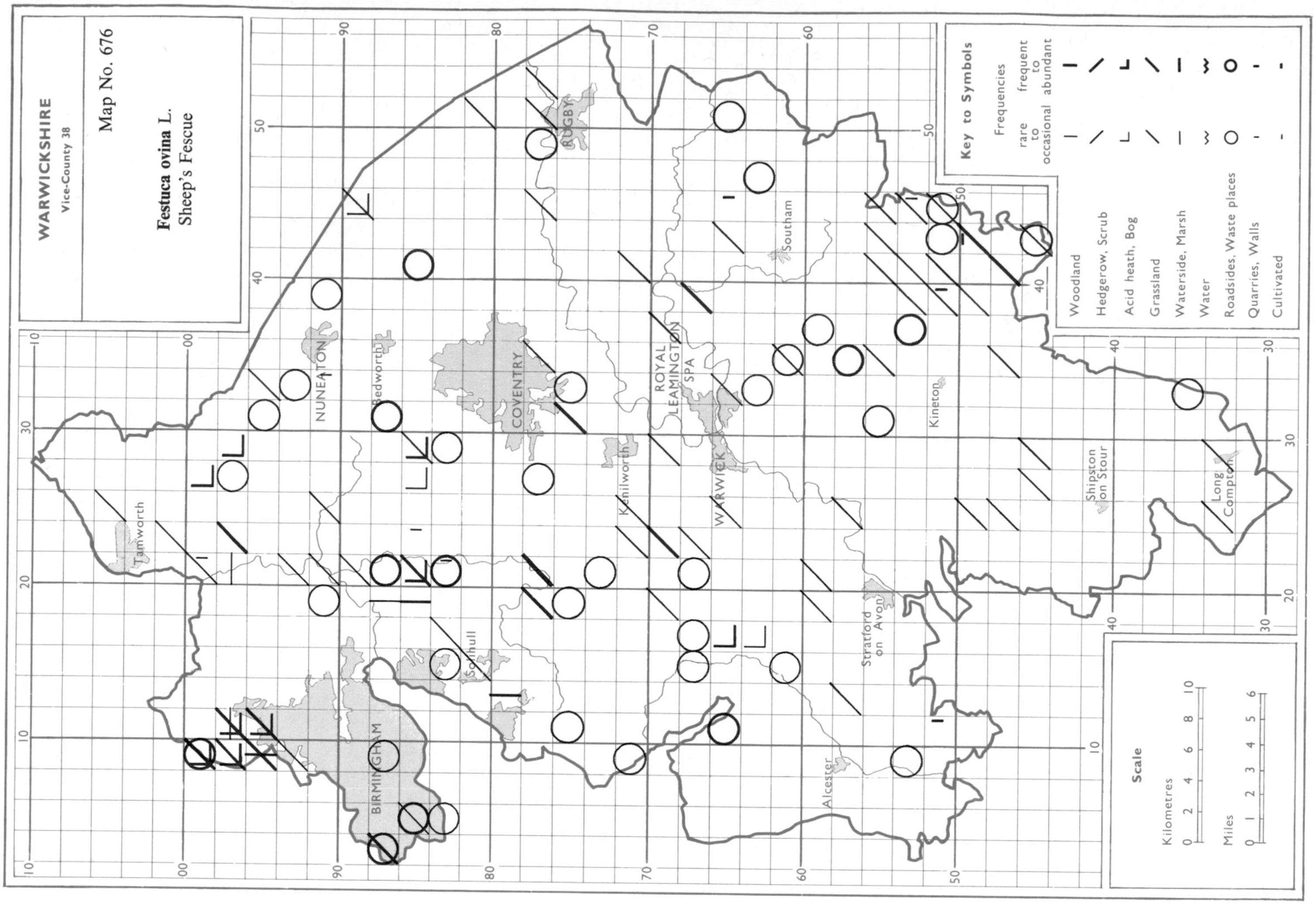
WARWICKSHIRE
Vice-County 38
Map No. 676
Festuca ovina L.
Sheep's Fescue
Key to Symbols
Frequencies
rare to occasional
frequent to abundant
Woodland
Hedgerow, Scrub
Acid heath, Bog
Grassland
Waterside, Marsh
Water
Roadsides, Waste places
Quarries, Walls
Cultivated
Scale
Kilometres
0 2 4 6 8 10
Miles
0 1 2 3 4 5 6
NUNEATON
Bedworth
COVENTRY
RUGBY
ROYAL LEAMINGTON SPA
Southam
Kineton
Tamworth
Kenilworth
WARWICK
Shipston on Stour
Long Compton
Solihull
Stratford on Avon
BIRMINGHAM
Alcester

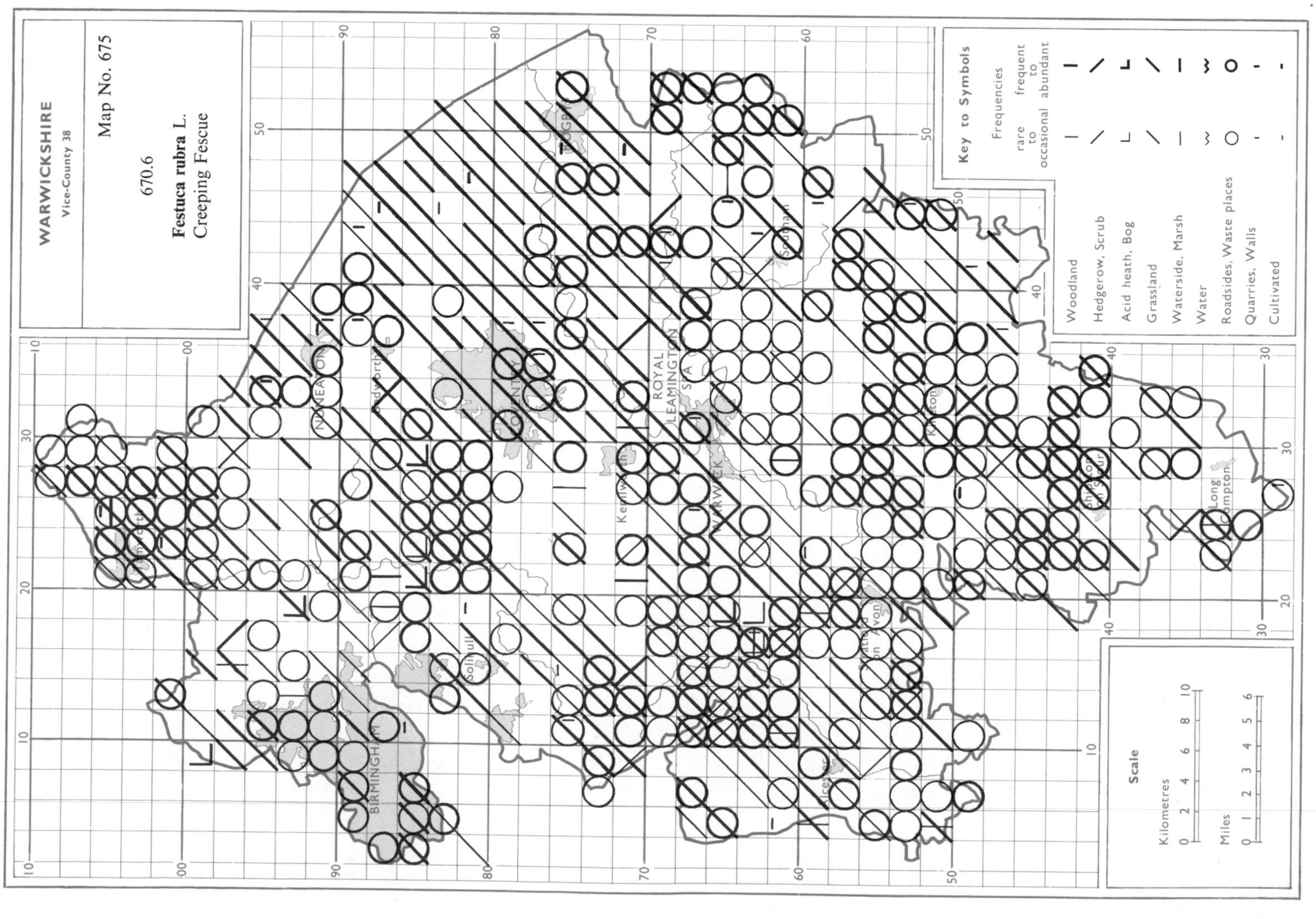
WARWICKSHIRE
Vice-County 38
Map No. 675
670.6
Festuca rubra L.
Creeping Fescue
Key to Symbols
Frequencies
rare to occasional
frequent to abundant
Woodland
Hedgerow, Scrub
Acid heath, Bog
Grassland
Waterside, Marsh
Water
Roadsides, Waste places
Quarries, Walls
Cultivated
Scale
Kilometres
0 2 4 6 8 10
Miles
0 1 2 3 4 5 6
ROYAL LEAMINGTON SPA
Long Compton

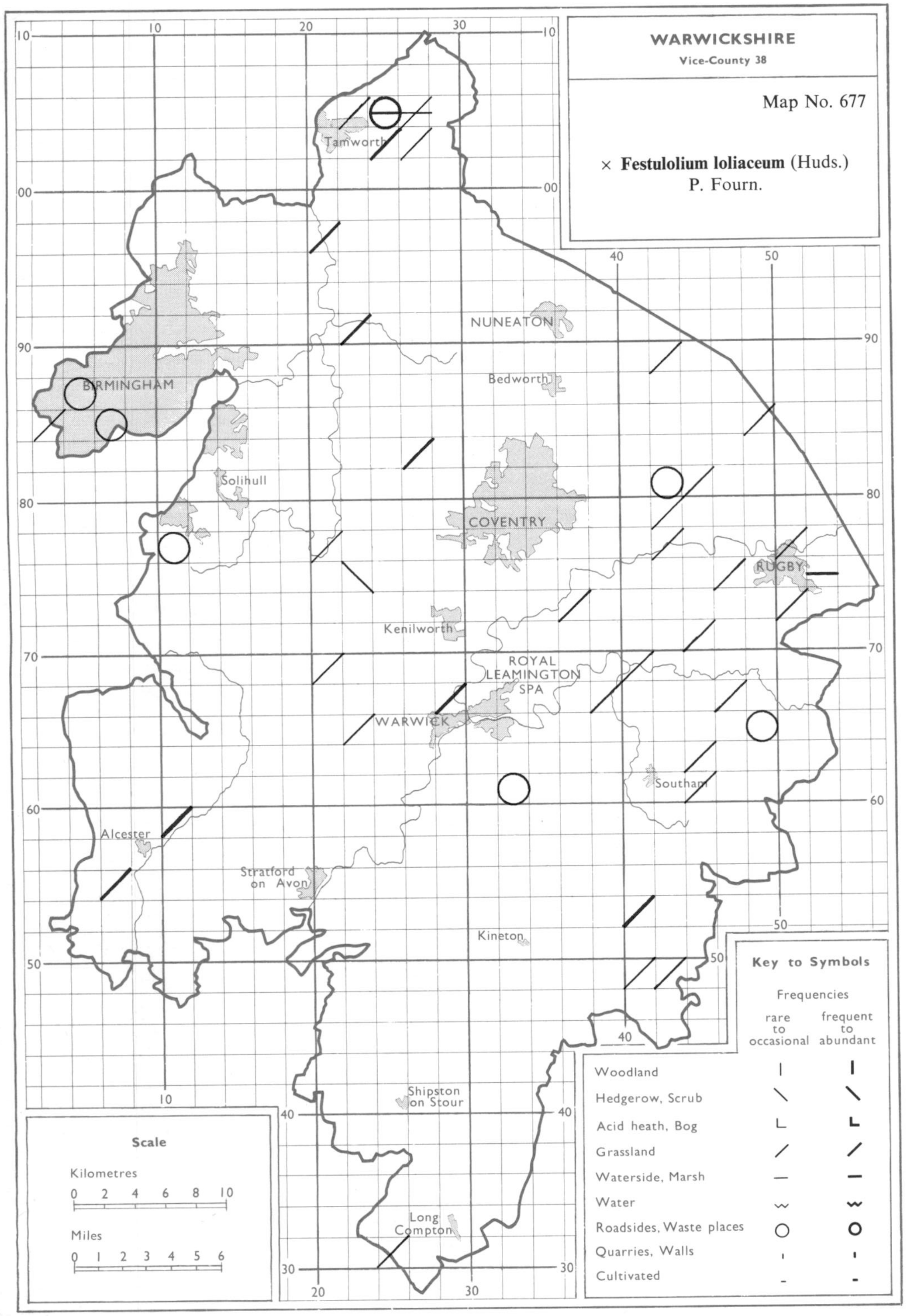
WARWICKSHIRE
Vice-County 38
Map No. 677
× Festulolium loliaceum (Huds.) P. Fourn.
Key to Symbols
Frequencies
rare to occasional
frequent to abundant
Woodland
Hedgerow, Scrub
Acid heath, Bog
Grassland
Waterside, Marsh
Water
Roadsides, Waste places
Quarries, Walls
Cultivated
Scale
Kilometres
Miles
Tamworth
Birmingham
Solihull
Nuneaton
Bedworth
Coventry
Rugby
Kenilworth
Royal Leamington Spa
Warwick
Southam
Alcester
Stratford on Avon
Kineton
Shipston on Stour
Long Compton

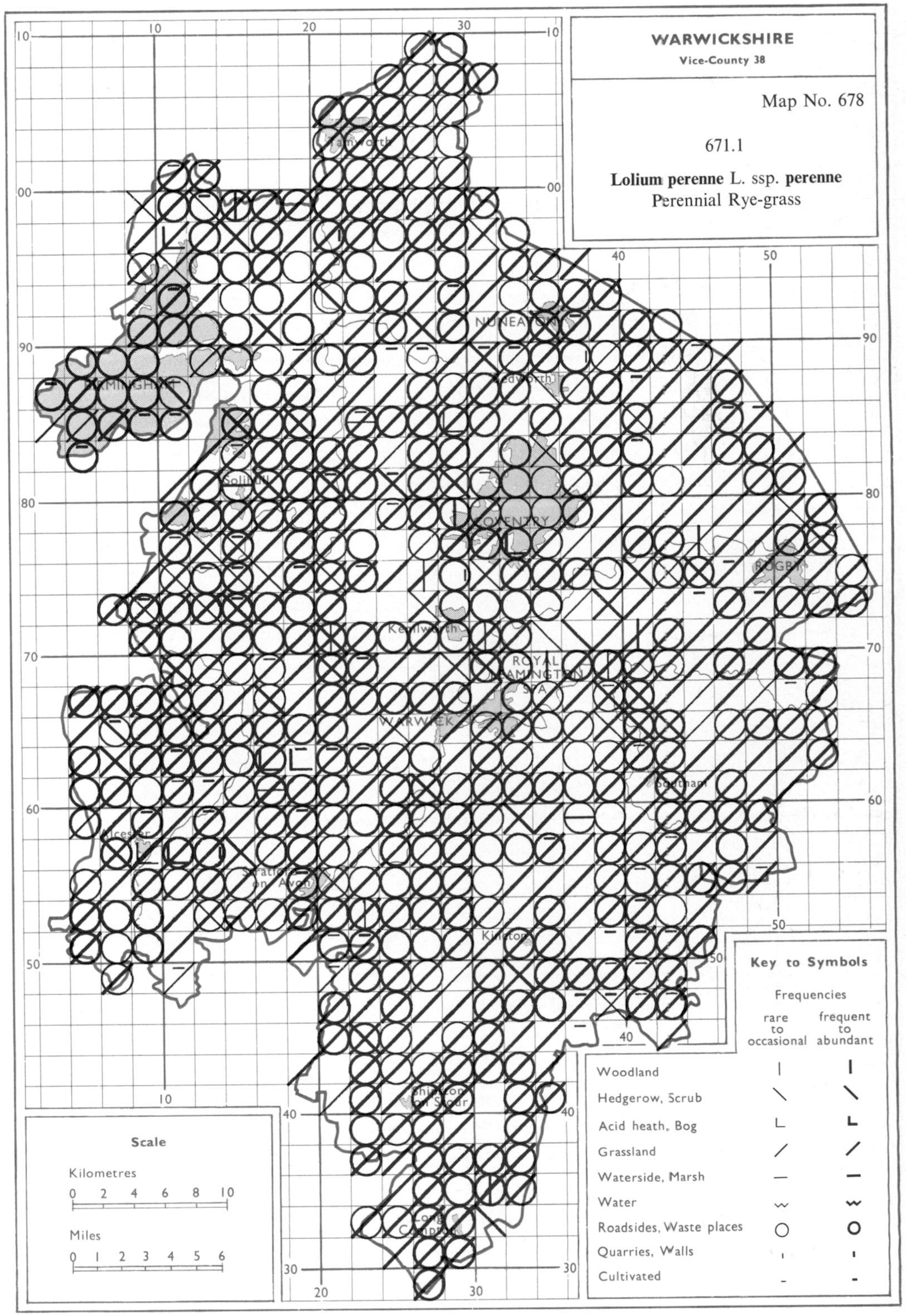
WARWICKSHIRE
Vice-County 38
Map No. 678
671.1
Lolium perenne L. ssp. perenne
Perennial Rye-grass
Key to Symbols
Frequencies
rare to occasional
frequent to abundant
Woodland
Hedgerow, Scrub
Acid heath, Bog
Grassland
Waterside, Marsh
Water
Roadsides, Waste places
Quarries, Walls
Cultivated
Scale
Kilometres
Miles

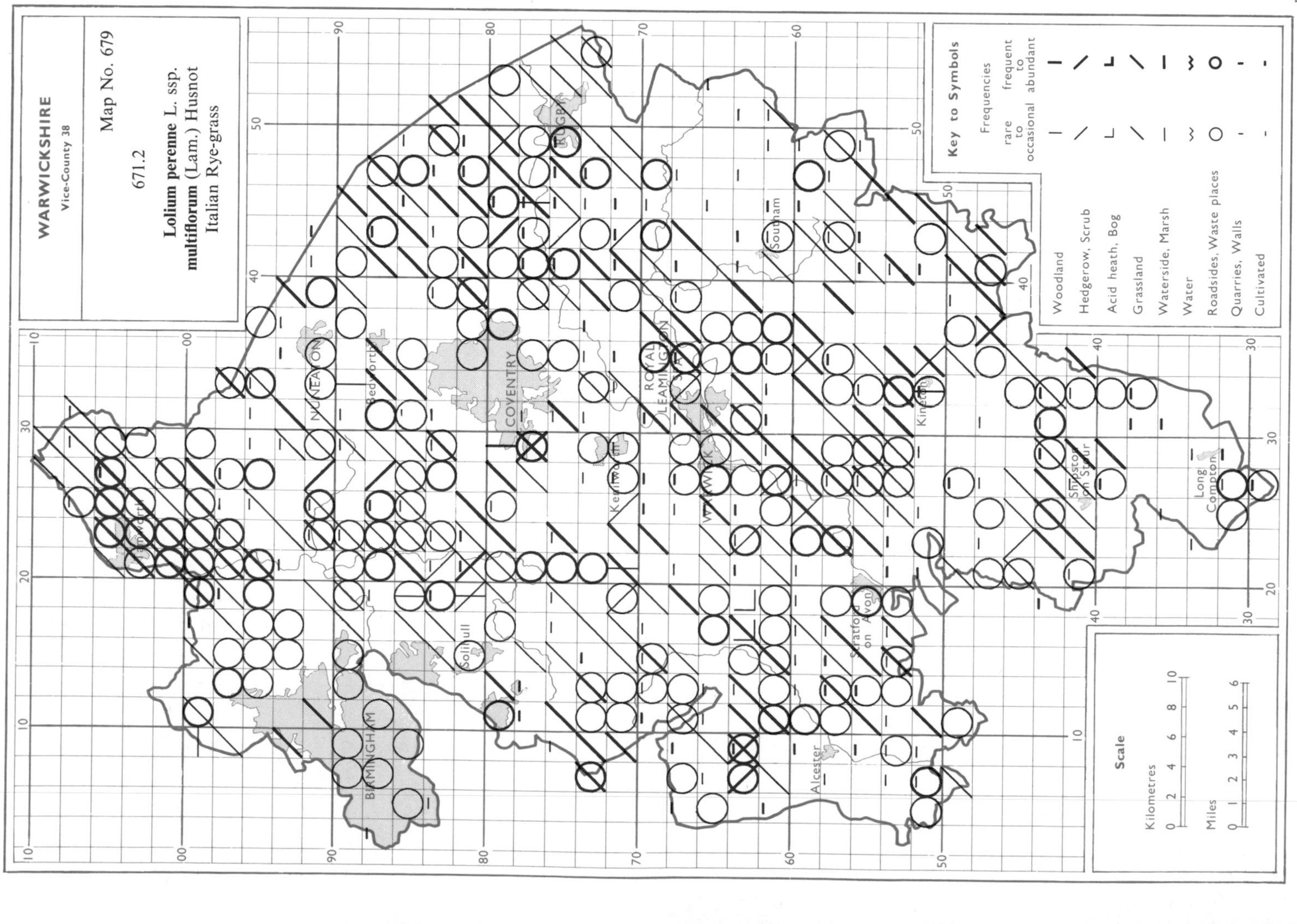
WARWICKSHIRE
Vice-County 38
Map No. 679
671.2
Lolium perenne L. ssp. multiflorum (Lam.) Husnot
Italian Rye-grass
Key to Symbols
Frequencies
rare to occasional
frequent to abundant
Woodland
Hedgerow, Scrub
Acid heath, Bog
Grassland
Waterside, Marsh
Water
Roadsides, Waste places
Quarries, Walls
Cultivated
Scale
Kilometres
0 2 4 6 8 10
Miles
0 1 2 3 4 5 6
NUNEATON
Bedworth
COVENTRY
RUGBY
ROYAL LEAMINGTON SPA
Kenilworth
WARWICK
Southam
Kineton
Shipston on Stour
Long Compton
Stratford on Avon
Alcester
Solihull
BIRMINGHAM
Tamworth

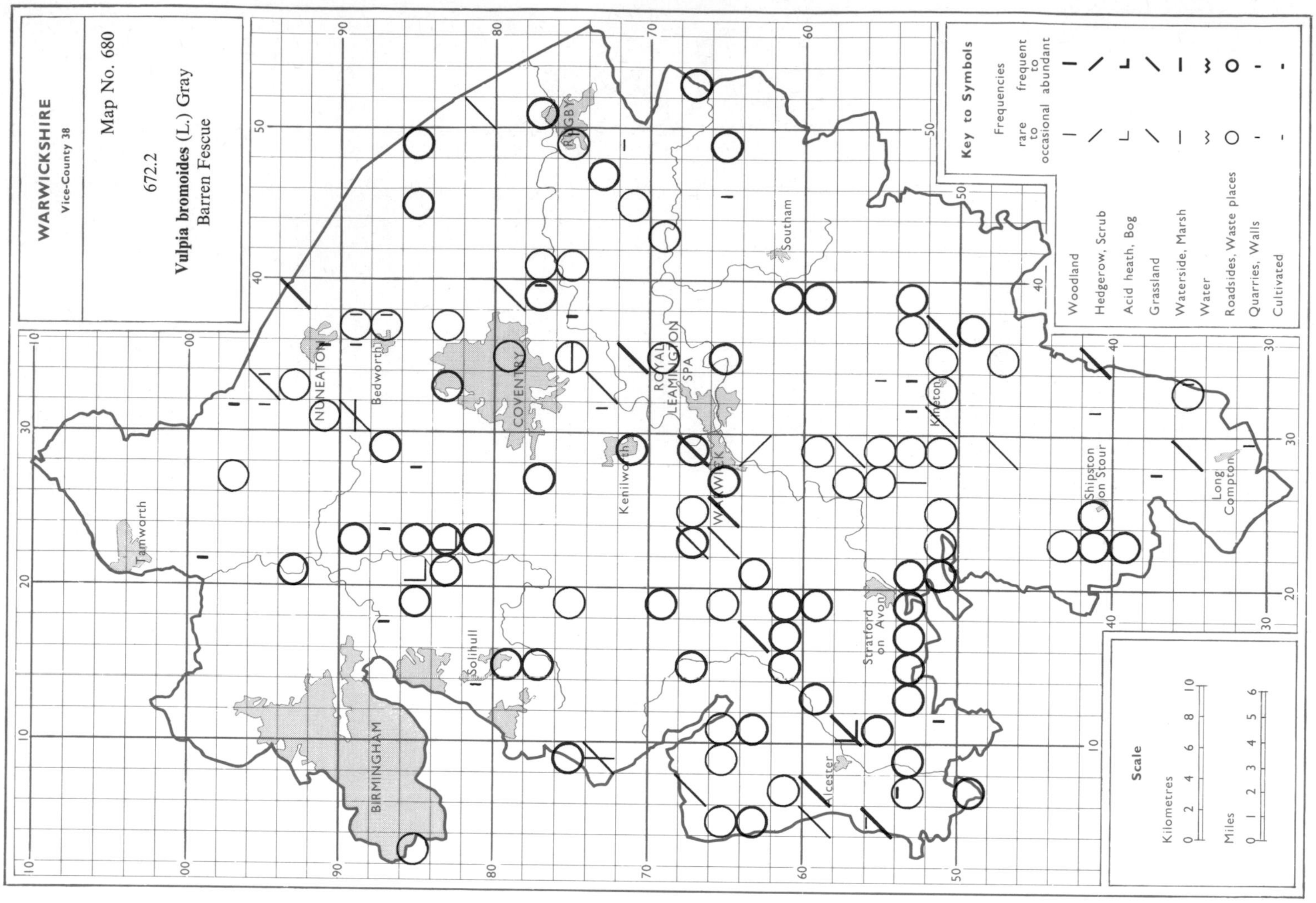
WARWICKSHIRE
Vice-County 38
Map No. 680
672.2
Vulpia bromoides (L.) Gray
Barren Fescue
Key to Symbols
Frequencies
rare to occasional
frequent to abundant
Woodland
Hedgerow, Scrub
Acid heath, Bog
Grassland
Waterside, Marsh
Water
Roadsides, Waste places
Quarries, Walls
Cultivated
Scale
Kilometres
0 2 4 6 8 10
Miles
0 1 2 3 4 5 6
NUNEATON
Bedworth
COVENTRY
RUGBY
ROYAL LEAMINGTON SPA
Kenilworth
WARWICK
Southam
Kineton
Shipston on Stour
Long Compton
Stratford on Avon
Alcester
Solihull
BIRMINGHAM
Tamworth

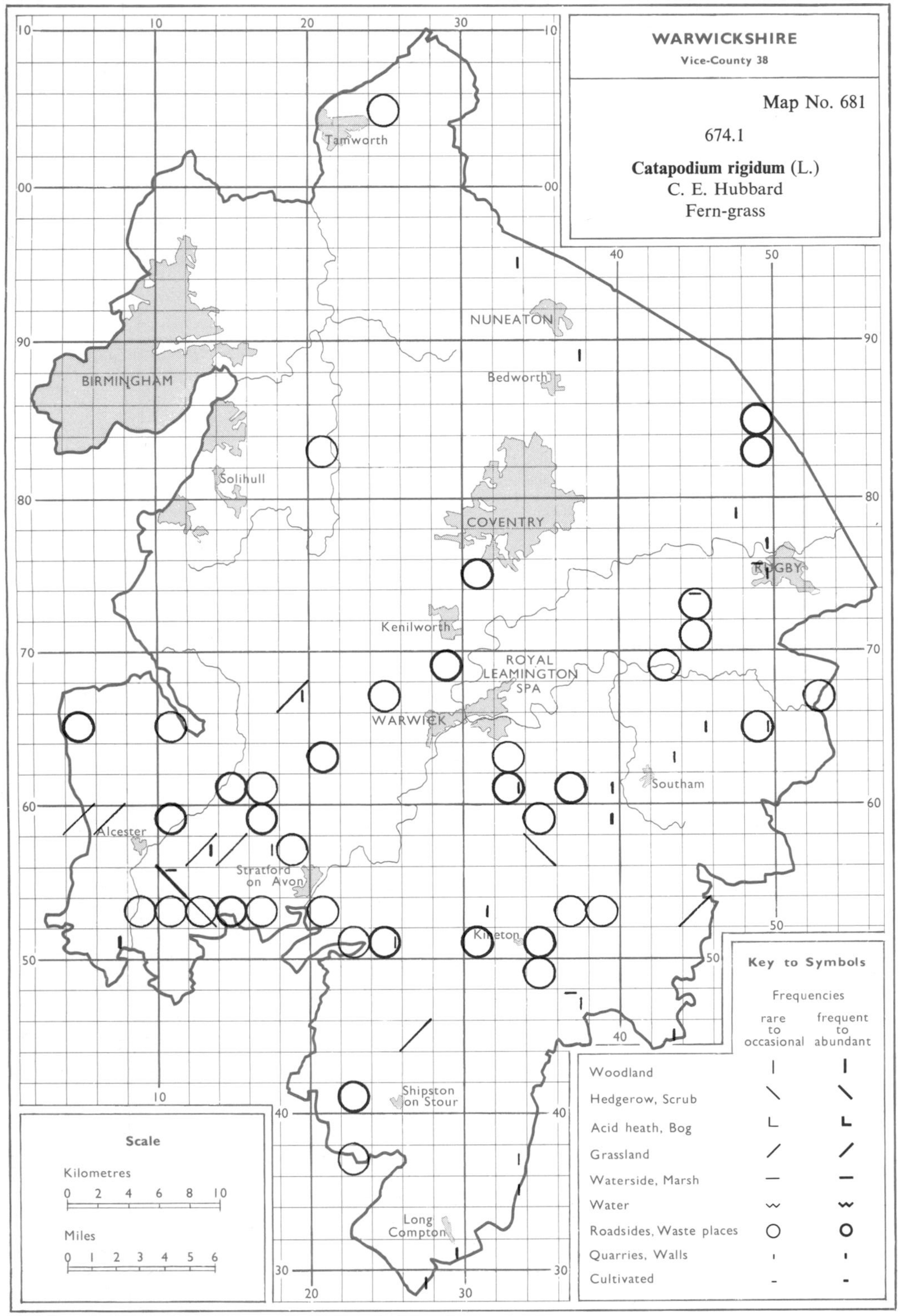
WARWICKSHIRE
Vice-County 38
Map No. 681
674.1
Catapodium rigidum (L.) C. E. Hubbard
Fern-grass
Tamworth
BIRMINGHAM
Solihull
NUNEATON
Bedworth
COVENTRY
RUGBY
Kenilworth
ROYAL LEAMINGTON SPA
WARWICK
Southam
Alcester
Stratford on Avon
Kineton
Shipston on Stour
Long Compton
Key to Symbols
Frequencies
rare to occasional
frequent to abundant
Woodland
Hedgerow, Scrub
Acid heath, Bog
Grassland
Waterside, Marsh
Water
Roadsides, Waste places
Quarries, Walls
Cultivated
Scale
Kilometres
0 2 4 6 8 10
Miles
0 1 2 3 4 5 6

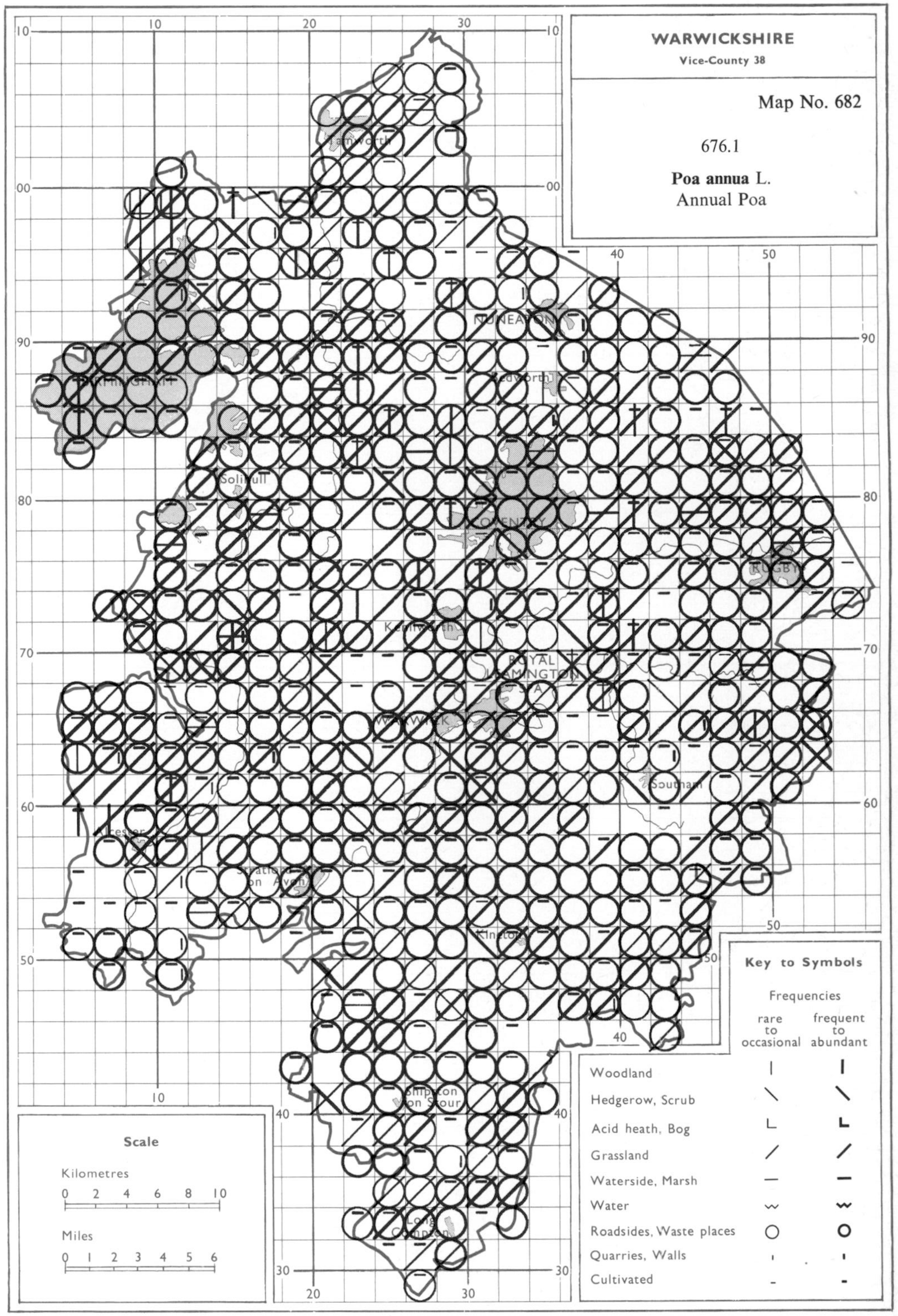
WARWICKSHIRE
Vice-County 38
Map No. 682
676.1
Poa annua L.
Annual Poa
Key to Symbols
Frequencies
rare to occasional
frequent to abundant
Woodland
Hedgerow, Scrub
Acid heath, Bog
Grassland
Waterside, Marsh
Water
Roadsides, Waste places
Quarries, Walls
Cultivated
Scale
Kilometres
0 2 4 6 8 10
Miles
0 1 2 3 4 5 6

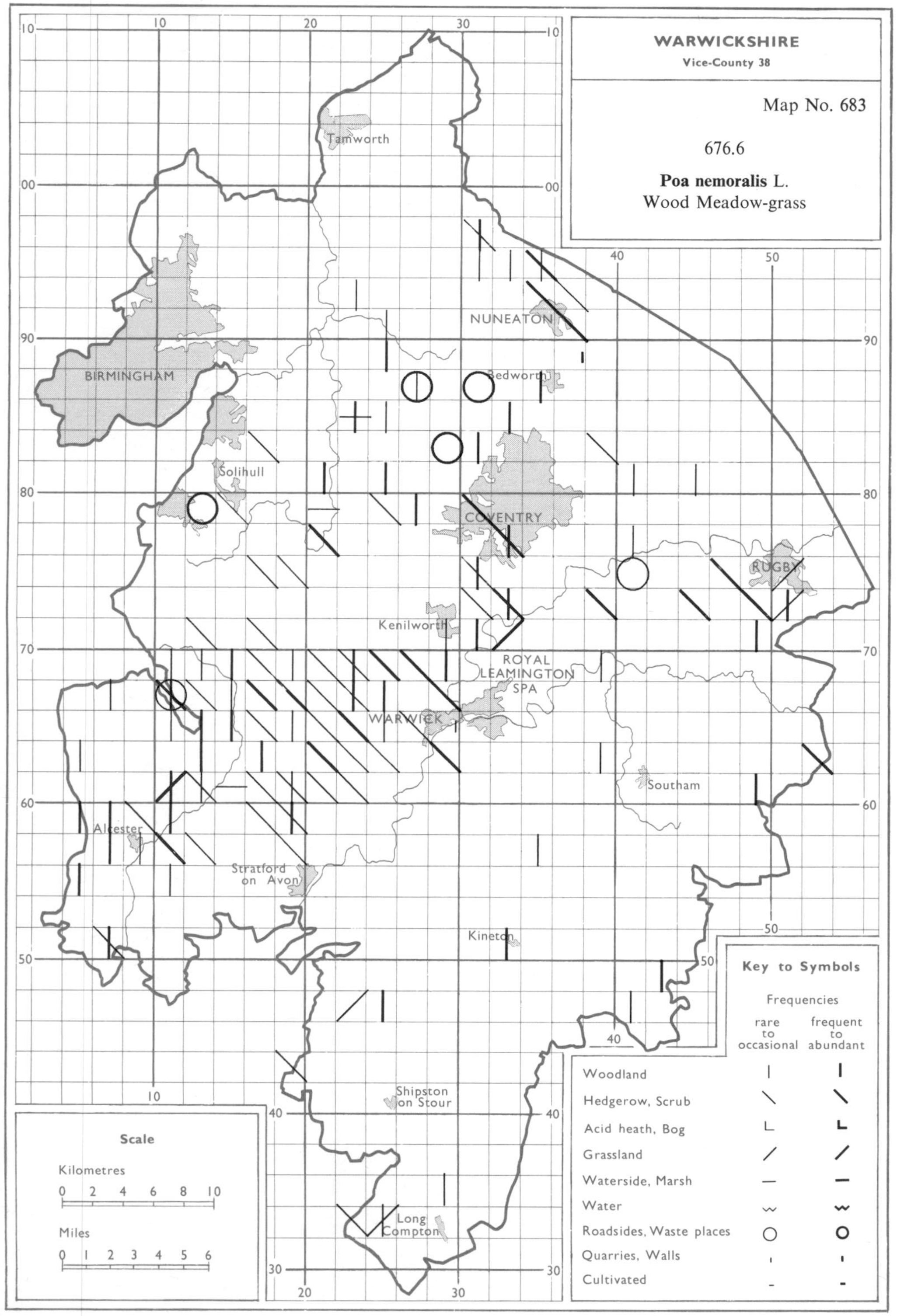
WARWICKSHIRE
Vice-County 38
Map No. 683
676.6
Poa nemoralis L.
Wood Meadow-grass
Tamworth
NUNEATON
Bedworth
BIRMINGHAM
Solihull
COVENTRY
RUGBY
Kenilworth
ROYAL LEAMINGTON SPA
WARWICK
Southam
Alcester
Stratford on Avon
Kineton
Shipston on Stour
Long Compton
Key to Symbols
Frequencies
rare to occasional
frequent to abundant
Woodland
Hedgerow, Scrub
Acid heath, Bog
Grassland
Waterside, Marsh
Water
Roadsides, Waste places
Quarries, Walls
Cultivated
Scale
Kilometres
0 2 4 6 8 10
Miles
0 1 2 3 4 5 6

WARWICKSHIRE
Vice-County 38
Map No. 684
676.9
Poa compressa L.
Flattened Meadow-grass
Tamworth
NUNEATON
Bedworth
BIRMINGHAM
Solihull
COVENTRY
RUGBY
Kenilworth
ROYAL LEAMINGTON SPA
WARWICK
Southam
Alcester
Stratford on Avon
Kineton
Shipston on Stour
Long Compton
Key to Symbols
Frequencies
rare to occasional
frequent to abundant
Woodland
Hedgerow, Scrub
Acid heath, Bog
Grassland
Waterside, Marsh
Water
Roadsides, Waste places
Quarries, Walls
Cultivated
Scale
Kilometres
0 2 4 6 8 10
Miles
0 1 2 3 4 5 6

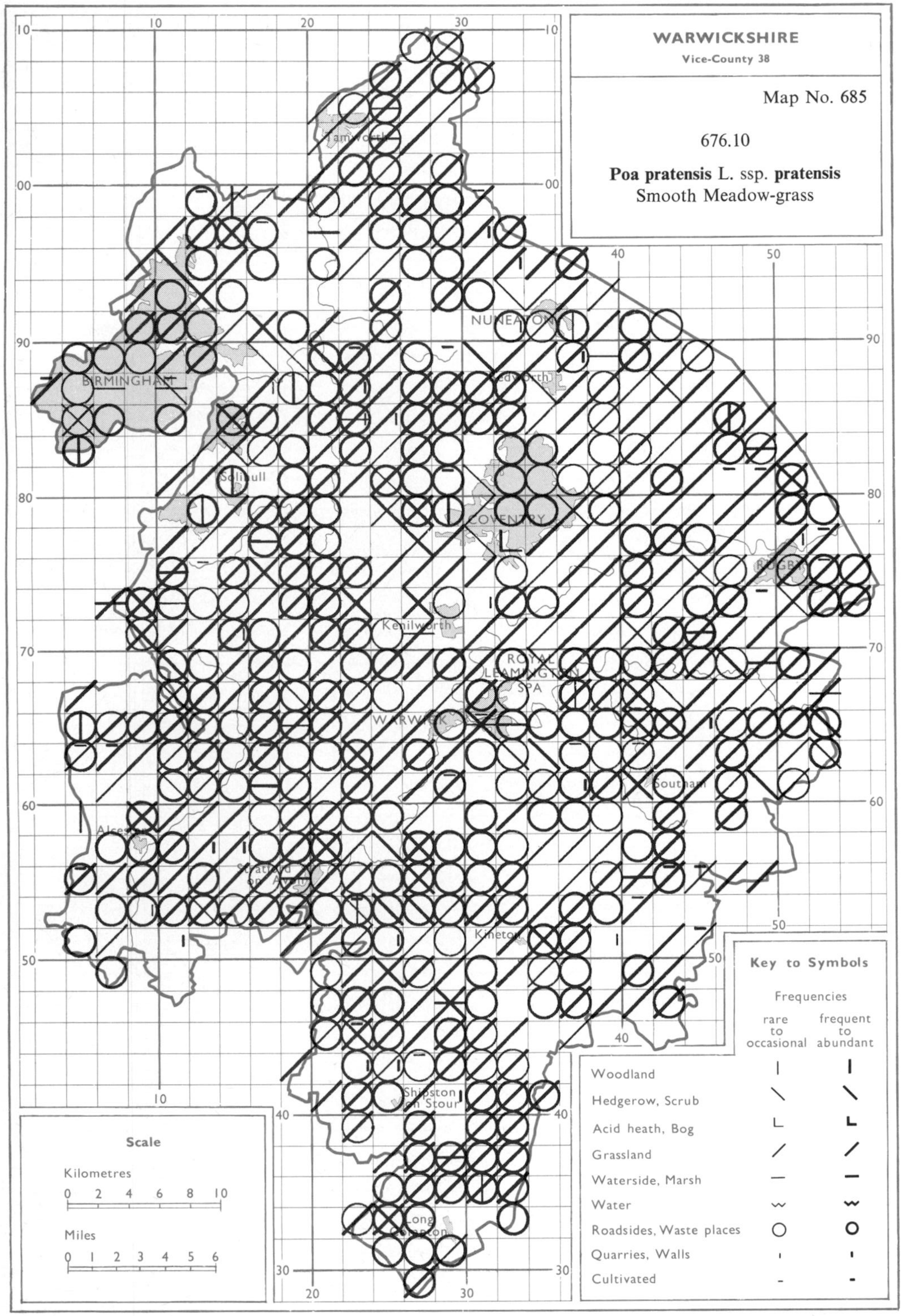
WARWICKSHIRE
Vice-County 38
Map No. 685
676.10
Poa pratensis L. ssp. pratensis
Smooth Meadow-grass
Tamworth
NUNEATON
Bedworth
BIRMINGHAM
Solihull
COVENTRY
RUGBY
Kenilworth
ROYAL LEAMINGTON SPA
WARWICK
Southam
Alcester
Stratford on Avon
Kineton
Shipston on Stour
Long Compton
Scale
Kilometres
0 2 4 6 8 10
Miles
0 1 2 3 4 5 6
Key to Symbols
Frequencies
rare to occasional
frequent to abundant
Woodland
Hedgerow, Scrub
Acid heath, Bog
Grassland
Waterside, Marsh
Water
Roadsides, Waste places
Quarries, Walls
Cultivated

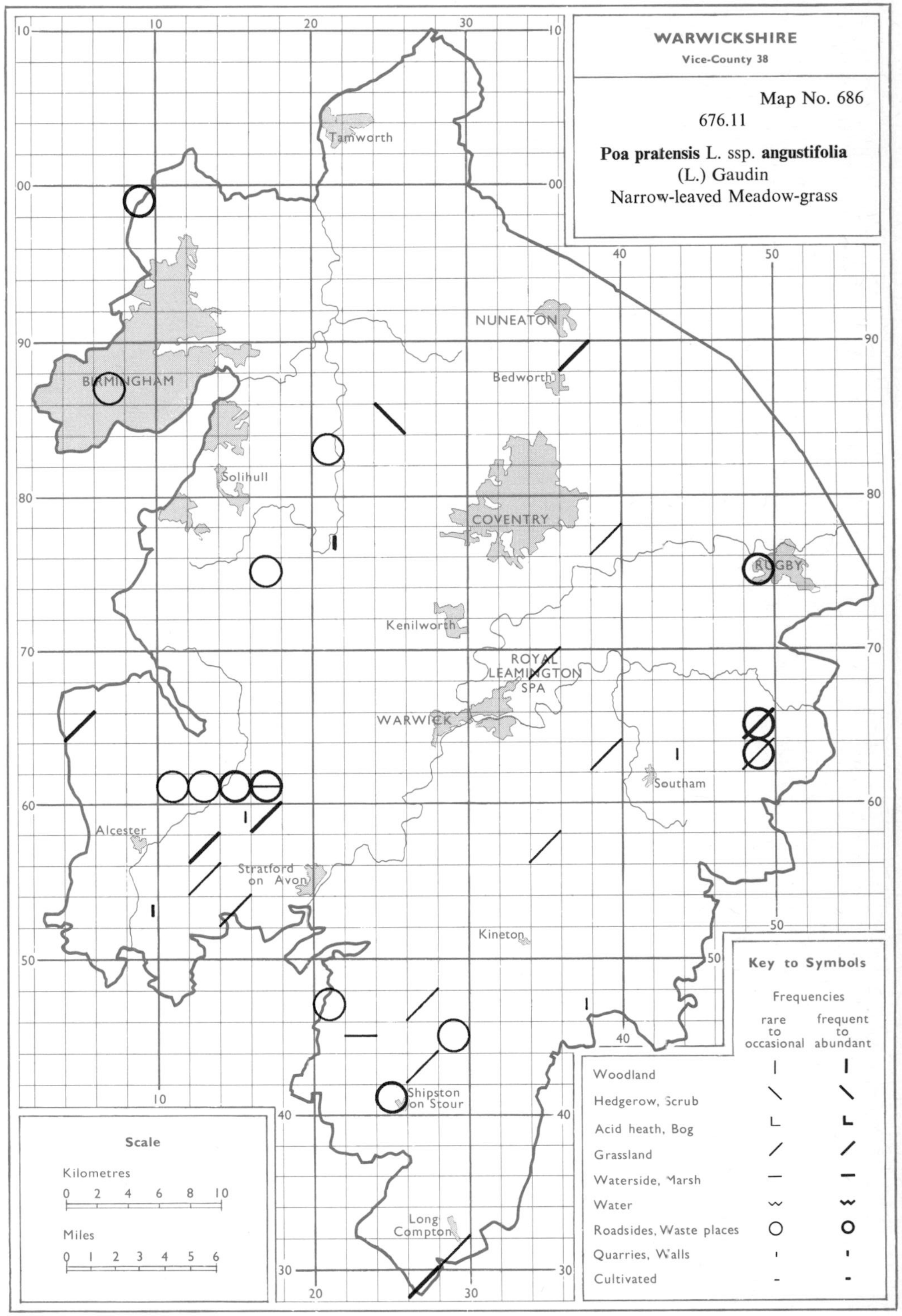
WARWICKSHIRE
Vice-County 38
Map No. 686
676.11
Poa pratensis L. ssp. angustifolia
(L.) Gaudin
Narrow-leaved Meadow-grass
Tamworth
NUNEATON
Bedworth
BIRMINGHAM
Solihull
COVENTRY
RUGBY
Kenilworth
ROYAL LEAMINGTON SPA
WARWICK
Southam
Alcester
Stratford on Avon
Kineton
Shipston on Stour
Long Compton
Scale
Kilometres
0 2 4 6 8 10
Miles
0 1 2 3 4 5 6
Key to Symbols
Frequencies
rare to occasional
frequent to abundant
Woodland
Hedgerow, Scrub
Acid heath, Bog
Grassland
Waterside, Marsh
Water
Roadsides, Waste places
Quarries, Walls
Cultivated

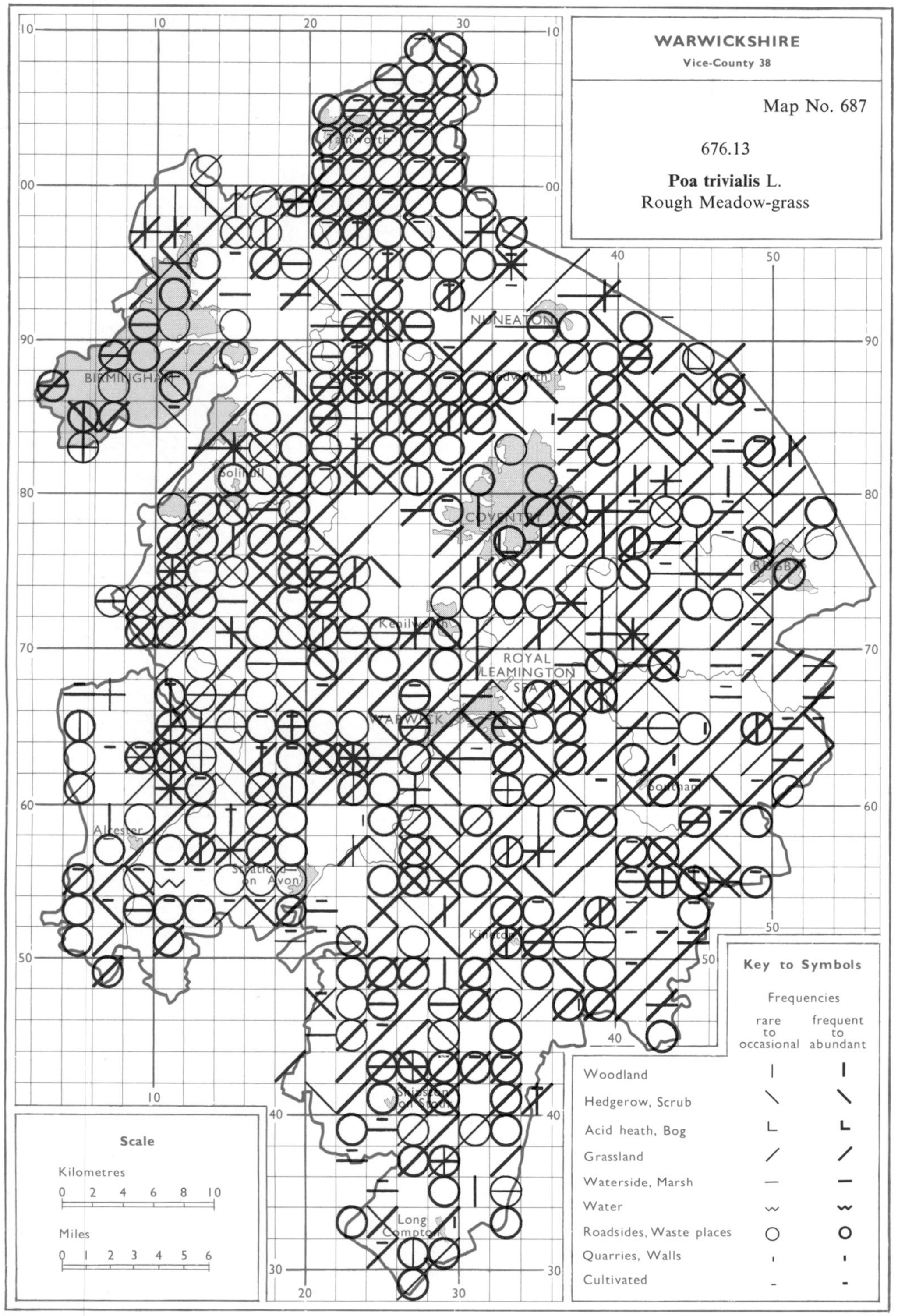
WARWICKSHIRE
Vice-County 38
Map No. 687
676.13
Poa trivialis L.
Rough Meadow-grass
NUNEATON
Bedworth
BIRMINGHAM
Tamworth
Solihull
COVENTRY
RUGBY
Kenilworth
ROYAL LEAMINGTON SPA
WARWICK
Southam
Alcester
Stratford on Avon
Kineton
Shipston on Stour
Long Compton
Key to Symbols
Frequencies
rare to occasional
frequent to abundant
Woodland
Hedgerow, Scrub
Acid heath, Bog
Grassland
Waterside, Marsh
Water
Roadsides, Waste places
Quarries, Walls
Cultivated
Scale
Kilometres
0 2 4 6 8 10
Miles
0 1 2 3 4 5 6

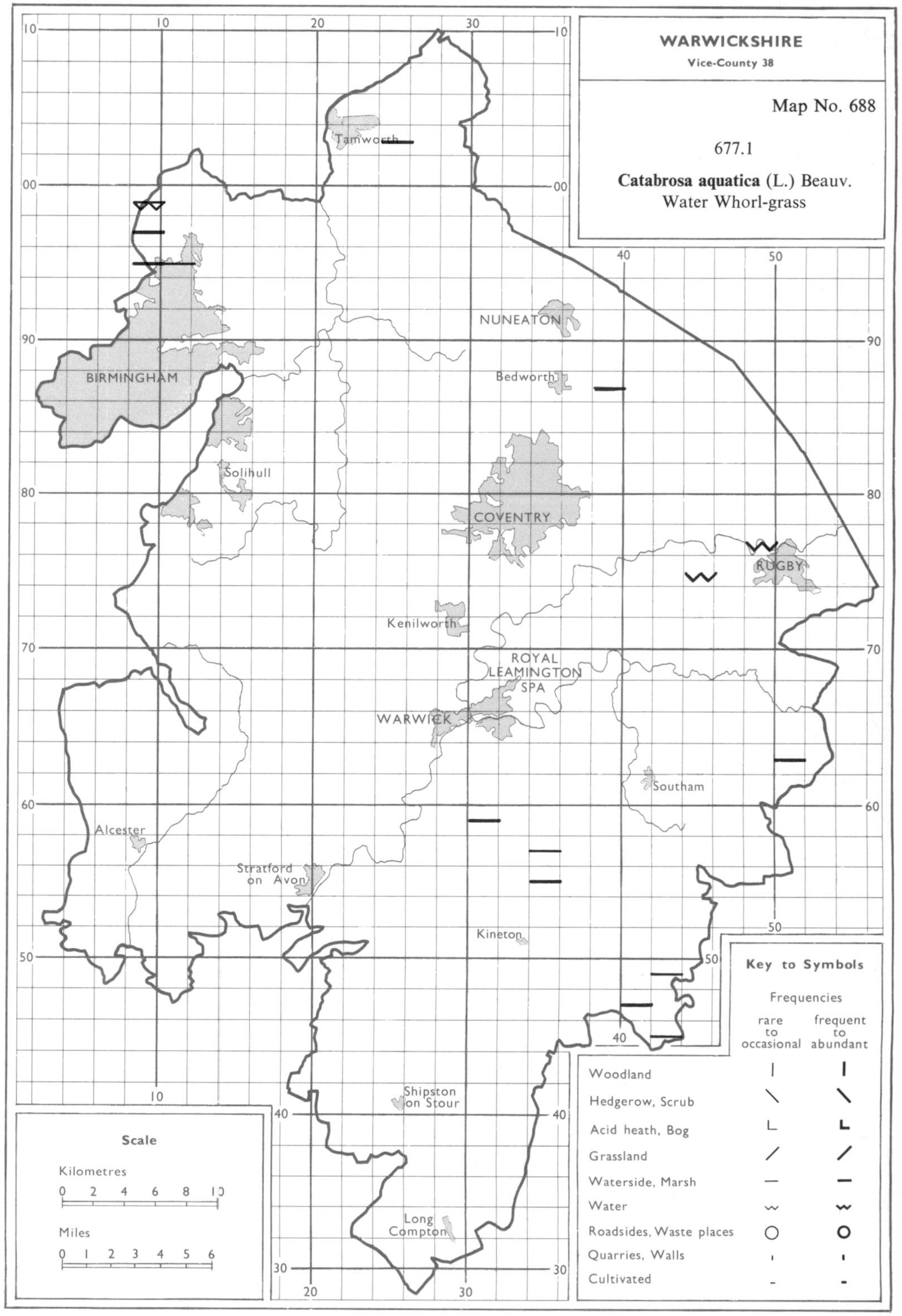
WARWICKSHIRE
Vice-County 38
Map No. 688
677.1
Catabrosa aquatica (L.) Beauv.
Water Whorl-grass
NUNEATON
Bedworth
BIRMINGHAM
Tamworth
Solihull
COVENTRY
RUGBY
Kenilworth
ROYAL LEAMINGTON SPA
WARWICK
Southam
Alcester
Stratford on Avon
Kineton
Shipston on Stour
Long Compton
Key to Symbols
Frequencies
rare to occasional
frequent to abundant
Woodland
Hedgerow, Scrub
Acid heath, Bog
Grassland
Waterside, Marsh
Water
Roadsides, Waste places
Quarries, Walls
Cultivated
Scale
Kilometres
0 2 4 6 8 10
Miles
0 1 2 3 4 5 6

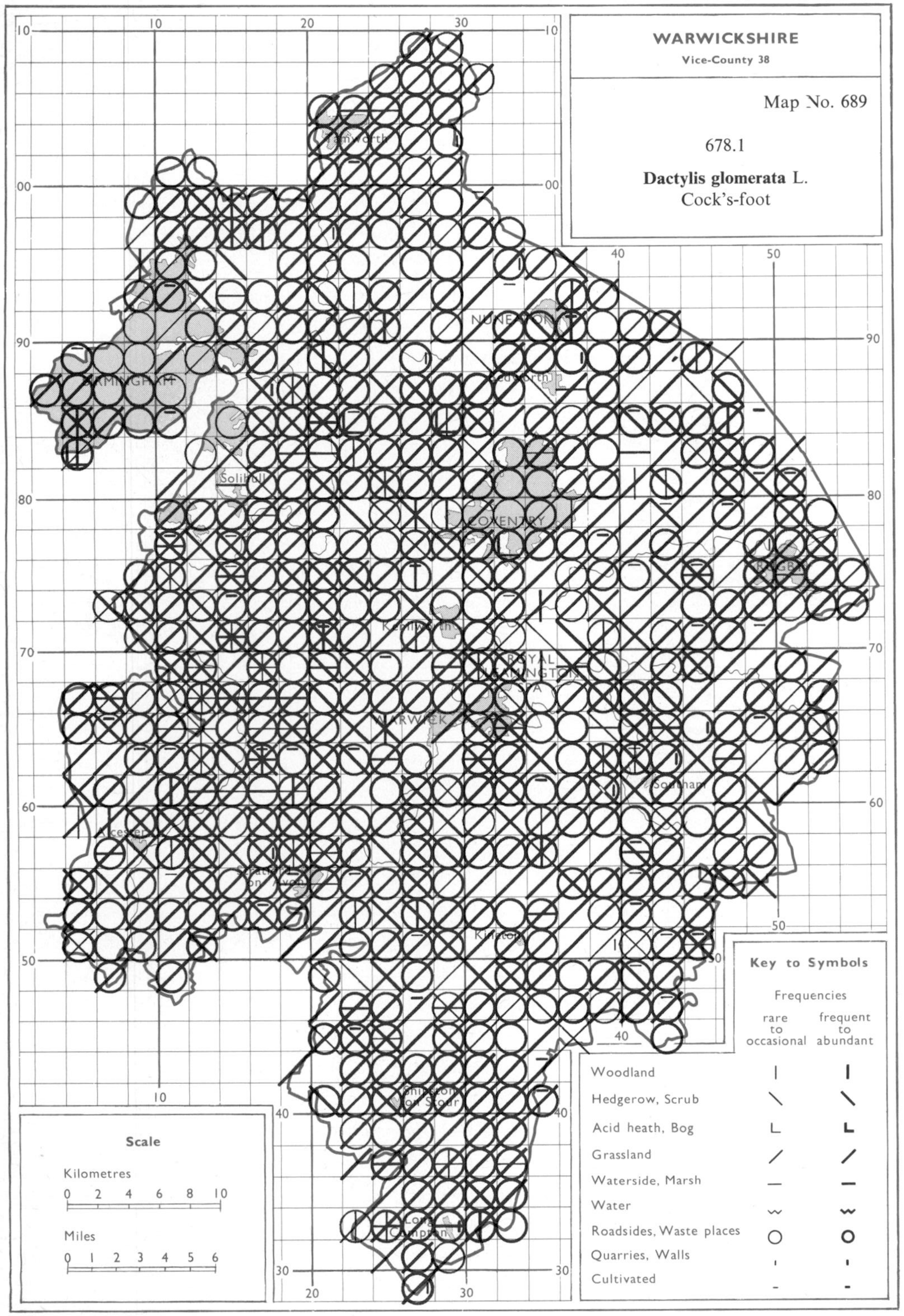
WARWICKSHIRE
Vice-County 38
Map No. 689
678.1
Dactylis glomerata L.
Cock's-foot
Tamworth
NUNEATON
Bedworth
BIRMINGHAM
Solihull
COVENTRY
RUGBY
Kenilworth
ROYAL LEAMINGTON SPA
WARWICK
Southam
Alcester
Stratford on Avon
Kineton
Shipston on Stour
Long Compton
Scale
Kilometres
0 2 4 6 8 10
Miles
0 1 2 3 4 5 6
Key to Symbols
Frequencies
rare to occasional
frequent to abundant
Woodland
Hedgerow, Scrub
Acid heath, Bog
Grassland
Waterside, Marsh
Water
Roadsides, Waste places
Quarries, Walls
Cultivated

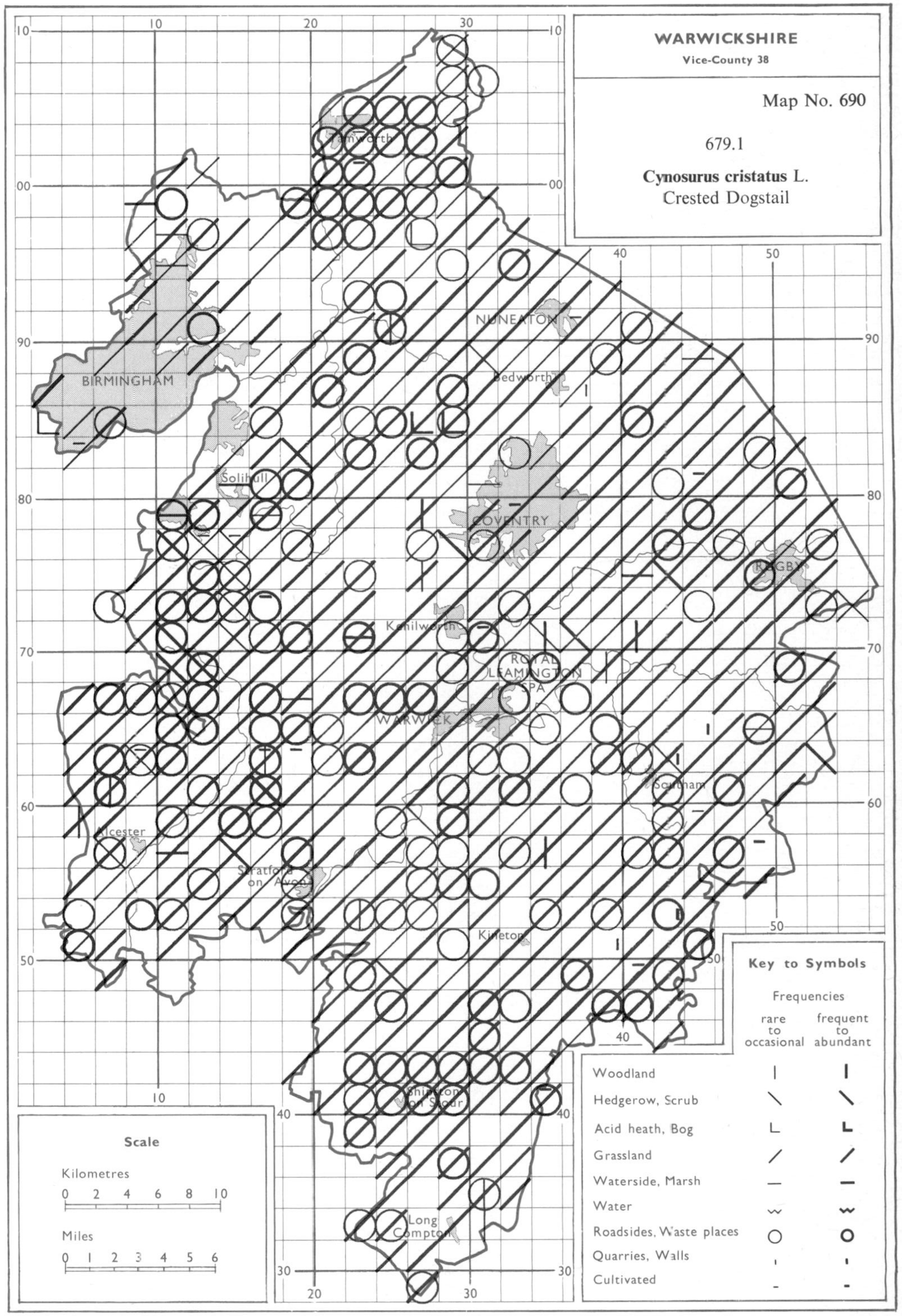
WARWICKSHIRE
Vice-County 38
Map No. 690
679.1
Cynosurus cristatus L.
Crested Dogstail
Tamworth
NUNEATON
Bedworth
BIRMINGHAM
Solihull
COVENTRY
RUGBY
Kenilworth
ROYAL LEAMINGTON SPA
WARWICK
Southam
Alcester
Stratford on Avon
Kineton
Shipston on Stour
Long Compton
Scale
Kilometres
0 2 4 6 8 10
Miles
0 1 2 3 4 5 6
Key to Symbols
Frequencies
rare to occasional
frequent to abundant
Woodland
Hedgerow, Scrub
Acid heath, Bog
Grassland
Waterside, Marsh
Water
Roadsides, Waste places
Quarries, Walls
Cultivated

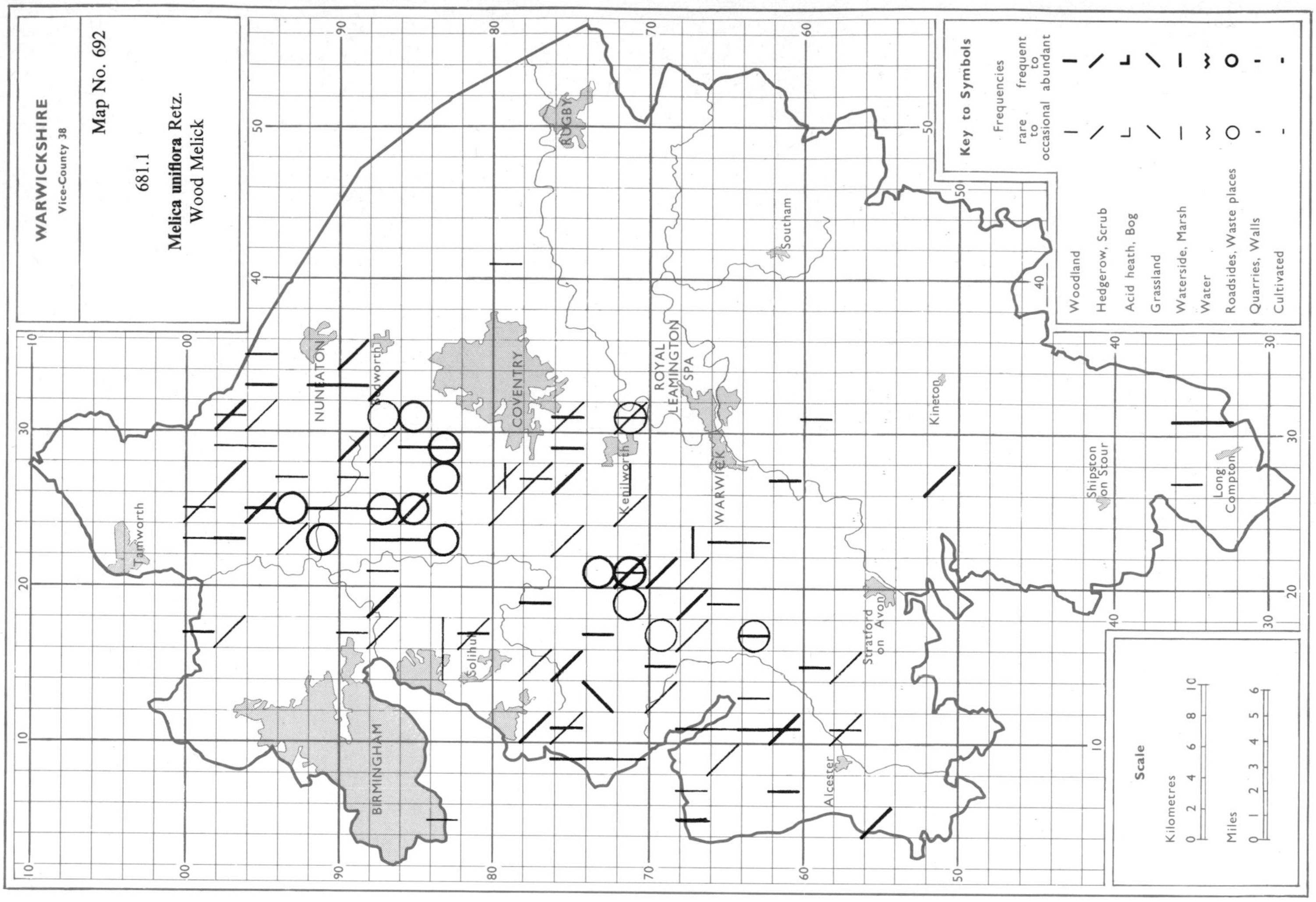
WARWICKSHIRE
Vice-County 38
Map No. 692
681.1
Melica uniflora Retz.
Wood Melick
Key to Symbols
Frequencies
rare to occasional
frequent to abundant
Woodland
Hedgerow, Scrub
Acid heath, Bog
Grassland
Waterside, Marsh
Water
Roadsides, Waste places
Quarries, Walls
Cultivated
Scale
Kilometres 0 2 4 6 8 10
Miles 0 1 2 3 4 5 6
NUNEATON
Bedworth
COVENTRY
RUGBY
ROYAL LEAMINGTON SPA
Kenilworth
WARWICK
Southam
Kineton
Shipston on Stour
Long Compton
Tamworth
Solihull
BIRMINGHAM
Stratford on Avon
Alcester

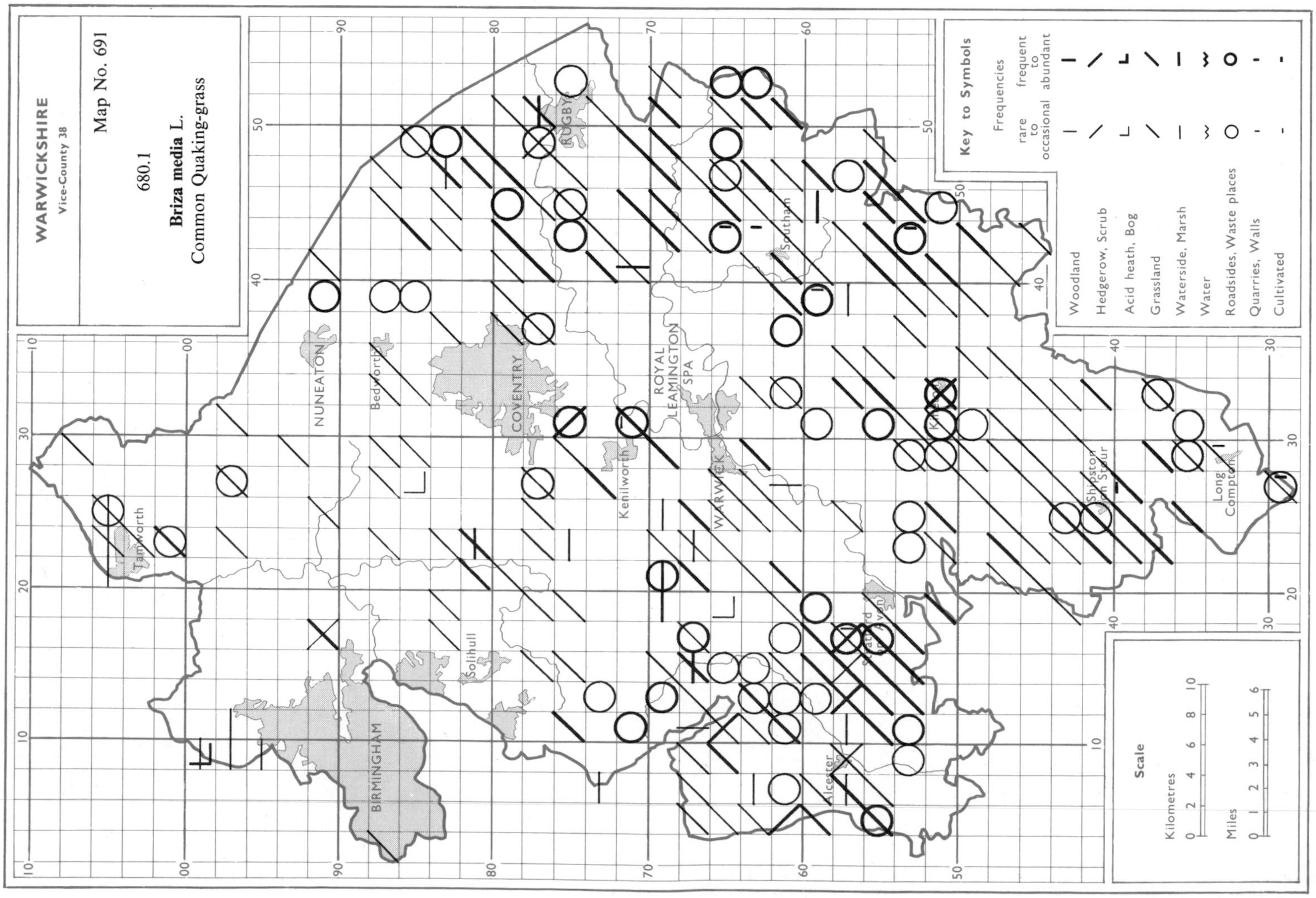
WARWICKSHIRE
Vice-County 38
Map No. 691
680.1
Briza media L.
Common Quaking-grass
Key to Symbols
Frequencies
rare to occasional
frequent to abundant
Woodland
Hedgerow, Scrub
Acid heath, Bog
Grassland
Waterside, Marsh
Water
Roadsides, Waste places
Quarries, Walls
Cultivated
Scale
Kilometres 0 2 4 6 8 10
Miles 0 1 2 3 4 5 6
NUNEATON
Bedworth
COVENTRY
RUGBY
ROYAL LEAMINGTON SPA
Kenilworth
WARWICK
Southam
Shipston on Stour
Long Compton
Tamworth
Solihull
BIRMINGHAM
Alcester

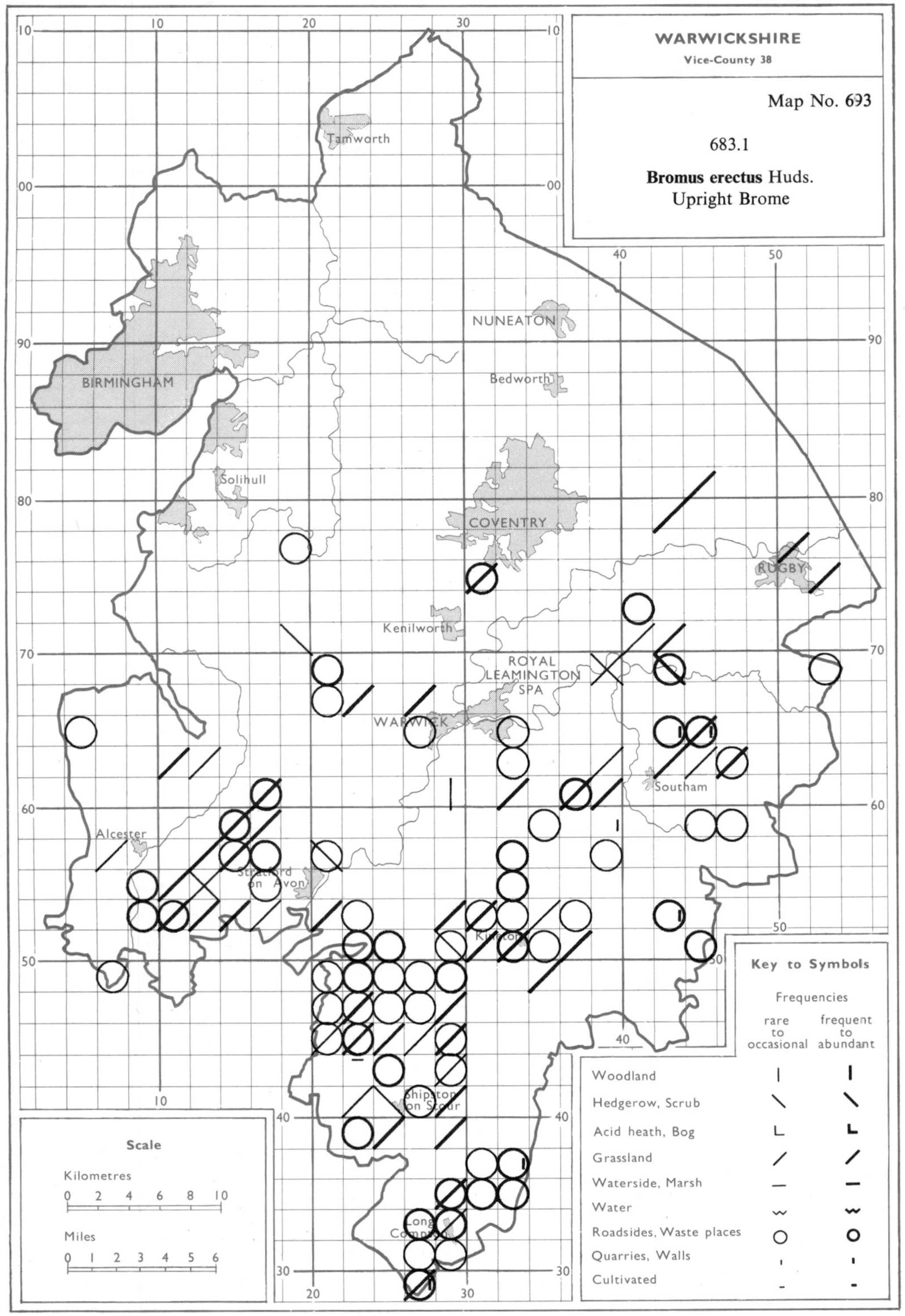
WARWICKSHIRE
Vice-County 38
Map No. 693
683.1
Bromus erectus Huds.
Upright Brome
Tamworth
BIRMINGHAM
NUNEATON
Bedworth
Solihull
COVENTRY
RUGBY
Kenilworth
ROYAL LEAMINGTON SPA
WARWICK
Southam
Alcester
Stratford on Avon
Kineton
Shipston on Stour
Long Compton
Key to Symbols
Frequencies
rare to occasional
frequent to abundant
Woodland
Hedgerow, Scrub
Acid heath, Bog
Grassland
Waterside, Marsh
Water
Roadsides, Waste places
Quarries, Walls
Cultivated
Scale
Kilometres
0 2 4 6 8 10
Miles
0 1 2 3 4 5 6

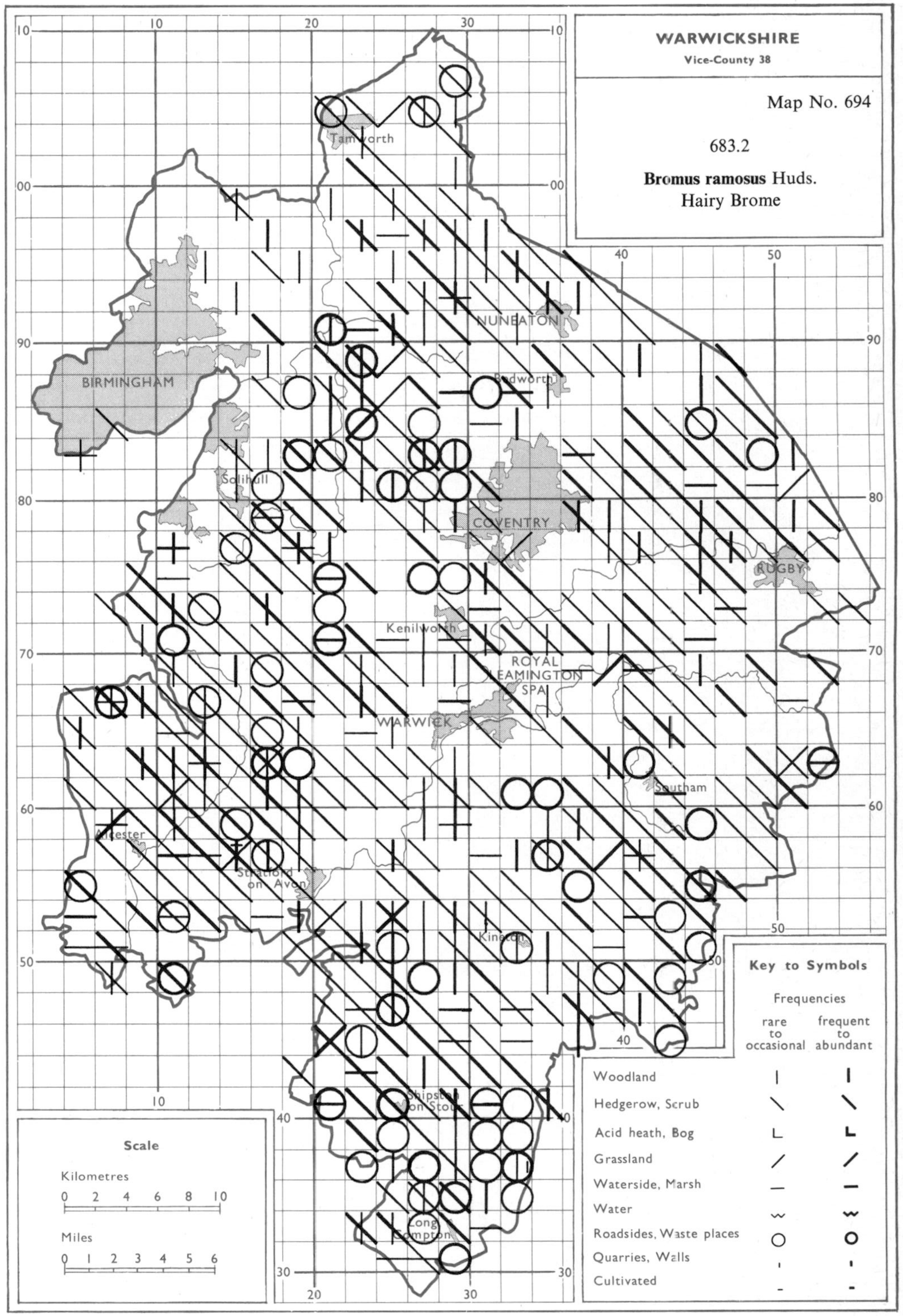
WARWICKSHIRE
Vice-County 38
Map No. 694
683.2
Bromus ramosus Huds.
Hairy Brome
Tamworth
BIRMINGHAM
NUNEATON
Bedworth
Solihull
COVENTRY
RUGBY
Kenilworth
ROYAL LEAMINGTON SPA
WARWICK
Southam
Alcester
Stratford on Avon
Kineton
Shipston on Stour
Long Compton
Key to Symbols
Frequencies
rare to occasional
frequent to abundant
Woodland
Hedgerow, Scrub
Acid heath, Bog
Grassland
Waterside, Marsh
Water
Roadsides, Waste places
Quarries, Walls
Cultivated
Scale
Kilometres
0 2 4 6 8 10
Miles
0 1 2 3 4 5 6

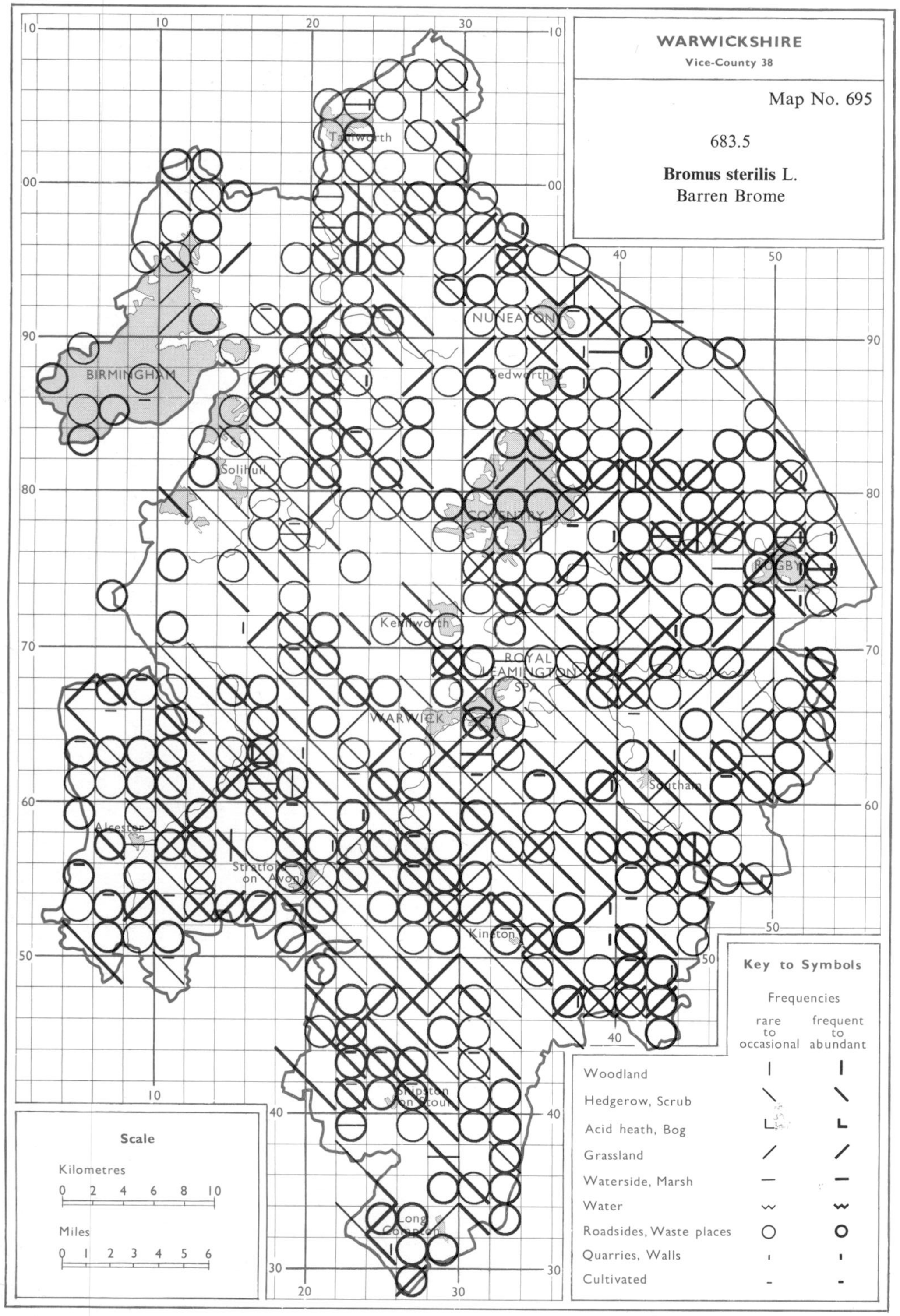
WARWICKSHIRE
Vice-County 38
Map No. 695
683.5
Bromus sterilis L.
Barren Brome
Key to Symbols
Frequencies
rare to occasional
frequent to abundant
Woodland
Hedgerow, Scrub
Acid heath, Bog
Grassland
Waterside, Marsh
Water
Roadsides, Waste places
Quarries, Walls
Cultivated
Scale
Kilometres
Miles
Tamworth
BIRMINGHAM
NUNEATON
Bedworth
Solihull
COVENTRY
RUGBY
Kenilworth
ROYAL LEAMINGTON SPA
WARWICK
Southam
Alcester
Stratford on Avon
Kineton
Shipston on Stour
Long Compton

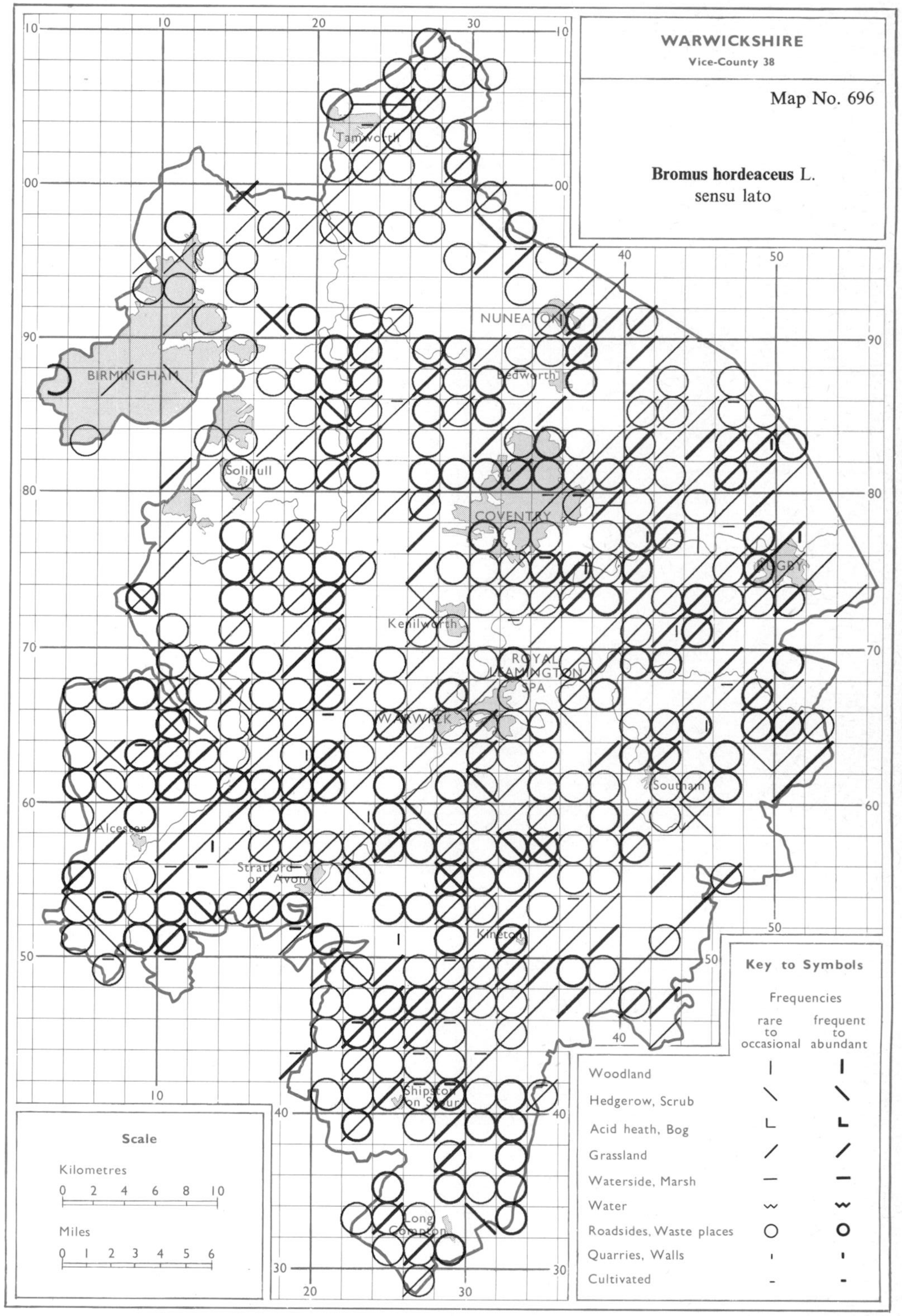
WARWICKSHIRE
Vice-County 38
Map No. 696
Bromus hordeaceus L.
sensu lato
Key to Symbols
Frequencies
rare to occasional
frequent to abundant
Woodland
Hedgerow, Scrub
Acid heath, Bog
Grassland
Waterside, Marsh
Water
Roadsides, Waste places
Quarries, Walls
Cultivated
Scale
Kilometres
Miles
Tamworth
BIRMINGHAM
NUNEATON
Bedworth
Solihull
COVENTRY
RUGBY
Kenilworth
ROYAL LEAMINGTON SPA
WARWICK
Southam
Alcester
Stratford on Avon
Kineton
Shipston on Stour
Long Compton

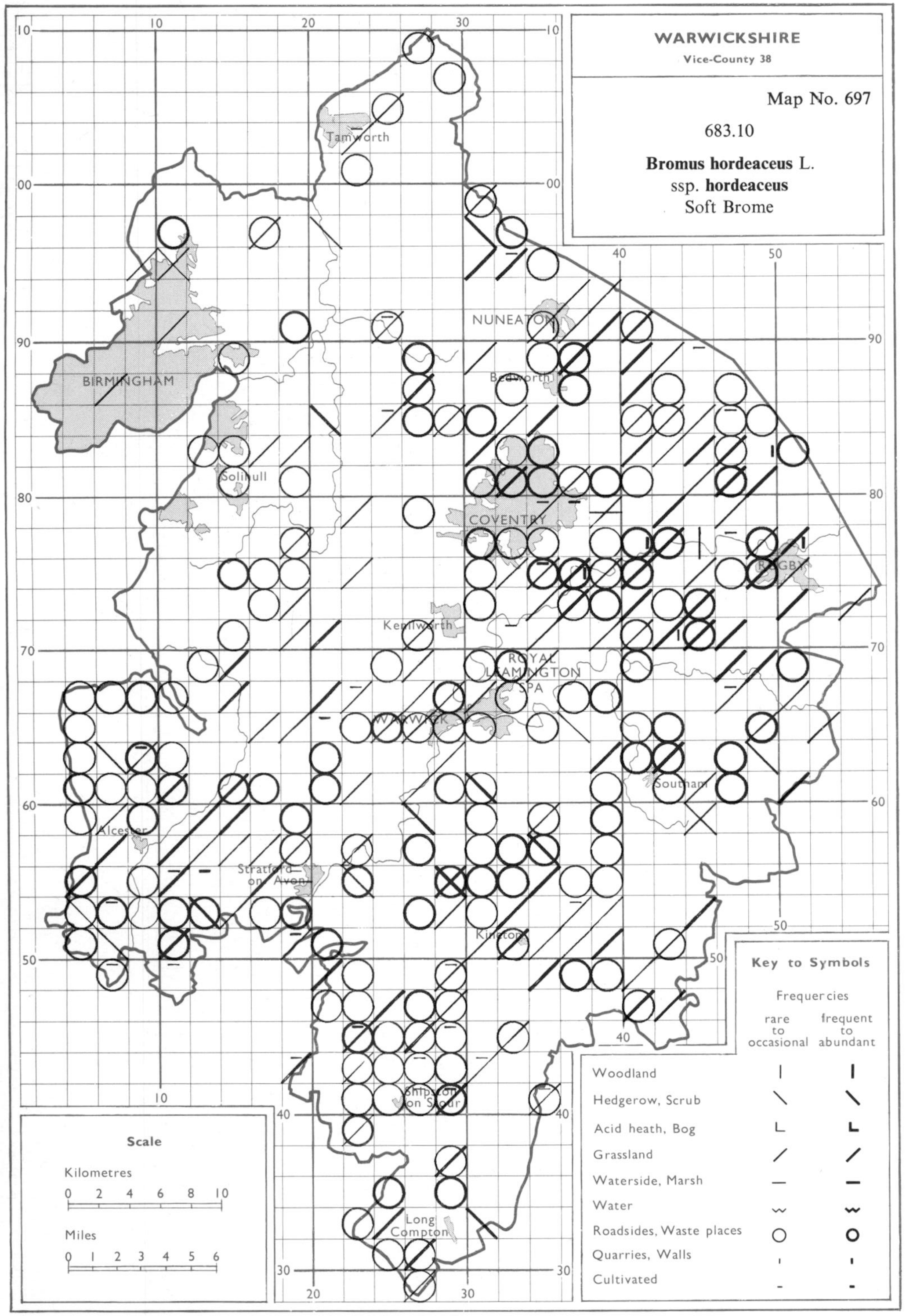
WARWICKSHIRE
Vice-County 38
Map No. 697
683.10
Bromus hordeaceus L.
ssp. **hordeaceus**
Soft Brome
Key to Symbols
Frequencies
rare to occasional
frequent to abundant
Woodland
Hedgerow, Scrub
Acid heath, Bog
Grassland
Waterside, Marsh
Water
Roadsides, Waste places
Quarries, Walls
Cultivated
Scale
Kilometres
0 2 4 6 8 10
Miles
0 1 2 3 4 5 6
Tamworth
NUNEATON
Bedworth
BIRMINGHAM
Solihull
COVENTRY
RUGBY
Kenilworth
ROYAL LEAMINGTON SPA
WARWICK
Southam
Alcester
Stratford on Avon
Kineton
Shipston on Stour
Long Compton

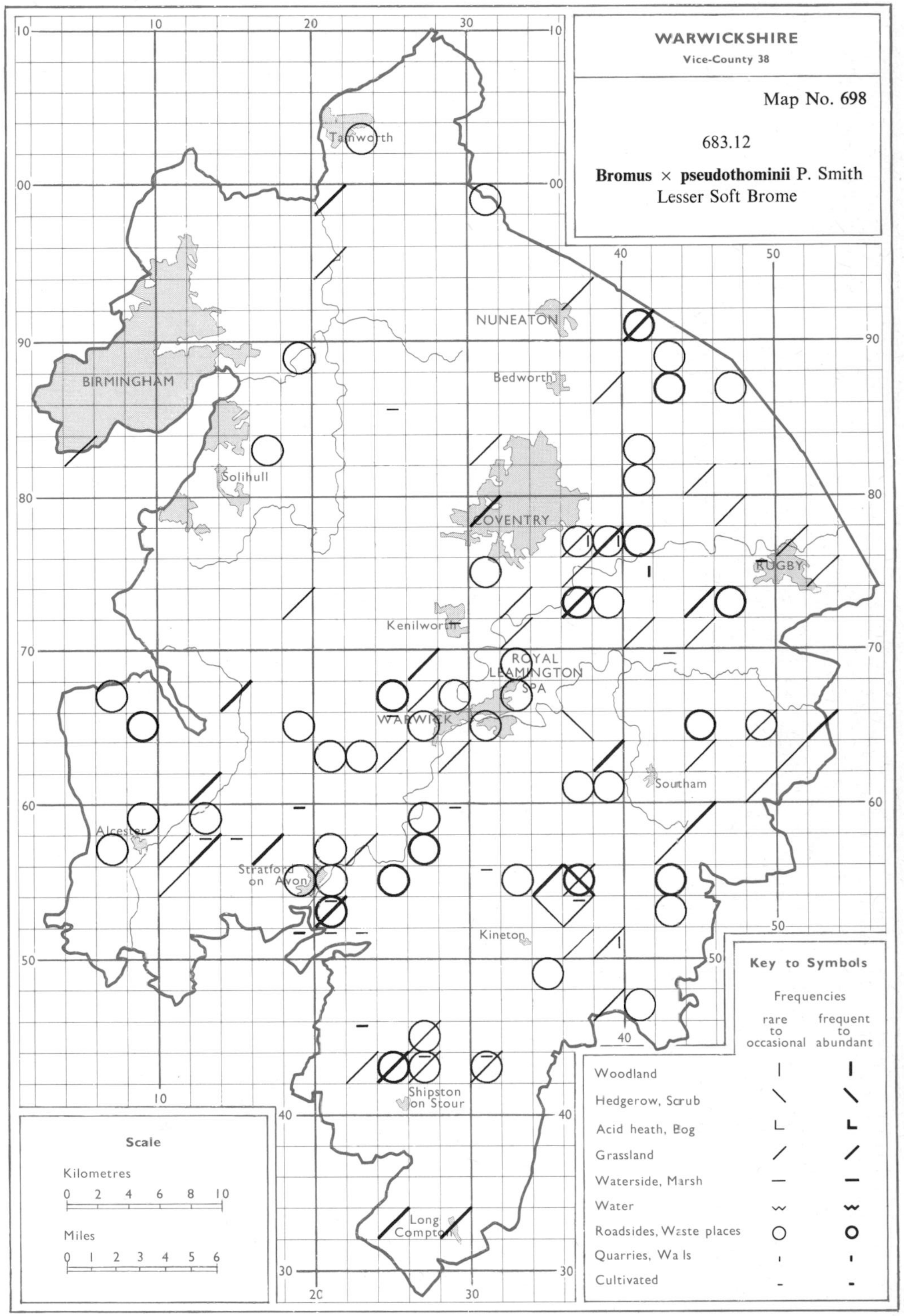
WARWICKSHIRE
Vice-County 38
Map No. 698
683.12
Bromus × pseudothominii P. Smith
Lesser Soft Brome
Key to Symbols
Frequencies
rare to occasional
frequent to abundant
Woodland
Hedgerow, Scrub
Acid heath, Bog
Grassland
Waterside, Marsh
Water
Roadsides, Waste places
Quarries, Walls
Cultivated
Scale
Kilometres
0 2 4 6 8 10
Miles
0 1 2 3 4 5 6
Tamworth
NUNEATON
Bedworth
BIRMINGHAM
Solihull
COVENTRY
RUGBY
Kenilworth
ROYAL LEAMINGTON SPA
WARWICK
Southam
Alcester
Stratford on Avon
Kineton
Shipston on Stour
Long Compton

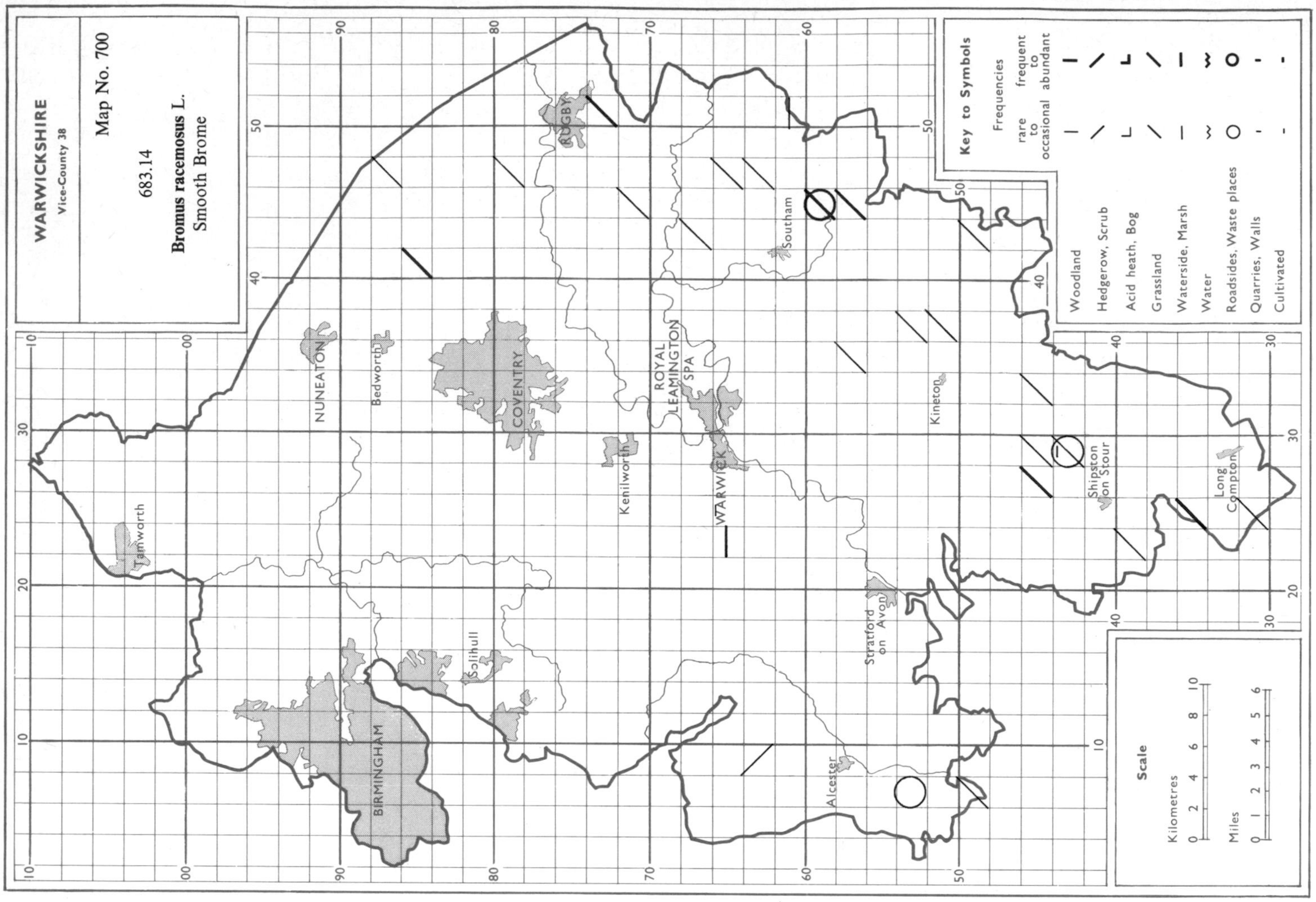

WARWICKSHIRE
Vice-County 38
Map No. 700
683.14
Bromus racemosus L.
Smooth Brome
Key to Symbols
Frequencies
rare to occasional
frequent to abundant
Woodland
Hedgerow, Scrub
Acid heath, Bog
Grassland
Waterside, Marsh
Water
Roadsides, Waste places
Quarries, Walls
Cultivated
Scale
Kilometres
0 2 4 6 8 10
Miles
0 1 2 3 4 5 6
NUNEATON
Bedworth
COVENTRY
RUGBY
Southam
ROYAL LEAMINGTON SPA
Kenilworth
WARWICK
Kineton
Shipston on Stour
Long Compton
Tamworth
Solihull
BIRMINGHAM
Stratford on Avon
Alcester

WARWICKSHIRE
Vice-County 38
Map No. 699
683.13
Bromus lepidus Holmberg
Slender Brome
Key to Symbols
Frequencies
rare to occasional
frequent to abundant
Woodland
Hedgerow, Scrub
Acid heath, Bog
Grassland
Waterside, Marsh
Water
Roadsides, Waste places
Quarries, Walls
Cultivated
Scale
Kilometres
0 2 4 6 8 10
Miles
0 1 2 3 4 5 6
NUNEATON
Bedworth
COVENTRY
RUGBY
Southam
ROYAL LEAMINGTON SPA
Kenilworth
WARWICK
Kineton
Shipston on Stour
Long Compton
Tamworth
Solihull
BIRMINGHAM
Stratford on Avon
Alcester

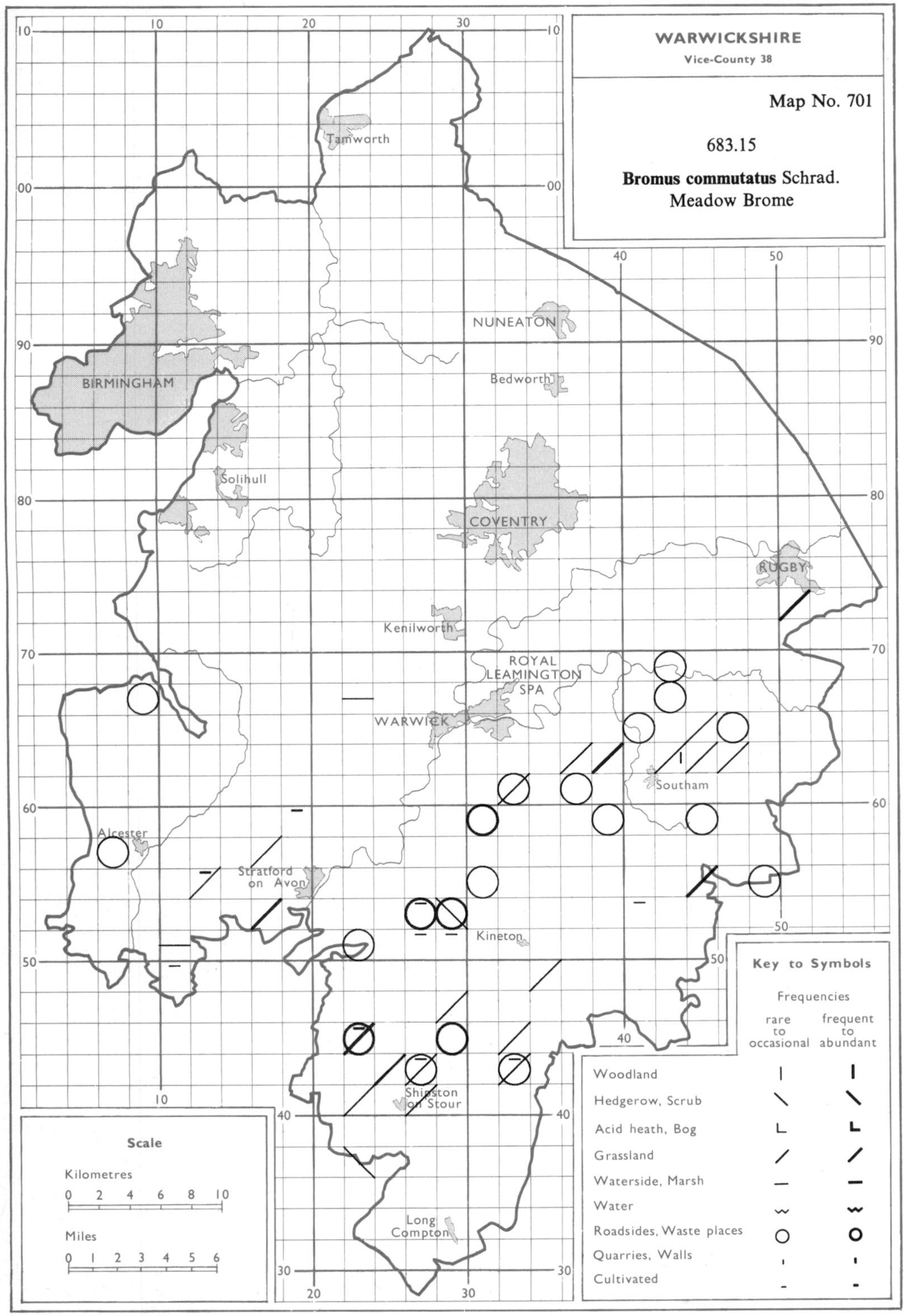
WARWICKSHIRE
Vice-County 38
Map No. 701
683.15
Bromus commutatus Schrad.
Meadow Brome
Tamworth
NUNEATON
Bedworth
BIRMINGHAM
Solihull
COVENTRY
RUGBY
Kenilworth
ROYAL LEAMINGTON SPA
WARWICK
Southam
Alcester
Stratford on Avon
Kineton
Shipston on Stour
Long Compton
Key to Symbols
Frequencies
rare to occasional
frequent to abundant
Woodland
Hedgerow, Scrub
Acid heath, Bog
Grassland
Waterside, Marsh
Water
Roadsides, Waste places
Quarries, Walls
Cultivated
Scale
Kilometres
0 2 4 6 8 10
Miles
0 1 2 3 4 5 6

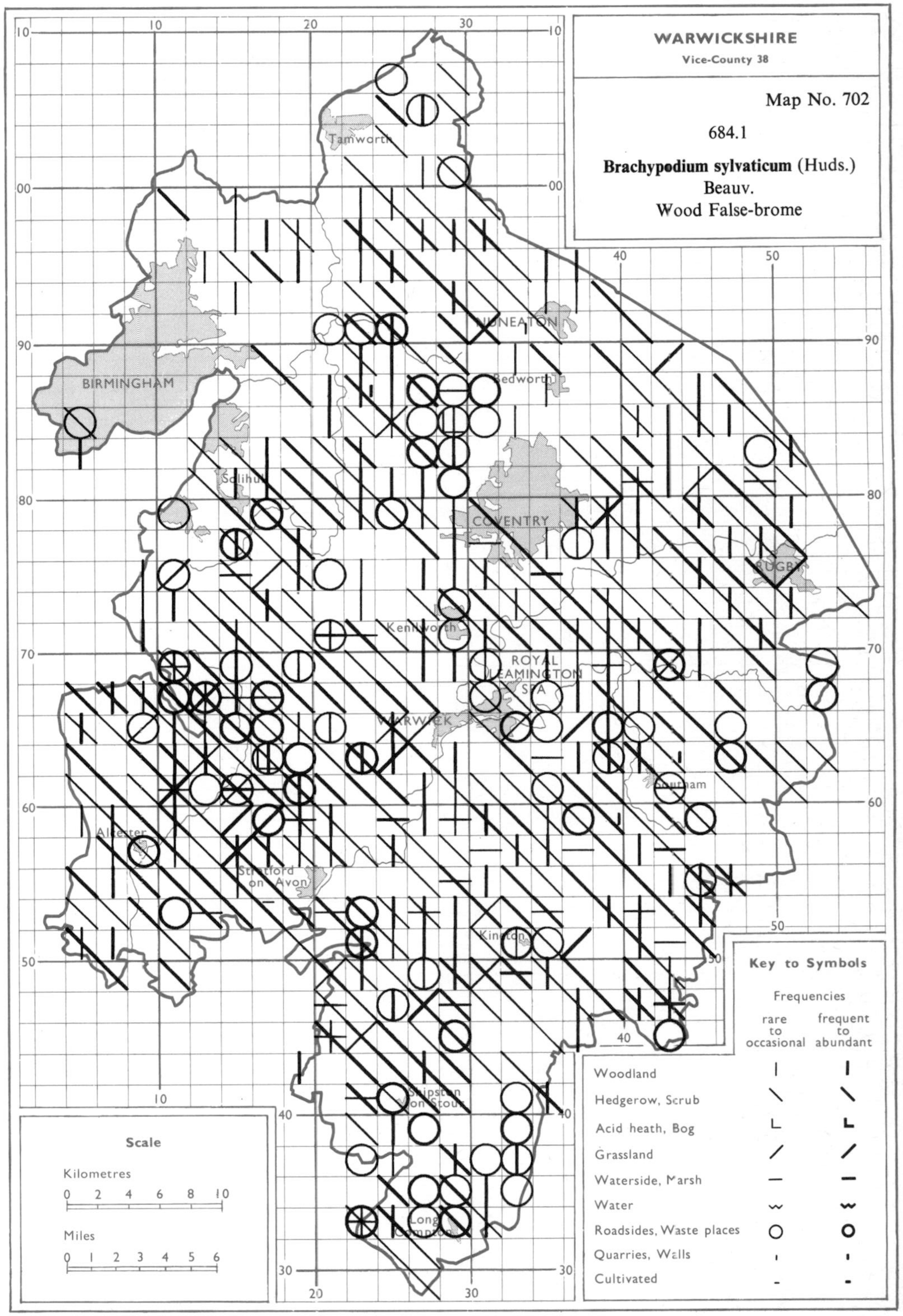
WARWICKSHIRE
Vice-County 38
Map No. 702
684.1
Brachypodium sylvaticum (Huds.) Beauv.
Wood False-brome
Tamworth
NUNEATON
Bedworth
BIRMINGHAM
Solihull
COVENTRY
RUGBY
Kenilworth
ROYAL LEAMINGTON SPA
WARWICK
Southam
Alcester
Stratford on Avon
Kineton
Shipston on Stour
Long Compton
Key to Symbols
Frequencies
rare to occasional
frequent to abundant
Woodland
Hedgerow, Scrub
Acid heath, Bog
Grassland
Waterside, Marsh
Water
Roadsides, Waste places
Quarries, Walls
Cultivated
Scale
Kilometres
0 2 4 6 8 10
Miles
0 1 2 3 4 5 6

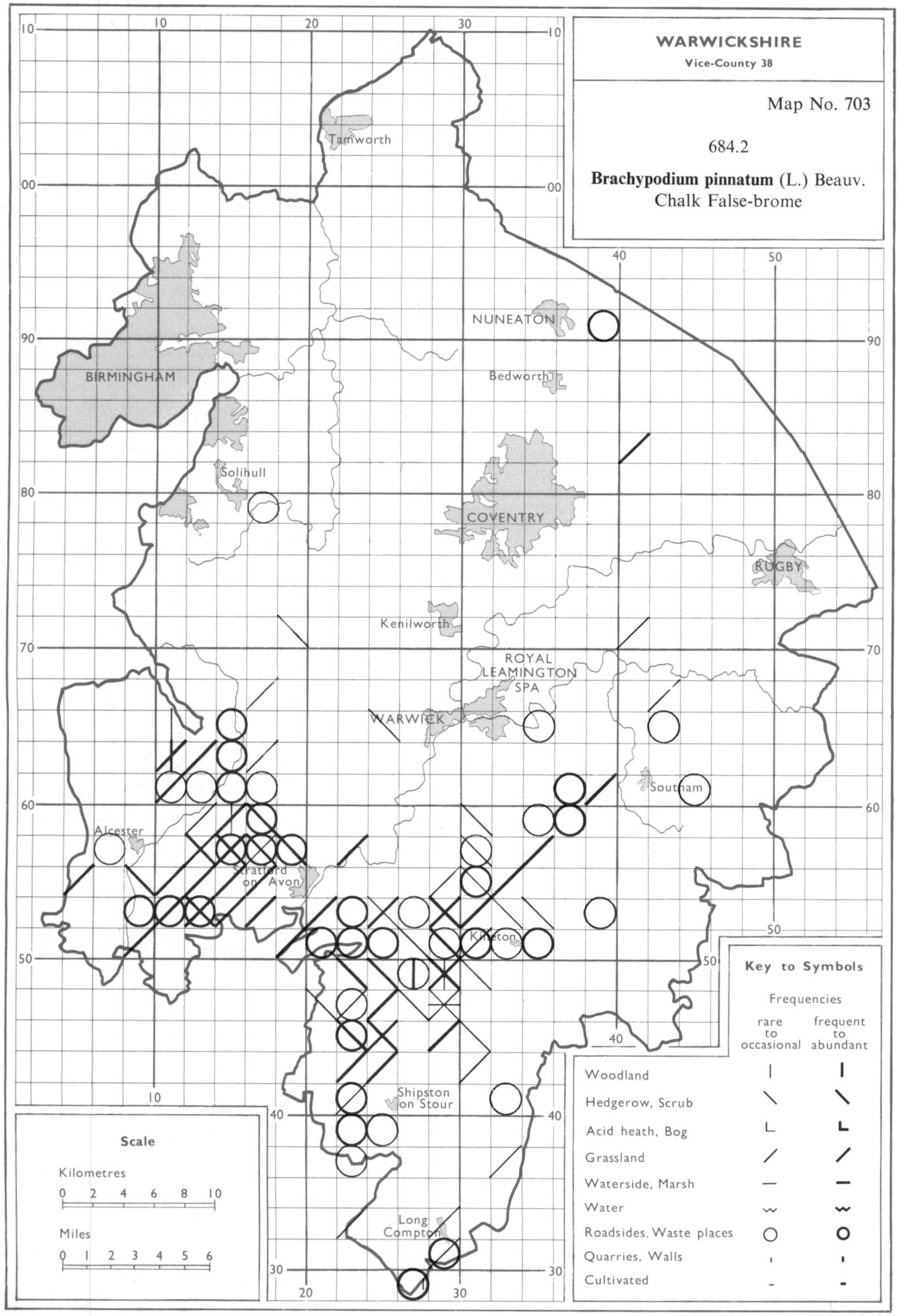
WARWICKSHIRE
Vice-County 38
Map No. 703
684.2
Brachypodium pinnatum (L.) Beauv.
Chalk False-brome
Tamworth
NUNEATON
Bedworth
BIRMINGHAM
Solihull
COVENTRY
RUGBY
Kenilworth
ROYAL LEAMINGTON SPA
WARWICK
Southam
Alcester
Stratford on Avon
Kineton
Shipston on Stour
Long Compton
Key to Symbols
Frequencies
rare to occasional
frequent to abundant
Woodland
Hedgerow, Scrub
Acid heath, Bog
Grassland
Waterside, Marsh
Water
Roadsides, Waste places
Quarries, Walls
Cultivated
Scale
Kilometres
0 2 4 6 8 10
Miles
0 1 2 3 4 5 6

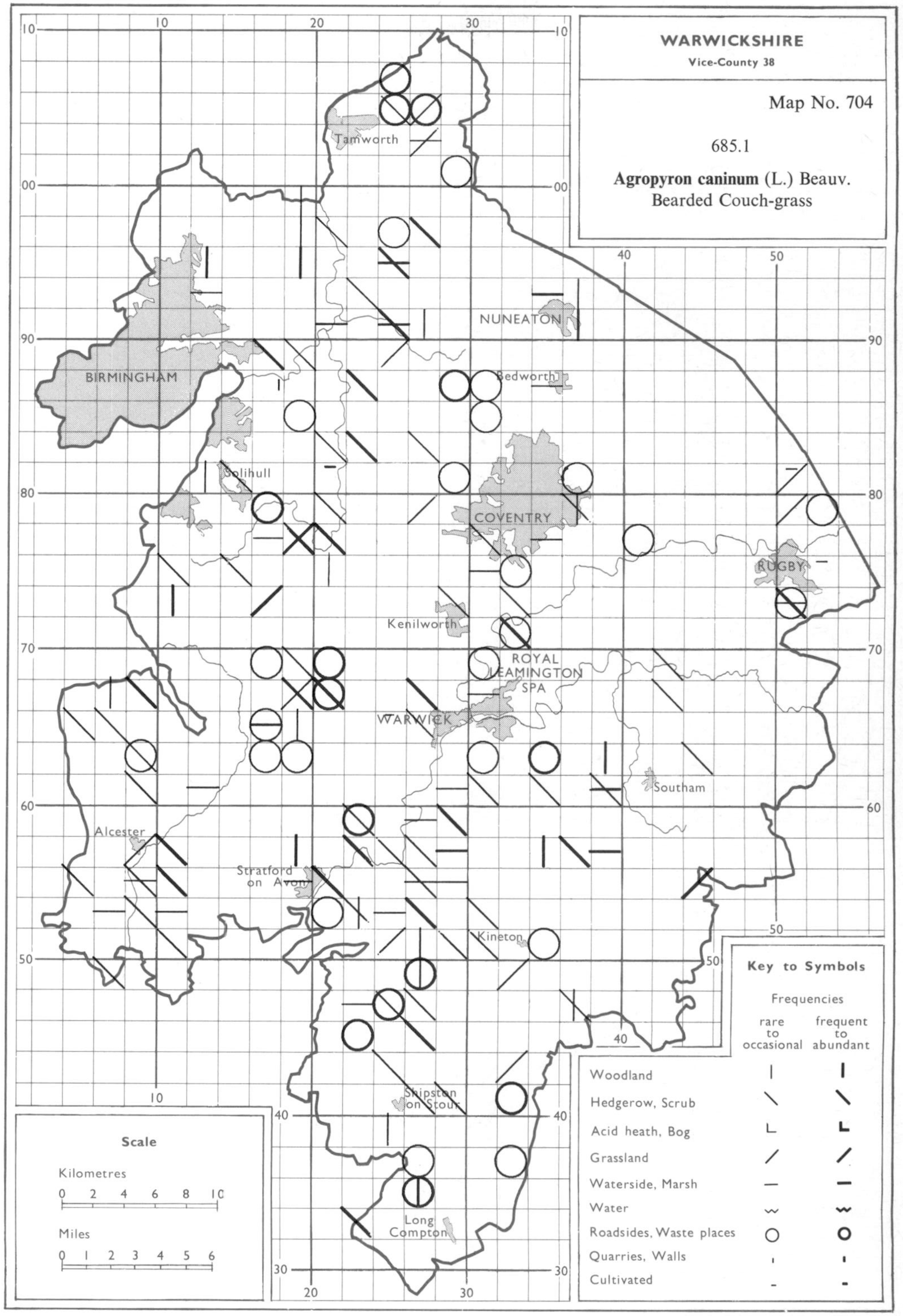
WARWICKSHIRE
Vice-County 38
Map No. 704
685.1
Agropyron caninum (L.) Beauv.
Bearded Couch-grass
Tamworth
NUNEATON
Bedworth
BIRMINGHAM
Solihull
COVENTRY
RUGBY
Kenilworth
ROYAL LEAMINGTON SPA
WARWICK
Southam
Alcester
Stratford on Avon
Kineton
Shipston on Stour
Long Compton
Key to Symbols
Frequencies
rare to occasional
frequent to abundant
Woodland
Hedgerow, Scrub
Acid heath, Bog
Grassland
Waterside, Marsh
Water
Roadsides, Waste places
Quarries, Walls
Cultivated
Scale
Kilometres
0 2 4 6 8 10
Miles
0 1 2 3 4 5 6

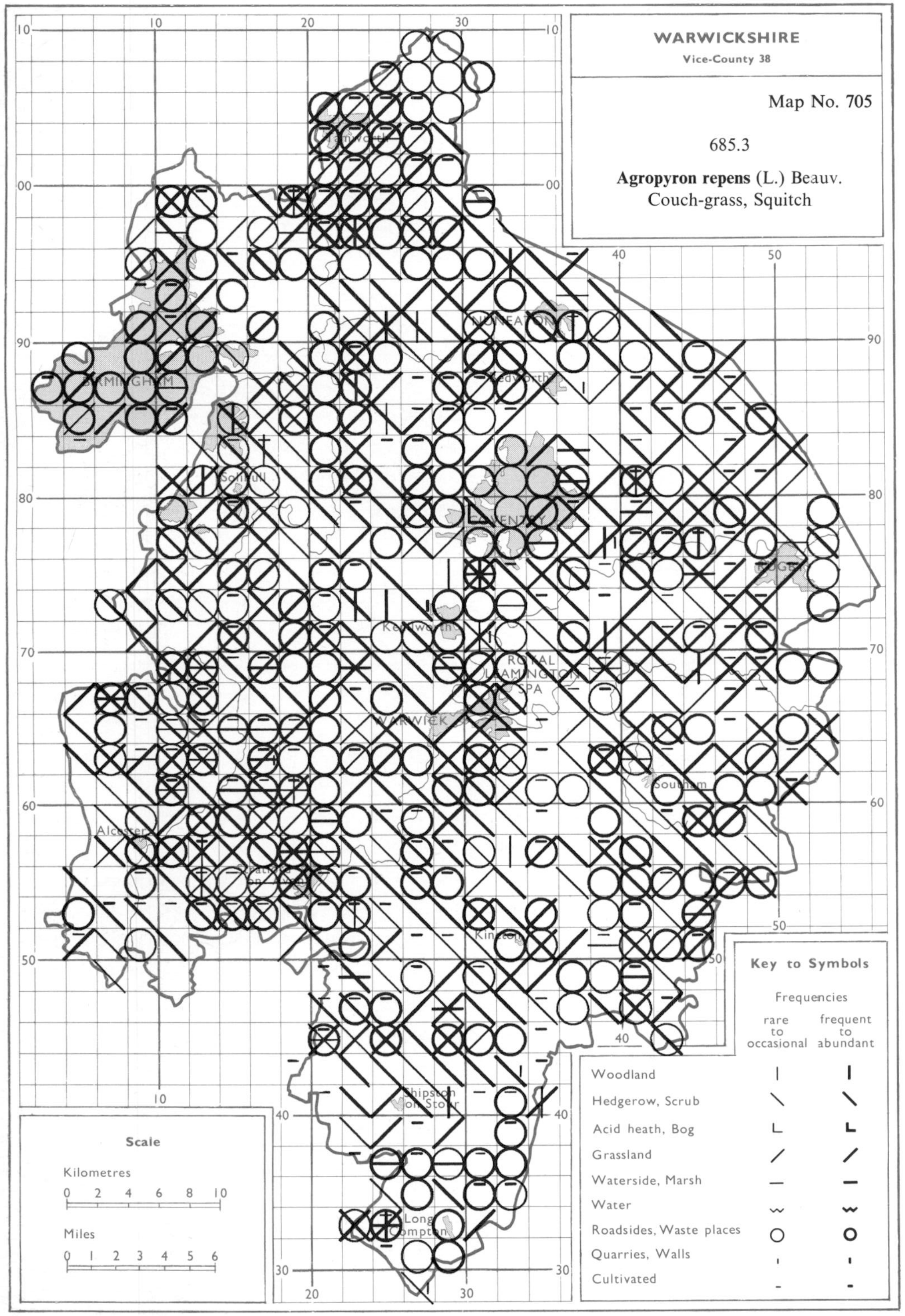
WARWICKSHIRE
Vice-County 38
Map No. 705
685.3
Agropyron repens (L.) Beauv.
Couch-grass, Squitch
BIRMINGHAM
Solihull
NUNEATON
Bedworth
COVENTRY
RUGBY
Kenilworth
ROYAL LEAMINGTON SPA
WARWICK
Southam
Alcester
Stratford on Avon
Kineton
Shipston on Stour
Long Compton
Key to Symbols
Frequencies
rare to occasional
frequent to abundant
Woodland
Hedgerow, Scrub
Acid heath, Bog
Grassland
Waterside, Marsh
Water
Roadsides, Waste places
Quarries, Walls
Cultivated
Scale
Kilometres
0 2 4 6 8 10
Miles
0 1 2 3 4 5 6

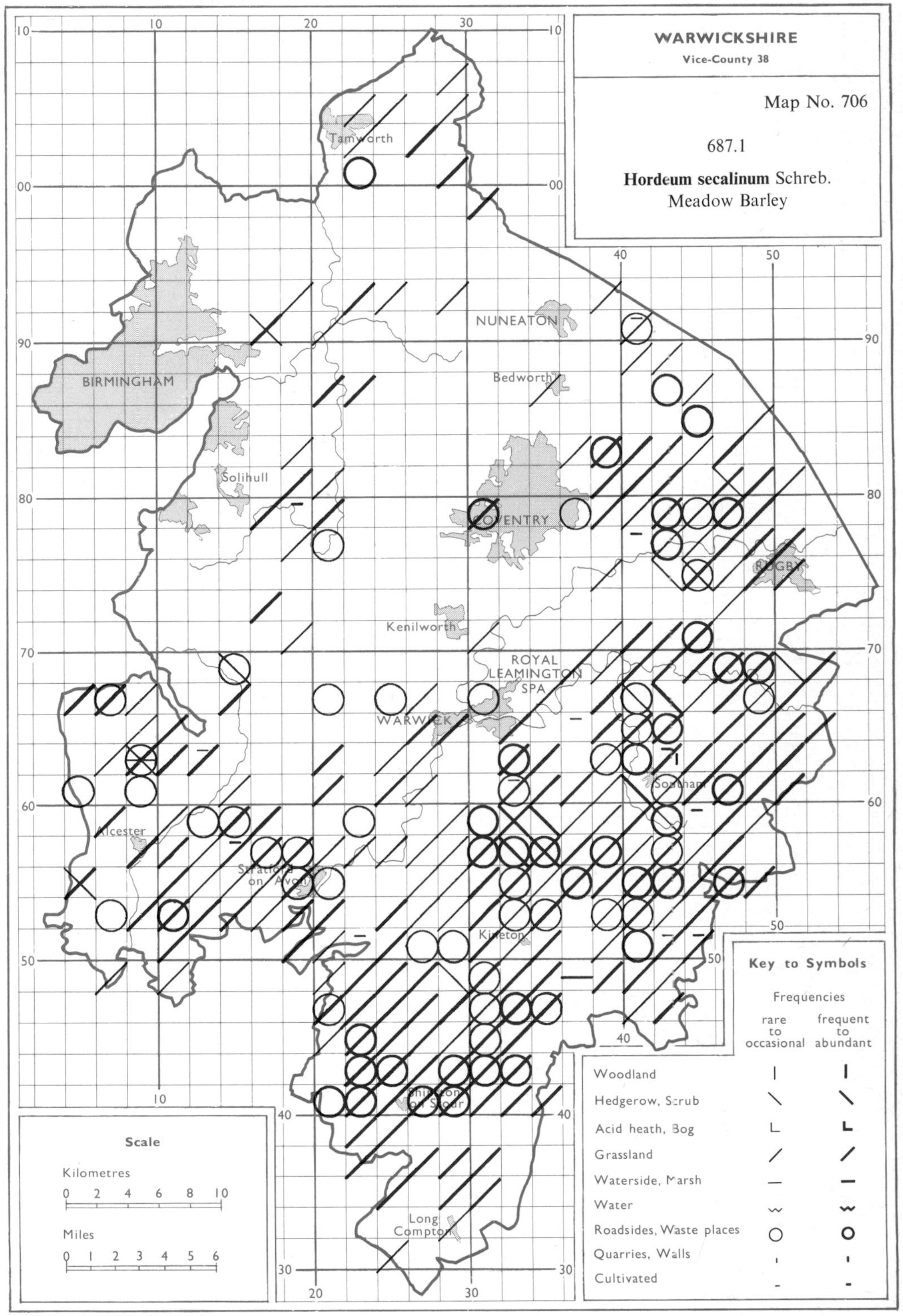
WARWICKSHIRE
Vice-County 38
Map No. 706
687.1
Hordeum secalinum Schreb.
Meadow Barley
Tamworth
BIRMINGHAM
Solihull
NUNEATON
Bedworth
COVENTRY
RUGBY
Kenilworth
ROYAL LEAMINGTON SPA
WARWICK
Southam
Alcester
Stratford on Avon
Kineton
Shipston on Stour
Long Compton
Key to Symbols
Frequencies
rare to occasional
frequent to abundant
Woodland
Hedgerow, Scrub
Acid heath, Bog
Grassland
Waterside, Marsh
Water
Roadsides, Waste places
Quarries, Walls
Cultivated
Scale
Kilometres
0 2 4 6 8 10
Miles
0 1 2 3 4 5 6

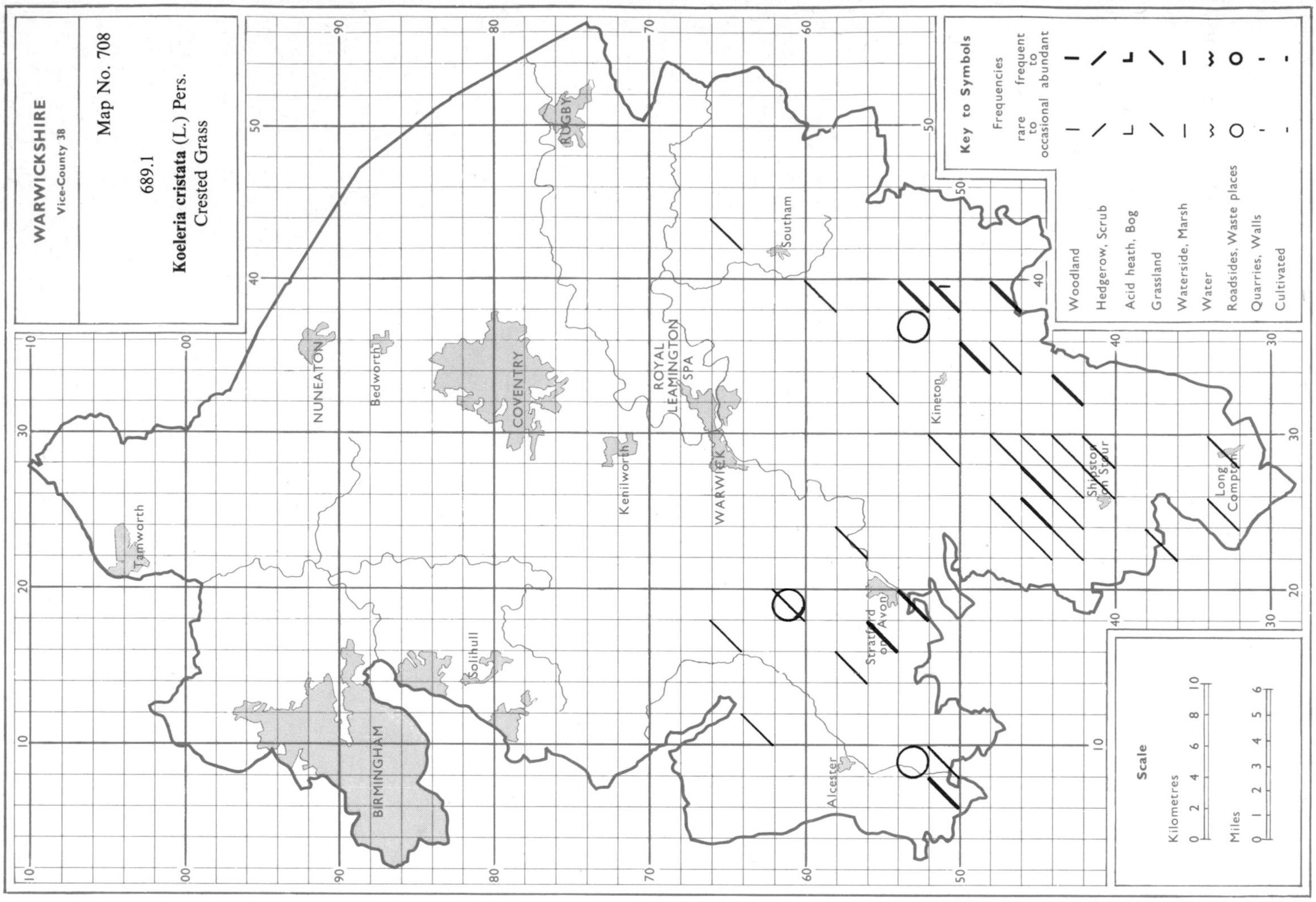
WARWICKSHIRE
Vice-County 38
Map No. 708
689.1
Koeleria cristata (L.) Pers.
Crested Grass
Key to Symbols
Frequencies
rare to occasional
frequent to abundant
Woodland
Hedgerow, Scrub
Acid heath, Bog
Grassland
Waterside, Marsh
Water
Roadsides, Waste places
Quarries, Walls
Cultivated
Scale
Kilometres 0 2 4 6 8 10
Miles 0 1 2 3 4 5 6
NUNEATON
Bedworth
COVENTRY
RUGBY
Southam
ROYAL LEAMINGTON SPA
Kineton
Kenilworth
WARWICK
Shipston on Stour
Long Compton
Tamworth
Solihull
BIRMINGHAM
Stratford on Avon
Alcester

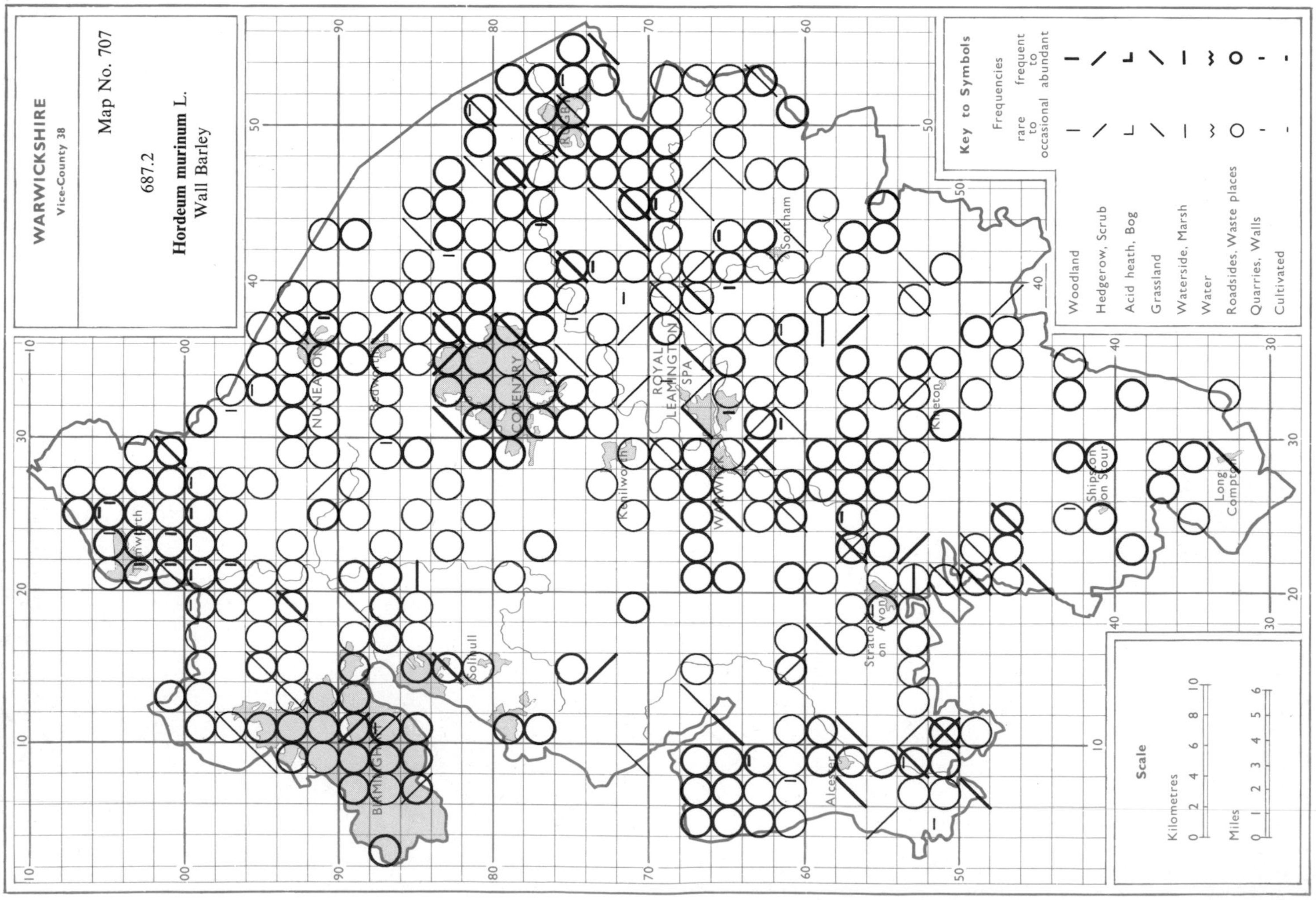
WARWICKSHIRE
Vice-County 38
Map No. 707
687.2
Hordeum murinum L.
Wall Barley
Key to Symbols
Frequencies
rare to occasional
frequent to abundant
Woodland
Hedgerow, Scrub
Acid heath, Bog
Grassland
Waterside, Marsh
Water
Roadsides, Waste places
Quarries, Walls
Cultivated
Scale
Kilometres 0 2 4 6 8 10
Miles 0 1 2 3 4 5 6
NUNEATON
COVENTRY
Southam
ROYAL LEAMINGTON SPA
Kineton
Kenilworth
Shipston on Stour
Long Compton
Solihull
Stratford on Avon
Alcester

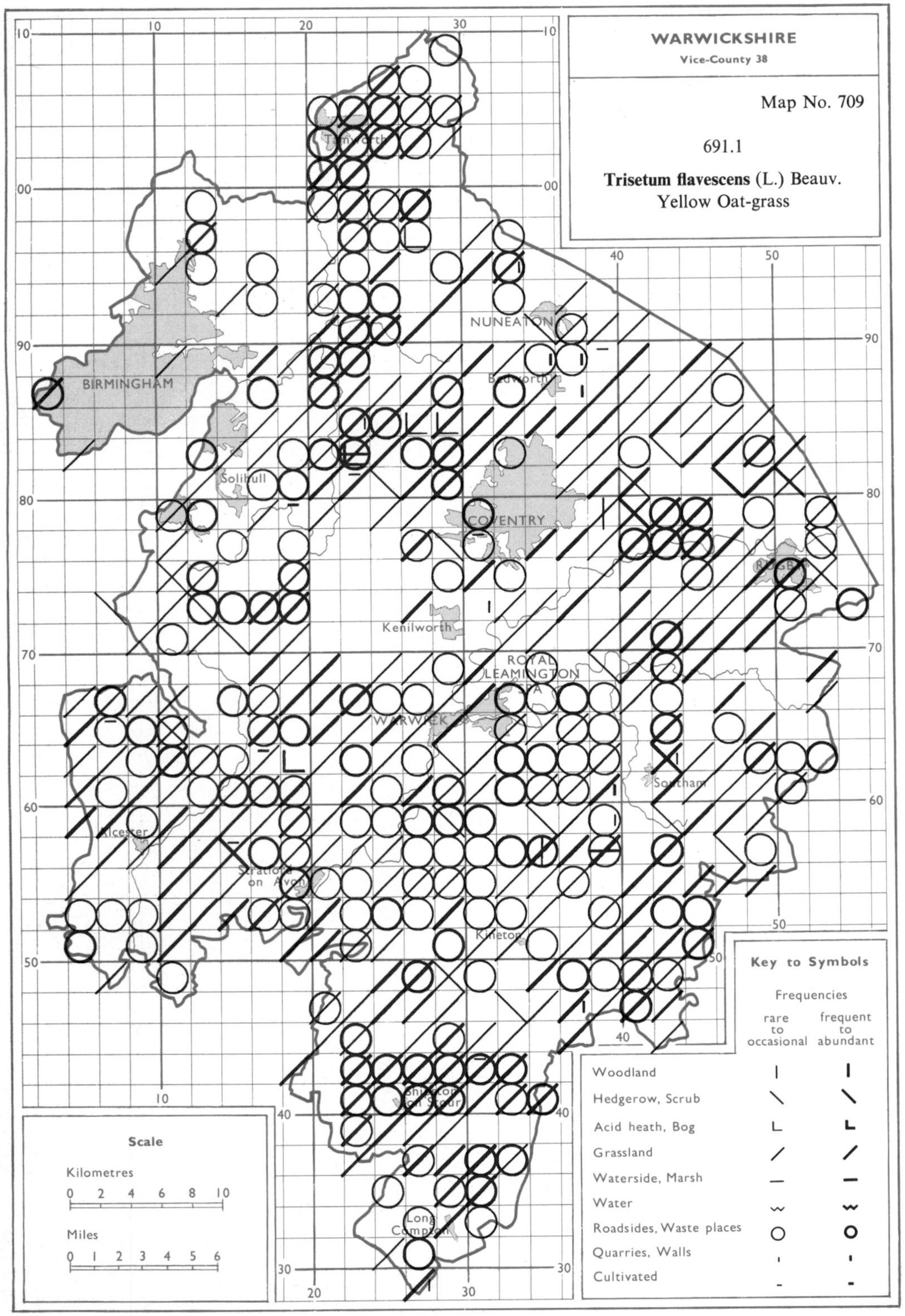
WARWICKSHIRE
Vice-County 38
Map No. 709
691.1
Trisetum flavescens (L.) Beauv.
Yellow Oat-grass
Tamworth
Birmingham
Solihull
Nuneaton
Bedworth
Coventry
Rugby
Kenilworth
Royal Leamington Spa
Warwick
Southam
Alcester
Stratford on Avon
Kineton
Shipston on Stour
Long Compton
Key to Symbols
Frequencies
rare to occasional
frequent to abundant
Woodland
Hedgerow, Scrub
Acid heath, Bog
Grassland
Waterside, Marsh
Water
Roadsides, Waste places
Quarries, Walls
Cultivated
Scale
Kilometres
0 2 4 6 8 10
Miles
0 1 2 3 4 5 6

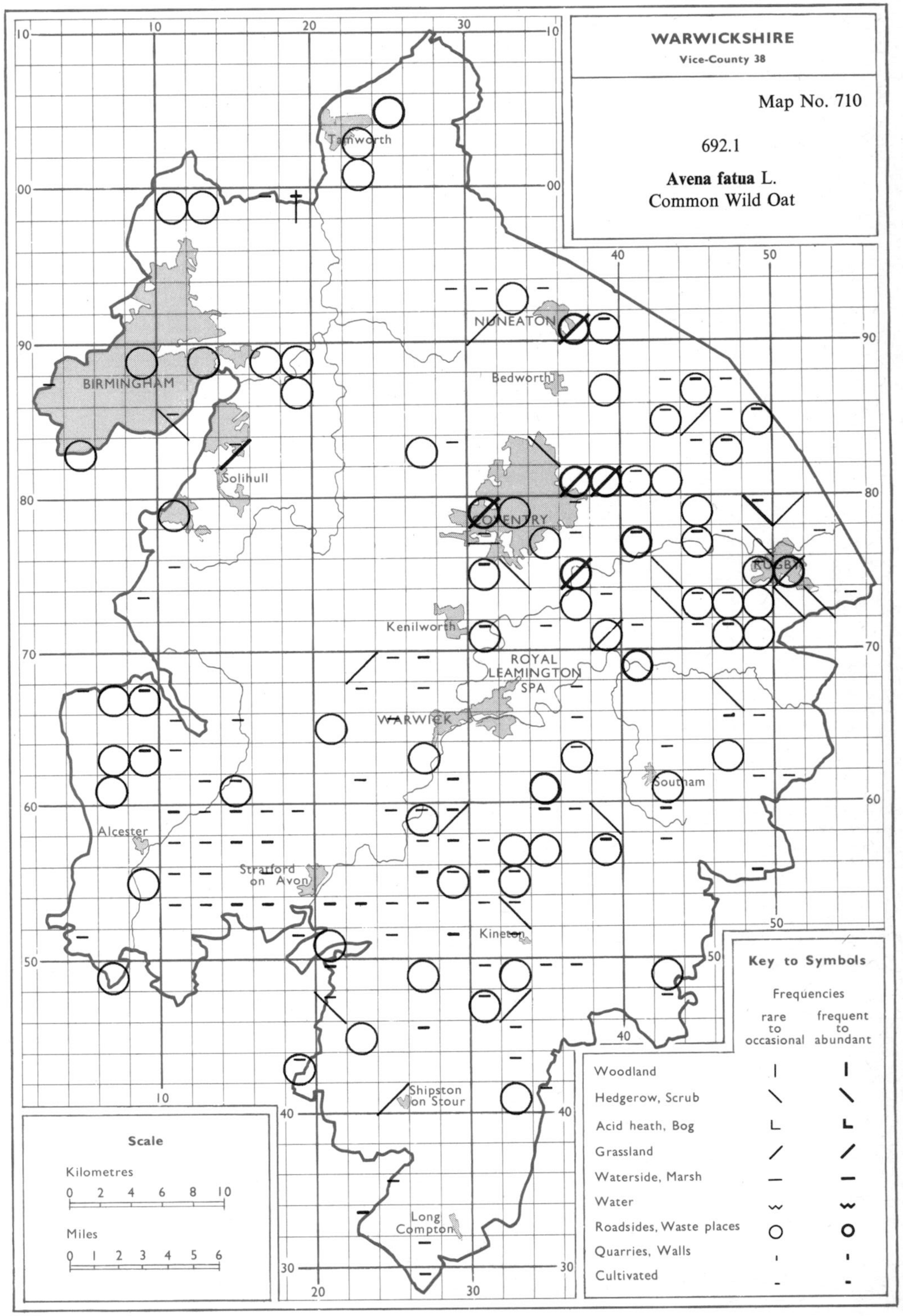
WARWICKSHIRE
Vice-County 38
Map No. 710
692.1
Avena fatua L.
Common Wild Oat
Tamworth
Birmingham
Solihull
Nuneaton
Bedworth
Coventry
Rugby
Kenilworth
Royal Leamington Spa
Warwick
Southam
Alcester
Stratford on Avon
Kineton
Shipston on Stour
Long Compton
Key to Symbols
Frequencies
rare to occasional
frequent to abundant
Woodland
Hedgerow, Scrub
Acid heath, Bog
Grassland
Waterside, Marsh
Water
Roadsides, Waste places
Quarries, Walls
Cultivated
Scale
Kilometres
0 2 4 6 8 10
Miles
0 1 2 3 4 5 6

WARWICKSHIRE
Vice-County 38
Map No. 712
693.1
Helictotrichon pratense (L.) Pilg.
Meadow Oat-grass
Key to Symbols
Frequencies
rare to occasional
frequent to abundant
Woodland
Hedgerow, Scrub
Acid heath, Bog
Grassland
Waterside, Marsh
Water
Roadsides, Waste places
Quarries, Walls
Cultivated
Scale
Kilometres
0 2 4 6 8 10
Miles
0 1 2 3 4 5 6
NUNEATON
Bedworth
COVENTRY
RUGBY
Southam
ROYAL LEAMINGTON SPA
Kineton
Kenilworth
WARWICK
Tamworth
Solihull
BIRMINGHAM
Stratford on Avon
Shipston on Stour
Long Compton
Alcester

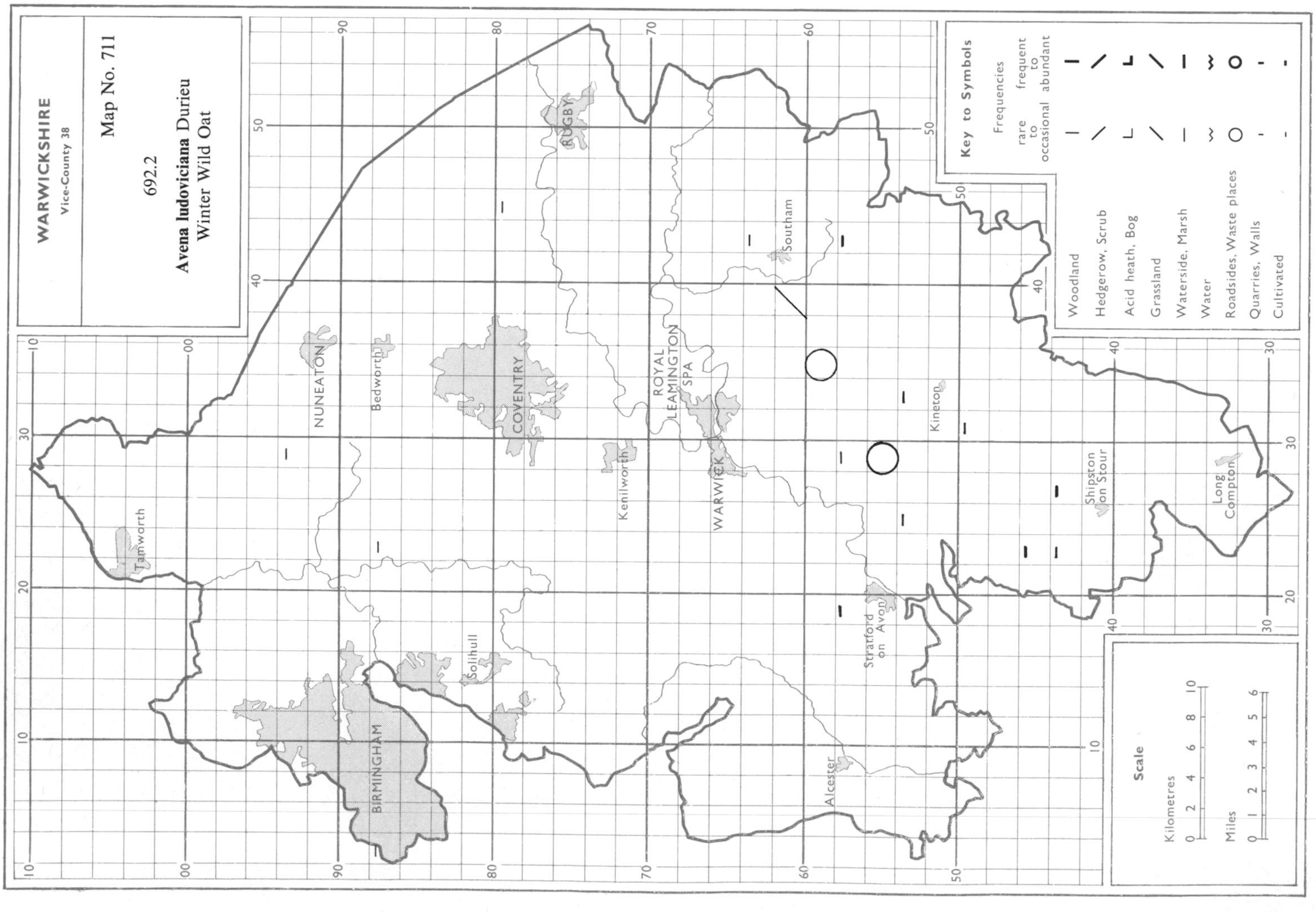
WARWICKSHIRE
Vice-County 38
Map No. 711
692.2
Avena ludoviciana Durieu
Winter Wild Oat
Key to Symbols
Frequencies
rare to occasional
frequent to abundant
Woodland
Hedgerow, Scrub
Acid heath, Bog
Grassland
Waterside, Marsh
Water
Roadsides, Waste places
Quarries, Walls
Cultivated
Scale
Kilometres
0 2 4 6 8 10
Miles
0 1 2 3 4 5 6
NUNEATON
Bedworth
COVENTRY
RUGBY
Southam
ROYAL LEAMINGTON SPA
Kineton
Kenilworth
WARWICK
Tamworth
Solihull
BIRMINGHAM
Stratford on Avon
Shipston on Stour
Long Compton
Alcester

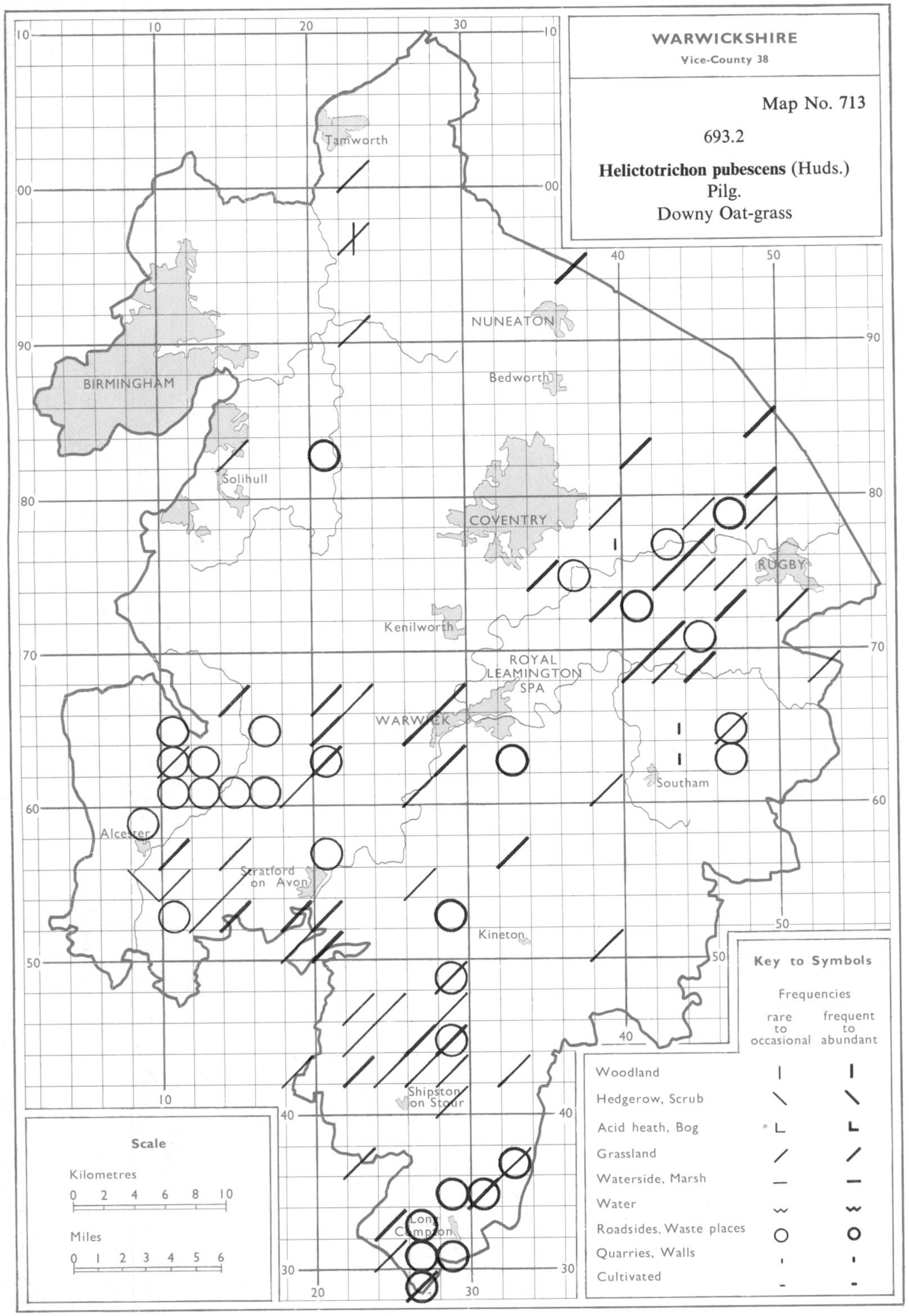
WARWICKSHIRE
Vice-County 38
Map No. 713
693.2
Helictotrichon pubescens (Huds.) Pilg.
Downy Oat-grass
Tamworth
NUNEATON
Bedworth
BIRMINGHAM
Solihull
COVENTRY
RUGBY
Kenilworth
ROYAL LEAMINGTON SPA
WARWICK
Southam
Alcester
Stratford on Avon
Kineton
Shipston on Stour
Long Compton
Key to Symbols
Frequencies
rare to occasional
frequent to abundant
Woodland
Hedgerow, Scrub
Acid heath, Bog
Grassland
Waterside, Marsh
Water
Roadsides, Waste places
Quarries, Walls
Cultivated
Scale
Kilometres
0 2 4 6 8 10
Miles
0 1 2 3 4 5 6

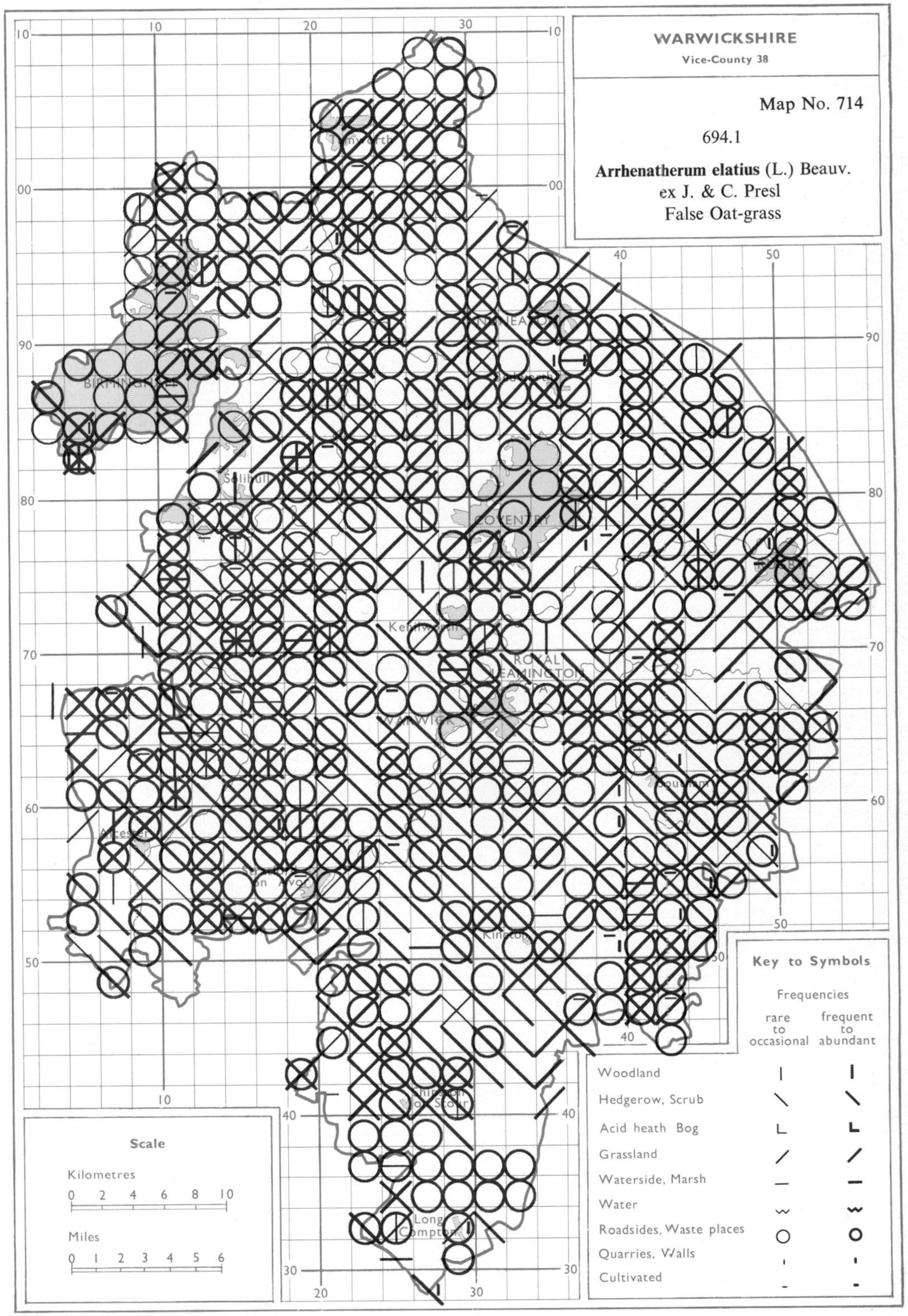
WARWICKSHIRE
Vice-County 38
Map No. 714
694.1
Arrhenatherum elatius (L.) Beauv. ex J. & C. Presl
False Oat-grass
Key to Symbols
Frequencies
rare to occasional
frequent to abundant
Woodland
Hedgerow, Scrub
Acid heath Bog
Grassland
Waterside, Marsh
Water
Roadsides, Waste places
Quarries, Walls
Cultivated
Scale
Kilometres
0 2 4 6 8 10
Miles
0 1 2 3 4 5 6

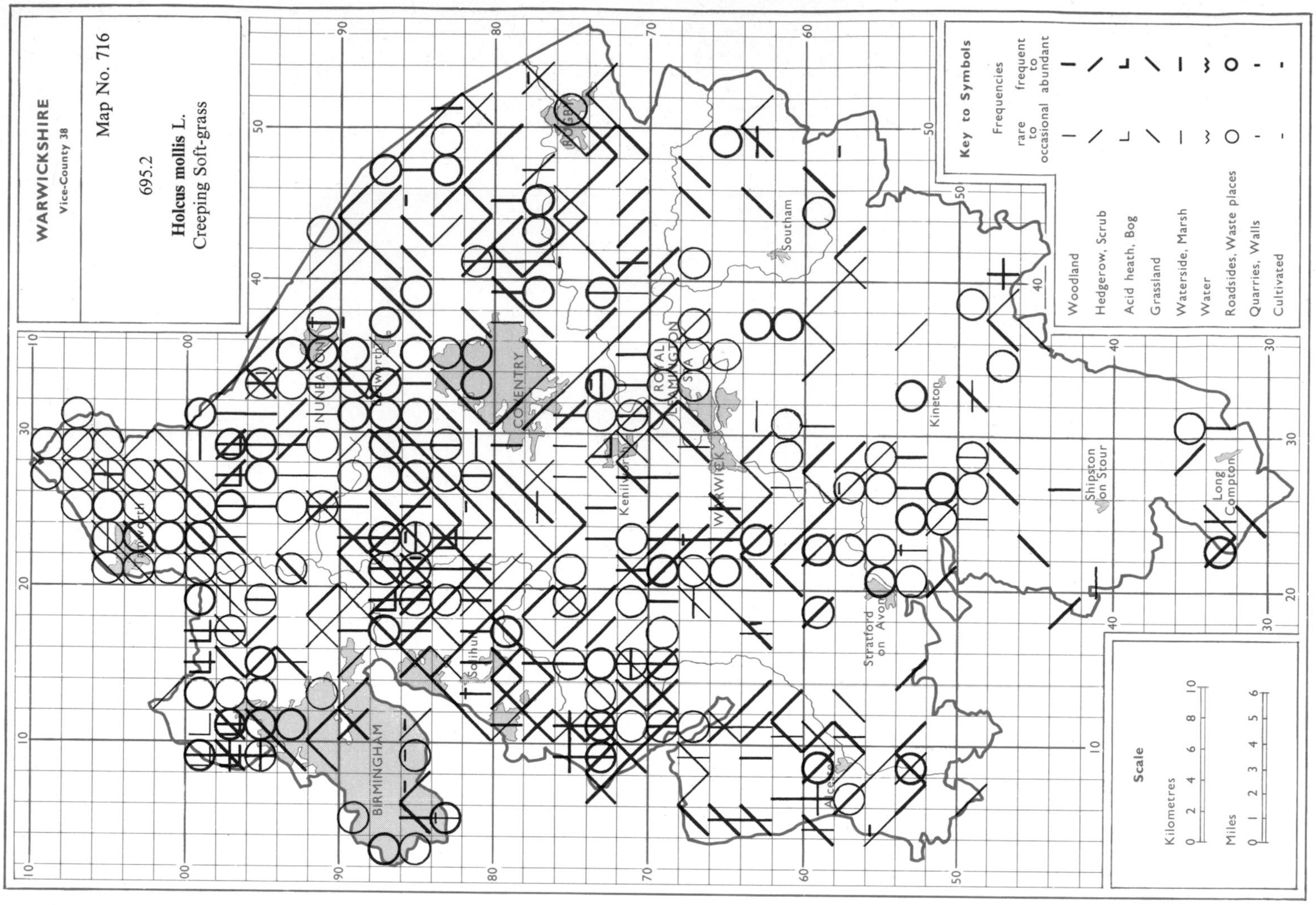
WARWICKSHIRE
Vice-County 38
Map No. 716
695.2
Holcus mollis L.
Creeping Soft-grass
Key to Symbols
Frequencies
rare to occasional
frequent to abundant
Woodland
Hedgerow, Scrub
Acid heath, Bog
Grassland
Waterside, Marsh
Water
Roadsides, Waste places
Quarries, Walls
Cultivated
Scale
Kilometres
Miles
BIRMINGHAM
COVENTRY
NUNEATON
RUGBY
ROYAL LEAMINGTON SPA
WARWICK
Kenilworth
Solihull
Tamworth
Southam
Kineton
Stratford on Avon
Shipston on Stour
Long Compton
Alcester

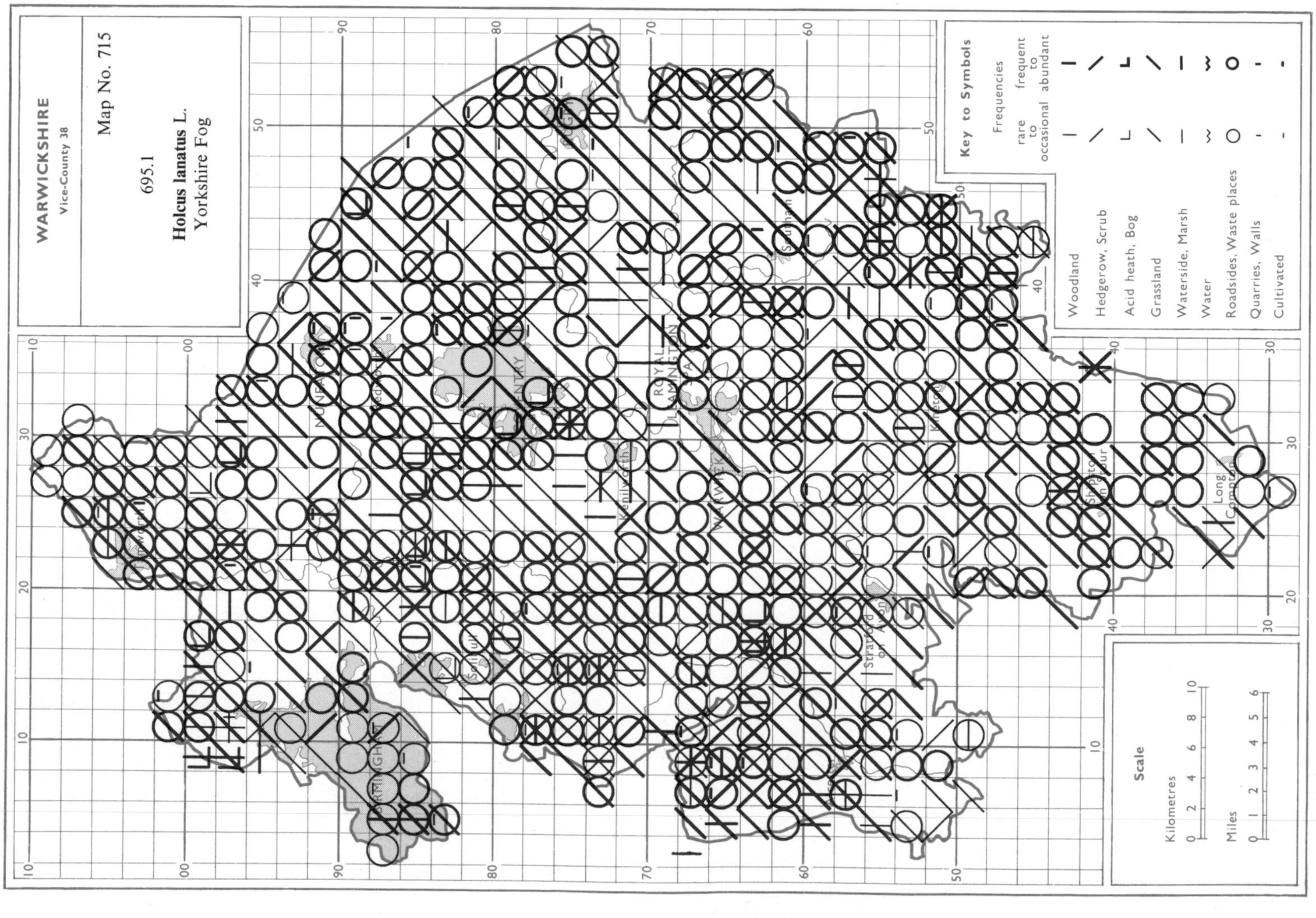
WARWICKSHIRE
Vice-County 38
Map No. 715
695.1
Holcus lanatus L.
Yorkshire Fog
Key to Symbols
Frequencies
rare to occasional
frequent to abundant
Woodland
Hedgerow, Scrub
Acid heath, Bog
Grassland
Waterside, Marsh
Water
Roadsides, Waste places
Quarries, Walls
Cultivated
Scale
Kilometres
Miles

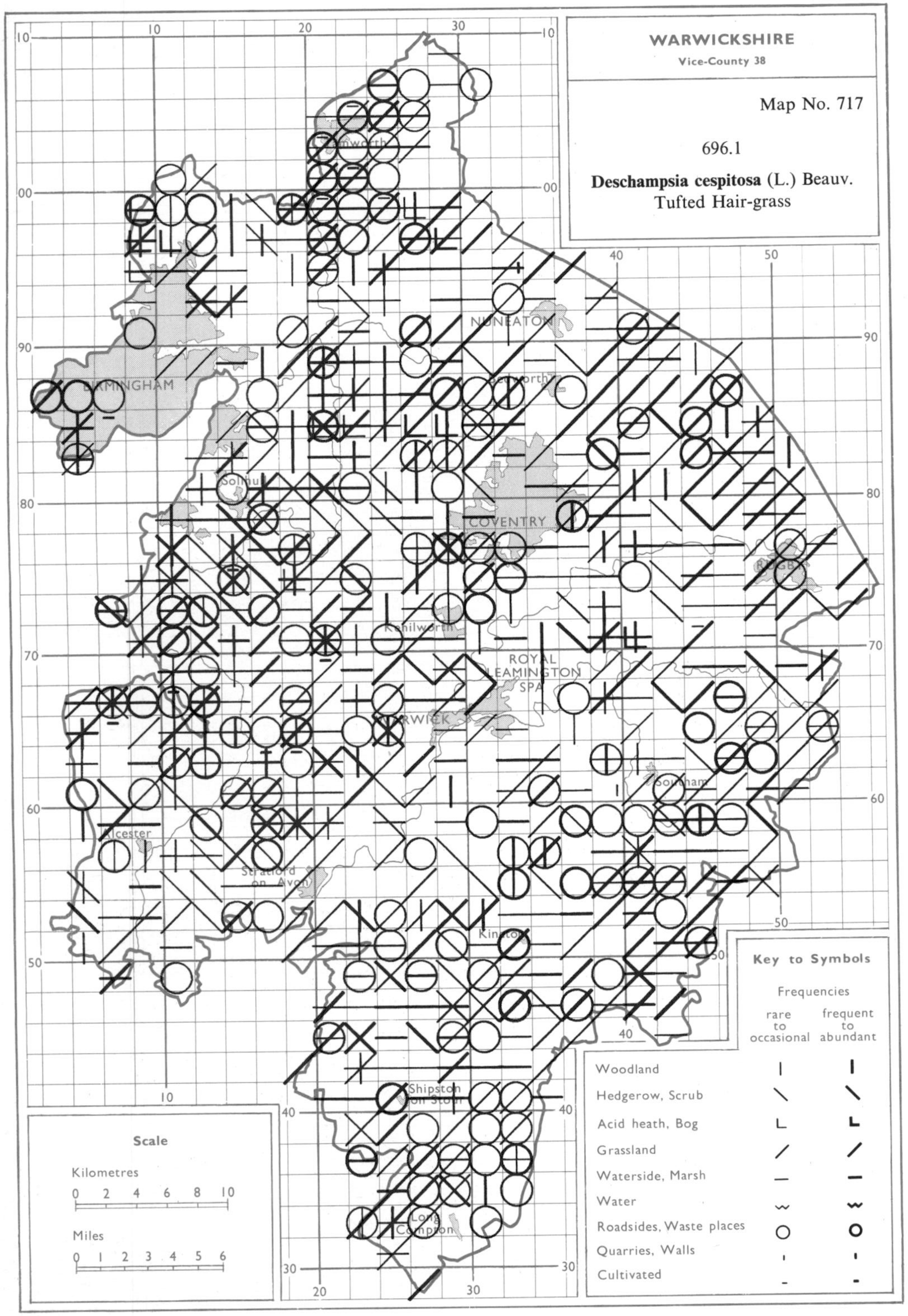
WARWICKSHIRE
Vice-County 38
Map No. 717
696.1
Deschampsia cespitosa (L.) Beauv.
Tufted Hair-grass
Tamworth
Birmingham
Nuneaton
Bedworth
Solihull
Coventry
Rugby
Kenilworth
Royal Leamington Spa
Warwick
Southam
Alcester
Stratford on Avon
Kineton
Shipston on Stour
Long Compton
Key to Symbols
Frequencies
rare to occasional
frequent to abundant
Woodland
Hedgerow, Scrub
Acid heath, Bog
Grassland
Waterside, Marsh
Water
Roadsides, Waste places
Quarries, Walls
Cultivated
Scale
Kilometres
0 2 4 6 8 10
Miles
0 1 2 3 4 5 6

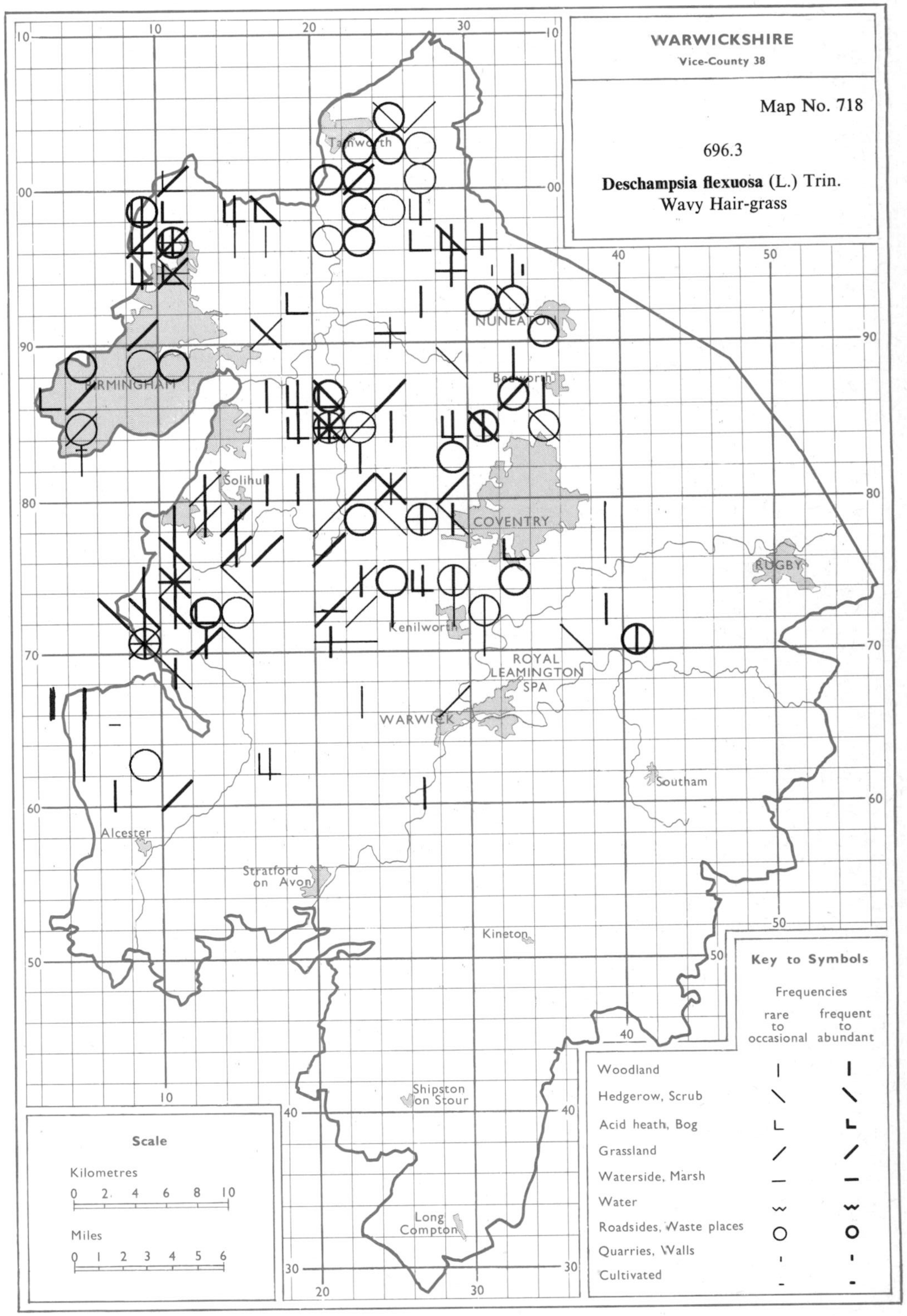
WARWICKSHIRE
Vice-County 38
Map No. 718
696.3
Deschampsia flexuosa (L.) Trin.
Wavy Hair-grass
Tamworth
Birmingham
Nuneaton
Bedworth
Solihull
Coventry
Rugby
Kenilworth
Royal Leamington Spa
Warwick
Southam
Alcester
Stratford on Avon
Kineton
Shipston on Stour
Long Compton
Key to Symbols
Frequencies
rare to occasional
frequent to abundant
Woodland
Hedgerow, Scrub
Acid heath, Bog
Grassland
Waterside, Marsh
Water
Roadsides, Waste places
Quarries, Walls
Cultivated
Scale
Kilometres
0 2 4 6 8 10
Miles
0 1 2 3 4 5 6

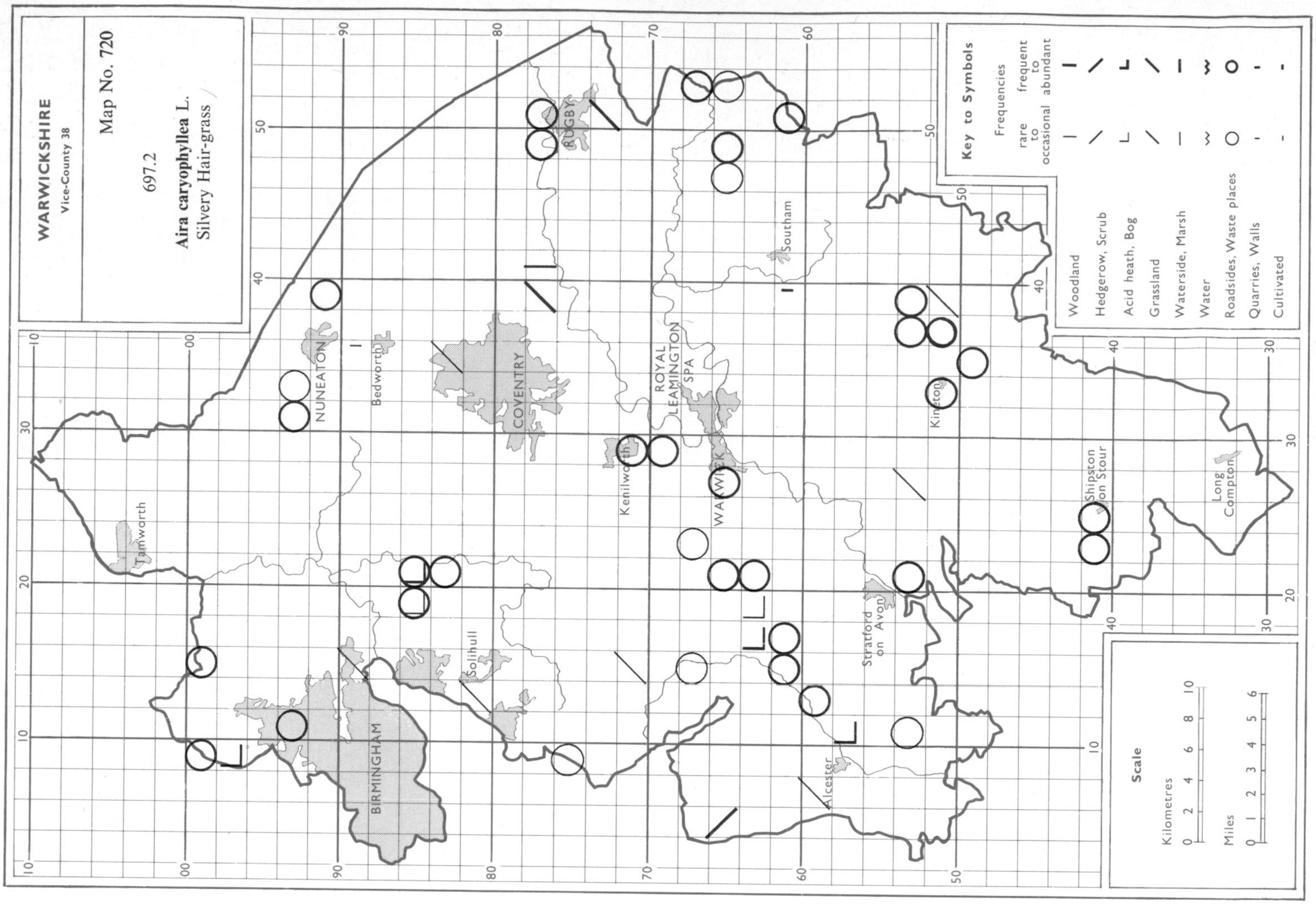
WARWICKSHIRE
Vice-County 38
Map No. 720
697.2
Aira caryophyllea L.
Silvery Hair-grass
Key to Symbols
Frequencies
rare to occasional
frequent to abundant
Woodland
Hedgerow, Scrub
Acid heath, Bog
Grassland
Waterside, Marsh
Water
Roadsides, Waste places
Quarries, Walls
Cultivated
Scale
Kilometres
0 2 4 6 8 10
Miles
0 1 2 3 4 5 6
Tamworth
NUNEATON
Bedworth
COVENTRY
RUGBY
Solihull
BIRMINGHAM
Kenilworth
ROYAL LEAMINGTON SPA
WARWICK
Southam
Stratford on Avon
Alcester
Kineton
Shipston on Stour
Long Compton

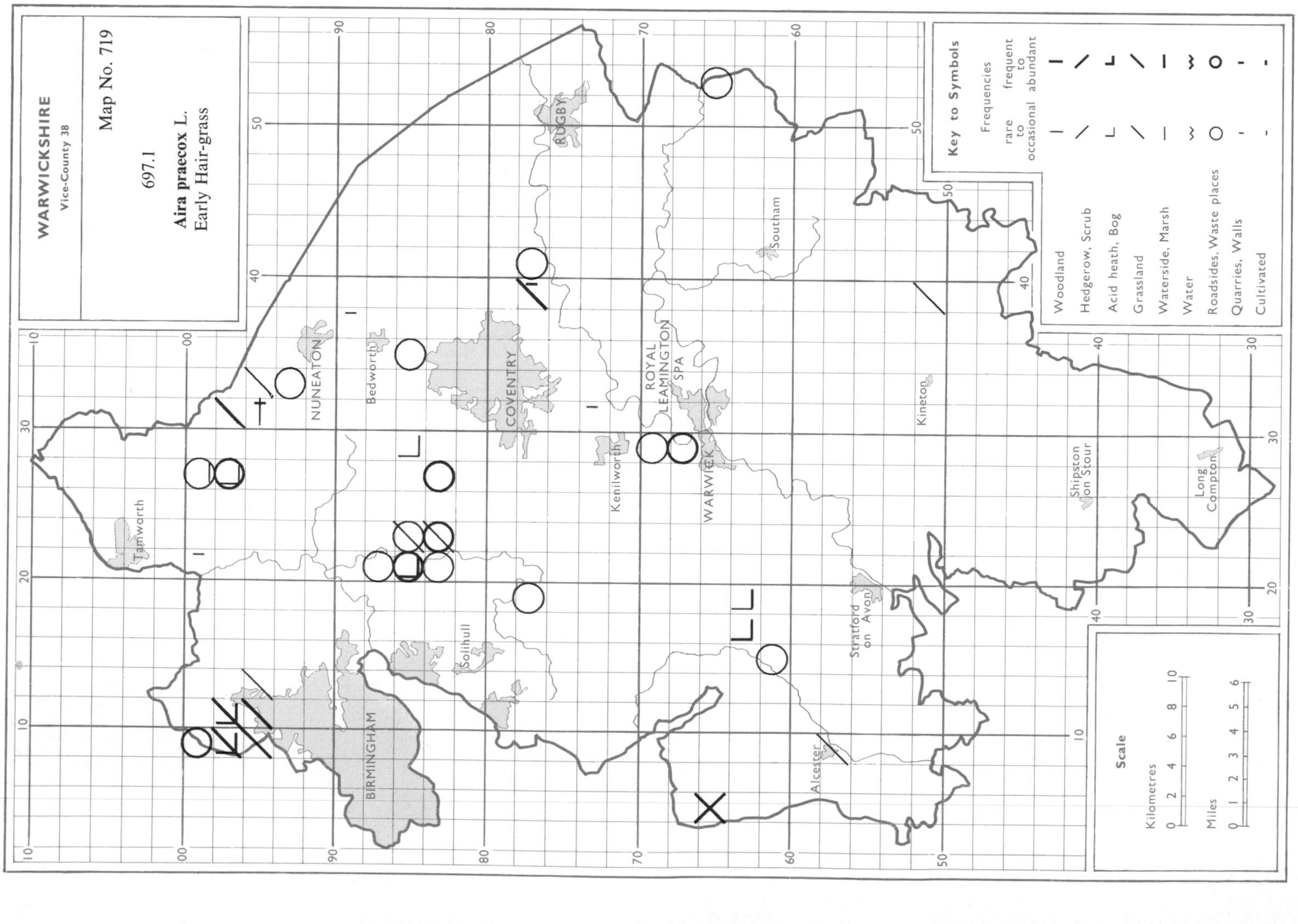
WARWICKSHIRE
Vice-County 38
Map No. 719
697.1
Aira praecox L.
Early Hair-grass
Key to Symbols
Frequencies
rare to occasional
frequent to abundant
Woodland
Hedgerow, Scrub
Acid heath, Bog
Grassland
Waterside, Marsh
Water
Roadsides, Waste places
Quarries, Walls
Cultivated
Scale
Kilometres
0 2 4 6 8 10
Miles
0 1 2 3 4 5 6
Tamworth
NUNEATON
Bedworth
COVENTRY
RUGBY
Solihull
BIRMINGHAM
Kenilworth
ROYAL LEAMINGTON SPA
WARWICK
Southam
Stratford on Avon
Alcester
Kineton
Shipston on Stour
Long Compton

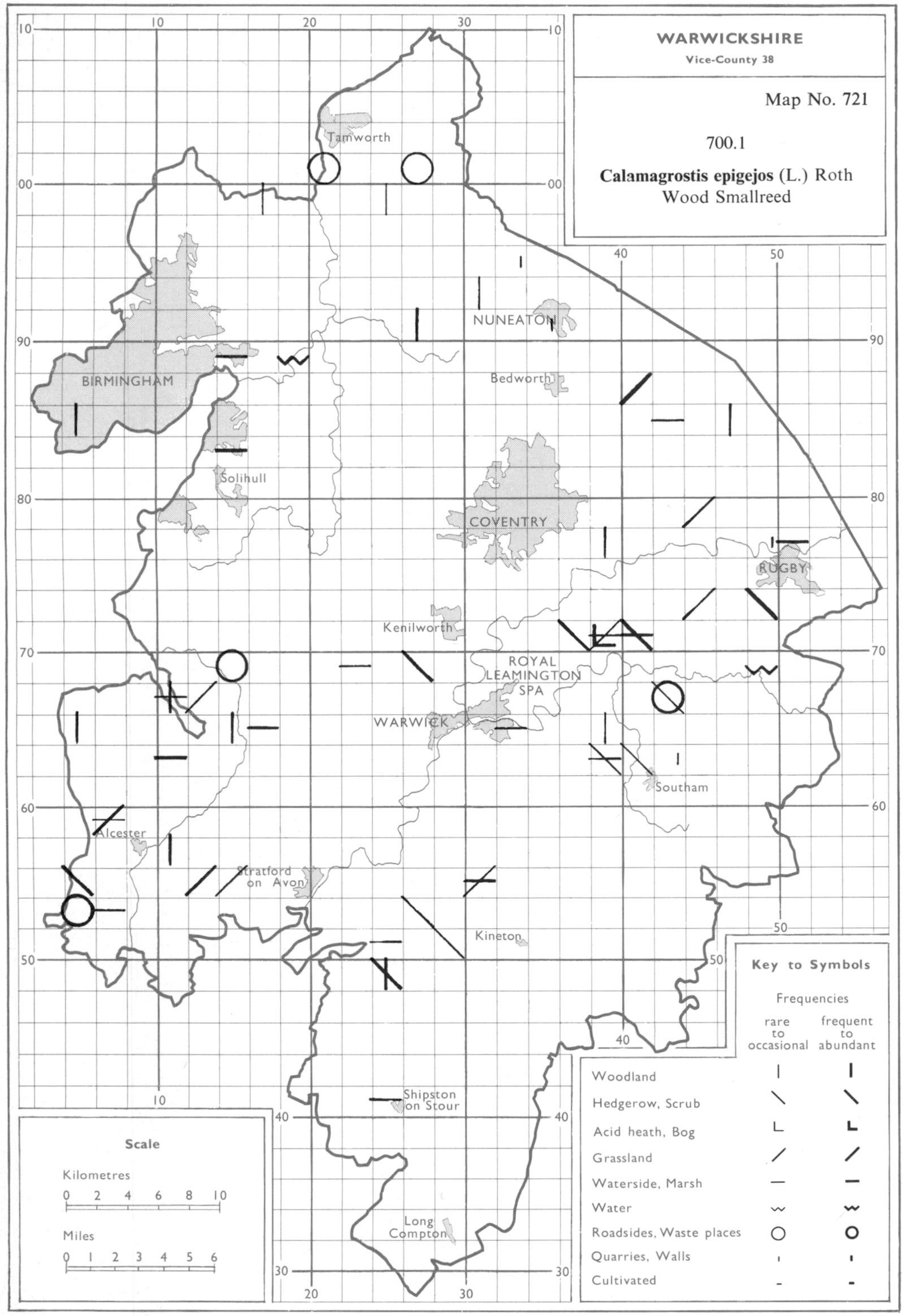
WARWICKSHIRE
Vice-County 38
Map No. 721
700.1
Calamagrostis epigejos (L.) Roth
Wood Smallreed
Tamworth
NUNEATON
Bedworth
BIRMINGHAM
Solihull
COVENTRY
RUGBY
Kenilworth
ROYAL LEAMINGTON SPA
WARWICK
Southam
Alcester
Stratford on Avon
Kineton
Shipston on Stour
Long Compton
Key to Symbols
Frequencies
rare to occasional
frequent to abundant
Woodland
Hedgerow, Scrub
Acid heath, Bog
Grassland
Waterside, Marsh
Water
Roadsides, Waste places
Quarries, Walls
Cultivated
Scale
Kilometres
0 2 4 6 8 10
Miles
0 1 2 3 4 5 6

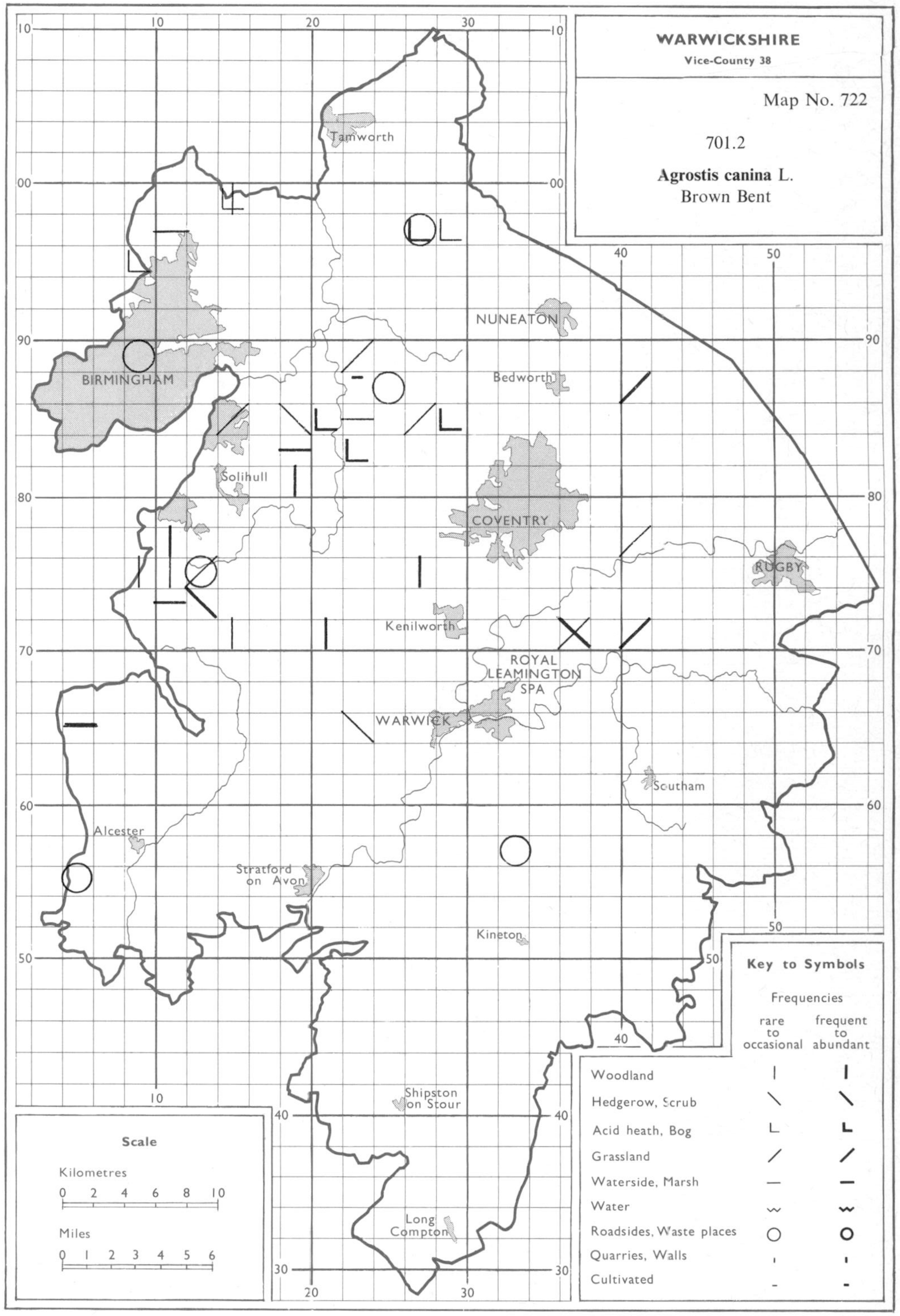
WARWICKSHIRE
Vice-County 38
Map No. 722
701.2
Agrostis canina L.
Brown Bent
Tamworth
NUNEATON
Bedworth
BIRMINGHAM
Solihull
COVENTRY
RUGBY
Kenilworth
ROYAL LEAMINGTON SPA
WARWICK
Southam
Alcester
Stratford on Avon
Kineton
Shipston on Stour
Long Compton
Key to Symbols
Frequencies
rare to occasional
frequent to abundant
Woodland
Hedgerow, Scrub
Acid heath, Bog
Grassland
Waterside, Marsh
Water
Roadsides, Waste places
Quarries, Walls
Cultivated
Scale
Kilometres
0 2 4 6 8 10
Miles
0 1 2 3 4 5 6

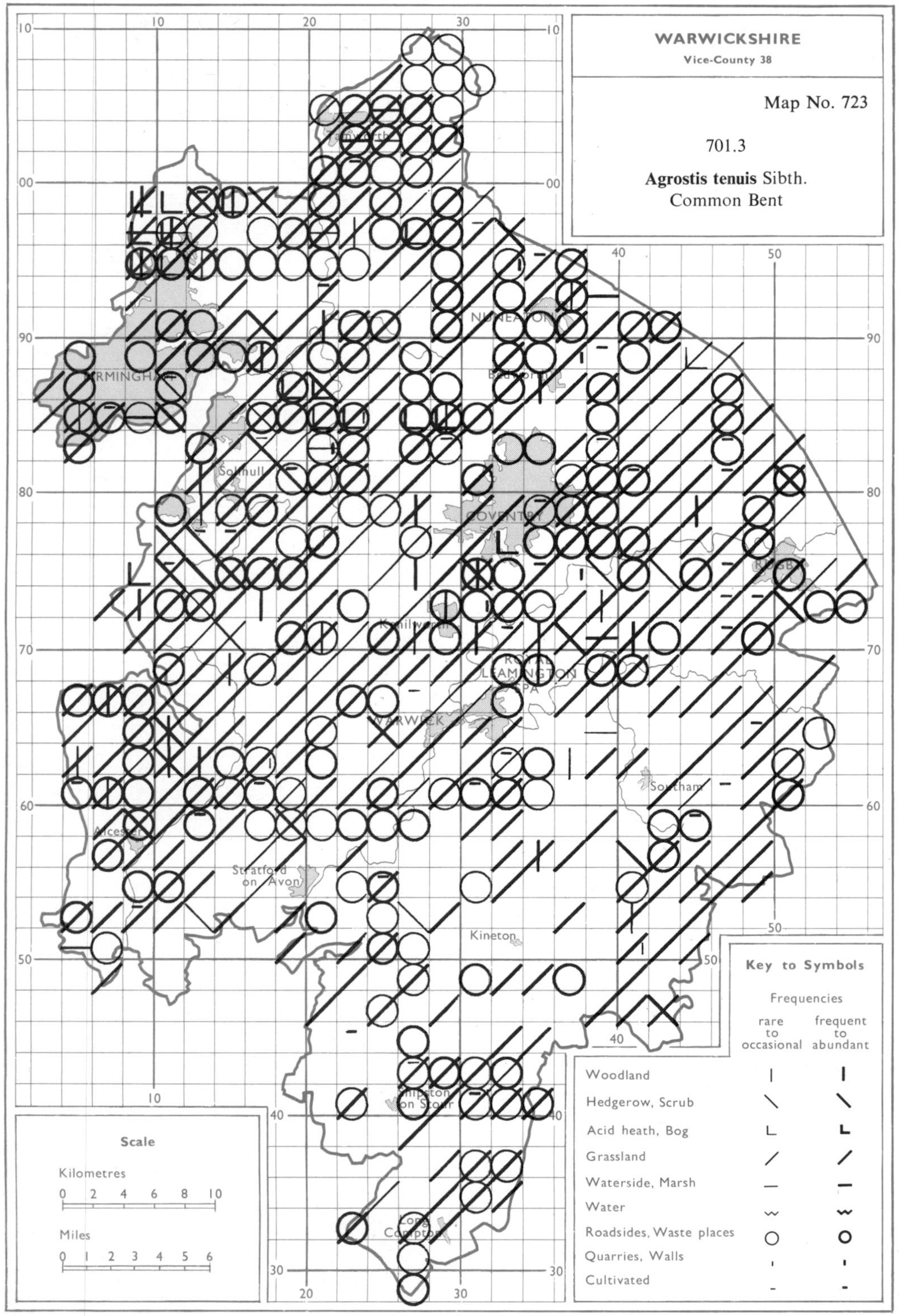
WARWICKSHIRE
Vice-County 38
Map No. 723
701.3
Agrostis tenuis Sibth.
Common Bent
Tamworth
Birmingham
Solihull
Nuneaton
Bedworth
Coventry
Rugby
Kenilworth
Royal Leamington Spa
Warwick
Southam
Alcester
Stratford on Avon
Kineton
Shipston on Stour
Long Compton
Key to Symbols
Frequencies
rare to occasional
frequent to abundant
Woodland
Hedgerow, Scrub
Acid heath, Bog
Grassland
Waterside, Marsh
Water
Roadsides, Waste places
Quarries, Walls
Cultivated
Scale
Kilometres
0 2 4 6 8 10
Miles
0 1 2 3 4 5 6

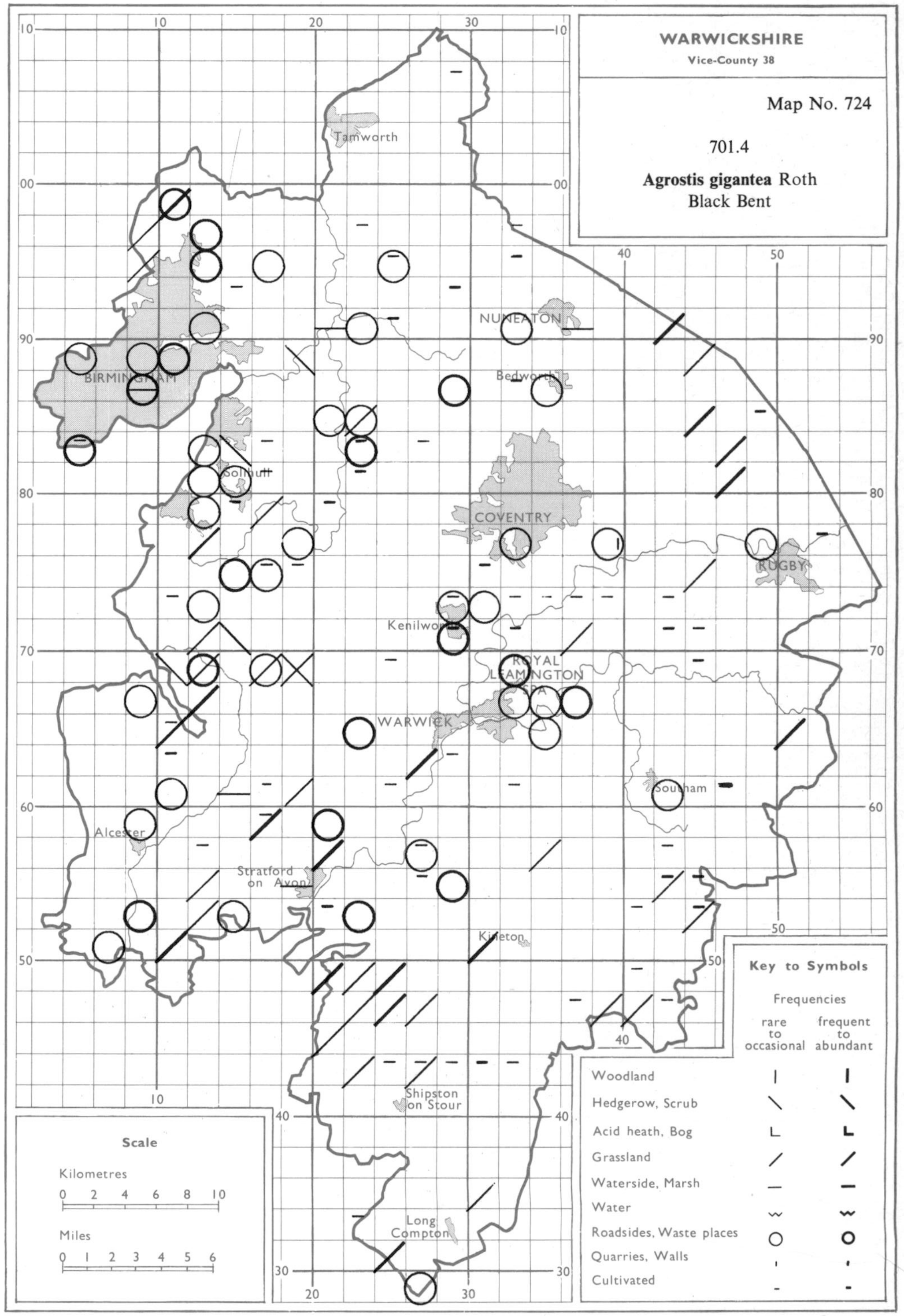
WARWICKSHIRE
Vice-County 38
Map No. 724
701.4
Agrostis gigantea Roth
Black Bent
Tamworth
Birmingham
Solihull
Nuneaton
Bedworth
Coventry
Rugby
Kenilworth
Royal Leamington Spa
Warwick
Southam
Alcester
Stratford on Avon
Kineton
Shipston on Stour
Long Compton
Key to Symbols
Frequencies
rare to occasional
frequent to abundant
Woodland
Hedgerow, Scrub
Acid heath, Bog
Grassland
Waterside, Marsh
Water
Roadsides, Waste places
Quarries, Walls
Cultivated
Scale
Kilometres
0 2 4 6 8 10
Miles
0 1 2 3 4 5 6

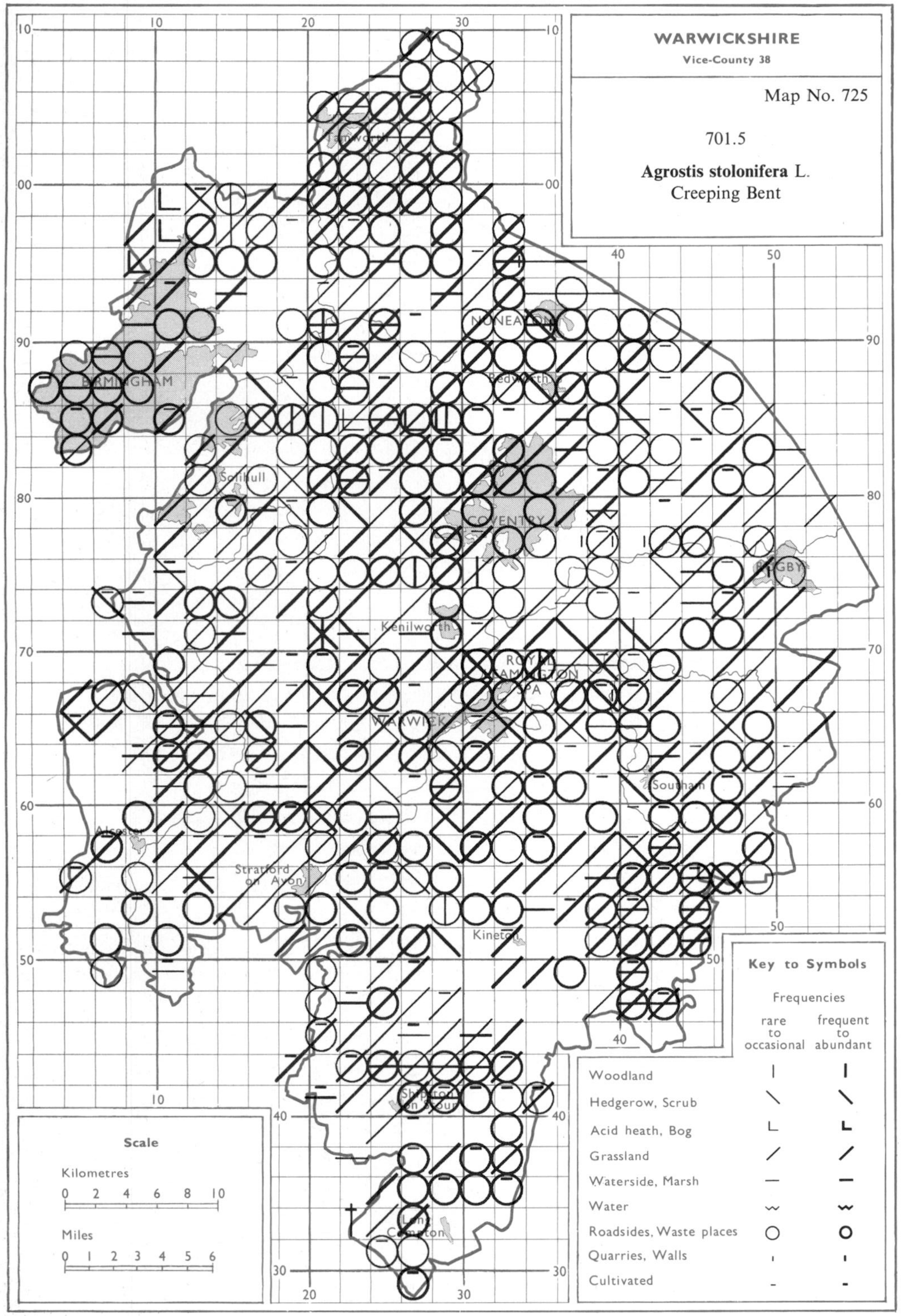
WARWICKSHIRE
Vice-County 38
Map No. 725
701.5
Agrostis stolonifera L.
Creeping Bent
Key to Symbols
Frequencies
rare to occasional
frequent to abundant
Woodland
Hedgerow, Scrub
Acid heath, Bog
Grassland
Waterside, Marsh
Water
Roadsides, Waste places
Quarries, Walls
Cultivated
Scale
Kilometres
0 2 4 6 8 10
Miles
0 1 2 3 4 5 6
BIRMINGHAM
Tamworth
Solihull
NUNEATON
Bedworth
COVENTRY
RUGBY
Kenilworth
ROYAL LEAMINGTON SPA
WARWICK
Southam
Alcester
Stratford on Avon
Kineton
Shipston on Stour
Long Compton

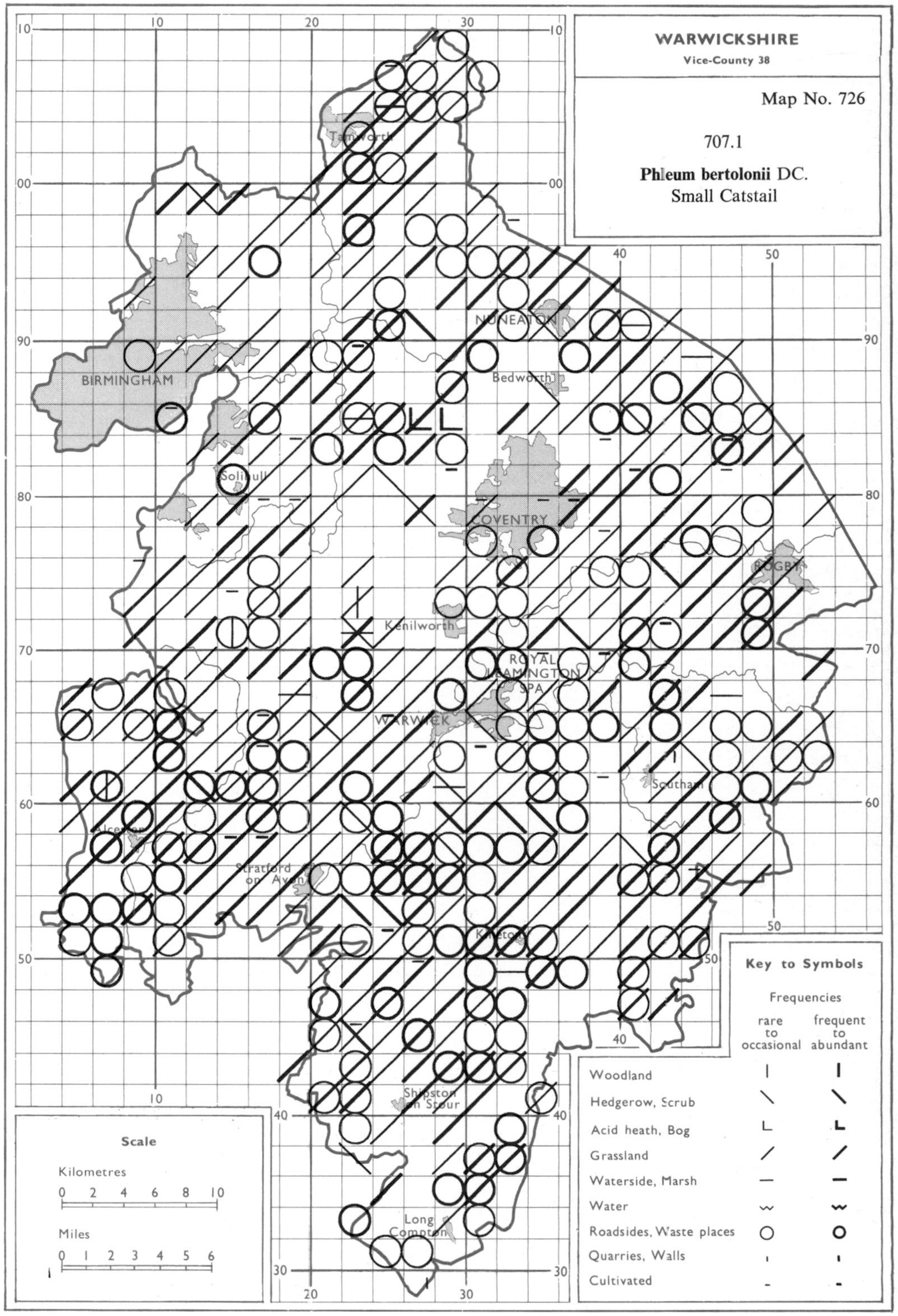
WARWICKSHIRE
Vice-County 38
Map No. 726
707.1
Phleum bertolonii DC.
Small Catstail
Key to Symbols
Frequencies
rare to occasional
frequent to abundant
Woodland
Hedgerow, Scrub
Acid heath, Bog
Grassland
Waterside, Marsh
Water
Roadsides, Waste places
Quarries, Walls
Cultivated
Scale
Kilometres
0 2 4 6 8 10
Miles
0 1 2 3 4 5 6
BIRMINGHAM
Tamworth
Solihull
NUNEATON
Bedworth
COVENTRY
RUGBY
Kenilworth
ROYAL LEAMINGTON SPA
WARWICK
Southam
Alcester
Stratford on Avon
Kineton
Shipston on Stour
Long Compton

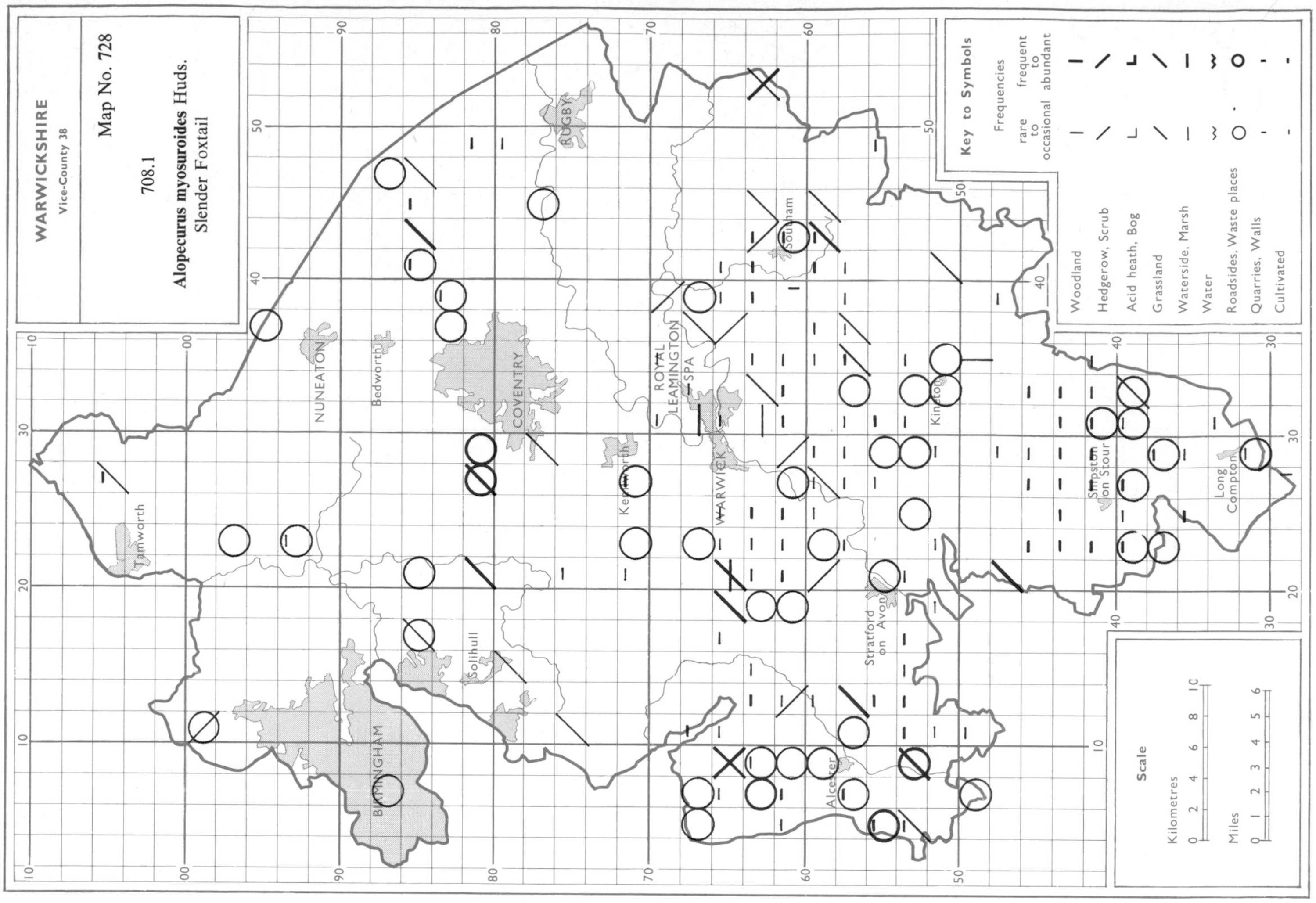
WARWICKSHIRE
Vice-County 38
Map No. 728
708.1
Alopecurus myosuroides Huds.
Slender Foxtail
Key to Symbols
Frequencies
rare to occasional
frequent to abundant
Woodland
Hedgerow, Scrub
Acid heath, Bog
Grassland
Waterside, Marsh
Water
Roadsides, Waste places
Quarries, Walls
Cultivated
Scale
Kilometres
Miles
NUNEATON
Bedworth
COVENTRY
RUGBY
ROYAL LEAMINGTON SPA
WARWICK
Kenilworth
Southam
Kineton
Shipston on Stour
Long Compton
Stratford on Avon
Alcester
Solihull
BIRMINGHAM
Tamworth

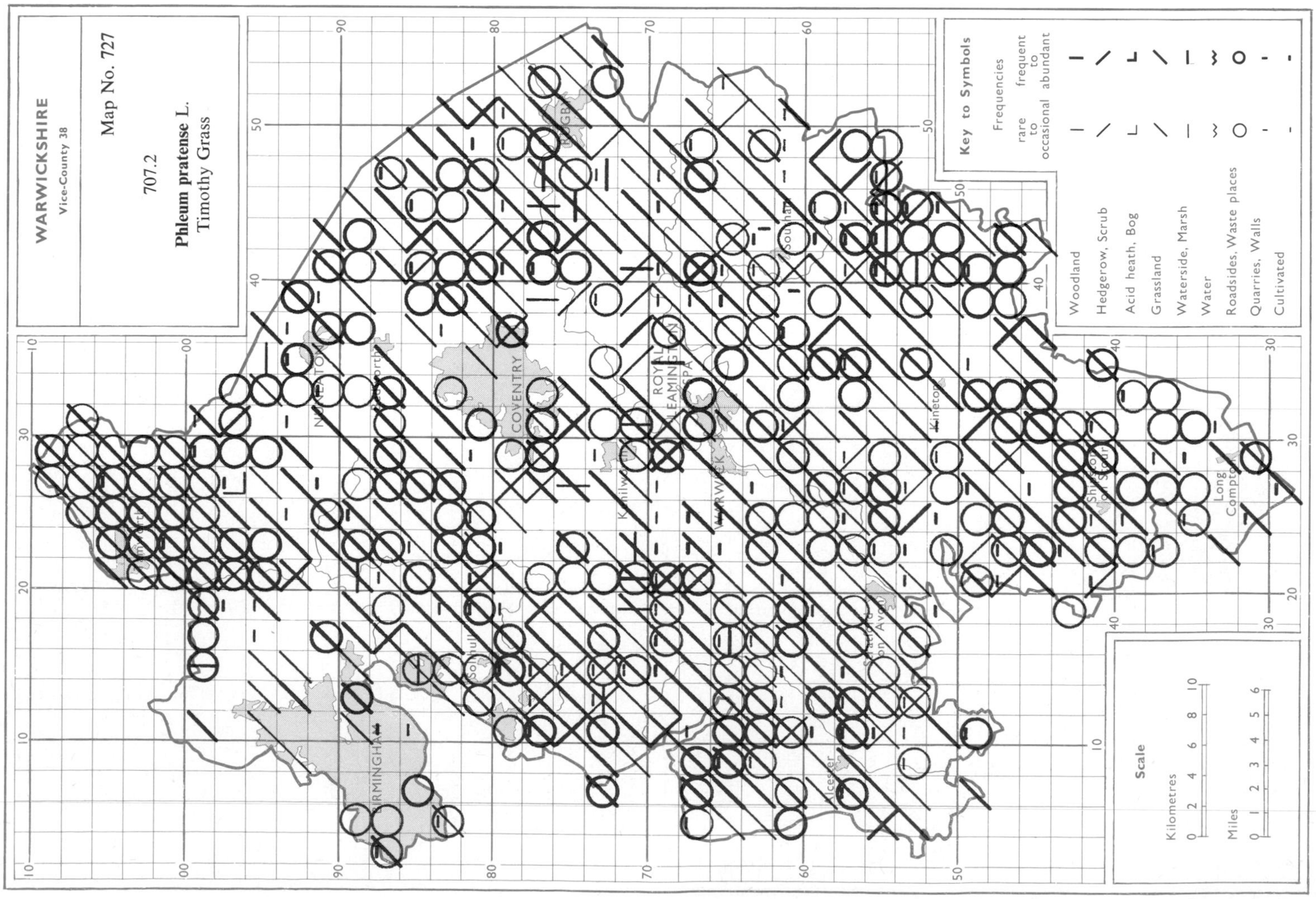
WARWICKSHIRE
Vice-County 38
Map No. 727
707.2
Phleum pratense L.
Timothy Grass
Key to Symbols
Frequencies
rare to occasional
frequent to abundant
Woodland
Hedgerow, Scrub
Acid heath, Bog
Grassland
Waterside, Marsh
Water
Roadsides, Waste places
Quarries, Walls
Cultivated
Scale
Kilometres
Miles
COVENTRY
RUGBY
ROYAL LEAMINGTON SPA
WARWICK
Southam
Kineton
Long Compton
Alcester
Solihull

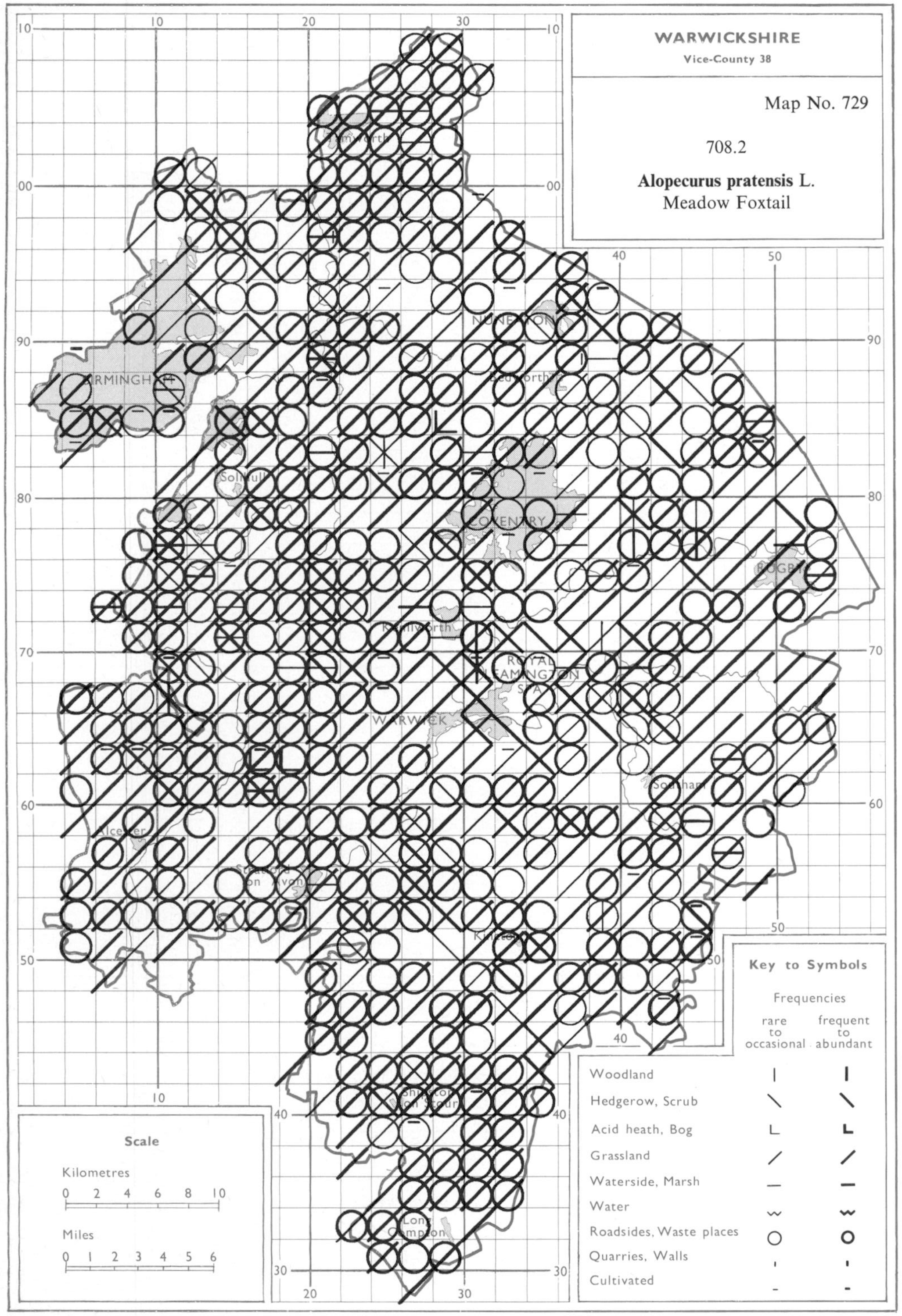
WARWICKSHIRE
Vice-County 38
Map No. 729
708.2
Alopecurus pratensis L.
Meadow Foxtail
Key to Symbols
Frequencies
rare to occasional
frequent to abundant
Woodland
Hedgerow, Scrub
Acid heath, Bog
Grassland
Waterside, Marsh
Water
Roadsides, Waste places
Quarries, Walls
Cultivated
Scale
Kilometres
0 2 4 6 8 10
Miles
0 1 2 3 4 5 6

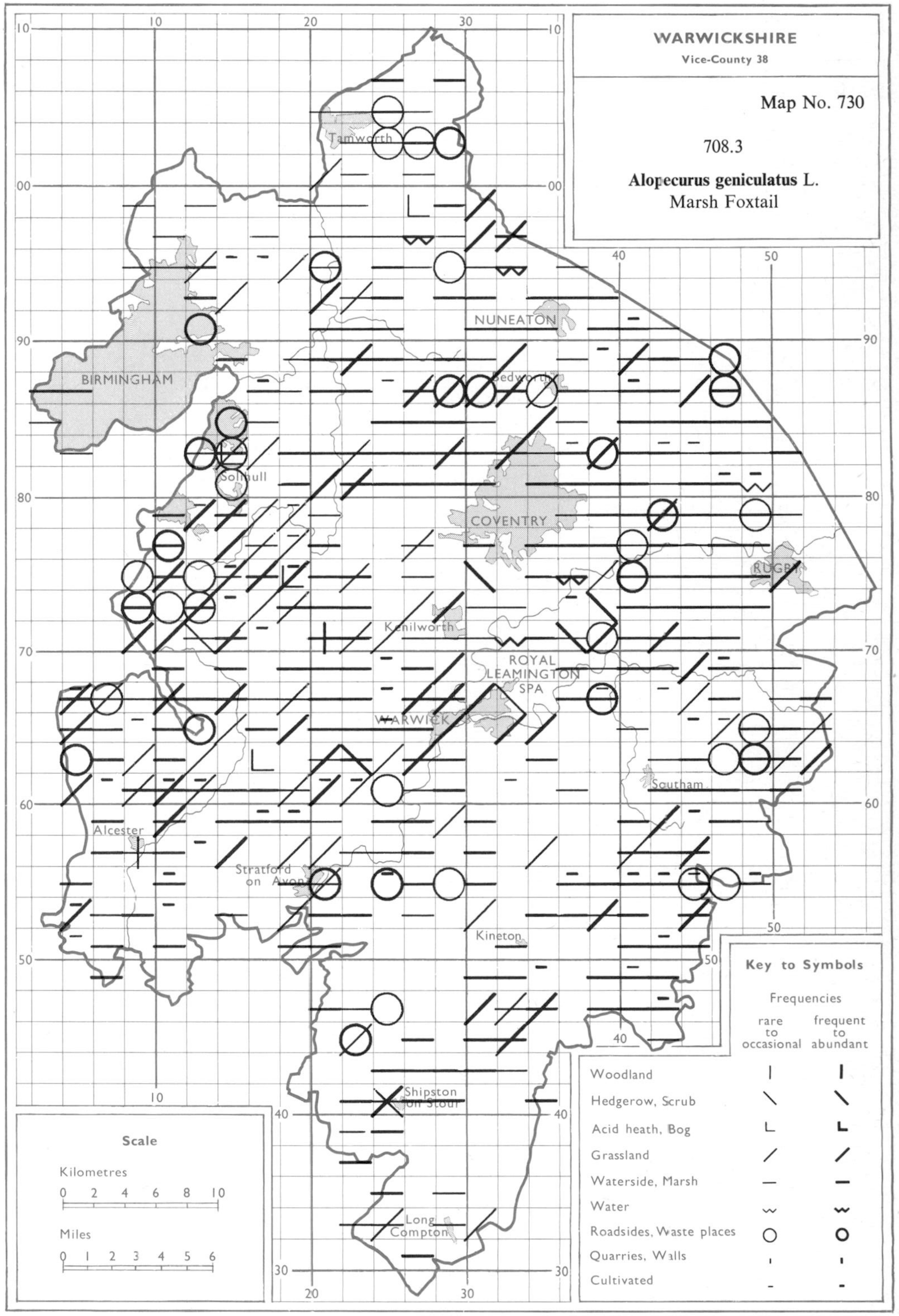
WARWICKSHIRE
Vice-County 38
Map No. 730
708.3
Alopecurus geniculatus L.
Marsh Foxtail
Tamworth
NUNEATON
BIRMINGHAM
Bedworth
Solihull
COVENTRY
RUGBY
Kenilworth
ROYAL LEAMINGTON SPA
WARWICK
Southam
Alcester
Stratford on Avon
Kineton
Shipston on Stour
Long Compton
Key to Symbols
Frequencies
rare to occasional
frequent to abundant
Woodland
Hedgerow, Scrub
Acid heath, Bog
Grassland
Waterside, Marsh
Water
Roadsides, Waste places
Quarries, Walls
Cultivated
Scale
Kilometres
0 2 4 6 8 10
Miles
0 1 2 3 4 5 6

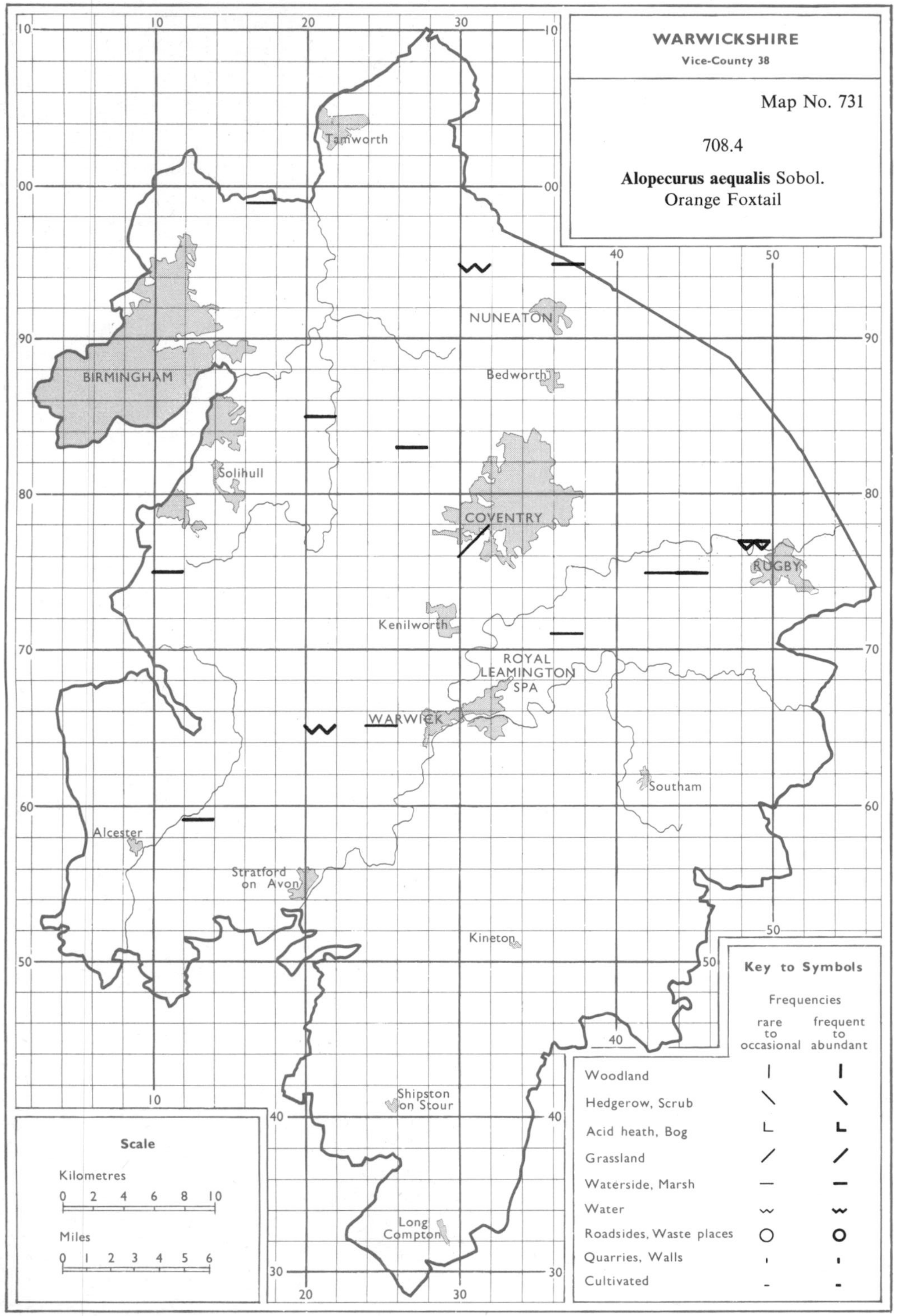

WARWICKSHIRE
Vice-County 38
Map No. 731
708.4
Alopecurus aequalis Sobol.
Orange Foxtail
Tamworth
NUNEATON
Bedworth
BIRMINGHAM
Solihull
COVENTRY
RUGBY
Kenilworth
ROYAL LEAMINGTON SPA
WARWICK
Southam
Alcester
Stratford on Avon
Kineton
Shipston on Stour
Long Compton
Key to Symbols
Frequencies
rare to occasional
frequent to abundant
Woodland
Hedgerow, Scrub
Acid heath, Bog
Grassland
Waterside, Marsh
Water
Roadsides, Waste places
Quarries, Walls
Cultivated
Scale
Kilometres
0 2 4 6 8 10
Miles
0 1 2 3 4 5 6

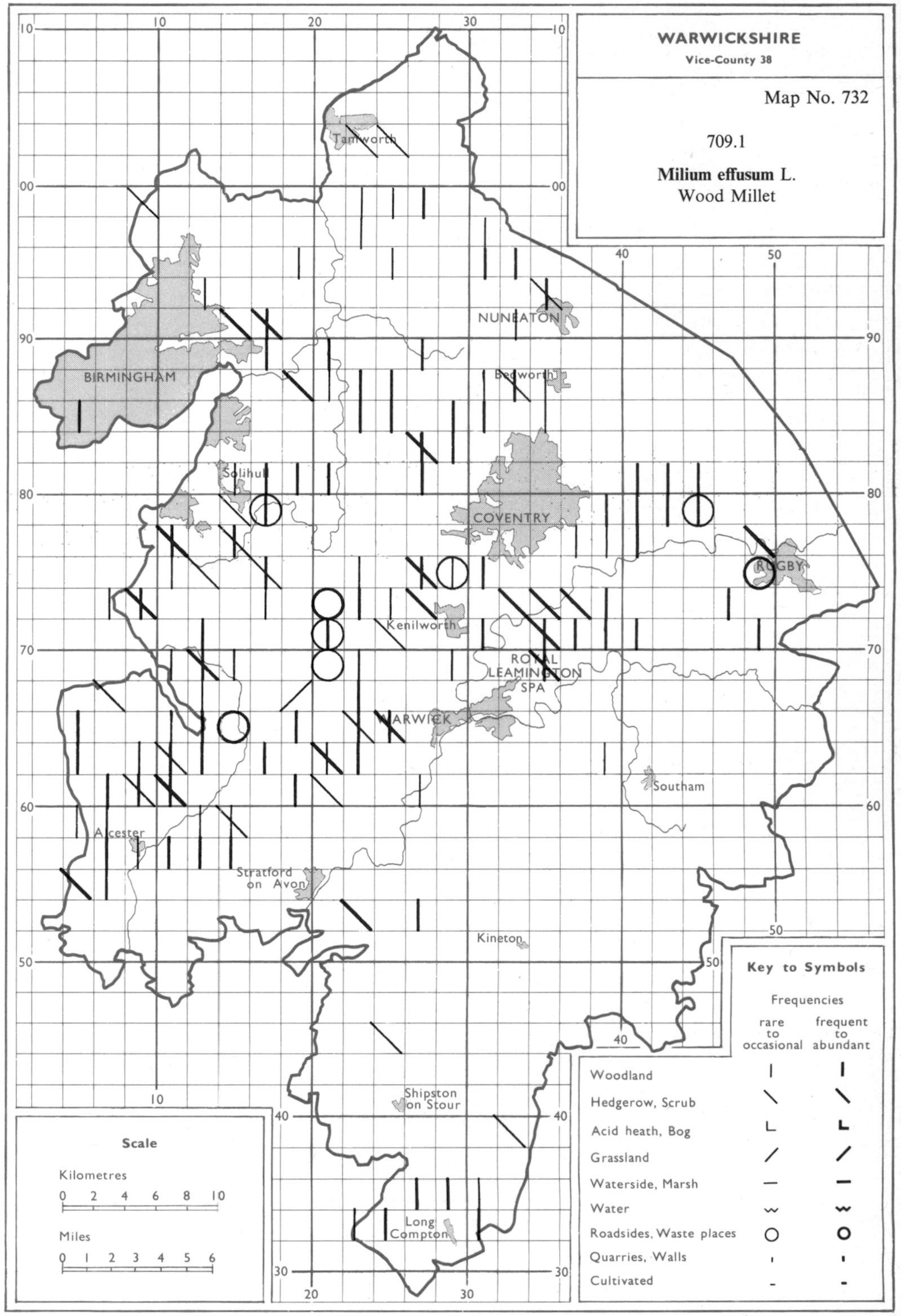

WARWICKSHIRE
Vice-County 38
Map No. 732
709.1
Milium effusum L.
Wood Millet
Tamworth
NUNEATON
Bedworth
BIRMINGHAM
Solihull
COVENTRY
RUGBY
Kenilworth
ROYAL LEAMINGTON SPA
WARWICK
Southam
Alcester
Stratford on Avon
Kineton
Shipston on Stour
Long Compton
Key to Symbols
Frequencies
rare to occasional
frequent to abundant
Woodland
Hedgerow, Scrub
Acid heath, Bog
Grassland
Waterside, Marsh
Water
Roadsides, Waste places
Quarries, Walls
Cultivated
Scale
Kilometres
0 2 4 6 8 10
Miles
0 1 2 3 4 5 6

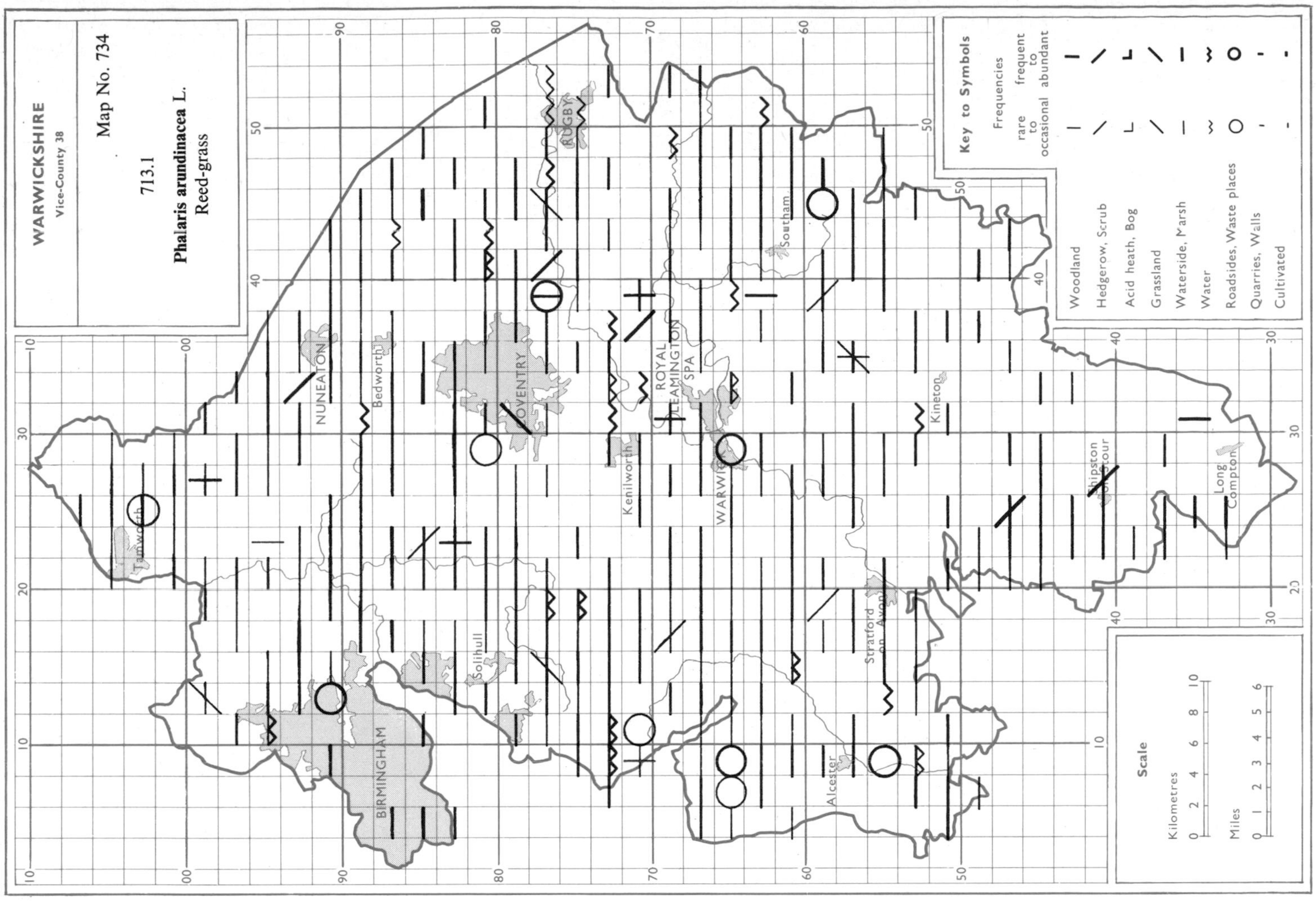
WARWICKSHIRE
Vice-County 38
Map No. 734
713.1
Phalaris arundinacea L.
Reed-grass
Key to Symbols
Frequencies
rare to occasional
frequent to abundant
Woodland
Hedgerow, Scrub
Acid heath, Bog
Grassland
Waterside, Marsh
Water
Roadsides, Waste places
Quarries, Walls
Cultivated
Scale
Kilometres 0 2 4 6 8 10
Miles 0 1 2 3 4 5 6
Tamworth
NUNEATON
Bedworth
COVENTRY
RUGBY
BIRMINGHAM
Solihull
Kenilworth
ROYAL LEAMINGTON SPA
WARWICK
Southam
Kineton
Stratford on Avon
Alcester
Shipston on Stour
Long Compton

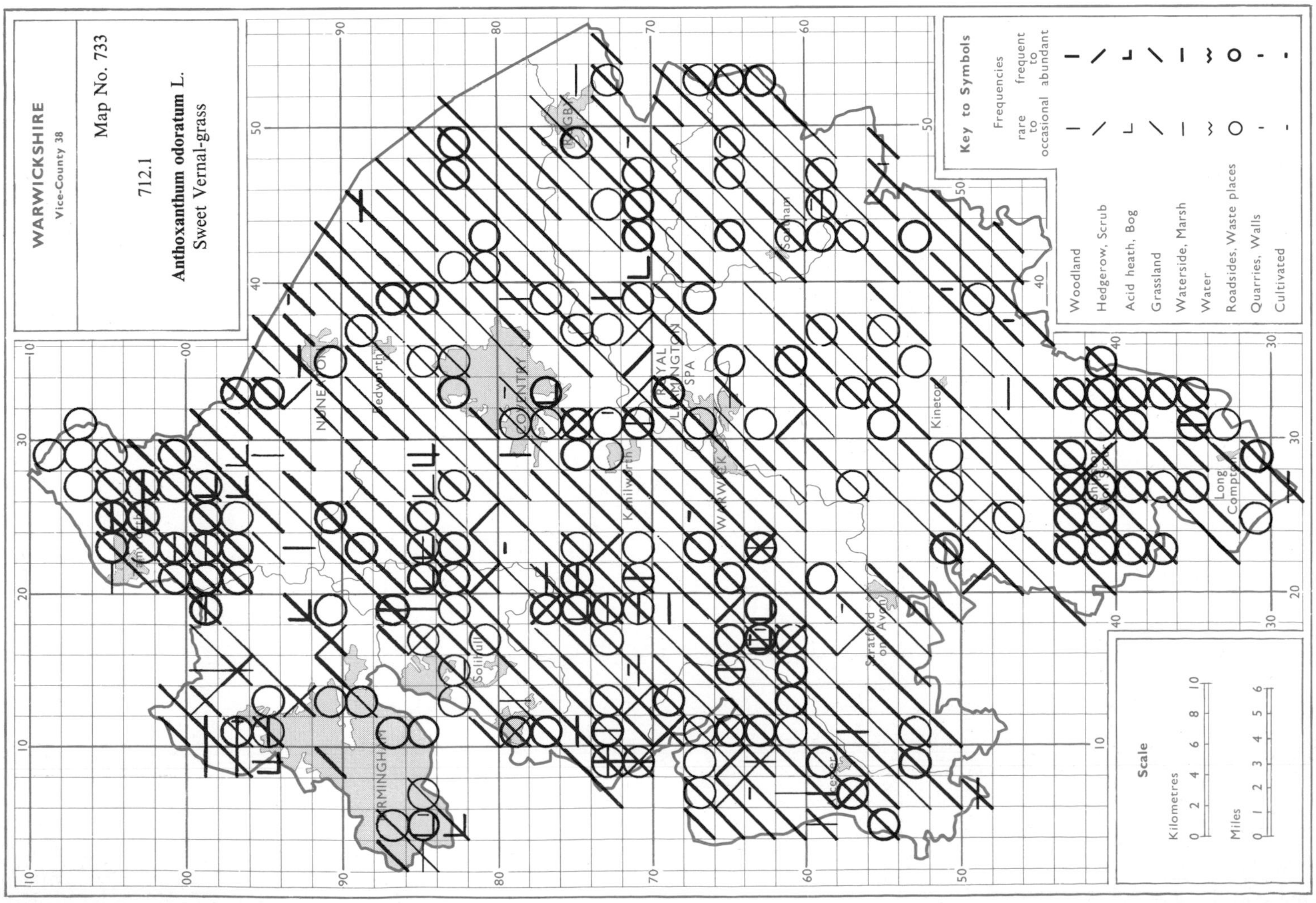
WARWICKSHIRE
Vice-County 38
Map No. 733
712.1
Anthoxanthum odoratum L.
Sweet Vernal-grass
Key to Symbols
Frequencies
rare to occasional
frequent to abundant
Woodland
Hedgerow, Scrub
Acid heath, Bog
Grassland
Waterside, Marsh
Water
Roadsides, Waste places
Quarries, Walls
Cultivated
Scale
Kilometres 0 2 4 6 8 10
Miles 0 1 2 3 4 5 6
NUNEATON
Bedworth
COVENTRY
RUGBY
Solihull
Kenilworth
ROYAL LEAMINGTON SPA
WARWICK
Southam
Kineton
Stratford on Avon
Long Compton

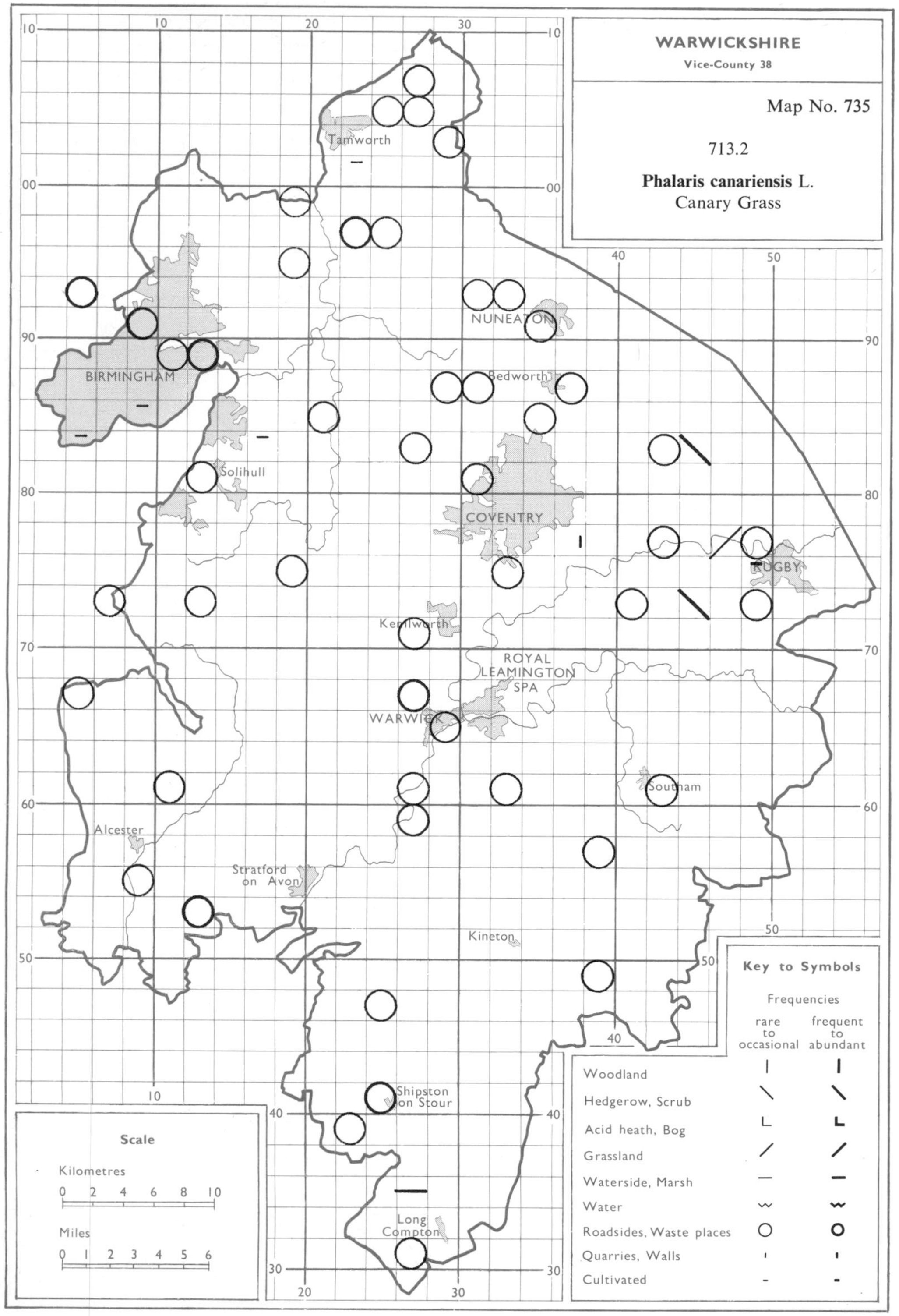
WARWICKSHIRE
Vice-County 38
Map No. 735
713.2
Phalaris canariensis L.
Canary Grass
Tamworth
NUNEATON
Bedworth
BIRMINGHAM
Solihull
COVENTRY
RUGBY
Kenilworth
ROYAL LEAMINGTON SPA
WARWICK
Southam
Alcester
Stratford on Avon
Kineton
Shipston on Stour
Long Compton
Key to Symbols
Frequencies
rare to occasional
frequent to abundant
Woodland
Hedgerow, Scrub
Acid heath, Bog
Grassland
Waterside, Marsh
Water
Roadsides, Waste places
Quarries, Walls
Cultivated
Scale
Kilometres
0 2 4 6 8 10
Miles
0 1 2 3 4 5 6

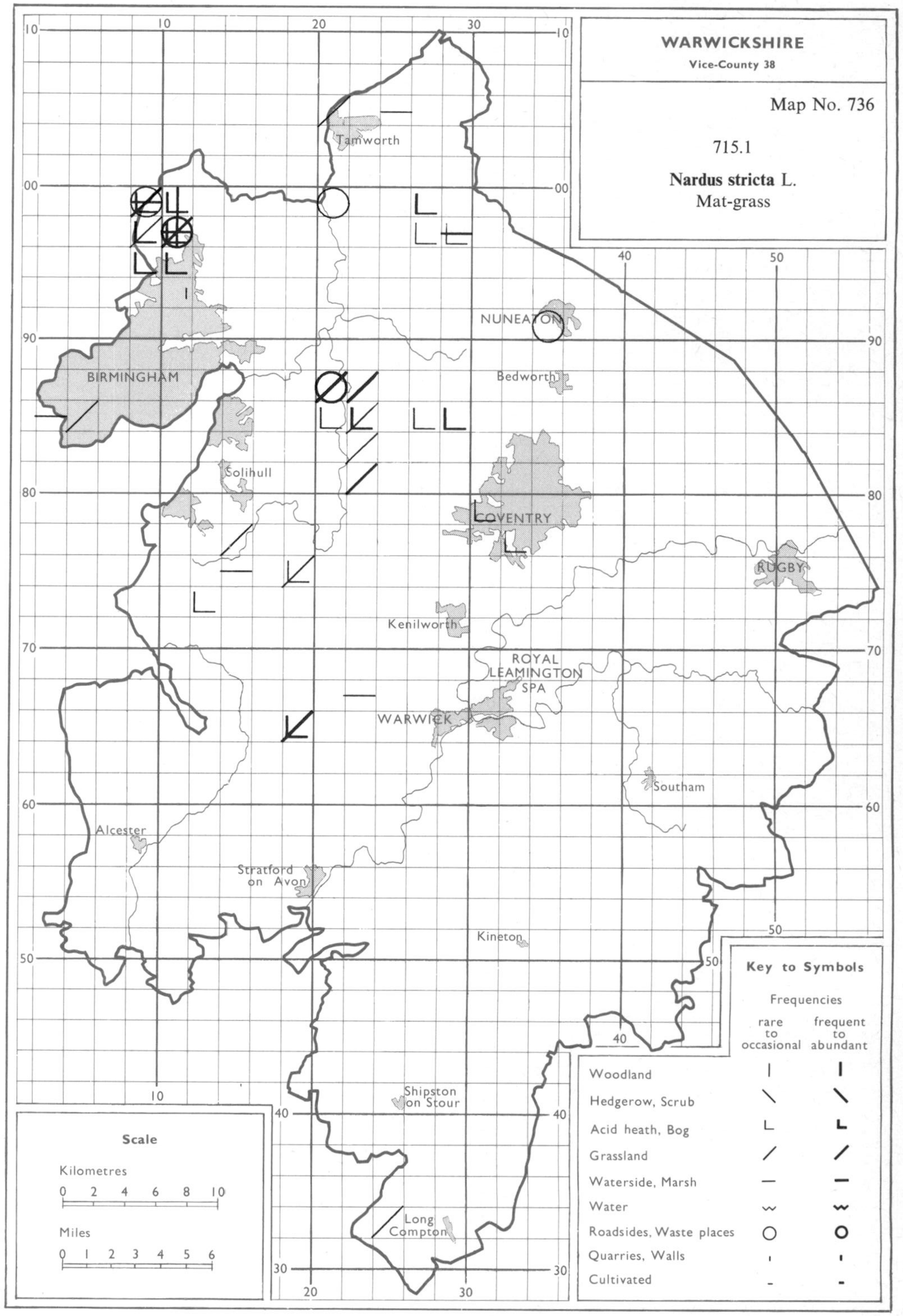
WARWICKSHIRE
Vice-County 38
Map No. 736
715.1
Nardus stricta L.
Mat-grass
Tamworth
NUNEATON
Bedworth
BIRMINGHAM
Solihull
COVENTRY
RUGBY
Kenilworth
ROYAL LEAMINGTON SPA
WARWICK
Southam
Alcester
Stratford on Avon
Kineton
Shipston on Stour
Long Compton
Key to Symbols
Frequencies
rare to occasional
frequent to abundant
Woodland
Hedgerow, Scrub
Acid heath, Bog
Grassland
Waterside, Marsh
Water
Roadsides, Waste places
Quarries, Walls
Cultivated
Scale
Kilometres
0 2 4 6 8 10
Miles
0 1 2 3 4 5 6

10. LINE PRINTED MAPS (Nos. 737-795)

This series of maps (4 per page) was produced on a line printer. Because of the physical limitations of the printer in which the space between lines is 4 mm and that between alternate symbols on the lines is 5 mm the resulting maps were slightly distorted by an increase of E–W to N–S dimensions in the ratio of 5 : 4. A double scale in the bottom left-hand corner indicates these differences graphically.

The symbols presented on these maps are very much simpler than those in the graph-plotted maps, and indicate only frequency, not habitat range. The symbol "×" indicates "abundant", "locally abundant", "frequent", or "locally frequent"; the symbol "+" indicates "occasional" or "rare". As with the graph-plotted maps, the computer has been programmed to select the "most frequent" reading in the sequence "a, la, f, lf" and "o, r", respectively.

Transparent overlays are available in a pocket at the back of the book to be used in conjunction with these maps, as follows:

Map IX Relief and drainage.
Map X Surface geology.
Map XI Land use, 1960 (including canals and railways—used and disused).

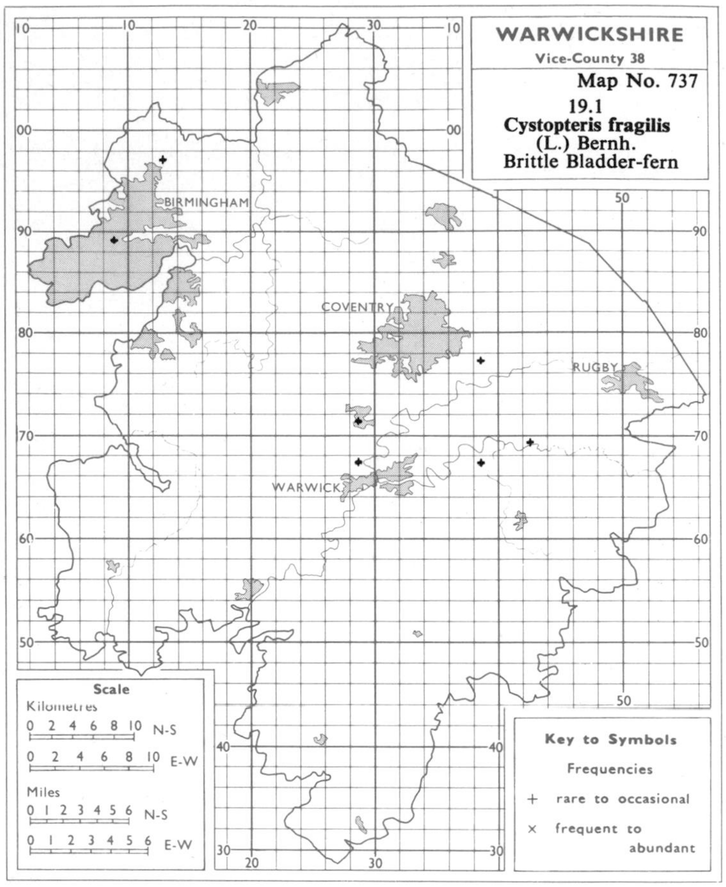
WARWICKSHIRE
Vice-County 38
Map No. 737
19.1
Cystopteris fragilis
(L.) Bernh.
Brittle Bladder-fern
BIRMINGHAM
COVENTRY
RUGBY
WARWICK
Scale
Kilometres
0 2 4 6 8 10 N-S
0 2 4 6 8 10 E-W
Miles
0 1 2 3 4 5 6 N-S
0 1 2 3 4 5 6 E-W
Key to Symbols
Frequencies
+ rare to occasional
× frequent to abundant

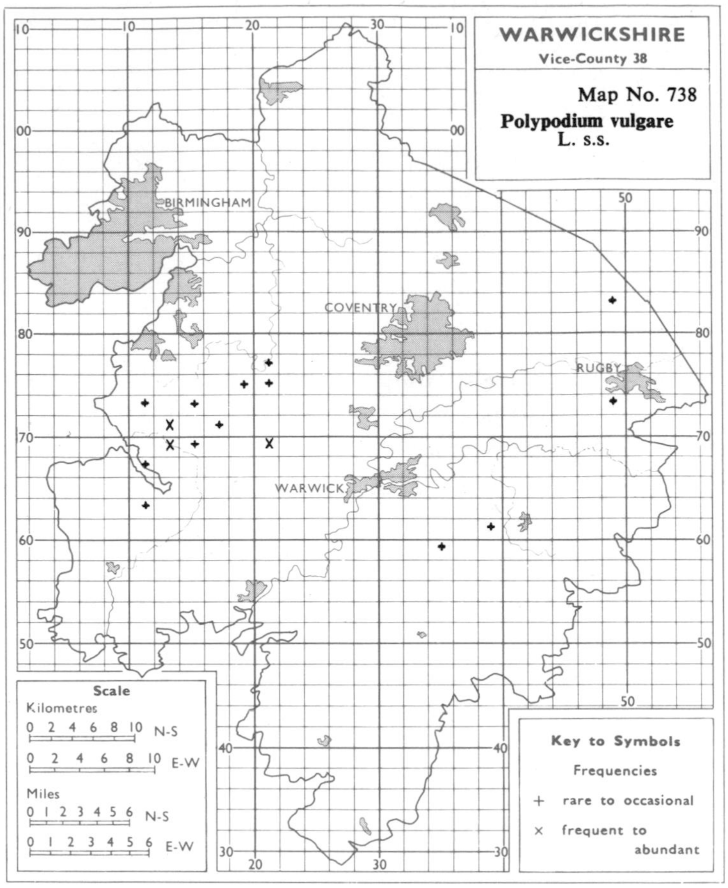
WARWICKSHIRE
Vice-County 38
Map No. 738
Polypodium vulgare
L. s.s.
BIRMINGHAM
COVENTRY
RUGBY
WARWICK
Scale
Kilometres
0 2 4 6 8 10 N-S
0 2 4 6 8 10 E-W
Miles
0 1 2 3 4 5 6 N-S
0 1 2 3 4 5 6 E-W
Key to Symbols
Frequencies
+ rare to occasional
× frequent to abundant

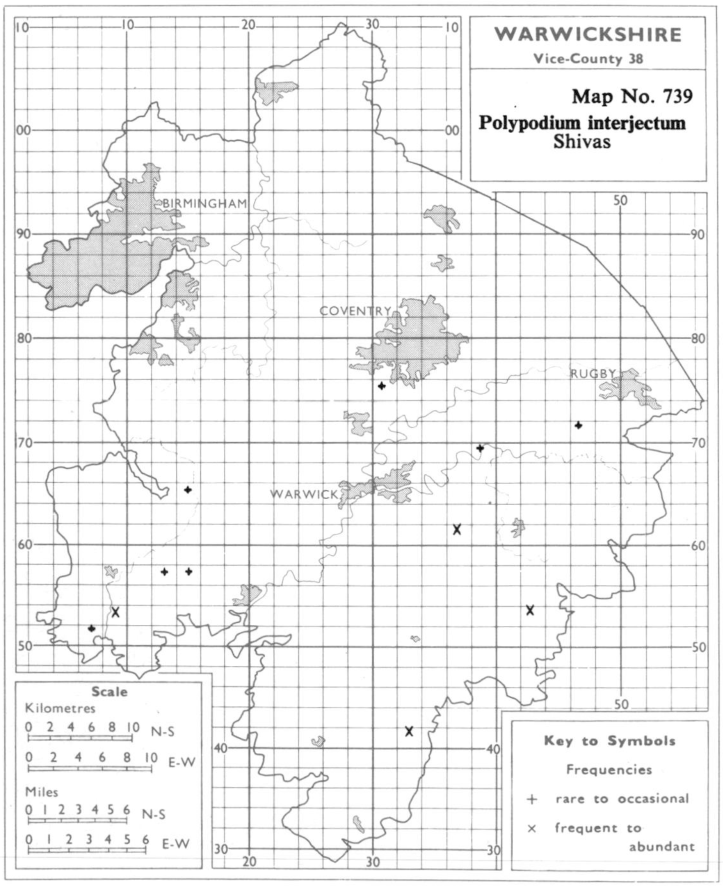
WARWICKSHIRE
Vice-County 38
Map No. 739
Polypodium interjectum
Shivas
BIRMINGHAM
COVENTRY
RUGBY
WARWICK
Scale
Kilometres
0 2 4 6 8 10 N-S
0 2 4 6 8 10 E-W
Miles
0 1 2 3 4 5 6 N-S
0 1 2 3 4 5 6 E-W
Key to Symbols
Frequencies
+ rare to occasional
× frequent to abundant

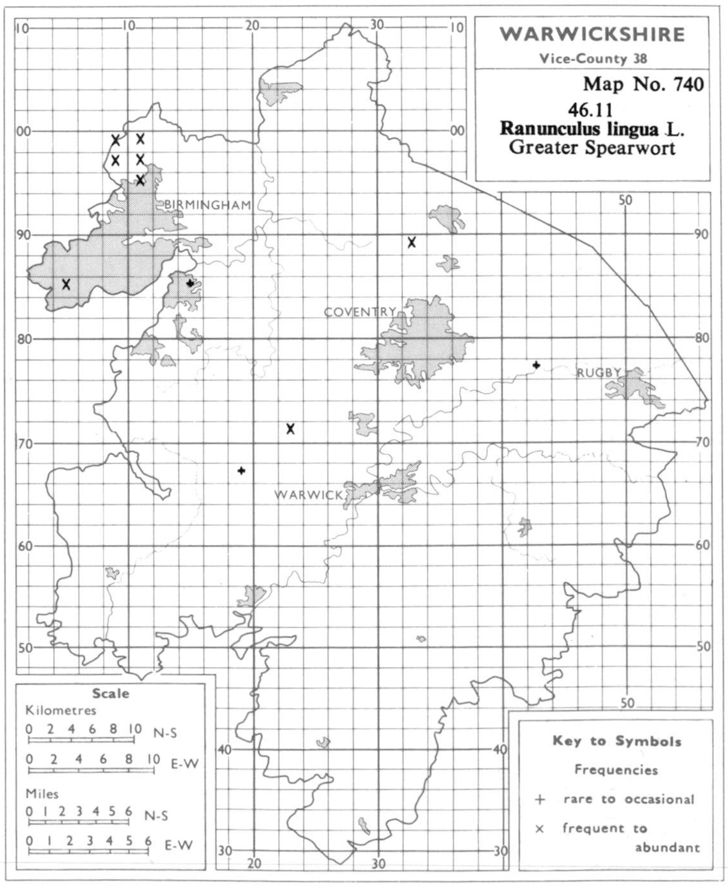
WARWICKSHIRE
Vice-County 38
Map No. 740
46.11
Ranunculus lingua L.
Greater Spearwort
BIRMINGHAM
COVENTRY
RUGBY
WARWICK
Scale
Kilometres
0 2 4 6 8 10 N-S
0 2 4 6 8 10 E-W
Miles
0 1 2 3 4 5 6 N-S
0 1 2 3 4 5 6 E-W
Key to Symbols
Frequencies
+ rare to occasional
× frequent to abundant

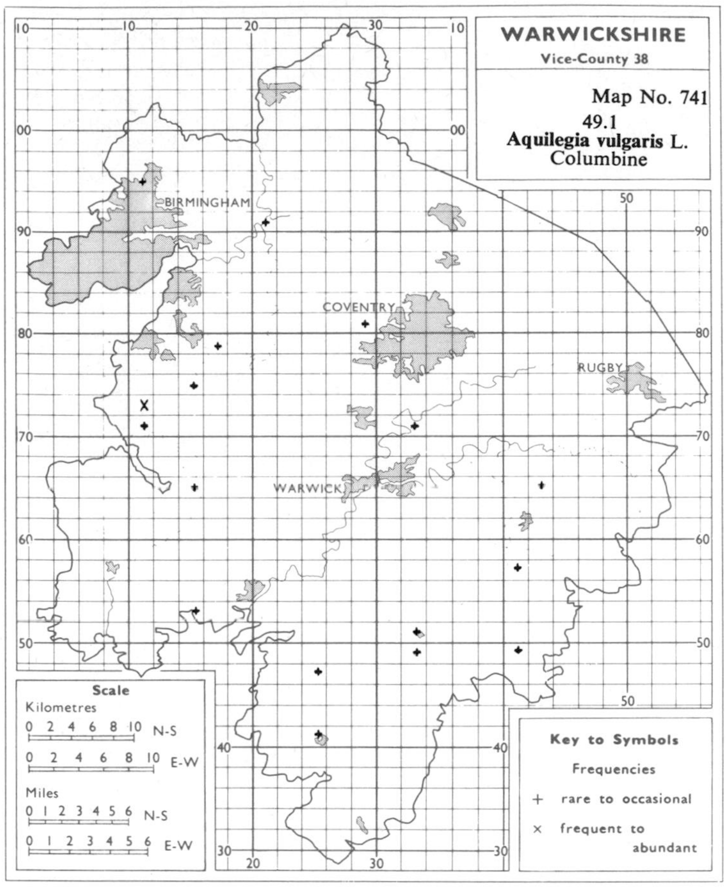
WARWICKSHIRE
Vice-County 38
Map No. 741
49.1
Aquilegia vulgaris L.
Columbine
BIRMINGHAM
COVENTRY
RUGBY
WARWICK
Scale
Kilometres
0 2 4 6 8 10 N-S
0 2 4 6 8 10 E-W
Miles
0 1 2 3 4 5 6 N-S
0 1 2 3 4 5 6 E-W
Key to Symbols
Frequencies
+ rare to occasional
× frequent to abundant

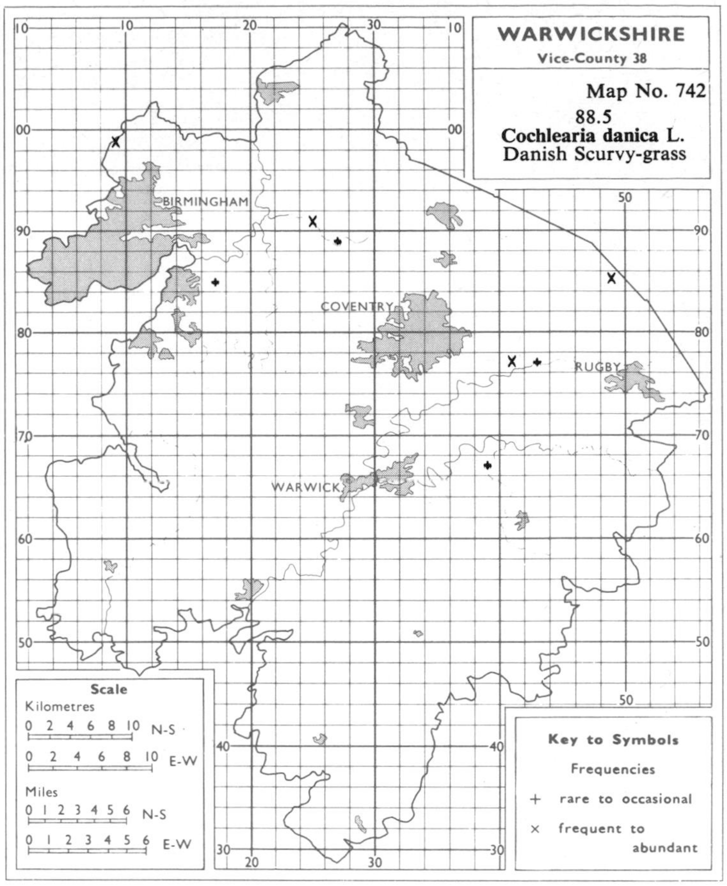
WARWICKSHIRE
Vice-County 38
Map No. 742
88.5
Cochlearia danica L.
Danish Scurvy-grass
BIRMINGHAM
COVENTRY
RUGBY
WARWICK
Scale
Kilometres
0 2 4 6 8 10 N-S
0 2 4 6 8 10 E-W
Miles
0 1 2 3 4 5 6 N-S
0 1 2 3 4 5 6 E-W
Key to Symbols
Frequencies
+ rare to occasional
× frequent to abundant

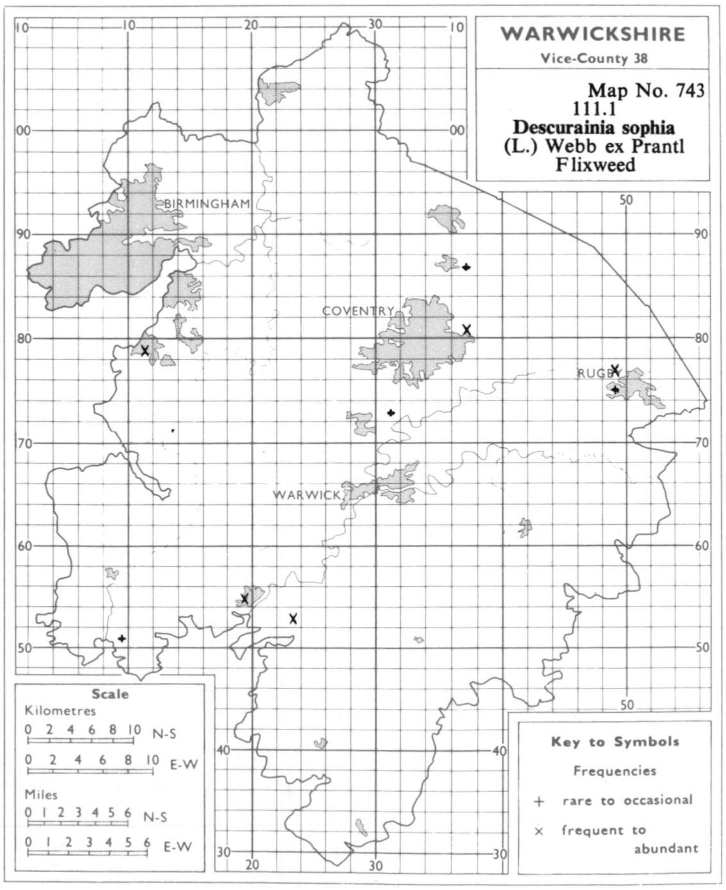
WARWICKSHIRE
Vice-County 38
Map No. 743
111.1
Descurainia sophia
(L.) Webb ex Prantl
Flixweed
BIRMINGHAM
COVENTRY
RUGBY
WARWICK
Scale
Kilometres
0 2 4 6 8 10 N-S
0 2 4 6 8 10 E-W
Miles
0 1 2 3 4 5 6 N-S
0 1 2 3 4 5 6 E-W
Key to Symbols
Frequencies
+ rare to occasional
× frequent to abundant

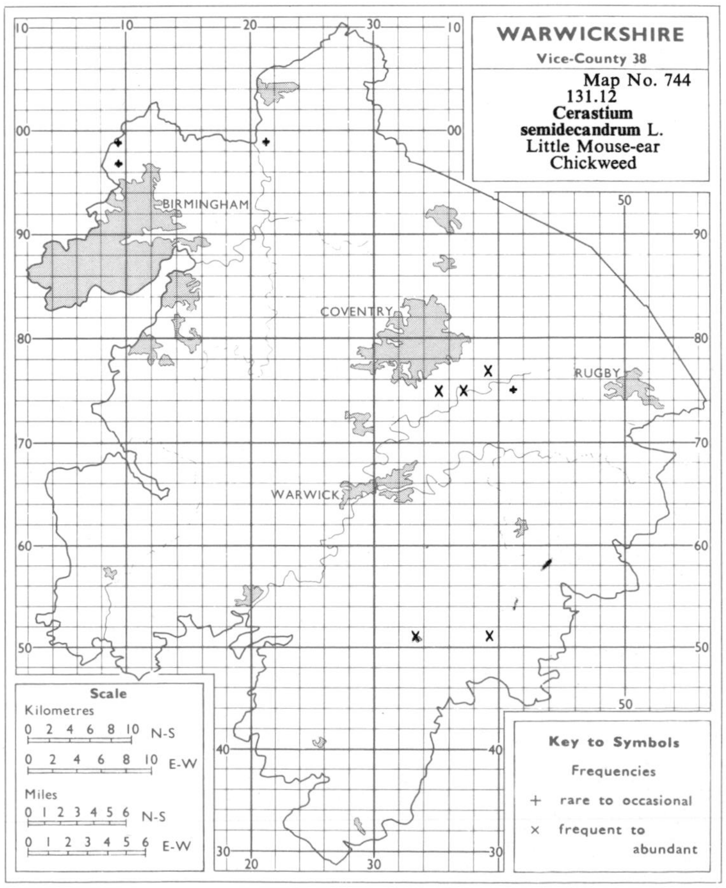
WARWICKSHIRE
Vice-County 38
Map No. 744
131.12
Cerastium
semidecandrum L.
Little Mouse-ear
Chickweed
BIRMINGHAM
COVENTRY
RUGBY
WARWICK
Scale
Kilometres
0 2 4 6 8 10 N-S
0 2 4 6 8 10 E-W
Miles
0 1 2 3 4 5 6 N-S
0 1 2 3 4 5 6 E-W
Key to Symbols
Frequencies
+ rare to occasional
× frequent to abundant

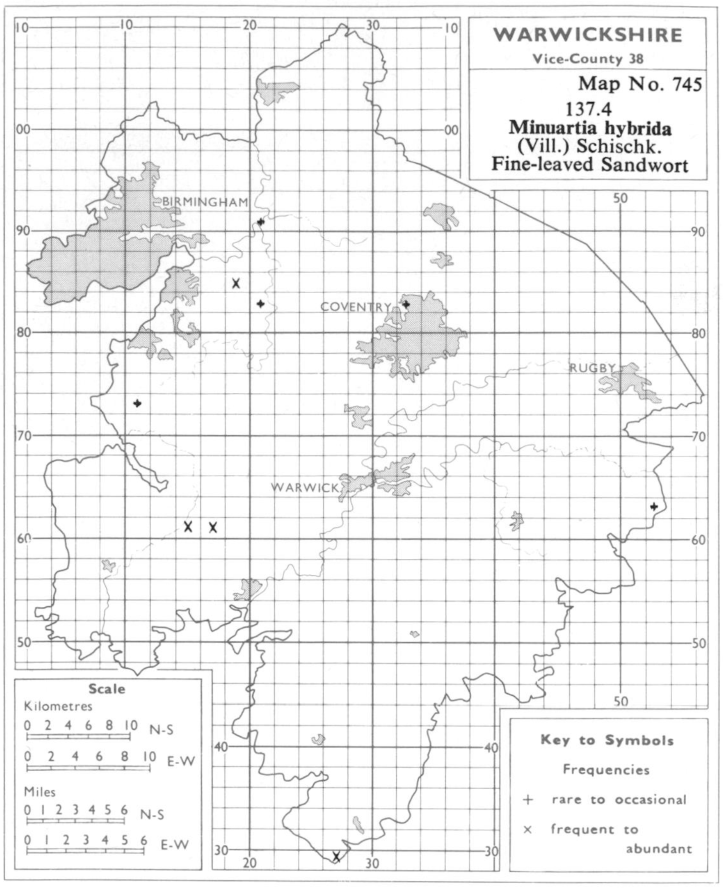
WARWICKSHIRE
Vice-County 38
Map No. 745
137.4
Minuartia hybrida
(Vill.) Schischk.
Fine-leaved Sandwort
BIRMINGHAM
COVENTRY
RUGBY
WARWICK
Scale
Kilometres
0 2 4 6 8 10 N-S
0 2 4 6 8 10 E-W
Miles
0 1 2 3 4 5 6 N-S
0 1 2 3 4 5 6 E-W
Key to Symbols
Frequencies
+ rare to occasional
× frequent to abundant

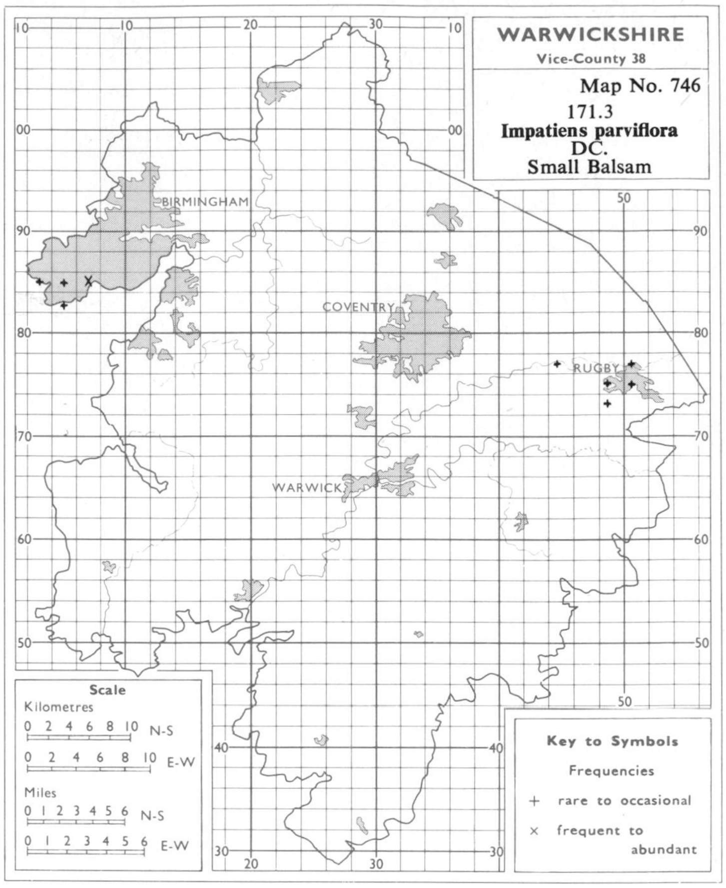
WARWICKSHIRE
Vice-County 38
Map No. 746
171.3
Impatiens parviflora
DC.
Small Balsam
BIRMINGHAM
COVENTRY
RUGBY
WARWICK
Scale
Kilometres
0 2 4 6 8 10 N-S
0 2 4 6 8 10 E-W
Miles
0 1 2 3 4 5 6 N-S
0 1 2 3 4 5 6 E-W
Key to Symbols
Frequencies
+ rare to occasional
× frequent to abundant

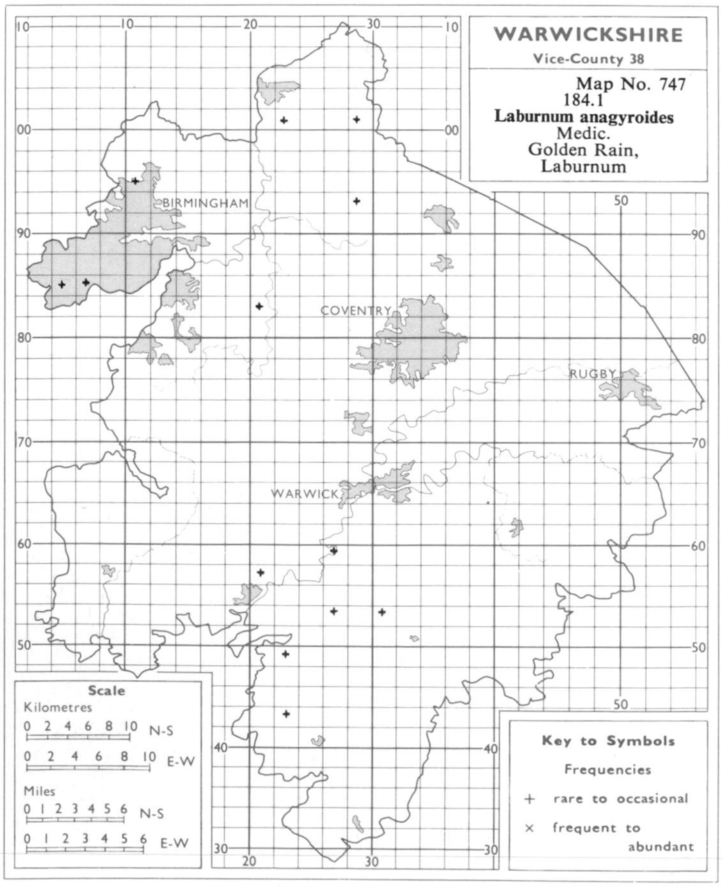
WARWICKSHIRE
Vice-County 38
Map No. 747
184.1
Laburnum anagyroides
Medic.
Golden Rain,
Laburnum
BIRMINGHAM
COVENTRY
RUGBY
WARWICK
Scale
Kilometres
0 2 4 6 8 10 N-S
0 2 4 6 8 10 E-W
Miles
0 1 2 3 4 5 6 N-S
0 1 2 3 4 5 6 E-W
Key to Symbols
Frequencies
+ rare to occasional
× frequent to abundant

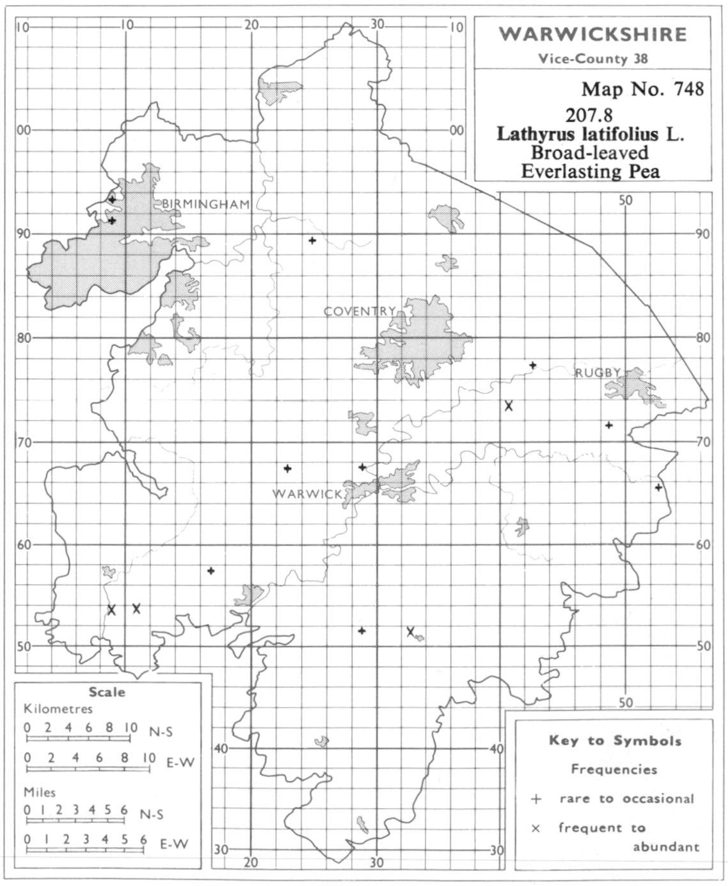
WARWICKSHIRE
Vice-County 38
Map No. 748
207.8
Lathyrus latifolius L.
Broad-leaved
Everlasting Pea
BIRMINGHAM
COVENTRY
RUGBY
WARWICK
Scale
Kilometres
0 2 4 6 8 10 N-S
0 2 4 6 8 10 E-W
Miles
0 1 2 3 4 5 6 N-S
0 1 2 3 4 5 6 E-W
Key to Symbols
Frequencies
+ rare to occasional
× frequent to abundant

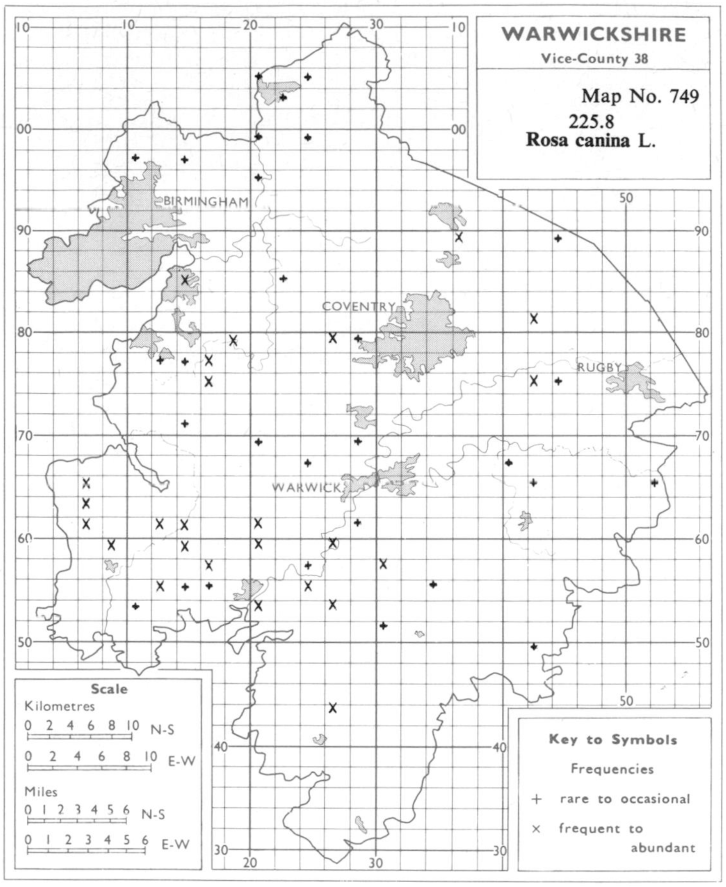
WARWICKSHIRE
Vice-County 38
Map No. 749
225.8
Rosa canina L.
BIRMINGHAM
COVENTRY
RUGBY
WARWICK
Scale
Kilometres
0 2 4 6 8 10 N-S
0 2 4 6 8 10 E-W
Miles
0 1 2 3 4 5 6 N-S
0 1 2 3 4 5 6 E-W
Key to Symbols
Frequencies
+ rare to occasional
× frequent to abundant

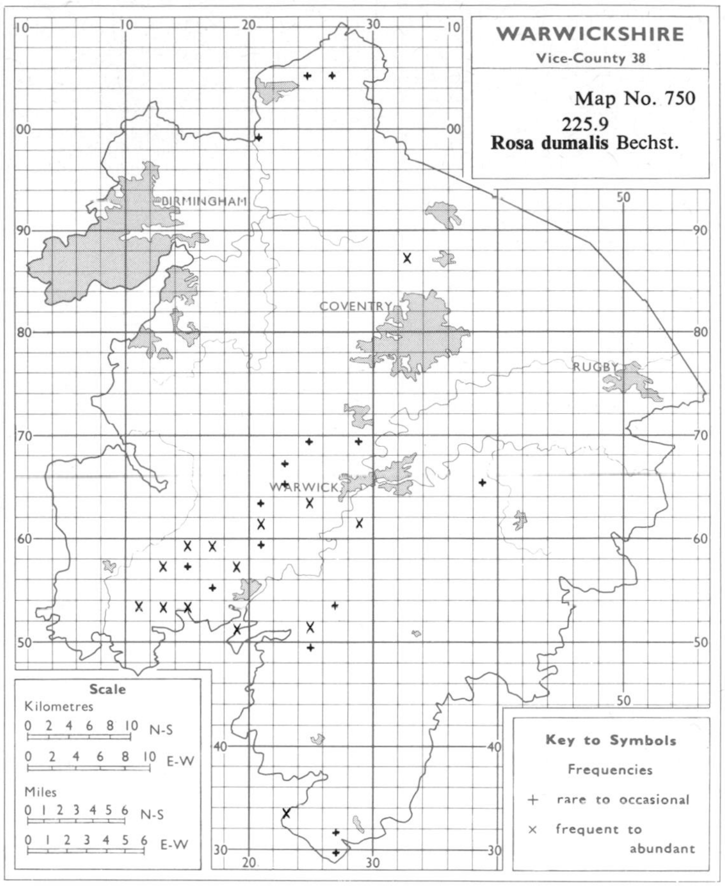
WARWICKSHIRE
Vice-County 38
Map No. 750
225.9
Rosa dumalis Bechst.
BIRMINGHAM
COVENTRY
RUGBY
WARWICK
Scale
Kilometres
0 2 4 6 8 10 N-S
0 2 4 6 8 10 E-W
Miles
0 1 2 3 4 5 6 N-S
0 1 2 3 4 5 6 E-W
Key to Symbols
Frequencies
+ rare to occasional
× frequent to abundant

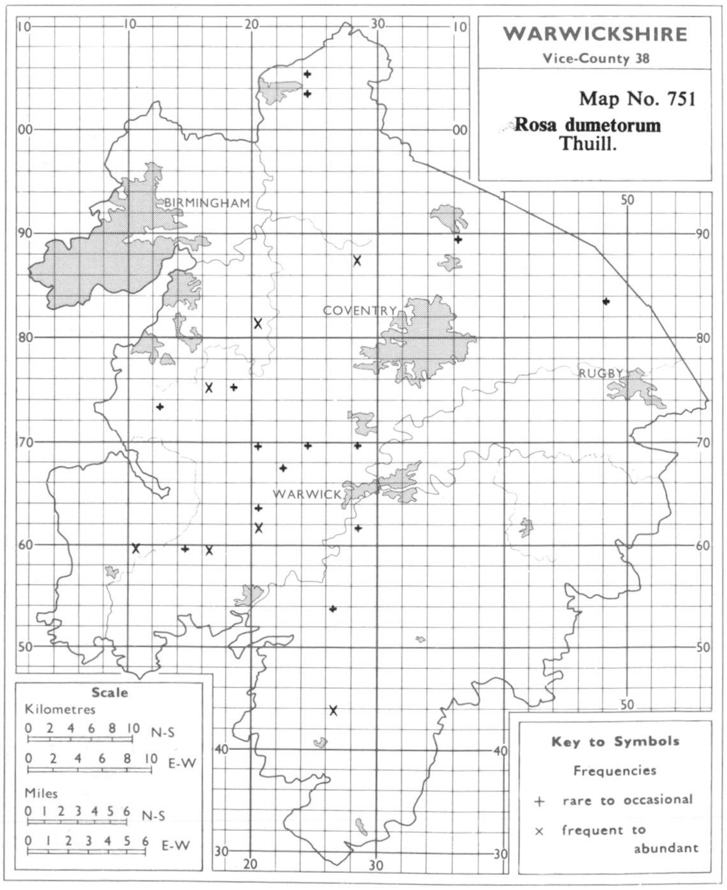
WARWICKSHIRE
Vice-County 38
Map No. 751
Rosa dumetorum Thuill.
BIRMINGHAM
COVENTRY
RUGBY
WARWICK
Scale
Kilometres
0 2 4 6 8 10 N-S
0 2 4 6 8 10 E-W
Miles
0 1 2 3 4 5 6 N-S
0 1 2 3 4 5 6 E-W
Key to Symbols
Frequencies
+ rare to occasional
× frequent to abundant

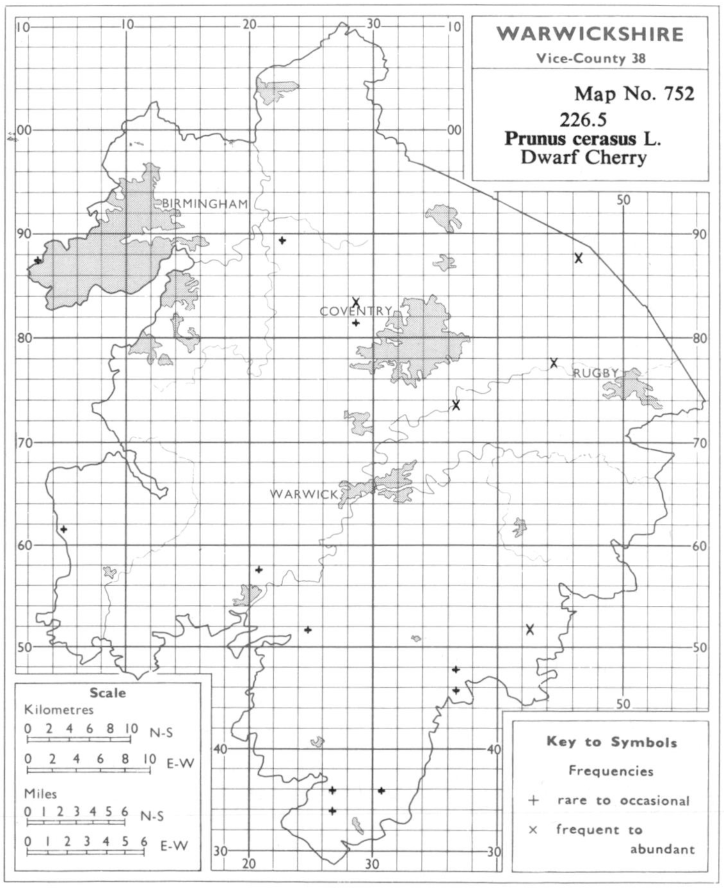
WARWICKSHIRE
Vice-County 38
Map No. 752
226.5
Prunus cerasus L.
Dwarf Cherry
BIRMINGHAM
COVENTRY
RUGBY
WARWICK
Scale
Kilometres
0 2 4 6 8 10 N-S
0 2 4 6 8 10 E-W
Miles
0 1 2 3 4 5 6 N-S
0 1 2 3 4 5 6 E-W
Key to Symbols
Frequencies
+ rare to occasional
× frequent to abundant

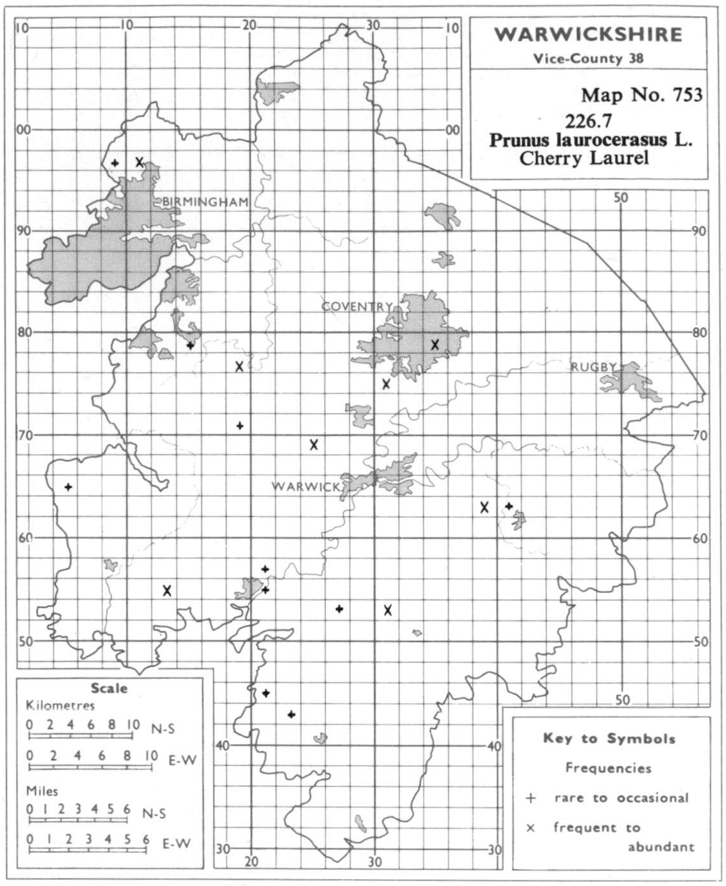
WARWICKSHIRE
Vice-County 38
Map No. 753
226.7
Prunus laurocerasus L.
Cherry Laurel
BIRMINGHAM
COVENTRY
RUGBY
WARWICK
Scale
Kilometres
0 2 4 6 8 10 N-S
0 2 4 6 8 10 E-W
Miles
0 1 2 3 4 5 6 N-S
0 1 2 3 4 5 6 E-W
Key to Symbols
Frequencies
+ rare to occasional
× frequent to abundant

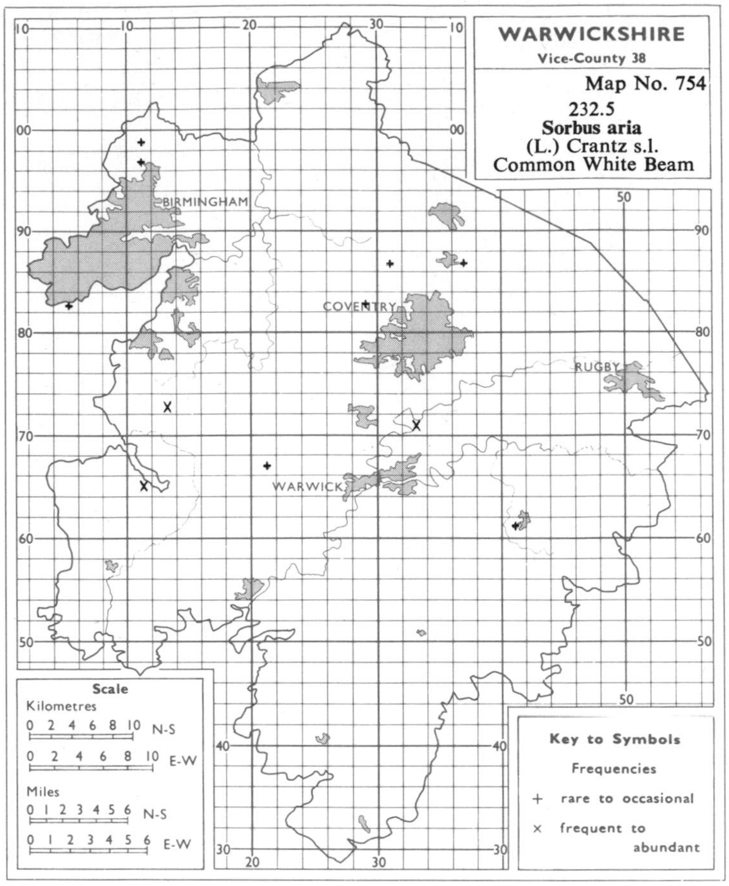
WARWICKSHIRE
Vice-County 38
Map No. 754
232.5
Sorbus aria
(L.) Crantz s.l.
Common White Beam
BIRMINGHAM
COVENTRY
RUGBY
WARWICK
Scale
Kilometres
0 2 4 6 8 10 N-S
0 2 4 6 8 10 E-W
Miles
0 1 2 3 4 5 6 N-S
0 1 2 3 4 5 6 E-W
Key to Symbols
Frequencies
+ rare to occasional
× frequent to abundant

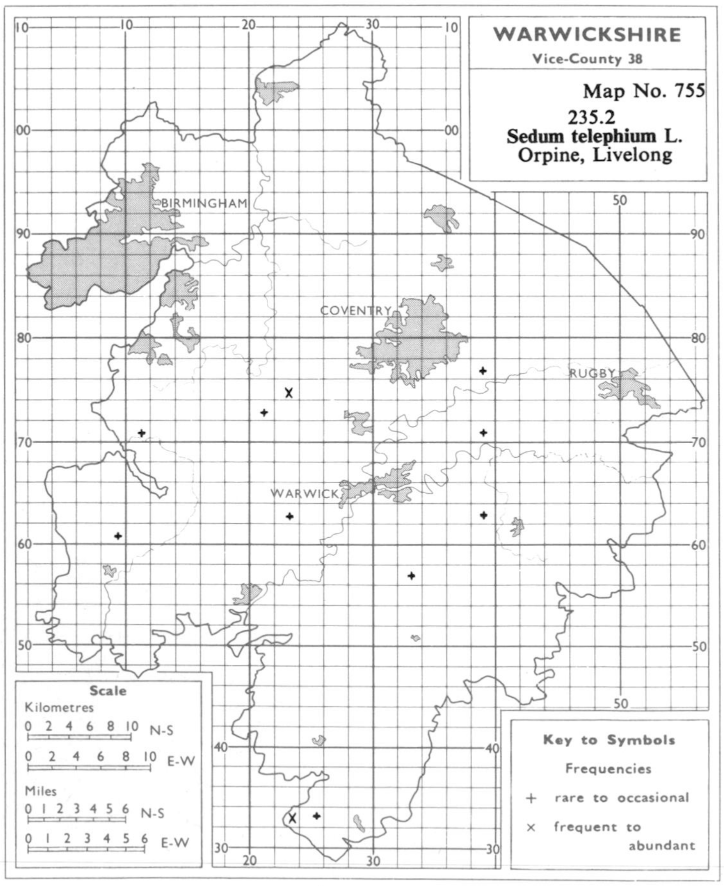
WARWICKSHIRE
Vice-County 38
Map No. 755
235.2
Sedum telephium L.
Orpine, Livelong
BIRMINGHAM
COVENTRY
RUGBY
WARWICK
Scale
Kilometres
0 2 4 6 8 10 N-S
0 2 4 6 8 10 E-W
Miles
0 1 2 3 4 5 6 N-S
0 1 2 3 4 5 6 E-W
Key to Symbols
Frequencies
+ rare to occasional
× frequent to abundant

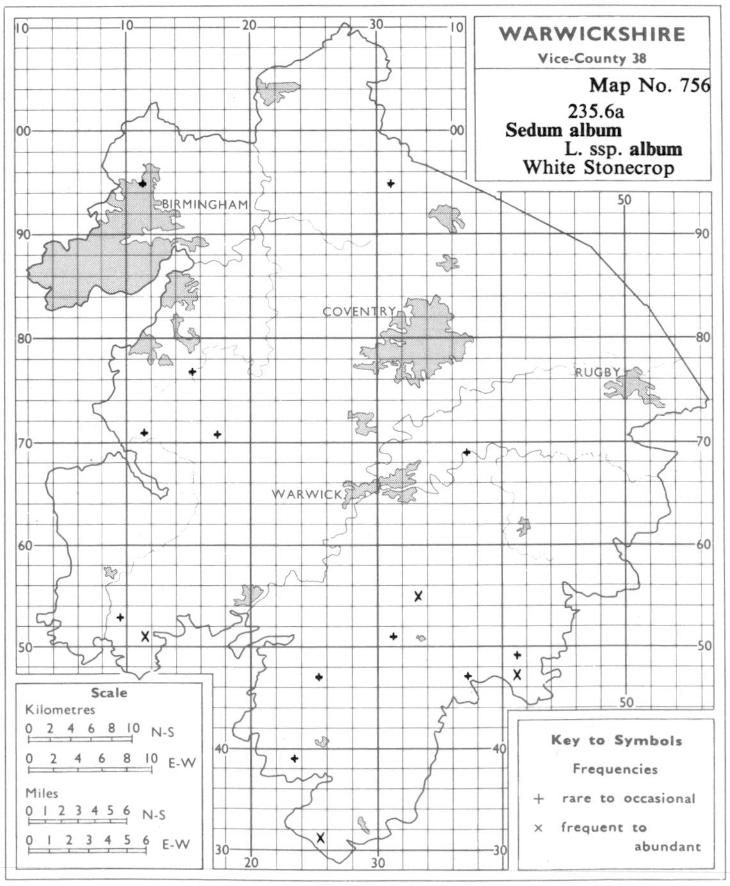
WARWICKSHIRE
Vice-County 38
Map No. 756
235.6a
Sedum album
L. ssp. album
White Stonecrop
BIRMINGHAM
COVENTRY
RUGBY
WARWICK
Scale
Kilometres
0 2 4 6 8 10 N-S
0 2 4 6 8 10 E-W
Miles
0 1 2 3 4 5 6 N-S
0 1 2 3 4 5 6 E-W
Key to Symbols
Frequencies
+ rare to occasional
× frequent to abundant

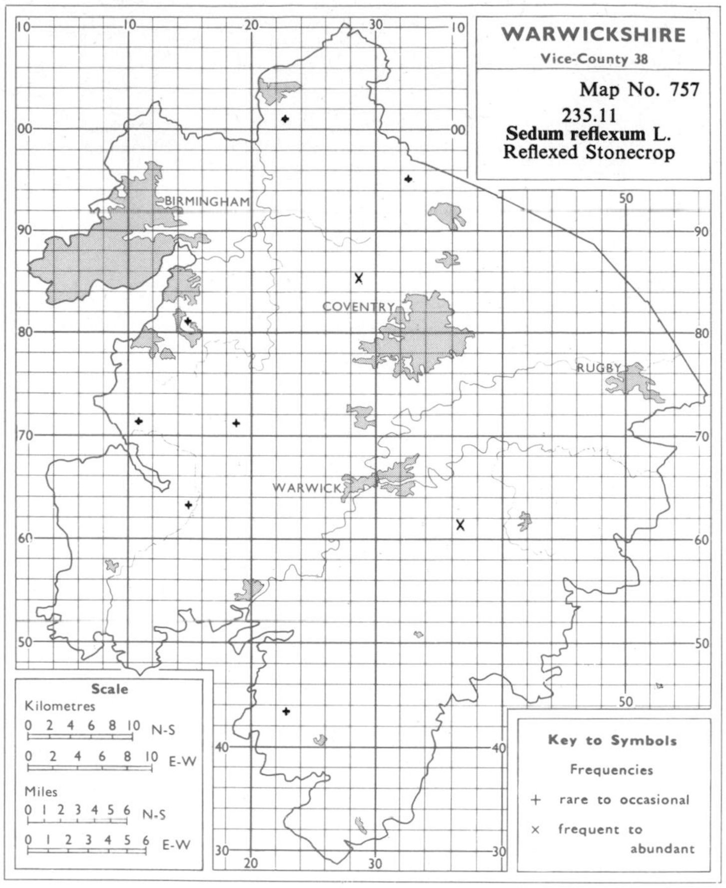
WARWICKSHIRE
Vice-County 38
Map No. 757
235.11
Sedum reflexum L.
Reflexed Stonecrop
BIRMINGHAM
COVENTRY
RUGBY
WARWICK
Scale
Kilometres
0 2 4 6 8 10 N-S
0 2 4 6 8 10 E-W
Miles
0 1 2 3 4 5 6 N-S
0 1 2 3 4 5 6 E-W
Key to Symbols
Frequencies
+ rare to occasional
× frequent to abundant

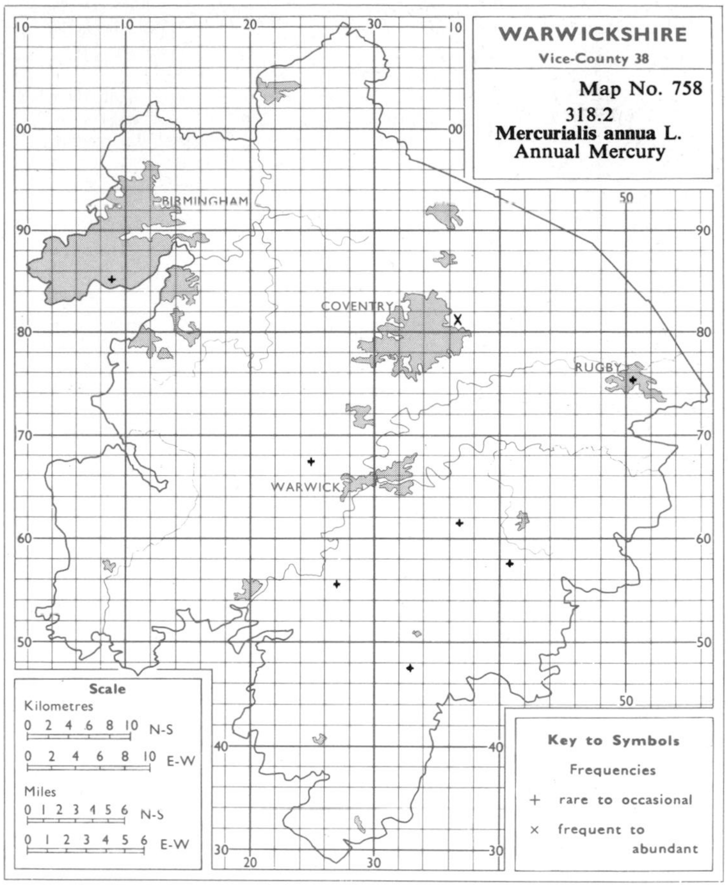
WARWICKSHIRE
Vice-County 38
Map No. 758
318.2
Mercurialis annua L.
Annual Mercury
BIRMINGHAM
COVENTRY
RUGBY
WARWICK
Scale
Kilometres
0 2 4 6 8 10 N-S
0 2 4 6 8 10 E-W
Miles
0 1 2 3 4 5 6 N-S
0 1 2 3 4 5 6 E-W
Key to Symbols
Frequencies
+ rare to occasional
× frequent to abundant

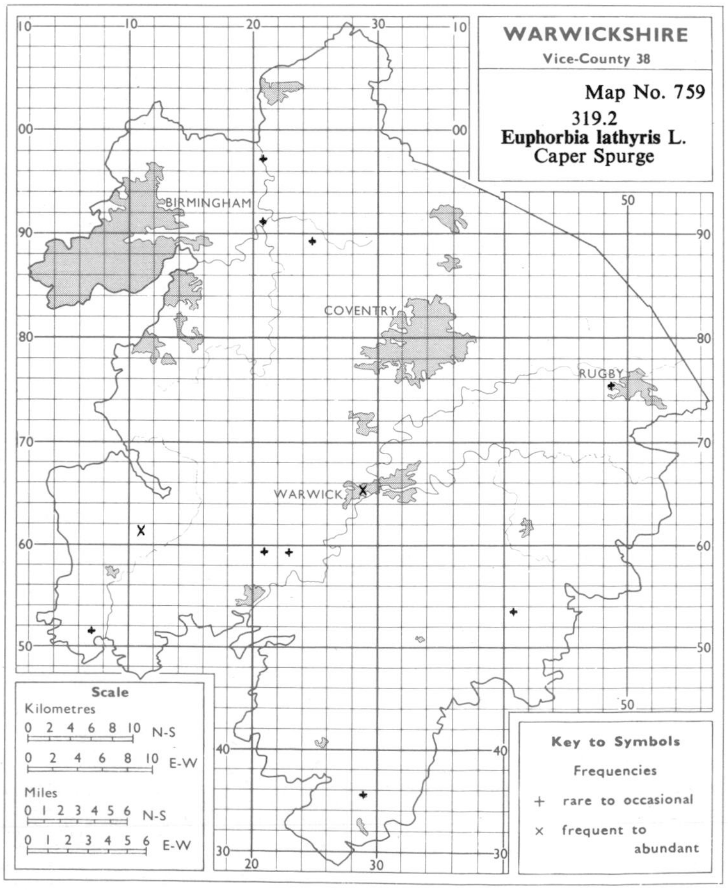
WARWICKSHIRE
Vice-County 38
Map No. 759
319.2
Euphorbia lathyris L.
Caper Spurge
BIRMINGHAM
COVENTRY
RUGBY
WARWICK
Scale
Kilometres
0 2 4 6 8 10 N-S
0 2 4 6 8 10 E-W
Miles
0 1 2 3 4 5 6 N-S
0 1 2 3 4 5 6 E-W
Key to Symbols
Frequencies
+ rare to occasional
× frequent to abundant

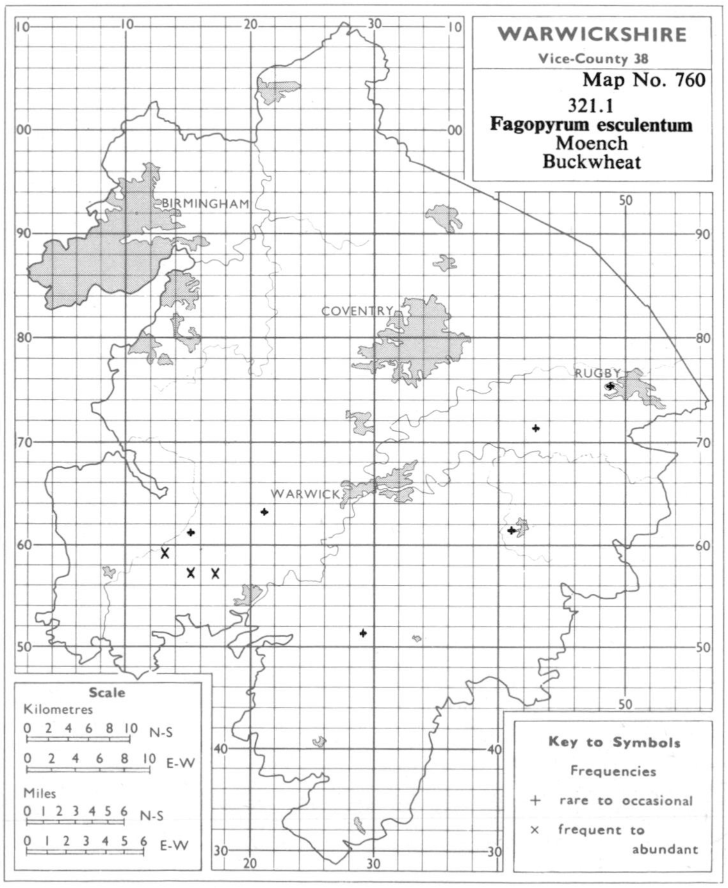
WARWICKSHIRE
Vice-County 38
Map No. 760
321.1
Fagopyrum esculentum Moench
Buckwheat
BIRMINGHAM
COVENTRY
RUGBY
WARWICK
Scale
Kilometres
0 2 4 6 8 10 N-S
0 2 4 6 8 10 E-W
Miles
0 1 2 3 4 5 6 N-S
0 1 2 3 4 5 6 E-W
Key to Symbols
Frequencies
+ rare to occasional
× frequent to abundant

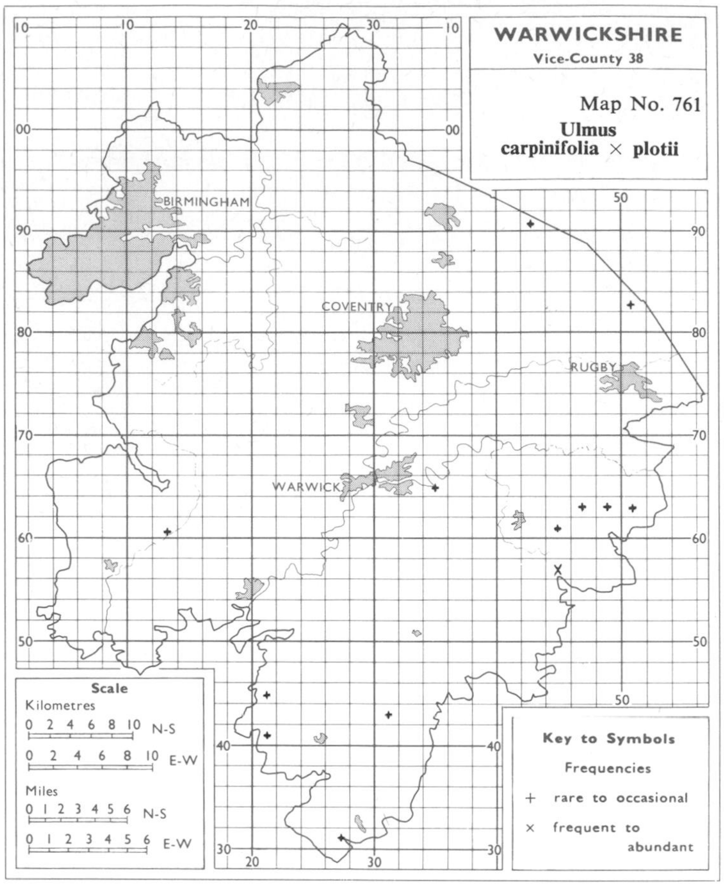
WARWICKSHIRE
Vice-County 38
Map No. 761
Ulmus carpinifolia × plotii
BIRMINGHAM
COVENTRY
RUGBY
WARWICK
Scale
Kilometres
0 2 4 6 8 10 N-S
0 2 4 6 8 10 E-W
Miles
0 1 2 3 4 5 6 N-S
0 1 2 3 4 5 6 E-W
Key to Symbols
Frequencies
+ rare to occasional
× frequent to abundant

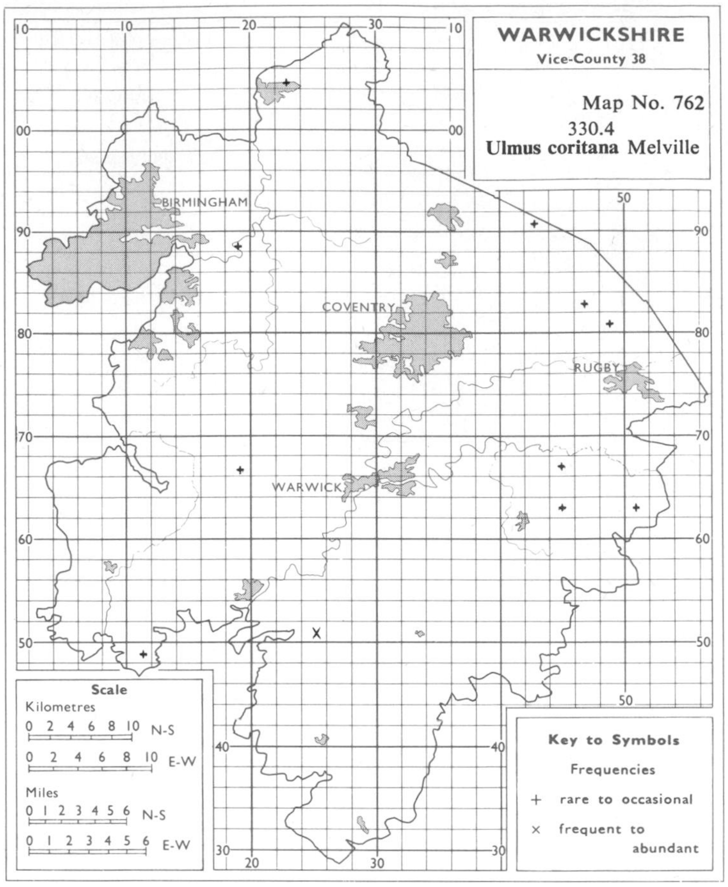
WARWICKSHIRE
Vice-County 38
Map No. 762
330.4
Ulmus coritana Melville
BIRMINGHAM
COVENTRY
RUGBY
WARWICK
Scale
Kilometres
0 2 4 6 8 10 N-S
0 2 4 6 8 10 E-W
Miles
0 1 2 3 4 5 6 N-S
0 1 2 3 4 5 6 E-W
Key to Symbols
Frequencies
+ rare to occasional
× frequent to abundant

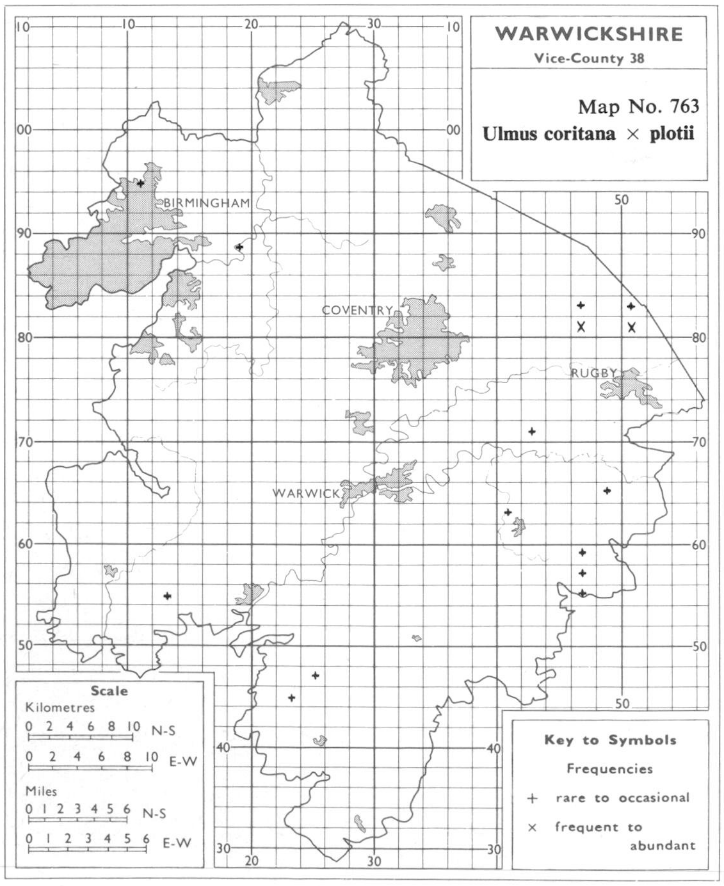
WARWICKSHIRE
Vice-County 38
Map No. 763
Ulmus coritana × plotii
BIRMINGHAM
COVENTRY
RUGBY
WARWICK
Scale
Kilometres
0 2 4 6 8 10 N-S
0 2 4 6 8 10 E-W
Miles
0 1 2 3 4 5 6 N-S
0 1 2 3 4 5 6 E-W
Key to Symbols
Frequencies
+ rare to occasional
× frequent to abundant

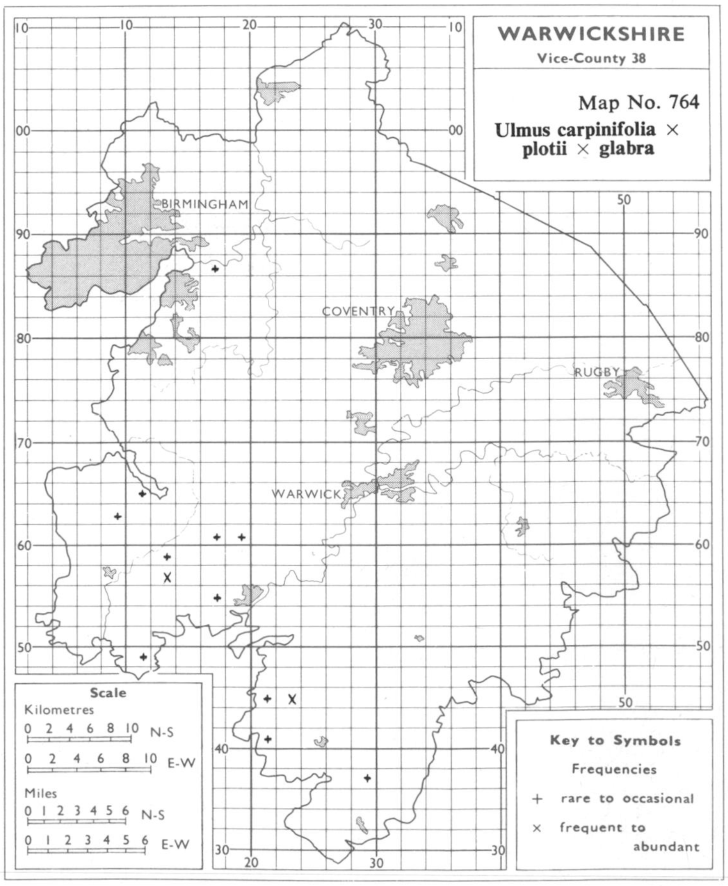
WARWICKSHIRE
Vice-County 38
Map No. 764
Ulmus carpinifolia × plotii × glabra
BIRMINGHAM
COVENTRY
RUGBY
WARWICK
Scale
Kilometres
0 2 4 6 8 10 N-S
0 2 4 6 8 10 E-W
Miles
0 1 2 3 4 5 6 N-S
0 1 2 3 4 5 6 E-W
Key to Symbols
Frequencies
+ rare to occasional
× frequent to abundant

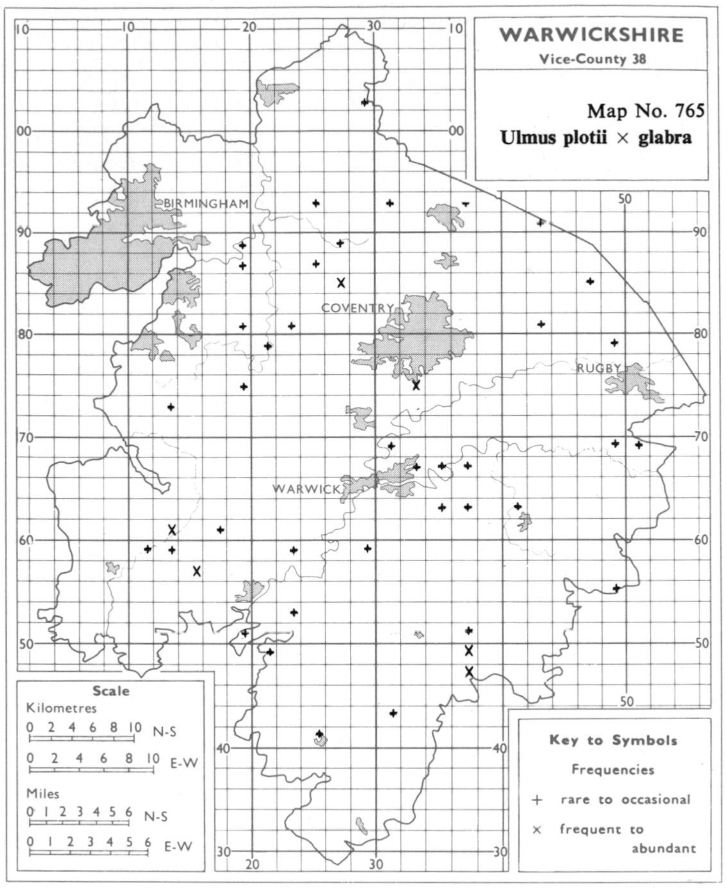
WARWICKSHIRE
Vice-County 38
Map No. 765
Ulmus plotii × glabra
BIRMINGHAM
COVENTRY
RUGBY
WARWICK
Scale
Kilometres
0 2 4 6 8 10 N-S
0 2 4 6 8 10 E-W
Miles
0 1 2 3 4 5 6 N-S
0 1 2 3 4 5 6 E-W
Key to Symbols
Frequencies
+ rare to occasional
× frequent to abundant

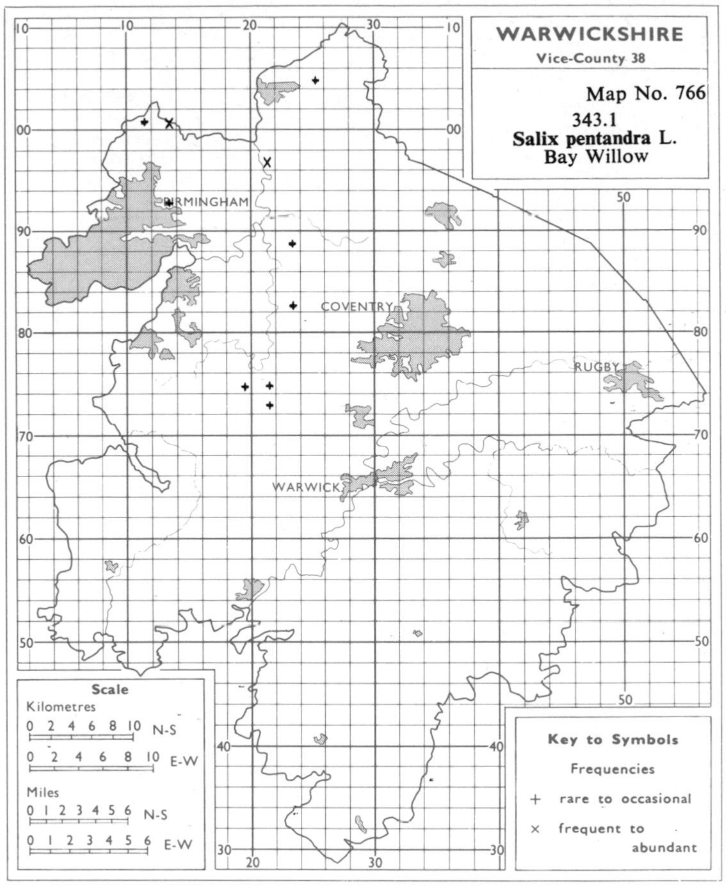
WARWICKSHIRE
Vice-County 38
Map No. 766
343.1
Salix pentandra L.
Bay Willow
BIRMINGHAM
COVENTRY
RUGBY
WARWICK
Scale
Kilometres
0 2 4 6 8 10 N-S
0 2 4 6 8 10 E-W
Miles
0 1 2 3 4 5 6 N-S
0 1 2 3 4 5 6 E-W
Key to Symbols
Frequencies
+ rare to occasional
× frequent to abundant

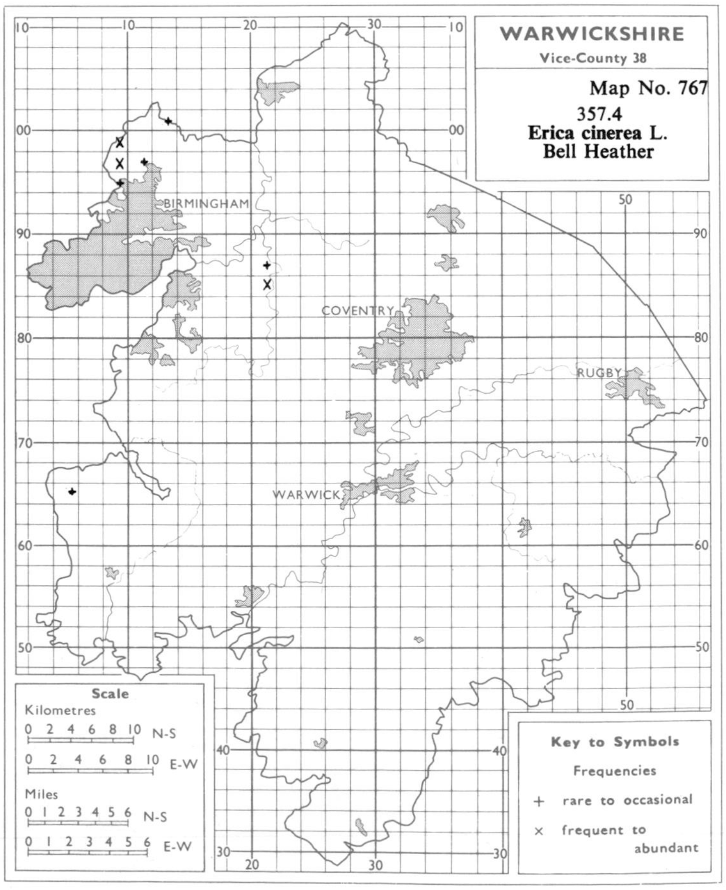
WARWICKSHIRE
Vice-County 38
Map No. 767
357.4
Erica cinerea L.
Bell Heather
BIRMINGHAM
COVENTRY
RUGBY
WARWICK
Scale
Kilometres
0 2 4 6 8 10 N-S
0 2 4 6 8 10 E-W
Miles
0 1 2 3 4 5 6 N-S
0 1 2 3 4 5 6 E-W
Key to Symbols
Frequencies
+ rare to occasional
× frequent to abundant

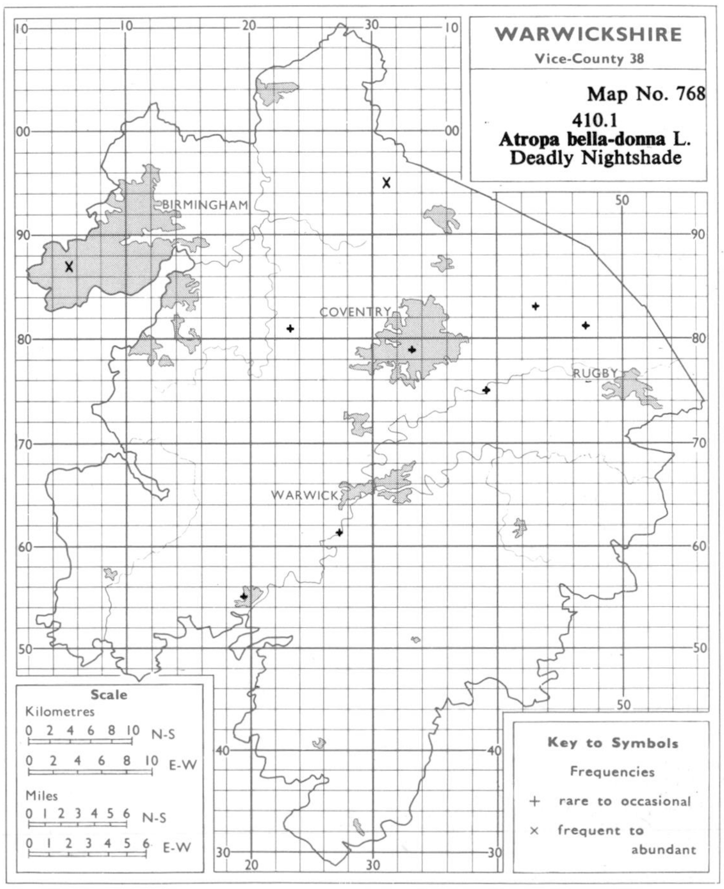
WARWICKSHIRE
Vice-County 38
Map No. 768
410.1
Atropa bella-donna L.
Deadly Nightshade
BIRMINGHAM
COVENTRY
RUGBY
WARWICK
Scale
Kilometres
0 2 4 6 8 10 N-S
0 2 4 6 8 10 E-W
Miles
0 1 2 3 4 5 6 N-S
0 1 2 3 4 5 6 E-W
Key to Symbols
Frequencies
+ rare to occasional
× frequent to abundant

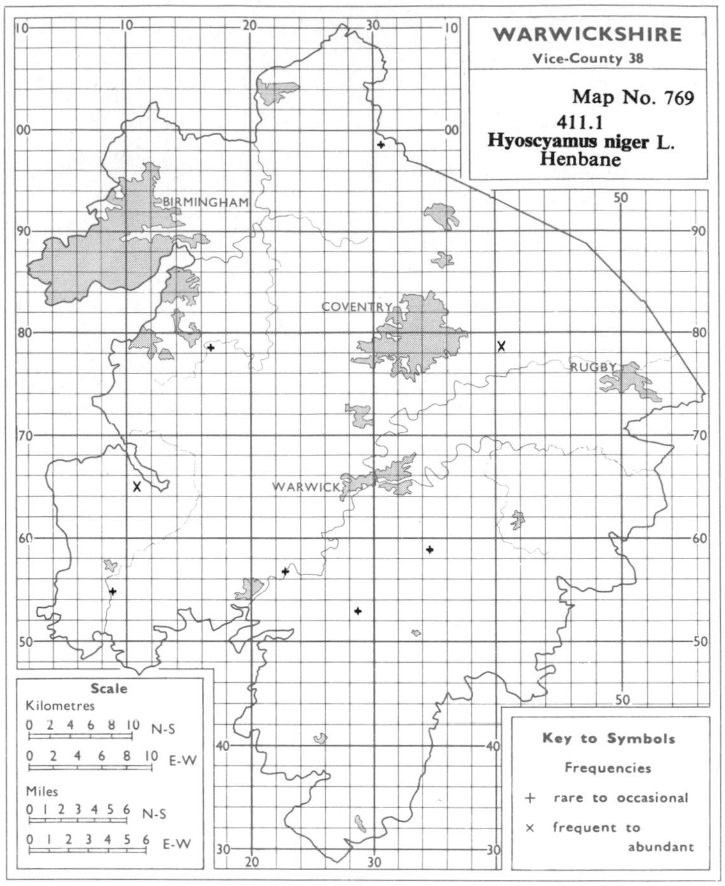
WARWICKSHIRE
Vice-County 38
Map No. 769
411.1
Hyoscyamus niger L.
Henbane
BIRMINGHAM
COVENTRY
RUGBY
WARWICK
Scale
Kilometres
0 2 4 6 8 10 N-S
0 2 4 6 8 10 E-W
Miles
0 1 2 3 4 5 6 N-S
0 1 2 3 4 5 6 E-W
Key to Symbols
Frequencies
+ rare to occasional
× frequent to abundant

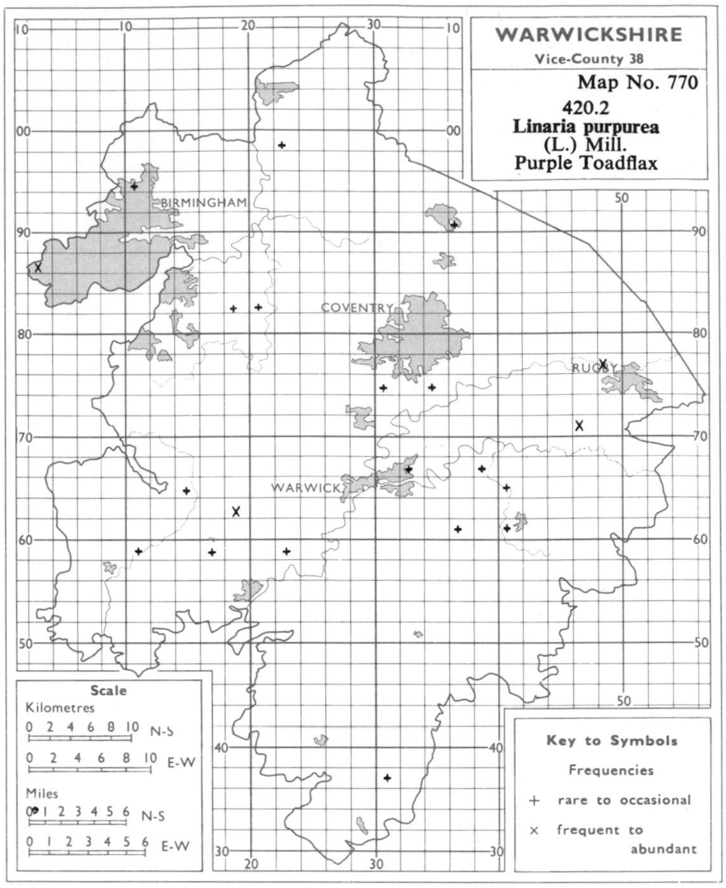
WARWICKSHIRE
Vice-County 38
Map No. 770
420.2
Linaria purpurea (L.) Mill.
Purple Toadflax
BIRMINGHAM
COVENTRY
RUGBY
WARWICK
Scale
Kilometres
0 2 4 6 8 10 N-S
0 2 4 6 8 10 E-W
Miles
0 1 2 3 4 5 6 N-S
0 1 2 3 4 5 6 E-W
Key to Symbols
Frequencies
+ rare to occasional
× frequent to abundant

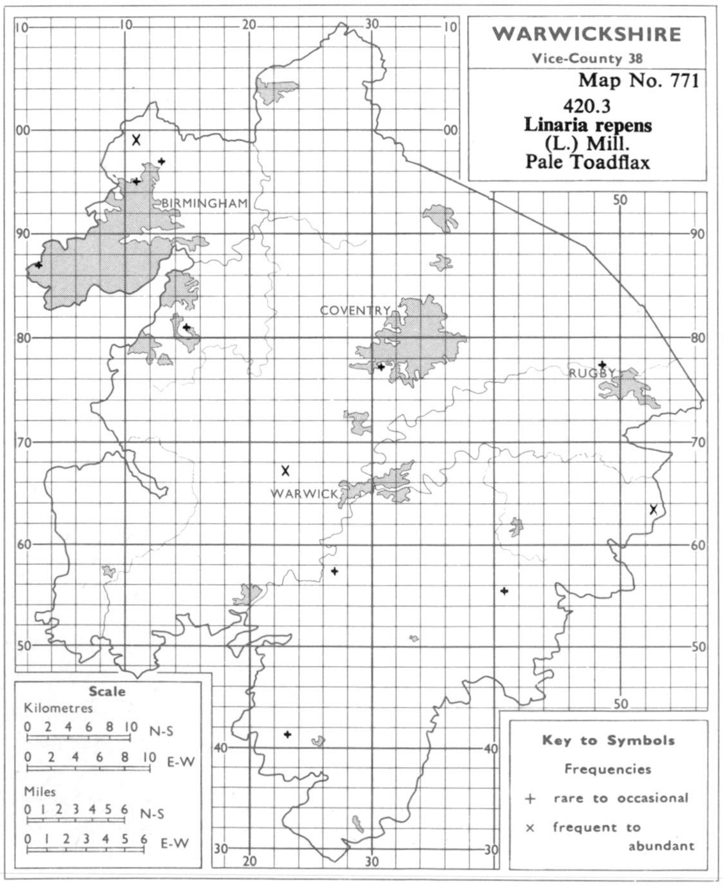
WARWICKSHIRE
Vice-County 38
Map No. 771
420.3
Linaria repens (L.) Mill.
Pale Toadflax
BIRMINGHAM
COVENTRY
RUGBY
WARWICK
Scale
Kilometres
0 2 4 6 8 10 N-S
0 2 4 6 8 10 E-W
Miles
0 1 2 3 4 5 6 N-S
0 1 2 3 4 5 6 E-W
Key to Symbols
Frequencies
+ rare to occasional
× frequent to abundant

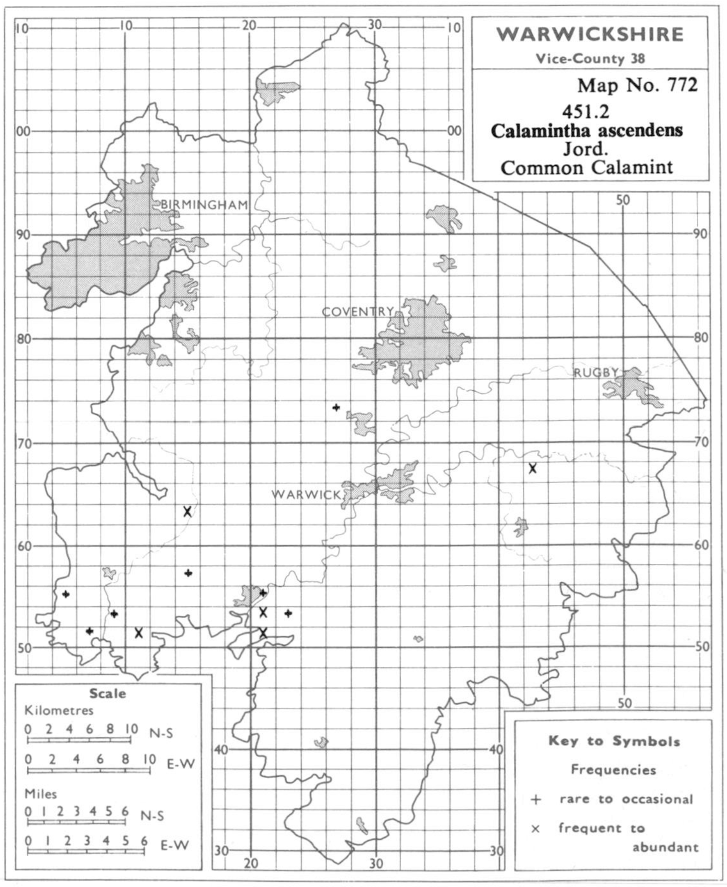
WARWICKSHIRE
Vice-County 38
Map No. 772
451.2
Calamintha ascendens Jord.
Common Calamint
BIRMINGHAM
COVENTRY
RUGBY
WARWICK
Scale
Kilometres
0 2 4 6 8 10 N-S
0 2 4 6 8 10 E-W
Miles
0 1 2 3 4 5 6 N-S
0 1 2 3 4 5 6 E-W
Key to Symbols
Frequencies
+ rare to occasional
× frequent to abundant

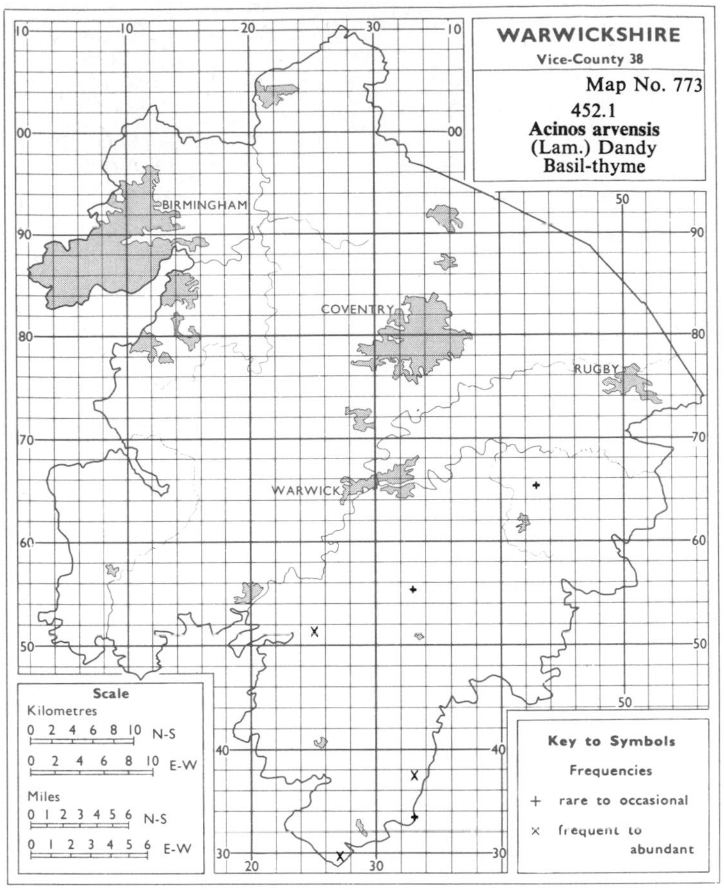
WARWICKSHIRE
Vice-County 38
Map No. 773
452.1
Acinos arvensis (Lam.) Dandy
Basil-thyme
BIRMINGHAM
COVENTRY
RUGBY
WARWICK
Scale
Kilometres
0 2 4 6 8 10 N-S
0 2 4 6 8 10 E-W
Miles
0 1 2 3 4 5 6 N-S
0 1 2 3 4 5 6 E-W
Key to Symbols
Frequencies
+ rare to occasional
× frequent to abundant

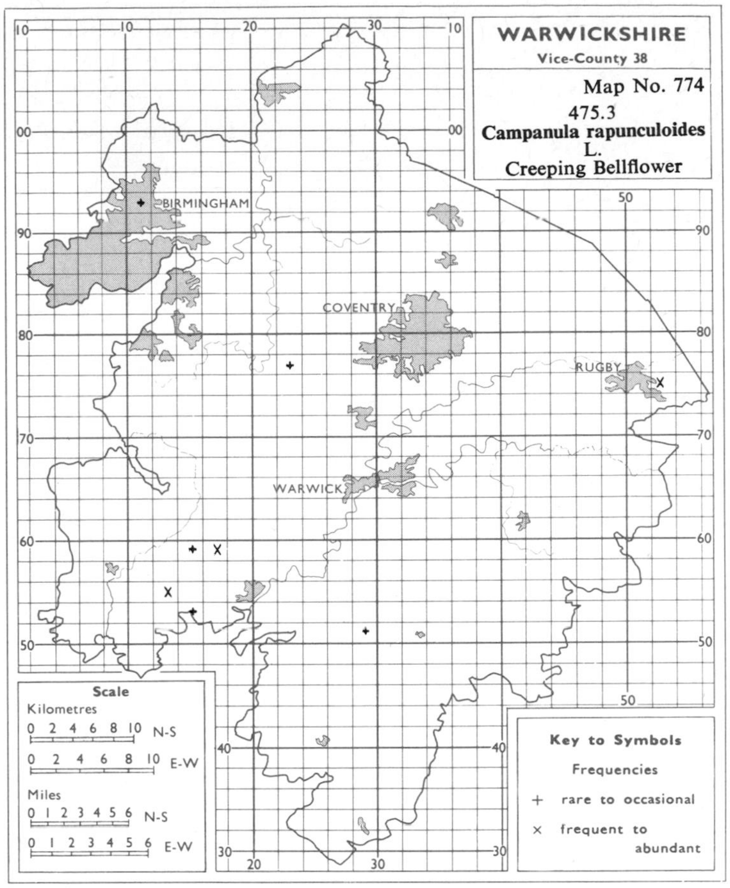
WARWICKSHIRE
Vice-County 38
Map No. 774
475.3
Campanula rapunculoides L.
Creeping Bellflower
BIRMINGHAM
COVENTRY
RUGBY
WARWICK
Scale
Kilometres
0 2 4 6 8 10 N-S
0 2 4 6 8 10 E-W
Miles
0 1 2 3 4 5 6 N-S
0 1 2 3 4 5 6 E-W
Key to Symbols
Frequencies
+ rare to occasional
× frequent to abundant

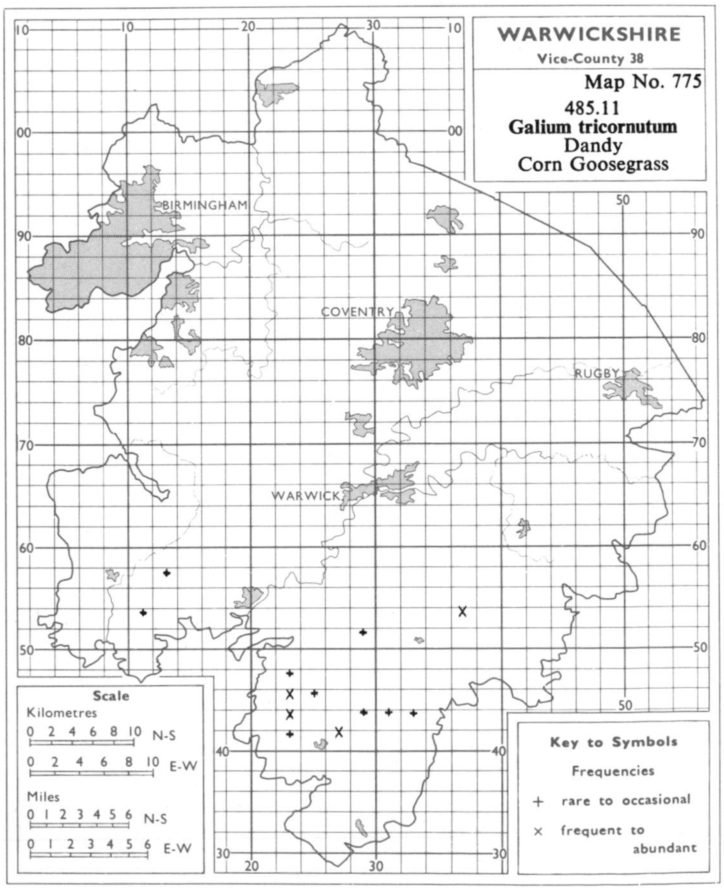
WARWICKSHIRE
Vice-County 38
Map No. 775
485.11
Galium tricornutum Dandy
Corn Goosegrass
BIRMINGHAM
COVENTRY
RUGBY
WARWICK
Scale
Kilometres
0 2 4 6 8 10 N-S
0 2 4 6 8 10 E-W
Miles
0 1 2 3 4 5 6 N-S
0 1 2 3 4 5 6 E-W
Key to Symbols
Frequencies
+ rare to occasional
× frequent to abundant

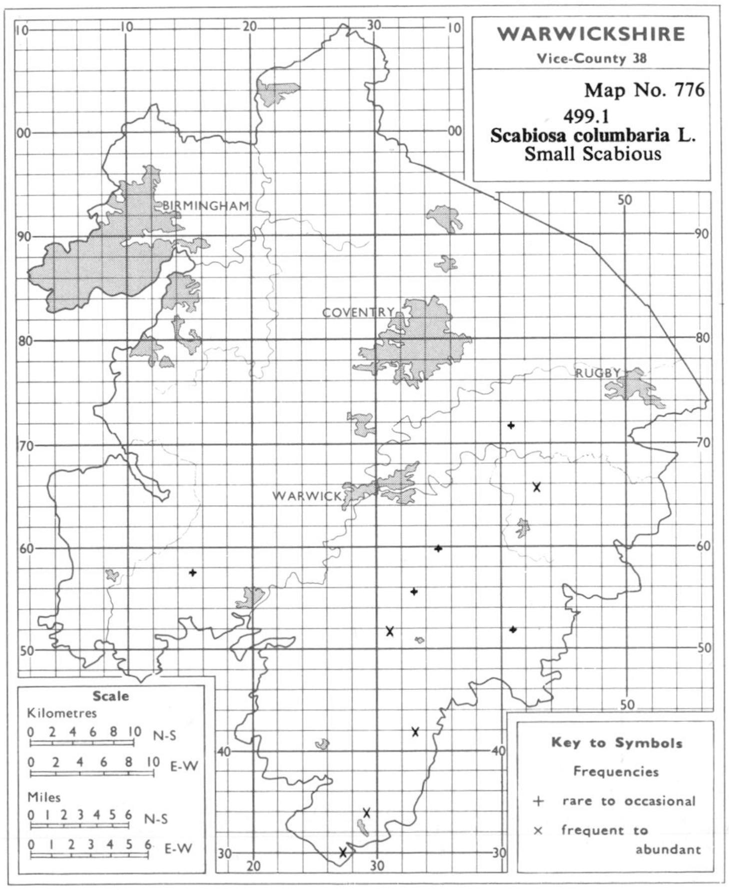
WARWICKSHIRE
Vice-County 38
Map No. 776
499.1
Scabiosa columbaria L.
Small Scabious
BIRMINGHAM
COVENTRY
RUGBY
WARWICK
Scale
Kilometres
0 2 4 6 8 10 N-S
0 2 4 6 8 10 E-W
Miles
0 1 2 3 4 5 6 N-S
0 1 2 3 4 5 6 E-W
Key to Symbols
Frequencies
+ rare to occasional
× frequent to abundant

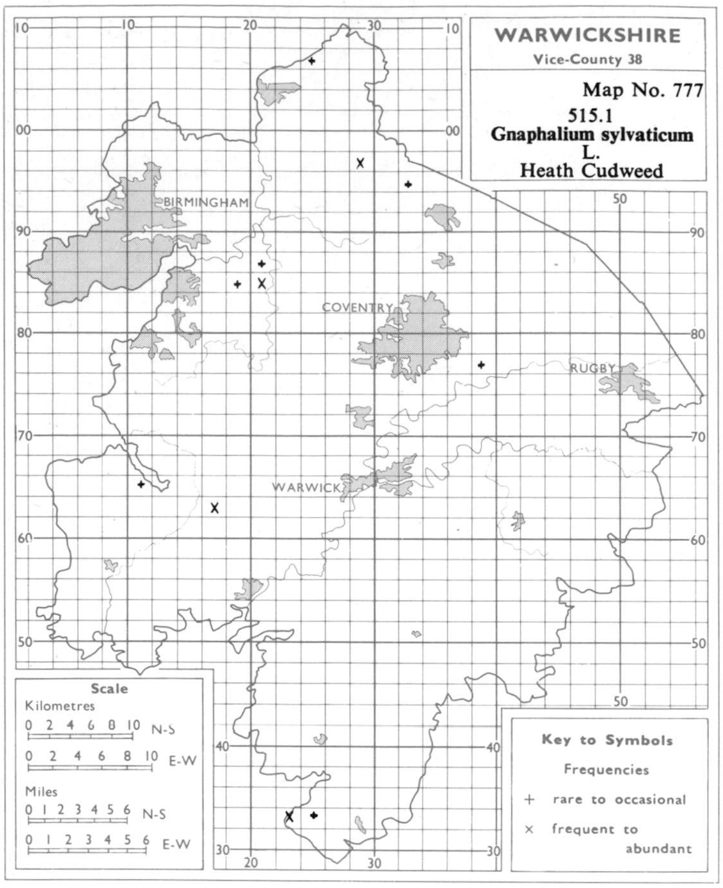

WARWICKSHIRE
Vice-County 38
Map No. 777
515.1
Gnaphalium sylvaticum L.
Heath Cudweed
BIRMINGHAM
COVENTRY
RUGBY
WARWICK
Scale
Kilometres
0 2 4 6 8 10 N-S
0 2 4 6 8 10 E-W
Miles
0 1 2 3 4 5 6 N-S
0 1 2 3 4 5 6 E-W
Key to Symbols
Frequencies
+ rare to occasional
× frequent to abundant

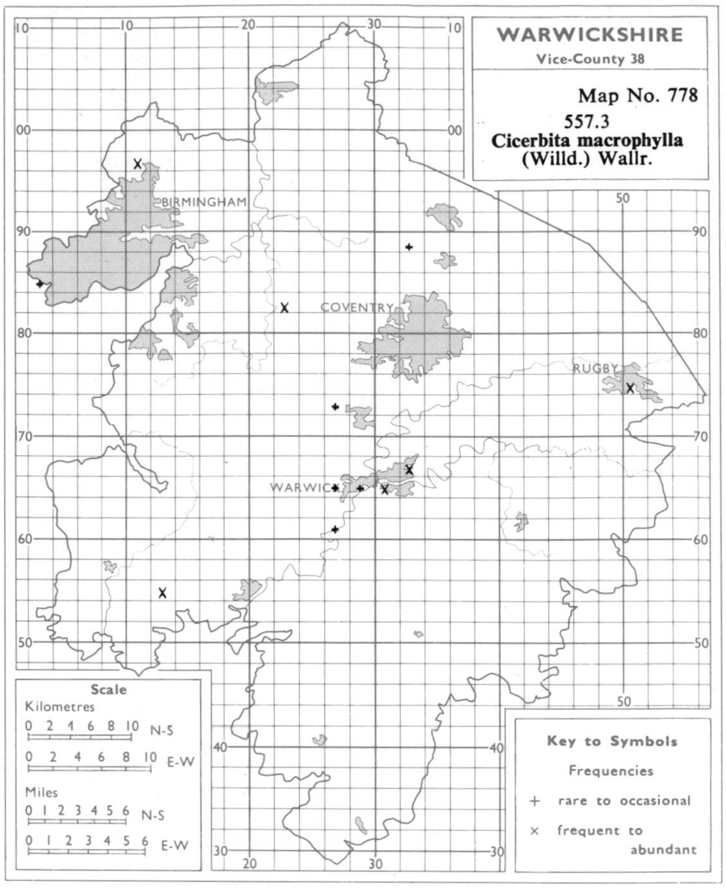

WARWICKSHIRE
Vice-County 38
Map No. 778
557.3
Cicerbita macrophylla (Willd.) Wallr.
BIRMINGHAM
COVENTRY
RUGBY
WARWICK
Scale
Kilometres
0 2 4 6 8 10 N-S
0 2 4 6 8 10 E-W
Miles
0 1 2 3 4 5 6 N-S
0 1 2 3 4 5 6 E-W
Key to Symbols
Frequencies
+ rare to occasional
× frequent to abundant

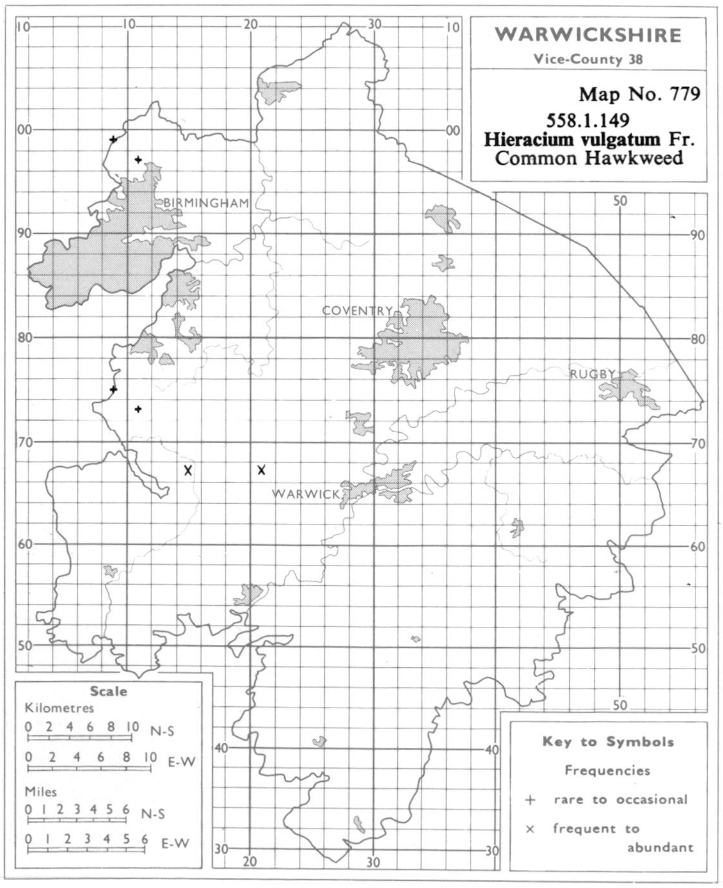

WARWICKSHIRE
Vice-County 38
Map No. 779
558.1.149
Hieracium vulgatum Fr.
Common Hawkweed
BIRMINGHAM
COVENTRY
RUGBY
WARWICK
Scale
Kilometres
0 2 4 6 8 10 N-S
0 2 4 6 8 10 E-W
Miles
0 1 2 3 4 5 6 N-S
0 1 2 3 4 5 6 E-W
Key to Symbols
Frequencies
+ rare to occasional
× frequent to abundant

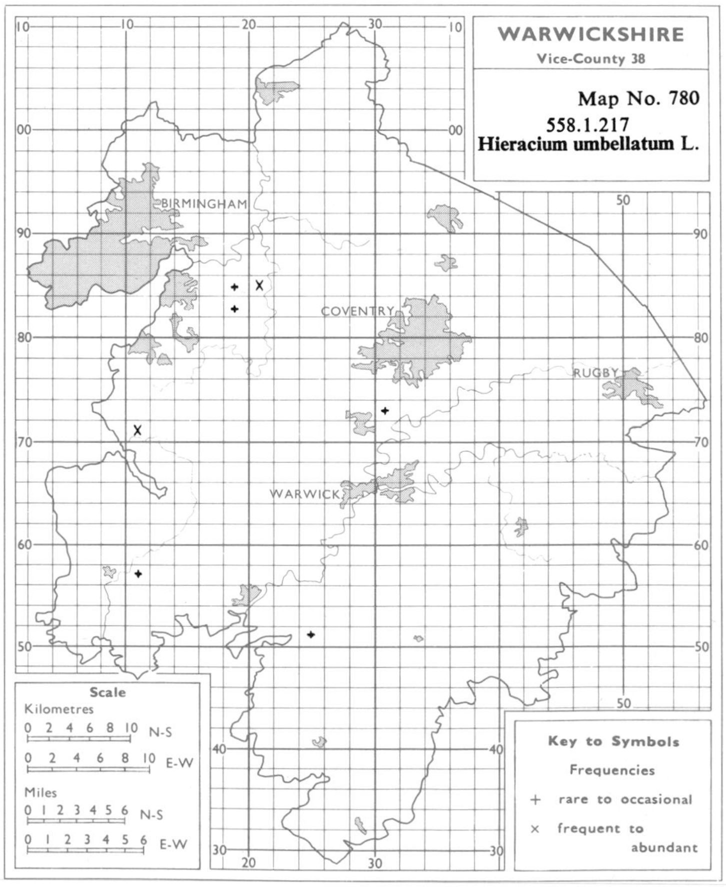

WARWICKSHIRE
Vice-County 38
Map No. 780
558.1.217
Hieracium umbellatum L.
BIRMINGHAM
COVENTRY
RUGBY
WARWICK
Scale
Kilometres
0 2 4 6 8 10 N-S
0 2 4 6 8 10 E-W
Miles
0 1 2 3 4 5 6 N-S
0 1 2 3 4 5 6 E-W
Key to Symbols
Frequencies
+ rare to occasional
× frequent to abundant

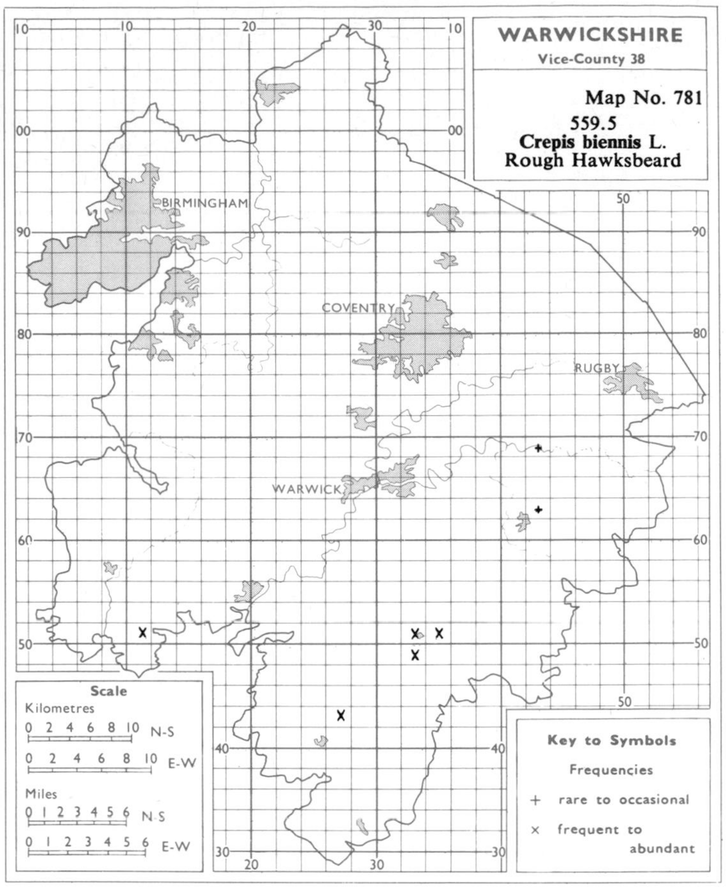
WARWICKSHIRE
Vice-County 38
Map No. 781
559.5
Crepis biennis L.
Rough Hawksbeard
BIRMINGHAM
COVENTRY
RUGBY
WARWICK
Scale
Kilometres
0 2 4 6 8 10 N-S
0 2 4 6 8 10 E-W
Miles
0 1 2 3 4 5 6 N-S
0 1 2 3 4 5 6 E-W
Key to Symbols
Frequencies
+ rare to occasional
× frequent to abundant

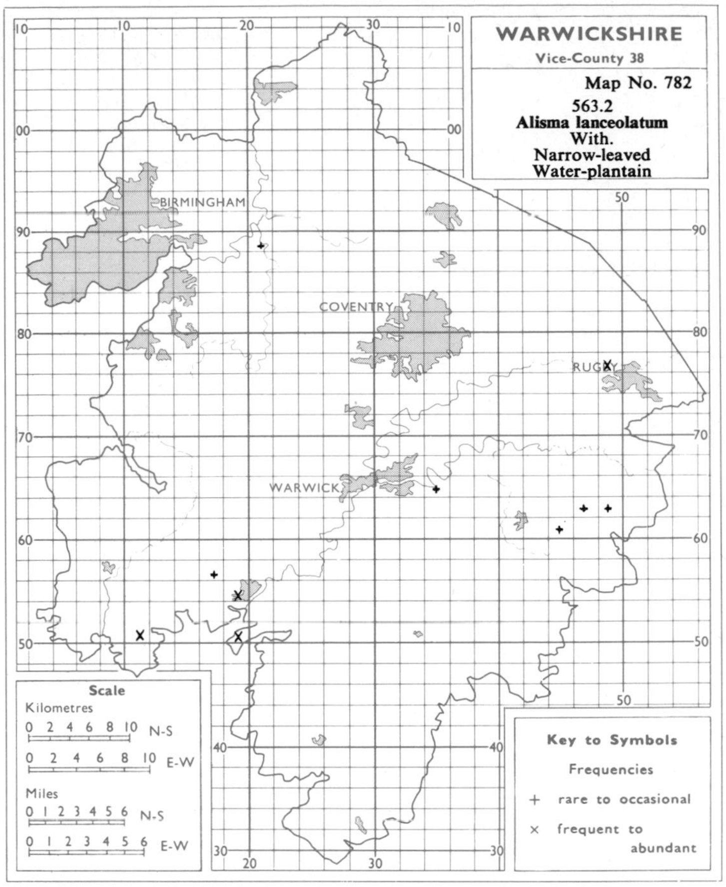
WARWICKSHIRE
Vice-County 38
Map No. 782
563.2
Alisma lanceolatum With.
Narrow-leaved Water-plantain
BIRMINGHAM
COVENTRY
RUGBY
WARWICK
Scale
Kilometres
0 2 4 6 8 10 N-S
0 2 4 6 8 10 E-W
Miles
0 1 2 3 4 5 6 N-S
0 1 2 3 4 5 6 E-W
Key to Symbols
Frequencies
+ rare to occasional
× frequent to abundant

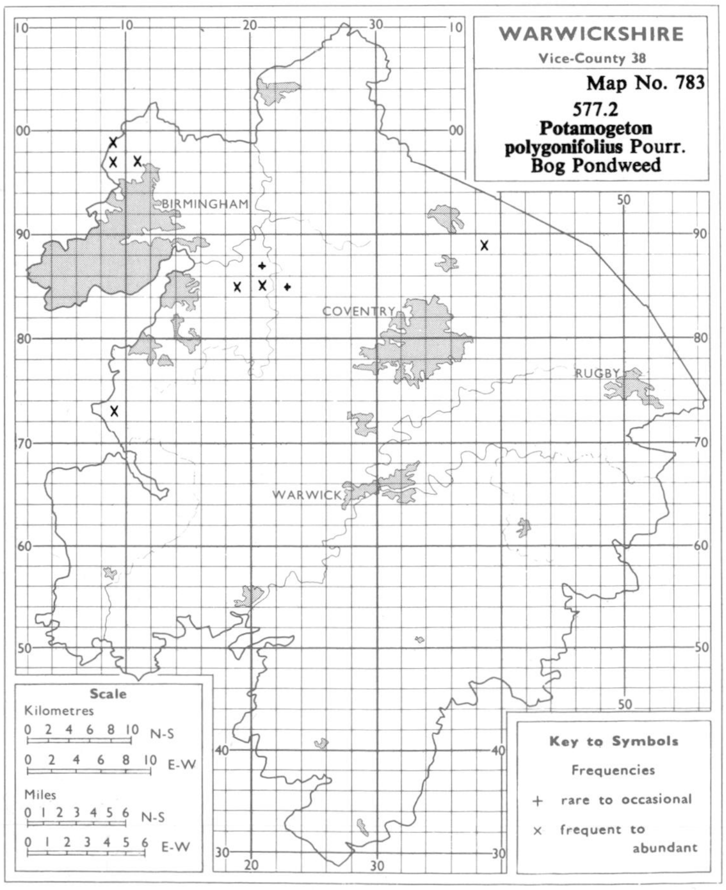
WARWICKSHIRE
Vice-County 38
Map No. 783
577.2
Potamogeton polygonifolius Pourr.
Bog Pondweed
BIRMINGHAM
COVENTRY
RUGBY
WARWICK
Scale
Kilometres
0 2 4 6 8 10 N-S
0 2 4 6 8 10 E-W
Miles
0 1 2 3 4 5 6 N-S
0 1 2 3 4 5 6 E-W
Key to Symbols
Frequencies
+ rare to occasional
× frequent to abundant

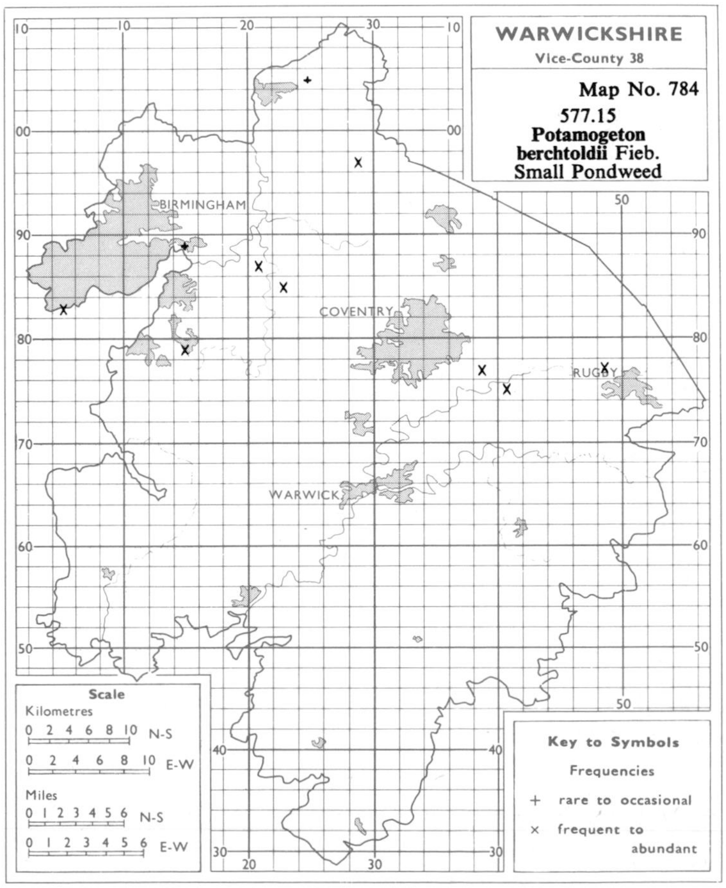
WARWICKSHIRE
Vice-County 38
Map No. 784
577.15
Potamogeton berchtoldii Fieb.
Small Pondweed
BIRMINGHAM
COVENTRY
RUGBY
WARWICK
Scale
Kilometres
0 2 4 6 8 10 N-S
0 2 4 6 8 10 E-W
Miles
0 1 2 3 4 5 6 N-S
0 1 2 3 4 5 6 E-W
Key to Symbols
Frequencies
+ rare to occasional
× frequent to abundant

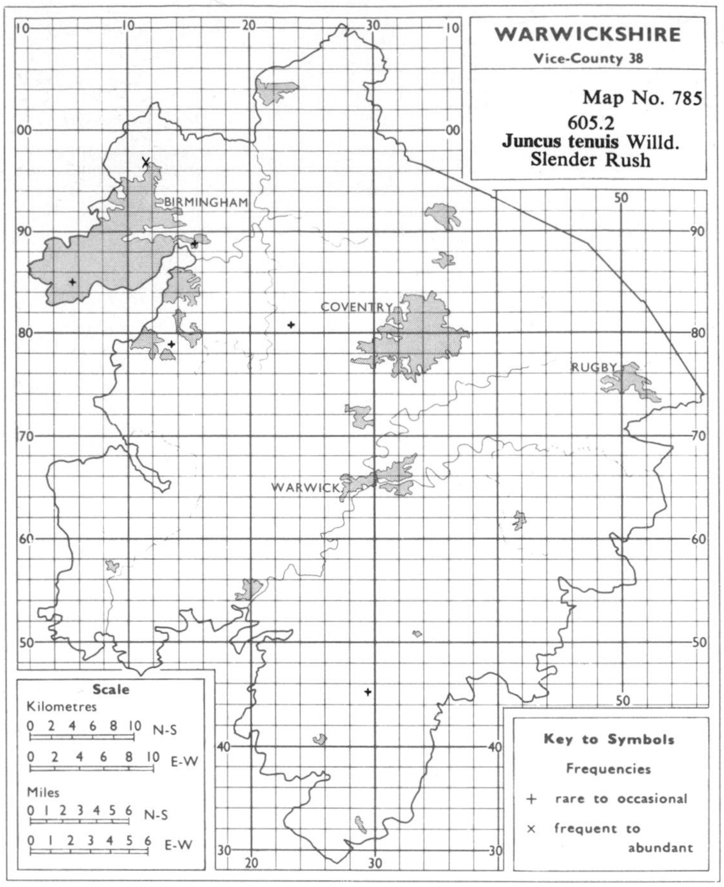
WARWICKSHIRE
Vice-County 38
Map No. 785
605.2
Juncus tenuis Willd.
Slender Rush
BIRMINGHAM
COVENTRY
RUGBY
WARWICK
Scale
Kilometres
0 2 4 6 8 10 N-S
0 2 4 6 8 10 E-W
Miles
0 1 2 3 4 5 6 N-S
0 1 2 3 4 5 6 E-W
Key to Symbols
Frequencies
+ rare to occasional
× frequent to abundant

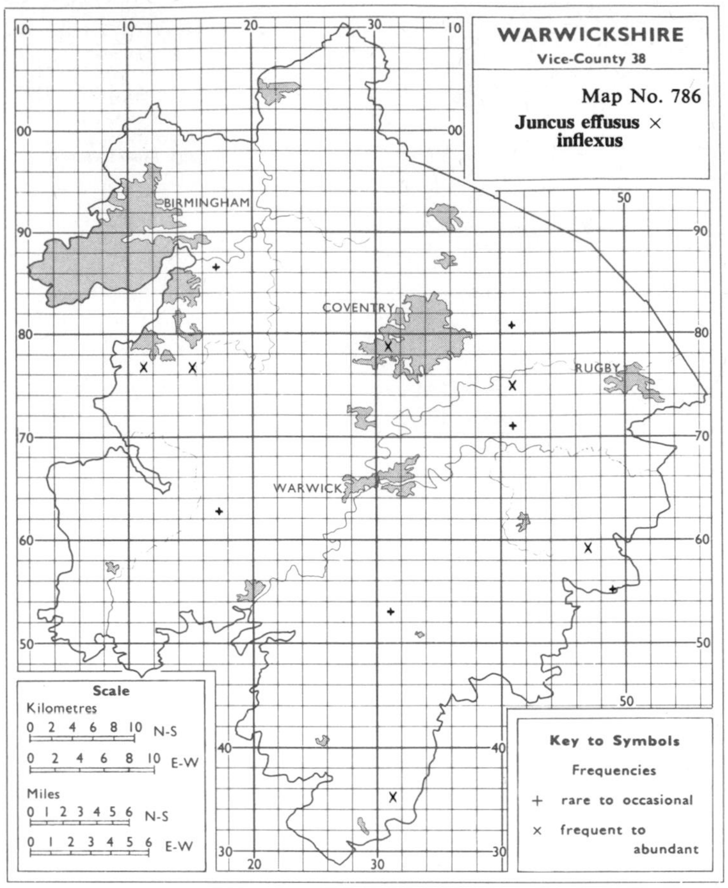
WARWICKSHIRE
Vice-County 38
Map No. 786
Juncus effusus × inflexus
BIRMINGHAM
COVENTRY
RUGBY
WARWICK
Scale
Kilometres
0 2 4 6 8 10 N-S
0 2 4 6 8 10 E-W
Miles
0 1 2 3 4 5 6 N-S
0 1 2 3 4 5 6 E-W
Key to Symbols
Frequencies
+ rare to occasional
× frequent to abundant

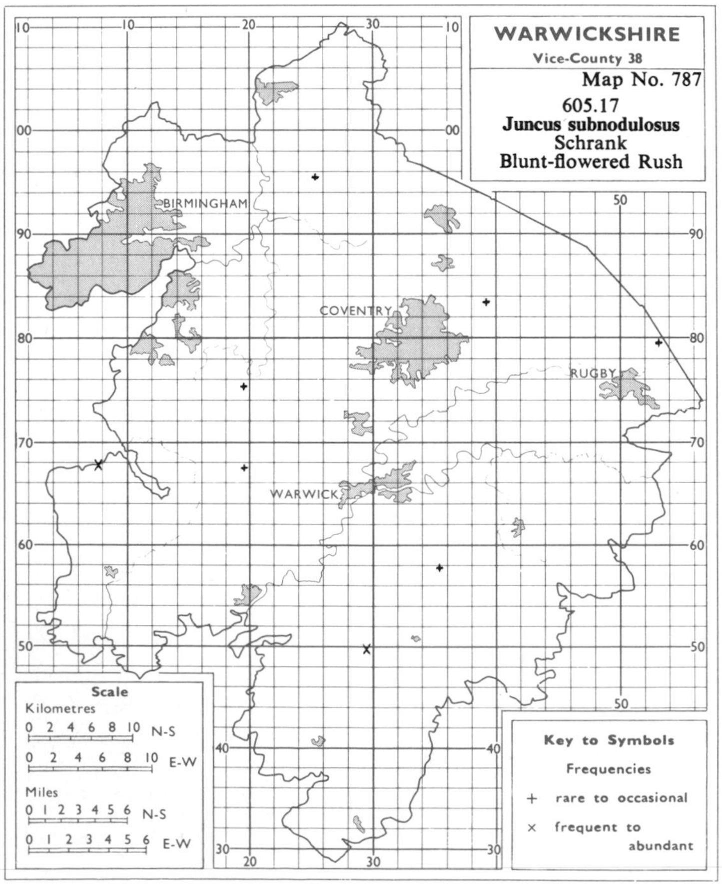
WARWICKSHIRE
Vice-County 38
Map No. 787
605.17
Juncus subnodulosus Schrank
Blunt-flowered Rush
BIRMINGHAM
COVENTRY
RUGBY
WARWICK
Scale
Kilometres
0 2 4 6 8 10 N-S
0 2 4 6 8 10 E-W
Miles
0 1 2 3 4 5 6 N-S
0 1 2 3 4 5 6 E-W
Key to Symbols
Frequencies
+ rare to occasional
× frequent to abundant

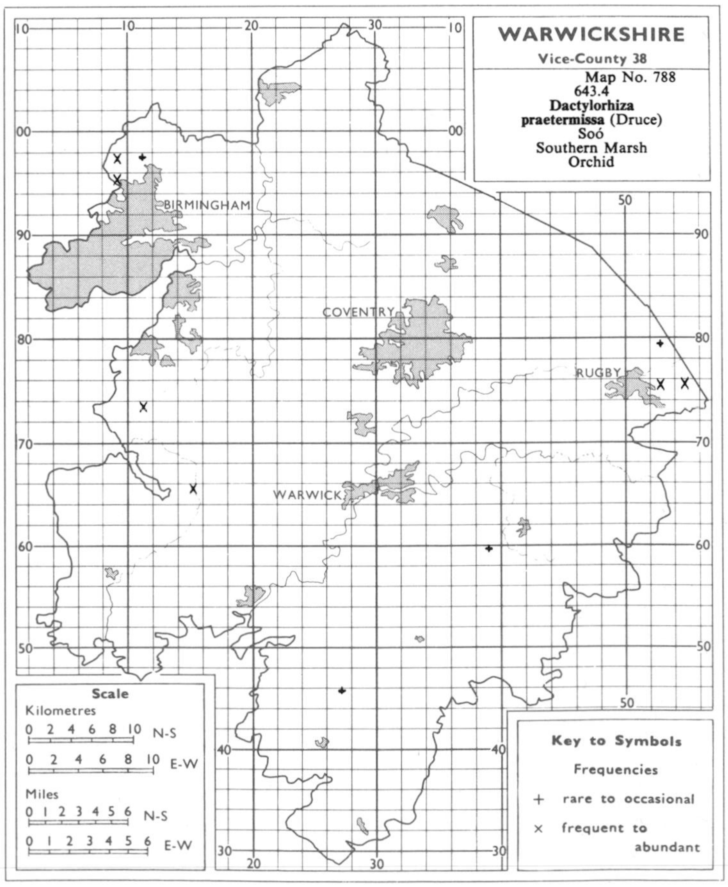
WARWICKSHIRE
Vice-County 38
Map No. 788
643.4
Dactylorhiza praetermissa (Druce) Soó
Southern Marsh Orchid
BIRMINGHAM
COVENTRY
RUGBY
WARWICK
Scale
Kilometres
0 2 4 6 8 10 N-S
0 2 4 6 8 10 E-W
Miles
0 1 2 3 4 5 6 N-S
0 1 2 3 4 5 6 E-W
Key to Symbols
Frequencies
+ rare to occasional
× frequent to abundant

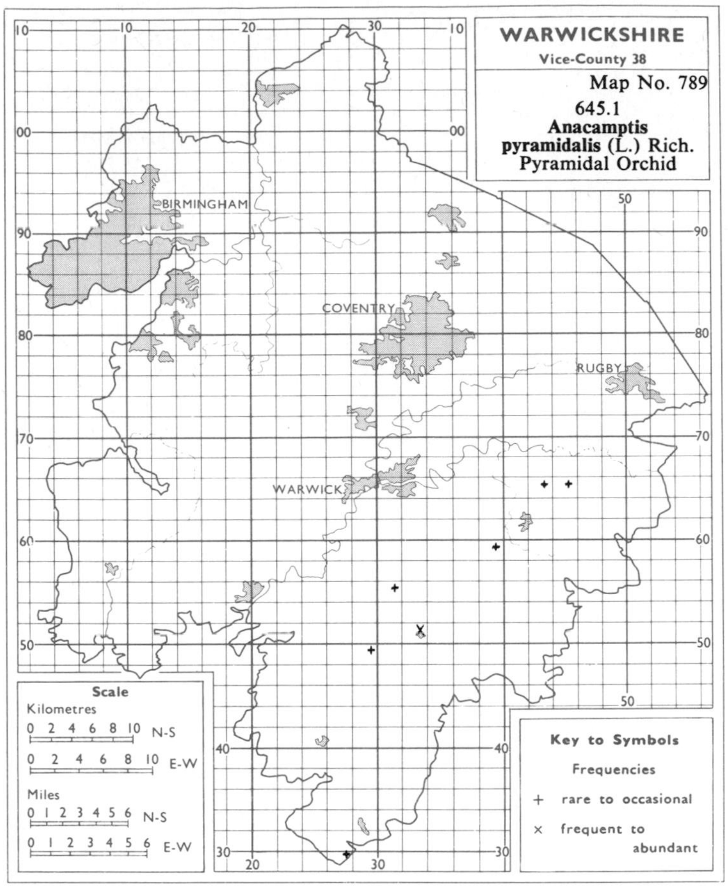
WARWICKSHIRE
Vice-County 38
Map No. 789
645.1
Anacamptis pyramidalis (L.) Rich.
Pyramidal Orchid
BIRMINGHAM
COVENTRY
RUGBY
WARWICK
Scale
Kilometres
0 2 4 6 8 10 N-S
0 2 4 6 8 10 E-W
Miles
0 1 2 3 4 5 6 N-S
0 1 2 3 4 5 6 E-W
Key to Symbols
Frequencies
+ rare to occasional
× frequent to abundant

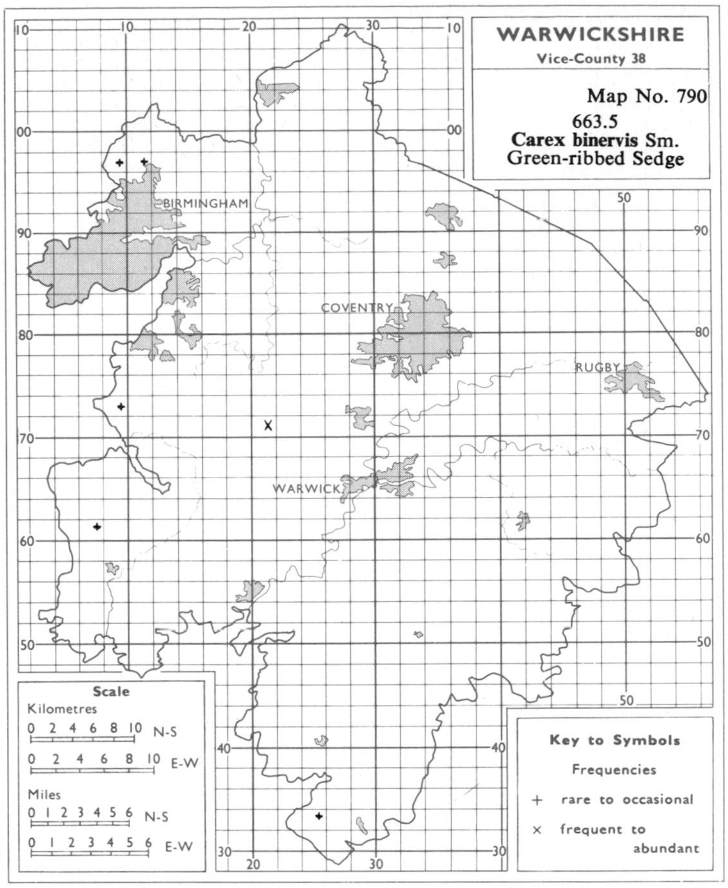
WARWICKSHIRE
Vice-County 38
Map No. 790
663.5
Carex binervis Sm.
Green-ribbed Sedge
BIRMINGHAM
COVENTRY
RUGBY
WARWICK
Scale
Kilometres
0 2 4 6 8 10 N-S
0 2 4 6 8 10 E-W
Miles
0 1 2 3 4 5 6 N-S
0 1 2 3 4 5 6 E-W
Key to Symbols
Frequencies
+ rare to occasional
× frequent to abundant

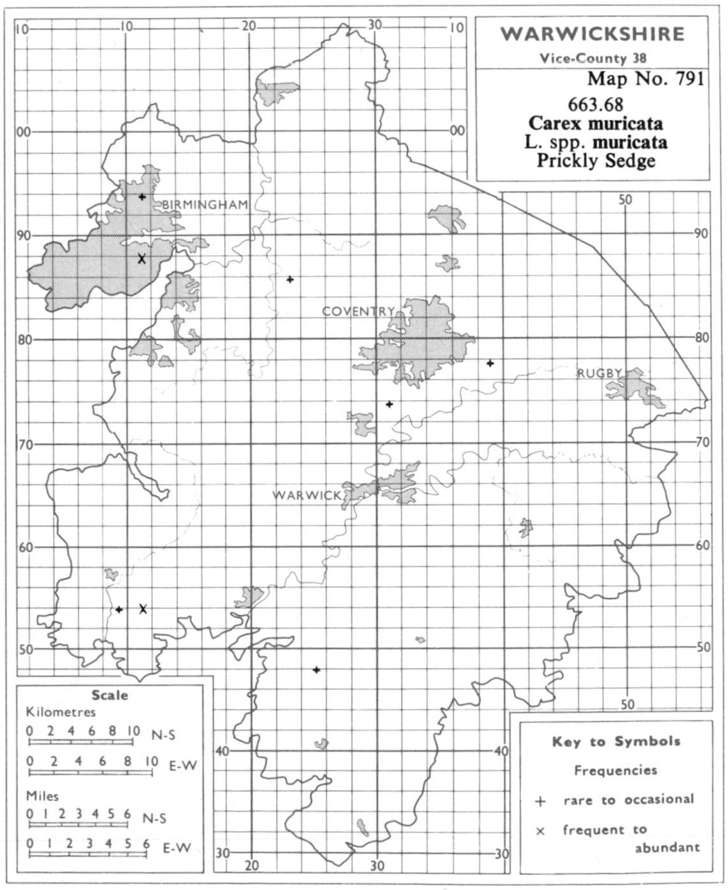
WARWICKSHIRE
Vice-County 38
Map No. 791
663.68
Carex muricata L. spp. **muricata**
Prickly Sedge
BIRMINGHAM
COVENTRY
RUGBY
WARWICK
Scale
Kilometres
0 2 4 6 8 10 N-S
0 2 4 6 8 10 E-W
Miles
0 1 2 3 4 5 6 N-S
0 1 2 3 4 5 6 E-W
Key to Symbols
Frequencies
+ rare to occasional
× frequent to abundant

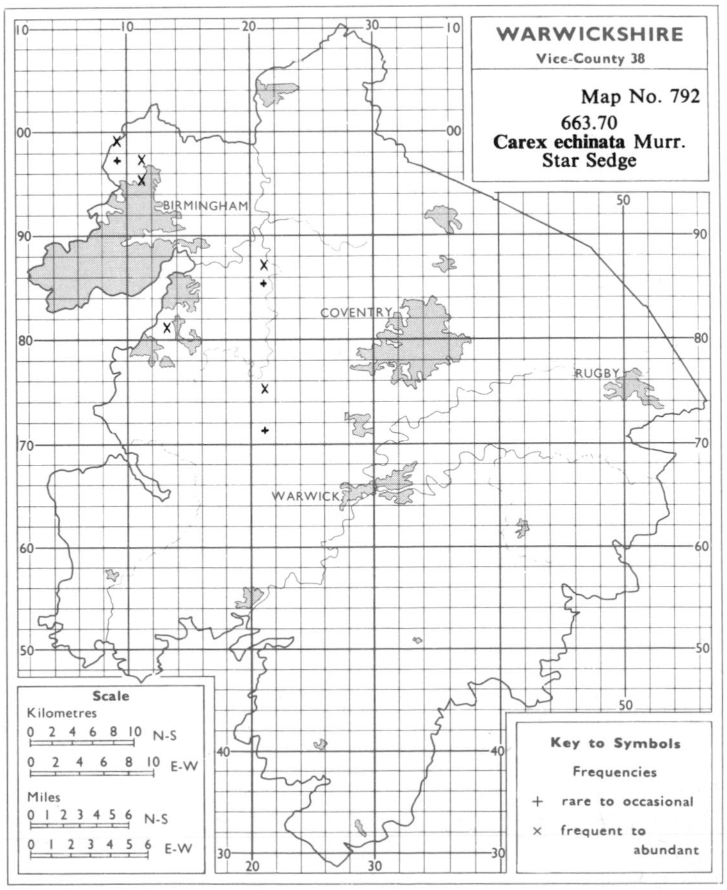
WARWICKSHIRE
Vice-County 38
Map No. 792
663.70
Carex echinata Murr.
Star Sedge
BIRMINGHAM
COVENTRY
RUGBY
WARWICK
Scale
Kilometres
0 2 4 6 8 10 N-S
0 2 4 6 8 10 E-W
Miles
0 1 2 3 4 5 6 N-S
0 1 2 3 4 5 6 E-W
Key to Symbols
Frequencies
+ rare to occasional
× frequent to abundant

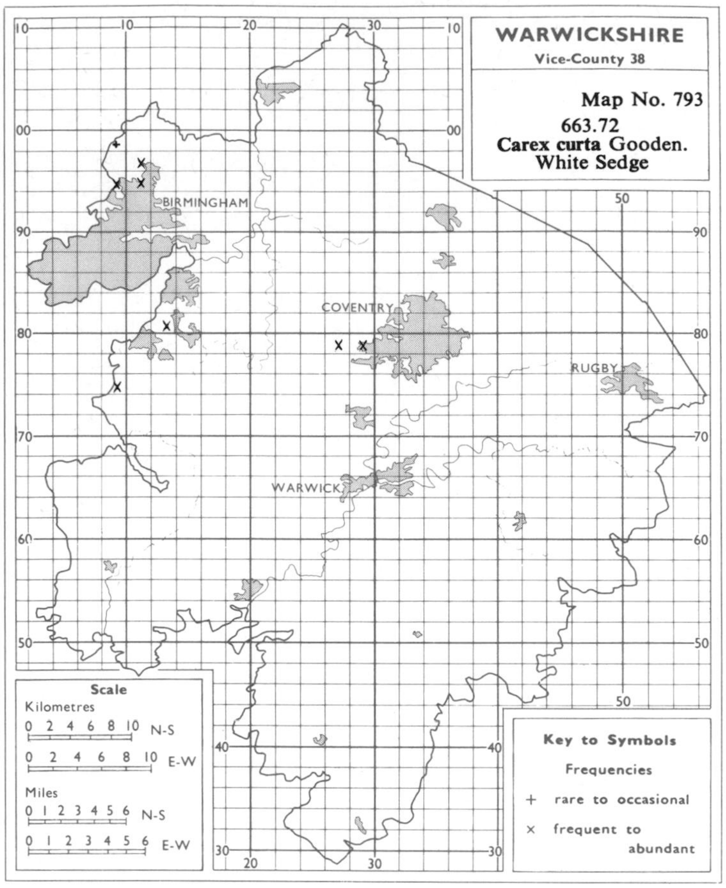
WARWICKSHIRE
Vice-County 38
Map No. 793
663.72
Carex curta Gooden.
White Sedge
BIRMINGHAM
COVENTRY
RUGBY
WARWICK
Scale
Kilometres
0 2 4 6 8 10 N-S
0 2 4 6 8 10 E-W
Miles
0 1 2 3 4 5 6 N-S
0 1 2 3 4 5 6 E-W
Key to Symbols
Frequencies
+ rare to occasional
× frequent to abundant

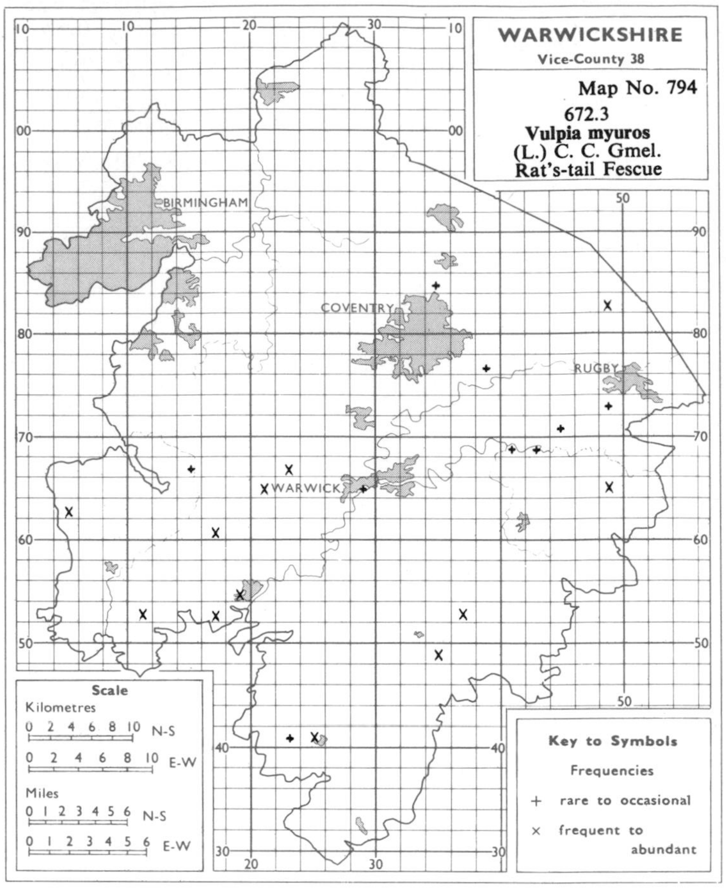
WARWICKSHIRE
Vice-County 38
Map No. 794
672.3
Vulpia myuros
(L.) C. C. Gmel.
Rat's-tail Fescue
BIRMINGHAM
COVENTRY
RUGBY
WARWICK
Scale
Kilometres
0 2 4 6 8 10 N-S
0 2 4 6 8 10 E-W
Miles
0 1 2 3 4 5 6 N-S
0 1 2 3 4 5 6 E-W
Key to Symbols
Frequencies
+ rare to occasional
× frequent to abundant

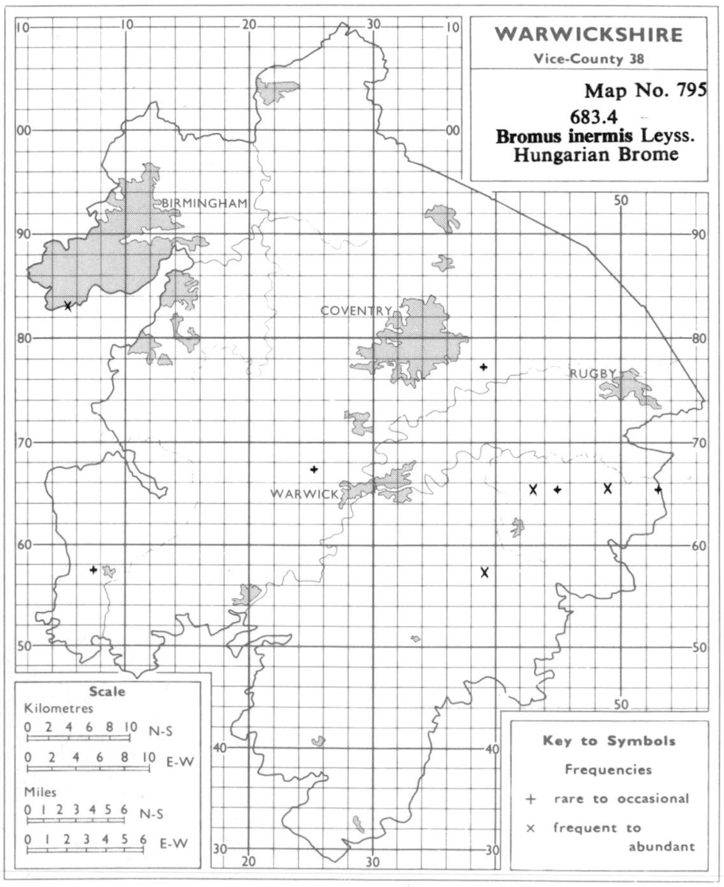
WARWICKSHIRE
Vice-County 38
Map No. 795
683.4
Bromus inermis Leyss.
Hungarian Brome
BIRMINGHAM
COVENTRY
RUGBY
WARWICK
Scale
Kilometres
0 2 4 6 8 10 N-S
0 2 4 6 8 10 E-W
Miles
0 1 2 3 4 5 6 N-S
0 1 2 3 4 5 6 E-W
Key to Symbols
Frequencies
+ rare to occasional
× frequent to abundant

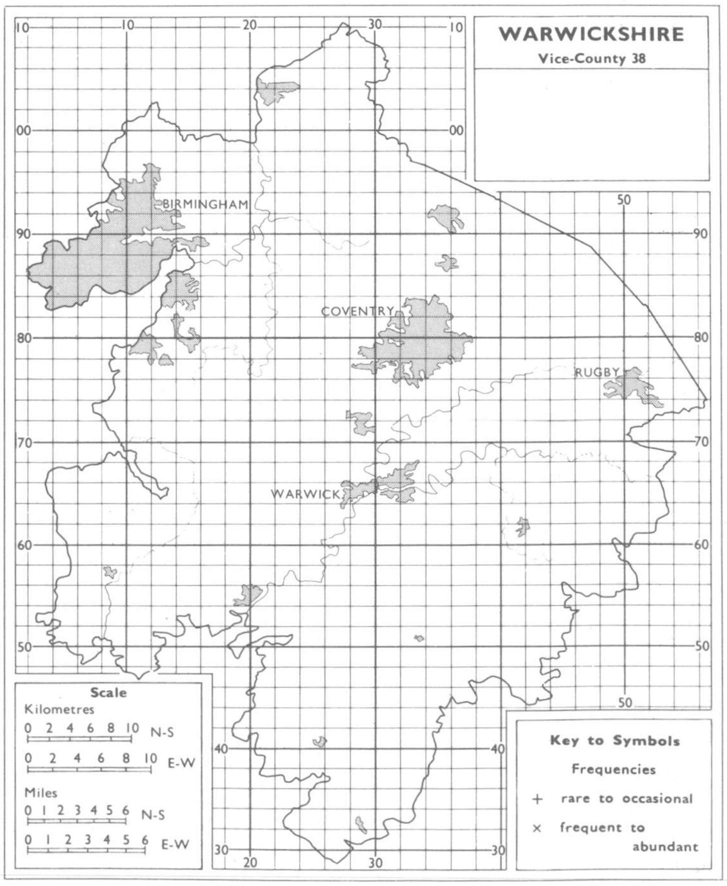
WARWICKSHIRE
Vice-County 38
BIRMINGHAM
COVENTRY
RUGBY
WARWICK
Scale
Kilometres
0 2 4 6 8 10 N-S
0 2 4 6 8 10 E-W
Miles
0 1 2 3 4 5 6 N-S
0 1 2 3 4 5 6 E-W
Key to Symbols
Frequencies
+ rare to occasional
× frequent to abundant

11. VASCULAR PLANT CHECK LIST AND ANALYSES OF HABITAT AND FREQUENCY RECORDS

This section is based on the list published by the present authors in the "Proceedings of the Birmingham Natural History Society", **20,** No. 4, 1965, from which the previous records have been corrected and brought up to date, and statistical information added.

Where a species has been recorded for VC 38 and the number of records is sufficient to make the figures significant we have added computer calculated data from the random square records as follows:

(a) the number of records in the nine major habitat groups together with the percentage values for each;
(b) the total number of records with percentages, classed as *abundant* (including also "locally abundant", "locally frequent" and "frequent") and *occasional* (including also "rare");
(c) the total number of records for each species, and
(d) the percentage of the total number of 613 random squares in which it has been recorded. In a few cases where an asterisk appears in the last column no calculated figures are available, but the percentages then would always be very low.

Most available records from the VCs surrounding VC 38 have been included in the Check List, whether or not the species in question have been recorded from VC 38. However, where a species has been recorded from only one of the surrounding VCs and from a habitat not found in VC 38, it has been omitted. A few rare aliens have also been left out, and, in general, hybrids have been included from other VCs only when recorded from VC 38. Again, the presence of *Rubus* species in other VCs is indicated only for those recorded from VC 38.

Vascular Plant Check List and Analyses of Habitat and Frequency Records

Species	Vice-County Records 38	23	32	33	37	39	55	57	Woodland No.	%	Hedgerow No.	%	Grassland No.	%	Heath No.	%	Waterside No.	%	Water No.	%	Ruderal No.	%	Quarries No.	%	Cultivated No.	%	Total Abundant No.	%	Total Occasional No.	%	Total Records No.	% of Random Squares
Acaena anserinifolia				★	★																											
Acer campestre	★	★	★	★	★	★	★	★	70	*10·9*	514	*80·1*	7	*1·1*			21	*3·3*			29	*4·5*	1	*0·2*			188	*29·3*	454	*70·7*	642	87
platanoides	★			★	★			★																								
pseudoplatanus	★	★	★	★	★	★	★	★	137	*23·6*	331	*57·0*	15	*2·6*	1	*0·2*	15	*2·6*			64	*11·0*	4	*0·7*	14	*2·4*	145	*25·0*	436	*75·0*	581	72
Aceras anthropophorum	★	★	★				★	★																								
Achillea millefolium	★	★	★	★	★	★	★	★	2	*0·2*	62	*5·9*	399	*38·1*	1	*0·1*	20	*1·9*			512	*48·9*	10	*1·0*	40	*3·8*	594	*56·8*	452	*43·2*	1046	97
ptarmica	★	★	★	★	★	★	★	★			3	*3·2*	37	*39·8*	2	*2·2*	35	*37·6*			14	*15·1*	2	*2·2*			23	*24·7*	70	*75·3*	93	14
Acinos arvensis	★	★	★	★	★	★	★	★													1	*50·0*	1	*50·0*			1	*50·0*	1	*50·0*	2	*
Aconitum anglicum	★	★		★	★	★	★	★																								
Acorus calamus	★	★	★	★	★	★	★	★									6	*60·0*	4	*40·0*							6	*60·0*	4	*40·0*	10	1
Actaea spicata	★	★					★																									
Adiantum capillus-veneris	★		★		★																											
Adonis annua		★	★	★	★	★	★	★																								
Adoxa moschatellina	★	★	★	★	★	★	★	★	12	*22·2*	18	*33·3*	2	*3·7*			13	*24·1*			8	*14·8*	1	*1·9*			34	*63·0*	20	*37·0*	54	6
Aegopodium podagraria	★	★	★	★	★	★	★	★	12	*2·5*	88	*18·3*	29	*6·0*			10	*2·1*			284	*59·2*	1	*0·2*	56	*11·7*	300	*62·5*	180	*37·5*	480	63
Aesculus carnea	★				★																											
hippocastanum	★	★	★	★	★	★	★	★	68	*20·9*	132	*40·5*	53	*16·3*			5	*1·5*			36	*11·0*	2	*0·6*	30	*9·2*	38	*11·7*	288	*88·3*	326	43
Aethusa cynapium	★	★	★	★	★	★	★	★			27	*4·4*	34	*5·6*							183	*30·1*	1	*0·2*	362	*59·6*	180	*29·7*	427	*70·3*	607	75
Agrimonia eupatoria	★	★	★	★	★	★	★	★	8	*1·5*	62	*11·9*	170	*32·5*			14	*2·7*			266	*50·9*	2	*0·4*	1	*0·2*	130	*24·9*	393	*75·1*	523	63
procera	★	★	★		★	★	★	★	1	*16·7*	2	*33·3*	1	*16·7*			1	*16·7*			1	*16·7*							6	*100·0*	6	1
Agropyron caninum	★	★	★	★	★	★	★	★	11	*8·3*	51	*38·6*	10	*7·6*			23	*17·4*			32	*24·2*	2	*1·5*	3	*2·3*	39	*29·5*	93	*70·5*	132	19
junceiforme	★																															
repens	★	★	★	★	★	★	★	★	18	*1·9*	263	*27·5*	196	*20·5*	1	*0·1*	34	*3·6*			283	*29·6*	4	*0·4*	157	*16·4*	707	*74·0*	249	*26·0*	956	11
Agrostemma githago	★	★	★	★	★	★	★	★																								
Agrostis avenacea	★				★																											
canina	★	★	★	★	★	★	★	★	3	*15·8*	1	*5·3*	7	*36·8*	3	*15·8*	1	*5·3*			4	*21·1*					5	*26·3*	14	*73·7*	19	3
ssp. *canina*	★																															
ssp. *montana*	★																															
gigantea	★	★	★	★	★	★	★	★			3	*2·3*	35	*26·9*			3	*2·3*			40	*30·8*			49	*37·7*	42	*32·3*	88	*67·7*	130	19
gigantea × *stolonifera*	★																															
semiverticillata	★			★	★	★																										
stolonifera	★	★	★	★	★	★	★	★	7	*0·9*	40	*5·0*	302	*38·0*	3	*0·4*	65	*8·2*	1	*0·1*	270	*34·0*	5	*0·6*	102	*12·8*	522	*65·7*	273	*34·3*	795	81
tenuis	★	★	★	★	★	★	★	★	20	*2·7*	24	*3·3*	417	*56·6*	12	*1·6*	6	*0·8*			220	*29·9*	6	*0·8*	32	*4·3*	608	*82·5*	129	*17·5*	737	76
Aira caryophyllea	★	★	★	★	★	★	★	★	1	*2·5*			5	*12·5*	3	*7·5*					30	*75·0*	1	*2·5*			26	*65·0*	14	*35·0*	40	6
praecox	★	★	★	★	★	★	★	★					6	*23·1*	3	*11·5*	1	*3·8*			11	*42·3*	5	*19·2*			10	*38·5*	16	*61·5*	26	4
Ajuga reptans	★	★	★	★	★	★	★	★	86	*25·7*	49	*14·7*	94	*28·1*	1	*0·3*	54	*16·2*			48	*14·4*	1	*0·3*	1	*0·3*	143	*42·8*	191	*57·2*	334	43
Alchemilla vulgaris	★	★	★	★	★	★	★	★																								
glabra						★	★	★																								
mollis	★			★																												
venosa	★																															
vestita	★	★	★	★	★	★	★	★	2	*2·8*	1	*1·4*	41	*57·7*			10	*14·1*			15	*21·1*	1	*1·4*	1	*1·4*	12	*16·9*	59	*83·1*	71	10
xanthochlora	★	★		★	★	★	★	★																								
Alisma gramineum					★																											
lanceolatum	★	★	★	★	★	★	★	★									1	*16·7*	5	*83·3*							2	*33·3*	4	*66·7*	6	*
plantago-aquatica	★	★	★	★	★	★	★	★			1	*0·3*					197	*55·0*	160	*44·7*							141	*39·4*	217	*60·6*	358	51
Alliaria petiolata	★	★	★	★	★	★	★	★	31	*4·1*	408	*53·5*	22	*2·9*			93	*12·2*	1	*0·1*	192	*25·2*	5	*0·7*	11	*1·4*	419	*54·9*	344	*45·1*	763	86

Species	Vice-County Records 38	23	32	33	37	39	55	57	Woodland No.	%	Hedgerow No.	%	Grassland No.	%	Heath No.	%	Waterside No.	%	Water No.	%	Ruderal No.	%	Quarries No.	%	Cultivated No.	%	Total Abundant No.	%	Total Occasional No.	%	Total Records No.	% of Random Squares
Allium carinatum				★	★																											
oleraceum	★		★	★	★	★	★	★																								
paradoxum		★		★	★		★																									
ursinum	★	★	★	★	★	★	★	★	17	*24·6*	18	*26·1*	1	*1·4*			32	*46·4*			1	*1·4*					38	*55·1*	31	*44·9*	69	9
vineale	★	★	★	★	★	★	★	★			6	*7·1*	25	*29·8*			1	*1·2*			45	*53·6*			7	*8·3*	28	*33·3*	56	*66·7*	84	11
Alnus glutinosa	★	★	★	★	★	★	★	★	32	*9·2*	50	*14·5*	2	*0·6*	2	*0·6*	254	*73·4*	1	*0·3*	5	*1·4*					123	*35·5*	223	*64·5*	346	45
Alopecurus aequalis	★	★	★	★	★	★	★	★					1	*8·3*			8	*66·7*	3	*25·0*							7	*58·3*	5	*41·7*	12	2
geniculatus	★	★	★	★	★	★	★	★	2	*0·4*	5	*1·0*	88	*17·4*			328	*64·7*	4	*0·8*	30	*5·9*			50	*9·9*	331	*65·3*	176	*34·7*	507	63
geniculatus × *pratensis*	★	★					★																									
myosuroides	★	★	★	★	★	★	★	★	1	*0·6*	9	*5·4*	25	*15·1*			3	*1·8*			38	*22·9*	1	*0·6*	89	*53·6*	56	*33·7*	110	*66·3*	166	23
pratensis	★	★	★	★	★	★	★	★	9	*0·9*	70	*6·8*	567	*55·3*	1	*0·1*	18	*1·8*			335	*32·7*			25	*2·4*	802	*78·2*	223	*21·8*	1025	94
Althaea hirsuta	★	★		★		★	★																									
Alyssum alyssoides	★	★	★	★	★	★	★	★																								
Amaranthus albus	★	★			★																											
deflexus	★				★																											
lividus	★																															
retroflexus	★	★	★	★	★		★	★																								
Ammi majus		★		★	★		★																									
visnaga	★			★	★		★																									
Ammophila arenaria	★																															
Anacamptis pyramidalis	★	★	★	★	★	★	★	★					3	*75·0*									1	*25·0*					4	*100·0*	4	*
Anagallis arvensis	★	★	★	★	★	★	★	★			7	*1·2*	31	*5·4*	1	*0·2*	4	*0·7*			120	*20·9*	4	*0·7*	408	*71·0*	217	*37·7*	358	*62·3*	575	75
arvensis × *foemina*	★																															
foemina	★	★	★	★	★	★	★	★			1	*8·3*									1	*8·3*	1	*8·3*	9	*75·0*	4	*33·3*	8	*66·7*	12	2
minima	★	★			★	★		★																								
tenella	★	★	★	★	★	★	★	★																								
Anaphalis margaritacea	★			★	★	★		★																								
Anchusa arvensis	★	★	★	★	★	★	★	★			2	*2·2*	5	*5·6*			1	*1·1*			33	*37·1*	2	*2·2*	46	*51·7*	26	*29·2*	63	*70·8*	89	13
Andromeda polifolia					★	★		★																								
Anemone nemorosa	★	★	★	★	★	★	★	★	99	*52·4*	49	*25·9*	14	*7·4*			5	*2·6*			21	*11·1*			1	*0·5*	96	*50·8*	93	*49·2*	189	25
Angelica archangelica					★	★	★																									
sylvestris	★	★	★	★	★	★	★	★	65	*12·3*	61	*11·5*	40	*7·5*			282	*53·2*			80	*15·1*			2	*0·4*	170	*32·1*	360	*67·9*	530	61
Antennaria dioica		★	★	★	★	★	★	★																								
Anthemis arvensis	★	★	★	★	★	★	★	★					1	*4·3*							6	*26·1*			16	*69·6*	14	*60·9*	9	*39·1*	23	3
cotula	★	★	★	★	★	★	★	★			8	*3·0*	17	*6·4*			2	*0·7*			108	*40·4*	2	*0·7*	130	*48·7*	116	*43·4*	151	*56·6*	267	34
tinctoria	★	★	★	★		★	★	★																								
Anthoxanthum odoratum	★	★	★	★	★	★	★	★	16	*2·2*	24	*3·3*	507	*69·7*	8	*1·1*	18	*2·5*			145	*19·9*	4	*0·6*	5	*0·7*	468	*64·4*	259	*35·6*	727	81
puelii	★	★	★		★	★																										
Anthriscus caucalis	★	★	★	★	★	★	★	★																								
cerefolium	★	★		★	★		★	★																								
sylvestris	★	★	★	★	★	★	★	★	54	*5·6*	416	*43·2*	123	*12·8*			39	*4·0*			314	*32·6*	5	*0·5*	13	*1·3*	759	*78·7*	205	*21·3*	964	96
Anthyllis vulneraria	★	★	★	★	★	★	★	★					3	*10·7*							21	*75·0*	4	*14·3*			10	*35·7*	18	*64·3*	28	4
Antirrhinum majus	★	★	★	★	★	★	★	★																								
Apera interrupta		★	★				★	★																								
spica-venti	★	★		★	★	★	★	★																								
Aphanes arvensis s.l.	★	★	★	★	★	★	★	★	1	*0·4*			22	*9·5*	1	*0·4*					44	*19·0*	8	*3·4*	156	*67·2*	86	*37·1*	146	*62·9*	232	31
arvensis	★	★	★	★	★	★	★	★					8	*6·8*	1	*0·9*					18	*15·4*	7	*6·0*	83	*70·9*	53	*45·3*	64	*54·7*	117	13
microcarpa	★	★	★	★	★	★	★	★					5	*31·2*							9	*56·2*			2	*12·5*	6	*37·5*	10	*62·5*	16	2

Vascular Plant Check List and Analyses of Habitat and Frequency Records (contd.)

Species	Vice-County Records								Wood-land		Hedge-row		Grass-land		Heath		Water-side		Water		Ruderal		Quarries		Culti-vated		Total Abun-dant		Total Occa-sional		Total Records	% of Random Squares
	38	23	32	33	37	39	55	57	No.	%	No.	%	No.	%	No.	%	No.	%	No.	%	No.	%	No.	%	No.	%	No.	%	No.	%	No.	
Apium graveolens	★	★	★	★	★	★	★	★																								
inundatum	★	★	★	★	★	★	★	★																								
inundatum × *nodiflorum*			★					★																								
nodiflorum	★	★	★	★	★	★	★	★							1	*0·3*	256	*68·8*	115	*30·9*							226	*60·8*	146	*39·2*	372	51
Aquilegia vulgaris	★	★	★	★	★	★	★	★			4	*30·8*					1	*7·7*			8	*61·5*							13	*100·0*	13	*
Arabidopsis thaliana	★	★	★	★	★	★	★	★			3	*1·5*	21	*10·7*							110	*55·8*	12	*6·1*	51	*25·9*	96	*48·7*	101	*51·3*	197	27
Arabis hirsuta	★	★	★	★	★	★	★	★																								
turrita		★					★																									
Arctium lappa	★	★	★	★	★	★	★	★	26	*8·4*	95	*30·6*	30	*9·7*			45	*14·5*			106	*34·2*	1	*0·3*	7	*2·3*	57	*18·4*	253	*81·6*	310	37
minus s.l.	★	★	★	★	★	★	★	★	50	*9·4*	157	*29·4*	66	*12·4*			35	*6·6*			209	*39·1*	1	*0·2*	16	*3·0*	96	*18·0*	438	*82·0*	534	61
minus ssp. *minus*	★	★	★	★	★	★	★	★																								
ssp. *nemorosum*	★					★		★																								
ssp. *pubens*	★							★																								
Arenaria balearica	★	★	★																													
serpyllifolia s.l.	★	★	★	★	★	★	★	★	1	*0·6*			19	*11·0*	1	*0·6*					95	*54·9*	44	*25·4*	13	*7·5*	81	*46·8*	92	*53·2*	173	24
leptoclados	★	★	★	★	★	★	★	★					5	*12·8*							22	*56·4*	8	*20·5*	4	*10·3*	16	*41·0*	23	*59·0*	39	6
serpyllifolia	★	★	★	★	★	★	★	★					3	*11·1*	1	*3·7*					11	*40·7*	11	*40·7*	1	*3·7*	14	*51·9*	13	*48·1*	27	4
Aristolochia clematitis		★	★		★	★	★	★																								
Armoracia rusticana	★	★	★	★	★	★	★	★			21	*6·6*	46	*14·4*			1	*0·3*			230	*72·1*	4	*1·3*	17	*5·3*	69	*21·6*	250	*78·4*	319	45
Arnoseris minima			★		★		★																									
Arrhenatherum elatius	★	★	★	★	★	★	★	★	26	*2·6*	262	*25·7*	264	*25·9*			15	*1·5*			423	*41·5*	10	*1·0*	19	*1·9*	907	*89·0*	112	*11·0*	1019	94
Artemisia absinthium	★	★	★	★	★	★	★	★			5	*5·2*	6	*6·2*							85	*88·5*					26	*27·1*	70	*72·9*	96	14
verlotorum		★			★																											
vulgaris	★	★	★	★	★	★	★	★	7	*1·5*	42	*9·3*	29	*6·4*							357	*78·8*	5	*1·1*	13	*2·9*	151	*33·3*	302	*66·7*	453	56
Arum maculatum	★	★	★	★	★	★	★	★	118	*19·2*	349	*56·8*	14	*2·3*			24	*3·9*			105	*17·1*	3	*0·5*	1	*0·2*	281	*45·8*	333	*54·2*	614	69
Asarum europaeum		★	★	★	★		★	★																								
Asparagus officinalis ssp. *officinalis*	★	★	★	★	★		★	★			5	*26·3*	4	*21·1*	1	*5·3*	1	*5·3*			7	*36·8*			1	*5·3*	2	*10·5*	17	*89·5*	19	3
Asperula arvensis	★	★	★	★	★	★	★	★																								
cynanchica	★	★	★	★	★		★	★																								
Asplenium adiantum-nigrum	★	★	★	★	★	★	★	★															27	*100·0*			11	*40·7*	16	*59·3*	27	4
ruta-muraria	★	★	★	★	★	★	★	★															60	*100·0*			26	*43·3*	34	*56·7*	60	10
trichomanes	★	★	★	★	★	★	★	★															34	*100·0*			14	*41·2*	20	*58·8*	34	6
viride	★				★	★		★																								
Aster novi-belgii	★	★		★	★						1	*5·3*					2	*10·5*			15	*78·9*	1	*5·3*			5	*26·3*	14	*73·7*	19	3
tripolium		★		★		★		★																								
Astragalus danicus		★	★	★	★		★																									
glycyphyllos	★	★	★	★	★	★	★	★	1	*7·1*	6	*42·9*	2	*14·3*							5	*35·7*					3	*21·4*	11	*78·6*	14	2
Athyrium filix-femina	★	★	★	★	★	★	★	★	23	*37·7*	12	*19·7*			1	*1·6*	22	*36·1*			2	*3·3*	1	*1·6*			23	*37·7*	38	*62·3*	61	8
Atriplex hastata	★	★	★	★	★	★	★	★			4	*1·7*	4	*1·7*			28	*12·1*			137	*59·1*	2	*0·9*	57	*24·6*	62	*26·7*	170	*73·3*	232	31
patula	★	★	★	★	★	★	★	★	3	*0·4*	10	*1·4*	18	*2·6*			26	*3·7*			306	*43·5*	3	*0·4*	338	*48·0*	318	*45·2*	386	*54·8*	704	84
Atropa bella-donna	★	★	★	★	★	★	★	★													5	*71·4*	1	*14·3*	1	*14·3*	2	*28·6*	5	*71·4*	7	*
Avena fatua	★	★	★	★	★	★	★	★			12	*6·2*	13	*6·8*			1	*0·5*			64	*33·3*			102	*53·1*	46	*24·0*	146	*76·0*	192	26
ludoviciana	★	★	★	★	★								1	*11·1*							1	*11·1*			7	*77·8*	2	*22·2*	7	*77·8*	9	1
sativa	★	★	★	★			★	★																								
strigosa	★	★	★	★	★	★	★	★																								
Azolla filiculoides	★	★		★	★																											

Species	Vice-County Records								Wood-land		Hedge-row		Grass-land		Heath		Water-side		Water		Ruderal		Quarries		Culti-vated		Total Abun-dant		Total Occa-sional		Total Records	% of Random Squares
	38	23	32	33	37	39	55	57	No.	%	No.	%	No.	%	No.	%	No.	%	No.	%	No.	%	No.	%	No.	%	No.	%	No.	%	No.	
Baldellia ranunculoides	★	★	★	★	★	★	★	★																								
Ballota nigra	★	★	★	★	★	★	★	★			70	*42·7*	8	*4·9*			1	*0·6*			80	*48·8*	3	*1·8*	2	*1·2*	36	*22·0*	128	*78·0*	164	24
Barbarea intermedia	★	★		★	★		★	★					1	*4·8*			5	*23·8*			3	*14·3*			12	*57·1*	8	*38·1*	13	*61·9*	21	3
stricta			★	★	★																											
verna	★	★	★	★	★	★	★	★					1	*14·3*							3	*42·9*			3	*42·9*	2	*28·6*	5	*71·4*	7	1
vulgaris	★	★	★	★	★	★	★	★			22	*7·0*	22	*7·0*			209	*66·1*			53	*16·8*	2	*0·6*	8	*2·5*	72	*22·8*	244	*77·2*	316	45
Bellis perennis	★	★	★	★	★	★	★	★	12	*1·2*	24	*2·4*	598	*59·1*	2	*0·2*	7	*0·7*			245	*24·2*	10	*1·0*	113	*11·2*	639	*63·2*	372	*36·8*	1011	98
Berberis vulgaris	★	★	★	★	★	★	★	★			16	*100·0*															1	*6·2*	15	*93·7*	16	3
Berteroa incana	★	★		★	★	★	★	★																								
Berula erecta	★	★	★	★	★	★	★	★							1	*2·1*	37	*77·1*	10	*20·8*							23	*47·9*	25	*52·1*	48	8
Beta vulgaris ssp. *maritima*				★	★																											
Betonica officinalis	★	★	★	★	★	★	★	★	11	*6·1*	41	*22·7*	83	*45·9*	2	*1·1*	15	*8·3*			25	*13·8*			4	*2·2*	61	*33·7*	120	*66·3*	181	25
Betula pendula	★	★	★	★	★	★	★	★	105	*40·2*	85	*32·6*	13	*5·0*	6	*2·3*	6	*2·3*			25	*9·6*	2	*0·8*	19	*7·3*	77	*29·5*	184	*70·5*	261	36
pubescens	★	★	★	★	★	★	★	★	75	*49·0*	54	*35·3*			7	*4·6*	4	*2·6*			9	*5·9*			4	*2·6*	67	*43·8*	86	*56·2*	153	21
Bidens cernua	★	★	★	★	★	★	★	★									38	*88·4*	5	*11·6*							19	*44·2*	24	*55·8*	43	7
frondosa	★			★	★																											
tripartita	★	★	★	★	★	★	★	★									59	*98·3*							1	*1·7*	25	*41·7*	35	*58·3*	60	9
Blackstonia perfoliata	★	★	★	★	★	★	★	★			2	*8·0*	10	*40·0*			1	*4·0*			9	*36·0*	3	*12·0*			10	*40·0*	15	*60·0*	25	4
Blechnum spicant	★	★	★	★	★	★	★	★	1	*14·3*	1	*14·3*			1	*14·3*	4	*57·1*									1	*14·3*	6	*85·7*	7	1
Blysmus compressus	★	★	★	★	★	★	★	★																								
Borago officinalis	★	★	★	★	★	★	★	★																								
Botrychium lunaria	★	★	★	★	★	★	★	★																								
Brachypodium pinnatum	★	★	★	★	★	★	★	★	2	*2·3*	22	*25·3*	33	*37·9*			1	*1·1*			29	*33·3*					47	*54·0*	40	*46·0*	87	12
sylvaticum	★	★	★	★	★	★	★	★	143	*24·4*	327	*55·7*	17	*2·9*			30	*5·1*			67	*11·4*	3	*0·5*			294	*50·1*	293	*49·9*	587	70
Brassica juncea		★			★																											
napus	★	★	★	★	★	★	★	★	1	*1·9*	1	*1·9*	2	*3·7*			6	*11·1*			17	*31·5*			27	*50·0*	9	*16·7*	45	*83·3*	54	8
nigra	★	★	★	★	★	★	★	★			5	*6·3*	6	*7·6*			11	*13·9*			32	*40·5*			25	*31·6*	23	*29·1*	56	*70·9*	79	11
rapa	★	★	★	★	★	★	★	★			1	*0·8*	10	*8·0*			22	*17·6*			37	*29·6*			55	*44·0*	29	*23·2*	96	*76·8*	125	18
Briza media	★	★	★	★	★	★	★	★	3	*1·3*	4	*1·7*	168	*71·2*	2	*0·8*	12	*5·1*			42	*17·8*	3	*1·3*	2	*0·8*	81	*34·3*	155	*65·7*	236	32
minor				★		★																										
Bromus arvensis	★	★	★	★	★	★	★																									
benekenii		★		★	★																											
carinatus	★	★			★																											
commutatus	★	★	★	★	★	★	★	★			1	*1·9*	24	*45·3*			1	*1·9*			18	*34·0*			9	*17·0*	12	*22·6*	41	*77·4*	53	6
var. *pubens*	★	★		★																												
commutatus × *racemosus*	★																															
diandrus				★	★		★	★																								
erectus	★	★	★	★	★	★	★	★	1	*0·9*	5	*4·6*	38	*35·2*							60	*55·6*	3	*2·8*	1	*0·9*	56	*51·9*	52	*48·1*	108	15
hordeaceus	★	★	★	★	★	★	★	★	1	*0·2*	24	*4·0*	222	*37·0*			3	*0·5*			316	*52·7*	9	*1·5*	25	*4·2*	227	*37·8*	373	*62·2*	600	69
ssp. *hordeaceus*	★	★	★	★	★	★	★	★	1	*0·3*	15	*4·2*	136	*38·5*			2	*0·6*			171	*48·4*	4	*1·1*	24	*6·8*	145	*41·1*	208	*58·9*	353	42
var. *leiostachys*	★	★		★	★																											
inermis	★	★					★						1	*16·7*							5	*83·3*					2	*33·3*	4	*66·7*	6	*
interruptus		★	★	★	★																											
lepidus	★	★	★	★	★	★	★	★			3	*4·6*	28	*43·1*							28	*43·1*			6	*9·2*	12	*18·5*	53	*81·5*	65	8
var. *micromollis*	★			★																												
madritensis				★	★																											
pseudosecalinus	★				★																											
pseudothominii	★	★	★	★	★	★	★	★			3	*2·7*	46	*41·4*							48	*43·2*	3	*2·7*	11	*9·9*	29	*26·1*	82	*73·9*	111	14
var. *hirsutus*	★																															

Vascular Plant Check List and Analyses of Habitat and Frequency Records (contd.)

Species	Vice-County Records								Wood-land		Hedge-row		Grass-land		Heath		Water-side		Water		Ruderal		Quarries		Culti-vated		Total Abun-dant		Total Occa-sional		Total Records	% of Random Squares
	38	23	32	33	37	39	55	57	No.	%	No.	%	No.	%	No.	%	No.	%	No.	%	No.	%	No.	%	No.	%	No.	%	No.	%	No.	
Bromus racemosus	★	★	★	★	★	★	★	★					16	*72·7*			2	*9·1*			3	*13·6*			1	*4·5*	5	*22·7*	17	*77·3*	22	3
ramosus	★	★	★	★	★	★	★	★	97	*19·4*	288	*57·7*	11	*2·2*			50	*10·0*			53	*10·6*					205	*41·1*	294	*58·9*	499	67
rigidus	★			★																												
secalinus	★	★	★	★	★	★	★	★																								
var. *hirtus*	★		★	★	★		★																									
squarrosus																																
var. *villosus*	★	★																														
sterilis	★	★	★	★	★	★	★	★	9	*1·3*	219	*31·3*	85	*12·1*			15	*2·1*			324	*46·3*	19	*2·7*	29	*4·1*	417	*59·6*	283	*40·4*	700	81
tectorum		★		★	★	★	★	★																								
willdenowii	★	★	★	★	★		★																									
Bryonia dioica	★	★	★	★	★	★	★	★	14	*4·2*	262	*78·2*	7	*2·1*			3	*0·9*			41	*12·2*	3	*0·9*	5	*1·5*	108	*32·2*	227	*67·8*	335	45
Buddleja davidii	★	★		★	★		★																									
Bunias orientalis		★	★	★		★	★	★																								
Bupleurum lancifolium	★	★		★		★	★	★																								
rotundifolium	★	★		★	★	★	★	★																								
tenuissimum			★		★																											
Butomus umbellatus	★	★	★	★	★	★	★	★									14	*41·2*	19	*55·9*	1	*2·9*					7	*20·6*	27	*79·4*	34	5
Buxus sempervirens	★	★	★	★	★	★	★	★	8	*25·8*	19	*61·3*									1	*3·2*			3	*9·7*	9	*29·0*	22	*71·0*	31	5
Cakile maritima	★																															
Calamagrostis canescens	★		★		★	★	★	★																								
epigejos	★	★	★	★	★	★	★	★	5	*12·2*	10	*24·4*	6	*14·6*			12	*29·3*	2	*4·9*	3	*7·3*	3	*7·3*			20	*48·8*	21	*51·2*	41	6
Calamintha ascendens	★	★	★	★	★	★	★	★			2	*25·0*	3	*37·5*							3	*37·5*					4	*50·0*	4	*50·0*	8	*
nepeta	★	★	★	★			★																									
Callitriche intermedia	★	★	★	★	★	★	★	★																								
ssp. *hamulata*	★	★					★	★																								
ssp. *pedunculata*	★	★					★																									
obtusangula	★	★	★	★	★	★	★	★																								
platycarpa	★	★	★ }				{ ★	★									3	*21·4*	11	*78·6*							8	*57·1*	6	*42·9*	14	2
				★	★	★																										
stagnalis	★	★	★ }				{ ★	★	6	*3·2*	2	*1·1*			1	*0·5*	68	*36·4*	109	*58·3*	1	*0·5*					130	*69·5*	57	*30·5*	187	27
Calluna vulgaris	★	★	★	★	★	★	★	★	11	*27·5*	5	*12·5*	4	*10·0*	12	*30·0*	1	*2·5*			7	*17·5*					23	*57·5*	17	*42·5*	40	5
Caltha palustris	★	★	★	★	★	★	★	★	8	*4·2*	1	*0·5*	15	*7·9*	2	*1·0*	161	*84·3*	4	*2·1*							85	*44·5*	106	*55·5*	191	26
Calystegia pulchra	★	★	★	★	★		★																									
sepium	★	★	★	★	★	★	★	★			319	*72·0*	12	*2·7*			12	*2·7*			77	*17·4*	5	*1·1*	18	*4·1*	193	*43·6*	250	*56·4*	443	64
silvatica	★	★	★	★	★	★	★	★	2	*1·0*	131	*68·2*	5	*2·6*			2	*1·0*			45	*23·4*	4	*2·1*	3	*1·6*	117	*60·9*	75	*39·1*	192	29
soldanella	★				★																											
Camelina sativa	★	★	★	★	★	★	★	★																								
Campanula glomerata	★	★	★	★	★		★	★																								
latifolia	★	★	★	★	★	★	★	★	9	*52·9*	5	*29·4*	1	*5·9*			1	*5·9*			1	*5·9*					4	*23·5*	13	*76·5*	17	2
patula	★			★	★	★	★	★																								
persicifolia	★	★		★	★			★																								
rapunculoides	★	★	★	★	★	★	★	★													1	*25·0*			3	*75·0*			4	*100·0*	4	*
rapunculus	★	★		★	★	★	★																									
rotundifolia	★	★	★	★	★	★	★	★	3	*1·9*	17	*10·6*	77	*48·1*	8	*5·0*					46	*28·7*	4	*2·5*	5	*3·1*	51	*31·9*	109	*68·1*	160	20
trachelium	★	★	★	★	★	★	★	★	6	*26·1*	8	*34·8*									6	*26·1*			3	*13·0*	9	*39·1*	14	*60·9*	23	3
Cannabis sativa	★	★	★		★		★	★																								
Capsella bursa-pastoris	★	★	★	★	★	★	★	★	2	*0·2*	20	*1·8*	105	*9·6*	1	*0·1*	2	*0·2*			549	*50·0*	9	*0·8*	409	*37·3*	598	*54·5*	499	*45·5*	1097	97

Species	Vice-County Records								Wood-land		Hedge-row		Grass-land		Heath		Water-side		Water		Ruderal		Quarries		Culti-vated		Total Abun-dant		Total Occa-sional		Total Records	% of Random Squares
	38	23	32	33	37	39	55	57	No.	%	No.	%	No.	%	No.	%	No.	%	No.	%	No.	%	No.	%	No.	%	No.	%	No.	%	No.	
Cardamine amara	★	★	★	★	★	★	★	★	1	2·1	2	4·2					43	89·6	2	4·2							21	43·7	27	56·2	48	7
bulbifera						★		★																								
flexuosa	★	★	★	★	★	★	★	★	27	7·7	19	5·4	17	4·8			275	78·1	1	0·3	13	3·7					130	36·9	222	63·1	352	47
hirsuta	★	★	★	★	★	★	★	★			1	0·6	7	4·0							104	59·4	17	9·7	46	26·3	89	50·9	86	49·1	175	24
impatiens	★		★	★	★	★		★																								
pratensis	★	★	★	★	★	★	★	★	13	2·2	22	3·6	226	37·4	2	0·3	294	48·7	2	0·3	41	6·8			4	0·7	194	32·1	410	67·9	604	69
Cardaria draba	★	★	★	★	★		★	★					5	14·7							20	58·8	4	11·8	5	14·7	15	44·1	19	55·9	34	5
Carduus acanthoides	★	★	★	★	★	★	★	★	6	2·7	73	33·3	42	19·2	1	0·5	41	18·7			46	21·0	4	1·8	6	2·7	59	26·9	160	73·1	219	27
acanthoides × *nutans*	★	★	★	★			★	★																								
nutans	★	★	★	★	★	★	★	★			2	5·0	14	35·0							18	45·0	4	10·0	2	5·0	11	27·5	29	72·5	40	6
tenuiflorus	★	★	★	★	★	★		★																								
Carex acuta	★	★	★	★	★	★	★	★					1	8·3			10	83·3	1	8·3							4	33·3	8	66·7	12	2
acutiformis	★	★	★	★	★	★	★	★							1	0·9	96	88·9	11	10·2							74	68·5	34	31·5	108	15
arenaria	★																															
binervis	★	★	★	★	★	★	★	★																								
caryophyllea	★	★	★	★	★	★	★	★	1	3·3			24	80·0	3	10·0	1	3·3			1	3·3					10	33·3	20	66·7	30	5
curta	★				★	★		★							1	33·3	2	66·7									2	66·7	1	33·3	3	*
demissa	★	★	★	★	★	★	★	★							1	25·0	3	75·0									1	25·0	3	75·0	4	1
demissa × *hostiana*	★			★				★																								
diandra	★				★	★	★	★																								
digitata				★	★			★																								
dioica	★	★	★	★	★	★	★	★																								
f. *isogyna*	★																															
distans	★	★	★	★	★	★	★						2	25·0			5	62·5			1	12·5					3	37·5	5	62·5	8	1
disticha	★	★	★	★	★	★	★	★			1	2·5	7	17·5	1	2·5	31	77·5									18	45·0	22	55·0	40	6
divulsa	★	★	★	★	★	★	★	★	2	13·3	6	40·0	4	26·7							3	20·0					4	26·7	11	73·3	15	2
echinata	★	★	★	★	★	★	★	★							1	33·3	2	66·7									2	66·7	1	33·3	3	*
elata		★	★			★	★	★																								
elongata	★				★																											
filiformis		★		★																												
flacca	★	★	★	★	★	★	★	★	8	2·8	13	4·5	130	45·3	1	0·3	74	25·8			58	20·2	2	0·7	1	0·3	135	47·0	152	53·0	287	35
hirta	★	★	★	★	★	★	★	★	4	0·8	11	2·2	164	33·0	3	0·6	221	44·5	1	0·2	89	17·9	2	0·4	2	0·4	201	40·4	296	59·6	497	62
hostiana	★	★	★	★	★	★	★	★																								
laevigata	★		★		★	★	★	★																								
lepidocarpa	★	★	★	★	★	★	★	★																								
montana					★			★																								
muricata																																
ssp. *muricata*	★	★	★	★	★	★	★	★													4	100·0					2	50·0	2	50·0	4	*
ssp. *leersii*	★	★	★	★	★	★	★	★																								
nigra	★	★	★	★	★	★	★	★					13	26·0	5	10·0	31	62·0			1	2·0					19	38·0	31	62·0	50	7
otrubae	★	★	★	★	★	★	★	★	11	2·7	14	3·5	27	6·7			318	79·5	8	2·0	21	5·2	1	0·2			140	35·0	260	65·0	400	51
otrubae × *remota*	★	★	★	★	★		★																									
ovalis	★	★	★	★	★	★	★	★	2	2·9	5	7·4	23	33·8	3	4·4	31	45·6			4	5·9					18	26·5	50	73·5	68	10
pallescens	★	★	★	★	★	★	★	★	2	15·4	2	15·4	2	15·4			6	46·2			1	7·7					6	46·2	7	53·8	13	2
panicea	★	★	★	★	★	★	★	★					21	44·7	1	2·1	23	48·9			2	4·3					23	48·9	24	51·1	47	7
paniculata	★	★	★	★	★	★	★	★	3	7·9	1	2·6	1	2·6	1	2·6	30	78·9	2	5·3							13	34·2	25	65·8	38	5
paniculata × *remota*		★	★		★																											
pendula	★	★	★	★	★	★	★	★	28	35·4	7	8·9					43	54·4	1	1·3							33	41·8	46	58·2	79	11

Vascular Plant Check List and Analyses of Habitat and Frequency Records (contd.)

Species	Vice-County Records								Wood-land		Hedge-row		Grass-land		Heath		Water-side		Water		Ruderal		Quarries		Culti-vated		Total Abun-dant		Total Occa-sional		Total Records	% of Random Squares
	38	23	32	33	37	39	55	57	No.	%	No.	%	No.	%	No.	%	No.	%	No.	%	No.	%	No.	%	No.	%	No.	%	No.	%	No.	
Carex pilulifera	★	★	★	★	★	★	★	★	2	*20·0*	1	*10·0*	2	*20·0*	5	*50·0*											3	*30·0*	7	*70·0*	10	1
pseudocyperus	★	★	★	★	★	★	★	★	1	*2·2*					1	*2·2*	39	*84·8*	5	*10·9*							23	*50·0*	23	*50·0*	46	7
pulicaris	★	★	★	★	★	★	★	★																								
remota	★	★	★	★	★	★	★	★	32	*25·8*	7	*5·6*					85	*68·5*									31	*25·0*	93	*75·0*	124	17
riparia	★	★	★	★	★	★	★	★									99	*78·0*	28	*22·0*							87	*68·5*	40	*31·5*	127	17
riparia × *vesicaria*	★	★																														
rostrata	★	★	★	★	★	★	★	★							2	*16·7*	10	*83·3*									9	*75·0*	3	*25·0*	12	1
serotina		★		★				★																								
spicata	★	★	★	★	★	★	★	★	3	*1·1*	20	*7·6*	106	*40·5*	1	*0·4*	43	*16·4*	1	*0·4*	86	*32·8*	2	*0·8*			52	*19·8*	210	*80·2*	262	36
strigosa	★	★	★	★	★	★	★	★																								
sylvatica	★	★	★	★	★	★	★	★	66	*72·5*	17	*18·7*	2	*2·2*			4	*4·4*			2	*2·2*					40	*44·0*	51	*56·0*	91	14
vesicaria	★	★	★	★	★	★	★	★									7	*100·0*									4	*57·1*	3	*42·9*	7	1
vulpina		★		★																												
Carlina vulgaris	★	★	★	★	★	★	★	★			2	*14·3*	6	*42·9*							5	*35·7*	1	*7·1*			6	*42·9*	8	*57·1*	14	2
Carpinus betulus	★	★	★	★	★	★	★	★	21	*51·2*	11	*26·8*	2	*4·9*			2	*4·9*			4	*9·8*			1	*2·4*	6	*14·6*	35	*85·4*	41	6
Carum carvi	★	★	★	★	★	★	★	★																								
Castanea sativa	★	★	★	★	★	★	★	★	51	*64·6*	16	*20·3*	7	*8·9*							5	*6·3*					8	*10·1*	71	*89·9*	79	12
Catabrosa aquatica	★	★	★	★	★	★	★	★									6	*66·7*	3	*33·3*							6	*66·7*	3	*33·3*	9	1
Catapodium rigidum	★	★	★	★	★	★	★	★			2	*4·2*	3	*6·2*							28	*58·3*	13	*27·1*	2	*4·2*	19	*39·6*	29	*60·4*	48	7
Caucalis platycarpos	★	★	★	★	★	★	★	★																								
Centaurea aspera	★																															
calcitrapa			★	★	★	★	★	★																								
cyanus	★	★	★	★	★	★	★	★					1	*12·5*							1	*12·5*			6	*75·0*	1	*12·5*	7	*87·5*	8	1
diluta	★				★																											
jacea	★	★		★		★																										
nigra	★	★	★	★	★	★	★	★	5	*0·6*	77	*9·3*	338	*40·7*	3	*0·4*	5	*0·6*			382	*46·0*	4	*0·5*	17	*2·0*	326	*39·2*	505	*60·8*	831	89
ssp. *nemoralis*	★	★	★	★	★	★	★	★																								
ssp. *nemoralis* × *jacea*	★																															
ssp. *nigra*	★	★	★	★	★	★	★	★																								
ssp. *nigra* × *jacea*	★																															
scabiosa	★	★	★	★	★	★	★	★			15	*14·0*	18	*16·8*							63	*58·9*	4	*3·7*	7	*6·5*	28	*26·2*	79	*73·8*	107	15
solstitialis	★	★	★	★	★		★																									
Centaurium erythraea	★	★	★	★	★	★	★	★	4	*4·3*	16	*17·0*	41	*43·6*	3	*3·2*					25	*26·6*	3	*3·2*	2	*2·1*	31	*33·0*	63	*67·0*	94	13
pulchellum	★	★	★	★	★																											
Centranthus ruber	★	★	★	★	★	★	★	★			1	*9·1*									3	*27·3*	7	*63·6*			3	*27·3*	8	*72·7*	11	2
Cephalanthera damasonium	★	★	★	★	★			★																								
longifolia	★	★		★	★	★		★																								
Cerastium arvense	★	★	★	★	★	★	★	★																								
diffusum	★	★	★	★	★	★	★						1	*5·9*							15	*88·2*	1	*5·9*			9	*52·9*	8	*47·1*	17	3
glomeratum	★	★	★	★	★	★	★	★	4	*1·4*	9	*3·0*	88	*29·7*			1	*0·3*			109	*36·8*	14	*4·7*	71	*24·0*	86	*29·1*	210	*70·9*	296	38
holosteoides	★	★	★	★	★	★	★	★	13	*1·4*	25	*2·6*	468	*49·5*	6	*0·6*	5	*0·5*			268	*28·3*	14	*1·5*	147	*15·5*	415	*43·9*	531	*56·1*	946	92
pumilum	★	★		★			★																									
semidecandrum	★	★	★	★	★	★	★	★					2	*40·0*							2	*40·0*	1	*20·0*			3	*60·0*	2	*40·0*	5	*
tomentosum	★	★	★	★	★		★	★																								
Ceratophyllum demersum	★	★	★	★	★	★	★	★											10	*100·0*							8	*80·0*	2	*20·0*	10	2
submersum		★	★	★	★		★	★																								
Ceterach officinarum	★	★	★	★	★	★	★	★															14	*100·0*			4	*28·6*	10	*71·4*	14	2

Species	Vice-County Records 38	23	32	33	37	39	55	57	Woodland No.	%	Hedgerow No.	%	Grassland No.	%	Heath No.	%	Waterside No.	%	Water No.	%	Ruderal No.	%	Quarries No.	%	Cultivated No.	%	Total Abundant No.	%	Total Occasional No.	%	Total Records No.	% of Random Squares
Chaenorhinum minus	★	★	★	★	★	★	★	★			1	*1·1*									71	*76·3*	5	*5·4*	16	*17·2*	44	*47·3*	49	*52·7*	93	15
Chaerophyllum temulentum	★	★	★	★	★	★	★	★	15	*3·3*	323	*70·2*	19	*4·1*			3	*0·7*			98	*21·3*	2	*0·4*			192	*41·7*	268	*58·3*	460	66
Chamaemelum nobile	★	★	★	★	★	★	★	★																								
Cheiranthus cheiri	★	★	★	★	★	★	★	★																								
Chelidonium majus	★	★	★	★	★	★	★	★	3	*1·8*	48	*28·6*	5	*3·0*			11	*6·5*			85	*50·6*	10	*6·0*	6	*3·6*	38	*22·6*	130	*77·4*	168	23
Chenopodium album	★	★	★	★	★	★	★	★	2	*0·2*	8	*1·0*	22	*2·6*	1	*0·1*	13	*1·6*			358	*42·9*	4	*0·5*	426	*51·1*	510	*61·2*	324	*38·8*	834	89
bonus-henricus	★	★	★	★	★	★	★	★					2	*8·3*							16	*66·7*			6	*25·0*	4	*16·7*	20	*83·3*	24	4
ficifolium	★	★	★	★	★	★	★	★																								
foliosum	★																															
glaucum		★		★		★	★	★																								
hircinum	★	★		★	★	★	★																									
hybridum	★	★	★	★	★	★	★														5	*55·6*			4	*44·4*	2	*22·2*	7	*77·8*	9	1
murale	★	★		★	★	★	★	★																								
opulifolium	★	★	★	★	★	★		★																								
polyspermum	★	★	★	★	★	★	★	★	1	*0·7*	3	*2·0*	5	*3·3*			7	*4·6*			40	*26·1*	1	*0·7*	96	*62·7*	46	*30·1*	107	*69·9*	153	22
rubrum	★	★	★	★	★	★	★	★					4	*2·9*			27	*19·7*			83	*60·6*	1	*0·7*	22	*16·1*	58	*42·3*	79	*57·7*	137	18
suecicum	★	★	★	★	★	★	★	★																								
urbicum	★	★		★	★	★	★	★																								
vulvaria	★	★	★	★	★	★	★	★																								
Chrysanthemum segetum	★	★	★	★	★	★	★	★			2	*2·9*	11	*15·7*							18	*25·7*			39	*55·7*	25	*35·7*	45	*64·3*	70	9
Chrysosplenium alternifolium	★	★		★	★	★	★	★																								
oppositifolium	★	★	★	★	★	★	★	★	11	*29·7*	1	*2·7*					25	*67·6*									21	*56·8*	16	*43·2*	37	6
Cicerbita macrophylla	★			★				★													6	*85·7*			1	*14·3*	3	*42·9*	4	*57·1*	7	*
Cichorium intybus	★	★	★	★	★	★	★	★			2	*4·3*	15	*31·9*			1	*2·1*			8	*17·0*	1	*2·1*	20	*42·6*	15	*31·9*	32	*68·1*	47	7
Circaea alpina agg.	★				★	★																										
× *intermedia*	★							★																								
lutetiana	★	★	★	★	★	★	★	★	139	*57·9*	59	*24·6*	2	*0·8*			8	*3·3*			21	*8·7*			11	*4·6*	136	*56·7*	104	*43·3*	240	35
Cirsium acaule	★	★	★	★	★	★	★	★			1	*1·1*	72	*82·8*							7	*8·0*	4	*4·6*	3	*3·4*	32	*36·8*	55	*63·2*	87	13
arvense	★	★	★	★	★	★	★	★	7	*0·5*	87	*6·8*	494	*38·5*	3	*0·2*	40	*3·1*			426	*33·2*	13	*1·0*	214	*16·7*	867	*67·5*	417	*32·5*	1284	98
dissectum	★	★	★	★	★	★	★	★					2	*40·0*	1	*20·0*	2	*40·0*									2	*40·0*	3	*60·0*	5	1
dissectum × *palustre*	★																															
eriophorum	★	★	★	★	★	★	★	★	1	*2·0*	6	*11·8*	25	*49·0*							13	*25·5*	5	*9·8*	1	*2·0*	17	*33·3*	34	*66·7*	51	7
heterophyllum						★		★																								
palustre	★	★	★	★	★	★	★	★	32	*5·3*	46	*7·6*	205	*34·0*	3	*0·5*	238	*39·5*	1	*0·2*	63	*10·4*	2	*0·3*	13	*2·2*	190	*31·5*	413	*68·5*	603	67
tuberosum				★																												
vulgare	★	★	★	★	★	★	★	★	22	*2·1*	145	*13·9*	432	*41·5*	6	*0·6*	39	*3·7*			329	*31·6*	8	*0·8*	61	*5·9*	318	*30·5*	724	*69·5*	1042	97
Cladium mariscus	★		★		★	★																										
Clematis vitalba	★	★	★	★	★	★	★	★	13	*14·4*	58	*64·4*	1	*1·1*			1	*1·1*			14	*15·6*	2	*2·2*	1	*1·1*	33	*36·7*	57	*63·3*	90	12
Clinopodium vulgare	★	★	★	★	★	★	★	★	3	*3·3*	27	*30·0*	19	*21·1*			4	*4·4*			35	*38·9*	1	*1·1*	1	*1·1*	22	*24·4*	68	*75·6*	90	13
Cochlearia danica	★		★	★		★	★														7	*100·0*					3	*42·9*	4	*57·1*	7	*
Coeloglossum viride	★	★	★	★	★	★	★	★																								
Colchicum autumnale	★	★	★	★	★	★	★	★	2	*33·3*	1	*16·7*	2	*33·3*			1	*16·7*											6	*100·0*	6	1
Conium maculatum	★	★	★	★	★	★	★	★	5	*2·1*	82	*33·9*	13	*5·4*			89	*36·8*			52	*21·5*			1	*0·4*	94	*38·8*	148	*61·2*	242	31
Conopodium majus	★	★	★	★	★	★	★	★	61	*11·5*	47	*8·8*	273	*51·3*	4	*0·8*	4	*0·8*			121	*22·7*			22	*4·1*	270	*50·8*	262	*49·2*	532	62
Conringia orientalis	★	★	★	★		★	★	★																								

Vascular Plant Check List and Analyses of Habitat and Frequency Records (contd.)

Species	Vice-County Records								Woodland		Hedgerow		Grassland		Heath		Waterside		Water		Ruderal		Quarries		Cultivated		Total Abundant		Total Occasional		Total Records	% of Random Squares
	38	23	32	33	37	39	55	57	No.	%	No.	%	No.	%	No.	%	No.	%	No.	%	No.	%	No.	%	No.	%	No.	%	No.	%	No.	
Consolida ambigua	★	★	★	★	★	★	★	★																								
Convallaria majalis	★	★	★	★	★	★	★	★	6	*85·7*	1	*14·3*															6	*85·7*	1	*14·3*	7	1
Convolvulus arvensis	★	★	★	★	★	★	★	★	4	*0·5*	104	*12·1*	120	*14·0*	1	*0·1*	4	*0·5*			406	*47·4*	12	*1·4*	206	*24·0*	515	*60·1*	342	*39·9*	857	84
Conyza canadensis	★	★	★	★	★	★	★	★													26	*96·3*			1	*3·7*	6	*22·2*	21	*77·8*	27	4
Coriandrum sativum	★	★	★	★	★	★	★																									
Corispermum leptopterum	★																															
Coronilla scorpioides	★	★				★	★																									
varia	★	★	★	★	★	★	★	★																								
Coronopus didymus	★	★		★	★	★	★	★													12	*75·0*	1	*6·2*	3	*18·7*	7	*43·7*	9	*56·2*	16	2
squamatus	★	★	★	★	★	★	★	★			2	*0·7*	38	*13·0*			18	*6·2*			190	*65·1*	1	*0·3*	43	*14·7*	99	*33·9*	193	*66·1*	292	40
Corydalis claviculata	★	★			★	★	★	★																								
lutea	★	★	★	★	★	★	★	★			3	*15·0*									6	*30·0*	11	*55·0*			5	*25·0*	15	*75·0*	20	3
solida	★	★	★	★	★	★	★																									
Corylus avellana	★	★	★	★	★	★	★	★	140	*23·1*	423	*69·9*	3	*0·5*			11	*1·8*			25	*4·1*	2	*0·3*	1	*0·2*	262	*43·3*	343	*56·7*	605	76
Cotoneaster microphyllus	★		★	★		★	★	★																								
simonsii		★	★	★																												
Crataegus laevigata	★	★	★	★	★	★	★	★	23	*8·9*	216	*84·0*	3	*1·2*			4	*1·6*			6	*2·3*	1	*0·4*	4	*1·6*	33	*12·8*	224	*87·2*	257	39
monogyna	★	★	★	★	★	★	★	★	95	*11·2*	618	*72·6*	33	*3·9*	2	*0·2*	38	*4·5*			59	*6·9*	2	*0·2*	4	*0·5*	703	*82·6*	148	*17·4*	851	97
Crepis biennis	★	★	★	★	★		★	★					3	*75·0*							1	*25·0*					2	*50·0*	2	*50·0*	4	*
capillaris	★	★	★	★	★	★	★	★	3	*0·5*	31	*4·9*	233	*37·2*	3	*0·5*	5	*0·8*			307	*49·0*	11	*1·8*	34	*5·4*	219	*34·9*	408	*65·1*	627	74
nicaeensis	★	★			★			★																								
paludosa	★				★	★	★	★																								
setosa	★	★	★	★	★		★																									
vesicaria																																
ssp. *taraxacifolia*	★	★	★	★	★	★	★	★			10	*4·7*	30	*14·2*			1	*0·5*			136	*64·2*	7	*3·3*	28	*13·2*	57	*26·9*	155	*73·1*	212	28
Crinitaria linosyris			★				★																									
Crocosmia × crocosmiflora	★				★			★																								
Crocus nudiflorus	★				★	★		★																								
purpureus	★	★		★	★	★		★																								
Cruciata laevipes	★	★	★	★	★	★	★	★	3	*2·1*	31	*22·0*	43	*30·5*			7	*5·0*			56	*39·7*	1	*0·7*			61	*43·3*	80	*56·7*	141	20
Cryptogramma crispa					★			★																								
Cuscuta cesatiana	★																															
epilinum	★	★	★		★		★																									
epithymum	★	★	★	★	★	★	★	★																								
europaea	★	★	★	★	★	★	★	★																								
suaveolens	★	★			★																											
Cyclamen hederifolium		★					★																									
Cymbalaria muralis	★	★	★	★	★	★	★	★			3	*5·3*									7	*12·3*	47	*82·5*			17	*29·8*	40	*70·2*	57	9
Cynoglossum germanicum	★	★	★	★	★																											
officinale	★	★	★	★	★	★	★	★			4	*25·0*	5	*31·2*			1	*6·2*			4	*25·0*	2	*12·5*			4	*25·0*	12	*75·0*	16	2
Cynosurus cristatus	★	★	★	★	★	★	★	★	7	*0·9*	14	*1·8*	603	*77·2*	2	*0·3*	9	*1·2*			134	*17·2*	4	*0·5*	8	*1·0*	553	*70·8*	228	*29·2*	781	89
echinatus	★	★	★	★	★	★	★	★																								
Cyperus longus		★	★	★																												
Cystopteris fragilis	★	★	★	★	★	★	★	★													1	*33·3*	2	*66·7*					3	*100·0*	3	*
Dactylis glomerata	★	★	★	★	★	★	★	★	31	*2·4*	127	*9·9*	552	*42·9*	2	*0·2*	30	*2·3*			497	*38·6*	9	*0·7*	38	*3·0*	1065	*82·8*	221	*17·2*	1286	99
Dactylorhiza fuchsii																																
ssp. *fuchsii*	★	★	★	★	★	★	★	★	34	*32·7*	12	*11·5*	21	*20·2*			15	*14·4*			18	*17·3*	4	*3·8*			35	*33·7*	69	*66·3*	104	14
ssp. *fuchsii*																																
× *praetermissa*	★	★	★	★			★	★																								

Species	Vice-County Records								Wood-land		Hedge-row		Grass-land		Heath		Water-side		Water		Ruderal		Quarries		Culti-vated		Total Abun-dant		Total Occa-sional		Total Records	% of Random Squares
	38	23	32	33	37	39	55	57	No.	%	No.	%	No.	%	No.	%	No.	%	No.	%	No.	%	No.	%	No.	%	No.	%	No.	%	No.	
Dactylorhiza incarnata																																
ssp. *pulchella*	★	★	★	★		★	★	★																								
maculata																																
ssp. *ericetorum*	★	★	★	★	★	★	★	★	1	*11·1*			2	*22·2*	2	*22·2*	3	*33·3*			1	*11·1*					3	*33·3*	6	*66·7*	9	1
praetermissa	★	★	★	★	★	★	★	★					1	*25·0*			3	*75·0*									2	*50·0*	2	*50·0*	4	*
purpurella			★			★		★																								
Daphne laureola	★	★	★	★	★	★	★	★	17	*39·5*	23	*53·5*									2	*4·7*			1	*2·3*	16	*37·2*	27	*62·8*	43	6
mezereum	★	★	★	★	★	★		★																								
Datura stramonium	★	★	★	★	★	★	★	★													4	*66·7*			2	*33·3*			6	*100·0*	6	1
Daucus carota ssp. *carota*	★	★	★	★	★	★	★	★			27	*10·2*	75	*28·3*			1	*0·4*			141	*53·2*	9	*3·4*	12	*4·5*	91	*34·3*	174	*65·7*	265	31
Deschampsia cespitosa	★	★	★	★	★	★	★	★	83	*10·5*	70	*8·8*	245	*30·9*	9	*1·1*	248	*31·2*			128	*16·1*	2	*0·3*	9	*1·1*	350	*44·1*	444	*55·9*	794	84
flexuosa	★	★	★	★	★	★	★	★	41	*30·4*	21	*15·6*	19	*14·1*	17	*12·6*	8	*5·9*			26	*19·3*	2	*1·5*	1	*0·7*	86	*63·7*	49	*36·3*	135	15
Descurainia sophia	★	★	★	★	★	★	★	★													5	*83·3*	1	*16·7*			3	*50·0*	3	*50·0*	6	*
Dianthus armeria	★	★		★	★	★		★																								
deltoides	★	★		★	★	★	★	★																								
gratianopolitanus	★	★	★	★	★																											
Digitalis purpurea	★	★	★	★	★	★	★	★	84	*21·3*	184	*46·6*	20	*5·1*	4	*1·0*	13	*3·3*			75	*19·0*	8	*2·0*	7	*1·8*	152	*38·5*	243	*61·5*	395	45
Digitaria sanguinalis	★	★			★	★																										
Diplotaxis erucoides					★	★																										
muralis	★	★	★	★	★	★	★	★													14	*73·7*	3	*15·8*	2	*10·5*	1	*5·3*	18	*94·7*	19	2
tenuifolia	★	★	★	★	★	★	★																									
Dipsacus fullonum	★	★	★	★	★	★	★	★	5	*2·0*	59	*23·4*	25	*9·9*			104	*41·3*			48	*19·0*	3	*1·2*	8	*3·2*	70	*27·8*	182	*72·2*	252	32
pilosus	★	★	★	★	★	★	★	★	3	*50·0*							3	*50·0*									1	*16·7*	5	*83·3*	6	1
Doronicum pardalianches	★	★	★	★	★	★	★	★																								
plantagineum		★	★		★		★																									
Draba incana						★		★																								
muralis	★	★		★		★	★	★																								
Drosera anglica						★		★																								
intermedia						★	★																									
rotundifolia	★	★	★	★	★	★	★	★																								
Dryopteris carthusiana	★	★	★	★	★	★	★	★	35	*53·8*	15	*23·1*	2	*3·1*	2	*3·1*	8	*12·3*			1	*1·5*	2	*3·1*			10	*15·4*	55	*84·6*	65	9
dilatata	★	★	★	★	★	★	★	★	111	*50·7*	47	*21·5*	1	*0·5*	3	*1·4*	35	*16·0*			15	*6·8*	6	*2·7*	1	*0·5*	49	*22·4*	170	*77·6*	219	30
filix-mas	★	★	★	★	★	★	★	★	164	*31·7*	176	*34·0*	6	*1·2*	1	*0·2*	71	*13·7*			60	*11·6*	30	*5·8*	10	*1·9*	137	*26·4*	381	*73·6*	518	65
pseudomas	★	★	★	★	★	★	★	★	5	*71·4*	2	*28·6*															1	*14·3*	6	*85·7*	7	1
Echinochloa crus-galli	★	★	★	★	★	★	★																									
Echinops sphaerocephalus	★	★				★																										
Echium lycopsis				★	★																											
vulgare	★	★	★	★	★	★	★	★													10	*90·9*	1	*9·1*			4	*36·4*	7	*63·6*	11	2
Elatine hexandra	★																															
hydropiper					★	★																										
Eleocharis acicularis	★	★	★	★	★	★	★	★																								
multicaulis	★	★	★	★	★		★	★																								
palustris	★	★	★	★	★	★	★	★	1	*0·8*	1	*0·8*	5	*4·1*	1	*0·8*	89	*72·4*	26	*21·1*							84	*68·3*	39	*31·7*	123	17
ssp. *palustris*	★	★		★	★																											
ssp. *vulgaris*	★	★	★	★	★	★	★	★																								
quinqueflora	★	★	★	★	★	★	★	★																								
uniglumis	★	★	★				★																									

Vascular Plant Check List and Analyses of Habitat and Frequency Records (contd.)

Species	Vice-County Records								Woodland		Hedgerow		Grassland		Heath		Waterside		Water		Ruderal		Quarries		Cultivated		Total Abundant		Total Occasional		Total Records	% of Random Squares
	38	23	32	33	37	39	55	57	No.	%	No.	%	No.	%	No.	%	No.	%	No.	%	No.	%	No.	%	No.	%	No.	%	No.	%	No.	
Elodea canadensis	★	★	★	★	★	★	★	★											46	*100·0*							37	*80·4*	9	*19·6*	46	7
Elymus arenarius							★	★																								
Empetrum nigrum	★					★	★	★																								
Endymion non-scriptus	★	★	★	★	★	★	★	★	220	*37·1*	272	*45·9*	34	*5·7*			10	*1·7*			53	*8·9*			4	*0·7*	329	*55·5*	264	*44·5*	593	64
Epilobium adenocaulon	★	★	★	★	★	★	★	★	15	*5·4*	30	*10·8*	23	*8·3*			112	*40·4*			73	*26·4*	4	*1·4*	20	*7·2*	82	*29·6*	195	*70·4*	277	37
adenocaulon × obscurum	★		★	★																												
adenocaulon × tetragonum	★																															
angustifolium	★	★	★	★	★	★	★	★	104	*11·7*	214	*24·0*	73	*8·2*	4	*0·4*	55	*6·2*			406	*45·6*	19	*2·1*	16	*1·8*	568	*63·7*	323	*36·3*	891	89
hirsutum	★	★	★	★	★	★	★	★	25	*2·7*	84	*9·2*	19	*2·1*	1	*0·1*	678	*73·9*	17	*1·9*	90	*9·8*			4	*0·4*	691	*75·3*	227	*24·7*	918	95
hirsutum × montanum	★	★	★	★			★	★																								
lanceolatum	★	★		★	★																											
montanum	★	★	★	★	★	★	★	★	59	*12·3*	87	*18·1*	18	*3·7*			99	*20·6*			155	*32·2*	14	*2·9*	49	*10·2*	112	*23·3*	369	*76·7*	481	63
montanum × parviflorum	★	★	★	★	★		★	★																								
montanum × roseum	★			★	★		★																									
nerterioides	★			★		★		★																								
obscurum	★	★	★	★	★	★	★	★	3	*3·2*	10	*10·8*	6	*6·5*			60	*64·5*			10	*10·8*			4	*4·3*	19	*20·4*	74	*79·6*	93	14
obscurum × palustre	★		★					★																								
obscurum × parviflorum	★	★	★	★	★		★	★																								
palustre	★	★	★	★	★	★	★	★									43	*95·6*	1	*2·2*	1	*2·2*					16	*35·6*	29	*64·4*	45	7
parviflorum	★	★	★	★	★	★	★	★	10	*3·4*	11	*3·7*	6	*2·0*			241	*81·1*	2	*0·7*	21	*7·1*	1	*0·3*	5	*1·7*	85	*28·6*	212	*71·4*	297	42
roseum	★	★	★	★	★	★	★	★			1	*3·6*					20	*71·4*	1	*3·6*	4	*14·3*	1	*3·6*	1	*3·6*	7	*25·0*	21	*75·0*	28	4
tetragonum ssp. *lamyi*	★	★	★	★	★		★																									
ssp. *tetragonum*	★	★	★	★	★	★	★	★	5	*4·0*	16	*12·9*	7	*5·6*	1	*0·8*	49	*39·5*			29	*23·4*	2	*1·6*	15	*12·1*	37	*29·8*	87	*70·2*	124	17
Epipactis atrorubens						★		★																								
helleborine	★	★	★	★	★	★	★	★	13	*72·2*	4	*22·2*											1	*5·6*			1	*5·6*	17	*94·4*	18	3
leptochila		★		★																												
palustris	★	★	★	★	★	★	★	★																								
phyllanthes	★	★		★			★	★																								
purpurata	★	★	★	★	★		★																									
Equisetum arvense	★	★	★	★	★	★	★	★	12	*1·4*	110	*13·0*	118	*13·9*	1	*0·1*	190	*22·4*	1	*0·1*	339	*40·0*	8	*0·9*	68	*8·0*	416	*49·1*	431	*50·9*	847	89
fluviatile	★	★	★	★	★	★	★	★									65	*64·4*	36	*35·6*							64	*63·4*	37	*36·6*	101	15
hyemale	★				★	★	★	★																								
palustre	★	★	★	★	★	★	★	★					1	*1·3*	3	*3·8*	71	*89·9*	4	*5·1*							46	*58·2*	33	*41·8*	79	11
sylvaticum	★	★	★	★	★	★	★	★																								
telmateia	★	★	★	★	★	★	★	★	3	*8·8*	6	*17·6*	1	*2·9*			15	*44·1*			8	*23·5*			1	*2·9*	25	*73·5*	9	*26·5*	34	4
Eranthis hyemalis	★	★	★	★	★	★	★	★																								
Erica cinerea	★	★	★		★	★	★	★	2	*40·0*					3	*60·0*											2	*40·0*	3	*60·0*	5	*
tetralix	★	★	★	★	★	★	★	★	4	*26·7*	1	*6·7*	1	*6·7*	8	*53·3*					1	*6·7*					7	*46·7*	8	*53·3*	15	1
Erigeron acer	★	★	★	★	★	★	★	★					3	*17·6*							9	*52·9*	5	*29·4*			9	*52·9*	8	*47·1*	17	2
Erinus alpinus		★		★	★			★																								

Species	Vice-County Records								Wood-land		Hedge-row		Grass-land		Heath		Water-side		Water		Ruderal		Quarries		Culti-vated		Total Abun-dant		Total Occa-sional		Total Records	% of Random Squares
	38	23	32	33	37	39	55	57	No.	%	No.	%	No.	%	No.	%	No.	%	No.	%	No.	%	No.	%	No.	%	No.	%	No.	%	No.	
Eriophorum angustifolium	★	★	★	★	★	★	★	★							4	*44·4*	5	*55·6*									6	*66·7*	3	*33·3*	9	1
latifolium		★	★	★	★		★	★																								
vaginatum	★				★	★	★	★																								
Erodium botrys	★				★																											
cicutarium																																
ssp. *cicutarium*	★	★	★	★	★	★	★	★					4	*6·9*							31	*53·4*	5	*8·6*	18	*31·0*	13	*22·4*	45	*77·6*	58	9
glutinosum	★																															
maritimum					★	★																										
moschatum	★	★	★	★	★	★	★																									
Erophila praecox	★	★	★		★		★	★																								
spathulata	★	★	★	★		★		★																								
verna	★	★	★	★	★	★	★	★					2	*2·6*							51	*66·2*	22	*28·6*	2	*2·6*	41	*53·2*	36	*46·8*	77	12
Eruca sativa	★	★	★	★			★	★																								
Erucastrum gallicum		★		★	★			★																								
nasturtiifolium	★			★	★																											
Eryngium campestre		★	★																													
Erysimum cheiranthoides	★	★	★	★	★	★	★	★					2	*16·7*							7	*58·3*	1	*8·3*	2	*16·7*	6	*50·0*	6	*50·0*	12	2
repandum	★	★		★																												
Euonymus europaeus	★	★	★	★	★	★	★	★	9	*20·5*	32	*72·7*	1	*2·3*			2	*4·5*									1	*2·3*	43	*97·7*	44	6
Eupatorium cannabinum	★	★	★	★	★	★	★	★	2	*7·1*	4	*14·3*			2	*7·1*	19	*67·9*			1	*3·6*					8	*28·6*	20	*71·4*	28	4
Euphorbia amygdaloides	★	★	★	★	★	★	★	★	11	*61·1*	2	*11·1*	2	*11·1*			1	*5·6*			2	*11·1*					4	*22·2*	14	*77·8*	18	2
corallioides		★	★																													
cyparissias	★	★	★	★	★	★	★																									
esula agg.	★	★	★	★	★	★	★	★																								
exigua	★	★	★	★	★	★	★	★			1	*0·7*	5	*3·3*			1	*0·7*			17	*11·3*	1	*0·7*	125	*83·3*	62	*41·3*	88	*58·7*	150	23
helioscopia	★	★	★	★	★	★	★	★	1	*0·2*	3	*0·7*	6	*1·4*			1	*0·2*			118	*28·4*	1	*0·2*	285	*68·7*	134	*32·3*	281	*67·7*	415	54
lathyris	★	★	★	★	★	★	★	★			1	*16·7*									3	*50·0*			2	*33·3*			6	*100·0*	6	*
peplus	★	★	★	★	★	★	★	★	1	*0·3*	3	*0·8*	5	*1·3*			1	*0·3*			108	*27·7*	6	*1·5*	266	*68·2*	149	*38·2*	241	*61·8*	390	50
platyphyllos	★	★	★	★	★		★																									
serrulata					★		★																									
Euphrasia anglica				★	★	★	★	★																								
borealis	★	★	★	★		★	★	★																								
confusa		★			★	★		★																								
micrantha	★																															
nemorosa	★	★	★	★	★	★	★	★	1	*3·7*	2	*7·4*	12	*44·4*			1	*3·7*			5	*18·5*	4	*14·8*	2	*7·4*	17	*63·0*	10	*37·0*	27	4
pseudokerneri		★	★	★																												
rostkoviana					★	★		★																								
tetraquetra		★		★				★																								
Fagopyrum esculentum	★	★	★	★	★	★	★	★													1	*25·0*			3	*75·0*	1	*25·0*	3	*75·0*	4	*
Fagus sylvatica	★	★	★	★	★	★	★	★	106	*37·9*	100	*35·7*	26	*9·3*			2	*0·7*			25	*8·9*			21	*7·5*	50	*17·9*	230	*82·1*	280	39
Falcaria vulgaris	★	★		★	★																											
Festuca altissima					★			★																								
arundinacea	★	★	★	★	★	★	★	★	4	*1·4*	29	*10·2*	96	*33·7*	1	*0·4*	48	*16·8*			104	*36·5*	1	*0·4*	2	*0·7*	90	*31·6*	195	*68·4*	285	38
gigantea	★	★	★	★	★	★	★	★	88	*26·8*	118	*36·0*	12	*3·7*			79	*24·1*			28	*8·5*			3	*0·9*	120	*36·6*	208	*63·4*	328	46
heterophylla	★	★																														
longifolia	★	★	★				★	★																								
ovina	★	★	★	★	★	★	★	★	3	*3·4*			43	*48·3*	6	*6·7*					29	*32·6*	7	*7·9*	1	*1·1*	22	*24·7*	67	*75·3*	89	13
ssp. *ovina*	★	★	★	★		★	★	★																								
ssp. *tenuifolia*	★	★	★	★		★	★	★																								

Vascular Plant Check List and Analyses of Habitat and Frequency Records (contd.)

Species	Vice-County Records								Wood-land		Hedge-row		Grass-land		Heath		Water-side		Water		Ruderal		Quarries		Culti-vated		Total Abun-dant		Total Occa-sional		Total Records	% of Random Squares
	38	23	32	33	37	39	55	57	No.	%	No.	%	No.	%	No.	%	No.	%	No.	%	No.	%	No.	%	No.	%	No.	%	No.	%	No.	
Festuca pratensis	★	★	★	★	★	★	★	★			11	2·3	314	64·6	2	0·4	16	3·3			133	27·4	1	0·2	9	1·9	276	56·8	210	43·2	486	60
rubra	★	★	★	★	★	★	★	★	7	1·0	19	2·8	364	53·9	4	0·6	2	0·3			255	37·8	15	2·2	9	1·3	418	61·9	257	38·1	675	76
ssp. *commutata*	★	★		★			★	★																								
ssp. *rubra*	★	★		★			★	★																								
× *Festulolium loliaceum*	★	★	★	★	★	★	★	★					24	82·8							5	17·2					6	20·7	23	79·3	29	5
Filago lutescens		★			★																											
minima	★	★	★	★	★	★	★	★																								
vulgaris	★	★	★	★	★	★	★	★																								
Filipendula ulmaria	★	★	★	★	★	★	★	★	30	4·1	82	11·1	62	8·4	2	0·3	493	66·7			68	9·2			2	0·3	459	62·1	280	37·9	739	78
vulgaris	★	★	★	★	★	★	★	★					23	79·3							6	20·7					9	31·0	20	69·0	29	5
Foeniculum vulgare	★	★	★	★	★	★	★	★			1	6·2	3	18·7							11	68·7			1	6·2	3	18·7	13	81·2	16	2
Fragaria ananassa	★	★	★	★			★	★																								
moschata	★	★	★	★	★	★	★	★																								
vesca	★	★	★	★	★	★	★	★	66	30·6	50	23·1	20	9·3	2	0·9	7	3·2			59	27·3	11	5·1	1	0·5	78	36·1	138	63·9	216	28
Frangula alnus	★	★	★	★	★	★	★	★	22	66·7	6	18·2	1	3·0	1	3·0	3	9·1									4	12·1	29	87·9	33	5
Fraxinus excelsior	★	★	★	★	★	★	★	★	155	18·6	560	67·2	21	2·5			27	3·2			61	7·3	3	0·4	6	0·7	326	39·1	507	60·9	833	98
Fritillaria meleagris	★	★	★	★	★	★	★																									
Fumaria bastardii	★	★			★																											
capreolata	★	★			★	★		★																								
densiflora	★	★			★																											
muralis	★	★		★	★	★	★	★																								
ssp. *boraei*	★	★		★	★	★	★	★																								
ssp. *muralis*					★	★																										
officinalis	★	★	★	★	★	★	★	★	1	0·3	8	2·6	5	1·6							112	35·9	1	0·3	185	59·3	85	27·2	227	72·8	312	44
parviflora		★				★		★																								
purpurea		★			★																											
vaillantii		★		★	★																											
Gagea lutea	★	★	★	★	★	★	★	★																								
Galanthus nivalis	★	★	★	★	★	★	★	★	2	14·3	1	7·1	2	14·3			1	7·1			6	42·9			2	14·3	5	35·7	9	64·3	14	2
Galega officinalis	★						★	★																								
Galeopsis angustifolia	★	★	★	★	★	★	★	★													1	11·1			8	88·9	1	11·1	8	88·9	9	1
bifida	★	★	★	★				★																								
speciosa	★	★	★	★	★	★	★	★																								
tetrahit s.l.	★	★	★	★	★	★	★	★	16	4·0	82	20·3	42	10·4	1	0·2	29	7·2			104	25·8			129	32·0	112	27·8	291	72·2	403	41
tetrahit s.s.	★	★	★	★	★	★	★	★																								
Galinsoga ciliata	★	★		★			★	★																								
parviflora	★	★	★	★	★	★	★	★																								
Galium aparine	★	★	★	★	★	★	★	★	99	9·1	508	46·9	51	4·7			49	4·5			224	20·7	5	0·5	148	13·7	844	77·9	240	22·1	1084	96
mollugo																																
ssp. *erectum*	★	★	★	★	★		★	★																								
ssp. *mollugo*	★	★	★	★	★	★	★	★	1	0·7	67	48·2	24	17·3			6	4·3			39	28·1	2	1·4			36	25·9	103	74·1	139	20
mollugo × *verum*	★	★	★	★	★		★	★																								
odoratum	★	★	★	★	★	★	★	★	24	72·7	5	15·2					1	3·0			3	9·1					19	57·6	14	42·4	33	5
palustre	★	★	★	★	★	★	★	★	18	5·0	6	1·7	3	0·8	2	0·6	329	91·1	3	0·8							174	48·2	187	51·8	361	50
ssp. *elongatum*	★	★	★	★	★	★	★	★																								
ssp. *palustre*	★	★	★	★	★	★	★	★																								
pumilum	★	★	★	★	★		★																									
saxatile	★	★	★	★	★	★	★	★	22	22·9	14	14·6	23	24·0	21	21·9	4	4·2			12	12·5					60	62·5	36	37·5	96	12
spurium	★	★		★		★	★	★																								

Species	Vice-County Records								Wood-land		Hedge-row		Grass-land		Heath		Water-side		Water		Ruderal		Quarries		Culti-vated		Total Abun-dant		Total Occa-sional		Total Records	% of Random Squares
	38	23	32	33	37	39	55	57	No.	%	No.	%	No.	%	No.	%	No.	%	No.	%	No.	%	No.	%	No.	%	No.	%	No.	%	No.	
Galium sterneri						★		★																								
tricornutum	★	★	★	★	★	★	★	★																	10	*100·0*	3	*30·0*	7	*70·0*	10	2
uliginosum	★	★	★	★	★	★	★	★	1	*3·1*			1	*3·1*	1	*3·1*	29	*90·6*									13	*40·6*	19	*59·4*	32	5
verum	★	★	★	★	★	★	★	★	4	*0·6*	65	*10·0*	308	*47·5*	2	*0·3*	6	*0·9*			245	*37·8*	10	*1·5*	8	*1·2*	268	*41·4*	380	*58·6*	648	76
Gastridium ventricosum	★			★	★																											
Genista anglica	★	★	★		★	★	★	★																								
tinctoria	★	★	★	★	★	★	★	★	1	*5·0*	5	*25·0*	11	*55·0*							3	*15·0*					6	*30·0*	14	*70·0*	20	3
Gentiana pneumonanthe							★	★																								
Gentianella amarella	★	★	★	★	★	★	★	★			1	*12·5*	3	*37·5*							3	*37·5*	1	*12·5*			5	*62·5*	3	*37·5*	8	1
anglica		★	★	★																												
campestris		★	★	★	★	★	★	★																								
germanica		★						★																								
Geranium columbinum	★	★	★	★	★	★	★	★			1	*25·0*	2	*50·0*							1	*25·0*					1	*25·0*	3	*75·0*	4	1
dissectum	★	★	★	★	★	★	★	★	2	*0·3*	50	*7·8*	198	*31·1*	1	*0·2*	3	*0·5*			231	*36·3*	7	*1·1*	145	*22·8*	164	*25·7*	473	*74·3*	637	72
endressii	★	★		★	★		★	★																								
lucidum	★	★	★	★	★	★	★	★			2	*18·2*									4	*36·4*	5	*45·5*			4	*36·4*	7	*63·6*	11	1
molle	★	★	★	★	★	★	★	★	4	*1·1*	32	*8·7*	80	*21·8*	1	*0·3*	1	*0·3*			150	*40·9*	7	*1·9*	92	*25·1*	81	*22·1*	286	*77·9*	367	45
phaeum	★	★	★	★	★	★	★	★																								
pratense	★	★	★	★	★	★	★	★	1	*0·6*	22	*12·8*	36	*20·9*			42	*24·4*			70	*40·7*			1	*0·6*	65	*37·8*	107	*62·2*	172	22
pusillum	★	★	★	★	★	★	★	★			11	*10·3*	19	*17·8*			2	*1·9*			39	*36·4*	7	*6·5*	29	*27·1*	30	*28·0*	77	*72·0*	107	15
pyrenaicum	★	★	★	★	★	★	★	★	1	*4·0*	5	*20·0*	7	*28·0*							10	*40·0*	1	*4·0*	1	*4·0*	8	*32·0*	17	*68·0*	25	3
robertianum	★	★	★	★	★	★	★	★	97	*13·0*	346	*46·3*	24	*3·2*			102	*13·7*			156	*20·9*	18	*2·4*	4	*0·5*	296	*39·6*	451	*60·4*	747	82
rotundifolium	★	★	★		★			★																								
sanguineum	★		★	★	★	★		★																								
sylvaticum	★			★	★	★	★	★																								
versicolor	★	★		★	★		★																									
Geum rivale	★	★	★	★	★	★	★	★	4	*40·0*	2	*20·0*					3	*30·0*			1	*10·0*					2	*20·0*	8	*80·0*	10	2
rivale × urbanum	★	★	★	★	★	★	★	★																								
urbanum	★	★	★	★	★	★	★	★	144	*19·6*	376	*51·3*	11	*1·5*			58	*7·9*			134	*18·3*	3	*0·4*	7	*1·0*	300	*40·9*	433	*59·1*	733	83
Glaucium corniculatum		★		★	★	★	★																									
flavum	★				★																											
Glaux maritima	★			★	★	★																										
Glechoma hederacea	★	★	★	★	★	★	★	★	115	*13·7*	358	*42·6*	78	*9·3*			44	*5·2*			212	*25·2*	11	*1·3*	22	*2·6*	498	*59·3*	342	*40·7*	840	85
Glyceria declinata	★	★	★	★	★	★	★	★									51	*85·0*	9	*15·0*							34	*56·7*	26	*43·3*	60	9
fluitans	★	★	★	★	★	★	★	★	2	*0·5*							348	*83·7*	66	*15·9*							293	*70·4*	123	*29·6*	416	57
maxima	★	★	★	★	★	★	★	★									112	*83·0*	22	*16·3*	1	*0·7*					106	*78·5*	29	*21·5*	135	20
× pedicellata	★	★	★	★	★	★	★	★									18	*90·0*	2	*10·0*							11	*55·0*	9	*45·0*	20	3
plicata	★	★	★	★	★	★	★	★					1	*0·6*	1	*0·6*	153	*85·5*	23	*12·8*	1	*0·6*					98	*54·7*	81	*45·3*	179	26
Gnaphalium sylvaticum	★	★	★	★	★	★	★	★			1	*14·3*			2	*28·6*	1	*14·3*			2	*28·6*			1	*14·3*	3	*42·9*	4	*57·1*	7	*
uliginosum	★	★	★	★	★	★	★	★	6	*1·9*	6	*1·9*	26	*8·3*	3	*1·0*	79	*25·1*	1	*0·3*	68	*21·6*	1	*0·3*	125	*39·7*	103	*32·7*	212	*67·3*	315	42
Groenlandia densa	★	★	★	★	★		★	★																								
Guizotia abyssinica	★																															
Gymnadenia conopsea	★	★	★	★	★	★	★	★																								
Gymnocarpium dryopteris	★	★			★	★	★	★																								
robertianum	★	★	★	★		★	★	★																								
Hammarbya paludosa						★		★																								
Hedera helix	★	★	★	★	★	★	★	★	164	*18·0*	538	*59·1*	7	*0·8*			19	*2·1*			111	*12·2*	59	*6·5*	12	*1·3*	573	*63·0*	337	*37·0*	910	95
Helianthemum nummularium	★	★	★	★	★	★	★	★					8	*80·0*							2	*20·0*					6	*60·0*	4	*40·0*	10	1

Vascular Plant Check List and Analyses of Habitat and Frequency Records (contd.)

Species	Vice-County Records 38	23	32	33	37	39	55	57	Wood-land No.	%	Hedge-row No.	%	Grass-land No.	%	Heath No.	%	Water-side No.	%	Water No.	%	Ruderal No.	%	Quarries No.	%	Culti-vated No.	%	Total Abun-dant No.	%	Total Occa-sional No.	%	Total Records No.	% of Random Squares
Helianthus rigidus	★																															
Helictotrichon pratense	★	★	★	★	★	★	★	★					14	*82·4*							3	*17·6*					2	*11·8*	15	*88·2*	17	3
pubescens	★	★	★	★	★	★	★	★	1	*1·3*	1	*1·3*	53	*69·7*							21	*27·6*					32	*42·1*	44	*57·9*	76	11
Helleborus foetidus	★	★	★	★	★	★	★	★																								
viridis	★	★	★	★	★	★	★	★																								
Heracleum mantegazzianum	★	★	★	★	★	★	★	★																								
sphondylium	★	★	★	★	★	★	★	★	45	*4·3*	262	*24·9*	220	*20·9*	1	*0·1*	43	*4·1*			449	*42·7*	5	*0·5*	26	*2·5*	640	*60·9*	411	*39·1*	1051	97
Herminium monorchis		★		★			★																									
Herniaria hirsuta	★				★	★		★																								
Hesperis matronalis	★	★	★	★	★	★	★	★	1	*3·8*	3	*11·5*	2	*7·7*			5	*19·2*			13	*50·0*			2	*7·7*	3	*11·5*	23	*88·5*	26	4
Hieracium aurantiacum				★				★																								
britannicum						★		★																								
brunneocroceum	★	★		★	★	★	★	★					8	*38·1*	1	*4·8*					10	*47·6*	1	*4·8*	1	*4·8*	6	*28·6*	15	*71·4*	21	3
calcaricola				★			★																									
caledonicum						★		★																								
cymbifolium						★		★																								
decolor						★		★																								
diaphanoides					★			★																								
diaphanum	★	★	★	★	★	★	★	★	1	*3·0*	3	*9·1*	5	*15·2*							19	*57·6*	5	*15·2*			13	*39·4*	20	*60·6*	33	5
dicella						★		★																								
eboracense	★				★			★																								
exotericum s.l.	★	★		★	★	★		★																								
flagellare	★	★	★				★																									
maculatum	★		★	★	★	★	★	★																								
perpropinquum	★	★	★	★	★	★	★	★	10	*14·5*	27	*39·1*	13	*18·8*			2	*2·9*			13	*18·8*	4	*5·8*			27	*39·1*	42	*60·9*	69	9
pilosella	★	★	★	★	★	★	★	★	3	*1·1*	13	*4·7*	107	*39·1*	4	*1·5*	1	*0·4*			119	*43·4*	20	*7·3*	7	*2·6*	135	*49·3*	139	*50·7*	274	37
rubiginosum						★		★																								
salticola	★		★		★	★	★						7	*24·1*	2	*6·9*					20	*69·0*					19	*65·5*	10	*34·5*	29	4
strumosum	★	★	★	★	★	★	★	★	5	*6·6*	10	*13·2*	14	*18·4*	1	*1·3*	1	*1·3*			38	*50·0*	7	*9·2*			31	*40·8*	45	*59·2*	76	10
subcrocatum						★		★																								
submutabile				★			★																									
umbellatum	★	★	★	★	★	★	★	★	2	*40·0*			1	*20·0*							2	*40·0*					2	*40·0*	3	*60·0*	5	*
vagum	★	★	★	★	★	★	★	★			3	*21·4*	3	*21·4*			1	*7·1*			6	*42·9*	1	*7·1*			4	*28·6*	10	*71·4*	14	2
vulgatum	★			★	★	★	★	★					3	*100·0*													2	*66·7*	1	*33·3*	3	*
Himantoglossum hircinum		★		★																												
Hippocrepis comosa	★	★	★	★	★	★	★	★																								
Hippuris vulgaris	★	★	★	★	★	★	★	★									1	*12·5*	7	*87·5*							6	*75·0*	2	*25·0*	8	1
Hirschfeldia incana				★	★		★																									
Holcus lanatus	★	★	★	★	★	★	★	★	45	*4·5*	82	*8·2*	439	*43·7*	9	*0·9*	30	*3·0*			358	*35·7*	6	*0·6*	35	*3·5*	713	*71·0*	291	*29·0*	1004	96
mollis	★	★	★	★	★	★	★	★	80	*16·5*	123	*25·4*	116	*23·9*	8	*1·6*	14	*2·9*			131	*27·0*	2	*0·4*	11	*2·3*	299	*61·6*	186	*38·4*	485	60
Hordelymus europaeus	★	★	★	★	★			★																								
Hordeum marinum			★	★	★	★																										
murinum	★	★	★	★	★	★	★	★			22	*5·1*	52	*12·0*			2	*0·5*			325	*75·1*	16	*3·7*	16	*3·7*	213	*49·2*	220	*50·8*	433	54
secalinum	★	★	★	★	★	★	★	★			13	*3·8*	249	*72·8*			1	*0·3*			71	*20·8*			8	*2·3*	213	*62·3*	129	*37·7*	342	42
Hornungia petraea						★		★																								
Hottonia palustris	★	★	★	★	★	★	★	★																								
Humulus lupulus	★	★	★	★	★	★	★	★	3	*2·4*	100	*81·3*	1	*0·8*			5	*4·1*			12	*9·8*			2	*1·6*	33	*26·8*	90	*73·2*	123	19
Hydrocharis morsus-ranae		★	★	★	★	★	★	★																								
Hydrocotyle vulgaris	★	★	★	★	★	★	★	★	1	*4·8*					3	*14·3*	16	*76·2*	1	*4·8*							17	*81·0*	4	*19·0*	21	2

Species	Vice-County Records								Wood-land		Hedge-row		Grass-land		Heath		Water-side		Water		Ruderal		Quarries		Culti-vated		Total Abun-dant		Total Occa-sional		Total Records	% of Random Squares
	38	23	32	33	37	39	55	57	No.	%	No.	%	No.	%	No.	%	No.	%	No.	%	No.	%	No.	%	No.	%	No.	%	No.	%	No.	
Hymenophyllum wilsonii						★		★																								
Hyoscyamus niger	★	★	★	★	★	★	★	★					2	*66·7*							1	*33·3*							3	*100·0*	3	*
Hypericum androsaemum	★	★	★	★	★	★	★	★																								
calycinum	★	★	★	★	★		★	★																								
× *desetangsii*		★		★				★																								
elodes	★				★	★	★	★																								
hircinum	★		★	★																												
hirsutum	★	★	★	★	★	★	★	★	19	*15·1*	36	*28·6*	29	*23·0*	1	*0·8*	7	*5·6*			33	*26·2*	1	*0·8*			36	*28·6*	90	*71·4*	126	17
humifusum	★	★	★	★	★	★	★	★	1	*4·0*	3	*12·0*	5	*20·0*	3	*12·0*	1	*4·0*			8	*32·0*	1	*4·0*	3	*12·0*	2	*8·0*	23	*92·0*	25	4
maculatum	★	★	★	★	★	★	★	★	5	*5·7*	9	*10·3*	20	*23·0*	1	*1·1*	11	*12·6*			41	*47·1*					26	*29·9*	61	*70·1*	87	13
montanum		★	★	★	★	★	★	★																								
perforatum	★	★	★	★	★	★	★	★	15	*4·1*	93	*25·1*	83	*22·4*	1	*0·3*	11	*3·0*			158	*42·7*	7	*1·9*	2	*0·5*	82	*22·2*	288	*77·8*	370	50
pulchrum	★	★	★	★	★	★	★	★	9	*23·1*	12	*30·8*	10	*25·6*	2	*5·1*	2	*5·1*			4	*10·3*					10	*25·6*	29	*74·4*	39	6
tetrapterum	★	★	★	★	★	★	★	★	9	*3·6*	17	*6·8*	24	*9·6*	2	*0·8*	173	*69·5*	1	*0·4*	23	*9·2*					36	*14·5*	213	*85·5*	249	36
Hypochoeris glabra	★	★	★		★		★																									
radicata	★	★	★	★	★	★	★	★	6	*0·9*	45	*6·8*	319	*48·3*	4	*0·6*	7	*1·1*			241	*36·5*	10	*1·5*	28	*4·2*	271	*41·1*	389	*58·9*	660	79
Iberis amara	★	★	★	★		★	★	★																								
Ilex aquifolium	★	★	★	★	★	★	★	★	124	*23·0*	360	*66·8*	5	*0·9*	1	*0·2*	3	*0·6*			25	*4·6*	2	*0·4*	19	*3·5*	124	*23·0*	415	*77·0*	539	66
Impatiens capensis	★	★	★	★	★	★	★	★			1	*2·6*					36	*94·7*	1	*2·6*							22	*57·9*	16	*42·1*	38	6
glandulifera	★	★	★	★	★	★	★	★									18	*78·3*	1	*4·3*	4	*17·4*					13	*56·5*	10	*43·5*	23	3
noli-tangere	★		★	★	★	★	★	★																								
parviflora	★	★	★	★	★	★	★	★			1	*20·0*									4	*80·0*							5	*100·0*	5	*
Inula conyza	★	★	★	★	★	★	★	★			3	*23·1*	2	*15·4*							5	*38·5*	2	*15·4*	1	*7·7*	2	*15·4*	11	*84·6*	13	2
helenium	★	★	★	★	★	★	★	★																								
Iris foetidissima	★	★	★	★	★		★	★	5	*38·5*	5	*38·5*	1	*7·7*							2	*15·4*					1	*7·7*	12	*92·3*	13	2
pseudacorus	★	★	★	★	★	★	★	★	3	*2·1*	1	*0·7*					104	*71·7*	35	*24·1*	2	*1·4*					66	*45·5*	79	*54·5*	145	21
Isatis tinctoria		★		★			★	★																								
Jasione montana	★	★	★		★	★	★	★																								
Juglans regia	★	★	★	★	★		★	★			5	*31·2*	7	*43·7*							2	*12·5*			2	*12·5*			16	*100·0*	16	3
Juncus acutiflorus	★	★	★	★	★	★	★	★	3	*4·3*	1	*1·4*	4	*5·7*	5	*7·1*	51	*72·9*	4	*5·7*	2	*2·9*					40	*57·1*	30	*42·9*	70	10
acutiflorus × *articulatus*	★				★																											
articulatus	★	★	★	★	★	★	★	★	3	*0·9*	3	*0·9*	33	*10·3*	1	*0·3*	260	*81·0*	10	*3·1*	7	*2·2*	3	*0·9*	1	*0·3*	143	*44·5*	178	*55·5*	321	46
bufonius	★	★	★	★	★	★	★	★	7	*2·3*	9	*3·0*	31	*10·2*	3	*1·0*	157	*51·8*	1	*0·3*	60	*19·8*	4	*1·3*	31	*10·2*	140	*46·2*	163	*53·8*	303	42
bulbosus s.l.	★	★	★	★	★	★	★	★	2	*16·7*					2	*16·7*	6	*50·0*			1	*8·3*	1	*8·3*			1	*8·3*	11	*91·7*	12	2
bulbosus s.s.	★																															
compressus	★	★	★	★	★	★	★	★					5	*38·5*			4	*30·8*			4	*30·8*					6	*46·2*	7	*53·8*	13	2
effusus	★	★	★	★	★	★	★	★	36	*5·3*	18	*2·6*	60	*8·8*	9	*1·3*	522	*76·5*	12	*1·8*	22	*3·2*	1	*0·1*	2	*0·3*	373	*54·7*	309	*45·3*	682	81
effusus × *inflexus*	★	★	★	★	★	★	★	★	1	*10·0*			3	*30·0*			5	*50·0*			1	*10·0*					5	*50·0*	5	*50·0*	10	*
gerardii				★	★	★																										
inflexus	★	★	★	★	★	★	★	★	14	*2·1*	10	*1·5*	108	*16·4*	3	*0·5*	467	*70·8*	5	*0·8*	46	*7·0*	7	*1·1*			340	*51·5*	320	*48·5*	660	78
kochii	★	★					★	★																								
squarrosus	★	★	★	★	★	★	★	★	3	*16·7*	1	*5·6*	1	*5·6*	10	*55·6*	3	*16·7*									10	*55·6*	8	*44·4*	18	2
subnodulosus	★	★	★	★	★	★	★	★									3	*75·0*	1	*25·0*							1	*25·0*	3	*75·0*	4	*
subuliflorus	★	★	★	★	★	★	★	★	20	*7·8*	14	*5·5*	39	*15·2*	8	*3·1*	156	*60·9*			16	*6·2*	2	*0·8*	1	*0·4*	86	*33·6*	170	*66·4*	256	37
tenuis	★	★		★	★	★	★	★							1	*33·3*	2	*66·7*											3	*100·0*	3	*
Juniperus communis		★	★	★	★	★	★	★																								
Kickxia elatine	★	★	★	★	★	★	★	★													3	*6·0*			47	*94·0*	22	*44·0*	28	*56·0*	50	8
spuria	★	★	★	★	★	★	★														2	*4·3*			45	*95·7*	25	*53·2*	22	*46·8*	47	8

Vascular Plant Check List and Analyses of Habitat and Frequency Records (contd.)

Species	Vice-County Records								Wood-land		Hedge-row		Grass-land		Heath		Water-side		Water		Ruderal		Quarries		Culti-vated		Total Abun-dant		Total Occa-sional		Total Records	% of Random Squares
	38	23	32	33	37	39	55	57	No.	%	No.	%	No.	%	No.	%	No.	%	No.	%	No.	%	No.	%	No.	%	No.	%	No.	%	No.	
Knautia arvensis	★	★	★	★	★	★	★	★			24	*14·4*	50	*29·9*							70	*41·9*	6	*3·6*	17	*10·2*	45	*26·9*	122	*73·1*	167	23
Koeleria cristata	★	★	★	★	★	★	★	★					21	*87·5*							2	*8·3*	1	*4·2*			4	*16·7*	20	*83·3*	24	4
Laburnum anagyroides	★	★	★		★	★	★		3	*27·3*	3	*27·3*	1	*9·1*							4	*36·4*							11	*100·0*	11	*
Lactuca serriola	★	★	★	★	★		★				2	*11·8*	1	*5·9*							13	*76·5*			1	*5·9*	2	*11·8*	15	*88·2*	17	2
virosa	★	★	★	★	★	★	★	★																								
Lamiastrum galeobdolon	★	★	★	★	★	★	★	★	54	*42·5*	40	*31·5*	5	*3·9*			6	*4·7*			22	*17·3*					64	*50·4*	63	*49·6*	127	16
Lamium album	★	★	★	★	★	★	★	★	20	*2·0*	302	*30·4*	90	*9·1*			23	*2·3*			478	*48·1*	15	*1·5*	65	*6·5*	583	*58·7*	410	*41·3*	993	97
amplexicaule	★	★	★	★	★	★	★	★			2	*2·4*									23	*27·4*	3	*3·6*	56	*66·7*	33	*39·3*	51	*60·7*	84	11
hybridum	★	★	★	★	★	★	★	★					2	*10·5*							5	*26·3*			12	*63·2*	4	*21·1*	15	*78·9*	19	3
maculatum	★	★	★	★	★	★	★	★			1	*5·6*	2	*11·1*							15	*83·3*					2	*11·1*	16	*88·9*	18	3
purpureum	★	★	★	★	★	★	★	★	1	*0·1*	76	*9·9*	48	*6·3*			6	*0·8*			297	*38·7*	7	*0·9*	332	*43·3*	243	*31·7*	524	*68·3*	767	81
Lappula myosotis	★	★	★				★																									
Lapsana communis	★	★	★	★	★	★	★	★	34	*3·7*	302	*32·9*	67	*7·3*	1	*0·1*	44	*4·8*			332	*36·2*	10	*1·1*	128	*13·9*	389	*42·4*	529	*57·6*	918	93
Larix decidua	★	★	★	★	★	★	★	★	83	*61·9*	36	*26·9*	5	*3·7*							5	*3·7*	1	*0·7*	4	*3·0*	28	*20·9*	106	*79·1*	134	21
Lathraea squamaria	★	★	★	★	★	★	★	★																								
Lathyrus aphaca	★	★		★	★	★	★	★																								
hirsutus	★	★		★		★		★																								
latifolius	★	★	★	★	★		★				3	*27·3*	2	*18·2*							5	*45·5*	1	*9·1*			3	*27·3*	8	*72·7*	11	*
montanus	★	★	★	★	★	★	★	★	8	*9·9*	35	*43·2*	21	*25·9*							17	*21·0*					18	*22·2*	63	*77·8*	81	11
nissolia	★	★	★	★	★	★	★	★			2	*10·0*	9	*45·0*							9	*45·0*					5	*25·0*	15	*75·0*	20	2
palustris					★		★	★																								
pratensis	★	★	★	★	★	★	★	★	8	*0·9*	121	*14·3*	316	*37·4*	1	*0·1*	44	*5·2*			338	*40·0*	3	*0·4*	13	*1·5*	396	*46·9*	448	*53·1*	844	91
sylvestris	★	★	★	★	★	★	★	★																								
tuberosus	★	★		★	★	★	★	★																								
Legousia hybrida	★	★	★	★	★	★	★	★													1	*9·1*			10	*90·9*	2	*18·2*	9	*81·8*	11	2
Lemna gibba	★	★	★	★	★	★	★	★									3	*8·1*	34	*91·9*							29	*78·4*	8	*21·6*	37	6
minor	★	★	★	★	★	★	★	★									39	*10·3*	339	*89·7*							289	*76·5*	89	*23·5*	378	56
polyrhiza	★	★	★	★	★	★	★	★																								
trisulca	★	★	★	★	★	★	★	★									5	*4·6*	103	*95·4*							76	*70·4*	32	*29·6*	108	18
Leontodon autumnalis	★	★	★	★	★	★	★	★	4	*0·6*	20	*3·0*	370	*55·2*	1	*0·1*	6	*0·9*			242	*36·1*	6	*0·9*	21	*3·1*	336	*50·1*	334	*49·9*	670	82
hispidus	★	★	★	★	★	★	★	★	2	*0·5*	7	*1·9*	210	*57·7*							140	*38·5*	3	*0·8*	2	*0·5*	120	*33·0*	244	*67·0*	364	48
taraxacoides	★	★	★	★	★	★	★	★	1	*0·6*	3	*1·8*	125	*76·2*			1	*0·6*			31	*18·9*	1	*0·6*	2	*1·2*	68	*41·5*	96	*58·5*	164	24
Leonurus cardiaca	★	★		★	★	★	★	★																								
Lepidium campestre	★	★	★	★	★	★	★	★			2	*2·4*	15	*18·1*							28	*33·7*			38	*45·8*	21	*25·3*	62	*74·7*	83	12
heterophyllum	★	★	★	★	★	★	★	★																								
latifolium			★	★	★	★	★																									
perfoliatum	★	★		★	★		★																									
ruderale	★	★	★	★	★	★	★	★													7	*100·0*					4	*57·1*	*3*	*42·9*	7	1
sativum	★	★			★		★																									
virginicum	★	★		★	★		★																									
Leucanthemum maximum	★						★																									
vulgare	★	★	★	★	★	★	★	★			14	*2·5*	313	*56·0*			4	*0·7*			186	*33·3*	9	*1·6*	33	*5·9*	189	*33·8*	370	*66·2*	559	68
Leucojum aestivum	★	★		★	★		★																									
Ligustrum ovalifolium	★	★			★																											
vulgare	★	★	★	★	★	★	★	★	66	*13·7*	367	*76·5*	3	*0·6*			5	*1·0*			29	*6·0*	2	*0·4*	8	*1·7*	169	*35·2*	311	*64·8*	480	66
Lilium martagon	★	★	★	★	★		★	★																								
Limosella aquatica	★	★	★	★	★	★	★	★																								

Species	Vice-County Records								Wood-land		Hedge-row		Grass-land		Heath		Water-side		Water		Ruderal		Quarries		Culti-vated		Total Abun-dant		Total Occa-sional		Total Records	% of Random Squares
	38	23	32	33	37	39	55	57	No.	%	No.	%	No.	%	No.	%	No.	%	No.	%	No.	%	No.	%	No.	%	No.	%	No.	%	No.	
Linaria pelisseriana	★																															
purpurea	★	★	★	★	★		★	★					2	13·3							13	86·7					3	20·0	12	80·0	15	*
repens	★	★		★	★	★	★	★					1	11·1							7	77·8	1	11·1			2	22·2	7	77·8	9	*
repens × vulgaris	★	★		★																												
vulgaris	★	★	★	★	★	★	★	★	1	0·5	18	9·6	20	10·7							137	73·3	3	1·6	8	4·3	71	38·0	116	62·0	187	25
Linum bienne	★	★		★	★	★	★	★																								
catharticum	★	★	★	★	★	★	★	★	2	1·4	5	3·6	79	57·2	1	0·7					42	30·4	7	5·1	2	1·4	66	47·8	72	52·2	138	19
perenne			★				★	★																								
usitatissimum	★	★	★	★	★	★	★	★																								
Listera cordata								★																								
ovata	★	★	★	★	★	★	★	★	23	45·1	6	11·8	7	13·7			1	2·0			11	21·6	3	5·9			13	25·5	38	74·5	51	7
Lithospermum arvense	★	★	★	★	★	★	★	★					2	6·9							3	10·3	1	3·4	23	79·3	5	17·2	24	82·8	29	4
officinale	★	★	★	★	★	★	★	★	1	5·0	10	50·0	3	15·0							3	15·0			3	15·0	2	10·0	18	90·0	20	3
purpuro-caeruleum		★	★			★		★																								
Littorella uniflora	★	★	★		★	★	★	★																								
Lobularia maritima	★	★		★	★		★																									
Lolium perenne																																
ssp. *multiflorum*	★	★	★	★	★	★	★	★	3	0·5	18	3·2	208	36·5	1	0·2					187	32·8			153	26·8	225	39·5	345	60·5	570	70
ssp. *perenne*	★	★	★	★	★	★	★	★	5	0·4	47	4·2	587	52·1	3	0·3	3	0·3			429	38·1	1	0·1	52	4·6	937	83·1	190	16·9	1127	98
temulentum	★	★	★	★	★	★	★	★																								
Lonicera caprifolium	★	★	★	★	★	★	★																									
periclymenum	★	★	★	★	★	★	★	★	140	30·0	306	65·5	3	0·6							14	3·0	3	0·6	1	0·2	163	34·9	304	65·1	467	58
xylosteum	★	★	★	★	★	★	★	★																								
Lotus corniculatus	★	★	★	★	★	★	★	★	9	1·1	37	4·6	430	53·3	2	0·2	8	1·0			293	36·3	12	1·5	16	2·0	335	41·5	472	58·5	807	92
tenuis	★	★	★	★	★	★	★	★			3	12·5	14	58·3							6	25·0	1	4·2			6	25·0	18	75·0	24	4
uliginosus	★	★	★	★	★	★	★	★	16	4·9	15	4·6	67	20·5	6	1·8	200	61·2			22	6·7	1	0·3			144	44·0	183	56·0	327	43
Lunaria annua	★	★		★				★																								
Lupinus polyphyllus	★																															
Luronium natans					★	★	★																									
Luzula campestris	★	★	★	★	★	★	★	★	12	2·2	6	1·1	422	77·6	6	1·1	6	1·1			81	14·9	3	0·6	8	1·5	269	49·4	275	50·6	544	71
forsteri		★		★	★		★																									
luzuloides	★			★	★																											
multiflora	★	★	★	★	★	★	★	★	19	35·2	7	13·0	14	25·9	5	9·3	3	5·6			6	11·1					23	42·6	31	57·4	54	7
var. *congesta*	★	★	★	★	★	★	★	★																								
pilosa	★	★	★	★	★	★	★	★	36	73·5	7	14·3	3	6·1							3	6·1					18	36·7	31	63·3	49	8
sylvatica	★	★	★	★	★	★	★	★	9	81·8	2	18·2															3	27·3	8	72·7	11	2
Lychnis flos-cuculi	★	★	★	★	★	★	★	★	24	11·9	11	5·5	32	15·9	1	0·5	126	62·7	1	0·5	6	3·0					66	32·8	135	67·2	201	27
Lycium barbarum	★	★	★	★	★	★	★	★			29	65·9	1	2·3			1	2·3			6	13·6	4	9·1	3	6·8	13	29·5	31	70·5	44	7
Lycopodium alpinum					★			★																								
clavatum	★	★	★		★	★	★	★																								
inundatum	★				★	★	★	★																								
selago	★	★		★	★	★	★	★																								
Lycopus europaeus	★	★	★	★	★	★	★	★	1	0·4					1	0·4	233	94·3	11	4·5	1	0·4					101	40·9	146	59·1	247	35
Lysimachia ciliata	★																															
nemorum	★	★	★	★	★	★	★	★	46	78·0	9	15·3					2	3·4			2	3·4					24	40·7	35	59·3	59	9
nummularia	★	★	★	★	★	★	★	★	16	8·6	15	8·1	20	10·8	1	0·5	107	57·5	1	0·5	24	12·9			2	1·1	76	40·9	110	59·1	186	26
punctata	★		★		★		★	★																								
vulgaris	★	★	★	★	★	★	★	★									5	62·5			3	37·5					3	37·5	5	62·5	8	1

Vascular Plant Check List and Analyses of Habitat and Frequency Records (contd.)

Species	Vice-County Records								Wood-land		Hedge-row		Grass-land		Heath		Water-side		Water		Ruderal		Quarries		Culti-vated		Total Abun-dant		Total Occa-sional		Total Records	% of Random Squares
	38	23	32	33	37	39	55	57	No.	%	No.	%	No.	%	No.	%	No.	%	No.	%	No.	%	No.	%	No.	%	No.	%	No.	%	No.	
Lythrum hyssopifolia	★	★	★		★	★	★	★																								
junceum	★																															
salicaria	★	★	★	★	★	★	★	★									63	*96·9*	1	*1·5*	1	*1·5*					20	*30·8*	45	*69·2*	65	10
Mahonia aquifolium	★	★	★	★	★	★	★	★	16	*25·4*	34	*54·0*	3	*4·8*							7	*11·1*	1	*1·6*	2	*3·2*	20	*31·7*	43	*68·3*	63	9
Maianthemum bifolium	★	★																														
Malus sylvestris	★	★	★	★	★	★	★	★	74	*12·5*	451	*76·3*	14	*2·4*	1	*0·2*	16	*2·7*			29	*4·9*	3	*0·5*	3	*0·5*	69	*11·7*	522	*88·3*	591	81
ssp. *mitis*	★	★	★	★	★	★	★	★	15	*14·2*	80	*75·5*	4	*3·8*			3	*2·8*			3	*2·8*			1	*0·9*	7	*6·6*	99	*93·4*	106	16
ssp. *sylvestris*	★	★	★	★	★	★	★	★	11	*12·1*	75	*82·4*	4	*4·4*							1	*1·1*					9	*9·9*	82	*90·1*	91	14
Malva moschata	★	★	★	★	★	★	★	★	3	*2·4*	27	*21·3*	35	*27·6*							56	*44·1*	3	*2·4*	3	*2·4*	23	*18·1*	104	*81·9*	127	19
neglecta	★	★	★	★	★	★	★	★			2	*2·2*	7	*7·6*			1	*1·1*			63	*68·5*	1	*1·1*	18	*19·6*	14	*15·2*	78	*84·8*	92	14
nicaeensis	★	★				★	★																									
parviflora	★			★	★	★	★	★																								
pusilla	★	★		★	★		★																									
sylvestris	★	★	★	★	★	★	★	★			35	*11·2*	32	*10·3*							225	*72·1*	5	*1·6*	15	*4·8*	51	*16·3*	261	*83·7*	312	41
Marrubium vulgare	★	★	★	★	★	★	★	★																								
Matricaria matricarioides	★	★	★	★	★	★	★	★	2	*0·2*	10	*1·1*	58	*6·4*			2	*0·2*			643	*71·0*	6	*0·7*	184	*20·3*	587	*64·9*	318	*35·1*	905	94
recutita	★	★	★	★	★	★	★	★			5	*1·2*	32	*7·8*			1	*0·2*			179	*43·8*	2	*0·5*	190	*46·5*	191	*46·7*	218	*53·3*	409	53
Matteuccia struthiopteris	★																															
Meconopsis cambrica	★	★	★		★	★	★	★																								
Medicago arabica	★	★	★	★	★		★	★			1	*4·3*	7	*30·4*							11	*47·8*			4	*17·4*	6	*26·1*	17	*73·9*	23	3
falcata		★	★	★	★	★	★	★																								
laciniata	★				★	★																										
lupulina	★	★	★	★	★	★	★	★	2	*0·3*	19	*2·8*	238	*35·3*			6	*0·9*			331	*49·0*	17	*2·5*	62	*9·2*	236	*35·0*	439	*65·0*	675	78
minima	★		★	★	★																											
polymorpha	★	★	★	★	★	★		★																								
sativa	★	★	★	★	★	★	★	★			5	*5·1*	20	*20·4*							51	*52·0*	1	*1·0*	21	*21·4*	30	*30·6*	68	*69·4*	98	14
Melampyrum cristatum			★				★																									
pratense	★	★	★	★	★	★	★	★	11	*78·6*	2	*14·3*									1	*7·1*					8	*57·1*	6	*42·9*	14	2
Melica nutans			★	★	★	★		★																								
uniflora	★	★	★	★	★	★	★	★	33	*48·5*	24	*35·3*					4	*5·9*			7	*10·3*					29	*42·6*	39	*57·4*	68	10
Melilotus alba	★	★	★	★	★	★	★	★			2	*16·7*									6	*50·0*	3	*25·0*	1	*8·3*	5	*41·7*	7	*58·3*	12	2
altissima	★	★	★	★	★	★	★	★	3	*3·8*	3	*3·8*	9	*11·5*							55	*70·5*	5	*6·4*	3	*3·8*	28	*35·9*	50	*64·1*	78	11
indica	★	★	★	★	★	★	★	★																								
officinalis	★	★	★	★	★	★	★	★			2	*6·2*	2	*6·2*							21	*65·6*	5	*15·6*	2	*6·2*	10	*31·2*	22	*68·7*	32	5
Melissa officinalis	★	★	★	★	★	★	★	★																								
Melittis melissophyllum				★	★		★	★																								
Mentha aquatica	★	★	★	★	★	★	★	★			2	*0·7*	1	*0·3*	1	*0·3*	248	*86·1*	33	*11·5*	2	*0·7*	1	*0·3*			171	*59·4*	117	*40·6*	288	40
arvensis	★	★	★	★	★	★	★	★	6	*6·1*	4	*4·1*	13	*13·3*			28	*28·6*	1	*1·0*	9	*9·2*			37	*37·8*	50	*51·0*	48	*49·0*	98	14
× *gentilis*	★	★	★	★	★	★	★	★																								
× *piperita*	★	★	★	★	★	★	★	★																								
pulegium	★	★	★	★	★	★	★	★																								
spicata	★	★	★	★	★	★	★	★			2	*5·7*	4	*11·4*			1	*2·9*			24	*68·6*	1	*2·9*	3	*8·6*	12	*34·3*	23	*65·7*	35	5
suaveolens	★				★	★		★																								
× *verticillata*	★	★	★	★	★	★	★	★			2	*12·5*					11	*68·7*	1	*6·2*					2	*12·5*	6	*37·5*	10	*62·5*	16	2
× *villosa* var. *alopecuroides*	★	★	★	★	★		★	★																								
Menyanthes trifoliata	★	★	★	★	★	★	★	★									4	*80·0*	1	*20·0*							5	*100·0*			5	1
Mercurialis annua	★	★	★	★	★	★	★	★			1	*20·0*									3	*60·0*			1	*20·0*	1	*20·0*	4	*80·0*	5	*
perennis	★	★	★	★	★	★	★	★	117	*30·5*	187	*48·8*	4	*1·0*			13	*3·4*			61	*15·9*	1	*0·3*			266	*69·5*	117	*30·5*	383	42

Species	Vice-County Records								Wood-land		Hedge-row		Grass-land		Heath		Water-side		Water		Ruderal		Quarries		Culti-vated		Total Abun-dant		Total Occa-sional		Total Records	% of Random Squares
	38	23	32	33	37	39	55	57	No.	%	No.	%	No.	%	No.	%	No.	%	No.	%	No.	%	No.	%	No.	%	No.	%	No.	%	No.	
Mespilus germanica	★	★	★	★			★																									
Milium effusum	★	★	★	★	★	★	★	★	81	*77·9*	21	*20·2*									2	*1·9*					65	*62·5*	39	*37·5*	104	16
Mimulus guttatus	★	★	★	★	★	★	★	★																								
luteus					★			★																								
moschatus					★	★	★	★																								
Minuartia hybrida	★	★	★	★	★	★	★	★													2	*100·0*					1	*50·0*	1	*50·0*	2	*
verna						★		★																								
Misopates orontium	★	★	★	★	★	★	★	★																								
Moehringia trinervia	★	★	★	★	★	★	★	★	116	*29·4*	207	*52·4*	11	*2·8*			23	*5·8*			35	*8·9*	2	*0·5*	1	*0·3*	155	*39·2*	240	*60·8*	395	54
Moenchia erecta	★	★	★		★	★	★	★																								
Molinia caerulea	★	★	★	★	★	★	★	★	4	*15·4*	2	*7·7*	1	*3·8*	14	*53·8*	5	*19·2*									16	*61·5*	10	*38·5*	26	3
Monotropa hypopitys s.l.	★	★	★	★	★	★	★	★																								
Montia fontana	★	★	★	★	★	★	★	★							1	*10·0*	6	*60·0*	2	*20·0*	1	*10·0*					3	*30·0*	7	*70·0*	10	1
ssp. *amporitana*	★			★	★	★	★	★																								
ssp. *chondro-sperma*	★			★	★																											
ssp. *variabilis*	★		★			★	★	★																								
perfoliata	★	★	★	★	★	★	★	★			1	*9·1*	1	*9·1*			2	*18·2*			2	*18·2*	2	*18·2*	3	*27·3*	4	*36·4*	7	*63·6*	11	2
sibirica	★				★	★	★	★																								
Muscari atlanticum		★	★	★			★																									
Mycelis muralis	★	★	★	★	★	★	★	★	5	*13·5*	17	*45·9*									7	*18·9*	8	*21·6*			6	*16·2*	31	*83·8*	37	5
Myosotis arvensis	★	★	★	★	★	★	★	★	43	*5·5*	156	*19·9*	100	*12·8*			58	*7·4*			211	*26·9*	11	*1·4*	205	*26·1*	225	*28·7*	559	*71·3*	784	83
caespitosa	★	★	★	★	★	★	★	★	1	*0·4*	2	*0·9*	1	*0·4*			216	*95·6*	5	*2·2*	1	*0·4*					87	*38·5*	139	*61·5*	226	34
caespitosa × *scorpioides*	★																															
discolor	★	★	★	★	★	★	★	★	3	*7·5*	2	*5·0*	13	*32·5*	2	*5·0*	1	*2·5*			10	*25·0*			9	*22·5*	16	*40·0*	24	*60·0*	40	6
ramosissima	★	★	★	★	★	★	★	★					3	*17·6*							9	*52·9*	5	*29·4*			9	*52·9*	8	*47·1*	17	3
scorpioides	★	★	★	★	★	★	★	★			1	*0·4*					234	*93·2*	16	*6·4*							120	*47·8*	131	*52·2*	251	35
secunda	★			★	★	★	★	★																								
sylvatica	★	★	★	★	★	★	★	★	12	*27·3*	13	*29·5*	6	*13·6*			6	*13·6*			7	*15·9*					13	*29·5*	31	*70·5*	44	7
Myosoton aquaticum	★	★	★	★	★	★	★	★	1	*0·7*	1	*0·7*	4	*2·9*			126	*91·3*			6	*4·3*					49	*35·5*	89	*64·5*	138	20
Myosurus minimus	★	★	★	★	★	★	★	★					2	*15·4*											11	*84·6*	8	*61·5*	5	*38·5*	13	2
Myriophyllum alterniflorum	★		★	★	★	★	★	★																								
spicatum	★	★	★	★	★	★	★	★											24	*100·0*							14	*58·3*	10	*41·7*	24	4
verticillatum	★	★	★	★	★	★	★	★																								
Myrrhis odorata	★	★		★	★	★	★	★																								
Narcissus × *incomparabilis*	★				★																											
majalis	★	★	★	★	★	★	★	★																								
× *medioluteus*	★	★	★	★	★	★		★																								
pseudonarcissus	★	★	★	★	★	★	★	★	11	*61·1*	1	*5·6*	4	*22·2*							2	*11·1*					5	*27·8*	13	*72·2*	18	3
Nardurus maritimus		★					★	★																								
Nardus stricta	★	★	★	★	★	★	★	★	1	*4·2*			3	*12·5*	12	*50·0*	3	*12·5*			4	*16·7*	1	*4·2*			11	*45·8*	13	*54·2*	24	3
Narthecium ossifragum	★				★	★		★																								
Neottia nidus-avis	★	★	★	★	★	★	★	★	5	*100·0*																			5	*100·0*	5	1
Nepeta cataria	★	★	★	★	★	★	★	★																								
Neslia paniculata	★	★	★	★																												
Nicandra physalodes	★	★	★		★		★																									
Nuphar lutea	★	★	★	★	★	★	★	★											89	*100·0*							62	*69·7*	27	*30·3*	89	14
Nymphaea alba	★	★	★	★	★	★	★	★											16	*100·0*							5	*31·2*	11	*68·7*	16	3

Vascular Plant Check List and Analyses of Habitat and Frequency Records (contd.)

Species	Vice-County Records								Wood-land		Hedge-row		Grass-land		Heath		Water-side		Water		Ruderal		Quarries		Culti-vated		Total Abun-dant		Total Occa-sional		Total Records	% of Random Squares
	38	23	32	33	37	39	55	57	No.	%	No.	%	No.	%	No.	%	No.	%	No.	%	No.	%	No.	%	No.	%	No.	%	No.	%	No.	
Nymphoides peltata	★	★	★	★	★	★	★	★																								
Odontites verna	★	★	★	★	★	★	★	★			8	*3·7*	68	*31·5*			3	*1·4*			81	*37·5*	1	*0·5*	55	*25·5*	80	*37·0*	136	*63·0*	216	29
ssp. *serotina*	★	★	★	★	★	★	★	★																								
ssp. *verna*	★	★	★	★	★	★	★	★																								
Oenanthe aquatica	★	★	★	★	★	★	★	★																								
crocata	★	★	★	★	★	★		★																								
fistulosa	★	★	★	★	★	★	★	★									19	*86·4*	3	*13·6*							13	*59·1*	9	*40·9*	22	3
fluviatilis	★	★	★	★			★	★																								
lachenalii	★	★	★	★	★		★																									
pimpinelloides				★	★																											
silaifolia	★	★	★	★	★		★																									
Oenothera biennis	★	★	★	★	★	★	★	★					1	*6·2*							12	*75·0*	1	*6·2*	2	*12·5*	3	*18·7*	13	*81·2*	16	3
erythrosepala	★	★	★	★	★		★	★													10	*90·9*			1	*9·1*	2	*18·2*	9	*81·8*	11	2
parviflora		★																														
perennis	★				★																											
stricta	★			★	★		★																									
Omphalodes verna		★		★	★																											
Onobrychis viciifolia	★	★	★	★	★	★	★	★					1	*12·5*							7	*87·5*					3	*37·5*	5	*62·5*	8	1
Onoclea sensibilis	★																															
Ononis repens	★	★	★	★	★	★	★	★			1	*1·1*	36	*40·4*							49	*55·1*	2	*2·2*	1	*1·1*	24	*27·0*	65	*73·0*	89	14
spinosa	★	★	★	★	★	★	★	★			3	*4·9*	40	*65·6*							17	*27·9*			1	*1·6*	27	*44·3*	34	*55·7*	61	8
Onopordum acanthium	★	★	★	★	★	★	★	★			1	*11·1*	1	*11·1*			1	*11·1*			5	*55·6*			1	*11·1*	2	*22·2*	7	*77·8*	9	1
Ophioglossum vulgatum	★	★	★	★	★	★	★	★	3	*9·1*	3	*9·1*	21	*63·6*							6	*18·2*					19	*57·6*	14	*42·4*	33	5
Ophrys apifera	★	★	★	★	★	★	★	★					4	*26·7*							6	*40·0*	5	*33·3*			4	*26·7*	11	*73·3*	15	2
insectifera	★	★	★	★	★	★		★																								
sphegodes		★	★																													
Orchis mascula	★	★	★	★	★	★	★	★	25	*71·4*	7	*20·0*	2	*5·7*							1	*2·9*					7	*20·0*	28	*80·0*	35	6
morio	★	★	★	★	★	★	★	★					14	*100·0*													3	*21·4*	11	*78·6*	14	2
ustulata		★	★	★	★	★	★	★																								
Origanum vulgare	★	★	★	★	★	★	★	★			2	*28·6*									5	*71·4*					1	*14·3*	6	*85·7*	7	1
Ornithogalum nutans	★	★	★	★	★		★	★																								
pyrenaicum		★		★																												
umbellatum	★	★	★	★	★	★	★	★	2	*12·5*	1	*6·2*	5	*31·2*							8	*50·0*					3	*18·7*	13	*81·2*	16	3
Ornithopus perpusillus	★	★	★	★	★	★	★	★					2	*20·0*	1	*10·0*					4	*40·0*	2	*20·0*	1	*10·0*	8	*80·0*	2	*20·0*	10	1
Orobanche elatior	★	★	★	★		★	★																									
hederae			★	★	★		★																									
minor	★	★	★	★	★		★	★																								
picridis		★			★																											
rapum-genistae	★			★	★	★	★	★																								
Osmunda regalis	★		★		★	★	★	★																								
Oxalis acetosella	★	★	★	★	★	★	★	★	90	*70·9*	26	*20·5*	2	*1·6*			5	*3·9*			4	*3·1*					59	*46·5*	68	*53·5*	127	19
corniculata	★	★	★	★	★		★	★																								
europaea	★	★		★	★		★	★																								
Panicum miliaceum	★	★	★	★	★	★	★	★																								
Papaver argemone	★	★	★	★	★	★	★	★	1	*2·6*											18	*47·4*			19	*50·0*	9	*23·7*	29	*76·3*	38	6
dubium	★	★	★	★	★	★	★	★			9	*2·3*	21	*5·5*	1	*0·3*					222	*57·7*	7	*1·8*	125	*32·5*	75	*19·5*	310	*80·5*	385	48
hybridum		★	★	★	★		★	★																								
lecoqii	★	★	★	★	★	★	★	★			2	*6·7*									12	*40·0*			16	*53·3*	10	*33·3*	20	*66·7*	30	5
rhoeas	★	★	★	★	★	★	★	★	1	*0·2*	17	*3·9*	30	*6·9*							199	*46·1*	4	*0·9*	181	*41·9*	98	*22·7*	334	*77·3*	432	54
somniferum	★	★	★	★	★	★	★	★			2	*2·8*	6	*8·5*							52	*73·2*			11	*15·5*	15	*21·1*	56	*78·9*	71	11

Species	Vice-County Records								Wood-land		Hedge-row		Grass-land		Heath		Water-side		Water		Ruderal		Quarries		Culti-vated		Total Abun-dant		Total Occa-sional		Total Records	% of Random Squares
	38	23	32	33	37	39	55	57	No.	%	No.	%	No.	%	No.	%	No.	%	No.	%	No.	%	No.	%	No.	%	No.	%	No.	%	No.	
Parapholis incurva	★																															
Parentucellia viscosa		★		★				★																								
Parietaria judaica	★	★	★	★	★	★	★	★					1	4·0							6	24·0	18	72·0			11	44·0	14	56·0	25	4
Paris quadrifolia	★	★	★	★	★	★	★	★	11	100·0																	4	36·4	7	63·6	11	2
Parnassia palustris	★	★	★	★	★	★	★	★																								
Pastinaca sativa	★	★	★	★	★	★	★	★	1	0·5	30	15·6	43	22·4			1	0·5			106	55·2	7	3·6	4	2·1	75	39·1	117	60·9	192	23
Pedicularis palustris	★	★	★	★	★	★	★	★																								
sylvatica	★	★	★	★	★	★	★	★					3	50·0			3	50·0									4	66·7	2	33·3	6	1
Pentaglottis sempervirens	★	★	★	★	★	★	★	★	3	6·8	16	36·4	5	11·4			1	2·3			18	40·9			1	2·3	15	34·1	29	65·9	44	6
Peplis portula	★	★	★	★	★	★	★	★	3	50·0	1	16·7					2	33·3									2	33·3	4	66·7	6	1
Petasites albus	★		★			★	★																									
fragrans	★	★	★	★	★	★	★	★			5	35·7									9	64·3					11	78·6	3	21·4	14	2
hybridus	★	★	★	★	★	★	★	★			4	8·9	3	6·7			27	60·0			10	22·2			1	2·2	28	62·2	17	37·8	45	6
japonicus	★			★																												
Petrorhagia nanteuilii				★	★																											
Petroselinum crispum	★	★	★	★	★	★	★	★																								
segetum	★	★	★	★	★	★	★																									
Peucedanum ostruthium					★	★		★																								
Phalaris arundinacea	★	★	★	★	★	★	★	★	4	1·1	5	1·4	4	1·1			328	90·1	18	4·9	5	1·4					240	65·9	124	34·1	364	52
canariensis	★	★	★	★	★	★	★	★			1	2·4	1	2·4			1	2·4			37	88·1	1	2·4	1	2·4	8	19·0	34	81·0	42	7
minor		★		★	★																											
paradoxa				★	★	★	★																									
Phleum arenarium	★			★																												
bertolonii	★	★	★	★	★	★	★	★	1	0·2	23	4·1	341	60·2	1	0·2	6	1·1			170	30·0			24	4·2	260	45·9	306	54·1	566	73
pratense	★	★	★	★	★	★	★	★	5	0·6	45	5·6	410	51·2	1	0·1	13	1·6			214	26·7	1	0·1	112	14·0	457	57·1	344	42·9	801	85
Phragmites australis	★	★	★	★	★	★	★	★			2	2·2					64	71·9	22	24·7	1	1·1					67	75·3	22	24·7	89	13
Phyllitis scolopendrium	★	★	★	★	★	★	★	★	3	6·5	3	6·5					8	17·4			3	6·5	29	63·0			5	10·9	41	89·1	46	7
Physalis alkekengi	★	★																														
Phyteuma spicatum	★					★																										
Picea abies	★	★			★		★		20	83·3	3	12·5													1	4·2	3	12·5	21	87·5	24	3
sitchensis	★				★																											
Picris echioides	★	★	★	★	★		★	★			7	13·5	15	28·8							16	30·8	1	1·9	13	25·0	13	25·0	39	75·0	52	7
hieracioides	★	★	★	★	★	★	★	★			4	12·9	8	25·8							18	58·1	1	3·2			7	22·6	24	77·4	31	4
Pilularia globulifera	★					★	★																									
Pimpinella major	★	★	★	★	★	★	★	★	3	4·4	13	19·1	13	19·1	1	1·5	5	7·4			31	45·6	1	1·5	1	1·5	32	47·1	36	52·9	68	8
saxifraga	★	★	★	★	★	★	★	★			9	4·5	91	46·0							96	48·5	2	1·0			57	28·8	141	71·2	198	27
Pinguicula vulgaris	★	★	★	★	★	★	★	★																								
Pinus nigra	★	★	★		★		★																									
sylvestris	★	★	★	★	★	★	★	★	114	48·5	70	29·8	34	14·5	2	0·9					14	6·0	1	0·4			36	15·3	199	84·7	235	31
Plantago coronopus	★	★	★	★	★	★	★	★																								
indica	★	★	★	★	★	★	★																									
lanceolata	★	★	★	★	★	★	★	★	4	0·4	46	4·1	433	38·6	1	0·1	2	0·2			554	49·3	12	1·1	71	6·3	813	72·4	310	27·6	1123	100
major	★	★	★	★	★	★	★	★	12	1·1	19	1·7	257	23·0			12	1·1			641	57·4	6	0·5	170	15·2	697	62·4	420	37·6	1117	97
maritima				★		★																										
media	★	★	★	★	★	★	★	★			6	2·8	110	50·7			1	0·5			94	43·3	5	2·3	1	0·5	81	37·3	136	62·7	217	28
Platanthera bifolia	★	★	★	★	★	★	★	★																								
chlorantha	★	★	★	★	★	★	★	★	15	71·4	2	9·5									3	14·3	1	4·8			3	14·3	18	85·7	21	3
Platanus × hybrida	★																															

Vascular Plant Check List and Analyses of Habitat and Frequency Records (contd.)

Species	Vice-County Records								Wood-land		Hedge-row		Grass-land		Heath		Water-side		Water		Ruderal		Quarries		Culti-vated		Total Abun-dant		Total Occa-sional		Total Records	% of Random Squares
	38	23	32	33	37	39	55	57	No.	%	No.	%	No.	%	No.	%	No.	%	No.	%	No.	%	No.	%	No.	%	No.	%	No.	%	No.	
Poa annua	★	★	★	★	★	★	★	★	23	*2·1*	32	*2·9*	243	*22·0*	1	*0·1*	8	*0·7*			527	*47·6*	12	*1·1*	260	*23·5*	869	*78·6*	237	*21·4*	1106	96
chaixii	★	★				★																										
compressa	★	★	★	★	★	★	★	★			1	*4·0*	1	*4·0*							6	*24·0*	17	*68·0*			11	*44·0*	14	*56·0*	25	4
nemoralis	★	★	★	★	★	★	★	★	47	*43·1*	53	*48·6*	3	*2·8*			1	*0·9*			5	*4·6*					49	*45·0*	60	*55·0*	109	16
palustris	★	★		★	★	★	★	★																								
pratensis	★	★	★	★	★	★	★	★																								
ssp. *angustifolia*	★	★	★	★	★	★	★	★			1	*3·7*	13	*48·1*			1	*3·7*			10	*37·0*	2	*7·4*			14	*51·9*	13	*48·1*	27	4
ssp. *irrigata*	★	★	★	★	★	★	★	★																								
ssp. *pratensis*	★	★	★	★	★	★	★	★	7	*0·9*	55	*6·7*	386	*46·9*	1	*0·1*	18	*2·2*			315	*38·3*	17	*2·1*	24	*2·9*	625	*75·9*	198	*24·1*	823	86
trivialis	★	★	★	★	★	★	★	★	58	*6·2*	99	*10·5*	304	*32·3*	2	*0·2*	101	*10·7*	2	*0·2*	293	*31·1*	4	*0·4*	79	*8·4*	739	*78·5*	203	*21·5*	942	86
Polemonium caeruleum	★	★		★	★	★	★	★																								
Polygala calcarea		★	★	★	★		★																									
serpyllifolia	★	★	★	★	★	★	★	★	2	*18·2*	1	*9·1*	2	*18·2*	2	*18·2*	1	*9·1*			3	*27·3*					2	*18·2*	9	*81·8*	11	1
vulgaris	★	★	★	★	★	★	★	★	1	*2·6*	5	*13·2*	20	*52·6*	1	*2·6*	1	*2·6*			10	*26·3*					12	*31·6*	26	*68·4*	38	6
Polygonatum x *hybridum*	★																															
multiflorum	★	★	★	★	★	★	★	★																								
odoratum	★	★		★		★		★																								
Polygonum amphibium	★	★	★	★	★	★	★	★					4	*1·6*	1	*0·4*	142	*57·3*	74	*29·8*	14	*5·6*	2	*0·8*	11	*4·4*	137	*55·2*	111	*44·8*	248	33
amplexicaule	★		★					★																								
arenastrum	★	★	★	★	★	★	★	★													16	*80·0*			4	*20·0*	15	*75·0*	5	*25·0*	20	3
arifolium	★																															
aubertii	★																															
aviculare s.l.	★	★	★	★	★	★	★	★	3	*0·3*	9	*0·9*	96	*9·5*			21	*2·1*			510	*50·3*	2	*0·2*	372	*36·7*	750	*74·0*	263	*26·0*	1013	95
bistorta	★	★	★	★	★	★	★	★	1	*4·5*	2	*9·1*	7	*31·8*			4	*18·2*			8	*36·4*					10	*45·5*	12	*54·5*	22	3
campanulatum	★																															
convolvulus	★	★	★	★	★	★	★	★	1	*0·2*	25	*4·4*	17	*3·0*			9	*1·6*			136	*24·0*	1	*0·2*	378	*66·7*	220	*38·8*	347	*61·2*	567	79
cuspidatum	★	★	★	★	★	★	★	★			16	*16·0*	6	*6·0*			3	*3·0*			60	*60·0*	4	*4·0*	11	*11·0*	35	*35·0*	65	*65·0*	100	14
dumetorum				★	★																											
hydropiper	★	★	★	★	★	★	★	★	12	*4·5*	4	*1·5*	15	*5·7*			188	*71·2*	7	*2·7*	11	*4·2*			27	*10·2*	122	*46·2*	142	*53·8*	264	36
lapathifolium	★	★	★	★	★	★	★	★	2	*0·7*	5	*1·7*	14	*4·7*			44	*14·9*			89	*30·2*			141	*47·8*	68	*23·1*	227	*76·9*	295	40
minus	★	★		★	★	★	★	★																								
mite	★	★	★	★	★	★	★	★																								
patulum	★					★																										
persicaria	★	★	★	★	★	★	★	★	6	*0·7*	18	*2·2*	62	*7·7*	1	*0·1*	130	*16·1*			259	*32·1*	1	*0·1*	330	*40·9*	357	*44·2*	450	*55·8*	807	87
polystachyum	★			★				★																								
rurivagum	★	★	★	★	★	★	★	★																								
sachalinense		★	★	★				★																								
Polypodium interjectum	★	★	★	★	★		★	★															7	*87·5*	1	*12·5*	2	*25·0*	6	*75·0*	8	*
vulgare s.l.	★	★	★	★	★	★	★	★	6	*14·0*	12	*27·9*									2	*4·7*	22	*51·2*	1	*2·3*	12	*27·9*	31	*72·1*	43	6
vulgare s.s.	★	★	★	★				★	1	*11·1*	6	*66·7*											2	*22·2*			2	*22·2*	7	*77·8*	9	*
Polypogon monspeliensis	★	★	★	★	★	★																										
Polystichum aculeatum	★	★	★	★	★	★	★	★	4	*19·0*	10	*47·6*					6	*28·6*			1	*4·8*					5	*23·8*	16	*76·2*	21	3
setiferum	★	★	★	★	★	★	★	★																								
Populus alba	★	★	★	★	★	★	★	★	8	*13·6*	34	*57·6*	7	*11·9*			3	*5·1*			7	*11·9*					10	*16·9*	49	*83·1*	59	9
× *canadensis*	★	★	★	★	★	★	★	★	28	*19·6*	69	*48·3*	13	*9·1*			16	*11·2*			14	*9·8*	3	*2·1*			19	*13·3*	124	*86·7*	143	21
canescens	★	★	★	★	★	★	★	★	17	*39·5*	18	*41·9*	1	*2·3*			5	*11·6*			2	*4·7*					14	*32·6*	29	*67·4*	43	6
gileadensis	★	★	★		★		★																									
nigra	★	★	★	★		★	★	★																								
var. *italica*	★	★	★				★		5	*10·9*	36	*78·3*	2	*4·3*							3	*6·5*					11	*23·9*	35	*76·1*	46	7

Species	Vice-County Records 38	23	32	33	37	39	55	57	Woodland No.	%	Hedgerow No.	%	Grassland No.	%	Heath No.	%	Waterside No.	%	Water No.	%	Ruderal No.	%	Quarries No.	%	Cultivated No.	%	Total Abundant No.	%	Total Occasional No.	%	Total Records No.	% of Random Squares
Populus tremula	★	★	★	★	★	★	★	★	56	33·1	80	47·3	4	2·4			19	11·2			10	5·9					42	24·9	127	75·1	169	25
trichocarpa	★																															
Potamogeton acutifolius	★		★																													
alpinus	★	★	★	★	★	★	★	★																								
berchtoldii	★	★	★	★	★	★	★	★											5	100·0							4	80·0	1	20·0	5	*
× *cadburyae*	★																															
compressus	★	★	★	★	★	★	★	★																								
× *cooperi*						★	★	★																								
crispus	★	★	★	★	★	★	★	★											75	100·0							47	62·7	28	37·3	75	12
friesii	★	★	★	★	★	★	★	★											5	100·0							3	60·0	2	40·0	5	1
gramineus			★				★																									
× *lintonii*	★				★	★	★	★																								
lucens	★	★	★	★	★	★	★	★											8	100·0							6	75·0	2	25·0	8	1
natans	★	★	★	★	★	★	★	★									9	5·0	170	95·0							136	76·0	43	24·0	179	28
nodosus	★	★																														
obtusifolius	★		★		★	★	★	★																								
pectinatus	★	★	★	★	★	★	★	★											84	100·0							56	66·7	28	33·3	84	13
perfoliatus	★	★	★	★	★	★	★	★											31	100·0							13	41·9	18	58·1	31	5
polygonifolius	★	★	★	★	★	★	★	★							2	40·0			3	60·0							2	40·0	3	60·0	5	*
praelongus	★	★	★			★		★																								
pusillus	★	★	★	★	★	★	★	★											5	100·0							2	40·0	3	60·0	5	1
× *salicifolius*	★	★	★				★																									
trichoides	★		★	★	★																											
× *zizii*	★					★	★																									
Potentilla anglica	★	★	★	★	★	★	★	★	3	6·8	4	9·1	25	56·8							11	25·0	1	2·3			14	31·8	30	68·2	44	7
anserina	★	★	★	★	★	★	★	★	7	0·9	37	4·6	145	17·9	1	0·1	96	11·9			485	59·9	6	0·7	33	4·1	409	50·5	401	49·5	810	88
argentea	★	★	★	★	★	★	★	★																								
erecta	★	★	★	★	★	★	★	★	31	11·7	16	6·0	113	42·6	14	5·3	15	5·7			71	26·8			5	1·9	118	44·5	147	55·5	265	34
intermedia		★	★	★	★	★	★																									
norvegica	★	★	★	★	★	★	★	★																								
palustris	★	★	★		★	★	★	★							1	9·1	8	72·7	2	18·2							11	100·0			11	1
recta	★	★	★	★	★		★	★																								
reptans	★	★	★	★	★	★	★	★	10	1·2	66	7·8	225	26·5	4	0·5	39	4·6			472	55·6	10	1·2	23	2·7	473	55·7	376	44·3	849	93
sterilis	★	★	★	★	★	★	★	★	32	16·5	47	24·2	45	23·2							68	35·1	2	1·0			73	37·6	121	62·4	194	27
tabernaemontani					★	★		★																								
Primula veris	★	★	★	★	★	★	★	★	36	7·9	93	20·4	216	47·5			8	1·8			97	21·3	3	0·7	2	0·4	155	34·1	300	65·9	455	50
veris × *vulgaris*	★	★	★	★	★		★	★	6	54·5	3	27·3	1	9·1							1	9·1					1	9·1	10	90·9	11	2
vulgaris	★	★	★	★	★	★	★	★	74	35·9	48	23·3	38	18·4			18	8·7			25	12·1			3	1·5	84	40·8	122	59·2	206	26
Prunella laciniata		★	★	★	★																											
vulgaris	★	★	★	★	★	★	★	★	32	4·2	33	4·3	426	55·7	4	0·5	47	6·1			158	20·7	8	1·0	57	7·5	307	40·1	458	59·9	765	83
Prunus avium	★	★	★	★	★	★	★	★	37	39·8	50	53·8	1	1·1			1	1·1			4	4·3					12	12·9	81	87·1	93	13
cerasifera	★	★	★	★	★		★	★																								
cerasus	★	★	★	★	★	★	★	★			11	84·6	1	7·7							1	7·7					3	23·1	10	76·9	13	*
domestica	★	★	★	★	★	★	★	★	11	7·5	126	86·3	2	1·4							5	3·4	1	0·7	1	0·7	35	24·0	111	76·0	146	23
ssp. *domestica*	★	★	★	★	★		★	★																								
ssp. *insititia*	★	★	★	★	★	★	★	★																								
laurocerasus	★	★	★	★	★				6	60·0	2	20·0									2	20·0					5	50·0	5	50·0	10	*
lusitanica	★																															
padus	★			★	★	★	★	★																								
spinosa	★	★	★	★	★	★	★	★	68	9·6	568	80·2	6	0·8	2	0·3	28	4·0			33	4·7	3	0·4			393	55·5	315	44·5	708	92

Vascular Plant Check List and Analyses of Habitat and Frequency Records (contd.)

Species	Vice-County Records								Wood-land		Hedge-row		Grass-land		Heath		Water-side		Water		Ruderal		Quarries		Culti-vated		Total Abun-dant		Total Occa-sional		Total Records	% of Random Squares
	38	23	32	33	37	39	55	57	No.	%	No.	%	No.	%	No.	%	No.	%	No.	%	No.	%	No.	%	No.	%	No.	%	No.	%	No.	
Pteridium aquilinum	★	★	★	★	★	★	★	★	106	*25·8*	170	*41·4*	25	*6·1*	5	*1·2*	2	*0·5*			77	*18·7*	21	*5·1*	5	*1·2*	246	*59·9*	165	*40·1*	411	49
Puccinellia distans	★	★	★	★	★	★	★	★																								
maritima				★	★	★																										
Pulicaria dysenterica	★	★	★	★	★	★	★	★			3	*4·2*	12	*16·7*			48	*66·7*			9	*12·5*					31	*43·1*	41	*56·9*	72	10
vulgaris	★	★			★		★																									
Pulmonaria longifolia				★		★																										
officinalis	★	★	★	★	★	★	★	★																								
Pulsatilla vulgaris		★	★	★			★																									
Pyrola minor	★	★		★	★	★	★	★																								
rotundifolia		★			★	★																										
Pyrus communis	★	★	★	★	★	★	★	★	2	*3·3*	50	*82·0*	5	*8·2*	1	*1·6*					3	*4·9*					2	*3·3*	59	*96·7*	61	10
Quercus cerris	★	★	★	★	★	★	★	★	23	*36·5*	28	*44·4*	7	*11·1*	1	*1·6*					4	*6·3*					7	*11·1*	56	*88·9*	63	9
ilex	★	★	★	★	★		★	★																								
petraea	★	★	★	★	★	★	★	★	17	*44·7*	13	*34·2*	5	*13·2*			2	*5·3*			1	*2·6*					5	*13·2*	33	*86·8*	38	6
petraea × *robur*	★			★	★		★	★																								
robur	★	★	★	★	★	★	★	★	195	*25·0*	479	*61·3*	48	*6·1*	2	*0·3*	4	*0·5*			46	*5·9*	5	*0·6*	2	*0·3*	297	*38·0*	484	*62·0*	781	91
Radiola linoides	★	★			★	★	★																									
Ranunculus acris	★	★	★	★	★	★	★	★	10	*1·0*	55	*5·4*	605	*59·1*			40	*3·9*			285	*27·8*	5	*0·5*	24	*2·3*	664	*64·8*	360	*35·2*	1024	96
aquatilis	★	★	★	★	★	★	★	★									21	*27·3*	55	*71·4*	1	*1·3*					58	*75·3*	19	*24·7*	77	12
aquatilis × *trichophyllus*	★																															
arvensis	★	★	★	★	★	★	★	★			1	*0·7*	4	*2·9*			2	*1·4*			22	*15·7*			111	*79·3*	40	*28·6*	100	*71·4*	140	21
auricomus	★	★	★	★	★	★	★	★	51	*24·8*	65	*31·6*	33	*16·0*			7	*3·4*			50	*24·3*					79	*38·3*	127	*61·7*	206	27
baudotii				★	★																											
bulbosus	★	★	★	★	★	★	★	★	2	*0·3*	22	*3·2*	445	*65·3*			9	*1·3*			182	*26·7*	4	*0·6*	17	*2·5*	392	*57·6*	289	*42·4*	681	78
circinatus	★	★	★	★	★	★	★	★											8	*100·0*							5	*62·5*	3	*37·5*	8	1
ficaria	★	★	★	★	★	★	★	★	138	*14·5*	238	*24·9*	187	*19·6*			213	*22·3*			164	*17·2*	3	*0·3*	11	*1·2*	558	*58·5*	396	*41·5*	954	81
ssp. *bulbifer*	★	★	★	★	★	★	★	★	19	*16·5*	40	*34·8*	20	*17·4*			23	*20·0*			12	*10·4*			1	*0·9*	75	*65·2*	40	*34·8*	115	14
ssp. *ficaria*	★	★	★	★	★	★	★	★	15	*12·6*	32	*26·9*	34	*28·6*			22	*18·5*			14	*11·8*	1	*0·8*	1	*0·8*	46	*38·7*	73	*61·3*	119	14
flammula ssp. *flammula*	★	★	★	★	★	★	★	★	3	*3·2*	4	*4·2*	8	*8·4*	4	*4·2*	74	*77·9*	2	*2·1*							38	*40·0*	57	*60·0*	95	13
fluitans	★	★	★	★	★	★	★	★																								
hederaceus	★	★	★	★	★	★	★	★							1	*5·0*	12	*60·0*	7	*35·0*							12	*60·0*	8	*40·0*	20	3
lingua	★	★	★	★	★	★	★	★									3	*75·0*	1	*25·0*							3	*75·0*	1	*25·0*	4	*
omiophyllus	★				★	★	★	★																								
parviflorus	★	★	★	★	★	★	★	★																								
peltatus	★	★	★	★	★	★	★	★									4	*13·3*	26	*86·7*							20	*66·7*	10	*33·3*	30	4
penicillatus	★	★	★	★	★	★	★	★											5	*100·0*							3	*60·0*	2	*40·0*	5	1
repens	★	★	★	★	★	★	★	★	37	*3·0*	83	*6·6*	368	*29·4*	2	*0·2*	258	*20·6*			398	*31·8*	6	*0·5*	98	*7·8*	936	*74·9*	314	*25·1*	1250	98
sardous	★				★	★	★	★																								
sceleratus	★	★	★	★	★	★	★	★	1	*0·2*	1	*0·2*					366	*91·0*	29	*7·2*	5	*1·2*					154	*38·3*	248	*61·7*	402	57
trichophyllus	★	★	★	★	★	★	★	★									8	*17·8*	37	*82·2*							27	*60·0*	18	*40·0*	45	7
Raphanus raphanistrum	★	★	★	★	★	★	★	★	2	*0·5*	3	*0·7*	28	*6·6*			3	*0·7*			165	*39·1*	3	*0·7*	218	*51·7*	109	*25·8*	313	*74·2*	422	57
Rapistrum rugosum	★	★	★	★	★	★	★	★																								
Reseda alba	★	★	★	★	★		★																									
lutea	★	★	★	★	★	★	★	★					1	*5·0*							16	*80·0*	3	*15·0*			4	*20·0*	16	*80·0*	20	3
luteola	★	★	★	★	★	★	★	★			5	*4·5*	11	*9·8*							86	*76·8*	9	*8·0*	1	*0·9*	33	*29·5*	79	*70·5*	112	16
Rhamnus catharticus	★	★	★	★	★	★	★	★	11	*6·0*	165	*90·7*	1	*0·5*			1	*0·5*			4	*2·2*					18	*9·9*	164	*90·1*	182	28

Species	Vice-County Records 38	23	32	33	37	39	55	57	Woodland No.	%	Hedgerow No.	%	Grassland No.	%	Heath No.	%	Waterside No.	%	Water No.	%	Ruderal No.	%	Quarries No.	%	Cultivated No.	%	Total Abundant No.	%	Total Occasional No.	%	Total Records No.	% of Random Squares
Rhinanthus minor																																
ssp. *minor*	★	★	★	★	★	★	★	★	2	*1·7*	5	*4·3*	73	*62·4*	1	*0·9*	3	*2·6*			23	*19·7*	3	*2·6*	7	*6·0*	57	*48·7*	60	*51·3*	117	16
ssp. *stenophyllus*		★	★	★			★	★																								
serotinus					★	★	★	★																								
Rhododendron ponticum	★	★	★	★	★	★	★	★	35	*70·0*	12	*24·0*	1	*2·0*							2	*4·0*					29	*58·0*	21	*42·0*	50	8
Rhynchosinapis cheiranthos	★	★		★	★																											
Rhynchospora alba	★	★			★	★	★																									
Ribes alpinum	★	★	★	★	★	★	★	★																								
nigrum	★	★	★	★	★	★	★	★	6	*23·1*	3	*11·5*	1	*3·8*			7	*26·9*	1	*3·8*	5	*19·2*	1	*3·8*	2	*7·7*	1	*3·8*	25	*96·2*	26	4
rubrum	★	★	★	★	★	★	★	★	29	*35·4*	24	*29·3*	2	*2·4*			23	*28·0*	1	*1·2*	2	*2·4*	1	*1·2*			6	*7·3*	76	*92·7*	82	12
uva-crispa	★	★	★	★	★	★	★	★	24	*25·3*	48	*50·5*					12	*12·6*			5	*5·3*	4	*4·2*	2	*2·1*	4	*4·2*	91	*95·8*	95	15
Robinia pseudoacacia	★	★		★	★			★																								
Roemeria hybrida						★	★																									
Rorippa amphibia	★	★	★	★	★	★	★	★									101	*85·6*	16	*13·6*	1	*0·8*					41	*34·7*	77	*65·3*	118	18
austriaca		★	★				★	★																								
islandica	★	★	★	★	★	★	★	★									85	*85·9*	2	*2·0*	11	*11·1*			1	*1·0*	26	*26·3*	73	*73·7*	99	15
microphylla	★	★	★	★	★	★	★	★									35	*79·5*	9	*20·5*							27	*61·4*	17	*38·6*	44	7
microphylla × *nasturtium-aquaticum*	★	★	★	★	★	★	★	★																								
nasturtium-aquaticum s.l.	★	★	★	★	★	★	★	★									244	*63·7*	139	*36·3*							192	*50·1*	191	*49·9*	383	50
nasturtium-aquaticum s.s.	★	★	★	★	★	★	★	★																								
sylvestris	★	★	★	★	★	★	★	★									11	*84·6*	1	*7·7*	1	*7·7*					8	*61·5*	5	*38·5*	13	2
Rosa arvensis	★	★	★	★	★	★	★	★	51	*11·8*	324	*75·2*	14	*3·2*			16	*3·7*			23	*5·3*	3	*0·7*			146	*33·9*	285	*66·1*	431	60
canina s.l.	★	★	★	★	★	★	★	★	45	*7·1*	486	*77·1*	17	*2·7*	2	*0·3*	26	*4·1*			47	*7·5*	3	*0·5*	4	*0·6*	281	*44·6*	349	*55·4*	630	82
canina s.s.	★	★	★	★	★	★	★	★			36	*83·7*	3	*7·0*							4	*9·3*					24	*55·8*	19	*44·2*	43	6
dumalis	★	★	★	★	★	★	★	★			23	*95·8*	1	*4·2*													13	*54·2*	11	*45·8*	24	*
dumetorum	★	★	★	★	★	★	★	★			14	*100·0*															5	*35·7*	9	*64·3*	14	*
obtusifolia	★	★	★	★	★	★	★	★																								
× *involuta*	★	★	★		★	★	★	★																								
pimpinellifolia	★		★	★	★	★	★	★																								
rubiginosa s.l.	★	★	★	★	★	★	★	★																								
agrestis	★	★			★		★	★																								
elliptica	★																															
micrantha	★	★	★	★	★	★	★	★																								
rubiginosa s.s.	★	★	★	★	★	★	★	★																								
rugosa	★				★																											
stylosa	★	★	★	★	★		★																									
villosa s.l.	★	★	★	★	★	★	★	★			8	*100·0*																	8	*100·0*	8	1
sherardii	★	★	★	★	★		★	★																								
tomentosa	★	★	★	★	★	★	★	★																								
mollis	★	★	★	★	★	★	★	★																								
Rubus caesius	★	★	★	★	★	★	★	★	5	*4·7*	50	*47·2*	8	*7·5*			21	*19·8*			21	*19·8*	1	*0·9*			47	*44·3*	59	*55·7*	106	15
chamaemorus						★		★																								
idaeus	★	★	★	★	★	★	★	★	73	*28·9*	84	*33·2*	14	*5·5*	2	*0·8*	16	*6·3*			60	*23·7*	2	*0·8*	2	*0·8*	86	*34·0*	167	*66·0*	253	34

Vascular Plant Check List and Analyses of Habitat and Frequency Records (contd.)

Species	Vice-County Records								Wood-land		Hedge-row		Grass-land		Heath		Water-side		Water		Ruderal		Quarries		Culti-vated		Total Abun-dant		Total Occa-sional		Total Records	% of Random Squares
	38	23	32	33	37	39	55	57	No.	%	No.	%	No.	%	No.	%	No.	%	No.	%	No.	%	No.	%	No.	%	No.	%	No.	%	No.	
Rubus saxatilis				★	★	★		★																								
fruticosus agg.	★	★	★	★	★	★	★	★	137	*14·5*	569	*60·2*	39	*4·1*	3	*0·3*	41	*4·3*			137	*14·5*	14	*1·5*	5	*0·5*	754	*79·8*	191	*20·2*	945	94
albionis	★	★																														
amplificatus	★	★		★	★	★	★	★																								
apiculatus	★	★			★		★	★																								
arrheniiformis	★																															
balfourianus	★					★			2	*18·2*	8	*72·7*									1	*9·1*					6	*54·5*	5	*45·5*	11	2
bloxamii	★	★		★	★	★	★																									
calvatus	★	★				★	★	★																								
cardiophyllus	★	★	★	★	★	★	★	★																								
carpinifolius	★					★	★	★																								
conjungens	★																															
crassifolius	★				★																											
dasyphyllus	★	★	★	★	★	★	★	★	1	*12·5*	5	*62·5*									2	*25·0*					1	*12·5*	7	*87·5*	8	1
diversus	★			★		★	★																									
drymophilus	★																															
echinatoides	★	★			★	★		★																								
echinatus	★	★	★	★	★	★	★	★																								
euryanthemus	★	★	★			★	★																									
falcatus	★	★	★	★		★	★	★																								
favonii	★			★																												
oliosus	★	★		★	★	★	★	★																								
fuscus	★	★		★	★	★	★																									
granulatus	★		★			★	★	★																								
gratus	★	★			★	★		★																								
hylonomus	★			★																												
insectifolius	★			★		★	★																									
laciniatus	★	★		★				★																								
leightonii	★	★	★	★		★	★	★																								
lindebergii	★					★	★	★																								
lindleianus	★	★	★	★	★	★	★	★																								
mercicus	★																															
var. *bracteatus*	★	★				★	★																									
mucronulatus	★					★	★	★																								
murrayi	★	★		★	★	★	★		3	*25·0*	8	*66·7*									1	*8·3*					4	*33·3*	8	*66·7*	12	2
myriacanthus	★			★		★		★																								
nemoralis	★	★			★	★	★	★																								
opacus	★				★	★	★																									
pallidus	★	★				★	★	★																								
plicatus	★				★	★		★																								
polyanthemus	★	★	★	★	★	★	★	★																								
pyramidalis	★		★			★																										
rotundifolius	★						★																									
rubritinctus	★	★			★	★																										
rufescens	★	★		★	★	★	★	★																								
scaber	★	★			★	★	★																									
scabripes	★				★	★	★																									
scabrosus	★					★																										
scissus	★	★			★	★	★	★																								

Species	Vice-County Records								Wood-land		Hedge-row		Grass-land		Heath		Water-side		Water		Ruderal		Quarries		Culti-vated		Total Abun-dant		Total Occa-sional		Total Records	% of Random Squares
	38	23	32	33	37	39	55	57	No.	%	No.	%	No.	%	No.	%	No.	%	No.	%	No.	%	No.	%	No.	%	No.	%	No.	%	No.	
Rubus fruticosus agg.																																
sprengelii	★	★	★		★	★	★	★																								
sublustris	★	★	★	★		★	★	★																								
tuberculatus	★	★		★	★	★		★																								
ulmifolius	★	★	★	★	★	★	★	★			22	*95·7*											1	*4·3*			14	*60·9*	9	*39·1*	23	4
vestitus	★	★	★	★	★	★	★	★	2	*28·6*	4	*57·1*									1	*14·3*					3	*42·9*	4	*57·1*	7	1
winteri	★		★					★																								
Rumex acetosa	★	★	★	★	★	★	★	★	7	*0·8*	39	*4·7*	456	*54·5*	3	*0·4*	17	*2·0*			275	*32·9*	5	*0·6*	35	*4·2*	416	*49·7*	421	*50·3*	837	91
acetosella s.s.	★	★	★	★	★	★	★	★	11	*2·3*	22	*4·6*	155	*32·1*	12	*2·5*	1	*0·2*			207	*42·9*	14	*2·9*	61	*12·6*	232	*48·0*	251	*52·0*	483	57
alpinus						★		★																								
conglomeratus	★	★	★	★	★	★	★	★	15	*4·1*	42	*11·4*	37	*10·1*			235	*64·0*	1	*0·3*	33	*9·0*			4	*1·1*	94	*25·6*	273	*74·4*	367	50
crispus	★	★	★	★	★	★	★	★	4	*0·5*	58	*7·3*	214	*27·0*	1	*0·1*	36	*4·5*			332	*41·8*	2	*0·3*	147	*18·5*	244	*30·7*	550	*69·3*	794	88
crispus × *obtusifolius*	★	★	★	★	★	★	★	★																								
crispus × *sanguineus*		★			★																											
hydrolapathum	★	★	★	★	★	★	★	★									67	*82·7*	14	*17·3*							27	*33·3*	54	*66·7*	81	12
longifolius						★		★																								
maritimus	★	★	★	★	★	★	★																									
obtusifolius	★	★	★	★	★	★	★	★	18	*2·0*	114	*12·6*	220	*24·4*			53	*5·9*			422	*46·7*	2	*0·2*	74	*8·2*	439	*48·6*	464	*51·4*	903	93
obtusifolius × *sanguineus*		★	★		★																											
palustris		★	★	★	★		★																									
patientia						★	★																									
pulcher	★	★	★	★	★	★	★	★																								
sanguineus	★	★	★	★	★	★	★	★	55	*8·8*	153	*24·6*	84	*13·5*	1	*0·2*	206	*33·1*			105	*16·9*	2	*0·3*	17	*2·7*	285	*45·7*	338	*54·3*	623	74
tenuifolius	★				★																											
Ruscus aculeatus	★	★	★	★	★		★	★																								
Sagina apetala	★	★	★	★	★	★	★	★					1	*2·0*			3	*6·1*			29	*59·2*	10	*20·4*	6	*12·2*	14	*28·6*	35	*71·4*	49	7
ciliata	★	★	★	★	★	★	★	★					1	*8·3*							7	*58·3*	4	*33·3*			3	*25·0*	9	*75·0*	12	2
nodosa	★	★	★	★	★	★	★	★																								
procumbens	★	★	★	★	★	★	★	★	4	*1·4*	9	*3·2*	15	*5·3*	2	*0·7*	27	*9·5*			137	*48·4*	47	*16·6*	42	*14·8*	103	*36·4*	180	*63·6*	283	38
subulata						★		★																								
Sagittaria sagittifolia	★	★	★	★	★	★	★	★									5	*8·5*	54	*91·5*							19	*32·2*	40	*67·8*	59	9
Salix alba	★	★	★	★	★	★	★	★	7	*3·0*	53	*22·7*	2	*0·9*			160	*68·7*	2	*0·9*	8	*3·4*	1	*0·4*			43	*18·5*	190	*81·5*	233	35
var. *coerulea*	★	★	★	★	★		★	★																								
var. *vitellina*	★	★	★	★	★	★	★	★																								
var. *vitellina* × *babylonica*	★																															
alba × *fragilis*	★	★	★	★	★	★	★	★			7	*20·6*					25	*73·5*			1	*2·9*	1	*2·9*			3	*8·8*	31	*91·2*	34	5
aurita	★	★	★	★	★	★	★	★	5	*13·2*	12	*31·6*			1	*2·6*	16	*42·1*			4	*10·5*					8	*21·1*	30	*78·9*	38	5
aurita × *cinerea* s.l.	★	★	★	★	★	★	★	★																								
× *cinerea* ssp. *oleifolia*	★																															
× *viminalis*	★																															
calodendron	★					★																										
caprea ssp. *caprea*	★	★	★	★	★	★	★	★	44	*13·7*	124	*38·7*	5	*1·6*	2	*0·6*	90	*28·1*	1	*0·3*	48	*15·0*	5	*1·6*	1	*0·3*	45	*14·1*	275	*85·9*	320	43

Vascular Plant Check List and Analyses of Habitat and Frequency Records (contd.)

Species	Vice-County Records								Woodland		Hedgerow		Grassland		Heath		Waterside		Water		Ruderal		Quarries		Cultivated		Total Abundant		Total Occasional		Total Records	% of Random Squares
	38	23	32	33	37	39	55	57	No.	%	No.	%	No.	%	No.	%	No.	%	No.	%	No.	%	No.	%	No.	%	No.	%	No.	%	No.	
Salix caprea × *cinerea* s.l.	★																															
× *cinerea* ssp. *oleifolia*	★																															
× *cinerea* ssp. *oleifolia* × *viminalis*	★																															
× *nigricans*	★																															
× *viminalis*	★		★			★	★	★																								
cinerea ssp. *cinerea*	★																															
ssp. *oleifolia*	★	★	★	★	★	★	★	★	43	*8·7*	175	*35·4*	6	*1·2*	3	*0·6*	216	*43·7*	3	*0·6*	45	*9·1*	3	*0·6*			110	*22·3*	384	*77·7*	494	64
cinerea × *viminalis*	★	★	★	★	★	★	★	★																								
cinerea ssp. *oleifolia* × *viminalis*	★																															
decipiens	★	★	★	★	★		★																									
× *forbyana*	★				★	★		★																								
fragilis	★	★	★	★	★	★	★	★	16	*2·8*	123	*21·6*	5	*0·9*			405	*71·1*	2	*0·4*	18	*3·2*			1	*0·2*	183	*32·1*	387	*67·9*	570	74
var. *latifolia*	★																															
var. *russelliana*	★	★			★	★	★	★																								
× *laurina*	★					★																										
nigricans	★	★																														
pentandra	★	★	★	★	★	★	★	★			2	*50·0*					2	*50·0*											4	*100·0*	4	*
purpurea	★	★	★	★	★	★	★	★			7	*24·1*	1	*3·4*			19	*65·5*			2	*6·9*					6	*20·7*	23	*79·3*	29	4
repens ssp. *repens*	★	★		★		★	★	★																								
triandra	★	★	★	★	★	★	★	★	1	*5·9*	4	*23·5*					10	*58·8*			2	*11·8*					3	*17·6*	14	*82·4*	17	3
triandra × *viminalis*	★	★					★	★																								
viminalis	★	★	★	★	★	★	★	★	4	*3·2*	27	*21·6*	2	*1·6*	1	*0·8*	76	*60·8*			14	*11·2*	1	*0·8*			23	*18·4*	102	*81·6*	125	19
Salsola kali	★	★	★	★		★																										
Salvia horminoides	★	★	★	★	★	★	★																									
pratensis	★	★	★	★	★		★	★																								
verticillata		★	★	★	★	★	★																									
Sambucus ebulus	★	★	★	★	★	★	★	★																								
nigra	★	★	★	★	★	★	★	★	148	*15·9*	604	*65·1*	8	*0·9*			35	*3·8*			116	*12·5*	9	*1·0*	8	*0·9*	463	*49·9*	465	*50·1*	928	98
racemosa						★		★																								
Samolus valerandi	★	★	★	★	★	★	★	★																								
Sanguisorba minor																																
ssp. *minor*	★	★	★	★	★	★	★	★			8	*5·3*	83	*54·6*			3	*2·0*			51	*33·6*	4	*2·6*	3	*2·0*	56	*36·8*	96	*63·2*	152	21
ssp. *muricata*	★	★	★	★	★	★	★	★																								
officinalis	★	★	★	★	★	★	★	★			14	*5·7*	142	*58·2*	1	*0·4*	38	*15·6*			44	*18·0*	1	*0·4*	4	*1·6*	91	*37·3*	153	*62·7*	244	34
Sanicula europaea	★	★	★	★	★	★	★	★	62	*70·5*	17	*19·3*	3	*3·4*			2	*2·3*			4	*4·5*					39	*44·3*	49	*55·7*	88	13
Saponaria officinalis	★	★	★	★	★	★	★	★			3	*23·1*	1	*7·7*							9	*69·2*					10	*76·9*	3	*23·1*	13	2
Sarothamnus scoparius	★	★	★	★	★	★	★	★	17	*6·9*	77	*31·4*	37	*15·1*	3	*1·2*	2	*0·8*			95	*38·8*	6	*2·4*	8	*3·3*	48	*19·6*	197	*80·4*	245	32
Saxifraga cymbalaria	★																															
granulata	★	★	★	★	★	★	★	★	3	*11·5*			17	*65·4*							6	*23·1*					14	*53·8*	12	*46·2*	26	4
hypnoides	★	★			★	★		★																								
tridactylites	★	★	★	★	★	★	★	★															16	*100·0*			3	*18·7*	13	*81·2*	16	3
Scabiosa columbaria	★	★	★	★	★	★	★	★					4	*80·0*							1	*20·0*					3	*60·0*	2	*40·0*	5	*
Scandix pecten-veneris	★	★	★	★	★	★	★	★					3	*5·5*							4	*7·3*			48	*87·3*	10	*18·2*	45	*81·8*	55	9

Species	Vice-County Records								Woodland		Hedgerow		Grassland		Heath		Waterside		Water		Ruderal		Quarries		Cultivated		Total Abundant		Total Occasional		Total Records	% of Random Squares
	38	23	32	33	37	39	55	57	No.	%	No.	%	No.	%	No.	%	No.	%	No.	%	No.	%	No.	%	No.	%	No.	%	No.	%	No.	
Schoenus nigricans		★	★	★	★	★	★	★																								
Scirpus cespitosus ssp. *germanicus*	★			★	★	★	★	★																								
fluitans	★	★	★		★	★	★																									
lacustris	★	★	★	★	★	★	★	★									25	*31·2*	55	*68·7*							55	*68·7*	25	*31·2*	80	12
maritimus	★			★	★	★																										
setaceus	★	★	★	★	★	★	★	★									5	*71·4*	1	*14·3*	1	*14·3*					4	*57·1*	3	*42·9*	7	1
sylvaticus	★	★	★	★	★	★	★	★									16	*100·0*									9	*56·2*	7	*43·7*	16	3
tabernaemontani	★	★		★	★	★	★	★																								
Scleranthus annuus	★	★	★	★	★	★	★	★					1	*2·0*							12	*23·5*	1	*2·0*	37	*72·5*	24	*47·1*	27	*52·9*	51	
Scorzonera humilis	★																															
Scrophularia auriculata	★	★	★	★	★	★	★	★	10	*2·6*	2	*0·5*	1	*0·3*			360	*94·0*	7	*1·8*	3	*0·8*					115	*30·0*	268	*70·0*	383	54
nodosa	★	★	★	★	★	★	★	★	48	*18·8*	54	*21·2*	15	*5·9*	1	*0·4*	83	*32·5*	1	*0·4*	50	*19·6*	3	*1·2*			30	*11·8*	225	*88·2*	255	36
umbrosa	★			★	★	★	★	★																								
vernalis		★	★		★																											
Scutellaria galericulata	★	★	★	★	★	★	★	★					1	*0·7*			138	*98·6*	1	*0·7*							68	*48·6*	72	*51·4*	140	22
galericulata × *minor*	★																															
minor	★	★			★	★	★	★																								
Sedum acre	★	★	★	★	★	★	★	★			1	*1·9*	1	*1·9*							18	*34·6*	32	*61·5*			13	*25·0*	39	*75·0*	52	8
album ssp. *album*	★	★	★	★	★		★	★													5	*38·5*	8	*61·5*			3	*23·1*	10	*76·9*	13	*
anglicum			★	★	★			★																								
dasyphyllum		★	★	★	★		★	★																								
forsteranum	★		★	★	★	★		★																								
reflexum	★	★	★	★	★	★	★	★			1	*25·0*									3	*75·0*					1	*25·0*	3	*75·0*	4	*
sexangulare	★	★		★	★																											
spurium		★	★	★																												
telephium	★	★	★	★	★	★	★	★	3	*33·3*	2	*22·2*					1	*11·1*			3	*33·3*					2	*22·2*	7	*77·8*	9	*
Sempervivum tectorum	★	★	★	★	★	★	★	★																								
Senecio aquaticus	★	★	★	★	★	★	★	★	2	*2·2*			4	*4·5*			83	*93·3*									26	*29·2*	63	*70·8*	89	13
erucifolius	★	★	★	★	★	★	★	★	3	*1·3*	26	*11·6*	79	*35·1*	3	*1·3*	1	*0·4*			110	*48·9*	2	*0·9*	1	*0·4*	60	*26·7*	165	*73·3*	225	28
fluviatilis		★						★																								
integrifolius		★	★	★			★																									
jacobaea	★	★	★	★	★	★	★	★	6	*1·7*	30	*8·5*	111	*31·4*	3	*0·8*	5	*1·4*			176	*49·9*	12	*3·4*	10	*2·8*	101	*28·6*	252	*71·4*	353	44
× *londinensis*	★					★																										
squalidus	★	★	★	★	★	★	★	★	3	*0·8*	14	*3·9*	35	*9·9*			2	*0·6*			261	*73·5*	24	*6·8*	16	*4·5*	151	*42·5*	204	*57·5*	355	42
squalidus × *vulgaris*	★	★		★	★																											
sylvaticus	★	★	★	★	★	★	★	★	3	*5·1*	23	*39·0*	7	*11·9*	2	*3·4*					17	*28·8*	3	*5·1*	4	*6·8*	17	*28·8*	42	*71·2*	59	8
sylvaticus × *viscosus*					★		★																									
viscosus	★	★	★	★	★	★	★	★			3	*3·3*	4	*4·4*							77	*85·6*	5	*5·6*	1	*1·1*	24	*26·7*	66	*73·3*	90	14
vulgaris	★	★	★	★	★	★	★	★	4	*0·4*	39	*3·6*	65	*6·0*			6	*0·6*			501	*46·5*	21	*1·9*	442	*41·0*	559	*51·9*	519	*48·1*	1078	96
vulgaris var. *hibernicus*	★	★		★	★		★	★																								
Serratula tinctoria	★	★	★	★	★	★	★	★	1	*3·1*	11	*34·4*	11	*34·4*			7	*21·9*			2	*6·2*					9	*28·1*	23	*71·9*	32	5
Setaria italica	★	★		★			★																									
lutescens	★	★	★	★	★	★	★																									
verticillata	★	★		★	★																											
viridis	★	★	★	★	★	★	★	★																								

Vascular Plant Check List and Analyses of Habitat and Frequency Records (contd.)

Species	Vice-County Records								Woodland		Hedgerow		Grassland		Heath		Waterside		Water		Ruderal		Quarries		Cultivated		Total Abundant		Total Occasional		Total Records	% of Random Squares
	38	23	32	33	37	39	55	57	No.	%	No.	%	No.	%	No.	%	No.	%	No.	%	No.	%	No.	%	No.	%	No.	%	No.	%	No.	
Sherardia arvensis	★	★	★	★	★	★	★	★			4	*5·3*	17	*22·7*							7	*9·3*	4	*5·3*	43	*57·3*	22	*29·3*	53	*70·7*	75	11
Sieglingia decumbens	★	★	★	★	★	★	★	★			1	*6·7*	4	*26·7*	6	*40·0*	3	*20·0*			1	*6·7*					2	*13·3*	13	*86·7*	15	2
Silaum silaus	★	★	★	★	★	★	★	★	1	*0·7*	9	*6·3*	92	*64·8*			3	*2·1*			35	*24·6*			2	*1·4*	24	*16·9*	118	*83·1*	142	20
Silene alba	★	★	★	★	★	★	★	★	13	*2·7*	151	*30·9*	50	*10·2*			1	*0·2*			194	*39·8*	5	*1·0*	74	*15·2*	95	*19·5*	393	*80·5*	488	58
alba × dioica	★	★	★	★	★	★	★	★			5	*26·3*	3	*15·8*			1	*5·3*			8	*42·1*			2	*10·5*	1	*5·3*	18	*94·7*	19	3
conica		★	★	★	★	★	★																									
cretica	★																															
dioica	★	★	★	★	★	★	★	★	104	*17·4*	244	*40·9*	26	*4·4*			80	*13·4*			132	*22·1*	3	*0·5*	7	*1·2*	296	*49·7*	300	*50·3*	596	64
gallica	★	★	★	★	★	★	★	★																								
maritima				★				★																								
noctiflora	★	★	★	★	★	★	★	★													2	*22·2*	1	*11·1*	6	*66·7*	3	*33·3*	6	*66·7*	9	1.
nutans	★				★	★	★	★																								
vulgaris	★	★	★	★	★	★	★	★			9	*14·1*	9	*14·1*							38	*59·4*	2	*3·1*	6	*9·4*	14	*21·9*	50	*78·1*	64	10
Silybum marianum	★	★	★	★	★	★	★	★																								
Sinapis alba	★	★	★	★	★	★	★	★					1	*3·6*							8	*28·6*			19	*67·9*	7	*25·0*	21	*75·0*	28	4
arvensis	★	★	★	★	★	★	★	★	2	*0·3*	13	*1·9*	53	*7·9*			9	*1·3*			241	*36·0*	1	*0·1*	350	*52·3*	206	*30·8*	463	*69·2*	669	82
Sison amomum	★	★	★	★	★	★	★	★	3	*1·9*	92	*59·7*	12	*7·8*			11	*7·1*			33	*21·4*	1	*0·6*	2	*1·3*	62	*40·3*	92	*59·7*	154	20
Sisymbrium altissimum	★	★	★	★	★	★	★	★													25	*92·6*			2	*7·4*	8	*29·6*	19	*70·4*	27	4
irio		★			★	★	★	★																								
officinale	★	★	★	★	★	★	★	★	3	*0·5*	66	*11·8*	44	*7·8*			14	*2·5*			353	*62·9*	9	*1·6*	72	*12·8*	135	*24·1*	426	*75·9*	561	67
orientale	★	★	★	★	★	★	★	★					2	*5·3*							31	*81·6*	1	*2·6*	4	*10·5*	7	*18·4*	31	*81·6*	38	6
Sisyrinchium bermudiana		★				★	★	★																								
Sium latifolium		★	★	★			★																									
Smyrnium olusatrum		★	★	★	★	★	★	★																								
Solanum dulcamara	★	★	★	★	★	★	★	★	41	*4·1*	415	*41·6*	13	*1·3*	2	*0·2*	370	*37·1*	14	*1·4*	116	*11·6*	10	*1·0*	16	*1·6*	389	*39·0*	608	*61·0*	997	96
nigrum	★	★	★	★	★	★	★	★	1	*1·1*	7	*7·8*	2	*2·2*			1	*1·1*			31	*34·4*	2	*2·2*	46	*51·1*	17	*18·9*	73	*81·1*	90	13
sarrachoides	★				★																											
Soleirolia soleirolii	★	★	★	★	★		★																									
Solidago canadensis	★	★	★		★		★		1	*2·5*	3	*7·5*	11	*27·5*							24	*60·0*			1	*2·5*	13	*32·5*	27	*67·5*	40	5
virgaurea	★	★	★	★	★	★	★	★			3	*50·0*									2	*33·3*	1	*16·7*					6	*100·0*	6	1
Sonchus arvensis	★	★	★	★	★	★	★	★	5	*0·9*	59	*10·5*	56	*9·9*	1	*0·2*	26	*4·6*			211	*37·5*	5	*0·9*	200	*35·5*	139	*24·7*	424	*75·3*	563	73
asper	★	★	★	★	★	★	★	★	9	*1·2*	55	*7·4*	78	*10·5*			41	*5·5*			324	*43·7*	8	*1·1*	227	*30·6*	195	*26·3*	547	*73·7*	742	86
oleraceus	★	★	★	★	★	★	★	★	9	*1·1*	61	*7·8*	47	*6·0*			15	*1·9*			386	*49·2*	10	*1·3*	257	*32·7*	230	*29·3*	555	*70·7*	785	86
palustris		★	★				★																									
Sorbus aria s.l.	★	★	★	★	★	★	★	★	2	*25·0*	4	*50·0*					1	*12·5*			1	*12·5*					2	*25·0*	6	*75·0*	8	*
aria × aucuparia	★																															
aucuparia	★	★	★	★	★	★	★	★	91	*42·5*	71	*33·2*	13	*6·1*	3	*1·4*	3	*1·4*			18	*8·4*	3	*1·4*	12	*5·6*	35	*16·4*	179	*83·6*	214	28
domestica					★			★																								
intermedia s.s.	★	★	★	★		★	★																									
rupicola						★	★	★																								
torminalis	★	★	★	★	★	★	★	★	7	*43·7*	8	*50·0*					1	*6·2*									1	*6·2*	15	*93·7*	16	3
Sparganium emersum	★	★	★	★	★	★	★	★									12	*28·6*	30	*71·4*							20	*47·6*	22	*52·4*	42	6
erectum	★	★	★	★	★	★	★	★									170	*56·5*	130	*43·2*	1	*0·3*					214	*71·1*	87	*28·9*	301	45
ssp. *erectum*	★	★					★	★																								
ssp. *micro-carpum*	★	★				★	★	★																								
ssp. *neglec-tum*	★	★			★		★	★																								
ssp. *oocar-pum*	★	★					★	★																								
minimum	★		★		★	★	★																									

Species	Vice-County Records								Woodland		Hedgerow		Grassland		Heath		Waterside		Water		Ruderal		Quarries		Cultivated		Total Abundant		Total Occasional		Total Records	% of Random Squares
	38	23	32	33	37	39	55	57	No.	%	No.	%	No.	%	No.	%	No.	%	No.	%	No.	%	No.	%	No.	%	No.	%	No.	%	No.	
Spergula arvensis	★	★	★	★	★	★	★	★	2	*0·6*	3	*0·9*	13	*3·9*			2	*0·6*			68	*20·2*	1	*0·3*	247	*73·5*	171	*50·9*	165	*49·1*	336	48
var. *sativa*	★	★	★	★	★	★	★	★																								
Spergularia marina				★	★	★																										
rubra	★	★	★	★	★	★	★	★			1	*3·1*									23	*71·9*	4	*12·5*	4	*12·5*	15	*46·9*	17	*53·1*	32	5
Spiraea menziesii	★																															
salicifolia	★	★	★	★	★	★	★	★																								
Spiranthes spiralis	★	★	★	★	★	★	★	★																								
Stachys annua	★	★	★			★	★																									
arvensis	★	★	★	★	★	★	★	★			3	*15·8*									6	*31·6*			10	*52·6*	9	*47·4*	10	*52·6*	19	3
germanica		★	★	★																												
palustris	★	★	★	★	★	★	★	★			3	*3·5*	2	*2·4*			75	*88·2*			2	*2·4*			3	*3·5*	28	*32·9*	57	*67·1*	85	12
palustris × *sylvatica*	★	★	★	★	★	★	★	★																								
sylvatica	★	★	★	★	★	★	★	★	99	*11·6*	492	*57·7*	32	*3·8*	1	*0·1*	46	*5·4*			173	*20·3*	3	*0·4*	7	*0·8*	524	*61·4*	329	*38·6*	853	95
Stellaria alsine	★	★	★	★	★	★	★	★	20	*8·0*	1	*0·4*	2	*0·8*	4	*1·6*	218	*87·6*			3	*1·2*			1	*0·4*	133	*53·4*	116	*46·6*	249	35
graminea	★	★	★	★	★	★	★	★	14	*2·6*	107	*20·0*	211	*39·5*	7	*1·3*	65	*12·2*			122	*22·8*	1	*0·2*	7	*1·3*	213	*39·9*	321	*60·1*	534	64
holostea	★	★	★	★	★	★	★	★	53	*12·4*	186	*43·4*	34	*7·9*			13	*3·0*			138	*32·2*	3	*0·7*	2	*0·5*	227	*52·9*	202	*47·1*	429	47
media	★	★	★	★	★	★	★	★	41	*3·7*	62	*5·5*	148	*13·2*	3	*0·3*	30	*2·7*			377	*33·7*	9	*0·8*	450	*40·2*	756	*67·5*	364	*32·5*	1120	95
neglecta	★	★		★	★	★	★	★	1	*2·9*	5	*14·3*	2	*5·7*			22	*62·9*			4	*11·4*			1	*2·9*	13	*37·1*	22	*62·9*	35	5
nemorum					★	★		★																								
pallida	★	★	★	★	★		★	★																								
palustris	★	★	★	★	★	★	★	★																								
Stratiotes aloides		★	★			★	★																									
Succisa pratensis	★	★	★	★	★	★	★	★	6	*5·1*	3	*2·5*	71	*60·2*	2	*1·7*	21	*17·8*			14	*11·9*			1	*0·8*	34	*28·8*	84	*71·2*	118	17
Swida sanguinea	★	★	★	★	★	★	★	★	39	*13·0*	235	*78·1*	2	*0·7*			9	*3·0*			15	*5·0*			1	*0·3*	62	*20·6*	239	*79·4*	301	44
sericea	★	★																														
Symphoricarpos rivularis	★	★	★	★	★	★	★	★	21	*12·4*	113	*66·9*	4	*2·4*			10	*5·9*			17	*10·1*	1	*0·6*	3	*1·8*	68	*40·2*	101	*59·8*	169	25
Symphytum asperum	★	★	★	★	★			★																								
caucasicum	★	★																														
grandiflorum	★		★	★	★																											
officinale	★	★	★	★	★	★	★	★					3	*21·4*			2	*14·3*			9	*64·3*					4	*28·6*	10	*71·4*	14	2
orientale	★	★	★		★		★	★																								
tauricum	★																															
tuberosum	★		★	★	★	★	★	★																								
× *uplandicum*	★	★	★	★	★	★	★	★	3	*1·6*	26	*14·2*	17	*9·3*			34	*18·6*			96	*52·5*	1	*0·5*	6	*3·3*	51	*27·9*	132	*72·1*	183	25
Syringa vulgaris	★	★	★	★	★	★	★	★																								
Tamus communis	★	★	★	★	★	★	★	★	52	*9·5*	471	*86·1*	2	*0·4*			3	*0·5*			16	*2·9*	3	*0·5*			202	*36·9*	345	*63·1*	547	79
Tanacetum parthenium	★	★	★	★	★	★	★	★	1	*0·7*	12	*8·0*	8	*5·3*			8	*5·3*			99	*66·0*	12	*8·0*	10	*6·7*	31	*20·7*	119	*79·3*	150	22
vulgare	★	★	★	★	★	★	★	★			10	*10·2*	12	*12·2*			7	*7·1*			65	*66·3*	1	*1·0*	3	*3·1*	24	*24·5*	74	*75·5*	98	14
Taraxacum laevigatum agg.	★	★	★	★	★	★	★	★			1	*2·4*	19	*46·3*							16	*39·0*	5	*12·2*			17	*41·5*	24	*58·5*	41	6
brachyglossum	★	★	★		★	★	★																									
excellens	★																															
fulviforme	★	★																														
glauciniforme	★	★																														
lacistophyllum	★	★	★		★			★																								
oxoniense	★	★			★			★																								
proximum	★																															

Vascular Plant Check List and Analyses of Habitat and Frequency Records (contd.)

Species	Vice-County Records								Woodland		Hedgerow		Grassland		Heath		Waterside		Water		Ruderal		Quarries		Cultivated		Total Abundant		Total Occasional		Total Records	% of Random Squares
	38	23	32	33	37	39	55	57	No.	%	No.	%	No.	%	No.	%	No.	%	No.	%	No.	%	No.	%	No.	%	No.	%	No.	%	No.	
Taraxacum officinale agg.	★	★	★	★	★	★	★	★	16	*1·3*	63	*5·2*	421	*34·6*	2	*0·2*	12	*1·0*			585	*48·1*	11	*0·9*	106	*8·7*	827	*68·0*	389	*32·0*	1216	97
alatum	★																															
duplidentifrons	★																															
hamatum	★																															
polyodon	★																															
subhamatum	★																															
spectabile agg.	★	★	★	★	★	★		★					9	*39·1*	1	*4·3*	12	*52·2*			1	*4·3*					8	*34·8*	15	*65·2*	23	3
britannicum	★																															
euryphyllum	★																															
faeroense	★												4	*36·4*	2	*18·2*	5	*45·5*									5	*45·5*	6	*54·5*	11	1
firmum	★																															
nordstedtii	★																															
unguilobum	★																															
Taxus baccata	★	★	★	★	★	★	★	★	42	*29·8*	32	*22·7*	15	*10·6*			7	*5·0*			10	*7·1*	1	*0·7*	34	*24·1*	16	*11·3*	125	*88·7*	141	22
Teesdalia nudicaulis	★	★	★		★	★	★	★																								
Tetragonolobus maritimus		★						★																								
Teucrium chamaedrys	★	★		★	★	★																										
scordium		★	★																													
scorodonia	★	★	★	★	★	★	★	★	54	*22·9*	107	*45·3*	12	*5·1*	7	*3·0*	2	*0·8*			52	*22·0*	2	*0·8*			133	*56·4*	103	*43·6*	236	26
Thalictrum flavum	★	★	★	★	★	★	★	★			1	*6·2*	1	*6·2*			14	*87·5*									4	*25·0*	12	*75·0*	16	2
minus				★		★	★	★																								
Thelypteris limbosperma	★	★	★	★	★	★	★	★																								
palustris	★	★	★			★																										
phegopteris	★				★	★		★																								
Thesium humifusum		★		★			★																									
Thlaspi arvense	★	★	★	★	★	★	★	★					6	*4·3*							43	*30·5*	1	*0·7*	91	*64·5*	45	*31·9*	96	*68·1*	141	20
perfoliatum		★		★			★																									
Thymus drucei	★	★	★	★	★	★	★	★			1	*3·8*	23	*88·5*									2	*7·7*			11	*42·3*	15	*57·7*	26	4
pulegioides	★	★	★	★	★	★	★	★																								
Tilia cordata	★	★	★	★	★	★	★	★	4	*14·8*	21	*77·8*	1	*3·7*			1	*3·7*									2	*7·4*	25	*92·6*	27	4
platyphyllos	★	★	★	★	★	★	★	★																								
tomentosa	★	★																														
× *vulgaris*	★	★	★	★	★	★	★	★	37	*20·0*	111	*60·0*	17	*9·2*			5	*2·7*			11	*5·9*			4	*2·2*	43	*23·2*	142	*76·8*	185	28
Torilis arvensis	★	★	★	★	★	★	★	★			2	*16·7*									1	*8·3*			9	*75·0*	4	*33·3*	8	*66·7*	12	2
japonica	★	★	★	★	★	★	★	★	11	*1·8*	327	*52·9*	80	*12·9*	1	*0·2*	20	*3·2*			167	*27·0*	3	*0·5*	9	*1·5*	259	*41·9*	359	*58·1*	618	77
nodosa	★	★	★	★	★	★	★	★			2	*12·5*	5	*31·2*							4	*25·0*	2	*12·5*	3	*18·7*	7	*43·7*	9	*56·2*	16	2
Trachystemon orientalis	★			★																												
Tragopogon porrifolius	★	★	★	★	★	★	★	★																								
pratensis	★	★	★	★	★	★	★	★			2	*0·4*	164	*33·3*			1	*0·2*			311	*63·2*	5	*1·0*	9	*1·8*	77	*15·7*	415	*84·3*	492	68
ssp. *minor*	★	★		★	★	★	★	★																								
ssp. *orientalis*	★	★																														
ssp. *pratensis*	★		★	★	★			★																								
Trifolium arvense	★	★	★	★	★	★	★	★					3	*12·0*	1	*4·0*					18	*72·0*	2	*8·0*	1	*4·0*	9	*36·0*	16	*64·0*	25	4
aureum	★	★	★	★			★	★																								
campestre	★	★	★	★	★	★	★	★	1	*0·4*	6	*2·5*	95	*40·1*			2	*0·8*			112	*47·3*	5	*2·1*	16	*6·8*	73	*30·8*	164	*69·2*	237	31
dubium	★	★	★	★	★	★	★	★	2	*0·3*	14	*2·1*	312	*46·4*	4	*0·6*	1	*0·1*			273	*40·6*	13	*1·9*	54	*8·0*	245	*36·4*	428	*63·6*	673	78

Species	Vice-County Records								Wood-land		Hedge-row		Grass-land		Heath		Water-side		Water		Ruderal		Quarries		Culti-vated		Total Abun-dant		Total Occa-sional		Total Records	% of Random Squares
	38	23	32	33	37	39	55	57	No.	%	No.	%	No.	%	No.	%	No.	%	No.	%	No.	%	No.	%	No.	%	No.	%	No.	%	No.	
Trifolium fragiferum	★	★	★	★	★		★	★					11	*45·8*			1	*4·2*			12	*50·0*					7	*29·2*	17	*70·8*	24	4
glomeratum				★	★																											
hybridum	★	★	★	★	★	★	★	★	1	*0·3*	4	*1·1*	157	*42·1*							91	*24·4*	1	*0·3*	119	*31·9*	96	*25·7*	277	*74·3*	373	52
incarnatum	★	★	★	★	★	★	★	★																								
medium	★	★	★	★	★	★	★	★			10	*4·6*	102	*46·6*	2	*0·9*					101	*46·1*	2	*0·9*	2	*0·9*	74	*33·8*	145	*66·2*	219	28
micranthum	★	★	★	★	★	★	★	★					11	*68·7*							4	*25·0*			1	*6·2*	10	*62·5*	6	*37·5*	16	2
ochroleucon			★	★	★		★																									
ornithopodioides					★			★																								
pratense	★	★	★	★	★	★	★	★	5	*0·5*	25	*2·3*	573	*53·7*	3	*0·3*	9	*0·8*			365	*34·2*	6	*0·6*	81	*7·6*	658	*61·7*	409	*38·3*	1067	98
repens	★	★	★	★	★	★	★	★	8	*0·7*	25	*2·2*	550	*47·3*	6	*0·5*	18	*1·5*			459	*39·5*	3	*0·3*	93	*8·0*	927	*79·8*	235	*20·2*	1162	98
resupinatum	★	★	★	★	★	★	★																									
scabrum	★	★	★	★	★		★	★																								
striatum	★	★	★	★	★	★	★	★					4	*66·7*							2	*33·3*					3	*50·0*	3	*50·0*	6	1
subterraneum	★	★		★	★		★	★																								
Triglochin maritima						★		★																								
palustris	★	★	★	★	★	★	★	★			1	*4·3*			1	*4·3*	21	*91·3*									6	*26·1*	17	*73·9*	23	4
Trigonella caerulea	★	★				★	★																									
Tripleurospermum maritimum ssp. *inodorum*	★	★	★	★	★	★	★	★	1	*0·1*	8	*1·0*	92	*11·2*			7	*0·9*			396	*48·4*	4	*0·5*	311	*38·0*	466	*56·9*	353	*43·1*	819	89
Trisetum flavescens	★	★	★	★	★	★	★	★	1	*0·2*	20	*3·7*	06	*57·2*	3	*0·6*	1	*0·2*			192	*35·9*	7	*1·3*	5	*0·9*	251	*46·9*	284	*53·1*	535	69
Trollius europaeus						★		★																								
Tulipa sylvestris	★	★	★	★	★	★	★	★																								
Turgenia latifolia		★	★	★	★	★	★	★																								
Turritis glabra	★	★		★	★	★	★	★																								
Tussilago farfara	★	★	★	★	★	★	★	★	12	*1·5*	35	*4·3*	81	*9·8*			64	*7·8*			504	*61·2*	15	*1·8*	112	*13·6*	360	*43·7*	463	*56·3*	823	88
Typha angustifolia	★	★	★	★	★	★	★	★									2	*40·0*	3	*60·0*							4	*80·0*	1	*20·0*	5	1
latifolia	★	★	★	★	★	★	★	★							1	*0·6*	87	*50·6*	83	*48·3*	1	*0·6*					118	*68·6*	54	*31·4*	172	25
Ulex europaeus	★	★	★	★	★	★	★	★	28	*6·9*	146	*36·0*	102	*25·1*	12	*3·0*	1	*0·2*			110	*27·1*	4	*1·0*	3	*0·7*	111	*27·3*	295	*72·7*	406	53
gallii	★	★		★	★	★	★	★	3	*11·1*	5	*18·5*	4	*14·8*	10	*37·0*					5	*18·5*					8	*29·6*	19	*70·4*	27	4
minor		★	★																													
Ulmus glabra	★	★	★	★	★	★	★	★	69	*21·0*	243	*73·9*	2	*0·6*			5	*1·5*			2	*0·6*	1	*0·3*	7	*2·1*	75	*22·8*	254	*77·2*	329	49
ssp. *glabra*	★																															
ssp. *montana*	★																															
× *hollandica* var. *hollandica*	★	★	★	★		★	★	★	2	*15·4*	11	*84·6*															4	*30·8*	9	*69·2*	13	2
var. *vegeta*	★	★	★	★																												
minor agg.	★	★	★	★	★	★	★	★	3	*9·7*	25	*80·6*					2	*6·5*			1	*3·2*					6	*19·4*	25	*80·6*	31	5
angustifolia	★			★			★																									
carpinifolia	★	★	★	★	★		★																									
carpinifolia × *plotii*	★	★									8	*88·9*					1	*11·1*									1	*11·1*	8	*88·9*	9	*
coritana	★	★					★		1	*12·5*	6	*75·0*					1	*12·5*									1	*12·5*	7	*87·5*	8	*
coritana × *plotii*	★								1	*11·1*	7	*77·8*									1	*11·1*					1	*11·1*	8	*88·9*	9	*
plotii	★	★	★	★		★	★	★	2	*40·0*	2	*40·0*											1	*20·0*			1	*20·0*	4	*80·0*	5	*
sarniensis	★	★	★	★			★																									
minor agg. × *glabra*	★			★	★	★	★	★	9	*18·0*	37	*74·0*					2	*4·0*			2	*4·0*					9	*18·0*	41	*82·0*	50	8
carpinifolia × *glabra*	★			★	★	★																										
carpinifolia × *plotii* × *glabra*	★							★	1	*14·3*	5	*71·4*					1	*14·3*									2	*28·6*	5	*71·4*	7	*

Vascular Plant Check List and Analyses of Habitat and Frequency Records (contd.)

Species	Vice-County Records								Wood-land		Hedge-row		Grass-land		Heath		Water-side		Water		Ruderal		Quarries		Culti-vated		Total Abun-dant		Total Occa-sional		Total Records	% of Random Squares
	38	23	32	33	37	39	55	57	No.	%	No.	%	No.	%	No.	%	No.	%	No.	%	No.	%	No.	%	No.	%	No.	%	No.	%	No.	
Ulmus minor agg. × *glabra*																																
coritana × *glabra*	★																															
coritana × *plotii* × *glabra*	★																															
plotii × *glabra*	★			★	★	★	★	★	5	*16·7*	24	*80·0*					1	*3·3*									5	*16·7*	25	*83·3*	30	*
minor agg. × *procera*	★																															
coritana × *procera*	★																															
plotii × *procera*	★																															
procera	★	★	★	★	★	★	★	★	2	*0·4*	521	*94·0*	23	*4·2*							6	*1·1*	1	*0·2*	1	*0·2*	372	*67·1*	182	*32·9*	554	83
Umbilicus rupestris	★	★	★	★	★	★	★	★																								
Urtica dioica	★	★	★	★	★	★	★	★	128	*8·9*	411	*28·6*	174	*12·1*	2	*0·1*	141	*9·8*			503	*35·0*	11	*0·8*	67	*4·7*	1150	*80·0*	287	*20·0*	1437	99
pilulifera	★	★	★			★	★																									
urens	★	★	★	★	★	★	★	★	4	*1·7*	8	*3·5*	6	*2·6*			9	*3·9*			87	*37·7*	3	*1·3*	114	*49·4*	69	*29·9*	162	*70·1*	231	32
Utricularia australis	★																															
minor	★	★	★		★	★																										
vulgaris agg.	★	★	★	★	★	★	★	★																								
Vaccaria pyramidata	★	★	★	★	★		★																									
Vaccinium myrtillus	★	★			★	★	★	★	10	*62·5*	2	*12·5*			3	*18·7*	1	*6·2*									10	*62·5*	6	*37·5*	16	2
myrtillus × *vitis-idaea*	★					★		★																								
oxycoccos	★				★	★		★																								
vitis-idaea	★				★	★	★	★																								
Valeriana dioica	★	★	★	★	★	★	★	★	1	*4·3*			1	*4·3*	1	*4·3*	20	*87·0*									15	*65·2*	8	*34·8*	23	3
officinalis	★	★	★	★	★	★	★	★	7	*8·1*	13	*15·1*	4	*4·7*	1	*1·2*	51	*59·3*	2	*2·3*	7	*8·1*			1	*1·2*	29	*33·7*	57	*66·3*	86	12
pyrenaica						★		★																								
Valerianella carinata		★	★	★	★	★	★	★																								
dentata	★	★	★	★	★	★	★	★																								
eriocarpa				★			★	★																								
locusta	★	★	★	★	★	★	★	★	1	*3·2*			3	*9·7*			1	*3·2*			22	*71·0*	1	*3·2*	3	*9·7*	19	*61·3*	12	*38·7*	31	5
rimosa	★	★	★	★	★	★		★																								
Vallisneria spiralis				★	★																											
Verbascum blattaria	★	★	★	★	★	★	★																									
lychnitis		★		★	★	★	★	★																								
nigrum	★	★	★	★	★	★	★	★			2	*20·0*	4	*40·0*							4	*40·0*					1	*10·0*	9	*90·0*	10	2
phlomoides		★		★				★																								
speciosum	★																															
thapsus	★	★	★	★	★	★	★	★	2	*3·3*	3	*4·9*	2	*3·3*							45	*73·8*	5	*8·2*	4	*6·6*	8	*13·1*	53	*86·9*	61	9
virgatum	★	★	★	★	★	★	★	★																								
Verbena officinalis	★	★	★	★	★	★	★	★																								
Veronica agrestis	★	★	★	★	★	★	★	★			1	*1·0*	8	*7·7*							16	*15·4*	2	*1·9*	77	*74·0*	42	*40·4*	62	*59·6*	104	15
anagallis-aquatica } *catenata*	★	★	★	★	★	★	★	★									67	*83·7*	13	*16·2*							32	*40·0*	48	*60·0*	80	12
catenata	★	★	★	★	★	★	★																									
arvensis	★	★	★	★	★	★	★	★	2	*0·7*	6	*2·1*	45	*15·6*							75	*26·0*	34	*11·8*	127	*43·9*	86	*29·8*	203	*70·2*	289	40
beccabunga	★	★	★	★	★	★	★	★	2	*0·4*	1	*0·2*	3	*0·6*	3	*0·6*	436	*82·0*	83	*15·6*	1	*0·2*	1	*0·2*	2	*0·4*	286	*53·8*	246	*46·2*	532	69
chamaedrys	★	★	★	★	★	★	★	★	61	*7·1*	164	*19·0*	261	*30·2*	2	*0·2*	15	*1·7*			282	*32·7*	9	*1·0*	69	*8·0*	323	*37·4*	540	*62·6*	863	90
filiformis	★	★	★	★	★	★	★	★	1	*3·6*	1	*3·6*	8	*28·6*			6	*21·4*			7	*25·0*			5	*17·9*	17	*60·7*	11	*39·3*	28	4
hederifolia	★	★	★	★	★	★	★	★	10	*2·5*	41	*10·4*	15	*3·8*			8	*2·0*			95	*24·1*	8	*2·0*	217	*55·1*	201	*51·0*	193	*49·0*	394	48

Species	Vice-County Records 38	23	32	33	37	39	55	57
Veronica montana	★	★	★	★	★	★	★	★
officinalis	★	★	★	★	★	★	★	★
persica	★	★	★	★	★	★	★	★
polita	★	★	★	★	★	★	★	★
scutellata	★	★	★	★	★	★	★	★
var. *villosa*	★				★		★	
serpyllifolia ssp. *serpyllifolia*	★	★	★	★	★	★	★	★
Viburnum lantana	★	★	★	★	★		★	★
opulus	★	★	★	★	★	★	★	★
Vicia bithynica				★	★			
cracca	★	★	★	★	★	★	★	★
hirsuta	★	★	★	★	★	★	★	★
hybrida	★					★	★	
lathyroides	★	★			★	★		★
lutea		★		★	★	★	★	★
sativa ssp. *angustifolia*	★	★	★	★	★	★	★	★
ssp. *sativa*	★	★	★	★	★	★	★	★
sepium	★	★	★	★	★	★	★	★
sylvatica	★	★	★	★	★	★	★	★
tenuifolia	★	★		★	★	★	★	
tenuissima	★	★		★	★			
tetrasperma	★	★	★	★	★	★	★	★
Vinca major	★	★	★	★	★	★	★	★
minor	★	★	★	★	★	★	★	★
Viola arvensis	★	★	★	★	★	★	★	★
canina	★	★	★	★	★	★	★	★
canina × *riviniana*	★							
hirta ssp. *hirta*	★	★	★	★	★	★	★	★
ssp. *calcarea*		★		★			★	
hirta × *odorata*	★	★	★	★	★		★	
lactea	★	★						
lutea						★		★
odorata	★	★	★	★	★	★	★	★
palustris	★	★			★	★	★	★
reichenbachiana	★	★	★	★	★	★	★	★
reichenbachiana × *riviniana*	★			★	★			★
riviniana ssp. *riviniana*	★	★	★	★	★	★	★	★
ssp. *minor*	★	★		★			★	★
tricolor	★	★	★	★	★	★	★	★
Viscum album	★	★	★	★	★	★	★	★
Vulpia bromoides	★	★	★	★	★	★	★	★
myuros	★	★	★	★	★	★	★	★
Wahlenbergia hederacea	★	★				★	★	★
Xanthium spinosum	★	★	★	★	★	★	★	
Zannichellia palustris	★	★	★	★	★	★	★	★

Species	Woodland No.	%	Hedgerow No.	%	Grassland No.	%	Heath No.	%	Waterside No.	%	Water No.	%	Ruderal No.	%	Quarries No.	%	Cultivated No.	%	Total Abundant No.	%	Total Occasional No.	%	Total Records No.	% of Random Squares
Veronica montana	35	*66·0*	14	*26·4*	1	*1·9*			3	*5·7*									16	*30·2*	37	*69·8*	53	8
officinalis	9	*14·5*	7	*11·3*	27	*43·5*			1	*1·6*			15	*24·2*	1	*1·6*	2	*3·2*	20	*32·3*	42	*67·7*	62	8
persica	1	*0·2*	7	*1·2*	48	*7·9*			3	*0·5*			130	*21·4*	7	*1·2*	412	*67·8*	274	*45·1*	334	*54·9*	608	75
polita					1	*3·2*							9	*29·0*			21	*67·7*	4	*12·9*	27	*87·1*	31	5
scutellata									8	*100·0*									3	*37·5*	5	*62·5*	8	1
var. *villosa*																								
serpyllifolia ssp. *serpyllifolia*	9	*2·6*	10	*2·8*	219	*62·4*	3	*0·9*	15	*4·3*			40	*11·4*	4	*1·1*	51	*14·5*	96	*27·4*	255	*72·6*	351	48
Viburnum lantana	23	*20·4*	83	*73·5*					2	*1·8*			4	*3·5*	1	*0·9*			15	*13·3*	98	*86·7*	113	16
opulus	62	*32·8*	78	*41·3*	2	*1·1*			41	*21·7*			5	*2·6*			1	*0·5*	24	*12·7*	165	*87·3*	189	27
Vicia bithynica																								
cracca	8	*1·5*	199	*37·7*	104	*19·7*	1	*0·2*	33	*6·2*			168	*31·8*	3	*0·6*	12	*2·3*	187	*35·4*	341	*64·6*	528	66
hirsuta	3	*0·6*	45	*9·4*	96	*20·1*	1	*0·2*	10	*2·1*			200	*41·9*	8	*1·7*	114	*23·9*	169	*35·4*	308	*64·6*	477	60
hybrida																								
lathyroides																								
lutea																								
sativa ssp. *angustifolia*	1	*0·7*	18	*11·8*	49	*32·2*			3	*2·0*			68	*44·7*	6	*3·9*	7	*4·6*	41	*27·0*	111	*73·0*	152	22
ssp. *sativa*	5	*0·9*	72	*12·7*	178	*31·4*	1	*0·2*	9	*1·6*			245	*43·3*	9	*1·6*	47	*8·3*	219	*38·7*	347	*61·3*	566	71
sepium	11	*2·8*	94	*24·2*	92	*23·7*			7	*1·8*			178	*45·8*	2	*0·5*	5	*1·3*	166	*42·7*	223	*57·3*	389	51
sylvatica																								
tenuifolia																								
tenuissima																								
tetrasperma	3	*1·8*	21	*12·8*	38	*23·2*			4	*2·4*			44	*26·8*			54	*32·9*	44	*26·8*	120	*73·2*	164	23
Vinca major	3	*6·7*	26	*57·8*	1	*2·2*			1	*2·2*			12	*26·7*			2	*4·4*	14	*31·1*	31	*68·9*	45	7
minor	3	*13·6*	8	*36·4*	2	*9·1*							7	*31·8*	1	*4·5*	1	*4·5*	12	*54·5*	10	*45·5*	22	3
Viola arvensis			8	*1·7*	29	*6·2*			2	*0·4*			76	*16·4*	5	*1·1*	344	*74·1*	149	*32·1*	315	*67·9*	464	65
canina																								
canina × *riviniana*																								
hirta ssp. *hirta*	8	*6·5*	32	*25·8*	32	*25·8*			2	*1·6*			46	*37·1*	1	*0·8*	3	*2·4*	35	*28·2*	89	*71·8*	124	16
ssp. *calcarea*																								
hirta × *odorata*																								
lactea																								
lutea																								
odorata	55	*16·0*	205	*59·8*	18	*5·2*			12	*3·5*			47	*13·7*	3	*0·9*	3	*0·9*	120	*35·0*	223	*65·0*	343	44
palustris	1	*7·7*					3	*23·1*	9	*69·2*									9	*69·2*	4	*30·8*	13	1
reichenbachiana	18	*46·2*	9	*23·1*	2	*5·1*			4	*10·3*			6	*15·4*					11	*28·2*	28	*71·8*	39	6
reichenbachiana × *riviniana*																								
riviniana ssp. *riviniana*	118	*28·6*	169	*41·0*	52	*12·6*	1	*0·2*	18	*4·4*			52	*12·6*	1	*0·2*	1	*0·2*	159	*38·6*	253	*61·4*	412	51
ssp. *minor*																								
tricolor					1	*4·3*							6	*26·1*			16	*69·6*	7	*30·4*	16	*69·6*	23	4
Viscum album																								
Vulpia bromoides	1	*1·0*	1	*1·0*	12	*11·8*	1	*1·0*	2	*2·0*			65	*63·7*	17	*16·7*	3	*2·9*	52	*51·0*	50	*49·0*	102	16
myuros													11	*91·7*	1	*8·3*			8	*66·7*	4	*33·3*	12	3
Wahlenbergia hederacea																								
Xanthium spinosum																								
Zannichellia palustris									1	*4·5*	21	*95·5*							17	*77·3*	5	*22·7*	22	*

12. BRYOPHYTES

T. LAFLIN

Introduction

The recent survey of bryophytes in Warwickshire was carried out between November, 1956, and August, 1967, and the maps are based on records made over this period. A few records of rare species, made subsequently, are included in the text. The basis for recording was the National Grid 1-km square, and the object was to cover evenly the whole vice-county and to record from all habitat types in every part of the county. The studies were not confined to any particular square within a tetrad, nor was every tetrad visited. In tetrads with a diversity of habitats more than one 1-km square was visited; in areas where only a few different habitat types are found over large tracts of country, e.g. in the urban area of Birmingham or some of the areas of uniformly farmed agricultural land, a smaller sample was studied. 438 tetrads (69%) were studied. Of the 2539 1-km squares in VC 38, records were made from 746 (29%). 674 squares were visited by me and I received lists from M. C. Clark (49 squares); G. A. and M. A. Arnold (29 squares); R. E. and Mrs. Evans (27 squares). Records from 23 squares were received from R. A. Austin (1), Miss B. M. B. Bailey (5), E. Bradford (4), Miss S. Coles (1), Miss G. Cooper (1), R. W. B. Dawson (3), J. H. Field (2), S. W. Greene (1), J. G. Hawkes (1), Mrs. M. D. G. Jones (3) and J. F. Lowe (1). Of the 746 squares studied, 40 were visited four or more times, 41 three times, 124 twice and 541 once. The actual squares surveyed are shown on overlay map XII.

Records included lists of species from different habitats, their frequency within the habitat and whether fruiting or sterile. Frequency records were subjective and followed a usual system of categories viz. abundant, frequent, occasional or rare. Frequencies were recorded for the habitat, and not for the square as a whole.

In addition to field work, I have searched national and local floras and botanical journals for earlier records and have examined the herbaria containing collections of the botanists who worked extensively in the county. A complete collection of species found during the present survey is in my herbarium. I also have the lists from the 1-km squares studied and lists of earlier records published and from herbaria.

The nomenclature follows the Census Catalogues; for the hepatics: Paton, J. A., Census Catalogue British Hepatics, 4th Edn., 1965; for the mosses: Warburg, E. F., Census Catalogue British Mosses, 3rd Edn., 1963. The seven-figure code numbers for the species are from these catalogues. The first figure is either 1 (hepatics) or 2 (mosses); the next three figures are the generic numbers; the next two are the species numbers; the last figure denotes the variety—1 for the "type" variety, then in numerical sequence for other varieties in their order in the catalogues. A few species new to Britain since the catalogues were published have been found in VC 38. These have been numbered to follow the varietal sequence of the species to which they are most nearly related.

Recorders and collectors of Warwickshire bryophytes

The following list gives all published papers and notes containing bryophyte records for VC 38 so far as I have been able to trace them. The study of herbaria has been confined to those which I know contain a substantial number of Warwickshire specimens, and names of collectors have been obtained from these. The dates immediately following the names indicate the periods over which active bryological study in VC 38 is known to have been carried out. Contributors of records to the present survey are listed in my Introduction. Their names are included only if they have published papers on Warwickshire bryophytes.

ARNOLD, G. A. and M. A., of Wilnecote (present day). List of bryophytes in "Alvecote Pools, Warwickshire Nature Reserve, 7th Annual Report for year ending 31 December, 1965" (1966).

BAGNALL, J. E., of Birmingham (1867–1909). Herbarium in City of Birmingham Museum; specimens scattered through many other herbaria. 18 papers, 62 short notes, and manuscript records; also records in Braithwaite, R. (q.v.).

A list of the mosses found within a 10-mile radius from Stephenson Place, Birmingham, 1867–70. *Proc. Bgham nat. Hist. microsc. Soc.* **1**, 97–101, and Addenda **1**, 110 (1869).

Report of the Botany Section, 1869–70. *Rep. Bgham nat. Hist. microsc. Soc.* **11**, 11 (1870).

The Moss Flora of Warwickshire. *J. Bot., Lond.* **12**, 18–22 (1874).

Tortula sinuosa in Warwickshire. *J. Bot., Lond.* **12**, 159 (1874).
Moss Flora of Sutton Park. *In* "Notes on Sutton Park", 19–22 (1876).
The Cryptogamic Flora of Warwickshire. *Midl. Nat.* **2**, 220–4, 253–6 and 278–80 (1879), and **3**, 10–12, 80–3 and 132–5 (1880).
Hepaticae of Warwickshire. *Midl. Nat.* **3**, 290–3 (1880).
Dicranum montanum; new Warwickshire habitat. *Midl. Nat.* **4**, 116–17 (1881).
Dicranum montanum. Midl. Nat. **5**, 187–8 (1882).
Notes on the Anker Valley and its Flora. *Midl. Nat.* **9**, 54–8, 69–73 and 89–92 (1886).
A half-day's ramble in the Arrow district. *Midl. Nat.* **9**, 117–19 (1886).
A new British moss. *Midl. Nat.* **10**, 182–3 (1887).
Notes on the Warwickshire Stour Valley and its Flora. *Midl. Nat.* **11**, 25–8, 67–71 and 98–103 (1888).
"Flora of Warwickshire." London and Birmingham (1891); Musci, 329–76; Hepaticae, 377–84.
Notes on the Flora of Warwickshire. *Midl. Nat.* **16**, 252–62 (1893).
"Musci" in *Vict. County Hist. Worcs.* **1**, 62–6 (1901).
"Botany" in *Vict. County Hist. Warws.* **1**, 51–6 (1904).
Musci. *In* Amphlett, J. and Rea, C., "The Botany of Worcestershire", 447–78. Birmingham (1909).
Notes in *Rep. Bgham nat. Hist. microsc. Soc.* **19**, 30 (1877); **20**, 8 (1878) and **21**, 18 (1879); *Midl. Nat.* **1**, 136 (1878); **2**, 104–5, 131, 190, 238 and 285 (1879); **3**, 70, 95, 122, 146–7, 148, 180 and 208 (1880); **4**, 17, 95, 119, 224 and 286 (1881); **5**, 210 (1882); **6**, 23, 71–2, 143, 167, 263, 282 and 284 (1883); **7**, 26, 87, 117, 145–6 and 173 (1884); **8**, 28, 87, 142, 174, 207, 238 and 270 (1885); **10**, 189 (1887); **11**, 134, 189, 215 and 238 (1888); **12**, 122–3 (1889); **14**, 239 (1891) and **16**, 167 (1893); *Rep. & Trans. Bgham nat. Hist. microsc. Soc.* for **1879**, xxv (1880); for **1881**, xx (1882); for **1882**, xxxi (1883) and for **1883**, xxiii–iv (1884); *Rep. Moss Exch. Club* for 1899, 1900, 1901, 1903, 1907 and (posthumously) 1921; *Rep. Br. bryol. Soc.* for 1925 and 1934.
MS records as annotations in his own copy of "Flora of Warwickshire", in the library of Oxford University.

BAXTER, W., of Rugby and Oxford (c. 1850–70). Records quoted in Bagnall's papers.

BLOOM, J. H., of Stratford on Avon (c. 1900–4). A few MS records in Bagnall's own "Flora" and specimens in City of Birmingham Museum.

BOLTON, S. P. (c. 1890–1910). Records quoted in Bagnall's papers and a few specimens in City of Birmingham Museum.

BRAITHWAITE, R. Records by Bagnall quoted in two papers: Recent additions to our Moss Flora, Part 4. *J. Bot., Lond.* **9**, 289–95 (1871); and Part 5. *J. Bot., Lond.* **10**, 193–9 (1872).
"British Moss Flora." 3 vols. London (1887–1905) gives records by Bagnall, Kirk, Mackay and Wait (q.v.).

BREE, W. T., of Allesley (c. 1810–40). Herbarium in Warwick County Museum; Records in Purton, **T.** (q.v.). One original paper:
The rarer plants of Warwickshire. *Loudon's Mag. nat. Hist.* **3**, 162–7 (1830).

BROMWICH, H. (1863). A few specimens in herbarium of Warwick County Museum.

CAMERON, D. A few records in Bagnall's "Flora of Warwickshire" (1891).

CHESHIRE, W., of Stratford on Avon (c. 1850–60). Specimen in Oxford University herbarium.

CLEMINSHAW, E. (c. 1900–10). Records in *Reps. Moss Exch. Club;* a few specimens in City of Birmingham Museum. His herbarium was at Rugby School, but I am informed that it is no longer there.

COLLINS, J. Record in Bagnall, 1893 (q.v.).

COOPER, Miss C. A., of Nuneaton (c. 1933–7). Only British record of *Gyroweisia reflexa* in *Reps. Br. bryol. Soc.* for 1933 to 1937 (1933–7).

CRUNDWELL, A. C. and **NYHOLM, E.** The European species of the *Bryum erythrocarpum* complex. *Trans. Br. bryol. Soc.* **4**, 597–637 (1964).

CUMMING, L. and **ELSEE, C.**, of Rugby (c. 1890–95).
List of mosses observed in 1894. *Rep. Rugby Sch. nat. Hist. Soc.* for 1894 (1895).

DALMAN, K. H. Bryophytes of the Nature Reserve, 1952–3. *Proc. Coventry Distr. nat. Hist. scient. Soc.* **2**, 195–7 (1953).

DIXON, H. N., of Northampton (c. 1900–10). Herbarium in Royal Botanic Gardens, Kew; a few specimens in City of Birmingham Museum; records in *Reps. Moss Exch. Club.*

DODD, A. J. Listed as contributing records from VC 38 in *Cens. Cat. Br. Mosses*, 2nd Edn. (1926), but I cannot locate records, nor find more information about his bryological activities.

DUNCAN, J. B. (c. 1920–6). MS records in B.B.S. Recorder's Notebook; records in Ingham, W., *Cens. Cat. Br. Hep.*, 3rd Edn. (1930).
"A Census Catalogue of British Mosses", 2nd Edn. (1926).

ELSEE, C., of Rugby (c. 1890–5). See Cumming and Elsee; Wait and Elsee.

FINCH, R., record in Crundwell and Nyholm, 1964 (q.v.).

GROVE, W. B. (c. 1890–5). Note in *Midl. Nat.* **16** (1893).

HOPKINS, Mrs. E. (c. 1870–80). Record in Bagnall, J. E., 1891 (q.v.).

HUGHES, V. R., note in *Proc. Coventry Distr. nat. Hist. scient. Soc.* **2**, 100 (1950).

INGHAM, W. "A Census Catalogue of British Hepatics", 2nd Edn. (1913).

JACKSON, A. B. (c. 1895–1905). Records in *Reps. Moss Exch. Club*, also Warwickshire mosses. *J. Bot., Lond.* **38**, 52 (1900).

JONES, E. W., of Kirtlington (present day). Records in *Trans. Br. bryol. Soc.*

KIRK, T., of Coventry (c. 1847–73). Records in several of Bagnall's papers and in Braithwaite, 1887–1905 (q.v.).
Herbarium at Oxford University; a few specimens in herbarium of City of Birmingham Museum.

KNIGHT, H. H. (c. 1920–6). MS records in B.B.S. Recorder's Notebook.

LAFLIN, T. Bryophytes in Check lists of the vascular plants and bryophytes of Warwickshire (VC 38), and surrounding vice-counties. *Proc. Bgham nat. Hist. phil. Soc.*, **20**, 45–64 (1965).

LOWE, H. E. (1835–7). Specimens in herbarium of Warwick County Museum.

MACKAY, —. Records in Braithwaite, 1887–1905 (q.v.).

MACVICAR, S. M. (1905). "A Census Catalogue of British Hepatics", 1st Edn.

NEWTON, Martha (present day). MS records from the Birmingham area.

NICHOLSON, W. E. (c. 1920–6). MS record in B.B.S. Recorder's Notebook.

PATON, J. A. *Riccia crystallina* L. and *Riccia cavernosa* Hoffm. in Britain. *Trans. Br. bryol. Soc.* **5**, 222–5 (1967).

PERRY, W. G., of Warwick. "*Plantae Varvicensis Selectae*" (1820). Repeats records from Withering (q.v.) and Purton (q.v.).

PURTON, T., of Alcester (c. 1800–21). His herbarium is reported to be in the City of Worcester Museum, but I could find no bryophytes there.
"A Botanical Description of British Plants in the Midland Counties" (1817).
"An Appendix to the Midland Flora" (1821).

RHODES, P. G. M. (c. 1914–28). Records in *Reps. Moss Exch. Club* and in *Reps. Br. bryol. Soc.*; a few specimens in herbarium of City of Birmingham Museum.

ROGERS, R. (c. 1875–90). Records in several of Bagnall's papers; a few specimens in herbarium of City of Birmingham Museum.

RUFFORD, W. S., of Salford (c. 1800–20). Records in Purton (q.v.).

RUSSELL, J. H. (c. 1890–1905). A few specimens in herbarium of City of Birmingham Museum.

SMITH, Miss K. E. (c. 1930–8). Record in *Rep. Br. bryol. Soc.* for 1938.

STONE, J. B., of Erdington (c. 1880–90). Specimens in herbarium of City of Birmingham Museum; records in several of Bagnall's papers.

THOMPSON, A. Records in *Rep. Br. bryol. Soc.* for 1940–3.

THOMPSON, J. F. S. (c. 1880–1905). Specimens in herbarium of J. B. Duncan.

THOMPSON, J. H. (c. 1865–70). Specimens in herbarium of City of Worcester Museum.

TINDALL, K. H., of Coventry (present day). Coventry Nature Reserve records: mosses and liverworts. *Proc. Coventry Distr. nat. Hist. scient. Soc.* **3**, 102–3 (1960).

TOWNSEND, C. C. (present day). MS list of bryophytes, mainly from Redditch area, 1954–5 (in my possession).

WADDELL, C. H. (c. 1900–7). Records in *Reps. Moss Exch. Club;* a few specimens in City of Birmingham Museum.

WAIT, W. O., of Rugby (c. 1885–95). Records in some of Bagnall's papers and in Braithwaite, 1887–1905 (q.v.). One original paper, with C. Elsee: List of Rugby mosses. *Rep. Rugby Sch. nat. Hist. Soc.* for 1893, 42–3 (1894).

WARBURG, E. F. *Pohlia pulchella* in Britain. *Trans. Br. bryol. Soc.* **4**, 760–2 (1965).

WATSON, E. V. *Pohlia lutescens* (Limpr.) Lindb. f. in Britain and Ireland. *Trans. Br. bryol. Soc.* **5**, 443–7 (1968).

WATSON, W. "*Weisia reflexa* in Britain", *Rep. Br. bryol. Soc.* for 1933 (1934).

WEBB, H., of Birmingham (c. 1850). A few records in Bagnall's papers; a few specimens in City of Birmingham Museum, all undated.

WEST, G. S. A biological investigation of the Peridineae of Sutton Park, Warwickshire. *New Phytol.* **8,** 181–96 (1909).

WILKINSON, W. H. Note in *Midl. Nat.*, **3** (1880).

WILSON, S. (c. 1890). A few records from the Harbury area in Bagnall's papers.

WITHERING, W., of Edgbaston.
"A Botanical Arrangement of British Plants", 2nd Edn., vol. 3 (1792).
"A Natural Arrangement of British Plants", 3rd Edn., vol. 3 (1796).

WITHERING, W., junr.
"A Systematic Arrangement of British Plants". 4th Edn., vol. 3 (1801); 5th Edn., vol. 3 (1812).
"An Arrangement of British Plants". 7th Edn., vol. 3 (1830).

Flora

For each species the Census Catalogue code number and name is given first. For those species recorded from ten or more squares the page number of the distribution map is given. This is followed by the first record from VC 38 and a summary of pre-survey records. Then follows a general description of habitats and distribution in the county, the number of tetrads from which the plant has been recorded during the present survey, this number as a percentage of the total number of tetrads surveyed and the percentage of single squares studied in which it has been found fruiting. For species not mapped, single square references are then given. Lastly are given abbreviated references of herbaria in which Warwickshire specimens may be found and details of occurrence in botanical vice-counties adjoining VC 38. As VC 57 so nearly joins Warwickshire this is also included. Species which have been erroneously or very doubtfully recorded, and those not yet found in the county, but likely to occur, are enclosed in brackets.

Abbreviations of herbaria locations: BIRA, City of Birmingham Museum; LAF, T. Laflin herbarium; OXF, University of Oxford; WAR, Warwick County Museum.

HEPATICAE

ANTHOCEROTACEAE

1001011 **Anthoceros punctatus** L.
Purton: Kinwarton, 1816, in "Midl. Flor." (1817).
Bagnall: near Old Park Wood, Leek Wootton; by Browns Wood, Solihull; by Maxstoke Castle; Over Green, Wishaw.
In acid fallow ground in stubble fields on marl and loamy sand, rare. 4; 1%; all fruiting. 0649, 0554, 1663, 2483.
BIRA, LAF. Recorded from all adjoining counties, but only doubtfully from 57.

1001021 **Anthoceros husnotii** Steph.
Kirk: Binley, 1859, in herb. Edinburgh Botanic Gardens, named *A. laevis* by Kirk.
In similar places to the preceding, very rare. 1; fruiting. 1663.
LAF. Recorded from 23.

1001031 **Anthoceros laevis** L.
Purton: tombstone in Arrow Churchyard, 1812, in "Midl. Flor." (1817).
Not found in present survey. Recorded from 33.

SPHAEROCARPACEAE

(1002011 **Sphaerocarpos michelii** Bell. and 1002021 **S. texanus** Aust. have been recorded from some adjoining counties and should be looked for in fallow loamy fields.)

MARCHANTIACEAE

1004011 **Reboulia hemisphaerica** (L.) Raddi
Bagnall: on banks of stream, Blythe Bridge, Solihull, 1880, in *Midl. Nat.* **3** (1880).
Not found in present survey. Recorded from 33, 37, 39, 55 and 57; doubtfully from 32.

1005011 **Conocephalum conicum** (L.) Underw. Map 796, p. 691
Rufford: ditch bank, Trent Lane to Hoo Mill, Alcester, in Purton, "Midl. Flor." (1817).
Bagnall; Stone; Townsend: many records from all over the county.
On wet banks, rocks and masonry by rivers, streams and lakes, most commonly by streams and small rivers, absent from banks which are scoured in times of flood; rarely away from water or creeping over low plants in marshes; throughout the county. 66; 15%; not found fruiting.
BIRA, LAF, WAR. Recorded from all adjoining counties and 57.

1006011 **Lunularia cruciata** (L.) Dum. Map 797, p. 691
Bagnall: Aston Waterworks Grounds, in *Midl. Nat.* **3** (1880); several other widely distributed records. Also recorded by Rogers.
On rocks, masonry and wood in and by water and at shady bases of walls and rock faces; on bare banks of rivers, streams, lakes and canals, rarely on banks in drier places; on compacted ground, e.g. gravelled or ashed paths; on soil in flower pots in glasshouses; always on or near to the ground; throughout the county. 64; 15%; not found fruiting.
LAF. Recorded from all adjoining counties and 57.

1008011 **Marchantia polymorpha** L. Map 798, p. 691
Kirk: near Coventry, 1849, in herb. Oxford University; Arbury Park.
Bagnall; Stone: several records, mainly from the north of the county.
At the margins of ponds and lakes, sometimes submerged (probably var. *aquatica* Nees); on wet ash or sewage deposits; in and around glasshouses, rarely on fallow ground; temporarily on heathy ground after burning; widely distributed. 17; 4%; not found fruiting.
BIRA, LAF, OXF. Recorded from all adjoining counties and 57.

RICCIACEAE

1010051 **Riccia glauca** L. Map 799, p. 691
Purton: stubble fields, Kinwarton, 1816, in "Midl. Flor." (1817); Salford.
Bagnall; Stone: several records from North Warwickshire.
In mildly acid to acid fallow ground in the centre and north of the county. 13; 3%; fruiting not recorded, but usually fertile.
BIRA, LAF. Recorded from all adjoining counties and 57.
19th century botanists did not record *R. sorocarpa* Bisch., which is the commoner of the two, and it is probable that some of their records of *R. glauca* L. belong to that species.

1010071 **Riccia beyrichiana** Hampe
Bagnall: on the new railway embankment, Sutton Park, 1879, in *Midl. Nat.* **3** (1880).
Not found in present survey Not recorded from any adjoining county, nor 57.

1010081 **Riccia sorocarpa** Bisch. Map 800, p. 692
Recorded for VC 38, without detail of locality or source of record, in the Supplement to the 3rd Edn. of the Census Catalogue, 1935.
In mildly acid to acid fallow ground in the centre and north of the county. 23; 5%; fruiting not recorded, but usually fertile.
LAF. Recorded from all adjoining counties and 57.

1010101 **Riccia fluitans** L.
Rufford: pond on Alne Hills, in Purton, "Midl. Flor." (1817).
Kirk: Griff; Arbury Park. J. H. Thompson: Coleshill Pool. Bagnall: between Shelfield and Wawensmoor; Baddesley Clinton; Balsall Common; near Corley Rectory; Coleshill Pool. Hughes: Tile Hill Wood.
Floating on ponds and lakes or on wet ground at margins, left after submergence or spreading out from the water; only in acid areas in the centre and north of the county. 9; 2%; not found fruiting. 1073, 1986, 1697, 1792, 2779, 2096, 2503, 2504, 4171.
BIRA, LAF, OXF, WAR. Recorded from 23, 33, 37, 55 and 57.

(1010131 **Riccia crystallina** L. was recorded for VC 38, but the records are now known to refer to *R. cavernosa* Hoffm., q.v.)

1010133 **Riccia cavernosa** Hoffm.
Bagnall: dry sediment of Rotton Park Reservoir, Birmingham, 1893, in herb. Birmingham Museum. J. F. S. Thompson: Edgbaston Reservoir, 1901, in herb. J. B. Duncan.
Not found in the present survey.
BIRA. Recorded from 32 and 55.
This species was referred to *R. crystallina* L. until 1967, when a critical examination of British material was made by Mrs. J. A. Paton, who saw both gatherings from VC 38 and referred them to *R. cavernosa* Hoffm.
See Paton, J. A. *Trans. Br. bryol. Soc.* **5**, (2) 222–5 (1967).

1011011 **Ricciocarpus natans** (L.) Corda
Bolton and Cleminshaw: ponds by footpath between Moat House and Rock Farm, Berkswell, 1903, in herb. Birmingham Museum.
Hughes: pond in Tile Hill Wood.
Floating on ponds and lakes, or on wet ground at margins, in acid soil areas in the north-east quarter of the county; rare. 4; 1%; not found fruiting. 2779, 4658, 4474, 4182.
BIRA, LAF, WAR. Recorded from 32 and 55.

RICCARDIACEAE

1012021 **Riccardia multifida** (L.) Gray
Purton: "common", but no localities given, "Midl. Flor." (1817).
Bagnall; Rogers: several records from the north-west quarter of the county; Wolford Wood.
On wet ground at the edges of ditches, canals, ponds and lakes, in marshes and, rarely, in fallow fields; uncommon in the northern half of the county. 8; 2%; not found fruiting. 1000, 2069, 2587, 2405, 2504, 3087, 3094, 4282, 5379.
BIRA, LAF. Recorded from counties to the north, but not from 23, 32 nor 33.

1012031 **Riccardia sinuata** (Dicks.) Trev. Map 801, p. 692
Kirk: wet ditch, Coventry Wood, Arbury, 1856, in herb. Oxford University.
Bagnall: several records from all over the county.
In varied wet bare ground habitats; floors, banks and spoil heaps of quarries and cuttings, wet woodland rides, ditch, stream and pond banks, wet heaths, marshes and bogs and, rarely, in fallow fields; widely distributed. 17; 4%; 22% fruiting.
LAF, OXF. Recorded from all adjoining counties and 57.

1012061 **Riccardia pinguis** (L.) Gray Map 802, p. 692
Purton: "common", but no localities given, "Midl. Flor." (1817).
Bagnall: several localities in north and central Warwickshire.
In bogs, marshes, flushes and springs, rarely in drier places on the ground; widely distributed, but mainly in the north. 9; 2%; not found fruiting.
BIRA, LAF. Recorded from all adjoining counties and 57.

PELLIACEAE

1014011 **Pellia epiphylla** (L.) Corda Map 803, p. 692
Purton: ditch bank between Aston Cantlow and Wilmcote, "Midl. Flor." (1817).
Bagnall; Bolton; Dalman: many records from all over the county.
On wet banks, sides of ditches, wet woodland rides and similar places on non-alkaline soils; rarely on sandstone; widespread. 89; 20%; 4% fruiting.
BIRA, LAF. Recorded from all adjoining counties and 57.

1014031 **Pellia neesiana** (Gottsche) Limpr.
Paton: 0995, marshy margin of Longmoor Mill Pool, Sutton Park, 1969.
1; fruiting. Possibly overlooked by me.
LAF. Recorded from 37 and 57.

1014041 **Pellia endiviifolia** Dicks. Map 804, p. 693
Bagnall: canal bank, Rowington, 1879, in herb. Birmingham Museum; several other widespread localities. Also recorded by Townsend.
On ditch and stream banks, wet rocks, in flushes and springs, on wet woodland rides and similar places; never in very acid places; throughout the county. 86; 20%; 3% fruiting.
BIRA, LAF, WAR. Recorded from all adjoining counties and 57.

METZGERIACEAE

1015011 **Metzgeria furcata** (L.) Dum. Map 805, p. 693
Purton: lane from Studley to Middletown, "Midl. Flor." (1817); Cookhill.
Kirk; Bagnall; Rogers: many records throughout the county and more common in north Warwickshire in the 19th century than it is now.
On trees and shrubs, particularly in hedgerows; rarely on rocks and walls; frequent in the south, rare in the north. 113; 26%; not found fruiting.
BIRA, LAF, OXF. Recorded from all adjoining counties and 57.

(1015021 **Metzgeria fruticulosa** (Dicks.) Evans has been recorded from 23, 32 and 33, and should be looked for in hedgerows in south Warwickshire.)

BLASIACEAE

1018011 **Blasia pusilla** L.
Laflin: old sandpits in Rough Hill Wood, Studley, 1963.
On banks in old sandpits, rare. 3; 1%; not found fruiting. 0563, 0564, 3876.
LAF. Recorded from all adjoining counties and 57.

FOSSOMBRONIACEAE

1020081 **Fossombronia pusilla** (L.) Dum. Map 806, p. 693
Purton: wheel rut in Oversley Wood, "Midl. Flor." App. (1821).
Bagnall: several records from the northern half of the county.
On bare acid ground in varied habitats; widespread in the north, rare in the south. 31; 7%; 59% fruiting.
BIRA, LAF. Recorded from all adjoining counties and 57.

1020091 **Fossombronia wondraczekii** (Corda) Dum.
Knight and Nicholson: in Supplement to 3rd Edn. Census Catalogue (1935); no details of locality.
In similar places to the preceding, rare. 1; fruiting. 3269.
LAF. Recorded from 23, 32, 33, 37, 39 and 57.

PTILIDIACEAE

1027011 **Ptilidium ciliare** (L.) Hampe
Laflin: heathy ground by Coleshill Pool, 1958.
Amongst *Calluna* on heaths, rare. 2; not found fruiting. 1985, 1986.
LAF. Recorded from counties to the north, but not from 23, 32 nor 33.

1027021 **Ptilidium pulcherrimum** (Web.) Hampe
Clark: on shrub, Spernall Park, 1960.
Epiphyte on small trees and shrubs, usually near to the ground, in woods and hedgerows, rare. 3; 1%; not found fruiting. 1062, 2332, 4246.
LAF. Recorded from 23, 32, 37, 55 and 57.

1028011 **Trichocolea tomentella** (Ehrh.) Dum.
Bagnall: marshy place above Blackroot Pool, Sutton Park, 1876, in herb. Birmingham Museum.
Refound in Sutton Park by J. H. Field, 1960; not precisely located in survey records.
BIRA. Recorded from 37, 39 and 57.

(1029011 **Blepharostoma trichophyllum** (L.) Dum. is recorded from adjoining counties except 32 and 55, and might be found on rotting stumps in some Warwickshire woods.)

LEPIDOZIACEAE

1031021 **Lepidozia reptans** (L.) Dum. Map 807, p. 693
Bagnall: Crackley Wood, Kenilworth, 1882, in herb. Birmingham Museum; Sutton Park; Trickley Coppice; Bentley Park; Hartshill Hayes.
On peaty banks in woods and on heaths, on rotting stumps and on siliceous rock outcrops; occasional in the northern half of the county. 10; 3%; not found fruiting.
BIRA, LAF. Recorded from all adjoining counties and 57.

Calypogeia Raddi

Early records of the *C. fissa—muellerana* group were all included under *C. trichomanis* (L.) Corda and, as such, were recorded by Purton, Kirk and Bagnall from many localities. Most were probably *C. fissa*, although some may have been *C. muellerana*. Except where specimens exist, their identity cannot be ascertained.

1033021 **Calypogeia muellerana** (Schiffn.) K. Müll.
Laflin: peaty ditch bank, Tile Hill Wood, 1958.
On peaty banks in the north of the county; not common. 5; 1%; not found fruiting. 0995, 0996, 1599, 2779, 2895.
LAF. Recorded from 23, 37, 39, 55 and 57.

(1033031 **Calypogeia trichomanis** (L.) Corda was frequently recorded by 19th-century botanists, but all were probably *C. fissa* or *C. muellerana*. *C. trichomanis*, as now recognised, has not been found in VC 38.)

1033041 **Calypogeia fissa** (L.) Raddi Map 808, p. 694
Kirk: Coventry Wood, Arbury, 1857, in herb. Oxford University.
Bagnall (specimens seen): Umberslade; Pool Hollies Wood; Bentley Park.
Townsend: by Bracebridge Pool, Sutton Park.
Dalman: Tile Hill Wood.
On loamy banks, in marshes and on sandstone faces; rarely on rotting wood; frequent on non-basic media throughout the county. 61; 14%; not found fruiting.
BIRA, LAF, OXF. Recorded from all adjoining counties and 57.

1033071 **Calypogeia arguta** Nees and Mont.
Clark: bank of stream, Bentley Park Wood, 1962.
On sandy and marly banks, woodland rides and sandstone faces; widely distributed in the north, but uncommon. 9; 3%; not found fruiting. 2488, 2790, 2797, 2995, 2996, 3067, 3071, 3671 and 3085.
LAF. Recorded from all adjoining counties and 57.

LOPHOZIACEAE

1034011 **Lophozia ventricosa** (Dicks.) Dum.
Bagnall: Sutton Park, in *Midl. Nat.* **3** (1880); three localities in or near Sutton Park. Townsend; Coughton Park.
Recorded by Kirk, 1857, but his specimen is *Lophocolea heterophylla* (Schrad.) Dum.
On peaty banks, banks over sandstone and sandstone rock outcrops; rare and only in the northern half of the county. 9; 3%; 25% fruiting. 0563, 0660, 1271, 1371, 2685, 2687, 2997, 3085, 3094.
BIRA, LAF. Recorded from 23, 32, 37, 39, 55 and 57.

(1034031 **Lophozia porphyroleuca** (Nees) Schiffn. is reported by Bagnall from VC 38, without locality, in Victoria County History Warws., vol. 1 (1904), and, on the strength of this record, included in the Census Catalogues from 1913 onwards. No specimen exists and, without better evidence, the record is suspect.)

1034061 **Lophozia excisa** (Dicks.) Dum.
Bagnall: Sutton Park, in *Midl. Nat.* **2** (1879); Wolford Wood.
Recorded by Kirk, 1857, but his specimen is *Lophocolea heterophylla* (Schrad.) Dum.
On peaty banks and woodland rides and on peaty accumulations over sandstone, rare. 8; 2%; all fruiting. 1994, 2284, 2199, 2304, 2504, 3866, 3876, 5173.
LAF. Recorded from 23, 33, 37, 39, 55 and 57.

(1034091 **Lophozia incisa** (Schrad.) Dum. has been recorded from counties to the north and might grow in Warwickshire in similar places to *L. ventricosa* (Dicks.) Dum.)

1034111 **Lophozia bicrenata** (Schmid.) Dum.
Bagnall: bank of Wormleighton Reservoir, 1887, in "Flor. Warws." (1891); near Rowton's Well, Sutton Park. Townsend; Rough Hill Wood.
On peaty ground and old sewage deposits, rare. 6; 1%; all fruiting. 0563, 1792, 2994, 2304, 2503, 3094.
LAF. Recorded from all adjoining counties and 57.

1035011 **Leiocolea turbinata** (Raddi) Buch
Map 809, p. 694
Bagnall: banks of the Avon near Barford, "Flor. Warws." (1891); heathland near Rowton's Well, Sutton Park (most unlikely); near Cornets End.
On calcareous clay banks and, rarely, on soft limestone; widespread on the Lower Lias, rare elsewhere. 14; 3%; 25% fruiting.
BIRA, LAF. Recorded from 23, 32, 33, 37, 39 and 57.

1035021 **Leiocolea badensis** (Gott.) Jörg.
Recorded by Duncan, without details of locality, in "Census Catalogue", 3rd Edn., 1930.
Not found in present survey. Recorded from 33 and 57.

1035041 **Leiocolea bantriensis** (Hook.) Jörg.
Laflin: 0998, boggy flush, Bracebridge Bog, Sutton Park, 1964; not fruiting.
LAF. Recorded from 57.

1044011 **Gymnocolea inflata** (Huds.) Dum. Map 810, p. 694
Bagnall: thatch of old outhouse, Boldmere Lane, Sutton Coldfield, in *Midl. Nat.* **3** (1880); Sutton Park; Baxterley Common.
On wet peaty ground, on colliery slag heaps and on sandstone rocks; widespread, but not common, in the north of the county. 24; 5%; 41% fruiting.
BIRA, LAF. Recorded from 32, 33, 37, 39, 55 and 57.

JUNGERMANNIACEAE

(1046011 **Solenostoma triste** (Nees) K. Müll., recorded from all adjoining counties, might be found on sandstone rocks in streams in the north of the county.)

(1046051 **Solenostoma cordifolium** (Hook.) Steph. is recorded from Shuttington Bridge in "Flor. Warws." (1891), but there is no specimen. It is most unlikely and is suspect. It is more likely to have been the preceding than this plant.)

(1046061 **Solenostoma sphaerocarpum** (Hook.) Steph. is another unlikely species recorded by Bagnall from several localities, first from Sutton Park in *Midl. Nat.* **2** (1879). There are no specimens in his herbarium now, but H. H. Knight, in a note in *Rep. Moss Exch. Club* for 1922, p. 297, reporting on his examination of the hepatics in Bagnall's herbarium, says "*Aplozia sphaerocarpa* is an error". All records are probably incorrect.)

1046081 **Solenostoma crenulatum** (Sm.) Mitt. Map 811, p. 694
Bagnall: Berkswell, in *Midl. Nat.* **3** (1880); several records from the north-west of the county; Newbold on Stour.
Townsend: Wolford Wood; Rough Hill Wood, Studley.
On wet woodland rides and floors of old quarries; rarely on stream banks; widespread on mildly acid to acid soils, but not common. The var. *gracillima* (Sm.) as common as the typical form. 14; 3%; 33% fruiting.
BIRA, LAF. Recorded from all adjoining counties and 57.

1047031 **Plectocolea hyalina** (Lyell) Mitt.
Bagnall: near Shirley Heath, 1880, in herb. Birmingham Museum.
Wet woodland rides on acid soils, rare. 1; not fruiting. 0563.
BIRA, LAF. Recorded from 23, 37 and 55.

1048021 **Nardia scalaris** (Schrad.) Gray Map 812, p. 695
Kirk: Stoke Heath, 1857, in herb. Oxford University.
Bagnall; Townsend: several records from the northern half of the county.
On wet ground on heaths, woodland rides, old pits and quarries and ditch and stream banks and on wet sandstone rocks; widespread in the north of the county, but not common; very rare in the south. 17; 4%; 5% fruiting.
LAF, OXF. Recorded from all adjoining counties and 57.

1048031 **Nardia geoscyphus** (De Not.) Lindb.
Arnold: old clay pit, Terra Cotta works, Glascote, 1966.
On wet acid compacted ground in old clay pits, rare. 2; both fruiting. 2303, 3094.
LAF. Recorded from 57.

PLAGIOCHILACEAE

1056021 **Plagiochila asplenioides** (L.) Dum. Map 813, p. 695
Purton: "common", but no localities given, "Midl. Flor." (1817).
Bagnall; Townsend: many localities throughout the county.
On the ground in woods, mostly on the heavier and base-rich soils, but not confined to them, frequent; occasionally by streams or on shaded rocks; throughout the county. 48; 11%; not found fruiting.
LAF, WAR. Recorded from all adjoining counties and 57.

1056022 var. **major** Nees Map 814, p. 695
Duncan: VC 38, without details of locality in "Census Catalogue", 3rd Edn. (1930).
In similar places to the type, probably under-recorded. 12; 3%; not found fruiting.
LAF. Recorded from 23, 32, 33, 37, 55 and 57.

HARPANTHACEAE

1057011 **Lophocolea bidentata** (L.) Dum. Map 815, p. 695
Purton: "common", but no localities given, "Midl. Flor." (1817).
Kirk; Bagnall; Dalman: many localities throughout the county.
In many varied habitats on all soil types; on the ground in woods, scrub, hedgebanks, roadside verges, pastures, rough grassland, heaths, bogs, marshes, on banks of rivers, streams, ditches, canals and lakes, very rarely submerged, on cutting banks, spoil heaps, rock exposures and blocks, masonry, stumps and fallen wood; common throughout the county. 186; 43%; 1% fruiting.
BIRA, LAF, WAR. Recorded from all adjoining counties and 57.

1057021 **Lophocolea cuspidata** (Nees) Limpr. Map 816, p. 696
Bagnall: Hartshill Quarries, 1880, in herb. Birmingham Museum; several other records from the northern half of the county.
Townsend: Rough Hill Wood, Studley.
On fallen wood and stumps and at tree bases, on banks and on rock outcrops and stonework, usually in shaded places; widespread, but not common, throughout the county. 22; 5%; 77% fruiting.
BIRA, LAF. Recorded from all adjoining counties and 57.

1057031 **Lophocolea heterophylla** (Schrad.) Dum. Map 817, p. 696
Kirk: foot of ash tree near the ponds, Berkswell Hall, 1857, in herb. Oxford University; Willenhall.
Bagnall: "in every wood I have visited". Also recorded by Stone and Wilkinson.
On trees, stumps and fallen wood in woods, scrub, hedgerows and open places, on fallow ground, bare ground of woodland rides and heathy clearings, banks, pits and quarries, and on rock exposures and blocks and masonry; abundant throughout the county. 287; 66%; 35% fruiting.
BIRA, LAF, OXF, WAR. Recorded from all adjoining counties and 57.

(1057051 **Lophocolea fragrans** (Mor. and De Not.) Mor. and De Not. was reported from Erdington by Bagnall in *Midl. Nat.* **3** (1880), but he had not seen the plant, nor does he name the finder. Undoubtedly an error.)

1058011 **Chiloscyphus polyanthos** (L.) Corda Map 818, p. 696
Bagnall: Sutton Park, in *Midl. Nat.* **2** (1879); several other localities in north Warwickshire. Also recorded by Stone and Russell.
On wet ground on acid soils, in woodland rides and on banks

of streams and lakes, sometimes submerged in shallow water, in marshes and on wet sandstone rocks in streams; widespread, but not common. 14; 3%; not found fruiting.
BIRA, LAF. Recorded from all adjoining counties and 57.

(1058012 **Chiloscyphus polyanthos** (L.) Corda, var. **rivularis** (Schrad.) Nees, was recorded by Bagnall, without locality, in Victoria County History Warws., vol. 1 (1904). Without better evidence the record is suspect.)

1058021 **Chiloscyphus pallescens** (Ehrh.) Dum.
Bagnall: Sutton Park, in *Midl. Nat.* **16** (1893).
In similar places to the preceding, but rare. 1; not fruiting; 2370.
Recorded from 23, 32, 33, 37, 39 and 57.

CEPHALOZIACEAE

(1062011 **Cephaloziella pearsonii** (Spruce) Douin was recorded by Bagnall from Hartshill Quarries in *Midl. Nat.* **16** (1893). There is no specimen in his herbarium and the record is probably an error.)

1062041 **Cephaloziella rubella** (Nees) Warnst.
E. W. Jones: Sutton Park, 1936, in *Trans. Br. bryol. Soc.* **4**, (3) (1963).
On heathy ground in the extreme north of the county, rare. 6; 1%; all fruiting. 1490, 1599, 1994, 1200, 2503, 3094.
LAF. Recorded from 23, 33, 37 and 39.

(1062051 **Cephaloziella hampeana** (Nees) Schiffn. is doubtfully recorded. *C. bifida* (Schreb.) Schiffn. was recorded for VC 38 in the "Census Catalogue", 2nd Edn. (1913), but the source of the record is not known. This was repeated as a doubtful record of *C. hampeana* (Nees) in the "Census Catalogue", 3rd Edn. (1930). It has been found in 23, 39 and 57, and could occur in Warwickshire.)

(1062071 **Cephaloziella stellulifera** (Spruce) Schiffn. was recorded by Bagnall from several localities in "Flor. Warws." (1891) and *Midl. Nat.* **16** (1893) but there are no specimens in his herbarium. The records are suspect, although it might be found on colliery slag heaps or similar places. Bagnall's records were from Sutton Park; Spring Wood, Hockley; Coleshill Heath; Baddesley Common; Bentley Park; Hartshill Quarries; Kenilworth Common and Wolford Heath.)

1062091 **Cephaloziella starkei** (Funck) Schiffn.
Map 819, p. 696
Bagnall: heathy footways, Mill Lane, Kenilworth, in *Midl. Nat.* **3** (1880); several other records from the northern half of the county; Wolford Heath.
On bare heathy ground in woodland rides, heaths, cutting banks, quarry spoil heaps and similar habitats, usually in dry, acid places; also on soil-covered rocks and rotting stumps; widespread, but not common. 26; 6%; 31% fruiting.
BIRA, LAF. Recorded from 23, 32, 33, 37 and 55; doubtfully from 39 and 57.

1062111 **Cephaloziella turneri** (Hook.) K. Müll.
Bagnall: ride in Wolford Wood, 1891, in herb. Birmingham Museum.
Not found in present survey.
BIRA. Recorded from 33.

1063021 **Cephalozia bicuspidata** (L.) Dum.
Map 820, p. 697
Purton: "common", but no localities given, "Midl. Flor." (1817).
Bagnall; Rhodes; Townsend: many records, mainly from the northern half of the county, but also a few from scattered localities in the south.
On peaty and heathy ground in woods and heaths, in peaty marshes, on spoil heaps and floors of pits and quarries, on sandstone rock faces and exposures, on rock-filled drainage channels of railway cuttings and sandstone rocks in streams; occasionally on rotting stumps; widespread, but not common and often in small quantity. 48; 11%; 62% fruiting.
BIRA, LAF. Recorded from all adjoining counties and 57.

1063022 var. **lammersiana** (Hüben.) Breidl.
Bagnall: without locality in Victoria County History Warws., vol. 1 (1904).
Rhodes: peaty ground, Sutton Park.
On wet peaty banks, rare. 1; fruiting. 1074.
LAF. Recorded from 39 and 57.

1063051 **Cephalozia connivens** (Dicks.) Lindb.
Purton: without locality, "Midl. Flor." (1817).
Bagnall: old tree stump above Blackroot Pool, Sutton Park.
On wet peaty ground, rare. 1; fruiting. 2503.
BIRA, LAF. Recorded from 23, 32, 39 and 57.

(1063071 **Cephalozia media** Lindb. was recorded by Bagnall from shaded bank, Arley Wood, in *Midl. Nat.* **16** (1893), but, without a specimen, the record is suspect.)

1065011 **Nowellia curvifolia** (Dicks.) Mitt.
Bree: Coleshill Heath, in Purton, "Midl. Flor." App. (1821).
On rotten fallen trunk in woodland, rare. 1; not fruiting. 1272.
LAF. Recorded from 33, 39 and 57.

(1066011 **Odontoschisma sphagni** (Dicks.) Dum., recorded from several adjoining counties, might be found in *Sphagnum* bogs in the extreme north of the county.)

SCAPANIACEAE

1069011 **Diplophyllum albicans** (L.) Dum.
Map 821, p. 697
Kirk: Arbury Park, 1859, in herb. Oxford University.
Bagnall: several records from the northern half of the county, but not in the north-east corner; Wolford Wood.
On sandy, loamy and peaty banks, on heaths, on colliery slag heaps and on sandstone rocks; widespread, but not common, in the northern half of the county. 11; 3%; not found fruiting.
BIRA, LAF, OXF, WAR. Recorded from all adjoining counties and 57.

(1070051 **Scapania curta** (Mart.) Dum. was recorded by Bagnall from Sutton Park in *Midl. Nat.* **3** (1880), and from Cornets End and Wolford Wood. There are no specimens and the records are suspect.)

1070081 **Scapania irrigua** (Nees) Dum.
Bagnall: Little Dickens, Earlswood, 1883, in herb. Birmingham Museum; several other records from the north-west quarter of the county; Wolford Wood.
Townsend: Sutton Park; Oversley Wood; Wolford Wood.
On wet, acid compacted ground, mainly in woodland rides, rarely on pit floors; in the central part of Warwickshire, very rare in the north. 8; 2%; not found fruiting. 0558, 0658, 0758, 0563, 1056, 2994, 3570, 3771, 3871.
BIRA, LAF. Recorded from 23, 33, 37, 39, 55 and 57.

1070131 **Scapania nemorea** (L.) Grolle
Bagnall: main drive of Oversley Wood, 1879, "Flor. Warws". (1891); Sutton Park; Arley Wood; Close Wood, Meriden; Cornets End; Fulford Heath; Martin's Wood Lane, Hockley; Waverley Wood; Wolford Wood (also found here by Townsend).
On wet banks of acid soil and on wet heaths in the

extreme north. 3; 1%; not found fruiting. 2798, 2995, 2996.
BIRA, LAF. Recorded from all adjoining counties and 57.

1070161 **Scapania undulata** (L.) Dum.
Bagnall: Four Ashes, Hockley in *Midl. Nat.* **3** (1880); Sutton Park; Marston Green; Salter Street; Holly Lane, Balsall; Cornets End; Oversley Wood; Wolford Wood.
On wet sandstone rocks in and by streams and in wet acid woodland rides, rare. 4; 1%; not found fruiting. 0956, 1056, 2969, 2996.
LAF. Recorded from 23, 32, 37, 39, 55 and 57.

RADULACEAE

1071011 **Radula complanata** (L.) Dum. Map 822, p. 697
Purton: "common", but no localities given, "Midl. Flor." (1817).
Bagnall: many records, mainly from the south and centre, but extending as far north as Shustoke and Arley; probably fairly frequent in the north central area, where it is no longer found.
On trees and shrubs, mainly ash, elm or maple, in hedgerows; widespread in the south and centre, absent from the north. 28; 6%; 4% fruiting.
BIRA, LAF. Recorded from all adjoining counties and 57.

PORELLACEAE

1073031 **Porella platyphylla** (L.) Lindb. Map 823, p. 697
Purton: "common", but no localities given, "Midl. Flor." (1817).
Bagnall: several localities from the south and centre of the county.
On trees and shrubs, particularly ash stools, in hedgerows and woods, and on limestone rocks and walls; widespread in the south, less common in the centre and absent from the north. 33; 8%; not found fruiting.
BIRA, LAF. Recorded from all adjoining counties and 57.

LEJEUNACEAE

1075011 **Lejeunea cavifolia** (Ehrh.) Lindb.
Duncan: without locality, in "Census Catalogue", 3rd Edn., 1930.
On ash stools in woods, rare. 3; 1%; not found fruiting. 1073, 2333, 2433.
LAF. Recorded from 23, 32, 33 and 57.

1082011 **Frullania tamarisci** (L.) Dum.
Purton: Ridgeway, in "Midl. Flor.", (1817).
Bagnall: lane near Shustoke; trees near Solihull; Morton Bagot; Yarningale Common; Rowington; Farnborough; Wormleighton; near Long Compton.
Not found in present survey. Recorded from all adjoining counties and 57.

1082051 **Frullania dilatata** (L.) Dum. Map 824, p. 698
Purton: "very common", but no localities given, "Midl. Flor." (1817).
Kirk; Bagnall; Townsend: many records, mainly from the south and centre of the county, but a few from the north.
On trees and shrubs, usually ash, elm or elder, in hedgerows, scrub and light woodland, not in very dense woodland, and on limestone rocks and walls; frequent in the south, decreasing northwards and rare in the north. 66; 15%; 3% fruiting.
BIRA, LAF, OXF. Recorded from all adjoining counties and 57.

MUSCI

SPHAGNACEAE

2001011 **Sphagnum palustre** L. Map 825, p. 698
Purton: Coughton Lane, in "Midl. Flor." (1817).
Bagnall; Wait and Elsee; West; Townsend: many records from the northern half of the county.
In bogs, at margins of ditches, ponds and lakes and on the ground in wet woods and woodland rides, on acid soils; widespread in the north of the county. 19; 4%; not found fruiting.
BIRA, LAF, OXF, WAR. Recorded from 23, 32, 37, 39, 55 and 57.

2001031 **Sphagnum papillosum** Lindb.
Bagnall: Sutton Park, 1874, in herb. Oxford University; several places in Sutton Park (also found by Townsend); Arley Wood; Hill Bickenhill; Coleshill Heath; Browns Wood, Solihull; Tile Hill Wood.
In bogs and boggy margins of lakes in very acid places, sometimes forming hummocks; rare, in the extreme north of the county. 3; 1%; not found fruiting. 0996, 2304, 2504.
BIRA, LAF, OXF. Recorded from 37, 39 and 57.

(2001051 **Sphagnum compactum** DC. is recorded for VC 38 in the "Census Catalogue", 2nd Edn., 1926, but the source of the record is not known.)

2001071 **Sphagnum teres** (Schimp.) Ångstr.
Bagnall: marsh by Windley Pool, Sutton Park, 1876, in herb. Oxford University.
Rhodes; A. Thompson: Longmoor, Sutton Park.
At boggy margins of lakes, very rare. 2; not fruiting. 1985, 1998.
LAF, OXF. Recorded from 57.

2001081 **Sphagnum squarrosum** Pers. Map 826, p. 698
Kirk: boggy place near Astley, 1849, in herb. Oxford University.
Bagnall: several places in Sutton Park (also found by Rhodes); Cornets End; Seckington.
In boggy places in woods, on heaths and by ditches, ponds and lakes, requiring less acid conditions than many species. 15; 3%; not found fruiting.
BIRA, LAF, OXF, WAR. Recorded from 23, 37, 39, 55 and 57.

2001121 **Sphagnum recurvum** P. Beauv. Map 827, p. 698
Bagnall: Sutton Park, in *J. Bot., Lond.* **12** (1874); several places in Sutton Park (also found by Rhodes and Townsend); Trickley Coppice; Seckington.
In bogs and boggy margins of ponds and lakes, sometimes in standing water; widespread in the north of the county. 12; 3%; not found fruiting.
BIRA, LAF, WAR. Recorded from 32, 37, 39, 55 and 57.

(2001131 **Sphagnum pulchrum** (Lindb.) Warnst. was recorded from Sutton Park by Bagnall in *Midl. Nat.* **6** (1883), and by Rhodes in *Rep. Moss Exch. Club* for 1919, and included in the "Census Catalogue", 2nd Edn. (1926). Rhodes' specimen is *S. recurvum* P. Beauv. and Bagnall's record probably also refers to this.)

(2001151 **Sphagnum tenellum** Pers. was recorded by Bagnall from Sutton Park in *Proc. Bgham nat. Hist. microsc. Soc.* **1** (1869), but this is probably an error. He also named a specimen from Coleshill Bog, 1874, as this species, but it is *S. rubellum* Wils.)

2001161 **Sphagnum cuspidatum** Ehrh.
Bagnall: Sutton Park, in *Proc. Bgham nat. Hist. microsc.* Soc. **1** (1869); several places in Sutton Park (also found by Rhodes and Townsend); Trickley Coppice; Tile Hill Wood.
At the margins of acid pools and lakes, sometimes submerged; frequent in Sutton Park, very rare elsewhere. 4; 1%; not found fruiting. 0563, 0998, 1096, 2304.
BIRA, LAF. Recorded from 37, 39, 55 and 57.

(2001171 **Sphagnum contortum** Schultz is recorded from several places by Bagnall, but this is a var. *contortum* Schimp. of *S. subsecundum* Nees, which is not now recognised and is not the same as Schultz's *S. contortum*. True *S. contortum* has not been found in VC 38.)

2001181 **Sphagnum subsecundum** Nees, agg. Map 828, p. 699
Bagnall: Sutton Park, in *Proc. Bgham nat. Hist. microsc. Soc.* **1** (1869); many other records from the northern half of the county.
Rhodes: Sutton Park. Wait and Elsee: Brandon Wood.
In bogs, at boggy margins of ditches, ponds and lakes, sometimes submerged, on wet heaths, wet woods and woodland rides; widespread in the north and in a few localities in the centre of the county. 19; 4%; not found fruiting.
BIRA, LAF, WAR. Recorded from 32, 33, 37, 39, 55 and 57.
The varieties were not recorded in detail during the survey, but the following were found.

2001182 var. **inundatum** (Russ.) C. Jens.
Bagnall: Hill Bickenhill, 1874, in herb. Birmingham Museum; several other localities in north Warwickshire.
Frequent over the range of the aggregate.
BIRA. Recorded from 33, 37, 55 and 57.

2001183 var. **auriculatum** (Schimp.) Lindb.
Bagnall: Sutton Park, 1875, in herb. Oxford University; many other records.
The common form in Warwickshire.
BIRA, LAF, OXF. Recorded from 32, 33, 37, 39, 55 and 57.

2001191 **Sphagnum fimbriatum** Wils. Map 829, p. 699
Bagnall: Sutton Park, in *Proc. Bgham nat. Hist. microsc. Soc.* **1** (1869); many other records from north-west Warwickshire.
Townsend: Bracebridge Pool, Sutton Park.
In bogs, at boggy margins of ditches, ponds and lakes, on wet heaths and in wet woodlands on acid soils, but requiring less acid conditions than many species; the commonest species of the genus; in the northern half of the county. 21; 5%; not found fruiting.
BIRA, LAF, WAR. Recorded from 23, 37, 39, 55 and 57.

2001201 **Sphagnum girgensohnii** Russ.
Bagnall: Arley Wood, 1884, in herb. Birmingham Museum.
Not found in the present survey.
BIRA. Recorded from 39 and 57.

2001241 **Sphagnum rubellum** Wils.
Bagnall: Sutton Park, in *Proc. Bgham nat. Hist. microsc. Soc.* **1** (1869); Coleshill Bog; Coleshill Heath.
Not found in the present survey.
BIRA. Recorded from 39, 55 and 57.

(**Sphagnum acutifolium** Ehrh. agg. was recorded by Kirk and Bagnall from several localities in the north of the county. Specimens in herbaria are mostly *S. plumulosum* Röll, although one is *S. capillaceum* (Weiss) Schrank.)

2001251 **Sphagnum capillaceum** (Weiss) Schrank
Bagnall: Coleshill Bog, 1878, in herb. Birmingham Museum, named *S. acutifolium* Ehrh.
Rhodes: Sutton Park.
At boggy margins of pools, rare. 1; not fruiting. 2404.
BIRA, LAF. Recorded from 32, 37, 39, 55 and 57.

(2001261 **Sphagnum quinquefarium** (Lindb.) Warnst. was recorded, without details of locality, by Bagnall in a MS note in his own copy of Flora Warws. and, on the strength of this record, included in the "Census Catalogue", 1st Edn., 1907. There is no specimen and the record is suspect.)

2001271 **Sphagnum plumulosum** Röll
Kirk: The Pools, near Binley Grange, undated, in herb. Oxford University.
Bagnall: Sutton Park; Coleshill Bog; Browns Wood, Solihull; Chalcot Wood.
Rhodes: Bracebridge Bog, Sutton Park.
All the above are based on specimens, named *S. acutifolium* Ehrh. In boggy areas on heaths and by acid lakes and pools, rare. 8; 2%; not found fruiting. 0584, 0998, 2779, 2284, 2798, 2303, 2404, 2504, 5278.
BIRA, LAF, OXF. Recorded from 23, 37, 39, 55 and 57.

POLYTRICHACEAE

2003021 **Atrichum undulatum** (Hedw.) P. Beauv. Map 830, p. 699
Purton: "common", but no localities given, "Midl. Flor." (1817).
Bree; Kirk; Bagnall; Stone; Wait and Elsee; Baxter; Dalman; Townsend: many records from all over the county.
On the ground in woods, coppices, scrub and shaded hedge or ditch banks, typically on brown earth soils and avoiding very acid or highly alkaline media; occasionally on less acid heaths where podsolisation is not or hardly evident and on banks and spoil heaps of sand and gravel pits, coal pits or acid stone quarries; rarely on soft sandstone, in arable fields or on silted tree trunks; common in the north and middle of the county and on Middle Lias loam in the south; less common on the Lower Lias clay, but present where leaching or leaf litter have reduced the base status of the surface soil. 173; 40%; 35% fruiting.
BIRA, LAF, OXF, WAR. Recorded from all adjoining counties and 57.

2003022 var. **minus** (Hedw.) Par.
Bagnall: railway siding, Four Oaks, 1897, in herb. Birmingham Museum; Solihull.
Not recorded in the present survey.
BIRA. Recorded from 32, 39, 55 and 57.

2003023 var. **haussknechtii** (Jur. and Milde) Frye
Bagnall: Cornets End, 1893, in herb. Birmingham Museum; Little Dickens.
Not recorded in the present survey.
BIRA. Recorded from 37.

2005011 **Polytrichum nanum** Hedw.
Purton: "common", but no localities given, "Midl. Flor." (1817).
Bree: Meriden Heath. Kirk: Stoke Heath.
Bagnall: several records from the northern half of the county. Bloom: Wolford Heath.
On banks of old sandpits, rare, 1; fruiting. 2381.
BIRA, LAF, OXF, WAR. Recorded from all adjoining counties and 57.

2005021 **Polytrichum aloides** Hedw.
Withering: Edgbaston Plantations, in *Bot. Arr. British Plants*, 2nd Edn., vol. 3, 1792.
Purton; Bree; Kirk; Bagnall; Wait and Elsee; Baxter; Townsend: several records from the north and centre of the county; Wolford Wood; Whichford Wood.
On sandy banks of hedgerows, old sand and gravel pits and sandy faces; in several widespread localities in north and central Warwickshire, but uncommon. 6; 1%; 66% fruiting. 0998, 1763, 2874, 2381, 2797, 3168.
BIRA, LAF, OXF, WAR. Recorded from all adjoining counties and 57.

2005022 var. **minimum** (Crome) Rich. and Wall.
Bagnall: Browns Wood, Solihull, 1874, in herb. Birmingham Museum; Shirley Heath; near Packington Park.
Not recorded in the present survey.
BIRA. Recorded from 23.

2005031 **Polytrichum urnigerum** Hedw.
Kirk: railway cutting, Coventry Park, 1856, in herb. Oxford University.
Bagnall: old sandpits, Cornets End.
On banks in old sandpits, rare. 2; not fruiting. 1763, 2381.
BIRA, LAF, OXF, WAR. Recorded from all adjoining counties and 57, although doubtfully from 23.

2005061 **Polytrichum piliferum** Hedw. Map 831, p. 699
Bagnall: Sutton Park, in *Proc. Bgham nat. Hist. microsc. Soc.*, **1** (1869).
Bagnall; Wait and Elsee; Townsend: several other records from the north of the county.
In dry sandy ground, on heaths, in open heathy woodland, on floors, banks and spoil heaps of old sand and gravel pits, coal pits and siliceous rock quarries and on peaty accumulations in drainage channels of railway cuttings; widespread, but not very common. 27; 6%; 3% fruiting.
BIRA, LAF, WAR. Recorded from all adjoining counties and 57.

2005071 **Polytrichum juniperinum** Hedw. Map 832, p. 700
Bree: Coleshill Heath, in Purton, "Midl. Flor.", App. (1821).
Kirk; Bagnall; Stone; Townsend: many records from the north and centre of the county; Wolford Heath.
On dry sandy ground, on heaths, in open places in heathy woodland, on floors, banks and spoil heaps of old sand and gravel pits, coal pits and siliceous rock quarries, on ashed tracks of railways, on peaty banks of railway cuttings, on soil-covered sandstone rocks and walls and on rotten stumps; widespread and fairly frequent. 70; 16%; 13% fruiting.
BIRA, LAF, OXF, WAR. Recorded from all adjoining counties and 57.

2005091 **Polytrichum aurantiacum** Sw. Map 833, p. 700
Bagnall: Sutton Park, in *Proc. Bgham nat. Hist. microsc. Soc.* **1** (1869); Trickley Coppice; New Park; Middleton Heath; Hartshill Hayes; Windmill Naps; Browns Wood, Solihull; Mockley Wood; Waverley Wood.
In open woodland on marly, sandy and heathy soil, on colliery slag heaps and on rotten stumps; widespread, but uncommon. 12; 3%; not fruiting.
BIRA, LAF. Recorded from all adjoining counties and 57.
A remarkable form, resembling an *Atrichum*, has been found twice during the present survey, in 2503 and 4080. These are the same as the plant found by Dixon near Northampton and described by him as *Catharinaea dixonii* Braithw., in *J. Bot., Lond.* **23**, 169 (1885).

2005101 **Polytrichum formosum** Hedw. Map 834, p. 700
H. E. Lowe: Edgbaston, 1837, in herb. Warwick County Museum, named *P. juniperinum*.
Bagnall; Stone; Wait: many records from the northern half of the county; Wolford Wood.
On the ground in mildly to strongly acid woodland, scrub, hedgerows and heaths and on shaded banks of cuttings, sand and gravel pits and siliceous stone quarries; rarely on railway tracks or rotten stumps or woodwork; throughout the county. 84; 19%; 4% fruiting.
BIRA, LAF, WAR. Recorded from all adjoining counties and 57.

2005111 **Polytrichum commune** Hedw. Map 835, p. 700
Purton: "common", but no localities given, "Midl. Flor." (1817).
Bree; Perry; Bagnall; Baxter; Wait, Cumming and Elsee: many records from the northern half of the county; Wolford Heath.
In bogs, at boggy margins of pools and lakes, on wet heaths and in heathy woodland and on wet compacted ground, floors of old sandpits, colliery slag heaps and wet woodland rides; on acid soils, widespread in the northern half of the county. 37; 8%; 5% fruiting.
BIRA, LAF, WAR. Recorded from all adjoining counties and 57.

2005112 var. **perigoniale** (Michx.) B. and S.
Bagnall: Sutton Park, 1881, in herb. Birmingham Museum; Hartshill Quarries.
Not recorded in the present survey.
BIRA. Recorded from 32, 37 and 39.

2005113 var. **humile** Sw.
Bagnall: Browns Wood, Solihull, 1874, in herb. Birmingham Museum; Sutton Park; Purley Park.
Not recorded in the present survey.
BIRA. Not recorded from any adjoining county, nor from 57.

FISSIDENTACEAE

2008011 **Fissidens viridulus** Wahlenb. Map 836, p. 701
Bagnall: Brownshill Green, Coventry, 1874, in herb. Birmingham Museum; near Knowle; Bentley Park.
Bloom recorded it from Haseley, but his specimen is *F. incurvus* Starke.
On bare acid ground on banks of rivers, streams and ditches, on clay faces, in woodland rides, in scrub and on hedgebanks, rarely on soft sandstone near water or in fallow fields; usually on clayey soils, widely distributed. 48; 11%; 35% fruiting.
BIRA, LAF. Recorded from all adjoining counties and 57.

2008021 **Fissidens bambergeri** Schimp.
Laflin: clay bank of railway cutting, Stretton on Fosse, 1963.
On bare clay of calcareous clay banks where disturbance is infrequent; rare. 2; both fruiting. 2239, 3952.
LAF. Recorded from 23 and 33.

(2008031 **Fissidens minutulus** Sull. Bagnall and Wait and Elsee recorded *F. pusillus* Wils. from a few scattered localities. Three specimens so named in herb. Birmingham Museum are all *F. bryoides* Hedw. and Bagnall's described habitat, "damp sandy banks", is unlikely. All are probably erroneous. It might be found on sandstone rocks in or by streams.)

(**Fissidens pusillus** Wils. var. **lylei** (Wils.) Braithw. was recorded by Bagnall from three localities. This variety is not now recognised and, without specimens, it is not possible to place the records.)

2008032 **Fissidens minutulus** Sull., var. **tenuifolius** (Boul.) Norkett Map 837, p. 701
Laflin: sandstone fragments, old quarry, Stareton, 1957.
On limestone or sandstone fragments and blocks on the ground in shade and on soft rock exposures or stone masonry in shade, near to the ground; widespread, but uncommon. 13; 3%; 80% fruiting.
LAF. Recorded from 23, 32, 33, 37 and 57.

(**Fissidens tamarindifolius** Brid. was recorded from a few localities by Bagnall and by Wait and Elsee. The species is not now recognised and, in the absence of specimens, it is not possible to place these records.)

2008041 **Fissidens bryoides** Hedw. Map 838, p. 701
Purton: "common", but no localties given, "Midl. Flor." (1817).
H. E. Lowe; Bagnall; Baxter; Cumming, Wait and Elsee; Townsend: many records throughout the county.
On acid sandy and marly ground on hedge and ditch banks, in woodland and scrub, on sandy heaths, on faces, banks and spoil heaps of sand and gravel pits, coal pits and siliceous stone quarries, by streams and ponds and in fallow ground; on soft sandstone; rarely on silted tree bases; throughout the county except on calcareous clay and limestone, frequent. 172; 39%; 76% fruiting.
BIRA, LAF, WAR. Recorded from all adjoining counties and 57.

2008051 **Fissidens incurvus** Starke Map 839, p. 701
Bagnall: Solihull, in *Proc. Bgham nat. Hist. microsc. Soc.* **1** (1869).
Bagnall; Wait and Elsee; Bloom; Townsend: many records from all parts of the county.
On mildly acid to calcareous clayey ground on hedge and ditch banks, by rivers, streams and ponds, in woodland and scrub, on banks and spoil heaps of clay pits and cuttings, rarely in fallow fields or marshes; frequent, but rarer in the north than in the south. 81; 18%; 99% fruiting.
BIRA, LAF, WAR. Recorded from 23, 32, 33, 37, 55 and 57.

2008071 **Fissidens crassipes** Wils. Map 840, p. 702
Bagnall: River Avon, Guys Cliffe Mill, 1877, in herb. Birmingham Museum.
On stonework of bridges, culverts, weirs and other masonry by water and on stone blocks in streams, rarely on woodwork by rivers; usually submerged in running water of rivers and streams, rarely on canal locks; widespread, but not common. 26; 6%; 50% fruiting.
BIRA, LAF. Recorded from 23, 32, 33, 37 and 57.

2008121 **Fissidens exilis** Hedw. Map 841, p. 702
Bagnall: Hay Wood, 1872, in herb. Birmingham Museum; several other records from the north and centre of the county; also recorded by Bloom.
On loamy and marly acid soil in woodland and scrub, rarely on hedge or ditch banks; not common and mainly in the centre of the county. 13; 3%; all fruiting.
BIRA, LAF. Recorded from all adjoining counties and 57.

2008131 **Fissidens exiguus** Sull.
Laflin: 3270, sandstone block in small stream running into River Avon, Bericote Wood, Ashow, 1957; not fruiting.
LAF. Not recorded from any adjoining county, nor from 57.

2008151 **Fissidens taxifolius** Hedw. Map 842, p. 702
Purton: "common", but no localities given, "Midl. Flor." (1817).
Bree; Bagnall: many records, throughout the county.
On the ground in woodland and scrub, on hedge banks, banks of ditches, streams, rivers, ponds and canals, on floors and banks of clay pits and cuttings, in fallow ground and in marshes, particularly on clay soil, but not confined to it; rarely on soft rock or stonework, both limestone and sandstone; common throughout the county. 288; 66%; 7% fruiting.
BIRA, LAF, WAR. Recorded from all adjoining counties and 57.

2008161 **Fissidens cristatus** Wils. Map 843, p. 702
Laflin: banks of disused oolite quarry, Little Compton, 1959.
On broken banks in overgrown limestone quarries, banky pastures and cuttings, on the Lias and oolitic limestones and calcareous clay, rare. 9; 2%; not found fruiting.
LAF. Recorded from all adjoining counties and 57.

2008171 **Fissidens adianthoides** Hedw.
Purton: on a strawberry bed, Studley Castle, in "Midl. Flor." (1817).
Bagnall: several places in Sutton Park; canal, Small Heath; Ballards Green; near Binley.
In bogs, marshes and marshy flushes, locally frequent in a very few places; one record from base of hawthorn in scrub. 5; 1%; 20% fruiting. 0996, 0998, 1096, 3150, 3859.
BIRA, LAF. Recorded from all adjoining counties and 57.

2009011 **Octodiceras fontanum** (La Pyl.) Lindb.
Duncan: "Avon and Stour", MS note in Br. Bryol. Soc. Recorder's Notebook; "Census Catalogue", Supplement to 1st Edn., 1923, and subsequent editions.
On masonry of weir, in sheltered water on River Stour. 1; not fruiting. 2049.
LAF. Recorded from 23, 33, 37 and 55.

ARCHIDIACEAE

2010011 **Archidium alternifolium** (Hedw.) Mitt.
Bagnall: shores of Coleshill Pool, 1868, in *Proc. Bgham nat. Hist. microsc. Soc.* **1** (1869); Gannaway Grove, Claverdon.
Cameron: Edgbaston.
On wet acid ground in woodland rides, rare. 6; 1%; not found fruiting. 0956, 1056, 2333, 2433, 3771, 3871.
BIRA, LAF. Recorded from 23, 33, 37, 39, 55 and 57.

DITRICHACEAE

2011011 **Pleuridium acuminatum** Lindb. Map 844, p. 703
Purton: "common", but no localities given, "Midl. Flor." (1817).
Kirk; Bagnall; Stone; Rogers; Wait, Cumming and Elsee: many records from the northern half of the county and from the Middle Lias in the extreme south.
In fallow acid sandy or marly ground in arable fields, floors and banks of sand and gravel pits, on ditch and pond banks, on heaths and on woodland rides; frequent on acid soil throughout the county. 62; 14%; all fruiting.
BIRA, LAF, OXF, WAR. Recorded from all adjoining counties and 57.

2011021 **Pleuridium subulatum** (Hedw.) Lindb.
Kirk: Styvichall Common, 1856, in herb. Oxford University.
Bagnall: fields near Marston Green Railway Station; clay pit by Erdington Railway Station.
Stone: Grange Garden, Erdington. Wait and Elsee: Little Lawford. Townsend: near Coughton Park.
On bare acid sandy or marly soil in arable fields, old sandpits, woodland rides and ditch banks; not common. 10; 2%; all fruiting. 1869, 2433, 2141, 2352, 2549, 2381, 2504, 3747, 3772, 3083.

BIRA, LAF, OXF, WAR. Recorded from all adjoining counties and 57.

2012011 **Ditrichum cylindricum** (Hedw.) Grout
Map 845, p. 703
Laflin: ride in Brandon Wood, 1957.
On bare acid ground in fallow fields, woodland rides, on floors of old sandpits, pond and ditch banks and marshes; throughout the county except on calcareous clay and limestone, 54; 12%; not found fruiting.
LAF. Recorded from all adjoining counties and 57.

2012031 **Ditrichum heteromallum** (Hedw.) Britt.
Knight: VC 38, without details of locality or date, in Brit. Bryol. Soc. Recorder's Notebook; "Census Catalogue", 2nd Edn., 1926.
Not recorded in the present survey.
Recorded from 32, 37, 39, 55 and 57.

2012071 **Ditrichum flexicaule** (Schwaegr.) Hampe
Bagnall: Marlcliff, 1872, in herb. Birmingham Museum; heathy lane by Stoopers Wood, Wawensmoor.
On banks and floors of disused limestone quarries, rare. 2; not found fruiting. 2729, 3336.
BIRA, LAF. Recorded from all adjoining counties and 57.

2015011 **Ceratodon purpureus** (Hedw.) Brid.
Map 846, p. 703
Purton: "common", but no localities given, "Midl. Flor." (1817).
Bree; H. E. Lowe; Bagnall; Stone; Wait and Elsee; Dalman; Townsend: many records, throughout the county.
On rock exposures and blocks, on masonry of all kinds, on bare ground in fallow fields, tracks, pit and quarry floors, banks and spoil heaps, garden paths, ashed and burnt ground, railway tracks, marshes, heaths, banks of ditches, streams, rivers, ponds and canals, woodland rides, open places in scrub, on hedgebanks and on trees, stumps and woodwork; very common throughout the county. 387; 88%; 53% fruiting.
BIRA, LAF, WAR. Recorded from all adjoining counties and 57.

2015012 var. **conicus** (Hampe) Husn.
Recorded for VC 38 in "Census Catalogue", 1st Edn., 1907, but the source of the record is not known; omitted from subsequent editions.
Jones: on thin, dry earth on stony path, stone pits, Edge Hill, 1964, in *Trans. Br. bryol. Soc.* **4**, (5), 1965.
Not found by me in present survey.
Recorded from 23, 32, 33, 37 and 39.

SELIGERIACEAE

2017021 **Seligeria pusilla** (Hedw.) B., S. and G.
Recorded from sandstone walls, Warwick Castle, by Mrs. E. Hopkins and included by Bagnall in "Flor. Warws.", 1891 and the "Census Catalogues". The specimen in herb. Birmingham Museum is undated and Bagnall has written on the label "I do not think the locality is correct". The locality is unlikely and the plant was probably collected in the neighbourhood of Bath, from where Mrs. Hopkins' other records come.
Laflin: 2333, oolite block by culvert by ride, Wolford Wood, 1964; fruiting.
LAF. Recorded from 23, 33, 37, 39 and 57.

DICRANACEAE

2021011 **Pseudephemerum nitidum** (Hedw.) Reim.
Map 847, p. 703
Bagnall: "frequent", but no localities given, in *Proc. Bgham nat. Hist. microsc. Soc.* **1** (1869); a few records from north Warwickshire.
On bare acid ground in fallow fields, on banks of ditches, streams, rivers and ponds, on paths, woodland rides, compacted floors of old sand, gravel and sandstone quarries, marshes and on rotten willow trunks by rivers; throughout the county. 90; 21%; 35% fruiting.
BIRA, LAF, WAR. Recorded from all adjoining counties and 57.

Since writing the above account and preparing the map, a new species, Dicranella staphylina *Whiteh. (q.v.) has been described. All sterile material named* Pseudephemerum nitidum *(Hedw.) Reim. in my herbarium proves to be this new species and it is probable that most of the 55 records of non-fruiting* P. nitidum *are also this species. This does not alter the distribution pattern of* P. nitidum, *but its frequency is reduced. Only fruiting records of this plant should be noticed.*

(2022011 **Dicranella palustris** (Dicks.) Crundw. was recorded by Bagnall, without details of locality, in a MS note in his own copy of the Flora of Warws. and, on the strength of this, it was included in the "Census Catalogue", 1st Edn., 1907. Without better evidence the record is suspect.)

2022021 **Dicranella schreberana** (Hedw.) Dix.
Map 848, p. 704
Bagnall: below Rowton's Well, Sutton Park, 1884, in *Midl. Nat.* **8** (1885); by Bracebridge Pool, Sutton Park.
On bare ground in fallow fields, on banks of ditches, streams, rivers and ponds, in marshes, on compacted floors of sand and clay pits and in woodland rides, widely distributed; most frequent on mildly acid marl, but also found on basic clay and acid sand, but not on very acid media. 85; 19%; 1% fruiting.
BIRA, LAF. Recorded from all adjoining counties and 57.

2022041 **Dicranella varia** (Hedw.) Schimp.
Map 849, p. 704
Bagnall: canal bank near Rowington, 1873, in herb. Birmingham Museum.
Bagnall; Wait and Elsee; Jackson: several records from all over the county.
On bare ground on banks and spoil heaps of limestone quarries and clay pits, cutting banks, banks of ditches, streams, rivers, ponds and canals, fallow fields, woodland rides, marshes and scrub; most frequent on basic media, but not confined to it; widespread in the south, much less frequent in the north. 132; 30%; 19% fruiting.
BIRA, LAF. Recorded from all adjoining counties and 57.

2022043 **Dicranella staphylina** Whitehouse
Laflin: fallow stubble field, Manor Farm, Luddington, 1958.
This species was first described by H. L. K. Whitehouse in *Trans. Br. bryol. Soc.* **5**, (4), 759 (1969).
The leaves and areolation are very like those of *Pseudephemerum nitidum* (Hedw.) Reim. (q.v.). During the survey I recorded that plant in the sterile state from 55 tetrads. All my herbarium specimens of this sterile material prove to be *Dicranella staphylina* Whiteh. and I have since found it in several widely dispersed stubble fields. It is obviously widespread and frequent in bare ground habitats on acid soils, particularly in the centre and north of the county.

2022051 **Dicranella rufescens** (With.) Schimp. Map 850, p. 704
Bagnall: near Knowle, in *J. Bot., Lond.* **12** (1874); Tythall Lane, Solihull.
Townsend: Wolford Wood.
On bare acid ground in woodland rides, rarely on floors of old sandpits or in fallow fields; uncommon and mainly in the centre of the county. 10; 2%; 70% fruiting.
LAF. Recorded from all adjoining counties and 57.

2022061 **Dicranella crispa** (Hedw.) Schimp.
Bagnall: Solihull, in *J. Bot., Lond.* **12** (1874); sandstone rocks, lane out of Sandy Lane, Milverton.
Mackay: Birmingham.
Not recorded in present survey.
Recorded from 39 and 57.

2022081 **Dicranella cerviculata** (Hedw.) Schimp.
Bagnall: Sutton Park, in *Proc. Bgham nat. Hist. microsc. Soc.* **1** (1869); several places in Sutton Park; Coleshill Heath; Merevale Park.
On peaty ground on banks of runnels, spoil heaps of coal pits, wet heaths and woodland rides; uncommon in the north of the county. 5; 1%; all fruiting. 0996, 2070, 2790, 2996, 2503.
BIRA, LAF. Recorded from all adjoining counties and 57.

2022091 **Dicranella heteromalla** (Hedw.) Schimp. Map 851, p. 704
Withering, junr.: Further Plantations, Edgbaston, in "Syst. Arr. Brit. Plants," 4th Edn., vol. 3, 1801.
Purton; Kirk; Bagnall; Wait and Elsee; Dalman; Townsend: many records from all over the county.
On bare ground in woods, scrub, heaths, hedgebanks, banks of ditches, streams, rivers and ponds, on banks and spoil heaps of sand and gravel pits, coal pits, ironstone quarries and siliceous stone quarries, on cutting banks, on soft sandstone exposures, blocks and masonry and on rotting stumps; rarely at tree bases and in fallow fields; widespread and common, but absent from calcareous media. 252; 58%; 37% fruiting.
BIRA, LAF, OXF. Recorded from all adjoining counties and 57.

(2022092 var. **sericea** (Schimp.) Schimp. was recorded by Bagnall from Corley Rock, 1888, in herb. Birmingham Museum, but his specimen appears to be the type. It probably grows on sandstone rock faces, but was not recorded in the present survey.)

2026011 **Dichodontium pellucidum** (Hedw.) Schimp.
Bagnall: Hardings Wood, Maxstoke, 1881, in herb. Birmingham Museum; several other records from the north-west of the county.
On sandstone rocks in and by streams, widespread but uncommon; mainly on the Coal Measures. 8; 2%; not found fruiting. 0583, 1598, 2969, 2974, 2488, 2489, 2996, 3069, 3270.
BIRA, LAF. Recorded from 33, 37, 39, 55 and 57.

(2026013 var. **flavescens** (With.) Husn. might be found in similar places to the type.)

2027011 **Dicranoweisia cirrata** (Hedw.) Lindb. Map 852, p. 705
Purton: "common", but no localities given, "Midl. Flor." (1817).
Bree; Bagnall; Wait and Elsee; Jackson; Townsend: many records from all over the county.
On trees and shrubs in woodland, scrub, hedgerows and by water, on woodwork, on fallen wood and on stumps, common everywhere; less frequent on stonework of all kinds, but most often on sandstone walls. 309; 71%; 50% fruiting.
BIRA, LAF, WAR. Recorded from all adjoining counties and 57.

2029051 **Dicranum montanum** Hedw. Map 853, p. 705
Bagnall: Lower Nuthurst, Sutton Park, 1867, in Notes on Sutton Park, 1876, the first British record; several other records from north Warwickshire.
In woods in the northern half of the county and on the Middle Lias in the extreme south; mainly on stumps, but also at the bases of oak trees. 27; 6%; not found fruiting.
BIRA, LAF, OXF, WAR. Recorded from all adjoining counties and 57.

2029061 **Dicranum flagellare** Hedw.
Laflin: West Grove Wood, Haselor, 1959.
On oak stumps in woodland, rare. 4; 1%; not found fruiting. 0660, 1256, 1356, 3034, 3357.
LAF. Recorded from 23 and 39.

2029071 **Dicranum strictum** Schleich. Map 854, p. 705
Laflin: Frogmore Wood, Fen End, 1957.
In woods, mainly on stumps or dead branches, but occasionally on living wood, rare. 10; 2%; not found fruiting.
LAF. Recorded from all adjoining counties and 57.

2029101 **Dicranum fuscescens** Sm.
Bagnall: oak tree, Birchley Stump Wood, Maxstoke, 1882, in herb. Birmingham Museum; oak tree, Morgrove Coppice, Spernall.
Not found in the present survey.
BIRA. Recorded from 37, 39 and 57.

2029111 **Dicranum majus** Sm. Map 855, p. 705
Kirk: Keresley, no date, in herb. Oxford University.
Bagnall: Bentley Park; Hartshill Hayes; Arley Wood; Browns Wood, Solihull; Rough Hill Wood (also found by Townsend); Coughton Park.
In oak woods on acid soil, rare. 8; 2%; not found fruiting.
BIRA, LAF, OXF, WAR. Recorded from all adjoining counties and 57.

2029121 **Dicranum bonjeanii** De Not.
Bagnall: Sutton Park, in *Proc. Bgham nat. Hist. microsc. Soc.*, **1** (1869).
Bagnall; Wait and Elsee; Jackson; Townsend: several other records from the northern half of the county; Wolford and Barton Heaths.
On wet and dry heaths, heathy grassland and in woods, rare. 5; 1%; not found fruiting. 0454, 0873, 1062, 2384, 3688.
BIRA, LAF, OXF. Recorded from all adjoining counties and 57.

2029131 **Dicranum scoparium** Hedw. Map 856, p. 706
Purton: "common", but no localities given, in "Midl. Flor." (1817).
Kirk; Bagnall; Rogers; Wait and Elsee; Townsend: many records from all over the county.
On the ground on wet and dry heaths, in rough grassland on acid or basic soils, in woods and scrub, on overgrown spoil heaps of stone quarries or coal pits and on stumps and at tree bases in woods; rarely on stumps in hedgerows, on stone blocks in drainage channels of railway cuttings and on sandstone wall tops; throughout the county. 90; 21%; 1% fruiting.
BIRA, LAF, OXF, WAR. Recorded from all adjoining counties and 57.

2029141 **Dicranum polysetum** Michx.
Bagnall: Wolford Heath, 1887, in herb. Birmingham Museum. Braithwaite, in "Brit. Moss Flor.", vol. 1, 1887, gives this as the first British record, but there is an earlier record from Scotland.
Not found in the present survey.
BIRA, OXF. Recorded from 55.

2029161 **Dicranum spurium** Hedw.
Webb: Coleshill Heath, undated, in herb. Birmingham Museum.
Not recorded in the present survey.
BIRA. Not recorded from any adjoining county nor from 57.

(2030031 **Dicranodontium denudatum** (Brid.) Britt. was included in the Supplement to the 2nd Edn., "Census Catalogue", 1935, based on a plant collected by Bagnall from Tile Hill Wood, 1880, and named by him *Campylopus pyriformis* (Schultz) Brid. Later bryologists named the plant *D. denudatum* (Brid.). I have sectioned leaves from plants from all tufts in the packet. The large empty cells lie in one or two rows on the upper face of the nerve, not in a median line. There is no doubt that Bagnall named it correctly.)

(2031041 **Campylopus fragilis** (Brid.) B., S. and G. was recorded by Bagnall from five places in north Warwickshire. A specimen in his herbarium, from Hampton in Arden, is *C. flexuosus* (Hedw.) Brid. This plant is unlikely to occur in VC 38, and the records are probably erroneous.)

2031051 **Campylopus pyriformis** (Schultz) Brid. Map 857, p. 706
Bagnall: Sutton Park, in *Proc. Bgham nat. Hist. microsc. Soc.*, **1** (1869); several other records from the northern half of the county; Wolford Heath.
Wait and Elsee: Coombe Woods.
On compacted and burnt ground on heaths and heathy woodland rides and clearings (one of the first colonists after burning) and on rotten stumps in woods on acid soils in the north and centre; locally frequent in a few places, but not generally common. 42; 10%; 8% fruiting.
BIRA, LAF. Recorded from all adjoining counties and 57.

2031061 **Campylopus flexuosus** (Hedw.) Brid. Map 858, p. 706
Purton: Ragley Woods, in "Midl. Flor." (1817).
Bagnall; Rogers; Townsend: a few records from scattered localities in the northern half of the county.
On heathy ground in woods and on heaths and on stumps in woodland; rarely on stumps in hedgerows or on the tops of soft sandstone walls; in the northern half of the county and in the extreme south; not common. 31; 7%; not found fruiting.
BIRA, LAF, WAR. Recorded from all adjoining counties and 57.

(2031101 **Campylopus polytrichoides** De Not. was recorded for VC 38, based on a plant collected by Arnold from Pooley Fields, 1965, in *Trans. Br. bryol. Soc.* **5**, (1), 193 (1966). Examination of further material from the original gathering, and from subsequent gatherings from the same locality, has shown it to be *C. introflexus* (Hedw.) Brid.)

2031111 **Campylopus introflexus** (Hedw.) Brid.
Laflin: rotten stump in hedgerow between Shotteswell and Mollington, 1961.
On bare or burnt ground on wet or dry heaths or in heathy woodland rides, or on colliery slag heaps; rarely on rotten stumps or on fallow sandy ground; in disturbed habitats on acid media; almost certainly a recent arrival and spreading fairly rapidly in suitable places. 9; 2%; not found fruiting. 2757, 2070, 2896, 2503, 2504, 3078, 3094, 3294, 4246.
LAF, WAR. Not recorded from any adjoining county nor from 57.

LEUCOBRYACEAE

2033011 **Leucobryum glaucum** (Hedw.) Schimp.
Bree: Coleshill Heath in Purton, "Midl. Flor.", App. (1821).
Bagnall: several places in Sutton Park; Coleshill Heath; Coleshill Bog; Marston Green; Ballards Green; Forshaw Heath.
Not recorded in the present survey.
BIRA. Recorded from all adjoining counties and 57.

ENCALYPTACEAE

2034021 **Encalypta vulgaris** Hedw.
Laflin: old quarries, Burton Dassett Hill, 1957.
On soft limestone rock exposures and walls and on rubbly limestone faces; rare in the south of the county. 5; 1%; all fruiting. 2629, 2729, 2530, 3553, 3951, 3952.
LAF. Recorded from all adjoining counties and 57.
Bagnall recorded this species from Earlswood in *J. Bot., Lond.* **12** (1874), but later corrected this as a mistake for the next species.

2034051 **Encalypta streptocarpa** Hedw.
Bagnall: small bridge near Earlswood, in *J. Bot., Lond.* **12** (1874), as *E. vulgaris* Hedw., later corrected; other records from limestone or diorite walls, i.e. Red Hill (also found by Townsend); Meriden Shafts; Fillongley; Arley; Atherstone.
On limestone or diorite walls and rock exposures, on rubble and on spoil heaps of coal pits or quartzite quarries, widespread but uncommon. 7; 2%; not found fruiting. 1356, 1073, 2199, 3952, 3760, 3095, 3096.
BIRA, LAF, WAR. Recorded from all adjoining counties and 57.

TORTULACEAE

2035011 **Tortula ruralis** (Hedw.) Crome Map 859, p. 706
Purton: "very common", but no localities given, "Midl. Flor." (1817).
Bagnall; Wait and Elsee; Stacey Wilson: many records from all over the county.
On rock exposures and blocks, wall tops, especially on canal locks and walls, canal towpaths, railway tracks, spoil heaps and floors of old limestone quarries and colliery slag heaps; widespread and frequent in the south, becoming rarer northwards. 72; 16%; 2% fruiting.
BIRA, LAF. Recorded from all adjoining counties and 57.

2035041 **Tortula intermedia** (Brid.) Berk. Map 860, p. 707
Bagnall: Binton, 1873, in herb. Birmingham Museum; several records from the south and centre of the county; Fillongley; Mancetter.
On limestone walls, rock exposures and blocks, widespread but not common in the south, absent from the north. 22; 5%; 8% fruiting.
BIRA, LAF. Recorded from all adjoining counties and 57.

2035061 **Tortula laevipila** (Brid.) Schwaegr. Map 861, p. 707
Kirk: by the River Sowe, Willenhall, 1859, in herb. Oxford University.
Bagnall; Wait and Elsee: many records from south and central Warwickshire; also from Olton; Hampton in Arden; Grendon; Griff; Wolvey; Shilton.
On trees and shrubs, mainly on isolated trees in hedgerows or by rivers and streams, occasionally in scrub or open woodland; very rare on masonry; only in the southern half of the county. 89; 20%; 35% fruiting.
BIRA, LAF. Recorded from all adjoining counties and 57.

(2035062 var. **laevipiliformis** (De Not.) Limpr. might be found in similar places to the type.)

2035081 **Tortula papillosa** Wils. Map 862, p. 707
Bagnall: Forge Mills, in *Proc. Bgham nat. Hist. microsc. Soc.* **1** (1869); several other records from all over the county; apparently more common and over a wider area than it is now.
On trunks of trees, usually elms, in hedgerows or by rivers and streams; occasional on the Lower Lias clay in the south. 13; 3%; not found fruiting.
BIRA, LAF. Recorded from all adjoining counties but not from 57.

2035091 **Tortula latifolia** Hartm. Map 863, p. 707
Kirk: by the River Sowe, Willenhall, 1859, in herb. Oxford University.
Bagnall; Wait and Elsee: many records from all over the county.
On trees, hedgerow shrubs and masonry by rivers and streams, particularly those silted by occasional flooding; less frequent by canals; rarely temporarily establishing itself in ruderal habitats away from water, e.g. one record from an ashed footpath, one from anthills on bank of cutting; throughout the county, but frequent only in the southern half and not found by the Rea, Cole or Tame or streams feeding them. 65; 15%; not found fruiting.
BIRA, LAF, OXF. Recorded from all adjoining counties and 57.

2035101 **Tortula subulata** Hedw. Map 864, p. 708
Purton: "not rare", but no localities given, "Midl. Flor." (1817).
Bree; Bagnall; Wait and Elsee: several records from all over the county.
On sandy hedgebanks, sandy tracks, canal towpaths, soft sandstone masonry, spoil heaps and banks of old quarries and sandpits, banks of ponds and ditches, on the ground under beech trees and on rotting stumps in hedgerows; widespread, but mainly in the south and centre; not common. 15; 3%; 67% fruiting.
BIRA, LAF, WAR. Recorded from all adjoining counties and 57.

2035121 **Tortula muralis** Hedw. Map 865, p. 708
Purton: "very common", but no localities given, "Midl. Flor." (1817).
Bagnall; Wait and Elsee; Townsend: many records from all over the county.
On rock exposures and blocks, both basic and siliceous, on masonry of all kinds and on dumped bricks and concrete, abundant throughout the county; rarely on trees or on marly banks. 326; 74%; all fruiting.
BIRA, LAF, WAR. Recorded from all adjoining counties and 57.

2035122 var. **aestiva** Hedw.
Bagnall: dam of Bracebridge Pool and by Powell's Pool, Sutton Park, in *J. Bot., Lond.* **12** (1874).
Stone: Erdington Grange garden.
Not recorded in the present survey, but I have not attempted to separate the variety from the type.
BIRA. Recorded from all adjoining counties and 57.

2035131 **Tortula marginata** (B. and S.) Spruce Map 866, p. 708
Bagnall: Sutton Park, in *Proc. Bgham nat. Hist. microsc. Soc.* **1** (1869); several other records from the Warwick area and a few from elsewhere in the northern half of the county.
On sandstone walls and rock outcrops, particularly on the Keuper sandstone around Warwick; not common. 17; 4%; 11% fruiting.
BIRA, LAF. Recorded from 23, 32, 33, 37 and 39.

2036021 **Aloina rigida** (Hedw.) Limpr.
Purton: on the side of the Ridgeway, in "Midl. Flor." (1817).
Bagnall: Ilmington; Newbold on Stour; Fenny Compton; Tysoe; Kineton; Harbury; Wilmcote.
Not found in the present survey, but should grow in similar places to, and in company with, the next two species.
BIRA. Recorded from all adjoining counties and 57.

2036031 **Aloina ambigua** (B. and S.) Limpr. Map 867, p. 708
Bagnall: wall tops, Binton, 1873, in herb. Birmingham Museum; several records from the southern half of the county; also from Arley; Astley; between Nuneaton and Hartshill.
On floors, spoil heaps and faces of old limestone quarries; rarely on marly faces or on banks in old sandpits; in the south and centre of the county, not common. 12; 3%; 62% fruiting.
BIRA, LAF. Recorded from all adjoining counties and 57.

2036041 **Aloina aloides** (Schultz) Kindb.
Kirk: clay bank, Radford Tollgate, near Coventry, 1858, in herb. Oxford University.
Bagnall: several records from all over the county and apparently more common than the last species in his time.
In similar places to the preceding and often growing with it. 10; 2%; all fruiting. 0950, 1763, 2730, 2830, 2545, 2450, 2550, 2466, 3255, 3951, 4150.
BIRA, LAF, OXF. Recorded from all adjoining counties and 57.

2037021 **Desmatodon convolutus** (Brid.) Grout
Bagnall: marly bank on the Alcester Road, Drayton Bushes, three miles from Stratford on Avon, in *J. Bot., Lond.* **12** (1874).
Not found in the present survey.
Recorded from 57.

2038011 **Pterygoneurum ovatum** (Hedw.) Dix.
Purton: "common", but no localities given, "Midl. Flor." (1817).
Bagnall: several records from the southern half of the county.
Wait and Elsee: Clifton.
On rubbly wall tops and on spoil heaps in calcareous clay pits, rare; this species was frequent in the 19th century, but has become rare since limestone walls are no longer capped with soil. 2; both fruiting. 2141, 4364.
BIRA, LAF. Recorded from 23, 32, 33, 37, 55 and 57.

2040011 **Pottia lanceolata** (Hedw.) C. Müll. Map 868, p. 709
Purton: "common", but no localities given, "Midl. Flor." (1817).
Bagnall; Wait and Elsee; Bloom: several localities in the southern half of the county; also Solihull; Tanworth; Arley Wood.
On floors, spoil heaps and faces of old limestone quarries and calcareous clay pits, on anthills on calcareous banks and on rubbly limestone walls in the southern half of the county; not common. 16; 4%; all fruiting.
BIRA, LAF. Recorded from all adjoining counties and 57.

2040031 **Pottia heimii** (Hedw.) Fürnr.
Laflin: 4460, muddy margin of brackish pool, salt spring, Southam Holt, 1957. 1; fruiting.
LAF. Recorded from 32 and 37.

2040041 **Pottia intermedia** (Turn.) Fürnr. Map 869, p. 709
Purton: "not rare", but no localities given, "Midl. Flor." (1817).
Kirk; Bagnall; Wait and Elsee: several records from all over the county.
On fallow ground, on floors, banks and faces of old pits and quarries and on bare earth patches on banks, on a variety of soils; widespread, but not common in the centre of the county. 10; 2%; all fruiting.
BIRA, LAF, OXF. Recorded from all adjoining counties and 57.

2040051 **Pottia truncata** (Hedw.) Fürnr. Map 870, p. 709
Purton: "common", but no localities given, "Midl. Flor." (1817).
Bagnall; Wait and Elsee; Townsend: many records from all over the county.
On fallow ground in arable fields, on tracks and paths, canal towpaths, woodland rides, banks of rivers, streams, ditches and ponds, floors and spoil heaps of pits and quarries and in marshes; common throughout the county; most frequent on acid or neutral soil, but also on basic media. 163; 37%; 94% fruiting.
LAF. Recorded from all adjoining counties and 57.

(2040071 **Pottia wilsonii** (Hook.) B. and S. was recorded from Claverdon by Bloom in a MS note in Bagnall's own copy of "Flor. Warws.", but this is almost certainly an error.)

2040091 **Pottia davalliana** (Sm.) C. Jens. Map 871, p. 709
Bagnall: Maxstoke, in *Proc. Bgham nat. Hist. microsc. Soc.* **1** (1869); several other records from all over the county.
Cumming and Elsee: Little Lawford. Townsend: Spernall.
On bare ground in fallow fields, on floors and spoil heaps of limestone quarries, calcareous clay pits and coal pits and on anthills in scrub; always on basic soil; rare, but widespread, mainly in the south of the county. 13; 3%; all fruiting.
BIRA, LAF. Recorded from all adjoining counties and 57.

(2040121 **Pottia bryoides** (Dicks.) Mitt. has been recorded four times: by Bagnall from near Coleshill; by Wait and Elsee from Bilton and Little Lawford; and by Bloom from Stratford on Avon. There is a specimen of the last in the herbarium at Birmingham Museum which is *P. lanceolata* (Hedw.) and all are probably erroneous.)

2041011 **Phascum curvicollum** Hedw.
Clark: 2530, rubbly wall top near Little Compton village, 1960, in herb. Br. Bryol. Soc. 1; fruiting.
Recorded from 23, 32, 33, 37 and 57.

2041021 **Phascum cuspidatum** Hedw. Map 872, p. 710
Purton: "common", but no localities given, "Midl. Flor." (1817).
Bagnall; Rogers; Wait and Elsee; Cumming; Townsend: many records from all over the county.
On bare ground in arable fields, compacted ground of paths and waysides, canal towpaths, floors of old pits and quarries, banks of rivers, streams, ditches and ponds, woodland rides and marshes on all soil types, but preferring heavy soils; common throughout the county. 119; 27%; all fruiting.
BIRA, LAF, WAR. Recorded from all adjoining counties and 57.

(2041022 var. **piliferum** (Hedw.) Hook. and Tayl. was recorded from Thurlaston by Cumming, Wait and Elsee in *Reps. Rugby Sch. nat. Hist. Soc.* for 1893 and 1894, but almost certainly an error.)

2041024 var. **curvisetum** (Dicks.) Nees and Hornsch. Bagnall: fallow field near Coleshill Pool, in *Proc. Bgham nat. Hist. microsc. Soc.* **1** (1869).
Not recorded in the present survey, nor from any adjoining county nor 57.

2042011 **Acaulon muticum** (Hedw.) C. Müll.
Purton: "common", but no localities given, "Midl. Flor." (1817).
Bagnall: Sutton Park; Coleshill Heath; Tilehouse Green, Knowle.
In sandy and marly acid bare ground of fallow fields and woodland rides; rare. 4; 1%; all fruiting. 0973, 1552, 2433, 2352.
LAF, WAR. Recorded from all adjoining counties and 57.

2043011 **Cinclidotus fontinaloides** (Hedw.) P. Beauv.
Purton: stone cistern, watering place at Binton, opposite Dr. Carlton's house, in "Midl. Flor." (1817).
Bree: mill wheel, Bidford Grange. Cheshire: weir, Luddington.
On masonry of weirs, sluices and bridges, submerged; along a limited length of the Stour and the Avon between Milcote and Binton; one record from the Oxford Canal, Hillmorton. 3; 1%; not found fruiting. 1652, 1853, 5473.
LAF, OXF. Recorded from all adjoining counties and 57.

2043031 **Cinclidotus mucronatus** (Brid.) Mach.
Map 873, p. 710
Kirk: near Stratford on Avon, 1860, in herb. Oxford University; by the River Sowe, Willenhall.
Bagnall: several records from the Avon, Itchen, Stour, Arrow and Alne.
On trees by rivers and streams, particularly when silted by periodic flooding; less frequently on masonry by rivers; by the lower Avon and its tributaries from the Leam down river; not common. 19; 4%; 11% fruiting.
BIRA, LAF, OXF. Recorded from 23, 32, 33, 37 and 39.

2044011 **Barbula convoluta** Hedw. Map 874, p. 710
Kirk: Binley, undated, in herb. Oxford University.
Bagnall; Cumming and Elsee; Townsend: many records from all over the county.
On bare ground, particularly where compacted, as on paths, tracks, canal towpaths and floors of pits and quarries; frequent on ashed ground, on railway tracks and spoil heaps of stone quarries and coal pits; occasional on rock exposures and blocks and on masonry of all kinds, in fallow arable fields, woodland rides, and banks of rivers, ditches, ponds and canals; on acid or basic media, common throughout the county. 195; 45%; 16% fruiting.
BIRA, LAF, OXF, WAR. Recorded from all adjoining counties and 57.

2044012 var. **commutata** (Jur.) Husn.
Kirk: old coal pits, Griff, 1857, in herb. Oxford University.
Not recorded in the present survey, but no attempt has been made to separate the variety from the type; probably still present.
OXF. Recorded from 23, 32, 33, 39 and 57.

2044021 **Barbula unguiculata** Hedw. Map 875, p. 710
Purton: "not rare", but no localities given, "Midl. Flor." (1817).
Bagnall; Wait and Elsee; Stacey Wilson; Townsend: many records from all over the county.
On bare ground in arable fields, on paths, tracks, canal towpaths, banks of rivers, streams, ditches, ponds and canals, on floors, spoil heaps and banks of pits and quarries, on hedge and cutting banks, woodland rides, bare ground in scrub, railway tracks and marshes; on soft rock exposures

and blocks and on masonry of all kinds, particularly where rubble has accumulated; on many kinds of media, but absent from very acid soils; common throughout the county. 269; 61%; 22% fruiting.
BIRA, LAF, WAR. Recorded from all adjoining counties and 57.

2044022 var. **cuspidata** (Schultz) Brid.
Bagnall: mortar coping of wall near Hartshill, in *Midl. Nat.* **2** (1879); Wixford; Yarningale Common; Arley; Fillongley; Tysoe; Avon Dassett.
Not recorded in the present survey, but no attempt has been made to separate the variety from the type; probably still present.
Recorded from 32, 37, 39 and 57.

2044031 **Barbula revoluta** Brid. Map 876, p. 711
Bagnall: canal bridge, Olton, in *Proc. Bgham nat. Hist. microsc. Soc.* **1** (1869); many other records from all over the county except the north-east quarter. Jackson: Radford Semele.
On limestone and sandstone masonry and on brickwork jointed with lime mortar, most common by canals and mainly in the centre of the county. 21; 5%; 8% fruiting.
BIRA, LAF. Recorded from all adjoining counties and 57.

2044041 **Barbula hornschuchiana** Schultz Map 877, p. 711
Bagnall: Kineton, 1874, in herb. Birmingham Museum; Bearley; Yarningale Common; Shirley Heath; Fillongley; Ballards Green; Whitley Abbey. Bloom: Whitchurch. Townsend: Rough Hill Wood.
On compacted and ashed ground of paths, tracks, canal towpaths, railway tracks and floors of limestone quarries or sandpits; rarely on rubbly masonry; widespread, but not very common. 27; 6%; 4% fruiting.
BIRA, LAF. Recorded from all adjoining counties and 57.

2044071 **Barbula fallax** Hedw. Map 878, p. 711
Bree: Allesley, in Purton, "Midl. Flor." App. (1821).
Bagnall; Wait and Elsee; Townsend: many records from all over the county.
On soft limestone rock exposures and masonry, on bare ground of tracks, canal towpaths, floors and spoil heaps of limestone quarries and calcareous clay pits, in scrub, on cutting banks, on banks of rivers, streams, ditches and ponds, woodland rides and marshes; almost always on basic media, but rarely on mildly acid marl; frequent over most of the county, but absent from the north-west quarter. 82; 19%; 7% fruiting.
BIRA, LAF. Recorded from all adjoining counties and 57.

2044081 **Barbula reflexa** (Brid.) Brid.
Bagnall: marly banks near Preston Bagot, in *Midl. Nat.* **16** (1893).
On bank in old limestone quarry, Traitors Ford; rare. 1; not fruiting. 3336.
LAF, Recorded from 23, 32, 33, 39 and 57.

2044091 **Barbula spadicea** (Mitt.) Braithw.
Bagnall: Henley in Arden, 1873, in herb. Birmingham Museum; 24 other records from widespread localities throughout the county, except the north-east quarter.
Not found in the present survey.
BIRA. Recorded from 33, 37, 39, 55 and 57.

2044121 **Barbula rigidula** (Hedw.) Milde Map 879, p. 711
Bagnall: canal bridge, Olton, 1872, in herb. Birmingham Museum; a few other widely distributed records.
Frequent on masonry by water, particularly on canal locks and retaining walls; occasional on limestone rock exposures, blocks and walls away from water; rarely on sandstone; throughout the county. 62; 14%; 1% fruiting.
BIRA, LAF. Recorded from all adjoining counties and 57.

2044131 **Barbula nicholsonii** Culm.
Laflin: masonry of canal lock, Grand Union Canal, Hatton, 1957.
On masonry of canal locks and retaining walls, occasional; rare on masonry or compacted ground by streams. 6; 1%; not found fruiting. 1652, 1663, 1867, 2758, 2466, 4150.
LAF, Recorded from 23, 32 and 37.

2044141 **Barbula trifaria** (Hedw.) Mitt. Map 880, p. 712
Bagnall: Hatton, 1884, in herb. Birmingham Museum; a few other records from widely dispersed localities all over the county.
Townsend: by pond in Old Park Wood, Arrow.
On rock blocks, exposures and masonry, usually near to the ground and often by water; frequent in the southern half of the county, rare in the north. 58; 13%; 1% fruiting.
BIRA, LAF. Recorded from all adjoining counties and 57.

2044161 **Barbula tophacea** (Brid.) Mitt. Map 881, p. 712
Bagnall: old clay pits, Erdington, in *Proc. Bgham nat. Hist. microsc. Soc.* **1** (1869).
Bagnall; Jackson: many other records from all over the county.
On rock exposures and blocks, on masonry of all kinds, particularly by water, on rock blocks in streams, on wet ground on paths, tracks, canal towpaths, floors, spoil heaps and faces of old pits and quarries, on banks of rivers, streams, ditches and ponds, ruts in wet woodland rides, in marshes, on tufa by springs and on silted trees by rivers; frequent throughout the county. 108; 25%; 10% fruiting.
BIRA, LAF. Recorded from all adjoining counties and 57.

2044171 **Barbula cylindrica** (Tayl.) Schimp.
Map 882, p. 712
Bagnall: Sutton Park, in *J. Bot., Lond.* **12** (1874).
Bagnall; Jackson; Townsend: many records from all over the county.
On rock exposures and blocks of all types, on masonry of all kinds, on blocks filling drainage channels in cuttings, on compacted ground, on floors, spoil heaps and faces of old pits and quarries, on silted trees by water, on trees and shrubs in hedgerows and on sandy and loamy hedgebanks and stream banks; frequent throughout the county. 175; 40%; 1% fruiting.
BIRA, LAF, WAR. Recorded from all adjoining counties and 57.

2044181 **Barbula vinealis** Brid. Map 883, p. 712
Bagnall: Bearley, 1872, in herb. Birmingham Museum; Ipsley; Milverton; Astley; Fillongley; Oxhill; Warmington.
Jackson: Leamington Spa. Townsend: near Wolford Wood.
On rock exposures, blocks and walls, particularly on soft sandstone masonry; widespread, but not common. 34; 8%; not found fruiting.
BIRA, LAF. Recorded from all adjoining counties and 57.

2044201 **Barbula recurvirostra** (Hedw.) Dix.
Map 884, p. 713
Kirk: Red Hill, 1855, in herb. Oxford University; Meriden.
Bagnall; Townsend: many records from all over the county.
On rock exposures and blocks, particularly limestone, but also on sandstone, on masonry of all kinds, on spoil heaps and faces of old quarries, on trees by water; rarely on shrubs in hedgerows and scrub, or on the ground under beech trees;

widespread and fairly common in much of the county, but rare in the north-west. 69; 16%; 59% fruiting.
BIRA, LAF, OXF, WAR. Recorded from all adjoining counties and 57.

2046011 **Gyroweisia tenuis** (Hedw.) Schimp.
Map 885, p. 713
Bagnall: stone walls, Harborne Road, Edgbaston, in *Proc. Bgham nat. Hist. microsc. Soc.* **1** (1869); a few other records from the northern half of the county.
On masonry and rock blocks in and by water; widespread, but in few localities. 12; 3%; 50% fruiting.
BIRA, LAF. Recorded from all adjoining counties and 57.

2046021 **Gyroweisia reflexa** (Brid.) Schimp.
Cooper: sandstone of rock garden, The Quarry, Lutterworth Road, Nuneaton, 1933. in *Rep. Br. bryol. Soc.* for 1933; refound periodically until 1938. The only British locality for a species which is mainly Mediterranean in distribution.
Not found in the present survey.
LAF.

2047011 **Eucladium verticillatum** (With.) B., S. and G.
Laflin: lock on Grand Union Canal, Lapworth, 1957.
On calcareous or soft sandstone masonry, usually in constantly dripping water, on tufa round springs, often encrusted with lime; very rarely in calcareous marshes; rare. 6; 1%; not found fruiting. 1870, 2450, 2167, 2966, 3336, 3051.
LAF. Recorded from all adjoining counties and 57.

2049021 **Tortella tortuosa** (Hedw.) Limpr.
Bagnall: wall of canal bridge, Olton, 1872, in herb. Birmingham Museum.
Not found in the present survey.
BIRA. Recorded from 23, 33, 37, 39, 55 and 57.

2051031 **Trichostomum sinuosum** (Mitt.) Lindb.
Map 886, p. 713
Bagnall: bridge near Wootton Wawen, in *J. Bot., Lond.* **12** (1874); Fenny Compton; Alderminster; Loxley; near Henley in Arden; between Birdingbury and Marton; Mancetter.
On limestone exposures, blocks and masonry, rarely on sandstone walls or blocks; occasional on brickwork or concrete; on hedgerow trees and shrubs; almost always near to the ground and often by water; frequent in the south, occasional in the centre, absent from the north. 71; 16%; not found fruiting.
BIRA, LAF, WAR. Recorded from 23, 32, 33, 37, 55 and 57.

2051041 **Trichostomum crispulum** Bruch
Laflin: 2729, on bank of disused oolite quarry, Little Compton, 1957. 1; not found fruiting.
LAF. Recorded from all adjoining counties and 57.

(2051051 **Trichostomum brachydontium** Bruch was included for VC 38 in the 2nd Edn., "Census Catalogue", 1926, but the source of the record is not known. Without better evidence it must be suspect.)

2052011 **Weissia controversa** Hedw. Map 887, p. 713
Bree: Allesley, in Purton, "Midl. Flor.", App. (1821).
H. E. Lowe; Bagnall; Townsend: many records from all over the county except the north-east quarter.
On bare ground of spoil heaps, banks and faces of old quarries, pits, cuttings and hedgebanks; rarely in fallow arable fields, woodland rides and pond and ditch banks; widespread, but not common. 31; 7%; all fruiting.
BIRA, LAF, WAR. Recorded from all adjoining counties and 57.

2052041 **Weissia rutilans** (Hedw.) Lindb.
Bagnall: canal bank, Olton, 1868, in herb. Birmingham Museum; Maxstoke; near Duke End.
Not found in the present survey.
BIRA. Recorded from 23, 37, 39 and 57.

2052061 **Weissia microstoma** (Hedw.) C. Müll.
Bagnall and Cameron recorded this plant from several widespread localities in the acid soil areas of the north-west and extreme north of the county, but I suspect that they were probably the var. *brachycarpa* (Nees and Hornsch.) C. Müll.
One specimen in his herbarium is in poor condition, with immature capsules, but the leaves suggest the variety.
Laflin: bank of old limestone quarry, Burton Dassett Hill, 1958.
On bare ground patches on calcareous banks of cuttings and limestone quarries, rare. 4; 1%; all fruiting. 0958, 1858, 2042, 3952.
LAF. Recorded from all adjoining counties and 57.

2052062 var. **brachycarpa** (Nees and Hornsch.) C. Müll.
Map 888 p. 714
Bagnall and Cameron: probably several records from the north-west and extreme north of the county (see notes under preceding).
On bare fallow acid marly or loamy ground in arable fields, woodland rides, marshes and ditch banks, in several widespread localities. 21; 5%; all fruiting. This variety is probably more common than the records suggest. I confused this with *W. squarrosa* (Nees and Hornsch.) C. Müll. until Mr. A. Crundwell pointed out my mistake. Only records supported by specimens have been included in the tables and map.
LAF. Recorded from 23 and 57.

2052071 **Weissia squarrosa** (Nees and Hornsch.) C. Müll.
Bagnall: fields, Kingsbury Wood, 1882, in herb. Birmingham Museum.
In acid marly fallow fields, rare. 2; both fruiting. 3885, 4282.
BIRA, LAF. Recorded from 32, 37, 55 and 57.

2052081 **Weissia rostellata** (Brid.) Lindb.
Bagnall: on bare sediment at dried margin of Alcester Reservoir, 1888, in herb. Birmingham Museum.
Not recorded in the present survey.
BIRA. Recorded from 23, 55 and 57.

(2052101 **Weissia multicapsularis** (Sm.) Mitt. was recorded by Bagnall from Sutton Park, but erroneously. His specimen is *W. crispa* (Hedw.) Mitt., var. *aciculata* (Mitt.) Dix. Braithwaite suggested to Bagnall that the plant was a form of *W. crispa* (Hedw.) (see Bagnall, "Notes on Sutton Park", 1876), but the record was included in the 1st Edn. of the "Census Catalogue" and has been repeated since.)

2052111 **Weissia crispa** (Hedw.) Mitt.
Recorded by Bagnall from Sutton Park in *Rep. Bgham nat. Hist. microsc. Soc.* **19** (1877), but this is the same plant as the preceding (see notes above), viz. var. *aciculata* (Mitt.) Dix.
Laflin: Lias clay bank opposite railway station, Binton, 1956.
On bare ground in fallow fields, on banks and on cutting faces, on calcareous clay, rare in south and central Warwickshire. 8; 2%; all fruiting. 1352, 1453, 1455, 1556, 1163, 2042, 2653, 3052.
LAF. Recorded from all adjoining counties and 57.

2052112 var. **aciculata** (Mitt.) Dix.
Bagnall: by Powell's Pool, Sutton Park, 1876, in herb. Birmingham Museum, variously named by Bagnall *W. crispa* (Hedw.) Mitt. and *W. multicapsularis* (Sm.) Mitt.
On bare acid sandy, gravelly or marly ground in fallow fields and on ditch banks, rare. 6; 1%; all fruiting. 0960, 1552, 1652, 2188, 3057, 4282.
BIRA, LAF, WAR. Recorded from 32, 33, 37 and 55.

2053011 **Leptodontium flexifolium** (Sm.) Hampe
Bagnall: Purley Park, Atherstone, 1885, in herb. Birmingham Museum.
On peaty ground on heaths, rare. 1; not fruiting. 1095.
BIRA, LAF. Recorded from 32, 37, 39, 55 and 57.

GRIMMIACEAE

2055021 **Grimmia apocarpa** Hedw. Map 889, p. 714
Purton: "not rare", but no localities given, "Midl. Flor." (1817).
Bagnall: many records from all over the county.
On limestone exposures, blocks and masonry and on brickwork with lime-mortar jointing, frequent; occasionally on sandstone walls and on rock blocks in drainage channels of cuttings; frequent in the south and centre, rare in the north. 98; 22%; 95% fruiting.
BIRA, LAF, WAR. Recorded from all adjoining counties and 57.

(2055041 **Grimmia stricta** Turn. was recorded four times by Bagnall as *G. apocarpa* Hedw., var. *gracilis* auct. There is one specimen in his herbarium which is *G. apocarpa* Hedw., and probably all were this species.)

2055072 **Grimmia alpicola** Hedw. var. **rivularis** (Brid.) Broth.
Bagnall: stones in stream out of large pool, Arbury Hall, in *Midl. Nat.* **2** (1879).
Not found in the present survey.
Recorded from 37, 39, 55 and 57.

2055091 **Grimmia crinita** Brid.
Bagnall: mortar of old bridge over Grand Union Canal, Hatton, 1872, in herb. Birmingham Museum. Found up to 1889, but not since. The only British record for a mainly southern European moss.

2055201 **Grimmia pulvinata** (Hedw.) Sm. Map 890, p. 714
Purton: "very common", but no localities given, "Midl. Flor." (1817).
H. E. Lowe; Bagnall; Wait and Elsee; Townsend: many records from all over the county.
On limestone rock exposures and blocks and on masonry of all kinds, more common on limestone and joints in brickwork than on sandstone; common throughout the county, but less so in the north than in the south and centre. 155; 36%; 95% fruiting.
BIRA, LAF, WAR. Recorded from all adjoining counties and 57.

(2055202 var. **africana** (Hedw.) Dix. was recorded by Bagnall as var. *obtusa* (Brid.) Hüben., from Ilmington, 1893, but his specimen is only the type.)

2055241 **Grimmia trichophylla** Grev.
Bagnall: wall of Lapworth Churchyard, 1873, in herb. Birmingham Museum; stone walls of farmyard between Cubbington and Chesford; Stoneleigh; Radford Semele; Avon Dassett.
On sandstone or ironstone masonry, mainly on walls, rare. 9; 2%; not found fruiting. 1869, 1879, 2656, 2865, 2967, 2386, 3371, 4961, 4982, 5172.
BIRA, LAF. Recorded from 32, 37, 39, 55 and 57.

2056021 **Rhacomitrium aciculare** (Hedw.) Brid.
Laflin: 4982, rock blocks filling drainage channel, railway cutting at north end of tunnel, Gills Corner, Monks Kirby, 1965. 1; not fruiting.
LAF. Recorded from 37, 39, 55 and 57.

2056041 **Rhacomitrium fasciculare** (Hedw.) Brid.
Laflin: tiled roof of old barn, facing north, Stareton, 1958.
On rock blocks filling drainage channels on railway cuttings and embankments; rarely on tiled roofs or sandstone masonry by canals; rare. 8; 2%; 11% fruiting. 1073, 3371, 4979, 4982, 4983, 4984, 5161, 5262, 5173.
LAF. Recorded from 32, 33, 37, 39, 55 and 57.

2056051 **Rhacomitrium heterostichum** (Hedw.) Brid.
Kirk: stone wall, Binley, undated, in herb. Birmingham Museum.
In similar places to the preceding and often growing with it; rare. 6; 1%; not found fruiting. 3371, 4982, 4983, 4984, 5162, 5263, 5172, 5173. Several gatherings were entirely without hair points to the leaves and this form appears to be as frequent as the typical plant.
BIRA, LAF. Recorded from 32, 37, 39, 55 and 57.

2056071 **Rhacomitrium canescens** (Hedw.) Brid.
Bree: shores of Coleshill Pool, in Purton, "Midl. Flor." App. (1821); (also found there by Kirk).
Bagnall: near Four Ashes, lane leading to Monkspath; lane from Solihull to Sharmans Cross; wayside, main road to Kenilworth, near Berkswell railway station.
Stacey Wilson: railway cutting, Harbury.
On rock blocks filling drainage channels in railway cuttings, old Great Central line; rare. 2; not found fruiting. 4982, 5268.
BIRA, LAF, OXF. Recorded from 23, 33, 37, 39, 55 and 57.

2056081 **Rhacomitrium lanuginosum** (Hedw.) Brid. Map 891, p. 714
Bagnall: Pinley, in *J. Bot., Lond.* **12** (1874); Kenilworth, on the road from Leamington; Chesford Bridge.
On rock blocks filling drainage channels in railway cuttings, rarely on sandstone walls; the most frequent species of the genus; not in the south of the county. 11; 3%; not found fruiting.
LAF, WAR. Recorded from 32, 33, 37, 39, 55 and 57.

FUNARIACEAE

2058011 **Funaria hygrometrica** Hedw. Map 892, p. 715
Purton: "common", but no localities given, "Midl. Flor." (1817).
Kirk; Bagnall; Wait and Elsee; Townsend: many records from all over the county.
On bare ground habitats of all kinds, but particularly on fire sites, burnt heaths, ashed ground and railway tracks, flower pots in glasshouses and canal towpaths, but also in fallow fields, on tracks, paths, floors and spoil heaps of pits and quarries, margins of ponds, woodland rides, marshes and banks of streams and ditches; on soft rock blocks and masonry of all kinds and on silted wood by water; common throughout the county and found on waste ground in the middle of big towns. 246; 56%; all fruiting.
BIRA, LAF, OXF. WAR. Recorded from all adjoining counties and 57.

(2058012 **Funaria microstoma** B., S. and G. was recorded by Bagnall from Oversley Wood in *Midl. Nat.* **16** (1893), but undoubtedly an error.)

2058041 **Funaria fascicularis** (Hedw.) Schimp.
Bagnall: Minworth, in *Proc. Bgham nat. Hist. microsc. Soc.* **1** (1869); several other records from the north-west quarter of the county; also recorded by Rogers.
In fallow fields on marly soil, usually when left long without disturbance; rarely on canal towpaths; rare. 5; 1%; all fruiting. 0858, 0960, 1058, 3057, 4561.
BIRA, LAF. Recorded from all adjoining counties and 57.

2059011 **Physcomitrium pyriforme** (Hedw.) Brid. Map 893, p. 715
Purton: bank bounding the millpond, Oversley, in "Midl. Flor." (1817).
Bree; Kirk; Bagnall; Stone; Cumming, Wait and Elsee; Rhodes: many records from all over the county.
On bare wet ground in marshes, on banks of rivers, streams, ditches and ponds, often where mud is thrown up onto the banks, and on canal towpaths; on both basic and acid soils; rarely on silted tree bases by water; frequent throughout the county. 57; 31%; all fruiting.
BIRA, LAF, OXF, WAR. Recorded from all adjoining counties and 57.

2060011 **Physcomitrella patens** (Hedw.) B., S. and G.
Bagnall: damp marly bank near New Fillongley Hall, in *Midl. Nat.* **2** (1879); Rotton Park Reservoir, Birmingham; Maxstoke; by Watling Street near Merevale Park; Alcester Reservoir; Wormleighton.
Wait and Elsee: Bourton on Dunsmore; Lawford Road, Rugby. Rhodes: Edgbaston Reservoir.
On drying mud at the edges of ponds and on banks of rivers and canals, rare. 3; 1%; all fruiting. 2843, 4656, 5181.
BIRA, LAF. Recorded from all adjoining counties and 57.

EPHEMERACEAE

2062011 **Ephemerum recurvifolium** (Dicks.) Boul.
Laflin: bare ground, ride in Alveston Pasture Wood, 1957.
On bare clayey ground in fields and woodland rides, rare. 2; both fruiting. 1858, 2352.
LAF. Recorded from 23 and 32.

2062061 **Ephemerum serratum** (Hedw.) Hampe Map 894, p. 715
Bagnall: Birmingham area, without precise locality, in *Rep. Bgham nat. Hist. microsc. Soc.* **2** (1870); several other records from the north-west quarter of the county.
On bare acid ground of fallow fields and woodland rides; widespread in the centre of the county and on the Middle Lias on the southern boundary; easily overlooked. 17; 4%; all fruiting.
BIRA, LAF. Recorded from all adjoining counties and 57.

(2062062 var. **minutissimum** (Lindb.) Grout has not been separated from the type in the survey. It probably occurs, but its relative frequency has not been noted.)

SPLACHNACEAE

2067021 **Splachnum ampullaceum** Hedw.
Hawkes: 0996, on horse dung in bog, Longmoor, Sutton Park, 1963. 1; fruiting.
LAF. Recorded from 23, 32, 39 and 55.

SCHISTOSTEGACEAE

2068011 **Schistostega pennata** (Hedw.) Web. and Mohr
Laflin: 2488, clefts of sandstone rocks in bank above stream, Dumble Wood, Maxstoke, 1966. 1; fruiting.
LAF. Recorded from 32, 37, 39, 55 and 57.

TETRAPHIDACEAE

2069011 **Tetraphis pellucida** Hedw. Map 895, p. 715
Bree: Allesley, in Purton, "Midl. Flor.", App. (1821).
Bagnall; Stone; Rogers; Dalman; Townsend: many records from the northern half of the county.
On stumps and tree bases in woods, less frequently in scrub and hedgerows, on alders in swamps and trees by rivers; occasionally on peaty banks of streams and ditches or in woods, and on sandstone rock exposures; frequent in the centre and north of the county. 83; 19%; 1% fruiting.
BIRA, LAF, WAR. Recorded from all adjoining counties and 57.

BRYACEAE

2071021 **Orthodontium lineare** Schwaegr. Map 896, p. 716
Laflin: tree stumps, Waverley Wood, 1957.
On rotting stumps, stools and tree bases in woods, less often in scrub and hedgerows or by water; occasionally on peaty ground on heaths or in heathy woodland; on acid soils in the north and centre of the county and on the Middle Lias in the south; widespread and frequent. 92; 21%; 90% fruiting. Well established and widespread when first found; it has probably spread rapidly since the 1930's.
LAF, WAR. Recorded from all adjoining counties and 57.

2072011 **Leptobryum pyriforme** (Hedw.) Wils. Map 897, p. 716
Bree: walls of Warwick Castle, in Purton, "Midl. Flor." App. (1821).
Kirk; Bagnall; Stone; Wait and Elsee: several records from the north-west and centre of the county.
On sandstone rock exposures, blocks and masonry, on brickwork by water, on ashed ground, on soil in flower pots and on staging in glasshouses; occasionally on bare earth in marshes and on stream banks, on colliery slag heaps; rarely in fallow ground in arable fields and woodland rides, on silted tree bases by water or on rotting stumps; widespread, but very rare in the south. 49; 11%; 23% fruiting.
BIRA, LAF, WAR. Recorded from all adjoining counties and 57.

2073051 **Pohlia nutans** (Hedw.) Lindb. Map 898, p. 716
Bree: Coleshill Heath, undated, in herb. Warwick County Museum.
Bagnall; Townsend: many records from all over the county except the Lias clay areas.
On peaty and sandy ground on heaths, in acid woodland, scrub and hedgerows, on peaty ditch banks, on tree bases and stumps, on woodwork of canal locks, on soft sandstone rock exposures and masonry, on peaty soil between rocks in drainage channels in cuttings, on ashed ground and tracks, on soil in pots in glasshouses and on floors and spoil heaps of siliceous stone quarries and sand, gravel and coal pits; frequent and widespread in acid soil areas, but only in artificial habitats in calcareous clay areas. 182; 42%; 42% fruiting.
BIRA, LAF, WAR. Recorded from all adjoining counties and 57.

2073091 **Pohlia rothii** (Corr.) Broth.
Laflin: wet ride, Oakley Wood, Bishops Tachbrook, 1957.
On wet acid soil in woodland rides or on compacted floors of sandpits and at peaty margins of pools; rare. 6; 1%; not found fruiting. 0563, 0564, 0998, 1056, 2572, 3059.
LAF. Recorded from 23, 37 and 57.

2073101 **Pohlia bulbifera** (Warnst.) Warnst.
Laflin: 0996, peaty bank of ditch, Longmoor, Sutton Park, 1964; not fruiting.
LAF. Not recorded from adjoining counties nor 57.
Bagnall recorded this from Shirley Street, 1880, but his specimen in herb. Birmingham Museum is *Pohlia annotina* (Hedw.) Loeske.

2073111 **Pohlia annotina** (Hedw.) Loeske Map 899, p. 716
Purton: Coughton Lane, in "Midl. Flor." (1817).
Bagnall: several records from the northern half of the county.
In wet, compacted, bare acid ground of woodland rides, heaths, tracks, paths, canal towpaths, floors of sand and gravel pits and colliery slag heaps, on soft sandstone rock exposures, blocks and walls and on peaty ditch banks or margins of pools; widespread, but not common, in the north and centre of the county. 32; 7%; not found fruiting.
BIRA, LAF. Recorded from 23, 32, 37, 39, 55 and 57.
The var. *decipiens* Loeske is the more common form than the type.

(2073121 **Pohlia proligera** (Lindb.) Limpr. was included in the 2nd Edn. of the "Census Catalogue", 1926, but the source of the record is not known. This species has been much confused with the var. *decipiens* Loeske of the preceding and it is likely that the record refers to that.)

2073122 **Pohlia pulchella** (Hedw.) Lindb.
Laflin: wet ride in Hay Wood, Baddesley Clinton, 1963.
In marly or sandy ground of woodland rides or clearings or on deep, shaded banks; rare. 3; 1%; not found fruiting. 2071, 2683, 3059.
LAF. Not recorded from any adjoining county nor from 57.

2073123 **Pohlia lutescens** (Limpr.) Lindb. f.
Field: wet clay in ditch, Earslwood, 1964, in Watson, *Trans Br. bryol. Soc.* **5** (3), 443 (1968).
On shaded marly or sandy banks, rare. 1 (in addition to above); female plants, but not fruiting. 2874.
LAF. Not recorded from any adjoining county nor from 57.

2073141 **Pohlia wahlenbergii** (Web. and Mohr) Andr. Map 900, p. 717
Kirk: Binley, undated, in Bagnall, *J. Bot., Lond.* **12** (1874). Bagnall: many records from all parts of the county, but mainly from the north-west and centre.
On wet soil in woodland rides, on floors of sandpits, on faces and banks of old quarries and cuttings and stream and ditch banks; usually on acid soil, but rarely on basic media; widespread, but not common. 11; 3%; not found fruiting.
BIRA, LAF. Recorded from all adjoining counties and 57.

2073151 **Pohlia delicatula** (Hedw.) Grout Map 901, p. 717
Kirk: Wyken Pumps, near Coventry, undated, in herb. Oxford University, as *Webera carnea* (Brid.) Schimp.
Bagnall; Stone; Wait and Elsee; Townsend: many records, all from the northern half of the county.
On wet bare ground, particularly on clayey soil, but also on other soil types except very acid media; on steep banks of rivers, streams, ditches and ponds, canal towpaths, marshes, woodland rides and faces and banks of clay pits; occasionally in fallow ground of arable fields, paths and tracks and sandpits, and on silted rocks in streams; common throughout the county. 217; 50%; 2% fruiting.
BIRA, LAF, OXF. Recorded from all adjoining counties and 57.

(2077041 **Bryum pendulum** (Hornsch.) Schimp. was recorded by Bagnall from nine localities. His plant from Gaydon in herb. Birmingham Museum is *B. inclinatum* (Brid.) Bland. and that from Rowington, although with immature fruit, is probably also that species. It is doubtful whether *B. pendulum* (Hornsch.) has been found in VC 38, although it is recorded from 23, 33, 37, 39 and 57.)

(2077071 **Bryum knowltonii** Barnes was recorded by Bagnall, as *B. lacustre* (Web. and Mohr) Bland., from Acorn Coppice, Tanworth, in *Midl. Nat.* **16** (1893), but almost certainly an error.)

2077081 **Bryum inclinatum** (Brid.) Bland.
Kirk: Griff, 1859, in herb. Oxford University.
Bagnall: near Stechford; near Ryton End; Shrewley Heath; Seckington; near Hemlingford Green; Gaydon Turnpike.
Jackson: Sutton Park; canal, Warwick.
On sandstone walls, ashed ground, railway tracks, spoil heaps and banks of siliceous stone quarries, coal pits and sand and gravel pits, peaty ground and rotting stumps and wood, throughout the county. As it is most frequently found with immature fruit or sterile, when it is not possible to determine it certainly, I have not mapped it. It is probably common and a number of plants in herbaria under other names have proved to be this species.
BIRA, LAF, OXF. Recorded from all adjoining counties and 57.

(2077121 **Bryum uliginosum** (Brid.) B. and S. was recorded by Bagnall from Ansty and Coventry and by Wait from Combe Fields, but probably erroneously. Bagnall's specimen from Coventry in herb. Birmingham Museum is *B. pseudotriquetrum* (Hedw.) Schwaegr.)

2077131 **Bryum pallens** Sw. Map 902, p. 717
Bagnall: Sutton Park, 1868, in herb. Birmingham Museum; several other records from the north of the county; Ilmington.
On wet compacted ground, particularly canal towpaths and at margins of pools, in marshes and marshy flushes and wet woodland rides; on silted rock blocks in streams and on compacted floors of limestone quarries; widespread, but not common and always in small quantity. 26; 6%; 4% fruiting.
BIRA, LAF, WAR. Recorded from all adjoining counties and 57.

(2077141 **Bryum turbinatum** (Hedw.) Turn. was doubtfully recorded by Bagnall from Sutton Park in *J. Bot., Lond.* **12** (1874), but almost certainly an error.)

2077191 **Bryum pseudotriquetrum** (Hedw.) Schwaegr. Map 903, p. 717
Kirk: Binley, undated, in herb. Oxford University.
Bagnall: several places in Sutton Park (also found by Townsend); Packington Park; canal near Coventry.
In bogs, marshes and marshy flushes, on wet banks of streams, canals and ponds and on masonry and woodwork of canal locks; on both basic and acid ground, widespread, but not common. 29; 7%; 28% fruiting.
BIRA, LAF, OXF. Recorded from all adjoining counties and 57.

2077192 var. **bimum** (Brid.) Lilj.
Kirk: Binley, 1859, in herb. Oxford University.
Bagnall; Wait and Elsee: many records from all over the county. Bagnall recorded it far more often than the type, but his determinations of *Bryum* material were not reliable, although the two specimens in his herbarium are correctly named.
On trees and woodwork by canals, rare. 3; 1%; with inflorescence, but not fruiting. 1869, 3664, 3763.
BIRA, LAF, OXF. Recorded from all adjoining counties and 57.

(2077201 **Bryum creberrimum** Tayl. was recorded twice by Bagnall; from Fillongley as *B. affine* Lindb. and Arn., and from Hemlingford Green as *B. bimum* Turn. var. *cuspidatum* B. and S. Neither is supported by a specimen and, in view of the many wrong determinations of *Bryum* spp. by Bagnall, the records must be suspect.)

(2077211 **Bryum pallescens** Schleich. was recorded by Bagnall from stone walls, New Fillongley Hall, 1882, in "Flor. Warws.", 1891. There is no specimen in his herbarium and the record must be suspect.)

2077221 **Bryum intermedium** (Brid.) Bland.
Kirk: Brandon, 1858, in herb. Oxford University.
Bagnall: Sutton Park; Erdington; Berkswell; Rowington; Warwick.
Townsend: Longmoor Mill Pool, Sutton Park.
On canal lock. 1; fruiting. 5374.
BIRA, OXF. Recorded from all adjoining counties and 57.

2077231 **Bryum caespiticium** Hedw.
Purton: "common", but no localities given, "Midl. Flor." (1817).
Bagnall; Stone; Wait and Elsee; Jackson: many records from all over the county.
Frequent on limestone masonry and brickwork jointed with lime mortar; also on sandstone, on railway tracks and on rotting stumps in hedgerows; throughout the county and probably common. It is often found with immature fruit or sterile, when it is not possible to determine it certainly, so I have not mapped it.
BIRA, LAF, WAR. Recorded from all adjoining counties and 57.

2077251 **Bryum argenteum** Hedw. Map 904, p. 718
Purton: "common", but no localities given, "Midl. Flor." (1817).
Kirk; Webb; Bagnall; Russell; Wait and Elsee; Townsend: many records from all over the county.
On many bare ground habitats such as paths, tracks, arable fields, ashed ground, railway tracks, towpaths of canals, floors and spoil heaps of pits and quarries, woodland rides and banks; on masonry of all kinds, rock exposures and blocks; frequent in the angles of pavements and similar places in towns; occasionally at tree bases or on woodwork; common throughout the county. 277; 63%; 25% fruiting.
BIRA, LAF, OXF, WAR. Recorded from all adjoining counties and 57.

2077252 var. **lanatum** (P. Beauv.) Hampe Map 905, p. 718
Bagnall: north Warwickshire, without precise locality, in "Victoria County Hist. Warws.", **1** (1904); Alderminster.
Townsend: Sutton Park.
On masonry of all kinds, on made-up paths and on railway tracks; throughout the county, but apparently not common, although it may have been under-recorded. 27; 6%; 10% fruiting.
LAF. Recorded from all adjoining counties and 57.

2077261 **Bryum bicolor** Dicks. Map 906, p. 718
Bree: shores of Coleshill Pool, in Purton, "Midl. Flor.", App. (1821).
Kirk; Bagnall; Wait and Elsee; Townsend: many records from all over the county.
On compacted ground of all kinds, paths, tracks, canal towpaths, railway tracks, floors and faces of quarries and pits, spoil heaps and woodland rides and on masonry of all kinds; occasionally on banks of rivers, streams, ditches and ponds; common throughout the county. 139; 32%; 44% fruiting.
BIRA, LAF, OXF, WAR. Recorded from all adjoining counties and 57.

2077271 **Bryum radiculosum** Brid.
Bagnall: Castle Bromwich, in *Proc. Bgham nat. Hist. microsc. Soc.* **1** (1869); several other records from central and north Warws.; also recorded by Jackson.
On lime-jointed brickwork of bridges and retaining walls and on compacted lime dumps; mainly in the eastern part of the county; probably under-recorded. 8; 1%; 57% fruiting. 2566, 2405, 3853, 3361, 4064, 4769, 4479, 5278.
BIRA, LAF. Recorded from all adjoining counties and 57.

2077281 **Bryum erythrocarpum** Schwaegr. Map 907, p. 718
Bagnall: near the railway, Whitacre, in *Midl. Nat.* **2** (1879); a few other records from widely distributed localities.
On bare ground on all types of soil, in fallow fields, compacted ground of paths, tracks, canal towpaths, woodland rides and clearings, railway tracks, floors, banks, faces and spoil heaps of pits, quarries and cuttings, in scrub, on hedgebanks, in marshes and on banks of rivers, streams, ditches and ponds; rarely on soft rock blocks or masonry or on silted trees by rivers; frequent throughout the county. 201; 45%; 2% fruiting.
Recorded from all adjoining counties and 57.
Crundwell and Nyholm (*Trans. Br. bryol. Soc.* **4**, (4), 1964) have shown that this species is a complex of a number of species, of which only the five following have, so far, been found in VC 38. As much of the recording in the survey was done before 1964 under the name of the aggregate, and knowledge of the distribution in Warwickshire of the species of Crundwell and Nyholm is incomplete, I have mapped *B. erythrocarpum* Schwaegr.

2077291 **Bryum rubens** Mitt.
Laflin: fallow ground, Luddington, 1957.
The common species of the group, found in similar places as are listed for *B. erythrocarpum* Schwaegr., on all except very acid soils. Since 1964 it has been found in 103 tetrads throughout the county and many of the pre-1964 records of *B. erythrocarpum* Schwaegr. probably refer to this. Frequent throughout the county. Not found fruiting.
LAF. Recorded from 23, 32, 33, 37 and 55.

2077292 **Bryum microerythrocarpum** C. Müll. and Kindb.
Laflin: fallow field, Luddington, 1957.
Widespread on mildly acid sandy and marly bare ground, but apparently much less common than *B. rubens* Mitt., although it may have been overlooked. Since 1964 it has been found in 22 tetrads, mainly in the centre and north of the county, but also on the Middle Lias in the extreme south. Not found fruiting.
LAF. Recorded from 23 and 32.

2077293 **Bryum ruderale** Crundw. and Nyh.
Laflin: ashed siding, railway station, Binton, 1956.
On compacted, often ashed, ground of made-up tracks, railway sidings and tracks, canal towpaths and similar places; apparently not common, but widespread. 8; not

found fruiting. 0960, 1453, 2672, 3859, 3783, 4656, 4182, 5064.
LAF. Recorded from 23, 32, 33 and 55.

2077294 **Bryum violaceum** Crundw. and Nyh.
Laflin: 2672, fallow field by Chase Wood, Kenilworth, 1969; not fruiting.
LAF. Recorded from 23 and 39.

2077301 **Bryum klinggraeffii** Schimp.
Laflin: 2572, fallow field south of Chase Wood, Kenilworth, 1969; not fruiting.
LAF. Recorded from 23, 39, 55 and 57.

(Of the other species of the *erythrocarpum* group, the most likely to be found in Warwickshire is probably 2077303, *B. bornholmense* Wink. & Ruthe, which might grow in woodland rides.)

2077371 **Bryum capillare** Hedw. Map 908, p. 719
Purton: loose sandy bank, lane from Coughton to Sambourne, in "Midl. Flor." (1817).
Kirk; Bagnall; Wait and Elsee; Townsend: many records from all over the county.
On rock exposures, rock blocks and masonry of all kinds; on trees, stumps and fallen wood in woods, scrub, hedgerows and by water; less frequently on railway tracks and spoil heaps of pits and quarries; rarely on sandy hedgebanks or on peaty ground under beeches; common throughout the county and very variable in form. 323; 74%; 27% fruiting.
BIRA, LAF, OXF, WAR. Recorded from all adjoining counties and 57.

(2077372, var. **torquescens** (Bruch) Husn. was recorded by Bloom from Shrewley New Pool as a MS note in Bagnall's own copy of "Flora Warws." and, on the strength of this record, included in the "Census Catalogues". Without better evidence it must be suspect.)

2077381 **Bryum obconicum** Hornsch.
Bagnall: wall of canal bridge near Rowington, in *Midl. Nat.* **7** (1884) and in herb. Birmingham Museum, 1886.
On sandy banks, rare. 1; fruiting. 3576.
BIRA, LAF. Recorded from 39.

2078011 **Rhodobryum roseum** (Hedw.) Limpr.
Bree: Allesley, undated, in herb. Warwick County Museum.
Bagnall: near Pool Hollies Wood, Sutton Park; Marston Green.
In well-drained grassy places, rare. 1; not fruiting. 3951.
BIRA, LAF, WAR. Recorded from all adjoining counties and 57.

MNIACEAE

2079011 **Mnium hornum** Hedw. Map 909, p. 719
Purton: Ragley Woods, in "Midl. Flor." (1817).
Bree; Kirk; Bagnall; Stone; Wait and Elsee; Jackson; Dalman; Townsend: many records from all over the county.
On the ground on acid soil in woods, scrub and hedgebanks; found also in woods on basic soils where a shallow humus layer accumulates on the soil surface; on acid banks of rivers, streams, ditches and ponds, on soft sandstone rock exposures, blocks and masonry, on hummocks in organic marshes, on non-podsolised heaths and on stumps and tree bases in woods, scrub, hedgerows and by water; common throughout the county except in the Lias clay areas, where it is found, but less frequently. 196; 45%; 11% fruiting.
BIRA, LAF, OXF, WAR. Recorded from all adjoining counties and 57.

(2079041 **Mnium marginatum** (With.) P. Beauv. was recorded, as *M. serratum* Brid., by Bagnall from the Birmingham area in *Rep. Bgham nat. Hist. microsc. Soc.* **2** (1870), but erroneously.)

2079061 **Mnium stellare** Hedw.
Bagnall: near Maxstoke Priory, in *J. Bot., Lond.* **12**, (1874); Shawberries Wood, Maxstoke; dripping rocks, Milverton.
On sandstone rocks in streams, rare. 1; not fruiting. 2488.
LAF. Recorded from all adjoining counties and 57.

2079071 **Mnium cuspidatum** Hedw.
Bloom: near Whitchurch, 1903, as MS note in Bagnall's copy of "Flora Warws."; wrongly attributed to Bloxham in Brit. Bryol. Soc's. Recorder's Notebook.
In banky pasture, in short turf round rock outcrops, rare. 1; not fruiting. 2042, 2142.
LAF. Recorded from all adjoining counties and 57.

2079091 **Mnium longirostrum** Brid. Map 910, p. 719
Purton: Oversley Mill, in "Midl. Flor." App. (1821).
Bagnall; Rogers; Wait and Elsee: several records from the north and centre of the county.
Amongst turf on banks in rough grassland, old pastures, cuttings and embankments, on gravelled or stoned tracks, on spoil heaps of pits and quarries, in woodland rides, on soil-covered rock exposures and walls, in marshes, on banks of rivers, streams and ditches, on the ground in woods and scrub, on hedgebanks and on silted trees by rivers; widespread and in many places, but usually in small quantity; throughout the county. 90; 21%; not found fruiting.
BIRA, LAF, WAR. Recorded from all adjoining counties and 57.

2079101 **Mnium affine** Bland.
Bagnall: Sutton Park, in *Proc. Bgham nat. Hist. microsc. Soc.* **1** (1869); several other records from the northern half of the county; also recorded by Cumming and Elsee, and Townsend.
Amongst grass in pastures, on railway cuttings, stream banks, canal locks and woodland floors, rare. 7; 2%; not found fruiting. 2437, 2352, 3154, 3967, 4877, 4286, 4982.
LAF. Recorded from 23, 32, 37, 39, 55 and 57.

2079111 **Mnium rugicum** Laur.
Newton: 0995, wet grassy margin of Longmoor Mill Pool, Sutton Park, 1969; not fruiting; in herb. Br. Bryol. Soc.
Recorded from 23, 32 and 57.

2079121 **Mnium seligeri** (Lindb.) Limpr.
Kirk: Whitley Common, undated, in herb. Oxford University as *M. affine* Bland. Cleminshaw: Sutton Park.
In boggy flushes in open ground and in organic marshes in alder swamps or loamy woods, rare. 5; 1%; not found fruiting. 0996, 0998, 1095, 1096, 2086.
BIRA, LAF, OXF. Recorded from all adjoining counties and 57.

2079131 **Mnium undulatum** Hedw. Map 911, p. 719
Purton: moist shady bank near Marsom's Gate, Dunnington, in "Midl. Flor." (1817).
Bree; Bagnall; Wait and Elsee; Townsend: many records from all over the county.
On the ground in woodland and scrub, particularly on heavy basic soils, but also on mildly acid soil where wet enough; on shaded hedgebanks, in marshes, particularly organic marshes in shade, on banks of rivers, streams, ditches, ponds and canals and in old grassland; occasionally on shaded walls or rock blocks; common throughout the county. 207; 47%; not found fruiting.

BIRA, LAF, WAR. Recorded from all adjoining counties and 57.

2079141 **Mnium punctatum** Hedw. Map 912, p. 720
Purton: Oversley Lodge, near the watering place, in "Midl. Flor." (1817).
Bagnall; Wait and Elsee; Townsend: many records from all over the county, but mainly in the north and centre.
On the ground, on stumps and fallen wood in wet oak woods and scrub, in alder swamps, in organic marshes in shade, in marshy flushes, on banks and wood by streams and ponds and on shaded sandstone rocks and walls by water; throughout the county. 92; 21%; 1% fruiting.
BIRA, LAF, WAR. Recorded from all adjoining counties and 57.

2079151 **Mnium pseudopunctatum** B. and S.
Bagnall: Sutton Park, in *Proc. Bgham nat. Hist. microsc. Soc.* **1** (1869); several places in Sutton Park (also found by Rhodes and Townsend); Black Font, Shirley Heath; Kingsbury Wood; Binley; drain near Waverley Wood; Wolford Wood.
In bogs and boggy flushes in Sutton Park, locally frequent there. 2; not found fruiting. 0996, 0998.
BIRA, LAF. Recorded from 37, 39 and 57.

AULACOMNIACEAE

2081011 **Aulacomnium palustre** (Hedw.) Schwaegr. Map 913, p. 720
Bree: Coleshill Heath, in Purton, "Midl. Flor.", App. (1821). (Also found by H. E. Lowe and Kirk.)
Bagnall: Sutton Park (also found by Webb and Townsend); Baddesley Common; Marston Green; Hill Bickenhill; Cornets End; Packington; canal siding, Ansty; near Brinklow; Barton Flat Heath; Wolford Heath.
In bogs, at boggy margins of pools, on wet heaths and in wet birch woods; in few localities, but usually frequent where found; only in the northern half of the county. 17; 4%; not found fruiting.
BIRA, LAF, OXF, WAR. Recorded from all adjoining counties and 57.

2081031 **Aulacomnium androgynum** (Hedw.) Schwaegr. Map 914, p. 720
Bagnall: Sutton Park, in *Proc. Bgham nat. Hist. microsc. Soc.* **1** (1869).
Bagnall; Wait and Elsee; Dalman: many records from the northern half of the county.
On stumps, fallen wood and shrubs in woods, scrub, hedgerows and by water, on soft sandstone rocks and walls, on shady hedgebanks and on woodwork and masonry by canals; rarely on compacted ground of woodland rides, spoil heaps of quarries, floors of sandpits, canal towpaths, ditch banks and peaty heaths; widespread and frequent except in calcareous clay areas. 208; 47%; not found fruiting.
BIRA, LAF, WAR. Recorded from all adjoining counties and 57.

MEESIACEAE

2084011 **Amblyodon dealbatus** (Hedw.) B. and S.
Bagnall: boggy heath, Longmoor, near Rowton's Well, Sutton Park, in *Proc. Bgham nat. Hist. microsc. Soc.* **1** (1869), also 1870 in herb. Birmingham Museum: found up to 1889, but not seen since.
BIRA. Recorded from 37 and 57.

BARTRAMIACEAE

2087021 **Bartramia pomiformis** Hedw.
Purton: in a lane leading from Spernall Ash to Middletown, in "Midl. Flor." (1817).
Bree: Meriden. Bagnall: near Westwood Coppice, Sutton Park; Middleton Heath; Curdworth; Marston Green; Browns Wood, Solihull; Little Dickens; near Bacons End; Holly Lane, Balsall. Stone: Erdington. Wait and Elsee: Dunchurch Road, Rugby.
On sandy and loamy hedgebanks and banks of cuttings, rare. 4; 1%; all fruiting. 2874, 3665, 3765, 3866.
BIRA, LAF, WAR. Recorded from 23, 32, 37, 39, 55 and 57.

2090021 **Philonotis fontana** (Hedw.) Brid.
Purton: Cook Hill, in "Midl. Flor." (1817).
H. E. Lowe; Bagnall; Collins; Townsend: several widespread records from the north-west of the county. Bagnall: Harborough Magna. Wait and Elsee: Birdingbury.
In bogs and boggy flushes, wet margins of pools and wet woodland rides, rare. 6; 2%; not found fruiting. 0996, 0998, 1095, 1097, 2572, 3059.
BIRA, LAF, WAR. Recorded from all adjoining counties and 57.

2090031 **Philonotis caespitosa** Wils.
Bagnall: Curdworth, 1883, in herb. Birmingham Museum; Studley; Spernall; Austy Wood; Tanworth; Knowle; Hockley; Grendon; Hartshill; Newbold on Stour.
At wet margins of lakes, boggy flushes and wet woodland rides, rare. 4; 1%; not found fruiting. 0758, 0998, 1097, 3871.
BIRA, LAF. Recorded from 37, 39 and 57.

2090041 **Philonotis capillaris** Lindb.
Recorded for VC 38 in "Census Catalogue", 2nd Edn., 1926, but the source of the record is not known.
In boggy flushes in Sutton Park, rare. 2; 1%; not found fruiting. 0996, 1096, 1097.
LAF. Recorded from 23 and 37.

2090061 **Philonotis calcarea** (B. and S.) Schimp.
Bagnall: Lapworth, 1882, in herb. Birmingham Museum; Austrey; Sutton Park; Ansty.
In boggy flushes in Sutton Park, rare. 1; not fruiting. 0998.
BIRA, LAF. Recorded from 23, 32, 33, 37, 39 and 57.

PTYCHOMITRIACEAE

2094011 **Ptychomitrium polyphyllum** (Sw.) Fürnr.
Kirk: stone wall near Binley, undated, in Bagnall, *J. Bot., Lond.* **12** (1874).
On stone blocks filling drainage channels in railway cuttings and embankments, rare. 6; 2%; 86% fruiting. 1073, 3860, 4982, 4983, 4984, 5161, 5173.
LAF. Recorded from 32, 33, 37, 39, 55 and 57.

ORTHOTRICHACEAE

2097011 **Zygodon viridissimus** (Dicks.) R. Br. Map 915, p. 720
Bagnall: Oakley Wood, in *J. Bot., Lond.* **12** (1874); many other records from all over the county. Wait and Elsee: Bretford.
On trees and shrubs in open woodland, scrub, hedgerows and by water; frequent in the south, becoming rarer northwards and rare in the north; occasional on limestone rock exposures and masonry and on brickwork near water; rare on sandstone walls and on rock blocks in drainage channels in railway cuttings. 88; 20%; not found fruiting.
BIRA, LAF, WAR. Recorded from all adjoining counties and 57.

2098011 **Orthotrichum rupestre** Schleich.
H. E. Lowe: Weston Park wall, 1835, in herb. Warwick County Museum.
Not found in present survey.
WAR. Recorded from 37.
Recorded from Wootton Wawen by Bagnall, but his specimen in Birmingham Museum is sterile and probably *O. anomalum* Hedw. Also recorded by Wait and Elsee from Stretton on Dunsmore, but probably an error.

2098021 **Orthotrichum anomalum** Hedw. Map 916, p. 721
Purton: "not rare", but no localities given, in "Midl. Flor." (1817).
Bree; Bagnall; Rogers; Wait: several records from south and central Warwickshire and a few from the north.
On limestone rock exposures and walls and on brickwork and concrete by water; rare on soft sandstone walls; frequent in the south and centre, rare in the north. 62; 14%; 93% fruiting.
BIRA, LAF, WAR. Recorded from all adjoining counties and 57.

2098031 **Orthotrichum cupulatum** Brid.
Bagnall: Whitchurch, 1888, in herb. Birmingham Museum; stone walls near Newbold on Stour.
On stonework of masonry by River Stour, rare. 2; both fruiting. 2347, 2447.
BIRA, LAF. Recorded from all adjoining counties and 57.

(2098032 var. **nudum** (Dicks.) Braithw. was recorded by Bagnall from by the Stour, Alderminster, in Amphlett and Rea, "Bot. Worcs.", 1909, but this was probably only the type.)

2098051 **Orthotrichum affine** Brid. Map 917, p. 721
Bree: Bidford Grange, undated, in herb. Warwick County Museum.
Kirk; Bagnall; Wait and Elsee; Dixon and Waddell: several records from all over the county, including north Warwickshire, where it is now very rare.
On trees and shrubs in hedgerows, scrub, open woodland and by water; rarely on limestone rock exposures and masonry; fairly frequent in the south of the county, decreasing northwards and very rare in the northern half of Warws. 68; 16%; 87% fruiting.
BIRA, LAF, OXF, WAR. Recorded from all adjoining counties and 57.

2098071 **Orthotrichum striatum** Hedw.
Bagnall: on poplars, near Rowington village, on the way to Kingswood, 1878, in herb. Birmingham Museum.
Not found in the present survey.
BIRA. Recorded from 23, 32, 33, 37, 55 and 57.
Purton, in "Midl. Flor." (1817), records *O. striatum* Hedw. as "common on trunks of old trees, but his descriptive notes suggest *O. affine* Brid., a much more likely plant.

2098081 **Orthotrichum lyellii** Hook. and Tayl. Map 918, p. 721
Bagnall: Old Park Wood, Arrow, 1872, in herb. Birmingham Museum.
Bagnall; Cumming, Wait and Elsee: several records, particularly from the south and centre, but also from the north, from where it has not recently been found.
On trees and shrubs, particularly ash, elm, maple and elder, in hedgerows, by water, and in open woodland, occasional in the south and centre of the county. 14; 3%; 19% fruiting.
BIRA, LAF. Recorded from all adjoining counties and 57.

2098091 **Orthotrichum rivulare** Turn.
Purton, in "Midl. Flor." (1817): "I do not exactly recollect where my specimen was gathered."
Bree: stones and mill wheel, River Avon near Bidford Grange. Bagnall: stream by Crab Mill, Preston Bagot; Wootton Wawen; stonework of canal, Ansty; River Stour, Alderminster.
Not found in the present survey.
BIRA, WAR. Recorded from 33, 37, 39, 55 and 57.

2098101 **Orthotrichum sprucei** Mont.
Duncan: River Avon, Alveston, 1926, in Br. Bryol. Soc's. Recorder's Notebook.
On trees by River Dene and Knee Brook, rare. 3; 1%; all fruiting. 2437, 2850, 2950.
LAF. Recorded from 32, 33, 37, 39, 55 and 57.

2098121 **Orthotrichum stramineum** Hornsch.
Bagnall: trees, Washford, Studley, 1886, in herb. Birmingham Museum; Ilmington; Wimpstone. Wait and Elsee: near Birdingbury.
On elders in open woodland, very rare. 1; fruiting. 3159.
BIRA, LAF. Recorded from 23, 33, 37, 39 and 57.

2098141 **Orthotrichum tenellum** Bruch
Bagnall: near Stratford on Avon, in *J. Bot., Lond.* **12** (1874); Little Wolford; ash trees, Tredington; Ilmington; Wimpstone; between Red Hill and Stratford on Avon; near Offchurch; near Old Fillongley Hall; near Weddington.
Not found in the present survey.
Recorded from 23, 32, 33, 37, 55 and 57.

2098151 **Orthotrichum pulchellum** Brunt.
Laflin: ash tree, Alveston Pasture, 1958.
On trees and shrubs (ash and hawthorn) in open woodland and by rivers, rare in the south of the county. 4; 1%; 50% fruiting. 2734, 2840, 2352, 3336.
LAF. Recorded from 23, 33, 55 and 57.

2098161 **Orthotrichum diaphanum** Brid. Map 919, p. 721
Withering, junr.: trunk of tree in the poultry yard at Edgbaston, in *Syst. Arr. Brit. Plants*, 4th Edn. vol. 3 (1801).
Purton; Bagnall; Cumming, Wait and Elsee; Rhodes: many records from all over the county.
On limestone rock exposures and blocks, on limestone and sandstone walls and on brickwork jointed with lime mortar, particularly by water; on rocks filling drainage channels in cuttings and embankments; on trees, shrubs, fallen logs and woodwork in open woodland, scrub, hedgerows and by water; frequent in the south and centre, occasional in the north. 159; 36%; 98% fruiting.
BIRA, LAF. Recorded from all adjoining counties and 57.

2098171 **Orthotrichum obtusifolium** Brid.
Bagnall: near Stratford on Avon, 1873, in herb. Birmingham Museum; 13 other localities south of National Grid horizontal line SP 600.
Not found in the present survey.
BIRA. Recorded from 23, 32, 33 and 37.

2099031 **Ulota crispa** (Hedw.) Brid. Map 920. p. 722
Bree: tree trunks, especially young oak poles, Allesley, in Purton, "Midl. Flor.", App. (1821).
Bagnall; Wait and Elsee: several records from all over the county; apparently more frequent and more widespread than it is now.
On trees in open woodland, scrub and by water, rare and stunted. 10; 2%; 64% fruiting.
BIRA, LAF, WAR. Recorded from all adjoining counties and 57.

(2099051 **Ulota bruchii** Hornsch. was recorded by Bagnall from Waverley Wood, 1884, but his specimen in herb. Birmingham Museum is *U. crispa* (Hedw.) Brid. This record is probably the basis for its inclusion for VC 38 in the "Census Catalogue", 2nd Edn., 1926.)

FONTINALACEAE

2100011 **Fontinalis antipyretica** Hedw. Map 921, p. 722
Purton: "common", but no localities given in "Midl. Flor." (1817).
Bree; Kirk; Bagnall; Wait and Elsee; Townsend: many records from all parts of the county.
On rocks, stones, masonry, woodwork and tree bases, submerged in rivers, streams, lakes, ponds and canals; fairly frequent throughout the county. 40; 9%; 2% fruiting.
BIRA, LAF, OXF, WAR. Recorded from all adjoining counties and 57.

2100012 var. **gigantea** (Sull.) Sull.
Bagnall: near Stratford, 1873, in herb. Birmingham Museum; cattle pool, Billesley.
In similar places to the type, apparently rare. 1; not fruiting. 1456.
BIRA, LAF. Recorded from 23 and 57.

(2100021 **Fontinalis dolosa** Card. was recorded by Bagnall from Birdingbury Wharf and from the canal near Rugby, as MS. notes in his own copy of "Flor. Warws.", and repeated in "Victoria County Hist. Warws.", **1**, 1904, but these are errors for the last species.)

CLIMACIACEAE

2101011 **Climacium dendroides** (Hedw.) Web. and Mohr
Purton: woods beyond Cold Comfort, Alcester, in "Midl. Flor." (1817).
Kirk: Allesley; Arbury. Bagnall: Sutton Park (also found by Townsend); Erdington; Shirley; Earlswood; by canal, Holywell.
In boggy and marshy places and in rough calcareous grassland; in very few localities, locally frequent in Sutton Park, rare elsewhere. 6; 2%; not found fruiting. 0996, 0998, 1095, 1096, 4675, 5278.
BIRA, LAF, OXF. Recorded from all adjoining counties and 57.

HEDWIGIACEAE

2102011 **Hedwigia ciliata** (Hedw.) P. Beauv.
Kirk: sandstone walls and stones near the Cascade, Arbury Hall, 1857, in herb. Birmingham Museum.
Not found in the present survey.
BIRA, OXF. Recorded from 37, 39, 55 and 57.

CRYPHAEACEAE

2103011 **Cryphaea heteromalla** (Hedw.) Mohr
Purton: root of tree, thicket at Alcester Mill, in "Midl. Flor." (1817).
Bree: Allesley. Bagnall: several records from the south and centre of the county.
Wait and Elsee: Birdingbury. Baxter: Hillmorton.
On elm, ash or elder in open woodland, scrub and hedgerow in the Lower Lias area of south Warwickshire; rare and much rarer than in the 19th century. 4; 1%; 50% fruiting. 1356, 2734, 2245, 2946.
BIRA, LAF. Recorded from all adjoining counties and 57.

LEUCODONTACEAE

2104011 **Leucodon sciuroides** (Hedw.) Schwaegr.
Map 922, p. 722
Purton: "not rare", but no localities given, in "Midl. Flor." (1817).
Kirk; Bagnall; Wait and Elsee: many records from the south and centre and a few from the north of the county.
On trees and shrubs, particularly ash, elm and maple, in open woodland, scrub, hedgerows and by water; rarely on limestone rock exposures or walls; widespread and occasional in the south, becoming rarer northwards and absent from the north; less common than in the 19th century. 26; 6%; not found fruiting.
BIRA, LAF, OXF, WAR. Recorded from all adjoining counties and 57.

2105011 **Antitrichia curtipendula** (Hedw.) Brid.
Laflin: 3057, on horizontal branch of ash in hedgerow near Ashorne village, stunted, 1969. 1; not fruiting.
LAF. Recorded from all adjoining counties and 57.

NECKERACEAE

(2109021 **Neckera crispa** Hedw. was imported on limestone into a garden at Stratford on Avon in 1963, where it persisted for a time.)

2109041 **Neckera complanata** (Hedw.) Hüben.
Map 923, p. 722
Purton: "common", but no localities given, in "Midl. Flor." (1817).
Bree; Bagnall; Wait and Elsee; Baxter; Townsend: many records, mainly from the south and centre, but several from the north.
On trees and shrubs, particularly elm, ash and maple, in woodland, scrub, hedgerows and by water; less frequently on limestone exposures, blocks and masonry; rarely on brickwork by water; frequent in the south, widespread but less common in the centre, rare in the north. 99; 23%; not found fruiting.
BIRA, LAF, WAR. Recorded from all adjoining counties and 57.

2110011 **Omalia trichomanoides** (Hedw.) B., S. and G.
Map 924, p. 723
Bree: about Allesley, in Purton, "Midl. Flor.", App. (1821).
Bagnall; Baxter: several records from all over the county.
On tree bases, stools and shrubs, particularly ash, near to the ground in open woodland, scrub, hedgerows and by water; most frequently on ash stools or protruding roots by ditches and streams near to water level; rarely on rock blocks on the ground or masonry by water. 51; 12%; 46% fruiting.
BIRA, LAF, WAR. Recorded from all adjoining counties and 57.

2111011 **Thamnium alopecurum** (Hedw.) B., S. and G.
Map 925, p. 723
Purton: upon the river bank leading from Oversley to the Mill, in "Midl. Flor." (1817).
Bagnall; Wait and Elsee; Townsend: many records from all parts of the county.
On the ground in woodland and scrub, particularly on heavy basic soils, and occasionally on river, stream and canal banks; on rock exposures and blocks on the ground and on stone masonry, most frequently on limestone, but also on sandstone; on rocks in streams and on tree bases and stools, particularly ash, in woods, scrub, hedgerows and by water; frequent in the south and centre, widespread but not common in the north. 66; 15%; 1% fruiting.

BIRA, LAF, WAR. Recorded from all adjoining counties and 57.

HOOKERIACEAE

2113011 **Hookeria lucens** (Hedw.) Sm.
Clark: 2995, stream bank in Bentley Park Wood, 1962. 1; not fruiting.
LAF. Recorded from 37, 39, 55 and 57.

LESKEACEAE

2118011 **Leskea polycarpa** Hedw. Map 926, p. 723
Bree: on tiles, Allesley, in Purton, "Midl. Flor.", App. (1821); alders in the River Arrow.
Bagnall; Cumming, Wait and Elsee; Townsend: several records from all parts of the county.
On trees, shrubs and protruding tree roots, occasionally on masonry and woodwork by rivers and streams; frequent in the south and centre, less common in the north and not recorded from the Rea, Tame, Anker and Sowe valleys. 69; 16%; 52% fruiting.
BIRA, LAF, WAR. Recorded from all adjoining counties and 57.

THUIDIACEAE

2122031 **Anomodon viticulosus** (Hedw.) Hook. and Tayl. Map 927, p. 723
Purton: "common", but no localities given, in "Midl. Flor." (1817).
Bagnall: several records, widespread throughout the county, but particularly from the south and west-central parts of the county.
On tree bases and shrubs, mainly ash and elm, in open woodland, scrub, hedgerows and by water; rarely on limestone rock exposures and walls; occasional in the south and centre, absent from the north. 14; 3%; not found fruiting.
BIRA, LAF. Recorded from all adjoining counties and 57.

2123031 **Thuidium tamariscinum** (Hedw.) B., S. and G. Map 928, p. 724
Purton: "common", but no localities given, in "Midl. Flor." (1817).
Bree; Bagnall; Cumming, Wait and Elsee; Jackson; Dalman; Townsend: many records, throughout the county.
On the ground in woods and scrub, occasionally on woodland rides, shaded hedgebanks, river banks, shaded rough grassland, marshes and heaths; widespread and fairly frequent throughout the county except in some areas in the north. 92; 21%; not found fruiting.
BIRA, LAF, WAR. Recorded from all adjoining counties and 57.

2123051 **Thuidium philibertii** Limpr.
Recorded for VC 38 in 2nd Edn. "Census Catalogue", 1926, but the source of the record is not known.
In rough grassland, grassy quarry or cutting banks and marshy ground on limestone or calcareous clay soils; occasional in the south of the county. 8; 2%; not found fruiting. 0956, 1455, 1557, 2042, 2940, 3150, 3951, 3860.
LAF. Recorded from all adjoining counties and 57.

AMBLYSTEGIACEAE

2125011 **Cratoneuron filicinum** (Hedw.) Spruce Map 929, p. 724
Bree: Allesley Pool Tail, undated, in herb. Warwick County Museum.
Bromwich; Kirk; Bagnall; Wait and Elsee; Jackson; Townsend: many records from all over the county.
In very varied, but usually wet, habitats, but occasionally in drier places; sometimes locally dominant in marshes and marshy flushes; on masonry by streams and canals, on rock blocks, both limestone and sandstone, in streams, at bases of walls, on wet compacted ground and tracks, particularly canal towpaths and wet woodland rides, on banks of rivers, streams, ditches, ponds and canals, sometimes growing in the water, in wet grassland, on wet ground in woodland and scrub and on tree bases and woodwork by water; common throughout the county. 201; 46%; not found fruiting.
BIRA, LAF, OXF, WAR. Recorded from all adjoining counties and 57.

2125013 var. **fallax** (Brid.) Roth
Bagnall: Wire Hill, Sambourne, MS record in his own copy of "Flora Warws."
Not recorded in the present survey, but no attempt has been made to separate it from the type.
Recorded from all adjoining counties and 57.

2125021 **Cratoneuron commutatum** (Hedw.) Roth
Bree: bog near Bidford Grange, in Purton, "Midl. Flor.", App. (1821).
Kirk: near Ryton Tollgate. Bagnall: several places in Sutton Park (also found by Townsend); Bentley Park; near Crab Mill, Preston Bagot; canal near Shrewley Common and Rowington.
In bogs and boggy flushes, locally frequent in Sutton Park, rare elsewhere. 4; 1%; 25% fruiting. 0996, 2995, 2996, 5278.
LAF, OXF. Recorded from all adjoining counties and 57.

2125022 var. **falcatum** (Brid.) Mönk.
Kirk: Bannerley Pool, 1857, in herb. Oxford University, named *Hypnum revolvens* Turn. by Kirk.
Bagnall: several places in Sutton Park (also found by Stone and Townsend); by Bannams Wood, Morton Bagot; near Greenhill Green.
In bogs and boggy flushes, locally frequent in Sutton Park, rare elsewhere. 4; 1%; not found fruiting. 0996, 0998, 1096, 5278.
BIRA, LAF, OXF. Recorded from all adjoining counties and 57.

(2125024 var. **sulcatum** (Lindb.) Mönk. was recorded by Bagnall from Sutton Park as "probably this", in *Midl. Nat.* **3** (1880), but undoubtedly an error.)

2126011 **Campylium stellatum** (Hedw.) J. Lange and C. Jens.
Bree: bog near Bidford Grange, undated, in herb. Warwick County Museum.
Kirk; Bagnall: several records from north and central Warwickshire, particularly in the western half of this area; also Wolford Wood and Wimpstone Fields in the south.
In marshes and boggy flushes, rare and obviously more rare than in the 19th century. 2; not found fruiting. 0998, 3094.
BIRA, LAF, OXF, WAR. Recorded from 23, 32, 37, 39, 55 and 57.

2126021 **Campylium protensum** (Brid.) Kindb.
Jackson: grassy bank near Rugby, in *J. Bot., Lond.* **38** (1900).
In basic marshy grassland around springs, rare. 3; 1%; not found fruiting. 1557, 1960, 2450.
LAF. Recorded from 23, 32, 33, 37, 55 and 57.

2126031 **Campylium chrysophyllum** (Brid.) J. Lange Map 930, p. 724
Bagnall: Yarningale Common, in *J. Bot., Lond.* **12** (1874); many other records from the centre and north of the county; also Wimpstone Fields in the south.
In rough calcareous grassland, open scrub and marshy grassland, on floors of old clay pits; rarely in woodland rides and in arable fields; mainly on calcareous clay and only in

the south and centre, in contrast to Bagnall's records, which are mainly from the north or the central marl areas. 32; 7%; not found fruiting.
BIRA, LAF. Recorded from all adjoining counties and 57.

2126041 **Campylium polygamum** (B., S. and G.) J. Lange and C. Jens.
Bagnall: marsh near Tythall Lane, Solihull, in *J. Bot., Lond.* **12** (1874); Sutton Park; Hardings Wood, Maxstoke; Earlswood Reservoir; near Stratford on Avon; near Wilmcote; Wimpstone Fields; Itchington Holt; Chadshunt.
In marshy places, rare. 3; 1%; not found fruiting. 1000, 2145, 2245, 3094.
BIRA, LAF. Recorded from 23, 37, 39 and 55.

2126051 **Campylium elodes** (Lindb.) Kindb.
Bagnall: marsh, Wimpstone Fields, 1887, in herb. Birmingham Museum.
In marshy places, rare. 4; 1%; 50% fruiting. 2142, 2797, 2404, 2504.
BIRA, LAF. Recorded from 55.

2126061 **Campylium calcareum** Crundw. and Nyh.
Laflin: bank of long-disused clay pit, Glascote, 1967.
In rough grassland on calcareous banks, rare and in small quantity. 3; 1%; not found fruiting. 2202, 3864, 4369.
LAF. Recorded from 23, 32, 33, 37, 55 and 57.

2127011 **Leptodictyum riparium** (Hedw.) Warnst.
Map 931, p. 724
Purton: "common", but no localities given, in "Midl. Flor." (1817).
Kirk; Bagnall; Stone; Wait and Elsee; Townsend: many records from all over the county.
On the ground, stonework, masonry, woodwork, trees, stumps, and fallen logs by rivers, streams, ditches, ponds and canals; common throughout the county; rarely in marshes, on tree bases in alder swamps or willow carr or in drier places on shrubs in hedgerows. The var. *longifolium* (Schultz) well distinct, usually submerged, less frequent but widespread. 189; 43%; 62% fruiting.
BIRA, LAF, OXF, WAR. Recorded from all adjoining counties and 57.

2128011 **Hygroamblystegium tenax** (Hedw.) Jenn.
Map 932, p. 725
Bagnall: Sutton Park, in *J. Bot., Lond.* **12** (1874).
Bagnall; Wait and Elsee: several records from widespread localities in all parts of the county.
On woodwork and masonry by streams and canals and on limestone or sandstone blocks in streams, widespread but infrequent. 12; 3%; 47% fruiting.
BIRA, LAF. Recorded from all adjoining counties and 57.

2128021 **Hygroamblystegium fluviatile** (Hedw.) Loeske
Bagnall: water wheel in Sutton Park, 1877, published with that date in "Notes on Sutton Park", 1876.
On sandstone rocks in stream, very rare. 1; not fruiting. 2488.
LAF. Recorded from 33, 37, 39 and 57.

2129011 **Amblystegium serpens** (Hedw.) B., S. and G.
Map 933, p. 725
Purton: "common", but no localities given, in "Midl. Flor." (1817).
Kirk; Bagnall; Wait and Elsee; Townsend: many records from all over the county.
On rock exposures and blocks and masonry of all kinds, on tree bases, shrubs, stumps, fallen wood and woodwork in woodland, scrub, hedgerows and by water, on compacted ground of tracks, paths and railway tracks, on floors and waste heaps of disused pits and quarries, in grassland, on banks of rivers, streams, ditches and canals and on the ground in woodland, scrub and hedgebanks; common throughout the county. 359; 82%; 78% fruiting.
BIRA, LAF, OXF, WAR. Recorded from all adjoining counties and 57.

2129021 **Amblystegium juratzkanum** Schimp.
Map 934, p. 725
Recorded for VC 38 in 2nd Edn. "Census Catalogue", 1926, but the source and details of the record are not known.
In similar places to the preceding but much less common, although perhaps under-recorded, since it is easily confused with that species; usually in shade or by water; widespread. 15; 3%; 41% fruiting.
LAF. Recorded from all adjoining counties and 57.

2129031 **Amblystegium kochii** B., S. and G.
Laflin: stone footbridge by ford over brook, Preston Fields, Preston Bagot, 1963.
On stonework by streams and canals, rare. 2; not found fruiting. 1766, 2800.
LAF. Recorded from 32.

2129041 **Amblystegium varium** (Hedw.) Lindb.
Kirk: wet ditch, Arbury Park, 1856, in herb. Oxford University.
Bagnall: Sutton Park; between Water Orton and Minworth; Hurley; Kingsbury; Preston Bagot; Wormleighton Reservoir.
On woodwork and fallen wood by rivers and canals, rare. 2; 50% fruiting. 2456, 4353.
BIRA, LAF, OXF. Recorded from all adjoining counties and 57.

2131011 **Drepanocladus aduncus** (Hedw.) Warnst.
Map 935, p. 725
Bree: bog by Bidford Grange, in Purton, "Midl. Flor.", App. (1821).
Webb; Bagnall; Townsend: several records from widespread localities all over the county.
In ponds, marshes and marshy pastures, often in standing water, less frequently in canals; most common on calcareous clay, but also on other soil types; tolerant of brackish water in salt spring; one record from masonry of canal lock and one from an arable field; widespread. 36; 8%; 3% fruiting.
BIRA, LAF, WAR. Recorded from all adjoining counties and 57.

2131021 **Drepanocladus sendtneri** (Schimp.) Warnst.
Bagnall: by Longmoor Mill Pool, Sutton Park, in *Proc. Bgham nat. Hist. microsc. Soc.* **1** (1869).
Wait and Elsee: pit near Newbold on Avon.
Not found in the present survey.
BIRA. Recorded from 37, 39 and 57.

2131022 var. **wilsonii** (Schimp.) Warnst.
Bagnall: Wimpstone Fields, 1874, in herb. Birmingham Museum; Earlswood Reservoir.
Not found in the present survey.
BIRA. Recorded from 23 and 32.

2131031 **Drepanocladus lycopodioides** (Brid.) Warnst.
Bagnall: Wimpstone Fields and Crimscote Downs, both 1887, in herb. Birmingham Museum.
Not found in the present survey; both of the above areas are now drained.
BIRA. Recorded from 32.

2131041 **Drepanocladus fluitans** (Hedw.) Warnst.
Map 936, p. 726
Bree: pit near Bidford Grange, in Purton, "Midl. Flor.", App. (1821).
Bagnall; Stone; West; Townsend: several widespread records from all parts of the county.
In and at the margins of acid pools, persisting on wet peat where the pools dry out in summer; rarely in peaty ditches

or peaty places in bogs and marshes; in a few localities in north and central Warwickshire. 15; 3%; 24% fruiting.
BIRA, LAF, WAR. Recorded from 32, 33, 37, 39, 55 and 57.

2131042 var. **falcatus** (B., S. and G.) Warnst.
Recorded for VC 38 in 2nd Edn. "Census Catalogue", 1926, but details of the record are not known.
In similar places to the type, but apparently rarer. 2; not found fruiting. 1986, 1097.
LAF. Recorded from 37, 55 and 57.

2131051 **Drepanocladus exannulatus** (B., S. and G.) Warnst.
Bagnall: Keepers Pool, Sutton Park, in *Proc. Bgham nat. Hist. microsc. Soc.* **1**, (1869); several places in Sutton Park (also found by Rhodes and Townsend); Baddesley Common; Packington Park; Clarksland Coppice, Tanworth; near Blythe Bridge, Solihull; Henley in Arden; Wimpstone Fields.
In acid bogs and at boggy margins of acid pools, rare; only in the north of the county. 4; 1%; not found fruiting. 0996, 1985, 1986, 2798.
BIRA, LAF. Recorded from 33, 37, 39, 55 and 57.

2131052 var. **rotae** (De Not.) Wynne
Recorded for VC 38 in 2nd Edn. "Census Catalogue", 1926, but the source and details of the record are not known.
In similar places to the type, rare. 1; not found fruiting. 1097.
LAF. Recorded from 37 and 57.

2131061 **Drepanocladus revolvens** (Turn.) Warnst.
Bree: bog near Bidford Grange, in Purton, "Midl. Flor.", App. (1821).
Kirk: Bannerley Pool. Bagnall: Sutton Park (also found by Townsend); Earlswood Reservoir; Wimpstone Fields; Wormleighton Reservoir.
In boggy flushes in Sutton Park, rare. 1; not fruiting. 0998.
BIRA, LAF, OXF. Recorded from 32, 37, 39 and 57.

2131062 var. **intermedius** (Lindb.) Rich. and Wall.
Kirk: Corley Moor, 1860, in herb. Oxford University, named *D. exannulatus* by Kirk; Coleshill Bog.
Bagnall: several places in Sutton Park (also found by Townsend); Hill Hook; Wimpstone Fields. Waddell: Crimscote Fields.
In boggy flushes in Sutton Park, rare. 1; not fruiting. 0996.
BIRA, LAF, OXF. Recorded from 23, 32, 33, 37, 39 and 57.

2131071 **Drepanocladus vernicosus** (Lindb.) Warnst.
Bagnall: near Windley Pool, Sutton Park, 1872, in herb. Birmingham Museum; Bracebridge Bog, Sutton Park (also found by Stone and Townsend).
In boggy flush in Sutton Park, rare. 1; not fruiting. 0998.
BIRA, LAF, OXF. Recorded from 37, 39 and 57.

2131081 **Drepanocladus uncinatus** (Hedw.) Warnst.
Kirk: railway cutting, Coventry Park, 1856, in herb. Oxford University.
Not found in the present survey.
OXF. Recorded from 33, 37, 39 and 57.

2132021 **Hygrohypnum luridum** (Hedw.) Jenn. Map 937, p. 726
Bree: mill wheel at Bidford Grange, undated, in herb. Warwick County Museum.
Kirk: Arbury. Bagnall: several records from by canals in central and north Warwickshire and from stones in streams in the north.
On masonry and woodwork of locks and retaining walls on all the canals, widespread; rarely on stonework by streams, on rock blocks in streams or on sandstone masonry in dry, but shaded, places. 18; 4%; 59% fruiting.
BIRA, LAF, OXF, WAR. Recorded from all adjoining counties and 57.

(2132022 var. **subsphaericarpon** (Schleich.) C. Jens. was recorded for VC 38, without details of date or locality, by Bagnall in a MS note in the Br. Bryol. Soc. Recorder's Notebook and included in the 1st Edn. "Census Catalogue", 1907. Without better evidence the record must be suspect.)

2134011 **Acrocladium stramineum** (Brid.) Rich. and Wall.
Bagnall: Sutton Park, in *Proc. Bgham nat. Hist. microsc. Soc.* **1**, (1869); (also recorded by Jones and Townsend).
In bogs and at boggy margins of pools and on the ground in wet birch wood, in a few localities in the north of the county. 8; 2%; not found fruiting. 0995, 0996, 0998, 1985, 1986, 1095, 1098, 2779.
BIRA, LAF. Recorded from 37, 39 and 57.

2134031 **Acrocladium cordifolium** (Hedw.) Rich. and Wall. Map 938, p. 726
Kirk: Binley, 1857, in herb. Oxford University; Yew Tree Wood, Arbury Hall.
Bagnall; Stone; Townsend: several records from the north of the county and Wimpstone Fields and Barton Flat Covert in the south.
At the margins of acid pools, often growing in water at the bases of reeds; less frequently in acid marshes; widespread in the north and centre of the county. 17; 4%; not found fruiting.
BIRA, LAF, OXF, WAR. Recorded from all adjoining counties and 57.

2134041 **Acrocladium giganteum** (Schimp.) Rich. and Wall.
Bagnall: by Windley Pool, Sutton Park, 1870, in herb. Birmingham Museum; several other places in Sutton Park; Acocks Green; Old Park, Warwick; Wimpstone Fields.
In boggy flush in Sutton Park, rare. 1; not fruiting. 0996.
BIRA, LAF. Recorded from all adjoining counties, but not 57.

2134061 **Acrocladium cuspidatum** (Hedw.) Lindb. Map 939, p. 726
Purton: "common", but no localities given, in "Midl. Flor." (1817).
Bree; Kirk; Bagnall; Wait and Elsee; Jackson; Townsend: many records from all over the county.
In marshes and wet grassland, in dry calcareous grassland and scrub, in wet compacted ground, particularly canal towpaths and wet woodland rides, in bogs and boggy flushes, wet woodland and scrub, on banks of rivers, streams, ditches, ponds and canals, sometimes growing in the water; frequent throughout the county; less often on wet heaths, on limestone masonry and blocks, on brickwork by water, as on canal locks, on hedgebanks and on tree stumps. 237; 54%; 1% fruiting.
BIRA, LAF, OXF, WAR. Recorded from all adjoining counties and 57.

BRACHYTHECIACEAE

2135011 **Isothecium myurum** Brid. Map 940, p. 727
Kirk: near Berkswell, undated, in herb. Oxford University.
Bagnall; Wait and Elsee; Townsend: several records from all over the county.
At the bases of trees and on stools and stumps in woodland, rarely in scrub, in several widespread localities in the southern half of the county, absent from the north; apparently less common than in the 19th century. 19; 4%; not found fruiting.
BIRA, LAF, OXF. Recorded from all adjoining counties and 57.

2135021 **Isothecium myosuroides** Brid. Map 941, p. 727
Bree: trunks of trees near the ground about Allesley, in Purton, "Midl. Flor.", App. (1821).
Bagnall: many records from all over the county.
On tree bases and on shrubs, stools and stumps in woodland, scrub and hedgerows, persisting well in cleared woodland; fairly frequent; occasionally on both limestone and sandstone rock exposures and blocks near to the ground; rarely on masonry by water; throughout the county, but rarer in the north than elsewhere. 92; 21%; 2% fruiting.
BIRA, LAF, WAR. Recorded from all adjoining counties and 57.

2137011 **Camptothecium sericeum** (Hedw.) Kindb. Map 942, p. 727
Purton: "very common", but no localities given, in "Midl. Flor." (1817).
Kirk; Bagnall; Wait and Elsee; Townsend: many records from all over the county.
On exposures and blocks of all types of rocks and on masonry of all kinds; on bases and trunks of trees and shrubs in open woodland, scrub, hedgerows and by water; common; rarely on spoil heaps of old limestone quarries or colliery slag heaps; throughout the county, but a little less frequent in the north than elsewhere and not surviving in very built-up areas. 236; 54%; 5% fruiting.
BIRA, LAF, OXF, WAR. Recorded from all adjoining counties and 57.

2137021 **Camptothecium lutescens** (Hedw.) B., S. and G. Map 943, p. 727
Bree: Bidford Grange, undated, in herb. Warwick County Museum.
Kirk; Bagnall; Jackson: several records from the southern half of the county.
In rough calcareous grassland, particularly on banks of old quarries and cuttings, and in open scrub on calcareous clay; rarely on canal banks; confined to the Lias soils, oolite and calcareous marl in the southern half of the county. 23; 5%; not found fruiting.
BIRA, LAF, OXF, WAR. Recorded from all adjoining counties and 57.

2138011 **Brachythecium albicans** (Hedw.) B., S. and G. Map 944. p, 728
Bagnall: Witton, in *Proc. Bgham nat. Hist. microsc. Soc.* **1**, (1869).
Bagnall; Townsend: many records from all over the county.
On compacted acid ground of tracks, paths, sidings, canal towpaths, railway tracks, woodland rides and floors and spoil heaps of siliceous stone quarries and sand and gravel pits; on the tops of walls, particularly of canal locks and in compacted areas of acid grassland or dry heaths; widespread, but not very common. 55; 13%; not found fruiting.
BIRA, LAF. Recorded from all adjoining counties and 57.

2138021 **Brachythecium glareosum** (Spruce) B., S. and G. Map 945, p. 728
Bagnall: Lapworth Street, in *J. Bot., Lond.* **12** (1874).
Bagnall; Wait and Elsee; Jackson: several records from all parts of the county.
In rough calcareous grassland, on calcareous banks and spoil heaps of old limestone quarries and calcareous clay pits, on limestone wall tops, on rock blocks in drainage channels of railway cuttings and on canal banks and towpaths; rarely on fallen wood in open woodland; occasional in the southern half of the county; absent from the north-west. 20; 5%; not found fruiting.
BIRA, LAF. Recorded from all adjoining counties and 57.

2138041 **Brachythecium salebrosum** (Web. and Mohr) B., S. and G.
Purton: "not rare", but no localities given, in "Midl. Flor." (1817).
Bagnall: field by Kingsbury Wood; Olton; Morton Bagot; sandpit between Crabs Cross and Ipsley; Ilmington.
On wall tops by streams, on railway tracks and on shrubs in hedgerows, rare; possibly under-recorded as it is easily overlooked as *B. rutabulum* (Hedw.) in the sterile state, but definitely uncommon. 7; 2%; 86% fruiting. 1960, 2239, 2868, 2504, 3336, 3760, 3860.
LAF. Recorded from 23, 32, 33, 37, 39 and 55.

2138051 **Brachythecium mildeanum** (Schimp.) Milde Map 946, p. 728
Bagnall: Ipsley Fields, 1886, in herb. Birmingham Museum; Snowford Bridge.
In marshes and marshy pastures, usually where bare ground is present and on wet compacted ground, widespread but uncommon, although easily mistaken for the next species when sterile. 11; 3%; 8% fruiting.
BIRA, LAF. Recorded from 23, 32, 37, 39, 55 and 57.

2138061 **Brachythecium rutabulum** (Hedw.) B., S. and G. Map 947, p. 728
Purton: "very common", but no localities given, in "Midl. Flor." (1817).
Bree; Kirk; Bagnall; Wait and Elsee; Townsend: many records from all over the county.
On bare compacted and wet ground, in grassland of all kinds, in marshes, woodland, scrub, hedgebanks, by rivers, streams, ditches, ponds and canals, on rock exposures, blocks and masonry near to the ground, on railway tracks, on floors, banks and spoil heaps of pits and quarries, at tree bases, stools, stumps and woodwork in woods, scrub, hedgerows and by water; rarely on heaths or in arable fields; very common throughout the county. 390; 89%; 28% fruiting.
BIRA, LAF, OXF, WAR. Recorded from all adjoining counties and 57.

2138071 **Brachythecium rivulare** B., S. and G. Map 948, p. 729
Kirk: The Cascade, Arbury Hall, undated, in herb. Oxford University.
Bagnall; Wait and Elsee; Townsend: many records from the centre and north of the county; also from the base of Edgehill and Whichford Wood in the south.
In marshes and boggy flushes, in wet pastures, on wood, rocks and masonry in streams, canal locks, on banks of rivers and streams and on tree bases by rivers where they are periodically inundated during flood periods; throughout the county; always where the water is moving, even if very slowly. 65; 15%; 1% fruiting.
BIRA, LAF, OXF. Recorded from all adjoining counties and 57.

2138111 **Brachythecium velutinum** (Hedw.) B., S. and G. Map 949, p. 729
Purton: "common", but no localities given, in "Midl. Flor." (1817).
Bree; Kirk; Bagnall; Wait and Elsee; Townsend: many records from all over the county.
On tree bases, shrubs, stools and stumps in open woodland, scrub, hedgerows and by water, on sandy hedgebanks, on spoil heaps of siliceous stone quarries and coal pits; occasionally on the ground in woodland on sand and marl or on rock exposures, blocks or masonry; frequent throughout the county. 271; 62%; 82% fruiting.
BIRA, LAF, OXF, WAR. Recorded from all adjoining counties and 57.

2138121 **Brachythecium populeum** (Hedw.) B., S. and G.
Kirk: railway bank near Binley, 1858, in herb. Oxford University.
Bagnall: Marston Green; canal bank, Olton; Tythall Lane, Solihull; Merevale Park.
On rock exposures, blocks and masonry of Middle Lias stone in the extreme south and on sandstone and diorite in the extreme north; rarely on rock blocks in drainage channels in railway cuttings, on masonry by water or on tree bases in woodland or hedgerow; uncommon. 11; 3%; 46% fruiting. 1062, 2076, 2996, 3748, 3848, 3097, 4364, 4089, 4982, 4984, 5161.
BIRA, LAF, OXF. Recorded from all adjoining counties and 57.

2138131 **Brachythecium plumosum** (Hedw.) B., S. and G.
Bagnall: stone coping of canal bridge near Olton, in *Midl. Nat.* **3**, (1880); tree roots near water, River Alne, Preston Bagot.
On sandstone blocks in streams and on masonry by streams or canals, rare. 4; 1%; 50% fruiting. 1879, 2488, 2996, 3290.
BIRA, LAF. Recorded from 37, 39, 55 and 57.

2139011 **Scleropodium caespitosum** (Wils.) B., S. and G. Map 950, p. 729
Bagnall: bank of pool by Middleton Park, 1873, in herb. Birmingham Museum.
Bagnall; Wait and Elsee; Rhodes: many records from all over the county.
On tree bases and protruding roots, stumps, shrubs, woodwork, masonry of all kinds and compacted ground by water, either near to water level or where occasionally inundated by flooding; less frequently on tree bases, rock blocks or masonry near to the ground in drier places and on compacted ground of gravelled or hardcored paths and tracks; rarely on dry compacted ground elsewhere; frequent and widespread. 113; 26%; 1% fruiting.
BIRA, LAF. Recorded from all adjoining counties and 57.

2139021 **Scleropodium tourretii** (Brid.) L. F. Koch Map 951, p. 729
Bagnall: bank near Spernall Ash, 1886, in herb. Birmingham Museum; near Grendon; Furnace End, Whitacre; near Fillongley Hall; between Crab Mill and Wootton Wawen; Coughton Court.
On banks of rivers, streams and canals and on tree bases, masonry and rock blocks by water; rarely in marshes or on walls or tree bases in dry places; widespread, but not common. 37; 8%; not found fruiting.
BIRA, LAF. Recorded from 23, 33, 37, 39 and 57.

2140011 **Cirriphyllum piliferum** (Hedw.) Grout Map 952, p. 730
Bagnall: canal bank, Olton, in *Proc. Bgham nat. Hist. microsc. Soc.* **1** (1869).
Bagnall; Wait and Elsee: many records from all parts of the county.
On the ground in woodland and scrub on heavy soils, particularly basic soils; occasionally in marshes, on shaded hedgebanks and banks of rivers, streams and ditches and damp grassland; rarely on tree bases; widespread and fairly frequent, but absent from the north-west part of the county. 91; 21%; not found fruiting.
BIRA, LAF, WAR. Recorded from all adjoining counties and 57.

2140031 **Cirriphyllum crassinervium** (Tayl.) Loeske and Fleisch. Map 953, p. 730
Laflin: walls of Kenilworth Castle, 1956.
On limestone rock exposures, blocks and masonry; rarely on dry sandstone or other masonry by water or tree bases and stumps; occasional in the south and centre of the county. 22; 5%; 5% fruiting.
LAF. Recorded from all adjoining counties and 57.

2141011 **Eurhynchium striatum** (Hedw.) Schimp. Map 954, p. 730
Purton: "common", but no localities given, in "Midl. Flor." (1817).
Kirk; Bagnall; Wait and Elsee; Jackson: many records from woods and thickets all over the county.
On the ground in woodland and scrub, most frequent on heavy basic soils and not found in very acid habitats; occasional on rock exposures and wall tops, shaded river banks and rough grassland; throughout the county, but infrequent in the north. 73; 17%; 1% fruiting.
BIRA, LAF, OXF. Recorded from all adjoining counties and 57.

2141041 **Eurhynchium praelongum** (Hedw.) Hobk. Map 955, p. 730
Purton: "common", but no localities given, in "Midl. Flor." (1817).
Bree; Bagnall; Wait and Elsee; Dalman; Townsend: many records from all over the county.
On the ground in woods, scrub, hedgebanks, banks of rivers, streams, ditches, ponds and canals, compacted and waste ground, floors, banks and spoil heaps of old pits and quarries, fallow ground, grassland, marshes and heaths; less often on rock exposures and blocks and masonry of all kinds, fallen wood, woodwork, tree bases and stumps, usually near to the ground; common throughout the county. 388; 89%; 2% fruiting.
BIRA, LAF, WAR. Recorded from all adjoining counties and 57.

2141042 var. **stokesii** (Turn.) Hobk.
Bagnall: Edgehill Woods, in "Flor. Warws.", 1891; a few other widespread records; also recorded by Townsend.
On the ground in woods and scrub, particularly on heavy basic soils, widespread and frequent, but not recorded in detail in the survey.
LAF. Recorded from all adjoining counties and 57.

2141051 **Eurhynchium swartzii** (Turn.) Curn. Map 956, p. 731
Kirk: Binley Bridge, 1859, in herb. Oxford University.
Bagnall; Wait and Elsee; Townsend: many records from all over the county.
On the ground in open woodland, scrub, hedgebanks, shaded grassland, banks of rivers, streams, ditches, ponds and canals, in arable fields (the most common pleurocarpous species in this habitat), in compacted ground, particularly canal towpaths, on floors and spoil heaps of old limestone quarries and calcareous clay pits and in basic and neutral marshes; occasionally on limestone rock exposures, blocks and walls and on masonry by water; rarely on sandstone; frequent throughout the county, but less common in the north than elsewhere. 243; 55%; not found fruiting.
BIRA, LAF, OXF, WAR. Recorded from all adjoining counties and 57.

(2141061 **Eurhynchium schleicheri** (Hedw. fil.) Lor. was recorded by Bagnall from banks, Little Wolford, 1887, but his specimen in herb. Birmingham Museum is *E. swartzii* (Turn.) Curn.)

2141071 **Eurhynchium speciosum** (Brid.) Milde
Bagnall: tree roots near water, coppice by Windley Pool, Sutton Park, 1870, in "Notes on Sutton Park", 1876; small bridge, lane from Water Orton to Minworth; Curdworth; canal siding, Lapworth Wharf.
Not found in the present survey.
BIRA. Recorded from 23, 32, 33, 37 and 55.

2141081 **Eurhynchium riparioides** (Hedw.) Rich.
Map 957, p. 731
Purton: Oversley Mill, in "Midl. Flor.", App. (1821).
H. E. Lowe; Kirk; Bagnall; Wait and Elsee; Townsend: many records from all over the county.
On rocks, masonry, woodwork, tree bases and roots and banks in and by water, submerged or near water level, in rivers, streams and canals; always in moving water, even if moving only slowly, and most frequent in swift-flowing, well-aerated water, as in weirs and sluices; frequent throughout the county. 154; 35%; 14% fruiting.
BIRA, LAF, OXF, WAR. Recorded from all adjoining counties and 57.

2141101 **Eurhynchium murale** (Hedw.) Milde
Map 958, p. 731
Bree: Allesley, in Purton, "Midl. Flor.", App. (1821).
Kirk; Bagnall; Jackson: many records from all over the county.
On shaded rock exposures, blocks and masonry of all kinds, but particularly on limestone and usually near to the ground; rarely on tree bases by water; widespread. 60; 14%; 83% fruiting.
BIRA, LAF, OXF. Recorded from all adjoining counties and 57.

2141111 **Eurhynchium confertum** (Dicks.) Milde
Map 959, p. 731
Bree: Bidford Grange, undated, in herb. Warwick County Museum.
Kirk; Bagnall; Wait and Elsee; Townsend: many records from all over the county.
On rock exposures, blocks and walls and masonry of all kinds, usually in shade; less frequently on tree bases, shrubs, stools and stumps in open woodland, scrub, hedgerows and by water; rarely on banks on the ground; throughout the county. 222; 51%; 93% fruiting.
BIRA, LAF, OXF, WAR. Recorded from all adjoining counties and 57.

2141121 **Eurhynchium megapolitanum** (Bland.) Milde
Bagnall: banks of River Tame near Curdworth, in *Midl. Nat.* **16** (1893); Merevale; Stivichall; Preston Bagot; Burton Dassett.
Amongst grass on banks and on rock exposures and wall tops, rare. 8; 2%; 13% fruiting. 0854, 1356, 2830, 2257, 2685, 3760, 3866, 4978.
LAF. Recorded from 23, 37 and 57.

2142011 **Rhynchostegiella pumila** (Wils.) E. F. Warb.
Map 960, p. 732
Kirk: Whitley, 1858, in herb. Oxford University; Stoneleigh.
Bagnall: several records from north and central Warwickshire.
On hedgebanks, on the ground in scrub and on shaded banks of streams and ditches; rarely on rock exposures in banks or retaining walls by streams; widespread, but not frequent. 23; 5%; 9% fruiting.
BIRA, LAF, OXF. Recorded from all adjoining counties and 57.

2142021 **Rhynchostegiella curviseta** (Brid.) Limpr.
Map 961, p. 732
Laflin: canal lock, Bishopton, 1957.
On rocks, stones, masonry and drains by water; rarely on wet ground; by streams and canals in the southern half of the county. 10; 2%; 92% fruiting.
LAF, WAR. Recorded from 23, 32, 33 and 37.

2142031 **Rhynchostegiella teesdalei** (Sm.) Limpr.
Purton: moist shady place between Oversley Green and the Mill, upon the upper bank, in "Midl. Flor." (1817).
Bagnall: stones in stream, Bentley Park.
Not found in the present survey.
BIRA. Recorded from 23, 33, 37 and 57.

2142041 **Rhynchostegiella tenella** (Dicks.) Limpr.
Map 962, p. 732
Bagnall: masonry of railway bridge near Marston Green, 1869, in herb. Birmingham Museum; Merevale Park; Bentley Park; Baddesley Ensor; Compton Verney.
On shaded rock exposures and blocks, particularly limestone, and on masonry; one record from base of shrub in scrub; widespread but uncommon in the southern half of the county. 19; 4%; 70% fruiting.
BIRA, LAF. Recorded from all adjoining counties and 57.

ENTODONTACEAE

2145011 **Entodon concinnus** (De Not.) Par.
Laflin: 0956, rough calcareous grassland, bank of old Lias limestone diggings, Primrose Hill, Oversley, 1961. Not fruiting.
LAF. Recorded from all adjoining counties and 57.

2146011 **Pseudoscleropodium purum** (Hedw.) Fleisch.
Map 963, p. 732
Purton: "common", but no localities given, in "Midl. Flor." (1817).
Bree; Bromwich; Bagnall; Wait and Elsee; Dalman; Townsend: many records from all over the county.
In rough grassland, poor grazed pastures, marshes, heaths, open woodland, scrub, woodland rides, heathy hedgebanks; rarely on wall tops, railway tracks or at wet margins of ponds and canals; one record from water at the base of reeds; widespread throughout the county. 153; 35%; not found fruiting.
BIRA, LAF, WAR. Recorded from all adjoining counties and 57.

2147011 **Pleurozium schreberi** (Brid.) Mitt.
Map 964, p. 733
Purton: wood beyond Cold Comfort, Alcester, in "Midl. Flor." (1817).
Perry; Kirk; Bagnall; Dalman: several records from the north-west and north-central parts of the county; also Wimpstone Fields and Wolford Heath in the south.
On heaths and in heathy grassland or heathy woodland on acid soils in central and north Warwickshire; Wolford Wood in the south; not common. 14; 3%; not found fruiting.
BIRA, LAF, OXF, WAR. Recorded from all adjoining counties and 57.

PLAGIOTHECIACEAE

(2148011 **Isopterygium seligeri** (Brid.) Dix. has been found in VC 32, and might occur on stumps in woods in S.E. Warwickshire.)

2148021 **Isopterygium depressum** (Bruch) Mitt.
Laflin: 2934, limestone block in hedgerow, lane leading to Whichford Wood, 1959. Not fruiting.
LAF. Recorded from 23, 33, 37 and 57.

2148041 **Isopterygium elegans** (Hook.) Lindb.
Map 965, p. 733
Bagnall: Sutton Park, in *Proc. Bgham nat. Hist. microsc. Soc.* **1** (1869); Bentley Park; Shustoke; Maxstoke; Shirley; Griff Hollow.
Wait and Elsee: Birtley Rough, Brandon. Dalman: Tile Hill Wood.
Townsend: Sutton Park; Rough Hill Wood.
On banks on acid soils in woods, scrub, hedgerows and by water; on dry sandstone rock exposures and masonry; rarely between rock blocks in drainage channels of railway cuttings or on stumps in woodland or hedgerows; frequent in acid soil areas in the north and centre and on the Middle Lias on the southern boundary. It is difficult to understand the small number of 19th-century records. 109; 25%; not found fruiting.
BIRA, LAF, WAR. Recorded from all adjoining counties and 57.

2149011 **Plagiothecium latebricola** B., S. and G.
Map 966, p. 733
Bagnall: Windley Pool, Sutton Park, 1868, in *Midl. Nat.* **3** (1880); Waters Wood, Maxstoke; Shawberries Wood; Berkswell.
On stumps and stools in woodland on acid soil; rarely on stumps in hedgerows or by water; in the centre and north of the county. 34; 8%; not found fruiting.
BIRA, LAF, WAR. Recorded from 23, 33, 37, 39, 55 and 57.

2149041 **Plagiothecium denticulatum** (Hedw.) B., S. and G.
Map 967, p. 733
Bree: Allesley, in Purton, "Midl. Flor.", App. (1821).
Bagnall; Cumming, Wait and Elsee; Dalman; Townsend: many records from the north and centre and a few from the south.
On stumps, stools, tree bases and low shrubs in woods, scrub, hedgerows and by water, less frequently on banks in these habitats; occasionally in marshes and at the margins of lakes and ponds, sometimes growing in the water at the bases of reeds; rarely on sandstone blocks and walls or on the ground on heaths; frequent in acid soil areas; only occasional in basic soil areas. 192; 44%; 49% fruiting.
BIRA, LAF, WAR. Recorded from all adjoining counties and 57.

(2149051 **Plagiothecium ruthei** Limpr. is recorded from Earlswood by Field, 1964 (see *Trans. Br. bryol. Soc.* **5**, (3), 1968). I cannot distinguish this species from some wet habitat forms of *P. denticulatum* (Hedw.) B., S. & G. I have found a large number of plants with one or other of the distinguishing features of *P. ruthei*, but typical auricles are more often associated with arched leaf margins than with those with one straight margin and asymmetrical leaves of all types may often be found on the same plant. Examination of more than a hundred Warwickshire gatherings of *ruthei* type makes me suspect strongly that British *P. ruthei* is no more than a wet habitat form of *P. denticulatum.*)

2149061 **Plagiothecium curvifolium** Schlieph.
Bagnall: New Park, Middleton, 1881, in herb. Birmingham Museum; Sutton Park; Bentley Park; near Bulkington.
Townsend: Coughton Park.
On stumps and stools, rarely on tree bases, in woodland on acid soils in central Warwickshire; not common. 9; 2%; not found fruiting. 0454; 0563, 0660, 2268, 2070, 2873, 3570, 3671, 4078.
BIRA, LAF. Recorded from 23, 37, 39 and 57.

(2149081 **Plagiothecium platyphyllum** Mönk. Bagnall gives three localities, Armscote, Packington and Corley, for *P. denticulatum* (Hedw.) var. *majus* Boul. There is one specimen, from Packington, so named in herb. Birmingham Museum, but this is *P. denticulatum* and all records probably refer to that plant.)

2149101 **Plagiothecium succulentum** (Wils.) Lindb.
Map 968, p. 734
Russell: Chelmsley Wood, 1891, in herb. Birmingham Museum, named *P. sylvaticum* (Brid.) B., S. and G.
On banks in woodland, scrub, hedgerows and by water on mildly acid and neutral soils, on stumps, stools and tree bases in similar places and also in basic soil areas; occasionally on siliceous rock exposures; frequent throughout the county. 110; 25%; 3% fruiting.
BIRA, LAF. Recorded from all adjoining counties and 57.

2149111 **Plagiothecium sylvaticum** (Brid.) B., S. and G.
Map 969, p. 734
Bagnall: near Powells Pool, Sutton Park, in *Proc. Bgham nat. Hist. microsc. Soc.* **1** (1869).
Bagnall; Jackson; Townsend: several records from central and north Warwickshire; also Wolford Wood in the south.
In similar places to the preceding but less common in most habitats, although more frequent on sandstone. 55; 13%; not found fruiting.
BIRA, LAF. Recorded from 23, 33, 37, 39 and 57.

2149121 **Plagiothecium undulatum** (Hedw.) B., S. and G.
Map 970, p. 734
Bree: Coleshill, in Purton, "Midl. Flor.", App. (1821); Allesley.
Bagnall; Dalman: several records from the north-west and west-central parts of the county.
In heathy woodland, on heaths and on long-disused spoil heaps of siliceous stone quarries, not common; in the northern half of the county. 11; 3%; not found fruiting.
BIRA, LAF. Recorded from 23, 32, 37, 39, 55 and 57.

HYPNACEAE

(2151011 **Pylaisia polyantha** (Hedw.) B., S. and G. was recorded by Bagnall from Frogmore Wood, Fen End, 1880, in *Midl. Nat.* **3** (1880), but there is no specimen in his herbarium. Confirmation is desirable.)

2154011 **Hypnum cupressiforme** Hedw. Map 971, p. 734
Purton: "common", but no localities given, in "Midl. Flor." (1817).
H. E. Lowe; Kirk; Bagnall; Wait and Elsee; Townsend: many records from all over the county.
On trees, shrubs, stumps, stools, fallen wood and woodwork in all types of habitat; on rock exposures and blocks and masonry of all kinds; occasional on spoil heaps of pits and quarries and on compacted ground of paths, railway tracks, banks and on the ground in woods; very common throughout the county. 389; 89%; 4% fruiting.
BIRA, LAF, OXF, WAR. Recorded from all adjoining counties and 57.

2154012 var. **resupinatum** (Wils.) Schimp. Map 972, p. 735
Bagnall: Sutton Park, in *Proc. Bgham nat. Hist. microsc. Soc.* **1** (1869); many other records; also recorded by Cumming, Wait and Elsee, and Townsend.
On trunks of trees, particularly oak and elm, mainly in woodland but occasionally in scrub and hedgerows; rather few records and probably under-recorded. 23; 5%; not found fruiting.
BIRA, LAF. Recorded from all adjoining counties and 57.

2154013 var. **filiforme** Brid.
Bagnall: Sutton Park, in "Notes on Sutton Park," 1876; several other records from widespread localities.
Dalman: Tile Hill Wood.
On tree bases or stools in mature oak woodland on acid

soils, rare. 5; 1%; not found fruiting. 1860, 1597, 2840, 3258, 3571.
BIRA, LAF. Recorded from all adjoining counties and 57.

2154015 var. **ericetorum** B., S. and G. Map 973, p. 735
Bagnall: Purley Park, 1885, in herb. Birmingham Museum; Wolford Heath and Wood.
Townsend: Rough Hill Wood; Oversley Wood.
On the ground on heaths, in heathy grassland and heathy woodland rides and clearings; rarely in acid boggy places; common in association with *Calluna* and *Agrostis* spp.; widespread in the centre and north; only from Wolford Wood in the south. 39; 9%; not found fruiting.
BIRA, LAF. Recorded from all adjoining counties and 57.

2154016 var. **tectorum** B., S. and G. Map 974, p. 735
Bagnall: walls near Meriden Shafts, 1888, in herb. Birmingham Museum; several other records from widespread localities.
On rock exposures, blocks and masonry and on wall tops; less frequently on tree bases in open woodland or hedgerow; usually growing on flat horizontal surfaces; widespread but not common. 26; 6%; not found fruiting.
BIRA, LAF. Recorded from all adjoining counties and 57.

2154017 var. **lacunosum** Brid.
Bagnall: Milverton, in *J. Bot., Lond.* **12** (1874); several other records from widespread localities.
Amongst short turf in rough grassland on calcareous clay, limestone or leached loam, rare. 5; 1%; not found fruiting. 1763, 3544, 3859, 3683, 4982.
BIRA, LAF. Recorded from all adjoining counties and 57.

2154071 **Hypnum lindbergii** Mitt. Map 975, p. 735
Bagnall: Sharmans Cross, Solihull, 1872, in herb. Birmingham Museum.
Bagnall; Townsend: several records, mainly from the west-central part of the county, but a few records from the north and Wolford Wood in the south.
On woodland rides on the heavier, mildly acid soils; rarely on wet floors of sandpits; mainly in the centre, but also in the extreme south; not common. 10; 2%; not found fruiting.
BIRA, LAF. Recorded from 23, 32, 37, 39, 55 and 57.

2156011 **Ctenidium molluscum** (Hedw.) Mitt.
Map 976, p. 736
Purton: Oversley Hill, in "Midl. Flor." (1817); Red Hill.
Bree; Kirk; Bagnall; Wait and Elsee; Townsend: many records from all over the county.
On the ground on basic to mildly acid soils in woodland, in rough calcareous grassland and open scrub, in marshes and boggy flushes and on limestone walls or blocks on the ground; mainly in the south and centre, but a few localities in the north. 37; 8%; not found fruiting.
BIRA, LAF, OXF, WAR. Recorded from all adjoining counties and 57.
None of the varieties has been recorded, but no attempt has been made in the present survey to recognise them.

HYLOCOMIACEAE

2160011 **Rhytidiadelphus triquetrus** (Hedw.) Warnst.
Map 977, p. 736
Purton: "very common", but no localities given, in "Midl. Flor." (1817).
Bree; Bagnall; Wait and Elsee: several records from all over the county.
On the ground in woodland, scrub and rough grassland, mainly on calcareous soils, but occasionally in acid woodland; rarely in marshes or on heaths; almost confined to the south and south-central parts of the county. 38; 9%; 2% fruiting.
BIRA, LAF, WAR. Recorded from all adjoining counties and 57.

2160021 **Rhytidiadelphus squarrosus** (Hedw.) Warnst.
Map 978, p. 736
Purton: "common", but no localities given, in "Midl. Flor." (1817).
Bagnall; Cumming and Elsee; Townsend: many records from all parts of the county.
In grazed and rough grassland, in marshes, on heaths, on compacted ground, particularly along woodland rides, on the ground in open woodland and scrub; occasionally by canals or at wet margins of ponds; frequent throughout the county. 129; 29%; not found fruiting.
BIRA, LAF. Recorded from all adjoining counties and 57.

(2160022 var. **calvescens** (Wils.) Mönk. was recorded by Bagnall from south Warwickshire, without details of precise locality, in "Victoria County Hist. Warws.", **1**, 1904, but probably erroneously.)

2160031 **Rhytidiadelphus loreus** (Hedw.) Warnst.
Bree: woods, Allesley, in Purton, "Midl. Flor.", App. (1821).
On the ground in oak woods, rare. 1; not fruiting. 2996.
LAF. Recorded from all adjoining counties and 57.

2161011 **Hylocomium brevirostre** (Brid.) B., S. and G.
Bree: Allesley, undated, in herb. Warwick County Museum; Meriden.
Bagnall: Wire Hill Wood; Wolford Wood; Whichford Wood.
Not found in the present survey.
BIRA, WAR. Recorded from all adjoining counties and 57.

2161041 **Hylocomium splendens** (Hedw.) B., S. and G.
Map 979, p. 736
Purton: "common", but no localities given, in "Midl. Flor." (1817).
Bree; Perry; Bagnall; Wait and Elsee; Jackson: many records from all over the county.
On the ground in rough grassland, particularly on banks, in open woodland and scrub and on dry heaths; only in the southern half of the county and less frequent now than in the 19th century. 24; 5%; not found fruiting.
BIRA, LAF, WAR. Recorded from all adjoining counties and 57.

Acknowledgements

I wish to thank a large number of people who have helped and advised in many ways and, in particular, the following:

For advice on the presentation of the material in the text, the preparation of the maps and more general advice: Miss D. A. Cadbury, Mr. M. C. Clark, Prof. J. G. Hawkes and Mr. R. C. Readett, members of the Editorial Panel.

For assistance with location of published records and loan of books and manuscripts: Mr. D. E. Allen, Mr. P. Falk and Mr. N. C. Kittermaster (Rugby School); Mr. H. E. E. Babb (journals in the Library of the Birmingham Natural History Society); Prof. J. G. Hawkes (Bagnall records and journals in the Library of the University of Birmingham); Mrs. M. D. G. Jones (Rugby Public Library); Miss J. M. Morris (Purton, Perry and several journals); Mrs. J. A. Paton (hepatic records in British Bryological Society's Notebook); Mr. A. J. Pettifer (journals in the British Bryological Society's Library) ;Mr. C. C. Townsend (a valuable MS list of his own records); Mr. E. C. Wallace (various); Dr. E. F. Warburg (Bagnall and moss records in British Bryological Society's Notebook).

For determination of critical species: Mr. A. C. Crundwell (*Weissia* and other mosses); Mr. J. C. Field (*Philonotis*); Dr. S. W. Greene (*Plagiothecium*); Dr. E. Lodge (*Drepanocladus*); Mr. A. H. Norkett (*Fissidens*); Mrs. J. A. Paton (Hepaticae); Dr. E. F. Warburg (*Pohlia annotina* group and other mosses); Dr. H. L. K. Whitehouse (*Dicranella*).

For permission to examine herbaria and for help in examination and extraction of records: Mr. L. Bilton, Mr. K. E. Davies and Mr. P. W. Hanney (City of Birmingham Museum); Miss U. K. Duncan (herb. J. B. Duncan); Miss J. M. Morris and Mr. R. W. B. Dawson (Warwick County Museum); Mr. C. Phipps and Mr. D. R. Shearer (City of Worcester Museum); Dr. E. F. Warburg and Mr. A. R. Perry (University of Oxford herbarium).

For help with field work and recording: those named in the Introduction and particularly Mr. G. A. and Mr. M. A. Arnold, Mr. M. C. Clark and Mr. and Mrs. R. Evans.

For reading the manuscript and the proofs: Mr. M. C. Clark, Mr. R. E. Evans, Prof. J. G. Hawkes and Mr. A. R. Perry.

13. DISTRIBUTION MAPS OF BRYOPHYTES

The following maps (four per page) show the recorded distribution of the 184 bryophyte species that were found in at least ten of the 1 km squares recorded. Details of records for species found in fewer than ten 1 km squares are given in the text.

The computer program for the bryophytes reduced the single square records to tetrads (groups of four 1 km squares), gave priority to "fruiting" records over "sterile" ones in the same tetrad, and computed total numbers and percentages of records for each species. These numbers are shown in the bryophyte text.

The distribution maps were produced on a line printer, as for the vascular plant maps in section 10. However, the two symbols here indicate fruiting (×) and sterile (+) records rather than frequencies.

The maps are slightly distorted (as are those in section 10) by an increase in E–W to N–S dimensions in the ratio of 5:4, because of the physical limitations of the line printer. A double scale in the bottom left-hand corner indicates these differences graphically.

Transparent overlays are available in a pocket at the back of the book showing

Map IX	Relief and drainage
Map X	Surface geology
Map XI	Land use, 1960 (including canals and railways—used and disused)
Map XII	1 km squares surveyed for bryophyte records.

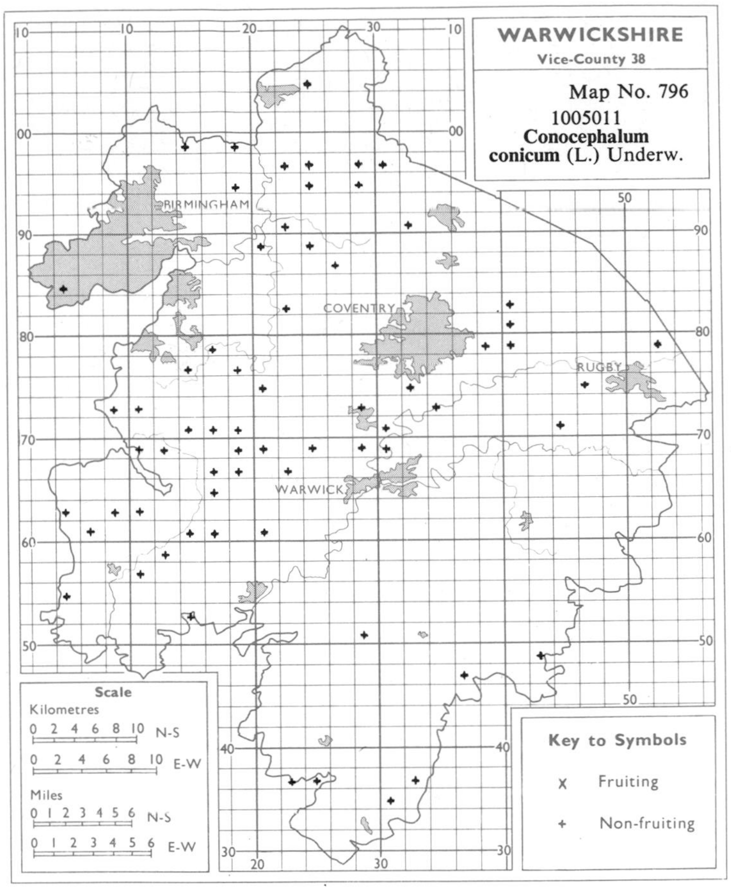
WARWICKSHIRE
Vice-County 38
Map No. 796
1005011
Conocephalum conicum (L.) Underw.
BIRMINGHAM
COVENTRY
RUGBY
WARWICK
Scale
Kilometres
0 2 4 6 8 10 N-S
0 2 4 6 8 10 E-W
Miles
0 1 2 3 4 5 6 N-S
0 1 2 3 4 5 6 E-W
Key to Symbols
X Fruiting
+ Non-fruiting

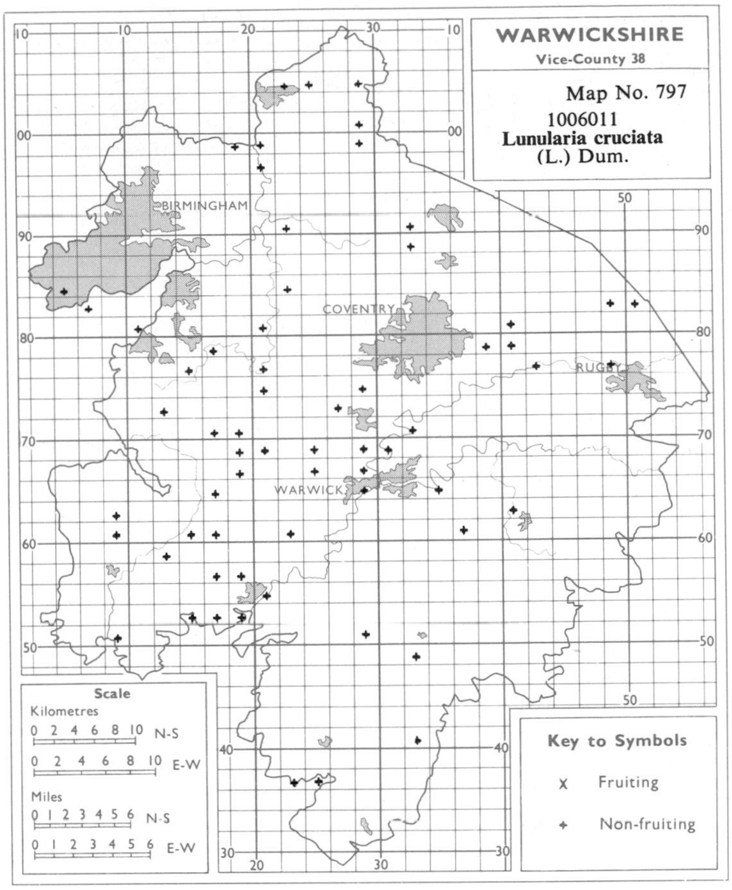
WARWICKSHIRE
Vice-County 38
Map No. 797
1006011
Lunularia cruciata (L.) Dum.
BIRMINGHAM
COVENTRY
RUGBY
WARWICK
Scale
Kilometres
0 2 4 6 8 10 N-S
0 2 4 6 8 10 E-W
Miles
0 1 2 3 4 5 6 N-S
0 1 2 3 4 5 6 E-W
Key to Symbols
X Fruiting
+ Non-fruiting

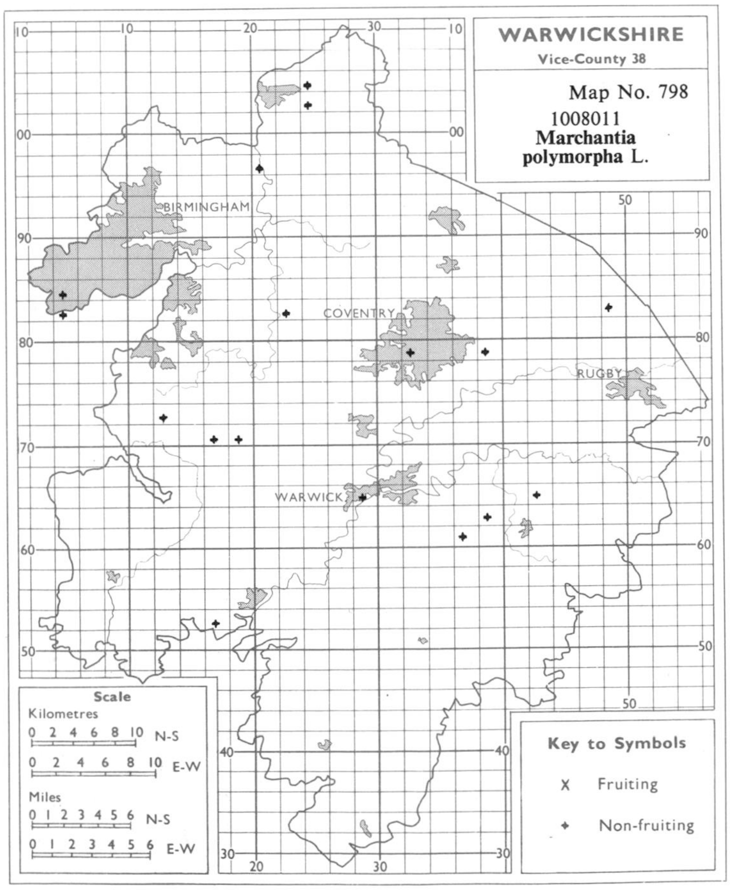
WARWICKSHIRE
Vice-County 38
Map No. 798
1008011
Marchantia polymorpha L.
BIRMINGHAM
COVENTRY
RUGBY
WARWICK
Scale
Kilometres
0 2 4 6 8 10 N-S
0 2 4 6 8 10 E-W
Miles
0 1 2 3 4 5 6 N-S
0 1 2 3 4 5 6 E-W
Key to Symbols
X Fruiting
+ Non-fruiting

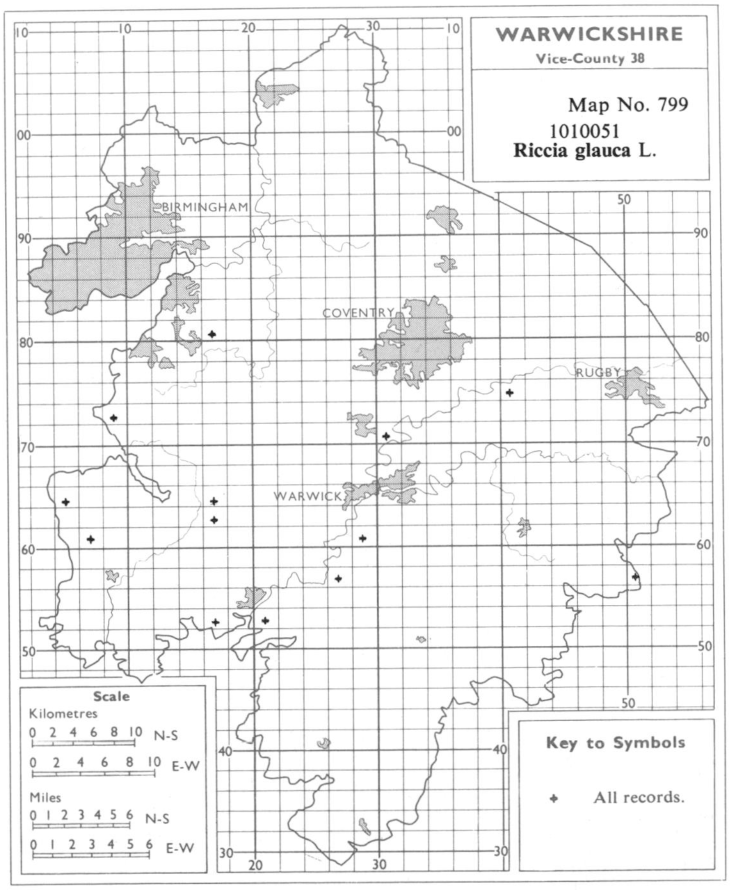
WARWICKSHIRE
Vice-County 38
Map No. 799
1010051
Riccia glauca L.
BIRMINGHAM
COVENTRY
RUGBY
WARWICK
Scale
Kilometres
0 2 4 6 8 10 N-S
0 2 4 6 8 10 E-W
Miles
0 1 2 3 4 5 6 N-S
0 1 2 3 4 5 6 E-W
Key to Symbols
+ All records.

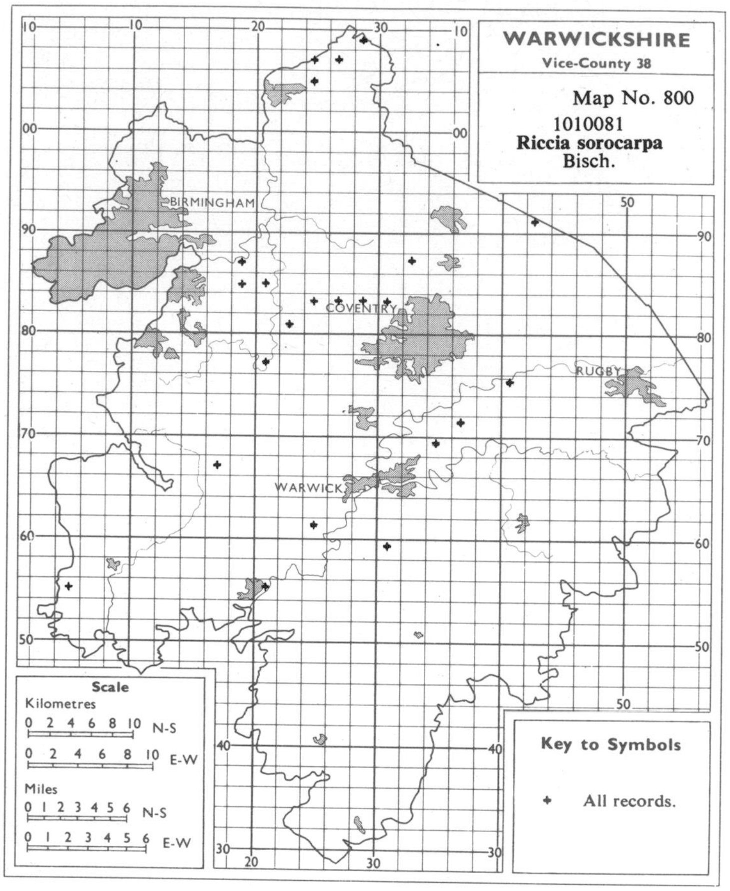
WARWICKSHIRE
Vice-County 38
Map No. 800
1010081
Riccia sorocarpa
Bisch.
BIRMINGHAM
COVENTRY
RUGBY
WARWICK
Scale
Kilometres
N-S
E-W
Miles
N-S
E-W
Key to Symbols
All records.

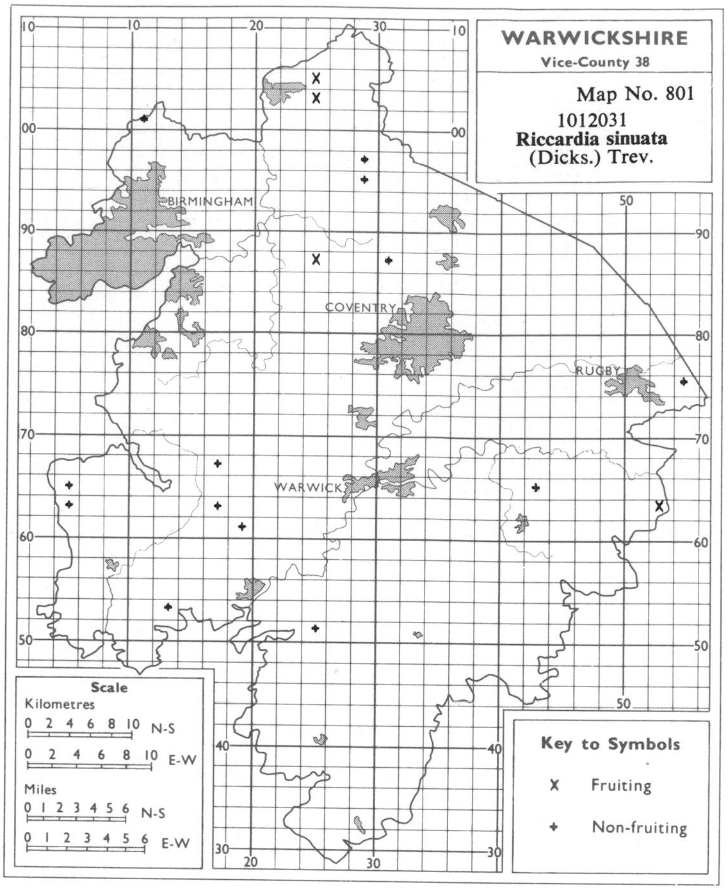
WARWICKSHIRE
Vice-County 38
Map No. 801
1012031
Riccardia sinuata
(Dicks.) Trev.
BIRMINGHAM
COVENTRY
RUGBY
WARWICK
Scale
Kilometres
N-S
E-W
Miles
N-S
E-W
Key to Symbols
Fruiting
Non-fruiting

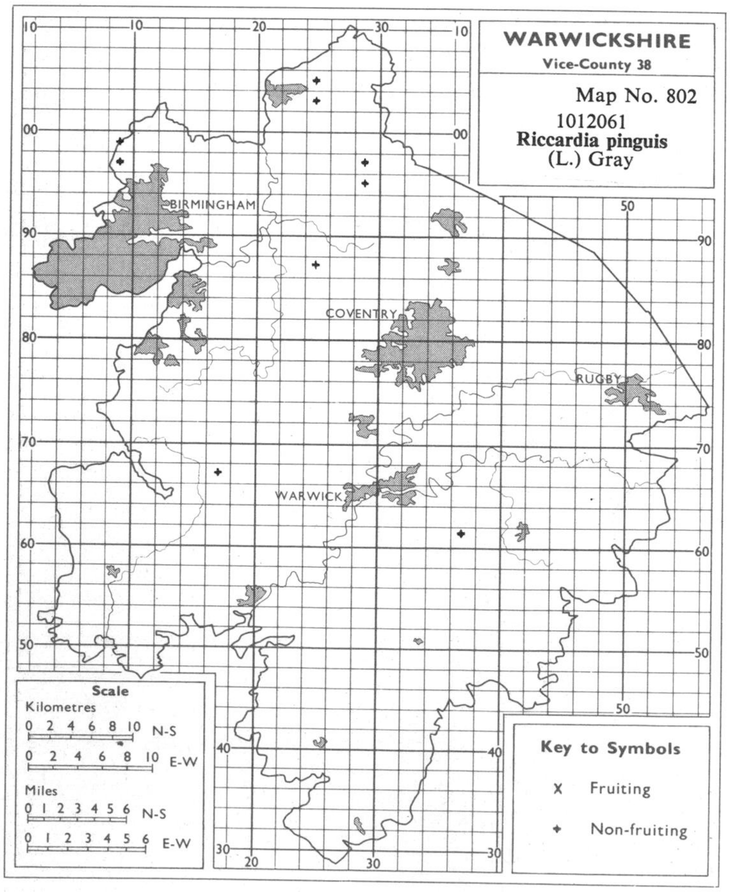
WARWICKSHIRE
Vice-County 38
Map No. 802
1012061
Riccardia pinguis
(L.) Gray
BIRMINGHAM
COVENTRY
RUGBY
WARWICK
Scale
Kilometres
N-S
E-W
Miles
N-S
E-W
Key to Symbols
Fruiting
Non-fruiting

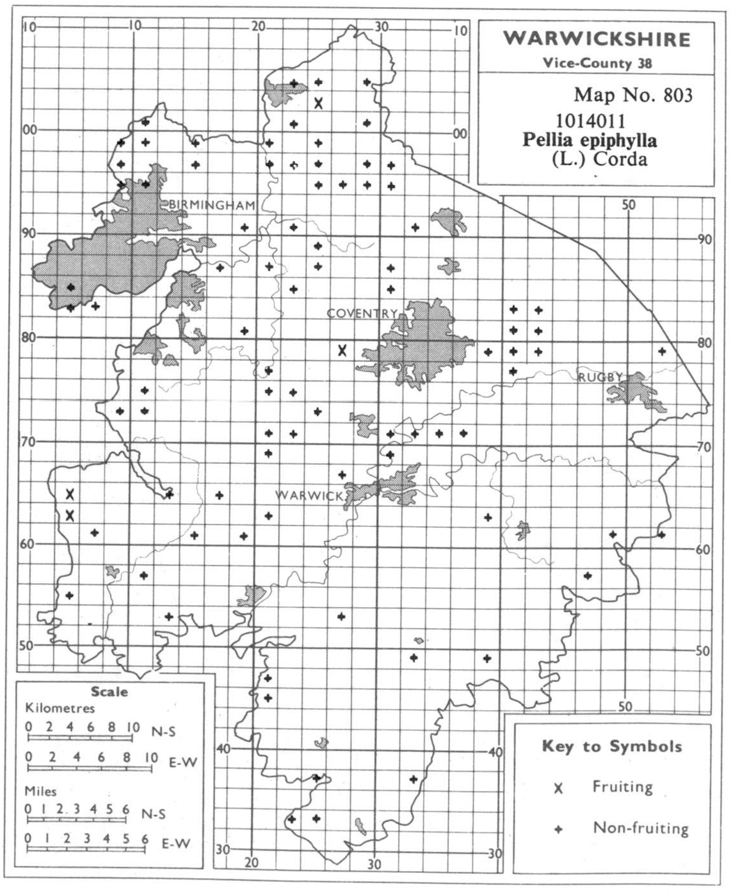
WARWICKSHIRE
Vice-County 38
Map No. 803
1014011
Pellia epiphylla
(L.) Corda
BIRMINGHAM
COVENTRY
RUGBY
WARWICK
Scale
Kilometres
N-S
E-W
Miles
N-S
E-W
Key to Symbols
Fruiting
Non-fruiting

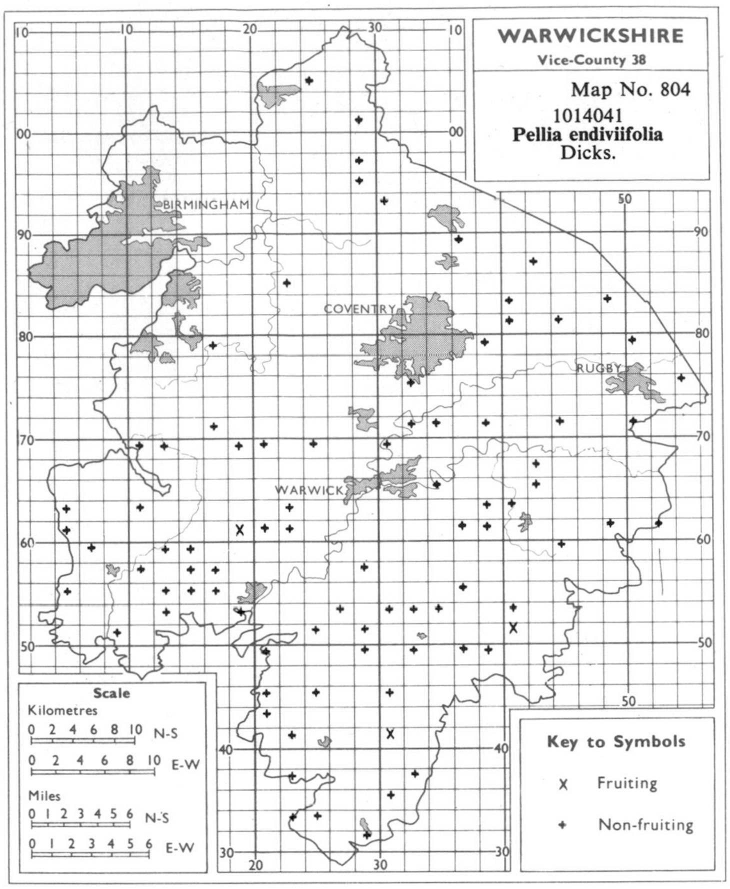
WARWICKSHIRE
Vice-County 38
Map No. 804
1014041
Pellia endiviifolia
Dicks.
BIRMINGHAM
COVENTRY
RUGBY
WARWICK
Scale
Kilometres
0 2 4 6 8 10 N-S
0 2 4 6 8 10 E-W
Miles
0 1 2 3 4 5 6 N-S
0 1 2 3 4 5 6 E-W
Key to Symbols
X Fruiting
+ Non-fruiting

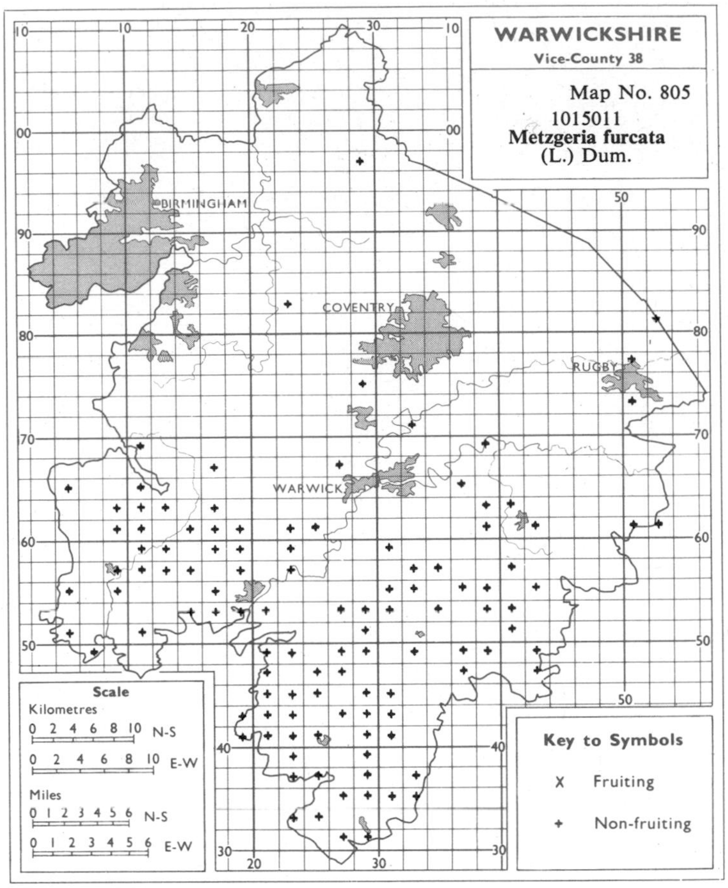
WARWICKSHIRE
Vice-County 38
Map No. 805
1015011
Metzgeria furcata
(L.) Dum.
BIRMINGHAM
COVENTRY
RUGBY
WARWICK
Scale
Kilometres
0 2 4 6 8 10 N-S
0 2 4 6 8 10 E-W
Miles
0 1 2 3 4 5 6 N-S
0 1 2 3 4 5 6 E-W
Key to Symbols
X Fruiting
+ Non-fruiting

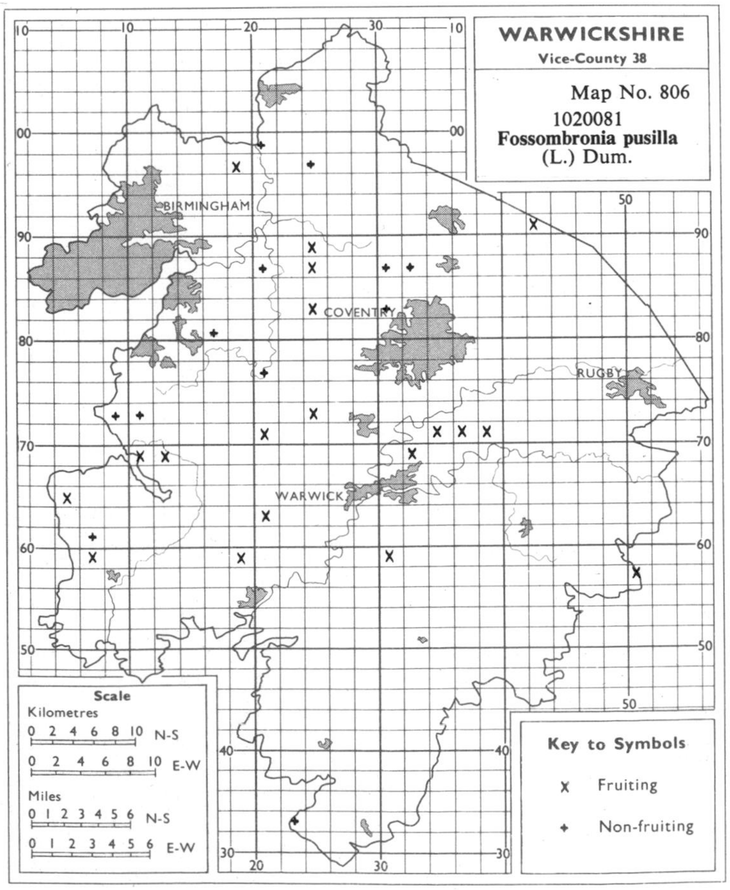
WARWICKSHIRE
Vice-County 38
Map No. 806
1020081
Fossombronia pusilla
(L.) Dum.
BIRMINGHAM
COVENTRY
RUGBY
WARWICK
Scale
Kilometres
0 2 4 6 8 10 N-S
0 2 4 6 8 10 E-W
Miles
0 1 2 3 4 5 6 N-S
0 1 2 3 4 5 6 E-W
Key to Symbols
X Fruiting
+ Non-fruiting

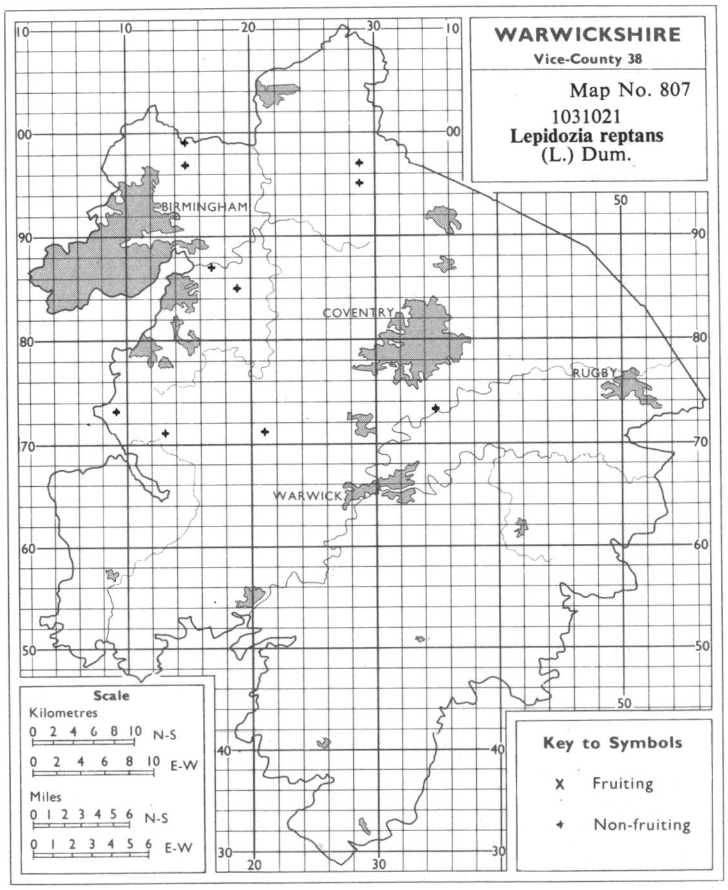
WARWICKSHIRE
Vice-County 38
Map No. 807
1031021
Lepidozia reptans
(L.) Dum.
BIRMINGHAM
COVENTRY
RUGBY
WARWICK
Scale
Kilometres
0 2 4 6 8 10 N-S
0 2 4 6 8 10 E-W
Miles
0 1 2 3 4 5 6 N-S
0 1 2 3 4 5 6 E-W
Key to Symbols
X Fruiting
+ Non-fruiting

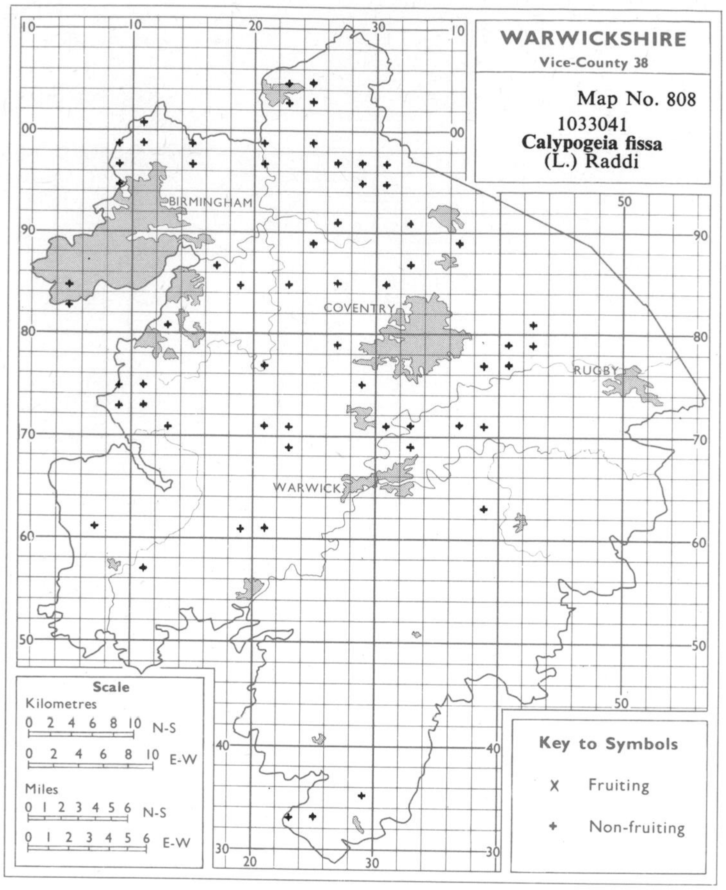
WARWICKSHIRE
Vice-County 38
Map No. 808
1033041
Calypogeia fissa
(L.) Raddi
BIRMINGHAM
COVENTRY
RUGBY
WARWICK
Scale
Kilometres
0 2 4 6 8 10 N-S
0 2 4 6 8 10 E-W
Miles
0 1 2 3 4 5 6 N-S
0 1 2 3 4 5 6 E-W
Key to Symbols
X Fruiting
+ Non-fruiting

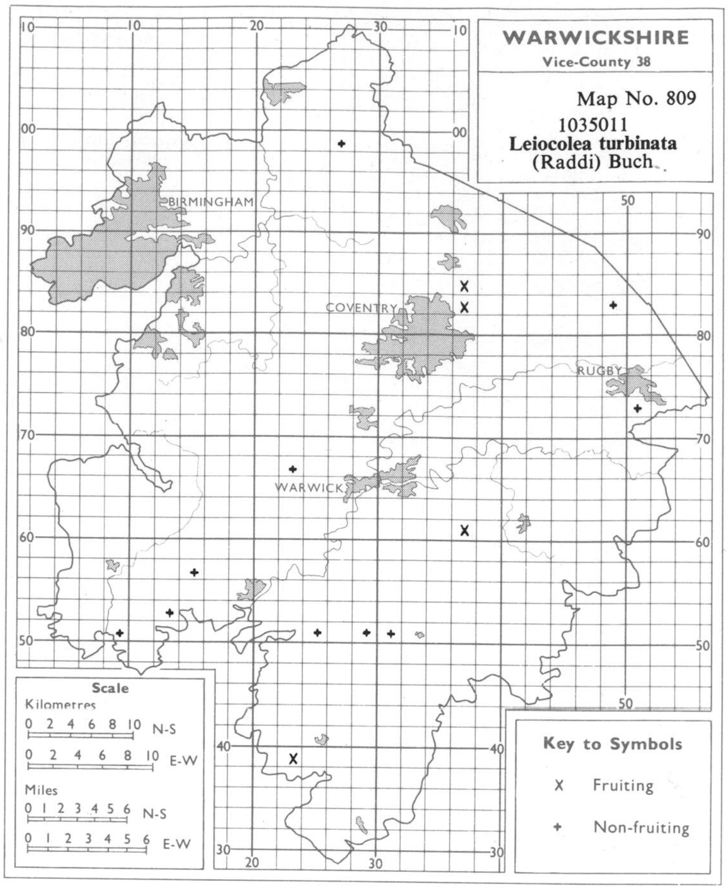
WARWICKSHIRE
Vice-County 38
Map No. 809
1035011
Leiocolea turbinata
(Raddi) Buch.
BIRMINGHAM
COVENTRY
RUGBY
WARWICK
Scale
Kilometres
0 2 4 6 8 10 N-S
0 2 4 6 8 10 E-W
Miles
0 1 2 3 4 5 6 N-S
0 1 2 3 4 5 6 E-W
Key to Symbols
X Fruiting
+ Non-fruiting

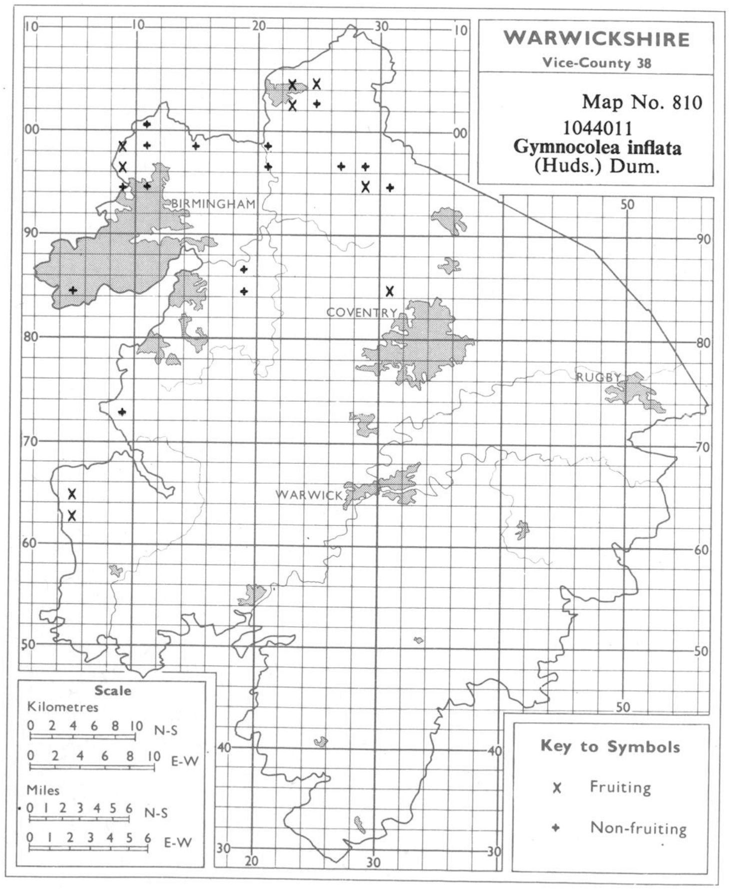
WARWICKSHIRE
Vice-County 38
Map No. 810
1044011
Gymnocolea inflata
(Huds.) Dum.
BIRMINGHAM
COVENTRY
RUGBY
WARWICK
Scale
Kilometres
0 2 4 6 8 10 N-S
0 2 4 6 8 10 E-W
Miles
0 1 2 3 4 5 6 N-S
0 1 2 3 4 5 6 E-W
Key to Symbols
X Fruiting
+ Non-fruiting

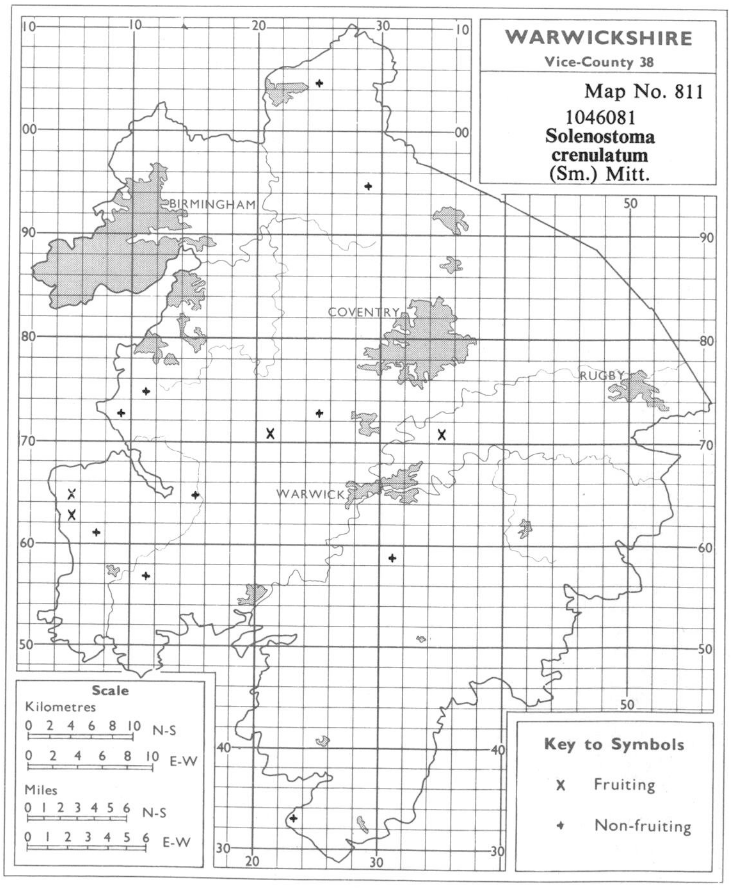
WARWICKSHIRE
Vice-County 38
Map No. 811
1046081
Solenostoma
crenulatum
(Sm.) Mitt.
BIRMINGHAM
COVENTRY
RUGBY
WARWICK
Scale
Kilometres
0 2 4 6 8 10 N-S
0 2 4 6 8 10 E-W
Miles
0 1 2 3 4 5 6 N-S
0 1 2 3 4 5 6 E-W
Key to Symbols
X Fruiting
+ Non-fruiting

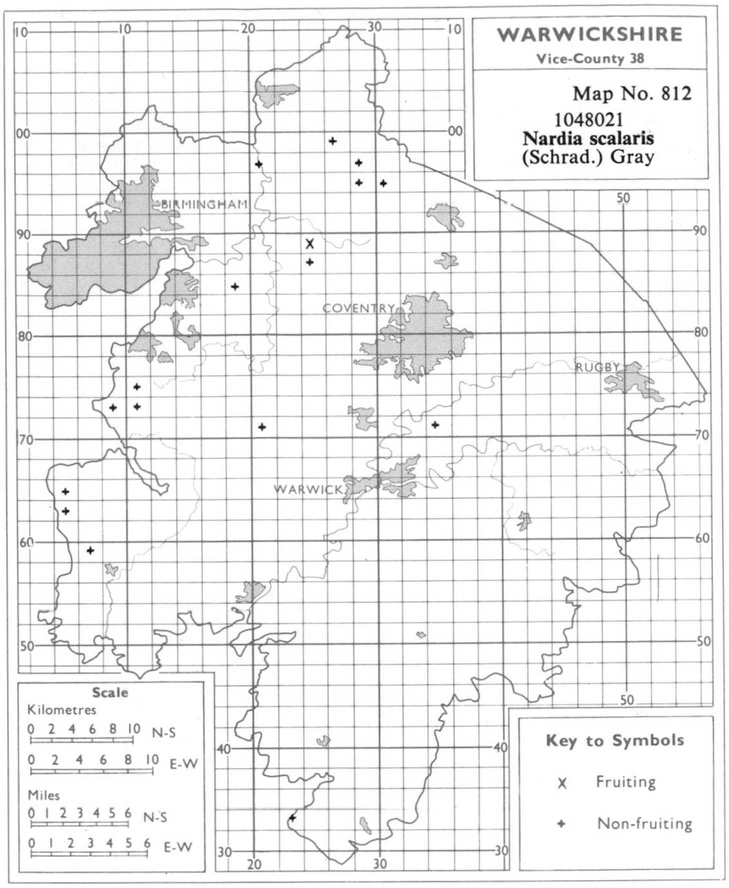
WARWICKSHIRE
Vice-County 38
Map No. 812
1048021
Nardia scalaris
(Schrad.) Gray
BIRMINGHAM
COVENTRY
RUGBY
WARWICK
Scale
Kilometres
0 2 4 6 8 10 N-S
0 2 4 6 8 10 E-W
Miles
0 1 2 3 4 5 6 N-S
0 1 2 3 4 5 6 E-W
Key to Symbols
X Fruiting
+ Non-fruiting

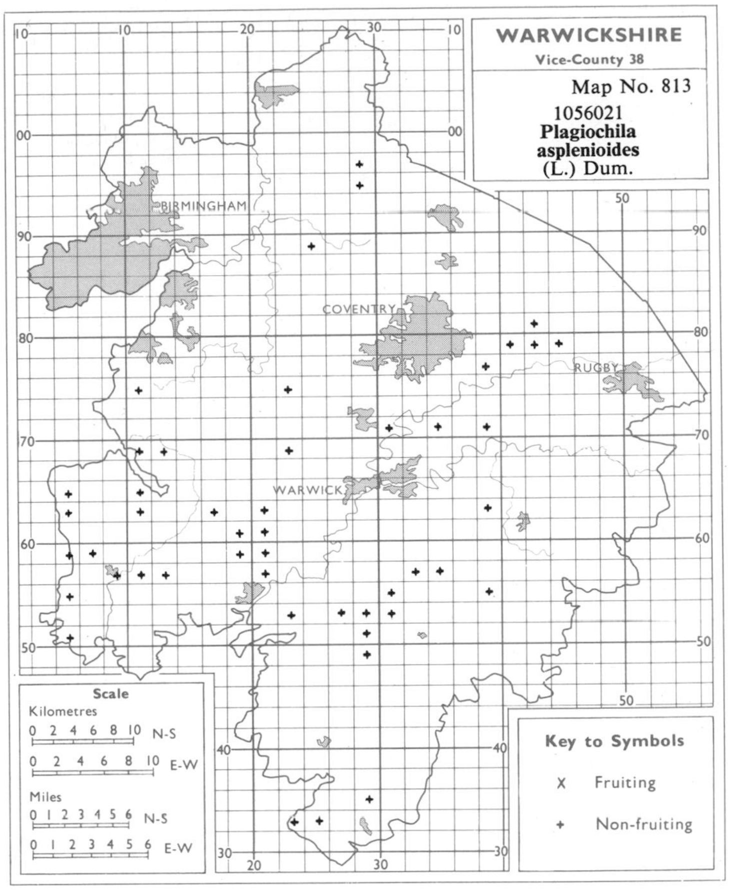
WARWICKSHIRE
Vice-County 38
Map No. 813
1056021
Plagiochila
asplenioides
(L.) Dum.
BIRMINGHAM
COVENTRY
RUGBY
WARWICK
Scale
Kilometres
0 2 4 6 8 10 N-S
0 2 4 6 8 10 E-W
Miles
0 1 2 3 4 5 6 N-S
0 1 2 3 4 5 6 E-W
Key to Symbols
X Fruiting
+ Non-fruiting

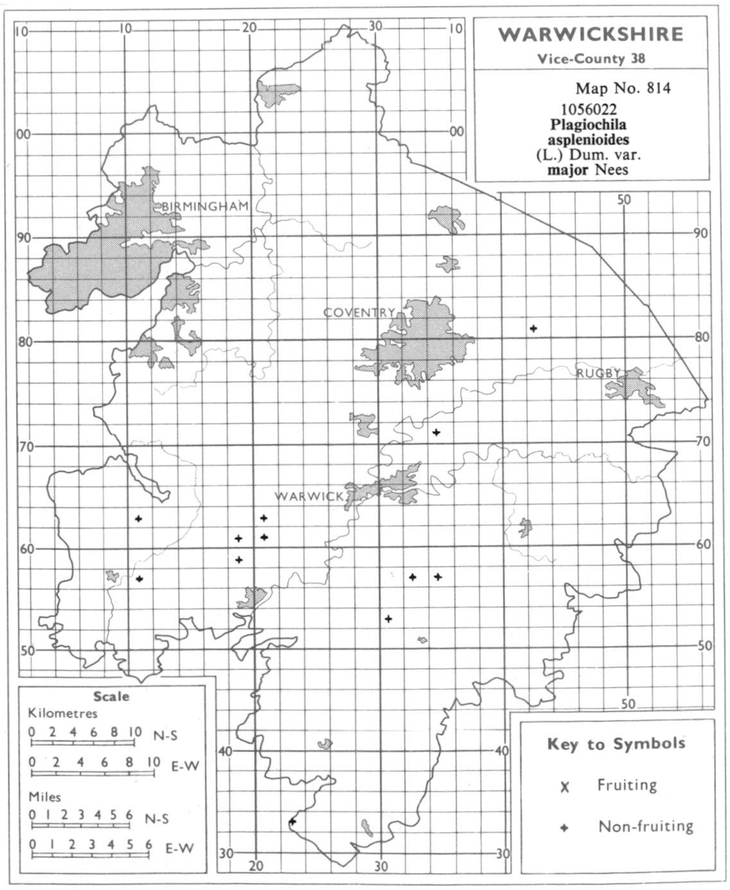
WARWICKSHIRE
Vice-County 38
Map No. 814
1056022
Plagiochila
asplenioides
(L.) Dum. var.
major Nees
BIRMINGHAM
COVENTRY
RUGBY
WARWICK
Scale
Kilometres
0 2 4 6 8 10 N-S
0 2 4 6 8 10 E-W
Miles
0 1 2 3 4 5 6 N-S
0 1 2 3 4 5 6 E-W
Key to Symbols
X Fruiting
+ Non-fruiting

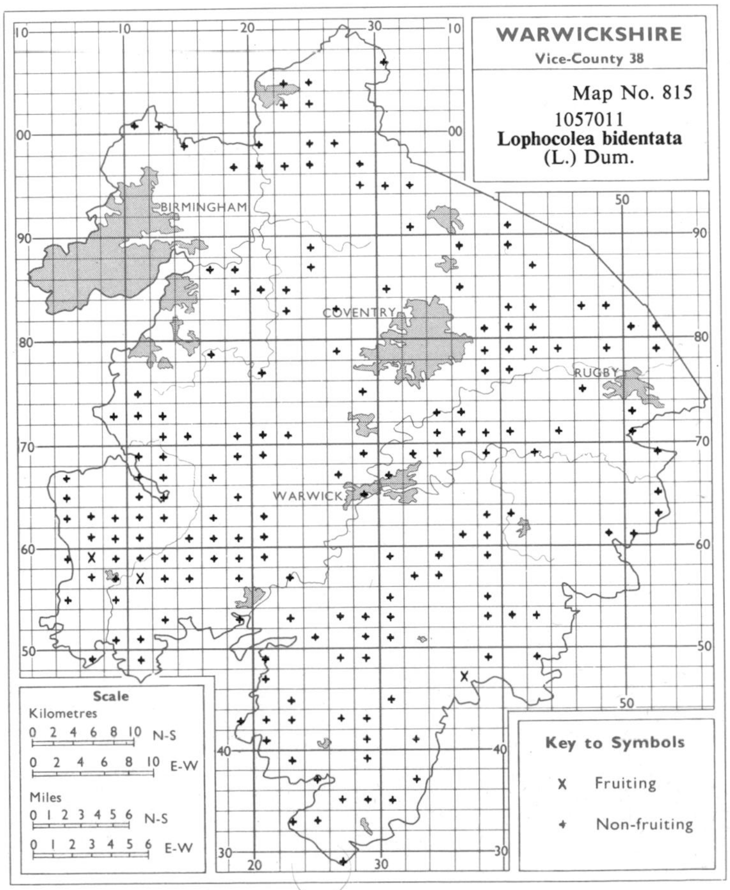
WARWICKSHIRE
Vice-County 38
Map No. 815
1057011
Lophocolea bidentata
(L.) Dum.
BIRMINGHAM
COVENTRY
RUGBY
WARWICK
Scale
Kilometres
0 2 4 6 8 10 N-S
0 2 4 6 8 10 E-W
Miles
0 1 2 3 4 5 6 N-S
0 1 2 3 4 5 6 E-W
Key to Symbols
X Fruiting
+ Non-fruiting

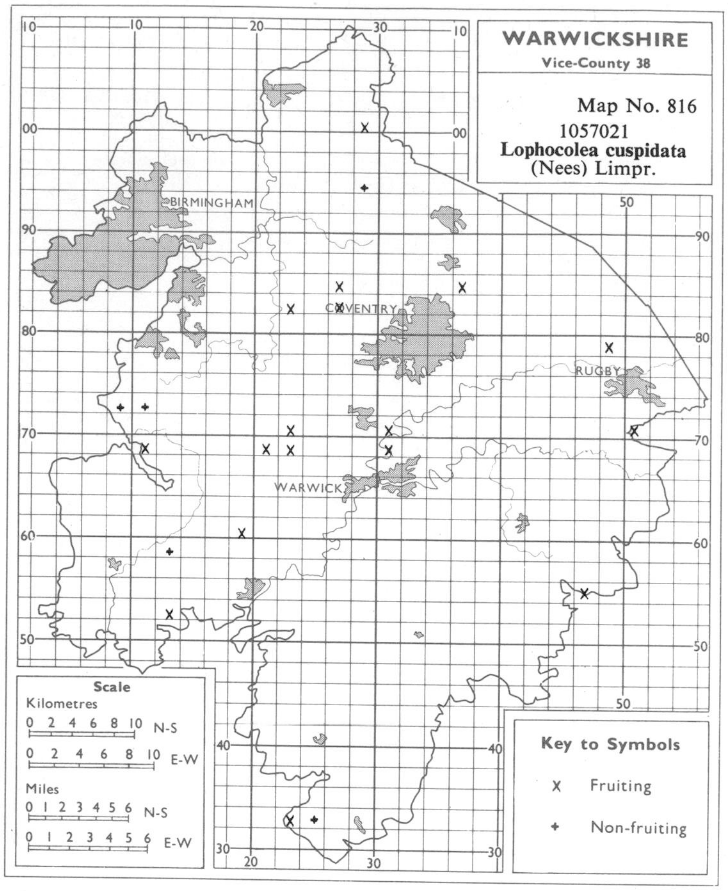
WARWICKSHIRE
Vice-County 38
Map No. 816
1057021
Lophocolea cuspidata
(Nees) Limpr.
BIRMINGHAM
COVENTRY
RUGBY
WARWICK
Scale
Kilometres
Miles
Key to Symbols
X Fruiting
+ Non-fruiting

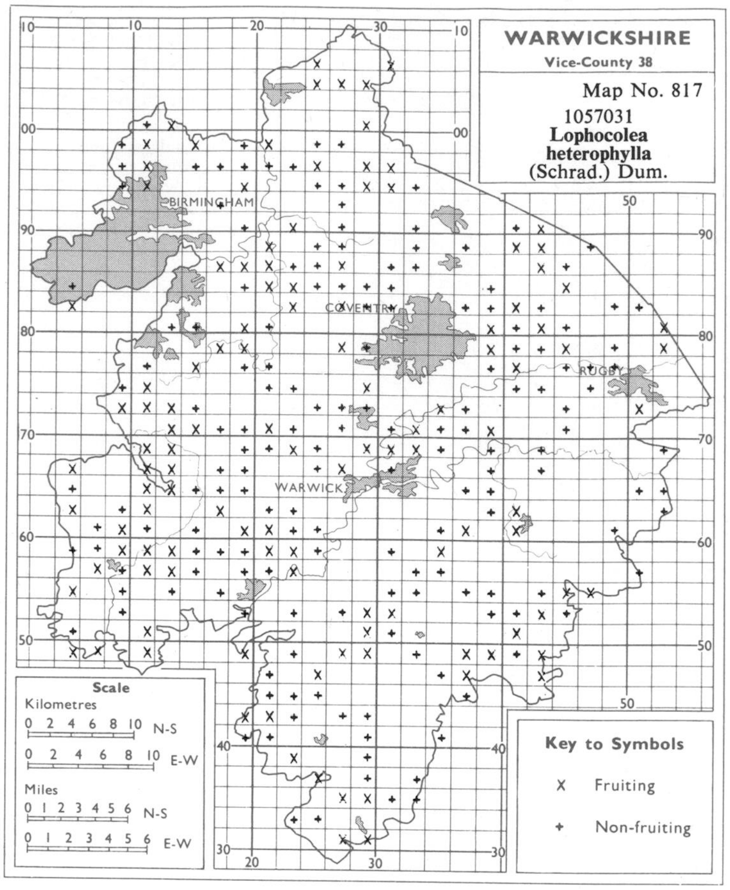
WARWICKSHIRE
Vice-County 38
Map No. 817
1057031
Lophocolea
heterophylla
(Schrad.) Dum.
BIRMINGHAM
COVENTRY
RUGBY
WARWICK
Scale
Kilometres
Miles
Key to Symbols
X Fruiting
+ Non-fruiting

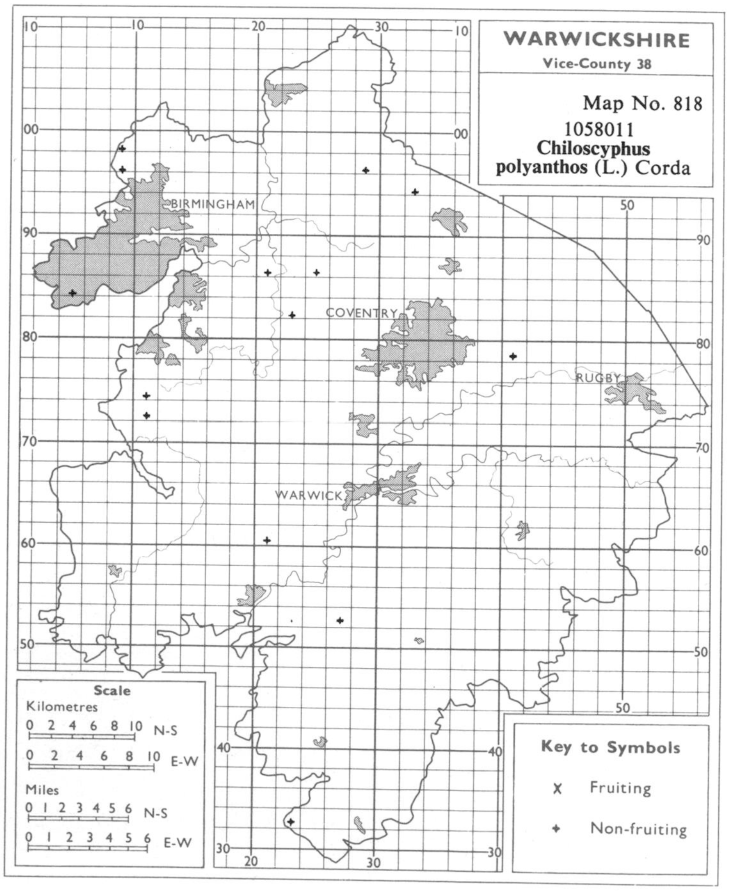
WARWICKSHIRE
Vice-County 38
Map No. 818
1058011
Chiloscyphus
polyanthos (L.) Corda
BIRMINGHAM
COVENTRY
RUGBY
WARWICK
Scale
Kilometres
Miles
Key to Symbols
X Fruiting
+ Non-fruiting

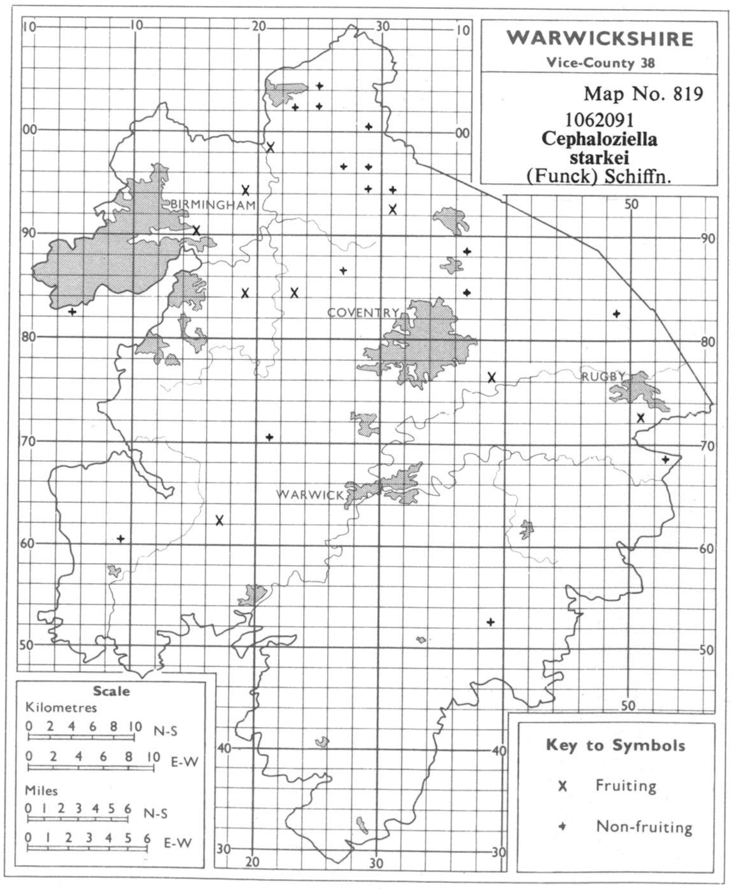
WARWICKSHIRE
Vice-County 38
Map No. 819
1062091
Cephaloziella
starkei
(Funck) Schiffn.
BIRMINGHAM
COVENTRY
RUGBY
WARWICK
Scale
Kilometres
Miles
Key to Symbols
X Fruiting
+ Non-fruiting

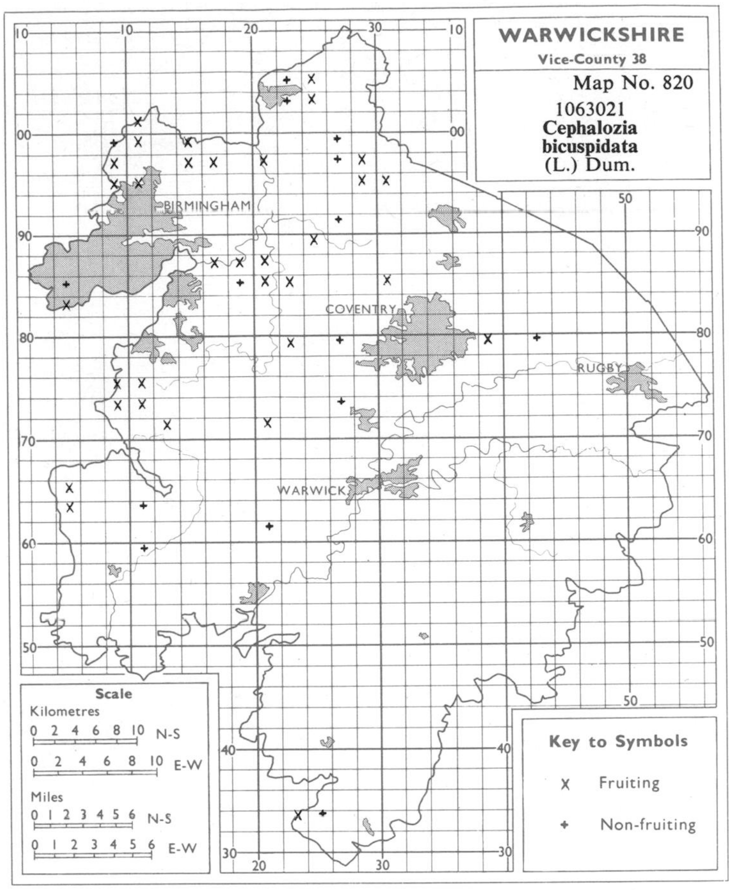
WARWICKSHIRE
Vice-County 38
Map No. 820
1063021
Cephalozia bicuspidata
(L.) Dum.
BIRMINGHAM
COVENTRY
RUGBY
WARWICK
Scale
Kilometres
0 2 4 6 8 10 N-S
0 2 4 6 8 10 E-W
Miles
0 1 2 3 4 5 6 N-S
0 1 2 3 4 5 6 E-W
Key to Symbols
X Fruiting
+ Non-fruiting

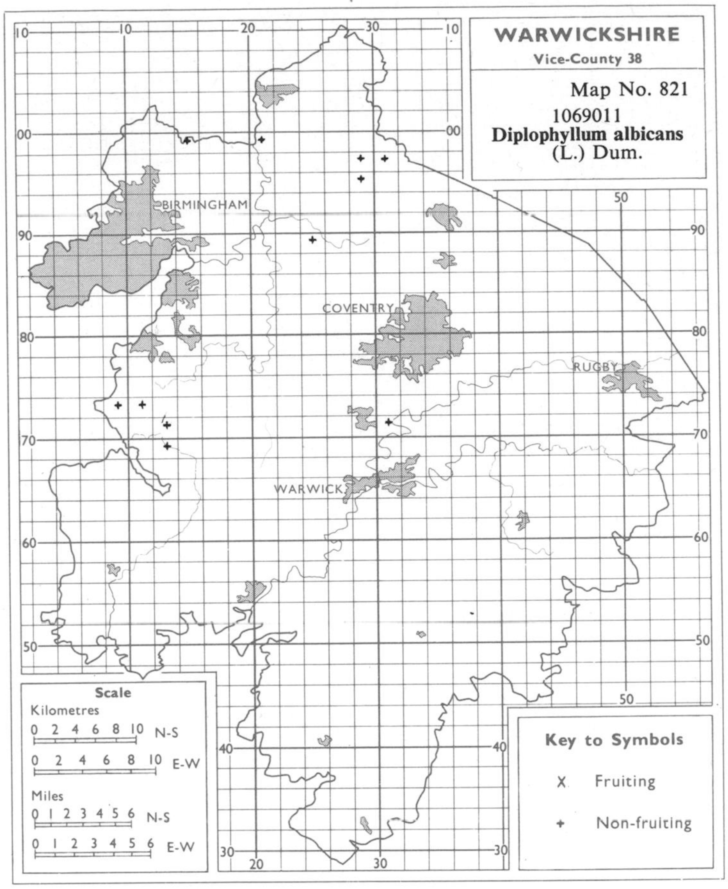
WARWICKSHIRE
Vice-County 38
Map No. 821
1069011
Diplophyllum albicans
(L.) Dum.
BIRMINGHAM
COVENTRY
RUGBY
WARWICK
Scale
Kilometres
0 2 4 6 8 10 N-S
0 2 4 6 8 10 E-W
Miles
0 1 2 3 4 5 6 N-S
0 1 2 3 4 5 6 E-W
Key to Symbols
X Fruiting
+ Non-fruiting

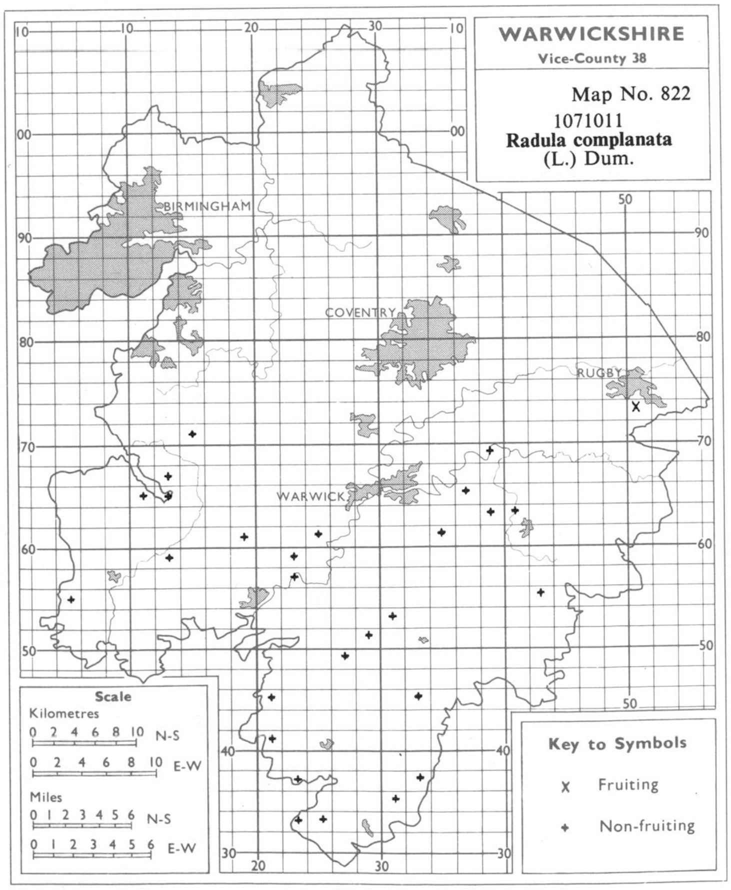
WARWICKSHIRE
Vice-County 38
Map No. 822
1071011
Radula complanata
(L.) Dum.
BIRMINGHAM
COVENTRY
RUGBY
WARWICK
Scale
Kilometres
0 2 4 6 8 10 N-S
0 2 4 6 8 10 E-W
Miles
0 1 2 3 4 5 6 N-S
0 1 2 3 4 5 6 E-W
Key to Symbols
X Fruiting
+ Non-fruiting

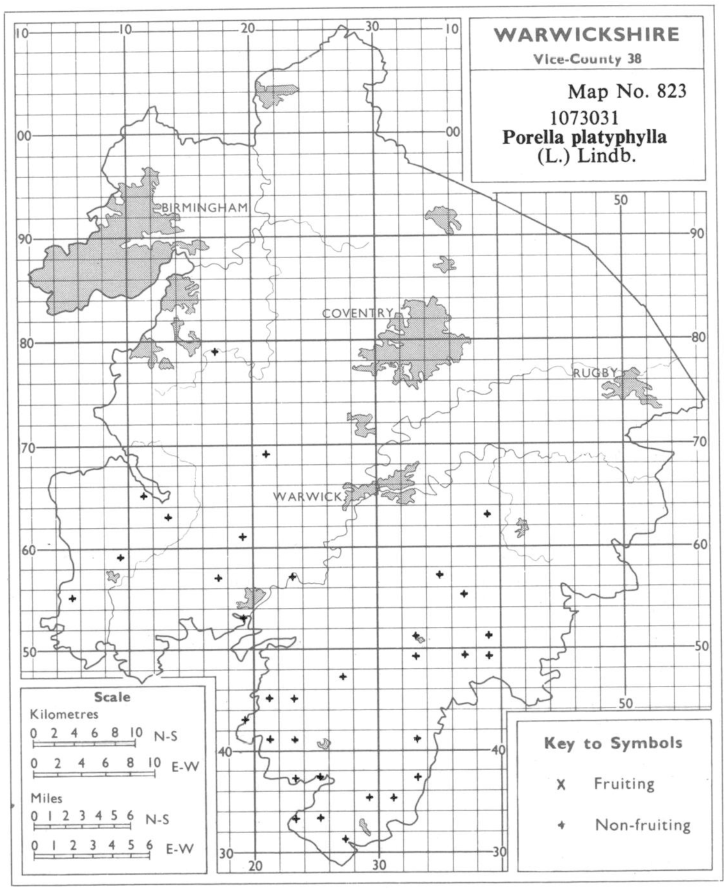
WARWICKSHIRE
Vice-County 38
Map No. 823
1073031
Porella platyphylla
(L.) Lindb.
BIRMINGHAM
COVENTRY
RUGBY
WARWICK
Scale
Kilometres
0 2 4 6 8 10 N-S
0 2 4 6 8 10 E-W
Miles
0 1 2 3 4 5 6 N-S
0 1 2 3 4 5 6 E-W
Key to Symbols
X Fruiting
+ Non-fruiting

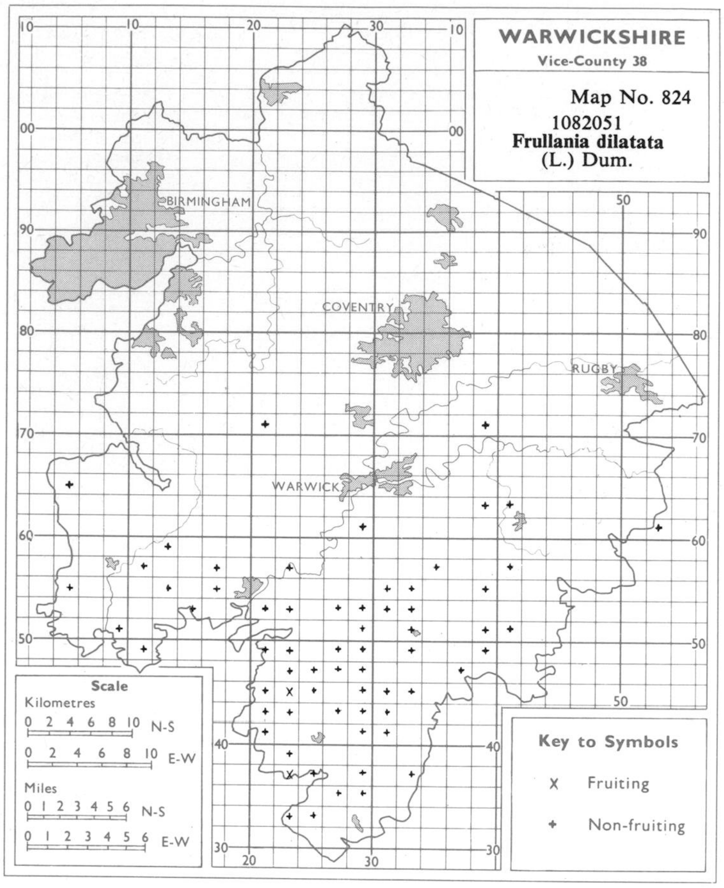
WARWICKSHIRE
Vice-County 38
Map No. 824
1082051
Frullania dilatata
(L.) Dum.
BIRMINGHAM
COVENTRY
RUGBY
WARWICK
Scale
Kilometres
0 2 4 6 8 10 N-S
0 2 4 6 8 10 E-W
Miles
0 1 2 3 4 5 6 N-S
0 1 2 3 4 5 6 E-W
Key to Symbols
X Fruiting
+ Non-fruiting

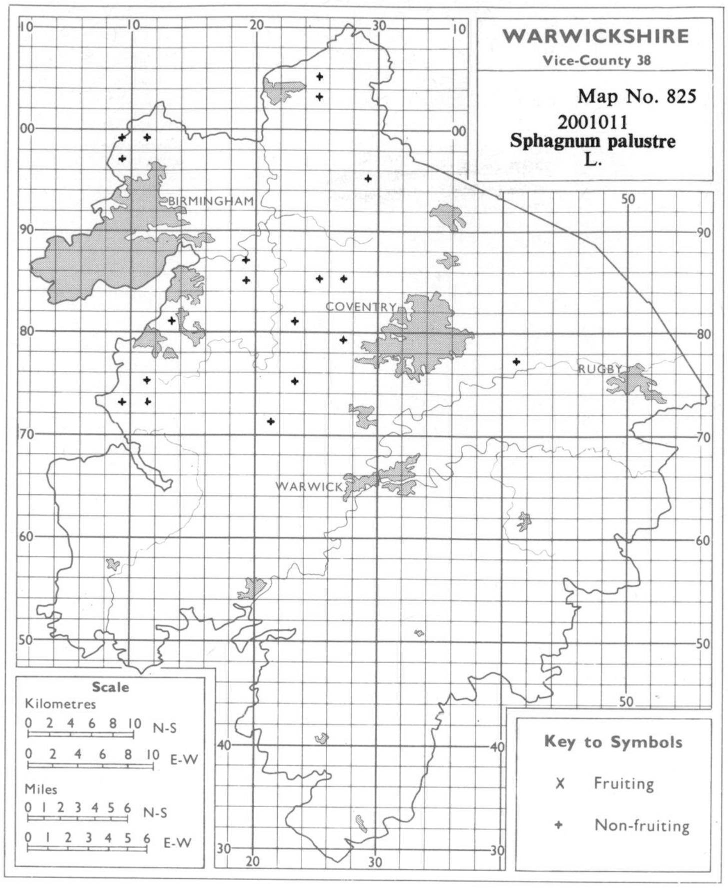
WARWICKSHIRE
Vice-County 38
Map No. 825
2001011
Sphagnum palustre
L.
BIRMINGHAM
COVENTRY
RUGBY
WARWICK
Scale
Kilometres
0 2 4 6 8 10 N-S
0 2 4 6 8 10 E-W
Miles
0 1 2 3 4 5 6 N-S
0 1 2 3 4 5 6 E-W
Key to Symbols
X Fruiting
+ Non-fruiting

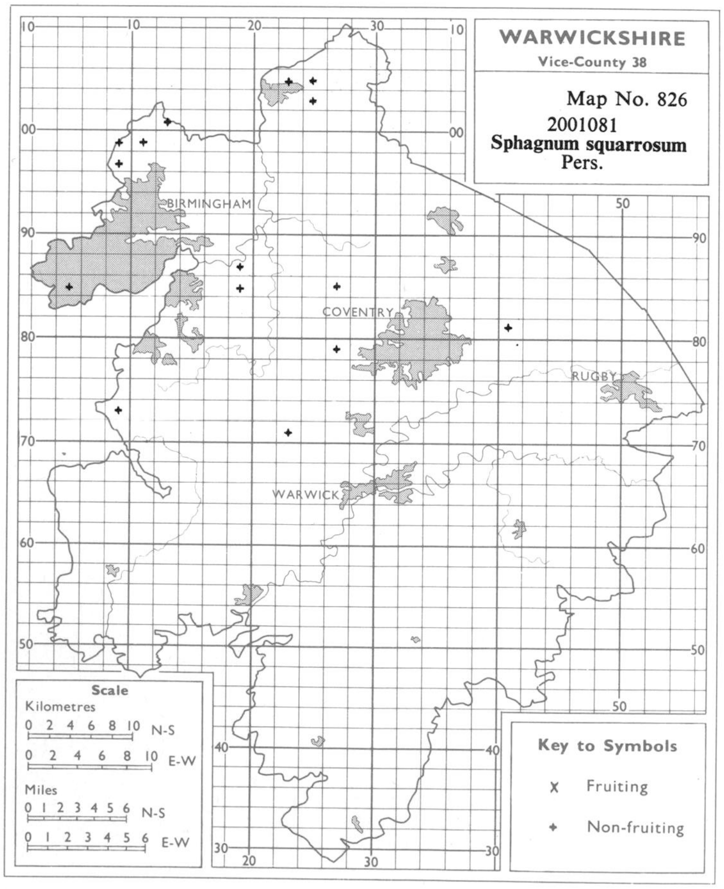
WARWICKSHIRE
Vice-County 38
Map No. 826
2001081
Sphagnum squarrosum
Pers.
BIRMINGHAM
COVENTRY
RUGBY
WARWICK
Scale
Kilometres
0 2 4 6 8 10 N-S
0 2 4 6 8 10 E-W
Miles
0 1 2 3 4 5 6 N-S
0 1 2 3 4 5 6 E-W
Key to Symbols
X Fruiting
+ Non-fruiting

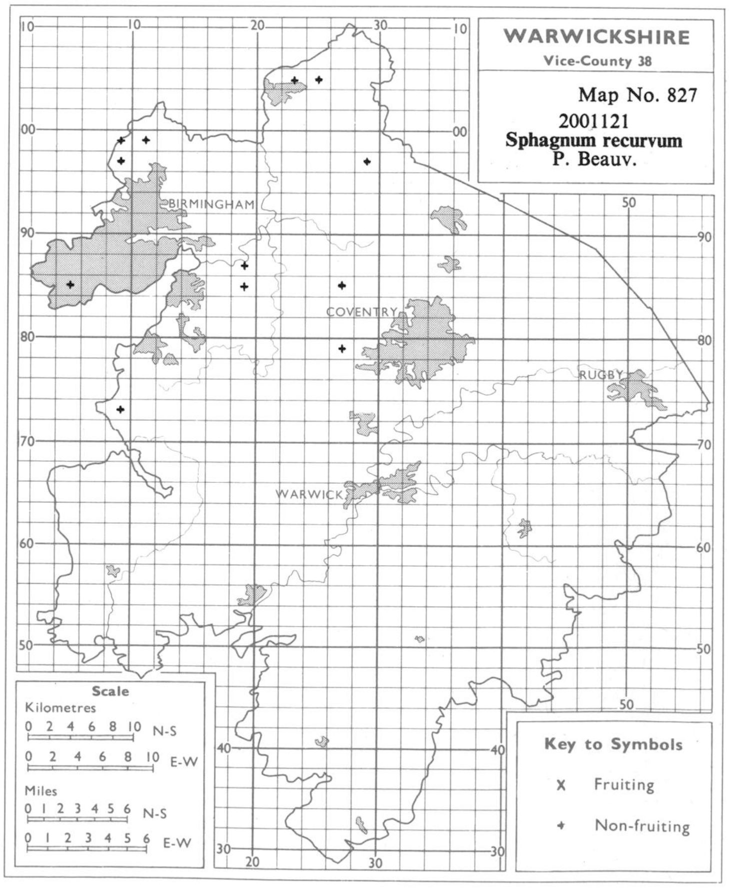
WARWICKSHIRE
Vice-County 38
Map No. 827
2001121
Sphagnum recurvum
P. Beauv.
BIRMINGHAM
COVENTRY
RUGBY
WARWICK
Scale
Kilometres
0 2 4 6 8 10 N-S
0 2 4 6 8 10 E-W
Miles
0 1 2 3 4 5 6 N-S
0 1 2 3 4 5 6 E-W
Key to Symbols
X Fruiting
+ Non-fruiting

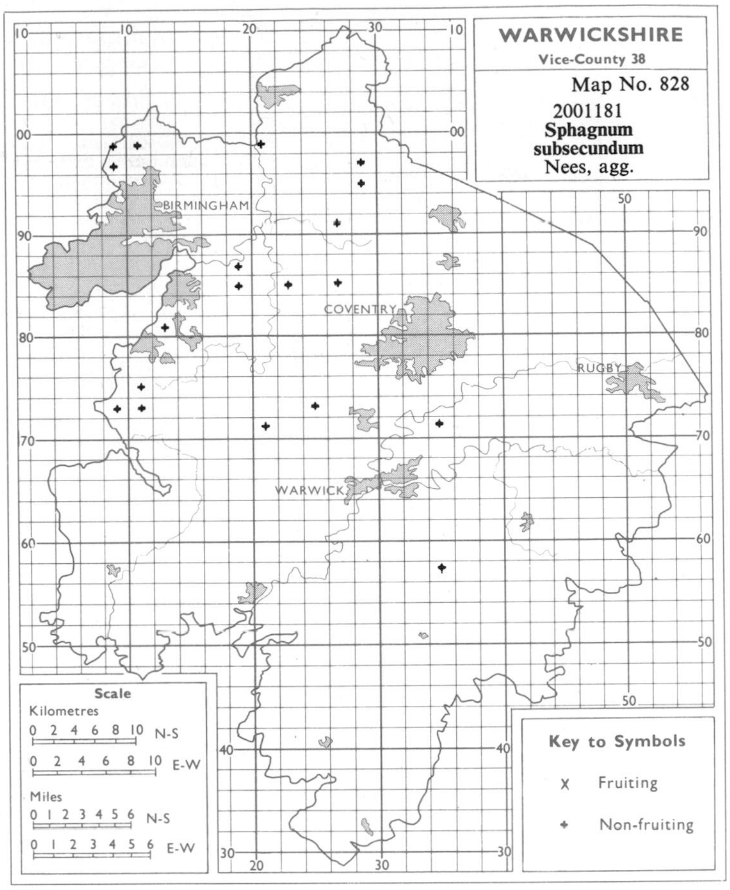
WARWICKSHIRE
Vice-County 38
Map No. 828
2001181
Sphagnum subsecundum
Nees, agg.
BIRMINGHAM
COVENTRY
RUGBY
WARWICK
Scale
Kilometres
0 2 4 6 8 10 N-S
0 2 4 6 8 10 E-W
Miles
0 1 2 3 4 5 6 N-S
0 1 2 3 4 5 6 E-W
Key to Symbols
X Fruiting
+ Non-fruiting

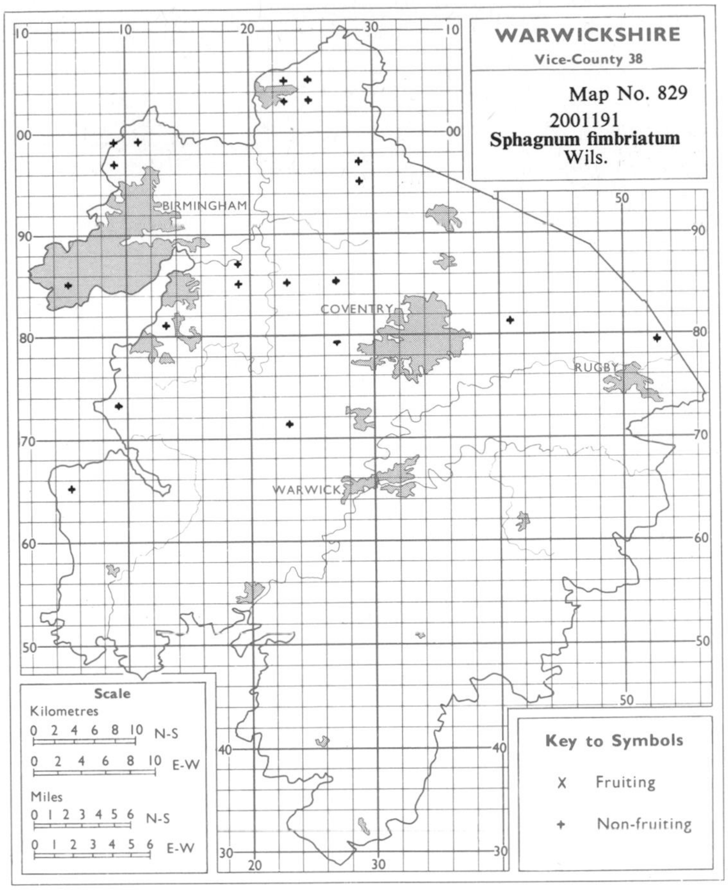
WARWICKSHIRE
Vice-County 38
Map No. 829
2001191
Sphagnum fimbriatum
Wils.
BIRMINGHAM
COVENTRY
RUGBY
WARWICK
Scale
Kilometres
0 2 4 6 8 10 N-S
0 2 4 6 8 10 E-W
Miles
0 1 2 3 4 5 6 N-S
0 1 2 3 4 5 6 E-W
Key to Symbols
X Fruiting
+ Non-fruiting

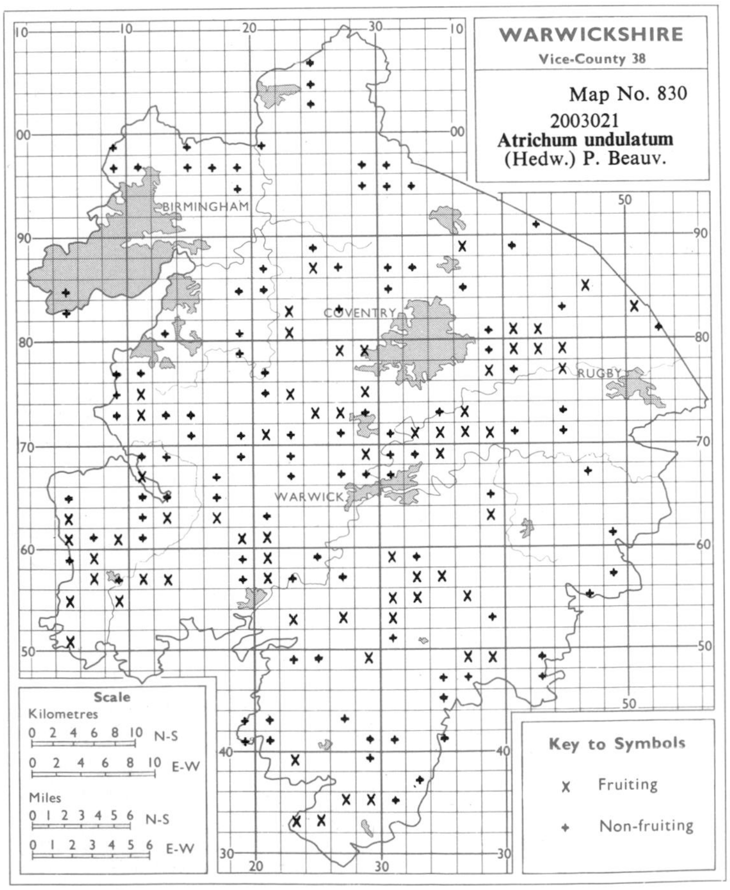
WARWICKSHIRE
Vice-County 38
Map No. 830
2003021
Atrichum undulatum
(Hedw.) P. Beauv.
BIRMINGHAM
COVENTRY
RUGBY
WARWICK
Scale
Kilometres
0 2 4 6 8 10 N-S
0 2 4 6 8 10 E-W
Miles
0 1 2 3 4 5 6 N-S
0 1 2 3 4 5 6 E-W
Key to Symbols
X Fruiting
+ Non-fruiting

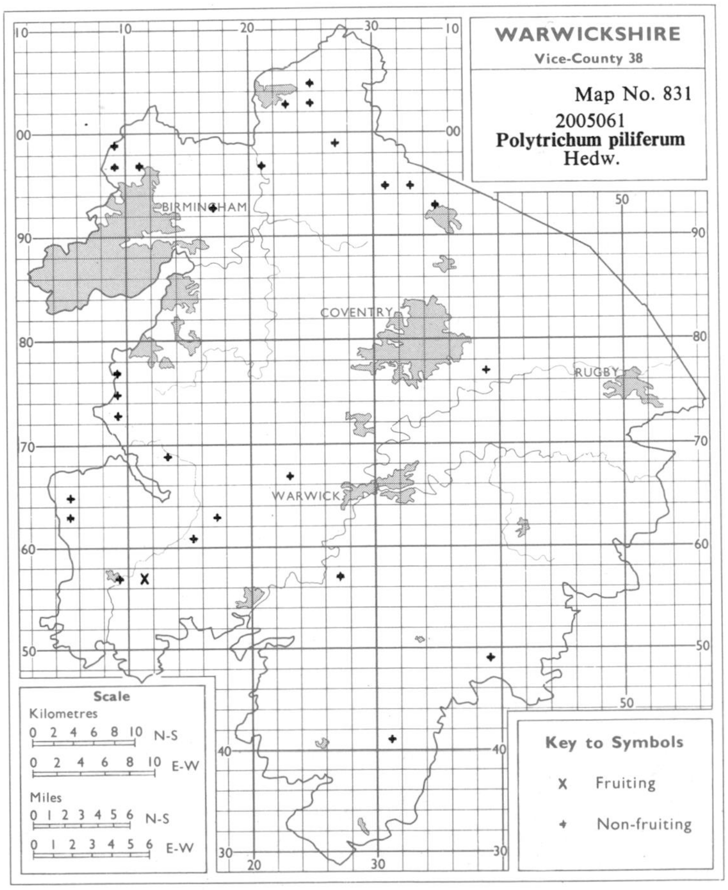
WARWICKSHIRE
Vice-County 38
Map No. 831
2005061
Polytrichum piliferum
Hedw.
BIRMINGHAM
COVENTRY
RUGBY
WARWICK
Scale
Kilometres
0 2 4 6 8 10 N-S
0 2 4 6 8 10 E-W
Miles
0 1 2 3 4 5 6 N-S
0 1 2 3 4 5 6 E-W
Key to Symbols
X Fruiting
+ Non-fruiting

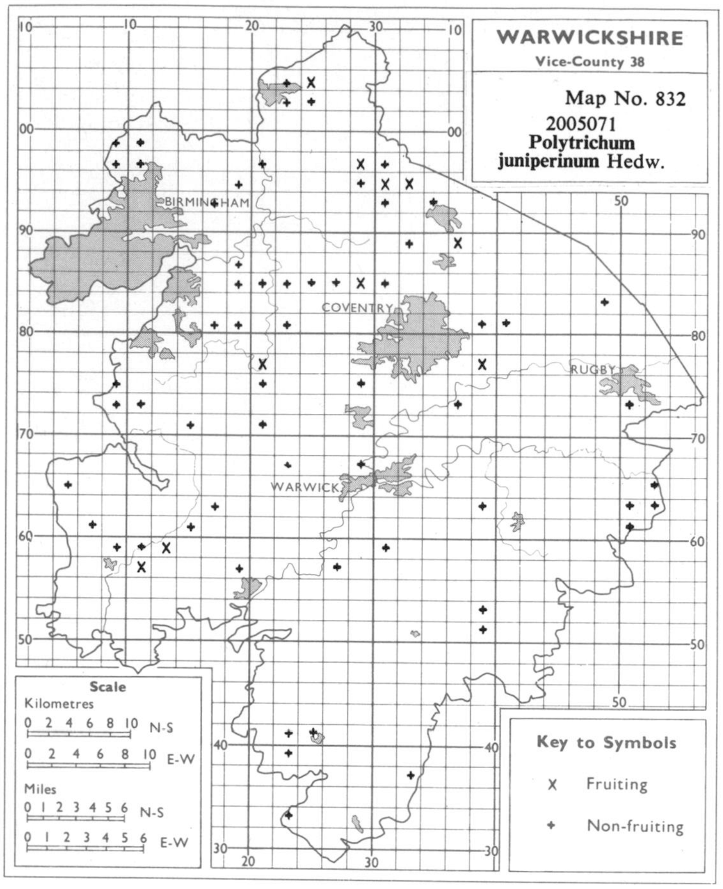
WARWICKSHIRE
Vice-County 38
Map No. 832
2005071
Polytrichum juniperinum Hedw.
BIRMINGHAM
COVENTRY
RUGBY
WARWICK
Scale
Kilometres
N-S
E-W
Miles
N-S
E-W
Key to Symbols
X Fruiting
+ Non-fruiting

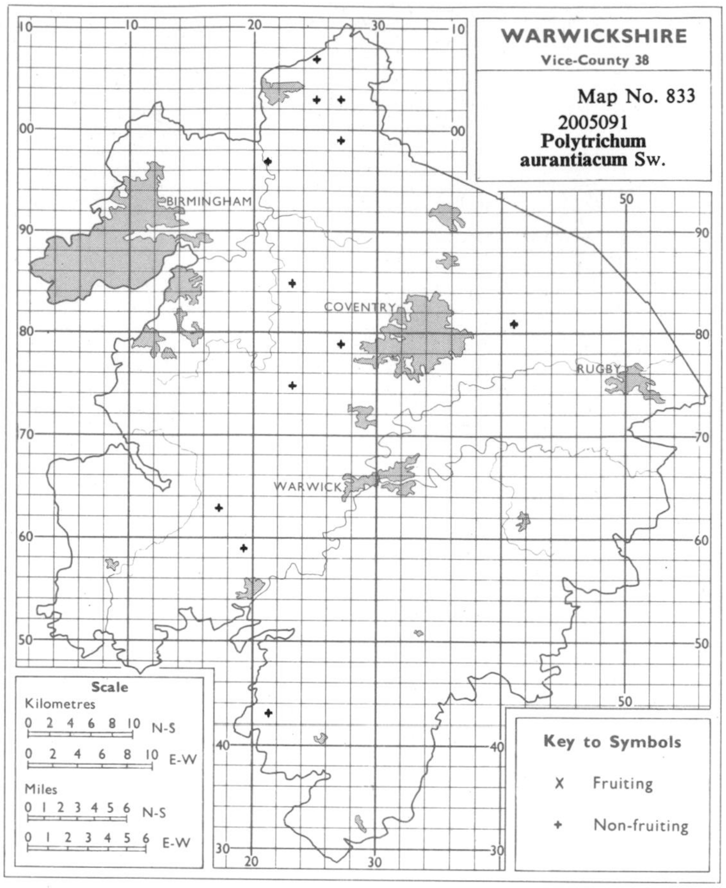
WARWICKSHIRE
Vice-County 38
Map No. 833
2005091
Polytrichum aurantiacum Sw.
BIRMINGHAM
COVENTRY
RUGBY
WARWICK
Scale
Kilometres
N-S
E-W
Miles
N-S
E-W
Key to Symbols
X Fruiting
+ Non-fruiting

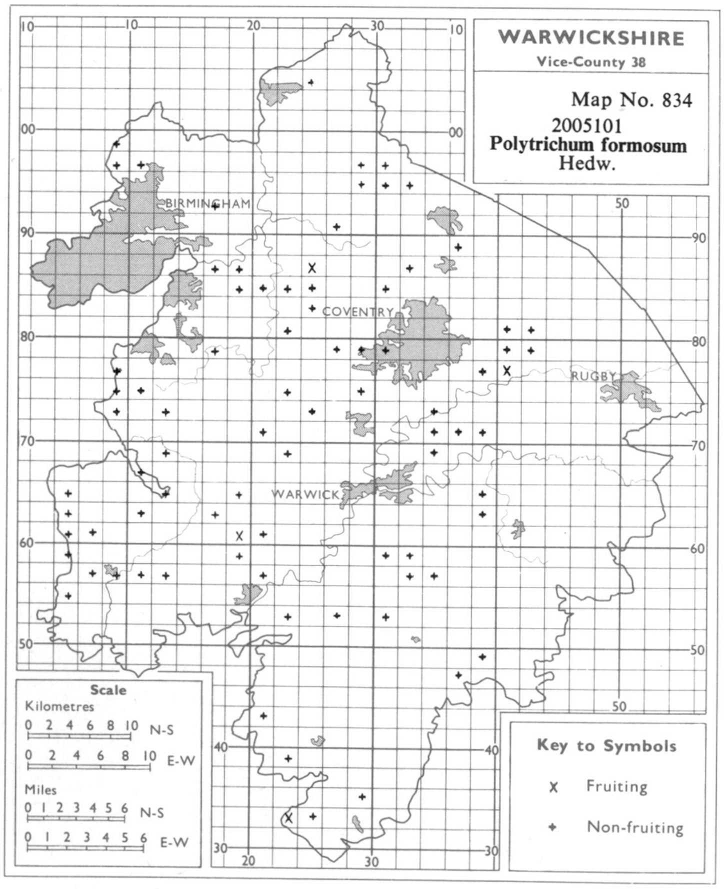
WARWICKSHIRE
Vice-County 38
Map No. 834
2005101
Polytrichum formosum Hedw.
BIRMINGHAM
COVENTRY
RUGBY
WARWICK
Scale
Kilometres
N-S
E-W
Miles
N-S
E-W
Key to Symbols
X Fruiting
+ Non-fruiting

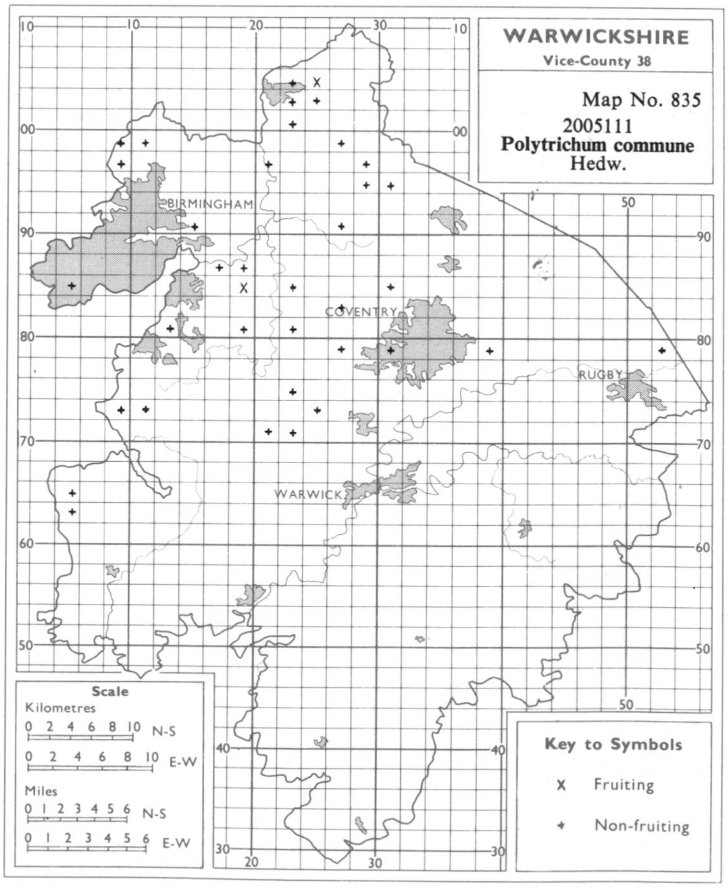
WARWICKSHIRE
Vice-County 38
Map No. 835
2005111
Polytrichum commune Hedw.
BIRMINGHAM
COVENTRY
RUGBY
WARWICK
Scale
Kilometres
N-S
E-W
Miles
N-S
E-W
Key to Symbols
X Fruiting
+ Non-fruiting

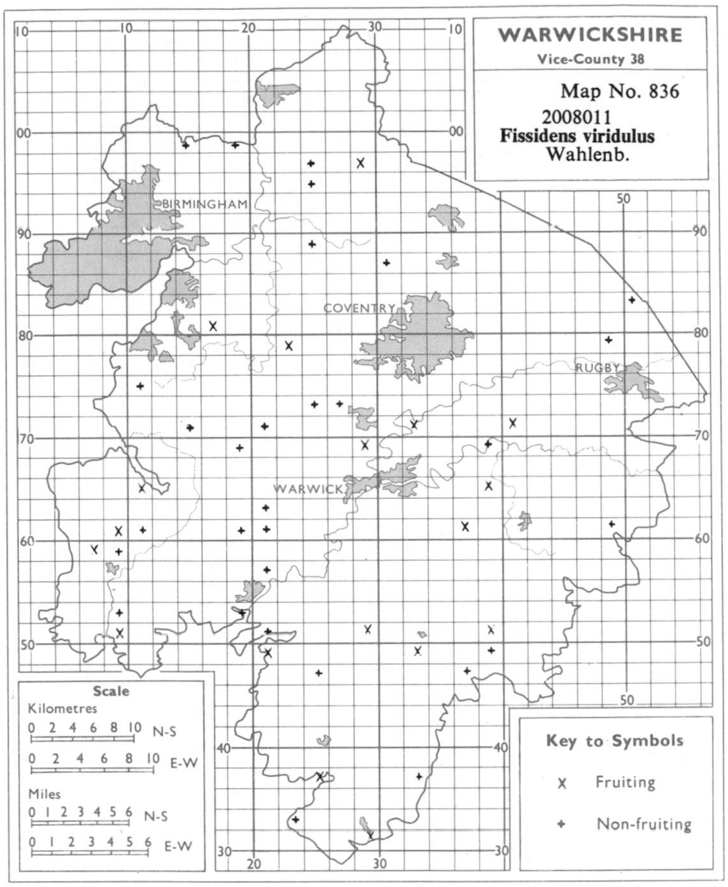
WARWICKSHIRE
Vice-County 38
Map No. 836
2008011
Fissidens viridulus
Wahlenb.
BIRMINGHAM
COVENTRY
RUGBY
WARWICK
Scale
Kilometres
N-S
E-W
Miles
Key to Symbols
X Fruiting
+ Non-fruiting

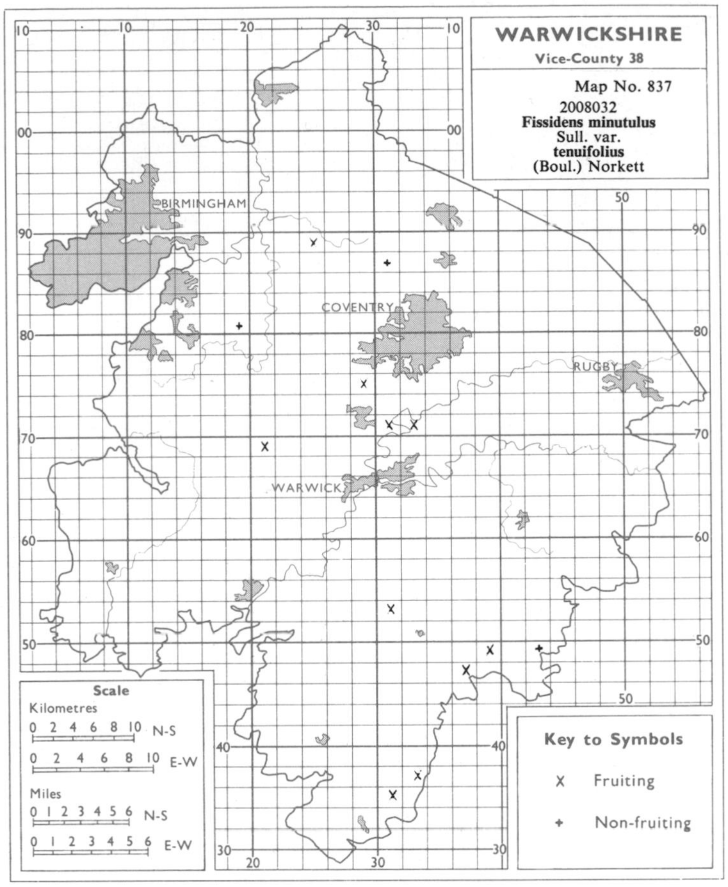
WARWICKSHIRE
Vice-County 38
Map No. 837
2008032
Fissidens minutulus
Sull. var.
tenuifolius
(Boul.) Norkett
BIRMINGHAM
COVENTRY
RUGBY
WARWICK
Scale
Kilometres
N-S
E-W
Miles
Key to Symbols
X Fruiting
+ Non-fruiting

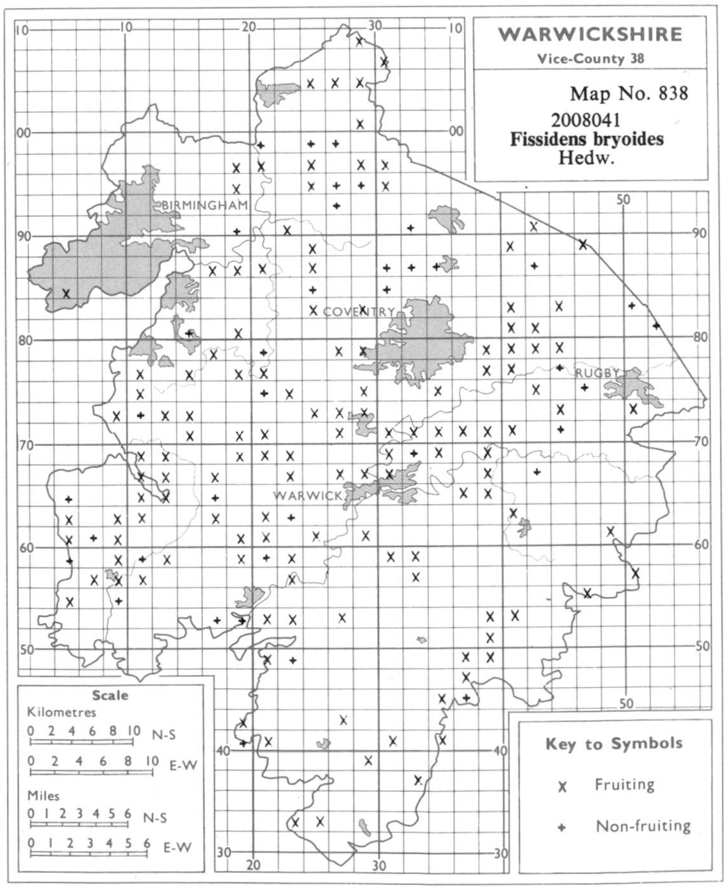
WARWICKSHIRE
Vice-County 38
Map No. 838
2008041
Fissidens bryoides
Hedw.
BIRMINGHAM
COVENTRY
RUGBY
WARWICK
Scale
Kilometres
N-S
E-W
Miles
Key to Symbols
X Fruiting
+ Non-fruiting

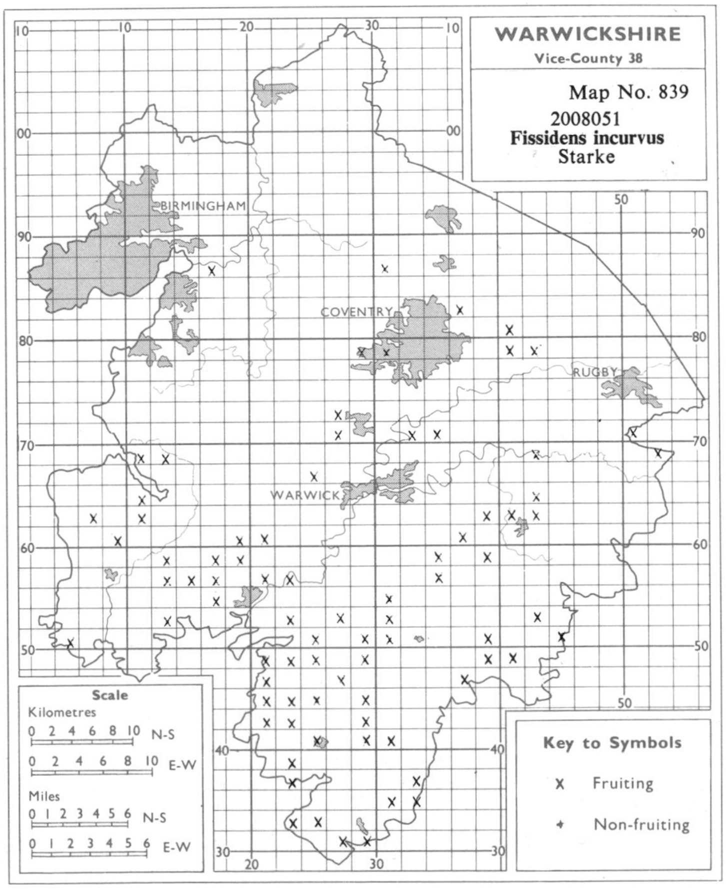
WARWICKSHIRE
Vice-County 38
Map No. 839
2008051
Fissidens incurvus
Starke
BIRMINGHAM
COVENTRY
RUGBY
WARWICK
Scale
Kilometres
N-S
E-W
Miles
Key to Symbols
X Fruiting
+ Non-fruiting

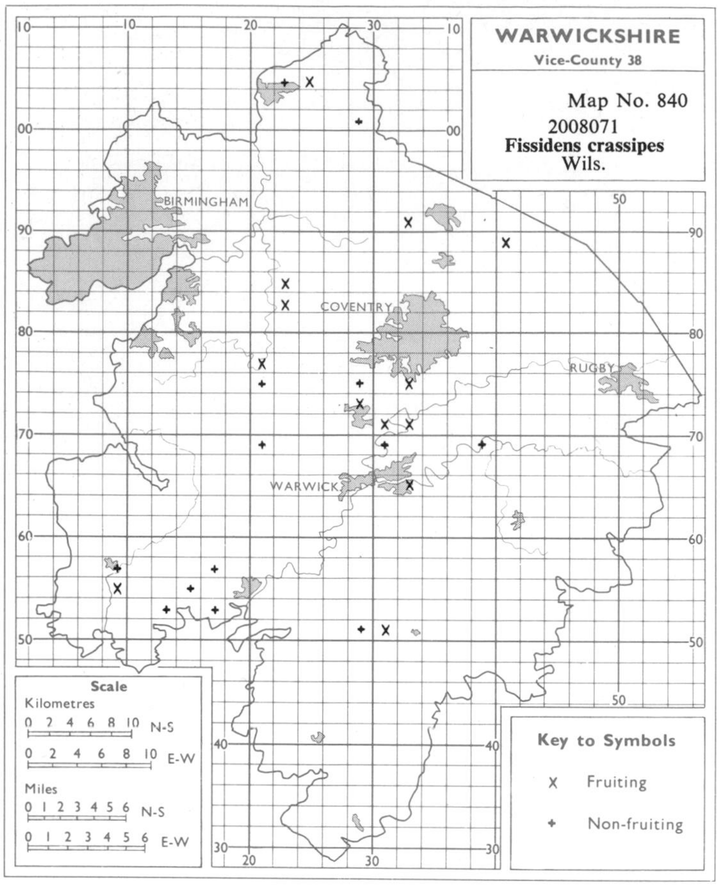
WARWICKSHIRE
Vice-County 38
Map No. 840
2008071
Fissidens crassipes
Wils.
BIRMINGHAM
COVENTRY
RUGBY
WARWICK
Scale
Kilometres
0 2 4 6 8 10 N-S
0 2 4 6 8 10 E-W
Miles
0 1 2 3 4 5 6 N-S
0 1 2 3 4 5 6 E-W
Key to Symbols
X Fruiting
+ Non-fruiting

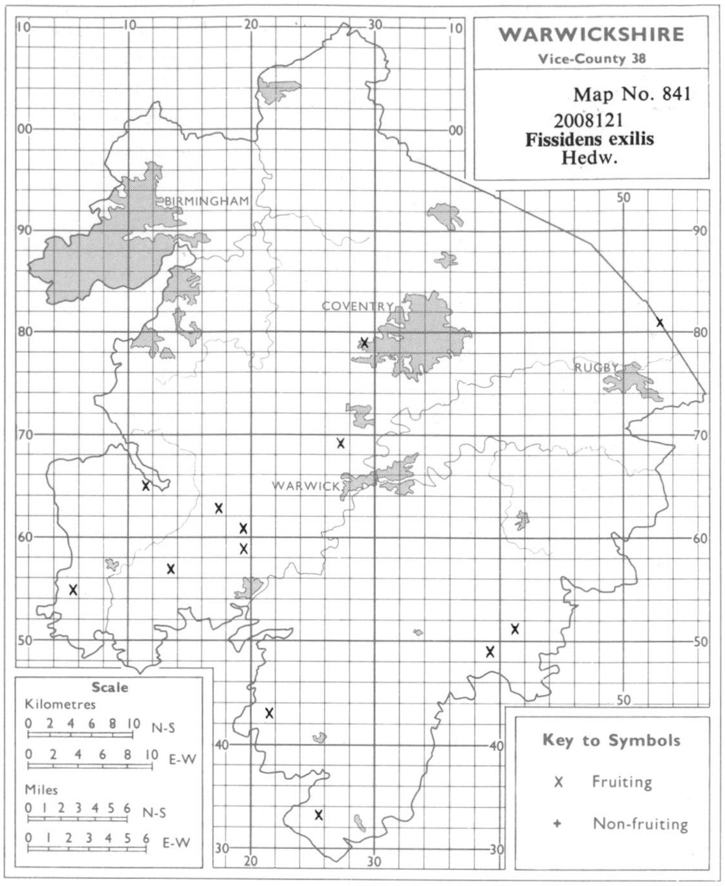
WARWICKSHIRE
Vice-County 38
Map No. 841
2008121
Fissidens exilis
Hedw.
BIRMINGHAM
COVENTRY
RUGBY
WARWICK
Scale
Kilometres
0 2 4 6 8 10 N-S
0 2 4 6 8 10 E-W
Miles
0 1 2 3 4 5 6 N-S
0 1 2 3 4 5 6 E-W
Key to Symbols
X Fruiting
+ Non-fruiting

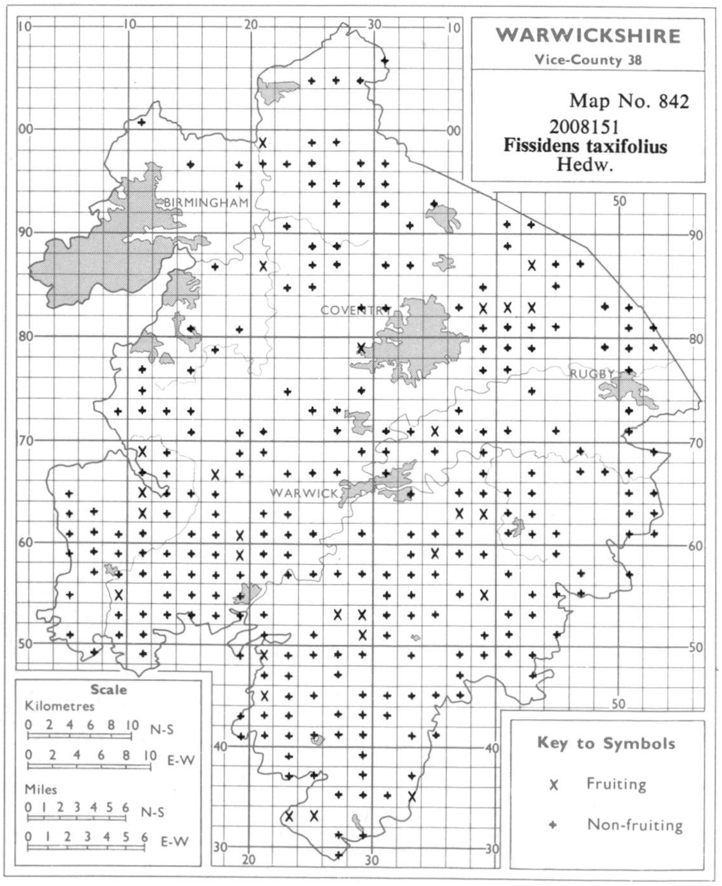
WARWICKSHIRE
Vice-County 38
Map No. 842
2008151
Fissidens taxifolius
Hedw.
BIRMINGHAM
COVENTRY
RUGBY
WARWICK
Scale
Kilometres
0 2 4 6 8 10 N-S
0 2 4 6 8 10 E-W
Miles
0 1 2 3 4 5 6 N-S
0 1 2 3 4 5 6 E-W
Key to Symbols
X Fruiting
+ Non-fruiting

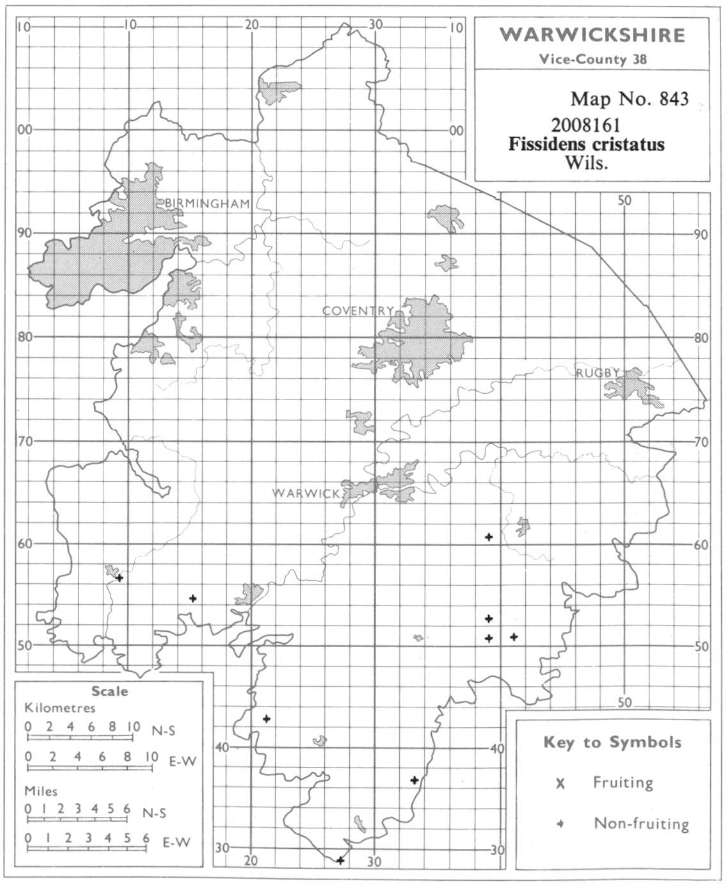
WARWICKSHIRE
Vice-County 38
Map No. 843
2008161
Fissidens cristatus
Wils.
BIRMINGHAM
COVENTRY
RUGBY
WARWICK
Scale
Kilometres
0 2 4 6 8 10 N-S
0 2 4 6 8 10 E-W
Miles
0 1 2 3 4 5 6 N-S
0 1 2 3 4 5 6 E-W
Key to Symbols
X Fruiting
+ Non-fruiting

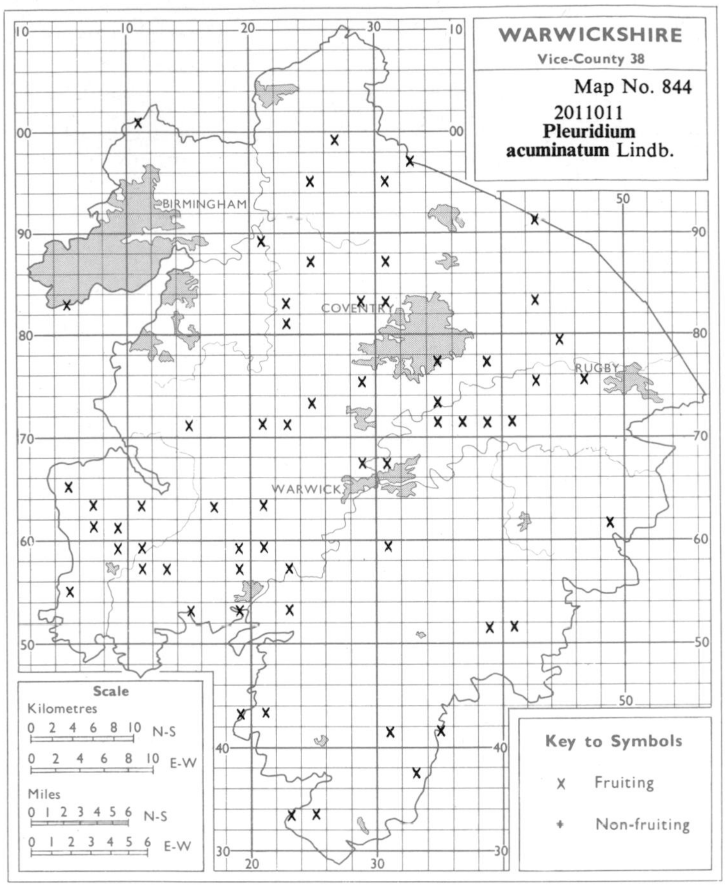
WARWICKSHIRE
Vice-County 38
Map No. 844
2011011
Pleuridium acuminatum Lindb.
BIRMINGHAM
COVENTRY
RUGBY
WARWICK
Scale
Kilometres
0 2 4 6 8 10 N-S
0 2 4 6 8 10 E-W
Miles
0 1 2 3 4 5 6 N-S
0 1 2 3 4 5 6 E-W
Key to Symbols
X Fruiting
+ Non-fruiting

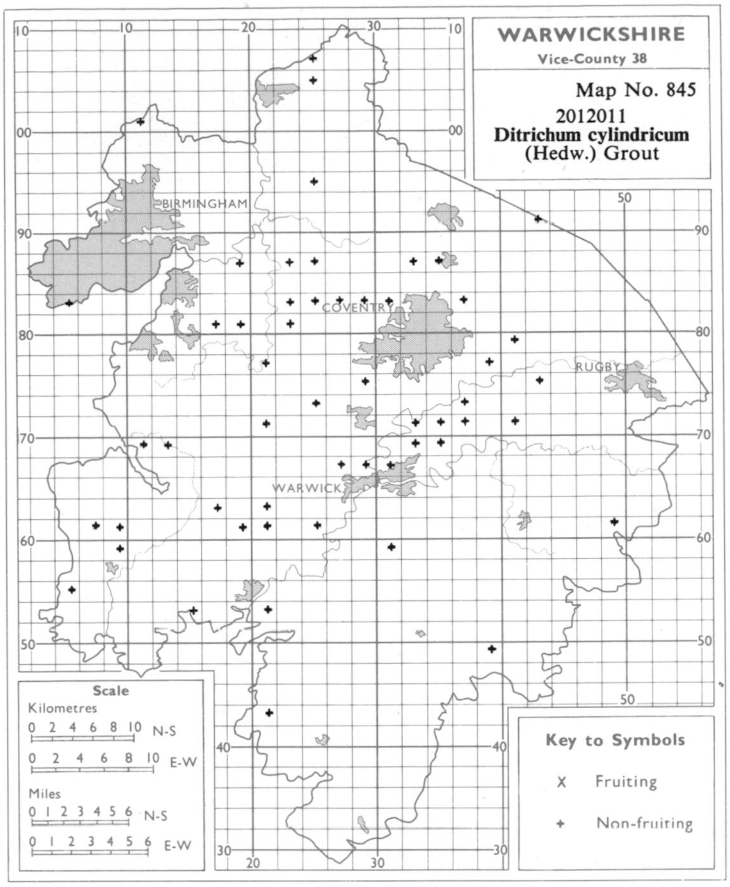
WARWICKSHIRE
Vice-County 38
Map No. 845
2012011
Ditrichum cylindricum (Hedw.) Grout
BIRMINGHAM
COVENTRY
RUGBY
WARWICK
Scale
Kilometres
0 2 4 6 8 10 N-S
0 2 4 6 8 10 E-W
Miles
0 1 2 3 4 5 6 N-S
0 1 2 3 4 5 6 E-W
Key to Symbols
X Fruiting
+ Non-fruiting

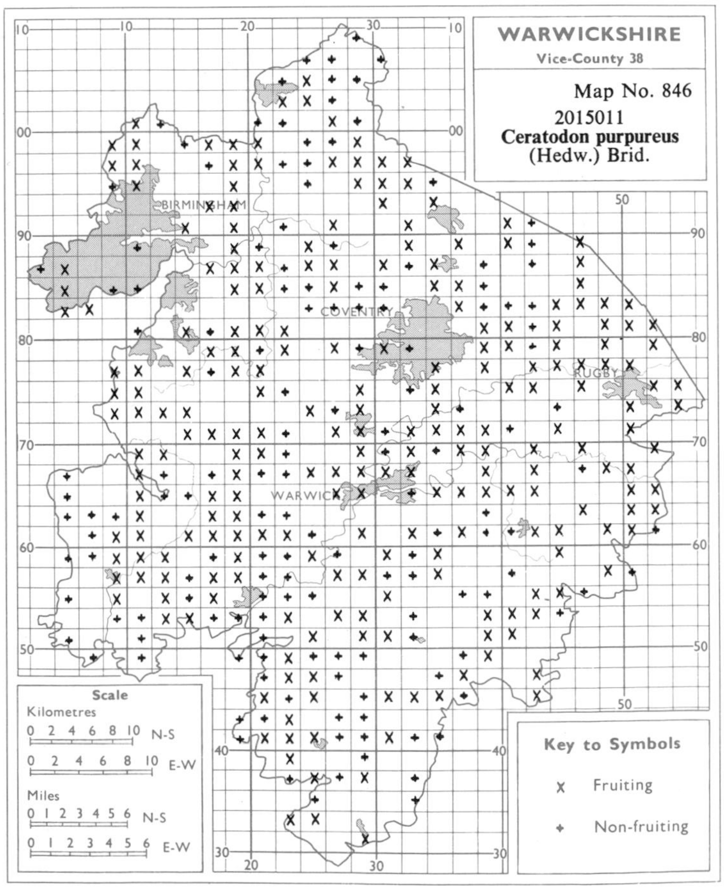
WARWICKSHIRE
Vice-County 38
Map No. 846
2015011
Ceratodon purpureus (Hedw.) Brid.
BIRMINGHAM
COVENTRY
RUGBY
WARWICK
Scale
Kilometres
0 2 4 6 8 10 N-S
0 2 4 6 8 10 E-W
Miles
0 1 2 3 4 5 6 N-S
0 1 2 3 4 5 6 E-W
Key to Symbols
X Fruiting
+ Non-fruiting

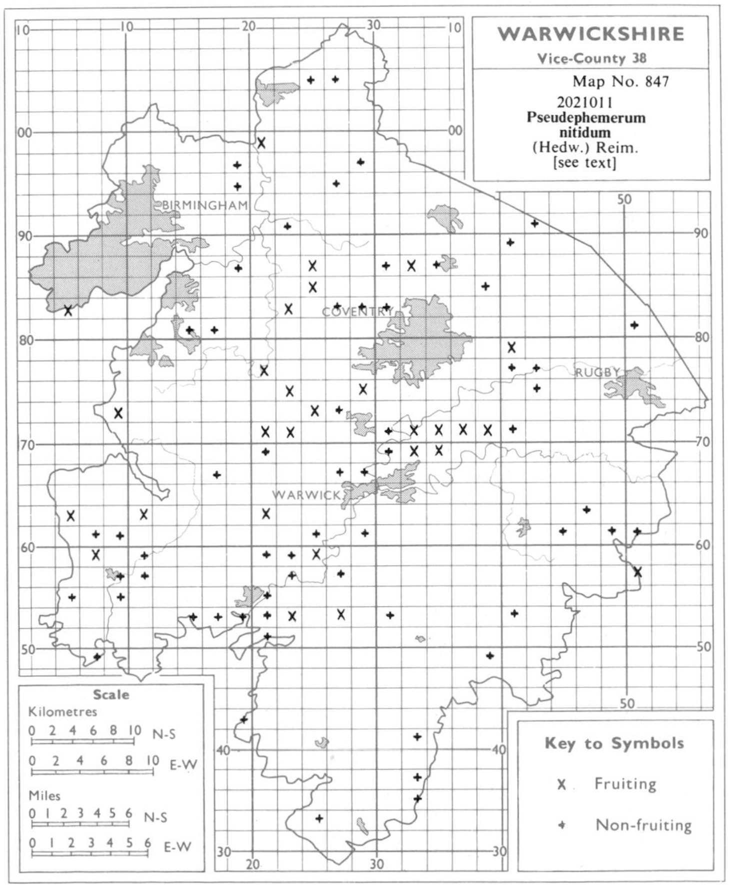
WARWICKSHIRE
Vice-County 38
Map No. 847
2021011
Pseudephemerum nitidum (Hedw.) Reim.
[see text]
BIRMINGHAM
COVENTRY
RUGBY
WARWICK
Scale
Kilometres
0 2 4 6 8 10 N-S
0 2 4 6 8 10 E-W
Miles
0 1 2 3 4 5 6 N-S
0 1 2 3 4 5 6 E-W
Key to Symbols
X Fruiting
+ Non-fruiting

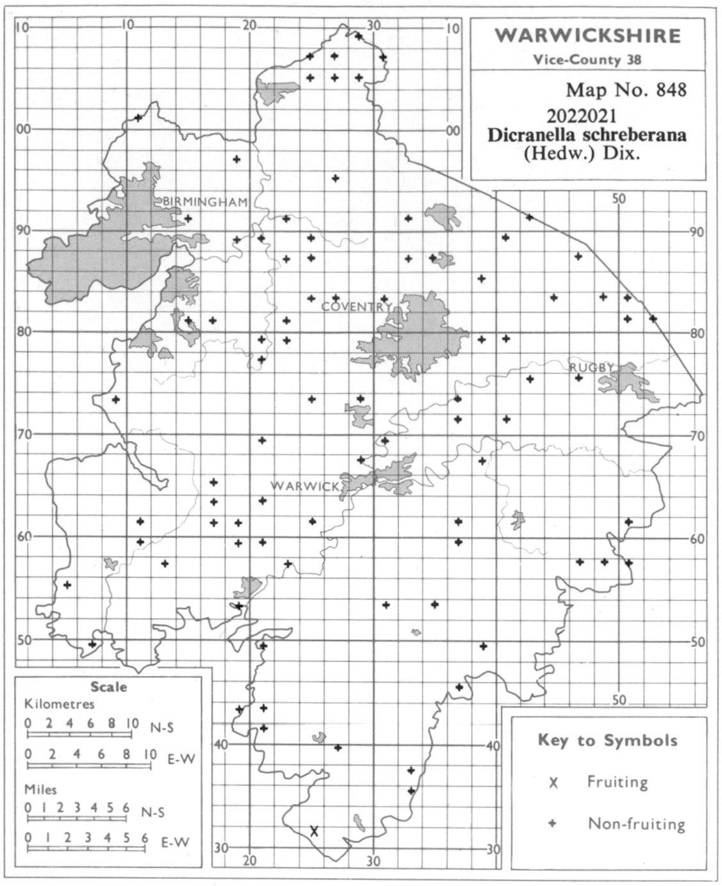
WARWICKSHIRE
Vice-County 38
Map No. 848
2022021
Dicranella schreberana
(Hedw.) Dix.
BIRMINGHAM
COVENTRY
RUGBY
WARWICK
Scale
Kilometres
0 2 4 6 8 10 N-S
0 2 4 6 8 10 E-W
Miles
0 1 2 3 4 5 6 N-S
0 1 2 3 4 5 6 E-W
Key to Symbols
X Fruiting
+ Non-fruiting

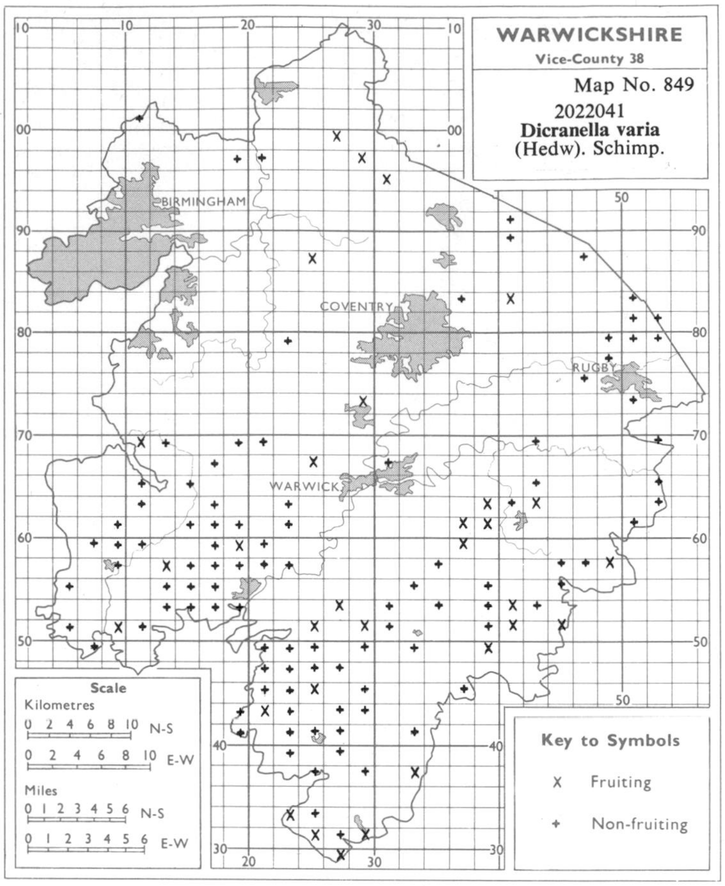
WARWICKSHIRE
Vice-County 38
Map No. 849
2022041
Dicranella varia
(Hedw). Schimp.
BIRMINGHAM
COVENTRY
RUGBY
WARWICK
Scale
Kilometres
0 2 4 6 8 10 N-S
0 2 4 6 8 10 E-W
Miles
0 1 2 3 4 5 6 N-S
0 1 2 3 4 5 6 E-W
Key to Symbols
X Fruiting
+ Non-fruiting

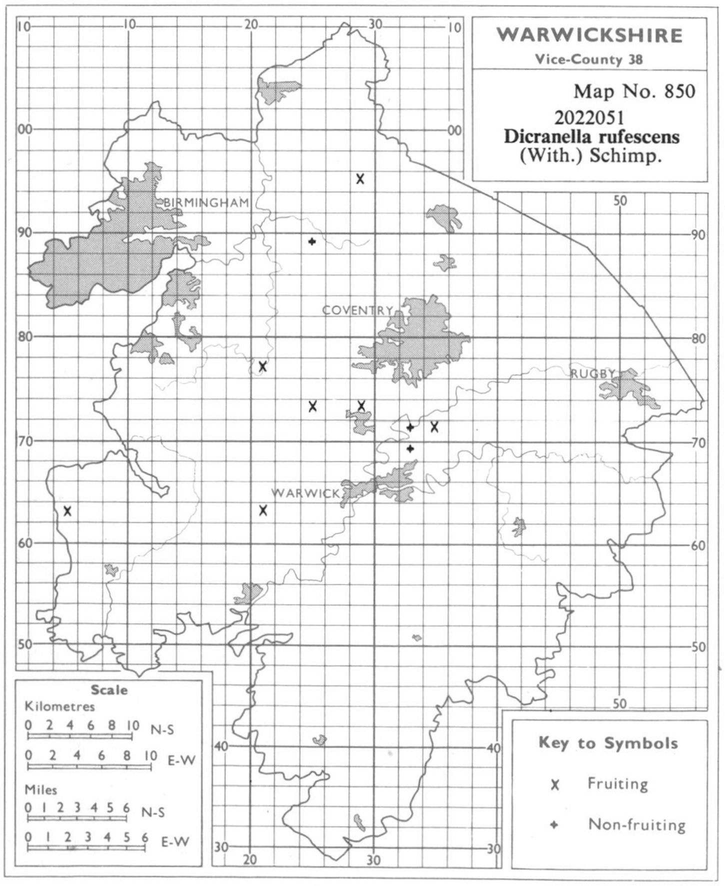
WARWICKSHIRE
Vice-County 38
Map No. 850
2022051
Dicranella rufescens
(With.) Schimp.
BIRMINGHAM
COVENTRY
RUGBY
WARWICK
Scale
Kilometres
0 2 4 6 8 10 N-S
0 2 4 6 8 10 E-W
Miles
0 1 2 3 4 5 6 N-S
0 1 2 3 4 5 6 E-W
Key to Symbols
X Fruiting
+ Non-fruiting

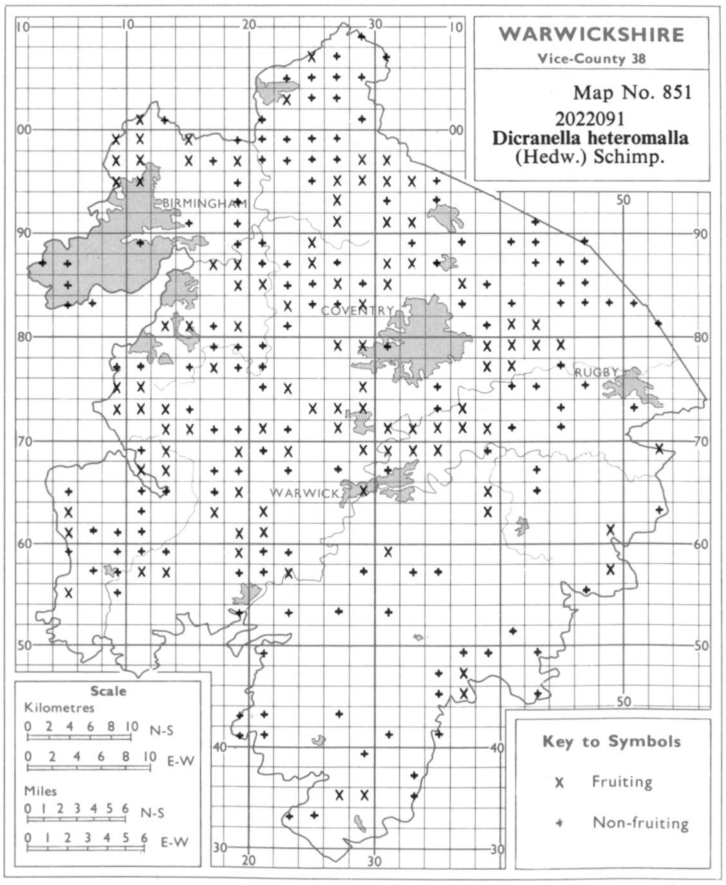
WARWICKSHIRE
Vice-County 38
Map No. 851
2022091
Dicranella heteromalla
(Hedw.) Schimp.
BIRMINGHAM
COVENTRY
RUGBY
WARWICK
Scale
Kilometres
0 2 4 6 8 10 N-S
0 2 4 6 8 10 E-W
Miles
0 1 2 3 4 5 6 N-S
0 1 2 3 4 5 6 E-W
Key to Symbols
X Fruiting
+ Non-fruiting

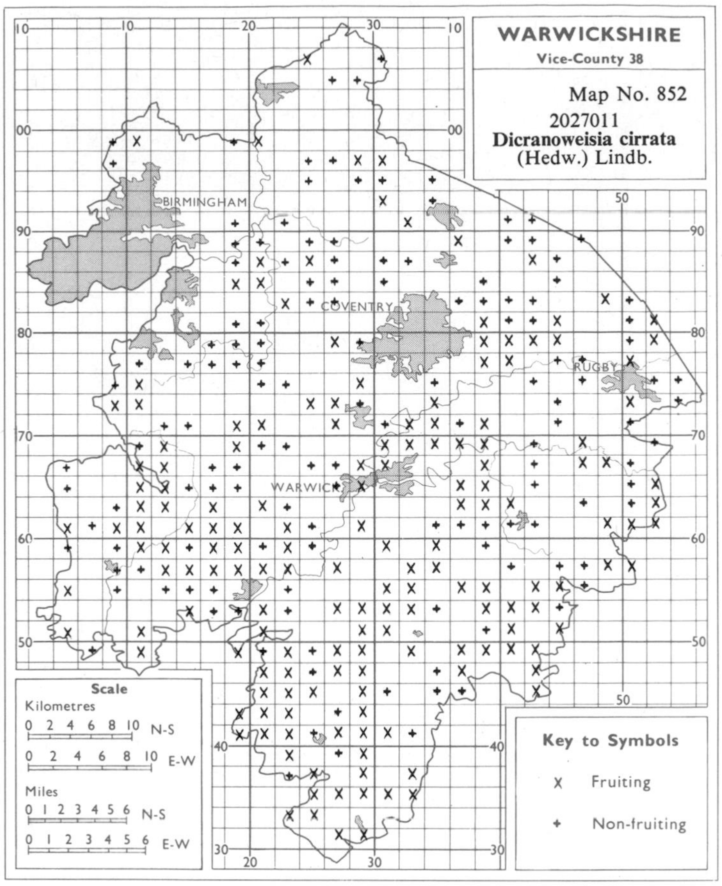
WARWICKSHIRE
Vice-County 38
Map No. 852
2027011
Dicranoweisia cirrata
(Hedw.) Lindb.
BIRMINGHAM
COVENTRY
RUGBY
WARWICK
Scale
Kilometres
0 2 4 6 8 10 N-S
0 2 4 6 8 10 E-W
Miles
0 1 2 3 4 5 6 N-S
0 1 2 3 4 5 6 E-W
Key to Symbols
X Fruiting
+ Non-fruiting

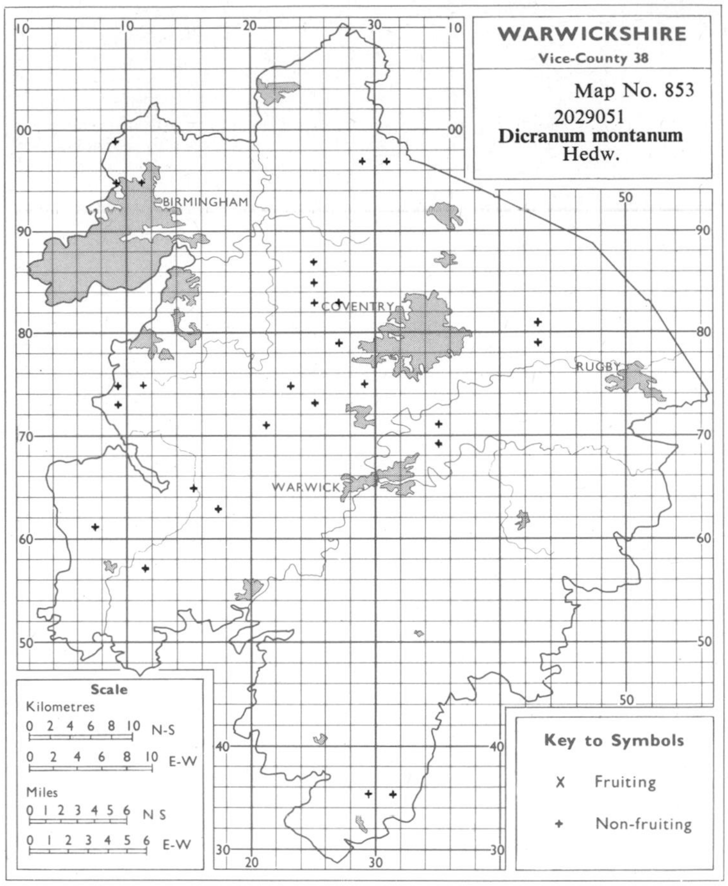
WARWICKSHIRE
Vice-County 38
Map No. 853
2029051
Dicranum montanum
Hedw.
BIRMINGHAM
COVENTRY
RUGBY
WARWICK
Scale
Kilometres
0 2 4 6 8 10 N-S
0 2 4 6 8 10 E-W
Miles
0 1 2 3 4 5 6 N S
0 1 2 3 4 5 6 E-W
Key to Symbols
X Fruiting
+ Non-fruiting

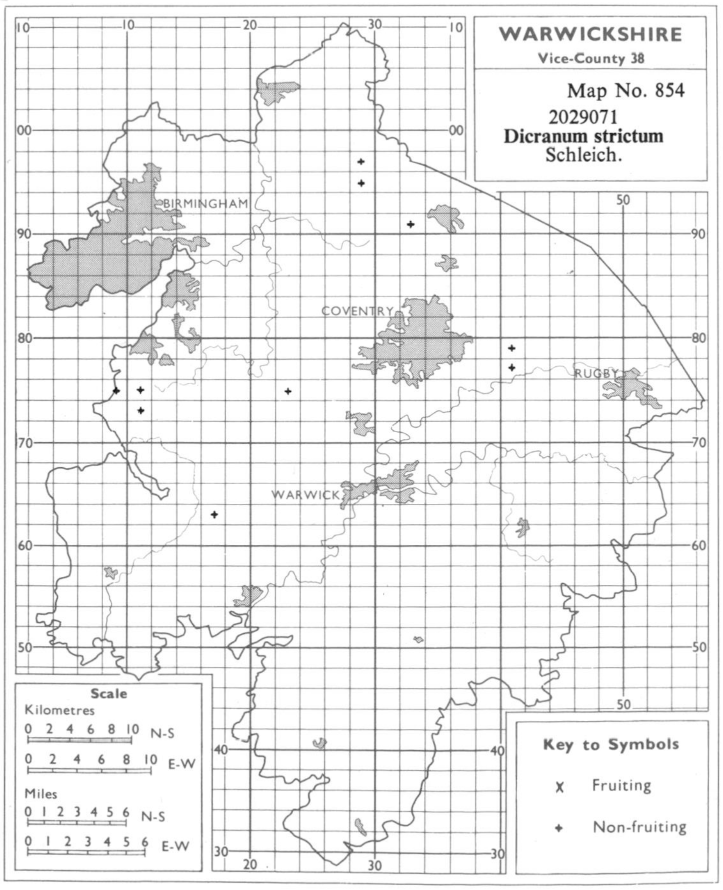
WARWICKSHIRE
Vice-County 38
Map No. 854
2029071
Dicranum strictum
Schleich.
BIRMINGHAM
COVENTRY
RUGBY
WARWICK
Scale
Kilometres
0 2 4 6 8 10 N-S
0 2 4 6 8 10 E-W
Miles
0 1 2 3 4 5 6 N-S
0 1 2 3 4 5 6 E-W
Key to Symbols
X Fruiting
+ Non-fruiting

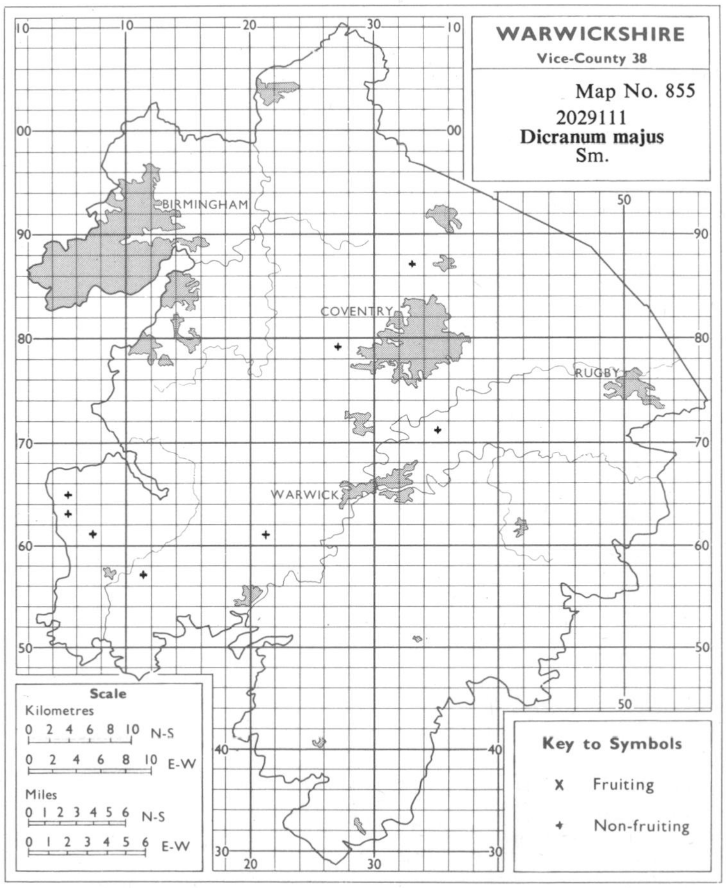
WARWICKSHIRE
Vice-County 38
Map No. 855
2029111
Dicranum majus
Sm.
BIRMINGHAM
COVENTRY
RUGBY
WARWICK
Scale
Kilometres
0 2 4 6 8 10 N-S
0 2 4 6 8 10 E-W
Miles
0 1 2 3 4 5 6 N-S
0 1 2 3 4 5 6 E-W
Key to Symbols
X Fruiting
+ Non-fruiting

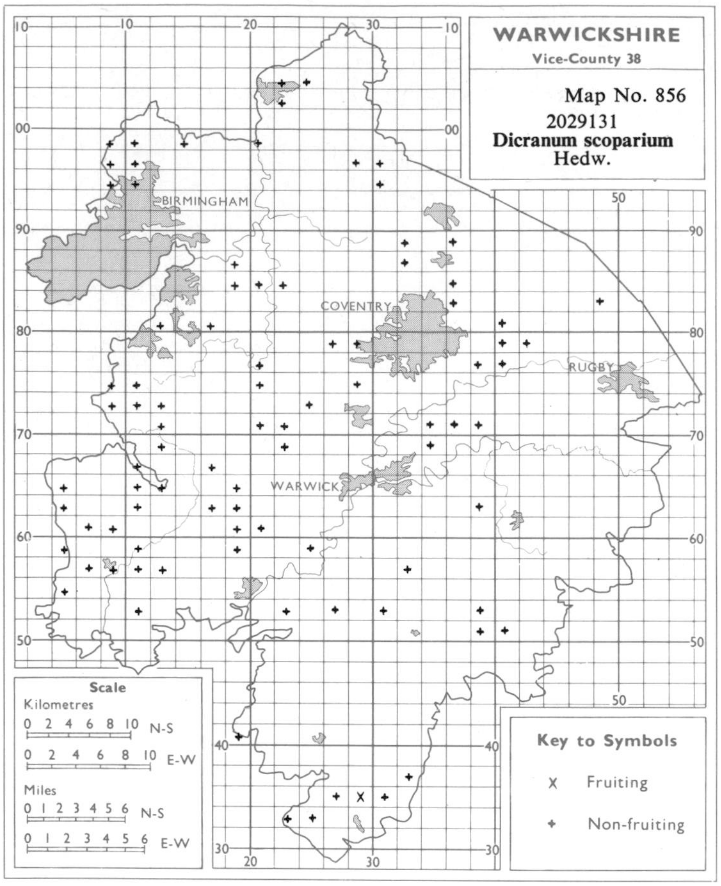
WARWICKSHIRE
Vice-County 38
Map No. 856
2029131
Dicranum scoparium
Hedw.
BIRMINGHAM
COVENTRY
RUGBY
WARWICK
Scale
Kilometres
0 2 4 6 8 10 N-S
0 2 4 6 8 10 E-W
Miles
0 1 2 3 4 5 6 N-S
0 1 2 3 4 5 6 E-W
Key to Symbols
X Fruiting
+ Non-fruiting

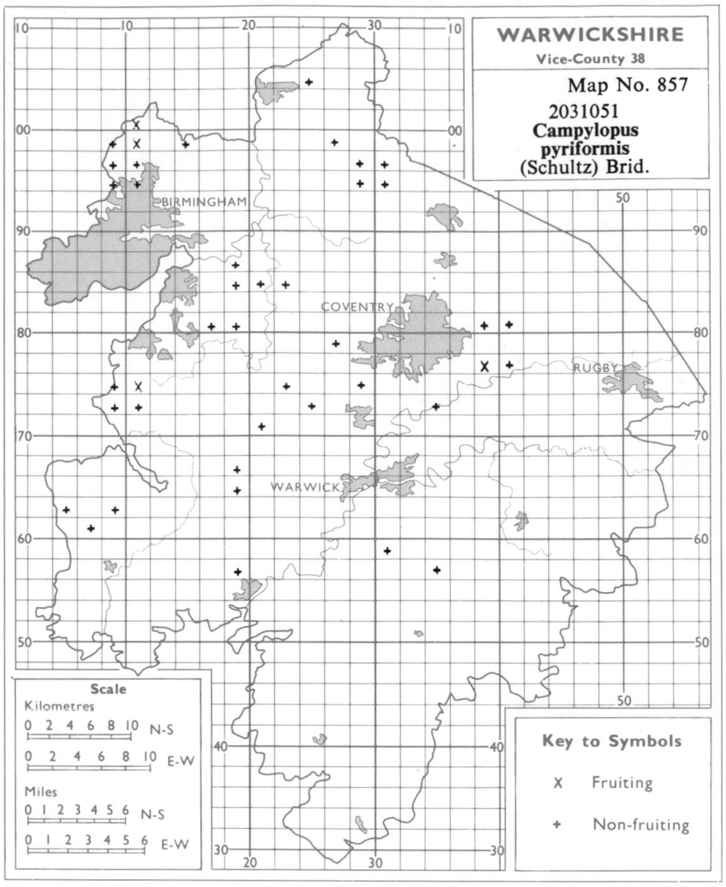
WARWICKSHIRE
Vice-County 38
Map No. 857
2031051
Campylopus
pyriformis
(Schultz) Brid.
BIRMINGHAM
COVENTRY
RUGBY
WARWICK
Scale
Kilometres
0 2 4 6 8 10 N-S
0 2 4 6 8 10 E-W
Miles
0 1 2 3 4 5 6 N-S
0 1 2 3 4 5 6 E-W
Key to Symbols
X Fruiting
+ Non-fruiting

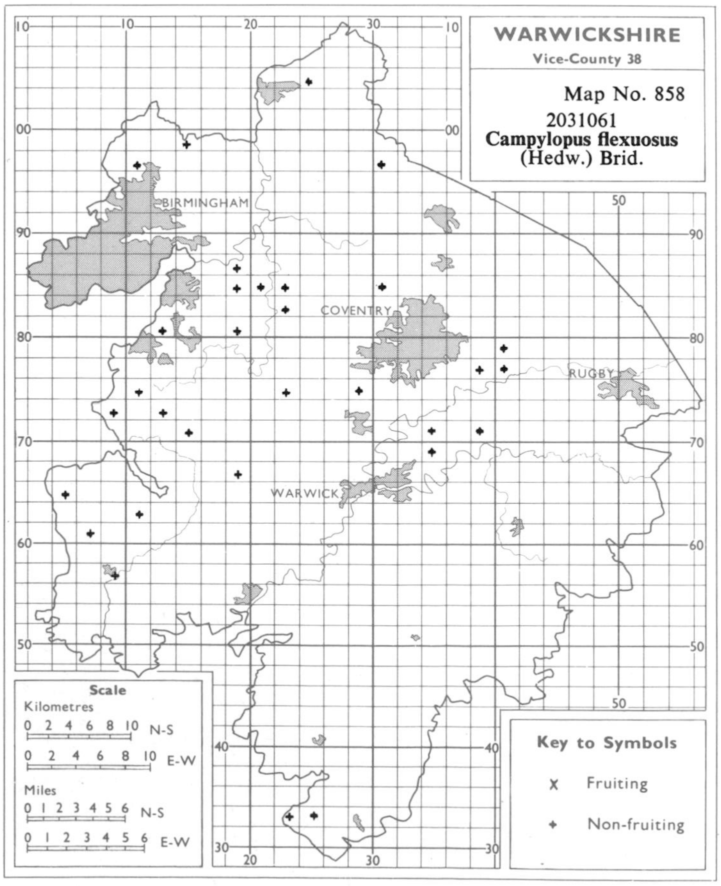
WARWICKSHIRE
Vice-County 38
Map No. 858
2031061
Campylopus flexuosus
(Hedw.) Brid.
BIRMINGHAM
COVENTRY
RUGBY
WARWICK
Scale
Kilometres
0 2 4 6 8 10 N-S
0 2 4 6 8 10 E-W
Miles
0 1 2 3 4 5 6 N-S
0 1 2 3 4 5 6 E-W
Key to Symbols
X Fruiting
+ Non-fruiting

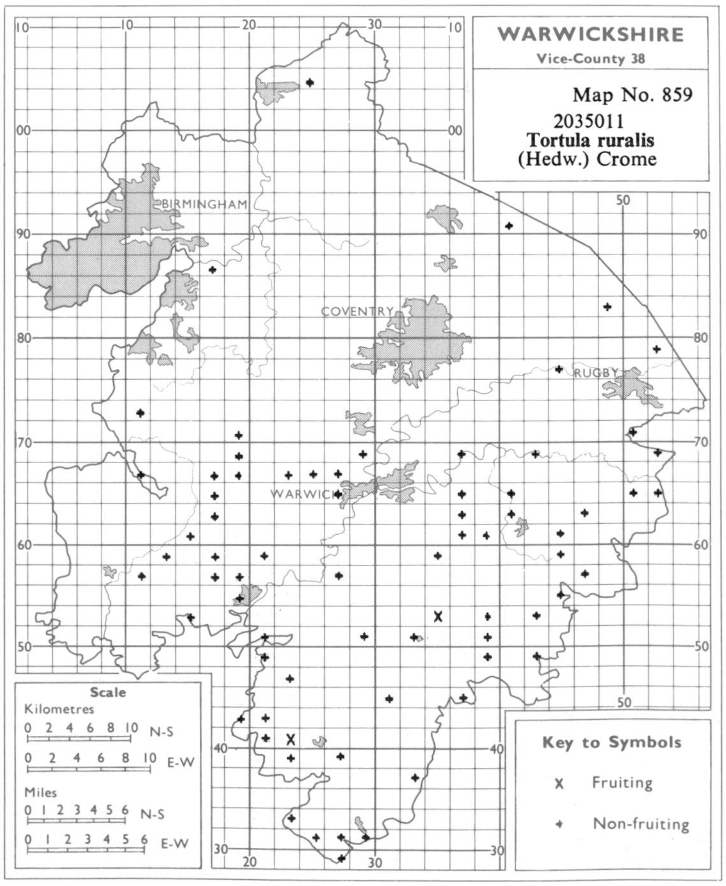
WARWICKSHIRE
Vice-County 38
Map No. 859
2035011
Tortula ruralis
(Hedw.) Crome
BIRMINGHAM
COVENTRY
RUGBY
WARWICK
Scale
Kilometres
0 2 4 6 8 10 N-S
0 2 4 6 8 10 E-W
Miles
0 1 2 3 4 5 6 N-S
0 1 2 3 4 5 6 E-W
Key to Symbols
X Fruiting
+ Non-fruiting

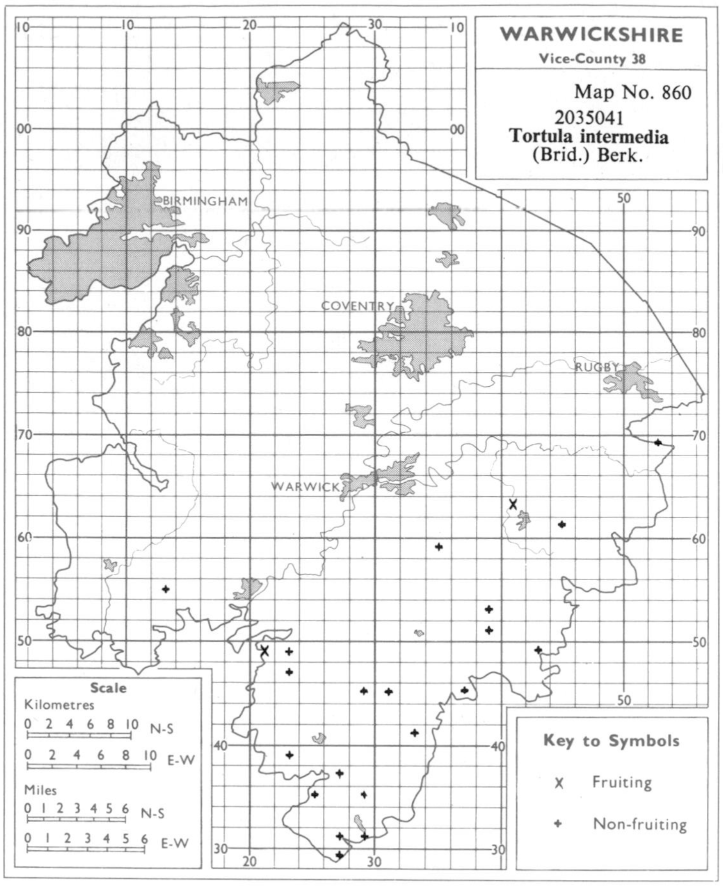
WARWICKSHIRE
Vice-County 38
Map No. 860
2035041
Tortula intermedia
(Brid.) Berk.
BIRMINGHAM
COVENTRY
RUGBY
WARWICK
Scale
Kilometres
0 2 4 6 8 10 N-S
0 2 4 6 8 10 E-W
Miles
0 1 2 3 4 5 6 N-S
0 1 2 3 4 5 6 E-W
Key to Symbols
X Fruiting
+ Non-fruiting

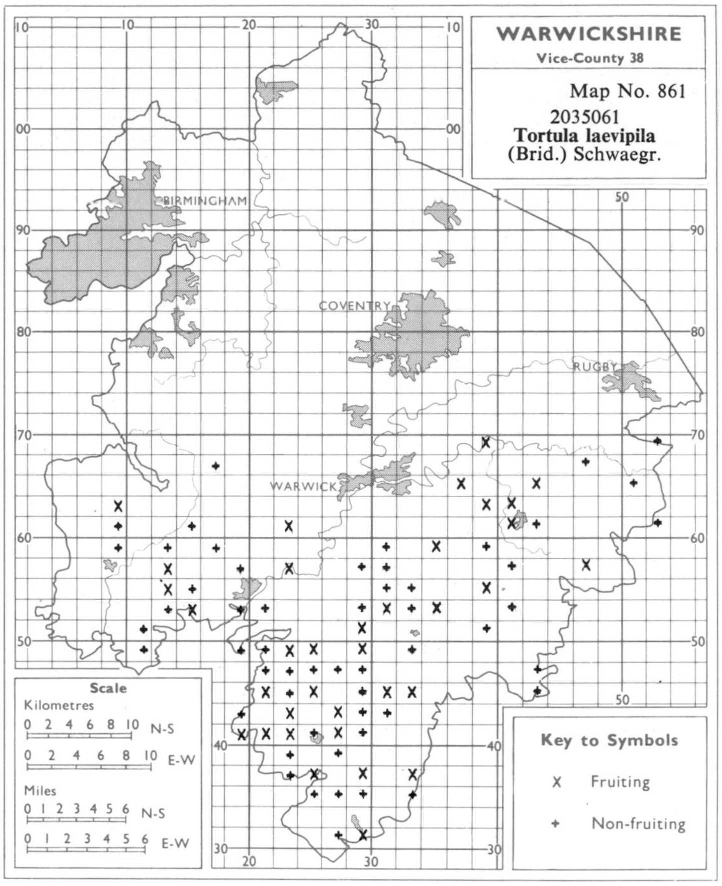
WARWICKSHIRE
Vice-County 38
Map No. 861
2035061
Tortula laevipila
(Brid.) Schwaegr.
BIRMINGHAM
COVENTRY
RUGBY
WARWICK
Scale
Kilometres
0 2 4 6 8 10 N-S
0 2 4 6 8 10 E-W
Miles
0 1 2 3 4 5 6 N-S
0 1 2 3 4 5 6 E-W
Key to Symbols
X Fruiting
+ Non-fruiting

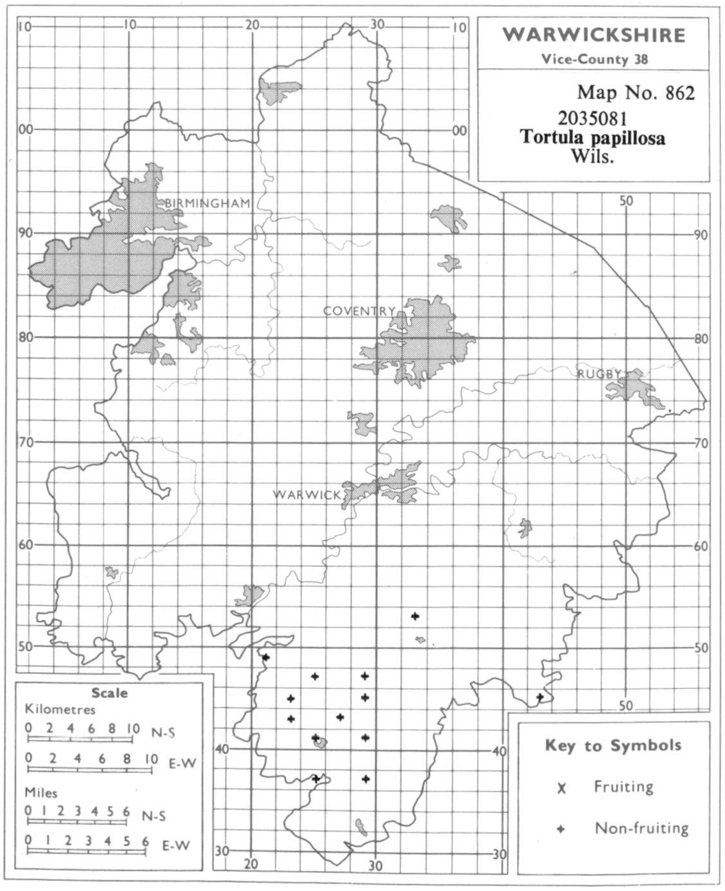
WARWICKSHIRE
Vice-County 38
Map No. 862
2035081
Tortula papillosa
Wils.
BIRMINGHAM
COVENTRY
RUGBY
WARWICK
Scale
Kilometres
0 2 4 6 8 10 N-S
0 2 4 6 8 10 E-W
Miles
0 1 2 3 4 5 6 N-S
0 1 2 3 4 5 6 E-W
Key to Symbols
X Fruiting
+ Non-fruiting

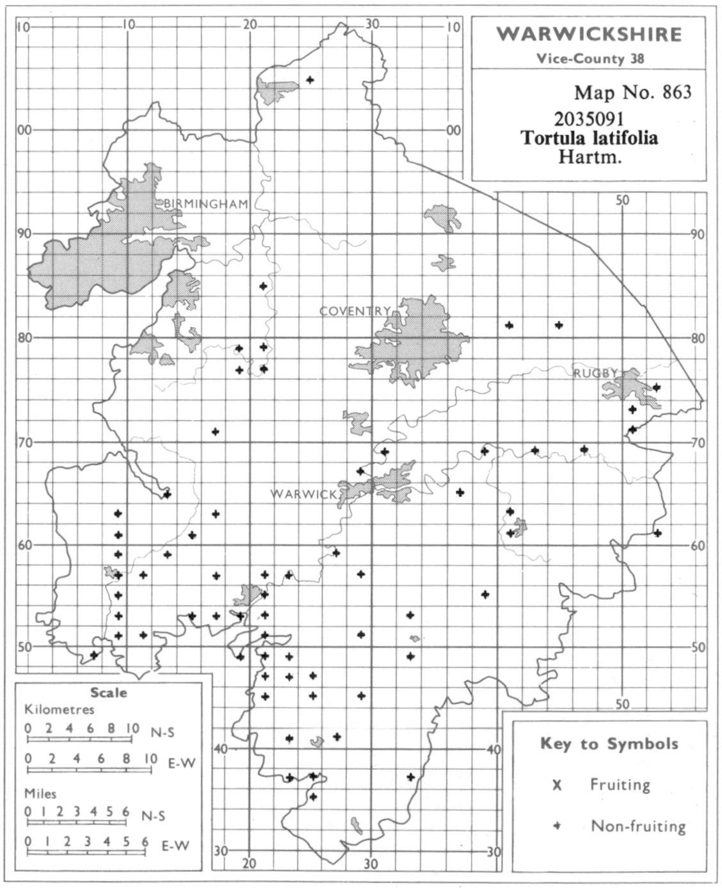
WARWICKSHIRE
Vice-County 38
Map No. 863
2035091
Tortula latifolia
Hartm.
BIRMINGHAM
COVENTRY
RUGBY
WARWICK
Scale
Kilometres
0 2 4 6 8 10 N-S
0 2 4 6 8 10 E-W
Miles
0 1 2 3 4 5 6 N-S
0 1 2 3 4 5 6 E-W
Key to Symbols
X Fruiting
+ Non-fruiting

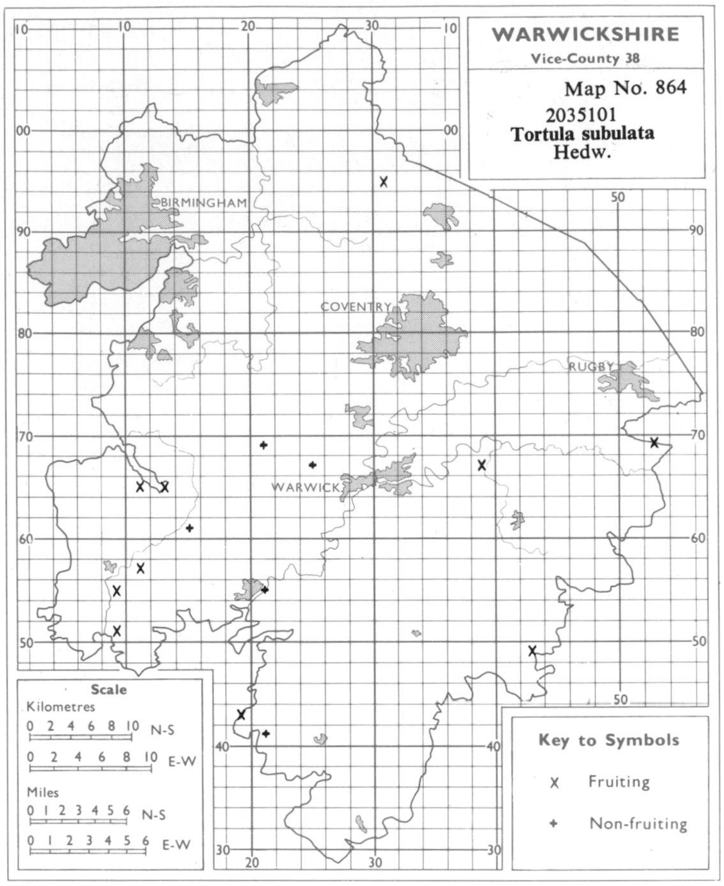
WARWICKSHIRE
Vice-County 38
Map No. 864
2035101
Tortula subulata
Hedw.
BIRMINGHAM
COVENTRY
RUGBY
WARWICK
Scale
Kilometres
0 2 4 6 8 10 N-S
0 2 4 6 8 10 E-W
Miles
0 1 2 3 4 5 6 N-S
0 1 2 3 4 5 6 E-W
Key to Symbols
X Fruiting
+ Non-fruiting

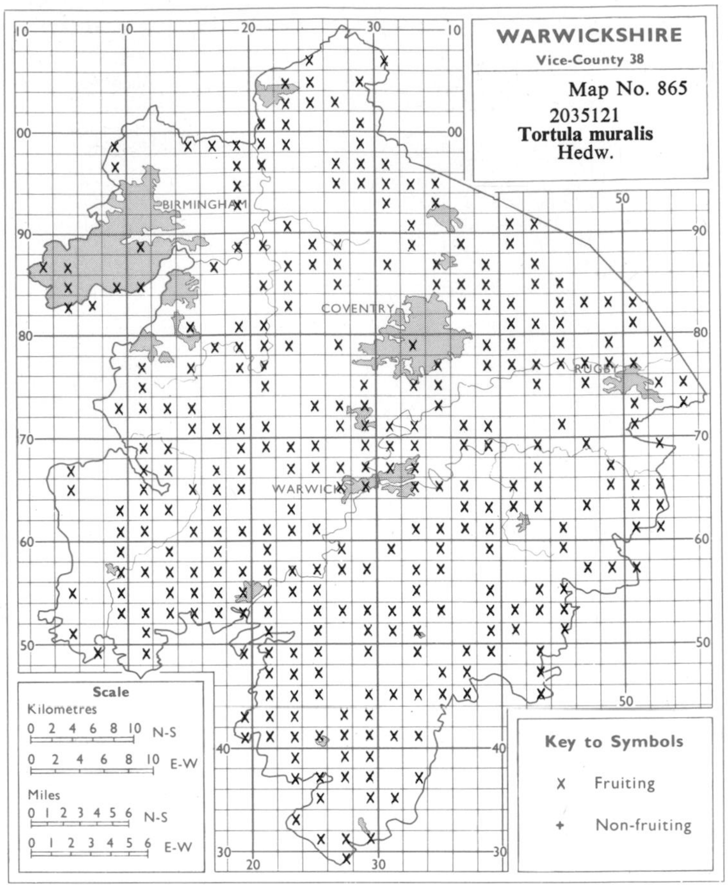
WARWICKSHIRE
Vice-County 38
Map No. 865
2035121
Tortula muralis
Hedw.
BIRMINGHAM
COVENTRY
RUGBY
WARWICK
Scale
Kilometres
0 2 4 6 8 10 N-S
0 2 4 6 8 10 E-W
Miles
0 1 2 3 4 5 6 N-S
0 1 2 3 4 5 6 E-W
Key to Symbols
X Fruiting
+ Non-fruiting

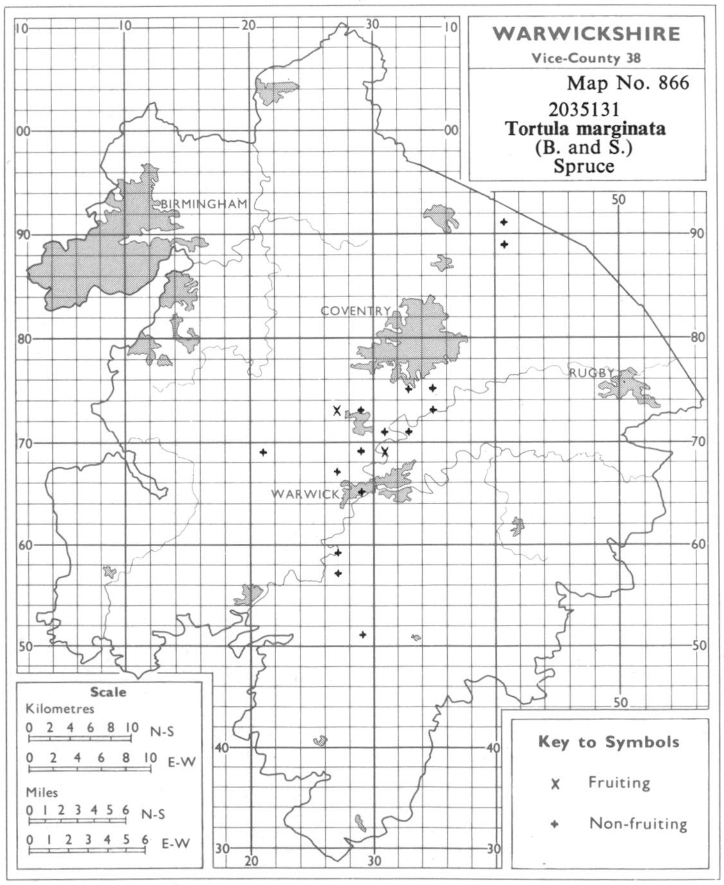
WARWICKSHIRE
Vice-County 38
Map No. 866
2035131
Tortula marginata
(B. and S.)
Spruce
BIRMINGHAM
COVENTRY
RUGBY
WARWICK
Scale
Kilometres
0 2 4 6 8 10 N-S
0 2 4 6 8 10 E-W
Miles
0 1 2 3 4 5 6 N-S
0 1 2 3 4 5 6 E-W
Key to Symbols
X Fruiting
+ Non-fruiting

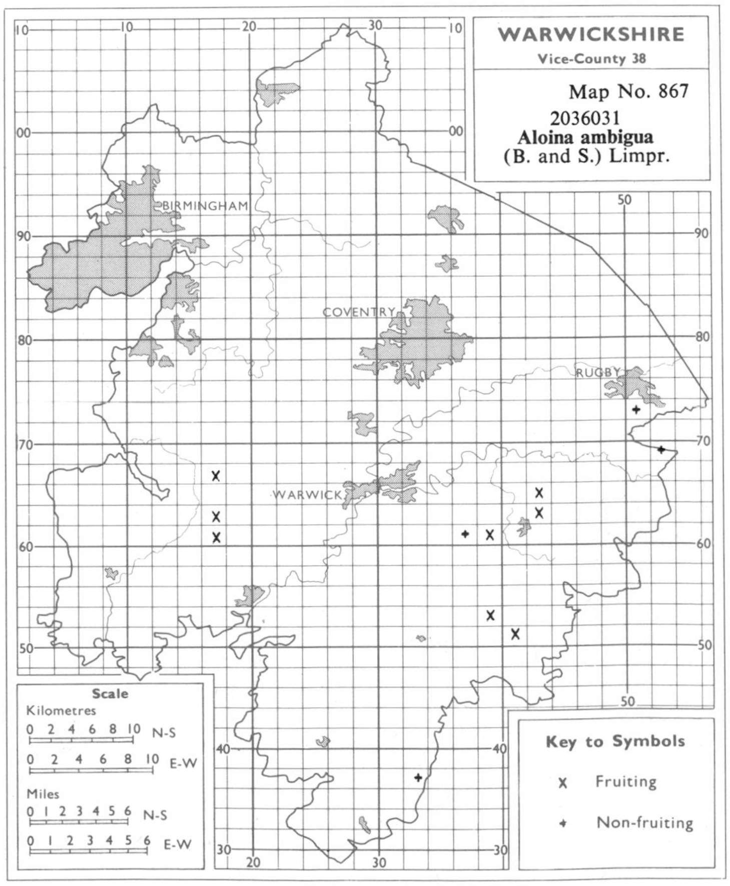
WARWICKSHIRE
Vice-County 38
Map No. 867
2036031
Aloina ambigua
(B. and S.) Limpr.
BIRMINGHAM
COVENTRY
RUGBY
WARWICK
Scale
Kilometres
0 2 4 6 8 10 N-S
0 2 4 6 8 10 E-W
Miles
0 1 2 3 4 5 6 N-S
0 1 2 3 4 5 6 E-W
Key to Symbols
X Fruiting
+ Non-fruiting

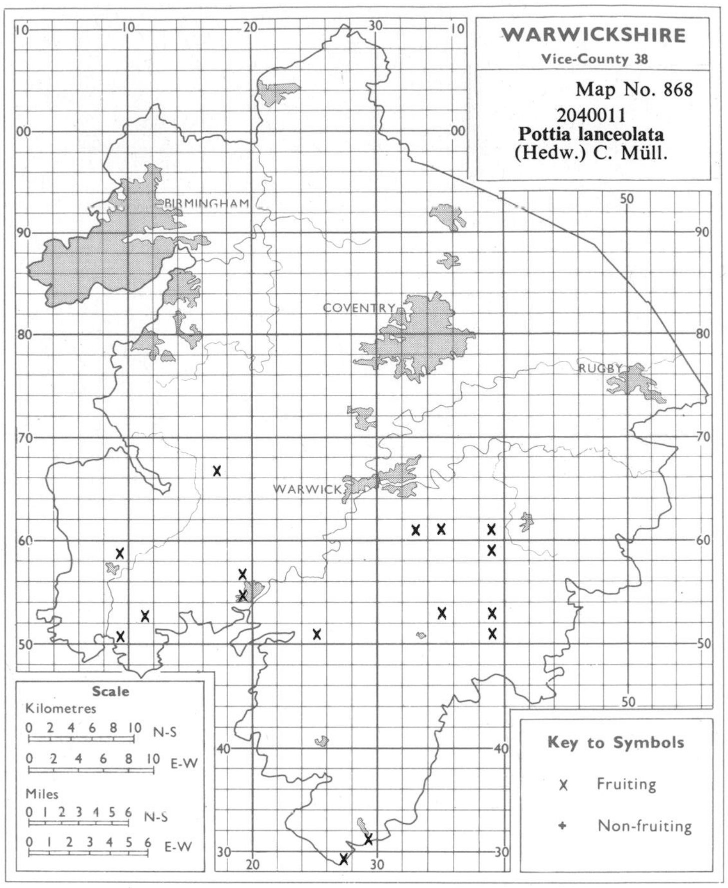
WARWICKSHIRE
Vice-County 38
Map No. 868
2040011
Pottia lanceolata
(Hedw.) C. Müll.
BIRMINGHAM
COVENTRY
RUGBY
WARWICK
Scale
Kilometres
N-S
E-W
Miles
N-S
E-W
Key to Symbols
X Fruiting
+ Non-fruiting

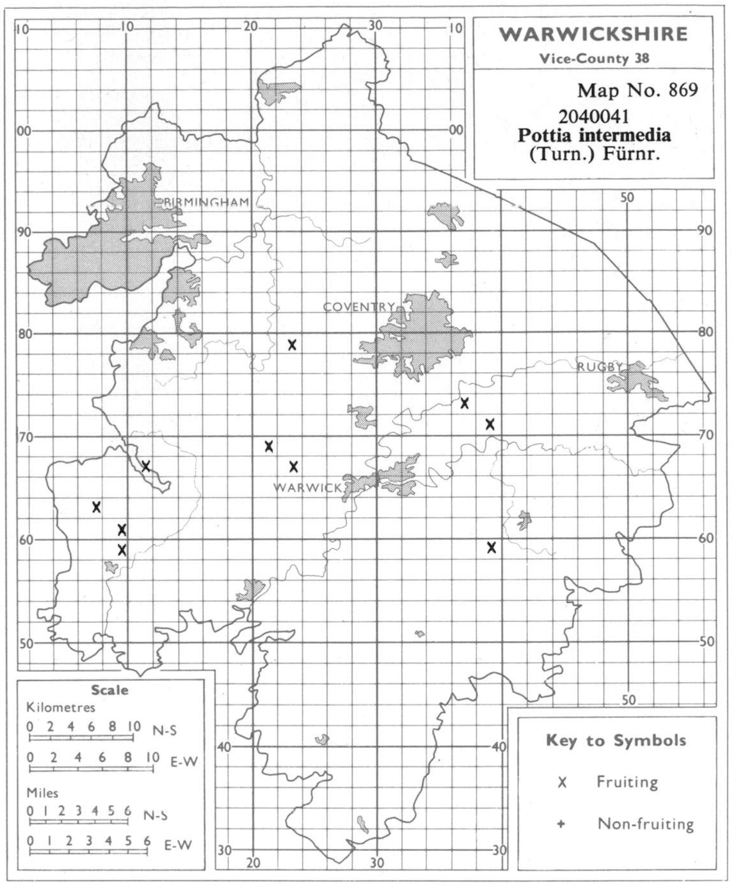
WARWICKSHIRE
Vice-County 38
Map No. 869
2040041
Pottia intermedia
(Turn.) Fürnr.
BIRMINGHAM
COVENTRY
RUGBY
WARWICK
Scale
Kilometres
N-S
E-W
Miles
N-S
E-W
Key to Symbols
X Fruiting
+ Non-fruiting

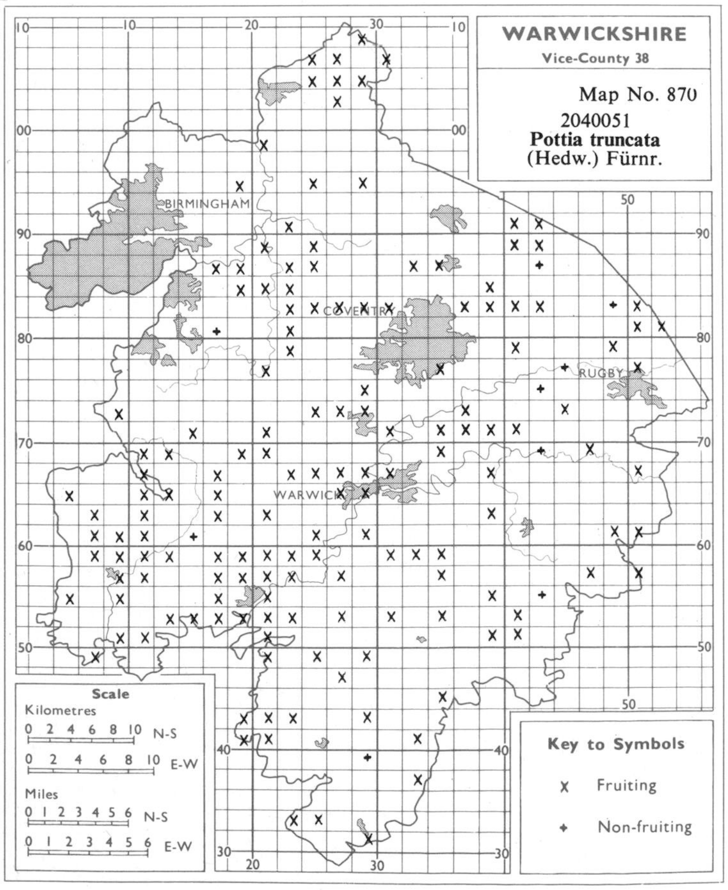
WARWICKSHIRE
Vice-County 38
Map No. 870
2040051
Pottia truncata
(Hedw.) Fürnr.
BIRMINGHAM
COVENTRY
RUGBY
WARWICK
Scale
Kilometres
N-S
E-W
Miles
N-S
E-W
Key to Symbols
X Fruiting
+ Non-fruiting

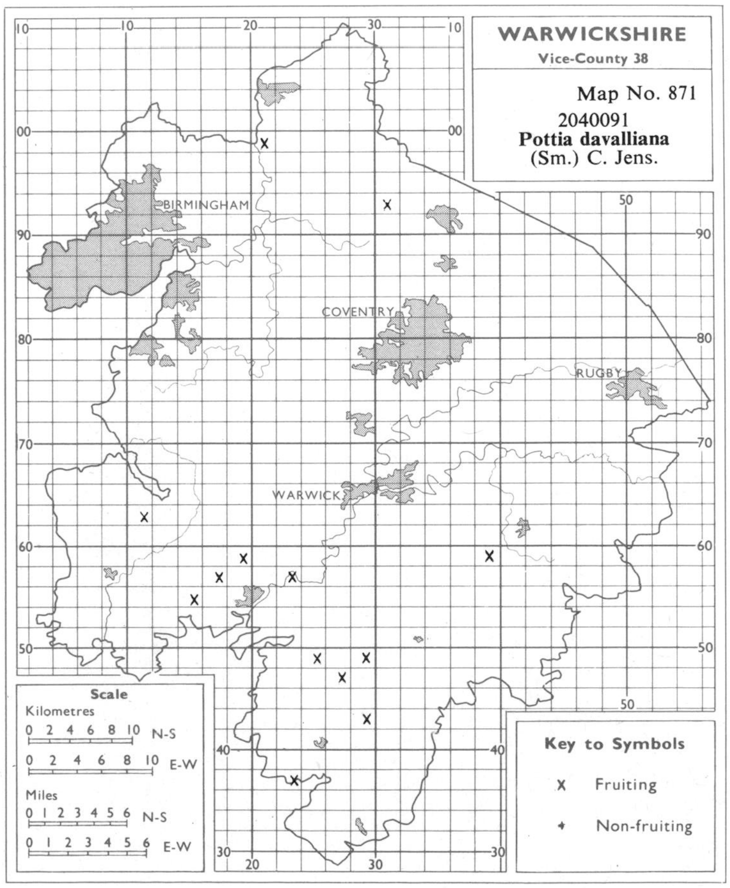
WARWICKSHIRE
Vice-County 38
Map No. 871
2040091
Pottia davalliana
(Sm.) C. Jens.
BIRMINGHAM
COVENTRY
RUGBY
WARWICK
Scale
Kilometres
N-S
E-W
Miles
N-S
E-W
Key to Symbols
X Fruiting
+ Non-fruiting

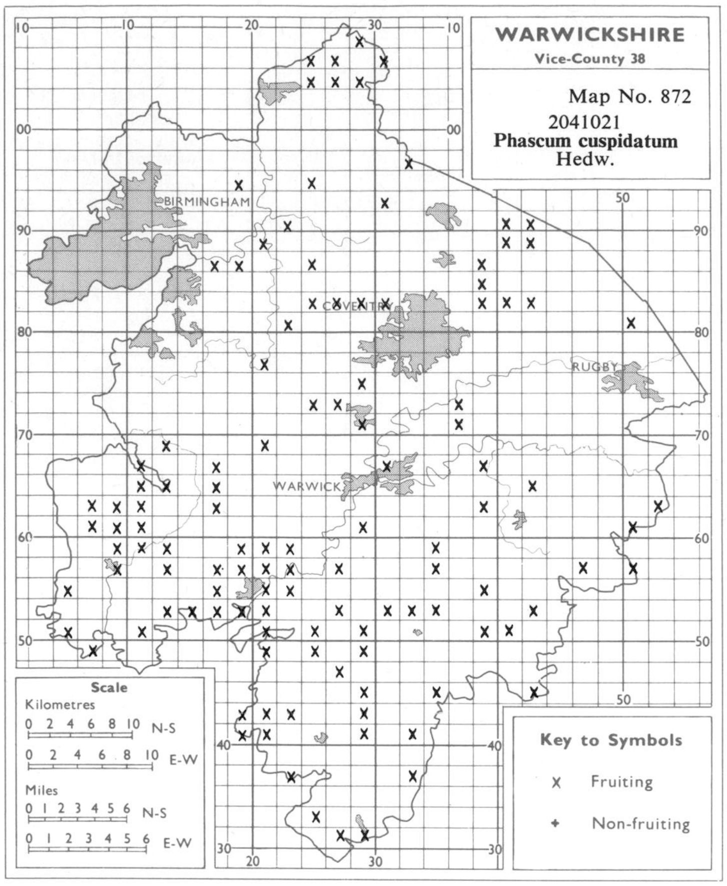
WARWICKSHIRE
Vice-County 38
Map No. 872
2041021
Phascum cuspidatum
Hedw.
BIRMINGHAM
COVENTRY
RUGBY
WARWICK
Scale
Kilometres
0 2 4 6 8 10 N-S
0 2 4 6 8 10 E-W
Miles
0 1 2 3 4 5 6 N-S
0 1 2 3 4 5 6 E-W
Key to Symbols
x Fruiting
+ Non-fruiting

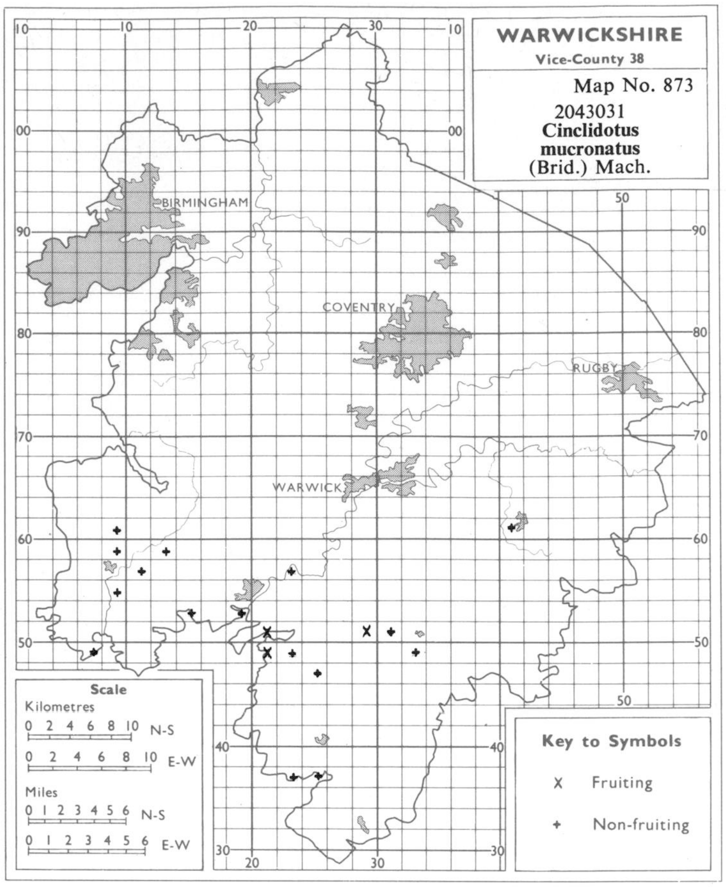
WARWICKSHIRE
Vice-County 38
Map No. 873
2043031
Cinclidotus
mucronatus
(Brid.) Mach.
BIRMINGHAM
COVENTRY
RUGBY
WARWICK
Scale
Kilometres
0 2 4 6 8 10 N-S
0 2 4 6 8 10 E-W
Miles
0 1 2 3 4 5 6 N-S
0 1 2 3 4 5 6 E-W
Key to Symbols
x Fruiting
+ Non-fruiting

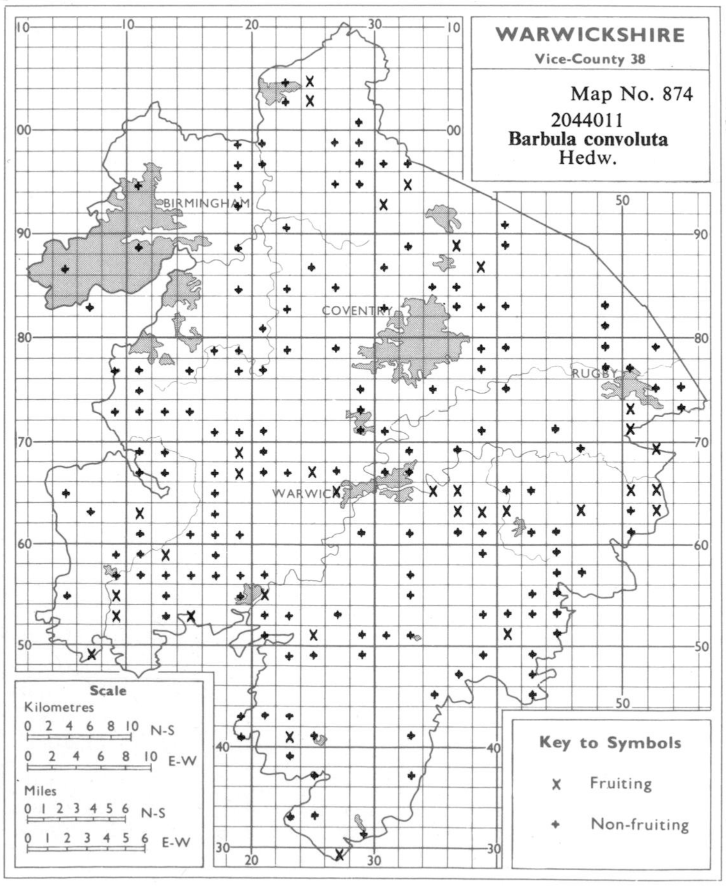
WARWICKSHIRE
Vice-County 38
Map No. 874
2044011
Barbula convoluta
Hedw.
BIRMINGHAM
COVENTRY
RUGBY
WARWICK
Scale
Kilometres
0 2 4 6 8 10 N-S
0 2 4 6 8 10 E-W
Miles
0 1 2 3 4 5 6 N-S
0 1 2 3 4 5 6 E-W
Key to Symbols
x Fruiting
+ Non-fruiting

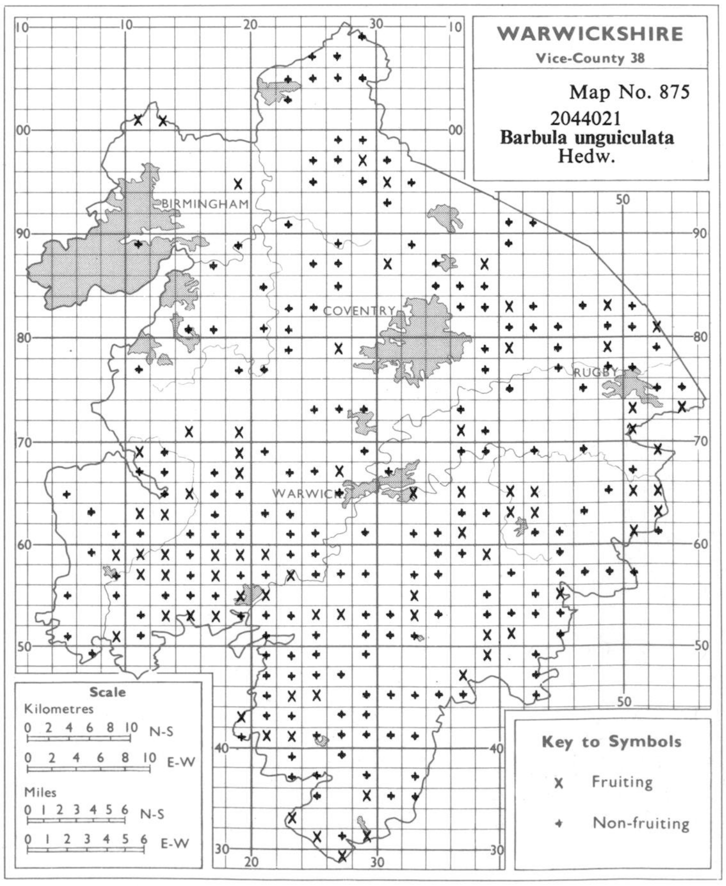
WARWICKSHIRE
Vice-County 38
Map No. 875
2044021
Barbula unguiculata
Hedw.
BIRMINGHAM
COVENTRY
RUGBY
WARWICK
Scale
Kilometres
0 2 4 6 8 10 N-S
0 2 4 6 8 10 E-W
Miles
0 1 2 3 4 5 6 N-S
0 1 2 3 4 5 6 E-W
Key to Symbols
x Fruiting
+ Non-fruiting

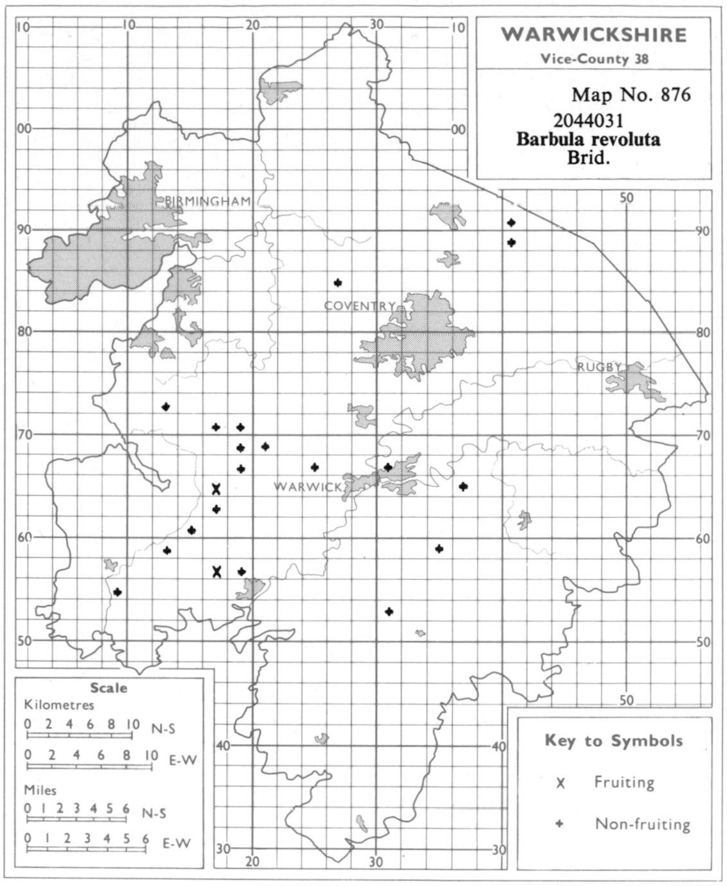
WARWICKSHIRE
Vice-County 38
Map No. 876
2044031
Barbula revoluta
Brid.
BIRMINGHAM
COVENTRY
RUGBY
WARWICK
Scale
Kilometres
0 2 4 6 8 10 N-S
0 2 4 6 8 10 E-W
Miles
0 1 2 3 4 5 6 N-S
0 1 2 3 4 5 6 E-W
Key to Symbols
X Fruiting
+ Non-fruiting

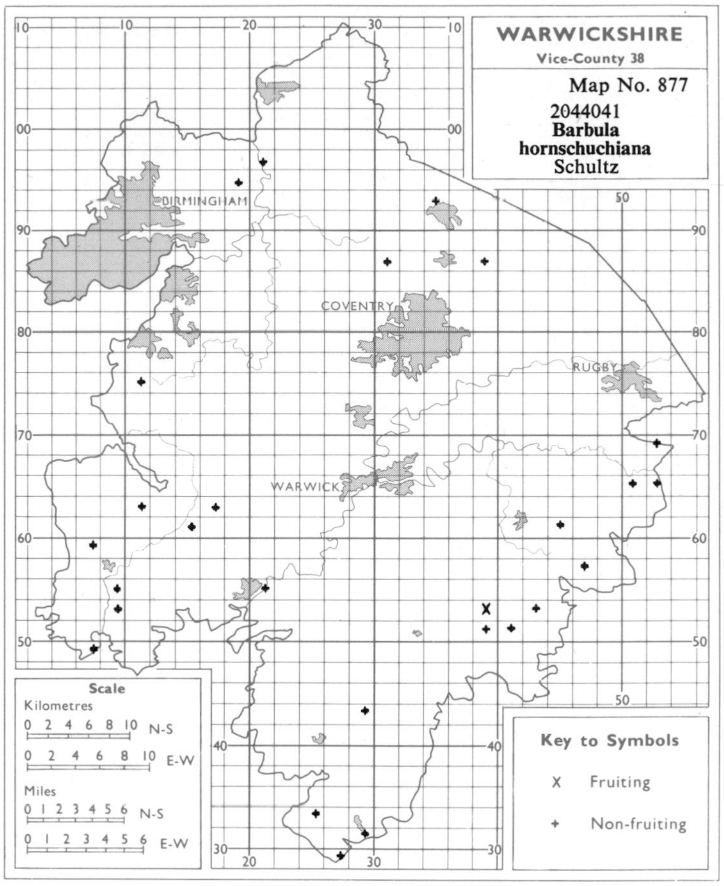
WARWICKSHIRE
Vice-County 38
Map No. 877
2044041
Barbula
hornschuchiana
Schultz
BIRMINGHAM
COVENTRY
RUGBY
WARWICK
Scale
Kilometres
0 2 4 6 8 10 N-S
0 2 4 6 8 10 E-W
Miles
0 1 2 3 4 5 6 N-S
0 1 2 3 4 5 6 E-W
Key to Symbols
X Fruiting
+ Non-fruiting

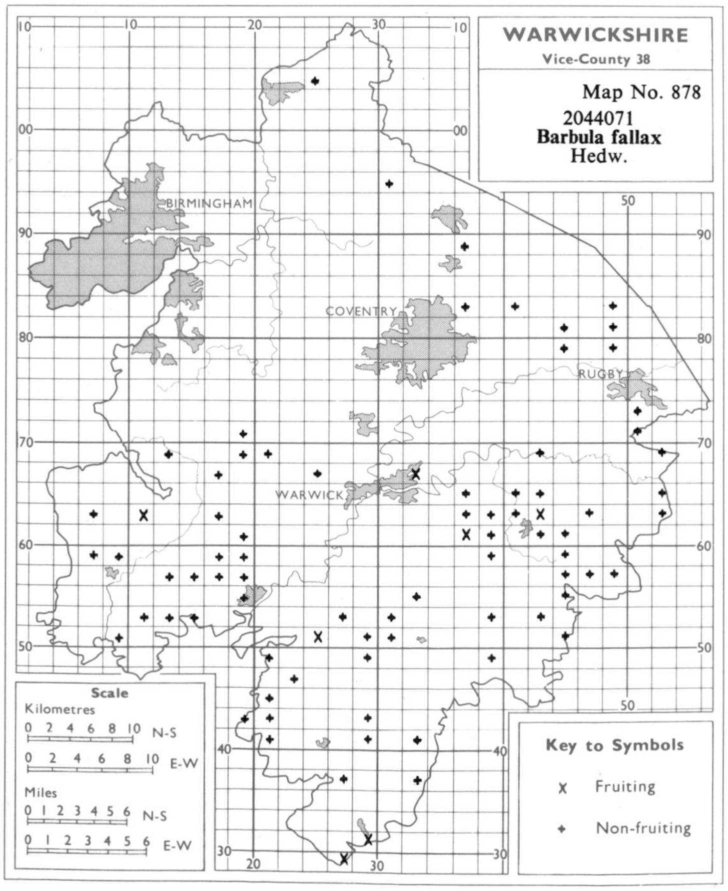
WARWICKSHIRE
Vice-County 38
Map No. 878
2044071
Barbula fallax
Hedw.
BIRMINGHAM
COVENTRY
RUGBY
WARWICK
Scale
Kilometres
0 2 4 6 8 10 N-S
0 2 4 6 8 10 E-W
Miles
0 1 2 3 4 5 6 N-S
0 1 2 3 4 5 6 E-W
Key to Symbols
X Fruiting
+ Non-fruiting

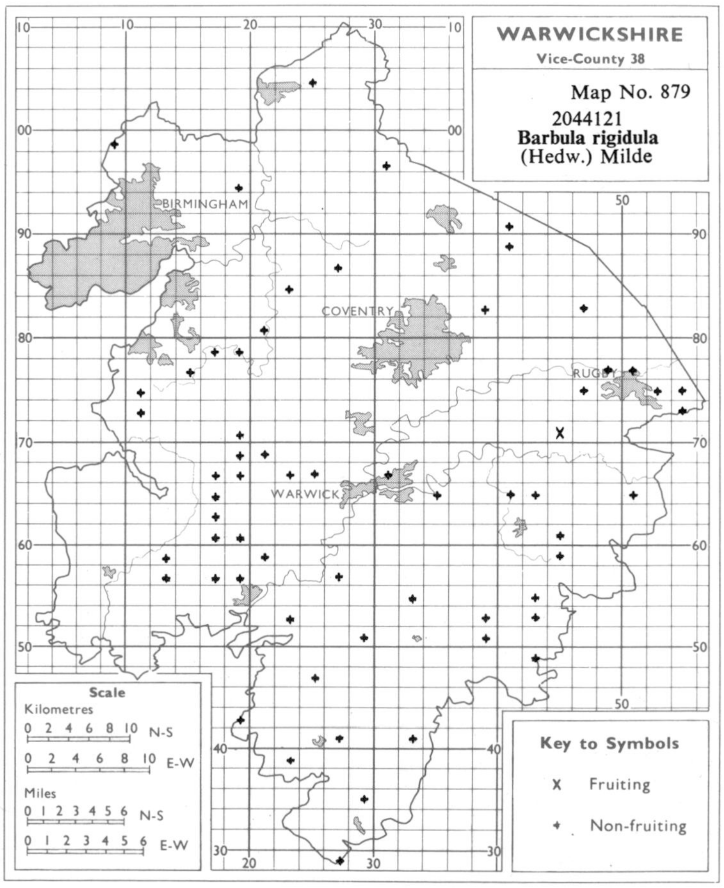
WARWICKSHIRE
Vice-County 38
Map No. 879
2044121
Barbula rigidula
(Hedw.) Milde
BIRMINGHAM
COVENTRY
RUGBY
WARWICK
Scale
Kilometres
0 2 4 6 8 10 N-S
0 2 4 6 8 10 E-W
Miles
0 1 2 3 4 5 6 N-S
0 1 2 3 4 5 6 E-W
Key to Symbols
X Fruiting
+ Non-fruiting

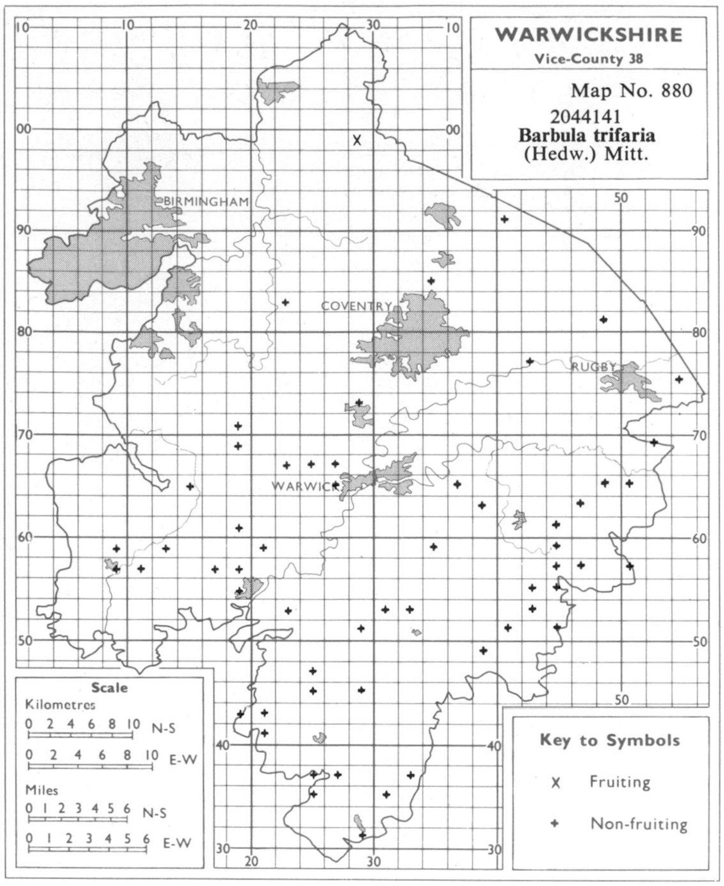
WARWICKSHIRE
Vice-County 38
Map No. 880
2044141
Barbula trifaria
(Hedw.) Mitt.
BIRMINGHAM
COVENTRY
RUGBY
WARWICK
Scale
Kilometres
N-S
E-W
Miles
Key to Symbols
X Fruiting
+ Non-fruiting

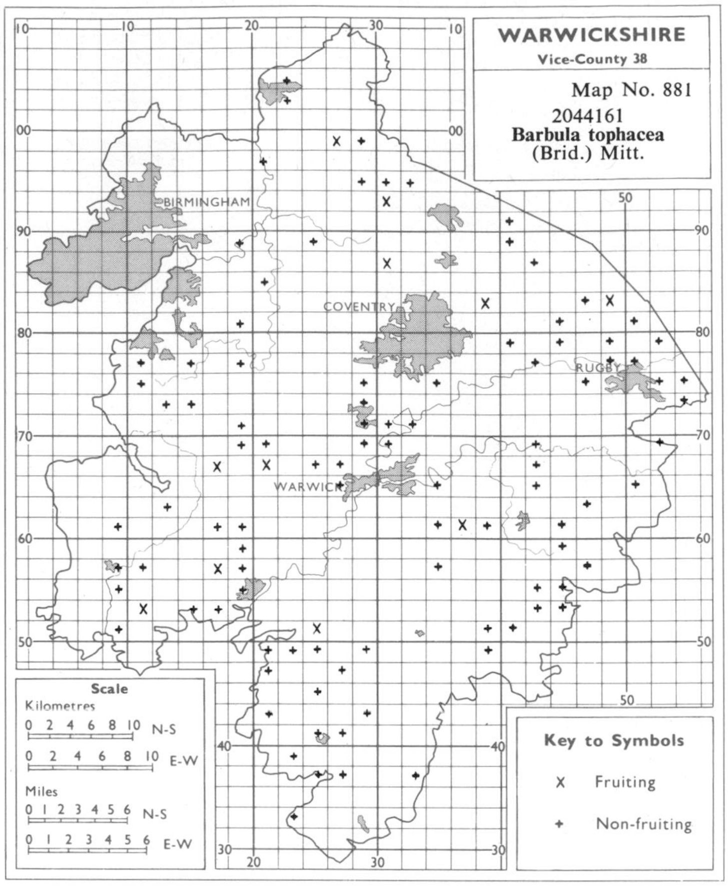
WARWICKSHIRE
Vice-County 38
Map No. 881
2044161
Barbula tophacea
(Brid.) Mitt.
BIRMINGHAM
COVENTRY
RUGBY
WARWICK
Scale
Kilometres
N-S
E-W
Miles
Key to Symbols
X Fruiting
+ Non-fruiting

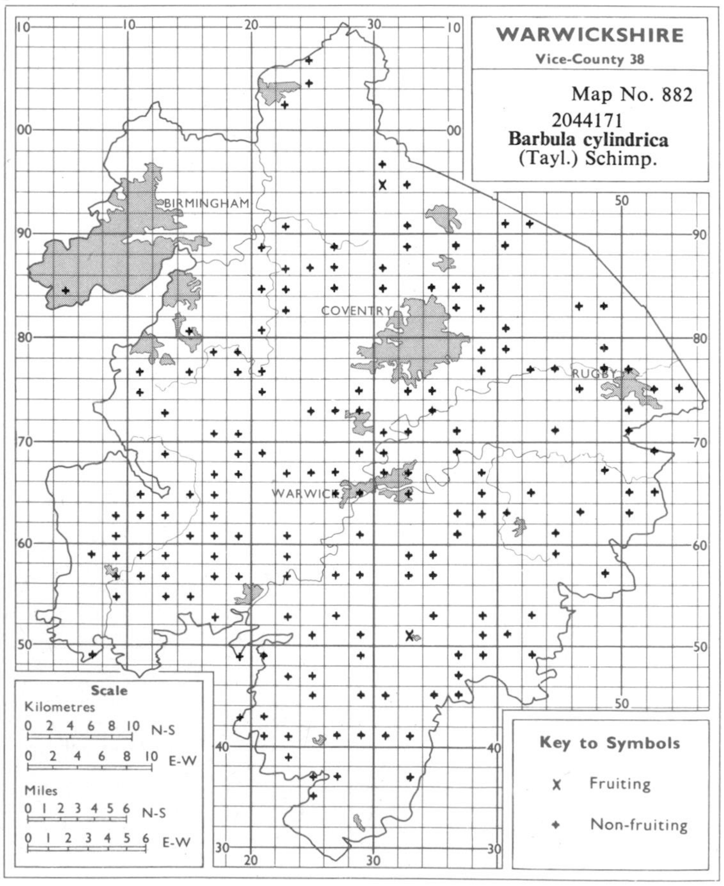
WARWICKSHIRE
Vice-County 38
Map No. 882
2044171
Barbula cylindrica
(Tayl.) Schimp.
BIRMINGHAM
COVENTRY
RUGBY
WARWICK
Scale
Kilometres
N-S
E-W
Miles
Key to Symbols
X Fruiting
+ Non-fruiting

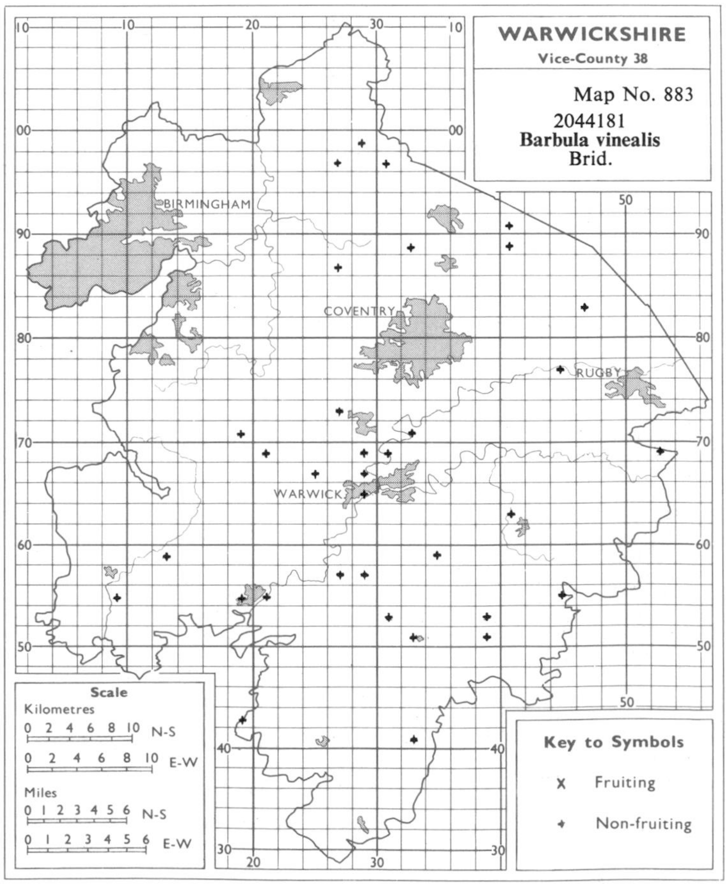
WARWICKSHIRE
Vice-County 38
Map No. 883
2044181
Barbula vinealis
Brid.
BIRMINGHAM
COVENTRY
RUGBY
WARWICK
Scale
Kilometres
N-S
E-W
Miles
Key to Symbols
X Fruiting
+ Non-fruiting

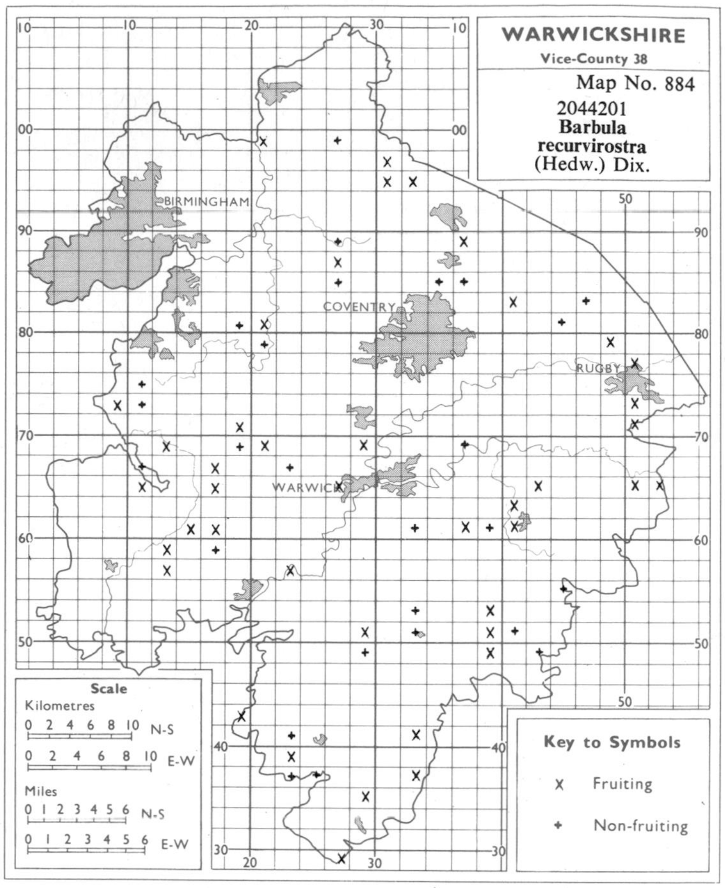
WARWICKSHIRE
Vice-County 38
Map No. 884
2044201
Barbula recurvirostra
(Hedw.) Dix.
BIRMINGHAM
COVENTRY
RUGBY
WARWICK
Scale
Kilometres
0 2 4 6 8 10 N-S
0 2 4 6 8 10 E-W
Miles
0 1 2 3 4 5 6 N-S
0 1 2 3 4 5 6 E-W
Key to Symbols
X Fruiting
+ Non-fruiting

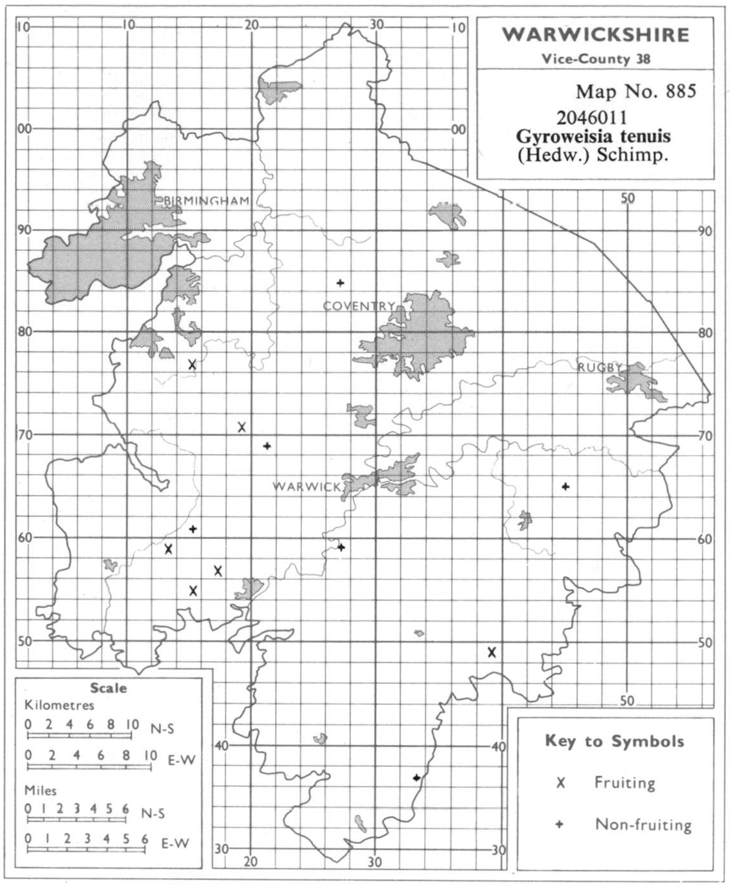
WARWICKSHIRE
Vice-County 38
Map No. 885
2046011
Gyroweisia tenuis
(Hedw.) Schimp.
BIRMINGHAM
COVENTRY
RUGBY
WARWICK
Scale
Kilometres
0 2 4 6 8 10 N-S
0 2 4 6 8 10 E-W
Miles
0 1 2 3 4 5 6 N-S
0 1 2 3 4 5 6 E-W
Key to Symbols
X Fruiting
+ Non-fruiting

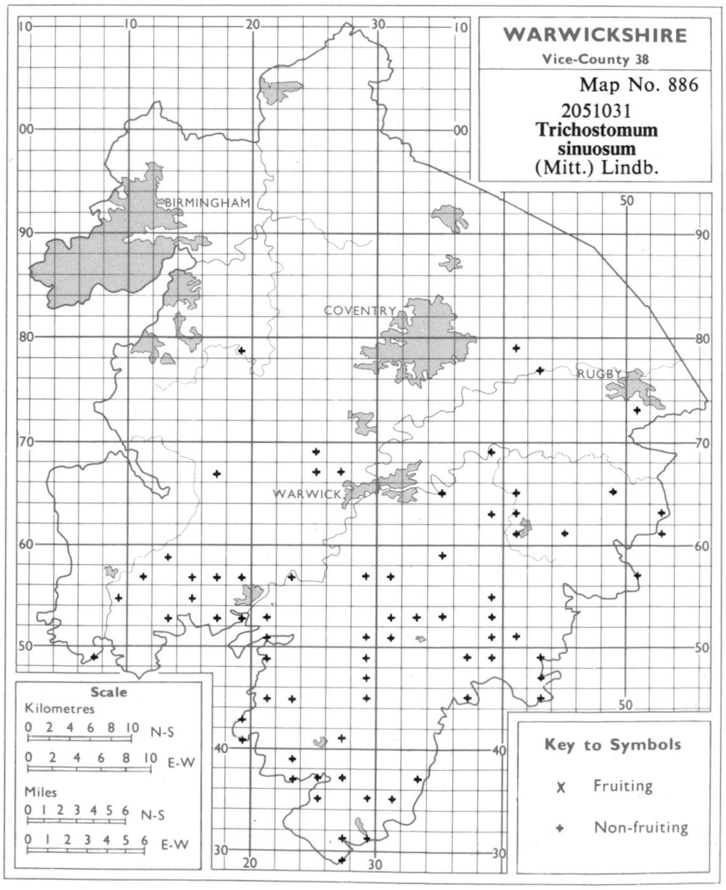
WARWICKSHIRE
Vice-County 38
Map No. 886
2051031
Trichostomum sinuosum
(Mitt.) Lindb.
BIRMINGHAM
COVENTRY
RUGBY
WARWICK
Scale
Kilometres
0 2 4 6 8 10 N-S
0 2 4 6 8 10 E-W
Miles
0 1 2 3 4 5 6 N-S
0 1 2 3 4 5 6 E-W
Key to Symbols
X Fruiting
+ Non-fruiting

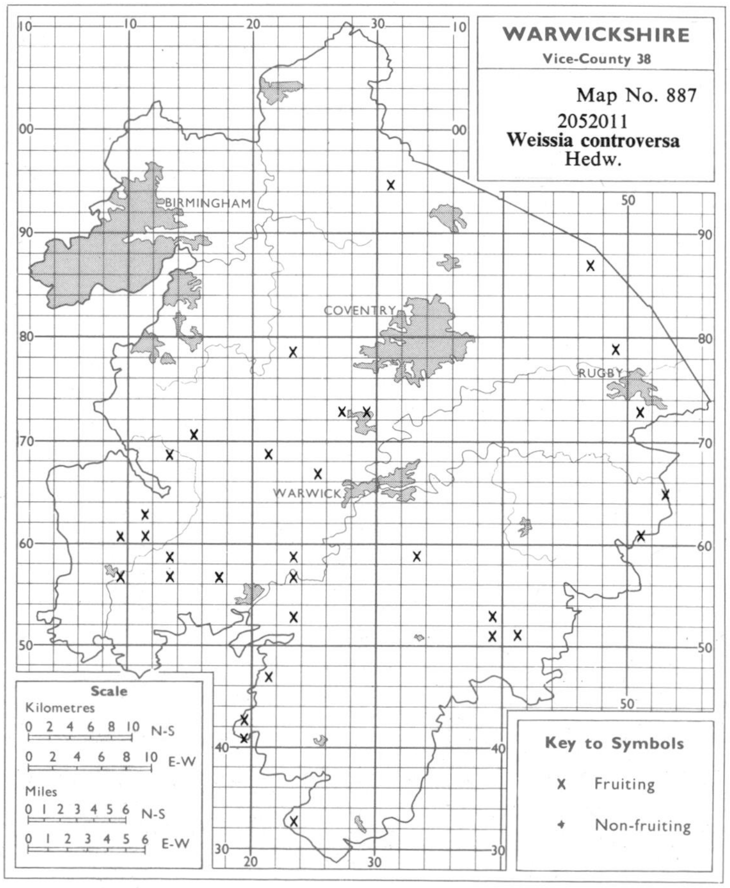
WARWICKSHIRE
Vice-County 38
Map No. 887
2052011
Weissia controversa
Hedw.
BIRMINGHAM
COVENTRY
RUGBY
WARWICK
Scale
Kilometres
0 2 4 6 8 10 N-S
0 2 4 6 8 10 E-W
Miles
0 1 2 3 4 5 6 N-S
0 1 2 3 4 5 6 E-W
Key to Symbols
X Fruiting
+ Non-fruiting

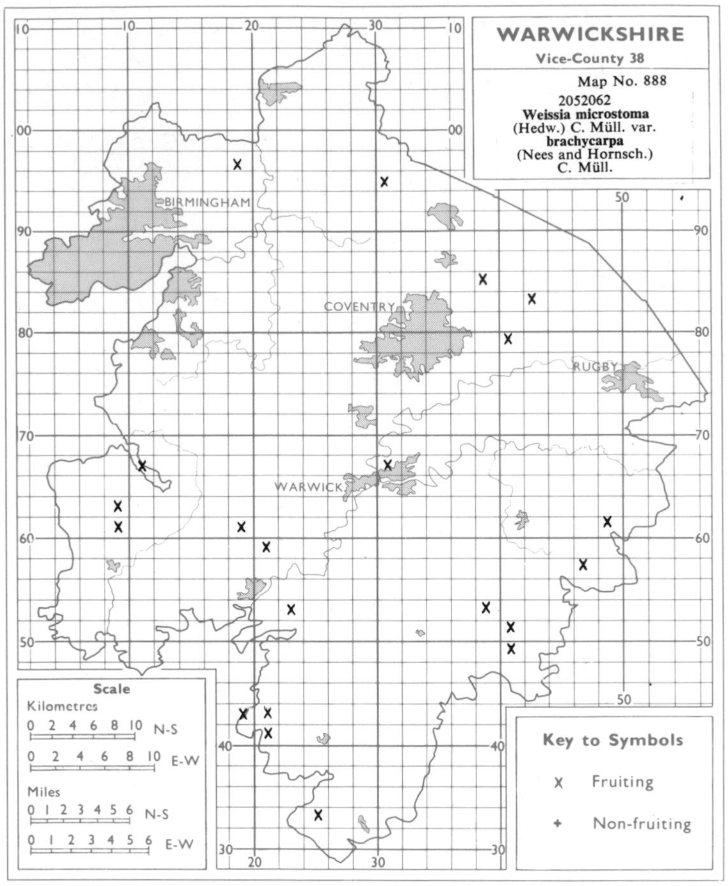
WARWICKSHIRE
Vice-County 38
Map No. 888
2052062
Weissia microstoma
(Hedw.) C. Müll. var.
brachycarpa
(Nees and Hornsch.)
C. Müll.
BIRMINGHAM
COVENTRY
RUGBY
WARWICK
Scale
Kilometres
0 2 4 6 8 10 N-S
0 2 4 6 8 10 E-W
Miles
0 1 2 3 4 5 6 N-S
0 1 2 3 4 5 6 E-W
Key to Symbols
x Fruiting
+ Non-fruiting

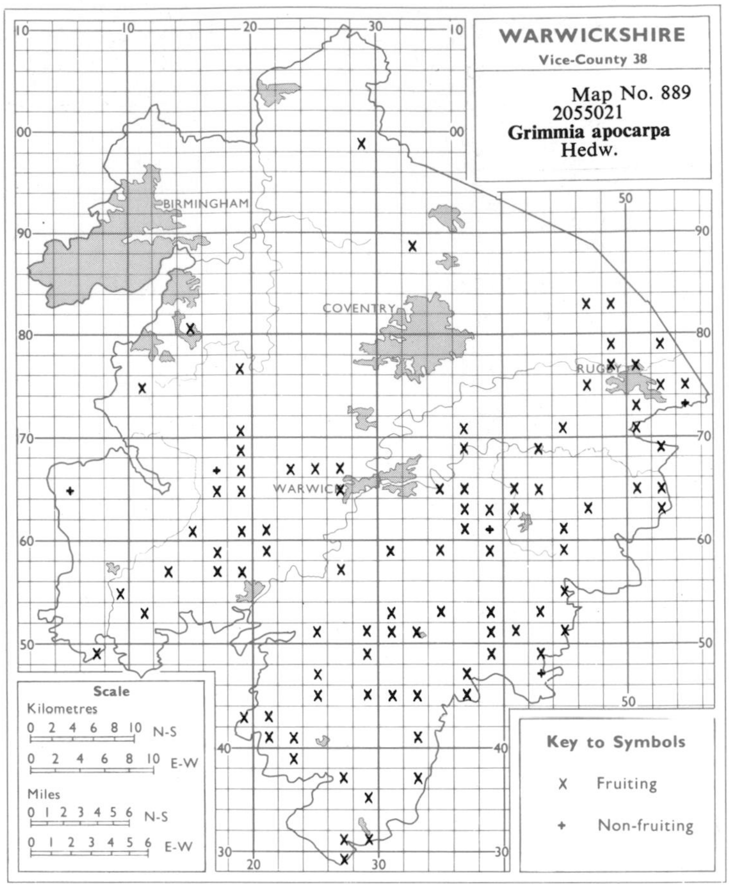
WARWICKSHIRE
Vice-County 38
Map No. 889
2055021
Grimmia apocarpa
Hedw.
BIRMINGHAM
COVENTRY
RUGBY
WARWICK
Scale
Kilometres
0 2 4 6 8 10 N-S
0 2 4 6 8 10 E-W
Miles
0 1 2 3 4 5 6 N-S
0 1 2 3 4 5 6 E-W
Key to Symbols
x Fruiting
+ Non-fruiting

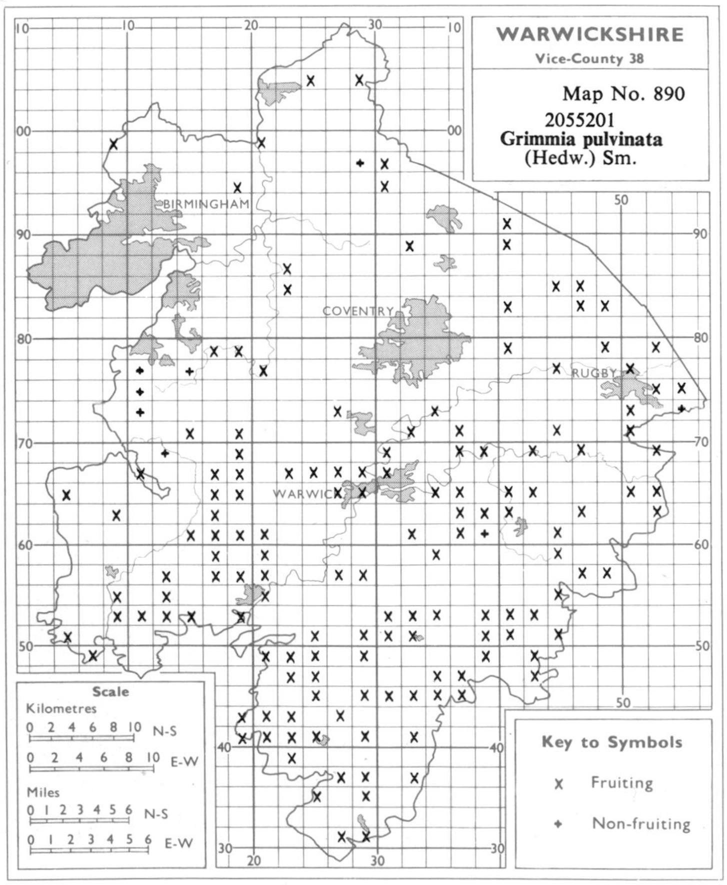
WARWICKSHIRE
Vice-County 38
Map No. 890
2055201
Grimmia pulvinata
(Hedw.) Sm.
BIRMINGHAM
COVENTRY
RUGBY
WARWICK
Scale
Kilometres
0 2 4 6 8 10 N-S
0 2 4 6 8 10 E-W
Miles
0 1 2 3 4 5 6 N-S
0 1 2 3 4 5 6 E-W
Key to Symbols
x Fruiting
+ Non-fruiting

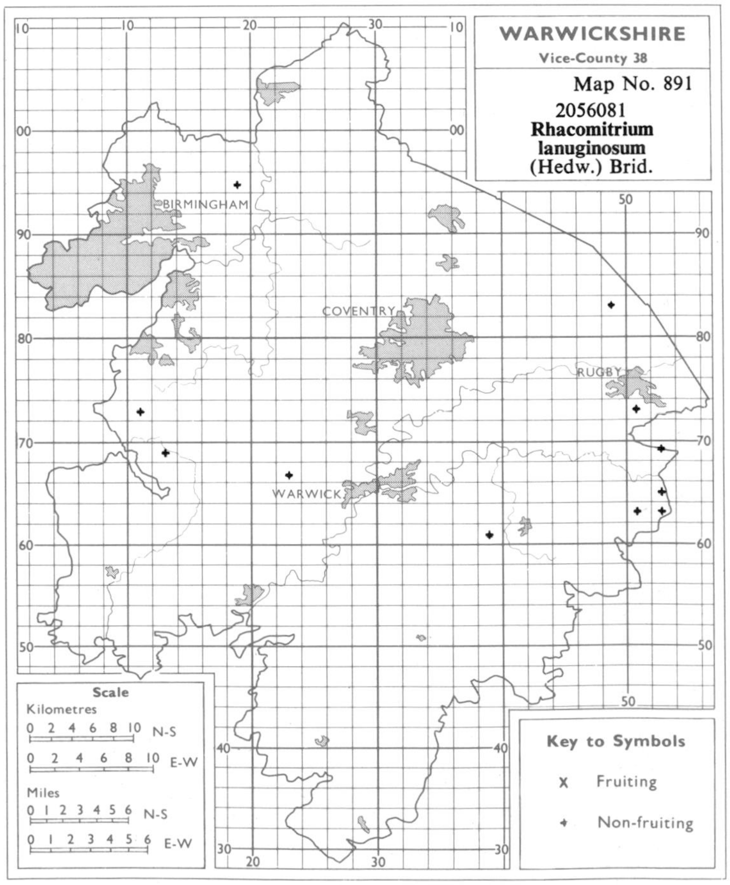
WARWICKSHIRE
Vice-County 38
Map No. 891
2056081
Rhacomitrium
lanuginosum
(Hedw.) Brid.
BIRMINGHAM
COVENTRY
RUGBY
WARWICK
Scale
Kilometres
0 2 4 6 8 10 N-S
0 2 4 6 8 10 E-W
Miles
0 1 2 3 4 5 6 N-S
0 1 2 3 4 5 6 E-W
Key to Symbols
x Fruiting
+ Non-fruiting

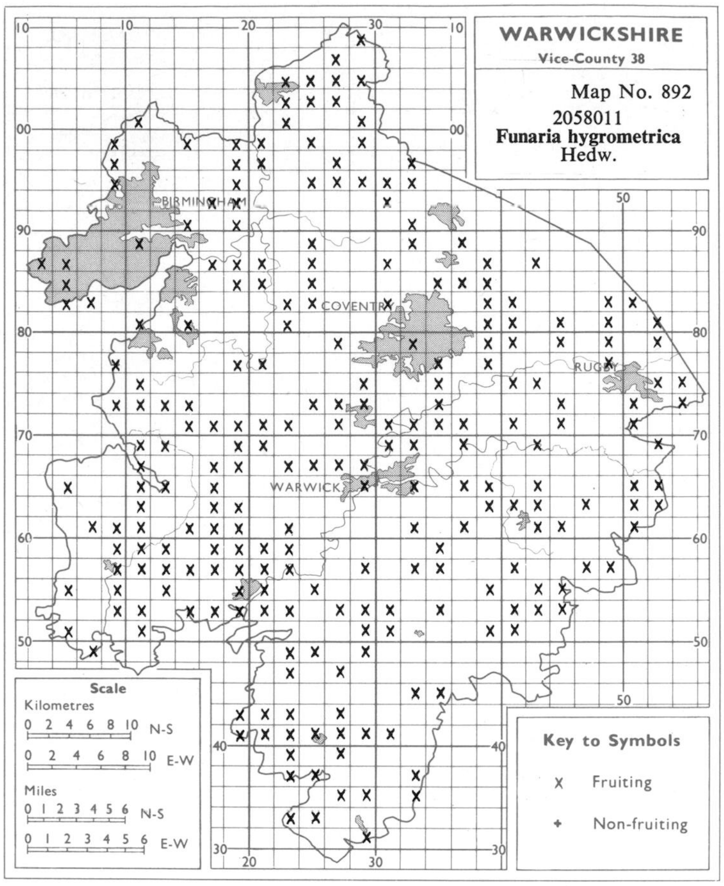
WARWICKSHIRE
Vice-County 38
Map No. 892
2058011
Funaria hygrometrica
Hedw.
BIRMINGHAM
COVENTRY
WARWICK
RUGBY
Scale
Kilometres
0 2 4 6 8 10 N-S
0 2 4 6 8 10 E-W
Miles
0 1 2 3 4 5 6 N-S
0 1 2 3 4 5 6 E-W
Key to Symbols
X Fruiting
+ Non-fruiting

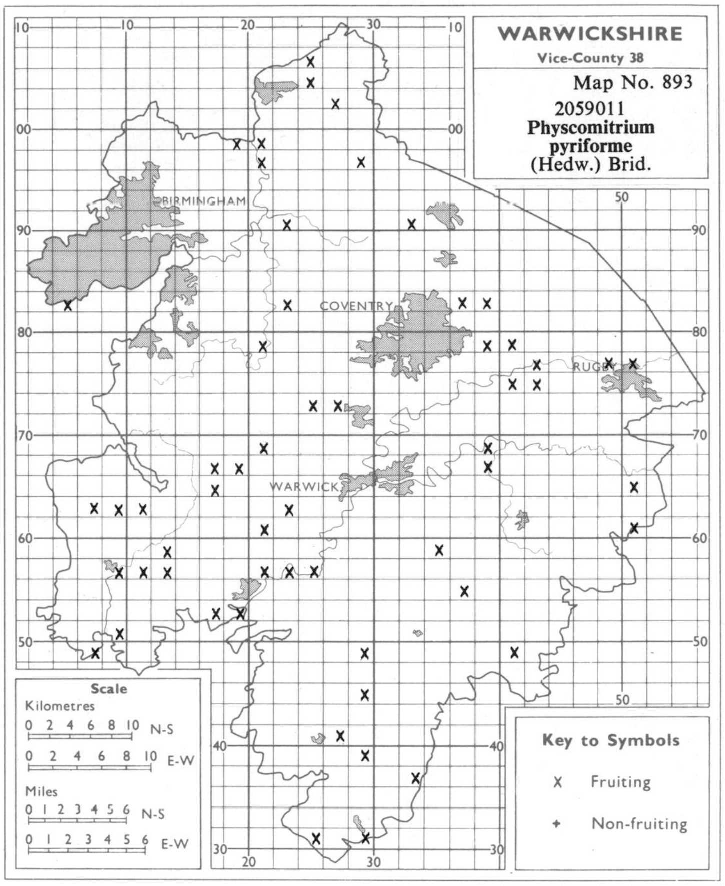
WARWICKSHIRE
Vice-County 38
Map No. 893
2059011
Physcomitrium
pyriforme
(Hedw.) Brid.
BIRMINGHAM
COVENTRY
WARWICK
RUGBY
Scale
Kilometres
0 2 4 6 8 10 N-S
0 2 4 6 8 10 E-W
Miles
0 1 2 3 4 5 6 N-S
0 1 2 3 4 5 6 E-W
Key to Symbols
X Fruiting
+ Non-fruiting

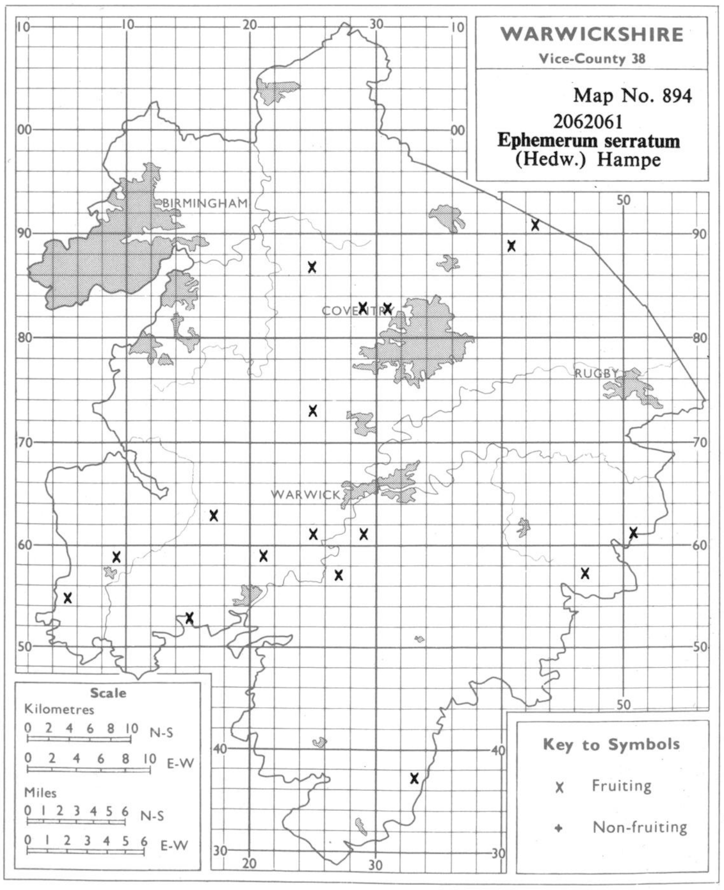
WARWICKSHIRE
Vice-County 38
Map No. 894
2062061
Ephemerum serratum
(Hedw.) Hampe
BIRMINGHAM
COVENTRY
WARWICK
RUGBY
Scale
Kilometres
0 2 4 6 8 10 N-S
0 2 4 6 8 10 E-W
Miles
0 1 2 3 4 5 6 N-S
0 1 2 3 4 5 6 E-W
Key to Symbols
X Fruiting
+ Non-fruiting

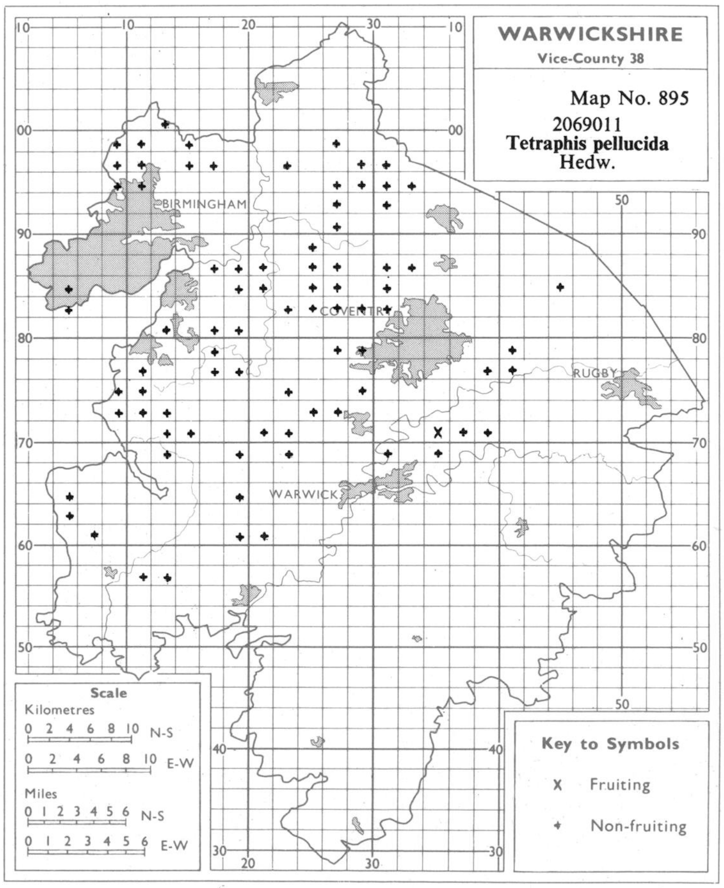
WARWICKSHIRE
Vice-County 38
Map No. 895
2069011
Tetraphis pellucida
Hedw.
BIRMINGHAM
COVENTRY
WARWICK
RUGBY
Scale
Kilometres
0 2 4 6 8 10 N-S
0 2 4 6 8 10 E-W
Miles
0 1 2 3 4 5 6 N-S
0 1 2 3 4 5 6 E-W
Key to Symbols
X Fruiting
+ Non-fruiting

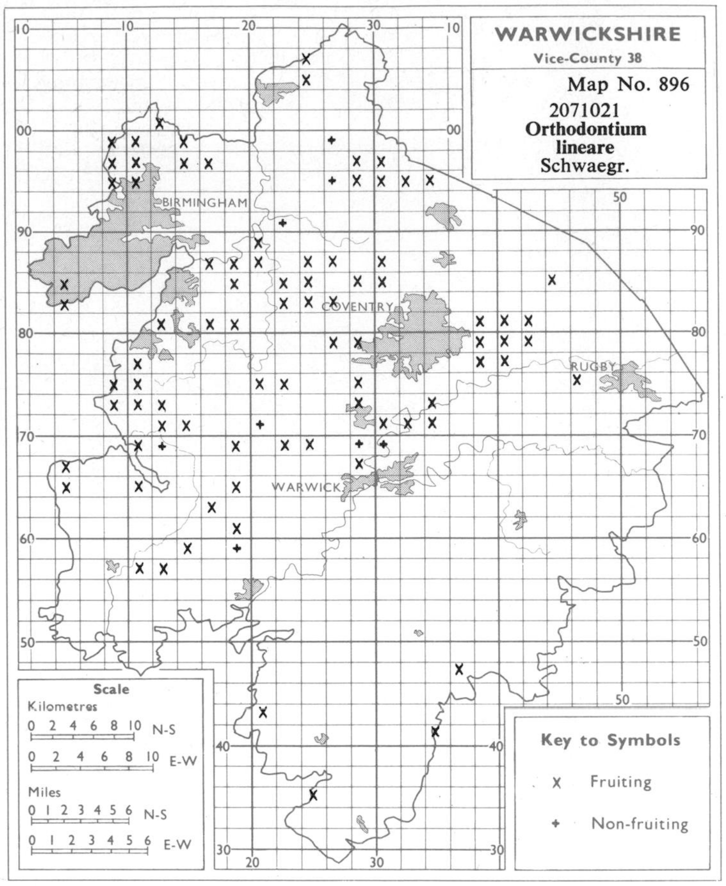
WARWICKSHIRE
Vice-County 38
Map No. 896
2071021
Orthodontium lineare
Schwaegr.
BIRMINGHAM
COVENTRY
RUGBY
WARWICK
Scale
Kilometres
0 2 4 6 8 10 N-S
0 2 4 6 8 10 E-W
Miles
0 1 2 3 4 5 6 N-S
0 1 2 3 4 5 6 E-W
Key to Symbols
X Fruiting
+ Non-fruiting

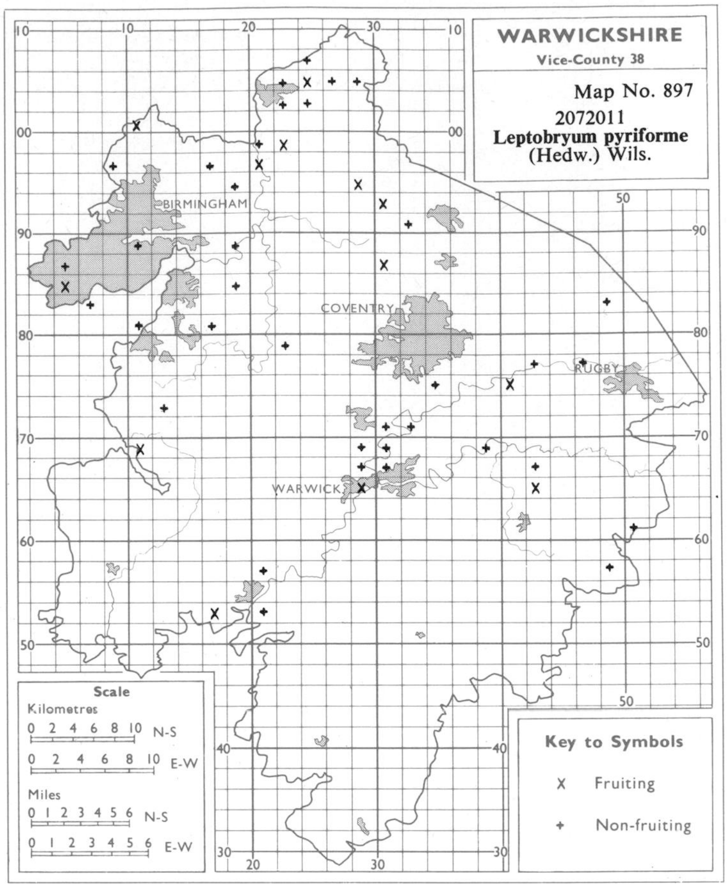
WARWICKSHIRE
Vice-County 38
Map No. 897
2072011
Leptobryum pyriforme
(Hedw.) Wils.
BIRMINGHAM
COVENTRY
RUGBY
WARWICK
Scale
Kilometres
0 2 4 6 8 10 N-S
0 2 4 6 8 10 E-W
Miles
0 1 2 3 4 5 6 N-S
0 1 2 3 4 5 6 E-W
Key to Symbols
X Fruiting
+ Non-fruiting

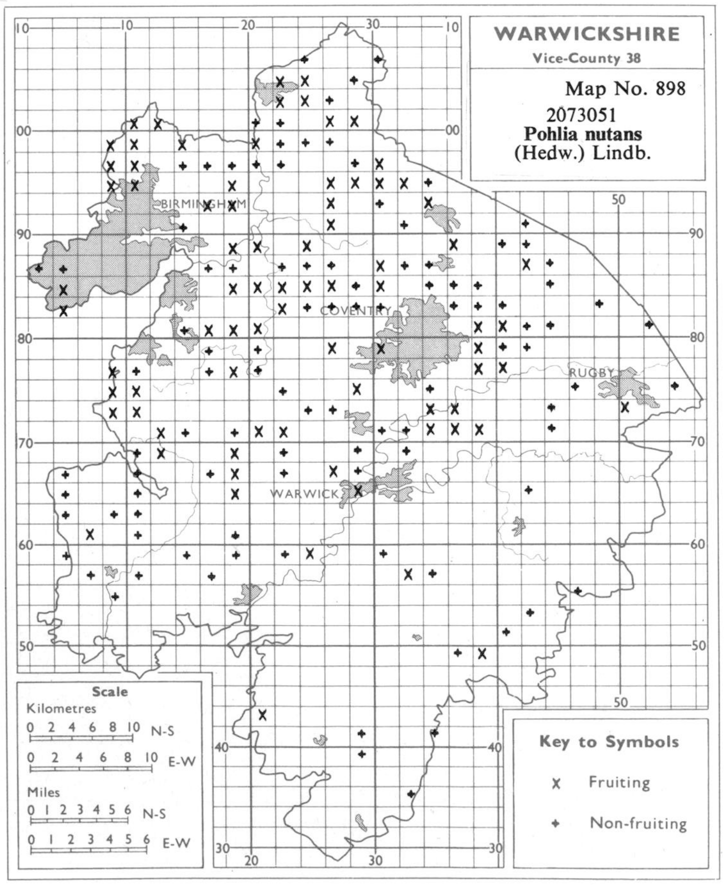
WARWICKSHIRE
Vice-County 38
Map No. 898
2073051
Pohlia nutans
(Hedw.) Lindb.
BIRMINGHAM
COVENTRY
RUGBY
WARWICK
Scale
Kilometres
0 2 4 6 8 10 N-S
0 2 4 6 8 10 E-W
Miles
0 1 2 3 4 5 6 N-S
0 1 2 3 4 5 6 E-W
Key to Symbols
X Fruiting
+ Non-fruiting

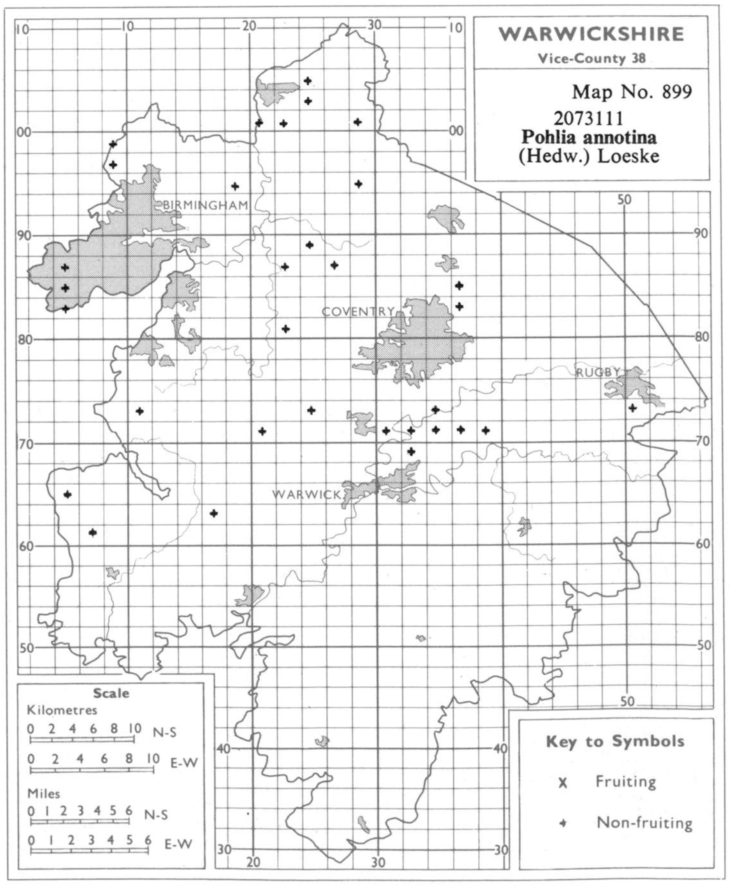
WARWICKSHIRE
Vice-County 38
Map No. 899
2073111
Pohlia annotina
(Hedw.) Loeske
BIRMINGHAM
COVENTRY
RUGBY
WARWICK
Scale
Kilometres
0 2 4 6 8 10 N-S
0 2 4 6 8 10 E-W
Miles
0 1 2 3 4 5 6 N-S
0 1 2 3 4 5 6 E-W
Key to Symbols
X Fruiting
+ Non-fruiting

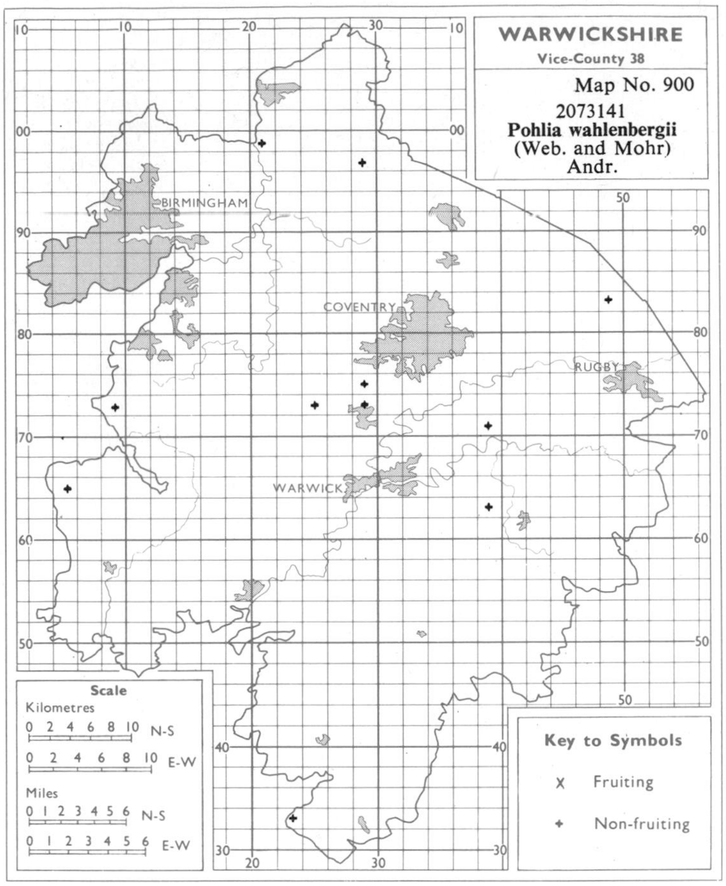
WARWICKSHIRE
Vice-County 38
Map No. 900
2073141
Pohlia wahlenbergii
(Web. and Mohr)
Andr.
BIRMINGHAM
COVENTRY
RUGBY
WARWICK
Scale
Kilometres
0 2 4 6 8 10 N-S
0 2 4 6 8 10 E-W
Miles
0 1 2 3 4 5 6 N-S
0 1 2 3 4 5 6 E-W
Key to Symbols
X Fruiting
+ Non-fruiting

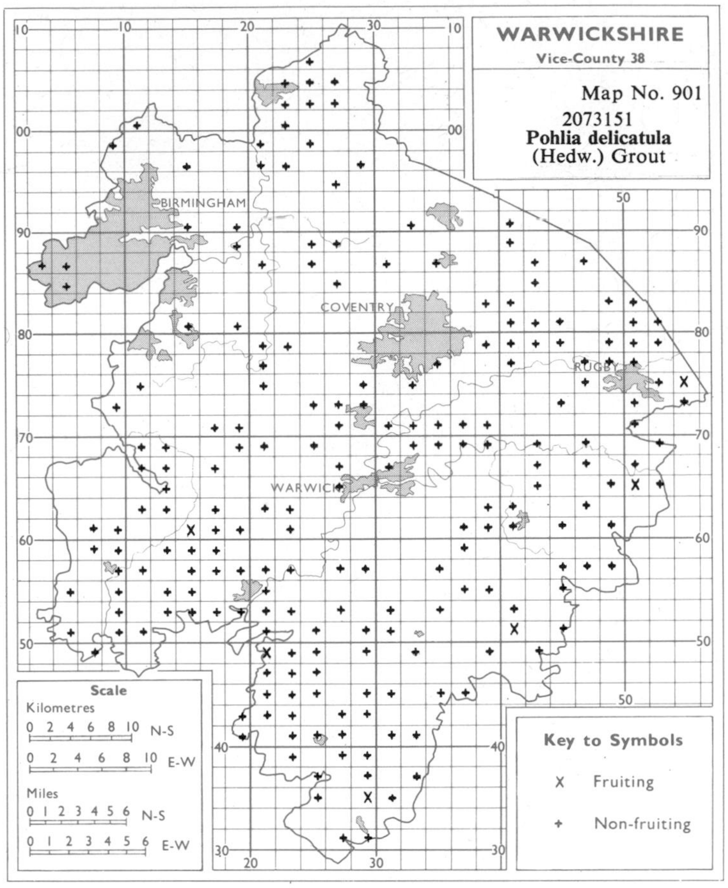
WARWICKSHIRE
Vice-County 38
Map No. 901
2073151
Pohlia delicatula
(Hedw.) Grout
BIRMINGHAM
COVENTRY
RUGBY
WARWICK
Scale
Kilometres
0 2 4 6 8 10 N-S
0 2 4 6 8 10 E-W
Miles
0 1 2 3 4 5 6 N-S
0 1 2 3 4 5 6 E-W
Key to Symbols
X Fruiting
+ Non-fruiting

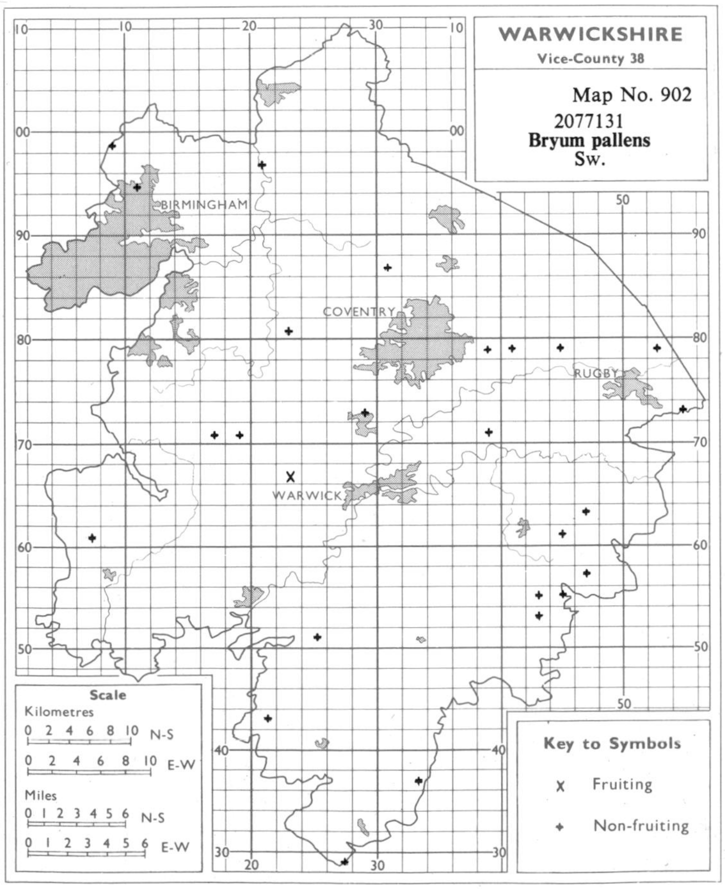
WARWICKSHIRE
Vice-County 38
Map No. 902
2077131
Bryum pallens
Sw.
BIRMINGHAM
COVENTRY
RUGBY
WARWICK
Scale
Kilometres
0 2 4 6 8 10 N-S
0 2 4 6 8 10 E-W
Miles
0 1 2 3 4 5 6 N-S
0 1 2 3 4 5 6 E-W
Key to Symbols
X Fruiting
+ Non-fruiting

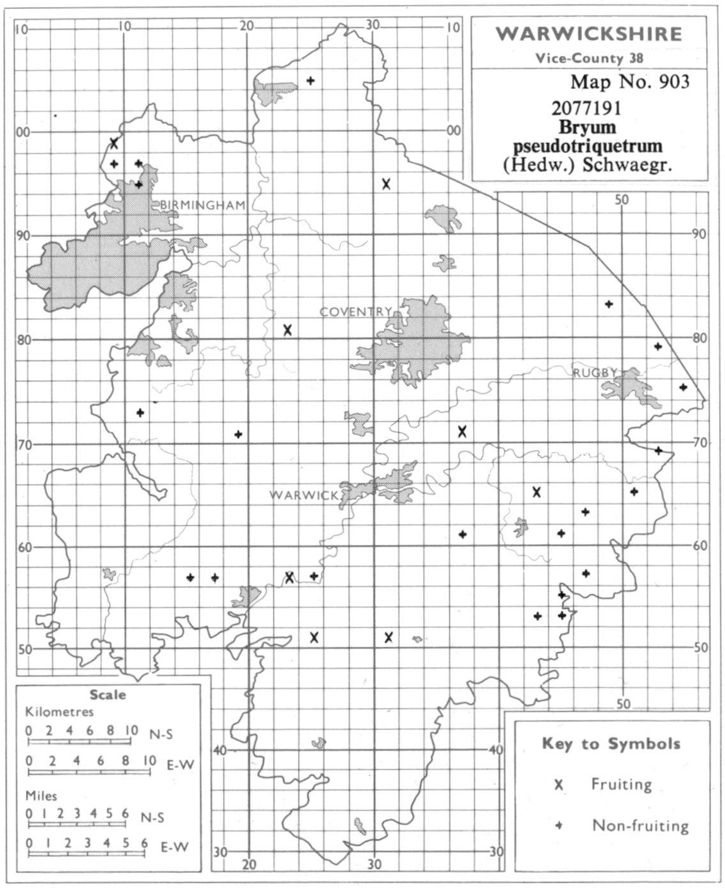
WARWICKSHIRE
Vice-County 38
Map No. 903
2077191
Bryum
pseudotriquetrum
(Hedw.) Schwaegr.
BIRMINGHAM
COVENTRY
RUGBY
WARWICK
Scale
Kilometres
0 2 4 6 8 10 N-S
0 2 4 6 8 10 E-W
Miles
0 1 2 3 4 5 6 N-S
0 1 2 3 4 5 6 E-W
Key to Symbols
X Fruiting
+ Non-fruiting

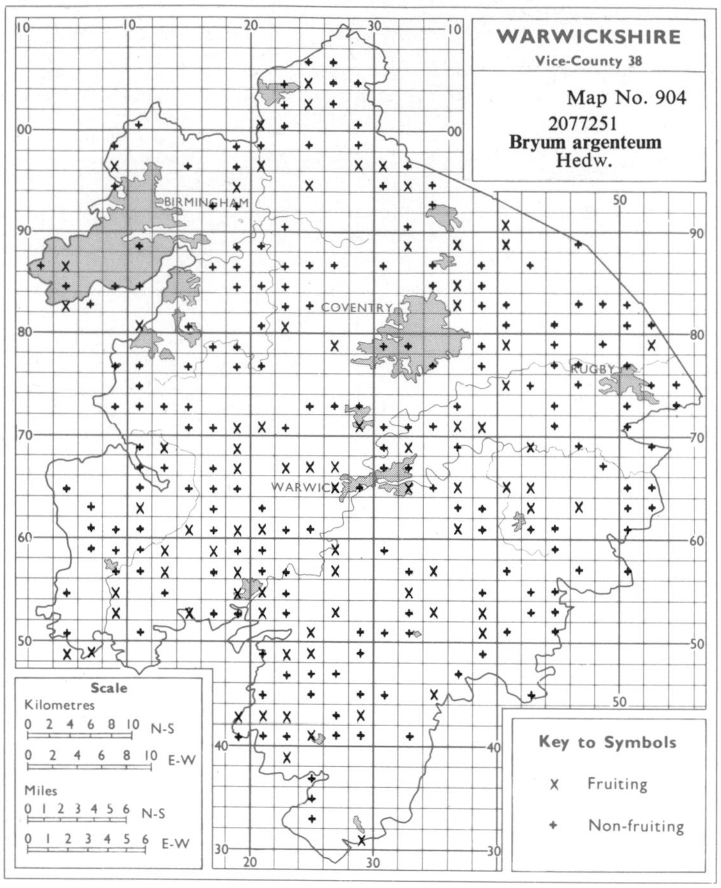
WARWICKSHIRE
Vice-County 38
Map No. 904
2077251
Bryum argenteum
Hedw.
BIRMINGHAM
COVENTRY
RUGBY
WARWICK
Scale
Kilometres
0 2 4 6 8 10 N-S
0 2 4 6 8 10 E-W
Miles
0 1 2 3 4 5 6 N-S
0 1 2 3 4 5 6 E-W
Key to Symbols
X Fruiting
+ Non-fruiting

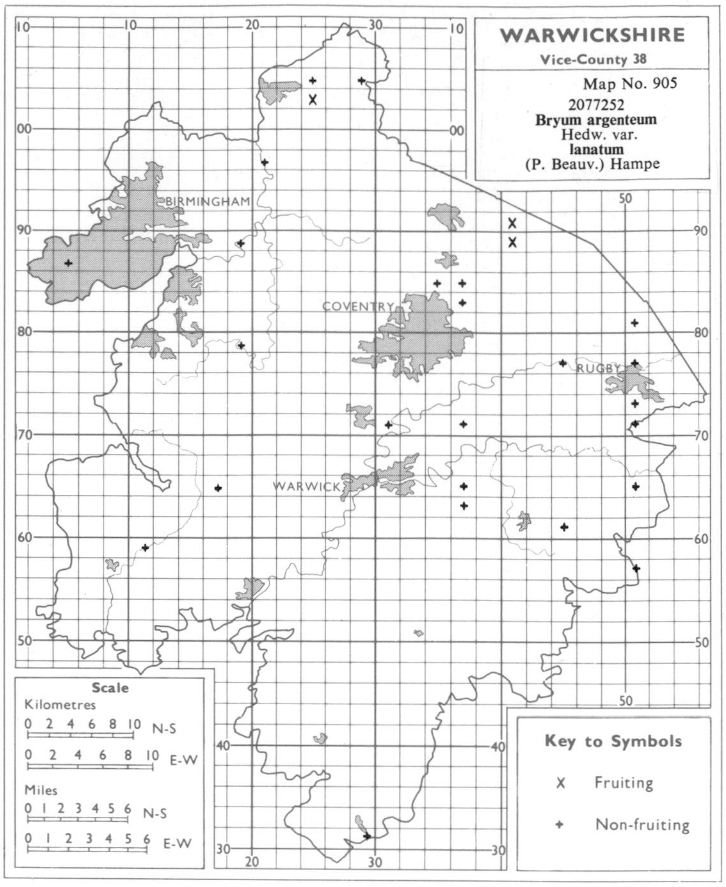
WARWICKSHIRE
Vice-County 38
Map No. 905
2077252
Bryum argenteum
Hedw. var.
lanatum
(P. Beauv.) Hampe
BIRMINGHAM
COVENTRY
RUGBY
WARWICK
Scale
Kilometres
0 2 4 6 8 10 N-S
0 2 4 6 8 10 E-W
Miles
0 1 2 3 4 5 6 N-S
0 1 2 3 4 5 6 E-W
Key to Symbols
X Fruiting
+ Non-fruiting

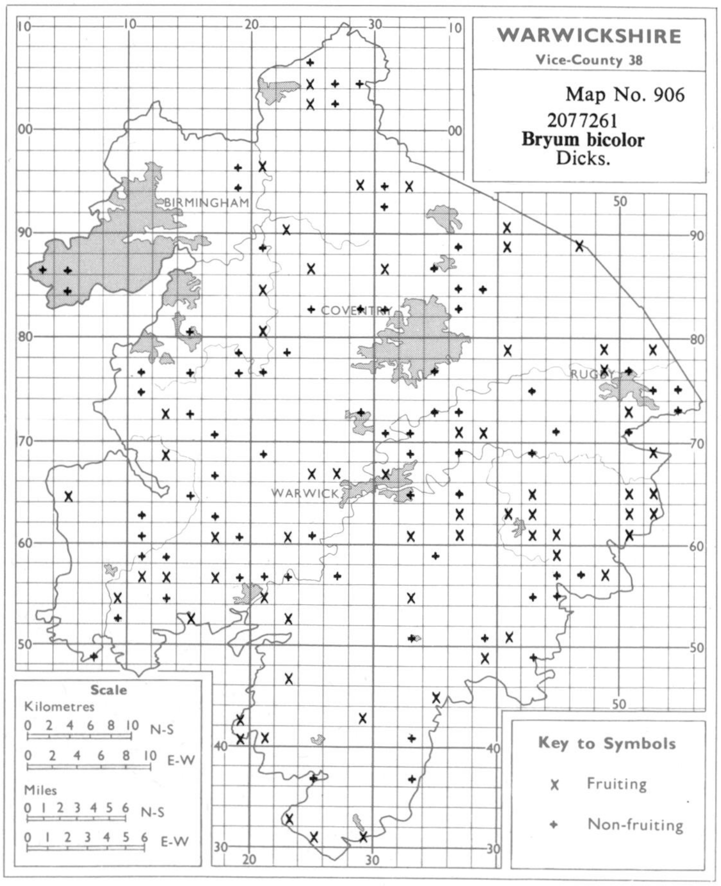
WARWICKSHIRE
Vice-County 38
Map No. 906
2077261
Bryum bicolor
Dicks.
BIRMINGHAM
COVENTRY
RUGBY
WARWICK
Scale
Kilometres
0 2 4 6 8 10 N-S
0 2 4 6 8 10 E-W
Miles
0 1 2 3 4 5 6 N-S
0 1 2 3 4 5 6 E-W
Key to Symbols
X Fruiting
+ Non-fruiting

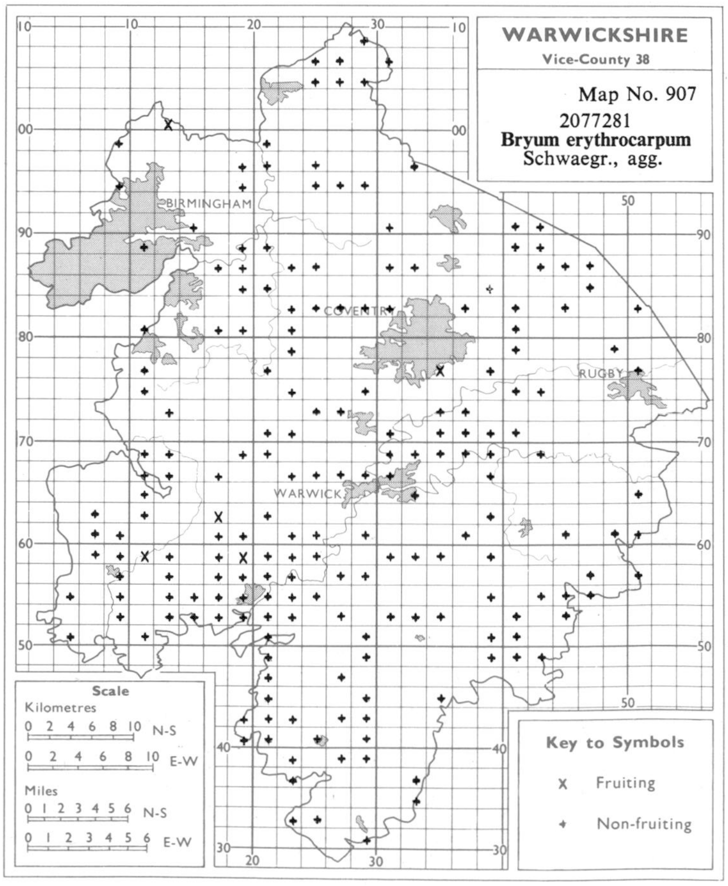
WARWICKSHIRE
Vice-County 38
Map No. 907
2077281
Bryum erythrocarpum
Schwaegr., agg.
BIRMINGHAM
COVENTRY
RUGBY
WARWICK
Scale
Kilometres
0 2 4 6 8 10 N-S
0 2 4 6 8 10 E-W
Miles
0 1 2 3 4 5 6 N-S
0 1 2 3 4 5 6 E-W
Key to Symbols
X Fruiting
+ Non-fruiting

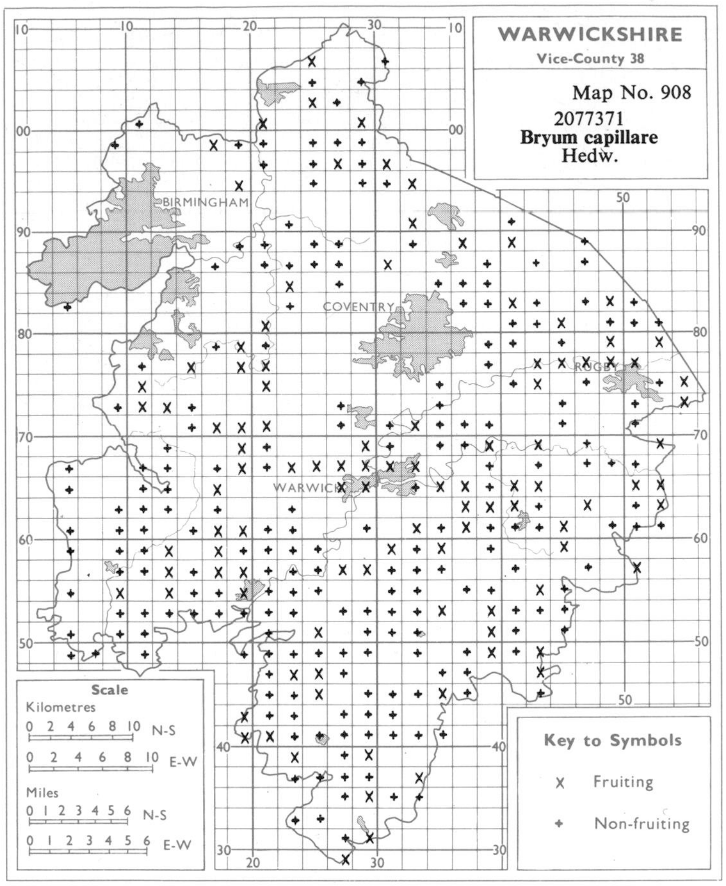
WARWICKSHIRE
Vice-County 38
Map No. 908
2077371
Bryum capillare
Hedw.
BIRMINGHAM
COVENTRY
RUGBY
WARWICK
Scale
Kilometres
0 2 4 6 8 10 N-S
0 2 4 6 8 10 E-W
Miles
0 1 2 3 4 5 6 N-S
0 1 2 3 4 5 6 E-W
Key to Symbols
X Fruiting
+ Non-fruiting

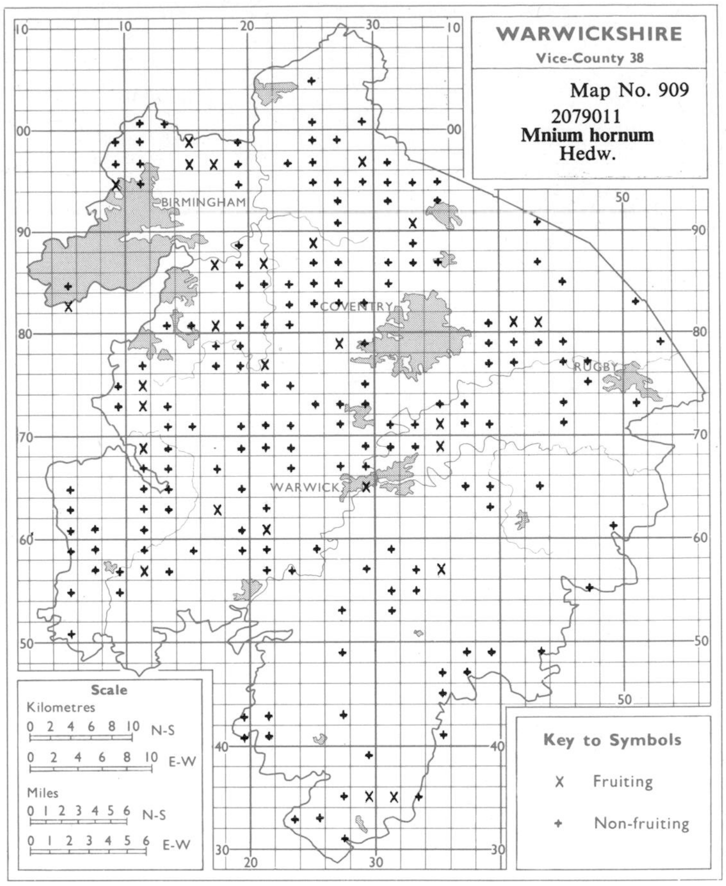
WARWICKSHIRE
Vice-County 38
Map No. 909
2079011
Mnium hornum
Hedw.
BIRMINGHAM
COVENTRY
RUGBY
WARWICK
Scale
Kilometres
0 2 4 6 8 10 N-S
0 2 4 6 8 10 E-W
Miles
0 1 2 3 4 5 6 N-S
0 1 2 3 4 5 6 E-W
Key to Symbols
X Fruiting
+ Non-fruiting

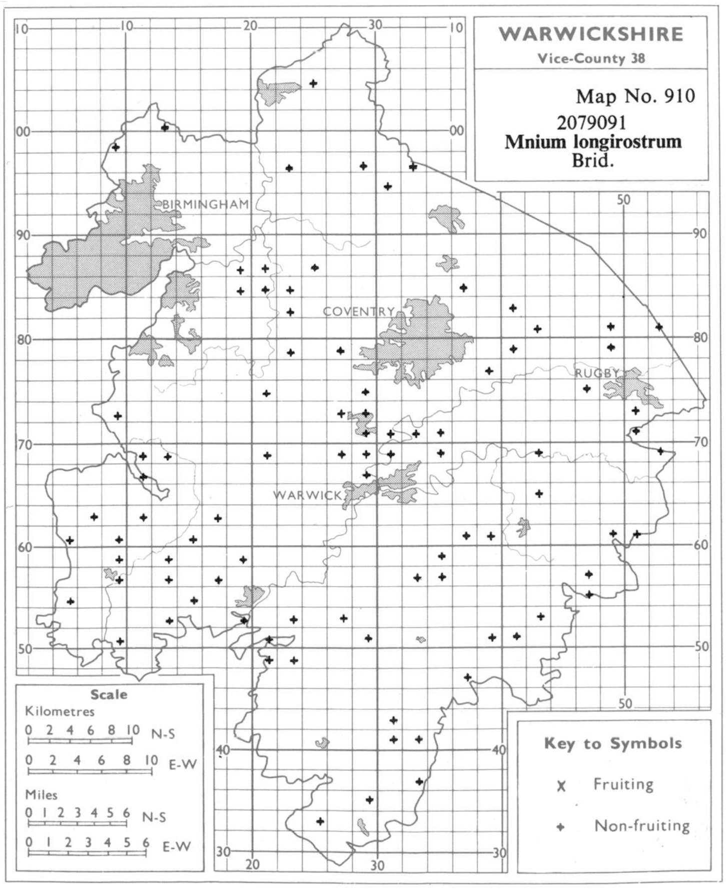
WARWICKSHIRE
Vice-County 38
Map No. 910
2079091
Mnium longirostrum
Brid.
BIRMINGHAM
COVENTRY
RUGBY
WARWICK
Scale
Kilometres
0 2 4 6 8 10 N-S
0 2 4 6 8 10 E-W
Miles
0 1 2 3 4 5 6 N-S
0 1 2 3 4 5 6 E-W
Key to Symbols
X Fruiting
+ Non-fruiting

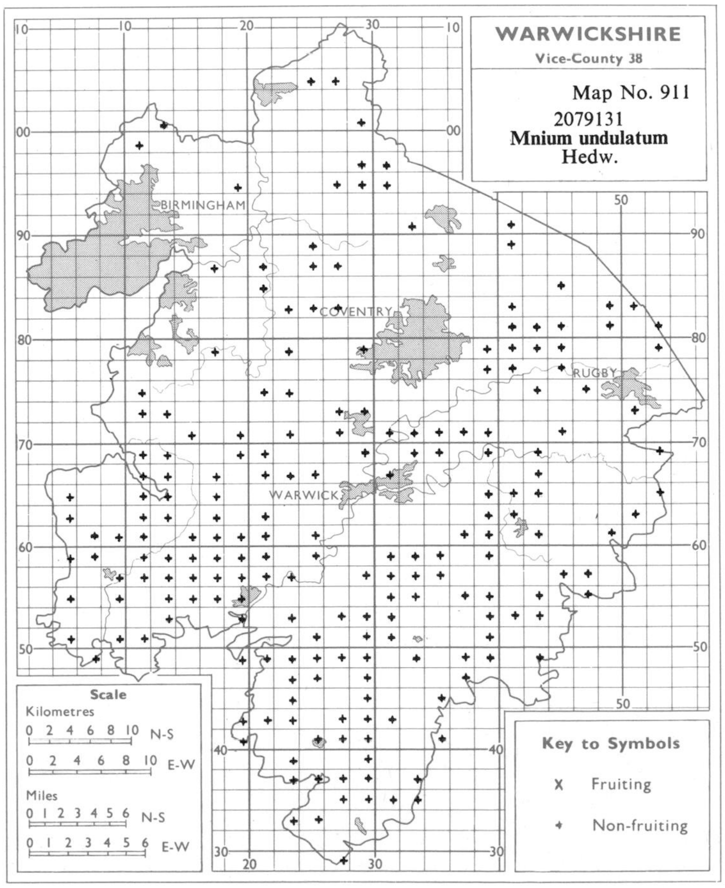
WARWICKSHIRE
Vice-County 38
Map No. 911
2079131
Mnium undulatum
Hedw.
BIRMINGHAM
COVENTRY
RUGBY
WARWICK
Scale
Kilometres
0 2 4 6 8 10 N-S
0 2 4 6 8 10 E-W
Miles
0 1 2 3 4 5 6 N-S
0 1 2 3 4 5 6 E-W
Key to Symbols
X Fruiting
+ Non-fruiting

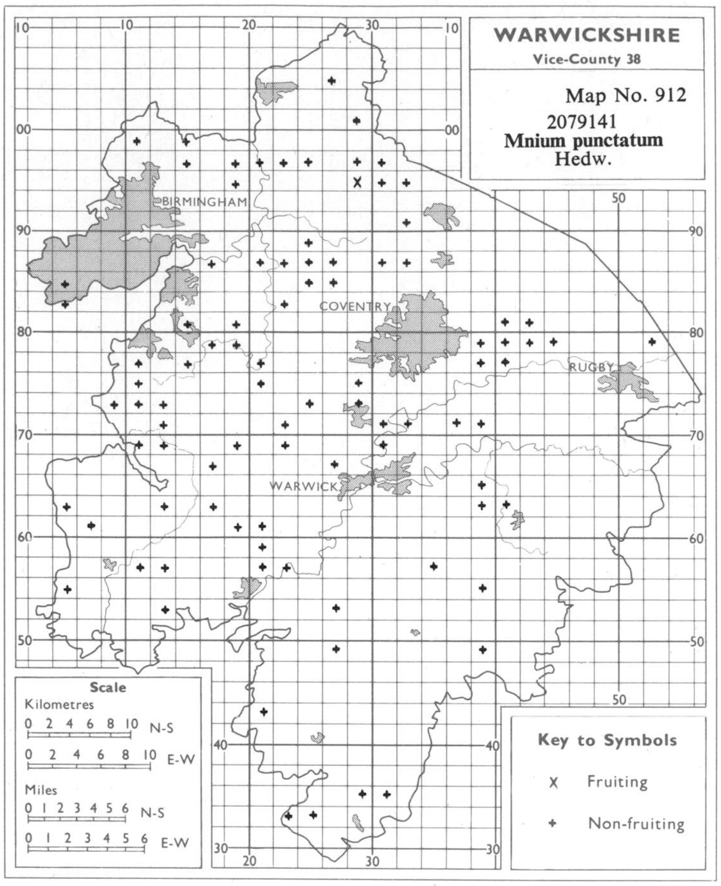
WARWICKSHIRE
Vice-County 38
Map No. 912
2079141
Mnium punctatum
Hedw.
BIRMINGHAM
COVENTRY
RUGBY
WARWICK
Scale
Kilometres
0 2 4 6 8 10 N-S
0 2 4 6 8 10 E-W
Miles
0 1 2 3 4 5 6 N-S
0 1 2 3 4 5 6 E-W
Key to Symbols
X Fruiting
+ Non-fruiting

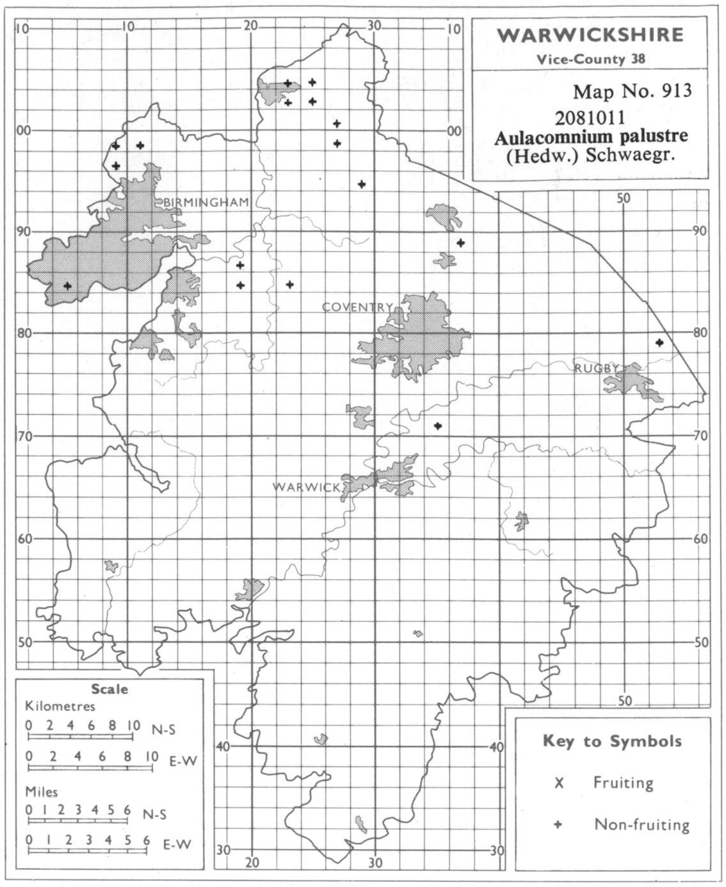
WARWICKSHIRE
Vice-County 38
Map No. 913
2081011
Aulacomnium palustre
(Hedw.) Schwaegr.
BIRMINGHAM
COVENTRY
RUGBY
WARWICK
Scale
Kilometres
0 2 4 6 8 10 N-S
0 2 4 6 8 10 E-W
Miles
0 1 2 3 4 5 6 N-S
0 1 2 3 4 5 6 E-W
Key to Symbols
X Fruiting
+ Non-fruiting

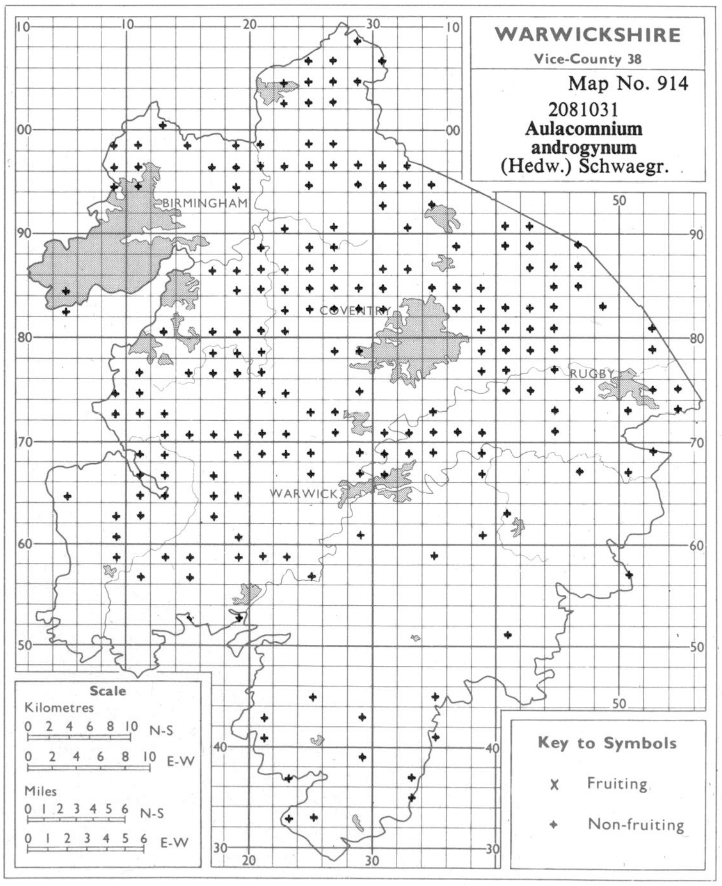
WARWICKSHIRE
Vice-County 38
Map No. 914
2081031
Aulacomnium
androgynum
(Hedw.) Schwaegr.
BIRMINGHAM
COVENTRY
RUGBY
WARWICK
Scale
Kilometres
0 2 4 6 8 10 N-S
0 2 4 6 8 10 E-W
Miles
0 1 2 3 4 5 6 N-S
0 1 2 3 4 5 6 E-W
Key to Symbols
X Fruiting
+ Non-fruiting

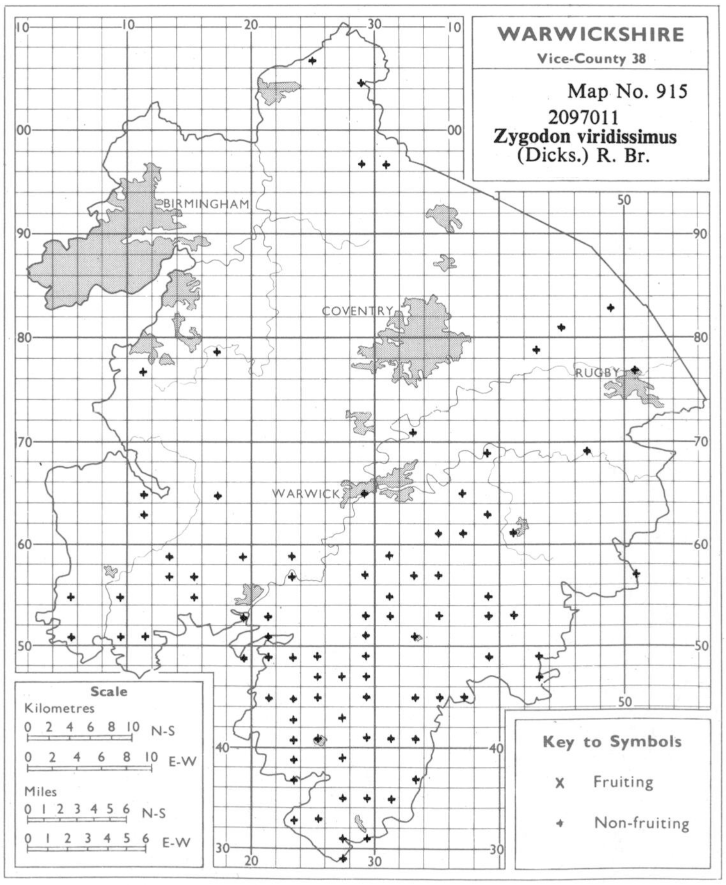
WARWICKSHIRE
Vice-County 38
Map No. 915
2097011
Zygodon viridissimus
(Dicks.) R. Br.
BIRMINGHAM
COVENTRY
RUGBY
WARWICK
Scale
Kilometres
0 2 4 6 8 10 N-S
0 2 4 6 8 10 E-W
Miles
0 1 2 3 4 5 6 N-S
0 1 2 3 4 5 6 E-W
Key to Symbols
X Fruiting
+ Non-fruiting

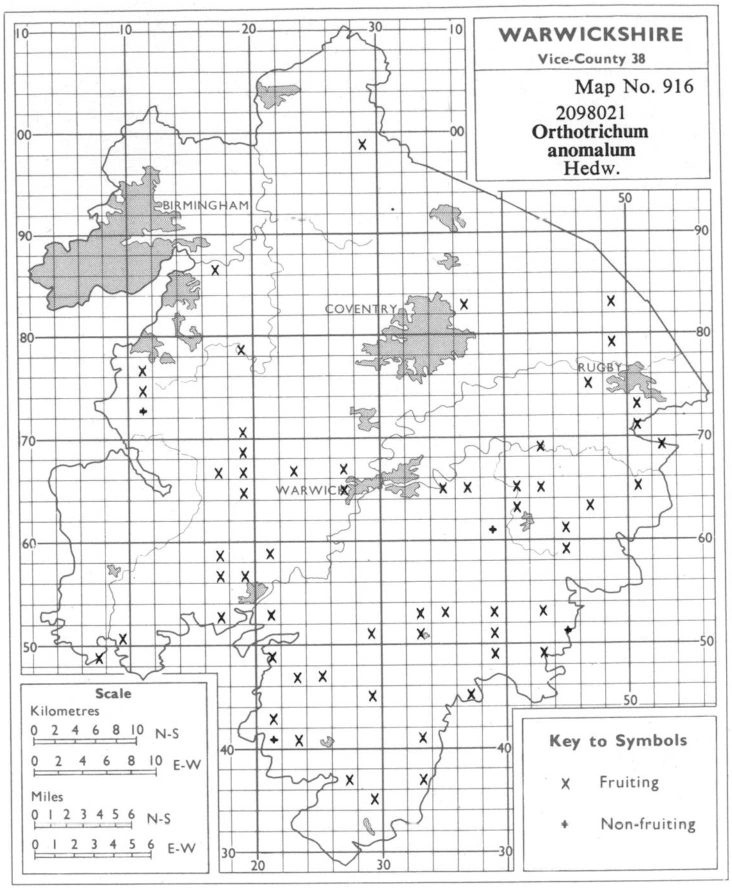
WARWICKSHIRE
Vice-County 38
Map No. 916
2098021
Orthotrichum anomalum Hedw.
BIRMINGHAM
COVENTRY
RUGBY
WARWICK
Scale
Kilometres
Miles
Key to Symbols
X Fruiting
+ Non-fruiting

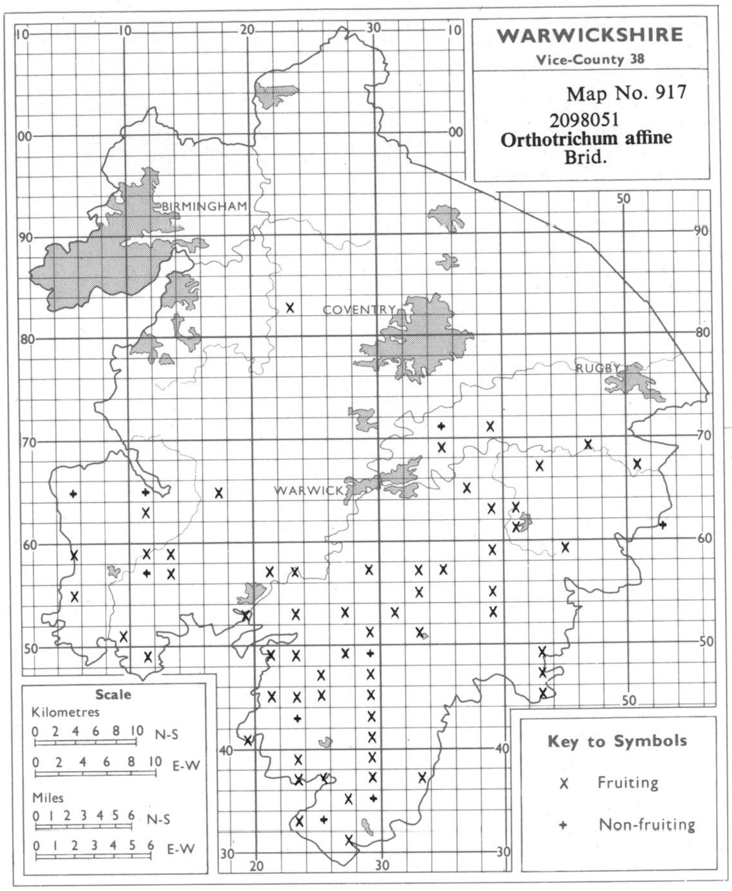
WARWICKSHIRE
Vice-County 38
Map No. 917
2098051
Orthotrichum affine Brid.
BIRMINGHAM
COVENTRY
RUGBY
WARWICK
Scale
Kilometres
Miles
Key to Symbols
X Fruiting
+ Non-fruiting

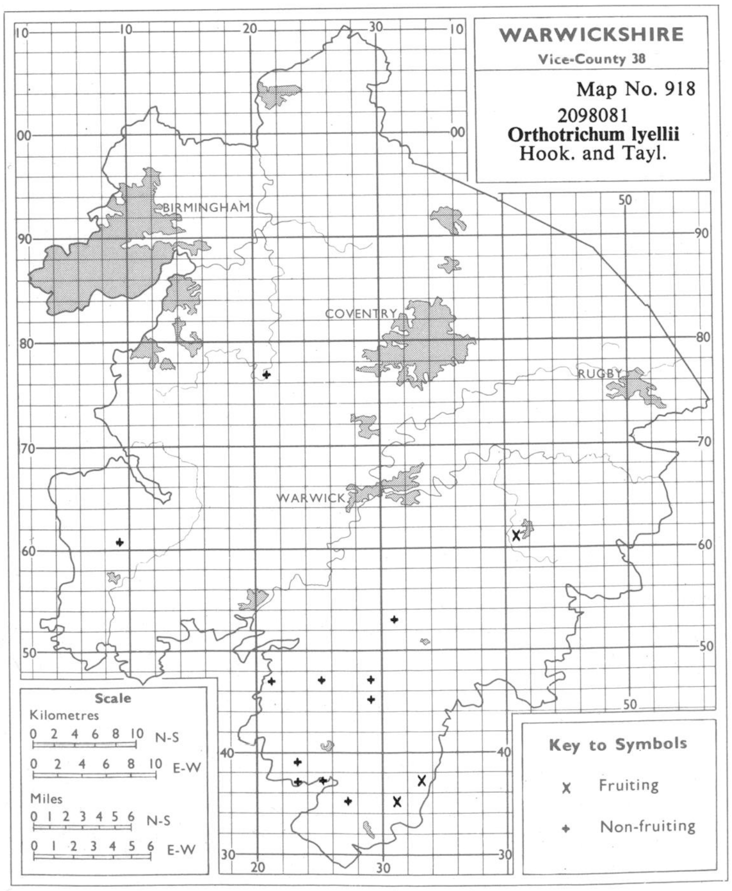
WARWICKSHIRE
Vice-County 38
Map No. 918
2098081
Orthotrichum lyellii Hook. and Tayl.
BIRMINGHAM
COVENTRY
RUGBY
WARWICK
Scale
Kilometres
Miles
Key to Symbols
X Fruiting
+ Non-fruiting

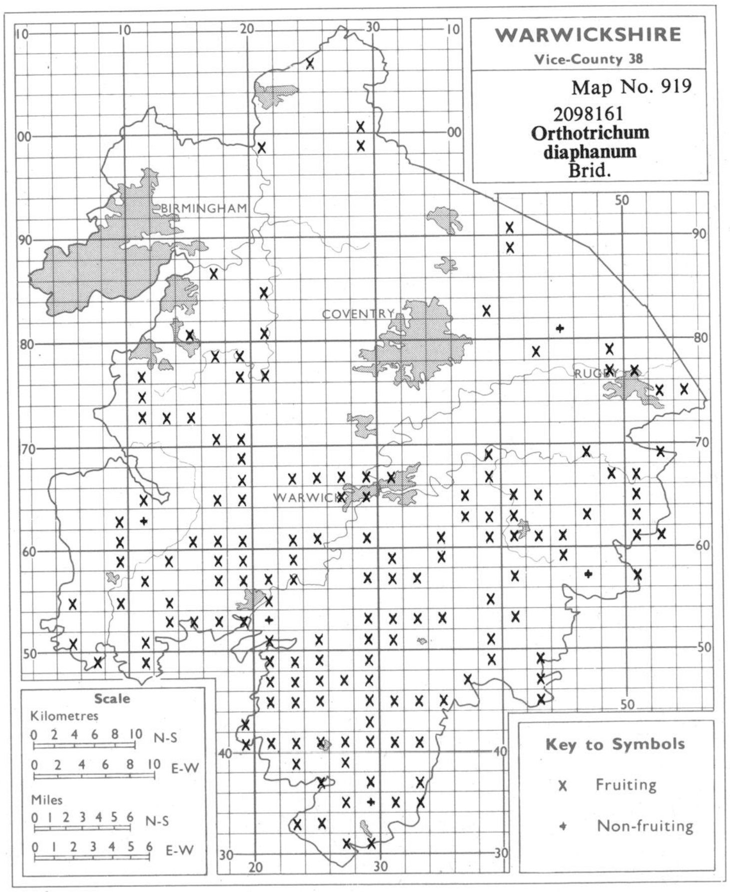
WARWICKSHIRE
Vice-County 38
Map No. 919
2098161
Orthotrichum diaphanum Brid.
BIRMINGHAM
COVENTRY
RUGBY
WARWICK
Scale
Kilometres
Miles
Key to Symbols
X Fruiting
+ Non-fruiting

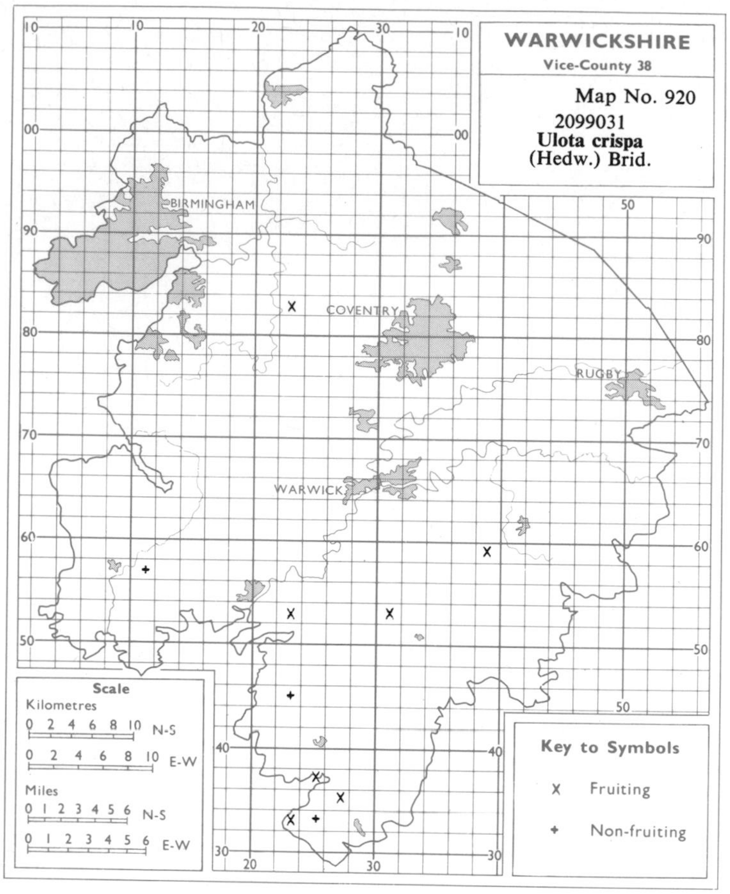
WARWICKSHIRE
Vice-County 38
Map No. 920
2099031
Ulota crispa
(Hedw.) Brid.
BIRMINGHAM
COVENTRY
RUGBY
WARWICK
Scale
Kilometres
Miles
Key to Symbols
X Fruiting
+ Non-fruiting

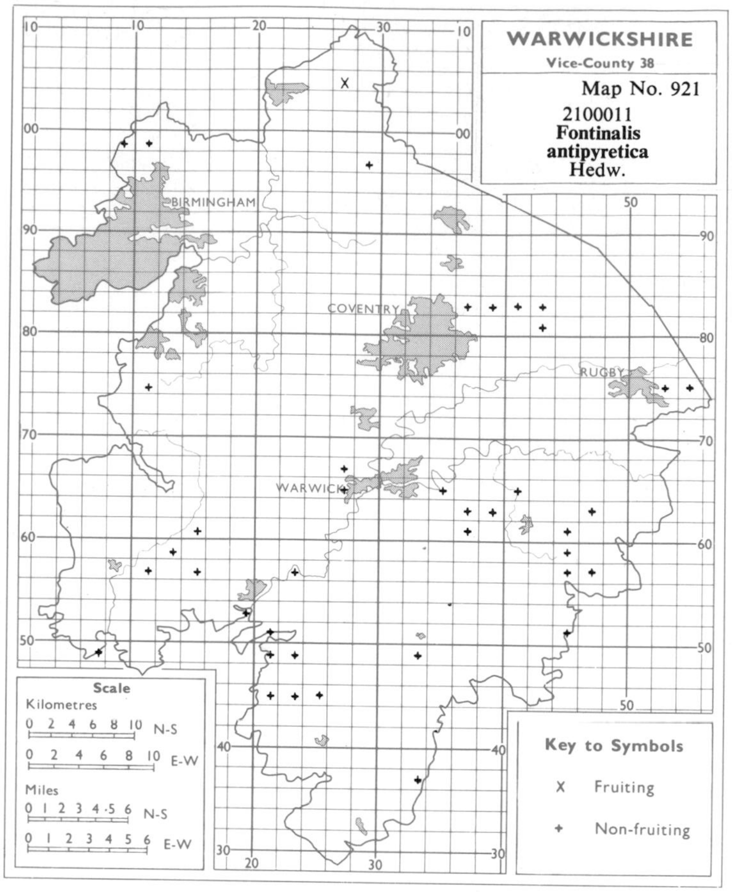
WARWICKSHIRE
Vice-County 38
Map No. 921
2100011
Fontinalis
antipyretica
Hedw.
BIRMINGHAM
COVENTRY
RUGBY
WARWICK
Scale
Kilometres
Miles
Key to Symbols
X Fruiting
+ Non-fruiting

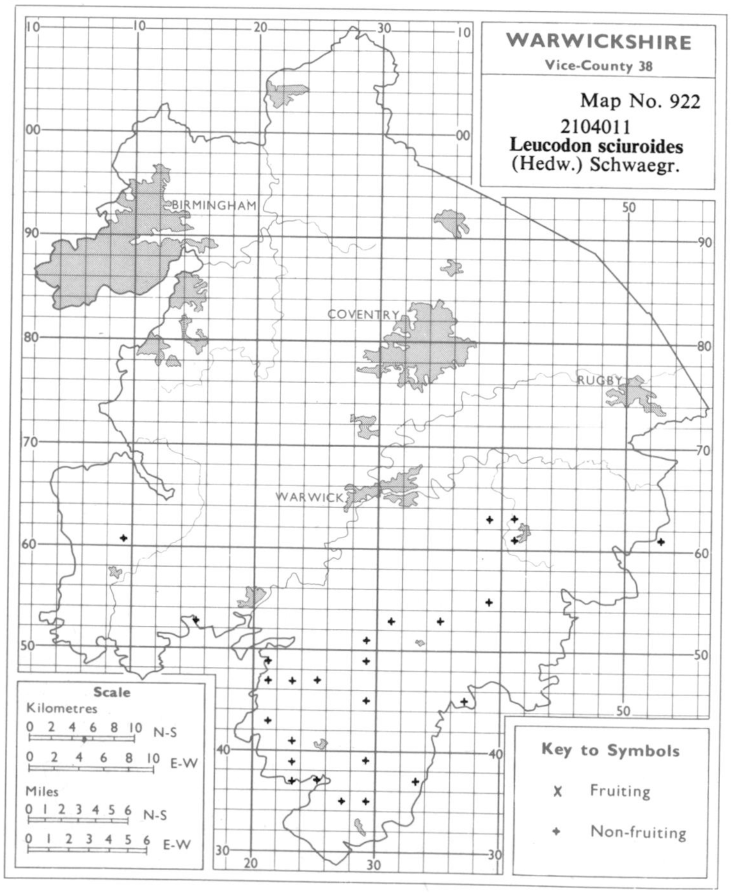
WARWICKSHIRE
Vice-County 38
Map No. 922
2104011
Leucodon sciuroides
(Hedw.) Schwaegr.
BIRMINGHAM
COVENTRY
RUGBY
WARWICK
Scale
Kilometres
Miles
Key to Symbols
X Fruiting
+ Non-fruiting

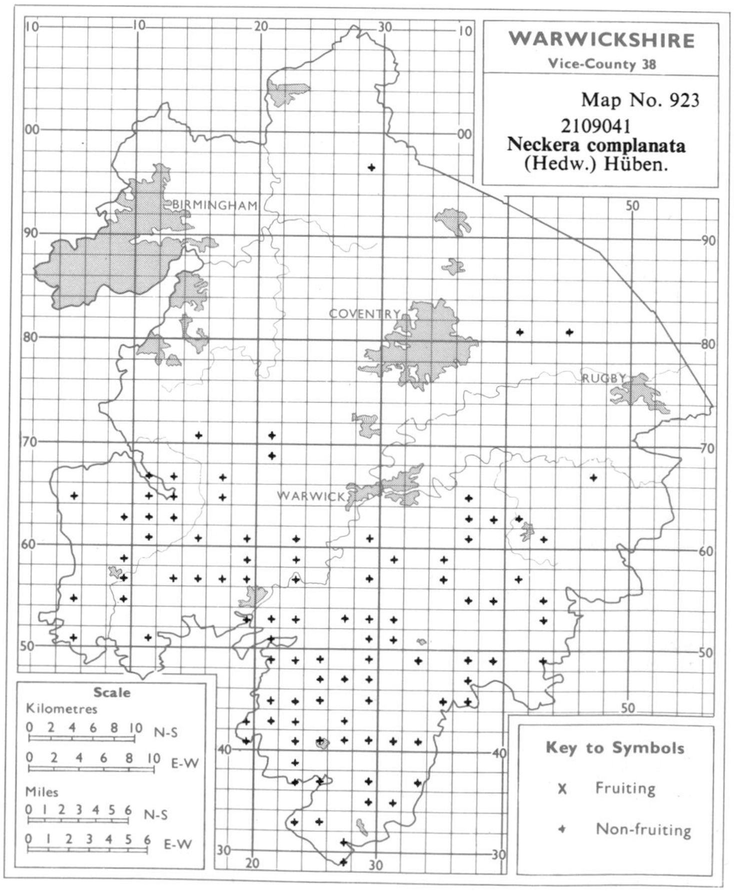
WARWICKSHIRE
Vice-County 38
Map No. 923
2109041
Neckera complanata
(Hedw.) Hüben.
BIRMINGHAM
COVENTRY
RUGBY
WARWICK
Scale
Kilometres
Miles
Key to Symbols
X Fruiting
+ Non-fruiting

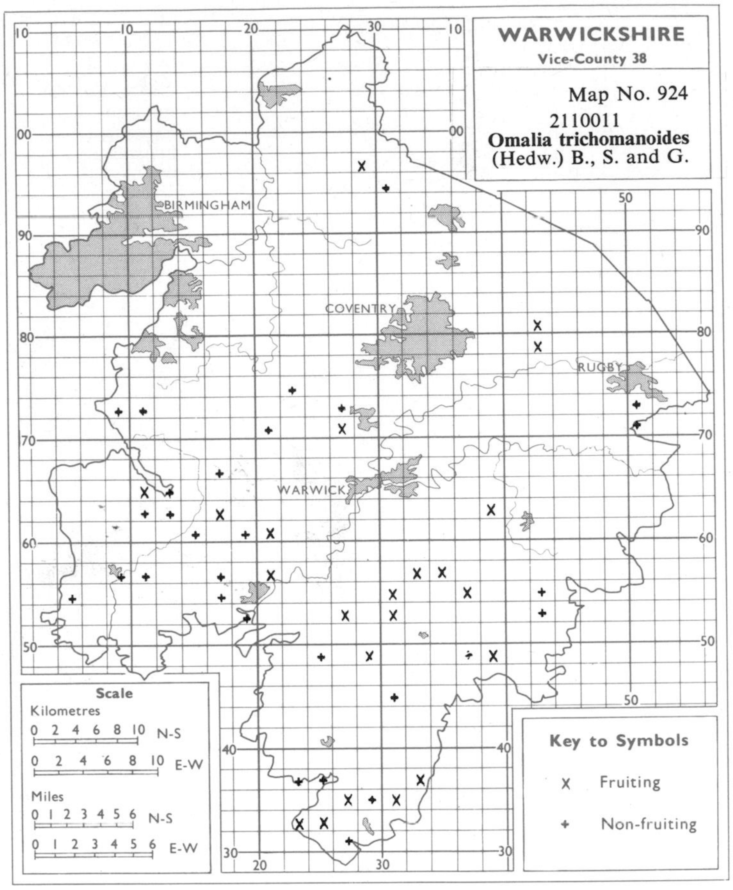
WARWICKSHIRE
Vice-County 38
Map No. 924
2110011
Omalia trichomanoides
(Hedw.) B., S. and G.
BIRMINGHAM
COVENTRY
RUGBY
WARWICK
Scale
Kilometres
0 2 4 6 8 10 N-S
0 2 4 6 8 10 E-W
Miles
0 1 2 3 4 5 6 N-S
0 1 2 3 4 5 6 E-W
Key to Symbols
x Fruiting
+ Non-fruiting

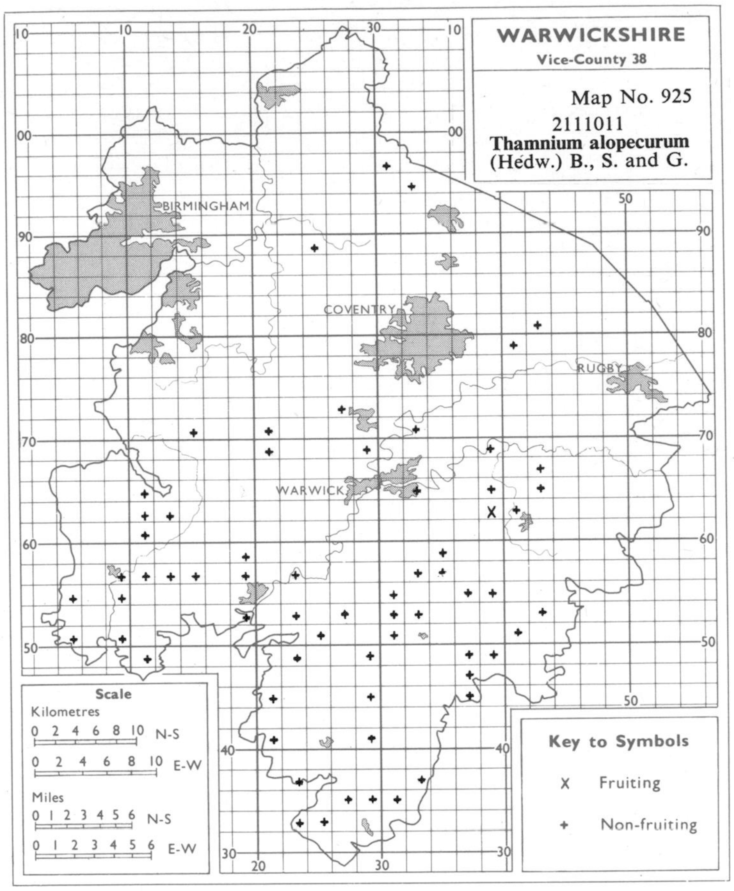
WARWICKSHIRE
Vice-County 38
Map No. 925
2111011
Thamnium alopecurum
(Hedw.) B., S. and G.
BIRMINGHAM
COVENTRY
RUGBY
WARWICK
Scale
Kilometres
0 2 4 6 8 10 N-S
0 2 4 6 8 10 E-W
Miles
0 1 2 3 4 5 6 N-S
0 1 2 3 4 5 6 E-W
Key to Symbols
x Fruiting
+ Non-fruiting

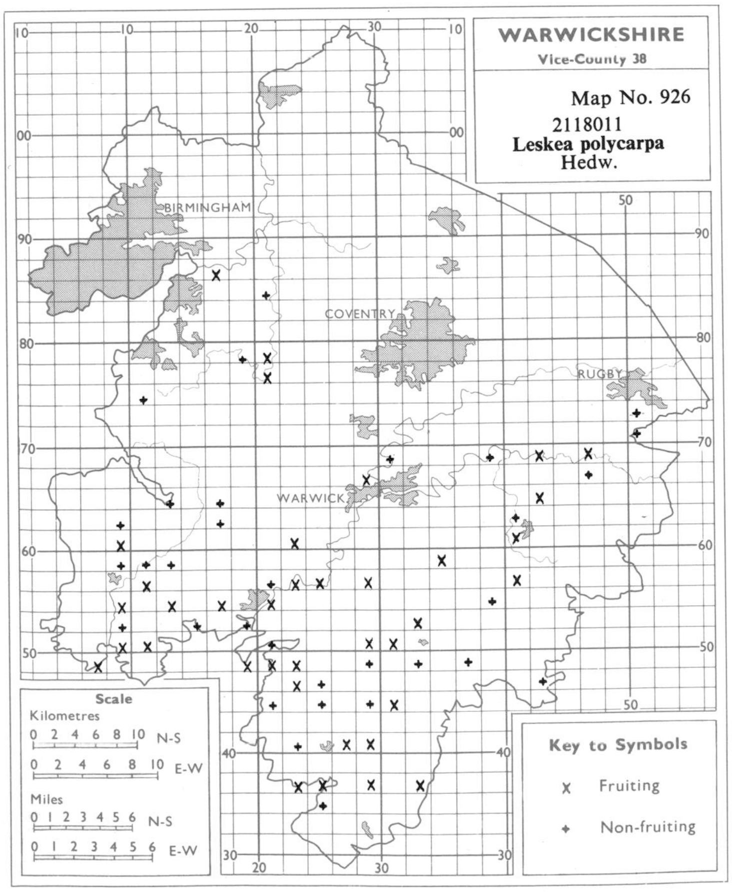
WARWICKSHIRE
Vice-County 38
Map No. 926
2118011
Leskea polycarpa
Hedw.
BIRMINGHAM
COVENTRY
RUGBY
WARWICK
Scale
Kilometres
0 2 4 6 8 10 N-S
0 2 4 6 8 10 E-W
Miles
0 1 2 3 4 5 6 N-S
0 1 2 3 4 5 6 E-W
Key to Symbols
x Fruiting
+ Non-fruiting

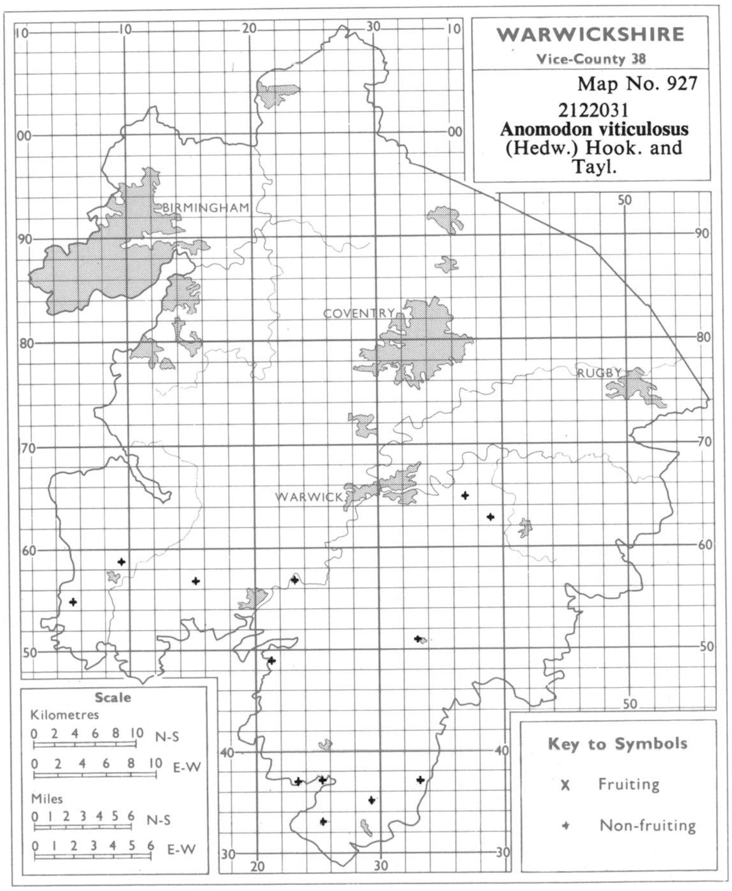
WARWICKSHIRE
Vice-County 38
Map No. 927
2122031
Anomodon viticulosus
(Hedw.) Hook. and
Tayl.
BIRMINGHAM
COVENTRY
RUGBY
WARWICK
Scale
Kilometres
0 2 4 6 8 10 N-S
0 2 4 6 8 10 E-W
Miles
0 1 2 3 4 5 6 N-S
0 1 2 3 4 5 6 E-W
Key to Symbols
x Fruiting
+ Non-fruiting

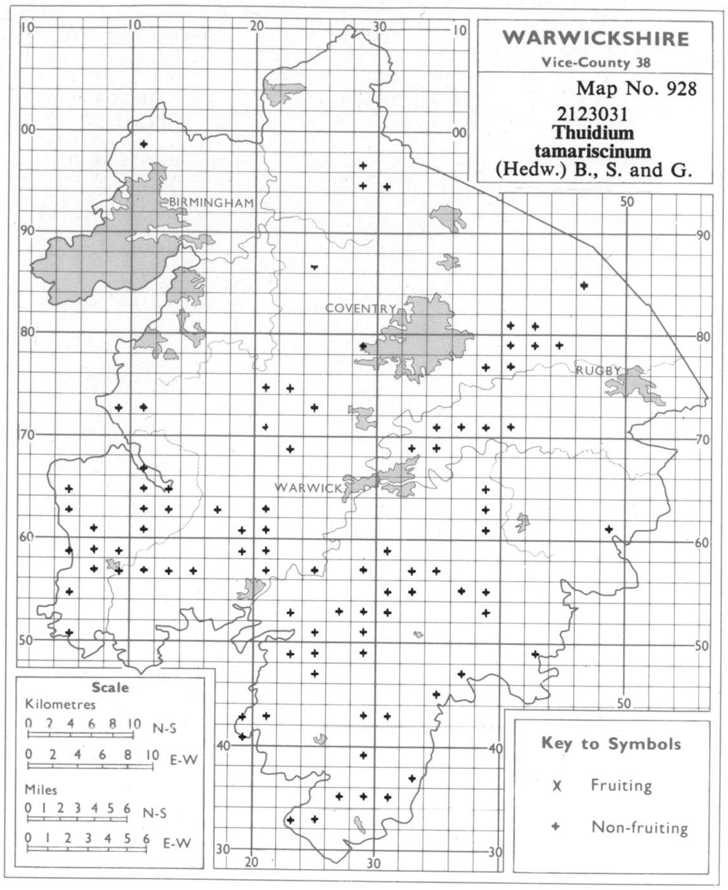
WARWICKSHIRE
Vice-County 38
Map No. 928
2123031
Thuidium tamariscinum
(Hedw.) B., S. and G.
BIRMINGHAM
COVENTRY
RUGBY
WARWICK
Scale
Kilometres
0 2 4 6 8 10 N-S
0 2 4 6 8 10 E-W
Miles
0 1 2 3 4 5 6 N-S
0 1 2 3 4 5 6 E-W
Key to Symbols
X Fruiting
+ Non-fruiting

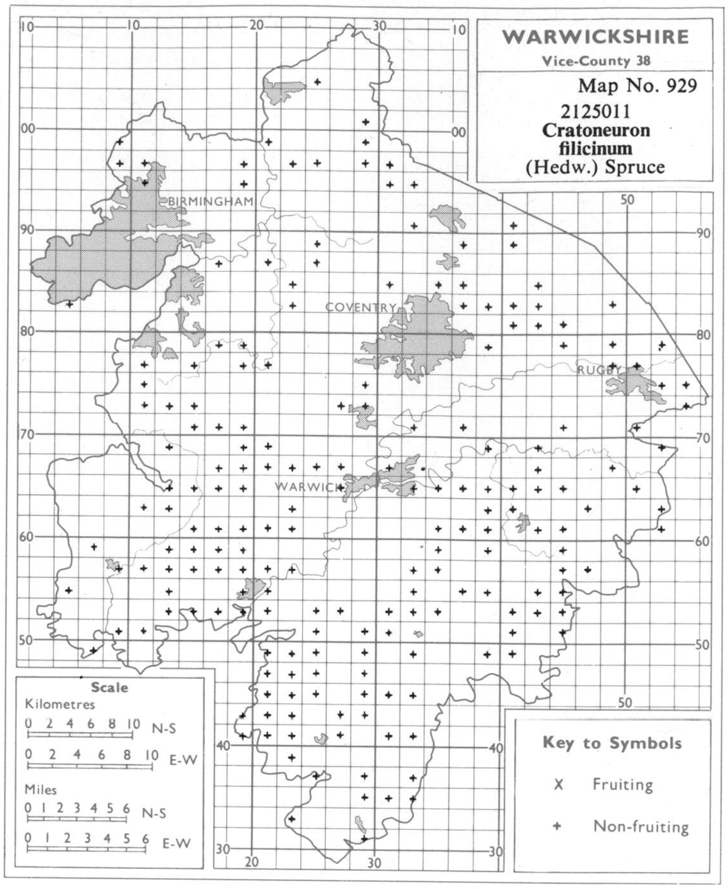
WARWICKSHIRE
Vice-County 38
Map No. 929
2125011
Cratoneuron filicinum
(Hedw.) Spruce
BIRMINGHAM
COVENTRY
RUGBY
WARWICK
Scale
Kilometres
0 2 4 6 8 10 N-S
0 2 4 6 8 10 E-W
Miles
0 1 2 3 4 5 6 N-S
0 1 2 3 4 5 6 E-W
Key to Symbols
X Fruiting
+ Non-fruiting

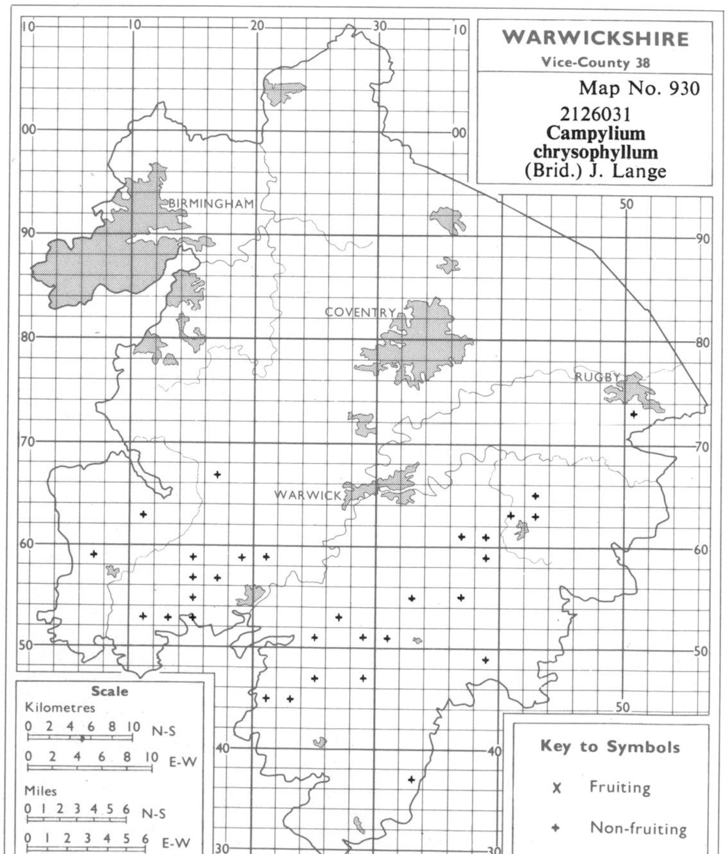
WARWICKSHIRE
Vice-County 38
Map No. 930
2126031
Campylium chrysophyllum
(Brid.) J. Lange
BIRMINGHAM
COVENTRY
RUGBY
WARWICK
Scale
Kilometres
0 2 4 6 8 10 N-S
0 2 4 6 8 10 E-W
Miles
0 1 2 3 4 5 6 N-S
0 1 2 3 4 5 6 E-W
Key to Symbols
X Fruiting
+ Non-fruiting

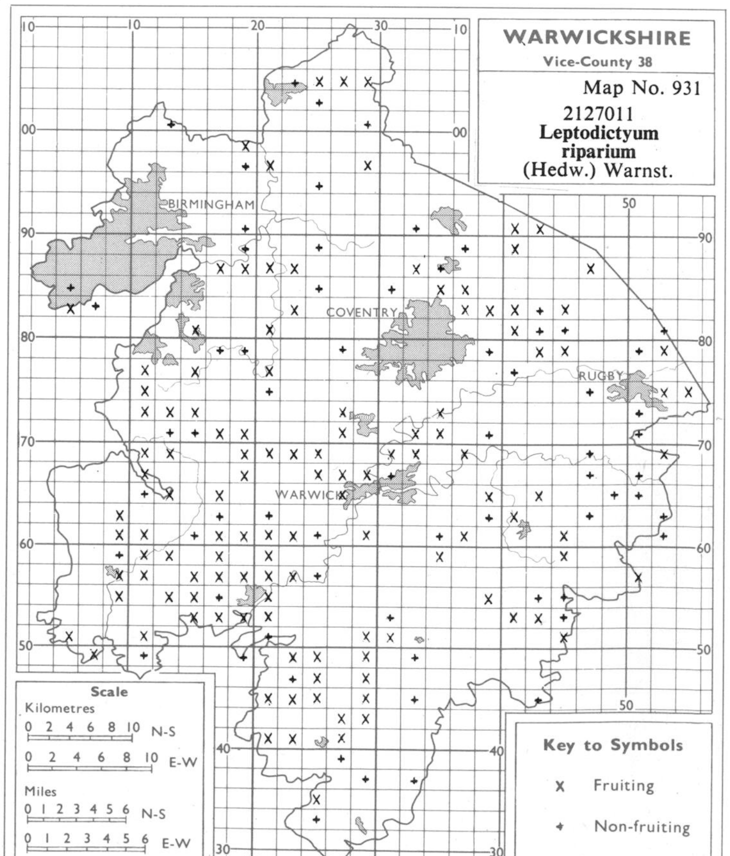
WARWICKSHIRE
Vice-County 38
Map No. 931
2127011
Leptodictyum riparium
(Hedw.) Warnst.
BIRMINGHAM
COVENTRY
RUGBY
WARWICK
Scale
Kilometres
0 2 4 6 8 10 N-S
0 2 4 6 8 10 E-W
Miles
0 1 2 3 4 5 6 N-S
0 1 2 3 4 5 6 E-W
Key to Symbols
X Fruiting
+ Non-fruiting

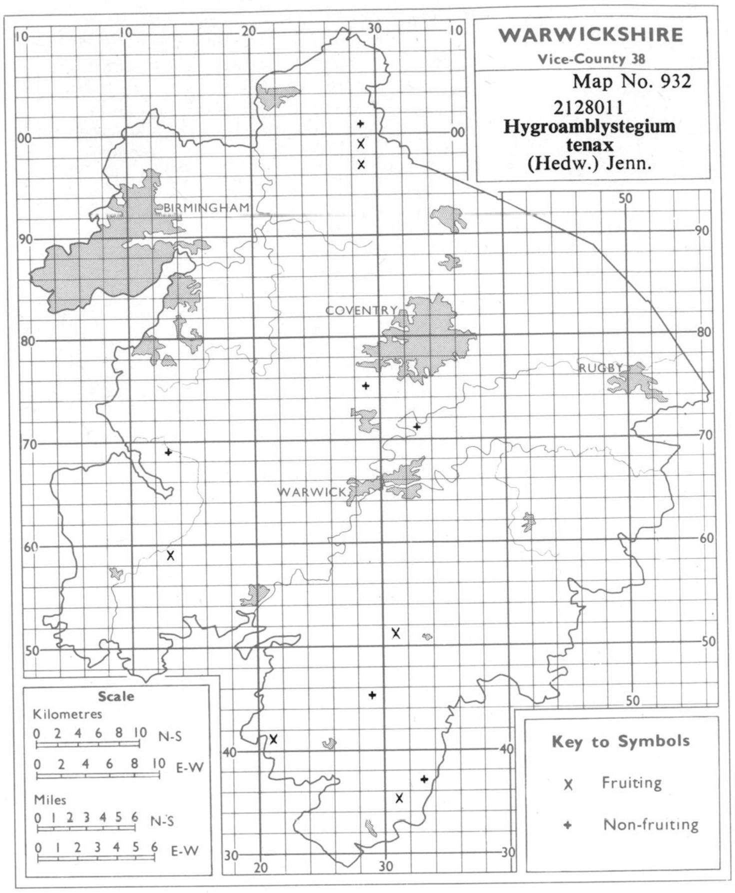
WARWICKSHIRE
Vice-County 38
Map No. 932
2128011
Hygroamblystegium tenax
(Hedw.) Jenn.
BIRMINGHAM
COVENTRY
RUGBY
WARWICK
Scale
Kilometres
N-S
E-W
Miles
N-S
E-W
Key to Symbols
X Fruiting
+ Non-fruiting

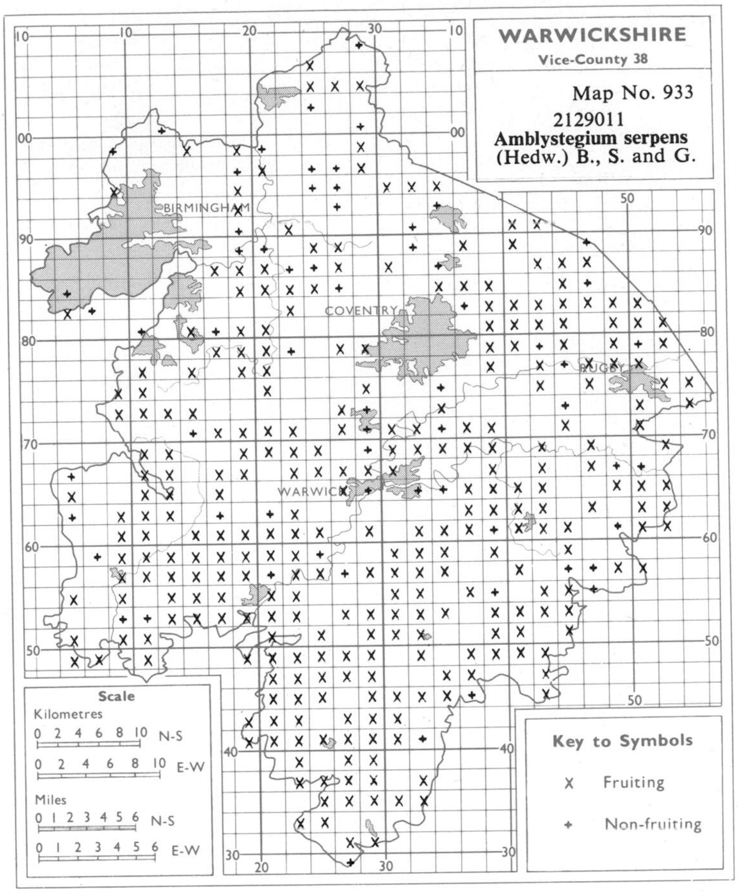
WARWICKSHIRE
Vice-County 38
Map No. 933
2129011
Amblystegium serpens
(Hedw.) B., S. and G.
BIRMINGHAM
COVENTRY
RUGBY
WARWICK
Scale
Kilometres
N-S
E-W
Miles
N-S
E-W
Key to Symbols
X Fruiting
+ Non-fruiting

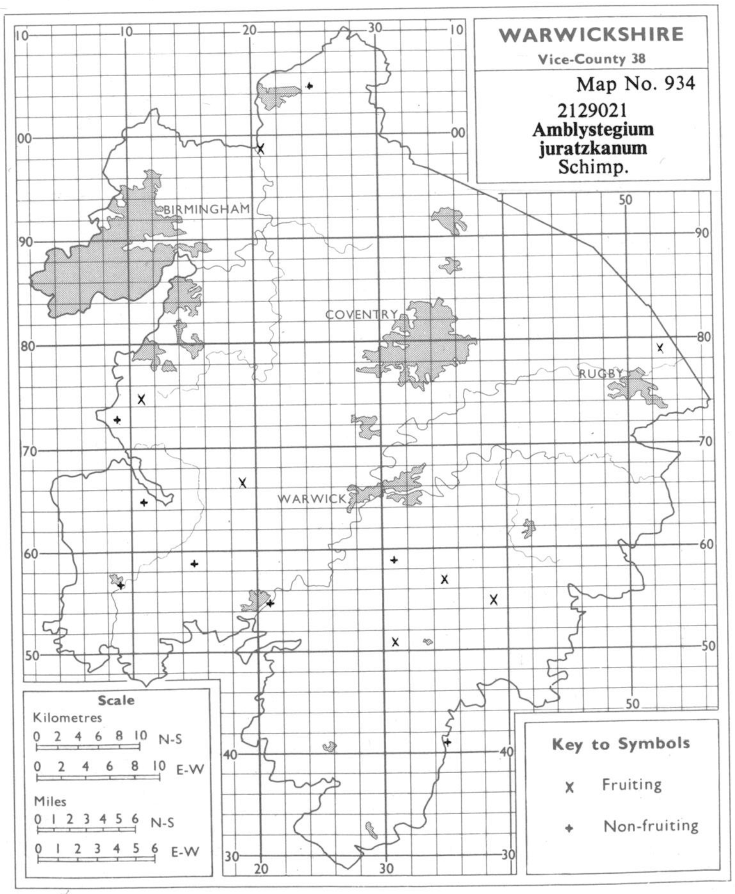
WARWICKSHIRE
Vice-County 38
Map No. 934
2129021
Amblystegium juratzkanum
Schimp.
BIRMINGHAM
COVENTRY
RUGBY
WARWICK
Scale
Kilometres
N-S
E-W
Miles
N-S
E-W
Key to Symbols
X Fruiting
+ Non-fruiting

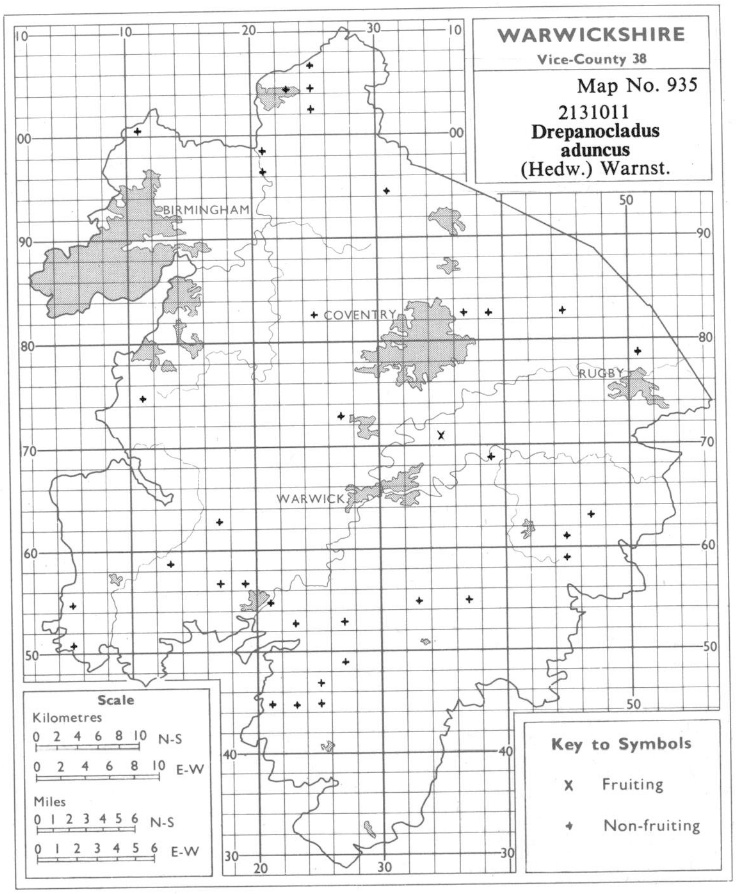
WARWICKSHIRE
Vice-County 38
Map No. 935
2131011
Drepanocladus aduncus
(Hedw.) Warnst.
BIRMINGHAM
COVENTRY
RUGBY
WARWICK
Scale
Kilometres
N-S
E-W
Miles
N-S
E-W
Key to Symbols
X Fruiting
+ Non-fruiting

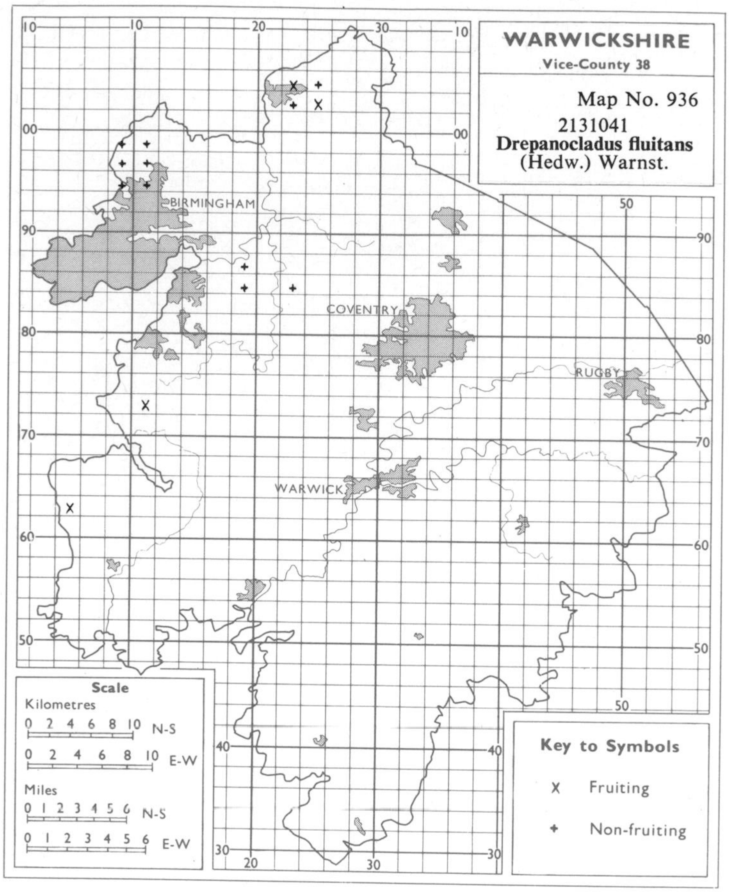
WARWICKSHIRE
Vice-County 38
Map No. 936
2131041
Drepanocladus fluitans
(Hedw.) Warnst.
BIRMINGHAM
COVENTRY
RUGBY
WARWICK
Scale
Kilometres
0 2 4 6 8 10 N-S
0 2 4 6 8 10 E-W
Miles
0 1 2 3 4 5 6 N-S
0 1 2 3 4 5 6 E-W
Key to Symbols
X Fruiting
+ Non-fruiting

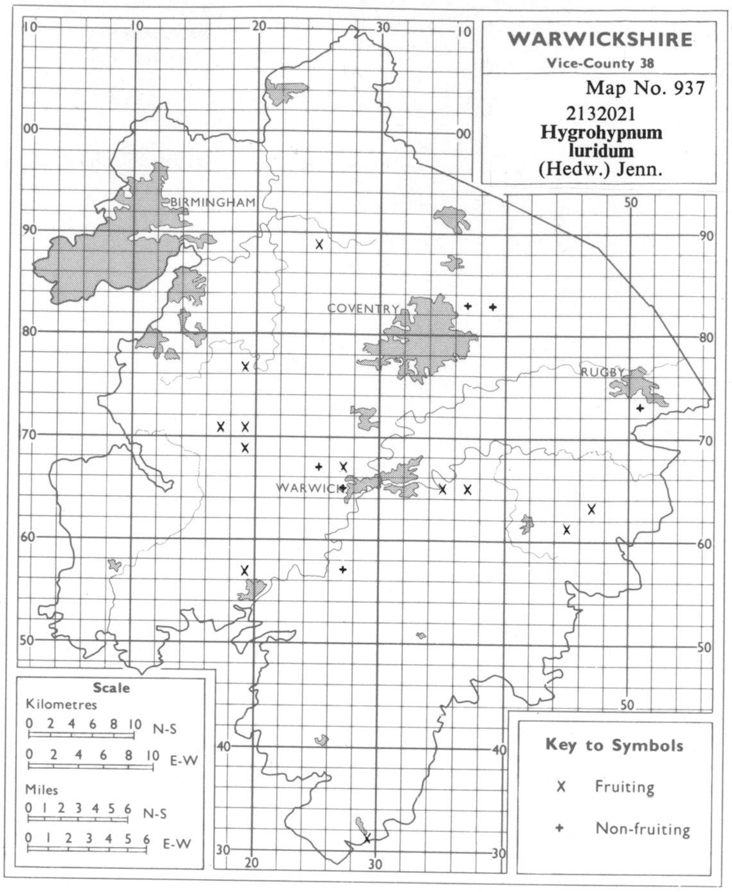
WARWICKSHIRE
Vice-County 38
Map No. 937
2132021
Hygrohypnum
luridum
(Hedw.) Jenn.
BIRMINGHAM
COVENTRY
RUGBY
WARWICK
Scale
Kilometres
0 2 4 6 8 10 N-S
0 2 4 6 8 10 E-W
Miles
0 1 2 3 4 5 6 N-S
0 1 2 3 4 5 6 E-W
Key to Symbols
X Fruiting
+ Non-fruiting

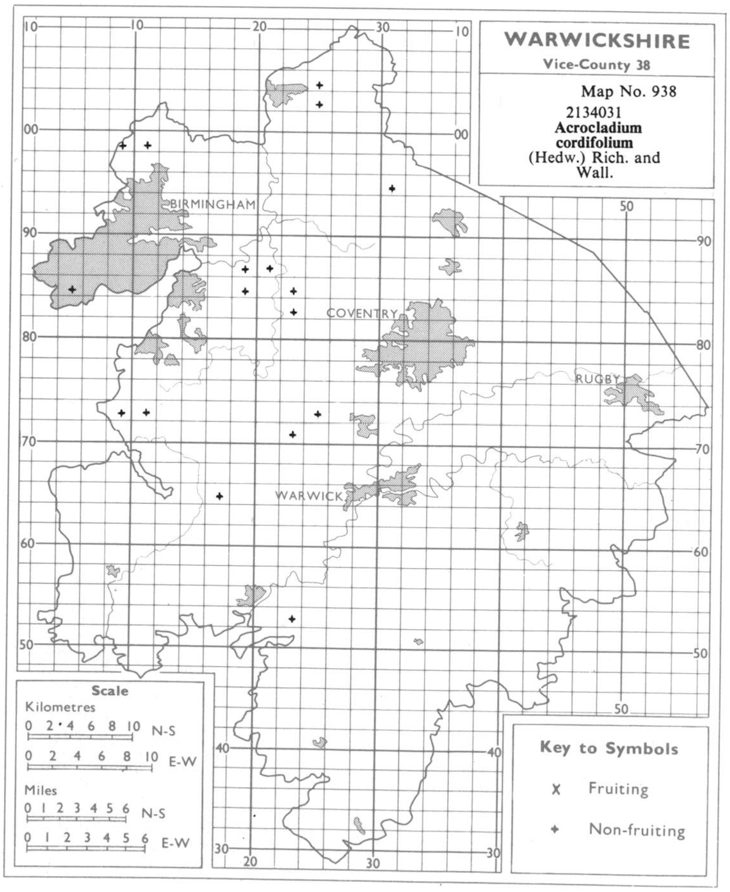
WARWICKSHIRE
Vice-County 38
Map No. 938
2134031
Acrocladium
cordifolium
(Hedw.) Rich. and
Wall.
BIRMINGHAM
COVENTRY
RUGBY
WARWICK
Scale
Kilometres
0 2 4 6 8 10 N-S
0 2 4 6 8 10 E-W
Miles
0 1 2 3 4 5 6 N-S
0 1 2 3 4 5 6 E-W
Key to Symbols
X Fruiting
+ Non-fruiting

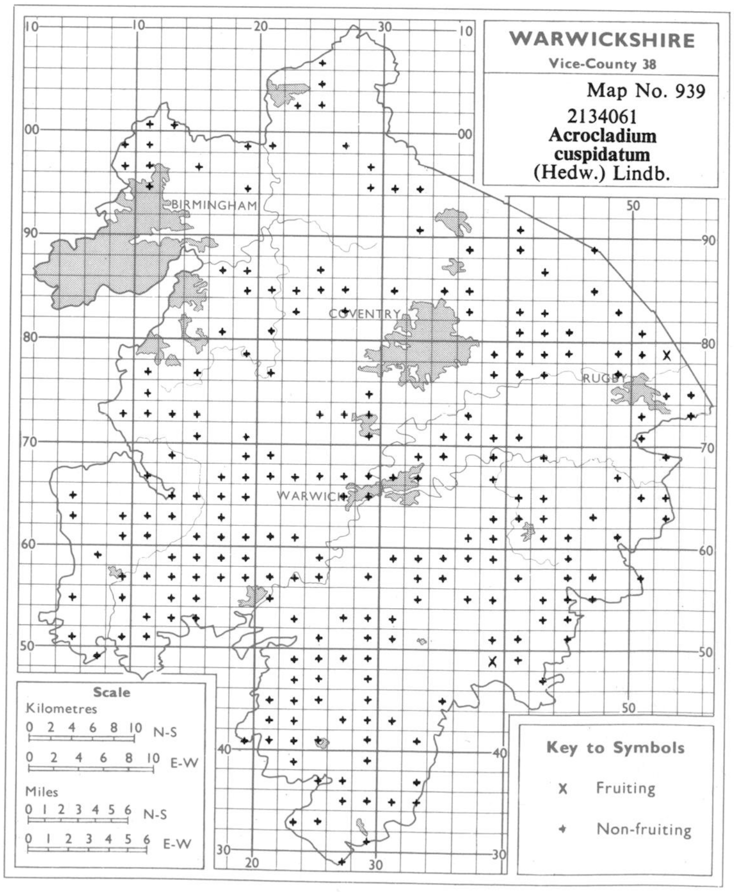
WARWICKSHIRE
Vice-County 38
Map No. 939
2134061
Acrocladium
cuspidatum
(Hedw.) Lindb.
BIRMINGHAM
COVENTRY
RUGBY
WARWICK
Scale
Kilometres
0 2 4 6 8 10 N-S
0 2 4 6 8 10 E-W
Miles
0 1 2 3 4 5 6 N-S
0 1 2 3 4 5 6 E-W
Key to Symbols
X Fruiting
+ Non-fruiting

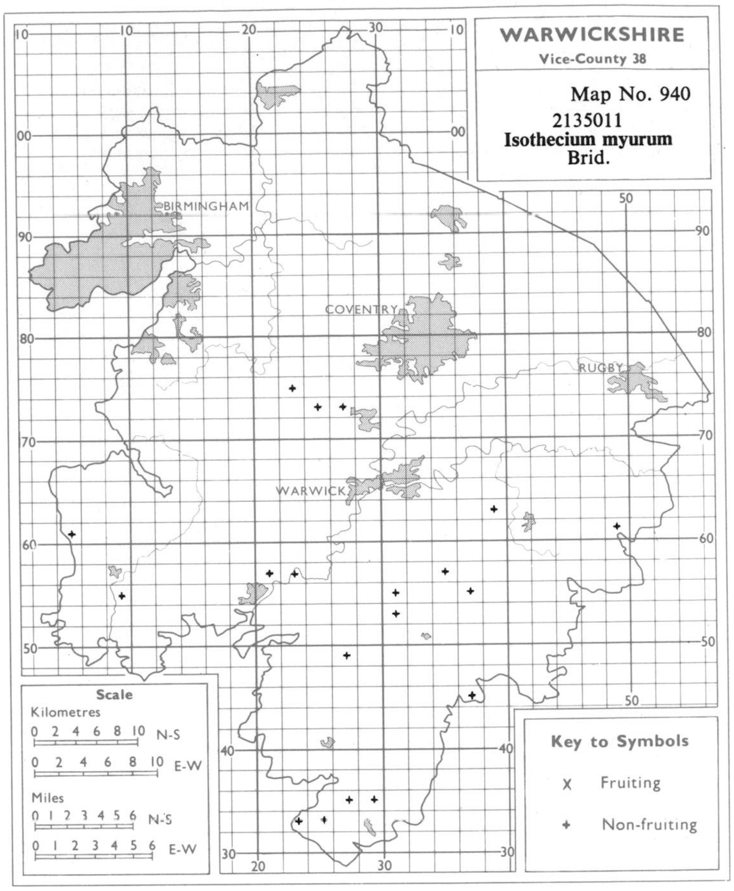
WARWICKSHIRE
Vice-County 38
Map No. 940
2135011
Isothecium myurum
Brid.
BIRMINGHAM
COVENTRY
RUGBY
WARWICK
Scale
Kilometres
0 2 4 6 8 10 N-S
0 2 4 6 8 10 E-W
Miles
0 1 2 3 4 5 6 N-S
0 1 2 3 4 5 6 E-W
Key to Symbols
X Fruiting
+ Non-fruiting

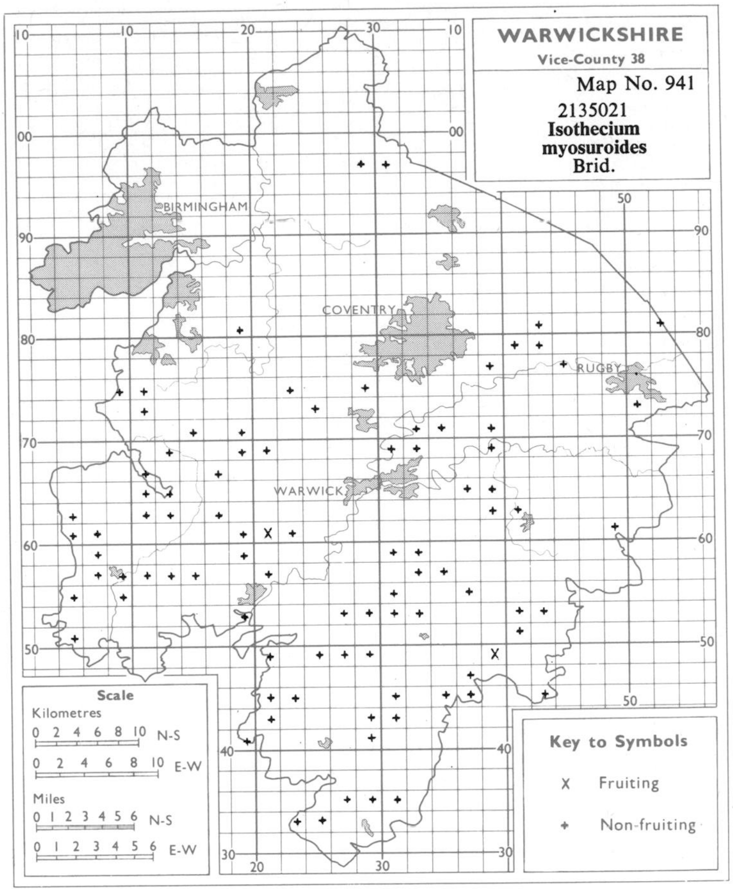
WARWICKSHIRE
Vice-County 38
Map No. 941
2135021
Isothecium
myosuroides
Brid.
BIRMINGHAM
COVENTRY
RUGBY
WARWICK
Scale
Kilometres
0 2 4 6 8 10 N-S
0 2 4 6 8 10 E-W
Miles
0 1 2 3 4 5 6 N-S
0 1 2 3 4 5 6 E-W
Key to Symbols
X Fruiting
+ Non-fruiting

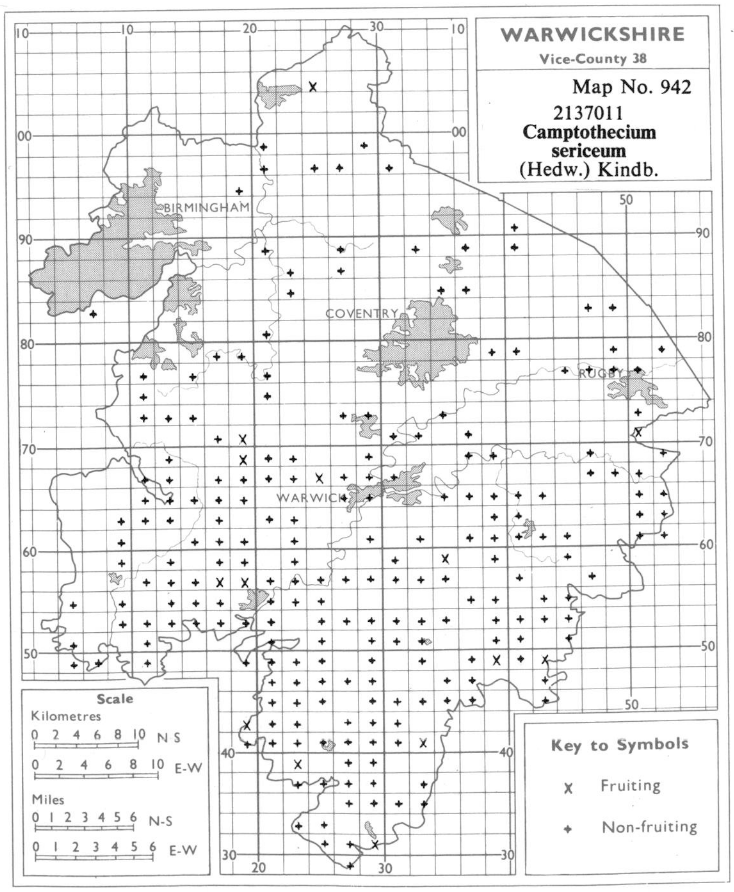
WARWICKSHIRE
Vice-County 38
Map No. 942
2137011
Camptothecium
sericeum
(Hedw.) Kindb.
BIRMINGHAM
COVENTRY
RUGBY
WARWICK
Scale
Kilometres
0 2 4 6 8 10 N S
0 2 4 6 8 10 E-W
Miles
0 1 2 3 4 5 6 N-S
0 1 2 3 4 5 6 E-W
Key to Symbols
X Fruiting
+ Non-fruiting

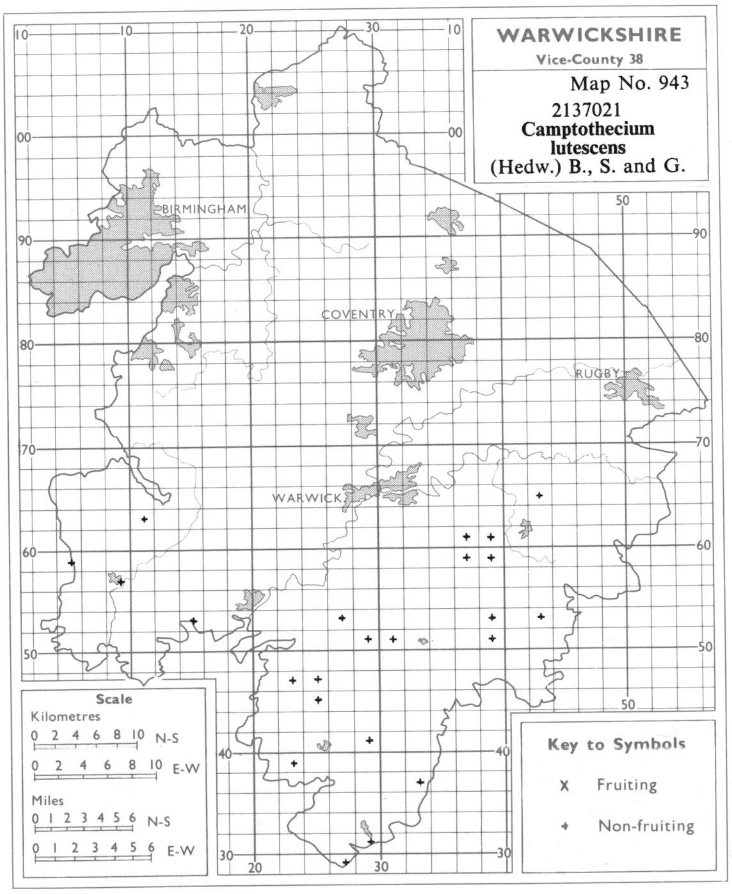
WARWICKSHIRE
Vice-County 38
Map No. 943
2137021
Camptothecium
lutescens
(Hedw.) B., S. and G.
BIRMINGHAM
COVENTRY
RUGBY
WARWICK
Scale
Kilometres
0 2 4 6 8 10 N-S
0 2 4 6 8 10 E-W
Miles
0 1 2 3 4 5 6 N-S
0 1 2 3 4 5 6 E-W
Key to Symbols
X Fruiting
+ Non-fruiting

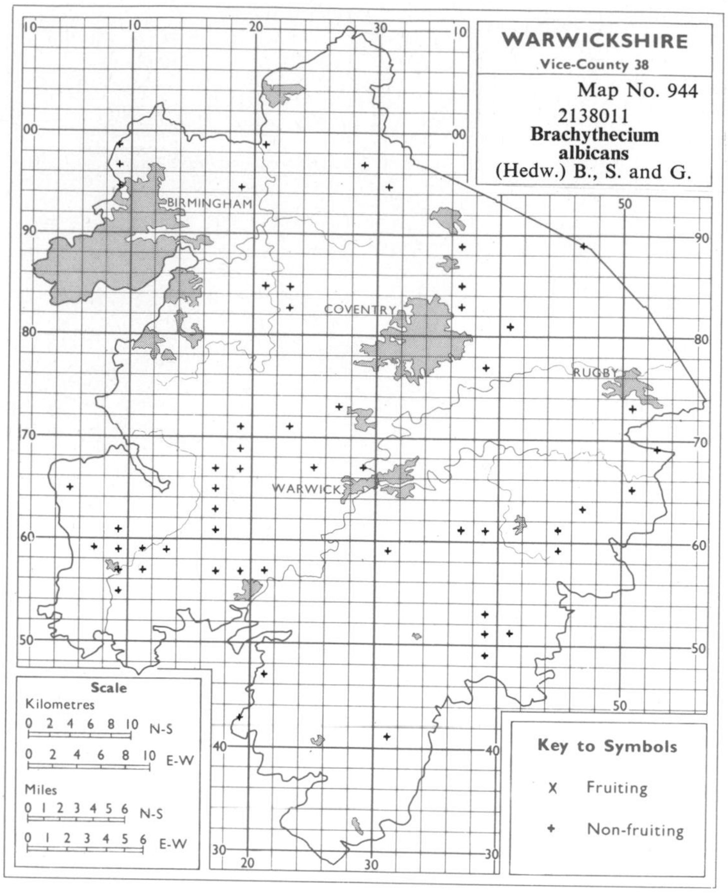
WARWICKSHIRE
Vice-County 38
Map No. 944
2138011
Brachythecium albicans
(Hedw.) B., S. and G.
BIRMINGHAM
COVENTRY
RUGBY
WARWICK
Scale
Kilometres
0 2 4 6 8 10 N-S
0 2 4 6 8 10 E-W
Miles
0 1 2 3 4 5 6 N-S
0 1 2 3 4 5 6 E-W
Key to Symbols
X Fruiting
+ Non-fruiting

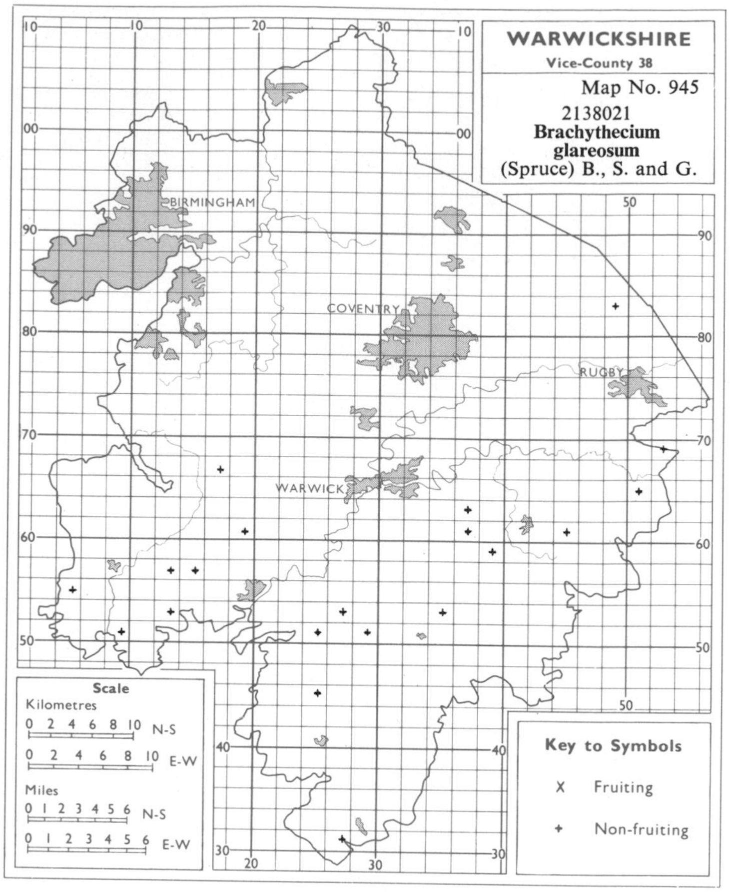
WARWICKSHIRE
Vice-County 38
Map No. 945
2138021
Brachythecium glareosum
(Spruce) B., S. and G.
BIRMINGHAM
COVENTRY
RUGBY
WARWICK
Scale
Kilometres
0 2 4 6 8 10 N-S
0 2 4 6 8 10 E-W
Miles
0 1 2 3 4 5 6 N-S
0 1 2 3 4 5 6 E-W
Key to Symbols
X Fruiting
+ Non-fruiting

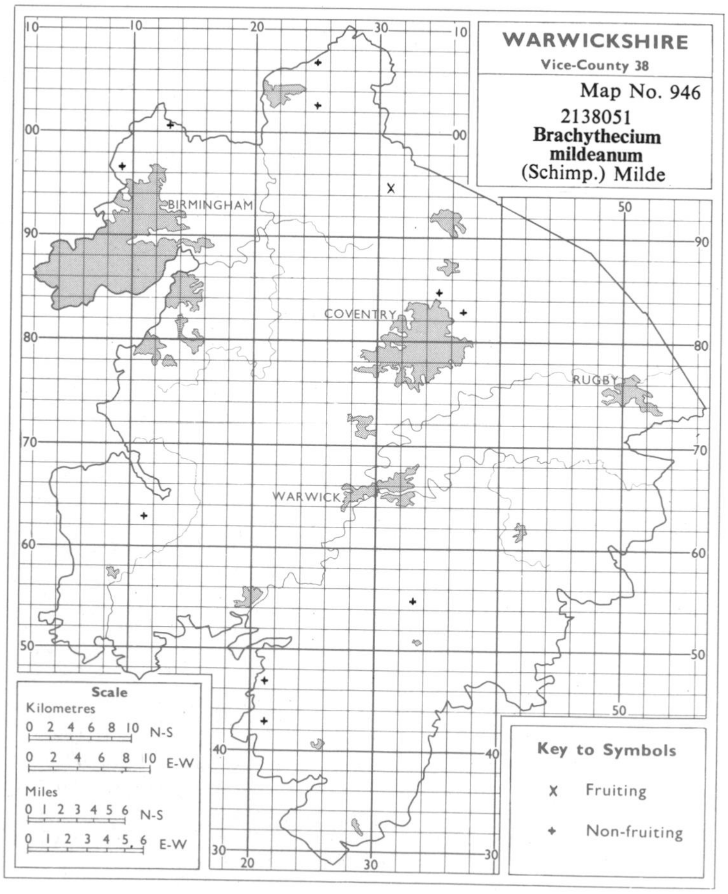
WARWICKSHIRE
Vice-County 38
Map No. 946
2138051
Brachythecium mildeanum
(Schimp.) Milde
BIRMINGHAM
COVENTRY
RUGBY
WARWICK
Scale
Kilometres
0 2 4 6 8 10 N-S
0 2 4 6 8 10 E-W
Miles
0 1 2 3 4 5 6 N-S
0 1 2 3 4 5 6 E-W
Key to Symbols
X Fruiting
+ Non-fruiting

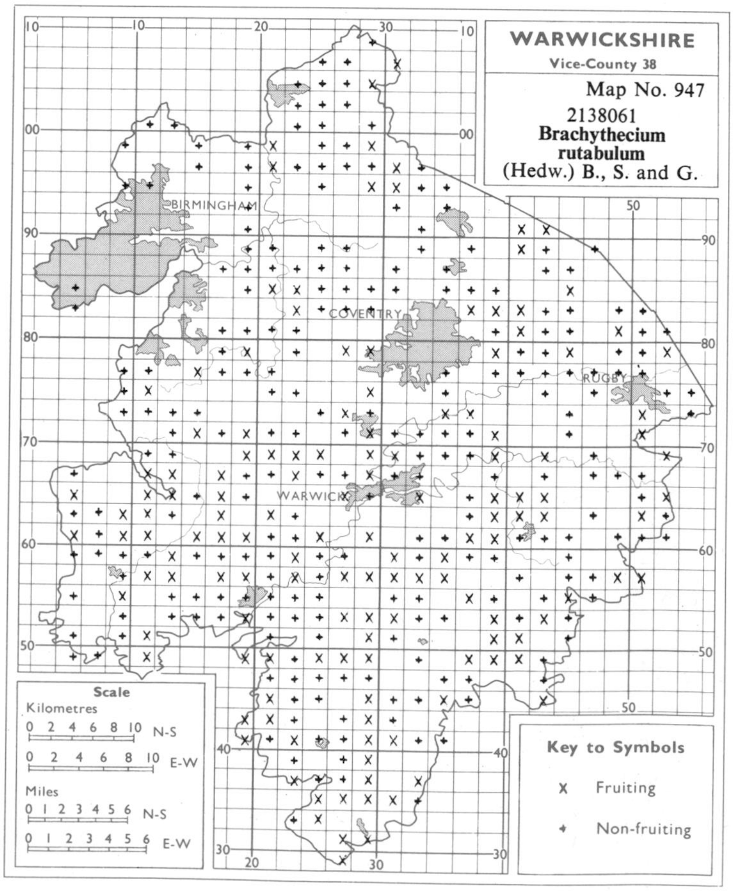
WARWICKSHIRE
Vice-County 38
Map No. 947
2138061
Brachythecium rutabulum
(Hedw.) B., S. and G.
BIRMINGHAM
COVENTRY
RUGBY
WARWICK
Scale
Kilometres
0 2 4 6 8 10 N-S
0 2 4 6 8 10 E-W
Miles
0 1 2 3 4 5 6 N-S
0 1 2 3 4 5 6 E-W
Key to Symbols
X Fruiting
+ Non-fruiting

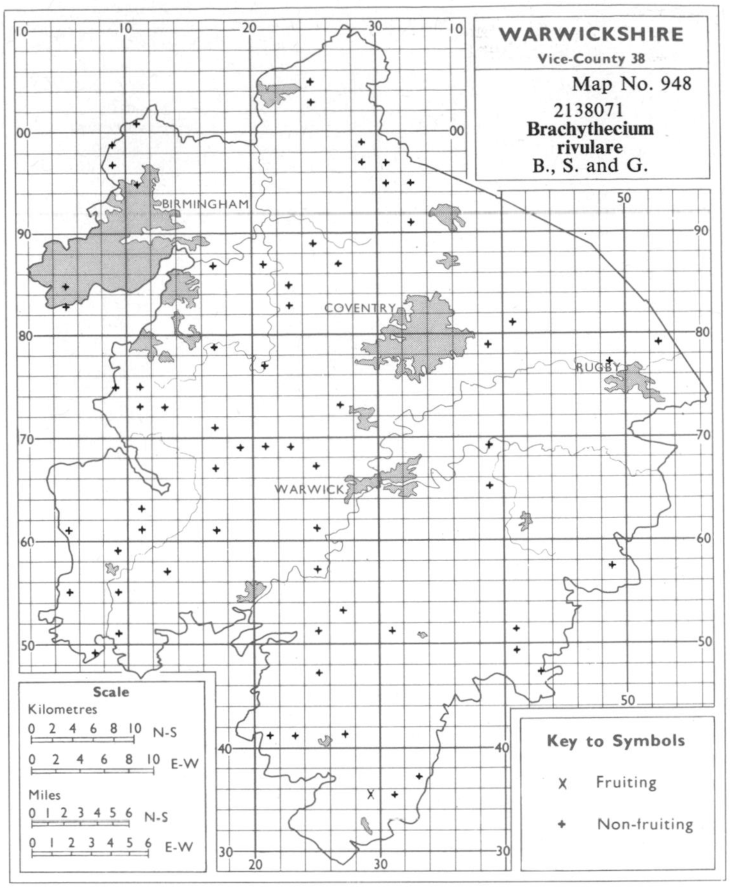
WARWICKSHIRE
Vice-County 38
Map No. 948
2138071
Brachythecium rivulare
B., S. and G.
BIRMINGHAM
COVENTRY
RUGBY
WARWICK
Scale
Kilometres
N-S
E-W
Miles
N-S
E-W
Key to Symbols
X Fruiting
+ Non-fruiting

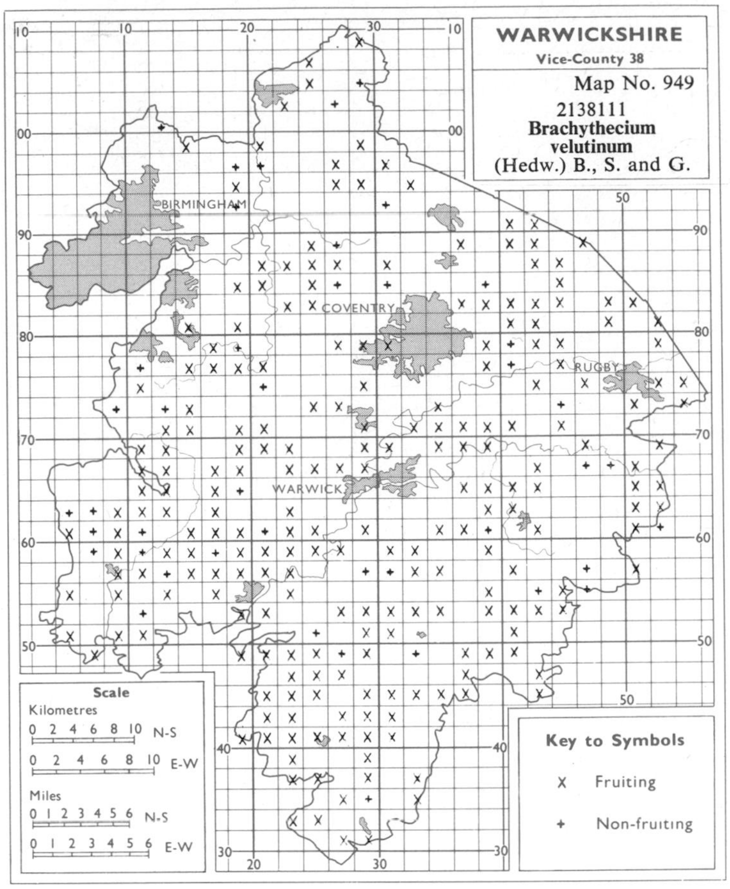
WARWICKSHIRE
Vice-County 38
Map No. 949
2138111
Brachythecium velutinum
(Hedw.) B., S. and G.
BIRMINGHAM
COVENTRY
RUGBY
WARWICK
Scale
Kilometres
N-S
E-W
Miles
N-S
E-W
Key to Symbols
X Fruiting
+ Non-fruiting

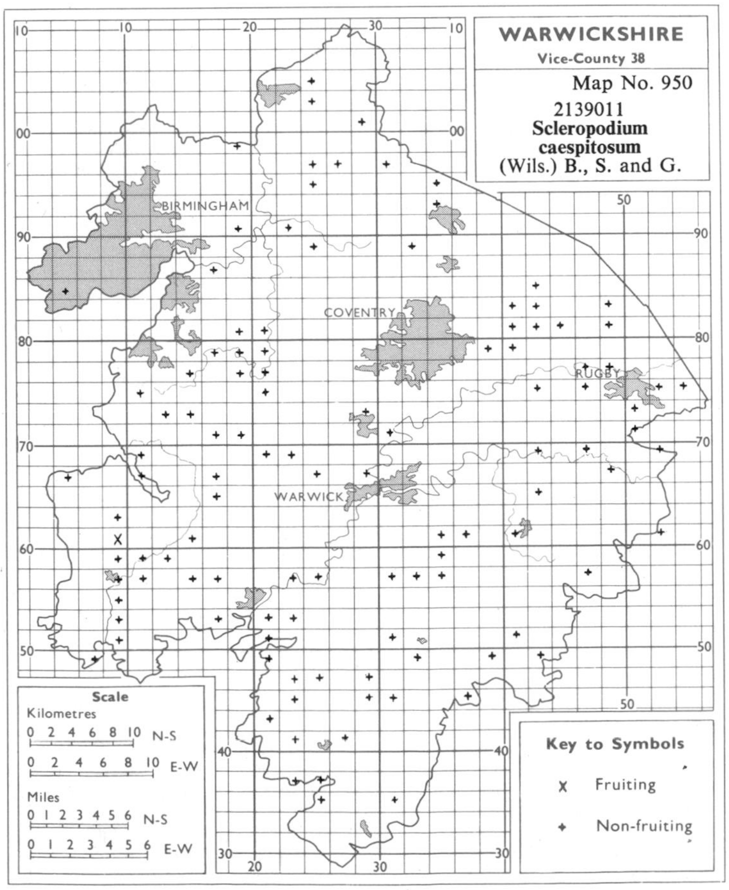
WARWICKSHIRE
Vice-County 38
Map No. 950
2139011
Scleropodium caespitosum
(Wils.) B., S. and G.
BIRMINGHAM
COVENTRY
RUGBY
WARWICK
Scale
Kilometres
N-S
E-W
Miles
N-S
E-W
Key to Symbols
X Fruiting
+ Non-fruiting

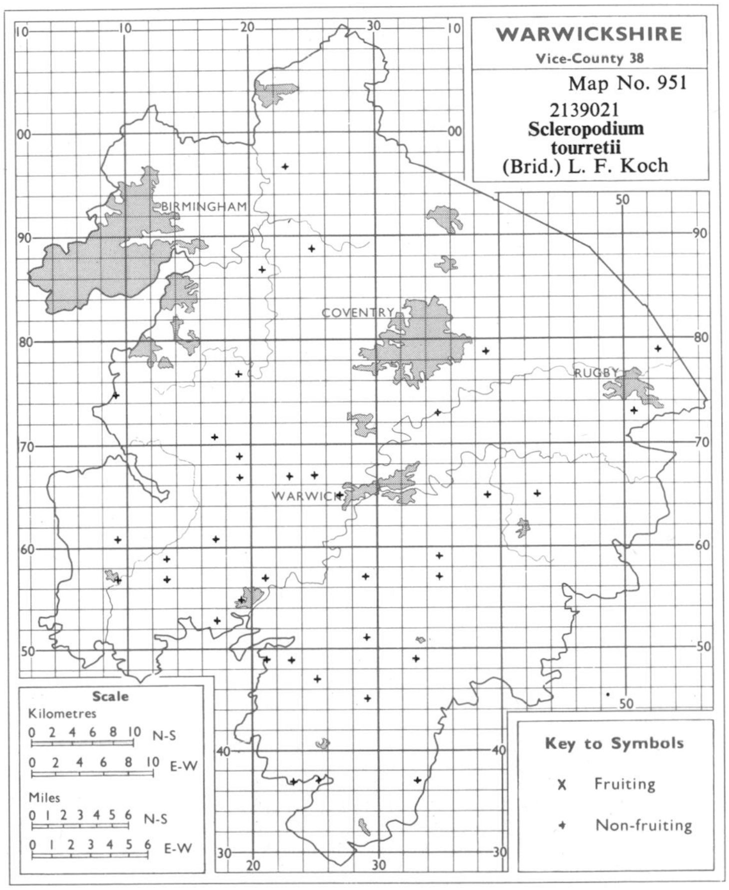
WARWICKSHIRE
Vice-County 38
Map No. 951
2139021
Scleropodium tourretii
(Brid.) L. F. Koch
BIRMINGHAM
COVENTRY
RUGBY
WARWICK
Scale
Kilometres
N-S
E-W
Miles
N-S
E-W
Key to Symbols
X Fruiting
+ Non-fruiting

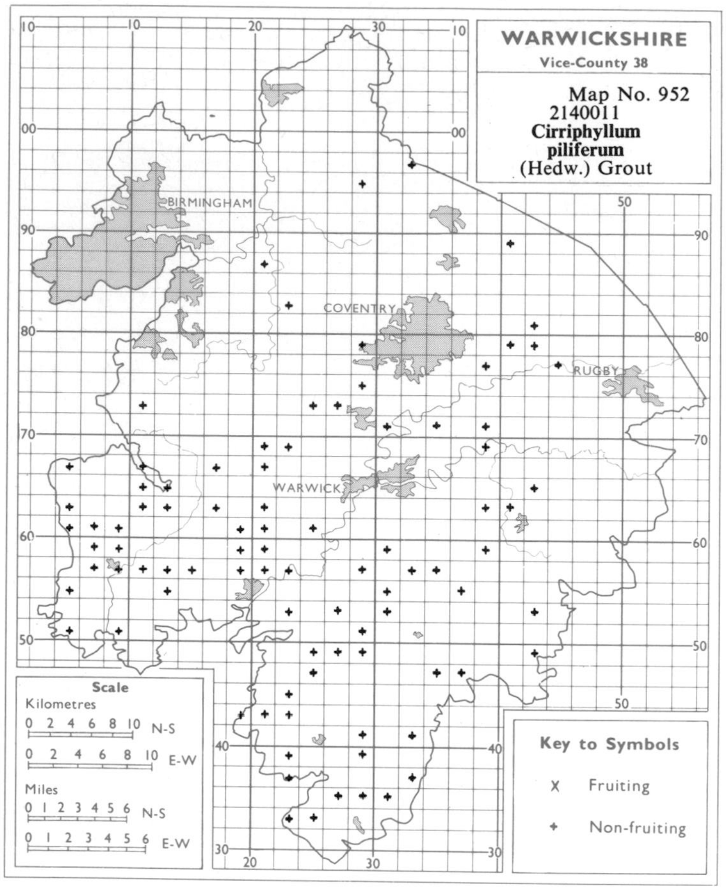

WARWICKSHIRE
Vice-County 38
Map No. 952
2140011
Cirriphyllum piliferum
(Hedw.) Grout
BIRMINGHAM
COVENTRY
RUGBY
WARWICK
Scale
Kilometres
N-S
E-W
Miles
N-S
E-W
Key to Symbols
X Fruiting
+ Non-fruiting

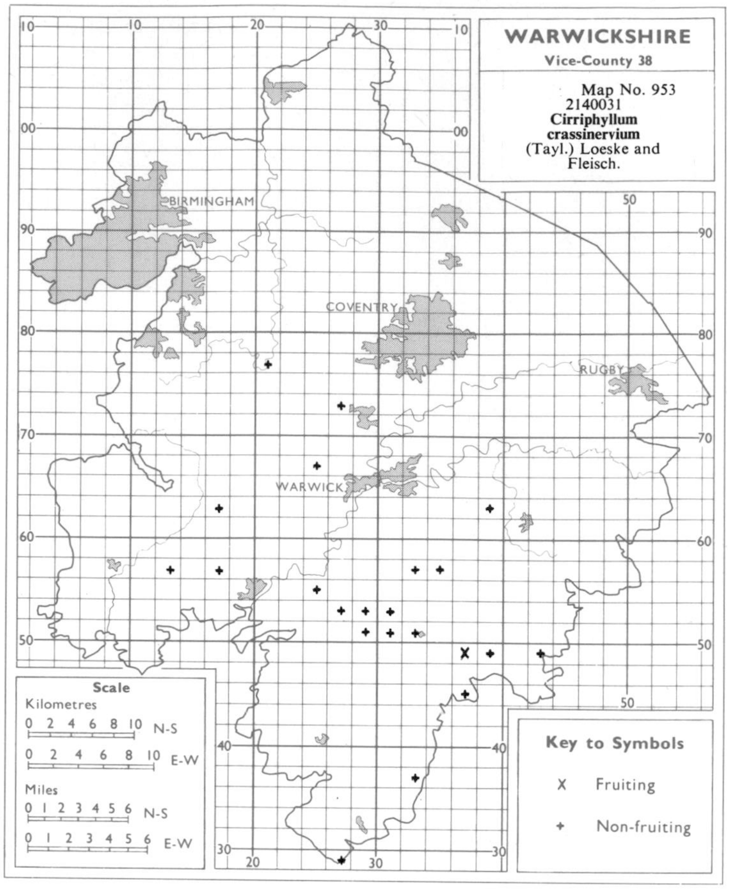

WARWICKSHIRE
Vice-County 38
Map No. 953
2140031
Cirriphyllum crassinervium
(Tayl.) Loeske and Fleisch.
BIRMINGHAM
COVENTRY
RUGBY
WARWICK
Scale
Kilometres
N-S
E-W
Miles
N-S
E-W
Key to Symbols
X Fruiting
+ Non-fruiting

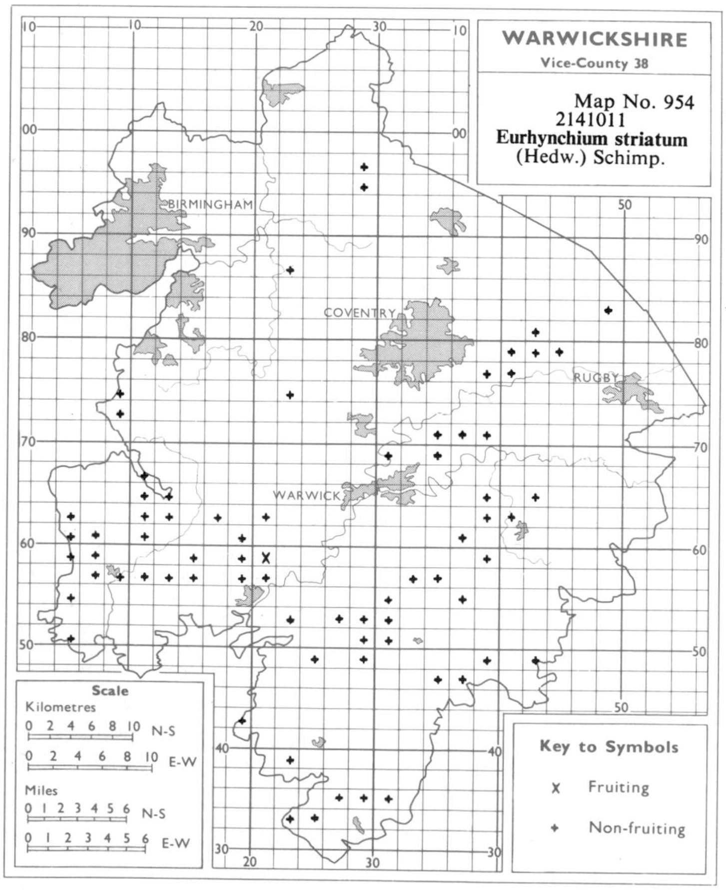

WARWICKSHIRE
Vice-County 38
Map No. 954
2141011
Eurhynchium striatum
(Hedw.) Schimp.
BIRMINGHAM
COVENTRY
RUGBY
WARWICK
Scale
Kilometres
N-S
E-W
Miles
N-S
E-W
Key to Symbols
X Fruiting
+ Non-fruiting

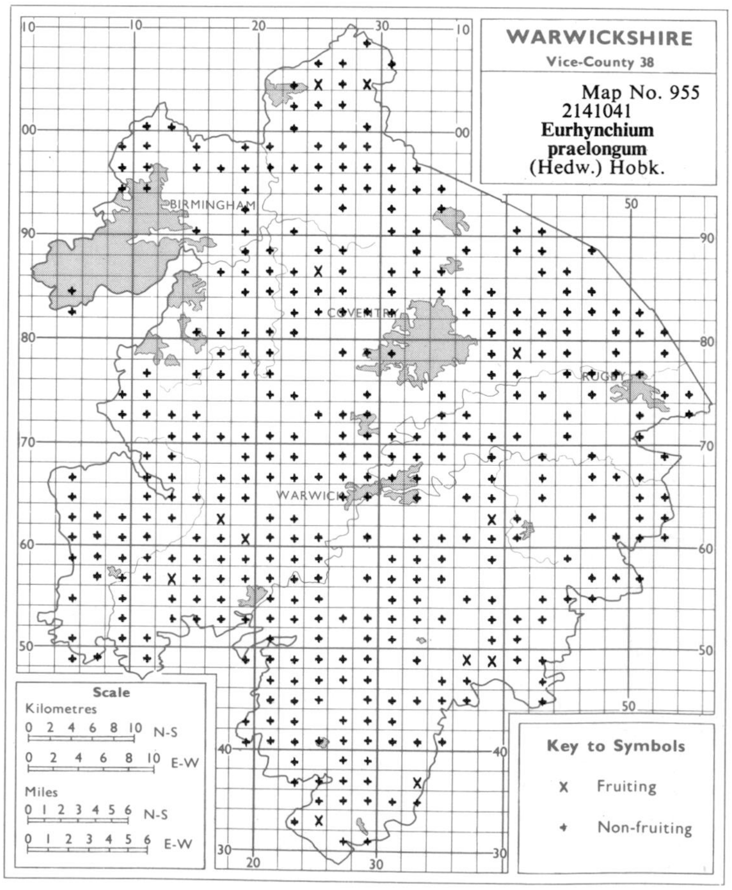

WARWICKSHIRE
Vice-County 38
Map No. 955
2141041
Eurhynchium praelongum
(Hedw.) Hobk.
BIRMINGHAM
COVENTRY
RUGBY
WARWICK
Scale
Kilometres
N-S
E-W
Miles
N-S
E-W
Key to Symbols
X Fruiting
+ Non-fruiting

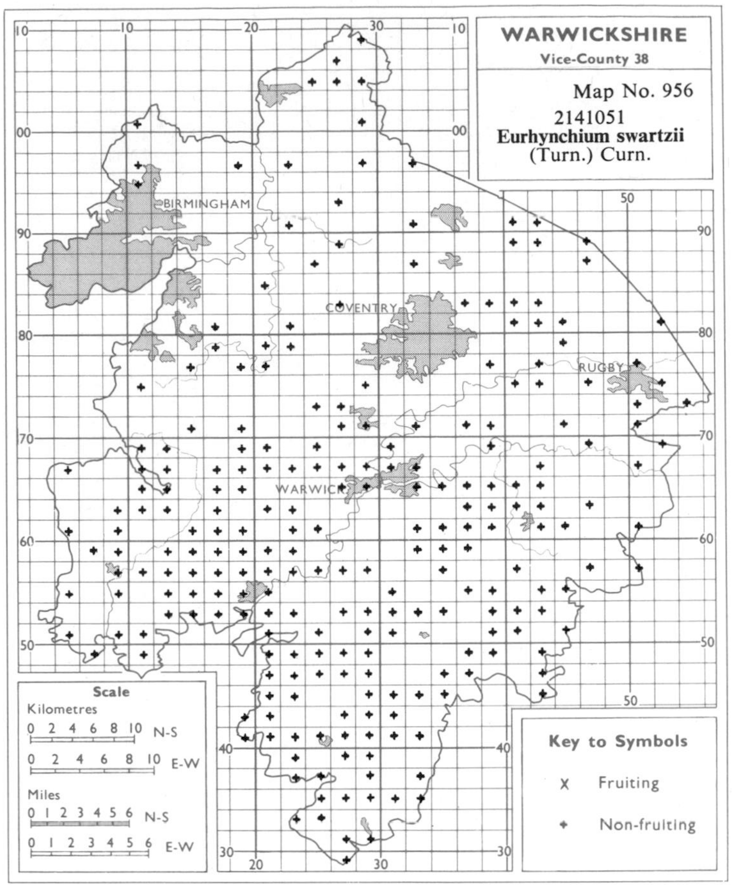
WARWICKSHIRE
Vice-County 38
Map No. 956
2141051
Eurhynchium swartzii
(Turn.) Curn.
BIRMINGHAM
COVENTRY
RUGBY
WARWICK
Scale
Kilometres
0 2 4 6 8 10 N-S
0 2 4 6 8 10 E-W
Miles
0 1 2 3 4 5 6 N-S
0 1 2 3 4 5 6 E-W
Key to Symbols
X Fruiting
+ Non-fruiting

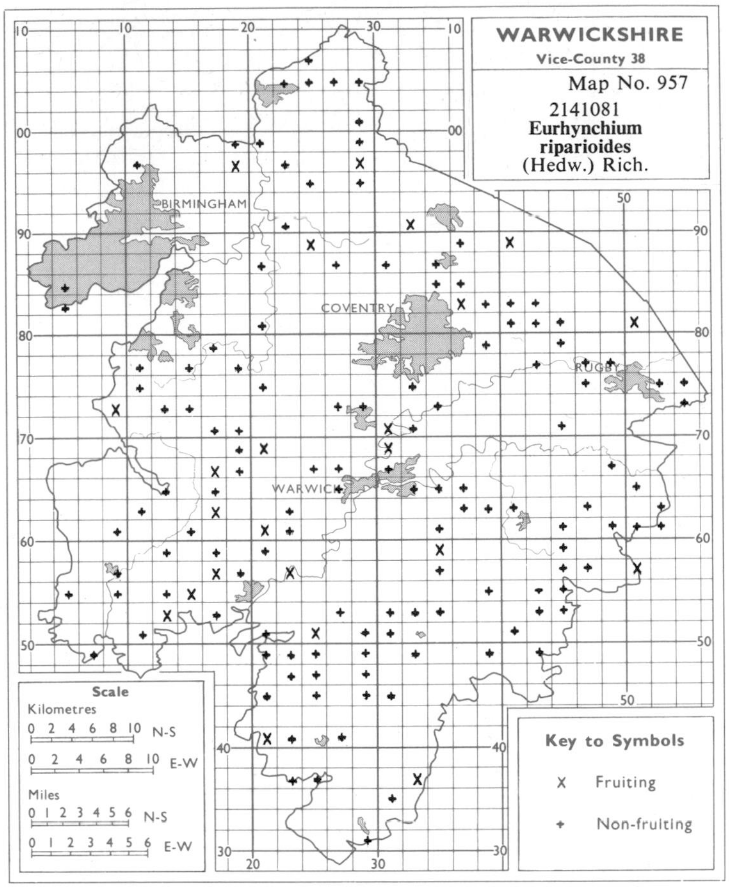
WARWICKSHIRE
Vice-County 38
Map No. 957
2141081
Eurhynchium
riparioides
(Hedw.) Rich.
BIRMINGHAM
COVENTRY
RUGBY
WARWICK
Scale
Kilometres
0 2 4 6 8 10 N-S
0 2 4 6 8 10 E-W
Miles
0 1 2 3 4 5 6 N-S
0 1 2 3 4 5 6 E-W
Key to Symbols
X Fruiting
+ Non-fruiting

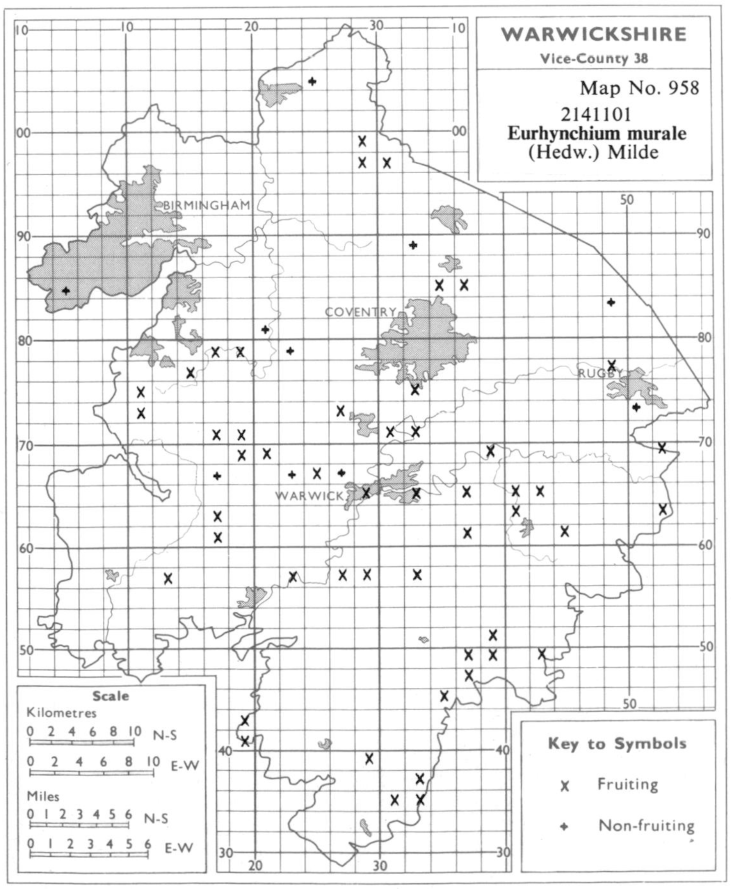
WARWICKSHIRE
Vice-County 38
Map No. 958
2141101
Eurhynchium murale
(Hedw.) Milde
BIRMINGHAM
COVENTRY
RUGBY
WARWICK
Scale
Kilometres
0 2 4 6 8 10 N-S
0 2 4 6 8 10 E-W
Miles
0 1 2 3 4 5 6 N-S
0 1 2 3 4 5 6 E-W
Key to Symbols
X Fruiting
+ Non-fruiting

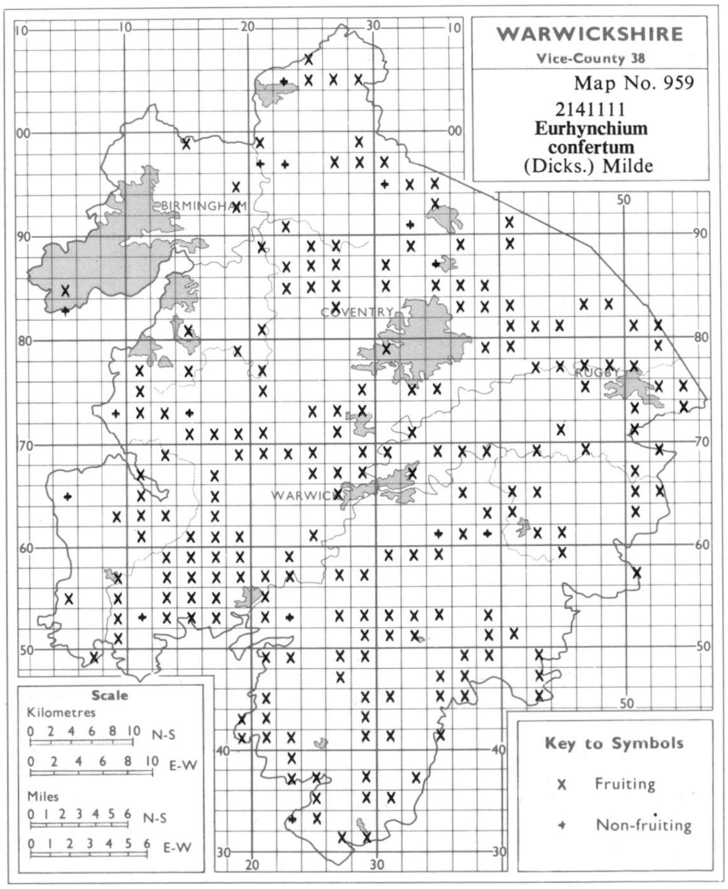
WARWICKSHIRE
Vice-County 38
Map No. 959
2141111
Eurhynchium
confertum
(Dicks.) Milde
BIRMINGHAM
COVENTRY
RUGBY
WARWICK
Scale
Kilometres
0 2 4 6 8 10 N-S
0 2 4 6 8 10 E-W
Miles
0 1 2 3 4 5 6 N-S
0 1 2 3 4 5 6 E-W
Key to Symbols
X Fruiting
+ Non-fruiting

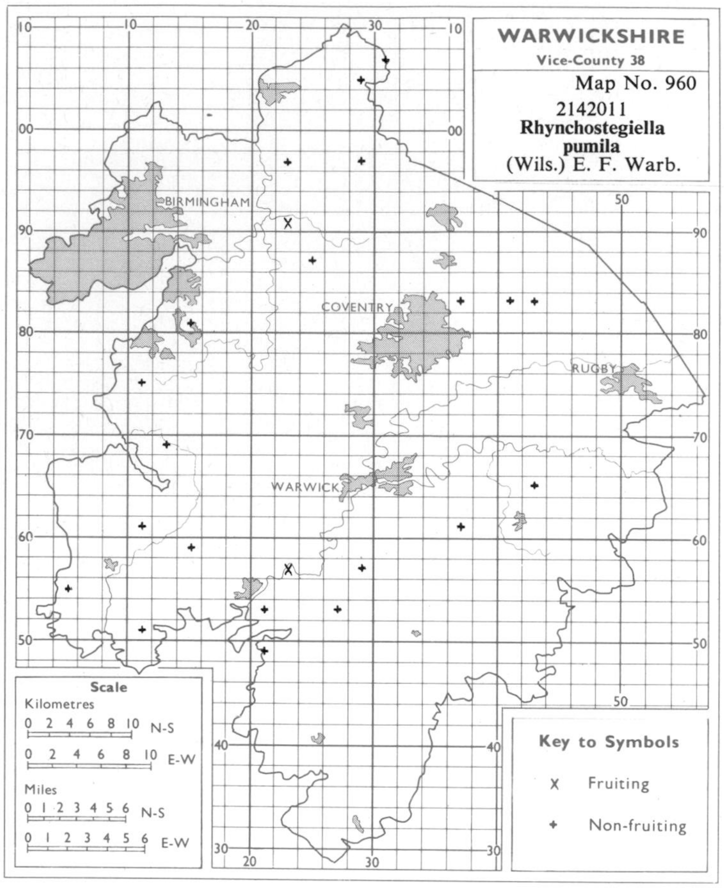
WARWICKSHIRE
Vice-County 38
Map No. 960
2142011
Rhynchostegiella pumila (Wils.) E. F. Warb.
BIRMINGHAM
COVENTRY
RUGBY
WARWICK
Scale
Kilometres
0 2 4 6 8 10 N-S
0 2 4 6 8 10 E-W
Miles
0 1 2 3 4 5 6 N-S
0 1 2 3 4 5 6 E-W
Key to Symbols
X Fruiting
+ Non-fruiting

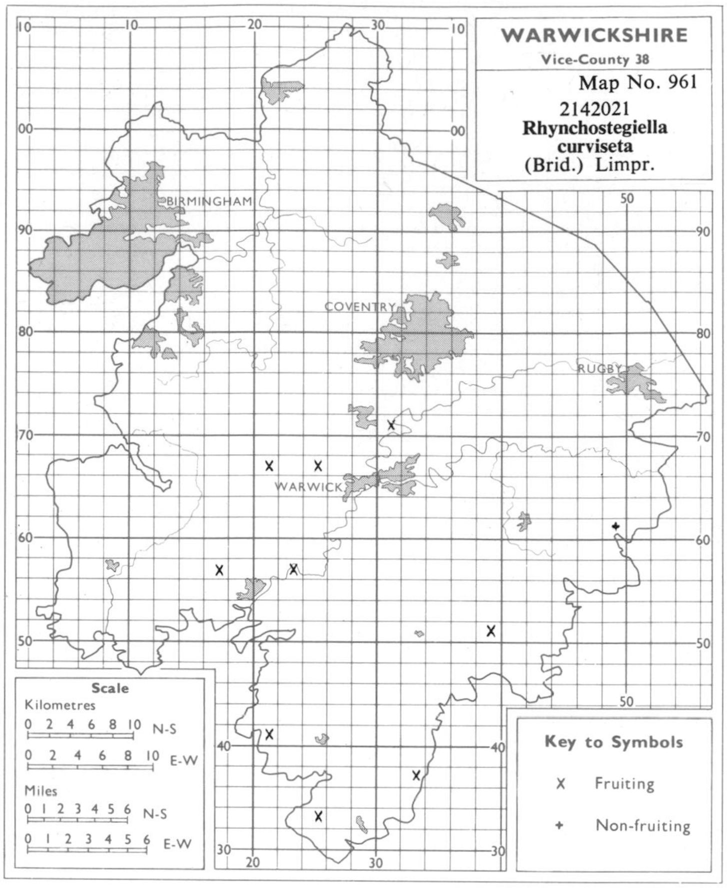
WARWICKSHIRE
Vice-County 38
Map No. 961
2142021
Rhynchostegiella curviseta (Brid.) Limpr.
BIRMINGHAM
COVENTRY
RUGBY
WARWICK
Scale
Kilometres
0 2 4 6 8 10 N-S
0 2 4 6 8 10 E-W
Miles
0 1 2 3 4 5 6 N-S
0 1 2 3 4 5 6 E-W
Key to Symbols
X Fruiting
+ Non-fruiting

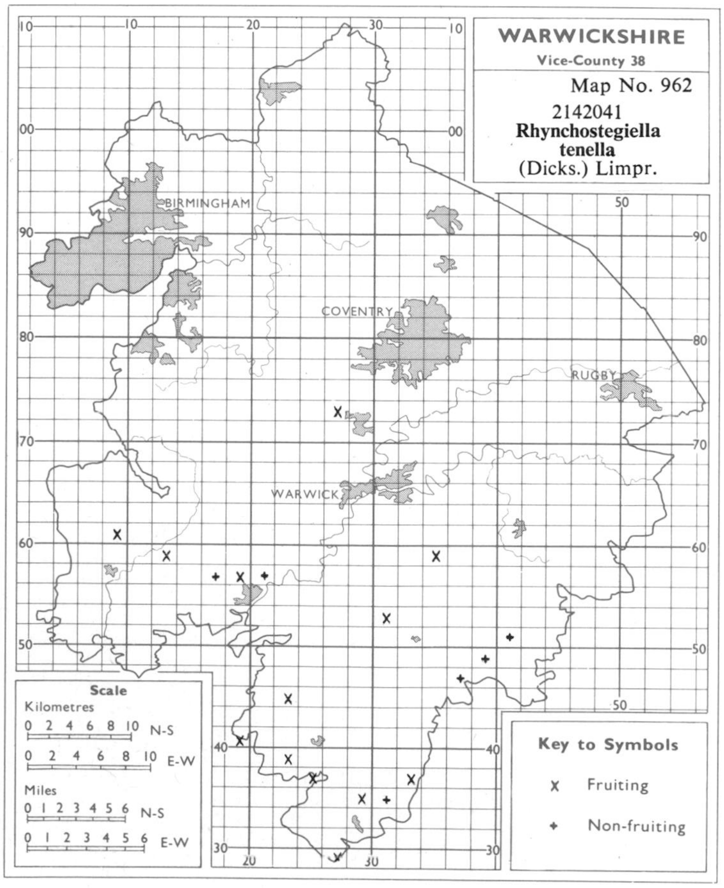
WARWICKSHIRE
Vice-County 38
Map No. 962
2142041
Rhynchostegiella tenella (Dicks.) Limpr.
BIRMINGHAM
COVENTRY
RUGBY
WARWICK
Scale
Kilometres
0 2 4 6 8 10 N-S
0 2 4 6 8 10 E-W
Miles
0 1 2 3 4 5 6 N-S
0 1 2 3 4 5 6 E-W
Key to Symbols
X Fruiting
+ Non-fruiting

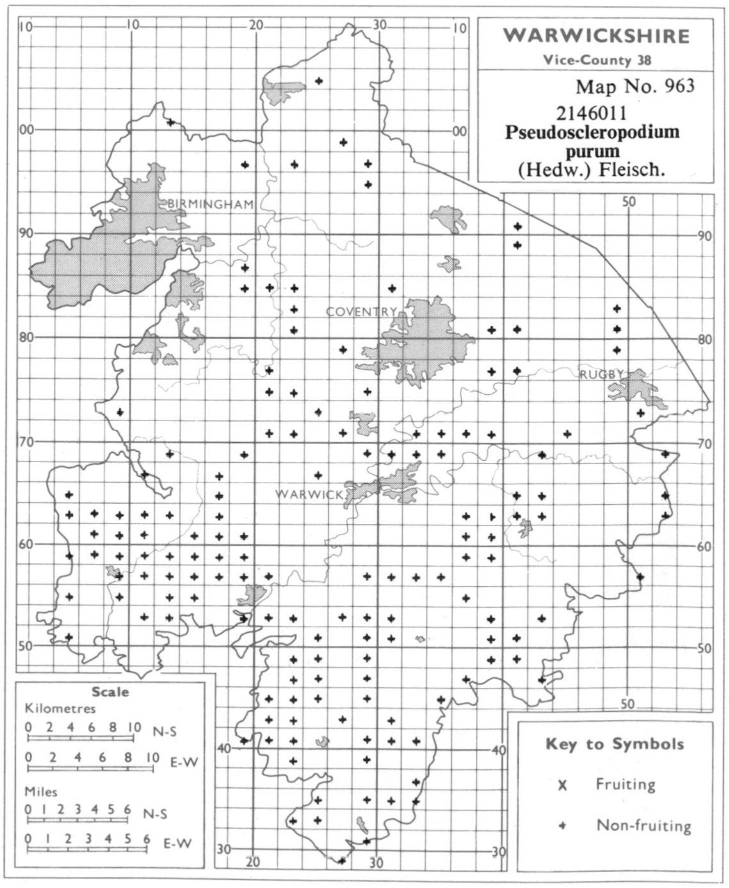
WARWICKSHIRE
Vice-County 38
Map No. 963
2146011
Pseudoscleropodium purum (Hedw.) Fleisch.
BIRMINGHAM
COVENTRY
RUGBY
WARWICK
Scale
Kilometres
0 2 4 6 8 10 N-S
0 2 4 6 8 10 E-W
Miles
0 1 2 3 4 5 6 N-S
0 1 2 3 4 5 6 E-W
Key to Symbols
X Fruiting
+ Non-fruiting

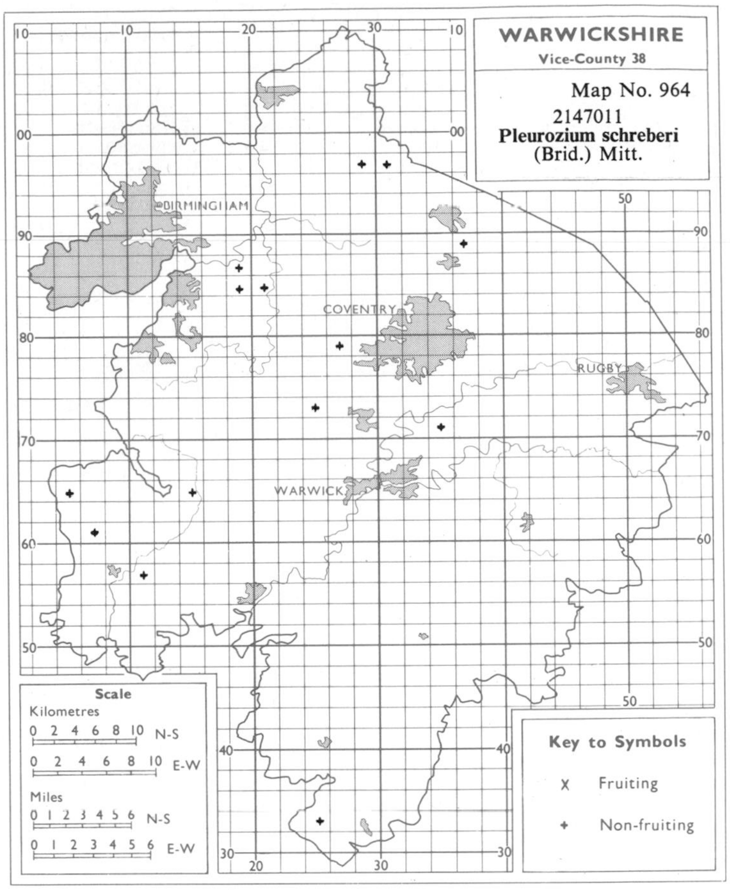
WARWICKSHIRE
Vice-County 38
Map No. 964
2147011
Pleurozium schreberi
(Brid.) Mitt.
BIRMINGHAM
COVENTRY
RUGBY
WARWICK
Scale
Kilometres
0 2 4 6 8 10 N-S
0 2 4 6 8 10 E-W
Miles
0 1 2 3 4 5 6 N-S
0 1 2 3 4 5 6 E-W
Key to Symbols
X Fruiting
+ Non-fruiting

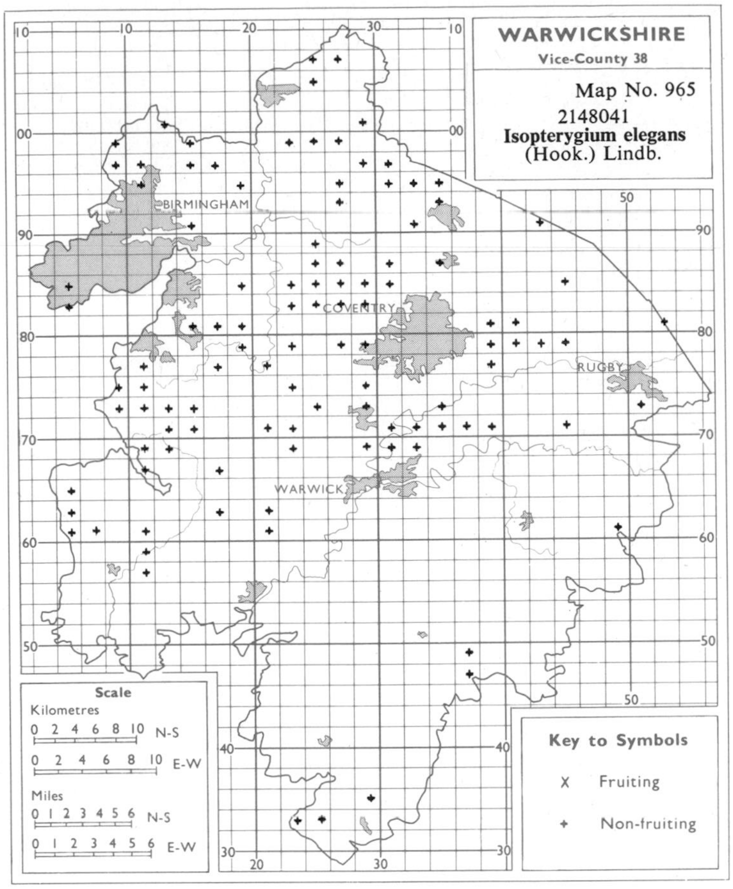
WARWICKSHIRE
Vice-County 38
Map No. 965
2148041
Isopterygium elegans
(Hook.) Lindb.
BIRMINGHAM
COVENTRY
RUGBY
WARWICK
Scale
Kilometres
0 2 4 6 8 10 N-S
0 2 4 6 8 10 E-W
Miles
0 1 2 3 4 5 6 N-S
0 1 2 3 4 5 6 E-W
Key to Symbols
X Fruiting
+ Non-fruiting

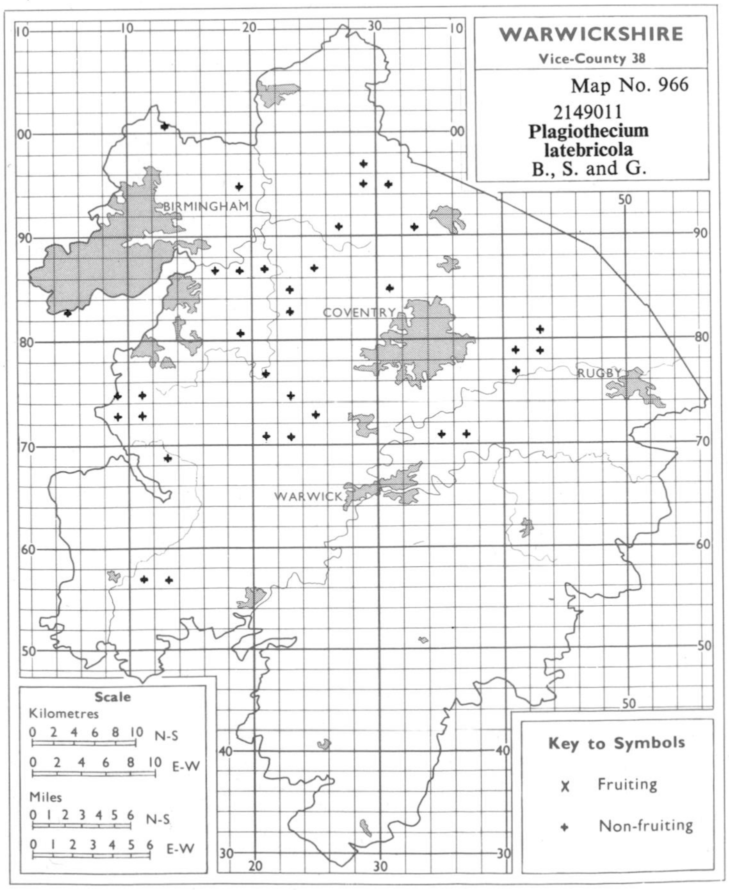
WARWICKSHIRE
Vice-County 38
Map No. 966
2149011
Plagiothecium
latebricola
B., S. and G.
BIRMINGHAM
COVENTRY
RUGBY
WARWICK
Scale
Kilometres
0 2 4 6 8 10 N-S
0 2 4 6 8 10 E-W
Miles
0 1 2 3 4 5 6 N-S
0 1 2 3 4 5 6 E-W
Key to Symbols
X Fruiting
+ Non-fruiting

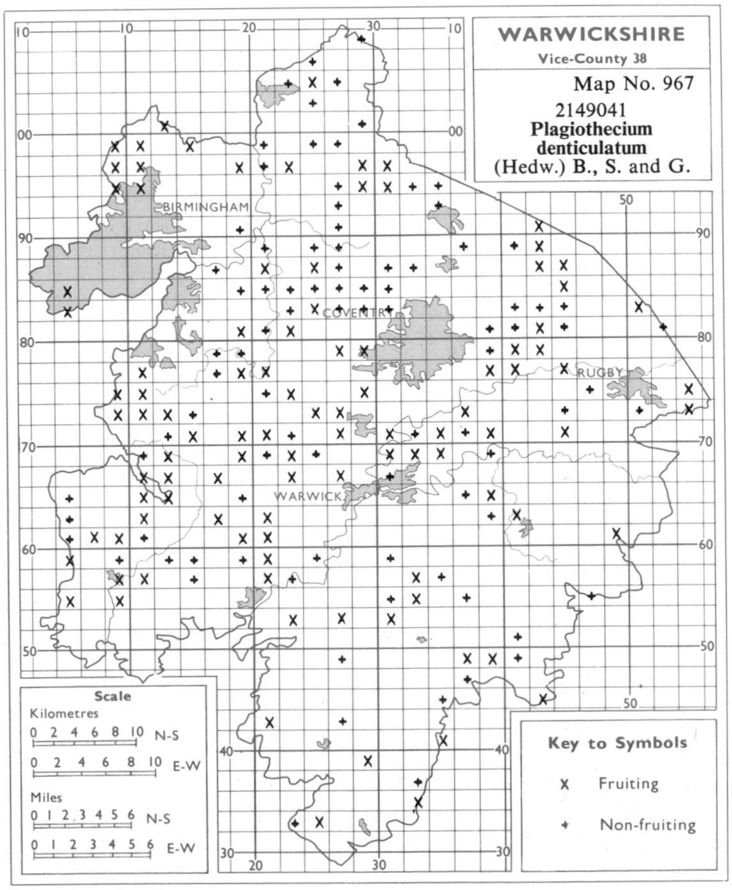
WARWICKSHIRE
Vice-County 38
Map No. 967
2149041
Plagiothecium
denticulatum
(Hedw.) B., S. and G.
BIRMINGHAM
COVENTRY
RUGBY
WARWICK
Scale
Kilometres
0 2 4 6 8 10 N-S
0 2 4 6 8 10 E-W
Miles
0 1 2 3 4 5 6 N-S
0 1 2 3 4 5 6 E-W
Key to Symbols
X Fruiting
+ Non-fruiting

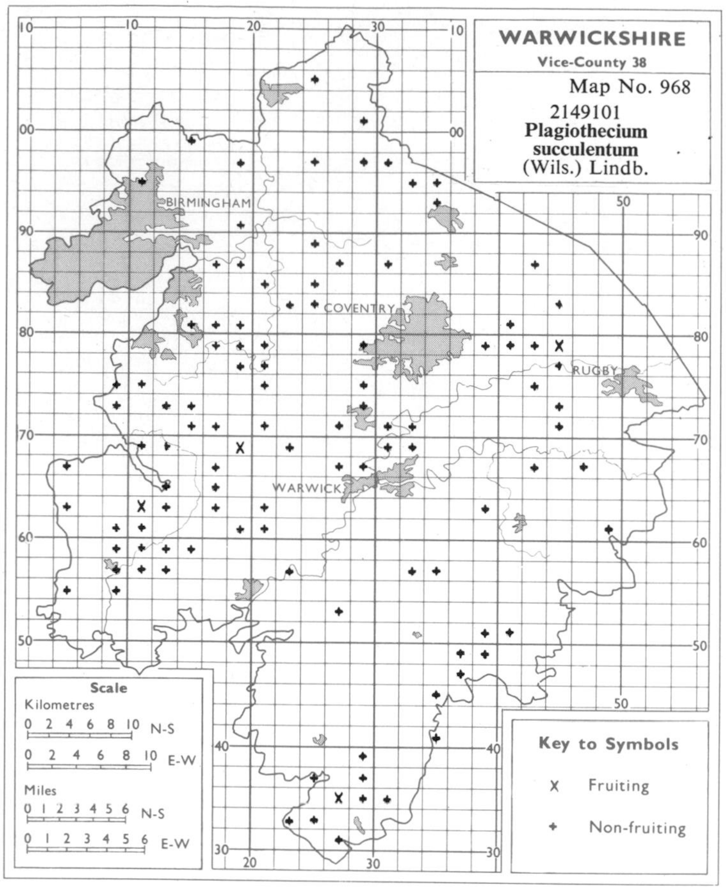
WARWICKSHIRE
Vice-County 38
Map No. 968
2149101
Plagiothecium succulentum (Wils.) Lindb.
BIRMINGHAM
COVENTRY
RUGBY
WARWICK
Scale
Kilometres
N-S
E-W
Miles
Key to Symbols
X Fruiting
+ Non-fruiting

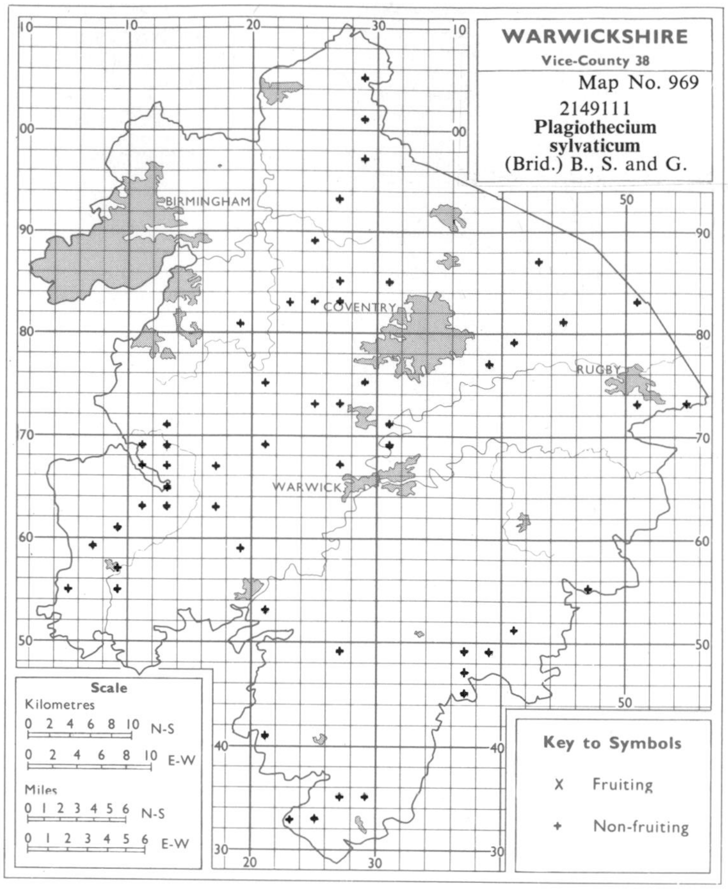
WARWICKSHIRE
Vice-County 38
Map No. 969
2149111
Plagiothecium sylvaticum (Brid.) B., S. and G.
BIRMINGHAM
COVENTRY
RUGBY
WARWICK
Scale
Kilometres
N-S
E-W
Miles
Key to Symbols
X Fruiting
+ Non-fruiting

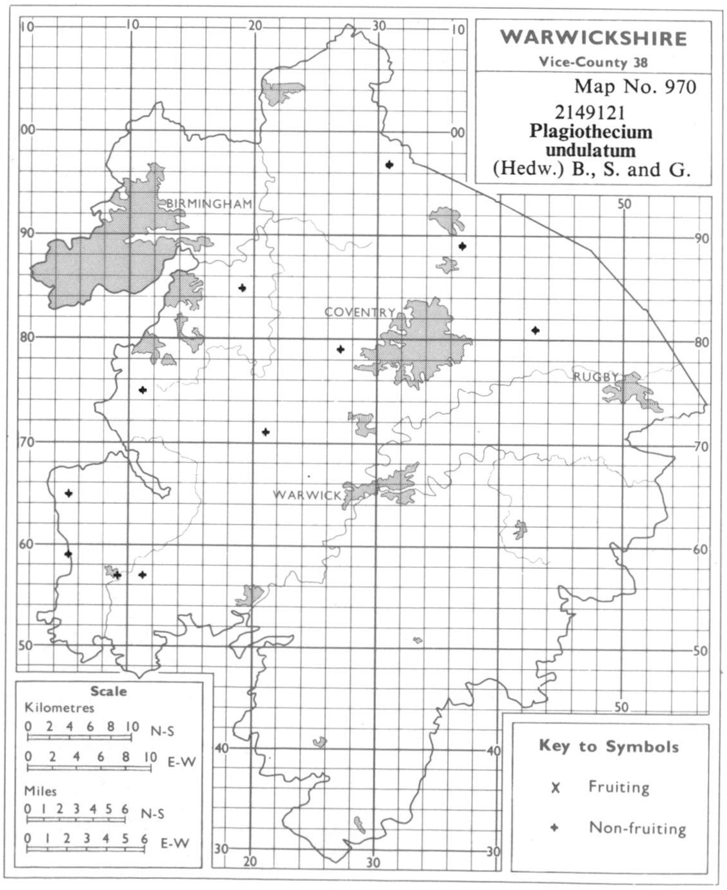
WARWICKSHIRE
Vice-County 38
Map No. 970
2149121
Plagiothecium undulatum (Hedw.) B., S. and G.
BIRMINGHAM
COVENTRY
RUGBY
WARWICK
Scale
Kilometres
N-S
E-W
Miles
Key to Symbols
X Fruiting
+ Non-fruiting

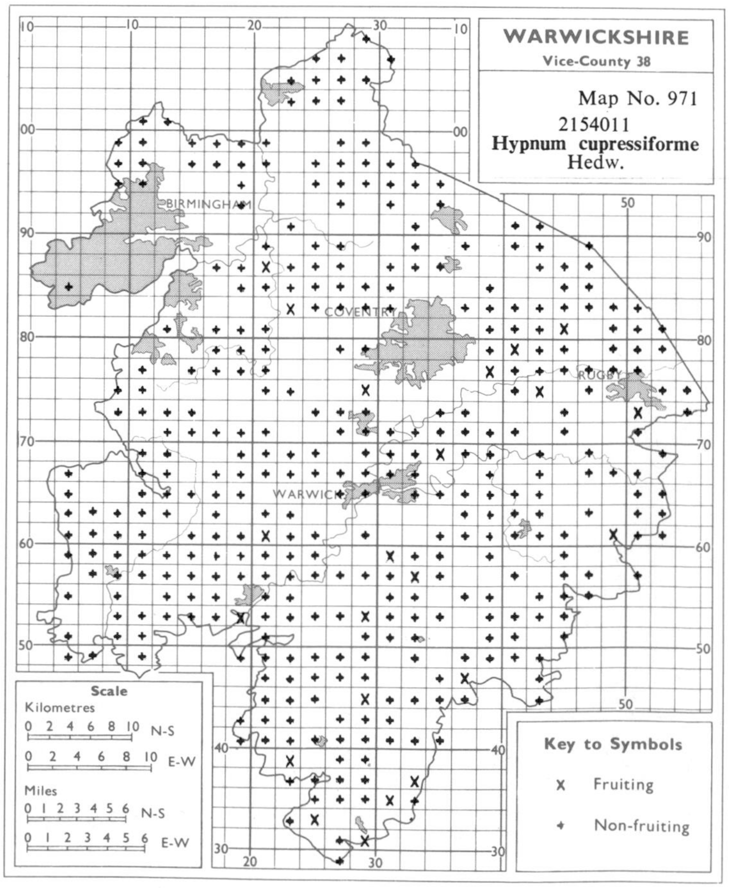
WARWICKSHIRE
Vice-County 38
Map No. 971
2154011
Hypnum cupressiforme Hedw.
BIRMINGHAM
COVENTRY
RUGBY
WARWICK
Scale
Kilometres
N-S
E-W
Miles
Key to Symbols
X Fruiting
+ Non-fruiting

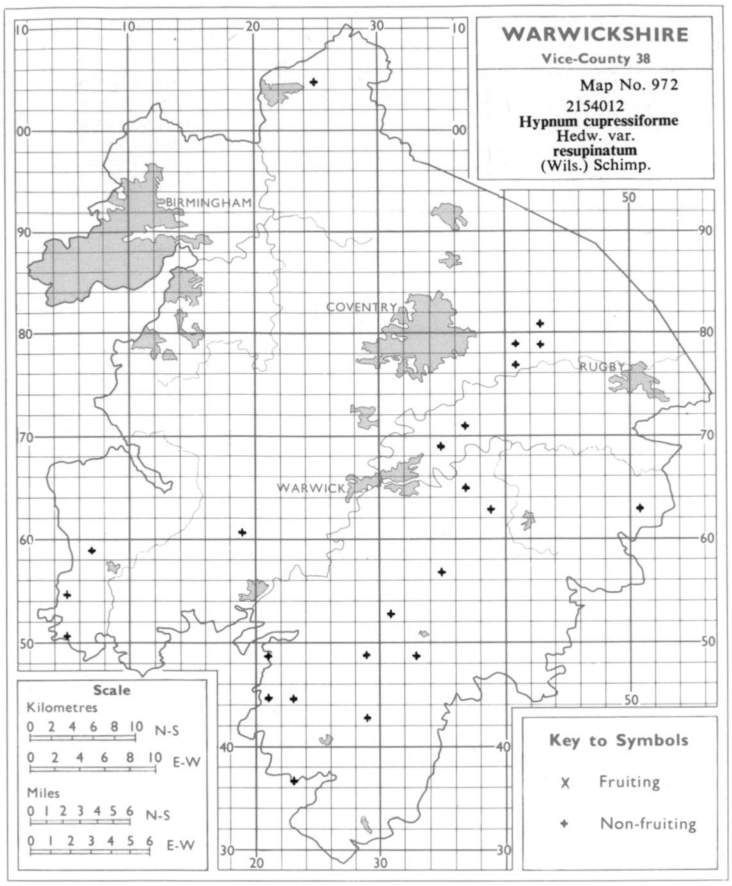
WARWICKSHIRE
Vice-County 38
Map No. 972
2154012
Hypnum cupressiforme
Hedw. var.
resupinatum
(Wils.) Schimp.
BIRMINGHAM
COVENTRY
RUGBY
WARWICK
Scale
Kilometres
0 2 4 6 8 10 N-S
0 2 4 6 8 10 E-W
Miles
0 1 2 3 4 5 6 N-S
0 1 2 3 4 5 6 E-W
Key to Symbols
X Fruiting
+ Non-fruiting

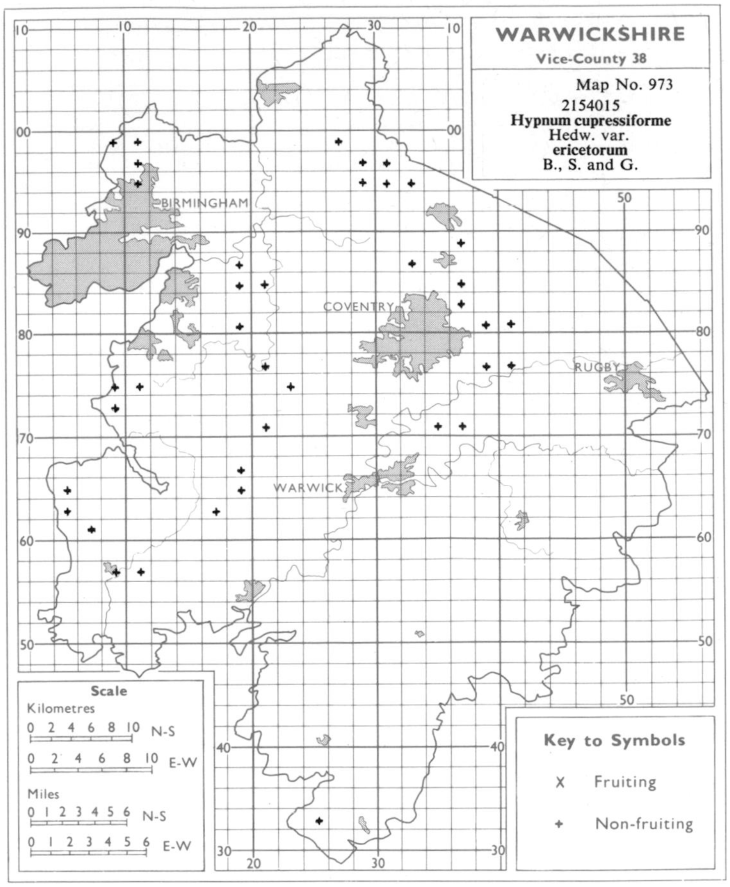
WARWICKSHIRE
Vice-County 38
Map No. 973
2154015
Hypnum cupressiforme
Hedw. var.
ericetorum
B., S. and G.
BIRMINGHAM
COVENTRY
RUGBY
WARWICK
Scale
Kilometres
0 2 4 6 8 10 N-S
0 2 4 6 8 10 E-W
Miles
0 1 2 3 4 5 6 N-S
0 1 2 3 4 5 6 E-W
Key to Symbols
X Fruiting
+ Non-fruiting

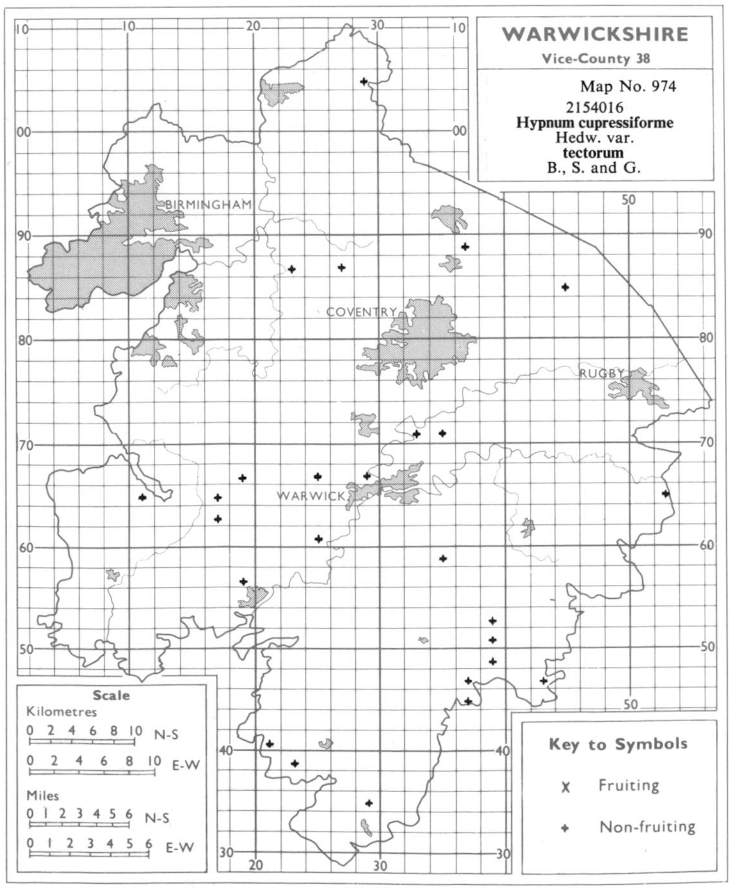
WARWICKSHIRE
Vice-County 38
Map No. 974
2154016
Hypnum cupressiforme
Hedw. var.
tectorum
B., S. and G.
BIRMINGHAM
COVENTRY
RUGBY
WARWICK
Scale
Kilometres
0 2 4 6 8 10 N-S
0 2 4 6 8 10 E-W
Miles
0 1 2 3 4 5 6 N-S
0 1 2 3 4 5 6 E-W
Key to Symbols
X Fruiting
+ Non-fruiting

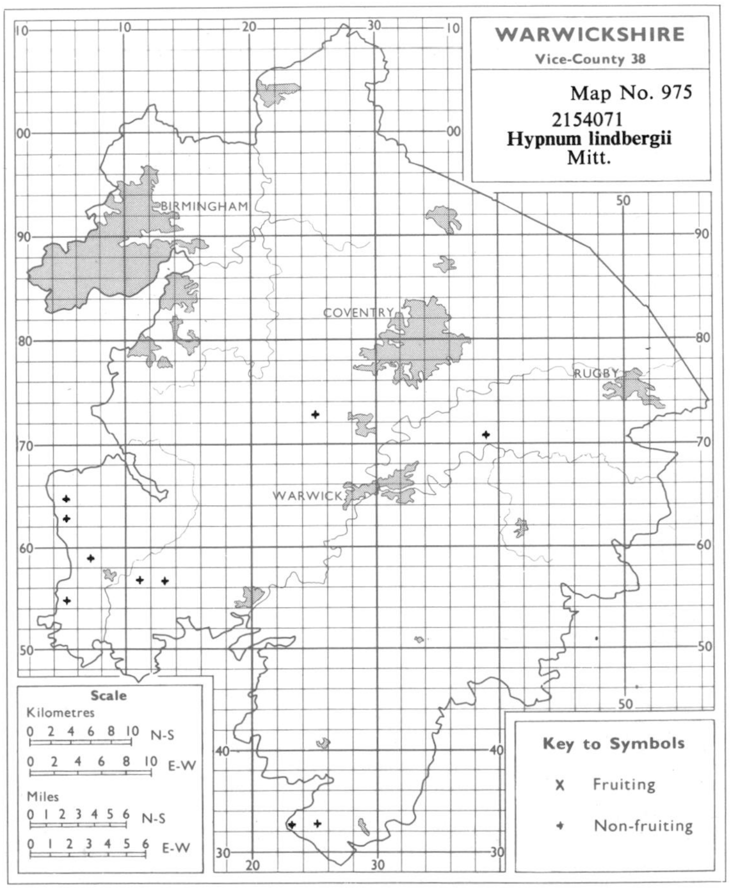
WARWICKSHIRE
Vice-County 38
Map No. 975
2154071
Hypnum lindbergii
Mitt.
BIRMINGHAM
COVENTRY
RUGBY
WARWICK
Scale
Kilometres
0 2 4 6 8 10 N-S
0 2 4 6 8 10 E-W
Miles
0 1 2 3 4 5 6 N-S
0 1 2 3 4 5 6 E-W
Key to Symbols
X Fruiting
+ Non-fruiting

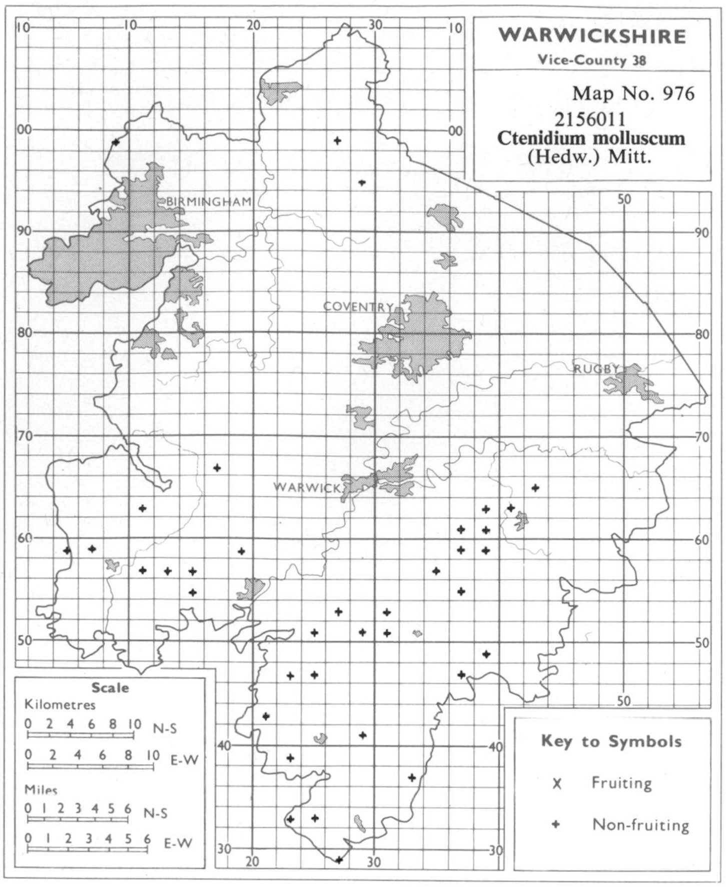
WARWICKSHIRE
Vice-County 38
Map No. 976
2156011
Ctenidium molluscum
(Hedw.) Mitt.
BIRMINGHAM
COVENTRY
RUGBY
WARWICK
Scale
Kilometres
0 2 4 6 8 10 N-S
0 2 4 6 8 10 E-W
Miles
0 1 2 3 4 5 6 N-S
0 1 2 3 4 5 6 E-W
Key to Symbols
X Fruiting
+ Non-fruiting

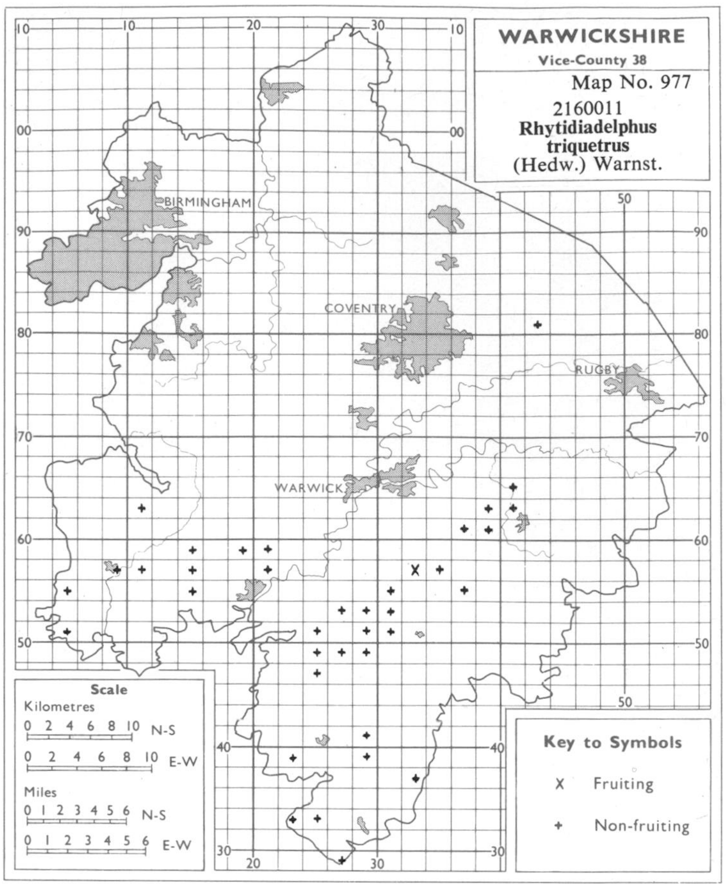
WARWICKSHIRE
Vice-County 38
Map No. 977
2160011
Rhytidiadelphus
triquetrus
(Hedw.) Warnst.
BIRMINGHAM
COVENTRY
RUGBY
WARWICK
Scale
Kilometres
0 2 4 6 8 10 N-S
0 2 4 6 8 10 E-W
Miles
0 1 2 3 4 5 6 N-S
0 1 2 3 4 5 6 E-W
Key to Symbols
X Fruiting
+ Non-fruiting

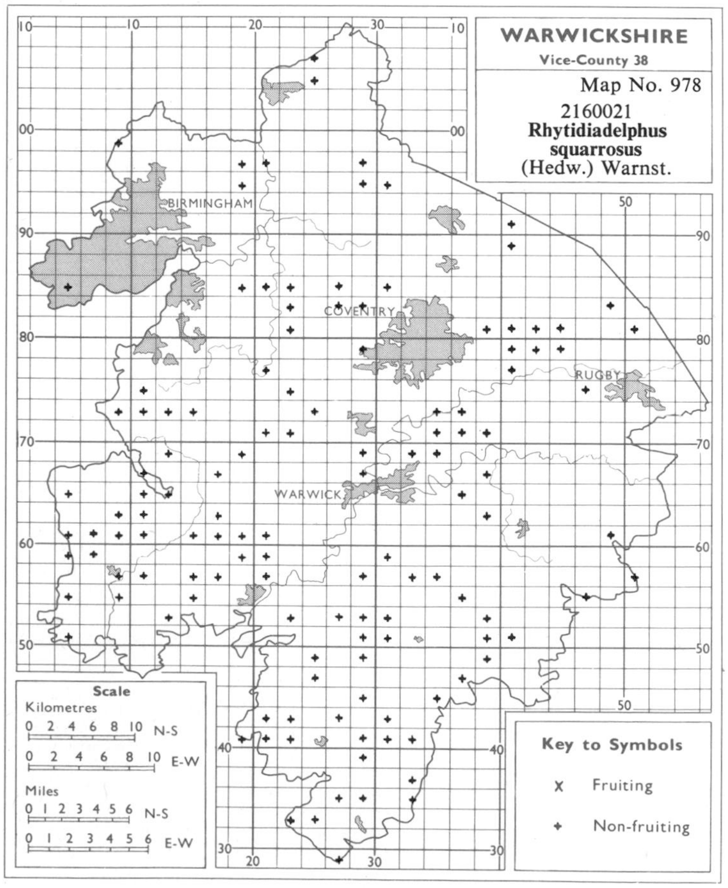
WARWICKSHIRE
Vice-County 38
Map No. 978
2160021
Rhytidiadelphus
squarrosus
(Hedw.) Warnst.
BIRMINGHAM
COVENTRY
RUGBY
WARWICK
Scale
Kilometres
0 2 4 6 8 10 N-S
0 2 4 6 8 10 E-W
Miles
0 1 2 3 4 5 6 N-S
0 1 2 3 4 5 6 E-W
Key to Symbols
X Fruiting
+ Non-fruiting

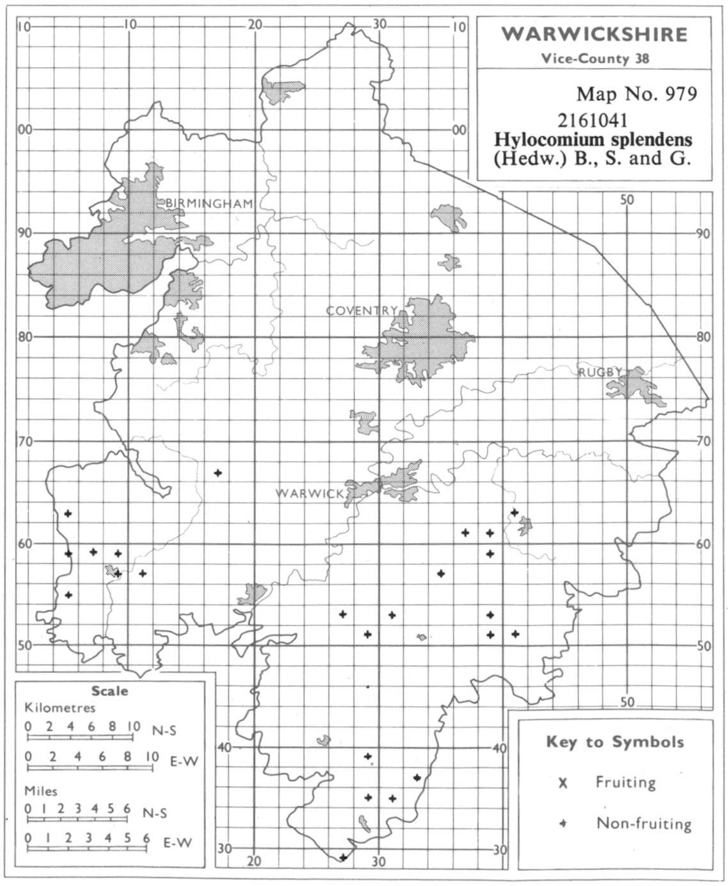
WARWICKSHIRE
Vice-County 38
Map No. 979
2161041
Hylocomium splendens
(Hedw.) B., S. and G.
BIRMINGHAM
COVENTRY
RUGBY
WARWICK
Scale
Kilometres
0 2 4 6 8 10 N-S
0 2 4 6 8 10 E-W
Miles
0 1 2 3 4 5 6 N-S
0 1 2 3 4 5 6 E-W
Key to Symbols
X Fruiting
+ Non-fruiting

14. CONTRIBUTORS TO THE SURVEY[1]

The following list gives the names of all those who contributed to the survey by observations in the field. However, by reason of the manner in which the survey was organised, there was wide variation in the extent and form of the contributions of individuals, so that a few others who helped in the work may have been inadvertently omitted; for any such omissions we tender our apologies.

The policy throughout was to encourage participation by as many people as possible, however little or however much time they were able to devote to the project. For instance, a series of special field meetings was arranged in under-recorded areas of the county in which many botany students wishing to get some experience of field work and a number of comparatively inexperienced amateur botanists took part, perhaps on only one or two occasions. Here they were usually paired with the more experienced workers, and records of observations were submitted jointly. Other helpers have contributed records of particular groups or limited areas or have merely submitted records of a few noteworthy finds.

A substantial number of individuals worked methodically in one or a number of 1 km squares and submitted complete lists. Some two-thirds of this methodical recording was done by the small number of contributors indicated by an asterisk (*), whose total lists ranged in number from 90 in the case of the most prolific recorder down to 14, with an average of about 35. Others who contributed fewer but at least one complete list are indicated by the symbol (†). A few contributors did a great deal of work in the later stages of the survey by making themselves responsible for back-checking doubtful records and filling in gaps over wide areas of the county; these are indicated by the symbol (B).

To all these contributors we should like to record here our very grateful thanks.

Allen, D. E.
*Allen, H.
Andrews, C. E. A.
*Arnold, M. A. and G. A. (B)
Ashton, Miss B.
Aslett, Mr. and Mrs. J. E.
Austen, Mrs. R.
Avery, K.
Bailey, Miss B.
†Baker, Mrs. H.
Baker, I.
Barnabas, Miss L.
†Bass, Miss E.
†Bates, Miss R. E.
†Batt, Mrs. J. M.
Benoit, P. M.
Bettridge, Mrs. M. M. H.
†Blackburn, Mrs. K.
Boardman, Mrs. M. E.
Bowers, G. M.
†Brierley, Mrs. J. N.
†Broady, Miss M. J.
Brooke, Miss A.
Brown, P. A.
Bullard, Mrs. R.
†Butler, Mrs. E.
Butler-Fleming, Miss A. M.
†Byles, Miss A.
†Bywater, Miss J.
*Cadbury, Miss D. A. (B)
*Clark, M. C. (B)
Clarke, R.
†Cockitt, R.

[1]See p. 654 for contributors to bryophyte survey.

Coles, Dr. S. M.
†Coley-Smith, Mrs. S.
†Cooper, Miss G.
Cornish, Miss B.
Cotes, R.
Cotton, B.
†Cotton, Mr. and Mrs. F.
*Coultas, P. G.
Cox, Mrs. J.
Darling, Miss D. F.
*Daulman, Miss B. M. (B)
†Daulman, Miss M.
Davies K. C.
†Dawson, R. W. B.
†Dean, Miss M.
Dengate, Miss A.
Diamond, B. C.
†Dix, H. M.
†Dony, Dr. J. G.
*Dony, Mrs. C. M.
Dudley, Miss K.
Duncan, Miss U. K.
Edees, E. S.
†Edwardson, A. L. G.
†Evans, A. W.
*Evans, Mr. and Mrs. R. E. (B)
Falk, P.
†Farmer, H. A.
†Farrand, E. A.
Field, J. H.
Fincher, F.
Ford, P.
Foster, Dr. W. D.
†Fountain, Mrs. A. M.
†Fowkes, H. H.
†Francis, Miss J.
†Franks, T. J.
Fremlin, Prof. J.
Gateley, M.
Ghaffar, A.
Gibby, Miss M.
†Goode, Miss P. M.
Grant, G. C.
†Green, P. S.
†Greene, Dr. S. W.
Grove, Miss J. H.
†Hackett, A.
Haggar, J. R.
†Hall, Mr. and Mrs. P. C.
Hancock, Miss M. B.
Hanson, F. D.
†Hanson, Mrs. H. E.
Hanson, M. K.
*Hardaker, W. H.
Harvey, M. J.
*Hawkes, Prof. J. G.
Hawkes, Miss P. D.
Heath, Rev. D. M.
†Henshall, T.
Heslop-Harrison, Prof. J.
Hill, J.
Hird, S.
†Hudson, Miss M.
Hull, R.
†Hulme, Mrs. B. A.
†Ibbotson, Miss E. M. H.
†Jacobs, V.
Jebbott, D. E.
†Jeffray, D. J.
Jessop, Miss A.
Jewsbury, Mrs. H.
*Jones, Mrs. M. D. G. (B)
†Kennedy, Mrs. J.
†Kiernan, Dr. J. A.
†Kilby, Dr. B. A.
*Laflin, T. (B)
Lapworth, H.
Latham, F. H.
Lay, E. J.
†Leonard, Miss S.
*Lester, Dr. R. N.
†Longton, Dr. R. E.
Lowe, Dr. C. R.
Lowe, J. H.
†Maddison, Miss S.
Marks, J.
Marvin, G. S.
Meikle, R. D.
Melville, Dr. R.
†Mitchell, Dr. G.
Moore, P. D.
Morley, B. D.
Morton, A. M.
†Murray, Rev. D. P.
†Neale, P. D.
Needham, Miss M.
Nelder, Mrs. M.
†Oakes, M.
Perry, D.
*Phipps, Dr. J. B.
†Pickering, B.
*Pickvance, Mrs. E.
Pickvance, S. M. J.
†Pinkess, L. H. W.
†Pitt, Miss L.
†Rainbow, M. J.
*Readett, R. C. (B)
*Roberts, H. A.
†Robertson, D. C.
†Robinson, C. G.
†Robson, R. J.
†Roe, Capt. and Mrs. R. G. B.
Rogers, Miss E.
†Rose A. C.
Rushforth, Miss M.
†Russell, Miss M. J.
Samuels, Mrs. K.
†Saunders, Miss M. G.
Savidge, Dr. J. P.
†Schofield, Miss M. M.
Scott, Mrs. G. E.
†Seeley, W. F. E.
†Shack, Dr. J.
Sharpe, J. R.
†Shotton, Prof. and Mrs. F. W.
†Skelding, Prof. A. D.
†Skull, A. J.
†Smart, Miss M. E.
Smith, F. C. C.
†Smith, Dr. and Mrs. P. (B)
†Smith, Dr. T. A.
†Spreadbury, Mrs. P. M.
†Stanley, P. E.
Stokes, Miss G.
Strachan, Dr. I.

Sutton, Miss A.
Tarn, Dr. R.
†Taylor, Mrs. H. H.
†Thomas, C.
Thorne, Miss B.
†Thorpe, Dr. M.
†Tindall, Mrs. K. H.
Tonks, E. S.
Townsend, C. C.
†Trought, T.
†Turner, G. J.
†Tye, Mrs. M. G.
*Tyrer, Dr. F. H. (B)
*Tyrer, Dr. P.
†Walker, I.
Ward, L. E.
†Warwick, Miss P. J.
Webster, Miss I. R.
Webster, Miss M. McCallum
†Weston, Miss R. L.
Westrup, A. W.
Wilkins, Dr. D. A.
†Wilson, Miss G. D.
†Witton, Miss R.
*Woolman, J. F.
†Wright, Miss P. M.
†Wright, Dr. S. T. C.
†Yardley, C.

15. LOCAL, NATIONAL AND INTERNATIONAL SPECIALISTS

We have mentioned already in the section on methods of recording (pp. 68 and 70) that our own Flora Panel carried out routine identifications of "critical" or difficult specimens whenever possible. At the beginning of the survey a wide range of critical material was submitted to local, national and international specialists and a voucher collection was built up of named specimens. With this collection on hand it was then found possible to identify the vast majority of critical specimens ourselves. Only particularly difficult specimens were then submitted to the specialists. However, with certain groups of microspecies and species hybrids (as, for example, in the genera *Salix*, *Ulmus*, *Rubus*, *Hieracium*, *Taraxacum*, etc.) all specimens were submitted for identification at this level, whilst identification at the aggregate level was made in Birmingham by the panel or by recorders in the field. A clear distinction is made in the species text when such cases are under consideration.

We give a list of specialists below, together with a note of the genera or species which we have submitted to them. We must reiterate that the specialists have seen only a small proportion of the total number of critical specimens collected, so that final responsibility for the critical groups must lie with us.

We should like to take this opportunity of renewing our thanks to these specialists for the time and care which they took with the specimens we sent them. Although space does not permit us to record the exact details in the species text of every specimen identified, all these details are kept in our card index and can be made freely available to anyone wishing to obtain further information.

Aellen, Dr. P. (Switzerland)	*Corispermum*
Alston, A. H. G.	*Dryopteris*
Andrews, C. E. A.	*Hieracium* (also contributed key and notes)
Ash, G. M.	*Epilobium*
Bradshaw, Dr. Margaret	*Alchemilla*
Brenan, J. P. M.	*Amaranthus*, *Atriplex*, *Chenopodium*
Brummitt, R. K.	*Calystegia*
Butcher, Dr. R. W.	*Ranunculus* sub-genus *Batrachium*
Cannon, J. F. M.	Umbelliferae
Cook, Prof. C. D. K.	*Ranunculus* sub-genus *Batrachium*, *Sparganium erectum*
Dandy, J. E.	*Potamogeton*
Edees, E. S.	*Rubus*
Graham, M. A.	*Mentha*
Harley, Dr. R. M.	*Mentha*

Harvey, Dr. J.	*Viola* sub-genus *Viola*
Heslop-Harrison, Prof. J.	*Dactylorhiza*, *Rosa*
Hubbard, Dr. C. E.	Gramineae
Hulme, Mrs. B. A.	*Atriplex*
Hyde, H. A.	*Aconitum*
Jeffrey, C.	Compositae
Jermy, A. C.	*Carex*, *Dryopteris*
Jones, Dr. B. M. G.	*Ranunculus ficaria*
Lewis, Dr. M. C.	*Arctium* (also contributed key and notes)
Lousley, J. E.	*Polygonum*, *Rumex*, *Senecio* etc.
Marsden-Jones, E. M.	*Anagallis*, *Centaurea*
Marshall, J. B.	*Crepis*
Meikle, R. D.	*Salix*
Melderis, Dr. A.	Gramineae
Melville, Dr. R.	*Ulmus*
Miles, B. A.	*Rubus*
Milne-Redhead, E.	*Cerastium*
Nelmes, E.	*Carex*
Pennington, T. D.	*Epilobium*
Pigott, Prof. C. D.	*Thymus*
Proctor, Dr. M. C. F.	*Ulex*
Raven, Prof. P. H. (California)	*Oenothera*
Richards, Dr. A. J.	*Taraxacum*
Richards, Prof. P. W.	*Juncus*
Sandwith, N. Y.	*Fumaria*
Savidge, Dr. J. P.	*Callitriche*
Sell, P. D.	*Hieracium*
Sledge, Dr. W. A.	*Carduus*
Smith, Dr. G.	*Erophila*
Smith, Dr. P.	*Bromus* (also contributed key and notes)
Styles, Dr. B. T.	*Polygonum*
Summerhayes, V. S.	Orchidaceae
Taylor, Sir George	*Potamogeton*
Taylor, P. G.	*Utricularia*
Timson, Dr. J.	*Polygonum*
Townsend, C. C.	*Galeopsis*, *Lepidium*
Turrill, Dr. W. B.	*Centaurea*
Tutin, Prof. T. G.	*Polygonum*
Valentine, Prof. D.	*Viola* sub-genus *Viola*
van Soest, Prof. J. L. (Netherlands)	*Taraxacum*
Wade, Dr. A. E.	*Myosotis*, *Symphytum*
Walters, Dr. S. M.	*Alchemilla*, *Aphanes*, *Calystegia*, *Eleocharis*, *Montia*, *Potentilla*, *Silene*
Warburg, Dr. E. F.	*Betula*, *Erodium*, *Euphrasia*, *Prunus*, *Viola hirta* etc.

West, Dr. C.	*Hieracium*
Woolman, J. F.	*Rubus* (also contributed key and notes)
Wright, F. R. Elliston	*Sagina*
Yeo, Dr. P. F.	*Euphrasia*
Young, Dr. D. P.	*Epipactis*

AUTHOR INDEX

(Including biographical references and contributors of literature and herbarium records)

Numbers in italics indicate those pages where references are given in full.

INDEX OF PLANT NAMES

Numbers in italics indicate the pages where first record, distribution and map reference are given. In Section 11 (pp. 617–653) the plants are listed in alphabetical order and references to those pages are therefore omitted here.

B

O

P

W

X

Y

Z

LIST OF MAPS IN BACK POCKET